TIME-SAVER STANDARDS
for
URBAN DESIGN

Dedication

Urban design is by definition an optimistic enterprise undertaken
as a legacy to the future, to preserve, celebrate and sustain
the values of a just, open and equitable
civil society and culture.

■

To those whose lives were lost and to those who lost loves ones,

September 11, 2001

and

to Judy.

TIME-SAVER STANDARDS for URBAN DESIGN

EDITORS

Donald Watson, FAIA, Editor-in-Chief
Alan Plattus
Robert G. Shibley

McGraw-Hill
New York ▪ Chicago ▪ San Francisco ▪ Lisbon ▪ London ▪ Madrid
Mexico City ▪ Milan ▪ New Delhi ▪ San Juan ▪ Seoul
Singapore ▪ Sydney ▪ Toronto

The McGraw-Hill Companies

Cataloging-in-Publication Data is on file with the Library of Congress

1 2 3 4 5 6 7 8 9 0 KGP/KGP 0 9 8 7 6 5 4 3

ISBN 0-07-068507-X

McGRAW-HILL PROFESSIONAL
Cary Sullivan, Editor

BOOK DESIGN & PRODUCTION
Susann Gierman / Suzani Design, Amherst, Massachusetts, USA
Todd S. Gordner

COVER DESIGN
Margaret Webster-Shapiro

Printed and bound by Quebecor/Kingsport
This book was printed on acid-free paper.

DISCLAIMER

The information in this book has been obtained from many sources, including publishers, government organizations, trade associations, manufacturers and professionals in research and in practice. The publisher, editors and authors have made every reasonable effort to make this reference work accurate and authoritative, but make no warranty, and assume no liability for the accuracy or completeness of the text, tables or illustrations or its fitness for any particular purpose. The appearance of technical data or editorial material in this publication does not constitute endorsement, warranty or guarantee by the publisher, editors or authors of any product, design, service, or process. It is the responsibility of users to apply their professional knowledge in the use of information and recommendations contained in this book, to consult original sources for more detailed information and to seek expert advice as required or as appropriate for the design and construction of buildings. Neither the authors, editors nor McGraw-Hill shall have any liability to any party for any damages resulting from the use, application or adaptation of information contained in *Time-Saver Standards*, whether such damages are direct or indirect, or in the nature of lost profits or consequential damages. The *Times-Saver Standards* is published with the understanding that McGraw-Hill is not engaged in providing architectural, engineering design or other professional services.

TABLE OF CONTENTS

1 CONTEXT OF URBANIZATION

2 CLASSIC TEXTS OF URBAN DESIGN

6 ELEMENTS OF THE CITY

7 URBAN DESIGN DETAILS

8 CASE STUDIES IN URBAN DESIGN

ABOUT THE EDITORS

DONALD WATSON, FAIA is an architect and author. He is editor-in-chief of *Time-Saver Standards for Architectural Design Data* (1997) and *Time-Saver Standards for Building Materials and Systems* (2000). He was formerly Visiting Professor at Yale School of Architecture and Professor of Architecture and Dean of the School of Architecture, Rensselaer Polytechnic Institute and is a 2002 recipient of the ACSA Distinguished Professor Award.

ALAN J. PLATTUS is Professor of Architecture at Yale University and founder and Director of the Yale Urban Design Workshop, a design center working with communities throughout Connecticut. His research interests include the history and development of cities. He is editor of the 1988 Princeton University Press reprint of Hegemann and Peets, *Civic Art*.

ROBERT G. SHIBLEY, AIA, AICP, is a professor of architecture and planning at the School of Architecture and Planning, University at Buffalo, State University of New York. He is the Director of the Urban Design Project and co-author of *Urban Excellence* with Philip Langdon and Polly Welsh (1990), *Placemaking: The Art and Practice of Building Communities* (1995) with Lynda Schneekloth, and *Commitment to Place: Urban Excellence and Community* (2000).

Acknowledgments

The editors are indebted to and acknowledge with appreciation the authors, contributors and publishers listed in credits and in About the Authors, and are grateful for the thoughtful and patient advice during formulation of this volume of Jonathan Barnett, Stephen W. Hurtt, Roger Lewis, Donlyn Lyndon, and Marguerite Villecco.

ABOUT THE AUTHORS

WAYNE ATTOE is an urban designer and author in Canberra, Australia. His books include *Skylines: Understanding and Molding Urban Silhouettes* (1981), *Architecture and the Critical Imagination* (1978), and *Urban Design: Reshaping Our Cities* with Anne Vernez Moudon.

EMILY AXEROD is a planner and the Executive Director of the Rudy Bruner Award for Urban Excellence and has worked in urban planning in both the public and private sectors in San Francisco and Boston.

EDMUND N. BACON, author of *Design of Cities* (1967), is known for his career-long focus upon one American city, Philadelphia, where he was Director of the Planning Commission from 1949 until retirement in 1970. His recognitions include the 1990 International Union of Architects' Abercrombie Prize for Town Planning. He is a senior fellow of the Urban Land Institute.

MICHAEL J. BEDNAR, FAIA, is a senior faculty member of Architecture at University of Virginia School of Architecture. He is a practicing architect in Charlottesville, Virginia, and former chair of the Charlottesville Planning Commission. His books include *The New Atrium* (1986) and *Interior Pedestrian Places* (1989).

MATTHEW J. BELL, AIA, is Associate Professor of Architecture at the University of Maryland School of Architecture and coordinator of the School's Graduate Program in Urban Design.

MICHAEL BERNICK is an attorney and codirector of the University of California National Transit Access Center. Since 1988, Mr. Bernick has served as Director and President of the Board of the San Francisco Bay Area Rapid Transit District (BART).

ALAIN BERTAUD is an independent urban planning consultant. He has worked for the World Bank as Principal Urban Planner in many large cities of Asia, Europe, and Latin America. He is currently working on a book on the spatial structure of world metropolis cities.

VIKRAM BHATT was educated in India and Canada. He is Professor of Architecture and Director of the Minimum Cost Housing Group at McGill University, Montreal. He has directed several award-winning international housing research projects. He is the author of *Resorts of the Raj: Hill Stations of India* (1997).

SHERI BLAKE, Ph.D., MCIP, is Assistant Professor, Department of City Planning, Faculty of Architecture, University of Manitoba, Winnipeg, Canada. She was 2000–2001 Fulbright Scholar at Pratt Institute Center for Community and Environmental Development, Brooklyn, NY.

ROBERT BRAMBILLA is former head of the New York-based Institute for Environmental Action and Director of the Italian Art & Landscape Foundation. He designed the 1980 exhibit and publication, "More Streets for People" followed by a series of books about pedestrian zones and planning, including *For Pedestrians Only: Planning, Design*, and *Management of Traffic-Free Zones* (1977), coauthored with Gianni Longo.

PETER CALTHORPE is principal of Calthorpe Associates, Berkeley, California, and is a cofounder of the Congress for the New Urbanism. He is co-author with Sym Van der Ryn of *Sustainable Communities* (1986) and author of *The Next American Metropolis* (1993) and *The Regional City: Planning for the End of Sprawl* (2001).

ROBERT CERVERO is professor in the Department of City and Regional Planning at University of California, Berkeley, and Co-Director of University of California National Transit Access Center. His books include *America's Suburban Centers: The Land Use Transportation Link* (1989), *The Transit Metropolis: a global inquiry* (1990), and *Paratransit in America: Redefining Mass Transportation* (1997).

MARK C. CHILDS is Director of the Design and Planning Assistance Center at the School of Architecture and Planning, University of New Mexico. He is currently completing a new book, *Parking Spaces and Square Design*.

NANCY CLANTON, I.A.L.D., is a registered professional engineer and president of Clanton & Associates, a lighting design firm based in Boulder, Colorado. She is currently chair of the IESNA Outdoor Environmental Lighting Committee.

NAHOUM COHEN is in private practice in architecture and town planning. He is Senior Lecturer at the Technion, Haifa, and at Hassadna College, Tel Aviv, and member of Israeli Institute of Architects and Town Planners.

BRUCE COLDHAM, AIA, is the founding principal of Coldham Architects in Amherst, Massachusetts. He is a graduate of Melbourne University School of Architecture and the Masters of Environmental Design degree program of Yale School of Architecture.

WALTER A. COOPER is a New York City-based consulting engineer with specialization in planning, design, and construction of telecommunications facilities throughout the United States and internationally.

CLARE COOPER MARCUS is Professor Emeritus of Landscape Architecture and Architecture at the University of California, Berkeley and principal of a consulting practice. She is coauthor with Wendy Sarkissian of *Housing as if People Mattered: Site Design Guidelines for Medium Density Family* (1986) and coeditor with Carolyn Francis of *People Places: Design Guides for Urban Open Space* (1990).

JAMES P. COWAN, I.N.C.E., is Senior Consultant with Acentech Incorporated, Cambridge, Massachusetts. He teaches university courses in acoustics on the Internet and is author of *Handbook of Environmental Acoustics* (1994) and *Architectural Acoustics Design Guide* (2000).

GORDON CULLEN (1914–1994) was a leader in development of British theories of urban design in the postwar period. He published *Townscape* in 1961. His work included research and planning studies based on the theories of Townscape, as well as a collaborative project for the unbuilt new town of Maryculter (1974).

He is subject of the monograph *Gordon Cullen: Visions of Urban Design* by David Gasling (1996).

ANDRES DUANY, FAIA, is a cofounder of the firm Duany Plater-Zyberk & Company (DPZ), recipient of the Thomas Jefferson Memorial Medal of Architecture. He is coauthor with Elizabeth Plater-Zyberk and Jeff Speck of *Suburban Nation* (2000). He served as editor of *The New Civic Art: Elements of Town Planning* (2001).

ROBERT REGIS DVOŘÁK is a painter, teacher, and lecturer on subjects of creativity in art, education, and business. He was formerly visiting professor of architecture at the University of Oregon and the University of California, Berkeley.

JAY FARBSTEIN, Ph.D., FAIA, is an architect and leads a consulting practice in San Luis Obispo, California, specializing in programmatic requirements for building projects as well as in post-occupancy evaluation.

GABRIEL FELD is a practicing architect and Associate Professor and Head of the Architecture Department, Rhode Island School of Design, Providence, Rhode Island.

CHAD FLOYD, FAIA, is an architect and urban designer and a principal of Centerbrook Architects and Planners, Centerbrook, Connecticut.

WERNER FORNOS is President of the Population Institute, Washington, DC, and is the representative of the International Union for the Scientific Study of Population, assigned to the United Nations International Council on Management of Population Programs.

JOHN J. FRUIN is former Research Engineer at the Port Authority of New York and New Jersey and consults on pedestrian facilities design for numerous agencies and organizations, including National Bureau of Standards, Occupational Safety and Health Administration, U.S. Department of Transportation, and U.S. Customs.

TONY GARNIER (1989–1948) was an architect and author of *Une Cité Industrielle* (1917), a fundamental treatise in the history of 20th century architecture and city planning.

ALEXANDER GARVIN is Adjunct Professor of Urban Planning and Management at Yale School of Architecture. He is a member of the National Advisory Council of the Trust for Public Land and a Fellow of the Urban Land Institute. In February 2002, Mr. Garvin was appointed as Vice President for Planning, Design and Development to coordinate urban development plans at the Lower Manhattan World Trade Center site and surroundings.

RAYMOND W. GASTIL is Director of the Van Alen Institute, New York City, and a coauthor with Robert Stern of *Pride of Place* (1986) and *Modern Classicism* (1988).

BARUCH GIVONI is Professor Emeritus of Architecture in the Graduate School of Architecture and Planning at UCLA and consultant to World Health Organization, World Meteorological Organization, the Israel Ministry of Housing, and numerous government organizations. He is author of *Man, Climate and Architecture* (1976) and *Climate Considerations in Building and Urban Design* (1994).

SUSAN GOLTSMAN, FSLA, is a founding principal of Moore, Iacofano, Goltsman, specializing in environmental design, planning, and education for children and youth. She currently serves on the Board of the Landscape Architecture Foundation.

WERNER HEGEMANN (1881–1936) was an architect, urbanist, and coauthor with Elbert Peets of *American Vitruvius*, first published in 1922. Hegemann, along with others of his generation, was interested in American developments and in monumental civic design in the classic European urban planning tradition.

DAVID HERZBERG is an historian and associate of the Urban Design Project at the University at Buffalo, State University of New York, and research contributor to *Rethinking The Niagara Frontier* (Shibley 2001).

JOHN W. HILL, FAIA, was founding dean of the School of Architecture, University of Maryland, where he taught for thirty years. His research interests include the morphological study of towns and neighborhoods and the conservation and preservation of rural and small town cultural landscapes.

LEWIS D. HOPKINS, AICP, is professor of urban and regional planning at the University of Illinois at Urbana-Champaign and chair of the Planning Accreditation Board. He was Fulbright Senior Scholar and former editor of the *Journal of Planning Education and Research.*

MICHAEL HOUGH is a landscape architect and founding partner with the firm Hough Stansbury Woodland Naylor Dance and Professor in the Faculty of Environmental Studies at York University, Ontario, Canada. His books include *City Form and Natural Process: Towards a New Urban Vernacular* (1984) and *Out of Place: Restoring Identity to the Regional Landscape* (1990).

STEVEN IZENOUR, FAIA, (1940–2001) was a principal of Venturi, Scott-Brown and Associates. In addition to his professional collaborations with Robert Venturi and Denise Scott-Brown, he was author of *White Towers* (1979).

ALLAN B. JACOBS is an urban designer and author of *Making City Planning Work* (1978), *Looking at Cities* (1985), and *Great Streets* (1993), and coauthor with Elizabeth Macdonald of *The Boulevard Book* (2001).

JANE JACOBS is author of numerous books, including *The Death and Life of Great American Cities* (1961), *Cities and the Wealth of Nations* (1984), and *The Nature of Economics* (2000).

HEATHER KINDAKE-LEVARO, ASLA, is Site Development Director for the Phoenix office of Qk4. She is a Ph.D. student in the Geography Department, Arizona State University.

JAMES R. KLEIN, ASLA, is a principal of Lardner/Klein Landscape Architects, of Alexandria, Virginia. He has completed numerous planning and design projects for urban greenways, bicycle trails and pedestrian safety projects throughout the United States.

RALPH L. KNOWLES is University Professor Emeritus of Architecture at the University of Southern California. He is recipient of the AIA Medal for Research and the ACSA Distinguished Professor Award.

His books include *Energy and Form* (1974) and *Sun Rhythm Form* (1981).

FRED KOETTER, FAIA, is Professor of Architecture and former Dean of the Yale School of Architecture and is architect and urban designer in private practice in Boston, Massachusetts, subject of the monograph *Koetter, Kim and Associates: Place/Time* (1977). He is coauthor with Colin Rowe of *Collage City* (1978).

SPIRO KOSTOF (1936–1991) was Professor of Architectural History at the University of California, Berkeley, California. His publications include *History of Architecture: Setting and Ritual* (1985), *America by Design* (1987) and *City Shaped Urban Patterns and Meanings Through History* (1991).

ROB KRIER is an architect, urban theorist and professor at Technical University of Vienna whose work is representative of new urban development in the European tradition. His books include *Architectural Composition* (1989) and *The Making of a Town*, coauthored with Christopher Kohl (2000). His work is subject of the monograph, Rob Krier Architecture and Urban Design (*AD* Vol. 30, 1993).

WALTER M. KULASH is a registered professional engineer and advocate of livable communities traffic solutions as consultant to dozens of cities. He is a member of the Institute of Transportation Engineers as well as The Congress of the New Urbanism.

KENNETH LABS (1950–1992) served as technics editor of *Progressive Architecture* in the 1980s. His publications include *Climatic Building Design*(1983), coauthored with Donald Watson, and *The Building Foundation Handbook* (1988).

JURG LANG is Professor of Architecture, University of California, Los Angeles, with research interests in urbanism and computer applications. He previously worked in the Athens office of C.A. Doxiadis.

ELIZABETH LARDNER ASLA, AICP, is a principal Lardner/Klein Landscape Architects of Alexandria, Virginia.

Le CORBUSIER (1887–1965), born Charles-Edouard Jeanneret, was among the foremost architects of 20th century early modernism and author of treatises on architecture and planning. *Vers Une Architecture* (1917–1923), published in English as *Towards a New Architecture* (1927), provided the first exposition of the modern movement in architecture by one of its principal prophets.

IAN M. LOCKWOOD is a registered professional engineer, and formerly Traffic Calming Director for the City of West Palm Beach. He served as chair of the Definitions Committee for the Institute of Transportation Engineers Traffic Calming Committee.

DONN LOGAN co-founded ELS/Elbasani & Logan Architects in 1967, an urban design firm responsible for downtown plans and major projects in the United States and internationally. Mr. Logan was Professor of Architecture at the University of California, Berkeley, from 1966 to 1986. In 1999 he inaugurated a new firm under his own name.

GIANNI LONGO is a founding principal of ACP-Visioning & Planning. He has worked with cities to involve citizens in urban planning decisions, including Chattanooga, New Haven, New York, and Washington, DC. His is coauthor with Roberto Brambilla of *For Pedestrians Only: Planning, Design, and Management of Traffic-Free Zones* (1977).

KEVIN LYNCH (1918–1984) was Professor of Urban Planning at Massachusetts Institute of Technology, author of *The Image of the City* (1960), *Site Planning* (1962) and *Managing the Sense of a Region* (1976).

ELIZABETH MACDONALD is an architect and urban designer. She is Assistant Professor of Urban Design at the University of California at Berkeley and is coauthor with Allan B. Jacobs of *The Boulevard Book* (2001).

SIR LESLIE MARTIN (1908–2000) was architect of the Royal Festival Hall (1951), Chair of Architecture at Cambridge University (1956–1972) and recipient of the Royal Gold Medal (1973). His work is documented in *Leslie Martin: Buildings and Ideas 1933–1983*, Cambridge University Press (1983).

MARY C. MEANS is principal of Mary Means & Associates, a community planning firm. She is past Vice President of the National Trust for Historic Preservation, where she created the National Main Street Program. She is founding chair of the National Coalition for Heritage Areas.

CHERYL A. O'NEILL is a principal with Torti Gallas and Partners–CHK Architects, in Silver Spring, Maryland. She won a 2002 national AIA urban design honor award for Pleasant View Gardens, a Hope IV project completed in Baltimore, Maryland.

ELBERT PEETS (1886–1968), an American architect, town planner, and writer, collaborated with Werner Hegemann on design of the town of Kohler, Wisconsin, in 1917. In 1922 they published *The American Vitruvius: An Architect's Handbook of Civic Art.*

CLARENCE ARTHUR PERRY (1872–1944) was a Chicago-trained sociologist and director of Russell Sage Foundation, which had developed Forest Hills Gardens in Queens, New York, begun in 1909, designed by Frederick Law Olmsted, Jr. and Grosvenor Atterbury, and where he lived. His approach made the concept of neighborhoods "axiomatic" in planning. His books included: *High School as a Social Center* (1914), *Neighborhood and Community Planning* (1929) and *Rebuilding of Blighted Areas: A Study of the Neighborhood Unit in Replanning and Plot* (1933).

ELIZABETH PLATER-ZYBERK is principal of Duany Plater-Zyberk & Company (DPZ), recipient of the Thomas Jefferson Memorial Medal of Architecture. She is Dean of Architecture, University of Miami, where she founded the Master of Architecture Program in Suburb and Town Design.

STEEN EILER RASMUSSEN (1898–1990) was a Danish architect, city planner and author of *Towns and Buildings* (1949) and *Experiencing Architecture* (1959). His first major book, *London: The Unique City* (1934, 1982), describes the British capital as the exemplar of what he calls the "scattered city," as opposed to the concentrated and concentric patterns of development that characterize European capitals such as Paris and Vienna.

JOHN S. REYNOLDS is Professor Emeritus of Architecture at the University of Oregon. He is coauthor with Benjamin Stein of *Mechanical and Electrical Equipment for Buildings* (1992).

ALDO ROSSI (1931–1997) was an architect, theoretician and artist and was recipient of the 1990 Pritzker Architecture Prize. He is author of *The Architecture of the City* (published in Italian in 1966 and in English in 1982). Other books include the *Aldo Rossi: Architecture, 1981–1991*, which won the 1992 AIA International Book Award.

COLIN ROWE (1920–1999) was Professor of Architecture at the University of Texas at Austin, at Cambridge University, and at Cornell University. In 1995, he was awarded the gold medal by the Royal Institute of British Architects. His publications include *The Mathematics of the Ideal Villa and Other Essays* (1976) and *Collage City*, co-authored with Fred Koetter (1978).

WITOLD RYBCZYNSKI is the Meyerson Professor of Urbanism at the University of Pennsylvania, and Professor of Real Estate at the Wharton School. He is author of numerous books, including the award-winning biography of Frederick Law Olmsted, *A Clearing in the Distance* (1999).

LYNDA H. SCHNEEKLOTH, ASLA, is a professor in the Department of Architecture, SUNY/Buffalo. She is co-editor of *Ordering Space: Types in Architecture and Design* (1994) with Karen Franck and coauthor of *Placemaking: The Art and Practice of Building Commun ities* (1995) with Robert Shibley.

LORING LaB. SCHWARZ has served as Director of New England Greenways for the Conservation Fund, a nonprofit land protection organization in Arlington, Virginia, and is currently deputy director of the Massachusetts Chapter of The Nature Conservancy.

DENISE SCOTT BROWN is an architect, urban designer, theoretician and author. She is a principal of Venturi, Scott-Brown & Associates, author of an AD Profile *Urban Concepts* (1990) and coauthor with Robert Venturi of *A View from Campidolio, Selected Essays* (1953–84).

CAMILLO SITTE (1843–1903) was an architect and educator in Vienna and gained international renown from the treatise, *City Planning According to Artistic Principles* (1889). Advocating a humanistic approach to planning and interested in vernacular objects, buildings and towns responsive to daily life, he devised a hierarchy of vernacular elements of anonymous, timeless development considered as the highest art.

ALISON SMITHSON (1928–1993) and **PETER SMITHSON** (1923–) were among the foremost theoreticians and practitioners of urban planning and design in the postwar period and were founders of Team 10.

ANN WHISTON SPIRN is Professor of Landscape Architecture and Planning, Massachusetts Institute of Technology and author of *The Language of Landscape* (1998) and *The Granite Garden: Urban Nature and Human Design* (1984).

PAUL D. SPREIREGEN, FAIA, is a Washington-based architect and planner. Author of numerous books and articles, he has been a commentator on design for National Public Radio and has conducted many design competitions throughout the United States.

CLARENCE STEIN (1882–1975) was a visionary architect and planner. Stein and Henry Wright planned Sunnyside Gardens in 1924 and Radburn, begun in 1928. Only partly completed because of the depression, Radburn was nonetheless influential as a model of planned communities and higher density housing in the United States. He is author of *Toward New Towns in America* (1951) and subject of a recent monograph *The Writings of Clarence S. Stein, Architect of the Planned Community* edited by Kermit Carly le Parsons (1998).

EDWARD STEINFELD, Ph.D. is Director of the Center for Inclusive Design and Environmental Access (IDEA) at State University of New York, Buffalo. He is a registered architect, design researcher, professor of architecture and adjunct professor of occupational therapy. He is a consultant to federal and state agencies, building owners and attorneys on issues of universal design and environmental access.

ROBERT A. M. STERN, FAIA, is Dean of the Yale School of Architecture and principal of Robert A. M. Stern Architects, recipient of numerous national design awards, and author of *Pride of Place* (1986), *New York 1900* (1983) and *New York 1930* (1987).

ROBERT VENTURI, FAIA, is principal of Venturi, Scott-Brown & Associates, author of *Complexity and Contradiction in Architecture* (1966), and architect for numerous award buildings. Among other recognitions, he is recipient of the Pritzker Architecture Prize, 1991.

POLLY WELCH, architect and formerly professor at the University of Oregon, directs design research and public policy development for the Massachusetts' Division of Capital Asset Management. She is editor of *Strategies for Teaching Universal Design* (1995) and other publications on universal design, building assessment, and urban design.

RICHARD WENER is an environmental psychologist and head of the Department of Humanities and Social Services at Polytechnic University in Brooklyn, New York. He has undertaken extensive research on the effects of built environments on individuals and communities.

WILLIAM H. WHYTE (1917–1999) was an author and urban research whose books include *The Organization Man* (1956), *The Last Landscape* (1970), *The Social Life of Urban Spaces* (1984), and *City: Rediscovering the Center* (1988). He is subject of a recent monograph, *The Essential William H. Whyte*, edited by Albert LaFrage (2000).

ROBERT YARO is President of Regional Plan Association in New York and Practice Professor in City and Regional Planning at the University of Pennsylvania.

PAUL ZUCKER (1889–1971). Trained as an architect in Berlin where he practiced before fleeing Nazi Germany in 1937, Paul Zucker became best known for his work as a theorist and historian. His publications include *Fascination of Decay* (1968) and *Town and Square from the Agora to the Village Green* (1959). Zucker taught as well as practiced in Berlin, and later taught at both Cooper Union and the New School in New York.

INTRODUCTION

Donald Watson, FAIA

Urban design and city building are surely among the most auspicious endeavors of this or any age, giving rise to a vision of life, art, artifact and culture that outlives its authors. It is the gift of its designers and makers to the future.

Increasingly apparent in the past decades, the global dimensions of urbanization and population growth have reached an unprecedented threshold. Heretofore, architecture, landscape design and urban planning were undertaken within the natural conditions of site, landscape and region. Not all cities were benign, the industrial city being often merely an efficient machine that consumed resources and lives. While many urban conglomerations are just that, those that are experiencing their maturation as post-industrial cities offer new opportunities for urban design, art and place making that reassert public culture and timeless civic values.

Vitruvius instructed the architect in building well, to take cues from opportunities of nature, site, and climate, expressing order and balance within the rules and proportions of the classic temple. At the beginning of the 21st century, the growth of cities and infrastructure approach an upper limit of size and reach—the way we build on the earth is changing the nature of the global climate. The urban designer is challenged to design not only for the health of citizens, but also for the well being of our global environment. The earth is now that temple.

Urban design is essentially an ethical endeavor, inspired by the vision of public art and architecture and reified by the science of construction. The "client" of urban architecture and design is not only the private or public sponsor, or the citizen in the street, but the biological system of all of life. The Vitruvian precepts of building well, *venustas, firmitas, and utilitas*—delight, firmness and commodity—are expressed in the Seattle Comprehensive Plan as principles of sustainable design: economic opportunity, social equity and environmental responsibility. These principles are evoked and enabled by the civic art of urban design.

Note on the composition of this volume

The idea for an inaugural volume of *Time-Saver Standards for Urban Design* began at an international conference on design education and practice of global sustainability, conveyed in Kenai, Alaska, in the late 1990s. Conferees affirmed that cities that are well planned, built and adapted over many generations provide the most convincing and enduring evidence of sustainable design. Without well-designed and vital cities, our civil society and life cannot be sustained. Cities deteriorating at their centers, or sprawling beyond recognizable boundaries, or exploding into unmanageable "mega-cities" are all evident despoilers of the world's natural and cultural resources.

At the Kenai conference, the question was asked, "Just what is the knowledge base and shared values, principles, and practices that define the discipline of urban design?"

This compendium is one answer to that question. An editorial board of advisors worked to define the key archival articles that are essential reading and reference for a student or practitioner of urban design. After some debate, the late 19th century writing of Camillo Sitte was selected as the earliest key reference for 20th century city design. As represented in the "Classic Texts of Urban Design" that begins Part 2 of this volume, Sitte promoted a choreographic perspective of urban composition that has been continuously elaborated upon by successive generations of urban designers.

An emphasis upon visual documentation is evident in this collection of articles. That is to say, urban design is defined here by its physical presence, its scale, dimension, and the perceived reality of plans, drawings and making of urban places. In addition to economic, social, and engineering data through which urban places are given quantitative measure, the defining quality of urban design is in the work at hand, revealed by visual documentation and experience.

A second characteristic of this compendium will be evident to the reader from the Table of Contents and the articles it represents. While a book of "Standards" is often taken to mean a compilation of details, dimensions and other technical data—one of our advisors said, "Wouldn't it be good to be able to look up in one book how wide a sidewalk should be and how high its curbs"—this notion of "urban design standards" may represent a misplaced search for certainty where none should exist. But what we do need and what these articles represent is the common ground of urban design, the shared principles and practices that are the basis of scholarship, research and practice. Quantitative standards—sidewalk widths and curb heights—are substantiated here in research on traffic, pedestrian ergonomics and behavior patterns. The qualitative standards of urban design are conveyed by debate and discussion, by historical scholarship, and by the positing of frameworks and theories of design that are then subject to the severe test of execution and built evidence.

In succeeding generations of work documented in this volume are common and continuously developing themes. The concept of "neighborhood" promoted in the 1929 writings of Charles Arthur Perry reappears in the contemporary *Lexicon of the New Urbanism* and in the environment/behavioral studies of Claire Cooper Marcus and Susan Goltsman. The discussion of hygiene, solar orientation and ventilation in Tony Garnier's *Une Cité Industrielle* is updated in contemporary environmentalism and bioclimatic design. Sitte's eye-level perspective of the cityscape is elaborated by Kevin Lynch and the sketchbook method by Gordon Cullen and Robert Dvorak, all of whom are represented in this volume.

Table A lists in order of frequency of citation the key words indexed in this volume, thus representing a cross-section of terms used by writers and practitioners of urban design in the past century. The tabulation of Elements of Urban Design reveals what one might expect as defining physical elements—the "nouns" that make up a city—or more accurately put, the city as conceived by urban designers, with the topic Roofscapes providing a reminder of the often forgotten "fifth façade" of a city. Some terms, such as Neighborhood and Block, have high frequency of use across the spectrum of 20th century writings. The biases of the writers (and of the editors who selected the articles) is revealed by frequent citation of terms such as Urban Square, Gardens, Plazas, Arcade, Atrium, Parks and Boulevard, which define the uniquely "urban" in "urban design."

Similarly, Design Issues and Principles—or the "verbs" of urban design—are represented in the foremost-cited term, Pedestrian Design, followed by Traffic, Trips. The most bothersome experience of cities might be traffic congestion, but for urban designers, places for the pedestrian, including streets for people and green infrastructure, are the antidote to the car-dominated city and thus deserve the detail of discussion found in this volume. Other frequently cited principles—Climate, Bioclimatic Design, Historic Preservation and Adaptive Reuse, Universal Design, and Citizen Involvement—reveal editorial bias but also, in terms of the literature and examples, the central and defining themes of urban design.

"Frequent citations" lists authors, theoreticians and practitioners most frequently referenced across the broad spectrum of articles in this book. Le Corbusier heads the list, although in this instance, his contribution to urban design is as often cited for criticism as for praise, perhaps the burden of any one person who dominates a generational discourse. Frederick Law Olmstead on the other hand is consistently cited for admirable and long-lasting precedents. The mid-20th century contributions of Kevin Lynch, Lewis Mumford, and William H. Whyte are highly cited for their contributions. In these citations, the principal contributions were made through authorship of influential books, rather than built work. Both Lynch and Whyte had intimate experience with built work through their research and consultative practices.

Robert Venturi/Denise Scott Brown and Andres Duany/Elizabeth Plater Zyberk are among the contemporary theoreticians and practitioners most frequently cited by (other) authors in this compilation, both cases representing work developed both in academic programs and in urban design practice. In these examples, the "seminal" role of academic-based research seminars and/or design studios in schools of architecture is significant, also evident in articles of Ralph Knowles on solar envelope, Vikram Bhatt and Witold Rybcynski on informal human settlements, John W. Hill on traditional Maryland towns, Colin Rowe/Fred Koetter on *Collage City*, and in the university-based Community Design Centers reported by Sheri Blake.

The list of cities most frequently cited as exemplars will not surprise, as much as reinforce the point that urban design principles and practices build upon the cultural inheritance of great cities, with Rome and European Renaissance cities mentioned most frequently. San Francisco and Portland, Oregon, are most often cited as recent examples of best practices in urban design, exemplifying the effective balance of civic leadership, multidisciplinary design, public art and citizen involvement.

These common terms and references show that the knowledge and value basis of urban design is conveyed by common and contrasting examples, best seen as a history of ideas in evolution. The editors hope that through this collection, the interested reader, student and practitioner will have ready access to the best references of urban design, so that these might inform and in turn be further refined in ongoing studies and design practices. ■

TABLE A: TOPIC CONTENT OF THIS BOOK listed by number of Index citations

ELEMENTS OF URBAN DESIGN

[56] Urban
[35] Square
  [5] St. Marks, Venice
[31] Land use, Landscape, Farming, Farmland
[31] Residential
[29] Street
[28] Architecture, building
[20] Gardens
[18] Plaza
[17] Roof, rooftops
[16] Arcade, Atrium
[16] Park
[14] Neighborhood
[14] Open space
[14] Suburb
[13] Path, Pathways
[12] Bikes, Bikeways
[10] Block
[10] Play, Playgrounds, Play spaces
[10] Rail, Railway, Railway station
[9] Automobile, Road, Roadway
[9] Shops, Shopping
[8] Boulevard
[8] Courtyard
[8] Market, Marketplace
[7] Public art
[6] Airports

DESIGN ISSUES & PRINCIPLES

[25] Pedestrian design
[22] Traffic, Trips
[20] Climate, Climatology, Bioclimatic design
[19] Density
[17] Environment
[15] Historic preservation and adaptive reuse
[14] Circulation
[13] Universal design
[12] Site planning
[11] Citizen involvement
[11] Population
[11] Regulation
[11] Visual character
[10] Region, Regional planning
[9] Safety
[8] Diversity
[8] Recreation
[7] New Urbanism

Analysis Method

[10] Visual
[3] Multidisciplinary
[2] Ian McHarg overlay
[2] Robert Yaro bird's-eye views
[2] William H. Whyte observational

DESIGN AND TECHNICAL DETAILS

[28] Graphics and interpretation
[23] Water
[23] Wind, Ventilation
[14] Parking, Parking lot, Parking spaces
[12] Plan, Planning
[12] Lighting
[9] Paving, Pavers
[9] Shading
[4] Air quality

CITATIONS

[17] Le Corbusier
[10] Olmstead, Frederick Law
[10] Wright, Frank Lloyd
[9] Lynch, Kevin
[7] Mumford, Lewis
[7] Whyte, William H.
[6] Duany, Andres and Elizabeth Plater Zyberk
[4] Venturi, Robert and Denise Scott Brown

CITIES

[13] Rome
[10] Renaissance cities
[10] New York
[9] Paris
[9] London
[6] Boston
[6] San Francisco
[6] Venice
[5] Portland, OR

1 • CONTEXT OF URBANIZATION

Werner Fornos, Desikan Thirunarayanapuram and Harold N. Burdett

Summary
While cities occupy only two percent of the earth's landmass, they contain 50 percent of world population, consume 75 percent of the world's resources and produce 75 percent of its waste. The overriding priority of 21st century planning and urban design will be to support the global agenda for environmental responsibility, social equity, and economic opportunity for all peoples. That challenge, in turn, is inexorably linked to the sustainability of the world's cities and urban areas.

Key words
demographics, developing countries, environmental resources, health, human settlements, life quality, megacities, migration, population statistics, poverty, sprawl, sustainability

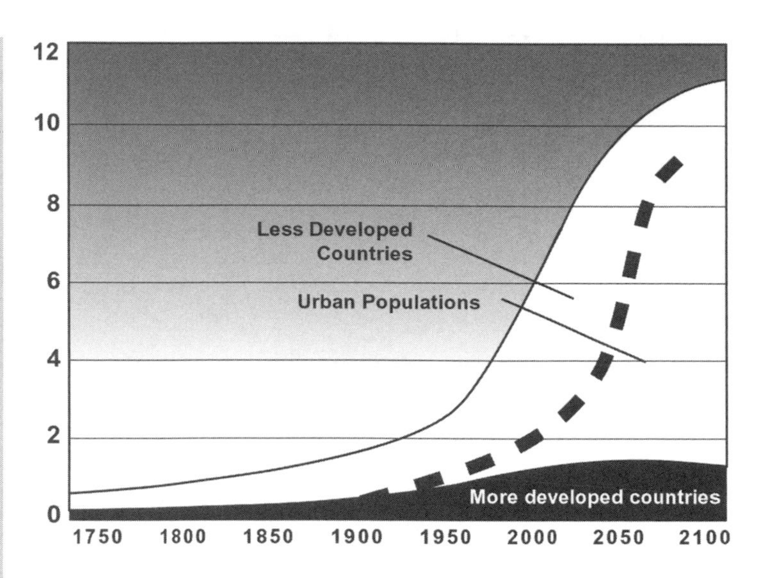

World population growth (in billions) Data: Population Reference Bureau.

Population and the urban future

The global scale and impact of urban settlements and cities will determine the course of the emergent 21st century. Cities are no longer isolated and responding only to local and regional influences. Very large cities—megacities—are exploding in size and growth, for the most part in newly developing regions of the globe. The impact of this growth, however, is not isolated. The urban explosion is international in its origins and in its influences, affecting every continent and country, as a result of globalization, multinational political affiliations, telecommunications, transport and commerce (Fig. 1).Urban design of human settlements and cities—what they are as a result of explosive growth and what they could be if guided by urban design instead of happenstance—is thus the foremost agenda for planners and design professionals throughout the world. Cities have shaped mankind. They now shape the quality of life for the entire biosphere of earth. But today, city building is subject to the unplanned and often chaotic flow of humankind undergoing unprecedented growth, migration, and an attendant diminution of natural resources, beyond the limits of sustainable life. Left to their own, these trends are aggravating if not destroying the hoped-for urban promise of employment, health, safety, equity and opportunity. The future of cities with a sustainable quality of life and values of civil society can result only by design.

From the beginning of history, civilization has been founded around great cities. Ancient Sumerians, Egyptians, Mayans, Chinese and inhabitants of the Indus Valley—in what is now Pakistan and parts of India—lived in urban settlements with amenities such as stone and brick construction and urban water and drainage systems. These traditional settlements balanced urban with rural living, never far removed from the natural and cultivated resources that sustained their populations.

In the eighteenth and nineteenth centuries, Europe and North America became increasingly urbanized and city and country more separated into disparate zones, impelled by the Industrial Revolution. London, Paris and New York were historic population magnets with the promises of wealth and a good life. Early city building benefited from the best of wealth and design with substantial construction of buildings and places that still endure as the world's cultural capitals. During these same centuries, cities in non-industrial countries remained small, still characterized by rural, herding and agricultural economies.

In 1900, only 160 million people—one-tenth of the world's population—lived in cities. In 2000, half of the population of the world—more than 3 billion people—are urban dwellers. In the last hundred years, world population has grown from about 1.6 billion to 6 billion. This near four-fold increase was accompanied by an alarming twenty-fold increase in the world's urban population.

In 1900, all of the world's most populous cities were in North America and Europe. By 2000, Tokyo, New York and Los Angeles were the only industrialized world cities left on the "top ten" list. By the year 2020, Dhaka, Karachi, and Jakarta will surpass New York and Los Angeles. By 2050, an estimated two-thirds of the world's population will live

Credits: This article is based upon a report by the authors in *The 21st Century Papers*, 1999, by kind permission of The Population Institute, Washington, DC. Additional sources and statistics are cited in tables and references.

Figs.1a–1d. Comparison of global urban infrastructure.
Source: Rimmer, Peter J. "Transport and telecommunications among world cities." (Lo, 1998).

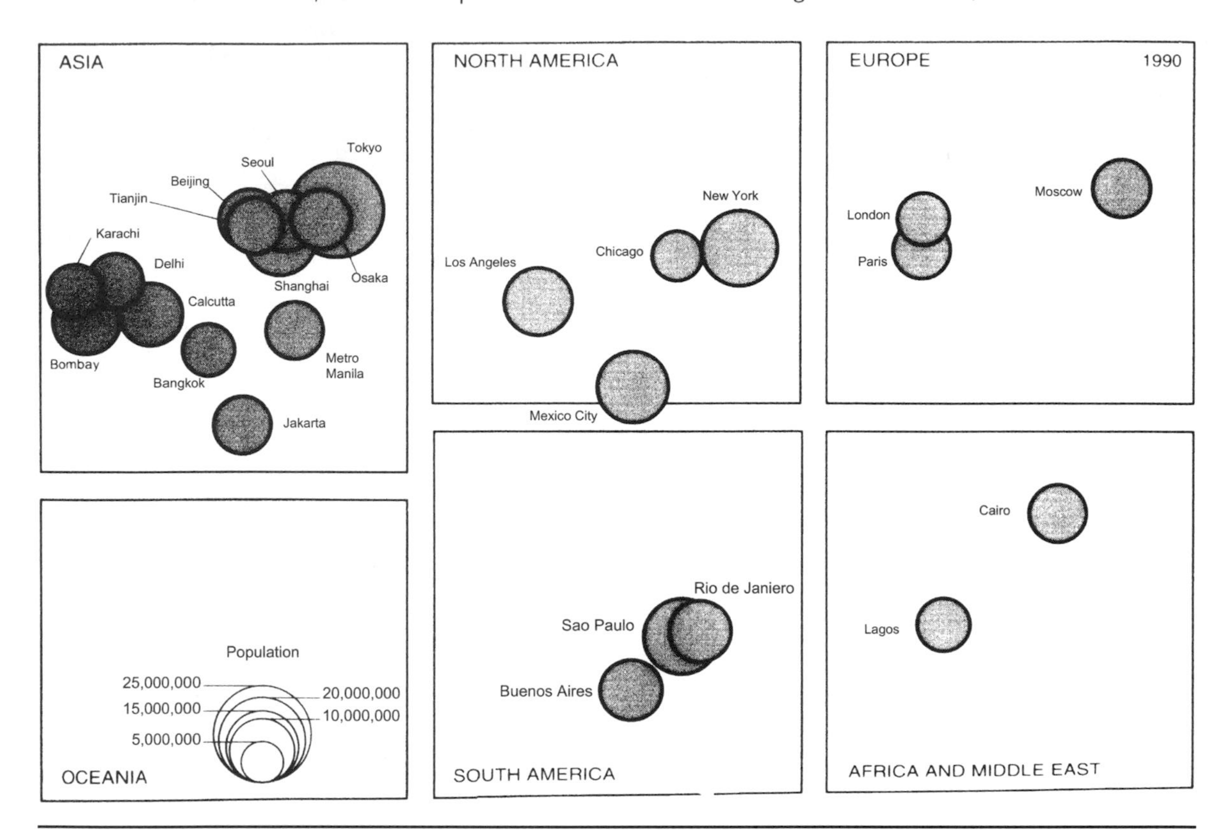

Fig. 1a. "Top 25" largest urban agglomerations, 1990.

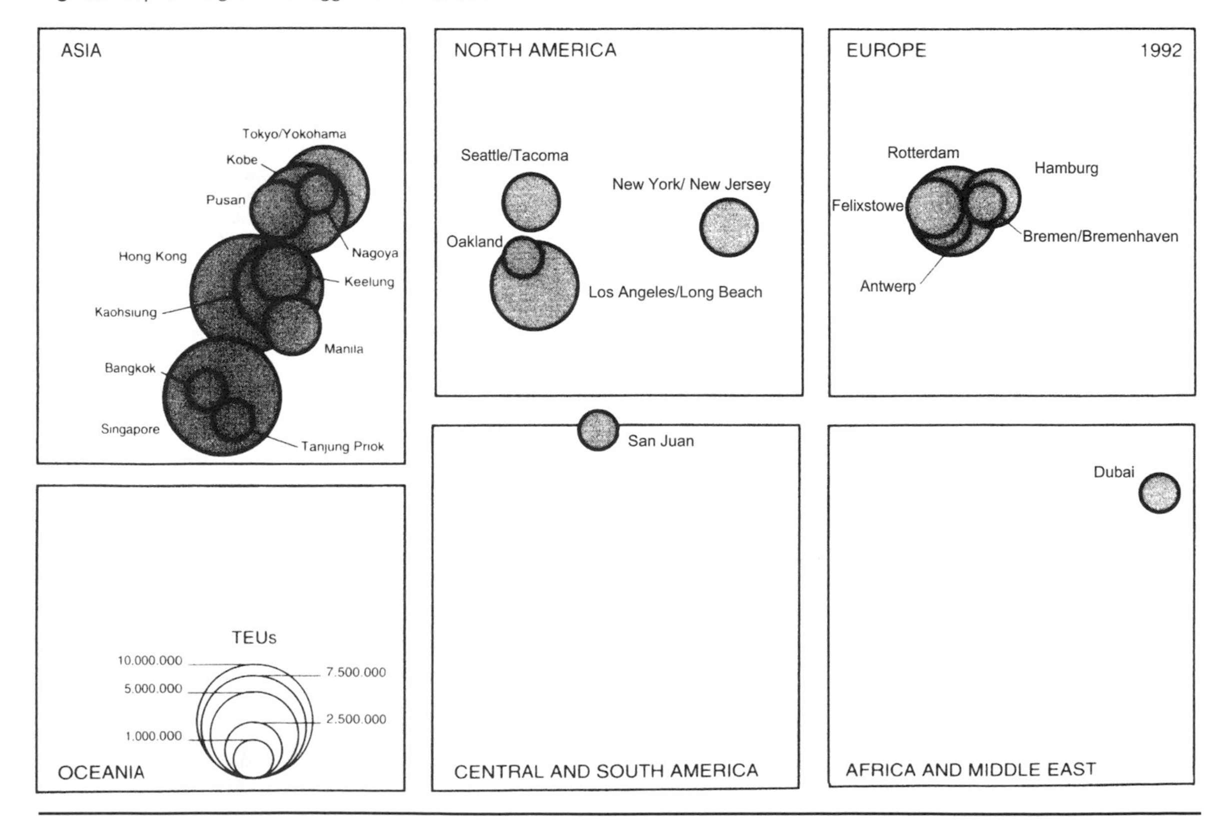

Fig. 1b. "Top 25" ports in the world container traffic league, 1992.

in urban areas. The most rapid urban growth over the next two decades is expected to occur in cities with populations from 250,000 to one million. The combined annual growth of such cities currently totals more than 28 million per year. Fifteen years from now, it will increase to 31 million per year.

Megacities and geopolitical consequences

The last half-century has witnessed the emergence of megacities—enormous metropolises with more than 8 million people each (see Tables 1–4). Most will be in newly developing economies of the developing world, making the urban explosion almost exclusively a developing world phenomenon. The impacts, however, will be evident internationally, both in terms of worldwide economic infrastructure but also global environmental quality and health and human culture influences that have no boundaries. The cities of the 21st century are by definition global cities.

By the year 2015, megacities will have a total of 400 million people, nearly 6 percent of the world's population. In the same year, there can expected to be thirty-eight cities with 5–10 million inhabitants—more than 4 percent of the world's population. While growth of cities in the developed nations has stabilized somewhat, most growth is occurring in the cities of poor, less developed nations that are ill equipped to accommodate it. This serves to aggravate the perception of a geopolitical and economic divide between the world's rich and the world's poor. Often as a result, the challenges of designing world cities are poorly understood by planners and designers, whose knowledge of urban design is called upon to improve urban conditions across the globe.

Migration and destabilization of the labor force

Population growth drives the increased pace of global urbanization and is accelerated by unplanned and unpredicted migration. In developing regions throughout the globe, there is an historic migration out of rural areas by workers and families attracted to cities by the promise of a better livelihood. In other regions subject to natural disasters and civil strife, migration is forced as people are displaced from their traditional homelands. Such migration offsets the best population stabilization efforts of cities, imposing added pressure on their infrastructures, and contributing to problems of health and sanitation, unemployment, crime and the cycle of poverty. In such cases, urbanization is part of the problem.

Of the global labor force of about 2.8 billion people, at least 120 million are unemployed and 700 million are classified as "underemployed," working long hours but not receiving enough money to cover even their basic needs. In many developing world countries and in Eastern Europe, both petty and violent crimes have increased. In Asia, over the period between 1975 and 1990, crime declined at a national level, but at the same time in cities of more than 100,000 inhabitants there was a considerable increase in crimes against property, organized crime and drug trafficking.

Poverty

More than half of the urban inhabitants of Asia, Africa and Latin America live in poverty. More than 3 billion people—half the world's total population—subsist on less than $2 a day. The number and proportion of those living in extreme poverty are rising. By 2025, six of every ten children in the developing world are expected to live in

Table 1. Population size of urban agglomerations with 8 million or more in 2000.

Agglomeration	Country	1950 (in millions)	1960	1970	1980	1990	2000
Bangalore	India	0.8	1.2	1.6	2.8	5.0	8.2
Bangkok	Thailand	1.4	2.2	3.1	4.7	7.2	10.3
Beijing	China	3.9	6.3	8.1	9.0	10.8	14.0
Bombay	India	2.9	4.1	5.8	8.1	11.2	15.4
Buenos Aires	Argentina	5.0	6.8	8.4	9.9	11.5	12.9
Cairo	Egypt	2.4	3.7	5.3	6.9	9.0	11.8
Calcutta	India	4.4	5.5	6.9	9.0	11.8	15.7
Dacca	Bangladesh	0.4	0.6	1.5	3.3	6.6	12.2
Delhi	India	1.4	2.3	3.5	5.6	8.8	13.2
Istanbul	Turkey	1.1	1.7	2.8	4.4	6.7	9.5
Jakarta	Indonesia	2.0	2.8	3.9	6.0	9.3	13.7
Karachi	Pakistan	1.0	1.8	3.1	4.9	7.7	11.7
Lagos	Nigeria	0.3	0.8	2.0	4.4	7.7	12.9
Lima	Peru	1.0	1.7	2.9	4.4	6.2	8.2
Los Angeles	USA	4.0	6.5	8.4	9.5	11.9	13.9
Manila	Philippines	1.5	2.3	3.5	6.0	8.5	11.8
Mexico City	Mexico	3.1	5.4	9.4	14.5	20.2	25.6
Moscow	Russia	4.8	6.3	7.1	8.2	8.8	9.0
New York	USA	12.3	14.2	16.2	15.6	16.2	16.8
Osaka	Japan	3.8	5.7	7.6	8.3	8.5	8.6
Paris	France	5.4	7.2	8.3	8.5	8.5	8.6
Rio de Janeiro	Brazil	2.9	4.9	7.0	8.8	10.7	12.5
São Paulo	Brazil	2.4	4.7	8.1	12.1	17.4	22.1
Seoul	Korea, Republic of	1.0	2.4	5.3	8.3	11.0	12.7
Shanghai	China	5.3	8.8	11.2	11.7	13.4	17.0
Teheran	Iran (Islamic Rep. of)	1.0	1.9	3.3	5.1	6.8	8.5
Tianjin	China	2.4	3.6	5.2	7.3	9.4	12.7
Tokyo	Japan	6.7	10.7	14.9	16.9	18.1	19.0

Notes:
Bangkok refers to Bangkok-Thonburi.
Cairo refers to Cairo-Giza-Imbâba.
Lima refers to Lima-Callao.
Los Angeles refers to Los Angeles–Long Beach.
Manila refers to Metro Manila.
New York refers to New York–North-eastern New Jersey.
Osaka refers to Osaka-Kobe.
Tokyo refers to Tokyo-Yokohama.
The population of Greater London exceeded 8 million in 1950 (8.7 million) and 1960 (9.1 million), but has been under 8 million since 1980.

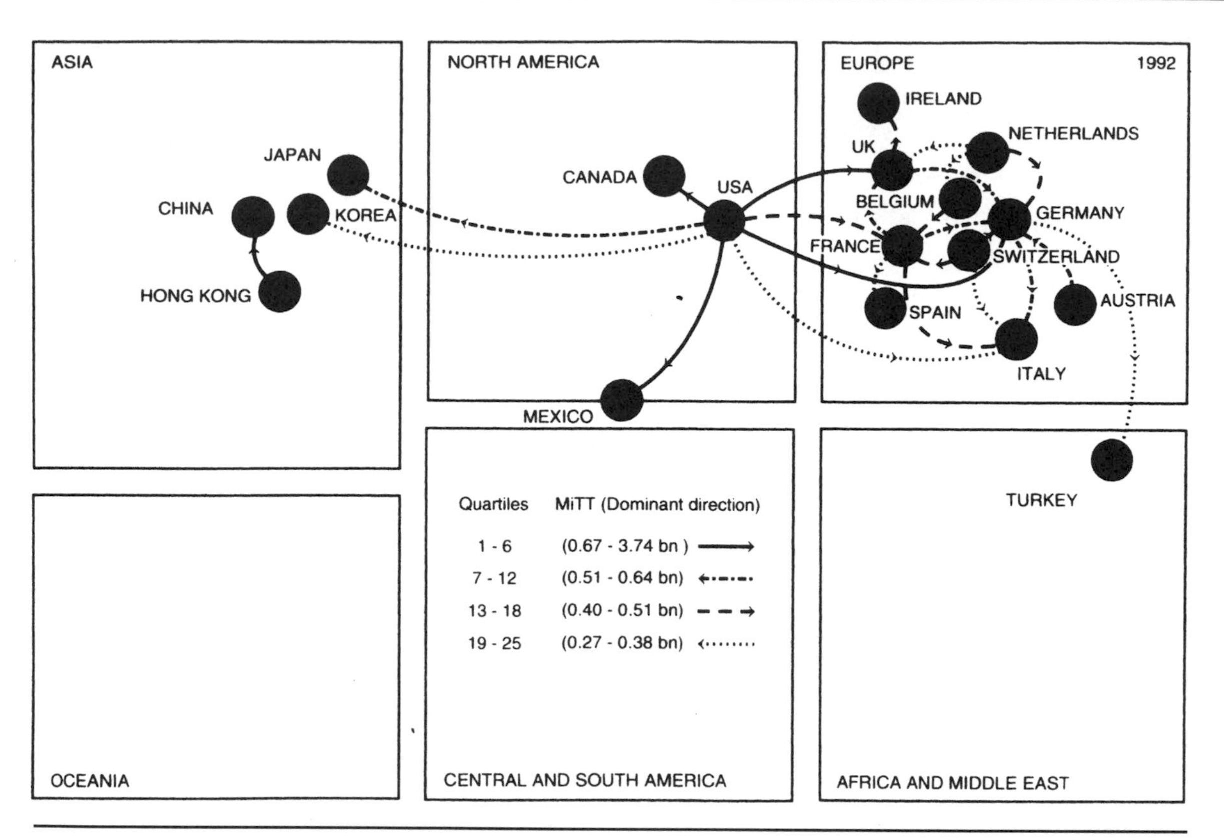

Fig. 1c. "Top 25" international routes with the largest volume of telecommunications traffic, 1992

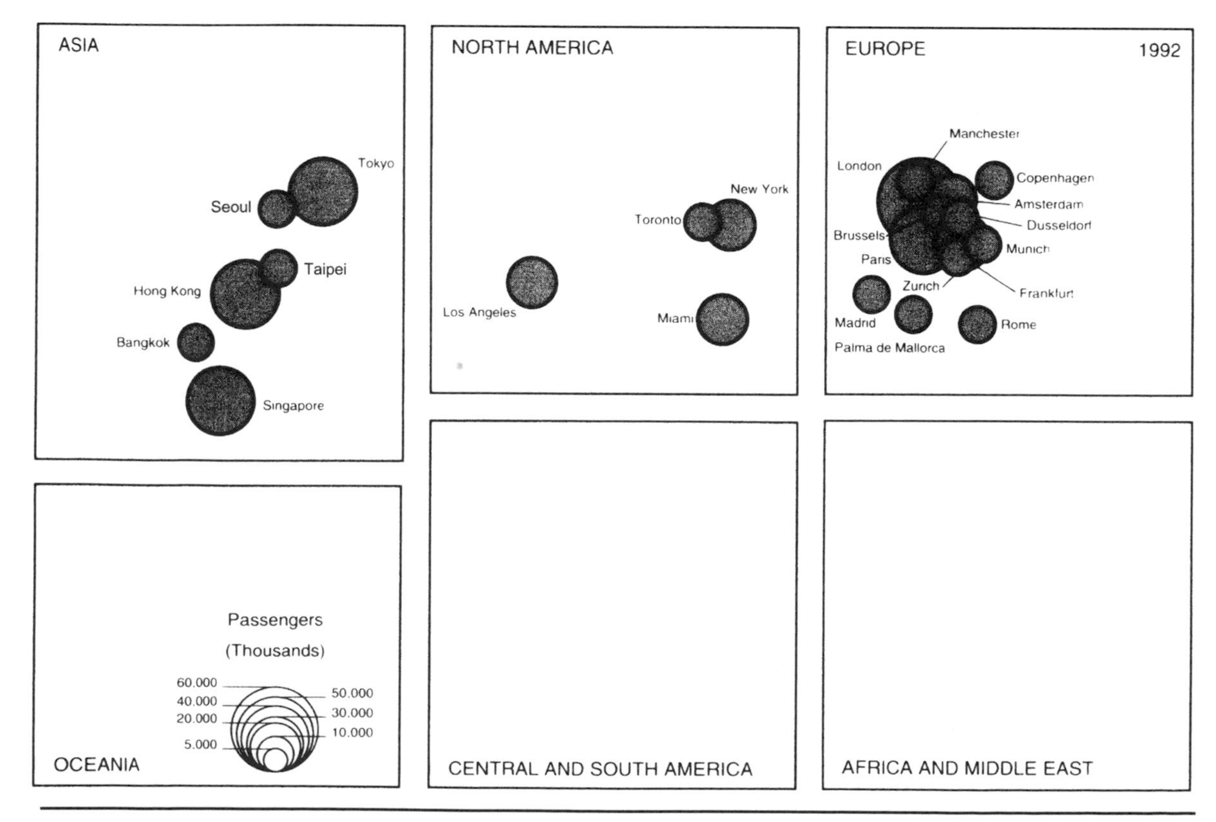

Fig. 1d. "Top 25" airports with highest international traffic, 1992.

cities and more than half will be poor. Even in more developed countries, poverty is concentrated in cities because, despite higher median incomes, more than 75 percent of poor live in urban areas.

In 1996, 28 percent of the urban population in less developed countries lived in poverty, including 41 percent in sub-Saharan Africa alone. The United Nations predicts that by 2025, the majority of the world's urban population will live in poverty. In less developed countries, there are conceptual and measurement problems in assessing urban poverty. But even conservative estimates show that urban poverty in the less developed world is high and growing rapidly.

The disparity of wealth aggravates the inequity, creating division in societies where civil society requires justice and fairness. The wealthiest 20 percent of world population benefit from close to 85 percent of combined global national product, compared to the poorest 20 percent who are without the benefits of global productivity almost entirely (Fig. 2).

Sanitation and health

Thousands of urban inhabitants in developing countries cannot afford the high costs of housing in the cities. The overstretched finances of city governments cannot subsidize housing for its millions of poor inhabitants. As a result, most of these people end up living in slums and shantytowns. It is estimated that 25 to 30 percent of the urban population lives in poor shantytowns, squatter settlements, or on the streets.

Squatter settlements lack running water and sanitation facilities. The World Bank estimates that in 1994, roughly 450 million urban dweller—or 25 percent of the less developed world's 1994 population—did not have access to the simplest latrines.

At least 220 million people in cities of the developing world lack clean drinking water, 600 million do not have adequate shelter, and 1.1 billion are exposed to elevated and unhealthy levels of air pollution. Such unsanitary living conditions have made city slums breeding grounds for various diseases. Waterborne diseases such as diarrhea, typhoid and gastroenteritis are rampant. Each year, 5 million children, mostly in cities, die from waterborne diarrheal diseases, due to a lack of proper sanitation and clean water. The pressures of urban life also lead to disproportionately high levels of sexually transmittal diseases, such as HIV/AIDS, in urban areas.

Good access to clean water has a proven impact in reducing waterborne diseases. The high child mortality rates in Africa, 12.6 percent for girls and 15.3 percent for boys, and low levels of access to water in African cities, where less than 50 percent of households are connected to water, shows the correlation between access to water and health.

Consumption and waste

Megacities are partly the result of an emergent international economic infrastructure, including the global trend to outsource jobs to developing countries where there is availability of skilled but cheaper labor. Cities cause extreme pressures on the natural resources of the rural areas and damage to their environments. Cities have very high population densities, with the result that city dwellers consume several times more food, energy and other resources than do the rural people.

Table 2. List of urban agglomerations of 8 million or more persons by development region: 1950, 1970, 1990, and 2000.

More developed regions

1950	1970	1990	2000
New York	New York	New York	New York
London	London	Tokyo	Tokyo
	Tokyo	Los Angeles	Los Angeles
	Los Angeles	Paris	Paris
	Paris	Moscow	Moscow
		Osaka	Osaka

Less developed regions

1950	1970	1990	2000
None	Mexico City	Mexico City	Mexico City
	São Paulo	São Paulo	São Paulo
	Shanghai	Shanghai	Shanghai
	Beijing	Beijing	Beijing
	Buenos Aires	Buenos Aires	Buenos Aires
		Calcutta	Calcutta
		Bombay	Bombay
		Jakarta	Jakarta
		Delhi	Delhi
		Tianjin	Tianjin
		Seoul	Seoul
		Rio de Janeiro	Rio de Janeiro
		Cairo	Cairo
		Manila	Manila
			Lagos
			Dacca
			Karachi
			Bangkok
			Istanbul
			Teheran
			Bangalore
			Lima

Rank of the megalopolises

Between 1985 and 1990, Mexico City surpassed Tokyo-Yokohama in population and became the largest megalopolis in the world. The United Nations estimates that Mexico City had 20.2 million inhabitants in 1990, exceeding the populations of Tokyo (18.1 million), Sao Paulo (17.4 million), and New York (16.2 million).

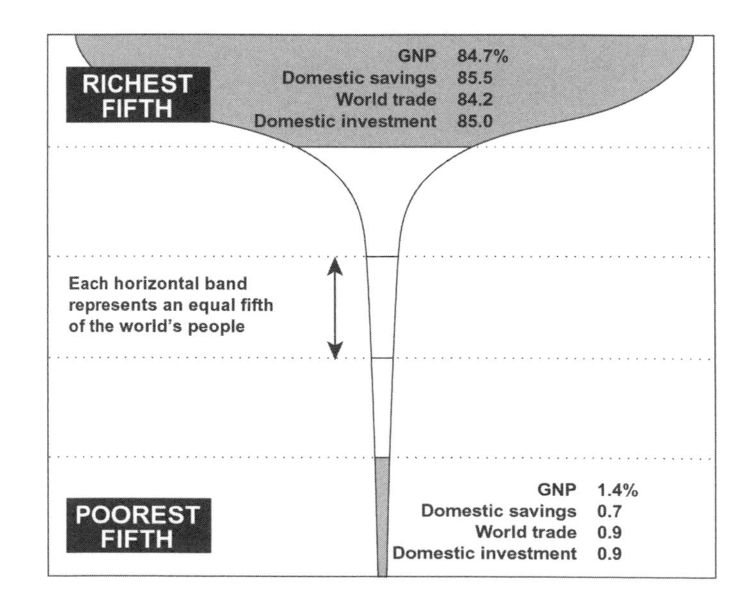

Fig. 2. Disparity of global wealth in terms of GNP and access to capital. Data: United Nations Development Program. 1991.

Table 3. World's ten largest urban agglomerations, ranked by population (in millions), 1950–2000.

1950		1960	
Rank, agglomeration	Population	Rank, agglomeration	Population
1. New York	12.3	1. New York	14.2
2. London	8.7	2. Tokyo	10.7
3. Tokyo	6.7	3. London	9.1
4. Paris	5.4	4. Shanghai	8.8
5. Shanghai	5.3	5. Paris	7.2
6. Buenos Aires	5.0	6. Buenos Aires	6.8
7. Chicago	4.9	7. Los Angeles	6.5
8. Moscow	4.8	8. Moscow	6.3
9. Calcutta	4.4	9. Beijing	6.3
10. Los Angeles	4.0	10. Chicago	6.0
1970		1980	
1. New York	16.2	1. Tokyo	16.9
2. Tokyo	14.9	2. New York	15.6
3. Shanghai	11.2	3. Mexico City	14.5
4. Mexico City	9.4	4. São Paulo	12.1
5. London	8.6	5. Shanghai	11.7
6. Buenos Aires	8.4	6. Buenos Aires	9.9
7. Los Angeles	8.4	7. Los Angeles	9.5
8. Paris	8.3	8. Calcutta	9.0
9. Beijing	8.1	9. Beijing	9.0
10. São Paulo	8.1	10. Rio de Janeiro	8.8
1990		2000	
1. Mexico City	20.2	1. Mexico City	25.6
2. Tokyo	18.1	2. São Paulo	22.1
3. São Paulo	17.4	3. Tokyo	19.0
4. New York	16.2	4. Shanghai	17.0
5. Shanghai	13.4	5. New York	16.8
6. Los Angeles	11.9	6. Calcutta	15.7
7. Calcutta	11.8	7. Bombay	15.4
8. Buenos Aires	11.5	8. Beijing	14.0
9. Bombay	11.2	9. Los Angeles	13.9
10. Seoul	11.0	10. Jakarta	13.7

Water is a precious resource that is often imported and consumed disproportionately by the world's cities.

Cities also generate disproportionate high levels of waste. Inadequate collection and unmanaged disposal present a number of problems for human health and productivity. Uncollected refuse dumped in public areas or in waterways contribute to the spread of disease. Disposing of waste has raised questions of space and logistics, besides health concerns. Even when municipal budgets are adequate for waste collection, safe disposal of collected waste often remains a problem in big cities. Dumping and uncontrolled landfills are sometimes the main disposal methods. Sanitary landfills are the norm in only a relative handful of cites.

Even prosperous cities in industrialized nations are not immune to the problems of waste disposal. Their problems are often even greater than those in developing nations because rich cities generate one hundred times more garbage per person. New York City, for instance, has been fast running out of landfill space. The vast Fresh Kills landfill in Staten Island that rises 176 feet (53.5 m) high—taller than the Statue of Liberty minus its pedestal—is proving to be inadequate. New York's plan to export its waste in barges to Virginia has faced opposition from environmental activists in that state.

Along with consumption and waste, is the despoliation of irreplaceable land, forest and other "natural capital" resources when reducing demand, reuse and recycling strategies could be more effective. Annual reports of the Worldwatch Institute (*e.g.*, Brown, 1999, and Worldwatch Institute, 2001) document the loss in the earth's natural resource base, much of it to make room for and support urban growth (Tables 5 and 6). Canadian ecologists, Rees and Wackernagel (1996) describe and calculate the extent of cities' environmental impact as the "ecological footprint" based on resource consumption levels. By this estimate, London's "footprint" extends to around 125 times its surface area, while Britain's footprint extends to eight times its actual surface area. With 12 percent of Britain's population, London requires the extent of the country's entire productive land. In these terms, if the needs of everyone in the world were held to the same standard as that of London, representative of the norm of the industrialized world, the equivalent of three more "earths" would be required. We aspire to live beyond our means.

Sprawl

As thousands of people move to cities in search of better opportunities or merely basic livelihood, cities face pressures upon limited space and infrastructure. The result is often a spreading out through sprawl, suburbanization and the creation of satellite cities. The horizontal expansion into green fields seems irresistible in spite of it dysfunctional results.

The trend is seen in both the industrialized and the developing world. Los Angeles has grown into a huge urban agglomerate of 4,000 square miles that includes cities such as Riverside, Anaheim, Santa Ana and Burbank, with a high dependence on cars. According to one estimate, suburban roads and houses replace 2.5 million acres of farmland each year in the United States, much of it on prime agricultural soil. In China, some 494,000 acres of arable land is lost each year for city streets and developments.

Sprawl also increases the need for transportation and other facilities. Travel time is a key performance measure of transportation systems.

Long transport time to work is a sign of urban dysfunction, associated with severe traffic congestion, uncontrolled mixes of traffic types, poorly operating public transport networks, lack of adequate local traffic management, accidents and general dissatisfaction of the travel-dependent population. Bangkok, Thailand, in 1959 was a compact city where people could walk from any one point to another. But within decades Bangkok's population soared and the city built for canal transport grew dependent on motor transport. Bangkok mushroomed from 41.6 square miles in 1953 to 265 square miles in 1990. Today, it can take three hours to cross Bangkok by car—due to long distances and traffic snarls.

Growing dependence on vehicles and the resultant traffic congestion make air pollution a primary urbanization concern. As much as 20 percent of the world's population is estimated to live in cities where the air is not safe to breathe.

Conclusion

In the next two decades, the world's urban population is expected to reach 4.2 billion with almost all of this growth occurring in developing countries. By 2050, cities will have to absorb between 2 and 4 billion more people than they have today. Though most of this increase will occur in developing world countries, industrialized nations will not be entirely free from the pressures of urbanization and population shifts.

The consequences will be international, evident in global loss of resources, as well as unsupportable inequities in life quality, risks to global health, and the instability that attends inequities wherever they unfairly exist. The design of world cities is a global agenda. Attention will have to focus more on the poorer countries facing more rapid population growth and urban migration. Most of these countries lack the finances, technology or the infrastructure to handle urban explosions. If the phenomenon of rural-to-urban migration continues, the urban areas of less developed countries will face social disintegration.

If urban explosion is an inevitable part of population growth, what should urban policy makers, planners and designer do to make the urban experience less painful for these millions, and to assert the historic role of cities as centers of civil society and culture?

Population stabilization: Whether rural or urban, nations and international aid organizations will have to redouble their efforts to stabilize population growth rates to sustainable levels. More and more developing countries have achieved remarkable progress in reducing their total fertility rates. Yet some countries in Africa, Asia, Latin America and the Caribbean still register fertility rates of more than five children per woman and have very high levels of infant mortality.

Sustainable development: Besides efforts to stabilize their populations, governments of countries with high levels of urban migration must initiate creative programs that will reduce the determination to migrate. These countries and the international industries that partner with them should give job incentives to rural people to keep them in their villages, disperse industries so the rural people can find jobs closer to home, and improve infrastructures in the rural areas. When they have better economic opportunities and other facilities at home, the people will feel little need to migrate to the cities.

Urban infrastructure: Even the most dramatic and large-scale effort to improve the quality of life of people in rural areas may not immediately stem the flow of migrants into the cities. So the problem will

Table 4. Average annual rate of change of urban agglomerations with 8 million or more in 2000.

Agglomeration	Country	1950–1960	1960–1970	1970–1980	1980–1990	1990–2000
Africa		*5.1*	*5.0*	*4.3*	*3.9*	*3.9*
Cairo	Egypt	4.3	3.6	2.6	2.7	2.7
Lagos	Nigeria	9.7	9.8	7.7	5.6	5.1
Latin America		*4.9*	*4.2*	*3.3*	*2.8*	*2.1*
Buenos Aires	Argentina	3.0	2.2	1.6	1.5	1.1
Lima	Peru	5.5	5.5	4.1	3.4	2.7
Mexico City	Mexico	5.4	5.5	4.3	3.3	2.4
Rio de Janeiro	Brazil	5.4	3.6	2.2	2.0	1.5
São Paulo	Brazil	6.6	5.4	4.1	3.6	2.4
Asia		*4.4*	*3.7*	*3.3*	*3.3*	*3.3*
Bangalore	India	4.3	3.2	5.5	5.7	5.0
Bangkok	Thailand	4.6	3.7	4.2	4.1	3.6
Beijing	China	4.7	2.5	1.1	1.8	2.6
Bombay	India	3.4	3.6	3.3	3.3	3.2
Calcutta	India	2.1	2.3	2.7	2.7	2.8
Dacca	Bangladesh	4.3	8.4	7.8	7.0	6.0
Delhi	India	5.0	4.4	4.5	4.6	4.1
Istanbul	Turkey	4.8	4.7	4.6	4.1	3.6
Jakarta	Indonesia	3.4	3.4	4.2	4.4	4.0
Karachi	Pakistan	5.9	5.2	4.6	4.4	4.1
Manila	Philippines	3.9	4.4	5.2	3.5	3.3
Seoul	Korea, Republic of	8.4	8.1	4.4	2.8	1.5
Shanghai	China	5.1	2.3	0.5	1.3	2.4
Teheran	Iran (Islamic Rep. of)	5.9	5.6	4.4	2.9	2.3
Tianjin	China	4.2	3.7	3.3	2.5	3.1
More developed regions		*3.1*	*2.1*	*0.7*	*0.7*	*0.5*
Los Angeles	USA	4.8	2.5	1.3	2.2	1.6
Moscow	Russia	2.6	1.2	1.4	0.8	0.2
New York	USA	1.4	1.3	−0.4	0.4	0.3
Osaka	Japan	4.1	2.8	0.9	0.2	0.1
Paris	France	2.8	1.4	0.2	−0.03	0.1
Tokyo	Japan	4.6	3.3	1.3	0.7	0.5
All agglomerations		*3.9*	*3.3*	*2.5*	*2.5*	*2.4*

Table 5. Human-induced land degradation worldwide, 1945 to present.

Region	Over-grazing	Defores-tation	Agricul-tural Misman-agement	Other[1]	Total	Degraded Area as Share of Total Vegetated Land
	(million hectares)					(percent)
Asia	197	298	204	47	746	20
Africa	243	67	121	63	494	22
South America	68	100	64	12	244	14
Europe	50	84	64	22	220	23
North & Cent. Amer.	38	18	91	11	158	8
Oceania	83	12	8	0	103	13
World	679	579	552	155	1,965	17

[1]Includes exploitation of vegetation for domestic use (133 million hectares) and bioindustrial activities, such as pollution (22 million hectares).
SOURCE: Worldwatch Institute, based on "The Extent of Human-Induced Soil Degradation," Annex 5 in L.R. Oldeman et al., *World Map of the Status of Human-Induced Soil Degradation* (Wageningen, Netherlands: United Nations Environment Programme and International Soil Reference and Information Centre, 1991).

Fig. 3. Squatter settlement upgrading project. Amman, Jordan. Financed by The World Bank with community development by Save The Children Federation. Water and sewer infrastructure services were "retrofitted" within a Palestinian refugee settlement, enabling phased self-help construction of homes.

have to be tackled at both ends. Countries will have to work to improve the lives of migrants in the cities so that they do not end up on the streets there, becoming a burden on the city's resources. In this respect, housing is a major issue. More and more migrants to cities end up living in slums and shantytowns, risking their health and the health of the cities. Building homes for the thousands in cities of the developing countries is a huge task that entails international involvement. Exemplary models exist, demonstrating that a combination of urban infrastructure and self-help strategies can create new communities and upgrading of former slums (Fig. 3).

There must be a focus on creating partnerships and participation for more equitable and sustainable development of the cities. The Istanbul Declaration from the United Nations Conference on Human Settlements II (United Nations, 1996) calls for "sustainable urban development" and "adequate shelter for all." At the Istanbul conference, the world's nations pledged "to ensure consistency and coordination of macro-economic and shelter policies and strategies as a social priority within the framework of national development programs and urban policies in order to support resource mobilization, employment generation, poverty eradication and social integration" (Table 7). Urban planners and designers from throughout the globe can help to fulfil the historic promise and opportunity of cities by helping to carry out this agenda. ■

Table 6. Population Size and Availability of Renewable Resources, circa 1990, with projections for 2010.

	Circa 1990	2010	Total Change	Per Capita Change
	(million)		(percent)	
Population	5,290	7,030	+33	–
Fish Catch (tons)[1]	85	102	+20	−10
Irrigated Land (hectares)	237	277	+17	−12
Cropland (hectares)	1,444	1,516	+5	−21
Rangeland and Pasture (hectares)	3,402	3,540	+4	−22
Forests (hectares)[2]	3,413	3,165	−7	−30

[1]Wild catch from fresh and marine waters; excludes aquaculture. [2]Includes plantations; excludes woodlands and shrublands.

SOURCES: Population figures from U.S. Bureau of the Census, Department of Commerce, *International Data Base,* unpublished printout, November 2, 1993; 1990 irrigated land, cropland, and rangeland from U.N. Food and Agriculture Organization (FAO), *Production Yearbook 1991* (Rome; 1992); fish catch from M. Perotti, chief, Statistics Branch, Fisheries Department, FAO, Rome, private communication, November 3, 1993; forests from FAO, *Forest Resources Assessment 1990* (Rome: 1992 and 1993)

Table 7. Habitat II Agenda: the Istanbul Declaration (United Nations, 1996)

1 Shelter

1.1 Provide security of land tenure.
1.2 Promote the right to adequate housing.
1.3 Provide equal access to land.
1.4 Promote equal access to credit.
1.5 Promote access to basic services.

2 Social development and eradication of poverty

2.1 Provide equal opportunities for a healthy and safe life.
2.2 Promote social integration and support disadvantaged groups.
2.3 Promote gender equality in human settlements development.

3 Environmental management

3.1 Promote geographically balanced settlement structures.
3.2 Manage the supply and demand for water in an effective manner.
3.3 Reduce urban pollution.
3.4 Mitigate effects of natural disasters with responsible preventive measures.
3.5 Promote effective and environmentally responsible transportation systems.
3.6 Support mechanisms to implement Agenda 21 (cf. United Nations, 1993).

4 Economic development

4.1 Strengthen small and micro-enterprises, particular hose developed by women.
4.2 Encourage public-private sector partnerships and stimulate productive employment

5 Governance

5.1 Promote decentralization of management and strengthen local authorities.
5.2 Encourage and support participation and civil participation.
5.3 Ensure transparent, accountable and efficient governance of towns, cities and metropolitan areas.

REFERENCES

Fuchs, Roland J., *et al.*, eds. 1999. *Mega-city Growth and the Future.* New York: United Nations University Press. pub # UNUP-820-6

Girardet, Herbert. 1992. *The Gaia Atlas of Cities: New directions for sustainable urban living.* London: Gaia Books, Ltd. Anchor/Doubleday.

Lo, Fu-Chen and Yue-Man Yeung, eds. 1998. *Globalization and the World of Large Cities.* New York: United Nations University Press. pub. # UNUP-999

Rees, William and Mathis Wackernagel. 1996. *Our Ecological Footprint: Reducing Human Impact on the Earth.* Philadelphia, PA: New Society Publishers.

United Nations. 1993. *Agenda 21: The Earth Summit—The United Nations Programme of Action from Rio.* New York: United Nations Publications. Pub. # E 93.1.11

United Nations. 1996. *Habitat Agenda and Istanbul Declaration: Second United Nations Conference on Human Settlements Istanbul, Turkey.* New York: United Nations Publications. Pub. # E DPI/1859

United Nations. 1985. *Estimates and projections of urban, rural and city populations, 1950-2025.* United Nations.

Worldwatch Institute. 2001. *Vital Signs 2001.* Washington, DC: Worldwatch Institute. www.worldwatch.org.

RELATED READING

Bongaarts, John. 1994. "Can the Growing Human Population Feed Itself." *Scientific American.*

Brown, Lester R., ed. 1999. *State of the World: Progress toward a Sustainable Society.* Worldwatch Institute Report. New York: W.W. Norton.

Douglas, Mary and Aaron Wildavsky. 1982. *Risk and Culture: An Essay on the Selection of Technical and Environmental Dangers* Berkeley, CA: University of California Press.

Ehlich, Paul R. and Anne H. Ehlich. 1991. *Healing the Planet: Strategies for Resolving the Environmental Crises.* New York: Addison-Wesley Publishing Company.

Jackson, Wes, Wendell Berry and Bruce Colman, eds. 1984. *Meeting the Expectations of the Land: Essays on Sustainable Agriculture and Stewardship.* San Francisco: North Point Press.

Lebel, Gregory G. and Hal Kane. 1989. *Sustainable Development: A Guide to Our Common Future—The Report of the World Commission on Environment and Development,* Washington, DC: The Global Tomorrow Coalition.

IUCN/UNEP/WWF. 1991. [The World Conservation Union-United Nations Environment Programme—World Wide Fund for Nature] *Caring For The Earth: A Strategy for Sustainable Living-Second World Conservation Strategy.* Gland, Switzerland.

Meadows, Donella H., Dennis L. Meadows, and Jørgen Randers. 1992. *Beyond the Limits: Confronting Global Collapse Envisioning a Sustainable Future.* Post Mills, VT: Chelsea Green Publishing Company.

Schumacher, E.F. 1973. *Small is Beautiful: Economics as if People Mattered.* New York: Harper & Row.

Thomas, William L. Jr., ed., 1956. *Man's Role in Changing the Face of the Earth.* Chicago: University of Chicago Press.

Turner, B. L. II, *et al.*, eds. 1990. *The Earth as Transformed by Human Action: Global and Regional Changes in the Biosphere over the past 300 Years.* New York: Cambridge University Press.

Wilson, Edward O. 1993. "Is Humanity Suicidal: We're flirting with extinction of our species." *New York Times Magazine.* May 30, 1993.

Alain Bertaud

Summary
A city is a very complex object, compounded by its constantly evolving shape and structure. To try to understand a city's inner mechanisms, urban planners and designers can develop analytic models that are simple enough to be easily understandable, but accurate enough to be useful. This article presents such a model that depicts the spatial distribution of population in seven large cities to provide an analytic tool that can be used to guide municipal development strategies.

Key words
census data, central business district (CBD), density, modeling, population, transport mode, trip patterns, urban planning

New York City

Metropolis: the spatial organization of seven large cities

Urban development strategies may address many concerns, such as the quality of the environment, the efficiency of infrastructure network, the growth of employment, or housing affordability. The task of the urban planner and designer is to identify the type of spatial organization that is compatible with the municipal strategy and the regulatory tools and infrastructure investments that will allow a city to evolve from its current spatial organization to the one implied by the strategy.

The complex economic and social relations that gave rise to the emergence of large cities produce a physical outcome—the urban built-up space—that can be mapped and measured. While one may never know with precision the nature of the forces that produced the built-up space, one is at least able to measure the end result. New technology developed during the last thirty years—satellite imagery, digital mapping and geographical information systems—enable us to have a much better knowledge of the urban shape than was the case in the past. At the same time, because the number of megacities in the world is increasing rapidly, monitoring and managing their spatial expansion is much more complex.

1 HOW TO DEFINE A CITY'S SPATIAL STRUCTURE

The spatial structure of a city can be defined by two complementary components; first, the spatial distribution of population as recorded by census data and second, the pattern of trips made by people when they go from their residence to their place of work, to schools, shops, social gatherings and to any other places where they will have a productive or social activity. The spatial distribution of population is therefore a static representation of the city when its people are at home, while the pattern of trips is a schematic view of the complex trajectories that these same people will follow during the time they are not at home.

Distribution of population

While some cities claim to be alive 24 hours a day, the reality is that most people are at home between midnight and 6 a.m., with relatively few exceptions such as night shift workers, etc. Where people are between midnight and 6 a.m. is the starting point of the daily trips toward the meeting places. The spatial distribution of population is therefore an image of the location of the majority of a population of a city between about midnight and 6 a.m. The urban planner's density maps are thus showing the densities around midnight, not the density during the day.

The spatial distribution of the population can be represented graphically by combining census data and land use maps to construct a 3-dimensional object, where the map of the built-up area is in the X-Y plane and the population density within the built-up area is shown in the Z dimension. To understand better a city spatial structure one must analyze the geometrical properties of this 3-dimensional graph. For instance, one can calculate the position of its center of gravity.

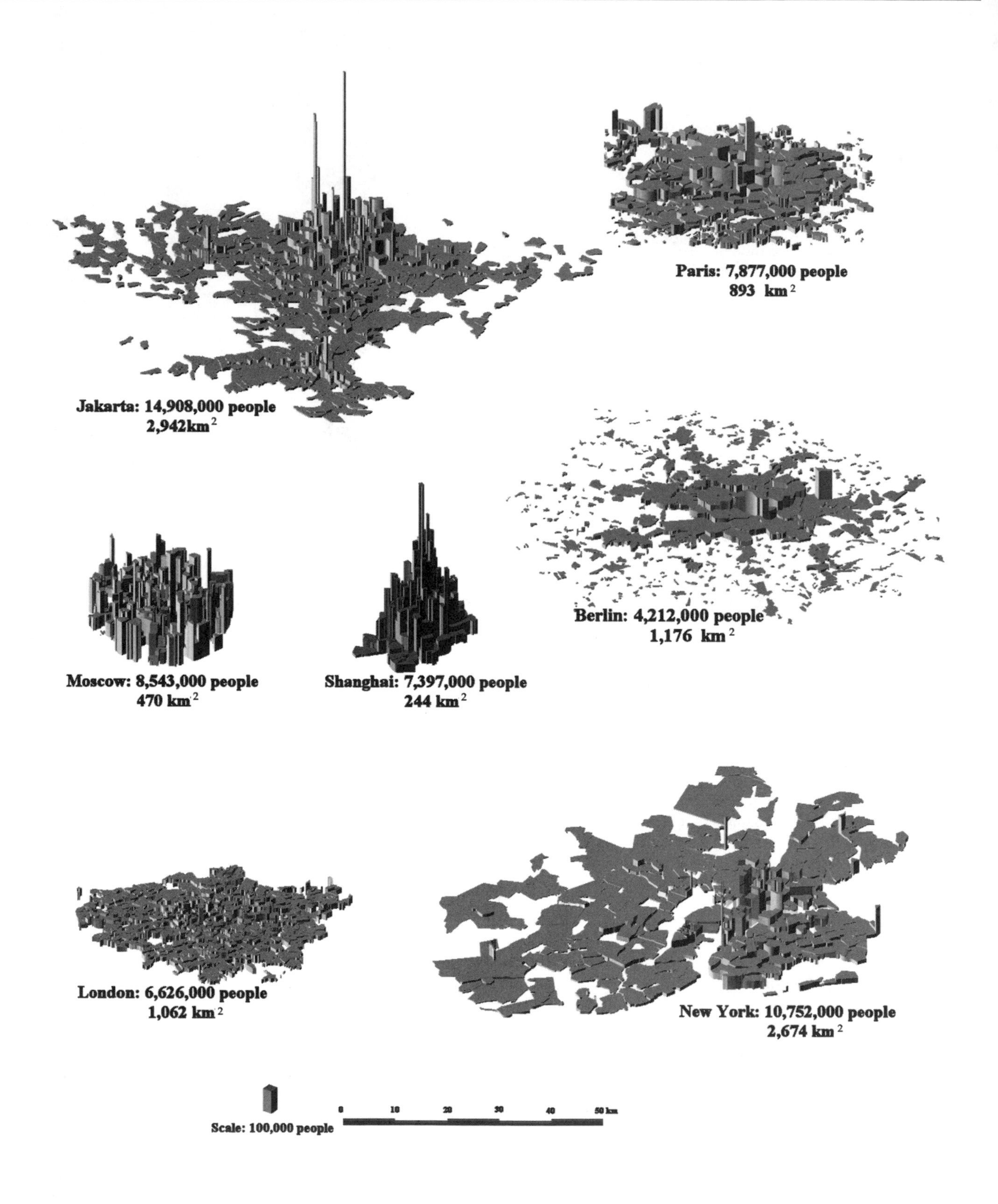

Fig. 1. The spatial structure of seven cities. This figure shows the spatial distribution of the population—at midnight—of 7 large world cities. The built-up area and the population of the 7 cities are represented at the same scale. The isometric view is from the South at a 30° angle with the horizontal plane.

The center of gravity will be the point to which the average distance per person is the shortest. One can also identify special areas of the city such as its central business district (CBD). One will be able then to calculate what is the average distance per person to the CBD and whether the CBD and center of gravity coincide. In other words, one is able to analyze the geometric property of this 3-dimensional graph and compare it to other similarly constructed ones representing different cities. The only constraint will be to make sure that the conventions used to build the 3-dimensional representation of population are consistent across cities.

Fig. 1 shows the spatial distribution of the population—at midnight—of seven large world cities. From east to west, they are: Shanghai, Jakarta, Moscow, Berlin, Paris, London and New York. The built-up area and the population of the seven cities are represented at the same scale. The isometric view is from the south at a 30° angle with the horizontal plane.

The spatial structure of the cities shown in Fig.1 appears complex. Each is dramatically different from the others. This article will discuss the indicators that help determine in what measurable way these spatial structures differ from one another and second, whether some structures might be performing better than others in meeting some simple criteria.

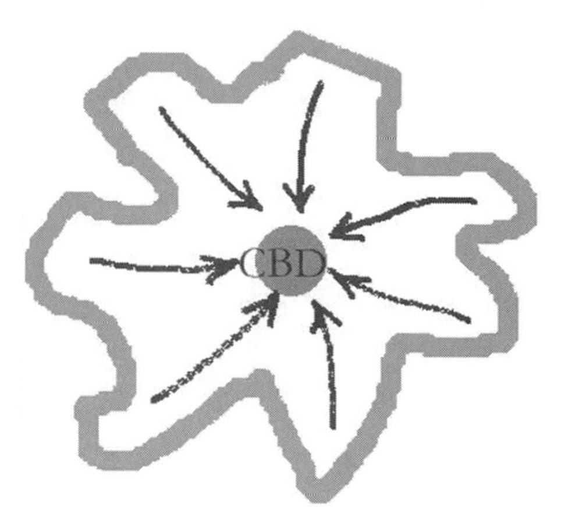

A. Monocentric model

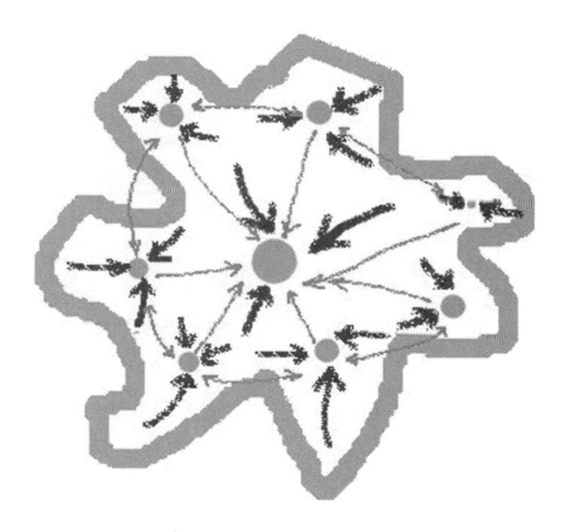

B. Polycentric model: the urban village version

Fig. 2. Pattern of trips in monocentric and polycentric cities.

2 PATTERN OF TRIPS: MONOCENTRIC AND POLYCENTRIC CITIES

Definition of trip pattern

Every day the inhabitants of a city move from their places of residence to other locations in the city that include places of work, shopping areas, cultural and social facilities. They also move to visit each other's residences. Many jobs involve more than one location within the same day. In a modern American city, the number of trips generated by commuting from residence to job is in most cases less than 50% of the number of total trips. One should resist the simplistic view that trip patterns are generated by daily return trip from residence to place of work. To avoid this confusion here, the term "meeting places" is used to refer to the places where people go when they leave their home. The definition of "meeting places" covers job locations, shopping places, cinemas, schools, other people's home, and so forth.

Trip patterns will thus depend on the relative location of residences and meeting places within the metropolitan area. In a *monocentric* city the location of the majority of these meeting places is heavily concentrated in a central area. In a *polycentric* city, most meeting places are distributed in clusters around the metropolitan areas. The number of clusters could range from three or four to several hundred.

To understand the pattern of trips in monocentric and in polycentric cities, it is necessary to look briefly at the economic justification of cities.

Labor and consumer markets as the base of the economic success of cities

A large unified labor and consumer market is the *raison d'être* of large cities whether they are monocentric or polycentric. A substantial literature treats cities as labor markets, such as Ihlandfeldt (1997) and the classic Goldner (1955). Prud'homme (1996) provides a convincing explanation for the growth of megacities in the last part of the twentieth century: large cities become more productive than small cities when they can provide larger effective labor markets. The capacity of megacities to maintain a unified labor market is the true long-term limit to their size. Market fragmentation due to management or infrastructure failure could therefore result initially in economic decay and eventually in a loss of population. Spatial indicators should therefore indicate if a city's current structure favors or disfavors the functioning of unified labor and consumer markets. The fragmentation of labor markets might be due to many different other non-spatial factors, for instance, the rigidity of labor laws or racial or sex discrimination.

Pattern of trips in monocentric and polycentric cities

A monocentric city can maintain a unified labor market by providing the possibility of moving easily along radial roads or rails from the periphery to the center (Fig. 2A) The shorter the trip to the CBD, the higher is the value of land. Densities, when market driven, tend to follow the price of land. Hence the negative slope of the density gradient from the center to the peripheries observed in most world cities.

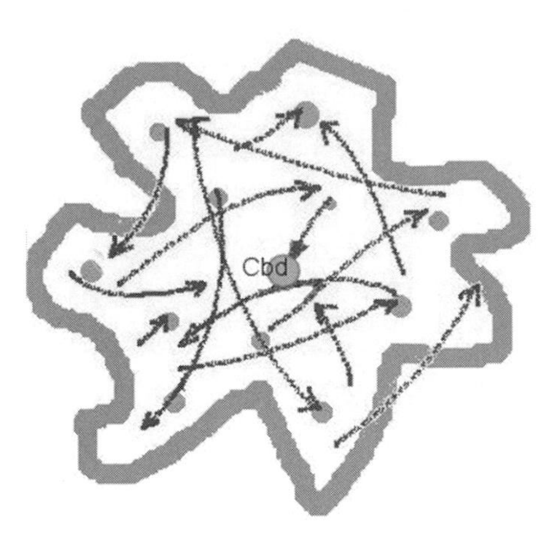

C. Polycentric model: random movement version

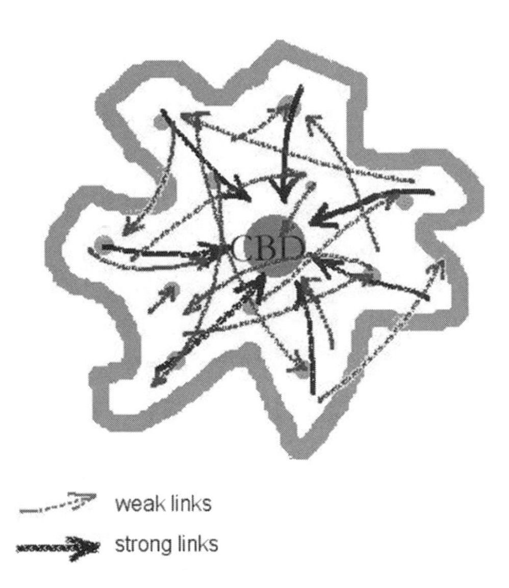

D. Mono-polycentricmodel: simultaneous radial and random movement

Fig. 2. Pattern of trips in monocentric and polycentric cities.

The growth of polycentric cities is also conditional on providing a unified labor and consumer market. Some urban planners often idealize polycentric cities by thinking that a self-sufficient community is likely to grow around each cluster of employment. According to this point of view, a number of self-sufficient "urban villages" would then aggregate to form a large polycentric metropolis (Fig. 2B). In such a large city, trips would be very short; ideally, everybody could even walk or bicycle to work. Nobody has ever observed this behavior in any large city. A metropolis constituted by self-sufficient "urban villages" would contradict the only valid explanation for the existence and continuous growth of large metropolitan areas: the increasing returns obtained by larger integrated labor markets. The urban village concept is the ultimate labor market fragmentation.

Although there are many polycentric cities in the world, there is no known example of one formed by an aggregation of small self-sufficient communities. The urban village concept flies in the face of common sense. It assumes that in a large metropolis most people will not look for work beyond a radius of few kilometers from their home, or that they would select a home only within the restricted boundaries limited by a given radius from their work.

In spite of not being encountered in the real world, the utopian concept of a polycentric city as a cluster of urban villages persists in the mind of many planners. For instance, in some suburbs of Stockholm urban regulations allow developers to build new dwelling units only to the extent than they can prove that there is a corresponding number of jobs in the neighborhood. The satellite towns built around Seoul and Shanghai are another example of the urban village conceit. The majority of the inhabitants of these towns work and commute daily to the main part of the city, and others who commute from the main city perform most jobs in the satellite towns.

In reality, a polycentric city functions very much in the same way as a monocentric city: jobs, wherever they are, attract people from all over the city. The pattern of trips is different, however. In a polycentric city each sub-center generates trips from all over the built-up area of the city (Fig. 2C). Trips tend to show a wide dispersion of origin and destination, appearing almost random. Trips in a polycentric city will tend to be longer than in a monocentric city, ceteris paribus. However, for a given point, the shorter the trip to all potential destinations, the higher should be the value of land. A geometrically central location will provide trips of a shorter length to all other locations in the city. Therefore, one should expect polycentric cities to also have a negatively sloped density gradient, not necessarily centered on the CBD but on the geometric center of gravity of the urbanized area. The slope of the gradient should be flatter, as the proximity to the center of gravity confers an accessibility advantage that is not as large as in a monocentric city. The existence of a flatter but negatively sloped density gradient in polycentric cities can be observed in cities that are obviously polycentric, like Los Angeles.

Monocentric cities tend to transform themselves as they grow into polycentric cities

Traditionally, the monocentric city has been the most widely used model to analyze the spatial organization of cities. The works on density gradients in metropolitan areas of Alonso (1964), Muth (1969), and Mills (1972) are based on the hypothesis of a monocentric city. It has become obvious over the years that the structure of many cities departed from the monocentric model and that many trip-generating activities were spread in clusters over a wide area outside the traditional CBD. Consequently, many have questioned whether the study of density gradients, which measures density variations from a central point located in the CBD, has any relevance in cities where the CBD is the destination of only a small fraction of metropolitan trips.

As it grows in size, the original monocentric structure of large metropolises tends with time to dissolve progressively into a polycentric structure. The CBD loses its primacy, and clusters of activities generating trips are spread out within the built-up area. Large cities are not born polycentric; they may evolve in that direction. Monocentric and polycentric cities are thus from the same specie, observed at a different time during their evolutionary process. No city is ever 100% monocentric and is seldom 100% polycentric (*i.e.*, with no discernable "downtown"). Some cities are predominantly monocentric, others predominantly polycentric and many are in between. Some circumstances tend to accelerate the mutation toward polycentricity, such as an historical business center with a low level of amenities, high private car ownership, cheap land, flat topography, or grid street design. Other circumstances tend to retard the trend, such as an historical center with a high level of amenities, rail-based public transport, radial primary road network, or difficult topography preventing communication between suburbs.

3 THE DEVELOPMENT OF SPATIAL INDICATORS

Can the same spatial indicators be used for monocentric and polycentric cities?

To compare and analyze the spatial organization of cities, one can use a number of spatial indicators. Some of these indicators require a central point, in this case the CBD or the center of gravity of the population. The measure of population distribution from a central point is usually associated with monocentric cities, *i.e.,* in cities where the CBD is the destination of the large majority of daily trips. The following section discusses how indicators based on central points are relevant to both monocentric and polycentric cities.

Density gradients, and other indicators linked to a central geometrical point, constitute very useful tools to reveal and compare the spatial structure of cities, whether they are monocentric or not. In many cities, the center of gravity and the historical CBD coincide, in particular in cities with few topographical constraints. When in a polycentric city these two points do not coincide, the center of gravity should be selected instead of the CBD to calculate the density gradient. In most large cities, some trips are following the monocentric mode from a random point to a central point. Others are following the polycentric mode from random points to random points (Fig. 2D). In this case one could select either the CBD or the center of gravity of the population as the reference point for density gradients.

In addition to the density gradient, the term "dispersion index" is used to compare the shape of various cities (Bertaud and Malpezzi 1999). All else being equal, a city shape which decreases the distance between people's residence and the main place of work and consumption will be more favorable to the functioning of labor and consumer markets. For a given built-up area, the shorter the average distance per person to the main place of work or to the main commercial areas, the better would be the performance of the city shape.

Location of the CBD and the center of gravity of the city shape

As we have seen, the average distance per person to the CBD would be a good measure of proximity for a monocentric city. However, the CBD is a real estate concept, not a geometric one. The location of the CBD within the city shape has an effect on the measure of proximity that is independent of the shape itself. One needs therefore first, to define a way to measure proximity within a shape independently from CBD location and then, to introduce the measure of the location of the CBD within a shape as an additional shape parameter.

Up to this point in the discussion, city shape is defined as a 3-dimensional graph in which the horizontal plane contains the base of the prisms corresponding to the outline of the different neighborhoods constituting the built-up area, while the population density in these same neighborhoods constitute the height of the prisms. The complex solid formed by the aggregations of neighborhood prisms possesses a center of gravity whose location can be calculated. This center of gravity is by definition the point from which the average distance to all the points of the solid is the shortest. Therefore, for a given shape, the optimum position of the CBD to maximize proximity coincides with the center of gravity of the shape. In the discussion that follows, another shape indicator is defined, relevant only for monocentric cities, using the measure of distance between CBD and the center of gravity.

Variations in distance to the CBD and in distance between random points for various spatial structures of identical area and average density

Before extending the discussion, consider as a test of the concept the variation in proximity for a fictional city of constant area and population (therefore constant average density) when the spatial structure varies. The example which follows measures the variations of the average distance per person to the CBD and between random points for a number of shapes while keeping the built-up area and population constant. Therefore, this measure could be applied to either a monocentric or a polycentric city. For this exercise, the assumption made is that the CBD coincides with the center of gravity of the shape.

Assume an imaginary city of one million people at an average density of 100 people per hectare, *i.e.,* a built-up area of 100 square kilometers (38.6 square miles). To limit the number of possible shapes the variations will be limited to those who are inscribed within a square of 12 by 12 kilometers (7.5 by 7.5 miles). To analyze the distance per person to the CBD and the average distance per person between random points for twenty variations of typical spatial structures, the average density, population and built-up area are assumed to remain as constants. The variables are the density of sub-areas, the location of sub-areas with different densities and the shape of the built-up area within the limit of a square of 12 kilometers (7.5 miles). The results are shown in Fig. 3.

The spatial organization types shown in Fig. 3 are presented by order of decreasing performance for average distance to the CBD. The results suggest three observations:

- *The variation of performance between types is large.* The distance to the CBD doubles between layout 1 to layout 20, that is, from 3 to 6 kilometers (1.9 to 3.7 miles), although the shape itself stays inscribed within a square of 12 by 12 km (7.5 by 7.5 miles). Between cities of identical average density, shape is therefore a very important factor in allowing markets to function and in affecting the length and the costs of networks.

- *The variation in the distance to the CBD is much larger between different spatial arrangement than between the distance to the CBD and the distance between random points for a given shape.* Shape itself is more important in city performance than whether the city function as a monocentric city or a random movement city.

- *While a poor performer for the distance to the CBD will generally be a poor performer for average distance between random locations, the correspondence is not linear.* Some types of spatial arrangements that are favorable to monocentric movements are not favorable to random movements. For instance, layout ranked 13 for distance to CBD performs better for random movement than the layout ranked 8.

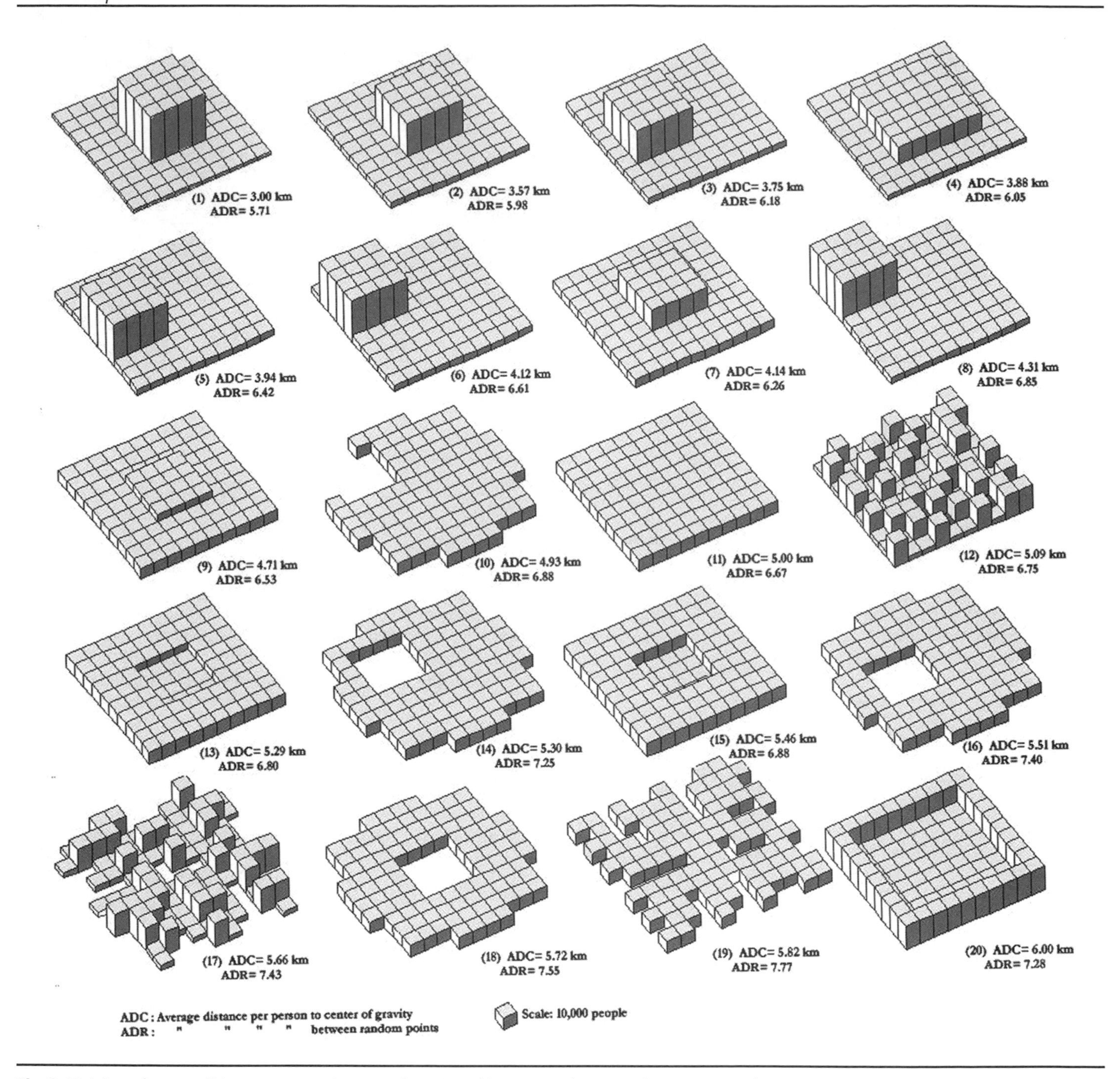

Fig. 3. Variation of average distance per person to center of gravity and between random points for various spatial structures.

Table 1. Selected Cities: Built-up Area, Population, and Average Density

	Built-up Area	Population	Average density in built-up area		Built-up land per person
	Km^2		people/Ha	People/sq.mile	m^2/person
Berlin	1,176	4,212,400	36	9,279	279
Jakarta Metropolitan Area (Jabotabek)	2,942	14,908,400	51	13,124	197
London	1,062	6,626,300	62	16,167	160
Moscow (Municipality)	503	8,497,200	169	43,711	59
New York Metropolitan area	2,674	10,752,900	40	10,414	249
Paris Metropolitan area	937	7,998,100	85	22,098	117
Shanghai (City proper)	244	7,397,200	303	78,483	33

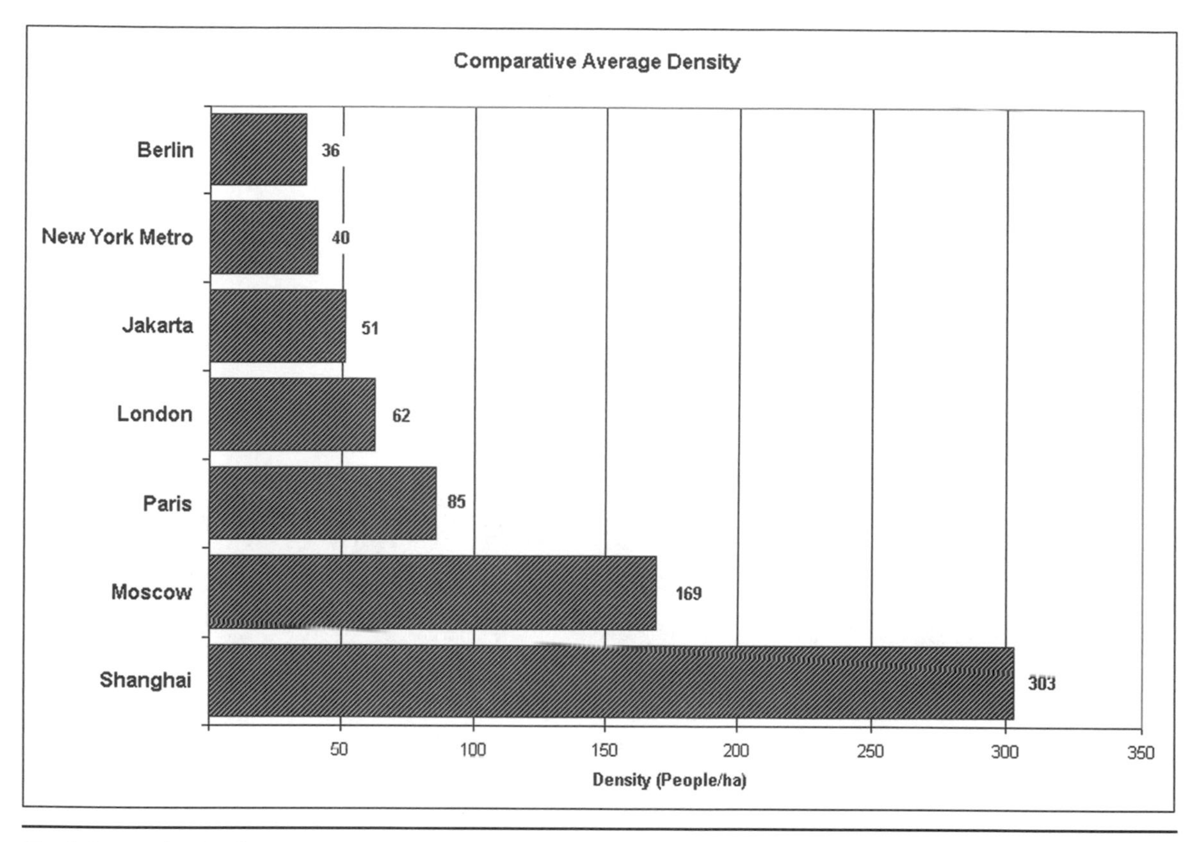

Fig. 4. Average density of seven cities.

The points made above can be summarized in the following manner:

Urban spatial structures have an impact on the functioning of labor and consumer markets. Because the length of trips—and hence their time and costs—from residence to meeting places varies with different type of spatial organization, spatial structures have therefore an impact on the economy of cities. Cities with a more favorable type of spatial organizations will possess a decisive economic competitive advantage over cities with a less favorable spatial structure.

Urban spatial structures have an important impact on the environment. Urban spatial structure has two ways to have an impact on the environment. First, because some spatial organizations shorten trips, they also reduce the need of cities for energy and therefore pollution. Second, because some spatial forms are more compact than others, they put less pressure on the natural environment surrounding cities.

The structures of cities evolve with time. A structure initially favorable to the economy and to the environment may eventually deteriorate into a less effective structure. Conversely, an inefficient urban structure might be guided to evolve into a more efficient one. It is therefore important that urban planners and designers constantly monitor the evolution of cities' structures with relevant indicators and establish targets for the future.

4 THE USE OF SPATIAL INDICATORS TO ANALYZE AND MONITOR SPATIAL ORGANIZTION

As discussed, Fig. 1 indicates the shape of seven large metropolises from different parts of the world, representing vastly different cultures, climate and economic systems. Selection of these cities has been neither random nor unbiased. The observations that follow should thus be considered as case studies and not as statistical evidence.

The basic data concerning the seven metropolises are presented in Table 1. The parameters of Table 1 are defined as follows:

Built-up area: expressed in square kilometers, measures the land area consumed by urban activities within a metropolitan area across administrative boundaries. It does not include large parks of more than 4 hectares, airports, agricultural land, and undeveloped vacant land and water bodies. However, the limits of a metropolitan area are themselves subjective. The boundary of the built-up areas is limited to where additional population corresponded to a density that was not significantly higher than the prevailing density in adjacent rural areas. While some might disagree in the way that these assumptions delimit the metropolitan built-up area, the advantage of this approach is its internal consistency. All the data presented here are derived from primary sources: census tract and land use maps or satellite imagery.

In the case of Shanghai, some settlements in the northwest suburbs have not been included due to a limitation in data availability. Shanghai data, however, corresponds to the area called "the city proper" by the municipality.

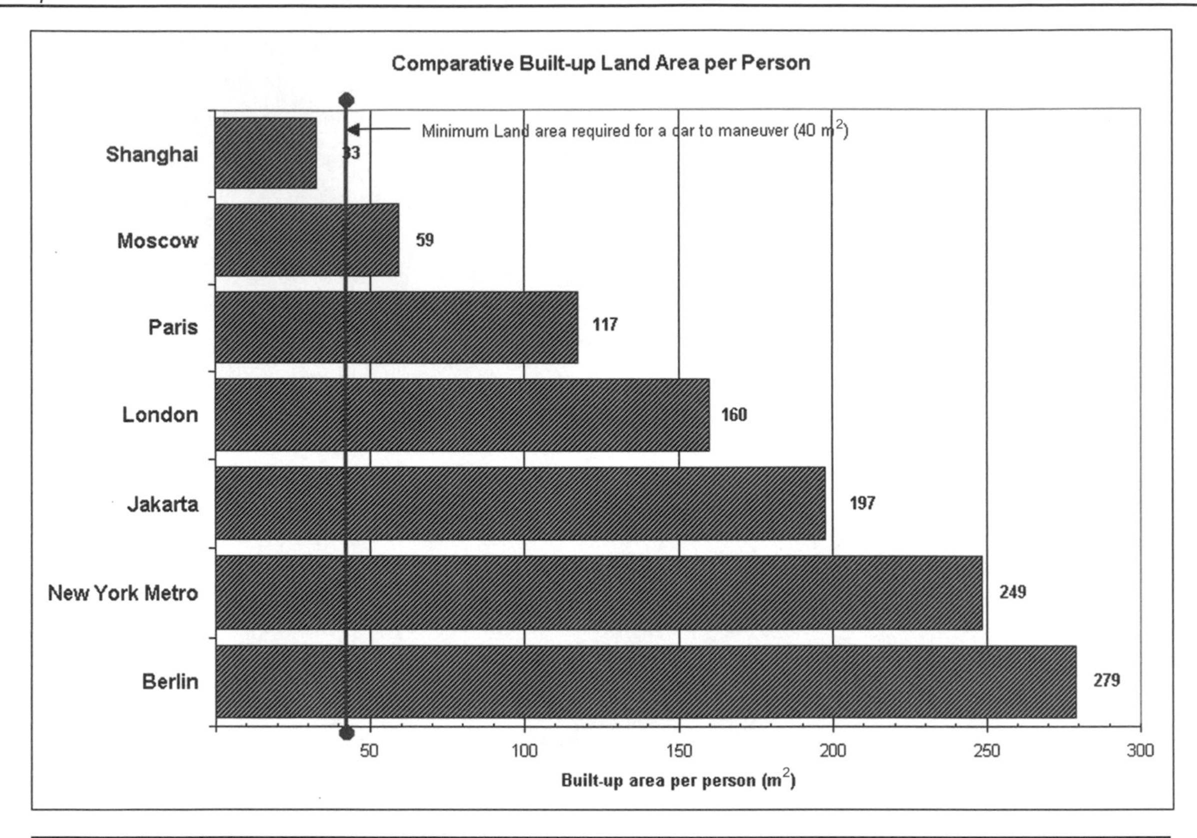

Fig. 5. Average land consumption per capita.

Population: Number of people within the built-up area defined above as given by census tracts within or intersecting the built-up area. All data are from the1990 census, except for Shanghai population data, which corresponds to a census taken in1987 by adding up data from street committees.

Average population density (expressed in people per hectare): Population divided by built-up area expressed in hectares (there are 257 hectares in a square mile).

Built-up area per person: Built-up area expressed in square meters divided by population.

Average density and land consumption

Fig. 4 indicates the wide range of density possible for a successful metropolis. The average density is an ambivalent parameter for the environment. In some instances, a high density may serve to preserve the environment outside the city but is not necessarily conducive to a pleasant environment inside the city.

One should note, however, that there is a lack of correlation between poverty and density or between crime and density. Jakarta is much poorer than Shanghai or Moscow but has a much lower density than either city. Shanghai and Paris have less violent crime per inhabitant than New York but both have a much higher density. As discussed below, average density *per se* is a rather crude spatial indicator. The way that density is spatially distributed within a metropolitan area is much more important for its shape performance than the average density. For instance, as seen above in Fig. 3, trip length is not directly related to average density but rather to the way densities are distributed across the metropolitan area.

Average land consumption per capita

Land consumption per capita is directly derived from the density. Fig. 5 shows the wide range of land consumption that is possible in a metropolitan area. There are no "good" or "bad" figures for land consumption. There are only trade-offs between spatial features and for instance transport options. The highlighted line in Fig. 5 indicated as "minimum land area" shows the amount of land space required by a car to park and to maneuver. The area required by a car is a small fraction of the land consumed per person in New York or in Berlin but it is a significant portion of it in Moscow and in Shanghai where a car consumes more space than a person!

The implication of this is clear: the smaller the consumption of land per capita, the more disruptive will be the use of the individual car as a means of transportation. While a decade ago, trips in Moscow were mostly by public transport (about 95%), by1996 public transport use had fallen to about 70% of trips. In Shanghai in 1987 about 65% of trips were using bicycle and about 30% buses. In both cities the rate of car ownership is growing very fast every year. The number of trips using private cars is following car ownership with a lag of a few years. It is clear that in cities with a very low land consumption per capita, such as Shanghai, Moscow and to a certain extent Paris, the car is directly encroaching on people space.

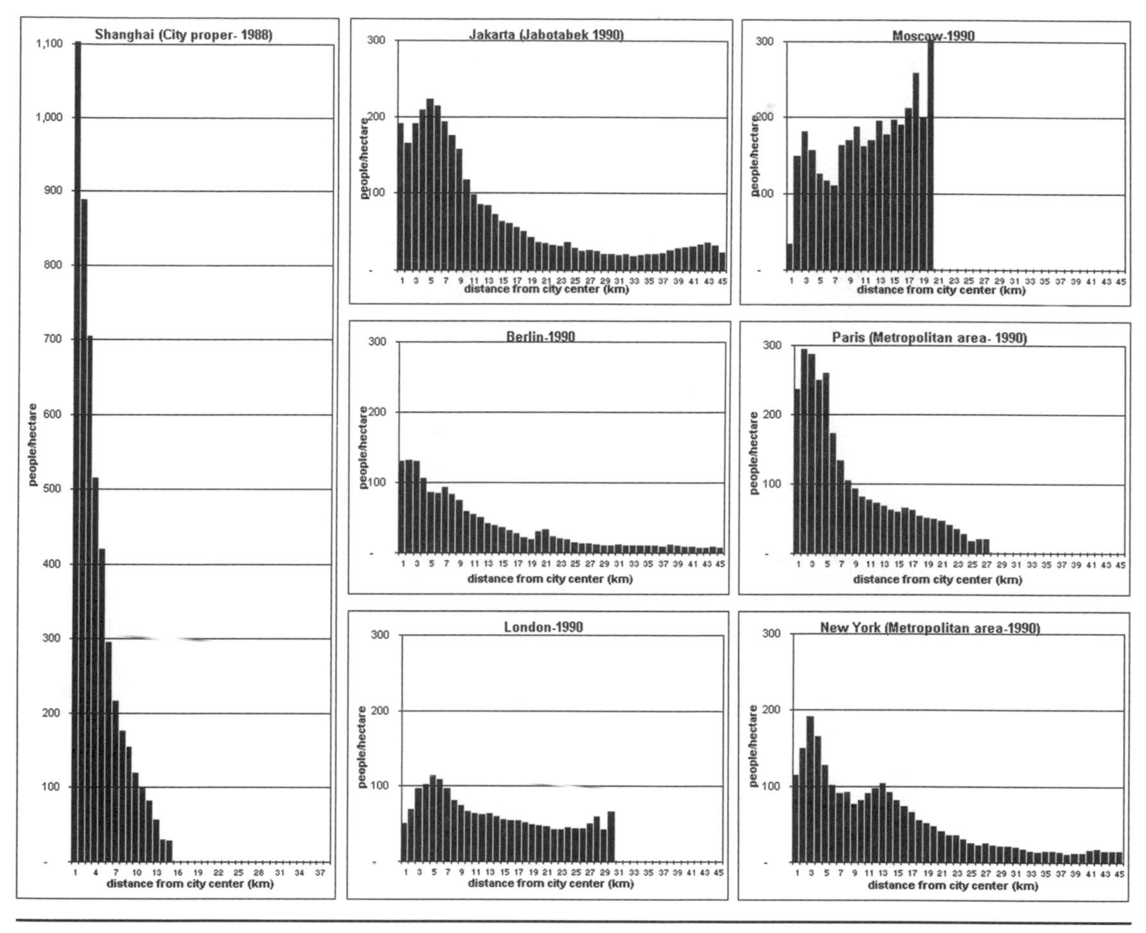

Fig. 6. Density profiles of seven cities.

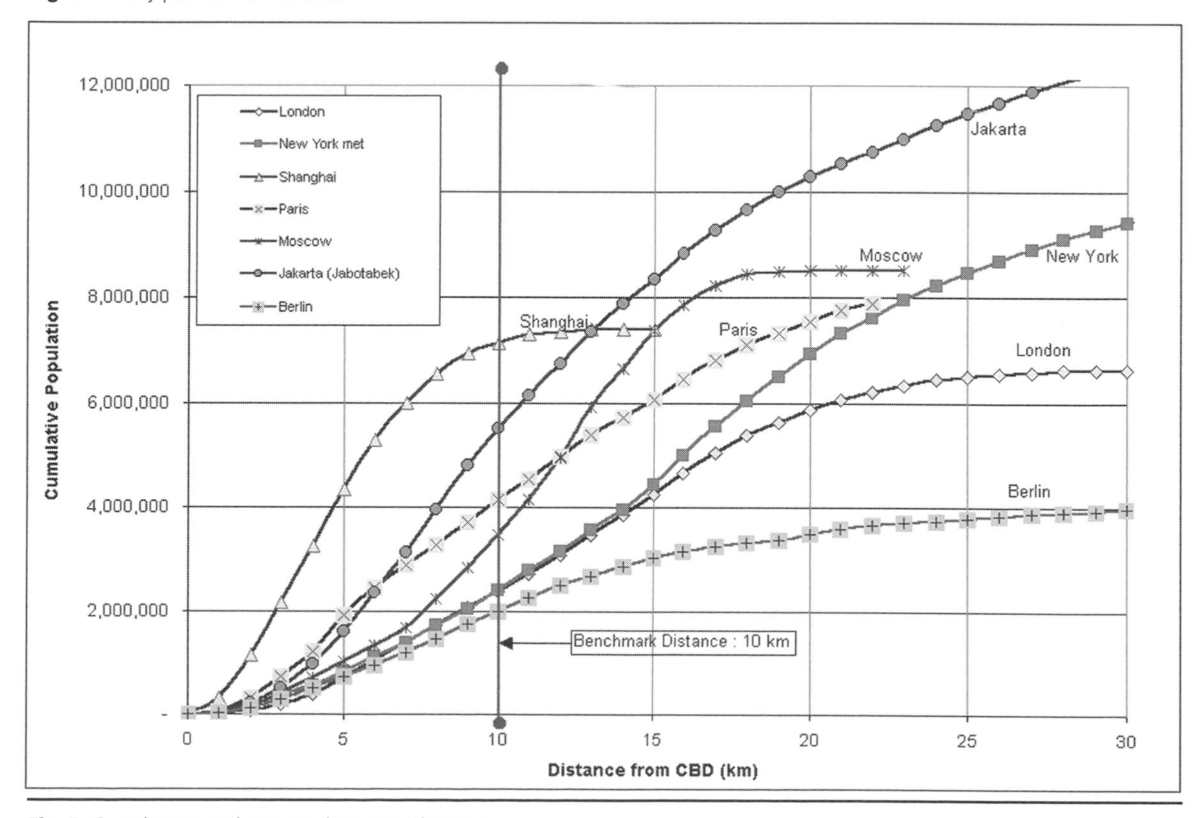

Fig. 7. Cumulative population per distance to the CBD.

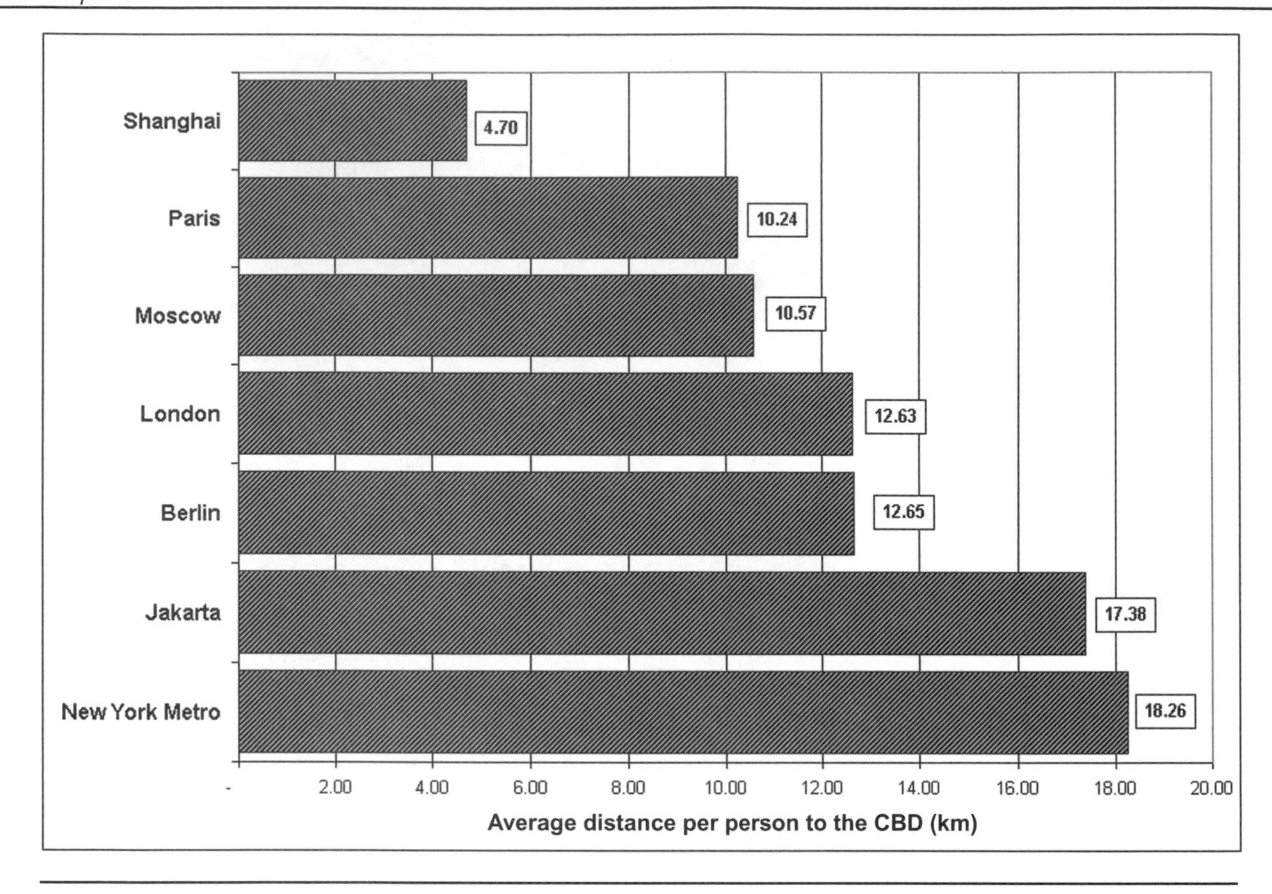

Fig. 8. Average distance per capita to CBD.

The impact of cars on the environment just in spatial terms (not including pollution) is much more disastrous in very dense cities. In the long run these cities will be confronted with a decision to either decrease densities—*i.e.*, destroy a significant portion of their city core—or to severely regulate the use of cars within the densest area of the city. Only Singapore, which by world standards is not particularly dense, has achieved serious measures to limit car traffic within the city core.

Density profile

Fig. 6 shows the density profile of the seven sample cities. The horizontal axis shows the distance from the city center at 1 kilometer interval from 0 to 45 kilometers. The vertical axis shows the population density expressed in people per hectare. The horizontal and vertical scale is the same on all graphs of Fig. 6. However, the vertical axis range from 0 to 300 people/hectare for all cities except for Shanghai where vertical axis range from 0 to 1,100 people per hectare.

This profile is established by calculating the number of people within each consecutive ring centered on the central business district (CBD), calculating the built-up area within the same rings, then for each ring dividing the population by the built-up area.

This graph shows that the spatial distribution of densities follows approximately the same pattern—a negatively sloped exponential curve—for all cities except Moscow. This is consistent with what urban economists Alonzo (1964), Muth (1969) and Mills (1972) present about the effect of land prices on densities. Moscow did not have a land market for seventy years so one should expect its density profile to be different from cities in market economies.

But what about Shanghai? This city did not have a land market for forty-five years but shows a density gradient much steeper than cities like Paris or New York where land markets exerted their pressure without interruptions for several hundred years. The only convincing explanation one can surmise for this puzzle is that the dominant use of the bicycle as a means of transportation during forty-five years shaped the density profile in the same way as the land market would have done it. Shanghai's density profile, as shown on Fig. 6, corresponds to a census taken in 1987. At that time, much of the city's housing stock had been built in pre-Revolutionary time. Very little new housing was added in the fifties because the national priority was the development of people's communes in rural areas. Then, from the mid-sixties to the end of the seventies, came the Cultural Revolution during which time many cities lost population to the countryside such that no new urban housing was built. The city absorbed additional population by densification and subdivision of the existing pre-revolution housing stock. New traditional socialist housing started to be built in significant numbers only in the eighties, a period too short to significantly affect the density profile. This might explain why the density does not rise in the Shanghai suburbs as it does in Moscow, but it does not explain the extremely steep negative gradient as compared to other cities in the sample. The dominant use of the bicycle as a means of urban transportation is the only explanation for the steep gradient. When people use a bicycle to commute in a big city

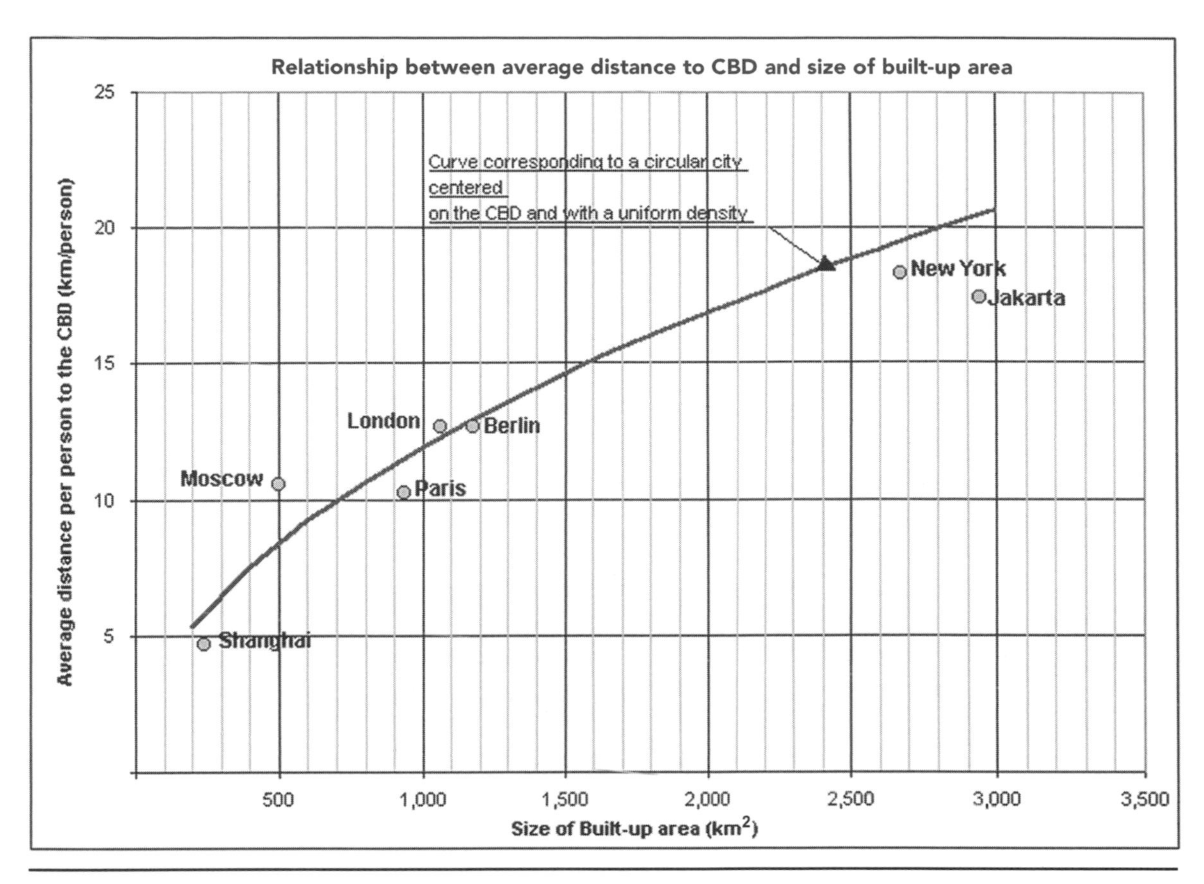

Fig. 9. Relationship between average distance and size of the built-up area.

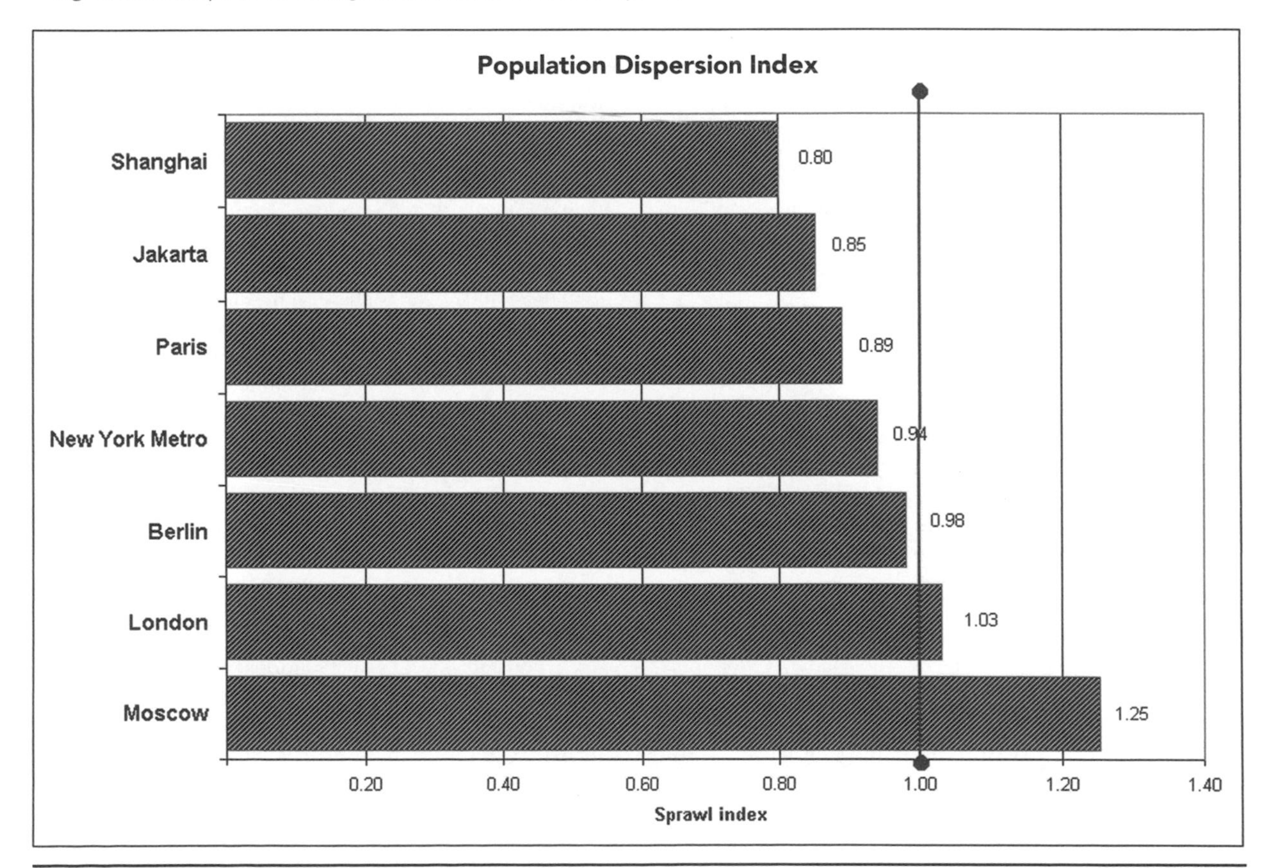

Fig. 10. Population dispersion index.

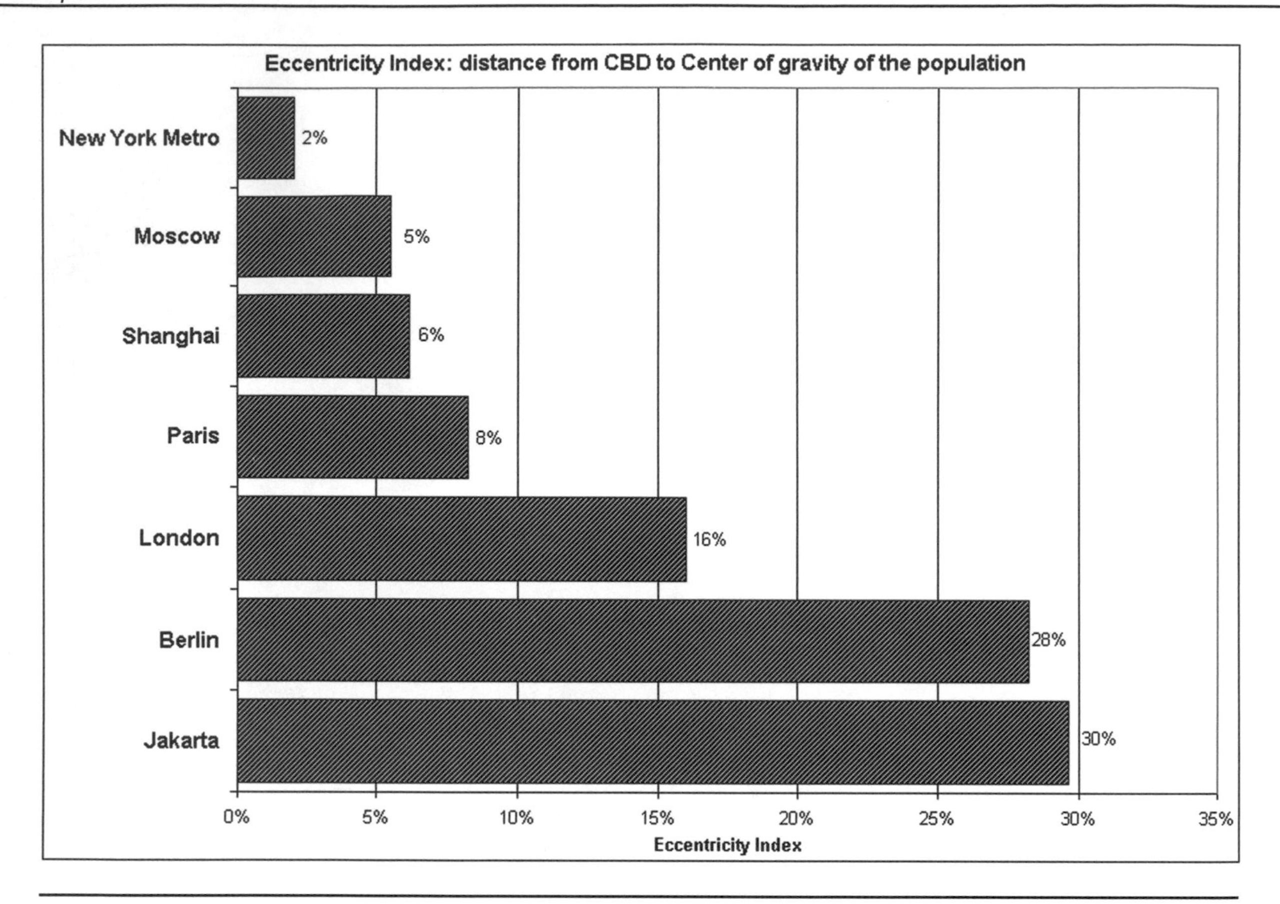

Fig. 11. Eccentricity index.

like Shanghai, "every kilometer counts." Houses located close to the center would have received a much higher pressure to densify than houses located in the periphery. Hence subdivisions and densification must have happened in a much more intensive way in the center than in the suburbs. The effort made to bicycle in the hot Shanghai summer and the cold and damp winter acted as a substitute for land price.

An additional observation could be made concerning the density profile of London. Between 26 and 30 kilometers (16 to 19 miles) from its center, London density rises slightly. This is the green belt effect. Although London's green belt is nowadays hard to detect on the ground, it did exist for a long time and it has the effect of raising density, first, because the reduction in land supply brought about by the green belt's regulatory restrictions and second, by the "amenity effect" offered by the green belt itself. Both factors contributed to an increase of land prices in the vicinity of the green belt, and consequently to an increase in density.

The density profile of a city in this sense is similar to the rings of a tree: it shows the history of a city and the markings of many events of the past that have affected the spatial structure.

Population by distance to center of gravity or to the CBD

The graph of Fig. 7 is particularly relevant for cities that are dominantly monocentric (the case for the seven cities in our sample). On the graph, the horizontal axis shows the distance from the CBD in kilometers and vertically the cumulative number of people. For instance, in a radius within 10 kilometers from the center Shanghai has more than 7 million people, Jakarta 5.5 million, Paris 4.1 million, Moscow 3.5 million, New York and London 2.5 million, and Berlin 2 million. The benchmark distance of 10 kilometers (6.2 miles) is interesting as it is a distance that, in urban areas, can easily be covered by bicycle in about 40 minutes and about 15 minutes with a tramway. We should note that the number of people that can have easy access to the center is very much related to the density profile shown on Fig. 6. The steeper the gradient of the density profile, the higher is the accessibility of the center. Note that while the Moscow average density (169 people/hecate) is about twice that of Paris, about 600,000 more people than in Moscow are located at less than 10 kilometers from Paris center. The spatial structure of Paris gives it a decisive advantage over Moscow in terms of accessibility. This advantage is easily translated into an advantage in the functioning of labor and consumer markets and in environmental cost.

Average distance per capita to the CBD

The average distance per person to the CBD is directly linked to the spatial structure. The distance is directly linked to the cost of getting to the city center and is proportional to the length and costs of random trip across the metropolitan area. It would be expected that the average distance per person would increase with the size of the built-up area. But one can see from the graph of Fig. 8 that this is not necessarily the case. The built-up area of Paris, (937 km^2 or 362 sq. mi.) is nearly twice the size of the built-up area of Moscow (503 km^2 or 117 sq. mi.), but the average distance per person to the center in Paris is slightly less than in Moscow. This apparent paradox is due to the distribution of densities within the built-up area, as seen on the density profiles for each city shown in Fig. 6. In this case the practical consequence of the positive density gradient of Moscow (density increasing with distance to the center instead of decreasing like in the other cities in our sample) is to lengthen the distance to the center, and this in spite of Moscow's compact aspect in Fig. 1.

Relationship between average distance per person to the CBD and built-up area

The graph of Fig. 9 shows horizontally the size of the built-up area expressed in square kilometers and vertically the average distance per person to the CBD. The curve on the graph shows the relationship between size of the built-up area and average distance per person for a fictitious city whose shape would be a circle and that would have a uniform density. This is useful to indicate how the average distance per person would vary when the shape stays constant but when the size of the built-up area becomes larger. On the graph, the relative vertical distance from the point representing each city and the curve gives an indication of shape performance. The farther below the curve that point that represents a city falls, the better is its particular shape performance. The farther it is above the line, the less favorable is the performance. For instance, we can see more clearly the relative performance of the spatial organization of Paris and Moscow as discussed above.

Population dispersion index

The measure of the average distance per person to the CBD in case of a monocentric city—or to the center of gravity in case of a polycentric city—provides a useful indicator of dispersion for a given city over time or between alternative spatial options. However, to have a comparative measure of shape performance between cities, it is necessary to have a measure of dispersion independent of the area of the city. All else being equal, in a city with a small built-up area the distance per person to the center will be shorter than in a city with a larger built-up area. To correct for the area effect, the index of population dispersion used in the calculations for Fig. 10 of this article is the ratio between the average distance per person to the CBD and the average distance to the center of gravity of a circle whose area would be equal to the built-up area.

The index of dispersion (Fig. 10) is thus independent from the area and from the density of a city. It reflects only the shape performance. It is thus possible to use the dispersion index to compare cities of very different sizes and of very different densities. A city of area "X" for which the average distance per person to the CBD is equal to the average distance to the center of a circle of area equal to "X" would have an index of dispersion of 1.

The dispersion index offers an interesting insight in the importance of a city spatial structure in reducing trip length. One can see that high density itself does not necessarily reduce trip length.

Eccentricity index

In the terms of this article, a city CBD may be termed as eccentric when it does not coincide with the center of gravity of the population. The eccentricity index shown in Fig. 11 is measured by calculating the percentage of the distance between the CBD and the center of gravity of the population over the average distance per person to the CBD. The larger that this percentage is, the more eccentric is its CBD. An eccentricity index below 10% is considered satisfactory. Between 10 and 20% the index indicates a mild eccentricity. If above 20% the city can be said to indicate a high eccentricity.

In a dominantly polycentric city, this indicator is not relevant, as only very few trips are directed to the CBD. By contrast, in a city that is predominantly monocentric, a large distance between the center of gravity and the CBD increases trip length significantly. On the graph of Fig. 11, only Berlin and Jakarta indicate a high eccentricity. Berlin up to 10 years ago, before the fall of the Berlin Wall, had in fact two CBDs, one in West Berlin, the other in East Berlin. The CBD used in this calculation is the CBD of West Berlin because land values are higher there than in the old East Berlin CBD. However the center of gravity of the city is in between the two centers in an area formerly occupied by the infamous wall, and precisely at the location of the new *Postdamerplatz* that is currently being built as the new business center of Berlin. When the new office buildings and shopping areas of the *Postdamerplatz* are completed and operating, the new CBD of Berlin will then have an eccentricity close to 0.

Jakarta's high eccentricity is due to its location along the sea. Most seaports are eccentric because the port area constituted originally at the time of the city creation the major center of activity. As the city develops, other services also develop and the port becomes an ancillary activity. Because of the asymmetry imposed by the proximity of the sea, the center of gravity of the population is constantly moving as the city area grows. In the case of Jakarta, the CBD has been moving away from the port since a number of years and has been following the center of gravity but with a lag of a few years. The CBD of Jakarta has been moving by 12 kilometers (7.5 miles) in the last fifteen years, always following the same path as the center of gravity, increasing thus the performance of the city shape. This displacement was not due to the initiative of urban planners but to market forces. Locations of higher accessibility became more valuable and attracted new businesses.

5 CONCLUSION: URBAN PLANNING AND DESIGN CAN MONITOR AND MODIFY URBAN SPATIAL STRUCTURE

Because of the impact of a city spatial structure on its economic and environmental performance, urban planners and designers should constantly monitor its evolution. Planners may use the basic indicators described above and additional ones more specific to the application. By mapping the location of new building permits and real estate investments, development trends can be easily monitored. The use of Geographical Information Systems (GIS) greatly facilitates this work.

Master planning usually includes broad municipal objectives that need to be translated into spatial terms. For instance, a municipal objective to increase the use of public transport implies a more compact city, maintaining a high degree of monocentricity and relatively high density. On the contrary, a policy aimed at increasing the consumption of housing by making it more affordable would imply an extension of the city and the opening of large peripheral areas for development. A policy aiming at improving the environment inside the city will require using more land for open space and for public facilities, thus increasing the size of the extended boundaries of the city at the expense of the agricultural or undeveloped areas in the periphery. By contrast, a policy aimed at protecting the natural environment outside the city would require higher density and a more compact development, probably implying a lower standard of land consumption and higher housing prices. There are no win-win urban spatial strategies. Most urban development policy involves painful trade-offs, which can be accomplished only through an accountable democratically elected municipality. No land use optimization can be achieved from a technical point of view alone. Political objectives are needed to set the strategy.

Urban planners and designers have to devise the tools that will help implement the spatial strategy derived from the municipal objectives. Planners have at their disposal only three types of tools to help shape an urban spatial structure:

- Land use regulations, part of which will be a zoning plan,
- Primary infrastructure investments and,
- Property taxation.

The impact of these tools will only be indirect. The real estate market, reacting to constraints and opportunities provided by regulations, infrastructure, and taxation, will in reality shape the city. Urban design and master plans will not have much direct influence. This is the reason why spatial indicators need to be constantly monitored to verify that the city is evolving in the spatial direction consistent with the municipal objectives. Many regulations and infrastructure investments have spatial side effects that are often unexpected. For instance, green belts have a tendency to increase the price of land and housing and to increase commuting distance as households and firms are looking for cheaper land on the other side of the green belt. Low density zoning designed to preserve more green areas have a tendency to generate sprawl by forcing land users to consume more land than would otherwise be at risk in the absence of zoning. New light rail and metro lines do not necessarily increase demand for public transport if densities are too low or the city structure is dominantly polycentric.

Urban design and planning is not an exact science but has to proceed by trial and error. That makes it all the more important to monitor and design for the constantly evolving spatial organization of cities. ■

REFERENCES

Alonso, William. *Location and land use.* Harvard University Press, 1964.

Bertaud, Alain. *Cracow in the Twenty First Century: Princes or Merchants? A city's structure under the conflicting influences of Land Markets, Zoning Regulations and a Socialist Past.* ECSIN Working Paper #8, The World Bank 1999.

Bertaud, Alain and Bertrand Renaud. Socialist Cities Without Land Markets. *Journal of Urban Economics*, 41, 1997, pp. 137–51.

Bertaud, Alain and Stephen Malpezzi, *The spatial distribution of population in 35 World Cities: the role of markets, planning and topography.* The Center for Urban Land and Economic Research, The University of Wisconsin, 1999.

Carlino, Gerald A. Increasing Returns to Scale in Metropolitan Manufacturing. *Journal of Regional Science,* 19, 1979, pp. 343–51.

Cervero, Robert. Jobs-Housing Balancing and Regional Mobility. *American Planning Association Journal,* 55, 1989, pp. 136–50.

Clark, Colin. Urban Population Densities. *Journal of the Royal Statistical Society,* 114, 1951, pp. 375–86.

Epple, Dennis, Thomas Romer and Radu Filimon. Community Development With Endogenous Land Use Controls. *Journal of Public Economics,* 35, 1988, pp. 133–62.

Fischel, William A. *Regulatory Takings: Law, Economics and Politics.* Harvard University Press, 1995.

Goldner, William. Spatial and Locational Aspects of Metropolitan Labour Markets. *American Economic Review,* 45, 1955, pp. 111–28.

Hall, Peter, *Cities of Tomorrow*, Blackwell, 1988, pp. 210–19.

Hamilton, Bruce W. Zoning and the Exercise of Monopoly Power. *Journal of Urban Economics,* 5, 1978, pp. 116–30.

Ihlandfeldt, Keith R. Information on the Spatial Distribution of Job Opportunities Within Metropolitan Areas. *Journal of Urban Economics,* 41, 1997, pp. 218–42.

Mills, Edwin. S. and Byong-Nak Song. *Urbanization and Urban Problems.* Harvard, 1979.

Muth, Richard F. *Cities and Housing.* University of Chicago Press, 1969.

Prud'homme, Remy, Managing Megacities, *Le courier du CNRS*, No. 82, 1996, pp. 174–176.

Sveikauskas, Leo. The Productivity of Cities. *Quarterly Journal of Economics,* 89, 1975, pp. 393–413.

Vikram Bhatt and Witold Rybczynski

Summary
This article describes an approach to improving housing and communities in developing world cities where there are large squatter settlements largely developed spontaneously and without planning, providing shelter for the "other half" of the world population estimated to be living in such subsistence conditions. An approach to analysis reveals existing living and building patterns as assets appreciated as viable social and economic benefits. The discussion suggests alternates to large scale "sites and services" approaches that permit greater community and individual choices—here called the "self-selection process"—within a framework of urban design for housing and communities.

Key words
housing, infrastructure, poverty, roads, self-help building, shelter, sites and services, slum upgrading, street, trees, workplace

How the other half builds

1 ELEMENTS OF INFORMAL SETTLEMENTS

Urban housing is a worldwide need, especially urgent in rapidly growing cities of the developing world, evident in unplanned and informal communities. These are examples of the informal sector housing—often described as squatter settlements or "slums"—that represent latent assets and the basis of urban solutions. It is moreover a solution that appears to deny conventional planning orthodoxy. In spite of its often spontaneous and improvised character, the informal sector, which maximizes self-help and mutual aid building, has been virtually the only group that has had any success in providing appropriate, low cost solutions to the shelter problems of the urban poor.

As a result of the rapid and extraordinary growth of cities, the shelter problems of the urban poor in less developed countries have increased in scale and in severity. Conventional resources are insufficient to deal with this situation, with the result that most low-income urban housing has been provided by the so-called informal sector, which exists and often thrives outside the traditional market economy. The informal sector, which can be found in every less-developed country, is characterized by decentralization and fragmentation, flexibility and by the small scale of its entrepreneurial activities. Understanding these as assets provides the basis of housing to revitalize informal settlements in the developing world. The same insights and approach can assist asset-based urban community development throughout the globe. The aim of the research reported in this article has been to understand this phenomenon, with a view to developing technologies and methods that will be suitable to application in the context of the developing world. The study is the result of site surveys carried out in four slum areas in the city of Indore, India.

There is nothing basic about "basic housing"—an inaccurate and misleading term. Present-day "standards" are a poor tool by which to evaluate this process. Standards that are based on "universal" measures, whether homogenized to the point that local assets are not appreciated, set false and unrealizable measures, too often unattainable in situations of greatest need. They reflect a view of solutions that is not only culturally inappropriate but also inadequate. A far more adaptive and appropriate set of settlement standards needs to be evolved. Housing and urban standards should seek to accommodate to and build up from a local situation, serving as guidelines to realizable improvements, rather than to reorganize. They should reflect the oftentimes harsh reality of the urban poor, but should respond to their special needs and capacities for improvement, not to an idealized set of criteria that may be inappropriate or unattainable.

The predominant housing planning methodology used in developing countries with international financing is to provide "sites and services," that is, to layout large tracts of small land plots and then installing an infrastructure of roads, water and sewerage. This methodology recognizes the distinction between house plots and circulation spaces but does not deal effectively with the diversity of the activities that take place in the "street." In addition to accommodating movement, the street is a place of work, shopping and commercial activity. It is the setting for social and religious functions. These activities and the spaces that they occupy are here described under the following element headings:

- House extensions
- Workplaces
- Small shops
- Trees
- Public structures
- Vehicles
- Access streets

Credits: The research described in this article was carried out by the Minimum Cost Housing group of the McGill School of Architecture, Montreal, Canada (see References).

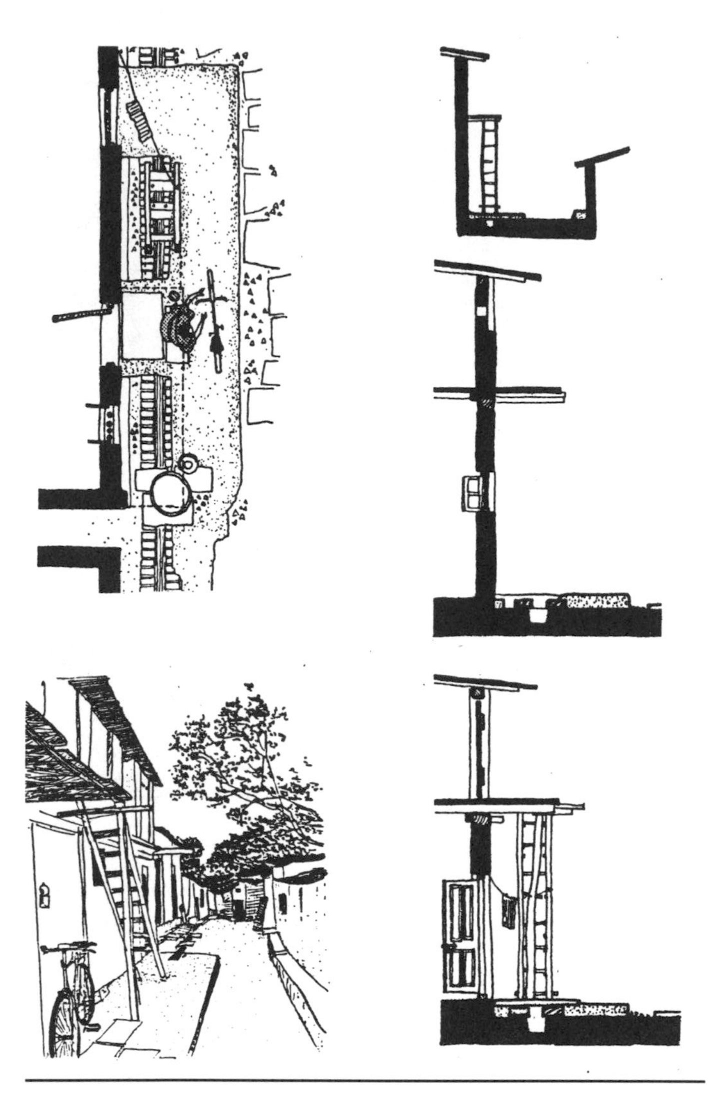

Fig. 1. Plan and sections: front stoop extensions at two levels.

Fig. 2. Basket-making.

House extensions take different forms, and represent a mediating zone between the house and the street. Another kind of space is the workplace. A wide range of commercial activities takes place in and around the home, and requires different sorts and sizes of space. Related to income-generation, likewise, are small shops, booths and kiosks, frequently a part of the dwelling or attached to it.

Streets and other public spaces are greatly influenced by the presence of *trees*, which have physical, social and sometimes religious significance. Streets are also marked by the presence of *public structures* such as public water taps, temples and plinths. A separate section of this study has been devoted to the *vehicles* that circulate in low-income urban settlements. Finally, we have studied the widths and character of the *streets* themselves.

House Extensions

There exists, in slum housing, a complex hierarchy of what we have called *house extensions*: spaces in front of the home that are nominally a part of the public realm, but that have acquired a private character through use, and through various physical modifications.

The simplest, and smallest, of these extensions is the stoop, often no more than an enlarged step, made out of beaten earth, stone or concrete. The stoop is usually less than 1 m (39 in.) wide, and is used as a step, as a seat, or as a workbench (Fig. 1).

A further elaboration of the porch is the outdoor room, a roofed platform that achieves a greater measure of privacy through the use of walls on one or two sides. The outdoor room can accommodate some fairly private activities, such as washing.

Why do people build house extensions? Plots are extremely small, and many activities cannot be accommodated inside the house. The porch or platform is a feature of rural housing that people understand, and which is easily integrated into everyday life. The public nature of this part of the house allows a greater contact with street life. And of course, platforms and porches are inexpensive.

Workplaces

It has long been known that slums and squatter settlements are not only places for living, but also places for working. The term "informal sector" precisely describes this phenomenon. Just as the urban poor take into their own hands the provision of inexpensive, appropriate shelter, so also do they participate in economic activities whose importance and complexity should be recognized.

These economic activities take many forms. Some of them are, in effect, service industries serving the slum directly. Inexpensive, usually recycled, construction materials and building components are produced for immediate, local consumption.

There are also many work activities that qualify as cottage industries, that is, the raw materials are provided by entrepreneurs who purchase the finished product from the slum worker (Fig. 2).

Slum workplaces are characterized by simple hand tools and primitive, although by no means crude, techniques. Most of the work activities require neither water nor electricity. As a result, the workplace is often mobile, moving easily in a fixed enclosure. This calls attention to the importance of streets and walkways as not just spaces for

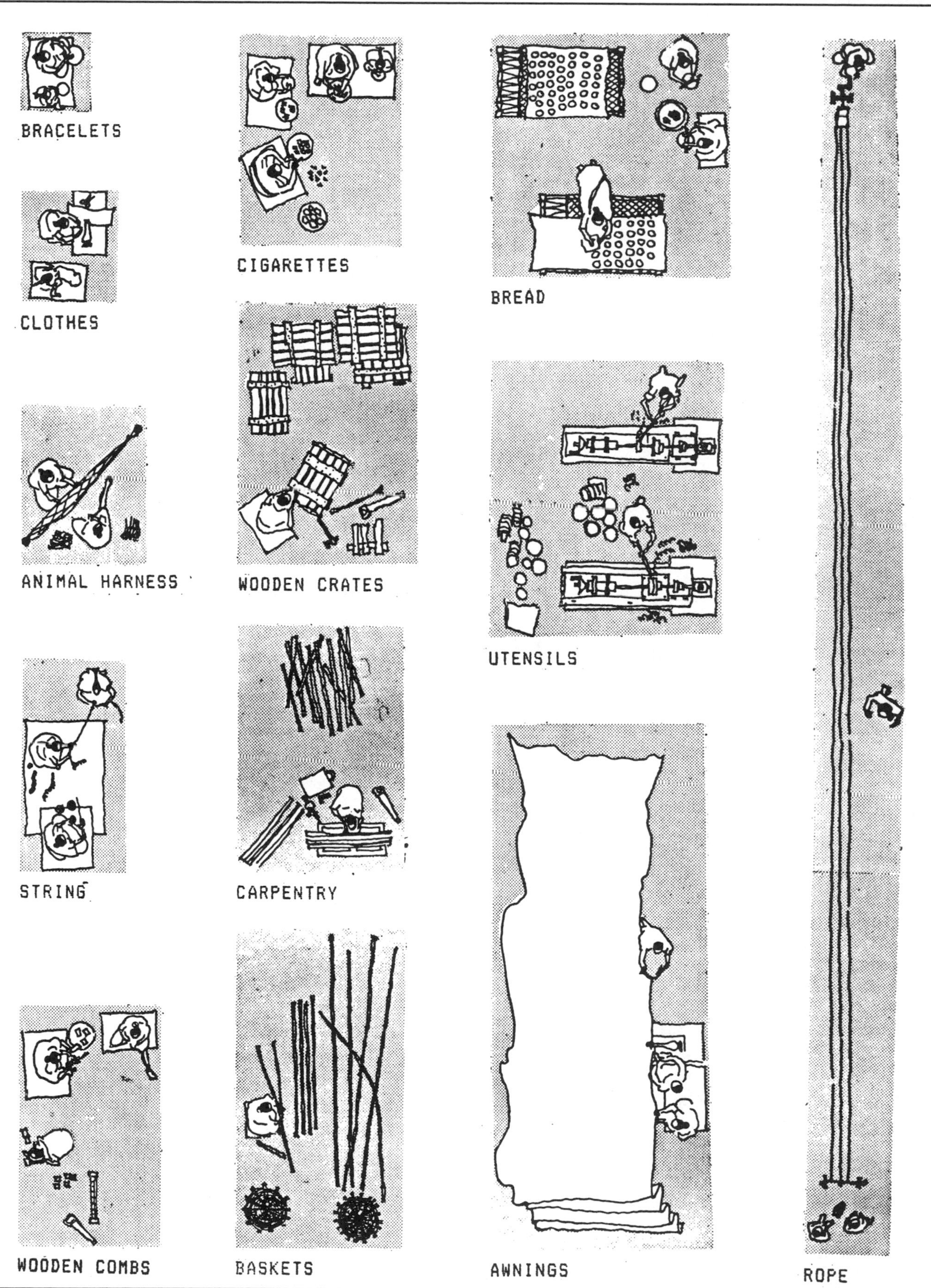

Fig. 3. Summary of workspaces typical of informal settlements.

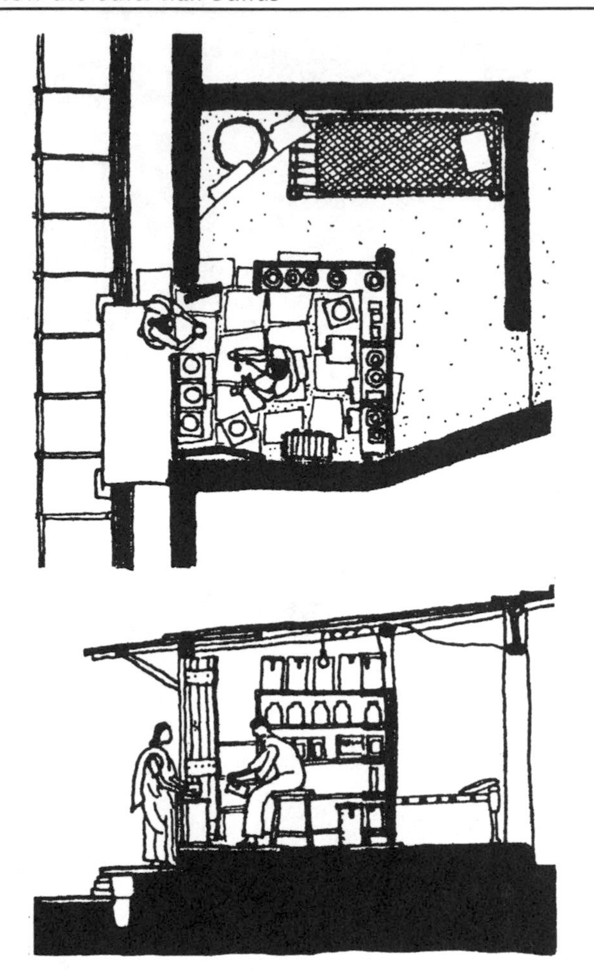

Fig. 4. Plan and section: groceries.

Fig. 5. Tree places.

circulation and socialization but also as workplaces. Unlike conventional housing, in which work and living are always physically separate, and distinct, low-income urban shelter requires the planner to be sensitive to a much richer mix of family, social and work activities (Fig. 3).

Small shops

Small neighborhood shops exist in addition to formal markets or more public shopping streets. They are usually distinguished by their extremely small size, the nature of their commerce, which is exclusively oriented to local needs, and their intimate proximity to the home (Fig. 4). The location of small shops follows sound commercial principles, that is, they tend to be located where there is the greatest exposure to passersby. This can be expressed as a hierarchy: along main streets, around squares and open spaces, at street corners, and, least desirably, along smaller streets.

Like small-scale workplaces, small shops are an important income generator in informal sector housing. They also distribute goods that are economically attuned to local resources and needs. This is especially true of repair shops. In the case of tea-shops, they function as neighborhood meeting places and informal social centers.

A close examination of small shops in marginal housing helps to explain why the "commercial centers" and "markets" that are planned for sites and services projects frequently stand empty and unused. Not only is the concept of centralized shopping inappropriate to the living patterns found in informal sector housing, it also fails to provide the economic and social benefits of "living over the store."

Trees

At first glance, landscaping seems an extraneous, if not irrelevant, issue in the context of low-income urban shelter. However, in existing slums and unplanned settlements, trees are conspicuously maintained, protected and planted by the inhabitants—without official assistance, and with some considerable labor. Why is this the case?

Trees, especially large species, not only provide shade but play the role of public buildings, and become a substitute for the arcades, porches and covered outdoor spaces that are a part of the normal urban fabric, but that are absent in most slums. Where large trees exist, they become the focus of public gathering places; where they do not, they are planted and cared for (Fig. 5).

Public squares, whose overall dimensions have been clearly dictated by the spread of the large trees located in their center, are a common feature of the slums included in this study. In almost all cases, the public square has grown up around, and chronologically later than, the tree. The space beneath such shade trees is used as an outdoor classroom, a meeting place, a workspace or a covered market. The central "meeting tree" which originates in the village (and not only in India) also reminds the urban immigrants of their rural roots.

Public structures

Traditional housing areas in Indian towns contain temples, sitting platforms, stairs, water fountains, bird feeders (Fig. 6), arcades, signs and entry gates. These public structures—as opposed to private house extensions—play an important role in establishing neighborhood identity and are important landmarks and visual reference points. They contribute to the rich texture of traditional Indian towns.

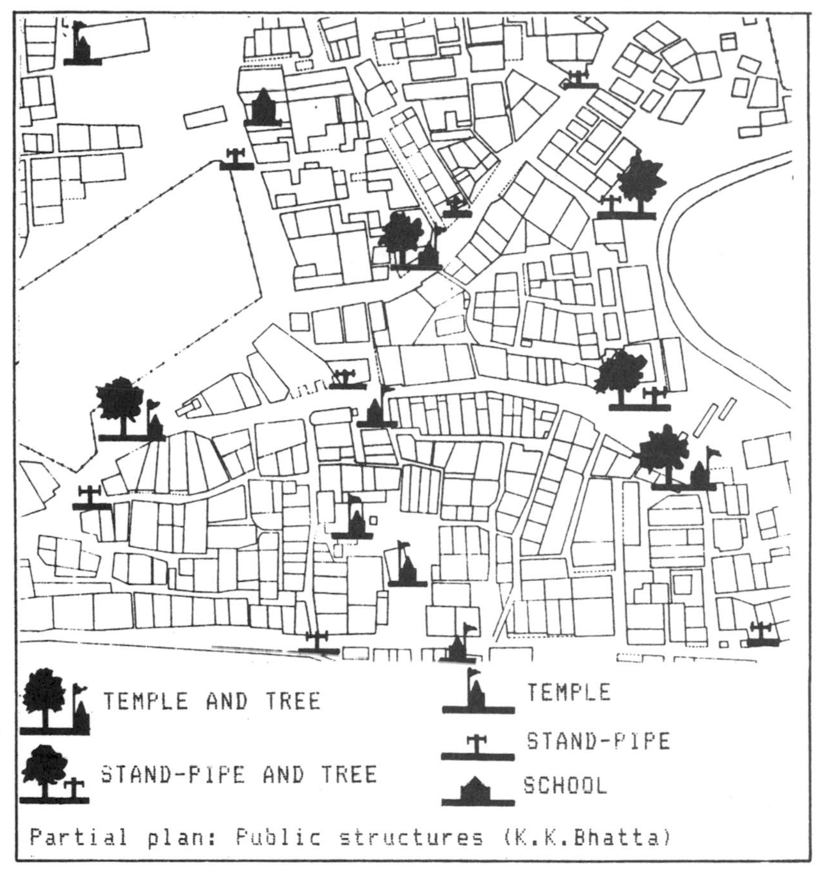

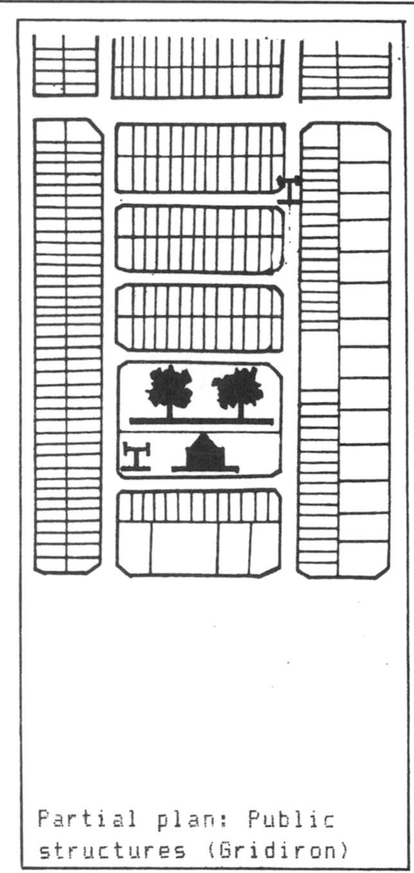

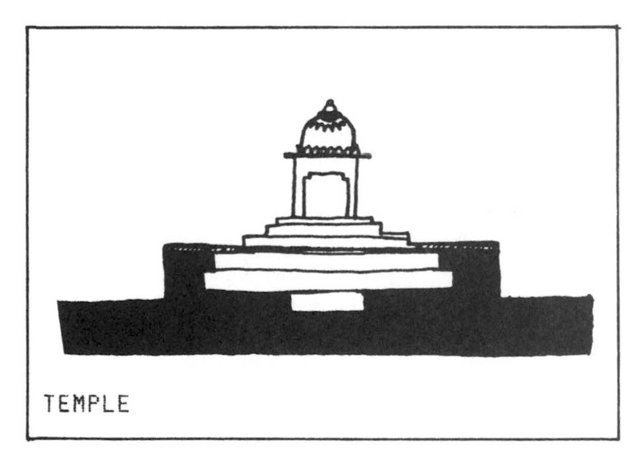

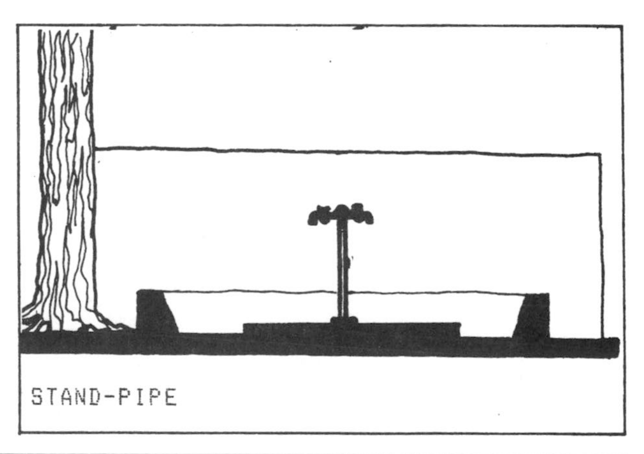

Fig. 6. Summary of public structures.

Fig. 7. Vehicles in informal settlements.

Fig. 8. Bullock cart.

Fig. 9. Narrow lane.

The public spaces of most planned low-income shelter projects are characterized by roads, drainage ditches, perhaps street lighting. . . and that is all. "Infrastructure" in this context is largely restricted to underground services. Planning is often barrack-like allocation of building plots and streets are simply movement spaces. It could be argued that in "basic" shelter, anything more would be a luxury and would not be required by the low-income inhabitants.

In informal settlements, on the other hand, there are evident signs of attempts to introduce identity into the public environment. These attempts may only be minimal. For one thing, beyond the neighborhood level, there is no "public" authority in these settlements. Whatever is done must be carried out by small groups of individuals, with minimal resources. But this only gives more importance to those public structures that do exist. They indicate the desire, and need, for personalizing and giving identity to the public space.

Vehicles

Automobile traffic in slums and squatter settlements is, naturally enough, extremely limited. This does not mean, however, that there is no vehicular traffic. If anything, there is a greater variety of vehicles in slums than in conventional housing.

The smallest vehicles include bicycles, mopeds, motor scooters and motorcycles. They are used for personal transportation; all the more important since informal settlements are usually some distance from the city (Fig. 7).

A second type of vehicle found in informal settlements is that which is used by the inhabitants for income generation. It includes pushcarts, pull carts, bicycle rickshaws, motorized rickshaws, and motorcycle-powered three-wheelers. Motorized rickshaws, especially, are the most common type of mass transit system, and can be found in all parts of the slum. Another, larger type of vehicle, is the animal-drawn cart used to move heavy materials (Fig. 8). Parking is a makeshift affair. Only the smallest vehicles can be stored inside houses, or in porches. The rest are simply left in the street, as close to the house as possible.

Access streets

Standards for the widths of streets vary considerably from place to place, but in most developing countries they are historically related to European codes. The general observation is often made that streets in planned housing areas tend to be extremely wide, considering the actual size and low frequency of vehicular traffic. This is part of a utopian and unachievable vision, which creates large voids that separate neighborhood and community elements.

In informal settlements, streets are generally narrower than streets with similar functions (shopping, for example) in planned housing areas (Figs. 9 and 10). While a main feeder in a planned project might be about 7 m (23 ft.) wide, the same street in the slums we surveyed was rarely wider than 3.5 m (11.5 ft.).

One should not imagine, however, that the widths of streets in informal sector housing settlements be reduced to minimal dimension. The largest number of access streets (in linear terms) found in long established squatter settlements and slums are *narrow streets* from 1.5 to 3 m (5 to 10 ft.). This reflects the large amount of social, work and domestic activity that takes place in the street and the need for

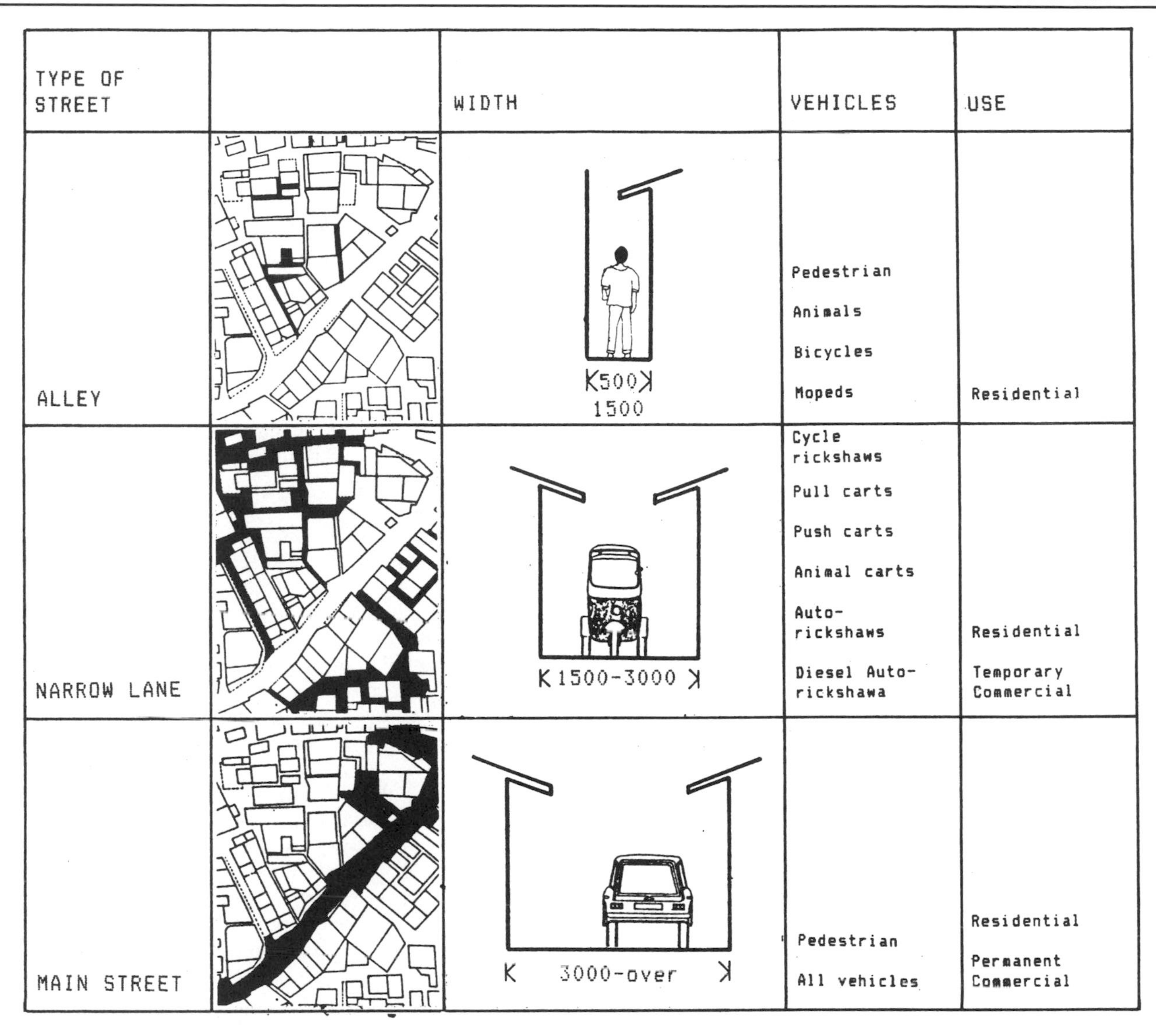

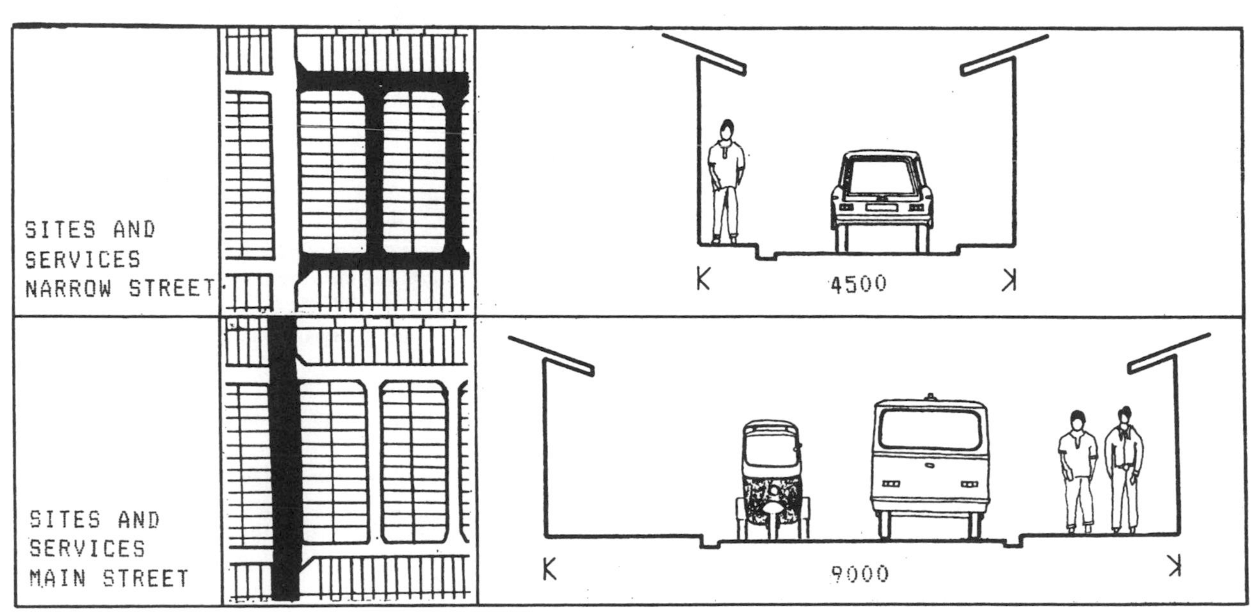

Fig. 10. Summary and comparison of access streets.

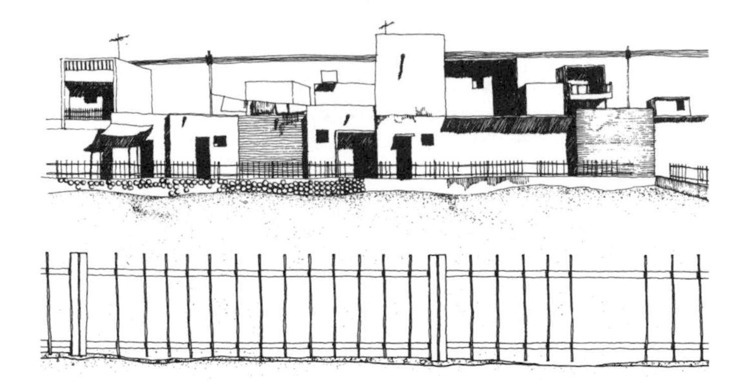

Fig. 11. Sites and services: spaces that are not well integrated remain empty and unused.

Fig. 12. Informal settlements: housing clusters formed around intimate family links.

streets wide enough to accommodate them. However, in the new, less-established slums, it is the wider *main streets* that predominate. This suggests that larger standards were established in the early life of the settlement and that streets become narrower as house extensions grow larger.

2 PROBLEMS ASSOCIATED WITH SITES AND SERVICES AND FORMAL HOUSING

One popular approach to the provision of low-cost housing for the urban poor is the so-called "sites and services" strategy. This has been advocated by several international aid agencies such as the World Bank and the United Nations and adopted by many housing authorities in developing countries since the 1970s. In a typical sites and services project, the serviced land is provided by the housing authority, while the actual house building is left to the homeowner. This change of role—from being a provider of housing to that of a facilitator that makes infrastructure and other related facilities available to users—represents a significant departure from the traditional attitude of housing agencies (Figs. 11 and 12).

The success of the sites and services strategy can, in large part, be explained by its economic rationale. Most developing countries are poor and cannot subsidize housing on a large scale for their growing populations. Limited resources can be usefully distributed among a larger number of beneficiaries, however, if housing authorities refrain from building complete dwellings and commit public funds only to the provision of land and infrastructure. Typical sites and services schemes are formally subsidized packages of shelter and related services that range in complexity from simple "surveyed plots," to an intermediate level of "serviced sites," to an upper level of "core housing" complete with utilities and access to community-based services (Mayo and Gross, 1987). Depending upon the capacity of the beneficiary to pay for the housing, the authority can choose the level of servicing. This flexibility permits the formal sector to target its projects toward very low-income groups.

Sites and services make housing available to the urban poor at a relatively low cost. However, there are several problems associated with this design approach. A survey of key sites and services projects in major cities of India (Bhatt 1984) and several other studies of completed projects identified the following problems inherent in this design approach:

- The bias of economics in planning typically discounts the social aspects of design
- Projects lack quality and variety of open spaces.
- During project planning incorrect assumptions are made concerning family income and plot sizes.
- Projects lack variety of plot sizes.
- Projects do not provide multi-family plots.
- They follow a blind plot allocation process.
- Project planning and implementation takes a very long time.

Towards a new approach

The problems associated with sites and services range from the lack of culturally appropriate housing to the creation of impersonal urban environments to wrong assumptions about the clients' needs, and so on. The same problems are common in most public housing projects. These problems are not technical, but more fundamental in nature, and are closely associated with the way sites and services or formal housing projects are conceived and implemented. Comparison of conventional planned housing and informal settlements reveal essential urban design principles and practices.

Conventional sites and services and other formally planned low-cost housing schemes do not take into account the actual user. Conventionally planned projects are developed in a few definite stages by a planner or an architect, and involvement of the project participants in the decision-making process is kept to a minimum. In a typical sites and services project, the level of user participation is greater but limited to the construction of individual houses. The project participants are still excluded from the crucial planning and design stages of the project, and thus from the actual process of development.

To address the problems associated with conventionally planned projects, a new design approach, radically different from the formal production of housing, is proposed. It is called the "self-selection" design process, a term borrowed from economics.

3 SELF SELECTION DESIGN PROCESS

It is proposed that the development process of a traditional or unplanned urban settlement should be adopted, but in such a way that it would also overcome the problems of poor and inadequate infrastructure that are associated with unplanned settlements (Box 1).

For the new design approach to increase the user input, the following development strategies are proposed: no pre-conceived development plan; progressive provision of major infrastructure; self-selection of plots; and a free choice in the selection of different plot sizes and shapes. The development strategies are such that they complement each other and will sustain the new approach (Figs. 13–16).

The hypothetical settlement was developed in seven stages, having two distinct parts, corresponding to the responsibilities of the two parties involved in the design process, the planner and the heads of households, and related to infrastructure and self-selection of plots.

Summary

The worked example demonstrates that the self-selection design process represents a viable alternative for providing appropriate living environments for the urban poor in developing countries. The self-selection process has demonstrated that it can produce a lively and user responsive built environment that can also be cost effective. It can therefore offer appealing features to both parties involved in the formal production of low-cost housing in developing countries—the users and the formal sector.

For the users, the self-selection design process offers a general control over the creation of their own living environments. Through an ample control over the location, size and shape of their plots, and the configuration of surrounding public spaces and circulations, the

BOX 1. Principles of the self-selection process

Two overriding principles inherent in the planning process of informal settlements are autonomous growth and continuous development. The increased use of these two concepts in planning new housing could produce a better living environment.

Autonomous growth. The idea behind autonomous growth suggests that the user should be involved in the housing process at every level of design. For a built environment to be socioculturally appropriate it should have, as a primary element, the contribution of its future residents. To achieve this, user participation in the decision-making process should be increased from the "micro level" of individual homes to the "macro level" of the settlement. Conversely, the duties of the design team should be decreased to the level of a general regulator of the settlement.

Continuous development. Conventional modern housing schemes are developed in a single or a few definite stages. Continuous development, on the contrary, assumes the development to take place in an unbroken cycle of events. This prevents the settlement from adopting an artificial or mechanical character. The organic nature of the urban fabric, representative of the traditional cities and unplanned settlements, can be attributed to such an autonomous and piecemeal growth process.

FIGS. 13–16. PRINCIPLES OF THE SELF-SELECTION PROCESS

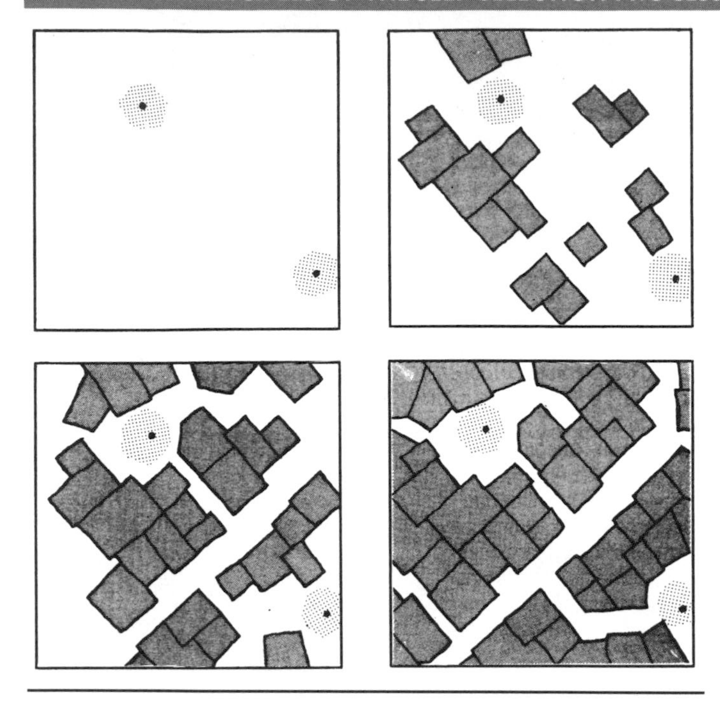

Fig. 13. No pre-conceived plan. There should be no pre-conceived plan to regulate the development and support the self-selection process. The location of streets, open areas and plots should take place in response to the requirements and aspirations of the project participants, to produce a culturally responsive environment. However, the creation of a set of rules is necessary to help the implementing team lead and ensure the development of a non-chaotic settlement.

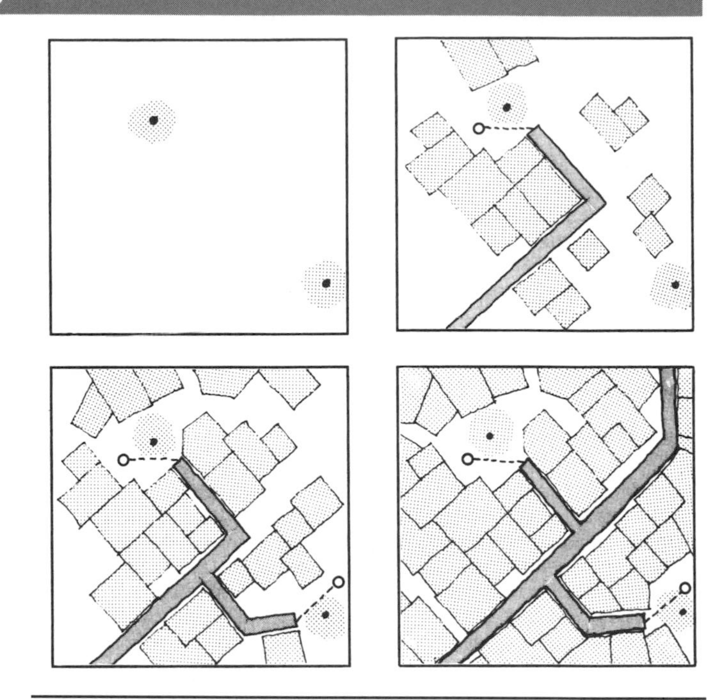

Fig. 14. Progressive infrastructure. The infrastructure should not be viewed as something to be made most efficient without any regard to the quality of the living environment. Rather, the infrastructure should be seen as a tool to assist the development of a new settlement, and also to serve the existing settlement. Instead of planning the entire infrastructure at once it should be introduced gradually. With the use of infrastructure—community water taps, public structures and paved roads—the designer can steer families in the desired direction and maintain a general control of the development.

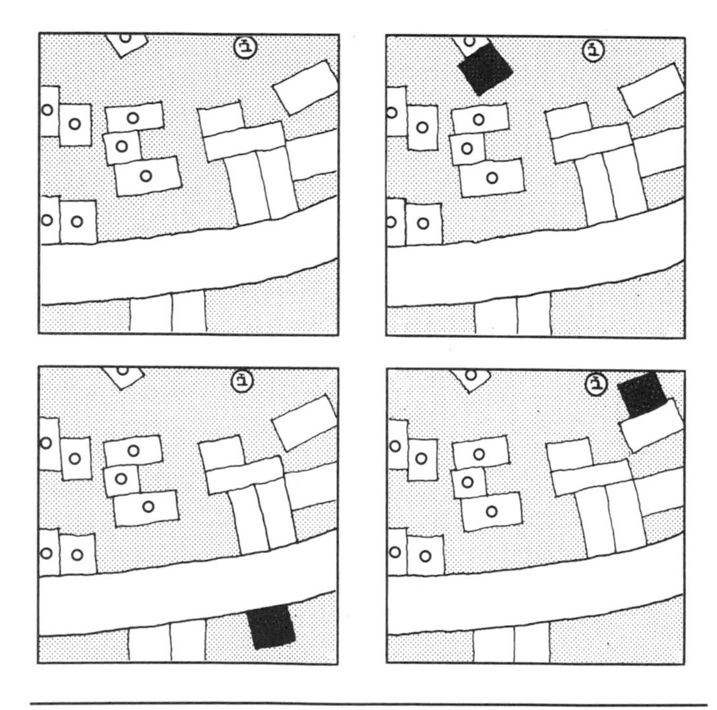

Fig. 15. Self-selection of plots. It is important to have an open plot-allocation system. In informal housing, families can select, according to their particular preferences, the location of their plots within the settlement. Under the self-selection design process, families will be free to choose the location of their plot.

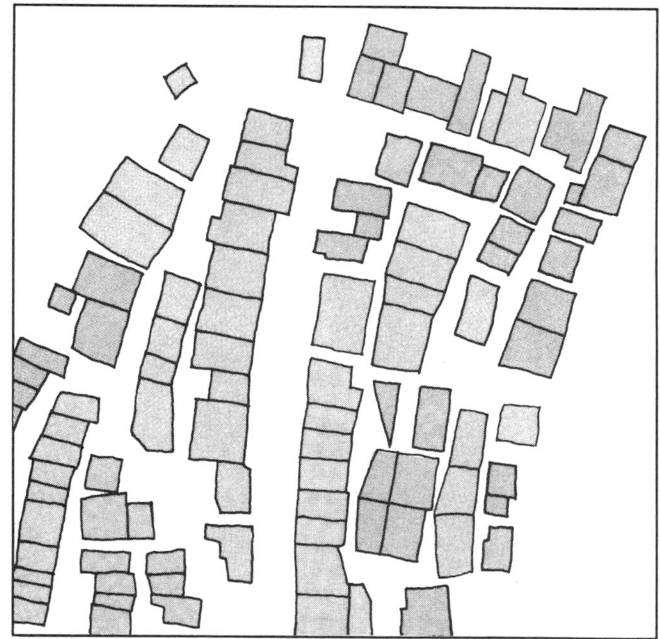

Informal Settlement, Indore, India: notice the variety of plot sizes.

Fig. 16. Variety of plot sizes. There should not be predetermined sizes or shapes of plots based on the economic classification of the beneficiaries. Plot sizes, shapes and proportions should be determined by the families themselves or self-selected. The plot size should be chosen according to individual family needs and its ability to pay for the plot.

self-selection process encourages the users to develop a built environment that matches their most particular needs, and supports a stronger community integration. More importantly, through an incremental upgrading of houses, infrastructure, and public amenities, the self-selection process offers the users the development of such a built environment according to their economic means (Fig. 17).

For the formal sector, the self-selection design process offers a lower initial investment on infrastructure, and a minimum involvement in the design, development, and maintenance of new residential projects. With such a lower investment, in both time and financial resources, the self-selection process offers the formal sector the possibility of developing a greater number of new residential projects. The self-selection design process, then, represents a viable and economic approach for coping with the growing housing demand in developing countries.

The principles behind the self-selection design process make it suitable for the development of new residential areas within the current formal housing context of developing countries. The self-selection process can be ideally suited for relocation projects where upgrading may not be possible, because of poor or dangerous location. Equally, the self-selection process can be used to develop new sites and services projects to accommodate the increased demand for low-cost housing. The flexibility inherent in the self-selection design process makes it also an effective planning method that can be easily applied to different contexts, since its basic rules may be modified to suit different cultural and regional needs. ■

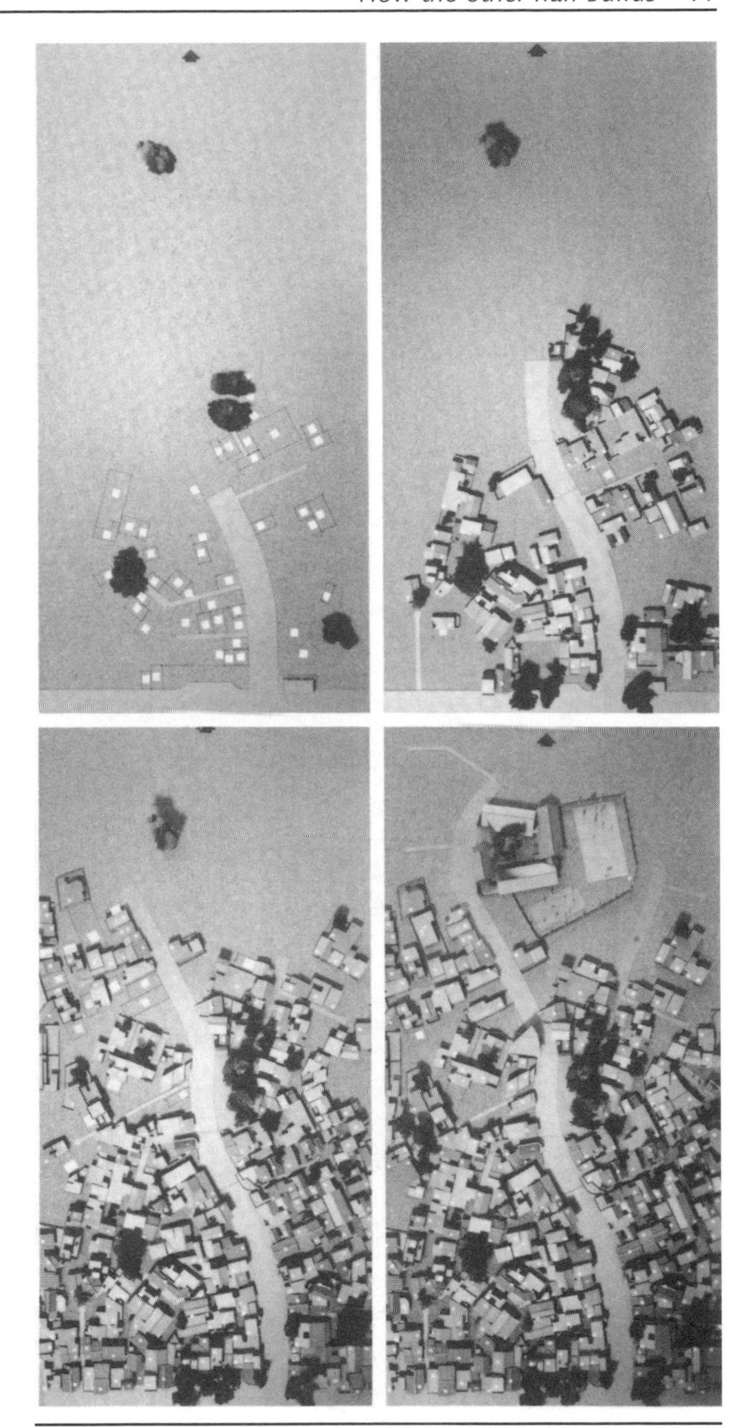

Fig. 17. Model for gaming and simulating possible growth alternatives based on self-selection process.

REFERENCES

Bhatt, Vikram, with Witold Rybczynski, Carlos Barquin, *et al., How the Other Half Builds.* A three-volume study, December, 1984–March 1990, undertaken at the Vastu-Shilpa Foundation, Ahmedabad and the Centre for Minimum Cost Housing, School of Architecture, McGill University, Macdonald-Harrington Building, 815 Sherbrooke Street West, Montreal, PQ, Canada H3A 2K6.

Volume 1: Space. Research paper No. 9. 1984. Witold Rybczynski, Vikram Bhatt, Mohammad Alghamdi, Ali Bahammam, Marcia Niskier, Bhushan Pathare, Amirali Pirani, Rajinder Puri, Nitin Raje, and Patrick Reid.

Volume II: Plots. Research paper No. 10. 1986. Carlos Barquin, Richard Brook, Rajinder Puri and Witold Rybczynski.

Volume III: The Self-Selection Process. Research paper No. 11. 1990. Vikram Bhatt, Jesus Navarrete, Avi Friedman, Walid Baharoon, Sun Minhui, Rubenilson Teixeira, and Stefan Wiedemann.

Other references:

Bertaud, Alain. 1985. *A Model for the Preparation of Physical Development Alternatives for Urban Settlement Projects.* Washington, DC: The World Bank.

Mayo, S.K. and David J. Gross, 1987. "Sites and Services and Subsidies: The Economics of Low-Cost Housing in Developing Countries." *World Bank Economic Review.* Vol. 1, No. 2: 301–335.

Oliver, Paul. ed., 1969. *Shelter and Society,* New York: Praeger.

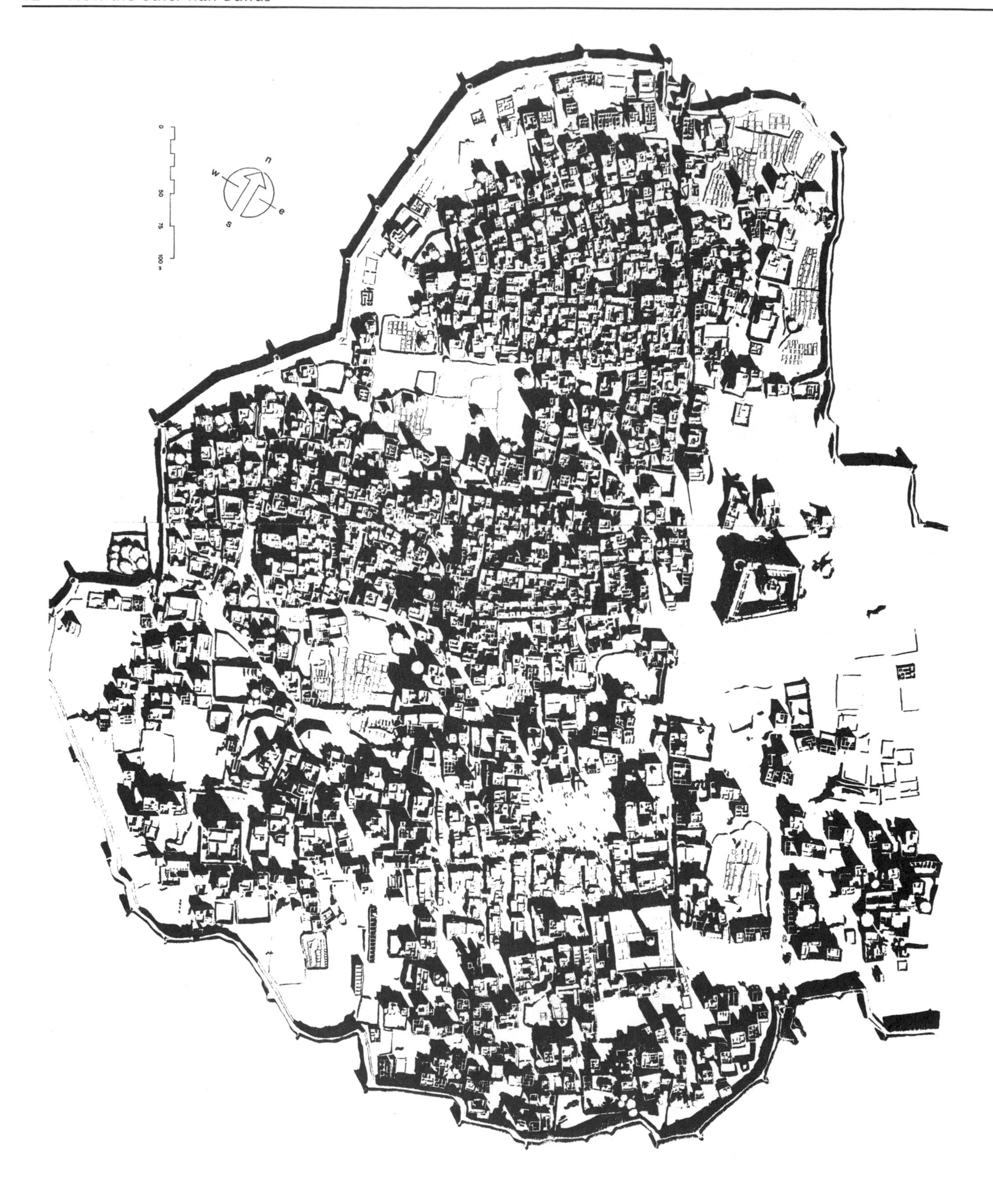

Fig. 18. Sa'dah, Yemen Arab Republic. Drawing reconstructed from aerial photo reveals the fabric of indigenous urban settlements. (Courtesy Alain and Marie-Agnes Bertaud).

Spiro Kostof

Summary

The discussion of cities, urban design and settlements freely moves across the man-made environment. In this context, it is worthwhile to emphasize that this physical canvas not be rent because of ideological or scholarly agendas. To see the interdependence of two landscapes and two ways of life—urban and rural—is an urgent scholarly strategy, to study the environment as one, not as village *versus* town or high style *versus* low. Tradition has no end: it cannot be superseded. The only enduring truth is in the seamless continuities of time and place.

Key words

architectural history, rural landscape, settlement patterns, *synoecism*, townscape

Fig. 1. Siena city wall of 1326.

Junctions of town and country

The prevailing sentiment, popular as well as scholarly, has always been to consider town and country one of the classic dichotomies of culture. In my own discipline, the history of architecture, the visual contrast of the two is ineluctable (Figs. 1 and 2). Primal images of the walled city, a densely packed structure of buildings and streets, are among our commonest documents. The open countryside with its patterns of field and cottages has never held the same interest, save for the architecturally distinguished villa and its landscaped setting. Even in the study of cities, architectural historians have been preoccupied with urban design in the sense of self-conscious and formal solutions of city form—a preoccupation that parallels our long-held exclusive claim to pedigreed buildings, to architecture as art. The more recent fascination with what urban designers like Gordon Cullen call "townscape" is equally conditioned by visual incident, albeit of a more informal, and anonymous sort.

My professional scrutiny for some time has been directed, instead, at physical continuities of time and place. I am interested in the built environment as a whole: in all buildings, the common place and the fancy, and their arrangements into landscapes of form subject to perennial change. For this effort, it has increasingly become evident, the disjunction of town and country is counterproductive, and the habit of viewing the city as a distinctive unit of analysis is quite possibly wrong-headed. We should be studying the history of settlement patterns, in which cities are merely accents, spontaneous or imposed by origin, that possess uncommon size and complexity. Physically, city-form is most beholden to prior systems of land division and settlement, farming practices and the disposition of common fields and pastures.

1 PREAMBLE: THE CASE OF SIENA

To introduce some of the themes I wish to address here, it might be useful to start in medias res, with a well-known and beloved specimen of urban form—the medieval commune of Siena.

Siena, at the end of the thirteenth century, was a powerful and well-run North Italian city-state, locked in fierce competition with its neighbor, Florence, to the north, and holding its own. Its territory stretched south and west almost to the edge of the Tyrrhenian Sea and the swampy coastal lowlands of the Maremma, in total an area of about a thirty-mile radius from the city. This hilly, poorly watered area included forests, good farmlands, pastures, and more than three hundred small towns, rural communities and feudal castles that recognized Siena's authority.

The city lay on the Via Francigena or Romea, a branch of the great Roman highway, the Via Emilia, connecting Parma with Rome. This tract ran right through the city, forming its north-southeast spine, and was dotted with inns and hospices, which served the crowds of pilgrims and other travelers who came down from the north and, under Sienese protection, headed toward the papal city. With another leg of this spine to the southwest, the city had the shape of an inverted

Credits: This article first appeared in *Dwellings, Settlements and Tradition: Cross-Cultural Perspectives* (1989) edited by Jean-Paul Bourdier and Nezar Alsayyad, and is reprinted by permission of the publisher, The International Association for the Study of Traditional Environments (IASTE), University of California, Berkeley, CA.

Fig. 2. Ambrogio Lorenzetti, *Effects of Good Government in the Town and Country*. Siena, Palazza Pubblico *ca.* 1340.

"Y" whose tips were marked by three hills—the domed Castlevecchio and San Martino, and the linear stretch of Camollia (Figs. 3 and 4). The story of the city of Siena begins with the merger of the communities on these three hills and the transformation of the harboring dip in their midst, the Campo, into a civic center—a process that started in the sixth or seventh century but was not formalized politically until the eleventh.

The walls were expanded several times as they grew. The last, begun in 1326, loosely hugged the city. There was a lot of agricultural land within, and even the built-up area was liberally punctuated with vegetable gardens and orchards. In prominent locations all over the hilly townscape, the principal families, many of them feudal nobility, had their *castellari*, fortified compounds with towers and other defensive appurtenances. These were rambling households with servant quarters, stables and warehousing facilities. The type was essentially the land-based feudal nucleus of the countryside, brought within the urban fabric by the magnates when the action moved from the countryside into the city and the agrarian economy of the earlier medieval centuries was superseded by an urban economy of banking and long-distance trade. To force these feudal lords to live within the walls, subject to the law of the city-state, was a main goal of the communal government. The defensible *castellare*, however, was a threatening unit for the city's self-government—an undigested lump in the urban body. The commune in time will run streets right through these enclaves, forcing them to open up and front the public space civilly with perforated facades.

The example of medieval Siena, and the few points I have selected to mention about its physical appearance, help to introduce several abiding lineaments of the mutual dependence of town and country.

First, both administratively and politically, the structure of human settlement has frequently engaged total landscapes. In this case, the Sienese commune itself is an extended pattern of townships, villages, and cultivated rural land. Second, the agricultural and pastoral uses of towns have always been important, especially in the so-called pre-industrial city. Third, the issue of topography is indeed central—not only in the primary sense that hills and valleys determine the configuration of settlements, which Siena's "Y" plainly demonstrates, but also in the additional and more significant sense that pre-extant rural order inevitably affects the developing city-form. In the case of Siena, urban morphogenesis is traceable to the coming together of three independent villages and their network of roads. The final lineament is that acclimation of rural housing patterns to city streets is an enduring theme of urban process for which the example of the Sienese *castellare* provides a dramatic illustration.

There are other aspects of my subject, coming to maturity in relatively recent times in the context of the modern city, which do not find prototypes in Siena. Chief among these is the history of Western experiments in urbanism that try to reconcile town and country, experiments that range from the Anglo-American picturesque suburb to the various formulations of the linear city.

2 RURAL/URBAN RELATIONSHIPS

It would be futile to attempt to deny the very real differences between cities and the countryside. The traditional labor of the farmer and the husbandman, set in the plains and pleats of the land and subject to seasonal rhythms, stands in millennial juxtaposition to the affairs of the city. The sociologist's distinctions must retain their qualified validity. It is surely not idle to recognize the informal social organizations of villages, with their static slow-changing ways, their low level of labor-division and their elemental sense of community of *Gemeinschat*, and to set this against the city's enterprise, its *Gesellschaft*, impersonal and dynamic, with a refined division of labor and a dependence on advanced technologies and industrial processes.

Long before sociology was born, these contrasts were eloquently articulated and enjoyed. The Renaissance in Italy, for example, set great store by the perfect life of the privileged classes balanced between *negotium* and *otium philosophicum*. That philosophical calm, for the rich and powerful Florentine merchants, could only be had in the country, where the villa engaged Nature, her flowers and trees and meadows, her secret springs and scurrying creatures. At Poggio, at Caiano, at Careggi and Caffagiolo, friends of Plato, lovers of his soul and students of his text, gathered to contemplate Truth. These were good men who had to be merchants and bankers and politicians in the city, but who slipped away here to a rustic, basic age, where pleasures were simple and thoughts deep. "Blessed villa," Alberti rhapsodized, "sure home of good cheer, which rewards one with countless benefits: verdure in spring, fruit in autumn, a meeting place for good men, an exquisite dwelling." Many cultures put store by this restorative balance between one's trade and one's sanity.

By the same token, the arguments of the present day also are well taken, which see a thorough interdependency between town and country, between agriculture and industry, and which point to the decline of the farming population, to the mechanization of farming processes and to the leveling influence of radio, television and tourism. And like Alberti's celebration of the country, they are also, for the purposes here, largely beside the point.

My concern here is man-land relations as manifested in physical planning, and my point is prismatic. How do we record the continuous processes of settlement and analyze its patterns? How do we resist seeing urban form as a finite thing, a complicated object, pitted against an irreconcilable, and allegedly inferior, rural context?

The architectural historian, I have already acknowledged, is congenitally handicapped in these matters. He or she is unwilling to accept that in the study of the built environment we are all recorders of a physicality akin to that of a flowing river or a changing sky.

Come the urban geographers, who are also fascinated with the city as an intricate artifact. Their methods and traditions are sufficiently different, however, to make their analysis of the form and internal structure of cities at once more comprehensive and more specialized. At the very least, their insistence that we pay attention to land parcels and plots and the particular arrangement of buildings within them—that the street system and plot pattern belong together—has enabled them to study urban fabric and its transformations with more thoroughness than the conventional approach of the architectural historian—to go beyond formal questions to a discussion of land use. And the steady interest they have shown in the distribution pattern of towns, and the flow of goods and people within that pattern, has led them to consider larger physical frames than the city itself. For the case I am pleading here, this of course is good.

But the emphasis is mistakenly on cities, and the preoccupation is with generating theory. The urban geographer is intent on discovering standard behavior, independent of particular historical circumstance. The historian of the rural landscape, on the other hand, has long been fascinated with methods of farming, enclosure, estate ownership and the like, all of which stop short at the city gates. As for other allied fields, it is symptomatic of what I propose we redress that there should be a "Rural Sociological Society" with a journal called *Rural Sociology*, even though, in the fifties at least, there were occasional papers in it on the "rural-urban continuum." In the fifties, that phrase meant primarily the rural-urban fringe in the parlance of the sociologists, or the "rurban" fringe, as they liked to call it. They inherited this interest from land economists who recognized the dynamic mixture of agricultural and urban (mostly residential) uses as something vital and worth studying.

The time frame here is recent. What is being observed is the result of changes in the American city since about the middle of the nineteenth century, specifically the history of suburbanization. Now the urban geographers were soon to seize on this urban fringe as a general principle of urban development throughout history. Their term was

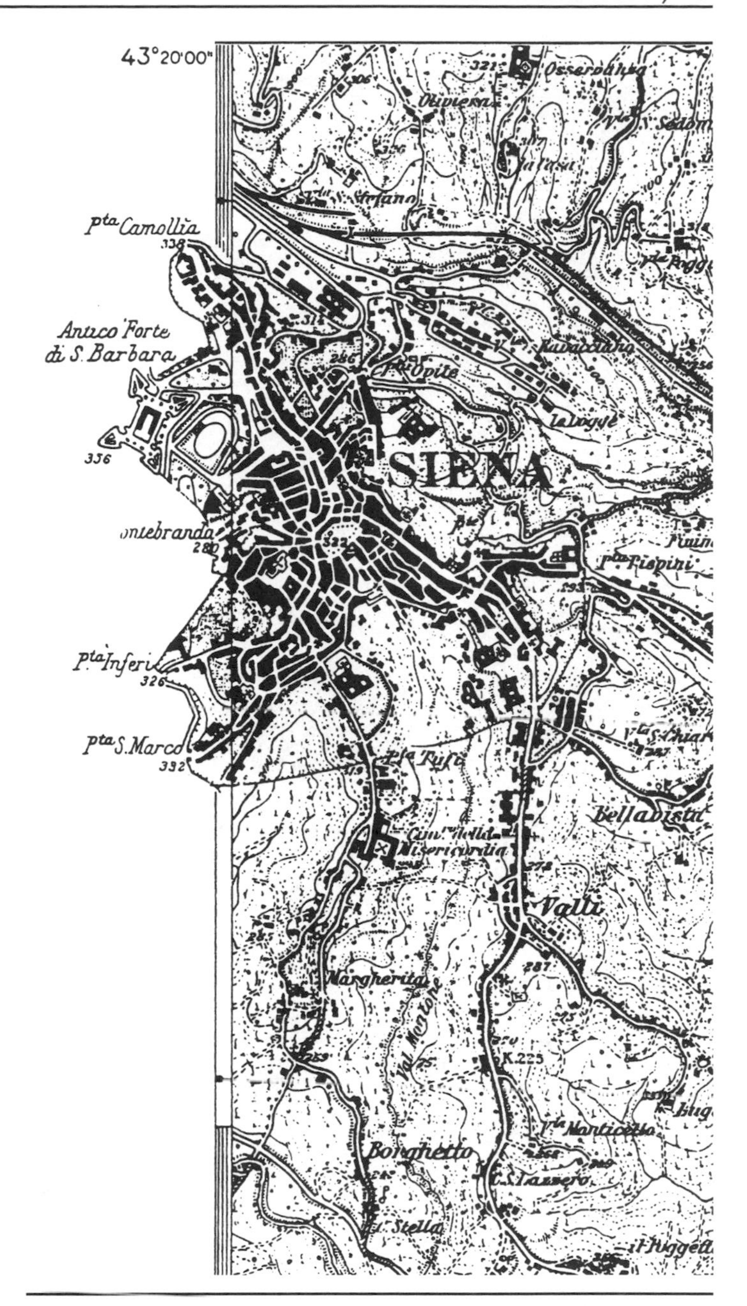

Fig. 3. Siena, plan of the city (Source: Instituto Geografico Militare, reprinted from Benevolo, *Storia della Citta*, 1976).

the "urban fringe belt." The German equivalent, applied first to Berlin by H. Louis, is *Stadtrandzone*. This area contains a heterogeneous collection of land uses, and shows a large-scale, low-density building pattern that contrasts with the thickly woven fabric of its core. What kind of uses? Well, horse and cattle markets, for example, noxious manufacturing processes like tanning, institutions deemed a health hazard like suburban leper houses of medieval London, Leicaster and Stamford, and religious houses like those of mendicant orders. In time, the city would incorporate this first, or inner, fringe belt, alter some of its character with an overlay of residential development, and give rise to a new fringe belt further out.

Alternating irregular rings of fringe belts and residential districts can be detected on the plans of many European cities. Each, according

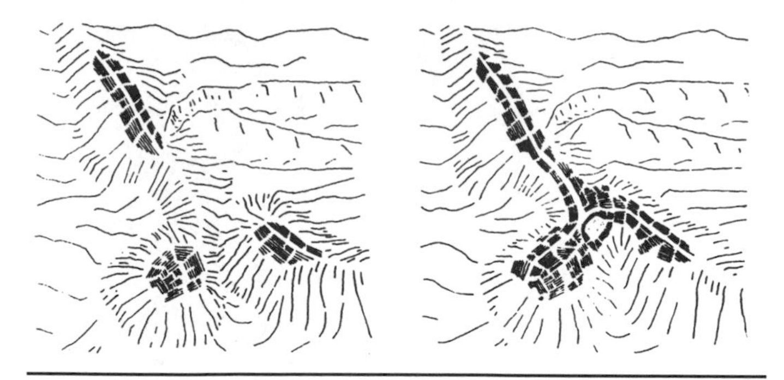

Fig. 4. The process of *synoecism*, illustrated by the merging of the villages of Castelvecchio (Citta), Camollia, and Castelmontone (San Martino) to form the city of Siena. (Drawing: R. Tobias.)

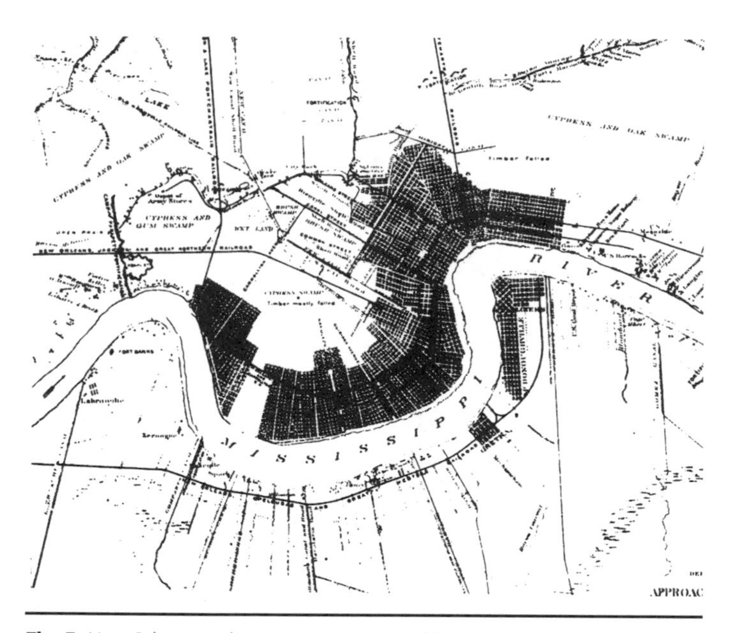

Fig. 5. New Orleans, Banks Map, 1863. Reprinted from Pierce Lewis. *New Orleans—The Making of an Urban Landscape.* 1976.

to its time, embodies distinct land uses. Parks do not predate the Baroque centuries; cemeteries, unless we are dealing with a distinct pre-Christian period like that of the Roman cities, are likely to start with the Enlightenment. More recently, we would find heavy industry, golf courses, universities, and in the outermost fringe, slaughterhouses, junkyards, sewage plants and oil refineries.

All of this is useful, but limited in application and method. The rural-urban continuum I am advocating as a perspective here is broader in scope, and intends to encourage the consideration of large social systems and urban-rural interdependence in the morphology of settlements. I am anxious to extend to our own professional domains of the built environment the seamless world that the United Nations presumed in terms of population when it asserted in its *Demographic Yearbook* of 1952 that, "there is no point in the continuum from large agglomerations to small clusters or scattered dwellings where urbanity disappears and rurality begins; the division between rural and urban populations is necessarily arbitrary."

3 SETTLEMENT SYSTEMS AND THEIR MUTATION

It is time to be more specific, and to elaborate the thematic lines I suggested at the beginning.

First, and at the broadest scale, we have to insist that human settlement is almost always continuous or concatenated, which means that towns and the countryside are subject to a responsive chain of design acts. If we think of cities alone, we tend to think of constellations based on some pretext or other, whether it is Walter Christhaller's or Auguste Losch's central place diagrams, or various systems of identifying urban spheres of influence like those advanced by A. E. Smailes, C. R. Lewis and others. There are other ways to group cities based on geographical logic: river or canal links, for example. Land routes and modern transportation means, like the railroad or the streetcar, are an equally effective basis of linkage. We could also cite political hierarchies of specific historical incident, for example, the designation of China of urban hierarchies by means of suffixes added to the names of towns—*fu* for a town of the first order, *chu* for a town of the second, *hieu* for a town of the third, and that is without counting elementary towns lower still.

But beyond the simple fact that a town can never exist unaccompanied by other towns, it is equally true, as Fernand Braudel put it, that "the town only exists as a town in relation to a form of life lower than its own. . .There is no town, no townlet without its villages, its scrap of rural life attached." And it is precisely this interdependence, as a physical phenomenon most of all, that has suffered scholarly neglect because of our persistent interest in a dualism of town and country.

We need to distinguish here between two kinds of processes. Let us call them *spontaneous* and *planned*, realizing of course that there is no aspect of human settlement that is not at least in part the result of premeditated action.

The planned process is easier to see. It is a common device of colonial enterprises when an alien land is readied for settlement by the colonizing power or agency. The city of course is the major vehicle of control and exploitation, but often the countryside is surveyed at the same time and distributed equitably and methodically. This is especially the case when the main colonial resource is agriculture rather than trade, say, or mining.

Both the Greeks and the Romans systematically divided the farming land at large, and matched urban plots for the settlers with corresponding extra urban allotments. The gridded order of the new cities was extended to the regional scale; or rather a standard matrix of all arable land provided the setting within which the cities themselves were accommodated. In the Roman system of centuriation, the module unit was 20 by 20 Roman *actus* (or 750 by 750 meters), further subdivided among farmer-colonists. An intersection of boundary lines for the *centuriae*, or square, could serve as the crossing point for the main axes—the *cardo* and *decumanus*—of the city.

The same possibilities for a uniform system of town and country planning existed for Spanish colonial rule in America, and later still for the opening up of the territories in the United States under the Land Ordinance instigated by Thomas Jefferson. And side by side with the *sitios* of New Spain and the townships of Jefferson's grid, we can recite the Japanese *jori* system, introduced in the seventh century, and the land division, or *polders*, applied by Dutch engineers to land reclaimed from the sea. In the Netherlands, where the very land is a result of human design, distinctions between town and country are particularly vacuous.

Once again, as students of environmental design, we have done very little with these comprehensive schemes, and much with the cities themselves which are only highlights within a larger coordinated design. We have written entire books on the ingenious 1732 plan of Savannah in colonial Georgia masterminded by James Oglethorpe, for example; but beyond a descriptive sentence or two we have left unexamined the extraordinary complement to this famous grid with its wards and squares—the outer zone of farming lots, the five-acre gardens further in, the "Common round the Town for Convenience and Air," and the town proper along the Savannah River. It is only because the orthogonal distribution applied uniformly, even when the garden squares might be shared by two colonists, each with a triangular lot, that Savannah could maintain the ward and square arrangement of the original town plat as it grew in the next century and moved into its cultivated land.

By spontaneous settlements, I mean a natural promotion of towns within a previously even, unaccented landscape. I am aware that this may sound very much like a return to the old favorite, the rise of cities, that has sustained those urban historians and geographers concerned with the pre-Classical world or the Middle Ages for several generations. But I am less interested here in the vast literature about when a town is a town—questions of size, density, economic activity, administrative function or occupational structure, and so forth—as I am in rare studies like those of Robert Adams concerning Mesopotamia (Adams, 1972, and Adams, 1981).

Adams' meticulous and demanding fieldwork to chart the ancient watercourses in the "birthplace of the city" (which, as we would expect, turned out to be totally different than the present pattern of rivers and canals) is as important for the physical history of human settlement as the more spectacular and photogenic archaeological discoveries of Leonard Wooley and his confreres have been for the history of Mesopotamian civilizations. I would also go so far as to claim, for our purposes, that the hesitant, tentative dotted lines of Adams' maps—carefully plotting ancient levees, variations of river discharges, and settlement patterns—are far more critical than the diagrams of Christhaller or Thyssen.

To my knowledge this sort of documentation and analysis does not as yet exist for that other period of nascent urbanization that has intrigued urban historians of Europe, namely the several centuries in the Middle Ages when, after the subsidence of Roman urban order around the Mediterranean and the northern provinces of the empire, new towns emerged out of non-urban cores in the rearranged countryside. We have a lot of theory, an increasing volume of case studies, especially for England, and promising new directions based on the archeology of early settlements. What we don't have is a scanning of regions, small or large, through that combined perspective of physical, political and social inquiry, to document and interpret transformations of the historic landscape of Europe.

Fieldwork may not be as helpful in writing the other aspects of the spontaneous process. *Synoecism*, the process behind the creation of medieval Siena, is one of these (Fig. 5). The term, according to Aristotle, describes the administrative coming together of several proximate villages to form a town. Translating his words: "When several villages are united in a single complete community, large enough to be nearly or quite self-sufficing, the *polis* comes into existence." This is how Athens was born and Rome and Venice and Viterbo and Novgorod and Calcutta, and a number of towns in Muslin Iran like Kazvin, Qum and Merv. In fact, *synoecism* is beginning to figure as one of the commonest origins of towns coming out of a rural context, and it is therefore unfortunate how little there is to read about the process.

The case of Islam is especially interesting in this context. Throughout its history, the Islamic city has given proof that it was conceived not so much as a tidy walled package contrasting with the open country side, but as a composite of walled units. Twin cities, where the two settlements slowly grew together across the intervening space, were not uncommon (*e.g.*, Isfahan, Raqqa). Ira Lapidus suggested some time ago that in some Iranian oases, entire regions might be considered composite cities, "in which the population was divided into non-contiguous, spatially isolated settlements." However you choose to categorize this constellation, it was a fully self-conscious system of settlement, in that the entire region would be surrounded by an outer wall, and "urban functions were not concentrated within the walls of the largest settlement, but were often distributed throughout the oasis" (Lapidus, 1969, p. 68).

The form of a synoecistic town absorbs the shapes of the original settlements, along with their road systems, and the open spaces that existed among the settlements are turned into marketplaces and communal centers. This is the origin of the Roman Forum and the Athens *agora* and Siena's Campo. In German lands during the Middle Ages, towns sometimes absorbed an adjacent rural parish, or *Landgemeinde*, in order to acquire common pasture in some cases "shifted their homes into the town and became fully privileged townsmen" (Dickinson, 1961, p. 331).

In Africa, traditional Black cities can be described as groups of village-like settlements with shared urban functions rather than a single center. These cities were very spread out. They consisted almost entirely of one-story structures arranged in residential compounds, no different an arrangement than the village. These compounds were often located with no particular care for alignment with the streets. Sudanese Muslim cities, Al Ubayyid, for example, have this same Black African pattern. In the nineteenth century Al Ubayyid was a city made up of five large villages, originally separated by cultivated areas which the Ottoman regime (*ca.* 1820–1884) partially filled in with barracks, mosques, a prefecture building and government worker's housing.

4 RUS IN URBE

The next two themes I set for myself at the start of this article follow logically from this discussion. One has to do with the agricultural and pastoral uses of urban land. The other has to do with the impact of rural land divisions on urban form in those spontaneous cases where an orderly regional matrix, like Roman centuriation or the American Land Ordinance grid, does not predetermine the pace and shape of suburban development.

I need not say much about the first of these themes. The urban accommodation of cattle has a history that stretches from Nineveh, where large open areas within the walls were set aside for the daily use of herds, down to the New England common. Little need be said, too, about agriculture in the city, except to point out again how radically an agricultural presence within the city challenges any strict

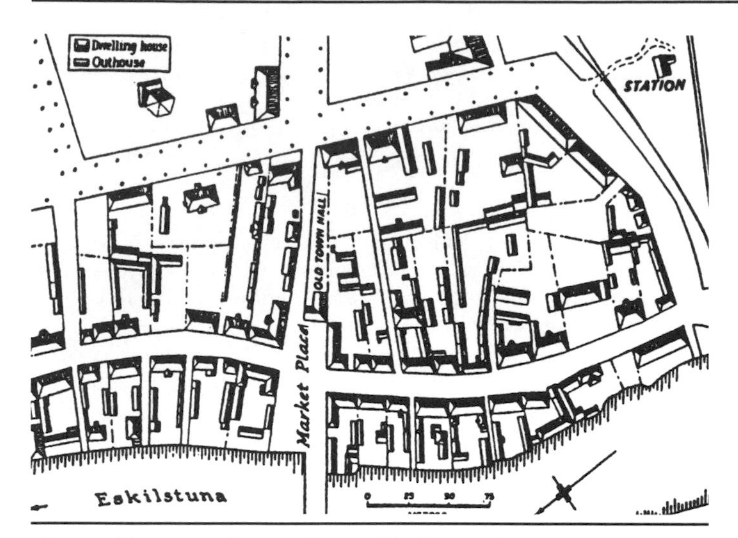

Fig. 6. Eskilstuna, Sweden. Distribution of buildings in the older quarter of town (Leighley, 1928).

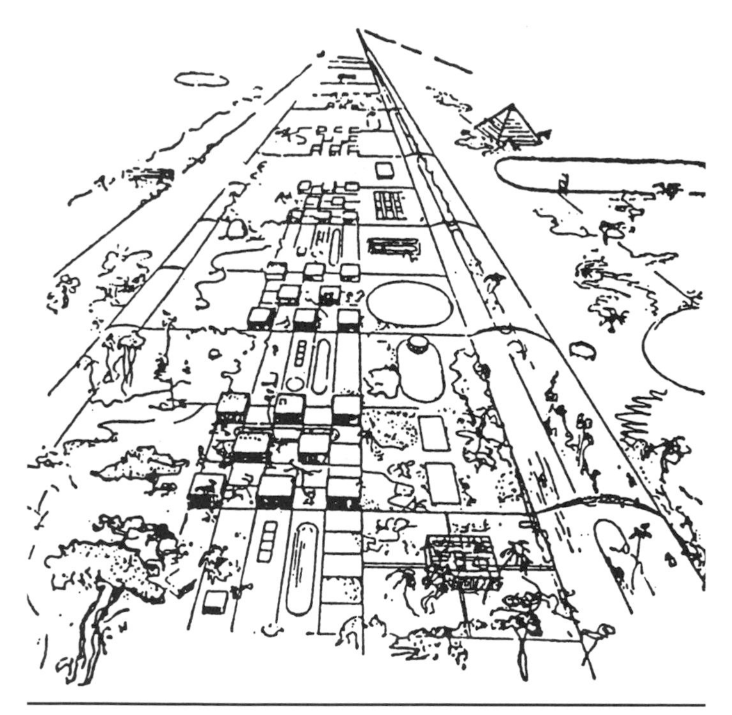

Fig. 7. Plan for Magnitogorsk, Ernst May, 1930, aerial view. A linear scheme that reflects Lenin's goals for Soviet development: "We must aim at the fusion of industry and agriculture ... by means of a more diffused settlement pattern for the people." Reprinted from D. Lewis, ed., *Urban Structure: Architects' Yearbook 12*, 1968.

separation of urban and rural domains. Sometimes a city would make its gardens and fields a walled component of the urban structure, as in the old nucleus of Cahors in a meander of the river Lot. In Yoruba cities, farmers and city-dwellers and the urban edge consists of a farm belt as much as fifteen miles wide. At other times, the producing garden is a regular component of houses throughout the city. In China, not only was intensive truck gardening found within the city, but also most houses devoted a small portion of their yard to gardening. We have only spottily written history of the transformation of producing gardens into idealized pleasure gardens. For Europe, the critical time for this was the sixteenth century (Jackson, 1980).

5 LAND DIVISION

The relation of rural land divisions to the urban form that supersedes them is a vast, critical subject, and we have not scratched much beyond the surface. I can point to some exemplary case studies to demonstrate how much we miss by neglecting this line of inquiry.

One instance is Pierce Lewis' classic study, *New Orleans—The Making of an Urban Landscape*, where we are shown the transformation of the narrow fan-shaped French plantation lots within the convex meander of the Mississippi into the radial boulevards of the expanding city (Fig. 5). Another instance is an article by Michael E. Bonine on "The Morphogenesis of Iranian Cities," where the loose grid of towns like Shiraz and Yazd, detectable through the so-called organic city form, and the long linear streets with their rows of courtyard houses, are derived by him from the channels of subterranean watercourses, or *qanat*, and a system of strip irrigation and rectangular field division (Bonine, 1979, pp. 208–224).

Now the documentation of this process whereby an antecedent rural landscape translates itself into urban form is exceedingly difficult. In most cases it is impossible to reconstruct this landscape except in relatively recent history. The English have started to rediscover, for example, how the common lands and open fields that surrounded towns were first alienated to individual ownership, beginning in the mid-nineteenth century, and were then transformed into a belt of urban extension. We can see there, if we know how to read the evidence, the medieval footpaths and the furlongs of the old open fields underneath the modern network of streets.

The English have also been able to take advantage of two invaluable field surveys of the nineteenth century. The tithe maps were produced after the Tithe Commutation Act of 1836 in connection with changing the tax system on land holdings from tithes to a money payment—that is, to a fixed rent on each holding. Most parishes in England and Wales were surveyed for this purpose, and the maps give us a precise picture of the configuration of fields. This one can superimpose on the famous ordinance survey maps of nineteenth century towns, drawn to a uniform scale. By matching the same segments in the tithe and ordinance maps of Leeds, for example, it is possible to show how the pre-urban *cadaster* determined the lines of the urban fabric.

We have still, within the context of the pre-industrial city, to comment on the acclimation of rural housing types to city streets. The evidence, never easy to gather, is steadily disappearing and scholarly interest remains marginal. The Burgerhaus of the Swiss town is a classic case of a farmstead being brought within the urban boundaries and changing over time under pressures from urban economy, lack of building space, and new architectural styles. In Sweden too, the adaptation of the rural farmstead to the city-form was slow and not complete until the late nineteenth century. An example, from the Malardalen towns of central Sweden, indicates the grouping of household buildings around a central yard before a single large house facing the street would absorb these scattered, individual sheltered functions (Fig. 6).

Let me site two other examples, one Western and the other Islamic. The medieval fortified court, or *curia*, has an urban apotheosis in more than one cultural sphere. The Islamic *haws*, on the other hand, was a large open courtyard with lodging on all four sides. The frequent location of this habitat on the outskirts of cities and the reported presence of cattle underscore the fact that we are dealing with the urban adaptation of a rural settlement form. Then there are those historical situations where no such adjustment was necessary, because the environmental order of the countryside and that of the city were made of the same cloth, both in terms of architecture and its arrangement into landscapes of form (Raymond, 1984, pp.86–87).

There are two points I want to pick up and emphasize here. The first is the self-conscious survival of village settlement patterns within the fabric of a city. The second is the seeming finality of city walls, and the contradiction this implies for the rural-urban continuum here espoused.

The first of these points entails both atavistic holdovers on the part of urbanized country folk and administrative control on the part of state authority. A striking modern example of the former is in squatter settlements like those of Zambian towns. The pattern may not be obvious at first, but the units are soon discovered—twenty huts or so around a common space and a physical grouping resembling a circle. In fact, squatter settlements would be a critical unit of study for those junctions of town and country under review in this essay. Are these ubiquitous formations best viewed as places in which rural people lose their traditional identity in preparation for city life, as it is so often maintained? Or do they, rather, represent a spontaneous opportunity for the city to regenerate its sense of tradition?

The administrative control has a number of rationales, all more or less coercive. In China, especially under the Han Dynasty, the aim was to integrate the lineage community into the administrative system. The same word—*li*—designated a village, a city quarter and a measure of length. The initial restructuring of the countryside may well have entailed a form of synoecism, where a number of adjacent villages were converted into market towns. The subsequent division of cities into *li* was probably intended to keep a check on the largely agricultural population by preserving village organization. How physical this affinity continued to be is hard to know. A recent study of some Florentine new towns in the fourteenth century showed how villages were forcefully de-mapped, and their inhabitants brought to live in separate corners of the new towns which, in their strictly gridded layouts, bore little physical relation to the original village nuclei.

As to the second point, note that the role of the walled edge as an emphatic divider of town and country has been overdrawn. Let me cite China. Despite the fact that any administrative town of consequence would be expected to be walled, the uniformity of building styles, and the layout and the use of ground space, carried one from city to suburb to open countryside without any appreciable disjunction, as Sinologists have persuasively argued. G. William Skinner writes, "The basic cultural cleavages in China were those of class and occupation . . .and of region. . .not those between cities and their hinterlands" (Skinner, 1977, p. 269). Indeed, we can go further and argue that the unchanging rural environment, not the city, was the dominant component of Chinese civilization.

6 MODERN PARADIGMS

Under its present system of government, the walls of Chinese towns would have come down for ideological reasons even when they were not functionally obsolete. It may have been Rousseau who first insisted that city walls artificially segregate crowds of urbanities from the peasants spread thinly over vast tracts of land. He urged that the territory be peopled evenly. But it was the Communists who gave their own gloss to this injunction of the Enlightenment. Through their Marxist forefathers, they would object to the idea of an urban/rural dichotomy: city walls artificially severed the mighty proletariat into an urban and a rural contingent, thus eviscerating its strength.

The fear of a conspiracy to weaken the masses by dividing them is behind Marxist doctrines of disurbanization. The Marx-Engels Manifesto of 1848 prescribed the "gradual abolition of the distinction between town and country, by a more equitable distribution of the population over the country." This was at the core of the great debate in Russia in the twenties between the urbanists—led by the planner L. Sabsovitch who advocated the construction of urban "agglomerations," vast communes that would hold four thousand people each in individual cells—and the disurbanists. N.A. Milyutin was chief among them, who proposed the abandonment of the old cities and the dispersal of the population by means of linear cities in free nature, first along the great highways that linked Moscow with its neighboring towns, and eventually across the whole extravagant spread of Russia (Fig. 7). The agricultural and industrial workers would live together, building a common proletariat, the new Communist aristocracy (Kopp, 1970).

The Russian debate thus appropriated one of the great settlement theories of modern times, which sought to erase the deprivations of the big city by bringing everyone close to nature along an open-ended transportation spine. The idea did not need Communism to support it. It was invented by a Spanish civil engineer in the 1880s, and was elaborated by the likes of Chambless, Richard Neutra and Le Corbusier for anyone who would buy it. We see its ultimate manifestation in the linear development along our freeways today, which, like the urban fringe with its suburbia, dissolved the city in open land.

The only other modern settlement concept of comparable power and seduction is the garden city, an English anti-urban dispersal fantasy of the turn of the century, which astonishingly was adopted as national policy after World War II. This is not the place to speak about Ebenezer Howard and Raymond Unwin, of Letchworth and Welwyn. In that circle, garden city and garden suburb were severely juxtaposed. Letchworth was a self-sufficient town of 30,000, a model for the future reconstruction of the capitalist/industrial environment; Hampstead Garden Suburb was a dependency of London, nothing more. For us, and for the case we are pleading, these elegances of dogma are unpersuasive. What is central to our argument is that from at least the middle of the nineteenth century, the history of the Anglo-American picturesque suburb, not to say suburbia in general, firmly established an intermediate environment between town and country, or as Frederick Law Olmstead was to phase it, "sylvan surroundings . . .with a considerable share of urban convenience" (Fig. 8). So durable has this intermediate environment proved that it alone should

Fig. 8. View of Glendale, Ohio, *ca.*1860. Planned by Robert C. Phillips in 1851, Glendale is probably the first American picturesque suburb designed in sympathy with the rural landscape ideals later popularized by Olmstead. (Source: Glendale Heritage Preservation.)

persuade us of the futility of ever seeing urban and rural as two distinct worlds, but rather as two aspects of a single continuum.

7 TOWN AND COUNTRY: BEYOND DUALISM

This article has a modest aim. In the context of a broad-based discussion of cities, urban design and settlements which freely moves across the man-made environment, it seemed worthwhile to emphasize that this physical canvas not be rent because of ideological or scholarly agendas. For too long we have extolled the city as a remarkable artifact, and urban life as an elevated form of engagement with the forces of progress, enterprise and an entire range of civilities. A polar opposite was needed, one that was easy-to-handle—the countryside. The village and its ways acquired many friends of its own in time, but also a heavy aura of sentiment that had to do with naturalness and honesty and enduring value.

Today, given the radical changes in the traditional landscapes of the world, we can hardly seek comfort in such antipodes. And yet the urge is irresistible to lament a paradise lost, a sad disjunction between a time-honored way of doing things and the arrogant disrespectful arrivism of the new.

To see the interdependence of two landscapes and two ways of life is an urgent scholarly strategy, to study the environment as one, not as village versus town or high style versus low. It could also be a healing thing that softens obstinate prejudices and eases anxieties rampant in those many parts of the world that are trapped between tradition and the present.

The land spreads out as one: time flows. The breaks, barriers, and divorces are of our own making. Our charge now, I venture to suggest, is to find tradition in the central business district of the metropolis, to see the old irrigation ditch beneath the fancy tree-lined avenue, to recognize the ancient process of synoecism that brought villages together to form cities at work still in our modern conurbations. Tradition has no end: it cannot be superseded. The only enduring truth is in the seamless continuities of time and place. ■

REFERENCES

Adams, R. McC. and H.J. Nissen. 1972. *The Uruk Countryside: The Natural Setting of Urban Societies.* Chicago: University of Chicago Press.

Adams, R. McC. 1981. *The Heartland of Cities: Surveys of Ancient Settlement and Land Use on the Central Floodplain of the Euphrates.* Chicago: University of Chicago Press.

Bonine, Michael E. 1979. "The Morphogenesis of Iranian Cities." *Annals of the Association of American Geographers.* 69.

Dickinson, R.E. 1961. *The West European City.* 2nd edition. London: Routledge and Kegan Paul.

Jackson, J.B. 1980. "Nearer than Eden," in *The Necessity for Ruins.* Amherst: University of Massachusetts Press.

Kopp, A. 1970. *Town and Revolution: Soviet Architecture and City Planning.* 1917–1935. New York: Brazilier.

Lapidus, I. 1969. *Middle Eastern Cities.* Berkeley: University of California Press.

Leighley, J. 1928. *The Towns of Malardalenin Sweden: A Study in Urban Morphology.* Berkeley: University of California Publications in Geography, 3.1.

Raymond, A. 1984. *The Great Arab Cities in the 16th–18th Centuries.* New York: New York University Press.

Skinner, G. William. 1977. *The City in Late Imperial China.* Palo Alto: Stanford University Press.

Peter Calthorpe

Summary

This article presents urban design guidelines for growth in our cities, suburbs and towns. Unlike typical "design guidelines," which deal primarily with aesthetic and architectural principles, the guidelines describe principles by which to integrate development of communities, neighborhoods, districts and regions. They follow three strategies: first, that the regional structure of growth should be guided by the expansion of transit and a more compact urban form; second, that single-use zoning should be replaced with standards for mixed-use and walkable neighborhoods; and third, that urban design should create an architecture oriented toward the public domain and human dimension rather than the private domain and auto scale.

Key words

corridor, greenway, infill, mixed use, neighborhood, pedestrian, transit-oriented development (TOD)

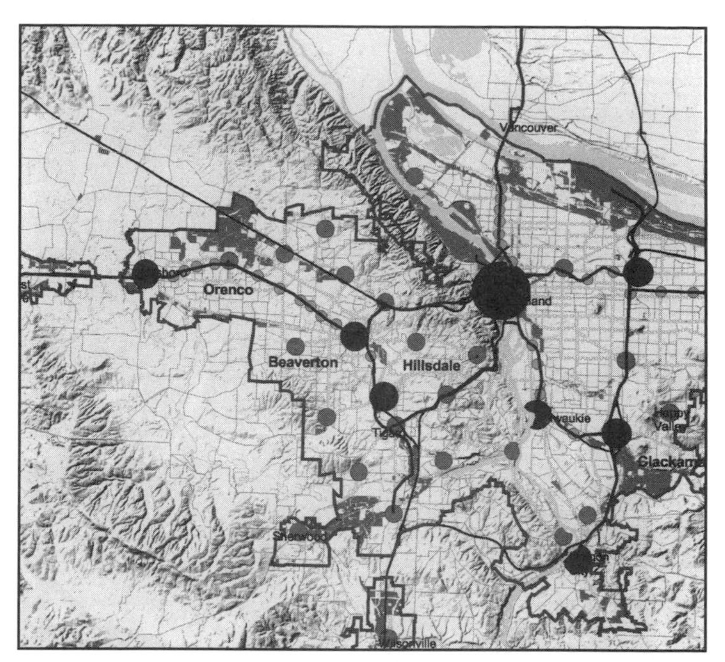

Metropolitan Portland, Oregon, centers, districts and corridors.

The regional city

1 THE BUILDING BLOCKS OF THE REGION

The Regional City is a concept developing in North America from three trends: the emergence of regionalism and bioregional planning initiatives, the maturation of the suburbs, and the revitalization of older urban neighborhoods. These are mutually interdependent: each prospect—region, suburb, and city—is needed to support the other. Taken together, these trends shape the outline of a new metropolitan form, what is best conceived of as "the Regional City." In the Regional City, networks of regional geography, demographics and growth, green space, and transit establish the basis of vital networks, through which transit-oriented development provides a coherent organizing system.

To facilitate the shift from "edge city" to the "regional city," the urban designer needs to reconceive the basic building blocks of the region and its jurisdictions. Rather than the twenty or thirty specific land-use designations found on most zoning maps, only four elements are sufficient to design complete regions, cities and towns.

Centers: the local and regional destinations at the neighborhood, village, town and urban scale.

Districts: the special-use areas, which are necessarily dominated by a single primary activity.

Preserves: the open-space elements that frame the region, protect farmlands and preserve critical habitat.

Corridors: the connecting elements based on either natural systems or infrastructure and transportation lines.

Centers are by definition mixed-use areas; they include jobs and housing as well as services and retail. Districts may be mixed use but are typically dominated by a single primary land use such as a university or an airport. Preserves may be productive agriculture or natural habitat. Corridors are the edges and connectors of the region's centers, neighborhoods and districts. They come in many forms, from roads and highways to rail lines and bikeways, from power-line easements to streams and rivers. Maps that use these four simple elements can help to redirect the fundamental quality of our regional habitat.

Centers: village, town and urban scale

Centers are the focal points and destinations within the Regional City. They gather together neighborhoods and local communities into the social and economic building blocks of the region. They are necessarily mixed use in nature: they combine housing of different scales, businesses, retail, entertainment and civic uses. There is a hierarchy from village center through urban center, but there are no hard-and-fast distinctions between them, only general qualitative differences.

Centers are distinct from neighborhoods but may include a neighborhood. Neighborhoods are primarily residential with some civic, recreational and support uses mixed in. Centers, on the other hand, are primarily retail, civic and workplace dominated with some residential uses mixed in. They are the destinations of several or many neighborhoods.

Credit: This article is adapted from publications by the author, including the *Next American Metropolis (1993)* and *The Regional City (2000)*, cited in the references.

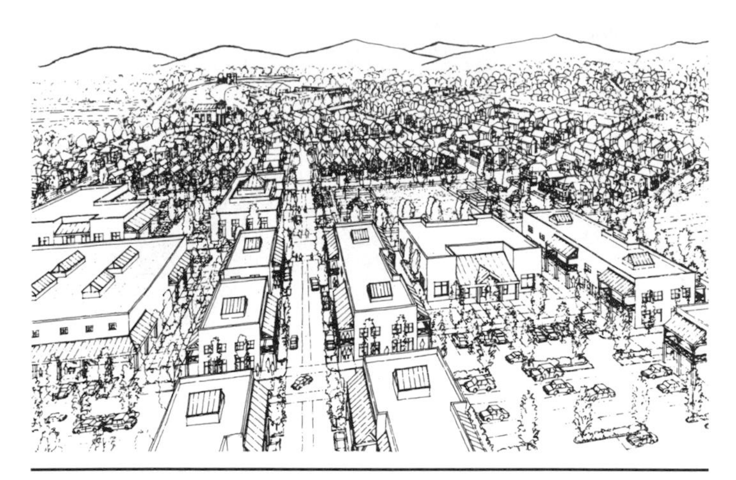

Fig. 1. The Regional City: interconnecting centers, districts, preserves and corridors.

The Regional City can and, in many cases, does have several urban centers. For example, the Bay Area has at least three: San Francisco, San Jose and Oakland. Either multiple or singular, urban centers form the prime structure of a region. They are the business, cultural and civic centers that provide the global identity and local focus for a region (Fig.1).

Districts

Districts are areas outside of neighborhoods and centers that accommodate uses not appropriate for a mixed-use environment. Not all uses can be of a scale, mix and character that fits within a neighborhood or a center. Examples of such uses are plentiful: light and heavy industrial areas, airports and major seaports, "big-box" retail and distribution centers, central business districts, military bases, and university campuses. These areas are critical to the economic and cultural life of a region but functionally must be separate from the fine grain of a neighborhood or the mix of a center.

However, some single uses, correctly segregated as districts, can be closely integrated with mixed-use areas and centers—and should be. Office parks are a prime example. Under current zoning, these primary work destinations are isolated and clustered into districts near highway interchanges. Through some bizarre identification with factories, offices are seen as a poor fit with village, town and urban centers. To the contrary, they should be integrated into our mixed-use centers. Such integration adds strength to the retail component of the center, reinforces the transit system supporting the center, and increases the value of any of its civic uses.

Other examples of important uses too often isolated from mixed-use centers are cultural and civic facilities. The ubiquitous suburban civic center or entertainment zone is another lost opportunity to complete and reinforce town and village centers. Our civic buildings along with our cultural facilities should be integrated into the fabric of our communities, mixed with employment, shopping and some housing. The modern equivalent of "courthouse square" should be a focal point of the Main Streets of the future. Theater districts and movie complexes should likewise be an essential part of the centers that draw our communities together.

Preserves

Preserves are perhaps the most complex and controversial building block of a regional design: complex because they include so many very different elements, locations and potential uses—controversial because the means of saving the land and the economic effects are hotly debated. Beyond those lands now protected by federal or state law (wetlands, critical habitat and so forth), the identification of which types of rural landscapes are appropriate for preservation is a central component of a regional vision. Such open-space preserves at the edge of a region are almost universally desired, as are the open-space corridors within the region. But their delineation and preservation is a political and economic challenge.

There are two distinct types of regional preserves: community separators and regional boundaries. Community separators function to create open-space breaks between individual communities within the region.

Preserving farmlands as regional boundaries is a different matter. The land values are not as high, and the need for preservation is justified by more than regional planning. Preservation is also needed because high-quality farmland is threatened in many areas of the country. According to the American Farmland Trust, if 1 million acres (40.5 million hectares) of farmland are to be lost to urbanization in California's fertile Central Valley, as much as 2.5 million (approximately 100 million hectares) will fall into this constrained zone at the edge of development.

Beyond the need to preserve our agricultural capacities is a large impulse among the electorate to preserve the rural heritage close to their urban areas, regardless of soil classification or ecological value.

Corridors

Although corridors come in many types and sizes, natural and human made, they always constitute a flow. Waterways, traffic flows and habitat movements define the unique corridors of each region. They can become either the boundary of a community or one of its unifying bits of common ground—a main street or riverfront are simultaneously destinations and passageways. Corridors are the skeletal structure of regional form and its connections. And they form the defining framework of its future.

Specific habitats, unique ecologies or larger watersheds can define natural corridors. In most cases they are a combination of all three. For this reason, a regional approach to open space is essential, and greenway corridors rather than segments are critical.

Each region has a watershed structure that is fundamental to its natural form. Every watershed is made up of catchment areas (hillsides), drainage areas (streams and rivers), wetlands (deltas and marshlands), and shorelines (beaches). There may be other natural corridors worth preserving in the region—specific habitats of endangered species, unique ecosystems or scenic corridors—but these four watershed domains are critical and contain many of the others. Continuity is more important than quantity in natural corridors.

Human-made corridors are as important to the quality of life within a region as are natural corridors. Roads and transportation systems have always provided the fundamental structure of human habitat in cities and regions. But it is time to balance our hyperextended highway system and road network with other types of mobility. Transit is foremost. It can form the regional armature for a different type of growth, one that naturally favors infill, walkability and human-scale development.

Light-rail, bus ways and streetcar lines, like historic boulevards and main streets, are pedestrian friendly and serve as catalysts to the formation of mixed-use neighborhoods and centers. They do not form barriers within communities; in fact, they often unify a place by creating a focus and a common destination.

Finally, bus routes and bikeways, like local streets, are the smallest elements of transit mobility. They are as essential to the larger forms of transit as pedestrians are to a main street. Feeder buses, safe sidewalks and comfortable routes for bikes are the "corridors" that knit together each neighborhood.

The unseen utility corridors are perhaps as important as the more visible road, transit and open-space corridors. Investments in water-delivery systems, sewers, drainage capacity and other utilities form the backbone of development. Designing these systems to be efficient, compact and responsive to the land-use vision of the region is essential.

Reusing and repairing old, underutilized and decaying corridors, either natural or human made, is an imperative for any regional strategy that includes significant infill and redevelopment. The strip commercial corridors of our older suburbs offer a chance, through redevelopment, to transform those places into mixed-use walkable districts. In these areas, the roads need to be redesigned and enhanced for pedestrian, bike and transit, and the infrastructure must be upgraded for higher densities and a mix of uses.

Perhaps the greatest opportunity for corridor reuse is in our underutilized railroad rights-of-way. Old and abandoned tracks can be reused for new transit links that run through the heart of a region's historic core and older suburbs.

In these design principles and regional building blocks, we are not proposing that the alternative to sprawl and inequity is a return to a small-town world of a historic culture or an acceleration of the fractured urbanism of many modern cities. A sustainable urban and regional form must be shaped out of the best of timeless traditions combined with the complexity and intensity of our contemporary world.

2 GUIDING PRINCIPLES OF TODS

The Transit-Oriented Development (TOD) concept is simple: moderate and high-density housing, along with complementary public uses, jobs, retail and services, are concentrated in mixed-use developments at strategic points along the regional transit system (Fig. 2). Recently, similar concepts have gone by many names: Pedestrian Pockets, Traditional Neighborhood Developments, Urban Villages and Compact Communities to name a few (Fig. 3). Although different in detail and emphasis, these concepts share a common perspective, design principles and set of goals: To add emphasis to the integration of transit on a regional basis, providing a perspective missing from strategies that deal primarily with the nature and structure of individual communities and neighborhoods. This regional perspective helps to define a meaningful edge for the metropolitan area, eliminating the danger of random growth in distant sites served only by highways. Such a larger

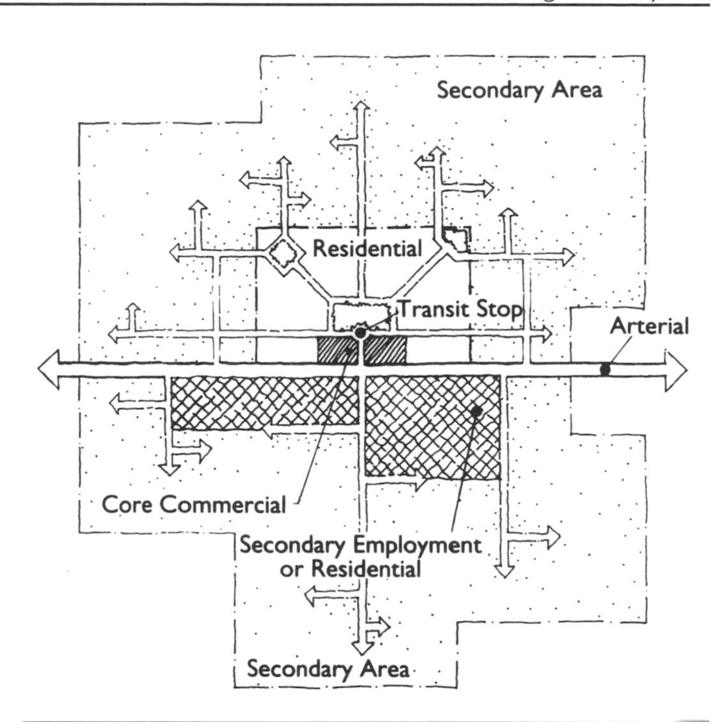

Fig. 2. Transit-oriented development.

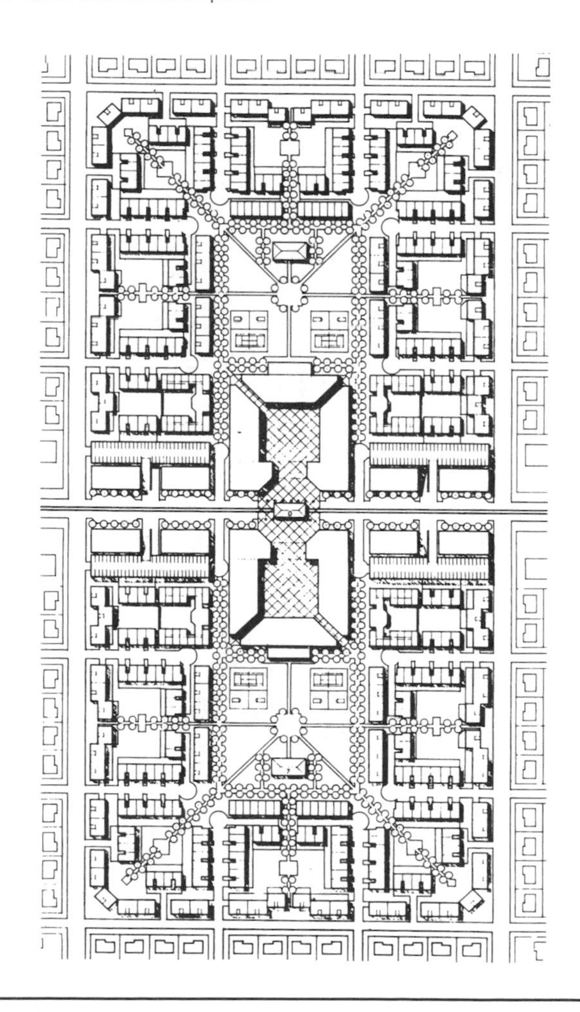

Fig. 3. Pedestrian pocket.

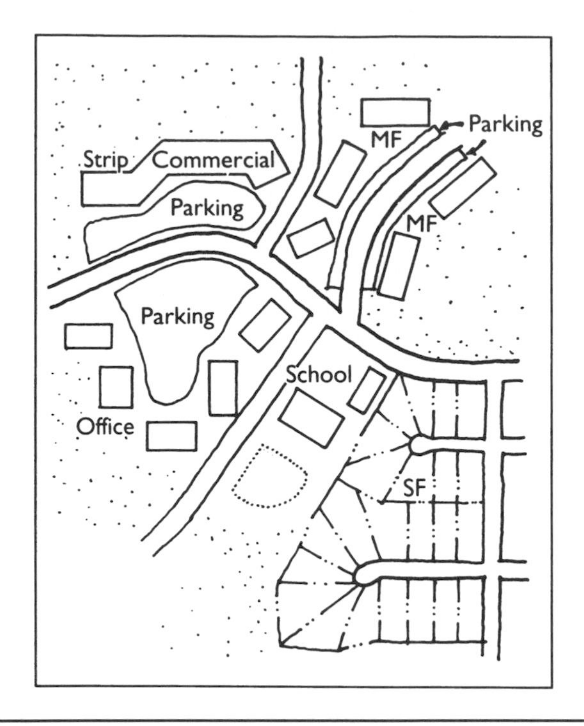

Fig. 4. Conventional suburban development.

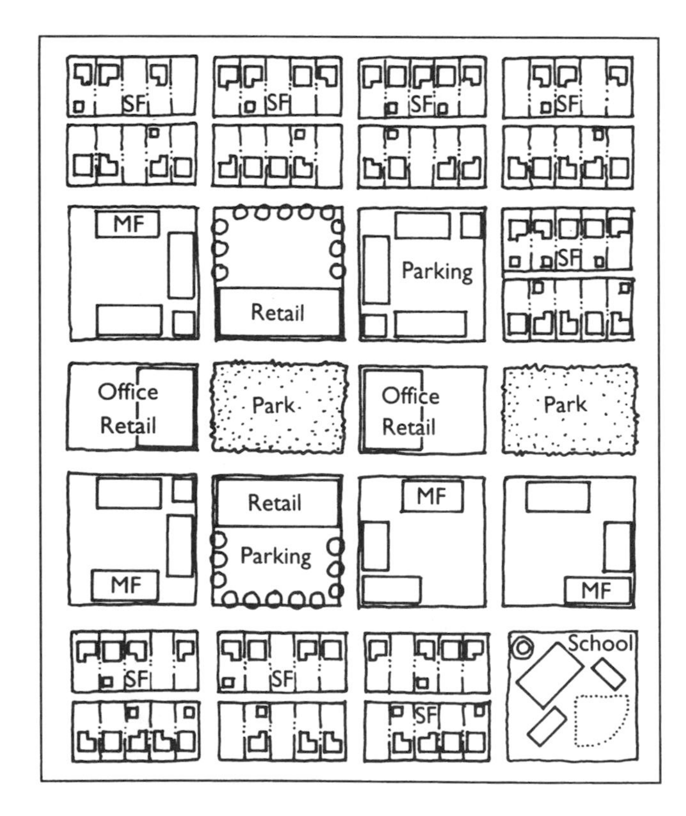

Fig. 5. Traditional neighborhood development.

view can help to order growth across balkanized metropolitan regions as well as to encourage infill and redevelopment efforts.

A "walkable" environment is perhaps the key aspect of the concept. In order to develop alternatives to drive-alone auto use, comfortable pedestrian environments should be created at the origin and destination of each trip. In contrast to conventional suburbs, walkable environments are characterized by streets lined by trees and building entries also help to make the TOD environment "pedestrian-friendly" (Figs. 4 and 5). Although focused on reinforcing transit, such land use configurations would equally support carpools and more efficient auto use.

The fundamental structure of the TOD is nodal—focused on a commercial center, civic uses and a potential transit stop. Defined by a comfortable walking distance, the TOD is made up of a core commercial area, with civic and transit uses integrated, and a flexible program of housing, jobs, and public space surrounding it.

TODs not only promote alternates to auto use, but can be a formula for affordable communities—affordable in many senses. Communities are affordable to the environment when they efficiently use land, help to preserve open space and reduce air pollution. They are affordable for diverse households when a variety of housing types, at various costs and densities, are encouraged in convenient locations.

The principles of Transit-Oriented Development are to:

- organize growth on a regional level to be compact and transit-supportive.
- place commercial, housing, jobs, parks, and civic uses within walking distance of transit stops.
- create pedestrian-friendly street networks that directly connect local destinations.
- provide a mix of housing types, densities, and costs.
- preserve sensitive habitat, riparian zones, and high quality open space.
- make public spaces the focus of building orientation and neighborhood activity.
- encourage infill and redevelopment along transit corridors within existing neighborhoods.

The following principles (Figs. 6–17) represent selected highlights of urban design principles for regional development (*cf.* Calthorpe, 1993 for fully detailed discussion). An example from the scale of region to civic plaza is shown in Figs. 18 and 19. ■

REFERENCES

Calthorpe, Peter and William Fulton. 2000. *The Regional City: Planning for the End of Sprawl.* Washington DC: Island Press.

Calthorpe, Peter. 1993. *The Next American Metropolis.* Princeton University Press.

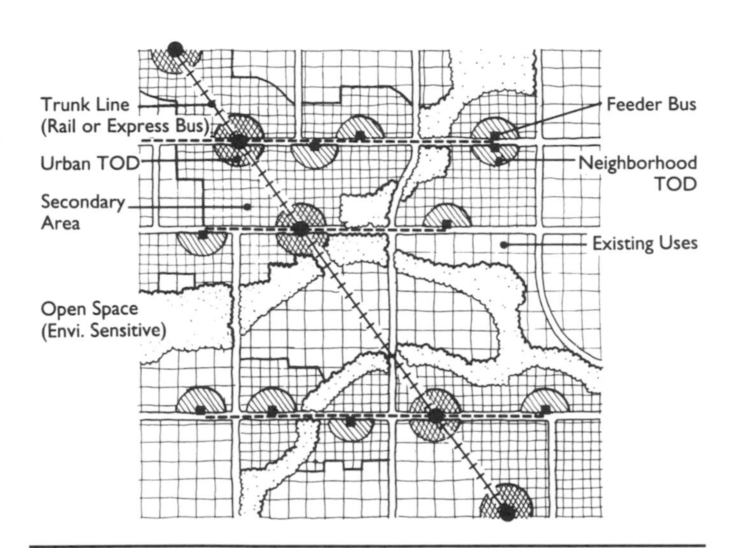

Fig. 6. Relationship to transit and circulation. The site must be located on an existing or planned trunk transit line or on a feeder bus route within 10 minutes transit travel time from a stop on the trunk line. Where transit may not occur for a period of time, the land use and street patterns within a TOD must function effectively in the interim.

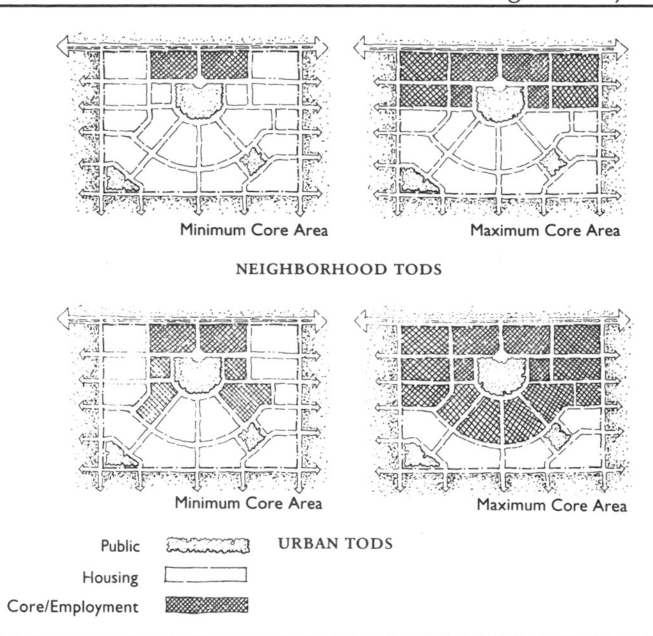

Fig. 7. Mix of uses. All TODs must be mixed-use and contain a minimum amount of public, core commercial and residential uses. Vertical mixed-use buildings are encouraged, but are considered a bonus to the basic horizontal mixed-use requirement. The following is a preferred mix of land uses, by percent of land area within a TOD.

USE	NEIGHBORHOOD TOD	URBAN TOD
Public	10% - 15%	5% - 15%
Core/Employment	10% - 40%	30% - 70%
Housing	50% - 80%	20% - 60%

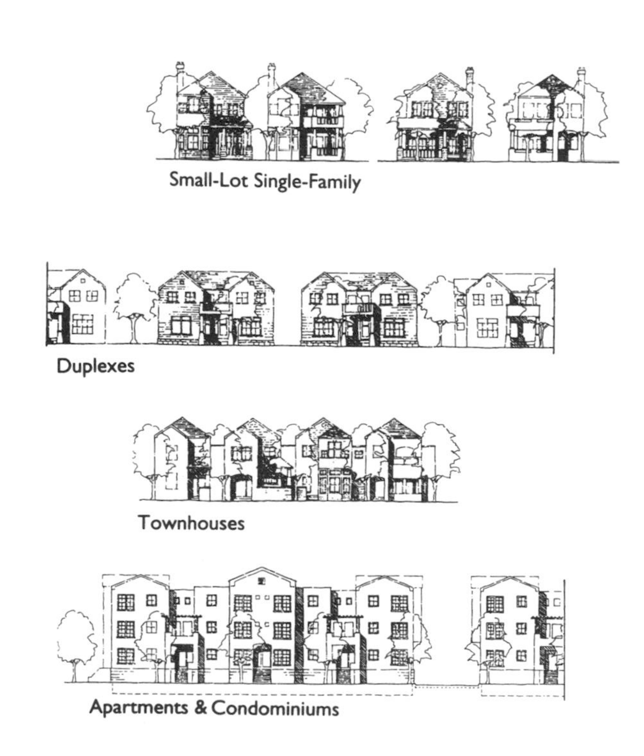

Fig. 8. Residential mix. A mix of housing densities, ownership patterns, price and building types is desirable in a TOD. Average minimum densities should vary between 10 and 25 dwelling units per net residential acre (.4 hectare), depending on the relationship to surrounding existing neighborhoods and location within the urban area.

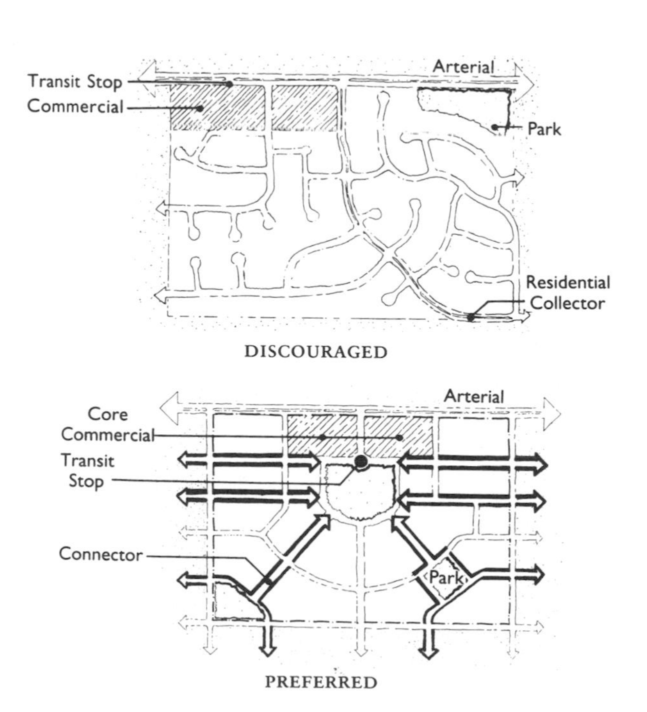

Fig. 9. Street and circulation system. The local street system should be recognizable, formalized and inter-connected, converging to transit stops, core commercial areas, schools and parks. Multiple and parallel routes must be provided between the core commercial area, residential and employment uses so that trips are not forced onto arterial streets.

Fig. 10. General design criteria. Buildings should address the street and sidewalk with entries, balconies, porches, architectural features and activities that help create safe, pleasant walking environments. Building intensities, orientation and massing should promote more active commercial centers, support transit and reinforce public spaces. Variation and human-scale detail in architecture is encouraged. Parking should be placed to the rear of buildings.

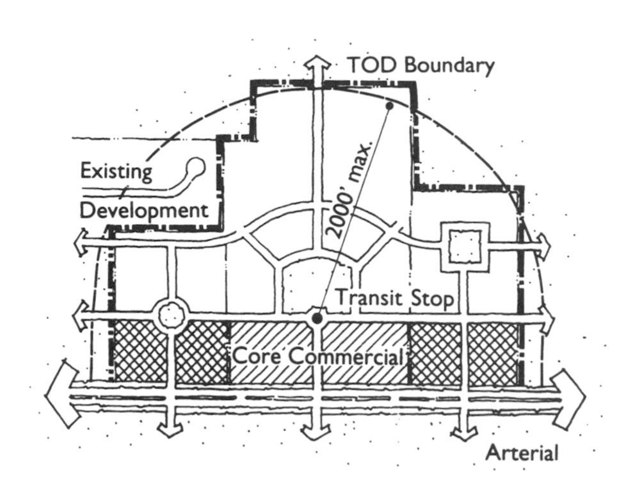

Fig. 11. Site boundary definition. The size of the TOD is variable depending on the ability to provide internal, local street connections. Parcels within an average 10-minute walking distance of the transit stop shall be included if direct access by local street or path can be established without use of an arterial. To allow for a basic mix of uses, the TOD area should be a minimum of 10 acres (4 hectares) for redevelopable and infill sites, and 40 acres (16 hectares) for New Growth Areas.

Fig. 12. Coordinated planning and specific area plans. Regardless of the number of property owners, development of a TOD must provide a coordinated plan for the entire site. This "Specific Area Plan" should be consistent with the Design Guidelines, coordinate development across property lines and provide strategies for financing construction of public improvements.

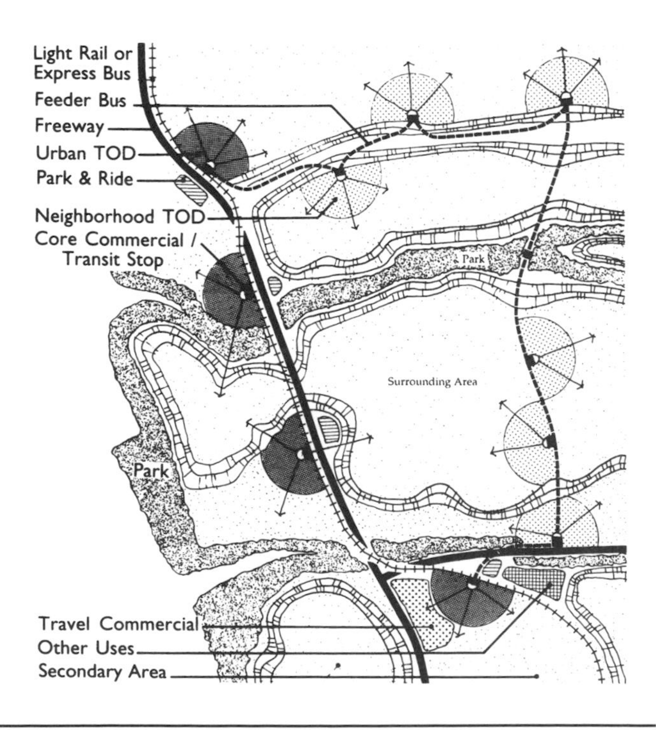

Fig. 13. Distribution of TODs. TODs should be located to maximize access to their Core Commercial Areas from surrounding areas without relying solely on arterials. TODs with major competing retail centers should be spaced a minimum of one mile apart and should be distributed to serve different neighborhoods. When located on fixed rail transit systems, they should be located to allow efficient station spacing.

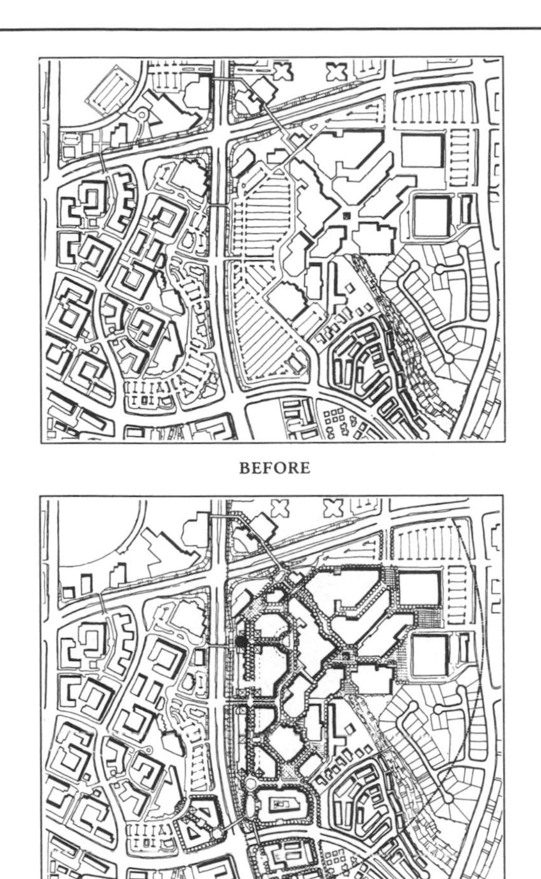

Fig. 14. Redevelopable and infill sites. Redevelopable and Infill Sites should develop underutilized parcels with new uses that allow them to function as walkable, mixed-use districts. Existing uses that are complementary, economical and physically viable should be integrated into the form and function of the neighborhood. Existing low-intensity and auto-oriented uses should be redeveloped to be consistent with the TOD's compact, pedestrian-oriented character.

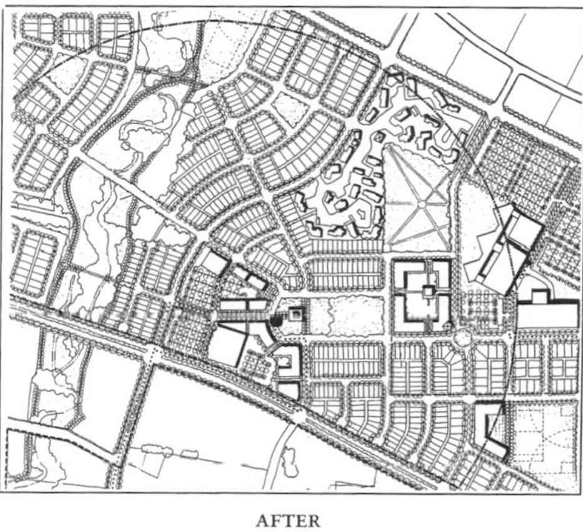

Fig. 15. New growth areas. New Growth Areas are typically located at the edge of the metropolitan region or on large sites that have been passed over. They may be large enough to create a network of Urban and Neighborhood TODs, as well as Secondary Areas, and should be planned in coordination with extensions of transit and an Urban Growth Boundary. New Growth Areas should not, however, be used to justify "leap frog" development or degrade sensitive environmental habitat.

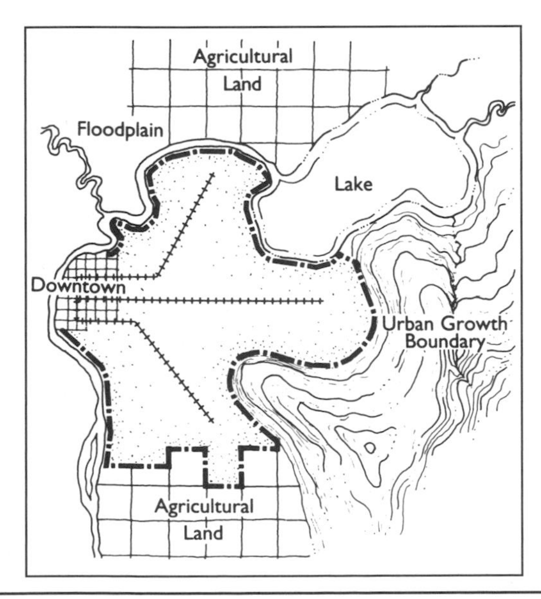

Fig. 16. Regional form. Regional form should be the product of transit accessibility and environmental constraints. Major natural resources, such as rivers, bays, ridge lands, agriculture and sensitive habitat should be preserved and enhanced. An Urban Growth Boundary should be established that provides adequate area for growth while honoring these criteria.

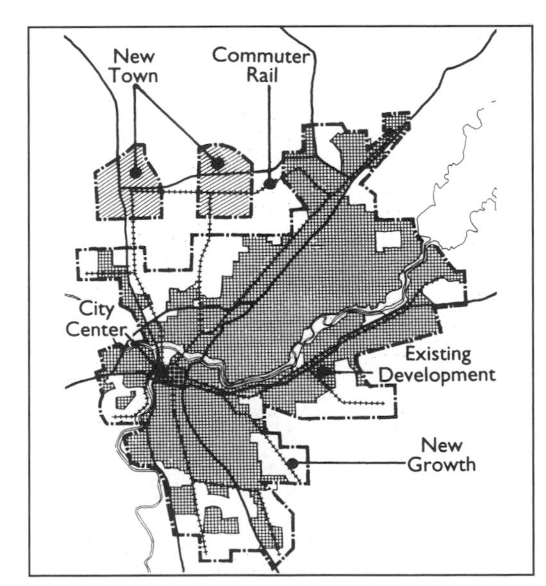

Fig. 17. Criteria for new towns. New Towns should only be planned if a region's growth is too large to be directed into Infill and adjacent New Growth Areas. They should be used to preserve the integrity of and separation between existing towns, as well as plan for a regional balance in jobs and housing. Appropriate sites should have a viable commuter transit connection and are not on environmentally sensitive lands.

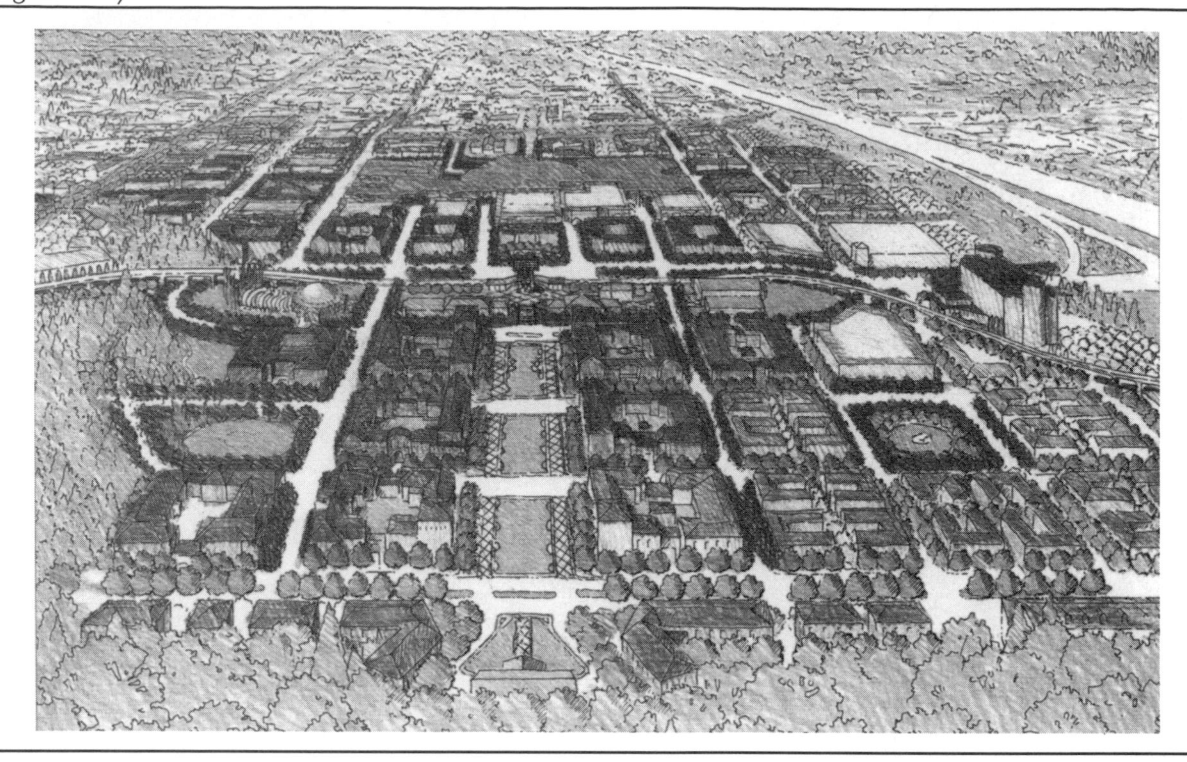

Fig. 18. Clackamas, Oregon. One of six Regional Centers in the Portland region, Clackamas Town Center is an example of the potential for redeveloping "greyfield" malls throughout suburbia. This site was one of the community urban-design cases conducted during development of the 2040 Plan. Here a traditional mall and a strip retail center are transformed into a mixed-use Town Center, complete with a large range of housing, parks, civic buildings, retail and offices. Calpthorpe Associates,1994.

Fig. 19. Clackamas, Oregon. The core of the old mall is retained and surrounded by blocks equal in size to those of downtown Portland. The parking is accommodated primarily in structures to the interior of each block. The existing arterial has been split into two one-way streets to better accommodate pedestrians. Between the two streets, the light-rail station and other civic buildings are central and more accessible. Since completion of this plan, many malls in the United States have looked to similar redevelopment plans to convert parking lots into more human-scaled, mixed-use environments.

Alexander Garvin

Summary
A new approach is needed to planning that explicitly deals with both public action and the probable private market reaction. Such change-oriented planning requires general acceptance of the idea that while urban designers are part of the urban improvement process, it is others who will make that change: civic leaders, interest groups, community organizations, property owners, developers, bankers, lawyers, architects, engineers, elected and appointed public officials. This article reviews successful examples and the role of planning, different forms of investment, regulation and incentives.

Key words
housing, historic preservation, investment, pedestrian, private partnerships, regulation, riverfronts, transit systems

Chicago Plan 1909 proposed extending Michigan Avenue across a new bridge north of the city's then existing business district.

A realistic approach to city and suburban planning

There is agreement neither on what to do to improve our cities and suburbs nor on how to get the job done. Some believe the answers are a matter of money; others believe they involve politics, or racial and ethnic conflict, or some other factor. One thing most people share, though, is disillusionment with planning as a way of fixing the American city.

This disillusionment with planning is far from justified. There are dozens of projects that are triumphs of American planning:

- Chicago would not have 23 miles (37 km) of continuous parkland along Lake Michigan if this land had not been included in the city's comprehensive plan of 1909.
- The glorious antebellum sections of Charleston, South Carolina, would not have survived if the city had not adopted zoning in 1931.
- Pittsburgh would not rank sixth in the nation as a major corporate headquarters center if it had not rebuilt its downtown during the 1940s and 1950s.
- Portland, Oregon, would not be a lively retail and employment center if during the 1970s and 1980s it had not enriched its pedestrian environment, built a light-rail system, and reclaimed its riverfront.

Such triumphs are easy to overlook. Once a problem is solved it disappears and is forgotten. Even local excitement over a successful project rarely spills over into national publications other than those with a narrow group of readers (preservationists, environmentalists, realtors, lawyers, architects, bankers, or some other group that is intimately involved with specific sets of city problems).

Many people are disillusioned with planning because so many of its promises are not kept. Usually these promises are made in good faith by planners who believe that their job is to establish municipal goals and provide blueprints for a better city. Too often the efforts of these planners end without much consideration of how they will obtain political support for their proposals, who will execute them, or where the money to finance them will come from.

Disillusionment with urban planning also develops when physical improvements fail to solve deep-seated social problems. This is not the fault of planning. After all, fixing cities does not fix people. The dis-

Credits: This article is adapted from the author's book, *The American City: What Works, What Doesn't*, McGraw-Hill, New York, 2002. Photos are by author unless otherwise noted.

illusionment is the product of false expectations. Crime, delinquency, and poverty are afflictions of city residents, not of the cities themselves. Such problems can be found in suburban and rural areas as well.

More realistic expectations are needed of what planning can accomplish. While it cannot change human nature and is therefore not a panacea for all urban ills, it surely can improve a city's physical plant and consequently affect the safety, utility, attractiveness, and character of city life. When Chicago began creating its waterfront parks, for example, large sections of the shoreline of Lake Michigan were being used as rail yards and garbage dumps. Simply removing these uses reduced hazards and made neighboring property more attractive.

A better understanding is needed of how effective planning is translated into a better quality of life. It is not accomplished by planners operating in a vacuum. By themselves, planners cannot accomplish very much. Improving cities requires the active participation of property owners, bankers, developers, architects, lawyers, contractors, and all sorts of people involved with real estate. It also requires the sanction of community groups, civic organizations, elected and appointed public officials, and municipal employees. Together they provide the financial and political means of bringing plans to fruition. Without them even the best plans will remain irrelevant dreams.

Finally, the planning profession itself needs to improve its understanding of the way physical changes to a city can achieve a more smoothly functioning environment, a healthier economy, and a better quality of life. For example, the restoration of Charleston's historic district generated substantial tourist spending, just as the reconstruction of the bridges and highways leading into downtown Pittsburgh reduced the cost of doing business and initiated an era of major corporate investment. These and other successful planning strategies are too frequently ignored in the search for more innovative prescriptions.

At its best, planning alters the very character of city life. During the 1970s and 1980s, Portland completely reorganized vehicular and pedestrian circulation. The business district was encircled by a ring road that greatly improved motor vehicle accessibility. A light-rail system provided transit service from the suburbs. Pedestrian precincts were established by transforming the old downtown highway into a riverfront park, by eliminating private motor vehicles from two downtown streets and repaving them as transit ways, and by acquiring several downtown blocks and converting them into new public parks. As a result, Portland became a safer, more convenient, more beautiful city. It also became a more attractive destination for the city's rapidly growing metropolitan region, drawing tens of thousands of additional weekday shoppers and weekend visitors.

Despite many remarkable successes, American planning has been plagued with continuing mistakes. These mistakes were and are avoidable. More than three decades have passed since Jane Jacobs observed that billions of dollars are spent for . . .

> *Housing projects that are truly marvels of dullness and regimentation. . .Civic centers that are avoided by everyone but bums, who have fewer choices of loitering places than others. Commercial centers that are lack-luster imitations of standardized suburban chain store shopping. Promenades that go from no place to nowhere and have no promenaders. Expressways that eviscerate great cities. This is not the rebuilding of cities. This is the sacking of cities. (Jacobs 1961, p. 4)*

Three decades and hundreds of billions of dollars later, her criticisms still ring true. Most cities continue to lack housing, civic and commercial centers, places to congregate and promenade, and traffic arteries. In too many cases, the attempt to remedy the situation constituted further "sacking of cities." These attempts may have been financially and politically feasible. However, they failed because they were conceived without proper consideration as to whether they would benefit the surrounding city.

Defining the planning process

Much of the nation's unsuccessful planning arises from the erroneous belief that project success equals planning success. Highways that are filled with automobiles, housing projects that are fully rented, and civic centers with plenty of busy bureaucrats may be successful on their own terms. The cities around them, however, may be completely unaffected. Worse, they may be in even greater trouble than they were prior to these projects.

Only when a project also has beneficial impact on the surrounding community can it be considered successful planning. Thus, planning should be defined as public action that generates a sustained and widespread private market reaction. That is precisely what has occurred wherever planning has been successful.

- When Chicago transformed its lake shore into a continuous park and drive, the real estate industry responded by spending billions to make it a setting for tens of thousands of new apartments.
- When Charleston preserved its "old and historic" district, it retained an extraordinary physical asset that, decades later, would attract a growing population and provide the basis of a thriving economy.
- When Pittsburgh cleared its downtown of the clutter of rail yards and warehouses; reduced air and water pollution; and built new highways, bridges, and downtown garages, businesses responded by rebuilding half the central business district.
- When Portland invested in a riverfront park, a light-rail system, and pedestrian streets, the private sector responded by erecting office buildings, retail stores, hotels, and apartment houses.

The scope of planning must be broadened. Over the past few decades, the areas of public concern and therefore of public action have expanded both substantively and geographically. Outraged citizens have demanded action to protect the natural environment, to preserve the national heritage, to provide a range of services that had never before been considered a public responsibility, and to deal with territory outside local political jurisdictions. The country should be deeply grateful to these activists for insisting that government fill important vacuums.

Similarly, the understanding of who generates public action must be expanded. Government agencies are not the only entities in the planning game. The public realm in many suburban areas is the creation of developers acting on the public's behalf. Non-profit organizations are responsible for many activities that have encouraged property owners to invest in neighborhood improvements. That is why the definition refers to public rather than government action that generates a sustained and widespread private market reaction.

Chicago, 1892 (Courtesy Chicago Historical Society)

Chicago, 1984

Pittsburgh, 1936 (Courtesy Carnegie Library of Pittsburgh)

Pittsburgh, 1984

Portland, OR, 1974 (Courtesy of Oregon Historical Society)

Portland, OR 1998

Too often, the response to their legitimate demands has been to create a set of protected special interests that are excluded from competition with other equally legitimate public concerns. As a result, large geographic areas are removed from active use without consideration of the social consequences. Buildings are declared landmarks without reference to economic impact. Services are provided to socially impaired individuals without any thought of the effect on the surrounding community. The situation can be rectified by including these new areas of public concern within the scope of planning and simultaneously including a far broader range of participants in the planning process.

The broad definition of planning suggested above highlights the fact that planning is about change: preventing undesirable change and encouraging desirable change. It may involve a tax incentive, a zoning regulation, or some other technical prescription, but only as a mechanism for instigating change. The important element is change itself. Planners obtain changes in safety, utility, and attractiveness of city life through strategic public investment, regulation, and incentives for private action.

Strategic public investment

Nineteenth-century planners were particularly enamored of strategic government investment. Just think of the many locations that were made more attractive for prospective developers through the installation of water mains, sewer pipes, or transit lines.

A more recent example is federal subsidization of the Interstate Highway System, created pursuant to the Federal Aid Highway Act of 1956. It vastly increased the amount of land within commuting distance of cities and, in the process, increased the attractiveness of suburban locations. Developers eagerly purchased the newly accessible land and built houses, shopping malls, and office parks. In the process millions of consumers were given the opportunity of owning a house in the country, close to shopping facilities and sometimes also near their jobs.

The 46,567 miles (78,800 km) of interstate highway that made up the system as of the year 2000 continue to alter land use patterns and local economies. At many exits "illuminated logos of McDonald's, Kentucky Fried Chicken, Waffle House, Phillips 66, Super 8 and Best Western have sprung up like lollipops on 100-foot (30 m) tall sticks. Just off the exits, often tucked out of sight, are its more muscular businesses—factories, meat-packing plants, warehouses and distribution centers." The effects are not just a matter of changing land use. For example, all but one of the thirteen counties bordering Interstate 40 in Arkansas increased in population during the decade between 1990-2000, while six of the eleven counties just beyond them lost population or failed to grow.

Capital investment can cause adverse impacts as well. Many interstate highways attracted motorists away from traditional urban arterials, thereby reducing demand in the retail establishments that had previously catered to the large market of automobile-oriented consumers. For decades after the highways had been built, cities were plagued with blighted retail streets, unable to replace the customers that previously had filled their no longer active stores.

The difference between routine capital spending and strategically planned investments lies in the use of these expenditures to spark further investment by private businesses, financial institutions, property owners, and developers. Governments regularly spend substantial sums of money to eliminate congestion and improve traffic flow. Such spending also can be used to spur development. One of the most dramatic examples occurred in Chicago when the city built a bridge extending Michigan Avenue north of the city's downtown "Loop." Since 1884, about half the traffic moving north and south across the Chicago River had squeezed into the 24 feet (7.3 m) of roadway and two 7 foot (2.13 m) wide sidewalks on the Rush Street Bridge. The congestion was sufficiently bad to generate a number of plans, culminating in 1909 with the proposals made by Daniel Burnham and Edward Bennett in The Plan of Chicago (Stamper 1991). They proposed both major street widening and a new bridge.

Pittsburgh, 1997

In 1913 the City Council decided to build an additional bridge connecting both sides of Michigan Avenue and to widen substantially the rights of way at either end of the new bridge. The section north of the river, then known as Pine Street, was nearly a mile long (1.6 km) and between 64-75 feet (19.5-23 m) wide. Michigan Avenue's new right-of-way was 150 feet wide. It was not entirely straight because the Council had insisted on preserving the city's landmark Water Tower & Pumping Station, which had survived the Fire of 1871 that had leveled so much of the city. Widening the street cost $14.9 million. It required the complete demolition of 34 buildings and the partial demolition of 33 others. The bridge and new avenue were opened to traffic in 1920. In the ensuing decade 31 major buildings were constructed or remodeled along this portion of North Michigan Avenue and property values increased six times. During the same period more valuable properties on State Street, south of the bridge, had only doubled in value. The new bridge and widened avenue had established North Michigan Avenue as the location of the future and set in motion the forces that by the end of the century would shift a substantial portion of downtown retailing from State Street to North Michigan Avenue, turning it into the miracle retailing mile of Chicago and one of the nation's most vibrant urban destinations.

Regulation

Regulation is most often used to alter the size and character of the market and the design of the physical environment. Perhaps the single most effective example occurred during the 1930s when the federal government restructured the banking system and in the process dramatically altered the housing market. Prior to that time few banks provided mortgage loans that covered more than half the cost of a house. These loans were extended for relatively short periods of time (two to five years) and involved little or no amortization.

The National Housing Act of 1934, which created the Federal Housing Administration (FHA), changed all that. It regulated the rate of interest and the terms of every mortgage that it insured. By 1938, a house could be bought for a cash-down payment equal to 10 percent of the purchase price. The other 90 percent came in the form of a 25-year, self-amortizing, FHA-insured mortgage loan. These new mortgage-lending practices greatly increased the number of people who could afford a down payment on a house as well as monthly debt service payments on a mortgage, and thereby also increased the size of the market for single-family houses. That is one of several reasons that the proportion of American households that owned their home increased from 44 percent in 1940 to 68 percent in 2000.

Not only did the FHA alter the size of the market, it also determined the design of the product. In order to be eligible for FHA mortgage insurance a house had to conform to published minimum property standards that included structure, materials, and room sizes. The effect of these regulations was to guarantee a minimum standard of quality on a national scale.

Regulation also can be used to alter the character of an entire area. This process usually begins with an attempt to prevent hazardous conditions. Local governments, for example, are usually interested in providing sufficient open land to permit natural drainage of rain and snow, to prevent waste from percolating through the ground to contaminate the water supply, and to ensure privacy. One way of achieving these objectives is to require a minimum lot size for any development, *e.g.*, no more than one house per acre (.4 ha.). The end result is the landscape of one-family houses on large lots that can be found throughout the nation.

As with strategic government investment, a regulation such as mandating minimum lot sizes also can produce an adverse impact. Since the amount of land in any community is finite, whenever a minimum lot size is adopted the future supply of house sites is reduced. This reciprocal relationship between the degree of regulation and the size of the market for the resulting product is inevitable. It was poignantly explained by Jacob Riis, who, in 1901, was already lamenting that the minimum construction requirements of "tenement house reform. . .

Portland, 1990. Pioneer Courthouse Square built at the intersection of the light-rail system and the city's two pedestrianized streets.

tended to make it impossible for anyone [not able] to pay $75 to live on Manhattan Island" (Riis, 1891).

Zoning regulations can be used to exclude the intrusive development incompatible with desired land-use patterns. By eliminating the possibility of such undesirable change, they reduce the risk of future problems (*e.g.*, traffic, pollution, and noise) and thereby increase the attractiveness of investing in real estate. In suburban communities zoning is routinely used to provide front and side yards that are large enough for residents to park their cars, thereby leaving room for visitors to park on the street. Rear yards have to be large enough to prevent unwarranted intrusions onto neighboring properties.

While regulations governing yard size can sustain real estate values, they rarely generate further improvements. But Santa Barbara, California, has demonstrated how land-use regulation can stimulate real estate activity by reducing the risk of developing property. Civic leaders were eager to spur economic growth and decided to do so by encouraging investment in tourist-oriented facilities. Not only did they need something with which to attract the growing tourist market, but also they needed to induce the real estate industry to build the necessary facilities.

At the beginning of the twentieth century, when this effort began, Santa Barbara was a dusty, wooden town so typical of those seen in western movies. It decided to reshape itself to conform to a California-Mediterranean image and adopted building laws that required property owners to develop in compliance with that image. By mandating design requirements, Santa Barbara increased its tourist appeal. More important, since property owners were assured of compatible neighboring buildings, the risk of failure was reduced and the likelihood of capturing the customers who had been attracted by the area's charming heritage was increased. It would be difficult to create a more auspicious climate for a tourist-based economy.

Incentives

Although the use of incentives is becoming more popular, the approach has been around a long time. One of the oldest examples is the incentive for people to own their homes. During the Civil War, the U.S. Congress allowed taxpayers to deduct interest payments and local taxes from the income that formed the basis of federal tax payments. The same deduction of mortgage interest and taxes was reintroduced in 1913, when the federal income tax was adopted. In both instances the incentives triggered an increase in home ownership. Yet neither incentive was intended to encourage development or redevelopment in specific communities.

There can be no serious change either in cities or suburbs without a favorable investment climate. In many instances government need only guarantee two things: intelligent spending on capital improvements and regulatory policies that provide stability and encourage market demand. Only when investment and regulation are insufficient to do the job should incentives come into play.

Santa Barbara, 1880. Aerial view of the city when it was a typical wooden western town. (Courtesy Santa Barbara Historical Society)

New York City faced such a situation during the mid-1970s. The city's fiscal crisis precluded most capital spending. Political gridlock prevented serious regulatory reform. At the same time the rate of housing deterioration and abandonment had reached alarming proportions. The city administration had to develop a strategy that would prevent further deterioration. The author, at that time Deputy Commissioner of Housing in charge of J-51 and other housing rehabilitation programs, proposed a strategy with one technique that seemed most likely to succeed: incentives that were sufficiently generous to induce private investment in the existing housing stock. Consequently, the Housing and Development Administration proposed to revise the city's little-known J-51 Program. It provided a twelve-year exemption from any increase in real estate tax assessment due to physical improvements and a deduction from annual real estate tax payments of a portion of the cost of those improvements.

The problem with the earlier J-51 Program was that it did not apply to three-quarters of the city's housing stock. Existing apartments that were not subject to rent control (because they were in structures that had been built after 1947, or had experienced a change in occupancy after 1971, or were owner-occupied) could not obtain these benefits unless they became subject to rent control. Nonresidential structures that had been converted to residential use were completely ineligible. Without J-51 benefits, any major investment in improvements resulted in punishment—a major increase in the real estate tax assessment. This was especially burdensome to the 770,000 apartments then subject to rent stabilization, New York City's second rent regulatory system. New York City regulates rents pursuant to two programs: rent control and rent stabilization. In 1975, 642,000 of New York City's 2,719,000 housing units were rent controlled. (Bloomberg, 1975).

During 1976, the Beame Administration persuaded the state legislature and the New York City Council to smash the rent control barrier by extending eligibility to rent-stabilized apartments. J-51 benefits were also provided for cooperative and condominium apartments in newly rehabilitated residential structures and to rental apartments in buildings converted from nonresidential to residential use, provided that they would become subject to some form of rent regulation.

These tax incentives completely altered the climate for investment in existing buildings. Banks increased their lending for housing rehabilitation, building owners increased their investments in building improvements, and developers began purchasing vacant structures for conversion to residential use. In fiscal 1977-1978, the first year in which the full impact of these incentives could be measured, more than 48,000 apartments were granted J-51 benefits.

J-51 provided an incentive that was sufficiently attractive to induce major investment in housing rehabilitation. However, there was another reason that so many property owners chose to apply for benefits. The administration of the program was made user-friendly. Until 1975 the program operated subject to unpublished regulations. Specific improvements that were eligible for benefits and the maximum

Santa Barbara, 1988. Aerial view of the city after it had been altered to conform to the Hispano-Mediterranean esthetic required by local zoning.

Charleston, 1991

New York City, 1991. Routine zoning regulations produced a routine landscape of identical front and side yards.

allowable expenditures for those improvements were listed on a typed schedule that was kept by the individual responsible for reviewing applications. Applicants had to file 26 separate forms. Program procedures were known to a few well-connected lawyers and developers, but had never been made public.

Within months of enactment of the revised J-51 Program, the administration published official regulations, made public a printed schedule of all allowable costs, and reduced the required filing to three one-page forms. Even unsophisticated property owners and poorly informed mortgage officers were now able to calculate probable J-51 benefits. As a result of these efforts, hundreds of property owners who had always had an aversion to government agencies were willing to seek the assistance they needed. In the process tens of millions of dollars were invested in improving the existing housing stock, demonstrating that properly conceived incentives can generate a desirable, sustained, and widespread market reaction. Between 1975 and 1990 many user-friendly characteristics of the program were eliminated. As a result, few property owners now apply for J-51 benefits without the assistance of a lawyer or expediter who specializes in agency processing.

A new approach to planning

A new approach is needed to planning that explicitly deals with both public action and the probable private market reaction. Such change-oriented planning requires general acceptance of the idea that while planners are in the change business, it is others who will make that change: civic leaders, interest groups, community organizations, property owners, developers, bankers, lawyers, architects, engineers, elected and appointed public officials—the list is endless.

Being entirely dependent on these other players, planners must concentrate on increasing the chances that everybody else's agenda will be successful. They may choose to do so by targeting public investment in infrastructure and community facilities, or by shaping the regulatory system, or by introducing incentives that will encourage market activity. But whatever they select, their role must be to initiate and shepherd often-controversial expenditures and legislation. More important, the public will be able to hold them accountable by evaluating the cost effectiveness of the private market reaction to their programs.

Only when this approach to planning takes hold will urban designers and planners get beyond the technical studies, needs analyses, and visions of the good city that currently masquerade as urban planning and get on with the business of fixing the American city. ■

REFERENCES

Bloomberg, Lawrence. 1975. (with Helen Lamale), *The Rental Housing Situation in New York City 1975,* New York: New York City Housing & Development Administration.

Garvin, Alexander. 1996. Revised 2002. *The American City: What Works, What Doesn't.* New York: McGraw-Hill.

Jacobs, Jane. 1961. *The Death and Life of Great American Cities,* New York: Random House.

Riis, Jacob. 1891. In a letter probably to Dr. Jane Robbins, October 10, 1891, quoted by Roy Lubove in *The Progressives and the Slums,* Pittsburgh: University of Pittsburgh Press, 1962 (p. 181).

Stamper, John W. 1991. *Chicago's North Michigan Avenue,* Chicago: University of Chicago Press (pp. 1–27 and 197–201).

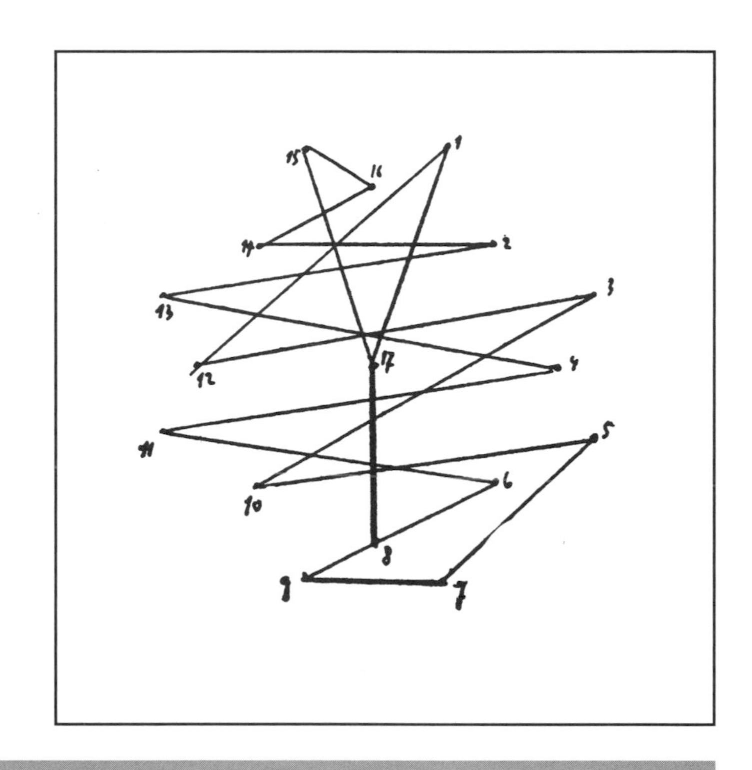

2 • CLASSIC TEXTS OF URBAN DESIGN

Camillo Sitte

Summary
Camillo Sitte (1843–1903) was an Austrian architect who through writing and practice sought to restore a humanistic approach to town planning, based on a rigorously analytical study of the perception of urban space, as opposed to what he saw as the increasingly technocratic attitude of engineers and traffic planners. In this he anticipated the incipient and continuing split between planning as a statistical, policy-based discipline and urban design as an aesthetically based field. He also shared the preoccupations of the academic field of art history in Austria and Germany, with its psychological and philosophical approach to the perception of form and space. The original edition of Sitte's text, first published in Vienna in 1889, was illustrated mainly with plans and contemporary prints and became what is arguably the founding text of modern urban design theory. The hand-drawn perspectives are from the 1902 French edition, to which Le Corbusier reportedly referred in characterizing Sitte's allegedly picturesque and medievalist approach as "the pack donkey's way."

Key words
arcade, city building, monuments, proportion, public square, urban scale, vista

Verona: *Piazza Erbe*

City building according to artistic principles

1 THE RELATIONSHIP BETWEEN BULDINGS, MONUMENTS AND PUBLIC SQUARES

In the South of Europe, and especially in Italy, where ancient cities and ancient public customs have remained alive for ages, even to the present in some places, public squares still follow the type of the ancient forum. They have preserved their role in public life. Their natural relationships with the buildings which enclose them may still be readily discerned. The distinction between the forum, or agora, and the market place also remains. As before, we find the tendency to concentrate outstanding buildings at a single place, and to ornament this center of community life with fountains, monuments, and statues which can bring back historical memories and which, during the Middle Ages and the Renaissance, constituted the glory and pride of each city.

It was there that traffic was most intense. That is where public festivals and theatrical presentations were held. There it was that official ceremonies were conducted and laws promulgated. In Italy, according to varying circumstances, two or three public places, rarely a single one, served these practical purposes.

The existence of two powers, temporal and spiritual, required two distinct centers: one, the cathedral square (Fig. 1) dominated by the campanile, the baptistry, and the palace of the bishop; the other, the

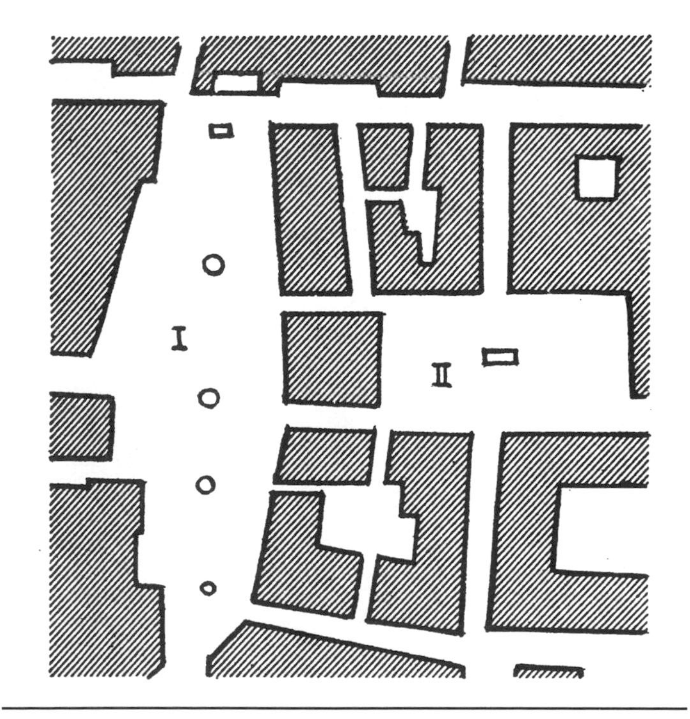

Fig. 1. Verona: I *Piazza Erbe,* II *Piazza dei Signori.*

Credits: This article is reprinted from Camillo Sitte, *The Art of Building Cities,* Reinhold Publishing Corporation, 1945, translated by Charles T. Stewart.

Signoria, or manor place, which is a kind of vestibule to a royal residence. It is enclosed by houses of the country's great and adorned with monuments. Sometimes we see there a *loggia*, or open gallery, used by a military guard, or a high terrace from which laws and public statements were promulgated. The *Signoria* of Florence is the finest example of this. The market square, rarely lacking even in cities of northern Europe, is the meeting place of the citizens. There stand the City Hall and the more or less richly decorated traditional fountain, the sole vestige of the past that has been conserved since the lively activity of merchants and traders has been moved within to iron cages and glass market places.

The important function of the public square in the community life of past ages is evident. The period of the Renaissance saw the birth of masterpieces in the manner of the Acropolis of Athens, where everything concurred to produce a finished artistic effect. The cathedral place at Pisa, an Acropolis of Pisa, is the proof of this. It includes everything that the people of the city have been able to create in building religious edifices of unparalleled richness and grandeur. The splendid cathedral, the campanile, the baptistry, the incomparable *Campo-Santo* are not depreciated by profane or banal surroundings of any kind. The effect produced by such a place, removed from the world of baseness while rich in the noblest works of the human spirit, is overpowering. Even those with a poorly developed sensitiveness to art are unable to escape the power of this impression. There is nothing there to distract our thoughts or to intrude our daily affairs. The esthetic enjoyment of those who look upon the noble façade of the cathedral is not spoiled by the sight of a modern haberdashery, by the cries of drivers and porters, or by the tumult of a café. Peace reigns over the place. It is thus possible to give full attention to the art work assembled there.

This situation is almost unique, although that of Saint Francis of Assissi and the arrangement of the *Certosa de Pavia* closely approach it. In general, the modern period does not encourage the formation of such perfect groupings. Cities, even in the fatherland of art, undergo the fate of palaces and dwellings. They no longer have distinct character. They present a mixture of motifs borrowed as much from the architecture of the north as from that of the southern countries. Ideas and tastes have been mingled as the people themselves have been interchanged. Local characteristics are gradually disappearing. The market place alone, with its City Hall and fountain, has here and there remained intact.

In passing we should like to remark that our intention is not to suggest a sterile imitation of the beauties spoken of as "picturesque" in the ancient cities for our present needs. The proverb, "Necessity breaks even iron," is fully applicable here. Changes made necessary by hygiene or other requirements must be carried out, even if the picturesque suffers from it. But that does not prevent us from examining the work of our forebears at close range to determine how much of it may be adapted to modern conditions. In this way alone can we resolve the esthetic part of the practical problem of city building, and determine what can be saved from the heritage of our ancestors.

Before determining the question in a positive manner, we state the principle that during the Middle Ages and Renaissance public squares were often used for practical purposes, and that they formed an entirety with the buildings which enclose them. Today they serve at best as places for stationing vehicles, and they have no relation to the buildings which dominate them. Our parliament buildings have no agora enclosed by columns. Our universities and cathedrals have lost their atmosphere of peace. Surging throngs no longer circulate on market days before our City Halls. In brief, activity is lacking precisely in those places where, in ancient times, it was most intense—near public structures. Thus, to a great extent, we have lost that which contributed to the splendor of public squares.

And the fabric of their very splendor, the numerous statues, is almost entirely lacking today. What have we to compare to the richness of ancient forums and to works of majestic style like the *Signoria* of Florence and its *Loggia dei Lanze*?

Buildings lay claim to so many statues that commissions are needed to find new subjects to be represented. It is often necessary to wait for years to find a suitable place for a statue although many appropriate places remain empty in the meantime. After long efforts we have reconciled ourselves to modern public squares as vast as they are deserted, and the monument, without a place of refuge, becomes stranded on some small and ancient space. That is even more strange, yet true. After much groping about, this fortunate result occurs, for it is thus that a work of art derives its value and produces a more powerful impression. Indifferent artists who neglect to provide for such effects must bear the entire responsibility of it.

The story of Michelangelo's *David* at Florence shows how mistakes of this kind are perpetrated in modern times. This gigantic marble statue stands close to the walls of the *Palazzo Vecchio*, to the left of its principal entrance, in the exact place chosen by Michelangelo. The idea of erecting a statue on this place of ordinary appearance would have appeared too modern or absurd if not insane. Michelangelo chose it, however, and without doubt deliberately; for all those who have seen the masterpiece in this place testify to the extraordinary impression that it makes. In contrast to the relative scantiness of the place, affording an easy comparison with human stature, the enormous statue seems to swell even beyond its actual dimensions. The sombre and uniform, but powerful, walls of the palace provide a background on which we could not wish to improve to make all the lines of the figure stand out.

Today the *David* is moved into one of the academy's halls under a glass cupola in the midst of plaster reproductions, photographs, and engravings. It serves as a model for study and an object of research for historians and critic. A special mental preparation is needed now to resist the morbid influences of an art prison that we call a museum, and to have the ability to enjoy the imposing work. Moreover, the spirit of the times, which believed that it was perfecting art, and which was still not satisfied with this innovation, had a bronze cast made of the *David* in its original grandeur and put it up on a vast plaza (naturally in its mathematical center) far from Florence at the *Via die Colli*. It has a superb horizon before it; behind it, cafes; on one side, a carriage station, a *corso*; and from all sides the murmurs of Baedeker readers ascend to it. In this setting the statue produces no effect at all. The opinion that its dimensions do not exceed human stature is often heard. Michelangelo thus understood best the kind of placement that would be suitable for his work, and, in general, the ancients were abler than we are in these matters.

The fundamental difference between the procedures of former times and those of today rests in the fact that we constantly seek the largest possible space for each little statue. Thus we diminish the effect that it could produce, instead of augmenting it with the assistance of a neutral background such as painters have used in their portraits. This explains why the ancients erected their monuments by the sides of public spaces, as is shown in the view of the *Signoria* of Florence. In this way, the number of statues could increase indefinitely without obstructing the circulation of traffic, and each of them had a fortunate background. Contrary to this, we hold the middle of a public place as the sole spot worthy to receive a monument. Thus no esplanade, however magnificent, can have more than one. If by misfortune it is irregular and if its center cannot be located geometrically we become confused and allow the space to remain empty for eternity.

2 OPEN CENTERS OF PUBLIC PLACES

It is instructive to study the manner in which the ancients located their fountains and monuments and to see how they were always able to make use of the conditions at hand. Ancient principles of art were applied anew in the Middle Ages, although less obviously. Only blindness can escape the observation that the Romans left the center of their forum free. Even in Vitruvius we may read that the center of a public place is destined not for statues but for gladiators. This question calls for a more attentive study of the following epochs.

In the Middle Ages the choice of placing fountains and statues together in many cases defies all definition. Some of the strangest arrangements were adopted. It is always necessary to recognize that, as for Michelangelo's *David*, this choice was guided by a fine sense for art, for the statue always harmonized admirably with its surroundings. Thus we face an enigma, the enigma of a natural art sense which, among the old masters, wrought miracles without the assistance of any esthetic rules. Modern technicians, who have succeeded them, armed with T-squares and compasses, have pretended to make fine decisions of taste through the coarseness of geometry.

Sometimes it is possible to catch a glimpse of the creative methods of our forebears and to find words that can explain the patterns of the successful effects that they attained. But since particular examples can vary so widely it seems difficult to wrest a general principle from the known facts. We shall try to see clearly in this apparent confusion, however, for our innate feeling has been lost for a long time. We can no longer get good results without deliberate intention. If then, we wish to rediscover the free inventiveness of the old masters and react against the inflexible geometrical principles of their successors, we must make an effort to allow the paths over which our ancestors went by instinct in ages that had a traditional esteem for art. The subject of this topic seems narrowly limited. Nevertheless, it is difficult to cover it in a few words. An example taken from everyday life which we hope is not shocking in its apparent triviality will make a useful substitute for an involved definition.

It is remarkable how children, working with no direction except their artistic instinct, often achieve the same results as primitive peoples in their crude designs. It may be surprising to say that one of their favorite games can teach us the principles of good monument location. As a matter of fact, the snowmen with which they amuse themselves in the winter are located in exactly the same manner as fountains and monuments were according to ancient practices. The explanation is quite simple. Consider the snow-covered plaza of a town. Here and there are snow paths—natural thoroughfares. There are large blocks of snow irregularly located between them, and the snowmen are built on these, for that is where the substance of which they are made is found.

It was at such points, similarly dispersed from traffic, that the ancient communities set up their fountains and monuments. It may be observed

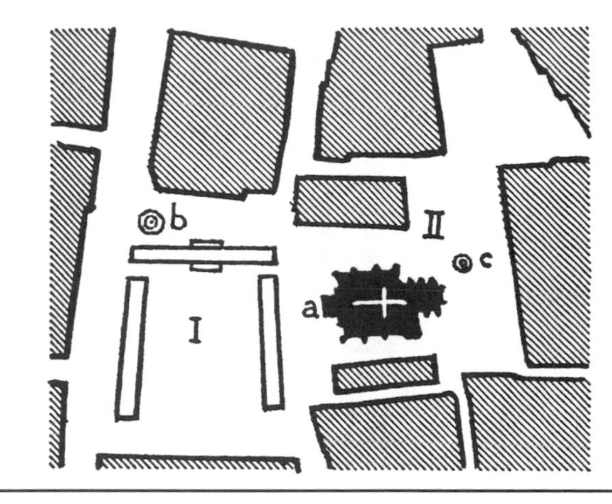

Fig. 2. Nuremberg: (I) Market Square, (II) *Frauen Plaza,* (a) *Marienkirche,* (b) Fountain.

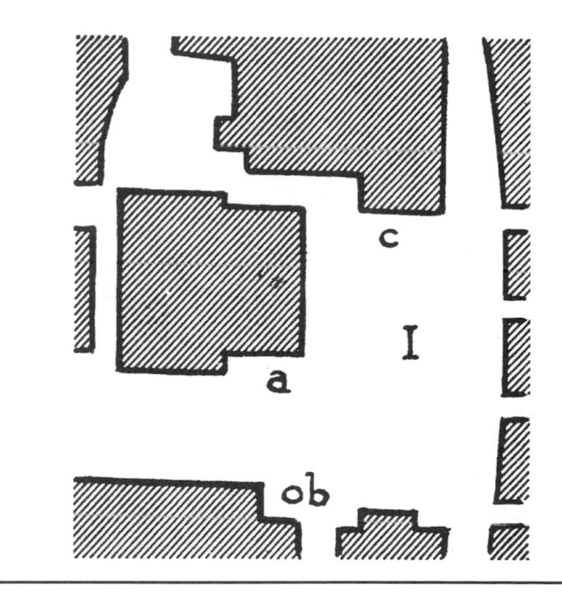

Fig. 3. Rothenburg ob der Tauber: (I) Market Square, (a) City Hall, (b) Fountain, (c) *Café.*

from old engravings that in ancient times the public places were not paved, nor even graded, but were furrowed in paths and gutters, as may still be seen in certain villages. If the building of a fountain were desired, it obviously would not be located on a thoroughfare, but rather on one of the island-like plots separated from traffic. When, with an increase in wealth, the community grew little by little, it had its public places graded and paved, but the fountain did not change its position. When it was desired to replace it by a similar, but more elaborate, structure, the new one was put up on the same spot.

Thus, each of these sites had its historical importance, and this explains why fountains and monuments are not located at places of intense traffic use, nor at the center of public places, nor on the axis of a monumental portal, but by preference to the side, even in the northern countries where Latin traditions have not had a direct influence. This also explains why in each city and in each public place the arrangement of monuments is different, for in each case streets open onto the square differently; traffic follows a different direction and leaves other points free. In short, the historical development of the public square varies according to the locality. It sometimes happens that the center of a

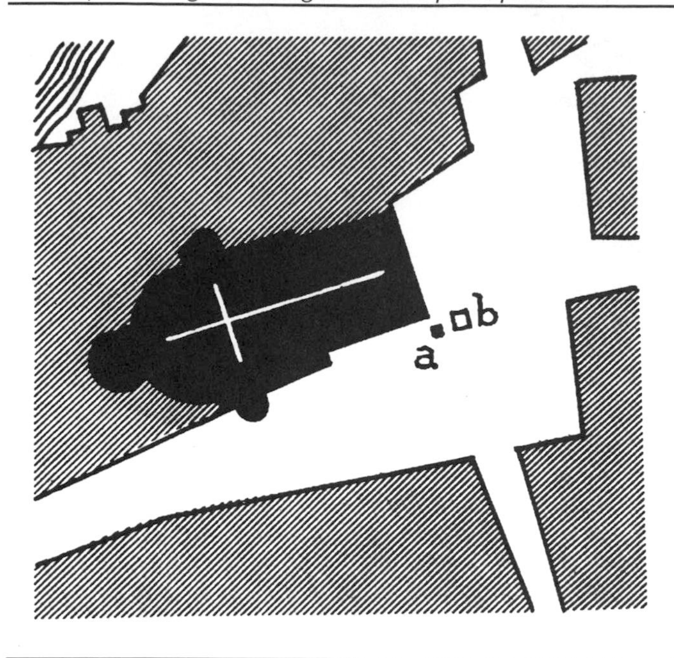

Fig. 4. Padua: *Piazza del Santo,* (a) Column, (b) Statue of *Gattamelata.*

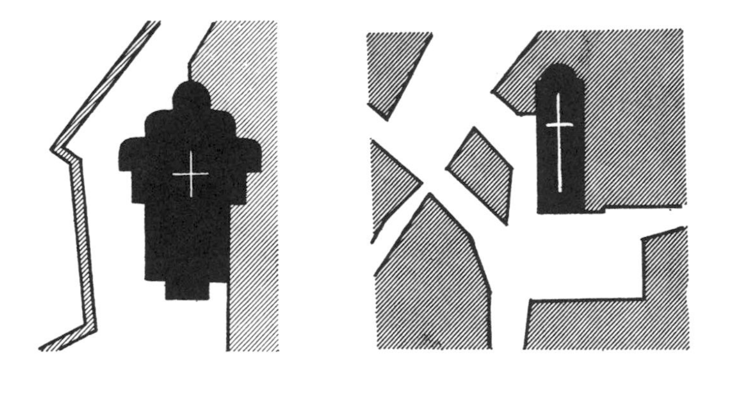

Fig. 5. Padua: *S. Giustina.*

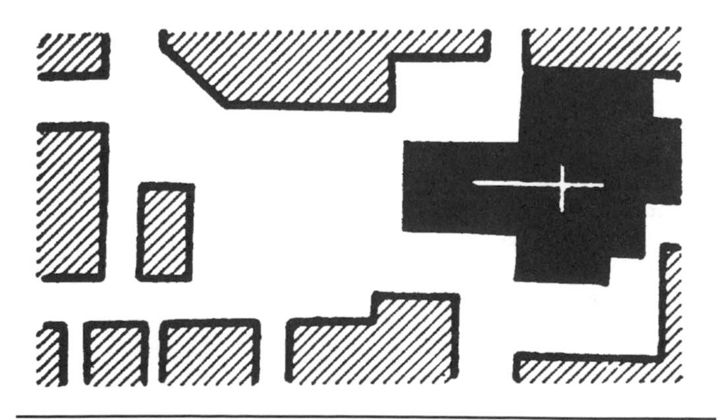

Fig. 6. Verona: *S. Fermo Maggiore.*

public place is selected for the placement of a statue. But this practice, preferred by modern architects, was never established as a principle by the ancients. They were not given to the excessive use of symmetry, for their fountains were built most frequently near the angle of a public square, where the principal street opened and where draft animals were brought for watering. The beautiful fountain of Nuremberg (Fig. 2) is a celebrated example of this. So also is the fountain of Rothenburg on the Tauber (Fig. 3).

In Italy, in front of the *Palazzo Vecchio*, on the *Signoria* of Florence, before the *Palazzo Communale* at Perugia, at the *Palazzo Farnese* in Rome, statues are located at the sides of the streets and not on the axis of the principal structure of the public square. In France the fountain of *Saint-Lazare* at Autun and the Fountain of the Innocents at Paris occupied the angle formed by the Rue aux Fers and the Rue Saint-Denis before 1786, instead of being aligned in the middle of the public square.

The location of the equestrian statue of *Gattamelata* by Donatello (Fig. 4) in front of Saint Anthony of Padua is most instructive. First we may be astonished at its great variance from our rigid modern system, but it is quickly and strikingly seen that the monument in this place produces a majestic effect. Finally we become convinced that removed to the center of the square its effect would be greatly diminished. We cease to wonder at its orientation and other locational advantages once this principle becomes familiar.

Thus to the ancient rule that prescribes the location of monuments at the edges of public squares may be added the principle followed during the Middle Ages, especially in cities of the north, according to which monuments and fountains were erected at points segregated from traffic. Now and then the two principles are put into practice simultaneously. They avoid common obstacles in seeking masterly artistic effects. Sometimes practical needs coincide with artistic requisites, and this is understandable, for a traffic obstacle may also interfere with a good view. The location of monuments on the axes of monumental buildings or richly adorned portals should be avoided for it conceals worthwhile architecture from the eyes; and, reciprocally, an excessively rich and ornate background is not appropriate for a monument. The ancient Egyptians understood this principle, for as Gattamelata and the little column stand beside the entrance to the Cathedral of Padua, the obelisks and the statues of the Pharaohs are aligned beside the temple doors. There is the entire secret that we refuse to decipher today.

The principle that we have just deduced applies not only to monuments and fountains, but to every type of construction, and especially to churches. Churches, which almost without exception occupy the center area of large sites, were not so located in former times. In Italy they are always set back with one or more sides against other buildings with which they form groups of open places that we shall now discuss.

The churches of Padua are classics in this respect. Only one side of San Giustina (Fig. 5) is set back against other buildings. Two sides of San Antonio and the Carmino are so set back, while all of one side with half of a second of the Jesuit church are so backed up. The bordering open space is quite irregular. This is also to be seen at Verona where a further tendency to preserve a square of great dimensions before the principal entrance to the church is observed; as in front of *S. Fermo Maggiore, S. Anastasia* and others. (Fig. 6). Each of these places has its particular history and creates a vivid impression. The facades and portals of the churches, which dominate them, assume their full splendor.

We rarely find an open place extending from the side of a church, as at *S. Cita* in Palermo (Fig. 7).

These few examples, in striking contrast to modern practices, are convincing enough to lead us into further study. No city could serve as a better subject for this study than Rome with its numerous religious edifices. A study of them from the point of view of location yields a surprising result. Of the 255 churches, 41 are set back with one side against other buildings; 96 with two sides against other buildings; 110 with three sides against other buildings; 2 set in with four sides to other buildings; and 6 stand free of other buildings.

It should be said that among the last six mentioned there are two modern buildings, the Protestant and Anglican chapels, and that the four others are surrounded by narrow streets. This is equally contrary to the modern practice that superimposes the center of the church upon the center of the site. We can definitely conclude that in Rome churches are never entirely free standing, and the same may be said of all Italy, for the principle is in use just as much in Pavia, in Venice, where only the Cathedral stands free on all sides; Cremona; Milan (with the exception of the Cathedral); Reggio (including the Cathedral here); Ferare; and many other places. The type seen in Lucca (Fig. 8) and Vicenza (Fig. 9) recalls the principles that we deduced on the subject of locating monuments. It is even more appropriate for monumental buildings to be situated on the sides of public squares of average spaciousness, for in that way alone can they be best utilized and looked at from a convenient distance.

The case of the Cathedral of Brescia (Fig. 10) is quite singular, but it is no exception to the rule, for the façade of the Cathedral serves as an enclosure for the open place. All of these observations indicate clearly that our modern systems are in direct opposition to established principles that were conscientiously observed in past epochs. We think it is impossible to place a new church anywhere except in the center of the site destined for it so that it will stand free on all sides, although this procedure is inconvenient and has few advantages. It detracts from the structure itself, for its potential effect cannot be concentrated but is scattered evenly around its circumference. Moreover, every organic relationship between the open place and its enclosure is made impossible, as all perspective effects require sufficient withdrawal. A cathedral requires a foreground to set off the majesty of its facade.

Location of a church in the middle of a site cannot even be defended in the name of the builder's interest, for it obliges him to extend, at great expense, all of the architectural features around the broad facades, the cornices, pedestals, and so on. In putting the building back with one or two sides against other buildings, the architect is spared all this expense, the free walls can be built of marble throughout, and there will still remain sufficient funds to embellish them with statues. Thus we should not have these monotonous side faces running continuously around the building, the perception of which cannot be appreciated from a single viewpoint. Furthermore, is it not often of advantage for a church to be joined to other buildings (cloisters or parsonages) in winter and bad weather? Besides, it is not only the building itself, but the public square as well, which suffers from the modern arrangement. Its name of place, square, or plaza is no longer any more than irony, for it is seldom more than a slightly widened street. In spite of all these inconveniences, and in spite of all the precepts of the history of ecclesiastical architecture, modern churches are located throughout the world almost without exception in the center of their sites. We have lost all discernment.

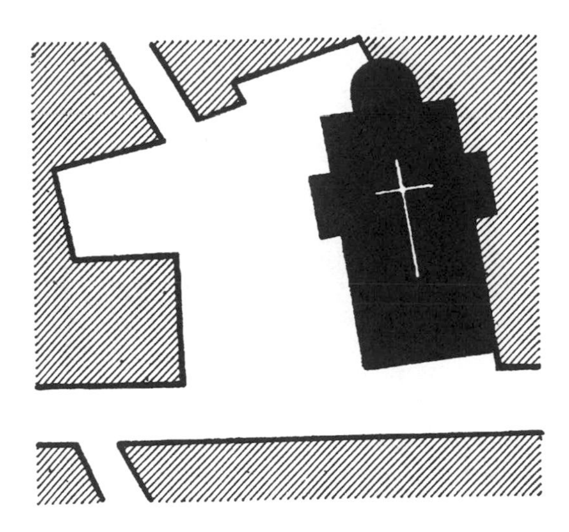

Fig. 7. Palermo: *S. Cita.*

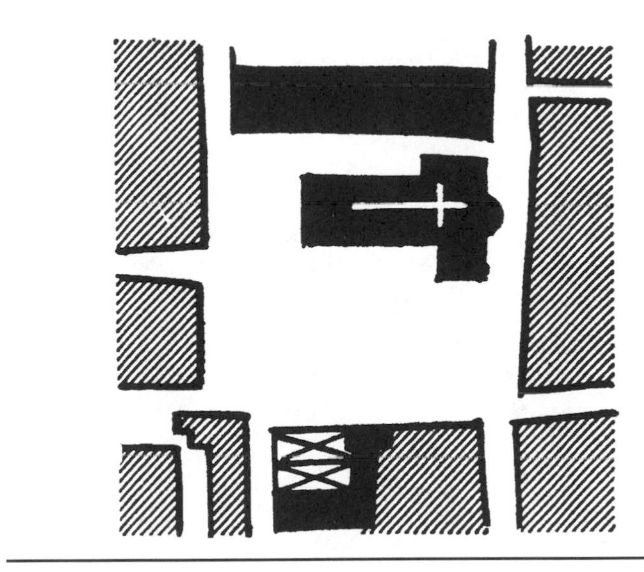

Fig. 8. Lucca: *S. Michele.*

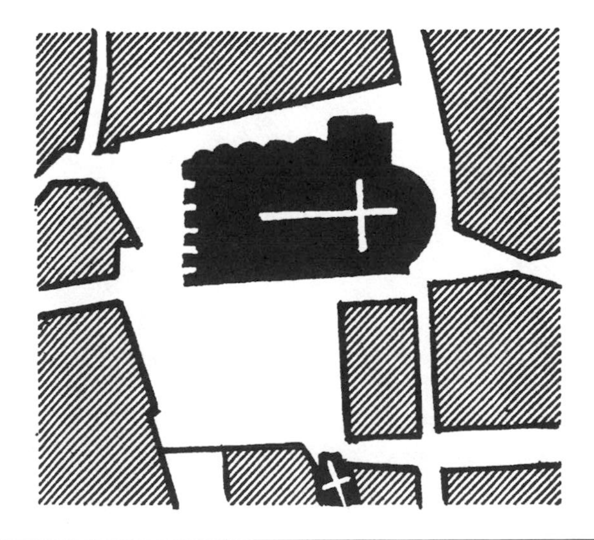

Fig. 9. Vicenza: *Piazza del Duomo.*

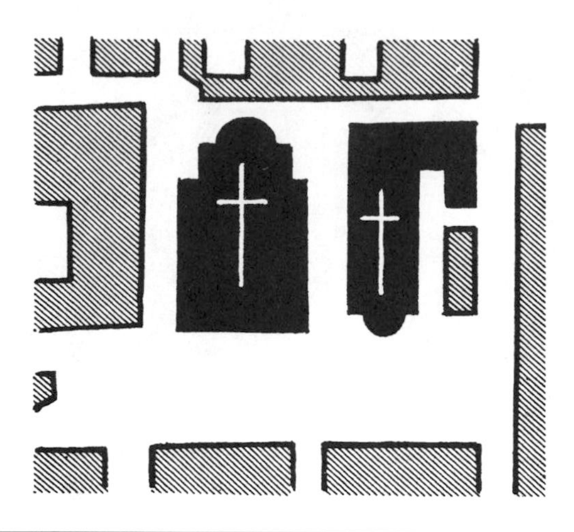

Fig. 10. Brescia: *Piazza del Duomo* (Cathedral Square) showing new (left) and old (right) cathedrals.

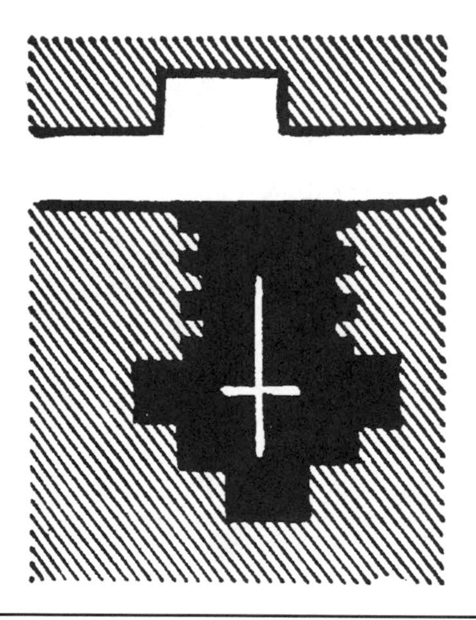

Fig. 11. Brescia: San Giovanni.

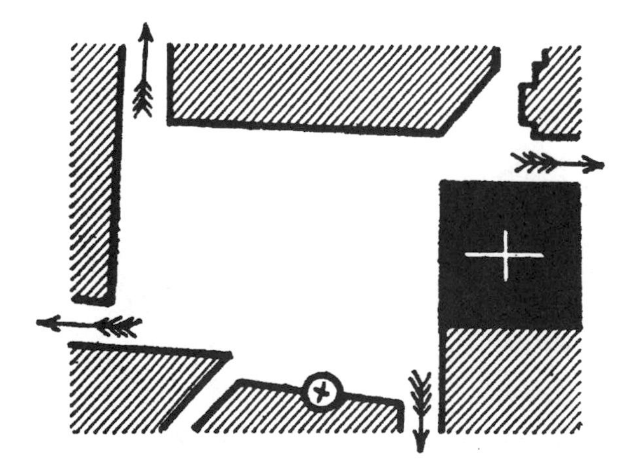

Fig. 12. Ravenna: Piazza del Duomo (Cathedral Square).

Theaters, city halls, and numerous other structures are also victims of this erroneous conception. Perhaps we hopefully believe in the possibility of seeing all sides of a building at once, or it may be thought that an interesting building is especially distinguished if its walls stand entirely free. Nobody imagines that putting a void around a building prevents it from forming, with its environs, various diverse scenes. The lordly square masses of the Florentine palaces present a picturesque appearance when seen from the adjacent narrow passages. In this way these buildings acquire a double worth, for they present different appearances from the *piazza* and the *vicolo*.

Modern taste is not satisfied with locating its own creations in the most unfavorable manner. It must also improve the works of the old masters by tearing them away from their surroundings. It does not hesitate to do so even when it is obvious that they have been designed to harmonize with the neighboring buildings, without which they would lose their worth. When a work of art is put in a place other than the one selected for it, part of its essential quality is taken away, and a great wrong is done to the artist who conceived it.

Performances of this kind are not rare. This rage for isolating everything is truly a modern sickness. R. Baumeister in his manual on city building even raises this to the status of a working principle. He writes, "Old buildings ought to be preserved, but we must, so to speak, peel them and preserve them." The object of this, then, is that by the transformation of surroundings the old buildings should be led to the midst of public places and in the axes of streets. This procedure is used everywhere and with special satisfaction in treating ancient city entrance portals. It is indeed a fine thing to have an isolated city gateway around which we may stroll instead of passing under its arches!

3 THE ENCLOSED CHARACTER OF THE PUBLIC SQUARE

The old practice of setting churches and palaces back against other buildings brings to mind the ancient forum and its unbroken frame of public buildings. In examining the public squares that came into being during the Middle Ages and the Renaissance, especially in Italy, it is seen that this pattern has been retained for ages by tradition. The old plazas produce a collective harmonious effect because they are uniformly enclosed. In fact, the public square owes its name to this characteristic in an expanse at the center of a city. It is true that we now use the term to indicate any parcel of land bounded by four streets on which all construction has been renounced.

That can satisfy the public health officer and the technician, but for the artist these few acres of ground are not yet a public square. Many things must be done to embellish the area to give it character and importance. For just as there are furnished and unfurnished rooms, we could speak of complete and incomplete squares. The essential thing of both room and square is the quality of enclosed space. It is the most essential condition of any artistic effect, although it is ignored by those who are now elaborating on city plans.

The ancients, on the contrary, employed the most diverse methods of fulfilling this condition under the most diverse circumstances. They were, it is true, supported by tradition and favored by the usual narrowness of streets and less active traffic movement. But it is precisely in cases where these aids were lacking that their talent and artistic feeling is displayed most conspicuously.

A few examples will assist in accounting for this. The following is the simplest. Directly facing a monumental building a large gap was made in a mass of masonry, and the square thus created, completely surrounded by buildings, produced a happy effect. Such is the *Piazza S. Giovanni* at *Brescia* (Fig. 11). Often a second street opens onto a small square, in which case care is taken to avoid an excessive breach in the border, so that the principal building will remain well enclosed. The methods used by the ancients to accomplish this were so greatly varied that chance alone could not have guided them. Undoubtedly they were often assisted by circumstance, but they also knew how to use circumstance admirably.

Today in such cases all obstructions would be taken down and large breaches in the border of the public place would be opened, as is done when we decide to "modernize" a city. Ancient streets would be found to open on the square in a manner precisely contrary to the methods of modern city builders, and mere chance would not account for this. Today the practice is to join two streets that intersect at right angles at each corner of the square, probably to enlarge as much as possible the opening made in the enclosure and to destroy every impression of cohesion. Formerly the procedure was entirely different. There was an effort to have only one street at each angle of the square. If a second artery was needed in a direction at right angles to the first, it was designed to terminate at a sufficient distance from the square to remain out of view from the square. And better still, the three or four streets which came in at the corners each ran in a different direction. This interesting arrangement was reproduced so frequently, and more or less completely, that it can be considered as one of the conscious or subconscious principles of ancient city building.

Careful study shows that there are many advantages to an arrangement of street openings in the form of turbine arms. From any part of the square there is but one exit on the streets opening into it, and the enclosure of buildings is not broken. It even seems to enclose the square completely, for the buildings set at an angle conceal each other, thanks to perspective, and unsightly impressions which might be made by openings are avoided. The secret is at right angles to the visual lines instead of parallel to them. Joiners and carpenters have followed this principle since the Middle Ages when, with subtle art, they sought to make joints of wood and stone inconspicuous if not invisible.

The Cathedral Square at Ravenna (Fig. 12) shows the purest type of arrangement just described. The square of Pistoia (Fig. 13) is in the same manner; as is the *Piazza S. Pietro* at Mantua (Fig. 14), and the *Piazza Grande* at Parma (Fig. 15). It is a little more difficult to recognize the principle in the *Signoria* of Florence (Fig. 16). The principal streets conform to the rule. The narrow strip of land of about a yard's width—at the side of the *Loggia dei Lanzi*—is much less noticeable in reality than it is on the map.

The ancients had recourse to still other means of closing in their squares. Often they broke the infinite perspective of a street by a monumental portal or by several arcades of which the size and number were determined by the intensity of traffic circulation. This splendid architectural pattern has almost entirely disappeared, or, more accurately, it has been suppressed. Again Florence gives us one of the best examples in the portico of the *Uffizi* with its view of the Arno in the distance. Every Italian city of average importance has its portico, and this is also true north of the Alps. We mention only the *Langasser Thor* at Danzig, the entrance portal of the City Hall and Chancellery at Bruges, the *Kerkboog* at Nimeguen, the great Bell Tower at Rouen, the monumental *Portals* of Nancy, and the windows of the Louvre.

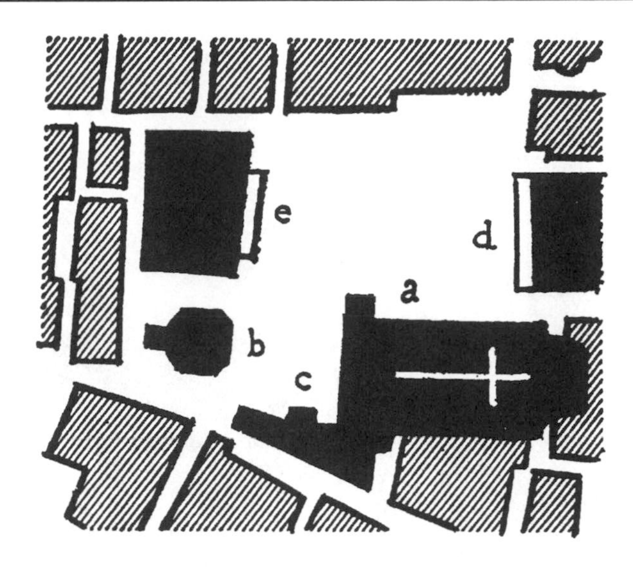

Fig. 13. Pistoia: *Piazza del Duomo* (Cathedral Square): (a) Cathedral, (b) Baptistry, (c) Bishop's Residence, (d) *Palazzo del Commune,* (e) *Pallazzo del Podesta.*

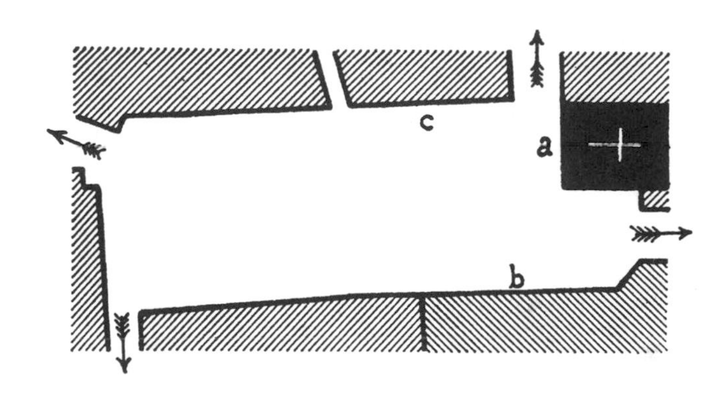

Fig. 14. Mantua: *Piazza S. Pietro:* (a) *S. Pietro,* (b) *Pal. Reale,* (c) *Vescoville.*

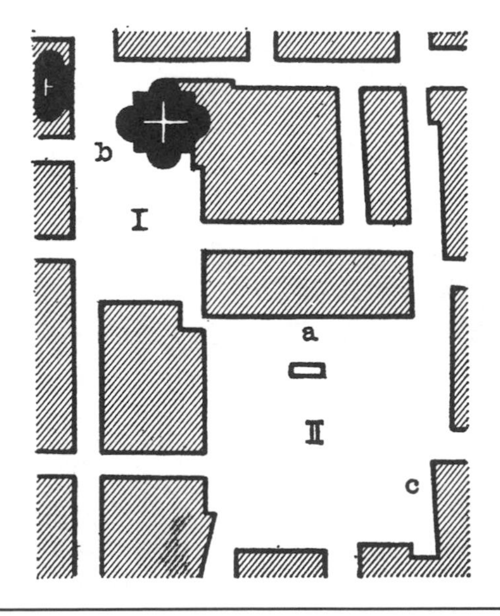

Fig. 15. Parma: (a) *Palazzo Communale (del Governatore),* (b) *Madonna della Steccata,* (c) *Palazzo della Podesteria (del Municipio),* (I) *Piazza d. Steccata,* (II) *Piazza Grande.*

Fig. 16. Florence: *Piazza della Signoria, (a) Palazzo Vecchio, (b) Loggia, (c) Fountain of Neptune, (d) Statue.*

Fig. 17. Verona: *Piazza dei Signori.*

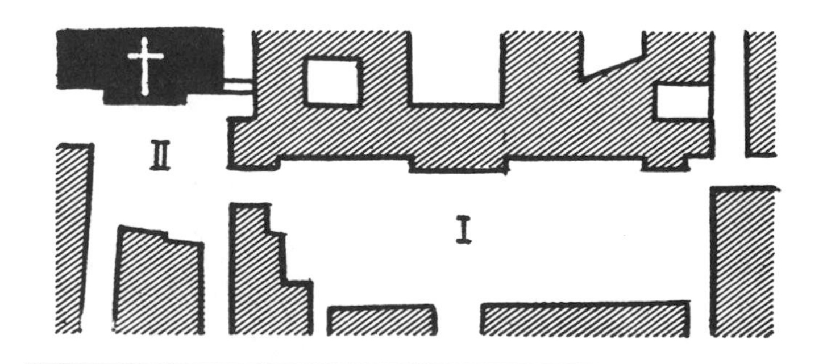

Fig. 18. Modena: (I) *Piazza Reale,* (II) *Piazza S. Dominico.*

More or less ornate portals like those that simply but effectively frame the *Piazza dei Signori* at Verona (Fig. 17) are to be found in all the royal residences, in the chateaux and city halls, and they are used as much for vehicular traffic as by pedestrians. While ancient architects used this pattern wherever possible with infinite variations, our modern builders seem to ignore its existence. Let us recall, to demonstrate the persistence of ancient traditions, that at Pompeii, too, there is an *Arc de Triomphe* at the entrance to the Forum.

Columns were used with porticos to form enclosures for public squares. Saint Peter's in Rome is the best example of this. In more modest proportions there is the hemicycle of the *Place de la Carriere* at Nancy. Sometimes portico and columns are combined as in the Cathedral Square at Salzburg. At *S. Maria Novella* in Florence the colonnade is replaced by a wall of rich architectural embellishment. At times public squares are completely surrounded by high walls opened by simple or monumental portals, as at the ancient episcopal residence of Bamberg (1591), at the City Hall of Altenbourg (1562–1564), at the old University of Fribourg-en-Brusgau, and other places.

Arcades were used to embellish monumental buildings more frequently in former times than at present, either on the higher stories, as in the City Halls of Halle (1548) and Cologne (1568), or on the ground level. Among the numerous examples, we call attention to arcades of the city halls of Paderborn, Ypres (1621–1622), Amsterdam, Lubeck, the Cloth Market at Brunswick, the City Hall at Brigue; arcades of market places, as at Munster and Bologna, or like the *Portico dei Servi* in the latter city. Let us recall also the beautiful portico of the *Palazzo Podesta* at Bologna, and at Brescia the superb arcade of *Monte Vecchio*, the fine loggias of *Udine* and *San Annunziata* at Florence. And finally, the arcade pattern was used in a thousand ways in the architecture of courts, cloister, and cemeteries.

All of these above mentioned architectural forms in former times made up a complete system of enclosing public squares. Today there is a contrary tendency to open them on all sides. It is easy to describe the results that have come about. It has tended to destroy completely the old public squares. Wherever these openings have been made the cohesive effect of the square has been completely nullified.

4 THE FORM AND EXPANSE OF PUBLIC SQUARES

We can distinguish between two kinds of public squares, those of depth and those of expanse. This classification has only a relative value, however, because it depends on the position of the observer and the direction in which he is looking. Thus, one square could have both forms at the same time, depending upon the observer's position with respect to a building, at the principal side or at one of the lesser sides of the square. In general, the character of a square is determined by one building of special importance.

For example, the *Piazza di S. Croce* at Florence is rather deep, because it is usually looked at while facing the church. Its bulk and its monuments are arranged to produce their best effect in a given direction. We readily see that a square having depth makes a favorable impression only when the dominant building is of rather slender form, as is the case with most churches. If the square extends from a building of exceptional breadth its contours should be modified accordingly.

If, then, church squares should generally be deep, the squares of city halls ought to be expansive. The position of monuments in either case should be determined by the form of the public square. The *Piazza Reale* at Modena (Fig. 18) is an example of a well-arranged expansive square, in form as well as in dimensions. The adjoining *Piazza di S. Dominico* is deep. Moreover, the manner in which the different streets open into the square is noteworthy. Everything is arranged to present a perfect setting. The street in front of the church does not detract from the general effect by breaking the enclosure, since it runs perpendicular to the direction of the observer's view. Neither do the two streets opening in the direction of the façade have a disturbing effect, for the observer turns his back on them in looking at the church. The projecting left wing of the chateau is not the work of pure chance. It serves to confine the view within the church square and to make a definite separation between the two squares.

The contrast between these two adjoining squares is striking. The effect of one is intensified by the opposing effect of the other. One is large, the other small; one expansive, the other deep; one is dominated by a palace, the other by a church.

It is truly a delight for the sensitive observer to analyze such a plan and to find the explanation for its wonderful effect. Like all true works of art, it continually reveals new beauties and further reason for admiring the methods and the resourcefulness of the ancient city builders. They had frequent occasion to find answers to difficult problems, for they took full account of contemporary necessities. That is why we seldom find "pure types" in their work, which developed slowly with the inspiration of a sound tradition.

It is difficult to determine the exact relationship that ought to exist between the magnitude of a square and the buildings which enclose it, but clearly it should be a harmonious balance. An excessively small square is worthless for a monumental structure. A square that is too big is even worse, for it will have the effect of reducing its dimensions, however colossal they may actually be. This has been observed thousands of times at Saint Peter's in Rome.

It is a delusion to think that the feeling of magnitude created by a square increases indefinitely with extensions of its size. We have already learned by experience in some places that continual expansion does not produce proportionate increases in impressiveness. It has been observed that the intensity of sound produced by a men's choir is increased at once by voices that are added to it, but there is a point at which the maximum effect is reached and at which the addition of more singers ceases to improve it. (This point is reached with 400 singers). This seems to be applicable to the impression of magnitude that can be created by certain public squares. If a narrow strip of a few yards is added to a small square the result is quite sensible and usually advantageous, but if the square is already large this increase is scarcely noticed, for the relative proportions of square and surrounding buildings remain about the same. Great esplanades are no longer found in modern cities except in the form of recreational areas. They can scarcely be considered as forming public squares since the buildings around them are like country houses in the open, or like villages seen from a distance.

Some places of this type are the *Champ de Mars* at Paris, the *Campo di Marte* at Venice, the *Piazza d'Armi* at Trieste, and the *Piazza d'Armi* at Turin. Although they do not come within the scope of our study, we mention them here because they have frequently been copied in the interior of cities with badly proportioned dimensions. Great buildings around them are reduced in appearance to ordinary rank, for in architecture the relationship of proportion plays a greater role than absolute size. Statues of dwarfs that are 6 feet (1.8 m) or more in height may be seen in public gardens. They stand in contrast to statuettes of Hercules of "Tom Thumb" size. Thus the greatest of the gods is made a dwarf and the smallest of men becomes heroic.

The person who is interested in city building should study the dimensions of some of the smaller squares and of a large square of his own city. It will convince him that the feeling of magnitude that they produce is often out of proportion to their actual dimensions. Above all it is important to achieve good relative proportion between the dimensions of a square and the enclosing buildings. This relationship, like all precepts of art, is difficult to establish precisely, for it must frequently submit to wide variations. A glance at the map of any great city demonstrates this. It is much easier to determine the proportions of a column and its entablature. It would be desirable to determine the relationship within a definite approximation, especially now when plans for city expansion are dashed off according to the whims of the draftsman and not gradually in accordance with the needs of the time. To assist in solving this important problem we have prepared the plans which accompany this study as far as possible on a common scale. It will be asserted that the variety of methods employed in past ages indicates almost arbitrary practice. However, it is possible to draw from an examination of them the following principles, which, in spite of platitudinous appearance, are far from being observed today.

- The principal squares of large cities are larger than those of small cities.

- In each city some principal public squares have expansive dimensions, while others must remain within confined limits.

- The dimensions of public squares also depend on the importance of the principal buildings that dominate them; or, put another way, the height of the principal building, measured from the ground to the cornice, should be in proportion to the dimension of the public square measured perpendicularly in the direction of the principal façade. In public squares of depth the height of the facade of the church should be compared with the depth of the square. In public squares of expanse the height of the facade of the palace or public building should be compared with the breadth of the square.

Experience shows that the minimum dimension of a square ought to be equal to the height of the principal building in it, and that its maximum dimension ought not to exceed twice that height unless the form, the purpose, and the design of the building will support greater dimensions. Buildings of medium height can be built on large public squares if, thanks to the few stories and massive architecture, they can be developed better in breadth.

It is also important to consider the relationship that should exist between the length and the breadth of a public square. In this, exact rules are of little worth, for the problem is not one of obtaining a good result on paper only but in reality as well. But the actual effect will depend largely on the position of the observer, and it may be said in passing that it is quite difficult to make accurate estimates of distance; and, too, we have but an imperfect perception of the relationship between length and breadth of a plaza. Let us say then only that public squares that are actually square are few in number and unattractive; that excessively elongated "squares" (where length is more than three times the width)

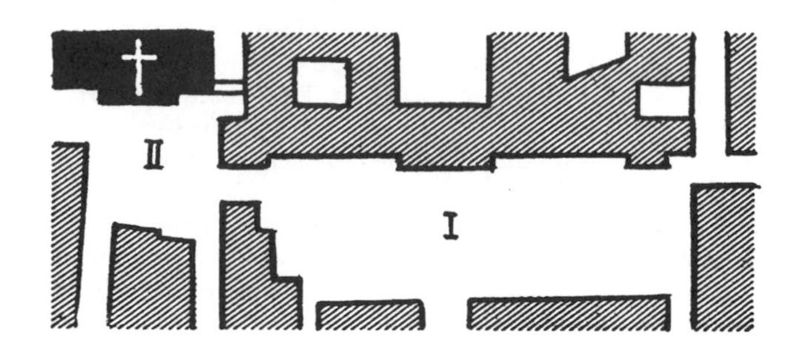

Fig. 19. Modena: (I) *Piazza della Legna,* (II) *Piazza Grande,* (III) *Piazza della Torre.*

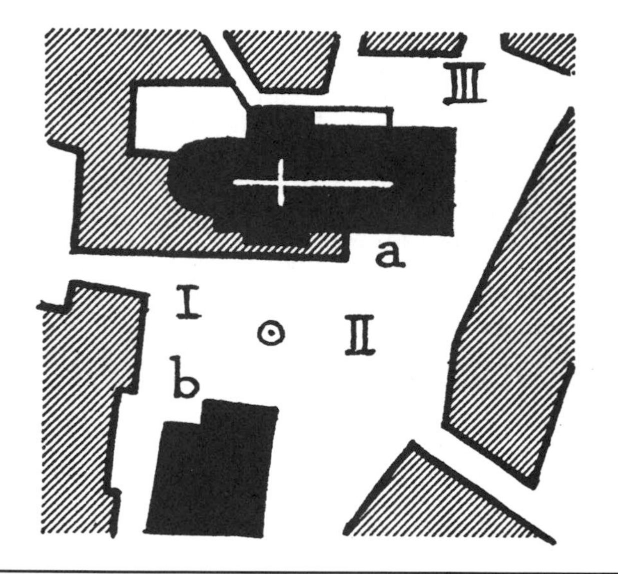

Fig. 20. Perugia: (I) *Piazza del Vescovato,* (II) *Piazza IV Novembre,* (III) *Piazza del Papa (Danti),* (a) *Cathedral,* (b) *Palazzo Comunale.*

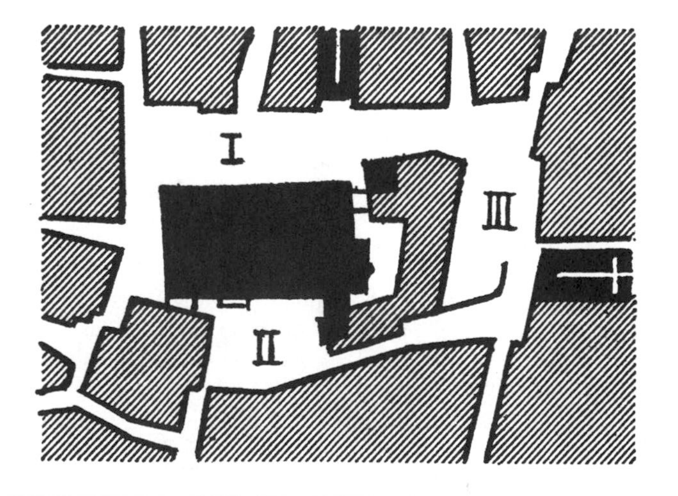

Fig. 21. Vicenza: (I) *Piazza dei Signori,* (II) *Pescheria (Piazza delle Erbe),* (III) *Piazza della Biava.*

have a scarcely better appearance. Expansive squares, in general, have much greater discrepancies between their two dimensions than squares which have depth. However, that depends on circumstances. The streets opening onto the square must also be considered. Narrow little streets of the old cities require squares of only modest dimensions, while today vast squares are needed to accommodate our streets of great width. Modern streets of medium width from 50 to 100 feet (15 to 30 m) would have been wide enough in past ages to form one of the sides of a typical, well-enclosed church square. Of course, that would only have been made possible by clever design and by the narrow width of old streets from 6-1/2 to 26 feet 2 to 8 m). What should be the dimensions of an adequate and well-proportioned square, located on a street of from 150 to 200 feet (46 to 60 m) wide? The *Ringstrasse* at Vienna is 186 feet (57 m) wide; the *Esplanade* at Hamburg is 150 feet (46 m); and the *Linden* at Berlin is 190 feet (58 m). Such dimensions are not even attained by the *Piazza di San Marco* at Venice, and there is still the *Avenue des Champs-Elysees* at Paris, which is 465 feet (142 m) wide. The average dimensions of the great squares of the old cities are 465 by 190 feet (142 by 58 m).

5 GROUPS OF PUBLIC SQUARES

We have already had occasion in our study to compare two squares located close together, and our illustrations have shown numerous other examples of similar groupings. They are so frequent, especially in Italy, that cities having the principal buildings grouped about a single square are rather exceptional. This is a result of the old practice of closing the frame of the square and setting churches and palaces back against other buildings.

Let us study another example from the plan of Modena (Fig. 19). The *Piazza Grande* is evidently intended to set off the lateral façade of the church. It is also of rather elongated shape and extends beyond the vault. This might be expressed theoretically by saying that a façade square and a vault square are joined together. The squares I and II are, on the contrary, quite distinct from each other. The *Piazza Grande* makes a complete entity by itself, and the *Piazza Torre* likewise has its individual character. Its purpose is to open up a perspective on the church tower, which thus produces its entire effect. Moreover, Square I, which commands the principal façade, is deep, conforming to the rule. The street opening there in the direction of the portal does not interfere with the harmony of the whole. At Lucca the *Piazza Grande* and the double square at the Cathedral, with one part in front of the church and the other at its side, establish a comparable rule. These examples, which could be multiplied indefinitely, demonstrate that the different facades of buildings have determined the form of the corresponding public squares in order to produce a fine work. In fact, it is not likely that two or three squares would have been created unless the various facades of a church could be readily adapted to it afterward. In any case, it is certain that this combination brings out all the beauties of a monumental building. We can scarcely ask for more than three squares and three different views, each forming a harmonious whole, around a single church.

This is new evidence of the wisdom of the ancients who, with a minimum of material resources, knew how to achieve great effects. It might almost be said that their procedures constitute a method for the greatest utilization of monumental buildings. In fact, each remarkable façade has a square to itself, and reciprocally, each square has its marble façade. That, too, has its importance, for those superb stone elevations, so desirable for giving a square enough character to save it from banality, are not found just everywhere.

This cleverly refined method can no longer be used, since its application requires the existence of well enclosed squares and buildings set back against other structures—two practices equally foreign to the style of the times, which prefers open breaches everywhere.

But let us return to the old masters. At Perugia the *Piazza di S. Lorenzo* (Fig. 20) separates the Cathedral from the *Palazzo Communale*. It is, then, both a cathedral square and a town hall square. Square III is given over to the Cathedral. At Vicenza the *Basilica* of Palladio (Figs. 21 and 22) is surrounded by two squares, each of which has its special character. Similarly, the *Signoria* at Florence also has its secondary square in the *Portico* of the Uffizi. From an architectural point of view this *Signoria* is the most remarkable square in the world. Every resource of the city building art has made its contribution here—the shape and dimensions of the square, contrasting with those of the adjoining square; the manner in which streets open; the location of fountains and monuments; all of this is admirably studied. Yet it required an unequaled quantity of labor. Several generations of able artists used centuries in transforming this site, of no special excellence in itself, into a masterpiece of architecture. We never tire of this spectacle, which is as pleasingly effective as the means of fabricating it are inconspicuous.

Venice also has a combination of squares that are remarkable in every way, the *Piazza di S. Marco* and the *Piazzetta* (shown as I and II, respectively, in Fig. 23). The first is a square having depth with respect to Saint Marks and an expansive square with respect to the Procuraties. Similarly, the second is expansive with respect to the Palace of the *Doges*, and primarily, deep with respect to the superb scene formed by the Grand Canal with the *Campanile S. Giorgio Maggiore* in the distance. There is a third small square before the lateral façade of Saint Marks. There is such an expanse of beauty here that no painter has ever conceived an architectural background more perfect than its setting. No theater ever created a more sublime tableau than the spectacle to be enjoyed at Venice. It is truly the seat of a great power, a power of spirit, of art, and of industry which has gathered the treasures of the world upon its vessels, which has thrust its supremacy over all the seas, and which has possessed its accumulated riches on this spot of the globe. The imagination of a Titian or of a Paul Veronese could not conceive of picture cities, in their backgrounds for great festival scenes, more splendid than these.

This unequaled grandeur assuredly was attained through extraordinary means: the effect of the sea, the great number of buildings embellished with sculpture, the magnificent coloring of Saint Marks, and the towering *Campanile*. But the exceptional affect of this assembly of marvels is due largely to skillful arrangement. We may be quite certain that these works of art would lose much of their value if they were located haphazardly by means of the compass and straightedge according to the modern system. Imagine Saint Marks isolated from its surroundings and transplanted to the center of a gigantic modern square; or the *Procuraties*, the library, and the *Campanile*, instead of being closely grouped, spread out over a wide area, and bordered by a 200 foot wide (60 m) boulevard. What a nightmare for an artist! The masterpiece would thus be reduced to nothing. The splendor of buildings alone is not enough to form a magnificent whole if the general arrangement of the square is not carefully worked out. The shape of St. Mark's Plaza, and of the squares that are subordinate to it, conforms to every principle that we have thus far discussed.

Fig. 22. Vicenza: *Piazza del Signori.*

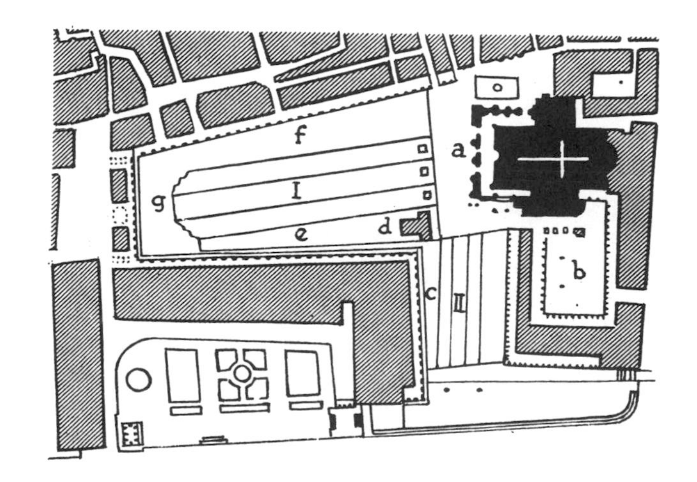

Fig. 23. Venice: (I) *Piazza San Marco,* (II) *Piazzetta.*

We should note especially the location of the *Campanile*, which, rising between the two squares, seems to stand guard.

What an impression is made by several grouped squares on the person who goes from one to the other! The eye encounters a new scene every instant, and we feel an infinite variety of impressions. This may be observed in photographs of St. Mark's and of the *Signoria* of Florence. There are more than a dozen popular views of each square, taken from various points. Each one presents a different picture, so much so that it is sometimes difficult to believe that they are all views of the same place. When we examine a modern square of strict right angle design, we can get only two or three views of different quality, for in general they express no artistic feeling. They are only surfaces of so much area. ■

Tony Garnier

Summary

Tony Garnier's (1869-1948) plans and sketches for an ideal "industrial city," *Une Cité Industrielle*, began as Ecole de Beaux Arts Prix de Rome studies or *envois* in 1901 and were eventually published as a loose-leaf folio in 1917, to become one of the formative theoretical proposals of 20th centruy urban design. This article is a translation of the Preface to *Une Cité Industrielle* with representative plates that indicate Garnier's prodigious drawing skill.

Key words

city planning, design standards, hospitals, hygiene, industrial zone, public garden, railway, residential district, solar orientation, town center, ventilation

Une Cité Industrielle, detail, dated 1917. Tony Garnier, Architect.

Une Cité Industrielle

Une Cité Industrielle portrays a utopian, modernist vision that incorporates functionalist principles a decade before they were advocated by any other architect (Miriani, 1990). Garnier envisioned an entire city in plan and in detail, including schools, hospitals, factories, residential quarters and recreational facilities. His generating concepts included a decentralized layout, traffic-free pedestrian zones, and residential districts with gardens to emphasize continuous pedestrian circulation and orientation and placement to follow local climatic design variables. In the Preface to *Une Cité Industrielle*, presented below, Garnier includes "regulations" to institute his design and plannng principles. The proposed materials and building techniques of reinforced concrete—up to then used only experimentally—would permit open plans and roof terraces, and glass windows disposed generously for sunlight and natural ventilation.

1 DISPOSITION AND LAYOUT

The architectural studies presented here with many plates focus on the establishment of a new city, Une Cité Industrielle. Most new towns that will be built from now on will be derived from industrial conditions. Therefore, the most generalized case is considered here. In such a scope of town planning, all possible architectural types will be required and all of these are examined here.

The town is assumed to be of average size, to have a population of 35,000 inhabitants. A generalized approach to research is adopted for this scale that would not have been applicable to the study of a smaller village or to a larger city. It is also assumed that the site includes an equal amount of hillside and level plain that is transected by a river. Although the town under study is fictitious, the existing towns of Rive-de-Gier, Saint-Étienne, Saint-Chamond, Chasse, and Givors that represent similar basic needs as the scheme we have imaged here. The setting is assumed to be southeastern France and the building materials proposed are indigenous to the region.

The reasons motivating the establishment of such a town could be assumed to be the availability of raw materials for manufacturing, the existence of a natural energy source available for industrial use, or the site's accessibility to transportation. In this case, the determining factor is taken to be a rushing stream that is advantageous for a dam and location of a hydroelectric power station to provide electricity for heating, lighting and power for factories and town. There are also mines nearby, although these could be assumed to be located farther away.

The main factory is situated in the plain, where the stream meets the river. A major railway line runs between the factory and the town, located above on a higher plain. Higher still are the well-spaced hospital buildings. Like the town itself, these are shielded from cold winds, oriented to the southern sun on terraces sloping facing the river. Each of these principal elements—factory, town, and hospital—is sited to allow for expansion, so that our study represents a more general longer term planning proposal.

To arrive at a design that completely fulfills the moral and material needs of the individual, a set of standards are established concerning traffic circulation, hygiene, and so on. The assumption is that a certain progress of social order would have already established such standards, thus insuring the adoption of such regulations although these are

Fig. 1. *Une Cité Industrielle,* general view, dated 1917. Tony Garnier, Architect.

completely unrecognized by current law. As such, it is assumed that there is the enabling public power of eminent domain, governance of uses of the land, distribution of water, food essentials and medicines, and the reutilization of refuse.

2 HOUSING

Many towns and cities have already enacted standards for hygiene according to local geographic and climatic conditions. In this city, it is assumed that the direction and conditions of the prevailing winds prompt particular practices represented in the following set of building regulations:

- In residences, each bedroom should have at least one south-facing window, large enough to illuminate the whole room and admit direct sunlight
- All spaces in residences, however small, should be illuminated and ventilated directly from outside and not rely upon internal shafts
- House interiors (walls, floors, and so forth) should be of a smooth surface with rounded corners

These standards, required for residential construction, will, whenever possible, also serve as guidelines for public buildings.

The area in residential quarters is subdivided into blocks measuring 150 meters in the east-west direction and 30 meters in the north-south. These blocks are then divided into lots of 15 by 15 meters, each one abutting directly onto the street. This partitioning ensures the best possible use of land and fulfills the standards cited above.

A residence or any building for a public function may occupy one or more lots. But the area of lot coverage for construction must always be less than half the entire site, with the remainder devoted to public garden accessible to pedestrian use: that is, each building lot must include a public pathway available from the street to the building behind.

This arrangement makes it possible for pedestrians to cross the city in any direction, independent of the street pattern. The land of the town as a whole is similar to a great park, free of enclosures and walls delimiting the terrain. The minimum distance between two houses in the north-south direction is equal to at least the height of the construction situated to the south. Due to these planning standards—which limit site coverage and prohibit the use of enclosures—and

Fig. 2. *Une Cité Industrielle,* factory view, dated 1917. Tony Garnier, Architect.

also because the land is graded for drainage, there is great variety in overall design.

The town is composed of a grid of parallel and perpendicular streets. Its main street originates at the railway station and runs east-west. The north-south roads, tree-lined on either side, are 20 meters wide and planted on both sides. The east-west roads are 13 or 19 meters wide; those of 19 meters are planted on the south side.

3 ADMINISTRATION—PUBLIC BUILDINGS

At the center of the city, an extensive area is reserved for public buildings. They form three groups:

I. Administrative services and assembly halls

II. Museums

III. Facilities for sports and entertainment

Groups II and III are situated in parks, bordered to the north by the main street and to the south by planted terraces which afford an open view of the plain, the river, and mountains beyond.

Group I: Administrative services and meeting halls

- A very open hall continuously accessible to the public, with a capacity of 3,000; the hall is equipped with public notice boards and a public address system to amplify meeting or musical entertainment; it is also used for large-scale meetings

- A second hall with amphitheater seating for 1,000 people, and two further amphitheaters for 500; all are equipped for conferences and film projection

- A large number of small meeting rooms (each with its own office and changing room) for unions, associations, and other groups

All these rooms are located beneath a vast portico that provides a covered promenade for the town center and a spacious area where people can meet, sheltered in case of inclement weather.

To the south of this portico is the clock tower, visible along the length of the main street. It is a landmark indicating the center of town. The administrative services include:

- A building containing municipal offices open to the public records (births, marriages, and deaths), and an arbitration tribunal; each of these will include rooms for the public, committees, and related offices

Fig. 3. Primary School (Pl. 38). Tony Garnier, Architect.

Fig. 4. Primary School Garden (Pl. 39). Tony Garnier, Architect.

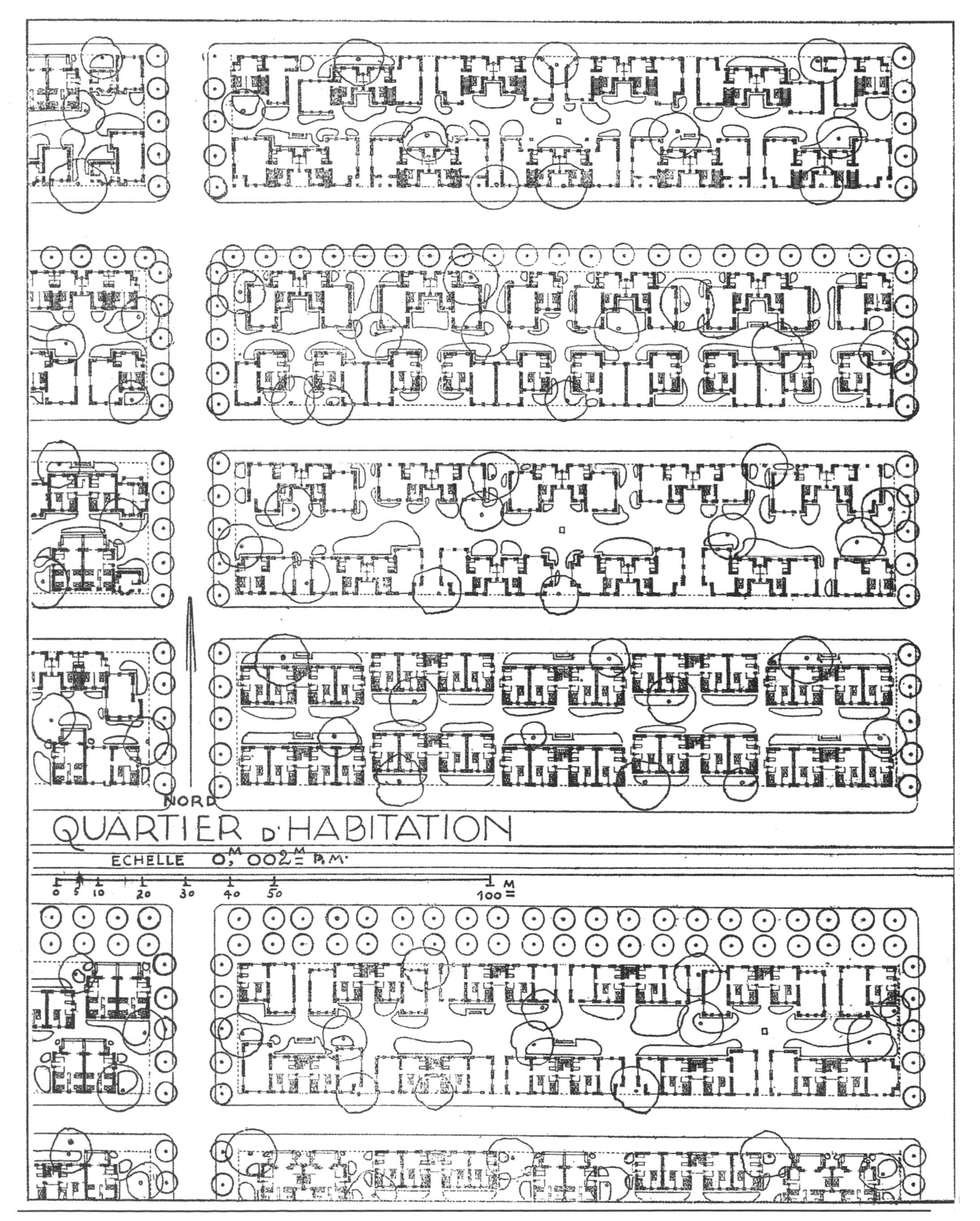

Fig. 5. Plan Residential Quarter (Pl. 66). Tony Garnier, Architect.

Fig. 6. View Residential Quarter (Pl. 72). Tony Garnier, Architect.

- An office building for all those branches of civic government that require at least one clerk in direct contact with the administration
- A building for social research
- A building for archives, sited near the fire station

There will also be an office housing the labor organizations, which include employment registry office; information offices; offices for trade union organizations and associations; temporary residences and cafeterias. There are also special advisory offices, including a building fitted out as a medical clinic, a pharmacy, and a center for hydrotherapy. Further south on the main street is the central post and telecommunications office, with complete mail, telex and telephone facilities.

Group II: Museum buildings

- Historical collections and important archaeological, artistic, industrial and commercial documents relating to the city; permanent monuments will be erected in the park surrounding the rooms containing the archives
- Botanical collections; in the garden and in a large greenhouse
- The library, including a spacious reading room (one side devoted to library volumes, the other to periodicals and newspapers) and a large map room (at its center a vast globe fitted with a stair to facilitate consultation). Located at the entrance to the library are service rooms for cataloging, book maintenance, book-binding, archiving, printing, a book loan office, and so forth.; surrounding these are the various storerooms
- A large separate hall for temporary exhibitions; with four entrances so that several small exhibits can be set up at once, or a large exhibition can utilize the entire hall

Group III: Public buildings for sports and entertainment

- A hall for entertainment and theater (1,900 seats), with all necessary support facilities; movable stage sets for quick scene changes (to eliminate equipment above and below the stage); green rooms for performers, orchestra and for theater sets; cloakrooms, toilets, foyer, and public restaurant
- A semi-circular amphitheater (after the ancient Greek theater) for open-air performances framed within a natural landscape
- Gymnasia
- A large public bath building with heated and unheated pools, changing cabins and bathing pools, shower rooms, massage and relaxation rooms, a restaurant, a fencing room, and tracks for athletic training
- Athletic fields (tennis courts, football pitches, and so forth), tracks for cycling and running; areas for high jump and discus throwing, and so forth; this area will be bordered by covered grandstands and grassy terraces screened by trees

Fig. 7. View Residential Quarter (Pl. 74). Tony Garnier, Architect.

As explained above, Groups II and III are situated in the midst of gardens, furnished with park benches, fountains, and so forth. All public buildings are constructed almost entirely in reinforced concrete and glass.

4 SCHOOLS

Conveniently located throughout the city's neighborhoods are primary schools for children up to approximately fourteen years of age. Schools will be coeducational; grouping of children will be by age, ability and advancement. A special landscaped street will separate the classes for smaller children from those of their elders, and will provide a play area for use between classes. Recreation areas will also include arcades and open porticoes. Schools will be equipped with projection theatres in addition to the necessary classrooms. The school principal and grounds attendants are housed nearby.

Secondary schools will be situated at the most northeastern point of the town. The curriculum will be addressed to the needs of an industrial town. For the majority of students, the education will involve general courses in vocational studies. A limited number of students will receive specialist training in administration and trade (that is, professional arts instruction). All children attend the secondary schools between fourteen and twenty years of age. Those qualified for further studies will enroll in professional schools or colleges.

The professional arts school is intended to prepare those who will engage in artistic production—in architecture, painting, and sculpture, as well as related areas of design such as furniture, fabrics, linen, embroidery, clothing, leatherwork; also in copper, tin, iron, glass, pottery, enamel, printing, lithography, photography, engraving, mosaic, poster art, and so forth.

The professional industrial school is concerned primarily with supporting the two major industries of the region, metallurgy and silk production, and will offer specialized courses devoted to the study of production and procedures.

5 HEALTH FACILITIES

The hospitals (715 beds) are situated on the hillside north of the city center. They are sheltered from the cold mountain winds by trees forming a screen to the east and west. The complex contains four main buildings.

- Hospital
- Heliotherapy center
- Hospital for contagious diseases
- Hospital for invalids

The plan as a whole as well as in detail has been designed according to current standards of medical science. Each section is disposed to accommodate future expansion.

6 RAILWAY STATION

The district around the railway station is mainly reserved for collective housing, such as hotels, and department stores, and so forth, so that the rest of the city is free of tall structures. The railway station square will face an open-air market.

The station is of average size and is sited at the intersection of the great artery leading out of town and the streets leading to the older developed area along the riverbanks. The main building opens onto the square and its clock tower is visible from all over town.

Public amenities are at street level and underground walkways are equipped with platforms and waiting rooms. The railway yard is situated farther to the east, with the sidings serving the factory to the west. The railway tracks are planned as straight lines, so that trains can move as rapidly as possible.

7 PUBLIC SERVICES

Certain basic services depend on the municipal administration and are subject to special requirements. These services include meat distribution, flour and bread production and storage, water supply, the control of pharmaceutical and dairy products. The administration is in charge of sewage and garbage disposal, and the recycling of refuse. It also controls the water supply, electrical power, and heating for industrial as well as for private consumption, and requires a centralized plant to provide such municipal services to all buildings and areas of the city.

8 FACTORIES

The main factory is a metallurgy works. Nearby mines supply raw materials. Energy is generated from the local hydropower site and power plant.

The factory produces steel rods and pipes, rolled-steel section, sheet metal, wheels, machine tools, and agricultural machinery. In addition, it fabricates metalwork for railway stock and naval equipment and bodywork for automobiles and airplanes.

The factory complex includes blast furnaces, steel mills, workshops with large presses and power hammers, assembly and repair shops, a dock for launching and repairing ships, a river port, workshops for outfitting automobile bodies, and workshops for refractories. It also includes vehicle testing tracks, numerous laboratories, and housing for engineering staff.

Support facilities will be distributed throughout the complex, including rest rooms, changing rooms, cafeteria, and first aid points.

Spacious roads with trees arranged in quincunx patterns will lead to the various areas of the industrial complex. Each department is arranged to allow for future expansion without curtailing other parts of the complex.

Around the center of the city, other manufacturing facilities may be added, including farmsteads for food production, silkworm production, spinning-mills, and so forth.

9 CONSTRUCTION

Materials used in building construction include concrete for foundations and walls, and reinforced concrete for the sills and roofing. Important structures are to be built in reinforced concrete. These two materials are highly plastic, and require specially prepared formwork; with simple forms, the installation is easier and construction costs are lower. This simplicity of means logically leads to simplicity of structural expression. Note that if the construction remains simple, without ornamentation or moldings and with sheer surfaces, the decorative arts can be effectively employed in all their forms, and each artistic object will maintain a cleaner and fresher expressiveness, due to its independence from the construction itself. Moreover, the use of concrete and cement makes it possible to obtain large horizontal and vertical surfaces, endowing the building with a sense of calm and balance in harmony with the natural contours of the landscape. Other construction methods and materials will without doubt contribute to other forms that will be equally interesting to study.

This concludes the summary of the planning of a city, an endeavor in which all can appreciate that work is a human law and that the cult of beauty and order can endow life with splendor. ■

REFERENCES

Miriani, Riccardo, ed.,1990. Tony Garnier: *Une Cité Industrielle,* New York: Rizzoli International Publications.

Wiebenson, Dora. 1969. Tony Garnier: *The Cité Industrielle*, New York: George Braziller.

Werner Hegemann and Elbert Peets

Summary

Werner Hegemann (1881–1936) was a leading German architectural journalist and urbanist, and Elbert Peets (1886–1968) an American architect, town planner, and writer. They published *The American Vitruvius: An Architect's Handbook of Civic Art* in 1922, describing their project as a visual "thesaurus." Indeed the eclectic, but fairly encyclopedic, scrapbook of illustrations has proven to be the book's most useful and influential aspect. Hegemann, along with others of his generation, was precocious in his interest in American developments, although he is less interested in their presumed modernity than in connecting the emergent interest in monumental civic design—fuelled by the City Beautiful movement and the accompanying classical revival—to a long European tradition, which he saw as directly relevant to both the practical and aesthetic challenges of the modern city.

Key words

campus plan, city form, civic center, plaza, rendering, Rome, streets, subdivision planning, Worlds fair

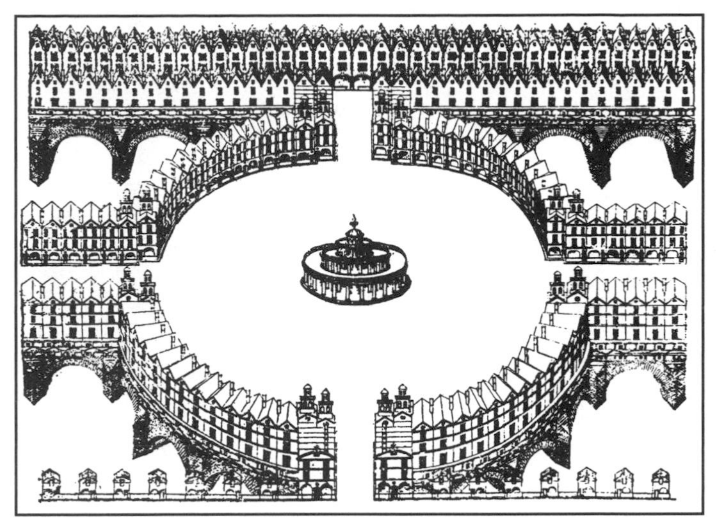

Paris. Drawing by Du Cerceau. Scheme for a circular plaza with a central building designed for the point of the island, in connection with the *Pont Neuf*.

Civic Art

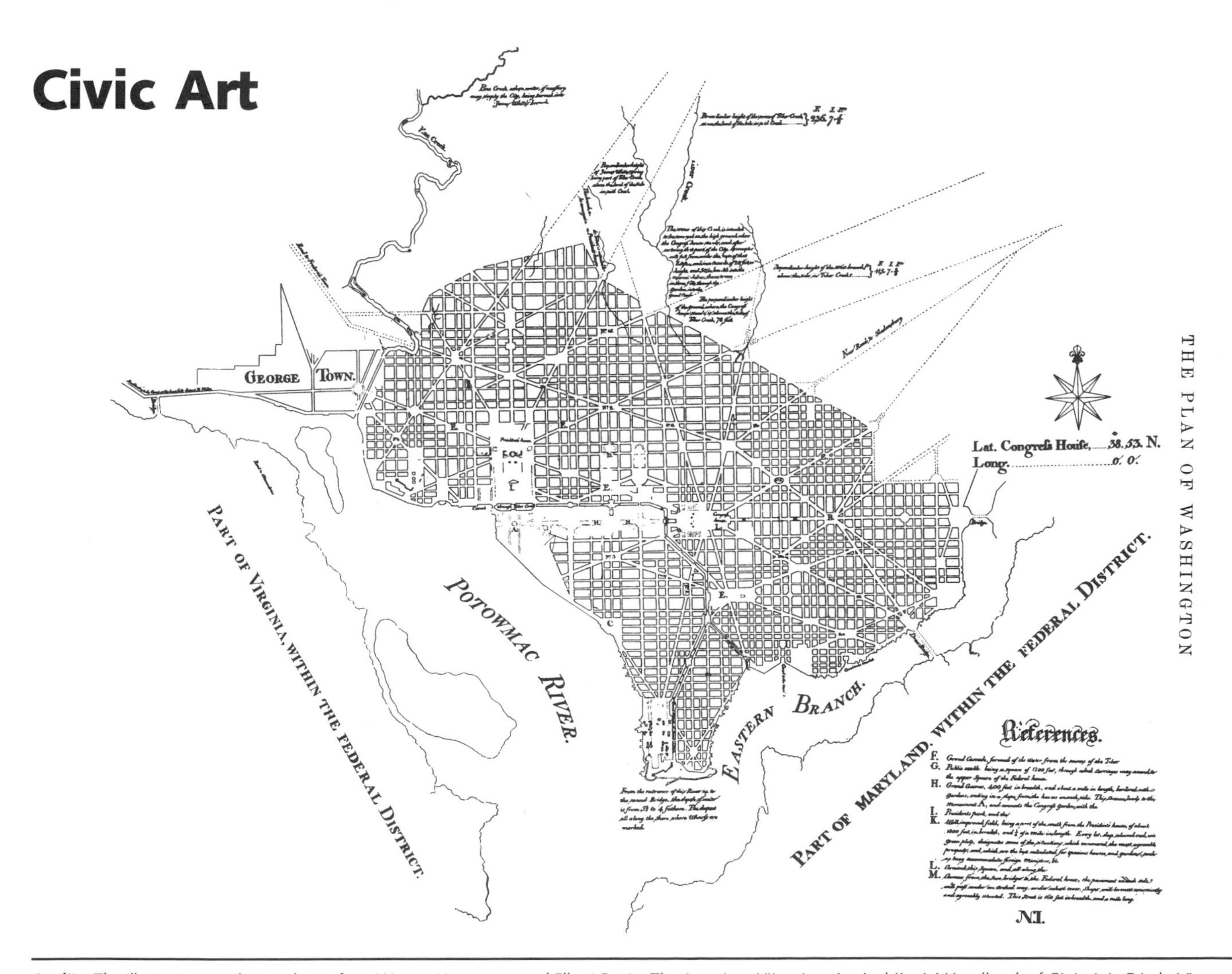

Credits: The illustrations in this article are from Werner Hegemann and Elbert Peets, *The American Vitruvius: An Architects' Handbook of Civic Art.*, Friedr. Vieweg & Sohn Braunschweig/Wesbaden.

THE SIZE OF RENAISSANCE PLAZAS

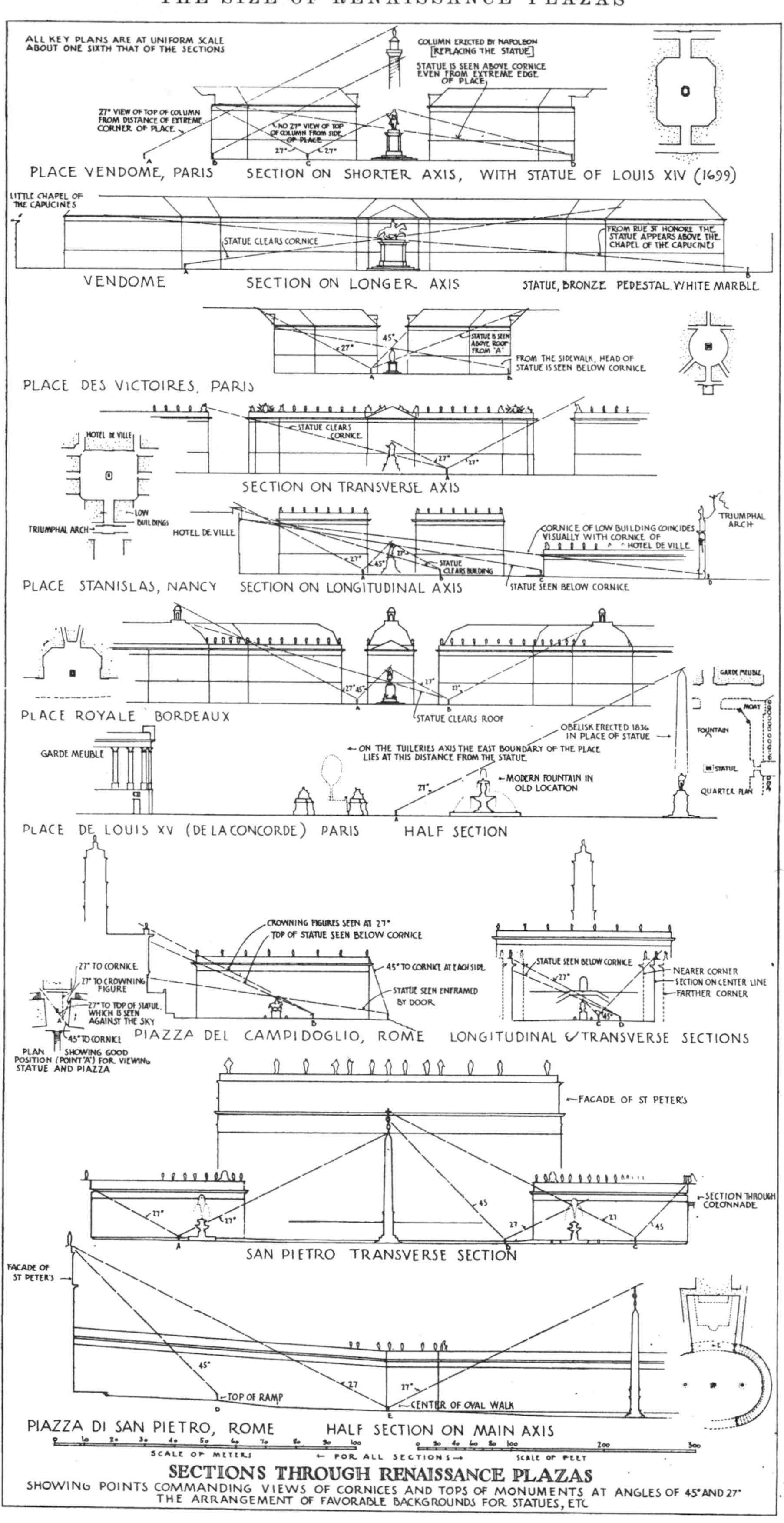

SECTIONS THROUGH RENAISSANCE PLAZAS
SHOWING POINTS COMMANDING VIEWS OF CORNICES AND TOPS OF MONUMENTS AT ANGLES OF 45° AND 27°
THE ARRANGEMENT OF FAVORABLE BACKGROUNDS FOR STATUES, ETC

THE SIZE OF RENAISSANCE PLAZAS

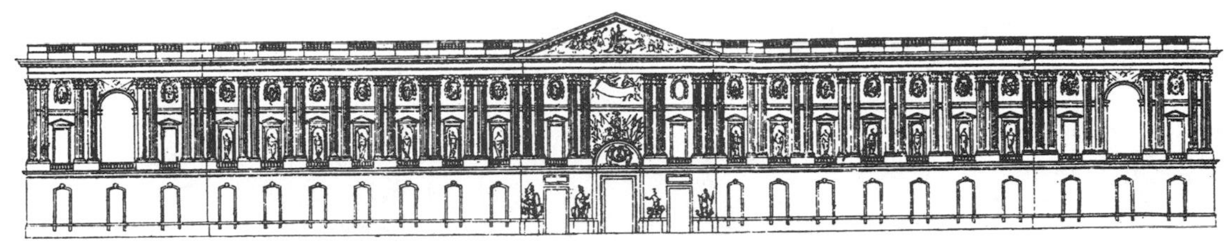

FIG. 221—PERRAULT'S DESIGN FOR THE EAST FRONT OF THE LOUVRE
Showing the niches and statues in place of windows. (From Durand.)

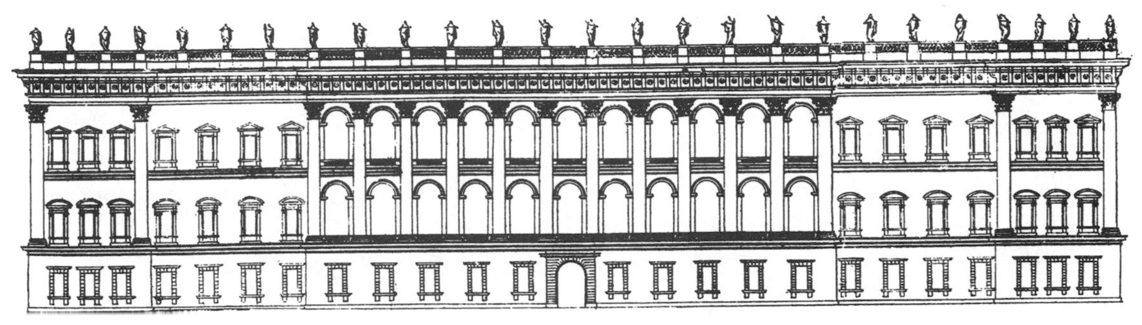

FIG. 222—BERNINI'S DESIGN FOR THE EAST FRONT OF THE LOUVRE
Same scale as Fig. 221 (From Durand.)

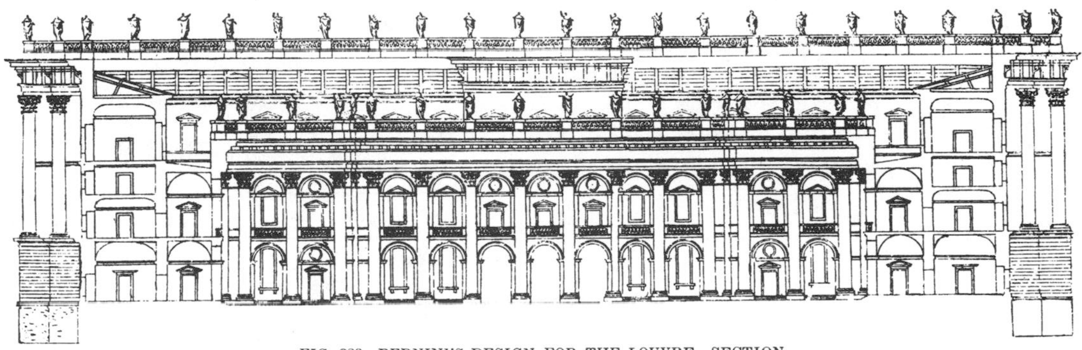

FIG. 223—BERNINI'S DESIGN FOR THE LOUVRE. SECTION
For plan see Fig. 213. (From Durand.)

FIG. 224—WEST FRONT OF THE LOUVRE IN 1660
A typical French palace front before the time of Perrault's boldly simplified colonnade. After a contemporary drawing. (From Babeau.)

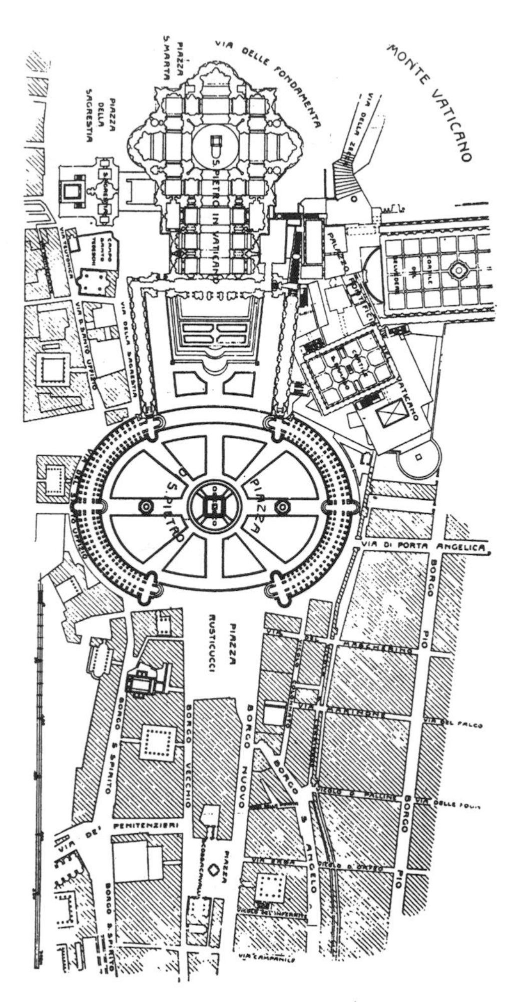

FIG. 246—ROME. VICINITY OF ST. PETER'S
Showing conditions in 1908. (Plan by Th. Hofmann.)

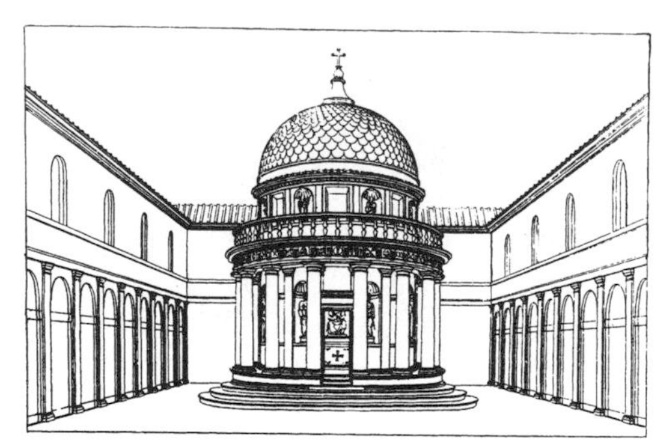

FIG. 241—THE TEMPIETTO AND SETTING, AS BUILT
(From J. A. Coussin.)

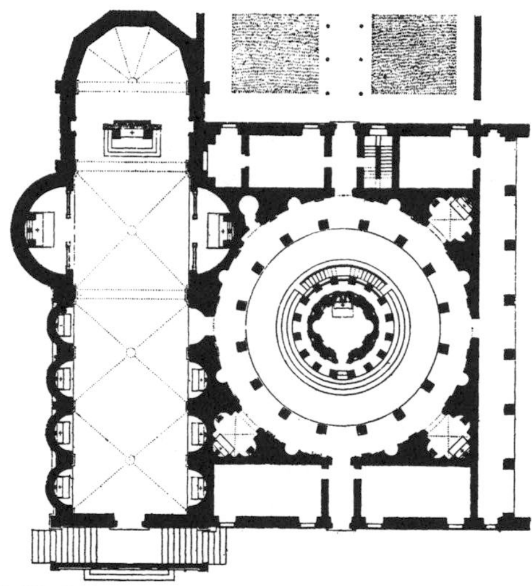

FIG. 240—ROME. BRAMANTE'S DESIGN FOR THE SETTING OF THE TEMPIETTO AT SAN PIETRO IN MONTORIO

The little temple is set in the center of a circular court of twice the diameter of the temple. The arcade around the court has the same number of columns as has the temple. The problem of the curved façade is interesting. (Compare Fig. 252.) To this project of Bramante's the following words by Wren might well be applied: "In this Court we have an Example of circular Walls; and certainly no Enclosure looks so gracefully as the circular: 'tis the Circle that equally bounds the Eye, and is every where uniform to itself." Plan from Serlio (Letarouilly).

FIG. 240A—PALLADIO. RESTORATION OF A TEMPLE ON THE VIA APPIA

The circular sepulchral temple of Romulus, part of the Circus of Maxentius, was built in the fourth century. Though the round temple has a radiating plan the extension at one side gives the whole building a definite orientation and makes reasonable the oblong shape of the court.

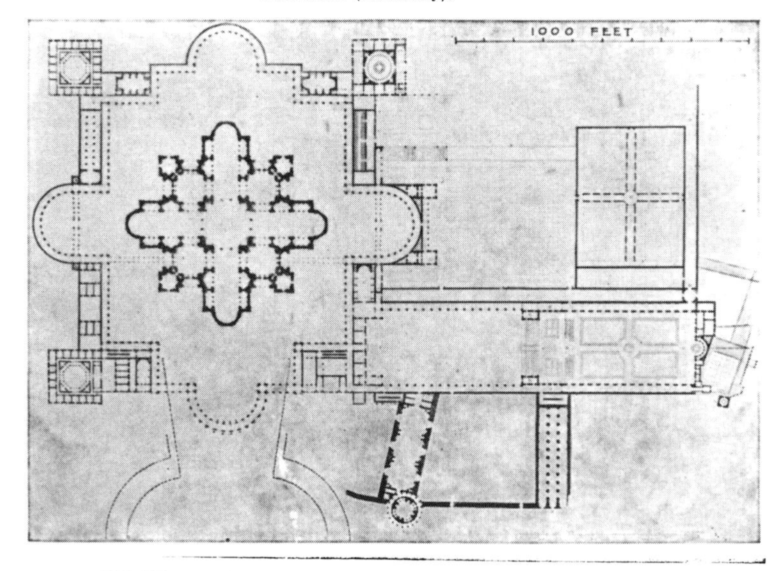

FIG. 242—BRAMANTE'S PLANS FOR ST. PETER'S AND THE VATICAN

A combination of Bramante's plans for the church, the plaza in which it was to be centered and for the court of the Belvedere, with part of Bernini's plaza shown in dotted lines. (From de Geymüller.)

CHURCHES FRONTING ON PLAZAS

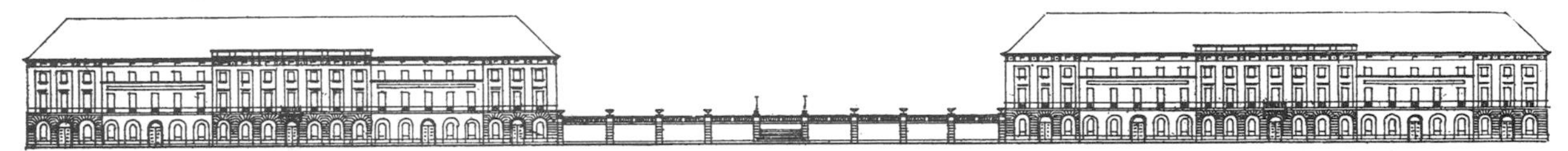

FIG. 285—PLAINE ST. PIERRE. ELEVATION OF SIDE OPPOSITE CHURCH

FIG. 286—PLAINE ST. PIERRE. VIEW TOWARD COURT

Note the arches closing the wall of the court.

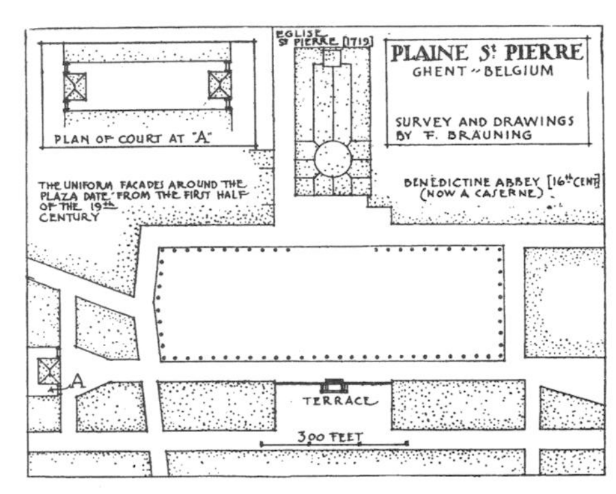

FIG. 287—GHENT. PLAINE ST. PIERRE

This fine plaza is a late flowering of Renaissance good taste. It was designed (perhaps by Roelandt) in connection with a reclamation scheme; the dignified buildings shown in Fig. 285 are rows of residences built with unified façades. The central area of the plaza is in gravel. (From "Der Staedtebau", 1918, with additions to the plan by the authors.)

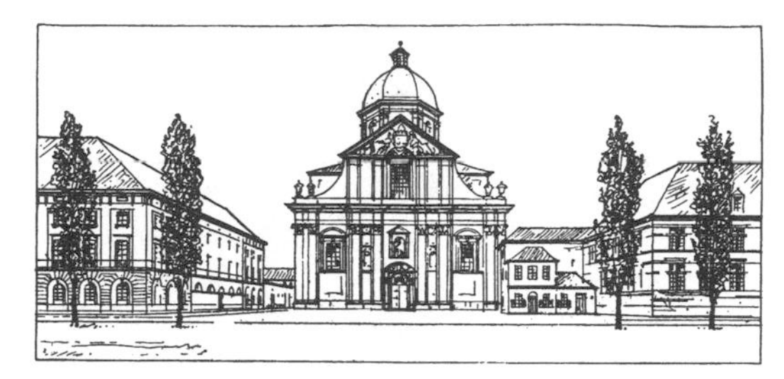

FIG. 288—PLAINE ST. PIERRE. THE CHURCH

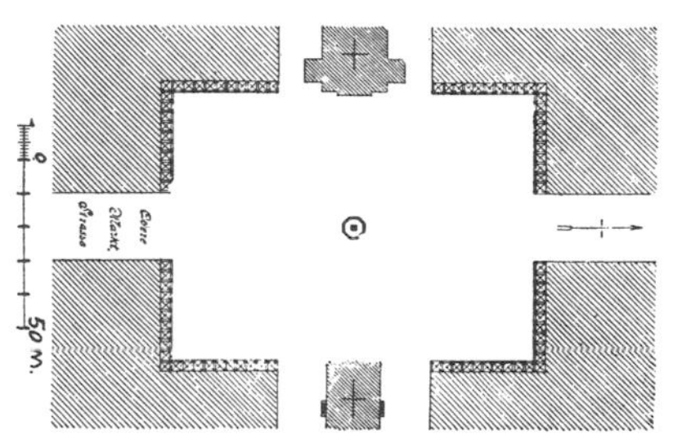

FIG. 289—LUDWIGSBURG. MARKET DEVELOPED AS FORE-COURT FOR TWO CHURCHES

The market place is surrounded by two-story houses. The churches are over two and a half times as high, and the contrast adds much to the impressiveness of the churches. The arcade arches are wider and less numerous than they are shown. (From Brinckmann.)

FIG. 290—CARLSRUHE. PROTESTANT CHURCH FACING THE MARKET PLACE

The church, which has no transepts, is centered on the tower in the rear and flanked by two courts which are separated from the market by double arches and framed by uniform buildings lower than the church. The portico of the church faces the entrance of the city hall; see plan of plaza Fig. 1000. Design of church and plaza by Weinbrenner, between 1801 and 1825. (From P. Klopfer.)

FIG. 291—PETROGRAD. COLONNADES OF KASAN CATHEDRAL

The façade of the Cathedral, built about 1800 by Wovonichin, is flanked by great colonnades which create a monumental setting somewhat resembling the Piazza del Plebiscito in Naples (Fig. 111), built some twenty years later.

PLAZA AND COURT DESIGN IN EUROPE

FIG. 335—RENNES. RUE NATIONALE

View looking east showing uniform façades framing the axis and the set-back of the old Palais de Justice to the left. See plan Fig. 339. (Drawing by Franz Herding.)

FIG. 337—RENNES. PLACE DU PALAIS DE JUSTICE

View showing the palace to the right and uniform façades framing it. The façades, originally inspired by Gabriel, were not completed before the end of the nineteenth century. See Fig. 339. (Drawing by Franz Herding.)

FIG. 338—RENNES. RUE DE BRILHAC

Looking west and past Gabriel's City Hall to the left. (Drawing by Franz Herding.) See Fig. 339.

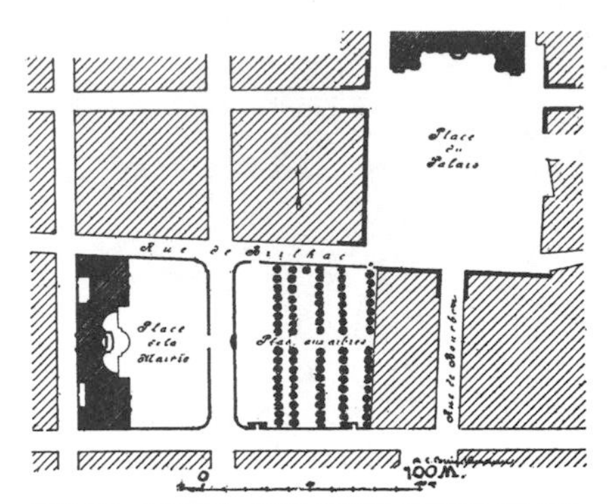

FIG. 336—RENNES. THE ROYAL PLAZAS IN 1826

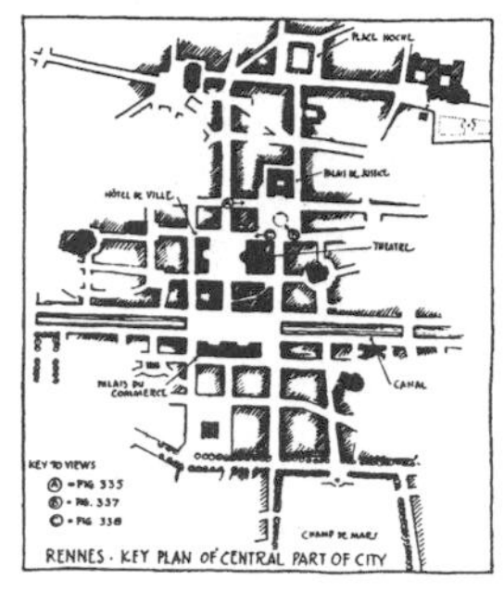

FIG. 336A—RENNES.

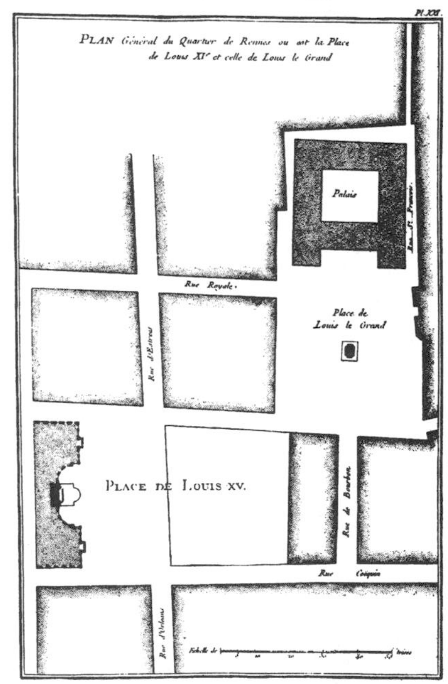

FIG. 339—RENNES. THE TWO ROYAL PLAZAS

(From Patte, 1765.) The heart of Rennes burned down in 1720 and the reconstruction was supervised by Gabriel, who later designed the Place de la Concorde. The design is specially interesting for American conditions because it largely fits into a gridiron. The old Palais de Justice, by Debrosse, 1618-54, (see Fig. 337), survived the fire. The new street plan lets it stand back a little to give it a more secure setting. The new building for the city hall (Mairie, see Fig. 336) is set still further back. In front of it is a court surrounded by balustrades. On the other side of the street a similar area is balustraded off and planted with trees in the back which, as Brinckmann suggests, may have been a temporary indication of another building by which Gabriel hoped to face the Mairie. A theater was later built there. The Mairie itself is divided into halves with separate roofs, with the tower in the middle and the monument of the king against the wall in the center.

PLAZA AND COURT DESIGN IN EUROPE

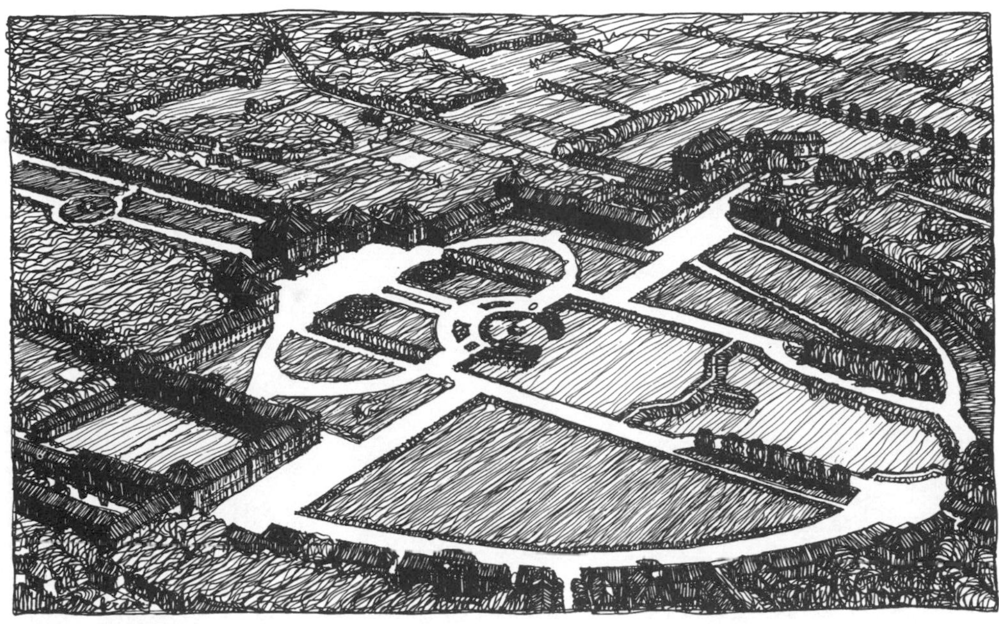

FIG. 415—MUNICH. ENTRANCE COURT TO THE NYMPHENBURG

Built 1663—1718 by Barelli, Viscardi, Zuccali, and Effner. Gardens designed by Girard, a disciple of Le Nôtre. There is a long canal, accompanied by quadruple allées, on axis of the main entrance. Thus a full axis view was not wanted, a case which has been compared with Bernini's sidewise entrances to the Piazza San Pietro in Rome. Approaching through one of the avenues one sees at its end a small framed picture which after one enters the forecourt suddenly broadens out. The contrast between the enclosed allée and the open forecourt makes the latter appear very large. The size of the central members of the palace appears large through being contrasted against the low members immediately to the right and left of it. The height of these low members corresponds again to the height of the other buildings one sees in the foreground immediately after entering the forecourt. As they frame the entire outer forecourt an optical scale is carried into the background from which the size of the main buildings can be appreciated. (From a drawing of Franz Herding, after a photograph published by Brinckmann.)

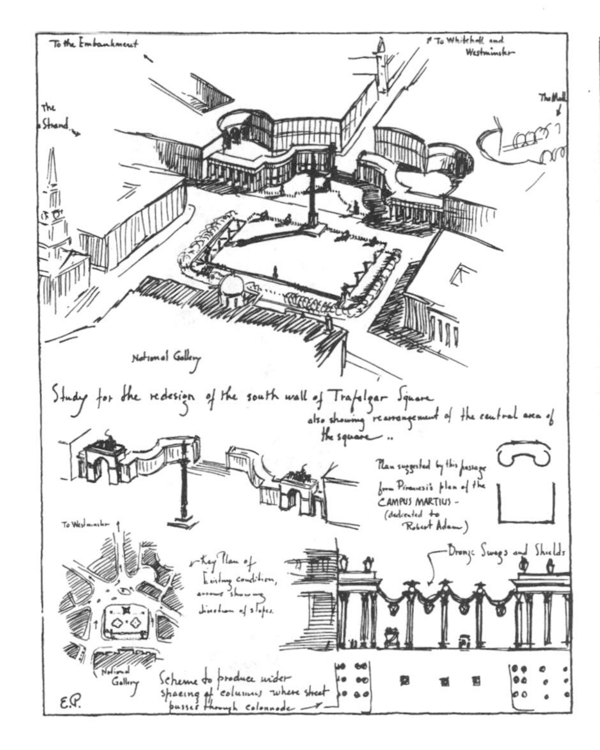

FIG. 417—LONDON. ADMIRALTY ARCH AND TRAFALGAR SQUARE

The square, which occupies the site of the "Royal Mews" or stables, was built in the thirties by Sir Charles Barry. The National Gallery, by Wilkins, and the Nelson Monument, by Baily, date from the same period. St. Martin's Church, by Gibbs, is a century older. The Admiralty Arch was built by Sir Aston Webb in 1910.

FIG. 416—LONDON. PAGE FROM TRAVEL SKETCH BOOK

FIG. 431—BERLIN—ZEHLENDORF. SEMICIRCULAR PLAZA

Designed by Schultze-Naumburg.

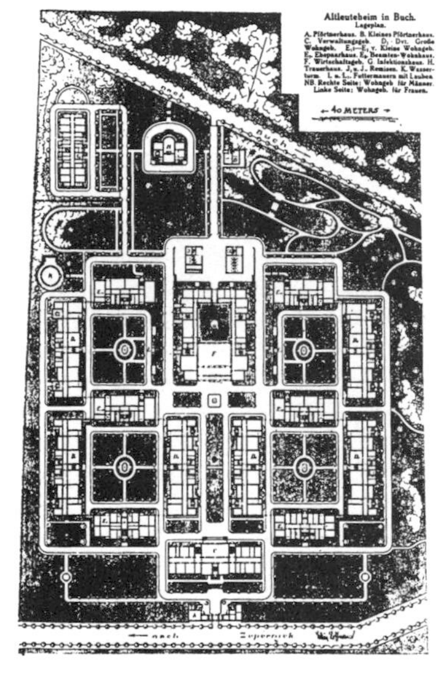

FIG. 432—BERLIN. OLD PEOPLE'S HOME

The buildings of this municipal institution are grouped around four plazas. Design by Ludwig Hoffmann. (Figs. 432-33 from Wasmuth's Monatshefte.)

FIG. 433—BERLIN. OLD PEOPLE'S HOME

View (going with Fig. 432) from the little fountain in the center looking south.

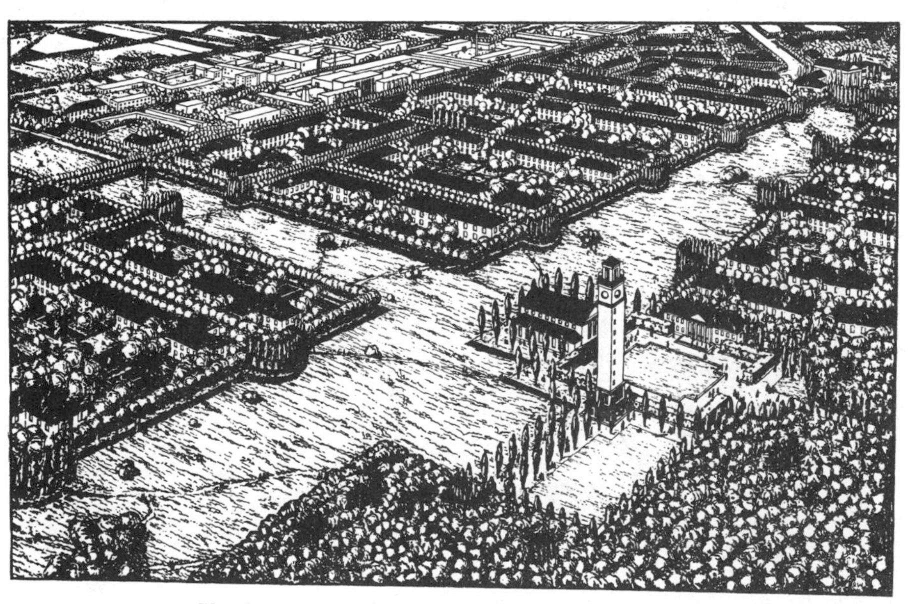

FIG. 434—BERLIN. DESIGN FOR SUBURBAN DEVELOPMENT

By Martin Wagner and Rudolf Wondracek.

THE GROUPING OF BUILDINGS IN AMERICA

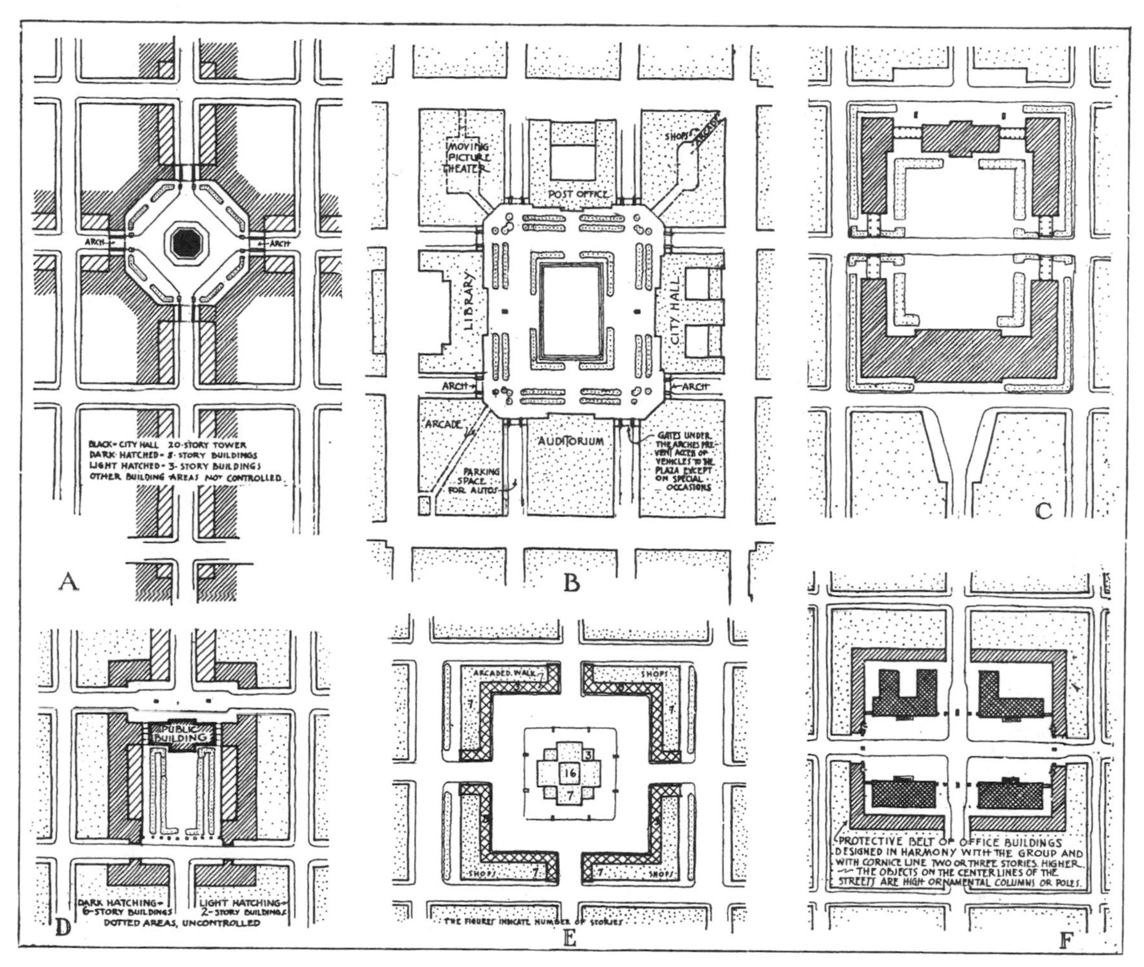

FIGS. 618-23—SIX PLANS FOR CIVIC CENTER GROUPS

These studies, by the authors, illustrate the adaptation of various Renaissance motives to modern conditions and gridiron street plans. For visualizations of these plans see Figs. 626-31.

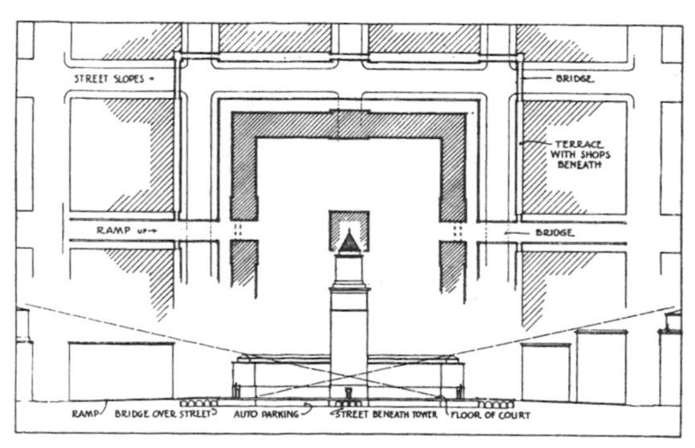

FIG. 624—CIVIC CENTER GROUP SURROUNDED BY A TRAFFIC CIRCLE AT A LOWER LEVEL

Regarding the two-level idea compare Fig. 614 . The intention is to raise the entire civic group upon a higher level surrounded by terraces. Since the interior court would be protected from traffic and by restricting buildings outside to heights below the lines of vision, the esthetic unit would remain uninterfered with. All traffic would stay on the lower level; the area under the civic group is reserved as parking space. The elevators of the tower would connect with the street beneath the tower and the court.

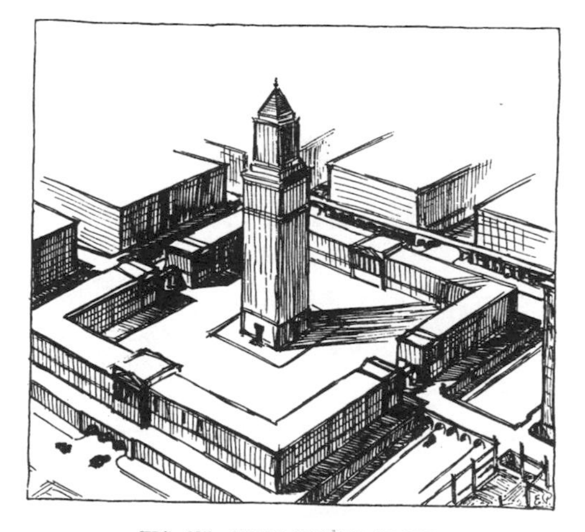

FIG. 625—CIVIC CENTER GROUP

View going with plan Fig. 624.

CIVIC CENTER SUGGESTIONS

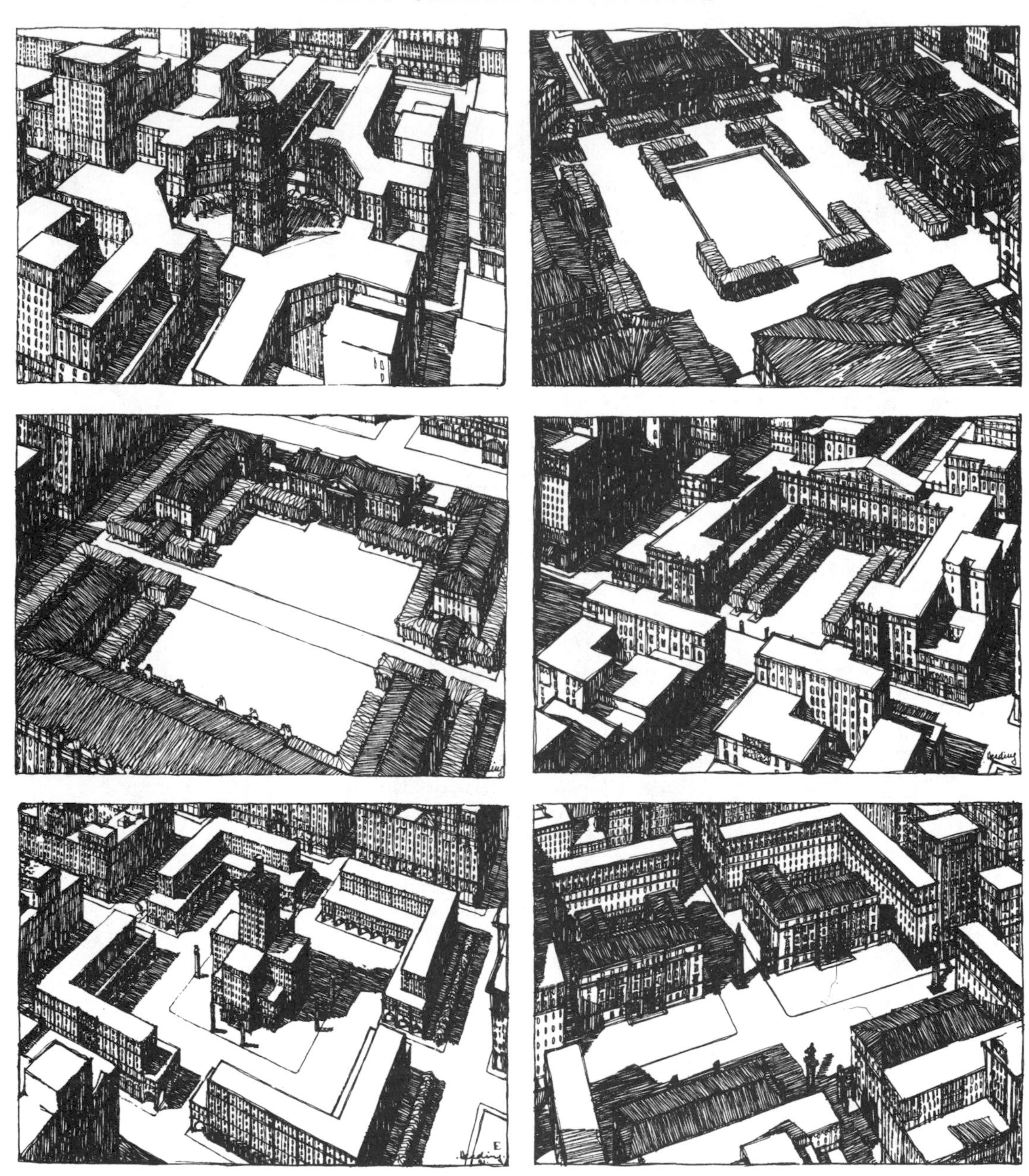

FIGS. 626-31—SIX CIVIC CENTER GROUPS

Bird's-eye views developed from the plans, Figs. 618-23. (From drawings by Franz Herding.)

FIGS. 632-33—PLAN AND SKETCH FOR CIVIC GROUP

The buildings fronting on the small oblong forecourt plazas would have to be simple and uniform, thus subduing the little left-over blocks at the ends of the plazas. Or these blocks might be the sites of specially designed pavilions, which would have to be high enough to conceal the buildings back of them on the diagonal streets.

THE WORLD'S FAIRS

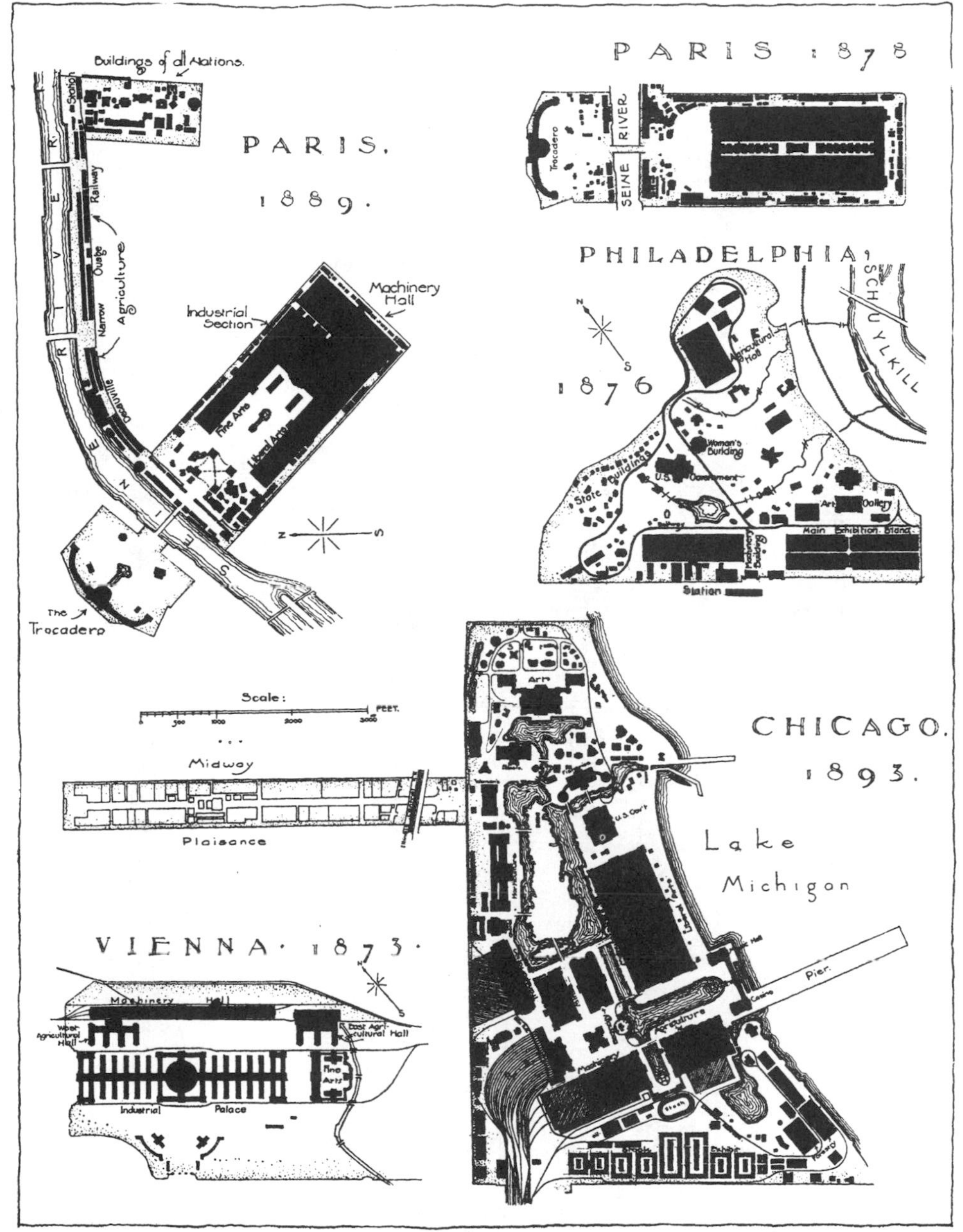

FIG. 441—FIVE WORLD'S FAIR PLANS (From the American Architect and Building News. 1893.)

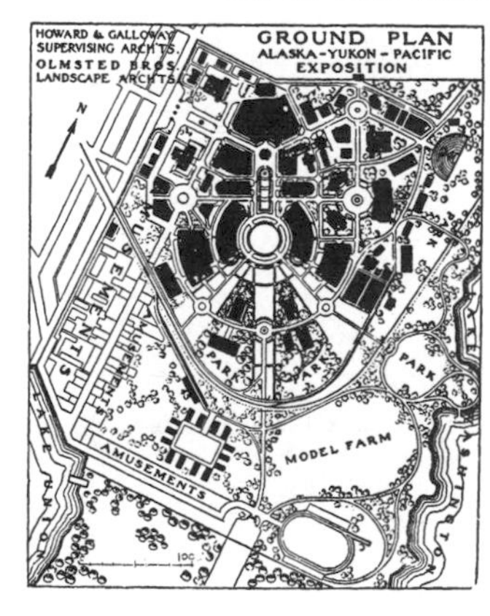

FIG. 442—ALASKA-YUKON-PACIFIC EXPOSITION, SEATTLE, 1910

(From Landscape Architecture.)

FIG. 443—LOUISIANA PURCHASE EXPOSITION, ST. LOUIS, 1904

(From the Architectural Review 1904.)

All plans on this page are brought to approximately the same scale.

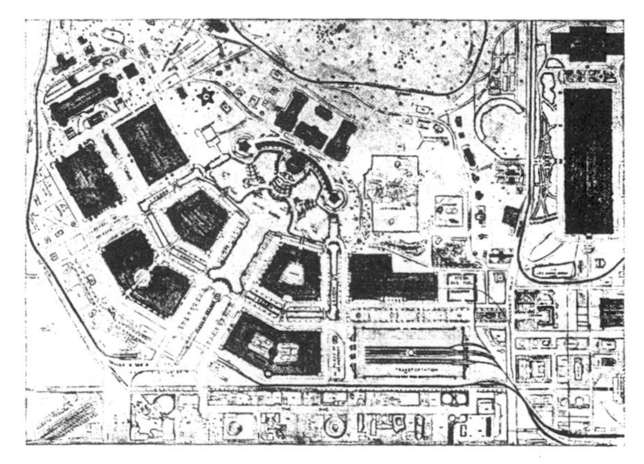

THE GROUPING OF BUILDINGS IN AMERICA

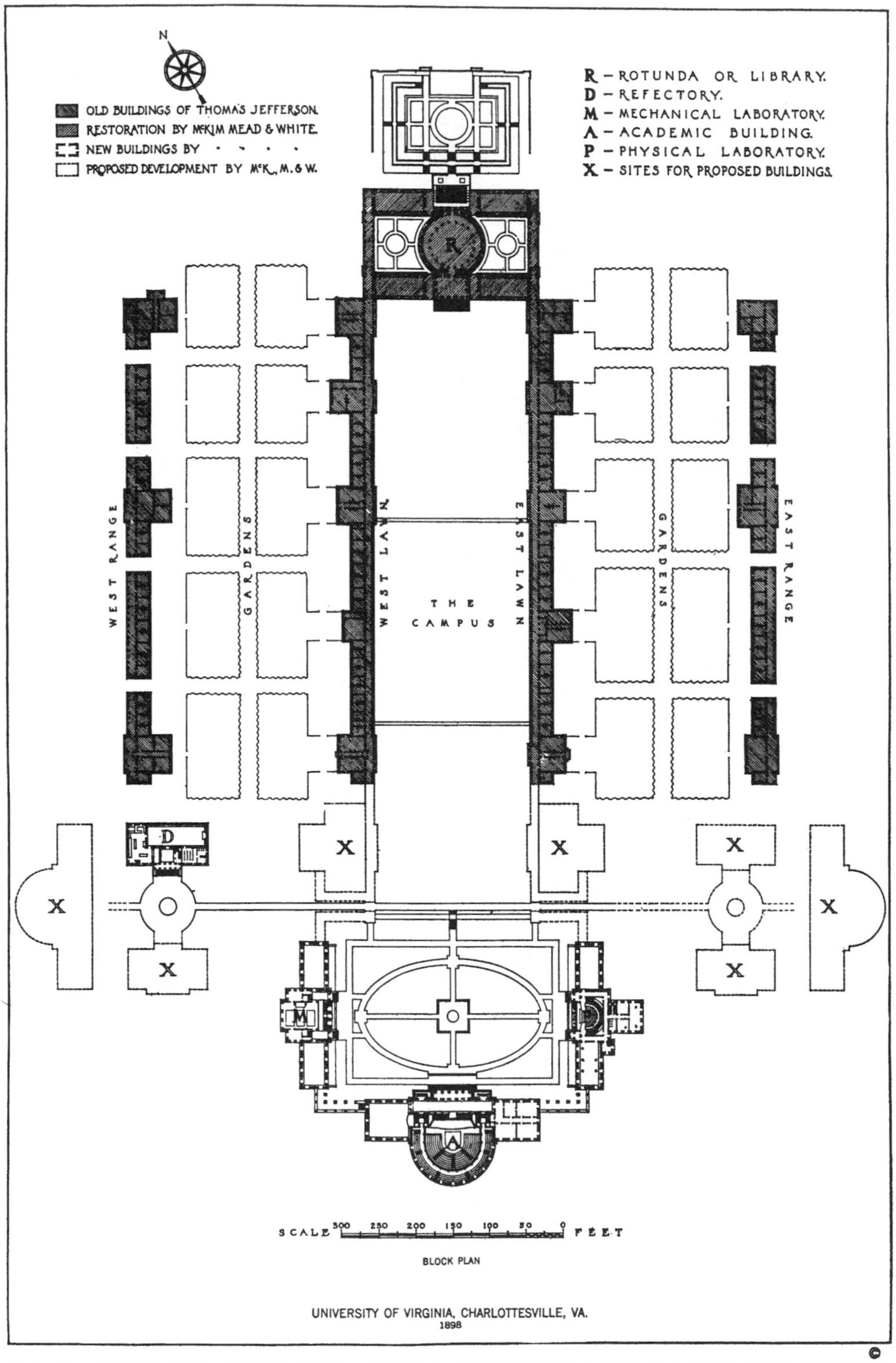

N

OLD BUILDINGS OF THOMAS JEFFERSON.
RESTORATION BY McKIM MEAD & WHITE.
NEW BUILDINGS BY · · · ·
PROPOSED DEVELOPMENT BY McK., M. & W.

R – ROTUNDA OR LIBRARY.
D – REFECTORY.
M – MECHANICAL LABORATORY.
A – ACADEMIC BUILDING.
P – PHYSICAL LABORATORY.
X – SITES FOR PROPOSED BUILDINGS.

WEST RANGE
GARDENS
WEST LAWN
THE CAMPUS
EAST LAWN
GARDENS
EAST RANGE

SCALE 300 250 200 150 100 50 0 FEET

BLOCK PLAN

UNIVERSITY OF VIRGINIA, CHARLOTTESVILLE, VA.
1898

©

FIG. 492—UNIVERSITY OF VIRGINIA. THOMAS JEFFERSON'S PLAN AS DEVELOPED BY McKIM MEAD AND WHITE

(From the Monograph of the work of McKim, Mead, and White.)

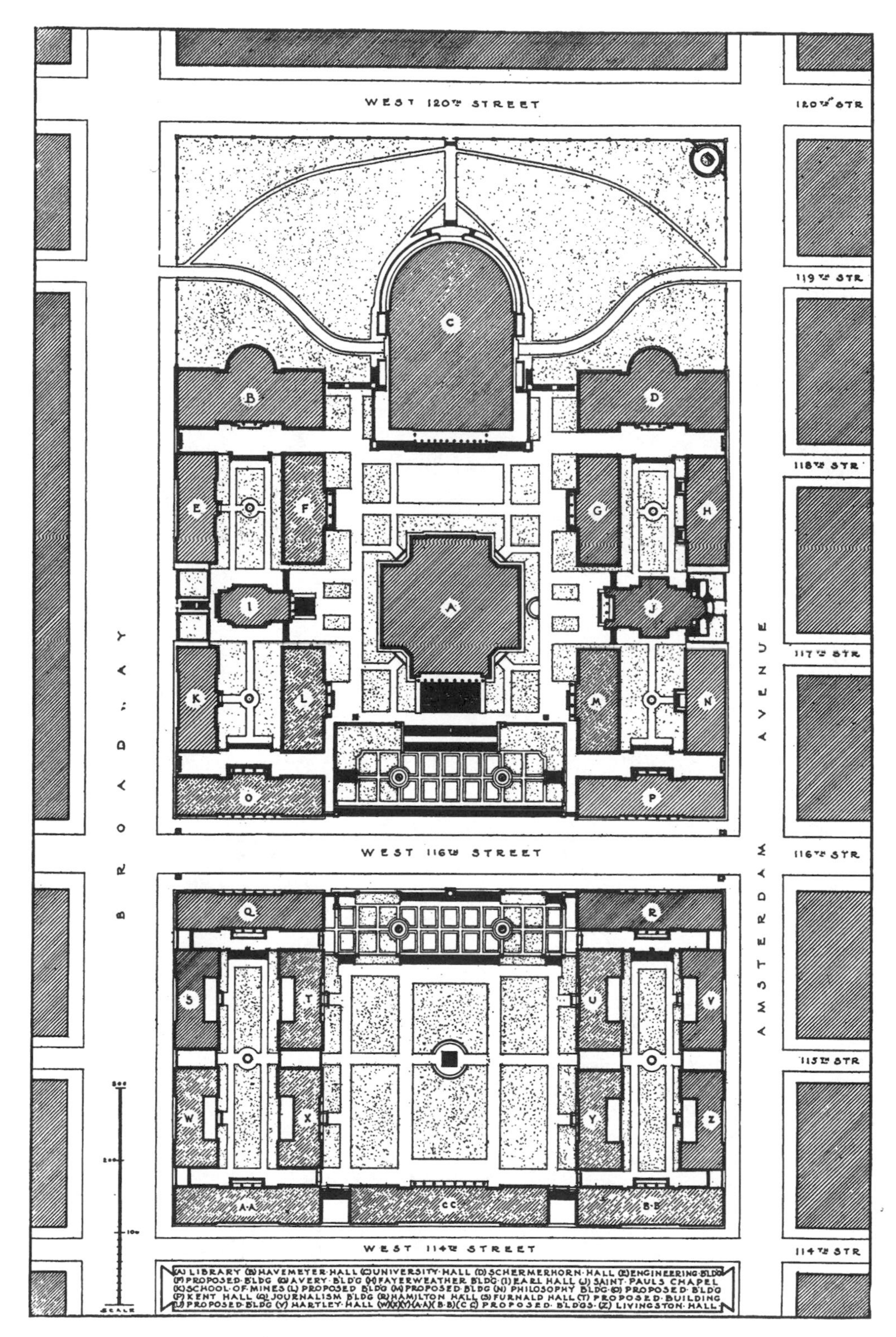

FIG. 510—NEW YORK. GENERAL PLAN OF COLUMBIA UNIVERSITY

By McKim, Mead, and White. The library, "A", stands in the center of a plateau formed by levelling the top of a hill. All the surrounding streets are well below the plateau, especially West 120th which is some thirty or forty feet below the central campus. The buildings stand on a high podium, as shown in Fig. 513A. (From the Monograph of the work of McKim, Mead, and White.)

CITY PLANS AS UNIFIED DESIGNS

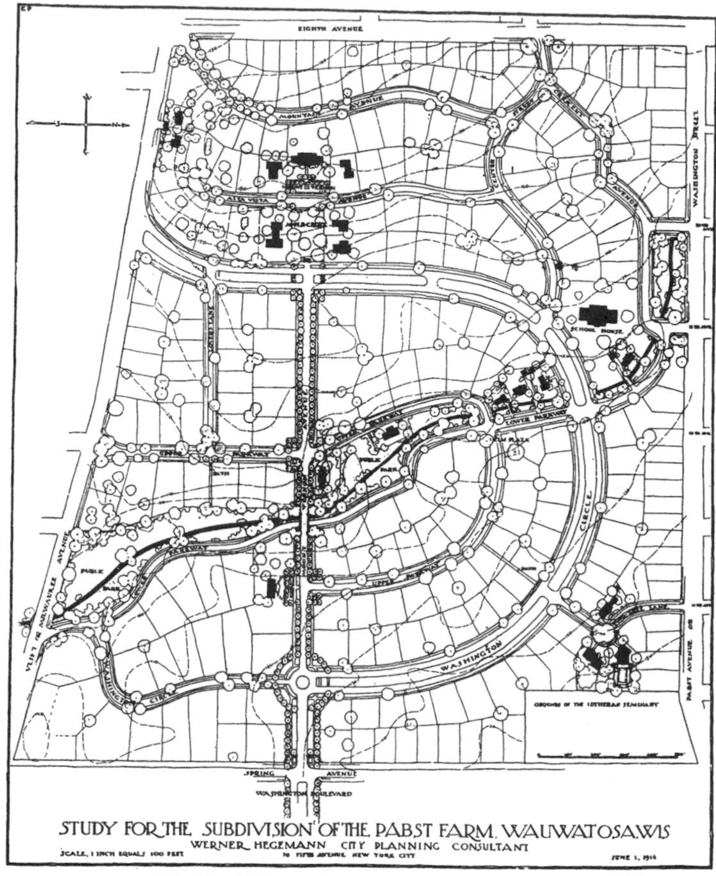

FIG. 1169—MILWAUKEE. WASHINGTON HIGHLANDS

The straight streeet which dominates the design is a continuation of a straight city boulevard running out from a large park. This formal axis, which drops ten feet to its middle point and then rises twenty, both in straight grades, bridges over a creek valley developed as an informal cross-axis. Above the end of the straight street (which in execution was tapered from a hundred feet at the east end to fifty-six at the west) the axis is carried up a steep hill (Apple Croft, an orchard park) and terminates in a group of large houses at the high point of the tract, a hundred feet above the valley park. Access to this high ground is had by streets adapted as closely as possible to the difficult topography. The wide Washington Circle connects the northern part of the property with the main axis and provides a route at easy grade for those who wish to avoid the steep ascent of Mount Vernon Avenue. Two Tree Lane and Elm Plaza are motived by fine existing trees. The entire property is bounded by a hedge and hedges line the principal axis street. All entrances are marked by uniform entrance posts and varying plantings of hedges and clipped lindens. The construction work on the subdivision was completed in 1920.

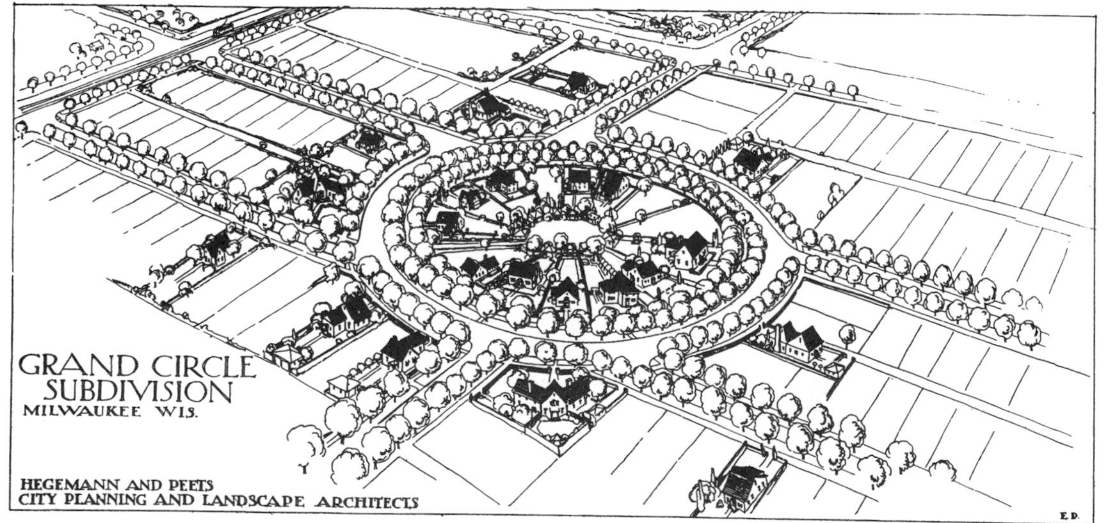

FIG. 1170—MILWAUKEE. GRAND CIRCLE

An effort to insert a pleasant variation into an existing gridiron. The plan facilitated the creation of a garage court in the center of the round block, reached by a lane from the street. The assorted doll-houses are due in part to the fact that the drawing had to be appropriate for use as a newspaper advertisement.

LAND SUBDIVISIONS AND RESIDENCE GROUPS

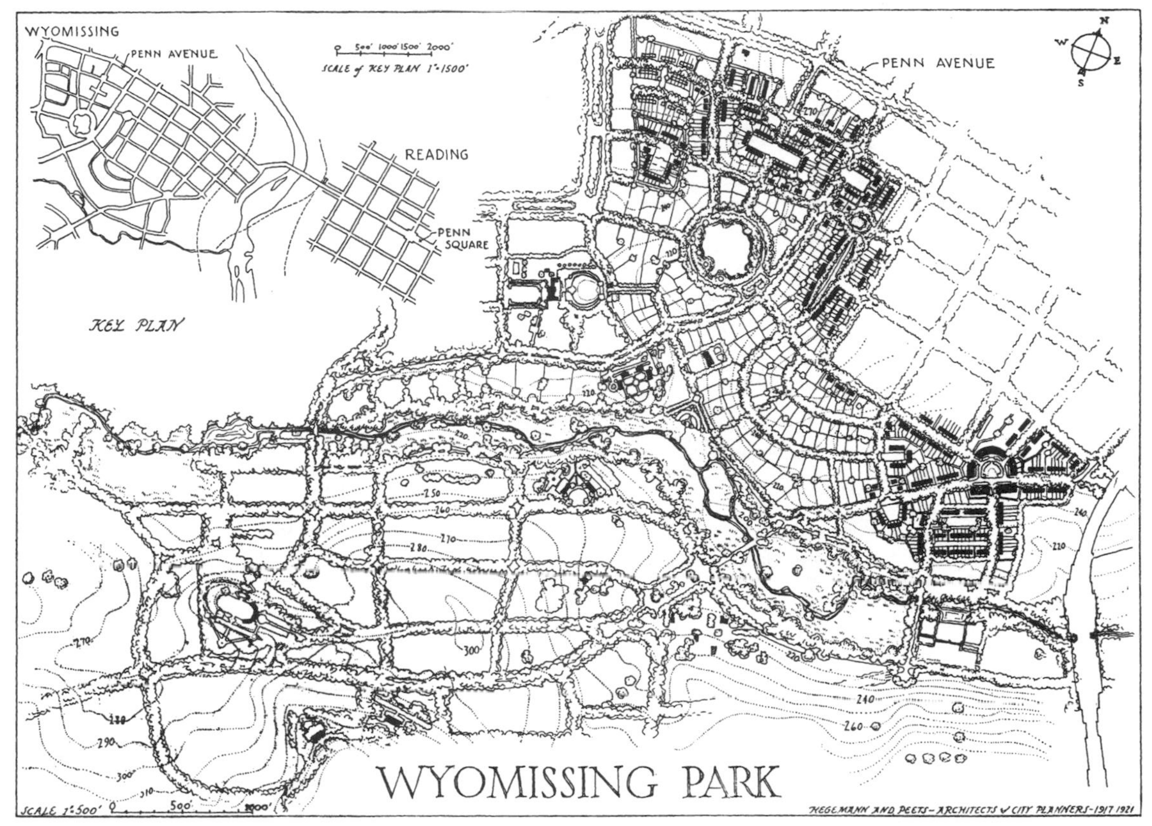

FIG. 1171—WYOMISSING, PA. GENERAL PLAN OF WYOMISSING PARK

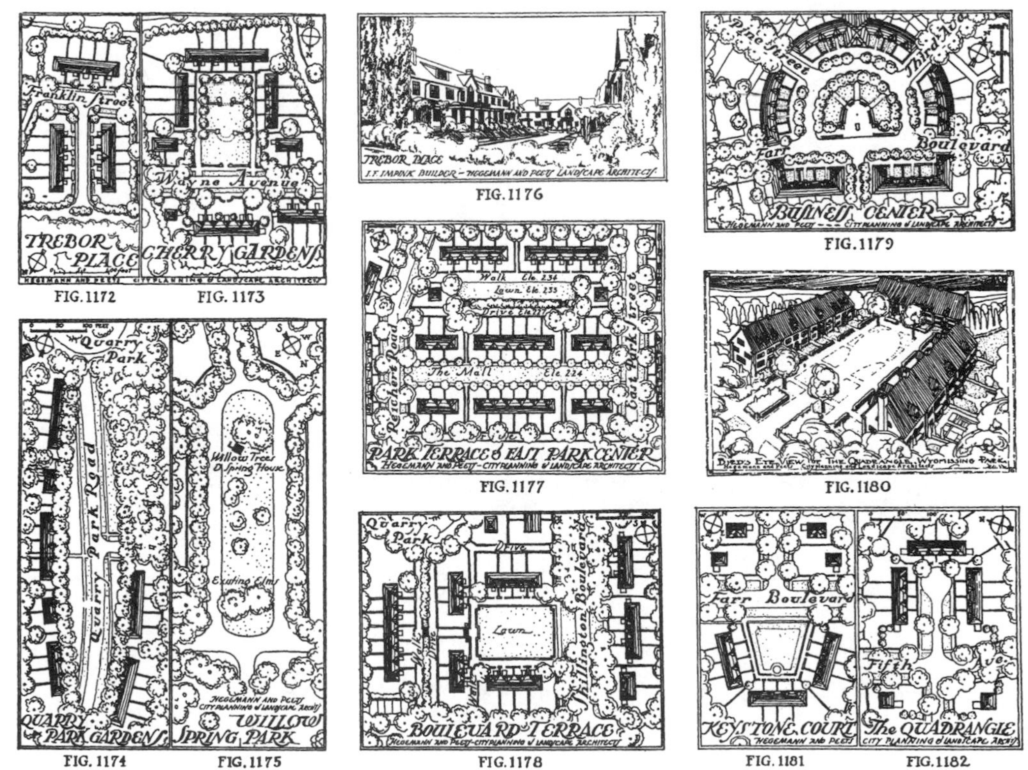

FIGS. 1172-82—WYOMISSING. DETAILS FROM THE PLAN OF WYOMISSING PARK

CITY PLANS AS UNIFIED DESIGNS

FIG. 1183—WYOMISSING. HOLLAND SQUARE. See Figs. 1186-87.

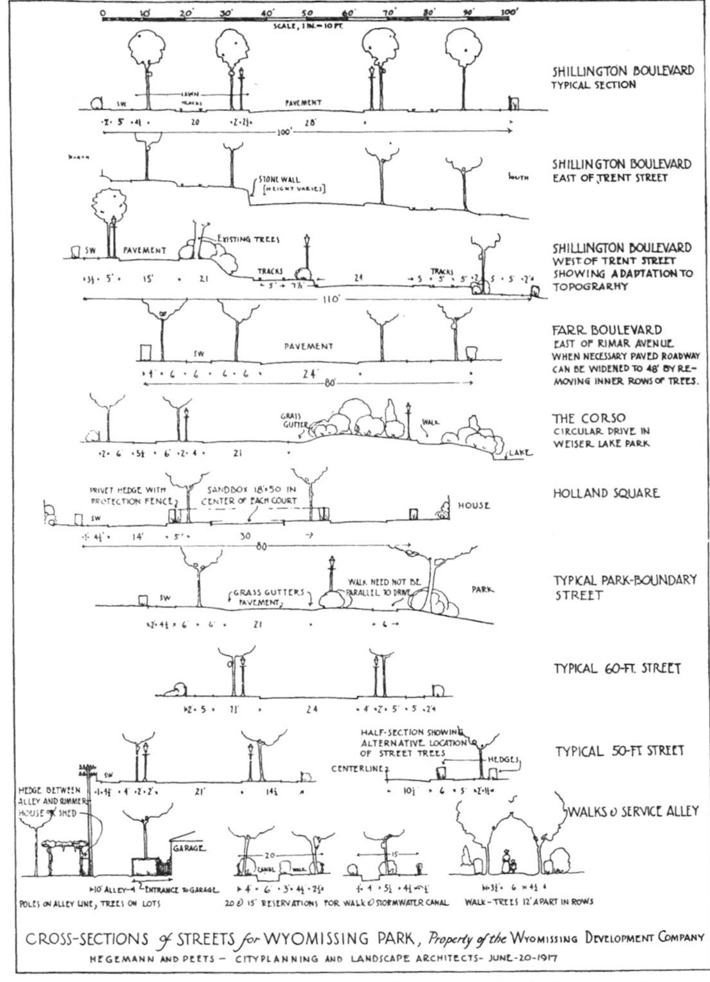

FIGS. 1184—WYOMISSING. STREET SECTIONS FOR WYOMISSING PARK

FIG. 1185—WYOMISSING. BUSINESS CENTER, WYOMISSING PARK

Wyomissing Park is a large land subdivision in Wyomissing, a suburb of Reading. (See the general plan, Fig. 1171.) In developing the plan the first principle fixed upon was to divide the tract roughly into three zones: the one nearest to the city and to the knitting mills in Wyomissing to be rather closely built up with grouped houses; the next, on the slopes toward Wyomissing creek, to be divided into fairly large lots for free-standing houses; the third, beyond the creek valley, to be divided into estates of various sizes. The building traditions of the region recognize the esthetic and economic value of row houses. It was therefore possible to plan the first zone mainly as a series of courts surrounded by rows of houses, in the style of the English garden cities. Two of these, Trebor Court and Holland Square, have been built and others are under way.

The Wyomissing Creek valley is a meadow fringed with elms, forming a pleasant natural park. The little lake in the northern part of the tract is the site of a mine from which iron was taken in colonial times.

LAND SUBDIVISIONS AND RESIDENCE GROUPS

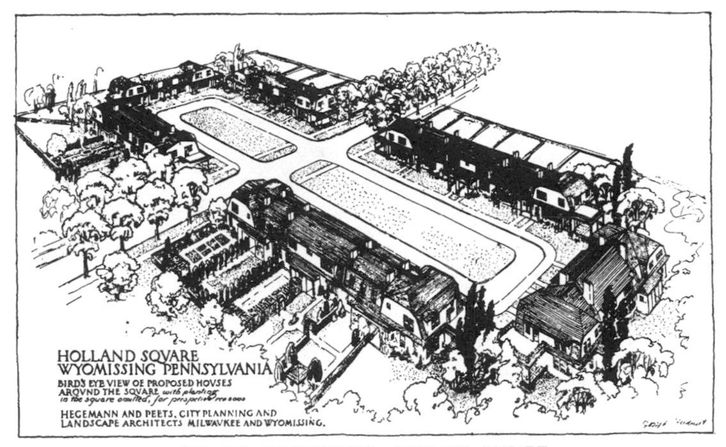

FIG. 1186—WYOMISSING. HOLLAND SQUARE

Comparison with Fig. 1183 will show that the suggestions embodied in this drawing were not followed very closely.

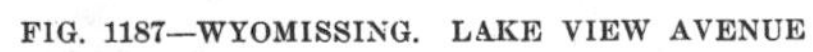

FIG. 1187—WYOMISSING. LAKE VIEW AVENUE

Holland Square constitutes a rest or pause in this sharply descending straight street which ends at the shore of a little lake. See plan, Fig. 1171.

FIG. 1188—WYOMISSING

An old stone house in Wyomissing Park. The farm buildings of the region are usually of stone, and the town houses of red brick.

FIG. 1189—WYOMISSING. LAKE VIEW PLAZA

A study for the uppermost unit in Lake View Avenue

FIG. 1190—MADISON. VISTA FROM THE HILL, LAKE FOREST

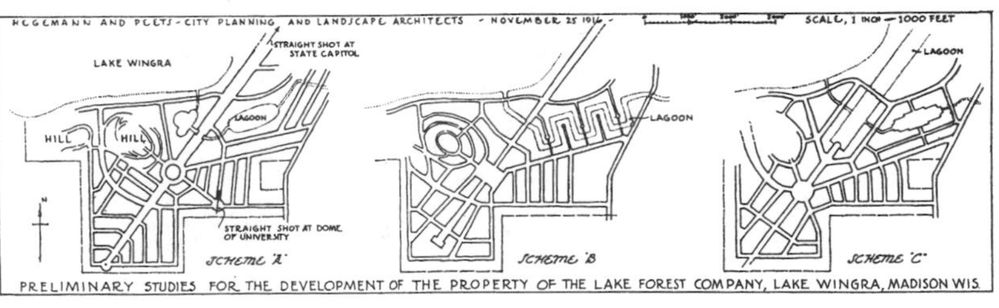

FIG. 1191—MADISON. STUDIES FOR THE PLAN OF LAKE FOREST

See Figs. 846 and 953 for other Lake Forest drawings.

SAN DIEGO, 1915. APPROACH OVER THE PUENTE CABRILLO
Bertram Grosvenor Goodhue, Consulting and advisory architect.

SAN DIEGO, 1915. VIEW OF EXPOSITION TAKEN FROM ACROSS THE CANYON
Cram, Goodhue and Ferguson, Architects; Bertram G. Goodhue, Consulting and advisory architect.

Clarence Arthur Perry

Summary
Traditional community, as opposed to the abstract and alienated social relations characteristic of modern urban industrial society, was a preoccupation of sociologists and social critics of the 19th and early 20th centuries. Chicago-trained sociologist Clarence Arthur Perry (1872–1944) became one of the principal theorists of and advocates for the traditional neighborhood as a basis for the planning of new towns and urban areas and for the redevelopment of blighted slums. His advocacy of the "neighborhood unit" as a principle element of planning was based not only on his academic interests, but also on his direct experience as sociologist-in-residence for the Russell Sage Foundation's model garden suburb of Forest Hill Gardens in New York, designed by Frederick Law Olmsted, Jr. and Grosvenor Atterbury (begun in 1909). His observations of that project led directly to the treatise on "Neighborhood and Community Planning" published as Volume VII of the *1929 Regional Plan of New York*, and excerpted here.

Key words
community plan, housing, neighborhood

The neighborhood unit. 160 acre tract for 6000 people with pedestrian ways for dwellings, schools, and shops with no transecting highways. Drawing: Chester B. Price, *House and Garden*, 1925.

The neighborhood unit

What is known as a neighborhood and what is now commonly defined as a region have at least one characteristic in common—they possess a certain unity that is quite independent of political boundaries. The area with which the Regional Plan of New York is concerned, for instance, has no political unity, although is it possessed of other unifying characteristics of a social, economic and physical nature. Within this area there are definite political entities, such as villages, counties and cities, forming suitable divisions for sub-regional planning, and within those units there are definite local or neighborhood communities which are entirely without governmental limits and sometimes overlap into two or more municipal areas. Thus, in the planning of any large metropolitan area, we find that three kinds of communities are involved:

- The regional community, which embraces many municipal communities and is therefore a family of communities;
- The village, county or city community;
- The neighborhood community.

Only the second of these groups has any political framework, although all three have an influence upon political life and development. While the neighborhood community has no political structure, frequently it has greater unity and coherence than found in the village or city and therefore is of fundamental importance to society.

1 THE NEIGHBORHOOD UNIT

The title "neighborhood unit" is a term of reference given, for purposes of the study described in this article, to the scheme of arrangement for a family-life community. Investigations have shown that residential communities, when they meet the universal needs of family life, have similar parts performing similar functions. In the neighborhood-unit system, those parts have been put together as an organic whole. The scheme is put forward as the framework of a model community and not as a detailed plan. Its actual realization as an individual real-estate development requires the embodiment and garniture which can be given to it only by the planner, the architect, and the builder.

The underlying principle of the scheme is that an urban neighborhood should be regarded both as a unit of a larger whole and as a distinct entity in itself. For government, fire and police protection and many other services, it depends upon the municipality. Its residents for the most part find their occupations outside of the neighborhood. To invest in bonds, attend the opera or visit the museum, perhaps even to buy a piano, they have to resort to the "downtown" district. But there are certain other facilities, functions or aspects that are strictly local and peculiar to a well-arranged residential community. They may be classified under four headings. Other neighborhood institutions and services are sometimes found, but these are practically universal:

Source: "Neighborhood and Community Planning" *Regional Survey—Volume VII New York City: Regional Plan of New York and its Environs,* 1929.

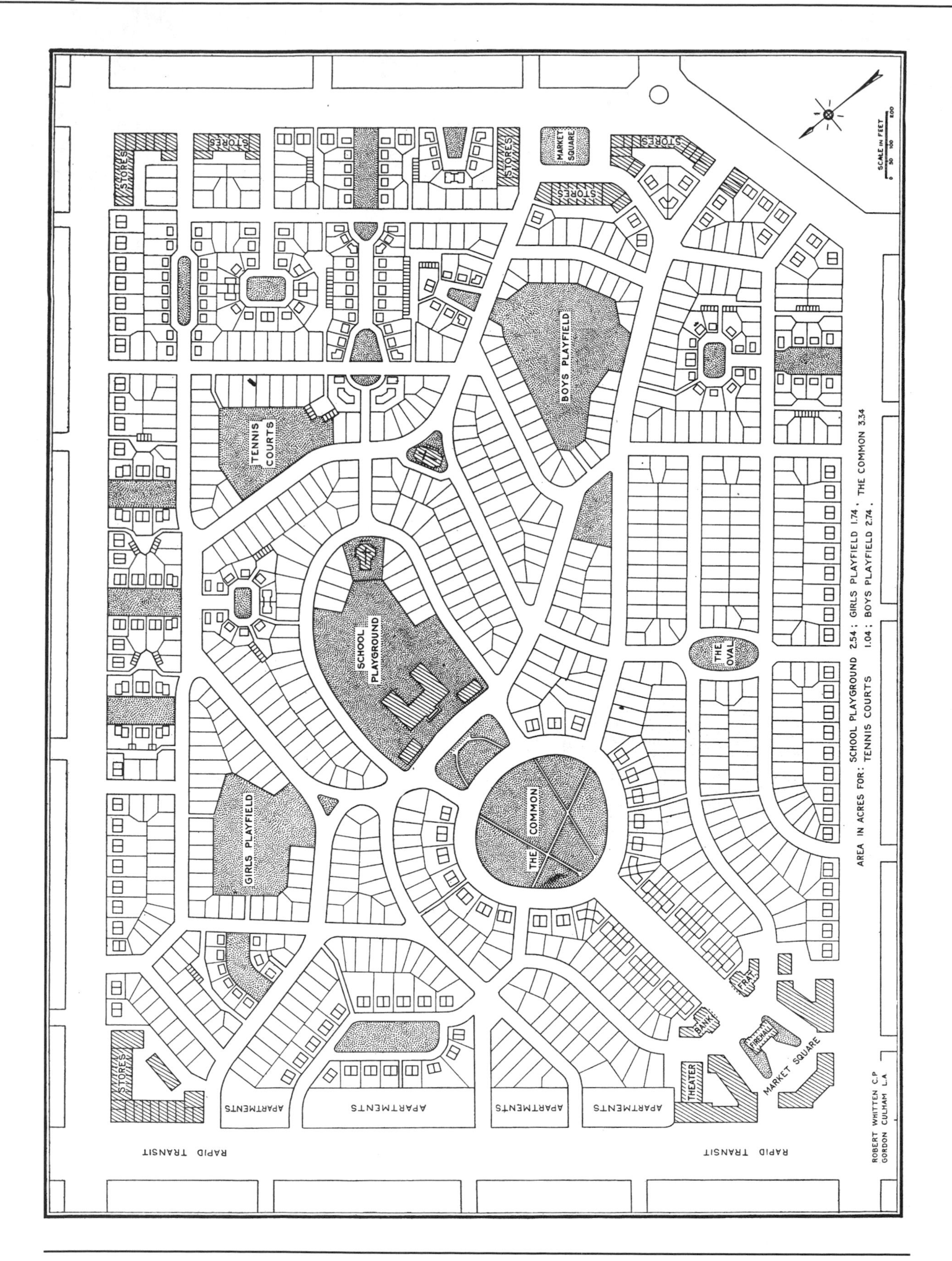

Fig. 1 Plan prepared in connection with research into "The Economics of Land Subdivision." It represents an attempt to apply the above principles in a layout suitable for a development of modest homes in the suburbs.

- Elementary school
- Small parks and playgrounds
- Local shops
- Residential environment

Parents have a general interest in the public school system of the city, but they feel a particular concern regarding the school attended by their children. Similarly, they have a special interest in the playgrounds where their own and their neighbors' children spend so many formative hours. In regard to small stores, the main concern of householders is that they be accessible but not next to their own doors. They should also be concentrated and provide for varied requirements.

Under the term "residential environment" is included the quality of architecture, the layout of streets, the planting along curbs and in yards, the arrangement and setback of buildings, and the relation of shops. Filling stations and other commercial institutions to dwelling places—all the elements which go into the environment of a home and constitute its external atmosphere. The "character" of the district in which people lives tells something about them.

It is with the neighborhood itself, and not its relation to the city at large, that this study is concerned. If it is to be treated as an organic entity, it logically follows that the first step in the conversion of unimproved acreage for residential purposes will be its division into unit areas, each one of which is suitable for a single neighborhood community. The next step consists in the planning of each unit so that adequate provision is made for the efficient operation of the four main neighborhood functions. The attainment of this major objective—as well as the securing of safety to pedestrians and the laying of the structural foundation for quality in environment—depends, according to our studies, upon the observance of the following requirements:

Neighborhood-unit principles

Size. A residential unit development should provide housing for that population for which one elementary school is ordinarily required, its actual area depending upon population density.

Boundaries. The unit should be bounded on all sides by arterial streets, sufficiently wide to facilitate its bypassing by all through traffic.

Open spaces. A system of small parks and recreations spaces should be provided, planned to meet the needs of the particular neighborhood.

Institution sites. Sites for the school and other institutions having service spheres coinciding with the limits of the unit should be suitably grouped about a central point, or common area.

Local shops. One or more shopping districts, adequate for the population to be served, should be laid out in the circumference of the unit, preferably at traffic junctions and adjacent to similar districts of adjoining neighborhoods.

Low-Cost Suburban Development
Area Relations of the Plan

Complete unit	160 acres	100 percent
Dwelling-house lots	86.5	54.0
Apartment-house lots	3.4	2.1
Business blocks	6.5	4.1
Market squares	1.2	0.8
School and church sites	1.6	1.0
Parks and playgrounds	13.8	8.6
Greens and circles	3.2	2.0
Streets	43.8	27.4

Internal street system. The unit should be provided with a special street system, each highway being proportioned to its probable traffic load, and the street net as a whole being designed to facilitate circulation within the unit and to discourage its use by through traffic.

To offer a clear picture of each of these principles, the figures illustrate plans and diagrams in which the principles have been applied.

2 LOW-COST SUBURBAN DEVELOPMENT

Character of district. The plan shown in Fig. 1 is based upon an actual tract of land in the outskirts of the Borough of Queens. The section is as yet entirely open and exhibits a gently rolling terrain, partly wooded. So far, the only roads are of the country type, but they are destined some day to be main thoroughfares. There are no business or industrial establishments in the vicinity.

Population and housing. The lot subdivision provides 822 single-family houses, 236 double houses, 36 row houses and 147 apartment suites, accommodations for a total of 1,241 families. At the rate of 4.93 persons per family, this would mean a population of 6,125 and a school enrollment of 1,021 pupils. For the whole tract the average density would be 7.75 families per gross acre.

Open spaces. The parks, playgrounds, small greens and circles in the tract total 17 acres, or 10.6 percent of the total area. If there is included also the 1.2 acres of market squares, the total acreage of open space is 18.2 acres. The largest of these spaces is the common of 3.3 acres. This serves both as a park and as a setting or approach to the school building. Back of the school is the main playground for the small children, of 2.54 acres, and near it is the girls' playfield of 1.74 acres. On the opposite side of the schoolyard, a little farther away, is the boys' playground of 2.7 acres. Space for tennis courts is located conveniently in another section of the district. At various other points are to be found parked ovals or small greens which give attractiveness to vistas and afford pleasing bits of landscaping for the surrounding homes.

Community center. The pivotal feature of the layout is the common, with the group of buildings that face upon it. These consist of the schoolhouse and two lateral structures facing a small central plaza. One of these buildings might be devoted to a public library and the other to any suitable neighborhood purpose. Sites are provided for

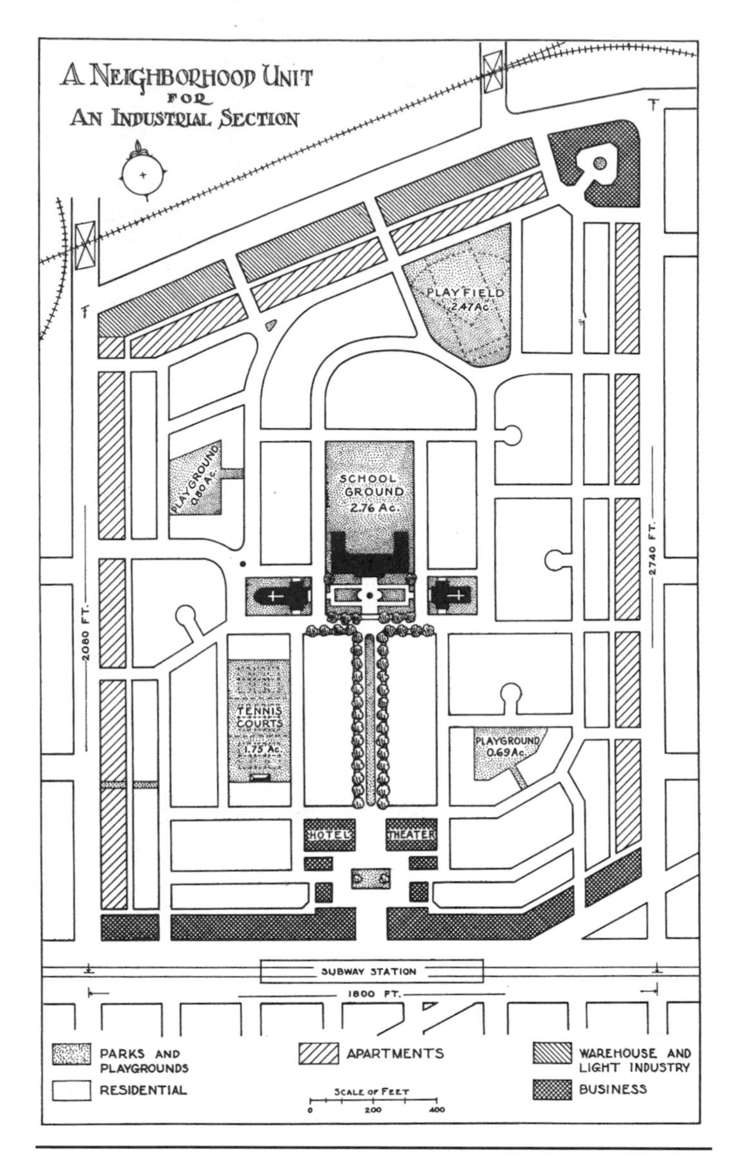

Fig. 2 Diagram suggesting the type of plan that might be devised for a more central area in the neighborhood of industry and business.

A Neighborhood Unit for an Industrial Section

Distribution of Area

Complete unit	101.4 acres	100 percent
Residences—houses	37.8	37.3
Residences—apartments	8.4	8.3
Parks and play spaces	10.8	10.6
Business	5.2	5.1
Warehouses	3.2	3.2
Streets	36.0	35.5

two churches, one adjoining the school playground and the other at a prominent street intersection. The school and its supporting buildings constitute a terminal vista for a parked main highway coming up from the market square. In both design and landscape treatment the common and the central buildings constitute an interesting and significant neighborhood community center.

Shopping district. Small shopping districts are located at each of the four corners of the development. The streets furnishing access to the stores are widened to provide for parking, and at the two more important points there are small market squares, which afford additional parking space and more opportunity for unloading space in the rear of the stores. The total area devoted to business blocks and market plazas amounts to 7.7 acres. The average business frontage per family provided by the plan is about 2.3 feet.

Street system. In carrying out the unit principle, the boundary streets have been made sufficiently wide to serve as main traffic arteries. One of the bounding streets is 160 feet wide, and the other three have widths of 120 feet. Each of these arterial highways is provided with a central roadway for through traffic and two service roadways for local traffic separated by planting strips. One-half of the area of the boundary streets is contributed by the development. This amounts or 15.3 acres, or 9.5 percent of the total area, which is a much larger contribution to general traffic facilities than is ordinarily made by the commercial subdivision, but not greater than that which is required by present-day traffic needs. The interior streets are generally 40 or 50 feet in width and are adequate for the amount of traffic that will be developed in a neighborhood of this single-family density. By the careful design of blocks, the area devoted to streets is rather lower than is usually found in a standard gridiron subdivision. If the bounding streets were not over 50 feet wide, the percent of the total street area would be reduced from the 27.4 percent to about 22 percent. It will be observed that most of the streets opening on the boundary thoroughfares are not opposite similar openings in the adjacent developments. There are no streets that run clear through the development without being interrupted.

3 NEIGHBORHOOD UNIT FOR AN INDUSTRIAL SECTOR

Fig. 2 is a sketch of the kind of layout, which might be devised for a district in the vicinity of factories and railways. Many cities possess somewhat central areas of this character, which have not been preempted by business or industry but which are unsuitable for high-cost housing and too valuable for a low-cost development entirely of single-family dwellings.

Economically, the only alternative use for such a section is industrial. If it were built up with factories, however, the non-residential area thereabouts would be increased and the daily travel distance of many workers would be lengthened. One of the main objectives of good city planning is therefore attained when it is made available for homes.

Along the northern boundary of the tract illustrated lie extensive railroad yards, while its southern side borders one of the city's main arteries, affording both an elevated railway and wide roadbeds for surface traffic. An elevated station is located at a point opposite the center of the southern limit, making that spot the main portal of the development.

Functional dispositions. The above features dictated the employment of a tree-like design for the street system. Its trunk rests upon the elevated station, passes through the main business district, and terminates at the community center. Branches, covering all sections of the unit, facilitate easy access to the school, to the main street stem, and to the business district.

Along the northern border, structures suitable for light industry, garages, or warehouses have been designated. These are to serve as a buffer both for the noises and the sights of the railway yards. Next to them, separated only by a narrow service street, is a row of apartments, whose main outlooks will all be directed toward the interior of the unit and its parked open spaces.

The apartments are assigned to sites at the sides of the unit that they may serve as conspicuous visible boundaries and enable the widest possible utilization of the attractive vistas which should be provided by the interior features—the ecclesiastical architecture around the civic center and the park-like open spaces.

Housing density. Fig.2 is intended to suggest mainly an arrangement of the various elements of a neighborhood and is not offered as a finished plan. The street layout is based upon a housing scheme providing for 2,000 families, of which 68 percent are allotted to houses, some semi-detached and some in rows; and 32 percent to apartments averaging 800 square feet of ground area per suite. On the basis of 4.5 persons in houses and 4.2 in suites, the total population would be around 8,800 people, and there would be some 1,400 children of elementary school age, a fine enrollment for a regulation city school. The average net ground area per family amounts to 1,003.7 square feet. If the parks and play areas are included, this figure becomes 1,216 square feet.

Recreation spaces. These consist of a large schoolyard and two playgrounds suitable for the younger children, grounds accommodating nine tennis courts, and a playfield adapted either for baseball or soccer football. In distributing these spaces regard was had both to convenience and to their usefulness as open spaces and vistas for the adjacent homes. All should have planting around the edges, and most of them could be seeded, thus avoiding the barren aspect so common to city playgrounds.

Community center. The educational, religious and civic life of the community is provided for by a group of structures, centrally located and disposed so as to furnish an attractive vista for the trunk street and a pivotal point for the whole layout. A capacious school is flanked by two churches, and all face upon a small square, which might be embellished with a monument, fountain, or other ornamental feature. The auditorium, gymnasium, and library of the school, as well as certain other rooms, could be used for civic, cultural and recreational activities of the neighborhood. With such an equipment and an environment possessing so much of interest and service to all the residents, a vigorous local consciousness would be bound to arise and find expression in all sorts of agreeable and useful face-to-face associations.

Shopping districts. The most important business area is, of course, around the main portal and along the southern arterial highway. For greater convenience and increased exposures a small market square has been introduced. Here would be the natural place for a motion picture theater, a hotel, and such services as a branch post office and a fire-engine house. Another and smaller shopping district has been placed at the northeast corner to serve the needs of the homes in that section.

Economic aspects. While this development is adapted to families of moderate means, comprehensive planning makes possible an intensive and profitable use of the land without the usual loss of a comfortable and attractive living environment. The back and side yards may be smaller, but pleasing outlooks and play spaces are still provided. They belong to all the families in common and the unit scheme preserves them for the exclusive use of the residents.

While this is primarily a housing scheme, it saves and utilizes for its own purposes that large unearned increment, in business and industrial value, which rises naturally out of the mere aggregation of so many people. The community creates that value and while it may apparently be absorbed by the management, nevertheless, some of it goes to the individual householder through the improved home and environment which a corporation, having that value in prospect, is able to offer.

The percentage of area devoted to streets (35.5) is higher than is usually required in a neighborhood unit scheme. In this case the proportion is boosted by the generous parking space provided in the market square and by the adjoining 200-foot boulevard, one-half of whose areas is included in this calculation. Ordinarily the unit scheme makes possible a saving in street area that is almost, if not quite, equal to the land devoted to open spaces.

The school and church sites need not be dedicated. They may simply be reserved and so marked in the advertising matter with full confidence that local community needs and sentiment will bring about their ultimate purchase by the proper bodies. If either or both of the church sites should not be taken, their very location will ensure their eventual appropriation for some public, or semi-public, use.

4 APARTMENT-HOUSE UNITS

Population. On the basis of five-story and basement buildings and allowing 1,320 square feet per suite, this plan would accommodate 2,381 families (Fig. 3). Counting 4.2 persons per family, the total population would number 10,000 individuals, of whom about 1,600 would probably be of elementary school age, a number which could be nicely accommodated in a modern elementary school.

Environment. The general locality is that section where downtown business establishments and residences begin to merge. One side of the unit faces on the principal street of the city and this would be devoted to general business concerns. A theater and a business block, penetrated by an arcade, would serve both the residents of the unit and the general public.

Street system. Wide streets bound the unit, while its interior system is broken up into shorter highways that give easy circulation within the unit but do not run uninterruptedly through it. In general they converge upon the community center. Their widths are varied to fit probably traffic loads and parking needs.

Open spaces. The land devoted to parks and playgrounds averages over one acre per 1,000 persons. If the space in apartment yards is also counted, this average amounts to 3.17 acres per 1,000 persons.

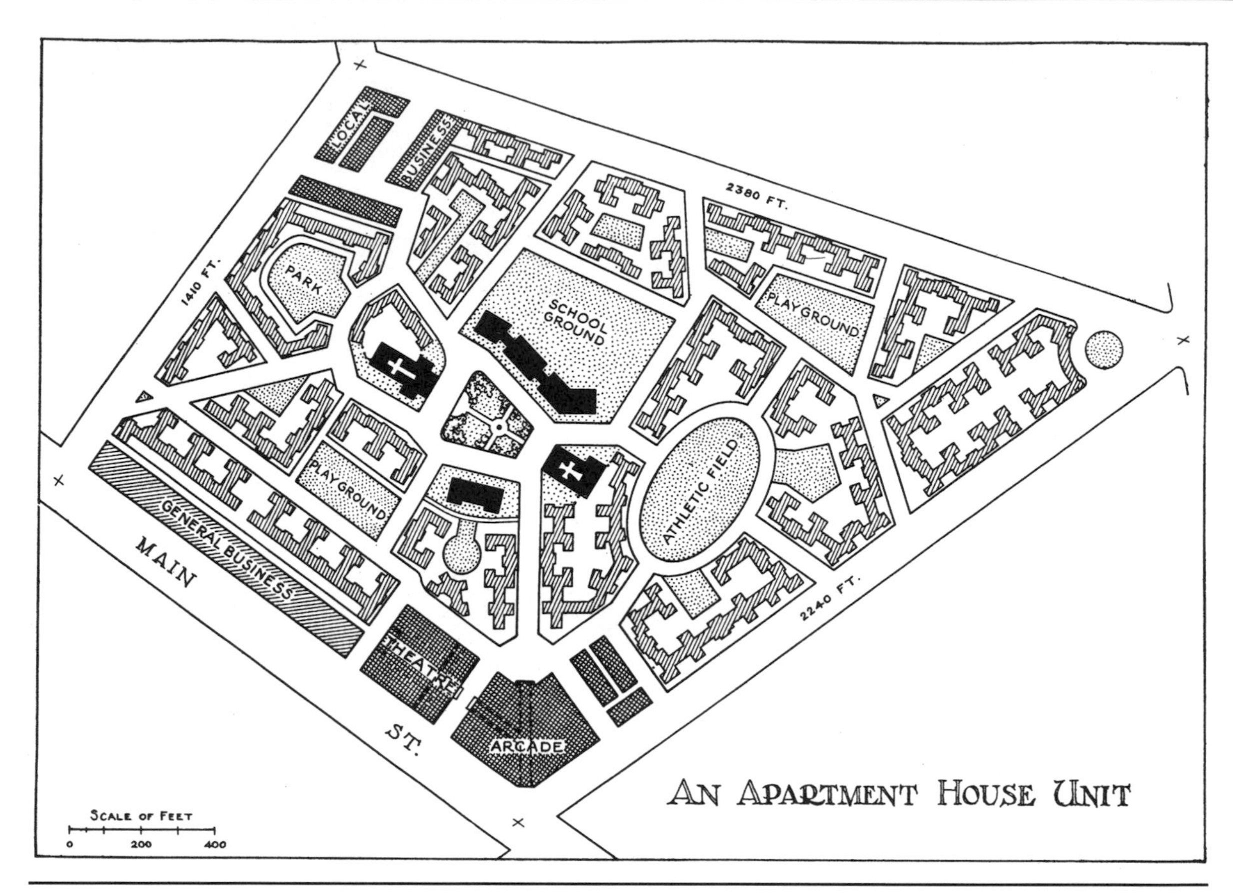

Fig. 3 Diagram of an apartment-house district such as might be laid out on the edge of a subsidiary business district and in a region in which the housing is a mixture of the single and multiple family types.

Apartment House Unit
Distribution of Area

Total area of unit	75.7 acres	100 percent
Apartment buildings	12.0	15.9
Apartment yards	21.3	28.0
Parks and playgrounds	10.4	13.8
Streets	25.3	33.4
Local business	4.9	6.5
General business	1.8	2.4

Area of Open Spaces

Kind	Acres
School grounds	3.27
Athletic field	1.85
Common	.81
Park	.61
Playground	1.03
Playground	.81
Circle	.18
Small greens	1.86
Total	10.42

For 1,600 children the space in the schoolyard provides an average of 89 square feet per pupil, which is a fair allowance considering that all the pupils will seldom be in the yard at the same time. The athletic field is large enough for baseball in the spring and summer, and football in the fall. By flooding it with a hose in the winter time it can be made available for skating.

On the smaller playground it will be possible, if desired, to mark off six tennis courts. The bottle-neck park is partly enclosed by a group of apartments, but it is also accessible to the residents in general.

The recreation spaces should be seeded and have planting around the edges, thus adding attractiveness to the vistas from the surrounding apartments.

Community center. Around a small common are grouped a school, two churches, and a public building. The last might be a branch public library, a museum, a "little theater," or a fraternal building. In any case it should be devoted to a local community use.

The common may exhibit some kind of formal treatment in which a monument and perhaps a bandstand may be elements of the

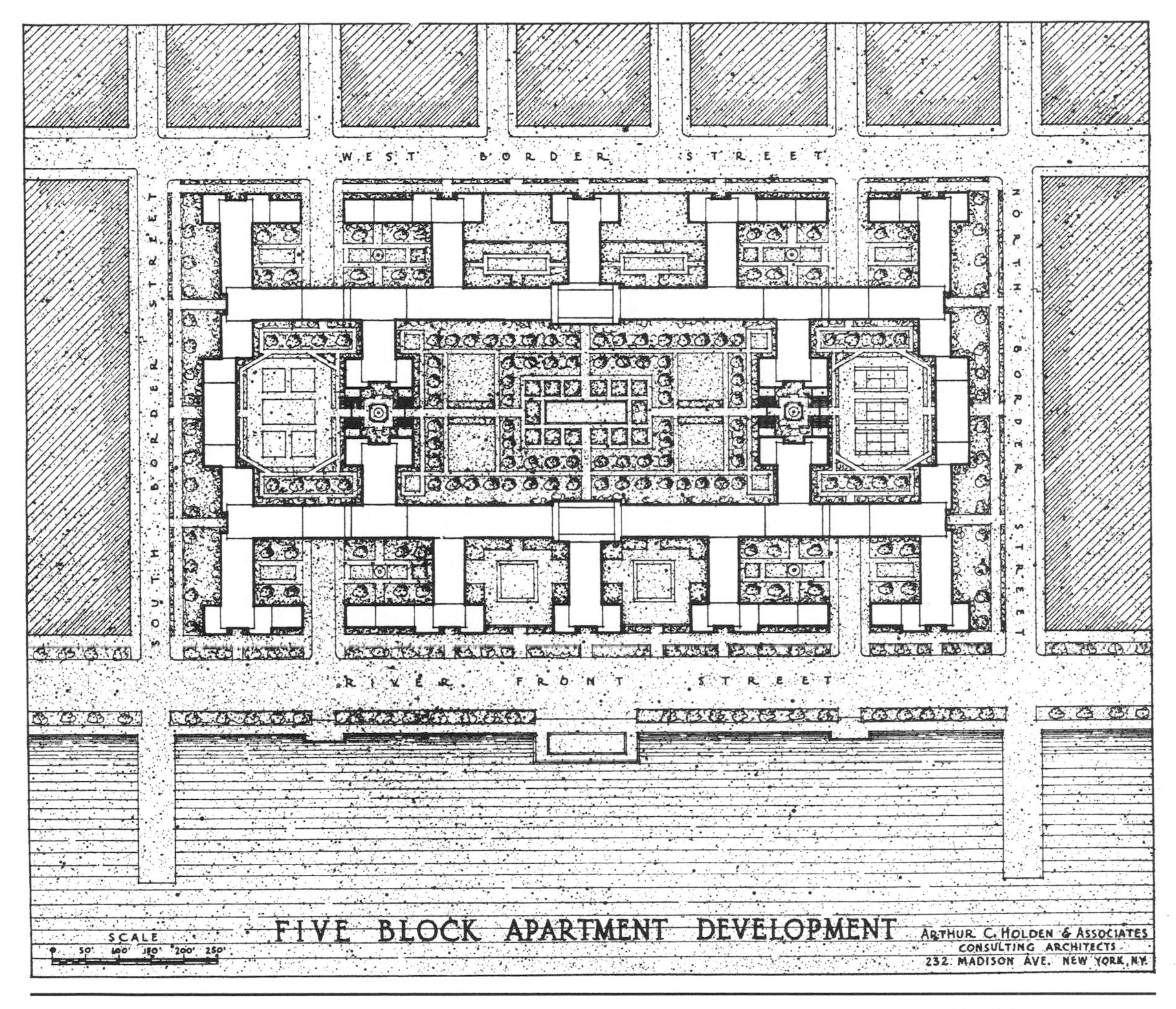

Fig. 4 Plan illustrating a five-block apartment unit suitable for a rebuilding operation in a central section that had suffered deterioration.

design. The situation is one that calls for embellishment, by means of both architecture and landscaping, and such a treatment would contribute greatly to local pride and the attractiveness of the development.

The ground plan of the school indicates a type in which the auditorium, the gymnasium and the classrooms are in separate buildings, connected by corridors. This arrangement greatly facilitates the use of the school plan by the public in general and permits, at the same time, an efficient utilization of the buildings for instruction purposes.

Apartment pattern. The layout of the apartment structures follows quite closely a design by Mr. Andrew J. Thomas employed for a group of "garden apartments" constructed for Mr. John D. Rockefeller, Jr., in New York City. The suites are of four, five, six and seven rooms and, in the case of the large ones, two bathrooms. Light comes in three sides of a room as a rule and, in some cases, from four sides. All rooms enjoy cross-ventilation.

In the Rockefeller plan every apartment looks out upon a central garden, which is ornamented with a Japanese rookery and a footbridge over running water. The walks are to be lined with shrubbery and the general effect will be park-like and refreshing.

Similar treatments could be given to the various interior spaces of the unit layout. Here, however, due to the short and irregular streets and the odd positions of the buildings, the charm of a given court would be greatly extended because, in many cases, it would constitute a part of the view of not merely one, but several, apartments.

5 FIVE-BLOCK APARTMENT-HOUSE UNIT

Locality. The plan shown above (Fig. 4) is put forward as a suggestion of the type of treatment which might be given to central residential areas of high land values destined for rebuilding because of deterioration or the sweep of a real estate movement. The blocks chosen for the ground site are 200 feet wide and 670 feet long, a length that is found in several sections in Manhattan. In this plan, which borders a river, two streets are closed and two are carried through the development as covered roadways under terraced central courts.

Ground Plan. The dimensions of the plot between the boundary streets are 650 feet by 1,200 feet, and the total area is approximately 16 acres. The building lines are set back from the streets 30 feet on the northern and southern boundaries. Both of the end streets, which were originally 60 feet, have been widened to 80 feet, the two 20-foot extra strips being taken out of the area of the development. The western boundary has been enlarged from 80 to 100 feet. The area given to street widening and to building set-back amounts to 89,800 square feet, or 11,800 square feet more than the area of the two streets which were appropriated.

It will be observed that the plan of the buildings encloses 53 percent of the total area devoted to open space in the form of central courts. The main central court is about the size of Gramercy Park, Manhattan, with its surrounding streets. Since this area would receive an unusual amount of sunlight, it would be susceptible to the finest sort of landscape and formal garden treatment.

Both of the end courts are on a level 20 feet higher than the central space and cover the two streets that are carried through the development. Underneath these courts are the service areas for the buildings. At one end of the central space there is room for tennis courts and, at the other, a children's playground of nearly one acre. By reason of the large open spaces and the arrangement of the buildings, the plan achieves an unusual standard as to light in that there is no habitable room that has an exposure to sunlight of less then 45 degrees. The width of all the structures is 50 feet, so that apartments of two-room depth are possible throughout the building, while the western central rib, being 130 feet from a 100-foot street, will never have its light unduly shut off by buildings on the adjacent blocks.

Accommodations. The capacity of the buildings is about 1,000 families, with suites ranging from three to fourteen rooms in size, the majority of then suitable for family occupancy. In addition there would be room for a hotel for transients, an elementary school, an auditorium, a gymnasium, a swimming pool, handball courts, locker rooms and other athletic facilities. The first floors of certain buildings on one or more sides of the unit could be devoted to shops. The auditorium could be suitable for motion pictures, lectures, little theater performances, public meetings, and possible for public worship. Dances could be easily held in the gymnasium. In the basement there might be squash courts.

Height. The buildings range in height from two and three stories on the boundary streets to ten stories in the abutting ribs, fifteen stories in the main central ribs, and thirty-three stories in the two towers. Many of the roofs could be given a garden-like treatment and thus contribute to the array of delightful prospects which are offered by the scheme.

This plan, though much more compact than the three others, nevertheless observes all of the unit principles. Neither the community center nor the shopping districts are conspicuous, but they are present. Children can play, attend school, and visit stores without crossing traffic ways. ■

Five-Block Apartment House Unit

Five blocks and four cross streets	19.07 acres
Two cross streets taken	78,000 sq. ft.
Given to boundary streets	50,800 " "
Area of set-backs	39,000 " "
Land developed	16.4 acres
Covered by buildings	6.5 "
Coverage	40.0 per cent
Three central courts	5.3 acres

Le Corbusier

Summary

The 1973 publication carried the following summary introduction: "The Athens Charter was first published in Paris—clandestinely, its author anonymous, with an introduction by Jean Giraudoux—in 1943, at the height of the Nazi occupation. Not until the second edition was published in 1957 did the name Le Corbusier illuminate the Charter. This great modernist manifesto, a mile-stone in the history of urban planning, was the result of the fourth meeting of the CIAM (International Congress for Modern Architecture), held in 1933 abroad the steamship Patris II, which cruised from Athens to Marseilles and back. The ninety-five clauses of the Charter are, remarkably, as meaningful and pertinent today, in our own search for solutions to the untended ills of the city, as when they were so compellingly documented by Le Corbusier after that historic conference nearly fifty years ago."

Key words

cities, density, health, heritage planning, housing, industrial zones, open space, recreation, traffic, urbanism

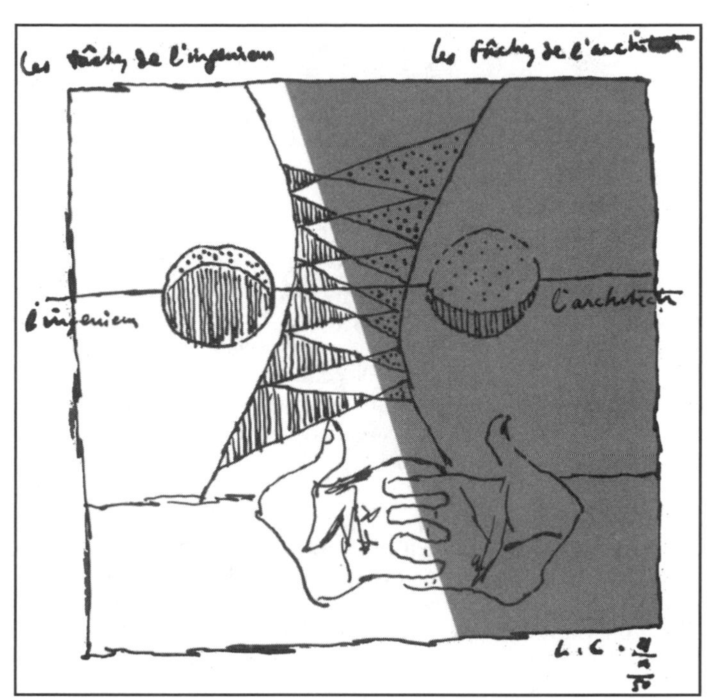

Le Corbusier's conception of the creative integration of design disciplines.

The Athens Charter

1 GENERALITIES

The city and its region

Observations

1. The city is only one element within an economic, social, and political complex that constitutes the region.

The political city unit rarely coincides with its geographical unit, that is to say, with its region. The laying out of the political territory of cities has been allowed to be arbitrary, either from the outset or later on, when, because of their growth, major agglomerations have met and then swallowed up other townships. Such artificial layouts stand in the way of good management for the new aggregation. Certain suburban townships have, in fact, been allowed to take on an unexpected and unforeseeable importance, either positive or negative, by becoming the seat of luxurious residences, or by giving place to heavy industrial centers, or by crowding the wretched working classes together. In such cases, the political boundaries that compartmentalize the urban complex become paralyzing. An urban agglomeration forms the vital nucleus of a geographical expanse whose boundary is determined only by the area of influence of another agglomeration. The conditions vital to its existence are determined by the paths of communication that secure its exchanges and closely connect it with its particular area. One can consider a problem of urbanism only by continually referring to the constituent elements of the region, and chiefly to its geography, which is destined to play a determining role in this question—the lines of watersheds and the neighboring crests that delineate natural contours and confirm paths of circulation naturally inscribed upon the earth. No undertaking may be considered if it is not in accord with the harmonious destiny of the region. The city plan is only one of the elements of this whole that constitutes the regional plan.

2. Juxtaposed with economic, social, and political values are values of a physiological and psychological origin which are bound up in the human person and which introduce concerns of both an individual and a collective order into the discussion. Life flourishes only to the extent of accord between the two contradictory principles that govern the human personality: the individual and the collective.

In isolation, man feels defenseless, and so, spontaneously, he attaches himself to a group. Left to his own devices, he would construct nothing more than his hut and, in that state of insecurity, would lead to a life of jeopardy and fatigue aggravated by all the anguish of solitude. Incorporated in a group, he feels the weight of the constraints imposed by inevitable social disciplines, but in return he is to some extent

Credit: The Athens Charter, translated from the French by Anthony Eardley, is from Le Corbusier, *The Athens Charter,* Grossman Publishers, New York, 1973, reprinted by permission of Penguin Putnam.

ensured against violence, illness, and hunger. He can think of improving his dwelling and he can also assuage his deep-seated need for social life. Once he has become a constituent element of a society that sustains him, he contributes, directly or indirectly, to the innumerable undertakings that provide security for his physical life and foster his spiritual life. His efforts become more fruitful and his more adequately protected liberty stops short only at the point where it would threaten the liberty of others. If there is wisdom in the undertakings of the group, the life of the individual is enlarged and ennobled by them. But if sloth, stupidity, and selfishness preponderate, the group—anemic and given over to disorder—brings its members nothing but rivalry, hatred, and disenchantment. A plan is well conceived when it allows fruitful cooperation while making maximum provision for individual liberty, for the effulgence of the individual within the framework of civic obligation.

3. These biological and psychological constants are subject to the influence of their environment—the geographical and topographical condition, the economic circumstances, the political situation. In the first place they are influenced by the geographical and topographical condition, the constitution of the elements, land and water, nature, soil, climate.

Geography and topography play a considerable role in the destiny of men. It must never be forgotten that the sun dominates all, imposing its law upon every undertaking whose object is to safeguard the human being. Plains, hills, and mountains likewise intermediate, to shape a sensibility and to give rise to a mentality. While the hillsman readily descends to the plain, the plainsman rarely climbs up the valleys or struggles over mountain passes. It is the crestlines of the mountain ranges that have delimited the "gathering zones" in which, little by little, men have gathered in clans and tribes, joined together by common customs and usages. The ratio of the elements of earth and water—whether it comes into play on the surface, contrasting the lake or river regions with the expanses of the steppes, or whether it is expressed as comparative rainfall, resulting in lush pasturelands here and heaths or deserts elsewhere—it also fashions mental attitudes which will be registered in mens' undertakings and which will find their expression in the house, in the village, and in the city. Depending on the angle at which the sun strikes the meridional curve, the seasons collide abruptly or succeed one another with imperceptible transitions; and although, in its continuous roundness, the Earth admits of no interruption from one parcel of land to the next, countless combinations emerge, each with its particular characteristics. Finally, the races of mankind, with their varied religions and philosophies, multiply the diversity of human undertakings, each proposing its own mode of perception and its own reason for being.

4. Secondly, these constants are affected by economic circumstances, by the resources of the region, by its natural and artificial contacts with the outside world.

Whether it be a circumstance of wealth or of poverty, the economic situation is one of the mainsprings of life, determining whether its movement will be in a progressive or a recessive direction. It plays the role of an engine which, depending on the power of its pulsations, brings prodigality, counsels prudence, or makes sobriety an imperative; it conditions the variations that delineate the history of the village, the city, and the country. The city that is surrounded by a region under cultivation is assured of its provisions. The city that has a precious substratum at its disposal becomes rich in substances that will serve it as exchange currency, especially if it is equipped with a traffic network ample enough to permit convenient contact with its near and distant neighbors. The degree of tension in the economic spring, though partly dependent on invariable circumstances, may be modified at any time by the advent of unexpected forces which chance or human initiative may render productive or leave inoperative. Neither latent wealth requiring exploitation nor individual energy has any absolute character. All is movement and, in the long run, economics is never anything but a momentary value.

5. Thirdly, the constants are affected by the political situation and the administrative system.

The political situation is a more unstable phenomenon than any other, the mark of a country's vitality, the expression of a wisdom that is approaching its apogee or is already on the slope of its decline. While politics is essentially unstable by nature, the administrative system, which is its outcome, possesses an inherent stability, which affords it a greater permanence over time and precludes too frequent modification. As a manifestation of changeable policy, its duration is assured by its own nature and by the very force of things. It is a system that, within somewhat rigid limits, administers the territory and the society consistently, imposes its ordinance upon them and, by bearing evenly on all the levers of control, determines uniform modes of action throughout the entire country. Yet, even in the merits of this economic and political framework have been confirmed by experience over a period of time, it can be shaken in a moment, whether in one of its parts or in the whole. Sometimes a scientific discovery is enough to upset the equilibrium, to reveal the discord between the administrative system of yesterday and the pressing realities of today. It may happen that communities, having managed to renovate their own particular framework, are crushed by the overall framework of the country—and this latter may, in turn, be immediately subject to the assault of major worldwide trends. There is no administrative framework that can lay claim to immutability.

6. Throughout history, specific circumstances have determined the characteristics of the city: military defense, scientific discoveries, successive administrations, and the progressive development of communications and means of transportation by land, water, rail, and air routes.

History is inscribed in the layouts and in the architecture of cities. Surviving layouts and architecture constitute a guideline which, together with written and graphic documents, enables us to recreate the successive images of the past. The motivations that gave birth to the cities were varied in nature. Sometimes it was a defensive asset—and a rocky summit or a loop of a river saw the growth of a fortified village. Sometimes it was the intersection of two roads, a bridgehead, or an indentation in the coastline that determined the location of the first settlement. The city had an uncertain form, most frequently that of a circle or semicircle. When it was a center of colonization, it was organized like a camp built on axes at right angles and girdled by rectilinear stockades. Everything was disposed according to proportion, hierarchy, and convenience. The highroads set out from the gates of the enclosure and threaded indirectly to distant points. One can still recognize in city plans the original close-set nucleus of the early market town, the successive enclosing walls, and the directions of divergent routes. People crowded together within the walls and, according to the degree of their civilization, enjoyed a variable proportion of well-being. In one place, deeply human codes dictated the choice of dispositions while, in another, arbitrary constraints gave rise to flagrant injustices. Then the age of machinism arose. To an age-old measure that one would have thought immutable—the speed of man's walking pace—was added to new measure, in the course of evolution—the speed of mechanized vehicles.

7. Hence the rationale governing the development of cities is subject to continual change.

The growth or decrease of a population, the prosperity or decline of the city, the bursting of fortified walls that become stifling enclosures, the new means of communication that extend the area of exchange, the beneficial or harmful effects of a policy of choice or submission, the advent of machinism, all of this is just movement. With the progression of time, certain values become unquestionably engrained in the heritage of a group, be they of a city, a country, or humanity in general; decay, however, must eventually come to every aggregation of buildings and roads. Death overtakes works as well as living beings. Who is to discriminate between what should remain standing and what must disappear? The spirit of the city has been formed over the years; the simplest buildings have taken on an eternal value insofar as they symbolize the collective soul; they are the armature of a tradition which, without meaning to limit the magnitude of future progress, conditions the formation of the individual just as climate, geographical region, race, and custom do. Because it is a "micro-cosmic motherland," the city admits of a considerable moral value to which it is indissolubly attached.

8. The advent of the machinist era has provoked immense disturbances in the conduct of men, in the patterns of their distribution over the earth's surface and in their undertakings: an unchecked trend, propelled by mechanized speeds, toward concentration in the cities, a precipitate and world-wide evolution without precedent in history. Chaos has entered the cities.

The use of the machine has completely disrupted the conditions of work. It has upset an ancient equilibrium, dealing a fatal blow to the craftsmen classes, emptying the fields, congesting the cities, and, by tossing century-old harmonies on the dung-hill, disturbing the natural relationships that used to exist between the home and places of work. A frenzied rhythm coupled with a discouraging precariousness disorganizes the conditions of life, impeding the mutual accord of fundamental needs. Dwellings give families poor shelter, corrupting their inner lives; and an ignorance of vital necessities, as much physical as moral, bears its poisoned fruits: illness, decay, revolt. The evil is universal, expressed in the cities by an overcrowding that drives them into disorder, and in the countryside by the abandonment of numerous agricultural regions.

2 THE PREVAILING CONDITION OF THE CITIES: CRITICAL EXAMINATION AND REMEDIAL MEASURES

Habitation

Observations

9. The population is too dense within the historic nuclei of cities, as it is in certain belts of nineteenth-century industrial expansion—reaching as many as four hundred and even six hundred inhabitants per acre.

Density—the ratio between the size of a population and the land area that it occupies—can be entirely changed by the height of buildings. But, until now, construction techniques have limited the height of buildings to about six stories. The admissible density for structures of this kind is from 100 to 200 inhabitants per acre. When this density increases, as it does in many districts, to 240, 320, or even 400 inhabitants, it then becomes a slum, which is characterized by the following symptoms:

- An inadequacy of habitable space per person;
- A mediocrity of openings to the outside;
- An absence of sunlight (because of northern orientation or as the result of shadow cast across the street or into the courtyard;
- Decay and a permanent breeding ground for deadly germs (tuberculosis);
- An absence or inadequacy of sanitary facilities;
- Promiscuity, arising from the interior layout of the dwelling, from the poor arrangement of the building, and from the presence of troublesome neighborhoods.

Constrained by their defensive enclosures, the nuclei of the old cities were generally filled with close-set structures and deprived of open space. But, in compensation, verdant space were directly accessible, just outside the city gates, making air of good quality available nearby. Over the course of the centuries, successive urban rings accumulated, replacing vegetation with stone and destroying the verdant areas—the lungs of the city. Under these conditions, high population densities indicate a permanent state of disease and discomfort.

10. In these congested urban sectors the housing conditions are disastrous, for lack of adequate space allocated to the dwelling, for lack of verdant areas in its vicinity and, ultimately, for lack of building maintenance (a form of exploitation based on speculation). This state of affairs is aggravated further by the presence of a population with a very low standard of living, incapable of taking defensive measures by itself (its mortality rate reaching as high as twenty percent).

The interior condition of a dwelling may constitute a slum, but its dilapidation is extended outside by the narrowness of dismal streets and the total absence of those verdant spaces, the generators of oxygen, which would be so favorable to the play of children. The cost of a structure erected centuries ago has long since been amortized; yet its owner is still tacitly allowed to consider it a marketable commodity, in the guise of housing. Even though its habitable value may be nil, it continues with impunity, and at the expense of the species, to produce substantial income. A butcher would be condemned for the sale of rotten meat, but the building codes allow rotten dwellings to be forced on the poor. For the enrichment of a few selfish people, we tolerate appalling mortality rates and diseases of every kind, which impose crushing burdens on the entire community.

11. The growth of the city gradually devours the surrounding verdant areas of which its successive belts once had a view. This ever-increasing remoteness from natural elements aggravates the disorder of public health all the more.

The more the city expands, the less the "conditions of nature" are respected within it. By "conditions of nature" we mean the presence, in sufficient proportions, of certain elements that are indispensable to living begins: sun, space, and verdure. An uncontrolled expansion has deprived the cities of these fundamental nourishments, which are of a psychological as well as physiological order. The individual who loses contact with nature is diminished as a result, and pays dearly, through illness and moral decay, for a rupture that weakens his body and ruins his sensibility, as it becomes corrupted by the illusory pleasures of the city. In this regard, all bounds have been exceeded in the course of these last hundred years, and this is not the least cause of the malaise with which the world is burdened at the present time.

12. Structures intended for habitation are spread out across the face of the city, at variance with the requirements of public health.

The first obligation of urbanism is to come into accord with the fundamental needs of men. The health of every person depends to a great extent of his submissions to the "conditions of nature." The sun, which governs all growth, should penetrate the interior of every dwelling, there to diffuse its rays, without which life withers and fades. The air, whose quality is assured by the presence of vegetation, should be pure and free from both inert dust particles and noxious gases. Lastly, space should be generously dispensed. Let us bear in mind that the sensation of space is of a psycho-physiological order, and that the narrowness of streets and the constriction of courtyards create an atmosphere as unhealthy for the body as it is depressing to the mind. The Fourth Congress of the CIAM, held in Athens, has proceeded from this postulate: sun, vegetation, and space are the three raw materials of urbanism. Adherence to this postulate will enable us to judge the existing condition of things and to appraise new propositions from a truly human point of view.

13. The most densely populated districts are located in the least favored zones (on badly oriented slopes, or in sectors invaded by fogs and industrial gases and vulnerable to floods).

No legislation has yet been effected to lay down the conditions for the modern habitation, not only to ensure the protection of the human person but also to provide him with the means for continual improvement. As a result, the land within the city, the residential districts, the dwellings themselves, are allocated from day to day at the discretion of the most unexpected—and at times the basest—interests. The municipal surveyor will not hesitate to lay out a street that will deprive thousands of dwellings of sunshine. Certain city officials will see fit, alas, to single out for the construction of a working-class district a zone hitherto disregarded because it is invaded by fog, because the dampness of the place is excessive, or because it swarms with mosquitoes. They will decide that some north-facing slope, which has never attracted anyone precisely because of its exposure, or that some stretch of ground reeking with soot, smoking coal slag, and the deleterious gases of some occasionally noisy industry, will always be good enough to house the uprooted, transient population known as unskilled labor.

14. Airy and comfortable structures (homes of the well-to-do) occupy the favored areas, sheltered from hostile winds, and are assured of pleasing views of the landscape—a lake, the sea, the mountains—and of abundant sunshine.

The favored areas are generally taken up by luxury residences, thus giving proof that man instinctively aspires, whenever his means allow it, to seek living conditions and a quality of well-being that are rooted in nature itself.

15. This biased allotment of habitation is sanctioned by custom and by the supposedly justified provisions of municipal administrations, namely, zoning resolutions.

Zoning is an operation carried out on the city map with the object of assigning every function and every individual to its rightful place. It is based on necessary differentiations between the various human activities, each of which requires its own specific space: residential quarters, industrial or commercial centers, halls or grounds intended to leisure hours. But while the force of circumstances differentiates the wealth residence from the modest dwelling, no one has the right to transgress rules that ought to be inviolable by allowing only the favored few to benefit from the conditions required for a health and well-ordered life. It is urgently necessary to modify certain practices. An implacable legislation is needed to ensure that a certain quality of well being is accessible to everyone, regardless of monetary considerations. It is necessary that precisely defined urban regulations forbid, once and for all, the practice of depriving entire households of light, air and space.

16. Structures built along transportation routes and around their intersections are detrimental to habitation because of noise, dust, and noxious gases.

Once we are willing to take this factor into consideration we will assign habitation and traffic to independent zones. From then on, the house will never again be fused to the street by a sidewalk. It will rise in its own surroundings, in which it will enjoy sunshine, clean air, and silence. Traffic will be separated by means of a network of footpaths for the slow-moving pedestrian and a network of fast roads for automobiles. Together these networks will fulfill their function, coming close to housing only as occasion demands.

17. The traditional alignment of habitations on the edges of streets ensures sunlight only for a minimum number of dwellings.

The traditional alignment of buildings along streets involves an inevitable arrangement of the built volume. When they intersect, parallel or oblique streets delineate square, rectangular, trapezoidal, and triangular areas of differing capacities which, once built up, from city "blocks." The need to admit light into the centers of these blocks gives birth to interior courtyards of varied dimensions. Unhappily, municipal regulations leave the profit-seekers free to confine these courts to utterly scandalous dimensions. And so we come to this dismal result: one façade out of four, whether it faces the street or the courtyard, is oriented to the north and never knows the sun, while the other three, owing to the narrowness of the streets and courts they face and to the resulting shadow, are half deprived of sunlight also. Analysis reveals that the proportion of city facades that get no sun varies from one-half to three-quarters of the total—and in certain cases, this ratio is even more disastrous.

18. Structures intended for collective use, as habitations, are arbitrarily distributed.

The dwelling shelters the family, a function that constitutes an entire program in itself and poses a problem whose solution—which in days gone by was sometimes a happy one—is nowadays most often left to chance. But outside the dwelling, and close to it, the family also requires the presence of collective institutions that could be considered actual extensions of the dwelling. These are: supply centers, medical services, infant nurseries, kindergartens, and schools, to which should be added the intellectual and athletic organizations that give adolescents an opportunity for the work or play suited to the particular aspirations of their age, and, to compete the "health-equipment," grounds and playing fields adapted to physical culture and daily sports activities for everyone. Although the benefit to be derived from these collective institutions is unquestionable, the masses are still badly in need of them. Their realization has barely been sketched out, and this in the most fragmentary manner, quite unrelated to overall housing needs.

19. Schools, particularly, are frequently situated on traffic routes and are too far away from housing.

Apart from any judgment as to their curricula and their architectural disposition, schools, as a general rule, are badly situated within the urban complex. Too far from the dwelling, they put the child in contact with the perils of the street. Moreover, they usually provide only instruction as such, so that the child under six and the adolescent over thirteen are consistently deprived of the pre-school and post-school organizations that would respond to the most imperative needs of those ages. The prevailing condition and distribution to the built domain is ill suited to the innovations that would not only shield childhood and youth from multifarious dangers but would also offer them the conditions that, alone, make a sound education possible, an education capable of guaranteeing them, in addition to instruction, a full physical as well as moral development.

20. The suburbs are laid out without any plan and without a normal connection to the city.

The suburbs are the degenerate progeny of the faubourgs, or "bastard boroughs." The borough was once a unit organized within a surrounding defensive wall. The faux bourg or false borough was backed against the wall from the outside and built out along an approach road deprived of protection. This was the outlet for the excess population, and the people had to accommodate themselves to its insecurity whether they liked it or not. At the time when the creation of a new fortified wall eventually embraced the false borough and its stretch of road within the bosom of the city, the first violence was done to the normal rules governing city layouts. The age of machinism is characterized by the suburb, a stretch of ground with no particular plan where all the dregs of society are dumped, where all the risky ventures are tried out, where the most modest working classes often live next to industries that are assumed, a priori, to be temporary—though some of them will experience enormous growth. The suburb is the symbol for waste, and, at the same time, for the risky venture. It is a kind of scum churning against the walls of the city. In the course of the nineteenth and the twentieth centuries this scum has become a flood tide, then an inundation. It has seriously compromised the destiny of the city and its possibilities of growth according to rule. The abode of an unsettled population enmeshed in numerous afflictions, the suburb is a culture medium of revolt, and is often ten times, even a hundred times, larger than the city. There are those who seek to turn these disordered suburbs, in which the time-distance function poses an ominous and unanswerable question, into garden-cities. Theirs is an illusory paradise, an irrational solution. The suburb is an urbanistic folly, scattered across the entire globe and carried to its extreme consequences in America. It constitutes one of the greatest evils of the century.

21. Attempts have been made to incorporate the suburbs into the administrative system.

Too late! The suburb has been belatedly incorporated into the administrative system. Throughout the entire area of the suburb an improvident code has allowed property rights to become established, and it declares them inviolable. The owner of a piece of vacant ground on which some shack, shed, or workshop has sprouted up, cannot be expropriated without multiple difficulties. The population density is very low and the ground is barely exploited; nevertheless, the city is obliged to furnish the suburban expanse with the necessary utilities and services: i.e., roads, utility mains, means of rapid communication, policing, street lighting and cleaning, hospital and school facilities, and the rest. The ruinous expense caused by so many obligations is shockingly disproportionate to the few taxes that such a scattered population can produce. On the day the Administration intervenes to redress the situation, it comes up against insurmountable obstacles and ruins itself in vain. To ensure the city the means for a harmonious development, the Administration must take responsibility for the management of the land surrounding the city before the suburbs spring up.

22. The suburbs are often mere aggregations of shacks hardly worth the trouble of maintaining.

Flimsily constructed little houses, boarded hovels, sheds thrown together out of the most incongruous materials, the domain of poor creatures tossed about in an undisciplined way of life—that is the suburb! Its bleak ugliness is a reproach to the city it surrounds. Its poverty, which necessitates the squandering of public funds without the compensation of adequate tax resources, is a crushing burden for the community. It is the squalid antechamber of the city; clinging to the major approach roads with its side streets and alleys, it endangers the traffic on them; seen from the air, it reveals the disorder and incoherence of its distribution to the least experienced eye; for the railroad traveler, excited by the thought of the city, it is a painful disillusion!

Requirements

23. Henceforth, residential districts must occupy the best locations within the urban space, using the topography to advantage, taking the climate into account, and having the best exposure to sunshine with accessible verdant areas at their disposal.

The cities, as they exist today, are built under conditions injurious to the public and private good. History shows that their founding and their development have resulted from a succession of deep-seated causes, and that not only have the cities been expanded, but they have also been often renewed over the centuries, and on the same site. By abruptly changing certain century-old conditions, the age of the machine has reduced the cities to chaos. Our task at this point is to extricate them from their disorder by means of plans that will provide for the staging of undertakings over a period of time. The problem of the dwelling, of habitation, takes precedence over all others. The best locations in the city must be reserved for it; and if they have been pillaged by greed or indifference, every effort must be made to recover them. Several factors contribute to the well being of the dwelling. We must seek simultaneously the finest views, the most healthful air (taking account of winds and fogs), the most favorably exposed slopes, and, finally, we must make use of existing verdant areas, create them if there are none, or restore them if they have been ruined.

24. The selection of residential zones must be dictated by considerations of public health.

The universally acknowledged laws of hygiene bring a grave indictment against the sanitary conditions of cities. It is not enough to formulate a diagnosis or even to discover a solution; the solution must be prescribed by the responsible authorities. In the name of public health, entire districts should be condemned. Some of them, the result of hasty speculation, merit only the pickaxe. Others should be spared in part, for the sake of their historical associations or for the elements of artistic value that they contain. There are ways of saving whatever deserves to be saved while relentlessly destroying whatever constitutes a hazard. But it is not enough to make the dwelling healthier; its outside extensions—places for physical education buildings and various playing fields—must be created and planned for by incorporating the areas that will be set aside for them into the overall plan ahead of time.

25. Reasonable population densities must be imposed, according to the forms of habitation suggested by the nature of the terrain itself.

The population densities of a city must be laid down by the authorities. They may vary according to the allocation of urban land to housing and may produce, depending on the total figure, a widespread or a compact city. To determine the urban densities is to perform an administrative act heavy with consequences. With the advent of the machine age, the cities expanded without control and without constraint. Negligence is the only valid explanation for that inordinate and utterly irrational growth, which is one cause of their troubles today. There are specific reasons for the birth of the cities and for their growth, and these must be carefully studied in terms of forecasts extending over a period of time: fifty years, let us say. A population figure can then be envisaged. It will be necessary to house this population, which involves anticipating which space will be used, foreseeing what "time-distance" function will be its daily lot, and determining the surface and area needed to carry out this fifty-year program. Once the population figure and the dimensions of the land are fixed, the "density" is determined.

26. A minimum number of hours of exposure to the sun must be determined for each dwelling.

Science, in it studies of solar radiations, has disclosed those that are indispensable to human health and also those that, in certain cases, could be harmful to it. The sun is the master of life. Medicine has shown that tuberculosis establishes itself whenever the sun fails to penetrate; it demands that the individual be returned, as much as possible, to "the conditions of nature." The sun must penetrate every dwelling several hours a day even during the season when sunlight is most scarce. Society will no longer tolerate a situation were entire families are cut off from the sun and thus doomed to declining health. Any housing design in which even a single dwelling is exclusively oriented to the north, or is deprived of the sun because it is cast in shadow, will be harshly condemned. Builders must be required to submit a diagram showing that the sun will penetrate each dwelling for a minimum of two hours on the day of the winter solstice, failing which, the building permit will be denied. To introduce the sun is the new and most imperative duty of the architect.

27. The alignment of dwellings along transportation routes must be prohibited.

The transportation routes, that is to say, the streets of our cities, have disparate purposes. They accommodate the most dissimilar traffic loads and must lend themselves to the walking pace of pedestrians as well as to the driving and intermittent stopping of rapid public transport vehicles, such as buses and tramcars, and to the even greater speeds of trucks and private automobiles. The sidewalks were created to avoid traffic accidents in the days of the horse, and only then after the introduction of the carriage; today they are absurdly ineffectual now that mechanized speeds have introduced a real menace of death into the streets. The present-day city opens its countless front doors onto this menace and its countless windows onto the noise, dust, and noxious gases produced by the heavy mechanized traffic flow. This state of things demands radical change: the speed of the pedestrian, some three miles an hour, and the mechanized speeds of thirty to sixty miles an hour must be separated. Habitation will be removed from mechanized speeds, which will be channeled into a separate roadbed, while the pedestrian will have paths and promenades reserved for him.

28. The resources offered by modern techniques for the erection of high structures must be taken into account.

Every age has used the construction technique imposed on it by its own particular resources. Until the nineteenth century, the art of building houses knew only bearing walls of stone, brick, or timber framing and floors made of wooden beams. In the nineteenth century, a transitional period made use of iron sections; and then, finally, in the twentieth century came homogeneous structures made entirely of steel or reinforced concrete. Before this completely revolutionary innovation in the history of building construction, builders were unable to erect premises exceeding six stories. The times are no longer so limited. Structures now reach sixty-five stories or more. What still must be resolved, through a serious examination of urban problems, is the most suitable building height for each particular case. As to the housing, the arguments postulated in favor of a certain decision are: the choice of the most agreeable view, the search for the purest air and the most complete exposure to sunshine, and finally, the possibility of establishing communal facilities—school buildings, welfare centers, and playing fields—within the immediate proximity of the dwelling, to form its extensions. Only structures of a certain height can satisfactorily meet these legitimate requirements.

29. High buildings, set far apart from one another, must free the ground for broad verdant areas.

Indeed, they will have to be situated at sufficiently great distances from one another, or else their height, far from being an improvement of the existing malaise, will actually worsen it; that is the grave error perpetrated in the cities of the two Americas. The construction of a city cannot be abandoned, without a program, to private initiative. Its population density must be great enough to justify the installation of the communal facilities that will form the extensions of the dwelling. Once this density has been determined, a presumable population figure will be adopted, permitting the calculation of the area to be reserved for the city. To determine the manner in which the ground is to be occupied, to establish the ratio of the built-up area to that left open or planted, to allocate the necessary land to private dwellings and to their various extensions, to fix an area for the city that will not be exceeded for a specified period of time—these constitute that important operation, which lies in the hands of the city authority: the promulgation of a "land ordinance." Thus, the city will henceforth be built in complete security and, within the limits of the rules prescribed by this statute, full scope will be given to private initiative and to the imagination of the artist.

Leisure

Observations

30. Open spaces are generally inadequate.

In certain cities, open spaces still exist. They are, for our day, the miraculously surviving remnants of reserves established during the course of history: parks surrounding princely mansions, gardens adjoining private houses, shaded promenades occupying the sites of a demolished fortification system. The last two centuries have greedily cut into these reserves—the authentic lungs of the city—covering them with buildings and setting masonry in place of grass and trees. At one time, open spaces had no other reason for existence than the pleasure and amusement of a privileged few. The social point of view, which today gives new meaning to the use of these spaces, had not yet emerged. Such areas may be the direct or the indirect extensions of the dwelling: direct if they surround the habitation itself, indirect if they are concentrated in a few large areas a little farther away. In either case, their assigned purpose will be the same, namely, to meet the collective activities of youth and to provide a favorable site for diversions, strolls, and games during leisure hours.

31. Even when open spaces are of an adequate size, they are often poorly located and are therefore not readily accessible to a great number of inhabitants.

When modern cities include a few sufficiently extensive open spaces, they are situated either on the city outskirts or in the midst of a particularly luxurious residential area. In the first instance, being remote from the working-class districts, they serve city-dwellers on Sundays alone and have no effect on their daily lives, which will continue to take place in trying conditions. In the second instance, they will actually be forbidden ground for the masses, and their function will consequently be reduced to that of embellishing the city without fulfilling their role as useful extensions of the dwelling. In either case, the severe problem of public health remains unimproved.

32. The remoteness of the outlying open spaces does not lend itself to better living conditions in the congested inner zones of the city.

Urbanism is called upon to devise the rules required to assure city-dwellers of living conditions that will safeguard not only their physical health but also their moral health and the joy of life that results from these. The working hours, often exhausting for the muscles or the nerves, should be followed every day by an adequate amount of free time. These hours of freedom, which machinism will unfailingly increase, will be devoted to a refreshing existence amidst natural elements. The maintenance and the establishment of open spaces are, therefore, a necessity, a matter of public welfare. This theme forms an integral part of the fundamentals of urbanism, and the city administrators should be compelled to give it their fullest attention. A just proportion of constructed volumes to open space—that is the only formula which resolves the problem of habitation.

33. In order to be accessible to their users, the few athletic facilities that are provided have generally been fitted out on a temporary basis, on sites destined for future housing or industrial districts. The result is precariousness and incessant upheaval.

A small number of athletic associations, eager to make use of their weekly leisure time, have found temporary shelter on the outskirts of cities but, since their existence is not officially recognized, it is, as a rule, extremely precarious. The hours of free or leisure time may be placed in three categories: daily, weekly, and yearly. The daily hours of free time should be spent close to the dwelling. The weekly hours of free time allow excursions out of the city and its vicinity. The yearly period of free time, that is to say, vacations and holidays, permit real travel, away from both the city and its region. Thus stated, the problems imply the creation of verdant reserves: 1) around the dwelling; 2) within the region; 3) throughout the country.

34. The sites that could be set aside for weekly leisure activities are often poorly connected to the city.

Once the sites close to the city that would make suitable centers for weekly leisure activities have been selected, the problem of mass transportation must be faced. This problem should be borne in mind from the moment the regional plan is first sketched out; it involves the investigation of various possible means of travel: roads, railroads, or rivers.

Requirements

35. Hereafter, every residential district must include the green area necessary for the rational disposition of games and athletic sports for children, adolescents, and adults.

This decision will have no effect unless it is supported by a genuine act of legislation: the "land ordinance." This ordinance will possess a diversity that corresponds to the needs to be satisfied. The population density, for instance, or the percentage of open area and built-up area may be varied, depending on functions, locales, and climates. Built volumes will be intimately blended with the green areas surrounding them. The built-up areas and the planted areas will be distributed on the basis of a reasonable amount of time needed to go from one to the other. In any event, the urban fabric will have to change its texture; the urban population centers will tend to become green cities. Contrary to what takes place in the "garden cities," the verdant areas will not be divided into small unit lots for private use but, instead, dedicated to the launching of the various communal activities that form the extensions of the dwelling. Kitchen gardening, the usefulness of which is actually the principal argument in favor of the garden cities, might very well be considered here: a percentage of the available ground will be allocated to it and divided into multiple individual plots, but certain collective gardening arrangements, such as tilling, irrigating, and watering, can lighten the labor and increase the yield.

36. Unsanitary blocks of houses must be demolished and replaced by green areas: the adjacent housing quarters will thus become more sanitary.

An elementary knowledge of the principal notions of health and sanitation is sufficient to detect a slum building and to discriminate a clearly unsanitary city block. These blocks must be demolished, and this should be an opportunity to replace them with parks which, at least in regard to the adjacent housing quarters, will be the first step toward improved health conditions. Some of these blocks, however, may happen to occupy sites particularly suitable for the construction of certain buildings indispensable to the life of the city.

In that event, intelligent urbanism will be able to assign to them the purpose that the overall regional plan and the city plan will have envisaged, in advance, as the most efficacious.

37. The new green areas must serve clearly defined purposes, namely, to contain the kindergardens, schools, youth centers, and all other buildings for community use, closely linked to housing.

The green areas that will have been intimately amalgamated with built volumes and integrated into the residential sectors will not have the embellishment of the city as their sole function. First, they must play a useful role, and it is facilities of a communal nature that will occupy their lawns: day nurseries, pre-school and post-school organizations, youth clubs, centers for intellectual relaxation or physical culture, reading rooms or game rooms, running tracks and outdoor swimming pools. These will be the extensions of the dwelling, and must be subject, like the dwelling itself, to the "land ordinance."

38. The weekly hours of free time should be passed in favorable prepared places: parks, forests, playing fields, stadiums, and beaches.

Nothing, or virtually nothing, has yet been provided for the weekly hours of leisure time. Vast spaces in the region surrounding the city will be reserved and equipped, and made accessible by sufficiently numerous and convenient modes of transportation. These spaces are no longer a matter of lawns around the house, more or less densely planted with trees, but of actual forests and meadows, natural and artificial beaches, which will constitute a vast and carefully tended preserve offering the city dweller numerous opportunities for health activity and beneficial relaxation. There are places on the outskirts of every city which are capable of fulfilling this program and which can become readily accessible provided there is a well-considered organization of the means of communication.

39. Parks, playing fields, stadiums, and beaches.

A program that will comprise every kind of relaxation must be decided upon: walking or hiking, alone or in groups, through the beauty of the landscape; every kind of sport—tennis, basketball, soccer, swimming, athletic exercises; staged entertainment—concerts, open-air theaters, and the various spectator sports and tournaments. Finally, specific facilities will have to be undertaken beforehand: the means of circulation, which will require rational organization; lodging-places—hotels, inns, and camping grounds; and one last, but not least important, provision—a supply of drinking water and food whose availability must be absolutely assured in all these places.

40. An assessment must be made of the available natural elements: rivers, forests, hills, mountains, valleys, lakes, and the sea.

Owing to the improvements in mechanized means of transportation, the question of distance is no longer a determining factor in this context. It is better to select appropriate natural elements, even though it may necessitate seeking them somewhat far afield. It is a matter not only of conserving the natural beauties that are still untouched but also of repairing the damage that certain of them may have suffered; in short, human industry will be called on to create in part the sites and landscapes to answer the program. This is another and very considerable social problem for which municipal officials are responsible—it is the problem of finding a counterpart to the exhausting labors of the week, of making the day of rest truly invigorating for physical and moral health, of never forsaking the population to the many disgraces of the streets. Putting the hours of leisure to fertile use will forge health and spirit in the inhabitants of cities.

Work

Observations

41. The places of work—factories, craft workshops, business and public administration offices, and commercial premises—are no longer rationally located within the urban complex.

In the past, the dwelling and the workshop, being linked together by close and permanent ties, were situated near one another. The unforeseen expansion of machinism has disrupted those harmonious conditions; in less than a century it has transformed the character of cities, shattered the age-old traditions of the craftsman classes and given birth to a new, anonymous labor force, which drifts from place to place. The rise of industry depends essentially on the means by which raw materials are supplied and on the facilities through which manufactured products are distributed. So industries rushed headlong to establish themselves along the railroad tracks introduced in the nineteenth century and on the banks of waterways, whose capacity was increased by steam navigation. But the founders of industry, by taking advantage of the immediate supply of food and lodging available in the cities, established their companies within the city or on its edges, heedless of the misfortune that might come of it. Implanted in the heart of residential districts, the factories fill them with noise and air pollution. Set on the outskirts and far removed from these residential districts, they force the workers to travel long distances every day in the tiring hustle and bustle of rush hour, needlessly causing them to lose part of their leisure time. This disruption of the former means of organizing work has led to an indescribable disorder and raised a problem that has, as yet, received only haphazard solutions. From this emerges the great ill of our time: the nomadism of the working population.

42. The connection between habitation and places of work is no longer normal; it necessitates the covering of inordinate distances.

From now on, normal relationships between those two essential functions of life—inhabiting and working—are disrupted. The faubourgs are full of workshops and mills, while the major industries, which continue to experience unlimited growth, are forced out into the suburbs. Since the city has reached its saturation point and is unable to accommodate any more inhabitants, suburban cities have been hastily thrown together, vast and densely packed batches of uncomfortable little rental apartments or needless housing developments. Morning, noon, and night, in summer and winter, the interchangeable labor force, which has no stable bond attaching it to industry, goes through its perpetual shifting in the depressing jostle of public transportation. Whole hours melt away in these disorganized displacements.

43. The rush hours betray a critical state of affairs.

The public transportation services—suburban trains, buses, and subways—are in full operation only four times a day. There is frenzied commotion in the rush hours, and the users pay dearly from their own pockets for an arrangement that adds hours of jostling and scurrying to the stresses of the workday itself. Running these transportation systems is a painstaking and costly business; the amount that the passengers pay is not enough to cover their operating expenses, so they have become a heavy public burden. To remedy such a state of affairs, two conflicting propositions have been upheld: to support the transportation, or to support the users of transportation? The choice must be made! The one implies an enlargement and the other a reduction in the diameter of the cities.

44. For lack of any program, there is unchecked expansion of cities, absence of forethought, speculation on land and other things, and industry, complying with no rule, establishes itself at random.

The land within the cities and in the surrounding areas is almost entirely privately owned. Industry itself is in the hands of private companies whose situation is sometimes unstable, being prone to all kinds of crises. Nothing has been done to subject industrial expansion to logical regulation; on the contrary, everything has been left to improvisation, which may occasionally favor the individual but which always over-burdens the collective.

45. Offices in the city are concentrated in business districts. Located on the best sites in town and provided with the most complete circulation systems, these business districts quickly fall prey to speculation. Since they are private undertakings, the organization necessary to their natural development is lacking.

The corollary of industrial expansion is the growth of business, private administration, and trade. Nothing in this area has been seriously estimated and planned. One must buy and sell, bring workshop and factory into contact with supplier and customer. These transactions require offices. Offices are premises requiring specific and critical facilities, indispensable to the efficient conduct of business. When they are isolated in separate offices, such facilities are costly. Everything points to a grouping of offices so that each is assured of optimum working conditions: ease of movement within, ready communication with the outside world, light, peace and quiet, high-quality air, heating and cooling systems, post office and telephone exchange, and radio.

Requirements

46. The distances between places of work and places of residence must be reduced to a minimum.

This implies a new distribution of all the places that are given over to work, in accordance with a carefully elaborated plan. The concentration of industries in belts around the large cities may have been a course of prosperity for certain firms, but the deplorable living conditions that have resulted for the masses must be denounced. This arbitrary disposition has given rise to an intolerable promiscuity. The time consumed in going back and forth between home and work bears no relationship to the daily course of the sun. Industries must be transplanted to the passageways for raw materials, along major waterways, highways, and railroads. A passageway is a linear element. Hence, instead of being concentric, the industrial cities will become linear.

47. The industrial areas must be independent of the residential areas, and separated from one another by a zone of vegetation.

The industrial city will extend along the canal, the highway, or the railroad, or, better yet, along all three traffic ways together. Once it has become linear instead of annular, the city will be able to align its own parallel band of habitation as it develops. A verdant zone will separate this band from the industrial buildings. The dwelling, which will thereafter stand in the open countryside, will be completely protected from noise and pollution and yet will be close enough to eliminate the long daily journeys to-and-fro; it will once again become a normal family organism. Thus recovered, the "conditions of nature" will help put a stop to the nomadism of the working population. The inhabitants will be able to choose from three available types of habitation: the individual house of the garden city; the individual house coupled with a small farm; and lastly, the collective apartment building furnished with all the service necessary to the well-being of its occupants.

48. The industrial zones must be contiguous to the railroad, the canal, and the highway.

The entirely new speed of mechanized transportation, whether utilizing road, rail, river, or canal, necessitates the creation of new traffic routes and the transformation of existing ones. This calls for a program of coordination that must take account of the new distribution of industrial establishments and the workers' dwellings that accompany them.

49. The craft occupations, closely bound up with the urban life from which they directly arise, must be able to occupy clearly designated places within the city.

The handicrafts differ from industry by their very nature and call for appropriate dispositions. They emanate directly from the cumulative potential of the urban centers. The crafts of bookmaking, jewelry, dressmaking, and fashion find the creative stimulus they need in the intellectual concentration of the city. They are essentially urban activities, whose work premises can be situated in the most intensely active points in the city.

50. The business city, devoted to public and private administration, must be assured of good communications with the residential quarters, as well as with industry or craft workshops remaining within or near the city.

Business has taken on so great an importance that the selection of the urban location to be reserved for it requires very special study. The business center must be located at the confluence of the traffic channels that serve the various sectors of the city: habitation, industry and craft workshops, public administration, certain hotels, and the different termini (railroad stations, bus stations, ports, and airports).

Traffic

Observations

51. The present network of urban streets is a set of ramifications that grew out of the major traffic arteries. In Europe, these arteries go back in time far beyond the Middle Ages, and sometimes even beyond antiquity.

Certain cities built for purposes of defense or colonization have had the benefit, since their origin, or a concerted plan. To begin with, a regularly formed fortification wall was laid down, against which the high roads came to a halt. The interior of the city was arranged with useful regularity. Other cities, greater in number, were born at the intersection of two cross-country high roads or, in some cases, at the junction of several roads radiating outward from a common center. These transportation arteries were closely linked to the topography of the region, which often forced them to follow a winding course. The first houses were established along their edges, and this was the origin of the principal thoroughfares, from which, as the city grew, an increasing number of secondary arteries branched out. The principal thoroughfares have always been the offspring of geography, and while many of them may have been straightened and rectified, they will nonetheless always retain their fundamental determinism.

52. The main transportation routes, originally conceived in terms of pedestrian and wagon traffic, no longer meet the requirements of today's mechanized means of transportation.

For reasons of security, ancient cities were surrounded by walls. Consequently, they were unable to expand as their population increased. If was necessary to practice economy in order to obtain the maximum habitable areas from the land. This accounts for the system of close-set streets and alleys that afforded access to the greatest possible number of front doors. Another consequence of this organization of cities was the system of city blocks, built perpendicularly above the street from which they took daylight and perforated with interior courtyards built for the same purpose. Later, when the fortified walls were expanded, the streets and alleys were extended beyond the initial nucleus as avenues and boulevards, while the nucleus itself retained its original structure. This system of building, which has long ceased to correspond to any need, still exists. Its facades give onto more or less narrow streets and interior courtyards. The traffic network that encloses it has multiple dimensions and intersections. Intended for other times, this network has not been able to adapt to the new speeds of mechanized vehicles.

53. The dimensioning of streets, ill adapted to the future, impedes the utilization of the new mechanized speeds and the orderly progress of the city.

The problem arises out of the impossibility of reconciling natural speeds, the pace of man or horse, with the mechanized speeds of cars, tramcars, trucks, and buses. The mixture of both is the source of countless conflicts. The pedestrian moves about in perpetual insecurity, while mechanized vehicles are obliged to brake incessantly and so are incapacitated—which does not, however, prevent them from being a continual source of mortal danger.

54. The distances between street intersections are too short.

Before reaching their normal cruising speed, mechanized vehicles have to start up and gradually accelerate. Sudden braking can only cause rapid wear and tear on major parts. A reasonable unit of length between the starting-up point and the point at which it becomes necessary to brake must therefore be gauged. Street intersections today, which occur at intervals of 100, 50, 20, and even 10 yards, are not suited to the proper operation of mechanized vehicles. They should be separated by intervals of from 200 to 400 yards.

55. The width of the streets is inadequate. Attempts to widen them are often very costly and ineffectual operations.

There is no uniform standard for street widths. It all depends on the number and type of vehicles they accommodate. The old thoroughfares, which were laid down by topography and geography from the very beginnings of the city and which form the trunks for an endless ramification of streets, have almost always maintained a heavy traffic flow. They are generally too narrow, but widening them is not always an easy or even an adequate solution. It is essential that the problem be investigated much more thoroughly.

56. Confronted with mechanized speeds, the street network seems irrational, lacking in precision, in adaptability, in diversity, and in conformity.

Modern traffic circulation is a highly complex operation. Traffic channels intended for multiple use must simultaneously permit automobiles to drive from door to door, pedestrians to walk from door to door, buses and tramcars to cover prescribed routes, trucks to go from supply centers to an infinite variety of distribution points, and certain vehicles to pass directly through the city. Each one of these activities requires a specific lane, geared to meet clearly distinguished requirements. Thus it is necessary to engage in a detailed study of the question, to consider its present state, and to seek solutions that really correspond to precisely defined needs.

57. Magnificent layouts, intended for show, may once have constituted awkward obstacles to traffic flow, and they still do.

What was admissible and even admirable in the days of horse-drawn carriages may now have become a source of constant disturbance. Certain avenues, which were conceived to ensure a monumental perspective crowned by a memorial or a public edifice, are a present cause of bottlenecks, of delays, and sometimes of danger. Such architectural compositions must be preserved from the invasion of mechanized vehicles, which they were not designed to accommodate and to whose speeds they can never be adapted. Traffic has now become a function of primary importance to urban life. It requires a carefully prepared program capable of providing whatever is needed to regulate its flow and to establish its indispensable outlets, thus doing away with traffic jams and the constant disturbance of which they are the cause.

58. In many cases, when the time comes for the expansion of the city, the track network of the railroad system proves a serious obstacle to urbanization. It hems in residential areas, depriving them of necessary contacts with the vital elements of the city.

Here again, time has flown too swiftly. The railroads were built before the prodigious industrial expansion that they themselves caused. By penetrating the cities, they arbitrarily cut off entire areas. The railroad track is a road one does not cross; it isolates certain areas from others which, having been gradually covered with dwellings, have found themselves deprived of contacts that are indispensable to them. In some cities, this situation has serious effects on the general economy, and urbanism is called upon to consider the modification and realignment of certain railroad systems in such a way as to draw them back into the harmony of an overall plan.

Requirements

59. The whole of city and regional traffic circulation must be closely analyzed on the basis of accurate statistics—an exercise that will reveal the traffic channels and their flow capacities.

Traffic circulation is a vital function whose present state must be expressed by graphic methods. The determining causes and the effects of its different intensities will then become clearly apparent, and it will be easier to detect its critical points. Only a clear view of the situation will permit the accomplishment of two indispensable improvements; namely, the assignment of a specific purpose to each traffic channel—to accommodate either pedestrians or automobiles, either heavy trucks or through traffic—and then the provision of each such channel with particular dimensions and features according to the role assigned it—the type of roadway, the width of the road surface, the locations and kinds of intersections and junctions.

60. Traffic channels must be classified according to type and constructed in terms of the vehicles and speeds they are intended to accommodate.

The single street, bequeathed by centuries past, once accepted both men on foot and men on horseback indiscriminately, and it was not until the end of the eighteenth century that the generalized use of carriages gave rise to the creation of sidewalks. In the twentieth century came the cataclysmic hordes of mechanical vehicles—bicycles, motorcycles, cars, trucks, and tramcars—traveling at unforeseen speeds. The overwhelming growth of certain cities, such as New York, for example, brought about an inconceivable crush of vehicles at certain specific points. It is high time that suitable measures were taken to remedy a situation that verges on disaster. The first effective measure in dealing with the congested arteries would be a radical separation of pedestrians from mechanized vehicles. The second would be to provide heavy trucks with a separate traffic channel. And the third would be to envisage throughways for heavy traffic that would be independent of the common roads intended only for light traffic.

61. Traffic at high-density intersections will be dispersed in an uninterrupted flow by means of changes of level.

Through vehicles should not be slowed down needlessly by having to stop at every intersection. Changes of level at each crossroad are the best means to assure them of uninterrupted motion. Laid out at distances calculated to obtain optimum efficiency, junctions will branch off the major throughways connecting them to the roads intended for local traffic.

62. The pedestrian must be able to follow other paths than the automobile.

This would constitute a fundamental reform in the pattern of city traffic. None would be more judicious, and none would open a fresher or more fertile era in urbanism. This requirement regarding the pattern of traffic movement may be considered just as strict as that which, in the area of habitation, condemns the northern orientation of any dwelling.

63. Roads must be differentiated according to their purposes: residential roads, promenades, throughways, and principal thoroughfares.

Instead of being given up to everyone and everything, roads must be governed by different rules, according to their category. Residential roads and the ground intended for collective uses require a particular atmosphere. So that dwellings and their "extensions" may enjoy the peace and calm that they need, mechanized vehicles will be channeled through special circuits. The avenues containing through traffic will have no contact with the local roads except at specified connecting points. The great principal thoroughfares, which are linked to the whole of the region, will naturally assert their predominance in the network. And promenades will also be envisaged where a reduced speed will be strictly imposed upon every type of vehicle so that pedestrians will at last be able to mingle with them without danger.

64. As a rule, verdant zones must isolate the major traffic channels.

Since the throughways or major roads will be quite distinct from the local roads, there will be no reason for them to come near either public or private structures. It would be advantageous to line them with dense screens of foliage.

The historic heritage of cities

65. Architectural assets must be protected, whether found in isolated buildings or in urban aggregations.

The life of a city is a continuous event that is expressed through the centuries by material works—lay-outs and building structures—which form the city's personality, and from which its soul gradually emanates. They are precious witnesses of the past which will be respected, first for their historical or sentimental value, and second, because certain of them convey a plastic virtue in which the utmost intensity of human genius has been incorporated. They form a part of the human heritage, and whoever owns them or it entrusted with their protection has the responsibility and the obligation to do whatever he legitimately can to hand this noble heritage down intact to the centuries to come.

66. They will be protected if they are the expression of a former culture and if they respond to a universal interest.

Death, which spares no living creature, also overtakes the works of men. In dealing with material evidence of the past, one must know how to recognize and differentiate that which is still truly alive. The whole of the past is not, by definition, entitled to last forever; it is advisable to choose wisely that which must be respected. If the continuance of certain significant and majestic presences from a bygone era proves injurious to the interests of the city, a solution capable of reconciling both points of view will be sought. In the case where one is confronted with structures repeated in numerous examples, some will be preserved as documents and the others will be demolished; in other cases, only the portion that constitutes a memorial or a real asset can be separated from the rest, which will be serviceably modified. Finally, in certain exceptional cases, complete transplantation may be envisaged for elements that prove to be inconveniently located but that are worth preservation for their important aesthetic or historical significance.

67. . . .and if their preservation does not entail the sacrifice of keeping people in unhealthy conditions.

By no means can any narrow-minded cult of the past bring about a disregard for the rules of social justice. Certain people, more concerned for aestheticism than social solidarity, militate for the preservation of certain picturesque old districts unmindful of the poverty, promiscuity, and diseases that these districts harbor. They assume a grave responsibility. The problem must be studied, and occasionally it may be solved through some ingenious solution; but under no circumstances should the cult of the picturesque and the historical take precedence over the healthfulness of the dwelling, upon which the well being and the moral health of the individual so closely depend.

68. . . .and if it is possible to remedy their detrimental presence by means of radical measures, such as detouring vital elements of the traffic system or even displacing centers hitherto regarded as immutable.

The exceptional growth of a city can create a perilous situation, leading to an impasse from which there is no escape without some measure of sacrifice. An obstacle can only be removed by demolition. But whenever this measure is attended by the destruction of genuine architectural, historical, or spiritual assets, then it is unquestionably better to seek another solution. Rather than removing the obstacle to traffic flow, the traffic itself can be diverted or, conditions permitting, its passage can be forced by tunneling beneath the obstacle. Finally, it is also possible to displace a center of intense activity and, by transplanting it elsewhere, entirely change the traffic pattern of a congested zone. Imagination, invention, and technical resources must be combined in order to disentangle even the knots that seem most inextricable.

69. The destruction of the slums around historic monuments will provide an opportunity to create verdant areas.

In certain cases, it is possible that the demolition of unsanitary houses and slums around some monument of historical value will destroy an age-old ambience. This is regrettable, but it is inevitable. The situation can be turned to advantage by the introduction of verdant areas. There, the vestiges of the past will be bathed in a new and possibly unexpected ambience, but certainly a tolerable one, and one from which the neighboring districts will amply benefit in any event.

70. The practice of using styles of the past on aesthetic pretexts for new structures erected in historic areas has harmful consequences. Neither the continuation of such practices nor the introduction of such initiatives will be tolerated in any form.

Such methods are contrary to the great lesson of history. Never has a return to the past been recorded, never has man retraced his own steps. The masterpieces of the past show us that each generation has had its way of thinking, its conceptions, its aesthetic, which called upon the entire range of the technical resources of its epoch to serve as the springboard for its imagination. To imitate the past slavishly is to condemn ourselves to delusion, to institute the "false" as a principle, since the working conditions of former times cannot be recreated and since the application of modern techniques to an outdated ideal can never lead to anything but a simulacrum devoid of all vitality. The mingling of the "false" with the "genuine," far from attaining an impression of unity and from giving a sense of purity of style, merely results in artificial reconstruction capable only of discrediting the authentic testimonies that we were most moved to preserve.

3 CONCLUSIONS: MAIN POINTS OF DOCTINE

71. The majority of the cities studies (by the Fourth Congress) today present the very image of chaos: they do not at all fulfill their purpose, which is to satisfy the primordial biological and psychological needs of their populations.

Through the efforts of the national groups within the International Congresses for Modern Architecture, some thirty-three cities were analyzed on the occasion of the Athens Congress: Amsterdam, Athens, Brussels, Baltimore, Bandung, Budapest, Berlin, Barcelona, Karlsruhe, Cologne, Como, Dalat, Detroit, Dessau, Frankfurt, Geneva, Genoa, The Hague, Los Angeles, Littoria, London, Madrid, Oslo, Paris, Prague, Rome, Rotterdam, Stockholm, Utrecht, Verona, Warsaw, Zagreb, and Zurich. These cities illustrate the history of the white race throughout the most diverse climates and latitudes. All of them bear witness to the same phenomenon: the disorder wrought by machinism in a situation that had previously allowed a relative harmony as well as the absence of any serious attempt at adaptation. In every one of these cities, man finds himself being molested. Everything that surrounds him stifles and crushes him. None of the things necessary for his physical and moral health has been preserved or introduced. A human crisis is raging in the major cities with repercussions throughout the land. The city is no longer serving its function, which is to shelter human beings, and to shelter them well.

72. This situation reveals the incessant accretion of private interests ever since the beginning of the machinist age.

The pre-eminence of private initiatives, motivated by self-interest and by the lure of profit, is at the root of this deplorable state of affairs. Not one authority, conscious of the nature and the importance of the machinist movement, has yet taken any step to avoid the damage for which no one can actually be held accountable. For a hundred years, every enterprise was left to chance. Housing and factories were constructed, roads laid out, waterways and railroads cut and graded, everything multi-piled in haste and in a climate of individual violence that left no room for any preconceived plan or premeditation. Today, the damage has been done. The cities are inhuman; the ferociousness of a few private interests has given rise to the suffering of countless individuals.

73. The ruthless violence of private interests provokes a disastrous upset in the balance between the thrust of economic forces on the one hand and the weakness of administrative control and the powerlessness of social solidarity on the other.

The sense of administrative responsibility and of social solidarity is daily driven to the breaking point by the keen and continually renewed forces of private interest. These diverse sources of energy are in perpetual conflict, and when one attacks, the other defends itself. In this unhappily uneven struggle it is generally the private interests that triumph, ensuring the success of the strong at the expense of the weak. But good sometimes comes from the very excess of evil, and the immense material and moral disorder of the modern city may ultimately result in the formation of new legislation for the city, a legislation supported by strong administrative responsibility, which will establish the regulations to the protection of human well being and dignity.

74. Although the cities are in a state of continuous transformation, their development is conducted without precision or control, and in utter disregard of the principles of contemporary urbanism, which have been laid down by qualified technical specialists.

The principles of modern urbanism, evolved through the labors of innumerable technicians—technicians in the art of building, technicians of health, technicians of social organization—have been the subject of articles, books, congresses, public and private debates. But they still must be acknowledged by the administrative agencies charged with watching over the destiny of cities, agencies that are often hostile to the major transformations proposed by the new data. The authorities must first be enlightened, and then they must act. Clear-sightedness and energy can salvage this dangerous situation.

75. On both spiritual and material planes, the city must ensure individual liberty and the advantages of collective action.

Individual liberty and collective action are the two poles between which the game of life is played. Any undertaking whose object is to improve the human condition must take these two factors into account. If it does not manage to satisfy the often contradictory requirements of both, it is inevitably doomed to failure. In any event, it is impossible to coordinate them in a harmonious way without preparing in advance a carefully studied program that leaves nothing to change.

76. The dimensions of all elements within the urban system can only be governed by human proportions.

The natural measurements of man himself must serve as a basis for all the scales that will be consonant with the life and diverse functions of the human being: a scale of measurements applying to areas and distances, a scale of distances that will be considered in relation to the natural walking pace of man, a time scale that must be determined according to the daily course of the sun.

77. The keys to urbanism are to be found in the four functions: inhabiting, working, recreation (in leisure time), and circulation.

Urbanism expresses the condition of an era. Until now, it has tackled only one problem, that of traffic circulation. It has been content to open up avenues or lay out streets, thus forming blocks of buildings whose purpose has been left to the haphazard ventures of private initiatives. This is a narrow and inadequate view of its mission. Urbanism has four principal functions. First, to assure mankind of sound and healthy lodging, that is to say places in which space, fresh air, and sunshine—those three essential conditions of nature—are abundantly available. Second, to organize places of work in such a way that instead of being a painful subjugation, work will once more regain its character as a natural human activity. Third, to set up the facilities necessary to the sound use of leisure time, making it productive and beneficial. And fourth, to establish links between these different organizations by means of a traffic network that provides the necessary connections while respecting the prerogatives of each element. These four functions, which are the four keys to urbanism, cover an enormous area, since urbanism is the outcome of a way of thinking, integrated into public life by means of a technique for action.

78. Plans will determine the structure of each of the sectors allocated to the four key functions and they will also determine their respective locations within the whole.

Since the CIAM Congress in Athens, the four key functions of urbanism have called for special measures offering each function the conditions most favorable to the development of its own activity so that they may be manifested in all their fullness and bring order and classification to the usual conditions of life, work, and culture. By taking account of this necessity, urbanism will transform the face of the city, break with the crushing constraint of practices that are no longer justified, and open an inexhaustible field of action to the creative. Each key function will have its own autonomy, based on circumstances arising out of climate, topography, and local customs; each will be regarded as an entity to which land and buildings will be allocated, and all of the prodigious resources of modern techniques will be used in arranging and equipping them. In this distribution, consideration will be given to the vital needs of the individual, not to the special interest or profit of any particular group. Urbanism must guarantee individual liberty at the same time as it must take advantage of the benefits of collective action.

79. The cycle of daily functions—inhabiting, working, recreation (recuperation)—will be regulated by urbanism with the strictest emphasis on time saving, the dwelling being regarded as the very center of urbanistic concern and the focal point for every measure of distance.

The desire to reintroduce the "conditions of nature" into daily life would seem, at first sight, to call for an ever greater horizontal expansion of cities; but the necessity of regulating the different activities in accordance with the duration of the sun's course goes counter to that idea, which has the disadvantage of imposing distances incommensurate with available time. The dwelling is the urbanist's central concern, and the interplay of distances will be governed by its location in the urban plan in conformity to the solar day of twenty-four hours, which dictates the rhythm of men's activity and gives correct measure to all their undertakings.

80. The new mechanical speeds have thrown the urban milieu into confusion, introducing constant danger, causing traffic congestion and paralyzing communications, and jeopardizing hygiene.

Mechanized vehicles should be liberating agents, and, with their speed, should bring about appreciable gains in time. But their accumulation and concentration at certain points has become both a hindrance to traffic and a source of continual danger. They have, moreover, introduced into urban life many factors that are injurious to health. The combustion gases with which they fill the air are harmful to the lungs, and their noise causes in man a state of permanent nervous irritability. The speeds that are now possible arouse the temptation to a daily exodus, away from it all and back to nature, they stimulate an intemperate and unbridled taste for mobility and foster ways of life which, by splitting up the family, profoundly disturb the basis of society itself. They condemn men to spend wearisome hours in all sorts of vehicles and, little by little, to abandon the practice of the healthiest and most natural function of all: walking.

81. The principle of urban and suburban traffic must be revised. A classification of available speeds must be devised. Zoning reforms bringing the key functions of the city into harmony will create natural links between them, in support of which a rational network of major traffic arteries will be planned.

By taking account of the key functions—housing, work, and recreation—zoning will introduce a measure of order into the urban territory. The fourth function, that of traffic movement, should have only one objective: to bring the other three into effective communication with one another. Major transformations are inevitable. The city and its region must be equipped with a road network that incorporates modern traffic techniques and is directly proportionate to its purposes and usage. The means of transportation must be differentiated and classified for each of them, and a channel must be provided appropriate to the exact nature of the vehicles employed. Traffic thus regulated becomes a steady function, which puts no constraint on the structure of either habitation or places of work.

82. Urbanism is a three-dimensional, not a two-dimensional, science. Introducing the element of height will solve the problems of modern traffic and leisure by utilizing the open spaces thus created.

The key functions—inhabiting, working, and recreation—develop within built volumes that are subject to three imperative necessities: adequate space, sun, and ventilation. These volumes are based not only on the ground and its two dimensions but also, and especially, on a third dimension: height. It is by making use of height that urbanism will recover the open land necessary for communications and for leisure spaces. A distinction must be made between sedentary functions, which develop inside volumes where the third dimension plays the most important role, and functions of traffic circulation, which, using only two dimensions, are tied to the ground; height plays a role only rarely and on a small scale—as, for instance, when changes of level are intended to regularize certain heavy flows of vehicular traffic.

83. The city must be studied within the whole of its region of influence. A regional plan will replace the simple municipal plan. The limit of the agglomeration will be expressed in terms of the radius of its economic action.

The particulars of a problem of urbanism are furnished by the sum of the activities carried out not only within the city itself but also throughout the region of which it is the center. The city's raison d'être must be sought and expressed in figures that will make it possible to forecast the stages of a plausible future development. The same operation applied to secondary population centers will provide a reading of the overall situation. Allocations, limitations, compensations can be determined, and these will provide each city, surrounded by its region, with its own character and destiny. Thus, each city will take its place and rank in the general economy of the country. The outcome will be the clear differentiation of regional boundaries. This is a total urbanism, capable of bringing equilibrium to each province and to the country as a whole.

84. Once the city is defined as a functional unit, it should grow harmoniously in each of its parts, having at hand the spaces and intercommunications within which the stages of its development may be inscribed with equilibrium.

The city will take on the character of an enterprise that has been carefully studied in advance and subjected to the rigor of an overall plan. Intelligent forecasts will have sketched its future, described its character, foreseen the extent of its expansions, and limited their excesses in advance. Subordinated to the needs of the region, assigned to provide a framework for the four key functions, the city will no longer be the disorderly result of random ventures. Its growth, instead of producing a catastrophe, will be a crowning achievement. And the increase in its population figures will no longer lead to that inhuman melee that is one of the afflictions of the big cities.

85. It is a matter of the most urgent necessity that every city draws up its program and enacts the laws that will enable it to be carried out.

Chance will give way to foresight, and program will replace improvisation. Each case will be written into the regional plan; sites will be measured and allocated to various activities: there will be clear rules governing the undertaking, which will begin tomorrow and proceed, little by little, in successive stages. The law will lay down the "land statute," endowing each key function with the means for its best self-expression, for its location on the most favorable sites at the most useful distances from other functions. The law must also make provisions for the protection and care of those areas that will one day be occupied. It will have the right to authorize—and to prohibit; it will encourage any carefully evaluated initiatives but will take care that they fit into the overall plan and are always subordinate to the collective interests that constitute the public good.

86. The program must be based on rigorous analyses carried out by specialists. It must provide for its stages in time and in space. It must bring together in fruitful harmony the natural resources of the site, the overall topography, the economic facts, the sociological demands, and the spiritual values.

The work will no longer be confined to the precarious plan of the land-surveyor who projects blocks of apartment houses and the dust of future building lots without a thought for the suburbs. It will be a true biological creation comprising clearly defined organs capable of fulfilling their vital functions to perfection. Soil conditions will be analyzed and the constraints they dictate identified; the general environment will be examined, and its natural assets arranged in hierarchical order. The major directions of traffic flow will be confined and placed in their proper positions, and the nature of their equipment determined according to their intended purposes. A growth curve will indicate the city's foreseeable economic future. Inviolable rules will guarantee the inhabitants good homes, comfortable working conditions, and the enjoyment of leisure. The soul of the city will be brought to life by the clarity of the plan.

87. For the architect occupied with the tasks of urbanism, the measuring rod will be the human scale.

After the downfall of the last hundred years, architecture must once again be placed in the service of man. It must lay sterile pomp aside, concern itself with the individual and create for his happiness the fixtures that will surround him, making all the movements of his life easier. Who can take the measures necessary to the accomplishment of this task if not the architect who possesses a complete awareness of man, who has abandoned illusory designs, and who, judiciously adapting the means to the desired ends, will create an order that bears within it a poetry of its own?

88. The initial nucleus of urbanism is a cell for living—a dwelling—and its insertion into a group forming a habitation unit of efficient size.

If the cell is the primordial biological elements, then the home, that is to say the family shelter, constitutes the social cell. After more than a century of subjection to the ruthless games of speculation, the construction of this home must now become a humane undertaking. The home is the initial nucleus of urbanism. It protects the growth of man and gives shelter to the joys and sorrows of his daily life. It is it to be filled with fresh air and sunshine inside, it must also be extended outside by various community facilities. So that dwellings can be more easily supplied with common services dealing conveniently with the supply of food, education, medical attention, and the enjoyment of leisure, it will be necessary to group them in "habitation units" of adequate size.

89. With this dwelling unit as the starting point, relationships within the urban space will be established between habitation, work places, and the facilities set aside for leisure.

The first of the functions that should engage the urbanist's attention is that of housing—and good housing. But people also have to work, and they must do so in conditions that demand a thorough revision of prevailing practices. Offices, workshops, and factories must be equipped in such a way as to guarantee the well being necessary to the accomplishment of this second function. Finally, the third function, which is recreation, the cultivation of one's body and mind, must not be neglected. The urbanist will have to make provision for the sites and premises required for this purpose.

90. To accomplish this great task, it is essential to utilize the resources of modern techniques, which, through the collaboration of specialists, will support the art of building with all the dependability that science can provide, and enrich it with the inventions and resources of the age.

The machinist era has introduced new techniques, which are one of the causes of the disorder and the upheaval of the cities. And yet it is to those very techniques that we must look for a solution to the problem. Modern construction techniques have established new methods, provided new facilities, made new dimensions possible. They have opened an entirely new cycle in the history of architecture. The new structures will be not only of a scale, but also of a complexity unknown until now. In order to fulfill the many-faceted task that has been imposed on him, the architect will have to join with many specialists at every stage of the undertaking.

91. The course of events will be profoundly influenced by political, social, and economic factors.

It is not enough to admit the necessity of a "land ordinance" and of certain principles of construction. To pass from theory to action still requires a combination of the following factors: a political power such as one might wish—clear-sighted, with earnest conviction, and determined to achieve those improved living conditions that have been worked out and set down on paper; an enlightened population that will understand, desire, and demand what the specialists have envisaged for it; an economic situation that will make it possible to embark upon and pursue building projects which, in certain instances, will be considerable. Yet it is possible, nonetheless, even at a time when everything is at a very low ebb, when the political, moral, and economic conditions are least favorable, that the necessity of building decent shelters will suddenly emerge as an overriding obligation, and that this obligation will provide politics, social life, and the economy with precisely the coherent goal and program that they were lacking.

92. . . . and it is not as a last resort that architecture will intervene.

Architecture presides over the destinies of the city. It orders the structure of the dwelling, that vital cell of the urban tissue whose health, gaiety, and harmony are subject to its decisions. It groups dwellings in habitation units whose success will depend on the accuracy of its calculations. It reserves in advance the open spaces in the midst of which will rise volumes built with harmonious proportions. It arranges the extensions of the dwelling, the places of work, the areas set aside for relaxation. It lays out the circulatory network that will bring the different zones into contact with one another. Architecture is responsible for the well being and the beauty of the city. It is architecture that takes charge of its creation or improvement, and it is architecture that must choose and allocate the different elements whose apt proportions will constitute a harmonious and lasting work. Architecture is the key to everything.

93. There are two opposing realities: the scale of the projects to be undertaken urgently for the reorganization of the cities, and the infinitely fragmented state of land ownership.

Works of major importance must be undertaken without delay, since all of the cities in the world, ancient or modern, reveal the same defects arising from the same causes. But no partial effort should be made unless it fits into the framework of the city and the region as they have been laid down by an extensive study and a broad overall plan. This plan will necessarily include parts that can be carried out immediately and others whose execution will have to be postponed indefinitely. Many pieces of land will have to be expropriated and will become subject to negotiations. It is at this point that we shall have to beware the sordid game of speculation, which so often smothers in the cradle great ventures animated by a concern for the public good. The problem of land ownership and possible land requisition arises in the cities and in their outskirts, and it extends throughout the more or less extensive area that constitutes their region.

93. The perilous contradiction indicated above raises one of the most hazardous questions of our day: the urgency of regulating the disposal of all usable ground by legal means in order to balance the vital needs of the individual in complete harmony with collective needs.

For years now, at every point on the globe, attempts at urban improvement have been dashed against the petrified law of private property. Ground—the territory of the country—must be made available at any time and at its fair market value, to be assessed before projects are worked out. The ground should be open to mobilization whenever it is a matter of the general interest. Countless difficulties have harassed people who were unable to gauge accurately the extent of technical transformations and their tremendous repercussions on public and private life. The absence of urbanism is the cause of the anarchy that prevails in the organization of cities and in the equipment of industries. Because we have misunderstood the rules, the fields have become empty, the cities have been filled beyond all reason, industrial concentrations have taken place haphazardly, workers' dwellings have become slums. No provision has been made for safeguarding man. The result is almost uniformly catastrophic in every country. It is the bitter fruit of a hundred years of undirected machinism.

95. Private interest will be subordinated to the collective interest.

Left to himself, a man is soon crushed by difficulties of every kind that he must overcome. If, conversely, he is subjected to too many collective constraints, his personality is stifled by them. Individual rights and collective rights must therefore support and reinforce one another, and all of their infinitely constructive aspects must be joined together. Individual rights have nothing to do with vulgar private interests. Such interests, which heap advantages upon a minority while relegating the rest of the social mass to a mediocre existence, require strict limitations. In every instance, private interests must be subordinated to the collective interest, so that each individual will have access to the fundamental joys, the well being of the home, and the beauty of the city. ■

EDITORS NOTE

The 1943 publication of *The Athens Charter* did not include illustrations. Many of its principles—rationalized planning based on modern functions, the dwelling as organizing unit of planning, exposure to sunlight and air, ample green space, *etc.*—are evident in Le Corbusier's early drawings and publications. Decrying both urban congestion and suburban sprawl, he espoused these principles in his work, writings and lectures throughout the world. The conception of *La Ville Radieuse*, envisioning a zoned city with superblocks and streamlined road systems, is widely cited as the theoretical model of the modern high rise apartment project and thus subject to urbanistic and sociological criticism based on the practical results. The drawings and captions below are from *Vers Une Architecture*, first published in French in 1917 and in English in 1927, which also included drawings by Tony Garnier of residential quarters from *The Industrial City* (*cf.* Article 2.2 earlier in this volume).

City of Towers. Le Corbusier. 1923. "The towers are place amidst gardens and playing fields. The main arteries, with their motor tracks built over them, allow of easy, or rapid, or very rapid circulation of traffic." *(Editions Crès)*

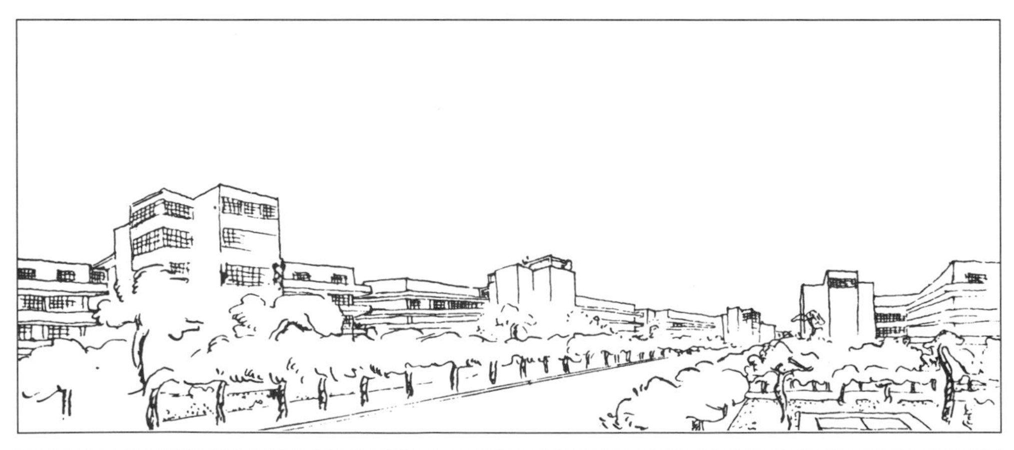

Streets with set-backs. Le Corbusier. 1920. "Vast airy and sunlit spaces on which all windows would open. Gardens and playgrounds around the buildings. Simple facades with immense bays. The successive projections give play of light and shade, and a feeling of richness is achieved by the scale of the main lines of the design and by the vegetation seen against the geometric background of the facades. Obviously we have here, as in the case of the City of Towers, a question of enterprise on a huge financial scale, capable of undertaking the construction of entire quarters. A street such as this would be designed by a single architect to obtain unity, grandeur, dignity and economy." *(Editions Crès)*

Steen Eiler Rasmussen

Summary
Steen Eiler Rasmussen (1898 –1990) was a Danish architect best known for his classic 1934 study of the planning and architecture of *London: The Unique City*, in which he describes the British capital as the exemplar of what he calls the "scattered city," as opposed to the concentrated and concentric patterns of development that characterize European capitals such as Paris and Vienna. In his 1949 survey of urban design history, *Towns and Buildings*, from which this article is reprinted, he sketches a direct comparison of the contrasting form and characteristic monumental urban spaces of London and Paris in the period before the French Revolution.

Key words
arcade, boulevard, London, monuments, open space, Paris, square, town planning

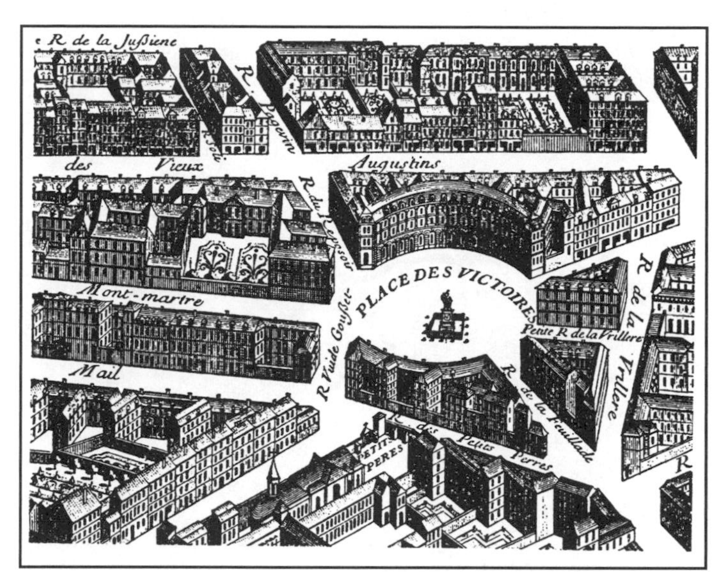

Place des Victoires, Paris. Section of Turgot's plan 1731. The circular plaza is composed in conjunction with straight streets that intersect it and determine the placing of the statue.

A tale of two cities

Paris and London represent two types of cities. Paris is the concentrated city in which many families live in each house. London is the scattered city in which one-family houses predominate, and where distances are great. One would naturally assume that the more a town grows, the greater becomes the necessity of crowding people together. But London, which is the second largest city in the world, proves otherwise. By and large, cities in England (but not in Scotland which, on this point, is very continental) and in America are the scattered type. Most of the cities on the Continent—though not all of them—are concentrated cities. The reasons for this are many. Here we have room only for a few indications which will help to characterize the two city types, Paris and London.

To a certain degree the special development of English towns can be attributed to the fact that England's best defense has always been her island location. Sine 1066 the country has never been invaded. Therefore, it has not been necessary to surround English towns with constricting rings of fortifications, as was so often the case on the Continent.

A city like Paris has expanded by placing one ring beyond the other, moving the line of defense further and further out. In the Middle Ages, in 1180, there was a wall around the island in the Seine—the *Isle de la Cité*—and small parts of the left and right banks. In 1370, the area of the city was enlarged by building a new wall on the right bank. The next expansion was due, not to the overcrowding of the city, but to the laying out of great royal gardens which broke through the city limits and formed a new boundary. This was carried further toward the northwest to protect the Tuileries gardens. Later, new rings were built around Paris, one in the 18th and one in the 19th centuries. The closed form continued to be regarded as absolutely necessary for a city. Building bans did not lead to a halt in the city's growth but only to the crowding of more people into each house.

London developed along quite different lines. Very early in its history the town within the Roman walls had become too small. At that time London was smaller than medieval Cologne or Paris. But new fortifications were not necessary. Instead, each village on the outskirts of old London became the nucleus of a new town. Together, they formed a cluster of towns, which gradually have grown into one—and now there are plans to separate them again. The names of these villages, some of them found in the *Domesday Book,* were the same as those designating the boroughs which now form the community we call London. London is not a city in the sense that Paris is. It is a collection of towns. At many places within London two such towns are separated only by a street, yet, when you go from one to the other there is a marked difference. The inhabitants speak another dialect, they have different political views, different municipal authorities and rates, different ideas about the propriety of children using the swings and seesaws in the parks on Sunday. Every one of these towns has maintained local government to an extent which, in many instances, borders on the ridiculous. The two dominating towns were London

Credits: This article is a chapter from Steen Eiler Rasmussen's *Towns and Buildings* described in drawings and words (first Danish edition 1949) reprinted by permission of MIT Press, Cambridge, MA, 1969 (1979).

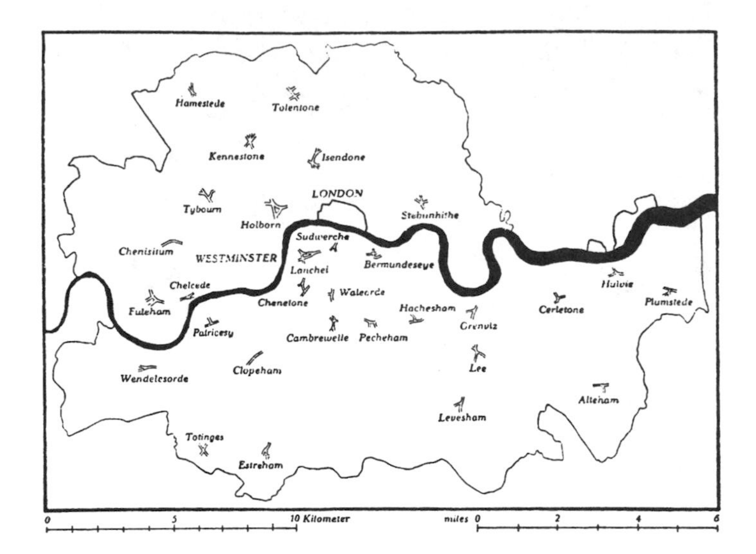

Villages near London mentioned in the *Domesday Book* from about the year 1080. Each village name signified a group of houses near a crossroad.

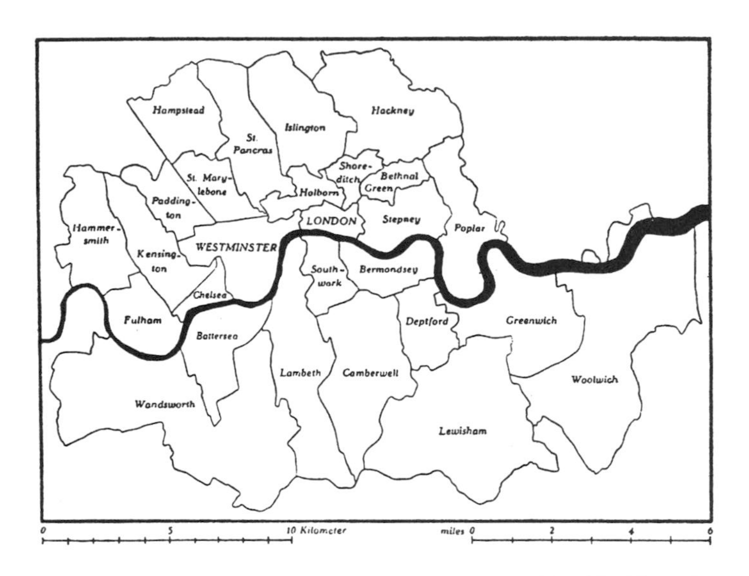

The boroughs that form the County of London today. The nucleus of each borough is one of the old villages, every one of whose names can be traced back to the *Domesday Book*.

proper, "the City of London," the seat of trade, and "the City of Westminster," the seat of government. The relations between these two have been decisive for England's history. The king (as well as the Government) does not reside in the City of London. When he comes there he is received with all the honors shown a foreign potentate visiting a free city. London's Lord Mayor comes to the spot where in the older days the gate of the city stood and, with great ceremony, hands over to the king the keys of the gate which no longer stands there.

When Henry IV of France, as a building speculator on the grand scale, built the *Place Royale*, now the *Place des Vosges*, it was a new and epoch-making idea. This was early in the 17th century. (Henry IV died in 1610.) At about the same time there lay a large, unbuilt area between London and Westminster where, earlier, a convent had stood. After the Reformation, Henry VIII had given the confiscated property to a nobleman who had been of great service to the royal house. Around 1630, this convent garden, or, as it is now called, Covent Garden, was ripe for exploitation and the fourth Earl of Bedford decided to utilize his land for a great building enterprise. But he wished to do it in just as stately a manner as the French king had carried out his building speculation. It was to be a monumental square with a church on its axis. The Earl employed the country's first architect, Inigo Jones, to design the church and the facades of the buildings and to plan the arcades that were to surround the square. What was to be hidden behind the facades was left to the tenants to decide. The project turned out to be a much more classical, more Italian *place* than the *Place des Vosges*. The church was lower than all the other buildings but it seemed large because it had the largest detail, a great portico of columns. But this monumentality did not last very long. While the *Place des Vosges* became the scene of knightly sports and tournaments, Covent Garden became a vegetable market which filled the coffers of the Bedford family. In Paris the court took over the square, in London, trade—which gives a very good idea of part of the difference between the two cities.

The arcades of Covent Garden, however, really took on something of the same significance of the arcades of ancient market places. They became a popular meeting place where friends strolled together, gossiping and discussing the news of the day. The arcade that led to some of the famous coffeehouses and to the Covent Garden Theatre became a London institution and has left many traces in English art and literature.

These two real estate projects, *Place des Vosges* in Paris and Covent Garden in London, had many traits in common. But as time went on, the development of the two cities greatly diverged. Paris became more and more a consumer city, a place where the enormous fortunes, made on the great manorial estates of the aristocracy, were spent on luxuries. While there was a general decree prohibiting construction on hitherto unbuilt land, the government encouraged all building which served to glorify the monarchy. Therefore, if one wanted to build on an empty site, all that was necessary was to fix upon a project which included a monumental place with a statue—and, lo! there was no longer any ban on building. It was even possible to obtain a subsidy from the government for the enterprise. This became the salvation of many a ruined nobleman, as for instance, the Duke de Vendome. In 1677 his creditors got together to find out if they could not make something out of his large holdings. The architect Mansard drew up plans for a great building enterprise around a monumental place with a statue. It turned out to be a long, troublesome undertaking. The plans were changed several times. In 1699 Girardon's equestrian statue of Louis

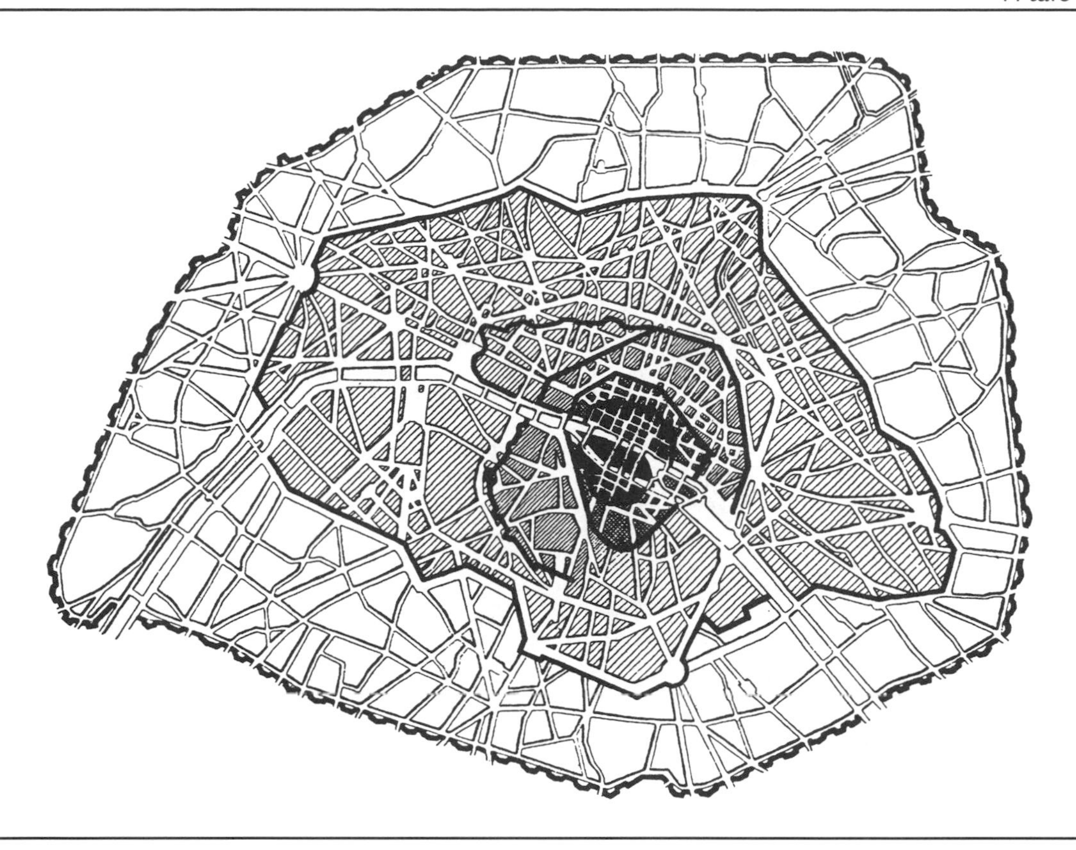

The development of Paris. *(North upward)* Central core early Middle Ages, around this heavy black lines show boundaries of c. 1180, 1370, 1676, 1784–91 and 1841–45.

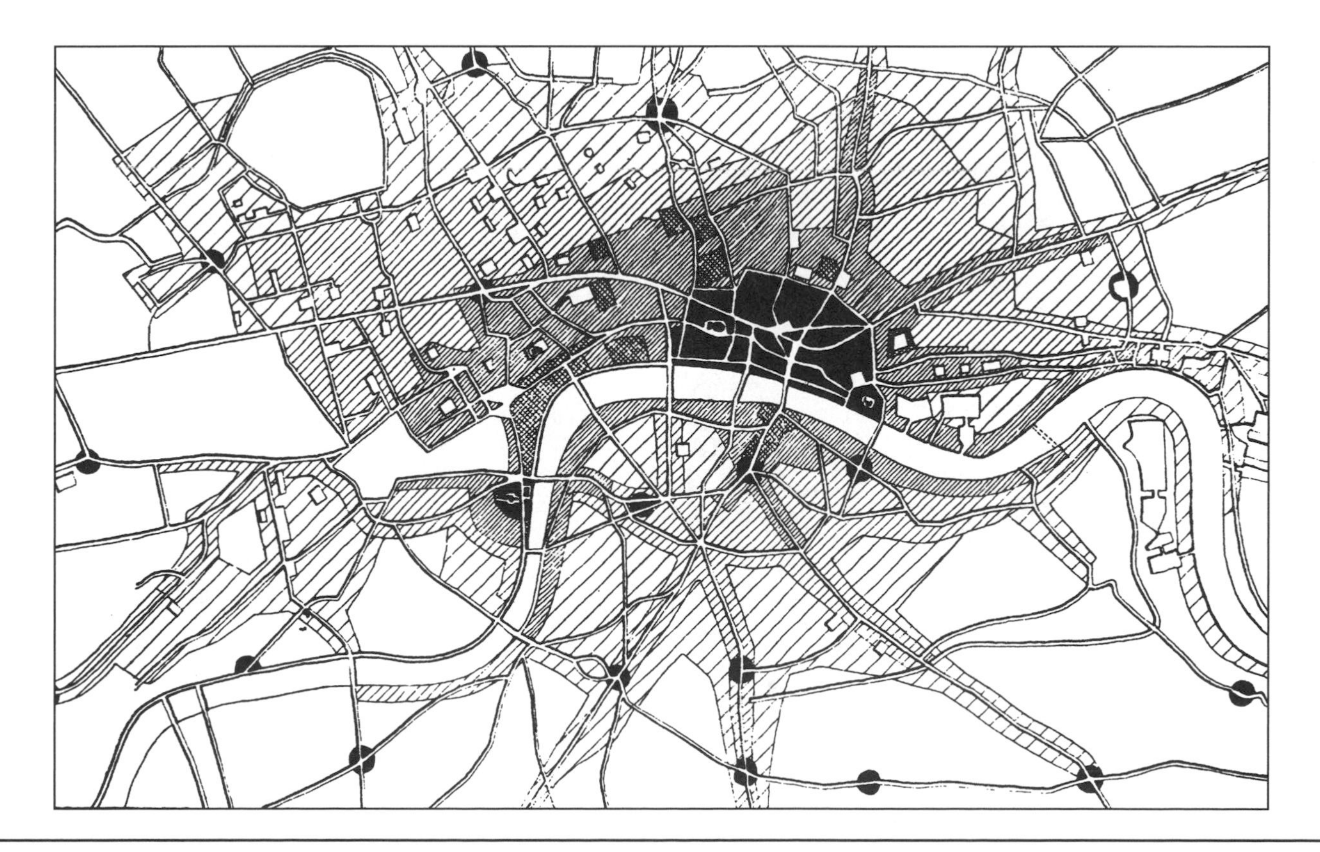

London's development. *(North upward)* Black denotes inhabited districts in early Middle Ages. Cross-hatching denotes later medieval settlements (convents, temple, buildings in Westminster and London) finely hatched around these. London c. 1660, thereafter c. 1790 and finally 1830.

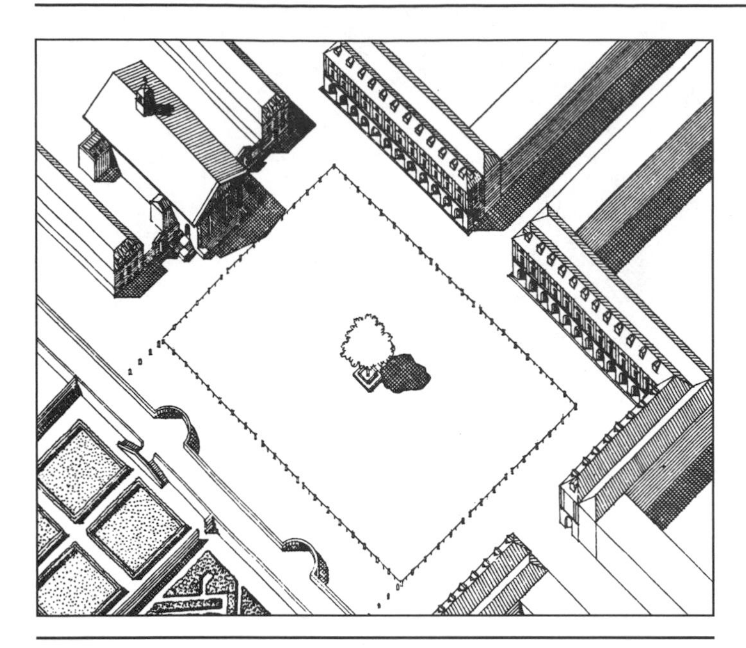

Covent Garden Square in London in its original form. In foreground, left, the Earl of Bedford's garden. Facing square St. Paul's Covent Garden, designed as a temple. The square was to be a classic forum with arcades and a public meeting place in the portico of the church.

Place Vendome, Paris. Section of Turgot's plan, 1731. In center, equestrian statue of Louis XIV with the magnificent, regular façade buildings around it which is in no way related to the buildings and courts behind.

XIV was erected. The property then passed into the hands of the municipality which carried out the project according to the final plan. In 1701 the facades were finished and not until then did the sale of the building lots behind them begin. There was no connection between the facades and the houses they hid. On the other hand, the height of the houses was carefully adapted to the 17-meter (56-feet) high monument so that the equestrian statue was seen rising above the cornices.

The Duke de la Feuillade also obtained permission to carry out a large building project around a circular plaza with a statue of Louis XIV. It was called the *Place des Victoires* and was laid out in 1697 just across the old city boundary, which had been pushed further out after the Cardinal's palace and the *Tuileries* had been built in the 16th century. Instead of the usual equestrian statue, this time the monument was a standing figure of the king being crowned with laurels by the goddess of victory. It exists no longer. Like other royal monuments, it was destroyed during the French Revolution.

The 18th century map of Paris shows other characteristic traits. Though the city's fortifications were no longer of vital importance, definite boundaries continued to be maintained. They were marked by the *Grands Boulevards.* The word "boulevard" is a corruption of the nordic *bulvirke* (bulwark) which means a palisade, a medieval form of defense work used before the employment of real walls and ramparts. The boulevard is the line of fortifications, itself, but when these were converted, in Paris, into broad, tree-lined wall streets, the designation *"boulevard"* was kept. And when, much later, under Napoleon III—as described in the chapter on Paris *Boulevards*—broad, radical thoroughfares, also planted with trees, were cut through the old city on all sides, these, too, were called boulevards. Today, the word simply means a broad, tree-lined avenue. However, in the 17th century the boulevards were actually boundary lines beyond which buildings could not be erected because uncontrolled expansion of the city was considered very dangerous. The result was, naturally, that the population within the walls became more and more dense.

There were building bans in London, also, but no clearly defined city boundaries because the town had spread beyond the Roman walls so early in its history. It expanded particularly toward the west, until London and Westminster had completely merged. In the new districts there were many open spaces. These were of two kinds, originating from different causes. Some of them were old village greens and fields, which from time immemorial had been set aside for the use for the inhabitants for sports, games and archery. Every form of custom and tradition has always been of great importance in England, where laws have never been collected into a logical system but have remained a simple record of rules and regulations naturally evolved from the daily life of the people. There are many accounts of the armed resistance of the inhabitants when building speculators attempted to exploit these old playing fields. They often became regular pitched battles with a number of wounded and even some dead. And in every case it was the defenders of the open spaces who held the field and won the support of the government. To this day there are still greens and commons spread all over London, where young people meet every Saturday all summer long, to play cricket, just as young people did when these green areas were parts of individual villages. In the center of the city the open spaces that have been preserved have become playgrounds, bandstands, public tennis courts, and other areas of recreation.

The other type of open spaces came into existence in the course of great building speculations. Covent Garden, the first real "square" in London, was such a great success that others followed it. West of London's "City" lay a number of old manorial estates and country houses. As the urban development approached nearer and nearer, the grounds of these estates were parceled out for building. The owners, however, desired to keep their old homes as long as possible, with sufficient open space around them, preferably toward the north where there was a splendid view of lovely, purpling hills crowned by the old villages of Hampstead and Highgate. Therefore, a large square was laid out in front of the house, which thus closed the south side of the square; new buildings were erected on the east and west sides, and the north side was kept open. Later, as the district grew, the north side was also built up, and a new London square had come into being.

In Paris they were just as much interested in open spaces as in London but these continued to be of a different type. They were statue *places*. Louis XIV or, as he was named on the monuments, Louis le Grand, had had his *places*. Louis XV, *le bien aimé*, must also have his. The object was not only to glorify the monarchy, but also to beautify the city and rid it of slums. The old districts, in which houses were crowded together and unsavory, were always present like a bad conscience.

In 1748 a great competition was held for the design of a monument *place* for Louis XV. The many plans that were sent in were reproduced in a large volume of engravings, published in 1765. But long before that year they had been spread all over Europe by newspapers and had been studied and copied even so far away as Denmark. The author of the stately work, Patte, had entered all the proposals on a map of Paris so that it resembled a city of royal places. And the proposals were by no means modest ones. One competitor submitted a plan for a new *Louvre* on the left bank of the Seine, duplicating that on the right bank, and with the entire western end of the *Isle de la Cité* turned into a monumental *place*, in keeping with great, new places on both banks. Some proposed the destruction of large numbers of houses to make way for circular, quadrangular or octagonal *places*. There was also a design for a complete system of squares, three monumental market places connected with each other by arcades. These many projects, however, did not lead to any slum clearance of the old districts. Instead, unbuilt land in front of the *Tuileries* was chosen as the site of the new *place*, bordered on one side by the Seine, on two others by rows of trees, and on the fourth side by new monumental buildings. In the centre a colossal equestrian statue of Louis XV was raised—now, long since vanished and replaced by a large obelisk.

In the 18th century London continued to spread out, adding new residential sections around open squares. The landlords were the great landowners who were not used to selling their property but only to leasing it out on long term. This had been the custom since the Middle Ages and it had been a good one for agriculture. Now, the same system was continued after the property had become urban. The realization of the appreciation of ground values took place only at long intervals, when a tenancy, which might run for 99 years, had terminated. But the owner could afford to wait. In London, which was a commercial city and where it was now possible to build as far out as one wished, speculators thought in terms of *building* speculation rather than *land* speculation, which are two very different things. Money was invested in firms of enterprising builders, and returns came as soon as the houses were finished and sold. The money was used to *produce* something, and the investor was not interested in building as many houses that were as attractive as possible. And as it was attractive to live in a house facing an open square, squares were naturally laid out.

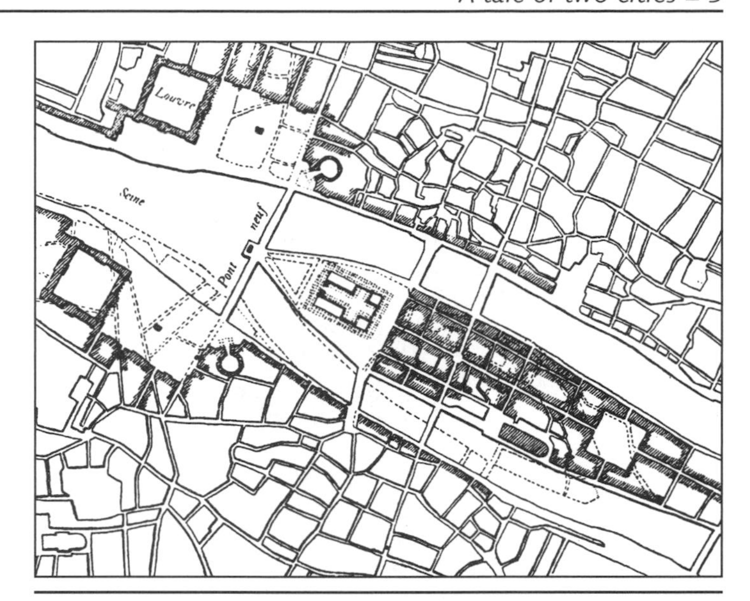

Examples of some of the unaccepted proposals for a monumental square in Paris in honor of Louis XV. This one is for a large square, to be built in front of the *Louvre*, containing an obelisk, and on the opposite bank a new *Louvre*, symmetrical with the old, and a new square. At the same time the entire *Isle de la Cité*, with the exception of *Notre Dame*, was to be regulated and rebuilt. Scale 1: 20.000. North upward.

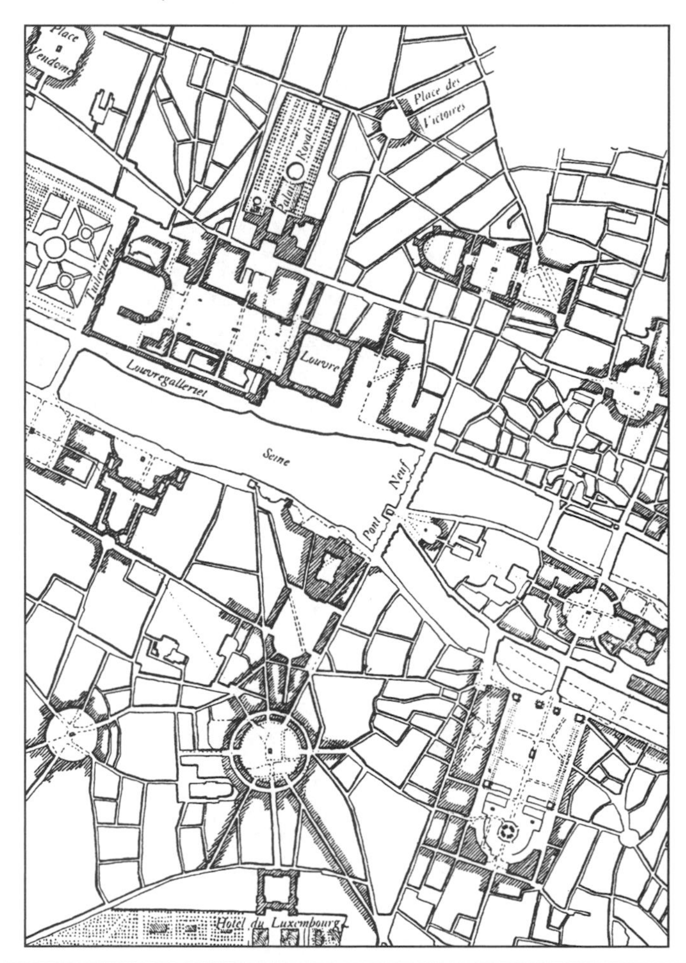

Proposals for monumental squares in Paris in honor of Louis XV; results of a competition, 1748—published 1765—and never carried out. At top are seen two squares from Louis XIV's time: *Place Vendome* and *Place des Victoires*. Scale 1: 20.000. North upward.

Section of London map, 1804. Reproduced on a scale of 1: 20.000. North upward. The map shows a number of the squares laid out in the 17th and 18th centuries.

When a district was no longer fashionable, the original residents moved to a new one, lying further out, which beckoned with modern houses, larger open spaces and gardens.

The monumental *places* of continental cities, on the one hand, and London's squares, on the other, were very different. The monumental *places* were great Baroque creations in which the house facades were of paramount importance and that which lay behind them quite unimportant. In the *Place Vendome*, as already noted, there was no connection between the subdivisions on the façades surrounding the *place* and the courtyards and rooms of the buildings behind them. The Baroque *place* was entirely dramatic in conception, forming an effective vista with entrance, approach and climax. Such effects were not found in London, where all four sides of the squares were generally the same. In the centre there was usually a fenced-in garden to which all the families living in the square had a key. The planting in them was informal and trees were allowed to grow naturally, becoming well formed and large. They were mostly plane trees. Neither Baroque nor Rococo found favor in these districts, where the houses were simple, anonymous brick buildings, their facades relieved only by sharply indented window-openings. In all the houses heating was done by coal fires on open hearths, which spread a layer of soot over the whole city. The houses became black. It was discovered that there were only two things to do about this. Either the brick walls could be covered with stucco and then oil painted, washed every year, and painted again when necessary; or you could make a virtue of necessity by painting the houses black from the start and, to relieve the gloom, draw up the brickwork joints with very fine white lines and paint the window casings a very light color. This was done in many cases and the charming effect became a characteristic London trait.

On each building lot there was only one house for one family. It might be a very large household with many family members and servants. In Paris there were usually many families in each house. At the entrance there was (and still is) a special Paris institution, *le concierge*. No one could enter or leave without passing him. He knew every inhabitant of his little kingdom on the other side of the entrance and saw to it that they received their mail and anything else brought to the door for them. . .

Under the influence of Carlyle, Charles Dickens wrote a book in 1859 which was very different from all his other books. Instead of describing his own time, he produced an historical novel, *A Tale of Two Cities*. It has not the documentary interest of many of his other books and there is, undoudtedly, some exaggeration in his description of monarchical Paris as compared to free London. But with amazing power he conjured up the two cities in unforgettable visions. As a symbol of Paris stands the minute and harrowing description of a staircase leading up through a tall tenement house, a steep and foul shaft with the doors of innumerable flats opening on to it. It symbolized the Paris that was tightly constricted within closed boundaries and which had to grow vertically because it could not spread out. At the very top of this dismal winding staircase was the miserable room where the noble and unhappy Dr. Manette had been brought after his mysterious release from the Bastille where he had been a life prisoner. Later in the book he is brought to London by friends and there we see him, sitting under a plane tree in his garden in Soho, a district of lovely squares where many emigrants found refuge. It is the London of the open spaces, with its air of humanity and with its green trees and black houses. ■

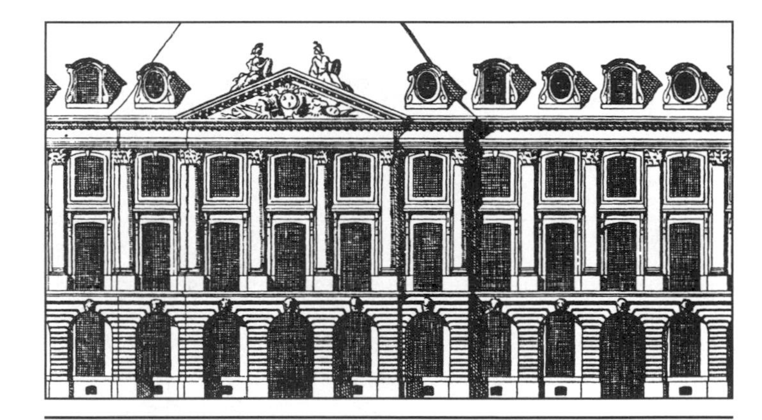

Section of façade, *Place Vendome,* Paris. Scale 1 : 500.

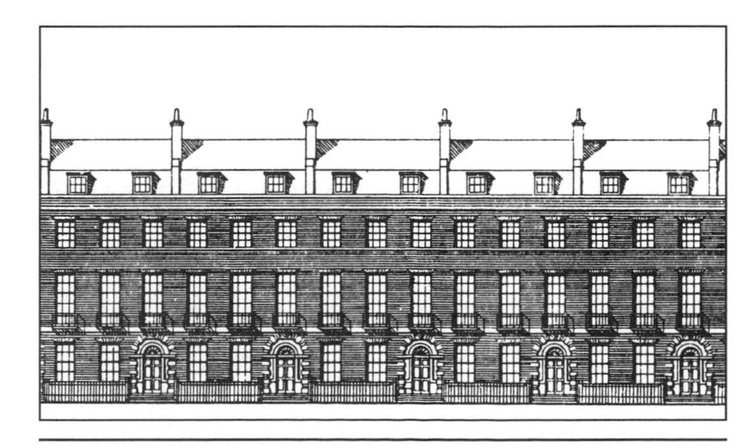

Façades on Bedford Square, London. Scale 1 : 500.

Clarence S. Stein

Summary
Clarence S. Stein (1882–1975) became one of the most influential urbanists of 20th century America, through his writing and extensive work as a town planner, much of it in collaboration with the architect Henry Wright, and his leading role in organizations such as the Regional Planning Association of America and the New York State Commission of Housing and Regional Planning (both founded in 1923). Early projects such as Sunnyside Gardens in Queens, New York (1924), led to Stein and Wright's design of Radburn, New Jersey (begun in 1928), where the principles of Stein's version of the garden city were clearly articulated. These include many key components of modernist city planning as well as what would become standard planning practice for suburban development: the superblock (traceable to Raymond Unwin's garden city work); the *cul-de-sac*; specialized roadways and separation of vehicular and pedestrian traffic (after the precedent of Olmsted's Central Park); and the neighborhood unit with an elementary school at its center (also developed by Clarence Perry).

Key words
automobiles, *cul-de-sac*, garden city, highways, neighborhood, new town planning, residences, superblocks

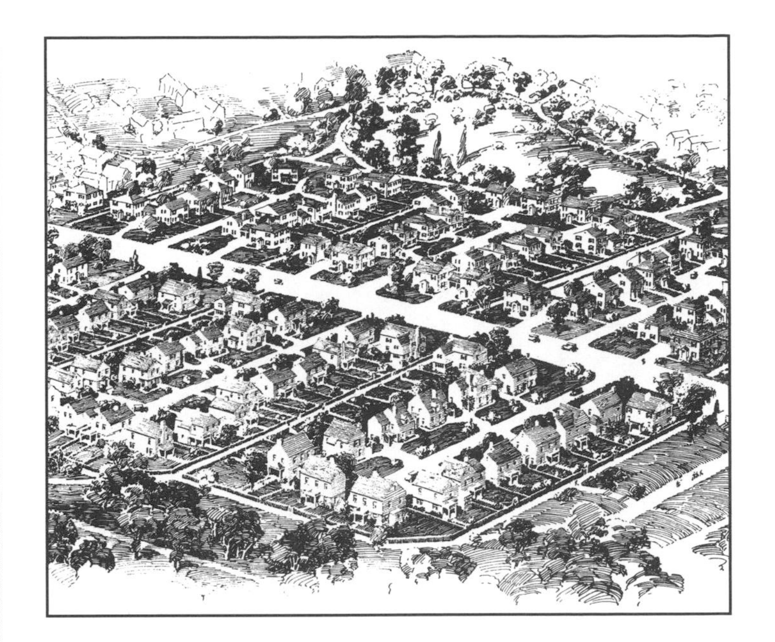

The Radburn Idea

Radburn's ultimate role was quite different from our original aim. It was not to be a Garden City. It did not become a complete, balanced New Town. Instead of proving the investment value of large-scale housing it became, as a result of the depression, a financial failure. Yet Radburn demonstrated for America a new form of city and community that fits the needs of present day urban living in America, and it is influencing city building throughout the world. We did our best to follow Aristotle's recommendation that a city should be built to give its inhabitants security and happiness.

The need for Radburn. American cities were certainly not places of security in the twenties. The automobile was a disrupting menace to city life in the United States—long before it was in Europe. In 1928 there were 21,308,159 automobiles registered (as compared with 5 in 1895). The flood of motors had already made the gridiron street pattern, which had formed the framework for urban real estate for over a century, as obsolete as a fortified town wall. Pedestrians risked a dangerous motor street crossing 20 times a mile. The roadbed was the children's main play space. Every year there were more Americans killed or injured in automobile accidents than the total of American war casualties in any year. The checkerboard pattern made all streets equally inviting to through traffic. Quiet and peaceful repose disappeared along with safety. Porches faced bedlams of motor throughways with blocked traffic, honking horns, noxious gases. Parked cars, hard grey roads and garages replaced gardens. It was in answer to such conditions that the Radburn plan was evolved. For America it was a revolution in planning; a revolution, I regret to say, which is far from completed.

ELEMENTS OF THE RADBURN PLAN

"The Radburn Idea"' to answer the enigma "How to live with the auto," or if you will, "How to live in spite of it," met those difficulties with a radical revision of relation of houses, roads, paths, gardens, parks, blocks, and local neighborhoods. For this purpose it used the following elements:

- **The Superblock in place of the characteristic narrow, rectangular block.**
- **Specialized roads planned and built for one use instead of for all uses:** service lanes for direct access to buildings; secondary collector roads around superblocks; main through roads, linking the traffic of various sections, neighborhoods and districts; express highways or parkways, for connection with outside communities. (Thus differentiating between movement, collection, service, parking, and visiting.)

Credits: This article is excerpted from C.S. Stein, *Towards New Towns for America,* copyright 1957 by Clarence S. Stein and is reprinted by permission of MIT Press, Cambridge, MA.

Fig. 1. Radburn: plan of the residential districts, 1929. Courtesy New York City Housing Authority.

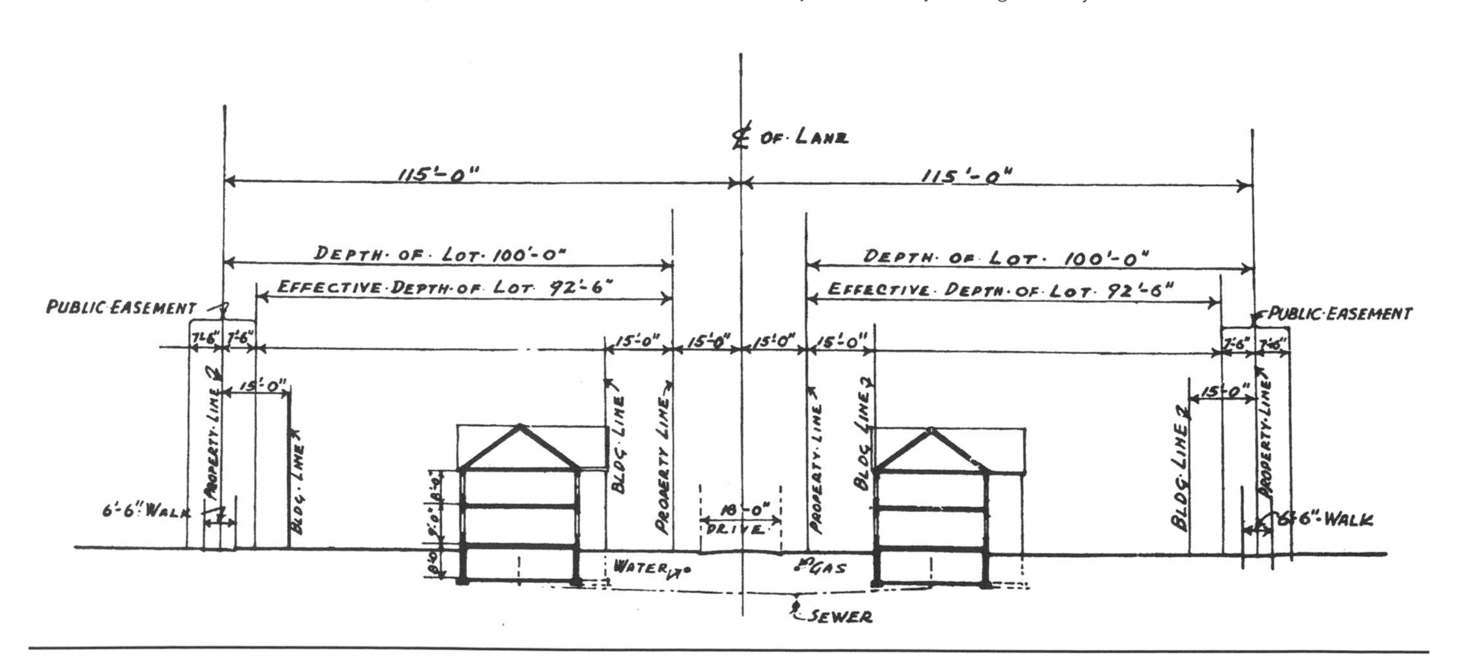

Fig. 2. Typical transverse section of a "lane" in the first unit of Radburn.

- **Complete separation of pedestrian and automobile, or as complete separation as possible.** Walks and paths routed at different places from roads and at different levels when they cross. For this purpose overpasses and underpasses were used.

- **Houses turned around.** Living and sleeping rooms facing toward gardens and parks; service rooms toward access roads.

- **Park as backbone of the neighborhood.** Large open areas in the center of superblocks, joined together as a continuous park. (Figs. 1–3).

Geddes Smith (Smith 1930) described Radburn compactly in 1929 as:

> *"A town built to live in—today and tomorrow. A town 'for the motor age.' A town turned out-side-in—without any backdoors. A town where road and parks fit together like the fingers of your right and left hands. A town in which children need never dodge motor-trucks on their way to school. A new town—newer than the garden cities, and the first major innovation in town-planning since they were built."*

Precedents

None of the elements of the plan was completely new. The distinctive innovations of Radburn were the integrating superblocks, specialized and separated means of circulation, the park backbone, and the house with two fronts. Radburn interwove these to form a new unity, as a practical and attractive setting for the realties of today's living.

There were precedents for all the elements.

Superblocks with great green interiors had been built in America. Before 1660, the Dutch in *Niece* Amsterdam (New York) built their homes around the periphery of large blocks, with farms behind and sometimes with a great garden core. However, throughout the nineteenth century and the early twentieth, most city growth was based on the repetitious geometric gridiron; a plan for facile plotting, surveying, legal recording—but not a plan for living. So Henry Wright and I went to Britain, on a special investigation to study superblocks with *cul-de-sac*, before we started planning Radburn. We concluded that, because of the greater use of the automobile in America, we were justified in increasing the size of superblocks over those at Welwyn, Letchworth and Hampstead Garden Suburb. The Radburn blocks were 30 to 50 acres (12 to 20 ha) in size. Their outlines were determined by their internal needs and by topography. Because of our heavier automobile traffic we faced fewer houses on main highways than most of the British examples. The English experiences helped us greatly, but if the superblock had not existed logic would have forced us to invent it. A rational escape from the limitations of the checkerboard plan in which all streets are through-streets, with the possibility of a collision between auto and pedestrian every 250 feet (76 m), compelled it.

Culs-de-sac. The dead-end lane had served in England for peacefulness and for economy of roads and utilities. *Culs-de-sac* had been used occasionally in our colonial villages. But the typical earlyAmerican arrangement of houses was along the main, and sometimes only, road. This was more neighborly, and it was easier to shovel snow away in winter. The costliness of through street pavement and main line utilities was not yet a factor of economic importance. Later the extravagance was not understood. Real estate and municipal engineering customs perpetuated obsolete forms.

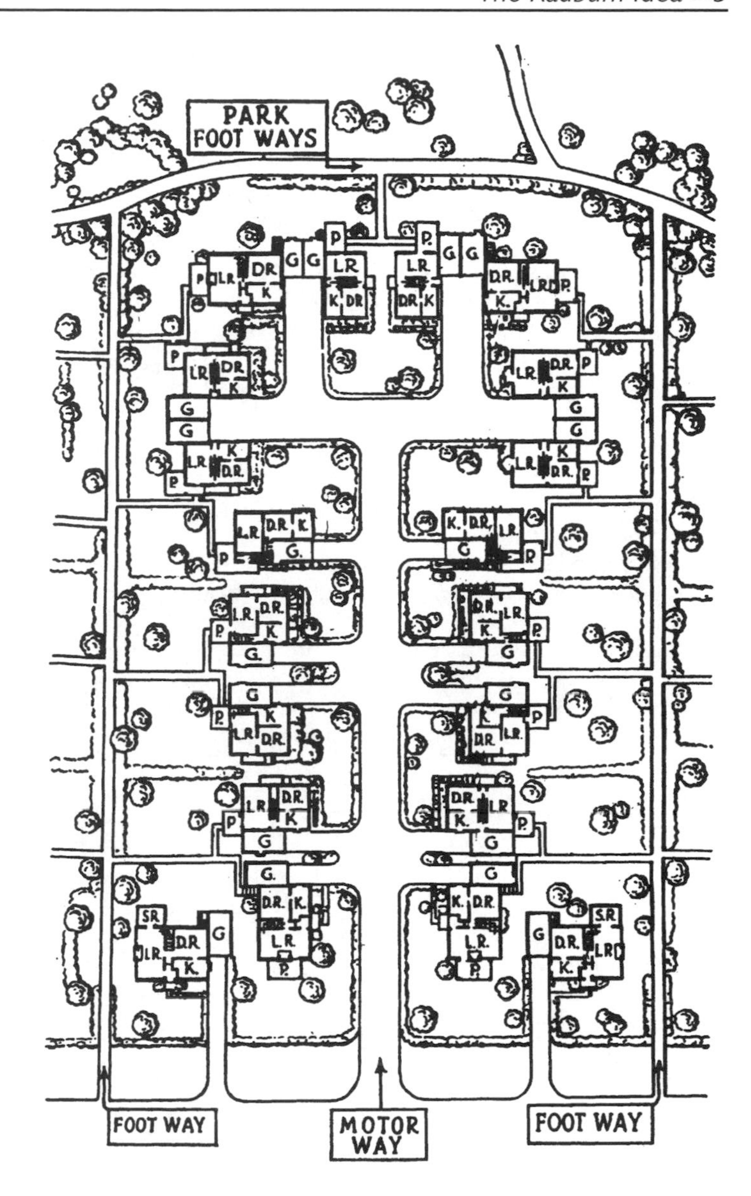

Fig. 3. Plan of a typical "lane" at Radburn. The park in the center of the superblock is shown at the top; the motorways to the houses are at right angles to the park.

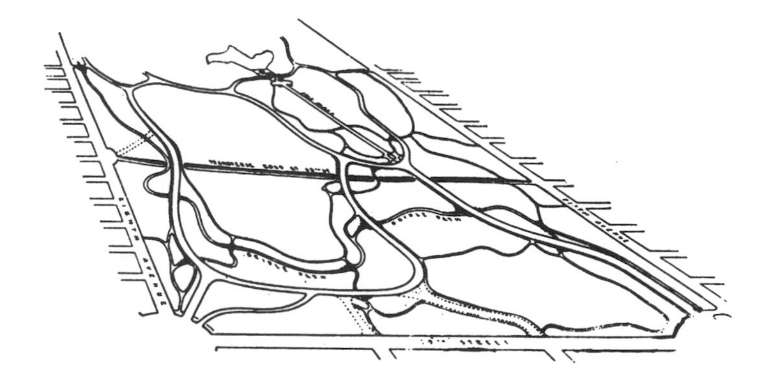

Fig. 4. Perspective sketch showing the separation of routes for vehicles, equestrians, pedestrians and outside traffic at the south end of Central Park. Greater comfort and safety is attained on all routes by the elimination of grade crossings, as planned by Frederick Law Olmstead.

I have already spoken of our experience with courts opening off streets at Sunnyside.

Separation of different means of communication had an excellent nearby precedent, Central Park in New York (Fig. 4). Here, almost half a century before the invention of the automobile, Frederick Law Olmsted and Calvert Vaux planned and executed what they described as:

> "*. . . A system of independent ways; 1st for carriages; 2nd, for horsemen; 3rd, for footmen; and 4th, for common street traffic requiring to cross the Park. By this means it was made possible . . . to go on foot to any district of the Park. . .without crossing a line of wheels on the same level. . .*" (Olmstead, 1851)

The automobile has multiplied the need of separating antagonistic uses of streets. The need is recorded in the statistics of automobile accidents—33,410 deaths in 1946, to say nothing of the million of more cripples. At Radburn we proposed to unscramble the varied services of urban streets. Each means of circulation would take care of its special job and no other: through traffic only on the main highways; with street intersections decreased about two-fold; most parking as well as garages, delivery, and other services, on the lanes; walks completely separated from autos by making them part of a park instead of a street, and by under or over passing the roads; finally, children's play spaces in the nearby park instead of in busy roads.

Specialized highways were in their infancy in the United States at the time that Radburn was conceived. There was not much more than the differentiation of parkways and pseudo-expressways from the ordinary city or town street. To plan or build roads for a particular use and no other use required a predetermined decision to make specialized use permanent or rather long-lived. That was contrary to the fundamentals of American real estate gambling, to serve which the pattern of ordinary highways had become the basis of city planning. I say this in spite of the fact that the 1920's were the heyday of zoning. None of the realtors, and few city planners who accepted zoning as their practical religion, seemed to have faith enough in the permanency of purely residential use to plan streets to serve solely that use. No, not even when the economy of so doing was clearly proved by Henry Wright and Raymond Unwin. Zone for dwellings? Yes, but don't give up the hope that your lot may be occupied some day by a store, gas station, or other more profitable use.

The Radburn Plan proposed to protect the residents, first, by planning and building for proposed use, and no other use; second, by private restrictions rather than by wishful zoning.

The house turned round. The creation of the Radburn Idea and of the Radburn Plan was a group activity. It was not merely the conception of its architect-planners. It took form out of actual experience at Sunnyside. It was influenced by the character and diversified abilities and experience of the technicians and the staff of the City Housing Corporation. But there can be no question that the seed from which the Radburn idea grew was conceived by that imaginative genius Henry Wright. Luckily we have in his own words "The Autobiography of Another Idea"—that is, the Radburn idea (Wright 1930).

> *"In 1902, as an impressionable youth just out of architectural school. . .at Waterford. . .Ireland, . . .I passed through an archway in a blank house wall on the street to a beautiful villa fronting upon spacious interior gardens. That archway was a passage to new ideas. . .I learned then that the comforts and privacy of family life are. . .to be found. . .in a house that judiciously related living space to open space, the open space. . .being capable to enjoyment by many as well as by few."*

From that time on Henry [Wright] started "to face kitchens and service rooms toward the street, and living rooms inward toward the garden." At Sunnyside we both wanted to turn all the houses that way, as we ultimately did at Radburn, but conservative opposition only permitted placing some of the porches on the lovely garden side.

Economy of the Radburn Plan. The parks that formed the interior core of the Radburn superblocks were secured without additional cost. Or rather the savings in expenditure for roads and public utilities at Radburn, as contrasted with the normal subdivision, paid for the parks. The Radburn type of plan requires less area of street to secure the same amount of frontage. In addition, for direct access to most houses, it uses narrower roads of less expensive construction, as well as smaller sized utility lines.

The superblock of 35 to 50 acres (14 to 20 ha) is surrounded by wide streets, but it replaces the greater number of wide broad streets of the normal checkerboard plan with service roads only 18 to 20 feet (5.5 to 6 m) wide. The use of these is limited to 15 to 20 families living on each *cul-de-sac*, and they carry no through traffic going elsewhere. Therefore they can be on lighter construction, and sewers and water lines are of lesser size and cost than the main lines on the through highways. In fact the area in streets and the length of utilities is 25 percent less than in the typical American street plan.

The saving in cost of these not only paid for the 12 to 14 percent of the total area that went into internal parks, but also covered the cost of grading and landscaping the play spaces and green links connecting the central block commons. The greater part of this expenditure was for improvement. The land itself—in spite of its value for spinach growing—cost only six cents a square foot (.09 sq. m). What makes subdivided land costly, even with the financing, carrying charges, taxes, and profits, is not the land itself. It is the roads and walks, sewers, water lines, electric, gas and other utilities that surround it. This land in lots along streets or lanes costs 6 cents gross or 10 cents per square foot, but an additional 25 cents must be added to pay its share of the improvement that lead to it. A park or playground in a regular town surrounded directly by improved streets would cost as much as it would with houses as a frontage. But not at Radburn—their land is just land (except for surrounding walks). There are no streets. So before landscaping the land, the cost of the parks was less than a fifth of what it would have been had dangerous highways encircled it.

The plan of Radburn

The time between the purchase of land in Fairlawn and the starting of construction was too short to develop a plan of Radburn as a whole (Fig. 5). This was vaguely in the back of our minds, to be given more definite form later. Our immediate problem was to relate the superblocks to the form of the land. We began with an area near the railroad station. As we did not want direct access to *culs-de-sac* from Fairlawn Avenue, which promised to become a main thoroughfare, we left a strip, an ordinary block 200 feet (61 m) wide, between it and our first superblock. If we had had time to study our whole plan carefully before deciding on the first superblocks, we probably would have eliminated all of the old forms of block and separated the superblock from Fairlawn Avenue merely by a parallel service road.

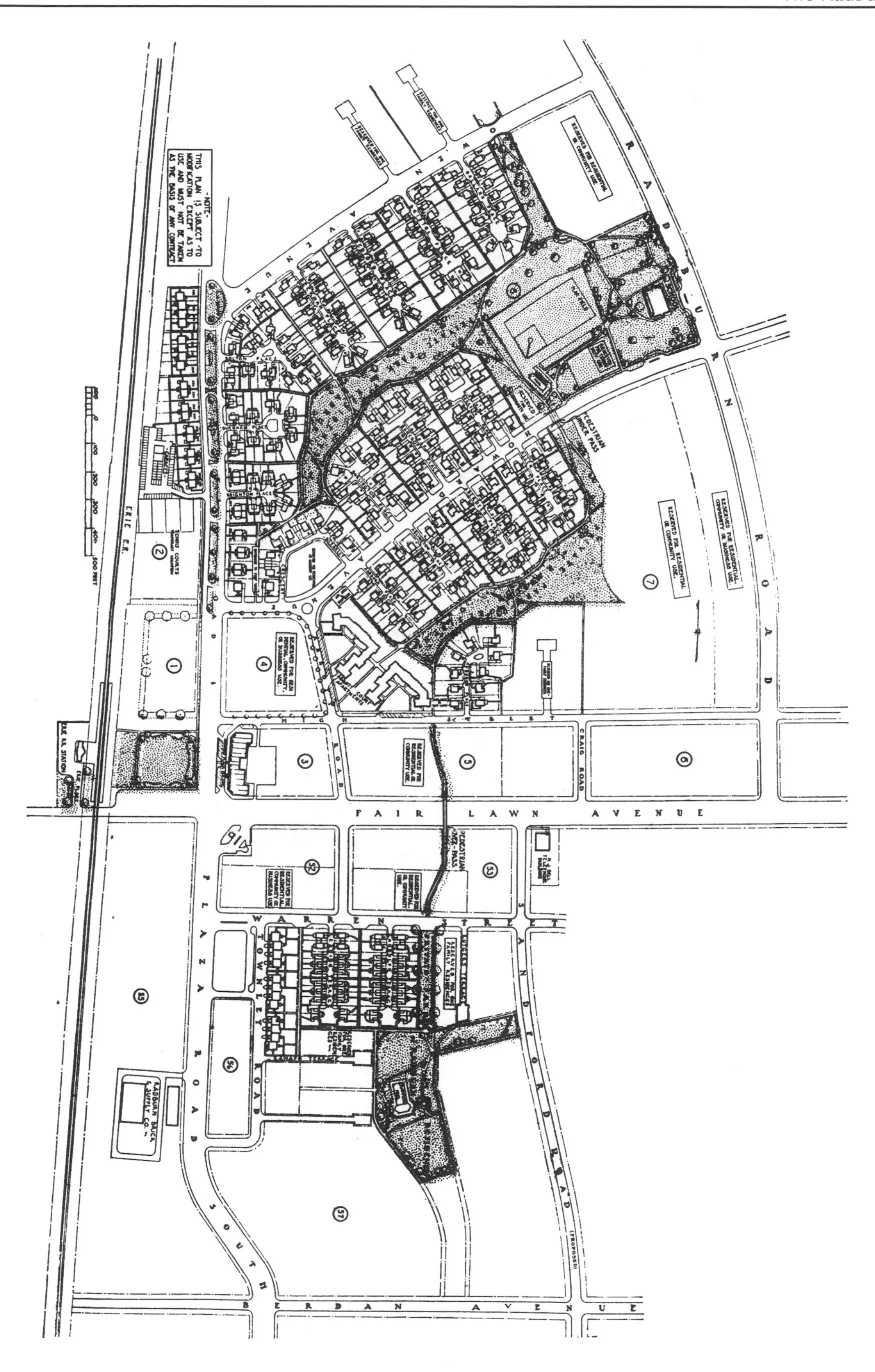

Fig. 5. Plan of Radburn completed by 1930.

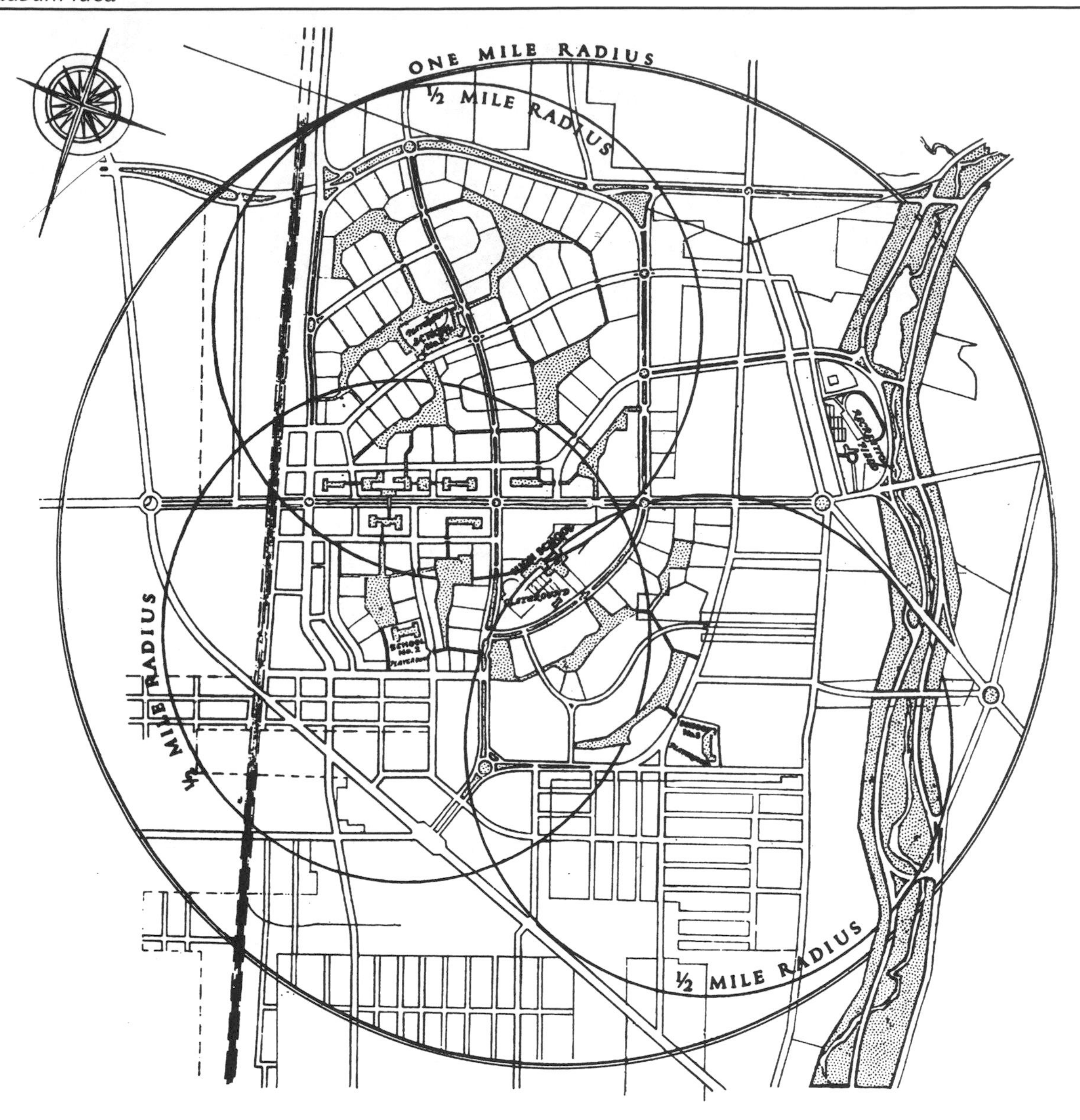

Fig. 6. General plan showing neighborhoods.

For these blocks have not lent themselves as well to practical development for modern living or shopping.

Neighborhoods. At Radburn, I believe, the modern neighborhood conception was supplied for the first time and, in part, realized in the form that is now generally accepted.

The neighborhoods were laid out with a radius of half a mile, centering on elementary schools and playgrounds. Each was to have its own shopping center. The size of the neighborhood was determined by the number of children cared for by a single school. So as to allow for flexibility in development, we tentatively overlapped our half-mile (.8 km) circles (Fig. 6). This left leeway for somewhat greater concentration of population in apartments or row houses, where it would be found most advisable to place these as building progressed. All parts of each neighborhood were to be connected by over and under passes.

The neighborhoods were planned for 7,500 to 10,000 people—this to depend on the most desirable number of pupils in a school—a matter that was then, and I believe still is, open to a wide diversity of opinions. Although a start was made in the building of two of the neighborhoods, ultimately neither was completed.

Town plan: the town as a whole. As a main educational and cultural center we chose a point nearby equidistant from the three proposed elementary schools, within a mile radius of all future houses. This was close to the intersection of the main north-south and east-west avenues. We planned to set the high school and town community building on a beautiful hill. Below was a low nearly marshy area. This, although not desirable for residential purposes, was excellent for the central recreational field, to serve both high school and town athletic needs.

The main commercial center might, we felt, serve as a regional market.

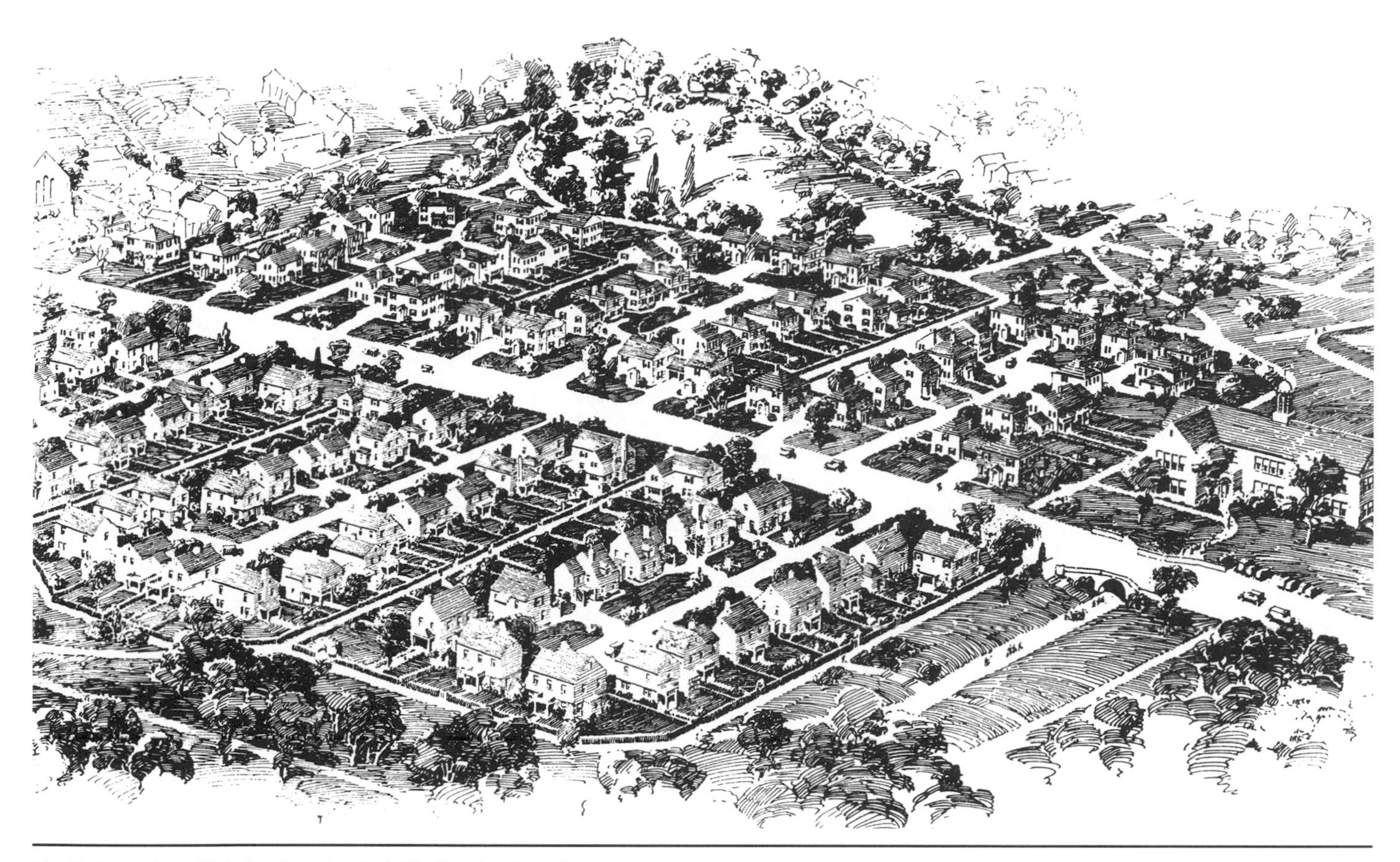

Fig. 7. Perspective, 1929. Courtesy New York City Housing Authority.

So we located it close to the proposed state throughway at the main entrance to Radburn, rather than in the physical center. We assumed that most of the regional market's clients would come by automobile. Therefore we planned superblocks, with an interior area of some 400,000 square feet (37,160 sq. m), to permit the parking of some 1,250 cars. This parking area was to be used in the evenings by the nearby Regional Theater.

For industry the section to the south of the State Highway was planned. This would have had direct access not only from the main entrance highways, but also from a spur from the railroad.

[Ed. Note: the original text is followed here by an extended discussion of "How the Radburn Plan Worked," including housing unit plans (Figs. 7 and 8) and evaluation commentary. The "Conclusions" section below summarizes the detailed discussion of Radburn].

CONCLUSIONS

I. The Radburn Plan serves present day requirements of good living in a more practical and pleasant way than does the conventional American city pattern.

- It is safer.
- It is more orderly and convenient.
- It is more spacious and peaceful.
- It brings people closer to nature.
- It costs less than other types of development with an equivalent amount of open space.
- Most people who live in Radburn prefer it. They enjoy the expansive nearby verdure; they appreciate the freedom from worry about their children's' safety.

Radburn works in practice as it was intended to function when it was only the Radburn Idea, twenty years ago.

II. A plan for living, in addition to an appropriate, flexible physical setting, requires an organization with vision, capable leaders and adequate finance for the operation of the physical plant.

- Until there are competent and well financed governmental agencies for this purpose, a private association is essential.
- Start to function when the New Town opens.
- Include in its membership all families in the community—both tenants and homeowners. All must pay for its services just as they pay rent and taxes.

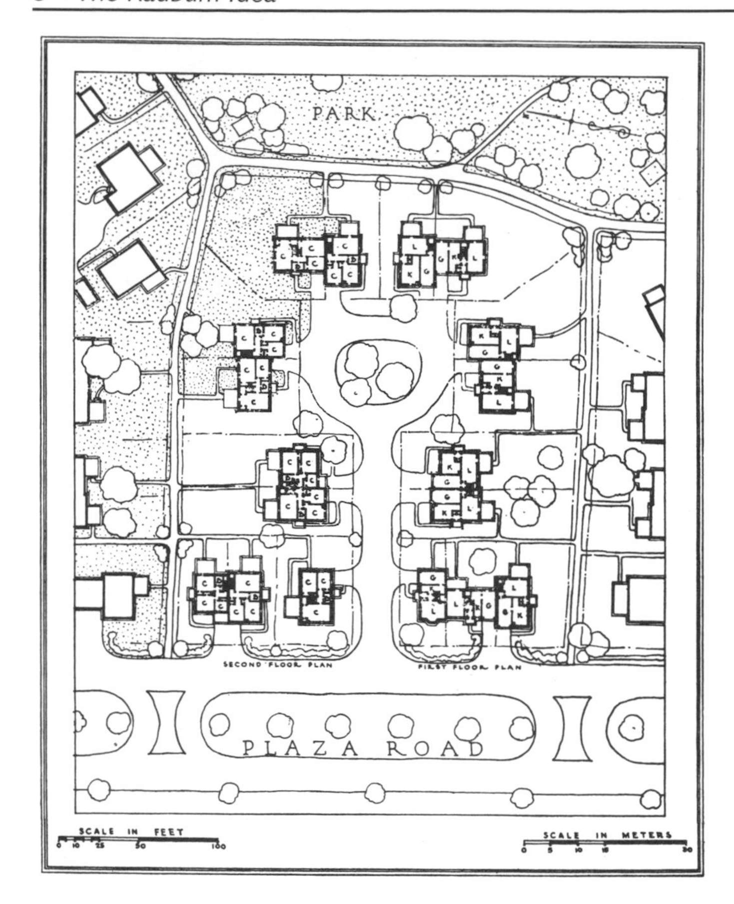

Fig. 8. Plan of Burnham Place. This, with is grouped houses and turning circle, is the most spacious *cul-de-sac* at Radburn. The turning circle allows vehicles to turn more easily and provides an island for planting.

- Be a single central organization rather than a group of separate sectional block associations.

III. A separate political entity is required by a New Town with a new form and advanced objectives, so that it may freely and clearly carry out its purposes.

- A private government within the borders of a political entity, which gives special services and privileges to its members, which are not available to the entire urban area, causes resentment and leads to disunion.

- All services for which people are taxed should be directed and operated by their elected representatives.

IV. The building of a new town requires large capital investment in land, utilities, highways and public buildings on which there can be little, if any, financial return for many years. Lacking governmental assistance, a private corporation (with the exception of organizations with large aggregates of capital such as insurance companies or endowed foundations) have small chance of more than temporary success under [uncertain] economic conditions [Ed. Note: author is referring here to the United States depression era that followed the construction of Radburn].

Governmental cooperation is required, at least in the following:

- Taking land—all the land that will be needed to complete the New Town.

- Holding the land until needed for construction; or financing the land cost at low rates for long periods.

- Financing the cost of main lines and central works of essential utilities and main highways, on low and long financial terms.

- Assisting the local government authorities in the construction of essential public buildings such as schools.

- Financial aid similar to that given to existing municipalities including subsidies, for housing low-income workers.

V. Continuous rapid growth of a New Town is imperative in the early years, so that overhead expenses do not devour all earnings.

VI. Conveniently placed and varied industry is an essential requirement of a New Town. Therefore industrial plans must be specific and realistic. Generalizations are valueless. Timing of industrial development must be synchronized with that of the building of homes and community equipment. ■

REFERENCES

Olmsted, Frederick Law, 1851. *Forty-Eight Years of Architecture. Vol. 2: Central Park.*

Smith, Geddes, 1930. *A Town for the Motor Age.* Survey Graphics reprint for the New York City Housing Corporation.

Wright, Henry. 1930. "The Autobiography of Another Idea." Regional Planning Association of America. Reprinted in *The Western Architect,* September 1930.

Paul Zucker

Summary

Paul Zucker focused on space and its representation as central to understanding architecture and urbanism. Along with scholars A.E. Brinckmann and Siegfried Giedion, he applied rigorous methods of spatial analysis to urban form. He shared with Giedion an interest in the emergent concept of space-time and its role in the perception of architecture and cities. *Town and Square* provides a comprehensive historical application of these ideas, attempting both a survey and a systematic categorization of different approaches to making of urban space.

Key words

architectural history, civic design, gardens, human scale, public parks, square, village green, vision angle

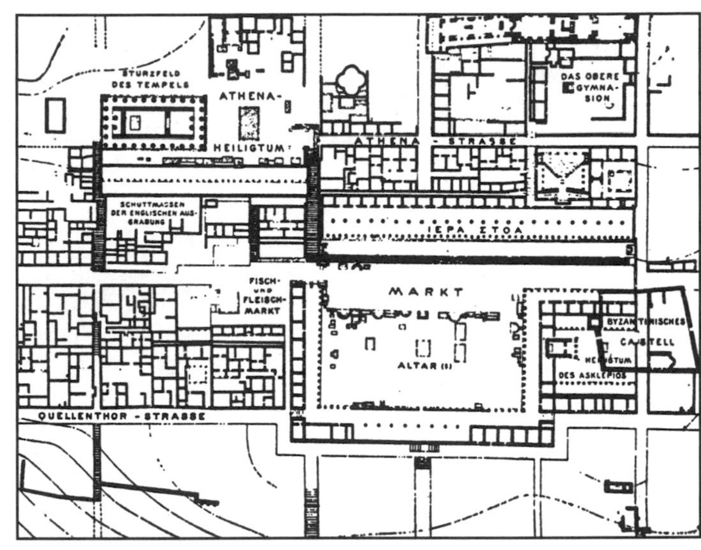

Priene

The square in space and time

St. Peter's Square in Rome, St. Mark's Square in Venice, and the Place Vendome in Paris are as generally known and admired as are Leonardo da Vinci's *Mona Lisa*, Michelangelo's *Moses*, and Rembrandt's *Night Watch*. These squares are undoubtedly as much "art" as any painting, sculpture, or individual work of architecture. The unique relationship between the open area of the square, the surrounding buildings, and the sky above creates a genuine emotional experience comparable to the impact of any other work of art. It is only of secondary importance for this effect whether and to what extent in each instance specific functional demands are fulfilled. Obvious as this statement may be, in our age of over-rationalization this fact must be emphasized. During the last decades, city planners have been primarily concerned with such problems as the use of land, the improvement of traffic and general communications, zoning, the relationship between residential and industrial areas, and so forth. These considerations have somewhat overshadowed the fundamental importance of the square as a basic factor in town planning, as the very heart of the city. Only now does interest turn toward this central formative element, "which makes the community a community and not merely an aggregate of individuals."

This physical and psychological function of the square does not depend on size or scale. The village green in a small New England town, the central square of a residential quarter within a larger city, the monumental plaza of a metropolis—all serve the same purpose. They create a gathering place for the people, humanizing them by mutual contact, providing them with a shelter against the haphazard traffic, and freeing them from the tension of rushing through the web of street.

The square represents actually a psychological parking place within the civic landscape. If one visualizes the streets as rivers, channeling the stream of human communication—which means much more than mere technical "traffic"—then the square represents a natural or artificial lake. The square dictates the flux of life not only within its own confines but also through the adjacent streets for which it forms a quasi estuary. This accent in space may make itself felt some blocks in advance—an experience shared by everyone who has ever driven a car in to an unfamiliar town.

This psychological function of the square is as true for the present and future as it has been for the past. As a matter of fact, the city planner of the past faced the same kind of problems as does the city planner of today: either the building of an entirely new settlement or—much more frequently—the reorganization of individual quarters within existing towns. Whether the curve of a street corner must be planned for the stoppage of a noble coach drawn by eight horses or for a line of Fords, bumper to bumper; whether an open area has to be shaped for royal spectacles or for political rallies, does not make any difference in principle. Such functional considerations have

Credits: This article is the opening chapter of Paul Zucher's monograph, *Town and Square: from the Agora to the Village Green*, first published in 1959 by Columbia University Press and, in paperback, by MIT Press, 1970. Reprinted by permission of MIT Press.

influenced width, length, and depth of streets and squares, their directions and their connections, then and now equally, so that one can say that planning in space today is hardly more functionalistic than it was in earlier centuries. The needs and demands of the past may have been fewer and less complex, but they were as basic for the determination of the final shape as they are now. Thus, our analysis of typical examples of the past need not remain a mere historical discussion, but should also stimulate some thoughts for town planning of today. While technical and socioeconomic conditions have changed completely since the Industrial Revolution, this should not deter us from applying lessons from the past to conditions and needs of the present. This does not mean, however, that the most impressive and convincing achievements of earlier centuries should simply be copied—a mistake to which nineteenth-century architects unfortunately succumbed only too often.

There exist today in towns and cities "squares" marked as such on maps which actually are no more than plain voids, empty areas within the web of streets. They differ from other areas of the town by the mere fact that they are bare of any structures. Artistically relevant squares, however, are more than mere voids; they represent organized space, and a history of the square actually means a history of space as the subject matter of artistic creation. For the consideration of the square as work of art it is not even important whether the confines of this "space" are real and tangible or whether they are partially imagined, if only the planner is able to guide and force our imagination in one distinct direction.

The term "space" has been used and misused so often and in so many contexts that it seems to have lost its intrinsic meaning. Therefore it may be helpful to define it anew in relation to city planning in order to give it freshness and usefulness. Here, "space," designating generally a three-dimensional expansion of any kind, is used more specifically. It means a structural organization as a frame for human activities and is based on very definite factors: on the relation between the forms of the surrounding buildings; on their uniformity or their variety; on their absolute dimensions and their relative proportions in comparison with width and length of the open area; on the angle of the entering streets; and, finally, on the location of monuments, fountains, or other three-dimensional accents. In other words, specific visual and kinesthetic relations will decide whether a square is a hole or a whole.

Exactly as towns in their totality either have grown naturally from villages, trading posts, military camps, castles, and monasteries, or were built following a preconceived design, so the individual square within a town either might have developed gradually out of certain existing conditions or might have been planned. Conditions which furthered the natural growth of a square were manifold: the intersection of important thoroughfares within the town; the open space at the approach of a bridge or in front of the west façade of a church, etc. In some instances, such sites became the nucleus for an artistically meaningful square of distinct three-dimensional form. However, most frequently a village green or an old market square expanded gradually into a definite spatial pattern through the successive erection of buildings around it. Such a development, typical for so many squares in Hellenistic and late medieval towns of Central Europe, needed, of course, centuries until the harmonious present-day appearance evolved.

In contrast to these organically grown squares, the planned square always appears as clearly defined as any individual piece of architecture. In some instances, generations of anonymous builders may have anticipated the underlying basic design in area and form before its actual realization; more often, individual artists have conceived the final scheme.

Planned squares, clearly recognizable as such, appeared in ancient Greece and her colonies from the fifth century B.C. on. They resulted indirectly from the gridiron scheme which then was introduced into Greece and Asia Minor by Hippodamus, although archaeological research has not yet definitely proved his personal part in these creations. In Hellenistic times, the majority of squares were planned. Here again, documentary evidence for the activities of individual architects is lacking. The same holds true for the most grandiose organization of squares in antiquity, the Imperial Fora in Rome. After the decline of ancient civilization with its Hellenistic and Roman creations, planned squares appeared again in the French and English bastides and the foundations by the Teutonic Knights in eastern Germany. In the early Renaissance, from the time of Leone Battista Alberti and Leonardo da Vinci onward, architects, in drawings and theoretical treatises, competed as fervently in the planning of whole towns and of individual squares as they did in creating churches, palaces, villas, and gardens. Later this architectural interest in the design and execution of squares reached its climax during the seventeenth and eighteenth centuries, the era to which posterity owes the majority of world-famous squares, especially in Italy and France.

The abundance of squares in Italy and France may be explained by a combination of climatic conditions and temperamental attitudes characteristic of the Romance peoples of Southern and Western Europe. These conditions led to a form of public life—and life in public—which made street and square the natural locale for community activities and representation. Not by chance then, Rome and Paris are the cities, which we associate primarily with the idea of the perfect public square, and it is therefore logical that in an analysis of generic types so many Roman and Parisian squares should be discussed.

However, almost identical climatic conditions in Spain and Greece, in the same Mediterranean area, have brought forth since ancient times neither a considerable number of squares nor any individual square comparable to the great creations in Rome and Paris. The reason for the lack of consciously shaped squares in Spain may be sought in the fact that even at the apex of Spain's political and economic power, in the sixteenth and seventeenth centuries, her specific societal structure and the psychological attitude of her population were not equally favorable for the development of a public life. And England, even during such a great era as the Elizabethan age, did not create any monumental solutions in town planning; nor did Holland in the seventeenth century, in a period of great commercial expansion and magnificent artistic activities. In these two countries it is the northern climate and the strong emphasis on domestic life that mostly prevented any desire for public spatial expression.

Being part of the living organism of a city with its changing socioeconomic and technical conditions, a square is never completed. In contrast to a painting or a sculpture, there is no last stroke of the brush or any final mark of the chisel. A painting or sculpture, once completed, never changes in itself but preserves its original appearance from the time of its creation except for the patina of age. Elements of the square, however, such as the surrounding structures, individual monuments, and fountains, are subjected to the flux of time; some may vanish, be destroyed or razed, others may be replaced and new ones added. Thus the original form of squares and streets may undergo fundamental changes, as the juxtaposition of old engravings and paintings with modern photographs of the same square and streets often shows. While at one time a square may have been primarily an accumulation of important individual buildings, the same square in another century may have developed into a comprehensible spatial form.

Frequently future spatial potentialities of a square were already latent in its initial stage. However, the basic spatial concept of a square is not always so strong that it prevails through centuries. Its appearance may change for two reasons: physically, through the erection of new buildings and the alternation or destruction of old ones, through a modification of the building line, and so forth; psychologically, through the different way in which each generation experiences and reacts to given proportions and distances, and through the new approach by which it interprets spatial relations. It is this combination of objective and subjective factors which makes the same square appear different to each generation. The large variety of impressions received from an identical reality is evidenced by descriptions in travel diaries. The *Piazzetta* in Venice or the *Piazza del Popolo* in Rome have stimulated the visual imaginations of travelers in the seventeenth, eighteenth, and nineteenth centuries in such different ways that it is sometimes hard to recognize the identity of the same locality. After all, our reaction toward nature presents a similar phenomenon: the same landscape is perceived quite differently by a painter of the seventeenth century and by an artist of the nineteenth century.

To mention another transformation of visual impression, closer to our days: in the nineteenth century, squares were no longer regarded as three-dimensional units. With the rise of classicism the awareness of the third dimension has vanished. Nor was it by chance that Impressionism, the ultimate fulfillment of the naturalistic development in painting since the Renaissance, never aimed at any conscious articulation of space—which was to be rediscovered by Cezanne and the twentieth-century Cubists. This lack of interest in space became even more obvious in nineteenth-century sculpture, architecture, and city planning. Squares simply provided opportunities for planting bushes and flowers, for placing statutes and comfort stations without any spatial ties—briefly, they represented a vain attempt to transplant the charm of a miniature English park in to the heart of the city.

With the twentieth century, as a consequence of new beginnings in architecture, the awareness of spatial relations and of the possibilities of their architectural expression returned. The increased kinesthetic sensitivity for three-dimensional forms, which leads from Thorwaldsen to Maillol, from Thomas Jefferson to Frank Lloyd Wright, distinguishes also the judgment of a Camillo Sitte from that of Raymond Unwin or Pierre Lavedan in their evaluation of famous squares and streets of the past. Shakespeare's *Hamlet* and Bach's fugues have been reinterpreted with each generation, as have Leonardo da Vinci's *Last Supper* and Michelangelo's Medici Chapel. Likewise the great creations of town planning of the past must be studied anew in our time. Such re-evaluation will stimulate and help, consciously or unconsciously, the creative work of our own generation.

The archetypes

Space is perceived by the visualization of its limits and by kinesthetic experience, *i.e.,* by the sensation of our movements. In the state of "visual tension," kinesthetic sensation and visual perception fuse most intensely—and the conscious enjoyment of townscapes as artistic experience produces just this visual tension. People in their movements are influenced and directed by three-dimensional confines and by the structural lines of such confines; in other words, the general tension becomes a specifically "directed" dynamic tension. If these confines are architectural structures, their volumes and their scale exert pressure and resistance and stimulate and direct our reaction to the space around us.

For the square, then, three space-confining elements exist: the row of surrounding structures, the expansion of the floor, and the imaginary sphere of the sky above. The forms of these three space-shaping elements—architectural frame, floor, and ceiling—are, of course, most decisively defined by the two-dimensional layout of the square.

These three factors that produce the final three-dimensional effect may vary in themselves: the surrounding structures may be of uniform height, proportion, and design, or they may differ; they may be more or less coherent. The floor, an equally important factor for the appearance of the square, may be homogeneous in expansion and texture (pavement), or it may be articulated by slopes, steps, different levels, and so forth. Its surface pattern may unify or isolate the framing vertical structures. The sky, the "ceiling" of the square, although distant, offers a visual boundary which in spite of its purely imaginary character confines aesthetically the space of the square just as definitely as to the surrounding houses or the pavement. The subjective impression of a definite height of the sky is caused by the interplay of the height of the surrounding buildings and of the expansion (width and length) of the floor. It is strongly influenced by the contours of eaves and gables, chimneys and towers. Generally the height above a closed square is imagined as three to four times the height of the tallest building on the square. It seems to be higher above squares which are dominated by one prominent building, whereas over wide-open squares, such as the *Place de la Concorde* in Paris, the visual distance of the sky is only vaguely perceived.

The correlation of these principal elements that confine a square is based on the focal point of all architecture and city planning: the contrast awareness of the human scale. As long as the size of the human body and the range of human vision are not recognized as the basic principles, any rules about absolute proportions, about design and composition of forms and motifs, or about symmetrical and asymmetrical organization, are meaningless. Experiments have proved that the human eye without moving perceives an expanse of a little more than 60 degrees in horizontal direction. In vertical direction, the expanse depends, of course, on the degree to which the eye opens. An angle of 27 degrees is most favorable for the perception of an individual work of architecture, and the beholder moves instinctively to a distance that allows this angle. However, in order to fuse various architectural units with their surroundings into a total impression, which effect is the genuine task of all city planning, the eye can employ an angle of 18 degrees only.

The appearance of each individual square represents a blend of intrinsic lasting factors (topographical, climatic, national) and of changing influences (stylistic, period-born), *i.e.,* of static and dynamic forces. Although squares of certain types prevail in certain periods, general space-volume relations are independent of particular historical forms. There exist definite basic types of squares which appear again and again. They show common characteristics in their spatial form, although the artistic expressions cannot be pressed into dogmatic categories. The specific function of a square, for instance, as a market square, as a traffic center, or as a *parvis*, never produces *automatically* a definite spatial form. Each particular function may be expressed in many different shapes.

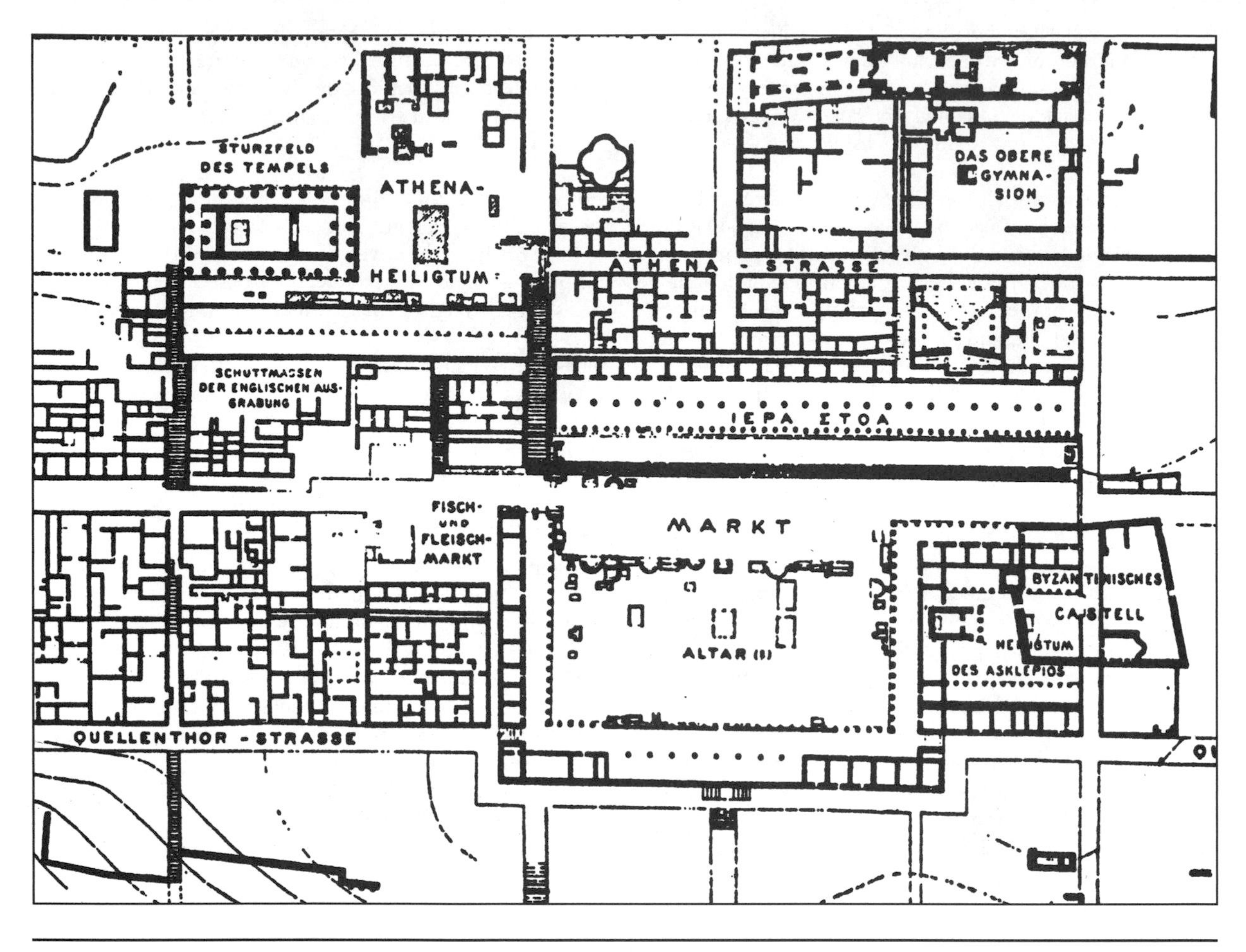

Fig. 1. Priene, plan of the agora (from Wiegand and Schrader, Priene).

Hence it is necessary to separate the various functions from the basic spatial concepts. On the one hand, many market squares may develop later on into monumental civic centers; on the other hand, grand, eloquently decorative plazas may sometimes be transformed amidst a changed neighborhood into mere recreational squares. Such developments prove that the archetypes are structural, that they are spatial, and not functionally, defined. These archetypes—not at all theoretical abstractions—become alive and real as soon as one visualized typical examples. They may be classified as follows:

- **Closed Square:** space self-contained
- **Dominated Square:** space directed
- **Nuclear Square:** space formed around a center
- **Grouped Squares:** space units combined
- **Amorphous Square:** space unlimited

This classification differs principally from the concepts of Camillo Sitte, Josef Gantner, Pierre Lavedan, and Sir Patrick Abercrombie; of course, it does not imply that any square represents only one *pure* type. Very often an individual square bears the characteristics of two of these types: it depends on the point of view, for instance, whether St. Mark's Square in Venice is regarded primarily as a closed square or as one element of the grouped squares of *Piazza* and *Piazzetta*; or whether New York City's Rockefeller Center is a closed square or a dominated square.

Moreover, it is quite obvious that even within one and the same category a variety of architectural expressions exists: the Forum of Trajan in ancient Rome was a closed square, as is the Place Royale in Brussels; but differences in function and sociological meaning in each instance created spatial units entirely different from each other. Likewise, the church-dominated medieval *parvis*, such as the original one of Notre Dame in Paris (now remodeled), differs widely in meaning and appearance from the baroque St. Peter's Square in Rome. And yet in both the dominating volume of a church structure directs the space.

Thus the above-outlined scheme of principal and basic categories should be taken rather as a starting point for aesthetic and historical analysis than as a rigid and dogmatic system.

The Closed Square. As the child imagines in its fantasy each mountain in the archform of a rising cone, looking like Fujiyama, so the average man thinks of a square primarily in the shape of the "closed" square (Fig. 1). Such a square would be visualized as a complete enclosure interrupted only by the streets leading into it. In terms of town planning, the closed square represents the purest and most immediate expression of man's fight against being lost in a gelatinous world, in

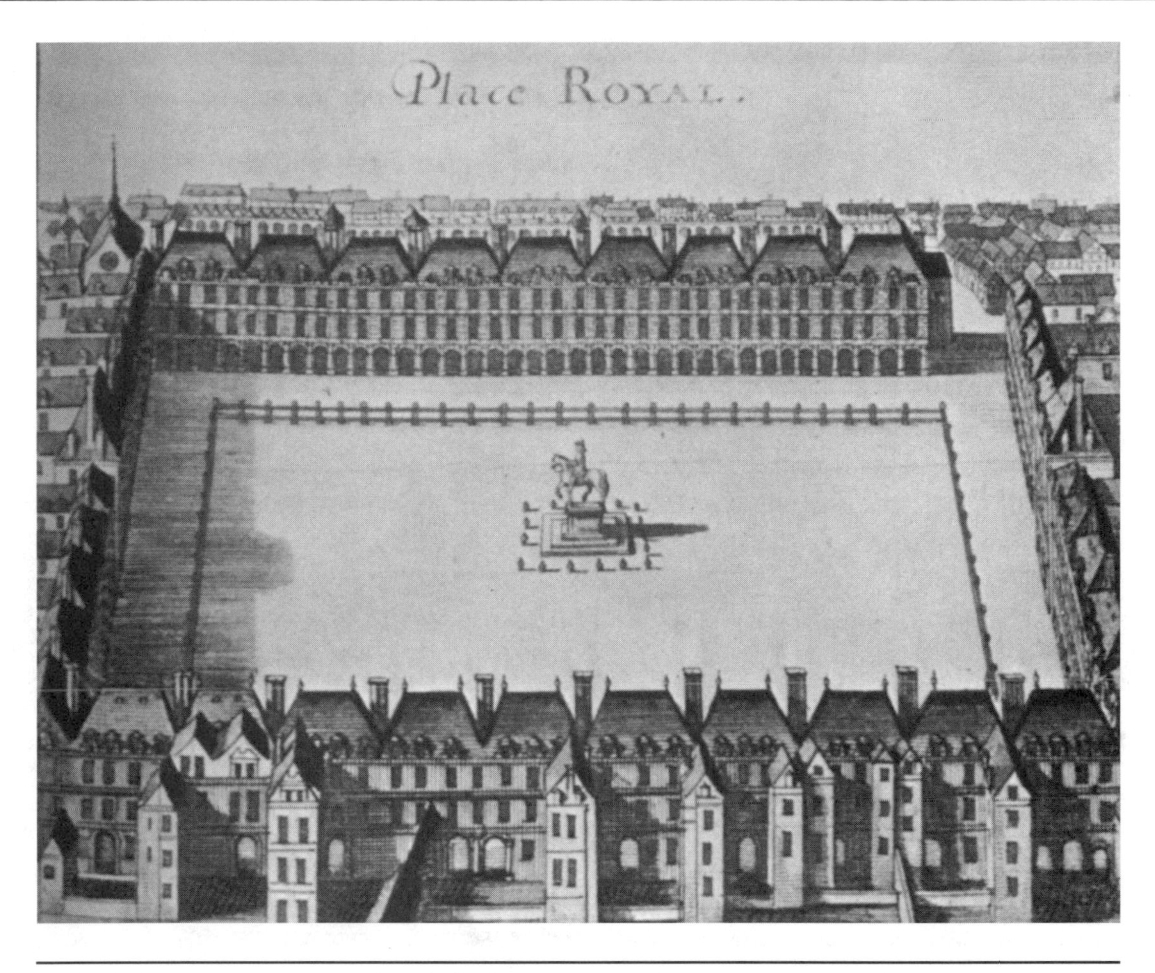

Fig. 2. Paris. Place Royale *(Place des Vosges)*. Copper engraving from Martin Zeiller, *Topographia Galliae*.

a disorderly mass of urban dwellings. Certainly the perfect realization of this form, in its Platonic purity so to speak, as, for instance, in the *Place des Vosges* in Paris (Fig. 2), will be encountered infrequently. This type, without being bound to specific periods or definite architectural styles, appears in its most perfect form in the Hellenistic and Roman eras and then again in the seventeenth and eighteenth centuries.

The primary element in the appearance of any closed square is its layout, be it a quadrangle, rectangle, circle, or any other regular geometrical form. Equally important is the repetition of identical houses or house types, facing the enclosed area, either with their broad fronts or with their gables. Such uniformity need not imply monotony, even when only one single type of structure is repeated all around the square. Mostly, however, a rhythmical alternation of two or more types is employed, the richer architectural accents concentrated on the corners or on the central parts of each side, or framing the streets running into the square.

The spatial impression of the square depends furthermore on differences in scale of the individual units, on the contrasts of higher and lower structures, on their relationships to the width and breadth of the horizontal area, on the location of monuments, fountains, *etc.*, and finally on variations in architectural decoration. Within this organization, the spatial balance of the square will always be achieved by the equation of horizontal and vertical forces. Each façade fulfills a dual function; on the one hand, it is part of an individual structure; on the other hand, it forms part of a common urban spatial order.

Differences in stylistic forms are of secondary importance. The single architectural elements may change. In antiquity, from the Hellenistic agora to the *Imperial Fora* in Rome, continuity and context of the framing structures were achieved by the porticus (colonnade), the rhythmical repetition of the vertical direction through columns. This was the main factor that tied the space together, as, for instance, on the agora in Priene. Later, not longer columns, but another architectural element served to bind the framing elements of the square: from the end of the Middle Ages through the seventeenth and eighteenth centuries, arched arcades connected the surrounding houses, as in the *Place des Vosges* in Paris. Sometimes the arcades of a closed square are continued into the adjacent streets, as, for instance, in Turin and Bologna in Italy, or in Madrid, Spain. This architectural element of the arcade is anticipated as an ornamental motif in the sculptural organization of Roman *sarcophagi*, in the arrangement of figures at the portals of early medieval churches, *etc.*

Yet, not every spatial enclosure, surrounded on all sides and thus creating a tightly knit architectural entity, can be considered as a closed square. The dimensions of the cloisters of a medieval monastery, such as Monreale or the *Certosa* in Pavia, Italy, or of an English collegiate square as in Oxford, or of the inner courtyard within a

Fig. 3. Paris. Notre Dame Cathedral and parvis before its remodeling in the nineteenth century. Copper engraving from Martin Zeiller, *Topographia Galliae*.

complex monumental structure, such as the Palace of the Doges in Venice or the *l'Hopital de Dieu* in Beaune, may be almost identical with the dimensions of a smaller Hellenistic agora or of a medieval arcaded town square. Nonetheless, those enclosures are not "squares" from the viewpoint of town planning. They represent rather an element of a comprehensive architectural organization and differ from the square in their sociological function by not serving public life and traffic.

The Dominated Square. The dominated square is characterized by one individual structure or a group of buildings toward which the open space is directed and to which all other surrounding structures are related. This dominating building may be a church—the medieval parvis represents the most obvious example of the dominated square—or any other monumental structure, a palace, a town hall, an architecturally developed fountain, a theater, a railway station. In any case, such a commanding volume directs the spatial relations of the open area. Expressed in terms of stage design, all perspectives lead toward a backdrop. Usually the direction of a main street which opens into the square established the axis toward the dominant building (*e.g.,* Versailles). The perspective of the surrounding buildings and the suction of the dominant structure create the spatial tension of the square, compelling the spectator to move toward and to look at the focal architecture. Thus the dominant square produces a dynamic directive of motion, whereas the closed square by careful proportioning creates a static equilibrium.

The most distinct relationship between the dominating building and the square exists in the *parvis*, originally the enclosed vacant areas before a church. The original medieval *parvis* of Notre Dame Cathedral in Paris is a good example of this relationship (Fig. 3).

Although the Renaissance and baroque squares before churches occasionally are given the name *parvis,* their spatial shape has an entirely different meaning. The aesthetic effect of the medieval *parvis* is characterized by its limited perspective, with the church façade, functioning merely as a two-dimensional wall. The later squares, however, are planned in such large dimensions that because of the greater distance the whole church building can be perceived in three dimensions. The cupola over the main body of the Renaissance and baroque churches becomes the center of attraction, and with it the development into depth becomes important, creating a spatial counterpoint (*e.g.,* St. Peter's, Rome; *Dome des Invalides,* Paris). Hence, the lateral houses, colonnades, or arcades represent merely framing elements for the dominating structure and a stage effect is achieved. Whereas the medieval *parvis* as space element is *subordinated* to the church, anticipating the spatial directions of its interior, the Renaissance and baroque, *parvis* creates a coordinated void in contrast to the controlling architecture. The same relation exists later on when the dominant building is no longer ecclesiastic but secular—a palace (Versailles), a theater (*Place de l'Odeon,* Paris), a railway station, or any other type of monumental architecture.

There is, however, a definite distinction between a dominated square in front of a palace and the court of honor of the palace. The latter is most intimately connected with all other architectural elements of the palace and is spatially a part of it. The square before the court of honor, however, is public and emphasizes the contrast to the architectural mass of the palace. Hence, here the usual sequence develops as follows: the street leading to the public square opposite the palace; the expanse of the public square proper; the more intimate court of honor; and finally the vestibule within the palace—the whole a gradual decrescendo in space (*e.g.,* Versailles).

The preponderant structure need not necessarily be voluminous. Very often is it merely a gate or an arch which may dominate a whole square, independent of its actual size. In such a case the imaginary continuation of the main thoroughfare toward the gate creates the main axis of the space between. All buildings surrounding such a square are subordinated to the street-gate axis, independently of their individual prominence, *e.g., Piazza del Popolo,* Rome; *Pariser Platz,* Berlin (Fig. 4). It was not before the Renaissance that the architectural form of a gate became monumentalized and could develop into the main accent of an area. For in medieval times, gates were more or less but articulated openings in the enclosing city walls and therefore could not visually dominate the square inside of those passageways.

A fountain may also dominate a square it if constitutes an entire front in combination with architecture, sculpture, and water. Then the falling waters compete with vertical elements of architecture in commanding our attention, especially in Mediterranean countries where the display of water in varied forms assumes such a decisive part in the townscape, *e.g., Fontana di Trevi,* Rome (Fig. 5).

It seems natural that a vertical structure like a church, a gate, or a fountain should direct the space of a square. Sometimes, however, even the adjacent horizontals of a bridge may have the same organizing quality. The direction of the main thoroughfare leading to a square is continued by the roadway of the bridge and vice versa. A firm axis is established along which the limited width of the street, the expansion of the square, and then again the relative narrowness of the bridge shape an alternating rhythm of space. Such a spatial sequence increases the visual effect by subordinating the square to a continuous axis. After the monumental solutions of antiquity (*Pons Mulvius,* before *Moles Hadriani* [the tomb of Hadrian], Rome), the motif was taken up again in the great axial organizations of the seventeenth and eighteenth centuries, *e.g.,* the bridge and square before the *Rue Royale,* Orleans; the bridge and entrance before the *Rue Nationale,* Tours; Andreas Schluter's project for the bridge and square before the Royal Palace, Berlin; the bridge over the Po River and *Piazza Vittorio Veneto,* Turin (Fig. 6).

Paradoxical as it may sound, the dominating element may also be a *void,* allowing a vista toward a mountain range in the distance (*Maria Theresienstrasse,* Innsbruck). It may also be a broad river (*Praca do Comercio,* Lisbon; Contant's eighteenth century project for the *Quai Malaquet,* Paris) or the open sea or a lagoon (*Piazzetta,* Venice). The view toward a distant landscape or the surface of bordering water dominates the direction of the square as if the opening into such an indefinite expansion were an actual structure. Seen from the water, in reversed direction, such a square appears as a closed area, resembling a stage set with its three walls. The same type of square is repeated in the innumerable enchanting small Italian towns around the north Italian lakes and along both Mediterranean coasts through the whole peninsula down to Sicily. The fourth wall of the stage has

PHOTO: Landesbildstelle Berlin

Fig. 4. Berlin. *Pariser Platz.*

Fig. 5. Rome. *Fontana di Trevi.* Etching by Giovanni Battista Piranesi. (Courtesy: Metropolitan Museum of Art).

PHOTO: Edizione S.A.C.A.T. Turin

Fig. 6. Turin. *Piazza Vittorio Veneto.*

been substituted by the edge of river, lakes, or sea, and nature and architecture melt into unforgettable vistas of unique spatial intimacy. These so-called marinas must be clearly distinguished from even the broadest quays and boardwalks whose directions parallel the shoreline or bank, without development into depth (*e.g., Quai aux Herbes,* Ghent; *Quai du Montblanc,* Geneva).

The Nuclear Square. The self-contained space of the closed square, shaped by the continuity of the surrounding buildings, is easily perceived. The space conception of the dominated square, although different in kind, is equally clear. It is directed through the visual magnetism of the governing structure or the dominant vista. More complex, although no less real, is the aesthetic sensation of what we would call *a nuclear square.*

The spatial shape of the nuclear square is of a definite order, although not so tightly knit as in both aforementioned instances—an entity, even without the frame of a continuous row of buildings or without the domination of a frontal structure. As long as there is a nucleus, a strong vertical accent—a monument, a fountain, an obelisk—powerful enough to charge the space around with a tension that keeps the whole together, the impression of a square will be evoked. As the pyramid in the vast expanse of the desert creates an aesthetically impervious space around it, with invisible walls and the sky as dome above, so the monument, the obelisk, or the fountain, or even an individual building, will tie the heterogeneous elements of the periphery into one visual unit. This spatial oneness is not endangered by any irregularity of the general layout or by the haphazard position, size, or shape of the adjacent buildings. Since the visual effect of the central monument, fountain, or other feature is naturally limited, the dimensions of such a nuclear square are consequently restricted. If the expansion of the square in relation to the size of the focal volume becomes too large, the square loses its unity.

The not too numerous examples of nuclear squares are confined to certain historical periods. The most typical nuclear squares belong to the Renaissance, exactly as the most outspoken closed squares originated in Hellenistic and Roman times and again in the seventeenth and eighteenth centuries in France, and as the most evident dominated squares appeared in the late Middle Ages and then again in baroque times. The *Piazza di SS. Giovanni e Paolo* with Verrochio's *Colleoni* monument in Venice and the *Piazza del Santo* in Padua with Donatello's equestrian figure of the Gattamelata show the artistic impact of the nuclear square most cogently.

However, it would be a mistake to believe that each fountain, each obelisk, or each monument within a square can become a space-shaping nucleus of three-dimensional orientation: the column on the *Place Vendome* in Paris; the obelisk on St. Peter's Square in Rome; the monument to King Stanislas Leszcynski on the *Place Stanislas* in Nancy—they do not create a nuclear square but merely contribute to the impression of an otherwise closed, dominated, or combined square.

Grouped Squares. The visual impact of a group of squares may be compared with the effect of a cycle of murals. In both instances, each unit, the individual square and the single fresco, represents an entity *per se*, aesthetically self-sufficient and yet part of a comprehensive higher order—"individuation and unity." An analogy on a more limited scale would be the relationship of successive rooms inside a baroque palace: the first room preparing for the second, the second for the third, and so forth, each room meaningful as a link in a chain, beyond its own architectural significance. Similarly, individual squares may be fused organically and aesthetically into one comprehensive whole.

A sequence of squares, different in size and form, develops in only one direction, thus establishing a straight axis, *e.g.,* the *Imperial Fora* in Rome, or the sequence of squares in Nancy.

Or, in a non-axial organization, a smaller square opens with one of its sides upon a larger square, so that the individual axes of each square meet in a right angle, *e.g.,* the *Piazza* and *Piazzetta* in Venice.

Or, a group of three or more squares of different shapes and proportions surround one dominant building, as in Salzburg, around the Cathedral, or in Bologna, around the *Palazzo Podesta.*

Or, finally, squares are related to each other without any direct physical connection. In other words, two individual squares fall into a coherent pattern although they are separated from each other by blocks of houses, thoroughfares, *etc.* These indirect spatial ties may differ in kind: in medieval times they are most likely irregular and the strongest connecting link, as, for instance, the mass of a church steeple, dominates both squares, *e.g.,* in Luneburg, the Johanniskirche and the Sand (Fig. 7) or short streets between both squares may function as passageways, *e.g.,* in Verona, *Piazza d'Erbe* and *Piazza dei Signori.* However, the most subtle relations are created in the planned organizations of the eighteenth century, *e.g.,* Rheims, the project for the *Place Royale;* Rennes, Place Louis XV and *Place Louis le Grand;* Copenhagen, Amalienborg and Frederikskirke.

In whichever way individual squares may be connected, the aesthetic effect of the whole depends on the mental registration of successive images of changing spatial relations. It is irrelevant whether this connection is direct or indirect. There are so many architectural means by which to achieve dynamic crescendos in the transition from one square to another—arcades, the position of monuments, and change in scale of the adjacent buildings—that the town planner is as free to mold the empty space of a group of squares as the sculptor is to mold the volume of a group of figures. Perspective potentialities and relative proportions are the decisive visual factors in reference to a system of grouped squares. Here the contrast of larger monumental buildings and smaller adjacent houses, of higher and lower eaves, of the location of monuments and fountains, of separating or connecting arcades and triumphal arches may increase or decrease the actual dimensions. The possibilities of illusionistic deception as to distance and expansion may come close to the effects of stage settings.

PHOTO: Hans Boy-Schmidt

Fig. 7. Luneburg, Johanniskirche and the Sand.

The Amorphous Square. It may seem odd to include the amorphous square in this discussion since by its very definition—amorphous, *i.e.,* formless, unorganized, having no specific shape—it does not represent any aesthetic qualities or artistic possibilities. However, if it shares at least some elements with the previously analyzed types of squares, it may appear at the first glance to be like one of them.

New York's Washington Square is laid out as a regular rectangle, framed by houses on all sides—and yet it is not a "closed" square. For its dimensions are so large, the proportions of many of its surrounding structures are so heterogeneous, so irregular, even contradictory, and the location and size of the small triumphal arch are so dissimilar to all the other given factors, that a unified impression cannot result. Disproportion in scale destroys all aesthetic possibilities.

Another factor spoils any aesthetic effect of Trafalgar Square in London: it could have developed into a "nuclear" square had not the tremendous façade of the National Gallery in contrast to the small adjacent blocks of houses and the irregular directions of the streets leading to the "Square" counteracted the effect of the Nelson Column as a space-creating element. But as it is, the column does not become a center of spatial relationships, a kernel of tension.

Correspondingly, the *Place de l'Opera* in Paris could not become a "dominated" square in spite of the monumental façade of the imposing opera house. The width of the Boulevard des Capucines running through it off-center and the presence of small structures like the entrance to the Metro, scattered all over its area ruin any spatial effect.

Not by change did the three aforementioned examples originate in the nineteenth century, in an era that had almost no feeling for three-dimensional qualities. However, these examples are at least "squares" from the surveyor's viewpoint, although without any artistic impact. But other metropolitan traffic centers, such as New York's Times Square, are "squares" in name only; they are actually mere cross-roads or *carrefours*.

Another type of amorphous square resulted from a typical misconception of the nineteenth century. The various eclectic revivals of that century tried to isolate, and in this way to emphasize the visual importance of, a church, a court of justice, a theater, and so forth, by surrounding it with a free area. However, a mere void as such does not create any specific three-dimensional impression. If in the Middle Ages the cathedral of a town, a monumental building which by its very mass and height dominated decisively all other structures, was surrounded by open areas and an irregular network of streets, it meant that the isolation of the building created the possibility of an overall vista for the spectator, and the volume of the edifice could be perceived in relation to the human scale. But then the awareness of the human scale was helped by small structures closely clustered around the cathedral. Most of these structures were removed closely during the nineteenth century's attempts at "stylistic purification." However, in the process, genuine external space was not shaped (*e.g.*, the remodeling of the space around the Freiburg and Ulm cathedrals).

The amorphous squares of the past. Although they neither unified nor confined the surrounding empty space, at least emphasized the volume. The nineteenth century, however, merely turned the void around a building into a tray or platter, on which the particular structure was presented.

Paradoxes of history

As was said before, the visual appearance of squares, in contrast to that of a painting, a sculpture, or even of an isolated individual work of architecture, cannot be understood or enjoyed as an expression of a single historical epoch. The square as a living organism changes continuously with varying socio-economic conditions and altered technological possibilities. Morphological differences of successive stylistic epochs are of minor importance. They mold the form of the surrounding buildings rather than the inartistic shape of the square proper. And yet the historical approach will clarify certain inner connections between seemingly heterogeneous solutions in the pattern of a town which could not be grasped by a mere aesthetic analysis. It will confirm the obvious fact that certain epochs brought forth, or at least preferred, certain types of squares. But it will also become evident that paradoxically such preferred types sometimes wandered, that they were taken over by epochs and countries where material conditions, sociological structure, and even functional needs were entirely different, sometimes contradictory. Thus the post-Hippodamic regular closed agora reappears after many centuries in the French bastides of the thirteenth century and somewhat later in the foundations of the Teutonic Knights in Germany. The axial organization of the Roman *Imperial Fora* is repeated more than one and a half millennia later in the sequence of squares in eighteenth-century Nancy. Certainly the political, economic, and spiritual realties had changed meanwhile as much as topographical and climatic situations differed. And yet the spatial structures are almost identical, although architectural details vary stylistically.

As the art forms of the Homeric epos, of the symphony, and of the altar-triptych gradually have become autonomous, existing *per se* as definite structural patterns, no longer time-bound, no longer specific but general, so have the three-dimensional archetypes of squares. The original motivation and the reasons for the development of each form are forgotten, are perhaps no longer existent, but the archetytpes remain as prime elements in the history of human society, of village, town, and city. ■

Kevin Lynch

Summary

In 1960, Kevin Lynch prefaced his signal book, *The Image of the City*, as follows, which serves both to introduce and succinctly encompass the content of his work: *"This [book] is about the look of cities and whether this look is of any importance, and whether it can be changed. The urban landscape, among its many roles, is also something to be seen, to be remembered, and to delight in. Giving visual form to the city is a special kind of design problem, and a rather new one at that."*

Key words

analysis method, districts, edges, image, landmark, nodes, paths, planning, urban design, way finding

Faneuil Hall Market Place, Boston, MA.

The city image and its elements

There seems to be a public image of any given city which is the overlap of many individual images. Or perhaps there is a series of public images, each held by some significant number of citizens. Such group images are necessary if an individual is to operate successfully within his environment and to cooperate with his fellows. Each individual picture is unique, with some content that is rarely or never communicated, yet it approximates the public image, which, in different environments, is more or less compelling, more or less embracing.

This analysis limits itself to the effects of physical, perceptible objects. There are other influences of imageability, such as the social meaning of an area, its function, its history, or even its name. These will be glossed over, since the objective here is to uncover the role of form itself. It is taken for granted that in actual design form should be used to reinforce meaning, and not to negate it.

The contents of the city images, which are referable to physical forms, can conveniently be classified into five types of elements: paths, edges, districts, nodes, and landmarks. Indeed, these elements may be of more general application, since they seem to reappear in many types of environmental images. These elements may be defined as follows:

1 Paths. Paths are the channels along which the observer customarily, occasionally, or potentially moves. They may be streets, walkways, transit lines, canals, railroads. For many people, these are the predominant elements in their image. People observe the city while moving through it, and along these paths the other environmental elements are arranged and related.

2 Edges. Edges are the linear elements not used or considered as paths by the observer. They are the boundaries between two phases, linear breaks in continuity: shores, railroad cuts, edges of development, walls. They are lateral references rather than coordinate axes. Such edges may be barriers, more or less penetrable, which close one region off from another; or they may be seams, lines along which two regions are related and joined together. These edge elements, although probably not as dominant as paths, are for many people important organizing features, particularly in the role of holding together generalized areas, as in the outline of a city by water or wall.

3 Districts. Districts are the medium-to-large sections of the city, conceived of as having two-dimensional extent, which the observer mentally enters "inside of," and which are recognizable as having some common, identifying character. Always identifiable from the inside, they are also used for exterior reference if visible from the outside. Most people structure their city to some extent in this way, with individual differences as to whether paths or districts are the dominant elements. It seems to depend not only upon the individual but also upon the given city.

4 Nodes. Nodes are points, the strategic spots in a city into which an observer can enter, and which are the intensive foci to and from which he is traveling. They may be primarily junctions, places of a

Credits: This article is an extracted digest of Chapter III of Kevin Lynch, *The Image of the City*, MIT Press, Cambridge, MA, 1960, and is reprinted by permission of the publisher. Photos of Boston are recent, to illustrate citations in the original text.

break in transportation, a crossing or convergence of paths, moments of shift from one structure to another. Or the nodes may be simply concentrations, which gain their importance from being the condensation of some use or physical character, as a street corner hangout or an enclosed square. Some of these concentration nodes are the focus and epitome of a district, over which their influence radiates and of which they stand as a symbol. They may be called cores. Many nodes, of course, partake of the nature of both junctions and concentrations. The concept of node is related to the concept of path, since junctions are typically the convergence of paths, events on the journey. It is similarly related to the concept of district, since cores are typically the intensive foci of districts, their polarizing center. In any event, some nodal points are to be found in almost every image, and in certain cases they may be the dominant feature.

5 Landmarks. Landmarks are another type of point-reference, but in this case the observer does not enter within them, they are external. They are usually a rather simply defined physical object: building, sign, store, or mountain. Their use involves the singling out of one element from a host of possibilities. Some landmarks are distant ones, typically seen from many angles and distances, over the tops of smaller elements, and used as radial references. They may be within the city or at such a distance that for all practical purposes they symbolize a constant direction. Such are isolated towers; golden domes, great hills. Even a mobile point, like the sun, whose motion is sufficiently slow and regular, may be employed. Other landmarks are primarily local, being visible only in restricted localities and from certain approaches. These are the innumerable signs, store fronts, trees, doorknobs, and other urban detail, which fill in the image of most observers. They are frequently used clues of identity and even of structure, and seem to be increasingly relied upon, as a journey becomes more and more familiar.

The image of a given physical reality may occasionally shift its type with different circumstances of viewing. Thus an expressway may be a path for the driver, an edge for the pedestrian. Or a central area may be a district when a city is organized on a medium scale, and a node when the entire metropolitan area is considered. But the categories seem to have stability for a given observer when he is operating at a given level.

None of the element types isolated above exist in isolation in the real case. Districts are structured with nodes, defined by edges, penetrated by paths, and sprinkled with landmarks. Elements regularly overlap and piece one another. If this analysis begins with the differentiation of the data into categories, it must end with their reintegration into the whole image. Our studies have furnished much information about the visual character of the element types. This will be discussed below. Only to a lesser extent, unfortunately, did the work make revelations about the interrelations between elements, or about image levels, image qualities, or the development of the image.

1 PATHS

For most people interviewed, paths were the predominant city elements, although their importance varied according to the degree of familiarity with the city. People with least knowledge of Boston tended to think of the city in terms of topography, large regions, generalized characteristics, and broad directional relationships. Subjects who knew the city better had usually mastered part of the path structure; these people thought more in terms of specific paths and their interrelationships. A tendency also appeared for the people who knew the city best of all to rely more upon small landmarks and less upon either regions or paths.

Particular paths may become important features in a number of ways. Customary travel will of course be one of the strongest influences, so that major access lines, such as Boylston Street, Storrow Drive, or Tremont Street in Boston, Hudson Boulevard in Jersey City, or the freeways in Los Angeles, are all key image features. Obstacles to traffic, which often complicate the structure, may in other cases clarify it by concentrating cross flow into fewer channels, which thus become conceptually dominant. Beacon Hill, acting as a giant rotary, raises the importance of Cambridge and Charles Streets; the Public Garden strengthens Beacon Street. The Charles River, by confining traffic to a few highly visible bridges, all of individual shape, undoubtedly clarifies the path structure. Quite similarly, the Palisades in Jersey City focus attention on the three streets that successfully surmount it.

Concentration of special use or activity along a street may give it prominence in the minds of observers. Washington Street is the outstanding Boston example: subjects consistently associated it with shopping and theaters. Some people extended these characteristics to parts of Washington Street that are quite different (*e.g.*, near State Street); many people seemed not to know that Washington extends beyond the entertainment segment, and thought it ended near Essex or Stuart Streets. Los Angeles has many examples—Broadway, Spring Street, Skid Row, 7th Street—where the use concentrations are prominent enough to make linear districts. People seemed to be sensitive to variations in the amount of activity they encountered, and sometimes guided themselves largely by following the main stream of traffic. Los Angeles' Broadway was recognized by its crowds and its streetcars; Washington Street in Boston was marked by a torrent of pedestrians. Other kinds of activity at ground level also seemed to make places memorable, such as construction work near South Station, or the bustle of the food markets.

Special façade characteristics were also important for path identity. Beacon Street and Commonwealth Avenue were distinctive partly because of the building facades that line them. Pavement texture seemed to be less important, except in special cases such as Olvera Street in Los Angeles. Details of planting seemed also to be relatively unimportant, but a great deal of planting, like that on Commonwealth Avenue, could reinforce a path image very effectively.

Proximity to special features of the city could also endow a path with increased importance. In this case the path would be acting secondarily as an edge. Occasionally, paths were important largely for structural reasons.

Where major paths lacked identity, or were easily confused one for the other, the entire city image was in difficulty. That the paths, once identifiable, have continuity as well, is an obvious functional necessity. People regularly depended upon this quality. The fundamental requirement is that the actual track, or bed of the pavement, go through; the continuity of other characteristics is less important. Paths which simply have a satisfactory degree of track continuity were selected as the dependable ones in an environment like Jersey City. They can be followed by the stranger, even if with difficulty. People often generalized that other kinds of characteristics along a continuous track were also continuous, despite actual changes.

Paths may not only be identifiable and continuous, but have directional quality as well: one direction along the line can easily be distinguished from the reverse. This can be done by a gradient, a regular change in some quality, which is cumulative in one direction. Most frequently sensed were the topographic gradients: in Boston, particularly on Cambridge Street, Beacon Street, and Beacon Hill.

People tended to think of path destinations and origin points: they liked to know where paths came from and where they led. Paths with clear and well-known origins and destinations had stronger identities, helped tie the city together, and gave the observer a sense of his bearings whenever he crossed them. Some subjects thought of general destinations for paths, to a section of the city, for example, while others thought of specific places.

This same kind of end-from-end differentiation, which is conferred by termini, can be created by other elements that may be visible near the end, or apparent end, of a path. The Common near one end of Charles Street acted this way, as did the State House for Beacon Street.

Once a path has directional quality, it may have the further attribute of being scaled: one may be able to sense one's position along the total length, to grasp the distance traversed or yet to go. Features which facilitate scaling, of course, usually confer a sense of direction as well, except for the simple technique of counting blocks, which is directionless but can be used to compute distances. Many subjects referred to this latter clue, but by no means all. It was most commonly used in the regular pattern of Los Angeles.

Given a directional quality in a path, we may next inquire if it is aligned, that is, if its direction is referable to some larger system. In Boston, there were many examples of unaligned paths. One common cause was the subtle, misleading curve. Most people missed the curve in Massachusetts Avenue at Falmouth Street, and confused their total map of Boston as a result.

At the same time more abrupt directional shifts may enhance visual clarity by limiting the spatial corridor, and by providing prominent sites for distinctive structures. The second common cause of misalignment to the rest of the city was the sharp separation of a path from surrounding elements. Paths in the Boston Common, for example, caused much confusion: people were uncertain which walkways to use in order to arrive at particular destinations outside the Common. Their view of these outside destinations was blocked, and the paths of the Common failed to tie to outside paths. In Los Angeles as well, the freeways were not felt to be "in" the rest of the city, and coming off an exit ramp was typically a moment of severe disorientation.

Research on the problems of erecting directional signs on the new freeways has shown that this disassociation from the surroundings causes each turning decision to be made under pressure and without adequate preparation. Even familiar drivers showed a surprising lack of knowledge of the freeway system and its connections. General orientation to the total landscape was the greatest need of these motorists.

The railroad lines and the subway are other examples of detachment. The buried paths of the Boston subway could not be related to the rest of the environment except where they come up for air, as in crossing the river. The surface entrances of the stations may be strategic nodes in the city, but they are related along invisible conceptual linkages. The subway is a disconnected nether world, and it is intriguing

Downtown Boston commercial district.

The State House, Boston.

The Boston Common.

to speculate what means might be used to mesh it into the structure of the whole.

Quite similarly, in the subway system, the successive branching of main lines was a problem, since it was hard to keep distinct the images of two slightly divergent branches and hard to remember where the branch occurred.

A few important paths may be imaged together as a simple structure, despite any minor irregularities, as long as they have a consistent general relationship to one another. A large number of paths may be seen as a total network, when repeating relationships are sufficiently regular and predictable. The Los Angeles grid is a good example. Almost every subject could easily put down some twenty major paths in correct relation to each other. At the same time, this very regularity made it difficult for them to distinguish one path from another.

2 EDGES

Edges are the linear elements not considered as paths: they are usually, but not quite always, the boundaries between two kinds of areas. They act as lateral references. They are strong in Boston and Jersey City but weaker in Los Angeles. Those edges seem strongest which are not only visually prominent, but also continuous in form and impenetrable to cross movement. The Charles River in Boston is the best example and has all of these qualities.

In Jersey City, the waterfront was also a strong edge, but a rather forbidding one. It was a no-man's land, a region beyond the barbed wire. Edges, whether of railroads, topography, throughways, or district boundaries, are a very typical feature of this environment and tend to fragment it. Some of the most unpleasant edges, such as the bank of the Hackensack River with its burning dump areas, seemed to be mentally erased.

While continuity and visibility are crucial, strong edges are not necessarily impenetrable. Many edges are uniting seams, rather than isolating barriers, and it is interesting to see the differences in effect.

Edges are often paths as well. Where this was so, and where the ordinary observer was not shut off from moving on the path, then the circulation image seemed to be the dominant one. The element was usually pictured as a path, reinforced by boundary characteristics.

The elevated railways of Jersey City and Boston are examples of what might be called overhead edges. Yet high overhead edges, which would not be barriers at the ground level, might in the future by very effective orientation elements in a city.

Edges may also, like paths, have directional qualities. The Charles River edge, for example, has the obvious side-from-side differentiation of water and city, and the end-from-end distinction provided by Beacon Hill. Most edges had little of this quality, however.

3 DISTRICTS

Districts are the relatively large city areas which the observer can mentally go inside of, and which have some common character. They can be recognized internally, and occasionally can be used as external reference as a person goes by or toward them. Many persons interviewed took care to point out that Boston, while confusing in its path pattern even to the experienced inhabitant, has a quality that quite makes up for it.

Subjects, when asked which city they felt to be a well-oriented one, mentioned several, but New York (meaning Manhattan) was unanimously cited. And this city was cited not so much for its grid, which Los Angeles has as well, but because it has a number of well-defined characteristic districts, set in an ordered frame of rivers and streets.

The physical characteristics that determine districts are thematic continuities which may consist of an endless variety of components: texture, space, form, detail, symbol, building type, use, activity, inhabitants, degree of maintenance, topography. In a closely built city such as Boston, homogeneities of façade—material, modeling, ornament, color, skyline, especially fenestration—were all basic clues in identifying major districts. Beacon Hill and Commonwealth Avenue are both examples. The clues were not only visual ones: noise was important as well. At times, indeed, confusion itself might be a clue, as it was for the woman who remarked that she knows she is in the North End as soon as she feels she is getting lost.

Usually, the typical features were imaged and recognized in a characteristic cluster, the thematic unit. The Beacon Hill image, for example, included steep narrow streets; old brick row houses of intimate scale; inset, highly maintained, white doorways; black trim; cobblestones and brick walks, quiet; and upper-class pedestrians The resulting thematic unit was distinctive by contrast to the rest of the city and could be recognized immediately. In other parts of central Boston, there was some thematic confusion. It was not uncommon to group the Back Bay with the South End, despite their very different use, status, and pattern. This was probably the result of a certain architectural homogeneity, plus some similarity of historical background. Such likenesses tend to blur the city image.

A certain reinforcement of clues is needed to produce a strong image. All too often, there are a few distinctive signs, but not enough for a full thematic unit. Then the region may be recognizable to someone familiar with the city, but it lacks any visual strength or impact.

Yet social connotations are quite significant in building regions. A series of street interviews indicated the class overtones that many people associate with different districts. District names also help to give identity to districts even when the thematic unit does not establish a striking contrast with other parts of the city, and traditional associations can play a similar role.

Districts have various kinds of boundaries. Some are hard, definite, precise. Such is the boundary of the Back Bay at the Charles River or at the Public Garden. All agreed on this exact location. Other boundaries may be soft or uncertain, such as the limit between downtown shopping and the office district, to whose existence and approximate location most people would testify. Still other regions have no boundaries at all, as did the South End for many of our subjects.

These edges seem to play a secondary role: they may set limits to a district, and may reinforce its identity, but they apparently have less to do with constituting it. Edges may augment the tendency of districts to fragment the city in a disorganizing way.

4 NODES

Nodes are the strategic foci into which the observer can enter, typically either junctions of paths, or concentrations of some characteristic. But although conceptually they are small points in the city image, they may in reality be large squares, or somewhat extended linear shapes, or even entire central districts when the city is being considered at a large enough level. Indeed, when conceiving the environment at a national or international level, then the whole city itself may become a node.

The junction, or place of a break in transportation, has compelling importance for the city observer. Because decisions must be made at junctions, people heighten their attention at such places and perceive nearby elements with more than normal clarity. This tendency was confirmed so repeatedly that elements located at junctions may automatically be assumed to derive special prominence from their location. The perceptual importance of such locations shows in another way as well. When subjects were asked where on a habitual trip they first felt a sense of arrival in downtown Boston, a large number of people singled out break-points of transportation as the key places. The transition from one transportation channel to another seems to mark the transition between major structural units.

The subway stations, strung along their invisible path systems, are strategic junction nodes. The stations themselves have many individual characteristics: some are easy to recognize, like Charles Street, others difficult, like Mechanics. A detailed analysis of the imageability of subway systems, or of transit systems in general, would be both useful and fascinating.

Major railroad stations are almost always important city nodes, although their importance may be declining. The same might have been said for airports, had our study areas included them. In theory, even ordinary street intersections are nodes, but generally they are not of sufficient prominence to be imaged as more than the incidental crossing of paths. The image cannot carry too many nodal centers.

The other type of node, the thematic concentration, also appeared frequently. Pershing Square in Los Angeles was a strong example, being perhaps the sharpest point of the city image, characterized by highly typical space, planting, and activity. The Jordan-Filene corner acts secondarily as a junction between Washington Street and Summer Street, and it is associated with a subway stop, but primarily it was recognized as being the very center of the center of the city. It is the "100 percent" commercial corner, epitomized to a degree rarely seen in a large American city, but culturally very familiar to Americans. It is a core: the focus and symbol of an important region.

Louisburg Square is another thematic concentration, a well-known quiet residential open space, redolent of the upper-class themes of the Hill, with a highly recognizable fenced park. It is a purer example of concentration than is the Jordan-Filene corner, since it is not a transfer point at all, and was only remembered as being "somewhere inside" Beacon Hill. Its importance as a node was out of all proportion to its function.

A strong physical form is not absolutely essential to the recognition of a node: witness Journal Square and Scollay Square. But where the space has some form, the impact is much stronger. The node becomes memorable.

Commonwealth Avenue, Boston.

Louisburg Square, Beacon Hill, Boston.

Piazza San Marco, Venice.

A node like Copley Square, on the contrary, which is of less functional importance and has to handle the angled intersection of Huntington Avenue, was very sharply imaged, and the connections of various paths were eminently clear. It was easily identified, principally in terms of its unique individual buildings: the Public Library, Trinity Church, the Copley Plaza Hotel, the sight of the John Hancock Building. It was less of a spatial whole than a concentration of activity and of some uniquely contrasting buildings.

Nodes, like districts, may be introvert or extrovert. Scollay Square is introverted, it gives little directional sense when one is in it or its environs. The principal direction in its surroundings is toward or away from it; the principal locational sensation on arrival is simply "here I am."

Many of these qualities may be summed up by the example of a famous Italian node: the *Piazza San Marco* in Venice. Highly differentiated, rich and intricate, it stands in sharp contrast to the general character of the city and to the narrow, twisting spaces of its immediate approaches. Yet it ties firmly to the major feature of the city, the Grand Canal, and has an oriented shape that clarifies the direction from which one enters. It is within itself highly differentiated and structured: into two spaces *(Piazza* and *Piazzetta)* and with many distinctive landmarks *(Duomo, Palazzo Ducale, Campanile, Libreria)*. Inside, one feels always in clear relation to it, precisely micro-located, as it were. So distinctive is this space that many people who have never been to Venice will recognize its photograph immediately.

5 LANDMARKS

Landmarks, the point references considered to be external to the observer, are simple physical elements that may vary widely in scale. There seemed to be a tendency for those more familiar with a city to rely increasingly on systems of landmarks for their guides—to enjoy uniqueness and specialization, in place of the continuities used earlier.

Since the use of landmarks involves the singling out of one element from a host of possibilities, the key physical characteristic of this class is singularity, some aspect that is unique or memorable in the context. Landmarks become more easily identifiable, more likely to be chosen as significant, if they have a clear form; if they contrast with their background; and if there is some prominence of spatial location. Figure-background contrast seems to be the principal factor. The background against which an element stands out need not be limited to immediate surroundings: the grasshopper weathervane of Faneuil Hall, the gold dome of the State House, or the peak of the Los Angeles City Hall are landmarks that are unique against the background of the entire city.

Spatial prominence can establish elements as landmarks in either of two ways: by making the element visible from many locations (the John Hancock Building in Boston, the Richfield Oil Building in Los Angeles), or by setting up a local contrast with nearby elements, *i.e.*, a variation in setback and height.

Location at a junction involving path decisions strengthens a landmark. Historical associations, or other meanings, are powerful reinforcements, as they are for Faneuil Hall or the State House in Boston. Once a history, a sign, or a meaning attaches to an object, its value as a landmark rises.

Distant landmarks, prominent points visible from many positions, were often well known, but only people unfamiliar with Boston seemed to use them to any great extent in organizing the city and selecting routes for trips.

Few people had an accurate sense of where these distant landmarks were and how to make one's way to the base of either building. Most of Boston's distant landmarks, in fact, were "bottomless;" they had a peculiar floating quality. The John Hancock Building, the Custom House, and the Court House are all dominant on the general skyline, but the location and identity of their base is by no means as significant as that of their top.

The gold dome of Boston's State House seems to be one of the few exceptions to this elusiveness. Its unique shape and function, its location at the hill crest and its exposure to the Common, the visibility from long distances of its bright gold dome, all make it a key sign for central Boston. It has the satisfying qualities of recognizability at many levels of reference, and of coincidence of symbolic with visual importance.

The *Duomo* of Florence is a prime example of a distant landmark: visible from near and far, by day or night; unmistakable; dominant by size and contour; closely related to the city's traditions; coincident with the religious and transit center; paired with its campanile in such a way that the direction of view can be gauged from a distance. It is difficult to conceive of the city without having this great edifice come to mind.

But local landmarks, visible only in restricted localities, were much more frequently employed in the three cities studied. They ran the full range of objects available. The number of local elements that become landmarks appears to depend as much upon how familiar the observer is with his surroundings as upon the elements themselves. Unfamiliar subjects usually mentioned only a few landmarks in office interviews, although they managed to find many more when they went on field trips. Sounds and smells sometimes reinforced visual landmarks, although they did not seem to constitute landmarks by themselves.

Landmarks may be isolated, single events without reinforcement. Except for large or very singular marks, these are weak references, since they are easy to miss and require sustained searching. The single traffic light or street name demands concentration to find. More often, local points were remembered as clusters, in which they reinforced each other by repetition, and were recognizable, partly by context.

A sequential series of landmarks, in which one detail calls up anticipation of the next and key details trigger specific moves of the observer, appeared to be a standard way in which these people traveled through the city. In such sequences, there were trigger cues whenever turning decisions must be made and reassuring cues that confirmed the observer in decisions gone by. Additional details often helped to give a sense of nearness to the final destination or to intermediate goals. For emotional security as well as functional efficiency, it is important that such sequences be fairly continuous, with no long gaps, although there may be a thickening of detail at nodes. The sequence facilitates recognition and memorization. Familiar observers can store up a vast quantity of point images in familiar sequences, although recognition may break down when the sequence is reversed or scrambled.

Element interrelations

The elements are simply the raw material of the environmental image at the city scale. They must be patterned together to provide a satisfying form. The preceding discussions have gone as far as groups of similar elements (nets of paths, cluster of landmarks, mosaics of regions). The next logical step is to consider the interaction of pairs of unlike elements.

Such pairs may reinforce one another, resonate so that they enhance each other's power; or they may conflict and destroy themselves. A great landmark may dwarf and throw out of scale a small region at its base. Properly located, another landmark may fix and strengthen a core; placed off center, it may only mislead, as does the John Hancock Building in relation to Boston's Copley Square. A large street, with its ambiguous character of both edge and path, may penetrate and thus expose a region to view, while at the same time disrupting it. A landmark feature may be so alien to the character of a district as to dissolve the regional continuity, or it may, on the other hand, stand in just the contrast that intensifies that continuity.

Districts in particular, which tend to be of larger size than the other elements, contain within themselves, and are thus related to, various paths, nodes, and landmarks. These other elements not only structure the region internally, they also intensify the identity of the whole by enriching and deepening its character. Beacon Hill in Boston is one example of this effect. In fact, the components of structure and identity (which are the parts of the image in which we are interested) seem to leapfrog as the observer moves up from level to level. The identity of a window may be structured into a pattern of windows, which is the cue for the identification of a building. The buildings themselves are interrelated so as to form an identifiable space, and so on.

Paths, which are dominant in many individual images, and which may be a principal resource in organization at the metropolitan scale, have intimate interrelations with other element types. Junction nodes occur automatically at major intersections and termini, and by their form should reinforce those critical moments in a journey. These nodes, in turn, are not only strengthened by the presence of landmarks (as is Copley Square) but provide a setting which almost guarantees attention for any such mark. The paths, again, are given identity and tempo not only by their own form, or by their nodal junctions, but by the regions they pass through, the edges they move along, and the landmarks distributed along their length.

All these elements operate together, in a context. It would be interesting to study the characteristics of various pairings: landmark-region, node-path, etc. Eventually, one should try to go beyond such pairings to consider total patterns.

Most observers seem to group their elements into intermediate organizations, which might be called complexes. The observer senses the complex as a whole whose parts are interdependent and are relatively fixed in relation to each other.

The method as the basis for design

Perhaps the best way of summarizing the method is to recommend a technique of image analysis developed as the basis of a plan for the future visual form of any given city.

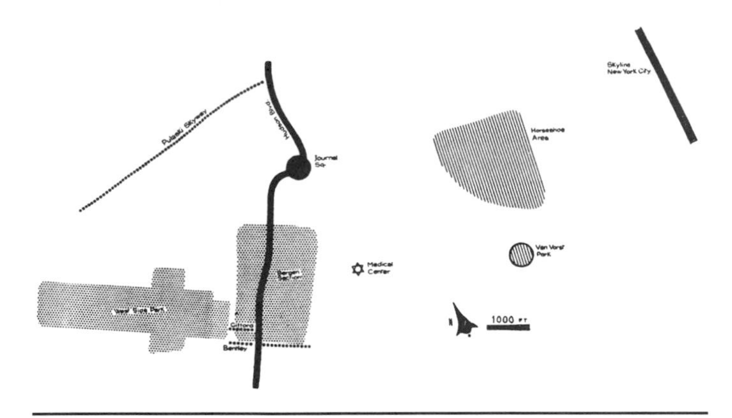

The distinctive elements of Jersey City.

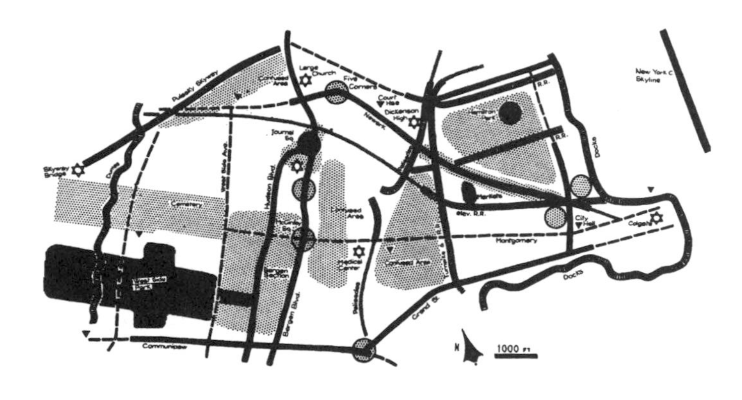

The visual form of Jersey City as seen in the field.

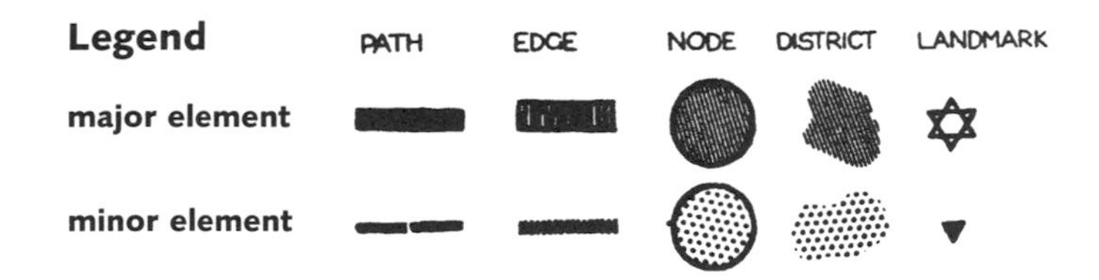

The procedure might begin with two studies. The first would be a generalized field reconnaissance by two or three trained observers, systematically covering the city both on foot and by vehicle, by night and day, and supplementing this coverage by several "problem" trips, as described above. This would culminate in a field analysis map and brief report, which would deal with strengths and weaknesses, and with general pattern as well as parts.

A parallel step would be the mass interview of a large sample, balanced to match the general population characteristics. This group, which could be interviewed simultaneously or in several parts, would be asked to do four things:

1. Draw a quick sketch map of the area in question, showing the most interesting and important features, and giving a stranger enough knowledge to move about without too much difficulty.

2. Make a similar sketch of the route and events along one or two imaginary trips, trips chosen to expose the length and breadth of the area.

3. Make a written list of the parts of the city felt to be most distinctive, the examiner explaining the meaning of "parts" and "distinctive."

4. Put down brief written answers to a few questions of the type: "Where is ________ located?"

The tests would be analyzed for frequency of mention of elements and their connections, for sequence of drawing, and for vivid elements, sense of structure, and composite image.

The field reconnaissance and the mass interview would then be compared for the relation of public image to visual form, to make a first-round analysis of the visual strengths and weakness of the whole area, and to identify the critical points, sequences, or patterns which are worth further attention.

Second-round investigation of these critical problems would then begin. Using a small sample, subjects would be asked individual interviews to locate selected critical elements, to operate with them in brief imaginary trips, to describe them, to make sketches of them, to discuss their feelings and memories about them. A few subjects might be taken out to these special locations, making brief field trips involving them, and describing and discussing them on the spot. Direction inquiries to the element from various origins might also be made of a random sample of persons in the street.

When these second-round studies had been analyzed for content and problems, equally intensive field reconnaissance of these same elements would then be carried out. Detailed studies of identity and structure under many different field conditions of light, distance, activity, and movement would follow. These studies would use the interview results, but would be by no means limited to them.

All this material would finally be synthesized in a series of maps and reports which would give the basic image of the area, the general visual problems and strengths, the critical elements and element interrelations, with their detailed qualities and possibilities for change. On such an analysis, continuously modified and kept up to date, a plan for the future visual form of the region could be based. ■

Jane Jacobs

Summary

Well before most professional and academic critiques of modernist planning, Jane Jacobs' passionate defense of her own neighborhood in *The Death and Life of Great American Cities* set the tone and articulated the themes of a new period in the development of urban design theory and practice, perhaps to as great an extent as Camillo Sitte's text of the late 19th century. Jacobs (1916) launched her attack on professional city planning doctrine and her defense of traditional urban neighborhoods from an explicitly and polemically amateur, common sense, and empirical position. Based on firsthand observation of the relationship between urban form and urban life in specific places, especially her own Greenwich Village, and grounded in an emergent movement of community-based activism and resistance. Her appreciation of the value of lively streets, crowded sidewalks and small blocks challenged official modernist urban theory descended from the Garden City movement, to Le Corbusier and CIAM, to the planners of European post-war redevelopment and American Urban Renewal.

Key words

city block, diversity, density, neighborhood, pedestrian, streets, urban scale

Court Street, New Haven, CT, a 1960s humanizing renovation of an urban streetscape.

The generators of diversity

Classified telephone directories tell us the greatest single fact about cities: the immense numbers of parts that make up a city, and the immense diversity of those parts. Diversity is natural to big cities.

"I have often amused myself," wrote James Boswell in 1771, "with thinking how different a place London is to different people. They, whose narrow minds are contracted to the consideration of some one particular pursuit, view it only through that medium. . .But the intellectual man is struck with it, as comprehending the whole of human life in all its variety, the contemplation of which is inexhaustible."

Boswell not only gave a good definition of cities, he put his finger on one of the chief troubles in dealing with them. It is so easy to fall into the trap on contemplating a city's uses one at a time, by categories. Indeed, just this—analysis of cities, use by use—has become a customary planning tactic. The findings on various categories of use are then put together into "broad, overall pictures."

The overall pictures such methods yield are about as useful as the picture assembled by the blind men who felt the elephant and pooled their findings. The elephant lumbered on, oblivious to the notion that he was a leaf, a snake, a wall, tree trunks and a rope all somehow stuck together. Cities, being our own artifacts, enjoy less defense against solemn nonsense.

To understand cities, we have to deal outright with combinations or mixtures of uses, not separate uses, as the essential phenomena. We have already seen the importance of this in the case of neighborhood parks. Parks can easily—too easily—be thought of as phenomena in their own right and described as adequate or inadequate in terms, say, of acreage ratios to thousands of population. Such an approach tells us something about the methods of planners, but it tells us nothing useful about the behavior or value of neighborhood parks.

A mixture of uses, if it is to be sufficiently complex to sustain city safety, public contact and cross-use, needs an enormous diversity of ingredients. So the first question—and I think by far the most important question—about planning cities is this: How can cities generate enough mixture among uses—enough diversity—throughout enough of their territories, to sustain their own civilization?

It is all very well to castigate the Great Blight of Dullness and to understand why it is destructive to city life, but in itself this does not get us far. Consider the problem posed by the street with the pretty sidewalk park in Baltimore. My friend from the street, Mrs. Kostritsky, is quite right when she reasons that it needs some commerce for its users' convenience. And as might be expected, inconvenience and lack of public street life are only two of the by-products of residential monotony here. Danger is another—fear of the streets after dark.

Credits: This article is a chapter from *The Death and Life of Great American Cities*, copyright 1961 by Jane Jacobs, and reprinted by permission of Random House, New York.

Some people fear to be alone in their houses by day since the occurrence of two nasty daytime assaults. Moreover, the place lacks commercial choices as well as any cultural interest. We can see very well how fatal is its monotony.

But having said this, then what? The missing diversity, convenience, interest and vitality do not spring forth because the area needs their benefits. Anybody who started a retail enterprise here, for example, would be stupid. He could not make a living. To wish a vital urban life might somehow spring up here is to play with daydreams. The place is an economic desert.

Although it is hard to believe, while looking at dull gray areas, or at housing projects or at civic centers, the fact is that big cities are natural generators of diversity and prolific incubators of new enterprises and ideas of all kinds. Moreover, big cities are the natural economic homes of immense numbers and ranges of small enterprises.

The principal studies of variety and size among city enterprises happen to be studies of manufacturing, notably those by Raymond Vernon, author of *Anatomy of a Metropolis*, and by P. Sargant Florence, who has examined the effect of cities on manufacturing both here and in England.

Characteristically, the larger a city, the greater the variety of its manufacturing, and also the greater both the number and the proportion of its small manufacturers. The reasons for this, in brief, are that big enterprises have greater self-sufficiency than small ones, are able to maintain within themselves most of the skills and equipment they need, can warehouse for themselves, and can sell to a broad market, which they can seek out wherever it may be. They need not be in cities, and although sometimes it is advantageous for them to be there, often it is more advantageous not to. But for small manufactures, everything is reversed. Typically they must draw on many and varied supplies and skills outside themselves, they must serve a narrow market at the point where a market exists, and they must be sensitive to quick changes in this market. Without cities, they would simply not exist. Dependent on a huge diversity of other city enterprises, they can add further to that diversity. This last is a most important point to remember. City diversity itself permits and stimulates more diversity.

For many activities other than manufacturing, the situation is analogous. For example, when Connecticut General Life Insurance Company built a new headquarters in the countryside beyond Hartford, it could do so only by dint of providing—in addition to the usual working space and rest rooms, medical suite and the like—a large general store, a beauty parlor, a bowling alley, a cafeteria, a theater and a great variety of games space. These facilities are inherently inefficient, idle most of the time. They require subsidy, not because they are kinds of enterprises that are necessarily money losers, but because here their use is so limited. They were presumed necessary, however, to compete for a working force, and to hold it. A large company can absorb the luxury of such inherent inefficiencies and balance them against other advantages it seeks. But small offices can do nothing of the kind. If they want to compete for a work force on even terms or better, they must be in a lively city setting where their employees find the range of subsidiary conveniences and choices that they want and need. Indeed, one reason, among many others, why the much-heralded postwar exodus of big cities from cities turned out to be mostly talk is that the differentials in cost of suburban land and space are typically canceled by the greater amount of space per worker required for facilities that in cities no single employee need provide, nor any one corps of workers of customers support. Another reason why such enterprises have stayed in cities, along with small firms, is that many of their employees, especially executives, need to be in close, face-to-face touch and communication with people outside the firm—including people from small firms.

The benefits that cities offer to smallness are just as marked in retail trade, cultural facilities and entertainment. This is because city populations are large enough to support wide ranges of variety and choice in these things. And again we find that bigness has all the advantages in smaller settlements. Towns and suburbs, for instance, are natural homes for huge supermarkets and for little else in the way of groceries, for standard movie houses or drive-ins and for little else in the way of theater. There are simply not enough people to support further variety, although there may be people (too few of them) who would draw upon it were it there. Cities, however, are the natural homes of supermarkets and standard movie houses plus delicatessens, Viennese bakeries, foreign groceries, art movies, and so on, all of which can be found coexisting, the standard with the strange, the large with the small. Wherever lively and popular parts of the cities are found, the small much outnumbers the large. Like the small manufacturers, these small enterprises would not exist somewhere else, in the absence of cities. Without cities, they would not exist.

The diversity, of whatever kind, that is generated by cities rests on the fact that in cities so many people are so close together, and among them contain so many different tastes, skills, needs, supplies, and bees in their bonnets.

Even quite standard, but small, operations like proprietor-and-one-clerk hardware stores, drug stores, candy stores and bars can and do flourish in extraordinary numbers and incidence in lively districts of cities because there are enough people to support their presence at short, convenient intervals, and in turn this convenience and neighborhood personal quality are big parts of such enterprises' stock in trade. Once they are unable to be supported at close, convenient intervals, they lose this advantage. In a given geographical territory, half as many people will not support half as many such enterprises spaced at twice the distance. When distance inconvenience sets in, the small, the various and the personal wither away.

As we have transformed from a rural and small-town country into an urban country, business enterprises have thus become more numerous, not only in absolute terms, but also in proportionate terms. In 1900 there were 21 independent non-farm businesses for each 1,000 persons in the total U.S. population. In 1959, in spite of the immense growth of giant enterprises during the interval, there were 26-1/2 independent non-farm businesses for each 1,000 persons in the population. With urbanization, the big get bigger, but the small also get more numerous.

Smallness and diversity, to be sure, are not synonyms. The diversity of city enterprises includes all degrees of size, but great variety does mean a high proportion of small elements. A lively city scene is lively largely by virtue of its enormous collection of small elements.

Nor is the diversity that is important for city districts by any means confined to profit-making enterprises and to retail commerce, and for this reason it may seem that I put an undue emphasis on retail trade. I think not, however. Commercial diversity is, in itself, immensely important for cities, socially as well as economically. Most of the uses of diversity on which I dwelt in Part I of *The Death and Life of Great American Cities* depend directly or indirectly upon the presence of plentiful, convenient, diverse city commerce. But more than this, wherever we find a city district with an exuberant variety and plenty in its commerce, we are apt to find that it contains a good many other kinds of diversity also including variety in its population and economic conditions that generate diverse commerce are intimately related to the production, or the presence, of other kinds of city variety.

But although cities may fairly be called natural economic generators of diversity and natural economic incubators of new enterprises, this does not mean that cities *automatically* generate diversity just by existing. They generate it because of the various efficient economic pools of use that they form. Wherever they fail to form such pools of use, they are little better, if any, at generating diversity socially, unlike small settlements, makes no difference. For our purposes here, the most striking fact to note is the extraordinary unevenness with which cities generate diversity.

On the one hand, for example, people who live and work in Boston's North End, or New York's Upper East Side or San Francisco's North Beach–Telegraph Hill, are able to use and enjoy very considerable amounts of diversity and vitality. Their visitors help immensely. But the visitors did not create the foundations of diversity in areas like these, nor in the many pockets of diversity and economic efficiency scattered here and there, sometimes most unexpectedly, in big cities. The visitors sniff out where something vigorous exists already, and come to share it, thereby further supporting it.

At the other extreme, huge city settlements of people exist without their presence generating anything much except stagnation and, ultimately, a fatal discontent with the place. It is not that they are a different kind of people, somehow duller or unappreciative of vigor and diversity. Often they include hordes of searchers, trying to sniff out these attributes somewhere, anywhere. Rather, something is wrong with their districts; something is lacking to catalyze a district population's ability to interact economically and help form effective pools of use.

Apparently there is no limit to the numbers of people in a city whose potentiality as city populations can thus be wasted. Consider, for instance, the Bronx, a borough of New York containing some one and a half million people. The Bronx is woefully short of urban vitality, diversity and magnetism. It has its loyal residents, to be sure, mostly attached to little blooming of street life here and there in "the old neighborhood," but not nearly enough of them.

In so simple a matter of city amenity and diversity as interesting restaurants, the 1,500,000 people in the Bronx cannot produce. Kate Simon, the author of a guidebook, *New York Places and Pleasures*, describes hundreds of restaurants and other commercial establishments, particularly in unexpected and out-of-the-way parts of the city. She is not snobbish, and dearly likes to present her readers with inexpensive discoveries. But although Miss Simon tries hard, she has to give up the great settlement of the Bronx as thin pickings at any price. After paying homage to the two solid metropolitan attractions in the borough, the zoo and the Botanical Gardens, she is hard put to recommend a single place to eat outside the zoo grounds. The one possibility she is able to offer, she accompanies with this apology:

> *"The neighborhood trails off sadly into a no man's land, and the restaurant can stand a little refurbishing, but there's the comfort of knowing that. . .the best of Bronx medical skill is likely to be sitting all around you."*

Well, that is the Bronx, and it is too bad it is so; too bad for the people who live there now, too bad for the people who are going to inherit it in future out of their lack of economic choice, and too bad for the city as a whole.

And if the Bronx is a sorry waste of city potentialities, as it is, consider the even more deplorable fact that it is possible for whole cities to exist, whole metropolitan areas, with pitifully little city diversity and choice. Virtually all of urban Detroit is as weak on vitality and diversity as the Bronx. It is ring superimposed upon ring of failed gray belts. Even Detroit's downtown itself cannot produce a respectable amount of diversity. It is dispirited and dull, and almost deserted by seven o'clock of an evening.

So long as we are content to believe that city diversity represents accident and chaos, of course its erratic generation appears to represent a mystery.

However, the conditions that generate city diversity are quite easy to discover by observing places in which diversity flourishes and studying the economic reasons why it can flourish in these places. Although the results are intricate, and the ingredients producing them may vary enormously, this complexity is based on tangible economic relationships, which in principle are much simpler than the intricate urban mixtures they make possible.

To generate exuberant diversity in a city's streets and districts, four conditions are indispensable:

1. The district, and indeed as many of its internal parts as possible, must serve more than one primary function; preferably more than two. These must ensure the presence of people who go outdoors on different schedules and are in the place for different purposes, but who are able to use many facilities in common.

2. Most blocks must be short; that is, streets and opportunities to turn corners must be frequent.

3. The district must mingle buildings that vary in age and condition, including a good proportion of old ones so that they vary in the economic yield they must produce. This mingling must be fairly close-grained.

4. There must be a sufficiently dense concentration of people, for whatever purposes they may be there. This includes dense concentration in the case of people who are there because of residence.

The necessity for these four conditions is the most important point this book has to make. In combination, these conditions create effective economic pools of use. Given these four conditions, not all city districts will produce a diversity equivalent to one another. The potentials of different districts differ for many reasons; but, given the development of these four conditions (or the best approximation to their full development that can be managed in real life), a city district should be able to realize its best potential, wherever they may live. Obstacles to doing so will have been removed. The range may not stretch to African sculpture or schools of drama or Romanian tea houses, but such as the possibilities are, whether for grocery stores, pottery schools, movies, candy stores, eating places, or whatever, they will get their best chance. And along with them, city life will get its best chances.

[Ed. note: In *The Death and Life of Great American Cities*, the author concludes this section with...] In the Chapters that follow, I shall discuss each of these four generators of diversity, one at a time. The purpose of explaining them one at a time is purely for convenience of exposition, not because any one—or even any three—of these necessary conditions is valid alone. All four in combination are necessary to generate city diversity; the absence of any one of the four frustrates a district's potential. ■

Edmund N. Bacon

Summary

Growth by Tension: At the beginning of the Baroque period, the ordering principle in the growth of the city of Rome was the establishment of lines of force which defined the tension between various landmarks in the old city. The interrelationship of these lines and their interaction with the old structures set into play a series of design forces which became the dominating element in the architecture work along them. Here the cohesive element is a line of force rather than a volumetric form.

Key words

composition, landscape, movement systems, nodal points, obelisk, pedestrian, plaza, scale, sequence, vista

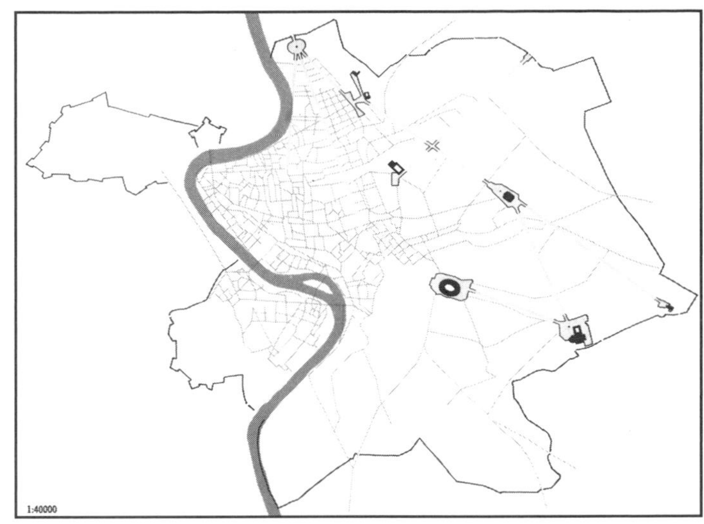

City growth by tension.

Design structure of Baroque Rome

Sixtus V, in his effort to recreate the city of Rome into a city worthy of the Church, saw clearly the need to establish a basic overall design structure in the form of a movement system as an idea, and at the same time the need to tie down its critical parts in positive physical forms which could not easily be removed. He hit upon the happy notion of using Egyptian obelisks, of which Rome had a substantial number, and erected these at important points within the structure of his design.

The power of this idea is demonstrated [by]. . .what actually happened to the Basilica of Saint Peter in various stages between the erection of the obelisk by Sixtus in 1586 and the completion of the Bernini colonnade. The design influence of the deliberate act of Sixtus was realized eighty years after his death, so it did not stem from any direct power he exercised during his lifetime. The point in space demarcated by the obelisk became the determinant in later construction because of the power of the idea in men's minds, transmitted over generations by the physical fact of the obelisk's existence. The continual criticism about the inapplicability of Sixtus's ideas in the present day, because his success was due to his despotic powers, which do not now exist, is absurd. Sixtus achieved far fewer actual architectural changes during the five years of his reign than any democratic city government could achieve today. It was the most inherent power to his idea, not his political influence, that caused the chain of events which followed.

That the actual physical accomplishment at the time of Sixtus's death was quite pathetic is shown by the two paintings (Figs. 1 and 2), from the Sistine Library in the Vatican. The upper fresco shows the confused and chaotic west façade of the venerable Saint Peter's Basilica before Sixtus started work, and the lower painting shows the appearance of the square at the time of his death, with the obelisk in

Fig. 1. Fresco of Saint Peter's Basilica before Sixtus started work.

Fig. 2. Painting shows the appearance of the square at the time of Sixtus death.

Credits: This article is an excerpt from Edmund N. Bacon, *Design of Cities,* 1976, reprinted by permission of Penquin Books.

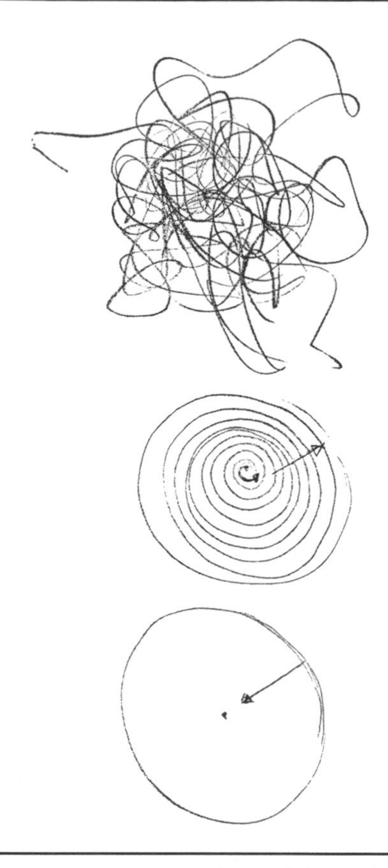

Fig. 3. A point in space as an organizing idea. Paul Klee sketch (*The Thinking Eye*).

Fig. 4. Engraving made by Israel Henriet *circa* 1640.

place. This is hardly an impressive civic achievement, but the idea of order has been implanted. As in the Klee drawings (Fig. 3), the single point in space can become a powerful design force, bringing order out of chaos.

The organizing force at work

The engraving made by Israel Henriet about 1640, shows the changes in Saint Peter's and the square around it after the placing of the obelisk (Fig. 4). The dome on Michelangelo's drum has been completed. The nave and façade of Saint Peter's have also been built, not, however, according to Michelangelo's or Bramante's ideas, but according to the new Baroque concepts expressed by the architect Carlo Maderna, who rejected the grandeur of the four-sided symmetry on the central dome. Considerable new work was carried out on the Vatican, shown to the right of the picture, but the palace still presents a rather raw appearance to the square. In the center is the two-story Bernini tower proposed by the architect to be one of the two symmetrical structures to rise above the Maderna façade. After its completion, cracks developed in the lower structure because of faulty engineering calculations and, when it became clear that the stability of the structure of a major part of Saint Peter's was threatened, the vast tower had to be torn down. The columns from this tower were transferred to the Piazza del Popolo to complete the great Baroque composition. They are now in place on the facades of the twin churches there.

A comment on the attitude toward professional practice of the day is the fact that the Roman Catholic Church, which commissioned Bernini to design the disastrous and costly Saint Peter's tower, recommissioned the same architect to design the great piazza before Saint Peter's some twenty years later, despite the display of engineering incompetence in the design of the defective tower. The world is the beneficiary of that remarkable decision.

The watercolor in Fig. 5, made in 1642 by Israel Silvestre of the view from the top of Saint Peter's dome, shows that, despite the ordering of Saint Peter's façade, the open space in the foreground was still chaotic and formless. The old fountain, the obelisk, and the square were unrelated, and the total effect was unimpressive, as it would continue to be for another twenty years.

The impulse is fulfilled

In this superb engraving by Giambattista Piranesi (Fig. 6), a portion of which has been removed, we see the obelisk on the site created for it by the great oval colonnade of Bernini. Yet it was the pre-existing obelisk that determined Bernini's design. The old fountain was moved to a new location, a new one was built to balance it, the columnar screen gave shape and definition to a magnificent extent of space, and order was achieved. In terms of mass the obelisk is only a tiny part of the whole; in terms of the idea, it dominates.

Throughout *Design of Cities,* the theme is stressed that we may look to the artist to help us see what is going on about us and to understand the essential nature of events. An excellent example is this engraving by Piranesi, in which, with superlative skill, the artist conveys the feeling of the obelisk as the organizing center of the entire complex. He shows it not simply as a static mass, but as a life force in the same sense that Klee conveys the feeling with a dot in the drawings on Fig. 3.

The influence of the point of space established here by the obelisk covers a small area, as it properly should do, in relation to the dominant basilica of Saint Peter. Its energies expire at the border of Bernini's oval, and it seems to me that Mussolini was mistaken in trying to extend them through his vulgar and ill-advised Via della Conciliazione. Bernini's original plan provided for an additional columned section defining the oval between the two end pavilions. It was his idea to complete, quite firmly, the self-contained character of this space and to preclude the very type of axial extension which Mussolini built.

In the rest of Rome, a chaotic, tangled area with the other six votive churches scattered about, Sixtus V saw the need for an entirely new design approach. During his five-year reign he placed the three remaining obelisks as points in space of his movement system and establishing the ends of the lines of force stretched taut between them. In this way the obelisks not only influenced the architecture in their immediate vicinity, but also extended their influence along the length of the connecting highways.

[Figs. 1–6 indicate] three stages of the development of Saint Peter's piazza simultaneously.

Fig. 5. Watercolor by Israel Silvestre, 1642.

Fig. 6. Engraving by Giambattista Piranesi.

Movement systems and design structure

"At that time Vasari was with Michelangelo [sic] every day: and one morning the Pope in his kindness gave them both leave that they might visit the Seven Churches on horseback (for it was Holy Year), and receive the Pardon in company. Whereupon, while going from one church to another, they had many useful and beautiful conversations on art and every industry."

These words written in 1560 by Giorgio Vasari, chronicler of the lives of the painters, architect of the Uffizi Palace in Florence, and painter in his own right, convey with fresh lucidity the feelings of the pilgrims as they moved from one votive church to another. The famous Antoine Lafrery engraving of 1575 (Fig. 7) shows quite accurately the ancient walls of Rome and some of its monuments and, on an enlarged scale, the seven churches that were the object of the pilgrimage. This is a graphic representation of the program for the Sixtus plan, of the phenomenon of undirected, undifferentiated movement. Because it is meandering, it would induce in the pilgrims no clearly organized sequence of purposeful architectural impressions, but rather a series of blunted visions of scattered houses, churches, and rolling countryside.

Apart from the uncompleted Renaissance drum of Michelangelo's design for the crossing of Saint Peter's, the architecture is Romanesque or Byzantine, with classical ruins remaining from earlier periods.

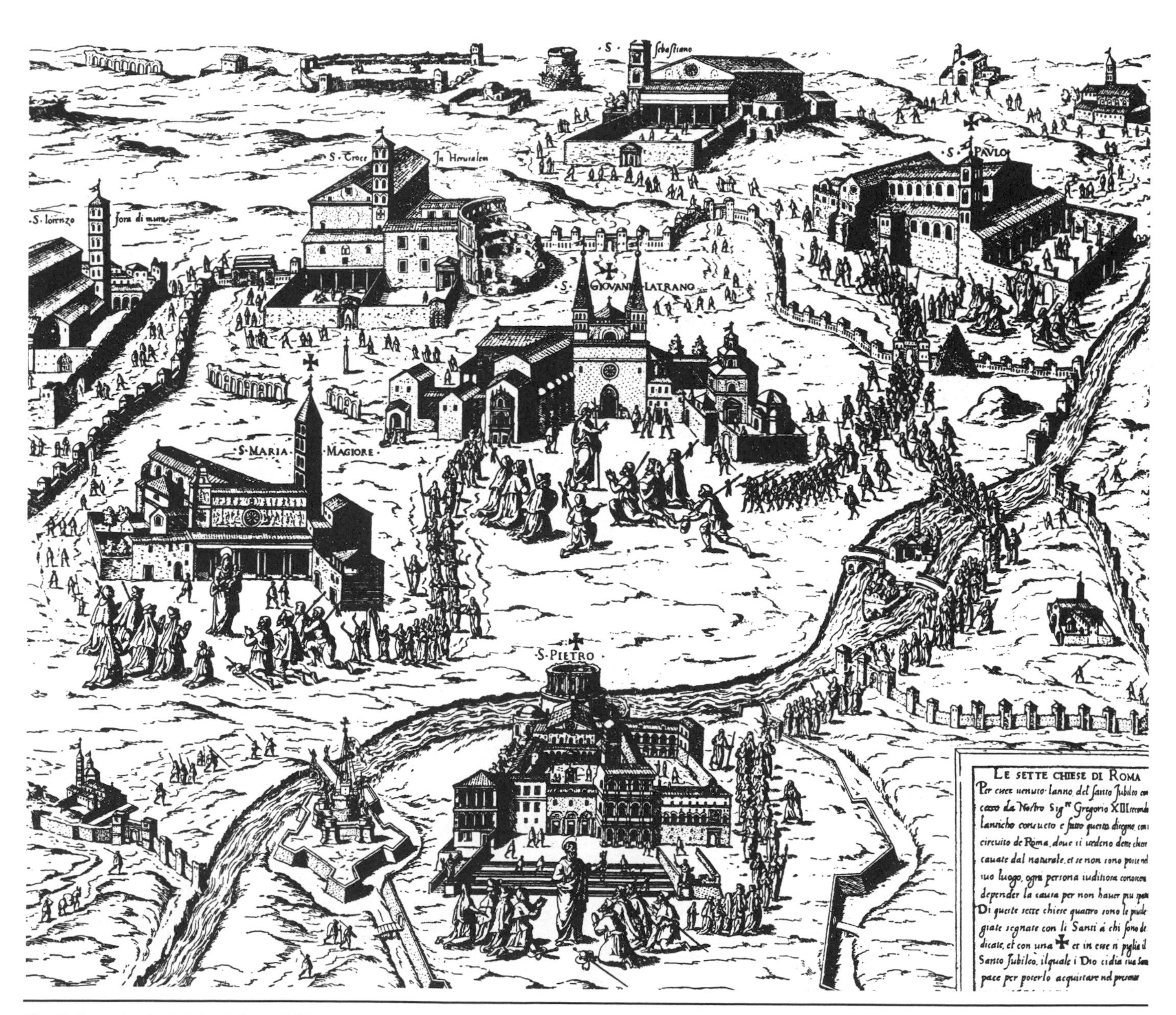

Fig. 7. Engraving by Antoine Lafrery, 1575.

Obelisks as points in movement systems

The remarkable engraving (Fig. 8), published in Rome in 1612, is evidently based on the one in Fig. 7 made by Lafrery thirty-seven years before. In the later example we see the architectural alterations that had been made during the intervening period, and, most specifically, the four newly placed obelisks of Sixtus V. A comparison of the two engravings provides a visual experience involving both the maturing of architectural form under the influence of new ideas and the appearance of the obelisks as design elements. The movement system emerges as a total design idea, symbolized by the obelisks positioned at its terminal points.

The obelisk at the extreme left is the one which Sixtus put near the Porta del Popolo, adjacent to which is the church of Santa Maria del Popolo, shown in small size in both engravings. The older fountain appears only in the later print, its existence evidently called to the engraver's mind by the new obelisk. This obelisk did not relate to any of the seven churches: it was a kind of greeting which confronted the visitor at the moment he entered the city through the principal gate.

The next obelisk is that on the west end of Santa Maria Maggiore, about two kilometers (1.2 miles) from the Porta del Popolo, which marked the terminus of Strada Felice originally intended to connect the two obelisks. Clearly visible are the twin domes, making a principal design interplay with the point of the obelisk. The dome on the right was built by Sixtus V.

The lower of the two central obelisks is, of course, in front of Saint Peter's, shown with the completed Maderna façade and the dome. The upper obelisk is the one erected by Sixtus V in front of San Giovanni in Laterano, now provided with a setting created by the newly built palace to the left of the church, and by the new two-tiered arcaded façade of the church itself. Both of these were the work of Domenico Fontana, Sixtus V's architect and adviser on building and city-planning projects.

Figs. 7 and 8 portray the metamorphosis of form over a period of time in response to a design idea.

Fig. 8. Engraving published in Rome in 1612.

Fig. 9. Painting by Taddeo di Bartolo, 1413.

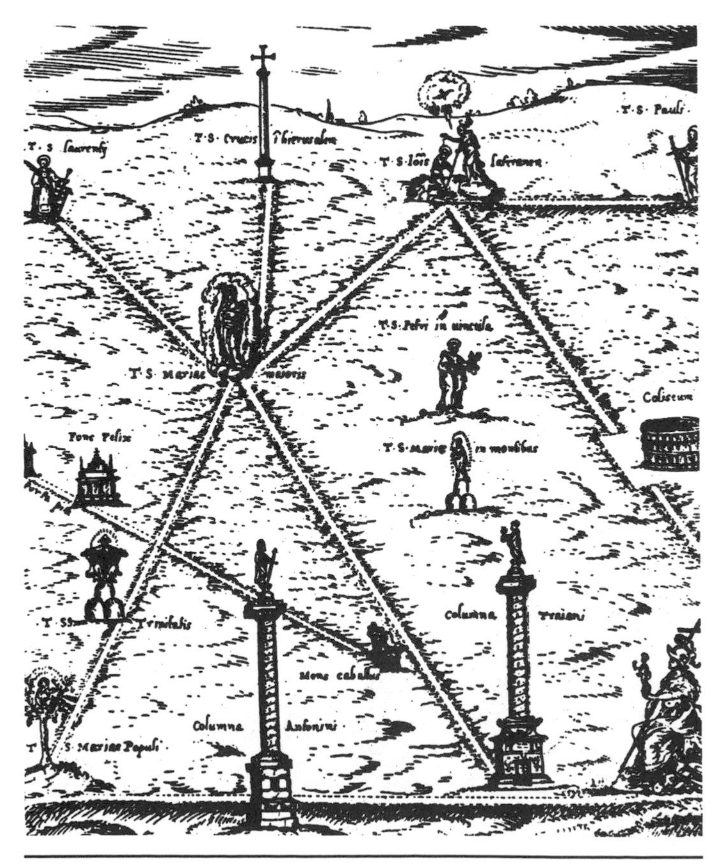

Fig. 10. Engraving by Giovani Francesco Bordino, 1413.

Nodal points dispersed

Again with the help of artists we see here the evolution of the popular image of Rome as it was affected by the larger design idea. Taddeo di Bartolo's painting of Rome (Fig. 9) was made in 1413, Giovanni Francesco Bordino's engraving (Fig. 10), was made in 1588, and Paul Klee's sketches from *The Thinking Eye* (Fig. 11) belong, of course, to the present century.

In the Bartolo painting in Fig. 9, Rome is seen strongly articulated by a series of ancient and medieval symbols associated with the various parts of the city: the Antonine and Trajan columns, the domed Pantheon and the Colosseum, the Dioscuri (statues of Castor and Pollux with their horses) on the Quirinal Hill, and, in the upper right, the equestrian statue of Marcus Aurelius before its removal from the San Giovanni in Laterano area to the Capitoline Hill. Interspersed among these remnants of classical Rome are medieval churches and towers. An early form of the Trevi Fountain with its three basins can be seen just below the Dioscuri. All these monuments are distributed in a general way in the appropriate parts of the city, but without any specific positioning or hint of design interrelationship. They may be compared with Klee's upper drawing showing points in space apparently distributed at random and producing no shape at all.

The Bordino engraving shown in Fig. 10, made only three years after the accession of Sixtus V to the papal throne, shows the astonishing rapidity with which his spatial-organization idea became known and understood. Here are some of the same symbols shown in Bartolo's painting. The two columns are clearly identifiable, as is the Colosseum. Santa Maria Maggiore is represented by a drawing of the Virgin, and the Dioscuri on the Quirinal Hill appear above the words *"Mons Caballus."* But each nodal image is precisely positioned and related to every other image by the design system of straight connecting streets. Although some of the streets had not been finished at this time, the idea of the movement system as a design-orienting element was so strong in the mind of Bordino that this became the organizing feature in his engraving and, by its very existence, served as a transmission agent, implanting the seeds of this idea in the mind of others.

Nodal points connected

When Paul Klee developed the idea shown in the Fig. 11 diagram, it is unlikely that he knew of the plans of Sixtus V for Rome, let alone that he made any connection with it. It is pleasant to accompany the rich representational drawings on the opposite page with the simple diagrams of Klee, because the fundamental design forces at work in Sistine Rome are as applicable today as they were in the sixteenth century if we consider their essential nature rather than their stylistic manifestations.

The establishments of points in space may be for emotional or spiritual associations with pre-existing monuments or structures (as was the case in Rome). Equally they may be points of production in regional economy, or centers of social regeneration in blighted areas. The concept of connecting these points by channels of energy, or lines of force, as demonstrated in the lower Klee drawing, may not only create an aesthetic physical design entity as happened in Sixtus's Rome, but produce an awareness of the structural relation of functions in what appeared to be a chaotic distribution of independent functions (upper diagram in Fig. 11).

Both the aesthetic design entity and the concept of a system of functional interrelationships are manifestations of the same underlying order, and the integration of the two is required if we are to solve contemporary problems on an urban scale. The fashion in contemporary architectural and planning thought of separating them by a "no-man's-land" to assure their continued individual identity—even to attach a whole professional vested interest separately to each one—has meant serious damage to efforts to solve the problems of the modern city.

Baroque Rome and Sixtus V

As Sixtus V cast his eye over the city of Rome after his election as Pope in 1585, he considered how he could make the sprawling, disorderly city into a fitting capital of Christendom. The only example in the entire city of a contemporary effort to relate more than one building to another in a design sense was Michelangelo's three-building "Capitoli," shown in Fig. 12, a 1561 engraving by Antonio Dosio. For the rest there was the crowded, jumbled medieval city, taking up about one-third of the space within the ancient Aurelian walls, the remainder being a few churches and ruined monuments scattered through vineyards and wasteland.

During the years of frustration which he spent in his villa near Santa Maria Maggiore as a neglected Cardinal ignored by hostile Popes, Sixtus formulated his ideas for the regeneration of Rome. Suddenly he was thrust into a position to do what he had so long desired, and his ideas took shape in the form of a clear plan for the city.

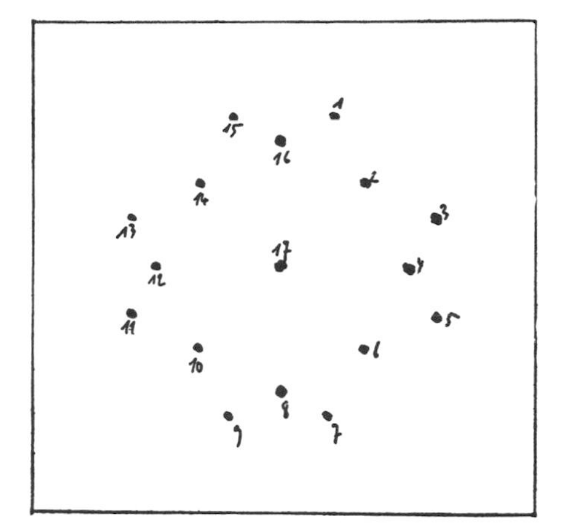

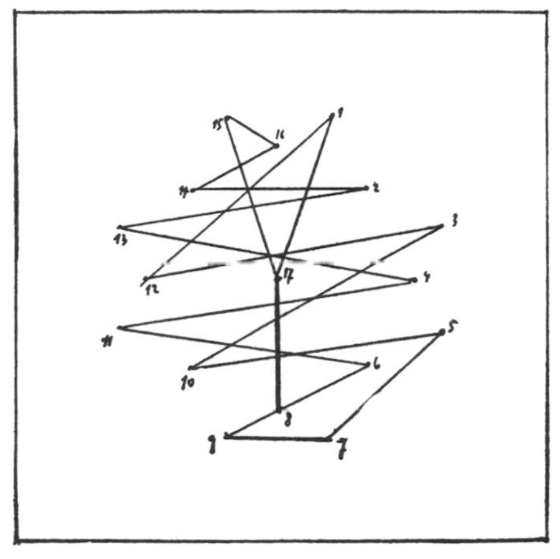

Fig. 11. Paul Klee sketch (The Thinking Eye).

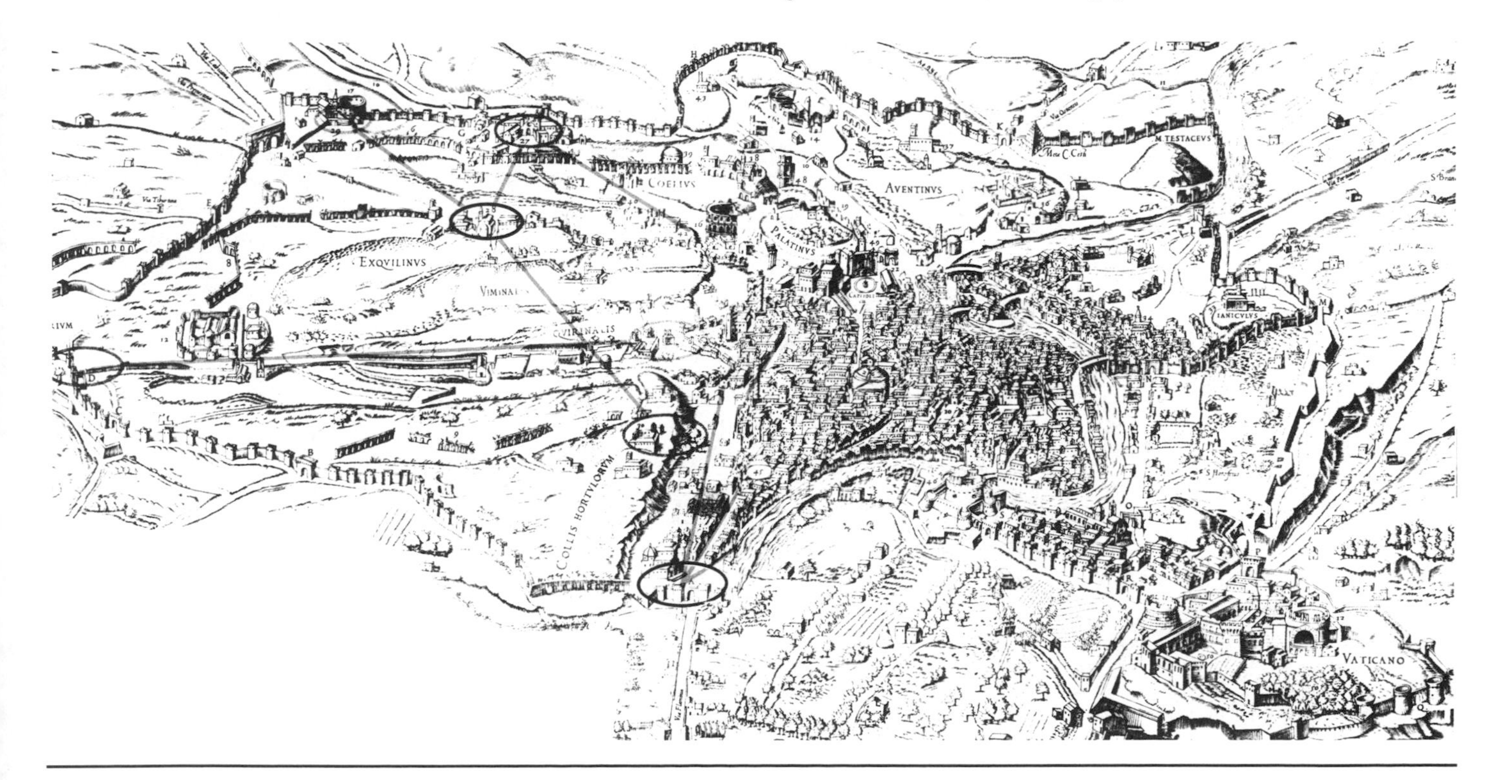

Fig. 12. Engraving by Antonio Dosio, 1561.

Extending from the Porta del Popolo in the northern wall of the city the foreground are three converging streets, the right-hand one leading to Porto di Ripetta at the Tiber River. To these streets Sixtus, in his mind's eye, added a fourth one, Strada Felice, extending directly to Santa Maria Maggiore. Only the portion from San Trinita dei Monti was built and is here shown as a yellow line intersected by the old Strada Pia. This connects with the Dioscuri clearly visible on the Quirinal Hill, and with Michelangelo's Porta Pia.

From Santa Maria Maggiore one new road branched off to Santa Croce and another one reached to San Giovanni in Laterano by a route indicated in yellow in this illustration. From here another route led to the Colosseum.

Here is demonstrated the seminal idea of the great plan for Rome, a colossal intellectual feat of an imposition of order on an environment of chaos.

[Ed. note: the original chapter in *Design of Cities* includes an extended discussion here on the chronology of the plan of Rome assisted by color maps.]

19th and 20th Century

The 1880 lithograph (Fig. 13) shows Rome in all its magnificence, a very different Rome from the one that Sixtus V knew when he became Pope three hundred years before indicated in Fig. 12.

Piazza del Popolo, with its obelisk, twin-domed churches, and the two semicircular extensions of Valadier, is prominent in the foreground. The street to the river had not yet been cut. To the left rises the series of ramps, loggias, and stairways that tie the Pincio Gardens and the piazza together. The Porta di Ripetta, the curving Baroque stairway to the river before Sixtus's San Girolamo degli Schiavoni, can be seen just at the bend of the Tiber to the right. The obelisk and the top of the Spanish Steps are visible in front of the twin towers of San Trinita dei Monti, and directly above is the obelisk before the twin-domed Santa Maria Maggiore, with the old Strada Felice stretching between. Just to the left, invading the horizon, is San Giovanni in Laterano, with the obelisk of Sixtus V in front of it.

This is a city in all its complexity, a city with an entirely different technological, sociological, and economic base from that which existed when Sixtus was Pope. Yet it is a city in which the quality of living, the joy of being there, and indeed the function of getting around are far more deeply influenced today by the vision and conviction of Pope Sixtus V than they were in his lifetime.

The great American landscape architect Frederick Law Olmsted, Sr., wrote,

> *"What artist so noble . . . as he who, with far-reaching conception of beauty and designing power, sketches the outlines, write the colors, and directs the shadows of a picture so great that Nature shall be employed upon it for generations, before the work he had arranged for her shall realize his intentions."*

Sixtus V succeeded in developing a similar kind of picture, employing the processes of city building rather than those of nature for the realization.

Fig. 13. Lithograph of Rome, 1880.

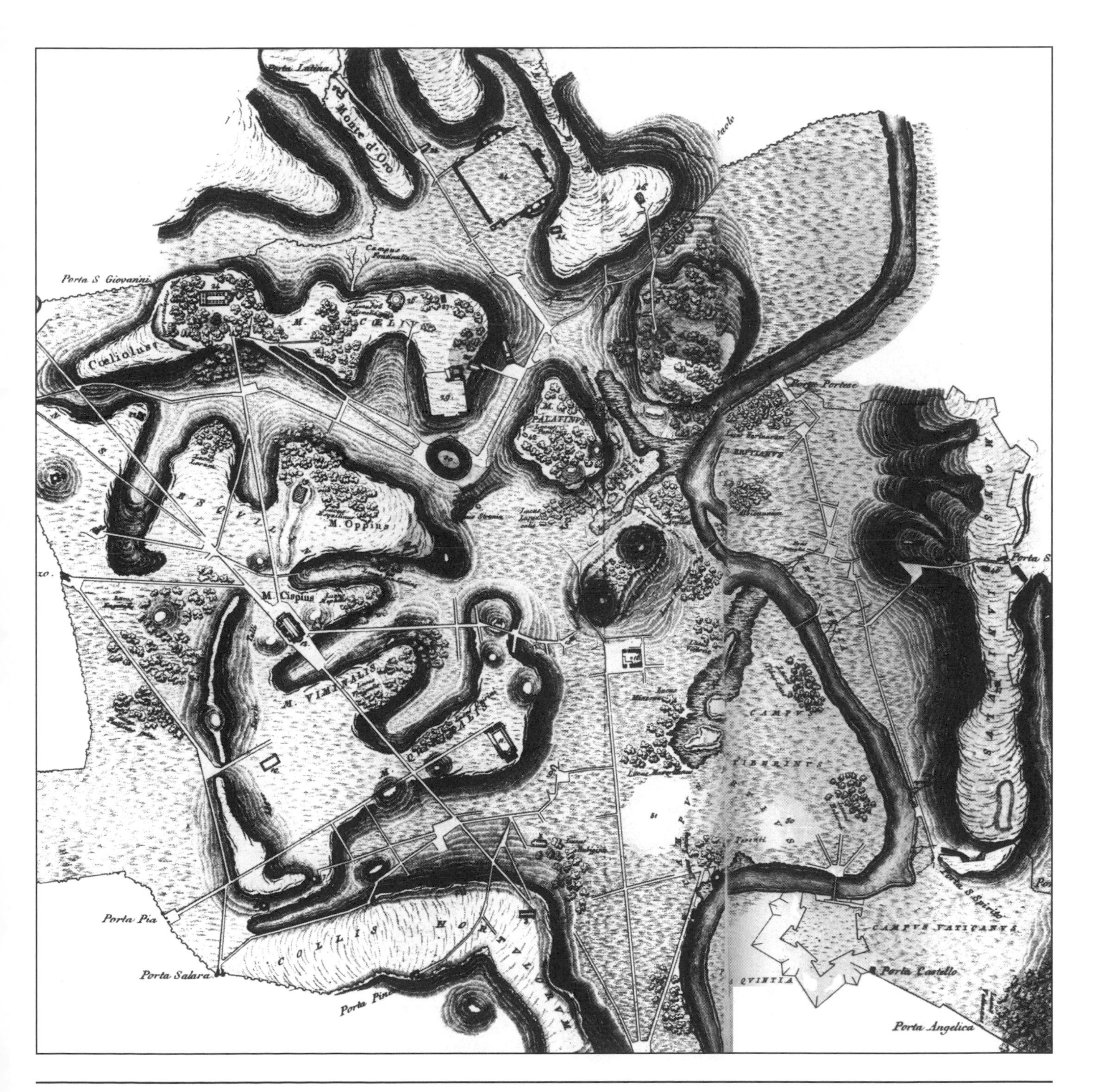

Fig. 14. Map of Rome. Giovanni Battista Brocchi, 1820.

Fig. 15. Fragment of the *Forma Urbis*, an incised marble map of the city set up by Septimius Severus early in the third century A.D.

Form and nature

There is more to the lesson of Rome than the interrelationship of planning and architectrure. There is the lesson of the relationship of planning and terrain. In this city of seven hills the problem of imposing a rational design network on such a rugged countryside is a formidable one.

During the classical period in Rome, when design concentrated on the establishment of self-contained building complexes, the topographical problem was principally that of hollowing out hills or filling in valleys to create exact geometrical planes for the formal, symmetrical buildings.

The design concept of Sixtus created entirely new topographical problems because he was concerned about a straight line, not a few hundred feet long as in the classical Roman *fora* or baths, but a few thousand feet long, as required for his vistas and movement system. The interaction between his design, the key buildings, and the topography is lucidly displayed on the map shown in Fig. 14, the work of Giovanni Battista Brocchi in 1820.

The movement system acts uncompromisingly across the countryside, tense and organic, moving directly to its goal but disturbing only what is necessary for the achievement of its purpose.

We can clearly see the hillock on the old Strada Pia that Sixtus had to cut away in order to connect Porta Pia visually with Piazza Quirinale. We see how Strada Felice went up and down hill in its straight course from San Trinita dei Monti to Santa Maria Maggiore, and in the very process of this rising and falling created a rhythmic experience, the impact of which would have been lost had it not been straight in plan. The branching roads to Santa Croce and to San Giovanni in Laterano also descend the valleys and ascend the hills, and it is the very purity of this counterpoint, the tense network of ways overlaid on the soft, rounded contours of the land, that contributes so greatly to the quality of Rome.

The quality of the land, made articulate by movement systems, is or should be a generating force in all architecture.

Ways of seeing the city

Classical Rome. Illustrated in Figs. 15 and 16 are two remarkable city plans, both of Rome, but of two widely separated periods. Fig. 15 is a fragment of the "Forma Urbis," an incised marble map of the city set up by Septimius Severus early in the third century A.D. on the wall of a building in the Forum. This map demonstrates an approach to design in which the exterior and the interior are integrated. Space flows freely throughout the entire area. Modulating the space is a unified system of columns rhythmically placed in disciplined rows, which give a sense of function and order.

The various fragments of this map present a great variety of forms, some rectangular, some semicircular, some circular, and some that are free and fantastic. But through the entire area the beat of the columnar structures is regulated to a common rhythm by the functional requirements of the masonry lintel. Even where vast vaults of space, the columns with their structural discipline were inserted as free screens across the spaces, recalling the rhythmic beat of the rest of Rome.

The Greek cities were of such a scale that only a few buildings, superbly rhythmic is within themselves but only modest in extent, could influence the extent of the entire city and, as symbols, dominate the less interesting sections. However, Roman ambition, and the concomitant scale of the city of Rome, demanded an entirely new principle of coherence and order. An individual building, or a series of individual buildings, without some binding element, would be swallowed up in the size of the great city. So, as in law and government, the Roman genius produced the ordering device on a broad enough scale to meet the demands of the new dimension of communal activity. And today the remnants of the classical city of Rome stand as a monument to the success of this prodigious effort.

Baroque Rome. The map in Fig. 16, drawn fifteen hundred years later, in a fragment of the Giambattista Nolli map of Rome in 1748. Implanted upon the formerly disciplined plans of classical Rome are the confused forms of the medieval city, which have been reordered by the architectural discipline of the Baroque. While this is a much tighter concept than that displayed in the map opposite (indeed the scale of the city was minuscule in relation to its former size), in the mind of Nolli and his contemporaries the exterior and interior public spaces were inextricably integrated into a singleness of thought and experience.

Here the rhythmic module is provided by the interior vaulted bay. The energy of the interior design, perfectly positioned and scaled in relation to the street and piazza on which it is placed, spills out into the void outside and animates the spaces of the city. The classical Pantheon is complemented by a Baroque fountain across the square: Vignola's Gesu Church debouches upon the square in front of it; Bernini's curve-fronted palazzo on Monte Citorio completes the ancient space movement and sets up tensions with the skillfully placed obelisk at the juncture of the broad and narrow squares. Most startling is the piazza in front of Sant' Ignazio, where the rhythmic forms of the nave and aisle vaults are extended across the square in the curved house walls which define the three interconnecting ovals in the plan. Here the inside spaces are an extension and completion of the experiences of the street, and at the same time extend their influence outward, molding the character, and in some cases the actual form, of the exterior spaces before them.

Again we have the phenomenon of the development of a common and powerful discipline of design, and, in the extraordinary map, a mode of representation commensurate with it. ■

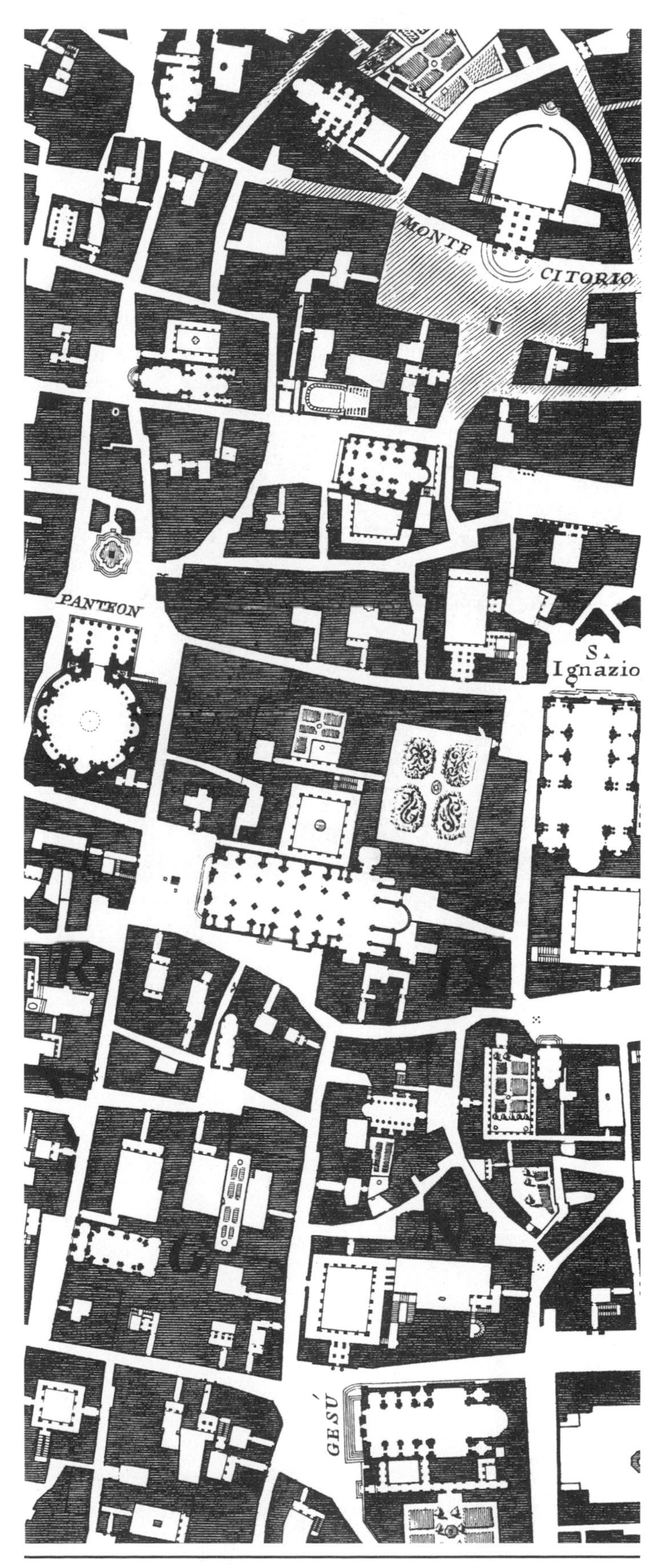

Fig. 16. Fragment of the Map of Rome by Giambattista Nolli, 1748.

Fig. 17. 1754 engraving by Piranesi of the Spanish Steps.

William H. Whyte

Summary
This article summarizes a three-year observational study undertaken in the 1970s by the author of how small urban plazas, parks and other outdoor gathering spaces are used, including the use and quality of design amenities for sitting, circulation, and related activities.

Keywords
circulation, density, entry, pedestrian, plaza, sidewalk, time-lapse photography

Ledge at Seagram's Plaza.

Social life of small urban spaces

INTRODUCTION

This is about city spaces, why some work for people, and some do not, and what the practical lessons may be. It is a by-product of first-hand observation.

In 1970, I formed a small research group, *The Street Life Project*, and began looking at city spaces. At that time, direct observation had long been used for the study of people in far off lands. It had not been used to any great extent in the United States city. There was much concern over urban crowding, but most of the research on the issue was done somewhere other than where it supposedly occurred. The most notable studies were of crowded animals, or of students and members of institutions responding to experimental situations—often valuable research, to be sure, but somewhat vicarious.

The Street Life Project began its study by looking at New York City parks and playgrounds and such informal recreation areas as city blocks. One of the first things that struck us was the lack of crowding in many of these areas. A few were jammed, but more were nearer empty than full, often in neighborhoods that ranked very high in density of people. Sheer space, obviously, was not of itself attracting children. Many streets were.

It is often assumed that children play in the street because they lack playground space. But many children play in the streets because they like to. One of the best play areas we came across was a block on 101st Street in East Harlem. It had its problems, but it worked. The street itself was the play area. Adjoining stoops and fire escapes provided prime viewing across the street and were highly functional for mothers and older people. There were other factors at work, too, and, had we been more prescient, we could have saved ourselves a lot of time spent later looking at plazas. Though we did not know it then, this block had within it all the basic elements of a successful urban place.

As our studies took us nearer the center of New York, the imbalance in space use was even more apparent. Most of the crowding could be traced to a series of choke points—subway stations, in particular. In total, these spaces are only a fraction of downtown, but the number of people using them is so high, the experience so abysmal, that it colors our perception of the city around, out of all proportion to the space involved. The fact that there may be lots of empty space somewhere else little mitigates the discomfort. And there is a strong carryover effect.

This affects researchers, too. We see what we expect to see, and have been so conditioned to see crowded spaces in center city that it is often difficult to see empty ones. But when we looked, there they were.

The amount of space, furthermore, was increasing. Since 1961, New York City has been giving incentive bonuses to builders who provided plazas. For each square foot of plaza, builders could add 10 square feet of commercial floor space over and above the amount normally permitted by zoning. So they did—without exception. Every new office building provided a plaza or comparable space: in total, by 1972, some 20 acres (8 ha) of the world's most expensive open space.

Credits: This article was first published in 1980 by the Conservation Foundation and is included in *The Essential William H. Whyte*, Albert LaFarge, editor, Fordham University Press, New York (2000). It is reprinted by kind permission of the Estate of William H. Whyte. Photos and captions are by the author.

We discovered that some plazas, especially at lunchtime, attracted a lot of people. One, the plaza of the Seagram Building, was the place that helped give the city the idea for the plaza bonus. Built in 1958, this austerely elegant area had not been planned as a people's plaza, but that is what it became. On a good day, there would be a hundred and fifty people sitting, sunbathing, picnicking, and schmoozing—idly gossiping, talking "nothing talk." People also liked 77 Water Street, known as "swingers' plaza" because of the young crowd that populated it.

But on most plans, we didn't see many people. The plazas weren't used for much except walking across. In the middle of the lunch hour on a beautiful, sunny day the number of people sitting on the plazas averaged four per 1,000 square feet (92.9 sq. m) of space—an extraordinarily low figure for so dense a center. The tightest knot central business district anywhere contained a surprising amount of open space that was relatively empty and unused.

If places like Seagram's and 77 Water Street could work so well, why not the others? The city was being had. For the millions of dollars of extra space it was handing out to the builders, it had every right to demand much better plazas in return.

I put the question to the chairman of the City Planning Commission, Donald Elliott. As a matter of fact, I entrapped him into spending a weekend looking at time-lapse films of plaza use and non-use. He felt that tougher zoning was in order. If we could find out why the good plazas worked and the bad ones didn't, and come up with hard guidelines, we could have the basis of a new code. Since we could expect the proposals to be strongly contested, it would be important to document the case to a fare-thee-well.

We set to work. We began studying a cross section of spaces—in all, sixteen plazas, three small parks, and a number of odds and ends. I will pass over the false starts, the dead ends, and the floundering arounds, save to note that there were a lot and that the research was nowhere as tidy and sequential as it can seem in the telling. Let me also note that the findings should have been staggeringly obvious to us had we thought of them in the first place. But we didn't. Opposite propositions were often what seemed obvious. We arrived at our eventual findings by a succession of busted hypotheses.

The research continued for some three years. I like to cite the figure because it sounds impressive. But it is calendar time. For all practical purposes, at the end of six months we had completed our basic research and arrived at our recommendations. The city, alas, had other concerns on its mind, and we found that communicating the findings was to take more time than arriving at them. We logged many hours in church basements and meeting rooms giving film and slide presentations to community groups, architects, planners, businessmen, developers, and real-estate people. We continued our research; we had to keep out findings up-to-date, for now we were disciplined by adversaries. But at length the City Planning Commissions incorporated our recommendations in a proposed new open space zoning code, and in May 1975 it was adopted by the city's Board of Estimate. As a consequence, there has been a salutary improvement in the design of new spaces and the rejuvenation of old ones. *[Ed. note: an abridged version of these Zoning Guidelines are reprinted as an Appendix in LaFarge (2000) cited in the credits].*

But zoning is certainly not the ideal way to achieve the better design of spaces. It ought to be done for its own sake. For economics alone, it makes sense. An enormous expenditure of design expertise, and of travertine and steel, went into the creation of the many really bum office-building plazas around the country. To what end? It is far easier, simpler to create spaces that work for people than those that do not—and a tremendous difference it can make to the life of a city.

The life of plazas

We started by studying how people use plazas. We mounted time-lapse cameras overlooking the plazas and recorded daily patterns (Fig. 1). We talked to people to find where they came from, where they worked, how frequently they used the place and what they thought of it. But, mostly, we watched people to see what they did.

Most of the people who use plazas, we found, are young office workers from nearby buildings. There may be relatively few patrons from the plaza's own building; as some secretaries confide, they'd just as soon put a little distance between themselves and the boss. But commuter distances are usually short; for most plazas, the effective market radius is about three blocks. Small parks, like Paley and Greenacre in New York, tend to have more assorted patrons throughout the day—upper-income older people, people coming from a distance. But office workers still predominate, the bulk from nearby.

This uncomplicated demography underscores an elemental point about good urban spaces: supply created demand. A good new space builds a new constituency. It stimulates people into new habits—al fresco lunches—and provides new paths to and from work, new places to pause. It does all this very quickly. In Chicago's Loop, there were no such amenities not so long ago. Now, the plaza of the First National Bank has thoroughly changed the midday way of life for thousands of people. A success like this in no way surfeits demand for space: it indicates how great the unrealized potential is.

The best-used plazas are sociable places, with a higher proportion of couples than you find in less-used places, more people in groups, more people meeting people, or exchanging good-byes. At five of the most used plazas in New York, the proportion of people in groups runs about 45 percent; in five of the least used, 32 percent. A high proportion of people in groups is an index of selectivity. When people go to a place in twos or threes or rendezvous there, it is more often because they have decided to. Nor are these sociable places less congenial to the individual. In absolute numbers, they attract more individuals than do less used spaces. If you are alone, a lively place can be the best place to be.

The most used places also tend to have a higher than average proportion of women. The male-female ratio of a plaza basically reflects the composition of the work force, which varies from area to area—in midtown New York it runs about 60 percent male, 40 percent female. Women are more discriminating than men as to where they will sit, more sensitive to annoyances, and women spend more time casting the various possibilities. If a plaza has a markedly lower than average proportion of women, something is wrong. Where there is a higher than average percentage of women, the plaza is probably a good one and has been chosen as such.

The rhythms of plaza life are much alike from place to place. In the morning hours, patronage will be sporadic. A hot dog vendor setting up his cart at the corner, elderly pedestrians pausing for a rest, a delivery messenger or two, a shoeshine man, some tourists, perhaps an odd type, like a scavenger woman with shopping bags. If there is any construction work in the vicinity, hard hats will appear shortly after 11:00 a.m. with beer cans and sandwiches. Things will start to liven up. Around noon, the main clientele begins to arrive. Soon, activity will be near peak and will stay there until a little before 2:00 p.m. Some 80 percent of the total hours of use will be concentrated in these two hours. In mid and late afternoon, use is again sporadic. If there's a special event, such as a jazz concert, the flow going home will be tapped, with the people staying at late as 6:00 or 6:30 p.m. Ordinarily, however, plazas go dead by 6:30 p.m. and stay that way until the next morning.

During peak hours the number of people on a plaza will vary considerably according to seasons and weather. The way people distribute themselves over the space, however, will be fairly consistent, with some sectors getting heavy use day in and day out, others much less. In our sightings we find it easy to map every person, but the patterns are regular enough that you could count the number in only one sector, then multiply by a given factor, and come within a percent or so of the total number of people at the plaza.

Off peak use often gives the best clues to people's preference. When a place is jammed, a person sits where he can (Fig. 2). This may or may not be where he most wants to. After the main crowd has left, the choices can be significant. Some parts of the plaza become quite empty; others continue to be used. At Seagram's, a rear ledge under the trees is moderately, but steadily occupied when other ledges are empty; it seems the most uncrowded of places, but on a cumulative basis it is the best used part of Seagram's.

Men show a tendency to take the front row seats, and, if there is a kind of gate, men will be the guardians to it. Women tend to favor places slightly secluded. If there are double-sided benches parallel to a street, the inner side will usually have a high proportion of women: the outer, of men.

Of the men up front, the most conspicuous are girl watchers. They work at it, and so demonstratively as to suggest that their chief interest may not really be the girls so much as the show of watching them. Generally, the watchers line up quite close together, in groups of three to five. If they are construction workers, they will be very demonstrative, much given to whistling, laughing, direct salutations. This is also true of most girl watchers in New York's financial areas. In midtown, they are more inhibited, playing it coolly, with a good bit of sniggering and smirking, as if the girls were not measuring up. It is all machismo, however, whether uptown or downtown. Not once have we ever seen a girl watcher pick up a girl, or attempt to.

Few others will either. Plazas are not ideal places for striking up acquaintances, and even on the most sociable of them, there is not much mingling. When strangers are in proximity, the nearest thing to exchange is what Erving Goffman has called "civil inattention." If there are, say, two smashing blondes on a ledge, the men nearby will usually put on an elaborate show of disregard. Watch closely, however, and you will see them give themselves away with covert glances, involuntary primping of the hair, tugs at the ear lobe.

Lovers are to be found on plazas. But not where you would expect them. When we first started interviewing, people told us we'd find lovers in the rear places. But they weren't usually there. They would be out front. The most fervent embracing we've recorded on film has usually taken place in the most visible of locations, with the couple oblivious of the crowd.

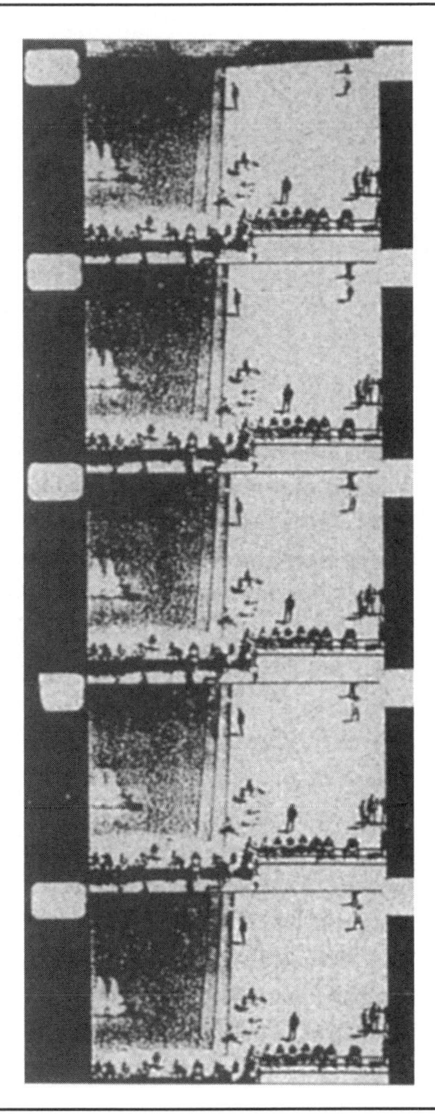

Fig. 1. Time-lapse documentation.

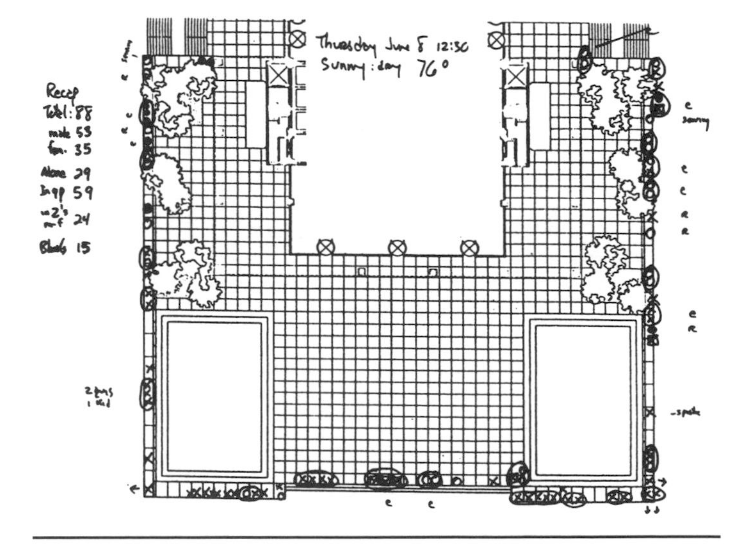

Fig. 2. A typical sighting map used to plot locations of sitters almost as quickly as a simple headcount. "X" and "O" represent male and female sitters, respectively.

Certain locations become rendezvous points for coteries of various kinds. For a while, the south wall of Chase plaza was a gathering point for camera bugs, the kind who like to buy new lenses and talk about them. Patterns of this sort last no more than a season—or persist for years. Some time ago, one particular spot became a gathering place for raffish younger people; since then, there have been many changeovers in personnel, but it is still a gathering place for raffish younger people.

What attracts people most, it would appear, is other people. If I belabor the point, it is because many urban spaces are being designed as though the opposite were true, and that what people liked best were the places that they stay away from. People often do talk along such lines; this is why their responses to questionnaires can be so misleading. How many people would say they like to sit in the middle of a crowd? Instead, they speak of getting away from it all, and use terms like "escape," "oasis," "retreat." What people do, however, reveals a different priority.

This was first brought home to us in a study of street conversations. When people stop to have a conversation, we wondered, how far away do they move from the main pedestrian flow? We were especially interested in finding out how much of the normally unused buffer space next to buildings would be used. So we set up time-lapse cameras overlooking several key street corners and began plotting the location of all conversations lasting a minute or longer.

People didn't move out of the main pedestrian flow. They stayed in it or moved into it, and the great bulk of the conversations were smack in the center of the flow—the 100 percent location, to use the real estate term. The same gravitation characterized "traveling conversations"—the kind in which two men move about, alternating the roles of straight man and principal talker. There is a lot of apparent motion. But if you plot the orbits, you will find they are usually centered around the 100 percent spot.

Just why people behave like this, we never have been able to determine. It is understandable that conversations should originate within the main flow. Conversations are incident to pedestrian journeys; where there are the most people, the likelihood of a meeting or a leave-taking is highest. What is less explainable is people's inclination to remain in the main flow, blocking traffic, being jostled by it. This does not seem to be a matter of inertia but of choice—instinctive, perhaps, but by no means illogical. In the center of the crowd, you have the maximum choice—to break off, to continue—much as you have in the center of a cocktail party, itself a moving conversation growing ever denser and denser.

People also sit in the mainstream. At the Seagram plaza, the main pedestrian paths are on diagonals from the building entrance to the corners of the steps. These are natural junction and transfer points and there is usually a lot of activity at them. They are also a favored place for sitting and picnicking. Sometimes there will be so many people that pedestrians have to step carefully to negotiate the steps. The pedestrians rarely complain. While some will detour around the blockage, most will thread their way through it.

Standing patterns are similar. When people stop to talk on a plaza, they usually do so in the middle of the traffic stream. They also show an inclination to station themselves near objects, such as a flagpole or a statue. They like well-defined places, such as steps, or the border of a pool. What they rarely choose is the middle of a large space.

There are a number of explanations. The preference for pillars might be ascribed to some primeval instinct: you have a full view of all corners but your rear is covered. But his doesn't explain the inclination men have for lining up at the curb. Typically, they face inwards, toward the sidewalk, with their backs exposed to the dangers of the street.

Foot movements are consistent, too. They seem to be a sort of silent language. Often, in a schmoozing group no one will be saying anything. Men stand bound in amiable silence, surveying the passing scene. Then, slowly, rhythmically, one of the men rocks up and down: first on the ball of the foot, then back on the heel. He stops. Another man starts the same movement. Sometimes there are reciprocal gestures. One man makes a half-turn to the right. Then, after a rhythmic interval, another responds with a half turn to the left. Some kind of communication seems to be taking place here, but I've never broken the code.

Whenever they may mean, people's movements are one of the great spectacles of a plaza. You do not see this in architectural photographs, which typically are empty of life and are taken from a persepctive few people share. It is a quite misleading one. At eye level, the scene comes alive with movement and color—people walking quickly, walking slowly, skipping up steps, weaving in and out on crossing patterns, accelerating and retarding to match the moves of others. There is a beauty that is beguiling to watch, and one senses that the players are quite aware of it themselves. You see this, too, in the way they arrange themselves on steps and ledges. They often do so with a grace that they, too, just sense. With its brown-gray monochrome, Seagram's is the best of the settings—especially in the rain, when an umbrella or two spots color in the right places, like Corot's red dots.

How peculiar are such patterns to New York? Our working assumption was that behavior in other cities would probably differ little and subsequent comparisons have proved our assumption correct. The important variable is city size. As I will discuss in more detail, in smaller cities, densities tend to lower, pedestrians move at a slower pace, and there is less of the social activity characteristic of high traffic areas. In most other respects, pedestrian patterns are similar.

Observers in other countries have also noted the tendency to self-congestion. In his study of pedestrians in Copenhagen, architect Jan Gehl mapped bunching patterns almost identical to those observable here. Matthew Ciolek studied an Australian shopping center, with similar results. "Contrary to 'common sense' expectations," Ciolek notes, 'the great majority of people were found to select their sites for social interaction right on or very close to the traffic lines intersecting the plaza. Relatively few people formed their gatherings away from the spaces used for navigation."

The strongest similarities are found among the world's largest cities. People in them tend to behave more like their counterparts in other world cities than like fellow nationals in smaller cities. Bit city people walk faster, for one thing, and they self-congest. After we had completed our New York study, we made a brief comparison study of Tokyo and found the proclivity to stop and talk in the middle of department store doorways, busy corners, and the like, is just as strong in that city as in New York. For all the cultural differences, sitting patterns in parks and plazas are much the same, too. Similarly, schmoozing patterns in Milan's Galleria are remarkably like those in New York's garment center. Modest conclusion: given the basic elements of a center city—such as high pedestrian volumes, and concentration and mixture of activities—people in one place tend to act much like people in another.

Indoor spaces

As an alternative to plazas, builders have been turning to indoor spaces. There are many variants: atriums, galleries, courtyards, through-block arcades, indoor parks, covered pedestrian areas of one shape or another. Some are dreadful. In return for extra floors, the developers provided space and welshed on the amenities. But some spaces have been very successful indeed, and there is enough of a record to indicate that the denominators are much the same as with outdoor space. Here, briefly, are the principal needs.

Sitting. Movable chairs are best for indoor parks. Most of the popular places have had excellent experience with them; some places, like Citicorp, have been adding to the numbers. In all cases the total amount of sitting space has met or exceeded the minimum recommended for outdoor spaces—one linear foot for every 30 square feet (2.8 sq. m) of open space. There is a tendency, however, to overlook the potentials of ledges and planters. Too many are by inadvertence higher than need be.

Food. Every successful indoor space provides food. The basic combination is snack bars and chairs and tables. Some places feature café operations as well.

Retailing. Shops are important for liveliness and the additional pedestrian flows they attract. Developers, who can often do better renting the space for banks or offices, are not always keen on including shops. They should be required to.

Toilets. If incentive zoning achieved nothing else, an increase in public toilets would justify it. Thanks to beneficent pressure, new indoor parks in New York are providing a pair or more, unisex-style as on airplanes. These facilities are modest, but their existence could have a considerable effect on the shopping patterns of many people, older ones especially.

One benefit of an indoor space is the through-block circulation it can provide for pedestrians. Planners believe this important, and developers have been allowed a lot of additional floor space in return for it. But walking space is about all that developers have provided, and it has proved no bargain. Unless there are attractions within, people don't use walkways very much, even in rainy or cold weather. At New York's Olympic Towers, which is taller by several million dollars worth of extra space for providing a through-block passage, the number of people traversing the passage is about 400 per hour at peak. On the Fifth Avenue sidewalk that parallels the passage, the flow is about 4,000 per hour.

Not so paradoxically, the walk-through function of a space is greatly enhanced if something is going on within it. Even if one does not tarry to sit or get a snack, just seeing the activity makes a walk more interesting. Conceivably, there could be conflict between uses. Planners tend to fret over this and, to ensure adequate separation, they specify wide walkways—in New York, 20 feet (6 m) at the minimum. But that is more than enough. As at plazas, the places people like best for sitting are those next to the main pedestrian flow, and for many conversations the very middle of the flow. Walkers like the proximity too. It makes navigation more challenging. At Minneapolis, one often gets blocked by people just standing or talking while there are others in crossing patterns or collision courses up ahead. The processional experience is all the better for the busyness.

Fig. 3. Crystal Court at the IDS Center in Minneapolis, designed by John Burgee and Philip Johnson, is one of the best interior spaces in the United States. Psychologically as well as visually, the Center has an excellent relationship with the street and surroundings, which are eminently visible. This helps make pedestrian flows easy. It has been a hospitable place, with plenty of seating—though a new owner bought the building and removed the open seating.

In an important respect, public spaces that are inside differ from public spaces that are outside. They're not as public. The look of a building, its entrances, the guards do have a filtering effect and the cross section of the public that uses the space within is somewhat skewed—with more higher-income people, fewer lower-income people, and presumably, fewer undesirables. This, of course, is just what the building management and shop owners want. But there is a question of equity posed. Should the public underwrite such space? In a critique of the Citicorp Building, Suzanne Stephens argues in *Progressive Architecture* that it should not. The suburban shopping mall, she notes, is frankly an enclave and:

> *"owes its popularity to what it keeps out as well as what it offers within. Whether this isolationism should occur in 'public space' created through the city' incentive zoning measures should be addressed at the city planning level. . .Open space amenities are moving from the true public domain blur. Thus this public space is becoming increasingly privatized."*

This is very much the case with most megastructures. They are exclusionary by design, and they are wrongly so. But buildings with indoor spaces can be quite hospitable if they are designed to be so, even rather large ones. The Crystal Court of the IDS Center is the best indoor space in the country, and it is used by a very wide mix of people. In mid-morning, the majority of the people sitting and talking are older people, and many of them are obviously of limited means (Fig. 3).

Inevitably, any internal space is bound to have a screening effect; its amenities, the merchandise lines offered, the level of the entertainment—all of these help determine the people who will choose to come, and it is not necessarily a bad thing if a good many of the people are educated and well-off. But there should be other kinds of people, too, and, if there are not, the place is not truly public. Or urban.

The big problem is the street. Internal spaces with shops can dilute the attractions of the street outside, and the more successfully they are, the greater the problem. How many more indoor spaces it might take to tip the scale is difficult to determine, but it is a matter the planning commissions should think very hard about. More immediate is the question of the internal space's relation to the street. If the space is underwritten by incentive zoning, it should not merely provide access to the public, it should invite it. A good internal space should not be blocked off by blank walls. It should be visible from the street; the street and its surroundings should be highly visible from it; and between the two, physically and psychologically, the connections should be easy and inviting. The Crystal Court of the IDS building is a splendid example. It is transparent. You are in the center of Minneapolis, no mistake. You see it. There is the street and the neighboring buildings, and what most catches the eye are the flows of people through doorways and walkways. It is an easy place to get in and out of.

Most places are not. Typically, building entrances are overengineered affairs centered around a set of so-called revolving doors. The doors do not of themselves revolve; you revolve them. From a standing start, this requires considerable foot pounds of energy. As does opening the swinging doors at the sides—which you are not supposed to use anyway. These doors are for emergency use. So there is frequently a sign saying "PLEASE USE REVOLVING DOOR" mounted on a pedestal blocking the center of the emergency door. Sometimes, for good measure, there is second set of doors 15 or 20 feet (4.6 or 6 m) inside the first.

All this is necessary, engineers say, for climate control and for an air seal to prevent stack-effect drafts in the elevator shafts. Maybe so. But on occasion revolving doors are folded to an open position. If you watch the entrances then, you will notice that the building still stands and no great drafts ensue. Watch the entrance long enough, and there is something else you will notice. The one time they function well is when they are very crowded.

I first noticed this phenomenon at *Place Ville Marie* in Montreal. I was clocking the flow through the main concourse entrance, a set of eight swinging doors. At 8:45 a.m., when the flow was 6,000 people per hour, there was a good bit of congestion, with many people lined up one behind another. Ten minutes later the flow was up to a peak rate of 8,000 people an hour (outside Tokyo, the heaviest I've ever clocked). Oddly, there was little congestion. People were moving faster and more easily, with little queuing. (Fig. 4)

The reason lies in the impulse for the open door. Some people are natural door openers. Most are not. Where there is a choice, they will follow someone who is opening a door. Sometimes they will queue up two or three deep rather than open a door themselves. Even where there are many doors, most of the time the bulk of the traffic will be self-channeled through one or two of them. As the crowd swells, however, an additional door will be opened, then another. The pace quickens. The headway between people shortens. In transportation planning, it is axiomatic that there should be a comfortable headway between people. In doorway situations, the opposite is true. If the interval between people shortens 1.2 seconds or less, the doors don't get a chance to close. All or most of the doors will open, and instead of bunching up at one or two of them, people will distribute themselves through the whole entrance.

One way to provide a good entrance, then, is to have big enough crowds. But there is another possibility. Why not leave a door open? This novel approach has been followed for the entrance of an indoor park. As part of the new Philip Morris building, architect Ulrich Franzen has designed an attractive space that the Whitney Museum will operate as a kind of sculpture garden. An entrance that invited people in was felt to be very important. Before the energy shortage an air door would have been the answer, and had been so specified in the zoning code for covered pedestrian areas. But this was out of the question now. So, at the other end of the scale, was the usual revolving-door barricade.

To check the potentials of an open door, I did a simple study of heavily used entrances. I filmed rush-hour flows with a digital stop watch recorded on the film, and then calculated how many people used which parts of the entrance. Happily, the weather was mild, and at several of the entrances one or two doors would be wedged open. As at *Place Ville Marie*, it was to the open door that most people went. This does not mean that the other doors were redundant; even if one doesn't choose to use them, having the choice to do so lessens one's sense of crowding. But for sheer efficiency, it became clear, a small space kept open is better than a wider space that is closed. At the main concourse entry to the RCA building, two open doors at one side of an eight-door entrance accounted for two thirds of the people passing through during the morning rush hour. At Grand Central Station, most of those using the nine-door entry at 42nd Street traversed open doors, and at any given time three doors accounted for the bulk of the traffic. The doors at Grand Central work well.

Franzen's design for the entrance to the Philip Morris indoor park incorporates these simple findings. Visually, the entrance will be a stretch of glass 20 feet (6 m) wide. At the center it will have a pair of automatic sliding doors. In good weather and at peak-use times, the doors will be kept open to provide a clear six-foot entry. This should be enough for the likely peak flows. For overflows, and people who like to open doors, there will be an option of swinging doors at either side. In bad weather, the sliding doors will open automatically when people approach. In effect, there will be an ever-open door. It is to be hoped that there will be many more.

Smaller cities and places

Will the factors that make a plaza or small space successful in one city work in another? Generally, the answer is yes—with one key variable to watch. It is scale, and it is particularly important for smaller cities. For a number of reasons, it is tougher for them to create lively spaces than it is for a big city. Big cities have lots of people in their downtowns. This density poses problems but it provides a strong supply of potential users for open spaces in most parts of the central business district. Where 3,000 people an hour pass by a site, a lot of mistakes can be made in design and a place may still end up being well used.

Smaller cities are not as compressed. True, some are blessed with a tight, well-defined center, with some fine old buildings to anchor it. But many have loosened up; they have torn down old buildings and not replaced them, leaving much of the space open. Parking lots and garages become the dominant land use, often accounting for more than 50 percent of downtown. This is true also of some big cities—Houston, for one. Houston has some fine elements in its downtown, but they are so interspersed with parking lots that they don't connect very well with one another.

Many cities have diffused their downtowns by locating new "downtown" developments outside of downtown, or just far enough away that one element does not support the other. The distances need not be great. If you have to get into a car and drive, a place six blocks away might as well be a mile or more. That is precisely the kind of trouble you have in a number of cities. Kansas City's Crown Center, for example, is only 11 blocks from the central business district, but the two centers still remain more or less unconnected.

Cities in the 100,000–200,000 range are not just scaled down versions of bigger cities. Relatively speaking, the downtowns of these smaller cities cover more space than the downtowns of bigger cities. Often their streets are wider, and their pedestrian densities much lower, with fewer people in any given area of the central business district. Sidewalk counts are a good index. If the number of passersby is under a rate of 1,000 per hour around noontime, a city could pave the street with gold for all the difference it would make. Something fundamental is missing: people. More stores, more offices, more reasons for being are what the downtown must have.

Some cities have sought to revitalize their downtowns by banning cars from the main street and turning it into a pedestrian mall. Some of these malls have worked well. Some have not. Again, the problem is diffusion. The malls may be too big for the number of people and the amount of activities. This seems to be particularly the case with the smaller cities—which tend to have the largest malls.

What such cities need to do is compress, to concentrate. Many of them were very low density to begin with; in some, most of the buildings

Fig. 4. *Place Ville Marie* sequence shows how heavier flows can make for less congestion in doorways. People tend to queue up behind open doors; when there are so many people that all doors are open, everyone moves faster.

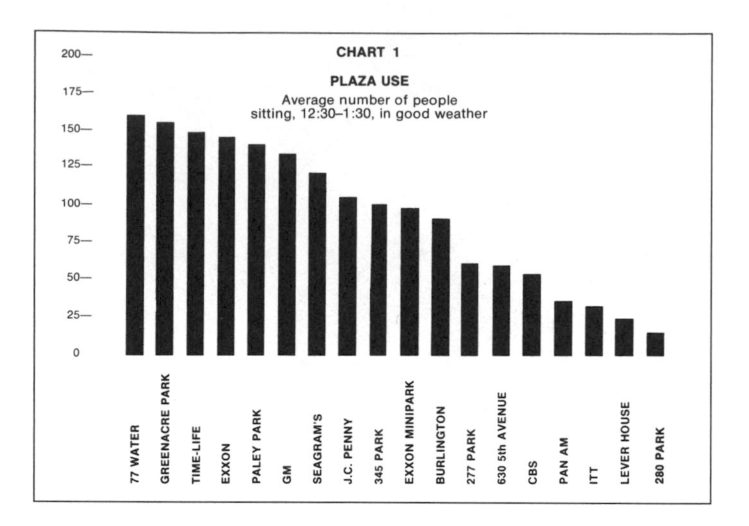

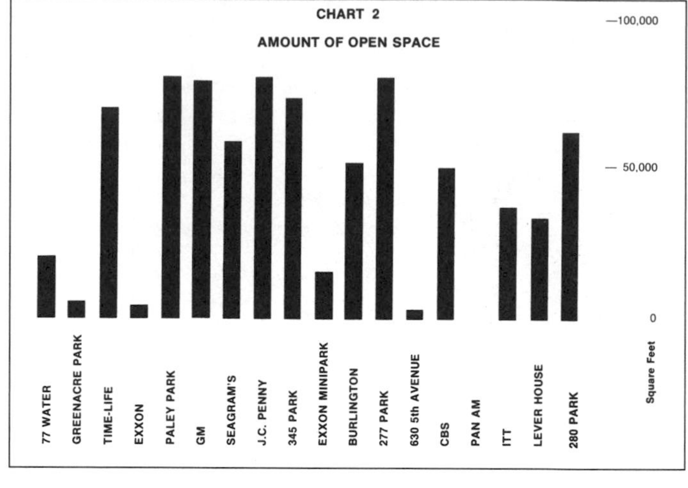

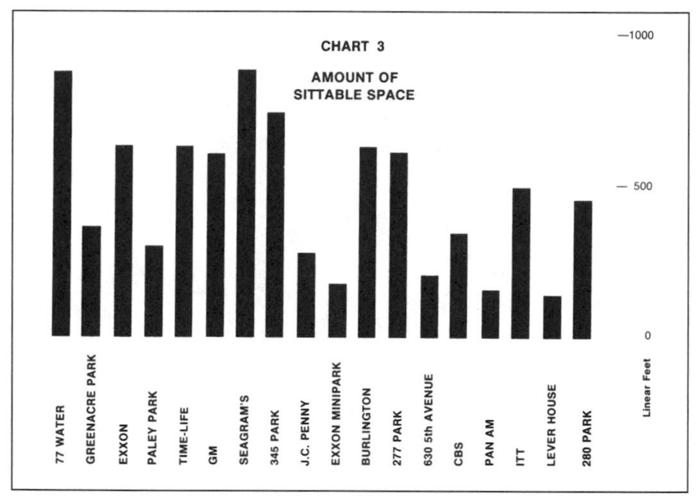

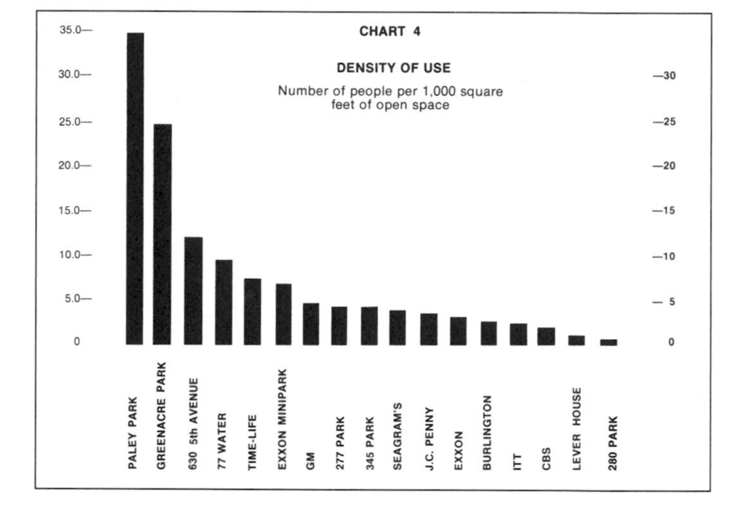

are only two or three stories high. Spread over many downtown blocks are activities and people that might have come together in a critical mass had they been compressed into two or three. Such places are sad to see. So many hopes, so many good intentions, so many fountains and play sculptures have gone into them. Yet they are nearly empty.

Smaller cities are also highly vulnerable to the competition of suburban shopping centers—in particular, the huge centralized ones going up next to interchanges. The suburban centers that do well are more urban in their use of space than the cities they are beating out. True, they are surrounded by a vast acreage of parking space, much of which is never used save on peak days. Unlike the earlier generation of linear shopping centers, however, the new ones are highly concentrated, one-stop places. You don't have to drive here to this and there for that. You enter an enclosed pedestrian system that is, in effect, a gigantic customer-processing machine.

A model for downtown? Some cities now think so. To beat suburbia at its own game, they have been inviting developers to put up shopping centers in downtown. The developers have responded with copies of their suburban models, with very little adaptation; concrete boxes, geared to people who drive to them, that have little relationship to the sidewalks or surrounding buildings of the city. These mini-megastructures may be an efficient setting for merchandising of the middle range; in suburbia, they provide something of a social center as well. But they are not for the downtown. They are the antithesis of what downtown should be.

Cities do best when they intensify their unique strengths. Salem, Oregon, for example, at one time thought its last, best hope would be a suburban-type shopping complex, complete with a skyway or two for razzle-dazzle. But somehow it didn't seem like Salem. The city decided on an opposite approach, filling in empty spaces with buildings to the scale of the place, putting glass canopies over sidewalks, converting alleys into shopping ways, tying strong points with pedestrian spaces and sitting areas. An old opera house has been converted into a complex of stores with felicitous results, and other old structures may be recycled, too. In sum, Salem has embarked on a plan that works with the grain of the city.

It is significant that the cities doing best by their downtowns are the ones doing best at historic preservation and reuse. Fine old buildings are worthwhile in their own right, but there is a greater benefit involved. They provide discipline. Architects and planners like a blank slate. They usually do their best work, however, when they don't have one. When they have to work with impossible lot lines and bits and pieces of space, beloved old eyesores, irrational street layouts, and other such constraints, they frequently produce the best of their new designs—and the most neighborly. ■

3 • URBAN DESIGN HISTORY AND THEORY

Gordon Cullen

Summary
Articles that formed the basis for *Townscape* (1961) first appeared in *The Architectural Review*. Like Jane Jacobs, William Whyte, Kevin Lynch, Robert Venturi and others in 1960s–70s, Gordon Cullen (1914–1994) approached urban design based upon the experience, perception, and particulars of specific places, as opposed to the rationalism of modernist urban theory, which was seen as abstract, formalist and disengaged. While there may be a tendency to see his position as picturesque and as the superficial use of "heritage" and cosmetic streetscapes, Cullen's seemingly lighthearted sketches were meant to illustrate a more rigorous theory based not on images but on underlying visual relationships that evoke a broad range of human response.

Key words
city, optics, place making, sketches, town square, viewscape, visual survey

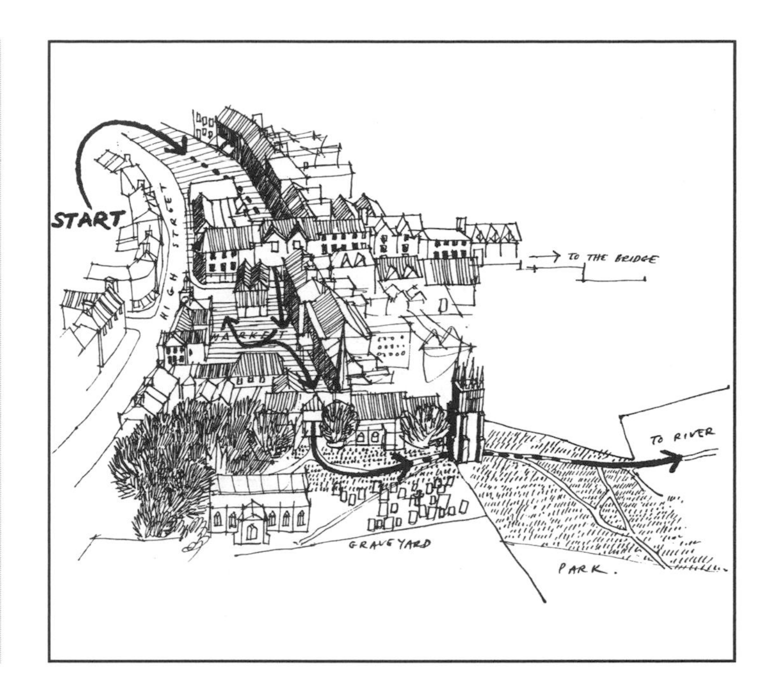

Introduction to townscape

1 INTRODUCTION

There are advantages to be gained from the gathering together of people to form a town. A single family living in the country can scarcely hope to drop into a theater, have a meal out or browse in a library, whereas the same family living in a town can enjoy these amenities. The little money that one family can afford is multiplied by thousands and so a collective amenity is made possible. A city is more than the sum of its inhabitants. It has the power to generate a surplus of amenity, which is one reason why people like to live in communities rather than in isolation.

Now turn to the visual impact, which a city has on those who live in it or visit it. I wish to show that an argument parallel to the one put forward above holds good for buildings: bring people together and they create a collective surplus of enjoyment; brings building together and collectively they can give visual pleasure which none can give separately.

One building standing alone in the countryside is experienced as a work of architecture, but bring a half dozen buildings together and an art other than architecture is made possible. Several things begin to happen in the group, which would be impossible for the isolated building. We may walk through and past the buildings, and as a corner is turned an unsuspected building is suddenly revealed. We may be surprised, even astonished (a reaction generated by the composition of the group and not by the individual building). Again, suppose that the buildings have been put together in a group so that one can get inside the group, then the space created between the buildings is seen to have a life of its own over and above the buildings which create it and one's reaction is to say "I am inside IT" or "I am entering IT." Note also that in this group of half a dozen buildings there maybe one which through reason of function does not conform. It may be a bank, a temple or a church amongst houses. Suppose that we are just looking at the temple by itself, it would stand in front of us and all its qualities, size, color and intricacy would be evident. But put the temple back amongst the small houses and immediately its size is made more real and more obvious by the comparison between the two scales. Instead of being a big temple it towers. The difference in meaning between bigness and towering is the measure of the relationship.

In fact there is an *art of relationship* just as there is an art of architecture. Its purpose is to take all the elements that go to create the environment: buildings, trees, nature, water, traffic, advertisements and so on, and to weave them together in such a way that drama is released. For a city is a dramatic event in the environment. Look at the research that is put into making a city work: demographers, sociologists, engineers, traffic experts; all cooperating to form the myriad factors into a workable, viable and healthy organization. It is a tremendous human undertaking.

And yet. . .if at the end of it all the city appears dull, uninteresting and soulless, then it is not fulfilling itself. It has failed. The fire has been laid but nobody has put a match to it.

Credits: This article is reprinted from *Townscape*, Architectural Press, London, 1961, with permission of the publisher.

Fig. 1. Here and There. Man-made enclosure, if only of the simplest kind, divides the environment into Here and There. On this side of the arch, in Ludlow, we are in the present, uncomplicated and direct world, our world. The other side is different, having in some small way a life of its own (a withholding). And just as the prow of a boat visible over a wall tells you of the proximity of the sea (vast, everlasting) so the church spire turns simple enclosure (left), into the drama of Here and There (right).

Firstly we have to rid ourselves of the thought that the excitement and drama that we seek can be born automatically out of the scientific research and solutions arrived at by the technical man (or the technical half of the brain). We naturally accept these solutions, but are not entirely bound by them. In fact we cannot be entirely bound by them because the scientific solution is based on the best that can be made of the average: of averages of human behavior, averages of weather, factors of safety and so on. And these averages do not give an inevitable result for any particular problem. They are, so to speak, wandering facts, which may synchronize or, just as likely, may conflict with each other. The upshot is that a town could take one of several patterns and still operate with success, equal success. Here then we discover a pliability in the scientific solution and it is precisely in the *manipulation of this pliability* that the art of relationship is made possible. As will be seen, the aim is not to dictate the shape of the town or environment, but is a modest one: simply to *manipulate within the tolerances.*

This means that we can get no further help from the scientific attitude and that we must therefore turn to other values and other standards.

We turn to the *faculty of sight*, for it is almost entirely through vision that the environment is apprehended. If someone knocks at your door and you open it to let him in, it sometimes happens that a gust of wind comes in too, sweeping round the room, blowing the curtains and making a great fuss. Vision is somewhat the same; we often get more than we bargained for. Glance at the clock to see the time and you see the wallpaper, the clock's carved brown mahogany frame, the fly crawling over the glass and the delicate rapierlike pointers. Cézanne might have made a painting of it. In fact, of course, vision is not only useful but it evokes our memories and experiences, those responsive emotions inside us which have the power to disturb the mind when aroused. It is this unlooked-for surplus that we are dealing with, for clearly if the environment is going to produce an emotional reaction, with or without our volition, it is up to us to try to understand the three ways in which this happens.

Concerning Optics

Let us suppose that we are walking through a town: here is a straight road off which is a courtyard, at the far side of which another street leads out and bends slightly before reaching a monument. Not very unusual. We take this path and our first view is that of the street. Upon turning into the courtyard the new view is revealed instantaneously at the point of turning, and this view remains with us whilst we walk across the courtyard. Leaving the courtyard we enter the further street. Again a new view is suddenly revealed although we are traveling at a uniform speed. Finally as the road bends the monument swings into view. The significance of all this is that although the pedestrian walks through the town at a uniform speed, the scenery of towns is often revealed in a series of jerks or revelations. This we call *Serial Vision.*

Examine what this means. Our original aim is to manipulate the elements of the town so that an impact on the emotions is achieved. A long straight road has little impact because the initial view is soon digested and becomes monotonous. The human mind reacts to a contrast, to the difference between things, and when two pictures (the street and the courtyard) are in mind at the same time, a vivid contrast is felt and the town becomes visible in a deeper sense. It comes alive through the drama of juxtaposition. Unless this happens the town will slip past us featureless and inert.

There is a further observation to be made concerning serial vision. Although from a scientific or commercial point of view the town may be a unity, from our optical viewpoint we have split it into two elements: the *existing view* and the *emerging view*. In the normal way this is an accidental chain of events and whatever significance may arise out of the linking of views will be fortuitous. Suppose, however, that we take over this linking as a branch of the art of relationship; then we are finding a tool with which human imagination can begin to mold the city into a coherent drama. The process of manipulation has begun to turn the blind facts into a taut emotional situation.

Concerning Place

This second point is concerned with our reactions to the position of our body in its environment. This is as simple as it appears to be. It means, for instance, that when you go into a room you utter to yourself the unspoken words, "I am outside IT, I am entering IT, I am in the middle of IT." At this level of consciousness we are dealing with a range of experience stemming from the major impacts of exposure and enclosure (which if taken to their morbid extremes result in the symptoms of agoraphobia and claustrophobia). Place a man on the edge of a 500-ft. (152-m) cliff and he will have a very lively sense of position, put him at the end of a deep cave and he will react to the fact of enclosure.

Since it is an instinctive and continuous habit of the body to relate itself to the environment, this sense of position cannot be ignored; it becomes a factor in the design of the environment (just as an additional source of light must be reckoned with by a photographer, however annoying it may be). I would go further and say that it should be exploited.

Here is an example. Suppose you are visiting one of the hill towns in the south of France. You climb laboriously up the winding road and eventually find yourself in a tiny village street at the summit. You feel thirsty and go to a nearby restaurant, your drink is served to you on a veranda and as you go out to it you find to your exhilaration or horror that the veranda is cantilevered out over a thousand-foot drop. By this devise of the containment (street) and the revelation (cantilever) the fact of height is dramatized and made real.

In a town we do not normally have such a dramatic situation to manipulate but the principle still holds good. There is, for instance, a typical emotional reaction to being below the general ground level and there is another resulting from being above it. There is a reaction to being hemmed in as in a tunnel and another to the wideness of the square. If, therefore, we design our towns from the point of view of the moving person (pedestrian or car-borne) it is easy to see how the whole city becomes a plastic experience, a journey through pressures and vacuums, a sequence of exposures and enclosures, of constraint and relief.

Arising out of this sense of identify or sympathy with the environment, this feeling of a person in street or square that he is in IT or entering IT or leaving IT, we discover that no sooner do we postulate a HERE than automatically we must create a THERE, for you cannot have one without the other. Some of the greatest townscape effects are created by a skillful relationship between the two, and I will name an example in India, where this introduction is being written: the approach from the Central Vista to the Rashtrapathi Bhawan in New Delhi. There is an open-ended courtyard composed of the two Secretariat buildings and, at the end, the Rashtrapathi Bhawan.

Fig. 2. Inside Extends Out. The corollary of this is the expression of inside volumes externally. In the case of the public house, the normal street façade is interrupted by the bulge which expresses the function. Again, the section through the shopping street shows how on one side, the left, we simply have shop windows whilst on the right the awnings and costers' barrows form an enclosure which transforms the whole street from an arid inside/outside statement to a comprehensive and dramatic linear market.

Fig. 3. Space Continuity. Similarly but on a larger scale, this view of Greenwich market, produces the effect of spatial continuity, a complex interlocking of volumes in which the quality of light and materials denies the concept of outside and inside.

Fig. 4. Public and Private. Emphasizing this difference are the various qualities attached to parts of the environment, qualities of character, scale, color, etc. In this case the change is from a public Here (Victoria Street) to a private or precinctual There (Westminster Cathedral).

All this is raised above normal ground level and the approach is by a ramp. At the top of the ramp and in front of the axis building is a tall screen of railings. This is the setting. Traveling through it from the Central Vista we see the two Secretariats in full, but the Rashtrapathi Bhawan is partially hidden by the ramp; only its upper part is visible. This effect of truncation serves to isolate and make remote. The building is withheld. We are Here and it is There. As we climb the ramp the Rashtrapathi Bhawan is gradually revealed, the mystery culminates in fulfillment as it becomes immediate to us, standing on the same floor. But at this point the railing, the wrought iron screen, is inserted; which again creates a form of Here and There by means of the screen vista. A brilliant, if painfully conceived sequence.

Concerning Content

In this last category we turn to an examination of the fabric of towns: color, texture, scale, style, character, personality and uniqueness. Accepting the fact that most towns are of old foundation, their fabric will show evidence of differing periods in its architectural styles and also in the various accidents of layout. Many towns display this mixture of styles, materials and scales.

Yet there exists at the back of our minds a feeling that could we only start again we would get rid of this hotchpotch and make all new and fine and perfect. We would create an orderly scene with straight roads and with buildings that conformed in height and style. Given a free hand that is what we might do...create symmetry, balance, perfection and conformity. After all, that is the popular conception of the purpose of town planning.

But what is this conformity? Let us approach it by a simile. Let us suppose a party in a private house, where are gathered together half a dozen people who are strangers to each other. The early part of the evening is passed in polite conversation on general subjects such as the weather and the current news. Cigarettes are passed and lights offered punctiliously. In fact it is all an exhibition of manners, of how one ought to behave. It is also very boring. This in conformity. However, later on the ice begins to break out of the straightjacket of orthodox manners and conformity and real human beings begin to emerge. It is found that Miss X's sharp but good-natured wit is just the right foil to Major Y's somewhat simple exuberance. And so on. It begins to be fun. Conformity gives way to the agreement to differ within a recognized tolerance of behavior.

Conformity, from the point of view of the planner, is difficult to avoid but to avoid it deliberately, by creating artificial diversions, is surely worse than the original boredom. Here, for instance, is a programme to rehouse 5,000 people. They are all treated the same, they get the same kind of house. How can one differentiate? Yet if we start from a much wider point of view we will see that tropical housing differs from temperate zone housing, that buildings in a brick country differ from buildings in a stone country, that religion and social manners vary the buildings. And as the field of observation narrows, so our sensitivity to the local gods must grow sharper. There is too much insensitivity in the building of towns, too much reliance on the tank and the armored car where the telescopic rifle is wanted.

Within a commonly accepted framework—one that produces lucidity and not anarchy—we can manipulate the nuances of scale and style, of texture and color and of character and individuality, juxtaposing them in order to create collective benefits. In fact the environment thus resolves itself into not conformity but the interplay of This and That.

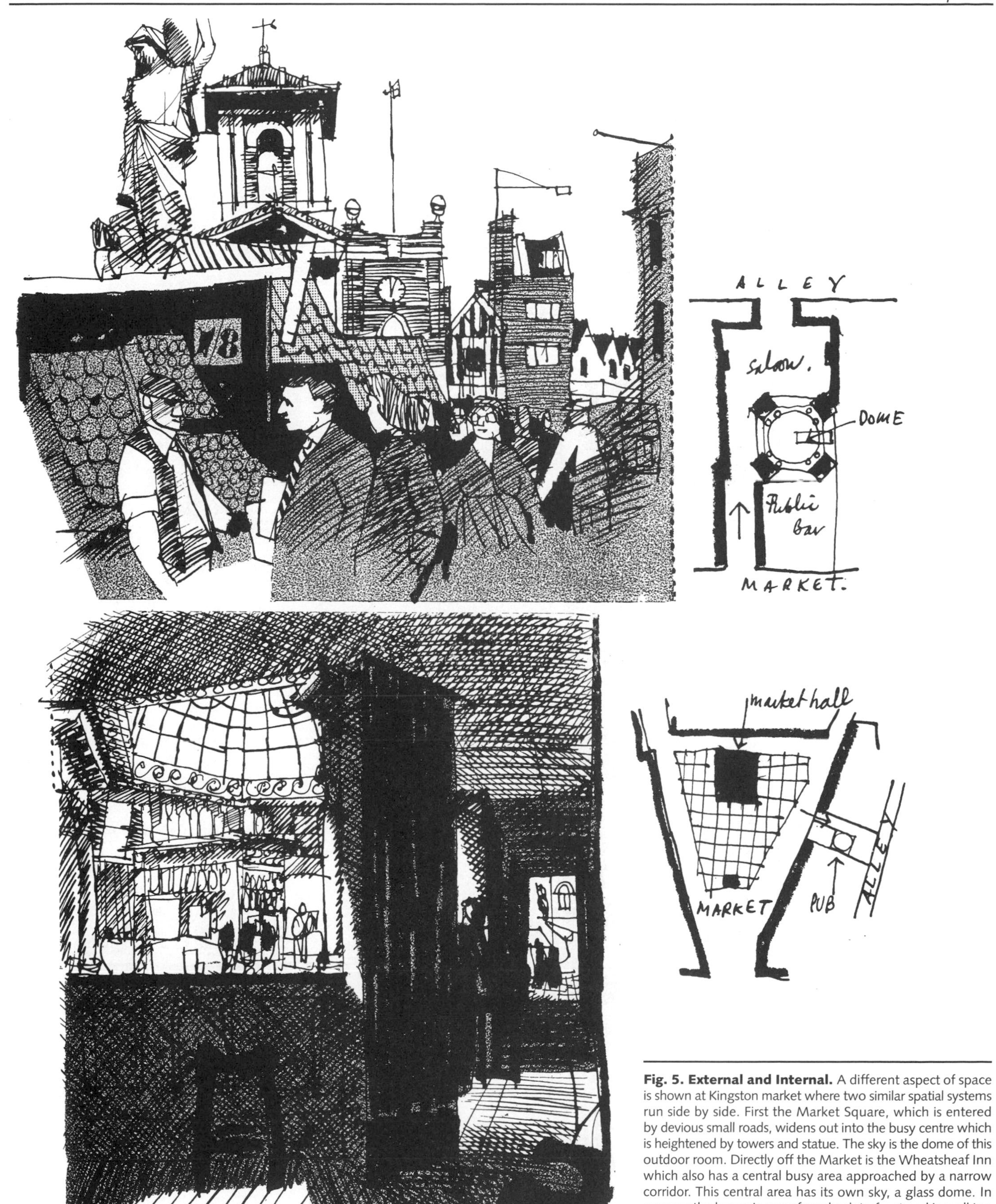

Fig. 5. External and Internal. A different aspect of space is shown at Kingston market where two similar spatial systems run side by side. First the Market Square, which is entered by devious small roads, widens out into the busy centre which is heightened by towers and statue. The sky is the dome of this outdoor room. Directly off the Market is the Wheatsheaf Inn which also has a central busy area approached by a narrow corridor. This central area has its own sky, a glass dome. In summer the house is open from back to front and in walking through one is struck by this unity of space sequence.

Fig. 6. Space and Infinity. The effect of infinity is not normally apparent in sky seen over rooftops. But if sky is suddenly seen where one might reasonably expect to walk, i.e., at ground level, then there is an effect of infinity or shock.

Fig. 7. Captured Space. The carved frets reach out and grip space, the slender rail and posts enclose it, the pierced wall reveals it. Behind, the louvered openings reveal the next dim layer of internal space and the windows complete it.

Fig. 8. Projection. Space, being occupiable, provokes colonization. This reaction may be exploited by placing space to achieve the desired results. In this view of the Bank of England the lofty portico, left, elevates the spirit more than a lofty solid building might.

It is a matter of observation that in a successful contrast of colors not only do we experience the harmony released but, equally, the colors become more truly themselves. In a large landscape by Corot, I forget its name, a landscape of somber greens, almost a monochrome, there is a small figure in red. It is probably the reddest thing I have ever seen.

Statistics are abstracts: when they are plucked out of the completeness of life and converted into plans and the plans into buildings they will be lifeless. The result will be a three-dimensional diagram in which people are asked to live. In trying to colonize such a wasteland, to translate it from an environment for walking stomachs into a home for human beings, the difficulty lay in finding the point of application, in finding the gateway into the castle. We discovered three gateways, that of motion, that of position and that of content. By the exercise of vision it became apparent that motion was not one simple, measurable progression useful in planning, it was in fact two things, the Existing and the Revealed view. We discovered that the human being is constantly aware of his position in the environment, that he feels the need for a sense of place and that this sense of identity is coupled with an awareness of elsewhere. Conformity killed, whereas the agreement to differ gave life. In this way the void of statistics, of the diagram city, has been split into two parts, whether they be those of Serial Vision, Here and There or This and That. All that remains is to join them together into a new pattern created by the warmth and power and vitality of human imagination so that we build the home of man.

That is the theory of the game, the background. In fact the most difficult part lies ahead, the Art of Playing. As in any other game there are recognized gambits and moves built up from experience and precedent. In the pages that follow an attempt is made to chart these moves under the three main heads as a series of cases, and later to show their application by means of town studies and planning proposals.

2 HERE AND THERE

On a flat plain a house is built. It is an object standing up on the flat surface. Inside the house there are rooms, volumes of space: but from the outside these are not obvious. All we see is the object. Many houses built together form streets and squares. They enclose space and thus a new factor is added to the internal volumes or spaces. . . the outside spaces. Whereas internal volumes, rooms, are justified in the purely functional sense of construction and shelter, there is no such forthright justification for external space/volume. It is accidental and marginal. Or is it?

In a purely materialistic world our environment would resemble a rock-strewn river, the rocks being building and the river being traffic passing them, vehicular and pedestrian. In fact, this conception of flow is false since people are by nature possessive. A group of people standing or chatting on the pavement colonize the spot and the passer-by has to walk round them. Social life is not confined to the interior of buildings. Where people forgather, in market place or forum, there will therefore be some expression of this to give identity to the activity. Marketplace, focal point, clearly defined promenade and so on. In other words, the outside is articulated into spaces just as is the inside, but for its own reasons.

We can therefore postulate an environment which is articulated; as opposed to one which is simply a part of the earth's surface, over which

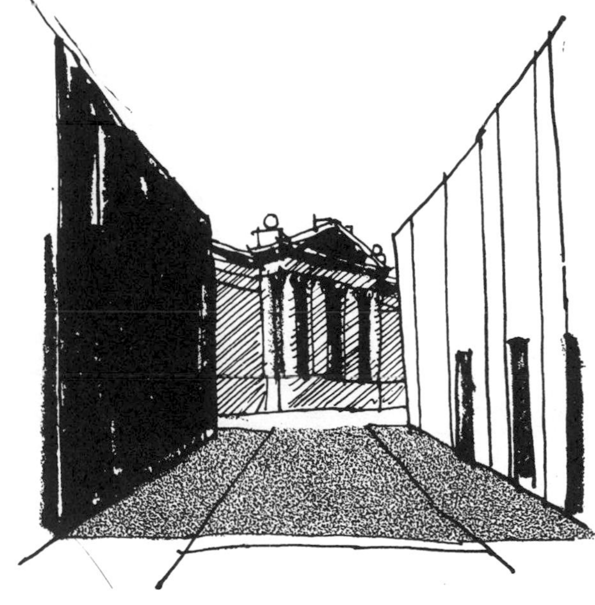

Fig. 9. Deflection. Where a view is terminated by a building at right angles to the axis then the enclosed space is complete. But a change of angle in the terminal building, as here in Edinburgh, creates a secondary space by implication. A space which you cannot see but feel must be there, facing the building.

Fig.10. Functional Space. What better way of emphasizing an event in the street such as a theater, than by giving this function its own space, which becomes alive and informed by sparkle and conversation and tension.

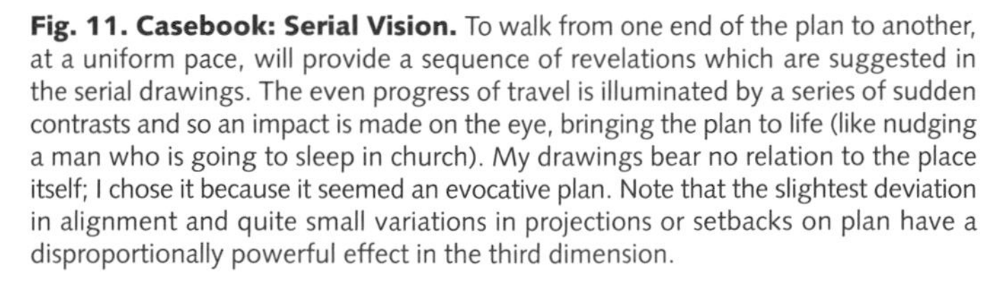

Fig. 11. Casebook: Serial Vision. To walk from one end of the plan to another, at a uniform pace, will provide a sequence of revelations which are suggested in the serial drawings. The even progress of travel is illuminated by a series of sudden contrasts and so an impact is made on the eye, bringing the plan to life (like nudging a man who is going to sleep in church). My drawings bear no relation to the place itself; I chose it because it seemed an evocative plan. Note that the slightest deviation in alignment and quite small variations in projections or setbacks on plan have a disproportionally powerful effect in the third dimension.

antlike people and vehicles are forever swarming and on to which buildings are plonked at random. Consequently, instead of a shapeless environment based on the principle of flow, we have an articulated environment resulting from the breaking up of flow into action and rest, into corridor street and marketplace, alley and square (and all their minor devolutions).

The practical result of so articulating the town into identifiable parts is that no sooner do we create a HERE than do we have to admit a THERE, and it is precisely in the manipulation of these spatial concepts that a large part of urban drama arises. The drawings in this article depict some points relevant to this use of space in urban scenery. ■

Leslie Martin

Summary
Some parts of *The grid as generator* were used in the Gropius Lecture at Harvard University in June 1966. The argument was developed later into the theme delivered at the University of Hull under the title, "The Framework of Planning," as the inaugural lecture by Leslie Martin as Visiting Ferens Professor of Fine Art. It is presented here in essentially that form. Also see reference "[various] 2000" for contemporary interpretations of the late Leslie Martin's contributions to 20th century urban planning and design theory.

Key words
density, grid, land development, planning, subdivision, street, theory

Leslie Martin (right) and Le Corbusier, University of Cambridge, 1959.

The grid as generator

The activity called city planning, or urban design, or just planning, is being sharply questioned. It is not simply that these questions come from those who are opposed to any kind of planning. Nor is it because so many of the physical effects of planning seem to be piecemeal. For example, roads can be proposed without any real consideration of their effect on environment; the answer to such proposals could be that they are just not planning at all. But is not just this type of criticism that is raised. The attack is more fundamental: what is being questioned is the adequacy of the assumptions on which planning doctrine is based.

What are those assumptions? To put this in the most general terms, they resolve themselves into two powerful lines of thought. The first, which stems from the work of the Viennese writer Camillo Sitte, whose book *City Planning according to Artistic Principles* was published in 1889, can be called the doctrine of the visually ordered city. To Sitte the total city plan is the inspired and the all-encompassing work of art. But Sitte went further: civic art must be an expression of the life of the community, and finally "works of art cannot be created by committee but only by a single individual" (Sitte 1898). The planner then is the inspired artist expressing in the total city plan the ambitions of a society. There are indeed many who, though not prepared to accept this total—it would not be inaccurate to say this totalitarian—role of the planner, have nevertheless been profoundly influenced by Sitte's doctrine of the visually ordered city. The doctrine has left its mark on the images that are used to illustrate high-density development of cities. It is to be seen equally in the layout and arrangement of Garden City development. The predominance of the visual image is evident in some proposals that work for the preservation of the past: it is again evident in the work of those that would carry us on, by an imagery of mechanisms, into the future. It remains central in the proposals of others who feel that, although the city as a total work of art is unlikely to be achieved, the changing aspect of its streets and squares may be ordered visually into a succession of pictures. The second line of doctrine is severely practical. It can be called the doctrine of the statistically ordered city. We know it well. It is the basis of those planning surveys in which uses are quantified, sorted out and zoned into particular areas; population densities are assessed and growth and change predicted. It is the raw material of the outline analyses and the town maps of the 1947 Act.

Now it is precisely these two aspects of planning (the first concerned with visual images and the second with procedure, and sometimes of course used in combination by planners), that were so sharply attacked by Mrs. Jane Jacobs in her book *The Death and Life of Great American Cities* (1961). For Mrs. Jacobs, both "the art of city planning

Credits: This article is the opening chapter of Leslie Martin and Lionel March, eds., *Urban Space and Structure (1972)*, reprinted with permission of Cambridge University Press.

and its companion, the pseudo-science of city planning, have not yet embarked on the effort to probe the real world of living." For her a city can never be the total work of art, nor can there ever be the statistically organized city. Indeed, to Mrs. Jacobs, the planning of any kind of order seems to be inconsistent with the organic development of cities which she sees as a direct outcome of the activities of living. Planning is a restrictive imposition: the areas of cities "in which people have lived are a natural growth. . .as natural as the beds of oysters." Planning, she says, is essentially artificial.

It is of course just this opposition between "organic" growth and the artificial nature of plans, between living and the preconceived system within which it might operate, that has been stressed so much in recent criticism. Christopher Alexander in a distinguished essay "A city is not a tree" puts the point directly when he says (Alexander 1966):

> *I want to call those cities that have arisen spontaneously over many many years "natural cities." And I shall call those cities or parts of cities that have been deliberately created by planners "artificial cities." Siena, Liverpool, Kyoto, Manhattan, are examples of natural cities. Levittown, Chandigarh and the British New Towns are examples of artificial cities. It is more and more widely recognized today that there is some essential ingredient missing in the artificial cities.*

Let us consider this. First of all would it be true to say that all old towns are a kind of spontaneous growth and that there have never been "artificial" or consciously planned towns in history? Leaving on one side ancient history, what about the four hundred extremely well-documented cases of new towns (deliberately planned towns) that Professor Beresford has collected for the Middle Ages in England, Wales and Gascony alone? (Beresford 1967). What about the medieval towns such as those built in Gascony between 1250 and 1318 on a systematic gridiron plan? All these towns were highly artificial in Alexander's sense. The planned town, as Professor Beresford observes, "is not a prisoner of an architectural past: it has no past." In it the best use of land meant an orderly use, hence the grid plan. In siting it and building it estimates had to be made about its future, about its trade, its population, and the size and number of its building plots. This contributes a highly artificial procedure.

But it is of course by no means uncommon. Indeed it is the method by which towns have been created in any rapidly developing or colonial situation. A book by John Reps, *The Making of Urban America* (1965) is a massive compendium of the planning of new towns throughout America, practically all of them based on highly artificial gridiron plans. He points out that there is a sense in which not merely cities but the whole of Western America is developed within an artificial frame: "the giant gridiron imposed upon the natural landscape by. . .the land ordinance of 1785."

The colonizer knows that the natural wilderness has to be transformed: areas must be reversed for agriculture as well as plots for building. The man-made landscape is a single entity: cities and their dependent agricultural areas are not separate elements. All these things are matters of measure and quantity. They are interrelated between themselves and numbers of people. The process demands a quality of abstract thought: a geometry and a relationship of numbers worked out in advance and irrespective of site. The 20-mile square plan for the proposed colony of Azilia, the plans of Savannah and Georgetown, are typical examples of this kind of thought. William Penn's plan for Philadelphia, the plans of such towns as Louisville, Cincinnati, Cleveland, New York City itself, Chicago and San Francisco, are all built on the basis on a preconceived frame.

In the case of the mediaeval towns described by Beresford, whilst some failed, a high proportion succeeded in their time. In a large number of American cities, the artificial grid originally laid down remains the working frame within which vigorous modern cities have developed. It is quite clear then that an artificial frame of some kind does not exclude the possibility of an organic development. The artificial grid of streets that was laid down throughout Manhattan in 1811 has not prevented the growth of those overlapping patterns of human activity which caused Alexander to describe New York as an organic city. Life and living have filled it out but the grid is there.

And this brings us closer to the centre of Alexander's main argument. What he is criticizing in the extended content of his essay, is the notion that the activities of living can be parceled out into separate entities and can be fixed forever by a plan. The assumption is common in much postwar planning. Consider an example. Housing is thought of in terms of density: 75, 100, 150 people per acre. That will occupy an area of land. Housing requires schools and they need open space: that will occupy another specific area. These areas in turn may be thought to justify another need: an area for recreation. That is one kind of thought about planning. But alternatively an effort may be made to see the needs of a community as a whole. It may be discovered that the way housing is arranged on the ground may provide so much free space that the needs of schools or recreation will overlap and may even be contained within it (Martin 1968).

In the first instance the uses are regarded as self-contained entities: Alexander equates this kind of thinking with an organization like that demonstrated by a mathematical tree. In the second instance the patterns of use overlap: the organization in this case is much closer to a far more complex mathematical structure: the semilattice. The illustration of the separate consideration of housing, schools and open space is elementary. But it is Alexander's argument that whole towns may be planned on this basis. And it is the attempt to deal with highly complex and overlapping patterns of use, of contracts and of communications in a way which prevents this overlap from happening that Alexander deplores. Hence the title of his paper: *"A city is not a tree."* In this sense of course he is correct. But the argument can be put in a different way. It can be argued that the notion (implied by Mrs. Jacobs) that elaborate patterns of living can never develop within a preconceived and artificial framework is entirely false. This can be developed by saying that an "organic" growth, without the structuring element of some kind of framework, is chaos. And finally that it is only through the understanding of that structuring framework that we can open up the range of choices and opportunities for future development.

The argument is this. Many towns of course grew up organically by accretion. Others, and they are numerous and just as flourishing, were established with a preconceived framework as a basis. Both are built up ultimately from a range of fairly simple formal situations: the grid of streets, the plots which this pattern creates and the building arrangements that are placed on these. The whole pattern of social behavior has been elaborated within a limited number of arrangements of this kind and this is true of the organized as well as the constructed town. Willmott and Young, studying kinship in the East End of London (1957), were able to show that everywhere elaborate patterns of living had been built up. All these elaborations, and a great variety of needs, were met within a general building pattern of terraces and streets. Change that pattern and you may prevent these relationships from developing or you may open up new choices that were not available in the original building form.

The grid of streets and plots, from which a city is composed, is like a net placed or thrown upon the ground. This might be called the framework of urbanization. That framework remains the controlling factor of the way we build whether it is artificial or accretion. And the way we build may either limit or open up new possibilities in the way in which we choose to live.

The understanding of the way the scale and pattern of this framework, net or grid affects the possible building arrangements on the land within it is fundamental to any reconsideration of the structure of existing towns. It is equally important in relation to any consideration of the development of metropolitan regions outside existing towns. The pattern of the grid of roads in a town or region is a kind of playboard that sets out the rules of the game. The rules outline the kind of game; but the players should have the opportunity to use to the full their individual skills whilst playing it.

How does the framework of a city work? In what way does the grid act as a generator and controlling influence on city form? How can it tolerate growth and change?

The answer to these questions is best given by historical examples, and in order to give the argument some point we can deliberately choose the most artificial framework for a city that exists: the grid as it has been used in the United States, and so well illustrated by Reps (1965).

We can start with the notion that to the colonizer the uncultivated wilderness must be tamed into a single urban-rural relationship. In the plan for the proposed Margravate of Azilia (the forerunner of the colony of Georgia) the ground to be controlled is 20 miles square, or 256,000 acres. Implicit in the subdivisions of this general square is a mile square grid; and out of the basic grid the areas for farmland, the great parks for the propagation of cattle and the individual estates are built up. At the centre is the city proper.

The Margravate was never built, but the concept of the single urban-rural unit and the principle of a grid controlled land subdivision with this remains. In the County map of Savannah, Georgia, made in 1735, a grid of (slightly less than) one mile square subdivides a rectangle nearly 10 miles long and 6 miles deep. Thirty-nine of these squares remain wooded areas: within this primary subdivision, further subdivisions create farms of 44 acres and 5-acre garden plots. These are the related grid systems of the city region. On the river front within this main system is the city itself.

Now it is the city grid of Savannah that can be used as a first example of a city grid. A view of Savannah in 1734 illustrated in John Reps's book describes the principle: the plots and streets of the embryo city are being laid out: some buildings are complete. The unit of the Savannah grid is square: it is called a ward and is separated from its neighbors by wide streets. Within each square (or ward) building plots for houses are arranged along two sides, the centre itself is open, and on each side of this open square are sites for shops and public buildings. Savannah grew by the addition of these ward units. In 1733 there were four units: in 1856 no less than twenty-four. The city became a checkerboard of square ward units, marked out by the street pattern. But within this again, the plaid is further elaborated. The central open spaces of each ward are connected in one direction by intermediate roads, in the other direction the central areas become a continuous band of open spaces and public buildings. Here is a unit grid with direction and orientation.

The second example of a grid is absolutely neutral. It lays down an extensive and uniform pattern of streets and plots. The whole process can be illustrated in one single large-scale example. In 1811 the largest city grid ever to be created was imposed upon a landscape. The unlikely site for this enterprise was an area of land between two geophysical provinces in which a succession of tilts, uplifts and erosions had brought through the younger strata two layers of crystalline rock. These appeared as rocky outcrops under a thin layer of soil and vegetation. Into their depressions sands and gravels had been deposited by glacial action to create swampy areas through which wandered brooks and creeks. Some of these still wander into the basements of the older areas of what is now Manhattan.

In 1613 the original Dutch settlement was limited to the tip of the island. In 1760 there was little expansion beyond this and contemporary illustrations depict to the north a rolling landscape. Taylor's plan of 1796 shows the first modest growth of a city laid out on a gridiron pattern. Surveys in 1785 and 1796 extending up the centre of Manhattan set out the basis for a grid, and in 1811 the special State Commissioners confirmed this in an 8 ft. long plan which plotted the numbered street system of Manhattan as far north as 155th Street. The plan showed 12 north-south avenues each 100 ft. wide and 155 cross streets each 66 ft. wide. The sizes of the rectangular building plots set out by this grid are generally 600 ft. by 200 ft. There were some public open spaces. (Central Park was of course carved out later.) And it is this framework that has served the successive developments of the built form from 1811 to the present day.

The third example of a city grid is of interest because of its dimensional links with the land ordinance, suggested by Thomas Jefferson and passed by Congress in 1785. Under that ordinance a huge network of survey lines was thrown across all the land north and west of the Ohio river (Robinson 1916). The base lines and principal meridians of the survey divided the landscape into square 36 miles each side. These in turn were subdivided into 6-mile squares or townships and further divided into 36 sections each one mile square. The mile squares are then subdivided by acreage: the quarter section 160 acres with further possible subdivisions of 80, 40, 20, 10 or 5 acres. The 5-acre sites lend themselves to further division into rectangular city blocks (not unlike those of Manhattan) and subdivision again into lots or building plots.

In 1832, according to Reps (1965), Chicago was not much more than a few log cabins on a swamp. The railway came in the mid-century and, by the seventies and eighties. A mile-square grid had been extended over a considerable area of the prairie and the city framework had developed within this through a plaiting and weaving of the subdivisions that have been described.

Here then are three types of grid, that of Savannah, the gridiron of Manhattan and that of Chicago. Each one is rectangular. Each one has admitted change in the form and style of its building. Each one has admitted growth, by intensification of land use or by extension. Savannah, as it grew, tended to produce a green and dispersed city of open squares (Fig. 1). In Manhattan, the small scale subdivision of the grid and the exceptional pressure to increase floor space within

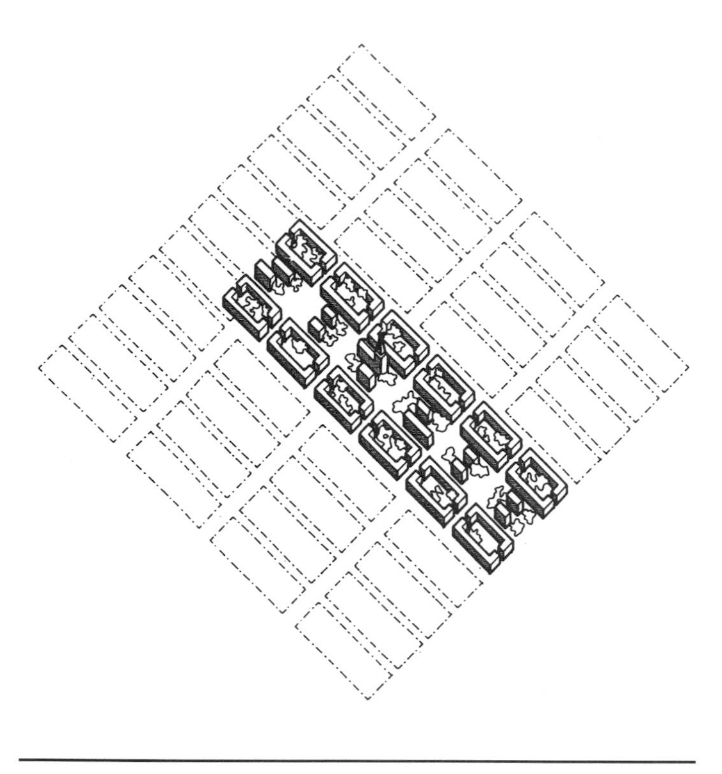

Fig 1. The basic plot layout of Manhattan is shown in the dotted lines. On this, four wards of the Savannah type of development have been superimposed. The example shows the effective way in which this layout opens up broad bands of green space and public buildings running across the developed areas.

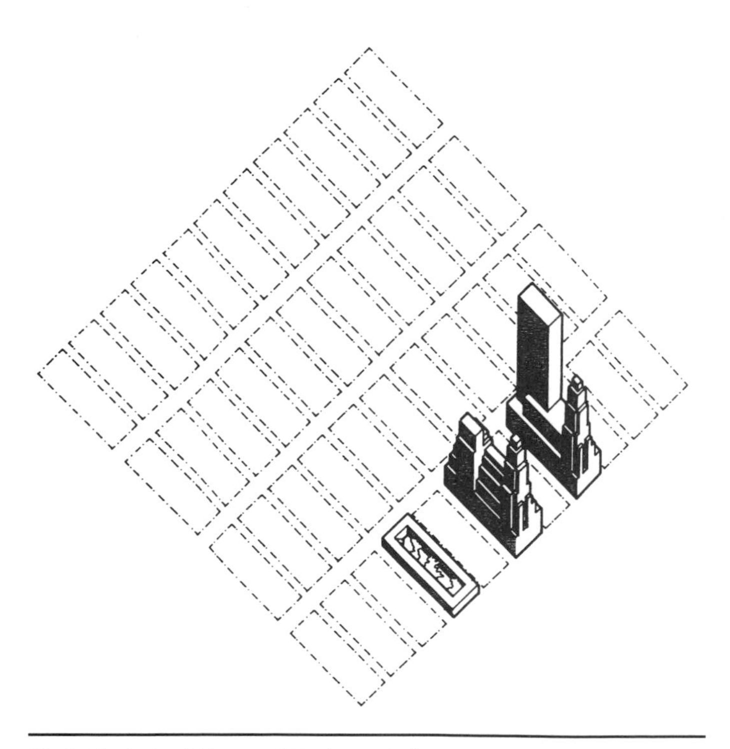

Fig 2. The basic plot layout of Manhattan is shown again in the dotted lines. The building forms show three stages of development including the original 4-6 story perimeter form with a garden at the centre, which was characteristic of the city in the 1850s, and two examples of the more intensive development during the present.

this, forced buildings upwards. Chicago spread, continually opening out the pattern of its grid. In each case the influence of the original grid remains: each one offers different possibilities and choices of building and of living.

In order to trace the influence of the grid, we can examine the building arrangement that developed within it in New York. We can identify at once what might be called the streets and the system that is established by the grid. If we now use the language of the urban geographers, we know that this defines the general plot pattern. The building arrangement develops within this (Conzen 1962).

The stages of this latter process can be traced in the early plans of Manhattan produced in 1850. The grid of roads is already built. Within this general plot pattern the separate building plots are being established. To the north, on the building frontier, there is a line of huts and shacks. Further south more permanent but separate buildings are being built. And in the most developed area further toward the tip of Manhattan the full building arrangement has solidified into connected terraces of four- to six-story houses arranged around the perimeter of the site and enclosing private gardens. Views of Manhattan in the 1850s show a city developed in this way: and this pattern of building arrangement can still be seen in many areas. At this point the building land is replete. A balance is maintained between the plot, the amount of building that it can reasonably support and the street system that serves this.

But as the pressure for floor space increases, the building form changes intensively at certain nodal points (Fig. 2). Deeper and high perimeter buildings first of all submerge the internal garden space. A process of colonization of the individual building plots begins, so that larger areas of the general plot are covered by higher buildings. In 1916 the first single building to occupy an entire city block rose a sheer 600 ft., its roof space almost exactly equaled the areas of its ground plan. It was this building that most clearly illustrated the need for the comprehensive zoning ordinances adopted that year, after arduous study and political compromise, to safeguard daylight in streets and adjoining buildings. But the grid now exerts a powerful influence; the limited size of the grid suggests the notion that increased floor space in an area can only be gained by tall buildings on each separate plot. The notion suggests the form; the regulations shape it into ziggurats and towers. Under the regulation that prevailed until recent years, if all the general building plots in central Manhattan had been fully developed, there would have been one single and universal tall building shape. And, to use an old argument by Raymond Unwin (1912), if the population of those buildings had been let out at a given moment, there would have been no room for them in the streets. The balance between area of plot, area of floor space and area of street has disappeared.

Now these descriptions of the grid, which have been used as a basis for the argument, have exposed the points at which it can be, and has been, extensively attacked for more than a century. A grid of any kind appears to be a rigid imposition on the natural landscape. It is this reaction against the grid that is voiced by Olmstead and Vaux writing in support of their design for Central Park in 1863: "The time will come when New York will be built up, when all the grading and the filing will be done and the picturesquely varied rocky formation of the island will have been converted into formations for rows of monotonous straight streets and piles of erect buildings" (Reps 1965).

In their opposition to the grid, the relief from its monotony became a specific aim. Central Park itself is an attempt to imitate nature and to recreate wild scenery within the grid.[1] The garden suburb with its curving streets is one form of attack on the grid system, and an attempt to replace it. And at the end of the century, the Chicago Fair (1893), Cass Gilbert's schemes in Washington (1900), and the plans for San Francisco (1905) and Chicago (1909) by Burnham are another attempt to transform the urban desert by means of vistas and focal points, into the "city beautiful." However, we recognize at once a contrast. The various types of grid that have been described opened up some possible patterns for the structure of a city but left the building form free to develop and change within this. The plans of the garden city designers or those concerned with making the "city beautiful" are an attempt to impose a form: and that form cannot change.

It is not possible to deny the force behind the criticisms of the grid. It can result in monotony: so can a curvilinear suburbia. It can fail to work: so can the organic city. What has been described is a process. It is now possible to extract some principles. Artificial grids of various kinds have been laid down. The choice of the grid allows different patterns of living to develop and different choices to be elaborated. The grid, unlike the fixed visual image, can accept reasons to growth and change. It can be developed unimaginatively and monotonously or with great freedom. There can be a point at which the original grid fails to respond to new demands (Fig. 3). As in Manhattan, it congeals. And it is at this point that we must try to discover from the old framework a new ordering principle that will open up new opportunities for elaboration by use.

It is precisely this that Le Corbusier underlined when he paid his first visit to New York in 1935 and made the comment: "What about the road?" (Le Corbusier 1939, 1947). The diagrams by which he illustrates this remark show the regenerative process that is necessary (Fig. 4). By increasing the size of the street net in Manhattan, Le Cobusier shows that the grid ceases to restrict. New building arrangements become possible and the balance between plot, building and street can be restored.

In the larger and more open mile square network of Chicago Frank Lloyd Wright had given a similar and vivid illustration of the capacity of the grid to respond to diversity and freedom. In 1913 a competition was held in order to "awaken interest in methods of dividing land in the interests of a community" (Yeomans 1916). The site was the standard section of the mile square grid. The standard subdivision of the grid, if rigidly applied, could divide it into 32 rectangles each 600-ft. (189 m) long by 250 ft. (76 m) deep. Mr. Wright accepts the established gridiron of the city "as a basis" for subdivision. He accepts the "characteristic aggregation of building. . .common to every semi-urban area of Chicago." The same number of people are housed. The business buildings, the factories, the heating plant, the utilities of the area are all there. But to use his words "they cling naturally to the main arteries of traffic. By thus drawing. . .all buildings of this nature into the location that they would prefer the great mass of the subdivision is left clear for residence purposes." Within this area parks (with their exhibition galleries and theaters), tree lined avenues and stretches of water diversify the layout. The range and choice of housing is wide. It is all natural, relaxed, capable of infinite variation and change as it develops within the framework of the grid. Mr. Wright's descriptive text includes these words: "in skilled hands the various treatments could rise to great beauty." It is prefaced by a quotation from Carlyle: "Fool! The deal is within thyself. Thy condition is the stuff thou shalt shape that same ideal out of."

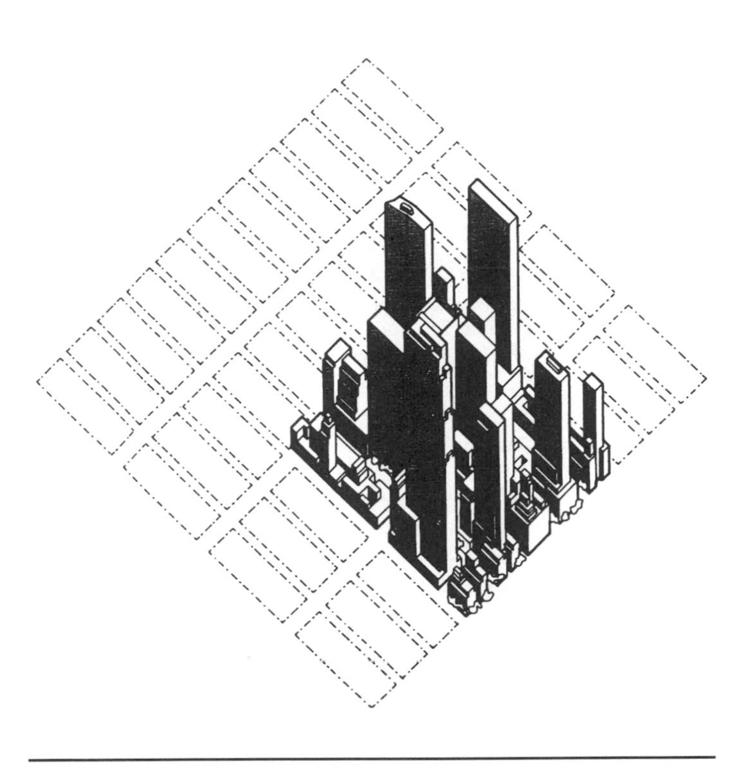

Fig 3. The illustration shows building plot development in its most intensive form.

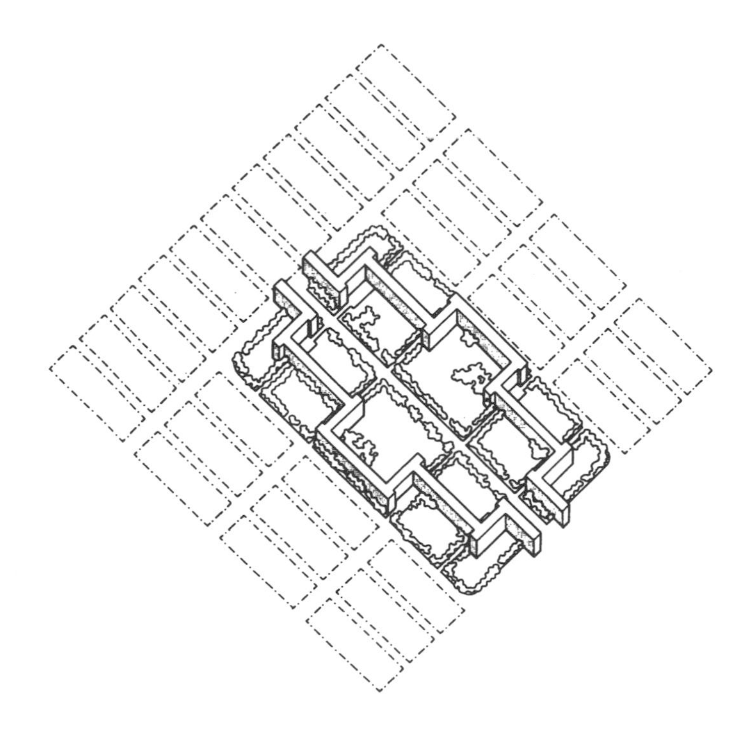

Fig. 4. Change in the scale of the grid. Le Corbusier's proposals for dwellings with setbacks (from his proposals for a city of 3 million people) are superimposed on the Manhattan grid and open up new possibilities in the building form.

In the case of these American cities the grid or framework can be regarded as an ordering principle. It sets out the rules of the environmental game. It allows the player the freedom to play with individual skill. The argument can now be extended by saying that the grid, which is so apparent in the American examples, is no less controlling and no less important in cities nearer home that would normally be called organic: London, Liverpool or Manchester. They too have a network of streets and, however much the grid is distorted, it is there. At a certain scale and under certain pressures, the grid combined with floor space limits and daylight controls is just as likely to force tall building solutions. And it is just as likely to congeal. It lends itself just as readily to regenerative action. The theoretical understanding of the interaction between the grid and the built form is therefore fundamental in considering either editing towns or the developing metropolitan regions.

The process of understanding this theoretical basis rests in measurement and relationships and it goes back certainly to Ebenezer Howard. Lionel March (1967) has pointed out a number of interesting things about Howard's book, *Tomorrow: A Peaceful Path to Real Reform*, first published in 1898. It is a book about how people might live in towns and how these might be distributed. But the important thing is that there is no image of what a town might look like. We know the type of housing, the size of plot, the sizes of avenues. We know that shopping, schools and places of work are all within walking distance of the residential areas. On the basis of these measurements we know the size of a town and the size of Howard's cluster of towns that he calls a City Federation. We know the choice that is offered and we know the measurements that relate to these. If we disagree with the choice, we can change the measurements. Lionel March took Howard's open centered city pattern linked by railways and showed that it could be reversed to a linear pattern linked by roads and that such patterns could be tested against the land occupied by our present stock of building and our future needs.

Now that is theory. It contains a body of ideas that are set down in measurable terms. It is open to rational argument. And as we challenge it successfully we develop its power. The results are frequently surprising and sometimes astonishingly simple. Ebenezer Howard's direct successor in this field was Raymond Unwin. The strength of his argument always rests in a simple demonstration of a mathematical fact. In an essay, "Nothing gained by overcrowding" (Unwin 1912), he presents two diagrams of development on 10 acres of land. One is typical development of parallel rows of dwellings: the other places dwellings round the perimeter. The second places fewer houses on the land but when all the variables are taken into account (including the savings on road costs), total development costs can be cut. From the point of view of theory, the important aspect of this study is the recognition of related factors: the land available, the built form placed on this, and the roads necessary to serve these. He demonstrated this in a simple diagram. Unwin began a lecture on tall buildings by a reference to a controversy that had profoundly moved the theological world of its day, namely, how many angels could stand on the head of a needle. His method of confounding the urban theologians by whom he was surrounded was to measure out the space required in the streets and sidewalks by the people and cars generated by 5-, 10- and 20-story buildings on an identical site. The interrelationship of measurable factors is again clearly demonstrated. But one of Unwin's most forceful contributions to theory is his recognition of the fact that, "the area of a circle is increased not in the direct proportion to the distance to be traveled from the center to the circumference, but in proportion to the square of that distance." Unwin used this geometrical principle to make a neat point about commuting time: as the population increase round the perimeter of a town, the commuting time is not increased in direct proportion to this.

The importance of this geometrical principle is profound. Unwin did not pursue its implications. He was too concerned to make his limited point about low density. But suppose this proposition is subjected to close examination. The principle is demonstrated again in Fresnel's diagram (Fig. 5) in which each successive annular ring diminishes in width but has exactly the same area as its predecessor. The outer band in the square form of this diagram has exactly the same area as the central square. And this lies at the root of our understanding of an important principle in relation to the way in which buildings are placed on the land. Suppose now that the central square and the outer annulus of the Fresnel diagram are considered as two possible ways of placing the same amount of floor space on the same site area: at once it is clear that the two buildings so arranged would pose totally different questions of access, of how the free space is distributed around them and what natural lighting and view the rooms within them might have. By this process a number of parameters have been defined which need to be considered in any theoretical attempt to understand land use by buildings.

This central square (which can be called the pavilion) and the outer annulus (which can be called the court) are two ways of placing buildings on the land. Let us now extend this. On any large site a development covering 50% of the site could be plotted as forty-nine pavilions, as shows in Fig. 6, and exactly the same site cover can be plotted in court form. A contrast in the ground space available and the use that can be made of it is at once apparent. But this contrast can be extended further: the forty-nine pavilions can be plotted in a form which is closer to that which they would assume as buildings (that is, low slab with a tower form over this). This can now be compared with its antiform: the same floor space planned as courts (Fig. 7). The comparison must be exact; the same site area, the same volume of building, the same internal depth of room. And when this is done we find that the antiform places the same amount of floor space into buildings which are exactly one third the total height of those in pavilion form (Martin and March 1966).

This brings the argument directly back to the question of the grid and its influence on the building form. Let us think of New York. The grid is developing a certain form: the tall building. The land may appear to be thoroughly used. Consider an area of the city. Seen on plan there is an absolutely even pattern of rectangular sites. Now assume that every one of those sites is completely occupied by a building: and that all these buildings have the same tower form and are 21 stories in height. That would undoubtedly look like a pretty full occupation of the land. But if the site of the road net were to be enlarged by omitting some of the cross streets, a new building form is possible. Exactly the same amount of floor space that was contained in the towers can be arranged in another form. If this floor space is placed in buildings around the edges of our enlarged grid then the same quantity of floor space that was contained in the 21-story towers now needs only 7-story buildings. And large open spaces are left at the centre.

Let us be more specific. If the area bounded by Park Avenue and Eighth Avenue, and between 42nd and 57th Street is used as a base and the whole area were developed in the form of Seagram buildings 36 stories high, this would certainly open up some ground space along the streets. If, however, the Seagram buildings were replaced by court forms (Fig. 8) then this type of development while using the same built volume would produce buildings only 8 stories high. But the courts thus provided would be roughly equivalent in area to Washington Square: and there could be 28 Washington Squares in this total area. Within squares of this size there could be large trees, perhaps some housing, and other buildings such as schools.

Of course no one may want this alternative.[2] But it is important to know that the possibility exists, and that, when high buildings and their skyline are being described, the talk is precisely about this and not about the best way of putting built space on to ground space. The alternative form of courts, taken in this test, is not a universal panacea. It suggests an alternative which would at once raise far reaching questions. For instance, the open space provided in the present block-by-block (or pavilion) form is simply a series of traffic corridors. In the court form, it could become traffic free courts. In this situation the question that needs answering is: at what point do we cease to define a built area by streets and corridors? At what point could we regard a larger area as a traffic free room surrounded by external traffic routes?

In all this the attempt has been simply to give a demonstration of procedure. The full repercussions of the questions are not obvious. They are highly complicated. But the factual aspect of the study establishes a better position from which to understand the nature of the complication and the limits of historical assumptions. What is left is something that can be built upon and needed decisions are brought back to the problem of the built form of an urban area not merely of a building. Here, the choice of the built form is critical in a number of ways, not least as a means of securing a new unity of conception.

Take for instance the question of the size of the road net. Professor Buchanan has looked at this from another angle (Ministry of Transport, 1963). Looking at cities in relation to traffic, he saw that most of them are built up from a collection of localities. He called these "environmental areas." These areas are recognizable working units. They are areas in which a pattern of related uses holds together: local housing, shopping, schools, etc., would be one obvious example. These areas are recognizable in Manhattan just as clearly as they are in London. They form, in Professor Buchanan's terms, "the rooms of a town." They need to be served by roads but they are destroyed when roads penetrate and subdivide them. His solution was to try to recognize and define these working areas and to place the net of roads in the cracks between them. By estimating the amount of traffic that might be generated by the buildings in such areas, Professor Buchanan was able to suggest some possible sizes for the networks. He had in fact by this procedure redefined the grid of a town in terms of modern traffic.

Here then is a proposition for a framework within which we can test out some possible arrangements of the built form. Professor Buchanan selected St. Marylebone as one of his test areas. This happens to adjoin the main London University site (already defined as a precinct in the London Plan) and this in turn is contiguous with the area around the Foundling Estate, which has been used in some Cambridge studies of the built form (Fig. 9). All three areas are approximately equal in size. The Foundling area (bounded on the north and south by Euston Road and Theobalds Road, and on the west and east by Woburn Place and Grays Inn Road) is about 3700 ft. from north to south and 2000 ft. wide. It developed a cohesion of its own. How did this happen?

This in turn can be related back to the main line of argument. In 1787 the whole of this area consisted of open fields: there were no controlling features. A plan of 1790 divides the land into building plots by its

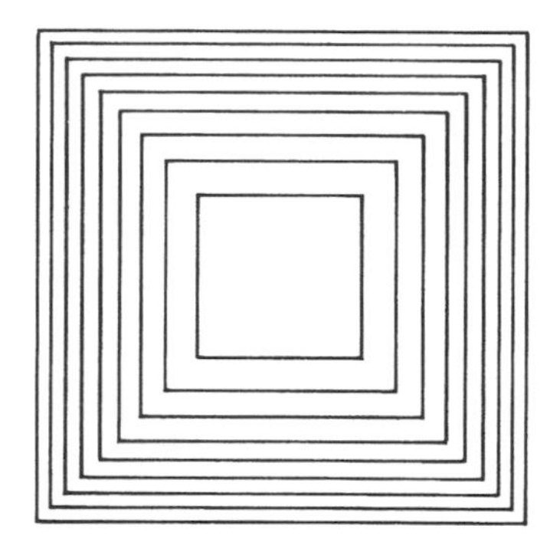

Fig. 5. Fresnel's diagram. The outer band has exactly the same area as the central square, an important principle in relation to the way in which buildings are placed on the land.

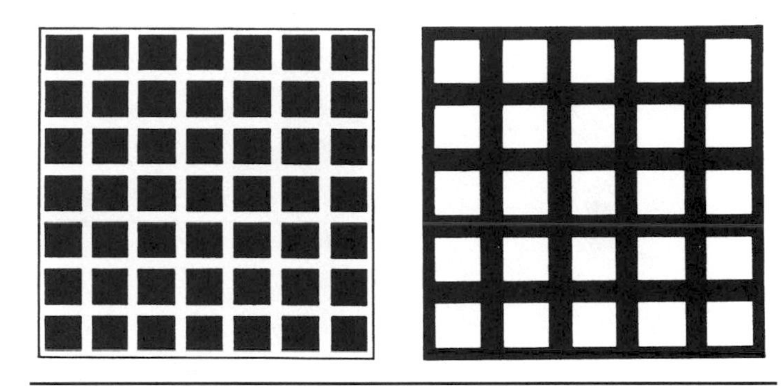

Fig. 6. Site covered with pavilions (left) or courts (right).

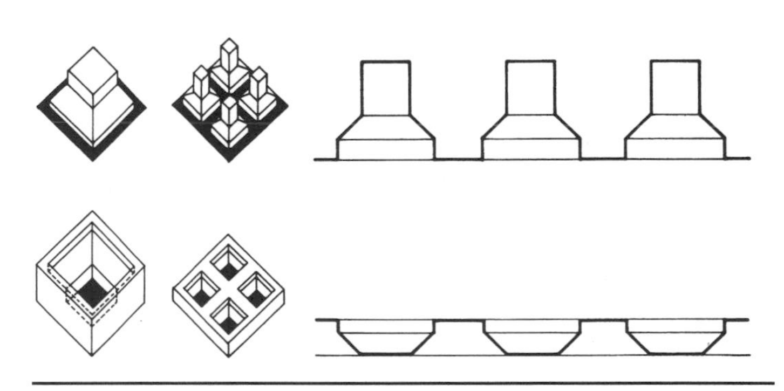

Fig. 7. Pavilions (upper) and counts (lower), their antiforms.

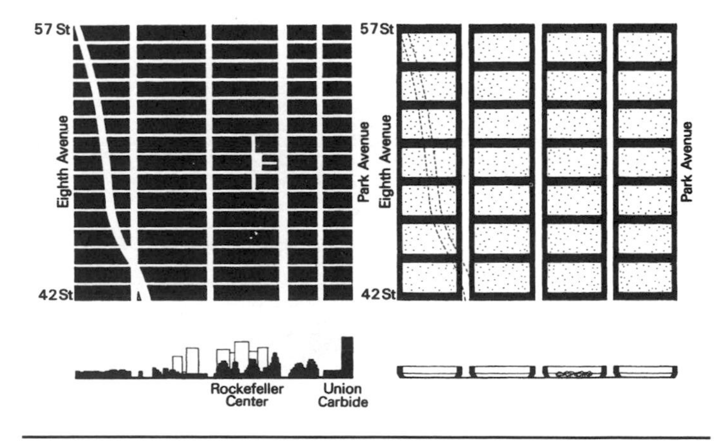

Fig. 8. Comparison of Manhatten landform coverage.

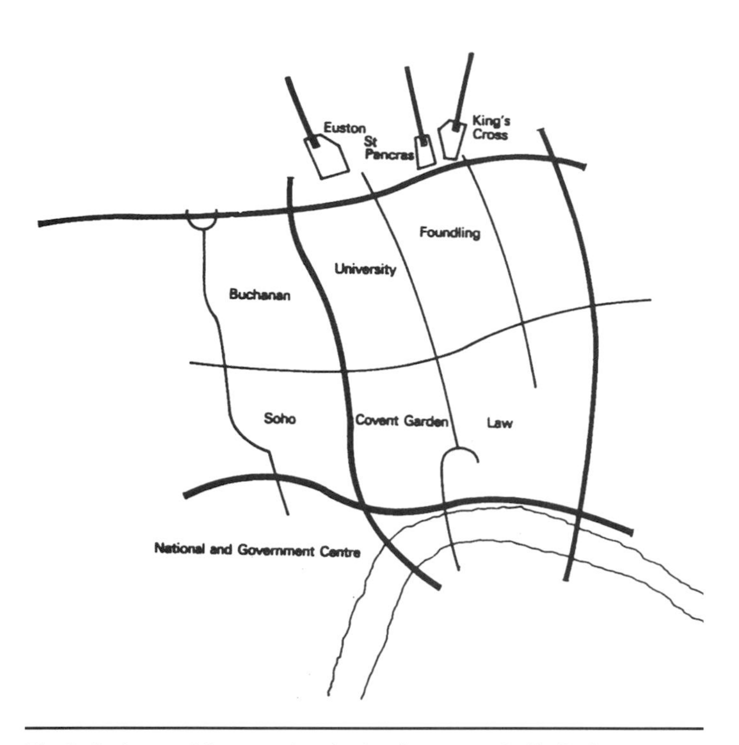

Fig. 9. Environmental areas and road networks as suggested by Buchanan.

network of streets and squares. The subsequent history, so well traced by Olsen (1964), shows the development and elaboration within this pattern. By 1900 the area could have been described by the language that Mrs. Jacobs applies to Greenwich Village. The intellectuals were there; so were the working Londoners; so were the Italians around their hospital in Queen Square. There were handsome houses, tenements and mews, hotels and boarding houses. The area had its own Underground station and its own shopping area along Marchmont Street. It served a complex community.

By 1960 the balance within the original pattern had radically altered. Fast moving traffic using the small scale grid of streets had subdivided the area. Site by site residential development at a zoned density of 136 people to the acre produces only one answer: tall blocks of flats. Redevelopment of sites for offices created taller and thicker buildings. The hospitals, which needed to expand, were hemmed in by surrounding development. The pattern congealed.

In this situation only a new framework can open up a free development. And if Professor Buchanan's surrounding road net is accepted as a basis for the development of the environmental area, the problem can be seen within a new unifying context. What sort of advantages could a rearrangement of the built form now create? Professor Buchanan in his study area outlines three possible solutions with progressive standards of improvement. The merit of this is that it sets out a comparative basis of assessment. But even his partial solution leads to an extensive road and parking system at ground level. From the point of view of the pedestrian the position is made tolerable by the use of a deck system to create a second level. Above this again, some comparatively tall buildings are required to rehouse the built space that is at present on the ground. This kind of image of the architecture of cities has a considerable history in modern architecture and has been

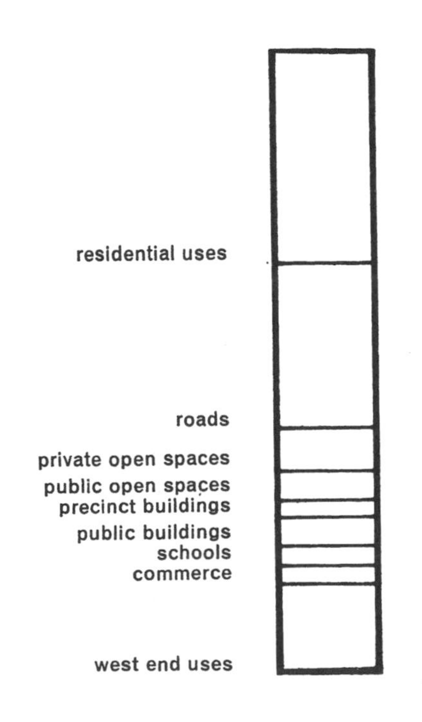

Fig. 10a. Quantities of built and open space in the Foundling Area.

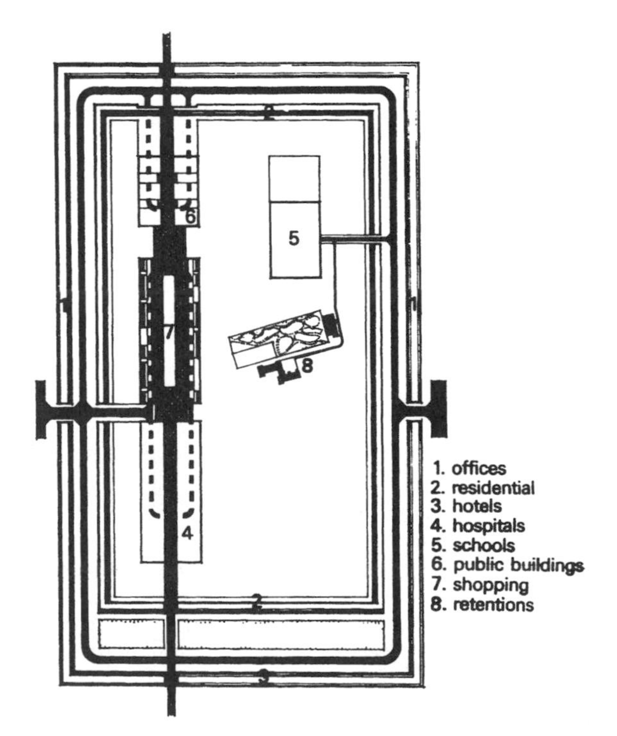

Fig. 10b. Possible geometric layout of the same quantities of built space in perimeter form.

much used as an illustration of central area reconstruction. But, as Professor Buchanan himself asks, what building complications does it produce and what sort of an environment does it create? Is it in fact worth building?

Professor Buchanan's range of choices could in fact be extended by applying some of the theoretical work which has been described. And when this is done the results are significantly different. The boundaries of the total area that are being considered have been defined by this new scale of the road network: the grid. Within this, the existing floor space can be assessed (Fig. 10): 34% of the site is occupied by housing: 25% by roads: 15% by office and commercial use: 12% is open space. In addition there is an important shopping street, a major hospital and several schools and educational buildings. With this information available it can be considered at a theoretical level how this might be disposed in a new building arrangement.

First, the shopping street, Marchmont Street, could be established as a north/south pedestrian route associated with the Underground and some housing. If all the office space which is at present scattered throughout the area could be placed in a single line of buildings around the perimeter of the area (where some of it already is), it need be no higher than eight stories. All the housing at present in the area could be placed within another band of buildings sited inside this and no higher than five stories. Of course, it could be arranged on the ground to include other forms and types of housing. But in theory, the bulk of the building at present covering the area could be placed in two single bands of building running around its edge, leaving the centre open, which would be a parklike area about the same size as St. James' Park (Fig. 11). Precisely the same amount of floor space would have been accommodated. There need be no tall buildings, unless they are specifically wanted. All the housing could look into a park. Buildings such as schools could stand freely within this. There would be a free site and a parklike setting for new hospital buildings.

All that may sound theoretical and abstract. But to know what is theoretically possible is to allow wider scope for decisions and objectives. We can choose. We can accept the grid of streets as it is. In that case we can never avoid the constant pressure on the land. Housing will be increasingly in tall flats. Hospitals will have no adequate space for expansion. Historic areas will be eaten into by new building. A total area once unified by use will be increasingly subdivided by traffic. We can leave things as they are and call development organic growth, or we can accept a new theoretical framework as an outline of the general rules of the game and work toward this. We shall know that the land we need is there if we use it effectively. We can modify the theoretical frame to respect historic areas and elaborate it as we build. And we shall also know that the overlapping needs of living in an area have been seen as a whole and that there will be new possibilities and choices for the future. ■

NOTES

1. This movement which began with gardens, was less appropriately applied to city layout. In Olmstead's words, "lines of roads were not to press forwards." Their curving forms suggest leisure and tranquillity. Compare this with the almost contemporary (1859) statements by Cerda in his plan for Barcelona in which there is "a reciprocal arrangement between that which is contained" (building plot and arrangement) and "that which contains" (grid and street system). "Urbanization is an appendix to universal movement: streets are for

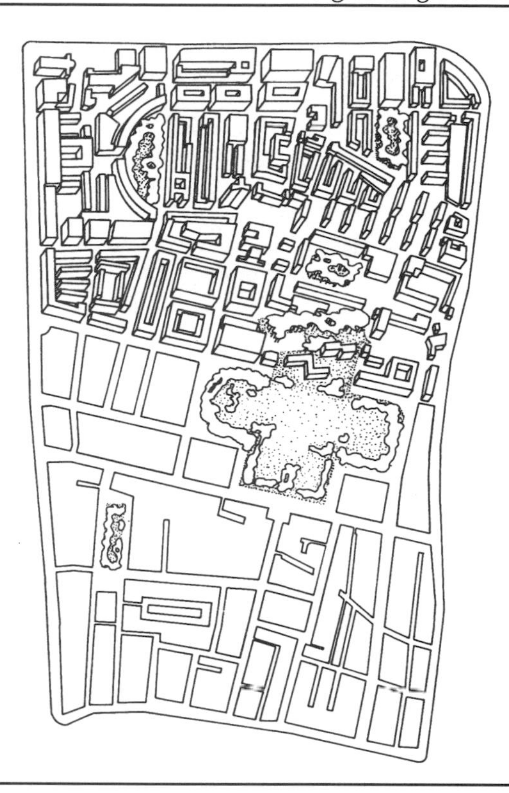

Fig. 11a. The existing plot layout and building development in an area of London that might be regarded as an environment room. But it is subdivided by roads and the limited size of the building plot increasingly forces development upward.

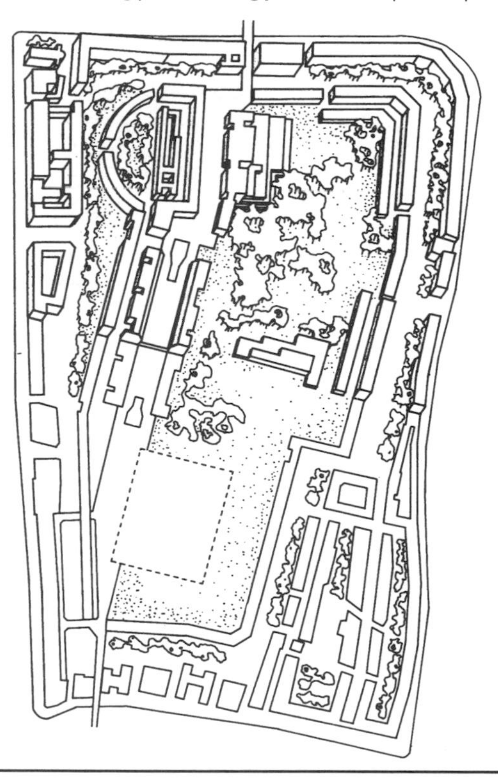

Fig. 11b. The same areas as that in Fig.11a. The road network is now enlarged and runs around the boundary of the area. Theoretically an entirely new disposition of buildings is possible and the illustration shows exactly the same amount of floor space in a new form. Tall buildings are no longer necessary: the buildings themselves have a new freedom for development and a considerable area of open space is discovered.

movement but they serve areas permanently reserved and isolated from that movement which agitates life" (the environmental area).

2. It is simply a demonstration of a possible choice within a general strategy, such as that, for instance proposed by the Goodmans (1960) in *Communitas,* Appendix A. The Goodmans also suggest a change in the scale of the grid by street closure (1960, p. 230) that with better layout on the residential acreage of Manhattan every room could face onto a Madison or Washington Square.

REFERENCES

Alexander, Christopher. 1965. "A city is not a tree." *Architectural Forum* magazine.

Beresford. M. 1967. *New Towns of the Middle Ages.* London: Lutterworth Press.

Consen, M.R.G. 1962. "The Plan Analysis of an English City Centre," in *Proceedings of the I.G.U. Symposium in Urban Geography.* Lund: Royal University of Lund.

Goodman, P. and Goodman, P. 1960. *Communitas.* New York: Vintage Books.

Howard, Ebenezer. 1898. *Tomorrow: A Peaceful Path to Real Reform.* Republished 1945 as *Garden Cities of Tomorrow.* Faber.

Jacobs, Jane. 1961. *The Death and Life of Great American Cities.* New York: Random House.

Le Corbusier. 1939. *Oeuvres Completes 1934-1938.* Zurich: Girsberger.

_______, 1947. *When the Cathedrals Were White.* London: Routledge.

Martin, Leslie. Aug. 1967. "Homes Beyond the Fringe." Royal Institute of British Architects *RIBA Journal.*

_______, Apr. 1968. "Education Without Walls." Royal Institute of British Architects *RIBA Journal.*

_______, and Lionel March, April 1966. Land Use and Built Forms. *Cambridge Research.*

_______, and Lionel March, eds. 1972. *Urban Space and Structure.* Cambridge, England: Cambridge University Press.

Ministry of Transport. 1963. *Traffic in Towns* (The Buchanan Report). London: Her Majesty's Stationery Office.

Olsen, D.J. 1964. *Town Planning in London.* New Haven: Yale University Press.

Reps, J. W. 1965. *The Making of Urban America.* Princeton: Princeton University Press.

Robinson, C. M. 1916. *City Planning.* New York: Putnam.

Sitte, Camillo. 1889. *City Planning according to Artistic Principles,* translated by Collins, G. R. and C. C. Collins. 1965. New York: Random House.

Unwin, R. 1912. "Nothing to be Gained by Overcrowding," in Cresse, W. L., ed. 1967, *The Legacy of Raymond Unwin: A Human Pattern for Planning.* Cambridge: MIT Press.

[various]. 2000. "Obituary—Leslie Martin 1908–2000" (commentaries on the professional contributions of Leslie Martin). *Arq, Architectural Research Quarterly* (4) 4-2000 (295-308).

Willmot, P. and M. Young. 1957. *Family and Kinship in East London.* London: Routledge.

Yeoman, A. B. 1916. *City Residential Land Development.* Chicago: University of Chicago Press.

Jurg Lang

Summary
C. A. Doxiadis, architect, planner and theoretician believed that problems of human settlements can be addressed only by a comprehensive and integrated approach. To this end he formulated ekistics, the science of human settlements. He defined five ekistic elements, man, society, shells, networks, nature and their interrelationships, and sought to relate any study of human settlements to units of the ekistic logarithmic scale, ranging from the individual (man, anthropos) to the entire world (ecumenopolis), extending from the past to the distant future.

Key words
cities, ekistics, global impact, metropolis, planning theory, regional planning, scale, taxonomy, transportation, urban planning

PHOTO: Fabian Bachrach

C. A. Doxiadis 1913–1975

C.A. Doxiadis and the Science of Human Settlements

1 CONSTANTINOS A. DOXIADIS

C. A. Doxiadis called himself a bricklayer and he talked with conviction about his visions of human settlements, which range from the individual room to the global community, extending from the past to the distant future. As a bricklayer he was connected to reality, in full command of his field, committed to implement his ideas and principles, operating from a basis of far-reaching visions and soaring aspirations. He left a formidible legacy, ekistics, proposed as a science of human settlements, consisting of a comprehensive framework for a new science, extensive research, proposals, visions and challenges and built projects, buildings and entire cities.

Constantinos Apostolou Doxiadis was born in 1913 in Bulgaria, the son of Greek parents. He lived in Athens most of his life and died June 28, 1975. Many of his ideas and convictions were influenced by childhood experiences and his early formative years. His father was a Greek Minister of Social Affairs during the time of the Asia Minor crisis in 1922, when millions of refugees came from the coastal areas of Asia Minor, particularly Smyrna. Doxiadis witnessed the hardship and suffering through his family's involvement in the relief programs. Much of his later work was based on practical experience, on real-life problems, on the urgency and seriousness of these problems and the need for immediate and effective answers.

In 1935 Doxiadis graduated from the Technical University of Athens as Architect-Engineer and he received his Doctorate Degree from the Berlin-Charlottenburg University in 1936. For his work and contributions in *ekistics* he was awarded Honorary Degrees by twelve universities, among them Swarthmore College, the University of Pittsburgh and the University of Michigan.[1]

Doxiadis experienced the misery of World War II and the German occupation of Greece. In his first major public role he served as Under-Secretary and Director General of the Ministry for Housing and Reconstruction for Greece between 1945 and 1948, and was Minister-Coordinator of the Greek Recovery Program and Under-Secretary, Ministry of Coordination from 1948 to 1951. The enormity and complexity of problems, scarcity of resources, urgency in guiding recovery and development and, afterwards, the complexity in allocating aid provided through the World Relief and Marshall Plan were decisive in Doxiadis' approach and his further involvement in the overall concept of a science of human settlements.

From these experiences in planning programs, Doxiadis realized that the basis of knowledge seemed to be lacking and what was undertaken was often ineffective. He formulated an agenda of priority issues for human settlements. It was then that he decided that the full spectrum of human knowledge would provide a better answer to the

Credits: The author acknowledges the contribution of Panayis Psomopoulos, President of the Athens Center of Ekistics and editor-in-chief of the journal *Ekistics*, who reviewed the manuscript and provided invaluable advice and material. Illustrations courtesy of the Doxiadis Archive at the Athens Center.

problems than any single discipline. He realized that the vocabulary of sociologists, economists, geographers were similar but were used with different meanings and that there was a need for them to sit down and define issues in a more objective and comprehensive way.

As early as 1950, Doxiadis postulated the need for a comprehensive science of human settlements, which he called *ekistics*. Throughout his life he worked with unending passion to formulate this science with a body of knowledge and a theory, and to disseminate these ideas and principles worldwide. Principles of ekistics guided Doxiadis in his professional projects, his research, his teaching and the foundation of institutions, symposia and societies, all with the purpose to use his ideas for the betterment of human settlements.

Doxiadis was a prolific writer and published several seminal books. His doctoral thesis *Raumordnung im Griechischen Staedtebau* was published as a book in 1937, later translated into English as *Architectural Space in Ancient Greece*, 1972.[2] In his book *Architecture in Transition*, 1963,[3] he addresses the role of the architect in a world of confusion and rapid change. Based on his broad experience and astute analysis of the increasing complexity and severity of the problems surrounding mankind, he postulated the comprehensive approach relying on scientific research, connecting the past with the future, architecture with the context. In his eloquent and passionate plea he joined the pioneers of the modern movement, and a generation later, he sought solutions not in the individual monument but in "simple, plain, human buildings" within the overall evolution of the environment and society. Ekistics, the science of human settlements, was summarized in his book *Ekistics* in 1968.[4] This new science was further elaborated in four more books,[5] *Anthropopolis* in 1974 (summary of a symposion on the City for Human Development), *Ecumenopolis* in 1975, *Building Entopia* in 1975, and *Action for Human Settlements* in 1976, published after his death.

In 1951 Doxiadis founded Doxiadis Associates, an international consulting firm for architecture, planning and development, headquartered in Athens, Greece, with affiliated offices worldwide. Doxiadis Associates provided opportunity for the application and testing of his ideas and theories. Projects ranged from architecture to rural and urban settlements, agricultural plans, public works, plans for tourism, transportation, housing, urban renewal and design, planning and development of new cities, such as Islamabad, the new capital of Pakistan, and Tema, in Ghana.

Doxiadis founded the Athens Technological Organization (originally the Athens Technological Institute) in 1958, consisting of distinguished members of Athenian and Greek society. The purpose of this organization was to help Greece to proceed from a developing country to a more advanced level by the spread of new technology, the education of young people as technicians and scientists, and the development of cultural programs and debates on burning issues of the country. Several technical schools were established and new educational programs were formulated and tested until 1967 when the schools and programs were taken over by the Ministry of Education and became public one year after Doxiadis' death.

The Athens Technological Organization offered a rich program of lectures, cultural events, exhibitions of art and performance of music. A new dimension was added in 1959 with the City of the Future Research project, initiated by Doxiadis with John G. Papaioannou as collaborator and project manager. Additional opportunities for research presented themselves immediately thereafter and the need grew to train experts in ekistics and have Doxiadis share his expertise with younger people. This lead in 1963 to the creation of the Athens Center of Ekistics, which became a major project of the Athens Technical Organization "to foster a concerted program of research, education, documentation, and international cooperation related to the development of human settlements."

Doxiadis designed and built an office and research complex on the slope of the Lykabettos hill in Athens as a sensitive blend of modern architecture, regionalism and urban integration. The building, housing his office, research facility and school grouped around a generous courtyard and meeting space, would soon become an international center for discourse. (Fig. 1)

Panayis Psomopoulos characterized Doxiadis and his work as follows:

> *Most of what Doxiadis has done in terms of ekistics is due to his exceptional intelligence, his immense thirst for knowledge, unabated energy and his very well organized mind, let alone his charming ability for communication. He started as a theoretician with his Ph.D. thesis, became a politician with enormous experience, a practitioner with Doxiadis Associates, and gradually shifted to research, documentation and education, so as to become a unique phenomenon in the first part of the second half of the 20th century.*[6]

While it is difficult to summarize Doxiadis' work within a short article, it is important that an attempt be made to provide a structure and entry to the body of his work, writings, projects and numerous creative contributions. The work was documented by Doxiadis himself in many books and papers, in contributions by colleagues, collaborators and students, but particularly by the journal *Ekistics*, which for over forty-five years has disseminated his ideas and principles.[7]

2 EKISTICS, THE SCIENCE OF HUMAN SETTLEMENTS

> *Human settlements are no longer satisfactory for their inhabitants. This is true everywhere in the world, in underdeveloped as well as in developed countries. It holds true both for the way of living of their inhabitants and for the forms we give to the shells of the settlements trying to satisfy their needs. And it is true whatever our aspect of the problem.*[8]

Doxiadis described our cities as urban nightmares. The irrational structure, clogged arteries and congested streets, pollution and environmental degradation, lack of sufficient housing, facilities and services are indications that cities no longer serve their inhabitants adequately. It is in our cities, where the problems of our societies manifest themselves most pointedly in an epoch of rapid growth and change.

While the fields of science and technology have experienced rapid advancement and progress, in dealing with our cities and settlements we face confusion, disorientation and isolated unconcerted action. There is plenty of research in focused aspects of cities and settlements, but there is a dearth of interdisciplinary work. Doxiadis therefore proposed the creation of a science, which integrates all aspects of human settlements including the implementation of ideas and solutions for a better habitat for mankind. He called this science *ekistics*, the science of human settlements, defined as follows:

Ekistics is the science of human settlements. The term derives from the Greek verb oikõ, *meaning "settling down," and denotes the existence of an overall science of human settlements conditioned by man and influenced by economic, social, political administrative, and technical sciences as well as the disciplines related to the arts.*[9]

There may be some dispute whether ekistics is a science at this stage, or whether it has to be developed more completely to be recognized as such. Doxiadis saw the purpose of his science to first provide a descriptive framework with increasing knowledge of facts and gradually evolving theories, and secondly a concerted effort to attain its goals and to implement ideas for better human conditions in improved settlements. The descriptive science requires scientific methodology, the prescriptive or implementation part requires scientists, technicians and artists to implement well-founded and creative ideas.

Cities and settlements are confronted with a great number of problems, some inherited from the past, some arising out of present conditions. According to Doxiadis there are critical conditions, which are common to all cities:

1. There is an unprecedented increase in population due to improved living conditions, accompanied by a migration to urban settlements. The result is growth of urban settlements at a tremendous scale.

2. We experience multiple impacts of machines in our lives. These impacts lead to higher productivity and new possibilities, but also bring unprecedented problems to the structure of cities and society, of resource use and environmental degradation.

3. There is a gradual socialization in the patterns of living, which allows the whole population to participate more and more in the city, its facilities and resources.

4. In the modern city, growth and change over time is a dominant feature, which must take precedence in all planning considerations.

Doxiadis stated that,

. . .we do not know enough about human settlements, and this is why we have failed to solve their problems and thus to create a better habitat for man. People are trying to learn more about them, but much more work is being undertaken in the creation of new settlements than in attempting to understand their functioning. Thus their creation is based on a very weak foundation.[10]

Scope and Organization

It is apparent that an integrated interdisciplinary science of the human settlement would have an enormous scope and complexity and Doxiadis in his early work concentrated on providing a framework of organization and emphasis. He started with an empiric study of settlements, which he pragmatically organized. All possible methods, all categories of studies, the findings of all disciplines related to human settlements had to be assembled and combined and their validity checked against the wealth of experience embedded in our settlements. He concluded:

Fig. 1. Courtyard of Doxiadis Associates and Athens Center of Ekistics.

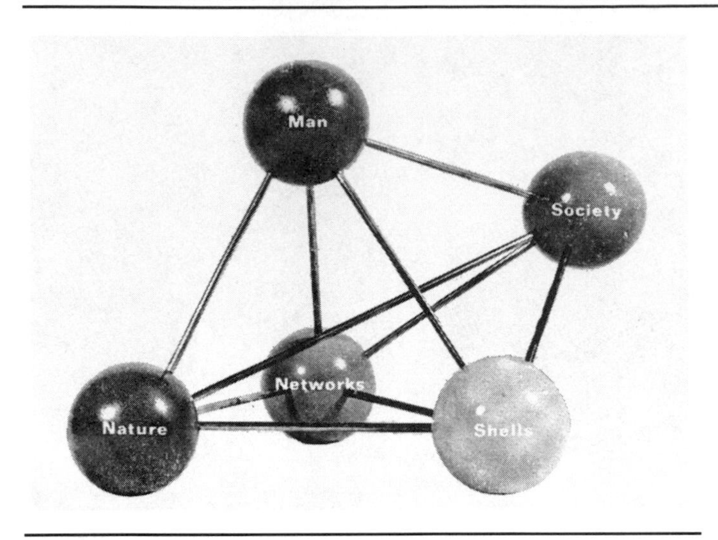

Fig. 2. Model of *ekistic* elements, used to demonstrate the importance of their interrelationship, or synthesis.

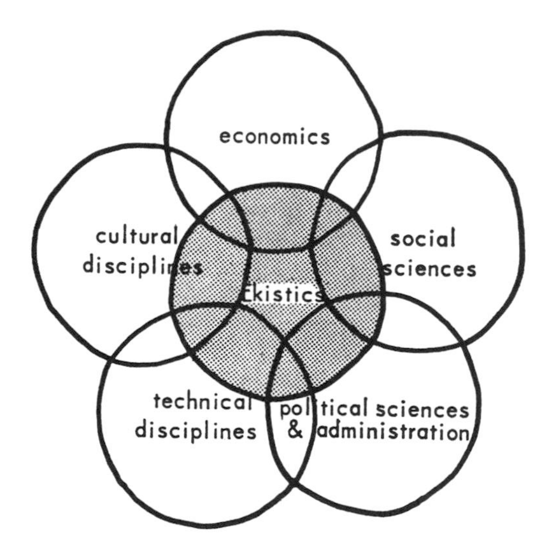

Fig. 3. *Ekistics* and related sciences.

These categories can be combined or studied in isolation. The studies can lead to conclusions or simply provide data to be fed into further studies. However, no studies can help us to illuminate the situation unless their findings are related to very specific categories and scales of human settlements.[11]

Doxiadis formulated a set of conceptual models, which formed the foundation of ekistics. Five of the most important ones are listed here[12]:

- The five elements of human settlements or ekistic elements: man, society, nature, shells and networks, and their relationships.
- The ekistic logarithmic scale (ELS) and the ekistic units.
- The ekistic grid, consisting of units of space and the five elements of human settlements.
- The model of satisfaction.
- The anthropocosmos model.

Ekistic elements

Doxiadis defined five elements of human settlements or ekistic elements, which are essential for an inclusive approach to human settlements. (Figs. 2 and 3)

- **Man.** In the center stands *man*, the individual human being. The generic term includes male and female. Later Doxiadis replaced man with the Greek term *anthropos* to be more inclusive.
- **Society** deals with people and their interaction with population trends, group behavior, social customs, occupation, income, and government. Of increasing importance is the preservation of values inherent in small communities after they have been absorbed by larger settlements.
- **Nature** represents the ecosystem within which man and society operate and cities and settlements are placed. The interrelation between man, machine, settlement and nature is of high importance, as is the carrying capacity of regions, continents and ultimately the entire planet.
- **Shells** are used as the generic term for all buildings and structures.
- **Networks** for transportation, communication and utilities support the settlements and tie them together with their organization and structure. Their changes profoundly affect urban patterns and often developments in networks have been portending new developments in cities and societies.

Individually the five elements have been well studied in their own scientific disciplines with man/*anthropos* in medicine and psychology, society in the social sciences, nature in geography and more recently in ecology, shells and networks in architecture, engineering and planning. Rather then creating a super-science through accumulation of every discipline, one shall concentrate on each discipline's aspects related to settlements and bring them together in an interdisciplinary approach. Over time, this will be supplemented with autonomous thought, methods, and approaches within ekistics itself. In essence, ekistics is an applied science, which needs to proceed from a solid scientific base to implementation and action within the most difficult and complex situations.

Rather than to consider the elements in isolation, Doxiadis urged an emphasis on their mutual interaction which he called *synthesis*.

Ekistic units and ekistic logarithmic scale

Any study of human settlements must be related to the scale. To this end, Doxiadis proposed 15 levels of settlement size or 15 ekistic units, ranging from the individual man or anthropos to the entire global settlement called ecumenopolis, with its ultimate size of 30–50 billion people. The 15 units are: man (*anthropos*), room, house, house group, small neighborhood, small city, city (*polis*), small metropolis, metropolis, small megalopolis, megalopolis, small eperopolis, eperopolis, ecumenopolis (for definition of the terms see glossary). The population figures for the first three units have been given with 1 for man/*anthropos*, 2 for room and 5 for house/dwelling. For the remaining 12 units, which are numbered as community classes (roman) I to XII, the population figures increase by a factor of roughly 7, which constitutes the *ekistic logarithmic scale*. The population figures range from 40 persons in the house group to 10,000 for the small city/*polis* (community class IV) to 4 million in the average metropolis to 30–50 billion for ecumenopolis, the city spanning the entire globe (Fig. 4).

Ekistic grid

> *Human settlements are so numerous and so different from each other that any attempt to study or understand them is meaningless unless we classify them in an orderly way. . . . All fields of knowledge which gradually become scientific pass through a state of effort towards a systematic classification in spite of the resistance which is sometimes made to this effort.*[13]

By combining the ekistic elements with the ekistic logarithmic scale, Doxiadis created the ekistic grid, an organizational matrix on which each study can be localized and categorized. This ekistic grid was henceforth to be included at the beginning of each publication. Thus it became a unique and most valuable element in the journal *Ekistics*. But it also became an indispensable research and policy tool in the establishment of the relation between research objects, and in the identification of overlaps or gaps in research topics and programs (Fig. 5).

Doxiadis was well aware of earlier uses of the grid as a tool for classification, or taxonomy, most notably the grids by the Scottish planner Patrick Geddes and Le Corbusier's CIAM grid. The CIAM grid with its elements housing, work, recreation and traffic undoubtedly had a built-in bias, which fundamentally influenced the understanding of the city and is at least partially responsible for the normative statements of the Athens Charter of 1933. Doxiadis' ekistic grid is more neutral in the choice of the elements, and the introduction of the ekistic elements, units and logarithmic scale constitutes a valuable and constructive innovation.

Goals and the model of satisfaction

Doxiadis derived the goals for human settlements from Aristotle's age-old saying that *the goals for our cities are to make man happy*

Ekistic unit	1	2	3	4	5	6	7	8	9	10	11	12	13	14	15
Com. class				I	II	III	IV	V	VI	VII	VIII	IX	X	XI	XII
Kinetic field	a	b	c	d	e	f	g	A	B	C	D	E	F	G	H
population range			3 - 15	15 - 100	100 - 750	750 - 5,000	5,000 - 30,000	30,000 - 200,000	200,000 - 1.5 M	1.5 M - 10 M	10 M - 75 M	75 M - 500 M	500 M - 3,000 M	3,000 M - 20,000 M	20,000 M and more
name of unit	Anthropos	room	house	housegroup	small neighborhood	neighborhood	small polis	polis	small metropolis	metropolis	small megalopolis	megalopolis	small eperopolis	eperopolis	Ecumenopolis
ekistic population scale	1	2	5	40	250	1,500	10,000	75,000	500,000	4 M	25 M	150 M	1,000 M	7,500 M	50,000 M

Fig. 4. *Ekistic* units.

COMMUNITY SCALE		i	ii	iii	I	II	III	IV	V	VI	VII	VIII	IX	X	XI	XII
EKISTIC UNITS		1	2	3	4	5	6	7	8	9	10	11	12	13	14	15
		ANTHROPOS	ROOM	HOUSE	HOUSE GROUP	SMALL NEIGHBORHOOD	NEIGHBORHOOD	SMALL POLIS	POLIS	SMALL METROPOLIS	METROPOLIS	SMALL MEGALOPOLIS	MEGALOPOLIS	SMALL EPEROPOLIS	EPEROPOLIS	ECUMENOPOLIS
EKISTIC ELEMENTS	NATURE															
	ANTHROPOS															
	SOCIETY															
	SHELLS															
	NETWORKS															
SYNTHESIS: HUMAN SETTLEMENTS																
POPULATION T (Thousands) M (Millions)		1	2	5	40	250	1,5 T	10 T	75 T	500 T	4 M	25 M	150 M	1,000 M	7,500 M	30,000 M

Ekistic Logarithmic Scale

Fig. 5. *Ekistic* grid with scale defined in terms of logarithmic scale of population (*x* axis).

5. aspects / 6. principles	desirability					feasibility				
	E	S	P	T	C	E	S	P	T	C
1. maximum of contacts										
2. minimum of effort										
3. optimum of protective space										
4. optimum of quality of the total environment										
5. optimum in the synthesis of all principles										

Aspects:

E economic
S social
P political
T technological
C cultural

Fig. 6. Model of Satisfaction.

and safe. Safety is not only limited to safety from wars, but safety from crime, pollution, and natural disasters (through location, codes).

> *The goal of ekistics is to achieve a balance between the elements of human settlements in order to guarantee happiness and safety of man.*[14]

Doxiadis followed the ancient Greek philosophers in asking what is the good life and, by referring to Aristotle, he gave his own position on what constitutes happiness. Doxiadis believed that to survive, to live and to achieve happiness, human beings built settlements, which always followed fundamental principles, and he defined five principles in man's quest for happiness:

1. **Maximum contacts.** Man is continuously reaching out for a greater number of contacts (material, aesthetic, intellectual) with nature and other people and elements. This maximizing of contacts leads to the expansion of cities.

2. **Minimum effort.** Man tries to expend minimum effort to achieve maximum contacts and to reduce energy, time and cost to a minimum. This leads to higher densities.

3. **Optimum space.** Man needs optimum (but not necessarily maximum) space, whether temporary or permanent, for man as an individual or as the member of a group, for the satisfaction of his needs.

4. **Quality of the environment.** The quality of the environment is determined by man's relation with nature, society, shells and networks, creating a balance of the ekistic elements. The relationships within the total environment need to be optimized.

5. **Optimum in the synthesis of all principles.** A balanced and beneficial synthesis of the preceding principles has to be created.

These principles were combined with desirability and feasibility of the economic, social, political, technological and cultural aspects to form the model of satisfaction. Doxiadis developed studies on maximizing the amount of time one spends in good and rewarding activities and minimizing idle waste of time. These efforts lead to time allocation studies and time management in urban activities as for instance the optimization of travel time (Fig. 6).

Anthropocosmos

Doxiadis elaborated further on the ekistic grid by adding the time scale and aspects of desirability and feasibility to the x-axis. On the y-axis the ekistic elements were expanded to include their interrelations. This formed the framework for the description of the total world (*cosmos*) of man (*anthropos*) leading to the term *anthropocosmos*. This model is intended to define the total system of life. It is described in *Ekistics 241*, December, 1975 and in a slightly longer version in *Ekistics 229*, December 1974 (Fig. 7).

The dimension of time

All too often settlements are only considered with their two-dimensional qualities. Doxiadis stressed again and again that we must deal with their three physical dimensions but also, and most importantly, include the fourth dimension of time. Settlements are no longer

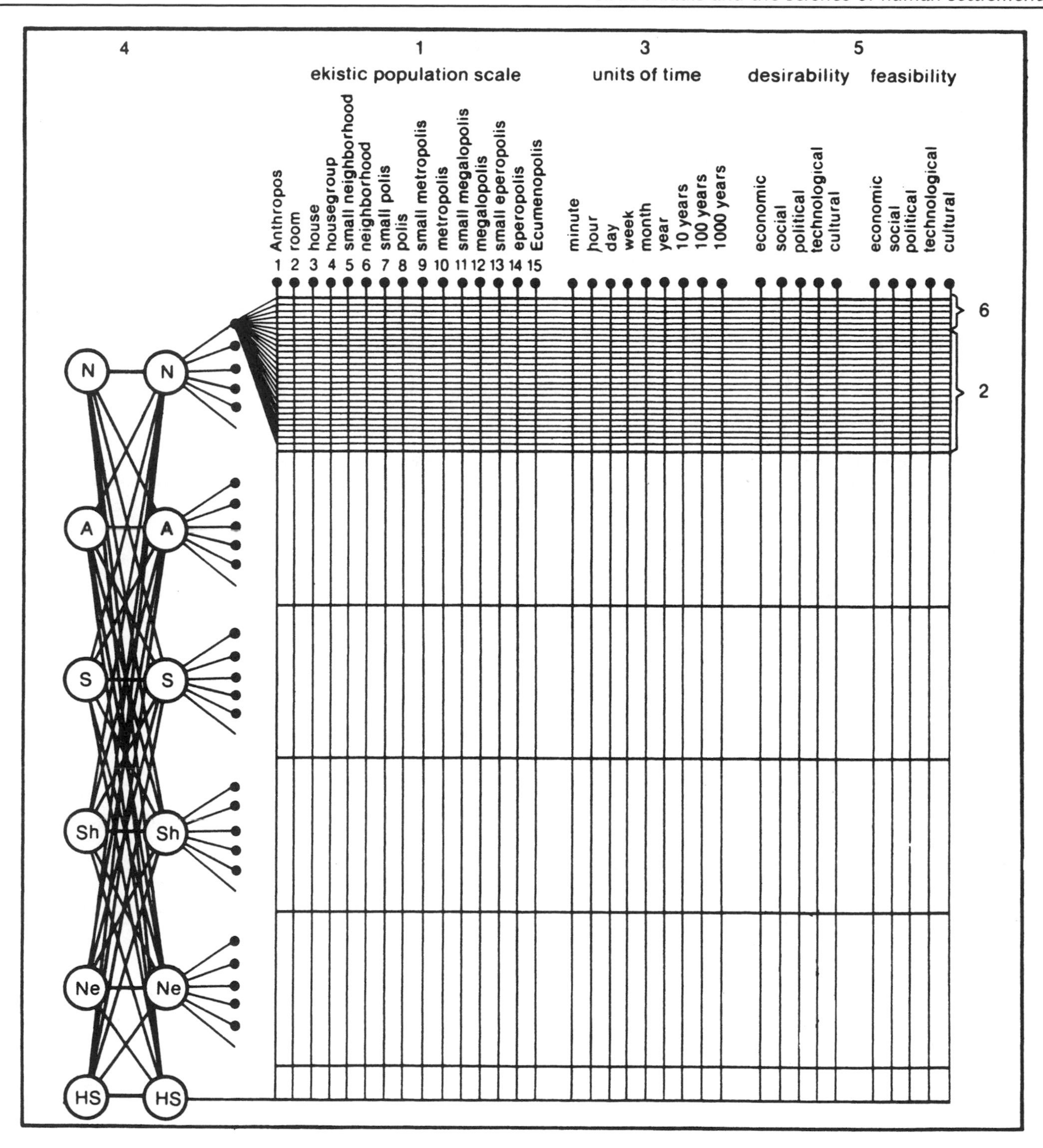

1. ekistic population scale
2. ekistic territorial scale
3. ekistic time scale — part of 1
4. ekistic elements — part of 2
5. aspects — part of 3
6. principles — part of 5

Fig. 7. *Anthropocosmos* model.

static but they are increasingly subject to rapid growth and change. By introducing the dimension of time in a bold way, we are encouraged to look at the past, which gives us an understanding of the evolution of settlements; by looking at the future we will anticipate the dynamic interplay of forces and be able to assess the complex implications. This anticipation of the future will enable us to plan for the future by acting in the present. The introduction of the dimension of time was a recurring theme and important part of Doxiadis' theory, which placed emphasis on the dynamic nature of cities.

In later years Doxiadis expanded from the thinking in four-dimensional space by adding more and more dimensions in a concept of the multi-dimensional or *n*-dimensional parameter space. This formed the basis for the conceptual model for the project for the Greater Detroit Region (Doxiadis 1966, 1967, 1970).

Concepts

Within the framework of Doxiadis' numerous models, several concepts emerged, which were developed and refined over time and served as recurring paradigms in many of his projects:

- Dynapolis is a term coined by Doxiadis in the early fifties meaning dynamic city or dynamic polis, recognizing the fact that contemporary city is a dynamic and not a static entity. He proposed a unique linear configuration to accommodate for growth and change.

- *Ecumenopolis* is a key concept of ekistics. As the largest unit of the ekistic logarithmic scale it is the inevitable city of the future, which will cover the entire earth as a continuous system forming a universal settlement (see below for further elaboration).

- The human community is the settlement unit designed at a human scale. It corresponds to unit 7 and community class IV on the ekistic logarithmic scale and is an indispensable building block within the changing and dynamic context of ecumenopolis, providing a stable spatial and functional unit to satisfy the needs and aspirations of our daily life.

- The functional classification of space postulates the complete allocation of land for human settlements consisting of the area of urban settlement, of cultivated areas and natural areas.

Search for ideal solutions

For thousands of years we have been developing static settlements, which can be categorized as natural and planned, radial and grid-iron, simple and monumental, ideal and utopian. Contemporary and future settlements are dynamic and as of today have not attained any clear forms to establish a morphology. Despite the uncertainties and difficulties, Doxiadis (1968) reminded that,

> *the search for the ideal is our greatest obligation. . .Only by searching for the ideal can we hope to face the real issues. As long as we start off with the narrow-minded view that the ideal is neither necessary nor attainable, we are compromised right from the beginning.*[15]

- **We have to guide our cities.** We cannot stop certain trends, but to avert adverse conditions in our cities, we must learn how to guide them.

- **We must define how much land we shall allocate for everything in its place.** This means we must define how much land is to be allocated for preservation of natural conditions, for cultivation and for settlements.

- **We must immediately designate common corridors** for networks of roads, communication and other infrastructure. Independent and uncoordinated networks lead to chaos.

- **We must create human communities** to enable us to live together with other people.

- **We must get rid of towers** and live in human dwellings or houses.

- **We must create new administrative systems,** one responsible for every large city and its region, another for every small city and another for every neighborhood.

Doxiadis developed the science of *ekistics* from a restricted framework of simple categories to an increasingly refined framework with a growing number of categories and relations. This in itself led to increasing complexity and sophistication. But it also encouraged increasingly complex and profound treatment of the subjects by others, filling the framework with knowledge and insights by top experts from many disciplines from all over the world. Doxiadis' evolution can be seen in his last article "Action for human settlements," *Ekistics, 241*, December 1975. In it he summarizes his position on the human settlement, the definition of the topic, the problems, the goals, policies and programs and finally a statement of the radical changes we need for our settlements to survive and become human.

3 ECUMENOPOLIS

> *The realistic view of the City of the Future*[16] *accepts that it will be a global city. This does not mean, that it will cover the whole globe—only a small part of the globe can and will be covered —but it will be a system of human settlements encircling the whole globe, made up of several types of cities and other settlements of all types interconnected into broader urbanized areas like the ones we today call megalopolises. It will consist of parts with very different densities from very high to very low, and of continuous built-up areas as well as separate areas interconnected by several types of transportation and communication lines. The global city of anthropos is the ecumenopolis or the city of the inhabited globe.*[17]

The unprecedented growth of metropolitan areas all over the world, the population explosion and migration to cities, the merging of metropolitan areas into continuous settlements such as the American eastern seaboard stretching from Boston to Washington, induced Doxiadis to conceive of a comprehensive model of urbanization which covered entire regions, countries and continents (Fig. 8). Studies of J. Gottmann (1961)[18] and others identified a trend of cities within large regions to become interdependent and to grow together into contiguous patterns along transportation networks, a settlement form, which they called *megalopolis*. Doxiadis predicted that this type of pattern would stretch across entire countries and continents and over time would form the unavoidable city of the future called *ecumenopolis*.

Doxiadis presented his concept of *ecumenopolis* in a 200-page

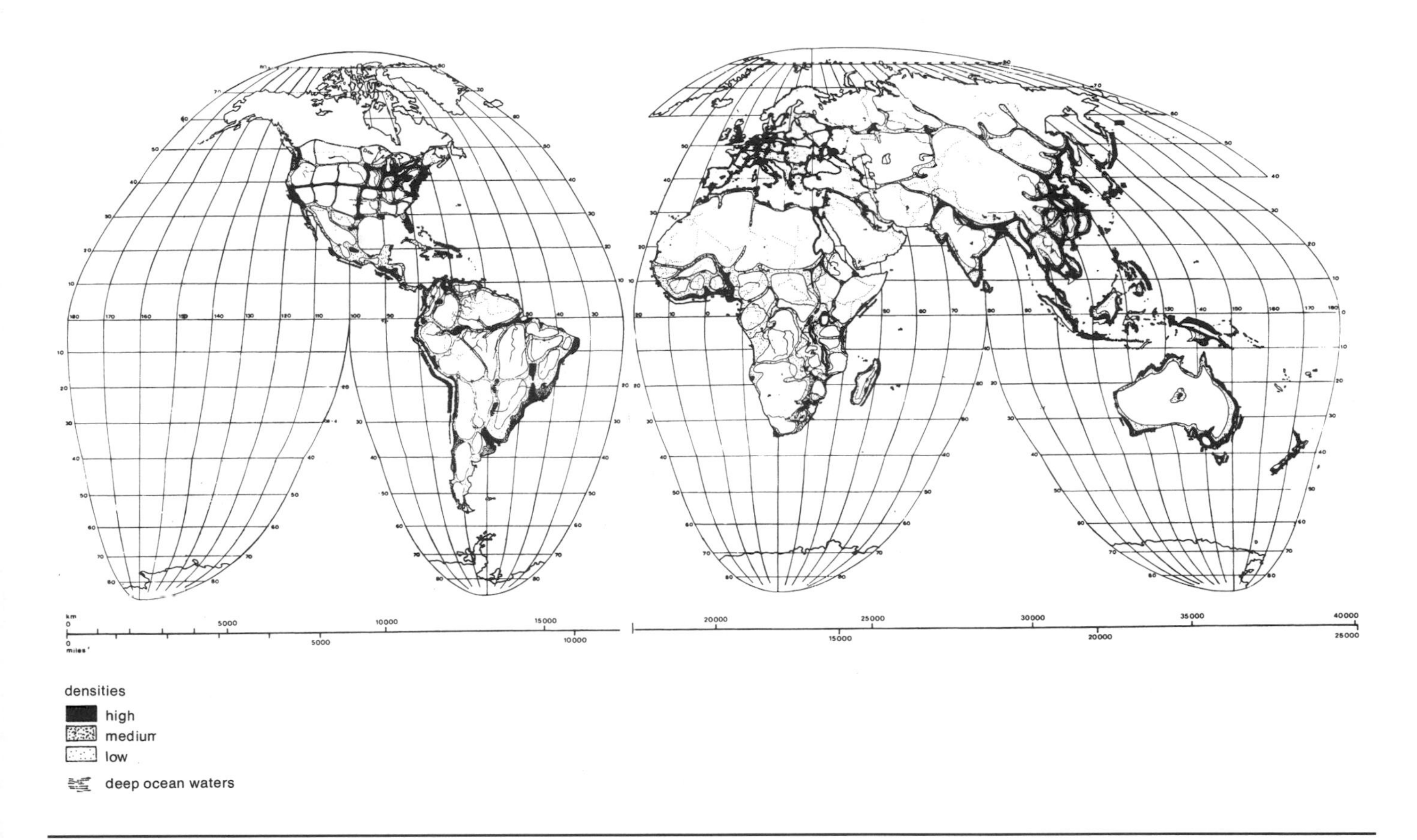

Fig. 8. *Ecumenopolis.*

report in 1959. This document already contained the major points of the project in a mature form with their conditions and far-reaching consequences, reflecting Doxiadis' many years of experience and thinking on the subject. In 1960 research began in a systematic way at the Athens Center of Ekistics under the leadership of John G. Papaioannou. The work was published in an issue of the journal *Ekistics* (June 1965)[19] devoted entirely to the subject. The book *Ecumenopolis, the Inevitable City of the Future* was written first by Papaioannou then rewritten by Chris Ripman and finally by Doxiadis himself. It appeared in 1975 as one of the so-called four red books to be presented at the Habitat Conference of the United Nations in Vancouver in 1976.

Projection of the dynamic settlement patterns proved to be surprisingly accurate, which goes a long way toward validating the concepts and methodology. The work on different parts of Ecumenopolis such as metropolitan areas and megalopolises in different regions of the world were projected to the year 2010 and 2100. After forty years (and nine years ahead of time) we can see that the 2010 projections will closely reflect reality. If there are discrepancies between projection and reality, they usually show that specific conditions are appearing ahead of time or to a greater extent than anticipated.

Why are these projections so accurate? Doxiadis did not merely follow the usual forecasting techniques such as trend extrapolations or independent parameter modeling. Rather, he defined and described a final condition of balance and saturation for the target date. From there he worked backwards connecting the future with the present and thus describing the development over time in reverse. This presupposes a state of balance and equilibrium as a final condition, in contrast to the explosive growth in population, in land and in resource consumption, and the increasing degradation of the environment, which would lead our planet toward disaster and cannot be sustained much longer. If mankind wants to survive, the trends have to be reversed. The limits we encounter force us to make critical choices.

One of the major difficulties in dynamically growing settlements is their unremitting change, which seems to affect the entire system. Yet an analogy to nature offers a solution. In a growing and changing organism, the individual cell usually maintains its size and structure despite growth and change of the system overall. Based on this example Doxiadis (1968) drew the conclusion that "the search for ideal solutions has to be geared towards [maintaining] static cells [within] the dynamic growth of the organism."[20] This condition can be achieved by removing the forces of change from a settlement or cell as, for example, by moving a transportation network from the center of a settlement to the periphery.

In the past the city block has been the basic unit of the city. With the increasing size of the settlement, the small town or community class IV with about 2,000 families or 10,000 inhabitants seems to be a better size for a recurring unit. The ideal building block should not exceed 30,000 inhabitants and a maximum size of 2,000 yards by 2,000 yards. It is important that the building block or cell is based on the human scale and the human experience. Growth is achieved by

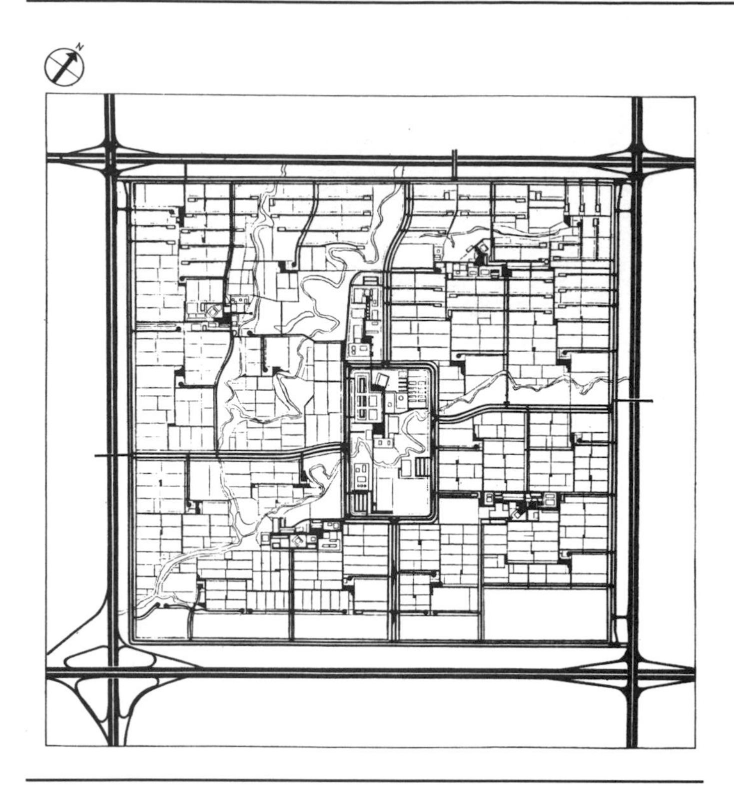

Fig. 9. The human community: four communities Class IV form a sector in Islamabad, the new capital of Pakistan.

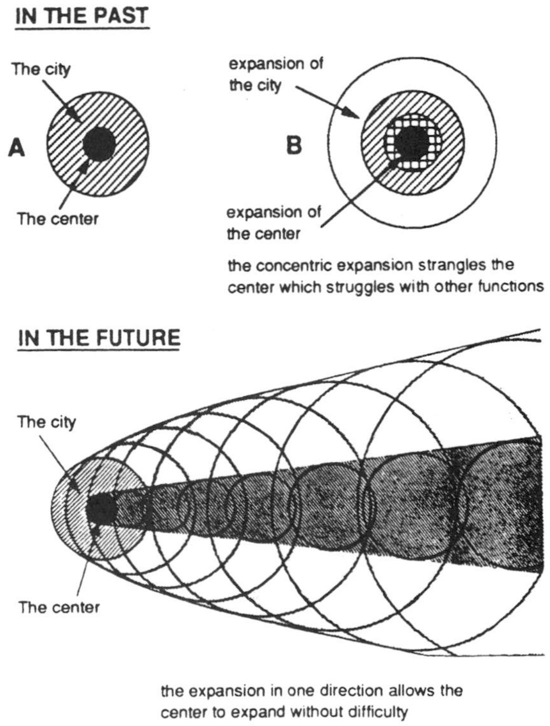

IN THE PAST

The city

A

The center

expansion of the city

B

expansion of the center

the concentric expansion strangles the center which struggles with other functions

IN THE FUTURE

The city

The center

the expansion in one direction allows the center to expand without difficulty

Fig. 10. The concept of *Dynapolis.*

adding units of a constant size rather than enlarging the units themselves. This is analogous to the evolution of the house, which, over time, grows not so much by the enlarging of one room but by adding more and more rooms. As the human city becomes bigger and bigger, it will consist of two categories of parts, the cells or building blocks and the networks organizing and serving them (Fig. 9).

"If we start building urban areas with static cells, it will be because of the need to create human conditions providing as much security against constant change as possible"(Doxiadis 1968).[21] But how is this achieved in dynamic settlements of the size of the metropolis, megalopolis and larger? One major problem of the rapid dynamic growth of settlements is the typical growth of their centers, expanding concentrically into their surrounding areas and having an impact on each subsequent concentric ring. The slum areas surrounding the downtowns are a typical result of the forces of transition from housing to office use and from lower to higher densities.

> *The only way to solve the problems of dynapolis [the dynamically growing metropolis] is to conceive a pattern which will permit its natural growth, especially that of its center, without allowing the new additions to destroy the existing patterns.*[22]

This led Doxiadis to the formulation of a compelling solution of his ideal *dynapolis*, which expands along a single axis emanating from the center, allowing growth by addition and leading to a unidirectional linear arrangement. Since empiric studies have shown that densities have declined over time, meaning newer cities show lower densities than older ones, this dynamic growth model can accommodate the lower densities in the future, accommodating increasing land consumption in subsequent growth along the growth axis. The requirements of efficient networks for traffic will likely superimpose a grid configuration over the theoretical diagram with the increasing circular areas (Fig. 10).

This diagram for the dynamic city is a creative invention for a settlement structure, which is based on a broad and carefully developed theory (Fig. 11). Doxiadis applied this model in many different contexts, most notably in Islamabad, Pakistan, 1960 (Fig. 12), and Tema, Ghana, 1964, both projects realized.[23]

Developing Urban Detroit Area Research Project

The Developing Urban Detroit Area (UDA) Research Project was one of Doxiadis major projects demonstrating his approach and the ekistic principles. It was conducted from 1965 to 1970 by Doxiadis Associates in collaboration with Detroit Edison Company and Wayne State University and resulted in a three-volume documentation *Emergence and Growth of an Urban Region—the Developing Urban Detroit Area Research Project* (Doxiadis 1966, 1967, 1970). The objectives of the study were:

- To analyze the nature and magnitude of the growth in the area
- To understand the problems
- To explore methods for finding solutions
- To assess the needs for electric energy and provide a framework to meet the demand. It should be a prototypical study for use in other areas.

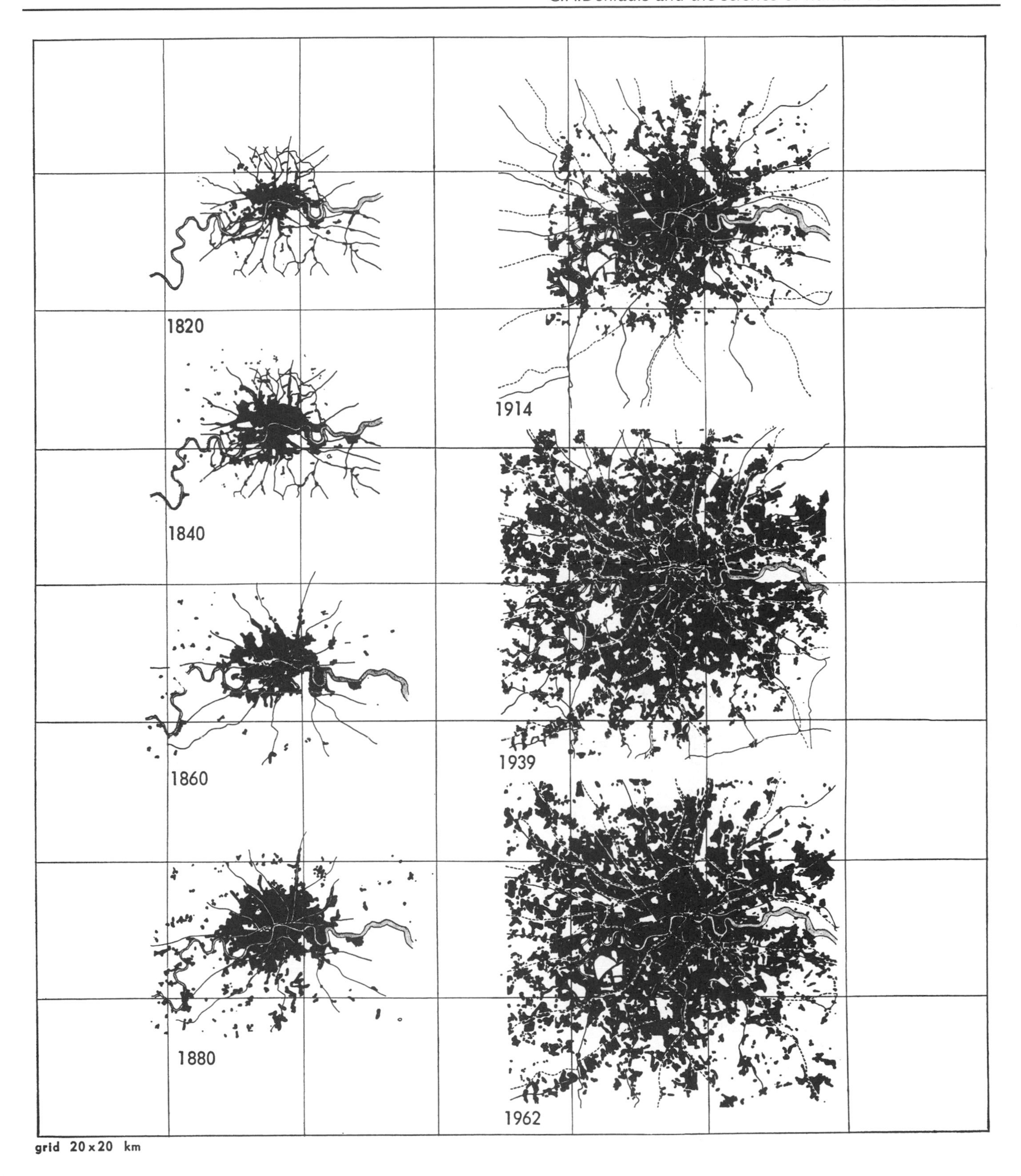

Fig. 11. Evolution of a *Dynapolis*. London England 1820–1962.

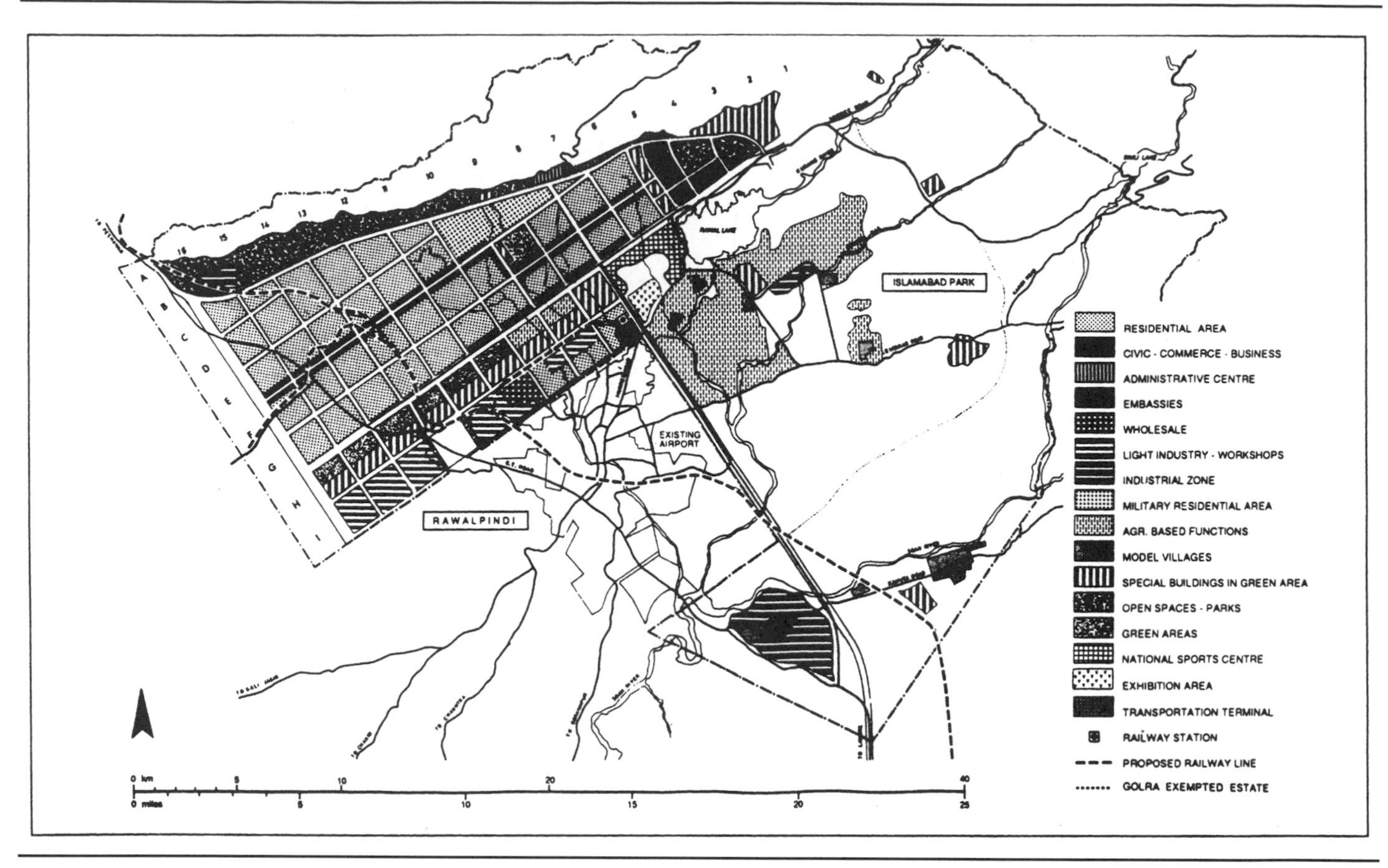

Fig. 12. Islamabad, the new capital of Pakistan.

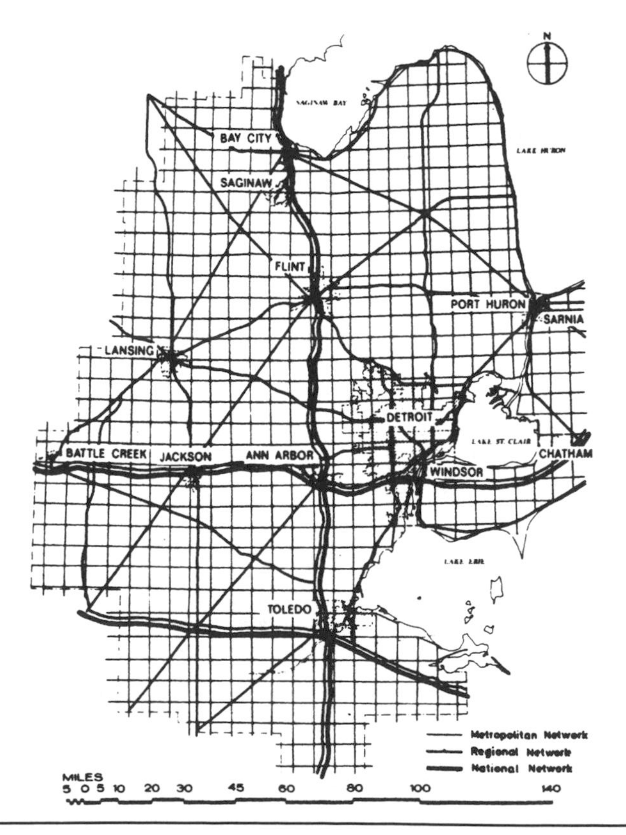

Fig. 13. Project for the Urban Detroit Area: Network Alternative.

Detroit with a population of 1.7 million inhabitants was classified as a metropolis, unit 10 on the ekistic logarithmic scale, and the scope of study was extended down to the scale of the town, unit 8, and up to the megalopolis, unit 12, and for some issues to ecumenopolis, unit 15. The immediate planning area was contained within a radius of roughly 100 miles (161 km) from the center of Detroit, stretching 75 miles (121 km) east and southeast and 120 miles (193 km) north (Figs. 13 and 14).

The time frame was also set up according to ekistic principles, starting in the present, going to the past (1900), extending into the distant future (2050), going back to the past and then to a near future (2000) and after a few iterations returning to the present to implement the solutions.

The Detroit region was considered to be a dynamic system undergoing growth and change. The form and structure of the future city was expected to be different from mere trend extrapolation, and Doxiadis tried to formulate a new method to be able to deal with these conditions. A variety of methods were used to predict the future and generate alternatives. These alternatives were then evaluated with a set of criteria and through successive rounds of elimination, the large number of alternatives was reduced to one by the IDEA–CID method.[24]

The main model used in the UDA project was an urban simulation model, with its centerpiece a transportation and land-use model derived from the Lowry model,[25] which simulated population distribution based on basic employment centers, transportation networks, speed, and a variety of behavioral parameters. The space of possible solutions or alternatives was defined by a number of

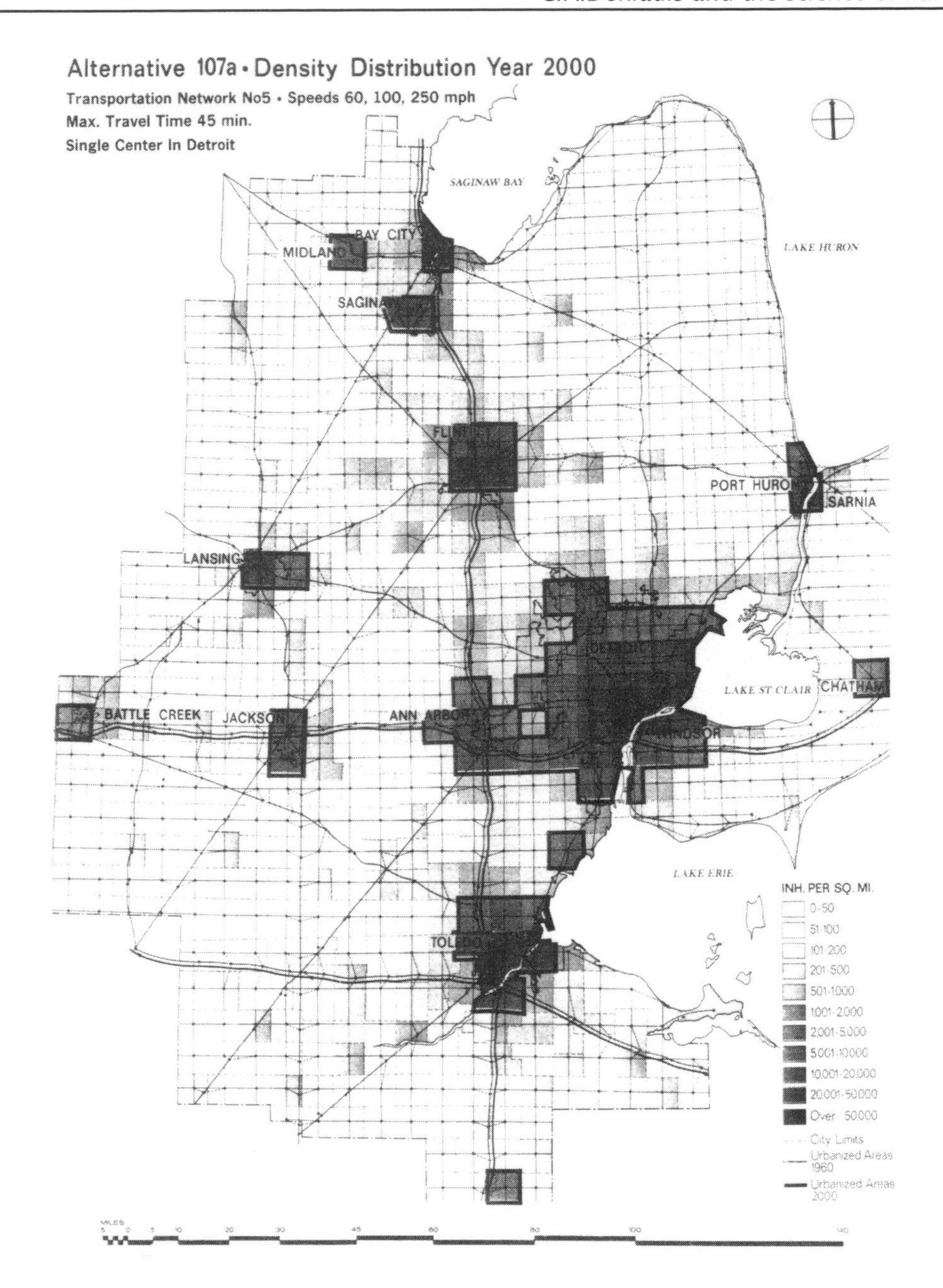

Fig. 14. Project for the urban Detroit area: Alternative of population densities based on 45-minute travel time and three projected transportation speeds.

relevant parameters or dimensions (n-dimensional parameter space). On each parameter a range of up to 16 values or options was selected and combining a set of values or points for each parameter created an alternative. This generated 49 million theoretical alternatives for the UDA project, which were then narrowed down by the IDEA–CID method, proceeding from the largest to the smaller scales.

The parameters considered in the UDA project were:

- Total population
- Basic employment concentrations
- Major center areas
- Education and research centers
- Industrial zones
- Port
- Airport
- Transportation networks
- Speed ranges
- Maximum travel times
- Behavioral parameters
- Growth parameters
- Parameters of employment structure

In the stepwise evaluation and elimination process, whole groups of alternatives were eliminated, whole groups of solutions were omitted. This type of model is by definition an abstraction of the real world and therefore incomplete. Whatever was included in the model was carefully considered; however, it was possible and even likely that whole parameters or ranges of parameter values were missing, which could limit the applicability of the model and its findings. Nevertheless, the UDA project was a pioneering application of a macro model which in its scope and theoretical rigor was unmatched at the time and in its clarity and strength of presentation is still unrivaled to the present time.

The project reflected Doxiadis' strong belief that the transportation networks are a decisive form determinant for the dynamic city of the future, particularly in the USA. He placed major transportation corridors across the planning area. Rather than designating specific transportation systems he assumed sets of speed as performance parameters, which can be expected for the target dates. His projections of speeds of 100, 250 and 400 mph (160, 400 and 640 km) reflected his future orientation and bold belief in technological progress but also were an anticipation that we had to transcend car transportation as we know it.

4 ACTIONS FOR HUMAN SETTLEMENTS

To implement his vision, Doxiadis embarked on many parallel and interrelated paths.

One of his strong arguments was that his theories and ideas were developed on the basis of a wealth of experience through projects and direct observation. His professional practice, Doxiadis Associates, was founded in 1951, and had influential projects in forty countries. At its peak Doxiadis Associates had 400 to 450 employees in a network of offices all over the world, including the United States (Washington, Philadelphia, Detroit). The mayor projects included Islamabad, the new capital of Pakistan (1960), the master plan of Accra-Tema, Ghana, in 1960, a housing program for Iraq with housing projects in Baghdad and other cities, study and plan for the Greater Detroit Area (1960–70), and housing in Eastwick, Philadelphia.

The Athens Center of Ekistics (ACE) was the research arm of the enterprise. Research started in 1959 with the initial projects the City of the Future(COF) (which later turned into the study of ecumenopolis), followed by the Human Community project (HUCO). In 1964 the project Capital of Greece was added to form the initial research program, covering three major units of the ekistic scale, namely small town, metropolis, and ecumenopolis.

The Athens Graduate School of Ekistics brought students together from all over the world to introduce them to ekistics, and with this to disseminate the ideas and principles worldwide through their future activities, projects, teaching and research. During their stay in Athens the students were also active participants in the research projects making valuable contributions from the viewpoint of their home countries. While the number of students was relatively small, the impact of ekistics is evident to the present day wherever they practice.

One of Doxiadis' most brilliantly conceived institutions was the *Delos Symposia*, which took place from 1963 to 1972. Doxiadis invited a group of approximately forty leaders from academia, politics, business and the professions related to human settlements to join in a cruise to the Greek islands. During the weeklong journey, the international experts discussed issues of human settlements, culminating in a joint postulation for action and further research presented in the ancient theater on the island of Delos. There could hardly have been any context more conducive to productive discourse than these cruises with visits to islands with indigenous villages, natural beauty and inspiring reminders of ancient Greek culture. After the formal and informal debates, the immersion into the local culture with eating, drinking, music and dancing served as a reminder that within the daunting problems of the future of settlements, there is at the center the human being with enjoyment of life and happiness.

The collaboration and interaction of the participants lasted well beyond the symposia. Doxiadis organized the Athens Ekistic Month at the headquarters in Athens to expand on the visit of the Delos participants and give exposure of the ekistic ideas to a wider interested public through lectures, meetings and discussion. At the third Delos Symposion in 1965, a decision was made to organize the participants and other personalities interested in the ekistic cause in the World Society for Ekistics, which was founded in 1967 and is active to the present time.

The Journal *Ekistics*

A most important element in the development and dissemination of the *ekistic* ideas was the existence of a journal. Initially Doxiadis supported a monthly bulletin to keep architects and planners in his numerous offices in developing countries informed of the relevant professional information elsewhere in the world. Mary Jaqueline Tyrwhitt, Professor at Harvard, became the first editor and directed the new journal also to the U.N. housing and development experts. The journal was called *Tropical Housing and Planning Monthly Bulletin* before assuming its name *Ekistics: Housing and Planning Abstracts* in October 1957, and its present name *Ekistics: The Problems and Science of Human Settlements* in January 1965. The first issue of October 1955 contained mainly reprints of articles from other journals, but over the years original articles started to dominate and give the journal a unique and highly respected position. The regular editorials and lead articles of Doxiadis himself gave a unique emphasis to the journal, and the recurring annual themes provided a reliable documentation on the development of the research and the evolution of the ekistic theory. Thus it became the major source for information on the Athens Center for Ekistics, but the articles on settlements in general set the ekistic work into the proper context. From 1978 to1985 the journal appeared bi-monthly and from 1986 it appeared four times a year under its present editor, Panayis Psomopoulos.[26]

A unique feature of the journal is the use of the ekistic index. The content of each article is identified in a matrix consisting of the five ekistic elements and the ekistic logarithmic scale, which allows for precise delineation of the topic and easy comparison. Annual overviews of all the published articles give precise indications of the emphasis of the journal and identify gaps in the existence of research or in the coverage of specific areas.

It is encouraging to see how the tools and methods of ekistics, and indeed ekistics as a whole, evolved over time. Most issues of *Ekistics* contained a detailed explanation of the ekistic index with an invitation for suggestions for improvements. While the principle of the index remained constant over the years, there were nevertheless several changes and refinements in the terminology and the ranges of the ekistic scale. These refinements occurred parallel to the development of the anthropocosmos model, which adds more dimensions to the

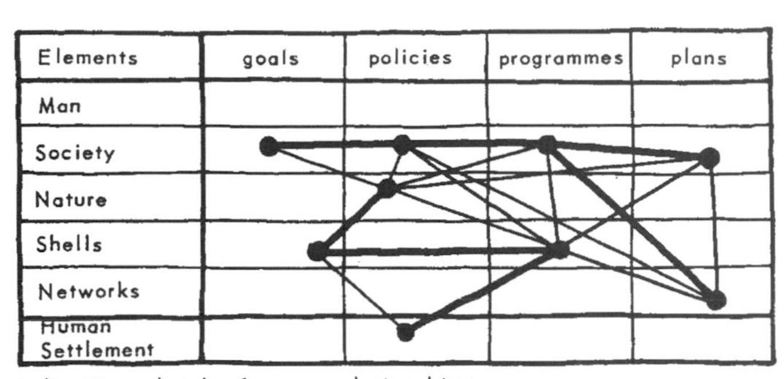

Fig. 15. Interrelationship of scales and elements.

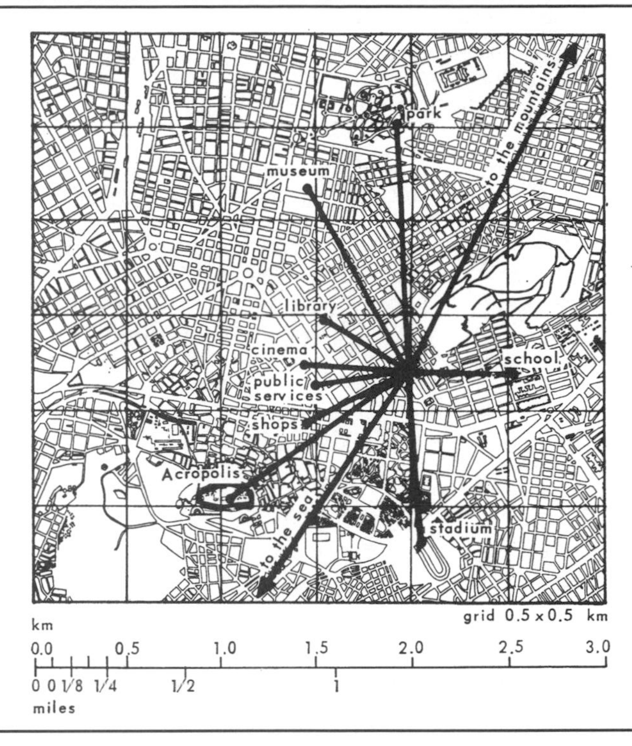

Fig. 16. The pedestrian realm of one individual (anthropos).

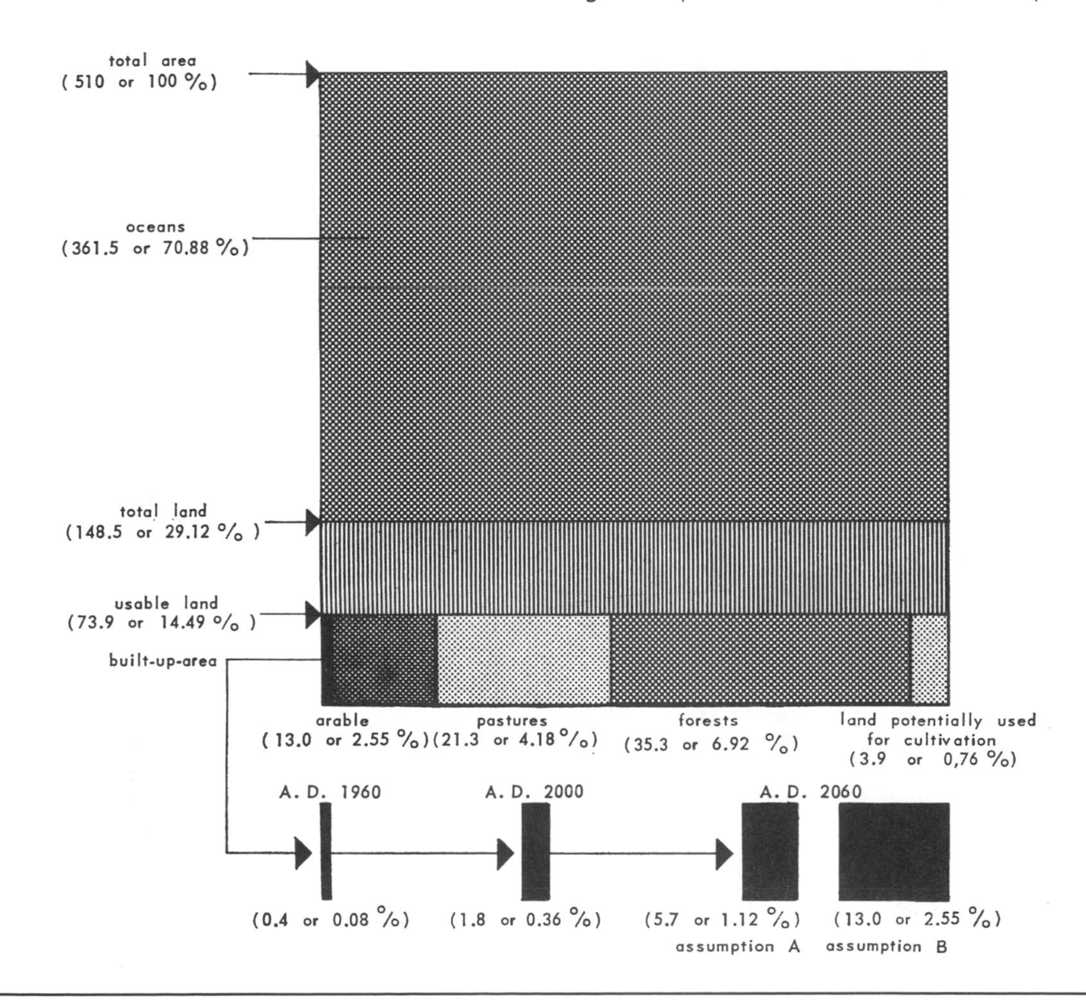

Fig. 17. The global impact of urban infrastructure (ecumenopolis).

diagram by introducing the interaction of the five ekistic units with each other as well as human settlements (formerly the synthesis) and the dimensions of time present, income, desirability and feasibility.

5 CONCLUSION

Doxiadis was masterful in distilling the most complex concepts into simple statements and concise summaries, allowing for easy comprehension and memorization, providing insight and inspiration and stimulating further action. In this spirit, a treatment of ekistics would not be complete without an attempt at a brief summary.

1. In dealing with human settlements we must be comprehensive, have an interdisciplinary scope and relate to the five ekistic elements: Man (or *anthropos*), society, nature, shells and networks, and their relationships (Fig. 15).

2. Any study of settlements shall refer to the ekistic units of scale from man to ecumenopolis, the fifteen levels in the ekistic logarithmic scale (Fig. 16).

3. The time dimension must be integrated in a bold manner in the analysis and design of human settlements from the past to the present to a distant future (Fig. 17).

4. The scientific method shall be used in a systematic treatment of human settlements, following the models, concepts, principles, values and postulations of ekistics. Each study shall be identified on the ekistic grid index.

5. Analysis must be followed by action.

6. The goal for our human settlements is derived from Aristotle's postulation for safety and happiness.

7. According to the model of satisfaction, increase to the maximum all possible contacts with people elements and functions; reduce energy, time and cost to a minimum; optimize protective space whether temporary or permanent, for man as an individual or as the member of a group; optimize relationships with nature, society, shells (buildings) and networks—in other words with the total environment; and finally, create a balanced and beneficial synthesis of the preceding principles.

8. Analysis, planning and design must take place within the larger vision of the concept of ecumenopolis.

9. The city must be treated as a dynamic settlement. The use of the concept of dynapolis allows for growth and change.

10. To deal with the overwhelming forces of growth and change, human communities must be created and preserved as stable building blocks, which can be replicated within the dynamically growing and changing city.

Ekistics started with Constantinos A. Doxiadis, one man and his vision, and blossomed into a framework for a science of human settlements of global scope and importance. Many contributors throughout the world worked on advancing the work and on closing the gap between ideals, aspirations and reality in creating a future, where balance and harmony between man, society, buildings, networks and nature prevail. Doxiadis offered a message of optimism and hope. He accepted the growth, complexity and interdependence of the modern settlements as inevitable and as a reflection of our civilization and progress, yet he vehemently rejected the dehumanizing conditions of the machine, the negative side effects of technology, wanton neglect of the environment and destruction of nature.

> *Doxiadis believed there was an urgent need to reconcile our increasingly complex scientific and technological achievements to our permanent, unchanging human nature and to the enduring prevalence of a "human scale" in our cities, whether the ones in which we reside today or those which we envisage for the future.*[27]

The ekistic movement grew in great part due to the charisma, personality and creative power of its principal innovator, Constantinos A. Doxiadis. His legacy, his ideas and principles are a rich resource to be applied, developed further and adapted to human settlements of the present and the future. ■

NOTES

1. For a detailed biography and listing of honorary degrees, honors and awards, see *Ekistics 373–375*, July–Dec. 1995, or the official website www.ekistics.org.

2. Doxiadis, Constantinos A. 1937. *Raumordnung im Griechischen Staedtebau.* Doctoral thesis Berlin-Charlottenburg University. Berlin: Vonwinckel. English Translation 1972. *Architectural Space in Ancient Greece.* Cambridge: MIT Press.

3. Doxiadis, Constantinos A. 1963. *Architecture in Transition.* London: Hutchinson.

4. Doxiadis, Constantinos A. 1968. *Ekistics, an Introduction to the Science of Human Settlements.* New York: Oxford University Press.

5. Doxiadis wrote the *Four Red Books* "to help us understand what will happen to our human settlements and what we can do to save them." Doxiadis, Constantinos A., 1974.

• *Anthropopolis* (summary of a symposion on the City for Human Development). Athens: Athens Publishing Center.

• and J. G. Papaioannou. 1975. *Ecumenopolis.* Athens: Athens Publishing Center.

• 1975. *Building Entopia.* New York: Norton.

• 1976. *Action for Human Settlements.* Athens: Athens Publishing Center. This last book was published posthumously as a key document to the United Nations First Conference on Human Settlements, *Habitat 1*, Vancouver, 1976.

6. Panayis Psomopoulos, letter to the author, April 27, 2001.

7. For detailed description of Doxiadis' background and work, see *Ekistics 373–375*, July–Dec. 1995, or the official website <www.ekistics.org.> Thomas W. Fookes. "Ekistics: An example of innovation in human settlement planning." *Ekistics 325–327*, July–Dec.

1987, gives an excellent overview over the concept of ekistics as an approach to planning. Suzanne Keller. "Planning at two scales: the work of C. A. Doxiadis." *Ekistics 282*, May–June 1980, summarizes basic themes in Doxiadis' work by an emphasis on the human dimension within a global framework. And, Panayis Psomopoulos. "The Apollonion in Attica: A bricklayer's dream." *Ekistics 312*, May–June 1985, describes how Doxiadis' vision found realization in one of his late projects for a new community near Athens "where human beings, not machines, can and must prevail and where the essential infrastructure is provided for the satisfaction of all the inhabitants' daily needs—regardless of age sex or ability."

8. Doxiadis, Constantinos A. 1968. *Ekistics, an introduction to the science of human settlements*. New York: Oxford University Press. p. 8.

9. Doxiadis, Constantinos A. *Ekistics 141*, Aug. 1967. p.131.

10. Doxiadis, Constantinos A. *Ekistics 140*, July 1967.

11. Doxiadis, Constantinos A. *Ekistics 140*, July 1967.

12. Fookes, *op. cit.*, includes two additional models, the force mobile and the IDEA-CID model (Isolation of Dimensions and Elimination of Alternatives – Continuously Increasing Dimensionality). The force mobile deals with the totality of forces and their impact on form and structure of human settlements. The IDEA – CID model was developed in connection with the study on the Detroit Region, see: Doxiadis, Constantinos A. 1966, 1967, 1970. *Emergence and Growth of an Urban Region, the Developing Urban Detroit Area. Vols. 1–3*. Detroit: Detroit Edison Company. It constitutes an important contribution to the generation and evaluation of large numbers of conceptual alternatives in urban modeling.

13. Doxiadis, Constantinos A. 1968. *op. cit.*, p. 31.

14. *Ibid*, p 56.

15. *Ibid*, p. 355.

16. The City of the Future was one of the major research projects of the Athens Center of Ekistics (ACE). "This project—a continuous one for over 40 years—examines the long term prospects, trends, developments and forecasts regarding networks of human settlements and networks of communication at a regional and global scale as well as in geopolitical systems of interdependency." *Ekistics 131*, July–Dec. 1998. Within this project, ecumenopolis is the largest scale element.

17. Constantinos A. Doxiadis. 1975. *Building Entopia*. New York: Norton, p. 2.

18. Gottmann, J. 1961. *Megalopolis: the Urbanized Northeastern Seaboard of the United States*. Cambridge: MIT Press.

19. *Ekistics 115*, June 1965.

20. Doxiadis, Constantinos A. 1968. *op. cit.*, p. 355.

21. *Ibid*, p. 364.

22. *Ibid*. p. 364.

23. The Dynapolis concept was also applied in conceptual proposals for Beirut, Lebanon, 1958, Caracas, Venezuela, 1960, Karachi, Pakistan, 1959. A proposal for Copenhagen, Denmark, 1963, argued for a limited number of expansion axes versus the official 5 Finger Plan, and a proposal for Washington, D.C., 1959, emphasizes the development along one axis towards south-west rather than the six official development directions in form of a star configuration. In both Copenhagen and Washington, Doxiadis predicted that the areas between the fingers would be filled in, which ultimately would choke the center.

24. The IDEA-CID model (Isolation of Dimensions and Elimination of Alternatives–Continuously Increasing Dimensionality) was developed in this context. For additional information on this model, see: Ulkuatam, Semih Rasid and Alexander N. Christakis "Algebraic representation of human settlement design." *Ekistics 325–327*, July–Dec. 1987.

25. Lowry, Ira S. 1964. *A Model of Metropolis. Santa Monica: Rand Corporation*. For additional background, see: Hansen, Walter G. 1959. "How Accessibility Shapes Land Use." *Journal of the American Institute of Planners 25.2*. p.73–76.

26. Psomopoulos, Panayis, ed. "Forty years of Ekistics on persisting priorities." Special issue of *Ekistics 373–375*, July–Dec. 1995, gives an overview of the evolution of the journal.

27. Psomopoulos, Panayis. "The Apollonion in Attica: A bricklayer's dream." *Ekistics 312*, May–June 1985, p. 274.

REFERENCES

Doxiadis, Constantinos A. 1968. *Ekistics, an Introduction to the Science of Human Settlements*. New York: Oxford University Press

_________ and J. G. Papaioannou. 1975. *Ecumenopolis*. Athens: Athens Publishing Center.

Fookes, Thomas W. "Ekistics: An example of innovation in human settlement planning." *Ekistics 325–327*, July–Dec. 1987.

Fig.18

metropolis

average overall density of metropolitan area
66 inh/ha (27 inh/acre)

Athens, Greece (1964)

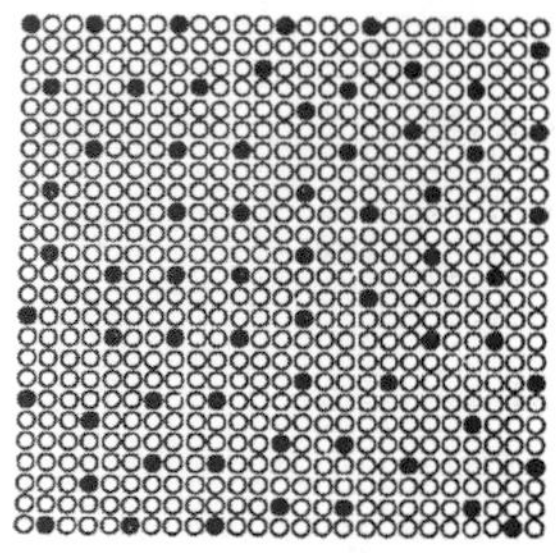

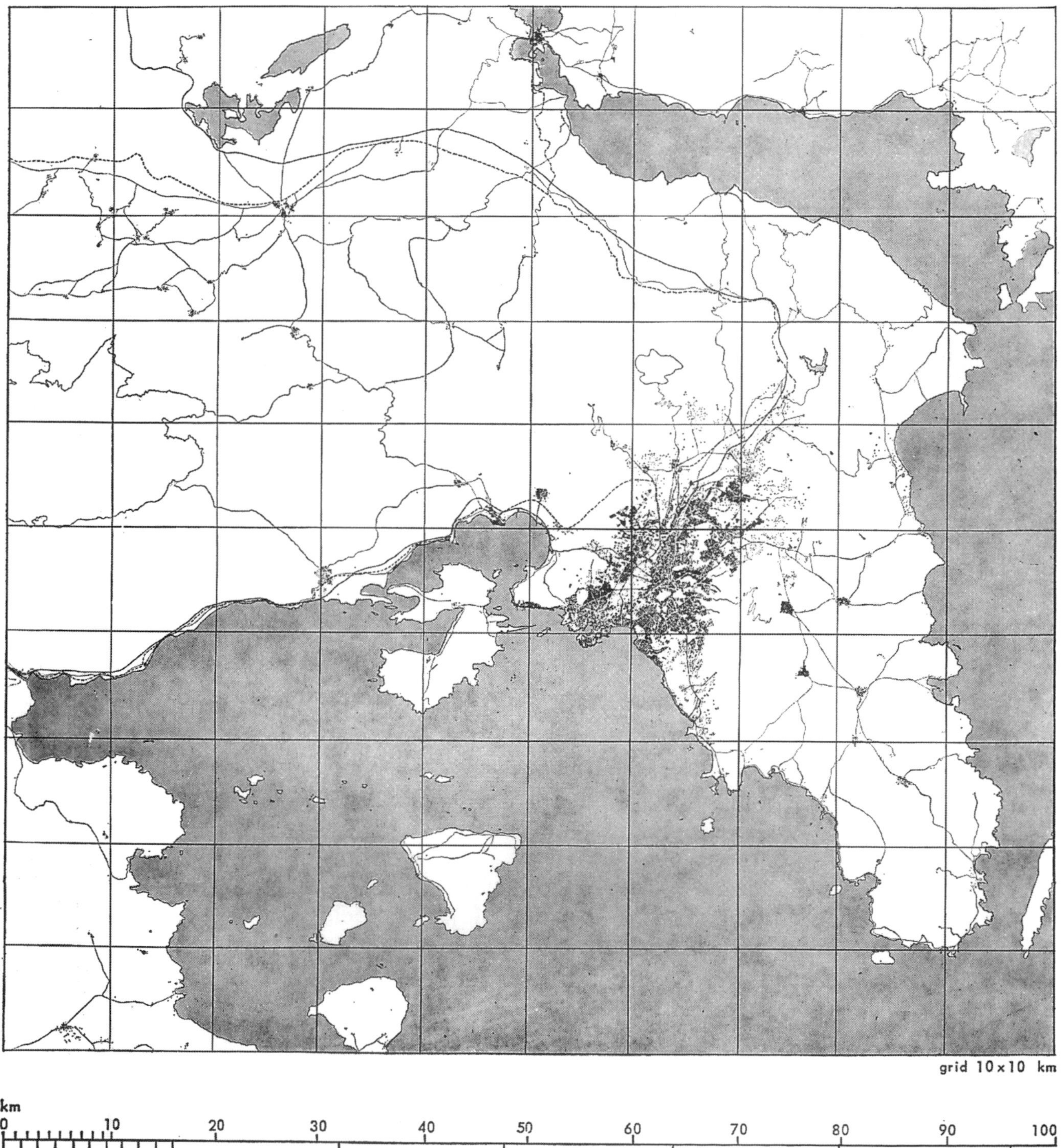

Fig.19

Dynametropolis

average overall density of metropolitan area
28 inh/ha (11 inh/acre)

London, England (1962)

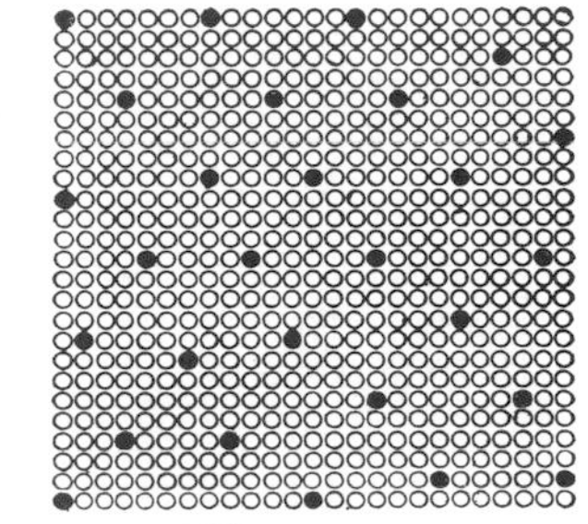

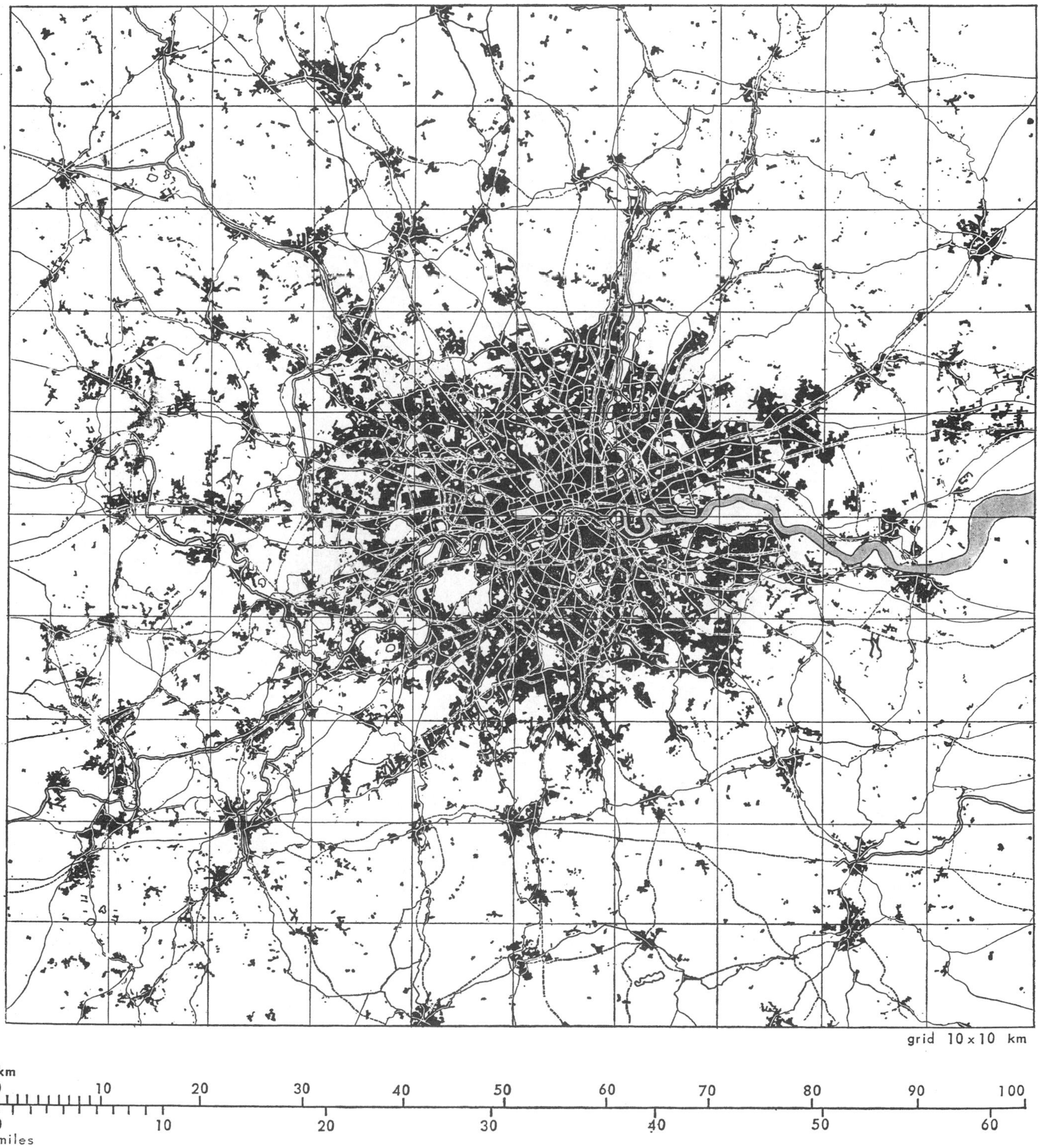

Fig. 20

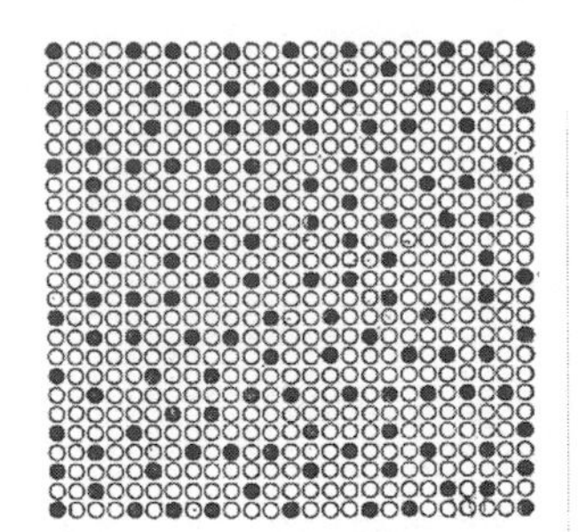

Dynamegalopolis

average density of built-up area
130 inh/ha (52 inh/acre)

New York, U.S.A. (1960)

Alison and Peter Smithson

Summary
Upon publication in 1968, the Preface to *The Team 10 Primer* read, "Here Team 10 tries to explain, in a similar form to the original Primer, what we stand for today. . .and why, because of our continually evolving attitude, the *Primer* is still a valid document for students of architecture to whom it was first directed in December 1962." This article indicates the development of the Smithsons' ideas of urban structuring via circulation systems, alongside the work of Jacob Bakema, Giancarolo de Carlo, Aldo van Eyck, Ralph Eskine, Shadrach Woods and others who loosely identified themselves as Team 10.

Key words
circulation, density, housing, mobility, motorway, parking, transportation center, urban infrastructure, way finding

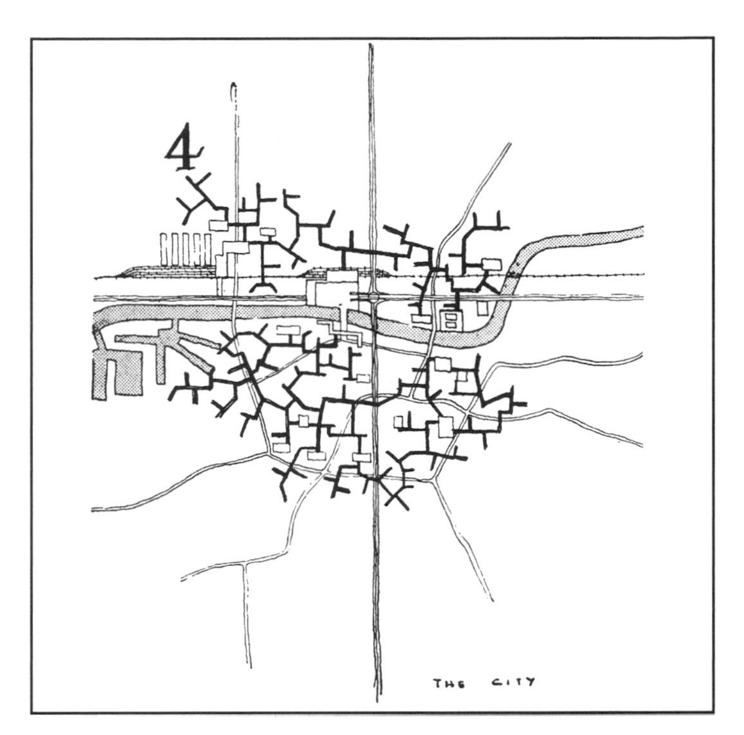

First cluster city diagram. Peter D. Smithson, 1952.

Urban infrastructure

Traditionally, some unchanging large scale thing—the Acropolis, the River, the Canal or some unique configuration of the ground—was the thing that made the whole community structure comprehensible and assured the identity of the parts within the whole. Today our most obvious failure is the lack of comprehensibility and identity in big cities, and the answer is surely in a clear, large scale, road system—the "Urban Motorway" lifted from an ameliorative function to a unifying function. In order to perform this unifying function, all roads must be integrated into a system, but the backbone of this system must be the motorways in the built-up areas themselves, where their very size in relationship to other development makes them capable of doing the visual and symbolic unifying job at the same time as they actually make the whole thing work.

The aim of urbanism is comprehensibility and clarity of organization. The community is by definition a comprehensible thing. And comprehensibility should also therefore be a characteristic of the parts. The community subdivisions might be thought of as "appreciated units"—an appreciated unit is not a "visual group" or a "neighborhood." but an in-some-way defined part of a human agglomeration. The appreciated unit must be different for each type of community. For each particular community, one must invent the structure of its subdivision (Fig. 1).

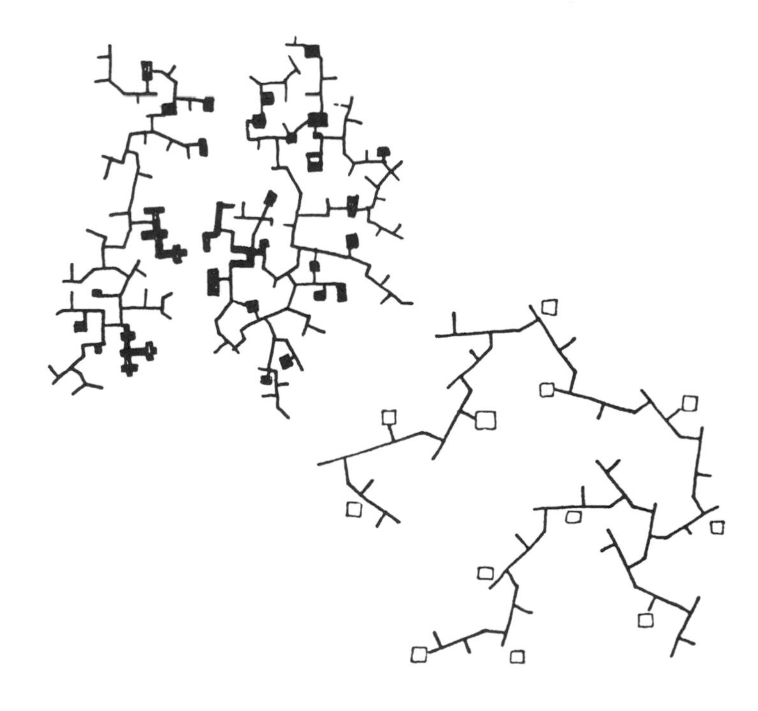

Fig. 1. "Appreciated units" as urban elements. (Alison and Peter Smithson, 1954.)

In most cases the grouping of dwellings does not reflect any reality of social organization. Rather they are the result of political, technical and mechanical expediency. Although it is extremely difficult to define

Credits: This article is excerpted from *Team 10 Primer*, 1968 (MIT Press), reprinted by kind permission of Peter Smithson.

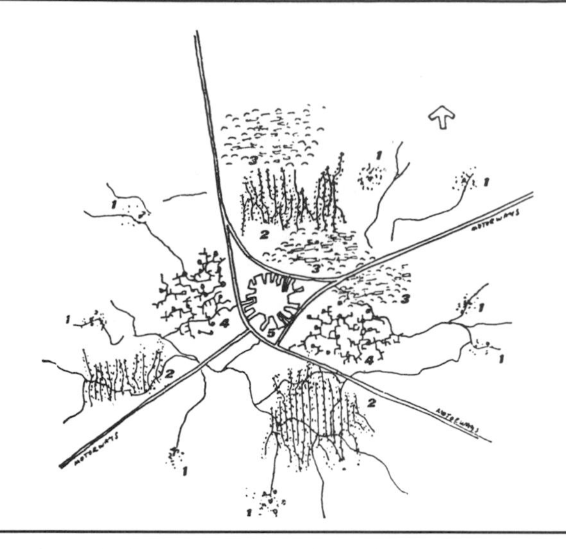

Fig. 2a. Cluster city diagram. (Peter D. Smithson, 1955.)

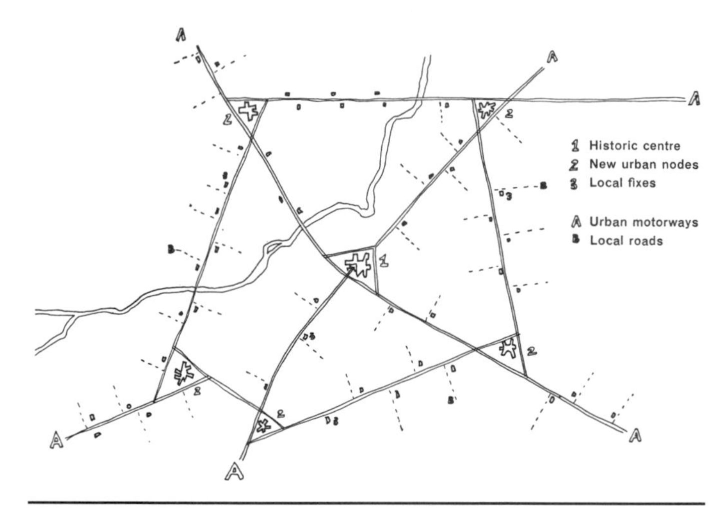

Fig. 2b. Roads as urban infrastructure, a basis for a pattern of growth. (Peter D. Smithson, 1955.)

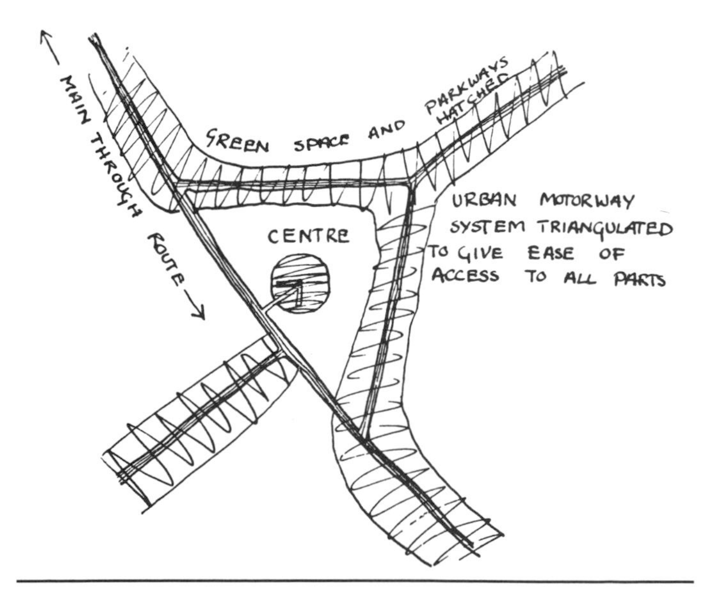

Fig. 2c. Equal flow and structure, basic diagram and first study, use of neutralized strip to define areas. (Peter D. Smithson, 1959.)

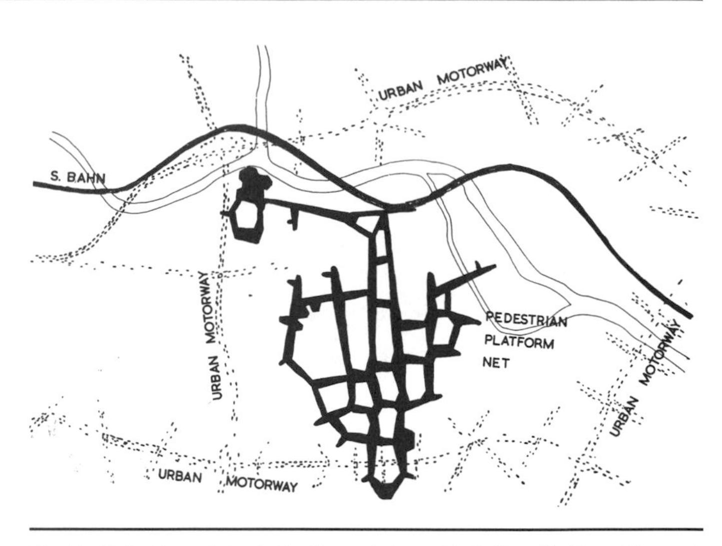

Fig. 3a. Pedestrian net structuring the central area. Berlin Plan. (Smithson/Sigmond, 1958.)

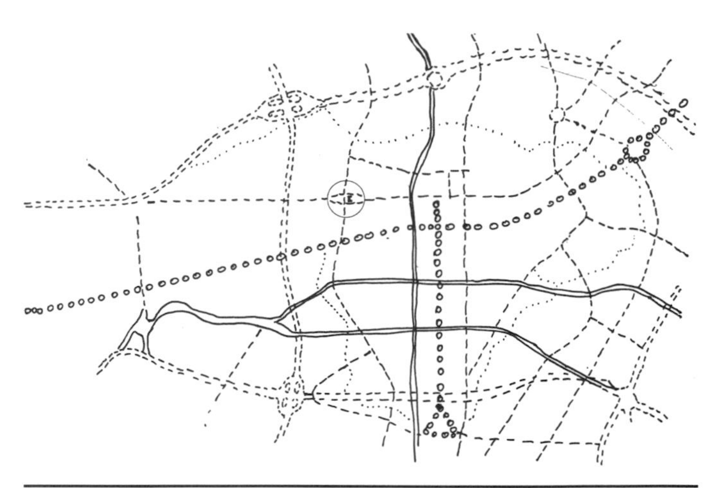

Fig. 3b. In order to perform this unifying function, all roads must be integrated into a system. Berlin Plan. (Smithson/Sigmond, 1958.)

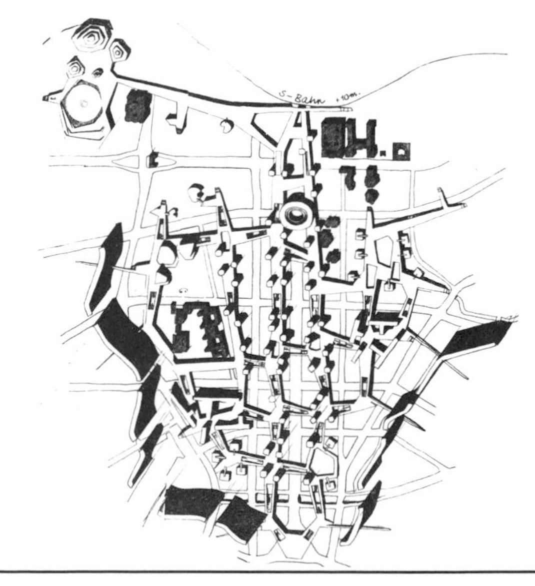

Fig. 3c. Definition of zones by superimposed movement nets, Berlin Plan Sketch. (Smithson/Sigmond, 1958.)

the higher levels of association, the street implies a physical contact community, the district an acquaintance community, and the city an intellectual contact community—a hierarchy of human associations.

In general, those town-building techniques that can make the community more comprehensible are:

1. To develop the road and communication systems as the urban infrastructure (motorways as a unifying force) and to realize the implication of flow and movement in the architecture itself.

2. To accept the dispersal implied in the concept of mobility and to rethink accepted density patterns and location of functions in relation to the new means of communication.

3. To understand and use the possibilities offered by a "throw-away"technology, to create a new sort of environment with different cycles of change for different functions.

4. To develop an aesthetic appropriate to mechanized building techniques and scale of operation.

5. To overcome the "cultural obsolescence" of most mass housing by finding solutions which project a genuinely twentieth-century technology image of the dwelling—comfortable, safe and not feudal.

6. To establish conditions not detrimental to mental health and well being. Past legislation and layout were geared to increasing standards of hygiene; in countries of higher standards of living this is no longer a problem. Criteria (for housing, etc.) have to be found to define the underlying environment. These might be: noise level, polluting and polluted environment, overcrowding, pressing and pushing, no space for the social gesture, all those demands made on the individual in societies inhabiting accumulated built forms.

The studies of association and identity led to the development of systems of linked building complexes which were intended to correspond more closely to the network of social relationships, as they now exist, than to the existing patterns of finite spaces and self-contained buildings. These freer systems are more capable of change, and, particularly in new communities, of mutating in scale and intention as they go along.

It was realized that the essential error of the English New Towns was that they were too rigidly conceived. In1956 we put forward an alternative system in which the "infrastructure" (roads and service) was the only fixed thing. The road system was devised to be simple and to give equal ease of access to all parts. This theme of the road system as the basis of the community structure was further explored in the Cluster City idea between 1957 and 1959, in the Haupstadt Berlin Plan 1958, and in the London Roads Study 1959 (Figs. 2 and 3).

Roads can be deliberately routed and the land beside them neutralized so that they become obviously fixed things (that is, changing on a long cycle). The routing of individual sections over rivers, through parks, or in relation to historic buildings or zones, provides a series of fixes or local identity points. The road net itself defines the zones identified by these fixes. Urban motorways thus designed form the structure of the community. In order to work they must be based on equal distribution of traffic loads over a comprehensive net, and this system is by its nature apparent all over the community, giving a sense of connectedness and potential release (Fig. 4).

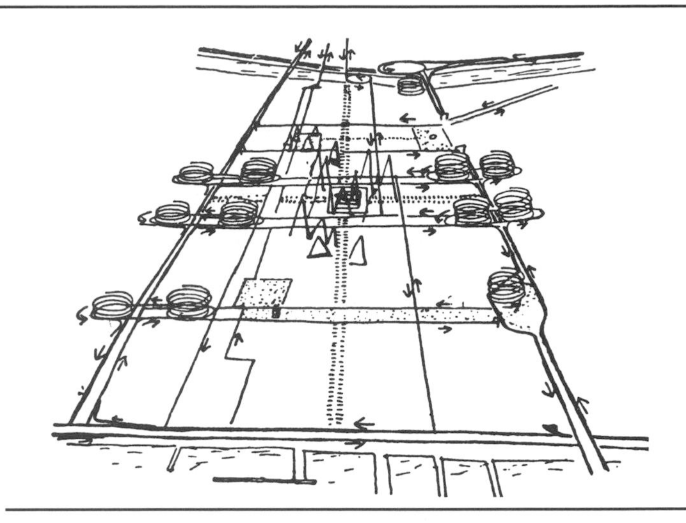

Fig. 4a. "To realize the implications of flow and movement in the architecture itself." (Louis Kahn, *Perspecta Yale Architectural Journal No. 2, 1953.)*

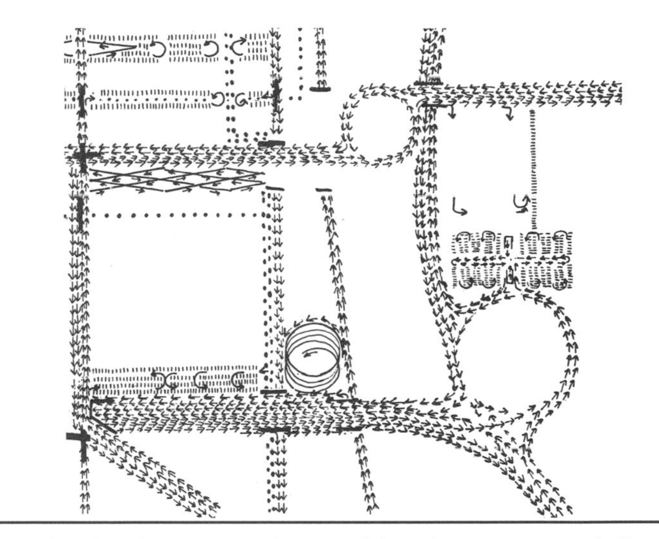

Fig. 4b. "The architect can control systems of physical communication and offer new concepts." (Louis Kahn, *Perspecta Yale Architectural Journal No. 2, 1953.)*

Fig. 4c. "A new sort of environment with different functions." (Louis Kahn, 1953.)

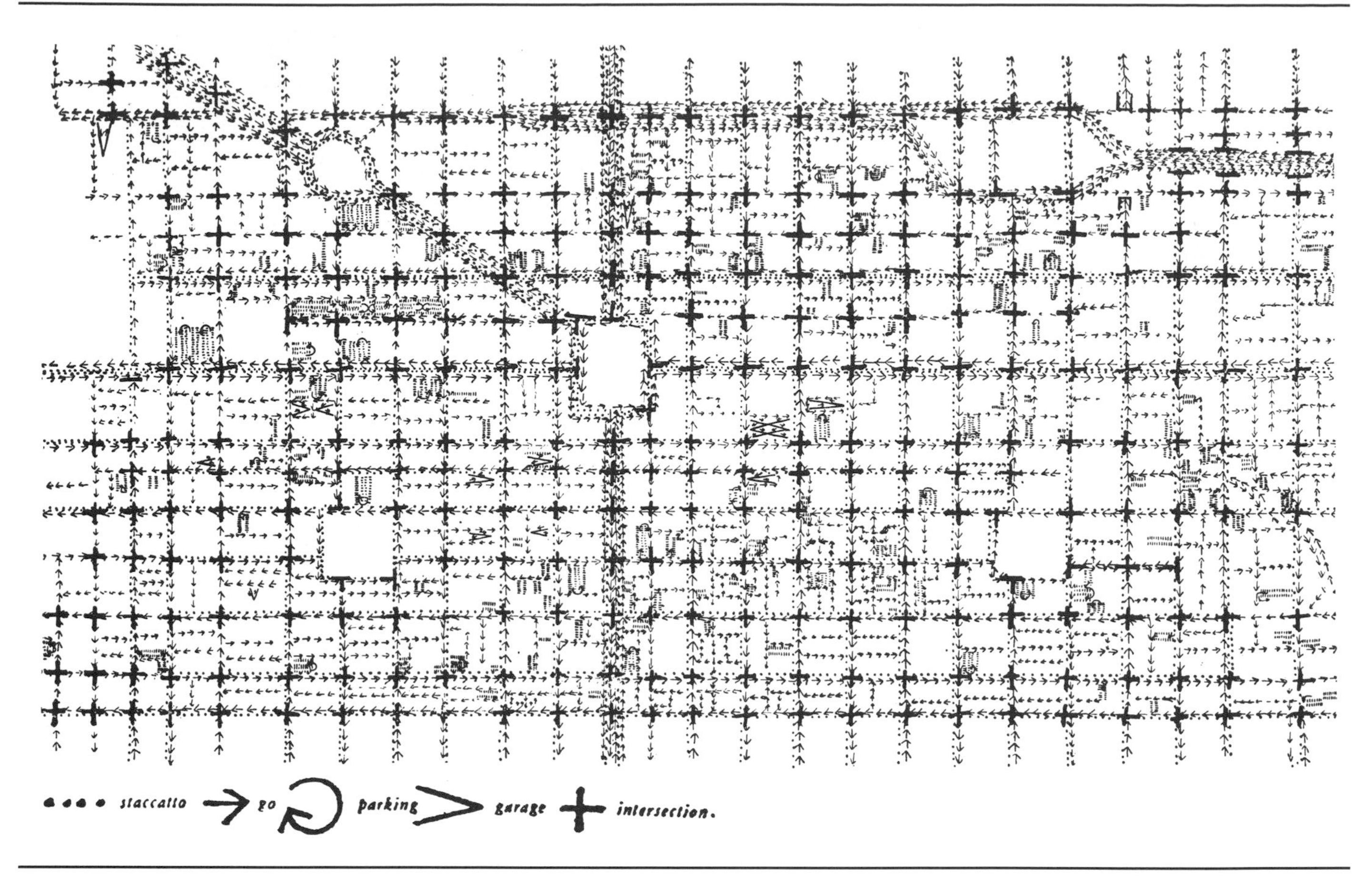

Fig. 5. Ideogram of existing traffic movement pattern, Philadelphia study. On the same streets trolleys, buses, trucks and cars with varying speeds, purposes and destinations. (Louis Kahn, *Perspecta Yale Architectural Journal No. 2, 1953.)*

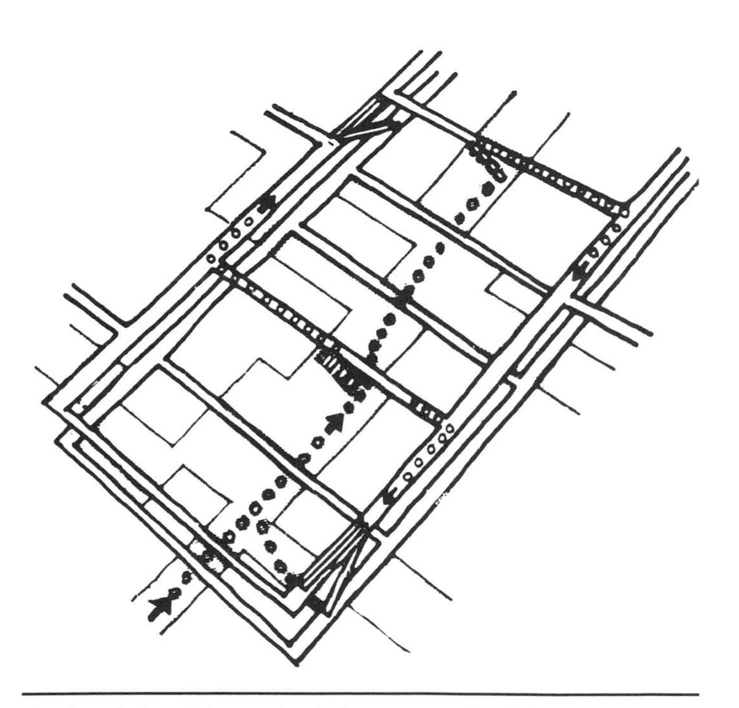

Fig. 6. Berlin Free University sketch showing types of multi-level and directional circulation built into the complex. Dotted track shows pedestrian free circulation and indicates random nature of pedestrian movement across open courts between teaching blocks. Direct pedestrian movement between faculties is via escalators. (Shadrach Woods, 1964.)

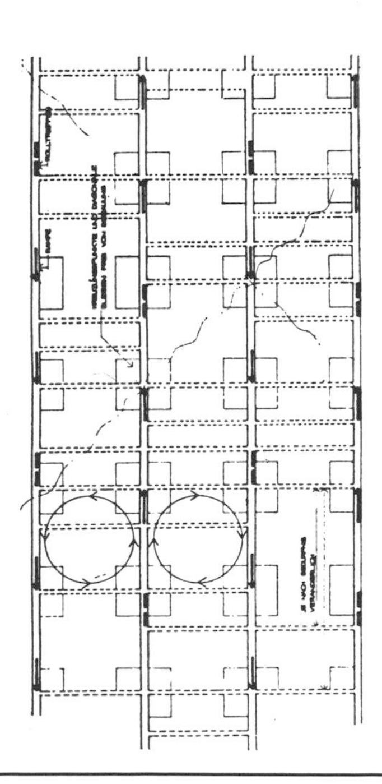

Fig. 7. Diagram showing the systems of orthogonal (grid-iron) and diagonal random cross circulation of Free University. (Shadrach Woods, 1964.)

Mobility has become the characteristic of our period. Social and physical mobility, the feeling of a certain sort of freedom, is one of the things that keep our society together, and the symbol of this freedom is the individually owned motorcar. Mobility is the key both socially and organizationally to town planning, for mobility is not only concerned with roads, but also with the whole concept of a mobile, fragmented, community. The roads (together with the main power lines and drains) form the essential physical infrastructure of the community. The most important thing about roads is that they are physically big, and have the same power as any big topographical feature, such as a hill or a river, to create geographical, and in consequence social, divisions.

We have to accept the dispersal implied in the concept of mobility and to rethink accepted density patterns and location of functions in relation to the new means of communication. In the dense, built-up areas of big cities the problems of movement are more complicated than those on out of town highways. Movement in cities must include the functions of parking and stopping. In general, national, inter-city, inter-sector and local (low-speed car, pedestrian) traffic should each have separate systems which offer no short cuts—all movement must proceed through each stage of the hierarchy—and the town building should respond to this hierarchy of movement. Louis Kahn's Plan for Midtown Philadelphia demonstrates how movement can be organically reorganized so that it is one with its inter-related functions of parking, and shopping (Fig. 5). The idea of relating the architecture to the type of movement can be shown most easily in examples uncomplicated by existing conditions, as in projects by Candilis, Josic, and Woods (Figs. 6–8).

The main case that is being presented here is that for a human agglomeration to be "a community" in the twentieth century, it is not necessary, practically or symbolically, for it to be a dense mass of buildings. But that does not necessarily mean that a bigger overall area need be covered than is covered at present. It is essentially a matter of regrouping of densities.

There is no need, for example, for low-density family houses to be excluded from the central areas of the city, nor is there any need to think conventionally that housing nearer the country should be at low densities. It depends on the life pattern of the people who live there what sort of environment is needed and what sort of density results. The overall pattern of the community is of clusters of varying densities, many parts with densities as high as 300 du per ac. (741 du per ha)—conceived as towers or "streets-in-the-air" (Fig. 9). Such concentrations will allow for the creation of the new road/green space system that compensates for family living development, without increasing the occupied area and without forcing people into an unwanted pattern. Dispersal of course must be disciplined so that any resultant development does not become absolutely structureless. Experience has shown that it is possible to maintain large-scale belts and parkway strips, and indeed the relating of the green areas to the major road system is an obvious way of providing the main urban structure.

It is the intention, by using the road system as the town structure, to keep the apparent level of mechanization under control. We are no longer in the position of needing to play up our devices, but rather to play them down, channeling mechanical noise and excitement and creating "pools of calm" for family living and regeneration.

Fig. 8. Competition project for Bochum University. Plan showing pedestrian routes and central public area. Dotted line shows monorail route running above spine of the whole university. (Candilis, Josic, Woods, 1962.)

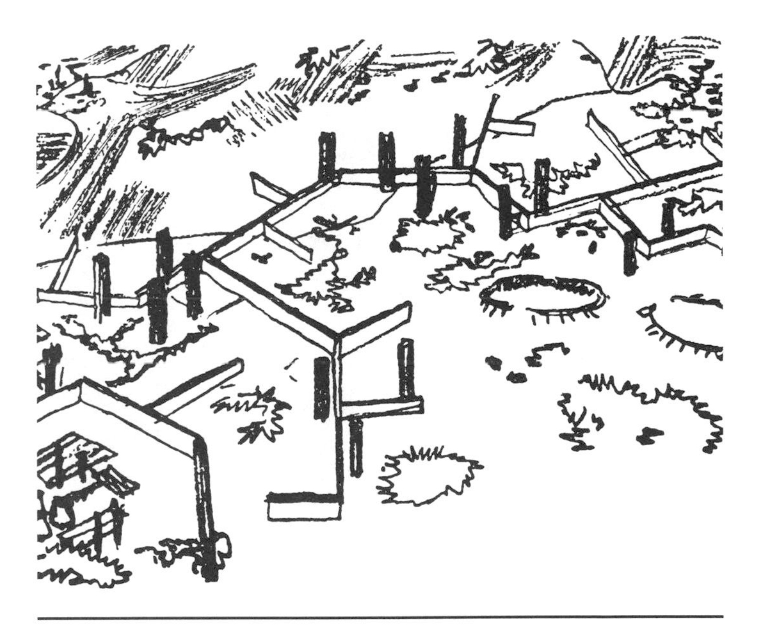

Fig. 9. Diagram, Greenspace compliments streets-in-the-air pattern for dwellings. (Alison Smithson, 1952.)

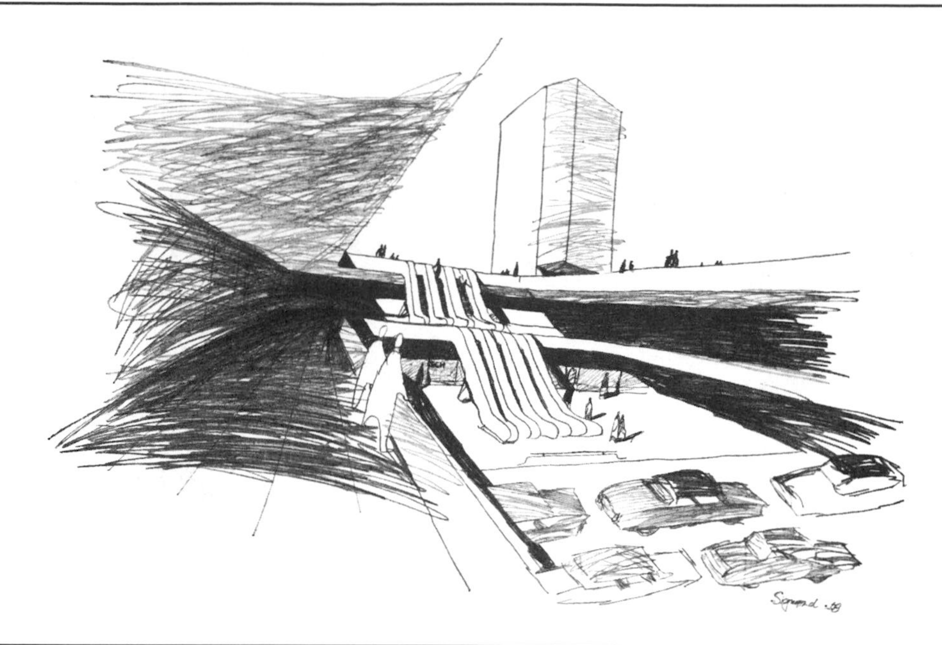

Fig. 10. Flow and speed. Berlin plan sketch. (Smithson/Sigmond, 1958.)

Density and intensity will be related to the function of the zone, resulting in communities of much greater genuine variation of living pattern. At some point there must be a place of maximum intensity. And somewhere there must be a place that not only allows for the contact of mind with mind, but also symbolizes it. It can only happen at the "center" (there can be only one place where the experience of the community reaches its maximum, if there were two, there would be two communities), but it follows that in a dispersed community it will be smaller and more intense than existing "city centers."

But it is in the question of social focus points that the difference between the Cluster City idea and what it is commonly compared with, Los Angeles, can be seen. Los Angeles is fine in many respects, but it lacks legibility—that factor which ultimately involves identity and the whole business of the city as an extension of oneself, and the necessity for comprehension of this extension. The layout of Los Angeles and the form of its buildings do not indicate places-to-stop-and-do-things-in. What form it has is entirely in its movement pattern, which is virtually an end in itself.

It is quite clear that in an ideal city at the present time, the communication set should serve (and indicate) places-to-stop-and-do-things-in. This is somewhat different from saying that every city needs a core. When Los Angeles is criticized for not being a city in the old European sense, it is not generally realized the colossal scatter of the places people go to; up to the mountains for a picnic; to the desert for a trip; to a far-off beach for a bathing party; or to Marshall Fields (in Chicago) for shopping. The social focus points are almost all outside the so-called downtown, and they are not really based on the sort of facility that can be readily molded to help the legibility of the town pattern.[2]

Just as our mental process needs fixed points (fixed in the sense that they are changing over a relatively long period) to enable it to classify and value transient information and thus remain clear and sane, so the city needs "fixes"—identifying points which have a long cycle of change by means of which things changing on a shorter cycle can be valued and identified. With a few fixed and clear things, the transient—housing, drug stores, advertising, sky signs, shops and at shortest cycle of all, of course, people and their extensions, clothes, cars and so on—are no longer a menace to sanity and sense of structure, but can uninhibitedly reflect short-term mood and need. If this distinction between the changing and the fixed were observed there would be less need for elaborate control and legislative energy could be concentrated on the long-term structure.

The establishment of an aesthetic of change (or transience) is in fact almost as important to the feeling for the structure as the maintenance of the inviolability of the road system (Fig. 10). An "aesthetic of change," paradoxically, generates a feeling of security and stability because of our ability to recognize the pattern of related cycles.[3]

This generation of architects must switch their focus to the problem of making the community structure more comprehensible; and this is not only a matter of "city planning" but must alter the nature of architecture itself—as least as far as the nature of architecture has been understood since the Renaissance. The time has come, we believe, to approach architecture urbanistically and urbanism architecturally. ■

REFERENCES

1. Peter Smithson, "London Road Study," *AD (Architectural Design)*, May 1960.

2. Alison and Peter Smithson, "Scatter," *AD (Architectural Design)*, April 1959.

3. Alison and Peter Smithson, *AR (Architectural Review)*, December 1960.

Gabriel Feld

Summary
The name of Shadrach Woods is associated with Team 10—the legendary group of young architects that redirected the work of the modern masters after World War II—and with the firm Candilis-Josic-Woods, best known for *Toulouse Le-Mirail* and other large housing and urban schemes from the fifties and sixties. Behind these few references is a remarkable set of projects and ideas that represent a major contribution to architecture, urbanism, and planning.

Key words
circulation, construction, grid, housing pedestrian, perimeter block, prefabrication, "stem," street, Team 10, urbanity

Shadrach Woods 1923–1973

Shadrach Woods and the architecture of everyday urbanism

Wood's career spanned twenty-five years, from his arrival at Le Corbusier's office in 1948 to his untimely death in 1973 at the age of 50. A turning point in the direction of architecture and urbanism, this period began with the undisputed triumph of modern architecture in the European reconstruction that followed the war, to finish in a vigorous process of critique and reassessment of the legacy of modernity.

Throughout those years, the work of Shadrach Woods focused on a few precise architectural and urban ideas, developing from project to project a set of principles of organization—what he often called "a search for systems." His ideas addressed the larger ideological issues involved in design of the physical environment, from the consequences of population growth, land use, transportation policies, allocation of natural resources and energy, to the underlying social circumstances resulting from inequities in the distribution of wealth.

From Le Corbusier's Office to Candilis-Josic-Woods

Shadrach Woods was born in Yonkers, New York, in 1923 into a family of Irish descent. He studied engineering briefly at New York University. During the war he was in the U.S. Navy, assigned to electronics and electronic maintenance. Upon the end of the war, he moved to Ireland where he studied literature and philosophy at Trinity College in Dublin. In 1948, stimulated by the challenges of the European reconstruction, he turned his attention to architecture. Too impatient to embark on a long course of formal education, he managed to land a job with Le Corbusier, reportedly begun with a discussion of their mutual interest in the literature of James Joyce. He spent a short time on the project for the Duval Factory at Saint-Dié and then moved on to the *Unité d'Habitation*, working first on the drawings in Paris and later in its construction supervision at Marseilles.

This initial encounter with architecture defined many of the concerns that would follow Woods for the rest of his career: a belief in the connection between architecture and social problems, a preference for large projects, particularly housing, an obsessive search for reductive diagrams and an emphasis on the close relation between form and construction. For the project at Marseilles he worked with two individuals who were to play an important role in his future work, Georges Candilis, the Greek architect who would become his partner, and Jean Prouvé, the engineer of his characteristic prefabrication solutions.

Upon completion of the *Unité*, Candilis and Woods moved to Morocco where they worked as part of *ATBAT-Afrique*. Le Corbusier along with some of his close collaborators founded *ATBAT (Atelier des Bâtisseurs*—"builders studio") after the war as a center for interdisciplinary research among architects, engineers, and planners. A major figure in this venture was Vladimir Bodianski, the Russian engineer who had been the team leader in the construction of the *Unité*.

With the arrival of Candilis and Woods, *ATBAT-Afrique* took on a number of projects, mostly low-cost housing schemes for the expanding urban populations. In these circumstances, Candilis and Woods found a fertile overlap between the experience at Marseilles and their observations of the North African traditional city, with its densely interwoven fabric of dwellings, courts and streets. The *ATBAT* projects

Credits: Research for this article was supported in part by a grant from the Graham Foundation for Advanced Studies in the Fine Arts. The author gratefully acknowledges review comments provided by Myles Weintraub and by Waltraude S. Woods along with original material from the Shadrach Woods archive.

Fig. 1. *Nil d'Abeilles* housing type, Casablanca. (*ATBAT-Afrique*,1953.)

Fig. 2. *Semiramis* housing type, Casablanca, (*ATBAT-Afrique*, 1953.)

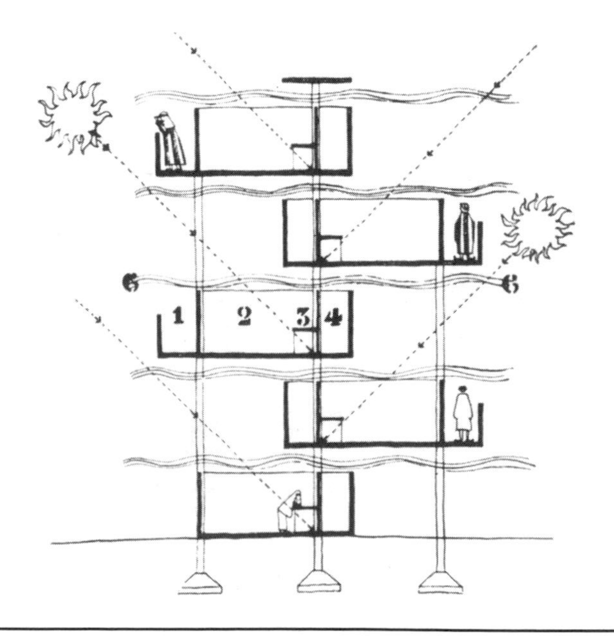

Fig. 3. *Semiramis* housing type. Diagrammatic cross-section.

were characterized by the regular geometry of their reinforced-concrete structure and an intricate articulation of rooms, voids and corridors. In less than five years, Candilis and Woods developed a wealth of housing types in response to specific site conditions and, in particular, to the habits of the different populations in the region. Some of the types were conventional exercises in the rationalization of plumbing cores and circulation; others experimented with vertical and horizontal aggregations of small units organized around exterior courtyards.

A 1953 project for the *Carriéres Centrales* section of Casablanca—100 dwelling units within a larger development under the direction of the French urbanist Michel Ecochard—included two of the most distinctive types: the south-facing *Nid d'Abeilles* or beehive, and the east-west facing *Semiramis*, both five-story blocks with dwelling units comprised of two rooms and an enclosed outdoor patio.

In the *Nid d'Abeilles* type, the rooms are between the corridor decks facing north and the patios facing south. These patios are displaced on alternate floors allowing light deep into the building through the resulting double height. The south elevation of the block is thus characterized by the texture of the individual outdoor spaces producing a vertical, highly geometricized version of the pattern of streets and passageways typically found in Arab cities; by contrast the north elevation appears almost like a veil draped over the block with long ribbon slits (Fig. 1).

The *Semiramis* type, perhaps the most radical of the experiments, placed the access galleries at the outside edges of both facades, the enclosed courts behind them, and the rooms at the center of the block. The corridors and patios alternate their orientation on every other floor, again allowing for light deep into the building and for cross ventilation. In this section, the access galleries and the patios appear as if they were weaving freely through the regularity of the reinforced-concrete columns; behind the bands of circulation, the enclosed rooms are organized in tower-like masses, a configuration highly reminiscent of the North-African casbah. With the corridors on the outside and the rooms deep inside the block, these buildings literally turned the section of the *Unité* inside out (Figs. 2 and 3).

The permutations of the North-African projects can be seen, variously, as responses to patterns of inhabitation, the climate of the region or the technology of reinforced concrete. As a body of work, it provided a formative experience, as much for the development of strategies of spatial or constructive organization, as for the attitude of an architecture tied to established social and cultural circumstances, anticipating many of the projects and realizations that were to come.

By the mid-fifties, working conditions may have deteriorated in the French colonies of North Africa while France was at the same time embarking on ambitious plans of government-subsidized housing construction. Candilis and Woods, now with the addition of the Yugoslav émigré Alexis Josic, seized the opportunities of a series of housing competitions and by 1955 had established a thriving practice in Paris. The projects of the firm drew from their *ATBAT* experience in the development of low-cost housing, articulation of spaces and enclosures, adaptability to changes in use, and rationalization of construction.

The first commissions of the period were part of "*l'opération million,*" a government program based on the standard cost of one million francs for a three-room dwelling unit. This was a program that gave municipalities who had been awarded government funding the choice to select among apartment plans submitted by several prized architects. Initial realizations included as little as forty or fifty units, but later ones were for developments of over three to four hundred units, organized in units of various heights. The massing of these projects betrayed a modern pedigree of diagrammatic dwelling masses dating back to Le Corbusier's *block à redents* of the early twenties. But unlike their precedent, these made a deliberate effort to define the space between masses and to recover the ground plane as the realm for social activity. The center of gravity for collective interaction went down from the high-rise to the ground. At the scale of the dwelling unit, these projects reformulated the spatial richness and flexibility of use from the North-African units and now addressed the more demanding expectations of the European context. The articulation of plumbing cores, bearing walls or frames, partitions, enclosures and fenestration became the subject of careful research; together with the development of access systems—corridors, stairs and elevators—they constituted a repertoire of architectural elements and solutions for a multitude of projects (Fig. 4).

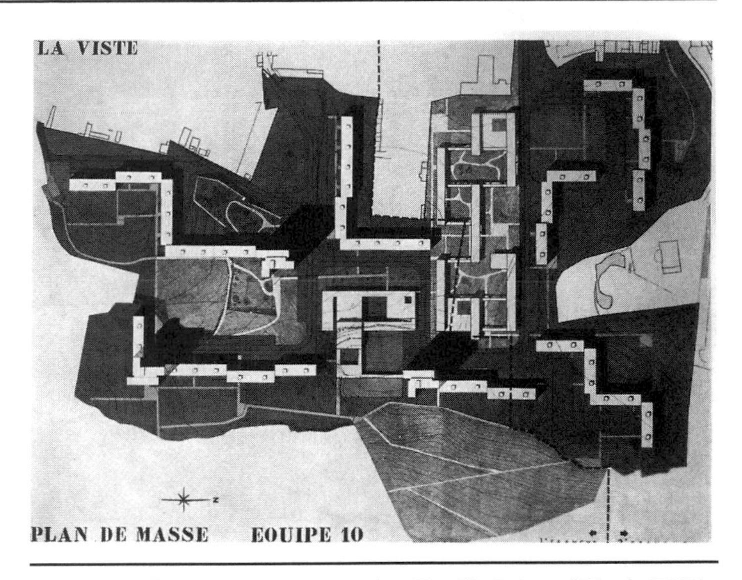

Fig. 4. Marseilles La Viste, competition plan. (Candilis, Josic, and Woods,1959.)

During this period in Paris, the firm continued more directly the line of exploration begun in North Africa, particularly the schemes of horizontal aggregation with central cores and internal open spaces. They were developed for Muslim countries such as Algeria and Iran, but some of the ideas were also extrapolated to other proposals for tropical environments. Both in the European and in the tropical projects, the role of prefabrication—either in concrete or steel—began to play a key role in the definition of architectural form. The signature "clip-on" panels with rounded-corner openings at Blanc Mesnil for example, were an early collaboration with Jean Prouvé (Fig. 5).

Fig. 5. Blanc Mesnil, elevation with "clip-on" panels by Jean Prouvé, 1956.

While housing represented the bulk of the commissions during the late fifties and was the assignment of their undisputed expertise, the firm also engaged projects involving other programs, including ancillary facilities for the housing projects, commercial centers, schools, and hotels.

Beyond the visual group: stem and web

By the end of the fifties, Candilis-Josic-Woods had developed an extensive experience of large-scale housing projects and urban planning. In the May 1960 issue of *Architectural Design* documenting the work of Team 10, the firm was represented by a large housing project for the French Atomic Energy Commission at Bagnols sur Cèze. Begun in 1956, this project of approximately 2,000 dwelling units included a variety of housing blocks and towers articulating open spaces of different scales and proportions. The Bagnols project capitalized on almost a decade of housing experience in which the architects had refined the planning of dwelling units around efficient plumbing and circulation cores and the aggregation of larger masses into characteristically modern ensembles. The Bagnols project also displayed a catalogue of plastic experiments in the prefabrication of facades, panels, and openings.

Within the context of the profusely illustrated projects and articles in the *Architectural Design* issue, Shadrach Woods introduced the idea of "*stem*" in a single, unassuming page. He presented the urban paradigm of a linear center of activities as an alternative to the urban strategies inherited from CIAM. He attacked the notion of urbanism based on the aggregation housing units into larger massing composed through visual or plastic notions, what he called with disdain "*le plan masse*." For Woods, this combination of "plastic art and plumbing in the search for self-expression," produced fixed structures, unable to

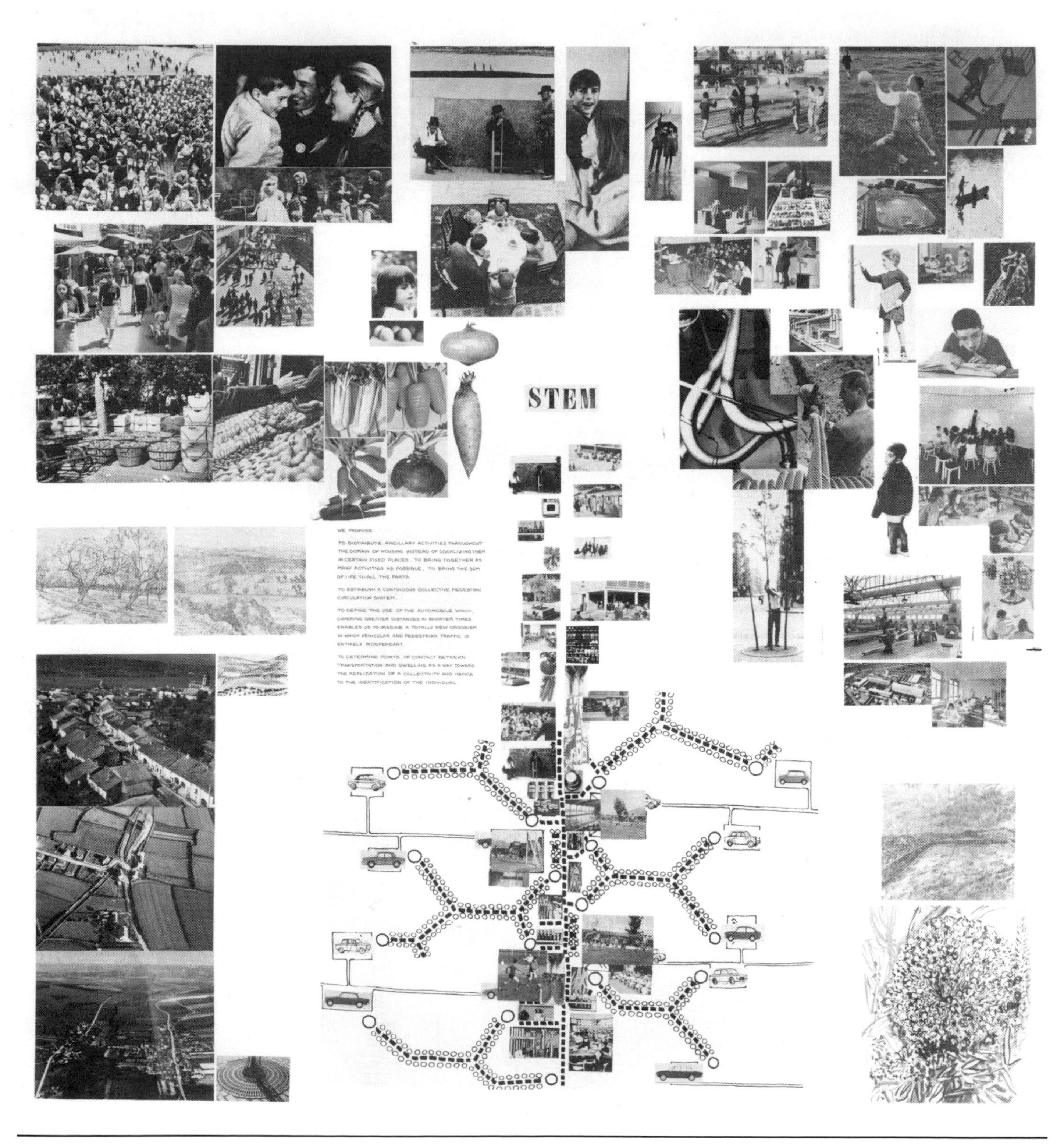

Fig. 6. Photomontage: the idea of "stem."

incorporate dynamic transformations, creating "an endless variation of geometric arrangements to make an endless series of virtually identical housing schemes, from Stockholm to Algiers and from Moscow to London."

In contrast to a form of urban organization proceeding from the dwelling unit to the arrangement of larger masses, Woods reversed the process of design starting with a basic structure containing all the collective programs of housing—commercial, cultural, educational, and recreational—together with circulation and services. He conceived a structuring device, the *stem*, as the place of social interaction. It included the collective facilities, the "*prolongements du logis*" or ancillaries, distributed throughout the housing rather than localized in the focal position established by the canon of Corbusier's *Ville Radieuse*. The density of housing became a function of the intensity of activity along the stem, with clusters of residential blocks attached to the stem at points. The structure was simple, a line without dimension that could respond to growth and change while addressing specific site conditions. This line was reserved for the pedestrian, with the vehicular circulation relegated to the periphery and serving the stem only at points. Woods expressed a clear bias toward the pedestrian, the active, engaged individual, the "Man in the Street" of his posthumous book. The stem was conceived as the domain of the pedestrian, "a street, not a road."

To reinforce a non-formal definition of the plan, the text dismissed dimensional characteristics in favor of time and speed: the 25-year cycle of the dwelling unit and the 5-year cycle of services (measures of validity), the 2 1/2 mi. (4 km) per hour of the pedestrian and the 60 mi. (97 km) per hour of the automobile. In a later photomontage, Woods moves further from formal definition, illustrating the idea of stem as an organic articulation of "the relationships between human activities" (Fig. 6).

Fig. 7. Caen-Hérrouville, model of general plan. Second stage. (Candilis, Josic, and Woods, 1961.)

This notion of organization was a clear response to the large urban complexes of modern pedigree built during the 1950s, even to Candilis-Josic-Woods's own earlier projects. Its aim was as clear as its choice of enemies: to recover "the street, which was destroyed by the combined assaults of the automobile and the *Charte d'Athénes*." The automobile had displaced the individual from center stage in the life of the city and the four functions of the Athens Charter had broken the conditions for interaction. This new notion of urban organization was an attempt to restore the lost urban intensity.

The 1960 text was a discussion of general principles with only two small cartoons contrasting the compositional arrangement of modern planning with the organization of the traditional city. A year later, the competition for Caen-Hérrouville, a housing project for 40,000 inhabitants, would give Woods the opportunity to test his new planning ideas (Fig. 7). The project was organized along two intersecting linear centers of activities, one running roughly north-south, and the other east-west. The pedestrian stems incorporated the various collective programs of the complex: commercial, cultural, educational, and leisure activities. The length of the stems was determined by a walking time of about twenty minutes. Their location and configuration followed the ridges in the topography of the site. Running at the highest level of the natural terrain, the pedestrian way was lined by retail stores and schools, rendered as diagrammatic low blocks or agglomerations of smaller constructions.

Behind these facilities, large recreational areas extended at the upper level with parking underneath. Structured parking served also the social centers, swimming pools, and other collective programs. Vehicular circulation ran at the lower level of the site and perpendicular to the pedestrian lines. The Caen-Hérrouville project organized the majority of the dwelling units into high-rise structures of 6, 10, and 14 stories, connecting the pedestrian paths at one vertical circulation node and placing the other vertical circulation nodes on the periphery where they were served by small open parking areas. The characteristic 120° angle configuration of the blocks responded to direct functional considerations: gradual change in corridor direction, reduction in views and shadows from one block to the other, and limit of access points to maintain open landscape. Low-rise housing clusters covered the edges of the site, removed from the pedestrian path and organized around parking lots and vehicular roads.

A series of small diagrams with hand-written notes—the expression of Wood's "word law and map law"—explained the generative arguments of the project tracing the path of decision making (Fig. 8). Here Woods deliberately attempted to demystify the design process through a sequence of simple incremental notions that informed the final configuration of the project. Caen-Hérrouville presented an open structure of organization that sustained social events and interaction, conceived without any preconceived formal idea, in Wood's words, "beyond the visual group." The linear center of activities is conceived as ". . .the exclusive domain of the pedestrian, without the car, it re-establishes the street: primordial and permanent function of urbanism." This distinction between pedestrian and vehicular realms can be traced to an earlier American precedent, the plan for Radburn by Clarence Stein and Henry Wright, while the configuration of

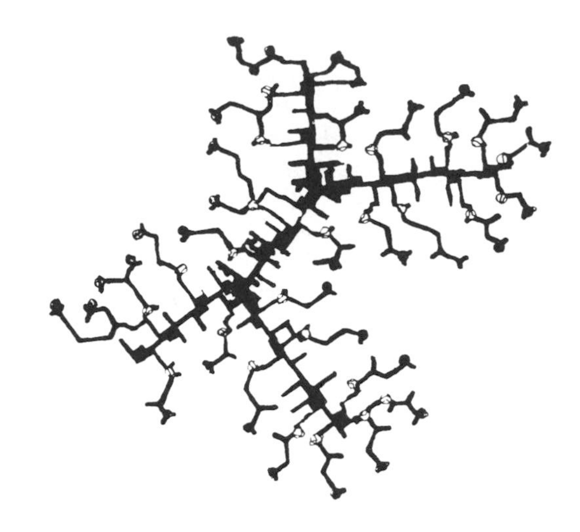

le centre linéaire est le domaine exclusif du piéton, il est desservi par la voiture et il rétablit la rue: fonction primordiale et permanente de l'urbanisme.

à partir des points d'arrêt des voitures l'homme dispose d'un réseau-piéton indépendant, soit au niveau du sol, soit aux différents niveaux desservis par les circulations verticales mécaniques -

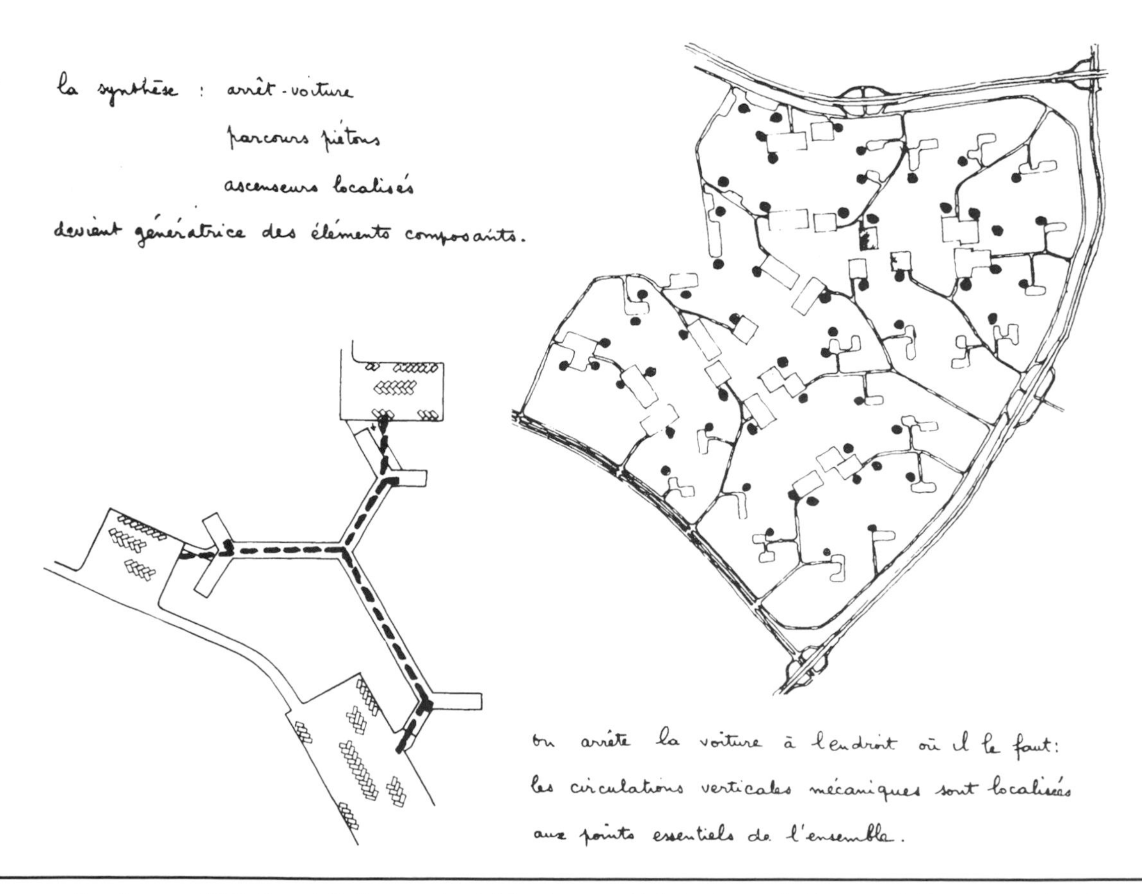

la synthèse : arrêt-voiture
parcours piétons
ascenseurs localisés
devient génératrice des éléments composants.
on arrête la voiture à l'endroit où il le faut:
les circulations verticales mécaniques sont localisées
aux points essentiels de l'ensemble.

Fig. 8. Caen-Hérrouville, generative diagrams, 1961. Upper left: Linear center. Building form would derive from pedestrian ways linking various points. Upper right: Principal built pedestrian network. Lower left: Principal pedestrian network. Lower right: Points of change from automobile to other circulation.

housing attached to services explicitly recognized a debt to Louis Kahn's notion of "servant" and "served" spaces.

The proposal for Caen-Hérrouville was submitted in April of 1961. It was selected for the second round of the competition and it was awarded third prize. In June of the same year Candilis-Josic-Woods submitted a proposal for *Toulouse le Mirail*, an ambitious competition launched by the socialist major of Toulouse. More than twice the size of Caen-Hérrouville, the Toulouse project was also organized along an open linear pedestrian center of activities. The proposal was selected for a second round. It was developed under the direction of Josic and was awarded first prize at the beginning of 1962. With this monumental commission, the idea of a linear center of activities, the concept of stem would finally find an opportunity for realization (Figs. 9 and 10).

Like Caen-Hérrouville, the *Toulouse le Mirail* project was termed a "Zone of Priority Urbanization" (ZUP), the largest of a series of *villes nouvelles* built in France after the war. The project was the result of decentralization policies based on the development of regional centers. It also responded to the influx of population from the former colonies in North Africa to the Midi region. The competition called for 23,000 dwelling units in an area of 2,000 acres, about the size of the historic core of the city. The first round of the competition received 150 entries of which the jury retained ten for the second round. Even a cursory look at the plans of the second round shows a marked distinction between the Candilis-Josic-Woods proposal and the other nine. A network of vehicular circulation dominates all the other plans. Some networks follow orthogonal patterns, others undulating gestures, a few lined by or enclosing large park areas. Some create superblocks with a variety of housing structures. Most display an identifiable center or concentration of activity. The winning scheme is the only one dominated by a pedestrian street, a *rue-centre*, with the collective equipment distributed throughout the project. Pedestrian and vehicular circulation routes are segregated, intersecting at points of high density of housing and collective facilities. Three-quarters of the dwelling units are grouped in high-rise blocks, following a characteristic hexagonal geometry and connected to the pedestrian deck. The remaining units are away from the spine, low-rise blocks and row houses served by a secondary vehicular network. At the edges, particularly to the south, the proposal included small factories and workshops.

At Caen-Hérrouville, the stem followed a natural ridge, establishing a sectional distinction between the upper pedestrian way and the lower vehicular movement. At Toulouse, the pedestrian street also followed a natural feature of the site, but in this case it was a low wetland that the project capitalized upon as a park running adjacent to the spine of the project. In order to establish a sectional distinction between the pedestrian and the automobile, the stem here was materialized as a series of large platforms with structured parking and vehicular movement below. The pedestrian deck of *Toulouse le Mirail* formalized the stem as a symbolic gesture of the project. Neither the contradiction of this monumentalization nor the spatial impact of the large platforms escaped Woods' critique: "[at] *Toulouse le Mirail*. . . the stem is a continuous platform over parking, an intimidating and expensive solution."

In the spring of 1962, while the office was frantically beginning to work on *Toulouse le Mirail*, Woods entered a competition for the Val d'Assua area of Bilbao, Spain. A housing competition for 85,000 inhabitants (about the scale of *Toulouse le Mirail*) Bilbao was organized along a series of smaller interconnected pedestrian strands that, like

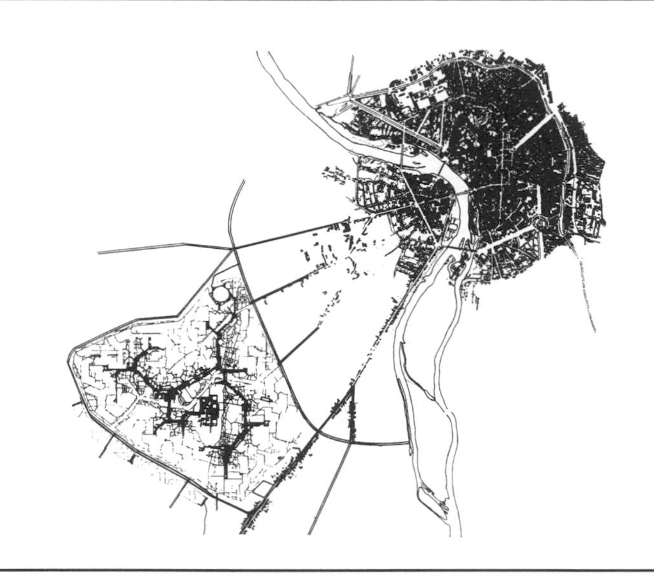

Fig. 9. *Toulouse le Mirail.* Plan in urban context. (Candilis, Josic, and Woods, 1961.)

Fig. 10. *Toulouse le Mirail*, aerial view.

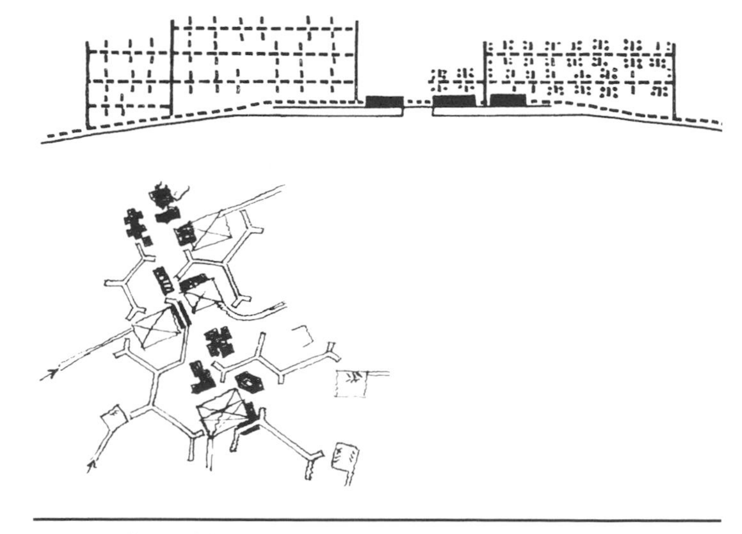

Fig. 11. Bilbao Val d'Assua Valley competition diagrammatic section and plan, 1962.

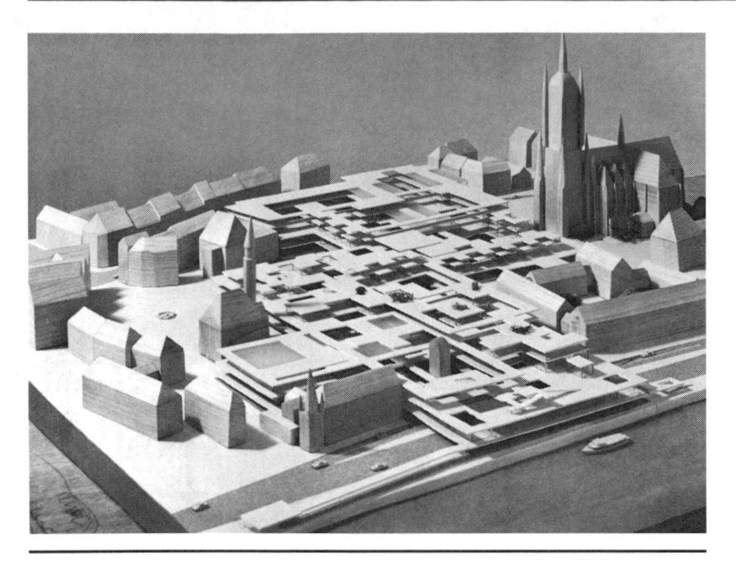

Fig. 12. Frankfurt Center competition model, 1963.

Fig. 13a. Frankfurt Center study model of right-of-way system.

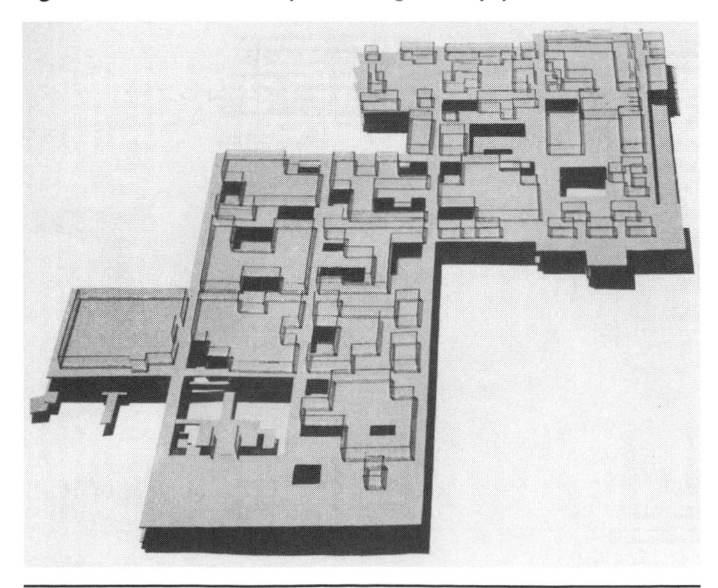

Fig. 13b. Frankfurt Center study model of second level.

at Caen, followed the ridges of the natural terrain (Fig. 11). This was an opportunity to develop the idea of stem in the most didactic and uncompromising manner. The pedestrian path again contained stores, schools, social centers. Reinforcing the integration of uses, each strand ended in light industry buildings. The canonical "separation of the four functions" of the Athens Charter had been completely dismantled. The automobile circulation maintained a secondary role and a third circulation was introduced: a *metro*, public transportation connecting the different pedestrian strands. The disposition of residential densities was reversed, with low-rise housing running parallel and adjacent to the stem and high-rise block extending farther back. This revision puts all the dwelling units in direct contact with the stem while displacing the need of parking to the periphery.

After the projects for Caen and Toulouse, the firm employed the idea of a linear organization device for diverse site conditions and programs: a mountain sky resort at Vallée de Belleville and a university at Bochum, Germany (*cf.* Smithson, Fig. 8, Article 3.4, this volume). By the end of 1962 Woods published a text titled "Web" [1] that expanded the notion of stem to that of a series of crossing strands forming a grid or web. He presented this notion as an extension of his linear structure of organization. The web may have been a direct response to his disappointment with Toulouse; against the broken geometries developed by Candilis and Josic, Woods proposed an uncompromising orthogonal geometry that relieved the plan from compositional gestures. It may have also responded to the limits of a linear organization that, in an extended field, left too many decisions to visual judgment.

In April 1963, a competition for the center of Frankfurt gave Woods the occasion to develop the idea. He proposed a regular matrix covering a large urban void, freely negotiating the edges in contact with the existing buildings. This multilevel grid provided an infrastructure of circulation and services where a variety of programs could find their place within the voids left by the structure. The grid also incorporates a few remaining historic icons such as the cathedral, as the only monuments in an otherwise non-descriptive building fabric. The result was a complex organism, different from, but comparable to the texture of the traditional city fabric, with an elaborate articulation of open spaces (Figs. 12 and 13).

This notion of "web" would eventually crystallize in the memorable project for the Berlin Free University. While Frankfurt is the most direct precedent for the Berlin competition, there was an earlier project that, at a much smaller scale, experimented with many of the spatial and constructive strategies that would inform the project for the university. In 1961, through the recommendation of Le Corbusier, Candilis-Josic-Woods received a commission for the extension of a convent at St. Julien L'Ars. Woods had always worked in vast projects; he took on this relatively small commission to test "on the ground" several of his larger urban ideas.

This project also offered Woods the opportunity to rekindle his professional collaboration with Jean Prouvé, the famous designer and builder of metal structures. Woods met Prouvé working at Marseilles. In the following years Woods turned to Prouvé for the design of housing facades, such as the characteristic metal panels with rounded corners at Blanc Mesnil. With the exception of the chapel, the entire enclosure of St. Julian was composed by a modular system of metal panels. This panel system designed by Prouvé defined and controlled the whole expression of the building. Woods even built a mock-up of the panel construction for the project on the terrace of his Paris apartment at *Rue Maturin Regnier*. The Notre Dame du Calvaire project can be seen as a "dress rehearsal" for the radical notion of the Berlin Free University: a building conceived as a matrix of organization, formally defined by its method of construction.

The competition for the extension of the Berlin Free University at the end of 1963 proposed a web as the organizational device for a large institution. But a dense urban area did not surround the site for the university in Dalhem. In fact, Dalhem was at that time one of a few remaining rural areas in the environs of Berlin. For the university, the amorphous network of Frankfurt is constrained to a clean rectangle. The grid or "*web*" is organized by four parallel main corridors or stems, 100 ft. (30.5 m) apart, connected by other, smaller transversal corridors. This simple framework receives a variety of spaces according to programmatic needs: auditoria, lecture halls and libraries located among the primary paths, while offices and seminar rooms were located along the secondary paths. They are configured in a localized way, generating an intricate pattern of open spaces. If the project attempts to recreate the richness of an urban fabric, the result was a large autonomous structure. The Free University has been criticized for its lack of connections to a surrounding urban context. After discussing the Frankfurt scheme as a milestone of the period, Kenneth Frampton argues that without the benefit of an established urban context, Berlin lost the conviction and intensity of events promised in the earlier project[2] (Figs. 14–17).

Again, the obvious reference to the intricate maze of Berlin is the Arab city, but there may have been two other historic models operating in Woods' mind. In *The Man in the Street*, he makes reference to the amphitheater turned square in Lucca and Diocletian's Palace in Split,[3] both large institutional structures that eventually gave way to city fabric. Woods is explicit about the relation between city and school:

> *The city itself, which is the natural habitat of Western man, is the school, college, university. We see the city as the total school, not the school as a "micro-community." Places of teaching and learning, when they can be identified as such, are an integral part of the structure of the city.*[4]

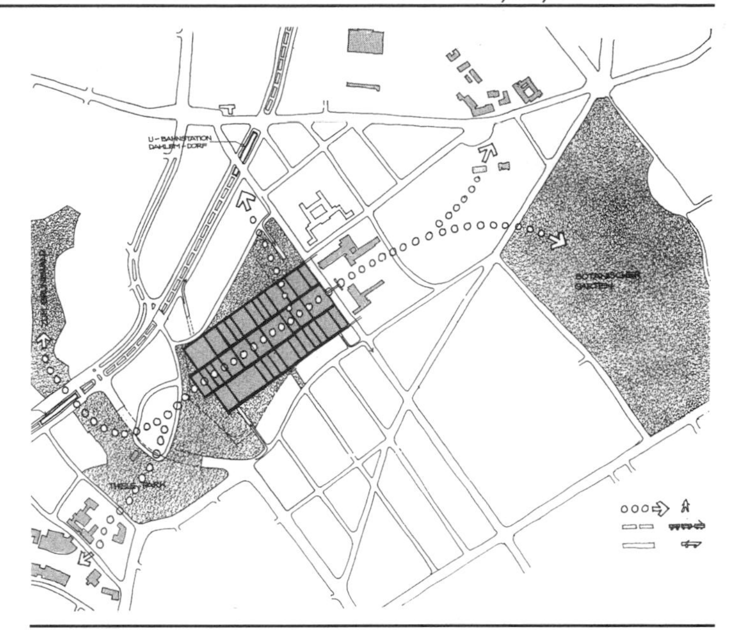

Fig. 14. Berlin Free University plan in urban context.

Fig. 15. Berlin Free University view of a courtyard.

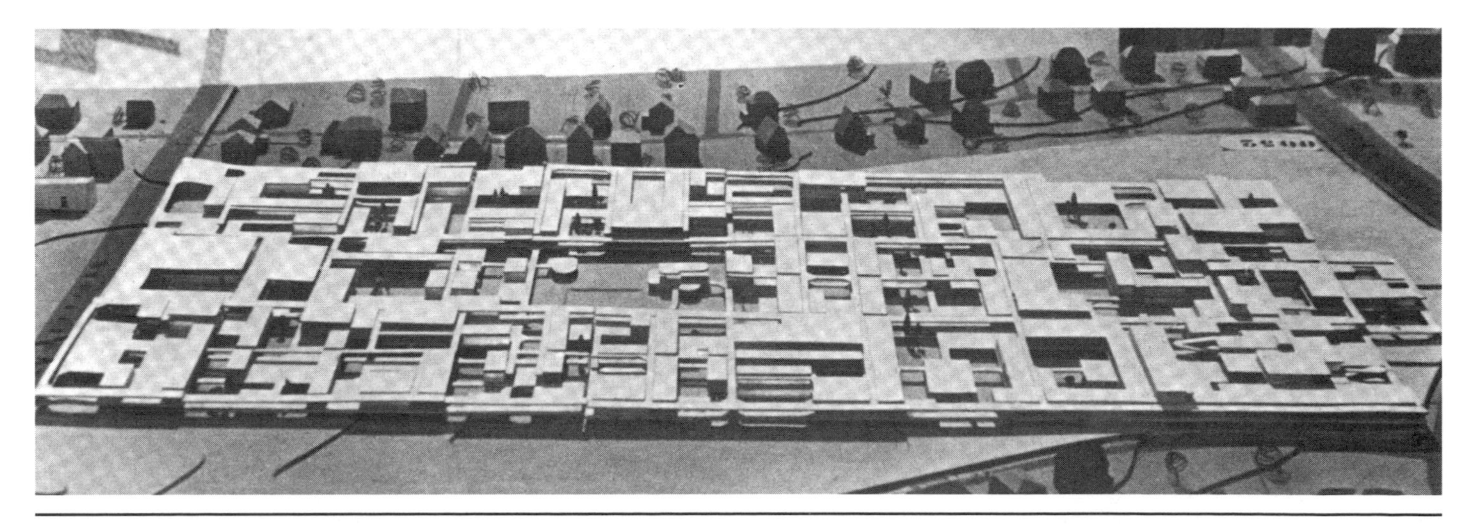

Fig. 16. Berlin Free University model. 1963.

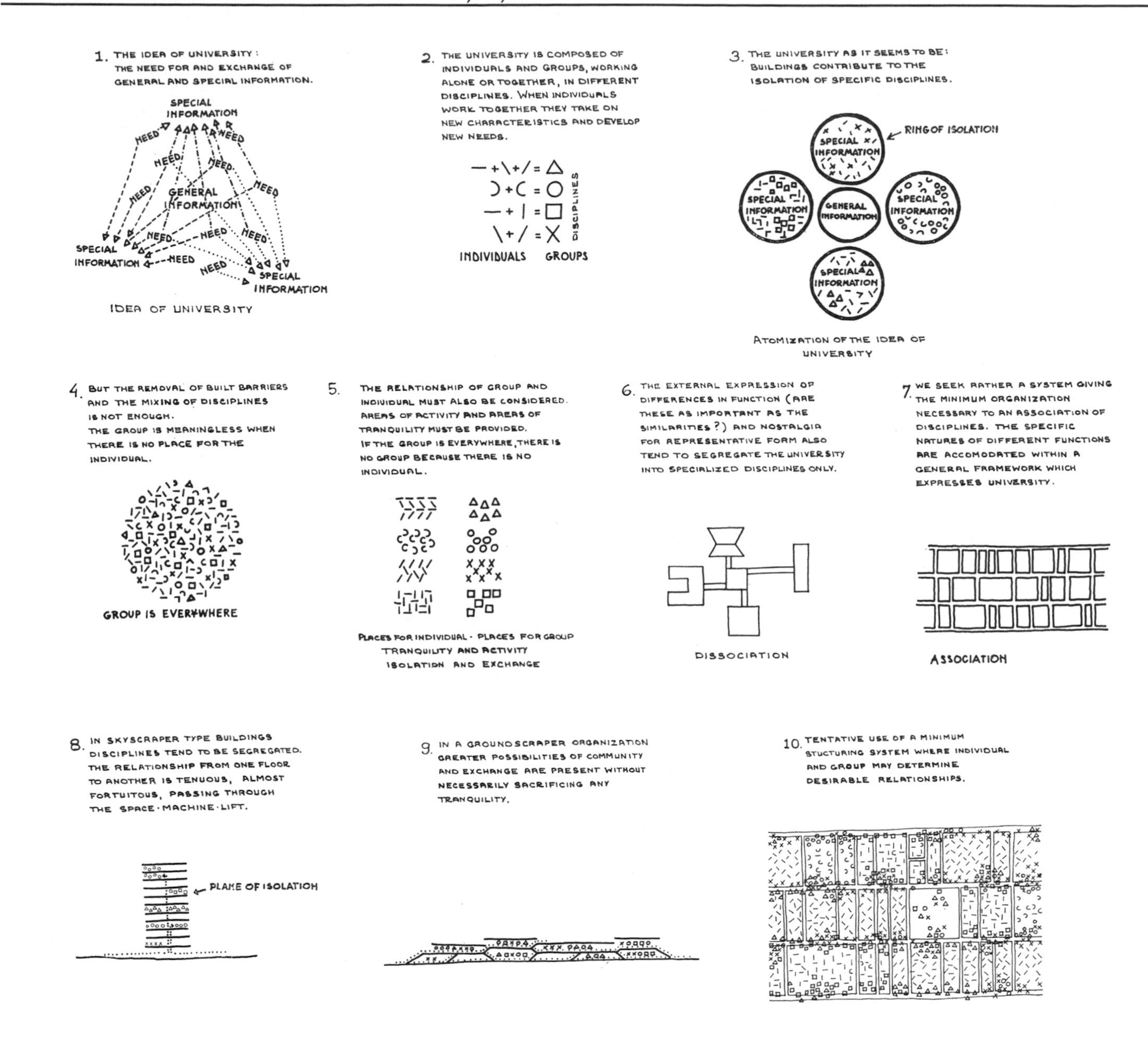

Fig. 17. Notes on the ideas underlying the concept of the Berlin and Dublin University proposals.

Such an ideological statement assumes the breaking of the boundaries between institution and everyday life of an open society. In a more direct way it anticipates movements such as the student revolts of the late 1960s. Perhaps the project for Berlin was conceived as a sort of "second foundation" for Dalhem. Over time, this structure would evolve into an elaborate urban fabric in which, as in the case of Diocletian's Palace, the rigid edges would dissolve. Undoubtedly, in the Berlin Free University we see Woods performing a balancing act at its riskiest. He could have simulated continuity or developed irregular edges—for example, by offsetting the grid half a module—but Woods is adamant in avoiding a formal gesture, quixotically waiting for time and the promise of a new society to take its course.

If the diagram is the generative force of the project, the choice of technology establishes the fundamental formal expression of the Berlin Free University. The construction system for the university reinforces the notion of latent transformation. In the earlier stages of design, Jean Prouvé developed a modular steel structure and system of clip-on weathering steel panels. The detailing of structure and panels was conceived so that the different building components could be disassembled and re-deployed to follow the evolving needs of the institution. Once again, Woods finds himself recreating the conditions of the historical city with the canonical means of modern architecture, prefabrication here attempting to replicate the transformations of the city over time.

Between Europe and America

The construction of the Berlin Free University would take ten years, from the 1963 competition to the opening of the first phase. By the mid-sixties, the Candilis-Josic-Woods team existed in name only. Josic established his own office not far from Paris. Candilis focused his practice on planning and design for leisure, undertakings in which he had been interested earlier and that in fact had created frictions with Woods. During the following decade, Shadrach Woods divided his time between Europe—mostly in the Berlin office that was developing the Free University under the direction of Manfred Schidhelm—and in the United States where he taught and lectured extensively. Woods compiled and wrote the book *Building for People*[5] that recorded the production of the firm. In his characteristic predilection for general ideas and principles, the content of the book is not organized by project or dates, but around issues—function, limits of space, space and massing, and public and private domains—illustrated by projects from their practice dating back to the *ATBAT* housing and culminating in the *stem* and *web* schemes, with more complete documentation of Frankfurt and Berlin.

In the Spring of 1962, Woods was invited to teach at Yale by Paul Rudolph, Chairman of the Department of Architecture at the time. At Yale, Woods made very strong connections with a group of graduating students—including Alex Cooper, Charles Gwathmey, Alex Tzonis, Myles Weintraub and Don Watson—that would eventually encourage him to come back to the U.S. The following year he was appointed professor at Harvard, where Jerzy Soltan, his friend from Le Corbusier's office and Team 10, was Chairman of the Architecture Department. Woods would continue to teach for the decade that followed, mostly at Harvard, but occasionally at other schools, including Washington University, Rice, and Cornell.

Now on his own, Shadrach Woods pursued a variety of projects from urban proposals to regional planning. In all these projects he developed the ideas of organization initially conceived in projects such as at Caen and Frankfurt, with increasing and more deliberate focus on the life of the street. He explored the notions of stem and web in a series of studies and competitions for the fabric of deteriorated by established urban cores. In proposals for Paris and New York, he deployed the characteristic 120° blocks mediated by linear spaces containing services at the lower levels. In Paris, the *Bonne Nouvelle* quarter—just north of *Les Halles*, a crowded neighborhood that had already been included in Le Corbusier's drastic *Plan Voisin* of the mid-twenties—is reconfigured as a pedestrian-oriented area organized through a matrix of multi-level glass-covered circulation in the tradition of Parisian arcades[6] (Figs. 18 and 19).

These arcades contain market-type shops and other educational, social and cultural facilities. They include an elevated mini-rail transit system and are connected to the metropolitan transit systems at points. The buildings above, rising to about ten stories, alternate office and residential levels in an intricate section of deep floor plates, maisonettes with double height terraces, external and internal corridors. The sections show the buildings described only schematically in their architecture (floor plates, corridors, connections to the *Metro*) but uses pictograms as symbols of people and their diverse activities. This project probably represents Wood's most deliberate effort to conceive a mixed-use urban environment. If the benefits of such environment are argued in terms of efficient use of public infrastructure, the overarching goal is "an effort to humanize the city." This humanization is predicated on the basis of opportunities for interaction, "Where people can at least see what others do, and ideally could meet and exchange views and opinions." Woods explicitly presents concrete forms of interaction such as those of children, parents and teachers; in his characteristic dry humor, he also includes the office workers

Fig. 18. Paris-Bonne Nouvelle. Plan in urban context, 1963.

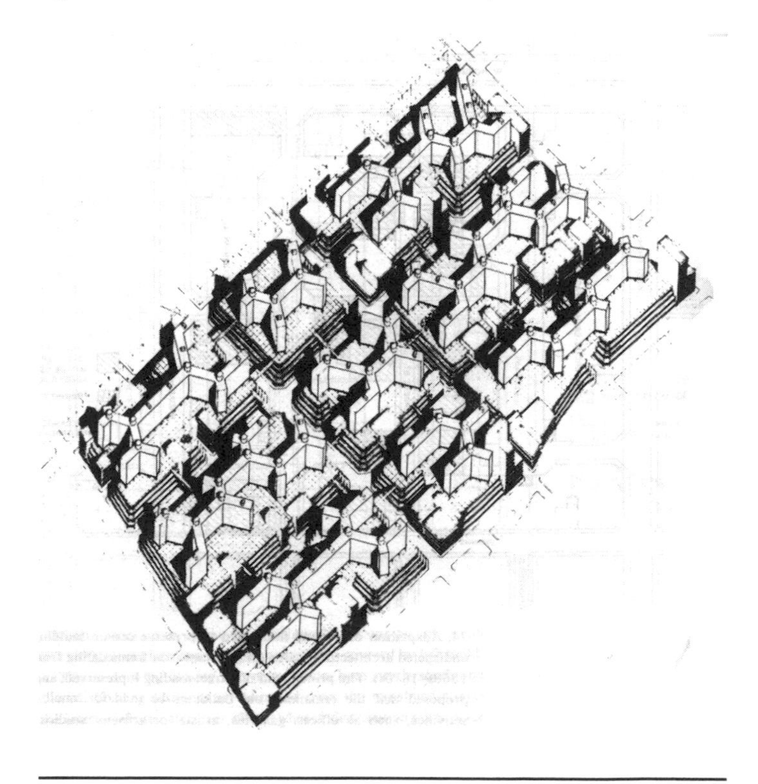

Fig. 19. Paris-Bonne Nouvelle. Bird's eye sketch, 1963.

Fig. 20. Douglas Circle proposal model. (Woods, Weintraub and Assoc. with Bond Ryder Associates, 1968.)

that should not "be exiled daily to the 9-to-5 mines, where nothing blooms but bureaucracy."

The *Bonne Nouvelle* proposal did not include parking since it deliberately rejected "private ownership of automobiles in a city well-endowed with public transportation systems." It included a secondary servicing road system. In an ensuing project for the SoHo district of Manhattan, the starting point is, in fact, the analysis of the service system. There is the same attention to detail of the people as in the *Bonne Nouvelle* schematic. The proposal reorganizes the street pattern closing half of the streets and creating center-of-the block loading areas with the remaining streets limited to moving vehicles and pedestrians. The blocks were proposed with large, loft-type bases incorporating existing cast-iron historical buildings with housing above and the extensive roof of the base conceived as a new "ground plane" with community services, schools and playgrounds. In general terms, both the SoHo and the *Bonne Nouvelle* projects deliberately challenge the modern assumption of segregation of uses with the most intense integration of diverse activities. Even if they betray a formal kinship with modern high-rise and super block patterns, the point of departure for both projects is not ideal forms of urban organization but existing conditions, traditional patterns and already emerging activities.

Shadrach Woods concludes his book *The Man in the Street* arguing that "the world is a city and urbanism is everybody's business." It would not be an exaggeration to say that for Woods the urban dimension was the measure of all things, both a way of understanding the physical environment at every scale and a mechanism to conceive social dynamics and political action. He was particularly concerned with the phenomenon of migration from rural areas to the city: as mechanized production takes over agriculture and mining operations, a series of physical and social dislocations. He cynically equates the "over-enrichment" of labor supply in the cities with the "over-enrichment" of chemicals in streams and rivers as the result of free market capitalism reduced to the goal of immediate profit. In his words, "the city grows, randomly, while the country, corporatized, grows barren of human habitat."

Urban and regional deterioration are thus linked. Two ensuing projects at regional scale gave Woods the opportunity to further develop issues such as urban space, density, private property of land, the role of public transportation and private cars, and zoning.

In 1965 the Paris region embarked upon a large study of future urban growth for an expected population of twelve million people. The areas surrounding the city were divided into four sectors and Candilis-Josic-Woods were assigned the study of the north area, south of Pontois. Woods' starting point for what amounted to a polemical counter-proposal is a conversation with the writer Roger Vaillard, in which they agreed that suburbs in the American fashion were a "*retour à la sauvagerie*"—reversion to the savage state—from the social environment of the city. They identified the private car as the enabling culprit for this process. This observation put the project in a collision course with the more expected planning mechanisms based on statistically defined populations and highway infrastructure. Instead, the proposal was based upon public transportation and pedestrian movement, organizing activities and exchange. Particularly radical in this proposal is a tentative arrangement based in only two "zones," one area of constructions and another free of constructions. The region was thus divided in equal zones of urban and rural stripes.

Two years later, the competition of ideas for the development of Bresse-Revermont, a region of France at the foothills of the Jura Mountains not far from Lyon, looks at the relation between urban and rural environments in an impoverished farmland region. The proposal advocates the stabilization of rural areas with viable agricultural economy and the regrouping of small populations into more viable communities with modern infrastructure. The goal of these ideas was to leverage the industrialization of agriculture and means of communication to "slow and eventually stop, the migration towards the cities."

As for many intellectuals at the time, 1968 was to be a watershed year for Shadrach Woods. Invited by fellow Team 10 member Giancarlo De Carlo, Woods designed and oversaw construction of a pavilion for the Milan *Triennale*. It was a two-story grid structure, based on the proportions of Le Corbusier's *modulor* and constructed of red painted steel. It displayed the stem and web concepts but also a series of large panels and photographs discussing general problems that challenged the design of the physical environment, from the degradation of the natural environment and the misallocation of energy resources to the inequities of wealth distribution. The *Triennale* became a casualty of the student revolts of the moment and the story goes that Woods invited a group of students to dismantle his pavilion.

Although Woods had been teaching regularly in American architecture schools since the early sixties, 1968 also marked his permanent return to the United States. Several of his former Yale students had administrative positions in the planning offices of New York City and encouraged him to move there. In 1970 Woods founded the office of Woods, Weintraub and Associates in partnership with his former student Myles Weintraub. However, at the time, only limited professional work came to the office and no built work resulted. In 1969 the Lindsay administration involved Woods as a consultant in the planning process of the controversial Lower Manhattan Expressway. Other projects included a variety of housing undertakings, most notably a project for Douglas Circle for the New York Housing Authority, designed in association with Bond Ryder Associates (Fig. 20).

Located across the northwest corner of Central Park, the project comprised approximately 780 dwelling units with a clear bias toward

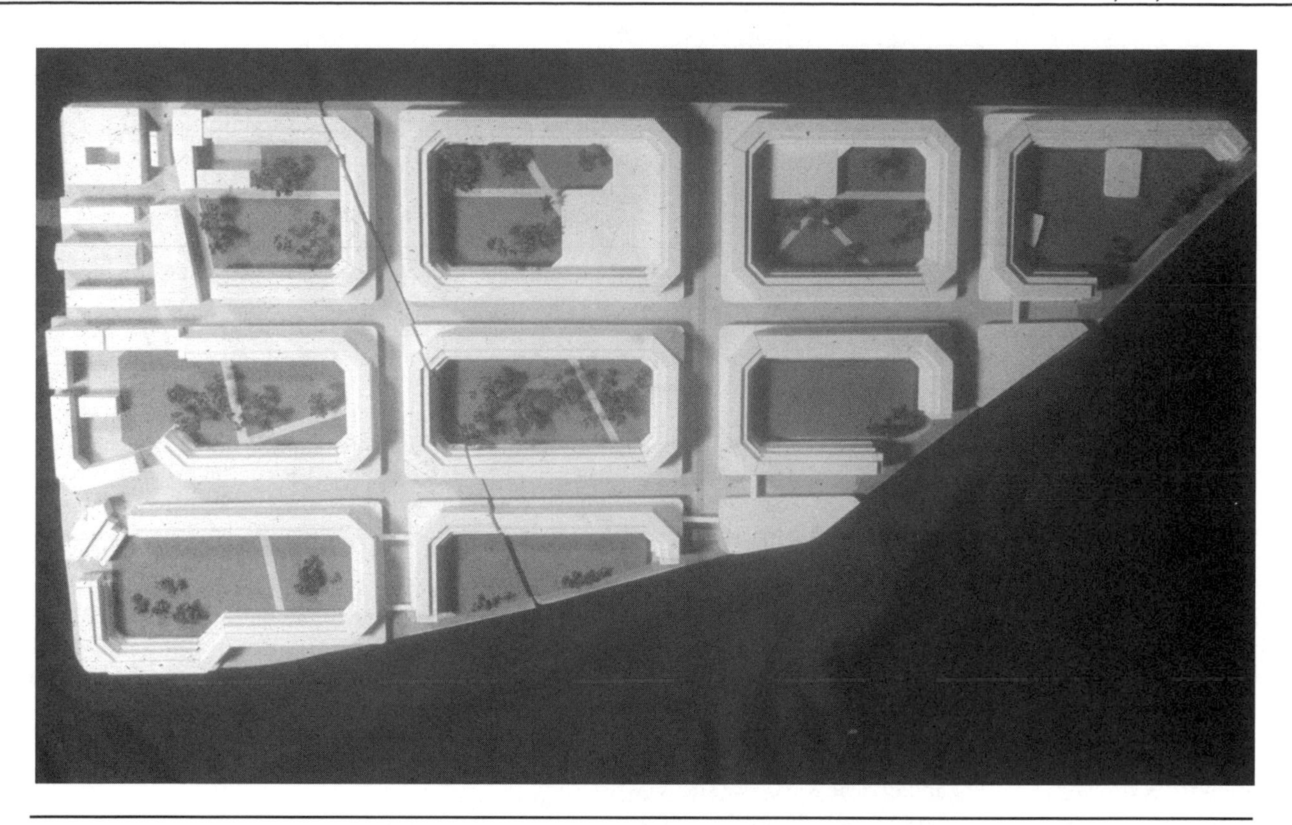

Fig. 21a. Karlsruhe competition model, 1971. (Woods, Weintraub and Assoc., Ilham Zebekoglu, Associate.)

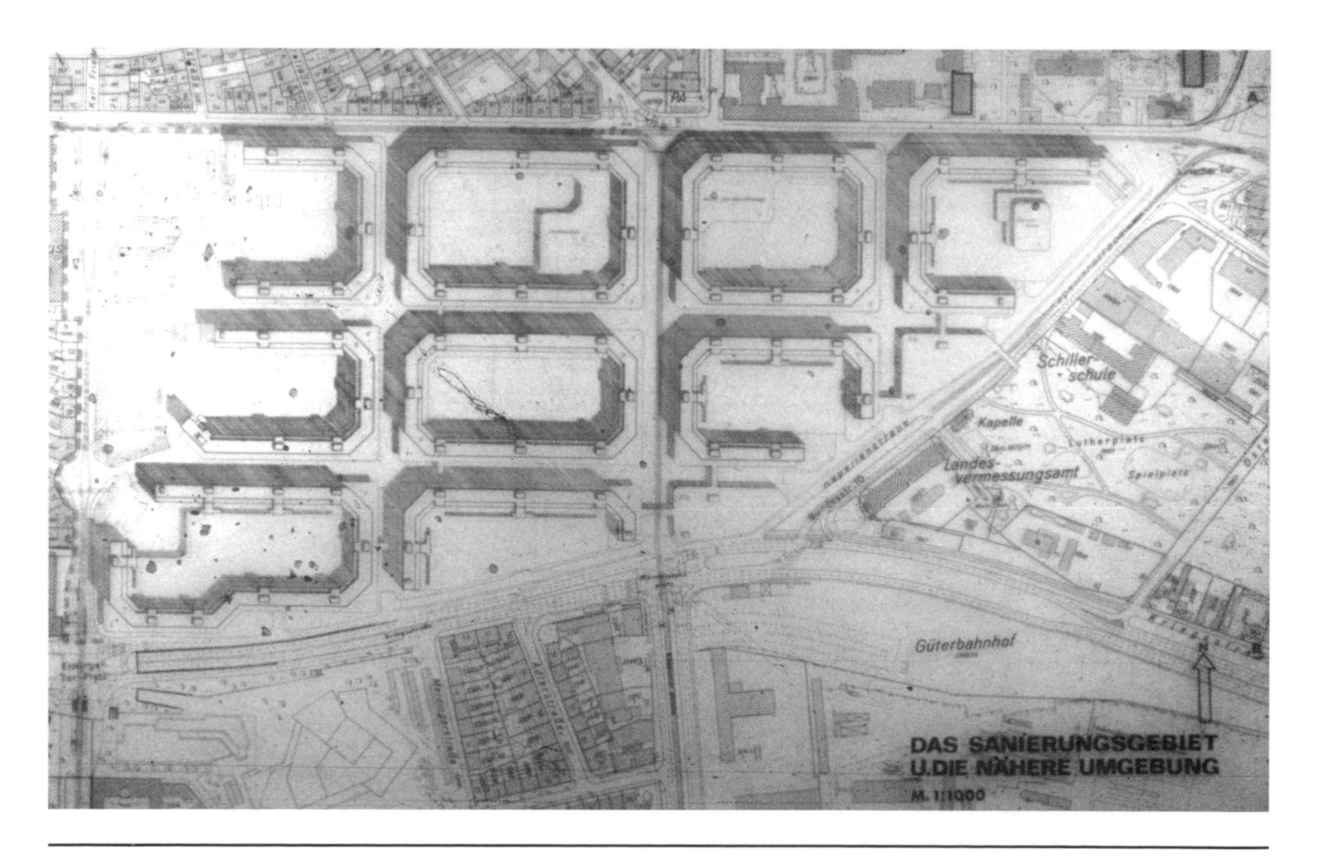

Fig. 21b. Karlsruhe competition. Plan in urban context, 1971.

pedestrian access and mixed programs and services. The site included the three corners adjacent to Central Park reconfigured to turn the traffic circle into a pedestrian-oriented cross intersection. The first two levels included a variety of public and semi-public uses (commercial, educational and cultural facilities) with overpasses connecting the three sites and Central Park above street level. The dwellings above this base were organized in high-rise blocks of characteristic angled geometry.

Other urban projects of these last few years moved away from the angled housing buildings in favor of more traditional rectangular blocks with perimeter housing complexes. The first of such projects was a 1966 officially accepted plan for Hamburg-Steilshoop wherein two linear spines are streets between rows of rectangular blocks, terminating at a commercial center; 6- to 14-story apartment buildings ring the perimeter block on three sides with a number of ground level openings to accommodate pedestrian paths for access to the interior playground courts. Within the context of *Schreber Garten*, the very lush existing gardens of the Hamburg site, Woods originally proposed a phased development of dense low 3- to 4-story terraced housing with small shops/studios and open-air markets at ground level. This was declared to be unsatisfactory because the city's planning administrators had determined that there must be high-rise construction of the site.

A similar organization to Hamburg-Steilshoop reappears in 1971 in a competition for *Karlsruhe* (Figs. 21a and b). The Karlsruhe site is arrayed as perimeter blocks ringing the rectangular spaces and the two lower levels are occupied by commercial and community facilities with housing on the upper levels. The scheme makes a clear distinction between the street outside the perimeter block as a place of activity and the inner part as an area of tranquility. The dwellings above this base were organized in high-rise units characteristic of affordable housing proposed for Cooper Square in New York for the Urban Development Corporation. This unbuilt project had a large interior court surrounded by a 6- to 10-story perimeter-building complex, with access to the apartments along the interior facing corridors located on every other floor. Based on his preliminary sketches, the project was developed by his colleagues Roger Cummings and Walteraude S. Woods, after Shadrach Woods' death in 1974.

Between utopia and fact

Near the end of his life Woods made one last trip to Europe. The first phase of the Berlin Free University, his most visible contribution to the culture of architecture, had just been completed. In its built form, the project could be seen both as a space for an alternative social order and, at the same time, an instance of the most ordinary spaces of the contemporary environment, almost like a curious cross between a phalanstery and an airport. Its deliberate excess of circulation connects the project to Fourier's phalanstery, where the corridors are the fundamental space of social interaction. Shadrach Woods explicitly recognizes the influence of Fourier.

In the first chapter of *The Man in the Street*, he presents the 19th-century French philosopher as an "illustration of how urbanism and architecture were incorporated into a social reformist movement" with an emphasis on social interaction.[7] Robert Filliou, a conceptual artist and good friend of Woods, also recognized the impact of Fourier. For him, Fourier represented an early effort to combine large economic and social transformations with a concern for sensual individual freedom. Filliou argues that Fourier "before Marx wrote and before Freud was born, succeeded in reconciling both."[8] Filliou and Woods see the phalanstery, a social order turned into building, as an institution capable of abolishing relationships of domination and replacing them with free circulation of individuals and ideas. Of course, this notion is at the core of the diagrams for Berlin, stressing the exchange of ideas, the association of disciplines, and "the use of a minimum structuring system where individual and group may determine desirable relationships."[9] If this preference betrays a connection with the utopian tradition, Woods' projects are connected also with a direct observation of the immediate built environment. They recognize that infrastructures of enormous scale and programs that are in constant transformation demand responsive forms of organization.

It is this combination of utopia and fact, two words that appear and reappear in the writings of Shadrach Woods, which may explain the renewed interest in his work. The current debates surrounding large projects of infrastructure, housing, public space, and production in a post-industrial society may point to the work of Shadrach Woods. It represents an effort to address emerging circumstances that—in introducing extraordinary disruptions to the fabric of the built environment—creates vast orders of their own. Without any trace of nostalgia, Woods attempted to maintain the cultural continuities of the city while addressing a large landscape of transformations. With all its contradictions, or perhaps because of its contradictions, this work may hold clues for the challenges of our contemporary environment. ■

NOTES AND REFERENCES

1. *le carré bleu*, no. 3, an issue of a small unassuming pamphlet that frequently published the work of Shadrach Woods.

2. Frampton, Kenneth, 1980. *Modern Architecture, A Critical History,* London: Thames and Hudson, p. 277.

3. Illustrated in Woods, Shadrach. 1975. *The Man in the Street: a Polemic on Urbanism,* Baltimore, MD: Penquin Books, pp. 86, 89.

4. Woods, Shadrach. 1969. "The Education Bazaar," *Harvard Educational Review,* vol. 39:4, 1969. *Architecture and Education,* p. 119.

5. Woods, Shadrach. 1968. *Candilis-Josic-Wood: Building for People.* New York: Praeger. Published in Europe as Joedicke, Jurgen 1968. *Candilis-Josic-Wood: A Decade of Architecture and Urban Design* Stuttgart/Bern: Karl Kramer Verlag.

6. Joachim Pfeufer was a collaborator on the later urban projects, including Bonne Nouvelle, Paris Nord, Bress-Revermont, Milan Trianelle, first scheme for Hamburg Steilshoop, SoHo, and Lower Manhattan expressway.

7. Woods, 1975. *op. cit.*, pp. 1–4.

8. Filliou, Robert. 1995. *From Political to Poetical Economy,* Vancouver, British Columbia: Morris and Helen Belkin Art Gallery, p. 58.

9. Donat, Kohn, ed. 1964. *World Architecture, 2.* New York: The Viking Press, p. 117.

Robert Venturi, Denise Scott Brown, and Steven Izenour

Summary

Based on a 1968 studio conducted at the Yale University School of Architecture and on research published as a provocative article in the March 1968 issue of *Architectural Design*, *Learning from Las Vegas* proposed a non-judgmental and rigorously analytic approach to the American commercial landscape. Following the opening salvo fired in Venturi's 1966 book, *Complexity and Contradiction in Architecture*, especially in its final chapter where Venturi poses the polemical question, "Is not Main Street almost alright?"— Venturi, Scott Brown and Izenour present the architecture and urbanism of the Las Vegas strip as a direct challenge to the core principles of modernist planning and architectural theory. The idea of "the decorated shed," the recognition of the dominance of sign and image over space, and the analysis of urban form and scale in relation to automotive, as opposed to pedestrian, experience all emerge from this study of the strip.

Key words

commercial strip, lighting, Nolli map, planning, sign graphics, theory

Fig. 1. The Las Vegas Strip, looking southwest.

A significance for A&P parking lots, or learning from Las Vegas

Substance for a writer consists not merely of those realities he thinks he discovers; it consists even more of those realities which have been made available to him by the literature and idioms of his own day and by the images that still have vitality in the literature of the past. Stylistically, a writer can express his feeling about this substance either by imitation, if it sits well with him, or by parody, if it doesn't. (Pourier 1967)

Learning from the existing landscape is a way of being revolutionary for an architect. Not the obvious way, which is to tear down Paris and begin again, as Le Corbusier suggested in the 1920s, but another, more tolerant way; that is, to question how we look at things.

The commercial strip, the Las Vegas Strip in particular—the example par excellence (Figs. 1 and 2)—challenges the architect to take a positive, non-chip-on-the-shoulder view. Architects are out of the habit of looking nonjudgmentally at the environment, because orthodox Modern architecture is progressive, if not revolutionary, utopian, and puristic; it is dissatisfied with *existing* conditions. Modern architecture has been anything but permissive: Architects have preferred to change the existing environment rather than enhance what is there.

But to gain insight from the commonplace is nothing new: Fine art often follows folk art. Romantic architects of the eighteenth century discovered an existing and conventional rustic architecture. Early Modern architects appropriated an existing and conventional industrial

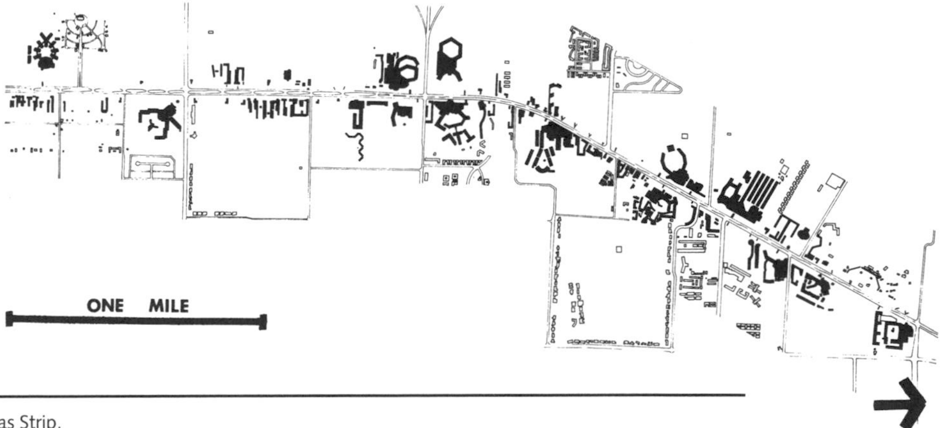

Fig 2. Map of the Las Vegas Strip.

Credits: This article is an extract of *Learning from Las Vegas* (Venturi, Scott Brown, and Izenour 1972), reprinted by permission of MIT Press. See endnote and References for full credits and citations.

vocabulary without much adaptation. Le Corbusier loved grain elevators and steamships; the Bauhaus looked like a factory; Mies refined the details of American steel factories for concrete buildings. Modern architects work through analogy, symbol, and image—although they have gone to lengths to disclaim almost all determinants of their forms except structural necessity and the program—and they derive insights, analogies, and stimulation from unexpected images. There is a perversity in the learning process: We look backward at history and tradition to go forward; we can also look downward to go upward. And withholding judgment may be used as a tool to make later judgment more sensitive. This is a way of learning from everything.

Commercial values and commercial methods

Las Vegas is analyzed here only as a phenomenon of architectural communication. Just as an analysis of the structure of a Gothic cathedral need not include a debate on the morality of medieval religion, so Las Vegas's values are not questioned here. The morality of commercial advertising, gambling interests, and the competitive instinct is not at issue here, although, indeed, we believe it should be in the architect's broader, synthetic tasks of which an analysis such as this is but one aspect. The analysis of a drive-in church in this context would match that of a drive-in restaurant, because this is a study of method, not content. Analysis of one of the architectural variables in isolation from the others is a respectable scientific and humanistic activity, so long as all are resynthesized in design. Analysis of existing American urbanism is a socially desirable activity to the extent that it teaches us architects to be more understanding and less authoritarian in the plans we make for both inner-city renewal and new development. In addition, there is no reason why the methods of commercial persuasion and the skyline of signs analyzed here should not serve the purpose of civic and cultural enhancement. But this is not entirely up to the architect.

Billboards are almost all right

Architects who can accept the lessons of primitive vernacular architecture, so easy to take in an exhibit like "Architecture without Architects," and of industrial, vernacular architecture, so easy to adapt to an electronic and space vernacular as elaborate neo-Brutalist or neo-Constructivist megastructures, do not easily acknowledge the validity of the commercial vernacular. For the artist, creating the new may mean choosing the old or the existing. Pop artists have relearned this. Our acknowledgment of existing, commercial architecture at the scale of the highway is within this tradition.

Modern architecture has not so much excluded the commercial vernacular as it has tried to take it over by inventing and enforcing a vernacular of its own, improved and universal. It has rejected the combination of fine art and crude art. The Italian landscape has always harmonized the vulgar and the Vitruvian: the *contorni* around the *duomo*, the *bortiere's* laundry across the *padrone's portone*, *Supercortemaggiore* against the Romanesque apse. Naked children have never played in our fountains, and I. M. Pei will never be happy on Route 66.

Architecture as space

Architects have been bewitched by a single element of the Italian landscape: the *piazza*. Its traditional, pedestrian-scaled, and intricately enclosed space is easier to like than the spatial sprawl of Route 66 and Los Angeles. Architects have been brought up on Space, and enclosed space is the easiest to handle. During the last forty years, theorists of Modern architecture (Wright and Le Corbusier sometimes excepted) have focused on space as the essential ingredient that separates architecture from painting, sculpture, and literature. Their definitions glory in the uniqueness of the medium; although sculpture and painting may sometimes be allowed spatial characteristics, sculptural or pictorial architecture is unacceptable—because Space is sacred.

Purist architecture was partly a reaction against nineteenth-century eclecticism. Gothic churches, Renaissance banks, and Jacobean manors were frankly picturesque. The mixing of styles meant the mixing of media. Dressed in historical styles, buildings evoked explicit associations and romantic allusions to the past to convey literary, ecclesiastical, national, or programmatic symbolism. Definitions of architecture were not enough. The overlapping of disciplines may have diluted the architecture, but it enriched the meaning.

Modern architects abandoned a tradition of iconology in which painting, sculpture, and graphics were combined with architecture. The delicate hieroglyphics on a bold pylon, the archetypal inscriptions of a Roman architrave, the mosaic processions in Sant'Apollinare, the ubiquitous tattoos over a Giotto Chapel, the enshrined hierarchies around a Gothic portal, even the illusionistic frescoes in a Venetian villa, all contain messages beyond their ornamental contribution to architectural space. The integration of the arts in Modern architecture has always been called a good thing. But one did not paint on Mies. Painted panels were floated independently of the structure by means of shadow joints; sculpture was in or near but seldom on the building. Objects of art were used to reinforce architectural space at the expenses of their own content. The Kolbe in the Barcelona Pavilion was a foil to the directed spaces: The message was mainly architectural. The diminutive signs in most Modern buildings contained only the most necessary messages, like LADIES, minor accents begrudgingly applied.

Architecture as symbol

Critics and historians, who documented the "decline of popular symbols" in art, supported orthodox Modern architects, who shunned symbolism of form as an expression or reinforcement of content: meaning was to be communicated, not through allusion to previously known forms, but through the inherent, physiognomic characteristics of form. The creation of architectural form was to be a logical process, free from images of past experience, determined solely by program and structure, with an occasional assist, as Alan Colquhoun (1967) has suggested from intuition.

But some recent critics have questioned the possible level of content to be derived from abstract forms. Others have demonstrated that the functionalists, despite their protestations, derived a formal vocabulary of their own, mainly from current art movements and the industrial vernacular; and latter-day followers such as the Archigram group have turned, while similarly protesting, to Pop Art and the space industry. However, most critics have slighted a continuing iconology in popular commercial art, the persuasive heraldry that pervades our environment from the advertising pages of *The New Yorker* to the superbillboards of Houston. And their theory of the "debasement" of symbolic architecture in nineteenth-century eclecticism has blinded them to the value of the representational architecture along highways. Those who acknowledge this roadside eclecticism denigrate it, because it flaunts the cliché of a decade ago as well as the style of a century ago. But why not? Time travels fast today.

The Miami Beach Modern motel on a bleak stretch of highway in southern Delaware reminds jaded drivers of the welcome luxury of a tropical resort, persuading them, perhaps, to forgo the gracious plantations across the Virginia border called Motel Monticello. The real hotel in Miami alludes to the international stylishness of a Brazilian resort, which, in turn, derives from the international style of middle Corbu. This evolution from the high source through the middle source to the low source took only thirty years. Today, the middle source, the neo-Eclectic architecture of the 1940s and the 1950s, is less interesting than its commercial adaptations. Roadside copies of Ed Stone are more interesting than the real Ed Stone.

Symbol in space before form in space: Las Vegas as a communication system

The sign for the Motel Monticello, a silhouette of an enormous Chippendale highboy, is visible on the highway before the motel itself. This architecture of styles and signs is antispatial; it is an architecture of communication over space; communication dominates space as an element in the architecture and in the landscape (Figs. 1–3). But it is for a new scale of landscape. The philosophical associations of the old eclecticism evoked subtle and complex meanings to be savored in the docile spaces of a traditional landscape. The commercial persuasion of roadside eclecticism provokes bold impact in the vast and complex setting of a new landscape of big spaces, high speeds, and complex programs. Styles and signs make connections among many elements, far apart and seen fast. The message is basely commercial; the context is basically new.

A driver thirty years ago could maintain a sense of orientation in space. At the simple crossroad a little sign with an arrow confirmed what was obvious. One knew where one was. When the crossroads becomes a cloverleaf, one must turn right to turn left, a contradiction poignantly evoked in the print by Allan D'Arcangelo (Fig. 4). But the driver has no time to ponder paradoxical subtleties within a dangerous, sinuous maze. He or she relies on signs for guidance—enormous signs in vast spaces at high speeds.

Fig. 3. Night messages. Las Vegas.

Fig. 4. Allan D'Arcangelo, *The Trip*.

The dominance of signs over space at a pedestrian scale occurs in big airports. Circulation in a big railroad station required little more than a simple axial system from taxi to train, by ticket window, stores, waiting room, and platform—all virtually without signs. Architects object to signs in buildings: "If the plan is clear, you can see where to go." But complex programs and setting require complex combinations of media beyond the purer architectural triad of structure, form, and light at the service of space. They suggest an architecture of bold communication rather than one of subtle expression.

The architecture of persuasion

The cloverleaf and airport communicate with moving crowds in cars or on foot for efficiency and safety. But words and symbols may be used in space for commercial persuasion (Fig. 5). The Middle Eastern bazaar contains no signs; the Strip is virtually all signs (Fig. 6). In the bazaar, communication works through proximity. Along its narrow aisles, buyers feel and smell the merchandise, and the merchant applies explicit oral persuasion. In the narrow streets of the medieval town, although signs occur, persuasion is mainly through the sight and smell of the real cakes through the doors and windows of the

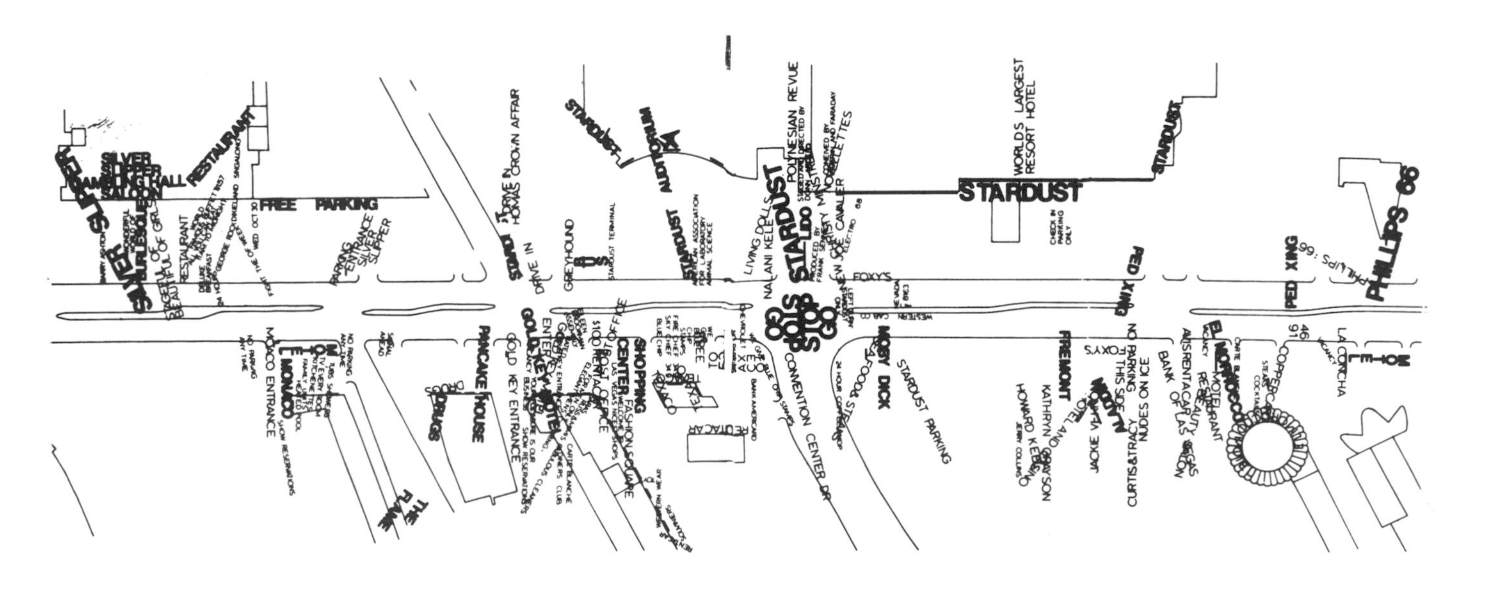

Fig. 5. Map of Las Vegas Strip (detail) showing every written work seen from the road.

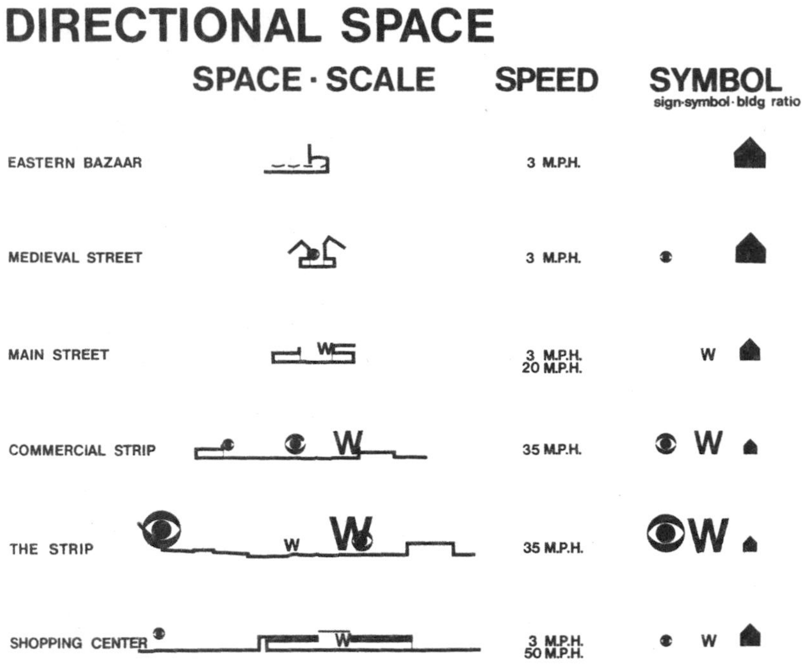

DIRECTIONAL SPACE

	SPACE · SCALE	SPEED	SYMBOL sign·symbol·bldg ratio
EASTERN BAZAAR		3 M.P.H.	
MEDIEVAL STREET		3 M.P.H.	
MAIN STREET	W	3 M.P.H. 20 M.P.H.	W
COMMERCIAL STRIP	W	35 M.P.H.	W
THE STRIP	W W	35 M.P.H.	W
SHOPPING CENTER	W	3 M.P.H. 50 M.P.H.	W

Fig. 6. A comparative analysis of directional spaces.

Fig. 7. Tanya billboard on the Strip.

Fig. 8. Lower Strip, looking north.

bakery. On Main Street, shop-window displays for pedestrians along the sidewalks and exterior signs, perpendicular to the street for motorists, dominate the scene almost equally.

On the commercial strip the supermarket windows contain no merchandise. There may be signs announcing the day's bargains, but they are to be read by pedestrians approaching from the parking lot. The building itself is set back from the highway and half hidden, as is most of the urban environment, by parked cars. The vast parking lot is in the front, not at the rear, since it is a symbol as well as a convenience. The building is low because air conditioning demands low spaces, and merchandising techniques discourage second floors; its architecture is neutral because it can hardly be seen from the road. Both merchandise and architecture are disconnected from the road. The big sign leaps to connect the driver to the store, and down the road the cake mixes and detergents are advertised by their national manufacturers on enormous billboards inflected toward the highway. The graphic sign in space has become the architecture of this landscape (Figs. 7 and 8). Inside, the A&P has reverted to the bazaar except that graphic packaging has replaced the oral persuasion of the merchant. At another scale, the shopping center off the highway returns in its pedestrian malls to the medieval street.

Vast space in the historical tradition and at the A&P

The A&P parking lot is a current phase in the evolution of vast space since Versailles (Fig. 9). The space that divides high-speed highway and low, sparse buildings produces no enclosure and little direction. To move through a *piazza* is to move between high enclosing forms. To move through this landscape is to move over vast expansive texture: the megatexture of the commercial landscape. The parking lot is the parterre of the asphalt landscape. The patterns of parking lines give direction much as the paving patterns, curbs, borders, and tapis vert give direction in Versailles; grids of lamp posts substitute for obelisks,

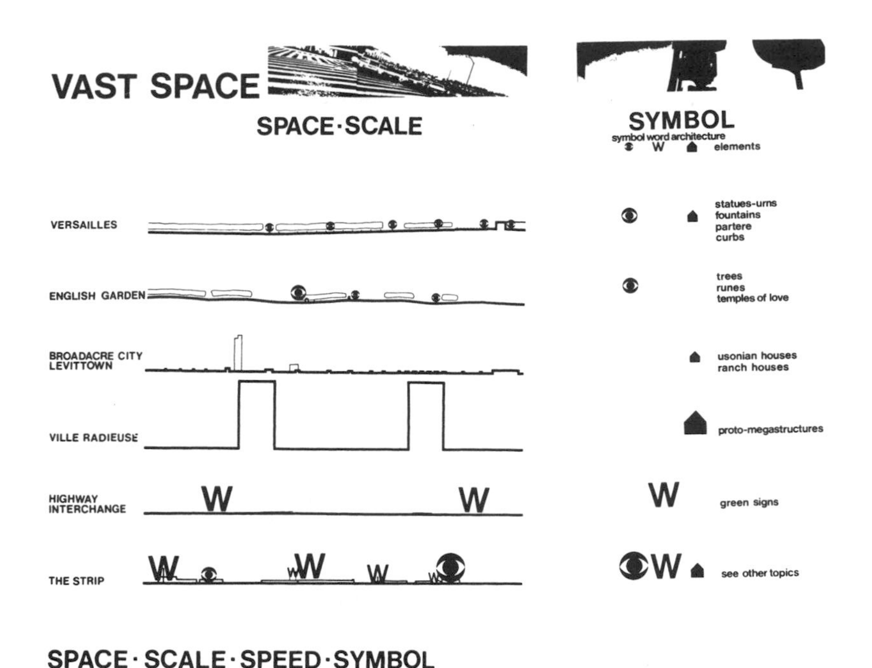

Fig. 9. A comparative analysis of vast spaces.

rows of urns and statues as points of identity and continuity in the vast space. But it is the highway signs, through their sculptural forms or pictorial silhouettes, their particular positions in space, their inflected shapes, and their graphic meanings, that identify and unify the megatexture. They make verbal and symbolic connections through space, communicating a complexity of meanings through hundreds of associations in few seconds from far away. Symbol dominates space. Architecture is not enough. Because the spatial relationships are made by symbols more than by forms, architecture in this landscape becomes symbolic in space rather than form in space. Architecture defines very little: The big sign and the little building is the rule of Route 66.

The sign is more important than the architecture. This is reflected in the proprietor's budget. The sign at the front is a vulgar extravaganza, the building at the back, a modest necessity. The architecture is what is cheap. Sometimes the building is the sign: The duck store in the shape of a duck, called "The Long Island Duckling" (Fig. 10) is sculptural symbol and architectural shelter. Contradiction between outside and inside was common in architecture before the Modern movement, particularly in urban and monumental architecture (Fig. 11). Baroque domes were symbols as well as spatial constructions, and they are bigger in scale and higher outside than inside in order to dominate their urban setting and communicate their symbolic message. The false fronts of Western stores did the same thing: They were bigger and taller than the interiors they fronted to communicate the store's importance and to enhance the quality and unity of the street. But false fronts are of the order and scale of Main Street. From the desert town on the highway in the West of today, we can learn new and vivid lessons about an impure architecture of communication. The little low buildings, gray-brown like the desert, separate and recede from the street that is now the highway as big, high signs. If you take the signs away, there is no place. The desert town is intensified communication along the highway.

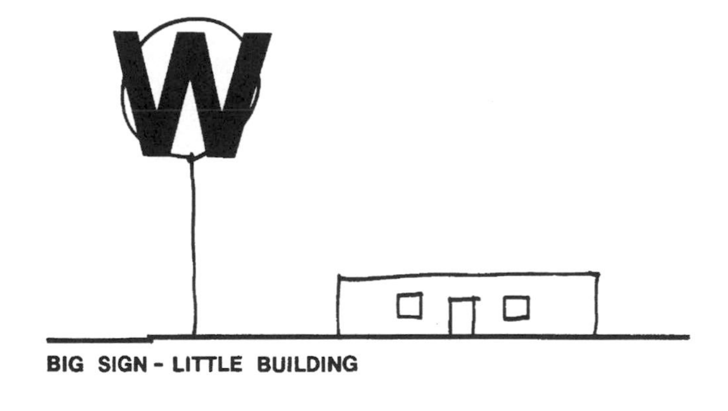

OR

Fig. 10. Big sign, little building or "building as sign."

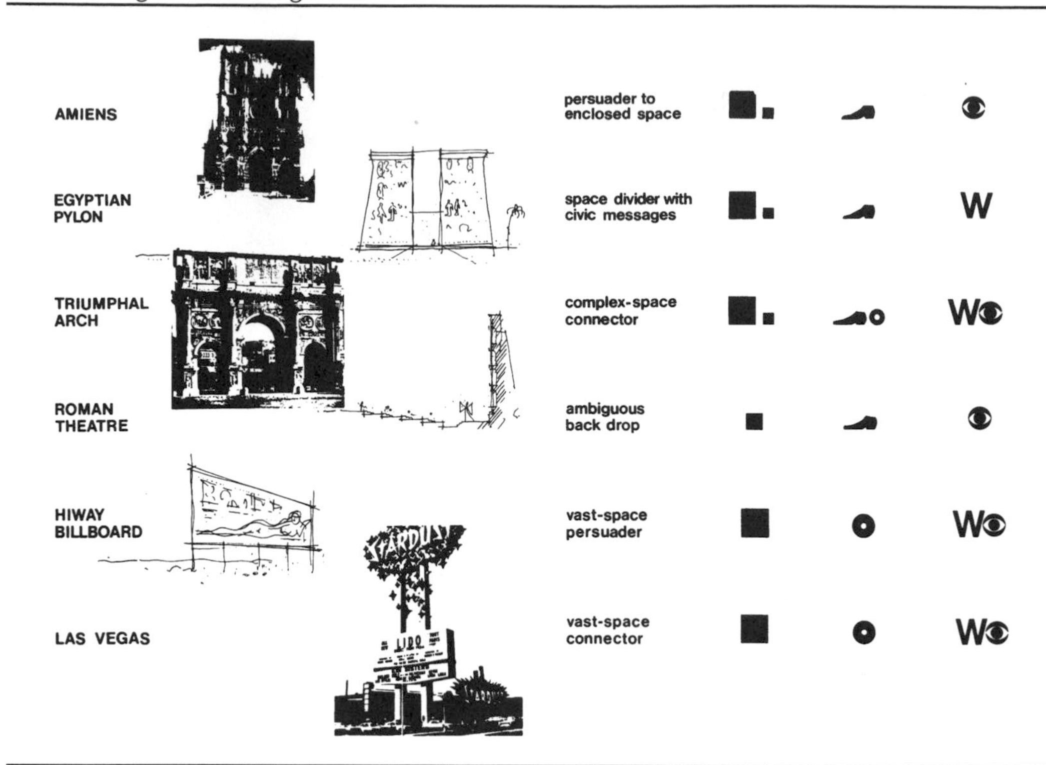

Fig. 11. A comparative analysis of "billboards" in space.

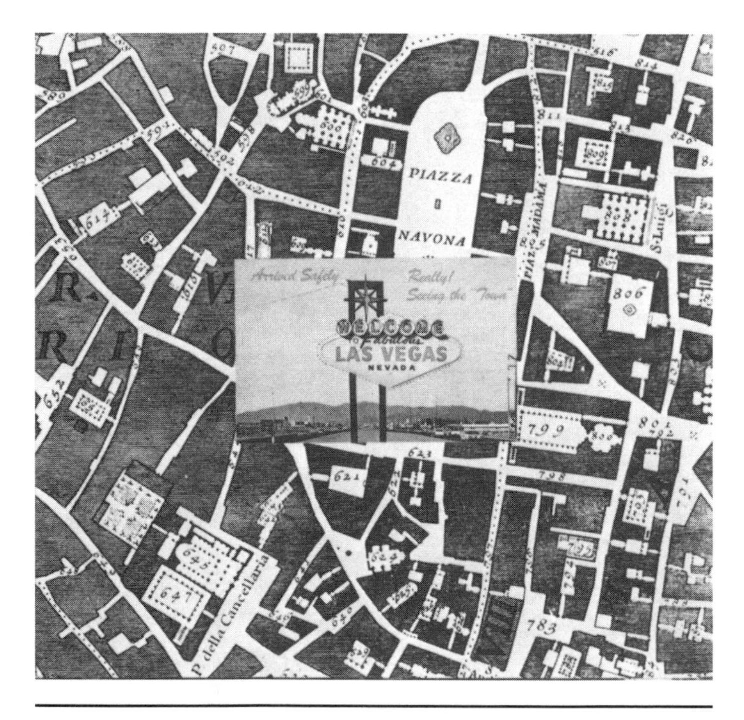

Fig. 12. Nolli's map of Rome (detail).

From Rome to Las Vegas

Las Vegas is the apotheosis of the desert town. Visiting Las Vegas in the mid-1960s was like visiting Rome in the late 1940s. For young Americans in the 1940s, familiar only with the auto-scaled, gridiron city and the anti-urban theories of the previous architectural generation, the traditional urban space, the pedestrian scale, and the mixtures, yet continuities, of styles of the Italian *piazzas* were a significant revelation. They rediscovered the *piazza*. Two decades later architects are perhaps ready for similar lessons about large open space, big scale, and high speed. Las Vegas is to the Strip what Rome is to the *piazza*.

There are other parallels between Rome and Las Vegas: their expansive settings in the Campagna and in the Mojave Desert, for instance, that tend to focus and clarify their images. On the other hand, Las Vegas was built in a day, or rather, the Strip was developed in a virgin desert in a short time. It was not superimposed on an older pattern as were the pilgrim's Rome of the Counter-Reformation and the commercial strips of eastern cities, and it is therefore easier to study. Each city is an archetype rather than a prototype, an exaggerated example from which to derive lessons for the typical. Each city vividly superimposes elements of a supranational scale on the local fabric: churches in the religious capital, casinos and their signs in the entertainment capital. These cause violent juxtapositions of use and scale in both cities. Rome's churches, off streets and *piazzas*, are open to the public; the pilgrim, religious or architectural, can walk from church to church. The gambler or architect in Las Vegas can similarly take in a variety of casinos along the Strip. The casinos and lobbies of Las Vegas are ornamental and monumental and open to the promenading public; a few old banks and railroad stations excepted, they are unique in American cities. Nolli's map of the mid-eighteenth century reveals

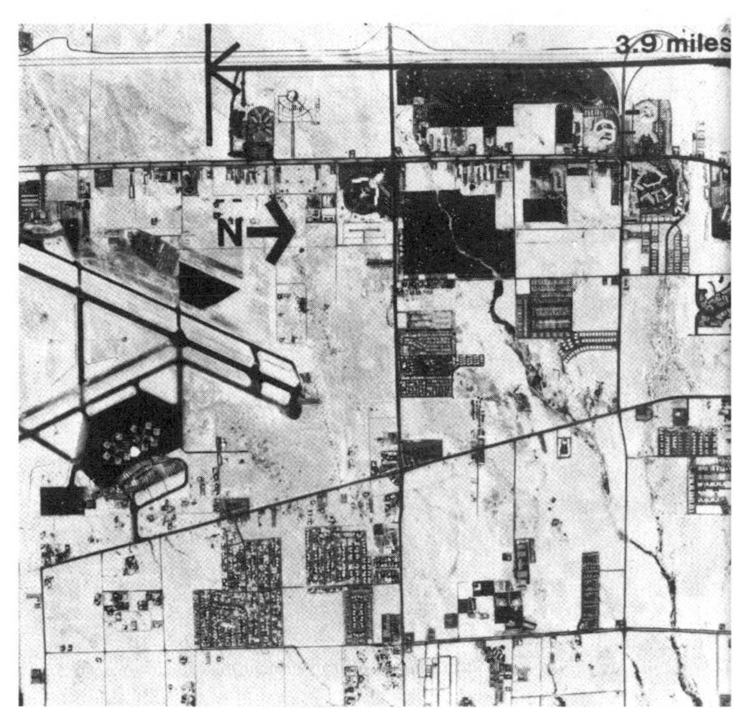

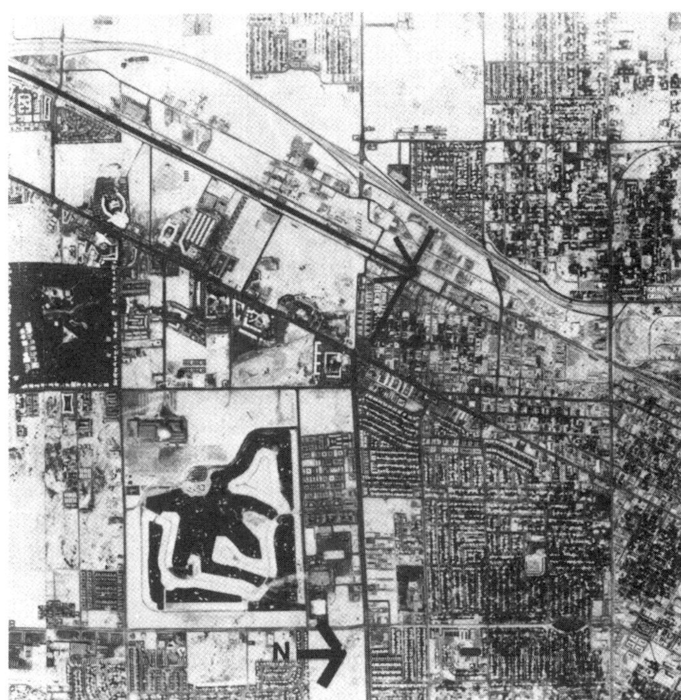

Fig. 13. Aerial photograph of upper Strip.

the sensitive and complex connections between public and private space in Rome (Fig. 12). Private building is shown in gray cross-hatching that is carved into by the public spaces, exterior and interior. These spaces, open or roofed, are shown in minute detail through darker poché. Interior of churches read like *piazzas* and courtyards of palaces, yet a variety of qualities and scales is articulated.

Maps of Las Vegas

A Nolli map of the Las Vegas Strip reveals and clarifies what is public and what is private, but here the scale is enlarged by the inclusion of the parking lot, and the solid-to-void ratio is reversed by the open spaces of the desert. Mapping the Nolli components from an aerial photograph provides an intriguing crosscut of Strip systems (Fig. 13). These components, separated and redefined, could be undeveloped land, asphalt, autos, buildings, and ceremonial space (Figs. 14 a–e). Reassembled (Fig. 14f), they describe the Las Vegas equivalent of the pilgrims' way, although the description, like Nolli's map, misses the iconological dimensions of the experience.

A conventional land-use map of Las Vegas can show the overall structure of commercial use in the city as it relates to other uses but none of the detail of use type or intensity. "Land-use" maps of the insides of casino complexes, however, begin to suggest the systematic planning that all casinos share (Fig. 15). Strip "address" and "establishment" maps can depict both intensity and variety of use (Fig. 16). Distribution maps show patterns of, for example, churches, and food stores that Las Vegas shares with other cities and those such as wedding chapels and auto rental stations that are Strip-oriented and unique. It is extremely hard to suggest that atmospheric qualities of Las Vegas, because these are primarily dependent on Watts (Fig. 17a), animation, and Iconology; however, "message maps," tourist maps, and brochures suggest some of it.

Main Street and the Strip

A street map of Las Vegas reveals two scales of movement within the gridiron plan: that of Main Street and that of the Strip. The main street of Las Vegas is Fremont Street and the earlier of two concentrations of casinos is located along three of four blocks of this street. The casinos here are bazaar-like in the immediacy to the sidewalk of their clocking and tinkling gambling machines. The Fremont Street casinos and hotels focus on the railroad depot at the head of the street; here the railroad and main street scale of movement connect. The depot building is now gone, replaced by a hotel, and the bus station is now the busier entrance to town, but the axial focus on the railroad depot from Fremont Street was visual, and possibly symbolic. This contrasts with the Strip, where a second and later development of casinos extends southward to the airport, the jet-scale entrance to town.

One's first introduction to Las Vegas architecture is a forebear of Eero Saarinen's TWA Terminal, which is the local airport building. Beyond this piece of architectural image, impressions are scaled to the car rented at the airport. Here is the unraveling of the famous Strip itself, which, as Route 91, connects the airport with the downtown.

System and order on the Strip

The image of the commercial strip is chaos. The order in this landscape is not obvious (Fig. 17b). The continuous highway itself and its systems for turning area absolutely consistent. The median strip accommodates the U-turns necessary to a vehicular promenade for casino crawlers

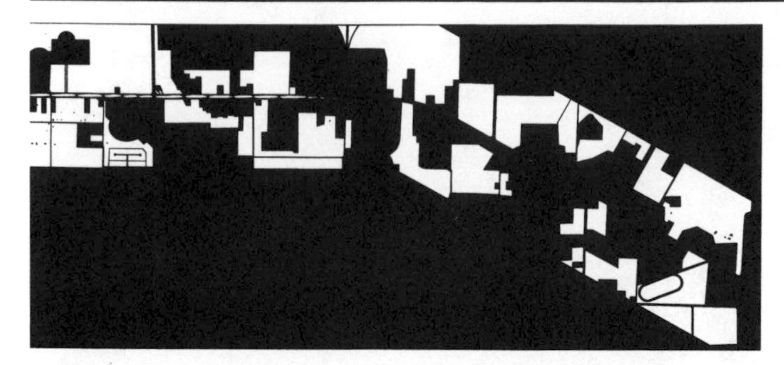

14a. Upper strip, undeveloped land.

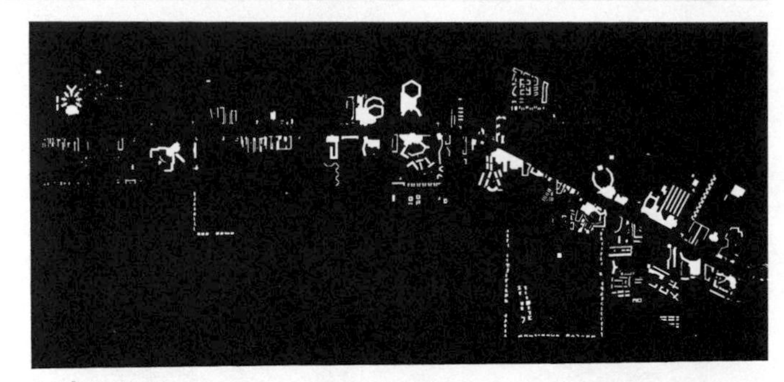

14d. Buildings.

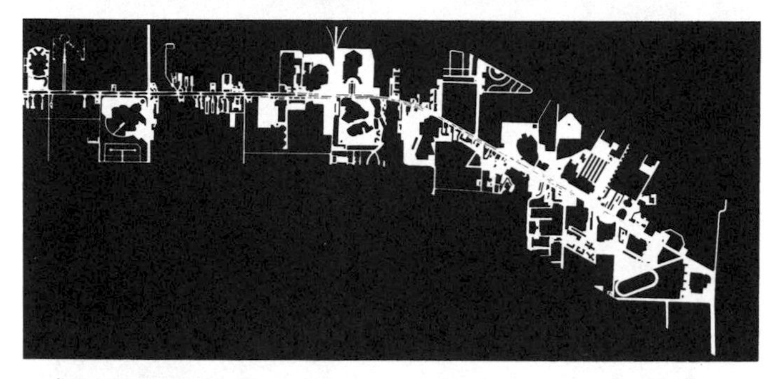

14b. Asphalt.

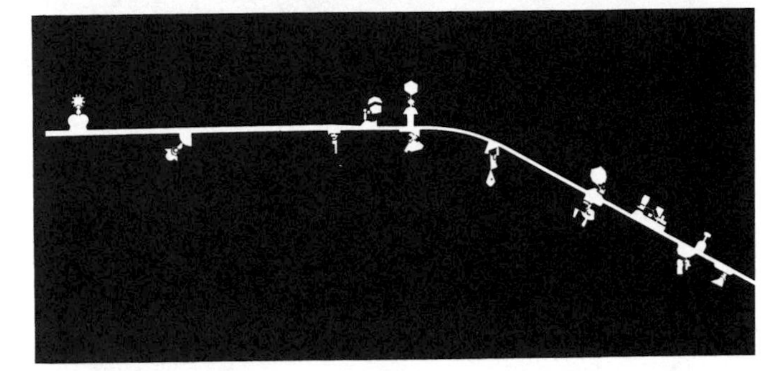

14e. Ceremonial space.

14c. Autos.

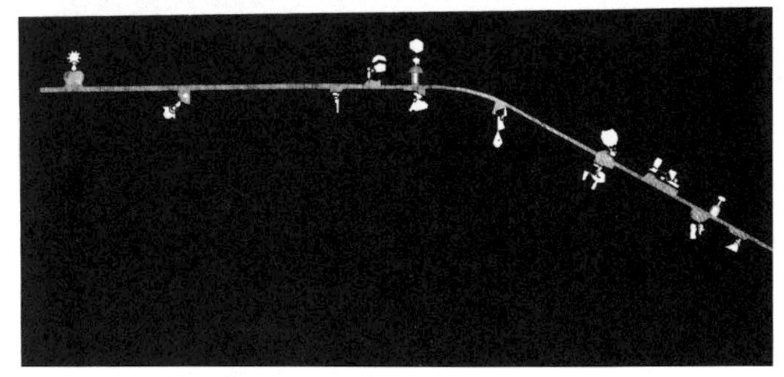

14f. "Nolli's Las Vegas."

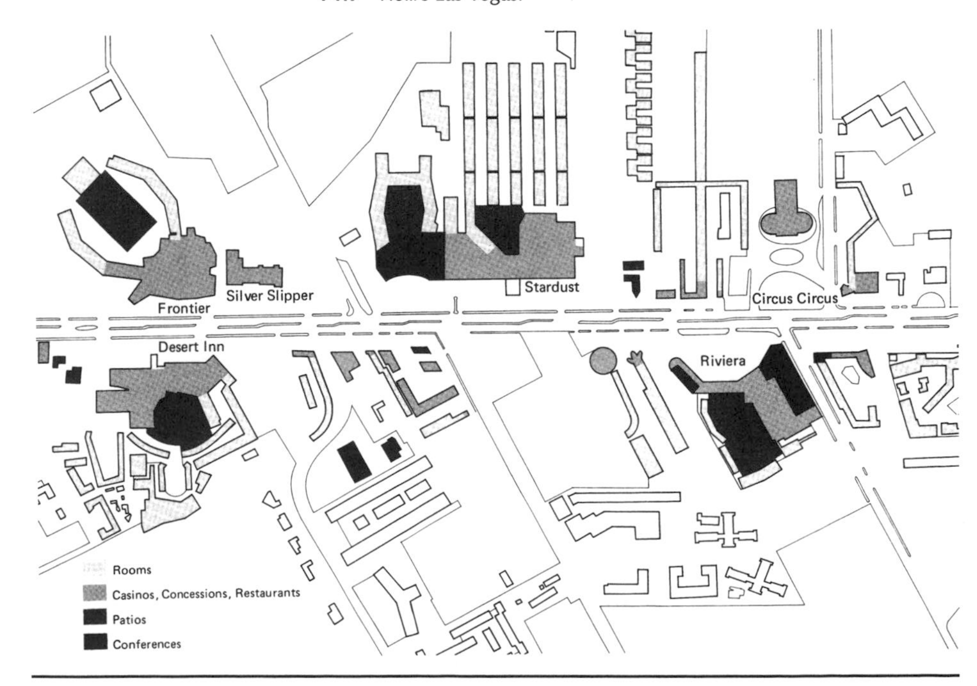

Fig. 15. Map of Las Vegas Strip (detail) showing uses within buildings.

as well as left turns onto the local street pattern that the Strip intersects. The curbing allows frequent right turns for casinos and other commercial enterprises and eases the difficult transitions from highway to parking. The streetlights function superfluously along many parts of the Strip that are incidentally but abundantly lit by signs, but their consistency of form and position and their arching shapes begin to identify by day a continuous space of the highway, and the constant rhythm contrasts effectively with the uneven rhythms of the signs behind.

This counterpoint reinforces the contrast between two types of order on the Strip: the obvious visual order of street elements and the difficult visual order of buildings and signs. The zone *off* the highway is a shared order. The zone *of* the highway is an individual order (Fig. 20). The elements of the highway are civic. The buildings and signs are private. In combination they embrace continuity and discontinuity, going *and* stopping, clarity *and* ambiguity, cooperation *and* competition, the community *and* rugged individualism. The system of the highway gives order to the sensitive functions of exist and entrance, as well as to the image of the Strip as a sequential whole. It also generates places for individual enterprises to grow and controls the general direction of that growth. It allows variety and change along its sides and accommodates the contrapuntal, competitive order of the individual enterprises.

There is an order along the sides of the highway. Varieties of activities are juxtaposed on the Strip: service stations, minor motels, and multimillion-dollar casinos. Marriage chapels ("credit cards accepted") converted from bungalows with added neon-lined steeples are apt to appear anywhere toward the downtown end. Immediate proximity of related cases, as on Main Street, where you *walk* from one store to another, is not required along the Strip because interaction is by car and highway. You *drive* from one casino to another even when they are adjacent because of the distance between them, and an intervening service station is not disagreeable.

Change and permanence on the Strip

The rate of obsolesce of a sign seems to be nearer to that of an automobile than that of a building. The reason is not physical degeneration but what competitors are doing around you. The leasing system operated by the sign companies and the possibility of total tax write-off may have something to do with it. The most unique, most monumental parts of the Strip, the signs and casino facades, are also the most changeable; it is the neutral, systems-motel structures behind that survive a succession of facelifts and a series of themes up front. The Aladdin Hotel and Casino is Moorish in front and Tudor behind.

Las Vegas's greatest growth has been since World War II. There are noticeable changes every year: new hotels and signs as well as neon-embossed parking structures replacing on-lot parking on and behind Fremont Street. Like the agglomeration of chapels in a Roman church and the stylistic sequence of piers in a Gothic cathedral, the Golden Nugget casino has evolved over thirty years from a building with a sign on it to a totally sign-covered building. The Stardust Hotel has engulfed a small restaurant and a second hotel in its expansion and has united the three-piece façade with 600 ft. (183 m) of computer-programmed animated neon.

The architecture of the Strip

It is hard to think of each flamboyant casino as anything but unique, and this is as it should be, because good advertising technique requires

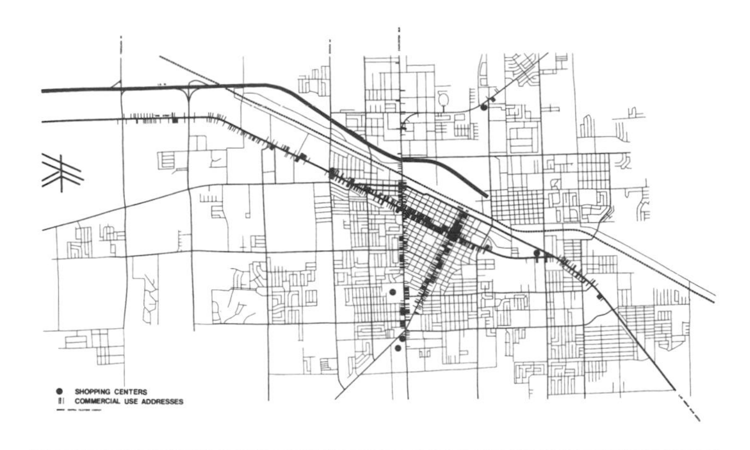

Fig. 16. Map showing location of ground floor commercial establishments (1961) on three Las Vegas strips.

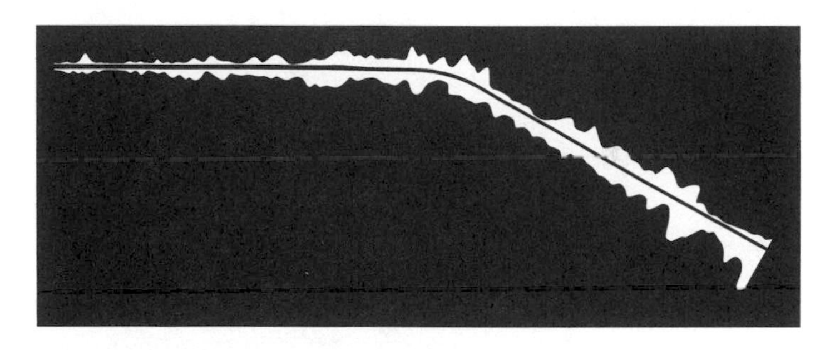

Fig. 17a. Illumination levels on the Strip.

Fig. 17b. The order in this landscape is not obvious.

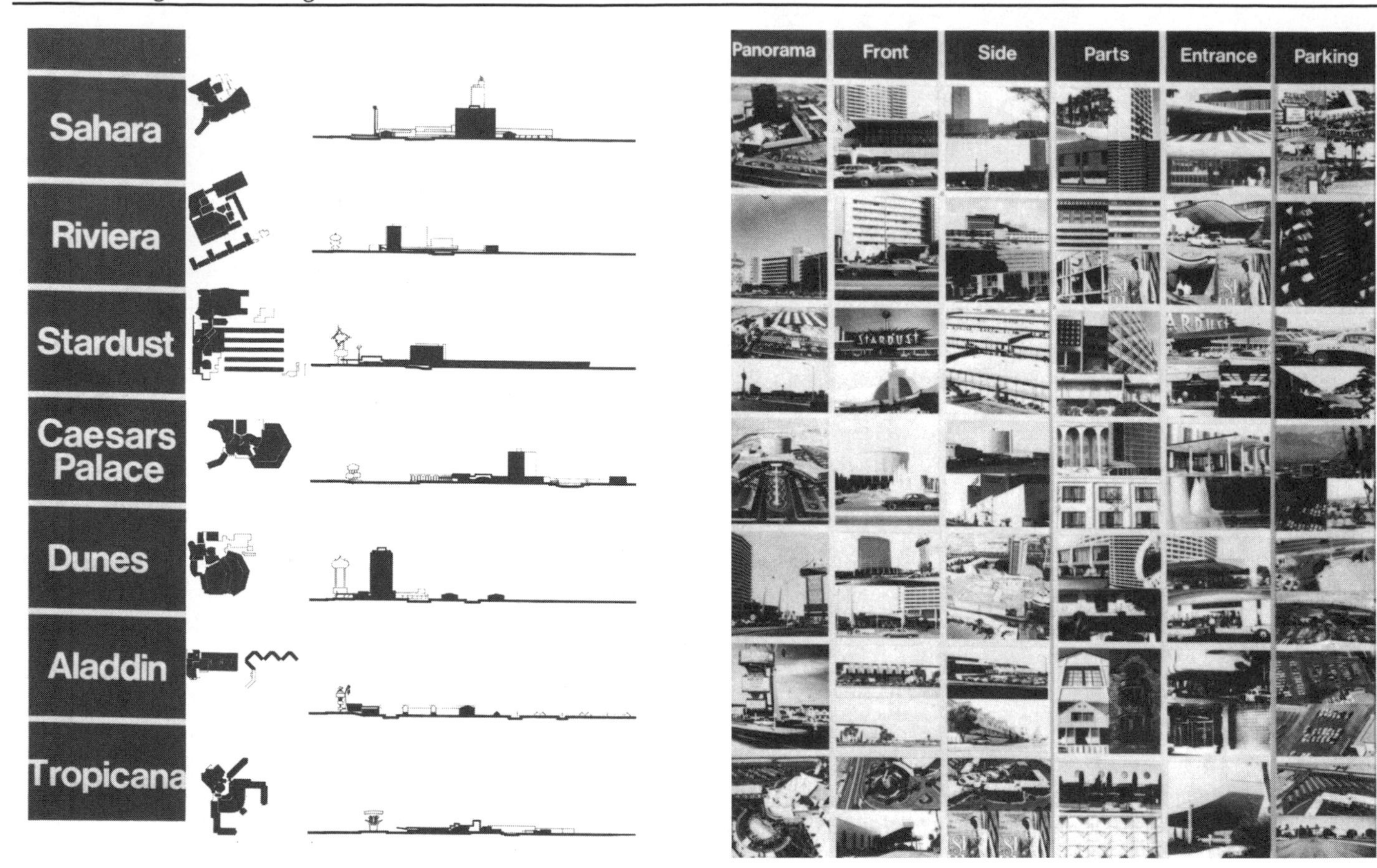

Fig. 18. A schedule of Las Vegas Strip hotels: plans, sections, and elements.

the differentiation of the product. However, these casinos have much in common because they are under the same sun, on the same Strip, and perform similar functions; they differ from other casinos—say, on Fremont Street—and from other hotels that are not casinos, as indicated in Fig. 18.

A typical hotel-casino complex contains a building that is near enough to the highway to be seen from the road across the parked cars, yet far enough back to accommodate driveways, turnarounds, and parking. The parking in front is a token: It reassures the customers but does not obscure the building. It is prestige parking: The customer pays. The bulk of the parking, along the sides of the complex, allows direct access to the hotel yet stays visible from the highway. Parking is seldom at the back. The scales of movement and space of the highway relate to the distances between buildings; because they are far apart, they can be comprehended at high speeds. Front footage on the Strip has not yet reached the value it once had on Main Street, and parking is still an appropriate filler. Big space between buildings is characteristic of the Strip. It is significant that Fremont Street is more photogenic than the Strip. A single postcard can carry a view of the Golden Horseshoe, the Mint Hotel, the Golden Nugget, and the Lucky Casino. A single shot of the Strip is less spectacular; its enormous spaces must be seen as moving sequences.

The side elevation of the complex is important, because it is seen by approaching traffic from a greater distance and for a longer time than the façade. The rhythmic gables on the long, low, English medieval style, half-timbered motel sides of the Aladdin read empathetically across the parking space and through the signs and the giant statue of the neighboring Texaco station, and contrast with the modern Near Eastern flavor of the casino front. Casino fronts on the Strip often inflect in shape and ornament toward the right, to welcome right-lane traffic. Modern styles use a porte cochere that is diagonal in plan. Brazilianoid International styles use free forms.

Service stations, motels, and other simpler types of buildings conform in general to this system of inflection toward the highway through the position and form of their elements. Regardless of the front, the back of the building is styleless, because the whole is turned toward the front and no one sees the back. The gasoline stations parade their universality. The aim is to demonstrate their similarity to the one at home—your friendly gasoline station. A motel is a motel anywhere. But here the imagery is heated up by the need to compete in the surroundings. The artistic influence has spread, and Las Vegas motels have signs like no others. Their ardor lies somewhere between the casinos and the wedding chapels. Wedding chapels, like many urban land uses, are not form-specific. They tend to be one of a succession of uses a more generalized building type (a bungalow or a store front) may have. But a wedding-chapel style or image is maintained in different types through the use of symbolic ornament in neon, and the activity adapts itself to different inherited plans. Street furniture exists on the Strip as on other city streets, yet it is hardly in evidence.

Fig. 18 cont. A schedule of Las Vegas Strip hotels: plans, sections, and elements.

Fig. 19. Upper Strip looking north.

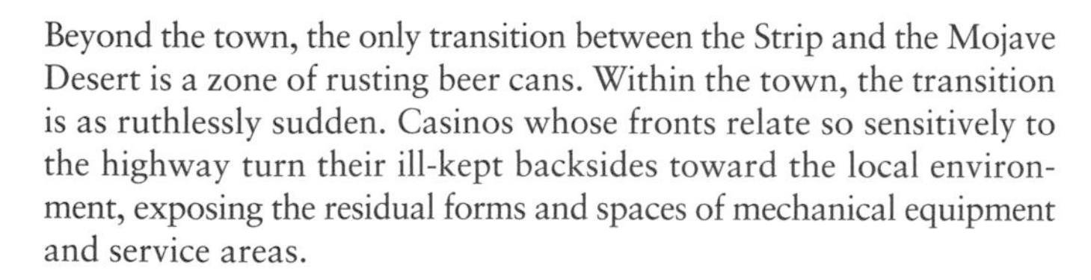

Beyond the town, the only transition between the Strip and the Mojave Desert is a zone of rusting beer cans. Within the town, the transition is as ruthlessly sudden. Casinos whose fronts relate so sensitively to the highway turn their ill-kept backsides toward the local environment, exposing the residual forms and spaces of mechanical equipment and service areas.

Inclusion and the difficult order

Henri Bergson called disorder an order we cannot see. The emerging order of the Strip is a complex order. It is not the easy, rigid order of the urban renewal project or the fashionable "total design" of the megastructure. It is, on the contrary, a manifestation of an opposite direction in architectural theory: Broadacre City—a travesty of Broadacre City, perhaps, but a kind of vindication of Frank Lloyd Wright's predictions for the American landscape. The commercial strip within the urban sprawl is, of course, Broadacre City with a difference. Broadacre City's easy, motival order identified and unified its vast spaces and separate buildings at the scale of the omnipotent automobile. Each building, without doubt, was to be designed by the Master or by his Taliesin Fellowship, with no room for honky-tonk improvisations. An easy control would be exercised over similar elements within the universal, Usonian vocabulary to the exclusion, certainly, of commercial vulgarities. But the order of the Strip *includes*; it includes at all levels, from the mixture of seemingly incongruous land uses to the mixture of seemingly incongruous advertising media plus a system of neo-Organic or neo-Wrightian restaurant motifs in walnut formica. It is not an order dominated by the expert and made easy for the eye. The moving eye in the moving body must work to pick out and interpret a variety of changing, juxtaposed orders, like the shifting configurations of a Victor Vasarely painting. It is the unity that "maintains, but only just maintains, a control over the clashing elements which compose it. Chaos is very near; its nearness, but its avoidance, gives. . .force" (Heckscher 1962, p. 289).

Image of Las Vegas: inclusion and allusion in architecture

Tom Wolfe used Pop prose to suggest powerful images of Las Vegas (Wolfe 1966, p. 8). Hotel brochures and tourist handouts suggest others. J. B. Jackson, Robert Riley, Edward Ruscha, John Kouwenhoven, Reyner Banham, and William Wilson have elaborated on related images. For the architect or urban designer, comparisons of Las Vegas with others of the world's "pleasure zones"—with Marienbad, the Alhambra, Xanadu, and Disneyland, for instance—suggest that essential to the imagery of pleasure-zone architecture are lightness, the quality of being an oasis in a perhaps hostile context, heightened symbolism, and the ability to engulf the visitor in a new role: for three days one may imagine oneself at the Riveria rather than a salesperson from Des Moines, Iowa, or an architect from Haddonfield, New Jersey.

However, there are didactic images more important than the images of recreation for us to take home to New Jersey and Iowa: one is the

Avis with the Venus; another, Jack Benny under a classical pediment with Shell Oil beside him, or the gasoline station beside the multi-million-dollar casino. These show the vitality that may be achieved by an architecture of inclusion or, by contrast, the deadness that results from too great a preoccupation with tastefulness and total design. The Strip shows the value of symbolism and allusion in an architecture of vast space and speed and proves that people, even architects, have fun with architecture that reminds them of something else, perhaps of harems or the Wild West in Las Vegas, perhaps of the nation's New England forebears in New Jersey. Allusion and comment, on the past or present or on our real commonplaces or old clichés, and inclusion of the everyday in the environment, sacred and profane—these are what are lacking in present-day Modern architecture. We can learn about them from Las Vegas as have other artists from their own profane and stylistic sources.

Pop artists have shown the value of the old cliché used in a new context to achieve a new meaning—the soup can in the art gallery—to make the common uncommon. And in literature, Eliot and Joyce display, according to Poirier (1967, p. 20), "an extraordinary vulnerability. . . to the idioms, rhythms, artifacts, associated with certain urban senvironments or situations. The multitudinous styles of Ulysses are so dominated by them that there are only intermittent sounds of Joyce in the novel and no extended passage certifiably in his as distinguished from a mimicked style." Poirier refers to this as the "decreative impulse." Eliot himself speaks of Joyce's doing the best he can "with the material at hand" (Eliot 1958, p. 125). Perhaps a fitting requiem for the irrelevant works of Art that are today's descendants of a once meaningful Modern Architecture are Eliot's lines in "East Coker" (Eliot 1943, p. 13):

That was a way of putting it—not very satisfactory:
A periphrastic study in a worn-out poetical fashion,
Leaving one still with the intolerable wrestle
With words and meanings. The poetry does not matter. . .

ENDNOTE

This article represents work from a Fall 1968 Masters of Architecture design studio, School of Art and Architectures, Yale University, directed by the authors. Students participating in the work included: Ralph Carlson, Tony Farmer, Ron Filson, Glen Hodges, Peter Hoyt, Charles Korn, John Kranz, Peter Schlaifer, Peter Schmitt, Dan Scully, Doug Southworth, Martha Wagner, Tony Zunino. The studio program, *A&P Parking Lots, or Learning From Las Vegas. A Significance for A Studio Research Problem*, and work topics were prepared by Denise Scott Brown. ■

REFERENCES

Colquhoun, Alan. 1967. "Typology and Design Method," *Arena, Journal of the Architectural Association.* June 1967, pp. 11–14.

Elliot. T. S. 1943. *Four Quartets.* New York: Harcourt Brace.

Elliot, T. S. 1958. *The Complete Poems and Plays, 1909–1950.* New York: Harcourt, Brace.

Heckscher, August. 1962. *The Public Happiness.* New York: Atheneum Publishers.

Pourier, Richard, 1967. "T.S. Eliot and the Literature of Waste," *The New Republic* May 20, 1967, pp. 21–22.

Venturi, Robert, Denise Scott Brown, and Steven Izenour.1972. *Learning from Las Vegas.* Cambridge: The MIT Press.

Wolfe, Tom. 1966. *The Kandy Colored Tangerine Flake Streamline Baby.* New York: Noonday Press.

Aldo Rossi

Summary

A leading architect and theorist of the postwar generation in Europe, Aldo Rossi first came to the attention of a wider audience through his participation in the Rationalist movement and exhibition of early projects in the Milan Triennale of 1973. He had published a major theoretical treatise in Italian in 1966, translated into English as *The Architecture of the City* in 1982, in which he derives the role of architecture from an analytic and historical account of urban form. Rossi develops a critique of modernist architecture and urbanism based upon a rejection of functionalist theory, especially as applied to the problem of the city. While Rossi's own work takes a distinctive direction—in part conditioned by his interpretation of the Enlightenment concept of "type"—he emphasizes urban themes such as memory and monumentality, public and private, that are characteristic of that critical watershed in urban design theory when the whole city, rather than the individual building, became the starting point for an account of urban form.

Key words

classification, form, functionalism, history of architecture, house, human geography, scale, typology, urban structure

Courtyard of the House of Diana, Ostia Antica, Rome.
Rendering by Italio Gismondi. c. 1940.

Problems of classification

We have indicated the principal questions that arise in relation to an urban artifact—among them, individuality, *locus*, memory, design itself. Function was not mentioned. I believe that any explanation of urban artifacts in terms of function must be rejected if the issue is to elucidate their structure and formation. We will later give some examples of important urban artifacts whose function has changed over time or for which a specific function does not even exist. Thus, one thesis of this study, in its efforts to affirm the value of architecture in the analysis of the city, is the denial of the explanation of urban artifacts in terms of function. I maintain, on the contrary, that far from being illuminating, this explanation is regressive because it impedes us from studying forms and knowing the world of architecture according to its true laws.

We hasten to say that this does not entail the rejection of the concept of function in its most proper sense, however, that is, as an algebra of values that can be known as functions of one another, nor does it deny that between functions and form one may seek to establish more complex ties than the linear ones of cause and effect (which are belied by reality itself). More specifically, we reject that conception of functionalism dictated by an ingenuous empiricism which holds that *functions bring form together* and in themselves constitute urban artifacts and architecture.

So conceived, function, physiological in nature, can be likened to a bodily organ whose function justifies its formation and development and whose alterations of function imply an alteration of form. In this light, functionalism and organicism, the two principal currents which have pervaded modern architecture, reveal their common roots and the reason for their weakness and fundamental ambiguity. Through them form is divested of its most complex derivations: type is reduced to a simple scheme of organization, a diagram of circulation routes, and architecture is seen as possessing no autonomous value. Thus the aesthetic intentionally and necessity that characterize urban artifacts and establish their complex ties cannot be further analyzed.

Although the doctrine of functionalism has earlier origins, it was enunciated and applied clearly by Bronislaw Malinowski, who refers explicitly to that which is man-made, to the object, the house: "Take the human habitation. . .here again the integral function of the object must be taken into account when the various phases of its technological construction and the elements of its structure are studied"(Malinowski 1944). From a beginning of this sort one quickly descends to a consideration solely of the purposes which man-made items, the object and the house, serve. The question "for what purpose?" ends up as a simple justification that prevents an analysis of what is real.

Credits: This article is excerpted from Chapter 1 of *The Architecture of the City*, Aldo Rossi, 1984, by permission of MIT Press (see References).

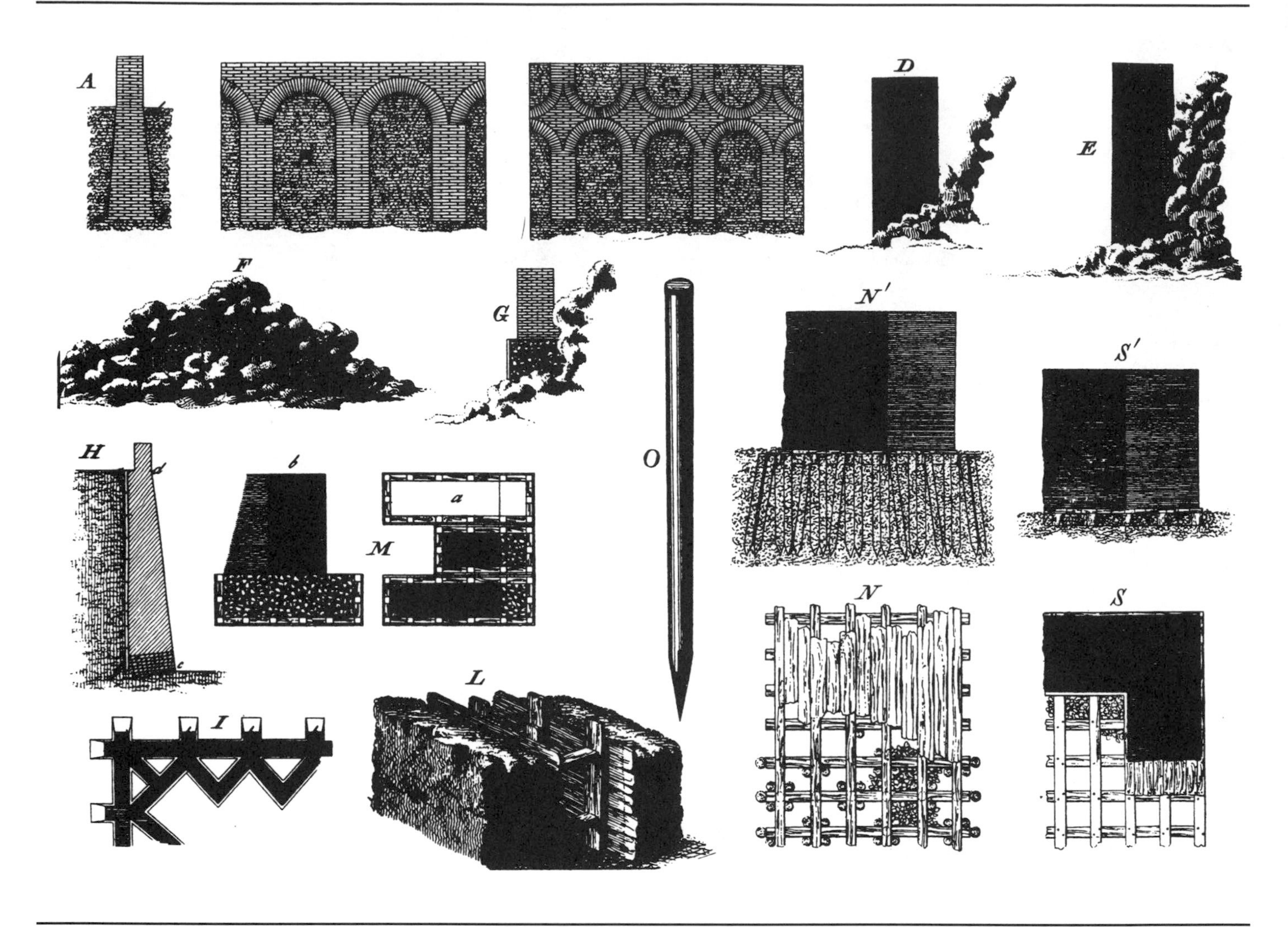

Various types of foundations. (From *Principj di Architettura Civile*. Francesco Milizia, 1832.)

This concept of function comes to be assumed as a given in all architectural and urbanistic thinking and, particularly in the field of geography, leads to a functionalist and organicist characterization of a large part of modern architecture. In studies of the classification of cities, it overwhelms and takes priority over the urban landscape and form; and although many writers express doubts as to the validity and exactitude of this type of classification, they argue that there is no other viable classification to offer as an alternative.

In such formulations, the city as an agglomeration is explained precisely on the basis of what functions its citizens seek to exercise; the function of a city becomes its *raison d'être*, and in this form reveals itself. In many cases the study of morphology is reduced to a simple study of function. Once the concept of function is established, in fact, one immediately arrives at obvious classifications: commercial cities, cultural cities, industrial cities, military cities, etc.

Moreover, even in the context of a somewhat general critique of the concept of function, it must be pointed out that there is already within this system of assigning functions a difficulty in establishing the role of the commercial function. In fact, as proposed, the concept of classification according to function is far too superficial; it assumes an identical value for all types of functions, which simply is not the case. Actually, the fact that the commercial function is predominant is increasingly evident.

This commercial function is the basis, in terms of production, of an "economic" explanation of the city that, beginning with the classical formulation offered by Max Weber (1956), has undergone a specific development, one to which we shall have to return later. Given a function-based classification of the city, it is only logical that the commercial function in both the city's formation and its development presents itself as the most convincing explanation for the multiplicity of urban artifacts and is tied to economic theories of the city.

Once we attribute different values to different functions, we deny the validity of naïve functionalism; in fact, using this line of reasoning we see that naïve functionalism ends up contradicting its own initial hypothesis. Furthermore, if urban artifacts were constantly able to reform and renew themselves simply by establishing new functions, the values of the urban structure, as revealed through its architecture, would be continuous and easily available. The permanence of buildings and forms would have no significance, and the very idea of the transmission of a culture, of which the city is an element, would be

The Doric order. (From *Principj di Architettura Civile*. Francesco Milizia, 1832.)

questionable. None of this corresponds to reality.

Naïve functionalist theory is quite convenient for elementary classifications, however, and it is difficult to see what can substitute for it at this level. It serves, that is, to maintain a certain order, and to provide us with a simple instrumental fact—just so long as it does not pretend that an explanation for more complex facts can be extracted from this same order.

On the other hand, the definition of type that we have tried to propose for urban artifacts and architecture, a definition which was first enunciated in the Enlightenment, allows us to proceed to an accurate classification of urban artifacts, and ultimately also to a classification based on function wherever the latter constitutes an aspect of the general definition. If, alternatively, we begin with a classification based on function, type would have to be treated in a very different way; indeed, if we insist on the primary of function we must then understand type as the organizing model of this function. But this understanding of type, and consequently urban artifacts and architecture, as the organizing principle of certain functions, almost totally denies us an adequate knowledge of reality. Even if a classification of buildings and cities according to their function is permissible as a generalization of certain kinds of data, it is inconceivable to reduce the structure of urban artifacts to a problem of organizing some more or less important function. Precisely this serious distortion has impeded and in large measure continues to impede any real progress in studies of the city.

For if urban artifacts present nothing but a problem of organization and classification, then they have neither continuity nor individuality. Monuments and architecture have no reason to exist; they do not "say" anything to us. Such positions clearly take on an ideological character when they pretend to objectify and quantify urban artifacts; utilitarian in nature, these views are adopted as if they were products for consumption. Later we will see the more specifically architectural implications of this notion.

To conclude, we are willing to accept functional classifications as a practical and contingent criterion, the equivalent of a number of other criteria—for example, social make-up, constructional system, development of the area, and so on—since such classifications have a certain utility; nonetheless it is clear that they are more useful for telling us something about the point of view adopted for classification than about an element itself. With these provisions in mind, they can be accepted.

Problems of classification

In my summary of functionalist theory I have deliberately emphasized those aspects that have made it so predominant and widely accepted. This is in part because functionalism has had great success in the world of architecture, and those who have been educated in this discipline over the past fifty years can detach themselves from it only with difficulty. One ought to inquire into how it has actually determined modern architecture, and still inhibits its progressive evolution today; but this is not an issue I wish to pursue here.

Instead, I wish to concentrate on the importance of other interpretations within the domain of architecture and the city, which constitute the foundations of the thesis that I am advancing. These include the social geography of Jean Tricart, the theory of persistence of Marcel Poète, and Enlightenment theory, particularly that of Milizia. All of these interest me primarily because they are based on a continuous reading of the city and its architecture and have implications for a general theory of urban artifacts.

For Tricart, the *social content* of the city is the basis for reading it; the study of social content must precede the description of the geographical artifacts that ultimately give the urban landscape its meaning. Social facts, to the extent that they present themselves as a specific content, precede forms and functions and, one might say, embrace them.

The task of human geography is to study the structures of the city in connection with the form of the place where they appear; this necessitates a sociological study of place. But before proceeding to an analysis of place, it is necessary to establish *a priori* the limits within which place can be defined. Tricart thus establishes three different orders or scales:

1. the scale of the street, including the built areas and empty spaces that surround it;
2. the scale of the district, consisting of a group of blocks with common characteristics;
3. the scale of the entire city, considered as a group of districts.

The principle that renders these quantities homogeneous and related them is social content.

On the basis of Tricart's thesis, I will develop one particular type of urban analysis which is consistent with his premises and takes a topographical point of view that seems quite important to me. But before doing so, I wish to register a fundamental objection to the scale of his study, or the three parts into which he divides the city. That urban artifacts should be studied solely in terms of place we can certainly admit, but what we cannot agree with is that places can somehow be explained on the basis of different scales. Moreover, even if we admit that the notion is useful either didactically or for practical research, it implies something unacceptable. This has to do with the *quality* of urban artifacts.

Therefore while we do not wholly deny that there are different scales of study, we believe that it is inconceivable to think that urban artifacts change in some way as a result of their size. The contrary thesis implies accepting, as do many, the principle that the city is modified as it extends, or that urban artifacts in themselves are different because of the size at which they are produced. As was stated by Richard Ratcliff,

> *To consider the problems of locational maldistribution only in the metropolitan context is to encourage the popular but false assumption that these are the problems of size. We shall see that the problems to be viewed crop us in varying degrees of intensity in villages, town, cities, and metropolises, for the dynamic forces of urbanism are vital wherever men and things are found compacted, and the urban organism is subject to the same natural and social laws regardless of size. To ascribe the problems of the city to size is to imply that solutions lie in reversing the growth process, that is, in deconcentration; both the assumption and the implication are questionable* (Ratcliff, 1959).

At the scale of the street, one of the fundamental elements in the urban landscape is the inhabited real estate and thus the structure of urban real property. I speak of inhabited real estate and not the house because the definition is far more precise in the various European languages. Real estate has to do with the deed registry of land parcels in which the principal use of the ground is for construction. The usage of inhabited land in large measures tends to be residential, but one could also speak of specialized real estate and mixed real estate, although this classification, while useful, is not sufficient.

To classify this land, we can begin with some considerations that are apparent from plans. Thus we have the following:

1. a block of houses surrounded by open space;
2. a block of houses connected to each other and facing the street, constituting a continuous wall parallel to the street itself;
3. a deep block of houses that almost totally occupies the available space;
4. houses with closed courts and small interior structures.

A classification of this type can be considered descriptive, geometric, or topographic. We can carry it further and accumulate other classificatory data relative to technical equipment, stylistic phenomena, the relationship between green and occupied spaces, etc. The questions this information gives rise to can lead us back to the principal issues which are, roughly speaking, those that deal with

1. objective facts;
2. the influence of the real-estate structure and economic data;
3. historical-social influences.

The real-estate structure and economic questions are of particular importance and are intimately bound up with what we call historical-social influences. In order to demonstrate the advantages of an analysis of this type, in the second chapter of this book we will examine the problems of housing and the residential district. For now, we will continue with the subject of real-estate structure and economic data, even if the second is given summary treatment.

The shape of the plots of land in a city, their formation and their evolution, represents a long history of urban property and of the classes intimately associated with the city. Tricart has stated very clearly that an analysis of the contrasts in the form of plots confirms the exis-

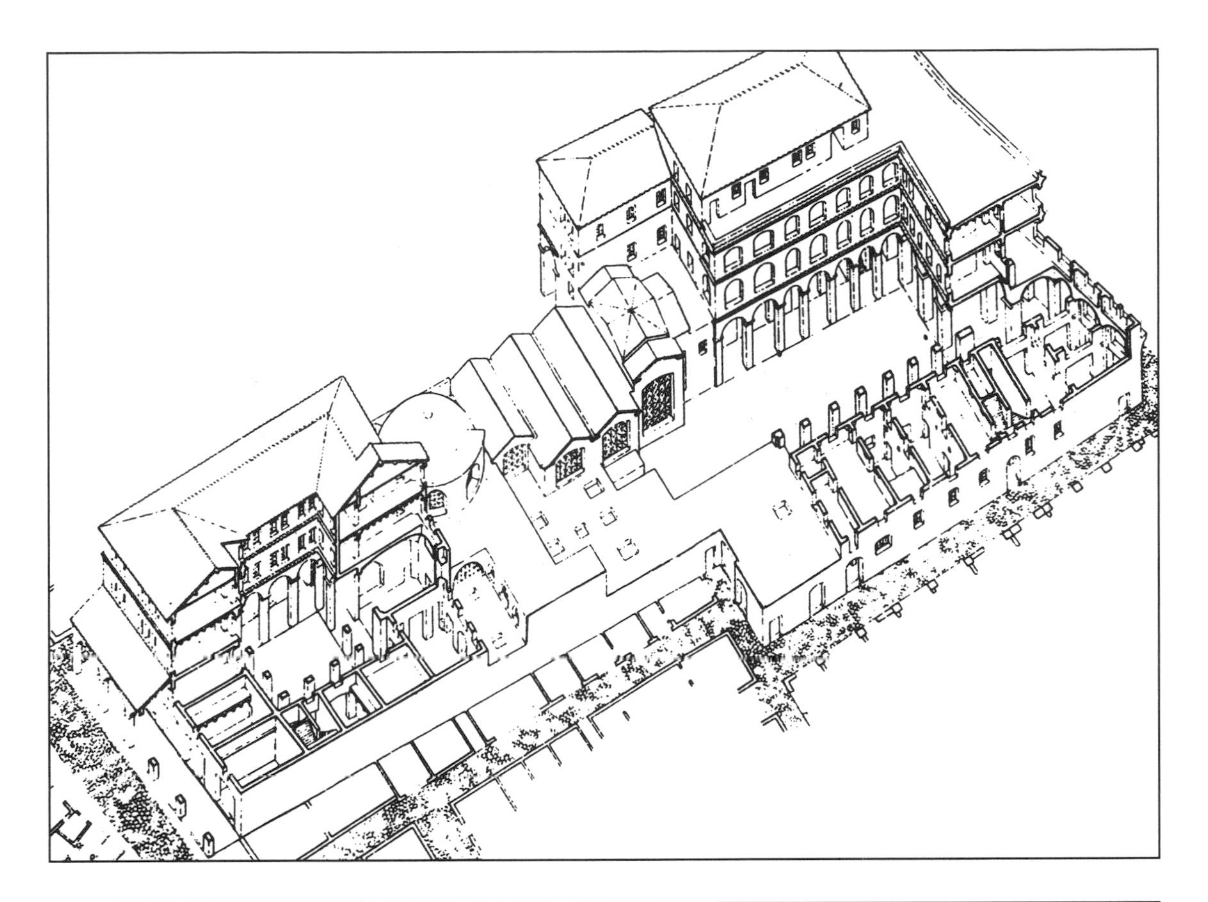

Ostia Antica, Rome. (upper) *Insula* (large lot) with the houses of Aurighi and Serapide, with bathhouses in the middle. (lower) Recostruction drawing by Italio Gismondi. c. 1940.

tence of a class struggle. Modifications of the real-estate structure, which we can follow with absolute precision through historical registry maps, indicate the emergence of an urban bourgeoisie and the phenomenon of the progressive concentration of capital.

A criterion of this type applied to a city with as extraordinary a life cycle as ancient Rome offers information of paradigmatic clarity. It allows us to trace the evolution from the agricultural city to the formation of the great public spaces of the Imperial age and the subsequent transition from the courtyard houses of the Republic to the formation of the great plebeians. The enormous lots that constituted the *insulae*, an extraordinary conception of the house-district, anticipate the concepts of the modern capitalist city and its spatial division. They also help to explain its dysfunction and contradictions.

Real estate, which we considered earlier from a topographic point of view, also offers other possibilities of classification when seen in a socio-economic context. We can distinguish the following:

1. the "precapitalist" house, which is established by a proprietor without exploitative ends;

2. the "capitalist" house, which is meant for rental and in which everything is subordinated to the production of revenue. Initially it might be intended either for the rich or the poor, but in the first case, following the usual evolution of needs, the house drops rapidly in class status in response to social changes. These changes in status create blighted zones, one of the most typical problems of the modern capitalist city and as such the object of particular study in the United States, where they are more evident than in Italy;

3. the "paracapitalist" house, built for one family with one floor rented out;

4. the "socialist" house, which is a new type of construction appearing in socialist countries where there is no longer private land ownership and also in advanced democratic countries. Among the earliest European examples are the houses constructed by the city of Vienna after the First World War.

When this analysis of social content is applied with particular attention to urban topography, it becomes capable of providing us with a fairly complete knowledge of the city; such an analysis proceeds by means of successive syntheses, causing certain elementary facts to come to light which ultimately encompass more general facts. In addition, through the analysis of social content, the formal aspect of urban artifacts takes on a reasonably convincing interpretation, and a number of themes emerge that play an important role in the urban structure.

From the scientific point of view, the work of Marcel Poète (1929) is without doubt one of the most modern studies of the city. Poète concerns himself with urban artifacts to the extent that they are indicative of the conditions of the urban organism; they provide precise information which is verifiable in the existing city. Their *raison d'être* is their continuity: while geographic, economic, and statistical information must also be taken into consideration along with historical facts, it is knowledge of the past that constitutes the terms of the present and the measure of the future.

Such knowledge can be derived from a study of city plans; these possess precise formal characteristics: for example, the form of a city's streets can be straight, sinuous, or curved. But the general form of the city also has a meaning of its own, and its needs naturally tend to be expressed in its built works, which beyond certain obvious differences present undeniable similarities. Thus in urban architecture a more or less clearly articulated bond is established between the shapes of things throughout history. Against a background of the differences between historical periods and civilizations, it therefore becomes possible to verify a certain constancy of themes, and this constancy assures a relative unity to the urban expression. From this develop the relationships between the city and the geographic region, which can be analyzed effectively in terms of the role of the street. Thus in Poète's analysis, the street acquires major significance; the city is born in a fixed place but the street gives it life. The association of the destiny of the city with communication arteries becomes a fundamental principle of development.

In his study of the relationship between the street and the city, Poète arrives at important conclusions. For any given city it should be possible to establish a classification of streets which should then be reflected in the map of the geographic area. Streets, whether cultural or commercial, should also be able to be characterized according to the nature of the changes that are effected because of them. Thus Poète repeats the Greek geographer Strabo's observation about the "shadow cities" along the Flaminian Way, whose development is explained as occurring "more because they were found situated along that road than for any inherent importance" (Poète, 1929, p. 60).

From the street, Poète's analysis passes to the urban land, which contains natural artifacts as well as civic ones and becomes associated with the composition of the city. In the urban composition, everything must express as faithfully as possible the particular life of the collective organism. At the basis of this organism that is the city is the persistence of the plan.

This concept of persistence is fundamental to the theory of Poète; it also informs the analysis of Pierre Lavedan (1936) one of the most complete analyses available to us, with its interposing of elements drawn from geography and the history of architecture. In Lavedan, persistence is the generator of the plan, and this generator becomes the principal object of urban research because through an understanding of it one can rediscover the spatial formation of the city. The generator embodies a concept of persistence which is reflected in a city's physical structures, streets, and urban monuments.

The contributions of Poète and Lavedan, together with those of the geographers Chabot and Tricart, are among the most significant offerings of the French school to urban theory.

The contribution of Enlightenment thought to a comprehensive theory of urban artifacts would merit a separate study. One objective of the treatise writers of the eighteenth century was to establish principles of architecture that could be developed from logical bases, in a certain sense independently of design; thus the treatise took shape as a series of propositions derived serially from one another. Second, they conceived of the single element always as part of a system, the system of the city; therefore it was the city that conferred criteria of necessity and reality on single buildings. Third, they distinguished form, as the final manifestation of structure, from the analytical aspect of structure; thus form had a "classical" persistence of its own which could not be reduced to the logic of the moment.

One could discuss the second argument at length, but more substantial knowledge would certainly be necessary; clearly, while this argument applies to the existing city, it also postulates the future city and the inseparable relationship between the constitution of an artifact and its surroundings. Yet Voltaire had already indicated, in his analysis of the *grand siècle*, the limits of such architectures, how uninteresting a city would be if the task of every constructed work was to establish a direct relationship with the city itself. The manifestation of these concepts is found in the Napoleonic plans and projects, which represent one of the moments of major equilibrium in urban history.

On the basis of these three arguments developed in the Enlightenment, we can examine the theory of Milizia (1832). The classification proposed by Milizia, an architectural essayist concerned with theories of urban artifacts, deals with both individual buildings and the city as a whole. He classified urban buildings as either private or public, the former meaning housing and the latter referring to certain "principal elements" which I will call *primary*. In addition, he presents these groupings as classes, which permits him to make distinctions within classes, distinguishing each principal element as a building type within a general function, or better, a general idea of the city. For example, villas and houses are in the first class, while in the second are public buildings, public utilities, storage facilities, etc. Buildings for public use are further distinguished as universities, libraries, and so on.

Milizia's analysis refers in the first place, then, to classes (public and private), in the second to the location of elements in the city, and in the third to the form and organization of individual buildings. "Greater public convenience demands that these buildings [for public use] be situated near the center of the city and organized around a large community square." The general system is the city; the development of its elements is then bound up with the development of the system adopted.

What kind of city does Milizia have in mind? It is a city that is conceived together with its architecture. "Even without extravagant buildings, cities can appear beautiful and breathe desire. But to speak of a beautiful city is also to speak of good architecture." This assertion seems definitive for all Enlightenment treatises on architecture; a beautiful city means good architecture, and vice versa.

It in unlikely that Enlightenment thinkers paused over this statement, so ingrained was it in their way of thinking; we know that their lack of understanding of the Gothic city was a result of their inability to accept the validity of single elements that constituted an urban landscape without seeing these elements relative to some larger system. If in their failure to understand the meaning and thus the beauty of the Gothic city they were shortsighted, this of course does not make their own system incorrect. However, to us today the beauty of the Gothic city appears precisely in that it is an extraordinary urban artifact whose uniqueness is clearly recognizable in its components. Through our investigation of the parts of this city we grasp its beauty: it too participates in a system. There is nothing more false than an organic or spontaneous definition of the Gothic city.

There is yet another aspect of modernity in Milizia's position. After establishing his concept of classes, he goes on to classify each building type within the overall framework and to characterize it according to its function. This notion of function, which is treated independently of general considerations of form, is understood more as the building's purpose than as its function *per se*. Thus buildings for practical uses and those that are constructed for functions that are not equally tangible or pragmatic are put in the same class; for example, buildings for public health or safety are found in the same class as structures built for their magnificence or grandeur.

There are at least three arguments in favor of this position. Most important is the recognition of the city as a complex structure in which parts can be found that function as works of art. The second has to do with the value ascribed to a general typological discourse on urban artifacts or, in other words, the realization that one can give a technical explanation by reducing them to their typological essence. The third argument relates to the fact that this typological essence plays "its own role" in the constitution of the model.

For example, in analyzing the monument, Milizia arrives at three criteria:

> *that it is directed toward the public good; that it is appropriately located; and that it is constituted according to the laws of fitness. . . . With respect to the customs governing the construction of monuments, no more can be said here generally than that they should be meaningful and expressive, of a simple structure, and with a clear and short inscription, so that the briefest glance reveals the effect for which they were constructed.*

In other words, insofar as the nature of the monument is concerned, even if we cannot offer more than a tautology—a monument is a monument—we can still establish conditions around it which illustrate its typological and compositional characteristics, whether these precisely elucidate its nature or not. Again, these characteristics are for the most part of an urban nature; but they are equally conditions of architecture, that is, of composition.

This is a basic issue to which we will return later: namely, the way in which principles and classifications in the Enlightenment conception were a general aspect of architecture, but that in its realization and evaluation, architecture involved primarily the individual work and the individual architect. Milizia himself scorned the builders who mixed architectural and social orders as well as the proponents of objective models of functional organization such as were later produced by Romanticism, asserting that, "to derive functional organization from beehives is to go insect-hunting. . ." Here again we find within a single formulation the two themes which were to be fundamental in the subsequent development of architectural thought, and which already indicated in their dual aspects of organicism and functionalism their anticipation of the Romantic sensibility: the abstract order of organization and the reference to nature.

With respect to function itself, Milizia writes,

> *. . . because of its enormous variety functional organization cannot always be regulated by fixed and constant laws, and as a result must always resist generalization. For the most part, the most renowned architects, when they wish to concern themselves with functional organization, mainly produced drawings and descriptions of their buildings rather than rules that could then be learned.* (Milizia 1832)

This passage clearly shows how function is understood here as a relationship and not a scheme of organization; in fact, as such it is rejected. But this attitude did not preclude a contemporaneous search for rules that might transmit principles of architecture. ■

REFERENCES

Lavenda, Pierre. 1936. *Géographic des villes*. Paris: Gallimard.

Malinowski, Bronislaw. 1944. *A Scientific Theory of Culture and Other Essays*. Chapel Hill, NC: University of North Carolina Press.

Milizia, Francesco. 1832. *Principi di Architettura Civile*. Quoted passages are from *"Parte Seconda. Della Comodita."* Milan: Gabrielle Mazzotta. 1972.

Poète, Marcel. 1929. *Introduction à l'urbanisme. L'évolution des villes, la lecon de l'antiquité*. Paris: Boivin & Cie.

Ratcliff, Richard Updegraff. 1959. "The Dynamics of Efficiency in the Locational Distribution of Urban Activities," in Herman Melvin Mayer and Clyde Frederick Kohn, eds. *Readings in Urban Geography*. Chicago: University of Chicago Press. p. 299.

Rossi, Aldo. 1984. *The Architecture of the City*. Cambridge, MA: MIT Press. pp. 46–55.

Tricart, Jean. 1963. *Cours de Geographié Humaines*. Paris: Centre de Documentation Universitaire.

Weber, Max. 1956. *Wirtschaft and Gesellschaft*. Türbingen: U.C.B. Mohr-Paul Siebeck.

Colin Rowe and Fred Koetter

Summary

Colin Rowe (1920–1999) and Fred Koetter developed *Collage City* out of the ideas that had guided the research and design work of Cornell's urban design program. *Collage City* provides an anti-utopian theory of urban design, based on the broad art historical erudition, rigorous formal analysis, and revisionist account of the Modern Movement that had characterized Rowe's earlier writing and teaching. The principal protagonist—as well as antagonist—of this argument is Le Corbusier, whose architectural and urban proposals are reproduced here as figure-ground images and dramatically contrasted—as he had contrasted his own work with the ostensibly inefficient fabric of older cities—with the rich and continuous texture of traditional urban districts and projects. Eclectic, hybrid juxtaposition and layering of historic cities are presented here as formal correlatives of an open and inclusive approach more characteristic of, and conducive to, urban life than the abstract purity of modernist proposals.

Key words

architectural theory, building, city, context, form, historical continuity, infill, perception, street, typology

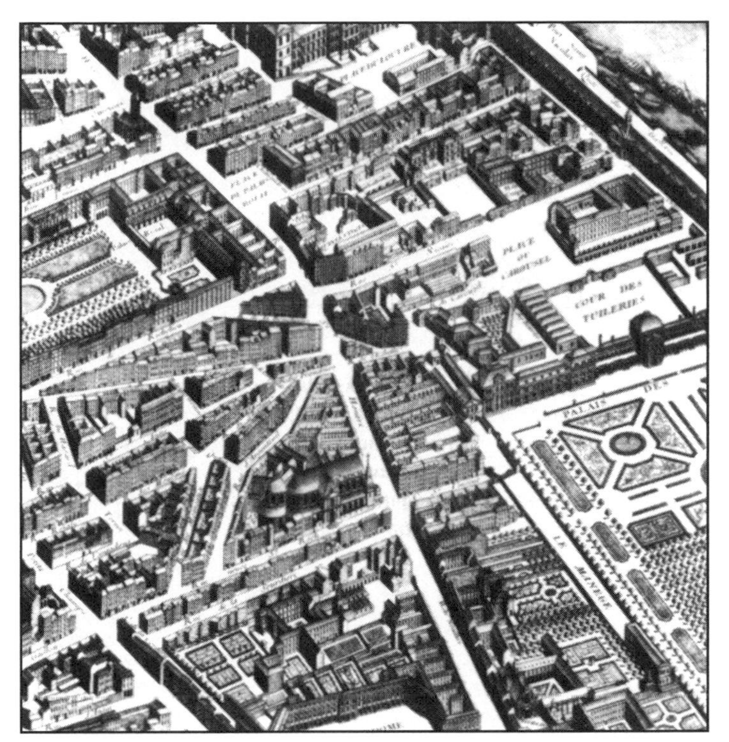

Paris, the Louvre, Tuileries, and Palais Royale, from the Plan Turgot, 1739.

Crisis of the object: Predicament of texture

Cities force growth and make men talkative and entertaining but they make men artificial. —Ralph Waldo Emerson

I think that our governments will remain virtuous as long as they are chiefly agricultural. —Thomas Jefferson

But. . .how can man withdraw himself from the fields? Where will he go, since the earth is one huge unbounded field? Quite simple; he will mark off a portion of this field by means of walls, which set up an enclosed finite space over against amorphous, limitless space. . . For in truth the most accurate definition of the urbs and the polis is very like the comic definition of a cannon. You take a hole, wrap some steel wire tightly around it, and that's your cannon. So the urbs or polis starts by being an empty space . . .and all the rest is just a means of fixing that empty space, of limiting its outlines. . .The square. . .This lesser rebellious field which secedes from the limitless one, and keeps to itself, is a space sui generis of the most novel kind in which man frees himself from the community of the plant and the animal. . .and creates an enclosure apart which is purely human, a civil space.

—Jose Ortega y Gasset

In intention the modern city was to be a fitting home for the noble savage. A being so aboriginally pure necessitated a domicile of equivalent purity: and, if way back the noble savage had emerged from the trees, then, if his will-transcending innocence was to be preserved, his virtues maintained intact, it was back into the trees that he must be returned.

One might imagine that such an argument was the ultimate psychological rationale of the *ville radieuse* or *Zeilenbau City*, a city which, in its complete projection, was almost literally imagined as becoming non-existent. Immediately necessary buildings appear, so far as possible, as delicate and unassertive intrusions into the natural continuum; buildings raised above the ground provide as little contact as possible with the potentially reclaimable earth: and, while there ensues a freedom-releasing qualification of gravity, we are perhaps also encouraged to recognize a commentary upon the dangers of prolonged exposure to any conspicuous artifact.

The projected modern city, in this way, may be seen as a transitional piece, a proposal which eventually, it is hoped, may lead to the re-establishment of an unadulterated natural setting.

Credits: This article is an excerpt from *Collage City*, MIT Press, 1978. Reprinted by permission of the publisher (see References).

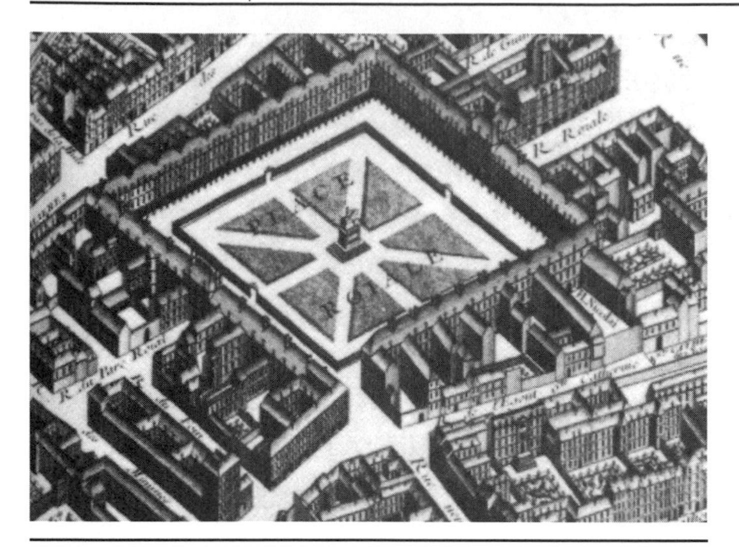

Fig. 1. Paris, *Place des Vosges (Place Royale)*. From the *Plan Turgot*, 1739.

Fig. 2. Le Corbusier: *Ville Radieuse*, 1930.

> *Sun, space, verdure: essential joys, through the four seasons stand the trees, friends of man. Great blocks of dwellings run through the town. What does it matter? They are behind the screen of trees. Nature is entered into the lease.*[1]

Such was the vision of an ever-evolving return to nature, a return that was (and is) evidently felt to be so important that, whenever possible, demonstrations of this vision have insisted on their absolute detachment, symbolic and physical, from any aspects of existing context which has been typically, envisaged as a contaminant, as something both morally and hygienically leprous. And thus Lewis Mumford on an illustration in *Culture of Cities*:

> *Rear of a handsome façade in Edinburgh: barracks architecture facing a catwalk: typical indifference to rear views characteristic of scene painting. An architecture of fronts. Beautiful silks, costly perfumes. Elegance of mind and small pox. Out of sight, out of mind. Modern functional planning distinguishes itself from this purely visual conception of the plan, by dealing honestly and competently with every side, abolishing the gross distinction between front and rear, seen and obscene, and creating structures that are harmonious in every dimension.*[2]

Which, allowing for a characteristically Mumfordian rhetoric, is all classically representative of the bias and the inter-war period. The prominent criteria are honesty and hygiene, the city of vested interest and impacted association is to disappear; and, in place of traditional subterfuge and imposition, there is to be introduced a visible and rational equality of parts—an equality which insists upon openness and is readily to be interpreted as both cause and effect of any condition of humane well-being.

Now, of course, the equation of the backyard with moral and physical insalubrity, which become the opposition of closure and openness and their investment with negative and positive qualities ("Elegance of mind and small pox"—as though the one automatically followed the other), could be illustrated from an abundance of other sources; and, in terms of that distinctively nineteenth century vision of the danse macabre, the human scarecrow in the cholera-infected courtyard, this style of augment should scarcely require reinforcement. Visually oriented architects and planners, preoccupied with the trophies and triumphs of culture, with the representation of the public realm and its public façades, had, for the most part, shamefully compromised not only the pleasurable possibilities but, worse than this, the essential sanitary bases of that more intimate world within which "real" people, people as deserving aspects of concern, actually do exist. And, if this statement were to be augmented to say something about pragmatically callous capitalists then its general substance would not be radically transformed.

But, if such was the one-time negative and necessary criticism of traditional metropolis, then if an overview of the nineteenth century Paris can be allowed to represent the evil, an overview of Amsterdam South may also be introduced to exhibit the initial conceptions of an alternative; and both illustrations derive from the accessible pages of Siegfried Giedion.[3]

The Hausmannesque situation, as witnessed by a bird or from a balloon, is so sufficiently comparable to the air photo of Berlagian Amsterdam as to need the minimum of comment. Both are subservient to the aesthetic of the French seventeenth century hunting forest with its *ronds-points* and *pattes-d'oie;* and, in being so, they both of them, by means of major arteries converging at a, hopefully, significant place, describe a triangular territory as subject for development or infill. But then it is here, with the infull, that resemblance ceases. For, if among the granduers and brutalities of Second Empire Paris, logical infill could be disregarded, if it could be reduced to the abstract volumetric status of trees in a garden by Le Notre, then in conscientious early twentieth century Holland such a highly visual universal matrix of "texture" was, emphatically, not available. And, because of the French prototype, the result is a Dutch embarrassment. In Amsterdam a genuine attempt has been made to provide a more tolerable theater of existence. Air, light, prospect, open space have all been made available; but, while one may sense that one is here on the threshold of the welfare state, one may still be overcome by the anomaly. The two big avenues, for all their ambitious protestation, are diffident and residual. They are lacking in the vulgar or the boring swagger and self-confidence of their Parisian prototypes. They are among the last pathetic gestures to the notion of the street; and their carefully edited concessions to *De Stijl* or to Expressionism do not conceal their predicament. They have become no more than the conservatively insinuated props to a dying idea. For, in the argument

Fig. 3. Paris, Boulevard Richard-Lenoir, 1861.

Fig. 4. Amsterdam South, c. 1961.

of solid versus void they have become redundant; and their references to a vision of classical Paris now have nothing to say. Simply these avenues are disposable. In no way do their facades designate any effective frontier between public and private. They are evasive. And much more than the façades of eighteenth century Edinburgh, they ineffectively conceal. For the important reality has not become what lies behind. The matrix of the city has become transformed from continuous solid to continuous void.

It goes without saying that both the failure and success of Amsterdam South, and of many comparable projects, could only activate the conscience; but, whatever may have been the doubts (the conscience is always more activated by failure than success), it probably remains true to say that logical skepticism was not able to digest the issue for at least some ten years. Which is to say that, until the late nineteen-twenties, the culturally obligatory street still dominated the scene and that, as a result, certain conclusions remained unapproachable.

In this sequence, the questions of who did what and precisely when and where are, for present purposes, irrelevant. The City of Three Million Inhabitants, miscellaneous Russian projects, Karlsruhe-Dammar-stock, etc., all have their dates; and the assignment of priority or praise or blame is not here an issue. Simply the issue is that, by 1930, the disintegration of the street and of all highly organized public space seemed to have become inevitable; and for two major reasons: the new and rationalized form of housing and the new dictates of vehicular activity. For, if the configuration of housing now evolved from the inside out, from the logical needs of the individual residential unit, then it could no longer be subservient to external pressures; and, if external public space had become so functionally chaotic as to be without effective significance, then—in any case—there were no valid pressures which it could any longer exert.

Such were the apparently unfaultable deductions which underlay the establishment of the city of modern architecture; but, around these primary arguments, there was evidently the opportunity for a whole miscellany of secondary rationalizations to proliferate. And thus the new city could achieve further justification in terms of sport or of science, in terms of democracy or equality, in terms of history and absence of traditional *parti pris*, in terms of private automobiles and public transport, in terms of technology and socio-political crisis; and, like the idea of the city of modern architecture itself, in some form or another, almost all of these arguments are still with us.

And, of course, they are reinforced (though whether reinforcement is the correct word may be doubted) by others. "A building is like a soap bubble. This bubble is perfect and harmonious if the breadth has been evenly distributed from the inside. The exterior is the result of an interior."[4] This debilitating half truth has proved to be one of Le Corbusier's more persuasive observations; but, if it is an impeccable statement of academic theory relating to domed and vaulted structures, it is also a dictum which could only lend support to the notion of the building as preferably free standing object in the round. Lewis Mumford intimates as much; but, if for Theo Van Doesburg and many others it was axiomatic that 'the new architecture will develop in an all sided plastic way,"[5] this placing of immensely high premia upon the building as "interesting" and detached object (that still continues) must now be brought into conjunction with the simultaneously entertained proposition that the building (object?) must be made to go away ("Great blocks of dwellings run through the town. What does it matter? They are behind the screen of trees"). And, if we have here presented this situation in terms of a typically Corbusian self-contradiction, there is obvious and abundant reason to recognize that one is confronted with this same contradiction any, and every, day. Indeed, in modern architecture, the pride in objects and the wish to dissimulate pride in this pride, which is everywhere revealed, is something so extraordinary as to defeat all possibility of compassionate comment.

But modern architecture's object fixation (the object which is not an object) is our present concern only insofar as it involves the city, the city which was to become evaporated. For, in its present and unevaporated form, the city of modern architecture that becomes a congeries of conspicuously disparate objects is quite as problematical as the traditional city which it has sought to replace.

Let us, first of all, consider the theoretical desideratum that the rational building is obliged to be an object and, then, let us attempt to place this proposition in conjunction with the evident suspicion that buildings, as man-made artifacts, enjoy a meretricious status, in some way, detrimental to an ultimate spiritual release. Let us further attempt to place this demand for the rational materialization of the object

Fig. 5. Harlow New Town, Market Square, 1950s.

Fig. 6. Le Corbusier. Project for city center of Saint-Dié, perspective, 1945.

and this parallel need for its disintegration alongside the very obvious feeling that space is, in some way, more sublime than matter, that, while the affirmation of matter is inevitably gross, the affirmation of a spatial continuum can only facilitate the demands of freedom, nature and spirit. And then let us qualify what became a widespread tendency to space worship with yet another prevalent supposition: that, if space is sublime, then limitless naturalistic space must be far more so than any abstracted and structured space; and, finally, let us upstage this whole implicit argument by introducing the notion that, in any case, space is far less important than time and that too much insistence—particularly upon delimited space—is likely to inhibit the unrolling of the future and the natural becoming of the "universal society."

Such are some of the ambivalences and fantasies which were, and still are, embedded in the city of modern architecture; but, though these could seem to add up to a cheerful and exhilarating prescription, as already noticed, even when realizations of this city, though pure, were only partial, doubts about it began very early to be entertained. Perhaps these were scarcely articulated doubts and whether they concerned the necessities of perception or the predicament of the public realm is difficult to determine; but, if, in the Athens Congress of 1933,[6] CIAM had spelled out the ground rules for the new city, then by the mid-forties there could be no such dogmatic certainty. For neither the state nor the object had vanished away; and, in CIAM's *Heart of the City*[7] conference of 1947, lurking reservations as to their continuing validity began, indecisively, to surface. Indeed, a consideration of the "city core," in itself, already indicates a certain hedging of bets and, possibly, the beginnings of a recognition that the ideal of indiscriminate neutrality or inconspicuous equality was hardly attainable or even desirable.

But, if a renewed interest in the possibilities of focus and hence of confluence seems, by this time, to have been developing, while the interest was there, the equipment to service it was lacking; and the problem presented by the revisionism of the late forties might best be typified and illustrated by Le Corbusier's plan for St. Dié, where modified standard elements of Athens Charter specification are loosely arranged so as to insinuate some notions of centrality and hierarchy, to simulate some version of "town center" or structured receptacle. And might it be said that, in spite of the name of its author, a built St. Dié would, probably, have been the reverse of successful; that St. Dié illustrates, as clearly as possible, the dilemma of the free standing building, the space occupier attempting to act as space definer? For, if it is to be doubted whether this "center" would facilitate confluence, then regardless of the desirability of this effect, it seems that what we are here provided with is a kind of unfulfilling schizophrenia—an acropolis of sorts which is attempting to perform as some version of an agora?

However, in spite of the anomaly of the undertaking, the re-affirmation of centralizing themes was not readily to be relinquished; and, if the "core of the city" argument might easily be interpreted as a seepage of townscape strategies into the CIAM city diagram, a point may now be made by bringing the St. Dié city center into comparison with that of the approximately contemporary Harlow new town which, though evidently "impure," may not be quite so implausible as, sometimes, has appeared to be the case.

At Harlow, where there is absolutely no by-play with metaphors of acropolis, there can be no doubt that what one is being offered is a "real" and literal marketplace; and, accordingly, the discrete aspects of the individual buildings are played down, the buildings themselves amalgamated, to appear as little more than a casually haphazard defining wrapper. But, if the Harlow town square, supposed to be the authentic thing itself, a product of the vicissitudes of time and all the rest, may be a little over-ingratiating in its illusory appeal, if one might be just a little fatigued with quite so enticing a combination of instant "history" and overt "modernity," if its simulation of medieval space may still appear believable as one stands inside it, then, as curiosity becomes aroused, even this illusion quickly disappears.

For an overview or quick dash behind the immediately visible set piece rapidly discloses the information that what one has been subjected to is little more than a stage set. That is, the space of the square, professing to be an elevation of density, the relief of an impacted context, quickly lends itself to be read as nothing of the kind. It exists without essential back up or support, without pressure, in built or human form, to give credibility or vitality to its existence; and, with the space thus fundamentally "unexplained," it becomes apparent that, far from being any outcropping of an historical or spatial context (which it would seem to be), the Harlow town square is, in effect, a foreign body interjected into a garden suburb without benefit of quotation marks.

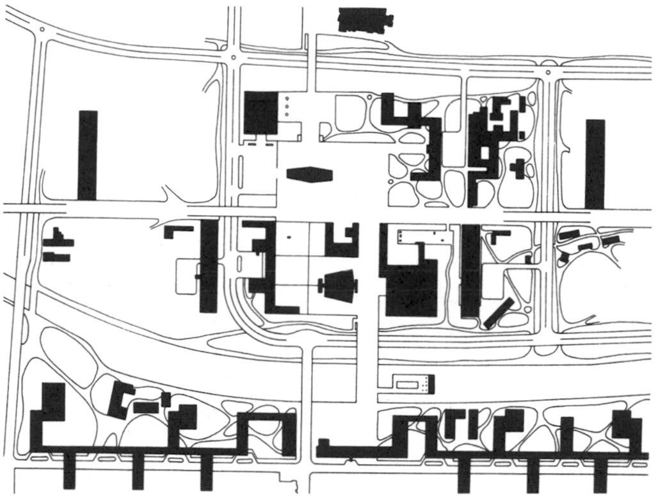

Fig. 7. Le Corbusier: project for Saint-Dié, figure-ground plan, 1945.

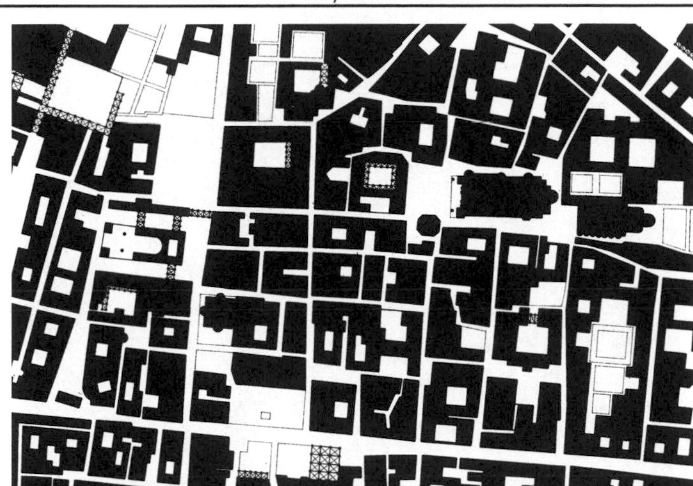

Fig. 8. Parma, figure-ground plan.

But, in the issue of Harlow *versus* St. Dié, one is still obliged to recognize a coincidence of intention. In both cases the object is the production of a significant urban foyer; and, given this aim, it seems perfectly fair to say that, whatever its merits as architecture, the Harlow town square provides a closer approximation to the imagined condition than ever St. Dié might have done. Which is neither to endorse Harlow nor condemn St. Dié; but is rather to allow them both, as attempts to simulate the qualities of "solid" city with the elements of "void," to emerge as comparable gestures of interrogation.

Now, as to the relevance of the questions which they propound, this might be best examined by once more directing attention to the typical format of the traditional city which, in every way, is so much the inverse of the city of modern architecture that the two of them together might, sometimes, almost present themselves as the alternative reading of some *Gestalt* diagram illustrating the fluctuations of the figure-ground phenomenon. Thus, the one is almost all white, the other almost all black; the one an accumulation of solids is largely unmanipulated void, the other an accumulation of voids in largely unmanipulated solid; and, in both cases, the fundamental ground promotes an entirely different category of figure—in the one *object*, in the other *space*.

However, not to comment upon this somewhat ironical condition; and simply, in spite of its obvious defects, to notice very briefly the apparent virtues of the traditional city: the solid and continuous matrix or texture giving energy to its reciprocal condition, the specific space; the ensuing square and street acting as some kind of public relief valve and providing some condition of legible structure; and, just as important, the very great versatility of the supporting texture or ground. For, as a condition of virtually continuous building of incidental make up and assignment, this is not under any great pressure for self-completion or overt expression of function; and, given the stabilizing effects of public façade, it remains relatively free to act according to local impulse or the requirements of immediate necessity.

Perhaps these are virtues which scarcely require to be proclaimed; but, if they are, everyday, more loudly asserted, the situation so described is still not quite tolerable. If it offers a debate between solid and void, public stability and private unpredictability, public figure and private ground which has not failed to stimulate, and if the object building, the soap bubble of sincere internal expression, when taken as a universal proportion, represents nothing short of a demolition of public life and decorum, if it reduces the public realm, the traditional world of visible civics to an amorphic remainder, one is still largely impelled to say: so what? And it is the logical, defensible presuppositions of modern architecture—light, air, hygiene, aspect, prospect, recreation, movement, openness—which inspire this reply.

So, if the sparse, anticipatory city of isolated objects and continuous voids, the alleged city of freedom and "universal" society will not be made to go away and if, perhaps, in its essentials, it is more valuable than its discreditors can allow, if, while it is felt to be "good," nobody seems to like it, the problem remains: what to try to do with it?

There are various possibilities. To adopt an ironical posture or to propound social revolution are two of them: but, since the possibilities of simple irony are almost totally pre-empted and since revolution tends to turn into its opposite, then, in spite of the persistent devotees of absolute freedom, it is to be doubted whether either of these are very useful strategies. To propose that more of the same, or more of approximately the same, will—like old-fashioned *laissez faire*—provide self-correction? This is just as much to be doubted as is the myth of the unimpaired capacities of self-regulation capitalism; but, all of these possibilities apart, it would seem, first of all, to be reasonable and plausible to examine the threatened or promised city of object fixation from the point of view of the possibility of its perception.

It is a matter of how much the mind and eye can absorb or comprehend; and it is a problem which has been around, without any successful solution, since the later years of the eighteenth century. The issue is that of quantification.

> *Pancras is like Marylebone, Marylebone is like Paddington: all the streets resemble each other. . .your Gloucester Places, and Baker Streets, and Harley Streets, and Wimpole Streets. . .all of those flat, dull, spiritless streets, resembling each other like a large family of plain children, with Portland Place and Portman Square for their respectable parents.*[8]

The time is 1847 and the judgment, which is Disraeli's, may be taken as a not so early reaction to the disorientation produced by repetition.

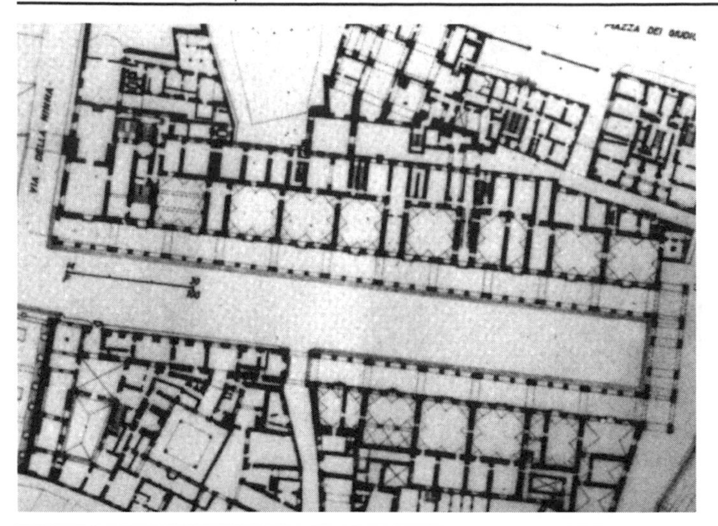

Fig. 9. Florence, Uffizi, plan.

Fig. 10. Le Corbusier: Marseilles, Unité d'Habitation, 1946, site plan.

But, if the multiplication of spaces long ago began to elicit such disgust, then what is there now to be said about the proliferation of objects? In other words, whatever may be said about the traditional city, is it possible that the city of modern architecture can sustain anything like so adequate a perceptual base? And the obvious answer would seem to be no. For it is surely apparent that, while limited structured spaces may facilitate identification and understanding, an intermediate naturalistic void without any recognizable boundaries will at least be likely to defeat all comprehension.

Certainly, in considering the modern city from the point of view of perceptual performance, by *Gestalt* criteria it can only be condemned. For, if the appreciation or perception of object or figure is assumed to require the presence of some sort of ground or field, if the recognition of some sort of however closed field is a prerequisite of all perceptual experience and, if consciousness of field precedes consciousness of figure, then, when figure is unsupported by any recognizable frame of reference, it can only become enfeebled and self-destructive. For, while it is possible to imagine—and to imagine being delighted by—a field of objects which are legible in terms of proximity, identity, common structure, density, etc., there are still questions as to how much such objects can be agglomerated and of how plausible, in reality, it is to assume the possibility of their exact multiplication. Or, alternatively, these are questions relative to optical mechanics, of how much can be supported before the trade breaks down and the introduction of closure, screening, segregation of information, becomes an experimental imperative.

Presumably this point has not, as yet, quite been reached. For the modern city in its cut-price versions (the city in the park become the city in the parking lot), for the most part still exists within the closed fields which the traditional city supplies. But, if, in this way—not only perceptually but also sociologically parasitic—it continues to feed off the organism which it proposes to supplant, then the time is now not very far remote when this sustaining background may finally disappear.

Such is the incipient crisis of more than perception. The traditional city goes away; but even the parody of the city of modern architecture refuses to become established. The public realm has shrunk to an apologetic ghost but the private realm has not been significantly enriched; there are no references—either historical or ideal; and, in this atomized society, except for what is electronically supplied or is reluctantly sought in print, communication has either collapsed or reduced itself to impoverished interchange of ever more banal verbal formulas.

Evidently, it is not necessary that the dictionary, whether *Webster's* or *OED*, need retain its present volume. It is redundant; its bulk is inflated: the indiscriminate use of its content lends itself to specious rhetoric; its sophistications have very little to do with the values of "jus' plain folks"; and, certainly, its semantic categories very little correspondence with the intellectual processes of the neo-noble savage. But, if the appeal, in the name of innocence, seriously to abbreviate the dictionary might find only a minimum of support, even though built forms are not quite the same as words, we have here sketched a program strictly analogous to that which was launched by modern architecture.

Let us eliminate the gratuitous; let us concern ourselves with needs rather than wants; let us not be too preoccupied with framing the distinctions; instead let us build from fundamentals. . . .Something very like this was the message which led to the present impasse; and, if contemporary happenings are believed (like modern architecture itself) to be inevitable, of course, they will become so. But, on the other hand, if we do not suppose ourselves to be in the Hegelian grip of irreversible fate, it is just possible that there are alternatives to be found.

In any case the question at this point is not so much whether the traditional city, in absolute terms, is good or bad, relevant or irrelevant, in tune with the *Zeitgeist* or otherwise. Nor is it a question of modern architecture's obvious defects. Rather it is a question of common sense and common interest. We have two models of the city. Ultimately, wishing to surrender neither, we wish to qualify both. For in an age, allegedly, of optional latitude and pluralist intention, it should be possible at least to plot some kind of strategy of accommodation and coexistence.

But, if in this way we now ask for deliverance from the city of deliverance, then in order to secure any approximation to this condition of freedom, there are certain cherished fantasies, not without final value, which the architect must be called upon to imagine as modified

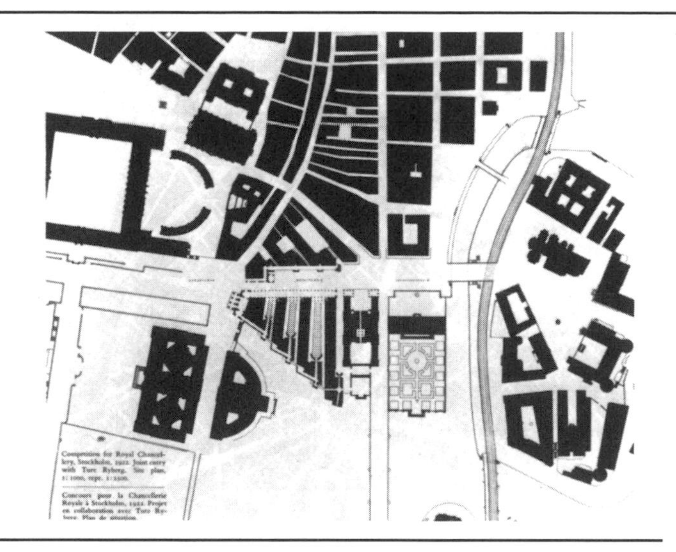

Fig. 11. Asplund: Chancellery, site plan.

Fig. 12. Le Corbusier: Paris, *Plan Voisin*, 1925. Partial excerpt of the figure-ground plan.

and redirected. The notion of himself as messiah is one of these; and, while the notion of himself as eternal proponent of avant gardeism is another, even more important is the strangely desperate idea of architecture as oppressive and coercive.[9] Indeed, particularly, this curious relic of neo-Hegelianism will require to be temporarily suppressed; and this in the interests of a recognition that "oppression" is always with us as the insuperable condition of existence "oppression" of birth and death, of place and time, of language and education, of memory and numbers, being all of them components of a condition which, as yet, is not to be superseded.

And so to proceed from diagnosis—usually perfunctory—to prognosis —generally even more causal—firstly there might be suggested the overthrow of one of modern architecture's least avowed but most visible tenets. This is the proposition that all outdoor space must be in public ownership and accessible to everybody; and, if there is no doubt that this was a central working idea and, has, long since, become a bureaucratic cliché, there is still the obligation to notice that, among the repertory of possible ideas, the inordinate importance of this one is very odd indeed. And thus, while its iconographic substance may be recognized—it meant a collectivized and emancipated society, which knew no artificial barriers—one may still marvel that such an offbeat proposition could ever have become so established. One walks through the city—whether it is New York, Rome, London or Paris who cares; one sees lights upstairs, a ceiling, shadows, some objects; but, as one mentally fills in the rest and imagines a society of unexampled brilliance from which one if fatally excluded, one does not feel exactly deprived. For, in this curious commerce between the visible and the undisclosed, we are well aware that we too can erect our own private hallucination which, however absurd it may be, is never other than stimulating.

This is to specify, in a particularly extreme form, a way in which exclusion may gratify the imagination. One is called upon to complete apparently mysterious but really normal situations of which one is made only partially aware; and, if literally to penetrate all these situations would be destructive of speculative pleasure, one might now apply the analogy of the illuminated room to the fabric of the city as a whole. Which is quite simple to say that the absolute spatial freedoms of the *ville radieuse* and its more recent derivatives are without interest; and that, rather than being empowered to walk everywhere—everywhere being always the same—almost certainly it would be more satisfying to be presented with the exclusions—walls, railings, fences, gates, barriers—of a reasonably constructed ground plane.

However, if to say so much is only to articulate what is already a dimly perceived tendency, and if it is usually provided with sociological justification[10] (identity, collective, "turf," etc.), there are more important sacrifices of contemporary tradition which are surely required; and we speak of a willingness to reconsider the object which allegedly nobody wants and to evaluate it not so much as figure but as ground.

A proposal which, for practical purposes, demands a willingness to imagine the present dispensation as inverted, the idea of such inversion is most immediately and succinctly to be explained by the comparison of a void and a solid of almost identical proportions. And, if to illustrate prime solid nothing will serve better than Le Corbusier's *Unité*, then, as an instance of the opposite and reciprocal condition. Vasari's Uffizi could scarcely be more adequate. The parallel is, of course, trans-cultural; but, if a sixteenth century office building become a museum may, with certain reservations, be brought into critical proximity with a twentieth century apartment house, then an obvious point can be made. For, if the Uffizi is Marseilles turned outside in, or if it is a jelly mold for the *Unité*, it is also void become figurative, active and positively charged; and, while the effect of Marseilles is to endorse a private and atomized society, the Uffizi is much more completely a "collective" structure. And, to further bias the comparison: while Le Corbusier presents a private and insulated building which, unambiguously, caters to a limited clientele, Vasari's model is sufficiently two-faced to be able to accommodate a good deal more. Urbanistically it is far more active. A central void-figure, stable and obviously planned, with, by way of entourage, an irregular back up which may be loose and responsive to close context. A stipulation of an ideal world and an engagement of empirical circumstance, the Uffizi may be seen as reconciling themes of self-conscious order and spontaneous randomness; and, while it accepts the existing, by then proclaiming the new, the Uffizi confers value upon both new and old.

Again, a comparison of a Le Corbusier product, this time with one by Auguste Perret, may be used to expand or to reinforce the preceding;

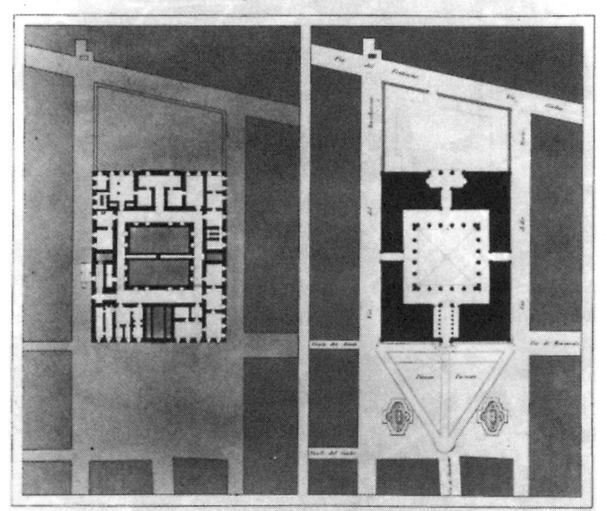

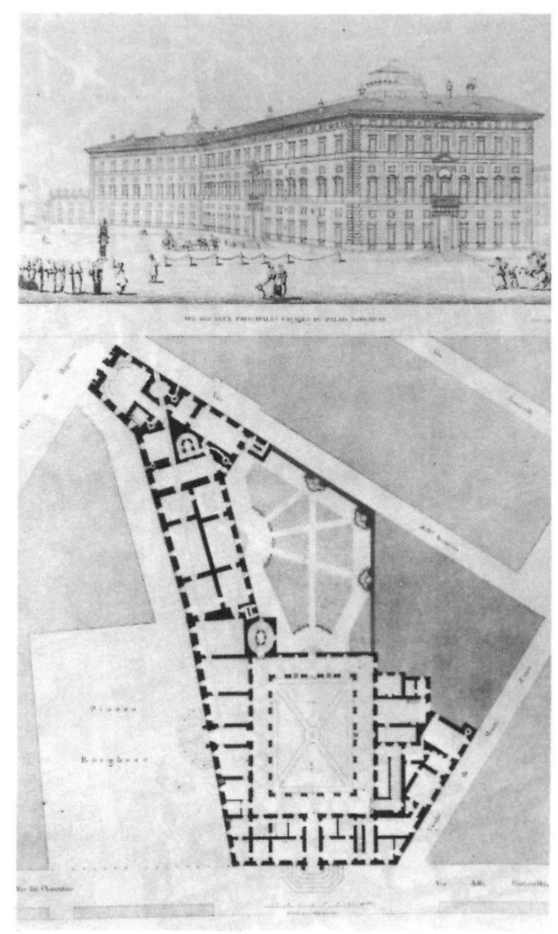

Fig. 13. Left above: Rome, *Palazzo Farnese*, view and plan; right above: Rome, *Palazzo Borghese*, view and plan; left below: *Todi, Santa Maria della Consolazione*; right below: Rome, *Sant' Agnese* in *Piazza Navona.*

and, since the comparison, originally made by Peter Collins, involves two interpretations of the same program, it may, to that extent, be considered the more legitimate. Le Corbusier and Perret's projects for the Palace of the Soviets which, the two together, might have been designed to confound the proposition that form follows function, could almost be allowed to speak for themselves. Perret gestures to immediate context and Le Corbusier scarcely so. With their explicit spatial connections with the Kremlin and the inflection of their courtyard toward the river, Perret's buildings enter into an idea of Moscow which they are evidently intended to elaborate; but Le Corbusier's buildings, which are apt to proclaim their derivation from internal necessity, are certainly not so much responsive to the site as they are symbolic constructs supposedly responsive to an assumed newly liberated cultural milieu. And if in each case, the use of site is iconographically representative of an attitude to tradition, then, in these two evaluations of tradition, it may be entirely fair to read the effects of a twenty-year generation gap.

But in one further parallel along these lines is no such gap that can be interposed. Gunnar Asplund and Le Corbusier were entirely of the same generation; and, if one is here and not dealing with comparable programs or proposals of equivalent size, the dates of Asplund's Royal Chancellery project (1922) and Le Corbusier's *Plan Voisin* (1925) may still facilitate their joint examination. The *Plan Voisin* is an outgrowth of Le Corbusier's *Ville Contemporaine* of 1922. It is the *Ville Contemporaine* injected into a specific Parisian site; and, however unvisionary it was professed to be—indeed however "real" it has become—it evidently proposes a completely different working model of reality from that employed by Asplund. The one is a statement of historical destiny, the other of historical continuity; the one is a celebration of generalities, the other of specifics; and, in both cases, the site functions as icon representative of these different evaluations.

Thus, as almost always in his urbanistic proposals, Le Corbusier largely responds to the idea of a reconstructed society and is largely unconcerned with local spatial minutiae. If the *Portes Saint-Denis* and *Saint-Martin* may be incorporated in the city center so far so good; if the Marais is to be destroyed no matter; the principal aim is manifesto. Le Corbusier is primarily involved with the building of a Phoenix symbol; and, in his concern to illustrate a new world rising above the ashes of the old, one may detect a reason for his highly perfunctory approach to major monuments—only to be inspected after cultural inoculation. And thus, by contrast, Asplund for whom, one might suppose, ideas of social continuity become represented in his attempt to make of his buildings, as much as possible, a part of the urban continuum.

But, if Le Corbusier simulates a future and Asplund a past, if one is almost all prophecy theater and the other almost all memory, and if it is the present contention that both of these ways of looking at the city—spatially as well as sentimentally—are valuable, the immediate concern is with their spatial implications. We have identified two models; we have suggested that it would be less than sane to abandon either; and we are, consequently, concerned with their reconciliation, with, at one level, a recognition of the specific and, at another, the possibilities of general statement. But there is also the problem of one model which is active and predominant and another which is highly recessive; and it is in order to correct this lack of equilibrium that we have been obliged to introduce Vasari, Perret and Asplund as purveyors of useful information. And, if there is no doubt about it that, of the three, Perret is the most banal and, maybe, Vasari the most suggestive, then, probably, Asplund may be felt to illustrate the

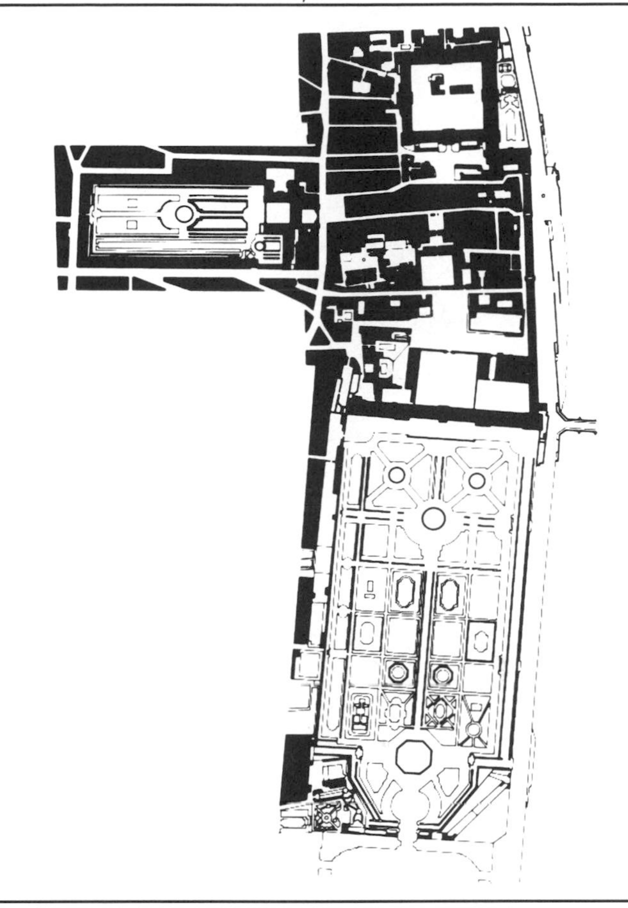

Fig. 14. Paris, the *Louvre*, *Tuileries*, and *Palais Royal*, c. 1780, figure-ground plan.

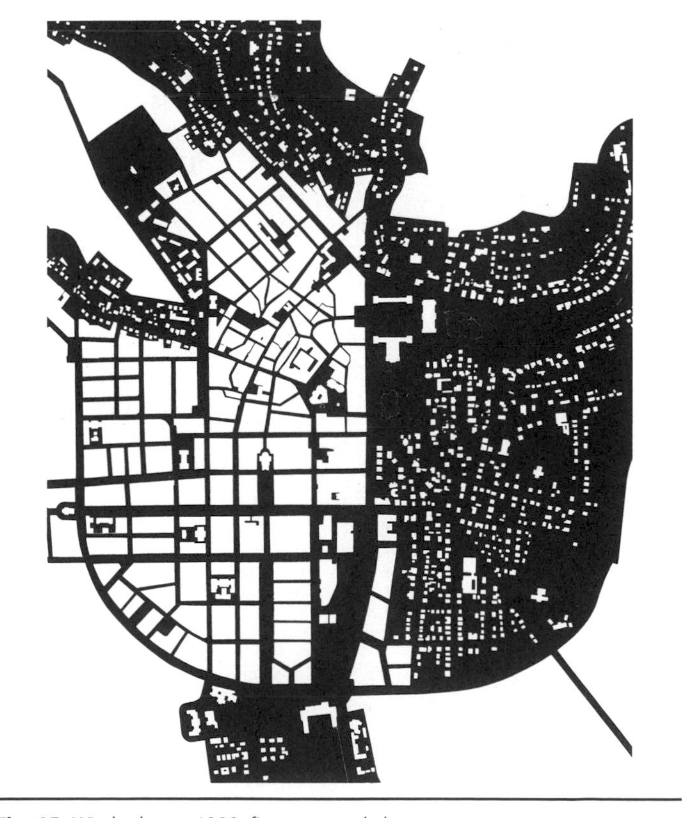

Fig. 15. Wiesbaden, c. 1900, figure-ground plan.

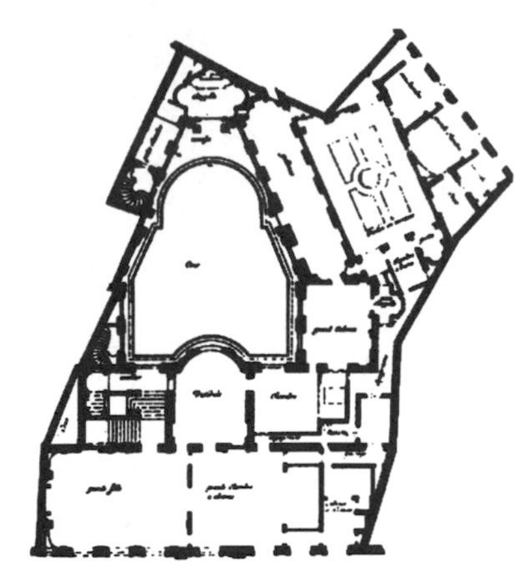
Paris, Hotel de Beauvais, plan

Hotel de Beauvais, elevation

Le Corbusier: Villa Savoye at Poissy

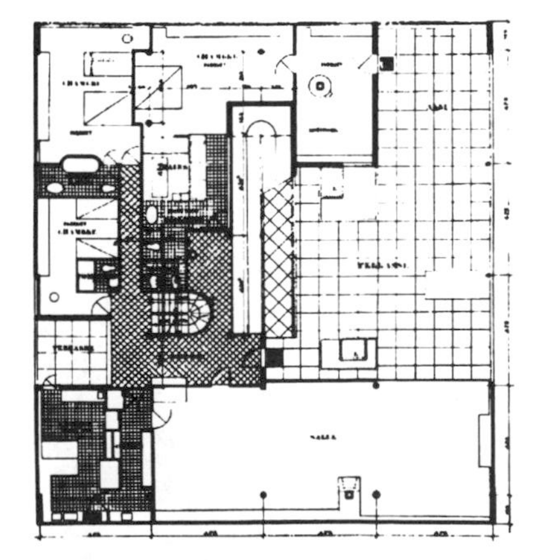
Villa Savoye, plan

Fig. 16. (a) Paris, *Hotel de Beauvais*, plan. *Hotel de Beauvais*, elevation; (b) Le Corbusier: *Villa Savoye* at Poissy.

most elaborate use of multiple design strategies. Simultaneously the empiricist reacting to site and the idealist concerned with normative condition, in one work he responds, adjusts, translates, asserts to be—and all at once—passive recipient and active reverberator.

However, Asplund's play with assumed contingencies and assumed absolutes, brilliant though it may be, does seem to involve mostly strategies of response; and, in considering problems of the object, it may be useful to consider the admittedly ancient technique of deliberately distorting what is also presented as the ideal type. And to take a Renaissance-Baroque example: if *Santa Maria della Consolazione* at Todi may, in spite of certain provincial details, be allowed to represent the "perfect" building in all its pristine integrity, then how is this building to be "compromised" for use in a less than "perfect" site? This is a problem which a functionalist theory could neither envisage nor admit. For though, in practice, functionalism could often become compounded with a theory of types, intrinsically it was scarcely able to comprehend the notion of already synthesized and pre-existent models being shifted around from place to place. But, if functionalism proposed an end to typologies in favor of a logical induction from concrete facts, it is precisely because it was unwilling to consider iconic significance as a concrete fact in itself, unwilling to imagine particular physical configurations as instruments of communication, that functionalism can have very little to say with reference to the deformation of ideal models. So Todi we know to be a sign and an advertisement; and, as we concede the freedom to use the advertisement wherever conditions may require it, we also infer the possibilities of sustaining, or salvaging the meaning while manipulating the form according to the exigencies of circumstance. And, in such terms, it may be possible to see *Sant' Agnese* in *Piazza Navona* as at Todi, which is simultaneously "compromised" and intact. The constricted site propounds its pressures; the piazza and the dome are the irreducible protagonists in a debate; the piazza has something to say about Rome, the dome about cosmic fantasy; and, finally, via a process of response and challenge, both of them make their point.

So the reading of *Sant' Agnese* continuously fluctuates between an interpretation of the building as object and its reinterpretation as texture; but, if the church may be sometimes an ideal object and sometimes a function of the piazza wall, yet another Roman instance of such figure-ground alternation—of both meanings and forms—might still be cited. Obviously not so elaborate a construct as *Sant' Agnese*, the *Palazzo Borghese*, located upon its highly idiosyncratic site, contrives both to response to this site and to behave as a representative palace of the Farnese type. The *Palazzo Farnese* provides its reference and meaning. It contributes certain factors of central stability, both of façade and plan; but, with the "perfect" cortile now embedded in a volume of highly "imperfect" and elastic perimeter, with the building predicated on a recognition of both archetype and accident, there follows from this duplicity of evaluation an internal situation of great richness and freedom.

Now this type of strategy which combines local concessions with a declaration of independence from anything local and specific could be indefinitely illustrated; but, perhaps, one more instance of it will suffice. Le Pautre's *Hotel de Beauvais*, with its ground floor of shops, is externally something of a minor Roman *palazzo* brought to Paris; and, as an even more elaborate version of a category of free plan, it might possibly prompt comparison with the great master and advocate of the free plan himself. But Le Corbusier's technique is, of course, the logical opposite to that of *Le Pautre*; and, if the "freedoms" of the *Villa Savoye* depend on the stability of its indestructible perimeter, the "freedoms" of the *Hotel de Beauvais* are derived from the equivalent stability of its central *cour d'honneur*.

In other words, one might almost write an equation: *Uffizi* : *Unité* = *Hotel de Beauvais* : *Villa Savoye*. As a simple convenience, this equation is of completely crucial importance. For on the one hand at the *Villa Savoye*, as at the *Unité*, there is an absolute insistence upon the virtues of primary solid, upon the isolation of the building as object and the urbanistic corollary of this insistence scarcely requires further commentary; and, on the other, in the *Hotel de Beauvais*, as at the *Palazzo Borghese*, the built solid is allowed to assume comparatively minor significance. Indeed, in these last cases, the built solid scarcely divulges itself; and, while unbuilt space (courtyard) assumes the directive role, becomes the predominant idea, the building's perimeter is enabled to act as no more than a "free" response to adjacency. On the one side of the equation building becomes prime and insulated, on the other the isolation of identifiable space reduces (or elevates) the status of building to infill.

But building as infill! The idea can seem to be deplorably passive and empirical—though such need not be the case. For, in spite of their spatial preoccupation neither the *Hotel de Beauvais* nor the *Palazzo Borghese* are, finally, flaccid. They, both of them, assert themselves by way of representational façade, by way of progression from façade-figure (solid) to courtyard-figure (void); and, in this context, although the *Villa Savoye* is by no means the simplistic construct which we have here made it appear (although it too, to some extent, operates as its opposite) for present purposes its arguments are not central.

For, far more clearly than at *Savoye*, at the *Hotel de Beauvais* and the *Palazzo Borghese* the *Gestalt* condition of ambivalence-double value and double meaning results in interest and provocation. However, though speculation may thus be incited by fluctuations of the figure-ground phenomenon (which may be volatile or may be sluggish), the possibilities of any such activity—especially at an urban scale—would seem very largely to depend upon the presence of what used to be called *poché*.

Frankly, we had forgotten the term, or relegated it to a catalogue of obsolete categories and were only recently reminded of its usefulness by Robert Venturi.[11] But if *poché* understood as the imprint upon the plan of the traditional heavy structure, acts to disengage the principal spaces of the building from each other, if it is a solid matrix which frames a series of major spatial events, it is not hard to acknowledge that the recognition of *poché* is also a matter of context and that, depending on perceptual field, a building itself may become a type of *poché*, for certain purposes a solid assisting that legibility of adjacent spaces. And thus, for instance, such buildings as the *Palazzo Borghese* may be taken as types of habitable *poché*, which articulate the transition of external voids.

So, thus far, implicitly, we have been concerned with an appeal for urban *poché*, and the argument has been primarily buttressed by perceptual criteria; but, if the same argument might, just as well, receive sociological support (and we would prefer to see the two findings as interrelated), we must still have a very brief question of how to do it.

It seems that the general usefulness of *poché*, in a revived and overhauled sense, comes by its ability, as a solid, to engage or be engaged by adjacent voids, to act as both figure and ground as necessity or circumstance might require; but with the city of modern architecture, of course, no such reciprocity is either possible or intended. But, though the employment of ambiguous resources might foul the cleanliness of this city's mission, since we are involved in this process anyway, it will be opportune again to produce the *Unité* and, this time, to bring it into confrontation with the *Quirinale*. In plan configuration, in its nimble relationship with the ground and in the equality of its two major faces the *Unité* ensures its own emphatic isolation. A housing block which, more or less, satisfies desired requirements in terms of exposure, ventilation, etc., its limitations with regard to collectively and context have already been noted; and it is in order to examine possible alleviation of these shortcomings that the *Palazzo del Quirinale* is now introduced. In its extension, the improbably attenuated *Manica Lunga* (which might be several *Unités* put end to end), the *Quirinale* carries within its general format all the possibilities of positive twentieth century living standards (access, light, air, aspect, prospect, etc.); but, while the *Unité* continues to enforce its isolation and object quality, the *Quirinale* extension acts in quite a different way.

Thus, with respect to the street on the one side and its gardens on the other, the *Manica Lunga* acts as both space occupier and space definer, as positive figure and passive ground, permitting both street and garden to exert their distinct and independent personalities. To the street it projects a hard, "outside" presence which acts as a kind of datum to service a condition of irregularity and circumstance (*Sant' Andrea, etc.*) across the way; but, while in this manner it establishes the public realm, it is also able to secure for the garden side a wholly contrary, softer, private and, potentially, more adaptable condition.

The elegance and the economy of the operation, all done with so little and all so obvious, may stand as a criticism of contemporary procedures; but, if a consideration of perhaps more than one building has here been implied, such an expansion may be carried a little further. To consider, for instance, the courtyard of the *Palais Royal*, admired but not "used" by Le Corbusier, as providing a clear differentiation between an internal condition of relative privacy and an external, less comprehensible world; to consider it not only as habitable *poché*, but as an urban room, perhaps one of many; and to consider then a number of towers, current specification—smooth, bumpy, with or without entrails, whatever—to be located as urban furniture, perhaps some inside the "room" and some outside. The order of the furniture is no matter; but the *Palais Royal* thus becomes an instrument of field recognition, an identifiable stabilizer and a means of collective orientation. The combination provides a condition of mutual reference, complete reciprocity, relative freedom. In addition, being essentially foolproof, it might almost "make the evil difficult and the good easy."

That all this is of no consequence. . . ? That between architecture and human "activity" there is no relationship. . . ? Such one knows to be the continuing prejudice of the "Let us evaporate the object, let us interact" school; but, if existing political structure—whatever one might wish—seems scarcely to be upon the threshold of impending dissolution and if the object seems equally intractable to important physico-chemical decomposition, then, by way of reply, it *might* be arguable that it could be justifiable to make at least *some* concessions to these circumstances.

To summarize: it is here proposed that, rather than hoping and waiting for the withering away of the object (while, simultaneously manufacturing versions of it in profusion unparalleled), it might be judicious, in most cases, to allow and encourage the object to become

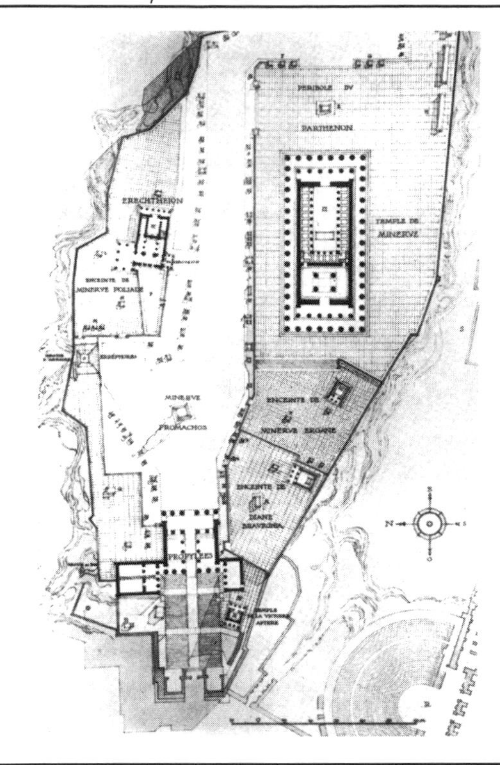

Fig. 17. Athens, the *Acropolis*.

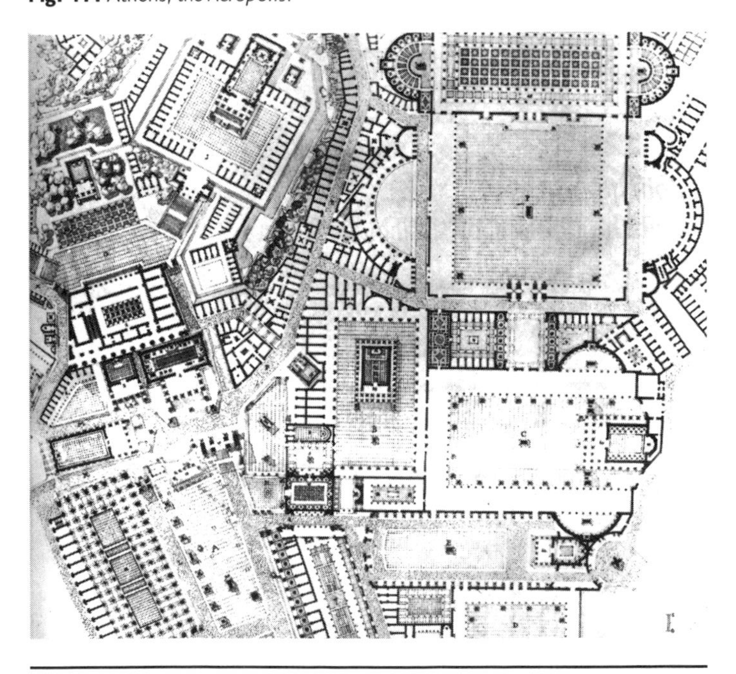

Fig. 18. Rome, the imperial *Fora*.

digested in a prevalent texture or matrix. It is further suggested that neither object nor space fixation are, in themselves, any longer representative of valuable attitudes. The one may, indeed, characterize the "new" city and the other old; but, if these situations which must be transcended rather than emulated, the situation to be hoped for should be recognized as one in which both buildings and spaces exist in an equality of sustained debate. A debate in which victory consists in each component emerging undefeated, the imagined condition is a type of solid-void dialectic which might allow for the joint existence of the overtly planned and the genuinely unplanned, of the set piece and the accident, of the public and the private, of the state and the individual. It is a condition of alerted equilibrium which is envisaged; and it is in order to illuminate the potential of such a contest that we have introduced a rudimentary variety of possible strategies. Cross-breeding, assimilation, distortion, challenge, response, imposition, superimposition, conciliation: these might be given any number of names and, surely, neither can nor should be too closely specified; but if the burden of the present discussion has rested upon the city's morphology, upon the physical and inanimate, neither "people" nor "politics" are, by now, clamoring for attention; but, if their scrutiny can barely be deferred, yet one more morphological stipulation may still be in order.

Ultimately, and in terms of figure-ground, the debate which is here postulated between solid and void is a debate between two models and, succinctly, these may be typified as *Acropolis* and *Forum*. ■

NOTES & REFERENCES

1. Le Corbusier. 1948. *The Home of Man*. London. pp. 91, 96.

2. Lewis Mumford. 1940. *The Culture of Cities*. London. p. 136.

3. Siegfried Giedion. 1941. *Space, Time and Architecture*. Cambridge, MA. p. 524.

4. Le Corbusier, 1927. *Towards a New Architecture*. London. p. 167.

5. Point XIV of Van Doesburg's Madrid lecture of 1930; but this statement was incipient in 1924: "In contrast to frontality sanctified by a rigid static concept of life, the new architecture offers a plastic wealth of multifaceted temporal and spatial effects." *De Stijl*. Vol. VI, No. 6-7, p. 80.

6. Le Corbusier, 1943. *La Charte d'Athenes*. Paris. English translation, Anthony Eardley. 1973. *The Athens Charter*. New York.

7. J. Tyrwhitt, J. L. Sert and E. N. Rogers (eds.).1951. *The Heart of the City: Towards the Humanization of Urban Life*. New York.

8. Benjamin Disraeli. 1847. *Tancred*. London.

9. Alexander Tsonis, 1972. *Towards a Non-Oppressive Environment*. Boston.

10. Oscar Newman, 1972: *Defensible Space*. New York and London. 1973: *Architectural Design for Crime Prevention*. Washington. Newman offers pragmatic justification for what in any case ought to be normative procedure; but his inference (surely correct) that spatial dispositions may operate to prevent crime is an argument distressingly far removed from the more classical supposition that the purposes of architecture are intimately related with the idea of the good society.

11. Robert Venturi. 1966. *Complexity and Contradiction in Architecture*. The Museum of Modern Art Papers on Architecture I. New York.

Rob Krier

Summary
The premise underlying this article is the conviction that in our modern cities we have lost sight of the traditional understanding of urban space. Of itself, this assertion is of no great service to town planning research. What has to be clearly defined is the term *urban space* and what meaning it holds within the urban structure, so that one can go on to examine whether the concept of urban space retains some validity in contemporary town planning and on what grounds. Space in this context is an actively debated concept. The intention here is not to generate a new definition but rather to bring its original meaning back into currency.

Key words
building section, courtyard, façade, morphology, scale, street, square, town planning, taxonomy, typology, urban space

Elements of the concept of urban space

Definition of the concept "urban space"

If we wish to clarify the concept of urban space without imposing aesthetic criteria, we are compelled to designate all types of space between buildings in towns and other localities as urban space. This space is geometrically bounded by a variety of elevations. It is only the clear legibility of its geometrical characteristics and aesthetic qualities which allows us consciously to perceive external space as urban space.

The polarity of internal-external space is constantly in evidence in this article, since both obey very similar laws not only in function but also in form. Internal space shielded from weather and environment is an effective symbol of privacy; external space is seen as open, unobstructed space for movement in the open air, with public, semi-public and private zones.

The basic concepts underlying the aesthetic characteristics of urban space will be expounded below and systematically classified by type. In the process, an attempt will be made to draw a clear distinction between precise aesthetic and confused emotional factors. Every aesthetic analysis runs the risk of foundering on subjective questions of taste. Visual and sensory habits, which vary from one individual to the next, are augmented by a vast number of socio-political and cultural attitudes, which are taken to represent aesthetic truths. Accepted styles in art history—for example, baroque town plans, revolutionary architecture, etc.—are both useful and necessary. However, my observations indicate that they are almost always identified with the social structure prevailing at the time in question. Certainly it can scarcely be proved that, because of the wishes of the ruling classes and their artists, the stylistic canons of the period in European art history between 1600 and 1730 appeared almost to be determined by fate. Of course for the historian every period of history forms a unit with its own internal logic, which cannot be fragmented and interchanged with elements of other periods at will.

The creative person, such as an artist, may use a completely different method of approach. The decisions he makes in deploying his aesthetic skills are not always based on assumptions which can be unequivocally explained. The artistic libido is of enormous importance here. The cultural contribution of an age develops on the basis of a highly complex pattern of related phenomena, which must subsequently be the subject of laborious research on the part of historians. This example throws us right into a complex problem which appears the same in whichever period of history we consider. We must discuss this example exhaustively before we start constructing our rational system. Each period in art history develops gradually out of the assimilated functional and formal elements which precede it. The more conscious a society is of its history, the more effortlessly and thoroughly it handles historical elements on style. This truism is important in as far as it legitimizes the artist's relationship with the universally accepted wealth of formal vocabulary of all preceding ages–this is as applicable in the 20th as in the 17th century.

Credits: This article is an excerpt from "Typological and Morphological Elements of the Concept of Urban Space," Chapter 1 of *Urban Design (Stradtraum)* published in German (1975) and in English by Academy Editions (1978), reprinted here by kind permission of the author.

I do not wish to rally support for eclecticism, but simply to warn against an all too naïve understanding of history, which has been guilty of such misjudgments as representing urban architecture amongst the Romans as markedly inferior to that of the Greeks, which from an historical point of view is simply not true. The same mistake persists today, as can be seen from attitudes to the architecture of the 19th century.

Our age has a remarkably distorted sense of history, which can only be characterized as irrational. Le Corbusier's apparent battle against *L'Academie* was not so much a revolt against an exhausted, aging school as the assumption of a pioneering stand in which he adopted its ideals and imbued them with a new and vigorous content.

This so-called "pioneering act" was a pretended break with history, but in reality was an artistic falsehood. The facts were these: he abandoned the tradition current until then that art supported by the ruling classes enjoyed the stamp of legitimacy and, being at an advanced stage of development, materially shaped the periods which followed. It was a revolt at one remove, so to speak, for the *Academie* lived on, and indeed came itself to share the same confused historical sense as the followers of the revolution.

I am speaking here about the modern age in general, and not about its exponents of genius who tower above the "image of the age." Rather than be indebted to elitist currents in art, the generation around the turn of the century sought new models. They found them in part in the folk art of other ages and continents, which had hitherto attracted little attention.

There began an unprecedented flurry of discovery of anonymous painting, sculpture, architecture, song and music of those peoples who were considered underdeveloped, and their contribution to culture was for the first time properly valued without regard to their stage of civilization. Other artists sought their creative material in the realm of pure theory and worked with the basic elements of visual form and its potential for transformation (the "abstracts"). Yet others found their material in social criticism and the denunciation of social injustice and carried out their mission using formally simple methods (the "expressionists"). The break with the elitist artistic tradition was identical to the artist's struggle for emancipation from his patron—the ruling class and its cultural dictatorship—which had been brewing even before the French revolution.

The example of the baroque town layout has already been mentioned, and the question raised of the identity of form, context and meaning. We must be more exact in asking:

1. Was the resulting from the free expression of the creative artist?

2. Alternatively, were the artistic wishes of the employing class imposed on the artist, and was he forced to adopt their notions of form?

3. Do contemporaneous periods exist, which on the basis of different cultural traditions in different countries or continents where similar social conditions prevail, produce the same artistic solutions?

4. Alternatively, are there noncontemporaneous periods which led to fundamentally different artistic solutions, each being a stage in the development of the same cultural tradition in the same country under the same conditioning social factors?

In this series of permutations, the following factors are relevant: aesthetics, artist, patron, social environment, leeway given to artistic expression, formal restrictions imposed by the patron, formal restrictions imposed by the social environment, fashion, management, level of development, technology and its potential applications, general cultural conditions, scientific knowledge, enlightenment, nature, landscape, and climate. We can conclude with a fair degree of certainty that none of these interrelated factors can be considered in isolation.

With this brief outline of the problem, a word of caution might be added about an oversimplistic undiscriminating outlook. It is certainly worth trying to establish why certain kinds of urban space were created in the 17th century which we now identify with that period. And it would be even more interesting to examine the real reasons by 20th century town planning has been impoverished and reduced to the lowest common denominator.

The following classification does not make any value judgments. It enumerates the basic forms which constitute urban space, with a limited number of possible variations and combinations. The aesthetic quality of each element of urban space is characterized by the structural interrelation of detail. I shall attempt to discern this quality wherever we are dealing with physical features of a spatial nature. The two basic elements are the street and the square. In the category of "interior space" one would be talking about the corridor and the room. The geometrical characteristics of both spatial forms are the same. They are differentiated only by the dimensions of the walls that bound them and by the patterns of function and circulation that characterize them (Figs. 1 and 2).

The square and the street

In all probability the square was the first way man discovered of using urban space. It is produced by the grouping of houses around an open space. This arrangement afforded a high degree of control of the inner space, as well as facilitating a ready defense against external aggression by minimizing the external surface area liable to attack. This kind of courtyard frequently came to bear a symbolic value and was therefore chosen as the model for the construction of numerous holy places (*agora*, forum, cloister, mosque courtyard). With the invention of houses built around a central courtyard or atrium this spatial pattern became a model for the future. Here rooms were arranged around a central courtyard like single housing units around a square (Figs. 3, 4).

The street is a product of the spread of a settlement once houses have been built on all available space around its central square. It provides a framework for the distribution of land and gives access to individual plots. It has a more pronouncedly functional character than the square, which by virtue of its size is a more attractive place to pass the time than the street, in whose confines one is involuntarily caught up in the bustle of traffic. Its architectural backdrop is only perceived in passing. The street layouts which we have inherited in our towns were devised for quite different functional purposes. They were planned to the scale of the human being, the horse and the carriage. The street is unsuitable for the flow of motorized traffic, whilst remaining appropriate to human circulation and activity. It rarely operates as an autonomous isolated space, as, for example, in the case of villages built along a single street. It is mainly to be perceived as part of a network. Our historic towns have made us familiar with the inexhaustible diversity of spatial relationships produced by such a complex layout.

Typical functions of urban spaces

The activities of a town take place in public and private spheres. The behavioral patterns of people are similar in both. So, the result is that the way in which public space has been organized has in all periods excerised a powerful influence on the design of private houses.

One might almost infer the existence of a kind of social ritual, which produces a perfect match between individual and collective. What concerns us above all here are those activities which take place in town in the open air: i.e., actions which a person performs outside the familiar territory of his own home and for which he utilizes public space, as, for example, traveling to work, shopping, selling goods, recreations, leisure activities, sporting events, deliveries, etc. Although the asphalt carpet that serves as a channel for the movement of cars is still called a "street," it retains no connection with the original significance of the term. Certainly the motorized transportation of people and goods is one of the primary functions of the town, but it requires no scenery in the space around it. It is different in the case of the movement of pedestrians or public transport vehicles which move at a moderate speed, like carriages. Today we have boulevards that apparently draw their life from the procession of flashy cars and pavement cafes that are visited despite the fact that the air is polluted by exhaust fumes. Looking at planning schemes of the turn of the century, one can appreciate that in cosmopolitan cities such as Paris, Rome or Berlin, the air was polluted in a different way: by horse manure, stinking sewage and uncollected refuse. A problem of urban hygiene, as old as the town itself, the only difference being that people can be poisoned by carbon monoxide but scarcely by horse manure.

On medical grounds we can no longer indulge in this kind of boulevard romanticism. While the automobile in its present form continues to occupy streets, it excludes all other users.

Fig. 1. The square.

Fig. 2. The street.

The square

This spatial model is admirably suited to residential use. In the private sphere it corresponds to the inner courtyard or atrium. The courtyard house is the oldest type of townhouse. In spite of its undisputed advantages, the courtyard house has now become discredited. It is all too easily subject to ideological misinterpretation, and people are afraid that this design may imply enforced conformity to a communal lifestyle or a particular philosophy.

Fig. 3. House.

A certain unease about one's neighbors has undoubtedly led to the suppression of this building type. Yet in the same way as communal living has gained in popularity for a minority of young people with the disappearance of the extended family, the concept of neighborhood and its accompanying building types will most certainly be readopted in the near future.

In the public sphere, the square has undergone the same development. Marketplaces, parade grounds, ceremonial squares, squares in front of churches and town halls, all relics of the Middle Ages, have been robbed of their original functions and their symbolic content and in many places are only kept up through the activities of conservationists.

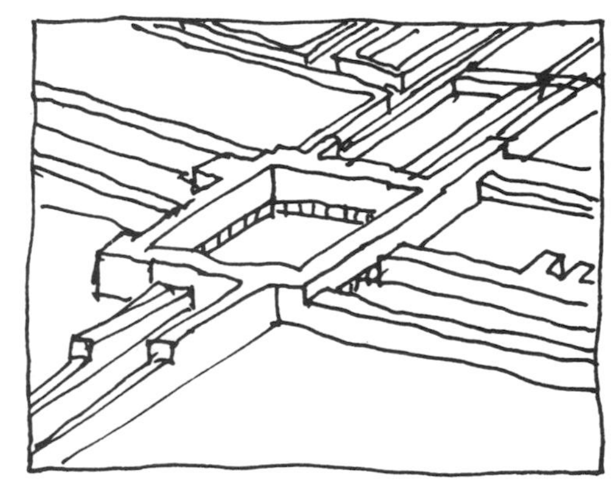

Fig. 4. Urban structure.

The loss of symbolism in architecture was described and lamented by Giedion in *Space, Time, Architecture*. The literary torch, which he carried for Le Corbusier in the '30s and for Jorn Utzon in the '60s, expressed his hope that this loss would perhaps be compensated by a powerful impetus toward artistic expression. He hoped for the same

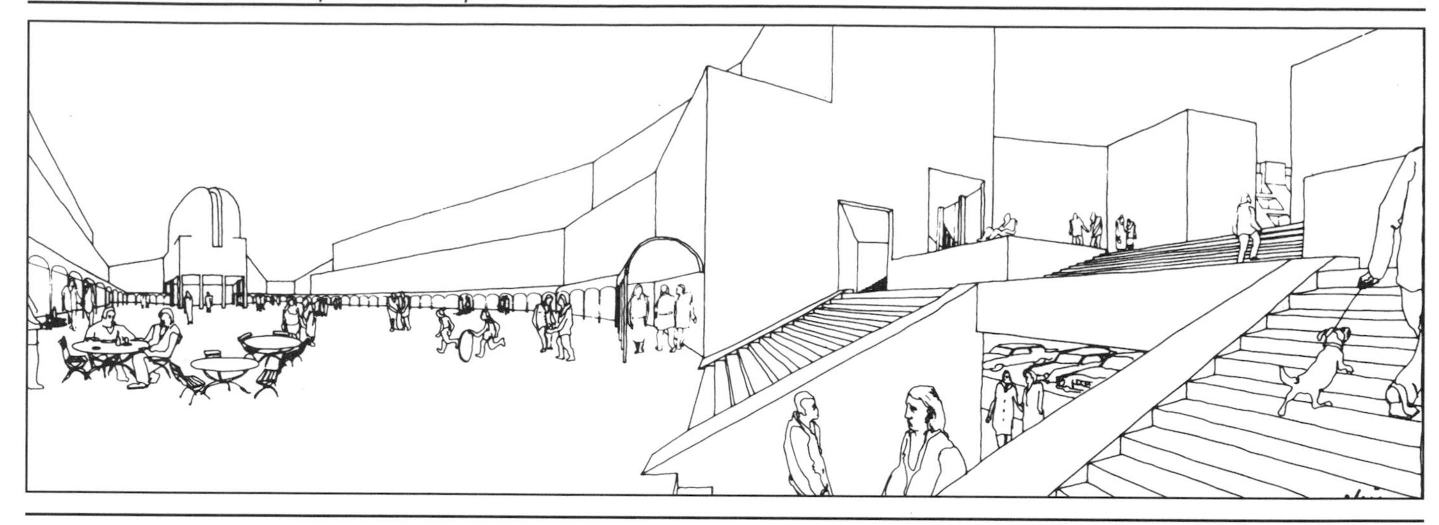

Fig. 5. The square as intersection of two roads, fixed point of orientation, and meeting place.

thing from new construction techniques. I have already stressed the importance of the poetic content and aesthetic quality of space and buildings. It is not my wish to introduce into this discussion the concept of symbolism, with all its ethical and religious overtones: and I would also like to warn against the arbitrary confusion of aesthetic and symbolic categories. If I maintain that the Louvre, instead of being a museum, might equally well be housing, a castle, or an office building etc., I am speaking of space- or building-type, not of external detailing or historical and sociopolitical factors which led to this structural solution. The aesthetic value of the different spatial types is as independent of short-lived functional concerns as it is of symbolic interpretations which may vary from one age to the next.

Another example to clarify this argument: The multistoried courtyard house, from the Middle Ages up to modern times, was the building type which acted as the starting point for the castle, the Renaissance and Baroque palace, etc. The Berlin tenements of the 19th century are also courtyard houses, but nowhere near being palaces. Anyone familiar with the architecture of Palladio should draw the right conclusion from this. The lavish use of materials certainly does not play the decisive role here. If that were the case, Palladio would long since have fallen into oblivion. So, even in the 20th century, one can construct a building with an inner courtyard without remotely aiming to imitate the palace architecture of the 16th century and the social class that produced it. There is no reason why the building types used by extinct dynasties to design their residences and show their material wealth should not serve as a model for housing today.

The early Christians were not afraid to adopt the building type of Roman judicial and commercial buildings, the basilica, as the prototype of their religious monuments. Le Corbusier took his row of *"redents"* from Baroque castles.

No contemporary public squares have been laid out which could be compared with urban squares like the *Grande Place* in Brussels, the *Place Stanislas* in Nancy, the *Piazza del Campo in* Siena, the *Place Vendome* and the *Place des Vosges* in Paris, the *Plaza Mayor* in Madrid, the *Plaza Real* in Barcelona and so forth. This spatial type awaits rediscovery. This can only occur, firstly, when it can be endowed with meaningful functions and, secondly, when planned in the right place with the appropriate approaches within the overall town layout (Fig. 5).

What functions are appropriate to the square? Commercial activities certainly are, such as the market, but above all, activities of a cultural nature. The establishment of public administrative offices, community halls, youth centers, libraries, theaters and concert halls, cafes, bars, etc. Where possible in the case of central squares, these should be functions that generate activity twenty-four hours a day. Residential use should not be excluded in any of these cases.

The street

In purely residential areas streets are universally seen as areas for public circulation and recreation. The distances at which houses are set back from the street, as regulations often demand, are so excessive that attractive spatial situations can only be achieved by gimmickry. In most cases, there is ample space available for gardens in addition to the emergency access required for public service vehicles. This street space can only function when it is part of a system in which pedestrian access leads off the street (Fig. 6).

This system can be unsettled by the following planning errors:

1. If some houses and flats cannot be approached directly from the street but only from the rear. In this way the street is deprived of a vital activity. The result is a state of competition between internal and external urban space. This characterization of space refers to the degree of public activity which takes place in each of these two areas.

2. If the garages and parking spaces are arranged in such a way that the flow of human traffic between car and house does not impinge upon the street space.

3. If the play spaces are squeezed out into isolated areas with the sole justification of preserving the intimacy of the residential zone. The same neurotic attitude toward neighbors is experienced in flats. The noise of cars outside the home is accepted, yet indoors children are prevented from playing noisily.

4. If no money can be invested in public open spaces, on such items as avenues of trees, paving and other such street furniture, given that the first priority is the visual appeal of space.

5. If the aesthetic quality of adjacent houses is neglected, if the facing frontages are out of harmony, if different sections of the street are inadequately demarcated or if the scale is unbalanced. These factors fulfill a precise cultural role in the functional coherence of the street and square. The need to meet the town's function of "poetry of space" should be as self-evident as the need to meet any technical requirements. In a purely objective sense, it is just as basic.

Can you imagine people no longer making music, painting, making pictures, dancing ...? Everybody would answer "no" to this. The role of architecture on the other hand is not apparently seen as so essential. "Architecture is something tangible, useful, practical" as far as most people are concerned. In any case its role is still considered as the creation of coziness indoors and of status symbols outdoors. Anything else is classed as icing on the cake, which one can perfectly well do without. A stage in history when architecture is not granted its full significance shows a society in cultural crises, the tragedy of which can scarcely be described in words. Contemporary music expresses it adequately.

The problems of the residential street touched on here apply equally to the commercial street. The separation of pedestrians and traffic carries with it the danger of the isolation of the pedestrian zone. Solutions must be carefully worked out which will keep the irritation of traffic noise and exhaust fumes away from the pedestrian, without completely distancing one zone from the other. This means an overlapping of these functions, to be achieved with considerable investment in the technological sphere, a price which the motorized society must be prepared to pay. This problem will remain much the same even when the well-known technical shortcomings and acknowledged design failings of the individual car have been ironed out. The number of cars, and their speed, remains a source of anxiety. With the ways things are going at the moment, there seems little hope of either factor being corrected. On the contrary, nobody today can predict what catastrophic dimensions these problems will assume and what solutions will be needed to overcome them.

It is absurd to labor under the misapprehension that one day the growing need to adopt new modes of transport will leave our countryside littered with gigantic and obsolete monuments of civil engineering. One is inclined to think that, considering the level of investment in the car and all that goes with it, a fundamental change is no longer feasible in the long term.

All this illustrates the conflict of interests between investments for the demands of machine/car and investments for living creature/man; it also indicates that there is a price to be paid for the restoration of urban space, if our society is to continue to value life in its cities.

Back to the problem of the commercial street which has already been outlined. It must be fashioned differently from the purely residential street. It must be relatively narrow. The passerby must be able to cast an eye over all the goods on display in the shops opposite without perpetually having to cross from one side of the street to the other. At least, this is what the shopper and certainly the tradesman would like to see. Another spatial configuration of the shopping street is provided by the old town center of Berne, in which pedestrians can examine the goods on display protected by arcades from the inclemency of the weather. This type of shopping street has retained its charm and also its functional efficiency up to the present day. The pedestrian is relatively untroubled by the road, which lies on a lower level. This street space can serve as an example to us.

Fig. 6. The street as artery and means of orientation.

The same can be said of the glass-roofed arcades or passages which originated in the 19th century. Strangely enough, they have fallen out of favor today. From the point of view of ventilation it was obviously disadvantageous then to lead the street frontage into a passageway. With today's fully air-conditioned commercial and office buildings, however, this building type could come back into fashion. Protection against the elements is a financially justifiable amenity for shopping streets, developed by the Romans from the colonnades that surrounded the Greek *agora*, has completely died out. The remains of such formal streets can still be found, *inter alia*, at Palmyra, Perge, Apameia, Sidon, Ephesus, Leptis Magna, and Timgad.

The appearance of this type of street is a fascinating event in the history of town planing. With the increased prosperity of Roman rule, a need arose for the uniform and schematic plan of the Greek colonial town to be modified, with emphasis being placed on arterial roads within the homogeneous network of streets, and this was achieved by marking them with particularly splendid architectural features. They certainly had important functional connotations which today can no longer be clearly surmised. Whatever these connotations were, they had an obviously commercial as well as symbolic character, in contrast to the *agora* and the *forum*, which were reserved primarily for political and religious purposes.

Weinbrenner, with his proposed scheme for the improvement of the *Kaiserstrassein Karlsruhe*, attempted to revive this idea. The *Konigsbau* in Stuttgart designed by Leins could be a fragment of the arcade street of Ephesus. The Romans were astoundingly imaginative in perfecting this type of street space. So, for example, changes in the direction of streets, dictated by existing features of the urban structure, were highlighted as cardinal points by having gateways built across them. In the *Galeries St. Hubert* in Brussels, this problem has been solved on the same principle. By this expedient, the street space is divided up into visually manageable sections, in contrast to the seemingly infinite persepctive of the remaining network of streets. It should equally be noted that in rare cases streets broaden out into squares directly without their articulation being marked by buildings. The street and the square were conceived as largely independent and autonomous spaces.

Such devices, used by Roman and Greek town planners to indicate spatial relationships, lapsed into oblivion with the decline of the Roman Empire in Europe. Isolated building types such as the forum and the basilica were adopted unchanged in the Middle Ages, for example, in monasteries. The forum was no longer employed as a public space. Not so in North Africa and the Near East, and to some extent in Spain, where these ancient types of urban space survived almost unchanged until the turn of the century using traditional construction methods.

Fig. 7. One type of urban space on three different scales.

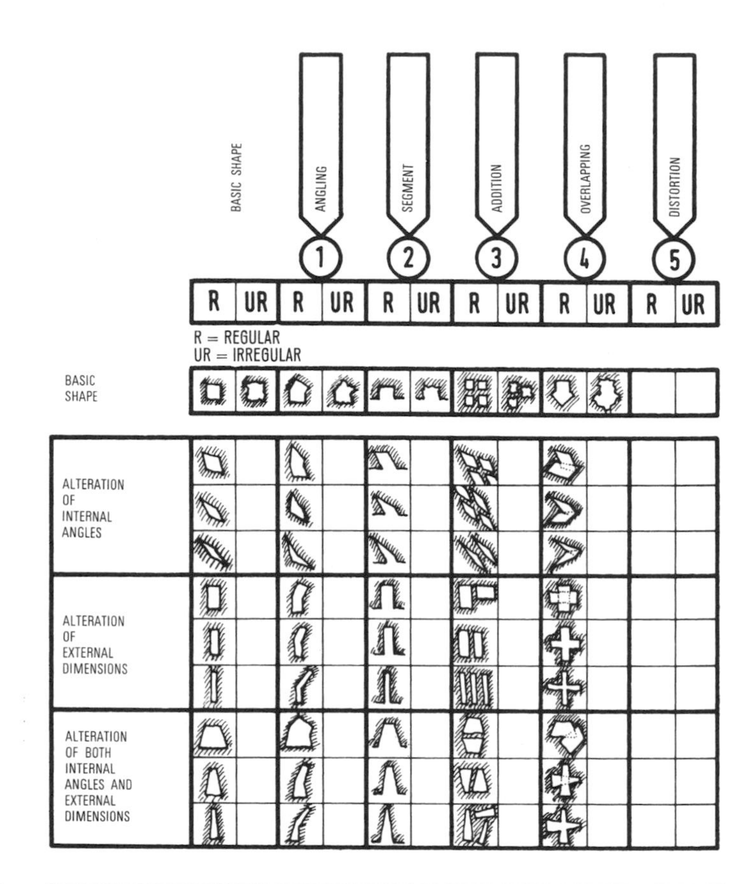

Fig. 8. Plan typology matrix.

Typology of urban space

In formulating a typology of urban space, spatial forms and their derivatives may be divided into three main groups, according to the geometrical pattern of their ground plan: these groups derive from the square, the circle or the triangle.

Without doubt the scale of an urban space is also related to its geometrical qualities. Scale can only be mentioned in passing in this typology. Scale and the significance of proportions in external space do not affect the arrangement of the typology (Fig. 7).

Modulation of a given spatial type

The plan typology matrix depicted in Fig. 8 shows, reading from top to bottom:

1. The basis element.

2. The modification of the basic element resulting from the enlargement of reduction of the angles contained within it, where the external dimensions remain constant.

3. The angles remain constant and the length of two sides changes in the same proportion.

4. Angles and external dimensions are altered arbitrarily.

Reading from left to right, the matrix illustrates the following stages of modulation:

1. Angled space. This indicates a space that is a compound of two parts of the basic element with two parallel sides bent.

2. This shows only a segment of the basic element.

3. The basic element is added to.

4. The basic elements overlap or merge.

5. The heading "distortion" includes spatial forms that are difficult or impossible to define. This category is intended to cover those shapes which can only with difficulty be traced back to their original geometric model. These shapes may also be described as species born out of chaos. Here the elevation of buildings may be distorted or concealed to such an extent that they can no longer be distinguished as clear demarcations of space, for example, a façade or mirror glass or one completely obscured by advertisements, so that a cuckoo-clock as big as a house stands next to an outside ice-cream cone, or an advertisement for chewing gum stands in place of the usual pierced façade.

Even the dimensions of a space can have a distorting influence on its effect, to such an extent that it ceases to bear any relation to the original. The column headed "distortion" has not been completed in this matrix, as these shapes cannot be diagrammactically expressed. All these processes of change show regular and irregular configurations.

How building sections affect urban space

The basic elements can be modified by a great variety of building sections. Fig. 9 depicts twenty-four different section types that substantially alter the features of urban space, as follows:

1. Standard traditional section with pitched roof.
2. With flat roof.
3. With top floor set back. This devise reduces the height of the building visible to the eye.
4. With a projection on pedestrian level in the form of an arcade or a solid structure. This device distances the pedestrian from the real body of the building and creates a pleasing human scale. John Nash in his Park Crescent, London, applied this type of section with particular virtuosity.
5. Half way up the building, the section is reduced by half its depth; this allows for extensive floors on the lower level and flats with access balconies on the upper level.
6. Random terracing.
7. Slopping elevation with vertical lower and upper floors.
8. Sloping elevation with protruding ground floor.
9. Stepped section.
10. Sloping section with moat or freestanding ground floor.
11. Standard section with moat.
12. Building with ground floor arcades.
13. Building on *pilotis*.
14. Building on *pilotis*, with an intermediate floor similarly supported.
15. Sloping ground in front of building.
16. A free-standing low building placed in front of a higher one.

17/18. Buildings with a very shallow incline, as, for example, arenas.

19. Building with arcade above ground level and access to pedestrian level.
20. Building with access balcony.
21. Inverted stepped section.
22. Building with pitched projections.
23. Building with projections.
24. Building with free-standing towers.

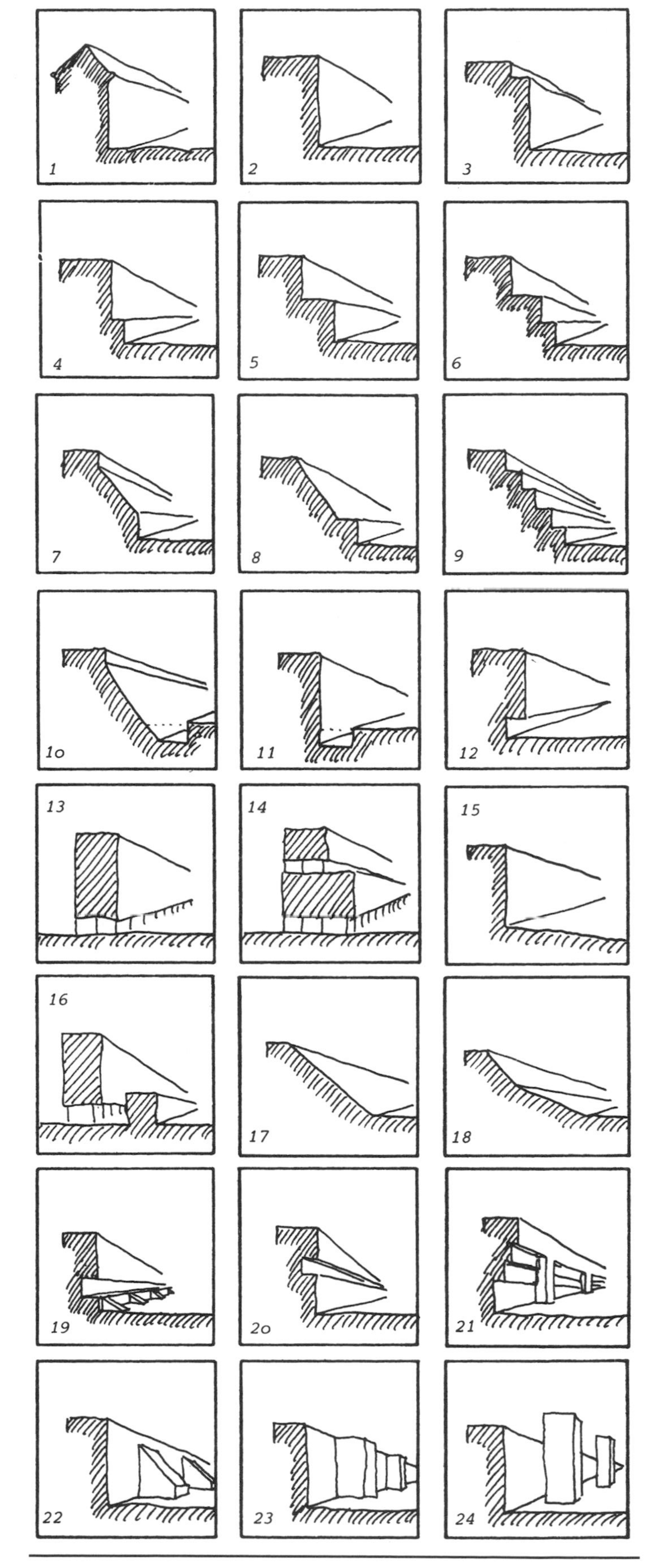

Fig. 9. Building sections.

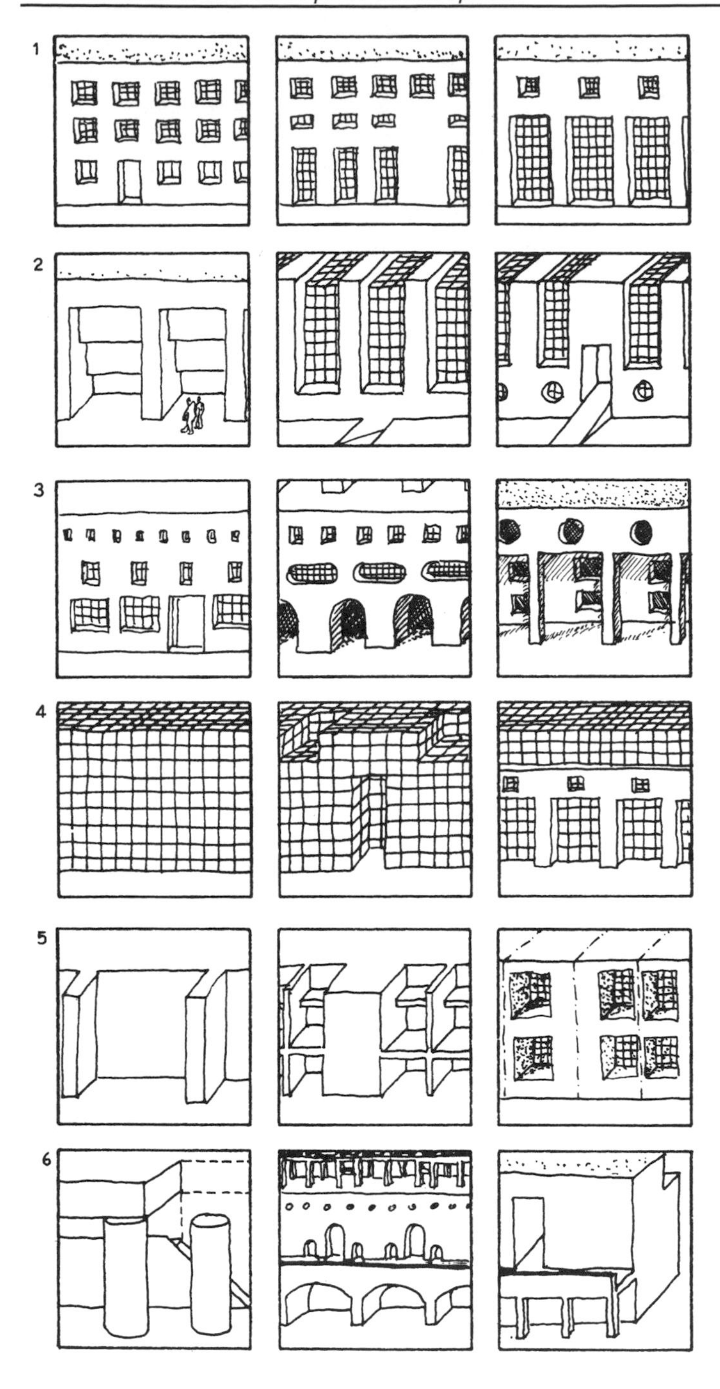

Fig. 10. Elevations.

Each of these building types can be given a façade appropriate to its function and method of construction. The sketches reproduced in Figs. 9 and 10 can only give some idea of the inexhaustible design possibilities. Each of these structures influences urban space in a particular way.

Elevations

Notes on sketches in Fig. 10 (reading left to right):

1. Pierced façade: the lowest level is more generously glazed in each sketch, reducing the solid area to a simple load-bearing structure.

2. The glazed area within the load-bearing structure can be modified according to taste. The following three pictures show a reverse of the design process portrayed in 1. A solid base forces the glazed area upward.

3. The window type can be modified horizontally and vertically according to the imagination of the designer.

4. Faceless modular façade as a theoretical (abstract) way in which the building might be enclosed. The modular façade can be adapted to all variations in the shape of the building. Solid sections of the building can be combined with the grid.

5. Windowless buildings: windows are placed in niches, etc., and the process starts again from the beginning.

6. Exploration of different geometries: a thematic interpretation of the elevation: lowest level = heavy; middle section = smooth with various perforations; upper part = light, transparent. One of the sketches of squares shows a variation on this theme on three sides of a square. Arcades are placed in front of houses, with different architectural styles juxtaposed.

Intersections of street and square

All spatial types examined up to now can be classified according to the types of street intersection laid out in Fig. 11. As an example, a set of permutations I set up for up to four intersections at four possible points of entry. This chart should only be taken as an indication of an almost unlimited range of possible permutations of these spatial forms. To attempt a comprehensive display here would conflict with the aim of this typological outline.

The vertical columns of this diagram show the number of streets intersecting with an urban space. Horizontally, it shows four spossible ways in which one or more streets may intersect with a square or street:

1. Centrally and at right angles to one side.

2. Off-center and at right angles to one side.

3. Meeting a corner at right angles.

4. Oblique, at any angle and at any point of entry.

5. Spatial types and how they may be combined.

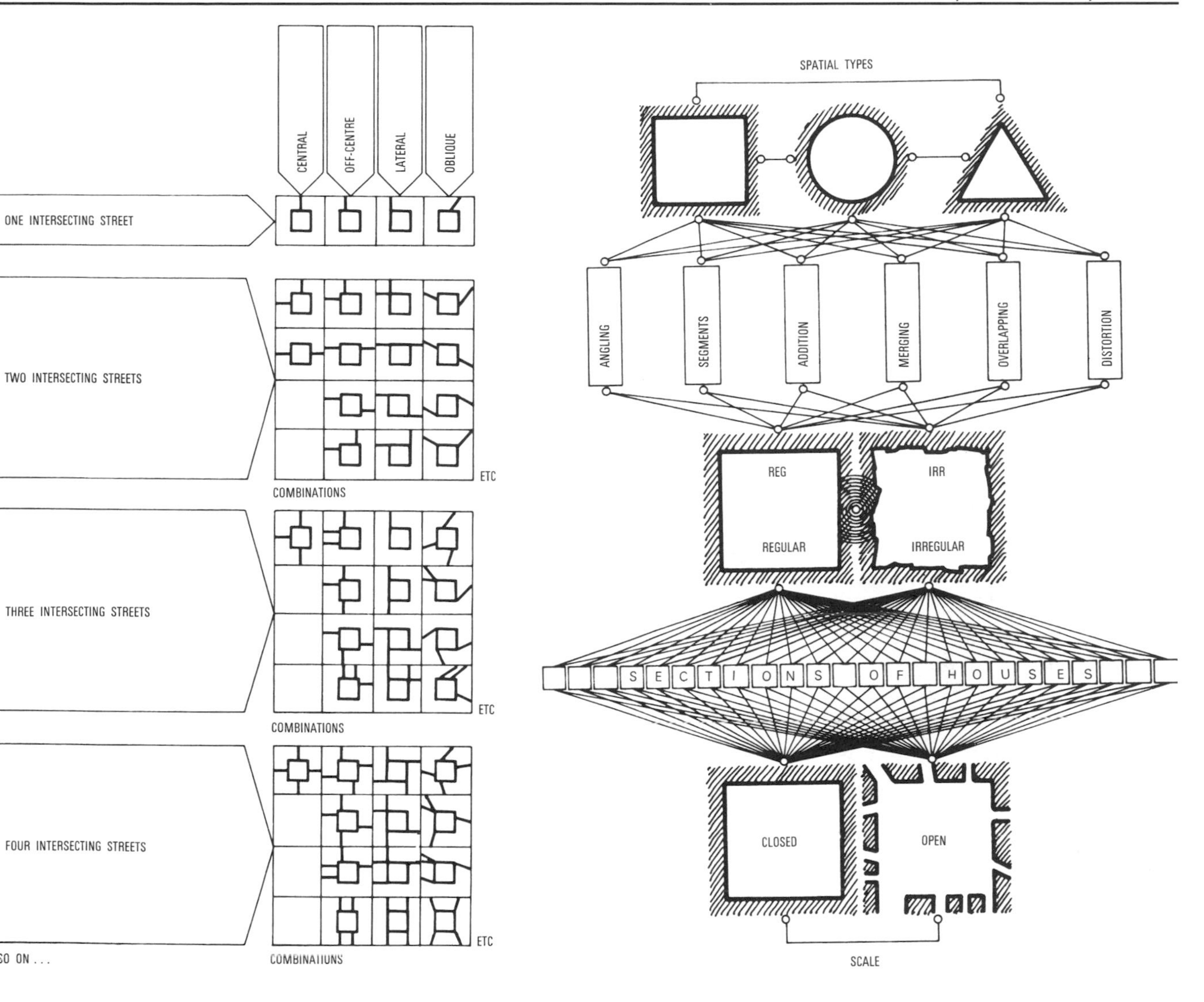

Fig.11. Intersections of street and square.

Fig. 12. Spatial types and how they might be combined.

The morphological classification of urban spaces may be summarized as follows:

The three basic shapes—square, circle and triangle—are affected by the following modulating factors: angling; segmentation; addition; merging; overlapping or amalgamation of elements; and distortion. These modulating factors can produce geometrically regular or irregular results on all spatial types (Fig. 12).

At the same time, the large number of possible building sections influences the quality of the space at all these stages of modulation. All sections are fundamentally applicable to these spatial forms. In the accompanying sketches the intention is to make clear as realistically as possible the effect of individual spatial types so that this typology can be more easily accessible and of practical use to the planner.

The terms "closed" and "open" may be applied to all spatial forms described up to now: that is, spaces which are completely or partially surrounded by buildings.

Finally, many compound forms can be created at will from the three spatial types and their modulations. In the case of all spatial forms, the differentiation of scale plays a particularly important role, as does the effect of various architectural styles on the urban space (Figs. 13 and 18).

Design exercises can be "played on the keyboard" that has just been described. Apart from this "formal" procedure, other factors also have their effect on space, and this effect is not insignificant. These factors are the rules governing building construction, which make architectural design possible in the first place, and above all else, determine the use or function or a building, which is the essential prerequisite for architecural design. The logic of this procedure would therefore demand this sequence: function, construction and finally the resultant design. ■

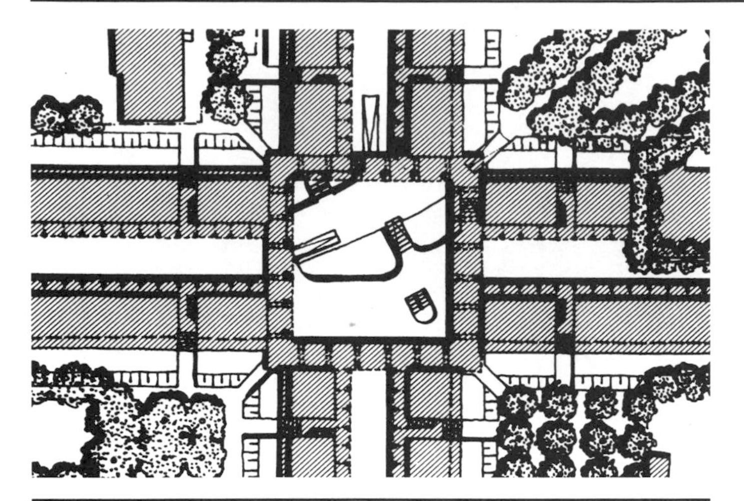

Fig. 13. The square as intersection.

Fig. 14. Arcade running around the square: high narrow columns.

Fig 15. Lower arcade.

Fig. 16. Low arcade, wide openings.

Fig. 17. Combination of three different façades.

Fig. 18. Combination as in Fig. 17, but overgrown with plants.

Congress of the New Urbanism

Summary

"New Urbanism is a relatively recent entry into the long-standing debate about sprawl. Beginning in 1993, the New Urbanism movement has grown to include urban designers, architects, planners, environmentalists, economists, landscape designers, traffic engineers, elected officials, sociologists, developers, and community activists, to start an incomplete list. It represents the interests of a broad coalition of environmentalists concerned with farmland preservation, habitat enhancements, and air quality as well as inner-city advocates concerned with urban reconstruction and social equity. It weds these groups and interests with a design ethic that spans from region to building."

Key Words

density, development practice, environment, housing, neighborhood, parks, revitalization, squares, streets, urbanism

Washington, D.C. North of Massachusetts Avenue (NoMA) Plan. (Calthorpe and Fulton, 2001.)

Charter of the New Urbanism

1 INTRODUCTORY COMMENT

Put simply, the New Urbanism sees physical design—regional design, urban design, architecture, landscape design, and environmental design—as critical to the future of our communities. While recognizing that economic, social, and political issues are critical, the movement advocates attention to design. The belief is that design can play a critical role in resolving problems that governmental programs and money alone cannot.

The "new" in New Urbanism has several aspects. It is the attempt to apply the age-old principles of urbanism—diversity, street life, and human scale—to the suburb in the twenty-first century. It is also an attempt to resolve the apparent conflict between the fine grain of traditional urban environments and the large-scale realities of contemporary institutions and technologies. It is an attempt to update traditional urbanism to fit our modern lifestyles and increasingly complex economics.

The Charter of the New Urbanism specifically structures its principles at three telescoping scales: the region, the neighborhood, and the building. But perhaps most important is its assertion that the three scales are interconnected and interdependent. The Charter is simply twenty-seven principles organized by these three scales. The three elements of this book—the emerging region, the maturing suburb, and the revitalized urban neighborhood—each benefit from the principles articulated in the Charter.

The regional section of the Charter posits principles similar to those described in this book as the foundation of the Regional City. Its neighborhood-scale principles go to an urban-design philosophy that reasserts mixed-use, walkable environments. Its principles of design at the scale of the street and building seek to recreate places in which continuity and public space are reestablished for the pedestrian.

Urbanism advances the fundamental policies and goals of regionalism: that the region should be bounded, that growth should occur in more compact forms, that existing towns and cities should be revitalized, that affordable housing should be fairly distributed throughout the region, that transit should be more widespread, and that local taxes should be equitably shared. Each of these strategies is elaborated in this book as fundamental to the Regional City. Each of these strategies has become central to the larger agenda of New Urbanism.

This larger agenda gives clarity to the precarious balance at the regional scale between inner-city investments, suburban redevelop-

Credits: Reprinted by permission of The Congress of the New Urbanism. The *Summary* and *Introductory Comment* appear in Calthorpe and Fulton (2001).

ment, and the appropriate siting of greenfield development. This balance is one of the last understood aspects of New Urbanism and one of its most important. It addresses the question of where development is appropriate at the regional scale.

New Urbanism is best known (and often stereotyped) for its work at the neighborhood and town scale. At this scale the Charter's principles describe a new way of thinking about and structuring our cities and towns. Rather than the simplistic single-use zoning of most contemporary city plans, the Charter proposes a structure of three fundamental elements—neighborhoods, districts, and corridors. The Charter does not sidestep the scale of modern business and retailing. It simply calls for their placement within special districts when they are not appropriate to the scale and character of a neighborhood. In this taxonomy, the special-use district and the corridor (natural, auto, or transit) provide complements to and connections for the basic urban tissue—complete and walkable neighborhoods.

It is at the scale of the city block, its streets, and individual buildings that the test of integrating the auto and the need for more pedestrian-friendly environments is resolved. The Charter does not call for the simplistic elimination of the car, but instead challenges us to create environments that can simultaneously support walking, biking, transit, and the car. It outlines urban design strategies that reinforce human scale at the same time that they incorporate contemporary realities. Jobs no longer need to be isolated in office parks, but their integration into mixed-use neighborhoods calls for sensitive urban design. Differing types of housing no longer need buffers to separate and isolate them, but they do need an architecture that articulates a fundamental continuity within the neighborhood. Retail and civic uses do not need special zones, but they do need block, street, and building patterns that connect them to their continuity.

The Charter calls for an architecture that respects human scale, respects regional history and ecology, and respects the need for modesty and continuity within a physical community. Traditional architectures has much to teach us about these imperatives without prescribing nostalgic forms. And these imperatives can lead to the use of historical precedents, especially in infilling and redeveloping areas that have a strong and reestablished character. On the other hand, climate-responsive design that honors the history and culture of a place, when combined with new technologies, can lead to innovative rather than imitative design. The "seamless" integration of new and old, and a respect for existing urban patterns and scale are the imperatives of the Charter.

Too often, New Urbanism is misinterpreted simply as a conservative movement to recapture the past while ignoring the issues of our time. It is not understood as a complex system of policies and design principles that operate at multiple scales. To some, New Urbanism simply means tree-lined streets, porch-front houses, and Main Street retail—the reworking of a Norman Rockwell fantasy of small-town America, primarily for the rich.

But nostalgia is not what New Urbanism is actually proposing. Its goals and breadth are much grander, more complete and challenging. Many of the misconceptions are caused by focusing only on the neighborhood-scale prescriptions of the Charter without seeing how they are embedded in regional structures or understanding that those neighborhoods are supports by design principles at the street and building scale that attend more to environmental imperatives and urban continuities than to historical precedent.

The Charter shares its central thesis with that of this book—sprawl and social inequity must be addressed comprehensively. A fundamental tenet of the Charter speaks to the critical issue of affordability and social integration through the principles of economic diversity and inclusive neighborhoods. Economic diversity calls for a broad range of housing opportunities as well as uses within each neighborhood—affordable and expensive, small and large, rental and ownership, single and family housing. This is a very radical proposition. It implies more low-income and affordable housing in the rich suburbs at the same time that it advocates more middle-class opportunities in urban neighborhoods. It advocates mixing income groups and ethnic groups in a way that is very frightening to many communities. It is a principle that is rarely realized in practice and, given the current political climate, is almost always compromised. But it is a central tenet of the Charter and The Regional City—and it sets a direction quite different from most new development in the suburbs and many urban renewal programs.

New Urbanism outlines a set of design and policy principles that provide the means to reintegrate the segregated geography of our cities and suburbs. In so doing, it raises a complex set of issues. When does "economic diversity" in a distressed inner-city neighborhood become gentrification? What is the appropriate mix of inclusionary housing in a suburban town? These are tough questions that only have local answers. Gentrification may be mitigated by more affordable housing at the regional level, but what of the coherence and identity of the old neighborhood and its unique culture? There are no simple solutions. Perhaps the appropriate amount of economic diversity for a low-income neighborhood is reached when success doesn't mean moving out. Perhaps the definition for a rich neighborhood is "when the schoolteacher and the fireman no longer have to drive in."

The Charter sees the physical design of a region—like the physical design of a neighborhood—as either fostering opportunities, sustainability, and diversity or inhibiting them. Such design cannot mandate a civil and vibrant culture, but it is a necessary framework. Much like healthy soil, the coherent design of a region and its neighborhoods can nurture a more equitable and robust society—or it can stunt them. This is not environmental determination. It is simply an attempt to find a better fit between our current realities and their physical armature.

2 THE CHARTER

The Congress for the New Urbanism views disinvestment in central cities, the spread of placeless sprawl, increasing separation by race and income, environmental deterioration, loss of agricultural lands and wilderness, and the erosion of society's built heritage as one interrelated community building challenge.

We stand for the restoration of existing urban centers and towns within coherent metropolitan regions, the reconfiguration of sprawling suburbs into communities of real neighborhoods and diverse districts, the conservation of natural environments, and the preservation of our built legacy.

We recognize that physical solutions by themselves will not solve social and economic problems, but neither can economic vitality, community stability, and environmental health be sustained without a coherent and supportive physical framework.

We advocate the restructuring of public policy and development practices to support the following principles: neighborhoods should be diverse in use and population; communities should be designed for the pedestrian and transit as well as the car; cities and towns should be shaped by physically defined and universally accessible public spaces and community institutions; urban places should be framed by architecture and landscape design that celebrate local history, climate, ecology, and building practice.

We present a broad-based citizenry, composed of public and private sector leaders, community activists, and multidisciplinary professionals. We are committed to reestablishing the relationship between the art of building and the making of community, through citizen-based participatory planning and design.

We dedicate ourselves to reclaiming our homes, blocks, streets, parks, neighborhoods, districts, towns, cities, regions, and environment.

We assert the following principles to guide public policy, development practice, urban planning, and design.

The Region: Metropolis, City and Town

1. Metropolitan regions are finites places with geographic boundaries derived from topography, watersheds, coastlines, farmlands, regional parks, and river basins. The metropolis is made of multiple centers that are cities, towns, and villages, each with its own identifiable center and edges.

2. The metropolitan region is a fundamental economic unit of the contemporary world. Governmental cooperation, public policy, physical planning, and economic strategies must reflect this new reality.

3. The metropolis has a necessary and fragile relationship to its agrarian hinterland and natural landscapes. The relationship is environmental, economic, and cultural. Farmland and nature are as important to the metropolis as the garden is to the house.

4. Development patterns should not blur or eradicate the edges of the metropolis. Infill development within existing urban areas conserves environmental resources, economic investment, and social fabric, while reclaiming marginal and abandoned areas. Metropolitan regions should develop strategies to encourage such infill development over peripheral expansion.

5. Where appropriate, new development contiguous to urban boundaries should be organized as neighborhoods and districts, and be integrated with the existing urban pattern. Noncontiguous development should be organized as towns and villages with their own urban edges, and planned for a jobs/housing balance, not as bedroom suburbs.

6. The development and redevelopment of towns and cities should respect historical patterns, precedents, and boundaries.

7. Cities and towns should bring into proximity a broad spectrum of public and private uses to support a regional economy that benefits people of all incomes. Affordable housing should be distributed throughout the region to match job opportunities and to avoid concentrations of poverty.

8. The physical organization of the region should be supported by a framework of transportation alternatives. Transit, pedestrian, and bicycle systems should maximize access and mobility throughout the region while reducing dependence upon the automobile.

9. Revenues and resources can be shared more cooperatively among the municipalities and centers within regions to avoid destructive competition for tax base and to promote rational coordination of transportation, recreation, public services, housing, and community institutions.

The Neighborhood, the District and the Corridor

1. The neighborhood, the district, and the corridor are the essential elements of development and redevelopment in the metropolis. They form identifiable areas that encourage citizens to take responsibility for their maintenance and evolution.

2. Neighborhoods should be compact, pedestrian friendly, and mixed use. Districts generally emphasize a special single use, and should follow the principles of neighborhood design when possible. Corridors are regional connectors of neighborhoods and districts; they range from boulevards and rail lines to rivers and parkways.

3. Many activities of daily living occur within walking distance, allowing independence to those who do not drive, especially the elderly and the young. Interconnected networks of streets should be designed to encourage walking, reduce the number and length of automobile trips, and conserve energy.

4. Within neighborhoods, a broad range of housing types and price levels can bring people of diverse ages, races and incomes into daily interaction, strengthening the personal and civic bonds essential to an authentic community.

5. Transit corridors, when properly planned and coordinated, can help organize metropolitan structure and revitalize urban centers. In contrast, highway corridors should not displace investment from existing centers.

6. Appropriate building densities and land uses should be within walking distance of transit stops, permitting public transit to become a viable alternative to the automobile.

7. Concentrations of civic, institutional, and commercial activity should be embedded in neighborhoods and districts, not isolated in remote, single-use complexes. Schools should be sized and located to enable children to walk or bicycle to them.

8. The economic health and harmonious evolution of neighborhoods, districts, and corridors can be improved through graphic urban design codes that serve as predictable guides for change.

9. A range of parks, from tot-lots and village greens to ballfields and community gardens, should be distributed within neighborhoods. Conservation areas and open lands should be used to define and connect different neighborhoods and districts.

The Block, the Street, and the Building

1. A primary task of all urban architecture and landscape design is the physical definition of streets and public spaces as places of shared use.

2. Individual architectural projects should be seamlessly linked to their surroundings. This issue transcends style.

3. The revitalization of urban places depends on safety and security. The design of streets and buildings should reinforce safe environments, but not at the expense of accessibility and openness.

4. In the contemporary metropolis, development must adequately accommodate automobiles. It should do so in ways that respect the pedestrian and the form of public space.

5. Streets and squares should be safe, comfortable, and interesting to the pedestrian. Properly configured, they encourage walking and enable neighbors to know each other and protect their communities.

6. Architecture and landscape design should grow from local climate, topography, history, and building practice.

7. Civic buildings and public gathering places require important sites to reinforce community identity and the culture of democracy. They deserve distinctive form, because their role is different from that of other buildings and places that constitute the fabric of the city.

8. All buildings should provide their inhabitants with a clear sense of location, weather, and time. Natural methods of heating and cooling can be more resource-efficient than mechanical systems.

9. Preservation and renewal of historic buildings, districts, and landscapes affirm the continuity and evolution of urban society.

Washington, D.C. North of Massachusetts Avenue (NoMa) Plan. "The city and neighborhood groups recognized the value of mixed-use urban environments, the need to preserve an existing arts community, and the need to diversify the economy of the city. Programs to support housing and high-tech industries were presently missing from the economic mix of the downtown area, as were urban design controls to allow lofts in converted warehouses and artist studios in surrounding infill locations. The gateway treatment of New York Avenue as it enters the central city is enhanced by a traditional traffic circle so unique to L'Enfant's plan of Washington. The structure of his famous radial boulevards is reinforced with new and rehabilitated buildings that orient to the streets." (Calthorpe and Fulton, 2001.)

REFERENCES

Calthorpe, Peter and William Fulton. 2001. *The Regional City: Planning for the End of Sprawl.* Washington, DC: Island Press. pp. 269–270.

Congress of the New Urbanism. 5 Third Street, Suite 725, San Francisco, CA 94103. www.cnu.org.

EDITORS NOTE

Like the Athens Charter of a half-century earlier, The Charter of the New Urbanism was published without illustrations. A sizable and increasing number of built projects forms the basis of its evolving theory and practice, promulgated by the Congress of the New Urbanism, whose founding members include Peter Calthorpe, Jean Driscoll, Andres Duany, Elizabeth Moule, Elizabeth Plater-Zyberk, Stephanos Polyoides, and Daniel Solomon.

The illustration and caption on this page illustrate application of New Urbanist principles to revitalization of an existing, and in this case renowned urban plan.

4 • PRINCIPLES AND PRACTICES OF URBAN DESIGN

Lewis D. Hopkins

Summary

This article considers how plans work and thus what plans are and how one can assess their success. Explanations of how plans work identify relationships between the attributes of plans and the effects that plans have. They thus identify what plans can achieve. Any one plan can work in one or several ways, which means that these are not categories for classification of plans but different mechanisms through which plans affect the world. Terms defined include *agenda, policy, vision, design* and *strategy*.

Key words

agenda, capital improvement program (CIP), design, maps, plans, planning, policy, regulation, strategy, vision

Vision and design for redevelopment of downtown Cleveland.
I.M. Pei and Associates.

How plans work

In the United States city planning is essentially a process of vision and survey, push and pull, barter and sell, education and exhortation, diplomacy and expediency, courts and juries.
—Walter D. Moody (1919), What of the City?

The plan here given is a program of improvements calculated to cover a period of many years. The order in which improvements are made, and when, is not so important as that each shall be so done as to fit into its place in the general plan.
—Harland Bartholomew (1924), The City Plan of Memphis, Tennessee

The free and easy meeting of problems as they arise will no longer suffice, and more than ever officials are looking for a solution of current problems in terms of the predictable future.
—Robert A. Walker (1950), The Planning Function in Urban Government

It is the intent of the Legislature that public facilities and services needed to support development shall be available concurrent with the impacts of such development.
—Florida Statutes 163.3177(10)h, adopted in 1985

How do plans work? Through what mechanisms or casual processes do plans affect actions? How can we explain why a particular plan is likely to have particular effects? As Moody, Bartholomew, Walker, and the Florida statutes quoted above demonstrate, plans can work in more than one way and planners have, for a hundred years, explained how they expected plans to work. These explanations do not provide precise predictions of what will happen in specific situations, but they do make sense of what we observe and enable us to talk about what we should do. This article considers how plans work and thus what plans are and how we can assess their success.

Agendas, policies, visions, designs, and strategies

Table 1 summarizes five different ways in which plans work: agendas, policies, visions, designs, and strategies. The definitions stipulated here are necessary to avoid the ambiguity of many meanings of "plan." For each distinct meaning, a word is used having a dominant connotation close to the narrower definition, even though the word may also be used to refer to other concepts. These words also have wider and richer meanings in related fields, such as the use of the term "policy" in policy analysis. To paraphrase Wildavsky (1973), if a word can mean everything, it can only mean nothing.

An agenda is a list of things to do.

An agenda works by recording a list to remind us what to do, or to share publicly a commitment to do these things. Agendas work when there are too many actions to remember or when there is benefit in gaining trust among people affected or legitimating actors as accountable. Publishing or publicly advocating an agenda serves both as a memory device and a commitment. We write down an agenda for a meeting so we will remember to discuss the intended issues. An agenda also implies repeated efforts to accomplish something. Agendas may merely list independent actions that only come together because someone chooses to focus on them at the same, or nearly the same, time. Once created, a Capital Improvements Program (CIP) or budget may function as an agenda. It keeps a record of a list too long to remember and known to be within the budget constraint set by projected

Credits: This article is excerpted from the author's book *Urban Development: The Logic of Making Plans* (Hopkins 2001) and is reprinted by permission of Island Press, Washington, DC.

revenues. Citizens who know that an item is on or not on the Capital Improvements Program list find some creditability in the assumption that it will or will not be built within a particular time. In this explanation, each citizen or each city council member is not concerned about any relationship among the projects or any interdependence among the decisions. Deciding on the CIP, however, is a focus of conflict among projects for different departments and different political wards and different interest groups as well as a competition for available budget. The process of creating a CIP is thus working in some way other than as an agenda.

Agendas differ from objectives. Agendas identify issues or actions; objectives identify valued attributes of outcomes. We can check off everything on our list of things to do, but still not accomplish the objectives that led to the list of things to do. We could create a list of measurable objectives but still have no ideas about what actions to take to achieve them. All explanations of plans must contend with the relationships of actions to outcomes and outcomes to objectives.

The items in the agenda of a meeting have in common the timing of decisions at the same meeting and perhaps a common decision maker's authority, but the choice for one item need have no relation to the choice for another. Agendas are of interest to planners because they are a tool that focuses the attention of a constituency, whether an individual, a legislature, a group, an electorate, or the public at large. Setting agendas and pursuing agendas are thus ways of affecting the decisions that will be made. Agendas keep our attention focused on important actions or issues rather then merely on what "comes across our desk" at the initiative of others. An agenda is one way to focus the attention of decision makers on some decisions rather than others.

A policy is an if-then rule.

A policy works by automating repeat decisions to save time or by ensuring that the same action is taken in the same circumstances, which yields fairness or predictability. Policies fit situations in which there are many repeat decisions and decisions are costly to make, consistency is viewed as fair, or predictability of repeat decisions is beneficial. For example, if the developer will pay for the cost of the sewer extension, then extend the sewer. This policy would save the costs of making this decision in each case, treat all developers alike, and make development actions predictable. Knowing the policies of other decision makers provides evidence for forecasting their decisions. Policies are distinct from regulations in that regulations change legally or administratively enforceable rights whereas policies identify standard responses for repeated instances of the same situation. If the policy is to grant tax incentives to new industrial firms, then when a new industrial firm proposes to locate in the community, tax incentives should be granted. The policy simplified decision making by deciding once on a decision rule to apply to all situations of the same class (Kerr 1976). Policies work in three ways: saving decision costs, ensuring consistency (fairness), and increasing predictability.

A vision is an image of what could be.

Visions compel action. Visions work by changing beliefs about how the world works (beliefs about the relationships between actions and outcomes), beliefs about intersubjective norms (peer group attitudes about good behaviors), or beliefs about the likelihood of success (raising aspirations or motivating effort). A vision could be interpreted as a normative forecast: a desired future that can work if people can be persuaded that it can and will come true. Visions, however, focus first on the outcome and then on the possibility of actions to attain this outcome. Henry David Thoreau expressed it this way in the concluding chapter of Walden: "If you have built castles in the air, your work need not be lost; that is where they should be. Now put the foundations under them." Visions are useful in situations in which they can change beliefs and thereby change investment actions, regulations, or activity patterns of residents. Visions are distinct from target designs, which are focused on a feasible solution to a complex problem of interdependencies. Visions work by their effect on beliefs, not by their feasibility of construction.

A vision can help overcome resilience in a system. Resilience dampens feedback that would give immediate responses to actions we might take. Lack of feedback makes intentional action both difficult and risky. If you are trying to change the attitudes of one ethnic group about another ethnic group, resilience is a hindrance. Even interventions that might change attitudes eventually with sufficient time or effort might not yield visible results in time to keep the effort going forward. The effort will then be stopped even though it might have succeeded eventually. A vision can help to motivate continued effort.

Guttenberg (1993) describes a "goal plan" approximately equivalent to this idea of vision: "The image is credible, it bears some relation to existing opportunities in the region, but apart from its ability to persuade, to move people by its attractiveness, it includes no explicit measures for ensuring that these opportunities will be realized" (p. 190). "The purpose of a goal plan is more to state a desired objective persuasively than to plot a course of action for the intervening years" (p. 193).

Graphic and verbal descriptions of future situations—social utopias or beautiful cities—have been developed for centuries, and "visioning" is a currently popular tool in urban planning. Visions can reframe problems by describing the present and its relationship to possible changes in a different way. Visions also describe what the world will look like after proposed changes occur. Literature on strategic planning for corporations—Bryson's (1995; Bryson and Crosby 1992) extension of this literature from hierarchical organizations to "shared power worlds"—use visions in all of these ways.

The Chicago Plan of 1909 is a familiar example of plan as vision. It included both graphic renderings of a physical vision and verbal descriptions of the characteristics of a great city.

In creating the ideal arrangement, everyone who lives here is better accommodated in his business and his social activities. In bringing about better freight and passenger facilities, every merchant and manufacturer is helped. In establishing a complete park and parkway system, the life of the wage earner and of this family is made healthier and pleasanter; while the greater attractiveness thus produced keeps at home the people of means and taste, and acts as a magnet to draw those who seek to live amid pleasing surroundings. The very beauty that attracts him who has money makes pleasant life of those among whom he lives, while anchoring him and his wealth to the city. The prosperity aimed at is for all Chicago (Burnham and Benett 1909, p. 8).

The Portland 2040 Plan also uses images of the implied future to sell its less palatable actions (Metro 2000). Planning for small towns often focuses on "visioning," a collaborative effort by a large portion of the town's citizens to follow a process fairly similarly to the corporate strategic planning process (Howe *et al.* 1997). The Atlanta 2020 Project used a visioning approach (Helling 1998).

Table 1

How Plans Work

Aspect	*Agenda*	*Policy*	*Vision*	*Design*	*Strategy*
Definition	List of things to do; actions, not outcomes	If-then rules for actions	Image of what could be, an outcome	Target, describes fully worked out outcome	Contingent actions (path in decision tree)
Examples	List of capital improvement projects	If developer pays for roads, then permit development	Social equality, picture of beautiful city	Building plan or city master plan	Road projects built depend on how much land development occurs when and where
Works by	Reminding; if publicly shared, then commitment to act	Automating repeat decisions to save time; taking same action in same circumstances to be fair	Motivating people to take actions they believe will give the imagined result	Showing fully worked out results of interdependent actions	Determining which actions to take when and where depending on situation when actions are taken
Works if	Many actions to remember and need trust among people affected	Repeated decisions should be efficiently made, consistent, and predictable	Can raise aspirations or motivate effort	Highly interdependent actions, little uncertainty about actions, and few actors involved	Interdependent actions by many actors over long time in relation to uncertain events
Measures of Effectiveness	Are actions on list taken?	Is rule applied without constant reconsideration, or is rule applied consistently?	Are beliefs changed as evidenced by beliefs elicited directly or revealed in actions?	Is design constructed or achieved?	Is contingent interdependence sustained in actions, and is information used in timely fashion?
Cases	Chicago Plan of 1909	Chicago Plan of 1967, Cleveland Policy Report of 1974, Lexington, Kentucky	Chicago Plan of 1909, Portland 2040, Washington, D.C. 2000	Chicago Plan of 1909	Lexington, Kentucky 1958, Integrated Action Plans, Nepal

A design is a fully worked out outcome.

Designs work by determining a fully worked out outcome from interdependent actions and providing this outcome as information before any action is taken. Designs fit situations in which there are highly interdependent actions, actions are easily inferred from information about the outcome, and there is little uncertainty about implementation of actions. We usually think of design as a process in which many ideas are tested and modified, but entirely in some simulated environment before any action is taken in the real world. Harris (1967) identifies design decisions as reversible at zero cost. All the decisions involved in design of a single building are tested as hypotheses in combination through diagrams and calculations to see how they fit together before any action is taken to construct the building. Designs usually focus on patterns of capital facilities rather than on the human activity patterns that will occur given these facilities. Measures of success should, however, access these human activity patterns.

Design works by figuring out a result for many interdependent actions before acting. It thus avoids the problems of interdependence, indivisibility, and irreversibility through a presumption of perfect foresight. There is no iterative adjustment; the result is determined first so that each action can immediately fit the solution. Bacon (1974, pp. 260–262) illustrates how the design concept breaks down over time in urban design but still results in somewhat coherent physical forms. A complete and coherent design for a section of a city is proposed. Some elements of the coherent design get implemented, but other elements do not because of citizen complaints, budget constraints, changes in government, or power relations. Then, situations change and new designs are proposed that in part relate to the elements of the previous design. Some of the elements of the new design are implemented. The realized urban form results from this sequence of dependent designs, none of which is implemented in its entirety.

As projects become more complex and more easily decomposed into actions that can be carried out separately (*e.g.*, more than one building, phased buildings to be constructed with long periods between each phase), they take on the character of a sequence of design projects linked by strategies about related decisions. Although any architect designing buildings will point out that in many cases the design may be modified during construction and that the cost of design changes is not zero, these costs are small and these modifications are minor relative to the whole design. In larger urban development situations, actions taken at different times are each of similar magnitude, such as building an interceptor sewer now and an expressway later. Modifying the expressway capacity or service area before it is built to complement a sewer system designed to absorb twenty-five years of growth is a different level of relationship from modifying details as a building is constructed.

A design approach solves problems before acting on any decision, whereas a strategy approach decides what action to take now cognizant of related future actions. We do not need to make all related decisions simultaneously, but we can consider potential future decisions before making a decision now. Note that the target creating of design is also different from agenda setting. An agenda is a list of things to do; a target is something to shoot at. A target might prompt an agenda. A strategy might be devised to achieve a target.

A strategy is a set of decisions that forms a contingent path through a decision tree.

Strategies work by determining what action should be taken now contingent on related future actions. Strategies fit situations in which there are many interdependent actions under the authority of many actors and occurring over a long time in relation to an uncertain environment. In sequential decision making, at the time action is taken on a current decision, the future decisions have been thought through for each outcome from that current decision. Saying that we plan to do something means that we will take certain actions under certain conditions when the time comes. Design and strategy represent the continuum sometimes described between synoptic or blueprint plans and incremental, decision-centered planning (*e.g.,* Faludi 1987). The crucial difference is the degree to which all decisions should or can be taken at once or only sequentially.

Strategy is arguable the most inclusive and thus fundamental notion of plans because it is the most explicit about the relationships among interdependent actions, their consequences, intentions, uncertainty, and outcomes. Strategies address most completely the problems of interdependence, indivisibility, irreversibility, and imperfect foresight. In contrast, designs focus primarily on outcomes. Visions, agendas, and policies are often joint effects of plans that also work as strategies or designs. Visions, policies, and agendas, as explained earlier, can also address situations that do not meet the strict criteria of interdependence, indivisibility, irreversibility, and imperfect foresight.

Plans address spatial phenomena, which is a direct result of interdependence among decisions in space. On the other hand, policy analysts tend to ignore spatial phenomena and focus on the impacts of individual programs or policies, not on plans for related actions. Analysis for a single decision or for repeat decisions of the same type may benefit from forecasts of impacts, but when interdependence actions can be taken sequentially, the relationships between decisions and forecasts become more complex. Plans working as strategies depend on functional, spatial, and temporal relationships among decisions themselves and their impacts. Policies are distinct from strategies, because policies apply to repeated decision situations of the same kind whereas strategies coordinate different but related decisions. Strategies may yield policies as statements of decision rules, such as "Allow development if the developer pays for the cost of sewer extensions." This policy might implement a strategy of providing sewer infrastructure over time concurrent with development. Plans may be hierarchically related. For example, under California planning legislation, area plans (or specific plans) are subject to policies and strategies set out in general plans (Olshansky 1996). The Chicago plan of 1966 (City of Chicago Department of Development and Planning 1966) set policies for area plans that were developed for each neighborhood.

In contrast to plans, regulations set the rights of a decision maker by identifying what decisions are permitted and by setting the range of discretion of choices and criteria in making these decisions. Regulations are enforced by the state through its monopoly on the use of force. For example, zoning restricts the range of uses, building height, and land coverage that may be undertaken on a particular parcel. A subdivision ordinance restricts the patterns by which land can be divided into building lots. Regulations may be created by private groups under the force of contracts, which are in turn enforced by the state. Thus a homeowners' association may impose design regulations for its members. Regulations affect decisions by restricting the set of choices, whereas plans affect decisions by providing information.

In contrast to regulations, none of the ways in which plans work is inherently binding on actors. Plans that work as strategies set forth contingent decisions that affect choices made now, but there is no current or future change in the range of alternatives from which the decision maker is permitted to choose. The effect on current decisions is only through the decision maker's own assessment of related decisions. Regulations define the set of future alternative from which a decision maker may choose, which can help to determine which decision is best for action now.

Plans also work as a focus of deliberation—discussion, argument, conflict, and resolution. Such work occurs both in the creation of plans and in their use to guide action.

Investments and regulations

Investments in physical infrastructures or facilities and regulations are widely recognized as the two major components of urban development plans (see, *e.g.*, Alexander 1992, 98ff; Neuman 1998). As in political interpretations, these different types of actions imply different tasks for plans (Table 2). Investments, whether by public agencies or private firms, change the capital stock of infrastructure or buildings. Regulations change rights, the range of discretion in making decisions. Plans often include recommendations for enabling legislation from a higher level of government to allow a lower level of government to take certain kinds of actions. Enabling legislation is thus analogous to regulation but is among levels of government rather than between governments and individuals. That we observe this pervasive focus of plans, whether made for governments or private firms or individuals, suggests that we should be able to explain why plans are made for investments in physical facilities and regulations rather than for other types of actions.

The simple explanation is that infrastructure investments, whether by public, private, or joint actors, are interdependent with other investments. They are partially indivisible and subject to significant economics of scale; they are durable, long lasting, and costly to reverse once action is taken; they are subject to imperfect foresight with respect to demand, technology, and related actions. When iterative adjustment does not work, plans that work as designs or strategies can yield improved outcomes because such plans consider other actions before taking an action now. Plans can yield such improvements not only from the perspective of a government, but also from the perspective of a private firm or individual.

Investments in physical facilities mediate between geographic space and people's behaviors. Thus two kinds of decisions matter: the decision to invest in infrastructure and the decisions to use the resulting infrastructure in particular ways. Indicators of quality of life depend on activities on populations, including their interactions with each other; the physical facilities in which they live and work, including the networks that connect these facilities; and the geographic locations in which these activities and facilities occur. Thus an indicator of vehicle miles traveled per person per day depend on where people who work downtown live and over what type of network they travel, which depends in turn on the geographic character of the site of the city. The important point is that investments occur in fixed locations and they create the physical context within which locational choice and daily behaviors occur. Whether investments are in buildings, housing, schools, treatment plants, or networks, roads, sewers, light-rail transit, they are fixed in place and cannot be moved without great cost. They are built with specific capacities, which cannot be changed without additional investment. Increments of capacity are subject to significant economies of scale. It is less costly per unit of treatment, for example, to build a larger rather than smaller sewer treatment plant. Although a treatment plant may take ten years to site, design, and build, it will still be expected to serve expected demand for a fifty-year life. Thus forecasts of demand for sixty years may be pertinent and must be precise enough to be useful. If demand occurs more slowly or at different densities than forecasted, however, contingent pipe sizes and construction timing should be available as strategy. It is very expensive to replace pipes to increase their capacity, however, so robust strategies for the major pipes in the network may be appropriate.

People choose to live or work in facilities that exist at particular locations because someone invested in the facility at that location. People choose transportation mode and route over a network of streets and transit based on investments made to link locations by roads or transit routes. The outcome of the investment is realized only when the location choice and travel choice behaviors occur. We must therefore estimate these behaviors for given investments rather than trying to estimate the effects of investments directly.

This logic of plans for investments also applies to capital investments by the private sector. Anas *et al.* (1998) give the example of the creation of new nodes in a multinucleated city. In many cases, no one developer has sufficient capital or land to build an entire new center alone. If several developers try to locate new subcenters when only one or two can be sustained, however, then some subcenters will fail. The capital invested will be lost and the underutilized land will displace other uses because of the high cost of conversion. Even the successfully established center will be slower in developing then necessary because some development and tenants will have to move from other failed centers. The private developers have much to gain from figuring out ahead of time which new center will succeed and building there initially. Public infrastructure providers and house buyers would also be affected by the uncertainty of location of new centers.

Table 2

Land Regulation and Implied Requirements of Plans

Regulation Type	*Regulation Logic*	*Implied Plan Logic*
Zoning	Externalities (positive and negative)	Strategy to address interdependence in advance because of irreversibility of investments and indeterminate adjustment process given imperfect foresight
	Infrastructure capacity	Strategy for capacity expansion and design for capacity at buildout because of irreversibility and indivisibility
	Fiscal objectives	Policy for consistent and fair repeated decisions for fiscal objectives
	Information costs or errors	Policy as means of providing information that is collective good or asymmetric between buyers and sellers
	Management of supply	Strategy to reduce infrastructure costs of spatial substitution of uses as technology changes given imperfect foresight
	Amenity protection	Target, permanent allocation yielding strategy of implementation to acquire rights
	Development timing	Strategy of zoning for non-urban uses until land is ripe for development
Official maps	Protect rights-of way	Strategy for rights-of-way because of irreversibility of investments
Subdivision regulations	External effects of design decisions	Policies to achieve design decisions by developer that have collective good external effects
Urban service areas (Urban growth boundaries)	Timing, resource lands protection, "optimal city size"—depending on how changes in area are managed over time	Strategy of efficient infrastructure provision and interaction costs over time; policy of consistent and fair resource land protection; target design of city
Adequate public facilities ordinances	Timing	Strategy of efficient infrastructure provision and interaction costs over time
Development rights (e.g., conservation easements, transferable development rights)	Permanent allocation of land to uses	Target design of pattern of uses among, e.g., resource lands and urban development
Impact fees	Timing, fiscal management, and distribution of costs among current and new residents	Policy for consistency and fairness and strategy for infrastructure financing

Regulations have a structure similar to investments with two kinds of decisions: decisions to regulate and decisions to act given the regulations. A decision to zone a municipality by land-use type and density is a decision to regulate. A decision to build a house in one of these zones is an action given the regulation. Usually the decision to regulate will be collective and the decision to act individual. In order to use regulations, decisions must be made about where to impose what regulations. These decisions are analogous to investments in that they face interdependence, indivisibility, irreversibility, and imperfect foresight. To implement a zoning regulation, we must consider a sufficiently large area to figure out a pattern of land uses that will reduce negative effects of adjacency of different uses and provide access to services. The area to be zoned must be considered in finite increments; it is indivisible. As with investments, regulations cannot, therefore, work by iterative adjustment. If a regulation is to reduce external effects of adjacent land uses, it will be effective only if it is imposed before the conflicting land uses invest in location next to each other. If a regulation is to match density with infrastructure capacity, it can only be effective before investments are made.

Investments and regulations are logical elements of plans that work as designs or strategies because they are likely to benefit from such plans. Social programs or other actions that are not interdependent, indivisible, irreversible, and subject to imperfect foresight are much less likely to benefit from such plans. For example, state-funded health care would affect quality of life and is worth careful analysis. A housing voucher program may be a valuable public program. Such programs may be on an agenda, may be implemented through policies, or be expressed as visions, but they are not likely to be the focus of a design or a strategy because they do not have the attributes of capital investments or spatially expressed regulations of interdependent actions.

This observation does not mean that such programs are unimportant; it means that instruments different from plans working as designs or strategies are likely to be more useful in achieving the intended outcomes. This observation also does not mean that equity goals or social purposes should not be criteria by which investments or regulations are judged. Regardless of the criterion of success, investments once made are costly to change. It is no more possible to iterate toward a social equity goal than toward an economic efficiency goal if the actions involved are irreversible investments.

Investments and regulations are likely to benefit from plans as designs or strategies because of the characteristics of these types of actions not because of characteristics of particular criteria for evaluating them. A full range of criteria is likely to be and should be considered.

Determining whether plans work

Do plans work? These explanations of how plans can work—as agendas, policies, visions, designs, and strategies—provide a means by which to assess whether plans do work. These explanations indicate what we can expect to observe if plans are working and how we can explain relationships among these observations. We can observe:

- **plan-making behaviors**—the things planners and their collaborators do when they make plans,
- **plans**—information available at particular times to particular people,
- **people using plans** while making decisions,
- **investments and regulations** that may have been affected by plans, and,
- **outcomes** in terms of activity patterns resulting from these investments and regulations

All of these observable phenomena provide opportunities for assessment. Here we focus on plans and whether they work, not on how they are made.

There are four broad criteria for assessing whether plans work:

- **Effect:** Did the plan have any effect on decision making, actions, or outcomes? For example, if it was intended to work as an agenda, how many of the listed actions were taken?
- **Net benefit:** Was the plan worth making and to whom? For example, if it was intended to work as strategy, were the gains in efficiency of infrastructure provision over time sufficient to compensate the costs of making the plans?
- **Internal validity (or quality):** Did the plan fulfill the logic of how it was intended to work? For example, if it was intended to work as strategy, did it address interdependence, indivisibility, irreversibility and imperfect foresight in appropriate ways?
- **External validity (or quality):** Did the outcomes intended or implied in the plan meet external criteria, such as claims for a just society? For example, if it was intended to work as a vision, did the vision include equity? Ethical acceptability is a crucial component of external validity.

Several authors have developed such typologies, none of which I follow completely, but some of which share common elements with this typology. Talen (1996) provides a thorough review of this literature and makes a strong case for the importance of assessing plans on the basis of whether they achieve objectives. Alexander and Faludi (1989) discuss the range of possibilities, including conformity of the actions to the plan, rationality of planning process, quality of the plan solution assessed before or after it has affected decisions, and whether the plan is utilized in the decision making process. Others (*e.g.*, Berke and French 1994; Dalton and Burby 1994) consider whether plans that meet standards in the literature or in state legislative mandates are more likely to result in a greater number of implementation tools being in place. More recently, Mastop and Faludi (1997) argue for a "performance" approach, which requires looking at how a plan affects decisions and how these decisions in turn affect outcomes. This casual chain links the plan to outcomes, which is consistent with my argument that we need explanations of how plans work. Connerly and Muller (1993) identify frequency of consultation of the plan by decision makers as a measure of plan quality, which highlights the necessary casual link but does not explain what the expected effects of use should be. Baer (1997) provides a checklist for assessing plans based primarily on the plan as document and the reported procedures by which it was made.

Some of the typologies focus more on what to assess and others on how to assess it. As most of these authors point out, it is very difficult to assess the effects of plans on outcomes and thus on measures of goal achievement. In urban development processes, it is almost impossible to say what would have happened without the plan and compare this to what did happen with the plan. Or, conversely, it is impossible to say what would have been different if there had been a plan. Calkins (1979) developed one of the most complete descriptions of the monitoring of plan accomplishment with respect to time and space. His key concepts are to recognize both underlying trends independent of the effects of a plan and trends caused by the plan.

Note that all of these assessment approaches are distinct from the question of evaluating a particular action in a plan, such as estimating the net benefits of a highway project or choosing between a transit-oriented or auto-oriented development pattern. That is, none of the above types of assessments addresses evaluating alternative plans in the process of choosing the content of a plan. Rather, they ask, "Did the plan work?"

Did the plan have any effect on decision making, actions, or outcomes?

Plans work by affecting actions, indirectly if not directly. Whether the actions taken yield intended outcomes is a distinct but important question. Good plans must not merely be more likely to affect actions. They must also be more likely to include actions that will yield intended outcomes. Does Chicago "look like" the Chicago Plan? Were the aspirations achieved based on some set of indicators? Does Cleveland "look like" the Cleveland Policy Report in the sense that indicators show an increase in choices for people who are least well off? Do the new towns of Reston, Virginia, or Milton Keynes in England look like the plans for them? To use this basis, we not only need to be able to measure outcomes pertinent to the plan, but also to provide an explanation of how the plan caused these outcomes. One difficulty is uncertainty in the relationship of actions to outcomes. Even if planned actions are taken, the intended outcomes may not occur. Even good choices in locating land uses relative to flooding or other natural hazards may yield larger losses over a given period than before the plan because of a particularly large flood or a cluster of hazard events. A plan that is based on the belief that people will use transit if they live in a transit-friendly environment may be used in decisions and affect actions but still not gain the outcomes that the plan sought because the belief about how the world works was wrong.

Talen (1996) argues strongly for the value of assessing plans directly in terms of the resulting patterns rather than in terms of actions taken. Rather than focus on the investment or regulation actions, she focused on whether the intent of the plan was achieved, in particular whether the equity distribution sought by a plan for city parks was achieved (Talen and Anselin 1998). If the objective of the plan is to provide parks for neighborhoods or types of households that are currently underserved, then a measure of whether the relative level of service for such neighborhoods and households improved is more pertinent than whether parks were built in the specific locations and sizes shown in the plan. This distinction returns us to explanations of how plans work. Whether the objective of equitable distribution of parks was achieved or not, the question of what effect a plan had on this remains to be shown. On the other hand, if the parks were built where the plan recommended and the plan recommended these locations in order to achieve equitable distribution, then just showing that the plan caused parks to be located in these places is also insufficient.

If the spatial diagram or map in the plan were meant merely to persuade constituencies of the possibility of action with respect to goals, then the specific locations of parks would not matter. The locations would be in the plan as an illusion of precision to achieve persuasion. If, on the other hand, the explanation of how the plan is

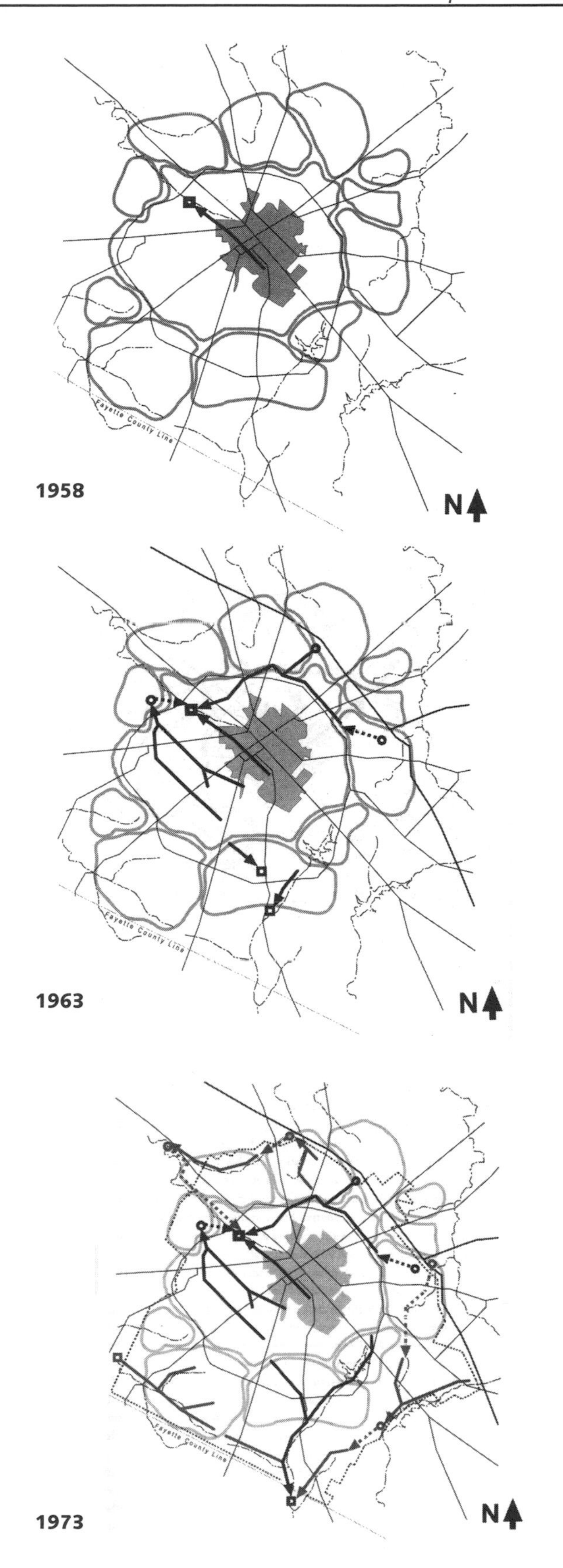

Policy plan and strategy plan for installation of sanitary sewer, contingent on land development. Lexington, Kentucky, (top) 1958, (middle) 1963; (bottom) 1973.

intended to work is as a set of fully worked out interdependent actions, then the particular locations of parks may be related to transit stops, dwelling unit densities, diagonal pedestrian access routes, and traffic calming street patterns. In this case it matters a great deal whether parks were built in particular places in conjunction with other actions. In this latter situation the assessment of equitable distribution with respect to demographic characteristics is insufficient. The substantive logic of the relationships among actions in the plan matters, and the logic of how the plan might affect action matters.

The 1929 Regional Plan of New York and Its Environs was several years in the making, involved many planners, and took a forty-year perspective. Johnson (1996) takes advantage of the resulting visibility of the planning process and the opportunity to track actions and outcomes to develop a thorough assessment of the effects of the plan. His analysis includes consideration of potential effects of the plan working as vision, agenda, policy, design, and strategy, though not based on the strict definitions and explanations presented here. He points out the difficulties of assessing whether a plan worked as an agenda.

> *It is difficult to separate forecasts of events that would have occurred, plan or no plan, from events whose occurrence is attributable to the Plan. And what of long-standing proposals that predated the Plan and were simply incorporated into it? To what extent can the fact of their being part of the Plan be credited with their realization? Each specific project or proposal needs to be analyzed as an individual case study if definitive judgments are to be made about the casual relationships of plan and reality (p. 244).*

Johnson also identifies difficulties in assessing the effects as policy or perhaps as vision.

> *For example, should decentralization be encouraged or discouraged? Or, should highways be emphasized over transit? General policy reveals itself in the making of specific decisions, but it is itself subject to modification and influence by plans, among other factors. But the extent to which plans as paradigms influence general policy is usually difficult to ascertain. The plan, if it embodies accepted public policy, can reinforce that policy, but the strength of that reinforcement can only be a matter of speculation. Where the plan breaks new ground or attempts to alter accepted policy assumptions, it may be a simpler task to estimate impact by reference to points at which policy changes. (p. 244)*

Johnson compares forecasts in the plan, such as population, to historical outcomes to interpret contingent strategies, though the plan itself did not identify such contingent strategies. He also reports which major projects were accomplished and which were not and computes percentages of open space projects completed by subregion, but as he argues, casual explanations linking these outcomes to the plan are difficult to construct.

Even a case study as detailed and thoroughly observed over several years as the traffic reduction scheme for Aalborg Denmark (Flyvbjerg 1998), however, still faces some of these difficulties. Flyvbjerg's interpretation centered on the power of certain actors to oppose parts of the plan on which he focuses his narrative. He interprets the inability to implement all of the interdependent elements of this scheme because of powerful opposition as a plan failing in the face of power. The proposed scheme in the plan he analyzes, however, contradicted the logic of interdependent actions of major capital investments made just before the plan was adopted. The plan's failure might be interrupted as the success of a previous plan that withstood the attempt to change it. Incomplete implementation of one plan is not generalizable as evidence that plans do not work.

Each of the ways in which plans work implies an explanation of how a plan affects the world and thus an assessment based on that particular explanation. The measure of effectiveness for an agenda is whether the tasks were accomplished. We may also be able to observe whether actors or citizens, to sustain the implied commitment to the list, referred to the agenda as a reminder. Such observations would be evidence that the actions occurred because of the plan and because it served as an external memory device.

For policies, there are distinct measures of success for its distinct purposes. For decision efficiency, the measure of effect is whether decisions were made by reference to the policy rather than by considering the next decision situation from scratch. Reference by decision makers to the policy may be observable. Or, the policy may become habit and therefore not be directly observable, even though conformance with policy can still be observed. For decision fairness or consistency, the measure of effectiveness is whether the policy was applied accurately in similar situations. This can be determined by assessing a sample of situations in which the policy should have been applied.

Observing beliefs of the plan's target audience before and after the plan and asking whether beliefs changed can assess the vision mechanism. Beliefs might be elicited directly or inferred or revealed in actions. To determine whether these changes in beliefs also changed actions as intended would require observations of actions. Without observation of changes in beliefs or inference of such changes, however, we could not tell whether a plan was working as a vision.

For designs, the measure of success is whether the design is constructed or achieved. This measure of conformance has been used in several plan effectiveness assessments (*e.g.,* Alterman and Hill 1978). It is generally not linked to a particular mechanism of how plans work, but rather a general notion of linking the plan directly to the outcome. Note that because the design mechanism is directly associated with the outcome, there is no intervening measure. The presumption is that we can recognize the outcome as resulting from the design because the design is sufficiently distinct that the outcome would not otherwise have occurred by chance. If design is not the mechanism by which a plan is expected to work, however, then conformance alone is not a sufficient measure of effect.

For strategies, the measure of success is whether the contingent strategy was pursued. Use of the strategy may or may not result in the most likely outcome being achieved. So for this explanation, the conformance measure is not directly pertinent.

Finally, it is important to distinguish between lack of plans and lack of action. In Kathmandu, Nepal, people lament the lack of planning, but there are actually many plans. There is a lack of action, in part because of severe budget constraints, and a lack of certain types of land development regulations. It is the lack of investments and regulations that people often mean when they say there is a lack of planning. These plans may have identified actions that were logically linked to good outcomes, but they are not good plans because they fail to consider whether any actor could take these actions. Or, if the plans are explained as visions, then they may be working, though slowly, by changing people's beliefs about how an urban settlement works and what other people believe is worth doing or feasible to do.

One can determine whether a plan worked by linking these observable phenomena:

- Was the plan used? Or, a plan is good because persons use it in choosing actions.
- Were the actions taken? Or, a plan is good because the actions implied by the plan were taken.
- Were the outcomes achieved? Or, a plan is good because the outcomes sought by the plan were achieved.

The combination of these three types of observations can yield a persuasive argument that a plan affected decision making, actions, and outcomes in turn. They can test an explanation of how plans work and thus provide generalizable implications for other similar circumstance. Whether the outcomes were and still are valued and ethical is a question of external validity discussed below.

Was the plan worth making and to whom?

Even if a plan is shown to have effects on decision making, or outcomes, it may not be worth the cost of making the plan. There is so little empirical evidence of the effects of plans, that it seems unnecessary to consider whether the effects compensate the costs. The question must be acknowledged, however, and effects should be identified in ways that might allow comparison to costs. Measuring the costs of making plans is conceptually straightforward. Measuring the benefits from the effects of plans, which might be negative, is a minefield of difficulties.

Helling (1998) reports a cost-effectiveness study of Atlanta 2020 collaborative visioning project. She assumes that the vision, the results of the process, and the process of creating the vision should somehow affect actions, which is consistent with explanations presented here. She concludes that the plan was relatively ineffective at anything other than increasing the interaction among participants, which might eventually have indirect effects on actions. She also estimated the costs for creating the vision at $4.4 million. These costs were carefully calculated as opportunity costs of resources used, including the opportunity cost of the time contributed by the twelve hundred populations. With no identifiable direct effects on actions, was the plan worth this cost? If it is intended to work as a vision, it might have changed beliefs, but not yet affected actions. Changes in beliefs are difficult to measure at best. In either case, costs matter. We can at least ask whether the benefits are plausibly greater than $4.4 million.

To value the effects of a plan requires considering all the ways in which the plan might work, distinguishing effects of the plan from what would have happened anyway, and estimating the value of these benefits. Uncertainty confounds these aspects further. Clearly the value of the benefits is different across individuals and groups and raises all the problems of assessing changes in social welfare if we take a collective perspective. Even from the perspective of a particular institution, such as a sanitary district planning for sewers, the estimate of benefits is problematic. In practice, the most practical way to ask

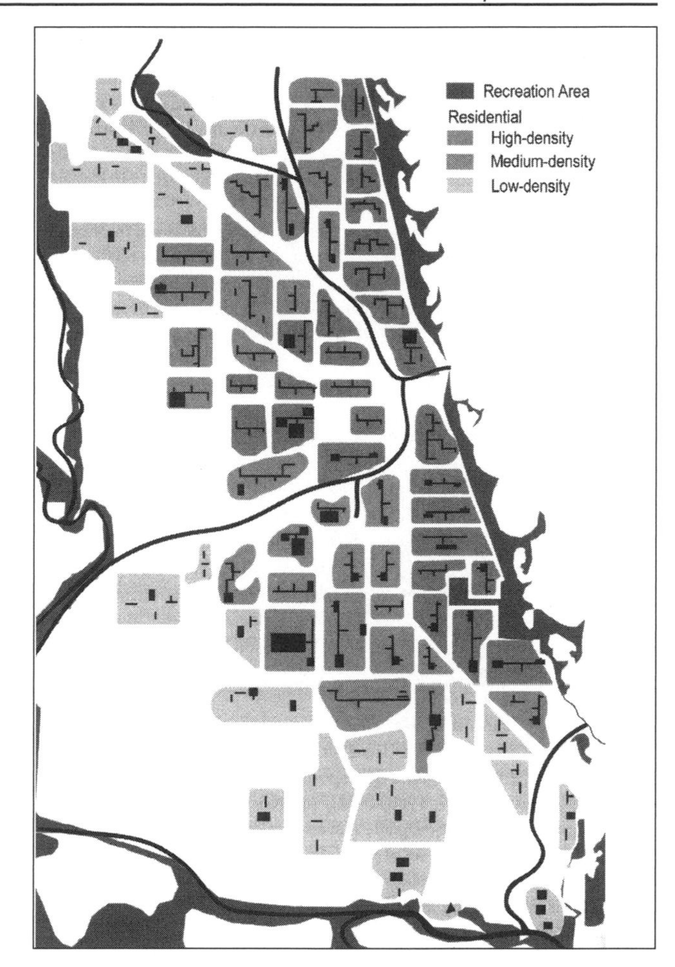

Policy plan diagram of proposed capital improvement projects providing park elements to serve and structure each neighborhood and to reconfigure the lake front. (Adapted from City of Chicago, Basic Policies for the Comprehensive Plan of Chigago, August, 1964, p.41.)

whether a plan was worth making is to estimate the costs of making it and then ask whether it is in rough terms plausible that the benefits could justify these costs. That is, it is unnecessary to estimate benefits any more precisely than whether they are greater or less than the costs. Thus Helling's example is an excellent model for addressing this question not just for the vision aspect of plans but for all aspects.

Was the plan internally consistent with the logic of how plans work?

Internal validity depends on attributes of the plan itself. The internal validity of a plan can be determined by looking only at the plan. As with any decision in the face of uncertainty, the question is whether a good plan was made given the information available when it was made, not whether the outcomes that resulted were good outcomes. The typical approach is to ask whether a plan contains a certain set of components, such as transportation and land use, or has a particular set of attributes, such as being organized for reference by decision makers. A more careful interpretation would ask whether a plan fulfills at least one of the logics of how plans work. For the strategy aspect of a plan: Are the actions linked together in contingent strategies that meet the logic of decision analysis? Or for the design aspect: Are the elements combined into a designed target configuration that works, in which the interdependent elements should function as intended?

Table 3

The Four I's

	Interdependence	*Indivisibility*	*Irreversibility*	*Imperfect Foresight*
Definition	Result of action A depends on action B.	Size of increment of action affects value of action.	No action available to return to previous state without cost.	More than one future is possible.
Examples	Value of land (or road) depends on road access (availability of land).	Road linking two locations must be complete and of width sufficient for vehicles.	Road cannot be relocated or resized without cost.	Jobs could increase at various rates and at various locations.
Implications	Actions are not separable.	Continuous marginal adjustment is not efficient or not possible.	History and dynamics matter.	Uncertainty cannot be eliminated.
Responses	Consider effects of combinations of actions.	Consider the sizes of changes.	Consider interdependent actions before taking action.	Consider uncertainty of actions, outcomes, and values.

The qualities of *interdependence, indivisibility, irreversibility* and *imperfect foresight* that characterize circumstances in which plans can improve outcomes.

Kent (1964, p. 91) identified the attributes of a general plan:

Subject-Matter Characteristics. The plan:

(1) Should Focus on Physical Development

(2) Should Be Long-Range

(3) Should Be Comprehensive

(4) Should Be General, and Should Remain General

(5) Should Clearly Relate Major Physical-Design Proposals to the Basic Policies of the Plan

Characteristics Relating to Governmental Procedures. The plan:

(6) Should Be in a Form Suitable for Public Debate

(7) Should Be Identified as the City Council's Plan

(8) Should Be Available and Understandable to the Public

(9) Should Be Designed to Capitalize on Its Educational Potential

(10) Should be Amendable

These are characteristics of a plan, not of the process by which it was created or the effects it had on the world. They are internal validity criteria. Some of these criteria can be derived from explanations of how plans work and thus argued to measure internal validity in this stronger sense. The "Four I's" argue that plans should focus on physical development (Table 3). A plan for physical development is sufficiently difficult and sufficiently independent from other municipal functions that it makes sense to have a plan that focuses on physical development only. Long range is probably too narrow an interpretation, but the concern with time horizons is pertinent because it recognizes that a set of interdependent actions may occur over time. The focus, however, should arguable be on multiple time horizons pertinent to particular sets of interdependent decisions. "Comprehensive" for Kent implies comprehensive across physical elements, comprehensive in scope of effects considered, and comprehensive in covering the entire municipality. Kent argues that it should be general in focusing on major policies and major physical design proposals rather then details. These claims are consistent with focusing on those projects that, because of the Four I's, are likely to benefit from the plans. Characteristics 6 through 10 increase the likelihood the plan will be used in making decisions, and thus link these internal validity criteria to the explanations of how plans affect actions and outcomes.

Plans funded by the federal government under Section 701 of the Housing Act of 1954 (Feiss 1985) and state-mandated local plans in several states must include particular elements, presumably because of a belief that good plans must have such elements. California, for example, requires land use, circulation, housing, conservation, open space, noise, and safety (Olshansky 1996). These requirements address both scope of decisions and scope of effects to be considered. Why states should mandate certain characteristics of plans raises a whole range of issues beyond the internal validity of plans. Such mandates demonstrate, however, that decisions about how to plan are made in part on the attributes of plans themselves, not on the way they are made or on the effects they have. Thus internal validity is an important category of criteria.

Did the plan seek outcomes that are ethically appropriate through means that are ethically appropriate?

A plan that seeks to achieve equity for the least well off is a better plan than one that seeks to increase the efficiency of urban development in a way that the efficiency gains accrue only to the most well off. Without elaborating ethical claims here, it is clear that a plan can affect decision making, actions, and outcomes, yield benefits sufficient to compensate its costs, be internally consistent in its logic, but still be a bad plan because of the goals it pursues or the means it employs. External validity calls a plan to the standards of ethics.

Keating and Krumholz (1991) assessed the effects of downtown plans by comparing six plans based in large part on whether they tended to accomplish what those who initiated and supported them intended to accomplish. Did the plans affect outcomes? They applied criteria from Sedway and Cooke (1983) who argued that plans for downtown development are worth making if there is support of major property owners and tenants, support and cooperation from all departments in city government, a citizens advisory committee, and a citywide plan within the downtown plan can be set. These criteria are predictive of whether a plan is likely to yield benefits to those who fund it that are sufficient to compensate its costs. Keating and Krumholz found that the six plans they studied all fit the Sedway and Cooke criteria. They were plans of a type that we expect to occur because they were initiated by landowners, business leaders, and local government, all of who can benefit from downtown development. These actors have incentives to produce these types of plans that are focused on decisions they can make and benefit from. None of the plans, however, dealt in a significant way with equity, which was predictable given who initiated them. The plans failed an external validity test on a prescriptive criterion of equity.

If one can explain situations in which plans are likely to be made and likely to work based on the first three broad criteria—effect, net benefit, and internal validity—then we can prescribe situations in which planners who measure success as plans affecting actions, and at costs that are compensated, should make plans. Such plans and planning are likely, however, to achieve what is easy and normal, not to accom plish unusual changes such as improvements in social equity. All four criteria—including external validity—are thus pertinent to evaluations of plans.

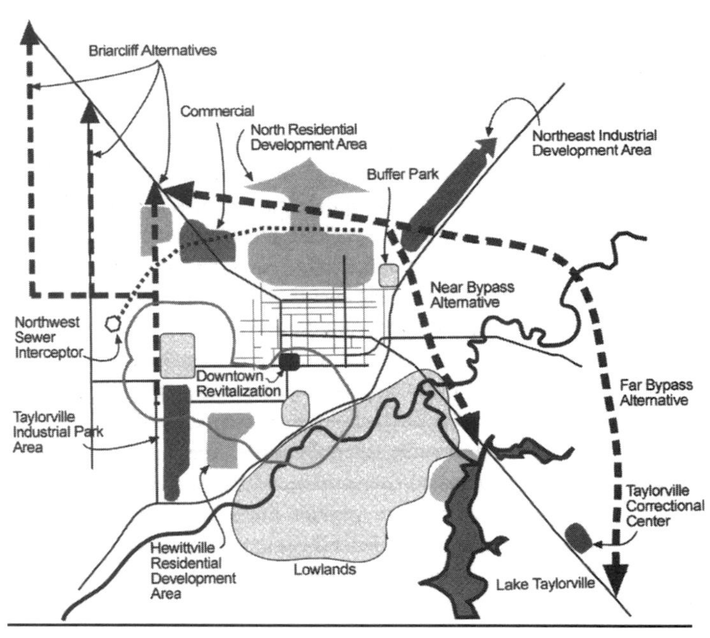

Strategy plan diagram for Taylorville, Illinois.

Summary: plans work in particular situations

Plans can work in more than one way. Given explanations of how plans work—explanations that link observable phenomena—it is possible to assess to what extent plans work in particular situations with respect to their effects, their net benefits, their internal validity, and their external validity. These explanations can also be used to predict that plans that meet these evaluation criteria will, in general, work in these ways in appropriate situations. They thus provide a basis for predicting what plans will be worth making. Plans for urban development often include agendas and policies as means of framing the actions implied by the plan. The vision, design, and strategy aspects of a plan are most pertinent, however, to figuring out the substantive logic of a plan for urban development and thus precede these agendas and policies. The fundamental reason for this precedence is that visions, designs, and strategies address interdependence among actions while agendas and policies do not. The strategy aspects of plans must also face uncertainty and thus forecasting, which leads us to an interpretation of plans through decision analysis. ■

Chicago Plan 1909. Daniel H. Burnham and Edward H. Bennett. The plan provided an agenda, a vision and a design.

REFERENCES

Alexander, Ernest R. 1992. *Approaches to Planning: Introducing Current Planning Theories, Concepts and Issues.* 2nd edition. Philadelphia: Gordon and Breach Science Publishers.

Alexander, Ernest R., and Andreas Faludi. 1989. Planning and Plan Implementation: Notes on Evaluation Criteria. *Environment and Planning B: Planning and Design* 16 (2):127–140.

Alterman, Rachelle and Morris Hill. 1978. Implementation of Land Use Plans. *Journal of the American Institute of Planners* 44 (3):274–285.

Anas, Alex, Richard J. Arnott, and Kenneth A. Small. 1998. Urban Spatial Structure. *Journal of Economic Literature 36* (3):1426–1464.

Bacon, Edmund N. 1974. *The Design of Cities.* Revised ed. New York: The Viking Press.

Baer, William C. 1997. General Plan Evaluation Criteria: An Approach to Making Better Plans. *Journal of the American Planning Association* 63 (3):329–344.

Berke, Philip, and Steven P. French. 1994. The Influence of State Planning Mandates on Local Plan Quality. *Journal of Planning Education and Research* 13 (4):237–250.

Bryson, John M. 1995. *Strategic Planning for Public and Nonprofit Organizations: A Guide to Strengthening and Sustaining Organizational Achievement.* Revised ed. San Francisco: Jossey-Bass.

Bryson, John M. and Barbara C. Crosby. 1992. *Leadership for the Common Good: Taking Public Problems in a Shared-Power World.* San Francisco: Jossey-Bass.

Burnham, Daniel H., and Edward H. Bennet 1909. *Plan of Chicago.* Chicago: The Commercial Club, 2–4.

Calkins, Hugh W. 1979. The Planning Monitor: An Accountability Theory of Plan Evaluation. *Environment and Planning* A 11 (7).

Connerly, Charles E., and Nancy A. Muller. 1993. "Evaluation Housing Elements in Growth Management Comprehensive Plans," in *Growth Management: The Planning Challenge of the 1990s*, J. Stein, ed. Thousand Oaks, CA: Sage.

City of Chicago Department of Development and Planning. 1966. *The Comprehensive Plan of Chicago.* Chicago: Department of Development and Planning.

Dalton, Linda C., and Raymond J. Burby. 1994. Mandates, Plans, and Planners. *Journal of the American Planning Association* 60 (4):444–461.

Faludi, Andreas. 1987. *A Decision-Centred View of Environmental Planning.* Oxford: Pergamon Press.

Feiss, Carl. 1985. The Foundations of Federal Planning Assistance: A Personal Account of the 701 Program. *Journal of the American Planning Association* 51 (2):175–184.

Flyvbjerg, Bent. 1998. *Rationality and Power: Democracy in Practice.* Trans. S. Sampson. Chicago: University of Chicago Press.

Guttenberg, Albert Z. 1993. *The Language of Planning: Essays on the Origins and Ends of American Planning Thought.* Urbana: University of Illinois Press.

Harris, Briton. 1967. The City of the Future: Problem of Optimal Design. *Papers and Proceedings of the Regional Science Association* 19:185–198.

Helling, Amy. 1998. Collaborative Visioning: Proceed with Caution. *Journal of the American Planning Association* 64 (3):335–349.

Howe, Jim, Ed McMahon, and Luther Propst. 1997. *Balancing Nature and Commerce in Gateway Communities.* Washington, DC: Island Press.

Hopkins, Lewis D. 2001. *Urban Development: The Logic of Making Plans.* Washington, DC: Island Press.

Johnson, David A. 1996. "Planning the Great Metropolis: The 1929 Regional Plan of New York and Its Environs" in G. E. Cherry and A. Sutcliffe, eds. *Studies in History, Planning and the Environment.* London: E & FN Spon.

Keating, W. Dennis, and Norman Krumhollz. 1991. Downtown Plans of the 1980s: The Case for More Equity in the 1990s. *Journal of the American Planning Association* 57 (2):136–152.

Kent, T. J. 1964. *The Urban General Plan.* San Francisco: Chandler Publishing Company.

Kerr, Donna H. 1976. The Logic of "Policy" and Successful Policies. *Policy Sciences* 7 (3):351–363.

Mastop, H. , and Andreas Faludi. 1997. Evaluation of Strategic Plans: The Performance Principle. *Environment and Planning B: Planning and Design* 24:815–832.

Metro. 2000. *Growth Management.* Growth Management Services Department, Metro, Portland, Oregon. September 27, 2000 [cited October 2, 2000]. Available from <http:www.multino mah.lib. or.us/ metro/gms.html>.

Neuman, Michael. 1998. Does Planning Need the Plan? *Journal of the American Planning Association.* 64 (2):208–220.

Olshansky, Robert. 1996. The California Environmental Quality Act and Local Planning. *Journal of the American Planning Association* 62 (3):313–330.

Pei and Associates. 1961. Erieview–Cleveland, Ohio: An Urban Renewal Plan for Downtown Cleveland.

Sedway, Paul H., and Thomas Cooke. 1983. Downtown Planning: Basic Steps. *Planning* 49 (12): 22–25.

Talen, Emily. 1996. Do Plans Get Implemented? A Review of Evaluation in Planning. *Journal of Planning Literature* 10 (3):248–259.

Talen, Emily, and Luc Anselin. 1998. Assessing Spatial Equity: The Role of Access Measures. *Environment and Planning A* 30 (4):595–259.

Wildavsky, Aaron. 1973. If Planning Is Everything, Maybe It's Nothing. *Policy Sciences 4* (2):137–153.

Matthew J. Bell and Cheryl A. O'Neill

Summary

City plans can be traced back to antiquity. Although invention has played a role today in how the city is represented, the urban designer is still confronted with the need to communicate a holistic vision of the designer's intentions over a greater area than a single building, and sometimes needing to assert a conception well beyond the life of the designer. The range of urban drawings and plans and the forms and values that they may carry with them are reviewed in a survey of representations of urban designs and plans.

Key words

cartography, drawing, grid, orthographic projection, survey, topography, United States Geological Survey (USGS), zoning

Fig. 1. 1911 plan for the city of Zagreb. Source: *Shaping the Great City: Modern Architecture in Central Europe 1890-1937.* (Prestel).

Representation and urban design

Urban design is an integral and essential part of architecture and the process of place making. Practitioners of urban design pursue an inclusive approach, often acting as "coordinator" of disciplines as unique as community planning, landscape architecture, civil engineering, urban forestry, and so forth. Seminal figures such as Bramante, Daniel Burnham, Vanbrugh, Otto Wagner and Frank Lloyd Wright were architects whose commissions often extended beyond the confines of the individual building. In their work, large landscapes, towns and cities, districts and neighborhoods could be designed with coherent principles of organization. All related disciplines, such as aspects of site engineering or landscape design could be related to design concepts by an inclusive tradition of drawing which represented, in urban design, the totality of the built environment. The 1911 plan for the city of Zagreb reveals such a tradition, including transportation systems, parks and open landscapes, zoning and the layout of streets and blocks all within the confines of one drawing. The city is represented as a system of inter-dependent and inter-related ideas, all brought together in the urban design plan (Fig. 1).

In the years since World War I, this tradition has been challenged by the rise of professional disciplines such as landscape architecture, transportation engineering, land-use law, and city planning to name but a few. The result has been both to distance design from urban design, and to render urban form the consequence of the actions by a separate set of specialists or as the result of policy decisions. Typically, these disciplines are less concerned with the city's physical reality and more with its social, economic, and political problems.

This shift has had an impact on the types of drawings used to represent the city. Each specialty has developed drawing types that describe conditions in their own field, but exclude ideas perceived to be outside their areas of expertise. Consider the typical zoning/land-use diagram utilized by planners or the traffic engineer drawing showing almost no buildings but every inch of curbs, concrete and guardrails. Concerns are maximized in one particular area of study with little information revealed about any other aspect of the city's form.

Drawings are useful to the urban designer in their capacity to provide information on the particulars of urban form and help to substantiate important regulating systems. The typical survey drawing almost routinely produced by most city planning departments or survey drawings by fire insurance companies provide significant information on a city's streets, blocks and fabric. This kind of drawing, traditionally a roof plan, provides accurate information about building heights, the nature of ground plane conditions and the location of parking and sidewalks. Baltimore, Maryland, like many cities, has an extraordinarily complete and accurate set of these drawing which are an indispensable source of site information (Fig. 2)

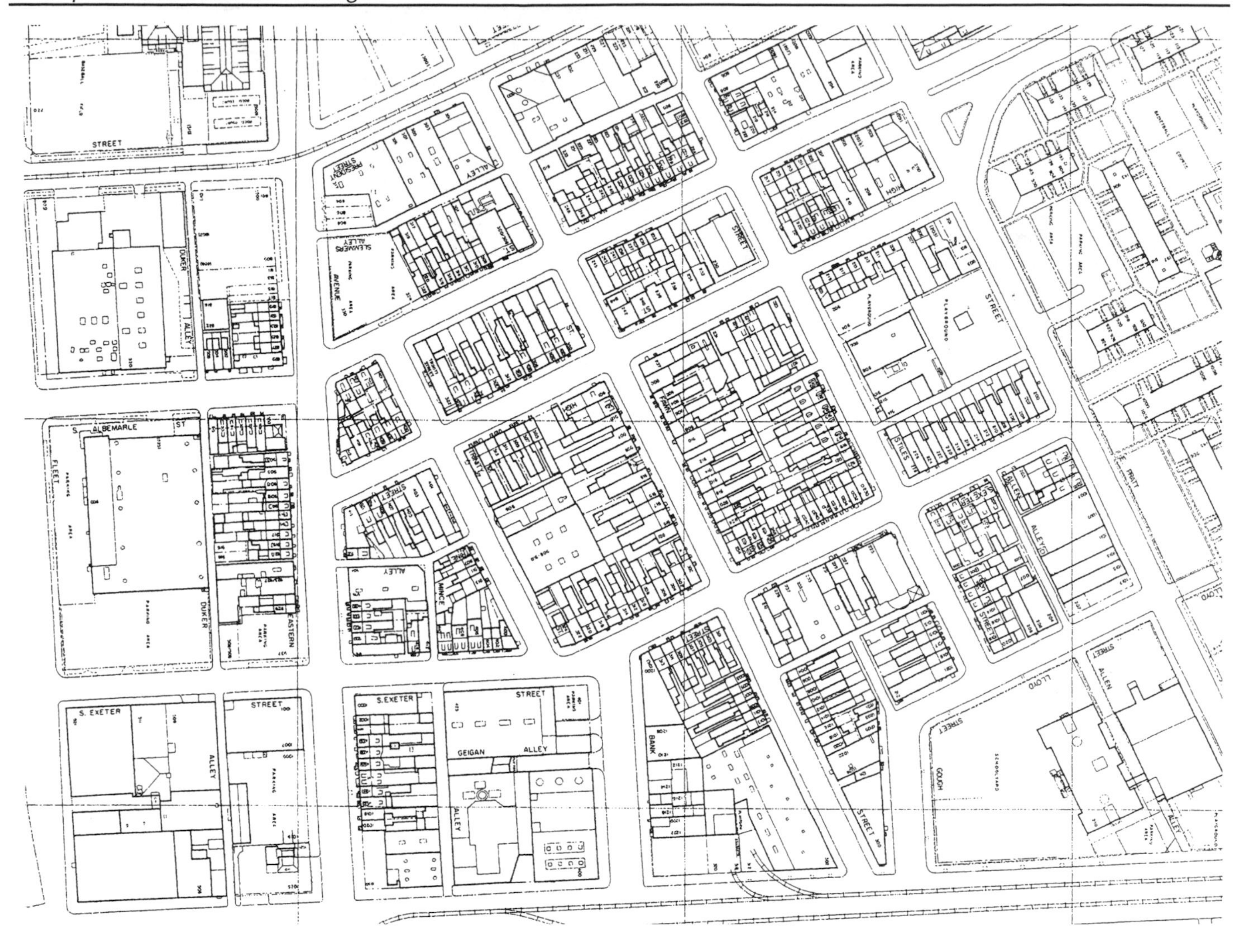

Fig. 2. Baltimore block plan. (Source: City of Baltimore.)

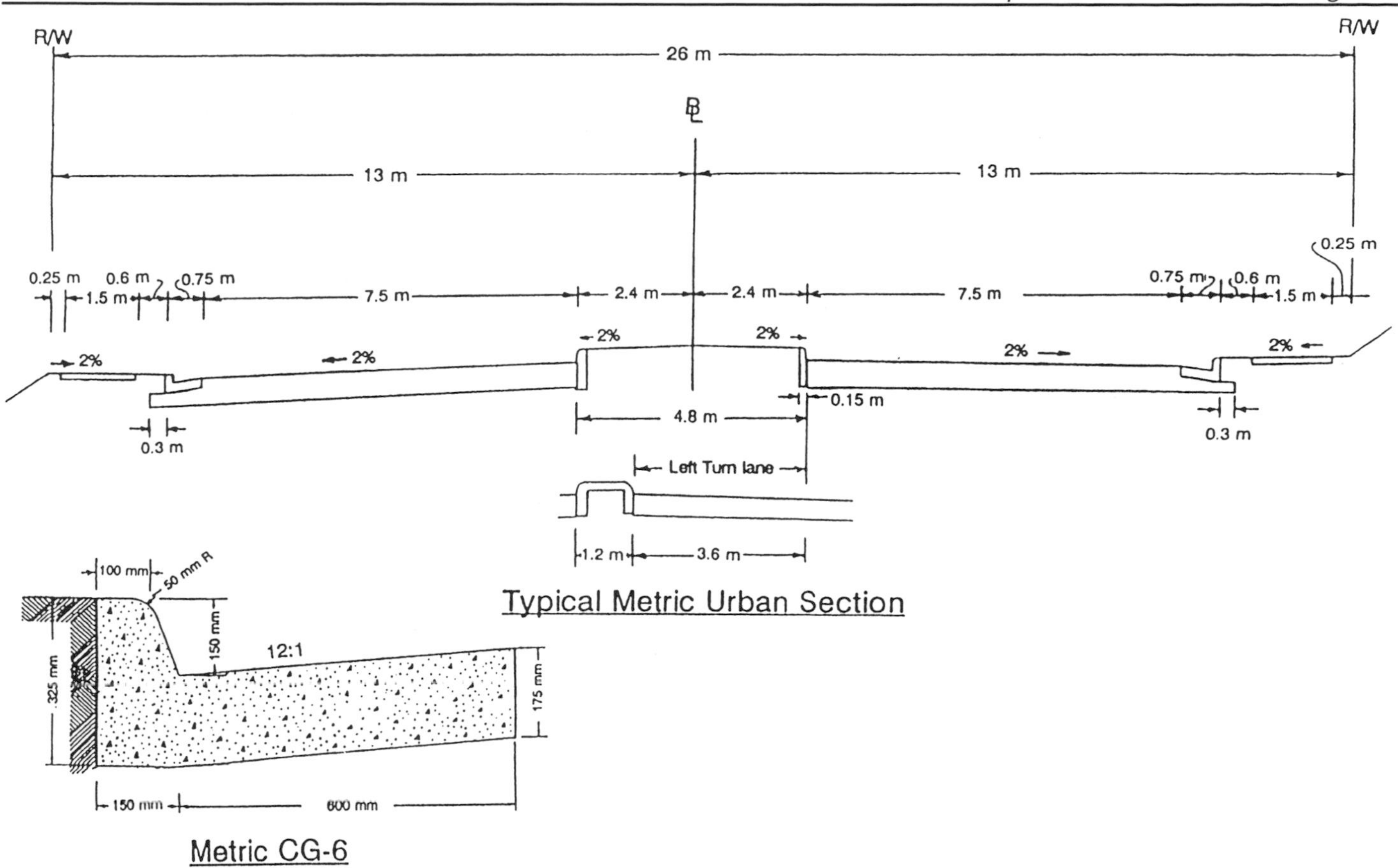

Fig. 3. Urban design street section. (Courtesy Torti Gallas and Partners.CHK.)

The narrow focus of specialized representations and their generally a spatial drawing conventions can, however, limit their broad applicability to the design task at the urban scale. Street sections produced by traffic engineering departments, for instance, accurately record the specifics of the cartway design of a given street typically with no information about adjacent conditions (Fig. 3). For the urban designer, the cartway is only a part of the street. Equally significant are the characteristics of front-yard setbacks, landscape conditions within the setback, the street right-of-way and perhaps most significantly the sectional profile of the buildings that establish the spatial volume of the street. Endowed with more holistic information about the spatial configuration of a street, the section can become a valuable drawing tool for urban design (Fig. 4). These types of representations, which have historically provided an integrated graphic approach at the urban scale, have become victims of the recent era of specialization. Yet they remain invaluable to the urban designer in the work of reconstructing our cities and building places of significant value.

Fig. 4. Traffic engineering road section drawing. (Source: Virginia Department of Transportation Road Standards.)

1 THE CITY MAP

The three-dimensional built reality and a city's conceptual organization are critical to the urban designer's understanding of the city. Urban design maps, generalized city plans rendered according to normative architectural graphic conventions, are an invaluable urban design tool with a lengthy tradition reaching back to the Renaissance. Two graphic conventions are important to this drawing type—first that the physical form of the city, its buildings and open spaces, is represented accurately and at some uniform level of specificity; and second, that a graphic

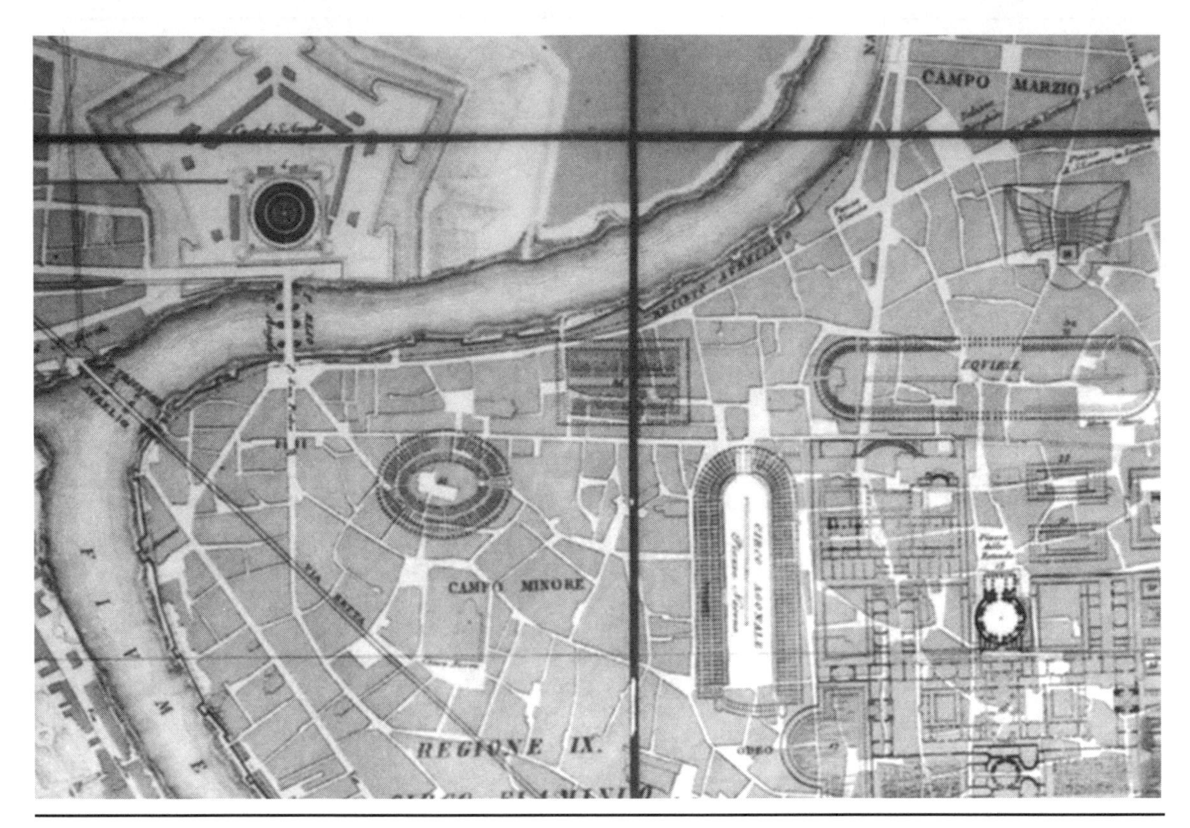

Fig. 5. Plan of modern and ancient Rome by Rodolfo Lanciani, 1902. (Source: Collection of Matthew Bell Caning.)

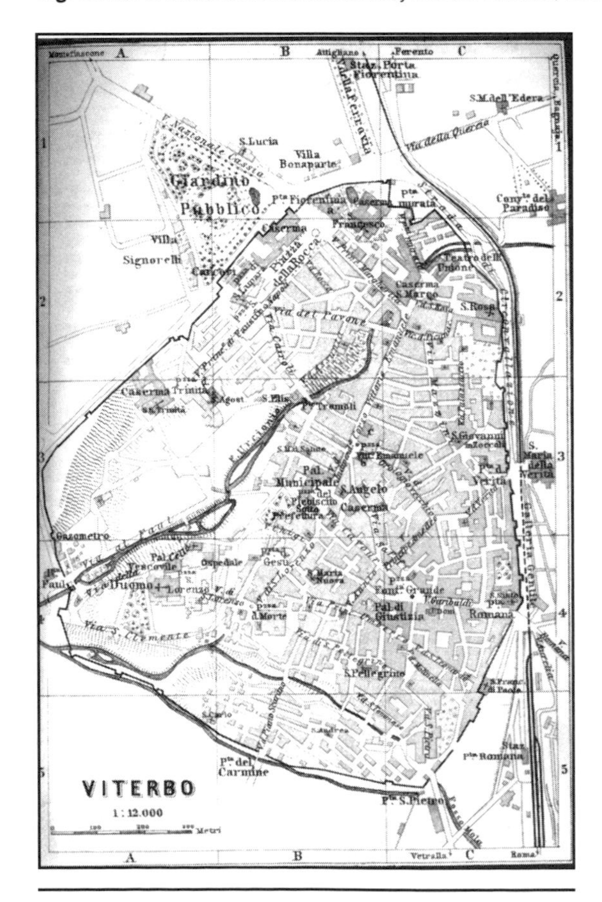

Fig. 6. Plan of Viterbo. (Source: Baedecker's Central Italy, 1904.)

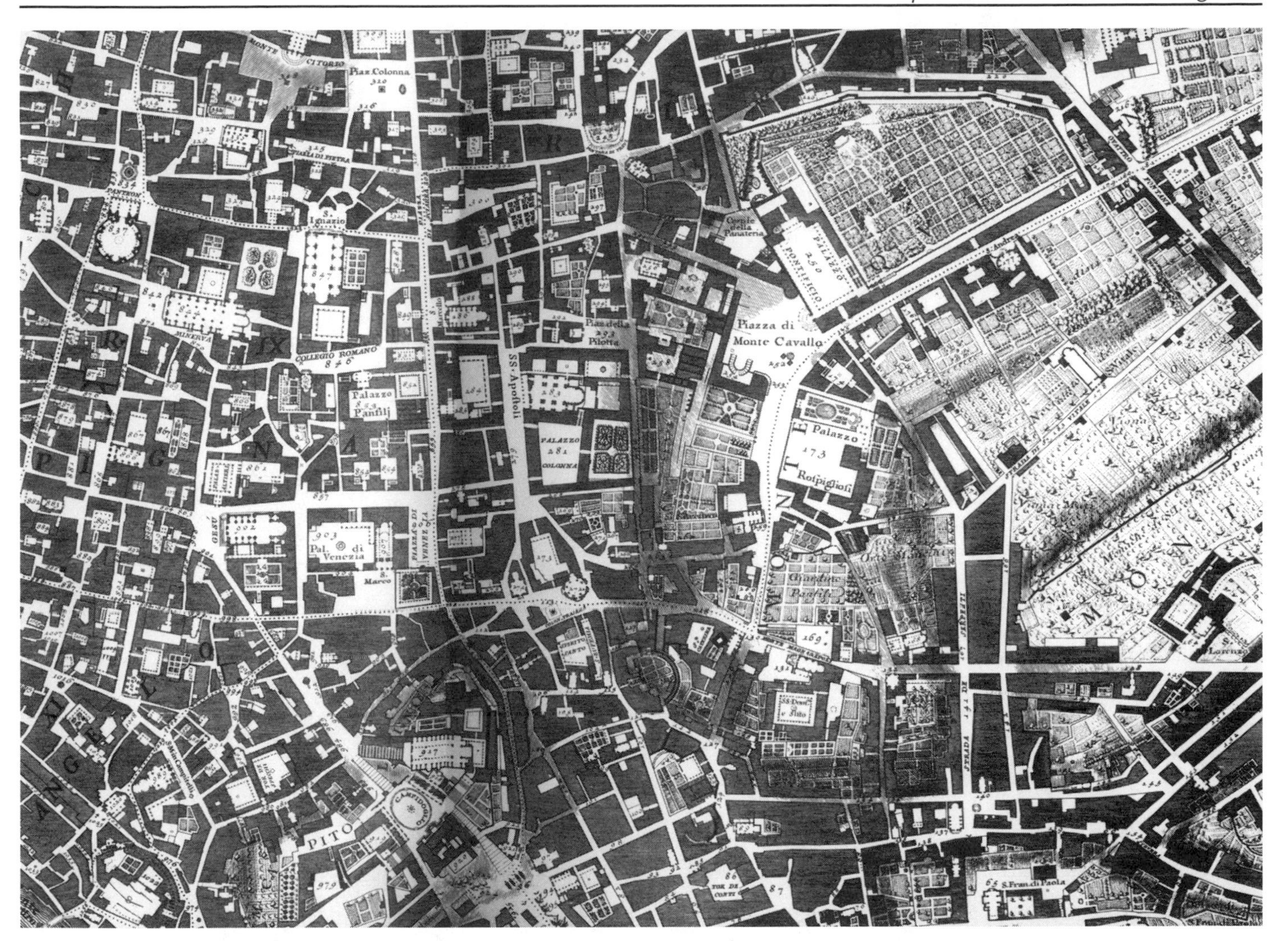

Fig. 7. Plan of the Campo Marzio, Rome. [Source: Plan of Rome by Giovanni Battista Nolli, 1748 (Officiana).]

distinction is made between the solid elements of the city or its buildings and its void elements (its streets, plazas, parks and other public open spaces). The contrast between a dark and a light hue, the one for buildings and the other for spaces, is critical in allowing these drawings to read spatially. The darker color is visually interpreted as a dense, weighty form while the paler hue, by contrast, asserts a lighter, airier visual presence. A visual analogy is thus made between the two color values and the role of buildings and spaces in creating three-dimensional urban form.

A cursory overview of this drawing type reveals a rich repository of urban maps. While the basic graphic system has remained consistent over the centuries, considerable variation has developed in drawing and media styles, both of which have evolved with the technology of drawing and reproduction. Levels of interpretation and the complexity of information conveyed by this map type have also varied. The famous Canina Plan of Rome, a 19th-century engraving of the city, utilizes this simple technique for the then modern-day city overlaying it with the known traces and speculations on the Roman one (Fig. 5). Maps as pedestrian as those produced for turn-of-the century tour guides, such as the well-known *Baedeker* guides, likewise use this graphic system to produce (what continue to be today) valuable urban representations. A map of the City of Viterbo, Italy, shows the basic conceptual organization of physical fabric including its winding medieval streets and irregular form adapted to the topography of its site (Fig. 6). The subtle rendering of significant buildings a darker hue allows them to be identified individually; the rest of the city's building, by contrast are shown in a highly generalized fashion. Blocks are rendered as completely solid, with no indication of either their subdivision into individual parcels or of exterior open space that may exist in the block's interior. The city is thus rendered at a uniform height and in a manner that emphasizes its composition in basic built solid and public voids.

The seminal 1748 plan of Rome by Giovanni Battista Nolli expanded on the information provided in the urban design map by further articulating the solid of building masses with the selective representation, according to conventional architectural plan graphics, of the primary public rooms of important civic and religious institutions (Fig. 7). The specificity of these interior space renditions captures their significant character and distinguishes between interior semi-public space and exterior public space. The presence of interior space in the plan also allows an identification of significant civic structures despite the fact that as an engraved work it is a black and white map.

Fig. 8. Plan of proposed group of municipal buildings or civic center at the intersection of Congress and Halsted Streets. [Sources: Burnham and Bennett Plan of Chicago, a reprint of 1909 editionn Daniel H. Burnham, Edward H. Bennett, published by Princeton Architectural Press (1999).]

For example, the rich correlation between the semi-public rooms of religious buildings and their counterparts in the many exterior open spaces, the *piazzas, piazzette* and streets of Rome is communicated in this simple graphic system. The complexity of urban spatial relationships captured here led to its frequent utilization by subsequent designers. Some of the students and later professionals trained at the *Ecole des Beaux-Arts* used techniques learned from the Nolli plan and others like it to make reconstructions of ancient sites. Daniel Burnham's Plan of Chicago also owes a great deal to the innovations initiated by Nolli two centuries earlier (Fig. 8).

Variations on this basic map type can also be found in the more recent work of Leon Krier and the American architects and town planners Andres Duany and Elizabeth Plater-Zyberk. Leon Krier's drawings for the new town of Poing in Germany renders his intervention similar in style to the *Baedeker* maps but operate at almost a diagrammatic level of generalization (Fig. 9). Only major streets, open spaces and civic buildings, which are colored a deeply saturated red, are included. The location of minor streets is inferred only by the inclusion of their intersections at major roadways. Like the *Baedeker* drawings, this representation emphasizes the organization of the city by its dominant public systems.

In another drawing, Krier highlights the role of civic buildings, similar to Nolli's Roman plan (Fig. 10). In this plan he marries two different drawing types, the plan and the orthographic projection, the latter utilized to precisely describe the three-dimensional forms of public structures. An alternative role of public buildings is here emphasized and juxtaposed with the spaces and defining fabric of the city. A two-toned shading system for the ground plane also allows a more specific rendition of block organization as well as a distinction between major and minor streets. Major streets are rendered in the whitest hue while minor streets and alleyways within the block become a pale orange-yellow. This technique suggests an organization of neighborhoods or quadrants within the overall plan as well as the composition in buildings of individual blocks without sacrificing the clarity of the overall organization.

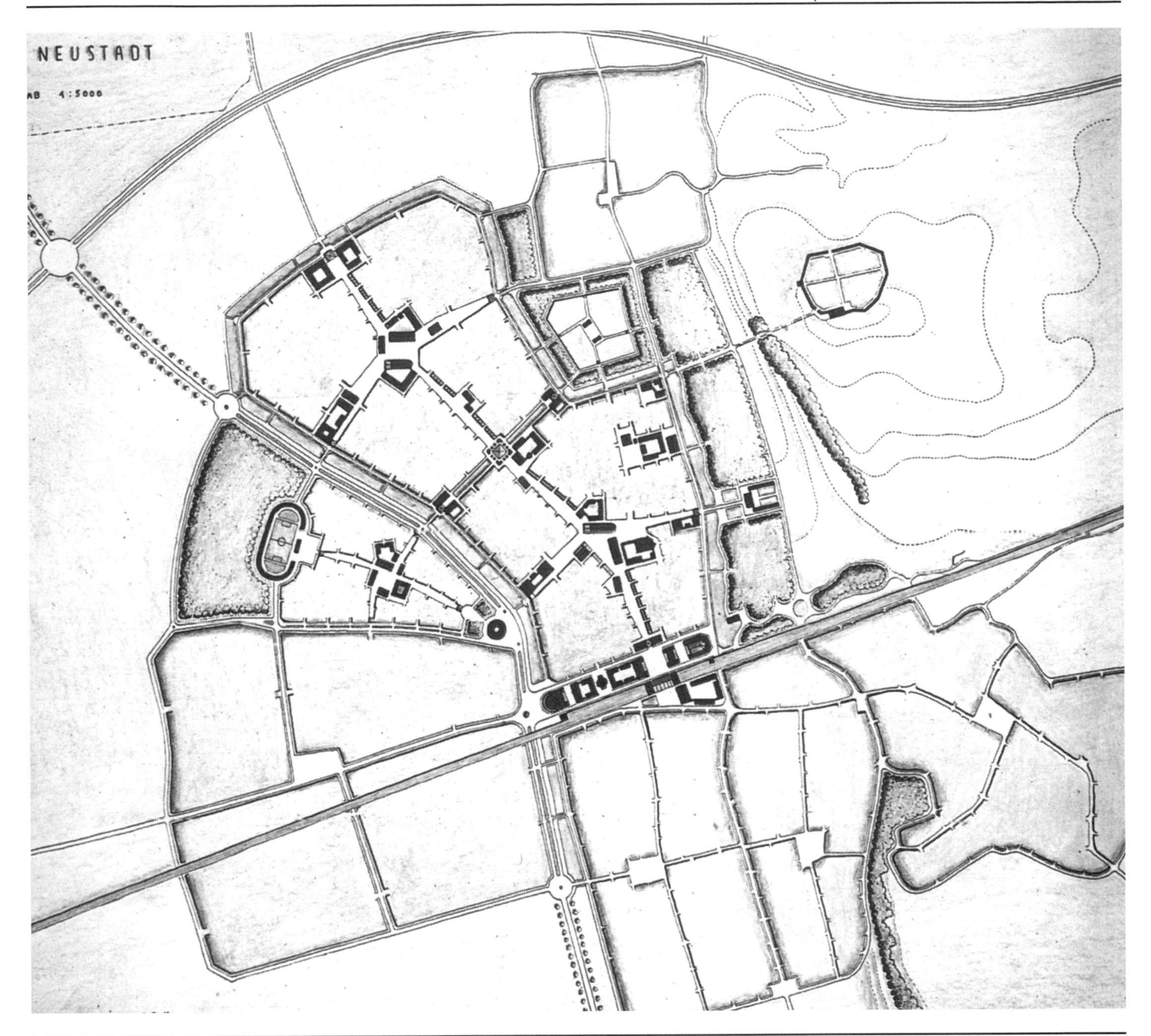

Fig. 9. Plan of Poing Nord New Town. [Source: Leon Krier: *Houses, Palaces, Cities* (Architectural Design).]

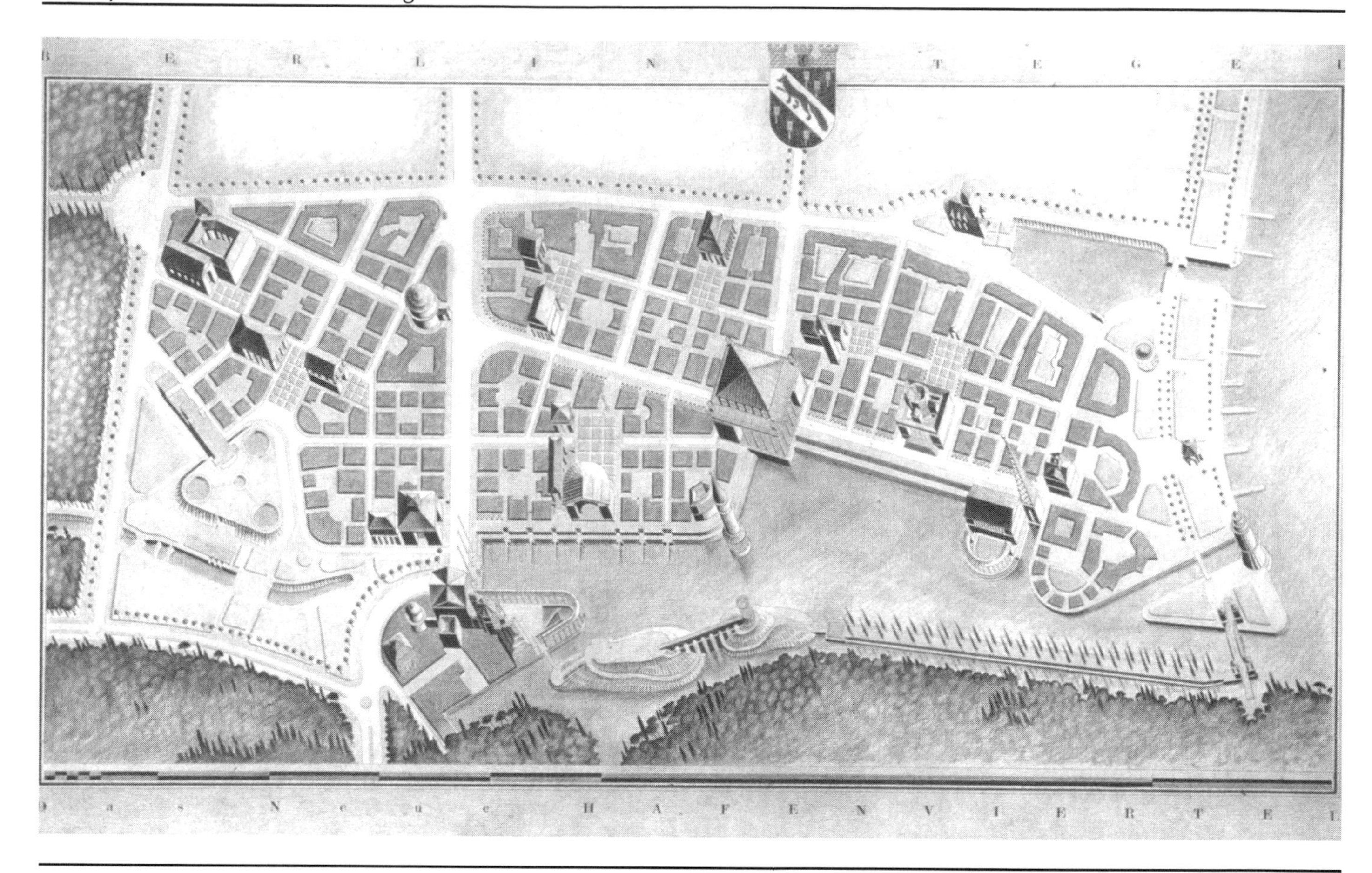

Fig. 10. Master plan for the New Hafenviertel, Berlin-Tegel. [Source: Leon Krier: *Houses, Palaces, Cities.* (Architectural Design).]

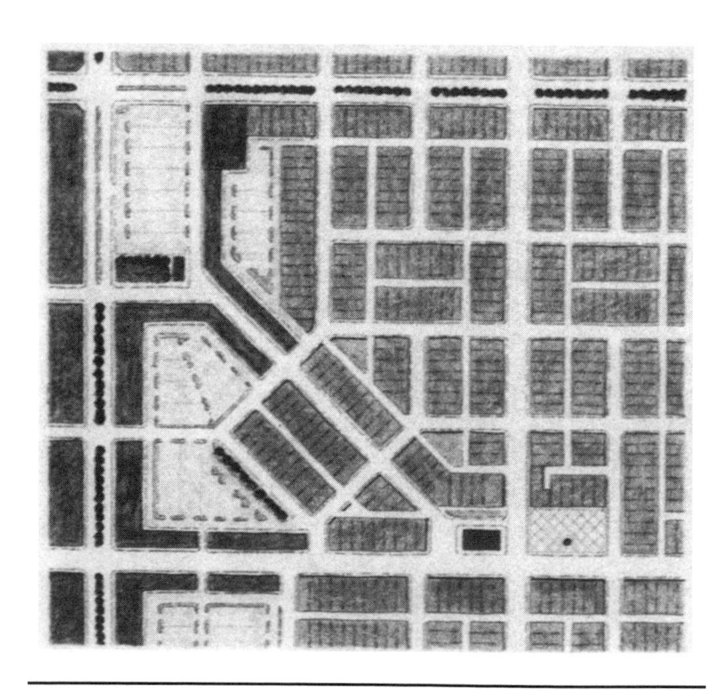

Fig. 11. City plan with lots and blocks. [Source: Andreas Duany and Elizabeth Plater-Zyberk.1991. *Towns and Town Making Principles* (Rizzoli).]

A unique invention of the many plans produced by Duany and Plater-Zyberk for their new towns is to substitute built form for the lotting system within the blocks (Fig. 11). This rendering system preserves the image of the lot as a solid while the inclusion of the subdivisions of the block give specific information on the organization of private space, in this case, housing. They also frequently marry their color rendering system to conventional land-use colors, which allows these drawings to distinguish the distribution of specific uses within the town.

A spectacularly rendered drawing by Eliel Saarinen for his project for Canberra, Australia, utilizes a more complicated rendering system to present the organization of urban form at multiple levels, including public buildings and spaces, normative fabric, and landscape (Fig. 12). Like the Krier drawing, a basic color distinction between black and various shades of pink and orange is used to distinguish the many public buildings of the new capital, shown in black, from normative city fabric. The variety of shades within the fabric, however, give more information as to it's character and organization, including it's relative density—three shades from light orange to deep pink are utilized indicating a gradation from less to more dense. Perimeter block buildings are shown with some degree of specificity while the rendering of interior space of the block is assigned an associated but less intense color value than attendant buildings. A distinction is thus made between the interior semi-public space of the block and public streets or piazza allowing for a clear reading, established in part by the black/white combination of public streets and buildings versus

Fig. 12. Plan for Canberra. [Source: Mirika Hausen, Kirmo Mikkola, Anna-Lisa Arnberg and Tytti Valto, 1990. *Eliel Saarinen Projects: 1896–1923* (MIT Press).]

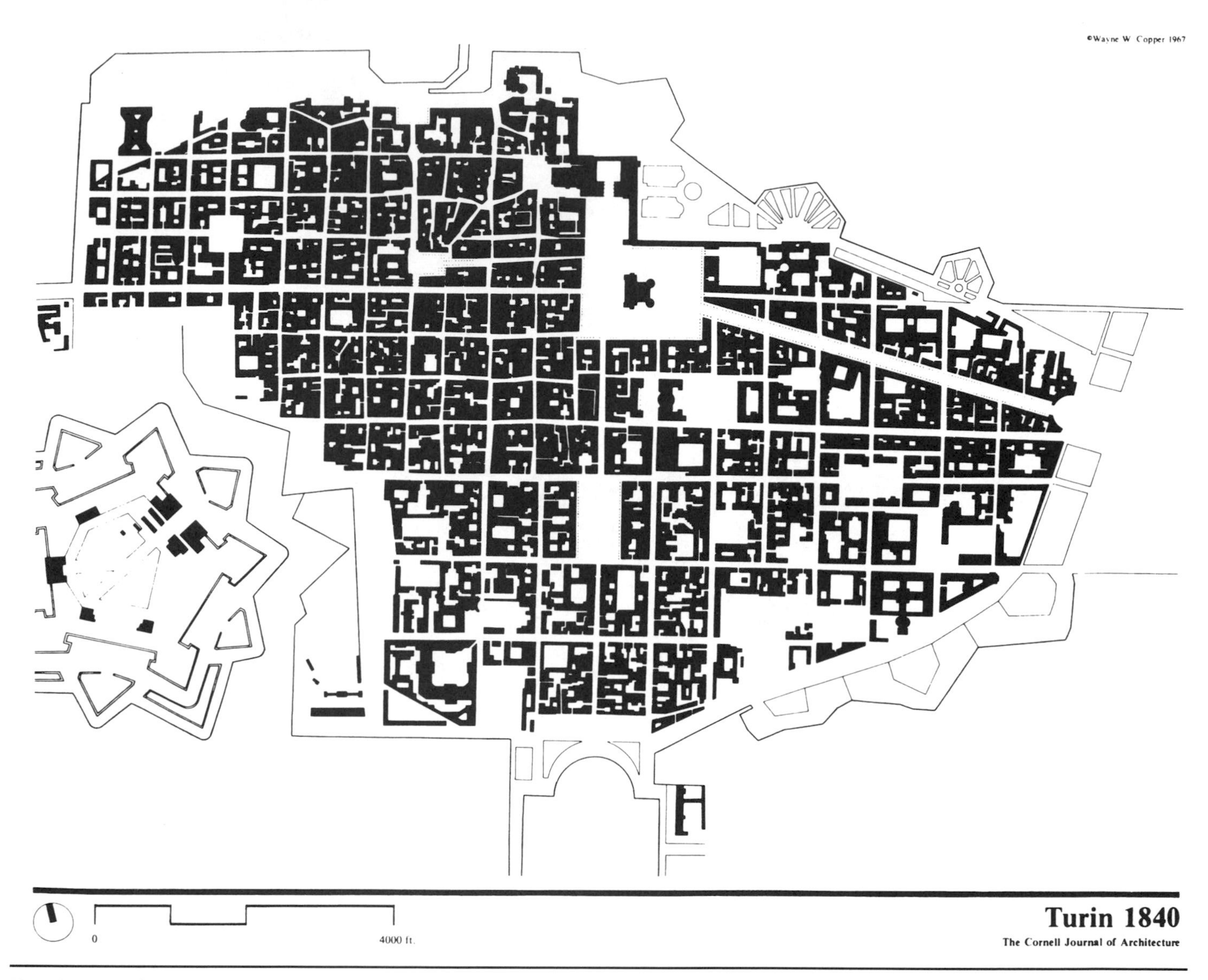

Fig. 13. Turin, Italy. [*Cornell Journal of Architecture 2* (Rizzoli).]

the warm colorized tones of the fabric, of the dominant public organization of the city.

Landscape is also rendered in shades of blue-green, a deep uniform blue for water and several grades of shade for trees. This technique clearly distinguishes the parks, canals and other open space systems from the buildings of the town. Trees rendered as a solid, dense mass, not unlike buildings, suggests a conceptual fit between the spatial organization of building and landscape components. Solids, be they buildings or landscape, become space-defining elements.

The figure-ground drawing type, invented in the Cornell Urban Design studio in the 1960s is perhaps the most elemental of representational types and has since become a standard part of the urban design drawing canon. While the spatial representation of solids and voids is similar to all previously discussed maps, the figure-ground uses a simple black and white color palette. Derived from Gestalt psychology this rendering system emphasizes the graphic inter-dependency of buildings and spaces, and their overall pattern to produce urban form. Unlike tonal plan drawings the figure-ground obliterates distinctions between building types as well as any information on the composition of the block. Only by virtue of scale, contour or location can important public structures be distinguished from others. A heightened level of abstraction, the figure-ground allows a morphological analysis of a city's form, particularly emphasizing the organization of texture into specific fields, hierarchies and the interdependency of space and solid.

A figure-ground of Turin, Italy, for instance, clearly depicts the simple organization of the city in a repetitive grid pattern oriented on the cardinal points of the compass (Fig. 13). Closer inspection reveals a distinction, discerned by the contrast in size and scale of the blocks, between the medieval fabric of the city, founded on the city's original Roman plan, and the later nineteenth-century extensions of that grid into larger city blocks. Significant public open spaces, the many *piazzas* that punctuate the simple grid are clearly represented and distinguished from the street grid by their shape and scale. The location of significant public buildings are suggested by virtue of their more unique contours—here the *Palazzo Madama*, a freestanding building, is clearly distinguished. Simple contrasts in shape suggests the chronology and associated patterns of urban growth—the many deformations of street grid in the old part of the city belie its largely medieval construction and stand in contrast to the precise grid of the nineteenth-century extensions. The one major nineteenth century street intervention, a broad avenue terminating in the main *piazza* is distinguished by the breadth of its street and its unique diagonal orientation.

2 THREE-DIMENSIONAL REPRESENTATIONS

All of the urban design maps discussed thus far are plan representations and so present physical form only in the manner in which it contributes to making public space such as street edges and plaza walls. An equally important aspect of urban space, however, is its three-dimensional qualities, including the height and volumetric composition of buildings as well as the character of street walls and building facades.

Topography, the rise and fall of the earth's surface, is another significant three-dimensional part of a city's form. Frequently this is represented, particularly in those places where it is significant, by the rendering of topographic lines inside and outside the city's plan. Two examples (Figs. 14 and 15) demonstrate this simple system. The extent of the landscape shown in a nineteenth century map of Gettysburg, Pennsylvania, clearly shows the position of the small town to its rural landscape. The color distinction between the brown contour lines and the black of the city's buildings powerfully communicates the difference between the natural and man-made conditions. The rendering of the hills surrounding the city in two shades of brown as though shadowed by the sun allow the contour of the land to read clearly, articulating the location of the village on a plateau whose shape has much to do with the town's form and extent. In today's context, the large-scale USGS map, linking large-scale orders of the natural landscape with the man-made can often be a critical reference for the urban designer.

A more contemporary drawing by Daniel Solomon for his Communication's Hill project in San Francisco (Fig. 16) uses a similar system to show its unique topography and how the form of the plan is influenced by and married to it. Solomon also utilizes a roof plan with rendered shadows to show new buildings. This simple system delineates height and mass of individual buildings within a block,

Fig. 14. Map of Gettysburg, PA, and environs. (Collection of Matthew Bell.)

Fig. 15. United States Geological Survey map (USGS).

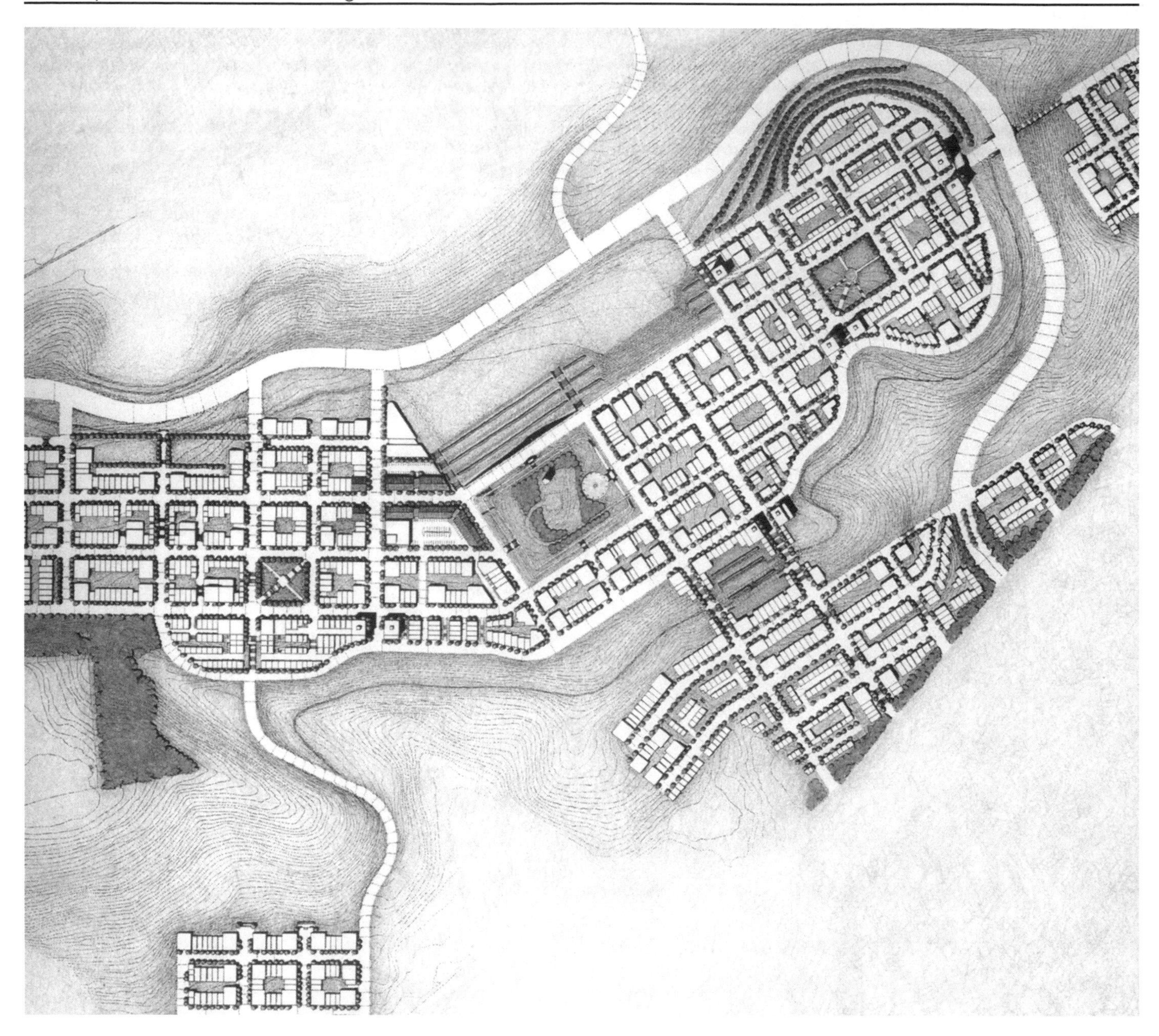

Fig. 16. Plan of Communications Hill Project, San Francisco. Daniel Solomon, architect and planner. [Source: Peter Katz and Vincent Scully, Jr. 1993. *An Architecture of Community* (McGraw-Hill).]

showing subdivisions as well as the major height distinctions between internal and corner block buildings.

The orthographic projection and the aerial perspective are two drawing types that marry the conceptual organization of plan representation with significant three-dimensional detail. The famous Turgot plan of Paris is an important representation of that city in the nineteenth century. An orthographic projection, the drawing presents, albeit from a single orientation, significant detail on building facades and roof compositions. Rather than a generalized solid mass of uniform height and composition, blocks can be understood in terms of their composition of individual buildings, whose specific character is suggested. The number of stories, the general pattern of windows and facade details and the composition of significant public buildings can all be discerned in this drawing. A portion of the plan around the *Place Royale,* for instance, shows the basic geometry and pattern of streets with the character of major public spaces including the *Place Royale* and adjacent gardens (Fig. 17).

The plan also articulates semi-private space within blocks and suggests the character of formal open spaces with elemental renderings of landscape. Roof detail, including features such as the massive mansards around the *Place* or the more unique gables, towers and cupolas of adjacent churches are shown as well. The basic composition of facades such as number of stories, window patterns, and ground floor conditions including primary entries can all be discerned. Because an orthographic projection preserves plan relationships, the conceptual organization of the city can be read without distortion between fore-

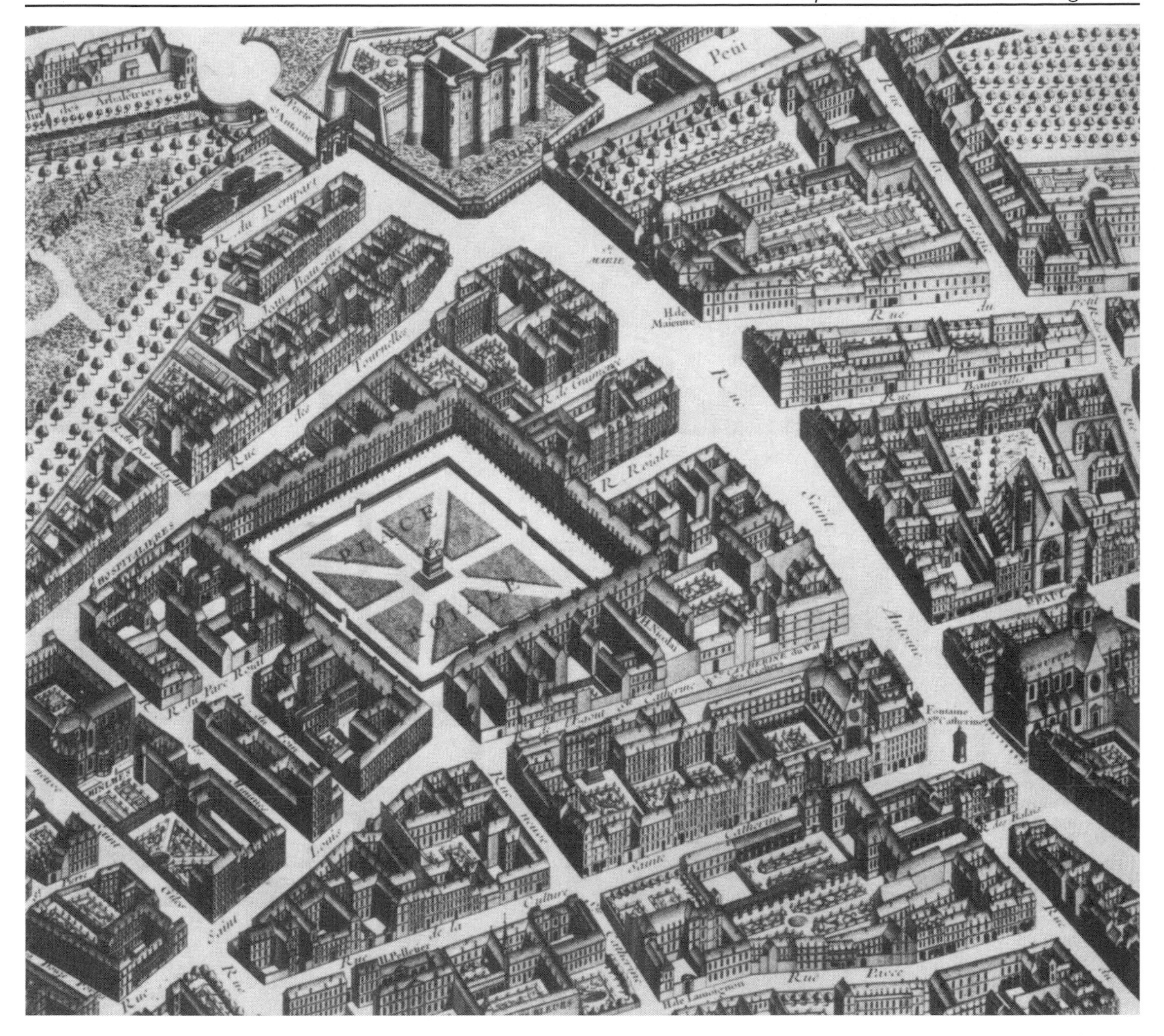

Fig. 17. *Place Royale* from Turgot Plan of Paris 1734–39. [Source: Colin Rowe and Fred Kotter. 1978. *Collage City* (MIT Press).]

ground and background images. The drawing allows an equivalency of detail for blocks and areas throughout the city attempting to provide a complete accounting of the three-dimensional form of the city's surfaces and volumes, whether solid or void.

The aerial perspective is, by contrast, a perceptual representation of a city view as though one was flying above it. Unlike orthographic projections, the same distortions present in normal sight—the shrinking of elements in the distance, the convergence to a horizon line—are present in this drawing. The virtue of these images lies in their ability to marry an experiential image, derived from the perspective construction, with the conceptual organization of the plan offered by the aerial vantage point. More than any others, the selection of vantage point, composition on the page and drawing media profoundly influence the reading of the image.

The dramatic height, centralized axial composition and abstract color system all emphasize the heroic aspects of Daniel Burnham's plan for the Chicago waterfront (Fig. 18). Leon Krier's drawing for a new quarter in *La Villette* has a similar composition but a significantly lower vantage point and is a simple line drawing (Fig. 19). The axial composition of Krier's plan is similarly emphasized by the location of the vantage point which is positioned on one of the two dominating axis of the plan. The new aerial view incorporates significant adjacent neighborhoods suggesting a clear vision of the integration of the new quarter with the urban patterns of existing development at its perimeter.

Fig. 18. Aerial perspective. [Source: Burnham and Bennett *Plan of Chicago,* a reprint of 1909 edition. Daniel H. Burnham, Edward H.Bennett, Princeton Architectural Press. (1990).]

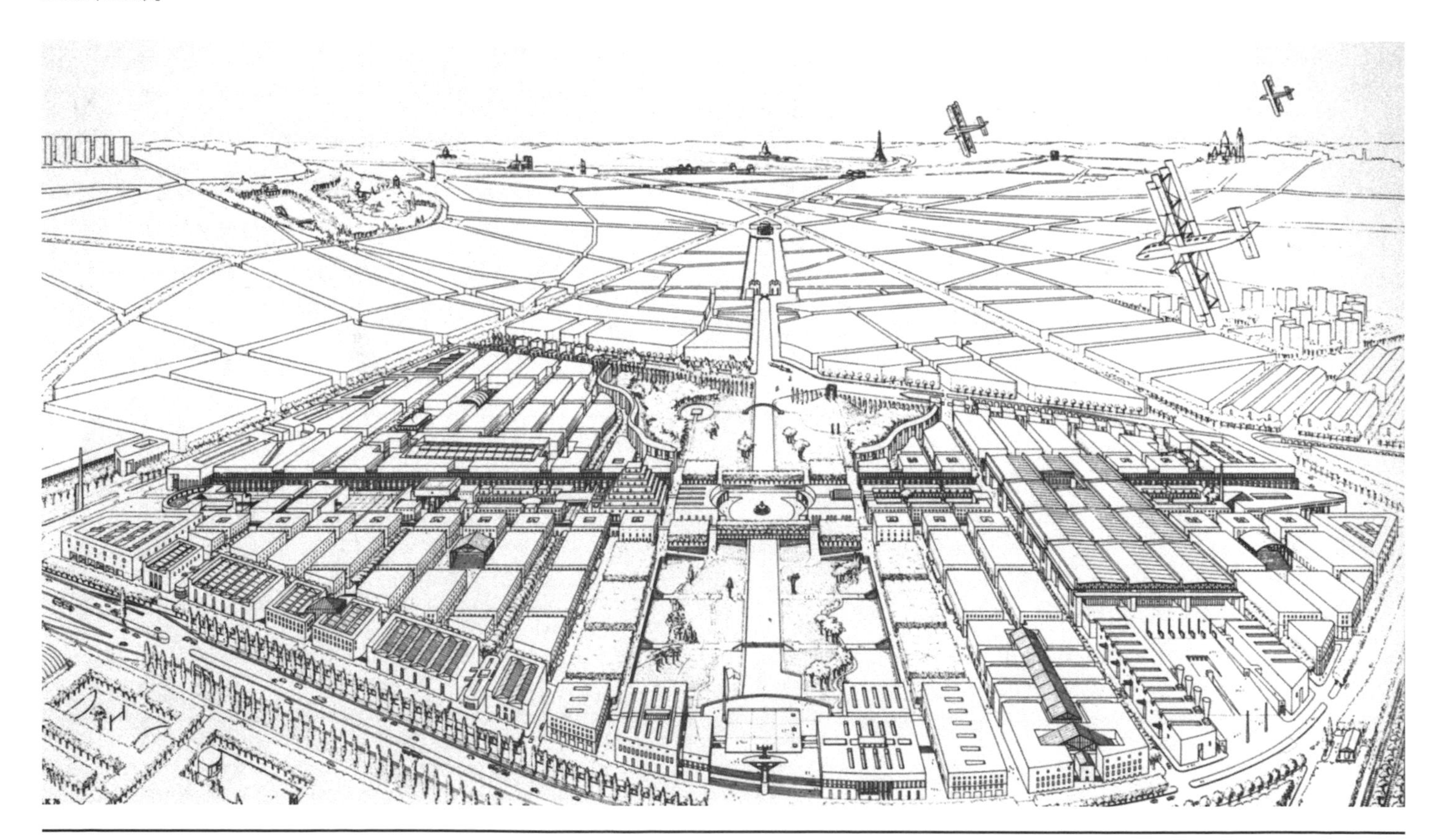

Fig. 19. Aerial view of La Villette (project) Leon Krier: *Houses, Palaces, Cities.* (*Architectural Design* Profile Series 54, 1984.)

Fig. 20. Aerial view of Kentlands, charrette drawing. [Source: Peter Katz and Vincent Scully, Jr., 1993. *The New Urbanism—Toward an Architecture of Community* (McGraw-Hill).]

Andres Duany and Elizabeth Plater-Zyberk use the aerial perspective extensively in their many town-planning projects (Fig. 20). The style of these drawings, generally watercolor or colored pencil, immediately strikes a warmer more familial tone, which is reinforced by the generally more intimate selection of views. A feature of their drawings is the aggressive inclusion of tree-lined streets and landscaped neighborhoods, evoking that most distinct American tradition of the suburban street.

In contrast to these, the ground-level perspective sacrifices the conceptual overview for the benefits of a perceptual, experiential understanding of the city—a fragment rather than a whole is presented. The development of this drawing type ran tandem with its use in architecture beginning with the Renaissance and figures such as Sebastian Serlio (Figs. 21a and b) with his famous representations of tragic and comic stage sets based on city streets. The medium of the rendering here tends to lend emphasis to those aspects of three-dimensional form that are to be experienced, with a particular emphasis on the focal point of the view. Many engraved images of Renaissance and Baroque Rome offer grand views with strong axial compositions whose intricate linework emphasize the rich classical detailing of the city's buildings.

Hugh Ferris' drawings (Fig. 22) of early skyscraper New York are by contrast highly atmospheric romantic renditions rendered in grays and whites emphasizing the dramatic solid and spatial volumes of the growing city. These can be contrasted with modern-day computer graphics (Fig. 23) which represent interventions in a style that mimics photography attempting to eliminate the fictive or perhaps more evocative aspect of a city or a building.

Fig. 21a. Comic Stage Set. (*Five Books of Architecture* by Serlio, 1611.)

Fig. 21b. Tragic Stage Set. (*Five Books of Architecture* by Serlio, 1611.)

Fig. 22. City Center perspective. (Hugh Ferris. *The Metropolis of Tomorrow*,1929.)

Fig. 23. Project for mixed-use building, Washington, D.C. (Courtesy of Torti Gallas and Partners.CHK and Interface Multimedia.)

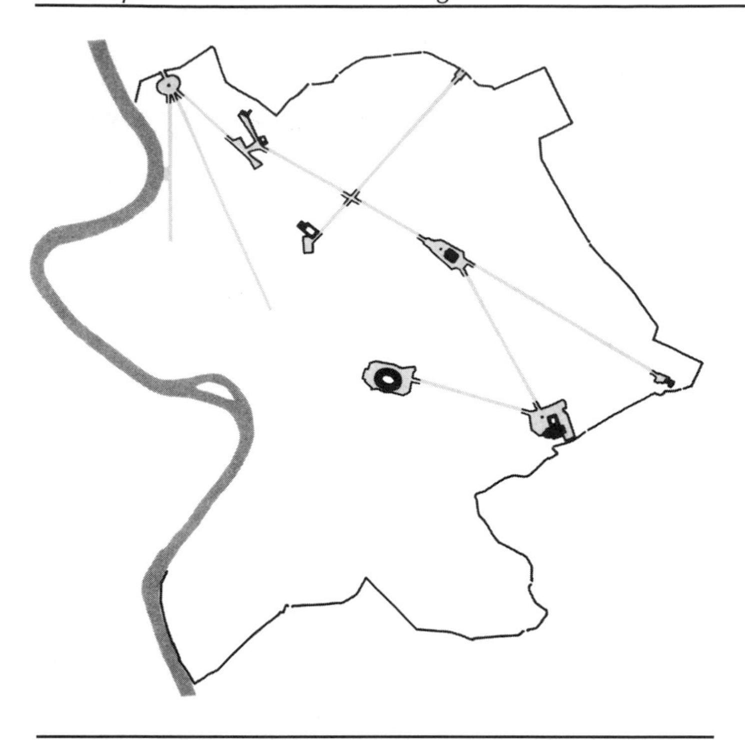

Fig. 24. City structures of Baroque Rome. [Source: Edmund Bacon. *Design of Cites* (Viking).]

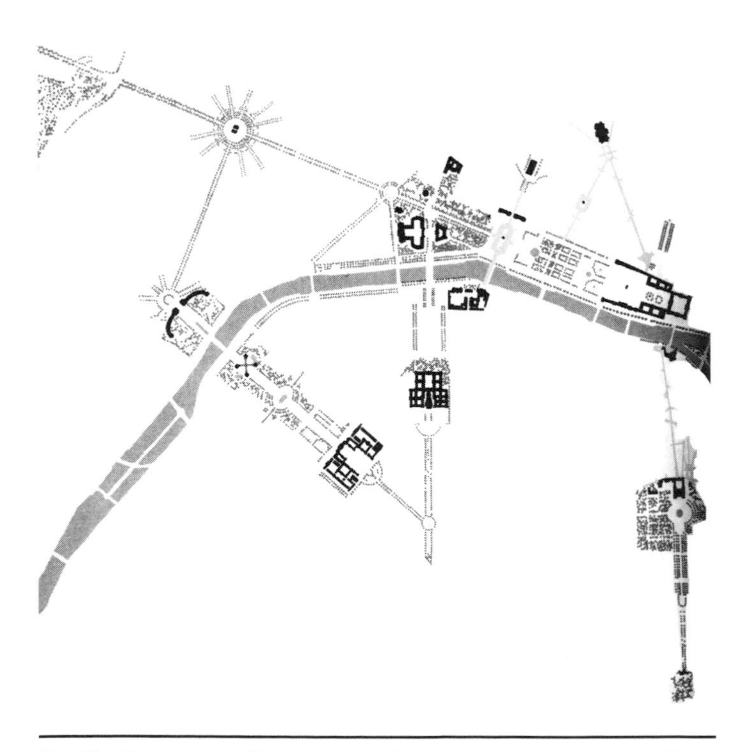

Fig. 25. City structures of Paris. [Source: Edmund Bacon. *Design of Cities* (Viking).]

3 DIAGRAMS AND REGULATING DRAWINGS

A final category of urban design drawings is that created to regulate its growth or particularize one aspect of its organization. This category of drawing is by nature more idiosyncratic and relies heavily on the use of the diagram. In contrast to the multitude of information all communicated at a similar level of specificity in the urban design map, the goal of the diagram is to isolate and communicate a single idea or aspect of a city's form. Eliminating information or making one aspect or idea more prominent in the drawing generally does this.

Edmund Bacon in his book *Design of Cities* produced a number of diagrams to illustrate his perceptions of urban morphology. Bacon uses colored plan diagrams of complex urban relationships, such as *Sixtus V* interventions in Rome, or an analysis of the palace and garden relationships in Paris (Figs. 24 and 25). The latter renders both the palace and garden in a figure-ground style. Buildings are colored black while the trees in the gardens, which similarly operate as solid masses, are green. This simple technique allows a distinction between built and landscape forms while emphasizing the similar role they have in establishing the primary axial sequences of the city of Paris. At times the pairing of a diagram with other types of drawings can provide a rich vision for a piece of a city.

A second set of diagrams are those associated with codes or other regulating systems and here again diagrammatic drawings are used to record only those elements of urban form to be governed by a code. An example from one of Duany and Plater-Zyberk's codes for the town of Wellington shows the types of drawings used to regulate various kinds of residential development (Fig. 26). These include highly diagrammatic sections as well as plan drawings of individual lots. The section drawings identify controls over building heights while the plan drawings layout open space controls, particularly mandatory setbacks and lot coverage.

A set of guidelines developed by Koetter Kim and Associates to govern the development of building facades for a project for Canary Wharf in London, England includes a variety of diagrammatic sections and elevations (Fig. 27). A section drawing highlights the three-dimensional profile of a building illustrating mandatory setbacks and height limitations while a schematic elevation demonstrates the impact of those setbacks on the facades of the space. These drawings are designed to regulate the basic architectural form of the buildings as they are to be developed within the framework of the urban design plan. Typically this type of drawing is employed when the project is to be developed by several different developers and/or architects and the diagram is a representation of an aspect of the regulating design guidelines or code. Since the project is to be built over the span of many years, such measures help the architecture to fill out the promise of the urban design plan.

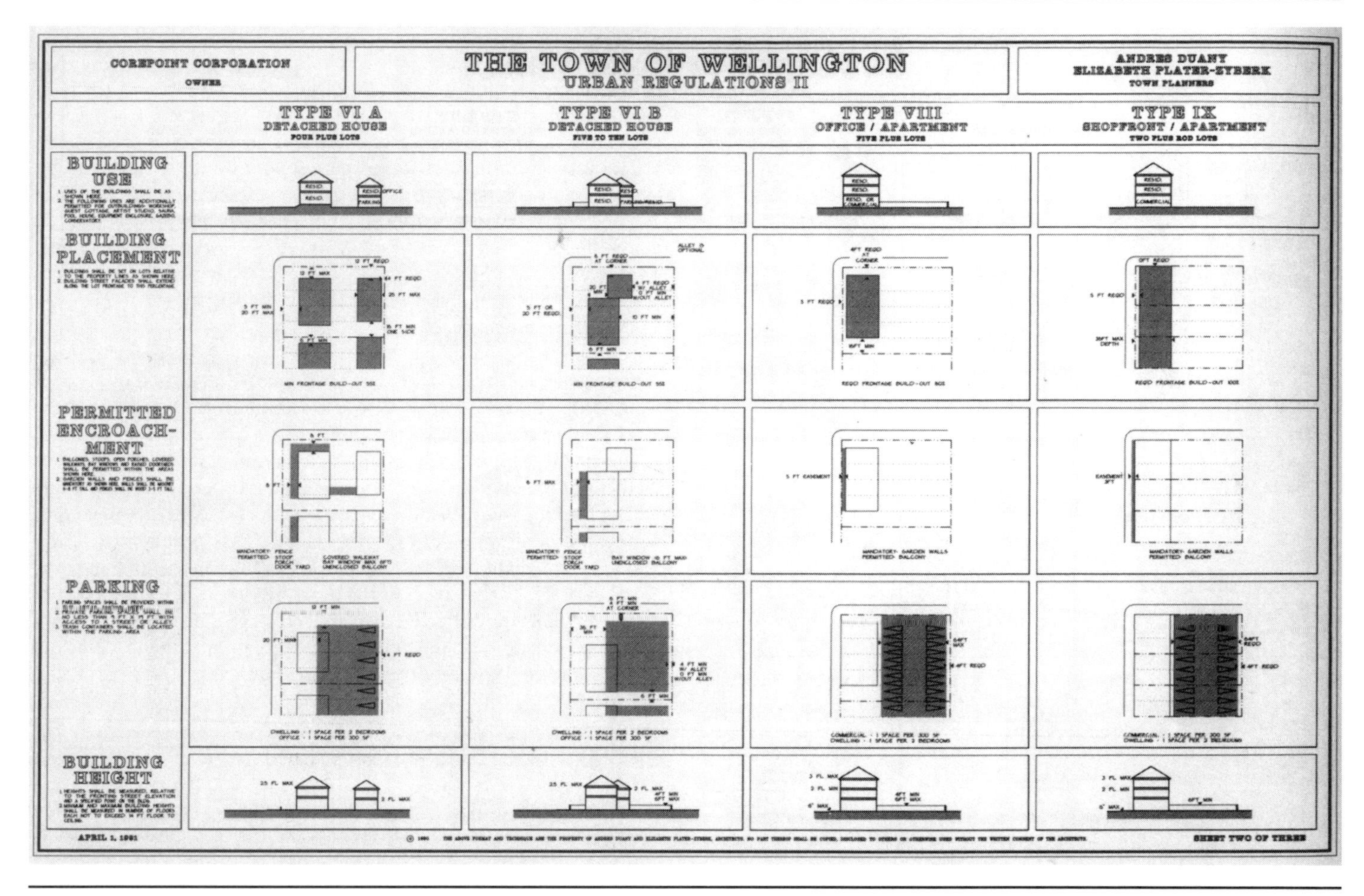

Fig. 26. Urban Code for Wellington. [Source: Peter Katz and Vincent Scully, Jr., 1993. *The New Urbanism—Toward an Architecture of Community* (McGraw-Hill).]

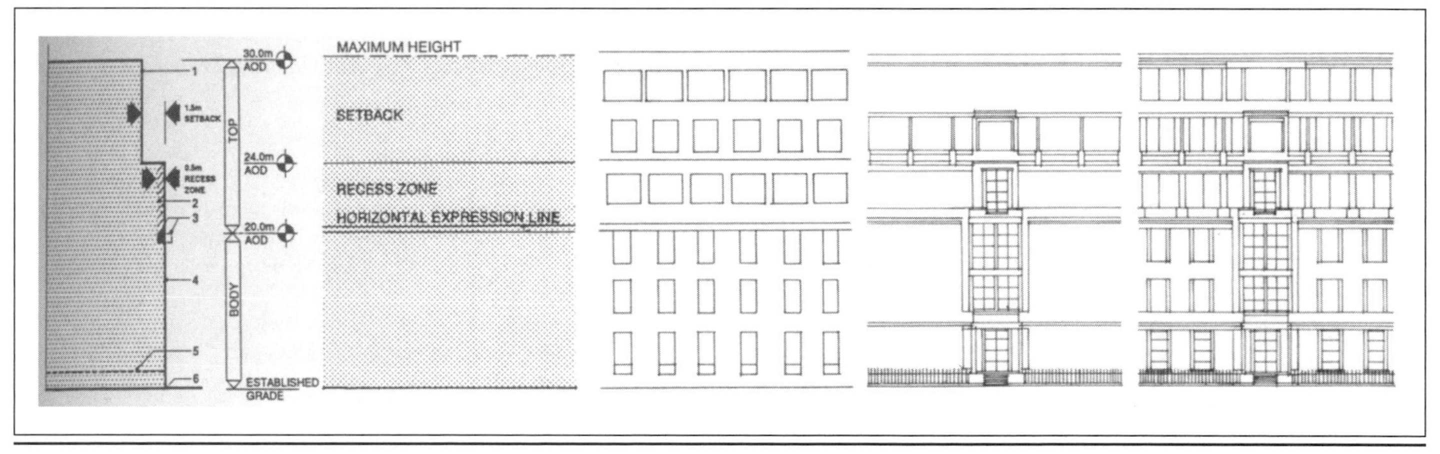

Fig. 27. Canary Wharf Façade Code. [Source: Koetter Kim and Associates, 1997. *Place Time* (Rizzoli).]

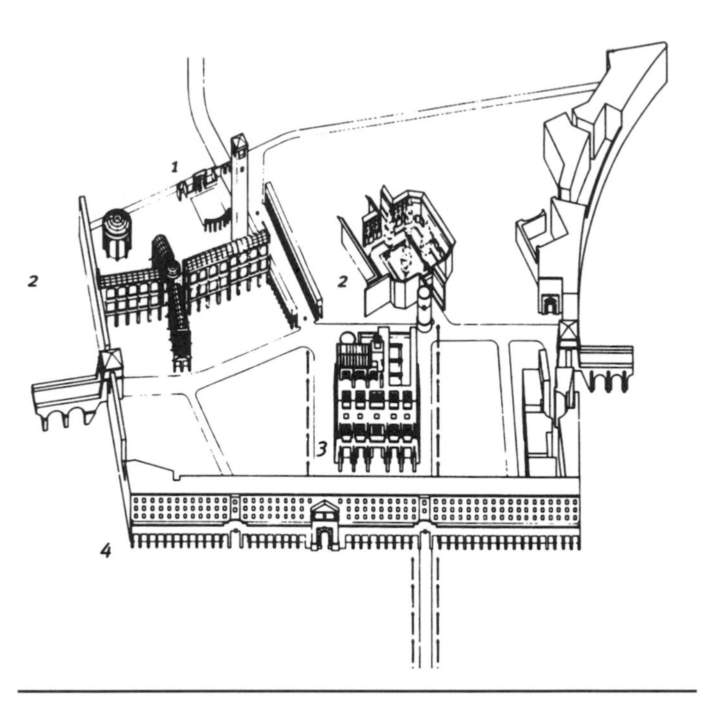

Fig. 28. Chinatown urban design isometric drawing. [Source: Koetter Kim and Associates.1997. *Place Time* (Rizzoli).]

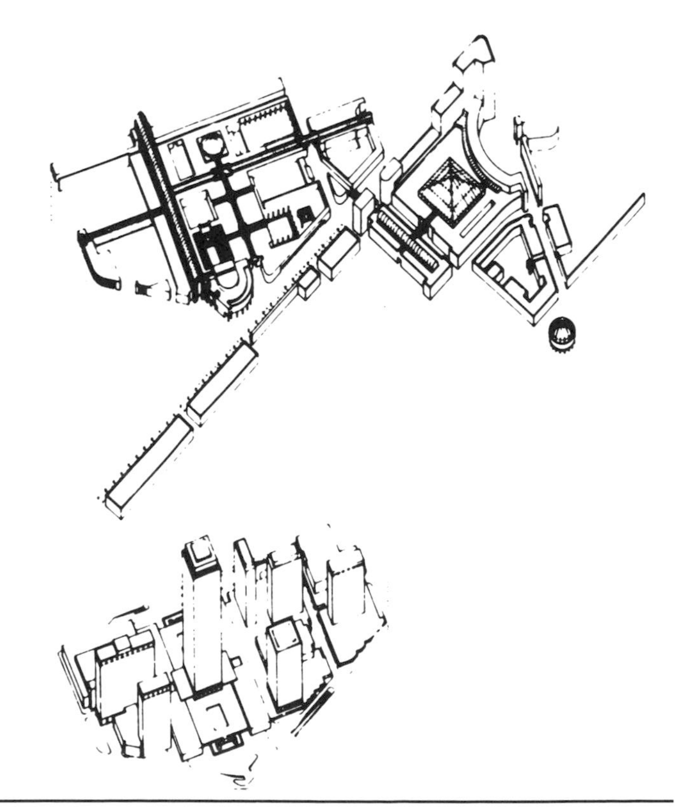

Fig. 29. Boston Back Bay urban design isometric diagram. [Source: Koetter Kim and Associates.1997. *Place Time* (Rizzoli).]

More complex diagrammatic drawings as illustrated by Koetter Kim Associates help to crystallize the conceptual underpinnings of an urban design scheme (Fig. 28). In a design conceived to "re-knit" the fragmented pieces of the Chinatown neighborhood in Boston, Koetter Kim use a series of selective isometrics of building and urban design elements to reveal the complex collage and interrelationships of the concept. Such an approach is successful at a smaller scale of urban design, such as a neighborhood or a district and when the existing urban fabric has many distinctive elements to be analyzed, explored, and extended. The technique also suggests a more complex interpretation of the city with convergent and divergent systems of order, at times overlapping or existing side-by side.

A more general kind of layered analytical drawing is used by Koetter Kim Associates to illustrate how the section of Boston around the Prudential Center might be retrofitted to make a more cogent urban environment (Fig. 29). The isometric diagram explodes the various themes of the design to reveal the conceptual order of the plan while keying into the isometric of the existing context. The drawing communicates the situation of the urban fabric "as is" while providing a critique, immediately visible with the proposed changes hovering over the plan. Each of the layers of the drawing can be read individually and together as they inform the overall concept. The technique is not a simple recording of concepts and features but rather a carefully edited diagram of overall intent informed by selected details.

CONCLUSION

It is conceivable that the history of representation in urban design is longer than the history of drawing used to delineate a single building. Why is this so? Perhaps this is because, until the modern age, craftsmen with a detailed knowledge of building construction constructed buildings. The architect was on-site directing construction. Buildings could be laid out in proportional relationships of materials and stylistic concepts. The architect provided a foundation for the appearance and working with experienced craftsmen designed the structure. It is interesting to note how few drawings exist for many of the most well-known buildings from the history of architecture prior to industrialization.

The converse may exist in urban design. City plans can be traced back to antiquity and although invention has played a role in how today the city is represented, the urban designer is still confronted with the needs to communicate a holistic vision of his or her intentions over a greater area than a single building, and sometimes needing to assert a conception well beyond the life of the designer, witness L'Enfant and Washington, DC.

The profusion of information in the so-called "Information Age" offers the urban design professional more choice of technique and the promise of a more comprehensive graphic system to understand cities and contexts. Techniques for representing specific detailed aspect of the city have become more sophisticated, such as Graphic Information Systems (GIS) and computer-aided design, yet the technological limitations of previous societies may indeed have made information more understandable and comprehensive. The challenge of today is to avoid the general tendency for specialization and to use techniques available for more comprehensive and interdisciplinary representations of the city to aid the designer. ■

Paul D. Spreiregen

Summary
Architectural design begins with the preparation of a building program and site analysis. So, too, do plans for designing or redesigning a portion of a city. In the case of a city, the analysis is a diagnosis of the city's component pieces, to see the relations between these pieces and to assess their condition. A visual survey in urban design is an examination of the form, appearance, and composition of a city—an evaluation of its assets and liabilities. A visual survey also enables the urban designer to see where the city needs reshaping.

Key words
activity, density, district, image, route, scale, skyline, traffic pattern, urban structure, visual survey, vocabulary of urban form

Fine-grain and uniform texture. Row houses in Philadelphia.

Making a visual survey

A Working Vocabulary of Urban Form

Architectural design begins with the preparation of a building program and site analysis. So, too, do plans for designing or redesigning a portion of a city. In the case of a city, the analysis is a diagnosis of the city's component pieces, to see the relations between these pieces and to assess their condition. A visual survey in urban design is an examination of the form, appearance, and composition of a city—an evaluation of its assets and liabilities. A visual survey also enables us to see where the city needs reshaping.

A visual survey can be made of any city or town, regardless of size. It can also be made at different scales—a neighborhood, the center, a suburban area, or a small group of buildings. Furthermore, it can be made for a built-up part of the city which is going to be altered very slightly or for a part of the city which is going to be rebuilt entirely. The process of making a visual survey is not complicated, nor need it be done with a high degree of precision. As a matter of fact, it is best done in general terms, for to deal with the city on a large scale we must think broadly.

To conduct a visual survey, one must have a basic idea of the elements of urban form. These necessitate a descriptive vocab-

A visual survey discerns a city's assets . . .

. . . and its liabilities.

Credits: This article is a chapter of the author's writings, *Urban Design*, published by the American Institute of Architects and McGraw-Hill, 1965, and first appeared in *AIA Journal,S* 1964. Drawings are by the author.

ulary. Next, one must examine the city and describe it in terms of this vocabulary. It is also necessary to relate the elements, in order to understand its workings, its form, and its consequent appearance.

While making a visual survey, it is important to constantly evaluate. Certain discordant elements must be noted as faults to be corrected; certain appropriate elements must be noted as assets to be protected. A good urban design survey will also disclose a number of specific ideas for improving, correcting, or replacing parts of the city, for a good survey leads to ideas for action.

The Image of the City

People's impressions of a building, a particular environment, or a whole city, are, of course, more than visual. Within the city lie many connotations, memories, experiences, smells, hopes, crowds, places, buildings, the drama of life and death, affecting each person according to his particular predilections. From his environment each person constructs his own mental picture of the parts of the city in physical relationship to one another. The most essential parts of an individual's mental image, or map, overlap and complement those of his fellows. Hence we can assume a collective image-map or impressions-map of a city: a collective picture of what people extract from the physical reality of a city. That extracted picture is the image of the city.

Every work of architecture affects the details and often the whole of the collective image. The collective mental picture—the image of the city—is largely formed by many works of architecture seen in concert or in chaos, but definitely seen together.

Several years ago, Prof. Kevin Lynch conducted a study of what people mentally extract from the physical reality of a city. He reported the results in a book called *The Image of the City,* and his findings are a major contribution to understanding urban form and to architecture as component parts of that form. Professor Lynch is one of the country's leading investigators of urban form. Many of the ideas in this book were derived from his studies. In his examination of the form of the city, Professor Lynch found that there are five basic elements which people use to construct their mental image of a city:

Pathways: These are the major and minor routes of circulation which people use to move about. A city has a network of major routes and a neighborhood network of minor routes. A building has several main routes which people use to get to it and from it. An urban highway network is a network of pathways for a whole city. The footpaths of a college campus are pathways for the campus.

Districts: A city is composed of component neighborhoods or districts; its center, uptown, midtown, its in-town residential areas, trainyards, factory areas, suburbs, college campuses, etc. Sometimes they are distinct in form and extent—like the Wall Street area of Manhattan. Sometimes they are considerably mixed in character and do not have distinct limits—like the midtown area of Manhattan.

Edges: The termination of a district is its edge. Some districts have no distinct edges at all but gradually taper off and blend into another district. When two districts are joined at an edge they form a seam. Fifth Avenue is an eastern edge for Central Park. A narrow park may be a joining seam for two urban neighborhoods.

Paths.

Districts.

Edges.

Landmarks: The prominent visual features of the city are its landmarks. Some landmarks are very large and are seen at great distances, like the Empire State Building or a radio mast. Some landmarks are very small and can only be seen close up, like a street clock, a fountain, or a small statue in a park. Landmarks are an important element of urban form because they help people to orient themselves in the city and help identify an area. A good landmark is a distinct but harmonious element in its urban setting.

Nodes: A node is a center of activity. Actually it is a type of landmark but is distinguished from a landmark by virtue of its active function. Where a landmark is a distinct visual object, a node is a distinct hub of activity. Times Square in New York City is both a landmark and a node.

These five elements of urban form alone are sufficient to make a useful visual survey of the form of a city. Their importance lies in the fact that people think of a city's form in terms of these basic elements. To test them, sketch a map of your own city, or better still, ask someone else to do it, taking only a few minutes. The result will be twofold: a picture of the most salient features of a city's form—its image—and a map of the sketcher's particular interests as they relate to the city. The result will also be akin to the cartoon maps of the United States as seen through the eyes of a Texan or a New Yorker. The features will be distorted and probably exaggerated, the degree of distortion reflecting the hierarchy of values of the sketcher.

The more "imageable" a city, the easier it is to find one's way about in it, even if its street pattern is not clear. In designing a city, it is important to consider how a new development will affect the total urban image. A new development can be made to tie visibly into a city's path system; to form or help reinforce a district; if on an edge, to strengthen the edge; and if at a seam, to maintain continuity. It can also become a good landmark and an active node.

Paths, landmarks, nodes, districts, and edges are the skeletal elements of a city form. Upon that basic framework hangs a tapestry of embellishing characteristics which all together constitute the personality of a city. To build a broader vocabulary upon this basic framework we must consider landform, natural verdure, climate, several aspects of urban form itself, certain details and several lesser facets of form.

Landform and Nature

Every city is built on a piece of land. The form of this land and its features are the foremost determinants of a city's form. In speaking of landform, we are speaking primarily of topography.

In looking at landscape, we are seeking its character. As urban designers we observe the form of the terrain—flat, gently rolling, hilly, mountainous—in relation to the architecture and the cities which are set in it. A flat site may suggest either vertical architecture or assertive horizontals. A slightly hilly site may call for vertical architecture at the summits with a flow of cubes on the slopes, or may suggest a termination of architecture just below the crests. A steep hillside or valley may lend itself to terracing, with orientation to the sun. In every case we must assess the qualities of the terrain, including the design relationships they express.

The prominent features of a landscape should be carefully noted—cliffs, mountain peaks, ranges of hills on the horizon,

Landmarks.

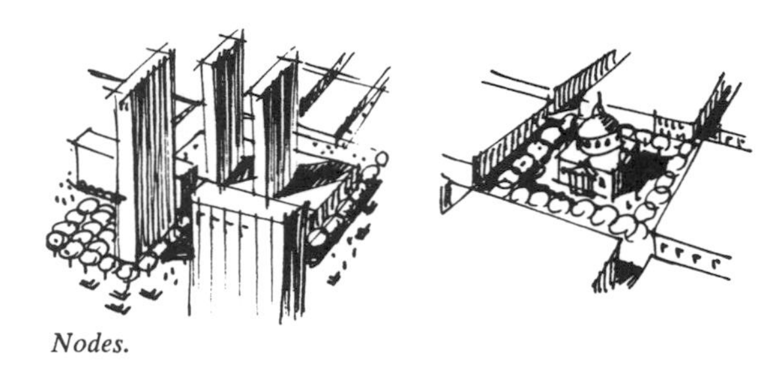

Nodes.

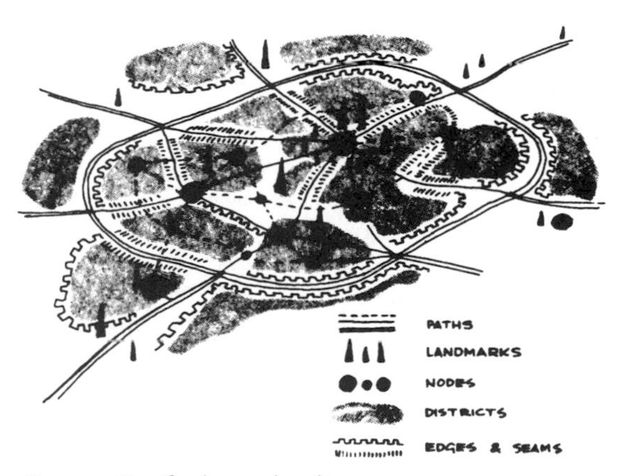

Representing the form of a city with abstract *symbols.*

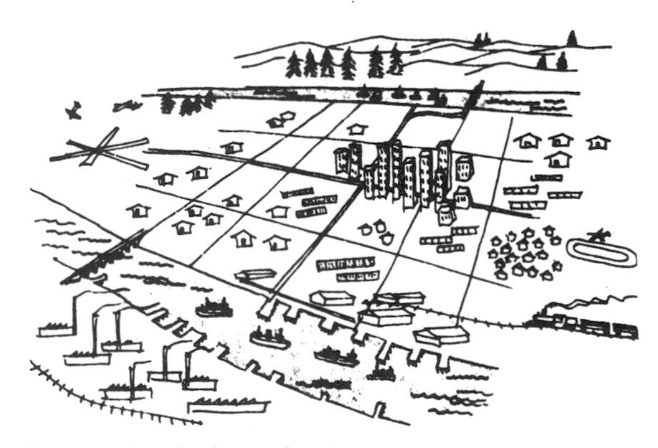

Representing the form of a city with representational *symbols.*

Topography.

Form and character.

Characteristic features.

Classification of native trees.

Architectural form in relation to terrain.

The city in nature; nature in the city.

plateaus, rivers, or lakes. These are accenting landscape features which can be employed actively as sites or passively as vistas, supplementing architectural and urban form. They can be used as major vista objectives from points within the city or as special sites for buildings. Some are better left in their natural state.

Indigenous greenery should be assessed in terms of shape, size, character, practicability, and seasonal change. An urban designer needs a working knowledge of the local flora and its suitability in various uses. A thickly foliated tree, formally shaped, might be proper for lining a road to shield the automobilist from a low sun. A spreading shade tree of informal shape, might be quite appropriate as a restful sitting place in the bustle of the city.

Characteristic detail of the landscape should be considered for possible use as architectural and urban design embellishment: a native rock or gravel, a characteristic earth color, the form of local streams, characteristic stands of trees. Indigenous architecture should also be noted, particularly in older towns. These are the result of evolution and may have achieved a mature relationship with their environment.

Certain areas of landscape should not be touched, but preserved in their natural state. A survey of the natural landscape may disclose areas which are better left as wilderness. These might well be chosen with relation to nearby cities and towns so that they are accessible as necessary complements to urban life, but not menaced by it.

Buildings and small towns can often be seen in their entirety in the framework of nature. As such, they are accents or counterpoints to their natural settings, elements of vitality in a setting of repose. A larger town, however, can seldom be seen in its entirety, but only in part, from various viewing places. Here we have a one-to-one relationship, nature being less a setting than a major component of the whole scene—balancing the sight of the city rather than acting as a setting for it. Raw nature sometimes exists within large cities in the form of streams, rivers, shore lines, cliffs, etc. Here the city is the setting for nature. Nature, in this circumstance, becomes a foil or counterpoint to the urban surroundings.

Thus, we might regard a small town as an object in the embrace of nature, a larger town as being hand-in-hand with nature, and finally, the large city as assuming the role of nature and becoming the embracer.

A visual survey of nature in relation to achitecture and urban design is threefold in scope. We first try to determine the character of the surrounding landscape to which our architectural and urban forms must respond esthetically and functionally. Second, we evaluate the degree to which our existing architecture and cities enhance nature. Third, we must decide what natural areas are to be left alone to act as complements to urban form. Throughout this process we search for assets and liabilities, preserving and enlarging upon the one and noting corrections to be made on the other.

Every work of architecture affects the natural landscape either positively or negatively; so does every structure and human settlement. Nature, in turn, as a setting for our constructions, is a visual framework to which all our constructions must respond.

Local Climate

While on the subject of the natural features of the terrain, it is wise to check on local climatological conditions. Local climate determines much of the character and appearance of the landscape and buildings. The following aspects of climate can be readily found in United States Weather Bureau publications.

Temperature: Seasonal temperature and humidity as averages and extremes which indicate the periods of relative comfort, the extremes which must be ameliorated, and which therefore determine architectural and urban form.

Light: The number of clear, partly cloudy, and fully cloudy days, which conditions the light affecting the appearance of the city and of buildings.

Precipitation: The amount of precipitation in the form of rain and snow.

Sun: The angles of the sun in different seasons, which affects viewing conditions and, thus, design. It is useful to make a simple three-dimensional model to study these angles.

Winds: The prevailing seasonal winds including the direction and intensity of cold winter winds, gentle or severe fall and spring gusts, and cooling summer breezes. These affect design considerably.

In addition to these quantitative factors there are a number of qualitative aspects of climate which are as important in urban design. Some cities are well oriented toward the rising sun or the setting sun. Some cities have forms that derive almost directly from their climates—arcaded cities in the sun, for example. Considerable research or experimentation might be done to determine how cold winter winds could be slackened and cooling summer breezes induced. The quality of light—sharp and clear or cloudy and dull—should be a determinant in the design of building facades including their degree of intricacy and their coloring. These are always a matter of artistic consideration but a careful appraisal of actual conditions can help a decision.

Shape

Every city has a general overall shape. There are several classifications of shape.

Radiocentric: The most frequently found urban form is the radiocentric, a large circle with radial corridors of intense development emanating from the center.

Rectilinear: A variation on radiocentric form is the rectangle, which usually has two corridors of intense development crossing at the center. This variant of the radiocentric form is found in small cities rather than large. It is the radiocentric form with right angles.

Star: A star shape is a radiocentric form with open spaces between the outreaching corridors of development.

Ring: A ring shape is a city built around a large open space. The San Francisco Bay is such an open space for the cities of the bay area. A ring and star may be found in combination, particularly where a loop road is built around the outskirts of an expanding metropolis.

Linear: The linear shape is usually the result of natural topography which restricts growth or the result of a transportation spine. Stalingrad in the Soviet Union was planned as a linear city. The megalopolis on the East Coast has become a vast metropolitan area with a linear configuration.

Branch: The branch form is a linear spine with connecting arms.

Sheet: A vast urban area with little or no articulation.

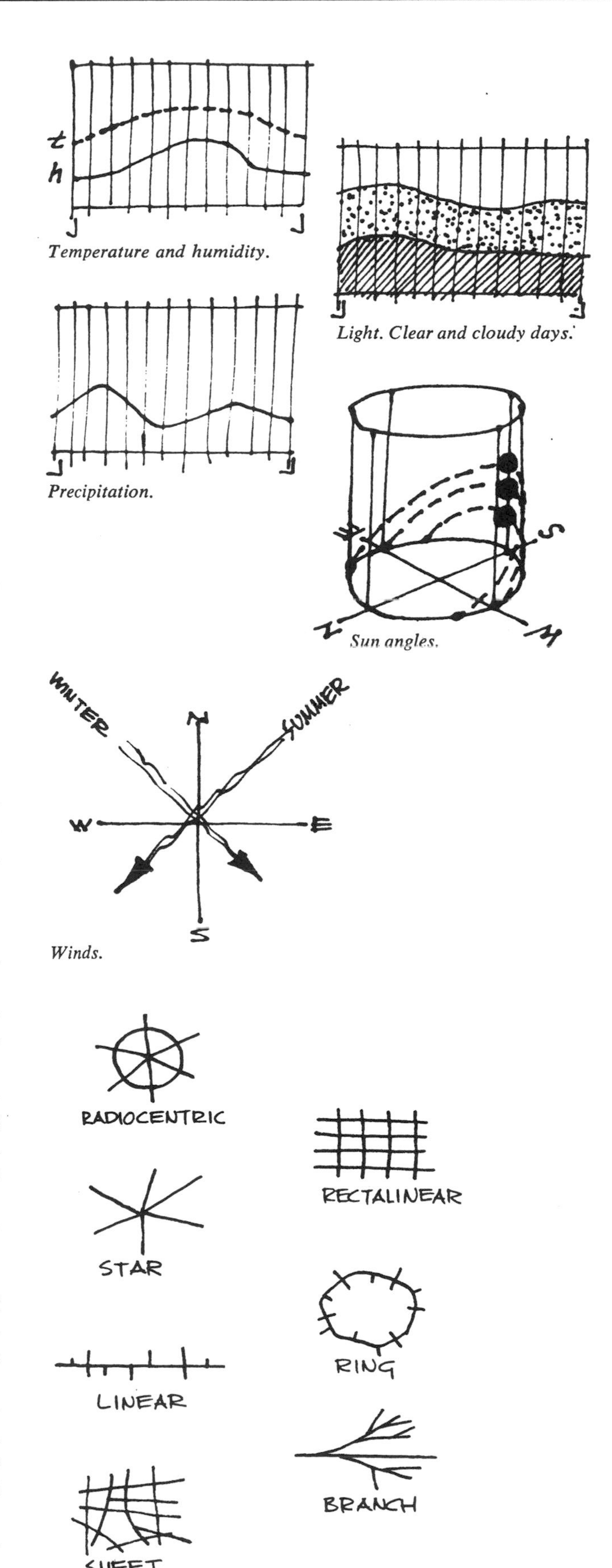

Temperature and humidity.

Light. Clear and cloudy days.

Precipitation.

Sun angles.

Winds.

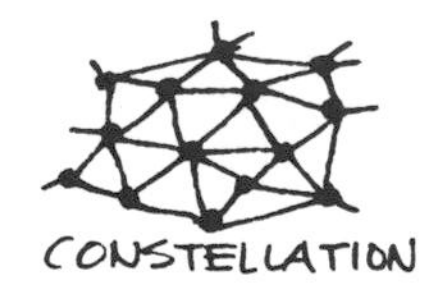

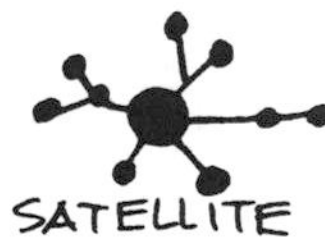

Articulated Sheet: The articulated sheet form is accented by one or more central clusters and several subclusters.

Constellation: The constellation is a series of nearly equal-size cities in close proximity.

Satellite: The satellite is a constellation of cities around a main center.

These classifications of form have definite implications for a city's function. They have advantages and disadvantages related to circulation, proximity to open space, and articulation of neighborhoods or districts. Further, these classifications may be applied to the city as a whole or to parts of the city, isolated for study, like open spaces or circulation. The open spaces of a city may be linear or branched; or they may form a radiocentric pattern. The circulation networks may likewise be described as one or another shape.

Size and Density

Closely related to a city's shape is its size, a quantitative aspect which can be approached several ways. We first of all think of the physical extent of a city: so many miles across or so many miles from center to outskirts. We can also describe size in terms of the number of inhabitants. The relation between size and density is important, for it indicates the distribution of people and the city's urban massing.

Density can be computed mathematically in several ways: the number of people per square mile; the number of houses per acre or square mile; or the amount of building floor area in a given section. It can also be expressed in terms of automobile population. In 1962 Los Angeles, the country's most auto-oriented city, had 2,220 cars per square mile and Washington, D.C., had 4,100 cars per square mile. This comes as a surprise to most people. One would think that those figures are reversed—which suggests a note of caution in judging aspects of quality from statistics of quantity.

The gross size of a city in terms of its population is also revealing. Classifications according to size alone are quite useful. A basic population of about 200,000 to 300,000 is necessary to support basic public cultural facilities. Amsterdam, Holland, with a population of about a million people, is of the maximum size that can be traversed on foot by a hearty walker, from center to outskirts.

Unless a city is evenly built-up, studies of density are best made on separate sectors of a city. Density figures indicate the relationship between built-up and open land; therefore they can describe almost graphically the image of a suburban residential area or an in-town row-house area. Densities have definite implications for various forms of transportation. In making a visual survey, it is helpful to determine the density of various areas and to relate the density figures to physical patterns of land and buildings and, hence, the visible form of the area.

Anchorage, Alaska. A rectilinear pattern on flat land in the lap of a mountain range.

Pattern, Grain, and Texture

Urban areas have distinct patterns. Usually these are seen in their block and street layouts. Most American cities have rectilinear block and street patterns. On rolling terrain, in outlying areas, curvilinear streets and blocks form another type of pattern. A cul-de-sac system forms a third pattern. Mixtures of open space and built-up space constitute still another pattern. A basic design pattern can be very helpful in planning a residential area or a campus area. An urban pattern is the geometry, regular or irregular, formed by routes, open spaces, and buildings.

Grain is the degree of fineness or coarseness in an urban area. Texture is the degree of mixture of fine and coarse elements. A suburban area with small houses on small plots has a fine grain and a uniform texture. With small houses on varying size lots, it could still have a fine grain but an uneven texture. In the city, large blocks with buildings of varying sizes could be described as having a coarse and an uneven texture. If the buildings are uniform in size, they could be described as having a coarse grain but a uniform texture.

Such distinctions are easily indicated on a sketch map. They are useful in evaluating an area's form and in making decisions about a design treatment for it. For example, a coarse-grained unevenly textured area may be impersonal and repellent and could be treated with some fine scale and unifying design elements. An extensive and uniformly-grained area might well be treated with relieving accents.

Urban Spaces and Open Spaces

Urban shape, pattern, grain, size, density, and texture are primarily aspects of solid form—the building masses of the city. In architecture it is rather helpful to conceive of a building not only as a solid but as spaces modeled by solids. It is also helpful to consider a city this way. The spaces of the city range from the space of the street to the space of a park system and, ultimately, to the vast space in which an entire city exists. It is helpful to think of these spaces as two generic types: formal or "urban spaces," usually molded by building facades and the city's floor; and natural or "open spaces," which represent nature brought into, and around, the city.

Basically an urban space must be distinguished by a predominant characteristic, such as the quality of its enclosure, the quality of its detailed treatment or outfittings, and the activity that occurs in it. An urban space should, ideally, be enclosed by surrounding walls, have a floor which suits its purpose, and have a distinct purpose to serve. If, however, any one of these qualities is sufficiently strong, it alone may establish the sense of urban space.

A group of office buildings may contain a space around a poorly designed plaza or a complex road intersection, the floor space being devoted entirely to traffic. This is an urban space which has a sense of place in the city. It is both a landmark and a traffic node, as well as an office node. An urban square may be beautifully landscaped as a restful urban park, but it may lack entirely the peripheral building facades which are needed for a sense of enclosure. Here we have a poorly enclosed space, but a space nevertheless. In another instance, a particular place in the city may function as the locale of an important activity while possessing neither physical enclosure nor appropriate floor. Times Square in New York is such an example.

In all these examples we have a sense of space. Such spaces are islands or oases in the city. But urban spaces can also be linear corridors. Avenues and streets are linear urban spaces if they are enclosed on two sides or have some element of unifying character—trees or uniform buildings. Corridor spaces are spaces for linear movement. Island or oasis spaces are stopping places. Of course the two can be interconnected. In fact, a spatial structure for an

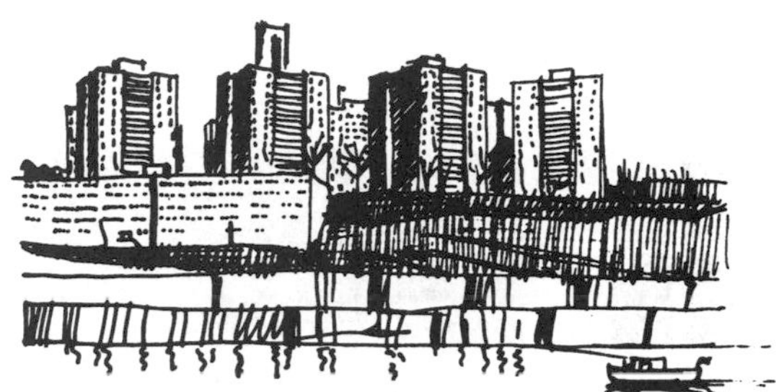

Coarse-grain and uniform texture. Yonkers, New York.

Coarse-grain and uneven texture. Lower Manhattan.

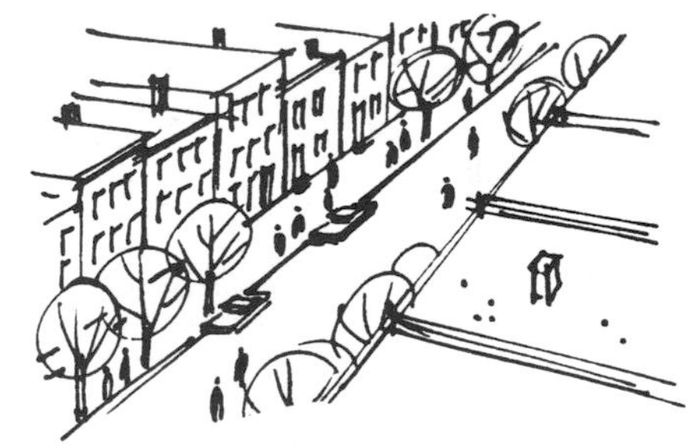

A corridor space.

An enclosed space. The University of Virginia Quadrangle, designed by Thomas Jefferson.

A spatial setting for a key building. A court house square in Missouri.

Urban spaces formed by building masses.

Routes traverse the countryside in many ways.

Routes can approach architecture or cities in many ways.

entire city is exactly such an arrangement at the city's total scale.

Open spaces, being nature brought into the city or open expanses allowed to remain in their original state, cannot be described in quite the same manner used for urban spaces. Their scale is given by the trees, shrubs, rocks, and ground surface rather than their gross width and length. Their appearance is characterized by the sight of natural verdure rather than surrounding buildings. However, a vista of a distant building may accent a particular spot and a bridge or pathway may complement nature's forms. Open spaces in the city have a wide variety of purposes. They are a complement and foil to urban form. They are also reservoirs of land for future use. For an urban design survey, one should study the spaces of the city as an overall structure. In doing this it is helpful to classify spaces according to their actual use and to consider formal urban spaces and the natural open spaces together.

For example, one could start by mapping all the recreational parks in the city, then the interconnected stream parks. The center city urban parks could be mapped, and the main corridor spaces that lead to them or connect them. The nodal spaces as well as the connector spaces all together would form the spatial network. Such a survey would disclose a need for creating spaces in certain areas, a need to improve existing spaces, and some possibilities for connecting all of them. The survey of spaces should disclose a hierarchy of spaces for rest and repose to spaces for meeting and bustling activity.

A city's entire system of public lands—roadways, schools, parks, civic buildings, libraries, etc.—could be thought of as an open space network possibly complemented here and there by public buildings. In an urban design survey we look for the location, quality, and amount of open space in relation to the city's built-up areas.

Routes

Landscape, architecture, and cities are seen as sequences as we travel along routes of movement. Routes of movement affect considerably the appearance of the landscape through which they pass and the architecture and cities which they serve. Routes of movement are a principal determinant of urban form. In making an urban design survey of the routes of a city, one should begin with the area well beyond the city limits, far out in the country. The primary function of a highway is to allow traffic to move, but a large part of that job depends on how clear the route is in relation to the city. This aspect of highway engineering—the "imageability" of the highway—is a matter of revealing its clarity of form and direction to the user. Too many highways have very poor physical relationships to the areas they serve. Rather than helping to define these areas, they often slash through them, actually acting as a blighting and disintegrating force.

Routes in the Countryside

In the open landscape, existing and proposed routes should be examined and assessed with a view to how well they relate to the natural terrain. How artfully or awkwardly do routes traverse the landscape, revealing its prominent features? Are vistas taken advantage of, or ignored? Some vistas might well be presented with dramatic suddenness; others might be introduced gradually, or be seen only in part.

Are there dull areas which require embellishment? Perhaps the introduction of a curve or rows of trees and shrubs could give more visual interest. Are there obstructions to the enjoyment of the prominent natural features? Does the road itself and its furniture mar the landscape or add beauty to it?

The outlying routes of your city are the first introductions which approaching visitors receive, giving them their major impressions. In making a visual survey of routes, the routes should be charted, noting the character of the terrain and the adaptation of the roads to it, the artful dramatization of landscape features, the quality of added features, the accenting of the route, its faults and possibilities for improvement or correction. Every new route should be examined and designed on these bases as a matter of sound road engineering.

What is the quality of the furnishings of a route?

Approach Routes and Surface Arteries

Approach routes present cities to us. They must satisfy the visual requirement of presenting architecture and cities in their best light, while enabling us to find our destination readily. The two requirements go hand in hand. An approach route must both inform us and conduct us.

The major routes through the city are surface arteries—high-volume traffic streets which carry buses and autos. They can be evaluated according to how they tie into the expressway pattern, their clarity of form, their relation to the cityscape, the shape of the building sites they pass by, and the way they pass through existing districts.

Modulating the approach route by screening a portal.

Another consideration is the street furnishings of the major surface streets. Can their design be improved and a program for improving signs and traffic furniture be started? How well do the through arteries tie into the pattern of slower-speed local streets? How well do they tie into major garages and parking areas? How easy is it to find a garage near your destination and to get into and out of it? Most important of all, we must examine the relation between a street's traffic and buildings. A good index is the degree to which street traffic is actually serving the buildings on the street, in contrast to traffic which is merely passing through on the way to some other destination.

A clear route with its own strength of character aids orientation.

Local Streets

The through arteries serve an intricate network of small streets, along which cars, buses, and delivery trucks stop and go. These streets carry a mixture of vehicles and people. In surveying them, we examine whether vehicular and pedestrian movement are in conflict with each other or aiding one another. Where do they belong together and where not? Are pedestrians forced to wait for long periods of time to cross streets, or are pedestrians free to cross streets anywhere? Is safety achieved by the use of stoplights or by grade-separated pedestrian crossings?

The foreground of a city should reinforce a view, not distract from it.

Is the vehicular traffic strictly local, or is much of it through traffic? Can this through traffic be relocated? How can existing small streets be protected against the intrusions of through traffic? What is the dimensional scale of the intimate local streets? How do they relate to the size of their districts? Can these patterns be improved and strengthened?

These comments illustrate that an inquiring survey raises many questions and stimulates many ideas. As we elaborate on the path-district-edge-landmark-node framework, we find that some of them require more attention than others. Degree of emphasis on one

Recording the visual sequence of a route.

A small town may have only a few districts; a large city, very many.

Districts may be expanding or shrinking. They may have clear edges or overlap. They may be uniform or complex.

Surface arteries may reinforce districts or slash them.

Strength of character—degree of identity —varies from one district to another.

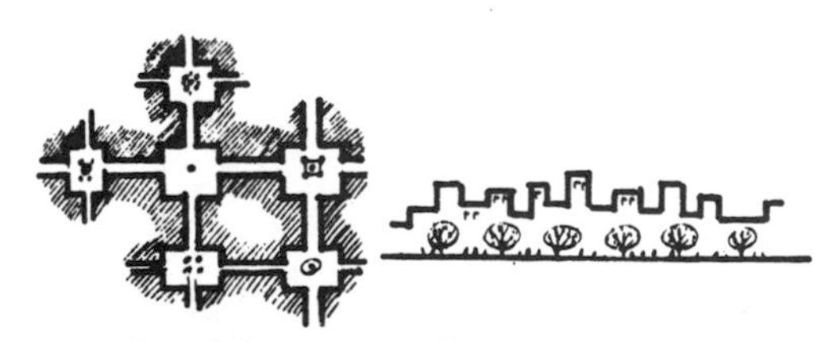

Regularly occurring elements give cohesiveness. Small squares or trees.

aspect or another will depend on the size of the city or the urban sector being surveyed. On the large-scale survey of the whole city, its paths, districts, and open spaces, for example, may be the predominant elements. However, at all levels, the examination of the *districts* of the city will probably require the greatest effort.

The Districts of a City

Every city consists of a series of parts which we refer to as districts or enclaves or sectors—or perhaps as quarters, precincts, or areas. They are distinguishable in that they have dominant and pervasive characteristic features. Our mental images of cities consist, to a large extent, of the arrangement of these parts. Some are distinct, some overlap others, some are uniform, some are very complex. Almost all are in a process of change, which further affects their appearance and their size. A very small town has at least several distinguishable areas; a metropolis may have fifty or a hundred.

The pattern of districts is closely related to the pattern of routes. The size of a district may be determined by the nature of the internal routes serving it. A commercial center, for example, can usually be traversed on foot or by a short cab ride. A residential section may often have local community facilities which can be reached on foot, although its gross size may be far beyond the limits of pedestrian traverse.

The districts of a city vary considerably in their strength of character. Districts which do have very strong characters often develop identifying names—Wall Street, Georgetown, Beacon Hill, Greenwich Village, the Loop. Other districts with less assertive character often bear names related to their historic origin—Market Street or Main Street, Foggy Bottom, Silver Spring, Brookline.

American cities, like most cities of the world, reflect their characteristics of culture, growth, and development in an urban nomenclature. In the United States this nomenclature includes: "downtown," the original center; "uptown," the enlargement of the original center; "midtown," an offshoot of both; "Chinatown" and "Harlem," ethnic areas; "the other side of the tracks," characterizing poor residential areas in the shadow of factories; "the waterfront"; "the outskirts."

Basically, there are two things to look for in discerning the various districts of a city: physical form and visible activity. For example, in a commercial center the types of buildings, the signs, the demolition and construction activity, the crowds of rushing people, the cabs and buses, the parking facilities—all these identify the place for us. On the other hand, in a residential area we have the houses, their spacing, the trees, the milk wagons, the parked cars, the children playing, the occasional neighborhood stores and schools. The sum impression of the individual parts and their relationships conveys to us the existence of a particular district of a city—a part in relation to a whole.

Few, if any, cities can be neatly compartmentalized in this way. The most prominent enclave may dissipate visually at its periphery. Most urban enclaves lack outstandingly prominent characteristics. Further, complexity in an urban enclave should not be mistaken for confusion. Urban complexity—the intense intermixture of complementary activities—is one of the major reasons for cities and the spice of urban life. One must also distinguish between uniformity amounting to dullness, and unifying architectural and landscape elements constituting visual cohesiveness, especially in the face of great variety. We should search for answers to the following:

Components: What are the principal component districts of the city? Where do they begin and end? What are their characteristics, physically and as defined by activity? How apparent are they?

Size: What is the size of a district—its shape, density, texture, landmarks, space?

Appearance: Regarding their physical appearance, what are the characteristics of building forms, building density, signs, materials, greenery, topography, route-pattern landmarks? What is the nature of the mixture of different building types?

Activity: Regarding visible activity, what are the principal clues of the activity of an area—the kinds of people, when and how they move about? What are the key visual elements—the things principally seen—which establish the character of a district?

Threats: What are the threats to a district? What external elements, such as a through road, threaten the health and survival of district? How is the district changing? Is it changing its position? Is an edge decaying? Is an edge advancing, perhaps into a peripheral district?

Emergence: Are there latent districts struggling to emerge, such as a new in-town residential section?

Relation: How do all these parts relate to each other and especially to the route patterns of the entire city? Finally, what are the areas in a city that cannot be classified easily, that lack cohesion in form and character? Are some of these targets for urban design work?

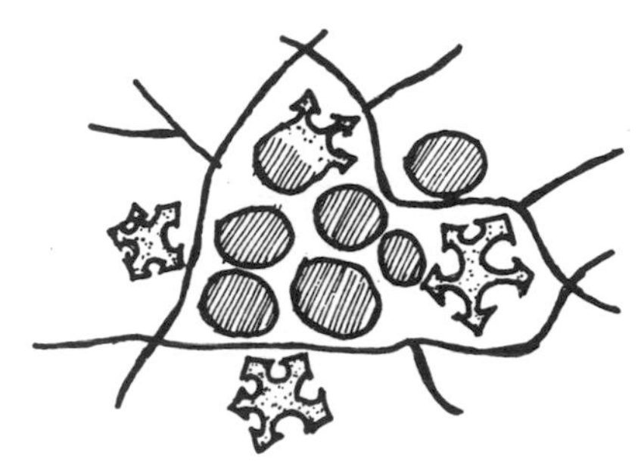

Are there new districts struggling to emerge or old ones struggling to enlarge?

Particular sights characterize particular districts.

The Anatomy of a District

Having distinguished the separate parts of the city, it remains to go one step further and survey the parts individually—to diagnose the districts, the parts which constitute the whole. In surveying the visual aspects of a district or enclave we should be asking:

Form: What is the physical form of the place—form and structure in three dimensions and in broad outline? What is the density and character of the buildings? What is the spacing of the buildings? How does it vary? What is the greenery of the place? How would you describe the paving, the signs, the night lighting? How uniform or how varied is the whole, or sections of it? Can a district be further dissected into meaningful places within it? What are these places like? What are the physical patterns of the place? What are the patterns and the linear and focal points or urban spaces within the district?

Activity: What do people do there? How well does architecture and the district serve people? What are the natural groupings of different activities within the district? How does the activity pattern change according to the time of day, week, or season? How lively are the central city areas? How does the local climate affect life in the areas? What are the detrimental aspects of the place?

Features: What are the features of the district—the major hubs or nodes, landmarks, and vistas? What are the major magnets, generators, and feeders? In a busy center-city area, what are the oases, the places of repose? In a quiet residential section, what are the hubs, the places of community focus?

Paths: What are the principal paths of movement in a district? How are they differentiated? How well do they serve the people

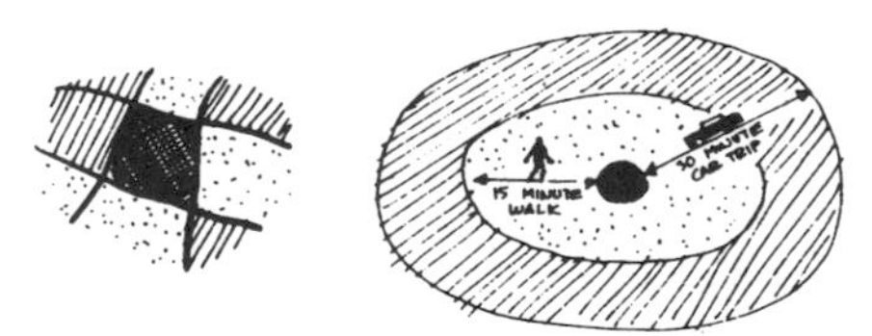

Patterns of routes help define a district; scales of distance help determine its size.

Some districts have unique unifying features.

What are the major vistas of a district?

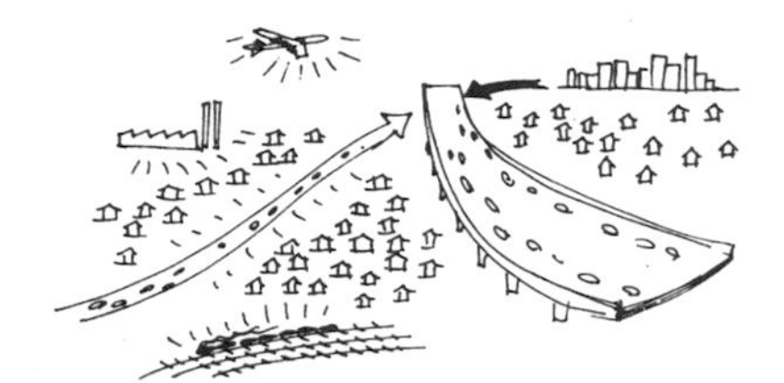

What are the blighting threats to a district?

How lively is the city core at different times of the day and week?

Active areas also need places of repose.

there? How well do they connect to the larger network of paths? Are the actual physical dimensions of the paths adequate or excessive? How do they determine the physical limits of the districts?

Centers: What are the features of a district that serve a symbolic civic role? What are these places like? Are they lively or lifeless? How can they be made lively? Are they integral parts of the areas around them? Are they part of the life of the community, or are they inanimate symbols?

Intrusions: What are the intrusions and detrimental features of a district? What are the blighting features? Here again one must be careful to distinguish between enlivening intermixtures and truly harmful elements. How much traffic can be tolerated on a street before it is impaired? How little before it is dead?

Change: How is the district changing, both in internal character and the adjustment of its periphery to change? Is there a direction of growth? In which direction is the center of gravity moving? Is the edge decaying? How can a decaying edge be invigorated? How can a district be stabilized?

Improvement: Finally, how can the formation of a new district be aided? What are the new elements of the city that are struggling to emerge? Which marginal districts can be protected and improved as part of the complementary complexity of the whole city? How would you analyze and depict the important districts in your city? What strengths do you discern, what weaknesses? What differences do you find between districts? Is there significance in their relative positions and character?

Activity Structure

An examination of districts and nodes reveals that there are certain spots in the city that have characteristic functions. Generally speaking, these districts fall into such categories as places of living, working, shopping, traveling, leisure, recreation, and learning. There is a logic to the location of these activities and there are definite visual results in their deployment and interrelationships. Density, topography, and transportation routes all affect an urban activity structure.

For example, a high-density residential area will have a central shopping cluster which many of its clients can reach on foot. A low-density residential area will more likely be served by a shopping center reached by automobile, its use shared with other low-density areas. Topography can dictate the location of routes and therefore the location of centers and subcenters. Topography can also dictate the location of hospitals and airports.

New transportation patterns can alter an existing urban structure by causing the relocation of facilities which depend on a high degree of public access. Shopping centers on a city's periphery are in large measure a consequence of circumferential expressways. Small neighborhood shopping centers on radial routes are approximate indicators of the centers of residential districts.

A large-scale study of activity structure will reveal the general centers of work and residence and a physical correspondence between activity and district. When tied into an examination of major routes of movement, the relation between activity and circulation access becomes clear. So do points of conflict and areas in transition.

Orientation

If there is logic in the arrangement of a city's anatomy and if that arrangement is visibly evident—articulated—the sense of orientation will be strong. If there is logic but little or no visible articulation, a city can be confusing even to the point where it arouses a high degree of frustration and anxiety, and the feeling of being lost.

Landmarks are a prime aid to orientation. On the overall scale of the city, prominent landmarks are tall verticals like central skyscraper groups, natural features such as rivers or shores, district edges, unique vistas, clear routes which lead to and from a known place, and districts with strong visual characteristics.

Orientation studies should be made on the scale of the whole metropolis as well as small enclaves such as shopping areas, commercial areas, or institutional groupings. Design programs to improve the sense of orientation are particularly important where there are many visitors, as at an airport, or downtown, or at a shopping center. The logic of arrangement and its visible evidence, achieved through design, is the prime device for improving orientation. Signs are a secondary device. Where signs are relied on too heavily, they may add to the confusion or go unheeded.

Orientation studies can be made by the urban design surveyor himself if he is unfamiliar with the area, but they are best made by an interview-map technique.

An old landmark. The Flatiron Building in New York.

Details

The appearance of small details, such as cracks in the pavement, parking meters, tree trunks, doorways, are major factors that characterize an area. They tell us of the area's age, purpose, upkeep, or decay. Signs are an important urban detail. A visual survey should examine the types of signs in an area: for advertising a product; for giving directions; and for marking a building, shop, theater, or hotel.

Consider the quality of details.

It is important to ascertain the intended audience of a particular sign. If the signs in a shopping area—store names, goods on sale, etc.—are scaled entirely to the pedestrian shopper, they will usually be appropriate. The signs on a highway should be designed for the fast-moving automobilist. On an expressway, signs other than traffic signs may be confusing, especially when seen together with traffic signs. The signs in a busy commercial area that relate to driving should be designed to be readily seen and to give information quickly and clearly. Confusion with signs arises when an area has too many conflicting uses. The solution to these conflicts lies not so much in trying to control the signs as it does in removing the reasons for their mixture.

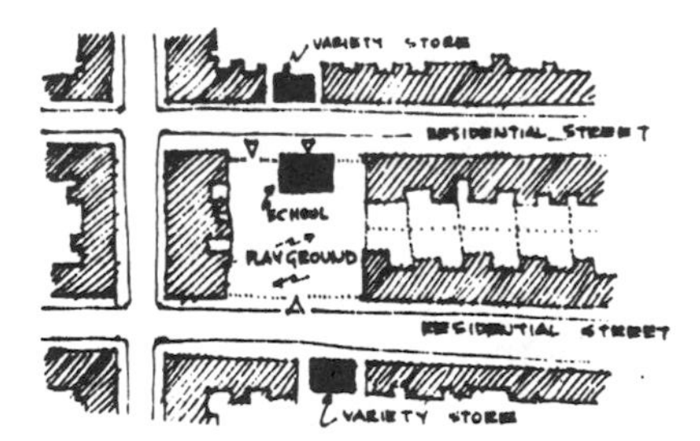

A very detailed analysis of a district can be helpful.

A visual survey of urban details should, therefore, include sign studies. More broadly, it includes the quality and conditions of park benches, wastebaskets, streetlamps, pavements, curbs, trees, fences, doorways, shopwindows, etc.—the street furniture and hardware of the city.

Pedestrian Areas

A large part of the difficulty in our cities arises because we have neglected the pedestrian. Walking will always remain a prime mode of transportation. Some areas of the city depend on it almost entirely as a means of communication and intermovement. Many new shopping centers and college campuses are models of design for pedestrian circulation. Older areas in the city need similar treatment.

We must be careful, however, in concluding that all the trouble

Central areas should be designed to accommodate pedestrians.

Bus terminals below, pedestrian area above. From the plan for Hook, *England.*

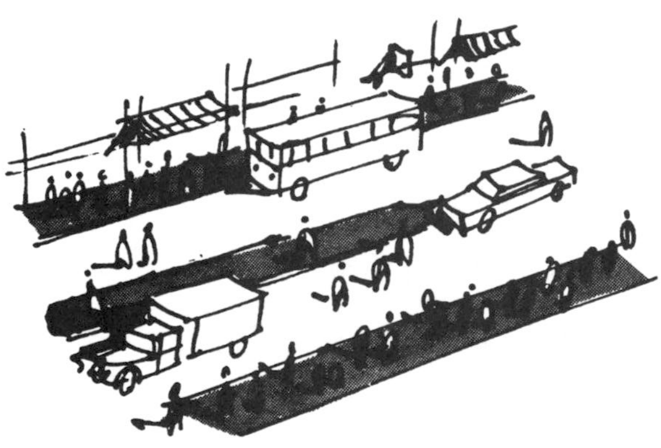

Pedestrian and vehicular circulation can be mixed if traffic is kept slow and islands provided.

In every city the pedestrian should be given primary consideration.

Vista from a tall building. Boston.

comes from the mixture of pedestrians and cars. Many city streets would be lifeless without cars. The problem comes when cars prevent the free flow of pedestrians. It is possible to have both—cars and pedestrians—in busy urban centers if the cars operate at very low speeds and if through traffic is reduced to the utmost.

A good way to check the quality of pedestrian movement in a busy area is first of all to examine the sidewalks for their adequacy—width, paving, condition, protection from rain and hot sun, and sidewalk outfittings such as benches. Second, one should walk through a pedestrian area taking several different paths to locate the main points of interrupted movement, generally speaking the intersections and crossovers. Too many intersections in city centers are designed to allow a maximum flow of traffic and to subjugate the pedestrian to long and annoying waiting periods. An answer to this problem is pedestrian safety islands and reduced-speed traffic. Pedestrian crossings should be frequent and convenient. A shopping street which is difficult to cross cuts the pedestrian-shop contact in half.

The ideal answer to this problem is the separation of cars and people onto different levels, the cars below and the people above. But this is impossible in most cities. In smaller towns it is possible to make the downtown area a pedestrian-oriented zone by providing a convenient bypass road for through traffic and by providing adequate parking garages around the downtown area itself. The essential approach to this problem lies in regarding a downtown area, or any center for that matter, as a stopping place and not a place for through traffic.

In outlying areas the same principles of pedestrian flow can be studied, to examine the pedestrian linkages between neighborhoods and their centers. Children should not have to cross busy streets, and the centers themselves should be at the center of a convenient walking as well as auto-shopping radius.

Vista and Skyline

Every city has a few striking vistas—of it and from it. Approaching Dallas, Texas, from the west, one sees a towering cluster of skyscrapers rising from the plains. The approach to Chicago along Lake Shore Drive is a dramatic urban entrance, as is the approach to New York City from the West Side Highway, and Salt Lake City through a pass in the Rockies. These are the major vistas of the city, and they must be protected from intrusion.

From the city, too, there are always a few dramatic outlooking vistas. Sometimes these vistas are modest, but still of great importance in characterizing the city. The slot views down the sloping streets of San Francisco afford fine vistas of the bay. Similar slot views down the side streets of Richmond, Virginia, afford glimpses of the surrounding countryside across the river.

The views into and out of a city are precious assets. They are an important part of an urban design plan. Some views of the city are in need of legal protection, like the shores of the Potomac across from Mount Vernon. Other views can be complemented by well-poised pieces of architecture, like the buildings of West Point on the palisades of the Hudson River.

An urban design survey should note the major views of the city and different points around the city, particularly points of approach. It should also note the major aspects of vista out of the city from points within. Evaluations should be made of improvements needed in both types of vista.

A further study can be made of a city's skyline. The city's skyline is a physical representation of its facts of life. But a skyline is also a potential work of art. An urban skyline is its collective vista. It is often the single visual phenomenon which embraces the maximum amount of urban form. Every building that alters the urban skyline should be studied for its effects on the overall view. Many skylines can be improved, particularly by adding a small counterpoint tower at an outlying location.

It is interesting to compare the visual effect of a cluster of towers with a single tower. A prominent single tower must be designed as a *chef d'oeuvre* if it is to be admired. It is too much on display to be mediocre. However, a cluster of towers, like a group of statues, can tolerate less distinguished design. If a tower is to be built at a prominent outlying location, it may be helpful to consider double or triple towers rather than a single shaft. Twins or triplets can be more modestly designed, and their profile as an ensemble can then be more assertive than the single shaft. On the other hand, an elegant profile may be easier to achieve with the single shaft.

Another aspect of vista and skyline is night lighting. Few cities are more dramatic as overall views than when seen at night. Twilight heightens the experience for it adds the drama of sunset to the unified view of the three-dimensional forms of the city. Colored beacons and the shafts of searchlights accent the scene with dots and thrusting lines. A new tower, particularly if it stands alone, should be studied from this aspect of appearance. The buildings of central Detroit are interesting examples of just such studies.

The night scene of the city, particularly its lights, is to a large extent within the control of the public. Every city that is building urban highways has a fine opportunity to add a ribbon of unifying illumination to the city's general appearance. The island of Manhattan is one of the most interesting cities to study from this point of view.

Manhattan can be seen on its two long sides and from the south across large expanses of water. It can be seen as a whole composition of masses and lights. At night the highways that ring the city are marked out by their evenly spaced and specially hued dots of highway lighting—ribbons of dotlike lights against a splash of electric stars. Further, Manhattan's several suspension bridges now have their catenary cables strung with lights. On a clear night several bridges can be seen simultaneously. The random lights, the even ribbons, and the gracefully curved bridge lights are one of the wondrous urban sights of the world. This illumination theme could well be used as a model in principle for every city. A visual survey of a city at night could suggest just when and how new lights could be added to achieve a similar composition.

Nonphysical Aspects

There are many nonarchitectural aspects of urban character: the New Year's Day parade in Philadelphia, the Rose Bowl parade in Los Angeles, Mardi Gras in New Orleans. These are a very large part of the image of a city and a large part of its personality. Architects can do much to improve the appearance and urban quality of cities by recognizing them and making better provision for them.

Every city has a history, linking it to its origin, and present in the minds of its population. Visible signs of that history can constitute a major aspect of its appearance. Architectural provision can be made for public ceremonies and events. In new areas the

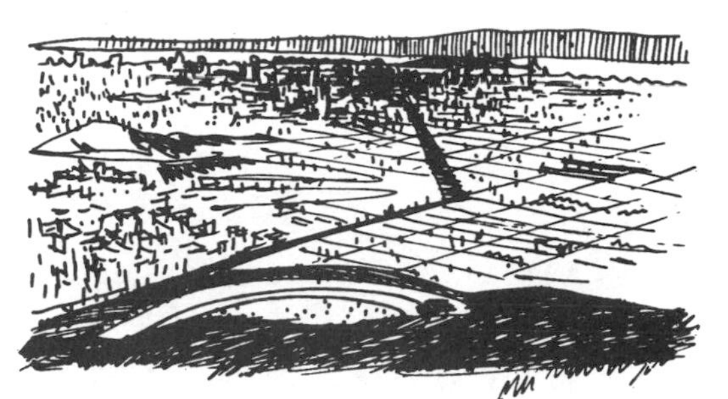

Vista from a mountain. Twin Peaks, San Francisco.

Manhattan at night. The outlined bridges are nocturnal landmarks.

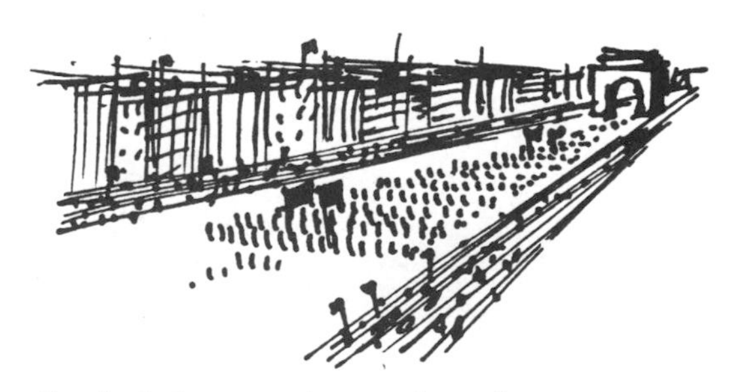

Nonphysical aspects. An annual parade.

inclusion of some visible symbols of the old city's personality give continuity and character to the new. Every city has a particular purpose which should be expressed architecturally. Boston is a center of learning as well as commerce. New York is a center of culture as well as finance. Miami is a center of leisure. Pittsburgh is a center of steel production. Detroit is our automobile manufacturing center.

A problem area.

Problem Areas

As a result of the visual survey, it is helpful to map the problem areas of a city alone. This map would stand as the urban design diagnosis of ills. It would show points of conflict between pedestrians and automobiles, areas with little or no sense of orientation, nondescript or gray areas, ugliness, communities lacking form and definition, areas with confusing signs, confusing circulation elements, incomplete routes, marred vistas, etc. As a diagnosis of ills, such a map would be a direct source of ideas for action programs.

Some Personal Techniques for Surveying

Anyone who attempts a visual survey of his city will undoubtedly develop his own technique and, very likely, a personal vocabulary. In our discussion of this subject we mean primarily to suggest the many different aspects of form that can be examined. But it is also important to be able to link all of the separate aspects of form into a chain of related aspects. One observer, therefore, might want to assemble his vocabulary of form elements into a coherent whole, such as the following: Paths, landmarks, nodes, districts, and edges are the skeletal elements of a city form. On that basic framework stand embellishing characteristics which all together constitute the personality of a city.

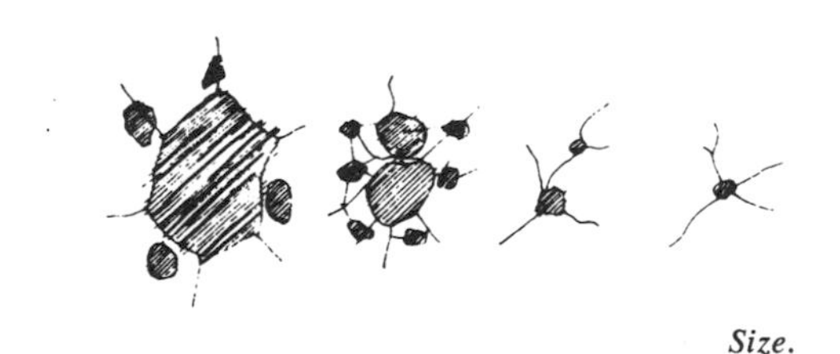

Size.

Suppose that we think of urban form in the following way: A city or town is generally thought of in terms of *size*—its population and physical extent. Size is closely linked to *shape*—the physical outline in horizontal plan form and vertical profile or contour. Size and shape are qualified by *pattern*—the underlying geometry of city form. Size, shape, and pattern are further modified by *density*—the intensity of use of land by people and buildings. Density is determined by urban *texture* and *grain*—the degree of homogeneity or heterogeneity of use by people or buildings.

Shape.

We can usually identify the parts of a city by their *dominant visible activities*. Often these activities are complementary, yet sometimes they are conflicting. It is important not to mistake complexity for conflict; complexity is the spice of urban life. The bustling urban centers are *magnets* of the city. People are the *generators* which require magnets around which to rally. *Feeders* are the links and paths which connect the two.

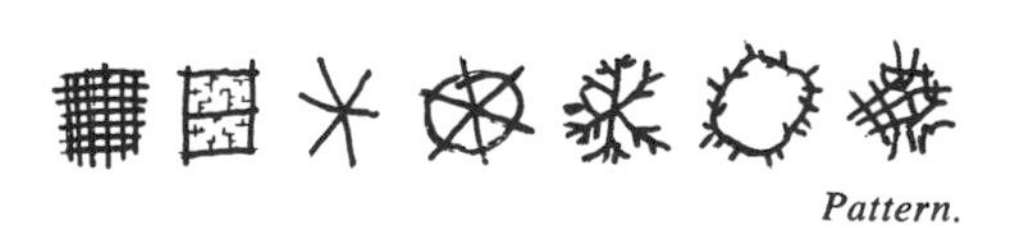

Pattern.

These areas of dominant visible activity exist in sequence as linked *accents*. The periodic occurrence of accents in sequence is rhythm. The disposition in a sequence has, of course, visible manifestations. Thus, accents in a sequence produce a *modulation of visual intensity*—varying degrees of richness of visual experience.

Density and grain.

Our use of the various parts of a city depends upon their degree of accessibility. Demands for accessibility produce channels of flow. Channels of flow vary in intensity, according to the time of day, week, or season, and thereby establish *patterns of movement*. Patterns of movement help define *districts* and act as links.

Visible activity, road signs, store signs, building signs, and symbolic objects are messages to us which convey purpose. They are *clues* to the organization of urban form.

The visual experience of a city is enriched by major *vistas*—views of large portions and major elements of the city, and of contrasting natural scenery. We are highly conscious of the nature of *land surface* (generally thought of as topography). We are aware of going up, going down, and the quality of the surface upon which we move. Natural landscape features form important *borders* or *edges* in cities.

Buildings are the immobile *masses* of a city. Arrangements of buildings form *patterns of mass.* Arrangements of buildings also form *urban spaces* which exist as *patterns of channels and reservoirs.* Entrances to a city can be accented by *portals,* the doorways to a city or a district in it. Pauses or relaxations in an intense area are *oases*—places inducing repose. They are passive accents which complement intense activity. Districts in a city are characterized by a pervading *continuity* of use, purpose, and appearance. Some districts are oriented to particular types of people or particular age groups. A fine distinction can even be made between masculine and feminine districts.

Our knowledge of these visible phenomena, the presence of visible landmarks, pattern, shape, etc., imparts a sense of *orientation*—a sense of where we are and where things are in relation to us. A sense of orientation is basic to our understanding, familiarity, and well-being in a city. We are conscious of the *age* of a city and its parts, the newness and oldness in buildings and places. We must avoid the danger of equating oldness with decay, or newness with amenity. In his work, the urban designer must transcend time and relate all parts of the city to each other. A major objective of urban design is to relate different kinds of buildings, regardless of differences in architectural style, age, or use.

Rhythmic sequence of accents.

Dominant visible activity.

Magnets and generators.

Recording the Results

Visual surveys are most readily recorded as simple maps accompanied by sketches, photographs, and brief notes. The maps can be base maps of the city, at the scale or scales of the survey. The sketches, photographs, and notes can be attached to the maps and the whole study put on display or published as a report. The maps and their notations are best done in a cartoon style and notations of certain features best indicated as a graphic symbol.

Routes of movement can be indicated by arrows, parking garages as a spiral, landmarks as large X's, vistas as sector lines, points of conflict in red, "gray areas" in gray, etc. One map should show the sum total of the general form of the city and its features. The remaining maps should complement this as a series of detailed aspects of the city's form.

A full set of survey maps might include the following:

1. Topography
2. Microclimate—sun, wind, and storm directions
3. Shape
4. Patterns, textures, and grains
5. Routes
6. Districts
7. Landmarks and nodes
8. Open spaces

9. Vistas
10. Magnets, generators, and linkages
11. Special activity centers and overall activity structure
12. Hubs of intense visual experience
13. Strong and weak areas of orientation
14. Sign areas
15. Points of conflict
16. Historic or special districts
17. Community structure
18. Areas for preservation, moderate remodeling, and complete overhaul
19. Places needing clarifying design elements
20. Sketch maps produced by the "man on the street" to discern the urban features and forms prominent in the public's eye.

Each of these maps should be illustrated by a few salient sketches or photos that show exactly what a map symbol represents and also a series of pictures which characterize the area.

Conclusion

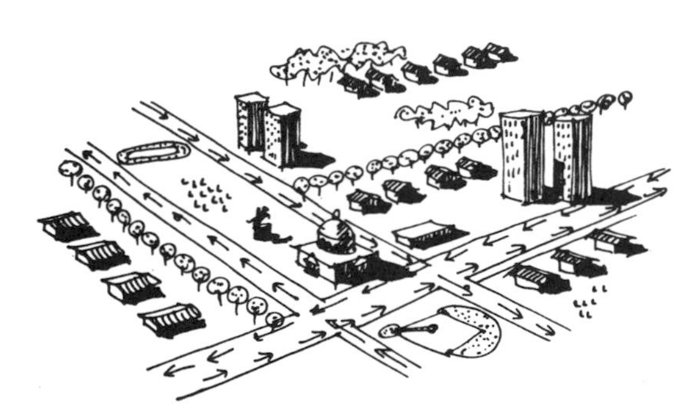

Whatever your survey technique, the results should be presented in everyday terms.

Music can be described and discussed with considerable precision and insight because it has a vocabulary and a body of literature. At present painting also enjoys this advantage. So does architecture, to an extent. Up to now the complex modern city has lacked a precise vocabulary for discussing its form and appearance. If we formulate such a vocabulary we will be able to discuss urban form with clarity. We will also be able to discuss the effects of various actions and policies that affect the city in terms of its buildings, parks, streets, and places. Therefore we will be better able to discuss its design.

While developing this language and engaging in conversations on urban form, we must avoid overly abstract terms. Unlike the complexities of modern music or painting, the city is familiar to everyone and can always be described in simple terms. That may be the best test of any vocabulary.

Robert Regis Dvořák

Summary
The *Piazza di Spagna* and the *Scala di Spagna* (Spanish Steps) are among the few and truly unique examples in urban design where stairs, by themselves, serve as a visual and spatial focal center. The sequence of space one experiences in this Late Baroque piazza and stairway (1721–25) by Alessandro Specchi and Francesco dé Santis, with its 137 steps, is a significant city space in central Rome. The sketchbook—using visual notation and drawing to explore formal, spatial relationships, details and activity of urban places—provides the urban designer a method of observation, documentation and understanding.

Key words
Rome, sketchbook, Spanish Steps, visual analysis

Sketchbook: Piazza di Spagna, Rome

The Spanish Steps provide an urban stage setting for innumerable activities, including a small flower market, a familiar meeting place in central Rome especially for foreign visitors, and a hub of pedestrian and vehicular traffic. There are a great variety of urban vistas, changes in elevation, and plenty of urban life, with a range of short and long vistas, views from below and above, details that bear close examination and light, shadow, texture and color.

In the process of making sketches, one becomes aware of the interconnectedness of all elements of urban life, climate, form and space. Here, the space was explored at all scales, using city maps, aerial photographs, historical information, and most of all experientially by making many sketches on site and from photographs. What makes this particular spatial and visual experience so unique is the way the stairs and the *piazza* are connected. The stair is the most important element and the *piazza* is relevant only as an adjunct to the stairway. City streets lead to the top and the bottom of the stairway. This outdoor city space has extraordinary richness and life, due to both circulation and the changes in level. It provides any number of choices to just sit—pick any stair tread—and read, talk, sketch, or just observe the urban scene.

Credits: The sketches are from studies made possible by a Rome Prize Fellowship in Environmental Design. 1970–72, The American Academy in Rome.

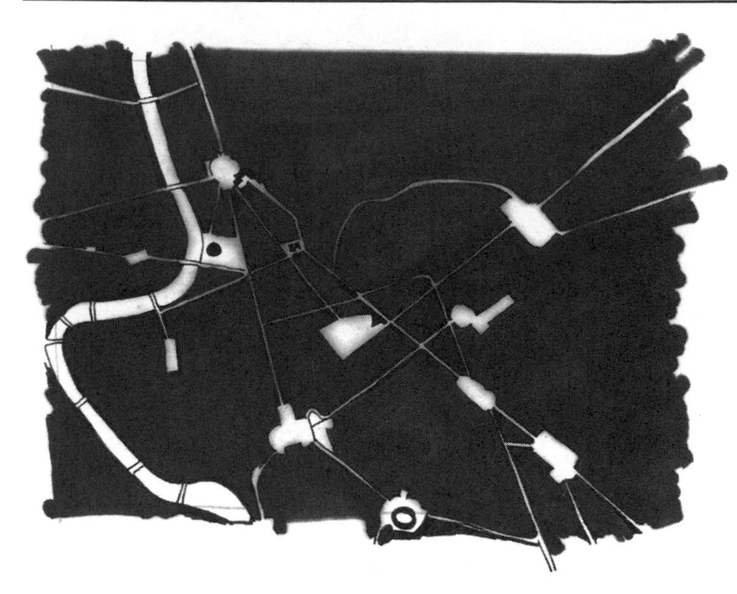

Most of these sketches were made with a soft pencil. Soft pencil sketches are difficult to keep from smudging and also more difficult to reproduce, so that I now work mostly with a pen. When I travel, I carry a sketchbook and usually a small camera. Sketching has provided many more hours of enjoyment and learning than with a camera. Any student of urban design will find in sketchbook drawing an entirely enjoyable method of note-taking, while at the same time sharpening one's observational skills and understanding of place. ■

REFERENCES

Dvořák, Robert Regis. *Drawing Without Fear, The Magic of Drawing,* Montara, CA: Inkwell Press. (www.youcreate.com)

Dvořák, Robert Regis. *Experiential Drawing,* Menlo Park, CA: Crisp Publications. (800-442-7477).

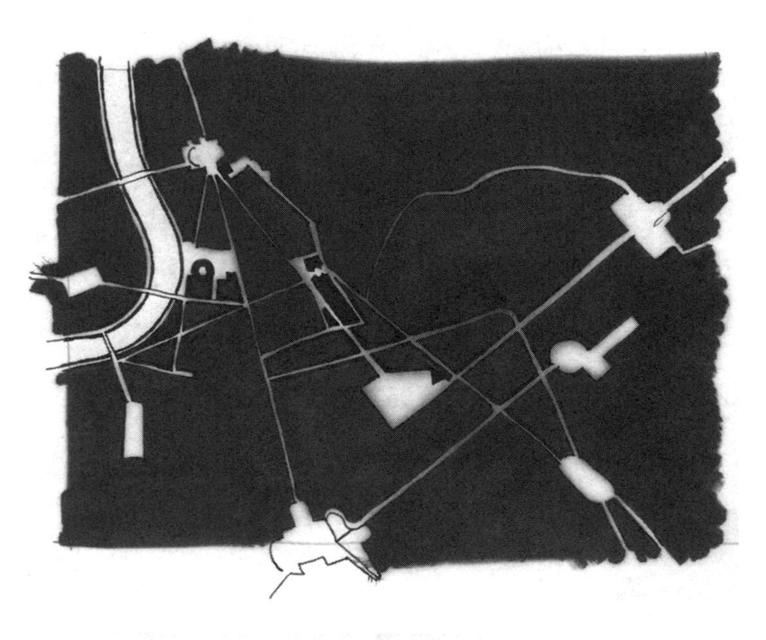

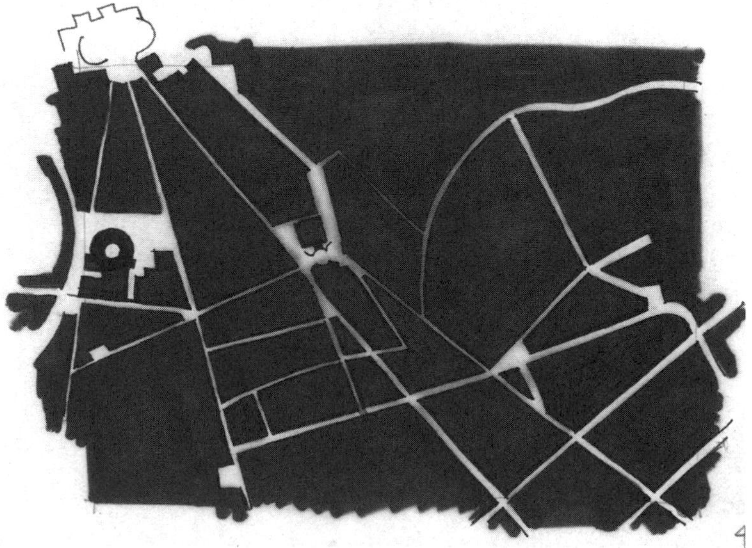

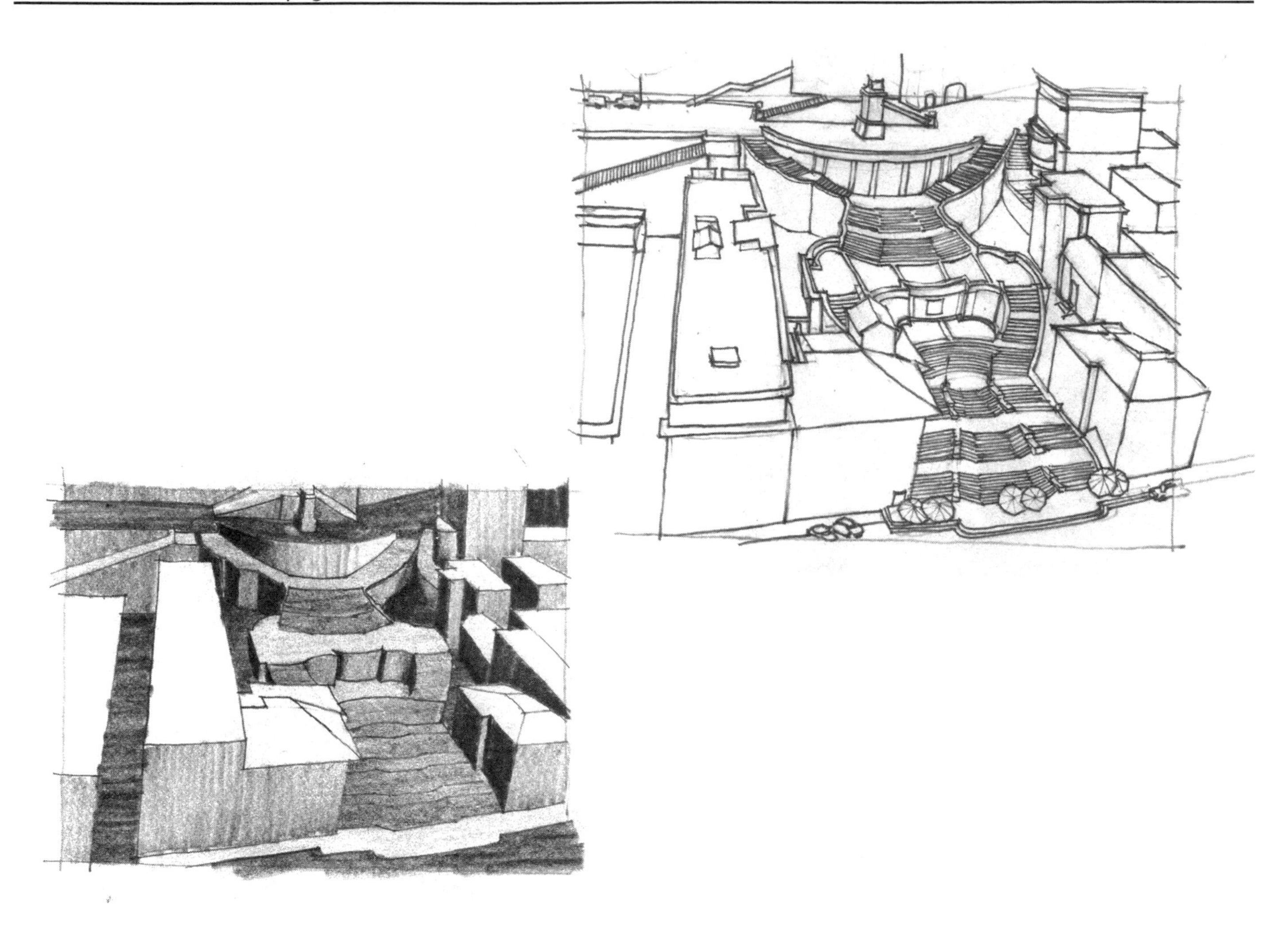

Nahoum Cohen

Summary
This article presents the case for comprehensive approaches to urban conservation. A methodology of assessing conservation potential is described and illustrated. Urban conservation provides a set of values and methods by which to preserve and renew the important elements of the city, from which it derives its unique sense of place. Planning for urban conservation assures that the cities of the world—with unique structures, places and districts—will be sustained in their irreplaceable role as the realm of vibrant life, culture and civil society.

Key words
conservation, district, historic preservation, infrastructure, methodology, public realm, regulation, street, urbanity

PHOTO: Donald Watson

Fig. 1. Naples, Italy. Galleria (l) and Opera House (r). The urban way of life is a unique human achievement. It enjoys a multinational consensus and reason for living in cities. Public space is sieved through the built volumes as to become an almost palpable entity, the defining quality of urbanity.

Planning for urban conservation

1 OVERVIEW

Historic city centers are made up of buildings, infrastructure and streets representing different periods that create various cultural and urban layers. Over time, their urban character is finely honed and offers irreplaceable qualities of urban culture absent from most new cities and suburbs.

Urban conservation is an approach to planning that seeks to preserve and as often necessary revitalize and transform the historic elements of cities into hubs of urban life and activity Piece-meal or "crisis driven" and reactive approaches to preserve structures only when they are threatened is always an inadequate response, even if resulting in one isolated success. Urban conservation is a long-term commitment to maintain a city's cultural and historic identity while also accommodating inevitable transition, growth and new uses (Figs. 1 and 2).

The aspects of historic preservation that address only a single building —focusing upon structural or architectural considerations dealing with its use, age or fitness—are only a small part, best seen as an outcome, of planning for urban conservation. Urban issues properly "begin" where the typical architectural considerations end. Urban conservation deals with questions of context that underlie a building's existence, including issues of ownership, land division, private and public property, the arrangement of urban space, including change of use, and their conservation by maintenance and adaptation through time.

Ownership

Public property and public policy considerations play a major role in the urban aspects of conservation. Public property, including urban infrastructure is a measure of cultural wealth, cannot be left to decay.

Fig. 2. London. Urban conservation in London (and throughout England) is the result of a long respect for building tradition. The concentration of the effort to conserve the common heritage is on a continuous scale, and not sporadically spread out in single examples. The conservation rules, regulating of heights, proportions and detail, allow adaptation.

Credits: The article is excerpted by the author from his publication. *Urban Conservation,* McGraw-Hill, 2002. Photos and illustrations are by the author, unless otherwise noted.

Fig. 3. Rome. The urban role of archeological conservation is almost self-evident by the Coliseum, Rome. In spite of some planning and circulation problems, there is no denying the importance of such a major historical presence to life in the city. At the same time, the city works around a building that cannot function too often or too well. (Photo courtesy of World Monuments Fund.)

Fig. 4. Jerusalem. The recognition and identification of an existing situation is exemplified in the conservation in Old Jerusalem. Urban conservation has established well-controlled development, with modern building restoration and adaptive use methods integrated with the existing urban web, developed inside existing city walls and having a somewhat anarchic structure. The established heights, materials, logic of urban form are preserved within existing squares.

The public entity of municipality or city normally shoulders a substantial portion of the financial burden of urban conservation.

Historical values

Conservation not only deals with modern urban factors, but delves into the past, often looking toward archeological findings and historic records for guidance and inspiration. In historic sites, archeological discoveries at sites influence the course that subsequent conservation and development may take.

Scope of preservation

The single building and protected archeological site both have limited scope in terms of planning for urban conservation. Conservation efforts will not be limited to small areas or individual neighborhoods, leading to an increase in the scope of conservation planning.

Coherent urban culture

The aim of urban conservation is to promote urban life characterized by a strong sense of historic and cultural vitality and continuity. Esthetic qualities, which will hopefully also be preserved, are not sufficient to achieve this goal. The values of day-to-day urban qualities, economic, social and environmental are the underpinning of the urban scene and must be rediscovered: the focus cannot be on monuments alone.

Issues of conservation

Urban analysis helps define the qualities that maintain the urban way of life. Urban conservation is an approach to aid in maintaining those cultural values. Analysis undertaken for urban conservation planning first seeks to gain an understanding of the urban web—the vital links of a city and its infrastructure—and how it came into existence (Figs. 3 and 4).

Analysis of the existing situation

- The urban situation must be thoroughly documented and analyzed accurately recording aspects that deserve and require preservation, and how these factors relate to other less obvious factors.

- When preserving a single building, it is important to understand what authority will preserve the urban elements, both for the building and then for its essential and supportive infrastructure. Where essential networks have been abridged, such as circulation, access, visibility, and surrounding uses, conservation planning seeks to identify these necessary factors and how they might be reestablished.

Urban space

The most evident feature of urban planning is space, usually public space. Urbanism may be defined as the result of a particular formation and interrelationship between public and accessible private spaces (in other words, what one may see from the street). Urban space creates reciprocity with the surrounding forms, activities, buildings and places, where all types of successful urban life spring forth and flourish.

Urban context

A commonly cited failure of mid-20th century. architecture has been overemphasis upon individual buildings as works of art, at the expense of the building's context. Urban conservation seeks to maintain and revitalize the weave or web of infrastructure, the functioning and arrangement of circulation, streets and public spaces, as the necessary context of urban life.

Urban regions

There are many variables involved in planning at urban and regional scale: the viability of regional transportation, circulation to the urban center and parking, the level of stability and permanence vs. growth and change of businesses, the proportion between public space, constructed volumes, building density and heights.

The level of conservation

Before approaching building renovations or physical reconstruction—called often the "restoration system"—it is essential that there be a clear conservation policy. The policy serves as the foundation on which other systems, such as restoration, rehabilitation or renovation (in this order) are based.

Conservation at the urban level and context is undertaken by identifying primary urban prototypes and establishing criteria to conserve these prototypes. This refers to urban elements, of which the individual building is a by-product. Urban conservation includes all the desired connections between these elements:

- The web and its external links
- The blocks within the local web
- The streets
- The pedestrian realm of walks, squares, etc.

Conservation on the architectural level is based on this context and identifies the architectural characteristics to prevent stylistic distortions and disruption of the existing architectural language. This can include structural renovation or rehabilitation, and in more extreme cases, restoration, generally more expensive.

2 URBAN STRUCTURE

Defining the underlying physical structure of a city (urban settlement) is prelude to planning for urban conservation. The urban conservation planner seeks to discern formative phases in the development of ancient cities, as a background to the formation of newer ones. It is worth noting, however, that the difference between old cities and new ones is not dramatically significant. Their differences lie in the level of complexity of new cities as well as in their physical size, the large number of various webs within the new city and the diverse specialization in the level of land designation and use (Figs. 5–9).

Successive phases of urban formation can be listed in somewhat inverse chronological order, despite the fact that throughout time they generally built up from small to large scale.

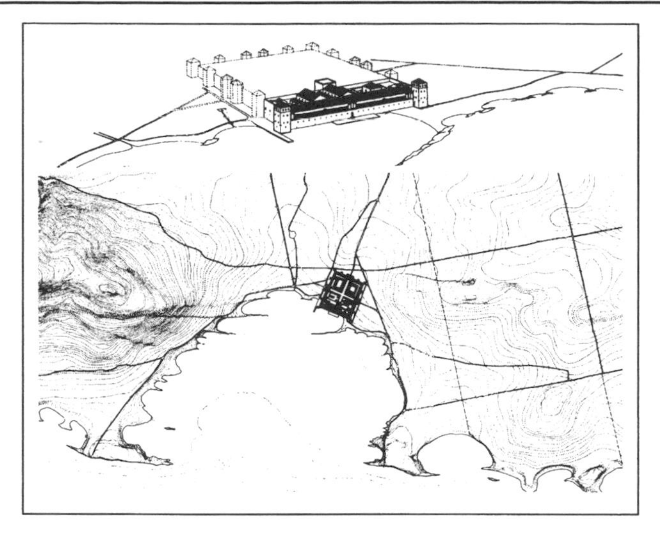

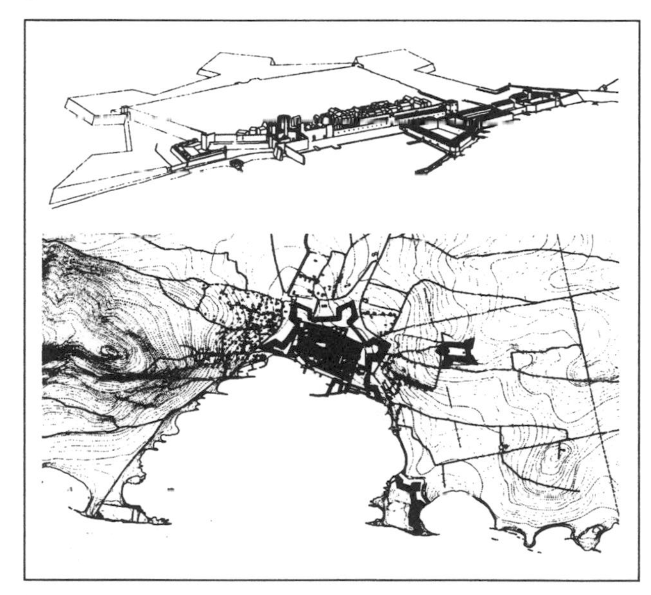

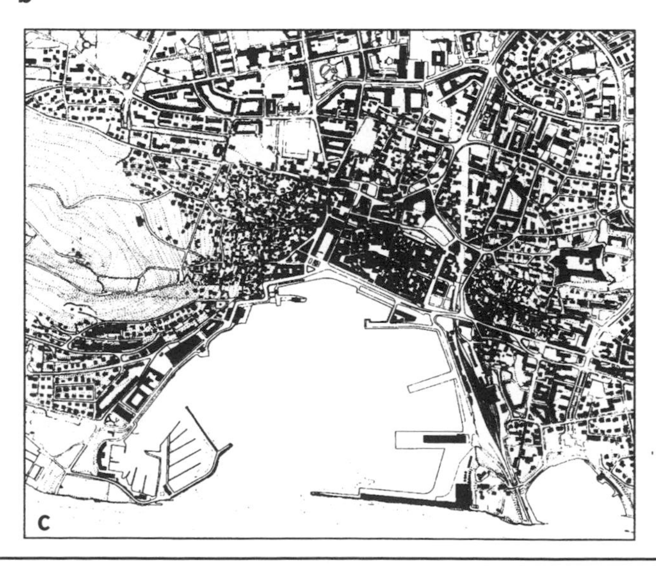

Fig. 5. City of Split, Croatia. (a) Spatial development, 4th century.; (b) Split Riva 1675; (c) spatial development 1985. (Illustration J. Marasovic in Armaly *et al.* 2001.)

Fig. 6. Carlovi Vari (near Prague) is an apt example of urban conservation and historic preservation. The vivacious clean colors render a somewhat sweet effect, oftentimes indicative of an overzealous conservation effort.

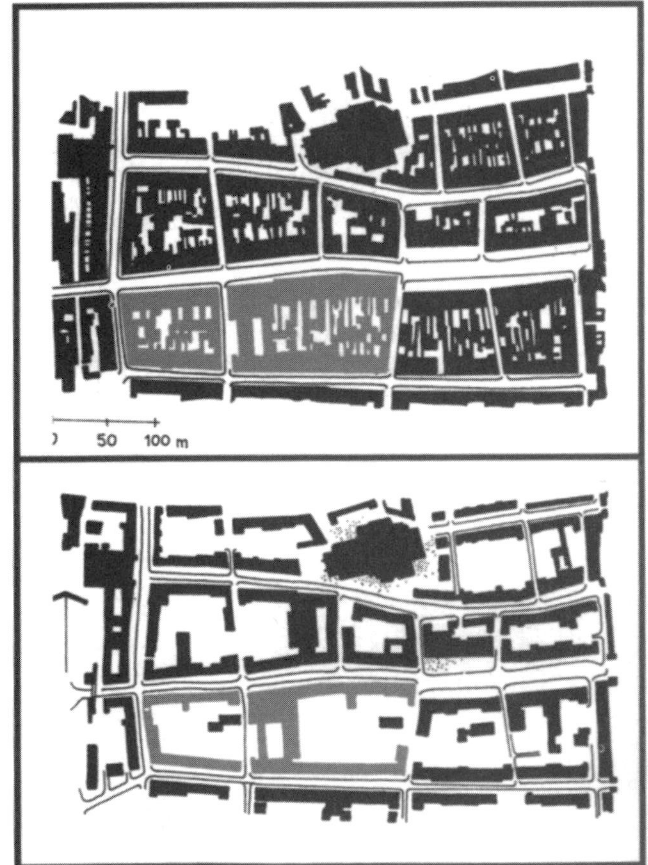

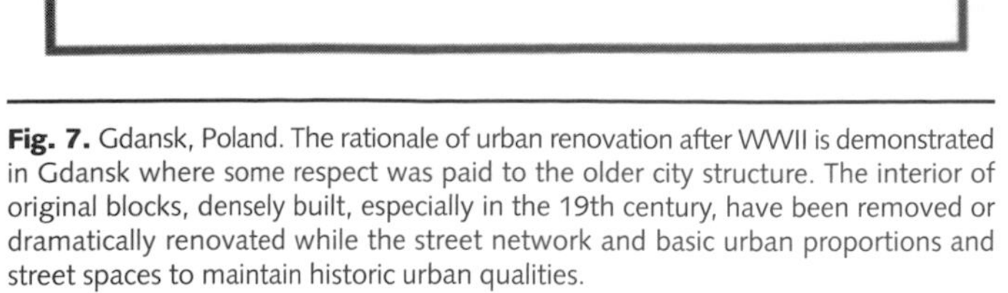

Fig. 7. Gdansk, Poland. The rationale of urban renovation after WWII is demonstrated in Gdansk where some respect was paid to the older city structure. The interior of original blocks, densely built, especially in the 19th century, have been removed or dramatically renovated while the street network and basic urban proportions and street spaces to maintain historic urban qualities.

Fig. 8. Milan, Italy. Milan's center contains a closely knit and slightly ill-defined city structure. As one moves farther from the city center, the blocks become more rational and orthogonal, resulting in a better-defined web.

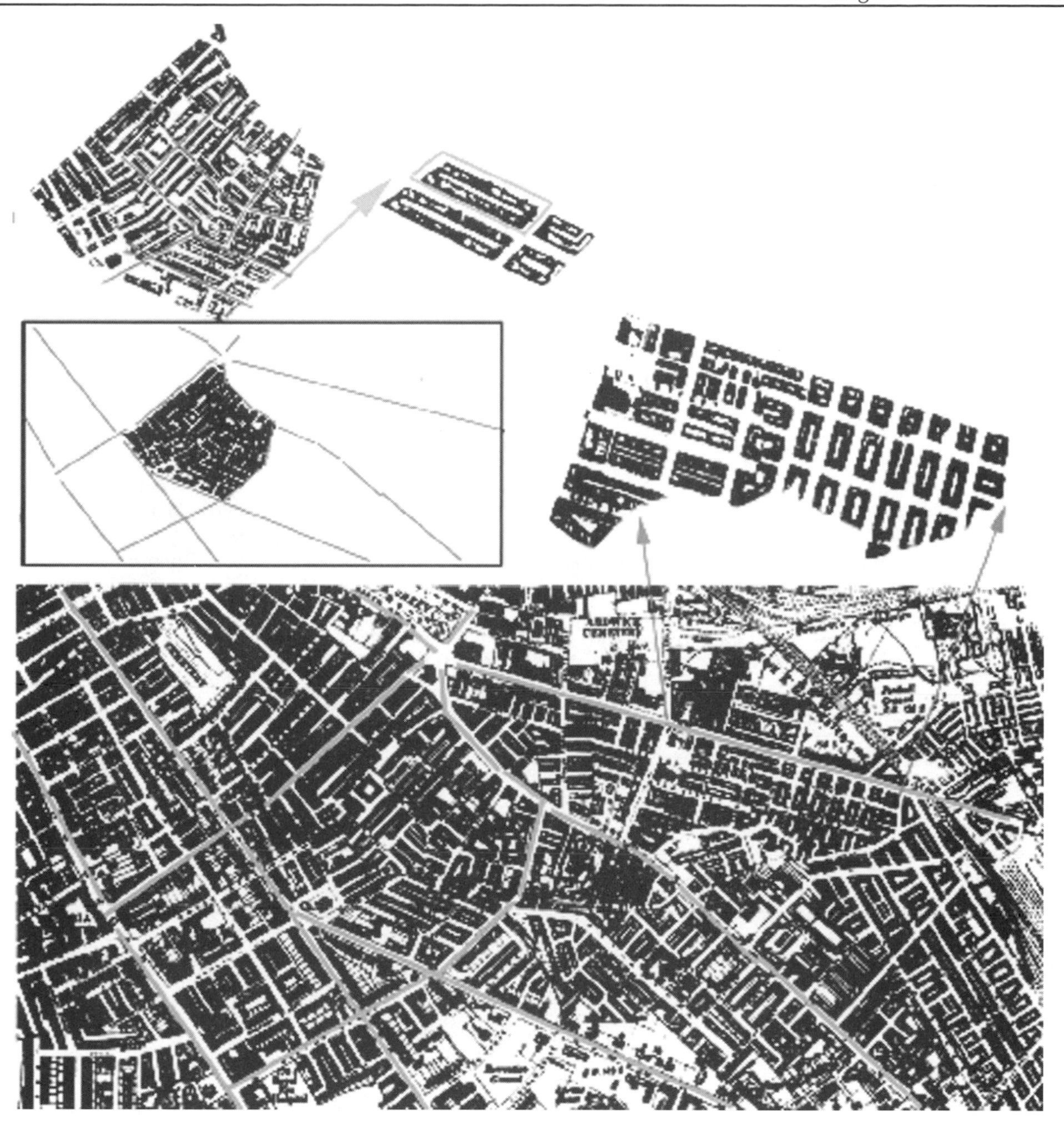

Fig. 9. Manchester, England. The urban patterns in Manchester do not offer such an abundance of features as Milan's (Fig. 8). Major and secondary roads create districts with common structural elements. The resulting definition of the internal blocks of the district can be simply understood. The block is defined by repetition, size and geometrical direction, and by regularity of plot dimensions.

A. Urban web

The urban web is the most general view of a settlement. The series of public passages create a unique format or structure that is physically and geometrically defined. The most prominent of those passages (roads and streets) can be seen as arteries that nourish the district, including its various buildings. Once the urban web is in view, the regularity of the emerging web comes into focus. The primary definition differentiates between groups of various geometrical webs (*e.g.*, orthogonal) and webs that lack a prominent geometry, appearing to be random. Other definitions relate to issues such as building density, repetition and heights.

The district

When approaching urban structures, one distinguishes features that are differentiated by internal qualities, despite the fact that they may merge. These differences are the formative nuclei that establish local features—variations in structure such as the size of the blocks and the distance between them. These formally different town structures can be considered "districts," definable as an uninterrupted accumulation of similarly featured groups (physical, restorable, symbolic, historical). Districts are in fact a type of neighborhood. Each district incorporates various properties within its structure. It is important to recognize the limits of each district, the links to its surroundings and city, as well as the extent of change or possible change in the future.

CONSERVATION POTENTIAL					URBAN MIX
CHARACTER DEFINITION	LOCALITY SENSE	INTERNAL RELATIONS	STYLE & DESIGN	METHODS & MATERIALS	
					LAND
					BUILDING
					USE

Fig. 10a Conservation potential. Matrix form for assessment of the urban conservation potential of a site.

URBAN PROFILE RELATED TO THE URBAN MIX						
SITE:						
URBAN MIX	STRUCTURAL ELEMENTS					
	WEB	DISTRICT	BLOCKS DIVISIONS	PRIMARY ELEMENTS	SECONDARY ELEMENTS	NATURE
LAND	regular					
BUILDING	3 stories					
USE	housing					

Fig. 10b. Structural potential. Matrix for structural assessment. "Primary" urban elements are streets and squares. Secondary ones are monuments, special districts, etc. "Nature" is comprised of open space elements such as parks and recreational areas.

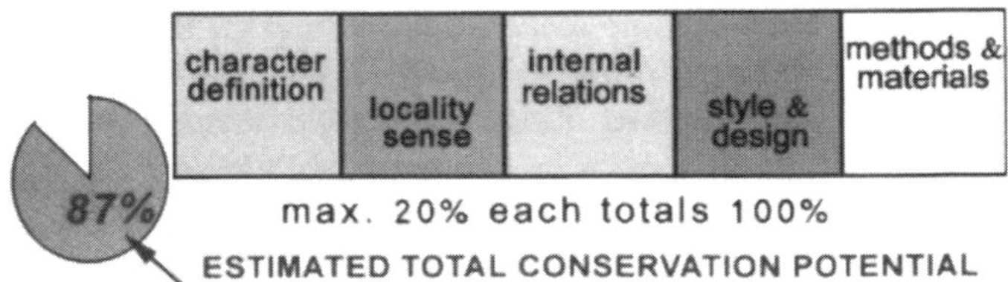

Fig. 11. Example of the pie chart summary, indicating how design style might be assessed in a particular case.

B. The block

A block is a coherent formation composed of smaller physical units, and a predecessor of the ones discussed. The physical continuity of parcels or lots, buildings and divisions, *i.e.*, the block, is surrounded and defined, by public passages *i.e.*, roads. This demarcation creates identifiable sections of land, which are, for the most part, privately owned (not by the city). These are the main characteristics of a block. Shared borders and relationships between the plots, *i.e.*, land divisions, as defined by law and planning codes, are created within the block.

The block is rarely, if ever, dissected by public passages. The block is more cohesive than a neighborhood, because it contains joint ownership and a proximal legal relationship between the factors that make it.

C. Division of the block

Block divisions create the smallest units of land division (into parcels or plots) defining building limits and often causing some repetition of building form. This arrangement or regularity is based on the needs, neighborhood relationships, laws and mutual agreement. It is deemed successful if it withstands the test of time, otherwise it disappears.

Improved and more sophisticated urban construction codes that do not ignore human and economic factors serve to guarantee the unit's existence. In the following discussion, the block is divided into the smallest possible unit—the land parcel.

The legal status of owner registration defines the long-term resilience of this basic formative urban unit. On the parcel of land, within a geometrical framework, buildings that house people and activities are constructed. With the passage of time, the building, as a unit, becomes more vulnerable than the parcel of land on which it rests and undergoes various permutations.

The chronological phases of urban formation can be summarized as follows:

- Humans settle within a land division framework, to create a building within it. This framework has its first land units attached to existing public service routes. Nearby neighbors will act similarly, forming the block.
- Neighborhoods establish a collection of blocks, containing unique internal relationships and shared borders. The block can be duplicated by similar, adjacent units, which are also nourished by even larger public roads. Concurrent growth of this type promotes urban formation.
- Sections of adjacent, similar blocks create the next formation—a type of neighborhood with a structured nature, a large unit containing internal urban wholeness. This is known as the district. Within the district operate advanced community functions, justifying the close proximity of the blocks.

The various blocks, each with its own reason for existing, move closer as they increase in size and create an urban language, though not of their own conscious doing. This urban language (a type of meta language) is capable of unifying the design style and character of the entire neighborhood while creating a generality, which appears as a

Fig. 12. Venice, Italy. A renowned example of urban conservation, with a score of 90%, indicating a very high conservation potential because of the city's uniqueness.

strong internal cooperation. These patterns have their sources of being and identity, including primary proportions, geometry of the section, distances and spaces, all of which represent a challenge in urban conservation. Such abstract issues, formative in nature, must be judged with the help of planning analysis of the underlying factors of urban structure, not only by appearance.

For this reason, urban conservation begins with planning and building codes. It is impossible to separate planning and conservation in establishment of a continual urban structure.

3 ASSESSING THE POTENTIAL FOR URBAN CONSERVATION

To guide urban conservation planning, a method for making judgments, while subjective, gives a basis of comparison to other, similar sites, as well as some rationale for aesthetic and historical value assessment.

The Conservation Potential Matrix (Fig. 10) provides a measure of the potential conservation quality present in a given site (Figs. 11–17). Five criteria and questions to ask in assessing the urban conservation potential of a particular site are:

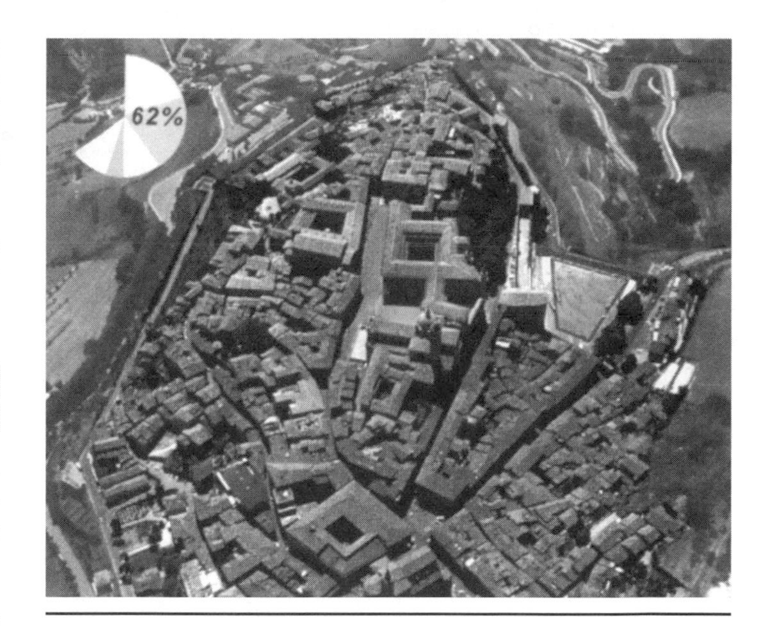

Fig. 13. Capua, Italy. Characteristics of the hill town—clear borders, good character definition, and a very good sense of locality—earn a positive assessment. Very restricted public areas, an overly dense and highly impacted web and internal volumes limit the potential for urban conservation beyond the building scale.

1 Definite character of the urban setting and clarity of the border of the site

This is a measure of the extent to which boundary and structure are recognizable as urban elements, such as city squares, parks, side streets and elements of nature. The region is given to definition by clear, physical borders that are well marked. If the physical definition has lent uniqueness to the site, it can be identified and qualified as the subject of a subsequent and detailed analysis.

2 Locality and sense of place

This is a measure of the site's regional and local character, atmosphere and its urban spaces, with links to the context of the city. Historical associations, scenic views, fusion with some defined issue, shade, comfort, feeling of local importance, vegetation and relation to the topography, are all contributing factors.

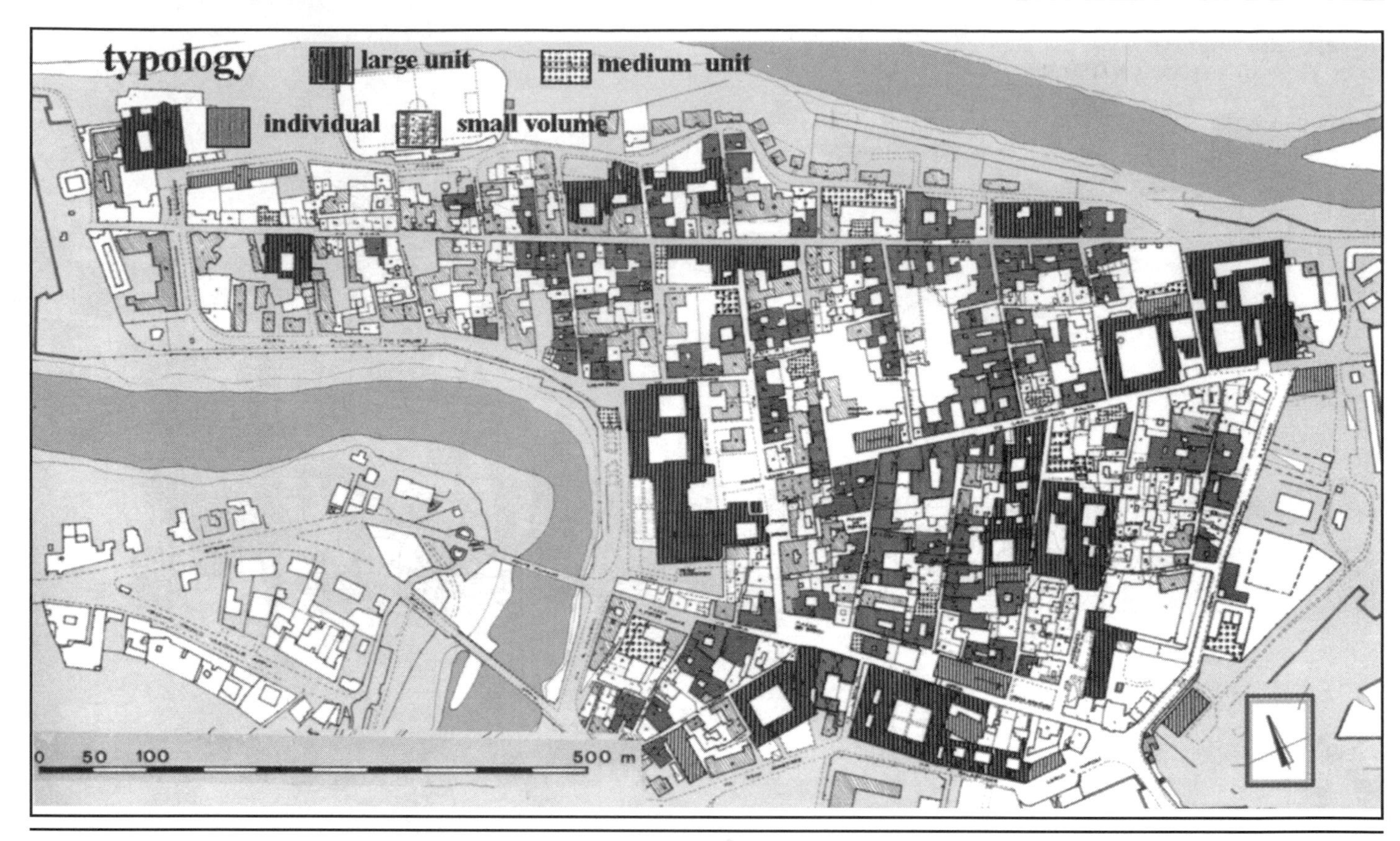

Fig. 14. Capua, Italy. Building typology, one of a number of factors considered in the analysis used as the basis of a building conservation program. (Brock, Giuliani, and Moisescu, 1973.)

URBAN PROFILE RELATED TO THE URBAN MIX

SITE: The town of Capua

CONSERVATION POTENTIAL					URBAN MIX	STRUCTURAL ELEMENTS					
CHARACTER DEFINITION	LOCALITY SENSE	INTERNAL RELATIONS	STYLE & DESIGN	METHODS & MATERIALS		WEB	DISTRICT	BLOCKS DIVISIONS	PRIMARY ELEMENTS	SECONDARY ELEMENTS	NATURE
3% +3%	5%	5%	1%	0%	LAND	geometrical !!		courts !!	streets		river
7%	6%	4%	4%	4%	BUILDING		similar method			churches	
4%	3% +3%	0% +5%	0%	2%	USE				0 mixed !!		ancient road
LEGEND: Total Potential — Present 48%, Future 59%; Present state %; Upgrading +%; Sub total 7% max						LEGEND: 0 ! !! !!! Degree of influence on conservation; STATE AT PRESENT; STATE IN FUTURE PLAN					

Fig. 15. Capua, Italy. Urban profile assessment of urban conservation related to the urban mix.

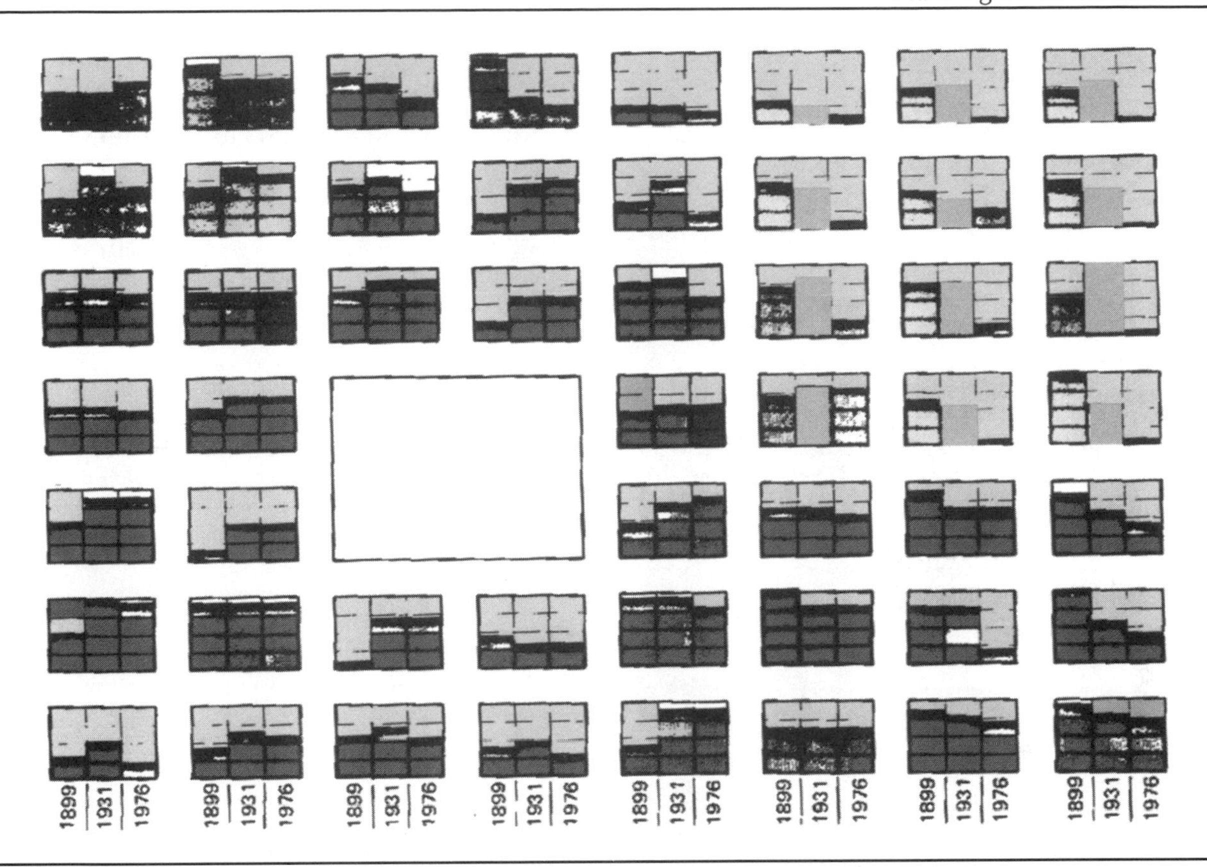

Fig. 16. San Francisco, CA Alamo Quarter. Analysis of long-term trends of the character of urban blocks. The tabulation indicates the decline between 1931 and 1976 of the number of dwelling units per block. High vacancy rates created a marked decline in the northeast (upper right) corner of the neighborhood. As individual houses were abandoned, these blocks were consolidated into single ownership, cleared of the traditional structures and subject to new out-of-scale projects. In spite of its inherent potential, urban renewal was not guided by conservation planning, allowing uncontrolled mixed use, with resulting decline and loss of historic character. (Anne Verdon Moudon, *Built for Change,* MIT Press.)

URBAN PROFILE RELATED TO THE URBAN MIX

SITE: Alamo, San Francisco

CONSERVATION POTENTIAL					URBAN MIX	STRUCTURAL ELEMENTS					
CHARACTER DEFINITION	LOCALITY SENSE	INTERNAL RELATIONS	STYLE & DESIGN	METHODS & MATERIALS		WEB	DISTRICT	BLOCKS DIVISIONS	PRIMARY ELEMENTS	SECONDARY ELEMENTS	NATURE
7%	6%	3% +4%	5%	5%	LAND	!!	! !!	0 !!			!!
6%	4%	0 +4%	4%	3%	BUILDING				! !!		
4%	1%	0 +3%	2%	–	USE					0	

LEGEND

Total Potential	
Present	Future
50%	59%

Present state: % — Upgrading: +% — Sub total 7% max

LEGEND

0 | ! | !! | !!! — Degree of influence on conservation

STATE AT PRESENT / STATE IN FUTURE PLAN

Fig. 17. San Francisco, CA Alamo Quarter. Urban profile related to the urban mix.

PHOTO: Middle East Centre

Fig. 18. Charleston, SC. The original city plat of narrow street front lots—a pecuniary response to property taxes calculated by length of street front—plus local building tradition and climatic responses have resulted in a coherent ordering of the city's urban structure and fabric.

PHOTO: Donald Watson

Fig. 19. Sana'a, Yemen. Multi-storied highly decorated houses are an architectural heritage and subject of urban conservation and preservation effort.

3 Internal space, proportions and relations

This is assessed by the connections of urban space created by the volumes of the built environment, but also internal continuity of function and uses. Sites examined will have to prove recognized relations between structures and spaces, guaranteeing the understanding of the environment, at a public and urban advantage. The internal relations imply special proportions between the various built volumes and the spaces thus created. In this criterion, the urban space has to be fully appraised.

4 Style and design

This provides a means to record evaluation of the overall design approach, character and style prevalent at the site, comprised of buildings, land and uses. The site will have to reflect a good level of mutual complementation as related to architectural style and design of structures. Here, due to overall considerations of design, the immediate esthetic rationale will come into play. Design involves proportions and silhouettes, not only details or artistry.

5 Construction methods and materials

This is a measure of the level of performance achieved by an authentic building technology. A high degree of processing and use of materials according to architectural specifications is apparent, with the ability to implement a defined building technology.

The summary matrix contains the full range of criteria and factors of an existing situation, in the form of an urban conservation profile. The right side contains the structural elements, reflected in the urban mix, and recognizable as factors in existing planning statutes, codes and zoning. The left side contains the relevant sum of the conservation components. On the right, decisions as to what urban components to keep, in a hierarchical order and in detailed urban mix. On the left, one can mark the qualities that need reinforcing, changing or elaborating upon.

The result of these deliberations will be marked by "change" (from present to future), which calls for a change in planning notions. The change may occur either in the structure of the site, which is a planning recommendation, or in the conservation makeup or upgrading, which has a more complex role. It is precisely these changes that become the input of conservation into the statutory planning procedures.

4 SUMMARY

The rarity and historical importance of historic building and urban values can serve as a lever for economic success and renewal. On its own, urban conservation cannot solve a city's economic hardships. This requires a great deal of public investment and combined private-public stewardship.

Urban conservation provides a set of values and methods by which to preserve and renew the important elements of the city from which it derives its unique sense of place and that often represent the world's multicultural heritage (Figs. 18 and 19). Planning for urban conservation assures that the cities of the world—with unique structures, places and districts—will be sustained in their irreplaceable role as the realm of vibrant life, culture and civil society. ■

REFERENCES

Armaly, Maha J., Stefano P. Paglioa, and Alain Bertaud. 2001. "Economics of Investing in Heritage: Historic Center of Split." in Serageldin, *et al.* 2001. pp. 165-179.

Brock, Giuliani, Moisescu, eds. 1973. *Il Centro Antico di Capua.* Italy: Marsilio Editori.

Kostof, Spiro. 1991. *The City Shaped: Urban patterns and meanings through history.* Boston: Bulfinch Press: Little, Brown & Co.

Serageldin, Ismail, Ephin Shluger and Joan Martin-Brown, eds. 2001. *Historic Cities and Sacred Sites.* Washington, DC: The World Bank.

Ralph L. Knowles

Summary
The concept of solar envelope provides architects and urban designers with a design analysis tool by which to understand and implement solar access to buildings for passive solar heating, solar control and daylighting. At the urban scale, the concept of *solar envelope* provides a means to regulate development within imaginary boundaries derived from the sun's relative motion. Buildings within this envelope will not overshadow their surroundings during critical periods of the day and year. The solar envelope provides zoning for low impact development and opens new aesthetic possibilities for architecture and urban design.

Key words
bioclimatic design, density, floor area ratio (FAR), solar access, solar envelope, sun rights legislation, zoning law

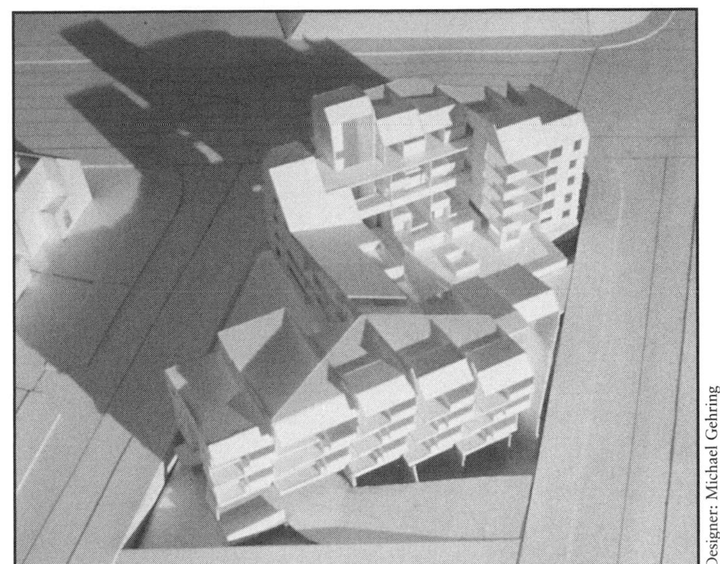

Designer: Michael Gehring

The solar envelope

1 SOLAR ACCESS

A 50-story office tower in Los Angeles casts a shadow 1,000 ft. (305 m) long between 1 and 2 PM in December. By 3 PM, that building's shadow is 1,800 ft. (329 m) long, with an area equivalent to two city blocks. Its leading edge cuts across the swimming pool of a popular downtown hotel, isolating a few sunbathers in a narrow strip of warm sunlight. The rest of the pool area is shadowed, cold and empty.

This picture, while perhaps a trivial case, raises ethical and legal issues of solar access. While I may choose to stand in shadow, I resist a developer's mandating it. And if I occupy a space or a building in the wake of that tower's shadow, I will resist that violation of my right to the sun's light and heat. It is this desire for sun rights, as well as the need to develop a sustainable architecture that relies upon sunlight for winter heat and daylighting, that argues for legally protected solar access (Knowles and Villecco 1980).

A continuing debate over the best way to guarantee solar access has left the consequences for urban form to be inadequately addressed. The notion that solar access is antiurban and against property rights has gained currency without a full exploration of its implications for the design and growth of cities. To say that solar access will destroy cities because it won't allow a tall building to be erected amid low ones is hardly an argument for urban quality.

Solar access, when achieved by using the solar envelope, does not automatically result in the elimination of tall buildings nor does it mandate suburban densities. Research reviewed in this article indicates that floor area ratios (FAR) can be achieved as high as 7.5 and housing densities in excess of 100 dwelling units per acre (.41 ha). These numbers exceed suburban densities and would be consistent with most urban areas in the U.S., with the exception of high-rise centers such as Manhattan.

The solar envelope does not abolish tall buildings but rather has a scaling impact on urban growth. Density can increase over time, according to public values, but violent disruptions of city scale are avoided. Where high-rise development already exists, the solar envelope can be used to protect rooftops and upper-floor solar access. New construction is shaped and proportioned with reference to the old.

The solar envelope carries an implied ethical obligation to use the sun and to relate to it formally. The designer is encouraged to differentiate building and urban form in graphic response to orientation. One side of a building will not look like another and one side of a street will not look like another. Development will tend to be lower on the south side of a street than on the north where a major southern exposure is thus preserved. Streets take on a directional character where orientation is clearly recognized.

The solar envelope calls for a design strategy based on natural rhythms. Sunlight is assured within the envelope's boundaries, hence designers can make use of the changing directions and properties of light without fear that a taller building will one day cancel their efforts. The potential exists to conceive of architecture in dynamic terms of form and space.

Credits: Work cited is the result of twenty five years of studio design and research at University of Southern California School of Architecture, supported in part by National Endowment for the Arts (NEA), The U.S. Solar Energy Research Institute (SERI) and the Lusk Center, University of Southern California (USC).

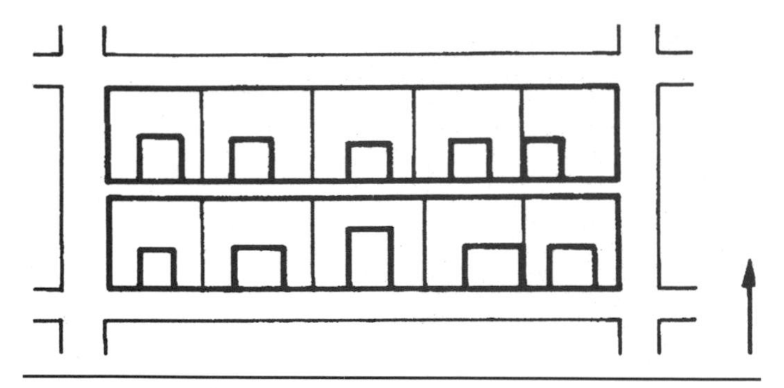

Fig. 1. The ancient Greek planners of Olynthus arranged houses to have two fronts: one to the sun and one to the street. (North up).

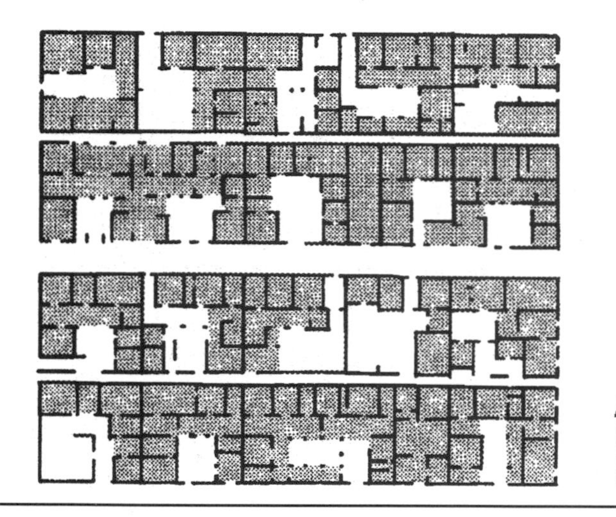

Fig. 2. Plan redrawn from excavations at Olynthus: The Hellenic House. (Robinson and Graham 1938.)

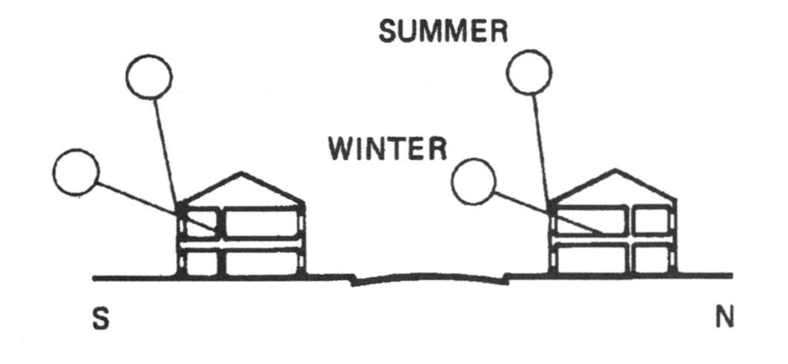

Fig. 3. Cross-section of conventional development, with major living spaces toward the street regardless of solar orientation thus losing the advantage of south exposure.

Fig. 4. Acoma Pueblo, NM. Drawing by Gary S. Shigemura. (Knowles 1974.)

Historical precedents

Solar access has become an international topic of discussion. Beginning in the 1970s, communities looked at sun rights primarily as a way to enhance energy efficiency for heating and lighting, a replacement for uncertain supplies of fossil fuels. Recently, environmental concerns are causing a renewed interest in sun rights. Solar access remains a legitimate area of public policy in which the aim is to regulate how and when neighbors may shadow one another.

The definition and implementation of solar access presently vary from one jurisdiction to another. In San Jose, CA, solar access is defined as the unobstructed availability of direct sunlight at solar noon on December 21. In Boulder, CO, sunlight must not be obstructed between 10 AM and 2 PM on that same date. Other definitions specify the percentage of wall area, glazing or roof that can be shaded by buildings or mature vegetation. The intent of such ordinances is clear enough, but a simple and effective means by which the developer or designer might achieve the objective is not set forth. Nor is it always certain how the community should go about proving compliance with the regulation. History shows prior instances of solar access can act as a model for today.

Olynthus: A model for solar access

The idea of solar access goes back at least 3000 years to the colonial cities of ancient Greece. Gridiron plans, attributed to Hippodamos of Militus, were arranged so that all houses faced to the sun for heat and light (Robinson and Graham 1938). Typical of this arrangement was the town plan of Olynthus where residential blocks ran long in the east-west direction and generally contained ten houses, five on each long side of the block (Fig. 1). The houses of Olynthus varied in size and layout but all contained a south-facing courtyard. Whatever conflicts may have resulted from a two-fronted house were resolved internally (Fig. 2). Consequently, while the house was always entered from the street, and in that sense had a street front, it was also oriented to the sun and had a sun front.

Internal differences of circulation resolved any conflicts resulting from this two-fronted plan. Those houses that faced south to the street and south to the sun were entered through the court. In contrast, houses that faced north to the street and south to the sun were entered through passageways that led from the street through the main body of the house and into the court where access was gained to all other spaces. This is in sharp contrast to most houses in the United States that tend to be oriented exclusively to the street, an arrangement that often conflicts with the solar orientation of major living spaces (Fig. 3).

Acoma: A model for the solar envelope

A thousand years ago in North America, settlements also respected solar access. Acoma Pueblo, located on a small plateau about 50 miles west of modern Albuquerque, NM, exemplifies such early planning (Fig. 4). Rows of houses step down to the south. Walls are of thick masonry. Roofs and terraces are of timber and reeds, overlaid with a mixture of clay and grass (Knowles 1974). House form at Acoma responds well to a high-desert climate (Fig. 5). The sun's low winter rays strike most directly south-facing masonry walls where energy is stored during the day, then released to warm inside spaces throughout the cold nights. In contrast, the summer sun passes high overhead, striking most directly the roof-terraces where the sun's energy is less effectively stored. East and west walls are covered by adjacent houses thus further reducing harmful summer-time effects.

The distance between rows of houses is just sufficient to avoid winter overshadowing of terraces and heat-storing walls (Fig. 6). It was the observation of this critical relationship of building-height to shadow-area that originally gave rise to the solar-envelope concept (Knowles 1981). But why the solar envelope, rather than some other legal device?

Legal background

Acute oil shortages in the United States during the 1970s prompted some American legal experts to call for a clarification or change of laws or even the formulation of new laws for solar access. This raised a question of legal precedents. The most commonly cited law outside the United States is the 19th-century. English Doctrine of Ancient Lights but there are problems with its application. Roughly, the doctrine states that if in living memory no one has overshadowed your property, they cannot now do so. However, U.S. courts have repeatedly disavowed this doctrine (Thomas 1976).

Some legal experts suggested that American water law, especially the Doctrine of Prior Appropriation, might offer a more useful precedent for sun rights (White 1976). Both sunlight and water are used rather than captured and sold; both may be consumed, but both are renewable. In addition, there is an equivalence between upstream and downstream in water law and the geometry of solar shadowing. Like the Doctrine of Ancient Lights, there are problems with the application of water law.

The Doctrine of Prior Appropriation is a formalizing of the general practice among early western settlers of appropriating available water according to who first put it to beneficial use. Simply put, "He who gets there first, gets the most." It was the American frontier's answer to the exigencies of pioneer settlement (Fig. 7).

The Doctrine of Prior Appropriation is not likely to apply very simply to solar allocation. Future access would not be assured for structures without present energy-conversion systems. Several permits acting on different, adjacent properties (as well as those on distant sites) may conceivably act to stop development completely on one of them. This point has been made abundantly clear in the writings of legal experts who point out serious weaknesses in any attempt to move directly from water law to solar law.

The difficulties in applying water law have led to arguments for straightforward zoning as a more appropriate approach to the problem (Hayes 1979). First, it offers the possibility of more local administration of rules affecting the allocation of sunlight. Second, zoning is traditionally applied to all properties in a district thus assuring future access and bypassing the problems of preference based on prior use. Finally, existing zoning limiting heights and setbacks is already based on the concept of an envelope of buildable volume. These reasons have been found compelling and have led to development of the *solar envelope* (Knowles 1980).

Street patterns

The street patterns of urban settlement greatly influence the solar envelope's size and shape. In the United States, those patterns are usually composed of subdivisions of the U.S. Land Ordinance of 1785

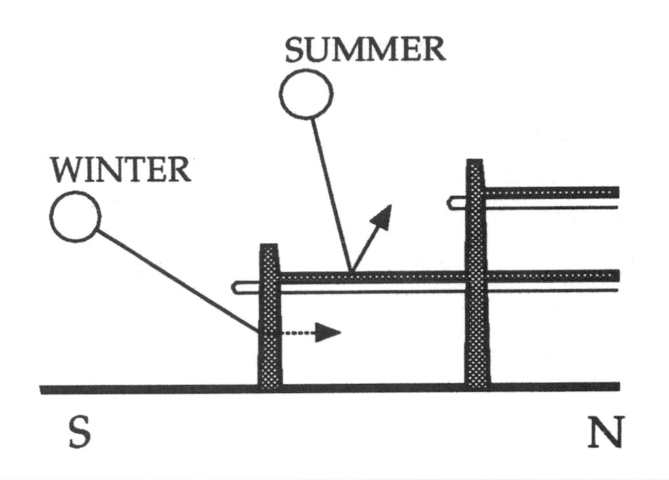

Fig. 5. Thick masonry walls and timber roof-terraces of Acoma Pueblo are thermally adapted to seasonal migrations of the sun.

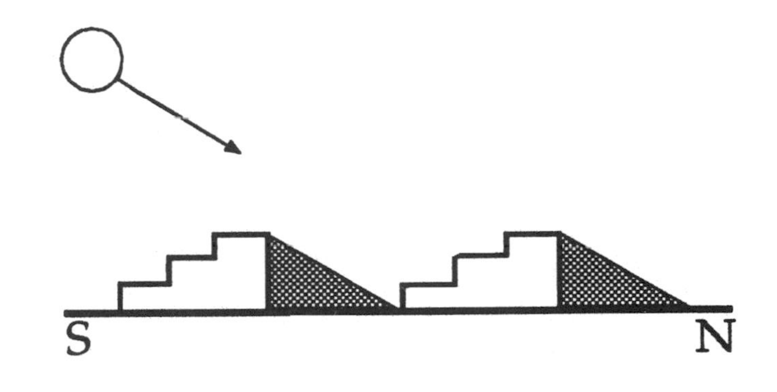

Fig. 6. Rows of houses at Acoma Pueblo are strategically spaced: no further apart than necessary to avoid winter shadows while conserving space on a small plateau site.

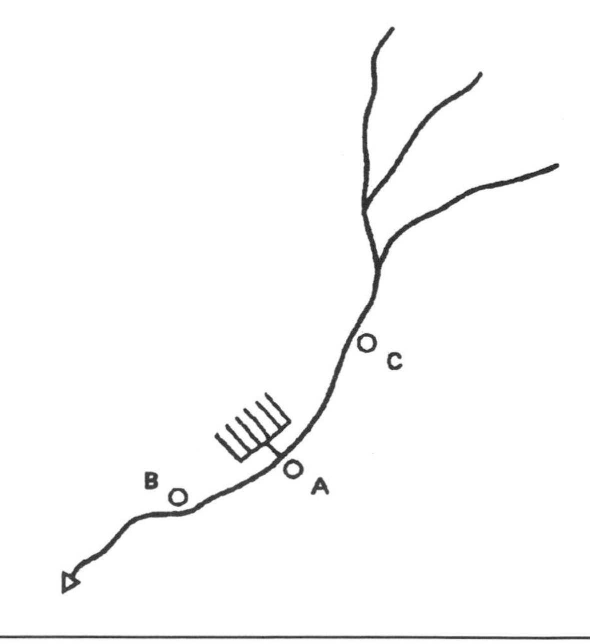

Fig. 7. The Doctrine of Prior Appropriation dictates that, after Settler A makes beneficial use of river water for irrigation, settlers B and C must acknowledge the prior claim without protest.

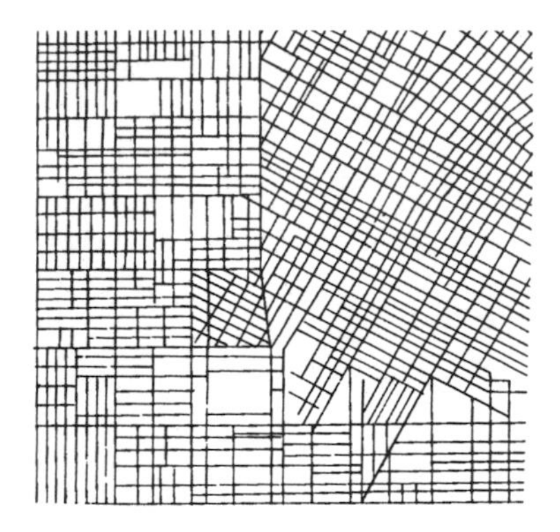

Fig. 8. Two street grids collide in Los Angeles. One follows the U.S. Land Ordinance of 1785; the other follows the diagonal Spanish grid.

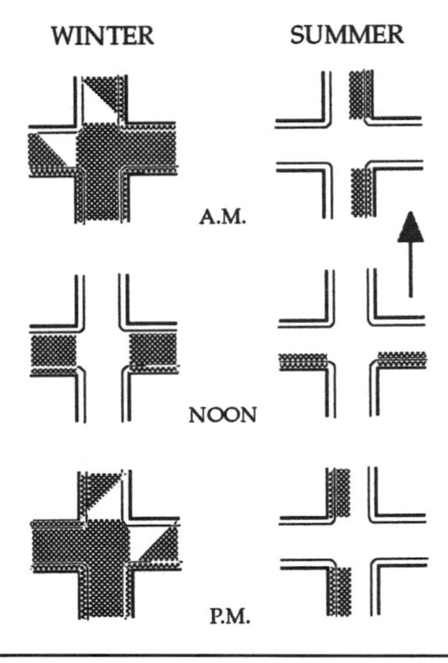

Fig. 9. Shadow patterns in streets laid out on the U.S. Land Ordinance indicate that E-W streets are especially uncomfortable: overshadowed and cold in winter, bright and hot in summer.

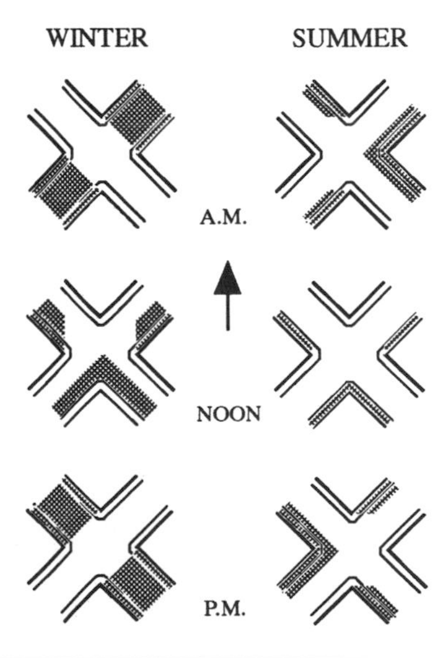

Fig. 10. Shadow patterns in streets laid out on the Spanish Grid are indication of greater comfort: warming winter sun and cooling summer shade in all streets.

that has ordered much of the land between Ohio and the Pacific Ocean. Typically, throughout the Midwest and the West, streets run with the cardinal points so that rectangular blocks extend in the east-west and north-south directions. But in Los Angeles, where most of the solar-envelope research has been done, there are two street grids (Fig. 8).

Most of Los Angeles follows the U.S. Land Ordinance but some streets run 26° off the cardinal points, following the original Spanish grid of *El Pueblo de la Reina de Los Angeles*. This diagonal orientation, an adaptation to sea breezes, is said to have been ordered by the King of Spain. It now extends from the original pueblo over the land that is modern downtown Los Angeles.

The effects of orientation on changing street qualities

Before discussing the street grid's influence on the solar envelope, there should be some mention of important temporal differences of streets themselves resulting from orientation. Everyone has experienced the discomfort of walking in a street too cold or too hot. Accordingly, many of us will walk on the sunny side of a street in winter, the shady side in summer. The choice may vary from morning to afternoon. This tendency to favor different paths at different times and seasons can in turn influence the actions of shoppers, sidewalk vendors, and store owners. A comparison of shadow patterns in the U.S. and the Spanish grids serves to demonstrate important differences that can profoundly effect the kind and quality of street life.

Urban streets running east-west are shadowed during winter. In Los Angeles at 34° N, the streets remain dark and cold. By contrast, streets that run north-south are lighted and warmed during the midday and are more pleasant during the busy noontime shopping period (Fig. 9).

Summer presents an entirely different picture. Unlike winter, when the sun's rays come only from the southern sky, the summer sun comes more directly from the east in the morning and the west in the afternoon. At midday, it is nearly overhead. Streets running east-west receive a little shadow at midday, but much less in the morning and afternoon—a critical factor especially on a hot afternoon. Streets that run north-south will be shadowed in the morning and afternoon, but will receive full sun for a brief period at midday.

The U.S. grid does not produce the best street qualities of solar light and heat. Its east-west streets are too dark and cold in winter, too bright and hot in summer. Its north-south streets, while pleasant in winter, lack any protective summer shadow at midday. In comparison, the older Spanish grid has advantages regarding street qualities of light and heat (Fig. 10). During the winter, every street on the Spanish grid receives direct light and heat sometime between 9 AM and 3 PM, the six hours of greatest insolation. It is true that at midday, all streets have shadows; but because of their diagonal orientation, more sunlight enters than if they ran due east-west

There is also a summer-time advantage to the Spanish grid. Except for a short period each morning and afternoon, when the sun passes quickly over first one diagonal street and then the other, some shadow appears in every street all day long. There is, consequently, almost always some cool place to walk. Real-estate experts, vitally concerned with the value of favored pathways, fully recognize these differences of street quality. Unfortunately, the spatial and temporal consequences of street orientation are rarely a basis for land-use planning or urban design.

Influence of street orientation on the solar envelope

Street orientation influences the solar envelope in two ways. The first of these has important consequences for development. The second relates more to issues of urban design.

The size of the solar envelope, and hence development potential, varies with street orientation. Generally, more envelope height is attainable at either of the two possible block orientations within the U.S. grid while less volume is possible within the Spanish grid. The street's gain in sunlight and shadow thus appears to be the developer's loss under the solar envelope when applied to the Spanish grid. This has made downtown Los Angeles a challenging problem from the viewpoint of solar-access zoning.

The shape of the solar envelope also varies with street orientation thus enhancing legibility. Kevin Lynch wrote, "To become completely lost is perhaps a rather rare experience. . .but let the mishap of disorientation once occur, and the sense of anxiety and even terror that accompanies it reveals to us how closely it is linked to our sense of balance and well-being" (Lynch 1960). Pathways, districts, and directions take on clear perceptual meaning when the solar envelope becomes a framework for urban design.

A comparison of three different block orientations shows how the solar envelope can enhance urban legibility (Fig. 11). Immediately evident are not only different envelope sizes but different shapes as well. These differences, when observed by developers and architects, will result in street asymmetries, district variety, and clear directionality along streets. Such differences would occur systematically, not randomly, thus serving as dependable cues to way finding through the city.

Space-time construct

The solar envelope is a construct of space and time: the physical boundaries of surrounding properties and the period of their assured access to sunshine. The way these measures are set decides the envelope's final size and shape.

First, the solar envelope avoids unacceptable shadows above designated boundaries along neighboring property lines; these boundaries can be referred to as *shadow fences* (Kensek and Knowles 1995). The height of shadow fences can be set in response to any number of different surrounding elements such as privacy fences, windows, or party walls. Their height may also be set by adjacent land uses with, for example, housing demanding lower shadow fences than commercial or industrial uses. Different heights of shadow fence will effect the shape and size of the solar envelope (Fig. 12).

Second, the envelope provides the largest volume within time constraints, called *cut-off times*. The envelope accomplishes this by defining the largest theoretical container of space that would not cast shadows off-site between specified times of the day. Greater periods of assured solar access will be more constraining on the solar envelope. Cut-off times that are specified very early in the morning and late in the afternoon will result in smaller volumes than would result from later times in the morning and earlier times in the afternoon (Fig. 13).

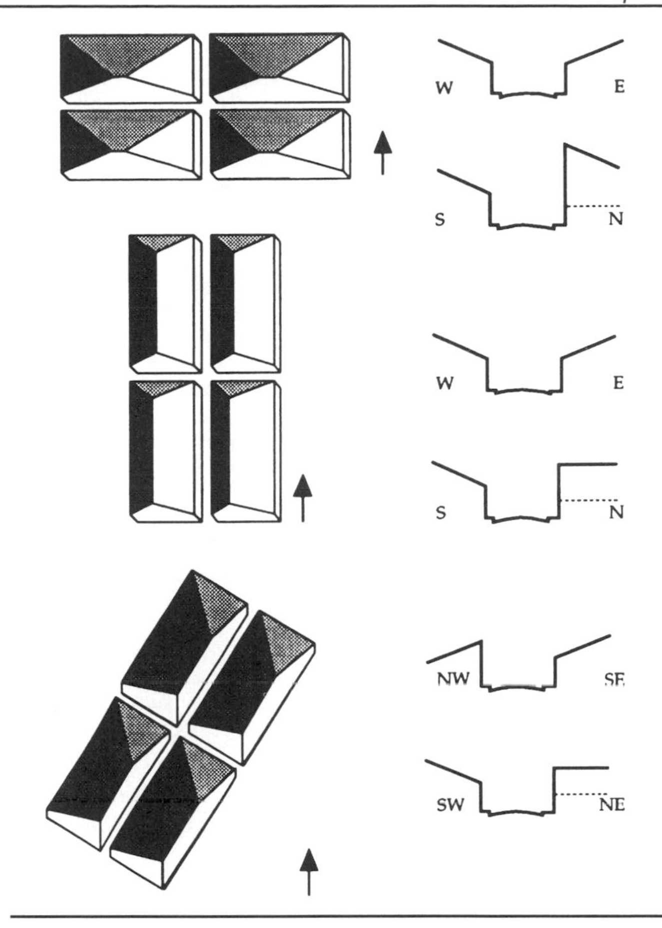

Fig. 11. Three different block orientations demonstrate the effect on size and shape of solar envelopes. Solar envelopes over E-W blocks have the most volume and the highest ridge, generally located near the south boundary. N-S blocks produce less volume and a lower ridge running length-wise down the middle. The diagonal blocks produce the least volume and a ridge along the south-east boundary.

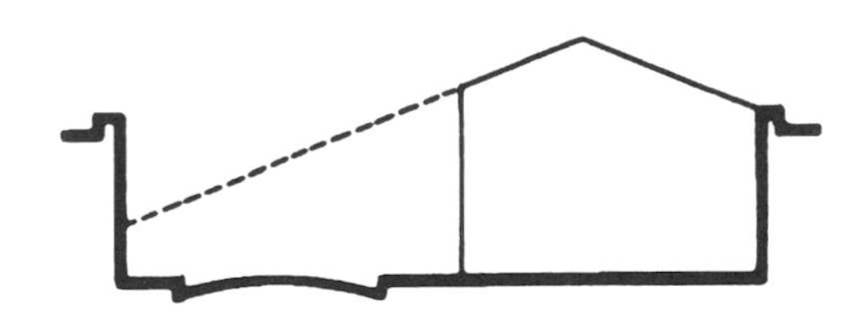

Fig. 12. Shadow fences may be specified at different heights on adjacent properties to avoid overshadowing such elements as windows or rooftops that may benefit from direct sunshine.

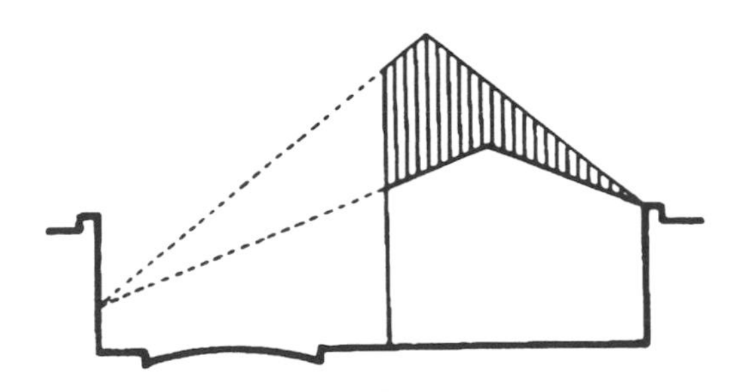

Fig. 13. Cut-off times can be specified to increase or decrease volume under the solar envelope.

The CRA Example

Model guidelines prepared in 1994 for the L.A. Community Redevelopment Agency (CRA) show how shadow fences influence the solar envelope. Shadow fences vary according to street character as set forth in a proposal by the CRA for Downtown Los Angeles. Shadowing is allowed up to 10 ft. (3.05 m) along *alleys*, up to 20 ft. (6.1 m) along *paseaos* and *avenidas*, and up to 45 ft. (13.7 m) along *boulevards*. Such differentiation anticipates varied land uses and street qualities (Fig. 14).

The same example shows how cut-off times also influence the solar envelope. Winter has the greater impact on volume because sun angles are low at that season; cut-off times are 10 AM and 2 PM thus providing four hours of direct access to sunshine, the minimum for good passive solar design in Los Angeles. Summer angles are much higher so have less impact on the envelope; the cut-off times are 8 AM and 4 PM, a longer period than winter but desirable in a mild, Mediterranean climate where people enjoy gardening and outdoor recreation.

The Saintes example

The CRA case typifies rectangular parcels but some building sites have complicated boundaries. A case in point involves a 1997 study in Saintes, France. The City of Saintes proposed a *Beaux Arts* Museum in the historical district. Existing on the site are some buildings intended to be removed, but others to be preserved. Surrounding the site is a medieval network of streets of irregular widths and angles, lined by existing structures of different heights and shapes.

The resulting solar envelope fits organically within its surround (Fig. 15). Rays of the sun, slanting across the site at winter cut-off times, determine envelope height changes. The form suggests an affinity of scale, an interdependence that lends a quality of nature. It does not clash with history.

Such complex sites, plus the need for an accurate legal description of sun rights, have led to the development of computer-based programs at USC (Schiler and Yeh 1993). *CalcSolar* is an interactive program used in the Saintes study. The site data and shadow constraints are drawn in AutoCAD, and the program is run integrally. The final envelope appears as a surface model that can be readily imported into other CAD programs to pursue the design process (Kesek and Knowles 1995, 1997).

CalcSolar provides visualization and evaluation. First, the formal impact of solar-zoning on a landscape can be quickly understood. Second, the program facilitates measures of density and floor area ratio (FAR) of contrasting space-time inputs. Prompt results help counter arguments that allowing solar access to surrounding properties diminishes chances to "improve" one's own property, when in fact opportunities are increased.

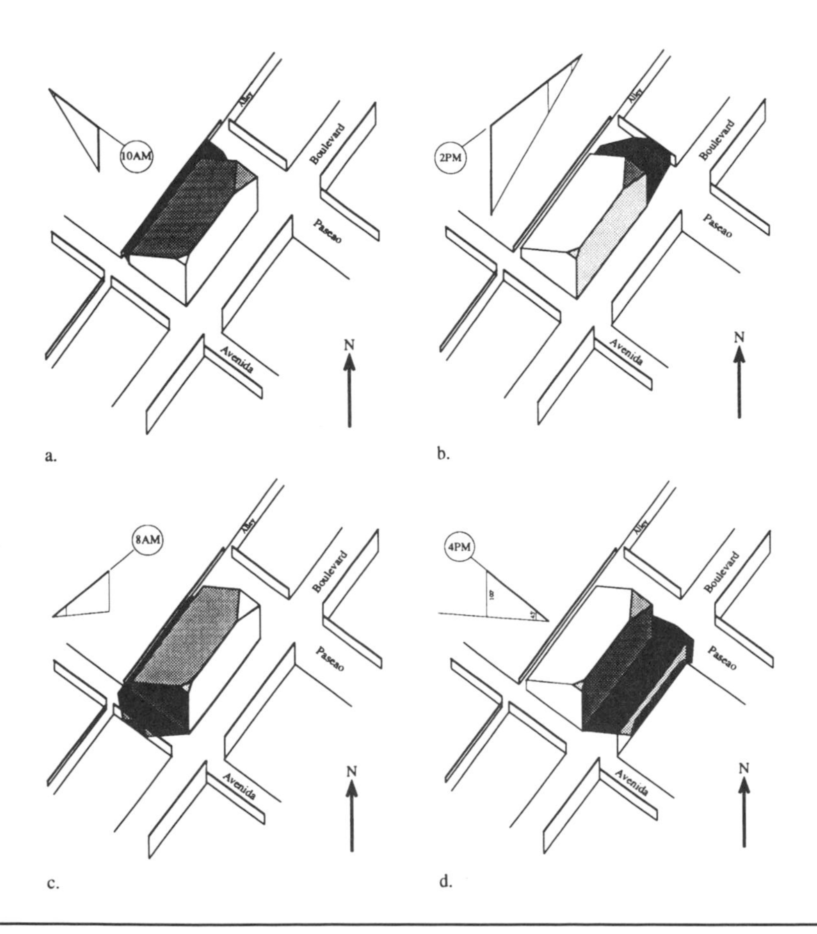

Fig. 14. The size and shape of the solar envelope are determined by different shadow fences acting at diverse cut-off times: (a) A shadow cast to the top of a 10' shadow fence across an alley at 10 AM on the winter solstice has by far the greatest impact on the envelope's form; (b & c) shadows cast to 20' at 2 PM winter and 8 AM summer have relatively little effect; (d) a shadow cast across a boulevard at 4 PM on the summer solstice has no impact at all because it doesn't reach the top of the 45' shadow fence.

Research protocol

Most tests of the solar envelope have concentrated on multiple housing. Other projects of a public or institutional sort have occasionally included commercial offices and, most recently, libraries; but the vast majority have focused on housing. One reason is the widening commitment to sustainable urban design (Steel 1997). Another reason is that research data are more easily accumulated and compared by adhering to a single use.

Each study begins by matching solar-envelope rules to actual land uses and economics. For example, in high-density urban areas, the period of guaranteed solar access is likely to be reduced and the shadow fences raised when compared to lower-density areas. Other variations include allowing the solar envelope to run continuously along streets rather than rising and falling at each property sideline. This last alteration of envelope rules may depend less on the sun and more on the provision of a clear passageway through a block for cooling summer breezes and cross-ventilation.

Following the generation of solar envelopes, the building-design process begins with each designer (architecture studio students) being required to follow realistic guidelines. For example, attention is paid to parking, circulation and fire and safety codes. While density is always a crucial issue, each dwelling unit in a multiple-housing design is required to have cross-ventilation and at least four hours of direct sunshine for daylighting and winter heating. Beyond these specific requirements, the student designers were free to explore their own notions of design. Two early housing studies of the aggregated type illustrate the typical transition from solar envelope to building design and the resulting collective image. The first presses for high densities in a built-up urban setting. The second looks at larger units and lower densities in a typical California subdivision. Both hint at development limits and architectural forms that later studies have verified.

Southpark Housing 1982

The first project, with resulting densities of 80-100 dwelling units per acre (du/ac) (197–247 du/ha) tests the solar envelope within the Spanish grid of downtown Los Angeles. Viewed from the east, the solar envelopes appear crystal-like while existing buildings are rectilinear blocks (Fig. 16a). The envelopes are generated to provide four hours of sunshine in winter and eight hours in summer; they slope downward to a 20 ft. (6.1 m) shadow fence at all property lines, thus accommodating a base of street-front shops under housing. The envelopes are consistently higher on the south than on the north with the exception of tower-like shapes that project upwards at some corners where shadows are allowed to extend further northward into streets, but not onto properties across the street.

When buildings replace the envelopes, design elements appear that typify later housing studies (Fig. 16b). Terraces occur where the rectangular geometry of construction meets the sloping envelope. Courtyards center many designs to achieve a proper exposure for light and air. Facades are elaborate, enriched by the porches, screens, clerestories, and other devices of solar design—all differentiated by orientation to the sun and wind. Buildings meet each other gently, across sloping spaces, not abruptly across property sidelines and alleys. The resulting spaces, not confining and dark but rather liberating and filled with light, allow view and the free flow of air through the city.

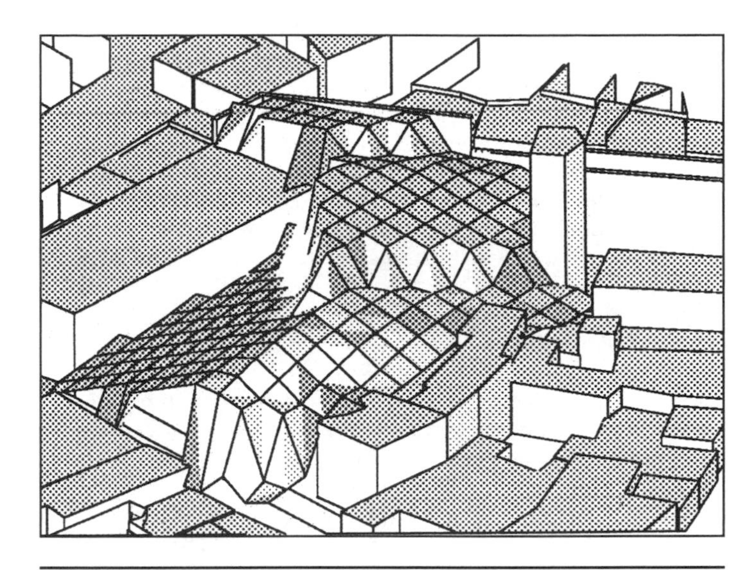

Fig. 15. The Saintes Project. The solar envelope must meet a great variety of boundary conditions within the context of a medieval street pattern. Viewed from the southeast.

Subdivision Housing — 1983

A second project replaces the typical suburban densities of 5–7 du/ac (12–17 du/ha) with higher densities of 25–45 du/ac (62–111 du/ha) under the solar envelope. Viewed from the south, the solar envelopes rise and fall with changes in street orientation and lot size (Fig. 17a). Envelope rules provide longer periods of sunshine than in the first project: six hours on a winter day, ten hours in summer. They are generated to a 6-ft (1.8 m) high shadow fence across streets at neighboring front yards and at rear property lines but they do not drop at property sidelines as in the first project. Instead, they form a continuous, undulating line along the street.

When building designs replace the envelopes, the result is remarkable innovation within harmony (Fig. 17b). The continuous envelopes result in a smooth flow of street facades. At the same time, building types range from town houses and courtyard clusters to apartments. Individual designers are clearly exploring separate formal ideas from one parcel to another. The consequence, if built, would be an enormous range of diversity and choice within a neighborhood.

2 DEVELOPMENT POTENTIAL

A rigorous 10-year housing study followed the two introductory projects to determine more precisely the solar envelope's development potential. Data were collected relating *density* (a count of dwelling units per acre, generally corresponding to land values) to V/S (volume-to-surface ratio, an energy-related measure of building form). At the end of the study, this relationship was taken as grounds for concluding that buildings of 3–7 stories generally represent the best size range for urban dwellings in Los Angeles. These figures can vary among cities, but the underlying suppositions of solar-access policy and design are broadly applicable to places of density everywhere.

Los Angeles zoning

Los Angeles zoning provides the urban housing reference for this study. First, the dwelling classifications are the actual ones used in the design studio. Second, they show in which density range the greatest variety of types is officially recognized by L.A. planners. Finally, each part of the range symbolizes not only different dwelling classifications, but also a separate grouping of possibilities for designers, developers, and users. To evaluate these possibilities, it is useful to establish the meaning and the method of calculating V/S and density (Fig. 18).

Calculations of volume/surface (V/S)

V/S, measured on the vertical axis of the graph, varies directly with building size. Symbols clustered at the low end of the graph represent small buildings up to 3 stories. Symbols at the high end signify large buildings of unlimited height. Between these extremes lies a range of mid-sized buildings, 4–7 stories.

V/S acts both as an energy-related descriptor of form and an expression of design choices. The low V/S of a small building means that energy must be expended mainly to overcome surface or "skin" loads; it also implies a strong architectural bond to sunshine, fresh air, and view. On the other hand, the high V/S of a large building means that more energy must be expended to handle the internal stresses of overheating; it also means less potential for the architect to design with nature.

Calculations for volume (cu. ft.) include only the space within dwelling units, not support facilities. This is consistent with the research protocol that excludes from a guarantee of sunshine all commercial space and shops below housing as well as major parking facilities either above or below grade. This method of calculating building volume also conforms to the design program that emphasizes natural light for housing.

Calculations for surface (sq. ft.) include exposed portions of the lot as well as the building's faces; this combination is used for three reasons. First, zoning codes usually list minimum yard and lot sizes together with building dimensions as a combined basis for classification. Second, energy is expended to maintain the lot as well as the building and when the lot is an acre or more, the proportion used for lawns and gardens can be enormous. Finally, when assuring solar access for winter heating and access to summer winds for cooling, the lot and the building must be seen as an integral set.

Calculations of density

Density (dwelling units per acre), measured on the horizontal axis of the graph, varies with housing classification. One-family dwellings are generally spread out compared with multiple dwellings, thus providing more space outside in the yards. Also, one-family houses tend to have more floor space than a unit within an apartment building.

Density, an indicator of land values, expresses development options. High densities correspond with inflated land values. Units, and even whole buildings, become compact and essentially repetitive. Low densities coincide with smaller land costs; developers concentrate on one-family houses multiplied over enormous tracts. But for urban housing on restricted sites in Los Angeles, developers usually try for the highest densities the market and zoning will support.

Calculations for density are based on net acreage only: the land within lot lines. It does not include streets, parkways, or such facilities as schools or shopping centers as do some density calculations for planned unit developments (PUDs). While density calculations based on gross acreage are commonly used, net acreage is here taken to be more consistent with the V/S calculations.

Exemplary housing projects

Four projects, covering a range of settings and densities, are exemplary. The design program calls for solar access and cross-ventilation to all dwelling units. The research protocol for solar envelopes has been adjusted to increase density in successive projects. As density increases and buildings become larger, solar access and cross-ventilation to individual dwelling units becomes progressively harder to achieve. The corresponding rise of V/S accurately measures this growing difficulty.

Low density

The first project achieves a density range of 7–18 du/ac (17–45 du/ha) with a corresponding V/S range of 2.5–6.0. Located on the N-S ridge of a low hill, it calls for replacing nearly all the existing one-family dwellings with multiple dwellings. Individual units are in the 2–4 bedroom range or about 1,350–2,500 sq. ft. (125–232 sq. m). The rules for generating solar envelopes call for guaranteeing 6 hours of sunshine on a winter day and 10 hours in summer for outdoor recreation and for gardening. Shadowing is allowed at any time below

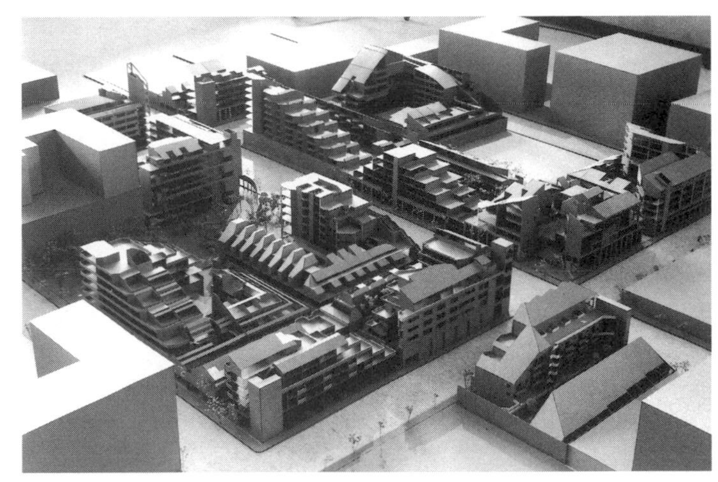

Fig. 16. Southpark housing, viewed from the east: (a) solar envelopes; (b) design studies.

17a

17b

Fig. 17. Subdivision housing, viewed from the south: (a) solar envelopes; (b) design studies.

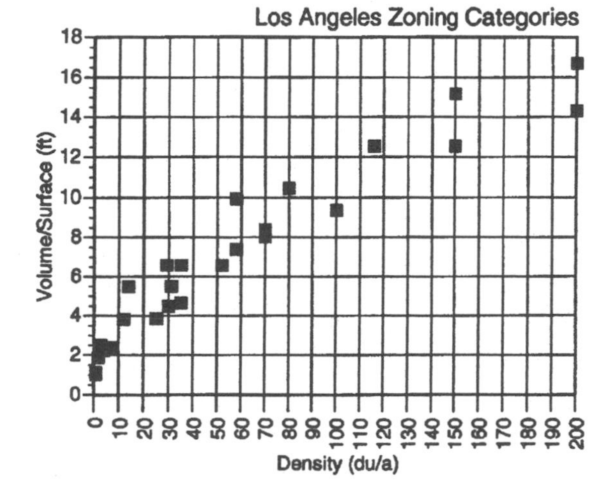

Fig. 18. A plot of V/S (volume-to-surface) against density (dwelling units per acre) for all dwelling classifications covered by Los Angeles Zoning Regulations from 1 du/a (A2) to 200 du/a (R5). (Two separate V/S values at 150 du/a and again at 200 du/a signify that building and lot configurations can vary significantly without changing density.)

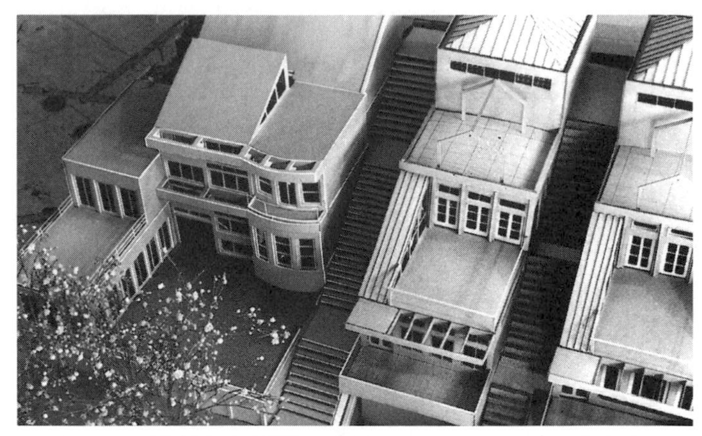

Fig. 19a & b. Project 1. Low density hillside housing with a density range of 7–18 du/a and a corresponding V/S range of 2.5–6.0: (a) A view from the southwest along the N-S ridge; (b) a close view of the west-facing slope.

Fig. 20 a & b. Project 2. Low-mid density housing with a density range of 14–28 du/a and a related V/S range of 3.9–6.2: (a) View from the south along east-facing slope; (b) view from the southwest on flat portion of site.

8 ft. (2.44 m) at front and rear property lines, but is unlimited at side property lines and on all public rights-of-way.

A view from the southwest shows several lots surrounding an intersection, each with a somewhat different condition of slope and solar access (Fig. 19a). The results accentuate the natural topography of the hill: high on the ridge, low on the slopes where the solar access of houses further downhill must also be protected. A detail view of the western slope shows two different siting strategies (Fig. 19b). One lot (right) contains two small houses separated by a stair for access to the rear of the lot. An adjacent lot (left) holds only one large house.

Both views show great architectural variety. Partly this results from different site conditions that affect the solar envelopes. But even within a parcel, relatively low values of V/S provide designers with especially rich possibilities to explore architectural form.

Low-mid density

The second project, located on a more gentle eastern slope of the same hill, nearly doubles the density to 14–28 du/ac (35-69 du/ha) with a V/S range of 3.9–6.2. All programmatic requirements for dwelling types, and also the solar envelope rules, are the same as for the first project. But here, as the hill flattens, solar envelopes can rise, providing more buildable volume than in the first project.

A view from the south (Fig. 20a) shows characteristics similar to those seen earlier on the west side of the hill. The solar envelope accentuates the downward tilt of the natural topography. Portions of otherwise buildable volume are cut away and clerestories installed over stairwells to capture south sun for daylighting and especially for winter heating. Downhill and on the flatter portion of the site, houses are taller and their shapes less forced by the solar envelope (Fig. 20b). A view from the southwest shows fairly typical three-story row houses lined up along a very deep lot with gardens and entry along one edge. Inside, a tall clerestory provides light and air to an atrium or garden center of each house.

Both views show somewhat less architectural variety than was seen in the first project. Windows tend to be more of a size and shape. Terraces and gardens are smaller and less visible. This is not the result of less desire on the designers' part for creative self-expression, but almost entirely the result of increased V/S that decreases the possibilities for identifying separate dwelling units.

Mid-High Density

The third project, located on the Spanish grid, raises densities still higher to 38–72 du/ac (94–178 du/ha) and the V/S range to 6.4–9.4. The pro-

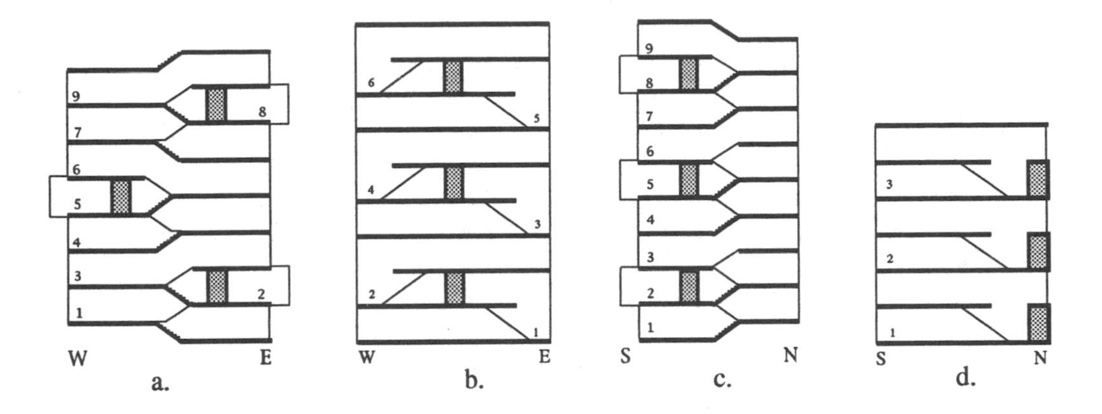

Fig. 21. Housing sections developed in Western Europe. Best for E-W exposures are (a) by Dutch architect Jacob Bakema and (b) by French architect Le Corbusier; (c) and (d) are adapted in the USC Solar Studio for N-S exposures.

Fig. 22 a & b. Project 3. Mid-high density housing with a density range of 38–72 du/a and a V/S range of 6.4–9.4: (a) Overview from the southwest showing tree-lined, central street; (b) a detail view from the northwest showing different ways to open outdoor spaces to winter sunshine.

gram calls for replacing dilapidated one-family dwellings, but not existing multiple-dwellings, with a market mix of units averaging 1000 sq. ft. (93 sq. m). Parking is below grade on some lots, but is naturally ventilated.

The rules for solar envelopes meet less generous time and space constraints thus allowing more building volume here than for the first two projects. While the earlier protocol guaranteed 6 hours of direct sunshine, the rules here guarantee only 4 hours—the minimum generally recommended for passive design in this "Mediterranean" climate. Shadowing is allowed at any time below 10 ft. (3.01 m) on residences and below 20 ft. (6.02 m) on surrounding commercial properties.

Two western European prototypes solve the problem of solar access and cross-ventilation in apartment buildings (Fig. 21 a-b). Higher densities in the U.S. generally depend on "double-loaded" corridors and mechanical systems. Hallways in the European designs systematically skip some floors, allowing units to pass freely both over and under for access to light and air in opposite directions.

The two European prototypes work well if site conditions suggest an E-W exposure, but need adjusting where the exposure is N-S (Fig. 21 c-d). Winter sunshine enters only the south-facing rooms, leaving those on the north relatively darker and colder. Since every dwelling unit is required to have a south exposure, the N-S building section becomes asymmetrical in its spatial organization.

The size of these sections depends on orientation. A building depth of 40–45 ft. (12.19–13.7 m) is about right for N-S exposures whereas the depth for an E-W exposure averages about 10 ft. (3.01 m) greater. This results from the fact that useful sunlight, especially in winter, can enter from only one side of a N-S section but two sides of an E-W section: 2-3 hours from the east in morning, another 2–3 hours from the west in the afternoon thus enlivening most of the space in the deeper unit. (Floor-to-floor height is maintained at 10 ft. for study.)

The use of these sections, along with tighter envelope rules, allows considerable building-height variation within a given land parcel (Fig. 22a). Along the central tree-lined street where the solar envelope is high, apartments can now rise 5–7 stories. Along the alley, where the envelope slopes downward, 2–3 story townhouses reflect more the scale of the first two projects. (Existing structures appear in the photos as simple blocks, 3–5 stories high and closely spaced.) A detail view from the northwest shows different ways to open outdoor spaces to winter sunshine (Fig. 22b). A broad south exposure allows one design to use a generous, plazalike courtyard. Another design, because of its narrow lot shape and the diagonal orientation of the Spanish grid, splits open along a true N-S axis for the midday sun to enter a streetlike court.

Fig. 23a & b. Project 4: High density housing with a density range 76–128 du/a and an accompanying V/S range of 7.9–10.5: (a) A view from the southeast showing three designs in the foreground sharing a continuous envelope; (b) a view from the northwest showing two well-integrated designs in the foreground, again sharing a continuous envelope.

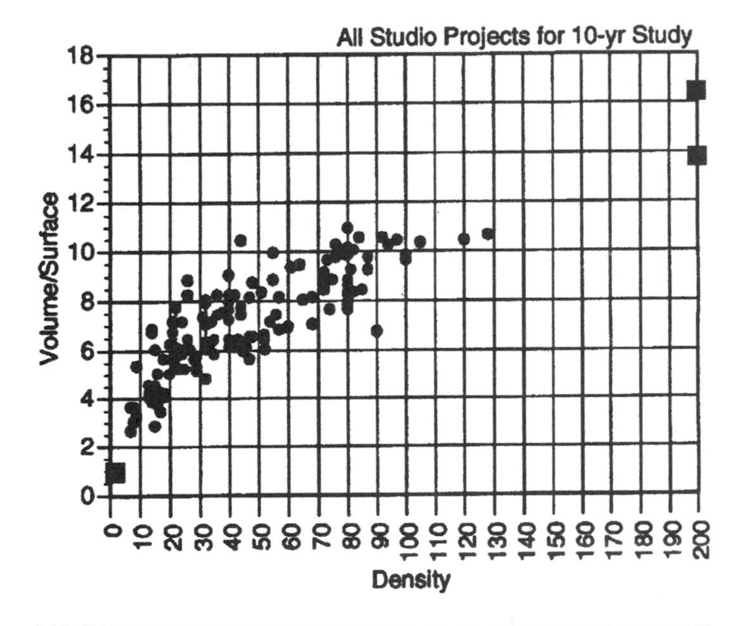

Fig. 24. The graph shows round symbols representing all 150 student housing designs clustering in a density range of 7–128 du/a corresponding to a V/S range of 2.5–10.5. (Square symbols represent the extent of L.A. zoning range.) In general, all round symbols can be seen to level off at a V/S of 10.0 corresponding to a maximum density of about 100 du/a. The few exceptions that rise fractionally above this plateau symbolize unusually tall buildings resulting from special site conditions.

Both views show more architectural repetition than was seen in either of the first two projects. Still, for the higher densities involved, there is remarkable design variety among parcels and unit diversity within any one lot.

High density

The fourth project, located on a hillside close to downtown, achieves a density range of 76–128 du/ac (195–316 du/ha) and a matching V/S range of 7.9–10.5. The site is inappropriate for very large commercial structures, but ideal for needed housing. Design requirements for unit size and parking are the same as for the prior example, and the building sections diagrammed there are used here as well. The solar-envelope rules for time constraints are the same, but the space constraints are different than the prior project. First, there are no setbacks. Second, the solar envelope runs continuously without dropping at property sidelines. Finally, overshadowing is purposely allowed on a north-facing slope that has been left open as a park. Combined, these three changes have the effect of providing more envelope volume than in any of the earlier projects.

A view from the southeast shows that, because the solar envelopes do not drop at sidelines as they do in Project 3, separate designs merge (Fig. 23a). In the foreground are actually three separate designs. The closest one is L-shaped. The middle one is very tall with a low building in the front yard facing the street. The last design in the row breaks from the envelope but holds the south facade line. A view from the northwest and across the park again shows designs merging under shared solar envelopes (Fig. 23b). In the background are two separate designs that divide one envelope. In the foreground are two other designs that take the division a step further by coalescing into a single, dynamic composition.

This fourth project not only has the highest density of the projects shown but it extends the V/S range to the highest values achieved in the 10-year study. These last results turn out to have the most significance for urban growth and form.

Fig. 25a & b. The Bunker Hill Project: (a) The solar envelope, viewed from the south, varies in height from 100' to 500' and contains a huge development potential (FAR = 20); (b) an exemplary design (FAR = 7.5) trades off development potential of the envelope for solar access to buildings and spaces within project boundaries. Designer, Randall Hong.

Research findings

A composite graph, representing all 150 designs, falls short of the full range of Los Angeles zoning but for two different and opposing reasons (Fig. 24). The lowest density of the study (7 du/ac or 17 du/ha)) was deliberate, the result of an initial decision to exclude from investigation one-family dwellings on very big lots as inappropriate for urban housing. On the other hand, the high end of the study's density range (128 du/ac or 316 du/ha) was not deliberate but the chance result of a step-by-step disclosure over the ten years of testing. Between these values, the study found a remarkable variety of ways to live in the city within a height range of 3–7 stories. The conclusion of the study is that ample opportunities do exist in this size range to provide both energy conservation and passive design and thus improved comfort and life quality—the goals of sustainable design—without overly limiting development options for urban growth.

The leveling of symbols at a plateau in Fig. 24 marks the most important discovery of the study. The consistent effort to achieve both energy efficiency and life quality, while striving for higher densities, produced a critical cut-off value of V/S = 10.0 corresponding with a maximum density of about 100 du/a. A few special circumstances, as a park or wide boulevard where longer shadows could be cast without harming a neighboring property, resulted in taller buildings with fractionally higher V/S values. Otherwise, for desired exposure to solar access and cross-ventilation in a compact and continuous urban fabric, the rule holds. Designers who break this rule lose the opportunity to utilize architectural design to sustain building comfort and must fall back to energy-intensive mechanical systems.

The cut-off value of V/S = 10.0 provides a simple but powerful design tool. Architects don't have to wait until a project is far advanced to evaluate its passive-design potential. Even at very early stages of planning, a simple calculation, performed on alternative massing schemes, provides an unequivocal basis for comparing the eventual character of their energy usage. If V/S is less than 10.0, designing with nature is a good option. On the other hand, if V/S rises above 10.0, reliance on energy-intensive systems is inevitable.

3 DESIGN IMPLICATIONS OF THE SOLAR ENVELOPE

Architects generally ignore the temporal dimension that nature adds to our perceptions of space. They acknowledge that buildings may be transformed or deteriorate over time but on the whole their artistic idea of space is complete and static, the final product of many imaginative decisions. By contrast, guaranteed access to sunlight opens the possibility that a spatial concept linking solar rhythms to life's rituals can give new life and freshness to design. Space is generated by flux itself.

One of the earliest revelations of new design possibilities under the solar envelope came in a 1979 study of Bunker Hill, a 9-acre site in a key downtown location of Los Angeles (Knowles 1981). The Community Redevelopment Agency (CRA), after expressing an interest in solar-access zoning for the site, made available to USC a preliminary version of a program that later became the basis for a competition among five developer-architect teams. In its July 1980 majority report, the CRA cited the winning entry of the competition for its attention to scale and to solar access with regard to neighboring properties.

Solar-envelope rules for the USC study of Bunker Hill made distinctions in the magnitude of acceptable shadowing of surrounding properties. Least shadowing was allowed on housing and most on commercial office buildings that, in this specific test of the envelope, were shadowed to a maximum of 33% of their window-wall areas at any given time. The final envelope that emerges from these rules is of variable height from 100 to 500 ft. (30.5 to 152.4 m). Its profile generally rises and falls along with that of the surrounding skyline (Fig. 25a).

A design that does not fully use up the floor-area potential of the solar envelope has instead made a skillful trade-off (Fig. 25b). The solar envelope has a latent FAR in excess of 20 based on a floor height of

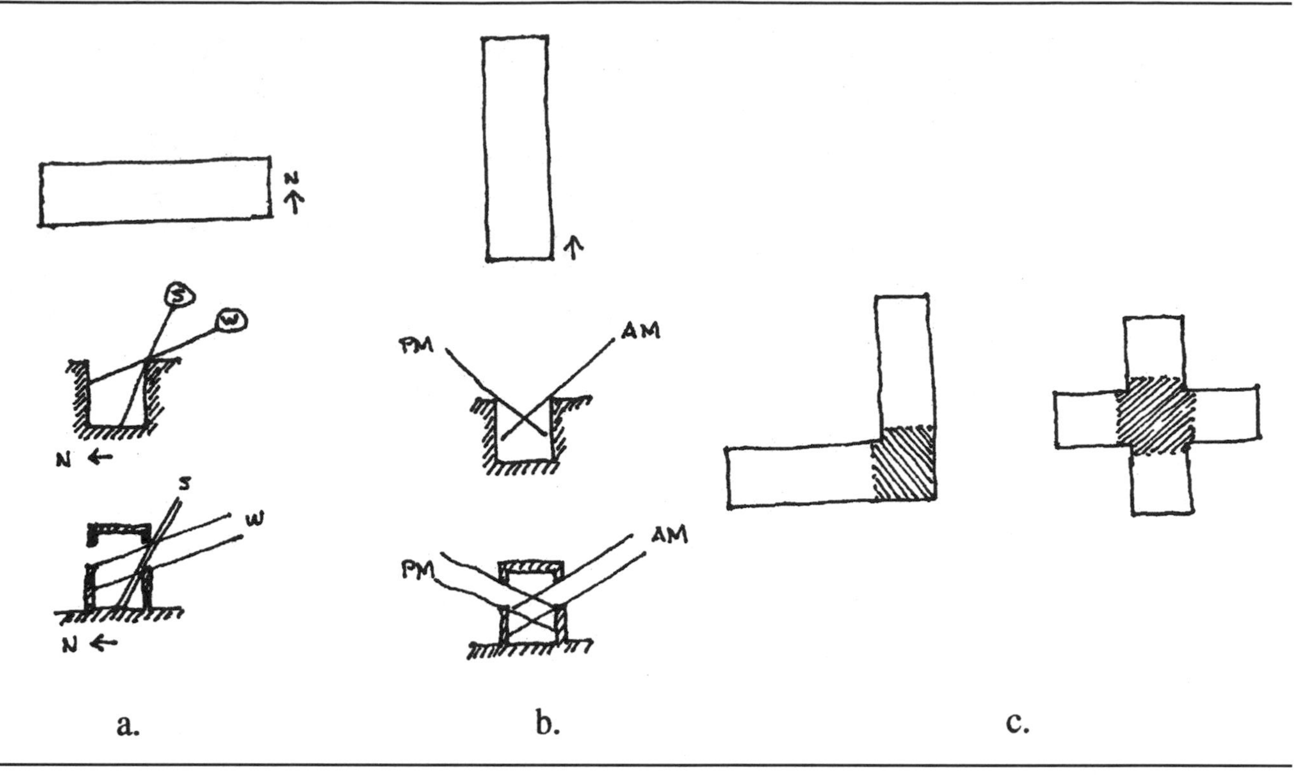

Fig. 26. The orientation of a space determines its solar rhythm: (a) A long E-W space accentuates a seasonal rhythm; (b) a long N-S space accentuates a daily rhythm; (c) where N-S and E-W spaces intersect, the rhythm is contrapuntal.

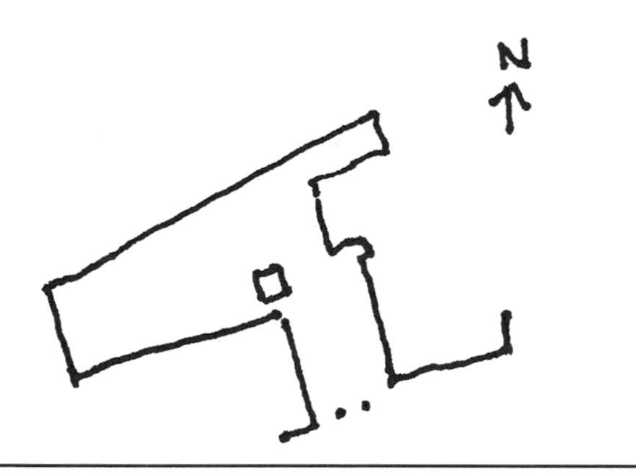

Fig. 27. San Marco Plaza, Venice: Its two parts accentuate different solar rhythms influencing when as well as where people choose to sit down.

Fig. 28. The cloister of Salsbury Cathedral, A.D. 1263: Monks of the Middle Ages divided each day into eight ceremonial periods corresponding to the liturgical phases of monastic life.

15 ft. (4.57 m). With such a huge volume, the CRA's programmed FAR of 8–10 could easily have been met by following standard development procedures. Rather, the designer chose to split the project open along a favorable N-S axis. The FAR is reduced to 7.5 but with great qualitative gains. The maneuver combines the best qualities of both the U.S. and Spanish grids, thus providing ample winter sunshine to housing and to public open space within a mixed-use project. Splitting the plan achieves an additional quality.

Rhythm

Rhythm itself is a mysterious fact of aesthetic experience (Gross 1966). This observation has generally been made about poetry or music. But rhythm has the quality of being actual in all our movements and in the events around us. Rhythm is the way our bodies and emotions respond to the passage of time. It is therefore a universal means of affecting experience and conveying meaning in our lives.

Consider how the rhythms of sunlight alter our understanding of space. Whether we occupy an inside or an outside space, two scales of time can be noted. One measures a day of experiences and actions; the other, a year. People cannot really know their possibilities, or the possibilities of a place, by passing through just once. They must remain awhile.

Any space that is oriented from east to west strengthens our experience of the seasons (Fig. 26a). One main wall is nearly always dark; on the other side of the space, a shadow line moves gradually up the wall and then down again. To experience the whole cycle takes exactly one year. The basic movement is always the same. As the sun's path drops lower in the sky during late summer and fall, the shadow moves up. As the sun's path rises in the sky during late winter and spring, the shadow moves down. Changing the orientation of the space will evoke a different cycle and rhythm.

Any space that is oriented from north to south sharpens our experience of a day (Fig. 26b). Both main walls are lighted, but at different hours. Every morning, light from the east will cast a shadow that moves quickly down the opposite wall and across the floor. Every afternoon, light from the west will cast a shadow that crosses the floor and climbs the opposing wall. To experience the whole cycle takes from before sunrise until after sundown.

Where east-west and north-south spaces pierce each other, we can experience two measures of time (Fig. 26c). The common volume intensifies both a seasonal and a daily cycle. It combines them, laying one over the other. The result is a crossing in space that proportions time. Its celebration has often lifted architecture out of the region of fact into the realm of art.

In great cathedrals, cycles of sunlight occupy a year in the nave, a day in the transept. In a monastic church of the Middle Ages as well as a *basilica* of Renaissance Rome, some sort of spatial crossing proportions time. A perpetually renewed experience of wonder is intensified by the changing qualities of light, holy observances measured by variations of hue and value.

San Marco Plaza in Venice is oriented so that its two parts manifest different solar cycles (Fig. 27). Some building surfaces are mostly subdued, even a bit gloomy. Others are mostly alive and radiant. Still others alternate, moving first to one mode, then to another. The places where people choose to direct their attentions and energies, especially where they find a cafe table to sit down, all vary with the day and season. A special place in *San Marco Plaza* celebrates these two levels of awareness; it is marked by a great campanile. Shadows from its towering height sweep from west to east by day, from north to south by year. The tower and its shadows mark a ritual passage in time.

Ritual

We human beings, in all our practices, engage in rituals everywhere, in all parts of the globe and in all types of society (Meyer and Smith 1994). Some rituals are construed as an attribute of culture, collective actions in common spaces. Other rituals seem to be personal inventions carried out in private settings. Whether an attribute of culture or personal invention, rituals very often emphasize the rhythm of actual experience in a place. Rituals have their origins in the things we keep on doing in a place. Everyone forgets his everyday life: washing the dishes, dusting the furniture, sitting down to eat. Yet if it weren't for the forgotten, the unforgettable wouldn't be able to grow. We would thus be denied a sense of who we are, of where and when we exist in a continuously repeating cycle of life.

Rituals tend to amplify nature's rhythms. For example, the sun comes into our garden or through our window. We can move either toward or away from it for relief. At first our movements may only rehearse a direct search for comfort. But eventually, simple movements may be embellished, expanded in detail to express our feelings about life in the place. It is this latent connection between rhythm and ritual that holds enormous promise for designers to act with imagination.

The Middle Ages provides elaborate examples of religious rituals related to solar rhythms (Fig. 28). Monks divided each day into eight ceremonial periods corresponding to the liturgical phases of monastic life.

Summer shadows on the back porch, morning, midday and afternoon.

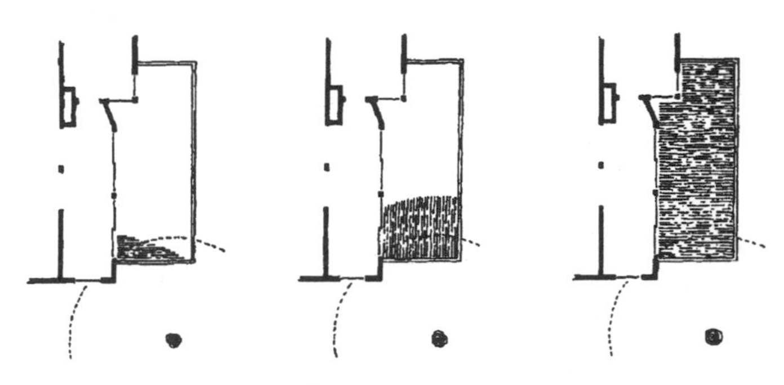

Winter shadows on the back porch, morning, midday and afternoon.

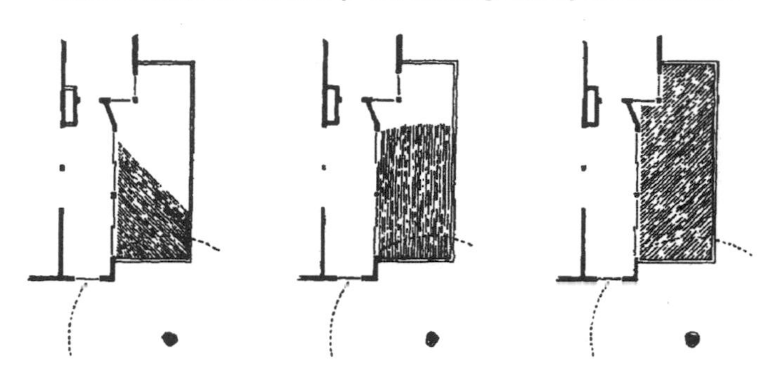

Fig. 29. Changing shadows on our back porch in Los Angeles: Summer and winter shadows; morning, midday and afternoon, (North, up).

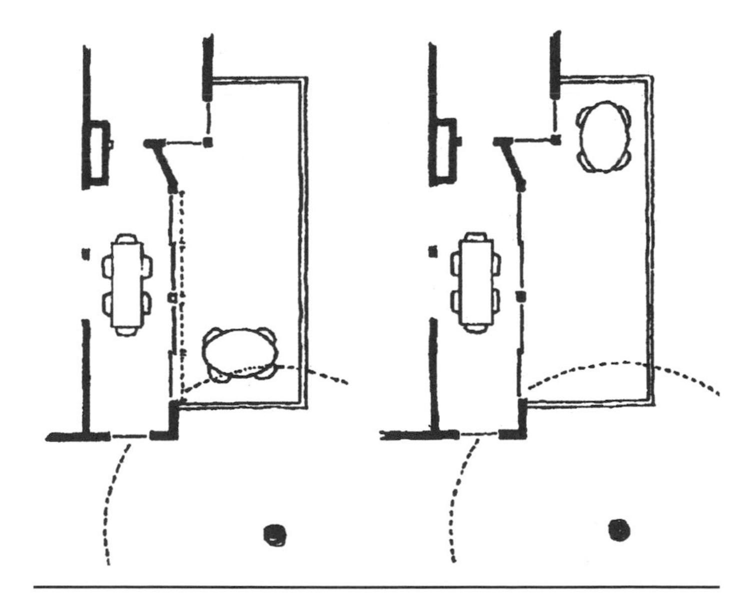

Fig 30. Changing locations of our back-porch table: (a) Southward under the shade of a neighbor's tree in summer; (b) northward to avoid the tree's longer winter shadow.

Fig. 31. Longhouse Pueblo, Mesa Verde, CO (AD 750–1100). Viewed from the southeast toward the western end of the cave. Drawing by Gary S. Shigemura, (Knowles 1974.)

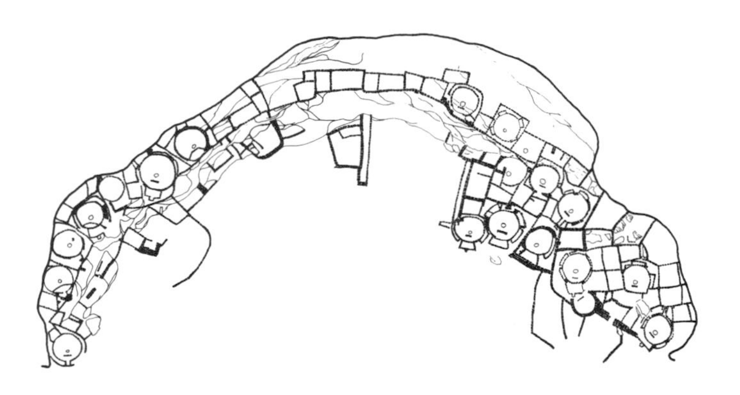

Fig. 32. Composite plan of Longhouse. Drawing based on a reconstruction by anthropologist Douglas Osborne, California State College, Long Beach, CA.

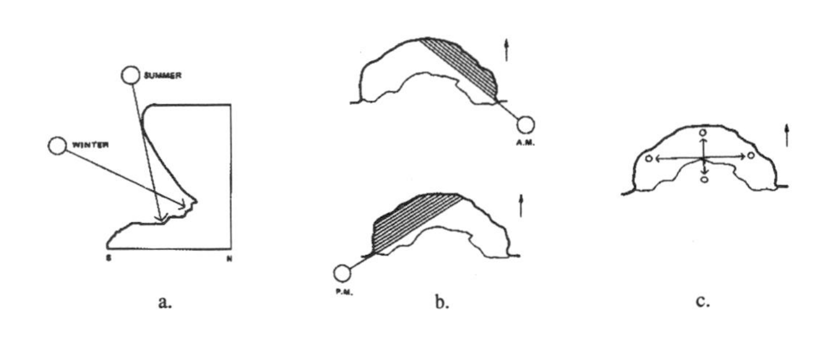

Fig. 33. Activities within Longhouse migrated with the sun's passages: (a) Seasonal passages of the sun; (b) daily passages of the sun; (c) contrapuntal migrations of activity.

From *Matins*, the darkness of early morning, to *Compline*, late evening twilight, appropriate actions were universally prescribed for each period. The exact meaning and hour varied according to place and season. The length of the same day was not equal in northern England and southern Italy. Summer days were different in each place as were those of winter. Changing place made time different; changing time made the qualities of a place different. Changing either changed meaning.

The connection between rhythm and ritual, while more intricate in the Medieval example, can also be found in modest circumstances. For example, my family sits down every day to eat, but instead of sitting in the same place, we sit at different tables (Knowles 1992). The way this happens has been gently guided by rhythmic changes of sunlight and shadow on our back porch (Fig. 29).

A neighbor's tree spreads over the south end of our porch casting shadows. To sit in the cool summer shadow, we move the table southward. To catch the warm winter sun, we move our table northward (Fig. 30). Back and forth, once each spring and again each fall, we carry the table across a shadow line. The moment we know it is spring is not exact. We could remember the calendar, but more often than not, on a warm and sunny morning, somebody will say, "Let's move the table." And then it is spring, no matter the date. Our daily passages to and from the porch are forgettable and yet they carry the seasonal changes that bring new possibilities. Moving the table shifts all our connections to the house, the view and each other. After 30 years, we are still thinking up new ways of arranging things.

Pueblos of the Southwest

The connection between rhythm and ritual exists everywhere in the way people have traditionally identified with their environments. Some of the most inspiring examples appear in our American southwest. Here, pressed by the harshness of a high-desert climate, native people recognized natural rhythms in their adaptations of location and form. Customary movements and corresponding rituals confirm the impact of natural cycles (Knowles 1974 and 1981).

Longhouse, Mesa Verde, Colorado

The location and form of Longhouse provided ancient dwellers with year-round comfort (Fig. 31). The settlement (c.1100 A.D.) is located within a large, south-facing cave, 500 feet across, 130 feet deep, and 200 feet high (Fig. 32). The brow of the cave admits warming rays of the low winter sun but shields the interior of the cave from the rays of the more northern summer sun (Fig. 33a).

A daily response complements the seasonal adaptation of Longhouse. As the sun moves from the eastern to the western sky, morning rays enter the west end of the cave and twilight rays enter the east end (Fig. 33b). The thermal mass of the cave itself, as well as the structure of buildings within, helped to mitigate extreme daily variations, but a main adaptation for comfort may be best understood in the rhythms of people's lives.

The dwellers of Mesa Verde tended to move in and out of the cave with the seasons. The way they worked suggests that they moved deeper into the cave for shelter during summer and spread further to the south in winter, using their terraces for work and play with full exposure to the warm solar rays. This north-south migration coincided with the shadows and thus the thermal variations of exactly one year's duration. It is possible to imagine a contrapuntal migration (Fig. 33c).

Fig. 34. The ruins of Pueblo Bonito (AD 919–1080). Viewed from the north rim of Chaco Canyon. Drawing by Gary S. Shigemura. (Knowles 1974.)

There is not much documentation of the tempo of daily life at Longhouse, but a transverse migration might very well have occurred through the cave. During much of the year, morning light entered the dark cave from the east and outlined those buildings at the western end. All families in the pueblo kept turkeys, but only the turkeys first struck by the morning rays of sunlight would awake and become noisy. The turkeys roused the children who, in turn, disturbed the parents. Those first pulses of activity would then echo 500 ft. (152 m) against the eastern walls of the cave. The day passed with fairly general activity distributed throughout the cave until evening. Twilight came first to the western end of the cave and subdued the turkeys, then the children. Their parents could breathe those last sighs of relief while parents in the eastern end of the cave still had time to go before their day was done. As their light disappeared, the last turkeys uttered a quiet gobble and the children of the eastern cave let that day finally go.

The foregoing sequences are imaginary but not without historical foundation. The fact is that such a scene could very well have taken place. And what is more to the point, it would have occurred every day with seasonal variations suggesting a strategic rhythm both for thermal comfort and for the emergence of ritual.

Longhouse depends for its adaptation to climate on a south-facing natural formation but other pueblos, also south-facing, are standing in the open. Some of these, like the living pueblo in Taos, New Mexico, are irregular and appear to be randomly assembled. Others, like the nearly square ruins at Aztec National Monument, are symmetrical and clearly planned. One of the most impressive of the planned type, 300 ft. (91.4 m) across its south edge and 3–5 stories high, is the semicircular Pueblo Bonito (Judd 1964).

Pueblo Bonito, Chaco Canyon, New Mexico

Bonito, in its original form (circa 919–1080 A.D.), was built in two stages to resemble an amphitheater with a stage surrounded by curving rows of seats too big for people—the Greek theater of Epidauros, transplanted and scaled for gods (Fig. 34). Even today, standing among the silent debris of Chaco Canyon, people often express deep feelings

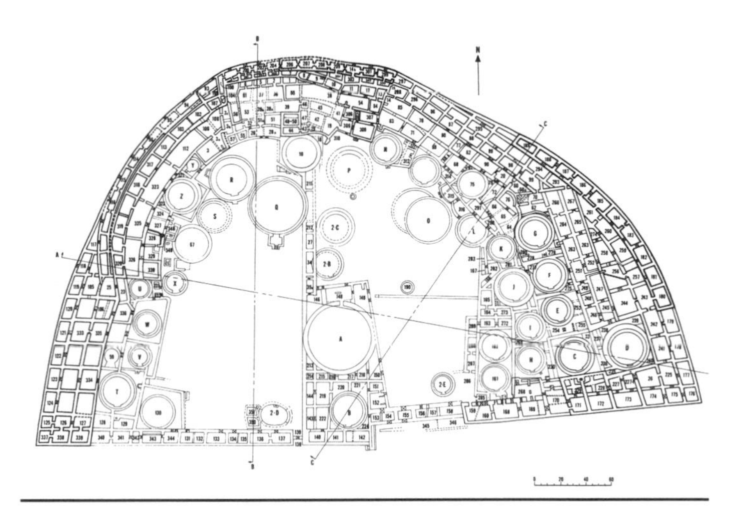

Fig. 35. Plan of Pueblo Bonito, Chaco Canyon, NM. Redrawn from *The Architecture of Pueblo Bonito* by Neil M. Judd. (Washington D.C.: Smithsonian Institution, 1964.)

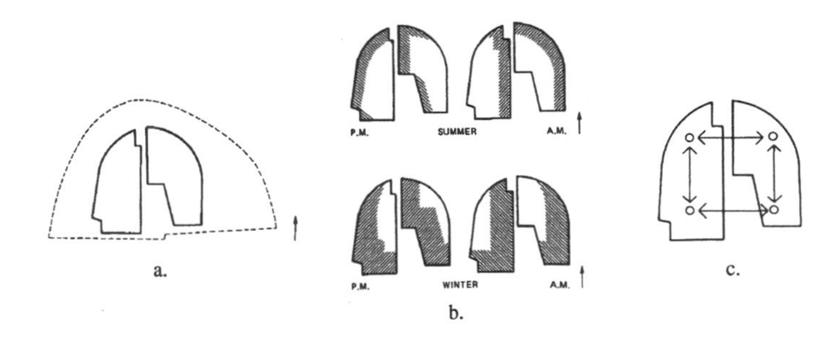

Fig. 36. Ritual dances within Bonito's courtyards very likely migrated with the sun's passages: (a) The courtyards outlined; (b) shifting shadows by day and season; (c) contrapuntal migrations of ritual dances.

after experiencing the passage of sunlight over the powerful half-forms of Bonito's ruins (Fig. 35).

There is much evidence for seeing Bonito as an astronomical observatory and perhaps a ritual temple to honor the sun (Knowles 1974; Aveni 1975). Daily and seasonal rhythms were not only a condition of pueblo life, they were a cause for sacred celebration. Some sense of this can be conveyed by simulating changes in shadow patterns within the original pueblo courtyards (Fig. 36a). The result of the sun's daily passage is a shift in the patterns of dark and light in the great courtyards: west to east by day; north-south by year (Fig. 36b). In the shifting spot-lighted areas, it is easy to imagine ceremonial dances taking place at different locations in the courts at different times of the day and year (Fig. 36c). Dances of the sort occur today on the terraces and in the courtyards of modern pueblos (Scully 1975).

Such observations, both on site and in the laboratory, are compelling when contrasted with modern architecture's avoidance of nature. Our machines automatically reduce natural variation to a narrow range of light, heat, and humidity. In the process, we have lost a sense of harmony that derives from feeling the complex rhythms of nature. And we have lost more.

We have lost an essential challenge to our imaginations. When we flatten and simplify nature, we lessen the need for many customarily repeated acts. Take the most banal example: putting on one's sweater or taking it off. The sweater itself—its age, color, texture—asserts our personality. So does the way we carry it on our bodies—tightly fitting or loose and baggy. The sweater and the wearing of the sweater can become a system of rites, a ritual observance of our environment and ourselves.

4 CONCLUSION

Not only does the solar envelope allow for potential development—increasing density while maintaining access to sunlight—but also opens new possibilities for architecture and urban design to design with nature and unique circumstance. The outcome of a research and design studies reported above shows that buildings under the solar envelope are neither too big for human scale nor too small for urban growth; their scale is consistent with some of the most admired cities in the world. These results suggest that the solar envelope can provide urban legibility while, at the same time, satisfying many conditions of a sustainable future. ■

REFERENCES

Aveni, Anthony F., ed. 1975. *Archaeoastronomy in Pre-Columbian America.* Austin: University of Texas Press.

Coleman, Kim and Ralph Knowles. 1992. "A Case Study in Design Studio Teaching: The Parallel Studio" *The Design Studio in the 21st Century: Proceedings of the 1992 ACSA West Regional Meeting.* California Polytechnic State University, San Luis Obispo, CA.

Gross, Harvey, ed. 1966. *The Structure of Verse.* Greenwich, Connecticut: Fawcett.

Hayes, Gail Boyer. 1979. *Solar Access Law.* Cambridge: Ballinger.

Judd, Neil M. 1964. *The Architecture of Pueblo Bonito.* Washington, DC: Smithsonian Institution.

Kensek, Karen M. and Ralph L. Knowles. 1995. "Work in Progress: Solar Zoning and Solar Envelopes," *ACADIA Quarterly.* (Spring): 11–17.

——. 1997. Solar Access Zoning: Computer Generation of the Solar Envelope," *Proceedings of the ACSA SW Regional Meeting.* University of New Mexico, Albuquerque, NM.

Knowles, Ralph L. and Richard D. Berry. 1980. *Solar Envelope Concepts: Moderate Density Building Applications.* Golden, CO: Solar Energy Research Institute.

Knowles, Ralph L. 1974. *Energy and Form.* Cambridge: MIT Press.

——. 1980. "The Solar Envelope," *Solar Law Reporter,* Vol. 2. (July/August): 263–297.

——. 1981. *Sun Rhythm Form.* Cambridge: MIT Press. [out of print].

——. 1992. "For Those Who Spend Time In a Place," *PLACES 8:2* (Fall): 11–14.

——. 1996."Rhythm and Ritual: A Motive for Design," *Proceedings: ACSA Western Regional Meeting.* University of Hawaii at Manoa, Honolulu, Hawaii.

Knowles. Ralph L. and Marguerite N. Villecco. 1980. "Solar Access and Urban Form," *AIA Journal* (February): 42–49 and 70.

Lynch, Kevin. 1960. *The Image of the City.* Cambridge: The Technology Press & Harvard University Press.

Meyer, Marvin and Richard Smith, eds. 1994. *Ancient Christian Magic: Coptic Texts of Ritual Power.* New York: Harper Collins Publishers.

Robinson, David M. and J. Walter Graham. 1938. *Excavations at Olynthus: The Hellenic House.* Baltimore: Johns Hopkins University Press.

Schiler, M. and P. Yeh. 1993. SolVelope: An Interactive Computer Program for Defining and Drawing Solar Envelopes, *Proceedings: American Solar Energy Conference,* Washington, D.C.

Scully, Vincent J. 1975. *Mountain, Village, Dance.* New York: Viking Press.

Steele, James. 1997. *Sustainable Architecture: Principles, Paradigms, and Case Studies.* New York: McGraw-Hill.

Thomas, William. 1976. "Access to Sunlight," *Solar Radiation Considerations in Building Planning and Design: Prodeedings of a Working Conference* (National Academy of Sciences, Washington, D.C. (14–18).

White, Mary R. 1976. "The Allocation of Sunlight: Solar Rights and the Prior Appropriation Doctrine" *Colorado Law Review,* vol. 47: 421–427.

Baruch Givoni

Summary

Many features of the physical structure of the city can affect the urban microclimate. Therefore it is possible to modify the urban climate through urban policies and design of neighborhoods and whole new cities. With such modifications it is possible to improve the comfort of the inhabitants outdoors and indoors and to reduce the energy demand for heating in winter and for cooling in summer.

The following urban design elements that affect the urban climate are discussed in this article, with reference to the research literature:

- Size of the city
- Orientation and width of the streets
- Density of the built-up area
- Height of buildings
- Building color
- Effect of parks and other green areas on the urban climate

Key words

albedo, climatology, cool roofs, density, evaporative cooling, heat island, landscape, orientation, shading, ventilation, wind protection

San Francisco. The microclimatic effects of urban areas are subject to complex interaction of regional meteorology, building configuration, orientation, surfaces and materials.

Urban design and climate

This article reviews literature in building climatology documenting experimental studies of the effects of building shape and orientation, landscape, ground and building surfaces, on the microclimate. The microclimate as defined by Geiger (1965) is the conditions at or immediately near surface levels as affected by its physical attributes. The first part discusses general affects of urban climatology. A second part discusses issues related to the effect of green roofs and parks.

1 URBAN CLIMATOLOGY

In big cities it is common to observe, at the late hours of the nights, air temperatures that are 3–5°C (5.4–9°F) higher than the surrounding areas, and in extreme cases, higher by up to 8–10°C (14.4–18°F) This describes the "Urban Heat Island" effect. Heat islands are urban areas whose temperatures exceed the surrounding suburban and rural temperatures because the increased amount of buildings and asphalt retain heat. The heat island intensity is defined as the maximum difference between the urban center and the open country obtained by automobile traverses, usually during clear calm nights. The main cause of this phenomenon is the faster rate of cooling of the open areas around cities compared with the rate of nocturnal cooling of densely built-up urban centers. The larger and denser the city is, the greater is the difference in air temperature commonly observed between the center of the city and the surrounding area during the nights. During the daytime hours, however, this difference in air temperature between the city and its surrounding area is smaller—only about 1–2°C (1.8–3.6°F) degrees—and often the daytime temperatures in a densely built area may be lower than in the open.

The attempts to describe this effect quantitatively met with difficulties in expressing the size and density of the city numerically. Part of the factors causing the heat island phenomenon depend upon the size and density of the population, as well as on its standard of living, such as vehicular traffic, intensity of heating in the winter and air conditioning in the summer, and industrial plants. Other factors depend upon the size of the built-up urban area and the building density. In most cases the density of buildings and energy producing activities in the center of cities also increase with the size of a city.

Two types of factors affect the differences between the urban and the "rural" temperatures. One, they are correlated with meteorological factors such as the cloud cover, humidity, and wind speed. Two, various features of the urban structure, such as the size of cities, the

Credits: This article is a summary by the author of literature search and studies reported in his publications cited in the References. Photos are by Donald Watson, unless otherwise noted.

density of the built-up areas, and the ratio of building heights to the distances between them can have strong effect on the magnitude of the urban heat island.

The size of a city is a parameter easy to define and to obtain. On the other hand, the density of buildings, although having more direct causal relationship to the nocturnal urban heat island, is a complex urban feature and in real cities, very difficult to be defined in a way meaningful to urban climatology. Therefore, it is convenient to substitute the city population size for its density.

It has been demonstrated (Oke 1982) that the maximum difference between the urban center and the open country can be statistically related to the size of the population of the city. The heat island decreases, however, as the regional wind is stronger. Oke has also found that for European cities of a given size, the heat island is weaker than in North America. He attributes this discrepancy to the fact that the centers of North American cities have taller buildings and higher densities than typical European cities. Jauregui (1984) has added to the work of Oke his own data on numerous cities located in low latitudes in South America and India. He has shown that the heat islands in South America's cities are weaker than even in the European ones. Jauregui suggests that this phenomenon can be attributed in part to the difference in morphology (physical structure) between the South American and the European cities.

Density of the built-up area

Building density affects both the urban temperature and the wind speed. A given urban density can result from independent design features, which affect the urban climate in different ways such as:

- Fraction of urban land covered by buildings,
- Distances between buildings, including streets' width,
- Average height of buildings.

Higher urban density means larger ratio of the height of the buildings to the distances between them (H/W) and less view of the sky (a smaller sky view). Several attempts have been made to express the intensity of the heat island as a function of those urban physical features, based upon the hypothesis that the urban heat island is caused mainly by reduced radiant heat loss to the sky from the ground level of densely built urban centers where the heat island phenomenon is measured.

However, during daytime hours the effect of the width of streets on temperature is often quite different. In July 1988 the author took air temperature measurements during two days, at about 1 m (3.3 ft.) height, in three streets of very different width, ranging from a wide avenue to a very narrow alley in Seville, Spain. The measurements were taken with a sling psychrometer between 6 a.m. and 11 p.m. (Givoni 1998). In the early morning, the temperature in the wide avenue was the lowest (in accordance with the urban heat island models) but during the rest of the day, especially around noon and in the afternoon hours, the temperature patterns were reversed. The wider the streets the higher were the temperatures. The highest temperature was measured in the wide modern avenue. The lowest temperature was in the very narrow alley, with a height-to-width ratio (H/W) of about 6. The largest differences between the temperatures in the avenue and in the alley occurred in the afternoons.

An extensive study of the effect of street width, and the resulting solar radiation reaching the ground, was conducted by Sharlin and Hoffman (1984). They took continuous measurements at nine "stations" in the Tel Aviv area, Israel, along two traverses equidistant from the sea, over periods of 21 days in the summer of 1979 and in the winter of 1980. The nine stations were characterized by various numerical indices, each expressed a specific characteristic to the plot area. The most significant were found to be the ratio of the buildings envelope (walls) area to the site area and the ratio of the permanently shaded area around the buildings to the plot area. Other characteristics included the total built and paved area, green area, and estimated population, but their effect was found to be statistically not significant.

In summer both the effects of the buildings exterior wall area and the permanently shaded areas were significant in affecting the maximum temperatures and the daily temperature range. In winter the effect of only the buildings envelope area was significant. Both studies, Sharlin and Hoffman's extensive research and Givoni's two days measurements, indicate that when higher urban density reduces the amount of sun reaching the street and increases the amount of mass that absorbs the radiation and its surface area, it can effectively lower the urban daytime maximum temperature.

Urban density and wind speed

A higher urban density generally reduces the wind speed in the urban area, as a result of increased friction near the ground. However, this effect depends mainly on the various physical details of urban space, including the orientation of the streets and the buildings with respect to the wind direction (Figs. 1–3). It is possible therefore to have a wide range of ventilation conditions for a given level of density. Distances between buildings, either across streets or within an urban block, greatly affect the ventilation conditions, both outdoors and indoors. With the same height of buildings, reducing the distances between them reduces the urban wind speed. With the same pattern of land coverage, higher buildings reduce the ground level wind speed.

When the wind, flowing over open land, approaches an urban area the first rows of buildings divert the approaching wind current upwards, and the rest of the buildings behind are left in the wind "shadow" of the buildings standing in front of them. In this situation, two separate airflow regimes are created. The regional air currents flow mainly over the tops of the buildings, creating a suction zone above the buildings that is strongest above the first row of buildings and decreases gradually as the wind flows over successive rows.

In between the buildings a secondary air current is created as a result of the friction between the upper air currents and the building. This secondary flow is strongest in the first street, as a result from the strongest suction above the buildings, but still much slower than the wind speed over the open area, and it decreases gradually in the successive streets downwind. As the distances between the buildings are smaller and/or the height of the buildings is higher (a higher H/W), the wind speed between the buildings at ground level is decreasing.

High-rise buildings located among lower buildings

One of the main factors determining the effect of building density on ventilation conditions in the city is the difference in height of buildings. Under a given density condition, high buildings with large open spaces between them will have better ventilation conditions than closely spaced low buildings—approaching in the extreme a pattern similar to a lone building with open space on all sides. However, more than the average building height, it is the difference between building heights that affects the ventilation conditions.

Individual buildings rising high above other buildings around them create strong air currents in the area. This phenomenon is due to the fact that the high-rise building is exposed to the main wind currents that flow above the "general" level of the urban canopy, and are stronger than those flowing through the urban canopy itself. Against the facade of the high-rise building which faces the wind, a high air pressure pocket is formed, which causes a strong downward current, and in this way mixes up the air layers near the ground between the lower buildings.

In cold countries, and during the winter months in many hot countries, this current is undesirable as far as the comfort of the local residents is concerned, although it is always helpful in dispersing air pollutants from traffic that are generated near the ground. During the summer, in warm humid regions, which often experience light winds, the stronger air currents may be welcomed for increasing the comfort level of the local residents.

Paciuk (1975) has conducted extensive wind tunnel measurements, under the direction of the author, on the effect of building height, distances between buildings and the effect of high-rise buildings on the wind speed in the streets. With uniform height, when the H/W was changes from 1:3, through 1:1 to 3:1, the respective airspeeds between the buildings were 28%, 21% and 14% of the wind speed in the open area upwind of the buildings, at the same height above ground.

Introducing tall buildings—which rise appreciably above the roof level of the neighboring buildings in specific locations within the urban fabric—can greatly increase the wind flow pattern in the streets at the "pedestrian level." When a high-rise building is placed, along with lower buildings, on the downwind side of a street running perpendicular to the wind direction, a high pressure is created on the windward side of the building, causing strong downward airflow. When another high-rise building is placed on the upwind side of the same street at some distance from the first building, a low pressure (suction) is created in the street on the leeward side of that building. A strong airflow is created in the street, flowing from the pressure point to the suction point and then upward.

As a general pattern, in all the configurations with the "towers" the highest speeds were measured in front of the towers (the pressure point), but the air speeds have been increased in the whole configuration, relative to the speeds with the same height buildings. While addition of high buildings increases the density of the built-up area, their impact can, in effect, be to increase substantially the overall urban wind speed. However, with inappropriate locations, such as placing concentration of high-rise buildings upwind of a neighborhood, the high-rise buildings can block the wind and reduce appreciably the wind speed in the affected area.

The shape of the upwind wall of a high-rise building can modify the flow pattern. A convex wall diverts more air to the sides and less upward and downward. It smoothes the deflection of the flow and therefore reduces the resulting turbulence at the sidewall and at the windward wall. On the other hand, a concave windward wall concentrates the flow along this wall, upward and downward. As a result, the turbulence increases.

Fig. 1. Wind tunnel testing of urban morphology.

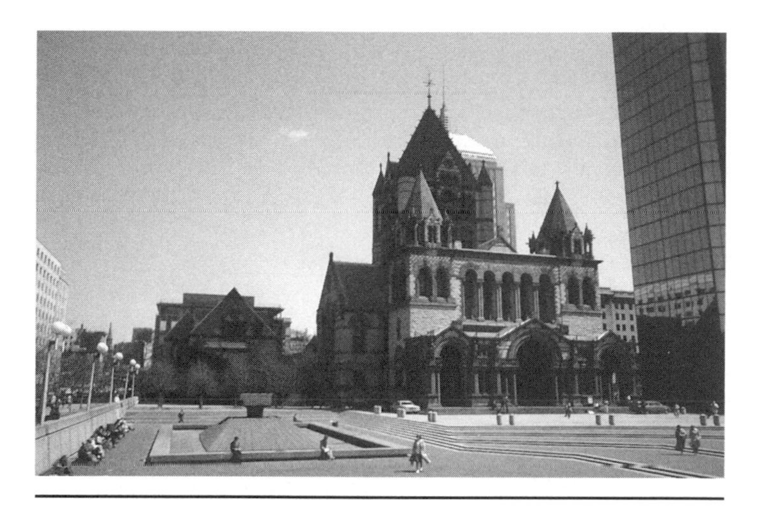

Fig. 2. Copley Plaza, Boston. Sun-exposed and wind-sheltered seating area provides tolerable microclimate in otherwise cool, wind-blown season.

Fig. 3. Sea Ranch. Landscape and building form reflect the dominant wind patterns of the North California Coast. (Lawrence Halprin, Landscape Architect.)

A setback of the tower, with respect to its "base," starting about 6–10 m (20–33 ft.) above street level, can eliminate most of the down flow at the street, where it affects the pedestrians. Such a design solution still maintains the positive effect of the high-rise building on the mixing rate of the street level polluted air with the clearer air from above.

Street orientation

In many urban situations, especially in commercial districts, the main facades of the buildings face the street. The orientation of the streets with respect to the sun may thus affect the solar impact on the buildings and the solar utilization potential. Street orientation with respect to the wind direction greatly determines the ventilation potential of the buildings, as well as the outdoor ventilation conditions. The width of the streets determines the distance between the buildings on both sides of the street, with impacts both on the ventilation and solar exposure of the buildings.

A north-south orientation of a street may result in an east-west orientation of buildings along and parallel to the street, which will cause unfavorable solar exposure for these buildings. From the solar exposure viewpoint an east-west street orientation is preferable. In dust-prone areas, common in hot, dry regions, wide streets parallel to the wind direction may aggravate the dust problem in the town as a whole. As the wind direction in many of the hot, dry regions is from the west, there is a conflict between the solar and the dust considerations with respect to street orientation.

Street orientation and urban ventilation

In built-up areas wind speeds vary in the streets, as well as around and between buildings, depending on the relationship between the wind direction and the orientations of both the streets and the buildings, respectively.

When city streets are perpendicular to the wind direction, and the buildings lining these streets are long row buildings, the principal air current flows above the buildings. The airflow in the streets is mainly the result of a secondary air current, caused by the friction of the wind blowing above the city against the buildings lining the streets. Under these conditions the ventilation of urban space is suppressed. The width of the streets has a minor effect on the urban ventilation, within the range encountered in urban areas.

When the city streets are parallel to the direction of the wind, they create obstacle-free passageways, through which the prevailing winds can penetrate into the heart of the urban area. In this case, as the streets are wider, the airflow encounters less resistance from the buildings on the sides of the streets, thus improving the general urban ventilation. A similar phenomenon occurs when the streets lie at a small angle to the prevailing winds

When the wind is oblique to the streets and the buildings along it (assumed parallel to the streets) then the wind pressure and speeds will be very different over the two sides of the buildings and along the walkways. Pedestrians on the sidewalk at the downwind side of the street will experience a much higher wind speed than the pedestrians on the sidewalk at the up-wind side of the street. The reason is that along the downwind buildings a strong downward airflow will be generated by the winds, especially near junctions with cross streets. This will directly affect the down-wind sidewalk down flow of air, while the other sidewalk will be sheltered from the wind.

In regions where higher near-ground wind speeds are desirable this situation can be modified to a great extent by suitable placement of high-rise buildings. Such buildings create zones of high and low pressures above the built-up area, and thus generate vertical currents stirring the urban air mass, as was discussed above.

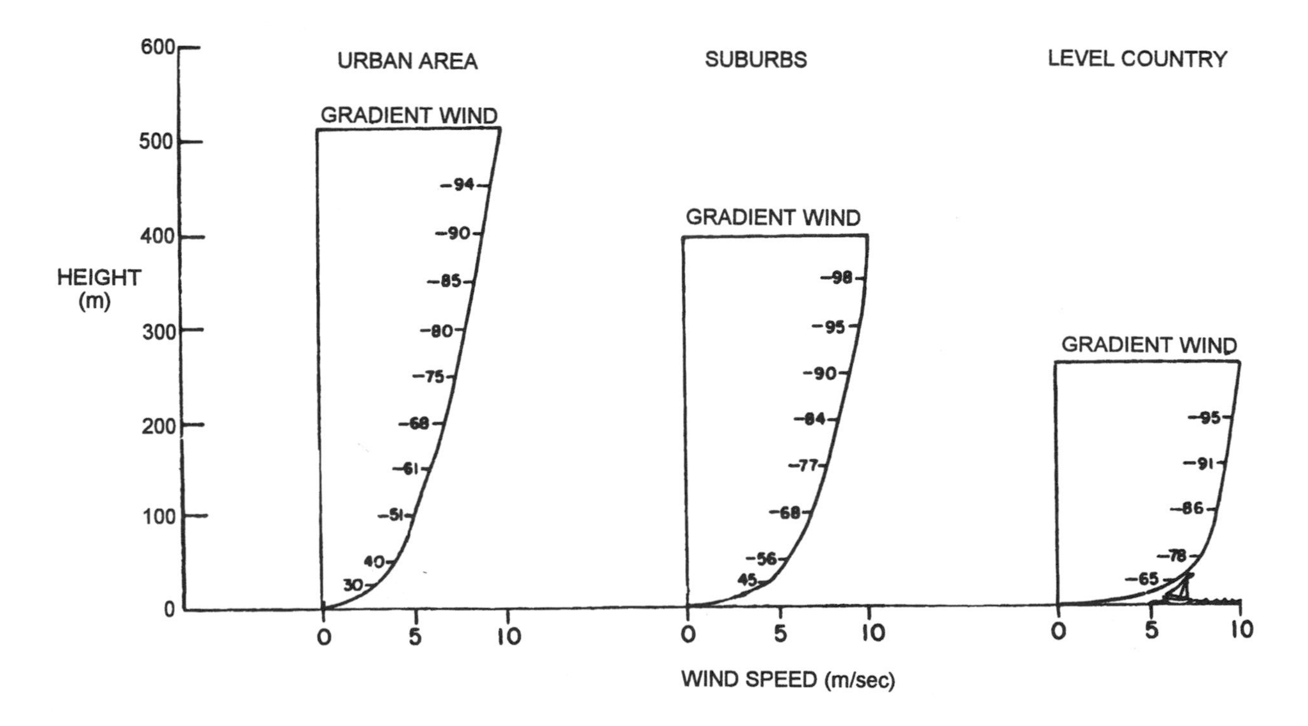

Fig. 4. Effect of terrain roughness on the wind profile. Diagrammatic wind speed profiles above urban, rural and sea surfaces; percentage of the gradient wind speed. (Givoni 1998.)

Urban albedo (color)

The urban energy balance and temperature depends on the amount of solar radiation absorbed within the urban fabric. This, in turn, depends on the average urban albedo. Albedo is defined (Taha *et al.* 1988) as the ratio of reflected-to-incident radiation at a particular surface over the whole solar radiation spectrum. It depends mainly on the color of the surface. Dark colors have low albedo while a white color has a very high albedo. The average urban albedo depends on the color of the roofs, walls and roads, parking areas and so forth. The land fraction covered by vegetation affects the urban temperature but this effect is not due to the albedo of the plants, as the solar absorptivity of leaves is rather high, but to the evaporation from the leaves (Fig. 5).

The urban albedo is the main factor determining the amount of solar radiation absorbed in the urban area. The color of the built urban elements, especially the building's roofs, is controllable by urban design. Because the roofs comprise a large fraction of the urban area in a densely built town, the radiation balance in such cases can be controlled and thus may have a pronounced effect on urban air temperature (Fig. 6).

Fig. 7 illustrates the effect of roof color on their surface temperature. Surface temperatures taken during three days, of two identical roofs as measured in Haifa, Israel, are shown in the figure. One roof was painted gray color (albedo of about 0.3) and the other roof was painted white (Albedo of about 0.75). The ambient air temperature has ranged from about 20°C (68°F) to 26–27°C (78–81°F). The maximum surface temperature of the gray roof was about 58°C (114°F) and that of the white roof was 26–28°C (78–82°F). The minimum temperatures of the two roofs were the same, about 15°C (59°F). The drop of the surface temperatures below the ambient air temperature during the night hours results from the longwave radiation to the sky, reflecting the identical emissivity of the two colors.

The average surface temperature of the gray roof was about 35°C (95°F), while that of the white roof was about 21°C (70°F). Thus, while the dark roof is a source of heating, the white roof serves as a

Fig. 5. Street scene in Portugal illustrates design responses to variable climatic conditions, overhands and shaded balconies, adjustable awnings, ventilating shutters, bright and light colors.

Fig. 6. Roofs of Bermuda dwellings are painted white, principally as a means of maintaining water collection, and have the effect of dramatically reducing solar heat gain.

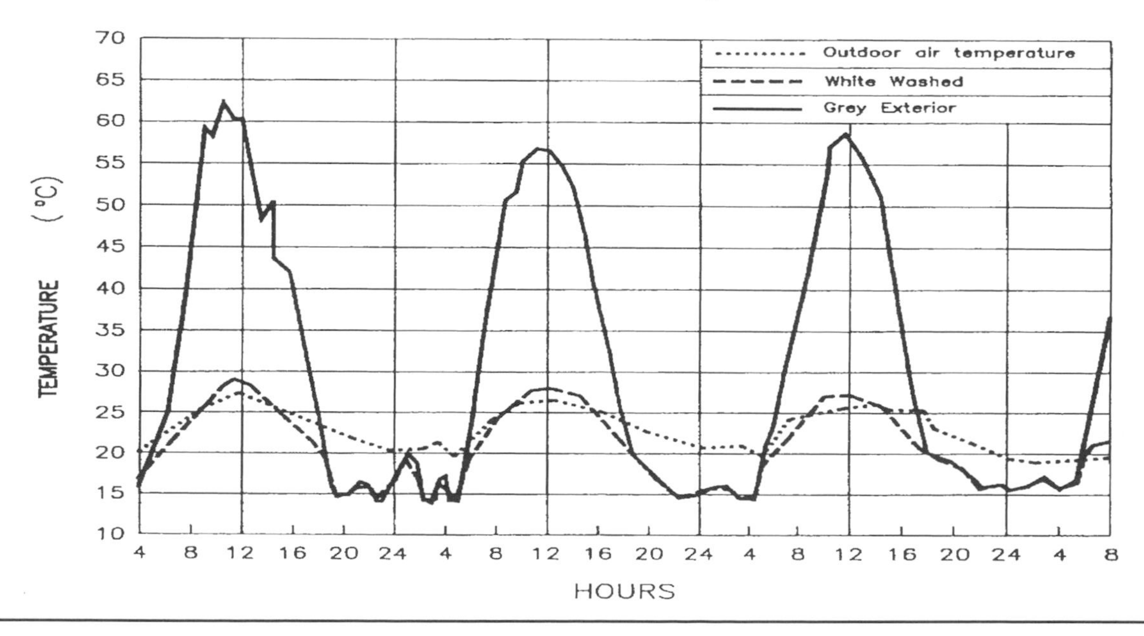

Fig. 7. Effect of roof color on surface temperatures. (Givoni 1994.)

cooling element with respect to the ambient air. In densely built cities, where roofs comprise a significant fraction of the urban area, ensuring white roofs can change the radiant balance of the city and reduce significantly its daytime temperatures.

Changing the color of a building and increasing its albedo reduces the amount of solar energy absorbed at the building's envelope and thus reduces its cooling load. Akbari *et al.* (1990) have simulated the saving in cooling energy and in peak power in July in Sacramento, California, by changing the albedo of a house and its surroundings from 0.25 to 0.40. They have calculated that the cooling energy would be reduced by 45% and the peak power by 21%.

COMMENTS ON URBAN DESIGN IN DIFFERENT CLIMATES

Hot, dry regions

The main objective of urban design in hot, dry regions, from the climatic aspect, is to mitigate the stresses imposed by the climate on people staying outdoors (working, shopping, playing, strolling). An additional objective is to improve the chances of the individual buildings to provide a comfortable indoor environment. To ameliorate the heat stress on summer days, neighborhoods should be planned so that distances for walking people and playing children are short. Sidewalks should be shaded as much as possible, either by trees or by the buildings along them. Shade is particularly desirable in places where people (and mainly children) congregate outdoors during daytime hours.

Urban density considerations in hot-dry regions

A high density of the built up area (land coverage by buildings) in hot, dry climate may have both positive and negative effects on human comfort outdoors and on the indoor climate of the buildings. With a given building's height increasing urban density means smaller open spaces between and around the buildings. The effect of the reduced distances in hot-dry regions depends to a large extent on the orientation of the walls in question and on the color (albedo), especially that of the roofs.

When the distance between the buildings is decreased along the east-west axis the mutual shading of the E/W walls of a given building by its neighbors increases. As long as adequate natural ventilation of the building can be achieved, the effect on the indoor climate can be beneficial because the solar impact on the walls in summer is reduced. The effect of distance between buildings along the north/south axis is quite different. In winter it may reduce the possibilities of utilizing solar energy for heating, which has great potential in hot, dry regions.

It is possible to infer from theoretical considerations, as well as from actual measurements (Givoni 1998) in streets and around buildings, that in hot, dry regions it might be possible to plan high-density residential neighborhoods in cities so that the ambient daytime air temperatures would be lower than in the surrounding country.

High land coverage by buildings means that a large part of the radiation exchange will take place at the roof surfaces and not at the ground level and at the walls. By assuring that all the roofs are colored white—by yearly repainting, for example—it is possible to achieve a negative radiation balance: the longwave radiant loss can substantially exceed the absorbed solar radiation even on a clear day in mid-summer. Under these conditions, the average temperature of the roof surfaces, and the air in contact with them, would be lower than the average regional air temperature. If the neighborhood is large enough, and built densely enough, it can be assumed that it will be possible to achieve a daytime air temperature at the street level that is than in the surrounding arid areas.

In the open areas between buildings, planting trees should be encouraged as much as water availability and financial resources enable. Solar radiation absorbed in leaves of vegetation increases the evaporation rate instead of raising the temperature. The resulting elevated humidity, in an arid region, can be welcomed from the comfort viewpoint. Thus, large coverage of the land surface by a combination of white roof and trees can result in significant lowering of the urban temperatures in hot, dry regions.

Hot, humid regions

The main urban design objectives in hot, humid regions involve two types of issues:

- Minimizing the hazards from tropical storms and floods.
- Enhancing comfort by providing good natural ventilation.

Minimizing flood hazards by urban design features

Floods in urban areas can be caused either by water flowing through the city originating in far away areas or by excess rainwater generated over the area of the city, which cannot be absorbed in the soil or discharged away fast enough. During rains, the city itself increases the excess of water run-off because the ability of the ground in an urban area to absorb water is reduced, as a result of the coverage of the land by buildings, roads, and parking areas. This factor increases the risks from floods in low-lying flat urban areas.

The risk of floods from excess rainwater within the boundary of the city itself can be minimized by details of urban design. The following design details can be applied to achieve this goal:

- Increasing rain absorption in soil in the urban area, thus reducing run-off,
- Preserving land features of natural drainage such as interconnected valley systems,
- Collecting excess run-off in urban reservoirs, such as minilakes.

A significant portion of the urban surfaced area is not subjected to heavy vehicular traffic, such as parking lots, and pedestrian areas. These areas can be surfaced with permeable pavement, such as open-grated concrete blocks, and special bricks. Soil can be laid and seeded with suitable grass to promote infiltration. A layer consisting of a mixture of sand and gravel under the blocks can increase the area of effective infiltration below the semihard surface, thus increasing the rate of water absorption in the ground.

An interesting planning policy for increasing absorption of rainwater in the ground was implemented in the city of Davis, California. All public parking areas are surfaced with perforated blocks. Instead of underground drainage pipes, the runoff discharge is through interconnected wide and shallow grassy minivalleys—called bioswales —functioning as natural drainage features of the area. These shallow "valleys" form an integral part of the urban open spaces system. The natural exposed drainage channels proved to be more effective than the conventional underground drainage pipes during extreme rainstorms.

Enhancing ventilation

An effective street layout from the urban ventilation aspect in a hot, humid region is when wide main avenues are oriented at an oblique angle to the prevailing winds, *e.g.*, at about 30°. This orientation still enables penetration of the wind into the heart of the town. The buildings along such avenues are exposed to different air pressures on their front and back facades. The upwind wall is at the pressure zone while the downwind wall is at the suction zone. This street orientation thus provides a good potential for natural ventilation of the buildings while at the same time provides also good ventilation within the streets. It is desirable mainly in high-density residential urban zones.

An urban configuration to be avoided as much as possible in hot, humid regions is that of high long buildings, of the same height, perpendicular to the prevailing wind direction. This configuration blocks the wind and creates poor ventilation conditions both in the streets and for the buildings, as the first row of buildings acts as a wind barrier. An urban profile of variable heights, where buildings of different heights are placed next to each other, and when the long facades of the buildings are oblique to the wind, actually enhances the urban ventilation.

At a given density level, the theoretically best urban climate conditions exists in a hot, humid climate when that density is obtained with high narrow buildings ("towers"), placed as far apart from each other as is consistent with the given density. Such configuration provides the best ventilation conditions for the given urban section as a whole, and especially for the occupants of the buildings. Several factors contribute to this effect:

- Narrow, "towerlike" buildings located far apart cause mixing of the airstream at higher elevations with the air near ground. Part of the wind momentum at the higher level is thus transferred to the lower level, increasing the wind speed near the ground. This effect improves the ventilation conditions for the lower floors, as well as for the pedestrians in the streets and the open spaces between the buildings (Fig. 8).

- With this configuration a greater proportion of the population lives and works at higher elevation above the ground. Both the temperature and the vapor pressure decrease with height, especially in a densely built-up urban area. Therefore the comfort conditions for the people living in the upper floors are improved.

- Thermal updraft and mixing of the air layers from different heights captured by the high buildings, lowers the air temperature at the pedestrian level. This occurs because the air temperature at a height of 30–40 meters is lower than the air near the ground, which is heated by the warmer ground.

Cold climate

In view of the harsh climate during the winter in cold climates towns planners may face a dilemma: whether to "escape" the winter in enclosed spaces or to develop an infrastructure enabling the inhabitants of the town to enjoy potential winter outdoor activities.

Pressman (1988) suggests the following principles for urban design in cold climates:

- High density in the residential, retail and commercial sectors of the city, in order to reduce the transportation needs and the space heating requirements. High urban density implies more intensive use of the land.

- Accommodating diverse uses in the same building: residences, offices, shops, in order to enable people to work and shop in the same building complex where they reside.

- Mixed land use on the urban scale: mixing residential, offices, retail, industrial, and recreation land uses within the same neighborhood. The objective is to reduce the need for commuting.

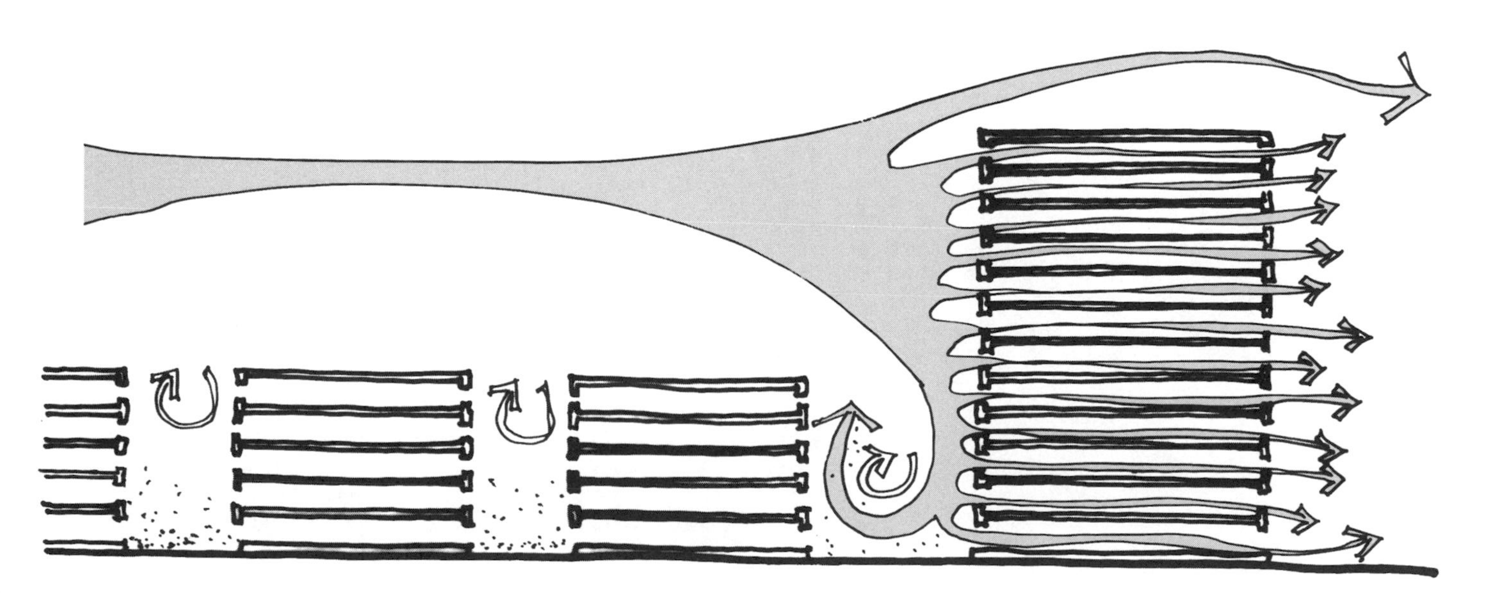

Fig. 8. Wind speed near the ground in front of a high-rise building is increased. (After Givoni 1998.)

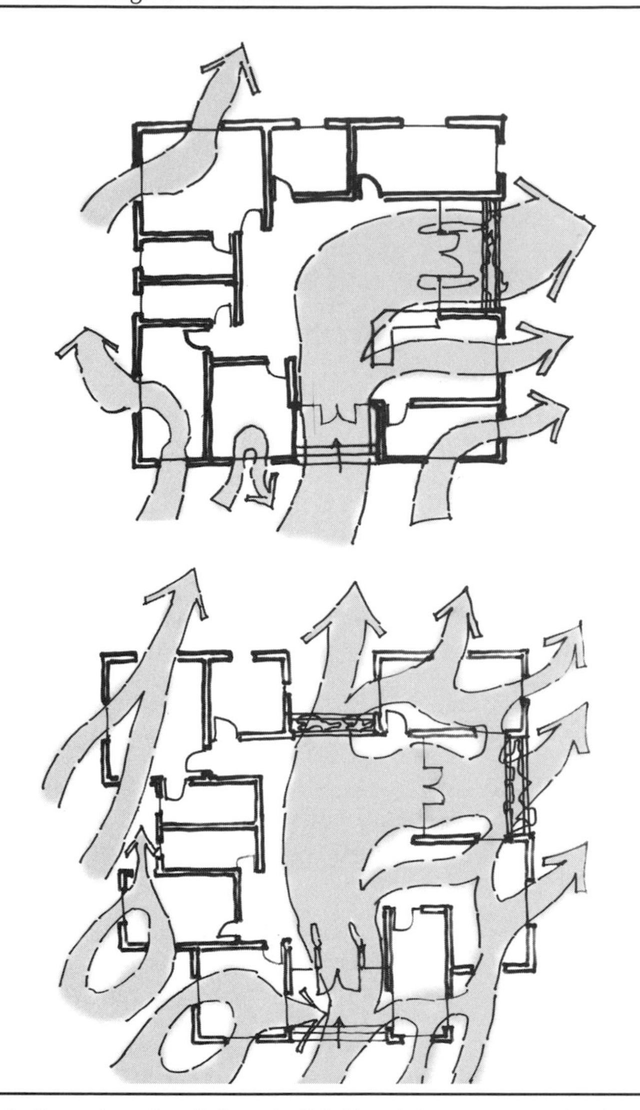

Fig. 9. Comparison of ventilation potential of two designs: one compact and one spread-out, with the same area and program. (After Givoni 1998).

Fig. 10. Charleston, SC. Traditional dwelling type (narrow street frontage) provides shaded portico facing interior courtyard garden for sun tempering and cross-ventilation.

Pressman suggests that such mixed land use will enhance self-sufficiency of the neighborhood because a broader range of services can be made economically viable with improved accessibility.

- Promoting public transit through urban planning, as this is the most energy efficient form of movement. To be cost effective it must serve a high-density area.
- Complementary location of various urban functions. Complementary functions should be grouped according to relative compatibility: residential and work areas linked by variety of movement modes, recreational zones accessible from both employment and dwelling zones.

Covered and glazed streets

In commercial urban sections, a street covered by a glazed roof ("passages") provides protection from the wind and the snow to a large volume of pedestrians, and thus enhances greatly the comfort of the customers in the winter. This, in turn, may attract more customers to the covered street and to the stores and offices located in those places, increasing their profitability. The owners may thus be inclined to cover the extra expenses involved in covering the street.

Covered highways, usually underground, can provide snow-free traffic between urban sections and across the city. In Oslo, a new three-lane tunnel highway has enabled snow-free car traffic under the city and reduced significantly the number of cars crossing the City Hall Square. It also opened the waterfront for pedestrian use and enjoyment.

2 GREEN AREAS AND THE URBAN CLIMATE

Urban "green" areas, in public open spaces like parks as well as in private planted areas around buildings, can have a marked effect on the climatic conditions to which an individual building is exposed and on the urban climate at large. The type and details of the plants around a building can affect its exposure to the sun and the wind, its indoor comfort conditions and energy use. The accumulated effect of trees and other plants in the urban area can modify the urban temperature, the urban wind conditions and the solar exposure of buildings and of pedestrians in the streets, in public parks and other open spaces (Figs. 9–12).

Leaves of plants absorb most of the solar radiation that strikes them but transform a very small part of the radiant energy into chemical energy by photosynthesis. Most of the absorbed solar radiation is transformed into latent heat by diffusion and evaporation of water from the leaves (a process known as evapotranspiration). The evaporation significantly cools the leaves and the air in contact with them and at the same time increases the humidity of the air. The importance and desirability of this factor depends on the local humidity and temperature conditions. As a result of the evapotranspiration process, the air near the ground in green areas is cooler than the air in built-up areas covered by asphalt or concrete or over land bare of plants. Furthermore, as a result of their lower temperature, the longwave radiation emitted from leaves is lower than that emitted from the surrounding walls and other hard surfaces.

The effect of vegetation on wind conditions depends to a great extent on the type of the plants and on the details of the planting pattern.

Grassy areas pose the least friction (resistance) to the wind and allow for the best ventilation conditions. Bushes impede the wind near the ground surface and above their foliage. The type and density of trees, in particular, have a noticeable impact on the wind speed near the ground. A densely planted row of trees may serve as a windbreak. A grove of trees lowers significantly the wind speed within it and downwind from the grove. Blockage of the wind by plants is desirable in cold regions and seasons, but is undesirable in hot climates and seasons, especially in hot, humid regions.

After general discussion of the climatic effects of plants, this section presents quantitative information obtained in experimental studies in which the effects of various plant types were investigated. This presentation is divided into three sections:

- Shading effects of plants.
- Effect of plants on ambient and surface temperatures.
- Effects of plants on the wind speed.

Climatic effects of plants upon comfort and energy use

Trees and other plants modify the climatic conditions to which buildings, as well as persons staying outdoors, are exposed. Trees and vines provide shade over windows and walls of buildings, as well as shading persons staying outdoors. All plants reduce the ambient air temperature and increase the moisture content of the air. Most plant types reduce the wind speed near the ground.

The effects of green spaces on the urban climate depends on the fraction of "green" areas, public as well as private, relative to the whole built-up urban area. The effect of urban plants can be different, under different planting schemes, especially with respect to solar exposure of pedestrians and the wind speed in the streets, depending on the particular choice of the plants and the details of the landscaping.

In high-density cities, where buildings and roads cover most of the land, and only very limited area is available for plants, the effect of plants on the air temperature may be rather small. Their main contribution to the urban climate in such cities may be in providing shade for pedestrians. On the other hand, in suburban areas where more land is available for vegetation and the fraction of the "green" area can be increased, the effect of plant can be more significant.

Plants can affect the indoor temperature and the cooling and heating loads of buildings in several ways:

- Trees with high canopy, and pergolas near walls and windows, provide shade and reduce the solar heat gain with relatively small blockage of the wind (shading effect) (Fig. 13).

- Vines climbing over walls and high shrubs next to the walls, while providing shade, also reduce appreciably the wind speed next to the walls (shading and insulation effects) (Fig. 14).

- Dense plants near a building can lower the air temperature next to the skin of the building, thus reducing the conductive and the infiltration heat gains. In winter they, of course, reduce the desired solar gain and may increase the wall's wetness after rains.

- Ground cover by plants around a building reduces the reflected solar radiation and the longwave radiation emitted toward the

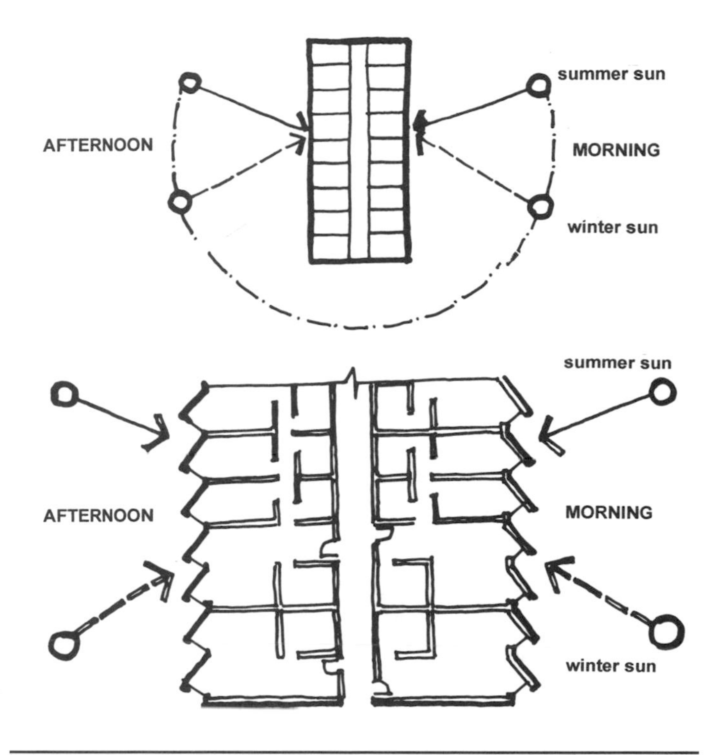

Fig. 11. Triangular bay windows in east- and west-facing walls, maximize solar gain in winter and minimize solar overheating in summer.

Fig. 12. Carpenter Center, Harvard University, Cambridge, MA. Le Corbusier. Angled *brise soleil* admit direct winter sun and reflected sunlight year-round.

walls from the surrounding area, thus lowering the solar and long-wave heat gain in summer.

- If plants lower the ambient temperature around the condenser of an air-conditioning unit of a building, the coefficient of performance (COP) of the cooling equipment would be improved.

- By reducing the wind speed around a building in winter, plants can reduce the infiltration rates and the heating energy use of the building (insulation effect). However, trees and shrubs can also direct the wind to a desired spot, *e.g.*, an opening serving as an inlet for ventilating a building, or to a bench in a park.

- Evergreen trees on the southern side of a building reduce its potential to use solar energy for heating. Trees and high shrubs at the western and eastern sides of a building can provide effective protection from solar gain in summer.

Quantitatively, the effects of plants depend on the density and thickness of the foliage layer and the type of the leaves of the plants. These properties change with the age of the plants and with the seasons. The seasonal changes are greatest, of course, in the case of deciduous plants.

Experimental data on the effect of plants and green areas

Although much is written on the shading effect of plants, not many controlled experimental studies are done on this subject. A selection of experimental studies that have been conducted are highlighted below. In different studies where air temperatures were measured, the measurement points were at different heights above the ground. Therefore the quantitative effects observed in the different studies are not directly comparable.

Al Hemiddi has measured the effect of surface treatments, including plant cover, on ambient temperatures (Al Hemiddi 1991). In this study the surface and air temperatures at a height of about 1 m (3.3 ft.) were measured above land areas with different ground treatments in the UCLA campus. The measurements were taken around noontime during different periods over an entire year for a total of about 70 days. The ground treatments included a shaded sidewalk, unshaded paved plaza, unshaded lawn, a space between a high and dense shrub fence and a building, and a parking lot.

Table 1 shows averages of air and surface temperatures measured during a sequence of several clear days in summer. The lowest temperatures were in a space between a line of high shrubs and a wall of a building. The highest temperatures were, as expected, in the parking lot. During the hottest period of the study, the parking lot surface reached about 50°C (about 120°F) while the surface of the unshaded lawn was about 29°C (88°F) and that of the shaded sidewalk about 13°C (73°F).

Parker (1983) has studied the effect of landscaping on walls' temperatures in Miami, Fl. No information was given on the color of the walls. On hot sunny late-summer days the average temperature of walls shaded by trees or by a combination of trees and shrubs was reduced by 13.5–15.5°C (24–27.5°F). Climbing vines reduced the surface temperature by 10–12°C (18–21.6°F). Parker also measured the effect of landscaping on cooling energy consumption of a test building with and without landscaping. The landscaping consisted of trees and shrubs around the building. Average daily rate of energy consumption for air conditioning on hot summer days was reduced by 59% as a result of landscaping.

Taha *et al.* (1989) have measured the air temperature and wind speed within the canopy of an isolated orchard and in the upwind and downwind open areas to the south and north of it, in Davis, California. The orchard size was approximately 307 m (1006 ft.) long N-S and 150 m (490 ft.) wide E-W. The canopy had covered about 30% of the land area and the soil was wet from irrigation. Empty fields stretched for 2 kms (1.2 miles) away from the canopy. To the south of the orchard was a stream lined with a strip of tall trees and shrubs. The daytime wind was mostly from the north and the evening and night winds from the south. Air temperature and wind speed were measured at 1.5 m (5 ft.) above ground. Measurements were reported for three "stations" within the orchard and for two stations in each the northern and the southern end of open fields (behind a strip of trees). The measurements period lasted for two weeks (October 12–25, 1986). The northernmost field station served as a control, from which deviations of the temperature and wind speed at the other stations were calculated as "effects."

The average maximum temperature within the orchard was 23.9°C (75° F) while in the upwind field it was 28.3°C (83°F). The average minimum temperature in the orchard was higher than in the open field, 2.7 vs. 1.8°C (37°F vs. 35°F). The average maximum wind speed in the orchard was much lower than in the open upwind field, 3.75 vs. 8.5 m/s (750 vs. 1,700 fpm).

Sandifer and Givoni have conducted experiments at UCLA since 1997, quantifying under controlled conditions the effects of vines on wall surface temperatures (Givoni and Sandifer 1997). Various plant varieties were grown in front of both west and south facing test panels, including evergreen jasmine and deciduous grape. One of the objectives of this study was to evaluate the interrelationship between the shading effects and wall's color. The jasmine has provided effective shading year round (average shading effectiveness about 63%). The shading effectiveness of the grape in summer was as high as that of the jasmine but in winter, from November through April, the shading by the bare branches of the grape fell to about 30%.

Potchter (2001) has measured the effect of shading a western concrete wall by a tree in Tel Aviv. The measurements included the ambient air temperature and the wall's surface temperatures at two points: one point exposed to the sun and the other point shaded by a nearby tree. With ambient air temperature of about 28.5°C (83°F) in the afternoon, the temperature of the wall exposed to the sun was 41.5°C (106.5°F) and that of the shaded wall was about 30.5°C (87°F).

Effect of plants on wind speed

In any urban environment the spatial distribution of wind speeds varies greatly over very short distances. Even the value of the average wind speed in a given area depends on the choice of the exact specific locations of the points of measurements. Several experimental studies in which the effect of plants on wind speed was also measured are summarized below.

DeWalle (1983) has measured air infiltration and heating energy use in a small mobile house in central Pennsylvania. The mobile house was first calibrated to evaluate its infiltration rate and then located at different places, either "open" or protected, at various distances from a pine tree windbreak, expressed as multiples of the windbreak height,

H. Air infiltration and heating energy use were expressed as function of the wind speed in the open and of the inside-outside temperature difference. Measured infiltration rates and heating energy were then compared with the "predicted" ones. At distances of 1 to 4H from the windbreak the air speed was reduced by 40 to 50% of the undisturbed wind. The infiltration reductions were from 55% (at 1*H*) to 30% (at 4*H* and 8*H*). Heating energy reductions were about 20% (at 1*H*) to about 10% (at 4*H*).

Hoyano (1988) has installed a vertical vine (dishcloth gourd) in front of a veranda of a house and compared the air speed at the center of the window in a room behind a screened and an unscreened veranda. The vertical vine screen has reduced significantly the wind speed through the window in the screened veranda and the cross ventilation of the room behind it, as compared with the unscreened veranda. Without the screen the air speed at the center of the window was on the average about 45% of the outdoor wind speed while with the vine screen in the veranda it was about 17%. Thus the overall effect of a vertical screen on comfort in a hot, humid climate may be negative, due to the reduction of the indoor air speed.

Effect of urban tree canopy on cooling energy use

McPherson *et al.* (1989) measured the effect of landscaping on the cooling energy consumption of three 1/4-scale models of a building. The building's floor size was 3.7 x 3 m (8.4 x 5.4 ft.) and the plot's size was 15.3 x 15.3 m (27.5 x 27.5 ft.). The landscape treatments consisted of: (a) Bermuda grass turf around the building and no shade, (b) rock mulch around the buildings and the walls shaded by shrubs, and (c) rock mulch with neither grass nor shade. The color of the buildings is not noted.

The surface temperature of grass turf around noontime was lower by about 15°C (27°F) as compared with that of the rocks. The air temperature, measured about 0.5 m (1.6 ft) above the turf, was lower by about 2°C (3.6°F) than above the rocks. The model with the rock mulch consumed between 20% and 30% more cooling energy than the models with the turf and with the shrub's shade. The turf apparently has reduced the longwave radiant heat gain and the ambient air temperature next to the building's skin, while the shrubs reduced the solar heat gain by shading part of the walls

Huang *et al.* (1987) have simulated by the DOE-2.1 program the effect of increasing the amount of urban tree canopy by 10% (one tree per house), 25% and 30% (three trees) on the energy use for cooling in four cities: Sacramento, Phoenix, Lake Charles and Los Angeles. The base case condition is the unmodified weather tapes of these cities. The building was a one-story detached house of wood frame construction with 143 m^2 (1,540 ft^2) floor area and 14.3 m^2 (154 ft^2) of windows. Simulations were done with the house at two levels of insulation and with the trees added randomly or with trees planted purposely for effective shading.

The simulation calculated the individual and combined effects of the additional trees on solar heat gain (shading effect), on the urban wind speed (effect on infiltration) and on changes in the urban temperature (conductive gain) and humidity (latent heat).

Table 2 shows the combined effects of the increased tree canopy by 10% and 30% on the reductions in the annual kWh, the peak kW and the number of cooling hours, when the house was at the higher level of insulation and the trees were planted specifically for effective shading.

Fig. 13. Alhambra Palace, Cordoba, Spain. Planted garden and shaded walks of courtyards of Spain.

Fig. 14. South Beach, Miami. Planted trellis of Parking Garage contributes to reduction of urban heat island. SITE (James Wines) Design Architect.

Table 1. Averages of air and surface temperatures measured during a sequence of several clear days in summer.

Location	Air Temp. F	SurfaceTemp. F	Air Temp. C	Surface Temp. C
Parking lot	79	122	26.1	50.0
Open plaza	78	107	25.6	41.7
Shaded sidewalk	76	80	24.4	26.7
Grass lawn	75	88	23.9	31.1
Behind shrubs	74	73	23.3	22.8

Fig. 15. The Cloisters. New York City. One of a series of courtyards, each with different orientations and exposure. Springtime flowering can be seen to occur sequentially, as a function of the resulting garden microclimates.

Table 2. Effect of increasing trees' canopies on the saving in annual and peak energy use for cooling and the number of cooling hours

Location		Base Case	10% (1 tree per house) Reduction	%	30% (3 trees) Reduction	%
Sacramento	Annual kWh	1420	343	24.2	757	53.3
	Peak kW	7.10	1.24	17.5	2.44	34.4
	Cooling hours	904	165	18.3	514	56.9
Phoenix	Annual kWh	6011	873	12.6	2289	33.1
	Peak kW	8.87	0.80	9.0	1.57	17.7
	Cooling hours	3647	157	4.3	619	17.0
Lake Charles	Annual kWh	3908	466	11.9	1354	34.7
	Peak kW	7.17	0.89	12.4	1.55	21.
	Cooling hours	2489	170	6.8	584	23.5
Los Angeles	Annual kWh	359	0	0	0	0
	Peak kW	4.46	0.9	20.2	1.96	43.9
	Cooling hours	65	43	66.2	55	84.6

Note: For Los Angeles the reductions are minimal because the base case is very low. For the other cities even a 10 % increase in the canopy caused sizeable reductions in the annual peak cooling energy use and in the number of cooling hours.

COMMENTS ON URBAN PARK DESIGN IN DIFFERENT CLIMATES

Hot, dry regions

In hot, dry regions the rate of evaporation from bare soil is small. However, in irrigated urban parks and gardens, the water evaporation from plants and from the soil increases substantially, due to the low humidity. Therefore, the effect of green areas on the climate within and near the parks can be significant, and the effect on comfort desirable. Urban public and private green areas, to the extent that they can be kept properly (mainly irrigated), are great assets in a hot, dry region. Such areas are more pleasant and thermally more comfortable than built-up and hard-surfaced unplanted outdoor areas.

In hot, dry regions parks and playgrounds should provide ample shade and protection from dust in summer, as well as protection from cold winds in winter. Large lawns and flowerbeds without shade contribute little to the recreation possibilities of the inhabitants to rest, relax, or play on a hot sunny day. Thus, plenty of places to sit in the shade should be provided along roads and trails in public parks, and in children's playgrounds.

Hot, humid regions

The impact of plants on human comfort in hot, humid areas can be a mixed bag. The shading provided by trees is always welcomed. However, the blockage of the wind and the contribution to the humidity level by evaporation from the leaves may increase human discomfort, especially during hours with very light winds.

Trees with a high trunk and wide canopy are the most effective plants in providing usable shade around buildings. If densely placed on the windward side of the house they, of course, block the wind. Therefore, the best strategy with such trees is to have them only at the spots where their shade will be utilized without blocking the wind, such as near the walls but not in front of windows. Pergolas of vines in front and above windows can also provide effective shading without wind blocking. If the trees and the vines are deciduous they enable daylight and solar gain in winter. However, one should be careful to prevent low growing trees and high shrubs in front of windows on the windward sides of the building. Such plants can act in effect as windbreaks and reduce greatly the ventilation potential.

- High shrubs block the wind and "contribute" to the humidity level without providing useful shade, except when they are placed alongside walls. Therefore their introduction should be minimal, especially at the windward parts of the site.
- A combination of grasses, low flowerbeds and shade trees with high trunks is thus the most appropriate plant combination in landscaping in hot, humid climate.
- Low-lying areas prone to floods can be grassed and planted with trees, which can withstand flooding. If the vegetation can withstand a given height of water for a few days, then such areas can be utilized as flood controls during, and immediately after, rainstorms. In the periods between storms such areas can again be useful public green spaces for recreation.

Cold regions

In cold regions the main climatic considerations in designing public park areas are to provide access to the sun and protection from the wind. Robinette (1972) provides information on several studies in the USSR and in Germany in which the effect of shelterbelts on the wind speed was measured.

In places where the wind direction in winter is from the northwest, high and dense lines of evergreen trees (*e.g.*, conifers) on the north and west borders of the open space can provide wind protection without blocking the winter sun. Belts of evergreen shrubs along the base of trees will prevent wind penetration below the tree canopy.

Provisions for wind protection in urban parks may be of particular importance in arid regions with cold winters. The availability of sunshine and the absence of much rain in winter can enhance greatly the attraction of settlements in these regions as winter resorts. This factor may be of significant economic value to the community. Protection from cold winds will help greatly in enjoying activities in the public urban parks.

- U-shaped belts of high evergreen shrubs around benches, open to the south and with deciduous trees with high trunks behind it, can provide in cold regions pleasant seating places year round, as they will be exposed to the sun and protected from the wind in winter while enjoying the shade of the trees in summer. Such design details of the benches can increase the prospects of their use during sunny days in winter.

- Wind protection with solar exposure around facilities for various forms of winter sports, such as ice skating on frozen ponds, etc., can also encourage their use of the parks in winter, thus enhancing year-round use of the parks. ■

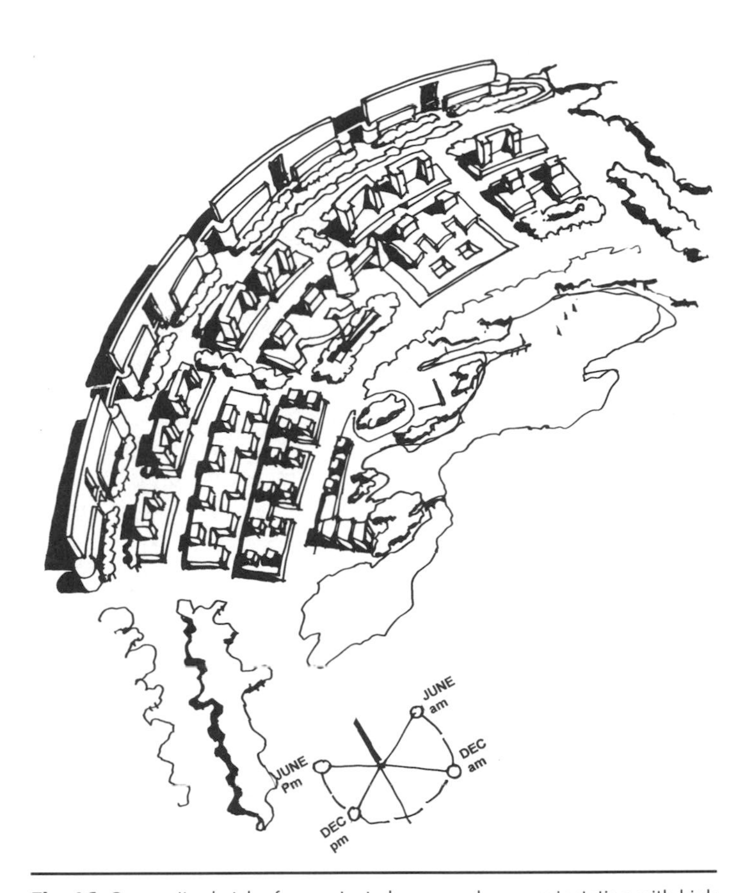

Fig. 16. Composite sketch of sun-oriented, summer breeze orientation with high buildings to the north for winter wind protection.

REFERENCES

Akbari, H., A.H. Rosenfeld and H. Taha. 1990. "Summer Heat Islands, Urban Trees, and White Surfaces." ASHRAE January 1990 Meeting. Atlanta, Georgia. Lawrence Berkeley Laboratory Publication LBL – 28308.

Al Hemiddi, N. 1991. "Measurements of Surface and Air Temperatures Over Sites with Different Land Treatments." PLEA '91 Conference in Seville, Spain.

Davis, I.R. 1984. "The Planning and Maintenance of Urban Settlements to Resist Extreme Climatic Forces" in *WMO* (1986) Pub. 277-310. Geneva: World Meteorological Organization.

DeWalle, D.R. 1983. Windbreak Effects on Air Infiltration and Space Heating in a Mobile House. *Energy and Buildings* 5, pp. 279–288.

Geiger, R. 1965. *The Climate Near the Ground*. Cambridge: Harvard University Press.

Givoni, B. 1979. *Man, Climate and Architecture*. Van Nostrand Reinhold. New York.

Givoni, B. 1989. *Urban Design in Different Climates*. WCAP–10, WMO/TD-No. 346. Geneva: World Meteorological Organization.

Givoni, B. 1994. *Passive and Low Energy Cooling of Buildings*. New York: Van Nostrand Reinhold.

Givoni, B. 1998. *Climate Considerations in Building and Urban Design*. New York: Van Nostrand Reinhold. Distributed by John Wiley and Sons.

Givoni, B. and S. Sandifer, 1997. "Thermal Effects of Vines on Wall Surface Temperatures," *Proceedings:* ISES 1997 Solar World Congress, South Korea. Vol. 4, pp. 258–265.

Hoyano, A. (1988). Climatological Uses of Plants for Solar Control and the Effects on the Thermal Environment of a Building. *Energy and Buildings, 11* (3): 181–199.

Huang, Y.J., H. Akbary, H. Taha and A.H. Rosenfeld. (1987). The Potential of Vegetation in Deducing Summer Cooling Loads in Residential Buildings," *J. Climate and Applied Meteorology, 26* (9), 1103–1116.

Jauregui, E. 1984. The Urban Climate of Mexico City. *WMO Proceedings*, Technical Conference on Urban Climatology and Its Application with Special Regard to Tropical Areas. WMO–No. 652.

McPherson, E.G., J.R. Simpson and M. Livingston. 1989. Effect of Three Landscape Treatments on Residential Energy and Water use in Tucson, AZ. *Energy and Buildings, 13* (2): 127–138.

Oke, T.R. 1981. "Canyon Geometry and the Nocturnal Urban Heat Island: Comparison of Scale Model and Field Observations." *Journal of Climatology, 1*: 237–254.

Paciuk, M. 1975. *Urban Wind Fields–An Experimental Study on the Effects of High Rise Buildings on Air Flow Around Them.* M.Sc. Thesis. Technion, Haifa, Israel.

Parker, J.H. 1983. The Effectiveness of Vegetation on Residential Cooling. *Passive Solar Journal.* 2(2): 123–132.

Potchter, O., Yaakov, Y. and Potchter K. 2001: The Influence of Vegetation Shading on Wall Temperatures. Fouier Systems Ltd & The Unit for Applied Climatology and Environmental Aspects. Tel Aviv University, Israel.

Pressman, N. 1988. "Developing Climate-Responsive Winter Cities." *Energy and Buildings*, 11: 11–22.

Robinette, G.O. 1972. "Plants/People/and Environmental Quality." U.S. Department of the Interior, Washington, D.C.

Sharlin, N. and M.E. Hoffman. 1984. "The Urban Complex as a Factor in the Air-Temperature Pattern in a Mediterranean Coastal Region." *Energy and Buildings*, 7: 149–158.

Taha, H., H. Akbari, A Rosenfeld and J. Huang. 1988. "Residential Cooling Loads and the Urban Heat Island—the Effect of Albedo." *Building and Environment* 2.3(4): 271–283.

Taha, H.T., H. Akbari and A. Rosenfeld. 1989. "Vegetation Canopy Micro-Climate: A Field Project in Davis California." Lawrence Berkeley Laboratory, LBL-24593.

Donald Watson and Kenneth Labs

Summary

Bioclimatic design—based on analysis of the climate and ambient energy represented by sun, wind, temperature and humidity—has developed out of sensitivity to ecological contexts and the need to conserve energy and environmental resources. This article demonstrates site planning approaches for wind protection in underheated conditions (typically winter) and for cooling in overheated periods (typically summer).

Key words

evaporative cooling, ground cover, land form, microclimate, passive heating and cooling, shading, solar radiation, temperature, vegetation, ventilation, wind shadow

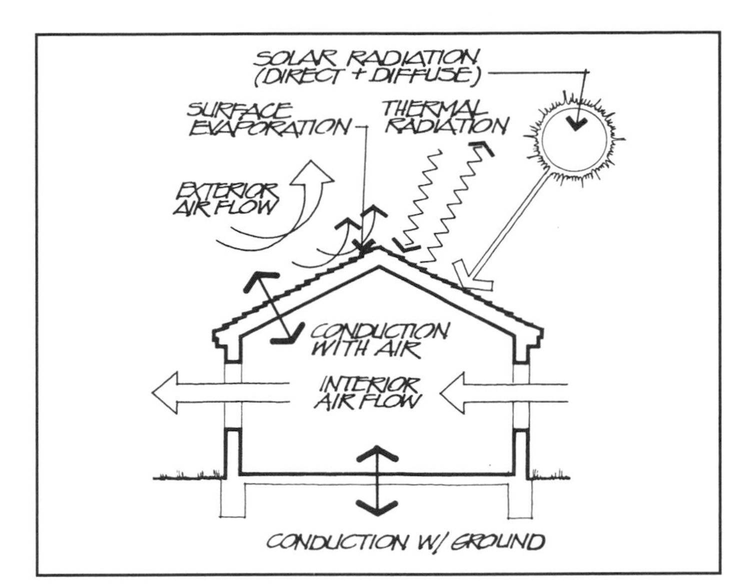

Paths of heating energy exchange at the building microclimate.

Bioclimatic design at the site planning scale

Urban designers can create favorable microclimatic conditions in and around buildings and outdoor spaces to dramatically increase comfort and reduce building energy requirements.

In winter (or underheated periods), the objectives of bioclimatic design at the site planning scale are to protect outdoor spaces, entryways and structures from the winter wind and to promote gain of solar heat. In summer (or overheated periods), these objectives are the reverse, to resist solar gain by shading and to promote cooling by ventilation. Evaporative cooling by ground cover and water can be made effective in temperate and hot dry climates.

Bioclimatic design techniques at the site planning scale include:

Wind breaks (winter)

- Use neighboring land forms, structures, or vegetation for winter wind protection.

Sun shading (summer): Because sun angles are different in summer than in winter, it is possible to shade spaces and building openings from the sun during the overheated summer period while allowing the sun's heat to reach spaces and window surfaces in winter. The concept to provide sun shading does not need to conflict with winter solar design concepts.

- Use neighboring land forms, structures, or vegetation for summer sun.

- Provide seasonal shading, including deciduous trees.

Natural ventilation (summer and seasonal): Natural ventilation is a simple concept by which to cool outdoor spaces and buildings.

- Use neighboring land forms, structures, or vegetation to increase exposure to summer breezes.

Plants and water (summer): Several landscaping techniques provide cooling by the use of plants and water near building surfaces and outdoor spaces for shading and evaporative cooling.

- Use ground cover and planting for site cooling.

- Maximize on-site evaporative cooling.

Each locale has its own bioclimatic profile of assets and liabilities, the responses to which are oftentimes evident in indigenous and local long-established building traditions. Recently developed energy design tools make it possible to utilize hourly weather data to accurately analyze climate. This enables the designer to apply sophisticated bioclimatic analysis to any location in North America (and other sites where hourly weather data is available), providing a systematic basis to guide design judgment. The site planning techniques illustrated in this article and other building bioclimatic design techniques can be selected as appropriate to the resource of local climate by analysis of local data using "Climate Consultant 2.0: A New Design Tool for Visualizing Climate" available via the Internet (Milne 1997).

REFERENCE

Milne, Murray. 1997. *Energy Design Tools.* Web Page, Department of Architecture and Urban Design. University of California Los Angeles (UCLA). http://www.aud.ucla.edu/energy-design-tools.

Credits: This article is excerpted from Donald Watson and Kenneth Labs, (1983, 1993). *Climatic Design*. New York: McGraw-Hill.

BIOCLIMATIC DESIGN AT THE SITE PLANNING SCALE

The following pages present representative microclimatic design details and data for site planning (Watson and Labs 1983). ■

Use neighboring land forms, structures, or vegetation for winter wind protection.

FIG. 2c
General Rules of Windbreak Design

1. **The range of protected area downwind is proportional to the height of the windbreak—the higher the barrier, the longer the "wind shadow". Angle of the windward edge is also important—the more vertical, the greater the effect.**

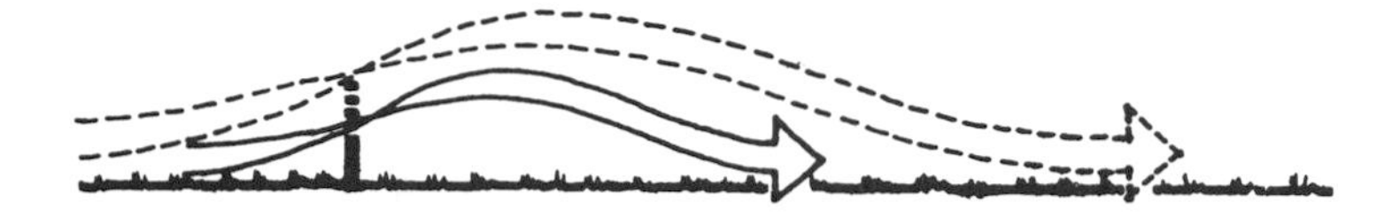

2. **The maximum length of wind shadow is developed only when the width of the windbreak is at least 11-12 times its height.**

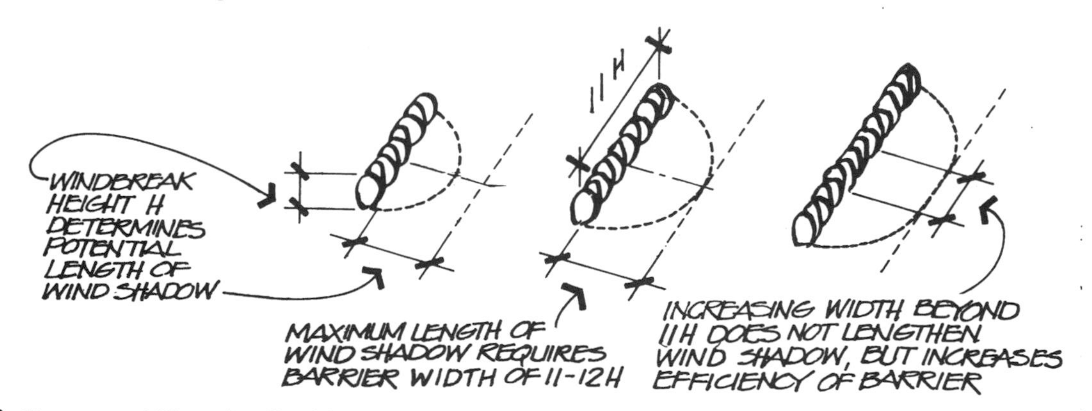

3. **The permeability or density of the barrier affects the length of the downwind protected zone—dense and solid barriers offer greatest reduction in wind speed, but only for a short distance immediately behind the barrier. Further downwind turbulence quickly restores wind speed. Wind permeable barriers pass some wind at reduced speed, creating an "air cushion" that reduces turbulence and extends the length of the sheltered zone downwind.**

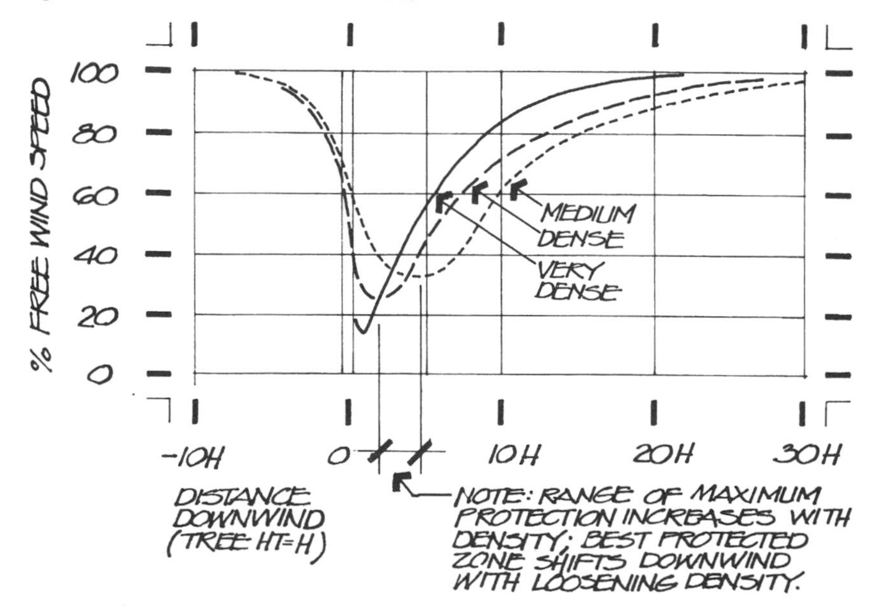

Use neighboring land forms, structures, or vegetation for winter wind protection.

An analysis of the building site should be made to determine if there are existing wind protected areas. In siting a house, the builder should avoid open areas, hilltops and valley floors that are directly exposed to prevailing winter winds. Existing hedgerows, hillocks and tree stands may be available as windshields (*FIG. 2a*). The arrangement of housing units in cluster plans and planned unit developments can be used to take advantage of these.

Cold air flows downhill just like water, so that although a valley may seem to be shielded from cross winds, it may actually be in the midst of a steady downstream nighttime cold air current. The best protected sites, occur on leeward slopes (*FIG. 2b*).

Windbreaks

In the absence of, or to augment, existing land forms and tree masses, barriers can be created for wind control. Trees and shrubs are the most common of these, but berms, fences and walls can also provide benefit. Usually, the higher the barrier, the larger the protective "wind shadow" (*FIG. 2d*). Solid or wind-impenetrable barriers are undesirable, as they create turbulent areas on the leeward side (*FIG. 2c*). The energy effectiveness of wind protection follows directly from the amount of windspeed reduction. The economic benefit of landscaping has been studied with the aid of computer simulation at the University of Wisconsin. Researchers have concluded: [Grist 1977]

> The most desirable windbreaks, from a heat loss point of view, are those that range in porosity from 25% to 60%. High porosity (about 50%) usually means more protection 5H to 20H downwind where velocity is reduced to about 30% of free stream velocity. Lower porosity (about 25%) usually indicates more protection up to 4H range, where velocity may be reduced to 10%, but less protection 4H to 20H downwind where velocity may be 60% of the free stream velocity.

Climatically planned landscaping can have significant energy benefits for homeowners, in addition to its aesthetic and property value. As a cautionary consideration, however, it should be noted that windbreaks can possibly impede summer ventilating breezes. This should be taken into account when planting plan is prepared.

FIG. 2a
Increasing the width of shelterbelt does not increase the length of its wind shadow. Contrarily, a wide belt (such as woods) consumes much of the area of its own protection.

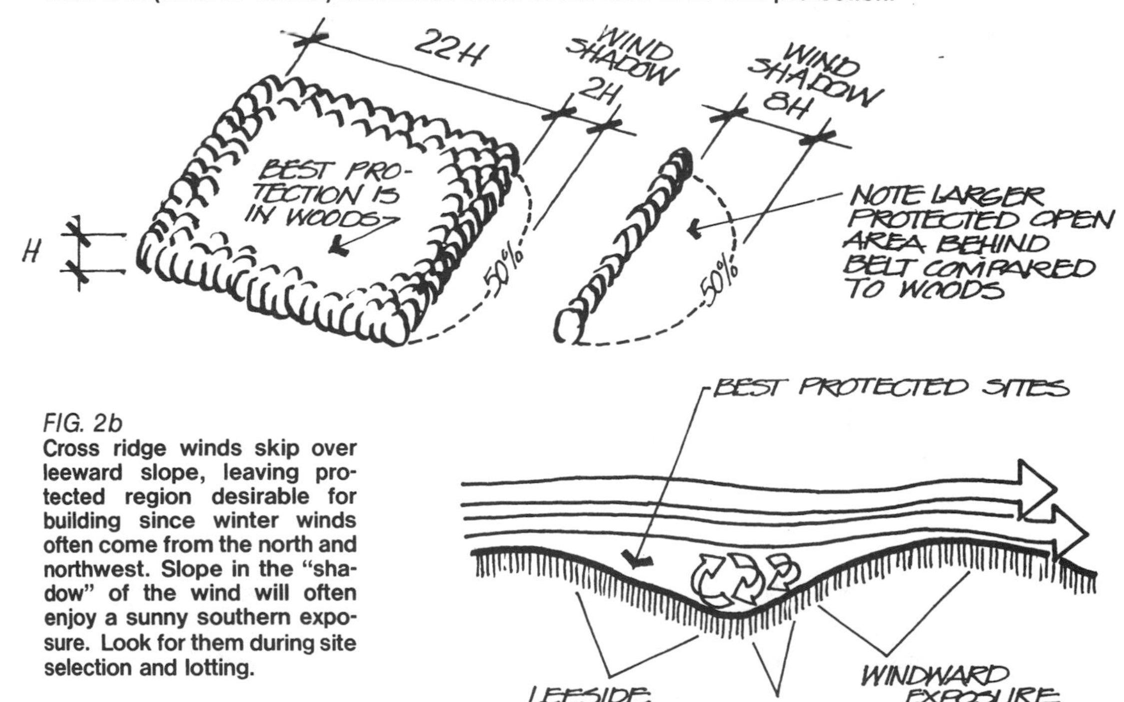

FIG. 2b
Cross ridge winds skip over leeward slope, leaving protected region desirable for building since winter winds often come from the north and northwest. Slope in the "shadow" of the wind will often enjoy a sunny southern exposure. Look for them during site selection and lotting.

Use neighboring land forms, structures, or vegetation for winter wind protection.

FIG. 2d.
Man-made Features: Fences & Walls

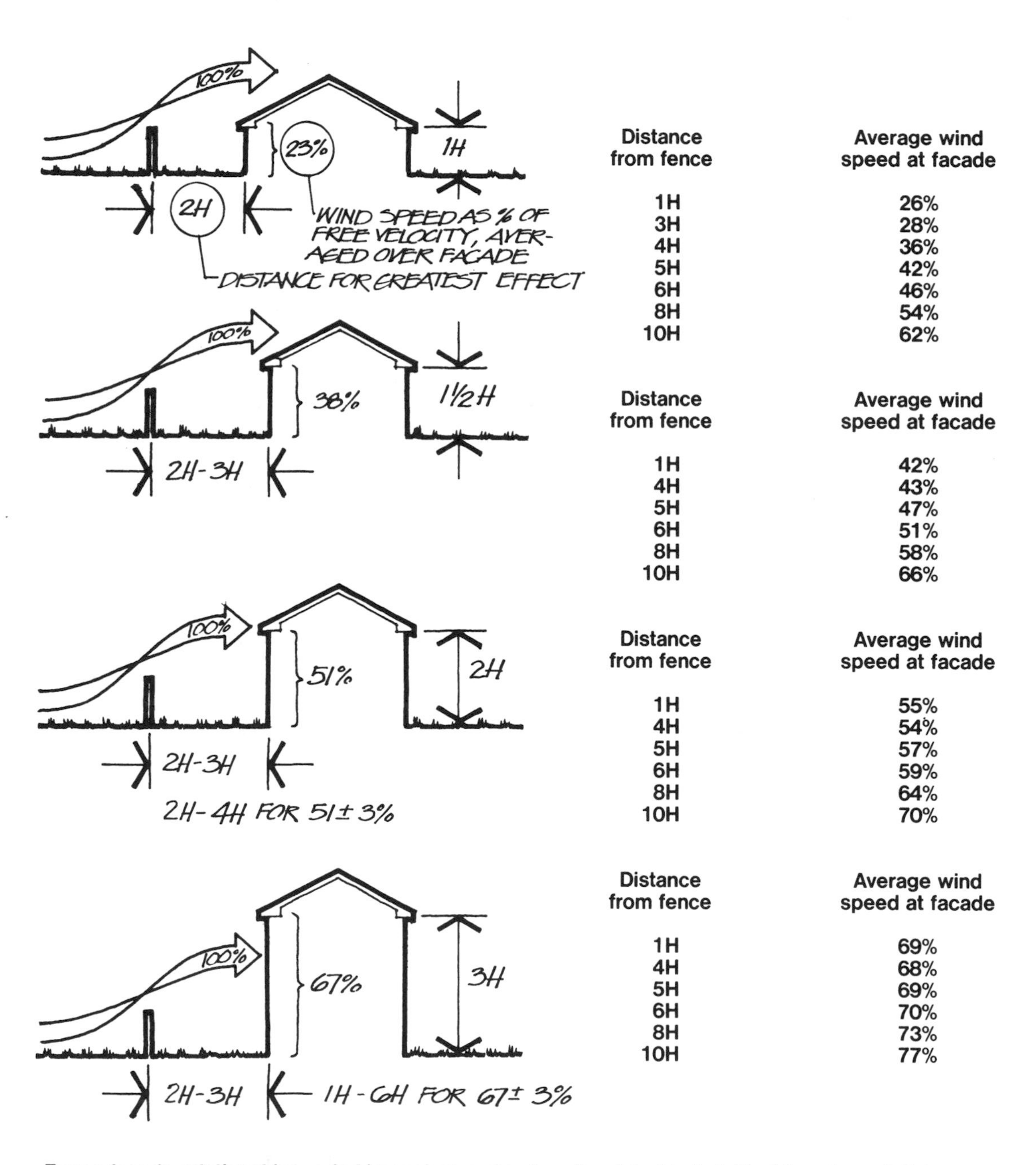

Distance from fence	Average wind speed at facade
1H	26%
3H	28%
4H	36%
5H	42%
6H	46%
8H	54%
10H	62%

Distance from fence	Average wind speed at facade
1H	42%
4H	43%
5H	47%
6H	51%
8H	58%
10H	66%

Distance from fence	Average wind speed at facade
1H	55%
4H	54%
5H	57%
6H	59%
8H	64%
10H	70%

Distance from fence	Average wind speed at facade
1H	69%
4H	68%
5H	69%
6H	70%
8H	73%
10H	77%

Fence-facade relationships: wind impact on a structure is related to height of facade and distance from windbreak. Recommendations given here are derived from observations in the open (not at a building surface) behind a windbreak of 15-25% permeability. Note that with decrease in relative fence height, the horizontal range of effectiveness (±3%) expands: although low fences provide less benefit, their location is less critical.

Use neighboring land forms, structures, or vegetation for summer shading.

KEEP UNDERSTOREY CLEAR
SO AS NOT TO DISRUPT
AIRFLOW FOR VENTILATION

FIG. 4b. **Plant tall canopy trees on south side of house to shade roof and walls.**

SHADE PLANTING ON WEST AND
NORTHWEST SIDES OFTEN
CAN DOUBLE AS WINTER WIND-
BREAK. CONSIDER EVERGREENS,
FENCES, AND WALLS.

FIG. 4c. **Plant dense trees, shrubs, hedges on west side of house to intercept afternoon sun.**

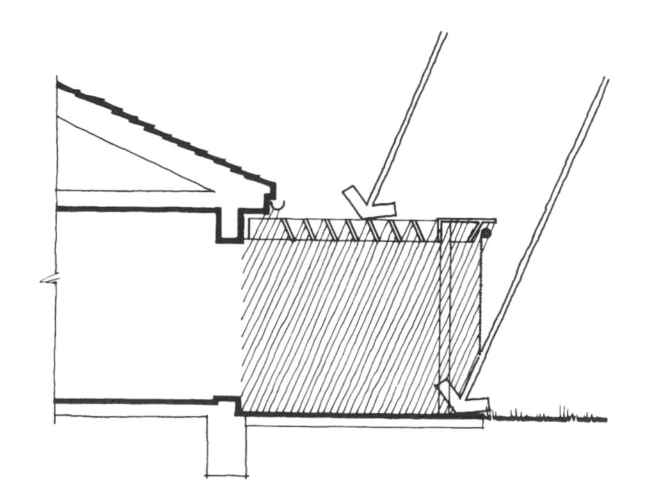

FIG. 4d. **Attached overhead shading structures can provide multiple benefits. Not only does this patio cover shade the wall, it also reduces reflected gain from loading on the wall.**

Use neighboring land forms, structures, or vegetation for summer shading.

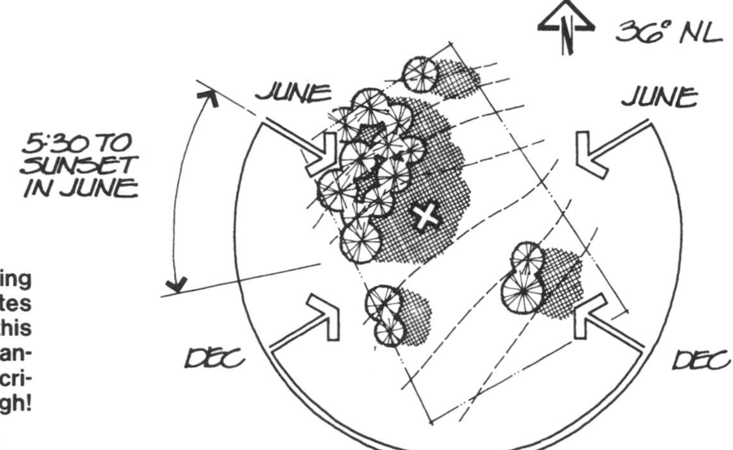

FIG. 4a **"X" marks the spot for existing on-site sun protection. Look for sites shrouded by trees on west side. In this example, trees are on upslope, enhancing their shading ability. Don't sacrifice winter southern exposure, though!**

Opportunities for sun control and shading exist in almost all aspects of site planning and development, from the initial selection of a building site or the plotting of the lots in a subdivision to the final selection of plant materials and details in landscaping.

Siting

For a custom house on property that is large enough to offer several building sites, the choices should be analyzed for natural shading from existing trees and land masses. These can be utilized to reduce solar gain from low afternoon summer sun by siting the building to the east of such features (*FIG. 4a*).

In plotting subdivisions, streets should be laid out to create lots having the best possible solar orientation. Generally, this means streets should run east-west; with this orientation, tightly clustered units will shade each other from afternoon summer sun, without obstructing desirable southern exposures.

On-lot development

Inasmuch as air temperatures usually peak during the end of July or in early August in most regions of the U.S., and since solar impact on *walls* is greater at this time than at the date of maximum insolation (June 21), the afternoons of this period can be considered as the selected design condition for determining a shade planting plan. When the site of the house has been determined, planning of other exterior shading devices can begin.

The best known, of course, are shade trees (*FIG. 4b*), which protect the house in summer and shed their leaves in winter to allow the house to receive solar gain. On the south side of the house, tall growing species should be selected that will shade the roof as well as the wall.

West elevations will benefit most from lower, more compact trees, and tall, dense shrubs that provide screening from the low afternoon summer sun (*FIG. 4c*). Because the winter sun does not reach around as far as it does in summer, shading vegetation for western walls may be evergreen. If winter winds are westerly or northwesterly, the planting may double as a windbreak.

Other shading devices for south and west walls include overhead trellises and shade walls and fences. Trellises can be built with louvers or slats to block the sun (*FIG. 4d*), or may consist of little more than a light framework to support climbing ivy.

The garage can be located on the western side of the house and may be sited to create a breezeway or shaded patio area between it and the house. There are many approaches to providing outdoor shading structures. The most suitable will depend on the interior organization of the plan, the intended areas for outdoor living space (patios, porches, etc.), the site of the lot and its relationship to other dwellings.

Use neighboring land forms, structures, or vegetation to increase exposure to summer breezes.

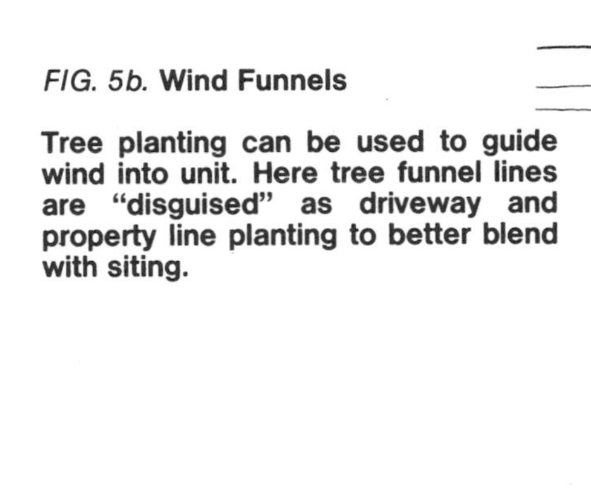

FIG. 5b. **Wind Funnels**

Tree planting can be used to guide wind into unit. Here tree funnel lines are "disguised" as driveway and property line planting to better blend with siting.

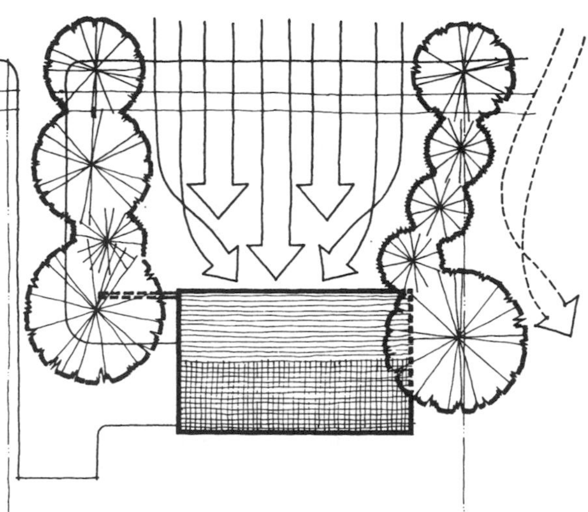

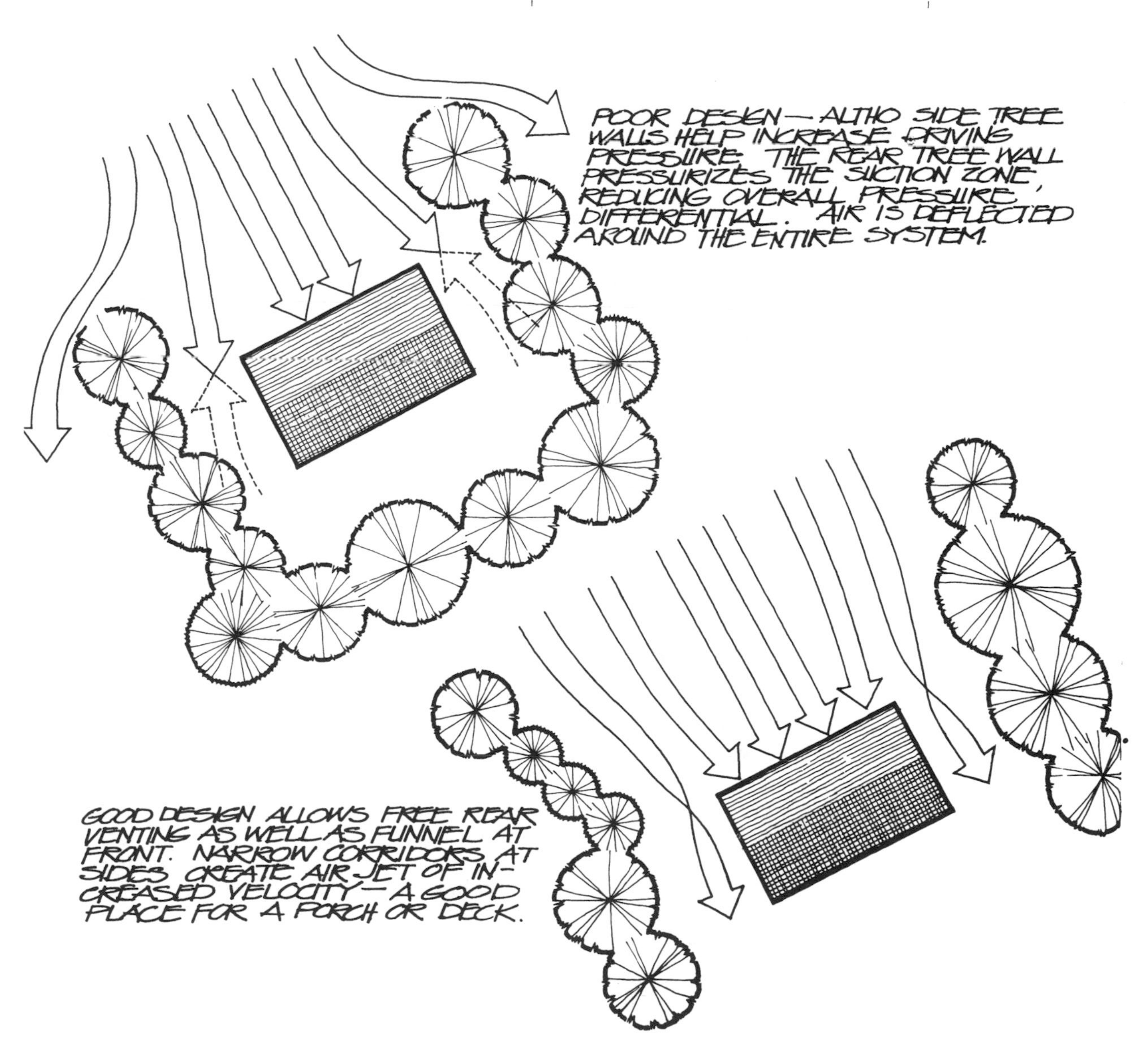

POOR DESIGN— ALTHO SIDE TREE
WALLS HELP INCREASE DRIVING
PRESSURE THE REAR TREE WALL
PRESSURIZES THE SUCTION ZONE,
REDUCING OVERALL PRESSURE
DIFFERENTIAL. AIR IS DEFLECTED
AROUND THE ENTIRE SYSTEM.
GOOD DESIGN ALLOWS FREE REAR
VENTING AS WELL AS FUNNEL AT
FRONT. NARROW CORRIDORS AT
SIDES CREATE AIR-JET OF IN-
CREASED VELOCITY— A GOOD
PLACE FOR A PORCH OR DECK.

Use neighboring land forms, structures, or vegetation to increase exposure to summer breezes.

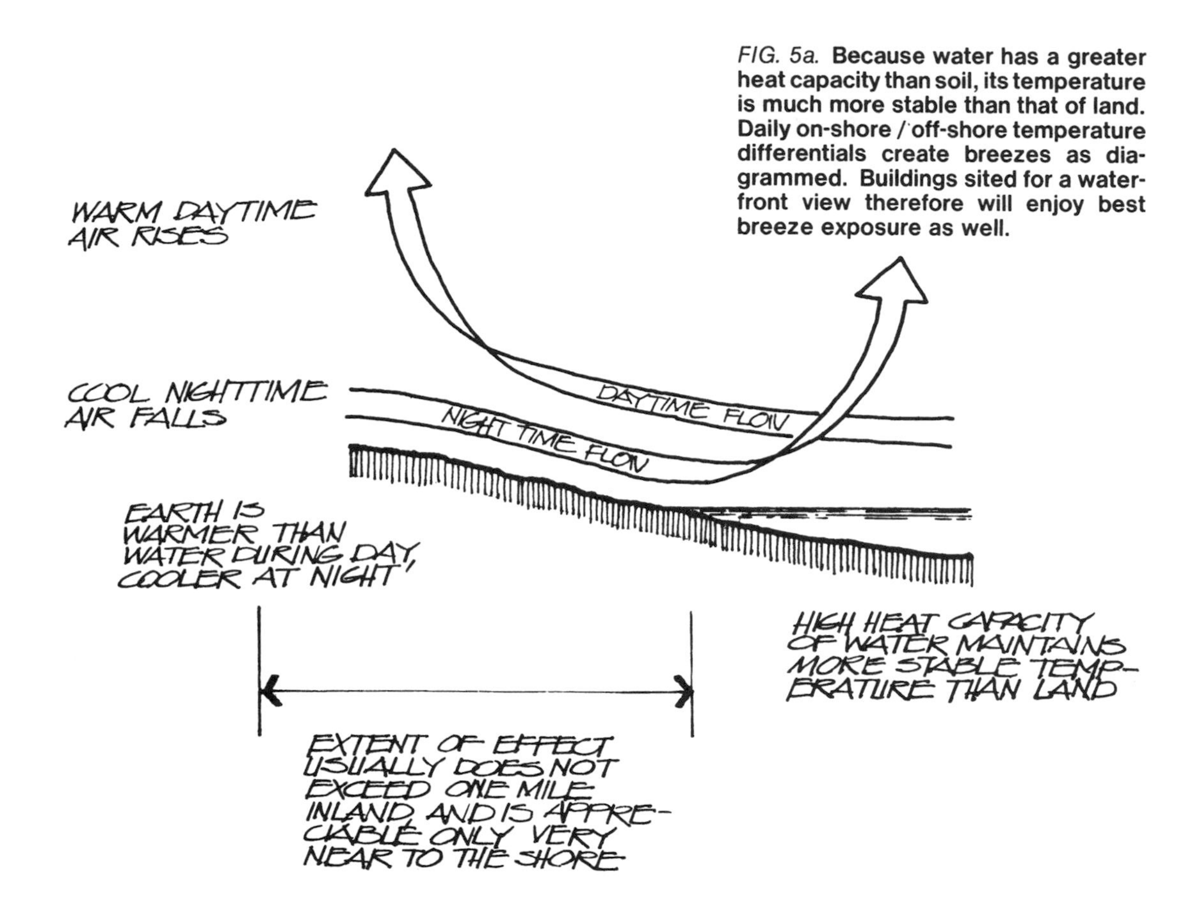

WARM DAYTIME AIR RISES
COOL NIGHTTIME AIR FALLS
DAYTIME FLOW
NIGHT TIME FLOW
EARTH IS WARMER THAN WATER DURING DAY, COOLER AT NIGHT
HIGH HEAT CAPACITY OF WATER MAINTAINS MORE STABLE TEMP-ERATURE THAN LAND
EXTENT OF EFFECT USUALLY DOES NOT EXCEED ONE MILE INLAND AND IS APPRE-CIABLE ONLY VERY NEAR TO THE SHORE

FIG. 5a. **Because water has a greater heat capacity than soil, its temperature is much more stable than that of land. Daily on-shore / off-shore temperature differentials create breezes as diagrammed. Buildings sited for a waterfront view therefore will enjoy best breeze exposure as well.**

The direction and velocity of flow of summer breezes are influenced considerably by local land forms, tree masses, and existing structures. The resulting air flow patterns may be quite different from that of prevailing breeze directions given in published (airport) climatic data.

Although building on the crest of a hill will maximize exposure to prevailing breezes, this practice also maximizes the structure's vulnerability to winter wind chill. Unless winters are mild and the ventilating season is long (as in the Gulf Coast States), it will be more advantageous to locate site development along hillsides, rather than at the ridgeline.

On slopes and in valleys, cool air flows downhill, washing along the slope and settling in depressions or following the valley downstream near large water bodies. Daily on-shore and off-shore breezes are created which are largely independent of regional air patterns (*FIG. 5a*). A topographic analysis of the area is thus necessary to determine probable on-site wind flow patterns and the most desirable building locations. (*FIG. 5e*).

Trees and shrubs can be used to channel air flow toward the structure, and may even be used to increase air velocity through the building by "funnelling" air into openings (*FIG. 5b*). Fences, walls, and adjacent structures can create air dams that increase the inflow pressures (*FIG. 5c*). Caution should be exercised in landscaping, however, since undesirable air dams can also be created that will cause air to pool rather than flow. Also, the location of planting outside of the house can either aid in deflecting air into the structure or hurt by deflecting it away from open windows (*FIG. 5d*).

Use neighboring land forms, structures, or vegetation to increase exposure to summer breezes.

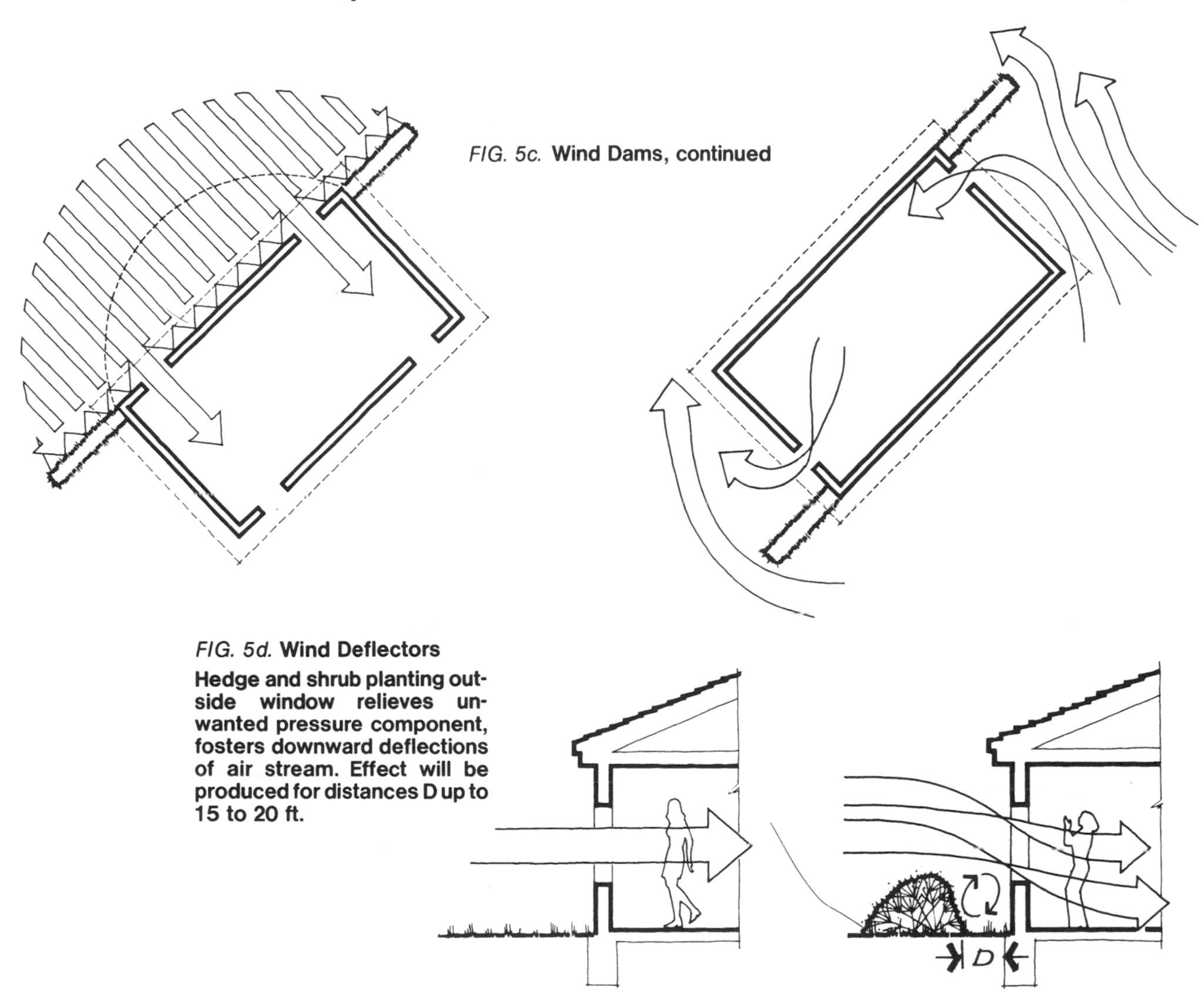

FIG. 5c. **Wind Dams, continued**

FIG. 5d. **Wind Deflectors**

Hedge and shrub planting outside window relieves unwanted pressure component, fosters downward deflections of air stream. Effect will be produced for distances D up to 15 to 20 ft.

Influence of tree canopy outside the window is to "lift" or warp the airstream upward by relieving downward pressure (opposite of shade-effect). If tree is immediately outside window it will produce a ceiling wash flow. At a distance from the house, canopy may warp the airstream sufficiently to miss the house altogether.

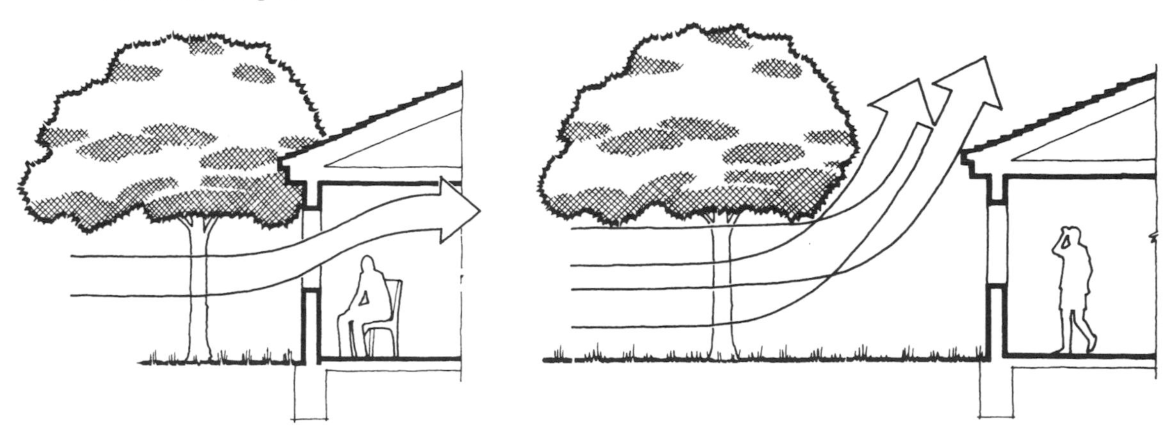

Use neighboring land forms, structures, or vegetation to increase exposure to summer breezes.

FIG. 5e.

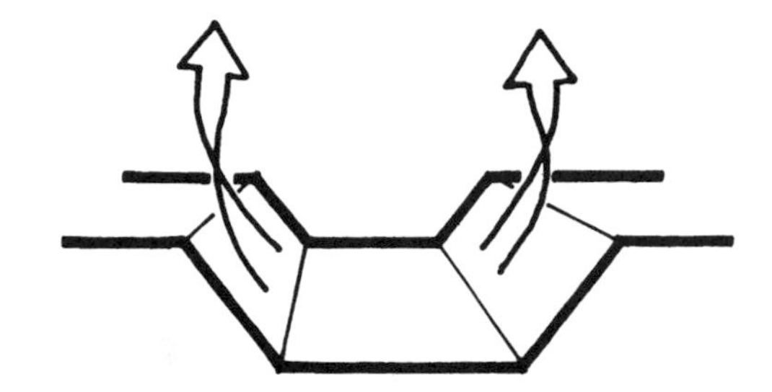

MORNING - BREEZE RISES UP FACE OF SLOPE DUE TO SURFACE WARMING

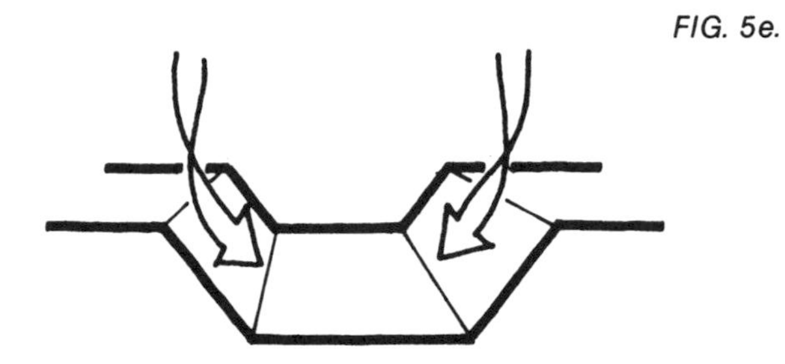

LATE EVENING - SURFACES COOL, START DOWNWARD FLOW OF COOL AIR

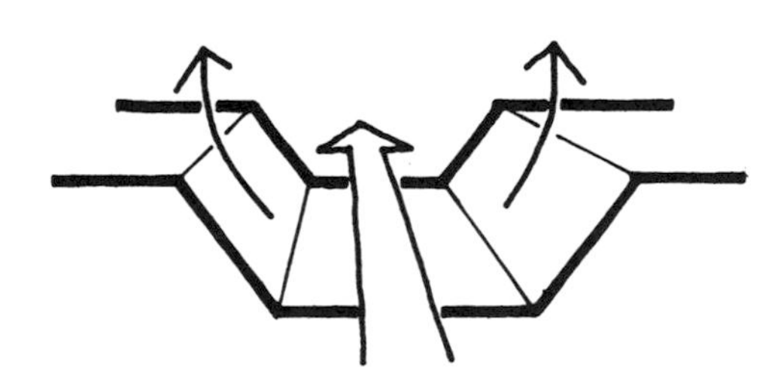

NOON - AIR POOL IN VALLEY HAS BEEN WARMED, CREATES UP-VALLEY BREEZE

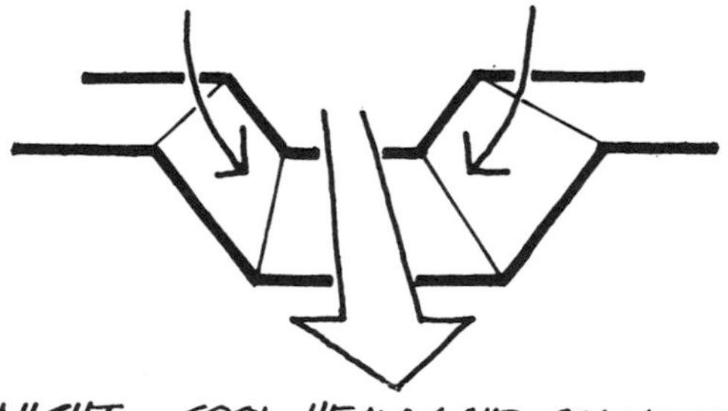

NIGHT — COOL, HEAVY AIR COLLECTED IN VALLEY BEGINS TO FLOW DOWNSTREAM

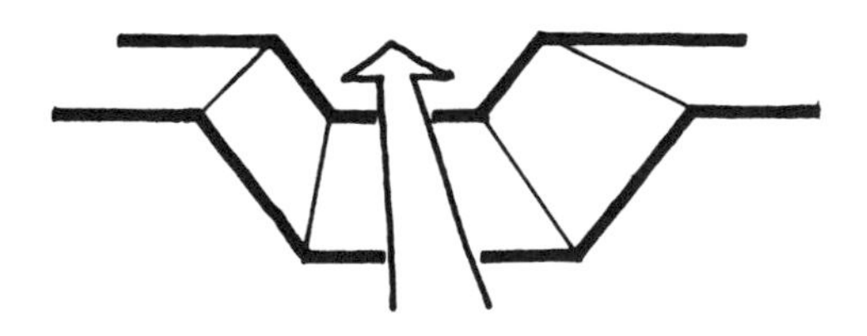

AFTERNOON - UP-VALLEY BREEZE PREDOMINATES AS SURFACES REACH & PASS MAXIMUM TEMPERATURE

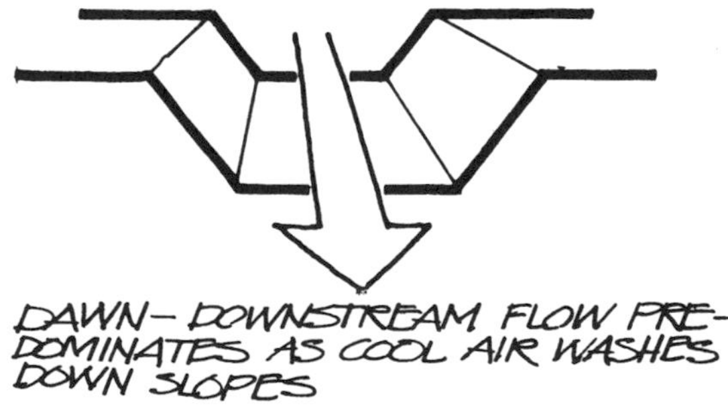

DAWN - DOWNSTREAM FLOW PRE-DOMINATES AS COOL AIR WASHES DOWN SLOPES

Use ground cover and planting for site cooling.

FIG. 6b **Neighborhood air temperatures can be kept low by minimizing the expanse of paving, and by shading paved areas. Many of the guidelines above save dollars or boost sales appeal.**

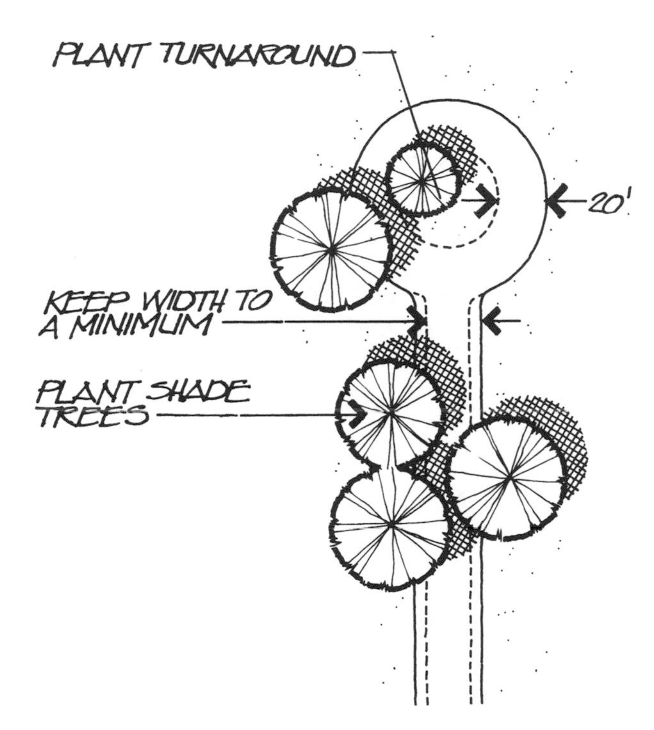

Site Planning Suggestions:

Keep paved area to a minimum—an 8ft. dia. turnaround with a 20 ft. ring road is recommended.

If spillover parking areas are required use a porous paving block instead of asphalt.

Plant shade trees to shade paving.

Use 18-20 ft. street width for large lot (34 acre or more) developments.

Use 26 ft. street width for 14 acre lots on cul-de-sacs and short loops.

Avoid 34-36 ft. street widths—these are never warranted in well planned new developments.

FIG. 6c. **Porous concrete paving can be precast or cast-in-place with forms made for this purpose ("grasstone," shown below, and "grasscrete" are examples) use it for stabilizing shoulders and for spillover parking spaces both on and off lot.**

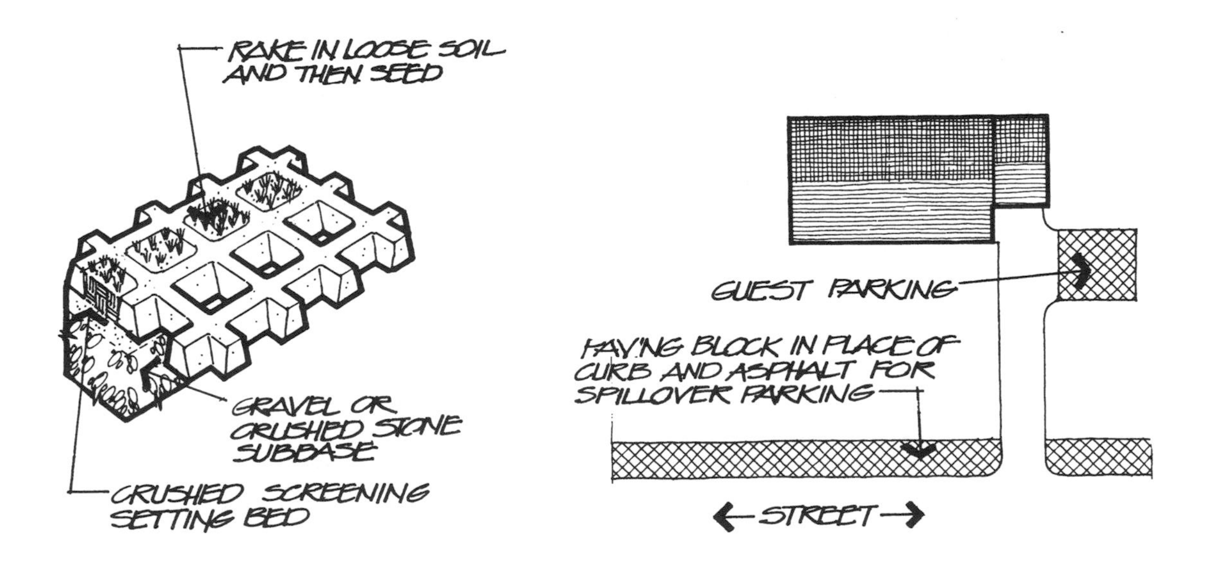

Use ground cover and planting for site cooling.

The climatic benefit of landscaped ground cover is seldom considered. On a sunny summer day, an acre of turf may evaporate about 2400 gallons of water. At this rate, the rear yard of a typical 1/4 acre lot will have the cooling effect of 2 million Btu per day. This has a significant influence on air temperature. In similar terms, the daily evaporation from a mature beech tree is said to provide an air cooling effect of one million Btu—the equivalent of 10 room-sized air conditioners operating 20 hours a day.

The difference in surface temperature between grass and asphalt can easily exceed 25°F (*FIG. 6a*). The air temperature in the "microclimate zone" (one to four feet) above these surfaces also differs appreciably, registering on the order of 10°F or more. The relationship of lawn and other living ground cover surfaces to non-evaporating surfaces (driveways, streets, roofs) will in part determine neighborhood air temperatures. These, in turn, will influence the cooling load on houses in the area, as well as the suitability of natural ventilation as a cooling strategy. Additionally, plants help create fresh air.

Stated a different way, vegetation should be maximized, and where possible, man-made surfaces such as streets and roofs should be shaded by trees (*FIG. 6b, 6c*).

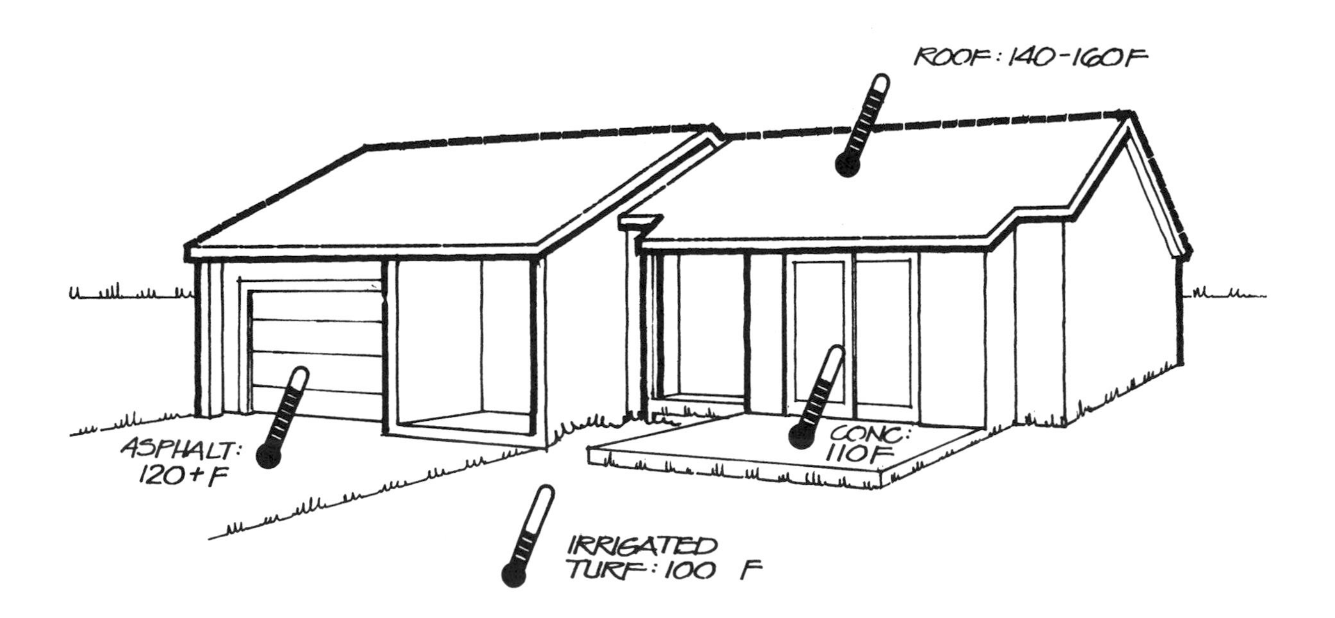

FIG. 6a. **Non-living surfaces are much hotter than grass (which would be cooler yet, if well irrigated) since they don't dissipate heat through evaporation. A black roof is hotter than an asphalt driveway, because the ground underneath the paving stores heat. The hottest roof will be one with insulation right under the roofing—having negligible mass. The coolest roofs will be sprayed, ponded, or covered with irrigated sod.**

Maximize on-site evaporative cooling.

Outdoor evaporative cooling mechanisms can help to provide outdoor comfort as well as to lower indoor cooling costs by lowering air temperature surrounding the building. This reduces the cooling load transmitted through the building shell and makes natural ventilation more desirable and more effective.

Since cool air is denser than warm air, it will tend to drain away, flowing downhill. The traditional response to retaining spray-treated air is to enclose the outdoor space with a wall or fence, in effect creating an open-topped "tank" of air. In some parts of the world, atrium or courtyard house design is standard practice, often with fountains or spray jets (*FIG. 7a*). In some cases, dual courtyard design has been employed, wherein a shaded, spray-cooled courtyard provides a cool ventilation air supply, while the heat trapping effect of a sunny courtyard on the other side of the unit propels an upward flow of warmed air, drawing the cool air through the house.

The same dual courtyard principle will work in many areas of the United States, but even simpler designs can achieve useful benefits. Small-lot patio houses can have cool yard areas with the aid of little more than a lawn sprinkler–provided that the fence or wall isn't too "leaky". Other low cost devices include running a perforated garden hose or pipe around the fencetop, creating a "wet wall". Spray-mist type area "foggers" can cool a large air mass instantly — and benefit the plants as well. The water that falls out of the air will help keep ground and patio surfaces cool by evaporation as well, so little water need be wasted (*FIG. 7b*).

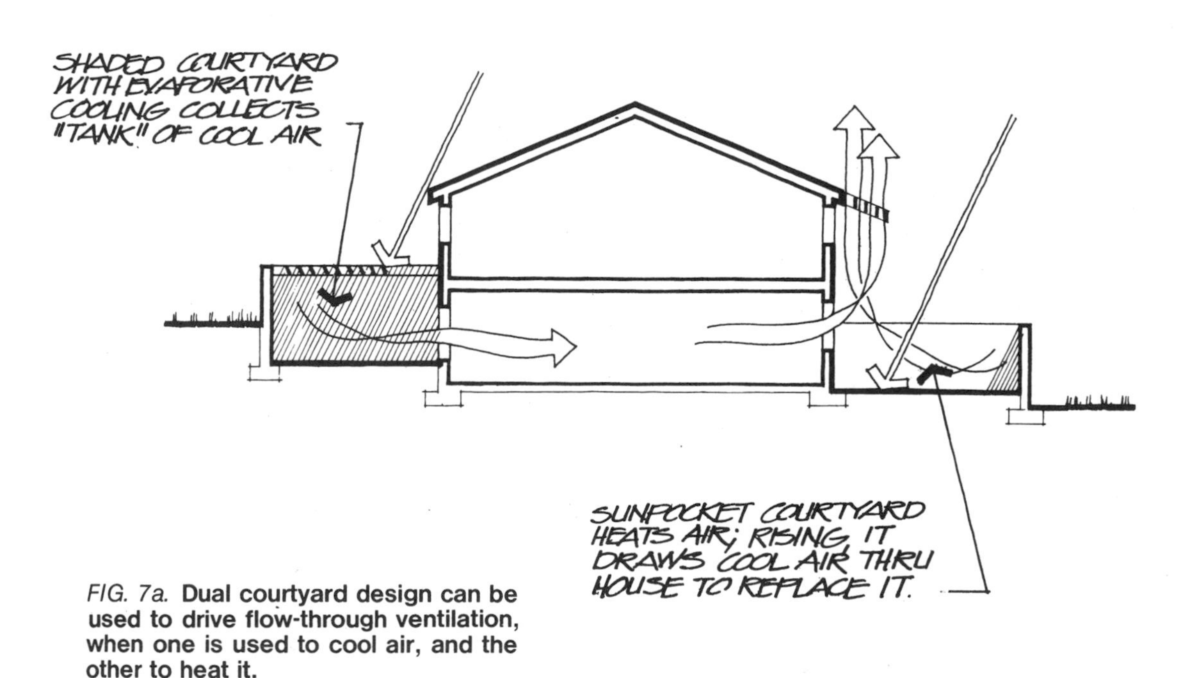
SHADED COURTYARD WITH EVAPORATIVE COOLING COLLECTS "TANK" OF COOL AIR
SUNPOCKET COURTYARD HEATS AIR; RISING IT DRAWS COOL AIR THRU HOUSE TO REPLACE IT.

FIG. 7a. **Dual courtyard design can be used to drive flow-through ventilation, when one is used to cool air, and the other to heat it.**

Temperature Scales

Temperature is defined as the thermal state of matter with reference to its tendency to communicate heat to matter in contact with it. Temperature is an index of the thermal energy content of materials, disregarding energies stored in chemical bonds and in the atomic structure of matter.

Fahrenheit temperature (F) refers to temperatures measured on a scale devised by G. D. Fahrenheit, the inventor of the alcohol and mercury thermometers, in the early 18th century. On the Fahrenheit scale, the freezing point of water is 32F and its boiling point is 212F at normal atmospheric pressure. It is said that Fahrenheit chose the gradations he used because it divides into 100 units the range of temperatures most commonly found in nature; the Fahrenheit scale, therefore, has a more humanistic basis than other temperature scales.

Rankine temperature (°R) refers to temperatures measured on the Rankine scale, which uses the same gradations as the Fahrenheit scale but which takes its origin at absolute zero (–459.69F); it is, therefore, an absolute scale of temperature.

Celsius temperature (°C) refers to temperatures measured on a scale devised in 1742 by Anders Celsius, a Swedish astronomer. The Celsius scale is graduated into 100 units between the freezing temperature of water (0°C) and its boiling point at normal atmospheric pressure (100°), and is, consequently, commonly referred to as the *Centigrade scale.*

Kelvin temperature (K) refers to temperatures measured on the Kelvin scale, which uses the same graduations as the Celsius scale but which takes its origin at absolute zero (–273.16°C); it is, like the Rankine, an absolute scale of temperature.

Temperature and Humidity Indicators

Dry-bulb temperature (T_d, DBT) is an indicator of sensible heat, or the heat content of perfectly dry air. It is the temperature measured by an ordinary (dry-bulb) thermometer, and is independent of the moisture content and insensitive to the latent heat of the air.

Wet-bulb temperature (T_w, WBT) is an indicator of the total heat content (or enthalpy) of the air, that is, of its combined sensible and latent heats. It is the temperature measured by a wet-bulb thermometer, a thermometer having a wetted sleeve over the bulb from which water is able to evaporate freely. A wet bulb thermometer is easily made by slipping a sleeve cut from a light cotton shoestring over the bulb of an ordinary outdoor or photo thermometer.

Dew point temperature (T_d, DPT) is an indicator of the moisture content of the air, with specific reference to the temperature of a surface upon which moisture contained in the air will condense. Stated differently, it is the temperature at which a given quantity of air will become saturated (reach 100% relative humidity) if chilled at constant pressure. Dew point temperature is not easily measured directly; it is conveniently found on a psychrometric chart if dry-bulb and wet-bulb temperatures are known.

Humidity refers to the water vapor contained in the air. Like the word "temperature," however, "humidity" must be qualified as to its type for it to have quantitative meaning.

Absolute humidity is defined as the weight of water vapor contained in a unit volume of air; typical units are pounds or grains of water per cubic foot. Absolute humidity is also known as the water vapor density (d_v).

Relative humidity (RH or r) is defined as the (dimensionless) ratio of the amount of moisture contained in the air under specified conditions to the amount of moisture contained in the air at saturation at the same (dry bulb) temperature. Relative humidity can be computed as the ratio of existing vapor pressure to vapor pressure at saturation, or the ratio of absolute humidity to absolute humidity at saturation existing at the same temperature and barometric pressure.

Specific humidity is the weight of water vapor contained in a unit weight of air. It may be expressed as a dimensionless ratio of pounds of water per pounds of (moist) air, or in terms of grains of moisture per pound of (moist) air.

Humidity ratio (W), or *moisture content* or *mixing ratio,* is defined as the (dimensionless) ratio of the weight of the water vapor to the weight of the dry air contained in a given volume of (moist) air.

Water vapor pressure (P_v) is that part of the atmospheric pressure ("partial pressure") which is exerted due to the amount of water vapor present in the air. It is expressed in terms of absolute pressure as inches of mercury (in. Hg) or pounds per square inch (psi). The vapor pressure of ambient air is conventionally associated with the dry bulb temperature at which the air would become saturated if it were chilled to; it is then termed the *water vapor saturation pressure* (analogous to dew point temperature).

Note: One pound contains 7000 grains

Donald Watson, editor

Summary
Sustainable design represents a set of principles of planning, design, and construction that endeavor to preserve and improve the environmental health of people and contingent natural systems. Sustainable design influence site design, rainwater harvesting, aquifer recharge, waste prevention and reclamation, and improved quality of air, water and vegetation by elimination of toxic chemicals. This article provides the context for sustainable design at the site, land planning and urban scale, with selected details and examples.

Key words
aquifer, bioclimatic design, bioregionalism, carrying capacity, conservation, health, infrastructure, nutrient recovery, rainwater reclamation, site planning, sustainabilility, utilities, waste treatment, water

Fig. 1. Center for Regenerative Studies.

Sustainable design

Sustainable design recognizes that human civilization is an integral part of the natural resources upon which all biological life of the planet depends. This places environmental understanding at the core of design of urban places and cities, conceived as part of their natural context that must be preserved and improved if the human community is to survive. The term *sustainability* has emerged in the past several decades as a broad set of principles that address economic, social and environmental development at all scales, local, regional and global. (See Box A *The concept of sustainability* on the following page).

Sustainable design dramatically enlarges the range of issues and opportunities that the design professions must address, in order to:

- preserve biological diversity and environmental integrity,
- contribute to the health of air, water, and soils,
- incorporate design and construction that reflect bioregional climatic conditions, and
- reduce and eliminate the deteriorating impacts of human use.

Sustainable design requires an understanding of environmental consequences of natural system requirements of the built environment. The Center for Regenerative Studies, California Polytechnic Institute, Pomona, CA (Figs. 1 and 2) is an example where an arid warm local climate has been impacted positively by construction of on-site water collection and water cleansing (through constructed wetlands)

Fig. 2. Center for Regenerative Studies, California State Polytechnic University, Pomona, CA. Recreated marshland and retention ponds. John Lyle, Landscape Architect.

Credits: Sections 1–4 of this article are adopted in part from *Guiding Principles of Sustainable Design*, National Park Service, 1993. Additional sources are cited in the figure captions and references. Photos are by the editor, unless otherwise noted.

Box A. The Concept of Sustainable Design

The term *sustainability* has emerged in the past several decades as a broad set of principles that address economic, social and environmental development at nearly any local, regional and global scale:

- The term was first applied to forestry and agricultural practices since the 1970s, to describe management policies to preserve their natural resource capacity.
- It was then enlarged to address any large-scale development policies and practices leading up to the United Nations 1992 Earth Summit held in Rio de Janeiro. Sustainability was given its most widely used definition, as "meeting the needs of the present without compromising the ability of future generations to meet their own needs."
- The term has been used by architects, landscape architects and urban designers, evident in the AIA/UIA World Congress of Architects June 1993 in Chicago, which adopted a "Declaration of Interdependence" to affirm a professional commitment to principles of sustainability.
- The term has appeared in urban and regional planning proposals, such as the 1994 Seattle Regional Plan where sustainability was referred to as a "three legged stool, combining economic opportunity, social equity, and environmental responsibility."
- The continuing effort to apply sustainability worldwide to urban design, architecture and planning is represented in the United Nations Conference on Human Settlements (Habitat II) held in Turkey, in 1996, which promulgated goals to "stop the deterioration of human settlement conditions, and, ultimately, to improve the living environments of all on a sustainable basis. . ." and to include "sustainable human settlements in an urbanizing world" and "adequate shelter for all."
- While architectural and building practices play a part in the technology of industrialization and use of resources, the greatest impact of population growth and demographics occurs in large-scale urban infrastructure, evident in unplanned growth of megacities throughout the world.
- This reinforces the importance of urban design, history, preservation and planning, and a commitment to sustainability in the life of our cities, defined comprehensively in economic, cultural, environmental and aesthetic terms.
- Sustainable design requires interdisciplinary teamwork in areas where economic, social and environmental issues have been previously considered separately.

The Land Ethic. The environmental and ethical basis of the concept of "sustainability" originates from ecology, as both a science and a set of values, evident in the early 20th century writing of Aldo Leopold. In a now classic essay "The Land Ethic" (Leopold 1949), Leopold defines the ecological approach to land and landscape in terms of the "biotic pyramid...[as] not merely soil; it is a fountain of energy flowing through a circuit of soil, plants and animals. Food chains are the living channels which conduct energy upward; death and decay return it to the soil. . .it is a sustained circuit, like a slowly augmented revolving fund of life."

Evolving "official" definitions of sustainability

The term "sustainability" now has an evolving "official" definition in international development, when global resource issues gained the attention of world leaders. Among the early advocates were Jimmy Carter, former President of the United States, and Gro Harlem Brundtland, former Prime Minister of Norway.

- **Global 2000: Entering the Twenty-First Century.** The 1980 Report Global 2000—The Report to the President: Entering the Twenty-First Century surveyed global demographic and environmental issues. Prepared during the presidency of Jimmy Carter and as stated in the preface, "the Global 2000 Study is the first U.S. Government effort to look at population, resources, and environment from a long-term global perspective that recognizes their interrelationships and attempts to make connections among them" (Barney 1980).
- **1987 Brundtland Commission Report.** The 1987 Report of the U.N. World Commission on Environment and Development Conference held in Stockholm, resulted in the Brundtland Commission Report, entitled "Our Common Future" (Lebel and Kane 1989) and offered a definition of "sustainability" that is the basis of the one offered at the Rio Conference.

Sustainable development means meeting the basic needs of all and extending to all the opportunity to satisfy their aspirations for a better life. But it also implies acceptance of consumption standards that are within the bounds of ecological possibility and to which all can aspire.

- **1991 World Conservation Strategy.** A fuller definition is offered in the Report Caring For The Earth—A Strategy for Sustainable Living: Second World Conservation Strategy (IUCN/UNEP/WWF 1991). The report defines sustainability as "a characteristic of a process or state that can be maintained indefinitely." First published in 1980, the report emphasizes three objectives:
- Essential ecological processes and life-support systems must be maintained,
- Genetic diversity must be preserved,
- Any use of species or ecosystems must be sustainable.

Quoting from the Report: If an activity is sustainable, for all practical purposes it can continue forever. . .A "sustainable society" lives by the nine principles:

(1) Respect and care for the community of life;
(2) Improve the quality of human life;
(3) Conserve the Earth's vitality and diversity;
(4) Minimize depletion of nonrenewable resources;
(5) Keep within the Earth's carrying capacity;
(6) Change personal attitudes and practices;
(7) Enable communities to care for their own environments;
(8) Provide an international framework for integrating development and conservation;
(9) Create a global alliance.

- **Rio Earth Summit's AGENDA 21.** The Rio Earth Summit produced a similar set of principles, stated as the six major themes of Agenda 21 (Daniel Sitarz, editor 1993):

In many portions of the world, the day-to-day quality of life is deteriorating due to a combination of poverty, malnutrition, unemployment, population growth, lack of health care and pollution. At the same time, a minority of humanity continues to sustain a lifestyle that is based on highly wasteful consumption patterns and pollution-generating production processes. To sustain and improve the quality of life on Earth requires:

- Efficient use of the world's natural resources,
- Protection and management of our global commons,
- Design and management of human settlements,
- Environmentally benign use of chemicals and management of human and industrial waste,
- Global economic growth based upon sustainability.

References

Barney, Gerald O., ed. 1980. Global 2000: *The Report to the President: Entering the Twenty-First Century.* U.S. Council on Environmental Quality and the Department of State Arlington, Virginia: Seven Lock Press.

Lebel, Gregory G. and Hal Kane. 1989. *Sustainable Development: A Guide to Our Common Future–The Report of the World Commission on Environment and Development.* Washington, DC: The Global Tomorrow Coalition.

Leopold, Aldo. 1949. *A Sand County Almanac.* New York: Oxford University Press.

IUCN/UNEP/WWF. 1991. (International Union Conservation Union–United Nations Environment Programme–World Wide Fund). *Caring For The Earth: A Strategy for Sustainable Living–Second World Conservation Strategy.* Gland, Switzerland.

Sitarz, Daniel, ed. 1993. *Agenda 21: The Earth Summit Strategy to Save Our Planet.* Boulder, CO: Earthpress.

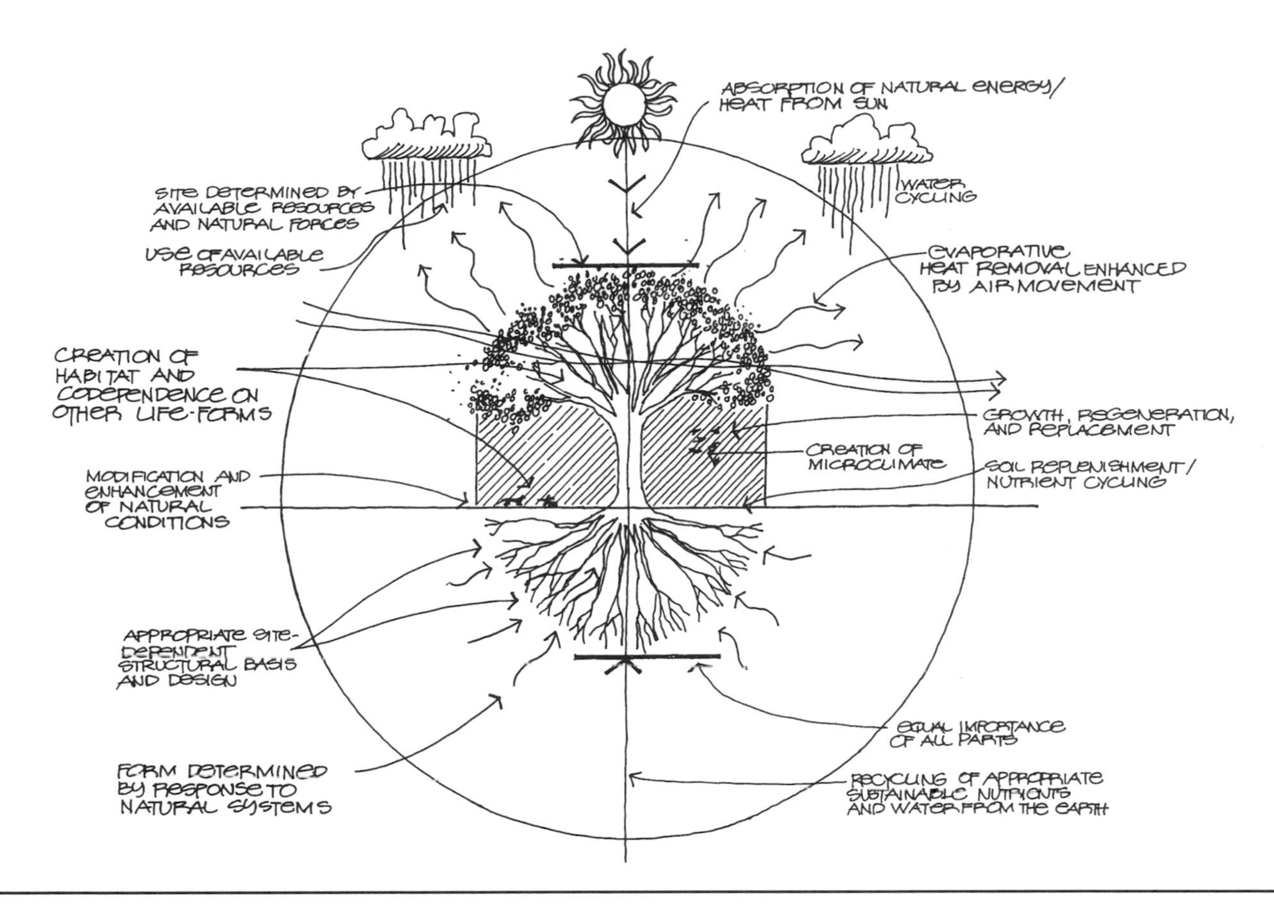

Fig. 3. Natural system model for design. (U.S Park Service 1995.)

supporting new vegetation and wetland zones that have attracted species of birds and other wildlife not normally resident in the locale.

Sustainable design is inspired by and learns from the lessons of nature. Consider the properties of a natural organism that utilizes sunlight and rainwater as sustenance, has mechanism to endure scarcity, produces nothing that is wasted, and coevolves with its surroundings to reproduce life. This natural model provides a helpful metaphor by which to inspire sustainable design (Fig. 3). For example:

- The natural organism makes use of immediately and locally available materials to construct itself, and does so with economy and efficiency. The same strategies when used in development can minimize global and local impacts on resources.
- The natural organism adapts to its environment through instinctive reaction and an evolutionary process of generations. Through the ability to rationalize and mechanize, humans have the ability to adapt psychologically and physically in a matter of hours, but often with little natural instinct or understanding of feedback and interrelationship with the environment.
- The natural organism maintains a sustaining relationship with its environment by a balance between its needs and available resources. Similarly, sustainable design adjusts demands, lifestyles, and technologies to evolve a compatible balance with the natural and cultural systems within its environment.

1 NATURAL RESOURCES

The operating premise of sustainable design is that built infrastructure and facilities must function within the ecosystem and its constraints, not only for foreseeable circumstances, but for a very long time. Ecosystems provide direct ecological services—what Amory Lovins and Paul Hawken have called *natural capital*—which for the built environment include passive solar heating, cooling and daylighting, vegetative screening, water/wastewater purification, and the physical and spiritual health we gain from natural resources (*e.g.*, beaches, forests, reefs, and wildlife).

The first level of sustainable design is precautionary, based on the principle of health to "Do no harm." Many negative impacts of design and construction practices are established by prevailing habits, conventions and even regulations. As a result, conscientious designers are led to assert and advocate for sustainable design policies, programs and regulations, well before a design project may fall within their direct professional purview. As a guide to designing to minimize possible negative impacts of urban development on the environment, Table A (following pages) indicates where negative impacts on the environment may occur. Potential impacts are arranged according to three categories—pollution, physical processes, and biological systems. These are further divided into specific impacts on the environment—that is, noise increased, erosion increased, vegetation altered. A solid black dot on the matrix indicates a negative impact.

Development Type/Activity	Pollution									Physical Processes									Biological Systems														
	Noise increased	Air quality deteriorated	Toxics released during construction	Toxic materials spilled or discharged	Pollutants introduced by vehicle transport	Petroleum products spilled	Toxics or sewage discharged by vessels	Odors released	Hot water discharged	Erosion increased	Sedimentation/siltation increased	Soil disturbed or compacted	Soil removed and disposed of	Surface water flow disrupted	Groundwater supply reduced	Groundwater supply depleted	Long-shore drift/beach dynamics altered	Dredging potentially required	Vegetation altered	Vegetation destroyed	Habitat altered	Habitat destroyed or fragmented	Coral reefs disturbed or destroyed	Barriers to wildlife movement created	Collisions or road kills on wildlife increased	Corridors for exotic species invasion created	Exotic/alien species directly introduced	Diseases introduced	Life cycles of wildlife disrupted	Nutrient flow/food chains altered	Nonnatural foods or habitat introduced	Nontarget species destroyed	Animal rights issues raised
Site Access																																	
Terrestrial																																	
Roads	●	●	●		●	●				●	●	●	●	●	●				●	●	●	●		●	●	●	●		●	●			
Trails	●									●	●	●		●					●	●	●					●							
Boardwalks	●		●																		●					●							
Aquatic																																	
Docks and piers						●	●										●	●					●										
Open anchorage						●	●												●	●	●	●	●										
Air																																	
Airport/strip	●	●	●		●	●				●	●	●	●	●					●	●	●	●		●									
Helicopter pad	●	●	●		●	●				●	●	●							●	●	●												
Seaplane dock	●	●			●	●																											
Remote/indirect																																	
Remote TV																			●														
Construction and Landscaping																																	
Site preparation																																	
Excavation	●	●			●	●				●	●	●	●	●					●	●	●	●											
Filling	●	●			●	●				●	●	●	●	●					●	●	●	●					●						
Draining														●	●				●	●	●	●		●						●	●		
Grading	●	●			●	●				●	●	●		●	●				●	●	●	●											
Foundation																																	
Slab												●		●						●		●											
Excavated	●	●										●	●	●						●		●											
Elevated			●																●		●												
Landscaping																																	
Shaping/planting	●	●								●	●	●							●	●	●		●	●		●	●	●	●	●	●		
Ponds													●	●						●		●									●		
Swimming pools													●	●	●					●		●											
Restoration	●	●																	●		●												
Energy																																	
Supply Source																																	
Wood-local		●																	●	●	●												
Fossil-imported		●		●	●	●																											
Biogas-local								●																									
Propane-imported		●			●																												
Solar-local																																	
Windmill-local			●																●														
Hydroelectric (with dam and storage)			●									●	●	●					●	●	●	●		●						●			
Hydroelectric (small-scale, water wheel or ram)												●							●		●												
Hydrothermal			●									●	●		●				●		●												
Electric-imported			●									●							●	●	●				●	●	●		●				
Electric-local	●	●	●		●	●		●				●							●		●												
Natural gas-imported			●	●								●							●		●												
Transmission of																																	
Powerlines																																	
Aboveground			●									●							●	●	●				●	●	●		●				
Buried			●									●							●		●												
Pipes																																	
Aboveground			●	●								●							●		●			●									
Buried			●	●								●							●		●												
Vehicle-bulk (also requires road access)	●	●		●	●	●																			●		●						
Boat-bulk (also requires docks, piers)	●	●				●	●											●					●										

Table A. Environmental impacts of development. (U.S Park Service 1995.)

Development Type/Activity	Pollution									Physical Processes									Biological Systems														
	Noise increased	Air quality deteriorated	Toxics released during construction	Toxic materials spilled or discharged	Pollutants introduced by vehicle transport	Petroleum products spilled	Toxics or sewage discharged by vessels	Odors released	Hot water discharged	Erosion increased	Sedimentation/siltation increased	Soil disturbed or compacted	Soil removed and disposed of	Surface water flow disrupted	Groundwater supply reduced	Groundwater supply depleted	Long-shore drift/beach dynamics altered	Dredging potentially required	Vegetation altered	Vegetation destroyed	Habitat altered	Habitat destroyed or fragmented	Coral reefs disturbed or destroyed	Barriers to wildlife movement created	Collisions or road kills on wildlife increased	Corridors for exotic species invasion created	Exotic/alien species directly introduced	Diseases introduced	Life cycles of wildlife disrupted	Nutrient flow/food chains altered	Nonnatural foods or habitat introduced	Nontarget species destroyed	Animal rights issues raised
Water																																	
Supply																																	
Wells															●	●																	
Cisterns			●																														
Impoundments												●	●	●					●	●	●	●		●						●			
Diversions														●					●	●	●									●			
Desalinization			●						●										●	●	●		●							●	●		
Recycling			●									●																●		●			
Importing (also requires road access)	●	●		●	●	●																			●		●						
Distribution																																	
Aboveground pipes			●									●							●		●			●									
Buried pipes			●									●							●		●												
Vehicle transport (also requires road access)	●	●			●	●																			●		●						
Waste Disposal—Storage																																	
Human/organic (secondary, onsite)			●	●				●				●		●	●				●		●							●		●	●		
Solid/trash (landfill, offsite)	●	●			●	●		●																	●		●	●					
Communication																																	
Radio/microwave transmission tower			●																●						●								
Satellite																																	
Telephone																																	
Lines aboveground			●									●							●	●	●				●	●	●		●				
Lines buried			●									●							●		●												
Walls and Fences																																	
Stone wall												●		●						●		●		●					●				
Cement/brick wall			●									●		●						●		●		●					●				
Wooden fence			●																			●		●					●				
Wire fence																						●		●	●				●				
Open trench												●	●	●						●	●	●		●									
Operations and Maintenance																																	
Machinery/vehicles	●	●			●	●	●																										
Routine recycling																																	
Fire management																																	
Fire breaks										●	●	●		●						●	●	●		●		●	●		●				
Controlled burns																			●		●												
Wildlife management																																	
Introduce predators																											●	●	●				
Trap/poison species				●																												●	●
Shoot species																																	●
Sterilization																													●				●
Natural controls																																	
Vegetation management																																	
Poisoning				●															●	●	●	●										●	
Cut/clear										●	●								●	●	●	●											
Natural controls																																	
Visitor activities (may require access roads, trails, docks, structures)																																	
Hiking	●									●					●																		
Boating	●					●	●																●		●								
Camping	●	●	●		●	●																											
Snorkel/SCUBA	●					●	●														●		●		●								
Horseback riding	●							●		●														●			●	●					
Nature study	●																																

Table A. (continued)

Design for integration of urban development with natural resources includes:

- **Natural behavior within an ecosystem.** A basic understanding of the natural behavior of an ecosystem is required before designing facilities to function within it, established by a resource inventory prior to a project design.

- **Links between ecosystems.** There are links between ecosystems that may be geographically separate, that is, between mountain forests and coastal mangroves, between mangroves and coral reefs. Changes in one ecosystem may have consequences in another.

- **Fragmentation of habitats.** Whether caused by constructing a specific facility or because of land-use decisions throughout an ecosystem, habitat fragmentation causes loss of biological diversity and must be minimize, and wherever possible, reversed by reconstitution of wildlife preserves and corridors.

- **Human demands on ecosystems.** The demands of human use on an ecosystem are cumulative. New proposals must account for the previous use of resources so that effects of past activity, proposed development, and anticipated future use do not exceed the ecosystem's capability. The scale and type of any potential development should be determined by the capability and resiliency of the ecosystem rather than by the physical capacity of the site.

- **Acceptable limits of change.** Change in the system is inevitable, but limits of acceptable environmental change—often called the carrying capacity—should be established before development begins. Acceptable change should not approach the upper limit of capacity because unpredictable events such as droughts and hurricanes may go beyond that limit and cause the entire system to collapse.

- **Ecosystem monitoring.** The effects on surrounding resources of developing and operating facilities should be routinely monitored and evaluated, and actions taken immediately to correct problems.

LAND VALUES

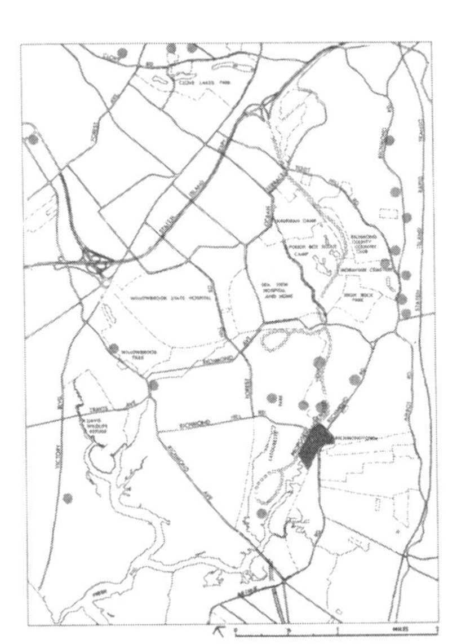

HISTORIC VALUES

TIDAL INUNDATION

WATER VALUES

LAND VALUES
ZONE 1 $3.50 a square foot and over.
ZONE 2 $2.50-$3.50 a square foot.
ZONE 3 Less than $2.50 a square foot.

TIDAL INUNDATION
ZONE 1 Inundation during 1962 hurricane.
ZONE 2 Area of hurricane surge.
ZONE 3 Areas above flood line.

HISTORIC VALUES
ZONE 1 Richmondtown Historic Area.
ZONE 2 Historic landmarks.
ZONE 3 Absence of historic sites.

SCENIC VALUES
ZONE 1 Scenic elements.
ZONE 2 Open areas of high scenic value.
ZONE 3 Urbanized areas with low scenic value.

RECREATION VALUES
ZONE 1 Public open space and institutions.
ZONE 2 Non-urbanized areas with high potential.
ZONE 3 Area with low recreation potential.

WATER VALUES
ZONE 1 Lakes, ponds, streams and marshes.
ZONE 2 Major aquifer and watersheds of important streams.
ZONE 3 Secondary aquifers and urbanized streams.

FOREST VALUES
ZONE 1 Forests and marshes of high quality.
ZONE 2 All other existing forests and marshes.
ZONE 3 Unforested lands.

WILDLIFE VALUES
ZONE 1 Best quality habitats.
ZONE 2 Second quality habitats.
ZONE 3 Poor habitat areas.

RESIDENTIAL VALUES
ZONE 1 Market value over $50,000.
ZONE 2 Market value $25,000-$50,000.
ZONE 3 Market value less than $25,000.

INSTITUTIONAL VALUES
ZONE 1 Highest value.
ZONE 2 Intermediate value.
ZONE 3 Least value.

Fig. 4. Site inventory Borough of Richmond, New York. Selected images. The entire set consists of 16 overlays. (McHarg 1969.)

2 SITE DESIGN

Site design of sustainable environments requires low-impact planning, construction and property maintenance, with strategies that do not alter or impair but instead help *repair* and *restore* existing site systems. Site systems such as plant and animal communities, soils, and hydrology must be maintained and improved as essential *processes of a healthy environment.*

Site selection for sustainable developments is a process of identifying, weighing, and balancing the attractiveness (natural and cultural environments, access) of a site against the costs inherent in its development (natural and cultural environments, access, hazards, operations). The characteristics of a region or site should be described spatially to provide a precise geographic inventory. The graphic overlay method was first proposed for site inventory and environmental design by Ian L. McHarg (1969) and evident in the work of environmental landscape architects (Figs. 4 and 5). When overlaid as a composite, the least sensitive area—and thus most amenable for low-impact development—are graphically identified.

The following general considerations apply to sustainable site design:

- Preserve and enhance the native landscape and its resources.
- Plan landscape development according to the unique features of the surrounding context rather than by overlaying standardized patterns and solutions.
- Understand the site as an integrated ecosystem with changes occurring over time in dynamic balance; the impacts of development must be confined within these natural changes.
- Allow simplicity of functions to prevail, while respecting basic human needs of comfort and safety.

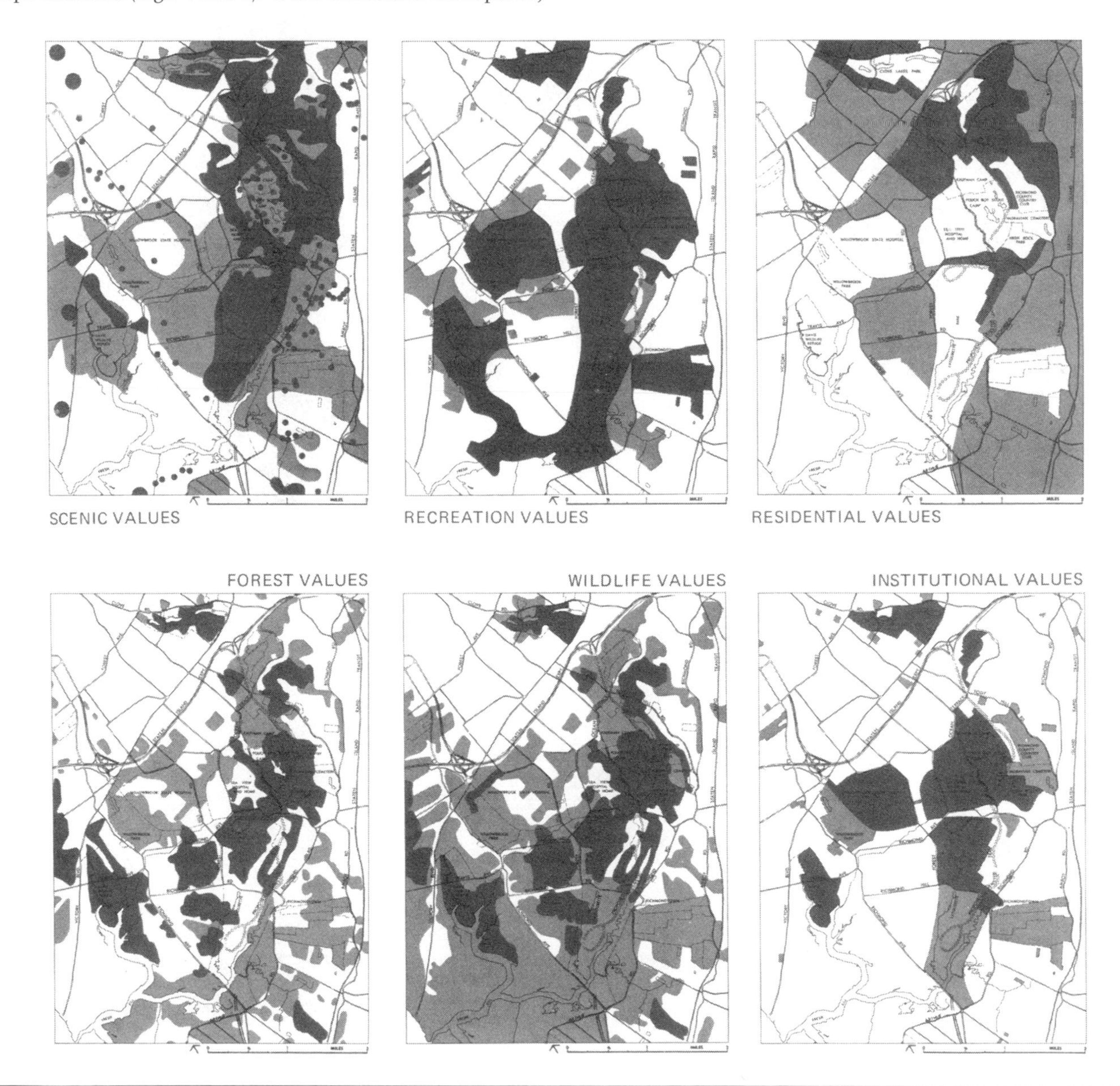

Fig. 4. Continued

LAND RESOURCE INVENTORY

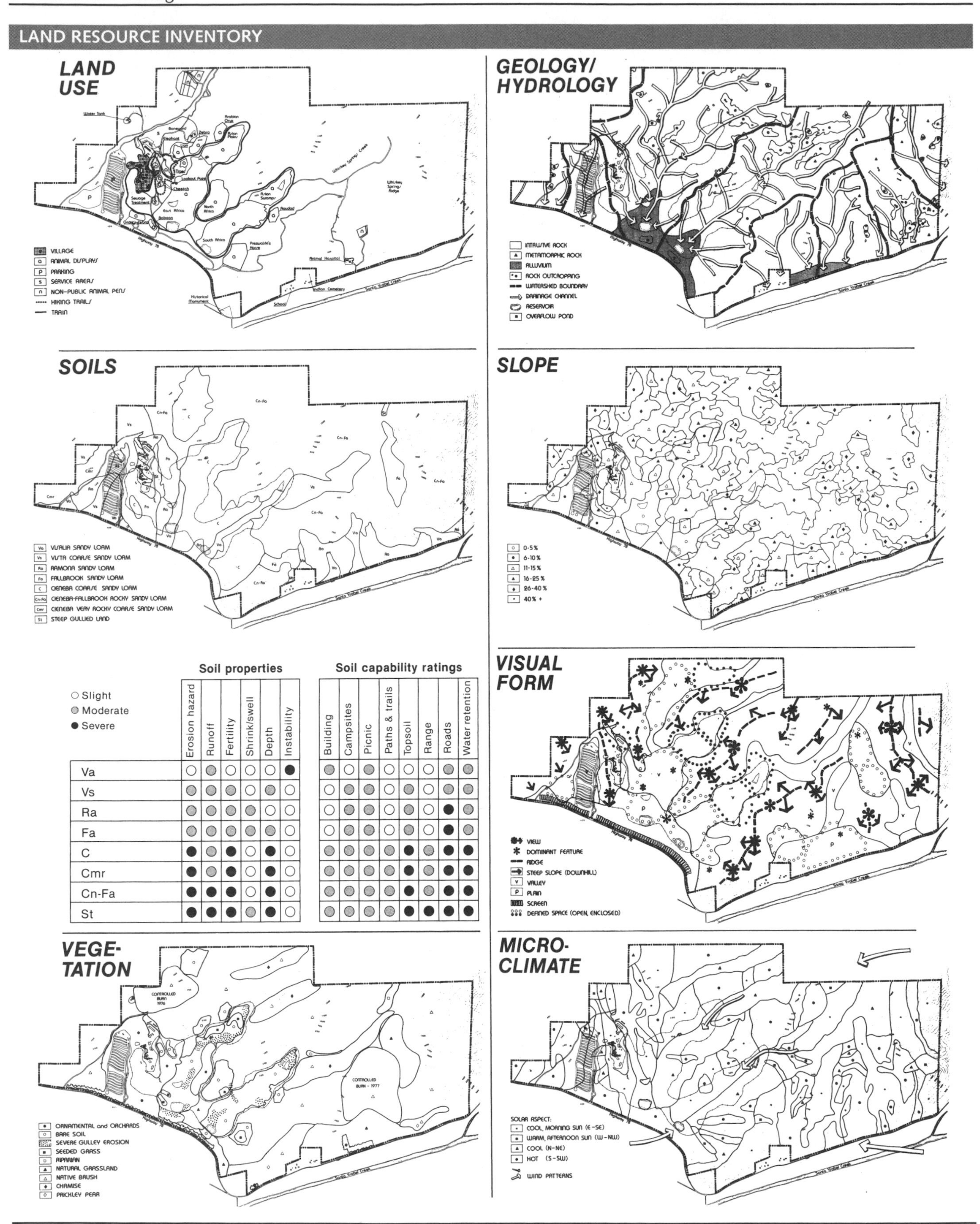

Soil properties / Soil capability ratings (○ Slight, ◐ Moderate, ● Severe)

	Erosion hazard	Runoff	Fertility	Shrink/swell	Depth	Instability	Building	Campsites	Picnic	Paths & trails	Topsoil	Range	Roads	Water retention
Va	Slight	Moderate	Slight	Slight	Slight	Severe	Moderate	Slight	Moderate	Slight	Slight	Slight	Moderate	Moderate
Vs	Moderate	Moderate	Moderate	Slight	Moderate	Slight	Slight	Moderate	Moderate	Slight	Moderate	Slight	Moderate	Moderate
Ra	Moderate	Moderate	Moderate	Moderate	Slight	Slight	Slight	Moderate	Moderate	Slight	Moderate	Slight	Severe	Moderate
Fa	Moderate	Moderate	Moderate	Moderate	Moderate	Slight	Slight	Moderate	Moderate	Slight	Moderate	Slight	Severe	Moderate
C	Severe	Moderate	Severe	Slight	Severe	Slight	Moderate	Moderate	Moderate	Moderate	Severe	Moderate	Severe	Severe
Cmr	Severe	Moderate	Severe	Slight	Severe	Slight	Moderate	Moderate	Moderate	Moderate	Severe	Moderate	Severe	Severe
Cn-Fa	Severe	Severe	Severe	Slight	Severe	Slight	Moderate	Moderate	Moderate	Moderate	Severe	Moderate	Severe	Severe
St	Severe	Severe	Severe	Moderate	Severe	Slight	Moderate	Moderate	Moderate	Moderate	Severe	Severe	Severe	Severe

Fig. 5. Land resource inventory and suitability analysis prepared for the San Diego Zoo Association by Studio 606, Department of Architecture, California State Polytechnic University, Pomona. (Lyle 1985.)

SUITABILITY MODELS

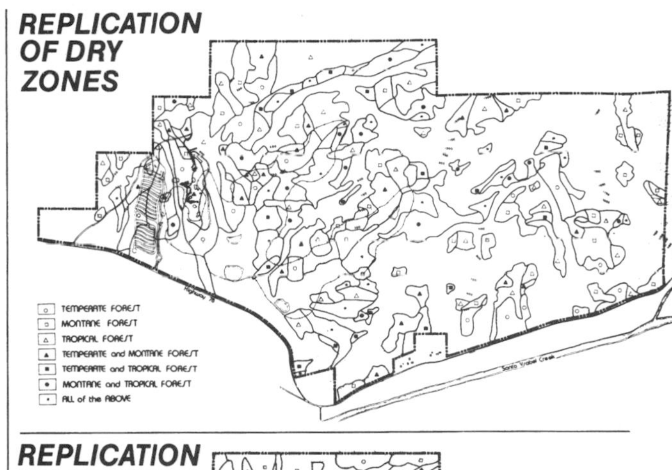

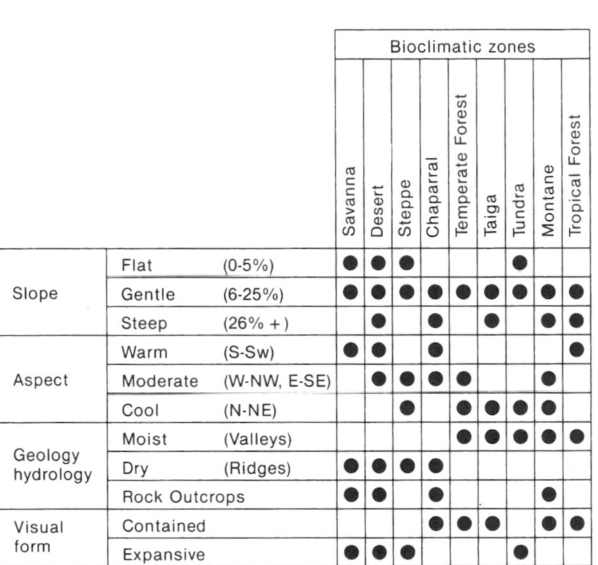

			Savanna	Desert	Steppe	Chaparral	Temperate Forest	Taiga	Tundra	Montane	Tropical Forest
			Bioclimatic zones								
Slope	Flat	(0-5%)	●	●	●				●		
	Gentle	(6-25%)	●	●	●	●	●	●	●	●	●
	Steep	(26% +)		●		●		●		●	●
Aspect	Warm	(S-Sw)	●	●		●					●
	Moderate	(W-NW, E-SE)		●	●	●	●			●	
	Cool	(N-NE)			●		●	●	●	●	
Geology hydrology	Moist	(Valleys)					●	●	●	●	●
	Dry	(Ridges)	●	●	●	●					
	Rock Outcrops		●	●		●				●	
Visual form	Contained					●	●	●		●	●
	Expansive		●	●	●				●		

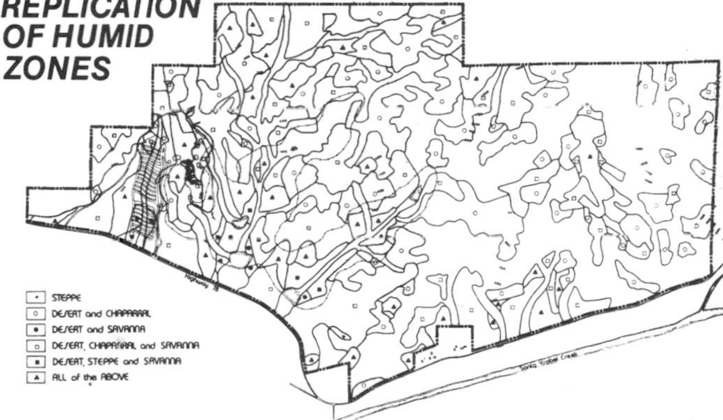

Severe constraints
- Steep slopes (over 40%) or
- Moderately steep slopes (25-40%) and erodible soils (Cl, Cn-Fa, Cmr) or
- Severely eroded soils (St) or
- Riparian vegetation

Moderate constraints
- Moderately steep slopes (25-40%) or
- Erodible soils

Minor constraints
- Moderate slopes (16-25%) or
- Fair soils (Fa, Ra) or
- Hot aspect (south, southwest)

No constraints
- Good soils (Va, Vs) or
- Gentle slopes (0-16%) or
- Rock outcroppings

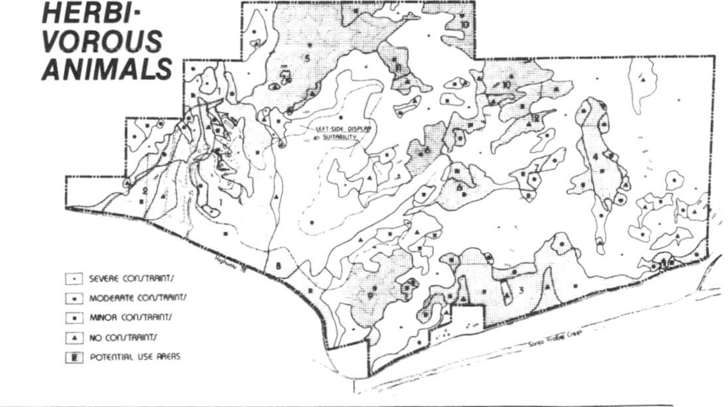

Severe constraints
- Steep slopes (over 40%) or
- Moderately steep slopes (25-40%) and unstable or erodible soils (Va, Cn-Fa, Cl, Cmr, St) or
- Moderately steep slopes and unstable or erodible soils and major drainage channels or
- Riparian vegetation or
- Rock outcroppings

Moderate constraints
- Moderately steep slopes (25-40%) or
- Moderate slopes (16-25%) and unstable or erodible soils (Va, Cn-Fa, Cl, Cmr, St) or
- Moderate slopes and unstable or erodible soils and major drainage channels or alluvial rock

Minor constraints
- Unstable or erodible soils (Va, Cn-Fa, Cl, Cmr, St) or
- Moderate slopes (16-25%) and fair soils (Fa, Ra) or
- Major drainage channels or alluvial rock

No constraints
- Good soils (Vs) and gentle to moderate slopes (0-25%) or
- Fair soils (Fa, Ra) and gentle slopes (0-15%)

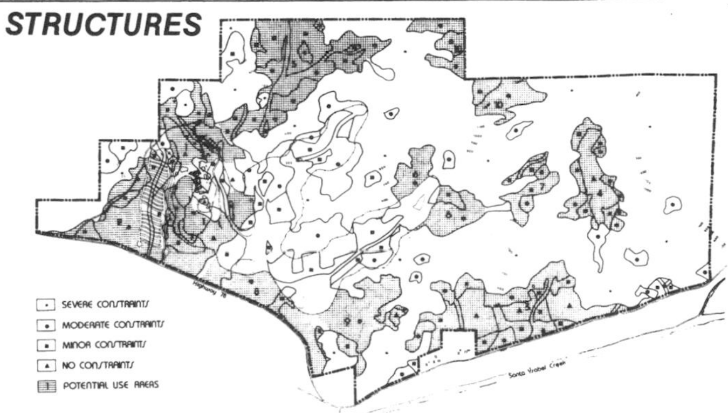

Fig. 5. Continued

Fig. 6. Visitor parking, Morris Arboretum, Philadelphia. The permeable parking surface feature is part of the interpretive design. Andropogan Associates, Landscape Architects.

Box B. Design Principles of Sustainable Site Design

The "Valdez Principles for Site Design," developed by Andropogon Associates, Ltd, to provide design and policy guidelines in site design for U.S. Park Service. They are applicable to any urban development.

- **Recognition of context.** No site can be understood and evaluated without looking outward to the site context.
- **Treatment of landscapes as interdependent and interconnected.** Conventional development often increased fragmentation of the landscape. The small remaining islands of natural landscape are typically surrounded by a fabric of development that diminishes their ability to support a variety of plant communities and habitats. Larger environmental systems must be created by reconnecting fragmented landscapes and establishing contiguous networks with other natural systems both within a site and beyond its boundaries.
- **Integration of the native landscape with development.** Even the most developed landscapes are not self-contained and should be redesigned to support some component of the natural landscape to provide critical connections to adjacent habitats.
- **Promotion of biodiversity.** The environment is experiencing extinction of both plant and animal species. Development itself affords the opportunity to emphasize the establishment of biodiversity on a site. Site design must protect local plan and animal communities, and new landscape plantings must deliberately reestablish diverse natural habitats in organic patterns that reflect the processes of the site.
- **Reuse of already disturbed areas.** Despite the declining availability of relatively unspoiled land and the wasteful way sites are conventionally developed, existing built areas are being abandoned and new development located on remaining rural and natural areas. This cycle must be reversed. Previously disturbed areas must be rehabilitated and restored to their natural integrity, especially urban landscapes.
- **Making a habit of restoration.** Where the landscape fabric is damaged, it must be repaired and/or restored. Every development project should have a restoration component. Effective restoration requires recognition of the interdependence of all site factors and must include repair of all site systems—soil, water, vegetation, and wildlife.

- Assess feasibility of development in long-term social and environmental costs, not just short-term construction costs.
- Analyze and model water and nutrient cycles prior to development intervention.
- Minimize areas of vegetation disturbance, earth grading, and water channel alteration.
- Locate structures to take maximum advantage of passive energy design and technologies to provide for human comfort.
- Provide space for processing all wastes created onsite (collection/recycling facilities, digesters, and lagoons) so that reusable/recyclable resources will not be lost and hazardous or destructive wastes will not be released into the environment.

Determine environmentally safe means of onsite energy production and storage in the early stages of site planning.

- Phase development to allow for the monitoring of cumulative environmental impacts of development.
- Allow the natural ecosystem to be self-maintaining to the greatest extent possible.

The following factors should be considered in site selection:

- **Capacity**—Every site has a carrying capacity for development and human activity. Site analysis should determine this capacity based on the sensitivity of site resources.
- **Density**—Siting of facilities should weigh the merits of concentration versus dispersal. Natural landscape values may be easier to maintain if facilities are carefully dispersed. Conversely, concentration of structure leaves more undisturbed natural areas.
- **Climate**—Environments for resource-related developments range from rain forest to desert. The characteristics of a specific climate should be considered when locating facilities so that human comfort can be maximized while protecting the facility from climatic forces such as violent storms and other extremes.
- **Slopes**—In many environments steep slopes predominate, requiring special siting of structures and costly construction practices. Building on slopes considered too steep can lead to soil erosion, loss of hillside vegetation, and damage to fragile wetland and marine ecosystems. Appropriate site selection should generally locate more intensive development on gentle slopes, dispersed development on moderate slopes, and no development on steep slopes.
- **Vegetation**—Retain as much existing native vegetation as possible to secure the integrity of the site. Natural vegetation is often an essential aspect of the visitor experience and should be preserved. Site selection should maintain large habitat areas and avoid habitat fragmentation and canopy loss. In some areas such as the tropics, most nutrients are held in the forest canopy, not in the soil—loss of canopy therefore causes nutrient loss as well. Plants live in natural associations (plant communities) and should remain as established naturally.

- **Views**—Views are critical and reinforce a visitor experience. Site location should maximize views of natural features and minimize views of visitor and support facilities.

- **Natural hazards**—Sustainable development should be located with consideration of natural hazards such as precipitous topography, dangerous animals and plants, and hazardous water areas. Site layout should allow controlled access to these features.

- **Access to natural and cultural features**—Good site design practices can maximize pedestrian access to the wide variety of onsite and offsite resources and recreational activities. Low-impact development is the key to protecting vital resource areas.

- **Energy and utilities**—Siting should consider possible connections to offsite utilities, or more likely, spatial needs for onsite utilities. The potential exists for alternative energy use in many places, particularly solar- and wind-based energy systems.

By way of summary, Box B and Box C provide guidelines by which landscape designers and architects incorporate these factors as principles of sustainable design

Site access, road design and construction

Site access refers to the means of physically entering a development and also the en route experience. Considerations for enhancing the experience of site access include:

- Select routes that minimize environmental impacts and that allow control of site development.

- Provide anticipation and drama by framing views or directing attention to landscape features along the access route and to provide a sense of arrival at the destination.

The need to construct a road into a site is the first critical decision to be made and one that has long-lasting implications. Building a road into a pristine site will change the site forever.

A curvilinear alignment can be designed to flow with the topography while adding visual interest. Crossing unstable slopes should be avoided. Steep grades should be used as needed to lay road lightly on the ground, and retaining walls should be included on cut slopes to ensure long-term slope stability. The road should have low design speeds (with more and tighter curves) and a narrower width to minimize cut-and-full disturbance.

Many soils are highly susceptible to erosion. Vegetation clearing on the road shoulders should be minimized to limit erosion impacts and retain the benefits of greenery. All fill slopes should be stabilized and walls provided in cut sections where needed. Exposed soils should be immediately replanted and mulched.

Unpaved surfaces are appropriate in areas of stable soils, lower slopes, and low traffic loads, but they require more maintenance. Permeable paved surfaces allow limited percolation of precipitation while providing better wear than unpaved surfaces. Permeable parking surfaces provide a means of recharging the local aquifer, rather than creating accelerated storm run-off (Fig. 6). Impermeable paved surfaces are needed for roads with the highest load and traffic requirements. Whenever possible, recycled materials should be used in the construction

Box C. Sustainable Design Guidelines

Bruce Coldham, Architect

Spatial analysis

1 Systematically record the natural resources of site and region.
2 Allocate land use according to productive potential and (human) social needs.
3 Understand bioclimatic design strategies for promoting winter heat gain and reducing summer overheating.
4 Use local resources to increase self-reliance.
5 Explore natural processes as the genesis of materials used and application of recycled and new materials to construction.
6 Value natural systems and processes.

Energy production and conversion

7 Use efficient equipment and appliances to reduce energy demand.
8 Match energy quality to end-use needs.
9 Promote daylighting in building design.
10 Consider biomass combustion.
11 Use annual cycle solar-thermal storage to balance seasonal excesses of heat and cold.
12 Generate electricity photovoltaically.
13 Install appropriate energy storage to buffer intermittent production of energy.
14 Harness wind/hydro/geothermal sources of power to balance winter slump in solar resource.
15 Provide grid connection to regional electric utility for storage.
16 Use metabolic energy to reduce capital energy intensity.

Water supply

17 Reduce water consumption by installing efficient fixtures and outlets.
18 Reduce water consumption by matching water quality to end use.
19 Collect rainwater for potable and process needs.
20 Collect surface runoff for process needs.

Nutrient/waste cycling

21 Collect "gray water" separately from "black water."
22 Use algal cultures/aquatic plants to remove nutrients, pollutants, and pathogens.
23 Compost organic materials.
24 Consider anaerobic digestion of organic material.

Food production

25 Match food production with human nutritional requirements.
26 Couple intensive food production to medium-density residential development to complete a nutrient/waste cycle.
27 Adopt organic and biodynamic farming techniques.
28 Maintain perennial polycultures for fruit and grain crops.
29 Use aquaculture techniques for fish and algal production.
30 Exploit the synergetic potential of a systematic approach to food production.
31 Use "bioshelters" to modify climate for increased production.
32 Provide food storage from one season to the next for seasonal produce.

Materials

33 Value the energy invested in the production of materials.
34 Design for long life and easy eventual recycling of material constituents.
35 Maintain recycling centers for the accumulation and marketing of source separated goods.
36 Value shape over mass in achieving strength and stability.

Shelter design

37 Locate buildings on sloping, "nonprime" agricultural land.
38 Construct low-rise dwelling units in attached clusters or rows.
39 Orient buildings and primary solar collectors for solar access.
40 Design structures for long life and adaptability.
41 Employ superinsulated building techniques to reduce heat gains and losses.
42 Landscape for microclimatic amelioration.

of the surfacing, *e.g.*, crushed glass, shredded rubber tires, or recycled aggregate.

Site utilities and waste systems

Utility systems. Early in the planning process, utility systems must be identified that will not adversely affect the environment and will work within established natural systems. After appropriate systems are selected, careful site planning and design is required to address secondary impacts such as soil disturbance and intrusion on the visual setting.

Utility corridors. When utility lines are necessary, they should be buried near other corridor areas that are already disturbed, such as roads and pedestrian paths. Overhead lines should not be located in desirable view sheds or over landform crests. Many utility lines can be concealed under boardwalks and thereby eliminate ground disturbance.

Night lighting. The nighttime sky can be dramatic. Light intrusion and overlighting glare can obscure what little night vision is available to humans. Care is required to limit night lighting to the minimum necessary for safety. Light fixtures should remain close to the ground to minimize eye level glare.

Storm drainage. In undisturbed landscapes, storm drainage is typically handled by vegetation canopy, ground cover plants, soil absorption, and streams and waterways. In a modified landscape, consideration must be given to the impacts of storm drainage on the existing natural system of drainage and the resulting structures and systems that will be necessary to handle the new drainage pattern. The main principles in storm drainage control are to regulate runoff to provide protection

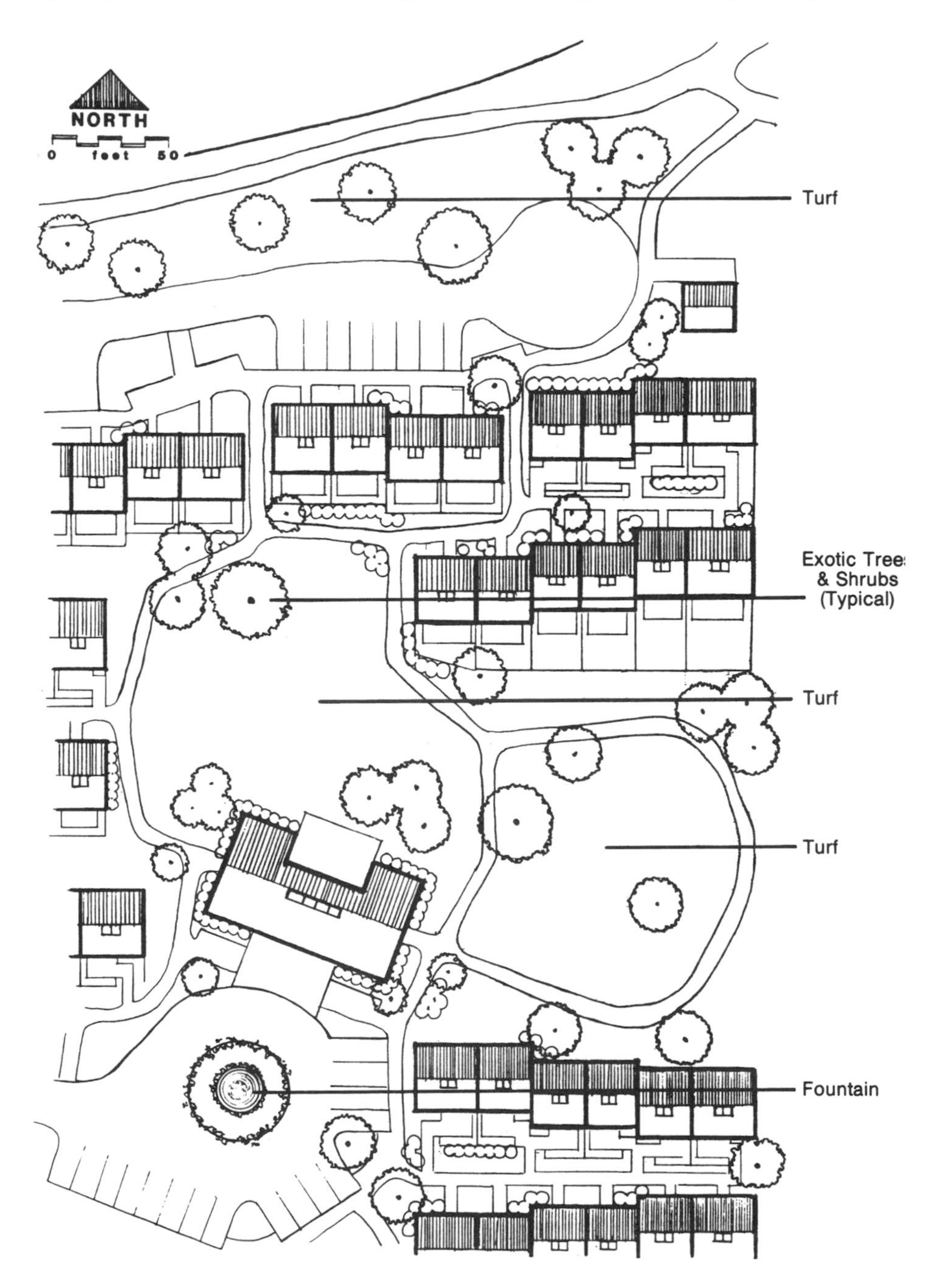

Fig. 7. Traditional development landscape with turf grass, widely placed trees and foundation shrubs, McPherson 1984. (Courtesy Landscape Architecture Foundation.)

from soil erosion and to avoid directing water into unmanageable volumes. Removal of natural vegetation, topsoil, and natural channels that provide natural drainage control should always be avoided. An alternative would be to stabilize soils, capture run-off in depressions—often called a bioswale and which have plants that help store and clean water and to help recharge local groundwater supply—and to revegetate areas to replicate natural drainage systems.

Irrigation systems. Low-volume irrigation systems are appropriate in most areas as a temporary method to help restore previously disturbed areas or as a means to support local agriculture and native traditions. Captured rainwater, recycled gray water, or treated effluent could be used as irrigation water.

Waste treatment. It is important to use treatment technologies that are biological, nonmechanical, and do not involve soil leaching or land disposal that causes soil disturbance. Constructed biological systems are being put to use increasingly to purify wastewater.

Site-adaptive design considerations

Natural Characteristics. When natural systems and climatic resources are incorporated into site designs, spaces can be more comfortable, interesting, and efficient (Figs. 7 and 8). They include:

- **Wind**—The major advantage of wind in recreational development is its cooling aspect. For example, trade winds in the Caribbean come from the northeast to the southeast quadrant, so many of the structures and outdoor gathering places of the native population are oriented to take advantage of this cooling wind movement, or "natural" air conditioning.

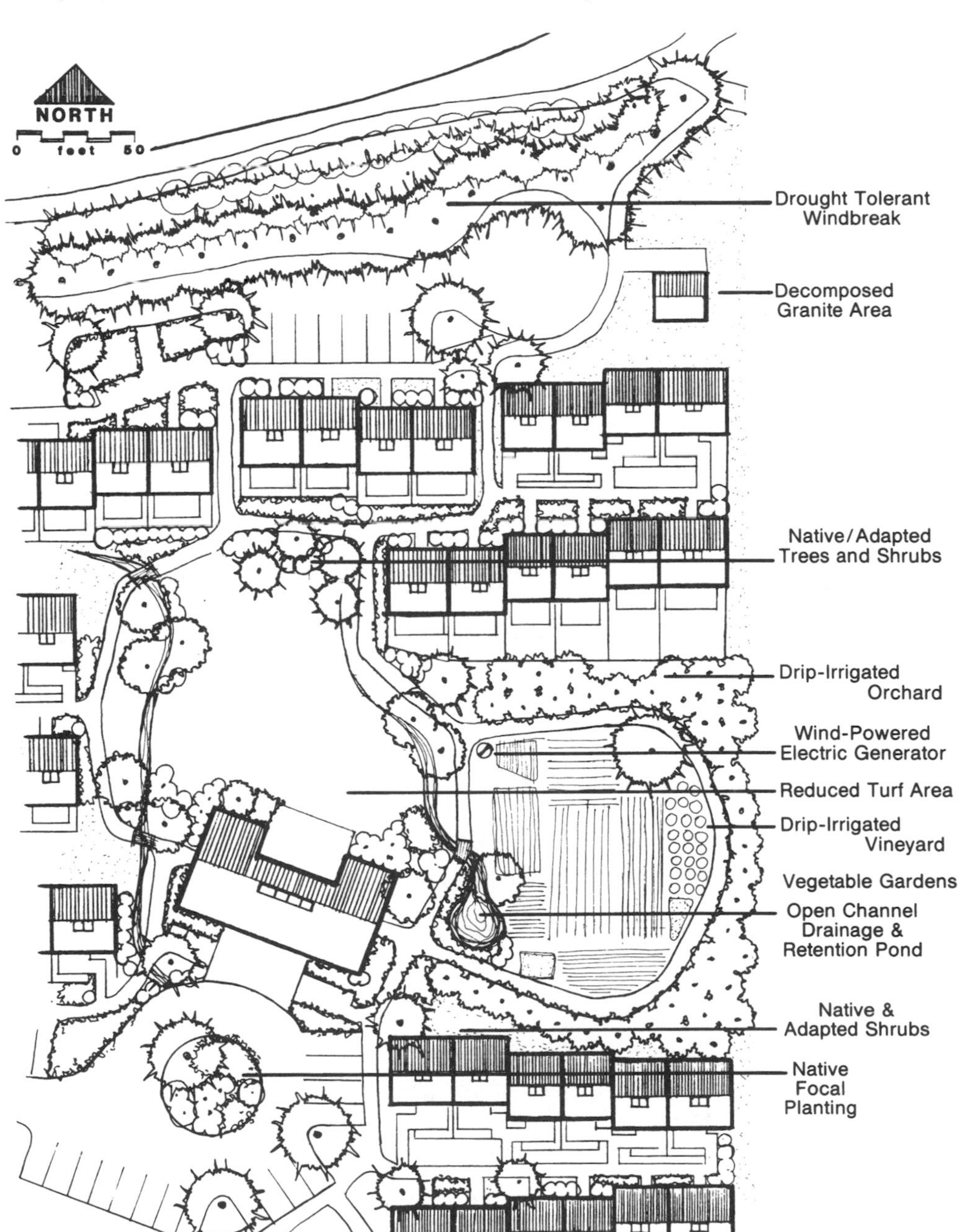

Fig. 8. Landscape modified to conserve water and reestablish indigenous plants. (McPherson 1984. Courtesy Landscape Architecture Foundation.)

- **Sun**—Where sun is abundant, it is imperative to provide shade for human comfort and safety in activity areas (pathways, patios). The most economical and practical way is to use natural vegetation, slope aspects, or introduced shade structures. Additional solar considerations for environmentally responsive site design include orientation of facilities to capitalize on daylighting and photovoltaic opportunities.

- **Rainfall**—Even in tropical rain forests where water is seemingly abundant, clean potable water is often in short supply. Many settings must import water, which substantially increases energy use and operating costs, and makes conservation of water important. Rainfall should be captured for a variety of uses (drinking, bathing) and this water reused for secondary purposes (flushing toilets, washing clothes).

- **Topography**—In many areas, flatland is at a premium and should be set aside for agricultural uses. This leaves only slopes upon which to build. Slopes do not have to be an insurmountable site constraint if innovative design solutions and sound construction techniques are applied. Protection of native soil and vegetation are critical concerns in high slope areas. Reducing the size of the footprint of development, eliminating the use of automobiles and their parking requirements, elevating walkways, and using point footings for structures are appropriate design solutions.

- **Vegetation**—Exotic plant materials, while possibly interesting and beautiful, are not amenable to maintaining healthy native ecosystems. Sensitive native plant species need to be identified and protected. Existing vegetation should be maintained to encourage biodiversity and to protect the nutrients held in the biomass of native vegetation.

- **Geology and soil**—Designing with geologic features such as rock outcrops can enhance the sense of place. Integrating rocks into the design of a deck or boardwalk brings people in direct contact with the character and uniqueness of a place.

- **Aquatic ecosystems**—Development near aquatic areas must be based on an understanding of sensitive resources and processes. In most cases, development should be set back from the aquatic zone and protective measures taken to address indirect environmental impacts, such as streamside vegetated buffers.

- **Wildlife**—Sensitive habitat areas should always be avoided. Encouraging wildlife to remain close to human activity centers enhances the visitor experience. This can be achieved by maintaining as much original habitat as possible.

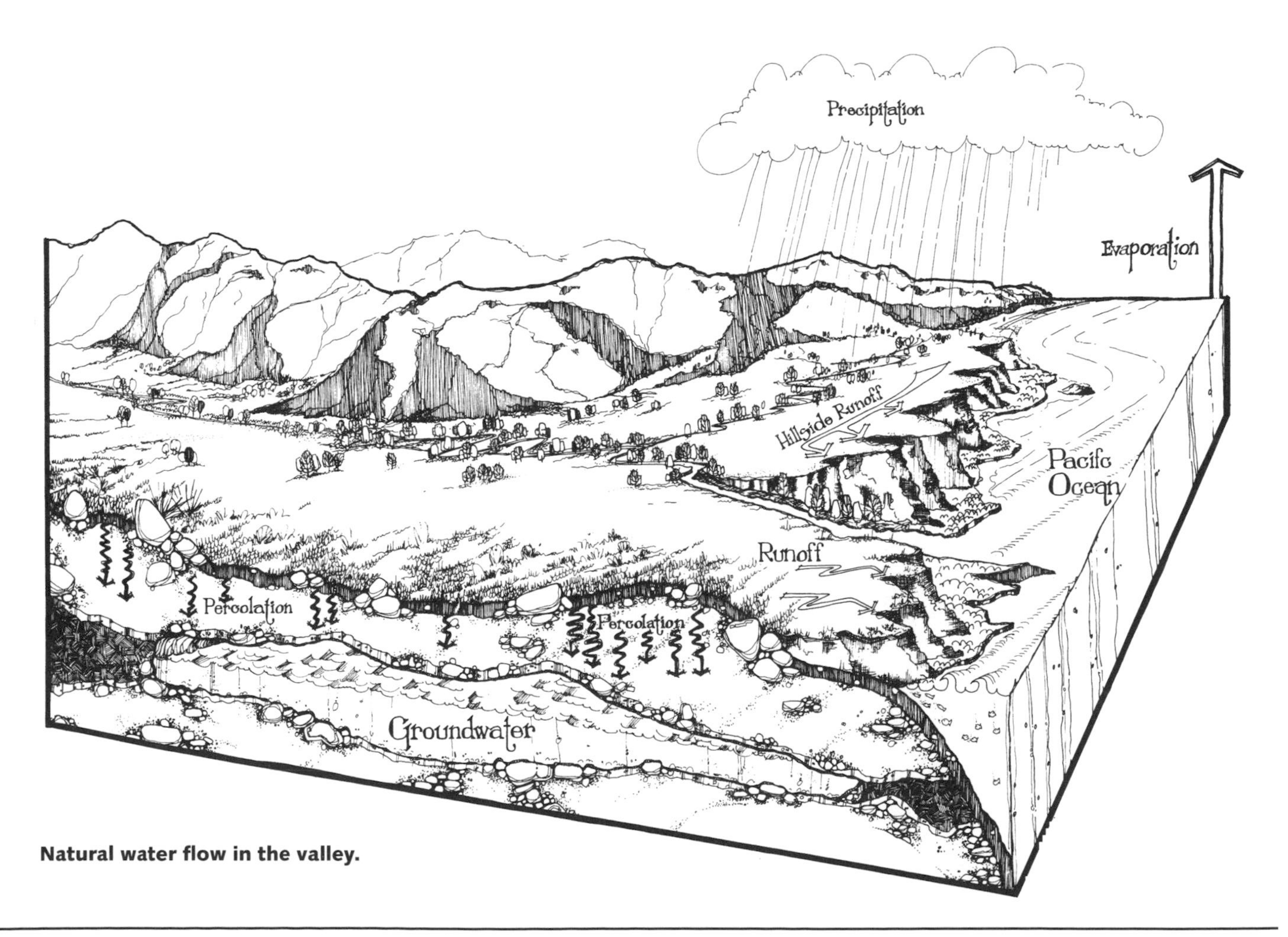

Natural water flow in the valley.

Fig. 9a. Aliso Creek Development Case Study. Studio 606, Department of Architecture, California State Polytechnic University, Pomona. (Lyle 1985.)

Native landscape preservation and restoration

Preservation of the natural landscape is of great importance during construction. It is much less expensive to retain it, rather than to remove and more ecologically sound than subsequent restoration. Preservation entails carefully defining the construction zone. Construction traffic can have the effect of compressing soils and making those areas incapable of absorbing water and holding oxygen necessary to support plant life in future. Avoid the specification to "clear and grub" any unnecessary soil areas because it encourages volunteer exotic growth in scarred areas.

Restoration of native planting patterns should be used when site disturbances are unavoidable. All native plants disturbed by the construction should be saved, healing them first in a temporary nursery. The site should be replanted with native materials in a mix consistent with that found in a natural ecosystem. In some instances, native materials should be used compositionally to achieve drama and visual interest for human benefit.

3 WATER SUPPLY

Water is a nourisher of plant and animal life, a bearer of food, a prime element of industrial processes, and a medium for transportation. It is an essential element of recreation, aesthetic and spiritual life. To ensure global, regional and local water resources can meet the demands of the future, all infrastructure, urban development and buildings require design for water conservation, collection, storage, treatment, and reuse.

The freshwater reservoir on which we are most dependent is the resultant runoff of a water cycle driven by the sun. Evaporation lifts purified water from the oceans and land, which then falls again as rain and snow. A reserve of freshwater is held in underground aquifers, but can be energy intensive to extract from and slow to replenish (Fig. 9). Certain land areas receive immense amounts of freshwater through precipitation, while others receive scant amounts. In some environments this imbalance is exhibited within a few hundred miles.

Water conserving and sustainable approaches to landscaping includes specification of indigenous planning, essentially accommodated to subsist on the water naturally available in the region, and, where needed, subsurface drip irrigation, ideally supported by on-site cisterns that use and stores local rainfall for even use throughout many seasons.

- **Regulations.** The World Health Organization and individual countries and associations all have health regulations governing drinking water. Parameters the maximum allowable contaminant

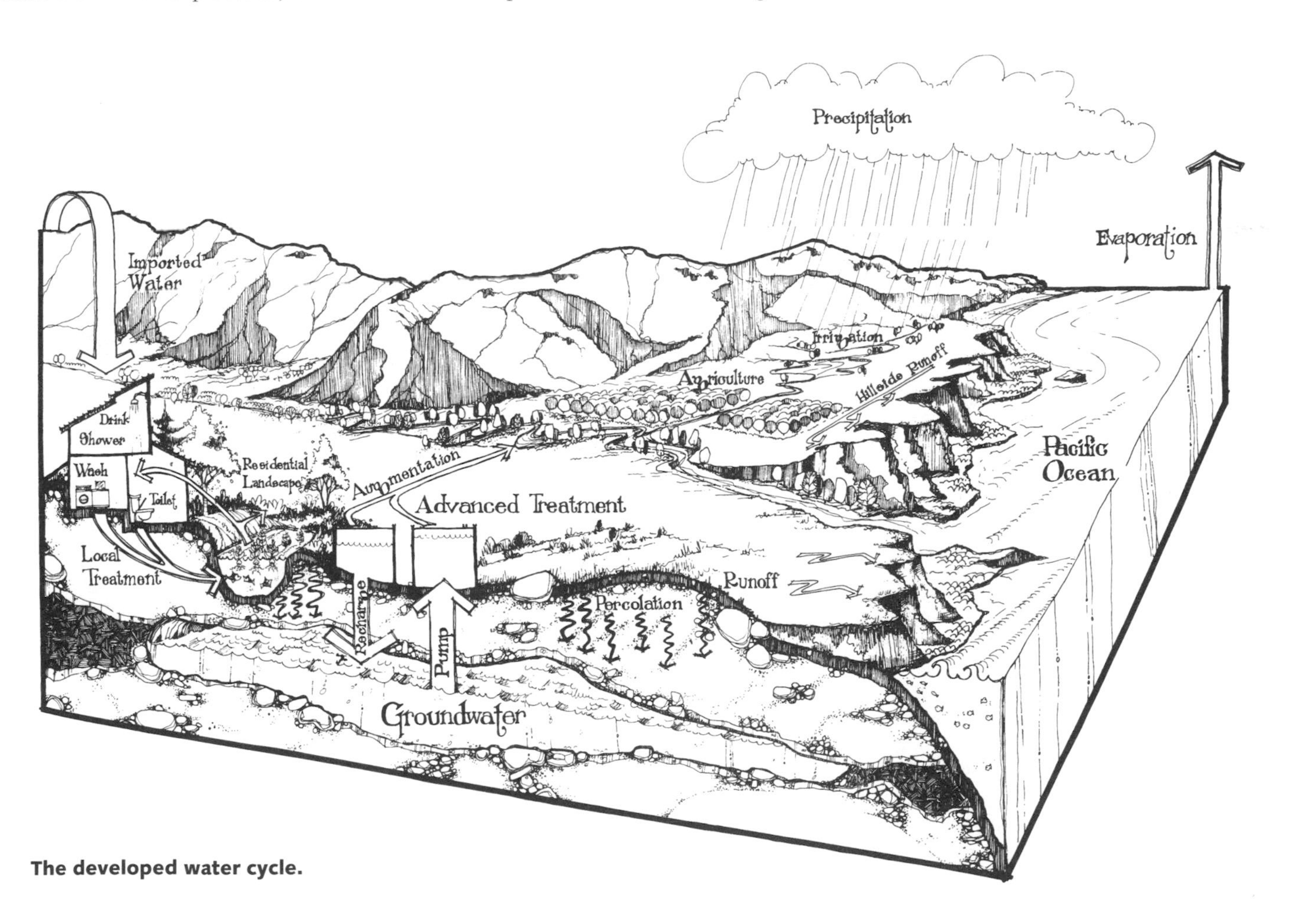

The developed water cycle.

Fig. 9b. Aliso Creek Development Case Study. Studio 606, Department of Architecture, California State Polytechnic University, Pomona. (Lyle 1985.)

levels are not identical. A resource-related development should determine applicable regulations prior to initiating a water supply program.

Water sources

- **Groundwater (wells and springs).** An uncontaminated groundwater source or spring usually requires the least input (energy, chemical, financial) to provide safe water for drinking, bathing, and cooking. Extreme efforts should be made to protect existing and potential groundwater sources from contamination.

  Use of groundwater is probably the least energy-intensive because renewable energy sources (wind, photovoltaic) can be used to pump the water to a hillside storage reservoir for distribution by gravity.

- **Surface water (fresh).** Fresh surface water can be used when groundwater is not available. Some locations have an abundance of fresh surface water such as streams, rivers, and lakes.

- **Lack of groundwater or surface water.** In those cases where there is a lack of water, rain catchment becomes an option as a standalone supply of water or a supplement to a limited ground or surface supply. Rainfall catchment from the roofs of structures is a recognized option for water supply, provided the necessary treatment processes are used prior to distribution. Care should be used in selecting a roofing material (*e.g.*, hard and smooth) that does not collect dirt. Metal roofs may release heavy metals into drinking water if the rainwater is acidic. Rainwater collected from ground surfaces can be used for secondary uses such as toilet flushing and irrigation of food crops, or groundwater recharge.

- **Extraction of freshwater from seawater, brackish water, or water vapor in the air.** Some areas have no readily available supply of freshwater and must rely on converting salt water to freshwater. Reverse osmosis, electrodialysis, distillation, and vapor compression are processes used. All are complex, extremely energy consumptive, costly, and difficult to operate and maintain, and present significant disposal problems caused by the brine concentrate.

Water treatment

The type(s) of treatment required will depend on the source of water and the quality of source water.

- **Groundwater.** Treatment of groundwater is accomplished by simple disinfection using sodium hypochlorite (laundry bleach). The sodium hypochlorite can be proportioned into the water being delivered to the storage tank using a water-powered or photovoltaic metering pump.

  An emerging water disinfection technology involves the use of liquid chlorine dioxide (Aqua Chlor). This technology provides excellent bacterial qualities while minimizing the formation of environmentally harmful disinfection by-products.

- **Surface water with low turbidity.** Before disinfection, surface water requires filtration. For resource-related developments, the recommended filtration processes would be slow sand filtration or cartridge filtration. Only the water used for drinking, washing, and cooking would need to be completely treated. Dual distribution systems are required—one for drinking water and one for lesser quality uses such as toilet flushing or garden and crop watering.

  The slow sand filter is an old technology that has recently been used in contemporary applications. An evenly graduated natural sand, approximately 3 ft. deep (.9 m) is placed in a constructed basin. The supply water is introduced into the top layer of sand and travels downward through the sand filter to perforated collection pipes on the bottom of the filter. Impurities in the water are removed in the top layer of the filter and accumulated for periodic removal by scraping. The removed impurities and top 1/2-in. (1.25 cm) of sand can be dried and used as a soil conditioner. No chemical additions or additional power are required. Operations and maintenance requirements are low. However, a certain land area is required for the filter basin. Disinfection with bleach is the final step.

  Cartridge filters using microporous filter elements (ceramic, paper, or fiber) with small pore sizes are suitable for low turbidity surface water. (Use a graduated series of cartridge sizes to prevent rapid clogging of filter.) Again, a dual distribution system is recommended to lessen the volume of high-quality water needed. Head loss through a cartridge filter is higher than through a slow sand filter, so a booster pump may be required to maintain adequate pressure in the water system. The paper and fiber filters are consumptive as they must be disposed of when full of sediment (disposal frequency depends on turbidity in supply water). The ceramic cartridge filter can be cleaned mechanically (scraped) and reused. Sediment cleaned from the ceramic cartridge can be dried and used as a soil amendment. Operations and maintenance is minimal. Disinfection with bleach is the final step.

- **Surface water with high turbidity.** If the source water has turbidity above 15–20 NTUs (nethelometric turbidity units), complete conventional treatment is required. This involves the addition of synthetic chemicals such as alum and polymer in a coagulation stage, followed by a flocculation stage before filtering in a rapid sand filter. The filter is hydraulically backwashed (usually once per day) to remove accumulated sediment from the filter. This backwash waste (containing the added chemicals) must be dried and disposed of in an approved manner. The complexity and cost of operation is high, maintenance costs are high, and chemical and power inputs are required. Dried waste sediments cannot be used as a soil amendment without further processing. The final step is disinfection with bleach.

Rainwater harvesting

Rainwater harvesting, collecting rainwater from building roofs, provides a means to collect water for seasonal and annual landscape watering and related water needs, including potable water, if treated. (See Box D on following pages.)

Gravity storage of any water product (raw, finished, reclaimed) should be used wherever possible. Gravity storage enables wind and photovoltaic pumping systems to be effective. Because these pumping systems work at relatively low pumping rates, the gravity storage tank acts as an accumulator to store water for heavy demand periods or for days when the wind does not blow or the sun does not shine.

The hydraulic ram provides another means of transferring raw water from a source to a storage tank at a higher elevation without electrical or hydrocarbon input. The hydraulic ram is a self-acting impulse pump that uses the momentum of a slight fall of water to force a part of the water to a higher elevation. A hydraulic ram is noisy, but the noise can be successfully mitigated with the use of sound-attenuating materials in an enclosure. It is practicable to operate a ram with a fall of only 18 in. (.46 m), but as the fall increases, the ram forces water to proportionately greater heights. The hydraulic ram is well suited for areas where electrical power is not available and where an excess supply of water is available.

As a gravity storage tank will be located in an elevated location, visual quality will be important. Multiple smaller tanks may be easier to screen than one tank. Multiple tanks also provide greater flexibility in operation. Tank materials should be noncorrosive and sectionalized for minimal transportation requirements to the tank site.

Water distribution

Most distribution systems are either buried or placed at grade. At-grade distribution systems have minimal effect on the site and vegetation during construction, but are subject to problems with accidental breakage, frost exposure, vandalism, and visual quality. Burying has the advantage of protecting against accidental breakage, but leaks are more difficult to locate on a buried distribution system. Leak detection and repair is imperative when dealing with such a precious resource as water.

Dual distribution systems are very effective in that different qualities of water can be delivered to different use points. Pipe contents should be color-coded so that cross-connection problems can be prevented.

Especially in environmental education facilities, water-related features provide a basis for distinctive design and may include indigenous landscaping (Fig. 10), water harvesting (Fig. 11), and gray water systems (Fig. 12).

4 WASTE PREVENTION

There is no completely safe method of waste disposal. All forms of disposal have negative impacts on the environment, public health, and local economies. Landfills have contaminated drinking water. Garbage burned in incinerators has poisoned air, soil, and water. Many water and wastewater treatment systems change the local ecology. Attempts to control or manage wastes after they are produced fail to eliminate environmental impacts.

The toxic components of household products pose serious health risks and aggravate the trash problem. In the U.S., about 8% in every ton of household garbage contains toxic materials, such as lead, cadmium, and mercury from batteries, insect sprays, nail polish, cleaners, and other products. When burned or buried, toxic materials also pose a serious threat to public health and the environment.

The only way to avoid environmental harm from waste is to prevent its generation. Pollution prevention means changing the way activities are conducted and eliminating the source of the problem.

Preventing pollution in a sensitive resource-related setting requires thinking through all of the activities and services associated with the facility and planning them in a way that generates less waste. Waste prevention leads to thinking about materials in terms of the three "R's," to *reduce, reuse, recycle*. The best way to prevent pollution is to avoid using materials that become waste problems. When such

Fig. 10. *Casa del Agua*, Tucson, AZ. Demonstration house that meets its water needs through rainfall harvesting. Indigenous landscaping is watered by the house gray water system.

PHOTO: Paul Bardagjy

Fig. 11. Rainwater collection and storage. Lady Bird Johnson National Wildflower Center, Austin, TX. Architects and Landscape Architects: Overland Partners.

Fig. 12. Gray water for plant watering utilizing wastewater from public toilets. National Audubon Sanctuary. Cork Screw Swamp, Florida. Designer: John Todd, Living Systems Technology, Inc.

Box D. Rainwater Harvesting Systems

Types

1. Passive: involves only two basic components of rainwater harvesting, catchment and conveyance.
2. Sophisticated: involves all six basic components of rainwater harvesting.

Components of Rainwater Harvesting

1. Catchment area: the surface upon which the rain falls. It may be a roof or another impervious pavement and may include landscape areas.
2. Conveyance system: transport channels from catchment area to storage.
3. Roof washing system: the systems that remove contaminants and debris.
4. Storage system: areas where collected rainwater is stored.
5. Delivery system: the system that delivers the treated rainwater, either by gravity or pump.
6. Water treatment system: includes filtering equipment and additives to settle, help filter, and disinfect the collected rainwater.

Levels of Commitment or Water Security for a Rainwater Harvesting System

1. Occasional: small storage capacity, one to two day supply of water after a rain, alternate water source required for dry weather.
2. Intermittent: small to medium storage capacity, partial year supply, alternate water source may be required.
3. Partial: medium to large storage capacity, majority of yearly supply, alternate water source may be required if rainwater is used for both landscape and potable water needs.
4. Full: large storage capacity, provides all of the water needed for all uses for whole year.

Water Balance Analysis

A water balance analysis, typically referred to as a water budget will initially allow a designer to determine how much rainwater can be collected by the project catchment area. A water budget will also provide a supply and demand analysis on a monthly basis and will help determine whether a cistern or storage area is needed and how much landscape water demand can be provided by the rainwater system. In addition, a water budget will determine how much, if any, supplemental water is needed to augment the landscape irrigation. If the landscape budget water requirements are larger than the rainwater system is able to provide, and supplemental water is not wanted or cannot be used, the water budget will help determine how much the landscape must be reduced to match the rainwater supplied. For landscape irrigation supply, rainwater plus a supplemental water source may be needed for a few years, until plants are established, because plants typically require more water to develop their root systems early in life or to establish a new root system when transplanted. Eventually, rainwater may be all that is needed for an established landscape. When the supplemental water is no longer required, it will serve as a security system for years with low rainfall or if a system needs servicing and the tank/cistern requires draining. Because a water budget is based on average rainfall and theoretical irrigation values, it will not be exact, but it should be used as a planning tool to help refine project goals. Four items are required to prepare a water budget: average rainfall data, a site plan, a landscape plan , and an irrigation plan.

Formulas for Calculating Rainwater Runoff and Landscape Irrigation Requirements

Formula 1 (Used for defined catchment areas)
(CA)*(R) *(E)*(7.48) = Catchment area runoff in gallons
Where:
CA = Catchment area (square feet)
R = Rainfall (percent of a foot)
E = Efficiency of catchment surface for runoff potential
7.48 = Number of gallons in a cubic foot

Formula 2 (Used for defining a catchment area size)
(TWR)*(365) = Total catchment area required in square feet
(AR)*(0.623)
Where:
TWR = Total water required/allowed per day for landscape irrigation
365 = Number of days in a year
AR = Annual rainfall in inches
0.623 = Converts inches of rain into gallons per square foot of area

Formula 3 (Landscape irrigation requirement in gallons)
(No.P)*(DU)*(DM) = Total monthly water requirement per plant type.
This needs to be calculated for each month for each plant type.
Where:
No.P = Number of plants per plant type
DU = Daily water usage of a single plant (changes per plant type)
DM = Days in the month being calculated

Box D. Rainwater Harvesting System for Landscape Irrigation (continued)

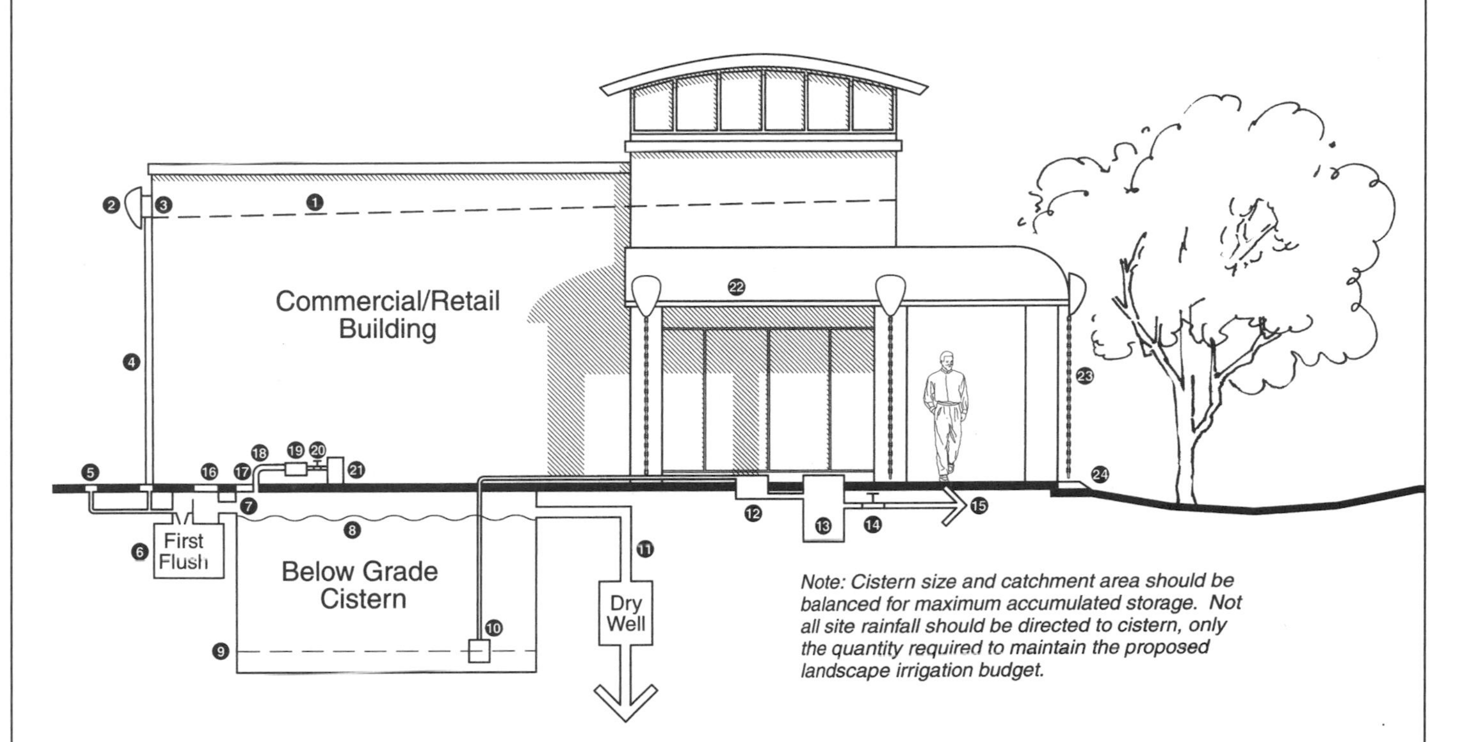
Commercial/Retail
Building
First
Flush
Below Grade
Cistern
Dry
Well
Note: Cistern size and catchment area should be balanced for maximum accumulated storage. Not all site rainfall should be directed to cistern, only the quantity required to maintain the proposed landscape irrigation budget.

1. Single plane or double plane "butterfly" rooftop collection.
2. Optional decorative scupper cover.
3. Scupper to downspout.
4. Downspout sized per local plumbing code, sediment trap at ground level.
5. Catch basin for paved/hard surface ground level runoff collection, with sediment trap.
6. Debris, sediment, and oil interceptor.
7. Rainwater inlet, inlet to cistern must be a minimum of ten inches below top of cistern. An inflow smoothing filter maybe appropriate at this location depending on proximity of rainwater inlet to irrigation supply filter. The smoothing filter will slow rainwater inlet turbulents that may disturb the fine sediment settled on the bottom of the cistern.
8. Maximum water level to be twelve inches below top of cistern.
9. Minimal level of water to maintain priming in landscape irrigation pump (approximately twelve inches), level of water to be determined by irrigation specialist.
10. Landscape irrigation supply filter with automatic shutoff to stop pump if water is below minimum level in cistern to maintain priming in pump, maintain minimum of six inches from cistern bottom to avoid settled fine sediment.
11. Cistern overflow (same size as inlet) to a dry well or gravity outlet to landscape basin if site conditions allow. An additional option would be to outlet to an adjacent flood retention underground storage pipe that is tied to a dry well. Cistern overflow must be a minimum of twelve inches below top of cistern to avoid contamination of alternate water supply.
12. Optional sand filter and pressure tank.
13. Landscape irrigation pump and pressure tank.
14. Typical valve.
15. Water supply line for irrigation system.
16. Thirty-two inch access for cleaning.
17. Twenty-four inch access for cleaning.
18. Alternate water supply, must not obstruct the twenty-four inch access. Alternate water supply may be proposed for a cistern manual fill option for droughts and plant establishment periods when additional/supplemental water is required.
19. Atmospheric vacuum breaker.
20. Typical valve.
21. Alternate water source, possibly domestic or municipal supply.
22. Gutter with leaf screen if building is adjacent to trees.
23. Rainchains or downspouts to splash pads and depressed landscape areas.
24. Splash pad.

Box D. Author: Heather Kinkade-Levario, ASLA.

materials must be used, they should be reused onsite. Materials that cannot be directly reused should be recycled.

Garbage/solid waste prevention

For projects of any scale, a comprehensive design strategy is needed for preventing generation of solid waste. An effective garbage prevention strategy would require that everything brought into a facility be recycled for reuse or recycled back into the environment through biodegradation.

Any resource-related development will have two basic sources of solid waste—materials purchased and used by the facility and those brought into the facility by visitors. The following waste prevention strategies apply to both, although different approaches will be needed for implementation:

- **Use products that minimize waste and that are nontoxic.**
- **Recovery nutrients from waste streams by composting, anaerobically digest biodegradable wastes, or constructed wastelands and solar-assisted aquaculture.**
- **Reuse materials onsite or collect suitable materials for offsite recycling.**
- **Use of products that minimize waste.**
- *Much of the growing volume of garbage is from the use of disposable consumer products and excess packaging.* Consideration must be given to materials or products that minimize waste disposal needs—purchase items with minimal packaging, buy in bulk, and replace disposable products with durable, reusable items. Use of plastics for packaging is increasing, thereby replacing recyclable products and materials. Plastics, which account for about 20% of solid waste by volume, are not biodegradable, difficult or impossible to recycle, have a high volume-to-weight ratio, and are toxic when burned.

  • *When selecting materials and goods, nothing should be purchased that will ultimately become toxic.* Nontoxic materials can often be substituted for products that cause contamination problems during disposal.

  • *Materials should be purchased locally whenever possible.* Locally produced goods needing less transport and less storage should have less packaging waste.

- **Nutrient recovery and biodegradation.** In the process of biodegradation, microorganisms break down the products of other living things and incorporate them back into the ecosystem. Biodegradable or bioconvertible material includes anything that is organic. Most of the organic components of garbage, such as paper and food wastes, can be eliminated through composting. Between 60 and 75% of the solid waste is bioconvertible.

  The biodegradable or bioconvertible percentage of the waste stream is large enough to consider at least two options for the conversion process. Two obvious options for conversion are composting and anaerobic digestion.

  • *Composting.* Composting is a familiar concept, and is used for handling yard waste and even sewage sludge. Both of these organic wastes require mixing of other materials to achieve a nutrient balance. Large chunks of relatively inert material (most commonly wood chips) add bulk and aeration to make the process work. This is typically done in open windrows or piles, with mixing done daily to provide aeration and homogeneity. This takes land space on a drainable surface, and a collection of any runoff for dispersal of the liquid to the process. It produces a quality soil amendment and reduces the bulk of the original material by approximately 40–50%. Composting does off-gas ammonia and carbon dioxide and produces offensive odors. It typically takes 50–60 days to process. Screening of the final product is necessary to remove the bulking material and provide granulation before use in the soil bed.

  The use of this product as a soil amendment is valuable, particularly in a tropical environment, because the soil is essentially sterile, with only about 2% organic content. Affected by humidity, rainfall, temperature, and normal soil activity, the organic material placed on the soil will typically last only 30–40 days in the tropics. In a temperate climate, that same material may last as long as six months.

  • *Anaerobic digestion.* Anaerobic digestion is used extensively worldwide for processing food waste, animal waste, the solids of human waste systems, and for the total array of solid waste such as waste paper, green waste, and landscape waste. This wet fermentation process converts the waste stream into three usable by-products: (1) biogas—an energy-rich gas stream, comparable to natural gas that can be used to offset the cost of energy utilities of the development, (2) a high-quality organic fertilizer solid that may be useful in landscaping efforts or crop production; and (3) a diluted liquid organic fertilizer that may be used in drip irrigation as an additive to any planting program, for feeding ornamentals, or in landscape plots for replenishing native or endangered species of plants.

- *Recycling.* A material doesn't become waste until it is thrown in the garbage can. If a material can be reused it is a resource, not waste. Reuse is the best form of recycling. Recycling can be maximized through the purchase of products for which there is a ready market as recycled materials.

  In circumstance where there is no available market for a given material, often a beneficial end use can be developed locally. Every effort should be made to work with the local community to determine if any of the materials generated by the facility can be used—*e.g.*, glass beverage containers can be ground up and incorporated into materials for construction and road building.

  Efficient recycling requires sorting of materials; convenient bins should be provided at the facility for the materials being recycled.

  • *Offsite disposal.* If a garbage prevention strategy has been fully executed, actual remaining waste should be minimal. Remaining residuals mean that the facility is not entirely environmentally sustainable. All residuals must be collected separately and disposed of offsite. In most cases residuals should be returned to their place of origin. Toxic material residuals must be segregated and disposed of separately.

5 THE CHALLENGE OF SUSTAINABLE DESIGN

The impetus for sustainable design came from outside of the design professions—essentially from international development community responding to global issues of population, poverty, threatened and diminishing natural resources and the imbalances of global development, evident in unattended growth of unplanned megacities. Sustainable design is thus central to the agenda of thought and action that responds to the root issues of global development. A number of significant ideas emerge from the concept of sustainability to shape an agenda for sustainable design, representing the contribution that can be made by the architecture, urban design and planning professions.

1 Bioclimatic design

Energy-efficient and environmentally responsible design has continuously evolved in architecture evident in the 1930s interest in solar design by early modernists such as the Kechs, but also Wright, Gropius, Breuer and Le Corbusier; the 1950s development of bioclimatic design by the Olgyays, 1970s research into energy-efficient heating, cooling and daylighting of buildings; and1980–90s concerns for human health, air quality, and environmental impact of buildings on the natural landscape. The microclimate provides opportunities to create comfort conditions in buildings by design strategies that include natural ventilation, daylighting, and passive heating and cooling (Watson and Milne 1997). In these approaches, architecture and environmental systems are conceived as integral to the microclimate of the site, modified by the design of healthy environments inside and outside spaces, which in turn create more favorable microclimates for gardening, wildlife and the restorative role of its impinging natural systems (Fig. 13).

2 Life-cycle and "cradle-to-cradle" materials reclamation

The "life-cycle" or "cradle-to-cradle" concept envisions all materials production as a continuous and sustainable process of use and reuse, essentially the recycleability of all materials design and production. The application of life-cycle thinking and materials reclamation to building suggests emphasis upon longevity, continuous preservation and renewal of building assets, adaptable systems and replaceable subcomponents, demountability, and reclaimed construction products and systems.

3 Sustainable community design

Sustainable community design combines architecture, landscape design and planning, in which towns and communities are conceived of in terms of environmental flows and resources (Van der Ryn and Calthorpe 1986). In addition to holistic environmental design approaches, community involvement is seen as essential, for which the design charrette and community design clinics provide models, as is the emphasis upon economic, social and community empowerment models that have been advanced in community development and advocacy disciplines.

4 Metro-regional planning

As an extension beyond the community scale, sustainability design issues can be best addressed by the inclusion of transportation, land use and metropolitan-scale environmental impacts of air and water, properly conceived as bioregional planning. This view is not beyond the architectural tradition, evidenced by the contributions to transit-

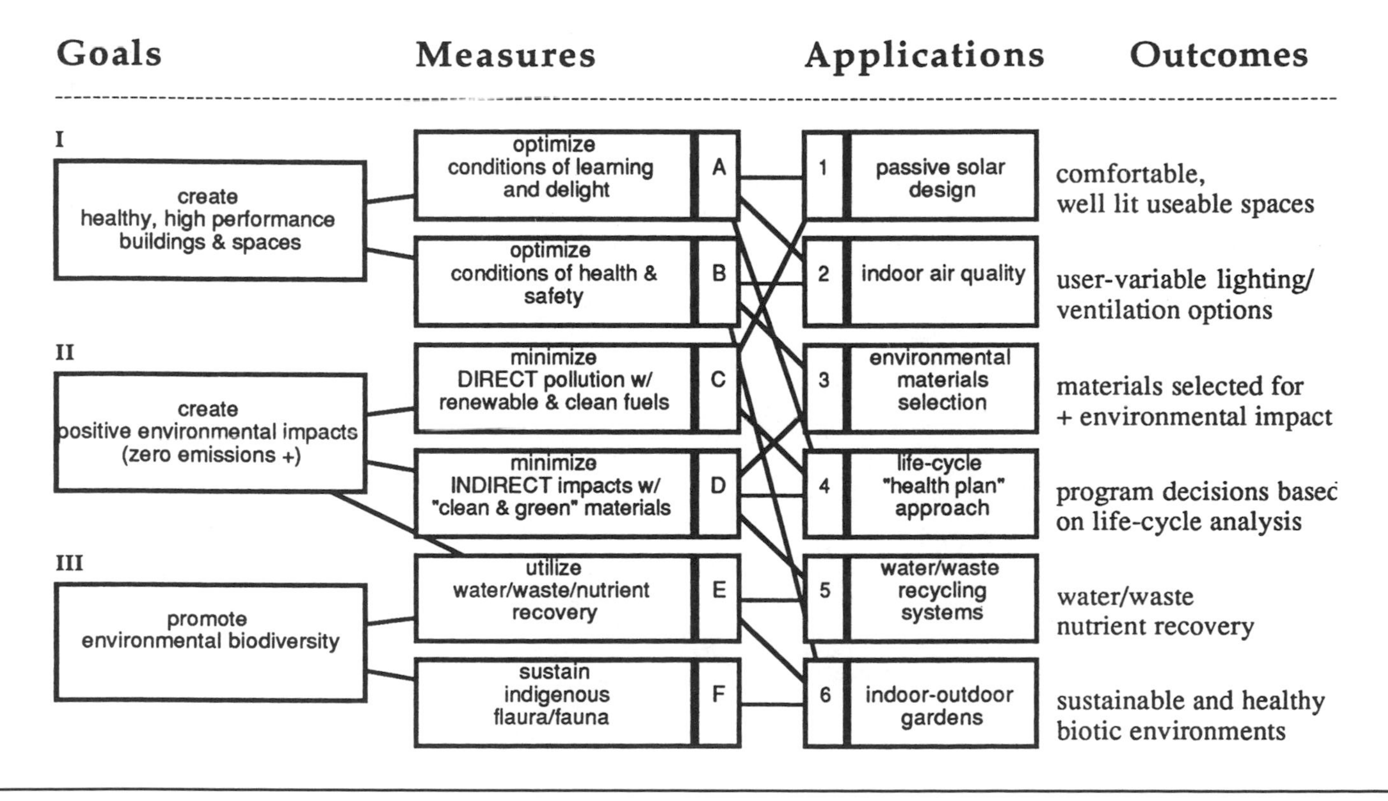

Fig. 13. Principles and practices of sustainable design. Donald Watson, FAIA.

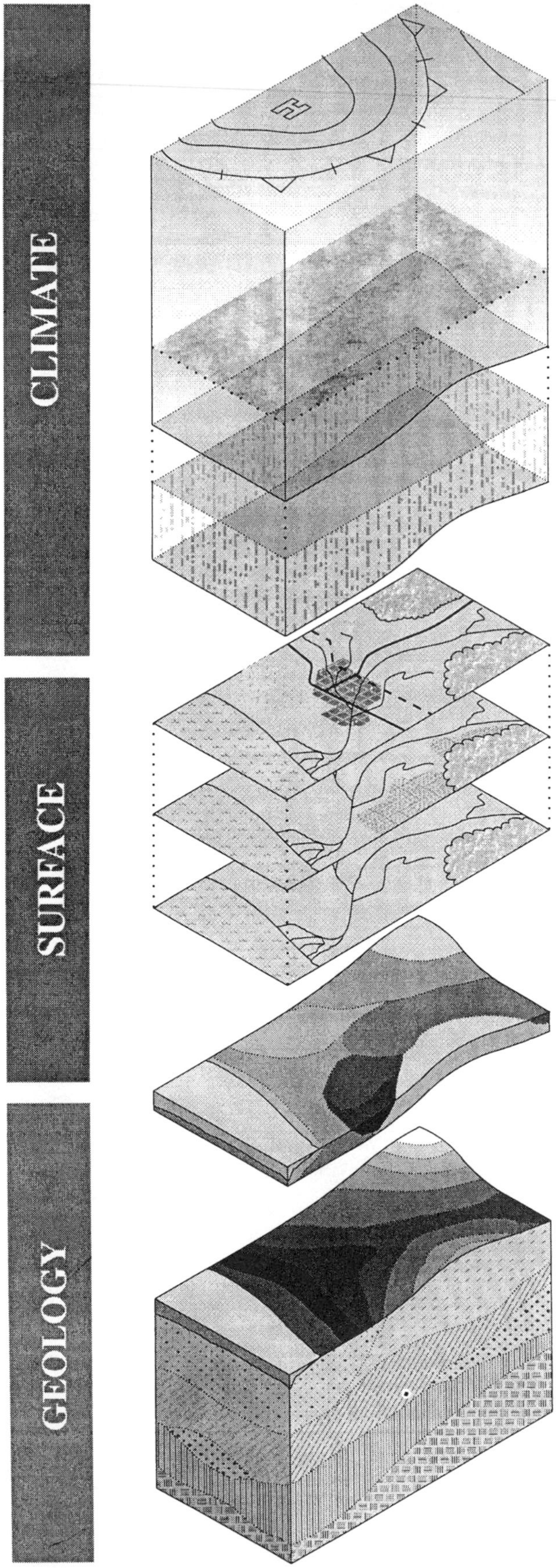

CLIMATE

SURFACE

GEOLOGY

AIRSHED

Quality:
- oxygen from plants
- pollution from manufacturing and industry
- gases from soil

Quantity:
- weather patterns (air currents/ winds)
- location of manufacturing and industry

PRECIPITATION

Quality:
- pollution from manufacturing and industry (acid rain)

Quantity:
- weather patterns (frequency and type)
- geographic location

LAND USE

Urban Systems
Agricultural Systems
Natural Systems

SOILS

- fertility (natural and agricultural plants)
- perviousness (type, depth, and location)
- water storage capability

WATERSHED

Geographic features:
- mountains
- valleys
- lake and sea basins

SUBSURFACE

Underground features:
- tectonics (stability)
  - earthquakes
  - sinkholes
  - caverns
- underground water storage (aquifers)
- underground water transportation

BOX. E. BIOREGIONAL LAYERS

by Dan Williams and Chris Jackson

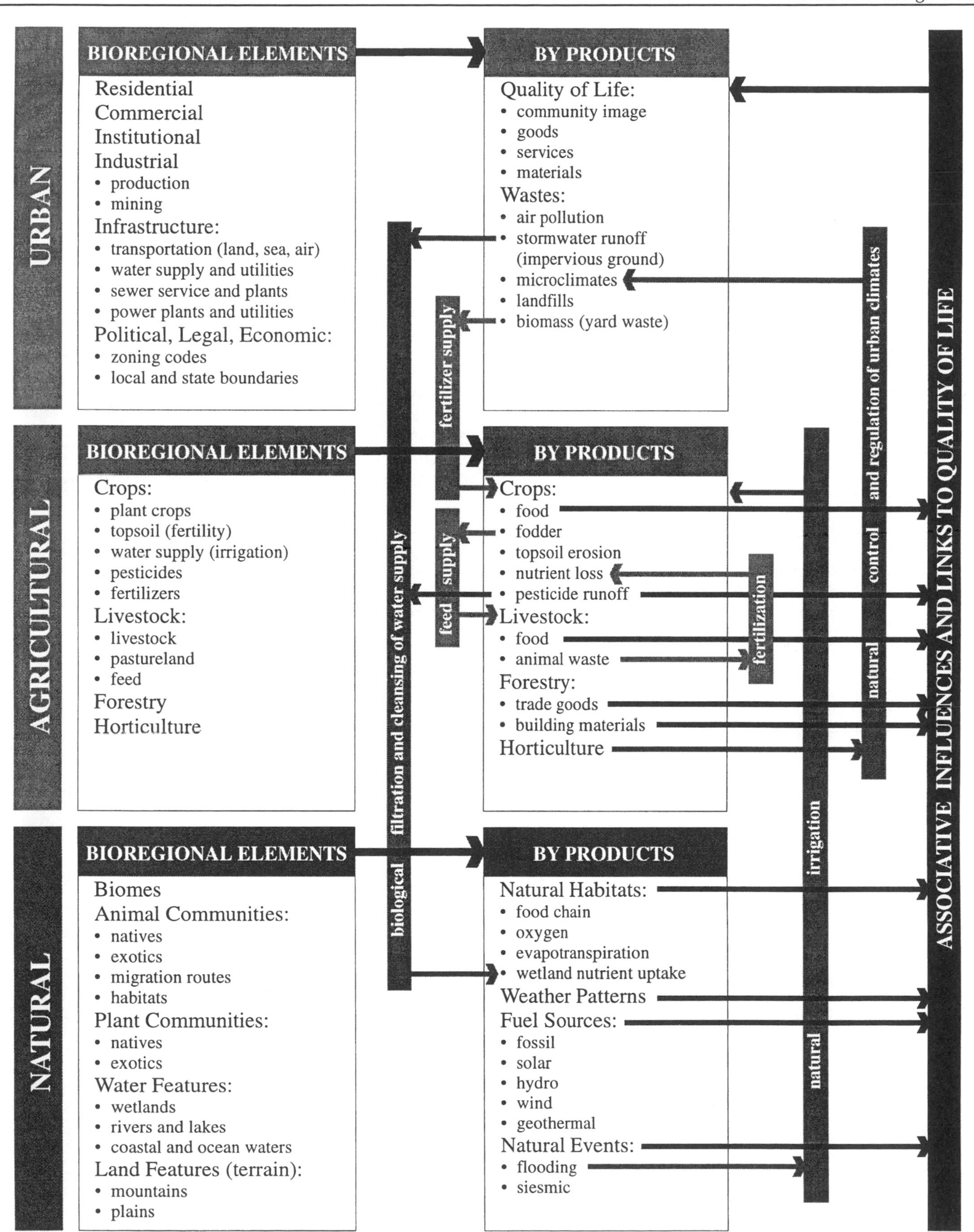

BOX E. cont. BIOREGIONAL LINKAGES by Dan Williams and Chris Jackson

oriented development approaches by Peter Calthorpe and to the town planning by Andreas Duany and Elizabeth Plater-Zyberk.

5 Bioregionalism

Bioregionalism is the approach to design urban places and infrastructure within the environmental context of their regions, defined by contingent landforms, vegetation and watersheds, and codependent living species, climate and resources (Box E). These issues are integrated in design by methods cited in this article pioneered by Ian McHarg (1969) and John Lyle (1994). Water conservation and waste nutrient recovery are also best conceived as regional strategies. Recharging of local aquifer through absorptive landscaping is a traditional but necessary alternative to conventionally engineered storm-water drainage, the neglect for which is now measured by the magnitude of major floods throughout the world. Sewer treatment that restores nutrients to topsoil is an economically viable and far more sustainable alternative to conventional disposal, demonstrated in biologically regenerative waste recovery systems at the municipal scale.

Bioregional design balances human needs with the carrying capacity of the natural and cultural environments. Designing within the carrying capacity of the land minimizes negative environmental impacts, importation of goods and energy, and seeks to restore and increase the carrying capacity by water and nutrient recovery, establishment of water courses and vegetated zones and the biotic systems they support.

6 Restoration of biological diversity and conservation

Beyond the enterprise of designing the built environment for human habitation, sustainability gives voice to the biological role of all living species in the web of life (Wilson 1992). The detrimental impact of building and land-use practices is now directly correlated to critical biological species decline in all areas of the world through habitat reduction, production of toxic chemicals and waste, combustion of fossil fuels, and related agricultural and industrial resource exploitation. Proposals that are part of bioregional planning at a regional and even continental scale call for the recovery of wilderness to preserve the range of endangered species as an international biological conservation strategy.

7 A global perspective

The discussion of sustainability that emerged from the Earth Summit in Rio has reconfigured the international view away from a geopolitical division of "first-, second-, and third-worlds" to "one world," increasingly interdependent in economic and environmental development. This aspiration has often become stalemated in political and ideological debate, surrounding economic issues of international aid and obligations of industrialized nations to support the economic development and conservation practices of developing nations. All the while, rapid industrialization continues apace especially in the developing world, largely uninformed by sustainable design practices. Regardless of these apparent expediencies, the sustainability discussion has given an unprecedented and undeniable perspective of the essential interconnectedness of all economies and environments that must be the framework of future design education and practice.

Implicit in all these discussions is a commitment and concern for the future well beyond our personal roles and realms—what author Robert Gilman has called "future fairness," offered as a succinct two-word definition of the concept of sustainability. Just as human impact has negative impact upon the global environment and thus upon future resources, the obverse can also be true, that human impact can have positive benefit through design intention. The role of stewardship through design conceives of human intelligence and creativity as integral in the evolution of life on earth. The capacity to design is our one best way to prepare for an unpredictable yet more sustainable future. While the interdependencies of global environmental health and biodiversity appear overwhelming, we do not yet know the upper limit of the human capacity for global education, stewardship, and sustainable design. ■

REFERENCES

Lyle, John Tillman. 1985. *Design for Human Ecosystems.* New York: Van Nostrand Reinhold.

McHarg, Ian L. 1969. *Design with Nature.* Garden City, New York: Doubleday/Natural History Press.

McPherson, E. Gregory. 1984. *Energy-Conserving Site Design.* Washington, DC: Landscape Architecture Foundation.

Olgyay, Victor. 1963. *Design with Climate.* Princeton: Princeton University Press.

National Park Service. 1995. *Guiding Principles of Sustainable Design.* Publication NPS D-902A. Denver: United States Department of the Interior. Denver Service Center.

Van der Ryn, Sym and Peter Calthorpe. 1986. *Sustainable Communities: A New Design Synthesis for Cities, Suburbs and Towns.* San Francisco: Sierra Club Books.

Watson, Donald and Murray Milne. 1997 "Bioclimatic Design," *Time-Saver Standards for Architectural Design*, 7th edition. New York: McGraw-Hill.

Wilson, Edward O. 1992. *The Diversity of Life.* Cambridge: Belknap Press/University Press.

Donald Watson with Chad Floyd

Summary
The capacity to help communities visualize preferred futures is a unique contribution that design professionals can make to community planning and decision making. Charrettes provide for an airing of views, possibilities and visions that can frame the terms to catalyze community-wide commitment to an urban design project. Charrettes offer a process for communities to envision and act upon what their neighborhoods might be and are thus vital to the process of "making democracy work." This article reviews elements of organizing and facilitating a design charrette, along with a description of utilizing television for community-wide involvement in urban design and planning.

Keywords
charrette, decision-making, facilitation, media, participatory design, planning, R/UDAT, television, zoning

Citizen design process, Dayton, Ohio

Community design charrettes

1 COMMUNITY INVOLVEMENT IN PLANNING

The term "community planning" refers to physical planning at the community scale, e.g., neighborhoods, urban scale community developments, and/or suburban communities, through a process involving community citizens and representatives.

Methods for involving communities in the physical planning process have been developed at least since the 1960s, sometimes referred to as "participatory design," and characterized by different degrees of involvement, or the "ladder of participation." The ladder of participation, a term that helps to define differences in how and when community stakeholders are invited to take part in the planning process and the degree to which they are part of the decision-making framework:

- ***low to modest involvement:***
  Participation in information and needs assessment.

Community members and representatives are interviewed as part of "needs assessment" or a "community area profile." Visioning workshops are an example, in which community members participate in describing needs and possible idealized outcomes.

- ***modest involvement:***
  Participation in advisory decision-making.

Community members and representatives are involved in an advisory role, providing input at several points in information gathering and assessment, including recommended courses of actions and/or feedback advice on planning and design proposals made by professionals. The RUDAT (Regional and Urban Design Assistance Team) process developed by the American Institute of Architects is a representative example, involving communities in a short-term (usually one week) intensive study of an urban area, with the results open to comment by community representatives.

- ***high involvement:***
  Participation in planning and design.

Community members and representatives are involved in the development of planning and design proposals, most often by participation in community design workshops or "charrettes." In a design workshop, community members provide the key information to guide professional designers, who in turn are asked to help the community group visualize options for future development. As community involvement more closely approaches "high involvement," community members and representatives are active not only in information and advisory guidance,

Credits: Sections 2-4 of this article are based on chapters in Donald Watson, 1996. *Environmental Design Charrette Workbook,* Washington, DC: AIA Publications. Unless otherwise noted, photos are courtesy of Centerbrook Architects and Planners.

Fig. 1. Community Design Charrette: separate breakout group design tables with work posted on walls for public walk-in review.

Fig. 2. Community Design Charrette: Design presentation and citizen forum.

but also by deciding amongst alternatives, thus providing the key value judgments and design decisions through either a consensus-based or a majority-voting process (Figs. 1, 2).

Differences between the charrette process and planning & zoning process

The community design processes, such as charrettes and vision workshops, are most often a preliminary and advisory phase utilized to explore options and to gain community input prior to more formal planning proposals. In a consensus-based decision process, the outcomes are developed by inclusive discussion, debate and agreement reached without any official or formal vote or an adversarial process. Expert option, such as environmental and other technical advice, is introduced by many experts, usually invited to work alongside of and as part of the planning and design process.

Most planning and zoning hearings, by definition, are quasi-legal proceedings, which adopt some form of receiving public comment, such as "pro and con" comments about a specific proposal. A public hearing is normally the only way that interested community members and citizens are able to comment in response to proposals already well developed. The process is thus often "reactive" to proposals already well formed and can easily lead to adversarial confrontation over debatable issues and design proposals. Opinions and judgments are expressed in order to convince a Planning and Zoning (P&Z) Board to make a regulatory and legally binding decision about a proposed plan. Technical input is presented as professional expert advice and has to be careful documented to have legal standing as evidence similar to submission and testimony at a legal hearing.

Strengths and weaknesses of community design process

Advantages:

- The process is "proactive." Enabling citizens to actively participate in planning.
- The process is open and informal, allowing a range of opinions to be heard and included.
- The process is undertaken early enough so that there is "low risk" and "low cost" and/or few barriers to public participation.
- The process involves citizens both as information courses and as evaluators, so that local community values are represented in decisions.
- The process allows for a diversity of opinion including extreme positions and, given the opportunity, these are moderated by the community itself.
- The process allows highly charged and divisive issues to be heard within a process of openness and fairness, thus enabling a process of conciliation.
- Expert opinion is introduced into the discussion in informal meetings where professionals are working alongside citizens. This helps to demystify professional expertise and to help educate the public regarding complex technical issues.

- Decision-making is relatively low-cost, often engaging local professionals on a volunteer, pro bono or reduced time and fee basis.

Disadvantages:

- The process is only loosely defined, and as such, can be manipulated and/or subject to criticism by participants and non-participants alike.
- The process requires early decision-making on key points, such as site conditions, property ownership, resources available, often before such data or decision commitments are available.
- The process is generally "advisory" and is sometimes overruled by authorities that do not agree with its recommendations.
- The process requires sensitive facilitation and broad community representation to avoid early frustration and resulting community resistance and/or apathy.
- The process takes time.
- The process is not widely known, requiring a "learning period" on the part of community stakeholders and the creation of trust in the process and its facilitation.

Strengths and weaknesses of planning and zoning process:

Advantages:

- The process is long-standing and defined by legal process, including precedents to establish a body of law related to planning decisions.
- Representatives on Planning & Zoning Boards and on Boards of Appeals are either elected or appointed, and thus representative and ultimately accountable to public interest.
- Public interest, as well as opposing private interests, is given legal standing by a defined process of public hearing, which vary according to each locality's Planning Board.

Disadvantages:

- The public interest is often represented only by opposition that is rallied in response to development proposals brought forward by private interests. Public often perceive such proposals as well financed and representing only private or commercial interests, against which public opinion is given unequal status and often without professional advice.
- Developers who make Planning & Zoning proposals have to invest a great deal in engineering studies prior to receiving permitting approval.
- Because local officials often make Planning & Zoning appointments, P&Z decisions can be seen as representative of the prevailing "political" interests.
- Public input is most often "reactive" that is, in response to proposals made by others, most often without any public comment period prior to the Planning and Zoning Public Hearing.

Note: Figs. 3-14 illustrate community design using TV discussed in Part 3 of this article.

Fig. 3. Roanoke, Virginia community design storefront office.

Fig. 4. Citizens involved in site assessment, Roanoke, Virginia. Local volunteers helped with the site analysis and evaluation, providing local information and expertise.

Fig. 5. Entertainment Break in daylong citizen site assessment, Dayton, Ohio. Combining the hard work of community analysis with "highlight" events, such as lunchtime entertainment by local musical groups.

- Public input, to be effective, has to be guided by legal counsel, thus creating the burden of the cost for professional and legal expense to represent broader public interests.

- Expert opinion is presented within the terms of a legal proceeding, and is thus costly to produce and often can be contradictory and open to interpretation, such as traffic studies, environmental impact studies, etc.

2 DESIGN CHARRETTES

The term "charrette" is adopted from the storied practice of *Ecole des Beaux Arts* architectural students in nineteenth century Paris who reputedly could be seen still drawing their projects until the last minute as they were carried "on the cart" or *en charrette* on the way to the academy's jury. In its modern-day adaptation, *charrette* refers to an intensive design workshop involving people working together under compressed deadlines.

In its use today, a "charrette" is a design and planning workshop held in a two- to three-day period in which architects and other design professionals, community leaders, public officials and citizens work together to envision alternatives for a local building program, neighborhood or regional community project, with an emphasis upon long-term economic, social and environmental sustainability (Watson 1996, Wates 1996).

The charrette process combines techniques familiar from brainstorming methods—letting ideas flow in an open way, each building upon the suggestions of all participants—as well as from "Future Search" processes—creating time-lines and issue maps and diagrams—all of which help individuals, groups and communities to visualize design alternatives and to discuss and evaluate best choices (Weisbord 1995). Design charrettes build upon the thirty-year history of the American Institute of Architects R/UDAT process (Regional / Urban Design Assistance Teams) in which expert design and planning professionals consult with communities about long-range strategies (Zucher 1990.)

A design charrette is the result of many months of planning, necessary to successfully convene a diverse set of community members and representatives, public leaders and outside "experts," each of whom by definition may represent conflicting agendas, diverse personalities, and cross-purposes. Meetings that are not well planned and facilitated can set community discussions back rather than to advance a hoped for community involvement proposal, due to miscommunication, misunderstanding or misuse of the initial good will that should otherwise prevail. Nevertheless, there are ways to help make such meetings successful. This section describes what some of the elements of success might be.

Any group or community meeting requires an organizational structure defined to a sufficient level of detail so that many people can work together, essentially "reading from the same page" to create a smooth running event. Decisions that need to be put into place include a charrette meeting location, sufficient planning time prior to the event, involvement of key stakeholders, and an organizational group or committee. The organizational roles require leadership, initiative, diplomacy, persistence and humor!

A charrette is typically a one- to two-day event to three-day event. In some cases, more time is needed, although this makes it more difficult to include a large number of people in the entire event. A typical size of group is between thirty and sixty people, although many charrettes have involved several hundred and more. Involving greater numbers is possible but should be considered "advanced level" in terms of organizational and facilitative capacity.

The charrette event in the context of community development

The following guidelines indicate characteristics that are recommended for successful community development that builds upon design charrettes:

1 Listen and learn: The charrette process provides for listening and understanding. It works if it facilitates mutual learning and capacity building among community groups, rather than perpetuating dependency upon outside "experts" and resources.

2 Combine and focus: Create focus by combining projects and programs within "high impact" areas strategically selected for demonstrable and replicable results.

3 Create community participation: Emphasize participatory decision-making that enables collaborative partnerships and encourages local initiative, volunteerism and community-based leadership.

4 Build upon local networking: Link local initiatives (bottom-up) to broad (horizontally-linked) networks, such as citywide collaborations.

5 Create multidisciplinary linkages: Link community- and university-based professionals from a range of disciplines to represent economic, social and community planning experience and knowledge.

6 Emphasize sustainable development: Emphasize the need to integrate economic, social and environmentally sustainable approaches to planning, programs and projects.

7 Invite scrutiny and evaluation: Experiment with and document different approaches, to report and disseminate lessons learned about action-based community development, establishing the basis for continuous professional and community learning.

8 Create youth initiatives: Include youth in charrette organization roles, providing "real-life" opportunities for youth leadership.

Preparing for a charrette

Generally, a minimum of three months (most typically six months) is needed to prepare for a successful event, with a longer time frame required for more complex events. The months prior to the charrette involve a series of key actions:

1 Identify a significant project

Everyone who has been involved in organizing charrettes usually has one message: Don't do it unless there is evident local commitment to use the charrette as part of a larger and longer range commitment to take action. It is important to choose a topic that will engage both positive and substantive community support. The initial proposal need not be perfect. It should be open to modification as discussions and planning proceeds. It has to be a project that is inspiring and at the same time feasible, that is, it does not raise false hopes.

Work at defining an issue or program focus that is meaningful, such as a local neighborhood area that deserves community discussion of alternatives and improvement. The size of project can range from building-scale to one well beyond the neighborhood scale.

Start with an initial proposal but let it develop with input from local constituents and stakeholders. Out of a number of preliminary discussions, key issues will emerge. Keep these and the goals of discussion relatively focused, while not ignoring larger and more complex implications.

It is critical to have all data maps and documentation assembled well ahead of time, prepared in the form of a briefing book, a process that sometimes takes months.

Early on in the planning, it will be essential to contact an experienced facilitator in order to add input and guidance to the charrette preparation. If a facilitator is not experienced in design charrettes (but is nonetheless a skilled group process facilitator), then a group of architects, landscape architects and/or planners will be needed to add the necessary design, planning and environmental experience. Additionally, university faculty at schools of architecture and planning may include individuals experienced in leading a charrette process.

2 Involve cosponsors who are stakeholders in the results

There are two general guidelines to involving stakeholders: The first guideline is, "Get stakeholders involved early on." Approach key stakeholders in a low-key way and in their terms, letting them know who you are and what you are about. Make their concerns yours.

A second guideline is "Don't leave anyone out." If the charrette project involves different groups or communities normally left out of the planning and decision process, organizers may find themselves from the outset dealing with a potentially disruptive situation. The advantage of a charrette is that the visioning exercise can be "low risk," that is, it can limit itself to proposing "unthought-of" alternatives and to illustrate new options, leaving it to others to evaluate and decide between competing options.

David Lewis, FAIA, one of the initiators of the AIA R/UDAT process, states, "You have to get to the point where the various goals and agendas are not in conflict. One never gets there by confrontation. . . it never works. Only through a public consensus building process do plans have a sustaining life." (Lewis 1996)

3 Establish preevent meetings that keep planning going forward

The overall questions in organizing a charrette are WHO, WHAT, WHEN, WHERE, WHY and HOW. Of all of these, the HOW is usually the most difficult question at the beginning. That is, there is a perceived need but the way to get there is not clear. In such cases, a broad-based and representative discussion and input from stakeholders is most helpful. A series of small-scale organizing meetings may be the best approach. Once established as regular meetings, they become "heartbeats" to the organizing process.

It is essential to gain "buy-in" of participants in the charrette goal. One of the initial event [or preevent] tasks is to develop a set of goals by

Fig. 6. Citizen involvement community map, Dayton, Ohio. Making the design process evident and participatory, in this case by large community maps to elicit comment and suggestions.

Fig. 7. TV charrette kick-off Roanoke, Virginia. The first of four televised programs in which the design charrette process was made accessible to more than 90,000 people in the viewing audience.

Fig. 8. TV charrette and call-in Dayton. Citizen involvement was sparked by featuring nationally known architects and planners, in this case, the late Charles Moore, Architect *(seated left)*.

consensus discussion. Communicating a clear goal is essential. Once the overall intent and purpose of the design charrette is agreed to, it is useful to state the goals or desired outcomes, keeping in mind that these too will evolve. Stating the goals as a desirable future vision makes it easier to capture the enthusiasm and support of participants.

Timing the event is important. Like any significant planning proposal, there is a right time and a wrong time. Without this groundwork, their event would not have worked so well. Preevent preparation, including training for facilitators and group leaders, is also crucial.

With a statement of intent and the program defined, organizers are then ready to move onto second-level decisions of implementing and preparing for the event, most typically through task group assignments that follow normal definitions:

- **program:** program definition and support materials
- **funding:** funding and/or contributions in kind
- **communications:** getting the word out and the press in
- **logistics:** particulars of space and support materials

4 Define the charrette program

There are at least two different meeting formats to provide the basic organization of groups participating in the event. The one to choose or the right combination depends upon the task at hand.

The first is to organize into generalized and integrated design teams, typically five to six people on each team who work together to develop a design, while "experts" roam between teams consulting with each team throughout the charrette. This option is appropriate where the predominant project goal is to come up with new design and planning visions.

An alternate is to organize specialized expert teams, in which case the number per team can be typically six or more, who work together to develop a set of recommendations related to a particular specialized topic, that is, lighting, building envelope, landscape, etc. This approach works well where there is an existing building or set of conditions that are preexisting and otherwise already designed that require a specialized set of environmental recommendations.

Combinations of these two options are most often adopted in events longer than one day. An obvious variation is to use both integrated design teams and specialized consulting groups, although limited time and the "getting used to" any particular organization cautions against anything overly complex. Each design project will suggest the nature of team composition, "division of labor" and integration of expertise. In any and all cases, individuals should have some choice in selecting the groups they work with, to allow for interpersonal choices and passionate interests.

5 Charrette introductory organizational meeting

The introductory meeting provides a crucial "kick-off." In many cases this began with a dinner followed by presentations intended to inspire, to inform and to set the stage for community creativity. The kick-off is an event to which one can invite local officials and others who would like to be briefed. . .some may be intrigued enough to change their schedules and stay on for the rest of the event. The kick-off or opening session is also typically used to provide the technical briefing to set the teams in place, get logistics out of the way and prepare groups for action.

In cases where the site is large, complex and not completely familiar to all participants, a tour of the site is appropriate and more immediate and informative than a slide show briefing.

6 Putting a funding strategy in place

Funding is required for a design charrette, to cover costs typical of space rental, food for participants, travel, honoraria and lodging for facilitators and group leaders, and printing and publications. Sources of such funds included local foundations, utilities, banks, businesses and Chambers of Commerce, with "contributions in kind" by restaurants, hotels, newspapers, television stations and art materials suppliers.

A likely source of funding support related to energy and environmental design goals are local utilities as well as municipal planning authorities. Both sources represent a vested interest in energy conservation, pollution prevention and waste elimination. Local and regional community and environmental agencies and associations are additional likely sources of endorsement and support funding.

7 Establish a news and communication plan

In most cases, the local press is easily involved in public communication and coverage, provided that notice is given (ideally, an informative and interesting press release) so that media reporters know about significant meetings and the event itself. Both the kick-off and the final public presentation of the charrette work can be organized to provide media coverage. Local newspapers are often a readily available means by which to publish the results, such as in a special "Sunday Supplement" printed as a community service. To assist in all of the media coverage, high quality reproducible graphics provide helpful visuals for articles. As described below in Section 3, local television stations are a source of media equipment, technical support and airtime, which demonstrate a capacity for the charity process to reach many thousands of people throughout the local region.

8 Logistics

There are several recommendations about the charrette workshop location itself. One recommendation is to hold the event at or near the actual project site, or at least to provide easy access so that site conditions can be visited, seen and discussed. An alternative recommendation is to hold the charrette in a publicly visible and accessible location. The ideal location combines both advantages, that is, proximity to the project site and centrally located.

The entire event should be held in one room, such as a gymnasium, in which many groups of five to six people can work. Each group should have access to flip charts and ample wall space. The charrette space should have at least one long wall that can be used to tape up flip-chart sheets and the maps and drawings that are produced during the event. Recording the ongoing discussions on flip charts is important, so that information can be inspected and in turn responded to by others. Flip-chart sheets, numbered appropriately, also become an important record of the event discussions that might otherwise be lost.

Standard folding tables can be placed throughout the space, to be easily moved by any of the groups. It is not necessary to be formal.

All that is required is a large space in which many discussion groups can occur. While part of the discussions will be "in plenary" with the entire group listening to one another, at least half the time and typically more is in small break-out groups. The number that makes up the "optimum size" of a small break-out group is debatable, but highly interactive groups, ones where all can have an active say, should be limited to five to six people per group. In large meetings, microphones placed informally around the room allow for plenary discussion to occur without a centralized podium layout.

9 Materials and resources

Charrette materials and supplies should include ample quantities of flip charts and pads of paper, removable masking tape, water-based markers and paper. Architect and engineering offices are a source of blueprint paper to be reused.

Other presentation tools that may be needed include overhead projectors, which allows both group and plenary presentations to be much simpler and easier to project and record. Convenient access to a copier is helpful for presentations. In some instances, availability of a one-hour photo processing shop may also be helpful, for example, for slides of views of the site for projection and enlargement to create overlay perspective drawings of design proposals.

It is useful to provide within the room a "community wall" and resource area for key information and exhibits. This normally includes the maps and resource information displays that are part of understanding the project context, such as overhead photos and maps from the local planning office. In addition to a resource board, some sort of interactive communication board is often helpful, especially in large groups. Computer interactive tools are increasingly available, to be used like a library resource during the charrette.

10 Charrette events schedule

Although the charrette event is relatively short, the overall process is extended in both directions, that is, from four- to six-months in preparation and an equal or longer time in implementation. The event itself needs to be scheduled within its allocated time, to capture the interest and focused energy of the participants.

There is a variety of experience and opinion related to the length of time or duration of the event itself. The AIA Committee of the Environment (COTE) October 1995 charrettes, including over a dozen different community charrettes, lasted essentially 48 hours over a three-day period, that is, beginning late Friday afternoon and running until mid- to late-Sunday afternoon (Watson 1996). Some participants considered that this was too short a period. In other instances, charrette events have run three entire days. R/UDATs are generally five days or longer. The proper length of time is a function of the number and complexity of the participating constituents. The greater the number of participants, the more difficult it is to keep the high-energy high-involvement pace. But, clearly even three days requires short cuts and may risk coming to an unsatisfactory conclusion before full consensus is reached.

At the conclusion of a charrette event, the summary is a critical point to have the work presented and debated in a public forum. Preparing and rehearsing for the final "public" presentation keeps the charrette schedule on track, working to deadline. In most cases, a spirit of

Fig. 9. Call-ins to the Roanoke Design Charrette encouraged by display of the ideas overlaid with the call-in phone number.

Fig. 10. Call-in suggestions to the Watkins Glen design charrette are immediately written on flip charts.

Fig. 11. Dayton TV charrette. Charles Moore, Architect (left) explains design proposals as they evolved in ways originally not anticipated (shown by improvised sheets added to the base map).

cooperation and participation develops, much like getting ready for a theatrical production. Ideally, everyone would feel that they can be part of the presentation, although not all can be "on stage," but it is important to present the work in the most representative manner possible. It is an ideal time for the "stakeholders" and/or implementors such as community leaders, youth and student participants to present the results, instead of the outside "experts." At least one dry run of the final presentation is recommended to make sure that timing and transitions are worked out.

Facilitation

The facilitator is given responsibility to direct the group process. The following provides a checklist of how a facilitator might best prepare for a charrette:

1 Audit existing conditions. Make sure the room will work and that all necessary materials and support functions are in place.

2 Discuss expectations with the organizing group in terms of goals and outcomes. Establish ways to measure success and consider an evaluation form to measure participant responses at the end of the event to find out what went well and what needs improvement.

3 Understand the nature of the group and community. The organizing group may or may not be representative of the community that is most impacted by the project or program being envisioned in the charrette. There is one rule of thumb: LISTEN, but then ECHO: reflect back what you have perceived to be the nature of the community issues.

4 Organize around goal statements. At any point in a group discussion, people will disagree on any and all points. An astute facilitator doesn't ask, "How do you feel?" but instead asks, "What will you do? How can we effect the outcome?" At the same time, the scope of the issue or problem area will expand or contract, depending on different views and styles of learning and action of those involved. Some think of parts while others think of wholes. Both are needed, but the best way to get everyone "reading from the same page" is to list goals and outcomes, and then detail how to get there.

5 Include all stakeholders. A stakeholder can be defined as anyone whose participation, energy, agreement and volunteerism will contribute to the success of the effort. This creates the "quandary" of trying to get some focus and action while listening to all constituents and stakeholders. Establish working relationships among the stakeholders early on in the charrette preparation and get the entire group to learn the habits of listening and echoing.

6 Prepare a schedule that anticipates variations in the discussion sequence. An experienced facilitator learns how to balance a fixed schedule with time and alternatives "built in" for discussions to take their own turn. At different and to some extent unpredictable times during a charrette, there are times for a "plenary" discussion and times for "break-out" group discussions. A good agenda has the flexibility to allow for both. A prepared agenda is especially important for the first half of a charrette by which the facilitator has in mind a "storyboard" of how the discussion might be undertaken. A "storyboard"—a term used in filmmaking, to refer to a set of steps that tell the story and show a sequence of events—is useful for both facilitator and organizing group to help anticipate the steps in the charrette and various checkpoints to be sure the event keeps on schedule. A facilitator could also set in place a "fast-response" team of advisors who are the "eyes and ears" of the process and can quickly meet to advise the facilitator on suggestions as the event proceeds.

Ground rules of facilitation

The facilitator might explain the following guidelines as "ground rules of discourse." The list is not complete. . .you can ask the group to add their own variations. Agreeing on ground rules helps to establish the setting for group discussion, listening and learning.

1 All ideas are valid.

2 One at a time. Only one person speaks at a time. . .listen to each other.

3 Get to "yes!" Emphasize "yes, and. . ." and discourage "no but. . ." statements.

4 Observe time frames.

5 Seek common group action, not problems and conflict.

A facilitator's task is made much easier by using flip charts, which the facilitator or a recorder (anyone who writes or prints clearly) uses to take notes of the discussion. By recording all ideas, everyone feels that they have been heard and recognized, that their idea is part of the record. An additional aid to communication and creative thinking is to use graphics, that is, to express ideas in both verbal and graphic terms. Often a capable cartoonist or graphic artist can express ideas more succinctly through a drawing and this helps visual imaging, which becomes more important as a charrette proceeds. The roles of discussion leader, facilitator, record and artist can be assumed by separate individuals or combined.

Evaluating the event

Most charrette events go quickly with a great degree of intensity and focus, so that when it is over, people leave quickly to catch up with things left undone for several days. However, without some evaluation process, valuable lessons learned may go unheeded. It is therefore recommended that some form of event evaluation be put in place, allowing time in the concluding session for evaluation forms to be completed before participants leave. This evaluation, whether in questionnaire form or otherwise, should allow for commentary to capture creative insights and suggestions for improvement.

If an evaluation questionnaire is simple and easily understood, participants will fill it out. It can be as simple as:

1 What worked?

2 What didn't work?

3 Suggestions for improvements?

4 Suggestions on next steps?

5 Other thoughts?

Ask people to fill out their responses in clearly written form, explaining that what they say is important and that answers can be transcribed as is and made available to all interested. Such responses will often yield an overall evaluation, along with very valuable insights and ideas for improvements and next steps. It is important to tabulate and publish the results. "Inviting scrutiny" can thus be shown to inform the entire process. In addition, the event might be monitored by an "evaluator" or "reporter" who is asked to summarize and evaluate results.

3 COMMUNITY DESIGN AND TV

Chad Floyd, FAIA

The community design process is suited to the television media as it evolved with experience in the 1970s and 1980s (see additional credits, also Crosbie 1984). They featured the following three elements:

1 Community office: A storefront office was located as close as possible to the downtown's "hundred percent corner," with a staff architects at a drawing desk in the front window. The office issued weekly press releases. The office included a walk-in interview area with comfortable chairs, tables, and maps to discuss ideas amid coffee and donuts.

2 Community committees: Several committees were established, starting with a small (15-person) Steering Committee, which brought together key public officials and community leaders. A second Citizens' Committee of 30 to 100 was representative of citizens, businesses, and special interests. It attended about a half dozen meetings, beginning with a "treasure hunt" exercise to record perceptions about places and ideas. The Committee process thus allowed community evaluation and testing and avoided the danger of us outsiders imposing our values and interpretations. In turn, the Citizens' Committee became a focus group to consider design proposals and eventually a representative mouthpiece for articulating proposals back to political leaders.

3 Television outreach: For greater outreach, local TV stations were used to carry the design message to a wider audience. To make this effective, the TV format was made entertaining and interactive. The shows paralleled the community design process over a three- to four-month period, each show airing for one hour, prime time on commercial stations wherever possible. TV costs were manageable, because commercial stations are required by the FCC to provide a few hours of community affairs programming every week. The stations provided camera time, airtime, and production facilities as in-kind contributions. With their help, the design charrette programs produced some pretty interesting television on prime-time air slots.

The sequence of TV shows, each one hour long and spaced about a month between shows, was as follows:

First Show: The objective of the typical first show was to get information from studio interviews and call-ins. We referred to specific geographic areas called out on "Idea Boards" arranged around the studio. In Roanoke, one of the early projects, we made the mistake of spreading ourselves out too thin around the studio. Intimacy of contact and some degree of control was compromised, but the show worked, nonetheless. We interviewed community representatives and experts in front of wall-sized posters summarizing information collected from prior community input. Call-in and interviewer ideas were sketched

Fig 12. Riverfront, Dayton, Ohio. General view BEFORE design charrette process was begun.

Fig 13. Dayton, Ohio Kiwanis Fountain. A "quick start" design to inaugurate and help promote the riverside revitalization project.

Fig. 14. Riverfront, Dayton, Ohio. General view AFTER design charrette process was completed.

(in Roanoke by Charles Moore who had the talent to make any idea look interesting). Their sketches captivated people. At Roanoke, we were able to obtain immediate opinions on economic feasibility from a financial analyst from the Rouse Company. In addition to call-ins, which provided real-time discussion and interest, we received responses from viewers after the show by mail. These were cut and pasted into issue categories, and ultimately were organized by topic and design response. Today, we would do it more easily by word processing.

Second Show: To provide visual focus for the second show, where our goal was to present planning options, we centered ourselves around a large map on a table. The map was colorfully painted for clarity of close-up camera detail. On it we showed a variety of 2-D design options. The map stimulated feedback from viewers, plus a few invited experts and municipal representatives. As we described options, we asked viewers to follow us with a questionnaire in the local newspaper. This enabled all who were interested to "vote" and even to add their own suggestions.

Third Show: By the third show we were illustrating favored ideas by means of three-dimensional models. The models addressed high-profile issues in detail. In Roanoke, we obtained a commitment from BlueCross-BlueShield to locate new offices downtown, and a top executive of the company made the announcement on the show in front of a model. We had listener call-ins and were able to get to points of consensus.

Fourth Show: The final show was intended to illustrate final planning proposals. No phone calls here. Just sketches and a detailed model allowing the camera to get down to eye level and bring our ideas to life. The show included the leaders who later would have to approve implementation. Discussions with them were moderated in a TV anchor desk format.

In Roanoke, we reached a viewing audience of 90,000 people, according to Nielsen ratings. Following each broadcast, our design issues achieved instant recognition, with phenomenal community awareness. Within three years of our TV programs, the community had voted for bond issues to fund all but seven of the 59 individual projects that made up the total community development plan, representing over $89 million in private funds and $41 million in combined local, state, and federal funds.

The Roanoke City Manager at the time, Bern Ewert, stated that "the television programs and the plan changed attitudes in Roanoke. . . initially they viewed the shows as purely entertaining, but soon large numbers of viewers tuned in, many from surrounding communities."

This process was employed in design projects in various communities, including downtown Roanoke, Virginia; Dayton and Springfield, Massachusetts (large and complex plans); Watkins Glen, New York (a relatively small focus), and Indianapolis (a large public project). In all, television helped to make the design process visible and community-wide consensus building easier. The TV experience developed into a planning approach we have come later to describe as "situation design," wherein each element must be capable of standing alone yet still relate to the plan as a whole. The goal is that if any particular proposal were to remain unimplemented, the plan is not at risk. This is the way cities naturally develop; progress occurs in discrete steps, each one responding to its situation. This approach seems to be well understood by laypersons, who are suspicious of the grand schemes.

TV certainly made the urban design work challenging, and it added an extra layer of coordination and detail. But we were pleased at how effective it made us as architects and urban designers. Most if not all community leaders and citizens were helpful in front of the "cool" TV camera. Many constructive comments were offered its unblinking eye, quite the opposite of the posturing that can occur in public hearings. Television helped us build a level of community interest and resolve that, as far as we have seen, is unmatched by conventional planning media. ■

ADDITIONAL CREDITS:

This article includes recommendations of many individuals who participated in the October 1995 AIA Environmental Design Charrettes, including Gregg Ander, Kirstine Anstead, Robert Bell, Bob Berkebile, Jestena Boughton, Robert Cevero, Brian Dunbar, Sue Ehrlich, Elizabeth Ericson, Pliny Fisk, Jim Franklin, Greg Franta, Kirk Gastinger, Harry Gordon, Chris Gribbs, Peg Howard, Chris Kelsey, Paul Leveille, Gail Lindsey, Andy Maurer, John B. Peers, David Sellers, and Lynn N. Simon.

The process of utilizing television media in community design charrettes is based on experience of Centerbrook Architects in the late 1970s and 1980s, Centerbrook undertook a number of community design projects, a half-dozen of which employed interactive television to involve the community.

REFERENCES

Crosbie, Michael J., 1984. "Television as a Tool of Urban Design," *Architecture Magazine,* November 1984.

Lewis, David, 1996. "Memory of the Future" pp. 87-93 in Watson, 1996.

Wates, Nick, ed., 1996. *Action Planning: How to use planning week-ends and urban design action teams to improve your Environment.* London: Prince of Wales's Institute of Architecture.

Watson, Donald. 1996. *Environmental Design Charrette Workbook.* Washington, DC: American Institute of Architects.

Weisbord, Marvin R., & Sandra Janoff, 1995. *Future Search: An Action Guide to Finding Common Ground in Organizations & Communities,* San Francisco: Berret-Koehler Publishers.

Zucher, Charles B., ed. 1990. *Creating a Design Assistance Team for Your Community,* Washington, DC: American Institute of Architects Regional Urban Design Committee.

Sheri Blake

Summary

The cumulative experience of Community Design Centers over a period of almost forty years is presented. The key goals of the field are defined, although these are not always realized, due to political, institutional or funding constraints. A history and description of types of centers are provided, which include university-based, independent nonprofit organizations and volunteer organizations. Centers respond directly to local contexts and community-based concerns and needs. Organizational issues of professional competition and the need for more comprehensive community design education are discussed.

Key words

capacity building, community design centers, economic development, nonprofit community development, participatory design, social justice, technical assistance, volunteerism

October 24, 1998. Neighborhood residents demonstrate to save the Knightsbridge Armory for adaptive reuse. (Northwest Bronx Community and Clergy Coalition and PICCED.)

Community design centers: an alternative practice

1 WHAT IS A COMMUNITY DESIGN CENTER?

Community Design Centers (CDCs) provide planning, design and technical assistance to low- and moderate-income urban and rural communities, many of which have limited resources. Those involved with CDCs thus work for social, economic and environmental justice, particularly serving local community-based development needs. CDCs plan and design with, not for, local community-based development organizations and residents. CDCs sustain relationships over a long period of time so that mutual trust can be established and shared learning can occur. They place themselves in opposition to institutional and corporate interests when a process or project may contribute to further decline of an already stressed environment or result in displacement of existing residents.

CDCs also build local knowledge and organizational capacity by assisting communities in leveraging the resources and tools necessary to meet a variety of challenges. CDCs held to educate community residents to participate effectively in the process of directing change. Community designers take on as a professional the responsibility to understand that communities are complex social systems and that physical design is only part of an integrated solution. Technical teams and solutions should remain flexible in response to specific communities and their unique conditions necessary to take economic control over their own assets. Community design practice is based on the three key tenets of affordability, accessibility (political, physical, economic, social and cultural) and aesthetics.

Why is design important in CDC practice?

Design, as a process, is an effective way for advancing the cause of social justice. It facilitates a wider range of community concerns and fosters a multidisciplinary approach. Better decisions are achieved by combining a community's experiential knowledge with academic and professional knowledge. Design solutions are, therefore, more contextual, inclusive and appropriate resulting in better use of resources. Process-oriented design translates human needs into achievable plans compared to traditional design which is often product oriented and devoted to institutional or corporate goals (Francis 1983). A specific style emerged in participation and environmental justice in landscape architecture. It was "highly personal, comfortably homemade and well-loved, indistinct edges that are more complex than the modern style, moving away from a preoccupation with joining materials to more emphasis on human movement and activity" (Hester 1983:53). "The resultant plans, being flexible and open-ended rather than fixed, anticipate and aid the incremental development that usually follows. Design became a continuous process of improvement."

If process and product are equally valued, the systematic and systemic ways in which various groups of people experience exclusion in their daily lives are more comprehensively addressed. Funders should recognize the value that CDCs bring in contributing to design quality within the framework of community organizing, Community Economic Development (CED) and capacity building. Designers should keep in mind that design is an important component of low-income and moderate-income communities because it demonstrates

Credits: The author would like to thank Terry Curry, Fiona Reid and Rex Curry for their comments on earlier versions of this article.

respect for the residents and contributes to further local investment. However, designers also should understand that community design is not solely about "fashioning more handsome buildings, interesting views, or attractive landscapes." It is about "empowering the citizens of local communities to shape their own preferred futures by acquiring and applying information and knowledge about their communities in a far more systematic, thoughtful, and democratic manner than current practice" (Mehrhoff 1999:122).

2 WHAT IS TECHNICAL ASSISTANCE?

Technical assistance provided by CDCs in the field of community development can be characterized and defined in three ways, helping, enabling, empowering:

Helping: Technical assistance helps in resolving the need or crisis but the person or organization being helped may not be in any position to avoid the same crisis in the future.

Enabling: The person or organization is encouraged to help themselves by being provided with the rudimentary tools, knowledge and sometimes money to resolve an immediate set of problems.

Empowering: The person or organization is provided with the knowledge to understand why they are in the situation that is contributing to local disinvestment. This type of technical assistance helps build local research, analytical and development skills moving the community toward organizing and taking responsibility for themselves and their neighborhood and toward developing their ability to change their conditions (that is, achieving social change).

Types of technical assistance provided by CDCs: CDCs serve a variety of clients (see Box 1: CDC Clients). The goal of effective CDCs is to empower nonprofit Community-Based Development Organizations (CBDOs) to be able to do comprehensive community building and development with and for existing residents. Services range from neighborhood planning, project development consultation, architectural design, graphic design and media services to a wide range of consultation services in planning, design, development and management. Activities such as information and referral assistance, community planning, envisioning, design, and the work leading to coalition building, program monitoring and evaluation, public policy development and analysis are all important aspects of a CDCs growth and development (ACD 1997). CDC services are responsive to their context, based on local needs, and available funding and staff skills and interests. Technical teams and solutions must remain flexible in response to specific communities and their unique conditions.

3 WHAT IS ORGANIZATIONAL CAPACITY BUILDING?

The core of CDC technical assistance that enables and empowers is based on providing education to residents and community-based development organizations or related agencies so that capacity is in place locally to continue work in the future. Examples include, but are not limited to, assistance in understanding what community building is and how to achieve it, providing technical courses in planning, development and management of local neighborhoods and their assets, or building knowledge of planning and design criteria during participatory design processes:

Community building: Recently, poverty alleviation in community development is more focused on the practices of community building, moving away from models of dependency to local self-reliance and responsibility. Community building is not new to CDC practice. It is defined by seven themes (Kingsley *et al.*1999):

- Focused around specific improvement initiatives in a manner that reinforces values and builds social and human capital;
- Community-driven with broad resident involvement;
- Comprehensive, strategic and entrepreneurial;
- Asset-based;
- Tailored to neighborhood scale and conditions;
- Collaboratively linked to the broader society to strengthen community institutions and enhance outside opportunities for residents;
- Consciously changing institutional barriers and racism.

CED Internship: In order to achieve self-reliance, CDCs have assisted in building development skills in CED. One of the premier training programs in organizational capacity building was collaboratively designed and delivered by Development Training Institute (DTI) in Baltimore and Pratt Institute Center for Community and Environmental Development (PICCED) in New York, the oldest CDC in the country. The CED internship included courses in the history of social movements and CED, strategic planning, organizational effectiveness, nonprofit real estate development and management, community economic development, accounting, law and tax, youth entrepreneurship, alternative financial institutions and practices and programs that empower. Once participants completed the course, consisting of six workshops over the period of nine months, they could access the second year of the graduate planning program at PICCED. The graduate program schedules all classes in the evening to allow opportunities for staff of local organizations to participate. This provides access to advanced education to many, particularly women and minorities. PICCED is now delivering related short technical workshops.

Participatory design and knowledge transfer: The inherent challenge is to move beyond participation in design that is limited to consultative processes or charrettes to building local capacity. Detroit Collaborative Design Center (DCDC) has been very effective in developing local knowledge during their participatory design process. Their former Director, Terry Curry, developed a set of eight workshops to assist participants in making informed decisions about building quality, project budget, programmatic requirements and building character, and spatial experience. For example, participants are provided with workshops on factors that affect decision making in housing design, regulatory rules that influence physical design and budget analysis (Fig. 1). The building is designed based on the results of a collaborative design process that includes both technical information and design evaluation knowledge (Fig. 2). Participants develop a greater understanding of key areas of planning bylaws, design criteria and process. They understand what to ask professional planners and designers on future projects and can collaborate more effectively rather than merely be consulted.

4 HISTORY OF COMMUNITY DESIGN CENTERS: FROM ADVOCACY TO EMPOWERMENT

CDCs emerged in the 1960s in opposition to urban renewal policy that encouraged removal and gentrification of neighborhoods. They also opposed policies that contributed to the destruction of the natural environment represented by suburban sprawl. In 1968 at the 100th Convention of the American Institute of Architects (AIA), Whitney M. Young, Jr., Executive Director of the Urban League, demanded greater accountability from the architecture profession. Referring to the "white noose around the central city," he indicated that architects shared the responsibility for the mess and that the profession had distinguished itself by its "thunderous silence" and "complete irrelevance" (Curry in Cary 2000). Architecture once had a strong care-taking tradition lost in the Master Architect model. CDCs began to rebuild civic engagement at the grassroots level (Mehrhoff 1999).

Studies and profiles produced between 1970 and 1977 documented between fifty to eighty CDCs in the U.S. But by 1987 only sixteen remained. Of these, only twelve were established in the early years (Curry in Cary 2000). By the 1980s, CDCs had to become more entrepreneurial, drawing their funding from a wider range of sources including government programs, foundations, local AIA chapters, fee-for-service work, universities, historic preservation programs, private philanthropists and corporations. They became less political and focused more on product than process. This evolution was influenced by a changing political and economic climate at the government level. Also community ownership and economic development initiatives were emerging at the grassroots level. However, justice, social change, community participation, empowerment and control of resources remained key to CDC practice (Comerio 1984).

As a result of this trend to an entrepreneurial model, many CDCs evolved into private practices. Some disappeared and some are occasionally "reactivated" to deal with a crisis. CDCs that evolved into community-based development organizations had difficulty managing the conflict between political intervention and specific project development. Licensed architects started many CDCs, but almost none of these centers survived, many pressured out of existence by the profession.

The ACSA Sourcebook (Cary 2000) documents forty-six university affiliated programs and twenty-six independent design centers. There has been a recent surge in the establishment of CDCs, in part due to recent government and foundation funding initiatives and a recognition by architecture schools of their value in education (see Box 2: SWOTS of Community Design Education).

A national networking organization, Community Design Center Director's Association was incorporated in 1978, renamed the Association for Community Design (ACD) in 1985. ACD holds an annual conference and runs a website and intranet forum (see References). ACD supports the formation of CDCs as capacity building professional vehicles with a broad range of architecture and planning professionals from universities, professional societies and the nonprofit sector. Members are dedicated to finding an alternate form of community planning and architectural practice that combats racism and policies that contribute to the persistence of poverty.

Box 1: CDC Clients

Examples of CDC clients include:

- Self-help neighborhood development organizations
- Unincorporated community-based associations
- Public officials/administrators of major metropolitan areas
- Nonprofit housing development corporations
- Private individuals/families eligible for special housing programs
- Nonprofit community development corporations
- Government (social welfare, housing, economic development)
- Community business assistance corporations
- Religious organizations providing services to community
- Senior citizen and youth organizations
- Block associations, civic organizations
- Rural and small town communities

Fig. 1. HOUSING DESIGN: Factors affecting decision making

Primary factors influencing the design of affordable housing are:

- organizational capacity
- marketability
- economic feasibility
- government requirements (codes and ordinances)
- project design
- stakeholder expectations

Source: Detroit Collaborative Design Center Affordable Housing and Community Design Forum

Fig. 2. HOUSING DESIGN: Factors affecting decision making

Critical questions:

- What is the occupancy and use group?
- What are the applicable construction classifications with and without sprinklers?
- What is the allowable number of occupants?
- What are the egress requirements that must be met?
- What are the fire separation distances required for site placement?
- What are the height and area options?
- What are the minimum levels of light and ventilation required?
- Are there any special requirements for handicap accessibility?

Source: Detroit Collaborative Design Center Affordable Housing and Community Design Forum

Box. 2: SWOTs of Community Design Education

Strengths

Design schools that teach the following subjects, for example, can bring real value to the community design process:

- Historic preservation;
- Sustainable planning and design;
- Research methods in planning and design;
- Design methods and programming;
- Landscape resource management;
- Landscape architecture;
- Building assessment;
- Affordable housing design;
- Placemaking;
- Small town conservation;
- Universal design;
- Graphic communication techniques.

Weaknesses

Designers should be able to articulate various alternatives in collaboration with the users and reflect on past projects and experience to inform future decisions. They should have the ability to translate the daily experiences of the users into not only plans and designs that can be built, but also policy. They need to understand that good environments are not designed but evolve and are only part of a comprehensive and integrated community development strategy. Multidisciplinary linkages are needed to integrate physical, social, economic, environmental and political concerns. Weaknesses are related to limited or no knowledge in design schools about the following areas:

- Community organizing and community building;
- Community economic development in community and downtown revitalization;
- Tools and techniques for effective community engagement combined with organizational capacity building;
- Nonprofit real estate development and management;
- Managing and marketing techniques in community design;
- Conflict mediation and resolution;
- Policy analysis;
- Evaluation criteria in planning and design.

Opportunities

CDCs that are connected to a design school can provide a commitment to both process and product. Communities are more than just statistics. As well as being about natural, social, economic and political environments, they are also about people and places. Wates (2000) provides a variety of tools, techniques and scenarios for collaboration in shaping various environments. He also explains the benefits of getting involved:

- Additional resources leveraged by including local resources;
- Better decisions by including local knowledge and experience;
- Building community by creating stronger relationships;
- Compliance with legislation that requires community participation;
- Democratic credibility when residents participate in decisions affecting them;
- Easier fundraising with grant organizations who encourage participation;
- Empowerment by building local capacity for addressing future problems;
- More appropriate contextual design solutions that use resources effectively;
- Professional education that results from mutual learning;
- Responsive environments that can be constantly tuned and refined over time;
- Satisfying public demand for participation;
- Speedier development by reducing time wasting conflicts;
- Sustainability, both environmentally, but also by reducing vandalism and neglect.

Threats

Universities have traditionally focused on research and evaluation. In turn they have devalued sustained implementation which is critical for mutual organizational learning and social change to occur. Studio and service-based learning models can only be mutually effective if they are a footnote in the longer term framework provided by established CDCs that function throughout the calendar year. The priority always has to be given to the community-based organizations and local residents, rather than students. Curry (2001) provides a few prerequisites for a service based learning studio:

- A neighborhood in socioeconomic distress is not a place for class experiments;
- It is necessary to define and redefine the problems collectively with all participants;
- The work is a full-time job with full-time responsibilities;
- Community engagement processes are not limited to one-time workshops and visioning processes, but are multiple planning and design techniques applied to shifting contexts throughout the timeframe of a process and product.

For a guide to evaluating implementation for tenure and promotion of academic staff, see Barry Checkoway. 1998. "Professionally Related Public Service as Applied Scholarship: Guidelines for the Evaluation of Planning Faculty." *Journal of Planning Education and Research.* 17, 4: 358-60.

Three Types of CDCs

Many CDCs are located in large U.S. cities and generally serve nonprofit community-based organizations in distressed urban and rural communities. In the United Kingdom, they are referred to as "community technical aid centers." Planning and design activities dominate. However some CDCs provide education and training programs for organizational capacity building, advocacy, referral and financial services. Most CDCs have more architects than planners on staff. Some have engineers, landscape architects, interior designers, graphic designers and other professionals, based on the nature of activities and opportunities for funding. Budgets range from a few thousand to a few million dollars.

CDCs are either university-affiliated, independent nonprofit or volunteer. The descriptions, below, provide examples of this range of CDCs. However, they vary extensively in their mission, types of services provided, the degree of project involvement, staffing, budget, size of project area, organizational structure and fee policy. They share a common interest in bringing together the integrative capacity of the various physical planning and design professions with the complex challenges of comprehensive CED. For the most complete description of CDCs, refer to the ACD website or The *ACSA Sourcebook of Community Design Programs* (Cary, 2000).

University Affiliated CDCs

The majority of university-affiliated CDCs are operated either as part of a university department or are nonprofits that have projects providing training opportunities for students. Some also have collaborative activities for community involvement and extension services. University affiliated CDCs are also involved in direct publication of academic journals, newsletters, or books. They may provide services for free, or at minimum charge or have a cost reimbursement policy based on available funding. University affiliated CDCs often evolved out of an individual staff person's initiative or are faculty- or university-wide initiatives.

The Pratt Institute for Community and Environmental Development (PICCED) in Brooklyn, New York emerged out of a collaboration between local residents, municipal authorities and university staff in planning and architecture in 1963 and eventually came under the administration of the President's office. PICCED, described on their website at www.picced.org., was founded on the triangulation of three key areas to support local community-based development initiatives—education, technical assistance and policy and advocacy, The Pratt Planning and Architectural Collaborative was established in 1975 to provide planning, design and development services, primarily for affordable housing production working from a base of community organizing (Fig. 3). In addition to planning and architecture design services, PICCED has recently developed a GIS team to compile and analyze data for assisting CBDOs in defining difficult problems and identifying resources for CED opportunities. PICCED has a strong relationship with the graduate planning program. Staff of PICCED teach in the program and students can work at the Center. Currently, full-time staff is about twenty, including five architects and several planners. Their operating budget is approximately two million dollars.

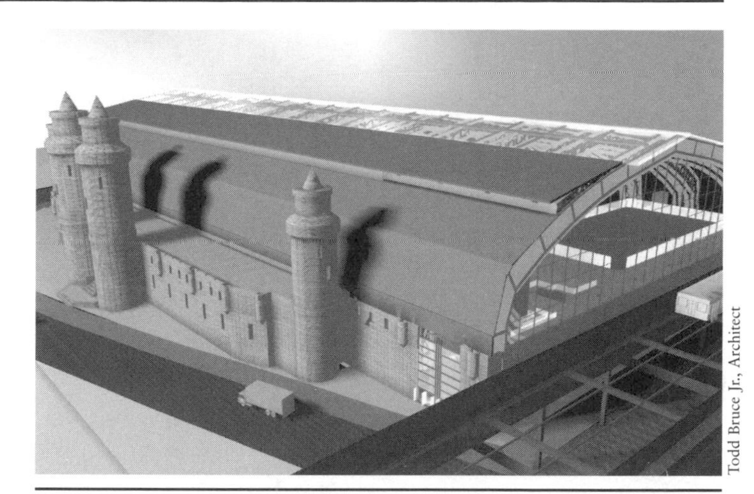

Todd Bruce Jr., Architect

Fig. 3. The Kingsbridge Armory. The Northwest Bronx Community and Clergy Coalition (NWBCCC) organized the community to save a historically significant building. NWBCCC and local residents believed that the Armory should remain a public asset. Given that local schools were overcrowded and vacant land scarce, the adaptive reuse design proposed by PICCED and NWBCCC provided room for three public schools, an indoor athletic field, retail frontage and community, recreational, cultural and parking facilities.

The mission of the Detroit Collaborative Design Center (DCDC)—established in 1994 as a nonprofit subsidiary of the School of Architecture, University of Detroit-Mercy—includes education, service and leadership. It focuses on educating responsible professionals through an undergraduate neighborhood design studio and providing internship opportunities for students at the center. It is committed to serving nonprofit organizations. Significantly, DCDC integrates its education and service mandates through an effective engagement technique that involves community-based organizations, funders, residents and other relevant parties in the design process and builds their organizational capacity at the same time. DCDC's goal for leadership is to set high standards for quality design solutions while building local capacity and taking local needs, budget limitations and program requirements into consideration. Their design work has been recognized by the American Institute of Architects and has won local, state and national awards. DCDC is staffed by five full-time employees, including two design fellows and one student intern (rotating position). Their yearly operating budget is $300,000. Further information can be accessed through the School of Architecture website at www.arch.udmercy.edu.

Independent Nonprofit CDCs

Independent nonprofit CDCs have been able to sustain themselves longer than most university-affiliated CDCs. They primarily do fee-for-service work based on ability to pay or direct project costs. Asian Neighborhood Design (AND), founded in 1973 as an independent nonprofit by Asian-American architecture students from the University of California, Berkeley, focused initially on providing design services

PHOTO: John Martine

PHOTO: Angeliki Georgiou

Figs. 4 and 5. Denny Row (before and after). A $6,000 recoverable Design Fund from CDCP enabled Allegheny West Civic Council (AWCC) to develop renovation plans for Denny Row, a row of six deteriorated historic homes. AWCC invested $500,000 in renovating the facades, designed by Integrated Architectural Services, to demonstrate the intrinsic value of the homes as a row. They then sold the shells to homebuyers, establishing a new development model that is being replicated elsewhere throughout the Pittsburgh area.

to nonprofit organizations, primarily serving the Chinese community. Their constituency and their organization evolved over time in direct response to the community development sector. Their activities—described on their website at www.andnet.org—include divisions in Architecture and Planning, Family and Youth Resources, Community Planning and Development, and Construction Management. AND becomes involved in a project before an architect is selected and stays involved through the completion. AND designs and builds housing, playground equipment, day care centers, youth clubs, and community service agency offices. They run job programs and developed a carpentry and cabinetmaking factory to train disadvantaged youth. This evolved into a very successful high-end furniture design and carpentry business. They publish, provide emergency repair services and lobby city officials on policy issues. Their focus has been on keeping housing affordable even if it means giving up the opportunity to do a major architectural design project (Comerio 1984). AND has evolved from a small community design center to a regional community development corporation with a primary focus on housing and community economic development. It has a staff of over one hundred and twenty, including nine architects and a budget of several million dollars (AND 1999).

The Community Design Center of Pittsburgh (CDCP) was originally established as the Pittsburgh Architects Workshop in the late 1960s, providing direct architectural services to organizations and individuals. It changed to its name and approach in 1987, becoming a combination independent nonprofit and broker of architectural services. Through its Design Fund, the CDCP provides recoverable grants plus technical assistance to help community-based organizations hire architects and professional planning assistance for early phases of revitalization projects (Figs. 4 and 5). The grants have been used for a wide variety of projects, including housing renovation and new construction, commercial and mixed-use development, open-space and community planning. The CDCP's Renovation Information Network provides opportunities for individuals to consult with volunteer or intern architects for advice and information. The organization's annual recreational event, Volkswagen Pedal Pittsburgh, has introduced over 10,000 riders to design landmarks and revitalization activities citywide. The CDCP is also engaged in broad Civic Stewardship initiatives that help strengthen the region's "quality of place." The CDCP offers a wide range of educational resources, including Design-In-Action! Workshops (Fig. 6), informal brownbag lunches and interactive visual presentations like "Negotiating with Retail Chains" and "Add Value to Your Home—by Design." Many educational resources are available at their website at www.cdcp.org. They have five full-time staff with a range of community development and design backgrounds, including one registered and one intern architect. Their budget is approximately $325,000 per year for operations plus approximately $60,000 for Design Fund grants. To date, the CDCP has focused primarily on distressed neighborhoods within the City of Pittsburgh, but is broadening their focus to include new clients and new geographical areas. Criteria for reviewing requests for support are based on neighborhood need, strategic impact, project feasibility and community-based organizational capacity.

Volunteer Organizations

Volunteer organizations act as resource centers that link professional service providers and community-based organizations, or coordinate activities of various professionals who volunteer their services to nonprofit community projects. Volunteer organizations often use a charrette process of intensive engagement over a short period of time.

The Minnesota Design Team (MDT), established in the early 1980s, is a volunteer model that coordinates professionals from a wide range of fields, including architecture, landscape architecture, city and regional planning, economic development, interior design, architectural history, anthropology, marine biology, agriculture, horticulture, forestry, and tourism. They primarily work with small towns and visit only those communities that demonstrate broad-based support, based on the results of a set of questions that the local community responds to before being selected. Once selected, a community is asked to contribute $3,500 to defray expenses and demonstrate their commitment to the process.

Mehrhoff (1999) describes the MDT process in detail. It involves several months of advance preparation by two team leaders and participating communities, collecting survey data, base maps, physical, economic, social and cultural data using action research. Several drafts of a community design framework are developed throughout the preparation and on-site process with the full team. Over a four-day period on site, the full team employs several key techniques and tools. These include slide presentations, briefing sessions, SWOT (strength, weaknesses, opportunities and threats) analysis, focus groups, visits to schools and other centers, bus and walking tours, visioning exercises, town meetings, democratic brainstorming, charrettes, storyboards, maps, drawings and networking. On the final day, community leaders are advised on ways to get started on key projects, additional resources they can call upon for help, and how they might coordinate public, private and volunteer efforts to move forward. An overall design framework that understands a community as a system of systems is crafted. The MDT team visits six months later to assess and evaluate progress. During the collaborative process, they post the on-going work at their website, www.minnesotadesignteam.org.

This volunteer process can be an effective way to galvanize public opinion, jump-start a revitalization process and generate initial excitement and participation. However, there are limitations to this model. A high level of citizen involvement is often produced in the short term, but underlying problems may be glossed over (Mehrhoff 1999). Too often, volunteer processes become limited to "step one," defining the problem and a visioning process. Although it is possible to connect the community vision to a plan of action, there may not be the necessary capacity, political will or democratic mechanism, locally, to sustain the process. Without longer-term technical assistance and willingness to politically intervene when necessary that an established CDC provides, allowing for mutual trust and learning to occur, there may not be opportunity for local community-based leadership and social change to be nurtured and alternative markets to be created. This volunteer process is important, but should not be considered the sum total of CDC practice. It still does not meet the challenge put forward by Whitney Young to the architecture and urban design professions. The goal of CDC practice is to generate a political process that involves plans, programs and projects. ■

PHOTO: Anne-Marie Lubenau

Fig. 6. Green Building Workshop. Design-in-Action workshops by CDCP, like this one on "Green Building" offer opportunities for participants to learn about issues influencing community development and practice what they have learned through hands-on exercises.

Box 3: On the Issue of Professional Competition

The issue of "unfair" competition for development projects is constantly raised in relation to fee structures, nonprofit status, and overall mission of the Community Design Center in serving the low- and moderate-income community and client. Whenever this issue reaches confrontation stages it has generally been between the private architectural community and architects working in the low-income community as members of community design centers. Design centers generally resolve these issues in the following ways:

First, the design center as a nonprofit has every right to bring public and private support funds into the process of community development. It is a legitimate strategy for project development, and it is in fact, the very essence of a policy statement made by the AIA that supported the formation of CDCs.

Second, where the support of the local architectural community is particularly strong on the provision of comprehensive high quality services in low-income areas, a support network is often established. CDCs provide the back-up support and continuity to the process in the form of a business partnership with volunteers and community-based clients. This partnership established the framework for training and education and the context for dialogue on public policy issues.

Third, volunteer systems are needed in every profession, but do not work without an administrative structure, often named "community design center."

Finally, ACD has documented the work of CDCs. In all cases the services of these organizations have facilitated hundreds of projects and joint ventures where they were most needed, yet least likely to occur without a commitment to advocacy, empowerment and access to technical assistance.

Operations and Policy Manual, ACD Inc., 1997. Also see this manual for minimum criteria for eligibility for CDC services, income eligibility guidelines used by CDCs, policy guidelines for free professional services, related policy issues regarding provision of services, fee structures, code of ethics and a policy statement developed by the AIA regarding CDCs.

REFERENCES

Association for Community Design (ACD). http://www.community-design.org. New York: ACD, Pratt Institute Center for Community and Environmental Development.

Association for Community Design (ACD). 1997. *Operations and Policy Manual.* New York: ACD, Pratt Institute Center for Community and Environmental Development.

Comerio, Mary C. 1984. "Community Design: Idealism and Entrepreneurship." *Journal of Architecture and Planning Research.* 1, 1: 227-43.

Cary Jr., John. Editor. 2000. *The ACSA Sourcebook of Community Design Programs at Schools of Architecture in North America.* Washington, D.C.: Association of Collegiate Schools of Architecture Press.

Curry, Rex. 2001. *Community Design Programs. Structures for Inclusion.* New York: Princeton Architectural Press. Forthcoming.

Comerio, Mary C. 1984. Community Design: Idealism and Entrepreneurship. *Journal of Architecture and Planning Research.* 1, 1: 227-243.

Francis, Mark. 1983. Community Design. *Journal of Architectural Education.* 3: 14-9.

Hardy, Flora. 1997. "Just Design?: The Role of Community Design Centers in Accommodating Difference." Cambridge: Newnham College. 10 pp. available from ACD, Pratt Institute Center for Community and Environmental Development.

Hester Jr., Randolph T. May 1983. "Process CAN Be Style: Participation and Conservation in Landscape Architecture." *Landscape Architecture,* p. 49-55.

Kingsley, G. Thomas, J.B. McNeely and J.O. Gibson. 1999. *Community Building: Coming of Age.* Washington D.C.: The Urban Institute.

Mehrhoff, Arthur W. 1999. *Community Design: A Team Approach to Dynamic Community Systems.* Thousand Oaks, CA: Sage.

Wates, Nick. 2000. *The Community Planning Handbook: How people can shape their cities, towns and villages in any part of the world.* London: Earthscan.

5 • REGIONAL AND URBAN-WIDE SCALE

John W. Hill

Summary

This article illustrates a method of formal analysis of traditional settlement patterns. It proposes that future development in states such as Maryland be informed by, and where appropriate be modeled after, the State's traditional, historic settlement patterns, and to show how village and hamlet development regulations might be written, based on those patterns.

Key words

alley, cultural preservation, farmland, regional planning, rural zones, settlements, street, town plans, Traditional Neighborhood Development (TND), zoning

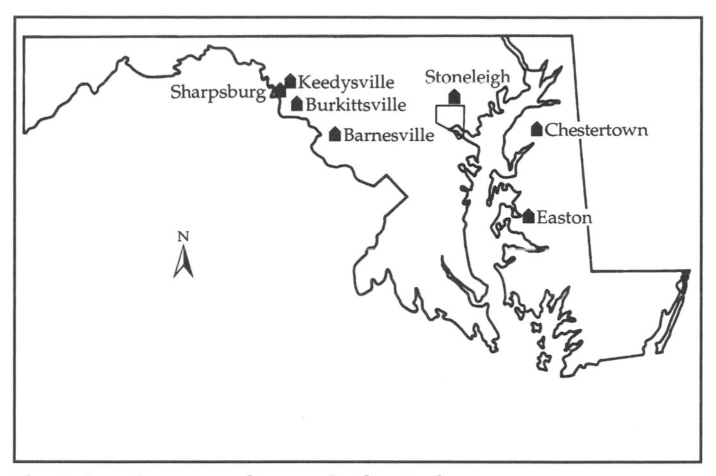

Fig. 1. Location map of towns in the study.

Design characteristics of Maryland's traditional settlements

1 BACKGROUND

Traditional land development patterns in Maryland tended, in the past, to produce cultural landscapes of great beauty and environmental quality in which compact settlements, farms and natural scenery coexisted in symbiotic harmony. In parts of the region, this cultural landscape remains intact and undamaged. In other areas, postwar development pressures and policies have produced a different environment, one in which the traditional cultural landscape has been replaced with suburban sprawl and strip highways. This article suggests a regionalist basis for shaping the future of our rural and ex-urban areas, one in which village and hamlet zoning, informed by a knowledge of the urban design characteristics of traditional regional settlements, will provide an alternative to large lot zoning as a more effective means of preserving rural character and providing better communities.

This article summarizes a study of the formal characteristics of a selected set of regional towns, villages, hamlets and neighborhoods (Fig. 1), undertaken in an attempt to understand the implicit rules that governed their visual organization. The communities were selected based on their visual character, their "intactness" and visual identifiability, and on the basis of a designed range of types and sizes. The methodology was as follows:

- The history of each of the communities and their overall plans were recorded, documenting each plan's growth over time.
- At a larger scale, the formal characteristics of selected "component sites" was documented within each community, permitting analysis of detailed layouts and visual characteristics of selected street corridors.

Analysis methods included archival investigation and interviews with officials and citizens. Fieldwork also included making on-site sketches and measurements. Later, low-level aerial color and black and white photographs were made of each site from a helicopter. From this vantage point, the effects of large lot subdivision development could be compared with traditional cultural landscape patterns across wide areas of the State. Finally, narrative reports were illustrated with graphic documentation showing the "urban design" characteristics of their study sites.

Aerial and ground level observation provided convincing evidence that sprawling large lot subdivision development has turned out to be a poor means of preserving rural character in Maryland and an inefficient use of land in designated growth areas. While the rationale for large lot development may have been the preservation of open space, the carpet of large lot sprawl in many areas has obliterated the character of Maryland's rural countryside, replacing traditional cultural landscapes of striking beauty with vast areas of suburban sprawl. The State's historic settlement patterns suggest a preferable way to shape future rural and urban growth. The future development in village and hamlet configurations based on these precedents should be encouraged, in a frame of reference valuing both growth and the conservation of the historic cultural landscape, economic development

Credits: This article is based on work conducted by the author at the School of Architecture, University of Maryland, in a seminar entitled "Regional Small Town Paradigms," 1993 (See References for complete citation).

Fig. 2. Barnesville base map.

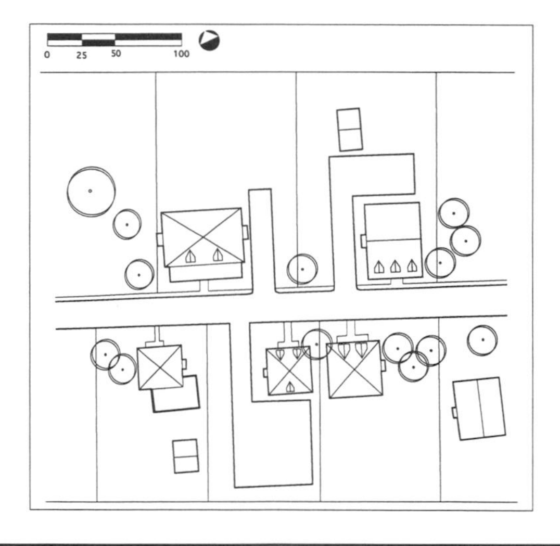

Fig. 3. Barnesville component site plan.

Fig. 4. Barnesville Road looking east: Corridor character.

and the conservation of agricultural land, and informed change and the preservation of our scenic and environmental legacy.

2 URBAN DESIGN CHARACTERISTICS OF HISTORIC COMMUNITIES

Traditional towns achieve their attributes in different ways, but all possess the following characteristics:

- The selected historic, paradigmatic rural settlements are compact and identifiable, and their boundaries are visually discernible.
- Their plans can be described as linear, crossroads or gridded, with variations designed to accommodate terrain or circumstance to achieve spatial hierarchy, or to enhance a localized "sense of place."
- They are visually coherent. Their character is established through consistent, subtle rules of formal organization and architectural language (conventions of composition, style, materials, use of component parts such as porches, ornament and detail).
- They possess a strong degree of spatial hierarchy. (For example, town centers are often marked with public spaces; local neighborhoods often have their own, less formal public open spaces.)
- Their street corridors are visually bounded, "layered" and intimate in feeling. The public realm is thus improved. At the same time a sense of privacy for individual houses is enhanced.
- Their street blocks can be understood as comprising their component neighborhoods, suggesting the role of the street as a "social channel" of neighborly interaction.
- They accommodate a mix of uses, even at the hamlet scale.
- They typically include a range of housing types.
- Parking is accommodated in a mix of on-street and off-street strategies. Large-scale parking lots are rare, and anomalous.
- Most important, the towns, their neighborhoods and their settings convey a strong "sense of place."

Some of the typical visual components of our traditional settlements are narrow roadways, street trees, sidewalks, "layered" front yard plantings, "layered" architectural designs, sometimes utilizing front porches, and relatively closely spaced structures on lots narrower than those conventional in current subdivision layouts. Traffic is controlled and managed through a variety of devices, including street width and discontinuous grid patterns. No *cul-de-sacs* are employed, however, except in areas developed after World War II.

Settlement types

For convenience, the study settlements are divided into four categories: hamlets, villages, towns, and a traditional suburb. Nearly all towns began as hamlets, so the first three of the categories represent "growth over time" morphology. Based on the examples, three working categories were defined as follows:

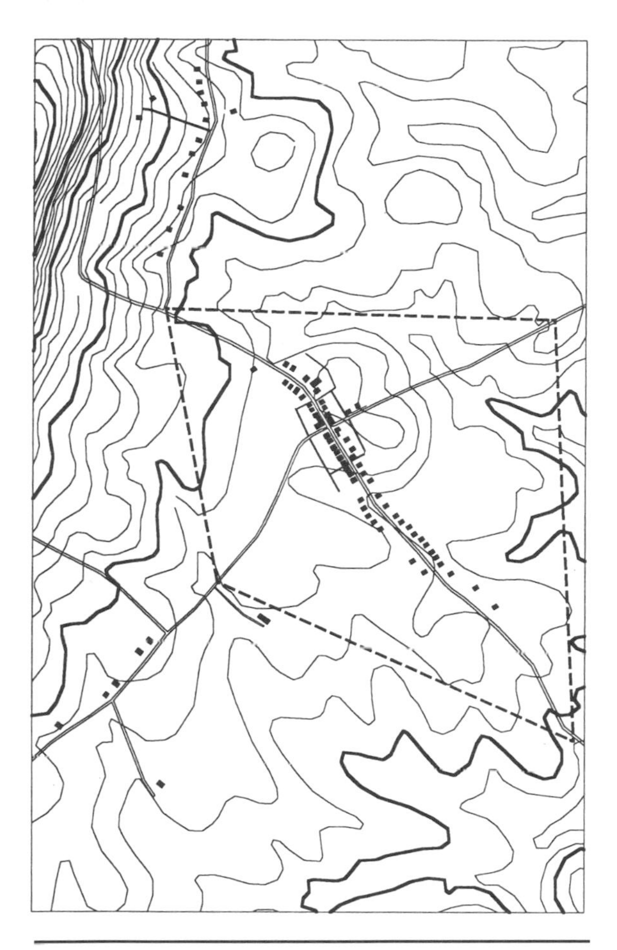

Fig. 5. Burkittsville town plan.

Fig. 6. Burkittsville component site plan, Main Street.

Fig. 7. Burkittsville Main Street, looking east.

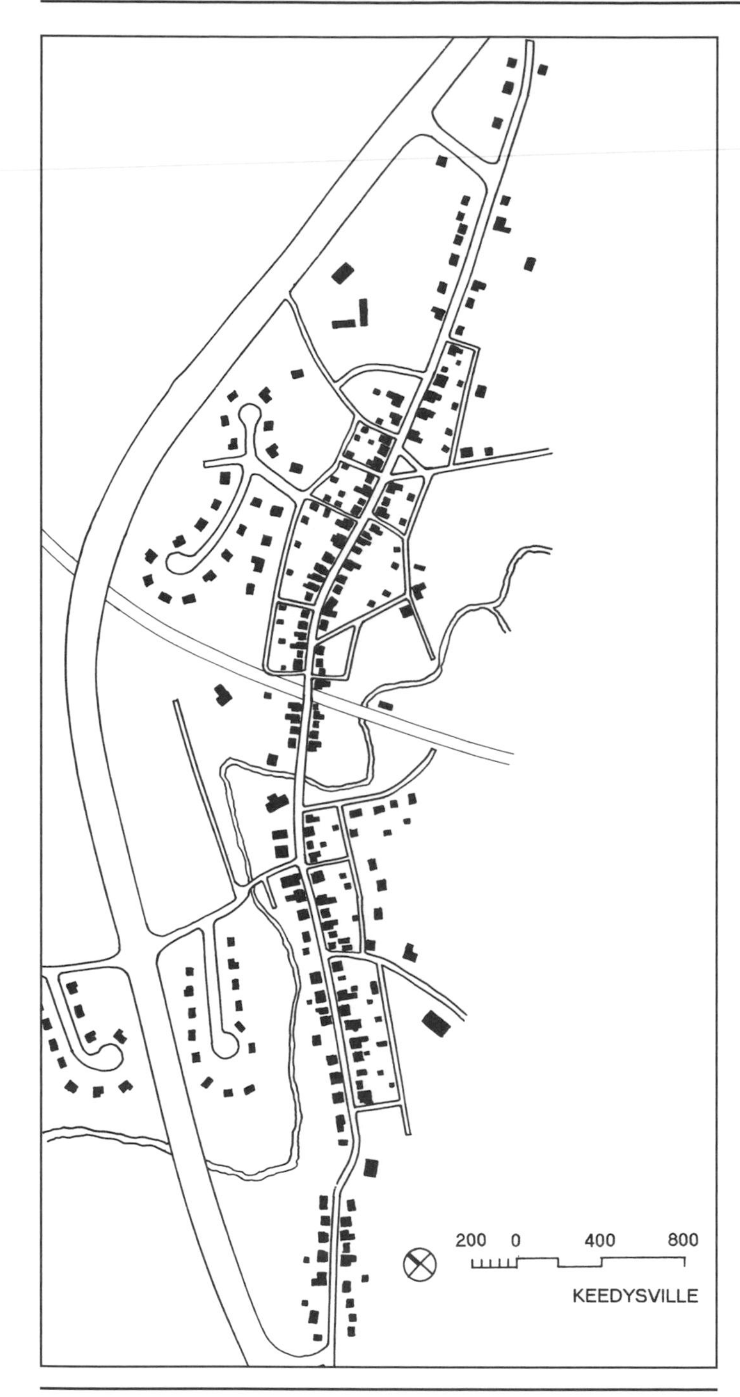

Fig. 8. Keedysville town plan.

Hamlets: compact, discernible settlements of 25 to 60 separate structures, with no, or a very small number of, commercial enterprises. (Example: Barnesville, Fig. 2.)

Villages: compact, discernible settlements of about 50 to perhaps 300 separate structures, accommodating a half dozen to several dozen commercial occupancies. (Examples: Burkittsville, Fig. 5 and Sharpsburg, Fig. 12.)

Towns: compact settlements larger than villages, containing several or a number of neighborhoods. Towns by definition have town centers (downtowns), and often play a role in governance as a jurisdictional center or subcenter. (Examples: Easton, Fig. 14 and Chestertown, Fig. 16.)

Linear plans

The village of Keedysville represents a clear example of a linear plan (Fig. 8). The road's traffic in the village's early decades created an economic growth opportunity. More recently, it threatened to overwhelm the settlement, and a highway by-pass was created.

Crossroads

Burkittsville and Barnesville are also mostly organized in linear plans, though one or more secondary roads cross both their main streets.

Pure linear plans, where there is no crossroad, do not have inherent centers where the potential locus of commercial and public activity is obvious.

When the opportunity presented itself, early roads were planned to follow ridgelines, or were located adjacent to a stream in a river valley. Good roadway drainage was thus provided to one or both sides of the road, making travel conditions in wet weather less muddy. Linear settlements built along such roads enjoy inherently good surface water drainage conditions, explaining one circumstantial advantage of their locations. (Sharpsburg is an exception. There, the main road through part of town follows a declivity in the terrain, turning the road into a surface water swale, not a recommended situation.)

Grids, distorted grids, and broken grids

A grid plan provides an "imageable" location map, and maximizes alternative circulation routes. The problem of grids is inherent: there is no implicit center or location hierarchy.

Classic Roman planned towns dealt with this lack of central focus by designating one central street the *cardo maximus* and the central crossing street the *decumanus maximus*. Their crossing provided a locational and hierarchical center, typically celebrated with a civic open space. Lesser hierarchical locations were usually provided along the streets of the grid by providing "exedral" spaces to one side of the road. Buildings located along the *cardo maximus* or *decumanus maximus* obviously enjoyed higher locational standing than those on other roads. Chestertown is planned in almost exact accordance with the classical Roman model.

Grids also possess the capacity for the accommodation of pattern distortions, such as circles, semicircles and curves, which can be employed to provide a sense of neighborhood location or spatial hierarchy.

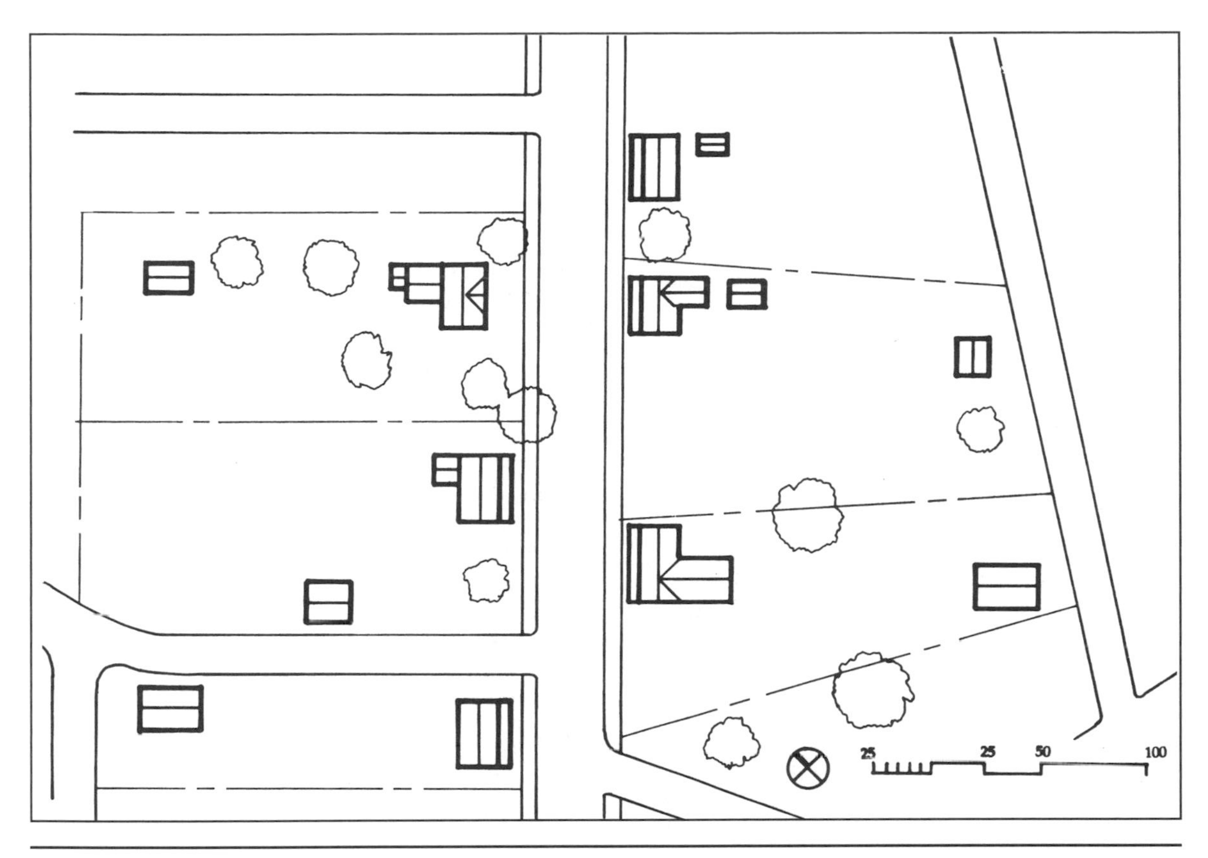

Fig. 9. Keedysville South Main Street architectural character.

Fig. 10. Keedysville North Main Street component site plan.

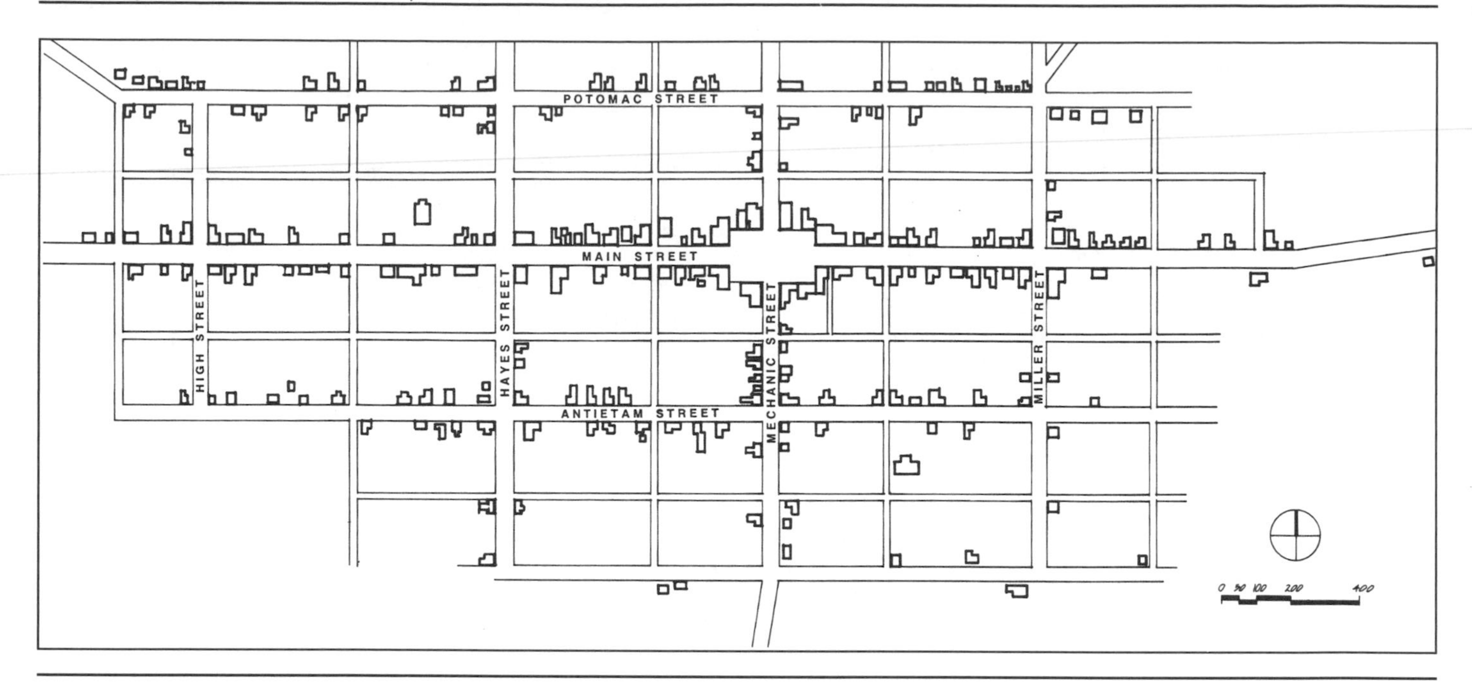

Fig. 11. Sharpsburg town plan, 1877.

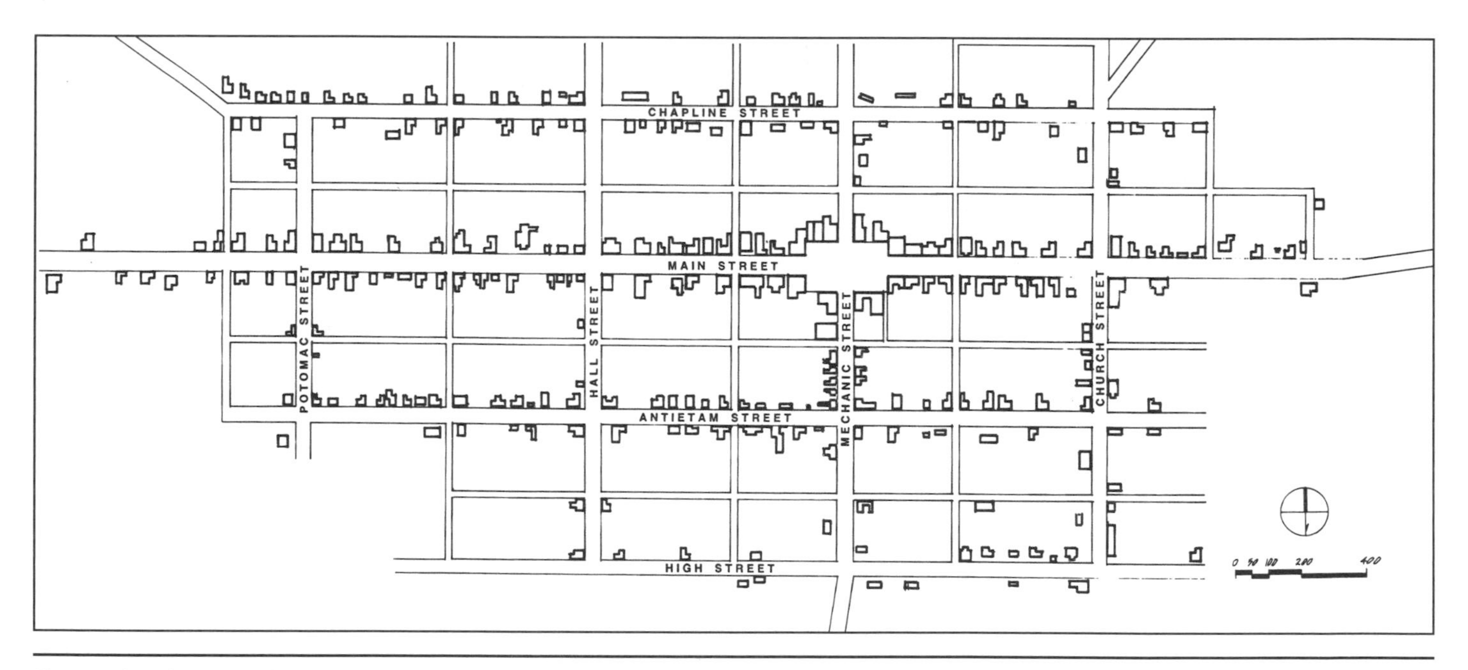

Fig. 12. Sharpsburg town plan, 1993.

Fig. 13. Sharpsburg West Main Street looking west.

A grid plan's inherent provision of numerous alternative circulation pathways can also pose a problem in the age of private cars and trucks. Every street can become a through street, unless distortions are introduced or the grid is broken. The discontinuous grid facilitates the local traffic's internal circulation while discouraging through traffic. The resulting visual character also contributes to a sense of visual closure in neighborhoods.

Hierarchy and open space

Villages have centers; towns have downtowns. Local neighborhoods need open space for recreation and to nurture a sense of community identity. Traditional towns, through plan disposition and distortions, and through the provision of centralizing and exedral figural space, achieve a sense of spatial and locational hierarchy.

Most of these spaces are square or rectangular. In neighborhoods, open space tends to be shaped more circumstantially, and casually.

Land use

Another important characteristic of traditional settlements, in contrast to conventional suburbs, is that there is an intimate mix of housing types and a presence of appropriately scaled commercial buildings. In traditional settlements, the range of housing types and commercial activities is housed in a built environment where continuity of scale seems to be the governing factor.

In all of the traditional towns presented here, the visually contained street corridor is the essential component. It constitutes the basic ordering devise of the traditional town, and provides its defining imagery.

Parking and planting

Neighborhood streets in these towns facilitate and control traffic circulation, function as social channels, and provide for parking. In traditional towns in Maryland, parking is accommodated through a variety of strategies, including on-street and off-street parking of the types described above, and small parking lots. However, the large fields of parking common to strip shopping centers and regional malls are almost never seen. The dominating value seems to be the preservation of the street corridor as a visual entity, obviating large-scale parking lots.

Visual character and identity

The characteristics of Maryland's traditional towns include an identity stemming from compactness, boundedness, visual coherence and memorable street corridors, which look like, and are used, as neighborhood open spaces. It's worth discussing the design components of these characteristics a little further.

Edges

Except for Stoneleigh, the towns presented in this publication have a major defining characteristic in common: they are visually bounded. Even Stoneleigh has "edge" definition as a community, stemming from the collector and arterial roads that surround it, and from its perceptible coherent visual character. Each of the other towns has a perceptible edge, where the fabric of the settlement meets natural or cultivated open space.

Formal coherence

Traditional towns achieve formal coherence with a variety of strategies,

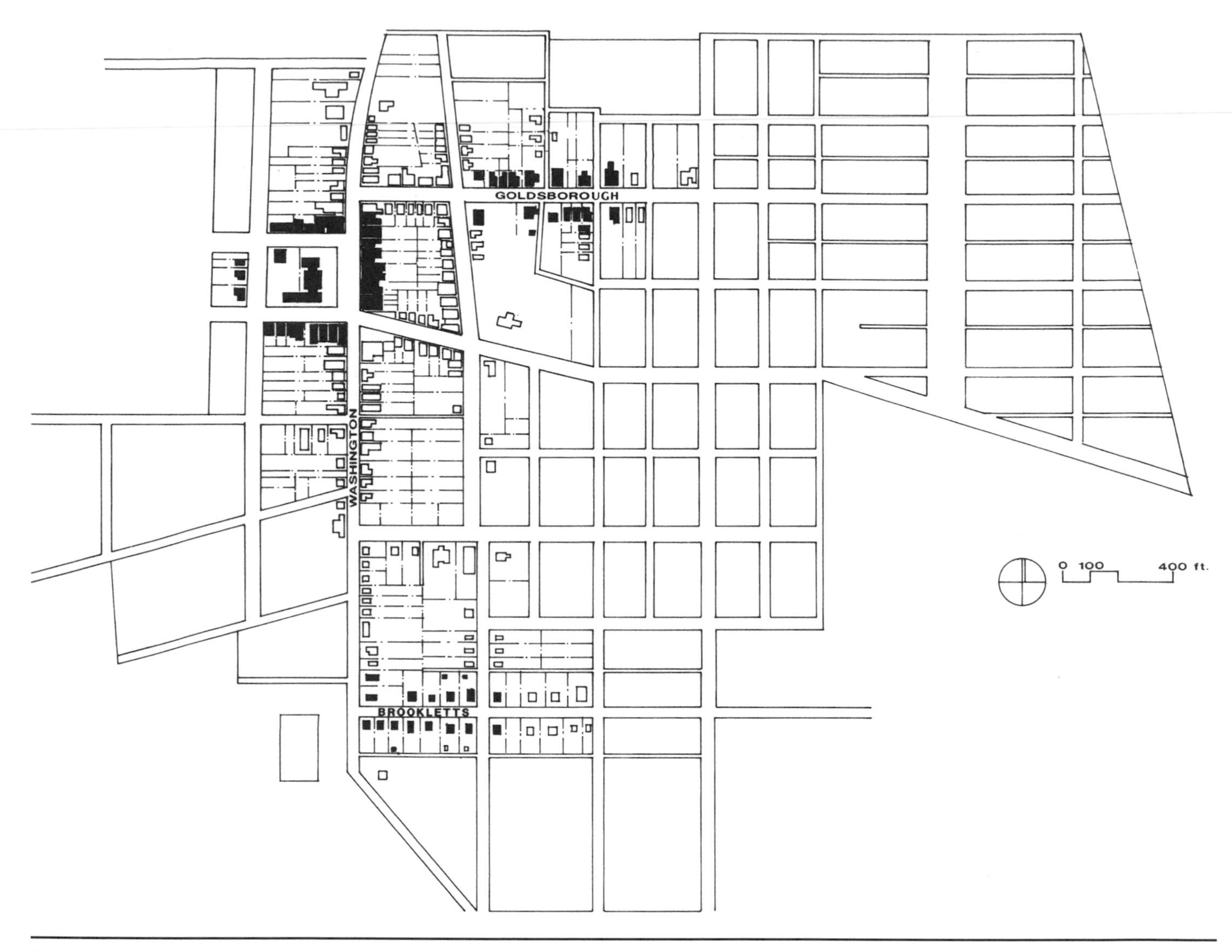

Fig. 14. Easton component site location map.

including house-lot-street relationships, building orientation and the use of consistent architectural language. All of the towns tend to be built with buildings situated on a common setback line, or in a close range of setbacks.

Architectural language

The formal coherence that typified traditional towns depends upon urban design constraints and the formal organization and orientation of buildings. But its achievement is also supported in the subtle and consistent architectural language of buildings. In traditional towns, coherence seems to have been originally achieved in an unwritten code of covenants, a sort of visual "social contract."

Visual closure

We've already talked about visual closure in the form of the "boundedness" of a settlement, and in the form of the visual definition of the street corridor. One more form of closure might be mentioned: the definition and closure afforded a neighborhood residential or commercial block through the "T" intersections of a discontinuous grid, or through the employment of grid anomalies such as curving streets. The grid of Stoneleigh and the view down Stoneleigh Road (Fig. 21) suggest the intimacy of the visual environment produced by such devices. Visual closure is one means of achieving the compact and intimate character common to traditional towns.

The following descriptive material comprises a virtual catalogue of traditional neighborhood elements, with no single rule of composition predominating. In general, however, in comparison with today's conventions, building lots in these traditional settlements are smaller, and setbacks, cartways and street corridors are narrower.

3 SEVEN MARYLAND EXAMPLES

A Hamlet

Barnesville

Barnesville, in Montgomery County, is a linear crossroads hamlet of some fifty residences, located at the crossing of Barnesville Road (its main street) and Md. Rt. 109. Once a farming community, its population hovers around 167 persons. A uniting element is the lining of

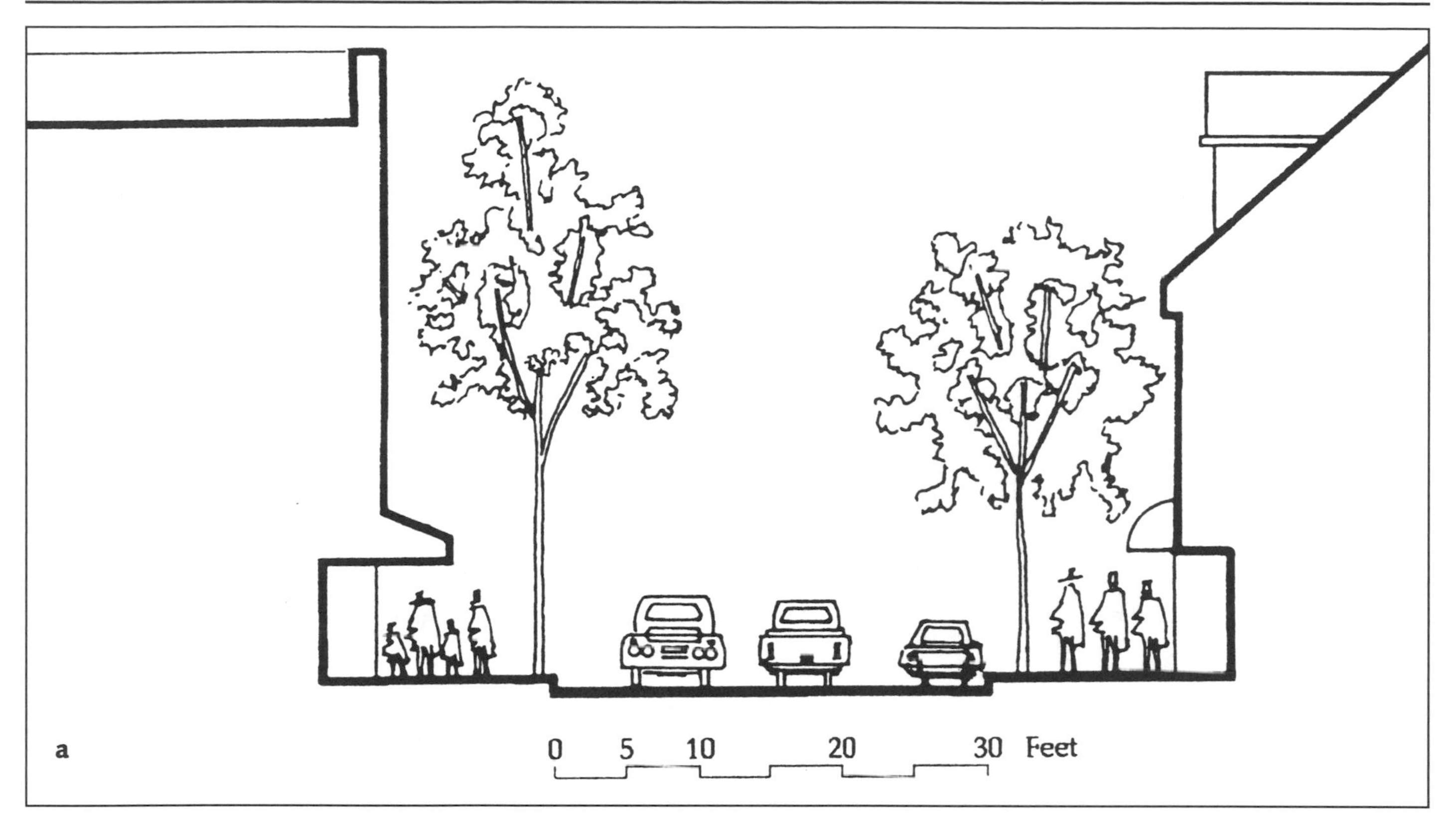

Fig. 15. Easton Washington Street (a) looking north,(b) architectural character.

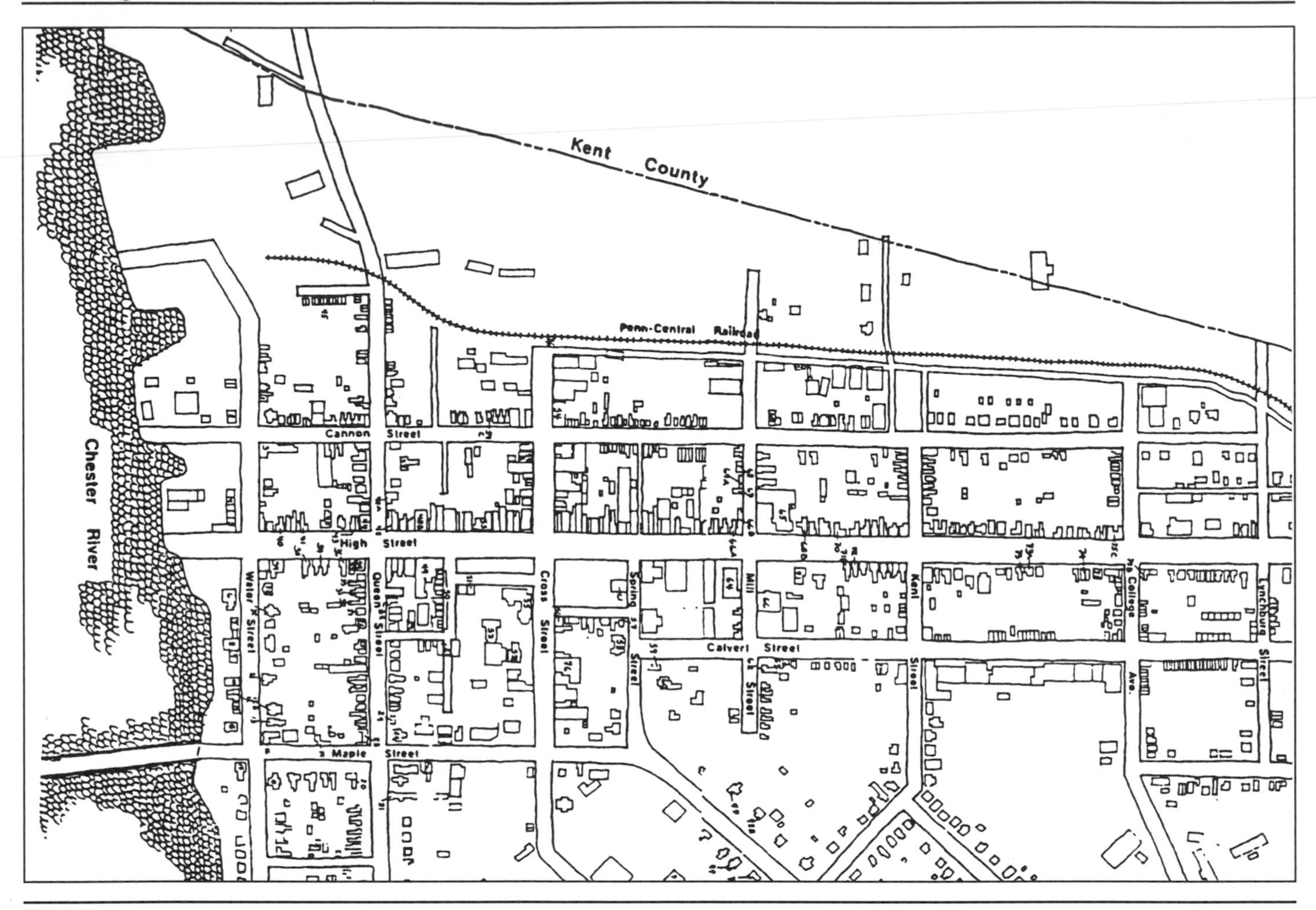

Fig. 16. Chestertown partial plan, 1993.

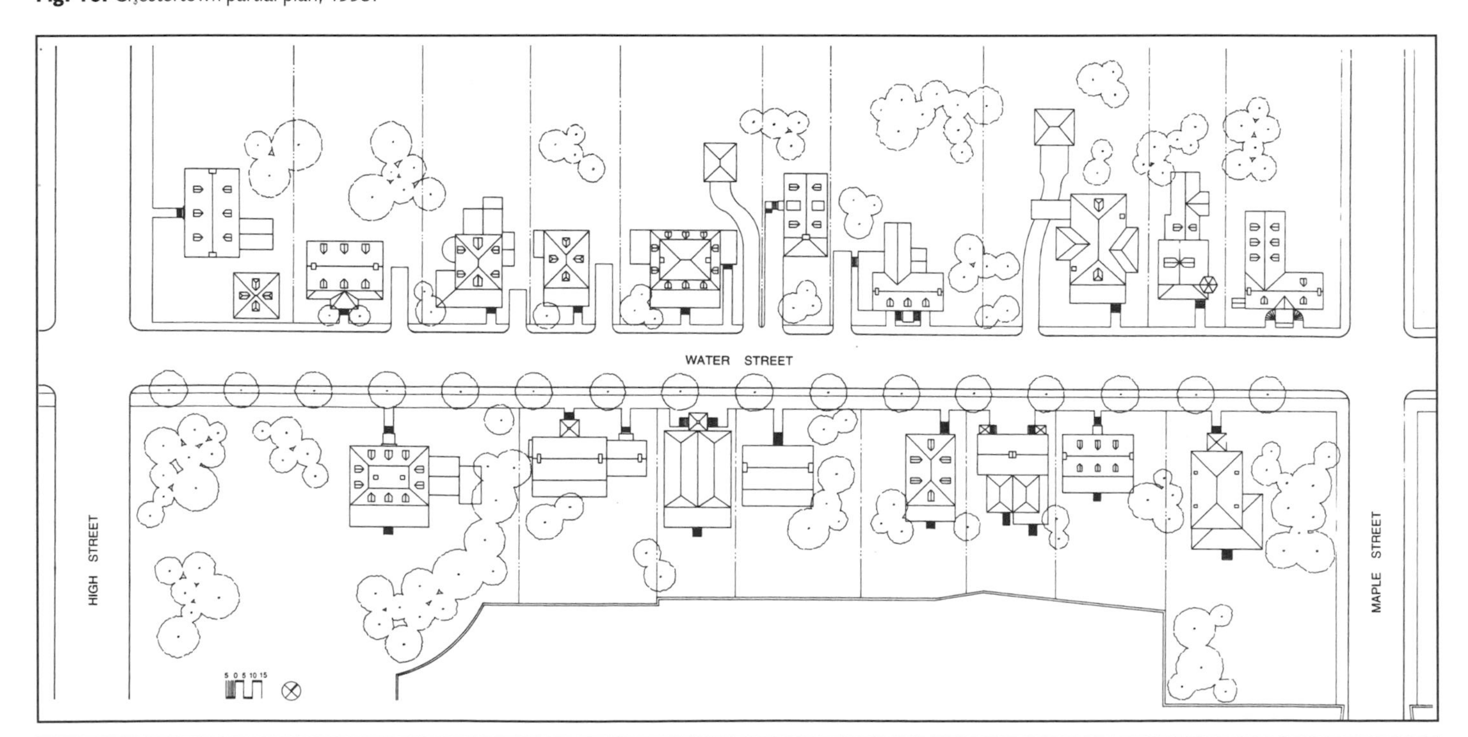

Fig. 17. Chestertown component site plan.

Fig. 18. Chestertown Water Street architectural character, south side.

the street corridor with front porches. Front yard depths and treatments vary. A number of houses are layered from the street with hedges and other plantings. With a varied palette of architectural languages and materials, the town's cohesion is established mostly by the narrow street corridor, the parallel orientation of roof ridges, and by the street trees, some of which are large enough to arch the roadway.

Street-corridor width varies, with a 42-ft. (12.8 m) minimum between house fronts. The occasional house is set back beyond the typical 11-ft. (3.35 m) front setback. Spacing between houses tends to be greater than setbacks from the roadway. Barnesville Road's pavement width is 22 ft. (6.7 m). There is a 40-in. (1 m) sidewalk along the north side of the main road. Street trees line the roadway at a spacing of about 10 ft. (3 m), closer-spaced than typical in contemporary practice, maintaining a feeling of definition and closure along the street corridor and contributing to the perception of the town as a coherent whole. Also contributing visual consistency in the townscape is the arrangement of most structures with their ridges paralleling the road.

Villages

Burkittsville

Burkittsville is a small early 19th century village located in Frederick County, at the intersection of Burkittsville Road (Maryland Route 17) and Main Street (called Gapland Road outside of town) in Burkittsville. Rolling countryside, with open meadows and fertile farm fields, beautifully surrounds the setting. The town center, with a tight cluster of structures, is situated at the crossroads. A mix of residential, commercial and institutional structures lines Main Street. Then, as at the time of its founding, Main Street backed up onto a rural landscape.

Keedysville

Keedysville is a narrow, linear village in Washington County located along the original main road from Boonsboro to Sharpsburg. Its character has been protected in recent years by the construction of a bypass along Maryland Route 34, directing through traffic around the town. The greatest impetus for Keedysville's growth came in 1867, with the construction of the Washington County branch of the B&O Railroad. The now abandoned rail line bisected Keedysville's Main Street: a small piece of track is still visible, marking the town center. Interestingly, the rail line and Main Street were perpendicular to each other, the latter having predated the former.

Sharpsburg

Sharpsburg is located three miles south of Keedysville. Its grid plan comprises eight streets, with a tiny centralized town square. The square, obvious in plan, is less evident from street level: the cartway width is unchanging, and the 8 foot setbacks which define the square are disguised by a partial planting of evergreens which continue the adjacent street walls. Each of the streets is the same width, 32 ft. (9.8 m).

The contemporary population is under seven hundred. Presently, the town enjoys some prominence as a tourist destination because of its location, centered on the site of one of the bloodiest battles of the Civil War, the Battle of Antietam. The town is very cohesive in character. A constant, narrow street corridor is everywhere maintained. Almost all buildings present their eaves to the street. Structures are all two or three stories in height. Most houses have porches, in various widths. Densities vary from about 2.5 to 2.8 dwelling units/acre (6.2 to 6.9 du/ha); building to building dimension, 65 ft. (19.8 m) except

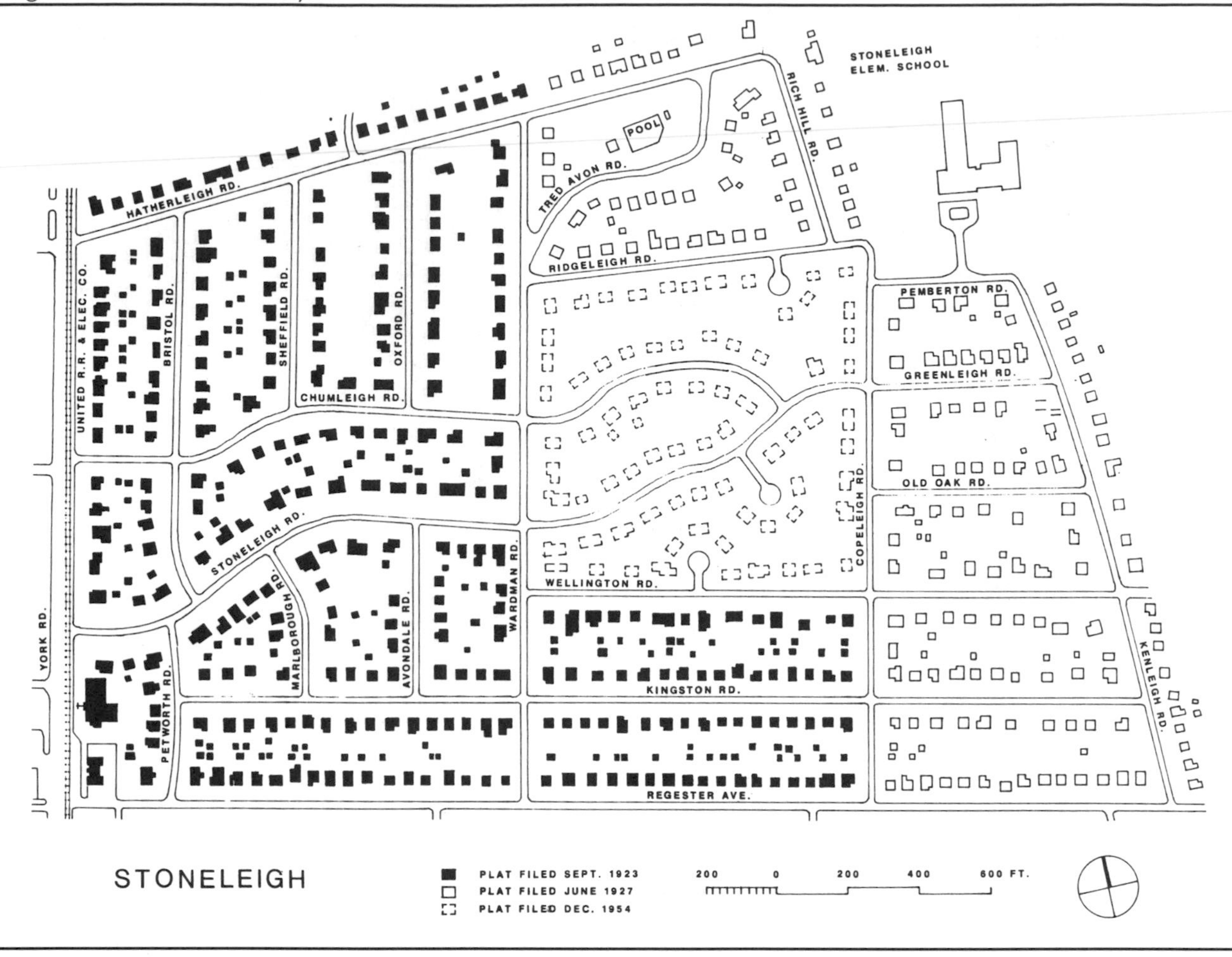

Fig. 19. Stoneleigh neighborhood plan showing growth over time.

at the town center, where it is 105 ft. (32 m); sidewalk width, 6 ft. (1.8 m); street tree planting (West Main Street), 30 ft. (9.14 m).

Towns

Easton

Easton is a Colonial Planted Town, laid out before the Revolution. Its history, however, antedates its planning. In 1684 Quakers built the Third Haven Meeting House, twenty-six years before the town was platted. About half a mile from the Meeting House, two acres were commissioned for a courthouse, and the true seed for the town was planted.

A noteworthy aspect of the original plan has been demonstrated in its adaptability to changing conditions; most of the ground floor spaces of facing structures are now occupied by commercial uses, with residential and office uses on second and third stories.

Chestertown

Chestertown, also a Colonial Planted Town, is situated on the Chester River, a little over 32 miles (52 km) north of Easton. The historic district of Chestertown is given over to commercial, office and residential use on High Street and Cross Street, with neighborhoods of relatively small houses on the other streets. An exception to this rule is the residential block along Water Street, where mostly three-story houses face the street to the north across narrow front yards, and the river, to the south, across wide lawns and gardens. The river facades of these houses, mostly three-story porches, give the town a noble face along the river, and dramatize the approach from the south.

A metropolitan suburb

Stoneleigh

Stoneleigh is the only metropolitan suburb considered in the study. Laid out in sections between 1923 and 1954, it comprises an archetypal traditional "edge city" suburb. The first streets were planned as a grid, except for Stoneleigh Road, which followed the curving path of the existing Stoneleigh House approach drive. Stoneleigh's second phase, platted in 1927, followed the gridded pattern. The final phase, platted in 1954, and occupying the site of newly demolished Stoneleigh House, was laid out in gently curving streets and three short cul-de-sacs.

AN ALTERNATIVE FUTURE FOR MARYLAND'S ENVIRONMENT

This study examined the relevance of historic models to the shaping of new, rural development in hamlet and village configurations. Their

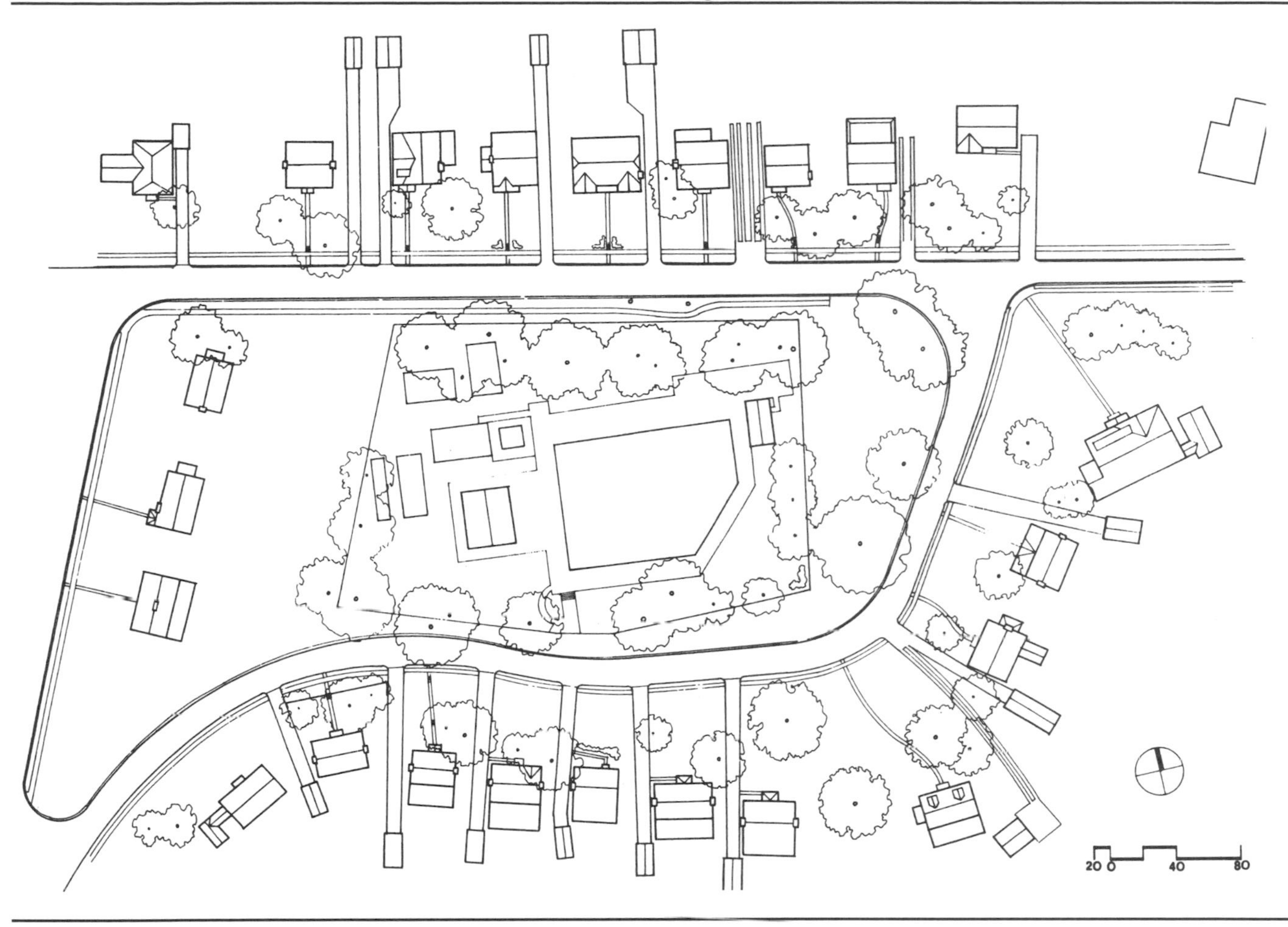

Fig. 20. Stoneleigh pool component site plan.

Fig. 21. Stoneleigh Road architectural character.

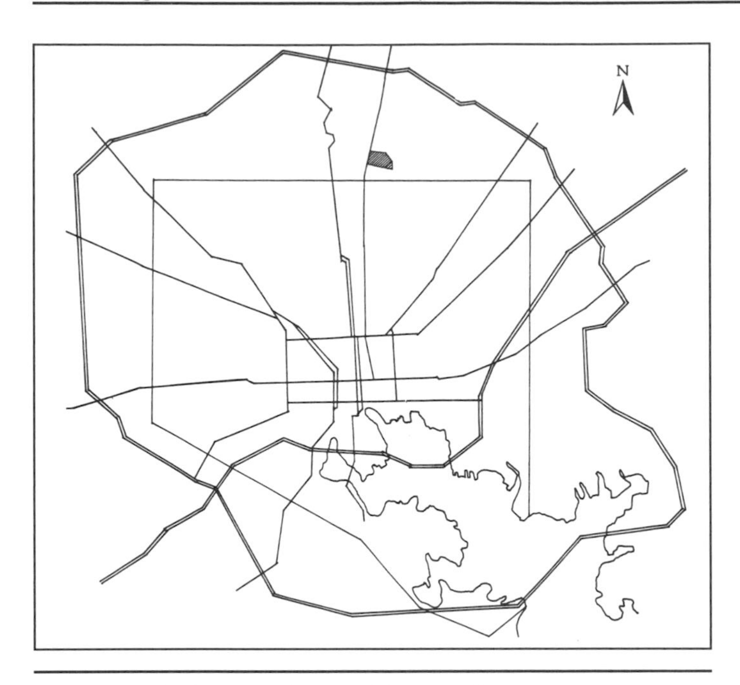

Fig. 22. Stoneleigh location relative to the City of Baltimore.

findings suggest that a carefully crafted hamlet and village development strategy, based on regional traditions and paradigms, can do much to protect existing historic cultural landscapes in Maryland. In compact hamlet and village settlements, new growth can be accommodated in a manner that will preserve the character of historic settlements and their settings.

In areas already impacted by suburbanization and spot development, a hamlet and village development strategy might be employed to give form, structure and a sense of hierarchy to what otherwise will inevitably become an environment of more or less continuous sprawl and strip highways. Judiciously placed new villages, buffered with open space would "center" otherwise amorphous communities, and provide opportunities for a localized concentration of services. Their development would expand the rage of housing accommodations in rural and suburbanizing parts of the State, and perhaps enhance the suburban population's sense of belonging to more comprehensible communities.

New growth around existing settlements, and freestanding new settlements located in important, surviving cultural landscapes, and even in suburbanizing areas, should be consistent, in their urban and architectural order, with the historic, regional context.

Summary

This study leads to the conclusion that a promising option for the long term preservation of the region's remaining rural, cultural landscapes and for the shaping of a better environment future for Maryland lies in the encouragement of new land development in hamlet and village configurations. (See Box A.)

One of several ingredients in making growth areas attractive is to use physical design elements to create functional neighborhoods—neighborhoods that work not only on a physical level, but on a social and aesthetic level as well.

Ten common visual design characteristics are identified in the case studies in this population. These characteristics are restated below as guidelines for creating quality neighborhoods:

- Neighborhoods should be compact and identifiable, and their boundaries visually discernible.
- Neighborhood plans should be comprehensible. For example, plans might be linear, crossroads or gridded, with variations to achieve spatial hierarchy, or to enhance local visual assets.
- Neighborhoods should be visually coherent. Character is established through consistent rules of organization and architecture.
- Neighborhoods should possess a strong degree of spatial hierarchy.
- Street corridors should be visually bounded, "layered," and intimate in feeling. Street trees, sidewalks, and front yard design elements can create visual layers and contribute to the intimacy of the streetscape.
- Street blocks should be understood to describe component neighborhoods, suggesting the role of streets and alleyways as a channel for neighborly interaction.
- Communities should accommodate a mix of uses, even at the hamlet scale.
- Communities should typically include a range of housing types.
- Parking should be accommodated in a mix of on-street and unobtrusive off-street strategies. Large-scale parking lots should be avoided.
- Most important, neighborhoods and their settings should convey a strong "sense of place." ■

REFERENCES AND CREDITS

Hill, John W., principal investigator. *Managing Maryland's Growth: Modeling Future Development on the Design Characteristics of Maryland's Traditional Settlements.* Jointly published by Maryland Office of Planning and the School of Architecture, University of Maryland, College Park, Maryland. 1992.

Faculty and students participating in the study and preparation of drawings are: Barnsville: Marc Burlinson, Ray Connor, Adam Hird, Brian Milnick; Burkittsville: Kyra Tallon, Daphne Quinn, Ron Shaeffer; Keedysville: Ted Strosser, Tim Denee, Mike Seibert; Sharpsburg: Sonja Shields, Rob Duckworth, Steve Teiler; Easton: Komal Bhatia, Chris Edsall, Olukayode Nejo; Chestertown: Andrew Hryniewicz, Paul Klee, John Wright; Stoneleigh: John Hill, Martin Towles. The comprehensive plan model for Traditional Neighborhood Development (Box A) was written by Bruce Bozman, Larry Duket and Mike Nortrup of the Maryland Office of Planning.

BOX A: Comprehensive plan model for Traditional Neighborhood Development

Goals Element. Any update or revision of the Comprehensive Plan should begin with the identification of goals, objectives, policies, and standards, since these statements establish the basic framework for the overall development philosophy of a jurisdiction. Following are examples of how Traditional Neighborhood Development (TND) principles might be incorporated into the Goals Element of the Plan:

1. Goals and objectives

To encourage the wide use of TND principles as one means of creating attractive living environments in our growth areas and in those rural population centers designated for growth.

To require the use of TND in growth areas and rural population centers where we recommend protecting a defined community character or seek to create a "traditional neighborhood."

To amend land development regulations to remove unwarranted obstacles to utilizing TND principle.

To foster a strong sense of community and other aspects that will make growth areas attractive to our citizens.

2. Policies

Encourage the use of narrow streets and alleys.

Encourage on-street parking to moderate vehicular speed and provide separation for pedestrian safety.

Encourage a grid street pattern.

Discourage the indiscriminate and random use of curvilinear street patterns and cul-de-sacs except as may be needed to avoid impacts to sensitive areas or to account for topography. Encourage the use of discontinuous street grids to control through traffic, rather than the use of numerous cul-de-sacs.

Allow narrow lots and shallow setbacks.

Promote a mixture of complementary land uses.

Encourage the creation of an environment that is "pedestrian friendly."

Encourage a wide range of housing types in an effort to promote socio-economic diversity and inclusiveness.

Permit higher densities in an effort to create a village atmosphere.

3. Standards

Since each traditional community is unique, it is not possible to recommend a uniform set of development standards that will serve as a model for every new TND development. The development standards recommended in the plan can be based on the standards used in an existing TND community. Typically, development standards in traditional neighborhoods would encourage shallow or no setbacks, narrow streets and alleyways, mixed uses, narrow lots, high densities, greater pedestrian activity, formal open spaces, and consistent architectural character. Generally, standards need to account for differences between rural and urban areas. For example, achievable densities would be largely controlled by whether public sewer and water exists; standards might also vary depending on aspects of community character.

Land Use Element. The Plan's Land Use Element, which discusses the major land use and development issues facing a jurisdiction and recommends the optimal future land use pattern, should be revised to incorporate appropriate references to TND. The traditional neighborhood development pattern (as described by the ten principles) should be recommended as a development technique in designated growth areas and in existing rural villages and towns.

The Land Use Element should consider whether there are areas where TND should be required, as opposed to merely "encouraged." This element could be used to address growth areas having unique character that could benefit from TND principles. The element should consider the issue from the perspectives of creating attractive new neighborhoods, protecting and expanding existing ones, and using TND along with other tools for unique issue—such as historic preservation and sensitive areas protection. Finally, this element should note that more flexible development standards are necessary in order to allow the narrow lots, higher densities, mixed uses and other features of TND.

Transportation Element. The Transportation Element of the Plan should be amended to incorporate standards that encourage the use of traditional neighborhood design principles. Street widths and minimum radii need to be reduced and provisions made to encourage on-street parking. The element should encourage the use of a grid street pattern and alleys. Multiple and redundant circulation and points of access are thus provided. Methods for ensuring pedestrian safety and circulation are needed, as well as means for creating character along the streetscape. The use of tree planting strips between sidewalks and travel lanes is a good method for addressing these issues.

Village street widths also represent a departure from typical suburban subdivision standards. While street widths differ greatly depending on local preferences, most sources recommend widths for local streets ranging around 20 ft. (6 m)—two travel lanes, no parking, or a one-way street with one parking lane. Even if parking is permitted on both sides, street width should not exceed 30 ft. (9.14 m) on-lot parking now required in all codes, on-street parking should be sporadic enough to permit oncoming cars to pass, even if some "weaving" is required. The objective should be to slow down and control vehicular traffic, not to increase its speed. On a street with commercial uses, however, where on-street parking is combined with larger traffic volumes, a four-lane width of 32 to 36 ft. (9.75 to 11 m)—two travel lanes, two parking lanes—may be needed.

Alleys are a key element in the local street pattern. Where lot widths are narrow, ranging from 40 to 60 ft (12.2 to 18.3 m), alleys are an alternative to multiple curb cuts for individual driveways, thereby providing more room for on-street parking on the main street. By removing the driveway from the front yard, alleys reduce the visual impact of the automobile; they can also be used to carry utility lines, to take trash collection activities off the main street, and to give children a sheltered play network removed from traffic.

Community Facilities Element. The Plan's Community Facilities Element should incorporate TND principles, which encourage additional open space through the creation of village greens, squares, and parks. The Plan should support the integration of these formal open spaces into development projects.

Community Character or Design Element. Since a number of the guiding principles of TND development involve architecture and design, a community may wish to prepare a separate Plan element that focuses on design guidelines. This element would address a number of design issues such as formal coherence, spatial hierarchy, layering, boundedness, edges, visual closure, and sense of place. This element should also establish policies and recommendations for the protection of historic character and historic structures that may be affected by new development. Traditional neighborhood design can be used to integrate historic structures into a project and to complement historic character.

Implementation Element. The Implementation Element of The Comprehensive Plan should recommend that the land development regulations (for example, the zoning ordinance and subdivision regulations) be revised to incorporate the TND guidelines recommended in the Plan. The Implementation Element should identify and recommend the removal of any unwarranted regulatory obstacles to the development of traditional neighborhoods. The Plan should promote the adoption of more flexible design standards and offer incentives to encourage the development of new traditional communities. This element should also follow through with the concept of mandatory TDN if the Land Use Element recommends this approach. It is possible that within a single jurisdiction there are areas where it should be encouraged and areas where it should be required.

Robert D. Yaro and Raymond W. Gastil

Summary
This article offers a methodology of presenting regional planning principles, utilizing aerial perspectives of significant regional and urbanizing locations in terms of existing conditions, and simulations or visualizations of future growth determined by present zoning and "build out" assumptions compared to alternatives utilizing compact and mixed use centers. Five examples from the New York Tri-State Region are shown, representing a range from semirural, suburban to urban conditions.

Keywords
beltway, commercial strip, neighborhood, rail transport, regional planning, suburb, urban revitalization, visualization

Visualization of alternative development futures (Yaro *et al.* 1988)

Visualizing a region's future

In the 1990s, New York based Regional Plan Association (RPA) undertook a project to establish regional design principles that would encourage compact and mixed-use centers in the next generation of growth in the States of New York, New Jersey, and Connecticut. Planning principles were developed through meetings that included land use law experts, community representatives, developers, architects, urban designers, and landscape architects. Regional design principles were visualized through "simulations" consisting of three drawings, or "triptychs," showing existing conditions, typical future development, and recommended future development. This article presents five illustrative examples of the method and the resulting regional planning recommendations.

The examples and simulations are based on real sites and typical conditions within the Tri-State Region. These include difference places, from dense urban locations to rural forest and farms, linked by their interdependent transportation, economic and natural systems, ranging from the commuter railroads and interstate highways to the Atlantic Shore and forested highlands—all of which cross municipal, country and state boundaries.

Each example provides an aerial perspective, as the place exists today, how it would appear in 2015 if typical trends continue, and how it would appear in 2015 if future development followed new regional design principles.

Five cases are presented as visualizations, or simulations (out of the full set of fifteen in the study). The following commentary offers a working definition of "simulation" and its relation to planning and design.

Visual simulation has been revived in the past two decades of planning and design practice, largely due to two factors:

• First, a renewed conviction that the public should have a role in the design of their communities, that their design judgments were a valid part of the process, and that design professionals have a responsibility to describe future design realistically and as engagingly as possible.

• Second, the belief that a new synthesis of photography, video, and computer imaging has the capacity to give the public a greater understanding of the future of their communities.

Simulation, as used here, attempts to imitate and represent visual experience as accurately as possible. To do so, it cannot communicate

Credits: This article is excerpted from "Visual Simulations: The Future of the Tri-State Region." *The New City*, winter issue 1993-94, the University of Miami School of Architecture. Dodson Associates executed the drawings. See endnote for additional credits.

Fig. 1. Rapidly developing urban fringe, existing conditions.

Fig. 2. Rapidly developing urban fringe, after typical development, 2015.

through the basic abstract tools of architecture and engineering, *i.e.*, plan, section, isometric and axonometric drawings. Instead, it uses the immediacy and relative accuracy of perspective, whether in drawing, photography, computer modeling, and ultimately virtual reality—to engage public understanding of the natural and built environment.

The visual simulations used by RPA can be compared to the landscape tradition of Humphrey Repton in the late 18th century, or to the more abstract renderings of urban alternatives by Pugin in the 19th century. What further distinguishes simulation as defined here is the conviction that the relationship between existing and future conditions should be rendered as directly as possible: the accuracy and meaning of the images of today, tomorrow and an alternative tomorrow benefit by the control of maintaining precisely the same view. In community and professional meetings, the triptychs gave a direct experiential reality to otherwise abstract zoning and development discussion.

Rapidly Developing Suburban Fringe

Existing Conditions (Fig. 1)

This prime rural farm and forest land is under extreme growth pressure, already sold to a developer and located less than five miles from a major commercial artery in a rapidly developing section of the region.

2015, After Typical Development (Fig. 2)

Houses on half-to-one-acre lots have been developed along new and wide (30 ft. (9 m) driving section or more) suburban roads. The pattern destroys the rural character and the environment that attracted buyers to the area in the first place, and urbanizes prime rural farm and forest land.

2015, After Recommended Development (Fig. 3)

The same housing type and lot sizes originally chosen by the developers have been reorganized along a narrower, more clearly defined road system. Lots keep the same total area as in the previous scenario, but have been reshaped to become narrower and longer. This creates more expansive backyards which can be organized into private spaces near the houses and into a common open space network in the center of the block.

Fig. 3. Rapidly developing urban fringe, after recommended development, 2015.

Fig. 4. Beltway Interchange, existing conditions.

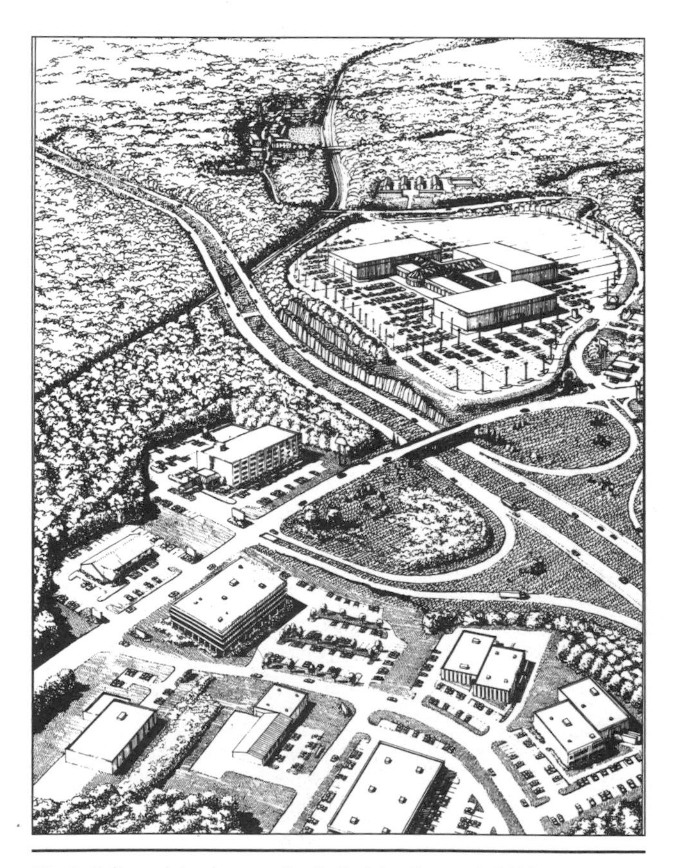

Fig 5. Beltway Interchange, after typical development, 2015.

The uniformity of lot sizes and building types in the previous scene is varied by creating lower density lots at the outskirts and higher density development in the center, including a new school, a small commercial district, and offices. Narrower streets and building setbacks help create a stronger sense of neighborhood.

Beltway interchange

Existing Conditions (Fig. 4)

A new freeway has recently been built through an outlying area of the metropolitan region. An interchange provides access to an older town center located along a major metropolitan rail line. The surrounding countryside consists of extensive forested lands, reservoirs supplying the city's drinking water, scattered suburban style homes, estates, and a few remaining farms.

2015, After Typical Development (Fig. 5)

Existing municipal zoning allows scattered commercial and industrial development to line both the freeway and nearby local roads. A regional shopping mall is built in the middle ground, while hotels, fast-food restaurants, offices, and industrial buildings are developed in the foreground. Forest land along the highway is replaced by development. Traffic congestion increases along the main access road leading to the interchange.

2015, After Recommended Development (Fig. 6)

The same amount of development as in the Typical Development scenario is located in a new town center located near the existing town along the railroad tracks. New land use regulations and strategies (including scenic easements) prevent and discourage development along the freeway, which becomes a "townless highway" as originally envisioned by planner Benton MacKaye in the 1920s. A direct and clearly marked access road links the interchange to the new commercial town center, allowing the new development to be accessible both to the automobile and the railroad.

A new and expanded railroad station services both the existing town as well as the new commercial center. A parking garage for commuters is located next to the station, allowing area residents to park and ride, bringing them into the downtown without destroying its character through large surface parking lots. The station is also linked to the new regional shopping mall, the central focus of the new town center. A large hotel is located at the roadway entrance to the town with office, retail, and light industrial buildings located around screened parking lots and structured parking. Residential units are located above commercial enterprises and at the periphery of the new town center. The existing town is also encouraged to expand out to the new rail station to connect with the new development.

Land use regulations as well as siting and design guidelines allow the new, private development to take shape around a network of attractively laid out streets and parks. The more intensive development adjacent to the rail line compensates for the preservation of the townless highway forested corridor.

Fig. 6. Beltway Interchange, after recommended development, 2015.

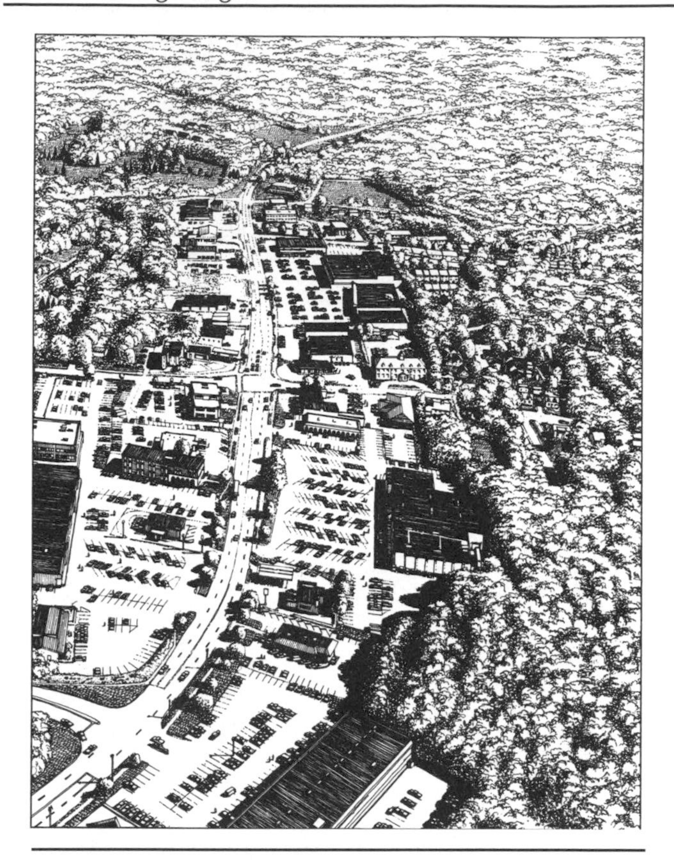

Fig. 7. Suburban commercial strip, existing conditions.

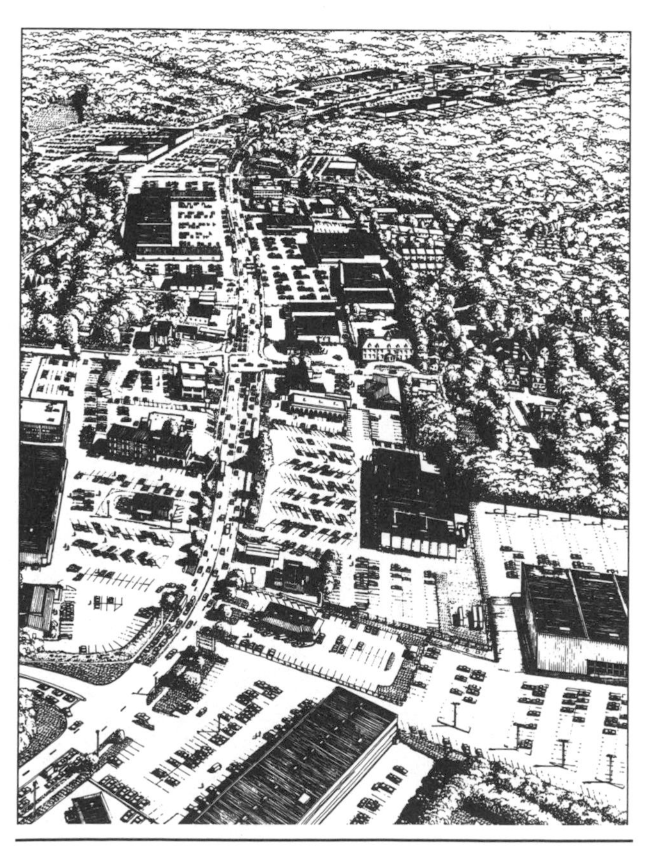

Fig. 8. Suburban commercial strip, after typical development, 2015.

Suburban commercial strip

Existing Conditions (Fig. 7)

The commercial strip is located along a suburban state highway on the outskirts of one of the Region's medium-sized traditional towns. Shopping centers, supermarkets, fast-food outlets, small office buildings, and gas stations compete loudly with neon signs and billboards for the dwindling business along the roadway. Traffic congestion is severe as a result of the new slew of curb cuts for parking lots and access roads. After its heyday in the 1970s, the strip is in decline.

2015, After Typical Development (Fig.8)

Even as the center of the strip begins to decline, new commercial development continues in adjacent underdeveloped rural and residential areas. The town extends the highway commercial zone along the entire length of Route 32 and enlarges the width of the zone extending commercial development into adjacent residential areas. The town's commercial zoning is amended to require improved landscaping around new buildings, but does little to prevent the effect of a solid band of asphalt parking extending on both sides of the road.

2015, After Recommended Development (Fig. 9)

A property-owner led task force developed a plan for shared parking that, with the cooperation of the State Department of Transportation, reduced the number of access roads and overall curb cuts, resolving congestion and safety problems. Through special district overlay zoning, property owners were encouraged to plant trees, rebuild closer to the lot lines, and provide active storefronts.

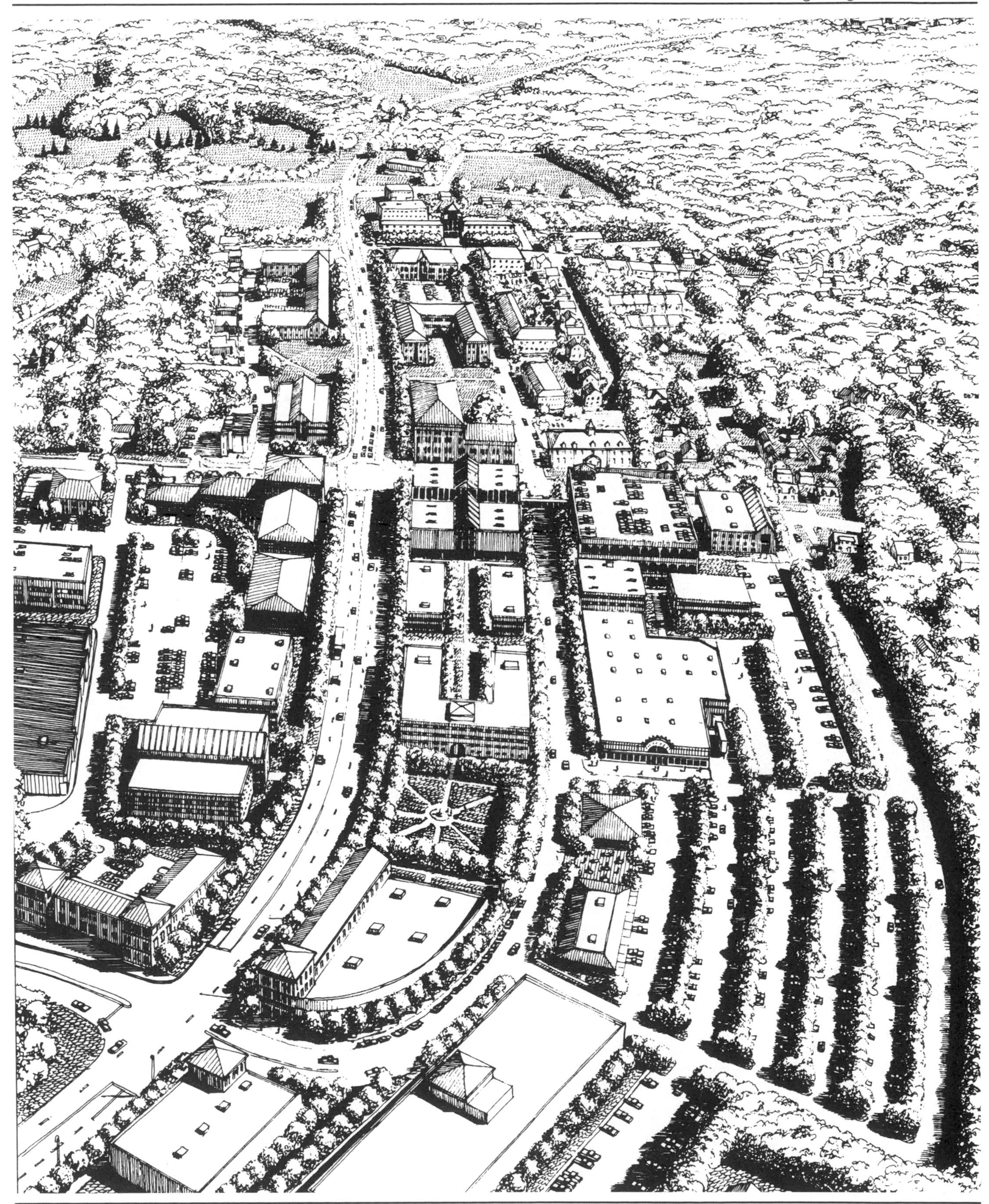

Fig. 9. Suburban commercial strip, after recommended development, 2015.

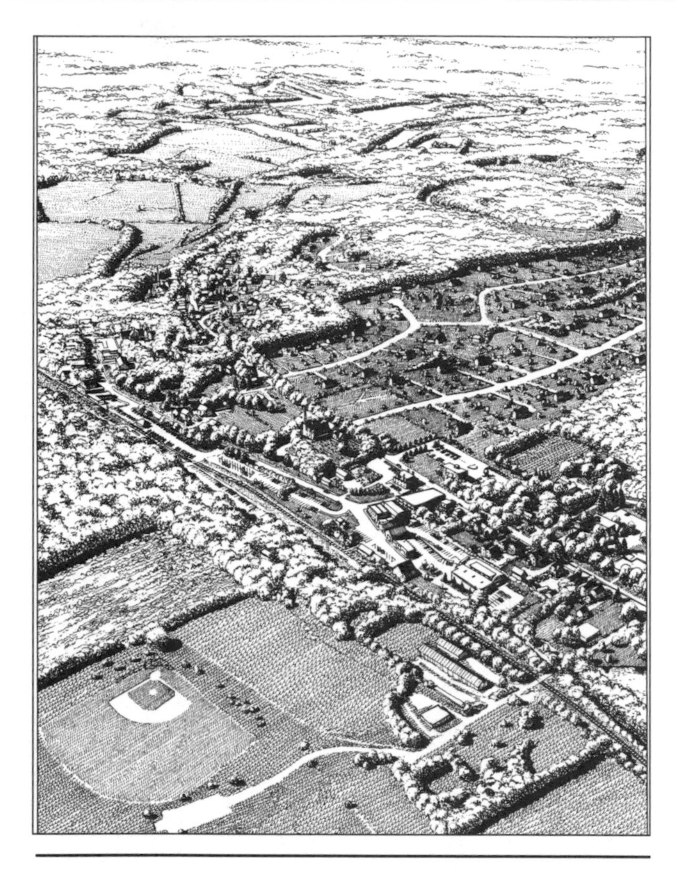

Fig. 10. Rail suburb, existing conditions.

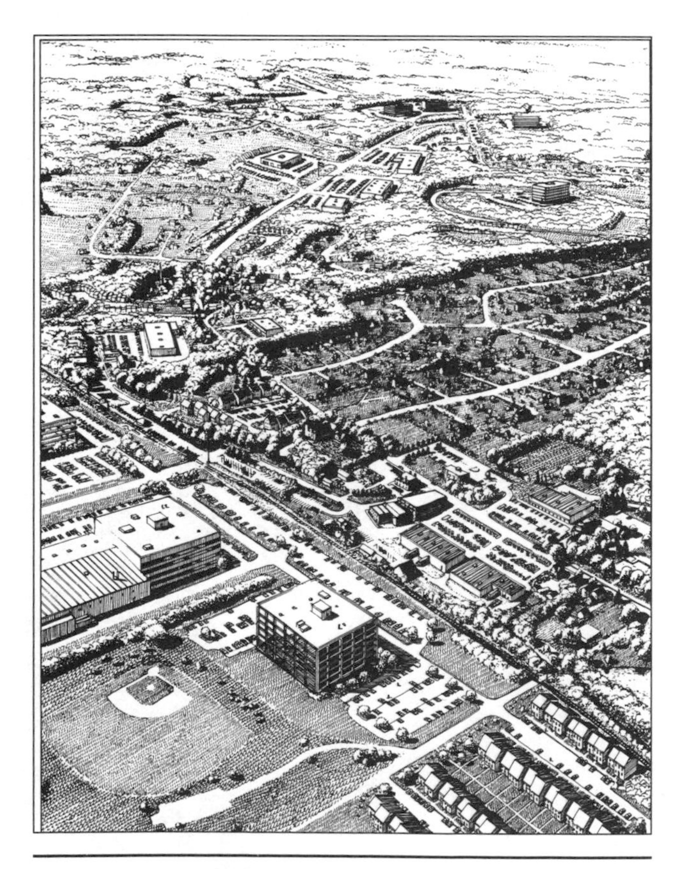

Fig. 11. Rail suburb, after typical development, 2015.

Rail suburb

Existing Conditions (Fig. 10)

An isolated farming community for over two hundred years, this town located in the middle portion of the metropolitan fringe developed as a rail suburb in the early part of this century. Because of its relative distance from the center of the region, it has preserved a large portion of its farmland and forestland. Suburban development reached the area during the boom of the 1980s and a large field north of town became the site of a large-lot housing development laid out according to the town's zoning and subdivision regulations.

2015, After Typical Development (Fig. 11)

During the development boom of the mid-1900s, suburbanization of the town increased. Office and shopping centers have sprung up along the major highways into town, which have been widened and straightened to handle increased traffic. The remaining woodland and farms were filled with large lot subdivisions, offices and retail outlets. The rail line continues to function but its new ridership, not located at the center of town, can only reach it by car. The fabric of the town center is eroded by the demolition of historic buildings, the widening of Main Street and construction of inappropriately scaled and designed new buildings and their associated parking lots. The town's zoning regulations have encouraged the separation of uses into distinct areas and have mandated the sprawling pattern of development.

2015, After Recommended Development (Fig. 12)

To avoid development of the surrounding countryside, the town channeled growth into the existing 1980s developments, north of town and other adjacent areas, by increasing allowable densities there from two to ? acre lots. Homeowner opposition to this new development was offset by the dramatic rise in their property values and the knowledge that new development would be carefully located and designed to enhance neighborhood character. A greater variety of building types, scales, and uses has been introduced in the redevelopment of the suburb to recreate the variety and scale of the existing village. A new mixed-use downtown has been created around the revitalized rail line. The same uses shown in the typical development have been incorporated in the more traditional center. Parking is located in lots behind buildings, or in structured parking with ground floor retail.

Fig. 12. Rail suburb, after recommended development, 2015.

Fig. 13. By-passed urban neighborhood, existing conditions.

Fig. 14. By-passed urban neighborhood, after typical development, 2015.

By-passed urban neighborhood

Existing Conditions (Fig. 13)

Near the center of a large city, a former urban neighborhood has been reduced to rubble. Highrise subsidized housing towers, lowrise apartment blocks, industrial buildings, recently built duplexes and rowhouses, as well as a few remaining nineteenth century commercial, residential, and religious buildings surround the vacant core. Recent market rate townhouse development in the middle ground has been well received, showing a demand for decent housing in the area. Recent slab highrise development outside the urban core extends into the distance as new centers of employment and housing in the suburbs replace the lost neighborhoods of the city center. High crime rates, drug addiction, and joblessness affect the neighborhood.

2015, After Typical Development (Fig. 14)

The mechanisms that created the previous scene have continued to work as more structures are abandoned and torn down, creating more rubble-strewn blocks and parking lots. Joblessness increases as business and industry continue to flee the area, reducing the demand for the newer housing, some of which is also torn down. Crime rates increase, creating even more of a fortress mentality among the remaining residents. Suburban sprawl and slab highrise towers in the distance mark the flight of residents and jobs from the abandoned urban core.

2015, After Recommended Development (Fig. 15)

In partnership with the community and city government, a major corporation has established a manufacturing research and administrative facility on the most distant group of empty blocks and has helped sponsor the reconstruction and future maintenance of the neighborhood. New market rate housing with basement parking garages occupies the foreground blocks, surrounding courtyards of protected and supervised open space, tree-lined streets, and small urban parks and squares. A recreation facility is located between the residential blocks and the corporate center. A regional park and golf course extend from the upper right into a new urban square at the intersection of the major streets. Suburban sprawl in the distance has been contained in well-defined centers while a regional greenbelt links major parks and open spaces to each other and the city center. Most of the subsidized housing towers have been renovated, while one has been removed to create a neighborhood park. Commercial and retail uses are located on the ground floors along the major streets.

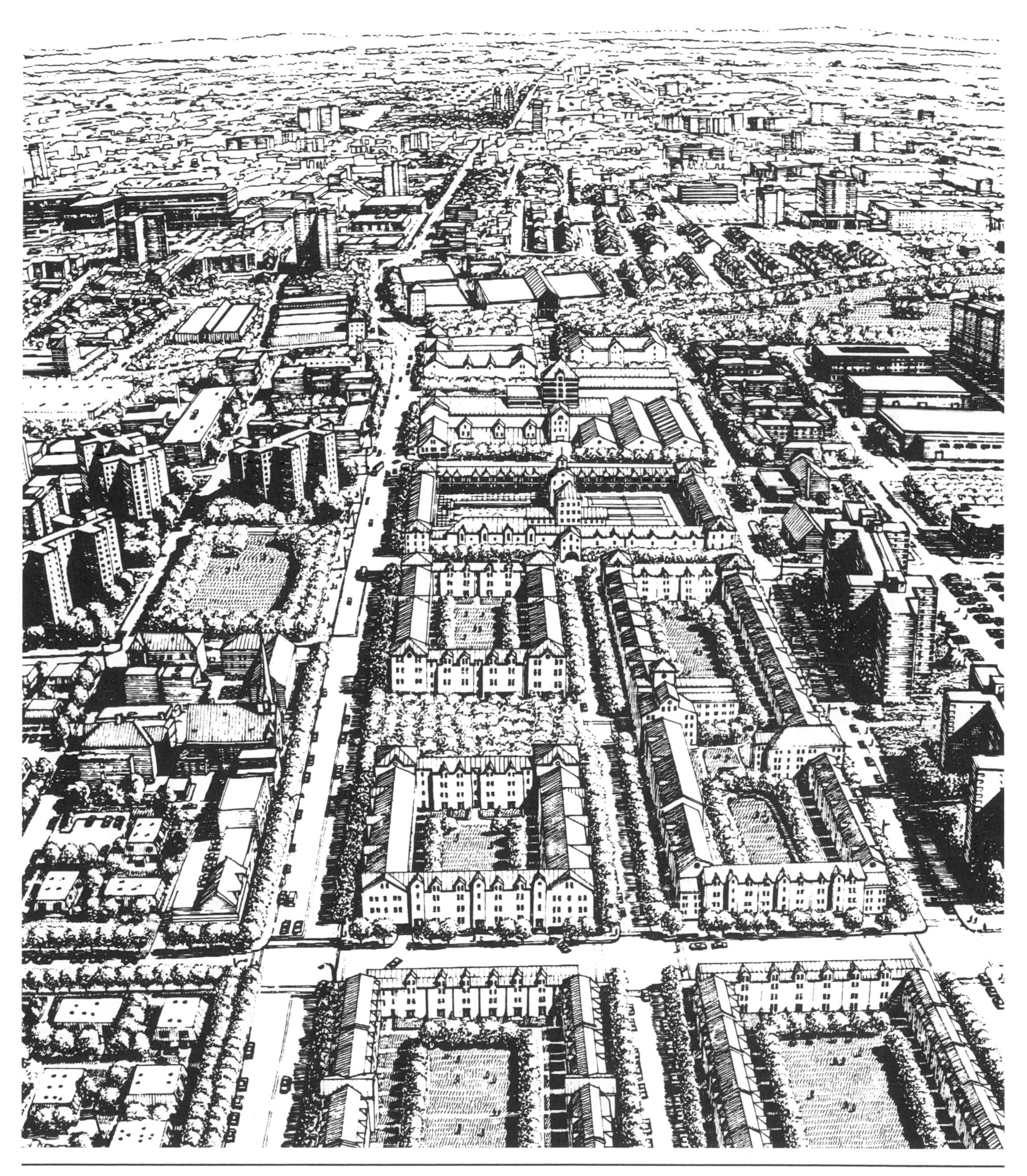

Fig. 15. By-passed urban neighborhood, after recommended development, 2015.

References and note on authorship and publication

In the 1990s RPA decided to use visual simulations as part of the broader effort to complete the Third Regional Plan, because the staff and colleagues had found it remarkably effective in earlier projects. Robert D. Yaro, before coming to RPA, had directed the Center for Rural Massachusetts at the University of Massachusetts, where with Randall G. Arendt, Harry L. Dodson, and Elizabeth A. Brabee, he completed *Dealing with Change in the Connecticut River Valley: A Design Manual for Conservation and Development,* published by the Lincoln Institute of Land Policy and the Environmental Law Foundation, Boston, MA (1988). This document used bird's eye perspective simulations, coupled with plans to dramatize the urbanized future faced by the towns of the Connecticut River Valley, unless they altered their zoning codes and practices.

Regional Plan Association's "Visual Simulations for the Region's Future" was developed as part of the Regional Design Program led by RPA Executive Director Robert D. Yaro, who collaborated with Jonathan Barnett, Harry L. Dodson and Robert L. Geddes. Funding for the first phase of the Regional Design Program included grants from the J. M. Kaplan Fund, the Andy Warhol Foundation for the Visual Arts, and The Vincent Astor Foundation. The full set of Visual Simulations are illustrated in *Tools for Shaping the Region*, published by the American Planning Association Press in 1994. ■

Michael Hough

Summary

Historically, the form of early towns and villages was dictated by their relationship to agricultural fields where food was produced. This symbiotic relationship between the fields that produced food for the town and the town that returned its refuse to enrich the fields was necessary to ensure survival. The creation of urban parks in the 18th and 19th centuries was based on spiritual and leisure needs rather than the necessities of food growing. Recreation is only one of the many functions that urban space must serve as we build an ecologically and socially viable future of our cities. Urban agriculture is a necessary urban land-use function that should be supported as long as poverty, hunger, multicultural traditions and land-connected recreational trends continue to influence city life and city form.

Key words

agriculture, animal husbandry, city farms, food security, gardens, integrated waste recovery, rooftop gardens, sewerage farming, soil, sustainability, water

PHOTO: Donald Watson

Terraced community garden created by International District Community Garden, Seattle, WA

Urban farming

1 URBAN ECOLOGY: A BASIS FOR SHAPING CITIES

Natural systems, including food production and waste recycling, are essential elements of the economic, engineering, political and design processes that shape cities. The often-unrecognized natural processes occurring within cities provide us with an alternative basis for urban design. The problems facing the larger regional context of the countryside have their roots in cities. Solutions must therefore also be sought there, by linking the concept of urbanism with nature. Sustainable agriculture and farming are inextricably linked to the processes of land planning and urban design.

At the beginning of the 21st century, environmental concerns and values have brought into focus an awareness of the earth's fragility as a natural system. We understand human beings as biological creatures immersed in vital ecological relationships within the biosphere, with the need that we live within its limits, sharing the planet with non-human life. Urban design can lead the transition to a society that gives priority to a more sustainable future through productive agriculture and arable land allocation and use.

The Bruntland Commission's definition of sustainability—"meeting the needs of the present without compromising the ability of future generations to meet their own" has helped to set the global framework for an ethic that recognizes the interdependence of all life forms and the maintenance of biological diversity (Lebel and Kane, 1989). Linkages between nature, cities and sustainability have profound implications for survival. World population is projected to be 10 billion by the year 2025, with 4.5 billion people in developing countries estimated to be living in urban areas, and with massive impacts of human activities on world ecosystems.

The elements of the self-perpetuating biosphere sustain life on earth give rise to the central determinants of planning physical landscape including rural and urban design. They determine the guiding principles that shape all human activities on the land:

- The interdependence of one life process on another.
- The interconnected development of living and physical processes of earth, climate, water, plants and animals.
- The continuous transformation and recycling of living and non-living materials.

When the urban landscape is considered in this context, some fundamental contradictions and paradoxes in the way that the city

Credits: This article is adopted by the author from his book, *Cities and Natural Process* (1995), cited in the References. Additional data are cited in the tables.

Table 1. Benefits of urban agriculture (after Koc *et al.*, 1999)

Agricultural production
- Marketed
- Nonmarketed

Indirect economic benefits
- Multiplier effects
- Recreational
- Economic diversity and stability
- Avoided disposal costs of solid waste

Social and psychological benefits
- Food security (available and affordable)
- Dietary diversity
- Personal psychological benefits
- Community cohesion and well-being

Ecological benefits
- Hydrologic functions
- Air quality
- Soil quality

Table 2. Costs of urban agriculture (after Koc *et al.*, 1999)

Inputs	Outputs
Natural resources • Land, rented or purchased • Land, vacant or donated • Water	Pollution and waste • Soil-quality impacts • Air-quality impacts • Water-quality impacts • Solid-waste and wastewater disposal
Labor • Wage and salary labor • Volunteer, unemployed, and contributed labor	
Capital and raw materials • Machinery and tools • Fertilizer and pesticides • Seeds and plants • Energy (fuel oil and electricity)	

Table 3. Energy efficiency of California Crops

Crops	Crop energy / input energy
Field crops (barley, corn, rice, sorghum, wheat)	3.90
Raw vegetables	0.77
Raw fruits	0.54
Average of all raw foods	1.36
Canned vegetables	0.25
Canned fruits	0.25
Frozen vegetables	0.22
Dried fruit and nuts	0.63
Average of all processed foods	0.47

Source: The New Alchemists Institute, "Modern Agriculture: A Wasteland Technology," *Journal of the New Alchemists*, 1974.

and the larger environment are perceived. In a world increasingly concerned with the problems of a deteriorating environment, including energy, pollution, vanishing plants, animals, natural or productive landscapes, there remains a marked propensity to bypass the environment most people live in—the city itself.

Conventional wisdom has regarded the modern city as the product of cheap energy, economic forces, high technology and a view of nature that is under our control. The disciplines that have shaped the city have often had little to do with the natural sciences or ecological values. If urban design can be described as that art and science dedicated to enhancing the quality of the physical environment in cities—providing civilizing and enriching places for the people who live in them—then the current basis for urban form must be re-examine in order to rediscover the nature of the places we live in and the role of urban farming within that environment (*cf.* Tables 1 and 2).

2 AGRICULTURE: PROCESS AND PRACTICE

To most urban people, the countryside is a recreational resource—a place to escape to, leaving the city behind. The connection between food and the land on which it is produced has become increasingly remote as an issue that directly concerns urban welfare. The food that appears in the supermarkets has little direct connection any more with the fields adjacent to the city. It is instead dependent on worldwide marketing and distribution networks operating on fossil fuels and based on international trade agreements.

However, patterns of consumption and environmental priorities are shifting. One of the indicators of change can be found in the concern for diet, the growth of health food stores, farmers' markets and allotment gardening, that suggest there are signs of a return to home-grown versus "factory-made" food. The links between people, food growing, "waste" and city space can be re-established as a central aspect of environmental and social values and incorporated into sustainable design of cities.

Agricultural systems are human-made communities of plants and animals, interacting with soils and climate. Unlike self-perpetuating natural systems, they are inherently unstable. Cultivation and harvesting and the biological simplicity of a few spaces inhibit the recycling of nutrients and make them susceptible to attack from pests. The degree to which agricultural systems can be stabilized depends on factors such as soil fertility, the extent to which animal nutrients are recycled and the diversity of the plant and animal species under cultivation. Traditional mixed farming practice—while varying widely in the type of cultivation and rapidly becoming extinct in most industrialized nations—has maintained a degree of ecological balance.

The enclosed field systems of European agriculture relied on nutrient input from farm animals, crop rotation and variety and the natural communities of hedges and woodlands to offset nutrient loss due to the harvest and the depredations of pests. Energy inputs to the system before the age of fossil fuels were limited to horses and men to pull the ploughs, sow the seed and thresh the grain. Fossil fuels provided the fundamental breakthrough. The rubber-tired tractor replaced human and animal labor. Chemical fertilizers and pest control agents replaced earlier and necessary biological methods of maintaining stability. Fossil fuels enabled agriculture to increase its efficiency, size and productivity, while decreasing its labor inputs.

The technologies that make this productivity and growth possible depend on large land areas and few farms for an efficient operation, a situation that has characterized farming in the industrialized countries. Relatively few, but highly productive, plant hybrids have replaced the more diverse but less productive plants of earlier farming. Worldwide, about 80 percent of human food supplies are dependent on eleven plant species.

Larger farms increasingly specialize in either crops or livestock and both are genetically engineered for maximum yield and performance. Animal feedlots and battery poultry have intensified farm output, converting feed into meat and eggs. New breeds of mechanical harvesters have been developed for harvesting genetically modified species of grains, fruits and vegetables with minimum labor. An entire agriculture industry has evolved of which farming is only a part. What is commonly known as "agribusiness" is based on three major components:

- An input-processing industry that produces seed, machines, fertilizers, fuel and related products required for large-scale farming
- The farm itself
- The food processing industry which transports farm products, processes food, markets and distributes products to wholesale and retail outlets.

This type of modern farming is dependent on fuel energy not only for growing food, but also for processing and distribution. Over the last century agriculture has in effect evolved from a labor-intensive, low-energy, small-scale and mixed farming operation to a vast industry that is capital- and energy-intensive and requires fewer and fewer people. But the benefits of high production farming—being able to feed more people cheaply and less grinding labor for the farmer—also bring considerable costs in environmental and social terms, for the countryside and for the city.

Environmental costs

The first and most basic issue is nonrenewable energy. Continued expansion of production, subsidized by an increasingly costly and diminishing fossil fuel resource, is clearly not sustainable. Barry Commoner pointed to the law of diminishing returns where, as cultivation becomes more intensive, greater amounts of energy subsidy must be used to obtain diminishing increments in yield (Commoner, 1971). The cyclical flow of energy through natural systems is simplified in industrial agriculture, which is sustained by nonrenewable energy at high environmental cost. High concentrations of fertilizers and chemicals used to maximize homogeneous crops threaten soil life and deplete the humus needed for the maintenance of biological health. Streams, rivers and groundwater receive nutrients and chemicals from "nonpoint source" pollution via run-off (that is, spread over very wide and diffuse areas) that lead to the destruction of aquatic species and threaten human health. Heavy machinery and tilling contribute to soil compaction and erosion and consequently the reduction of its fertility.

The complementary benefits to the soil of an animal/crop relationship disappear when crops and livestock become separate industries. Concentrating animals into feedlots involves energy, the disposal of wastes and chemical agents to prevent disease in constricted spaces. This type of industrialized agriculture has been shown to have built-in inefficiencies. Feed must be transported to the site. Enormous quantities of manure produced by the animals—estimated at 18 to 32 kilograms (40 to 70 pounds) per day per animal—are often uneconomical to transport back to the fields and must, therefore, be disposed of. Once fattened, the animals have to be transported long distances to urban markets.

The replacement of manpower by industrialized food production decreases the energy value of various food crops. Estimates of the energy efficiency of different crops grown in California indicate the ratios of crop energy over input energy illustrated in Table 3. Despite high yields generated by industrialized agriculture systems, there does not appear to be a viable net energy return to society. That is, the amount of energy that goes into growing, shipping, packaging and marketing the food is greater than the energy that we get out of it. Research conducted on farmland in upland areas of Britain concluded that, as farms get larger, they tend to produce less good per hectare on average rather than more. This suggests that policies aimed at amalgamating smaller farms into bigger ones in the interests of efficiency and increased production may, in fact, be counterproductive.

Social costs

Biological stability and the sustainability of agriculture are directly related to social issues. The development of modern agriculture has been accompanied by a progressive replacement of human labor with a capital-intensive agriculture industry. The labor-intensive farming of an earlier time maintained viable rural populations. But the relatively low cost technologies, fertilizers, chemicals and machinery of modern farming have become an effective substitute for labor. An increasing number of people have moved from the rural areas to the cities in search of urban jobs. The reasons for declining numbers have been credited to increased competition, more capital to run economic operations, amalgamation of farms into large units, the centralization of agriculture, increased mechanization needing fewer people to manage more land and influences of worldwide trading blocs. These factors work against the maintenance of economically healthy rural communities. Many once flourishing rural areas have been brought to economic ruin and have been unable to support or house their remaining population.

The steady disappearance of farmland in the face of urban growth has decreased the capacity of the rural areas to supply their local urban regions. City dwellers rely largely on the major food producing regions of the world for their food. A glance at the origins of the produce displayed in the supermarkets of most western cities tells us much of the distance (often many thousands of kilometers) that fruit, vegetables, and meat must travel, refrigerated, before the produce reaches the buyer. In Ontario between 1981 and 1986, 4,145 hectares (10,240 acres) of these rural lands were converted from agriculture use to meet the needs of the province's expanding populations. Many other regions are undergoing similar development pressures and the consequent loss of farmland. In the face of such land losses and rising energy and food costs over time, the reliance on imports from distant sources becomes an increasingly shortsighted method of feeding people.

This is particularly evident as a global problem. Since 1973 the global economy is reported to have experienced an economic slowdown associated with the loss of momentum in agriculture. There is a strong relationship between rapid population growth and declining capita grain production in many regions. Fig. 1 illustrates how different population growth rates are driving grain production trends in

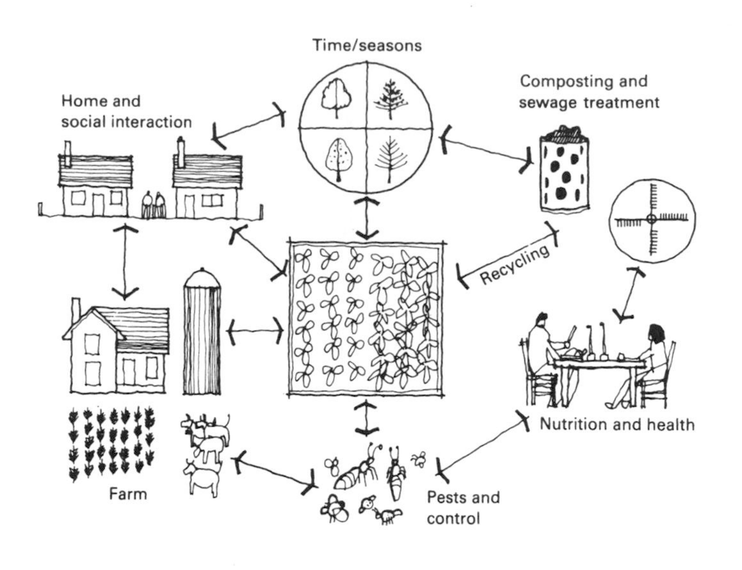

Fig. 1. Per capita grain production in Western Europe and Africa, 1950–1986. Source: Lester R. Brown and Jodi L. Jacobson. "Our Demographically Divided World." Worldwatch Paper 74. Washington, DC: Worldwatch Institute, December 1986.

Fig. 2. Community garden. Family garden allotments in common area of Pine Street Co-Housing, Amherst, MA. Bruce Coldham, Architect.

opposite direction, seen by comparing trends in Western Europe, the region with slowest population growth, and in Africa, which has the fastest population growth.

Crises and shortages lead to the adoption of alternative strategies for survival. Since the end of the Second World War, however, an affluent urban society has not had to concern itself with these basic needs as food has become readily available regardless of season or distance. But with the shift of societal values towards a better relationship with nature, an interest in "organic" farming, the primacy of human health and diet has emerged.

One indicator of these changing values is the rebirth of the farmers' market. Although farmers' markets had never completely disappeared, pressures from the supermarkets and large-scale farming during the 1950s and 1970s had reduced their competitiveness, and suburban expansion replaced the farmland close to the cities that had been the mainstay of their economy. Since the mid-1970s, farmers' markets have revived and by 1993 were reported to number some 2,000 in the United States. In the planning context, farmers' markets are an effective way of revitalizing downtowns.

Thus the opportunities for re-establishing constructive links with the land are tied to the food we eat and it is in the cities that these links can be best re-established. They are the places where farming alternatives can be explored and demonstrated through direct experience with soil productivity, the recycling of nutrient and material resource and urban metabolic processes. We can learn about energy and food production, farming practices, market gardening techniques, rural affairs as well as urban ones, while turning "waste" into useful products. To do so requires that farming becomes an integral part of the city's open space functions. It also requires that the overall physical and social structure of parks and open spaces be reshaped so that their productive value is recognized.

3 URBAN FARMING: RESOURCE AND OPPORTUNITIES

Urban sites have significant potential for small-scale agriculture, so it is here that the task of building a richer and more productive city environment can be achieved. This is borne out in cities in the developing nations that must produce food for large populations, often with limited space and energy for transportation and with minimal financial resources to import food. Design of urban farming is the process of seeking opportunities to match land availability, water and waste as nutrient sources, and available small-scale enterprise and residential and community markers, as natural connections to diversify and improve the quality of land, water and food (Figs. 2, 3).

A need, especially in developing nations where poverty is a major problem, is to link land, food production and waste recycling with employment and income. The exploitation of urban organic wastes and sewage for food production are to be found in the eastern fringe and hinterland of Calcutta. Sewage nourishes fishponds and paddy fields; productive vegetable farms are located on the garbage dumps created by the city in the 1860s, and use the natural compost generated by them. The practice of sewage farming, the growing of grass and other crops on municipal land with liquid wastes was brought from England to India in the 19th century and has survived in some twenty-five cities and towns.

The disposal of garbage is a universal problem of expanding cities worldwide and is often linked to recycling and income by scavenger groups. Recyclable plastic, after being cleaned, is in demand by plastic factories where it is reprocessed and used again. Similarly, iron scraps can also be sold for recasting after being melted down. In the U.S., biologist John Todd and others have developed practical applications of "living systems," wherein greenhouse agriculture is part of an integrated biological process that recovers nutrients from sewerage, produces crops and cleanses water (Fig. 4).

Such examples illustrate the basis for developing an "integrated recovery" approach as a dispersed system of waste processing, nutrient recovery and food production. It can be developed incrementally as market opportunities permit, and as a socially and environmentally viable alternative to conventional, large-scale waste management. It serves as an alternative to dumping unwanted materials into the local rivers, an endemic environmental problem in Indonesia. In effect, the two approaches—the formal and the informal—represent two different views of waste. The former considers it as a health and environmental hazard; the latter considers it as an economic resource of nutrient-rich material from which marketable products can be derived.

The integrated waste recovery approach thus achieves a number of objectives:

- reduces volumes to be dumped
- reduces the need for financing and subsidizing waste management
- creates jobs and income opportunities
- creates social and community cohesion; and it has significant environmental benefits.

The economic, social and environmental benefits can be considerable when developed as an integrated approach in urban planning and management. While they are the consequence of formal urban growth, integrating food production, water, waste and nutrient recycling can be seen as an essential process of urban design. Opportunities for recapturing productive and arable land include areas that are normally disregarded, such as land around electrical transmission lines, highway interchanges and also building rooftops.

The design potential of an integrated approach can be conceived in an inclusive view of the following urban design elements.

Physical and energy resources

Modern urban tradition has been molded by a combination of economics, technological and aesthetic influences and values. This is reflected in the fact that vast amounts of nutrient energy and land are conceived of as "waste." The idea that the city itself should be required to contribute to the production of such basic essentials as food involves a shift in social and ecological values, essentially to see "waste" as "food," as it is in natural systems, where nothing is wasted. The city's spaces, in effect, will be seen as having value beyond the recreational and aesthetic purposes ascribed to them.

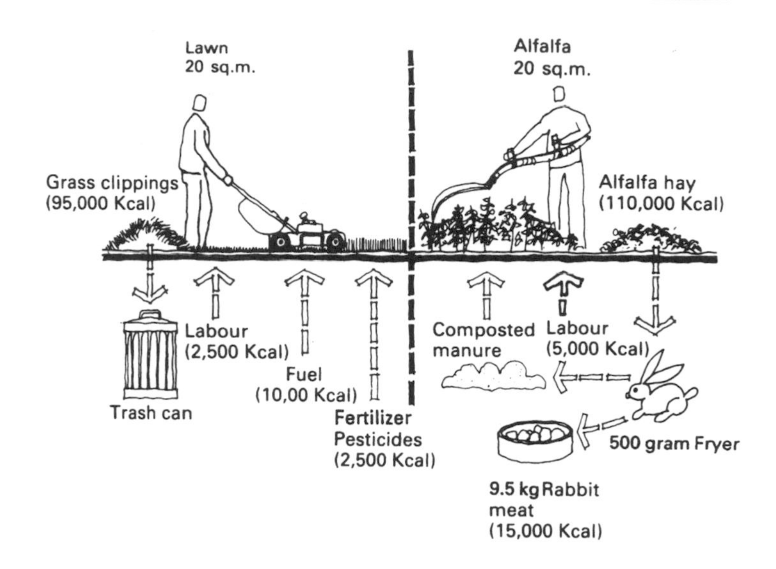

	Energy input Kcal	Energy yield plant	Kcal: Human food
Lawn	15,000	95,000	–
Alfalfa	5,000	110,000	15,000

Fig. 3. A comparison of two equally sized residential yards of 20 square meters (215 square feet), one producing lawn grass, and the other alfalfa. For every unit of energy invested (human labor only) in the production of alfalfa, 22 units of energy were returned in crop production. In the case of the lawn, the net production efficiency (assuming the grass clippings are discarded) is zero. It has been calculated that the rate of energy use for the maintenance of some 6.5 million hectares (16 million acres) of lawn in America exceeds the rate for the commercial production of corn on an equivalent amount of soil. Note: All notations represent annual totals. Source: The Farallones Institute, *The Integral Urban House: Self-Reliant Living in the City*, San Francisco: Sierra Club Books, 1978.

Fig. 4. Solar aquatic "Living Machine." Hydroponic gardening (for horticulture) part of nutrient waste recovery system utilizing sewerage waste from an office building, also cleaning water for landscaping. John Todd, Designer.

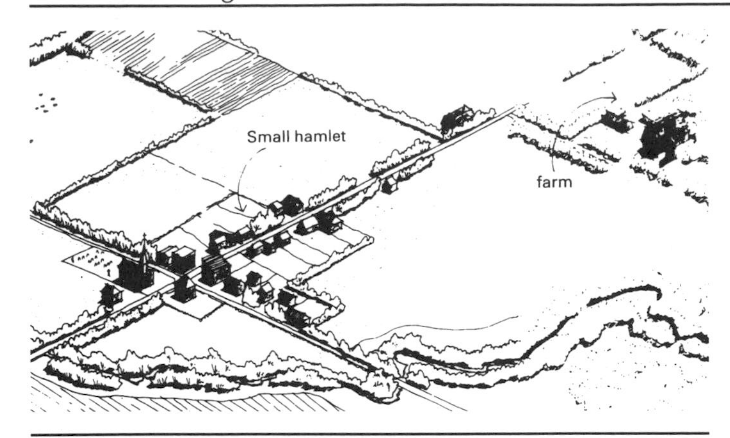

Fig. 5a. The traditional rural village and surrounding countryside.

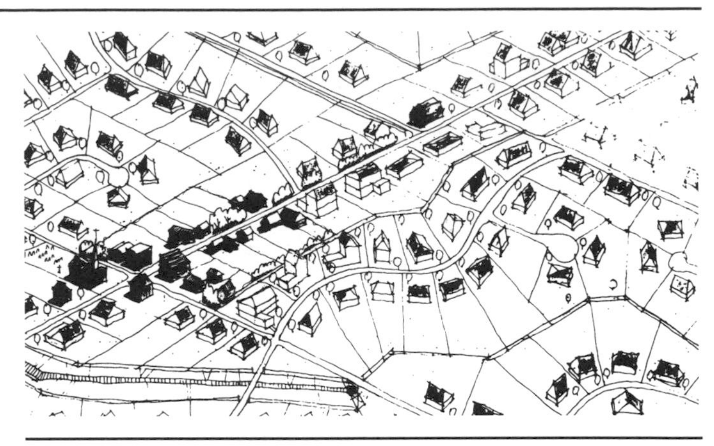

Fig. 5b. Conventional subdivisions typically swallow up the traditional rural village and countryside.

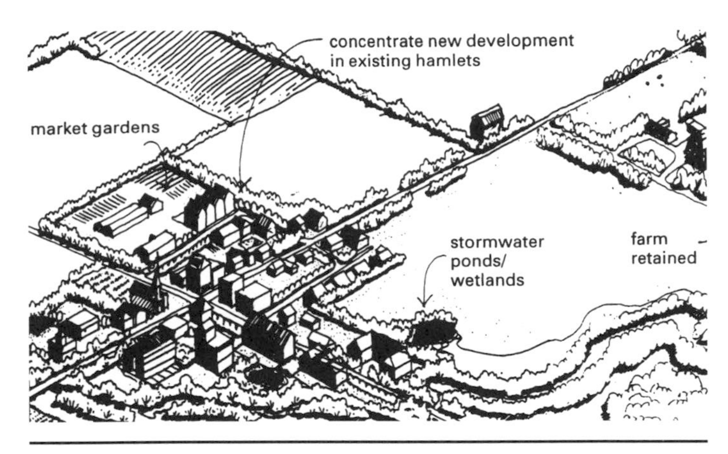

Fig. 5c. Clustering new development around the village leaves many rural functions, streams and woodland intact.

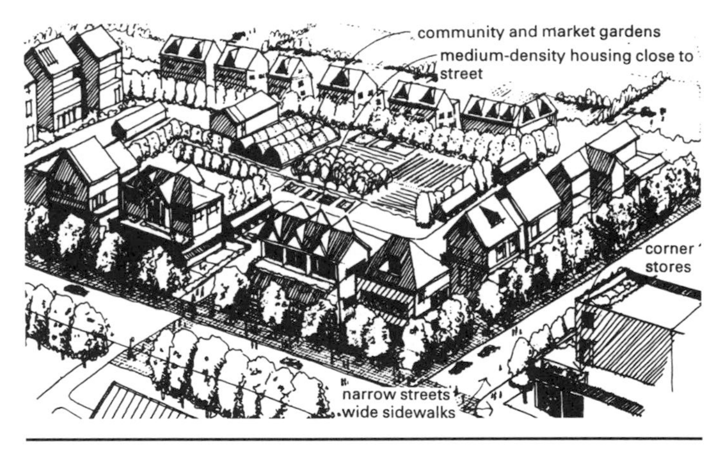

Fig. 5d. Mixed used development can be associated with small-scale market gardens and other rural-based occupations such as plant nurseries, pottery and crafts and recycling operations.

Figs. 5. The need to protect productive soils and natural features and processes as cities grow involves issues of energy and compact urban form, ecology, soil productivity and rural traditions, and finding new ways of establishing a new kind of urban that is an integration of city and countryside. Source: Royal Commission on the Future of the Toronto Waterfront Regeneration, Toronto: Minister of Supply and Services, 1972.

Land

The availability of urban land is potentially enormous in almost all major western cities. Railway, public works and public utility properties, vacant lots, cemeteries and industrial lands from a major proportion of unbuilt land that has been, and remains, sterilized or ineffectively used. The amount and type of land varies depending on the peculiar political, economic and urban renewal conditions of each place.

Residential property

At the smallest personal scale, the front or backyard provides some of the best opportunities for food growing in terms of energy, efficiency and direct benefit. With respect to diet, the food produced on a residential lot is a small scale, human energy-intensive operation. Thus, residential resources for urban farming are considerable and include any space where individuals or groups can grow a few tomatoes or lettuce. Private food growing does not affect public space directly, but the city as a whole becomes richer and more productive because of it (Fig. 5).

Wastelands and allotments

The demand for allotment gardens in or near the cities has been rapidly growing for many years, particularly among people who live in apartments in locations where outdoor space is at a premium. Increased leisure time, early retirement and unemployment are credited with the increasing demand for a plot of land to produce organically grown food. Allotments can be fitted to any shape of property.

Other urban wastelands include the hundreds of hectares of rooftops that for the most part lie desolate and forgotten in every city. Yet many present open space opportunities that could be turned to productive use. In Canada, the Toronto Food Policy Council is encouraging community gardens in allotment gardens and in the attempt to extend beyond that, also on apartment rooftops that exist in that city (Fig. 6).

Heat

A prodigious amount of heat energy is pumped into the city atmosphere from heating and cooling systems and industry, and this has a

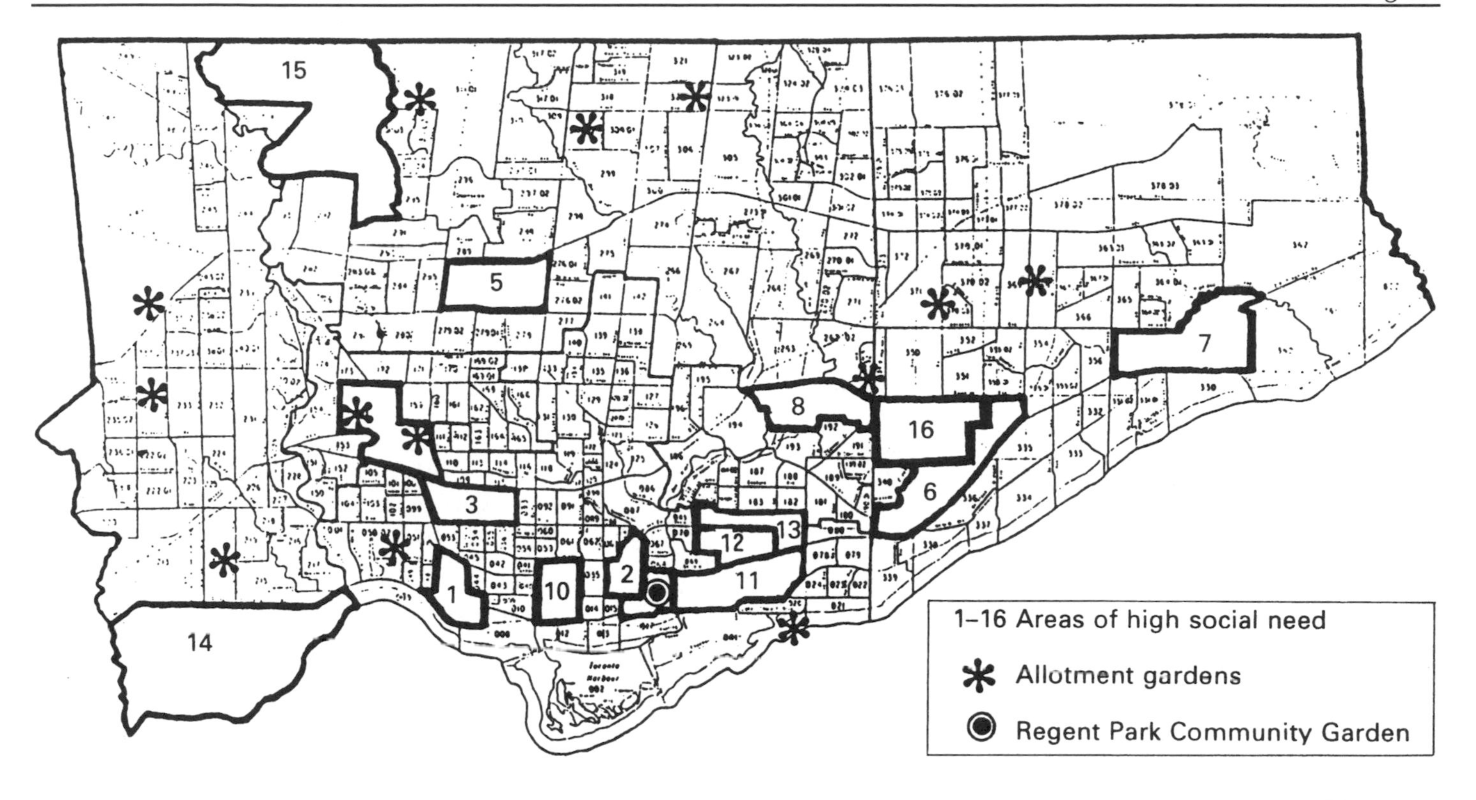

Fig. 6. Map showing areas of high social need and officially designated allotment gardens in Metro Toronto. Note that the opportunities rarely coincide with the areas of greatest need. The municipal parks are not designated as food-growing areas. Source: Social Planning Council of Metropolitan Toronto, 1985, cited in Michael Hough and Suzanne Barrett, *People and City Landscapes*, Toronto: Conservation Council on Ontario, 1987.

major impact on urban climate. This same energy must, however, be regarded as a resource rather than a problem in the context of urban farming. The wasted heat energy from buildings and generating stations may be considered in several ways. In the urban context the heat pumped out from industrial processes may be connected directly to good growing industries. The use of flat rooftops of many industrial buildings, using lightweight hydroponic techniques (water pond agriculture without soil), could begin to make more efficient use of both space and energy resources that are normally discarded. The industrial parks of many North American cities could become agricultural as well as industrial producers, combining functions that have traditionally been separated.

Nutrients

A key factor in urban farming is the question of soil enrichment and management. Many urban soils have been sterilized through constant disturbance and pollution. Consequently they have little fertility. But the move in many countries to a thriving agricultural industry during the Second World War was made possible by the immense inherent wealth of nutrient resources that is available in the city.

European allotment garden, located on the edge of the city on railway land and any other piece of ground that can be leased, is a precious resource. High demand and limited land have made it so. The garden plot is the citizen's summer cottage, often within cycling distance from home. Each plot is laid out with care, with a shed that includes storage for tools and often living space.

4 IMPLICATIONS FOR DESIGN

Historically speaking, the form of early towns and villages was dictated by their relationship to the agricultural fields where food was produced. This symbiotic relationship between the fields that produced food for the town, and the town that returned its refuse to enrich the fields, was necessary to ensure survival. The role of open space within the town, as Lewis Mumford observed, was primarily a functional one; it was important to citizens for growing useful produce or for livestock (Mumford, 1961). The agricultural cooperative settlements (kibbutzim) in Israel show a similar pattern of a direct relationship between habitat and food-producing land.

But new towns are rare and have little impact on the problems of existing cities. It is with the reshaping of the existing city landscape—with recovering the productivity of its soils—that we are concerned. The breakdown between people and direct connection with the land bean as a consequence of industrialization, growth ad mass migration into the cities for nonrural work—a phenomenon that continues in developing countries.

The creation of urban parks in the 18th and 19th centuries was based on spiritual and leisure needs rather than the functional necessities of food growing. This perception of open space—to provide recreational and aesthetic amenity at public expense, thereby satisfying the soul rather than the stomach—has persisted as the main objective of the parks. While recreation is today an essential facet of urban life, it is only one of the many functions that urban space must serve as we build an ecologically and socially viable future of our cities.

Urban agriculture will increasingly become a necessary urban land use function that should be publicly supported as long as poverty, hunger, multicultural traditions and land-connected recreational trends continue to influence city life and city form.

The examples that follow offer insights into the potential or urban agriculture to make productive use of derelict land while providing fortuitous alternatives to the treatment of unused land.

Farming on industrial land—a commercial venture

Examination of the city's open spaces will usually indicate that industrial lands take up very large areas, particularly on its fringes. These neglected places contribute to the city's visual blight, particularly when they become encircled by development. The stimulus to turn formerly vacant land to productive use of this kind lies in land tax laws. In Canada under the Ontario Assessment Act, land that is in use for agriculture is exempted from full land taxes levied by the local municipality, introduced by the province to protect farmland from the skyrocketing taxation that always follows urban development. As a result, the market garden concept has continued to flourish. Such cases illustrate ways that tax laws can assist in upgrading the general quality of the urban environment and have direct benefits to the surrounding community. The example points to the fact that creative policies are needed to create the productive landscape that will help provide cheap sources of food and point the way to an alternative approach to urban design.

Community action and urban form

The commonly held belief that parks should be provided at public expense by authorities and with little direct public involvement is a legacy of the past that is being re-examined in many cities. This belief dates from the growth of the industrial city in the nineteenth century and the development of the public park by indivuduals of social conscience who saw parks as an essential component of urban reform (Laurie, 1979 and Fein, 1968). In England, J.C. Louden's early 19th century writings advocating public parks contributed to the support by the middle classes of the concept of gardens for the less fortunate. Andrew Jackson Downing and Frederick Law Olmsted in the United States both saw contact with nature as a source of pleasure and benefit to society and a necessary way to improve the cities.

Today, we recognize that the nineteenth century romantic view of the park as a piece of natural scenery, separated from the city, for contemplation and spiritual renewal no longer has the validity it may have once had. The whole physical, technological and social structure of urban life has changed. In cities around the world, concerned citizens have initiated tree planting projects in valleys, and degraded lands, and the ecological restoration of urban rivers and their watersheds has provided the basis of cooperation between the public and government in achieving environmental and social goals.

The city farms

The determination among people at the grass roots to be participants and initiators in decision making, where it affects them and their neighborhoods, and the increasing incapacity on the part of public authorities to provide the amenities that have traditionally been their responsibility, are creating new conditions in cities. Experience in Britain has shown that those who have their roots, families and futures vested in the neighborhood and local community are best suited to turn around the physical decay and address the social needs prevalent in cities. The development of an alternative type of park, the city farm, was begun in the early 1970s. At that time it aimed to bring derelict land back into use for the benefit of local communities. The city farms have provided a basis for community revival in depressed urban areas and an educational link with rural occupations by enabling practical framework demonstrations to take place within urban neighborhoods.

5 SUMMARY

The implications for urban design and management are clear. A new approach to the urban landscape is needed, requiring radical conservation measures to ensure the future environmental and social viability of cities. Urban farming brings the activities of daily living closer to the land and the biological systems from which urban people have been alienated and gives the practical tools with which to sustain us in the future. The principles of productivity and diversity, the integration of environment and cultural diversity in the planning, design and management of urban landscapes flow from this philosophy. What logically follows is a view of urban land as functionally necessary to the biological health and quality of life in the city. Standards and design criteria for parks must be revised to permit a wider view of their agricultural role.

A policy for urban land as a whole is needed that encourages the creation of both commercial and community gardens, makes productive use of currently wasted energy and land resource, encourages the perpetuation of self-sustaining urban spaces, provides real economic benefits to the needy in times of economic depression and high unemployment, and contributes to the maintenance of cohesive and stable neighborhoods.

For many, urban farming is an alternative way of renewing contact with the land and nature through therapeutic and healthy work. For others it is an increasingly necessary method of obtaining food at a reasonable cost. This is particularly significant in development countries where the imperatives of employment, food, recycling of every scrap of material and economic survival in the burgeoning cities have forced immigrants on to government owned or wasteland. At the same time it can be argued that problems of poverty and homelessness are common to all cities, irrespective of how "developed" they have become.

People everywhere should be familiar with the broad scientific principles of the production, handling and use of the food that appears on the table. It teaches through direct experience something of rural occupations and the basis for alternative sustainable value. The concept of productivity in urban design terms has wide implications. Drawing its inspiration from ecologically and socially based management practice and understanding of human aspirations, urban farming deals with the maintenance of diverse natural and cultural environments. From this its authentic aesthetic inspiration is derived. This inherent variety and richness of purpose can shape a new vernacular and provide the foundation for the alternative urban design language we seek. ■

REFERENCES

Commoner, Barry. 1971. *The Closing Circle: Nature, Man & Technology.* New York: Alfred A. Knopf.

PHOTO: Donald Watson

Fig. 7. Rooftop gardens as part of integrated community design. Center for Regenerative Studies. California State Polytechnic University. Pomona, CA. John Lyle, Project Director.

Fein, Albert. 1968. *Landscape into Nature*. Ithaca, NY: Cornell University Press.

Jackson, Wes, Wendell Berry and Bruce Colman, eds. 1984. *Meeting the Expectations of the Land: Essays on Sustainable Agriculture and Stewardship*. San Francisco: North Point Press.

Koc, Mustafa , Rod MacRae, Luc J.A. Mougeot, and Jennifer Welsh 1999. *For Hunger-Proof Cities: Sustainable Urban Food Systems*. International Development Research Center, Toronto, Canada in association with the Centre for Food Security, Ryerson Polytechnic Institute, Toronto, Ontario, Canada.

Laurie, Michael. 1979. "Nature and City Planning in the Nineteenth Century" in Ian C. Laurie, ed., *Nature in Cities*. New York: John Wiley. 1979.

Lebel, Gregory G., and Hal Kane. 1989. *Sustainable Development: A Guide to Our Common Future – The Report of the World Commission on Environment and Development*. Washington, DC: The Global Tomorrow Coalition.

Lyle, John Tillman. 1994. *Regenerative Design for Sustainable Development*. New York: John Wiley.

Mumford, Lewis. 1961. *The City in History*. New York: Hartcourt, Brace and World.

Todd, John, and Nancy Jack Todd. 1986. *Bioshelters, Ocean Arks, City Farming*. Falmouth, MA: Ocean Arks International.

Food scarcity and the concept of food security

In recent decades, demographic and economic growth has challenged the limits of economic, social, and ecological sustainability, giving rise to questions about food security at the global level. The continuing reality of hunger and the unsustainability of current practices, both locally and globally, make food security an essential concern.

CONCEPT OF FOOD SECURITY

Despite technological advances that have modernized the conditions of production and distribution of food, hunger and malnutrition still threaten the health and well-being of millions of people around the world. According to the United Nations Food and Agriculture Organization's (FAO) widely accepted definition:

"Food security" means that food is available at all times; that all persons have means of access to it; that it is nutritionally adequate in terms of quality, quality and variety; and that it is acceptable within the given culture. Only when all these conditions are in place can a population be considered "food secure."

To achieve lasting self-reliance at the national and household levels, initiatives must be founded on the principles of economic feasibility, equity, broad participation, and the sustainable use of natural resources. In recent years, most of the research initiatives for food security have focused on four key components of the FAO's definition:

Availability: Providing a sufficient supply of food for all people at all times has historically been a major challenge. Although technical and scientific innovations have made important contributions focused on quantity and economies of scale, little attention has been paid to the sustainability of such practices.

Accessibility: The equality of access to food is a dimension of food security. Within and between societies, inequities have resulted in serious entitlement problems, reflecting class, gender, ethnic, racial, and age differentials, as well as national and regional gaps in development. Most measures to provide emergency food aid have attempted to help the disadvantaged but have had limited success in overcoming the structural conditions that perpetuate such inequities.

Acceptability: As essential ingredients in human health and well-being, food and food practices reflect the social and cultural diversity of humanity. Efforts to provide food without paying attention to the symbolic role of food in people's lives have failed to solve food-security problems. This dimension of food security is also important in determining whether information and food-system innovations will be accepted in a country, given the social and ecological concerns of its citizens.

Adequacy: Food security also requires that adequate measures are in place at all levels of the food system to guarantee the sustainability of production, distribution, consumption, and waste management. A sustainable food system should help to satisfy basic human needs, without compromising the ability of future generations to meet their needs. It must therefore maintain ecological integrity and integrate conservation and development.

Food security and urban populations

- Local food systems offer long-term sustainable solutions, both for the environment and for local and regional economic development. By linking the productive activities in the surrounding bioregion to the consumers in metropolitan centers, local food systems can reduce greenhouse gases and other pollutants caused by long-distance transportation and storage. They can reduce the vulnerability of food-supply systems to the impacts of weather and market-related supply problems of distant producers, offer greater choice through regional variations in biodiversity, provide fresher and more nutritious products in season, allow for more effective regional control of quality and chemical inputs, and create the potential for local development and employment opportunities. A regional or national network of local food systems does not necessarily diminish the possible advantages of the global food system for food security; rather, it would enhance these advantages.

- Cities need to encourage urban and peri-urban agriculture, aquaculture, food forestry, and animal husbandry, as well as safe waste recycling, as elements of more self-reliant local food-system initiatives. Food and nonfood production can tap idle resources and, through income and savings, improve food security, local employment, and urban resource management. Future plans for the flexible, creative, and combined use of urban space and form need to include permanent and temporary food production within metropolitan regions and to create land reserves for productive green space.

- Cities and metropolitan regions need to give priority to the availability and accessibility of food and develop their own food-security plans as part of their social and economic planning. Food-policy councils should be formed to advise local governments and planners.

- Food banks and other community assistance programs should only be relied on as emergency measures, rather than being institutionalized as permanent mechanisms for food access.

- No single solution will solve the problem of food insecurity. What is needed is a list of choices and a commitment to these principles of food security.

Development trends influencing future food production

- Lower-density urban expansion will increase land available for interim or permanent urban farming.

- Urban food production will continue to compete and outrun rural production in certain crops as urban production techniques improve.

- Following promotion programs and projects in the 1970s and 1980s, more national and local governments and specific public sectors will support urban food production in the South, for its food security, job, and environmental benefits, and in the North for its provision of a healthier product.

- Urban food production will be accepted and implemented more systematically as an intervention in food and social security programs (environmental agencies and programs will also increasingly include urban agriculture).

- Urban waste will be more commonly used as a production input because home and community-based treatment of waste will outperform massive and nonselective sewerage and landfill system.

- Information and communication technologies will enable small producers and processors to access and share prompt and reliable technical and market information, have access to credit, and organize themselves in virtual corporations.

- Community and civic organizations will increasingly support urban food production, and women will continue to dominate the industry in production, processing, and marketing (urban agriculture will grow with women's inexorable achievement of greater legal and financial rights).

- Public-private partnerships are accelerating, and national and local urban agricultural organizations appear destined to come together in regional networks.

- Food markets in many of the world's low-, medium-, and high-income countries will carry an increasing share of products grown and raised in urban areas (informal food markets will behave more like today's formal ones, and formal and informal markets will be better interrelated).

- Urban planning and design will more widely incorporate agriculture as a viable land use and urban-space allocation system.

Credit: Adapted from Koc, 1999 (*cf.* full citation in References).

James R. Klein and Elizabeth B. Lardner

Summary

Improving bicycling conditions for an urban bicyclist requires supportive planning and urban design policy and technical capability. There are competitive claims for the use of existing rights-of-way, a zone that is difficult to expand in urban areas. Urban bicyclists are continually presented with barriers that make their trips more difficult and dangerous. This article highlights some essential urban design principles and practices that improve safety and reduce the barriers to bicycle travel.

Key words

bicyclist, citizen involvement, crossways, greenway, path, pedestrian, safety, trailway

Fig. 1. Designated bikeways: Elevating the outer edge and widening the curve help bicyclists negotiate small-radii turns (Harris and Dines 1998).

Urban bikeways

1 INTRODUCTION

This section describes general information about the bicycle, bicycle user, and types of bicycle facilities that are common to all types of bicycle planning and design efforts (Figs. 1, 2). In Section 2, simple and inexpensive approaches to improving bicycling conditions in urban environments are described, followed in Section 3 by description of design principles and processes for new bicycle facilities in urban areas. Also see Harris and Dines (1998) for additional technical data on bikeway design.

The bicyclist

According to the AASHTO "Guide for the Development of Bicycle Facilities" (AASHTO 1999), bicyclists usually are categorized in three ways:

- **Advanced or experienced riders** that use a bicycle in the same way they would a motor vehicle. These riders prefer the street, using the most direct route available. They prefer and need ample shoulder room on the existing street system, choosing the road rather than a separated path.

- **Basic adult riders** may also use their bicycles for transportation purposes, but would rather avoid more heavily traveled or higher speed roads unless there is a designated (striped) bicycle lane. Basic riders will utilize shared-use paths and on-street routes through residential neighborhoods.

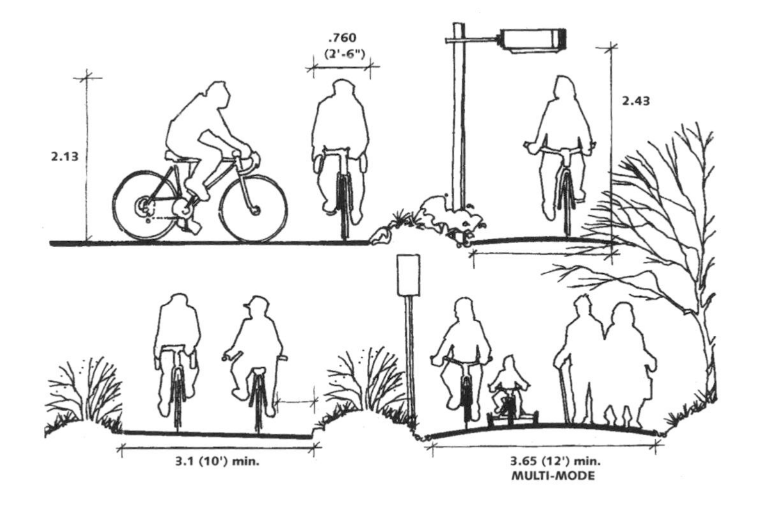

Fig. 2. Bikeway design criteria. A bicyclist uses a minimum of 30 in. (.76 m) of space. The 1999 AASHTO Guideline suggests a 10 ft. (3.1 m) trail width plus an additional 2 ft. (.61 m) maintenance clearance zone on either side of the trail.

- **Children and inexperienced riders** typically do not travel as fast or as far, but do benefit from direct access between schools, convenience stores, parks, and their homes. Children will usually ride on neighborhoods streets that connect to separate, shared-use paths. Crossing busy streets is not encouraged unless accompanied by a parent.

Credits: Photos are by the authors, unless otherwise noted.

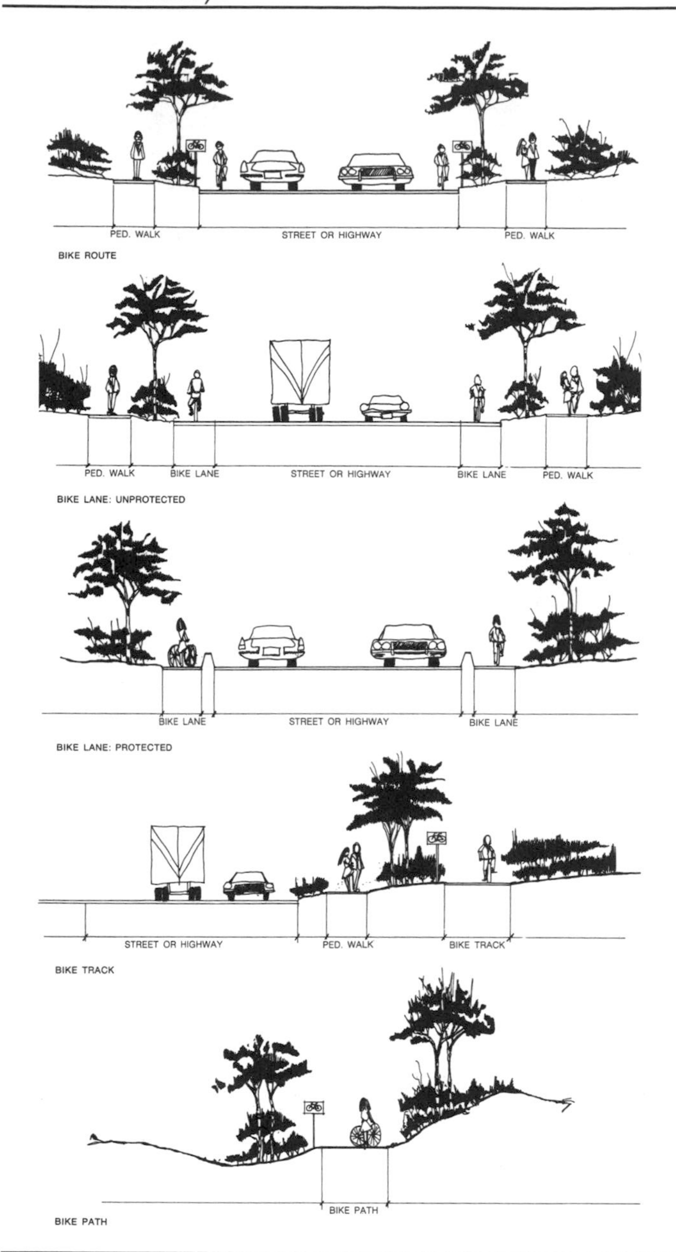

Fig. 3. Bikeway types.

Fig. 4. On-street signed bicycle route requires wider travel lanes.

Bicycle facilities

The 1999 AASHTO Guide defines four classifications or categories of bikeways (Fig. 3):

Shared roadway (no bikeway designation). Bicyclists will use these types of streets if there is enough room—usually about 14 ft. (4.27 m) wide paved travelways in each direction—if there is a relatively low volume of traffic, or if there is no other choice. These routes generally are not signed for bicycle use as a way to discourage a nonexperienced rider from using them until they gain enough experience to feel confident about handling unexpected conditions.

Signed shared roadway. This type is a signed route with the sign indicating that the road is generally suitable for advanced and basic adult riders.

On-street, signed bicycle lanes. This type of facility allocates a dedicated portion of the road surface for bicycle use by striping and signing an approximately 5-ft. (1.5 m) path between the curb and vehicular travel lane in both directions (Fig. 4)

Separated, multiuse paths. This type of facility is a dedicated hard and graded surface used by all forms of nonmotorized travel. In some cases separate travel lanes are dedicated for a pedestrian to use to minimize conflicts. The minimum width for a Multiuse Pathway as recommended by the 1999 AASHTO Guide is 10 ft. (3.05 m) (Fig. 5).

How much room does a bicycle need?

Although a typical bicycle takes approximately 30 in. (.76 m) of space (including the rider's elbows and knees) additional space is needed to operate a bicycle. The 1999 AASHTO Guide suggests four feet as the minimum width for a single bicycle lane, with five feet suggested for streets with multiple types of traffic. For separated, multiuse paths with two travel lanes, 10 ft. (3.05 m) is recommended. In addition to the space for the trail, minimum clearances should be maintained.

2 IMPROVING BICYCLING CONDITONS IN URBAN AREAS

Provision of safe and convenient bicycle travel in an urban area usually requires the construction of bicycle facilities. Most of the time, this means trying to "retrofit" a bicycle facility into the existing transportation system. Facilities typically include bicycle lanes, completely separated multiuse paths, or bicycle/pedestrian bridges to overcome major urban barriers such as rivers and streams, railroads, or major arterials. This is often a time-consuming process that requires implementation through a capital improvement program or securing outside funding, such as through the Federal Highway Administration's Transportation Enhancement Program (TE) of the Transportation Equity Act for the 21st Century (TEA-21).

Opportunities for new facilities

Opportunities for creating bicycle facilities in urban areas are plentiful, if there is demonstrated community support for capturing them. Opportunities for urban bicycle facilities include:

- Abandoned rail right-of-ways or the lands adjoining them.

- Underutilized industrial areas.
- Urban waterfront, riverfront, and stream corridors.
- Urban arterials undergoing expansion or reconstruction.
- Low-volume residential streets and minor collectors.
- Utility corridors. (Fig. 6)

By focusing too closely on the construction of facilities, urban designers may be overlooking a more cost-effective way to improve bicycling conditions in urban areas. By carefully examining routine transportation practices in a community, it may be possible to remove barriers and improve bicycling conditions in a cost-effective manner.

Bicycle improvements for urban streets

Most bicycle riders use city streets on at least a portion of their trip. According to the Federal Highway Administration, a typical bicycle ride is less than two miles in length. At a minimum, bicycle riders need to access neighborhood streets to get to a dedicated bikeway facility. The following summarizes some of the basic principles for integrating bicycle use into the everyday traffic mix in an urban environment.

Fix or replace dangerous drain grates. Parallel-bar drainage grates that may trap a bicycle wheel are obviously dangerous to the bicyclist. At a minimum, existing drainage grates can be retrofitted by turning the bars or welding flat steel bars across the grate so they are perpendicular to the flow of travel across the grate. Even this technique however, may jar the bicyclist. A better solution is to use a curb-face inlet or a "Vane Grate." The vane-grate has openings that are small enough to prevent a bicycle tire from entering, but large enough to prevent debris from collecting on its surface. Standard details should be permanently changed so that all grates installed that are safe or "bicycle friendly" (Fig. 7)

Pay attention to the roadway edge. Narrow bicycle tires can also get caught on a pavement edge that is higher than the gutter pan, or on pavement patches that form ridges on the surface. This is an especially problematic during utility work. Specifications for utility work should require a smooth match with the existing surface and a guarantee period should the patch fail.

Keep the street clean. Street sweeping is another simple technique that removes debris that collects at the side of the road. This is particularly important in flood-prone areas where overburden builds up in the area where bicyclists typically ride.

Improve railroad crossings. Tracks that cross the street at an angle of less than 45 degrees can catch a bicycle tire in the same manner as a drainage grate. Two options for solving this problem include using a smooth rubberized crossing surface or, if there is enough room, curving the path so that it comes across the tracks at a right angle.

Widen curb lanes. Providing additional space on standard streets sometimes is all that is needed to make it easier for cars and bicycles to share the travel surface. On four-lane arterial roads with existing 12-ft. (3.66 m) travel lanes, the inside lane can be narrowed to 11-ft. (3.35 m) and the outside curb lane widened to 13-ft. (3.96 m). The Maryland DOT found this to be effective in a 1985 study. On two-lane roads, increasing the lane width to a minimum of 14 ft. (4.27 m)

Fig. 5. Separated multiuse path.

Fig. 6. Bicycle path in Arlington, Virginia located adjacent to Interstate 66 sharing right-of-way with a high-voltage transmission line.

Fig. 7. Hazards of on-street cycling include drainage grates (upper sketch) that can be helped by a "Vane Grate" design (lower sketch). (Harris and Dines 1998).

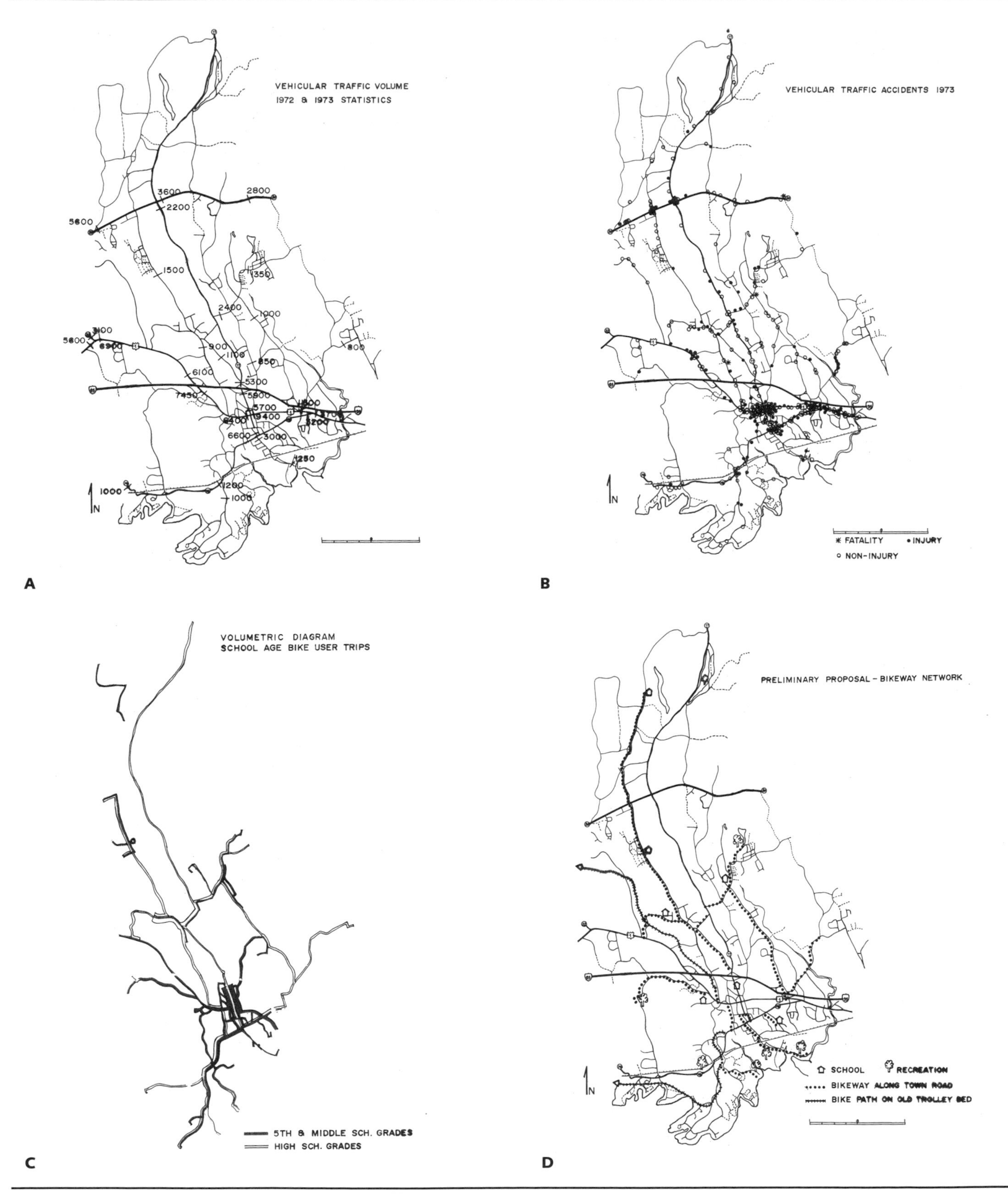

Box A: Guilford, CT Bikeway Plan. 1973. Analysis included: A-mapping of existing vehicular traffic volume; B-locations of accidents; C-circulation from residences to schools; D-composite recommendation based on citizen input (*cf.* Box B). Students and teachers of the entire school system participated in the documentation. The 1975 National Urban Bikeways Competition Award recognized the plan for citizen involvement in its planning and implementation. (Watson, Storer, and Waggoner 1974)

provides enough space (exclusive of the gutter pan). Montgomery County, Maryland has adopted this standard County-wide.

Install bike sensitive traffic signals. Bicycles often have problems triggering the sensors designed for vehicles to trip a signal. This can be very frustrating for a cyclist, who then proceeds to ignore the signals. This problem can be solved by using bicycle friendly detector loops, such as the "CalTrans Type D diagonal quadropole loop detector" or by marking the location in the bicycle lane where a bicyclist is most likely able to trigger the signal.

Consider traffic calming measures. Improved bicycle safety can also be achieved by reducing vehicular travel speeds. Care must be taken to make sure that pavement narrowing, lateral shifts, modern roundabouts and other types of traffic calming measures are bicycle friendly.

3 NEW BICYCLE FACILITIES IN URBAN AREAS

Creating new facilities in urban areas is primarily accomplished by "retrofitting" existing linear features such as an abandoned rail corridor or urban stream corridors to accommodate bicycles (and often other types of users).

Most of the decisions about bicycle routing are addressed at a community-wide scale. Most state Department's of Transportation have a statewide bicycle plan that defines priorities for major bicycle facilities. Metropolitan Planning Organizations (MPOs) are responsible for coordinating the budgets for transportation improvements. Many county and city governments also have bicycle facility plans that are part of the community's comprehensive plans. In order to get federal funding for a new bicycle facility, it must be included in at least the local comprehensive plan and the MPOs transportation improvement program. Some communities fund new projects through capital improvement projects at a local level. Demonstrating how a proposed facility fits in with an overall bicycle and pedestrian path network is a crucial part of gaining priority status for future funding and implementation. (See Boxes A and B for examples of citizen participation in the process.)

Process

Taking a look at the "big picture" is an important first step in creating bicycle facilities in urban areas. Some important questions to raise at the start of any bicycle planning exercise include:

- Who are the persons that have an interest in the bikeway: the various types of users, neighbors, regulating agencies; major institutions and employers along the route; and other key stakeholders?
- What are the existing conditions found along the proposed route and are there any "red flags" that might either prohibit construction or significantly raise the cost?
- Are there any major capital improvement projects or large-scale development or expansion projects in the corridor?
- Are there any significant environmental or cultural resources along the corridor?
- Are there any nearby community facilities that would benefit from improved bicycle and pedestrian access?

THIS IS YOUR BALLOT

1. (Check One)
   ____A. I would like bikeways in Guilford.
   ____B. I would **not** like bikeways in Guilford because (complete) ____________
2. (Check One)
   ____A. My family does not own bikes, but I think that bikeways would be desirable (Skip to Q. 5C.)
   ____B. My family owns bikes. (Answer i, ii, and iii)
   ____i. The adults own ______ (No. of bikes)
   ____ii. High school and younger family members own ____(No. of bikes)
   ____iii. Our bikes are located at (address or approximate location) ____________

3. The number of days each month I ride my bike is closest to (Circle One) (1) (5) (10) (15) (20) (25) (30)
4. I use my bike (Check One - complete if necessary)
   ____A. Less than I want to because (Complete) ____________
   ____B. More than I want to because (Complete)____________
   ____C. As often as I want to (Complete) ____________
5. (The Guilford Bike Committee would like your description of three types of bike trips. First, the trip you actually made most frequently during the past twelve months; second, the bike trip you make less frequently than you would like (or not at all) because of unsafe conditions, grade, lack of a place to leave bike, etc.; and last, the bikeway you think would be most beneficial to Guilford as a whole.)
   ____A. The bike trip I made most frequently during the past twelve months...
   Starts at ____________
   Ends at ____________
   ____B. A bike trip I would like to make more often...
   Starts at ____________
   Ends at ____________
   ____C. The bikeway that would be of the most beneficial to Guilford would...
   Start at ____________
   End at ____________
6. My age is ____________(Circle One)

Under 18	A	45-54	E
18-20	B	55-64	F
21-24	C	65-74	G
35-44	D	75 & Over	H

Although it isn't necessary for a valid ballot, the committee would appreciate the name and telephone number of anyone wishing to help.

Name ____________**Tel. No.**____________

If more than one family member desires to vote, facsimiles may be used or extra ballots may be obtained (and deposited) at the Guilford Libraries or the Tax Collectors Office in the Guilford Town Hall. Ballots from members of the same family should be clipped together. Ballots may be mailed to

Box B. Guilford, CT Bikeway Plan Survey Form distributed in schools and news media helped to establish priorities and increase community awareness and support. (Watson, Storer, and Waggoner 1974)

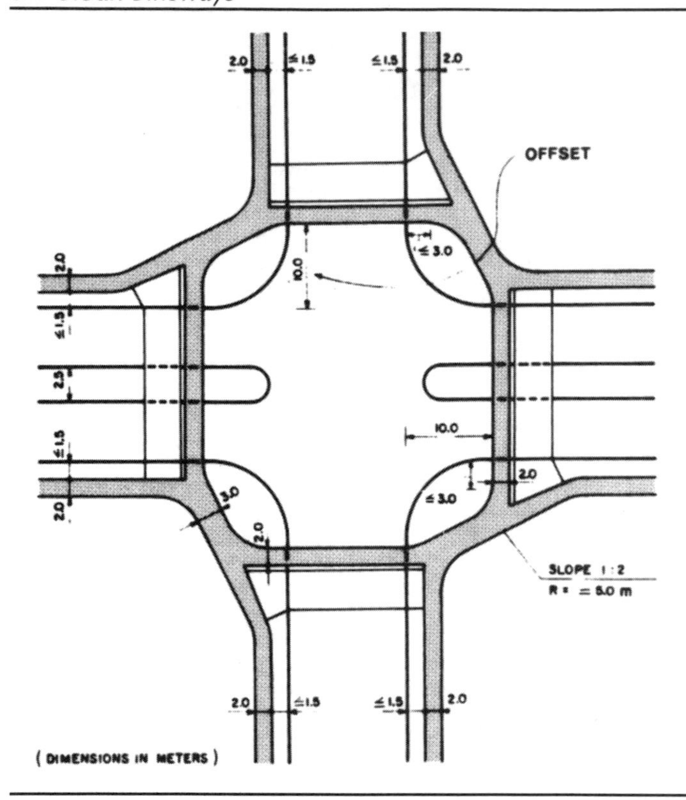

Fig. 8. Offset crossway. A separated bikeway loop is created around a vehicular intersection. The crossings, offset from the immediate intersection, place conflicting motor vehicular traffic in the cyclist's forward field of vision and moves the conflict area to a point where the turning motorist is no longer preoccupied with cross traffic from the driver's left. (FHA 1974).

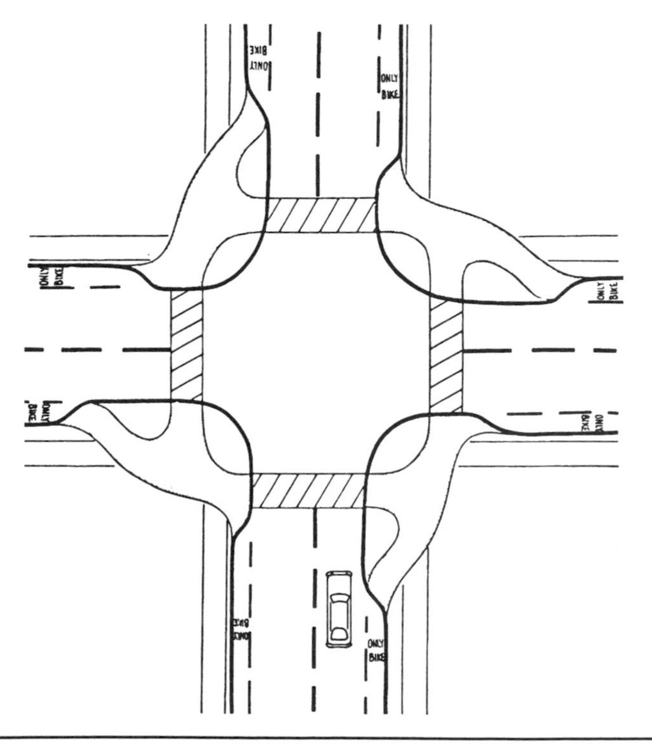

Fig. 9. Modified offset crossway. A modification of the offset crossing shown in Fig. 8, conforming to more restricted right-of-way conditions and on-street lane approaches. The principal concern with crossway design is that it inherently treats bicyclists as pedestrians, rather than as vehicle operators, and increases potential conflicts with pedestrians. These objections can be helped by adequate space for pedestrians and bicyclists. (FHA 1974).

Involving the persons with a keen interest in the bikeway is a critical first step in project planning. It is often useful to create a Project Advisory Group that includes representatives from each of the adjoining neighborhoods, agencies, institutions, user groups, and other key stakeholders. This approach will ensure that each of the groups with an interest in the project will be involved in the design process from the beginning rather than waiting until the "public meeting" to have an opportunity to provide input. The Advisory Group should walk the proposed route early in the process to identify important issues and other stakeholder groups.

In any bikeway planning project, it is important to clearly delineate the decision-making process. The following ten steps describe one possible approach that can lead to a successful agreement about the location of a multi-use pathway in an urban area:

- Step 1: Organize the project and form an advisory group
- Step 2: Identify "design user"
- Step 3: Identify and review existing and available data/identify "red flags"
- Step 4: Meet with community groups, user groups and neighbors
- Step 5: Assess the types of regulatory issues that will need to be addressed and avoid locations where regulations may be burdensome to project implementation
- Step 6: Assess the proposed alignment alternatives
- Step 7: Evaluate and select a preferred alignment
- Step 8: Develop the preferred alignment with options for design treatment
- Step 9: Public meetings and presentations
- Step 10: Prepare the final facility plan with implementation and phasing plan

The project advisory group should assist with the decision-making at each of the steps described above so that any potential problems and pitfalls have been ironed out well before the public meeting.

Existing conditions

Developing a strong technical basis for making location and design decisions is a critical step in the process. A summary map showing the existing conditions helps to explain the rationale for making location and design decisions and to facilitate the sharing of information with community groups, officials and all decision-makers. The following elements are helpful:

- Existing topography and planimetric features.
- Existing vegetation (trees, shrubs, woodland, natural features).
- Hydrology (wetland, floodplain, drainage features).
- Soil conditions that affect trail development.

- Existing paved surfaces and structures (roads, sidewalks, curb cuts, bridges, culverts, tunnels, etc.).
- Utility Easements (streetlights, water, sewer, storm drains, manholes gas, telephones, electric and traffic control devices).
- Location of active rail lines and safety zones associated with them.
- Existing pedestrian circulation and access, including flow restrictions and safety problem areas.
- Existing and potential adjacent land uses and planned projects .
- Historic sites (listed and eligible properties).
- Existing connecting trails and public lands.
- High-use destinations for bicyclists (*e.g.* schools, parks, other recreational venues).
- Documentation of accident locations, especially involving bicyclists and vehicles (as indication of danger points).

A photographic record of the route is also helpful to document specific issues and concerns. This is particularly important to identify situations where exceptions to design guidelines (such as AASHTO's 1999 guidelines may be needed).

Alignment alternatives and design options

A first level of screening is often used to avoid potential constraints to trail development. For example, a slope that is too steep to accommodate a bike path should be avoided (5% or greater requires ramps under the Americans with Disabilities Act). Alignment alternatives are then developed to avoid the constraints.

Design options should also be developed to incorporate decisions about trail width, trail surface, and the intended design user. In heavy use areas, trail widths should be widened to accommodate the anticipated volume of users. Changes to the trail surface can also be used to accommodate a wider variety of users. Skateboarders, roller bladers, and other small wheel users prefer asphalt to concrete. Joggers prefer softer surfaces such as crushed stone.

Overcoming obstacles

At the planning stage, it is important to determine whether or not a particular constraint or obstacle can be overcome and at what cost. In urban areas obstacles may include:

Major urban street intersections with no space for pedestrian and bicycle users. (*cf.* Figs. 8–10 for designs for intersections. Also see Kulash, this volume, Fig. 11).

- Rail lines
- Stream crossings
- Narrow bridges with little or no room for bicycles
- Lack of available right-of-way

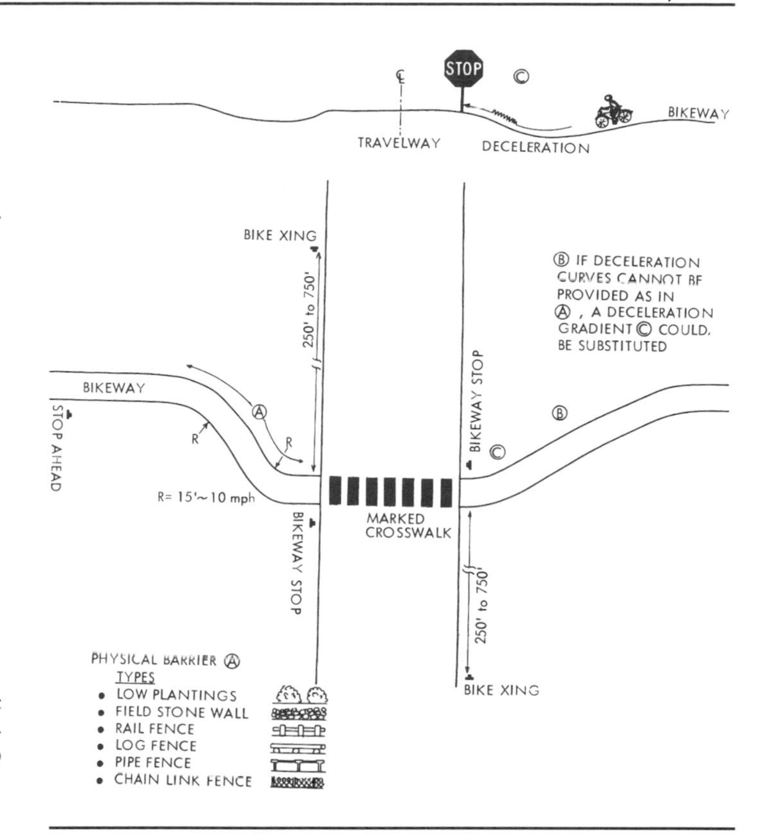

Fig. 10. Isolated crossings. Deceleration zones, created by offsets in direction and grade depressions, address concerns with vehicle/bicyclist conflicts and accidents at intersections of bike paths and roadways. Placement of physical barriers that require a bicyclist to slow down, to stop and/or dismount are appropriate at high hazard intersections.

Fig. 11. "Bicyclists Must Dismount" sign used along the Mt. Vernon Trail in Alexandria, Virginia

Fig. 12. Crossing a private drive and a busy intersection on the W&OD Trail in Arlington, Virginia.

Design exceptions. To overcome obstacles, it may be necessary to seek an exception to published design guidelines, and to sign the route to warn trail users of upcoming areas of concern that could not be avoided. For example, where a separated path must cross a driveway, it is important to place a sign that clearly indicates who has the right-of-way. Asking bicyclists to dismount may seem like an unreasonable request, but it makes it clear that if cyclists in both directions must cross a driveway, then the automobile should be given right-of-way (Fig. 11). If automobiles are asked to give up the right-of-way, bicyclists should be warned of upcoming vehicular crossings.

Bridges. Bicycle bridges are often needed to overcome major barriers. A bicycle bridge must be wide enough to accommodate a bicyclist plus a clear zone on either side of the path. 14 ft. (4.27 m) is a reasonable bicycle bridge width.

Clearance requirements for bridges often create additional design difficulties. This often necessitates the construction of custom made bridges. Bridges over railroads must meet minimum clearances by the rail owner. This can often by met by creating a subtle arch to the bridge. Less costly bridges can be manufactured off-site and assembled on-site. This is often appropriate for smaller bridges. Simple wooden structures can be constructed to cross smaller streams and creeks.

In the United States, ramps approaching the bridge must meet Americans with Disabilities Act (ADA) requirements. The maximum slope for an approaching pathway is 5%. Where slopes greater than 5% are required, then a ramp must be constructed. The maximum slope of the ramp is 1 ft. (.31 m) rise for every 12 ft. (3.66 m) horizontal distance. Level landings are required for every 32 ft. (9.75 m) in ramp length.

Signals. Crossing major urban arterials can often be accomplished through the use of bicycle friendly traffic signals. Having a dedicated phase for bicycle and pedestrian use is the safest option. The dedicated phase must have enough time to cross the street, and be frequent enough to discourage jaywalking. In some cases, providing a dedicated pedestrian/bicycle phase is not possible without causing dangerous vehicular backups. In this case, it is possible to signal an intersection so that pedestrians and bicyclists can move with the direction of travel (Fig. 12). "Yield to pedestrian" signs are needed for turning traffic, as are warning signs for bicyclists.

Warning signs. Where uncertain conditions cannot be avoided, warning signs need to be placed to give trail users a chance to slow down or in some cases dismount. The Manual of Uniform Control Devices has a complete chapter on bicycle signs (FHA 2000). It is important to utilize MUTCD signs so that trail users will learn to use the signs from one location to another. In some cases, special signs will be needed to alert motor vehicles to unusual and/or unexpected conditions.

Community linkages. Linking trails to community facilities is another important principle in urban bicycle path design. Bicycle paths are more likely to be utilized if they link homes with often-visited locations such as parks, schools, and neighborhoods. ■

REFERENCES

AASHTO (American Association of State Highway and Transportation Officials) Task Force on Geometric Design, 1999. *Guide for the Development of Bicycle Facilities.* Washington, D.C.: American Association of State Highway and Transportation Officials.

FAA (Federal Highway Administration). U.S. Department of Transportation. 1974. *Bikeways: State of the Art.* Pub. No. FHWA-RD-74-56. Washington, DC: U.S. Department of Transportation, Federal Highway Administration.

FAA (Federal Highway Administration). U.S. Department of Transportation. 1998. *Implementing Bicycle Improvements at the Local Level.* Publication No. FHWA-98-105. Washington, DC: U.S. Department of Transportation, Federal Highway Administration.

FAA (Federal Highway Administration). U.S. Department of Transportation. 2000. *Manual of Uniform Traffic Control Devices (MUTCD), December 2000 Millennium Edition.* Washington, DC: U.S. Department of Transportation, Federal Highway Administration.

Harris, Charles W. and Nicholas T. Dines, eds. 1998. *Time-Saver Standards for Landscape Architecture.* 2nd edition. p. 342–2.

Watson, Donald, Dorothy Storer, Barbara Waggoner. 1974. "A Bikeway Plan for Guilford, Connecticut." [unpublished project report].

Loring LaB. Schwarz, editor

Summary

Greenways are natural corridors that may traverse an urban landscape that has been otherwise transformed by development. They offer a low-cost means to help maintain the environmental integrity of an urban infrastructure. Greenways represent a relatively new form of urban park, conceived as a connecting network, often utilizing abandoned rail beds or restored river edges. They provide unique forms of outdoor recreation in waterways, bikeways and trails, community gardens, farm plots and pedestrian connections to schools and other public destinations.Greenways also serve as uninterrupted wildlife corridors and preserves and provide for necessary vegetated buffer zones along streams and rivers.

Key words

accessibility, bikeway, biodiversity, greenway, interpretation, rail beds, surface materials, trails, tread, vegetated buffer zones, wildlife corridor

PHOTO: Karl Gierman

Bridge of Flowers, Shelburne Falls, MA. Trailway constructed on abandoned railroad bridge.

Greenways

1 WHAT IS A GREENWAY?

The term "greenway" suggests two separate images: "green" suggests natural amenities such as forests, riverbanks, wildlife; "way" implies a route or path. Together they describe a corridors or trails crisscrossing a landscape that has been otherwise transformed by development.

Within the developed urban landscape, greenways serve a dual function: they provide open space for pubic access and recreational use, and they protect and improve remaining natural resources. In a broader sense, "greenway" is a generic term for a wide variety of linear open spaces that provide connections and thereby foster movement of some sort within an urbanized fabric. These may include bicycle routes, wildlife corridors, revitalized waterfronts, or tree-shaded footpaths along a stream or estuary that extends far from the city (Table 1).

Greenways respond to an increasing interest in outdoor recreation—walking, jogging, and bicycling. Greenways are an inexpensive way to meet this need because potential sites consist of relatively narrow corridors of land, often within available bottomlands, floodplains, abandoned railbeds, and other otherwise undevelopable locations. Greenways cross many properties and jurisdictions. They challenge greenways advocates to protect land through zoning, voluntary registration, and easements, along with traditional acquisition techniques. Greenways provide a means to protect natural, cultural, and historic resources which are accessible by greenway planning. Ecological greenways are valuable as movement corridors for wildlife and plant species. The provide buffer zones to serve as vegetative filters of sediment and pollution from runoff along stream edges. As greenscape, they are effective moderators of air pollution and temperature extremes in cities.

Defining a greenway corridor

Greenway design brings together the skills of urban design, technical engineering, the biological sciences, and community involvement in new ways.

- Minimize human-wildlife conflicts by siting recreation paths where there will be the least interference with natural systems (Fig. 1).
- Establish a comprehensive inventory of assets and liabilities (Table 2).
- Plan for native planting species and landscape that is self-maintaining and without chemical applications or excess watering. Parks along greenways should not require mowed turf grass. Plan for removing exotic or noxious species that invade the corridor.
- Limited the points of access and activities near and within sensitive areas.

Credits: This article is based in part on data from *Greenways: A Guide to Planning, Design, and Development*, 1993, a report of the Conservation Fund, edited by Loring LaB. Schwartz and reprinted by permission of Island Press (see References).

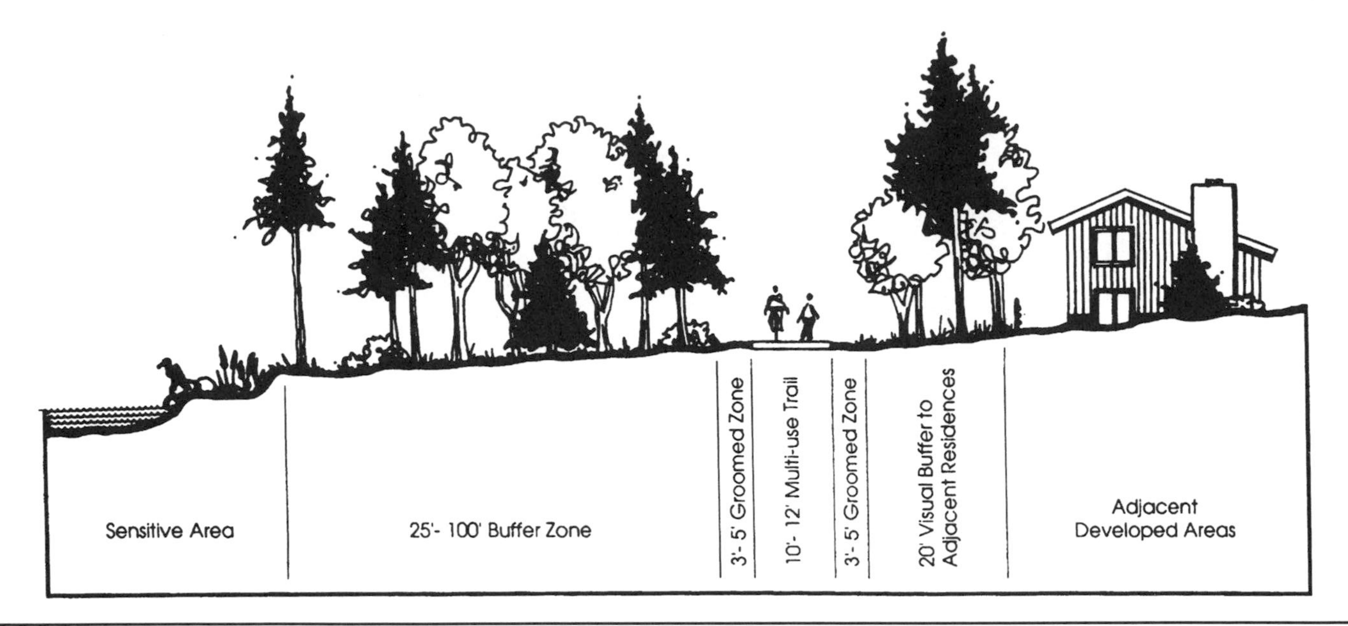

Fig. 1. Typical cross-section of a trail near sensitive areas.

Table 1. Types of Greenways

from *Greenways for America* (Little, 1990)

1. Urban riverside (or other waterbody) greenways, usually created as part of (or instead of) a redevelopment program along neglected, often run-down city waterfronts.

2. Recreational greenways, featuring paths and trails of various kinds, often relatively long distance, based on natural corridors as well as canals, abandoned railbeds, and public rights-of-way.

3. Ecologically significant natural corridors, usually along rivers and streams and less often ridgelines, to provide for wildlife migration and species interchange, nature study, and hiking.

4. Scenic historic routes, usually along a road, highway or waterway, the most representative of them making an effort to provide pedestrian access along the route or at least places to alight from the car.

5. Comprehensive greenway systems or networks, usually based on natural landforms such as valleys and ridges but sometimes simply an opportunistic assemblage of greenways and open spaces of various kinds to create an alternative municipal or regional green infrastructure.

- Restore and protect habitat along corridors where traditional animal migration routes have been identified. Identify and provide for species of plants, animals, bird life that may need a mosaic of different habitat types.

- Monitor changes in habitat quality of adjacent lands, which often provide necessary buffer zones. Promote sound land conservation practices and best practices by neighboring landowners and land managers.

- Establish a multi-agency review process that sets standards and guidelines for any actions affecting the corridor.

- Provide for community input at all phases of vision, design, implementation and maintenance. Address ongoing stewardship of the land with a management plan.

Other design guidelines include:

- Define a corridor large enough to allow flexibility, multiple connections and continuity.

- Focus on an overriding vision to prioritized and focus a set of goals. This could be to save an endangered plant, to provide a place where people can walk, to link regional parks to residential neighborhoods, or to prevent development from encroaching into floodplains.

- Keep a broad greenways concept in mind—integrated with other present and future possibilities—when defining the primary area.

- Define points of origin and destination that makes logical sense and that build upon other urban assets and points of interest.

- Base the size of the project on what the sponsoring organization and community can realistically undertake immediately, with phases to create a longer-term implementation plan.

Table 2. Greenway Inventory Checklist

Greenway Inventory Checklist

Project Name:	Date:

Description:

Land Ownership:
- ☐ Current use of land
- ☐ Zoning/types of use permitted
- ☐ Location of property lines
- ☐ Impact of greenway - current land use, ownership, activity
- ☐ Contact made with property owner

Environmental Assessment:
- ☐ Vegetation - species variety, size, growth habit, age, health
- ☐ Geology - landform type, composition, suitability
- ☐ Soils - type, composition, suitability
- ☐ Hydrology - drainage, water table, streams, tributaries, ponds, wells, springs, urban influence, wetlands
- ☐ Topography - steepness or flatness, longitudinal slope vs. cross-sectional slope
- ☐ Significant natural features - rare/endangered, unique, landmark
- ☐ Wildlife - species, nesting grounds, migratory routes, food sources
- ☐ Climate - wind, sun angle, exposure, rainfall

Access and Transportation:
- ☐ Existing access - motorized, nonmotorized, water-based
- ☐ Desired access - location, type
- ☐ Existing transportation within site - type, purpose, size
- ☐ Existing transportation available to & from site - types, purposes, sizes, relationships to each other
- ☐ Intersections - types, level of safety/danger, crossing type needed
- ☐ Relationship of greenway to existing transportation corridors - physical dimensions, traffic volume, noise, views
- ☐ Future transportation plans - expansions, modifications, additions
- ☐ Mass transit - linkage, mutual benefits

Socioeconomic Analysis:
- ☐ Political subdivision/jurisdictions regulating greenway
- ☐ Governing laws/regulations
- ☐ Organization of support - primary support, supplemental support
- ☐ Organizations in opposition - concerns, critical problems, & issues
- ☐ Fiscal resources/constraints - who, how much, additional resources
- ☐ Community events - types, times, additional desired events

Historic & Cultural Resources:
- ☐ Historic - National Register listings, components of historic districts
- ☐ Cultural - local, regional, national

Community Recreation:
- ☐ Inventory of community recreation activities and programs
- ☐ Providers - local government, private for-profit, nonprofit groups
- ☐ Facilities - types, location, capacity, usage
- ☐ Recreation needs assessment - critical needs greenway can satisfy
- ☐ Facility needs - develop greenway in conjunction w/ other facilities

Public/Private Infrastructure:
- ☐ Existing utilities - both within & around greenway: water, sanitary sewer, electricity, cable TV, fiber optic, telephone, natural gas, storm sewer
- ☐ Future utility plans - expansions, additions, possibility for installment or right-of-access fees
- ☐ Utility providers & operators - supporters or opponents in terms of maintenance, financial support, access
- ☐ The greenway as a utility - easement widths, ability for joint use

Community Impact:
- ☐ Physical impact - type, location, affect on community
- ☐ Cultural/social impact - opportunities, problems
- ☐ Economic opportunities - types, dollars, timeframe, who will benefit

2 GREENWAY TRAIL DESIGN

Trail design objectives

The most common feature of many greenways is a trail, which principally defines a land-based path but can also include waterways, navigable as a "trail." In all cases, design must meet specific purposes, which may include some or all of the following objectives:

- **Safety:** to indicate where it is safe to pass, and indicating whether or not marked trails have been inspected to remove obvious hazards, including falls, rock slides, overhanging limbs, management of poison ivy, *etc.*

- **Way-finding:** often assisted by trail maps, and best introduced by an easily understood map at the entry trailhead. Ideally, trails and choices of paths and direction should be identified in terms of length and time (average adult). It may help to mark distances on the trails. North arrows help at key orientation points and help to teach orienteering. Way-finding systems should be very easily understood, because they are most needed by the first-time visitor who can become disoriented. Where trails have multiple paths, path intersections should indicate recommended directions and/or shortest distance to "entry/exit/parking."

- **Interpretation:** "Interpretation" is the art of telling stories about a place. Interpretive trails are normally marked with numbers that help identify things to see. Often the most effective guide is a map that is at the trailhead or simply, a one-pager trail guide that explains highlights related to a numbered trail. Signage approaches to interpretation must be tested for comprehension, content and interest. Don't overdo these. Nature is best appreciated in silence and left as is. A good rule for interpretation is "Can you improve on silence?" (For guidelines for design of interpretive exhibit signage, see Trapp *et al.* 1996.)

- **Universal design:** Interpretive trails and stations that invite public use should be designed according to principles of universal design, accessible and welcoming to all people of all ages and abilities. Providing an accessible "sensory trail"near parking is ideal to encourage easy use. It should be hard surfaced and incorporate sounds and fragrances of nature for orientation using all senses. To fulfill the spirit of universal design, greenways can offer the entire range of outdoor experiences to persons with disabilities. Persons with disabilities may welcome the opportunity to be involved in the design of a sensory trail.

- **Mystery and delight:** The experience of the natural world provides a unique experience of physical and spiritual renewal. Greenway trail design can complement and heighten the experience and appreciation of nature by artfully revealing the variety of the landscape, by highlighting large trees and boulders, by planning vegetation for diversity of density, texture and pattern, by creating short and long vistas, by using meandering curves, and by diverting the trail for quiet places of respite. Consider seating, using sreadily available natural features, such as logs and boulders.

Trail design must address the above objectives with the following design elements: the types of users; the type of trail (land- or water-based, single or multiple users); how the trail fits into existing natural landscape; type and width of the trail tread; type of tread surface (Tables 3–5 at the end of this article).

The Trail User

There are many different trail users: walkers, hikers, joggers, dog walkers, equestrians, bicyclists, cross-country skiers, snow-mobilers, and so on. Trail users can be characterized by six categories of use: pedestrian, nonmotorized vehicular, nonmotorized water, pack and saddle animal users, motorized, and motorized water trail users.

Pedestrian trail users move along a trail at a leisurely to slow pace of 0 to 5 mi. (8 km) per hour. Nonmotorized vehicular trail users are primarily bicyclists, but also people on rollerblades, skates, and skateboards, and others who use self-propelled, wheeled nonmotorized equipment. They move through a trail at a wide range of speed; the average or most comfortable speed for most users ranges from 5 to 20 mi. (8 to 32 km.) per hour.

Saddle animal and pack animal trail users are principally equestrians. Tread width is generally similar to that for other land-based users, although the clearance height for horseback riders is greater. The surface of the trail tread should be soft, yet a firmly established base is needed for easy maintenance.

Motorized vehicular trial users include snowmobilers, backcountry jeep users, and those in all-terrain vehicles or on motorbikes. Perhaps the most controversial of all trail users, motorized users are likely to conflict with others because they move much faster and are propelled by noisy engines that emit exhaust fumes. Used unsafely, they can be involved in dangerous accidents. Unless their use is designated by design, they inflict potential damage to trail bases, accelerating erosion and degradation.

Nonmotorized water trail users include canoeists, kayakers, inner tubers or rafters, fishers, scuba or snorkel divers, sail boaters, pedal-power boaters, sailboarders, and wind surfers. For a waterway, the width of the trail may not dictate movement but rather the rate of water flow, wind speed, the size and weight of the water craft, the depth to the bottom of the water course, and the physical ability of the user to direct the craft or his or her own body through the water course.

Motorized water trail users include those in ski boats, electric motor craft, motorboats or varying size, shape, and function, and fishing boats. This group of use present problems of conflicting function similar to those of land-based motorized trail users.

Trail layout

The following steps are part of designing a greenway trail layout:

- Select a configuration.
- Position the trail within the greenway corridor.
- Determine the longitudinal and cross slopes of the corridor landscape that affect trail routing and adjusting trail location and alignment accordingly.
- Define the width of the trail tread based on the projected user groups.
- Compute the design load and traffic load for the trail based on projected use.
- Identify the dominant soils along the selected route.
- Accommodate drainage patterns and wetlands existing within the greenway corridor.
- Delineate the vegetative clearing limits necessary to accommodate trail and uses.

Selecting a layout configuration

Trails must be more than the shortest distance between two points. There are various layout strategies that may be considered (Fig. 2):

Linear layout satisfies the shortest distance criterion, and because of the narrowness of most greenway parcels, is one of the most common layouts.

Loop layouts, in which the origin and the destination are the same, offer more versatility for a variety of trail users. This layout is most effective around lakes and reservoirs, through planned community settings, and for interpretive trails that begin and end at one entry point.

A stacked loop layout consists of two or more loops stacked onto a single loop located close to the trailhead. Stacking the loops adds distance and variety to the route. This can be particularly useful if greenway land varies in elevation to provide optional levels and distances for users to explore and enjoy. Orientation can be provided by clearly designating the different trails.

The satellite loop layout is a series of looped and linear trails that radiate from a central collector loop. It provides users with a primary loop for origin-to-origin walking and offers subloops or connectors to other facilities. Orientation signage is important for extended versions of this layout.

A spoked-wheel loop layout includes a series of linear trails that radiate from a central core to a main loop trail. The core could be a smaller circular trail or simply a trailhead. Orientation is accomplished easily, due to the easily understood trail pattern.

The maze layout offers a great number of alternative routes through a series of interlinked looped and linear trails. It also has the greatest variety of distances and number of intersections. This layout will require a large area and the development of a sophisticated sign system so that trail users do not get lost.

Positioning trails

Greenway trails should be located with a greenway corridor to be compatible with the natural landscape and its functions (Fig. 3). It should blend with the natural contours of the land, accommodate all designated uses without straining the carrying capacity of the landscape, and provide safe access and passage for the users. It should stimulate all of the human senses and heighten awareness of the environment. Further, it has to be built and maintained in a cost-effective way.

Trail slopes

Longitudinal slopes are changes in elevation that occur along the running length of the trail. Generally, all trails should flow with, not against, the natural shape of the land within the greenway corridor. Ideally, trails that ascend a hill should do so gradually, at slopes that do not exceed eight percent, in compliance with recommended or mandated accessibility standards.

Cross-sectional slopes cross the width of the trail tread from one side to another to accommodate the intended user and drain water from the surface of the tread. The cross slopes for all trails, regardless of type and use, should fall between one and four percent. The most commonly used cross slope is two percent. Proper corridor selection and subgrade preparation will result in economy of surfacing materials.

Tread width

In trail design, the term "tread" refers to the area of the pathway (*cf. Definitions* at the end of this article). Typically, the width of a tread will depend on the amount of available land within the boundaries of the project, the type of users who will be permitted access to the trail, the potential for conflict among users, environmental sensitivity, and the cost of constructing a trail type. A trail width minimum of 5 ft. (1.5 m) is required to provide adequate room for passersby. Individual treads can be as narrow as 20 in. (50 cm) for hikers in remote wilderness terrain. Fig. 4 depicts ergometric considerations for urban pedestrian path and trails.

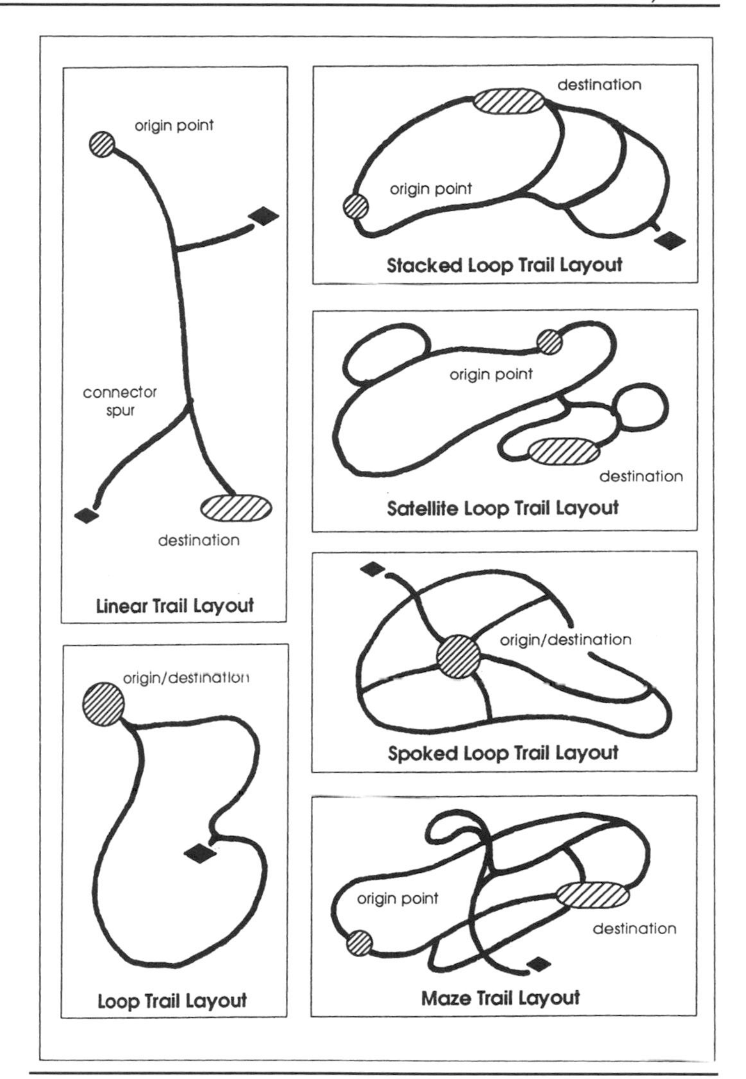

Fig. 2. Types of trail layouts.

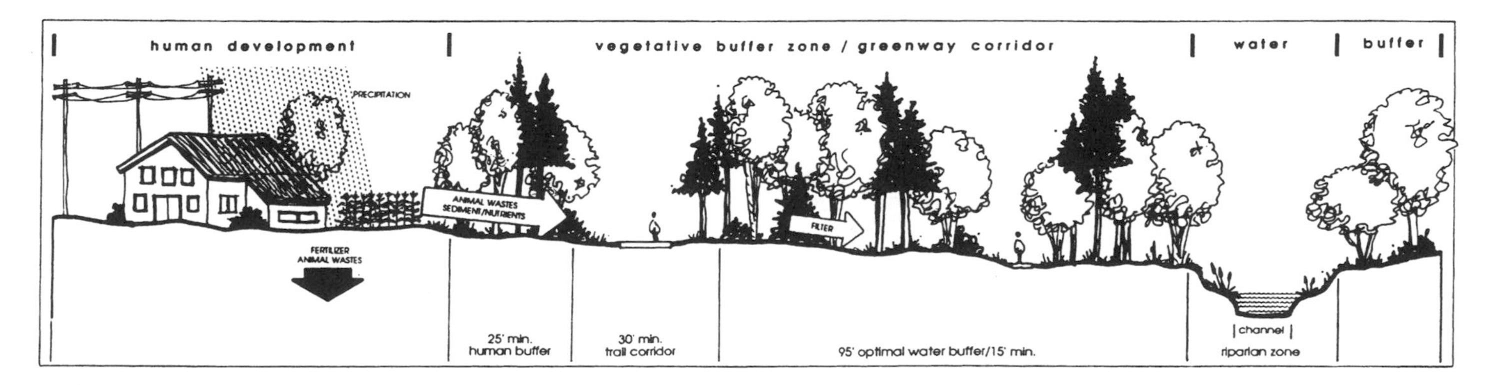

Fig. 3. Trail location within a corridor.

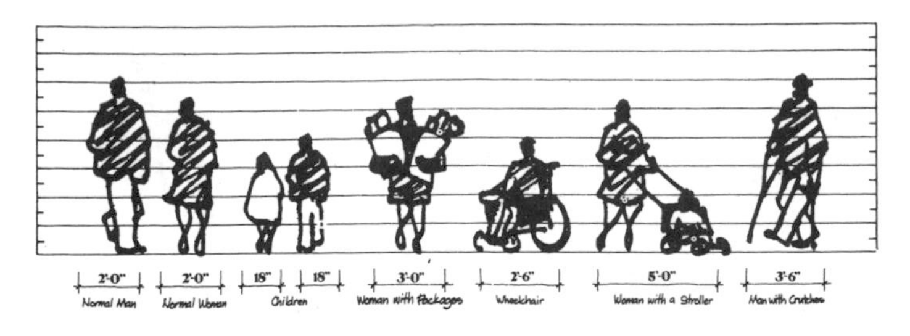

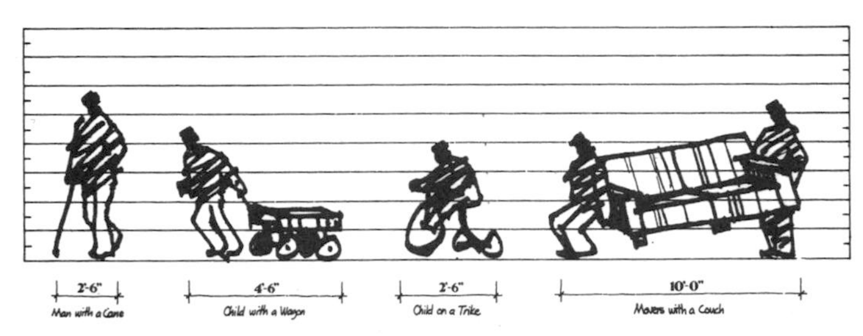

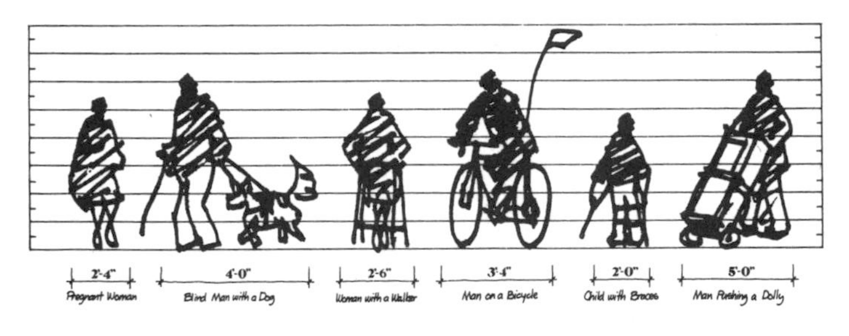

Fig. 4. Ergometric dimensions: people in outdoor settings and activities.

Low impact trail treads

Trail design must consider:

- **Suitability of the soil mass:** if the soil is seasonally under water, you will need to install an elevated walk.
- **Fragility of the eco-system:** some landscapes are too fragile to permit on-grade human access; and
- **Type of users** who are allowed access to the trail.

Not all environments can be modified to accept a multi-use trail. A low-impact trail is best suited to a specific landscape where certain uses (bicycles, horses) must be restricted. There are three factors to consider: environmental suitability of a particular use; the intensity of use; and the design and development of the trail tread.

Intense trail use of any kind can damage an ecologically sensitive landscape. A sensitive landscape can be protected and made accessible with a boardwalk. If providing a trail might encourage overuse and compromise sensitive environmental areas, it may be better to consider another route.

Trail tread standards

Following are six types of treads or pathways for land-based trails. The type that the designer selects depends on the number of different users groups to be accommodated. (Figs. 5–6)

Single-tread, single-use trails can include trails within street, road, or highway rights-of-way, off-road trails and sidewalks outside of rights-of-ways, and other pedestrian ways and trails that meander through a variety of urban, suburban, rural or wilderness landscapes.

Single-tread, multiple-use trails are the most common type in the nation. These trails vary in width from eight to more than twenty feet. Some have usage-control features such as signs or striping to separate trail users.

Single-tread, time of use trails—a relatively new concept—permits different groups who may represent conflicting uses to access single-tread trails at different times of the day, week, month, or year. This concept allows all trail users to enjoy the same facility by removing potential conflict between them.

Single-tread, zoning for multiple-use trails, also relatively new, opens different segments of a trail to different uses.

Multiple-tread, multiple-use trails allow for multiple uses within the same right-of-way but on separate treads. This generally requires a wider right-of-way to accommodate the diversity of users. All of the treads could be developed in parallel along the entire corridor or a portion of it.

Multiple-tread, single-use trails most often accommodate different skills levels or the speeds of different uses. This trail type is rare and usually developed by an organization that directly supports the single-user group, such as a walking or jogging club.

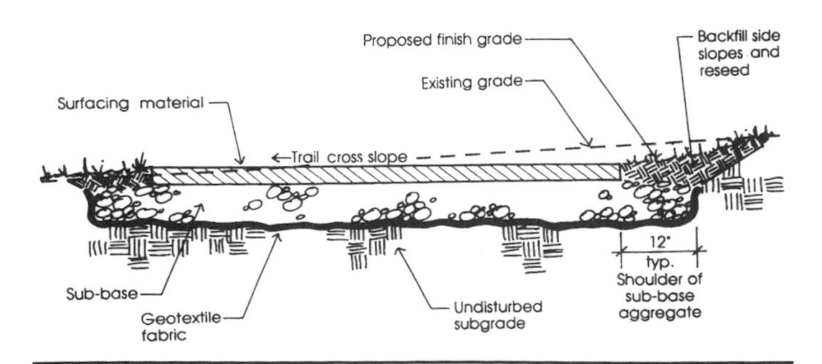

Fig. 5. Trail tread detail.

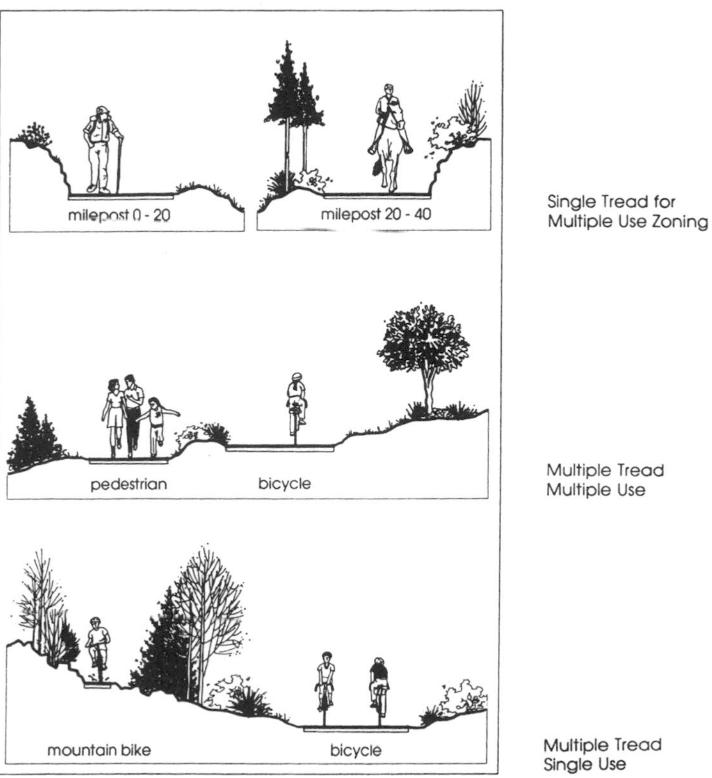

Fig. 6. Types of trail treads.

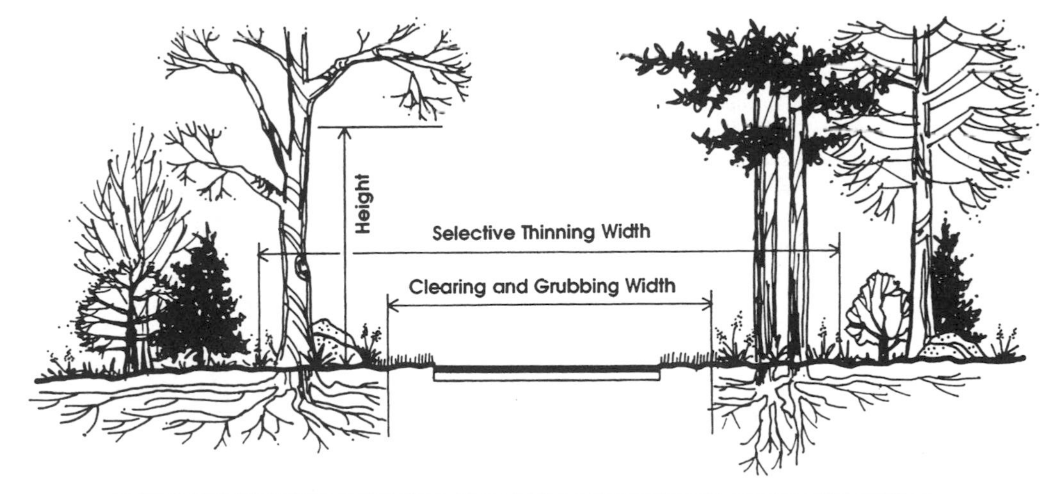

Trail Type	Clearing & Grubbing Width	Selective Thinning Width	Clearing Height
6-foot hiking only	10 feet	20 feet	8 feet
8-foot pedestrian only	14 feet	24 feet	8 feet
10-foot pedestrian only	16 feet	26 feet	8 feet
8-foot bicycle only	16 feet	26 feet	10 feet
10-foot bicycle/pedestrian	18 feet	28 feet	10 feet
6-foot horse only	12 feet	22 feet	12 feet
10-foot horse/pedestrian	16 feet	26 feet	12 feet
8-foot cross-country ski only	12 feet	22 feet	8 feet
12 foot snowmobile only	20 feet	30 feet	10 feet
18-foot ski/snowmobile	26 feet	36 feet	10 feet

Fig. 7. Vegetative clearing.

Trail tread cross-section

Constructed land-based trails consist of subgrade, subbase, geotextile fabrics, and surfacing material (Tables 6–10 and Fig. 8).

Subgrade: The subgrade is the undisturbed earth of the corridor, the primary foundation for the greenway trail. Subgrade suitability refers to the subgrade's ability to accommodate a greenway trail without drastic alternations or overly expensive development solutions. A highly suitable subgrade is one that contains moderate slopes, good drainages, and firm, dry soils. Because the subgrade ultimately receives all of the load or weight on the surface of the trail, it is important that the subgrade be structurally capable of supporting the trail design load. The subgrade must be sloped by crowning the tread or establishing a cross slope to provide property draining of surface and subsurface waters.

Subbase: The subbases' primary function is to transfer and distribute the load or weight from the surface of the trail to the subgrade. The subbase also serves vital drainage functions, preventing water from migrating up from the subgrade into the surface of the trail and allowing natural cross drainage to flow through the trail cross section. The subbase can be either hand- or machine-placed and should be compacted, smooth and level.

Geotextiles: are woven and nonwoven fabric mats used to strengthen the subgrade, subbase, and surface of a trail, especially in areas where soft or unsuitable soils are present. They ensure the strength and long life of the trail by maintaining the composition and integrity of the subbase material and preventing it from migrating into the subgrade.

Trail surfaces

The surface material should be selected on the basis of the type and intensity of trail use. Surfacing materials should give a firm, even, and dry trail tread, capable of supporting the designated users. Materials that can be used include asphalt, concrete, wood, soil hardeners, gravel, limestone, crusher fines, sand, wood chips or bark, brick or masonry, cobblestone, and grass. Surface materials are categorized as either hard or soft. The primary distinction between the two is based on the ability of the material to absorb rather than repel moisture. Advantages and disadvantages of trail surface materials are summarized in Table 11.

Soft surface materials

Soft surfaces are the most common trail materials. They are the least expensive type to install and the most compatible with the natural environment. A soft surface may not be the most practical, however, as most do not hold up well under intensive use and require continuing maintenance. There are only a few options within this category:

- **Natural surface "unprepared" tread trails**—essentially compacted earth, with occasional gravel infill in wet areas—are the oldest and most common type. They must be maintained to be free of tripping hazards. Proper drainage and elimination of mud holes is the key to optimal functioning of this trail type. If maintained, this is an appropriate surface for ecologically sensitive greenways.

- **Shredded wood fiber** is composed of mechanically shredded hardwood and softwood pulp, pine bark chips or nuggets, chipped wood pieces, or other by-products of tree trunks and limbs. This type of surface is soft on the feet and deters intrusion of growth within the trail area. It decays rapidly, and requires maintenance or replacement every several years.

Hard surface materials

Hard surfaces, although more costly, are more practical for multi-use urban and suburban trails. Most hard surface materials, while not inherently compatible with the native environment, can blend well with certain landscapes. Hard surfaces require less maintenance because they better absorb the impact of repeated use.

- **Soil cement** is a mixture of pulverized native soil and measured amounts of Portland cement. It is usually mixed at the project site and applied immediately after mixture. The mixture is then rolled and compacted into a dense surface by machinery. Soil cement will support most user groups, though use by bicyclists and horseback riders should be restricted.

- **Graded aggregate stone** is mined from natural deposits or formations. The variety of aggregate stone materials suitable for trail surfacing includes colored rock, pea gravel, river rock, washed stone and coarse sand. Installed in a loose manner, it can be considered a "soft" trail surface. This is a good surface for greenway trails because it is locally available, easy to install, maintain, and replenish, and accommodates the needs of a wide range of users within a single tread width. Aggregate surfaces are not normally recommended for bicycle trails or for use on wheelchair-accessible trails. However, granular stone is a good surface for greenway trails because it can be densely compacted, holds up well, and is compatible with the natural environment. If properly constructed in a very dense and compacted form (such as compacted cinder), it can support bicycle and accessible trail development, as long as it is properly maintained.

- **Asphalt concrete** is a hard surface material composed of asphalt cement and graded aggregate stone. Asphalt concrete is a flexible pavement, which means that it will form itself to the shapes and contours of the subbase and subgrade. Asphalt trails require little maintenance. Concrete surfaces are most often used for intensive urban applications. Concrete is a heavy, nonporous pavement. Of all surface types, it is the strongest and has the lowest maintenance requirement. It holds up well against the erosive action of water, root intrusion, and subgrade deficiencies such as soft soils.

- **Wood surfaces** are usually composed of sawn wood planks or lumber that form the top layers of a bridge, boardwalk, or deck. Wood is a preferred surface type for special applications because of its high strength to weight ratio, its aesthetic appeal, and its versatility.

5 OTHER ISSUES

The trail is the key feature in any greenway, but it is not the only consideration. There are often other structures and facilities associated with the greenway, including bridges, shelters, signage, parking, lighting, boardwalks, plantings and public art. Each of these elements requires planning, design, and engineering coordination with all urban infrastructure (Figs. 9–12).

Further, there are legal issues of liability, safety, security and rules of public conduct. Organizations and agencies that own land open to the public automatically assume a measure of responsibility, risk and liability. The owner of a greenway—whether public sector, private

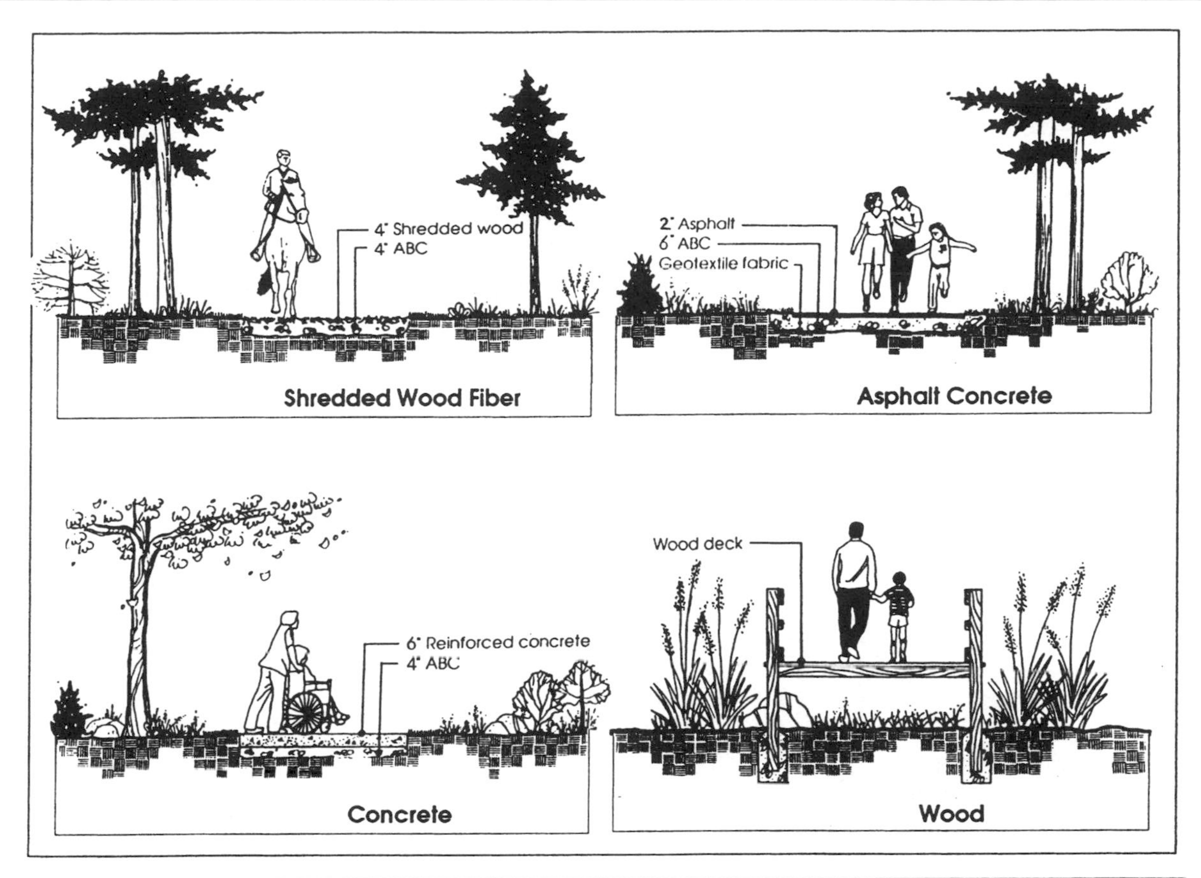

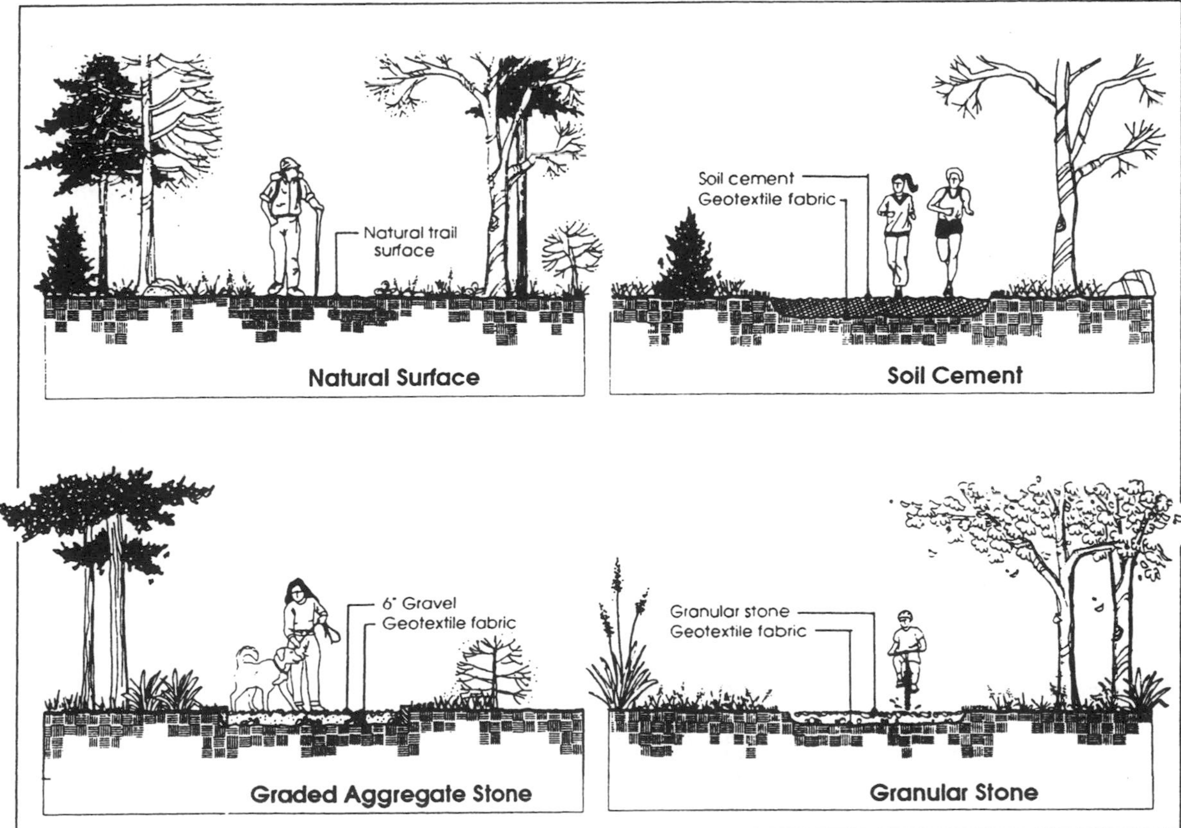

Fig. 8. Trail tread cross-sections.

PHOTO: Donald Watson

Fig. 9. Public Garden, Halifax, Nova Scotia, Canada.

Fig. 10. Trailhead. Hoadly Creek Preserve, Guilford, CT.

Fig. 11. Orientation and interpretive station marker.

sector, or nonprofit—must provide a safe facility for the full use and enjoyment of those who have access to it.

A well-conceived safety program often provides the user with a clear code of conduct for the greenway. This is especially important for heavily used multipurpose greenways. In some critical cases, such as blind spots where there is risk of conflict of use (such as pedestrians and bicyclists), painting a centerline or separated paths along those critical parts of the trail may be the simplest safety precaution. It provides a signal to users that people will be traveling in both directions, whether on bike, foot, or other modes of transportation.

It is also helpful to adopt a trail user ordinance. Although an ordinance does not in itself resolve multi-user conflict, it helps in promoting uniform and fair trail use while clarifying the expectations placed on all types of users.

Lastly, all greenways need a Management Plan. Greenways are long-term investments and with proper care, they can last for many generations. Some of the key components for a Greenway Management Plan include: user safety and risk management; maintenance; patrol and emergency procedures; administration; programming and events; stewardship and enhancement; and funding.

Good maintenance is required. Greenway planners must bring a special resourcefulness to the issue of maintenance. In many exemplary cases, community volunteers provide continuing maintenance and commitment to use of a greenway plan. ■

REFERENCES

Little, Charles. 1990. *Greenways for America*. Baltimore: Johns Hopkins University Press.

Mertes, James D., and James R. Hall. 1996. *Park Recreation, Open Space and Greenway Guidelines*. Washington, DC: National Recreation and Park Association.

Proudman, Robert D., and Reuben Rajala. 1993. *Trail Building and Maintenance,* 2nd edition. Gorham, NH: Appalachian Mountain Club.

Schwarz, Loring LaB. ed., with Charles A. Flink and Robert M. Searns, 1993. *Greenways: A Guide to Planning, Design and Development.* The Conservation Fund, Arlington, Virginia. Washington, DC: Island Press.

Smith, Daniel, and Paul Hellmund, eds. 1993. *The Ecology of Greenways*. Minneapolis: University of Minnesota Press.

Trapp, Suzanne, Michael Gross and Ron Zimmerman. 1994. *Signs, Trails, and Wayside Exhibits: Connecting People and Places*. Stevens Point, WI: University of Wisconsin Stevens Point Foundation Press.

USDA (United States Department of Agriculture) Forest Services.1996. *Standard Specifications for Construction and Maintenance of Trails.* ISBN 0-16-048802-8. Washington, DC: U.S. Government Printing Office.

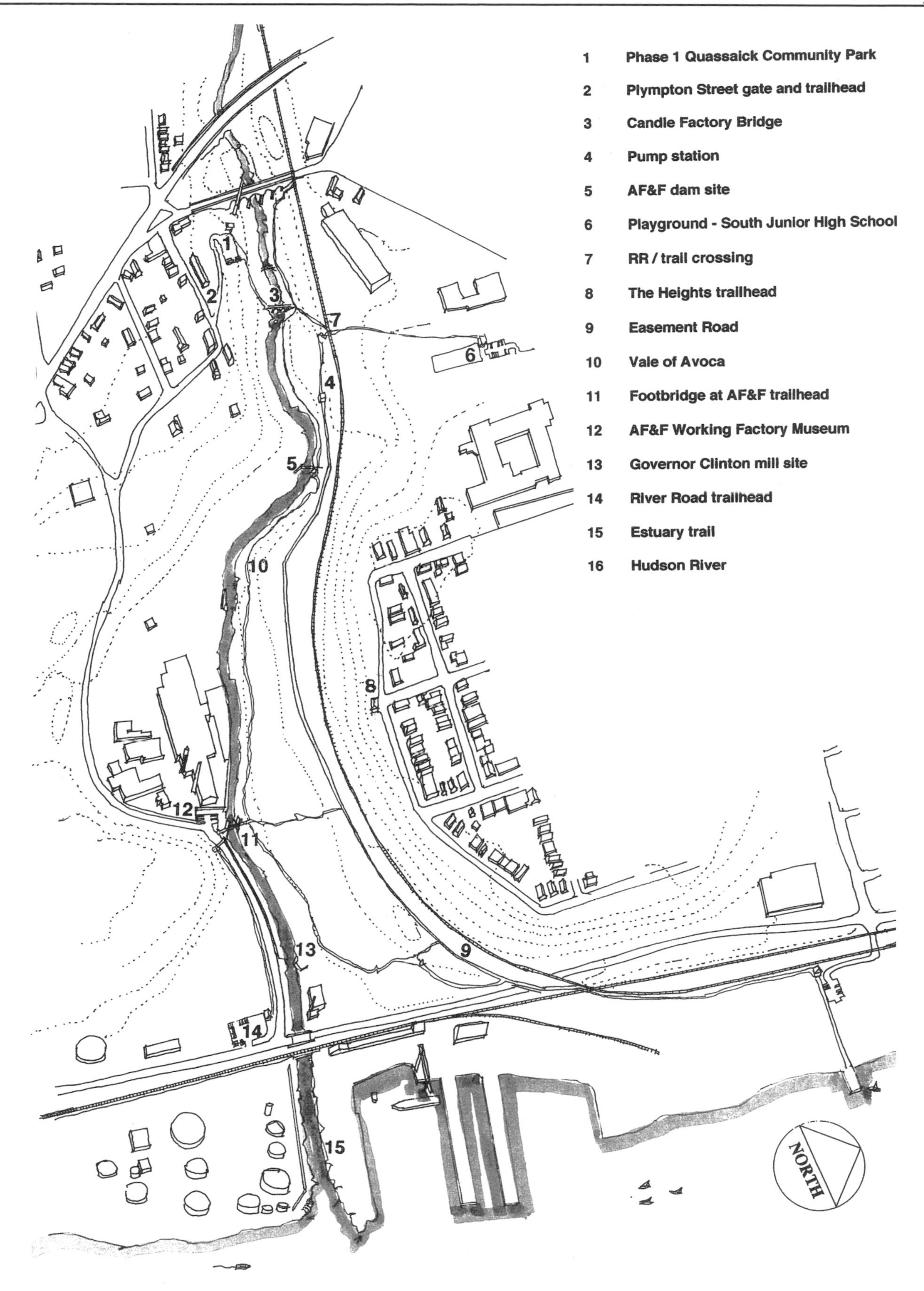

Fig. 12. Greenway Trail Quassaick Creek Estuary Preserve, Newburgh, New York. EarthRise Initiatives and Hudson & Pacific Design.

Table 3. International grading system of walking difficulty

International Grading System of Walking Difficulty

Level of Difficulty	Walking Time	Distance Traveled	Change in Elevation
Easy	Less than 3 hr	0–6 miles	Start to 1000 ft
Moderately strenuous	3–5 hr	6–12 miles	Start to 2000 ft
Strenuous	5–8 hr	12–15 miles	Start to 3000 ft
Very strenuous	8–10 hr	15–18 miles	Start to 4000 ft
Extremely strenuous	10–12 hr	18–25+ miles	Start to 5000 ft

Source: European Ramblers Association, 1987.

Table 4. Guidelines for developing accessible trails

Checklist: Access to Trails	Checklist: Accessibility of Trails
☐ Accessible parking—federal compliance	☐ Trail is at least 5 ft wide
☐ Accessible passenger loading zones	☐ Trail provides adequate passing lanes
☐ Trail is accessible from parking areas	☐ Surface is stable, firm, slip resistant
☐ Trailhead and parking area are appropriately signed	☐ Running slope of trail is 1:20 to 1:12
☐ Information provided in audio/print format	☐ Trail cross slopes are 1:50
☐ All support facilities are accessible	☐ Trail contains one distinct curb edge
☐ Activity areas are provided for all people	☐ Clear head room is at least 100 inches
☐ Summary information about trail is provided at trailhead—describes trail length, width, slopes, and surfaces	☐ Rest areas provided at 200-ft intervals
	☐ Rest areas are 5 ft by 5 ft in area
	☐ International Symbols of Access used

Source: Architectural and Transportation Barriers Compliance, Washington, D.C., 1981.

Table 5. Intensity of trail use

User Type	Low-Intensity Use	Medium-Intensity Use	High-Intensity Use
Pedestrians/walkers	2 PFM[a] or less	10–15 PFM	25 PFM
Bicyclists	50 ADBT[b] or less	250 ADBT	1000 ADBT
Equestrians	4 HTD[c] or less	20 HTD	100 HTD
Snowmobiles	100 ADT[d] or less	200 ADT	1000 ADT

[a] PFM, pedestrians per foot width or walkway/minute; source: Timesaver Standards for Landscape Architects.
[b] ADBT, average daily bicycle trip; source: Rhode Island Department of Transportation.
[c] HTD, horses per 6 foot trail/day; source: American Horse Council.
[d] ADT, average daily trips; source: International Association of Snowmobile Administrators.

Table 6. Trail design recommendations for longitudinal and cross slopes

Trail User	Average Speed (mph)	Longitudinal Slopes	Cross Slope
Hiker	3–5	No restriction	4% max.
Disabled pedestrian	3–5	2% prefer, 8% max.	2% prefer
Bicyclist	8–15	3% prefer, 8 % max.	2–4%
Horseback rider	5–15	5% prefer, 10% max.	4% max.
Cross-country skier	2–8	3% prefer, 5% max.	2% prefer
Snowmobiler	15–40	10% prefer, 25% max.	2–4% max.

Table 7. Standard tread width for bicycle-only trails (AASHTO)

AASHTO Standards	Recommended Minimum Width
One-way bicycle travel, single lane	5-ft
Two-way bicycle travel, dual lanes	10-ft
Three lanes of bicycle travel	12.5-ft

Table 8. Recommended trail tread widths for user-specific trails

Trail User Type	Recommended Tread Width
Bicyclist	10 ft (2-way travel)
Hiker/walker/jogger/runner	4 ft rural; 5 ft urban
Cross-country skier	8–10 ft for 2-track trail
Snowmobile	8-ft groomed surface for 1-way traffic; 10-ft goomed surface for 2-way traffic
Off-road vehicle	5-ft for 2-wheeled vehicles; 7-ft for 3- or 4-wheeled vehicles
Equestrian	4-ft tread; 8-ft cleared width
Wheelchair accessible	5-ft (1-way travel)

Courtesy of the State of Iowa Department of Natural Resources.

Table 9. Minimum recommended tread widths for multiuse trails (in feet)

Tread Type	Urban	Suburban	Rural
Single-tread, multiple use			
Pedestrian/nonmotorized	12	10	10
Pedestrian/saddle and pack animal	16	12	10
Pedestrian/motorized	22	22	16
Nonmotorized/saddle and pack animal	16	16	—
Motorized/saddle and pack animal (not recommended for simultaneous use			
Motorized/nonmotorized	22	16	16
Multiple tread, multiple use (each tread)			
Pedestrian only, 2-way travel	8	8	6
Nonmotorized only, dual travel	10	10	10
Saddle and pack animal, dual travel	8	8	8
Motorized use only, dual travel	16	16	6

Table 10. Environmental suitability of use within sensitive environments

	Type of Use					
Ecological Sensitivity	No Access	Walking	Bicycling	Equestrian	Snowmobile	OHV
Low (upland terrace, stable-rocky soils and mature forest)	O	●	●	●	●	◒
Moderate (lowland terrace, river valley, suitable soils, successional forest)	◒	●	◒	◒	O	O
High (wetlands, primary dunes, pocosins, steep slopes, hydric soils)	●	◒	O	O	O	O

Closed circles, most suitable; half-filled circles, suitable; open circles, not suitable.

Table 11 . Advantages and disadvantages of trail surface materials

Surface Material	Advantages	Disadvantages
Native soil	Natural material, lowest cost, low maintenance, can be altered for future improvements	Dusty and dirty, ruts under heavy use, not an all-weather surface, limited use
Soil cement	Uses natural materials, supports more usage than native soils, smoother surface, low cost	Surface wears unevenly, not a stable all-weather surface, costly, erodes, difficult to achieve correct mix.
Graded aggregate stone (washed stone, gravel)	Hard surface supports heavy use, moderate costs, natural material, accommodates multiple use	Angular stones can be sharp, continuous maintenance required, uneven surface, erosion, ruts
Granular stone (limestone, cinders)	Soft but firm surface, natural material, moderate costs, smooth surface, accommodates multiple use	Surface can wash away, ruts, erodes, constant maintenance to keep smooth surface, replenish stone—long-term expense, not for steep slopes
Shredded wood fiber	Soft, spongy surface—good for walking, moderate cost, natural material	Decomposes under high temperature, moisture, and sunlight, requires replenishment—long-term expense
Wood (boardwalks, bridge decking)	Pliable surface—excellent for multi-use; natural material blends with native landscape, spans streams, ecologically sensitive areas, and soft soils; only surface that places trail user above surrounding grade	High installation cost, easy to damage and vandalize, expensive to maintain, deteriorates with exposure to sun, wind, and water, susceptible to fire damage. Can be slippery when wet
Asphalt concrete	Hard surface, supports most types of use, all weather, does not erode, accommodates most users simultaneously, low maintenance	High installation cost, costly to repair, not a natural surface, leaches toxic chemicals, freeze and thaw can crack surface, access of heavy construction vehicles
Concrete	Hardest surface, easy to form to site conditions, supports multiple use, lowest maintenance, resists freezing and thawing the best, can be colored, all weather	Joints result in bumpy surface, high installation cost, costly to repair, not a natural looking surface, access of construction vehicles
Recycled materials	Good use of trash, surface can vary depending on materials, good life expectancy	High purchase and installation cost, aesthetics

DEFINITIONS

Terms commonly used in greenway and trail design (USDA Forest Service 1996).

Base Course. The layer or layers of specified material of designed thickness placed on a trailbed to support surfacing.

Batter. A backward and upward slope of the face of a wall.

Berm. The ridge of material formed on the outer edge of the trail that projects higher than the tread.

Borrow. Suitable materials taken from approved sources designated on the drawings or on the ground, to be used for embankments and backfilling.

Bridge. A structure, including supports, erected over a depression or stream, and having a deck for carry traffic.

Cap Rock. Rock placed in the top or uppermost layer in a constructed rock structure, such as a talus or rubble rock section or rock retaining wall.

Catch Point. The outer limits of a trailway where the excavation and/or embankment intersect with the ground line.

Clearing Limit. The area over and beside the trail that is cleared of tress, limbs, and other obstructions.

Climbing Turn. A reverse in direction of trail grade without a level landing used to change elevation on a steep slope.

Compacted. Consolidation that is obtained by tamping or rolling suitable material until no noticeable displacement of material is observed.

Culvert. A drainage structure composed of rock, metal, or wood that is placed approximately perpendicular to and under the trailway.

Cushion Material. Native or imported material, generally placed over rocky section of unsurfaced trail to provide a usable and maintained traveled way.

Danger Tree. An unstable tree (5 in./125 mm) or greater in diameter at breast height that is likely to fall across the trail.

Designated on the Ground. The location of materials, work areas, and construction items, including lines and grades, marked on the ground with stakes, flagging, tags, or paint.

Drawings. Documents showing details for construction of a facility, including but not limited to straight-line diagrams, trail logs, standard drawings, construction logs, plan and profile sheets, cross-sections, diagrams, layouts, schematics, descriptive literature, and similar materials.

Duff. Organic material overlying rock or mineral soil.

Embankment. A structure of suitable material placed on the prepared ground surface and constructed to the trailbed elevation.

Excess Excavation. Material in the trailway in excess of that needed for construction of designed trailways.

Ford. A water-level stream crossing constructed to provide a level surface for safe traffic passage.

Full Bench. Trailbed constructed entirely on undisturbed material.

Grade. The vertical distance of ascent or descent of the trail expressed as a percentage of the horizontal distance.

Header Rock. Rock laid with the narrow end toward the face of the wall.

Inslope. Where the trail is sloped downward toward the backslope.

Interpretation. An educational activity which aims to reveal meanings and relationships through the use of original objects, by firsthand experience and by illustrated media.

Mineral Soil. Soil or aggregate that is free from organic substances and contains no particles larger than (2 in./50 mm) at their greatest dimension.

Outslope. Where the trail tread is sloped downward toward the embankment or daylight side of the trailway.

Sideslope. The natural slope of the ground, usually expressed as a percentage.

Slough. The material from the backslope or the area of the backslope that has raveled onto the trailbed.

Slump. Where the trailbed material has moved downward, causing a dip in the trial grade.

Special Project Specification. Specifications that detail the conditions and requirements peculiar to an individual project, including additions and revisions to the standard specifications.

Surfacing. Material placed on top of the trailbed or base course that provides the desired tread.

Suitable Material. Rock that can be accommodated in the trail structure, and soil free of duff with a recognizable granular texture.

Switchback. A reverse in direction of trial grade with a level landing used to change elevation on a steep slope, usually involving special treatment of the approaches, barriers, and drainages.

Trailbed. The finished surface on which base course or surfacing may be constructed. For trials without surfacing the trailbed is the tread.

Trailway. The portion of the trail within the limits of the excavation and embankment.

Tread. The surface portion of the trail upon which traffic moves.

Turnout. A short section of extra trail width to provide for passage of trail users.

Waterbar. A structure used for turning water off the trail, usually made of logs or stones.

Water Courses. Any natural or constructed channel where water naturally flows or will collect and flow during spring runoff, rainstorms, etc.

Mary C. Means

Summary

Heritage areas form an armature upon which the urban design of cities and towns reside. Issues of density, zoning, and community character central to urban design are also part of a larger regional pattern. Relating the patterns of the regions to those of its municipalities and to in-between spaces demands a hybrid approach to planning and urban design. It needs the best of planning and urban design together with economic development, tourism, cultural interpretation, transportation and recreation planning. Several problems plague the heritage planner/ designer, including the coordination of multiple partners, developing the needed leadership for implementation, and defining stopping rules. Even so, heritage planning offers a kind of de facto regional planning base for urban design. The heritage base often can provide the much-needed vision for urban design interventions at several scales.

Key words

Adaptive reuse, heritage area, heritage corridor, National Park Service, regional planning, tourism, way finding graphics

PHOTO: Mary Means

Canal Locks, Lockport, Illinois. Along the Illinois & Michigan Canal National Heritage Corridor, the towpath trail is a popular visitor amenity, linking the many canal towns.

Heritage areas as an approach to regional planning

1 BRIEF HISTORY OF HERITAGE AREAS

In a scattering of places across the United States, a new hybrid approach to regional planning seems to be finding support among an unlikely range of interests. Heritage areas, heritage corridors, and regional heritage tourism efforts bear examination for what might be emerging as a promising platform for regional planning. Where did it begin, how does it happen, what results are emerging, and how does it relate to urban design?

Some twenty years ago, a visionary group of activist planners in northern Illinois, led by Gerald Adelmann of the nonprofit Open Lands Project, saw the potential for harnessing historic preservation to ecological conservation to create a new regional approach to community development. Their logic was based on the principles of National Parks, modified to work in living, working communities. Implementation would be accomplished via collaborative partnerships with landowners, local governments, state and Federal agencies. The birthplace of this new "heritage area" concept was the highly diverse landscape—river, historic towns, urban neighborhoods, steel mills, highways, tank farms, pristine natural areas, state parks, wetlands

PHOTO: Mary Means

Lancaster, Pennsylvania. The historic Watt & Shand department store in downtown Lancaster will house a luxury hotel and conference center and is an anchor project for the Lancaster-York Heritage Region, Pennsylvania's newest heritage area.

Credits: This article is based on Means, Mary. 1999. "Happy Trails," *Planning*, Chicago: American Planning Association, August, pp. 4–9.

Slatersville Mill, Massachusetts. In the 24 towns that comprise the Blackstone River Valley NHC, historic textile mills are the most prominent architectural landmarks. They present re-use challenges that are gradually being met. Slatersville was the first planned community in the U.S. Its handsome mill is used for warehousing by a manufacturer who intends to renovate it.

Mendon, Rhode Island. Always near the mills were villages of worker housing, such as these examples in the Blackstone River Valley National Heritage Corridor, in continuous use since the 1830s.

and shreds of prairie—along the 140-mile (225 km) route of the Illinois & Michigan Canal from its mouth in Chicago southwest to Peru, Illinois.

In the fifteen years since the Illinois & Michigan Canal National Heritage Corridor was designated by Congress, much has happened in the many other heritage corridors and heritage areas undertaken in the wake of I&M's success. A number of states have embraced the idea, setting up and funding their own heritage area programs, sometimes congruent with a nationally designated area, sometimes not.

By 2001, the U.S. Congress had designated and recognized twenty-three such efforts. Add to these the state heritage areas, and places that are actively working towards some formal designation, and the list would grow to well over 100. At first the National Park Service (NPS) was reluctant to acknowledge the hybrid heritage area approach. Now, however, NPS has become an active and supportive partner, although senior officials remain concerned about proliferation and potential budget impact on established National Parks.

Why is there such grassroots interest in this complicated, seemingly messy way of dealing with community development and historic preservation? How does heritage development differ from traditional regional planning? Does the heritage area movement hold promise for urban designers, planners and community development leaders? It seems timely to step back and take a look.

2 HERITAGE AREAS TODAY

Heritage areas range in size from the 500-square mile (1,300 sq. km) Essex National Heritage Area in Massachusetts, to Silos & Smokestacks America's Agricultural Heritage Partnership National Heritage Area, which encompasses 30+ counties in northeastern Iowa. Geographically, they include rural landscapes like much of the South Carolina National Heritage Corridor, as well as urbanized areas like Bethlehem and Allentown in Pennsylvania's Delaware & Lehigh Canal National Heritage Corridor. Heritage area planners share a belief that economic development, regional action, and cultural development are inextricably related. Their premise is that if new economic activity followed typical modern patterns in these special areas, development could obliterate much of what makes up the distinctive character of the place.

Typically, heritage planners are frustrated by the constraints of traditional economic development plans. They are concerned with the limitations of working within the rigid lines of municipal jurisdictions and see their work as offering a platform for crossing boundaries and working on a regional scale.

Urban designer and planner Tom Gallaher, AICP, is convinced that heritage areas are more than conventional regional planning. Gallaher, who headed one of the Congressionally designated areas (Silos & Smokestacks, America's Agricultural Heritage Partnership) through its formative years, visited many heritage corridors during a four-month field trip in 1998. "They are also places where community leaders are acting as convenors of ideas and dialogue, helping to put existing organizations together in new partnerships to accomplish projects," says Gallaher. "Project plans in heritage areas tend to be holistic and flexible, they welcome appropriate growth and manage change in an orderly manner. They really are addressing regional issues in new ways." Gallaher is assisting the Alliance of National

Heritage Areas, whose members are the twenty-three federalldesignated heritage areas, with a business plan.

The cultural forces that shape a region are common threads that do not recognize town lines. People can get together around the heritage area issue and overcome traditional conflicts—at least temporarily—to achieve "a lot of little," as Gallaher puts it. "Small, quick, positive and memorable events or programs where people can meet, work together, and develop the trust and confidence to undertake more challenging projects."

Virtually every successful heritage area involves leadership from business, government, and nonprofit groups. New York's Mohawk Valley Heritage Corridor Commission benefited from the leadership of F. X. Matt, head of the historic Matt Brewery in Utica. He and other business leaders saw the need to work together if their region was to survive. They viewed the heritage corridor as a natural vehicle for regional cooperation in a part of the country where such cooperation is rare.

Government is an active player in heritage areas, generally participating by providing technical assistance or small grants for planning and project feasibility. In most of the nationally designated heritage areas, federal funds help support a staffed management and coordination office, often governed by a commission. The federal presence is coordinated by the National Park Service, and in the case of Silos & Smokestacks, where agriculture is the central focus; the U.S. Department of Agriculture is also a partner.

States' contributions are growing. Pennsylvania's heritage parks program is entering its second decade of providing the supporting framework and flexible grant funding for ten designated "partnership parks." Maryland's new Heritage Preservation and Tourism Program complements that state's smart growth policies and programs, providing incentives for collaboration and reinvestment in heritage areas. New York's Urban Cultural Parks Program, the first of the statewide initiatives, has been renamed the New York Heritage Areas Program and continues to provide matching grants and a support system for some twenty recognized heritage areas, including the Mohawk Valley Heritage Corridor. Louisiana's Department of Culture, Recreation & Tourism is providing leadership for the thirteen-parish Atchafalaya Trace Heritage Area, which is the heart of Cajun country.

Nonprofit groups are vital partners in heritage development. The work (and funding) of the Federal I & M Canal NHC Commission is augmented significantly by the nonprofit Canal Corridor Association, which has become expert in matching corridor projects and programs with foundations and corporations. Among the strengths of nonprofits is their familiarity with philanthropic sources of funding for heritage activities, including land conservation. Nonprofit land trusts in Connecticut, Illinois, Maryland, Massachusetts and other states are ready partners in open space and stewardship projects, stepping in with money to buy key parcels at risk of inappropriate development.

3 WHAT GOES ON IN HERITAGE AREAS?

In terms of focus, most heritage areas blend education, revitalization, growth management, cultural conservation, and recreation. A premium is placed on raising local consciousness and pride through story-telling—making buildings and landscapes come alive through interpretive programs, walking tours, folk life festivals, museum

PHOTO: Mary Means

Providence & Worcester Railroad Station, Woonsocket, Rhode Island. The splendidly renovated railroad station houses the offices for the Heritage Corridor, and sparked several blocks of downtown revitalization.

PHOTO: Blackstone River Valley Heritage Corridor

The Blackstone River Valley Heritage Corridor blends natural and industrial heritage as the Chamber of Commerce pitches, "a local tradition of business development as old as the industrial revolution."

PHOTO: Mary Means

Grovesnordale, Connecticut. In the Quinebaug-Shetucket River National Heritage Corridor, a rural housing cooperative has been handsomely renovated mill housing. Architect: ICON Architecture, Boston.

displays and trailside exhibit systems. These educational efforts are vital ingredients in restoring community pride—an important first step for places hard-hit by postindustrial decline.

In Pennsylvania's once-thriving anthracite region, school children in the Lackawanna Heritage Valley each year produce a highly successful youth heritage festival, highlighting the continuing presence of the valley's many distinctive ethnic groups.

Both the Blackstone River Valley NHC and the Delaware & Lehigh Canal NHC have invested in well-designed, unified way finding graphic systems. Panels with the corridor's distinctive graphic image include the stories of noteworthy events or provide direction to visitors. "When the signs were finally installed all over the valley, it felt different, more like we were a bigger place," one local official said with pride.

Heritage areas seek a balance between preservation and economic development. Main Street programs are a natural part of the heritage development strategy, since historic downtowns are an ideal setting for such visitor services as specialty shopping, restaurants, and lodging. State Main Street programs in Iowa, Illinois, Louisiana, Pennsylvania, Massachusetts, and Maryland have been key contributors to heritage area efforts.

Adaptive use of historic buildings is widespread. A case in point is Peninsula, Ohio, a tiny crossroads village halfway between Cleveland and Akron that would likely have been obliterated by urban sprawl, but for the Cuyahoga National Recreation Area and the Ohio & Erie Canal National Heritage Corridor. Village buildings now house galleries, restaurants, and bike rental shops serving the growing traffic on the corridor's popular thirty-mile bike trail and scenic railroad.

More and more heritage areas are working as partners with industrial development agencies and chambers. The Blackstone Valley Chamber of Commerce proudly uses the heritage corridor as an example of its progressive regionalism and high quality of life. The Chamber's promotional materials pitch "a local tradition of business development as old as the industrial revolution."

Designers know about resistance to land use planning in rural areas. In this respect, heritage areas are no different. It can take a while to build the relationships, trust, and awareness necessary to take on land use and growth management issues. But taking a less-technical, more heritage-rooted approach can prove effective in motivating actions to conserve farmland, distinctive towns, and historic places.

Since its inception in 1986, the Blackstone River Valley NHC Commission has quietly and steadily advocated that towns become more active in planning. During the late 1980s, through a series of low-key workshops, heritage corridor staff helped several Corridor towns do vision plans and set goals, supported by funding from the Economic Development Administration. In the late 1990s, the need to deal seriously with the issue of land use was given greater urgency as the region's robust economy added to residential development pressures. Michael Creasey, executive director of the Commission, rallied local leaders. "For an area without regional powers," says Creasey, "the Corridor Commission, along with our partner, the Blackstone Valley Chamber of Commerce, comprise the regional response team." The Corridor and Chamber have started the new Blackstone Valley Institute, "a 'virtual institute' with a pot of 'hot-spot' money," as Creasey describes it. "When an opportunity or situation arises, BVI can act quickly by providing technical assistance and small feasibility grants to help towns engage in growth management and land use planning."

"Our work is beginning to show results," says Creasey. "As an example, a prime farm in Grafton, Massachusetts was likely to be developed. The town approached us and it led to a plan to do cluster development on a small part of the land, producing enough revenue to set aside the remainder to be leased and farmed. We brokered 'contingent commitments' that led to the action plan. Amazingly, the town is actually buying the farm and acting as the developer—an act that was strongly supported by town meeting!"

Creasey and his Corridor colleagues have also partnered with the regional chamber of commerce and the Central Massachusetts Regional Planning Commission on an urban design plan for appropriate industrial and business park development along Route 146, the heritage corridor's main highway. "A new interchange will soon connect Route 146 to the Massachusetts Turnpike, and we wanted to get ahead of the expected pressure for sprawl," explains Creasey.

Here's another example of back door planning: in Connecticut, scenic roads have become opportunities to build support for enhanced land use policies. With funding from the state department of transportation, a scenic byway management plan was developed for Route 169, the picturesque thirty-mile long spine of the Quinebaug & Shetucket Rivers NHC. The planning team's collaborative approach engaged property owners, elected officials, and business and civic groups from five towns in problem solving that led to heightened support for action. The public's strong support to "keep it just the way it is" prompted regional leaders to explore zoning overlays, tree preservation ordinances, and the purchase of development rights to protect threatened views. As a result, each town incorporated the scenic byway plan in its comprehensive plan. So far, two of the towns have passed protective ordinances and others are considering them.

Some projects are quite small. In the South Carolina Heritage Area, rural leaders became engaged when planners pitched the Savannah River Scenic Highway as a cultural experience. Rather than advocate building scenic overlooks, planners and town leaders proposed innovative pull-offs into town centers, encouraging visitors to discover a real variety store, snack at a boiled peanut stand, or take a walking tour. Communities are now implementing these plans, and cleaning up town gateways.

Many heritage areas incorporate cultural programs. In the Silos & Smokestacks NHA, nearly everyone in Spillville is of Czech descent. The tiny (pop. 387) Iowa community organized an international music festival to celebrate the 100th anniversary of the death of Anton Dvorak, who composed the New World Symphony after spending a summer there as church organist. More than 100,000 people came to the festival. With success under their belts, local leaders are now taking on other progressive initiatives.

Heritage areas often encompass brownfields, strip-mined lands, and other vestiges of our industrial past. Properly cleaned up, reforested or developed, sometimes with hiking and biking trails, these lands regain value and add to regional competitiveness.

Interviewed recently for an update of the Lackawanna Heritage Valley plan, long-time regional leader Robert Mellow, a Pennsylvania state senator, saw a direct connection between quality of life and economic value. He observed, "You would never have had a developer laying

PHOTO: Blackstone River Valley Heritage Corridor

Moffet Mill in Lincoln, Rhode Island (1812) is one of very few examples of early water powered machine shops.

PHOTO: Steven Schukraft

Plum Branch South Carolina. Along the Savannah River Scenic Highway, designers envision "pull-offs" in tiny towns, similar to scenic waysides along more traditionally scenic roads. Here, visitors are encouraged to get out and walk around the virtual ghost town to experience the rural south.

PHOTO: Holly Hanney

St. Martinville, Louisiana. In Louisiana's Atchafalaya Trace Heritage Area some eight unique towns are engaged in "Main Street" revitalization programs, which include building rehabilitation. St. Martinville, home of the legendary Evangeline, is the center of Cajun culture.

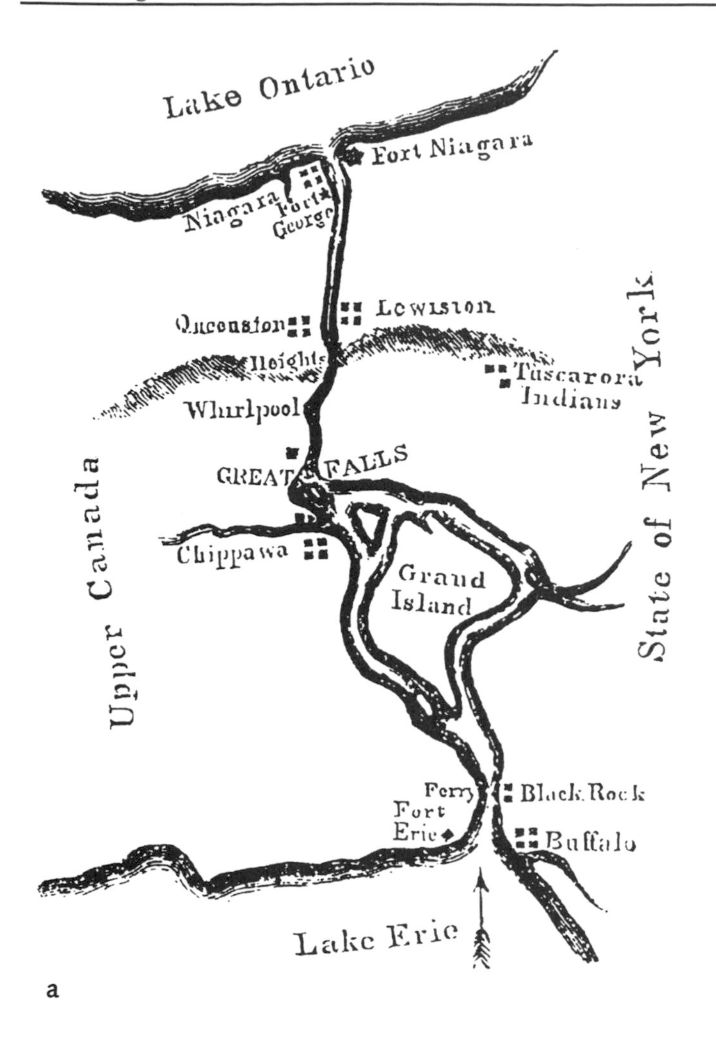

The tourist map of 1818–1819 of the Buffalo-Niagara region. (a) Illustrates a historical effort to redefine a region and its relationship to the world. The map was drawn by Johann George Kohl in (1923) and is part of the Toronto Metropolitan Reference Library archive. (b) The map is the aerial view looking from Lake Ontario towards Lake Erie depicting the region in a drawing from the same time period. Aerial drawing from the Buffalo and Erie County Historical Society.

out a subdivision and selling $25,000 lots on former culm banks (mine tailings) in Jessup, Pennsylvania without the adjacent Lackawanna Heritage Valley Trail and Riverside Park."

In western Pennsylvania, where acid mine drainage plagues streams long after coal mining has declined, the Allegheny Heritage Development Corporation, coordinating the large Allegheny Ridge NHA, has catalyzed efforts to tackle the problem at a watershed level, forming the eighteen-municipality Allegheny Ridge Heritage Coalition. The Coalition is sanguine about the challenge. "This is not a short-term turnaround," says planner David Sewak. "It took 100 years to get to where we are, we have to chip away as best we can."

Sewak is a "circuit rider," who assists grassroots efforts, including the Keystone Off Road Riders, a dirt-bikers group that ordinarily would be at odds with environmental groups. With the Coalition's encouragement, the bikers are purchasing abandoned mine sites, cleaning them up and using them as off-road recreation areas. "They aren't bad guys," says Sewak. "They love the outdoors too, they just love it in a different way."

Not surprising, tourism leaders are active proponents of heritage area development. They recognize heritage areas and corridors as ideal ways of packaging and organizing a region's historical, cultural and recreational attractions.

"The heritage area experience satisfies several core needs underpinning people's travel requirements, including the desire for authenticity," points out Elaine van S. Carmichael, a planner and economic analyst very active in tourism and leisure markets. "Moreover, heritage areas accommodate different travel modes and styles, ranging from intrepid explorers to outdoor recreation enthusiasts to group tour patrons."

4 NAGGING PROBLEMS

Heritage area planning and development touch on many fields of endeavor, from economic development to transportation and recreation. But with so many small projects, it is sometimes difficult to attract the attention of established constituencies.

Coordinating multiple partners can be wearing, but it is a vitally important task. Finding funding for coordination—such essential functions as recruiting progressive leaders, brokering contingent commitments and forging partnerships—is a lot harder than getting capital projects funded. "Elected officials seem to have no problem understanding brick and mortar projects," said one frustrated manager whose Congressman has been particularly effective in this regard. "But, the important programmatic elements—the noncapital parts—are less easily sold via traditional legislative paths."

Other funding sources include foundations and corporate grants, as savvy heritage developers are increasingly learning. Illinois' nonprofit Canal Corridor Association has been a model of success in attracting philanthropic and corporate support for activities in the I&M Canal Corridor. "The value of the Corridor in terms of quality of life, our work in environmental stewardship, recreation, and community pride make pitching it to funders less difficult than one might expect," says Gerald Adelmann. Gallaher agrees with this from his Silos & Smokestacks experience, where regional foundations made multi-year commitments for general operating funds, and viewed their participation as that of venture capitalists backing a new process through the formative stages.

Some heritage areas have a problem with leadership. It takes a special blend of talents to work in heritage area development. Not surprisingly, good planning professionals possess many of the right qualities. However, attracting and keeping talented staff is a challenge, especially as the more mature heritage areas navigate the transition from the first passionate spirited leader.

But sometimes heritage areas bring out the best in people. "Working with other forward-thinking people who share a love of the region's heritage is a very positive experience, especially if small projects succeed and lead to larger ones," says Dr. T. Allan Comp, a historian and planner who has worked with several heritage areas. "With a few successes, they are poised to take on more challenging issues."

5 WHEN TO STOP?

How many is too many? The National Park Service and key Interior Department worry that the national designation may be abused. While supportive of effective partnerships to achieve preservation and stewardship goals, the feelings about "pork barrel" politics still run beneath the surface. "I like the results I see, but I still wish we had some form of omnibus legislation to provide clear guidance about standards and roles," said one Congressional staffer.

Far more important is the realization that the creation of heritage areas has given us a kind of de facto regional planning. Think about it. By designating a heritage corridor and engaging progressive leaders from all walks of life in a web of activist projects, we meet many of the goals of regionalism, including growth management, farmland conservation, and economic revitalization. Heritage areas offer an important low-key platform for achieving many of the intentions of regionalism, including growth management, farmland conservation and environmental regeneration.

Heritage areas may be messy and hard to describe, but at the same time they seem to possess a tenacious spirit and grassroots vitality that is very healthy. ■

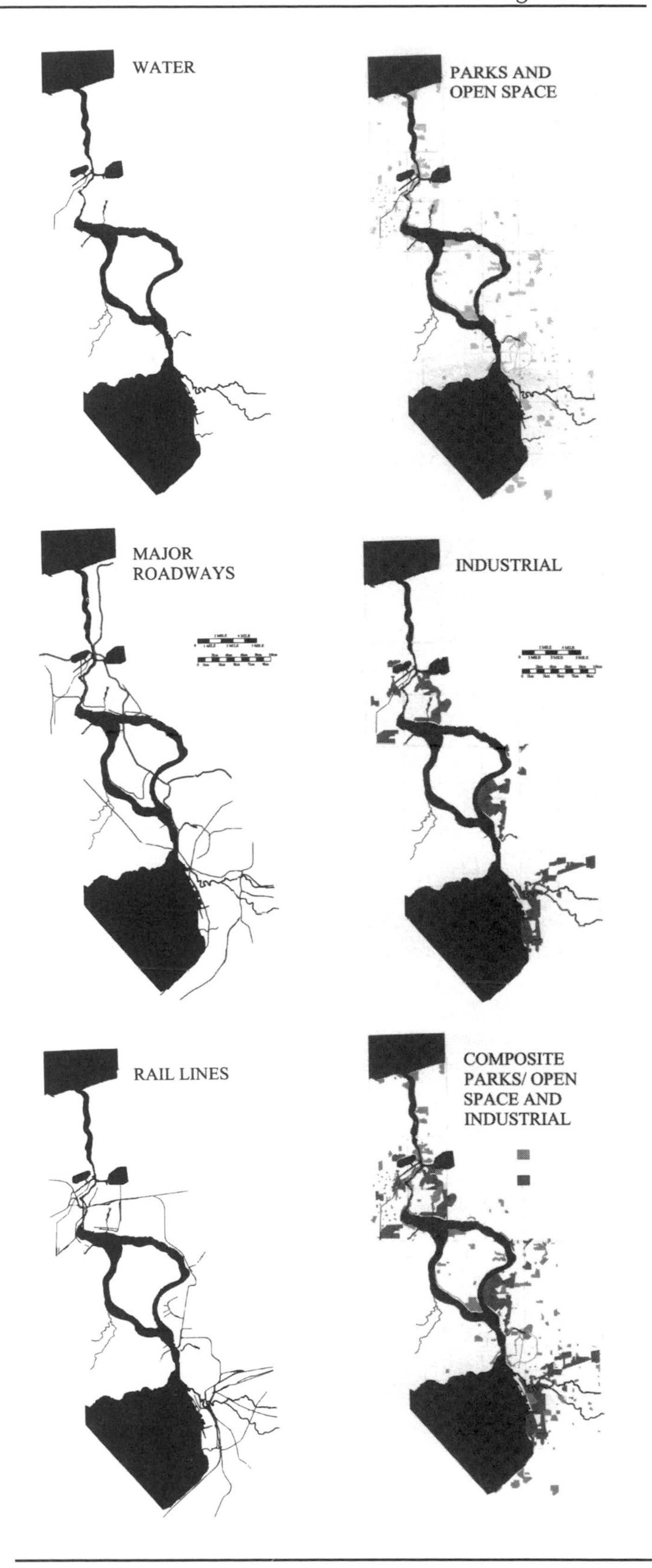

Fig. 11. Mapping key elements of the Niagara River on both the Canadian and U.S. sides was an initial part of the emerging heritage development program called Rethinking Niagara being developed by the Urban Design Project at the University at Buffalo, State University of New York and the Waterfront Regeneration Trust in Toronto, Canada. Maps: The Urban Design Project.

Aerial view of submerged village, Guerrero Viejo, B. Parvin from Sanchez, 1994.

REFERENCES

Alliance of National Heritage Areas Internet website at www.cofc.edu/~heritage

Blackstone Valley National Heritage Corridor Internet website at www.nps.gov/blac/index.htm

Sanchez, Mario L. 1994. *A Shared Experience.* Laredo: Los Caminos del Rio Heritage Project and the Texas Historical Commission.

Shibley, Robert G., and Bradshaw H. Hovey, eds. 2001. *Rethinking the Niagara Frontier.* The Urban Design Project. Buffalo: University at Buffalo, State University of New York.

Shibley, Robert G., and Bradshaw H. Hovey, eds. 2001. *A Canal Conversation.* The Urban Design Project. Buffalo: University at Buffalo, State University of New York.

Mary C. Means

Summary
Town centers are special places and require special effort if they are to remain vital in community life. Downtown revitalization calls for sensitive urban design, well executed and managed over time. Other essential elements include some form of collaborative management and public-private cooperation. The Main Street Approach is a time-tested framework that has been adapted successfully by more than 1000 communities in North America. Its lessons are significant for urban design practitioners.

Key words
Business Improvement Districts, downtowns, pedestrian zones, revitalization, traffic calming

Rutland, Vermont. Center Street and Merchant's Row. The scale of downtown Rutland typifies the image of "Main Street" held by many people. It was not always this way. Due to a strong community plan, and a decade-long commitment to downtown management and marketing, Rutland's many independent shops and restaurants now thrive—across the street from the nation's only downtown Wal-Mart.

Main Street—two decades of lessons learned

1 INTRODUCTION AND HISTORICAL OVERVIEW

Main Street is one of a handful of images imbedded in the American identity. For many, the term conjures up memories, real or imagined, of hometown friendliness, bustling activity, celebration and commerce—a people-scaled environment in the center of things. Born in an era where "urban design" was determined by the distance a person could walk or the turning radius of a horse and buggy and streets were framed by buildings constructed before elevators and steel made great heights possible, Main Street's pedestrian-friendly public environment has become an intuitive standard for quality place making.

The challenges facing Main Street became more complicated during the last quarter of the 20th century. Change has been hard on downtowns in modern times. Automobiles require different environments. Traffic-clogged streets were widened or made one-way, highways bypassed downtowns, and, later, entire towns were ringed by interstates. Commerce needs traffic to survive. Businesses responded to the new movement patterns by moving out to the highway. Older downtowns experienced enormous changes as the retail action shifted to strips and malls, mail order and Internet. Often, professional services and even government offices moved away from the core.

In addition to the shifts caused by the departure of downtown's traditional activity generators, there was a subtler, but ultimately equally devastating loss: leadership. Once the downtown had been "where it was at." The community's top business leaders owned buildings, ran businesses, worked in offices and interacted over lunch there. Big-budget advertising by department stores attracted customers to the whole downtown, bringing business to smaller shops, too. Downtown sort of "took care of itself." As department stores and major office tenants migrated, so did downtown's traditional "leaders." Disinvestment was more than in financial terms. If downtowns were to return as vital places, the traditional forces that "took care" of downtown had to be replicated by organized revitalization efforts.

Beginning in the 1960s, several waves of downtown design ideas gained fashion:

- **Ban cars.** The Nicolett Mall in Minneapolis was highly influential and stimulated Richmond, Indiana and scores of other small cities to remove traffic and pedestrianize the downtown (Fig. 1).

- **Enclose downtown.** Rockville, Maryland and Middletown, Ohio were two well-publicized examples of the move to compete with the mall for fair-weather shoppers.

- **Look modern.** Unified modernized storefront programs were pitched to progressive downtown merchants. A town in Nebraska "slip covered" every building on its town square in pastel shades of aluminum.

Credits: Photos are by the author unless otherwise noted.

PHOTO: Courtney Fisher

Fig. 1. Richmond, Indiana was an early adopter of pedestrianizing Main Street, and has struggled to retain retail activity.

Fig. 2. Cazenovia, New York. Preservationists encouraged restoration of fine 19th century commercial buildings through an early façade improvement program.

PHOTO: Steven Schukraft

Fig. 3. Saturday Farmers Market, Alexandria, Virginia. Historic Alexandria has had a farmers market on this location for more than 200 years. Many other downtowns have re-established farmers markets, sidewalk art shows, and festivals of all kinds to bring people back.

- **Look historic, theme it up.** Festival marketplaces, façade improvement programs, Victorian gaslights and streetscaping were attempts to recapture a sense of the past (Fig. 2).

- **Tear it down.** Urban renewal programs brought wholesale clearance of "blighted" downtowns, too often leading to decades of vacant land in the former center of town.

Smaller cities emulated larger ones. Unfortunately, the trickle-down "solutions" were too often out of scale, and seldom went beyond streetscaping—brick-tex sidewalks and honey locusts. Few downtowns seriously tackled the deeper, less visible issues: economic competitiveness, business reinvestment, and effective marketing. After all, it was easy to blame downtown merchants for downtown ills.

By the mid-1970s, activists in the preservation movement had become alarmed by the loss of important downtown buildings in towns across the country. Frustrated by the lack of a workable approach, the National Trust for Historic Preservation's Midwest Regional Office launched the Main Street Project, a pilot initiative to demonstrate "economic development within the context of historic preservation." In partnership with three towns (Madison, Indiana; Galesburg, Illinois; Hot Springs, South Dakota), the Trust worked with local leaders to find out what actually worked and to see if it could be replicated elsewhere. Trust staff served as project managers on-site for three years. Professionals in urban design, marketing and management, and economic development worked with the Trust and the pilot towns to develop practical strategies and implement them. The success of the Main Street Project—vacant buildings rehabilitated, new business started, small-scale but dramatic physical improvements, strong downtown management organizations—rekindled confidence and brought life back to all three downtowns.

2 THE MAIN STREET PROGRAM

Along the way, the Main Street Project team carefully documented what was happening, and produced publications, conferences, and a popular film that became a staple on the chamber of commerce/Rotary Club circuit. Hundreds of towns found the Trust's 4-point Main Street Approach to be a practical conceptual framework for their own downtown revitalization programs.

Derived from the pilot town successes, informed by the failures, and tested further as the initial program expanded to involve thirty towns in six states, the Main Street Approach is predicated on a program of conscious attention to the overall image of downtown, with an orchestrated balance of civic effort in four areas:

- **Organization.** Reaching consensus and building a strong, staffed downtown management group to orchestrate many individual players, and partners with government.

- **Promotion.** Marketing downtown as a destination, the center of community life; producing events that draw people, re-establishing a positive image for downtown (Fig. 3).

- **Design.** On-going attention to everything that contributes to downtown's overall image: buildings old and new, signs, sidewalks, streets, parking lots, connections, landscaping, lighting—seven window display. Design also addresses functional concerns, too, like circulation and the location of parking.

Economic Restructuring. Few older downtowns are still the center of their retail market, but strong businesses do remain and new uses for old buildings—residential, arts, culture, entertainment, and e-commerce—can capitalize on a downtown setting, bring new life, and enhance property values. Business development is a key activity.

As the original pilot program evolved into a national program, the National Trust formed the National Main Street Center in 1980. The Center continues to provide technical assistance and a support network for some 1,200 participating communities in the U.S. and Canada, including a number of urban neighborhoods where the commercial corridor is their "main street."

Also during the last two decades, many cities have discovered the value of the business improvement district (BID) as a mechanism to provide the sustainable funding required for downtown management and marketing. BIDs are publicly designated districts where property owners are assessed for additional funding to support additional services, including marketing, security, maintenance and management (Fig. 4).

Downtown areas generally are more complex and need a higher level of attention than most other parts of the city. Through BIDs downtown property owners make sure top-flight services are in place. The International Downtown Association (IDA) has provided leadership in helping both the Main Street Approach and BIDs gain widespread application. Both the National Main Street Center and IDA provide essential information and services for downtown practitioners and urban designers.

Fig. 4. Santa Monica, California. Bicycle patrols, parking management and the other centralized services provided by Santa Monica's BID make the Third Street Promenade a regional magnet for people seeking an enjoyable and entertaining experience.

3 LESSONS LEARNED

What can one learn from Main Street? Two decades of experience by countless downtowns can be summarized in ten lessons:

- **The four-point Main Street Approach**—organization, promotion, design and economic restructuring—really works. When orchestrated in an integrated manner, the process remains strongly relevant as a framework for coordinated action. Overemphasis on design or any of the other points, to the neglect of the others, is usually much less successful. Where there has been a sustained, balanced, well-led Main Street revitalization effort, property values and activity are up, as is community pride.

- **Design plays a vital role in restoring and revamping the overall public image of downtown.** Design efforts can be big and small, addressing the many ways downtown's image is projected. Celebrating small but important design changes—a new sign, a rehabilitated building—sends a constant positive message and emphasizes the fact that downtown is constantly changing, never finished (Fig. 5).

- **Investment in conscientious management pays off in business terms:** higher property values and a stronger downtown economy. Shopping center developers know that management and marketing are essential. So do smart Main Streets (Fig. 6).

- **Leadership will not come from the merchants.** Their job is to run successful businesses that attract people downtown. Instead of waiting for rescue, those with a strong vested interest in revitalization—

Fig. 5. Salem, Massachusetts. Business signs add a lot to the overall image downtown projects. Revitalization programs sometimes provide free design assistance or financial incentives to encourage better-designed storefronts and signs. This example shows that signs do not always have to be "old fashioned" to fit in to historic environments, if they are well designed.

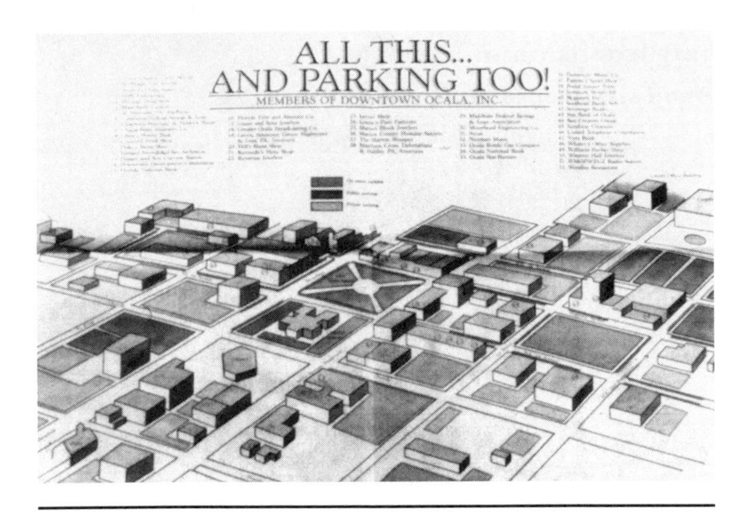

Fig. 6. Athens, Georgia. Well-designed presentations of downtown, such as the one the Downtown Development Authority has produced with this newspaper supplement, signal the overall image of quality—and let consumers know someone is making sure they can park conveniently.

Fig. 7. Holland, Michigan. Downtown is very much the heart of Holland, where healthy shops include such traditional businesses as hardware, clothing and pharmacies, as well as newcomers like a popular outdoor outfitter. Holland's sidewalks invite strolling, well-placed benches and bike racks are used seven days a week by office workers, shoppers, and students from Hope College, whose campus is on downtown's edge.

property owners, lenders, and city government—need to take the lead, reaching out and including merchants.

- **Sustainable funding of the downtown management operation is essential.** Struggles for money have killed otherwise effective programs. The best mechanism available is the business improvement district, which provides additional funding to support revitalization priorities.

- **Successful Main Street programs recognize the importance of retail shopping, but also know that downtown is much more than shops and restaurants.** Successful revitalization programs see downtown as the desirable location for many community activities—arts, cultural and educational institutions. Where would downtown Akron, Ohio be without its minor league ballpark, right on the 100 per cent corner? Or Savannah's Broughton Street minus the students and faculty of Savannah College of Art and Design—many of whom live downtown in renovated buildings?

- **Planning downtown with people in mind is essential.** Sociologist and urban planner William H. Whyte taught us to watch what people do, and make design decisions that reflect actual behaviors. Whyte's work has helped retrofit formerly sterile parks and plazas, and filled them with activity. Simple things like putting seating in the shade in hot climates (obvious? not necessarily in practice) can make a difference. Towns that work for people work well for business (Fig. 7).

- **Cars are not going to go away, but they do not have to dominate downtown environments.** Pedestrian malls failed because they removed the traffic retailers' need to show off merchandise, and in truncating traditional movement patterns, closed streets caused people to bypass downtown. Taking out pedestrian malls, instituting traffic calming, converting underused and overengineered one-way streets back to two-way movement, and managing parking are all promising urban design trends in downtowns (Fig. 8).

- **Downtown does not exist in a vacuum.** Downtown rebirth can and should catalyze the revitalization of near-in older neighborhoods, too. Many communities are seeing property values rise in older neighborhoods as road congestion eats up more and more time for busy families. The convenience of living near it all is more and more attractive to young people and empty nesters.

- **There is no formula for revitalization.** The four points—organization, promotion, design and economic restructuring—are a loose framework, not stone tablets. Evaluate, revisit, refresh. Trends change, markets shift, opportunities arise. Successful communities periodically evaluate their progress, keep their eyes and ears open to "best practices" and adapt them to local needs. One factor will always be present: the inevitability of change. Downtowns are remarkably adaptable when the community cares.

A final word about urban design and downtown. Design can be a noun or a verb. It can also be an adjective when used as a modifier, as in "design process" or "design management," an important and often neglected aspect of urban design. Orchestrating the many, many individual actions and changes that make up how downtown looks and works takes conscious design management. Much of the most important downtown planning and urban design work being done today is spearheaded by business improvement districts and Main Street programs. Helping guide their work is the body of experience known as the Main Street Approach. Two decades of practice and adaptation has enriched its value as a design management vehicle for historic town centers and urban commercial corridors.

Fig. 8. Nantucket, Massachusetts. Short-term on-street parking spaces are essential for retail in downtown. Parallel or diagonal parking provides convenient "errand running" locations.

4 IDEAS THAT WORK

- Provide Main Street building owners and merchants with professional design assistance for free or at low cost. Building improvements, store window display, signs and graphics are all part of downtown's image. A few "before and after" examples can make a big difference in momentum (Fig. 9).
- Use small matching grants to stimulate owner investment. Faced with a hideous out-of-scale and out-of-date sign? Provide the owner with a new design, pay to remove the old one and give a grant towards the new one—if it is a well-designed one.
- Promote a unified image for downtown through well-designed graphics and joint advertising programs.
- Make public investments contingent on private commitments: the city will step in with some of the hardscape and trees when building owners step up to the plate and form a business improvement district.
- Do a lot of little rather than bet the ranch on a mega-project. A lot of little can be accomplished incrementally, and can lead to mega-improvement in overall image.
- Document, document, document. Take photos, monitor economic indicators, and use both to tell the story, to promote the progress, to re-imagine a downtown on the move. ■

PHOTO: Norman Mintz

Fig. 9. Corning, New York. Urban design professionals from the nonprofit Market Street Restoration Agency worked with building owners and transformed a once-shabby downtown into a thriving retail and office environment. "Before and after" photographs from Corning inspired many other towns to undertake similar programs.

REFERENCES

Houstoun, Jr., Lawrence O. 1997. *BIDs: Business Improvement Districts*. Washington D.C.: Urban Land Institute/International Downtown Association.

International Downtown Association. www.ida-downtown.org

National Main Street Center. www.mainst.org

Paumier, Cyril B. 1988. *Designing the Successful Downtown*. Washington D.C.: Urban Land Institute.

Whyte, William H. 1980. *The Social Life of Small Urban Spaces*. Washington D.C.: Conservation Foundation.

Robert Cervero and Michael Bernick

Summary
This article outlines the transit village as a paradigm for creating attractive and sustainable communities, both in the city and the suburbs, where rail transit systems are or will be in place. The most important physical elements of the transit village—civic plazas near train entrances, pleasant walking environs, diversity in housing, compactness—are identified along with general principles of transit village design.

Key words
light rail transit, mixed uses, neighborhood revitalization, pedestrian planning, station-area development, transit agency, transportation

Terminus of pedestrian corridor and transit center, Bellevue, WA

Transit-centered urban villages

America's transit village movement is in many ways a reaction to the perceived declining quality of urban and suburban living. Traffic jams, faceless sprawl, and disconnected land uses are among the many reasons more Americans are looking for new and different paragons of suburbia. Failed public housing and inner-city entitlement programs are among the reasons that different approaches to urban revitalization are sought.

The transit village model is not new, having been successfully put into practice a good century ago in a number of U.S. cities. They provide important historical lessons as we consider, plan, and build transit villages of the 21st century.

1 THE CASE FOR THE TRANSIT VILLAGE

The transit village brings together ideas from the disciplines of urban design, transportation, and market economics. It is partly about creating a built form that encourages people to ride transit more often. However, equally important, it embraces goals related to neighborhood cohesion, social diversity, conservation, public safety, and community revitalization.

At its core, the transit village is a compact, mixed-use community, centered around the transit station that, by design, invites residents, workers, and shoppers to drive their cars less and ride mass transit more. The transit village extends roughly a quarter mile from a transit station, a distance that can be covered in about five minutes by foot. The centerpiece of the transit village is the transit station itself and the civic and public spaces that surround it. The surrounding public spaces or open grounds serve the important function of being a community gathering spot, a site for special events, and a place for celebrations—a modern-day version of the Ancient Greek agora.

Transit villages are not just physical entities. There are important social and economic dimensions behind transit village design. By creating an attractive built environment, complete with a civic core and prominent transit node, people are more likely to feel a sense of belonging and an attachment to the community.

Transit villages must also be economically viable and financially self-sustaining. Creating attractive urban environments that have good transit access to the rest of the region should, by definition, produce economic benefits. By creating better quality neighborhoods in areas with superior transit services, private investors will return to these areas, putting them on a road to financial recovery.

Elements of Transit Village design

The elements of Transit Village design are:

1. Enhanced mobility and environment
2. Pedestrian friendliness
3. Alternative suburban living and working environments
4. Neighborhood revitalization
5. Public safety
6. Public celebration

Enhanced mobility and environment: The primary transportation benefit of congregating housing, jobs, shops, and other activities

Credits: This article is adopted by the authors from their recent book, *Transit Villages in the 21st Century*, McGraw-Hill, 1997.

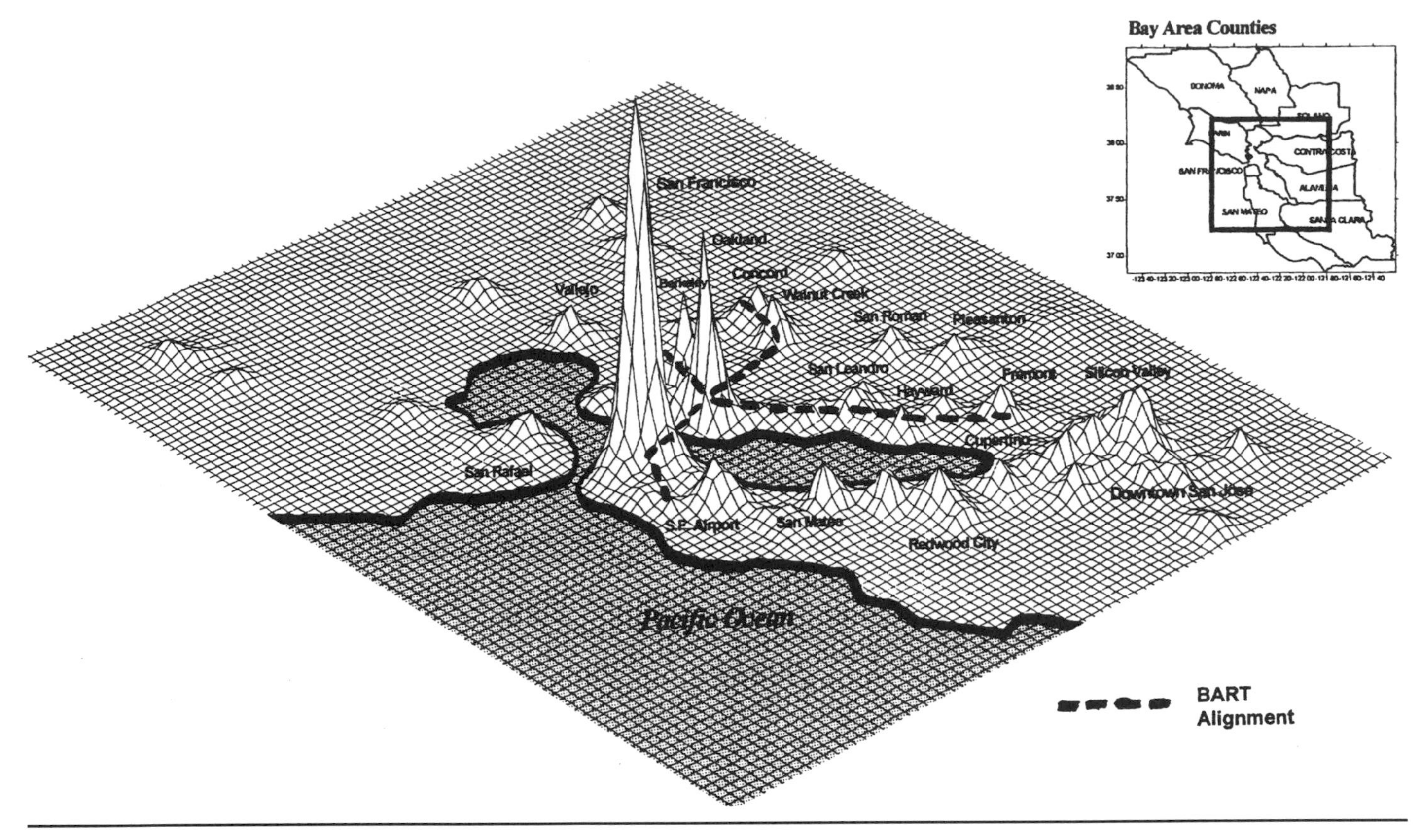

Fig. 1. Bay Area Employment Densities correlate closely with BART Authority Alignment (1990 data).

around transit stations is that transit ridership is likely to increase as a consequence. Replacing auto trips with train rides can help relieve traffic congestion along corridors served by rail. An important spin-off benefit is improved air quality, especially to the extent that park-and-ride trips are converted to walk-and-ride or bike-and-ride.

Pedestrian friendliness: By definition, transit villages should be inviting places for pedestrians. Mixes of land uses, with some housing built above ground-floor shops, can encourage walking, as can the narrow tree-lined streets, wide sidewalks, and an absence of large surface parking lots and long building setbacks.

Alternative suburban living and working environments: The transit village offers the opportunity to live in the suburbs without being entirely dependent on the automobile and with the rich variety of activities and services usually associated with cities. Transit lines can provide the location for intensifying commercial space and employment opportunities convenient to residences without increasing automobile dependence (Fig. 1). Transit villages should relieve pressures to intensify existing suburban neighborhoods, enabling them to maintain their cultural hegemony and low-density qualities. As metropolitan areas continue to grow and development pressures mount, transit villages provide a kind of safety valve. They produce additional housing that minimizes impacts on local and regional roads, does not contribute to sprawl, and enables existing neighborhoods to remain intact.

As an alternative suburban community, the transit village vision calls for a mix of housing suited to a range of incomes and lifestyle preferences: condominiums, duplexes, apartments, and single-family detached units. The diversity of housing in transit villages can mean a much-needed increase in the stock of affordable housing.

Neighborhood revitalization: The transit village offers a fresh, new approach to stimulating economic growth in inner-city neighborhoods served by rail. It invites private investment by creating the conditions for financial gain—from the foot traffic of the commuters regularly heading to stations, from the value added of siting commercial buildings near viable transit nodes, and from the benefits of a well-planned urban milieu.

Public safety: Few issues are more important to attracting people to transit villages than public safety. Residents, in particular, must regard the transit village, with its many activities and offerings, as a secure and safe place in which to live. Residents themselves are the most valuable asset in this regard. A transit village populated by residents, workers, and shopkeepers is a place where there is a continual security presence. Building a police substation, a common feature in many transit village plans, can further enhance public security and safety.

Public celebration: A public plaza that leads into the village's station entrance provides a natural spot for community gathering—a place for celebrations, parades, and performances. It is important that the transit station, functioning as the window to the rest of the region, is physically tied to and associated with the village's major gathering place.

The idea of a transit metropolis borrows from the visions of early city planners such as Sir Ebenezer Howard in England, and Frederick Law Olmstead and Edward Bellamy in America, who advanced the idea of building pedestrian-oriented "garden cities." Howard's vision

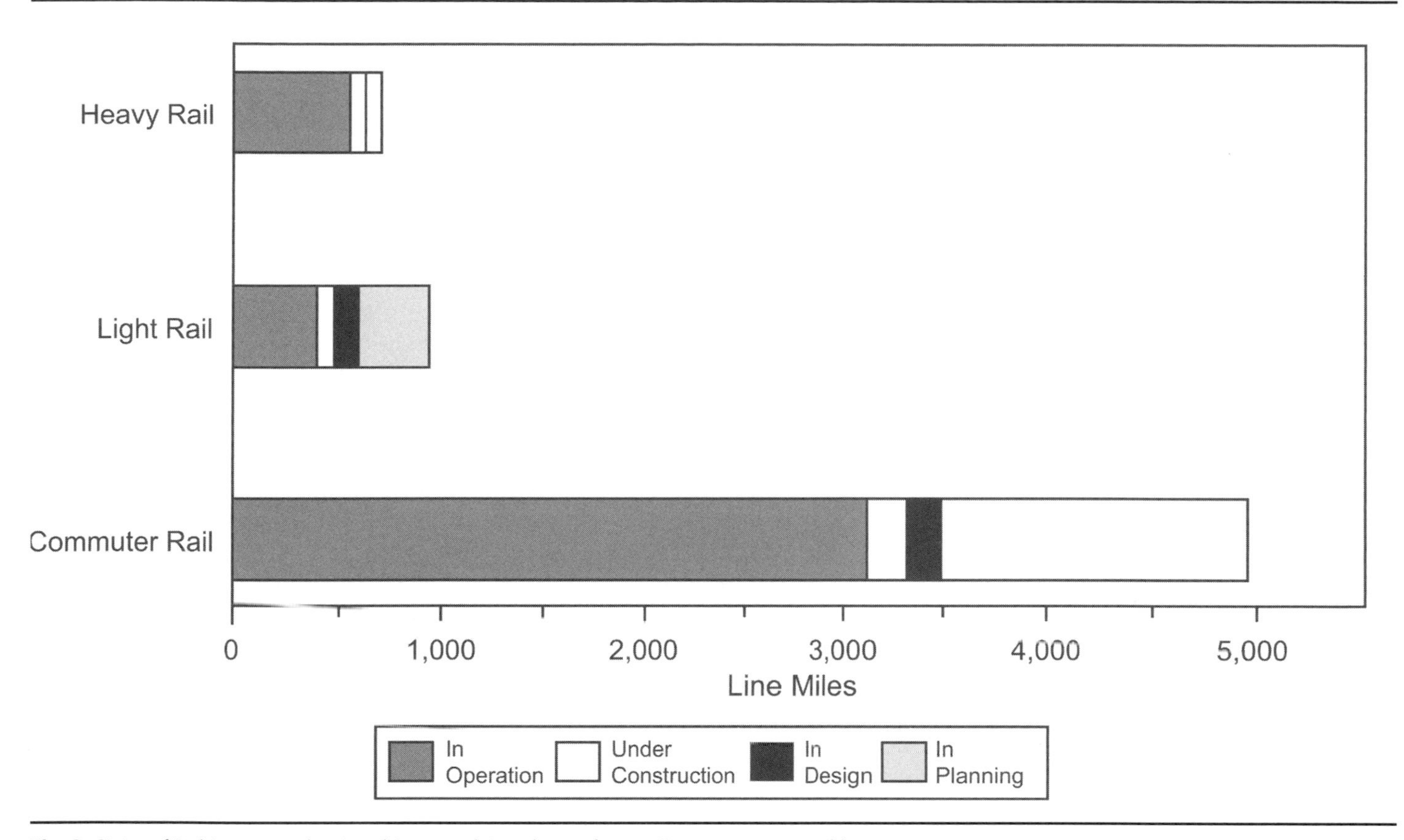

Fig. 2. Status of Rail Systems in the United States and Canada, as of 1995. (Source: American Public Transit Association, *Transit Fixed Guideway Inventory,* April 10, 1995).

was to build self-sufficient satellite communities of around 30,000 inhabitants that would orbit London, separated by protected greenbelts and connected by intermunicipal railways. Some vestiges of transit villages survive in the former "streetcar suburbs" of turn-of-the-century America, such as Shaker Heights in Cleveland, Chestnut Hill in Boston, Riverside near Chicago, Roland Park in Baltimore, and Country Club Plaza in Kansas City.

America's early rail-served neighborhoods featured a range of housing from large estates to small cottages, had distinctive gridiron street patterns, and focused on prominent civic areas near rail stops to instill a sense of community.

As a vision of urban development, the notion of a transit metropolis cannot be ignored if for no other reason than billions of dollars are today being invested in rail transit throughout the United States. In nearly every major metropolitan region of America, rail transit systems are being built, revived, or expanded.

2 CONTEMPORARY POLICY CONCERNS AND MASS TRANSIT

Fig. 2 and Table 1 document the current status of rail systems in North America. Continuing financial and political support for mass transit is rooted in many of the pressing urban policy dilemmas we face in the United States. Reasons for backing transit vary, but draw largely upon growing concerns over the sustainability—environmentally, economically, and socially—of a highly automobile-dependent society. The following are some of the escalating policy concerns that are behind the continuing support of mass transit in the United States.

Traffic congestion: Good quality transit services can induce shifts in traffic use modes and intensity. Concentrated development around transit nodes can induce spatial shifts in travel. In this sense, transit village development is a form of transportation demand management.

Air quality: Few other issues today drive transportation policy as much as concerns over air quality. Photochemical smog remains a serious problem in more than 100 U.S. cities, with the worst conditions in California and industrial areas of the northeast. At extreme levels, smog can impair visibility, damage crops, dirty buildings, and most troubling, threaten human health.

Other environmental concerns: Currently, automobiles and trucks are the two largest sources of carbon dioxide emissions in the United States, responsible for 20 percent of the total emissions. Some methane, in the earth's atmosphere, will eventually induce changes in precipitation, ocean currents, and seasonal weather patterns and cause crop damage, rising sea levels, and possibly the extinction of plant and animal species.

Environmentalists often attribute other costs to the automobile and the spread-out cities it has helped create. Roadways and parking lots consume over 30 percent of developed land in most U.S. cities and as much as 70 percent of downtown surface areas. In 1988, the United States averaged 82 feet (25 meters) of roadway per capita compared to about half this amount in Western Europe. Additionally, traffic noise is increasingly objectionable to many city dwellers. Although sound walls help attenuate noise pollution (at the expense of visual obstructions), depressed property values near busy highways remind us that no road can be made totally soundproof. One estimate places the noise damage to residential properties from cars and trucks at about $9 billion (in 1989 dollars) per year.

Energy conservation: Currently, transportation accounts for around three-quarters of the petroleum used in the United States, and about two-thirds of this amount is burned by motor vehicles.

Transit villages could conserve energy in two ways. Firstly, more compact development, in theory, should shorten trip lengths. And secondly, conversion of some motorized trips to mass transit should cut down on per capita consumption.

Social equity: Among the most pernicious and troubling effects of an increasingly auto-dependent society are the social injustices and economic disparities that result from physically and socially isolating significant segments of society. While America might be able to reengineer the car and better manage traffic flows to solve pollution, energy, and congestion problems, nothing can be done to the car or road system to reduce the social isolation and inequalities in access to jobs, clinics, and shops that many people experience.

Among the encouraging initiatives in recent times to rejuvenate poor transit-served neighborhoods and make them more accessible is Federal Transportation Administration's Livable Communities program. Many Livable Communities demonstration projects emphasize economic revitalization and land-use initiatives that bring people back to urban centers and strengthen community-based transportation services. The fusion of transit-oriented development and community rebuilding can be found in such projects as Baltimore's Reistertown Metro Station enhancement that will site a large child-care center on an underutilized parking lot; construction of new housing, retail shops, and pedestrian walkways near the 35th Street Station on Chicago's Green Line; and the rehabilitation of the Windemere Station in East Cleveland to incorporate a Head Start educational facility.

Quality of life: Urban transit villages are exactly what the term states: village style proximity of public and private services that are accessible and affordable and that offer choice, diversity, and richness of lifestyle.

Rebirth of rail transit in the United States: Partly in response to concerns over traffic congestion, air quality, energy dependence, and social inequities, a number of U.S. cities have sought to bolster mass transit in recent times by building new rail systems or extending existing ones. Rightly or wrongly, America is in the midst of a rail renaissance.

Fig. 3. Light Rail Transit, Portland, OR. Conveniently placed in pedestrian zone.

3 RAIL TRANSIT OPTIONS

Light rail transit: Light rail transit (LRT) is essentially a modern-day version of turn-of-the-century electric streetcars. Among its advantages are: light rail is relatively quiet, thus environmentally less obtrusive; it is electrically propelled, thus less dependent than buses on the availability of petrochemical fuels; it can operate effectively along available railroad rights of way and street medians, and is thus far cheaper, less disruptive, and easier to build than heavy rail; and it can be developed incrementally, a few miles at a time, eliminating the need for the long lead times associated with heavy rail construction. LRT's lack of exhaust fumes and its comparatively slow speeds make it particularly compatible with pedestrian settings like downtown malls.

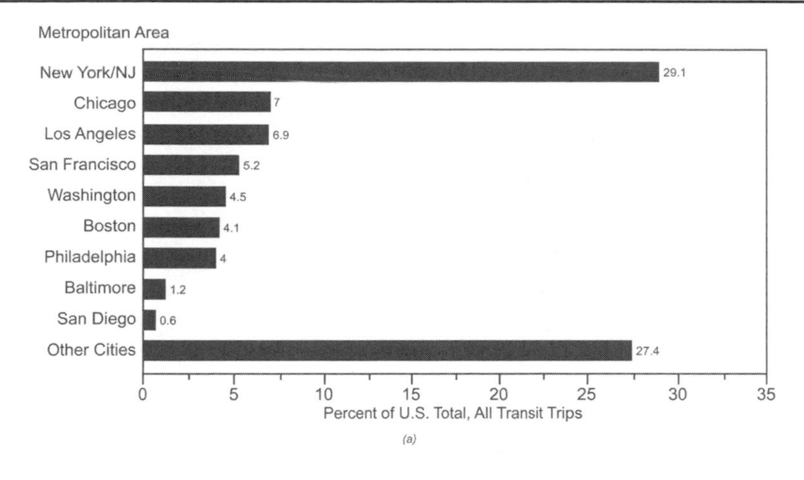

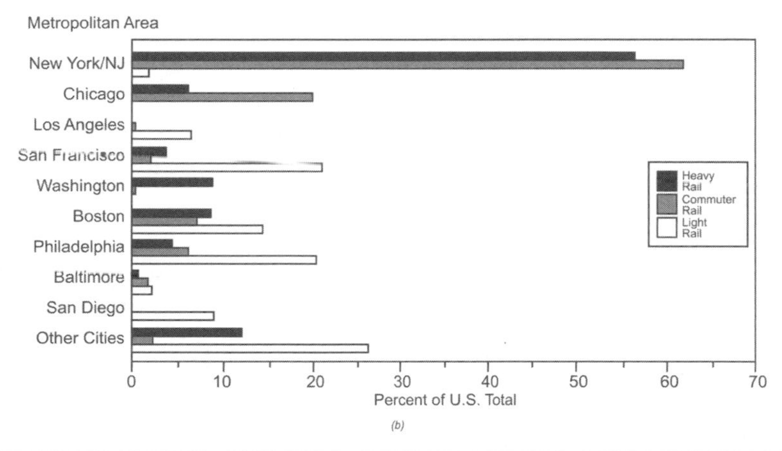

Fig. 4. Share of Transit Trips in Large Metropolitan Areas, All Modes (a) and Rail Modes Only (b), 1993. (Source: American Public Transit Association, 1994-1995, *Transit Fact Book*, Washington, DC, 1995). "

Commuter rail transit: Commuter rail services typically link outlying towns and suburban communities to a region's downtown. They are characterized by heavy equipment (e.g., locomotives that pull passenger coaches), wide station spacing, and high maximum speeds that compete with cars on suburban freeways, though slow in acceleration and deceleration. Services tend to be of a high quality, with every passenger normally getting a comfortable seat and ample legroom. Routes are typically 25 to 50 miles (40 to 80 km) long and lead to a stub-end downtown terminal. Because commuter transit is used by professional suburbanites to reach core-area jobs, commuter rail ridership tends to be highly concentrated in the peak.

Rapid rail transit: Sometimes called heavy rail or metros, rapid rail transit services are high-speed, high-performance systems within urbanized areas that connect neighborhoods and major activity centers (e.g., sports stadia, airports) to downtowns. In the city core, rapid rail lines almost always operate below ground, whereas outside of downtowns they are typically elevated or at grade. Stations tend to be about a mile or so apart, except in downtowns where they might be three or four blocks away. Heavy rail systems are electrically propelled, usually from a third rail, and each cart has its own motor. Since contact with the third rail (usually 600 volts) can be fatal, rapid rail stations usually have high platforms, and at-grade tracks are fenced.

In contrast to light rail and commuter rail systems, rapid rail transit is entering a slow-down rather than an expansion phase, mainly for fiscal reasons. Overall, the greatest prospect for building transit villages around new rail stations lies with LRT and commuter rail services.

Urban rail stations: The possibilities of station-area redevelopment are suggested by simple tabulations of existing rail station counts across the United States. Today, most stations are in older metropolitan areas with long-established rail networks (Figs. 3 and 4). For new station-area development, however, the greatest opportunities in terms of sheer numbers lie with light rail.

4 TRANSIT-SUPPORTIVE DESIGN

An essential element of the supportive physical characteristics of transit villages is urban design itself. Transit villages should encourage walking and transit riding. Since all transit trips involve some degree of walking, it follows that transit-friendly environments must also be pedestrian-friendly. The following provides a checklist of design criteria.

Commonly accepted transit-supportive designs often include the following types of treatments:

- Continuous and direct physical linkages between major activity centers; siting of buildings and complementary uses to minimize distances to transit stops.

Fig. 5. Compact versus spread-out development around a transit station. Are land uses complementary and within walking distance?

Fig. 6. Direct versus disconnected sight lines to a transit station. Are walk paths direct and separated from parking?

- Street walls of ground-floor retail and varied building heights, textures, and facades that enhance the walking experience; siting commercial buildings near the edge of sidewalks.

Compact versus spread-out development around a transit station. Are land uses complementary and within walking distance?

- Integration of major commercial centers with the transit facility, including air rights development (Fig. 5), which shows both poor and good examples.

- Gridlike street patterns that allow many origins and destinations to be connected by foot; avoiding cul-de-sacs, serpentine streets, and other curvilinear alignments that create circuitous walks and force buses to meander or retrace their paths; direct sight lines to transit stops (Fig. 6), which shows both poor and good examples.

- Minimizing off-street parking supplies; where land costs are high, tucking parking under buildings or placing it in peripheral structures; in other cases, siting parking at the rear of buildings instead of in front.

Building detachment versus building integration. Do buildings fit with and complement the transit system?

- Providing such pedestrian amenities as attractive landscaping, continuous and paved sidewalks, street furniture, urban art, screening of parking, building overhangs and weather protection, and safe street crossings.

- Convenient siting of transit shelters, benches, and route information.

- Creating public open spaces and pedestrian plazas that are convenient to transit.

Transforming suburban neighborhoods into more pedestrian-friendly, transit-supportive environments might occur over a number of stages. Fig. 7 shows a typical auto-oriented commercial district with a vast expanse of parking that separates buildings from the main street, numerous driveways and curb cuts, no internal or curbside sidewalks, exposed pathways, and minimal landscaping. Over time, the rather hostile environment for walking and transit riding could be redesigned, modified, and retrofitted so that it is more human in scale, compact,

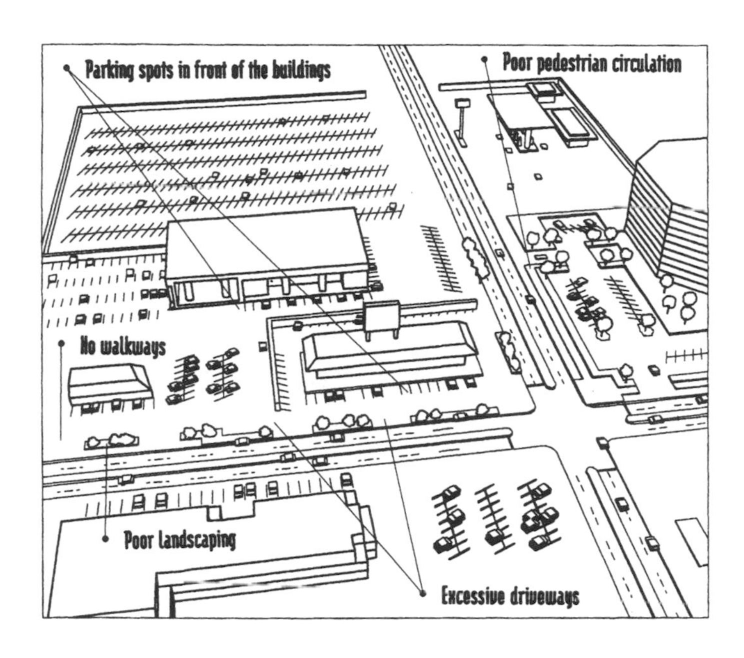

Fig. 7. Auto-oriented commercial district unfriendly to pedestrians and transit users.

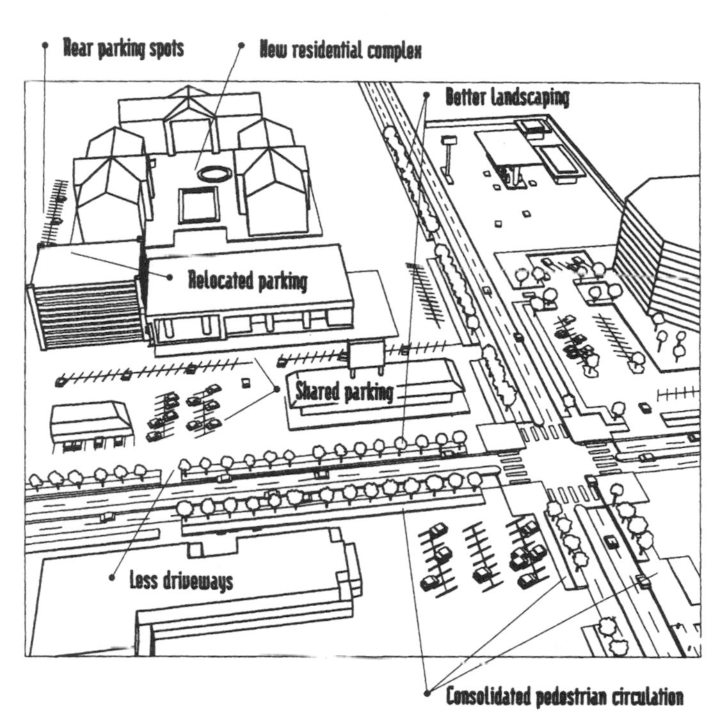

Fig. 8. Initial improvements friendly to pedestrians and transit users.

and attractive to pedestrians. In the early stages, less expensive things could be done: installing sidewalks and street lighting, improving pedestrian crossings, and consolidating driveways. The public improvements ideally would be enough to increase property values and spark a renewed interest in the area. This might lead to the intensification of uses, including the addition of housing. Fig. 8 portrays how the setting might look after such measures as relocating parking, consolidating driveways, integrating walkways, improving the landscape, and filling in the main street with more neighborhood-oriented uses like restaurants and specialty retail shops are accomplished. The final stage of transformation is depicted in Fig. 9. A light rail line penetrates the neighborhood. Flanking it is a public plaza that ties into a community complex. Courtyards, tree-canopied walkways, and further landscaping improvements enhance the setting. Additional housing increases the density of the neighborhood even more. The end result is the transformation of an auto-oriented commercial strip into a mixed-use neighborhood more conducive to walking and transit riding.

5 PRINCIPLES FOR TRANSIT VILLAGE IMPLEMENTATION

New rail investments, by themselves, do not automatically translate into significant land-use changes.

Among metropolitan areas in the United States, the introduction of a regional rail network has not by itself significantly affected urban form and property values. This is in good part because rail has been added during an era of high automobile accessibility and freeway development. Many localized factors can affect land-use outcomes. Among these are a healthy local real estate environment, community support, and attractive physical and social environment, and pro-development public policies.

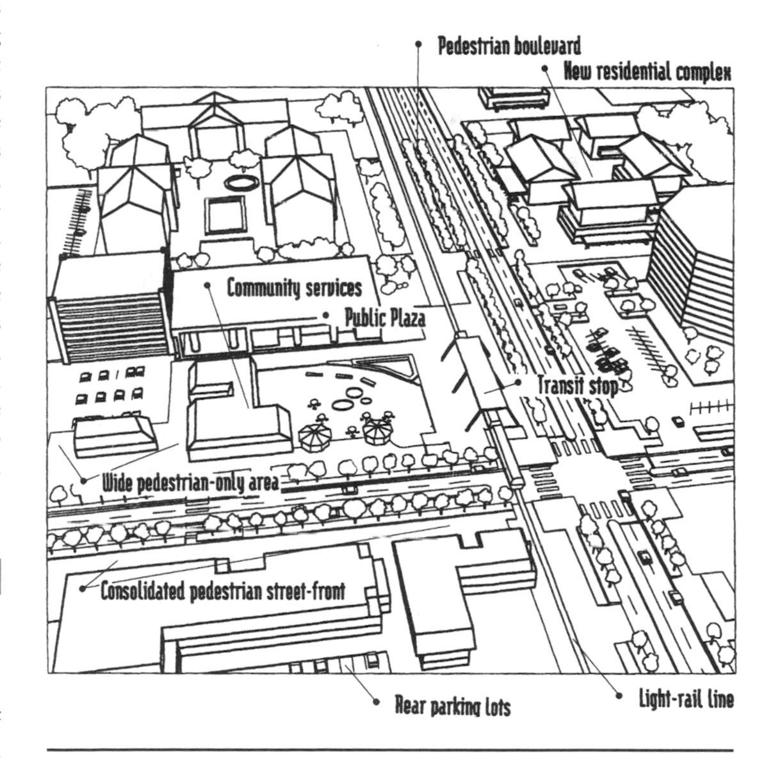

Fig. 9. Transformation into a transit-oriented neighborhood.

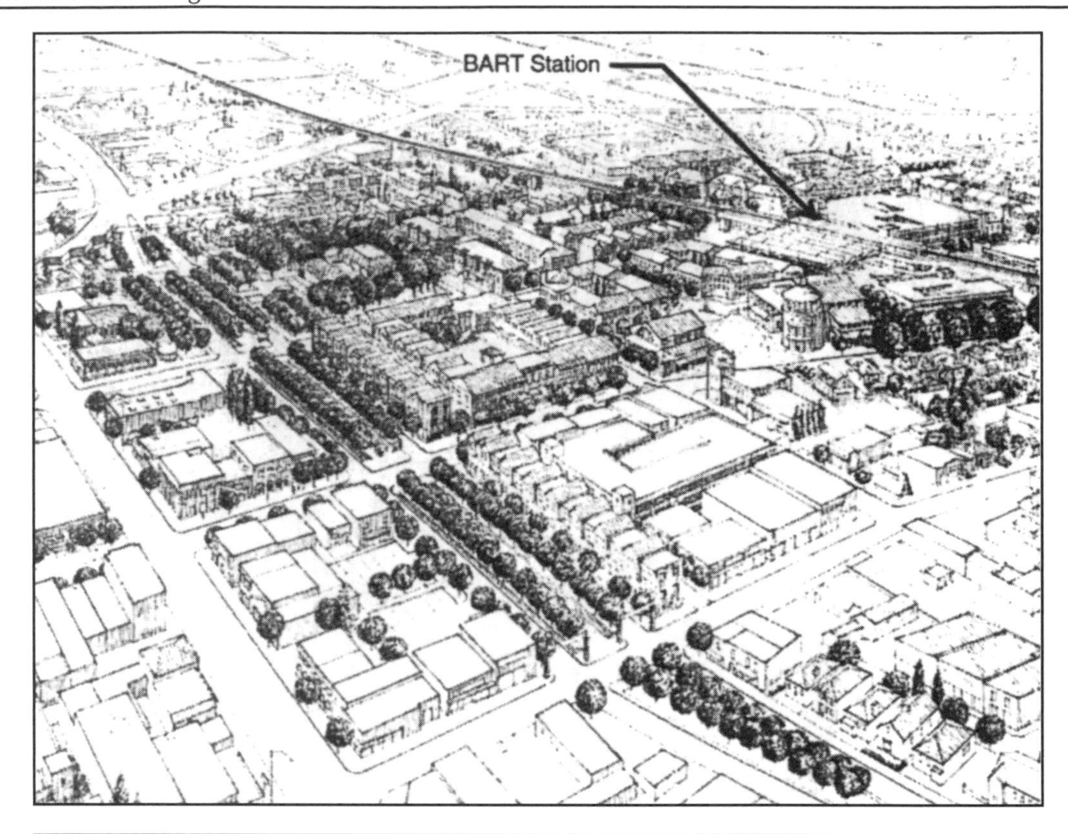

Fig. 10. Hayward Transit Village, San Francisco, CA. Mix of housing, stores, cultural activities and public spaces flanks the station. (Architect: Daniel Solomon)

Fig. 11. Emergence of a transit village: Pleasant Hill Station, San Francisco, CA.1994. The low-density coverage in the countryside is maintained, with new development concentrated at the transit station.

Transit-oriented development as long-term commitment.

The most successful examples of rail influencing urban form and guiding growth have been direct products of careful strategic planning. Cities like Stockholm, Singapore, and Tokyo offer the best examples. In these places, an overall vision of the future settlement pattern of the region was first established. Through selective and judicious station-area planning, these regions have successfully guided urban growth while creating world-class transit networks. Importantly, a stronger emphasis was placed on long-term development activities to making leasing and cost-sharing deals that have often yielded minimal financial benefits. Successful systems, on the other hand, have created station-area plans, formed joint development offices, banked land, and strategically introduced development incentives. In effect, they have controlled station-area development. Their payoffs have been over the long term in the form of high ridership and fare box returns associated with station-area development.

Critical mass in suburban-inner-city community building.

Suburban stations, especially ones at which the transit agency or other public entities own most of the land within a one-quarter to one-half mile (.4 to .8 km) radius of the station, offer opportunities for building new suburban communities that are oriented toward rail versus cars and highways. These communities might feature a civic and commercial core flanked by housing with varying densities and prices, interlaced by sidewalks and parks. Mixed land uses are critical to ensuring balanced, bi-directional flows on a rail system with comparable on and off counts at stations each morning. This has been an important part of the financial success of rail systems in Stockholm, Singapore, and Japan.

For inner-city stations, new community building also is possible through a critical mass of good quality and attractive development and redevelopment within a quarter-mile (.4 km) radius of a station. Successful inner-city community rebuilding emphasizes private businesses and co-investment (versus only public-led development and entitlements); home ownership (rather than predominantly subsidized rental housing); empowerment through such actions as establishing community-based business enterprises and neighborhood associations; and the establishment of a visible public safety presence.

Concentration of resources in achieving the first built projects.

Contemporary rail systems usually feature a diversity of station settings, each with varying levels of development promise. Planning effort should be devoted initially to developing or redeveloping a handful of station areas at most. This allows resources to be effectively targeted. Demonstrating that positive land-use changes are possible in conjunction with a rail investment. It is important for producing good "models" that the larger development community can emulate, as well as convincing banks and lenders that investing in station-area projects can be financially remunerative. Still, planning should be flexible, allowing for contingencies, to respond appropriately to shifting market conditions.

Proactive role of the transit agency and local government: assumption of risk by the public sector.

Rail investments can be important agents of economic growth, creating new forms of suburban development, enhancing commercial districts that lack weekend or evening activities, revitalizing otherwise stagnant

Fig. 12. Rapid rail transit, Seattle WA.

urban districts, and even regenerating depressed inner-city neighborhoods. Such changes, however, require a proactive public sector, one which takes the lead in preparing specific plans that win the consent of neighborhood and community groups, land banking and assembling land into developable parcels, writing down the cost of land in return for participation in project revenues, providing the infrastructure necessary for new development either through direct investment or through tax-increment financing, creating development incentives such as density bonuses, and underwriting early phases of housing retail development to generate private-sector interest in later phases.

A key player is the rail transit agency. The rail agency sometimes owns much of the land around stations and perhaps has the most to gain from good quality development. Other public-sector participants are also needed—the housing agency, the redevelopment agency, the regional planning authority—but none is in the position to spearhead positive chance as much as the transit agency. The private sector needs to be brought into the development process early, along with neighborhood groups. At times, the transit agency or local government must be prepared to assume some degree of financial risk for a project to move ahead. ■

Distance Measured in Traveling Time

Distance Traveled in 30 minutes

Conveyance	Miles	Km
Pedestrian walking leisurely	1.5	2.4
Pedestrian walking briskly	2.0	3.2
Bicycle at normal pace	5.0	8.0
Bicycle in one-hour race	15.0	24.0
Bus (in dense city traffic)	3.0	4.8
Bus (on suburban streets)	8.0	12.8
Bus (express)	15.0	25.0
Streetcar (in mixed traffic)	4.0	6.4
Light Rail 8.0 12.8 Subway (regular service)	12.0	19.2
Train (local service)	18.0	30.5
Train (regional express)	22.0	37.2
Train (Metroliner)	45.0	76.2
High-Speed Train (French TGV Train a Grande Vitesse)	80.0	135.4
Automobile (moving at normal urban speed limit)	12.0	20.3
Automobile (moving at 55 miles per hour)	27.0	45.7

Source: Material prepared by Sigurd Grava.

Top 20 U.S. Cities in Terms of Percentage of Workers Commuting by Mass Transit, 1990

City	% Commuting by mass transit	Types of transit services
New York, NY	53.4	HR, CR, LR, B, O
Hoboken, NJ	51.0	HR, CR, B, O
Jersey City, NJ	36.7	HR, CR, B, O
Washington, DC	36.6	HR, CR, B
San Francisco, CA	33.5	HR, CR, LR, B, O
Boston, MA	31.5	HR, CR, LR, B, O
Chicago, IL	29.7	HR, CR, B
Philadelphia, PA	28.7	HR, CR, LR, B, O
Atlantic City, NJ	26.2	B, O
Arlington, VA	25.4	HR, CR, B
Newark, NJ	24.6	HR, CR, LR, B, O
Cambridge, MA	23.5	HR, CR, B
Pittsburgh, PA	22.2	LR, B, O
Baltimore, MD	22.0	HR, CR, LR, B
Evanston, IL	20.9	HR, CR, B
Atlanta, GA	20.0	HR, B
White Plains, NY	19.1	CR, B
Camden, NJ	18.1	HR, CR, B
Oakland, CA	17.9	HR, B, O
Hartford, CT	17.1	CR, B

Key: HR = Heavy Rail; CR = Commuter Rail; LR = Light Rail; B = Bus (motor and trolley); O = Other, including ferry and legal general-public paratransit services.

Sources: American Public Transit Association, 1994-1995 *Transit Fact Book*, Washington, DC, 1995; U.S. Bureau of the Census, 1990 Census, *Journey to Work Characteristics of Workers in Metropolitan Areas*, Washington, DC, 1993.

Donn Logan and Wayne Attoe

Summary

A catalyst is an urban element that is shaped by the city and then, in turn, shapes its context. Catalytic urban design works not from a master plan, but from a master program. It sets out intentions and methods but not solutions. It uses zoning as combined tools of control, guideline and incentive. It is characterized by sequencing of development. It thus calls for both idealism and pragmatism: idealism about the specialness of the place and pragmatism about making that place work in relation to contemporary traffic needs and local culture and values. This dual need calls for a unique vision for each urban place.

Key words

central business district, development, financial incentives, mixed use, parking, skywalk, transportation, urban theory, zoning

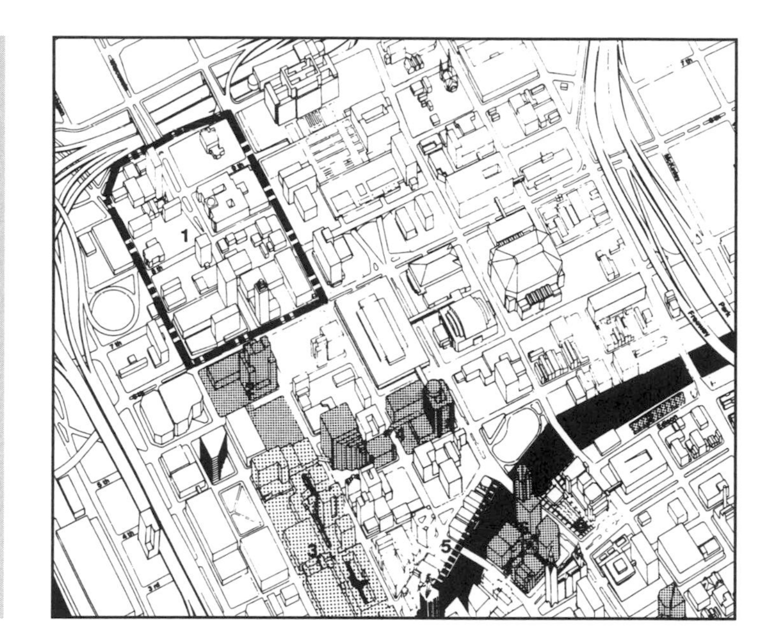

The concept of urban catalysts

Changing stances in European urban design theory have offered guidance for urban design that is not fully adequate to the American context. Most of the theories seem to assume a central government with the political and economic power to implement the envisioned development. We do not argue with existing European-based concepts. In fact, we recommend, pragmatically adopting many European urban values. But it is the values and not the forms associated with them that we commend. The following design principles derived from European cities and European-based urban design theories constitute the basis of good urbanism, not only in Europe but also in America.

1. Mixed activities are basic to cities.

2. Buildings (and the spaces they form) are the natural increments of urban growth.

3. New urban growth must recognize the context provided by past construction.

4. A major goal of urban design is the shaping of public open space, including meaningful street space.

5. Streets must accommodate various forms of transit and enhance pedestrian activity and movement.

6. Transportation systems should be rational.

7. Urban places should be varied to enhance the activities associated with them: housing, neighborhood shopping, major retail, civic, and so forth.

8. Citizens should have a role in shaping urban settings.

Urban design for center cities, instead of being conceived as the process of implementing one or another ideal image of the city, using various available tools, is more appropriately thought of as a process of arranging catalytic reactions. There should be no ultimate vision for the urban center, either functionalist, humanist, systemic, or formalist. Rather, there should be a sequence of limited, achievable visions, each with the power to kindle and condition other achievable visions. Visions for the new urban center should be modest and incremental, but their impact should be substantial, in contrast to the large visions that have been the rule, with their minimal or catastrophic impact.

A catalyst is an urban element that is shaped by the city (its "laboratory" setting) and then, in turn, shapes its context. Its purpose is the incremental, continuous regeneration of the urban fabric. The important point is that the catalyst is not a single end product but an element that impels and guides subsequent development.

Credits: This article is adapted by the authors from their book *American Urban Architecture: catalysts in the design of cities*, University of California Press, 1989.

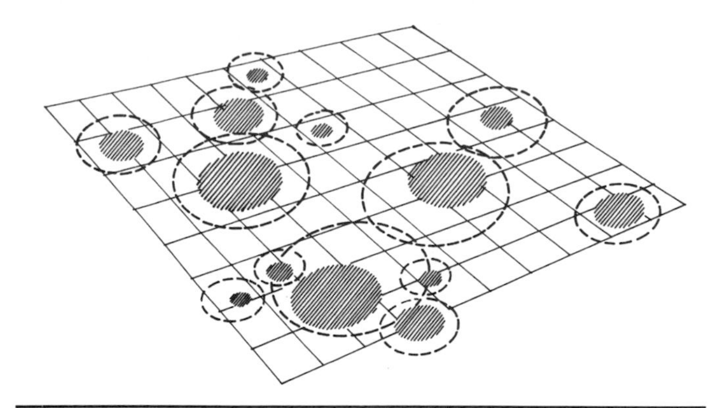

Fig. 1. Diagrammatic representation of the catalytic process. Actions (represented by hatching), whether developments, restorations, reports, or whatever, catalyze other actions, which in turn lend impetus to others. Each action is constrained too, so that the reaction does not destroy the city. The moderating aspect of the process is represented by the broken lines around the hatching.

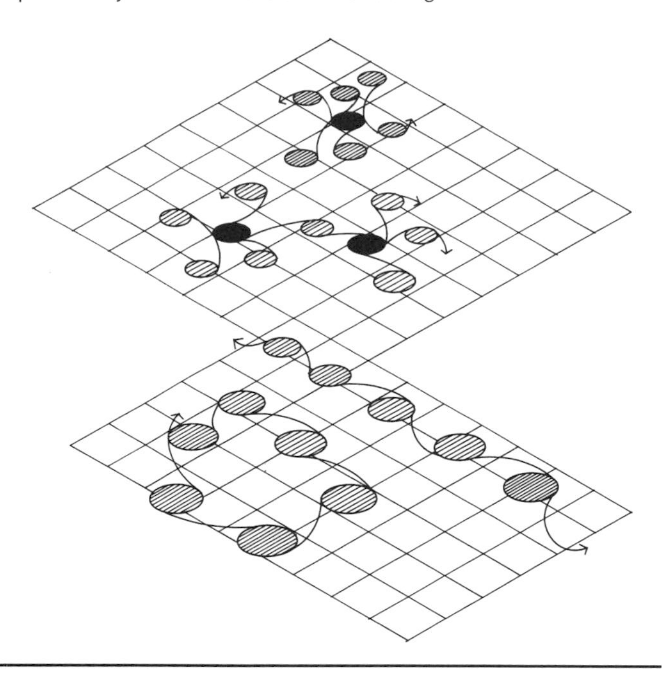

Fig. 2. Catalytic reactions can take several forms: nuclear (top), multinucleated, serial, and "necklace" (lower left).

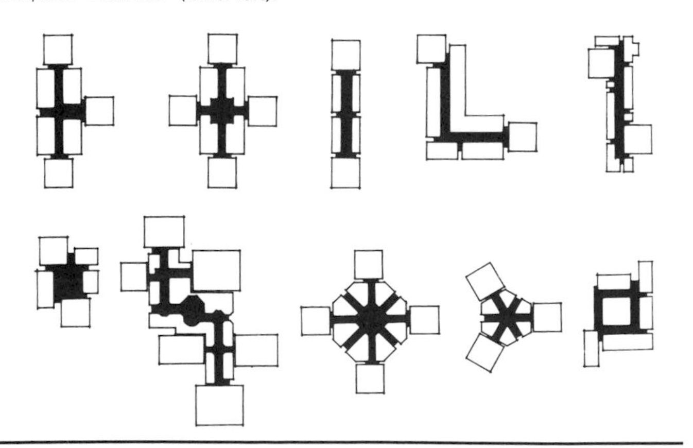

Fig. 3. Nodal, linear, and spread-form shopping centers. Stores in a cluster are the equivalent of shops around a square; stretched out, they are like shops along a "main street." These configurations can be extended to create spread-form variations.

Although renewal and revitalization schemes for cities are often touted as catalysts, many of these schemes remain inert and have little impact. They do not cause the promised urban reactive change. Sometimes the term catalyst refers to economic processes, typically an infusion of funds that leads to other infusions of funds, or, at a gross scale, it means that one development makes additional developmental projects look like good investment risks.

Architecture, too, is catalytic. Not only infusions of capital that incidentally produce new buildings and reconstructed streets but buildings themselves can be catalysts, ensuring the high quality of urban redevelopment. Urban design quality is determined at the scale of buildings.

Catalysis involves the introduction of one ingredient to modify others. In the process, the catalyst sometimes remains intact and sometimes it is modified. Adapted to describe the urban design process, catalysis may be characterized as follows:

The introduction of a new element (the catalyst) causes a reaction that modifies existing elements in an area. Although most often thought of as economic (investments beget investments), catalysts can also be social, legal, political, or—and this is our point—architectural. The potential of a building to influence other buildings, to lead urban design, is enormous.

Existing urban elements of value are enhanced or transformed in positive ways. The new need not obliterate or devalue the old but can redeem it.

The catalytic reaction is contained; it does not damage its context. To unleash a force is not enough. Its impact must be channeled.

To ensure a positive, desired, predictable catalytic reaction, the ingredients must be considered, understood, and accepted. (Note the paradox: a comprehensive understanding is needed to produce a good limited effect.) Cities differ; urban design cannot assume uniformity.

The chemistry of all catalytic reactions is not predetermined; no single formula can be specified for all circumstances.

Catalytic design is strategic. Change occurs not from simple intervention but through careful calculation to influence future urban form step by step. (Again, a paradox: no one recipe for successful urban catalysis exists, yet each catalytic reaction needs a strategic recipe.)

A product better than the sum of the ingredients is the goal of each catalytic reaction. Instead of a city of isolated pieces, imagine a city of wholes.

The catalyst need not be consumed in the process but can remain identifiable. Its identity need not be sacrificed when it becomes part of a larger whole. The persistence of individual identities—many owners, occupants, and architects—enriches the city.

Catalytic theory does not prescribe a single mechanism of implementation, a final form, or a preferred visual character for all urban areas. Rather, it prescribes an essential feature for urban developments: the power to kindle other action. The focus is the interaction of new and existing elements and their impact on future urban form, not the approximation of a preordained physical ideal (Figs. 1–3).

Catalyst theory: Case study of Milwaukee

The story of catalytic redevelopment in Milwaukee begins with a 1973 study commissioned by the Greater Milwaukee Committee. It offered a vision for a new downtown, which in turn could change attitudes. The study recommended the formation of a development corporation and the creation of a retail core with related uses.

Studies also indicated that Milwaukee needed risk capital for renewal, the Milwaukee Redevelopment Corporation (MRC) was formed in 1973 as a limited-profit, blue-chip grouping of large firms. Conceived by the MRC and the city in 1976, the Grand Avenue concept was reviewed in 1977, and negotiations with a developer, the Rouse Company, began in 1978. Private investment amounted to $18 million, with the Rouse Company contributing $19.5 million. The $39-million investment of the city of Milwaukee took the form of an Urban Development Action Grant and a tax-increment bond issue. No new tax dollars were involved.

The Milwaukee Redevelopment Corporation then took three steps. First, it proposed the construction of a retail complex called the Grand Avenue, which would both recall the former Grand Avenue, Milwaukee's historical retail/commercial artery (now Wisconsin Avenue), and offer an interior place, a semipublic realm better than that found in any suburban shopping center. Second, the MRC listed and responded to reactions to the idea. Third, it became a leading partner in the development and a link between private and public interest and investments in the project (Figs. 4–9).

From the point of view of catalytic urban design, the Grand Avenue project goes far beyond the bold vision, money, and political muscle that brought it into being. The value of Grand Avenue is only partly itself; it is equally valuable for its subsequent effects, the way it was able to catalyze other development.

Among the reasons that Grand Avenue became a success cited by both municipal planners and officials were the old vision, funds to turn a vision into reality, and willingness of government and business leaders to work toward a common goal. But any such statement fails to represent fully the sustained commitment and action on the part of private and public leaders. The measure of an urban design should therefore include its capacity to enable the imagination and commitment to significant urban leadership. Catalyst approaches, being made up of strategically conceived elements, is thus more likely to facilitate such commitments, compared to overly ambitious longer term master planning.

Principles of catalyst urban design

Catalyst theory offers the following new principles to traditional approaches to urban design, illustrated by reference to the Grand Avenue case study example:

1. The new element modifies the elements around it.

This is represented in the Milwaukee Development by the local and wider impacts of the improvements represented by the Hyatt Hotel and the Federal Plaza, the skywalk system which rapidly extended over many blocks, the positive development of nearby East Town, the riverwalk developed as an urban amenity, the theatre district and other nighttime uses that resulted from the development, and other local extensions of Grand Avenue, downtown to the west, and the warehouse district.

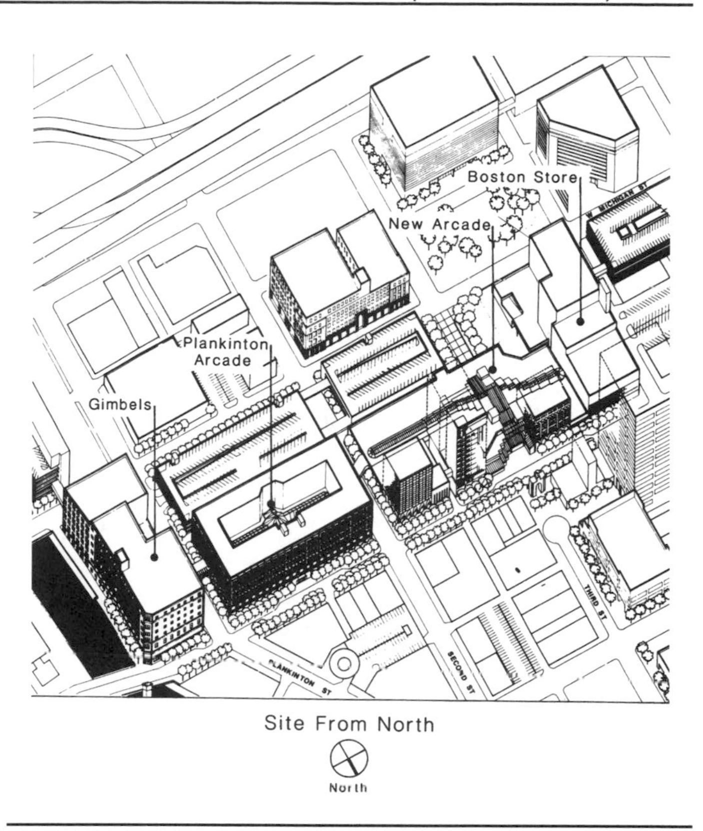

Fig. 4. Grand Avenue, Milwaukee. Aerial view, looking south, shows how Grand Avenue links the two existing department stores at either end and incorporates the existing Plankinton Arcade building. The tall buildings above Boston Store in this view predate the construction of Grand Avenue.

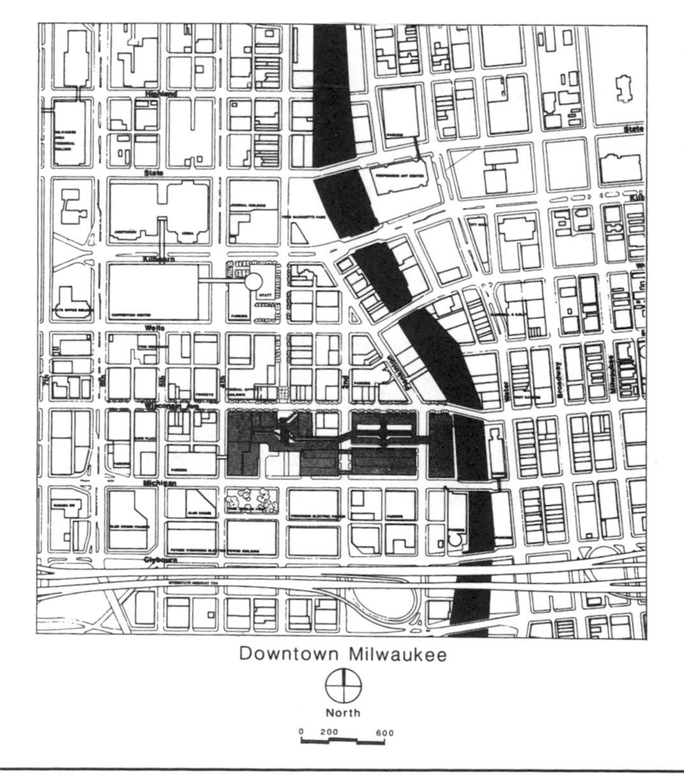

Fig. 5. Downtown Milwaukee. The scale of the Grand Avenue development is evident in this plan of downtown Milwaukee.

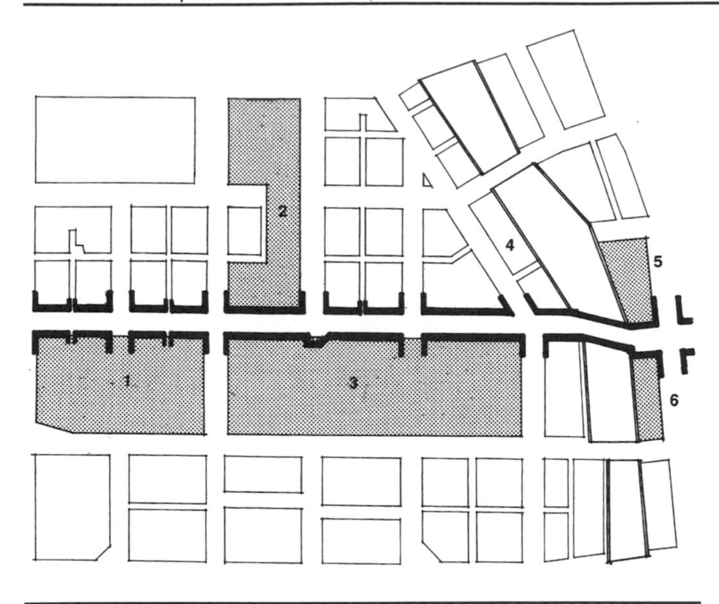

Fig. 6. Wisconsin Avenue, Milwaukee, as a frame and edge to various developments: (1) the Grand Avenue extensions, (2) Federal Building and Hyatt Hotel group, (3) the Grand Avenue retail development, (4) Riverwalk, (5) River Place, a mixed-use development, and (6) bank.

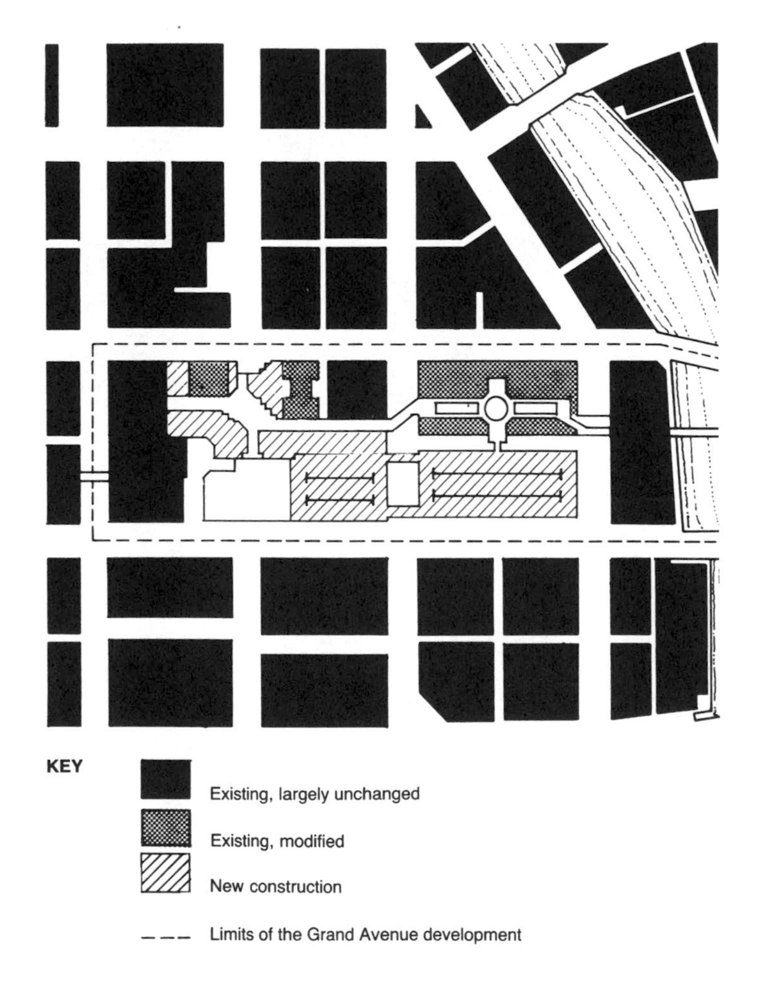

Fig. 7. The Grand Avenue skywalk system.

2. Existing elements are enhanced or transformed in positive ways.

This principle is manifested in two ways, in buildings and in people's behavior. In the case of Grand Avenue, the Plankinton Arcade, which had withered in both its role and its physical condition, was refurnished and given a new life as the centerpiece of the Grand Avenue development. In less that two years following the development of Grand Avenue, downtown was once again the primary center in the Milwaukee retail market. Other local structures, such as the Woolworth Building were restored or adaptively reused.

3. The catalytic reaction does not damage its context.

Although they are not a characteristic of all urban catalysts, the benign edges of Grand Avenue moderate its impact on its context. Its end points and most of its frontage (on Wisconsin Avenue) remain as they were before. It is noteworthy that a radical reorganization of pedestrian space could be accomplished without radically affecting the architectural character of downtown.

4. A positive catalytic reaction requires an understanding of the context.

The analysis of urban context is a complex mix of architectural and urban character, people, image and key urban functions such as transportation and parking. In each case, these are the basis of unique urban designs, which avoid formulaic responses. In relation to this concern for responding to the existing urban context, the most questionable decision at Grand Avenue was the closing of Third Street. Did this disrupt traffic in and violate the perceptions of downtown? There is no evidence that the street closure has impeded the flow of traffic and the closure does no hamper pedestrian movement. In fact, it invites pedestrians, for this is the principal entrance to the Grand Avenue. Although in this instance, modification in street pattern seems not to have been excessive, violations of the character of an existing setting remain a concern in any urban project that begins to restructure downtown in dramatic ways.

5. Not all catalytic reactions are the same.

The chemistry for urban revitalization in Milwaukee depended on several elements, first among them the configuration of traffic and traditional uses of the setting. The linear character of the Wisconsin Avenue commercial district is unlike that of cities whose shopping activities have developed around an intersection or a square. And in this case, parking tends to be concentrated. The center city developments must therefore be conceived as unique collections of existing ingredients needing to be customized to satisfy new requirements

6. Catalytic design is strategic.

Although much urban development is opportunistic ("Take advantage of tax credits; buy when prices are low; build what's profitable wherever you can"), better guarantees of profitability and urban quality can be had from strategic rather than opportunistic thinking. Opportunists think of the short term; strategists, of the longer term.

Catalytic urban design works not from a master plan, but from a master program. And it sets out intentions and methods but not solutions. So, for example, a classic master plan establishes transportation, zoning, and land-use patterns years or even decades hence and is typically inflexible

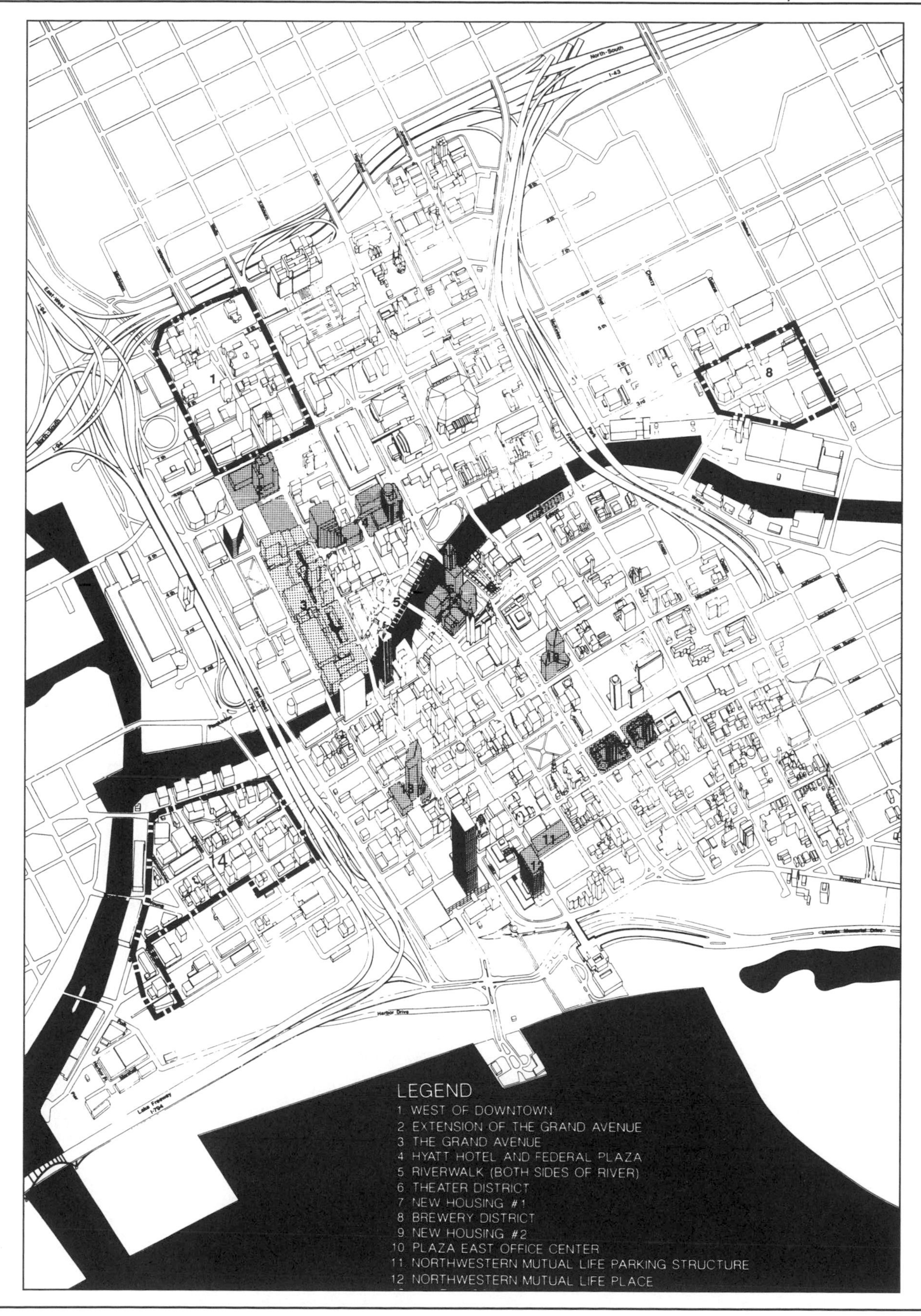

Fig. 8. Downtown Milwaukee, showing Grand Avenue and location of other developments that it has influenced.

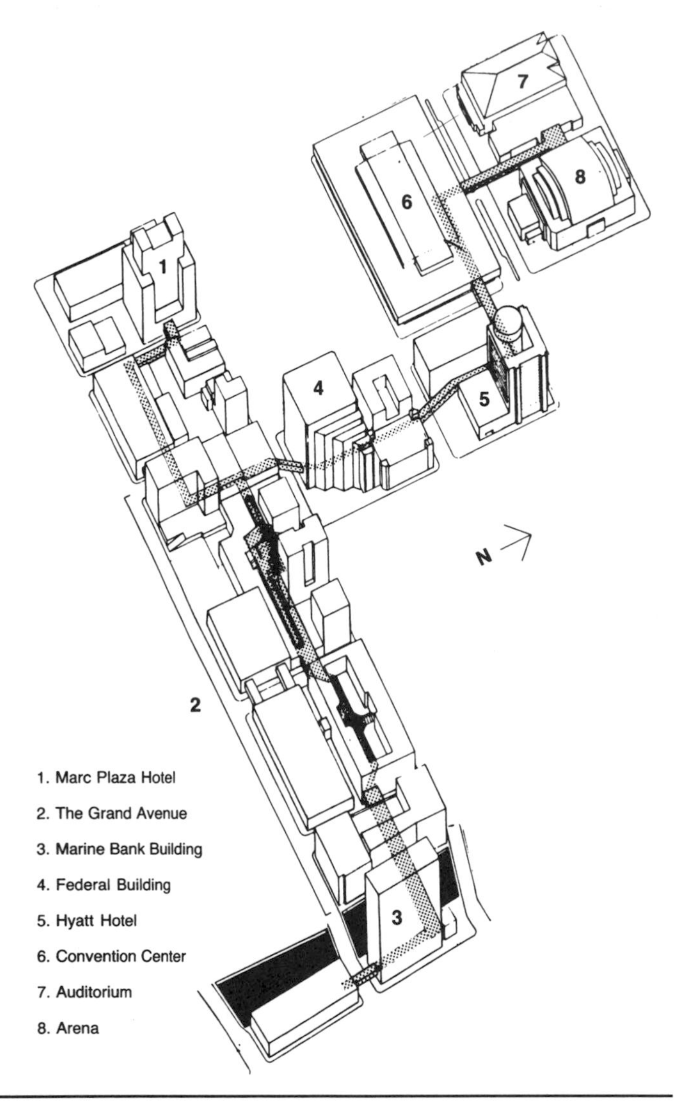

Fig. 9. The Grand Avenue development: New construction respects and incorporates existing buildings.

in responding to changing circumstances; a master program sets out to stimulate and control development in a way more responsive to the exigencies and opportunities that appear in potentially volatile American cities. In short, a master program is flexible. The key to keeping strategic and catalytic design malleable is to have multiple rather than single-minded views of the future. It uses zoning as combined tools of control, guide and incentive. It is characterized by sequencing of development.

7. A product better than the sum of the ingredients.

Before the Grand Avenue project was undertaken in Milwaukee, the auditorium and convention center was an island. It is now tied tightly to the hotel and to downtown by the skywalk system. The connections are practical, but the conceptual linkage of the disparate parts may be more important than mere practicality. The goal of any catalytic reaction should be not a collection of developments—so often the case in revitalization schemes—but integrative urbanism in which parts reinforce one another.

8. The catalyst can remain identifiable.

In chemistry the catalyst often disappears or is transformed in the course of a reaction, but this is usually not the case with urban chemistry. Instead, the ingredients of rejuvenation remain and contribute to the city's unique character and sense of depth. The layers of urban experience and urban history, the collage of styles and uses characteristic of a vital center city are the essence of urbanity.

Urban catalysis was necessary to initiate what has been accomplished in downtown Milwaukee and what is continuing to happen there and that unilateral, univalent renewal could not have met the challenge. The accomplishment is considerable:

- A unique place is emerging, a place composed of and responsive to what was there before. It is both old and new.
- It is for many a new gateway to a city they had abandoned, thus helping to restructure the image of downtown.
- It establishes precedents for other developments—precedents in design quality, precedents for thinking about existing buildings, precedents for using the city, precedents for relating interior-oriented architecture to existing streets and street life, and precedents for an integrative urban architecture that is new in the experience of most Milwaukeeans. ■

REFERENCE

Attoe, Wayne and Donn Logan. 1989. *American Urban Architecture: Catalysts in the Design of Cities.* Berkeley: University of California Press.

Robert A. M. Stern

Summary

There will be no new ideas about cities and their suburbs until our thinking frees itself from the biases and orthodoxies of our recent architectural and urban theories, especially those peculiar cultural biases and cultural prejudices which have encouraged us to see old cities and old buildings—not to mention traditions and recognizable forms—as worthless and wrong. Nor will we be able to deal confidently with the suburb until we free ourselves from the belief that new suburban ideas (or, in fact, new suburbs) can only grow on virgin land beyond the edges of existing development. Suburbs will not go away, nor should they. They may well hold the key to the solution of urban problems that were hitherto deemed insoluble.

Keywords

garden city, industrialization, planning theory, residential development, suburb, workers' housing

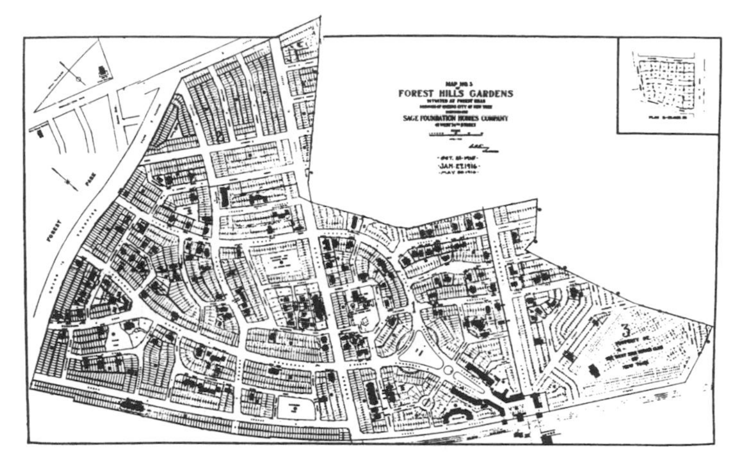

Forest Hills Gardens, New York, designed by Grosvenor Atterbury and the Olmsted Brothers, 1912.

The Anglo-American suburb

Little boxes on the hillside
Little boxes made of ticky-tacky.
Little boxes on the hillside.
Little boxes all the same. . .
Little people in the houses,
All went to the University,
Where they were put in little boxes,
And they came out just the same.
And there's doctors,
And there's lawyers,
And business executives,
And they look just the same.[1]

There is a story that Frank Lloyd Wright once took Alvar Aalto for a drive in the countryside around Boston to show him the American suburban landscape. Wright majestically gestured to the surrounding scene and said, "None of this could've been accomplished without me." And Aalto, telling the story later, commented on suburbia, "You know, I couldn't see it."[2]

Aalto's apparent myopia notwithstanding, the Anglo-American suburb is a remarkable achievement, not the degraded form of city planning that so many have called it. An important representation of our culture's traditions and aspirations, the suburb has nevertheless been spurned by modernist architects, theorists, and historians, who preferred the development of new dwelling types for the "brave new world" they promoted. Sigfried Giedion neatly summed up the modernist position in *Space, Time and Architecture*:

> *Contemporary architecture takes its start in a moral problem. . . . Contemporary architects have been. . .willing to anticipate public understanding. They too have refused to wait until they could be sure of universal approbation for their work. Following an impulse which was half ethical, half artistic, they have sought to provide our life with its corresponding shell or framework. And where contemporary architecture has been allowed to provide a new setting for contemporary life, this new setting has acted in its turn upon the life from which it springs. The new atmosphere has led to change and development in the conceptions of the people who live in it.*[3]

In the light of the postmodernist devolution we see that the modernist architects' *Neue Haus* and *Neue Baukunst* housing models were more than anticipation of public understanding: they were a part of modernist sensibility common to all the arts that opposed traditional and bourgeois culture in whatever form it took, a movement that Lionel Trilling has called the "adversary culture."[4]

"The intellectual bourgeois. . .has proved himself unfit to be the bearer of a German culture," said Gropius, "New, intellectually undeveloped levels of our people are rising from the depths. They are our chief hope."[5] Gropius's interest in the proletariat was perhaps only aesthetic or fashionable—"somewhat like the interest of President Rafael Trujillo of the Dominican Republic in republicanism," as Tom Wolfe put it in an issue of *VIA*[6]—but his antibourgeois prejudices were sincere. The aesthetic that he and other early modernist architects were developing at times took on the nature of a puritancial witch-

Credits: This article was prepared in conjunction with the exhibit "Suburbs," sponsored by the Cooper-Hewitt Smithsonian Institution National Museum of Design and previously published as a special issue of *AD Profile*, 1981.

hunt or crusade. From small scale to large, from ornament to building type, those architectural elements most intimately involved with traditional values were banished. New churches looked like factories, new forms of housing like office buildings.

Early modernist architects were inevitably influenced by the adversary culture already flourishing in the other arts. In literature, for example, it is virtually taken for granted that the modernist viewpoint is an adversary, subversive position intended to judge, condemn and perhaps revise the society that produced it. In his essay "On the Teaching of Modern Literature," first published in 1961, Trilling traces this modernist tradition back to Goethe, but writes that it had reached its apogee in the first quarter of the 20th century when the bright stars of early modernism—James Joyce, Pablo Picasso, Igor Stravinsky, *et al.*—had all adopted it. By the time that Trilling came to teach at Columbia University in the 1930s, several generations of college students accepted the adversary intentions of modernism as the norm, a paradoxical situation that left the adversary culture an important and perhaps dominant part of the culture it supposedly opposed.[7]

Since that time we have been through the exuberant "counter-cultural revolution" of the 1960s and the more lackadaisical cooling-off period of the 1970s that has been labeled the "Me Decade."[8] Our traditional values have been overwhelmingly challenged, leaving us all to some degree a part of what was once the adversary culture. The adversary position has lost its vitality in the arts: its tradition always insisted that the artist stands in inspired isolation against society, but who can claim fundamentally to oppose society when its universities, galleries, media—the cultural establishment—all support him?

Now many of the traditional values of our society are attractive again. Artists in all fields are rediscovering bourgeois virtues denied to the avant-garde as recently as ten years ago. Although Pop artists, for example, stated emphatically at the time that the content of their work was unimportant, the painting that followed Pop art has been more influenced by Pop's content than its form.

Architects, too, have rapidly progressed from an acceptance of "the dumb and the ordinary," ideas which seemed radical when Robert Venturi first proposed them, to an enthusiastic embrace of the simple, if not necessarily discrete, charms of the bourgeoisie, the obvious, and the familiar.[9] This *AD Profile* [the original publication of this article] necessarily written from an American perspective is a part of that movement. Unlike the Venturis' "Learning from Levittown"—a study of a suburban development one suspects they never really liked but felt obligated to learn from[10]—"The Anglo-American Suburb" is a look not at the kitschiest, most commercial suburbs, but at the tradition of planned suburbs and planned suburban enclaves which flourished between about 1790 and 1930 as the best and most comprehensively designed of their type. The sprawling suburbs we build now, based on the mobility of the automobile, are not the ones that our culture idealizes: Forest Hills Gardens and not Levittown rings the bell of status of Long Island; Roland Park is preferred in Maryland to the new town of Columbia. Even Lewis Mumford—the most articulate and in many ways most astute of the modernist urban critics and architectural historians and one who for years has railed against the automobile-dominated suburban environment—grudgingly acknowledges the important achievements of the planned suburbs of the preauto age. In his book *The City in History* he states emphatically that the builders of those suburbs "evolved a new form for the city."[11]

Forest Hills Gardens is in the New York City borough of Queens, while Roland Park is a part of Baltimore. Like their counterparts in London, Bedford Park and Hampstead Garden Suburb, they raise the question of what a suburb is. Though it is clearly a planning type, the suburb is perhaps most importantly a state of mind based on imagery and symbolism. Suburbia's curving roads and tended lawns, its houses with pitched roofs, shuttered windows, and colonial or otherwise elaborated doorways all speak of communities which value the tradition of family, pride of ownership and rural life. That symbolic imagery, discussed by Denise Scott Brown in her article "Suburban Space, Scale and Symbols,"[12] can be equally effective in rural or non-rural situations, with scattered single-family houses or at a higher density, so long as the imagery of the freestanding house on a tree-lined street is maintained as the dominant impression of the community.

While it is true that we cannot ignore the limitations of suburbia, it is also true that many of those limitations are the result of widespread social issues and not the result of the suburb as an architectural or planning type. But it should be pointed out that other problems of the suburbs—such as the dependency on the automobile for virtually any social or economic intercourse outside of the family, or the banality of the houses many of those families live in—may be as much the result of architects' and planners' neglect of the spread of the post-World War II automobile suburb as of the "crass commercialism" which is usually assigned the sole blame. Considering the disastrous consequences of the housing built on the model of Le Corbusier's *Ville Radieuse* and other modernist social experiments that failed to capture the support of the market (we might call that "benign neglect"), we are now reopening the issue: businessmen and theorists alike are coming to believe once again that the architect must attempt to reflect society as least as much as to reform it, and it is in this light that we believe it is time for a re-examination of the suburb.

The suburban ideal and the house

Suburban imagery is familiar to us all—as American as apple pie, as English as a pint of bitter—yet its role in culture is little studied, and even less understood. The word *suburb* is itself evolved from the Latin *suburbium*, most likely adapted by the English from the Old French *suburbe* during the period of Gallic influences during the 14th century. Chaucer's casual use of the term in 1386 in the *Canterbury Tales* suggests that it had long acquired a definite meaning.[13] But the suburb as we know it, the dependent dormitory town, could not exist without convenient transportation to carry the commuter into and out of his work in the city. Thus the origins of the modern suburb might be traced to the booming expansion of London under George III, when horse-drawn stages for the newly prosperous merchant class, aided by the building of an extensive paved highway system, fostered the development of country estates into new towns and the rapid growth of small, once remote villages that lay along the highways.

These early suburbs were popular because of their associations with life in the country. The merchants built small houses in emulation of the gentry's country estates, setting a pattern for the future suburban imagery. The romantic movements of the early 19th century contributed to the growth of suburbia. John Nash's rustic Blaise Hamlet, as well as the later twin villages and freestanding houses which he included in the predominately urban Regent Park's development, are seminal models for suburbia. But it is doubtful that they or any of the prototypical suburbs of the early 19th century had any direct effect on the American side of the Atlantic where the imagery of the New England villages was combined with the notion of Thomas

Jefferson's gentleman farmer to focus the drive towards the establishment of homogeneously populated towns that were also sound real estate investments. Regardless of the early sources of imagery in either country, however, it was not until the rise of industrialism in both that the suburb flourished for the general populace.

Industrialization contributed four factors to the development of the suburb: on the one hand, it brought increased prosperity for many; second, it brought better public transportation (particularly in the development of the railroad and the streetcar which in turn allowed workers new freedom of choice of where to live); third, although these advances were not of themselves exceptionable, they were often at the expense of unprecedented environmental and moral problems in the cities which were in the long run very disruptive to the urban core; and fourth, at least in the minds of some, it was damaging to family and spiritual life. The American Congregationalist minister Horace Bushnell gained popularity for his sermons and lectures addressing the changes brought on the domestic realm by industrialization and the problems of dealing with them. Bushnell did not mourn the passing of the pre-industrial age, in which he said everyday life was difficult, but he felt that the former existence had an "old simplicity" characterized by "severe virtues." Once the trails of what he called the "Age of Homespun" were removed, he feared that civilization might revert to barbarism. But to avert this, he preached the virtues of education and homelife in the raising of children, and proposed that "The home, having a domestic spirit of grace dwelling in it, should become the church of childhood, the table and hearth of a holy rite." Moreover (we continue to quote for fear, by paraphrasing, of losing the lovely flower of Bushnell's encomium), "the manners, personal views, prejudices, practical motives, and spirit of the house is an atmosphere which passes into all and pervades all as naturally as the air we breath." But when he spoke of the "spirit of the house," Bushnell meant not only the moral influence of the parents but the impact of the physical surroundings as well. He therefore advised parents to create pleasant homes, to make the "house no mere prison, but a place of attraction."[14]

Other ministers of the middle of the 19th century went so far as to describe the architecture of the suitable home, and often published their ideas in the form of stories and poems in the popular magazines of the day.[15] These edifying tales were little different from the secular literature which appeared in new journals like *Mother's Magazine*, *Family Circle*, *Happy Home* and *Parlor Magazine*, or from sentiments expressed in the widely read novel *Home* and the most popular song of its day, "Home, Sweet Home."

It was Andrew Jackson Downing who gave these ideas their fullest architectural expression. The author of *A Treatise on the Theory and Practice of Landscape Gardening* (published in 1841), *Cottage Residences* (1842), and *The Architecture of Country Houses* (1850), as Handern observes, was not a highly original thinker, but was virtually a personification of the domestic idea of the mid-19th century. Philosophically, Downing was a Romantic Rationalist, with practical ideas relating to the landscape painters of the Hudson River School. Architecturally, many of his theories were derived from the work of the English author John Claudies Loudon, who wrote *An Encyclopedia of Cottage, Farm and Villa Architecture* in 1839.[16] But the eclectic designs Downing published in his books were the most influential houses of the middle of the 19th century.

Downing decisively established the principles of the asymmetrical, picturesque house. Moreover, in popularizing the image of the house

Fig. 1. Welwyn Garden City advertisment.

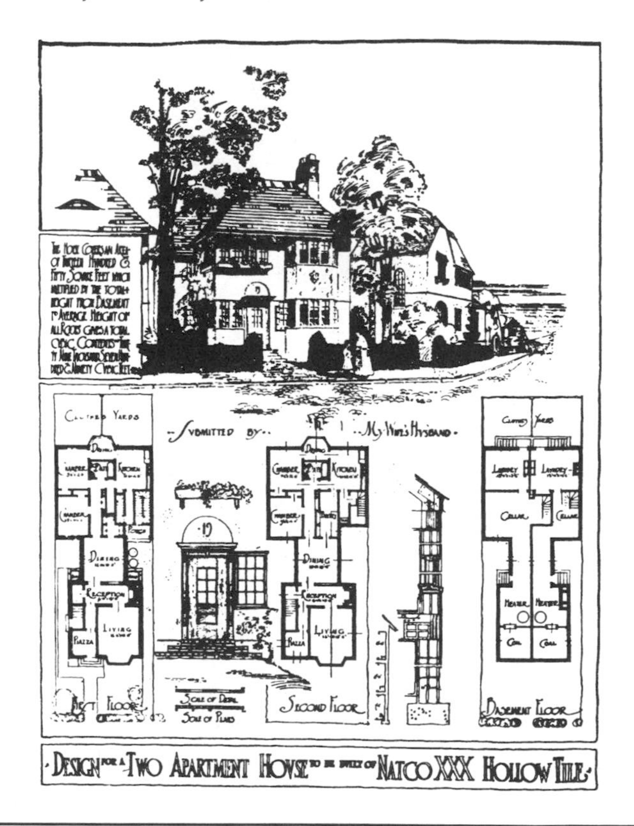

Fig. 2. Model Suburban House, Natco Hollow Tile Competition, Richard J. Shaw, 1914.

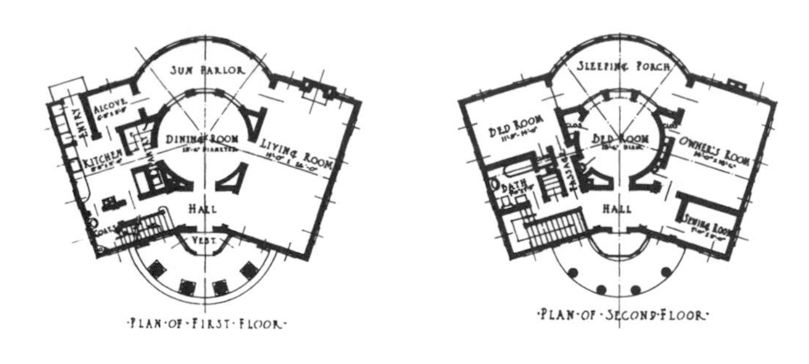

Fig. 3. Petrie House, Scottwood, Fiske Kimball, 1918.

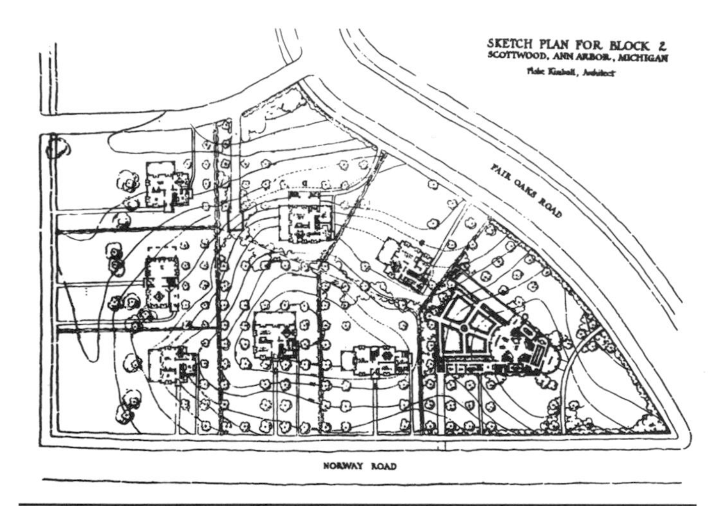

Fig. 4. Plot Plan, Scottwood, Ann Arbor, Michigan, Fiske Kimball, 1918.

Fig. 5. Eliel Saarinen, Munkkiniemi, Helsinki, Finland, 1914, perspective.

in the country—"Architectural beauty must be considered conjointly with the beauty of the landscape"[17]—he crystallized what would become the Anglo-American suburban ideal. Character was important to Downing, who stated that a house could be useful and beautiful but still not satisfactory, because "the intellect (must) approve what the senses relish and the heart loves."[18] What was needed was a house with feeling, particularly of domestic virtue.

After Downing the specific associations of the English country squire or the New England merchant with the freestanding house were replaced by the more general qualities of domestic and spiritual virtue, associations available to the members of any social class or background. From that time on, our best architects have concerned themselves with the problem of the small private house. Lewis Mumford has written, "From H.H. Richardson to Frank Lloyd Wright the most graciously original expressions of modern form were achieved in the suburban house."[19] Regrettably, if understandably, the standard architectural and urban histories have relied on the most elaborately conceived examples of suburban design when exploring Mumford's thesis, because these usually offer the fullest expression of architectural ideas. Nonetheless, it should be pointed out that many remarkable designs for modest suburban houses abound in the work of our leading architects. The majority of the suburbs illustrated in this issue were built for middle- or lower-class residents; in our market economy that means that they were necessarily simple, but the thoughtfulness of their design is evident in even the most cursory examination. Moreover, the importance of the inexpensive suburban house for traditionalist English and American architects is shown by their commitment to model houses. McKim, Mead & White's model houses in Short Hills may have been the developer's idea as much as their own, but Ernest Flagg's series of small suburban houses on his own estate in Staten Island and Frank Lloyd Wright's early experiments with Usonian houses were developed at their architects' expense.[20]

The modern suburban house is a specific type in its own right, not merely a stripped-down manor-house nor a farmhouse, both of which involve a lifestyle economically connected to the land. The suburban house is a direct response to the requirement of efficient, servantless domesticity and the need to reconcile the scale of the house with forms brought about by suburban transportation—whether that be the small lots of the streetcar suburb or the importance of the two-car garage. The Anglo-American suburb offers its users a comprehensible image of independence and privacy while accepting the responsibilities of the community, which is precisely what front, back, and side yards are all about.

The most successful suburban houses have addressed these issues within a recognizable cultural context. In developing the suburban house, American and English architects have drawn on examples of the past in order to establish a continuity and a sense of place in the open countryside where until recently new suburbs have traditionally been built. Mid-19th-century suburban architects tended to design in an 'associational' manner. Virtually all of A. J. Davis's or Samuel Sloan's villas were intended to evoke an earlier architectural style; there was the implication that the style carried with it a mood or characteristic that the prospective occupant could seize upon as emblematic of his own nature.

Later in the century, the suburban house was seen as a principal mechanism for the establishment or re-establishment of appropriate national styles. In England, C. F. A. Voysey's beautifully crafted suburban houses, typified by his own house, The Orchard, abstracted traditional village architectural imagery. Charles Rennie Mackintosh's Hill House was a commentary on vernacular Scottish form; Sir Edwin Lutyen's Salutation, a true revival of the Queen Anne Style. In America, the Shingle Style was in large part an interpretation of the salt-box architecture of pre-Georgian New England, and the Colonial Revival a reinterpretation of the Adamesque Georgian of the late Colonial and early Republican periods. In each of these cases, it goes without saying, the styles invoked were those of the pre-industrial age. And in virtually all the cases, it should be emphasized, the process was one of eclectic evocation and not one of archaeological reproduction.

The 20th century brought an enormous increase in the amount of suburban development with a number of interesting results. In England, the two "vernacular" styles of the 1890s—the Tudoresque free-cottage style and the "true" Queen Anne or Georgian—took root. In America, larger and more culturally diverse, the Shingle and Colonial styles were rejected in many parts of the country as too closely associated with the East Coast to meet the cultural needs of the rapidly developing communities of the Midwest, South and West. An associational design strategy based on regionalism emerged, even in the areas where no regional styles existed. Southern California, under the influence of Bertram Goodhue, adopted a loosely Hispanic Style. In the Midwest Frank Lloyd Wright forged a prairie style which attempted to connect back to New England while simultaneously involving the rude houses of the pioneers and the seemingly limitless landscape in which they found themselves and which was rapidly being transformed.

This regionalism took root not only in the design of single houses but also of suburban developments. Often, references to a local vernacular style were combined with references to a European prototype, as, for example, in the suburb of Chestnut Hill, located within the city limits of Philadelphia. Chestnut Hill was largely developed by one man, George Woodward, in the 1910s and 1920s. A number of very good architects built houses there, including Robert Rodes McGoodwin and the firm of Mellor, Meigs and Howe, and they acknowledged a common style based on the French farmhouses of Normandy and the local vernacular stone house architecture that could be seen in abundance in the adjacent village of Germantown. Thus Chestnut Hill today seems not only romantically evocative but contextually responsive, a grouping of new buildings that continue regional tradition. Later architects building in Chestnut Hill such as Kenneth Day, Oscar Stonorov, Louis Kahn and Romaldo Giurgola, chose not to continue or even acknowledge the Chestnut Hill vernacular that was once so firmly and successfully established.

Contrary to the modernist polemic, the pursuit of regional styles has not thwarted technological innovation or confounded the move towards abstraction that characterized much of the best traditionalist work of the 1910s and 1920s. Irving Gill's work at La Jolla, California, for example, though it refers to the Spanish Colonial tradition exhibits many characteristics considered modernist and is technologically innovative in its justly famous concrete tilt-up wall.

No architect has had more influence on the modern suburb than Frank Lloyd Wright. The centralized massing of Wright's most resolved works, such as the Cheney or Heurtley houses, led to the abandonment of the traditional gable-or-temple-fronted building in favor of a very low hip-roof characteristic of Jefferson, or to the gable form characteristic of the pre-Georgian New England model with its ridge line running parallel to the street.

Fig. 6. Walter Gropius and Marcel Breuer, New Kensington, Pennsylvania, 1942.

Fig. 7. Irving Gill, Women's Club, La Jolla, California, 1913.

It was not until the 1930s that Wright, with his Usonian houses, established the model for the "ranch" house which has characterized suburban development since World War II. In order to do this Wright abandoned the cross-axial plan of his early maturity and at the same time adopted single-story plan types with such features as low hipped roofs, carports and generous amounts of glass. Wright's impact on the suburb was even greater, however, in the area of landscape and townscape than it was on house design. With the introduction of the fully developed cross-axial plan in the Ward Willetts House of 1902 at then rural Highland Park (which was not a suburban house in the sense that his work in Oak Park was) Wright began a gradual transformation of the traditional suburban streetscape: the narrow but relatively deep lot characteristic of 19th-century suburban development was not suitable for Wright's new "prairie" house type, which de-emphasized the traditional, static relationship front, back and sides in favor of a new relationship based on the dynamism resulting from the simultaneous inward and outward focus of interior spaces and the composition of volumes in accord with principles of centrality and rotation. In order to accommodate Wright's Prairie House, suburban lots became square in plan.

Nevertheless, Wright's formal intentions were acceptable and accessible to an educated though not necessarily cultured clientele because they were in many ways quite traditional: their sheltering roofs and chimneys announced home and hearth to all but those critics too involved with an anti-sentimental modernism and who felt threatened in the 1930s and 1940s by Wright's claims to leadership as our most "modern" architect, more "modern," in fact, than the European modernists whom he dismissed as International Stylists. As a result of this split with modernism, as much as because of the loss of conviction on the part of traditionalists, serious pursuit of the design issues of the suburban house was abandoned. Just when the suburb burgeoned to unprecedented size our very best talents were pursuing such issues as architectural mass production or the joys of building "one-off" houses as monuments that would establish reputations leading to careers as designers of museums, or office buildings, or both. The subject of the suburb was left to the ordinary practitioner who felt largely confused and alienated in the crossfire of modernist polemics and to the speculative builder who reveled in the opportunities offered by an architectural ideology premised on the proposition that less could be more.

The social suburb: the planned industrial village

In his introduction to Walter Gropius's *Rebuilding Our Communities*, published in 1945, Paul Theobald writes:

> *Even 46 years ago no self-respecting architect would have endangered his reputation by designing low-cost housing projects or factories. The art of architecture was exclusive, a privilege jealously guarded by those who could afford it. But what seemed the "art" of architecture was actually a tragic nihilism, representing the compromise and business expediency of an architect's office instead of urgent human needs. It demanded heroism—the fearless determination of self-sacrifice—to break this spell... It has been the significant contribution of Walter Gropius to have demonstrated the indivisibility of social responsibility and structural soundness where needed most: in settlement housing projects, factory construction, and—above all—in education. Fearlessly and uncompromisingly he defended the thesis on which the Bauhaus was built: that art and architecture which fail to serve for the betterment of our environment are socially destructive by aggravating instead of healing the ills of an inequitable social system.*[21]

One of modernism's most persistent myths has been its insistence that because the academic and *Beaux-Arts* architects ignored what Sigfried Giedion called the "constituent facts" of the industrial age, those architects were ignorant or even contemptuous of the urgent social and economic problems that are inextricably linked in our time with the issue of housing. Even today, when many socially concerned architects see the irrelevance of Giedion's constituent facts for the majority of social problems, the feeling remains that the traditional, historicist architects were somehow guilty of social irresponsibility and elitism. But while the practice of architecture may have once been predominantly a rich man's profession, its American and English members as a group were neither elitist nor irresponsible: in fact, they were an important part of the progressive movement which, on both sides of the Atlantic, fought for better housing for the working and middle classes, and helped to implement the legal reforms which led to that housing.

Along with middle-class suburbs such as Bedford Park and Riverside, there were many industrial villages that contributed to the evolution of the Garden City ideal often assumed to be the first serious manifestation of a socially responsible conception of suburban planning. Although we often do not know whether it was the architects involved or their philanthropic clients who were responsible for the formal design decisions involved in the early industrial villages such as Lowell, Massachusetts or Saltaire, those towns were the start of a reform tradition for the housing of industrial workers carried on in the work of leading historicist architects at major American industrial sites between 1880 and 1920. McKim, Mead & White; George B. Post & Sons; Bertram Grosvenor Goodhue; Delano & Aldrich; York & Sawyer; Trowbridge & Livingston; and Walker & Gillette are in their way as important as Ebenezer Howard and Parker & Unwin in the history of the garden suburb. Moreover, the major American exponent of the Garden City ideas, Clarence Stein, was both a graduate of the Ecole des Beaux-Arts and a former employee in Goodhue's office, where he got his first taste for town planning while working on the new copper-mining village of Tyrone, New Mexico, Moholy-Nagy may not have known the American industrial villages, but he was aware, of course, of the Garden City and the contemporary work of Clarence Stein. But while the Garden City was widely acknowledged by modernists to be a primary source for their new theories of urbanism (Le Corbusier went so far as to describe his city for 3 million as a "Vertical Garden City"), the forms of its architecture were resoundingly rejected because they were sentimental, historical—familiar. Modernists, unlike Stein, abandoned the Garden City style and fixed on certain aspects of its compositional principles – mainly the idea of isolated wedges set in vast open spaces, flooded in sunlight and awash in greenery. They also shared the spirit of reform that Parker & Unwin inherited from John Ruskin and William Morris, a debt Parker acknowledged when he received the Gold Medal of the Royal Institute of British Architects in 1937:

> *One who was privileged to hear the beautiful voice of John Ruskin declaiming against the disorder and degradation resulting from the laissez-faire theories of life; to know William Morris and his work. . .could hardly fail to follow after the ideals of a more ordered form of society, and a better planned environment for it, than that which he saw around him in the seventies and eighties of the last century.*[22]

The forces that led to that spirit of reform were some of the same problems of industrialization instrumental in the reformation of family life. They were a reaction to the overcrowded, polluted and badly

Fig. 8. Frank Lloyd Wright, "A Home is a Prairie Town," *Ladies Home Journal* Project, 1900.

Fig. 9. Broadacre City, Frank Lloyd Wright, architect. Project 1935, is the apotheosis of the American suburban ideal at a vast scale adjusted to the motorcar. It was Wright's answer to the problem of the urban crisis. It is evident that Wright, far more than Clarence Stein or Henry Wright, clearly saw the planning potential of the automobile. The direct impact on the suburb of Wright's planning is difficult to document, but the influence of this architectural ideas is clear. The house types devised for Broadcare City had their nomenclature usually derived from the number of cars parked under their broad ubiquitous post-World War II ranch house. The modest "one-car" dwelling, with its low-hipped roofs, broad chimney, open interior plan and inclusion of workshop facilities, has surely become a constituent American type.

built quarters which grew up as industrial workers streamed into the cities. Three educated visitors to the English industrial cities of the early 19th century described "the social incoherence of these towns, their cold unhappiness, the class division of interests and pleasures, the concentration on a limited and limiting purpose." One of the visitors reported that little thought was given to "any comprehensive, any attention to general convenience, or to beauty and architectonic art. Capital is employed solely in the creation of new capital. What is not calculated to promote this end is regarded as useless and superfluous."[23]

Enlightened and benevolent employers set out to improve conditions by building model industrial towns out in the clean air and beauty of the open countryside. Unlike the earlier American mill towns such as Lowell, Massachusetts or Manchester, New Hampshire, where boarding houses were built to provide decent accommodation for workers who moved to the sparsely populated areas where sufficient water power was available to operate the mills, the English industrial villages were responses to the degradation of city life caused by the advance of industrialization: while the mill towns were only quite natural reflections of the surrounding New England communities the architects and mill owners were familiar with, the later company towns in England such as Saltaire, Port Sunlight, and Bournville were conscious idealizations of pre-industrial villages; the planned industrial village because a symbol of lost virtues and an affirmation of a new humanism. These new villages had two essential qualities: they were healthy environments, much healthier than the cities the industries fled, and they provided a romantic vision for both the workers and the benefactors that the dynamic processes for the industrial age were controllable. By virtue of their architectural character and the political and economic dependency on larger entities, they can be claimed as *suburban* industrial villages.

In America the process was somewhat different. The single-family workers' houses built by the S. D. Warren Company at Cumberland Mills in Westbrook, Maine, or the similar development in Willimantic, Connecticut, known as Oakgrove and built by the Willimantic Linen Company in 1865, are a departure from the rowhouses and group houses in Lowell. But like the houses at Lowell they are a straightforward reflection of the prevailing middle-class houses of their day. So too was Pullman, Illinois, built a full 30 years before Saltaire, but in the same decade as Port Sunlight. While Pullman shared—and in many ways anticipated—the English ideal of a fully planned model suburban community as a reaction to the city, its version was far less romantic. Its houses were not based on pre-industrial cottages, but were well-planned, if severe urban dwellings. The same cannot be said of other company towns such as Echota, a development of the Niagara Power Company in Niagara Falls, New York in 1892 where McKim, Mead & White built a reasonably convincing version of typical American village housing.[24]

Frederick Law Olmstead's romantically planned industrial village of Vandergrift, Pennsylvania, built in 1890 for the employees of the Apollo Steel Company, is perhaps the first of the American industrial towns to rival in appearance the English villages, although the houses on its winding streets were built to the tastes of individual owners and were somewhat erratic in quality. After the turn of the century, however, several company towns were built with romantic imagery. The Draper Company's large development at Hopedale, Massachusetts, planned in 1910 by Arthur A. Shurtleff with duplex houses designed by Robert A. Cook, was one of the best; also good were Kohler, Wisconsin and Tyrone, New Mexico. Others include: Goodyear Heights, Akron, Ohio; Allwood, New Jersey; Eclipse Park, Beloit, Wisconsin; Garden Suburb, Billerica, Massachusetts; Overlook Colony, Claymont, Delaware; Morgan Park, Duluth, Minnesota; Civic Building Association, Flint, Michigan; Framingham, Massachusetts; Hershey, Pennsylvania; Leclaire, Illinois; Marcus Hook, Pennsylvania; Midland, Pennsylvania; Waterbury Homes Corporation, Waterbury, Connecticut; Indian Hill, Worcester, Massachusetts; Loveland Farms and East Youngstown, Youngstown, Ohio.[25]

A lack of suitable housing for the increased numbers of workers needed during World War I in the vicinity of shipyards and other war plants prompted the United States government to plan model industrial villages along the lines of the company towns. Congress formed the United States Housing Corporation and a housing division of the Emergency Fleet Corporation; the former planned 128 developments for 19,100 dwellings at an estimated cost of $112 million, while the latter built 27 towns with 8,841 houses for a total expenditure of $71 million.[26] These projects, sold after the war at a loss, benefited the nation by the existence of such outstanding workers' housing as Atlantic Heights in Portsmouth, New Hampshire; Union Gardens at Wilmington, Delaware; Yorkship Village, Camden, New Jersey; and several subdivisions in Bridgeport, Connecticut. Other developments were not as good although when Congress investigated the projects in the 1920s it complained that the work had been too good; the houses were too well constructed.

Unfortunately, the industrial villages built during the First World War were the swan song of their type. After the war only a few industrial towns were built, and their planning failed to meet the standards achieved by the wartime work. One of those towns, Kingsport, Tennessee, was planned by John Nolen, the designer of several of the war developments, but it compares poorly with his best work. In the Second World War there was again a rush to provide housing for industrial workers, but by that time the impact of European modernism began to alter and even contradict prevailing standards of accommodation and taste. One of the most notable examples of the Second World War Defense Housing projects in the modernist mode, Walter Gropius and Marcel Breuer's Aluminum Terrace Housing near New Kensington, Pennsylvania, met with widespread user hostility at its inception. Though the criticism eventually subsided, it opened up the question of the ability or willingness of the new generation of architects to work within the suburban tradition.

At New Kensington and at Windsor Locks, Connecticut, designed by Hugh Stubbins, virtually all sense of the street and the hierarchy—or at least variety—of the houses is abandoned. The new values were best expressed by Howe, Stonorov & Kahn in their description of their "ground free" unit at Carver Court:

> *"In essence, the scheme consists of nothing more than a transportation of the basement to the ground level, with one storey of living space above. When developed into a typical four-family unit, some interesting features appear. There is open space between the utility blocks which serves very conveniently as a carport, a shelter for the entrance or a covered play area. These openings also eliminate the distinction between "front" and "back," as there is free circulation through, rather than around, each building. . ."*[27]

The impact of the Defense Housing program was widespread: the projects built across the country proposed design features which later became built into the federal loan programs for returning veterans, thereby setting the standard for postwar development. But their designs, often related to the Bauhaus *Siedlungen*, reflected the urban preferences of the modernist architects. Ironically, however, the same architects would usually build for themselves small, freestanding houses in the suburbs. The development of eleven houses at Six Moon Hill by the Architects' Collaborative, Gropius's firm, is a modernist Llewellyn Park; each house was sited on its own lot according to solar orientation as well as view, with sidewalks eliminated, and a communal park.[28]

The growth of the planned suburb

The suburb as we know it evolved from the combination of imagery already surfacing in the urban expansion projects such as Regent's Park, and the development of modern transportation. Christopher Tunnard and Henry Hope Reed argue that America's first suburb was Brooklyn Heights, made possible by the introduction of a steam-ferry sservice between Brooklyn and Manhattan in 1814, though that dormitory development's conventional urban rowhouses keep us from including it in this incomplete and introductory survey.[29] Moreover, the evolution of the suburb as a widespread phenomenon could not have happened without the railroad, commercially inaugurated in two places at once in 1830: in England with the opening of the Liverpool and Manchester Railway, and in the United States by the Baltimore and Ohio Railroad.[30] It was the rapid spread of the railroad throughout England and America which allowed the growth of the suburb as a popular concept, because it was the railroad which enabled the businessman and laborers of every major city to travel conveniently in and out from the country. While relatively little has been written on the immediate effects of the railroad and the development of the suburb, we know that as early as 1831, when the New York State Legislature granted its first railroad charter to the New York and Harlaem Railroad to run from 23rd Street through what was then open countryside to the village of Harlaem (where Harlem stands today), the *Morning Courier and Enquirer* speculated that Harlaem would soon become a suburb where commuters would be able to enjoy a comfortable house on an acre or two of land with "a garden, orchard, dairy, and other conveniences."[31] Five years later the tone of a prospectus for the speculative development of New Brighton on Staten Island made it clear that New Yorkers were familiar with suburbs.[32]

The commonly stated claim that Llewellyn Park is America's first suburb is clearly wrong. Founded by Llewellyn Haskell in 1853, its position as a fully planned, ideal community—for "long-haired men and short-haired women," one original resident wrote[33]—distinguishes it from earlier suburbs, and marks the start of the tradition of carefully planned suburbs that was to be a central concern of American and British architects and planners for the next 70 years, until it was interrupted by the polemics of modernism. Frederick Law Olmsted, better known for his great parks than his plans for suburban developments in Riverside, Illinois; Berkeley, California; Tarrytown Heights, New York; and Newport, Rhode Island, was one of the first designers to seriously consider the suburb as a development type. In 1868 he wrote that although many of the burgeoning suburbs. . .

> *". . . .are as yet little better than rude over-dressed villages, or fragmentary half-made towns, it can hardly be questioned that, already, there are to be found among them the most attractive, the most refined and the most soundly wholesome forms of domestic life, and the best application of the arts of civilization to which mankind has yet attained. It would appear then, that the demands of suburban life, with reference to civilized refinement, are not to be a retrogression from, but an advance upon, those which are characteristic of town life, and that no great town can long exist without great suburbs. It would also appear that whatever element of convenient residence is demanded in a town will soon be demanded in a suburb, so far as is possible for it to be associated with the conditions which are the peculiar advantage of the country, such as purity of air, umbrageousness, facilities for quiet out-of-door recreation and distance from the jar, noise, confusion, and bustle of commercial thorough-fares."*[34]

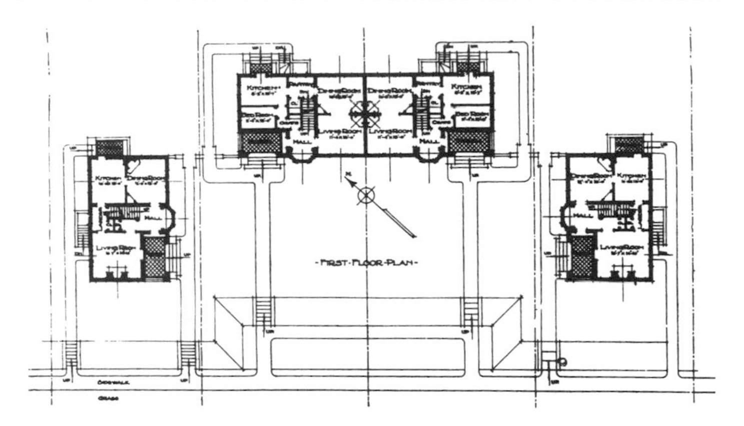

Fig. 10. Forest Hills Gardens, designed by Grosvenor Atterbury and the Olmsted Brothers, was built in 1912 by the Russell Sage Foundation as a model suburban residential town, 15 minutes by rail from Manhattan. The Sage Foundation intended Forest Hills gardens as a village of lower-income housing, but its nearness to Manhattan made the land cost too high, and the development quickly became the upper middle-class enclave it remains today.

To improve on the "fragmentary half-made towns" Olmsted and other suburban planners established building standards and design controls. Rules governing lot size, building placement, and property rights were first used in planned suburbs but became common in other suburbs long before zoning existed in the cities. The establishment of communal architectural styles in planned suburbs, as at Chestnut Hill or North Oxford, also influenced the form of the small villages which after the arrival of the railroad had grown into large suburbs. Of the hundreds of village improvement societies founded in the second half of the 19th century, many were interested in the romantic goal of making their town more village-like. Some went so far as to reform their town in the ideal image of a New England town or European village, depending on the background of the town's new residents. The most notable of these schemes was Jarvis Hunt's plan to cover the various Victorian buildings along Wheaton, Illinois' main street, in stucco and half-timber designs patterned on Elizabethan architecture, and E. D. Libby's gift of money to any citizen who would cooperate with his plan to transform Ojai, California, into a Spanish village.[35]

The railroad station was usually placed at the heart of the planned suburban villages. Before the invention of the automobile, commuters had to walk to and from the station every day, giving an additional reason for a compact plan. Moreover, the wives and children who were left at home during the day needed neighbors to socialize with. The same pattern developed along the streetcar and subway lines which after the mid-19th century gradually spread out from the compact cities. First with horse-drawn trolleys and later with electric trams, the areas between the old city cores and the railroad suburbs were brought within easy access to city centers, and although their relation to each other along the streetcar line might be continuous, they were often developed on the same distinct centers model as the earlier and usually grander railroad suburbs. Even residential resorts, where families would move for months at a time, were developed along lines similar to the commuter suburbs. After all, the summer and winter residents wanted the same sense of community and socialization as the suburban residents, and their houses allowed them to express their individuality within a framework which society appreciated.

In March 1913 the progressive City Club of Chicago sponsored a competition for the development of a tract of land connected to downtown Chicago by streetcar. The square site was undivided, without trees or buildings, but it was surrounded by the prevailing grid of Chicago. The competition rules stipulated only that a maximum of 1,280 families be housed on the site, but asked the contestants to show "the essentials of good housing in its broadest sense."[36] Included as a suggestion of what that might be glowing descriptions of the garden city movement.

The first and second prize schemes, by Wilhelm Bernard and Arthur Comey, were based on Frederick Law Olmsted Jr.'s and Grosvenor Atterbury's 1909 plan for Forest Hills Gardens. Both plans introduced focal community centers and curving streets of the sort often used in planned suburbs to define the neighborhoods. But only two designs attempted to connect the new enclave to the surrounding neighborhoods. One was William C. Drummond's, an interesting proposal with a grid rooted in the vast expanses of the prairie. The other, Frank Lloyd Wright's noncompetitive entry, also accepted the grid and the prairie. A reworking of Wright's quadruple block project, the entry was to prove important for the next development of the suburb, that in which the automobile became the dominant means of transportation.

A 1913 article entitled, "The Automobile and Its Mission," summarized the astounding progress of the auto age: in 1908 it was still a "transcendent play thing—thrilling, seductive, desperately expensive," but by 1913 it was opening up a new pattern of residential settlement.[37] The mobility of the car and other developments which reduced dependency on city centers—the cinema, radio, telephone and television—combined to produce a freedom, which never existed with the railroad or the streetcar. Suburban centers no longer had to be located at railroad stations or along trolley lines, and houses no longer needed to be within walking distance of the town center; if indeed there was a town center. In 1916 President Woodrow Wilson signed the Federal Roads Act, initiating a national road building program. Four years later more than nine million Americans owned cars, and the pattern of the suburb had been drastically altered. By 1941, when the Bureau of Public Roads prepared for World War II by undertaking a survey of the nation's daily commutation patterns, it found that 2,100 communities with populations ranging from 2,500 to 50,000 had dispensed with or had developed without inter-urban mass transportation systems and were almost entirely dependent on the private automobile for travel within their own boundaries.[38]

Only a few of our leading City Beautiful and City Functional planners, most notably John Nolen, appreciated the special characteristics of the automobile suburb. As the car became a central feature of American life, architects increasingly treated it as a problem rather than a virtue and abandoned the planning issues of the suburb to the developer. Even designers such as Clarence Stein, who clearly saw the advantages of the car's freedom to enable travel wherever it was desired, regarded the relationship of the car to the house as one to be hidden and subverted. In that way he was in accord with the urban preferences of the modernist architects who built World War II defense projects. Needless to say, these ideas were never popularly accepted or incorporated in suburban development.

Frank Lloyd Wright is perhaps the only significant architect of the mid-20th century to make a lasting contribution to the form of the auto suburb. Though his proposal for Broadcare City can be dismissed as an extreme vision of the Arcadian ideal such as a Detroit mogul might dream of, it did articulate principles of a new kind of planning based on the car which have come to govern suburban development sine 1945. Wright's aptly titled "clover-leaf" site plans of the late 1930s and early 1940s have had a tremendous impact on suburban land planning, because his characteristically American pragmatism went with the grain of Anglo-American traditions. Thus if one were to measure the success of an idea by its effect on built form and its theoretical insight rather than by its influence on professionals, Wright must be acknowledged as a 20th-century planner equal to or greater than the modernist heroes such as Le Corbusier, Gropius or the members of the Team Ten.

Without the intervention of the majority of our best architects and planners, the auto suburb has developed in two forms. One has been the enlargement of older suburbs, as at Greenwich, Connecticut; Lake Forest, Illinois; or countless other locations. The second has been the thoughtless surrounding of our cities with blankets of low-density housing, only interrupted by strips of commercial development along the main roads. As architects come to rediscover the positive virtues of the suburb as a type, it is clear that with the proper persepctive the first model, that of the rail suburb, can be made to work, and work well, although one suspects that in the future there will be less and less of that type of development. But we would suggest that precisely because of the popularity of that model it should be used for the rebuilding of our cities blighted by the postwar flight to the suburbs, a flight ironically helped by the urban redevelopment efforts of the 1940s and 1950s, when urban planners pulled down the worst sections of our slums without adequately providing new housing, so that the ghetto expanded into adjacent neighborhoods, and simultaneously built highways from the hearts of cities to the outer edges, affording easy routes to flee those newly declining neighborhoods. As a result, the areas between our downtowns and our suburbs now lie fallow and abandoned in cities all across the country. But these burnt-out wastelands, such as the Charlotte Street section of the South Bronx, the Barbican of New York's urban renewal blitz, while empty for block after block, still contain a network of utilities and mass transportation that make them extremely valuable in a time of diminishing resources.

New suburbs along the lines of Forest Hills and Baldwin Hills should be built where they are really needed, within walking distance of

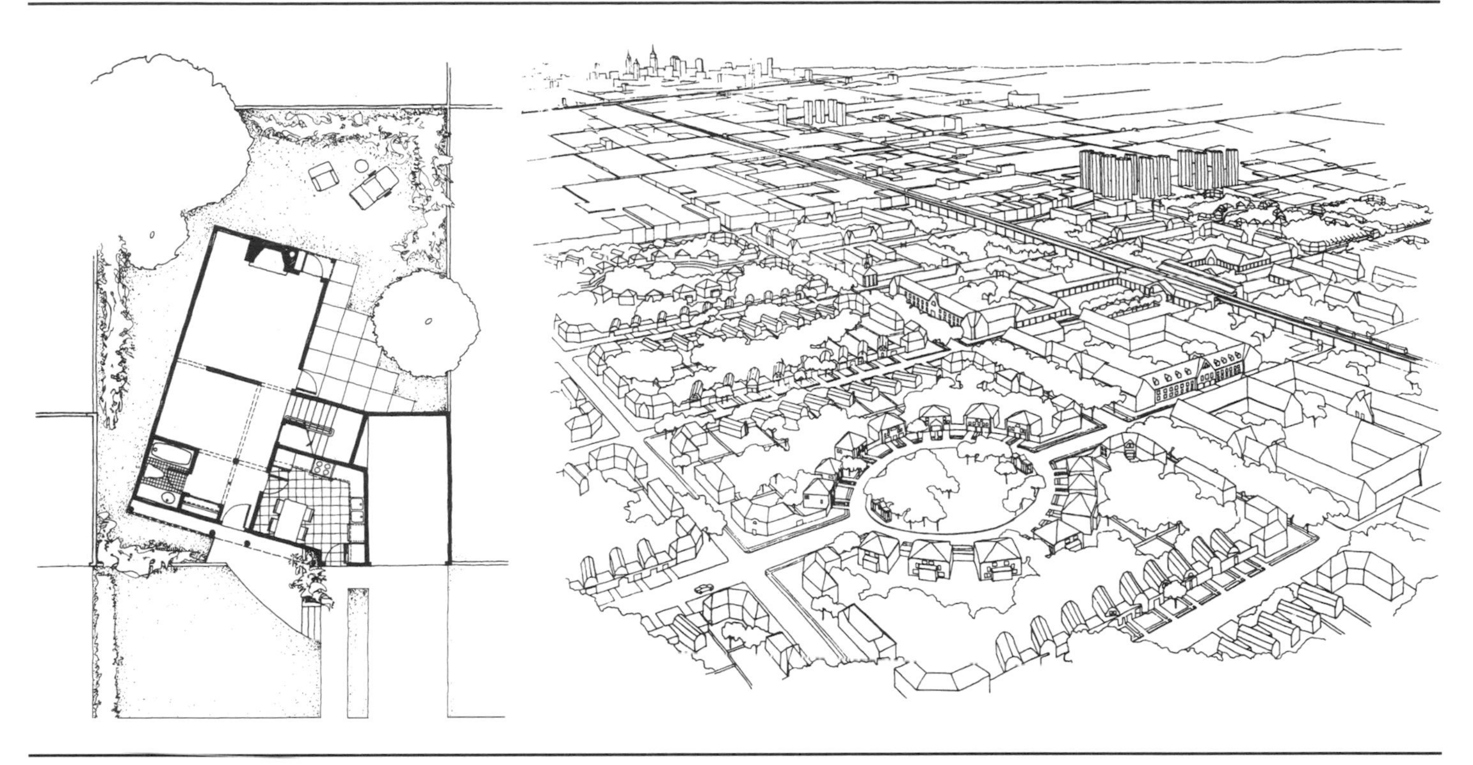

Fig. 11. Subway Suburb. Robert A. M. Stern, architect. Project 1976. Prepared for the Venice Biennale, the project is an attempt to define a new kind of suburb to be built within the legal confines of the city, relatively close to its center, utilizing urban land that has been abandoned and has no apparent higher value. This project suggest uncharacteristic ways to develop the land in blighted, marginal areas of the city, that will utilize existing street and utility grids to offset development costs, and take advantage of existing rapid transit services while also accommodating the automobile.

convenient transportation to business and cultural centers. In proposing the suburban model for the South Bronxes of our cities, we realize that we are taking on two sacred cows. The first is the notion that the history of cities is and has always been one of increasing population and therefore population density. The second is that the suburb is the particular fief of the middle classes, leaving the other forms of housing to those less well off economically and to the rich, who not only can choose what they want but who also are able to spend enough money to make their choices work. In challenging the former view, one merely contradicts dearly held theories of some physical and economic planners. But in challenging the later, one strikes at the jugular of the Anglo-American value system. Still it is important to point out that, though home ownership is a sine qua non for most Anglo-Americans, a badge honoring individual success, in reality such ownership for the past 30 years or more has in effect been subsidized by the government as a result of socialization in England, and the FHA, The GI Bill, and many other similar programs in the U.S.

Many English and Americans are uncomfortable in large, densely built cities. Our urbanism is shaped by the prejudices and preferences of this shared cultural heritage. Even London is a city of vast land area and low population density—a collection of loosely connected villages that set the stage for the kinds of urbanism that the automobile has made possible in Los Angeles and Houston. Even the actual density of a city like New York—once one goes beyond the dense core of Manhattan—is far lower than the city's image in literature and films would begin to suggest. Outside Manhattan, much of New York is a city of attached and semiattached one-two-and three family houses, interspersed with apartment blocks usually no more than six-stories high. For this reason, it can be argued that New York and most other American cities are, like London, collections of small towns—let us call them suburbs—united not by a uniform street grid or by a superhighway system, but by a system of roads which generally preceded urbanization and by underground and elevated rail systems that even now can make the suburbanization of our cities feasible.

There will be no new ideas about cities and their suburbs until our thinking frees itself from the biases and orthodoxies of our recent architectural and urban theories, especially those peculiar cultural biases and cultural prejudices which have encouraged us to see old cities and old buildings—not to mention traditions and recognizable forms—as worthless and wrong. Nor will we be able to deal confidently with the suburb until we free ourselves from the belief that new suburban ideas (or, in fact, new suburbs) can only grow on virgin land beyond the edges of existing development. Suburbs will not go away, nor should they. They may well hold the key to the solution of urban problems that were hitherto deemed insoluble. ■

NOTES AND REFERENCES

1. From "Little Boxes," a song made famous by Peter Seeger, written by Malvina Reynolds (copyright 1962 Schroder).

2. Anthony C. Antoniades, "Architecture from Inside Lens" *[sic]*, *A + U* 106, July 1979, pp. 3–22.

3. Third Edition, Harvard University Press, Cambridge, Mass., 1954, p. 607.

4. See Trilling's book *Beyond Culture*, uniform edition, Harcourt Brace, New York, 1979, *passim*. Daniel Bell also discusses the adversary culture in *The Cultural Contradictions of Capitalism*, Basic Books, New York, 1976.

5. Quoted by Tom Wolfe in his introduction to "Culture and the Social Vision," *VIA 4*, 1979, pp. 8–11.

6. Wolfe, *op cit.*

7. "On the Teaching of Modern Literature," first published as "On the Modern Element in Modern Literature" in the *Partisan Review*, is the first chapter in *Beyond Culture*, pp. 3–27.

8. Although the "Me Decade" was coined by Tom Wolfe, it has passed into general usage, respectable enough for *The New York Times.*

9. Romaldo Giurgola discussed the marriage of the International Style and bourgeois concerns in the work of the "New York Five" in his article, "Discreet Charm of the Bourgeoisie," *Architectural Forum.*

10. "Learning from Levittown" was a study made at Yale that developed into the Bicentennial exhibition "Signs of Life: Symbols in the American City," held at the Renwick Gallery in Washington, DC from February to October 1976, and an article by Denise Scott Brown, "Suburban Space, Scale and Symbols," *VIA III,* 1977. pp. 41–47.

11. Harcourt Brace, New York, 1961, p. 497.

12. See note 10.

13. *Oxford English Dictionary,* Oxford University Press, New York, 1971, p. 71.

14. David P. Handlin discussed Horace Bushnell in *The American Home*, Little Brown, Boston, 1979. pp. 5–11.

15. Handlin, *op cit.* pp. 11–12.

16. Handlin discussed the relationship of Downing to Loudon, *op cit.* pp. 31-38.

17. *Landscape Gardening*, p. 341. The modern pioneer study of Downing is in Vincent J. Scully Jr. *The Shingle Style and Stick Style: Architectural Theory and Design from Downing to the Origins of Wright,* Revised Edition, Yale University Press, New Haven, 1971, pp. xxiii-xiviii.

18. *Country Houses*, p. 47.

19. *The City in History*, p. 490.

20. Flagg's houses are published in his book, *Small Houses: Their Economic Design and Construction*, Scribner's, New York, 1922.

21. Paul Theobald, Chicago, 1945, np. (L. Moholy-Nagy, in his introduction to Walter Gropius's book, *Rebuilding our Communities.*

22. Quoted by Creese, p. 159.

23. J. L. Hammond and Barbara Hammond, *The Age of the Chartists: 1832–1854*, Longmans Green, London 1930, pp. 35-36. Quoted by Creese, pp. 2-3.

24. "Industrial Villages in America," *Garden City 1* ns. August 1906, pp. 49–50.

25. John Nolen, "Industrial Village Communities in the United States," *Garden Cities and Town Planning 7*, October 1918, pp. 6–9.

26. See *Report of the United States Housing Corporation*, Government Printing Office, Washington 1919; also Morris Knoniles, *Industrial Housing*, McGraw-Hill, New York, 1920.

27. George Howe, Oscar Stonorov, Louis Kahn, "Standards Versus Essential Space," *Architectural Forum 76*, May 1942, pp. 308–311.

28. For an introduction to these towns see "Defense Houses at New Kensington, Pa." *Architectural Forum*, October 1941, pp. 218–220; Isabel Bayley, "New Kensington Saga," *Task 5*, spring 1944, pp. 28–31; "U.S. Wartime Housing," *Architectural Review 96*, August 1944, pp. 30–60; "Aluminum City Terrace," *Journal of Housing 10*, October 1953, pp. 330–332; also see "Six Moon Hill," *Architectural Forum 92*, June 1950, pp. 112–127 and Gropius, *Rebuilding our Communities,* 1945.

29. In their book *American Skyline*, p. 62.

30. Carroll Meeks, *The Railroad Station,* Yale University Press, New Haven, 1956, p. 26.

31. William D. Middleton, *Grand Central: The World's Greatest Railway Terminal,* Golden West, San Marino, Cal. 1977, p. 12.

32. Barnett Shepherd, *Staten Island: An Architectural History,* exhibition catalogue, Staten Island Institute of Arts and Sciences, New York, 1979, p. 5.

33. Letter of Edward D. Page, March 24, 1916. Quoted by Richard Guy Wilson, "Idealism and the Origin of the First American Suburb, Llewellyn Park, New Jersey," *American Art Journal 11*, Fall 1979, pp. 79–90.

34. S. B. Sutton (ed.), *Civilizing American Cities: A Selection of Frederick Law Olmsted's Writings on City Landscape,* MIT Press, Cambridge, Mass. 1971, pp. 294–295.

35. Frederick Jennings, "Civic Improvement, Ojai, California, How an Old, Uninteresting Town Was Made Beautiful," *Architect and Engineer 58,* August 1919, pp. 39–48. Also S. Mays Ball, "The Suburb Beautiful, the 'Before and After' of a Progressive Idea," *House and Garden 15*, May 1909, pp. 166–171.

36. Alfred B. Yeomans (ed.), *City Residential Land Development: Competitive Plans for Subdividing a Typical Section of Land in the Outskirts of Chicago*, City Club, Chicago, 1916, p. 96.

37. Hubert Ladd Towle in *Scribner's Magazine 53*, 1913, cited by Handlin, p. 149.

38. Mel Scott, *American City Planning*, University of California, Berkeley, 1969, p. 540.

Andres Duany and Elizabeth Plater-Zyberk

Summary

These diagrams, reprinted from *The Lexicon of the New Urbanism*, illustrate the planning principles developed in the authors' professional practice, with emphasis upon common terms of reference and codifiable planning and design standards. The work—in-progress over many years—is grounded in the firm's experience in design of post- suburban development based upon traditional models of town planning. The text and illustrations provide a coherent summary of a neighborhood-based approach to design with its roots early in the 20th c. and now motivated by environmental and civic concerns. With articulation of the "transect," these planning principles are extended to a full spectrum of urban design practices.

Key words

density, mixed use, neighborhood, network, New Urbanism, pedestrian, rural, street, town, transect, urban

Albacoa Town Center, Palm Beach County, Florida

Lexicon of the New Urbanism

Our premises of the Lexicon are:

- the design of community directly affects human well-being and indirectly the continued viability of nature.
- communities are complex systems, most useful to designers when expressed as conventions.
- the conventions are supportive of the premises stated in the *Charter of the Congress of the New Urbanism*. [*cf.* article 3.10 in this volume].
- the criteria for inclusion of a convention is its long-term social and ecological success.
- the conventions take care to balance the order of community with the freedom of the individual.
- the standing and efficacy of the design profession requires a conventional design and planning language held in common.
- the conventions are integrated in order to resist their appropriation by specialists.
- the conventions are not immutable, but able to incorporate the evolution of practice.
- the conventions expand the range of possibilities currently available to the planning profession.
- the authority of the conventions is neither imposed nor protected, but confirmed through their empirical success.

Albacoa residential pattern. Duany Plater-Zyberk & Company, Calthorpe Associates, Moule and Polizoides, Town Planners

REFERENCES

Andres Duany and Elizabeth Plater-Zyberk. n.d. *The Lexicon of the New Urbanism*. Miami: Duany Plater-Zyberk & Company

Publications. Congress of the New Urbanism. 5 Third Street, Suite 725, San Francisco, CA 94103 <www.cnu.org>

Credits: The excerpts and plates are reproduced by permission from Duany Plater-Zyberk & Company *The Lexicon of the New Urbanism* (1998).

THE TRANSECT

- The Transect: a system of classification deploying the conceptual range rural-to-urban to arrange in useful order the typical elements of urbanism. The transect is a natural ordering system, as every element easily finds a place within its continuum. For example, a street is more urban than a road, a curb more urban than a swale, a brick wall more urban than a wooden one, an allee of trees more urban than a cluster.

The continuum of the Transect, when subdivided, forms the basis of the zoning categories: Rural, Sub-Urban, General Urban, Urban Center and Urban Core.

The Transect technique is derived from ecological analysis where it is applied to present the sequences of natural habitats such as shore-dune-upland or wetland-woodland-prairie.

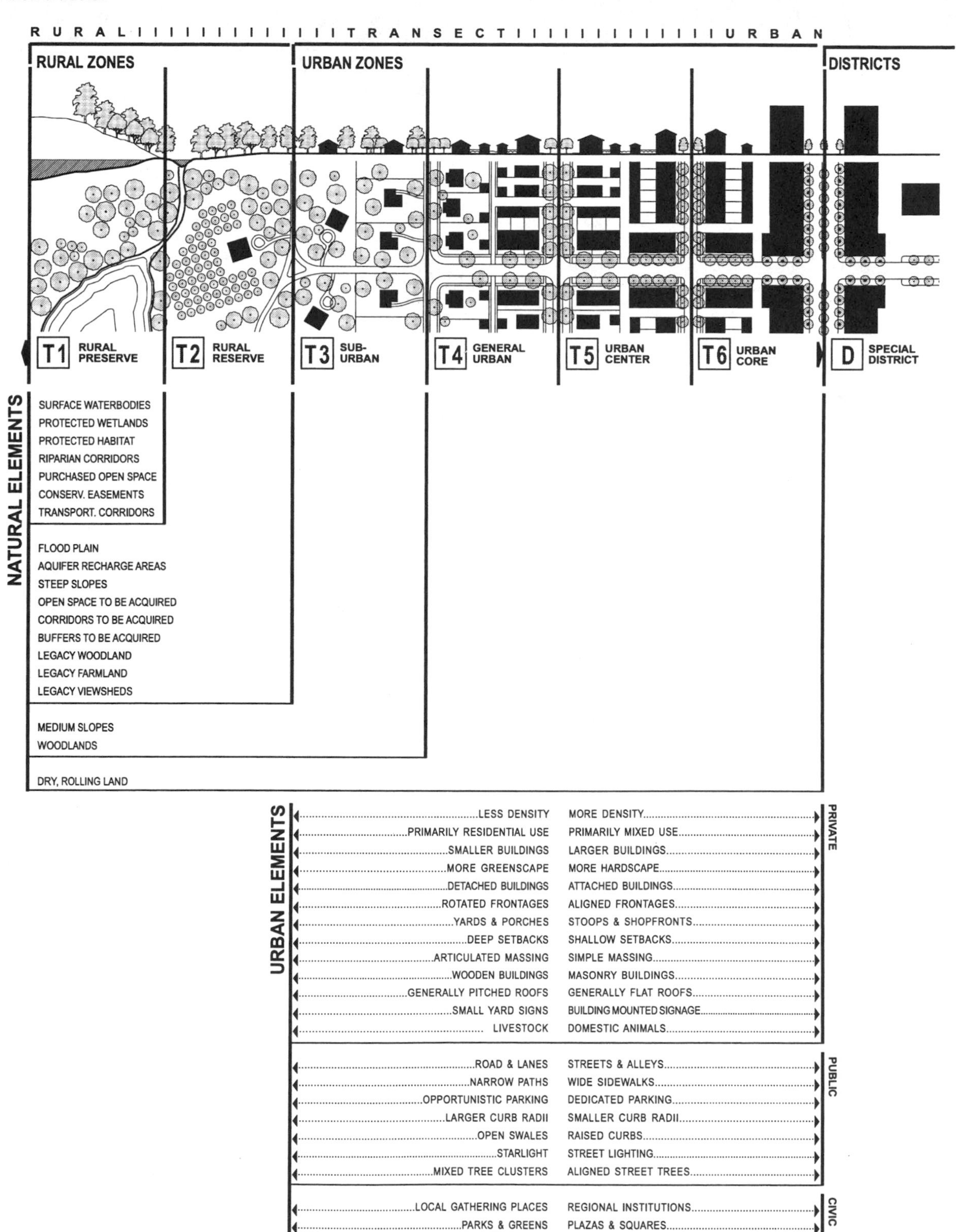

REGIONAL STRUCTURE

- **Regional Planning:** an armature that serves to structure the elements of the region. There are the Rural Areas, the Corridors, the Neighborhoods, and the Districts.

  In its short history as a discipline, regional planning has appeared to generate a great number of patterns for assembling these elements. These are usually presented as diagrams. When these diagrams are redrawn in a standardized graphic form *(taken from Sir Ebenezer Howard's Garden Cities of Tomorrow [1902])*, the options are reduced to three fundamental ones. Most regional plans make use of all of these three.

  As cities expand incrementally, these diagrams tend to be distorted due to circumstances which are both natural and man-made. The diagram of each model is therefore accompanied by an example of its application to an actual place.

- **Urban Boundary / Rural Boundary:** alternative tools of regional planning used to control and direct urban growth. There are two models, each with its own physical, political, and transportation implications.

  These boundaries have been metaphorically depicted as the Lake & Dam model, and the Stream & Levee model. The Urban Boundary restrains the flood of urban growth by surrounding the city with a single, continuous line, as a dam contains a rising lake. The Rural Boundary surrounds the open space with multiple lines, as levees protect valuable areas while allowing the urban flood to stream past.

  The urban boundary was conceptualized from the point of view of the city by the urbanist Ebenezer Howard, around 1900. The rural boundary was conceptualized from the point of view of nature by the environmentalist Benton McKaye, around 1920.

URBAN BOUNDARY MODEL

A regional planning method using statistical projection to delineate the urbanized area, beyond which is a rural area. The city is limited in its geographical extent, but not its density. Growth outside the boundary is envisioned as freestanding villages based on the Greenfield TND Code.

In a timeless pattern, a clearly defined core city, composed of neighborhoods, is surrounded by towns and villages connected by rail and separated by greenbelt. Ideally, each element is relatively self-sufficient.

This pattern emerged organically and was reinforced by the advent of the railroad. Moving along a fixed rail, but unable to stop frequently, the railroad created nodal points of settlement.

This pattern is the result of the Urban Boundary Model which limits the urbanized areas to protect the countryside, creating independent new towns. This pattern was proposed by Ebenezer Howard.

This is the pattern most influenced by the social structure of community.

Syn.: **Town and Country Pattern, Garden City, Towns and Villages, Railway Suburb.**

Pattern

Application: Madison, Wisconsin 1993

RURAL BOUNDARY MODEL

A regional planning method using cultural ecologically based criteria to protect certain open space from urbanization. The city is channelized past these boundaries, without limit to its geographical extent. Urban growth thus forms corridors between wedges of open space. The corridors are structured by the Greenfield TND Code. Densification occurs at designated intersections and transit stops along the corridors. Leapfrog development is prevented by a system of temporary **reserves** in addition to the permanent **preserves.**

Takes the form of urban corridors between wedges of natural preserves developed with the advent of the streetcar which could stop frequently. Moving along defined axes, the rails extended the boundaries of the core city, creating corridors of a width limited by the walking distance to the tracks. The advent of the automobile subsequently undermined the disciplined edges of the corridors.

By intentional planning, the urban fabric is channeled along the transportation corridors while residual wedges of open country are preserved between. The wedges are an irregular but continuous system, formed by an agglomeration of valuable natural features.

This pattern is the result of the implementation of a Rural Boundary which safeguards the valuable open space and allows the urbanized area to stream past. The rationalization of this model was by Paul Wolf and Benton McKaye.

This is the pattern most influenced by ecological concerns.

Syn.: **Linear City, Streetcar Suburb, Stream & Levee Model.**

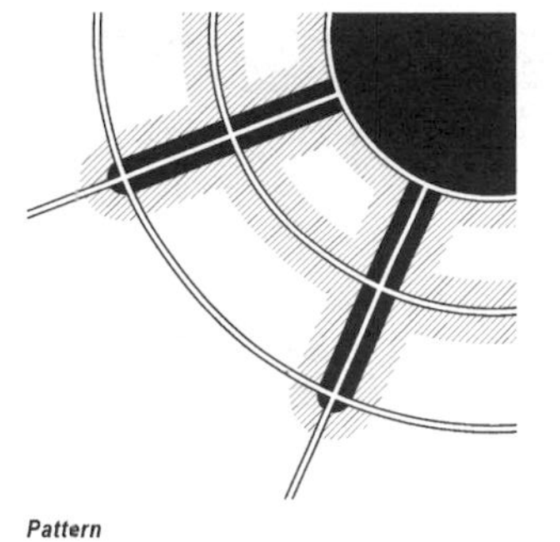

Pattern

Application: Baltimore, Maryland 1950

TRANSIT ORIENTED DEVELOPMENT (TOD)

This is a remedial pattern. Within a loose urbanized area, the TOD structure creates nodes at an efficient spacing for light rail. These nodes are mixed-use areas limited in extent by walking distance to the transit stop.

Called Urban TODs, these nodes are usually surrounded by a residential hinterland, structured as Neighborhood TODs, connected by a feeder bus system. Each Urban TOD may be specialized, with only the system as a whole being functionally complete.

This model was conceived and rationalized by Peter Calthorpe and Douglas Kelbaugh.

This is the pattern most influenced by the requirements of transportation.

Syn.: **Pedestrian Pocket**

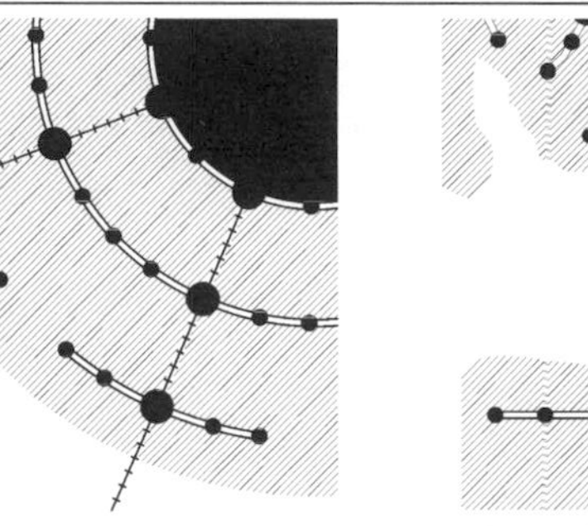

Pattern

Application: Portland, Oregon 2020

COMMUNITY NOMENCLATURE

Community Nomenclature: Various community concepts take the neighborhood as a model. Variations are due to a particular emphasis on density, spatial definition, transportation, or implementation. They have in common that they are socially and functionally variegated communities that are walkable and that manifest an urban gradient from urban center to rural edge.

Neologisms: The **Urban Village**, formulated by Patrick Gueddes early in the 20th century, is used both in the U.K. and Seattle. **The Quarter,** a transnational European term, was rationalized by Leon Krier. The **Neighborhood Unit**, the most influential U.S. proposal, was formulated by Clarence Perry in 1929 for the New York Regional Plan. The **Cell** proposal of Team X is influential in the British New Town movement and permeates the former colonies. **Traditional Neighborhood Development (TND)**, **Transit-Oriented Development (TOD)**, the latter with its synonym **Pedestrian Pocket,** and the Australian **Livable Neighborhood** are New Urbanist models.

Traditional Terms: A **Hamlet** is a neighborhood in the making. Standing free in the countryside, by virtue of its location away from transportation, the hamlet has a weak center. A **Village** is a complete neighborhood standing free in the countryside. The strong center of a village can usually be attributed, not to the population, but to its location on a transportation corridor. A **Town** is an assemblage of several neighborhoods, sharing a substantial center. A **City** is similar to a town in its neighborhood structure but has a strong core supported by the surrounding region.

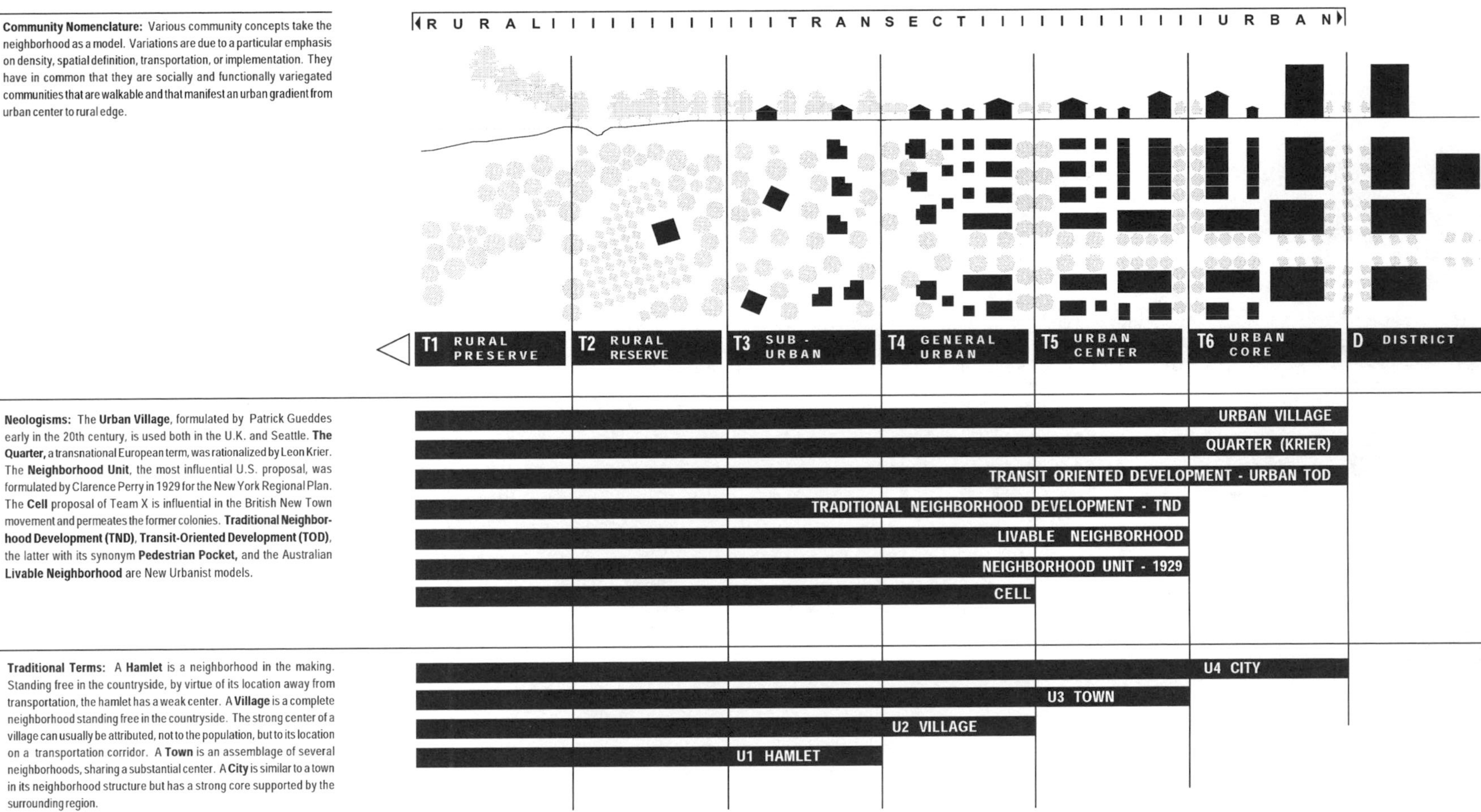

NEIGHBORHOOD STRUCTURE

- **Pedestrian Shed:** a determinant of urban size, defined as the distance which may be covered by a five-minute walk at an easy pace from the outer limit of the neighborhood proper to the edge of the neighborhood center. This is the distance that most persons will walk rather than drive, providing the environment is pedestrian-friendly. This distance is an axiomatic component of the neighborhood unit. It also defines the extent of the quartier, the TND, and the TOD. The pedestrian shed is conventionally one quarter of a mile or 1,320 feet. By variance, this dimension may be adjusted to accommodate site conditions: a) For TNDs of low density, by extension to a median distance of a half mile or 2,640 feet (this in order to increase the population catchment). b) For TNDs having an eccentrically located center, by calculating an average of the various edge-to-center distances. In CSD practice, the extent of parking lots and the length of shopping malls is similarly disciplined by walking distance.

NEIGHBORHOOD UNIT 1927

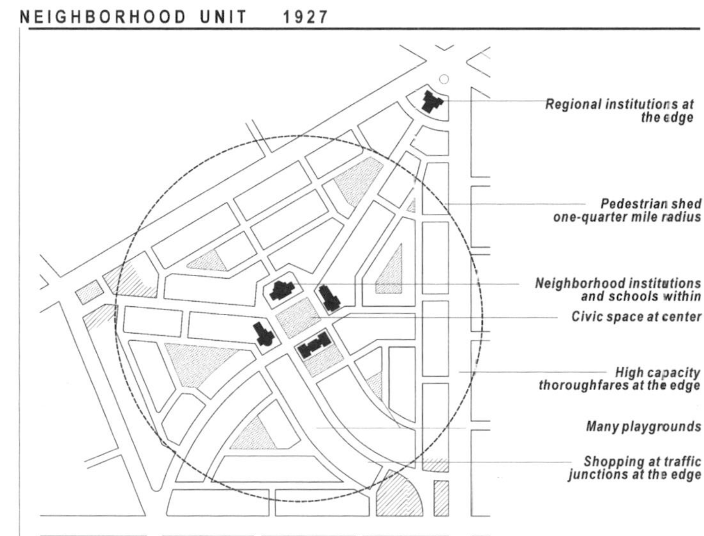

- **Neighborhood Unit:** A diagram and description from the First Regional Plan of New York (1927) which conceptualizes the neighborhood as the fundamental element of planning.

Size is determined by the walking distance of five minutes from center to edge, rather than by number of residents. Density is determined by the market. A community coalescing within a walkable area is the invariant.

An elementary school is at the center, within walking distance of most children. This is the most useful civic building, providing a meeting place for the adult population as well.

Local institutions are located within the neighborhood. Regional institutions are placed at the edges so that their traffic does not enter the neighborhood.

There is a civic open space at the center of the neighborhood, and several smaller playgrounds, one in close proximity to every household.

A network of small thoroughfares within the neighborhood disperses local traffic.

Larger thoroughfares channel traffic at the edges.

Retail is confined to the junction having the most traffic, accepting the realities of the automobile.

TRADITIONAL NEIGHBORHOOD DEVELOPMENT 1997

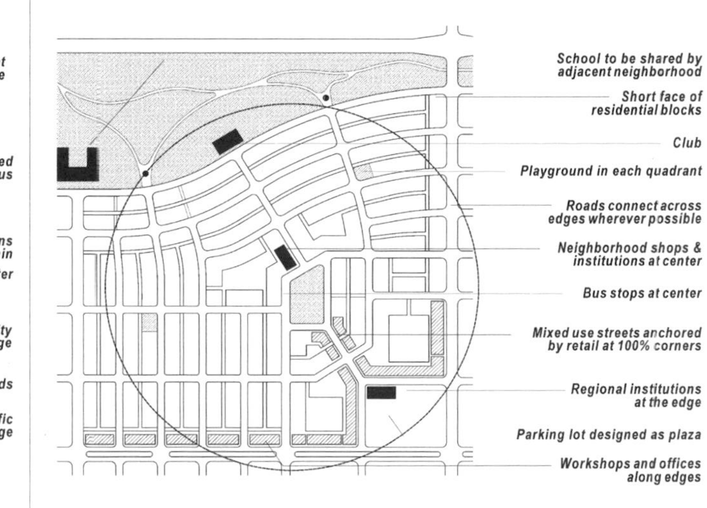

- **Traditional Neighborhood Development:** A diagram that updates the Neighborhood Unit and reconciles the current models.

The school is not at the center but at an edge, as the playing fields would hinder pedestrian access to the center. The school at the edge can be shared by several neighborhoods, mitigating the problem created by the tendency of neighborhoods to age in cohorts generating large student age populations that then drop off sharply.

There are few sites reserved for local institutions at the center and more for regional institutions at the edge. Ease of transportation has made membership in institutions a matter of proclivity rather than proximity.

The shops at the busiest intersections have been modified to accommodate larger parking plazas for convenience retail and extended by an attached main street for destination and live-work retail.

More service alleys and lanes have been added to accommodate the increased parking requirements.

The minor thoroughfares are connected with those outside the neighborhood in order to increase permeability and disperse traffic. This modification, however, increases the possibility of shortcuts.

The thoroughfare types support a transect from rectilinear streets at the urban center to curvilinear roads toward the rural edge.

The traffic along the boulevards at the edges is more unpleasant than originally envisioned. Three mitigating strategies are proposed: the provision of an end-grain of blocks at all edges, a green buffer shown along the bottom edge, and the location of resilient building types, such as office buildings, shown along the bottom edge.

The traffic along the highway shown at the top is assumed to be hostile and therefore buffered within a parkway.

CURRENT NEIGHBORHOOD MODELS

The neighborhood is the elemental building block of the regional plan. The neighborhood model may be structured by a variety of criteria, and there are social implications to each of the variants.

There are three neighborhood models currently proposed. They are very similar, differing primarily in the conception of the pedestrian shed: the location of its centroid, and its extent. These differences manifest secondary consequences regarding the density of the required model and the social quality of the center.

The alternatives can be easily compared when all are overlaid on the standard mile-square grid of the Continental Survey of the United States.

Although each of the models proposes a comprehensive regional strategy, their optimal application varies. All three should therefore be considered available for the appropriate circumstance.

T.N.D. PATTERN

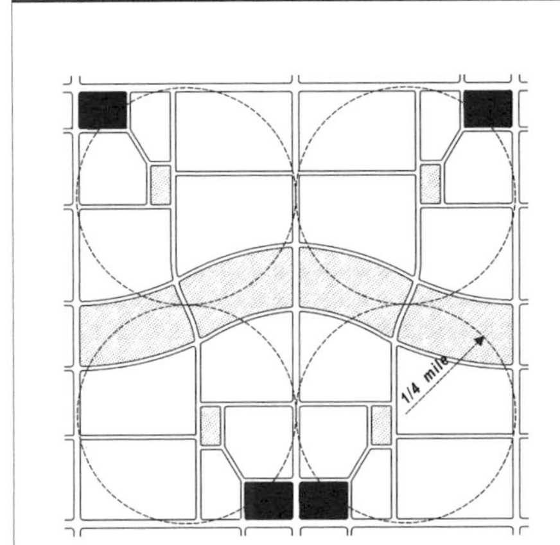

- **Traditional Neighborhood Development (TND)** is similar to the American Neighborhood Unit of 1927 and the European Quartier. It has its pedestrian shed centered on the centroid of the neighborhood proper, not necessarily coinciding with a major thoroughfare. The pedestrian shed is a walk of 5 minutes from edge to center. It is calculated as a circle, or multiple circles in the case of larger sites.

An advantage of the TND model is the high ratio of the neighborhood area that is within the pedestrian catchment. Taking the mile-square as a comparative matrix, the shed includes 70% of the developable area. Because a substantial proportion of the inhabitants are within walking distance of the center, bus transit will tend to be efficient, even at relatively low densities. Another advantage is that, because the center is not bisected by a high capacity thoroughfare (these remain at the edges), its spatial quality as a social condenser is not degraded by excessive traffic.

A disadvantage of the TND is that the commercial use at the center may only sustain neighborhood retail, as it does not benefit from the traffic straddling a main thoroughfare. This model tends to have only neighborhood institutions at the center, with regional institutions and commercial use at the edges shared by other neighborhoods.

T.O.D. PATTERN

- **Transit Oriented Development (TOD)** is similar to the railway suburb of the 19th century. Its pedestrian shed is centered on a rail transit station which, if possible, coincides with a major thoroughfare. This center is often at the edge of the centroid of the neighborhood area. Note: the pedestrian shed of the TOD model is traditionally drawn as a semicircle, although there is no intrinsic reason why this should be so.

An advantage of the TOD model is that rail is the most efficient form of transit. As it is also the most expensive, this model provides for its support by a high population density within the pedestrian catchment of each station -a minimum of 14 dwelling units per acre. Another advantage is that institutional as well as commercial uses are concentrated around a transportation node. This is likely to create retail that is well-supported by pedestrian and automobile traffic. The regional character of this transit center, however, may warrant the creation of local centers internal to the neighborhood, similar to the TND model. Another potential problem, is the spatial degradation resulting from of the traffic and parking requirement of a transit station at the center. This is mitigated by the dilution of the traffic by a one-way pair of principal thoroughfares at one block's spacing.

A disadvantage of the TOD is that the density required to support transit use may not be acceptable in certain markets. This is exacerbated by the low net ratio of area that is within a five minute pedestrian shed: taking the mile-square as a comparative matrix, such a shed includes only 7% of the gross developable area. The credible argument is made that the advantages of rail transit (as opposed to the bus mode of the other models) are sufficiently compelling that an effective pedestrian shed can be increased to a 10 minute/half-mile radius. This raise the catchment to 40% of the developed area.

LIVABLE NEIGHBORHOOD PATTERN

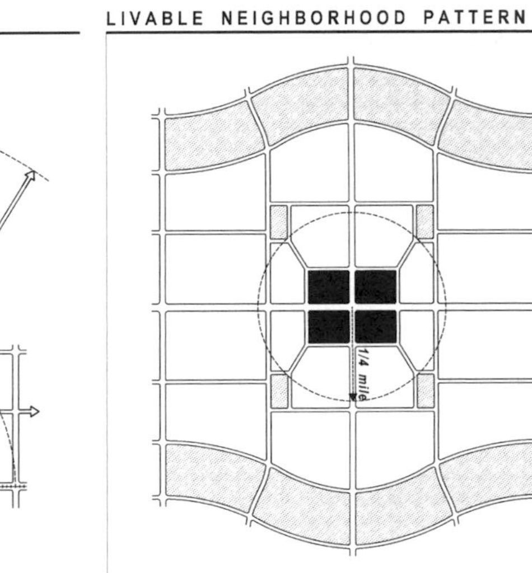

- **The Livable Neighborhood** combines aspects the TND and the TOD. It was conceptualized as a correction of the British Cell model, particularly as it was applied at Milton Keynes, a community held to have failed as the cells failed to coalesce into the greater social scale of a city, despite having the population and all the necessary elements (statistically) of one.

The Australian Livable Neighborhood has a pedestrian shed that appears to be eccentrically on a major edge thoroughfare, like the TOD; but actually, the neighborhood itself is centered on the regional thoroughfare. As with the TND and unlike the TOD, its pedestrian shed (this term itself derives from Australian usage) is conceptualized as a quarter-mile circle.

Like the TND, an advantage of this model is the high ratio of the neighborhood area that is within the pedestrian catchment. Taking the mile-square as a comparative matrix, the shed includes 70% of the developable area. Because a substantial proportion of the inhabitants are within walking distance of the center, transit will tend to work, even at relatively low densities. Also, the trajectory of bus transit is more direct than that of the TND.

The Australian Livable Neighborhood has the disadvantage that, because the center of the neighborhood is bisected by what is a high-capacity thoroughfare, its spatial quality as a social condenser may be degraded. A strategy to minimize this negative impact is the careful design of the thoroughfare as a boulevard. The strategy of the one-way pair proposed by the TOD may also apply. Note: in a repeated pattern of neighborhoods, with even dispersal of traffic, not all the neighborhood centers would have the traffic intensity to warrant either of these mitigating strategies.

NEIGHBORHOOD STRUCTURE – ELEMENTS

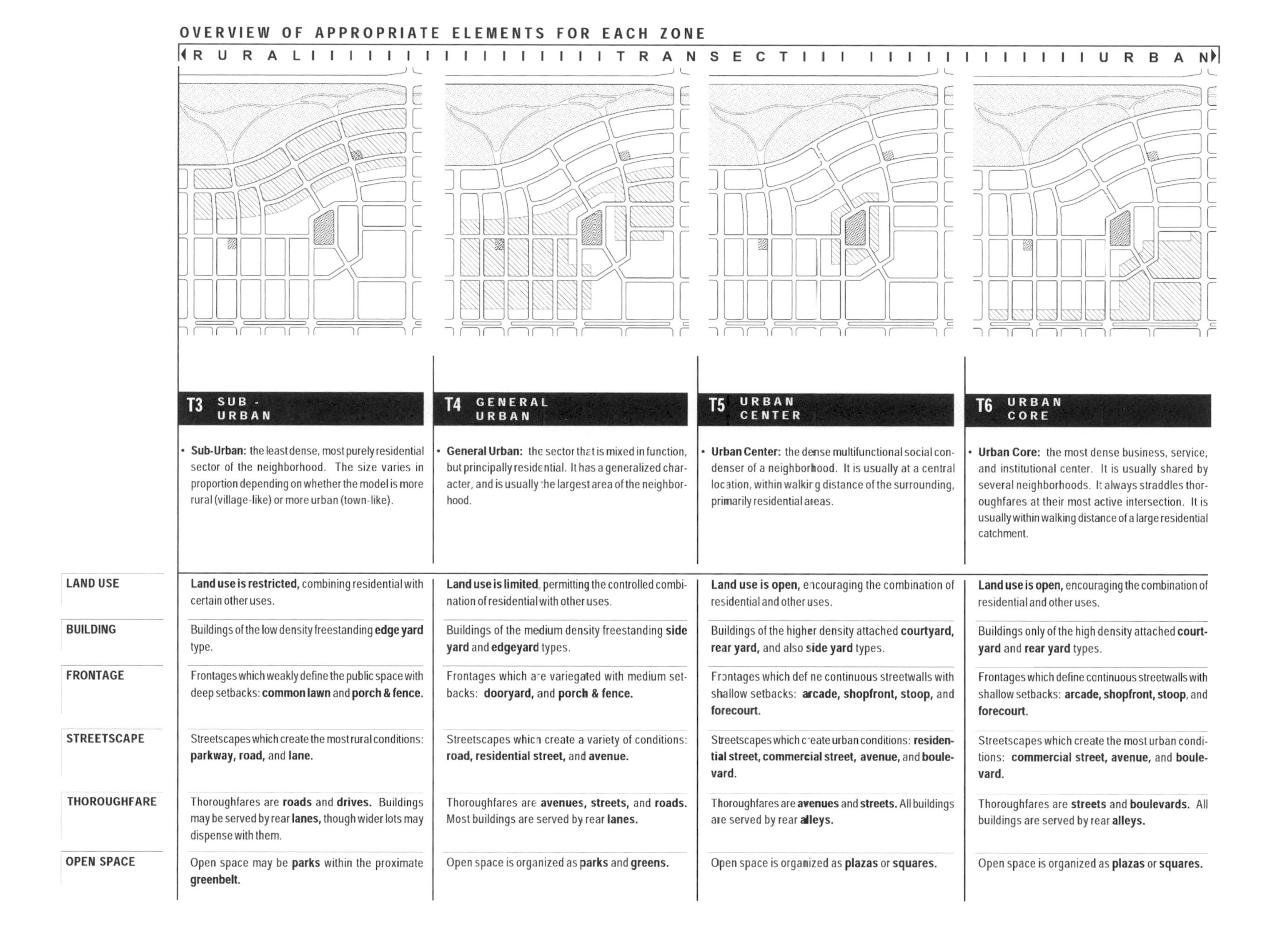

	T3 SUB-URBAN	T4 GENERAL URBAN	T5 URBAN CENTER	T6 URBAN CORE
	• **Sub-Urban:** the least dense, most purely residential sector of the neighborhood. The size varies in proportion depending on whether the model is more rural (village-like) or more urban (town-like).	• **General Urban:** the sector that is mixed in function, but principally residential. It has a generalized character, and is usually the largest area of the neighborhood.	• **Urban Center:** the dense multifunctional social condenser of a neighborhood. It is usually at a central location, within walking distance of the surrounding, primarily residential areas.	• **Urban Core:** the most dense business, service, and institutional center. It is usually shared by several neighborhoods. It always straddles thoroughfares at their most active intersection. It is usually within walking distance of a large residential catchment.
LAND USE	**Land use is restricted,** combining residential with certain other uses.	**Land use is limited**, permitting the controlled combination of residential with other uses.	**Land use is open,** encouraging the combination of residential and other uses.	**Land use is open,** encouraging the combination of residential and other uses.
BUILDING	Buildings of the low density freestanding **edge yard** type.	Buildings of the medium density freestanding **side yard** and **edgeyard** types.	Buildings of the higher density attached **courtyard, rear yard,** and also **side yard** types.	Buildings only of the high density attached **courtyard** and **rear yard** types.
FRONTAGE	Frontages which weakly define the public space with deep setbacks: **common lawn** and **porch & fence.**	Frontages which are variegated with medium setbacks: **dooryard,** and **porch & fence.**	Frontages which define continuous streetwalls with shallow setbacks: **arcade, shopfront, stoop,** and **forecourt.**	Frontages which define continuous streetwalls with shallow setbacks: **arcade, shopfront, stoop,** and **forecourt.**
STREETSCAPE	Streetscapes which create the most rural conditions: **parkway, road,** and **lane.**	Streetscapes which create a variety of conditions: **road, residential street,** and **avenue.**	Streetscapes which create urban conditions: **residential street, commercial street, avenue,** and **boulevard.**	Streetscapes which create the most urban conditions: **commercial street, avenue,** and **boulevard.**
THOROUGHFARE	Thoroughfares are **roads** and **drives.** Buildings may be served by rear **lanes,** though wider lots may dispense with them.	Thoroughfares are **avenues, streets,** and **roads.** Most buildings are served by rear **lanes.**	Thoroughfares are **avenues** and **streets.** All buildings are served by rear **alleys.**	Thoroughfares are **streets** and **boulevards.** All buildings are served by rear **alleys.**
OPEN SPACE	Open space may be **parks** within the proximate **greenbelt.**	Open space is organized as **parks** and **greens.**	Open space is organized as **plazas** or **squares.**	Open space is organized as **plazas** or **squares.**

IMPLEMENTATION

In the implementation process Traditional Neighborhood Developments should be vested, which is to say, permitted administratively. All other types of development are not precluded, but are required to undergo the conventional public process of justification, for denial or permit as a district.

Vesting is a strong incentive for development of the neighborhood model, which is assumed to be socially and environmentally benevolent. This system, however, is vulnerable to abuse by false or incomplete neighborhoods. A checklist of criteria can address this problem by forming a basis for acceptance.

The checklist enumerates the many qualities that distinguish TNDs from conventional suburban sprawl. While there are always exceptions, TNDs embody the majority of the principles that follow. All principles have a significant impact on the quality of a development, but those marked with an asterisk (*) are essential and non negotiable.

This list was compiled for the development of greenfield sites. The principles do not apply to smaller projects nor to infill projects.

The checklist can serve in different ways. For developers, the list allows them to review their plans to determine whether they can expect to realize the market premium which has been demonstrated to accrue to TNDs. For planning officials, the list allows them to determine whether submitted plans are likely to provide the social benefits associated with TNDs, to qualify for increased density allocation and for vesting.

LONG CHECKLIST

Regional Context

- ☐ Is the TND located within a comprehensive regional plan with a transit and an open space preservation strategy?
- ☐ Is the TND connected in as many locations as feasible to adjacent developments and thoroughfares?
- ☐ Do highways approaching the TND either pass to its side, or take on low-speed geometrics when entering it?*

Site Context

- ☐ Are most wetlands, lakes, streams, and other water amenities retained?*
- ☐ Are significant natural amenities at least partially fronted by thoroughfares rather than hidden behind back yards?*
- ☐ Is the site developed in such a way as to preserve as many trees as possible, with emphasis on saving specimen trees ?*
- ☐ Does the plan develop greens, squares, and parks, located at significant tree-save areas and other natural amenities?*
- ☐ Does the plan work with the topography to minimize the amount of grading necessary?*
- ☐ Are significant high points reserved for public tracts and/ or civic buildings?

Plan Structure

- ☐ Is the plan broken down into neighborhoods?
- ☐ Is each neighborhood roughly a ten-minute walk from edge to edge? (one half a mile)*
- ☐ Is the greatest density of housing toward the center?*
- ☐ Is the center the location of retail space (a corner store is required) and, ideally, employment, located in mixed-use buildings?* (Centers can be peripherally located in response to site conditions.)
- ☐ Is there a dry, dignified place to wait for transit at the center?*
- ☐ Is there a public space such as a square, plaza, or green at the center of the neighborhood?*
- ☐ Are buildings zoned, not by use, but by compatibility of building type?*
- ☐ Do zoning changes occur at mid-block rather than mid-street so that streets are coherent on both sides?
- ☐ Are there small playgrounds distributed evenly through the neighborhood, roughly within one eigth of a mile of every dwelling?*
- ☐ Is there an elementary school located within two miles of the T.N.D., and sized accordingly?
- ☐ Does each neighborhood reserve at least one prominent site for a meeting hall?*
- ☐ Are the large areas of open space between neighborhoods connected into continuous corridors?

Thoroughfare Network

- ☐ Are cul-de-sacs avoided where not absolutely necessary due to natural conditions?*
- ☐ Are streets organized in a network, where the average perimeter of blocks is less than 2000 feet?*
- ☐ Does the network vary in the character of the streetscape to support the urban-to-rural structure of the neighborhood?*
- ☐ Are most street vistas terminated either by a building carefully sited, or deflected by an angle in the street?
- ☐ Do most roads that curve maintain their cardinal orientation over their entire trajectory?

Streetscape

- ☐ Is there a main street, approximately 36 ft wide, with marked parking both sides;
- ☐ Are there through streets, approximately 28 ft wide, with parking one side?
- ☐ Are there local roads, approximately 26 ft wide, parking both sides?
- ☐ Are there local roads, approximately 20 ft wide, parking one side?
- ☐ Are there alleys, approximately 12 ft wide, with a 24 ft min right-of-way?
- ☐ Does every street have a sidewalk on at least one side, 4 to 5 ft in width, and three times that wide on the main street?* (Rural roads do not need sidewalks.)
- ☐ Does every thoroughfare have a tree planter 4 to 10 ft in width, of indigenous shade trees planted on average at approximately 30 ft on center?*
- ☐ Is the curb radius at intersections a maximum of 15 ft, with a typical measurement of 10 ft at main streets?*
- ☐ Are buildings placed relatively close to the street, such that they are generally set back the equivalent of one-quarter the width of the lot?*
- ☐ Do the building setbacks permit the encroachment of semi-public attachments, such as galleries, porches, bay windows, stoops, and balconies?

Buildings

- ☐ Is there a wide range of housing types located within close proximity to each other?
- ☐ Is there at least a minimum of 5% representation from each of the following categories:
  - live/work buildings
  - apartment buildings
  - rowhouses
  - sideyard houses or duplexes
  - cottages on small lots (30 ft- 40 ft wide)
  - houses on standard lots (40 ft-70 ft wide)
  - houses on large lots (70 ft and above)
- ☐ Are most lots smaller than 70 ft wide served by a rear alley to access garages?*
- ☐ Are all garages not served by an alley set back a minimum of 20 ft from the front of the building.
- ☐ Are parking lots located behind streetwalls or buildings, such that only their access is visible from streets?*
- ☐ Do townhouses have privacy fences on shared side property lines?*
- ☐ Do all commercial buildigs front directly on the sidewalk, with parking lots to the side or the rear?*
- ☐ Is each house permitted to have a small ancillary dwelling unit in the rear?
- ☐ Do commercial buildings have a second story (or more) for other uses?
- ☐ Is all subsidized housing:
  - distributed in ratios of no more than one unit in five?
  - similar in architecture from the other units?
- ☐ Do most buildings have a minimum of two stories and a maximum of four?
- ☐ Do most buildings have flat facades, with wings and articulations occurring to the rear?

Architectural Syntax

- ☐ Is a regional architectural syntax provided as a source of ecological responsibility?
- ☐ Are all windows and other openings either square or vertically proportioned?
- ☐ Are pitches within a limited range for the principal roofs?
- ☐ Are colors and materials limited to a harmonious range?

SHORT CHECKLIST

A neighborhood includes most of the following:

- ☐ There is a discernible center. This is often a plaza, square or green, and sometimes a busy or memorable intersection. A transit stop should be located at this center.
- ☐ Buildings at the center are placed close to the sidewalk and to eachother, creating an urban sense of spatial definition. Buildings towards the edges are placed further away and further apart from eachother, creating a more rural environment.
- ☐ Most of the dwellings are within a five-minute walk from the center. This pedestrian shed averages one- quarter of a mile.
- ☐ There is a variety of dwelling types. These take the form of houses, rowhouses, and apartments, such that younger and older, singles and families, the poorer and the wealthier, can find places to live.
- ☐ There are places to work in the form of office buildings or live-work units.
- ☐ There are shops sufficiently varied to supply the ordinary needs of a household. A convenience store, a post office, a teller machine, and a gym are the most important among them.
- ☐ A small ancillary building should be permitted within the backyard of each house. It may be used as a rental apartment, or as a place to work.
- ☐ There should be an elementary school close enough so that most children can walk from their dwelling. This distance should not be more than one mile.
- ☐ There are playgrounds near every dwelling. This distance should not be more than one-eighth of a mile.
- ☐ Thoroughfares within the neighborhood form a continuous network, providing a variety of itineraries and dispersing traffic. The thouroughfares connect to those of adjacent cities as often as possible.
- ☐ Thoroughfares are relatively narrow and shaded by rows of trees that slow traffic and create an appropriate environment for pedestrian and bicyclist.
- ☐ Parking lots and garage doors rarely end or front the thoroughfares. Parking is relegated to the rear of buildings and usually accessed by alleys or lanes.
- ☐ Certain prominent sites are reserved for public buildings. A building must be provided at the center for neighborhood meetings.
- ☐ The neighborhood should be self governing, deciding on matters of maintenance, security, and physical evolution.

STREET PATTERN TYPES

SAVANNAH PATTERN

Advantages
Excellent directional orientation
Controllable lot depth
Provides end grain of blocks for fast traffic
Even dispersal of traffic through the web
Straight lines enhance rolling terrain
Efficient double-loading of alleys and utilities

Disadvantages
Monotonous unless periodically interrupted
Does not easily absorb environmental interruptions
Unresponsive to steep terrain

Syn.: **Orthogonal Grid, Gridiron**

NANTUCKET PATTERN

Advantages
Hierarchy with long routes for through traffic
Even dispersal of traffic through web
Responsive to terrain
Easily absorbs environmental interruptions
Monotony eliminated by terminated vistas
Follows traces on the landscape

Disadvantages
Uncontrollable variety of blocks and lots

Syn.: **Sitte Model, Townscape**

MARIEMONT PATTERN

Advantages
Hierarchy with diagonals for through traffic
Even dispersal of traffic through the web
Monotony interrupted by deflected vistas
Diagonal intersections spatially well-defined

Disadvantages
Tends to be disorienting

Syn.: **Unwin Model, Spider Web**

WASHINGTON PATTERN

Advantages
Hierarchy with diagonals for through traffic
Even dispersal of traffic through the grid
Diagonals focus on terrain features
Diagonals interrupt monotony of the grid

Disadvantages
Uncontrollable variety of lots
High number of awkward lot shapes
Diagonal intersections spatially ill-defined

Syn.: **City Beautiful, Haussmann Model**

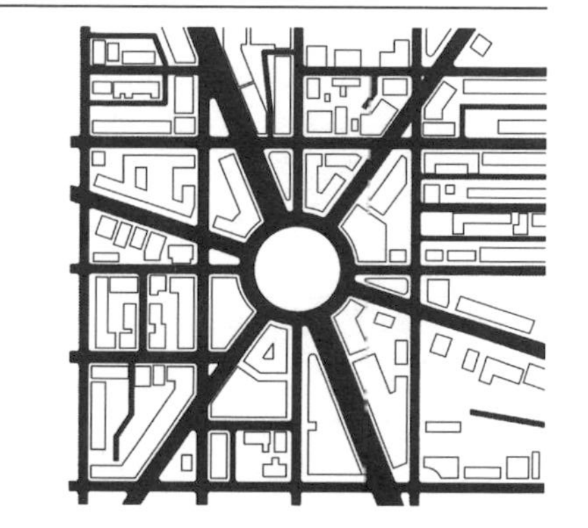

RIVERSIDE PATTERN

Advantages
Monotony interrupted by deflected vistas
Easily absorbs environmental interruptions
Highly responsive to terrain
Even dispersal of traffic through the web

Disadvantages
Highly disorienting
Uncontrollable variety of lots
No intrinsic hierarchy

Syn.: **Olmstedian**

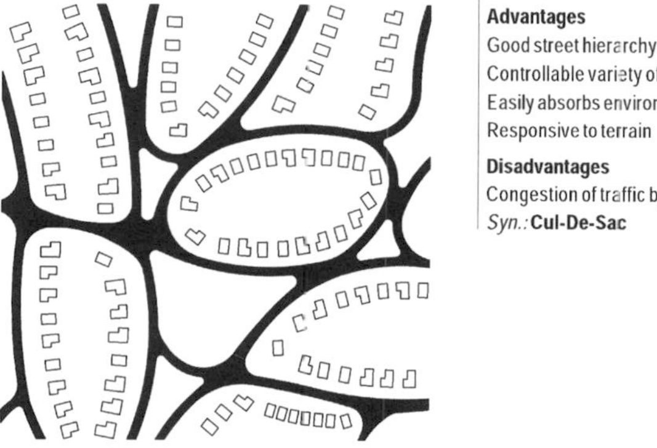

RADBURN PATTERN

Advantages
Good street hierarchy for locals and collectors
Controllable variety of blocks and lots
Easily absorbs environmental interruptions
Responsive to terrain

Disadvantages
Congestion of traffic by absence of web

Syn.: **Cul-De-Sac**

BLOCK TYPES

- **Block:** the aggregate of lots and tracts, circumscribed by thoroughfares.

The block is the middle scale of town planning. While it is not the determinant of the network nor of the building type, it strongly affects both.

There are a large number of block forms as implied by the six models of network; however, analysis reduces the variety to three categories: square, elongated, and irregular.

Each block type has distinct technical implications, and all types are useful even within a single neighborhood. For example, the square block accommodates the additional parking of a civic building within itself, useful at the Center Zone. The General Zone usually requires the normative lot sizes easily provided by the elongated block. The rural aspect, desirable at the Suburban Zone is supported by the picturesque qualities of the irregular block.

SQUARE BLOCK

The Square Block was an early model for planned settlements in America. It was sometimes associated with agricultural communities with four large lots per block, each with a house at its center. When the growth of the community produced additional subdivision, the replatting inevitably created irregular lots (Figure 1).

While this may provide a useful variety, it is more often regarded as a nuisance by a building industry accustomed to standardized products.

A disadvantage is that discontinuous rear lot lines prevent double-loaded alleys and rear-access utilities. Despite these shortcomings, the square block is useful as a specialized type. The forced variety of platting assures a range of lot prices. When platted only at its perimeter with the center open (Figure 2), it can accommodate the high parking requirements of civic buildings. The open center may also be used as a common garden or a playground, insulated from traffic.

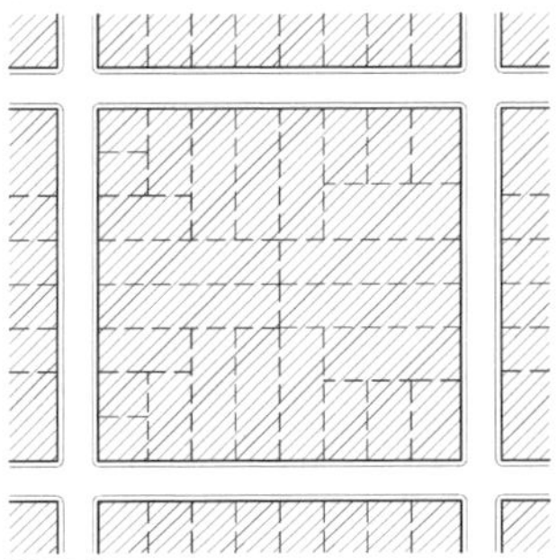

Figure 1

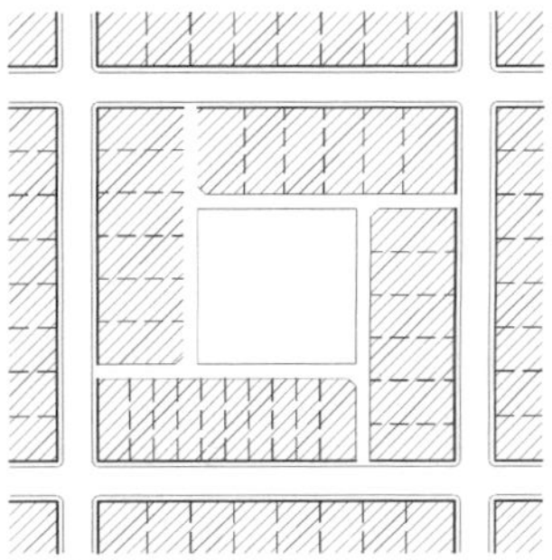

Figure 2

ELONGATED BLOCK

The Elongated Block is an evolution of the square block which overcomes some of its drawbacks. The elongated block eliminates the uncontrollable variable of lot depth, while maintaining the option of altering the lot width. Elongated blocks provide economical double-loaded alleys with short utility runs. The alley may be placed eccentrically, varying the depth of the lot (Figure 3-1). By adjusting the block length, it is possible to reduce cross-streets at the rural edges and to add them at the urban centers. This adjustment alters the pedestrian permeability of the grid, and controls the ratio of street parking to the building capacity of the block.

The elongated block can bend somewhat along its length, giving a limited ability to shape space and to negotiate slopes (Figure 4). Unlike the square block, it provides two distinct types of frontage. With the short side or end grain assigned to the higher traffic thoroughfare, most buildings can front the quieter long side of the block (Figure 3-2). For commercial buildings, the end grain can be platted to take advantage of the traffic while the amount of parking behind is controlled by the variable depth (Figure 3-3).

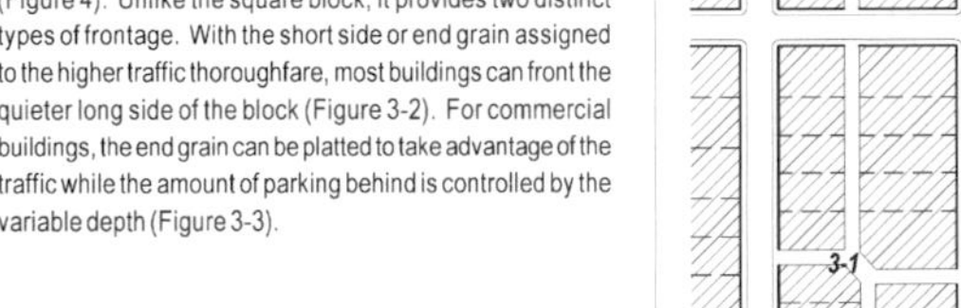

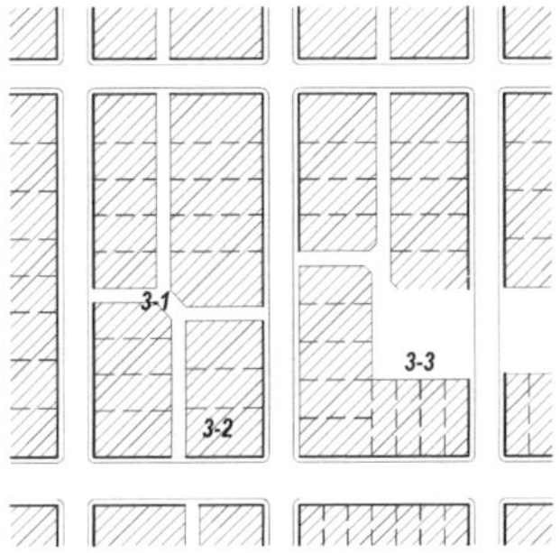

Figure 3

Figure 4

IRREGULAR BLOCK

The Irregular Block is characterized by its unlimited variations. The original organic block was created by the subdivision of land residual between well-worn paths.

It was later rationalized by Sitte, Cullen, Krier, and Olmsted to achieve a controllable picturesque effect and to organically negotiate sloping terrain. An important technique in the layout of irregular blocks is that the frontages of adjacent blocks need not be parallel (Figure 5).The irregular block, despite its variety, generates certain recurring conditions which must be resolved by sophisticated platting. At shallow curves, it is desirable to have the facades follow the frontage smoothly. This is achieved by maintaining the side lot lines perpendicular to the frontage line (Figure 6-1). It is important that the rear lot line be wide enough to permit vehicular access (Figure 6-2). At sharper curves, it is desirable to have the axis of a single lot bisect the acute angle (Figure 6-3). In the event of excessive block depth it is possible to access the interior of the block by means of a close (Figure 6-4). *Syn.:* **Organic Block** *(note: discuss topography)*

Figure 5

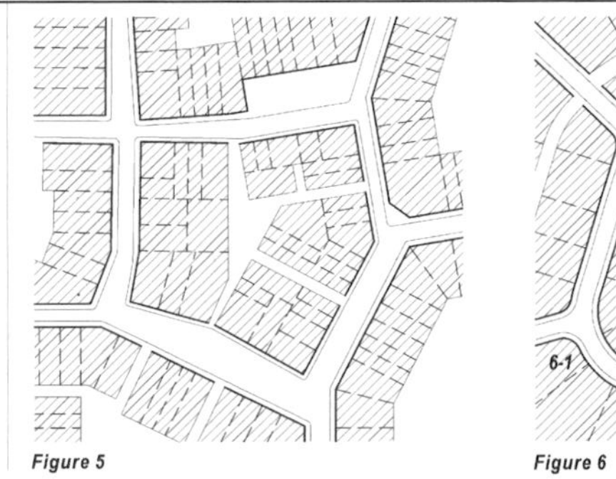

Figure 6

INFRASTRUCTURE TYPES

- **Infrastructure Cost:** the cost of service improvements on a given site, including utilities, streetscapes, and thoroughfares but excluding common amenities and buildings. The cost of infrastructure of Traditional Neighborhood Development should be equal to or less than that of Conventional Suburban Development for the following reasons:
  - the elimination of CSD front-loaded driveways compensates for the rear lanes, providing the latter are built to driveway standards.
  - the TNDs narrower thoroughfare widths compensate for the shorter length of cul-de-sacs.
  - TNDs use of simple open sections for drainage wherever roads are appropriate.
  - TNDs on-street parking consumes substantially less pavement than off-street parking by double-use of the moving lane as access lane.
  - since the TNDs network system has much greater connectivity than the CSDs stem system, there is a near-complete elimination of costly collectors that do not provide developable frontage;
  - TNDs increment of phasing is much smaller, as all market segments are accommodated within a single neighborhood as opposed to carrying the infrastructure costs of many homogeneous pods;
  - lot width based on an off-street parking module (12, 24, 36, 48, 60, 72 etc.) increases the density of TNDs by eliminating slivers of wasted land (parking controls density).

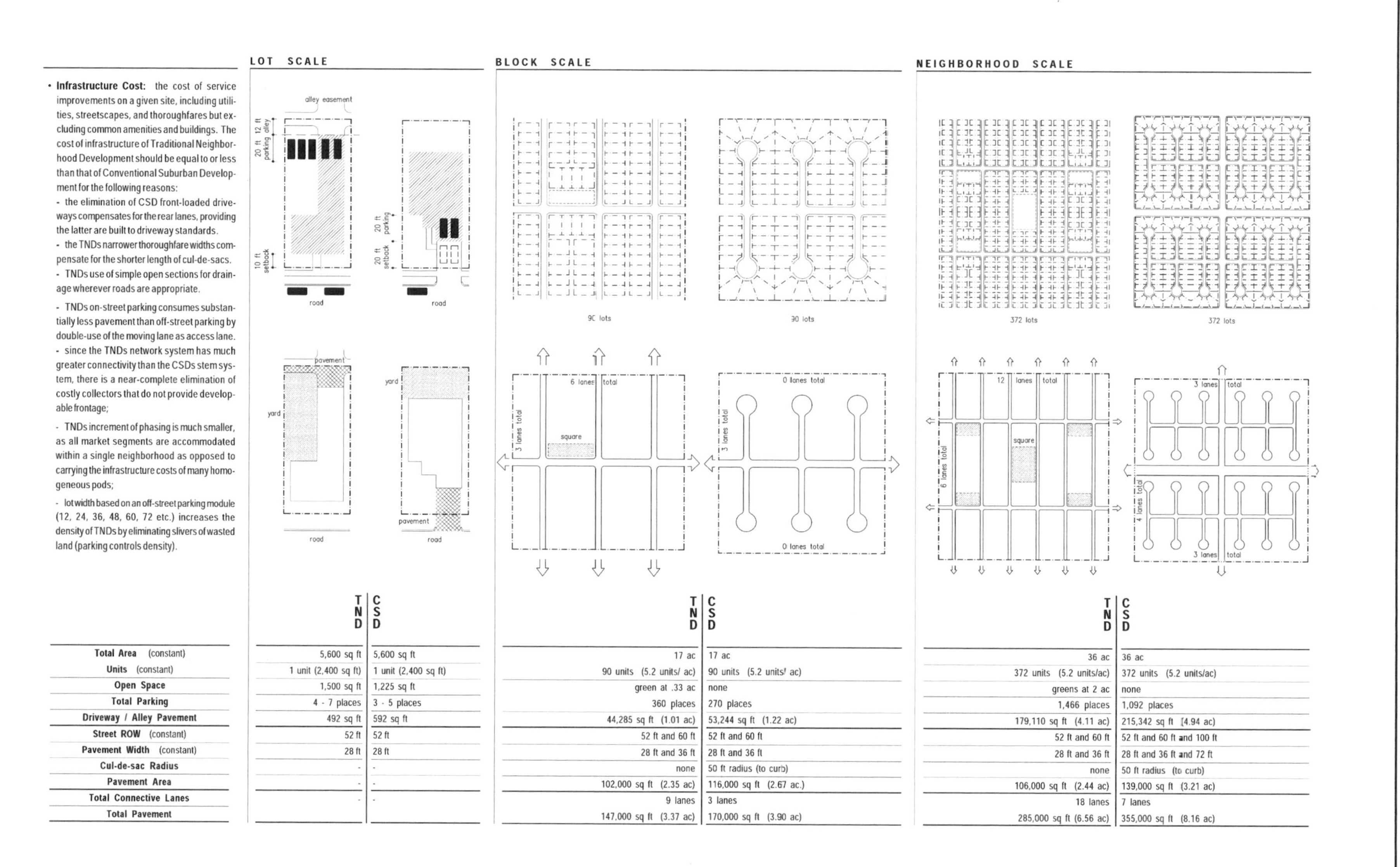

	Lot Scale TND	Lot Scale CSD	Block Scale TND	Block Scale CSD	Neighborhood Scale TND	Neighborhood Scale CSD
Total Area (constant)	5,600 sq ft	5,600 sq ft	17 ac	17 ac	36 ac	36 ac
Units (constant)	1 unit (2,400 sq ft)	1 unit (2,400 sq ft)	90 units (5.2 units/ ac)	90 units (5.2 units/ ac)	372 units (5.2 units/ac)	372 units (5.2 units/ac)
Open Space	1,500 sq ft	1,225 sq ft	green at .33 ac	none	greens at 2 ac	none
Total Parking	4 - 7 places	3 - 5 places	360 places	270 places	1,466 places	1,092 places
Driveway / Alley Pavement	492 sq ft	592 sq ft	44,285 sq ft (1.01 ac)	53,244 sq ft (1.22 ac)	179,110 sq ft (4.11 ac)	215,342 sq ft (4.94 ac)
Street ROW (constant)	52 ft	52 ft	52 ft and 60 ft	52 ft and 60 ft	52 ft and 60 ft	52 ft and 60 ft and 100 ft
Pavement Width (constant)	28 ft	28 ft	28 ft and 36 ft	28 ft and 36 ft	28 ft and 36 ft	28 ft and 36 ft and 72 ft
Cul-de-sac Radius	-	-	none	50 ft radius (to curb)	none	50 ft radius (to curb)
Pavement Area	-	-	102,000 sq ft (2.35 ac)	116,000 sq ft (2.67 ac.)	106,000 sq ft (2.44 ac)	139,000 sq ft (3.21 ac)
Total Connective Lanes	-	-	9 lanes	3 lanes	18 lanes	7 lanes
Total Pavement			147,000 sq ft (3.37 ac)	170,000 sq ft (3.90 ac)	285,000 sq ft (6.56 ac)	355,000 sq ft (8.16 ac)

STREETSCAPE TYPES

- **Public Streetscapes:** the section of the right-of-way between the lot line and the vehicular lanes. The public streetscape should be conceived integrally with the private frontage, sharing a continuous landscape and, in the case of the commercial street, a contiguous, seamless sidewalk. The following stretscape illustrations are based on a constant 12 ft dimension from curb to lot line.

Correlation: There are many types of frontages and streetscapes. Only certain of these serve to effectively define a public realm. When culled by the discipline of urbanism, the great number drops to a very few. While this determination may seem to represent an unnatural reduction of possibilities, these few are sufficient to the great cities, towns and villages of the world in all their variety. When embedded as options in a code, these few frontages represent an expansion of the available options which in the practice of conventional suburbia is usually limited to the 25 ft front yard.

Note: Frontages are independent of building type. For example, a row house type may have as its frontage a stoop, a dooryard, or a porch. However, discretion is necessary in the combination. As discretion cannot be assumed in the design process, the acceptable combinations should be controlled by code.

While thoroughfares would seem to be primarily vehicular and open space primarily for pedestrian use, such categorizations tend to erode the social function of the public realm. Thoroughfares are the most common public space and as such, require careful detailing of the streetscape; for open space to be effective, it must be designed with the appropriate landscape.

TYPICAL STREETSCAPE ASSEMBLAGES

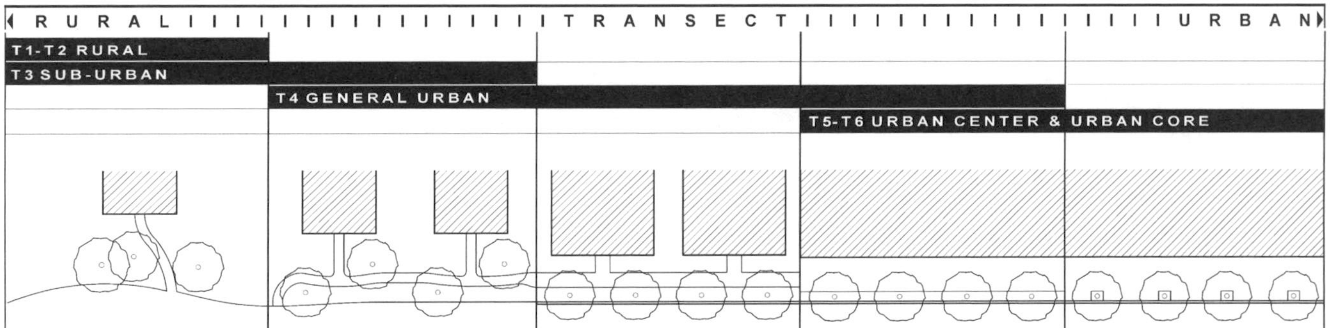

RURAL I I I I I I I I I I I I I I I I I TRANSECT I I I I I I I I I I I I I I I I URBAN

T1-T2 RURAL

T3 SUB-URBAN

T4 GENERAL URBAN

T5-T6 URBAN CENTER & URBAN CORE

- **Rural Road:** a very rural condition which comprises a road with open swales drained by percolationand no separate pedestrian path. Street trees consist of multiple species composed in clusters. This type is suitable within Rural and Edge Zones, especially when serving estate lots.
- **Road:** the next most rural condition which comprises a road with open swales drained by percolation and an informal walking path or bicycle trail along one side. The street trees consist of multiple species composed in clusters. This type is suitable within Edge Zones.
- **Residential Street:** a generalized condition which comprises a street drained by inlets in raised curbs. Narrow sidewalks along both sides are separated from the thoroughfare by a wide, continuous planter. The street trees consist of single or alternating pairs of species aligned in a regular allée. The trees of the front yard should be of compatible species. This type is suitable within General Zones, especially when enfronting house and cottage lots.
- **Residential Street:** a typical urban condition which comprises a street with raised curbs drained by inlets. A narrow, continuous planter separates wide sidewalks along both sides from the thoroughfare. The street trees consist of a single species aligned in a regular allée. This type is suitable within Center Zones, especially when enfronting rowhouse and apartment lots.
- **Commercial Street:** a very urban condition which comprises a street with raised curbs drained by inlets. Wide sidewalks along both sides are separated from the thoroughfare by small separate treewells. The street trees consist of a single species aligned in a regular allée. This type is suitable within Center and Core Zones, especially when serving shopfront lots. The tree spacing should be irregular, to stay clear of the entrances of the shops.

FRONTAGE TYPES

- **Frontage:** the privately held layer between the facade of a building and the lot line. The variables of frontage are the dimensional depth of the front yard and the combination of architectural elements such as fences, stoops, porches, and colonnades. *See:* **Streetscape**

The combination of the private frontage, the public streetscape and the types of thorougfare defines the character of the majority of the public realm. The combination of elements constitutes the layer between the private realm of buildings. It ranges in character from urban to rural as a function of the composition of their elements. These elements influence social behavior.

SECTION LOT ROW PLAN LOT ROW

T6 URBAN CORE | T5 URBAN CENTER | T4 GENERAL URBAN | T3 SUB-URBAN | T2 RURAL | D DISTRICT

- **Gallery & Arcade:** a facade of a building overlaps the sidewalk above while the ground story remains set back at the lot line. This type is indicated for retail use, but only when the sidewalk is fully absorbed within the arcade so that a pedestrian cannot bypass it. An easement for private use of the right-of-way is usually required. To be useful, the arcade should be no less than 12 ft wide.

- **Shopfront & Awning:** a facade is aligned close to the frontage line with the entrance at sidewalk grade. This type is conventional for retail frontage. It is commonly equipped with cantilevered shed roof or an awning. The absence of a raised ground story precludes residential use on the ground floor, although this use is appropriate above.

- **Stoop:** a facade is aligned close to the frontage line with the ground story elevated from the sidewalk, securing privacy for the windows. This type is suitable for ground-floor residential uses at short setbacks with rowhouses and apartment buildings. An easement may be necessary to accommodate the encroaching stoop. This type may be interspersed with the shopfront.

- **Forecourt:** a facade is aligned close to the frontage line with a portion of it set back. The forecourt created is suitable for gardens, vehicular drop offs, and utility off loading. This type should be used sparingly and in conjunction with the two frontage types above, as a continuous excessive setback is boring and unsafe for pedestrians. Trees within the forecourts should be placed to have their canopies overhanging the sidewalks.

- **Dooryard & Light Court:** a facade is set back from the frontage line with an elevated garden or terrace, or a sunken light court. This type can effectively buffer residential quarters from the sidewalk, while removing the private yard from public encroachment. The terrace is suitable for restaurants and cafes as the eye of the sitter is level with that of the standing passerby. The light court can give light and access to a basement.

- **Porch & Fence:** a facade is set back from the frontage line with an encroaching porch appended. The porch should be within a conversational distance of the sidewalk, while a fence at the frontage line maintains the demarcation of the yard. A great variety of porches is possible, but to be useful, none should be less than 8 ft wide

- **Common Lawn:** a facade set back substantially from the frontage line. The front yard thus created should remain unfenced and be visually continuous with adjacent yards. The ideal is to simulate buildings sitting in a common rural landscape. A front porch is not warranted, as social interaction from the enfronting throughfare is unlikely at such a distance. Common Lawns are suitable frontages for higher speed thoroughfares, as the large setback provides a buffer from the traffic.

- **Slip Lane:** a facade no more than 80 ft from the right-of-way. Parking is placed within the first layer. Private sidewalks are provided between the public sidewalk and the building entrances. The parking and private sidewalk system are landscaped to provide shade and shelter and a streetwall buffer. Appropriate transit stops are provided along the frontages, directly linked to the private sidewalk system.

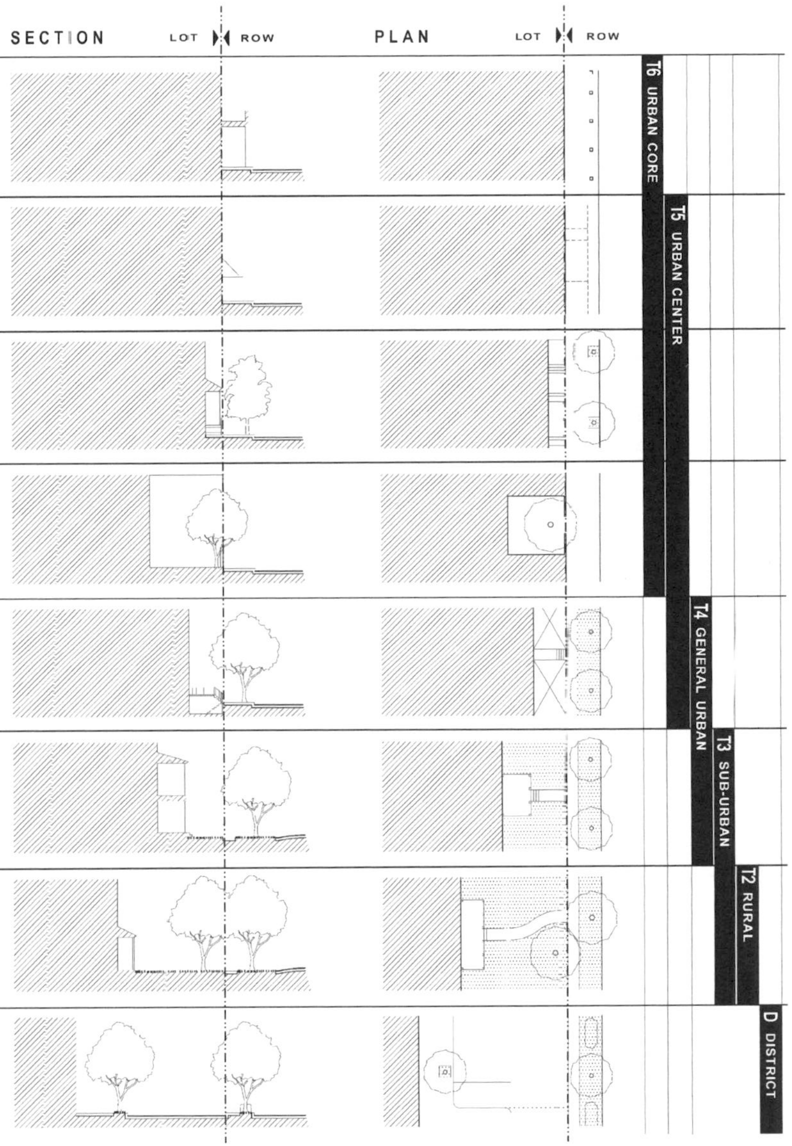

BUILDING TYPES

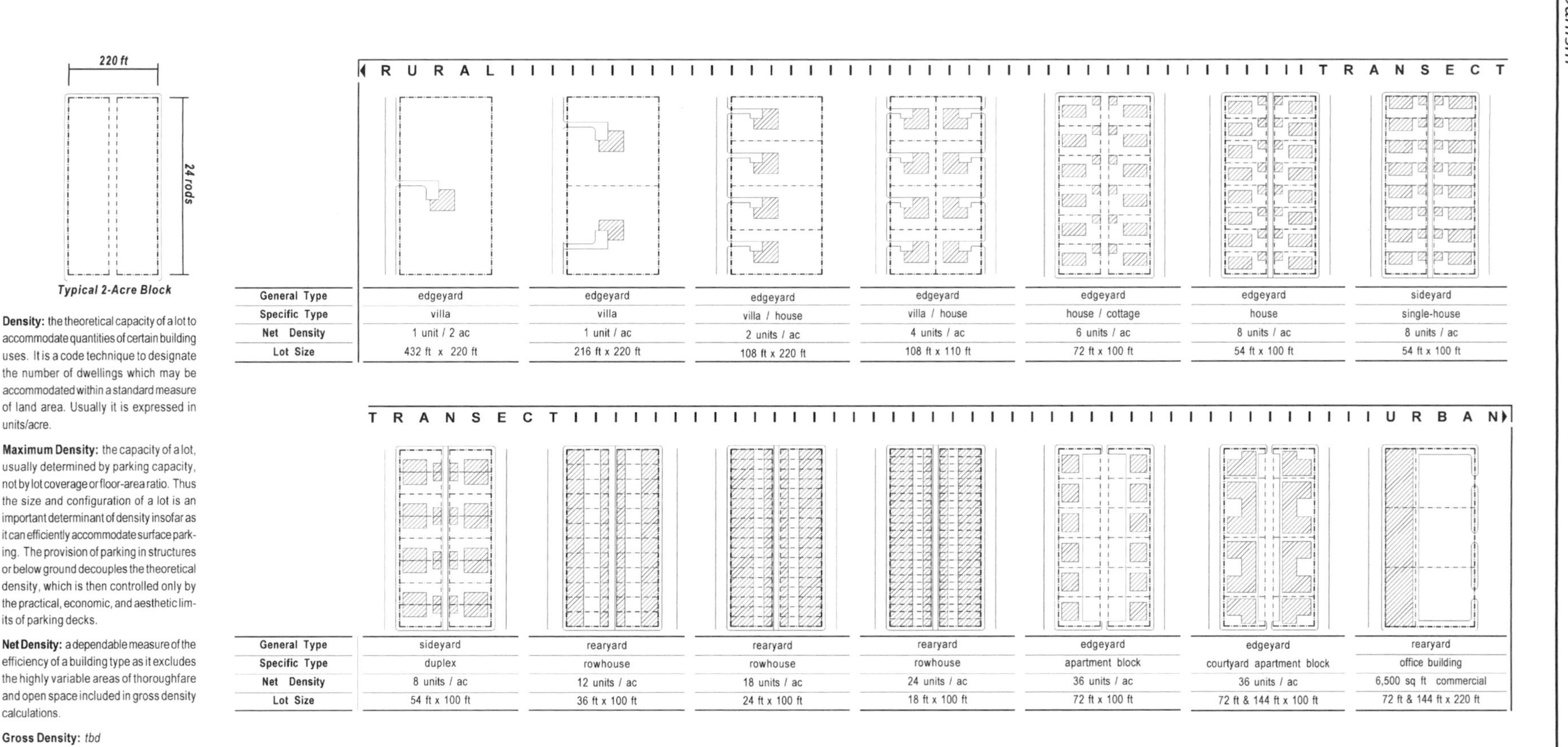

Typical 2-Acre Block

General Type	edgeyard	edgeyard	edgeyard	edgeyard	edgeyard	edgeyard	sideyard
Specific Type	villa	villa	villa / house	villa / house	house / cottage	house	single-house
Net Density	1 unit / 2 ac	1 unit / ac	2 units / ac	4 units / ac	6 units / ac	8 units / ac	8 units / ac
Lot Size	432 ft x 220 ft	216 ft x 220 ft	108 ft x 220 ft	108 ft x 110 ft	72 ft x 100 ft	54 ft x 100 ft	54 ft x 100 ft

General Type	sideyard	rearyard	rearyard	rearyard	edgeyard	edgeyard	rearyard
Specific Type	duplex	rowhouse	rowhouse	rowhouse	apartment block	courtyard apartment block	office building
Net Density	8 units / ac	12 units / ac	18 units / ac	24 units / ac	36 units / ac	36 units / ac	6,500 sq ft commercial
Lot Size	54 ft x 100 ft	36 ft x 100 ft	24 ft x 100 ft	18 ft x 100 ft	72 ft x 100 ft	72 ft & 144 ft x 100 ft	72 ft & 144 ft x 220 ft

Density: the theoretical capacity of a lot to accommodate quantities of certain building uses. It is a code technique to designate the number of dwellings which may be accommodated within a standard measure of land area. Usually it is expressed in units/acre.

Maximum Density: the capacity of a lot, usually determined by parking capacity, not by lot coverage or floor-area ratio. Thus the size and configuration of a lot is an important determinant of density insofar as it can efficiently accommodate surface parking. The provision of parking in structures or below ground decouples the theoretical density, which is then controlled only by the practical, economic, and aesthetic limits of parking decks.

Net Density: a dependable measure of the efficiency of a building type as it excludes the highly variable areas of thoroughfare and open space included in gross density calculations.

Gross Density: *tbd*

6 • ELEMENTS OF THE CITY

Michael J. Bednar

Summary

Urban atriums are public or semipublic pedestrian spaces partly or fully enclosed from the exterior climate. Atriums provide places for pedestrian circulation, gathering and socialization, with many secondary purposes such as exhibition, dining, retail, and performance. Their unique urban design contribution is to create multipurpose gathering spaces, to foster pedestrian-friendly connections between streets, plazas, concourses, and discrete buildings, and to provide climate-moderated environments for plants and for people.

Key words

arcade, atrium spaces, bioclimatic design, conservatory, galleria, pedestrian, plaza atrium, shopping mall, winter garden

PHOTO: Joachim W. Glassel

Galleria Vittorio Emanuele II, Milan

Urban atriums

True to its Latin derivative for "heart," atriums create central concourses and meeting places for urban life. Most often defined with large transparent enclosures that admit outdoor light and air, they include gallerias, conservatories, exhibition halls, train sheds, indoor markets, and winter gardens. Founded upon nineteenth century precedents, the public atrium represents a significant development of urban architecture and design in the past half-century. Enclosed pedestrian environments offer protection from the weather, a valuable attribute at some times and seasons in virtually all climates. Since the overhead enclosure is usually transparent, changes in daylight and season and the bioclimatic benefit are readily available. The protected climate allows pedestrians to utilize atrium spaces and their amenities freely throughout the day and evening. Such enclosure can also have the unwarranted consequence of increased energy costs if not carefully designed. Based on management, but also design limitations, atrium zoning also may lead to increased privatization of the urban realm, resulting in overly controlled use and limited access.

Precedents of covered outdoor city streets and arcades date from previous centuries. The new opportunities for large glazed spaces came with the introduction of glass-and-cast iron and later glass and steel enclosure systems in the nineteenth century. The first European examples began with "retrofitting" glass enclosures over exiting courtyards in hotels, institutional buildings and shopping streets. Early examples are the Burlington Arcade (1819) in London and the Galerie Vivienne and Galerie Colbert (1825–26) in Paris. The arcade concept was adopted in Brussels with the expansive Galerie St. Hubert (1846–47).

Following Joseph Paxton's realization of the Crystal Palace (1851), large glazed atrium structures advanced throughout Europe. In Italy, the concept assumed monumental proportions with grand gallerias in Turin, Naples, Genoa and Milan. The Galleria Vittorio Emanuele II in Milan (1865-77) is perhaps the best known example. The covered passage extending from the Piazza del Duomo to the Piazza della Scala has a great domed space at its center.

Adaptations in North America became evident in the 1880s, including merchandise markets in Providence and Cleveland, hotels such as the Brown Palace, Denver (1892) and the Isabella Stewart Gardner Museum, Boston (1902) which set a precedent for similar museums, and Wanamaker's Department Store, Philadelphia (1902). The Bradbury Building, Los Angeles (Fig. 1) is cited by architect/developer John Portman whose hotel developments of the late 1960s helped make the atrium concept a common organizing element of up-scale hotels and resorts.

1 URBAN DESIGN OPPORTUNITIES

Successful urban design and development result from integration between old and new urban context, between exterior and interior places, and between the public and the private realms. Because they create complementary connections, atriums are also serving an important design role in historic preservation. Additional amenities and features are often introduced into the public pedestrian zones, including landscaping, artwork, fountains, and exhibitions.

Credits: This article is based on the author's research and publications, *The New Atrium* (1986) and *Interior Pedestrian Places* (1989), cited in the references.

PHOTO: Michael Bednar

Fig. 1. Bradbury Building, Los Angeles. George Wyman, Architect. 1893

Fig. 2a. Paseo del Rio, San Antonio.

Fig. 2b. Hyatt Regency, San Antonio. Thompson, Ventulett, Stainback Assoc., 1981.

Atriums add to the inventory of public spaces available to the pedestrian, joining inside and outside, and in inhospitable climates, a continuous connection and extended urban experience. These connections can be to subways, and concourses. Atriums are also useful in creating access control for public, semipublic and private building users. The design of access and circulation is important to the success of atriums as public amenities.

Atriums can be designed as private, semipublic, or public spaces in many variations. Classically, the primary courtyard or atrium in a building was a private space created for exclusive use. In some instances, the atrium has been a semipublic space for controlled public use, commonly seen when private offices or hotel rooms overlook one or two floors of public areas. Atrium designs in public buildings can provide more extensive public use and access, with levels of public access defined by zoning of space and vertical separation.

Coordination between an atrium design and a significant urban pedestrian area is exemplified in Paseo del Rio (river walk), San Antonio, Texas, located one level below the street with a new connection to the historic Alamo (Fig. 2a). The Hyatt Regency San Antonio (Fig. 2b) is located where the new "spur" joins the old river loop. The running water and pedestrian way pass through the hotel atrium, connecting the river walk to hotel lobby, restaurants, and guestroom elevators.

Multi-usage ensures a steady flow of foot traffic—a kind of captive market of office workers, urban dwellers, or conventioneers. One of the earliest projects of this type was Citicorp Center in Manhattan, completed in 1978 (Fig. 3). The center consists of seven levels of shops and restaurants formed around an atrium at the base of a corporate office tower. The main attraction is the piazzalike quality of the atrium and its pedestrian connections to a subway station and the surrounding streets.

In some cities, atrium spaces are connected through systems of pedestrian bridges and tunnels, open arcades and covered plazas. A well-developed system exists in downtown Houston, where a series of pedestrian bridges and tunnels has evolved over a number of years to connect the partial atriums and lobbies of many buildings. At Pennzoil Place, a floor well, elevators and escalator relate the street and tunnel levels (Fig. 4). The street level system is being developed with interconnecting plazas. The proliferation of urban atriums has created the need to direct or control access for public, semipublic, and private users. One way to achieve access control is through vertical separation. At Houston Center, public access is limited to the street level, semipublic circulation is via bridges and tunnels, and private access is via high-rise elevators. Escalators, open stairs, or low-rise elevators control connections between public and semipublic floors.

Of interest from an urban design point of view are private entities that have created atriums as an amenity open to and shared with the public. An early and influential example, the Ford Foundation in New York provides an indoor garden atrium as the feature of a twelve-story office building for access at ground levels by staff, visitors, and public (Fig. 5).

Fig. 3a. Citicorp Center. (now "Citigroup Center") New York City. Hugh Stubbins and Associates, 1978.

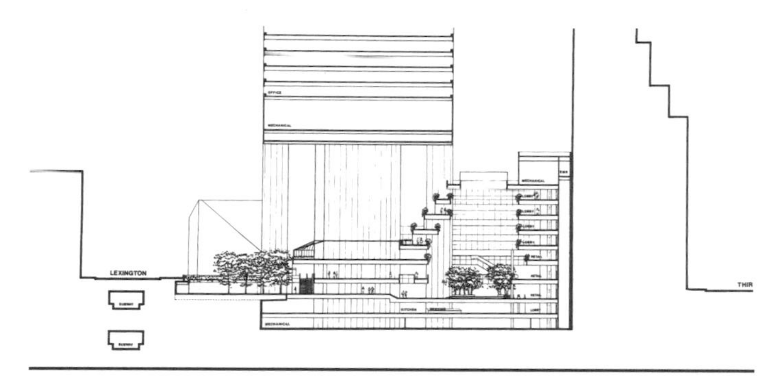

Fig. 3b. Citicorp Center, (cross-section through atrium). Drawing courtesy of Hugh Stubbins and Associates.

Fig. 4. Pennzoil Place, Houston, Texas. Philip Johnson and John Burgee, Architects,1976.

Fig. 5. Ford Foundation Headquarters, New York. Kevin Roche and John Dinkerloo, Architects with Dan Kiley, Landscape Architect, 1967.

PHOTO: Michael Bednar

Fig. 6. Pension Building. Engineer: Montgomery Meigs, 1882–87, now the National Building Museum, Washington, DC.

PHOTO: Joachim W. Glassel

Fig. 7. Erie County Community College, Buffalo, New York. Cannon Design, 1982.

Historic preservation and adaptive reuse

Atrium design offers options in historic preservation, adaptive reuse, and integration of new with old structures. The Bradbury Building in Los Angeles and the Cleveland Arcade were recently restored, and maintained in their original use. The Pension Building in Washington, DC, inspired by precedents such as Pallazzo Farnese in Rome, has been preserved as the National Building Museum (Fig. 6).

An advantage of the atrium concept is to enable restoration and protection of the exterior of an historic building, which is used as part of an atrium enclosure, thus maintaining the urban fabric while giving the building a new interior. This can allow a change of use, interior vertical circulation, and an improved energy strategy, as well as the installation of new building services. Two approaches are commonly utilized in this process: covering existing courtyards and carving out atriums from within existing structures.

Providing a glazed cover is a relatively easy means of converting a courtyard that is already enclosed on all sides. This strategy was used to create Erie County Community College in the adaptive reuse of the former Post Office and Federal Building, Buffalo, NY (Fig. 7).

A popular strategy, judged by frequent use, is to carve out an atrium space within an old building by removing portions of each floor form top to bottom. This can be accomplished in a wood- or steel-frame building by removing some of the internal frame structure, such as in the heavy timber-frame building that is now Butler Square, Minneapolis (Fig. 8). The structural frame is left exposed to express the original framing method and the subsequent atrium formation.

Among the most ambitious preservation projects to have utilized atriums to unite old and new structures is 2000 Pennsylvania Avenue in Washington, DC, where a new eleven-story office building has been placed behind a block of late nineteenth-century townhouses in order to preserve them (Fig. 9). The skylighted space between is a linear atrium formed by the juxtaposition of the new concrete and glass office building and the irregular brick facades of the old townhouses.

Plaza atrium

The "plaza atrium" is not contained within a given building but exterior to and connected to them to interrelate several buildings by enclosing a large shared space. Analogous to a covered square, the plaza atrium space is generous in scale enough to take on a character of its own. It can be varied in function and commonly intertwined with multileveled circulation systems.

Crystal Court at IDS Center in Minneapolis marks the modern day inauguration of this atrium type in a major U.S. city (Fig. 10). The IDS Center is truly a city center, at the nexus of an active commercial zone and multilevel pedestrian circulation. It draws together the covered walkways from surrounding blocks, connected by a system of second-level bridges, or skyways. Four bridges converge in the Crystal Court. Under each bridge is also a street-level entry. The roof structure is a steel space-frame with balcony walkways suspended from it. The ground floor is open and unencumbered, to create an open indoor plaza.

Fig. 8. Butler Square, Minneapolis. Phase 1: Miller, Hansen, Westerbeck, Bell, 1974. Phase II: Arvid Elness, Architects, 1980.

Fig. 9. 2000 Pennsylvania Avenue, Washington, DC. HOK Architects.

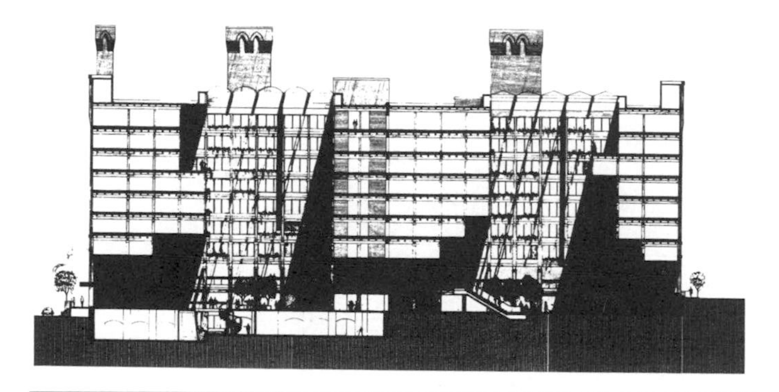

Fig. 9b. Butler Square, longitudinal cross-section.

Fig. 10. Crystal Court at IDS Center, Minneapolis.

PHOTO: Michael Bednar (3)

Fig. 11. The Cleveland Arcade. John Eisenmann and George Smith, Architects,1888. A prototype merchandise mall.

Fig. 12. Galleria at Crocker Center in San Francisco. Skidmore, Owings & Merrill, 1982.

Fig. 13. Eaton Centre. Bregman & Hamann and the Zeidler Partnership, 1978–79.

Arcades and galleria

More than any other purpose, atrium spaces are now used in a majority of shopping malls, devoted to commercial use. The precedents are found among the earliest examples of public atriums. Cleveland Arcade, 1888 (Fig. 11), is one of the great examples built in the United States and recently converted into a Hyatt Hotel. It represents the culmination of the Victorian commercial arcade and provided a model for the twentieth century of a naturally lit and naturally ventilated urban interior space.

Built between two of Cleveland's main shopping streets, the arcade is fronted by two nine-story office buildings located on these streets. The entry to the arcade is through these buildings, which, by their presence, create an indirect relationship between the arcade and the streets. This large scale has five levels, each with continuous galleries that step outward in three places so that the skylight roof provides an abundance of direct daylight. The elaborate roof structure consists of girders supported on arches that rest on columns with griffin-headed brackets. The bottom two floors are filled with shops and restaurants, and the upper floors reserved for small offices.

A contemporary project that recalls the Cleveland Arcade is the Galleria at Crocker Center in San Francisco (Fig. 12). The three-level shopping arcade, located mid-block, extends in a straight line between two adjoining streets. The center's shops and restaurants open off continuous galleries located on three levels, which are served by escalators obtrusively, positioned in the center. A glazed, barrel-vaulted skylight covers the arcade space. The Galleria at Crocker Center knits together the buildings and streets of this dense urban block by providing a generous system of pedestrian access.

The shopping center mall

In 1956, architect Victor Gruen designed the first fully enclosed shopping mall in North America, the Southdale Center in Edina, Minnesota, outside Minneapolis. Due to the severe Minnesota winters, a climate-controlled mall has a competitive advantage over open-air shopping centers. The Southdale Center has a primary, two-story mall space joining the department stores; and short mall spaces extend from it to the exterior. The mall's compact plan resulted in shorter walking distances for customers and reduced both the construction and land costs for the developer. Installing a centralized mechanical plant to supply conditioned air to the mall space reduced the operating costs. From there the air was drawn into the shops and then exhausted. The lightweight steel-roof structure of the mall space was raised above the stores' roof to provide daylighting from both clerestories and skylights, thus further reducing energy costs. Parking was arranged so that customers could enter at either level, with truck servicing occurring via ramps to the basement. The mall, here termed the Garden Court, has its own identity and presence. It became a feature unto itself, a basis for attracting customers and promoting the shopping center.

The concept of the shopping center as an arcade follows from the historic model of discrete shops lining a linear, skylit street. The design strategy is additive: the arcade is first created as a discrete space and then shops are attached to its sides. The important characteristic is the contrast between the two. The arcade is a public space with hard pavement, a distinctive roof form, bright daylight, and street furnishings. Energy conservation, an important consideration since the 1970s, seeks to balance higher levels of daylighting and light control with lower levels of cooling and heating in the public areas, with opportunities to utilize the atrium zone for bioclimatic moderation. This result can be obviated and increased energy costs can result, however, if careful attention is not paid to bioclimatic design principles and requirements. (See Box: "Bioclimatic Design Role of Atriums.")

Urban shopping centers can be described within two categories or market types. Anchor centers, like the Southdale precedent, rely upon large department stores as "magnets" to attract customers. Anchor centers are generally designed to attract people who live in the city or those who still come to the city to shop. Specialty centers rely upon their own ambiance, most often a function of design, to attract shoppers. Specialty centers attract the other constituencies: people who work in the city, tourists, travelers, and conventioneers.

The grandest version to date of an urban shopping center based on the anchor-store concept is Eaton Centre in Toronto (Fig. 13). Phases one and two of the project, which were completed in 1978 and 1979 (totaling 3.2 million square feet) include two office towers, Eaton's department store, a five-story atrium, 302 shops, two parking garages, and connection to the existing Simpson's department store. Eaton Centre is integrated with the surrounding context. Multiple entrances connect directly to surrounding developments and office buildings. Along the long façade is a four-level parking garage located above the shops, hidden by a constructivist steel screen that has a recessed, second-level walkway. The central atrium is colossal in scale. Its extensive dimensions give the space a strong linearity reinforced by the continuity of the barrel-vaulted skylight, which serves as a horizontal datum. Although Eaton Centre lacks the spatial aura of Milan's Galleria Vittorio Emanuele II, it is in fact some 20 percent larger in every one of its dimensions.

Union Station, Washington, DC, a turn-of-century Beaux-Arts railroad station designed by Daniel Burnham and completed in 1907, was recently renovated to include one of the grand specialty shopping centers in the world (Fig. 14). The monumental scale of the baths of imperial Rome inspired the building. The station has vaulted ceilings and coffers decorated with gold leaf. Through the station's main hall passed presidents and kings, for it was once the historic gateway into the capital city. The resurrected Union Station is a combination railroad station, Amtrak headquarters, transportation node, retail specialty center with day and evening functions, and parking garage. Shoppers enter through the majestic colonnades facing the Capitol; they then cross the main hall—passing by vendors, cafes, and kiosks on the way to the primary shopping area. Parking is at the rear, along with the railroad station and a lower-level connection to a Metro subway station. Care was taken to preserve and restore the existing building but then to make alterations and additions appear distinctively contemporary—including the new arched openings cut into the roof of the main hall to visually relate it to the concourse. The plaza in front of the Station is a celebratory urban space, with a line-of-sight connection to the new entry atrium of the Thurgood Marshall Judiciary Building, 1990 (Fig. 15).

After decades of retail migration to the suburbs, the return of retail to urban centers is a hopeful sign of urban revitalization. The success of enclosed suburban shopping malls is due in part to their amenities and environment. Designers of recent urban shopping centers are learning from this experience. The atrium space is the critical design element. The accompanying danger is introversion; these new urban centers must not become exclusive and inward-oriented. They must be good pieces of

Fig. 14. Union Station Plaza, Washington, DC, Interior renovation: Benjamin Thompson Associates, 1988, with view towards Thurgood Marshall Judiciary Building.

Fig. 15. Thurgood Marshall Judiciary Building. Interior atrium. 1990. Edward Larrabee Barnes, Architect.

urban design tied into exiting pedestrian, transport, and service systems, while acknowledging historic contextual forces.

Winter gardens

The creation of a winter garden—essentially a structure intended to create favorable climate for plants as well as for people—was originally motivated by the changing relationship between humans and nature in nineteenth-century Europe. Burgeoning urbanization was destroying vast natural areas, and the by-products of expanding industrialization were polluting the air and water. At the same time, scientists were gaining insights into the biological processes that led to a better understanding of plant growth. Winter gardens preserved the vision of a natural paradise, albeit in a controlled setting, and exhibited discovery of exotic plants and their cultivation. Another motivation was the enjoyment of nature as an art form—for its beauty and aesthetic appeal. These motivations—combined with the development of iron structural members and the ability to manufacture large sheets of glass—enabled the original greenhouse designers, architects and engineers to create wondrous winter gardens of great variety.

The first examples in the United States of winter gardens or conservatories were built towards the end of the nineteenth century, primarily for collection and educational purposes as botanical museums. Usually, they were located in large urban parks—as was the recently restored Enid A. Haupt Conservatory of New York Botanical Garden in the Bronx, designed in 1899 as a "cathedral devoted to nature." The structure is composed of eleven separate greenhouses joined together in a C-shaped plan, which defines an entry plaza and two courts (Fig. 16).

The history of winter gardens as an atrium type is of interest because it illustrates the genesis of an important architectural concept for urban atrium spaces. The notion of keeping plants and trees in an appropriate enclosed environment has had a pervasive influence on architecture since the nineteenth century. Greenhouse spaces have been incorporated into almost every building type—particularly residences, offices, restaurants, stores, and museums.

From the perspective of design integrity and pedestrian use, one of the most successful of these contemporary winter gardens is the IBM Garden Plaza at the back of the IBM Building in New York (Fig. 17a, b). It provides a visual and pedestrian-way linkage to the former AT&T Building across 56th Street, by means of a through-block passage. Direct spatial connections to the lobby and gallery of the IBM Building and visual connections from the third-floor cafeteria relate this space to all elements of its context. Architecturally, the IBM Garden Plaza might be compared to a nineteenth-century conservatory—a greenhouse brought into the city. A distinctive interior is created by use of one planting type throughout, 45-foot-high bamboo trees set in eleven groves. Although the structure is monumental, the bamboo trees create intimate zones of privacy. Movable chairs and tables add a noninstitutional and casual touch, as do the somewhat incongruous nineteenth-century street lamps and benches. The atmosphere resembles a European piazza, where individuals can be alone or in groups in an ennobling public domain.

Fig. 16. Enid A. Haupt Conservatory, New York Botanical Garden in the Bronx. William R. Cobb, Architect, 1899.

The concept of the winter garden is important to the development of atrium spaces because of its amenity and imagery. An interior garden is a conductive setting for respite and meeting within the context of the frenzied city. To be inside but visually outside has always been an experience of magical appeal. Here one can be in the presence of tropical plants on a bitterly cold winter day. Suffused in daylight and filled with strolling people, the winter garden presents a compelling attraction to every urban pedestrian.

2 DESIGN CRITERIA

The following criteria for designing successful atrium spaces are drawn from analysis of many historic and contemporary case studies, cited in the references.

- The spaces should be readily accessible from existing exterior places. This can be accomplished though direct physical linkage and/or visual transparency. Entrances and exits are the critical points whether located at street level or above, or whether related to a sidewalk or urban square.
- Successful atrium spaces should be legible and easily understood for purposes of way-finding circulation and orientation. Coherent spatial form and simple geometry will aid in achieving comprehensible circulation patterns. A strong spatial concept executed with consistent structural and architectural expression will aid in orientation.
- Atrium spaces should serve a public pedestrian purpose. They should contain pedestrian amenities and provide opportunities for socializing and public occasion.
- Technical requirements such as heating, ventilating, and air conditioning, smoke exhaust, fire suppression, and electric lighting should be integrated into a coherent architectonic expression that does not overwhelm the public space.

In the nineteenth-century precedents, the buildings were designed to be frameworks for civic and also retail activity, with clearly articulated structures and well-defined pedestrian spaces. In shopping streets, the shops, kiosks, stalls, and counters were destined to change as merchandising changed, while the building's structure remained permanent. Thus, there was an architectonic integrity to these permanent structures—a quality of daylight and an enduring character of space that made these buildings memorable.

The twentieth-century counterparts have not always been true to their nineteenth-century precedents in terms of architectural integrity and quality. For example, there has been considerable vacillation in the definition of the pedestrian space relative to the shops. In some cases, the shop fronts disappeared altogether; in others, the pedestrian space was treated as residual—as space between shops that did not need a character of its own. The design of twentieth-century commercial places has also suffered because of the increased horizontal and vertical size of shopping centers, which renders them overscaled for pedestrian use and orientation. Perhaps the greatest single change in architectural character has been caused by the influx of numerous mechanical and electrical systems. This plethora of technical requirements either has been concealed behind surfacing materials or left exposed and painted. Careful study of nineteenth century prototypes will both inspire and guide the contemporary urban designer. ■

Fig. 17a. IBM Garden Plaza interior, New York. Edward Larrabee Barnes, Architect.

Fig. 17b. IBM Garden Plaza, New York. Edward Larrabee Barnes, Architect.

Bioclimatic Design Role of Atriums (Watson 1997)

	COLD/CLOUDY Seattle Chicago Minneapolis	COOL/SUNNY Denver St. Louis Boston	WARM/DRY LosAngeles Phoenix Midland TX	HOT/WET Houston New Orleans Miami
HEATING				
H1 To maximize winter solar heat gain, orient the atrium aperture to the south.	●	❑	▼	
H2 For radiant heat storage and distribution, place interior masonry directly in the path of the winter sun.	▼	❑	●	
H3 To prevent excessive nighttime heat loss, consider an insulating system for the glazing.	●	❑		
H4 To recover heat, place a return air duct high in the space, directly in the sun	❑	●	▼	
COOLING				
C1 To minimize solar gain, provide shade from the summer sun.		❑	❑	●
C2 Use the atrium as an air plenum in the mechanical system of the building.	❑	❑	❑	❑
C3 To facilitate natural ventilation, create a vertical "chimney" effect with high outlets and low inlets.	❑	❑	❑	●
LIGHTING				
L1 To maximize daylight, use a stepped section (in predominantly cloudy areas).	❑	▼		
L2 To maximize daylight, select skylight glazing for predominant sky condition (clear and horizontal in predominantly cloudy areas).	❑	❑	❑	❑
L3 Provide sun- and glare-control	❑	❑	●	❑

Key: ● = Very important; ❑ = positive benefit; ▼ = discretionary

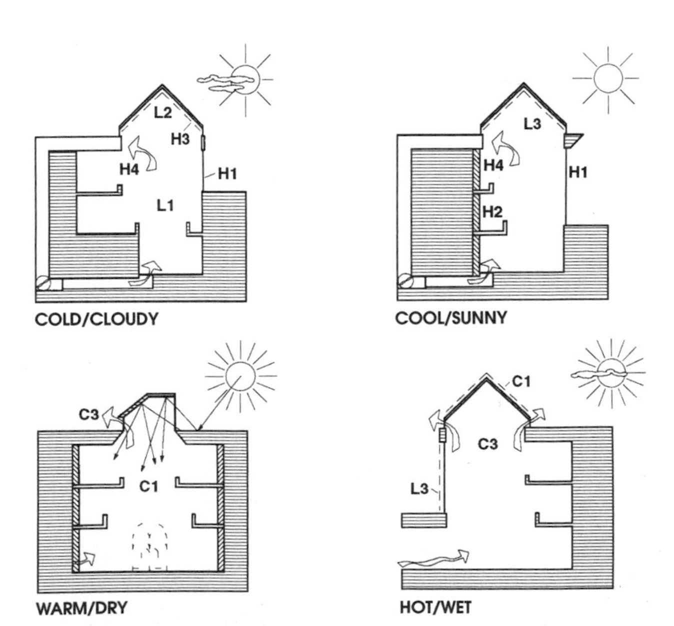

Atrium designs for solar daylighting, heating, gardens, and natural cooling.

REFERENCES

Bednar, Michael J. 1986. *The New Atrium*. New York: McGraw-Hill.

Bednar, Michael J. 1989. *Interior Pedestrian Places*. New York: Whitney Library of Design.

Geist, Johann Friedrich. 1983. *Arcades*. Cambridge: MIT Press.

Kohlmaier, George and von Sartory, Barna. 1986. *Houses of Glass*. Cambridge: MIT Press.

Maitland, Barry. 1985. *Shopping Malls*. London: Construction Press.

Saxon, Richard. 1983. *Atrium Buildings*. New York: Van Nostrand Reinhold.

Watson, Donald. 1997. "Design of Atriums for People and Plants." *Time-Saver Standards for Architectural Design*. Seventh Edition. New York: McGraw-Hill.

Zeidler, Eberhard H. 1983. *Multi-Use Architecture in the Urban Context*. Stuttgart: Karl Kramer Verlag.

John S. Reynolds

Summary
The courtyard building type has a long history. The word for *atrium* in Greece aithrius *ouvranós*, means "clear sky," and referred originally to open courtyards. Courtyards and atrium spaces are often used as the organizing element of buildings and urban designs throughout the world, adapting to different climates, cultures and urban functions. The terms *courtyard buildings* and *atrium spaces* describe a variety of buildings, all of which are examples of the courtyard type, but are hardly copies of one original model. This article presents design guidelines for new courtyard buildings derived from observations in Mexico and Spain.

Key words
aspect ratio, atrium, courtyard, daylighting, entryway, shading, plaza, radiative cooling, sun path charts, *zaguán*

Fig. 1. *Plaza de Armas*, Cuzco, Peru. This central square acts as a huge courtyard, symbolizing the civic center.

Courtyards: Guidelines for planning and design

1 CITY AND COURTYARD

The public plaza can become an example of a courtyard on a grander scale.

In many cities, arcades on the front facades of buildings only occur around the central plaza (Fig. 1). Thus the presence of a plaza can be anticipated from some distance, as an open sidewalk becomes an arcade, several blocks ahead.

The more unique the value of the plaza, the more unique should be its characteristics.

Surrounding arcades, width of the streets and their traffic patterns, plants within the plaza, and features such as a central bandstand or fountain, can be varied according to the relative importance of the plaza.

Some plazas are in fact courtyard forms, that is, entirely surrounded by walls (Fig. 2). In Spain, both Madrid and Salamanca feature a *plaza mayor*, walled plazas accessible by high arched openings for pedestrians on ground level, but continuous surrounding walls and windows with a visible sloped roof above. These large outdoor spaces, safe from cars and bright with umbrella tables, become the center of civic celebrations.

A typical zoning regulation for courtyard-type neighborhoods is that 25% of the site must be open to the sky, whether in one large or several smaller courtyards.

Fig. 2. *The Plaza Mayor,* Madrid. A large civic courtyard enclosed by buildings and free of automobiles.

Credits: This article is adapted by the author from his recent book, *Courtyards: Aesthetic, Social and Thermal Delight,* John Wiley & Sons, 2002. Photographs are by the author unless otherwise noted.

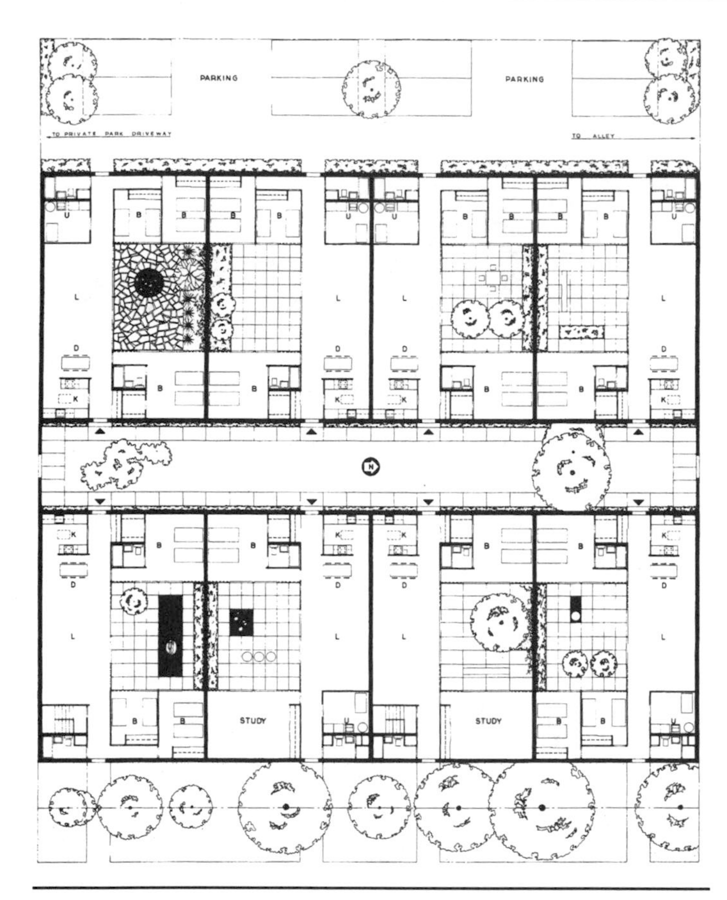

Fig. 3. Chicago courtyard townhouses. Humid summers and savage winters did not deter architect Y. C. Wong from his courtyard scheme for eight dwellings. These are located in a downtown residential district (Madison Park), with parking off an alley. Designed in 1961 with one (remote) parking space per dwelling, each courtyard serves its house on three sides. Only two openings, each a door, penetrate each row house's exterior walls. Half these houses place the living room north of the courtyard, where it can receive winter sun. (From Blaser 1985.)

Fig. 4. *Calle La Palma.* Córdoba, Spain. The narrower the street, the wider must be the garage door to accommodate the turning radius of even a small automobile.

For urban planners, tightly packed courtyard buildings promise a rather high density, compared to the individual house-in-the-garden approach typical of North America. The courtyard building requires much less land per building, because it wraps its rooms around a relatively small outdoor space (Fig. 3). This leads to smaller "footprints" than a similar sized building that is surrounded by outdoor space. Yet the building surrounded by open space can reach many floors before the quality of light and access to breeze is seriously compromised on its lower floors. For one- and two-family, or 1- to 3-story buildings, greater densities are achieved with courtyard buildings. At higher concentrations on smaller sites, courtyards become too deep to serve effectively as a conduit to light and air. The influence of aspect ratio on daylight and on cooling is detailed in later guidelines.

Consider more than one courtyard to serve differing needs.

A large courtyard on a large, multifamily site brings air and light into the surrounding building just as successfully as a similarly proportioned but smaller courtyard, since their aspect ratios would be the same. The difference is social; the smaller courtyard is more personal, and within but one family's control. Therefore, where sites allow, several smaller courtyards serving individual families may be preferable to one larger, but less private, one.

2 CARS AND COURTYARDS

Resolve the location of parked cars relative to the courtyard as an initial design decision.

The traditional hot-dry climate courtyard-type neighborhood has very narrow streets that encourage shading of buildings by those across the street. This leaves little room for sidewalks or for parked cars. These older traditional buildings rarely provided room for vehicles of any kind with access to the street. Sometimes, alleys allowed horses and carriages room to park at the rear. But today's older courtyard neighborhoods are choked with parked cars, and when garages are carved into these buildings, they displace the more formal ground-floor rooms that stood between street and courtyard. Also, some streets are so narrow that the garage doors must be quite wide to provide a proper turning radius, requiring even more space from the building's ground floor. A two-car-sized garage is needed to park a single car (Fig. 4).

Streets today are made wide by the requirements of traffic, fire-fighting equipment, and (sometimes) daylight and winter solar access. So the street width and orientation are set by considerations independent of climate, and are not usually an issue for the designer of a courtyard building. Thus the most challenging street-related issue may be storing the cars that accompany the courtyard.

Consider putting the carport behind or below the courtyard, rather than between courtyard and street.

Residential single-family courtyard buildings today are most likely to be built in a row-house configuration, where zoning allows common sidewalls. There may be access from a rear alley, but street frontage is common. Space planning suggests that the car should be stored as close to the street or alley as possible, to consume a minimum of a site. Yet, at minimum, covered or indoor parking may be expected, especially in harsh climates. Most residential neighborhoods mandate some setback from the street. The usual minimum would allow for a parked car in a driveway between the sidewalk and the building's front facade.

When the car enters the site from a rear alley, the car does not compete with the more public rooms such as living rooms that expect a close relationship with the front yard and sidewalk. But when the car enters the site from the street, this puts a garage door very close to the sidewalk. Planners often disparage the "snout houses" that put a wide garage door farthest forward, dominating the street facade of contemporary suburbs. Courtyard houses fall easily into this trap; but several alternatives are offered by the traditional urban courtyard building type, by taking advantage of the *zaguán* (entryway) as suggested in the following examples.

The entryway (zaguán) might serve as a carport.

One alternative is to use the *zaguán* as the carport. For a single-car family, this is fairly easy, necessitating a wide enough *zaguán* to allow comfortable passage beside the parked car. For a two-car family, this suggests a long enough *zaguán* to accommodate two car lengths, and the inconvenience of moving one car to allow another's departure. Another variation is a wider *zaguán* to accommodate side-by-side autos. This potentially exposes a great deal more of the courtyard to the sidewalk, with greater loss of privacy.

The entryway might serve as a ramp.

Another alternative is to combine within the *zaguán* a driveway down to a below-grade garage, and steps (or ramp) up to a courtyard above the garage (Fig. 5). The garage can be large enough for several cars, and the covering courtyard might be of crude construction, allowing minor water leaks onto the cars below. The ramps might be quite steep, especially if no front yard setback is involved. The raised courtyard might not be handicapped-accessible, and there will be little or no earth below the courtyard to sustain major plants such as vines and trees. But the *zaguán* would not be excessively wide, and the car storage would not unduly dominate the facade.

The entryway might serve as a driveway.

Although it is perhaps the least desirable alternative, it may be acceptable to use the *zaguán* as the driveway and the courtyard as the parking area (Fig. 6). Cars and other activities can co-exist in a courtyard especially in cases where the car may be away for most of the daytime hours. An alternative is to drive across the courtyard to a garage at the rear of the building. Either of these alternatives preserves the *zaguán* as an open passageway to the courtyard, while making a driveway of the courtyard.

3 COURTYARDS AND NEIGHORHOODS

Keep the building in touch with life on the street.

The quality of street life suffers when courtyard owners become discouraged about dirt and vandalism outside. Bumpers, rear-view mirrors, or other appendages on trucks and cars scratch walls on narrow streets. This may result in all attention turning inward, lavishing on the courtyard that share of the care that formerly went to washing off sidewalks, whitewashing exterior walls, watering and tending plants at windows and so forth. It is true that, by contrast, the less pleasant the street, the more spectacular will appear even an ordinary

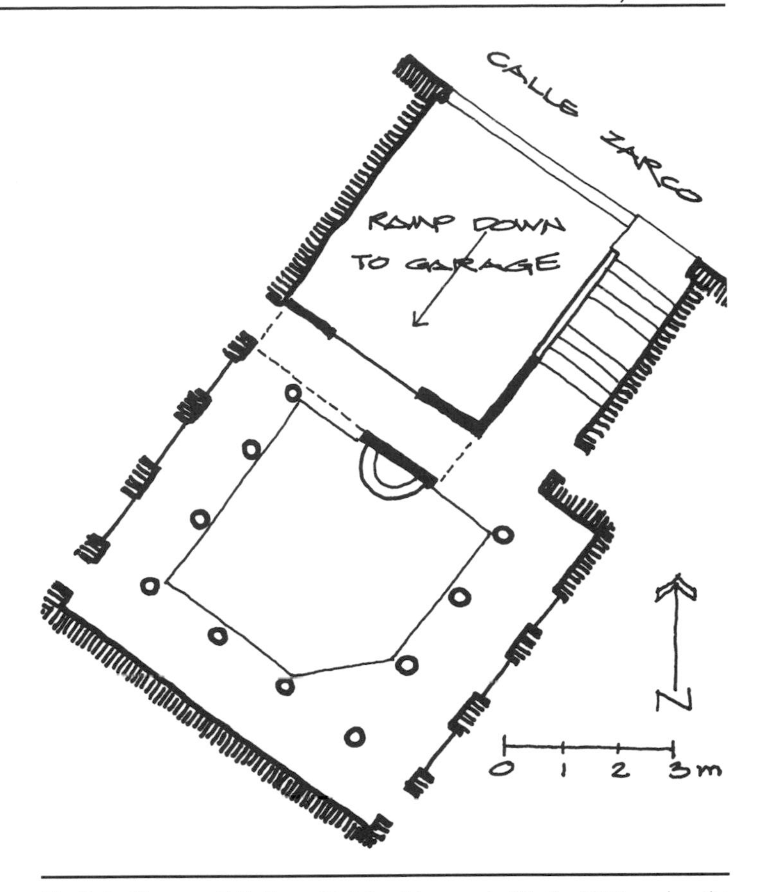

Fig. 5. Residence on *Calle Zarco*, Córdoba, A house rebuilt in the 1980s resolves the car and the courtyard, with ramp down to a lower level garage with stairway and courtyard above.

Fig. 6. Courtyard in Carmona, Spain. Cars parked in courtyard.

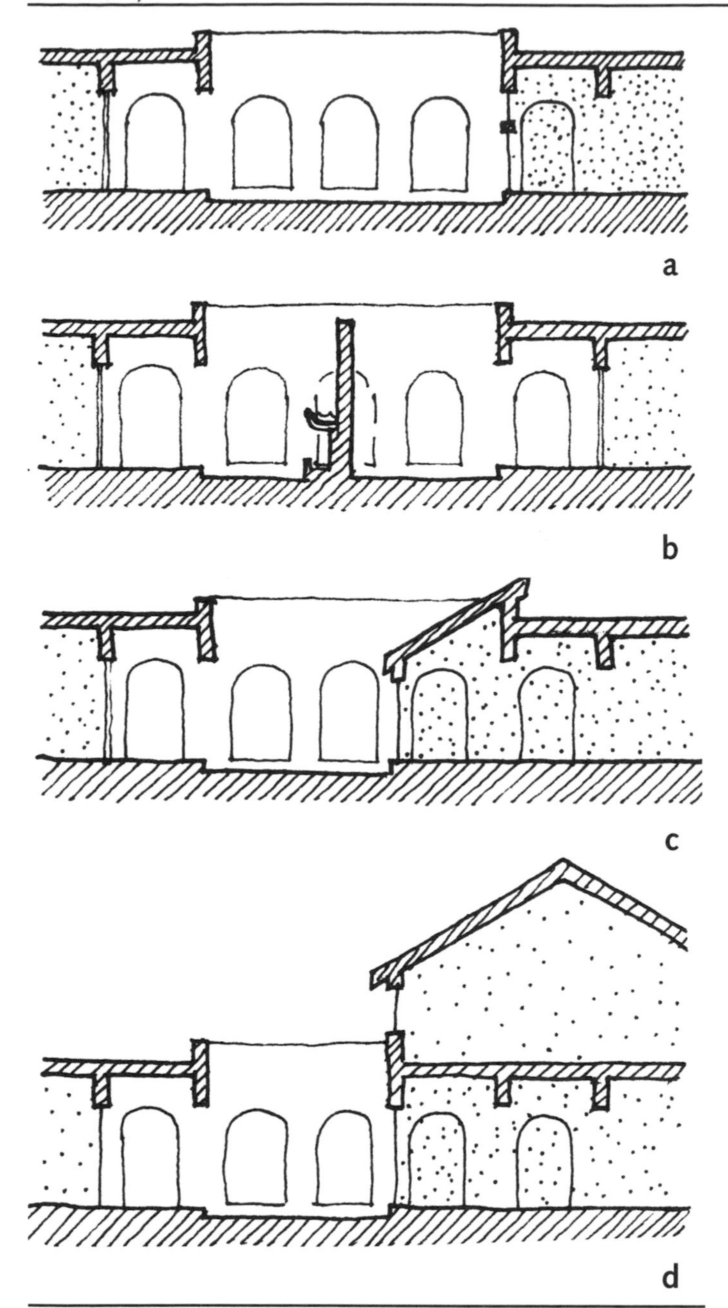

Fig. 7. Courtyards may be compromised by building expansion in several ways, for example: (a) Filling in one side of the arcade. (b) Dividing the courtyard. (c) Encroaching on the courtyard's open space. (d) Adding another floor. (Drawing: Michael Cockram.)

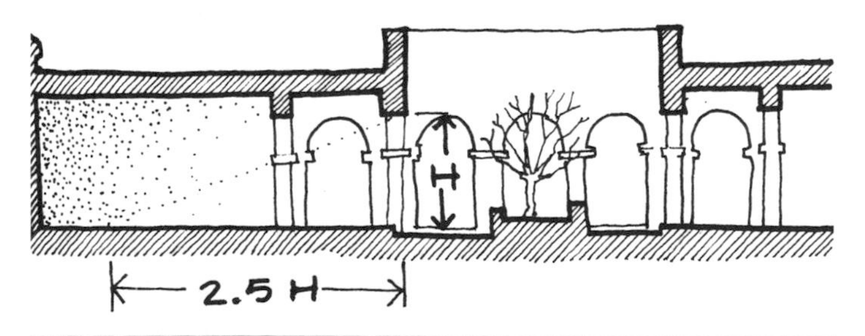

Fig. 8. Daylight penetration from windows or doors is usually adequate in a zone that extends to a maximum distance of 2.5H from the opening, where H is the opening's height above the floor. (Based on Stein and Reynolds, 2000. Drawing: Michael Cockram.)

courtyard. Unfortunately, this can lead to a downward cycle for the streetscape. The following guidelines suggest ways to provide privacy of dwellings while keeping the building in touch with life on the street.

Within the entryway, solid doors and iron grille rejas *are "switches" that give occupants degrees of control over privacy.*

The closer the facade to the street, the more urgent the need for "switches," that is, features which in closed position should be capable of blocking sound and sight. In the traditional examples, exterior wooden doors to the *zaguánes*, if closed, were cited as strong signals that the occupants are not to be disturbed. But when only the *reja* or iron grille gate is closed, it provides a view to the courtyard while barring passage. And, when these gates are left standing open, the message is "everyone's invited."

Where the facade can or must be set back from the street, more of a privacy screen can be developed within this setback.

Such a screen might include shade trees for the street and ramps down to parking or up to a courtyard. The indoor spaces at ground level that open onto the front yard/sidewalk are usually the most public spaces of the building; they are often living rooms or home offices in residences. These spaces can greatly enliven the sidewalk, especially when they can display some degree of their function to the passerby. Again, switches on all openings are crucial when the public realm passes so close to the private one.

The zaguán *can be the public eye on the courtyard, and vice versa.*

For a great number of Hispanic courtyards, the *zaguán* provides a controlled visual corridor into the center of a courtyard center. This important transition space also connects the courtyard to a view of life on the street. The *zaguán* may well be as important as the fence in North American suburbs, "good *zaguánes* build good neighbors."

4 COURTYARDS AND THEIR BUILDINGS

Make the courtyard floor initially as large as possible.

Perhaps the most obvious lesson is that a courtyard will likely be encroached on over the life of the building. Therefore, the designer might anticipate these future encroachments by "oversizing" the courtyard. This might translate as "one large courtyard is better than two smaller ones," yet the designer needs to remember that a first-encountered "formal" courtyard is often followed by a rear "working" courtyard. So many activities benefit from an occasional move outdoors, particularly when surrounded by the beauty and privacy of a courtyard. Consider possible changes in building function, to provide versatility for future adaptations that will not diminish its integrity (Fig. 7).

Whereas a shallow courtyard can have deeper arcades and rooms, a deeper courtyard should have more shallow arcades and rooms.

If the form of the courtyard can be entirely independent of the size and shape of the site, the designer's first task is to determine its proportions: length, width, and height. The choice of deeper or shallower is fundamental, affecting the courtyard's social role (introverted or extroverted), technical performance ("deeper" means "cooler and darker"), and aesthetics (sunny or shady). The choice of deep or shallow proportions also affects the arcades and rooms around the courtyard. The less daylight at the courtyard floor, the less penetration of daylight is possible within these adjacent spaces.

Generalized daylighting guidelines for courtyard proportions are based on the daylight factor (Reynolds 2002; also Millet and Bedrick 1980). The daylight factor (DF) is defined as the ratio between the daylight available at some point within a room, and the daylight available outdoors at the same moment. Because daylight coincides with the sun's heat, rooms in colder climates can benefit from higher daylight factors, while rooms in hotter climates must limit the solar gain that accompanies daylight, and thus are designed with lower daylight factors. The following six design guidelines address these issues.

Rooms that face the courtyard ideally are wider along the courtyard than they are deep.

A typical proportion is three times as wide as the depth. In such designs, daylight from the courtyard can fill the room more evenly. Obviously, this limits the number of such "ideal" rooms that can face the courtyard.

For good daylighting, the higher the top of the window above the floor, the deeper the daylight penetration into a room: the larger the window relative to the floor area, the higher the daylight factor (DF).

Beyond a distance into a room that exceeds 2.5 times the height (H) of the daylight opening (Fig. 8), there will be so little daylight relative to the daylight just inside the opening that electric lights will probably be routinely used. In courtyard buildings with arcades, this 2.5H distance must be measured from the face of the arcade at the courtyard edge. Clearly, the arcade intercepts daylight that would otherwise serve the room beyond.

Aspect ratio influences the available DF.

Fig. 9 charts the relationship between buildings with a square, white-walled "atrium" of given aspect ratios, and the expected DF in surrounding rooms. Three floor-level positions are listed: top floor, middle floor, and bottom floor. For each position, there is a range from most light (0% windows) and least light (50% windows); these refer to the opposite walls of the atrium that act as light reflectors toward the windows being evaluated. A light-colored wall without windows reflects more daylight, since any openings in such a wall will be darker in color, especially if they are arches with walls in shadow behind them.

Using this graph requires some additional judgment. Fig. 9 assumes the contemporary office building atrium, with continuous horizontal strip windows that begin at the ceiling and take up 50% of the wall area. The typical arcade, however, reaches toward but rarely touches the ceiling, and then only at the center of the arch. On the other hand, the arches typically constitute well over 50% of their wall area. The graph predicts the DF in the arcades rather well. But for spaces beyond the arcades, even using the "bottom floor" position might still over-predict the DF available. The aspect ratio $(L \times W)/H^2$ is a convenient measure of relative sky exposure.

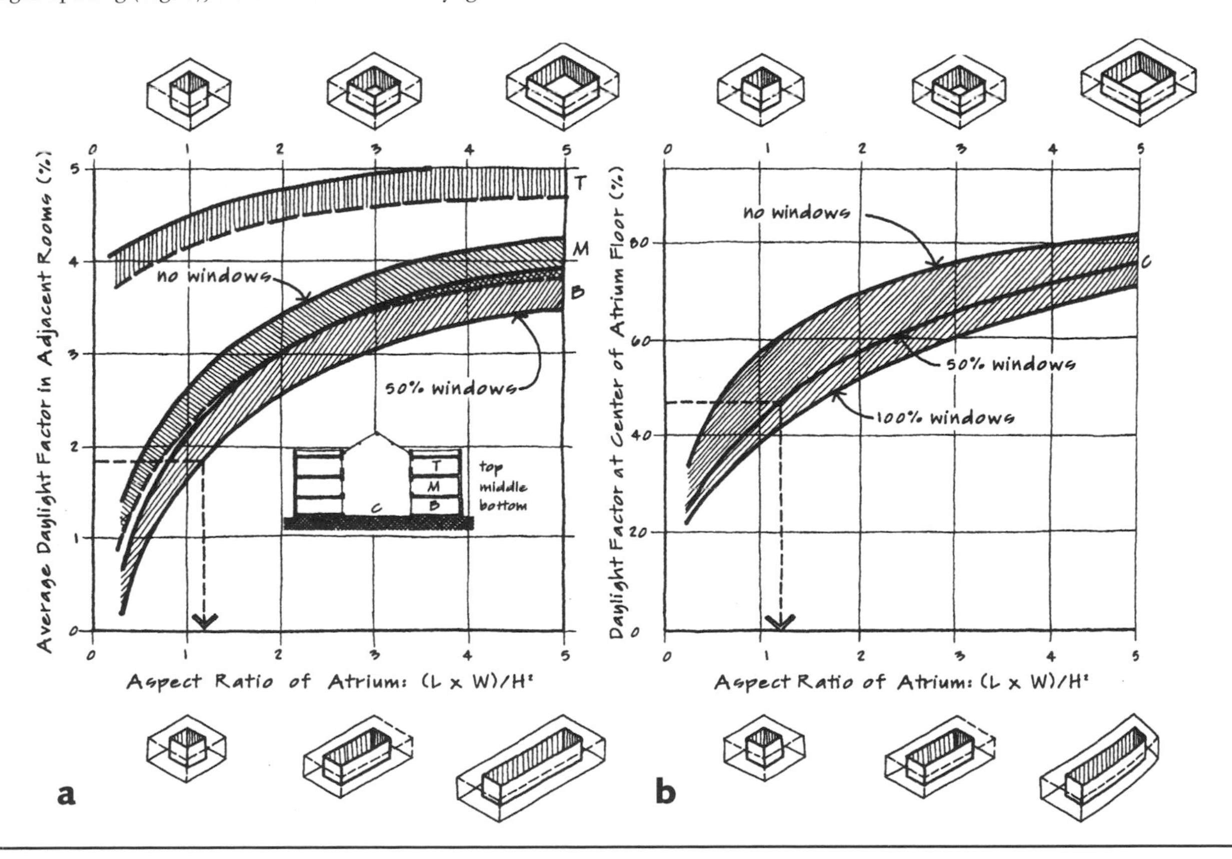

Fig. 9. Daylighting and the aspect ratio. (a) Daylighting factor (DF) estimates for spaces adjacent to courtyards. (b) Daylighting factor (DF) estimates for courtyard floors. (Reprinted by permission from Brown and DeKay 2001.)

PHOTO: Cesar Pelli

Fig. 10. Alhambra courtyards.

Light-colored courtyard surfaces diffuse daylight to the surrounding arcades and spaces.

The courtyard admits, absorbs and reflects daylight according to its proportions and surface colors. When the courtyard floor is very light in color, a substantial amount of reflected light is cast onto walls and ceilings of the arcade. Thus, elaborately carved ceilings can be appreciated for their subtlety and complexity, as in the courtyards of the Alhambra (Fig. 10).

Provide daylight openings with filtering options.

When a light floor surface is also specular (as with white marble), the sheen of light can become glare. White walls quickly become glare when direct sun strikes them, because they are directly in one's field of vision. An ancient glare-mitigation technique common in Islamic architecture, the *mashrabiya*, or window screen, breaks the sheet of glare into many sparkling components. It also provides privacy along with the filtered view, and the wooden screen construction conducts heat slowly, tending to keep the heat on the outer surface where winds might better carry it away.

The interior surfaces around daylight openings can ameliorate glare.

Glare is excessive contrast, as between a bright window and dark surfaces in a room. One strategy for glare reduction is to work with the surfaces immediately adjacent to the bright window. In the thick-walled courtyard buildings, most windows and doors are surrounded by surfaces placed diagonally in the wall; that is, the opening on the interior surface is much wider and taller than the opening in the exterior surface. Such surfaces not only help "spread" the daylight to the interior but become a "brightness mediator" between the window and the interior wall surface.

5 COURTYARDS AND COOLING

If radiation were the primary means of courtyard cooling, then a relationship between aspect ratio and courtyard temperatures, cooler than the official "design condition" temperatures, should become evident. (The "design condition" is the temperature used when sizing mechanical cooling equipment. It typically represents a maximum—but not extreme—temperature, such as one that is surpassed by only a few percent of hours in a typical season for a given city.)

In Fig. 11, these differences in temperature are shown for five courtyards in Andalucía. The courtyard plans are shown in Fig. 12. In each case, three consecutive hot-dry days at summer design conditions were monitored. The air temperatures at (or near) the center of each courtyard are compared with the "official" high and low temperatures for those same three days. The following preliminary design guidelines, at summer design conditions, emerge (shown as dashed lines in Fig. 11):

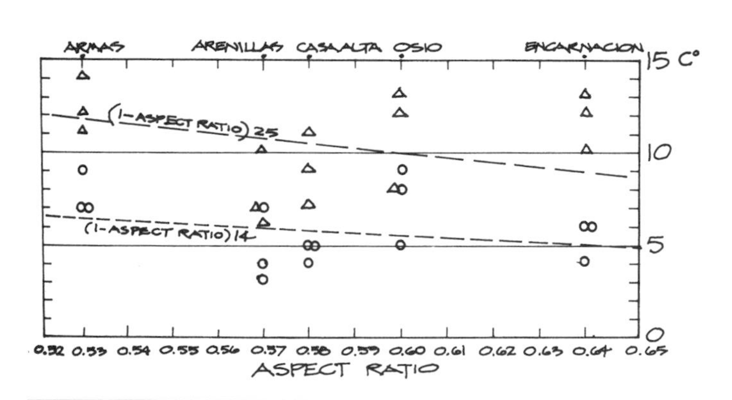

Fig. 11. Relating courtyard aspect ratio to differences between courtyard and official temperatures over three hot-dry days in Andalusian courtyards. Two kinds of temperature differences are indicated:

- Δ The C° difference in daily temperature range of the courtyard, compared to official daily temperature daily range, and
- **O** The C° difference in the highest daily temperature of the courtyard, compared to official highest daily temperature.

The C° difference in daily temperature range of courtyard, compared to official temperature daily range, is approximately = 25 (1-aspect ratio) C°.

(Theoretical maximum difference = official daily range; in such a case the courtyard range is zero, that is, no temperature change at all.)

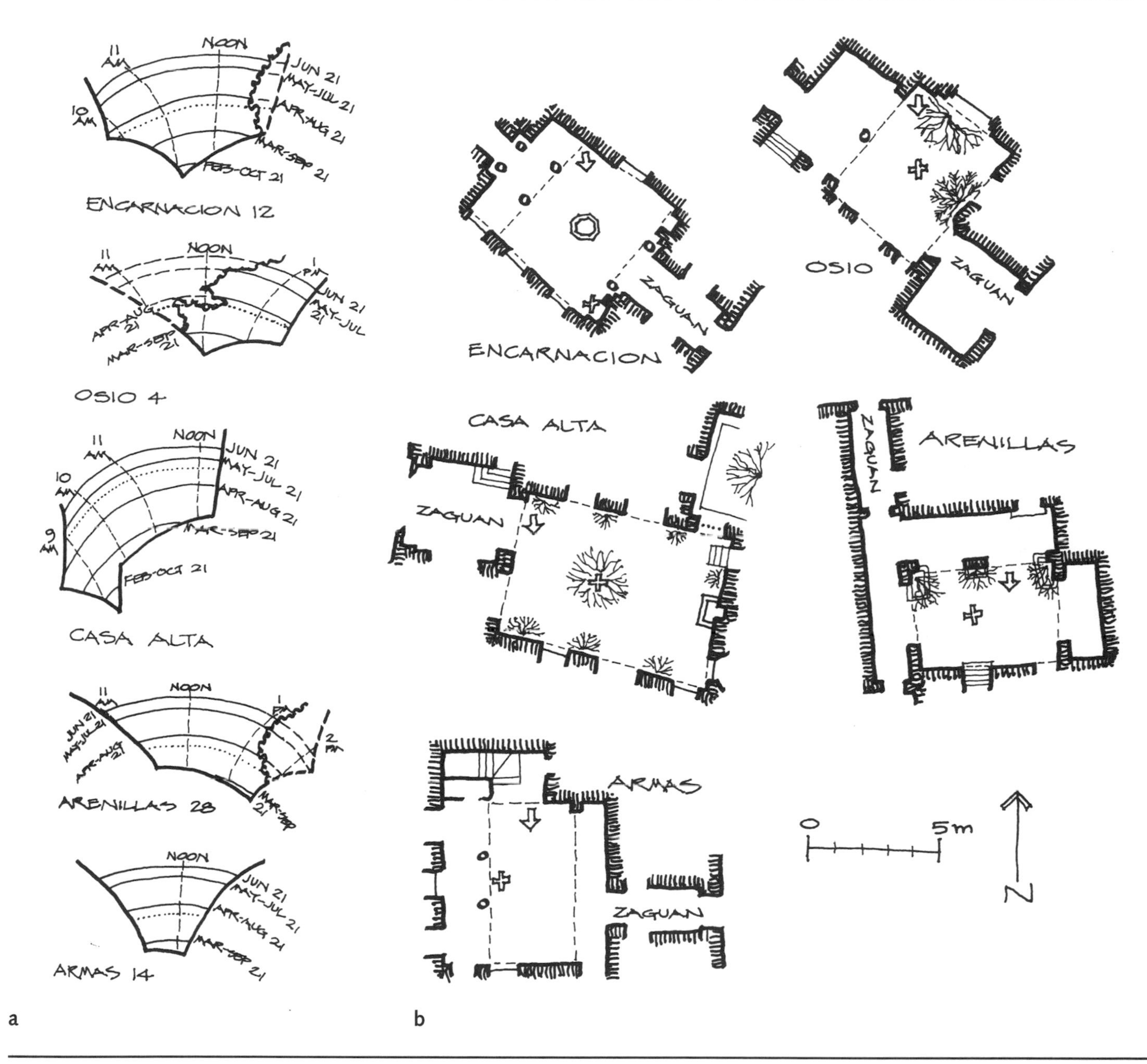

Fig. 12. The sun-path skycharts and plans of five Andulusian courtyards documented in Fig. 11.

The C° difference in the highest daily temperature of the courtyard, compared to official highest daily temperature, is approximately = 14 (1–aspect ratio) C°.

(Theoretical minimum courtyard high = average official; in such a case the courtyard's highest temperature only reaches the daily average official temperature.)

The five courtyards that were measured have similar aspect ratios, so the resulting design guideline must be considered quite preliminary, and subject to these cautions:

- There are far too few examples, over too small a range of aspect ratio, to draw final, firm conclusions for all courtyards.
- An aspect ratio of more than 1.0 suggests higher courtyard temperatures than the official temperatures. Consider that summer direct sun readily enters such courtyards and sol-air temperatures are higher than air temperatures.
- This is only applicable to hot-dry climates at about 36° north (or south) latitude. (This includes northern New Mexico and Arizona, and southern Nevada.)
- Official temperatures are usually from a more rural location, outside the urban heat island.
- Aspect ratio is a simple number, but sky exposure is more complex. Fig. 12 compares the sun-path sky charts for these five courtyards.

Provide for thermal sailing: watering, shading, and ventilation.

Courtyard cooling is even more complex than sky exposure, especially given the influence of evaporation in the hot-dry climate of southern Spain. Opening windows can facilitate ventilative cooling as well. Further, radiation can be significantly influenced by overhead movable screens (toldos): of these five courtyards, two courtyards (Encarnación #12 and Arenillas #28) did not have a toldo.

Evaporative cooling through watering of plants and surfaces, moveable shading, and opening and closing windows all depend on human intervention. This pro-active approach to cooling is also called thermal sailing. It may well be more influential than courtyard proportion in cooling a building.

Ventilation nearly always increases courtyard comfort in hot humid climates, but in hot arid ones, it is far more beneficial at night when temperatures are much lower. (Cf. Brown and DeKay 2001, p. 209 for estimates of wind speeds as a percentage of free, unobstructed incident wind, for a variety of courtyard proportions.)

Consider adjustable shading for the entire courtyard and for its individual openings.

A horizontal toldo can cover all or part of the courtyard. A toldo is used most often in hot dry climates, to keep sun out during the hours when the sun is highest in the sky. It is left open all night, and as long as possible each day, to maximize radiant cooling exposure to the cool evening summer sky.

Moveable shades are particularly useful, allowing considerable choice of how and when any part of an arcade might be used. Awnings or roll-down shades are sometimes used in arcades and are more common at windows. These smaller, individual shading devices brighten the courtyard, adding color and variety.

Theoretically, such an overall courtyard shade might be used to ward off rain, or to keep in heat on a cold winter night, because when closed it cuts off both the courtyard's radiant heat loss to the cold sky, and wind. But rain, along with wind, are a toldo's liabilities, possibly causing it to collapse with accumulated weight, or to be torn apart.

Fig. 13. Emerald People's Utility District Offices, Eugene, OR. Architect: John S. Reynolds, Equinox Design and Richard Williams, WE Group, PC.

An absorbent floor helps cool the courtyard.

In a hot, dry climate, evaporation is a powerful cooling technique. A damp floor, wicking up moisture from the soil below, can provide cooling throughout the hottest hours of the day or night.

6 COURTYARDS AND WINTER SUN

The skyline chart (as in Fig. 12) tells a more detailed story about winter solar access, but requires considerably greater effort to obtain.

In colder climates, provide increased winter solar gain by lowering the courtyard's south wall.

For courtyard buildings that need winter heating, limiting the sun-facing (more universally described as equatorial-facing) wall to minimum height is strongly recommended. Not only will more winter sun enter the courtyard, but more daylight as well. Except on heavily (uniformly) overcast days, the area near the sun will be a bit brighter on cloudy days. More summer sun will also enter over this lower equatorial wall, but shading over or within the courtyard can mitigate this potential disadvantage.

In wet-winter, dry-summer climate of Eugene, Oregon, the Emerald People's Utility District offices occupy three sides of a courtyard (Fig. 13). On the east side is the employee's lunchroom, which also acts as a meeting room available to the public after office hours. The courtyard serves as an outdoor break area for employees by day, and as an entry forecourt for the meeting room on evenings and weekends. The courtyard, and the deciduous vines that shade its two-story north side windows, are so inviting that the meeting room is fully booked months in advance. This 44° north latitude courtyard was designed to allow winter sun to nearly fill the courtyard's north wall.

7 COURTYARDS AND ARCADES

The arcade can be both "path" and "place."

To address the question of why the arcades, rather than the rooms around a courtyard, are so often chosen as places for work and social contact. Might it be as simple as avoiding glare? Rooms with daylight openings on only one side (typical of urban courtyard buildings) often suffer from the "cave effect," with an uncomfortably strong contrast between light levels in the room and the light streaming in through the rather small openings in only one wall. The lighter the courtyard surfaces, the more pervasive this daylight. The arcades are not as deep as the rooms beyond, and their daylit side usually has more opening area than opaque surface area. Thus the contrast between the arcade and the courtyard is much lower than between the room and the courtyard.

Design for migration: each arcade may be used at a different time of day.

Arcades modify the extremes of light, wind, and temperature in the courtyard; they mediate between inside and outside. This is another reason why they often seem the most desirable places to be. As the day progresses, different facades and their arcades are exposed to sun (Fig. 14). The parts still in shade are coolest; the parts now shaded but recently sunny will have considerable residual warmth. Depending

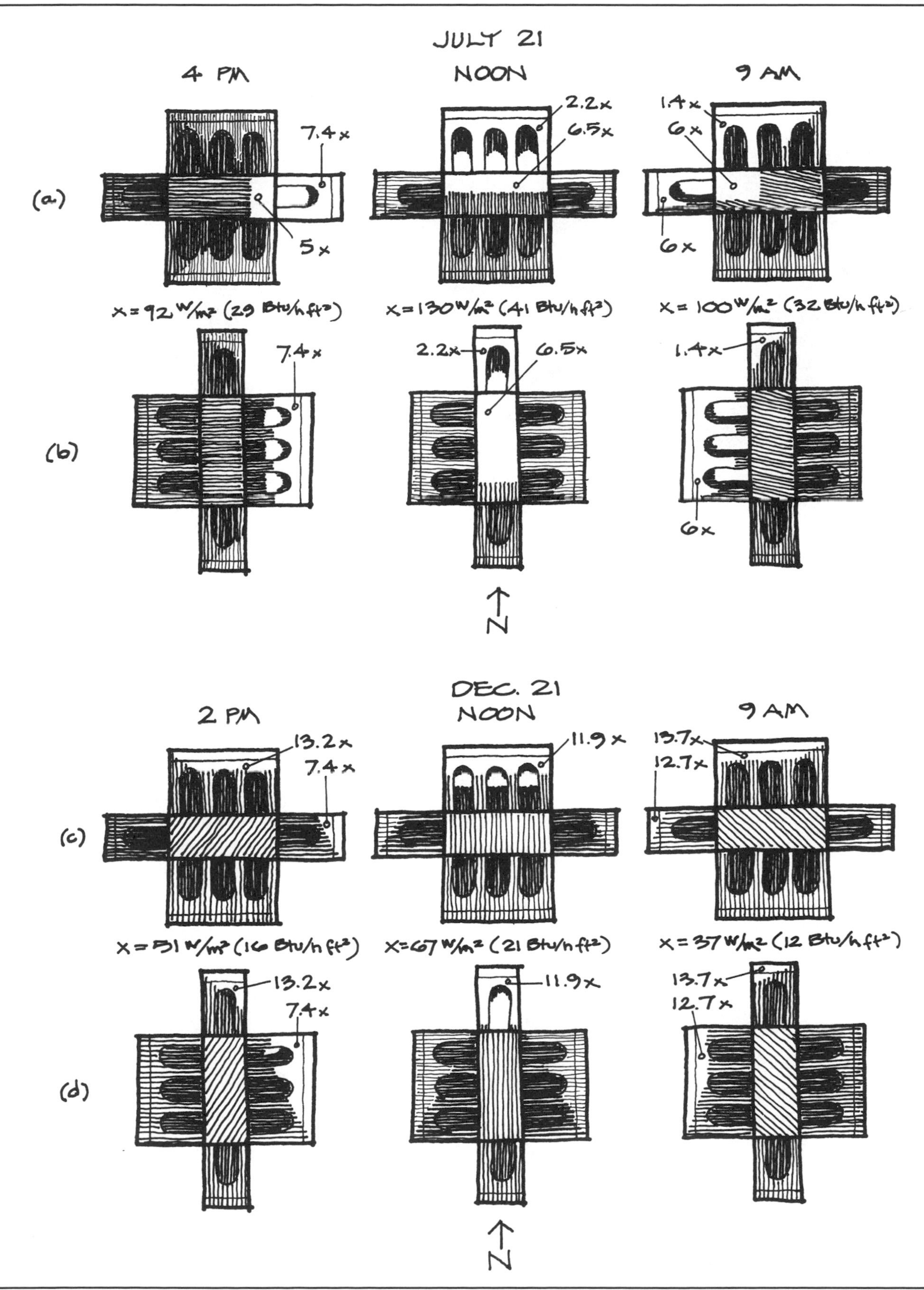

Fig. 14. Relative hourly solar heat gains on sunny courtyard surfaces (on a clear day at 36° north latitude including Cordoba, Spain). "X" represents the approximate (square meter) hourly gains. W/M² represents approximate (watts/square meter) hourly gains.

Fig. 15. The *Cabildo*, Salta, Argentina. Where the sky meets the courtyard, roof tile channels rains, funnels cool night air, and brings a sense of the sky itself into the courtyard.

on the desired use and the air temperature, almost any part of an arcade might be useful at a particular time.

Occupy the center.

Many traditional courtyards feature some object at or near their center; fountain, well, pond, tree, statue, large potted plant, table with chairs, hammock—something that draws the eye, and that may invite repose. This is particularly important when the *zaguán* aims the visitor's eye at this center. It also helps the courtyard to fulfill its role as the social center of a building.

Consider the role of the courtyard's silhouette.

The frame for the courtyard's opening to the sky presents another design decision: roof eave or parapet (Fig. 15). When the eave provides this frame (often in shallow courtyards), the sky contrasts with a rippled edge of many roof tiles. The wider the courtyard, the more of the actual roof tile surface can be seen from the opposite side of the courtyard. This kind of frame emphasizes both the expanse of the sky and the role of the roof as a shelter. It also promotes the collection of rainwater within the courtyard and allows night air, cooled by the roof surface radiating to the cold sky, to flow slowly into the courtyard. It connects the courtyard to nature.

When the walls provide the frame (often in deeper courtyards), parapets may be sculpted to emphasize one or more walls, or to accent a major axis at the center of a wall. This frame emphasizes the courtyard as a stage set, where the contents of the courtyard, not the sky, are the center of attention. It is a more inward-looking design.

Parapets might result from other design decisions, especially when the roof over the arcades is used as a terrace. The parapet then serves both as a safety railing and a privacy screen for summer night sleeping. Parapet walls also intercept some of the sun that would otherwise heat the roof surface; this intercepted heat can then be radiated or carried away by the wind.

The courtyard and stair as a multilevel stage for activities, displays and planting.

A stair within a courtyard invites a display of plants along its rising diagonal, and offers a potential stage at the landing. It also invites climbing, which may be welcome in a hotel or office building, but unwelcome in a residence where the upstairs rooms are the most private. Stairs within courtyards take up space, but in return provide a wealth of views from multiple levels, and chances to smell and touch vines that would otherwise be overhead, out of reach. Fountains, fires, stages for performers, blooming and fruiting plants in courtyards can enrich sight, smell, sound, taste and touch. In such a controlled landscape, sounds and aromas linger. At least two walls are always in shadow, the better against which to display a sunlit centerpiece. Very few courtyards contain no potted plants; it almost seems that a lack of potted plants indicates a lack of concern.

Large, rooted plants that provide significant shade require careful design attention. The sky is important to the courtyard's role; trees and vines both frame and filter the sky. Color and scent and swaying leaves add to summer delight, while maintenance and winter shade are likely less welcome. In temperate climates, deciduous vines and trees provide a dramatic seasonal change, often including a change in the color, as well as the quantity, of daylight through their leaves and branches.

Give water a place.

Water is often (but not always) at the center of the courtyard, where exposure to the sky (and its rain) is at maximum. Even if valued only for its evaporative cooling potential, water is a courtyard treasure.

Masking sounds (sometimes called acoustic perfume) are intended to cover up conversation or other unwanted sounds that echo in the hard surfaces of the courtyard. Masking sound sources include trickling water, birdsong, and wind through leaves, wind chimes; of these, only trickling water is reliably continuous. Water also conveys a message about cooling, most welcome in summer when windows are wide open by night.

Collect and use rainwater.

Rainwater is good for plants because it contains no chlorine or other chemical additives. Sizing the cistern depends both upon the monthly rainfall and the daily usage, as well as upon the total collection area available. Where the monthly rainfall is about the same year-round, a smaller cistern will suffice.

In dry climates, the roofs are often arranged to drain towards the courtyard, which then becomes the collecting place for all the site's rainwater (Fig. 16). The opening to a cistern below the courtyard might be as simple as a grate in the floor, or may be the basin around a central fountain. When eaves are the edges of the courtyard's aperture, gutters and downspouts are prominent and often painted in bright colors. When parapets are the edges, horizontal spouts, cannales, shoot water out and down to the courtyard during a storm. ■

REFERENCES

Blaser, Werner. 1985. *Atrium: Five Thousand Years of Open Courtyards.* Translated by D.Q. Stephenson. Basel, Switzerland: Wepf & Co.

Brown, G.Z., and Mark DeKay. 2001. *Sun, Wind and Light,* 2nd edition. New York: John Wiley & Sons.

Millet, Marietta, and James Bedrick. 1980. *Manual Graphic Daylighting Design Method.* Department of Architecture. University of Washington, Seattle. [out of print.]

Stein, Benjamin, and John Reynolds. 2000. *Mechanical and Electrical Equipment for Buildings.* 9th edition. New York: John Wiley & Sons.

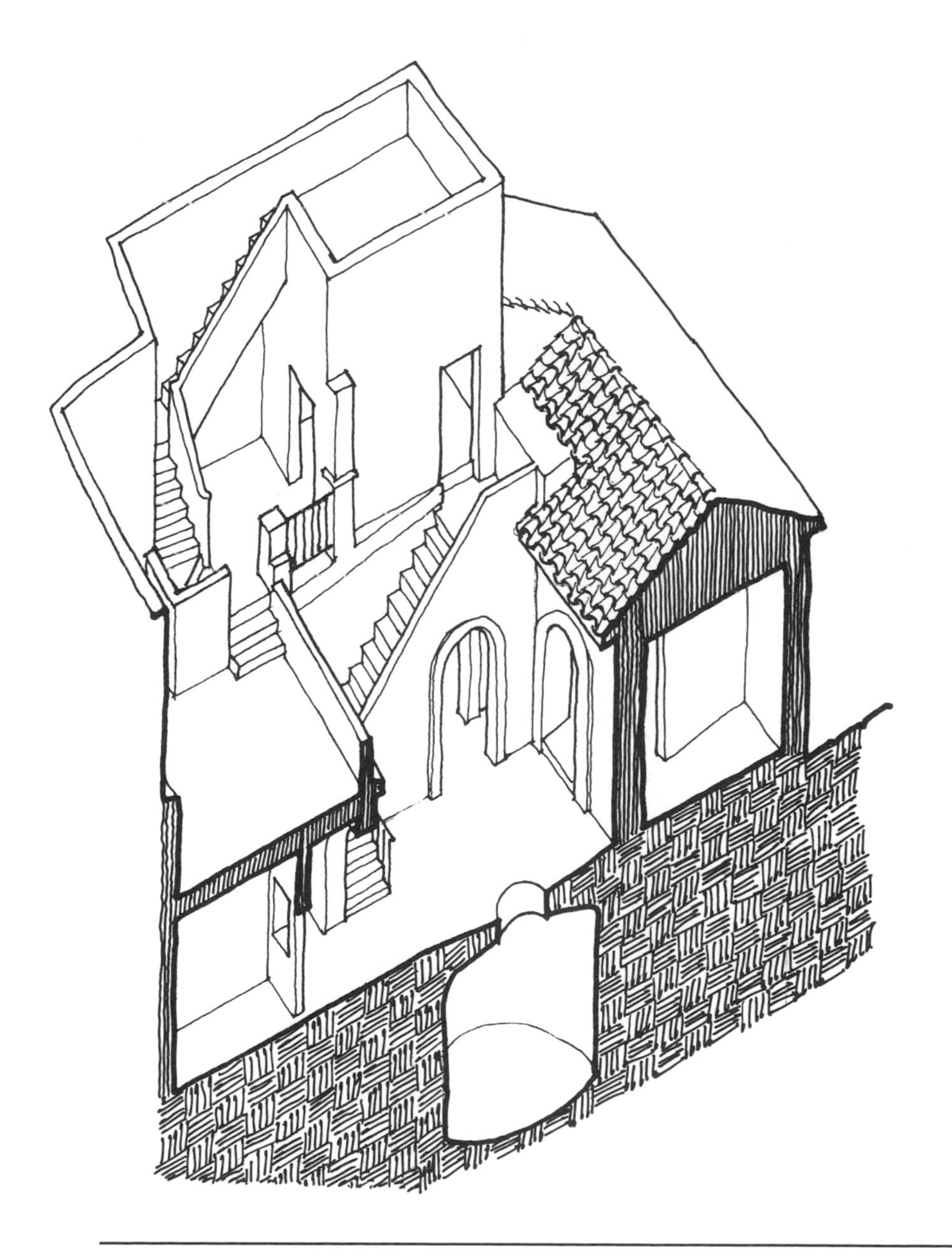

Fig. 16. Courtyard residence takes advantage of multiple levels. (after Matthew Wilson)

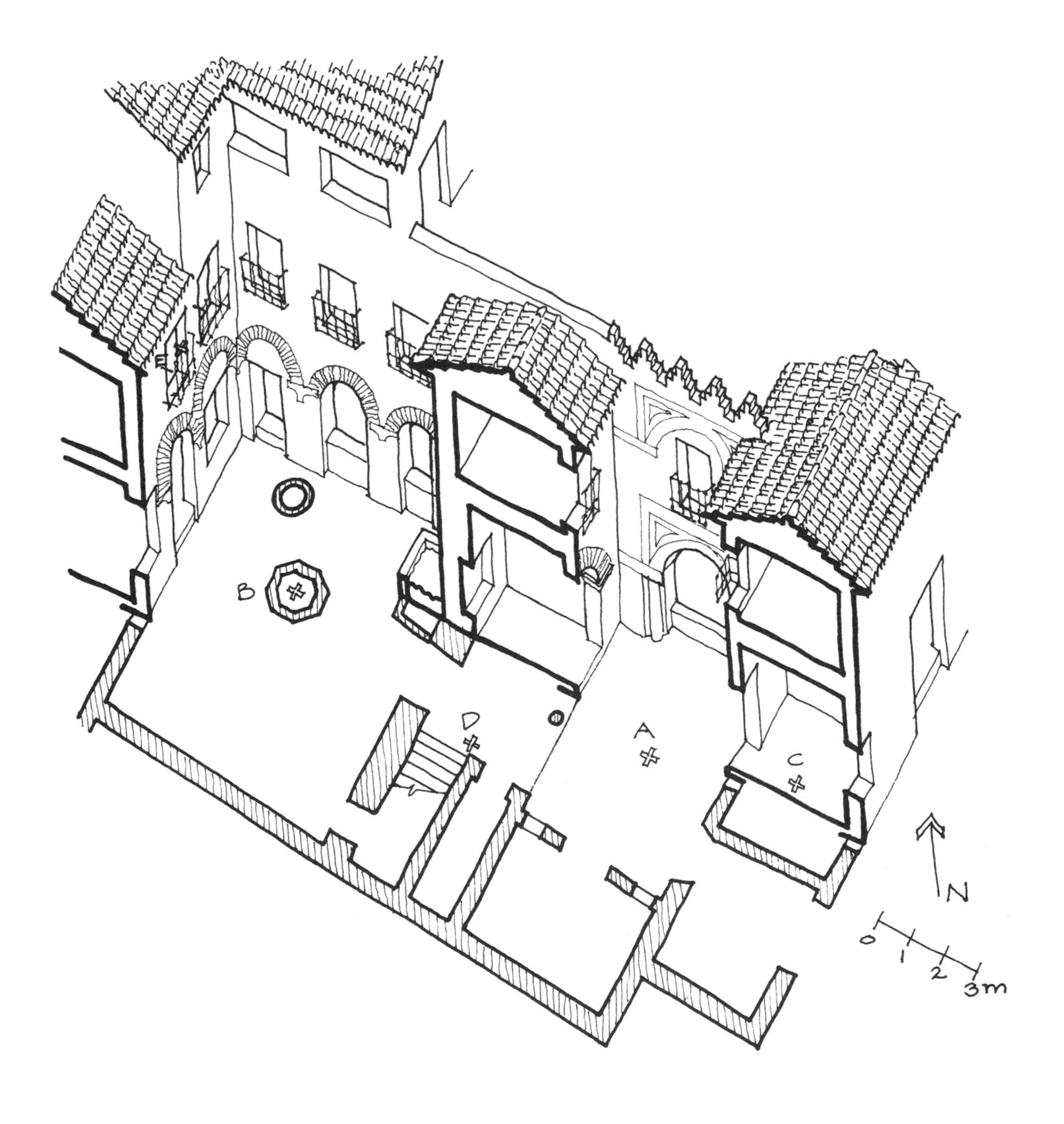

Fig. 17. Axonometric section through the *zaguán* at Osio #4. Points A, B, C, D, and E are locations of temperature sensors. (Reprinted with permission from Reynolds and Lowry 1996.)

Allan B. Jacobs

Summary
The article describes the requirements for great streets and qualities that contribute to their making. It declares that, first and foremost, a great street should help make community, should facilitate people acting and interacting to achieve in concert what they might not achieve alone. It states that a great street is physically comfortable and safe and describes that the best streets are those that encourage participation. The best streets are those that can be remembered. They leave strong, long continuing positive impressions. Finally, the truly great street is one that is representative: it is the epitome of a type; it can stand for others; it is the best.

Key words
accessibililty, parking, pedestrians, perspectives, requirements for street design; scale, street plans, urban quality

Along the Via dei Giubbonari in Rome.

Making great streets

1 REQUIREMENTS FOR GREAT STREETS

Certain physical qualities are required for a great street. All are required, not one or two. They are few in number and appear to be simple, but that may be deceptive. Most are directly related to social and economic criteria having to do with building good cities: accessibility, bringing people together, publicness, livability, safety, comfort, participation, and responsibility. The designable qualities of the best streets are the subject of this article.

A major objective of this work is to provide knowledge of the best streets for designers and urban decision makers, a reference for current work. The intent is not to provide formulae or recipes, but to provide knowledge as a basis for designs of future great streets.

Not unexpectedly, the course is full of caution signs. The connections between what can be built on a street and the socioeconomic criteria for a fine street are not always easy to make. If you cannot walk along a street or go from one side to the other, then you aren't likely to meet anyone on it. At the same time, it remains difficult to isolate physical features from social and economic activities that bring value to our experiences. To what extent is it the experiences that we have on a street rather than the physical setting that make it memorable in a positive way?

By themselves, as a group, the required qualities will not assure a great street, but they are necessary. Overall, though, a final ingredient—perhaps the most important—is necessary, and I call it "magic"—the magic of design. All of the parts, all of the requirements have to be put together into a whole street, and the ways of doing that, at least in detail, are infinite. It is hard to believe that all of them have been tried. There are new great street designs to come. Whatever is designed, though, will share with other great streets a few critical physical qualities.

Places for people to walk with some leisure

The point of view and interest in this inquiry has mainly to do with the best streets for people, mostly on foot. Vehicles often share public rights-of-way with people on foot; no problem, that has been and can be accounted for in design. But you do not meet other people while driving in a private car, nor often in a bus or trolley. It is on foot that you see people's faces and statures and that you meet and

Credits: This article is excerpted from Part 4, "Making Great Streets" in Jacobs, Allan B. (1993) *Great Streets*, Cambridge, Mass: MIT Press. Illustrations are by the author.

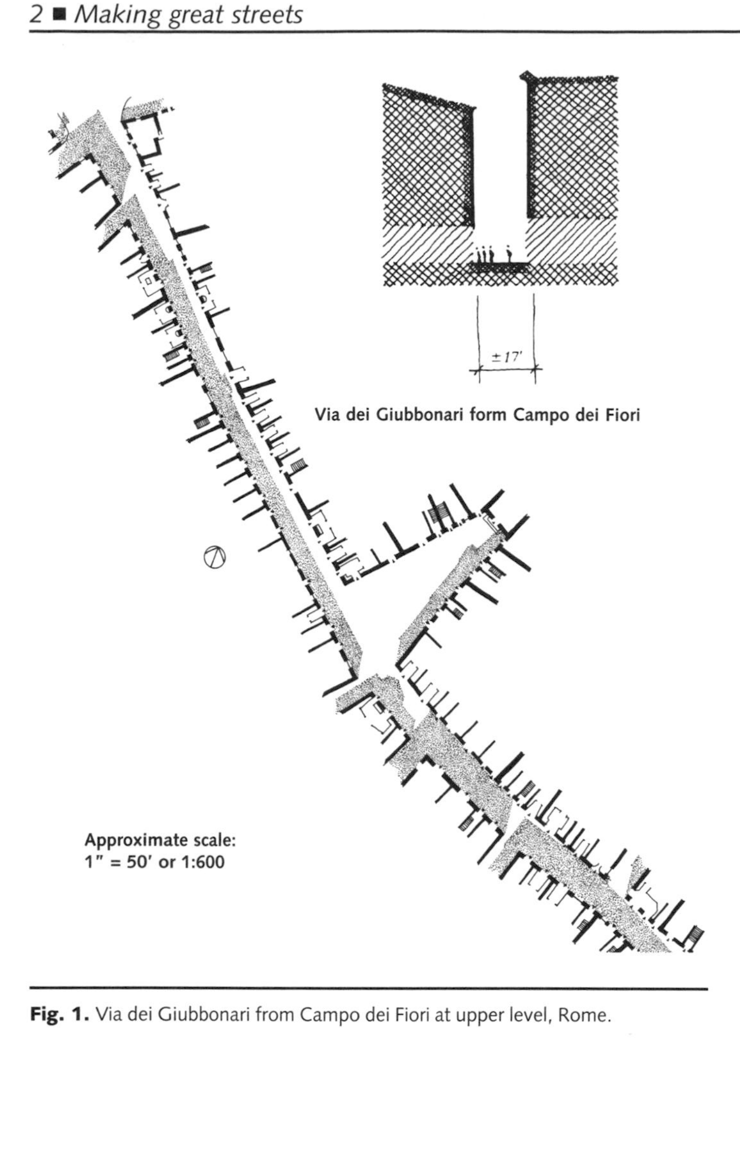

Fig. 1. Via dei Giubbonari from Campo dei Fiori at upper level, Rome.

experience them. That is how public socializing and community enjoyment in daily life can most easily occur. And it is on foot that one can be most intimately involved with the urban environment—with stores, houses, the natural environment, and people. As Marshall Berman (1982) explains, "[t]he essential purpose of this street, which gives it its special character, is sociability: people come here to see and be seen, and to communicate their vision to one another, not for any ulterior purpose, without greed or competition, but as an end in itself. . .This is the one place where people don't show themselves because they have to, where they aren't driven by the necessary and commercial interest that embraces the whole of St. Petersburg. . .The Nevsky is the common meeting ground and communication line of St. Petersburg" (p. 272). Great urban streets are often great streets to drive along as well as great public places to walk, but walking is the focus here.

Every fine street invites leisurely, safe walking. It sounds simple, and basically it is. There have to be walkways that permit people to walk at varying paces, including most importantly a leisurely pace, with neither a sense of crowding nor of being alone. They also must be safe, primarily from vehicles.

Walkways never seem crowded at three or four people per minute per meter, but at less than two they may seem empty. Any kind of walking is possible with up to eight people per minute per meter. Then, it seems, speed picks up even though leisurely walking still takes place. Crowding starts at perhaps thirteen people per minute per meter, when overall speed slows. People, it seems, make do in walking just as in so many other things; some can shut out distractions and walk at a leisurely pace even among large crowds. But at the high counts, about seventeen people per minute per meter, they must at times move to avoid people. And when people are forced off of walkways into streets because of too little space, then safety as well as leisure becomes an issue.

Curbs and sidewalks are the most common ways of separating and thereby protecting pedestrians from vehicles. They may physically separate but do not necessarily offer a sense of safety or tranquillity. Trees added at the curb line, if close enough to each other, create a pedestrian zone that feels safe. An auto-parking lane at the curb also creates separations, but while some great streets have parked cars, which is not one of the things you think of to create a great street. No physical separation at all between vehicle and pedestrian paths, that is, no curbs, can be a better solution, particularly on crowded, small streets; let cars and people mix. The Via dei Giubbonari is such a street and there are many others. The auto is forced to move at the pedestrian's pace. In part, the Netherlands and other countries, notably Denmark, use this principle successfully to achieve safe residential streets (Appleyard 1981, pp. 248–251; Eubank-Ahrens 1989, pp. 63–79; Homberger 1989, pp. 49–78).

Physical comfort

There are streets and places that we avoid because we know them to be physically uncomfortable. The best streets are comfortable, at least as comfortable as they can be in their settings (Arens and Bosselmann 1989, pp. 315–320; Bosselmann 1984, pp. 203–220). They offer warmth or sunlight when it is cool and shade and coolness when it is hot. They offer reasonable protection from the elements without trying to avoid or negate the natural environment. This is a contextual requirement for good streets: it is too much to expect a street in an Alaskan city to be warm in the winter, but it can be as warm as possible under its circumstances and not colder than it need be. A good urban street gives shelter from the wind. On urban streets winds will measure 25 to 40% of the winds outside the city in an open field, unless placement and height of buildings are such that winds are accelerated.

Climate-related characteristics of comfort are reasonably quantifiable and there is every reason that they should be a part of great streets. Sensitive designers of the past understood the requirements in planning their streets, often intuitively. It is possible to do better than that now, through measurement and pretesting of future street environments.

Definition

Great streets have definition. They have boundaries, usually walls of some sort or another, that communicate clearly where the edges of the street are, that set the street apart, that keep the eyes on and in the street, that make it a place. But what does that mean in operational terms? What does it take to define a street? If building facades or walls are the answer, how big do they have to be or how small can they be? What ought to be their spacing? These are difficult but important questions.

Streets are defined in two ways: vertically, which has to do with the height of buildings or walls or trees; and horizontally, which has most to do with the length of and spacing between whatever is doing the defining. There is also definition that may occur at the ends of a street, which is both vertical and horizontal. Usually buildings are the defining elements, sometimes walls, sometimes trees, sometimes trees and walls together. It is always the floor.

Vertical definition is a matter both of proportion and of absolute numbers. The wider a street gets, the more mass or height it takes to define it, until at some point the width can be so great that real street definition, not necessarily space definition, stops, regardless of height. For example, Blumenfeld (1967) observes that when the small dimensions of places exceed 450 feet (137 meters), spatial definition is weak and becomes "more of a field than of a plaza, despite the great height of the structures" (pp. 8). At another extreme, most buildings along the southern edge of the Giudecca Canal in Venice are only two or three stories high, yet they certainly define the Giudecca, at a distance of about 1,100 feet (335 meters) from the Zattere, across the canal. But that is not street definition or plaza definition. Rather, it is a clear definition of an urbanized shoreline.

Harmonious proportion has, at least since 1784, been a major objective of regulations of building height along Paris streets. The traditional proportion of two units street width to three units height to the cornice line was then formalized (Loyer 1988, pp. 121). Later, Haussmann would change to a square section for streets but without changing cornice height, although height above the cornice lines became greater for the city as a whole. Sunlight may well have been a factor in these height limits, but not, it seems, the idea of height to achieve street definition.

Hans Blumenfeld, relying heavily upon work by H. Maertens, is concerned with urban scale and principally with defining what can be meant by "human scale," but he does, indirectly, get to street definition (Blumenfeld 1967 pp. 216–234). Basing their work on physiological optics and experience, Maertens and then Blumenfeld use distances at which they report it is possible to recognize people (human scale) and distances at which facial expressions can be perceived (intimate human scale), together with angles at which objects can be perceived

Fig. 2. Via die Giubbonari from Campo dei Fiori.

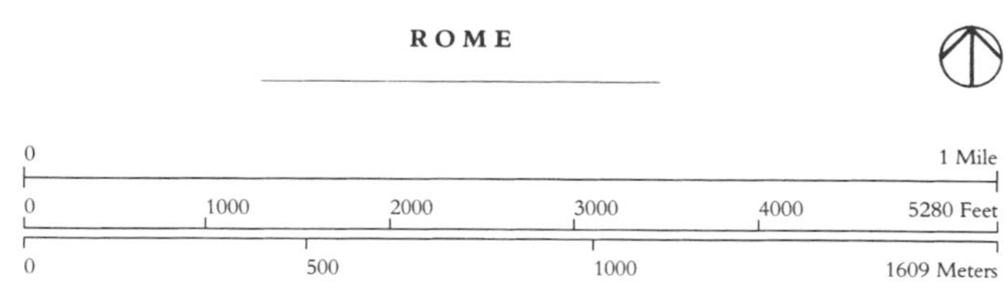

Fig. 3. Rome street pattern.

Fig. 4. Via dei Greci, Rome.

Fig. 5. Outer Market Street, San Francisco.

clearly, to judge the scale of buildings. They conclude that a building height of three stories—approximately 30 ft. (9.14 m)—and width of 36 ft. (11 m) with a street width of 72 ft. (22 m), are the maximum dimensions for a building of *human scale*. The smaller *intimate scale* requires a building height of 21 ft (6.4 m), a facade width of 24 ft. (7.3 m), and a street width of 48 ft. (14.6 m).

For our purposes two points need to be made: human scale and street definition are not necessarily the same thing; and, applied to a street, these conclusions would apply mostly to looking directly across a street, not along one. Also, while it may well be correct that a twenty-seven degree angle is the maximum at which an object may be perceived clearly, people are ornery and keep moving their heads and eyes, so the dimensions that are the result of the analysis may be fine for defining human scale but perhaps not for streets. More to the point of street definition, Maertens and Blumenfeld (1967) attest that:

> *At an angle of 27 degrees (height-distance ration 1:2) the object appears. . .as a little world in itself, with the surroundings only dimly perceived as a background; at an angle of 18 degrees (1:3) it still dominates the picture, but now its relation to its surroundings becomes equally important. At angles of 12 degrees (1:4) or less, the object becomes part of its surroundings and speaks mainly through its silhouette (p. 219).*

Noticeably, many fine streets are lined with trees, and these may be as important as the buildings in creating street definition.

Is there some point, some proportion or absolute height, at which the buildings are so high in relation to street width that the building wall becomes oppressive? Is there an upper limit to street definition as well as a lower? Maybe not. It may be that the upper limits are more appropriately determined by the impact of height on comfort and livability of the street, as measured by sunlight, temperature, and wind, than by absolute or proportional height. At the same time, it may be observed, none of the very best streets have tall buildings. The height of buildings along the best streets tends to be less than 100 ft (30.5 m).

There is another factor important to street definition: the spacing of buildings along a street. Buildings can be far enough apart so that, looking directly across a street or walking along one, it is normal to see beyond buildings to rear yards or to buildings on the next street. In the end, tighter spacing is more effective than looser in achieving street definition.

Qualities that engage the eyes

The eyes move. There is no stopping them, no keeping them still, unless there is nothing to see. Great streets require physical characteristics that help the eyes do what they want to do, must do: move. Every great street has this quality.

Achieving streets that prompt eye movement does not seem to be difficult. Generally, the constant movement of light over many different surfaces keeps the eyes engaged. Separate buildings, many separate windows or doors can achieve this effect, or surface changes. Or it can be the surfaces themselves that move and therefore attract the eye, if only for a split second, before something else—people, leaves, signs—gains momentary ascendancy. Visual complexity is the key, but not to the point of creating chaos or disorientation. Hong Kong street signs, for example, can be so demanding and insistent as to negate the street entirely and make the environment disorienting, even where the street plan is straightforward. Complexity within some holistic context, on the other hand, permits orientation.

Beyond helping to define a street, separating the pedestrian realm from vehicles, and providing shade, trees are special because of their movement. The constant movement of their branches and leaves, and the ever-changing light plays on, through, and around them. The leaves move and the light on them constantly changes, creating thousands and thousands of moving, changing surfaces. If light filters through them, casting moving shadows on walks and walls, so much the better. Branches of deciduous trees may not move much in winter, but the light plays over their uneven surfaces, changing them always, challenging the eyes if ever so subtly.

Buildings do not move. Light, though, moves over them, and as it does the surfaces change, in lightness, darkness and shadow, and therefore in color. The changes may be slow but they are real, and the eyes, ever sensitive, are happy to respond. Complex building facades over which light can pass or change make for better streets than do more simple ones.

What of streets at night? Much that was visible during the day disappears. The eyes may have less to look at. They become more focused, usually at a lower level than during the day, where the lights are: streetlights, signs, and store windows. Streets are different at night. They may almost cease to exist, like the Grand Canal in Venice, or they may exist only at night, because of the light and what it does to the eyes. The Paseo de Gracia, Boulevard Saint-Michel, and Strøget in winter stand out at night.

Transparency

The best streets have about them a quality of transparency at their edges, where the public realm of the street and the less public, often private realm of property and buildings meet. One can see or have a sense of what is behind whatever it is that defines the street; one senses an invitation to view or know, if only in the mind, what is behind the street wall.

Usually it is windows and doors that give transparency. On commercial streets, they invite you in, they show you what is there and, if there is something to sell or buy, they entice you. On the best shopping streets there may be a transition zone between the street and the actual shop doorways, a zone of receding show windows and space for outside displays that are welcoming attention getters. The best streets are replete with entryways, as little as 12 feet (3.7 meters) apart.

There are subtle ways to achieve transparency; it need not be all windows and doors. There are blank-walled passageways in Venice, less than three feet wide, with no windows and very few doorways, but with the branches and leaves of a tree overhanging the wall, that offer a kind of transparency. The branches, the leaves, the vine take you over the wall, into the garden beyond.

Complementarity

Overwhelmingly, the buildings on the best streets get along with each other. They are not the same but they express respect for one another, most particularly in height and in the way they look. It is not necessarily time of building or similarity of style that accounts for the design complementarity of buildings along the best streets. Rather, it is a series of characteristics, all of which are rarely present on any one street, but enough of which are always there to express regard

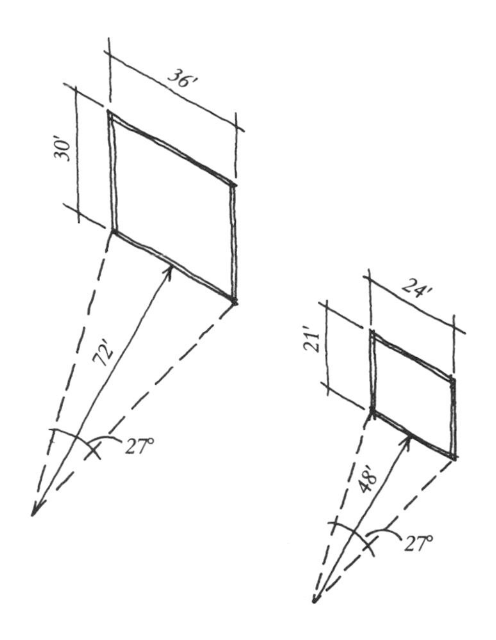

Fig. 6. Human scale (left) and intimate human scale (right), according to Blumenfeld and Maertens (1967).

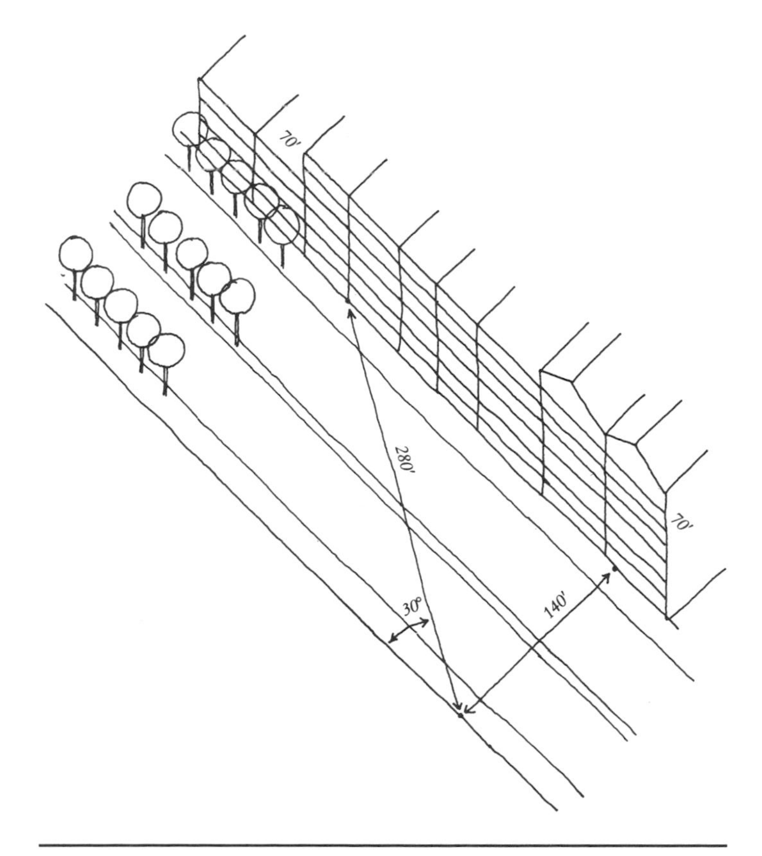

Fig. 7. Generally, buildings are likely to provide a sense of definition when height-to-horizontal-distance ratios are 1:4 when the viewer is looking at a 30-degree angle to the street direction.

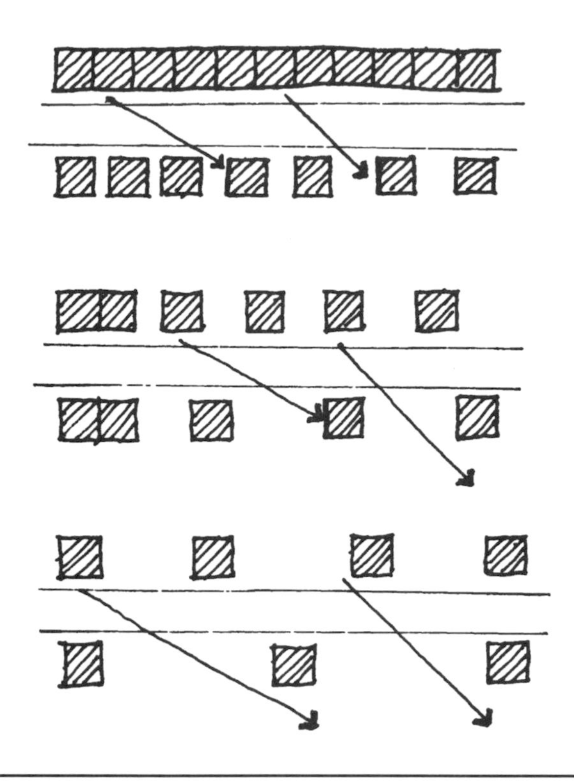

Fig. 8. Greater horizontal spacing and less definition.

Fig. 9. Hong Kong street signs.

and respect, one for another and for the street as a whole. The variables are materials, color, cornice lines and belt courses, building sizes, window openings and their details, entrances, bay windows, porches, overhangs and shadow lines and details like downspouts. A common architectural style is not to be discarded just because it may result in sameness of buildings. Formulae and prescriptions, however, are hard to come by. Caution and individual assessment are better ways of determining what it is that holds the buildings together. A generally similar building height, an overall building geometry dictated by the diagonal corners, and bay windows are the critical pieces.

Great streets are not generally characterized by standout, individual architectural wonders. That may be a part of fitting in. The architects of palaces along the Grand Canal built to the canal line, and there seems to have been no need to set them apart from adjoining buildings. Gaudi and his colleagues also found it possible to build within height standards and norms, and to respect street lines and the scale of other buildings along the Paseo de Gracia. Complementarity was both desirable and possible.

Maintenance

Ask a pedestrian on a San Francisco street what physical, buildable characteristics are most important to achieving a great street and the answers are very likely to include words like "cleanliness," "smooth," and "no potholes." (In a survey of approximately 100 people on San Francisco streets in 1989–1990, at four different locations, "cleanliness and good maintenance," and "smoothness and absence of potholes," were physical characteristics given second and third most often as most desirable to have on the best streets.) It is a point well taken. Care of trees, materials, buildings, and all the parts that make up a street is essential. Given a choice, and there usually are choices, people would prefer to be on well-maintained rather than poorly-maintained streets.

Physical maintenance is as important as any of the other requirements for great streets. It is more than a matter of keeping things clean and in good repair. It involves the use of materials that are relatively easy to maintain and street elements for which there is some history of caring.

Quality of construction and design

In the context of great streets, "quality" of construction and design is difficult to get a handle on. Mostly, it has to do with workmanship and materials and how they are used. There are streets that have all the characteristics we find present on the very best streets, and yet do not make the grade. Quality, or rather the lack of quality, is often the reason.

All the best maintenance in the world will not make a wobbly line straight or a skewed line vertical. Nor will it cure a sloppy putty seal, make a muddy color come to life, nor make right a wrong tree. These are matters of materials, workmanship, and design, all with the word "high-quality" before them when it comes to great streets.

2 QUALITIES THAT CONTRIBUTE

Many of the best streets have trees, but not all of them. Many but not all of the best streets have special public places to sit or stop along the way. Gateways, fountains, obelisks, and streetlights are among the physical, designable characteristics on great streets, but not always. Some physical qualities, then, contribute mightily to making great streets but are not required. On particular streets they can be as compelling and interesting as the necessary qualities, or they can add the salt and pepper, the spice or difference that turns a good street into a great one. Some factors, like accessibility and topography, are ever present. Other variables, most notably density and land uses, though not directly part of street design, are so intimately related to physical place that they cry out for discussion.

Trees

Given a limited budget, the most effective expenditure of funds to improve a street would probably be on trees. Assuming trees are appropriate in the first place and that someone will take care of them, trees can transform a street more easily than any other physical improvement. Moreover, for many people trees are the most important single characteristic of a good street. (In the survey of approximately 100 people on San Francisco streets in 1989–1990 mentioned earlier under maintenance, trees were the most frequently mentioned characteristic essential to good streets.)

Trees can do many things for a street and city, not the least of which is the provision of oxygen, and of shade for comfort. Green is a psychologically restful, agreeable color. Trees move and modulate the light. In terms of helping streets to work functionally, when planted in lines along a curb or even in the cartway they can effectively separate pedestrians from machines, machines from machines, and people from people. The trunks and branches create a screen, sometimes like a row of columns that gives a transparent but distinct edge. They can be a safety barrier between pedestrian and auto paths. Put a line of trees one lane into a street to make a parking lane, for example, as has been done on many European streets, and that lane becomes a part of the pedestrian realm while still functioning as a place to park cars. Even a few trees along the curb of a busy traffic street can have an impact if they are close enough together.

Which trees to use, their placement, their planting, and their maintenance are all important matters. Deciduous trees are more often appropriate than evergreens. Deciduous trees permit sunlight to reach the street in winter when it is either most needed or least a problem. Their leaf patterns are almost always less dense than those of non-deciduous trees and the leaves move more, subject to even slight wind changes; they permit light—mottled, moving light—to penetrate to the pedestrian.

To be effective, street trees need to be reasonably close together. They must be near enough to each other to create a line of columns that separates visually and psychologically one pathway from another, and if a further objective is to provide a canopy of branches and leaves to walk under. In practice, the most effective tree spacing is from 15 to 25 feet (4.5 to 7.6 meters) apart. On streets where the spacing reaches 30 feet (9 meters) or more, there are likely to be four rows of trees, or two to a side. If there is a rule of thumb to be learned from the best streets, it would be that closer is better.

Conventional wisdom holds that one should avoid street corners by forty or fifty feet (twelve to fifteen meters), for reasons of sight lines and therefore of auto safety. Nonetheless, tree planting along the best streets either preceded or has otherwise managed to avoid such dictums; it comes as close as possible to street corners. In fact, one reason why street trees are often not effective is a combination of imposed spacing and corner distance rules. Assuming a 400-foot (122 m) block (twice that of a north-south block in New York), a

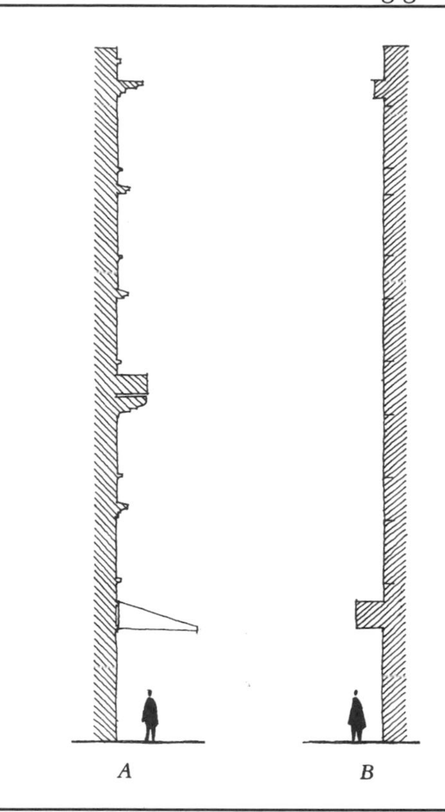

Fig. 10. Building wall sections of two buildings on Via Cola di Rienzo, Rome; the more complex façade, A, offers more surfaces, more opportunities for shadow and light changes than does façade B.

Fig. 11. Amsterdam canal.

Fig. 12. Stroget, Copenhagen: poster-filled windows of a vacant store.

Fig. 13. Trees along Viale Manlio Gelsomini, Rome.

50-foot (15 m) corner distance requirement, and a 50-foot (15 m) spacing standard, there will be seven trees along a block and then a 150-foot (45 m) gap for the intersection—not very many trees. A 200-foot (61 m) block would have only three trees.

The same spacing requirements help to explain why trees along center-of-street traffic medians are seldom effective. First there are the spacing requirements, and then even greater distances required at intersections to allow for left turn lanes. The results are even fewer trees and larger gaps.

Beginnings and endings

Every street starts and ends somewhere, and those locations are usually not too hard to fix. Perhaps in some perverse way it is the obviousness of the observation that keeps this from being an always-present requirement for great streets. And yet, though the entry to Roslyn Place in Pittsburgh is marked ever so subtly by two wrought iron gateposts, they are insignificant to that street's special character. It does not seem reasonable that every great street has to have something special, a physical thing, to mark its beginning or end, or that the start or finish should be crucial to making it what it is. Nonetheless, most great streets have notable starts and stops—not always fine, but notable. It could be argued that, since they have to start and stop somewhere, these points should be well designed. Some physical qualities that denote ends are helpful. They say, in effect, that one has arrived, or left, or they give boundaries. They are places to meet, or reference points.

Diversity of buildings

Generally, more buildings along a given length of street are better than fewer buildings. At the very least, there will be a vertical line between buildings where one ends and the other begins, and that line adds interest. The lines are also reference points, like markings on a ruler, that give a sense of scale. The more buildings, the more vertical lines. Diversity, or at least the greater likelihood of diversity, also comes with more rather than fewer buildings.

Having more rather than fewer buildings may or may not result in greater diversity of uses and activities. The different buildings can, however, be designed for a mix of uses and destinations that attract mixes of people from all over a city or neighborhood, which therefore helps build community: movies, different-sized stores, libraries. Yes, all of those uses can be designed into one single building—but those buildings are not truly public. People can be and are regularly excluded, and even when such buildings are on a public street, they lack the interest and diversity that comes with a variety of owners, buildings, stores, and designers. None of them can match the great streets.

Details

Details contribute mightily to the best streets. Gates, fountains, benches, kiosks, paving, lights, signs, and canopies can all be important, at times crucially so. At the same time, some contribute less than might be thought. The most important of them deserve special attention. According to Haussmann (quoted in Loyer 1967):

> *A gas lamp that is placed too high up will project light further but will not give adequate light to the immediate area around it. Obviously, that was not our goal. The higher a lamp, the greater the unlighted area at its base. By reducing the height of street lamps and the distance between them, and decreasing the intensity of*

the flame in each lamp so as not to use more gas, we were able to light the city's streets better. Extremely bright lights are useless; they blind people more than they light their way (p. 312).

Street designers give a great deal of attention to special paving and to *paving patterns*. Special paving can cost a lot of money, even though it rarely makes a significant difference. Generally, one wonders how often the results of such efforts are worth the expense, particularly if there is not likely to be a commitment to regular maintenance and an understanding that replacements of the special paving will have to be purchased and stored.

Benches help people stay on the street; they invite our presence by permitting rest, conversation, waiting for a friend, passing the time. They help to make community. They are less expected on residential streets and less present there than on commercial streets.

The small fountains along the Cours Mirabeau, the special circular seating at corners along the Paseo do Gracia, the drinking fountain and the bird stands on the Ramblas, the statues along Monument Avenue—all of these contribute to their streets. They can be joyful, magic-making things to look at. Signs and awning canopies can be too. The best signs are artfully conceived and executed store logos, like the old umbrella announcement along the Ramblas, which is hard not to look at. Such signs are public art in the best sense. Overhead awnings do something else; they create intimate spaces along streets, shady when it is sunny, protected and comfortable when it is not.

Understandably, people become excited about design features on streets that are special, existing ones they have seen and that are forever remembered, or design ideas that they just know will make a particular street great. But it is best not to depend on them. Alone, great fountains, or gates, or paving, or lights, are not enough. Details are the special seasonings of a great street.

Places

Somewhere along the path of a fine street, particularly if it is long, there is likely to be a break. More than just intersections, breaks are small plazas or parks, widenings, or open spaces. They are most important on narrow streets and long streets and streets that bend and turn. On those streets particularly they provide stopping places, pauses, reference points along the path.

Accessibility

One cannot forget that a major purpose of streets is to enable one to get from one place to another, not only to a location on the street but to and from areas beyond it. What sets the great streets apart is that they take people along their ways, from one part of the city to another, whether on foot or in a vehicle, with grace and at a reasonable pace. It is more pleasant to move along with deliberation, taking in what is at hand for the eyes and imagination. Even where speed is possible there are buffers to protect those who move more slowly.

There is another kind of street accessibility to consider: people must be able to get to the street with ease. In part it is a matter of location, particularly for major city-making or community-making streets. Notably, it is not difficult—it is even easy—to find and get to the best streets, within the city as a whole or in their more local context. Besides being places one can walk to, great streets are generally accessible by public transit, whether crossing them or along them or

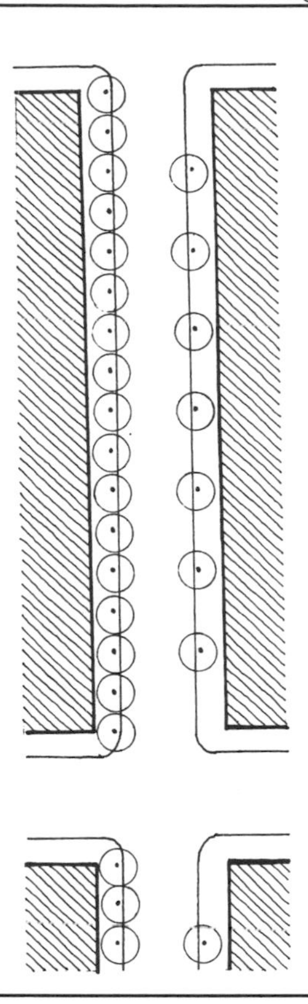

Fig. 14. Twenty-five-foot (7.6 m) versus fifty-foot (15 m) tree spacing, the former starting at the corners, the latter removed from the corners.

Fig. 15. Column that starts, or ends, the Ramblas in Barcelona

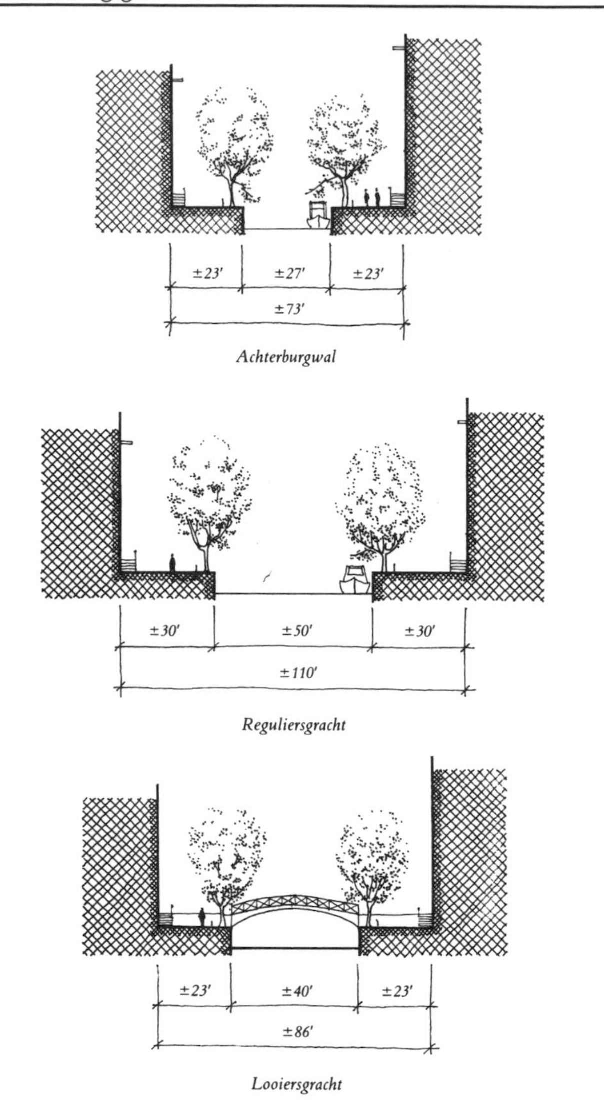

Fig. 16. Amsterdam's Voorburgwal.

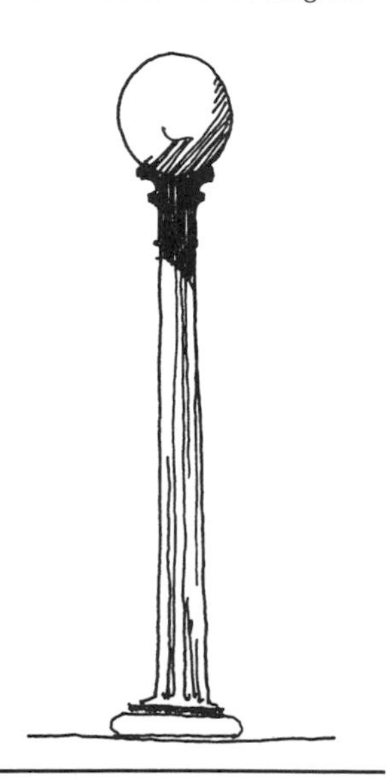

Fig. 17. Streetlight on Orange Grove Boulevard, Pasadena.

Fig. 18. Barcelona street light.

under them. Accessibility is also a matter of public access at places along the street, by intersecting or crossing streets or public ways. Streets with one entry for every 300 feet (90 meters) are easy to find, and some of the best streets approach that figure, but there are more entries on the busiest streets. (In these calculations an entry point is any public intersection, including those at the ends or beginnings of the streets.)

Still another type of access to be considered is for handicapped people. None of the identified great streets were designed with handicapped people in mind. And yet it is surprising that many of them accommodate wheelchairs with ease; the Via dei Giubbonari in Rome, the Strøget in Copenhagen, the Ramblas in Barcelona all do. Many of the best streets have places to rest. Handicapped people regularly use the boats along the Grand Canal. There is regularity about most of the best streets that one imagines to be relatively decipherable for the visually handicapped. Access for the handicapped, in a most elemental sense, is not too hard to achieve. Witness the ramps that have been added to so many streets around the world.

Density

Residential density and activities—the consequence of what urban planners call land uses—have been largely avoided thus far, despite admonitions that it is people and activities, more than what is physical and buildable, that make the best streets—an argument we accept. In physical design terms, however, one does not design or build density or land uses. One can design for a certain number of people for a given unit of land, or at a density of such and such, and those are physical relationships, but they are more a matter of urban policy made by urban policy makers, though designers may influence them. Similarly with activities and land uses, designers may influence buying and selling, or playing, or working, or interacting, or shopping, but they do not design them. The determination having been made that one space shall have a theatre, another sell cheese, and another be for people working at desks, a designer can design those spaces as well as spaces that can accommodate multiple uses; but the initial determinations are more matters of policy, often enacted into laws, that designer-builders will respond to, and not necessarily their basic task. Conversely, land use policy will certainly result in physical building, as it is intended to, but it is not physical in and of itself.

Whether or not they are directly designable and buildable, density and land use matters are important to streets. The best streets are wonderful places to be even if there aren't many or even any people on them. Void of human activity, streets soon cry out for people, they need people at the same time as they are for them. They are activated by people at the same time as they contribute to making a community for them. And that is achieved in considerable measure by having many people live along them or nearby—a matter of density.

Time and again in the 1980s and early 1990s, community groups responsible for planning or replenishing their central areas have called for streets, particularly main streets, to be what they call "24-hour" streets, areas populated and with a human presence all the time. It is density that achieves that objective. Elsewhere it has been observed that a minimum net residential density of fifteen dwelling units per acre can achieve active urban communities, and that fifty dwellings per net acre are possible without going above four stories or requiring overly wide streets (Jacobs and Appleyard 1987, p. 112–120). Mixed with other uses, particularly in central areas, the densities might be lower, though not on an individual building site. It is difficult for streets to help make community if there are not people to get to them easily.

On most great streets, there exist many different kinds of buildings designed for their uses—cinemas, theaters, or schools—or for earlier uses no longer present but adapted to present occupants—movie houses that became restaurants or stores—all of which add to interest and activity. Varieties, activity, liveliness of physical place are likely effects of diversity of uses.

Length

Great streets, it seems, come in all lengths. Why not? Yet at some point it can become difficult to sustain visual interest, diversity, eye-catching and thought-provoking images. Enough can become enough, or too much. If something special continues long enough it may no longer be special. The gentle turns and curves of the Grand Canal and the ever-changing buildings and light maintain interest, but even they have a limited power to attract; somewhere between the Rialto Bridge and the railway station one may be excused for wondering what is in his or her daily newspaper. Though we cannot specify just how long is too long, we can hypothesize that at some points along a long street changes are necessary if interest is to be sustained. They may consist in some special focal point like the statues along Monument Avenue, or a special building like the theater on the Ramblas, or a park like that along the Ringstrasse in Vienna, or perhaps a change in the street section.

Slope

More often than not, the best streets have noticeable changes in elevation, albeit none very steep. There would seem to be no reason why great streets could not have much greater elevation changes than those used as examples in this chapter. Certainly such streets must exist. The limit would be slopes so steep as to be difficult or uncomfortable for major population groups—the elderly, handicapped, mothers with young children. Otherwise, slope helps.

Parking

Automobile parking is a pervasive issue. Prepare a plan for an individual street or neighborhood, or for a central area, and parking is certain to be a major subject—a bone of contention—more time and energy consuming than housing. It has to do with accessibility. People with automobiles would like to park as close as possible to their destinations—directly in front is best. Merchants want them to. Parking standards and programs abound. So do guidelines and case studies for how best to landscape at-grade off-street parking lots, or where to locate and how to design and sometimes hide garages.

These standards can impact streets mightily and have done so, particularly in the United States. For example, if you put all of the parking in lots or structures behind the stores, then those stores will reorient to the rear, deadening the street. Large ground-level lots along a street leave gaps in street definition and activity. Garages on the street have a hard time fitting in, but can be made to do so, although they may be too expensive on a really busy street.

On-street auto parking is permitted and provided for along many of the best streets, far more often than not, but almost certainly in amounts that are far below demand or what any contemporary standard would require. At best, drivers seem to desire, at the very

Fig. 19. Monument Avenue, Richmond.

Fig. 20. Along Stroget in Copenhagen.

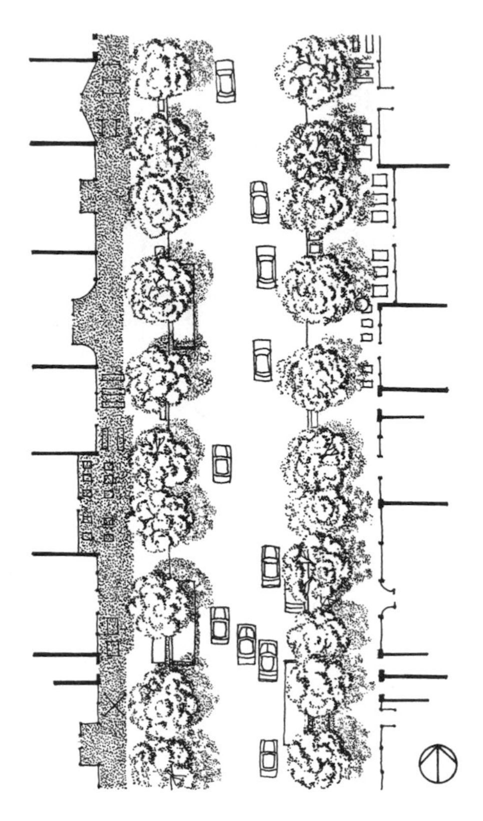

Approximate scale: 1" = 50' or 1:600

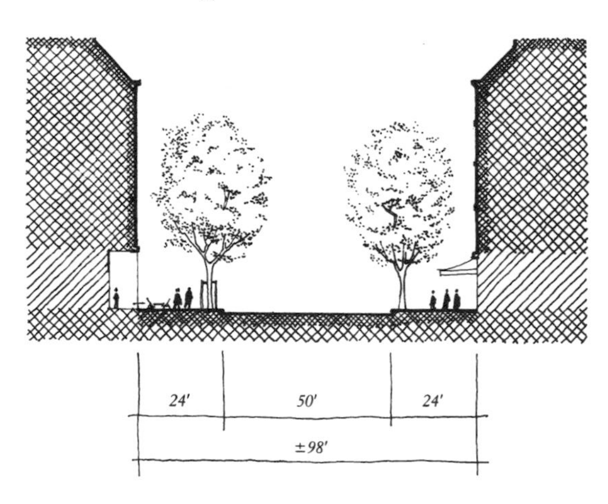

Fig. 21. Boulevard Saint-Michel: plan and section.

least, a long shot at finding a space in the block they are destined for; they take the chance, usually lose, then look elsewhere nearby for a place to park. That may be enough: a chance. The boulevard access streets are among the best in terms of handling cars, not in the numbers of cars accommodated. Driveways off of the best streets, or garage entrances for access to parking or for service, are rare, even on a fine residential street. Though present on more streets than not, auto parking in great amounts, to any contemporary standard, is not a characteristic of great streets. They seem to do well without "enough."

Contrast

Contrast in design is what sets one street apart from another, and ultimately what makes one great and another less so. Contrast in shape or length or size, or to the pattern of surrounding streets, is another matter. For many streets one or another of these qualities sets them apart from other streets; the Champs-Elysees is wider and longer than any other street in Paris; Regent Street in London is different in shape and regularity from the streets of Soho or Mayfair; the Ramblas stands visibly apart from the narrow, short streets to either side; Roslyn Place is shorter and narrower than most streets in Shadyside or in Pittsburgh as a whole; the five streets of Bologna that focus on the two towers at Piazza di Porta Ravegnana form in themselves a distinct pattern that sets them apart from others. For some of these streets it might not be too difficult to pick out the most special one or two from an otherwise unmarked map. But that is not always the case. Strøget or the Avenue Montaigne are not all that different in size or shape from other nearby streets.

In the end, shape or size or regularity within an urban physical context may set one street apart from others, may make it more noticeable, may give it a head start toward being special, but that is not likely to be enough and may not be a critical factor in determining a great street. It is the design of the street itself that makes the difference.

Time

Between the extremes of age and evolution, there are many variations of old or new streets and rapidly or slowly developed ones. If one criterion for being outstanding is that a street stand a test of time, be long continuing, then the likelihood is that the examples for study will be older rather than newer.

Some say that time is needed to make a great street, presumably to achieve diversity and change and a sense of history that comes with years. This patina of age—very different from grime—relates to the buildings more than the public way itself. But time is not necessarily required for a great street. There are many fine streets along which the buildings were developed in a relatively short time span, many along the French streets of the nineteenth century, the Boulevard Saint-Michel of Paris and Cours Mirabeau in Aix-en-Provence among them. On others, like Monument Avenue in Richmond, VA or the Paseo de Gracia in Barcelona the buildings took longer, about 100 years, and on Strøget the time period is longer still.

To the extent that incremental building and change do bring the diversity and sense of history that can give body and substance to a street, it may be argued that smaller rather than larger building parcels help. Diversity is likely to be greater initially as well as over time, as building decisions can be made incrementally.

If it is history and age that we want to see on a street, then there is nothing like time to gain them, remembering that many less than lovely streets have also been built over time. For the futurists and the impatient, however, it is encouraging to know that great streets can be built now. ■

REFERENCES

Appleyard, Donald. 1981. *Livable Streets*, Berkeley: University of California Press.

Arens, Edward, and Peter Bosselmann. 1989, "Wind, Sun and Temperature: Predicting the Thermal Comfort of People in Outdoor Spaces," *Building and Environment 24*, no. 4 (pp. 315-320).

Berman, Marshall. 1982. *All That Is Solid Melts Into Air*, New York: Viking Penguin.

Bosselmann, Peter *et. al.*1984, *Sun, Wind and Comfort*, Berkeley: Institute of Urban and Regional Development, University of California, Berkeley.

Blumenfeld, Hans. 1967. *The Modern Metropolis: Its Origins, Growth, Characteristics, and Planning,* Paul Spreiregan, ed. Cambridge: MIT Press.

Eubank-Ahrens, Brenda. 1987. "A Closer Look at the Users of Woonerven" *Public Streets for Public Use,* Anne Vernez Moudon, ed. New York: Van Nostrand Reinhold.

Homberger, W., *et al.* 1989. *Residential Street Design and Traffic Control,* Englewood Cliffs: Prentice Hall.

Jacobs, Allan and Donald Appleyard.1987. "Toward a New Urban Design Manifesto," *Journal of the American Planning Association* 53, no.1 (Winter), pp. 112-120.

Loyer, Francois 1988. (trans. Charles L. Clark). *Paris Nineteenth Century: Architecture and Urbanism*, New York: Abbeville Press.

Fig. 22. Princes Street, Edinburgh.

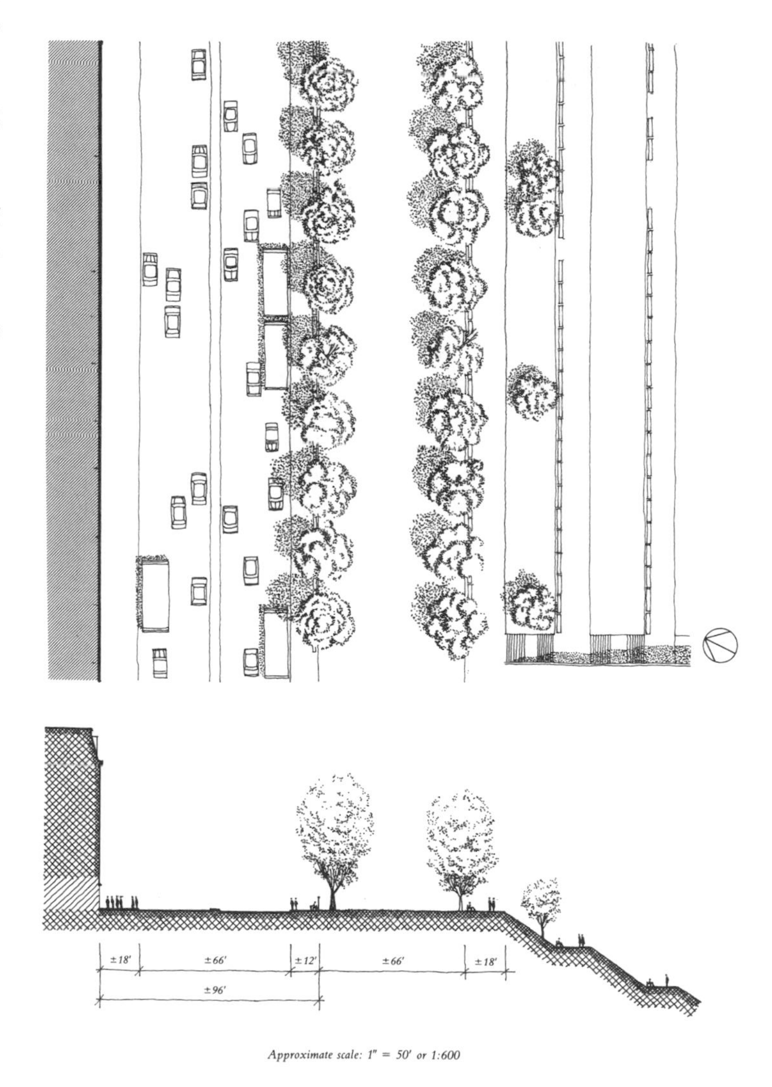

Fig. 23. Princes Street: plan and section.

Fig. 24. Along the Ramblas.

Fig. 25. Along the Ringstrausse.

Fig. 26. Strøget at Hojbro Plads: plan and section.

Elizabeth Macdonald

Summary

Multiway boulevards are complex urban streets that allow slow-moving local vehicular and pedestrian traffic to coexist alongside fast-moving through traffic within the same right-of-way. The design of the boulevard accomplishes this by providing specialized roadways separated by tree-lined medians. This article discusses the history of this street type and its usefulness in contemporary cities. It concludes with design guidelines.

Key words

arterial, boulevard, multimodal street, pedestrian realm, street design, urban trees

Ocean Parkway in Brooklyn, NY

The multiway boulevard

Multiway boulevards are complex urban streets that allow the close coexistence of different travel modes and travel speeds within the same right-of-way. The basic cross section consists of a central roadway, generally at least four lanes wide, for fast-moving through traffic, and narrow access roadways along either side, for slow-moving local traffic. The roadways are separated by tree-lined medians that often serve as pedestrian promenades. The configuration of parallel roadways serving distinctly different traffic functions is unique to multiway boulevards. The street form addresses the problems caused by the need for both through movement and local access to abutting property on urban streets. It can handle large amounts of through traffic while still allowing—and, if well designed, encouraging—active pedestrian activity along its edges.

Fig. 1. The *Grands Boulevards.*

1 A BRIEF HISTORY OF MULTIWAY BOULEVARDS

Early boulevards

The earliest boulevards were the tree-lined promenades built on the raised earthen ramparts of European cities after the defensive function of town walls became obsolete in the seventeenth century. The most significant were those built on the Paris ramparts beginning in 1670, which consisted of separate promenades on eleven rampart segments. Known as the Grands Boulevards, they were eventually connected to form an elevated route around the northern part of the city. The cross section of the boulevards varied, but they were generally between 100 to 125 feet (30 to 38 meters) wide—very wide in comparison with the average Parisian street at the time, only 24 feet (7.3 meters)—and lined with two, four, or five rows of trees. Their use was restricted: the center was for moving carriages and riders, while the side paths under the trees were for pedestrians. As the city grew to encompass the boulevards, uses built along them were entertainment-oriented. In the early nineteenth-century the boulevards were regraded lower and integrated into the general city street system (Fig. 1).

Nineteenth century boulevard systems

The mid- to late-19th century witnessed a second period of boulevard building. It began with Haussmann's reconstruction of Paris in the 1850s and continued in major cities in Europe and the United

Credits: This article is abstracted from Jacobs, Allan, Elizabeth Macdonald, and Yodan Rofé. 2001. *The Boulevard Book: History, Evolution, Design of Multiway Boulevards*, MIT Press, Cambridge, MA. It is reproduced by permission of the publisher. Illustrations are by the authors.

States into the early 1900s, and in Asia, India, and Central and South America into the 1940s (Fig. 2).

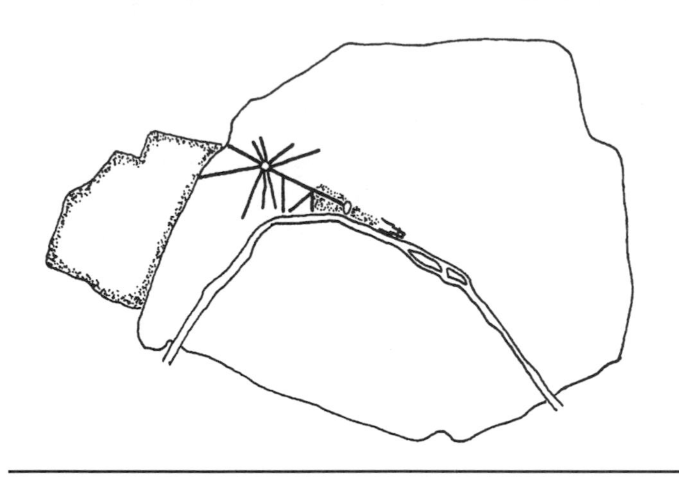

Fig. 2. Haussmann's multiway boulevards near the Etoile.

Nineteenth-century boulevards were generally built as part of large-scale city planning efforts and were designed to have many of the characteristics of the earlier rampart boulevards: generous width, lines of trees, separated pathways for carriages and pedestrians, and an association with leisure and entertainment. Typically their purpose was to open up areas adjacent to the city for new urban development. The physical form of the boulevard was also refined and modernized. Haussmann added paved roadway surfaces and raised curbs, and developed three distinct types: one with a central median, another with widened sidewalks, and the multiway boulevard (Fig. 3).

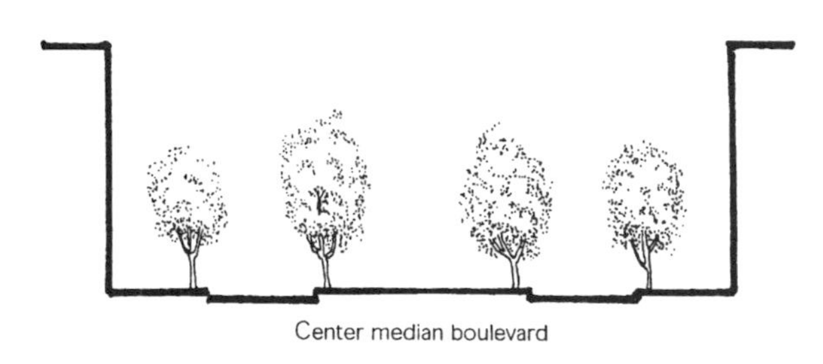

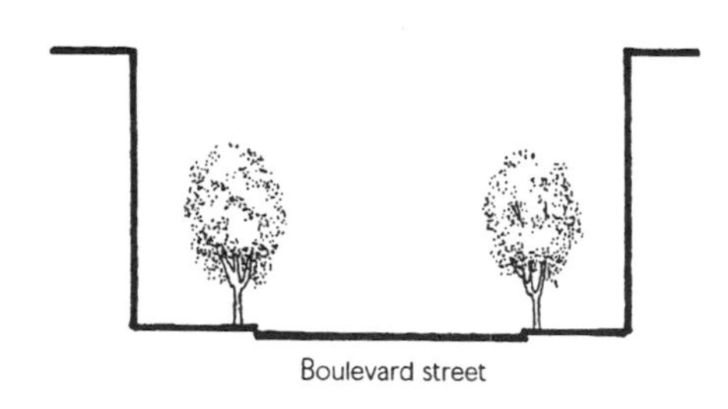

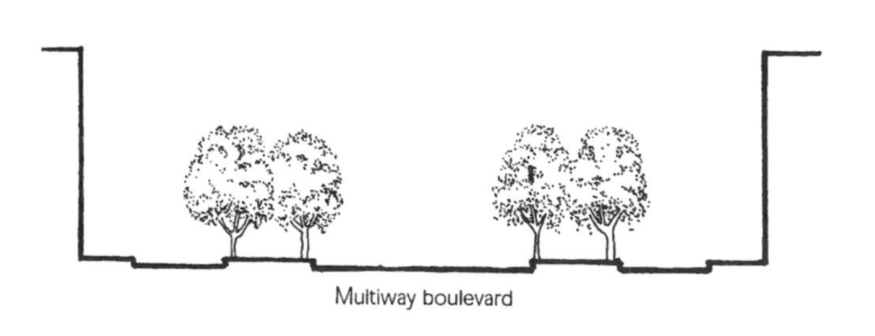

Fig. 3. Three types of boulevards.

Notable nineteenth-century city expansions structured with multiway boulevards include Haussmann's plan for the northwestern periphery of Paris, around the *Place d'Etoile*, Ildefons Cerdà's plan for Barcelona, and Frederick Law Olmsted and Calvert Vaux's plan for Brooklyn (Fig. 4). (Olmsted and Vaux called their boulevards "parkways.")

Twentieth century demise

After the turn of the century, few multiway boulevards were built in the United States. They fell out of favor with traffic engineers grappling with ways to rationalize automobile traffic and speed its flow. American engineers developed new kinds of street and highway systems designed to separate fast-moving through traffic from slow local traffic and pedestrians. Multiway boulevards, designed to handle all types of traffic, were deemed old-fashioned, inadequate, and unsafe, especially because of the complex intersections that resulted from the multiple roadways. Many existing ones were reconfigured.

In the 1930s, traffic engineers developed a system of classifying streets called Functional Classification, which is still in use. Streets are separated into types according to the vehicle movement and property access functions they are supposed to perform. Since movement and access functions are inversely correlated, the possibility of streets like multiway boulevards, which have both characteristics, was effectively excluded.

2 BOULEVARDS REBORN

The importance of the pedestrian realm

Research involving observation and analysis of existing streets suggests that multiway boulevards are in fact not less safe than normally configured streets carrying similar amounts of traffic—if they are well designed. A key factor that distinguishes safe multiway boulevards from unsafe ones is the presence or lack of an extended "pedestrian realm" along the edges of the boulevard (Fig. 5). This realm—which stretches to the outside edge of the median, including the sidewalk, the access roadway, and the planted median—is not reserved solely for pedestrian use. The access roadways are shared with vehicles, but their configuration encourages drivers to go very slowly (Figs. 6 and 7).

Necessary and conducive conditions for emergence of an extended pedestrian realm are:

- Uninterrupted median strips between the through lanes and the access lanes.

- A strong line of densely planted trees along the medians, continuing all the way to intersections.
- Relatively narrow access roadways that allow only one lane of traffic and are controlled by stop signs at every intersection.
- Placement on the medians of transit stops, kiosks, or benches—to encourage people to cross from the sidewalk to use them.
- Access ways that are further distinguished from the central through-way realm by a slight change of level and/or a different paving material.

Existing boulevards

Multiway boulevards exist the world over in different land use and socioeconomic contexts and have surprising physical variety—large and small, long and short, curvilinear and straight. While they tend to be wide and elegant and occupy prominent locations within the city structure, some particularly delightful ones are more modest.

Most Parisian multiway boulevards, such as the *Avenue Montaigne* and *Avenue Marceau*, are 120 to 135 feet (36 to 41 meters) wide, while several are much wider, like the *Avenue de la Grand Armée,* which is 230 feet (70 meters) wide. Barcelona boulevards range from the 165 foot (50 meters) wide *Avinguda Diagonal* to the 200 foot (61 meters) wide *Passeig de Gracia*. Eastern and Ocean Parkways in Brooklyn are 210 feet (64 meters) wide and in addition have 30-foot (9 meters) front yard setbacks on both sides. *Pasteur Boulevard,* in Ho Chi Minh City, at approximately 62 feet (19 meters), is perhaps the narrowest boulevard. Its very narrow access roads are dedicated to bicycle travel. At the other extreme is the *Avenida 9 de Julio*, in Buenos Aires, which has a right-of-way of 450 feet (137 meters) that includes 16 traffic lanes in the center roadway. It is more of an expressway than a boulevard (Figs. 8–10).

The wider boulevards—over 150 feet (46 meters) wide—generally have wide medians that often contain two lines of trees, pedestrian paths, benches, pedestrian-scaled light fixtures and sometimes fountains and special planting. The medians on the *Passeig de Gracia* contain particularly delightful tile benches and ornate lamps designed by Antonio Gaudi. The medians on Ocean Parkway contain an almost unbroken line of wood-slat benches facing toward the center roadway. Narrower boulevards generally have narrow medians—five to ten feet wide (1.5 to 3 meters) is common—containing a single line of trees.

Fig. 4. Ocean Parkway, Brooklyn, at the turn of the 20th Century (drawn from a photograph, New York City Photo Archive).

Fig. 5. The pedestrian realm on *Via Nomentana*, Rome.

3 DESIGN GUIDELINES

The success of multiway boulevards hinges on two words: "appropriate design." Do them well and multiway boulevards can be great streets. Do them poorly and, like other ill-designed streets, they will cause problems.

The unique feature of the multiway boulevard is its ability to accommodate many uses in a balanced way, not allowing any one use or mode of travel to dominate. It is therefore important to design them as integrated wholes. One must beware of reducing them to a series of issues for which specific guideline dimensions and solutions are given; the vision of a whole street, and the ways in which its different aspects interact, will be lost.

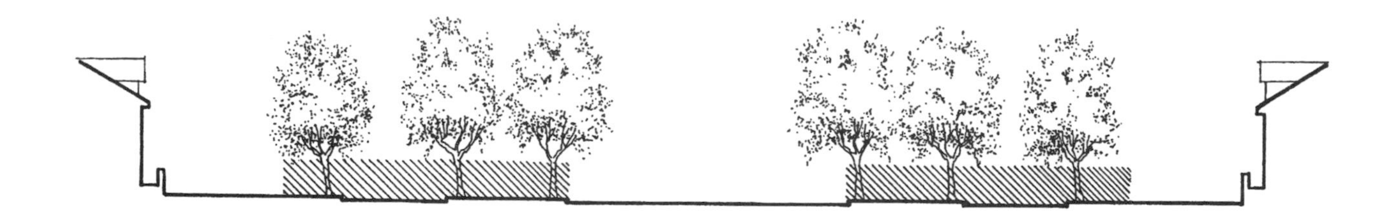

Fig. 6. Extended pedestrian realm on Ocean Parkway.

Fig. 7. Access roadway on *Via IV Novembre*, Ferrara.

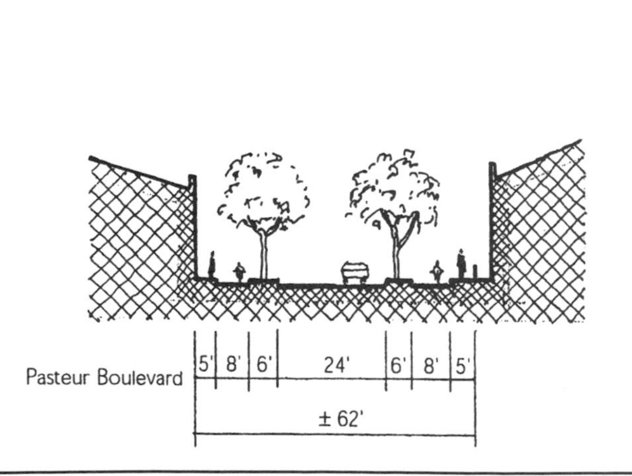

Fig. 8. *Pasteur Boulevard*, Ho Chi Minh City.

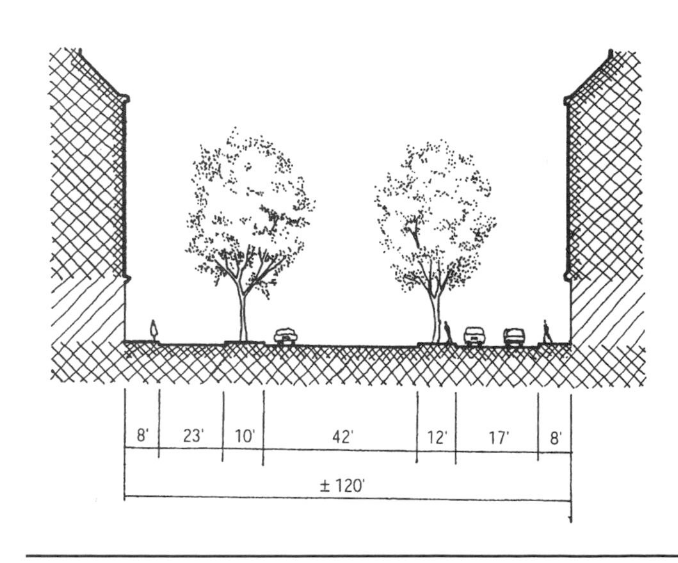

Fig. 9. *Boulevard de Courcelles*, Paris.

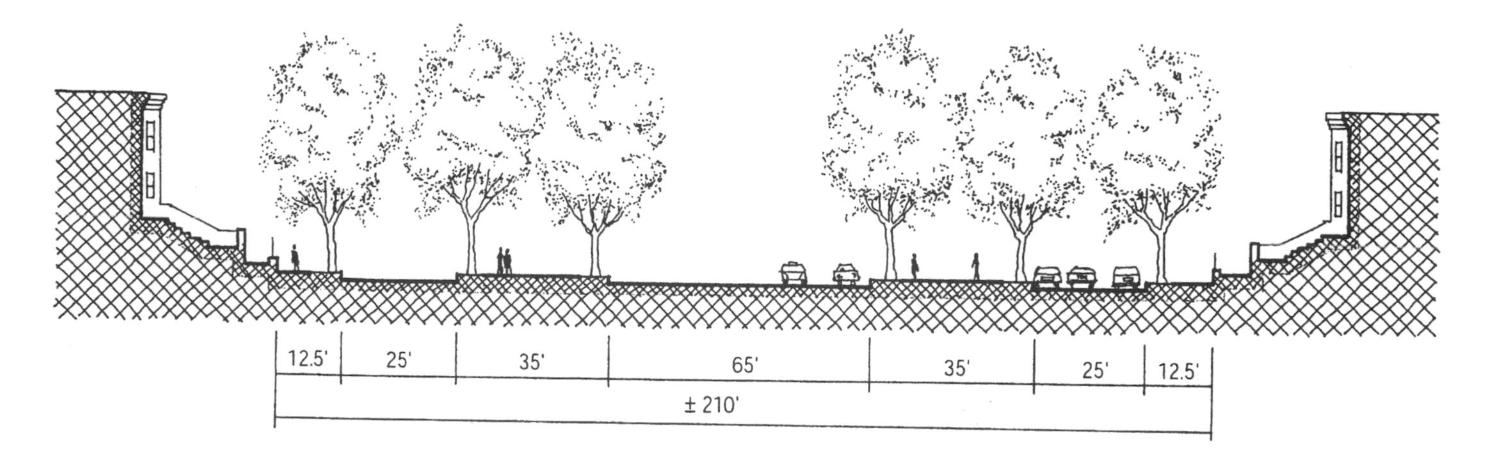

Fig. 10. Eastern Parkway, Brooklyn.

The following are a series of qualities that are crucial to the construction of a good boulevard. But as important as each of them is, the connections and the relationships among the qualities are even more important. The solution to any particular problem, the achievement of any one quality in the design of a boulevard, has to amplify and reinforce earlier decisions, and, in turn, be respected and strengthened by subsequent decisions.

Location, context, and uses

- Boulevards exist, and have a potential to exist, in very different contexts. They work well as residential, commercial, or mixed-use urban streets.
- Boulevards are appropriate where there is a need to carry both through traffic and local traffic, and where there is real or potential conflict between the two traffic types.
- They are appropriate for streets that, by virtue of their size and/or location, can become significant elements in the city (Fig. 11).
- They are appropriate where there is either a significant volume of pedestrians who need to cross the street or a potential desire to do so. Commercial streets, streets with high residential density, streets that incorporate public transit, or streets with a significant presence of public institutions, are examples of such streets.

Buildings that face the street

- Boulevards only make sense where adjacent buildings face the street (Fig. 12).
- Buildings along boulevards should have direct pedestrian access from sidewalks.
- Where a boulevard borders a public park or a major institution like a museum or civic center, then it can be one-sided, with an access roadway street only on the side with ordinary buildings and a wider sidewalk on the other side.

Fig.11. Eastern Parkway in winter.

Fig. 12. *Boulevard Courcelles,* Paris.

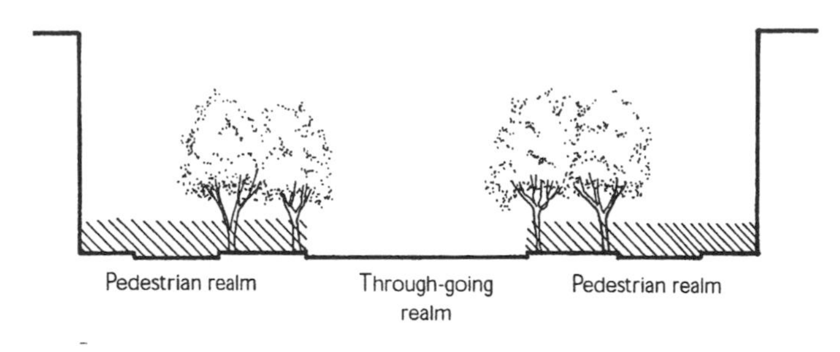

Fig. 13. Boulevard realms.

Boulevard realms and overall size

- Multiway boulevards are made up of two realms: a through-going realm and a pedestrian realm. The central roadway is devoted to relatively fast through-going traffic and may include a median separating the opposing traffic streams. On the outer sides of this roadway, and separating it from the abutting buildings, are pedestrian realms, which include the sidewalk, a narrow access roadway, and a continuous tree-lined median.

- A right-of way of 100 feet (30.48 meters) is the feasible minimum for a boulevard, assuming four lanes of travel in the central roadway. Narrower dimensions are possible if less through traffic is provided for or if the side roads serve bicycles instead of autos.

- Right-of-way dimensions of between 125 to 230 feet (38 to 70 meters) allow for more flexibility in the design.

- On the best boulevards, the pedestrian realm is never less than fifty percent of the total width of the right-of-way, and often approaches seventy percent (Fig. 13).

Fig. 14. Central Realm on Ocean Parkway.

The through-going central realm

- The boulevard's function as a carrier of relatively rapid and non-local traffic is just as important as its provision of local and pedestrian access. Multiway boulevards can form connections over the city and allow for easier and calmer through travel than on other city streets with direct access to abutting property, because the center lanes are subject to less interference from parking movements and from delivery vehicles (Fig. 14).

- Three lanes in each direction allow more flexibility in traffic arrangements and for the possibility of devoting an entire lane to public transit.

- The overall width of the center realm should be determined by balancing considerations of available right-of-way, traffic capacity desired, and need for safe crossings by pedestrians. Widths of more than about 80 feet (25 meters) can be difficult for pedestrians to cross and should be avoided.

- It is advisable to provide a refuge for pedestrians in the center of the boulevard between directional traffic flows.

The pedestrian realm

- An extended pedestrian realm is absolutely necessary to creating a safe boulevard. It provides space for parking, slow vehicles, and pedestrian movement; it makes street crossings shorter and easier; it provides the city and local inhabitants with an open space amenity; and it buffers abutting properties form the pollution, noise, and psychological impact of heavy traffic.

- The edge of the pedestrian realm should be defined strongly by a continuous median, planted with at least one uninterrupted, closely spaced line of trees.

- It is critical to allow parking on the access road. The friction caused by cars moving in and out of parking spaces discourages drivers in search of speed from moving into the pedestrian realm. Parking also gives access to buildings on the street (Fig. 15) .

- It is important to have only one travel lane in the pedestrian realm.

- Access to the pedestrian realm by vehicles is best achieved at intersections. Breaks in the median to allow vehicle access at mid-block locations may create more conflict points with through traffic and disturb the continuity of pedestrian use of the median (Fig. 15).

Continuous tree-lined medians

- On boulevards, continuous medians bound the center roadway and the pedestrian realms, separating them and joining them together at the same time. To a great extent their design determines the form and character of the boulevard.

- Medians may range from 5 to 50 feet (1.5 to 15 meters) wide.

- The most important element in the median, its defining characteristic, is the line or lines of trees: closely spaced, uninterrupted, and reaching all the way to the intersection.

- Medians should contain amenities to encourage pedestrian use. (Fig. 12)

- If there are buses or streetcars on the boulevard, stops should be located on the medians, as should subway access points. (Fig. 16)

Rows of trees and tree spacing

- Trees mark the boundary between the central, fast-moving realm and the slower pedestrian realms. It is extremely difficult to create a strongly defined, extended pedestrian realm with a treeless median. Trees also break down the visual scale of wide rights-of-way, and create a pleasant environment for pedestrians and drivers alike. (Fig. 17)

- Trees should have a maximum spacing of 35 feet (10.5 meters)—25 feet (7.5 meters) is better. A minimum spacing as low as 12 feet (3.5 meters) apart is possible.

- Deciduous trees are preferable. They provide shade in the summer yet allow sun into the street in winter.

- To maintain the integrity of the street as one whole, trees with dense foliage below eye level should not be used.

- Trees can be arranged in various ways: in medians 5 to 10 feet (1.5 to 3 meters) wide they are best planted in the center; in medians 10 to 25 feet (3 to 7.5 meters) wide, one row of trees may be best planted closer to the center roadway, so that most of the median's width is protected from fast traffic, while two rows of trees may be planted in a staggered pattern; on medians wider than twenty feet, two rows of trees make the most sense.

- Trees should be planted in a regular pattern to enhance the linearity and wholeness of the boulevard.

Fig. 15. Pedestrian realm along *Boulevard Courcelles*.

Fig. 16. Ocean Parkway, Brooklyn.

Fig. 17. *Avenue Grand Armée,* Paris.

Public transport

- Multiway boulevards are a natural location for public transportation.
- If buses are used, they should travel in the central roadway where curb lanes may become designated bus lanes(Fig. 18).
- If light-rail is incorporated into the street, it can run on the curb lane of the center roadway, or on the median, toward the center edge.
- Buses or light-rail can also run in dedicated lanes in the center of the street.

Parking

- Parking should be provided on the access roads. It slows traffic, increases the number of pedestrians moving along a street, and encourages street-oriented development. Moreover, parked vehicles act as a physical barrier between pedestrians and moving cars and thus provide a sense of safety. Parking should not dominate the pedestrian realm but should be balanced with other uses. Parking along the outer edge of the median, within the center roadway, should be avoided (Fig. 19).
- Access ways can include one or two rows of parallel parking.
- Parking lanes should be narrow. A width of 6 or 7 feet (1.8 or 2.1 meters) is possible and sufficient; 8 or 9 feet (2.4 or 2.7 meters) is the maximum.
- Angled parking can be incorporated into wide medians.
- Where an access way has two parking lanes, widening either the median or the sidewalk at intersections may help pedestrians. The presence of such a bulb or "neck" makes it easier for pedestrians to cross the access lane, and it slows cars that are entering or leaving it.
- **Parking may also be provided in underground parking garages beneath the central roadway, with entries and exits for cars from the access road and from the medians for pedestrians. These access points should not disrupt the pedestrian character of these spaces.**

Lane widths

- The ability to implement boulevards in limited space—between 100 and 140 feet (30.5 to 43 meters) depends on accepting narrow lanes. Narrow lanes work well in boulevards and may increase pedestrian safety by making crossing the street easier and slowing cars on the access way. A concern, of course, is access for emergency vehicles, particularly fire trucks, and for garbage and recycling trucks. This issue comes up in relation to narrow access ways, but is not acute. The width of most access ways is not less than that of many normal residential streets and medians may be designed with mountable curbs. Fire trucks can also operate from the center roadway, across a narrow median and access roadway.
- Lane widths on the access roadway and center roadway should be governed by different criteria.

Fig. 18. The Esplanade, Chico.

Fig. 19. Along the access roadway on The Esplanade.

Fig. 20. The Esplanade, Chico.

Fig. 21. Ocean Parkway median.

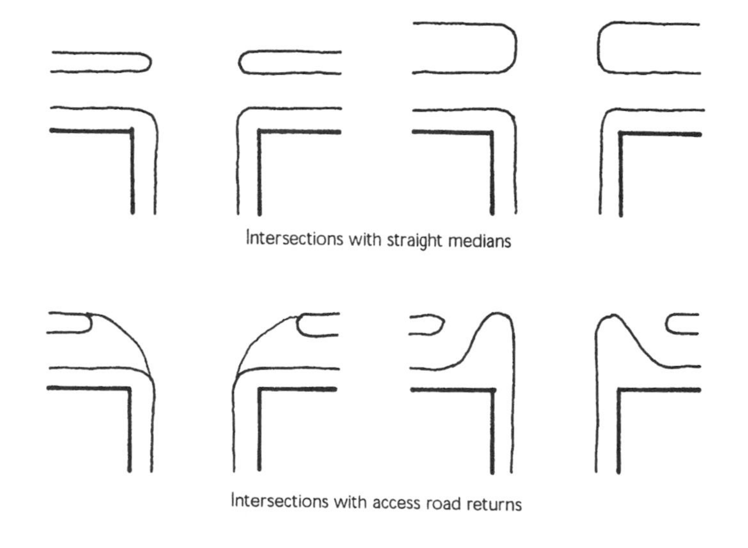

Intersections with straight medians
Intersections with access road returns

Fig. 22. Intersection configurations.

- Travel lanes on the access roads can be as narrow as seven feet and as wide as eleven feet. Where the access road is configured with two parking lanes its overall width should not exceed twenty-four feet.
- In the center roadway, the travel lane at the median curb should be a minimum of 9 feet and a maximum of thirteen feet wide.
- Inside lanes of the central roadway can range from 9.5 to 12 feet (2.9 to 3.7 meters) wide.

Bicycles

- Boulevards easily accommodate bicycle travel, either for local movement or for commuting and sport. These two types of movement have different characteristics and may be accommodated in different ways.
- Local bicycle traffic can easily be incorporated on the access roadway within the pedestrian realm (Fig. 20).
- Designated bicycle lanes for faster-moving commuter bicycles can be incorporated into a wide median on a designated path, or as a narrow first lane in the center roadway, next to the median (Fig. 21).

Intersection design

- Boulevard intersections should be designed to help people negotiate potential conflicts in consonance with an easily understood set of priorities, and with pedestrian safety in mind.
- All turning and weaving movements can be allowed at boulevard intersections unless there is a compelling reason to prohibit them. The presumption should be to allow them.
- Priority is given first to center through traffic, then to crossing traffic, then to movement on the access road.
- Turning radii and the configuration of medians are determined primarily to allow pedestrians easier crossing of intersections. Ease of turning for cars and large vehicles are secondary considerations.
- Different physical configurations are possible for intersections.
- The most straightforward intersection arrangement, which works for both narrow and wide medians, has straight medians that extend more or less the same distance into the intersection as the sidewalk edge.
- Alternatively, access roads may be designed with entries to and exits from the central roadway just before and just after intersections. This simplifies intersections but can make merging into the center difficult. It also forces local traffic to enter through-going traffic at every block. Moreover, it shortens the medians near every intersection, interrupting the continuity of the tree lines (Fig. 22).

Traffic controls

- Traffic controls should enhance the ability for both drivers and pedestrians to achieve their aims. They should be designed to acknowledge potential conflicts but provide clear and safe ways of negotiating them.

- Through traffic on the center roadway is given first priority, then traffic on cross-streets, and lastly, traffic on the access ways.

- The central lanes should be either uncontrolled at intersections, or controlled with a traffic light, with cross streets controlled by stop signs or lights accordingly.

- The access lanes should be controlled with a stop sign at every intersection (Fig. 23).

- On boulevards with narrow medians, the stop sign or signal controlling the cross street may be at the sidewalk edge or on the median. When it is on the sidewalk, the intersection remains clear of waiting cars and drivers on the access road can continue across it. When, however, the cross-street control is on the median, through travel on the access road may sometimes be partially blocked by waiting cars. This is not necessarily a problem, as drivers are likely to leave a gap (Fig. 24).

- On wide median boulevards with signaled intersections, the cross-street signal is usually placed at the sidewalk edge. Access road drivers wishing to enter or cross the center can then move into a waiting position in the protected space provided by the wide median.

- On wide median boulevards where the cross-street traffic is controlled by a stop sign, it is placed at the outer edge of the median. Access road drivers wishing to turn left under these controls must make two stops, along the access road and then at the median.

- Placing the stop sign or traffic light at the sidewalk edge on wide-median boulevards allows for generous pedestrian crosswalks in line with the medians, emphasizing the continuity of the median and encouraging people to walk or cycle along them (Figs. 23, 24).

Fig. 23. Intersection controls at The Esplanade.

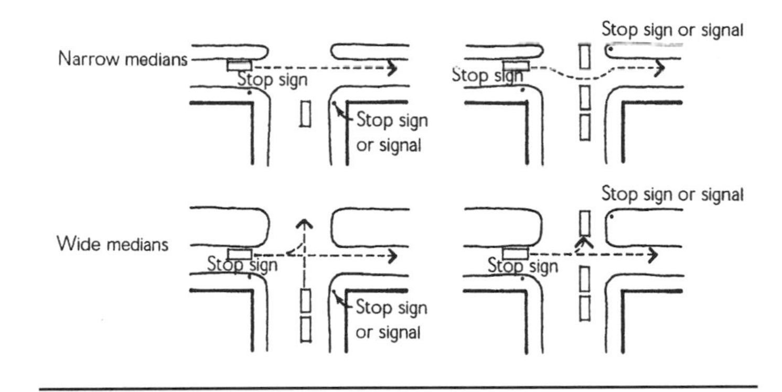

Fig. 24. Traffic control diagram.

Discouraging mid-block jaywalking

- Long blocks can create an unsafe situation in which jaywalking is frequent. It is possible to discourage jaywalking without defacing the street with barriers. Continuous benches or dense planting along the medians can form a barrier to jaywalking while enhancing the protection of the pedestrian realm (Fig. 25).

- Run benches or planters without interruption between intersections on the side of the median closer to the central roadway. Benches should face inward, toward the access road.

- Plants must be tall enough and dense enough to discourage walking through them (Fig. 25).

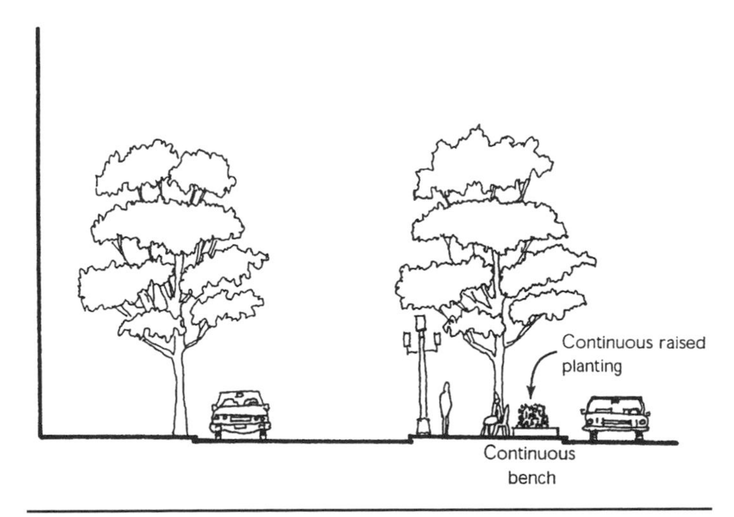

Fig. 25. Median barriers to jay walking.

Fig. 26. Raised access way on *Avenue George V*, Paris.

Differentiating the roadways

- Design details that differentiate the roadways and require drivers to moderate their speed on the access lanes can reinforce the boundary between the through-going realm and the pedestrian realm.
- A slight rise (about one inch) at the entrance to the access way makes cars slow down on entering or exiting the pedestrian realm. A raised crosswalk surface can have the same effect (Fig. 26).
- The access way can be differentiated by its paving. If the material looks similar to that used on the sidewalks and medians, it clearly suggests that all three are part of a single realm. ■

REFERENCES

Bosselmann, P., E. Macdonald. 1999. "Livable Streets Revisited." *Journal of the American Planning Association* 65:2, pp. 168–180.

Jacobs, A. B., Y. Rofé, E. Macdonald. 1995. "Another Look at Boulevards." *Places, A Forum of Environmental Design* 10:1, pp. 72–77.

Jacobs, A. B., Y. Rofé, E. Macdonald. 1995. *Multiple Roadway Boulevards: Case Studies, Designs, Design Guidelines.* (Working Paper #652). Berkeley, CA: University of California at Berkeley Institute of Urban and Regional Development.

Jacobs, A. B., Y. Rofé, E. Macdonald. 1994. *Boulevards: A Study of Safety, Behavior, and Usefulness.* Working Paper #625. Berkeley, CA: University of California at Berkeley Institute of Urban and Regional Development.

John J. Fruin

Summary
This article describes the documentation and analysis of pedestrian circulation as the basis of standards for pedestrian planning and design. The objectives of pedestrian improvement programs and analysis techniques are reviewed. The "Level of Service" concept is described by which to assess the convenience and safety of pedestrian areas, including walkways, stairs and queues.

Key words
data collection, density contours, design standards, entrances, pedestrians, perception, queuing, safety, stairs, traffic volume, universal design, walkways

Pedestrian Bridge, Boston Common

Planning and design for pedestrians

1 OBJECTIVES OF PEDESTRIAN PLANNING

The primary objectives of an improvement program for pedestrians are safety, security, convenience, continuity, comfort, system coherence and attractiveness.

• **Pedestrian safety:** The first means to improve pedestrian safety is reduction or elimination of pedestrian vehicle conflicts by space separation, either horizontal or vertical, or by time separation. Horizontal separation can be accomplished by a pedestrian precinct, street or mall, where vehicles are restricted or eliminated.

Vertical separation is attained through pedestrian underpasses or overpasses. There have been discouraging results where these were located off the normal and visible route of pedestrian trip desire lines, or where their use causes increased time and energy expenditures to climb stairs, or even the perception of increased effort. Pedestrian curbs and grade separations cannot be considered as sufficient safety improvements in and of themselves, but must be incorporated into a larger pedestrian network.

Engineering for pedestrian safety involves the provision of physical improvements to reduce pedestrian accident exposure. This would include standardization of signs and signals, distinctive crosswalk delineation particularly aimed at driver recognition of crosswalk zones, improving motorist lines of sight, upgraded street lighting, and any other physical features that contribute to pedestrian safety.

• **Pedestrian security:** Building and street configurations should be arranged to enhance clear observation by other pedestrians and policing personnel. High lighting levels, unobstructed lines of sight, and avoidance of building or landscaping configurations that provide concealment will assist in this objective. Television surveillance is increasingly used in buildings, transit stations, and major street locations.

Street lighting is an important aspect of pedestrian security, safety and favorable perception of the urban image. In addition to its safety and security value, lighting can be used to improve the pedestrian's perception of an urban space. Building facades, fountains and other attractive natural or architectural features may be floodlit to improve their appearance at night. Colored lighting can provide added visual variety.

• **Pedestrian convenience:** Sidewalk obstructions, such as mailboxes, telephone booths, newsstands, refuse cans, and planters, can be relocated to improve pedestrian flow at practically no cost. Other more stationary items, such as traffic light standards, fire hydrants, and fire alarm boxes, could be moved under normal replacement schedules. Ramped curb cuts provide convenience for handicapped pedestrians in wheelchairs, persons wheeling baby carriages, and for others who have difficulties with high curbs. Tactile trails for the blind can be installed in and around buildings and urban places that are frequented by the visually impaired.

Credits: This article is updated by the author from *Pedestrian Planning and Design,* originally written by the author with the support of The Port of New York Authority and the Howard Cullman Fellowship and is reprinted by permission of Elevator World (see References). Illustrations are by John Petrik Watson.

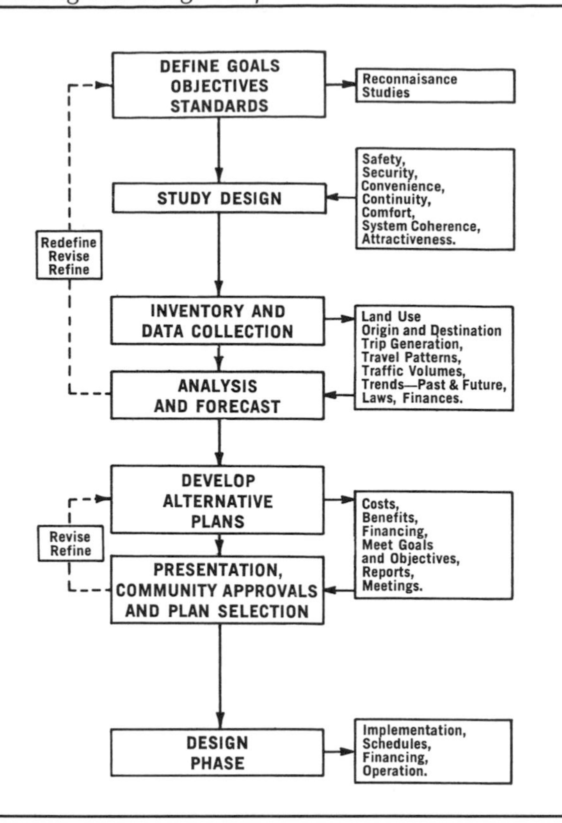

Fig. 1. Design process: analysis and design of a pedestrian improvement program.

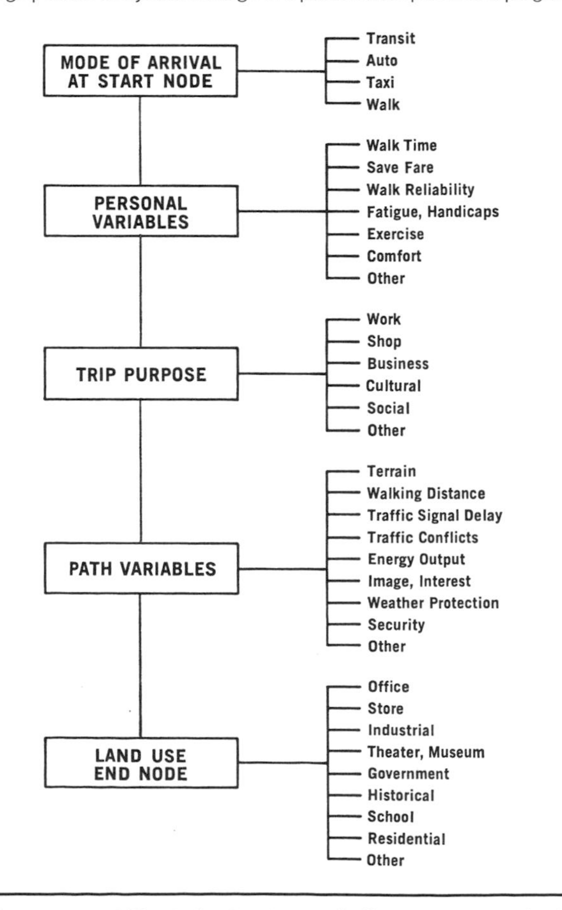

Fig. 2. Pedestrian trip variables. Pedestrian trips are also likely to show considerable variation because of the large number of trip variables.

Pedestrian amenities include the provision of bus shelters, covered arcades, benches and features for universal design and access. Ramped curbs assist the elderly, women with baby carriages and persons in wheel chairs, and ample provision of seating and rest areas.

Pedestrian traffic and circulation improvements include sidewalk widening, multiple building entrances, access to transit or parking garages, subsurface concourse or overhead bridge connections to other buildings or transit facilities, arcades, malls, plazas, pedestrian connectors bisecting long blocks, off-street taxi and bus passenger loading areas, off-street truck berths, restrictions on cross sidewalk freight operations, and traffic signals timed for pedestrians rather than for auto traffic.

• **Continuity:** The importance of system continuity cannot be overemphasized. That is, the pedestrian improvements must be direct, accessible and in the common pedestrian pathway of "perceived least effort." Developers of Chase Manhattan Plaza, in the downtown financial district of Manhattan, have built well-designed pedestrian oriented building service facilities, consisting of an elevated, attractively landscaped plaza, and generous belowground connections to the subway. This development serves, with great efficiency, the two large office buildings abutting the plaza. However, the plaza is an island without linkages to other buildings or transit stations in the area. The complex therefore provides little contribution to the systemwide deficiencies that exist in this crowded district.

• **System coherence:** Visual and functional coherence is a necessary element of pedestrian design if the full utility of the space is to be realized. All elements of the urban core, including street systems, transit facilities, office buildings, civic center and theater complexes, and shopping areas, should have clear visual statements that convey their direction, function and purpose. When pedestrians are able to satisfy their need for orientation and direction, their alertness of sensory gradients and changes in color, light, ground slope, smells, sound and textures, is increased.

System coherence is also an important element of building design, particularly for transportation terminals. An incoherent passenger terminal configuration multiplies the number of directional signs, thus decreasing individual sight effectiveness. Signs should be considered as a supplementary message to confirm the visual statements expressed by the legibility of the building design itself.

• **Attractiveness:** Landscaping, pavement color and texture, well-designed street furniture, fountains, and plazas increase the visual variety of the cityscape. Opportunities for introducing elements of surprise, through suddenly revealed vistas and panoramic views, should not be overlooked.

An aesthetic audit by a selected group of local artists and architects will pinpoint many of the more glaring violations of aesthetic design. Aesthetic improvements may also involve restoration of buildings with distinctive architectural features or historical significance, additions of landscaping, creating small parks or plazas on unused lots, variations in sidewalk pavement colors and textures, variations in lighting, and cultural innovations such as downtown art exhibits or concerts.

These criteria determine the factors to be established by pedestrian planning and design (Fig. 1). The extent to which pedestrian circulation and way finding is made convenient, safe and aesthetically satisfying determines first and lasting impressions of urban places and pathways, public transport and building entryways.

2 INVENTORY AND DATA COLLECTION

The inventory phase of the planning process requires collection of basic data about the study area, including physical features, land use, and characteristics of daily pedestrian population. Accurate classification of land use is important because of the extreme variability of pedestrian demand and design treatment. The inventory of land use should include the location and classification of all buildings, streets, and transportation services. Building classification includes type of use, total net usable area, estimated building value, building condition, and employee and daily transient populations. Hourly transit patronage, and auto parking accumulations on streets and in garages, should also be determined for the study area. Bus transit cash recording tapes, turnstile counts, and parking garage check stubs provide the best source for this data.

The physical inventory of the study area includes the plan configuration and dimensions of each street and sidewalk. Street inventories should include all traffic regulations, signs, signal locations and their cycle lengths, and vehicular traffic volumes. The sidewalk inventory should show the location and dimensions of building and transit system entrances, bus stops and the location of sidewalk furniture and other impedimenta that restrict sidewalk efficiency.

Ideally, all pedestrian walking trips should be determined by inventory (Fig. 2), including their origins and destinations, trip purposes, time of day, and volumes. In larger pedestrian networks this information is difficult, if not impossible to obtain, so that combinations of various sampling and analytic techniques are used to develop this data. These techniques include cordon counts, origin and destination surveys, pedestrian density surveys by aerial photography, and mathematical modeling.

- **Cordon counts** of pedestrians generally cover small areas, and are usually obtained by manual field counting, or more recently, by aerial time-lapse photography. The manual field count is a simple procedure that can provide much useful data. During light, off-hour pedestrian traffic, observers may be stationed on the roof of a tall building, to record pedestrian activity for several buildings or streets. Counts should be made on "typical" days, free from the distortions of weather and other seasonal effects.

- **Manual field count** procedures usually consist of stationing survey personnel at mid-block locations, away from any large traffic generators. Standard hand tally recorders are used for the count, with the observer tallying passing pedestrians by direction, and then recording the running totals for each time interval, usually every five or ten minutes.

- **Special entrance counts** may be made at large traffic generators, such as office buildings, theaters and department stores. These generator counts may be correlated to net floor area by classification type, and used to develop a pedestrian trip generation model.

- **Time-lapse photography** has been effectively used for pedestrian counting, but usually when detailed study data is required. Field procedure for time lapse photography surveys is dependent on the availability of vantage points high enough to provide a view of the study area. In order to conduct a complete study of a typical intersection, a vantage point of six stories or more in height may be required.

Fig. 3. Sky Count Survey. Pedestrian Density Contours presents a contour plot of pedestrian densities obtained in a downtown Manhattan sky count survey. The population of each of these intersections was obtained by counting pedestrians within a half block of the intersection on each street.

- **Sky count** is a data collection method adapted from the aerial photography and intelligence techniques (Fig. 3). It can be used to collect, and permanently record, a considerable variety of important transportation and land use data. For example, a single data collection flight over an airport can determine parking lot occupancy, length of checkout lines, roadway traffic volume, speed and composition, aircraft apron occupancy and other incidental data.

- **Origin and destination survey** is the classic method of obtaining transportation trip data. Depending on methodology and scope, the survey can determine the socio-economic characteristics of the trip maker, trip frequency, route and purpose. It also provides information on modal choice and the location and trip attraction rates of traffic generating centers.

- **Postcard mailback survey** is the most popular form of Origin and Destination survey, because it is convenient and gives a reasonably good return if properly publicized. Dimensional control of the sampling populations and questionnaire distribution is important since the sample data will be expanded and treated as representative of the whole trip population. This requires that the total number of persons entering the study zone by each transportation mode be accurately determined, and that the survey be conducted on a "typical" weekday with favorable weather conditions.

• **Analytic modeling and computer simulation** is used in transportation studies to provide capability of handling large amounts of data, especially to:

- Determine the relative value of the various elements of the system and to test the impact of changes,
- Forecast future system use,
- Synthesize existing or theoretical future systems where it is not practical or possible to collect sample data.

3 IMPROVEMENT PROGRAM FOR PEDESTRIANS

Improvement programs range from a remedial program with very little capital expenditure to a major program involving a complex and costly grade-separated network.

Pedestrian malls and streets are special street-grade pedestrian precincts in which vehicular intrusion has either been reduced or eliminated. In many respects, the pedestrian mall is the modern equivalent of the medieval plaza, with the most successful malls duplicating the qualities of human interest, interaction and communication provided by the plaza. The successful mall program must be conceived as a total system improvement which includes:

- Exclusion of all but emergency vehicles from the mall area;
- Development of an adequate perimeter street system to replace street circulation and capacity lost by the closing;
- Provision of adequate peripheral access for transit, private autos, emergency and service vehicles;
- Provision of adequate nearby parking, sufficient to replace all spaces lost by the street closing, plus additional parking generated by the mall;
- Promotional program based on building improvements, aesthetic landscaping, increased lighting, pedestrian amenities, coordinated advertising and special events.

Types of pedestrian precincts classified as malls

• **Service street or transitway:** limited access street where cross-street traffic is allowed, but through traffic is limited to transit buses and emergency vehicles only. If service cannot be provided through rear lots, commercial vehicles must enter and leave the system by the nearest cross-street. This is not a true mall, but it results in improvements in transit running times and reduced pedestrian conflicts.

• **Interrupted mall:** exclusive pedestrian street sections where cross traffic is allowed, but no vehicles are permitted on the mall sections except for emergencies. Sidewalks may be eliminated and exclusive pedestrian plazas are provided between cross-streets. Service by commercial vehicles is through rear lots, or by cross-street access only.

• **Continuous mall:** an exclusive pedestrian street with no cross-street traffic, and provision for only emergency vehicles on the mall. Service is by commercial vehicles through rear lots or side streets only.

• **Bonus zoning** provides an additional incentive and is an approach taken by a number of cities in zoning ordinances. Bonus zoning trades added development rights in exchange for the construction of desirable pedestrian improvements which otherwise might not be economically feasible for the developer. The types of amenities for which development bonuses are being awarded include special uses, such as theater and historical districts, pedestrian traffic and circulation improvements and open space, and aesthetic enhancement.

Design of street spaces

The street network is both the spinal column and the circulatory system of urban space, providing it with its structure and viability. The functional components of the traffic design of street spaces for pedestrians include sidewalks, street furniture, and street lighting.

• **Sidewalks** are the pedestrian's portion of the street space. Pedestrians must share this space with the countless pieces of hardware and furniture, much of which is required for the traffic control of vehicles. In addition to this intrusion into the pedestrian's space, vehicular conflicts and signal systems that are timed to favor vehicles reduce continuity of the sidewalk system further.

• **Street furniture** can be an amenity, such as proving welcomed seating, but is often an impediment. When designing or evaluating sidewalks, deductions must be made for the incursions by street furniture and time interruptions by signal cycles. Other considerations are involved in providing effective pedestrian space on sidewalks. Many new buildings have taken advantage of zoning bonus provisions, and have provided pedestrian plazas. When plazas are properly integrated with the sidewalk, they can be positive additions to effective pedestrian space. However, there are examples where the plaza has been separated from the sidewalk by walls, decorative planters, or grade discontinuities, which negate potential pedestrian space advantages.

Other design features limit the effectiveness of sidewalks from the human factor standpoint. Curb heights may be set too high for convenient human locomotion. Curb height should be no greater than the normal stair riser height maximum of 8 in. (20 cm). Preferably, curbs should either be set at 7 in. (17 cm) or less, or ramped, to facilitate use for universal accessibility. Because of sight limitations, locomotion handicaps, or convenience, many pedestrians can be observed bypassing raised curb safety islands, walking instead into active traffic roadways, to avoid stepping up and down. Walk through islands channelize pedestrian flow, and eliminate this practice. Sidewalk gratings similarly present a pedestrian inconvenience because of their slippery surfaces and tripping hazards.

• **Street lighting** is one of the most important design components of the pedestrian's environment after dark. High levels of illumination have been found to reduce pedestrian accidents, and to improve pedestrian security and area image.

Balanced light distribution is based on the use of luminaries with the most efficient lighting patterns for the particular space and by spacing lighting poles for efficient overlapping of lighting patterns. If light poles are spaced without this overlap, an uneven brightness occurs. Ideally, light poles should be spaced so that the illumination from one complements the other, to produce an even brightness level. Street lighting can be enhanced with floodlighting of pedestrian crosswalk areas and special interest features, such as statues, or architecturally attractive buildings.

Pedestrian walking distances

Walking distances, are important because they are a factor in plan configuration, and a measure of design serviceability. Tolerable distances are a subjective human variable. Based upon pedestrian studies at Port of New York Authority bus terminals, the practical limit of human walking distance appears to be related more to the context and situation than human energy. For most persons, the maximum tolerable walk distance is in the range of a normal 5-to-7 minute walk. The anxiety connected with meeting schedules, making the trip, and negotiating an unfamiliar building, however, tend to make these distances appear to be much longer. The tolerable walking distance for a given design situation is related to such factors as the trip purpose of the individual, available time, and the walking environment, rather than energy consumption. This suggests that improvement of the design environment to reduce negative psycho- logical factors is as important as reducing pedestrian walking distances.

Universal design and accessibility

The objective of universal design is to make pedestrian facilities accessible to, and usable by, individuals of all abilities and ages, provides direct benefits in improved utility and service to all. Pedestrians who are limited physically report that the dense, irregular, and usually hurried pedestrian traffic in most travel situations provides even greater challenge and difficulty. Higher level of service design standards for walkways would help to relieve many of the apprehensions of the handicapped, by providing sufficient area for all pedestrians to select their desired pace. Reduction of riser height provides a stair that is negotiated with a greater degree of traffic efficiency, ease and convenience.

The qualitative design of a pedestrian environment requires an understanding of related human characteristics and capabilities. The physical dimensions of the body determine working widths of doorways and passageways and affect the practical capacity of moving stairs and walkways. Psychological preferences of avoiding body contact with others are a determinant of interpersonal spacing in queues and other crowded pedestrian environments.

• **Human body dimensions: The Body Ellipse.** Body depth and shoulder breadth are the primary human measurements used by designers of pedestrian spaces and facilities. (Fig. 4.) Shoulder breadth is a factor affecting the practical capacity of doorways, passageways, stairways and mechanical devices such as escalators and moving walks. A compilation of body dimensions from a large number of human factors studies shows a shoulder breadth of 20.7 in. (52.5 cm) for the 99th percentile (99 percent are less than this) of civilian men, with a recommended addition of 1-1/2 in. (3.8 cm) for heavy clothing. In plan view, the average adult human body occupies an area of about 1-1/2 square feet (1.4 square meters). A larger ellipse of 18 in. by 24 in. (46 cm by 61 cm), equivalent to an area of 2.3 square feet (2 square meters), allows for the fact that many pedestrians are carrying personal articles, natural psychological preferences to avoid bodily contact with others, and body sway. This determines a practical standard for pedestrian design as an ellipse of 24 in. by 18 in. (61 cm by 46 cm).

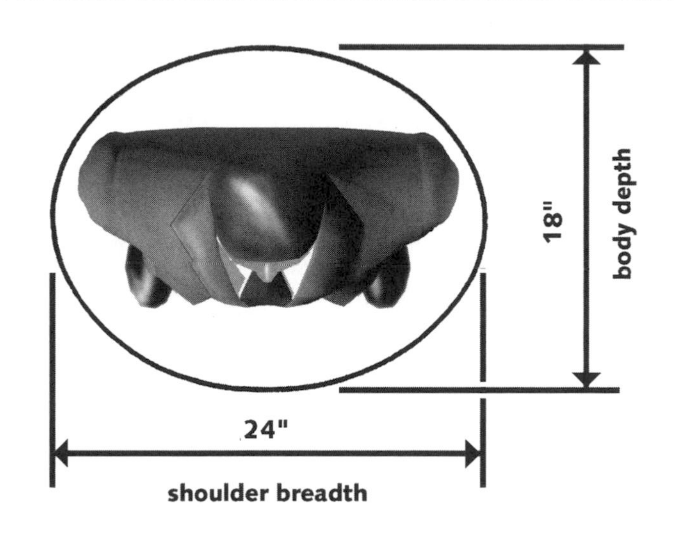

Fig. 4. Body ellipse assumed for pedestrian design purposes. An ellipse 18 in. by 24 in. (46 cm by 61 cm) equivalent to an ellipse area of 2.3 sq. ft. (2 sq. m).

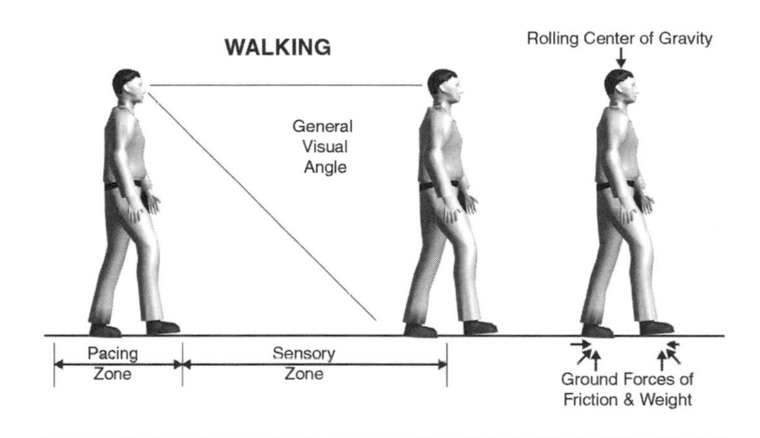

Fig. 5. Human Pacing and Walking Zones. horizontal surfaces.

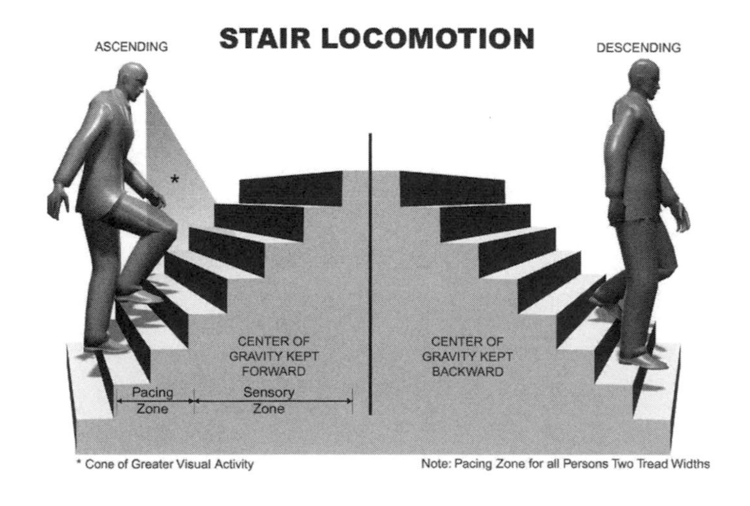

Fig. 6. Human Pacing and Walking Zones: stairs.

• **Perception of personal space: Body Buffer Zone.** Humans value person space. If freedom of choice exists, pedestrians will adopt personal spacing that avoids contact with others, except in special circumstances such as a crowded elevator, where this unwritten law may be temporarily suspended. Pedestrian contact on sidewalks and other walkways is governed by personal body buffer zone concepts and social and cultural politesse.

In the 1966 book *The Hidden Dimension*, Edward T. Hall observed that there are different cultural attitudes towards personal space throughout the world. Hall establishes four general categories: public distance, social distance, personal distance, and intimate distance, with close and far phases in each category. Hall's findings support the convention for public distance in public spaces in keeping with at least an "arms reach" separation. As a point of reference, an opened umbrella is about 43-in. (109 cm.) diameter with an area of about 10 sq. ft. (.9 sq. m).

The space required for locomotion may be divided into a pacing zone, the area required for foot placement, and the sensory zone, the area required by the pedestrian for perception, evaluation and reaction. For personal safety, or social conventions of avoiding brushing others, or just the joy of sightseeing, the pedestrian is constantly monitoring a whole range of sensory stimuli. The capabilities of human vision and distance judgment can have a significant effect on pedestrian activities. J. G. Gibson in *The Perception of the Visual World* (1960) uses the term "locomotor vision" to describe the specialized visual characteristics connected with judging the velocity, distance and direction of others during walking. Pedestrians, through use of vision and their own internal calculation, are able to keep track of the varying speeds and angles of oncoming pedestrians and to accurately adjust their pace and speed to avoid collision (Fig. 5).

• **Locomotion on stairs.** Stair climbing and descent is quite different from walking. Locomotion on stairs is necessarily more restricted because of safety considerations and the restraints imposed by tread and riser configurations. Human energy consumption for climbing stairs is about ten to fifteen times the energy needed for walking the equivalent horizontal distance, and surprisingly, about one-third greater for descent (Fig. 6).

• **Perception of urban space.** Kevin Lynch in *The Image of the City* (1960) has defined many of the elements that comprise the human perception of urban space. These elements were determined by cataloging the physical compositions of three distinctly different cities, and then questioning the inhabitants about their subjective place images. The components that define urban space were then classified into five broad categories: paths, nodes, landmarks, edges and districts. The path, or linkage between nodes, is considered to be the most dominant visual element of space because it is the unifying force upon which the other elements depend. The path provides cohesiveness, identity and the means of expression. While moving along the path, the observer is exposed to the kinesthesia of constantly changing relationships with all the visual elements that comprise the space. Movement enhances the sensory gradients of variations in smell, heat, light, color, texture and gravity, that all combine to produce the total human perception of a space. Nodes are the focal points and intersecting junctions of paths and supporting systems such as transit. Direct connections to node points would be a characteristic of a well-defined system. Landmarks are statements or points of reference that provide the observer with a continual sense of orientation and relationship with the space. The edges are the linear descriptors that define an urban space or district. Edges may be natural linear features, such as a river or lake, or man-made, such as a highway cut or railroad embankment. The configuration of these edges may describe an easily identifiable space entity or district, or it may be so disruptive that no space identity exists. A district is an identifiable space entity or precinct with a common identifying character.

The section that follows extends these principles with reference standards for pedestrian system design.

5 PEDESTRIAN TRAFFIC ANALYSIS AND DESIGN

This section discusses pedestrian traffic volumes and queuing relationships on the basis of average pedestrian area occupancy, providing easily understood measures for design. It presents the analysis and principles used to develop the "Level of Service" Design Standard detailed herewith.

• **Flow Volume** is the number of traffic units passing a point in a unit of time. In pedestrian design, flow is expressed as pedestrians per unit of width (in feet) of the walkway or stairway per minute, or Pedestrians/Foot/Minute (PFM). Flow is the most important traffic characteristic, because it determines the width of the pedestrian way (pedway). An inadequate width restricts flow, resulting in pedestrian inconvenience. Volume has been designated as P.

• **Speed.** Pedestrian locomotion speed, expressed in distance per unit of time, generally in feet per minute. When related to design of a pedway section, speed is the average speed of all pedestrians passing through the section during the design interval. Speed is designated as S.

• **Density.** The number of traffic units per unit of area. In pedestrian design, density would be expressed in tenths of a pedestrian per square foot, a difficult unit to visualize. A more manageable unit, the reciprocal of density, or the sq. ft. of area per pedestrian, is used consistently in this article. The reciprocal of density has been designated as M, the Pedestrian Area Module.

• **Headway.** The time and distance separation between traffic units (pedestrians, in the context of this article). For example, a two-second headway is the equivalent of one unit (pedestrian) passing a point each two seconds, or a flow volume of 30 traffic units per minute. Headway has not been commonly referred to in the design of pedestrian facilities, but it has useful applications.

• **Queue.** Is defined as one or more traffic units—and in this discussion pedestrians—who are waiting for service. If the pedestrian areas or service facility have insufficient capacity, a pedestrian queue will develop. Queue lengths and duration will vary according to traffic flow characteristics. In crowded systems, queues may be generated intermittently, due to random variations in traffic intensity.

The flow equation, the classic relationship of traffic design, derived from an analogy to fluid flow in channels, is expressed as follows:

Flow Volume = Average Speed x Average Density

or, $P = S \times D$

As mentioned above, application of the term "density" to pedestrian traffic results in the unwieldy unit of tenths of a pedestrian per square foot, so that the reciprocal of density, sq. ft. area per pedestrian (M), is more useful. The equation for pedestrian flow volume, (P), in pedestrian per foot width of pedway section, per minute, (PFM) is expressed as follows:

$$\text{Ped Volume} = \frac{\text{Average Ped Speed, feet per minute}}{\text{Average Ped Area, sq. ft. per ped.}}$$

$$\text{or,} \quad P = \frac{S}{M}$$

In this formulation, the designer has a clearer concept of relative design quality, since the units are easier to understand and manipulate. For example, the next section will show that a nearly normal average walking speed of 250 ft. (76 m) per minute is attained with an approximate average pedestrian area of 25 sq. ft. (2.3 m sq.) person. The simple division of area occupancy into average speed gives an equivalent design volume of 10 pedestrians per foot width of walkway, per minute.

Time and distance headways can be determined from flow volumes by assuming a specific pedestrian lane width. For example, if a 3-ft. (.9-m) wide pedestrian lane width has a flow of 10 PFM, as cited above, this is equal to a pedestrian volume per lane of 30 persons per minute, or the equivalent of an average time spacing of two seconds between pedestrians. This time spacing is translated into an average distance spacing of about 8 feet between following pedestrians, by multiplying by the 250 ft. (76 m) per minute walking speed.

Walking speeds

When unimpeded by crowd density or other traffic frictions, pedestrians may vary their walking speeds over a wide range. Fig. 7 illustrates the distribution of free flow walking speeds obtained in surveys of about 1000 non-baggage carrying pedestrians. On the basis of these surveys, average free flow walking speed for all males, females, and the combination of all pedestrians in the surveys, were from 270, 254 and 265 ft. (82, 77 and 80 m) per minute, respectively. Average speeds were found to decline with age, but slow and fast walkers were observed in all age groups. Factors as grade, and the presence of baggage or packages, have been found to have no appreciable effect on free flow walking speed.

The remaining determinant of pedestrian walking speed is traffic density. As traffic density increases, pedestrian speed is decreased, because of the reduction in available clear area for locomotion. Fig. 8 shows the effect of increased traffic density, or decreasing area occupancy, on pedestrian walking speeds for one-directional commuter traffic flow. Similar studies of two directional commuter and multi-directional flows of shoppers resulted in only small variations from this curve, confirming its more general applicability.

Traffic flows on walkways

Pedestrian volume, or the number of persons passing a given point in a unit of time, is the most important walkway design parameter. If traffic demand exceeds the capacity of a walkway section, crowding, uncomfortable shuffling locomotion and delay will result, producing a poor pedestrian environment.

Studies of pedestrian traffic flow on walkways have established flow volume relationships for three categories of pedestrian traffic (Fig. 9). These relationships, representing the average conditions of three

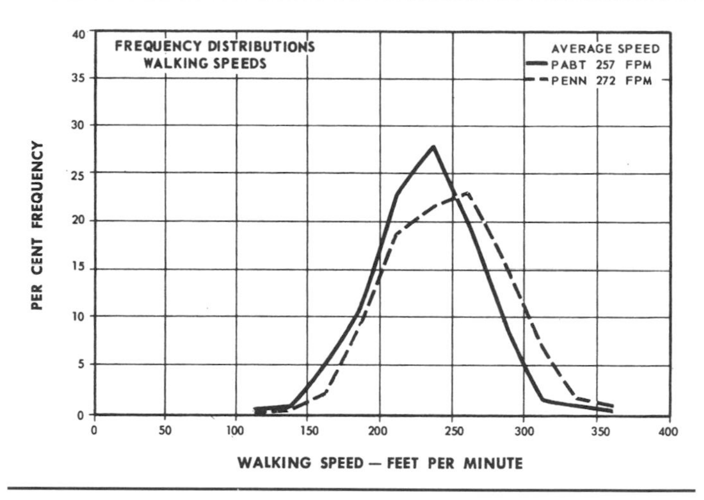

Fig. 7. Pedestrian speed unimpeded. Data obtained in surveys of 1000 pedestrians (without baggage) at The Port of New York Authority Bus Terminal and Pennsylvania Station.

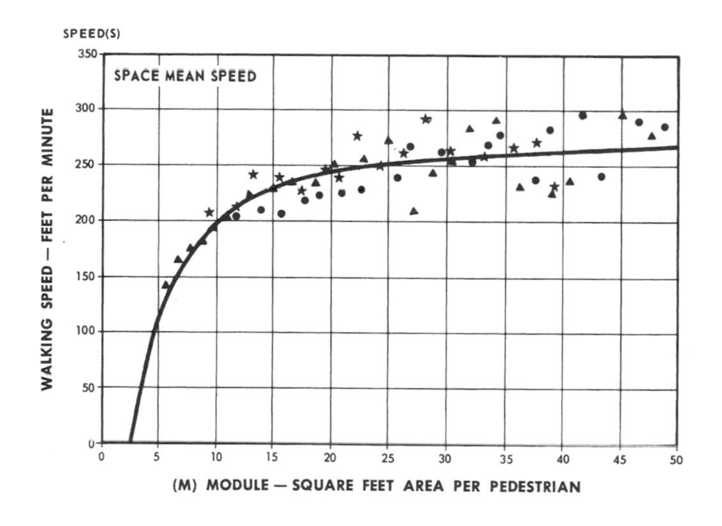

Fig. 8. Pedestrian speed impeded one way. Data obtained from time-lapse photography analysis.

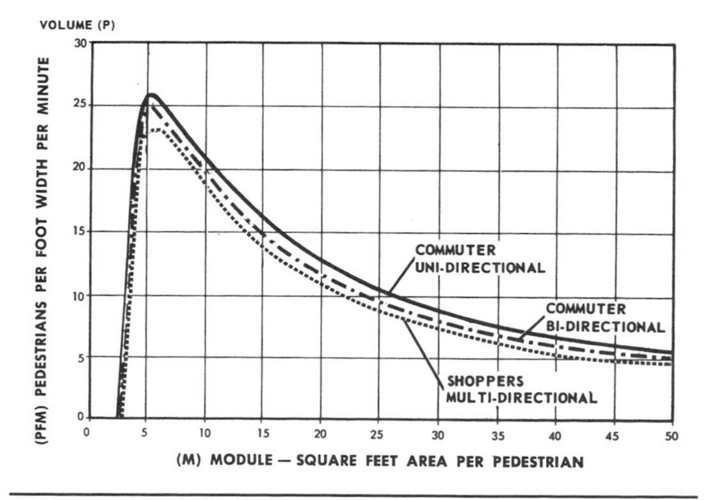

Fig. 9. Pedestrian flow volume and area occupancy. Data obtained from time-lapse photography analysis.

TABLE I. Observed headways at entrances

A summary of observed average headways for a number of entrance devices and portallike situations. While this summary is useful for comparison purposes, the designer is encouraged to examine entrance design problems from the standpoint of traffic headways. This would provide a more qualitative insight into system adequacy and Level of Service.

TYPE OF DEVICE	OBSERVED AVERAGE HEADWAY (Seconds)	EQUIVALENT PEDESTRIAN VOLUME (Persons per minute)
DOORS		
Free-swinging	1.0 – 1.5	40 – 60
Revolving – one direction	1.7 – 2.4	25 – 35
REGISTERING TURNSTILES		
Free Admission	1.0 – 1.5	40 – 60
With Ticket Collector	1.7 – 2.4	25 – 35
COIN OPERATED (LOW)		
Single slot	1.2 – 2.4	25 – 50
Double slot	2.5 – 4.0	15 – 25
BOARDING BUSES		
Single Coin Fare	2 – 3	20 – 30
Odd Cash Fares	3 – 4	15 – 20
Multizone Fares	4 – 6	10 – 15

TABLE II. Pedestrian speeds on stairs

	HORIZONTAL TIME-MEANS-SPEEDS (Feet Per Minute)			
	DOWN DIRECTION		UP DIRECTION	
	(1)	(2)	(1)	(2)
Age – 29 or under				
Males	163	83	110	120
Females	117	132	106	110
Group Average	**149**	**160**	**108**	**115**
Age – 30 to 50				
Males	136	160	101	116
Females	100	128	94	107
Group Average	**127**	**153**	**99**	**114**
Age – Over 50				
Males	112	118	85	81
Females	93	111	77	89
Group Average	**108**	**117**	**83**	**83**
Average – all ages, sexes	**132**	**152**	**100**	**113**

(1) Indoor Stair: 7 in. (18 cm) riser, 11.25 in. (27 cm) tread, 32 degree angle

(2) Outdoor Stair: 6 in. (16 cm) riser, 12.0 in. (30 cm) tread, 27 degree angle

distinctive types of pedestrian traffic, show a relatively small range of variation, which strongly suggests that reverse and cross flow traffic conflicts do not drastically reduce either pedestrian traffic volume or speed. This characteristic makes these curves applicable to a wider range of different design conditions.

A finding worth noting in the sidewalk study is that pedestrians tend to keep 6 to 8 in. (14 to 18 cm) lateral clear distance between themselves and walls. This suggests that this lateral distance should be deducted from the walkway dimension when determining effective walkway width.

The maximum average peak flow volumes of 26.2, 24.7 and 23.3 persons per foot of walkway width per minute (PFM), shown in Fig. 9, are representative of design values in use by a number of authorities.

Pedestrian spacing and conflicts

The previous discussion established the relationships of walking speed and flow volumes to average pedestrian area occupancy. The ability to select near normal walking speed, a qualitative measure of pedestrian flow, was found to require average areas of 25 sq. ft. (2.3 sq. m) per person or more. However, other measures are required to obtain a clearer understanding of walkway quality. Pedestrians must leave additional space for maneuvering within the traffic stream to bypass slower moving pedestrians and to avoid oncoming and crossing pedestrians. This additional space is required to sense the speed and direction of others, and to react without conflict or hesitation.

Pedestrian conflicts are thus a function of walking speed and pedestrian spacing in the traffic stream. Although wider pedestrian spacing provides larger crossing gaps, a corresponding increase in pedestrian speed tends to continue to make crossing the main stream difficult. The probability of conflicts due to crossing main stream traffic thus exists over a wide range of pedestrian densities.

Entrances

Entrances are, in effect, walkway sections in which pedestrians are channeled into equal, door-width traffic lanes. In addition to imposing restricted spacing in the traffic stream, entrances may require the pedestrian to perform some time-consuming function such as opening a door or turnstile. The earlier discussion of headways introduced the concept that pedestrians have different average time and distance separations in a traffic stream, dependent on the flow volume. The headway concept is a useful one for evaluating the design of doors, turnstiles, and other entrance devices. When a pedestrian opens a door, there must be a sufficient time and headway separation between that pedestrian and the following pedestrian, to allow for the performance of this function. Because of the added time required for the door opening, entrances will be the weak links in the pedway system, and therefore require added design attention. Observed headway distances at entrances and similar functions are summarized in Table I.

Traffic flow on stairways

Pedestrian locomotion on stairs is much more restricted than walking. The dimensions of the stairs themselves determine many of the aspects of locomotion that the pedestrian may more freely choose on a level surface. A summary of results of the surveys, which involved almost 700 pedestrians, is tabulated by differences in direction, stair angle, sex, and age, in Table II.

Great attention by the designer on stair dimensioning and factors of safety is needed as part of the qualitative design of stairways (Templer 1997).

Stairway volumes remain as the most significant design parameter. As with walkway volumes, maximum stairway flow occurs in the region of minimum pedestrian area occupancy, about at the point of a two tread length and one shoulder breadth area, or approximately 3 sq. ft. (.28 m) per person. In this confined area, movement of the pedestrian ahead determines forward progress. The maximum flow volume of 18.9 pedestrians per minute per foot of stairway width is representative of design values used by many authorities.

Studies conducted by the Regional Plan Association of New York disclose that stair use was reasonably fluid, without queuing, up to about 12 PFM (Pedestrian/Foot/Minute). Beyond this point, queuing developed as pedestrian spacing and speed were reduced. This value corresponds roughly to the level of average stair locomotion speed developed above. It is also compatible with the area occupancy of about 7 to 8 sq. ft. (.65 to .74 m sq.) per person, suggested as the minimum area required for normal stair locomotion. While this provides some minimum space and maximum volume criteria, it is not a free-flowing traffic stream where passing and selection of individual speed is possible.

Pedestrian queuing

Queuing is broadly defined as any form of pedestrian waiting that requires standing in a relatively stationary position for some period of time. Queues may be of two general types: a *linear* or *ordered queue*, with conventional first-come, first-served priority, or a *bulk queue*, which would be unordered and without an established discipline. Bulk queuing areas may be divided into those devoted only to standing and waiting, with limited movement within the queuing area, or those devoted to waiting combined with some need for reasonably free internal circulation through the queuing area. There are no known standards for the design of queuing spaces. As a result, this aspect of pedestrian design is often overlooked.

Lines of waiting pedestrians, caused by service stoppages or capacity restrictions of any type, are an aspect of design quality that should be carefully evaluated. The pedestrian holding capacity of all public spaces should be known.

Studies by the Otis Elevator Company to determine the practical capacity of elevators have determined the approximate upper limit of human occupancy in a confined space. The upper limit of mixed female and male occupancy was determined to be about 1.8 sq. ft. (.17 m sq.) per person. Observations of dense bulk queues, at escalators or crosswalks, show that pedestrian area occupancies will average about 5 sq. ft. (.46 m sq.) per person in these less confined circumstances.

Linear queues, such as those that occur on ticket lines, are remarkably consistent with the spacing observed in psychological experiments. Bus commuters have been found to select interpersonal spacings at about 19 to 20 in. (48 to 51 cm), with very little variation, for both ticket purchase and bus waiting lines.

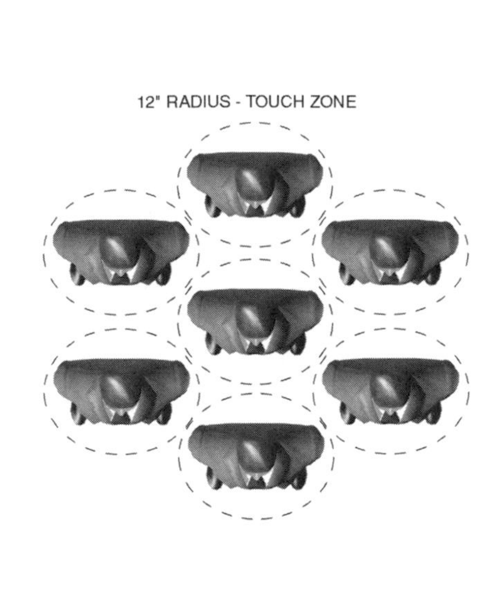

Fig. 10. Boundary of the touch zone: 3 sq. ft. (.28 m sq.) per person.

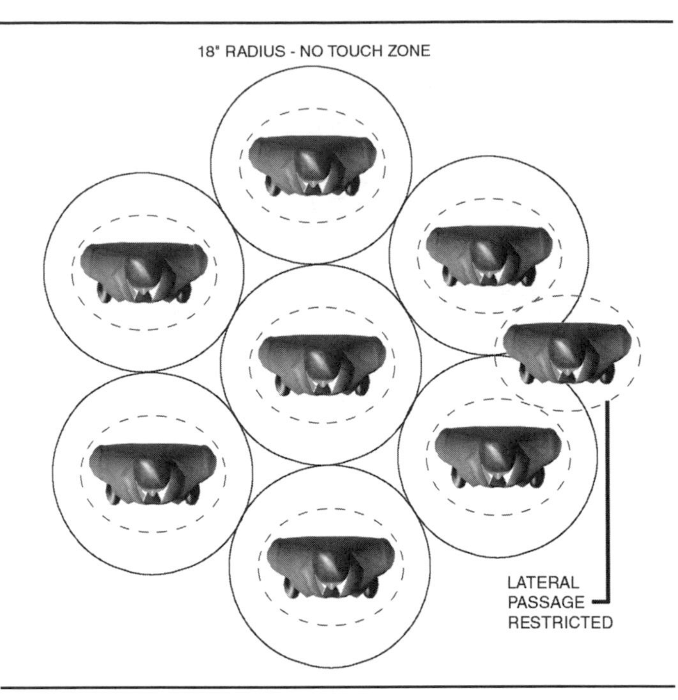

Fig. 11. Boundary of the no touch zone: 7 sq. ft. (.65 m sq.) per person.

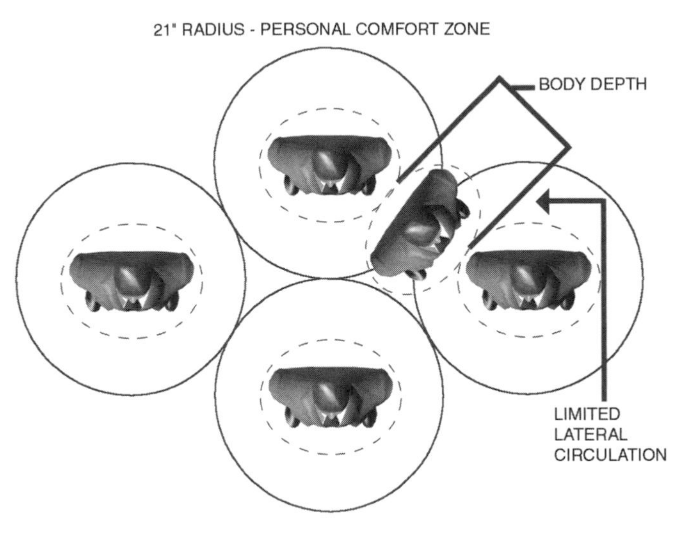

Fig. 12. Boundary of the personal comfort zone: 10 sq. ft. (.93 m sq.) per person.

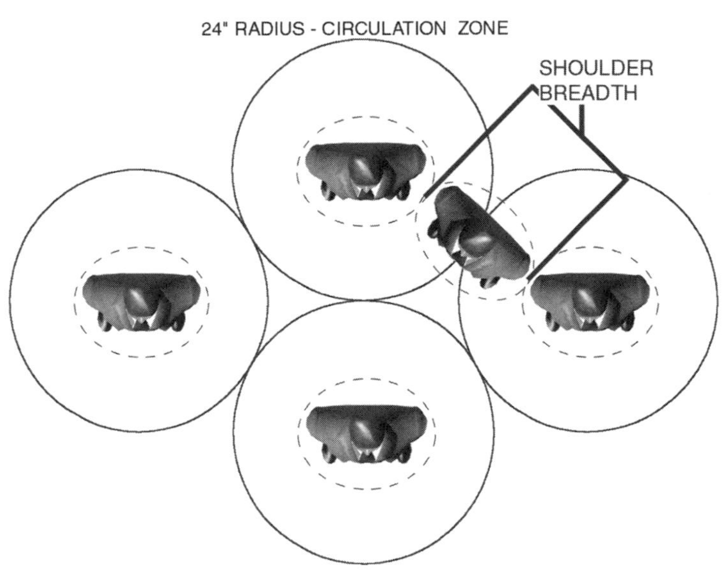

Fig. 13. Boundary of the circulation zone: 13 sq. ft. (1.2 m sq.) per person.

Figs. 10–13 illustrate the various levels of pedestrian area occupancy by which design criteria can be established, assuming uniform interpersonal spacing and circular body buffer zones. Fig. 10 depicts a group of pedestrian equally spaced in individual 2 ft. (61 cm) diameter buffer zones. This assumption results in a queuing area of 3 sq. ft. (.28 m sq.) per person and could be called the *boundary of the touch zone*. Below this area occupancy, frequent unavoidable contact between pedestrians is likely to occur.

In Fig. 11, the body buffer zone has been expanded to a spacing of 3 ft. (.91 cm) and a 7 sq. ft. (.65 m sq.) area. This might be called the *boundary of the no touch zone*, because contact with others can be avoided, as long as movement within the queuing area is not necessary.

Fig. 12 illustrates the expansion of the body buffer zone to a 42 in. (1.07 m) diameter and a 10 sq. ft (.93 m sq.) area. This might be termed the *boundary of the personal comfort zone*, within the range of spatial separation and area occupancy that people have selected in studies emphasizing the comfort criterion.

In Fig. 13, the body buffer zone has been expanded to a 4 ft. (1.2 m) diameter and a 13 sq. ft. (1.2 m sq.) area. This might be termed the *boundary of the circulation zone*, since circulation within a queuing area with an average occupancy that would be possible without disturbing others.

Notes on arrival processes

Arrival processes are an important determinant of the characteristics of use of all pedestrian facilities and spaces, indoors and outdoors. The arrival sequence should be clearly visualized and accommodated by the designer before qualitative design standards are applied. There are two basic types of pedestrian arrival processes: *bulk process* and *intermittent process* arrivals. An example of a bulk process arrival pattern occurs after a sporting event, when there is an immediate mass exodus of spectators, or at a railroad terminal platform, when a fully loaded passenger train discharges in a few minutes.

Intermittent arrival and departure patterns are characteristic of a typical pedestrian facility such as a large transportation terminal, or office building, served by multiple sources of demand. Intermittent arrival facilities tend to have more regular overall traffic patterns, but are subject to short term surges of traffic volume considerably higher than the average, as well as short-term gaps in which traffic volume falls far below the average.

6 LEVEL OF SERVICE DESIGN STANDARDS

Design of pedestrian spaces involves application of traffic engineering principles plus consideration of human convenience and the design environment. Many authorities use maximum capacity ratings for dimensioning pedestrian space. Too often, little or no consideration of human convenience has been made in adopting the ratings. In such cases, the maximum capacity of a pedestrian traffic stream is attained only when there is a dense crowding of pedestrians. This in effect is "planned congestion." Significant reductions in pedestrian convenience will result, as pedestrians are restricted and unable to maneuver. Since human convenience and amenity should be the primary consideration in environmental and urban design, pedestrian design standards should reflect the more generous space allocations. Level of Service standards permit a very close approximation for planning pedestrian zones that are safe, convenient and functional to a specified design standard.

The Level of Service Concept was first developed in the field of traffic engineering in recognition of the fact that capacity design was, in effect, resulting in planned congestion. The Level of Service concept provides a useful standard for the design of pedestrian spaces as well. Pedestrian service standards should, similarly, be based on the freedom to select normal locomotion speed, the ability to bypass slow moving pedestrians, and the relative ease of cross- and reverse-flow movements at traffic concentrations.

Level of Service standards are not a substitute for judgment. The designer must examine all elements of pedway design, including such traffic characteristics as the magnitude and duration of peaks, surging or platooning caused by traffic-light cycles or transit arrivals, and the economic ramifications of space utilization.

The Level of Service standards detailed below are based on a range of pedestrian area occupancies. Design volumes for walkways and stairways are presented as a range. Walkway and stairway Levels of Service are illustrated by drawings of one directional flow at the approximate pedestrian area occupancy representing that service level. The drawings are supplemented by a written description of the qualitative aspects of each Level of Service.

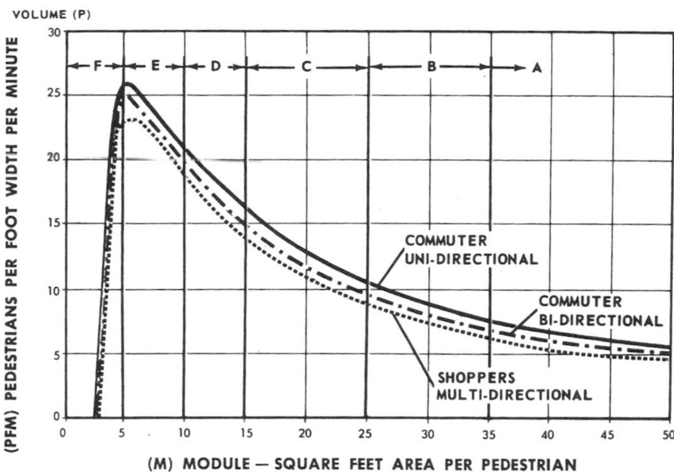

Fig. 14. Level of Service Standards of Walkways: Pedestrian flow volume and area relationships.

Walkway standards

Breakpoints that determine the various levels of service have been determined on the basis of the walking speed, pedestrian spacing, and the probabilities of conflict at various traffic concentrations. The standards provide the means of determining the design quality of corridors, sidewalks and entranceways. The effect width of corridors must be reduced by 18 in. (46 cm) on each corridor side, to account the human propensity to maintain this separation from stationary objects and walls, except under the most crowded conditions. Where there is a tendency for window-shopping or viewing of exhibits, net width should be reduced by an additional equal amount, to allow for standing pedestrians. When designing sidewalks, the effective walkway width must be reduced by an additional 2 ft. (61 cm) or more, to account for the constricting effects of street impediments such as parking meters, light standards, fire hydrants and receptacles. Traffic signals at corners also interrupt sidewalk flow, increasing sidewalk densities.

Illustrations of Walkway Levels of Service are shown in Fig. 15. The correspondence with pedestrian flow volume and area relationships is indicated on the curve in Fig. 14.

Walkways: Level of Service A

Average Pedestrian Area Occupancy: 35 sq. ft. (3.25 m sq.) per person, or greater.
Average Flow Volume: 7 PFM (Pedestrian Foot per Minute), or less.

At Walkway Level of Service A, sufficient area is provided for pedestrians to freely select their own walking speed, to bypass slower pedestrians, and to avoid crossing conflicts with others. Designs consistent with this Level of Service would include public buildings or plazas without severe peaking characteristics or space restrictions.

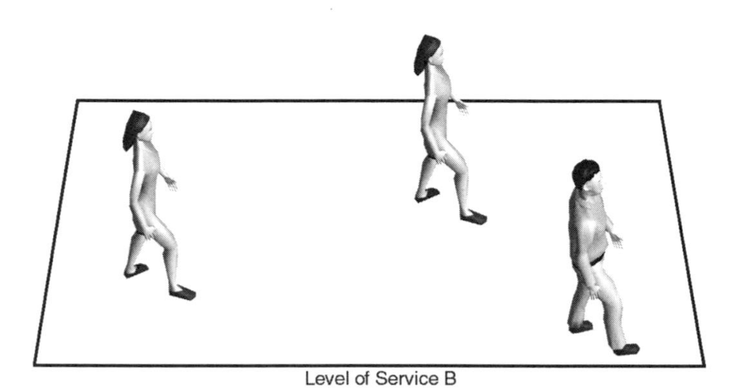

Walkways: Level of Service B

Average Pedestrian Area Occupancy: 25-35 sq. ft. (2.3-3.25 m sq.) per person.
Average Flow Volume: 7-10 PFM.

At Walkway Level of Service B, sufficient space is available to select normal walking speed, and to bypass other pedestrians in primarily one-directional flows. Where reverse directions or pedestrian crossing movements exist, minor conflicts will occur, slightly lowering mean pedestrian speeds and potential volumes. Designs consistent with this Level of Service would be of reasonably high quality, for transportation terminals and buildings in which recurrent, but not severe, peaks are likely to occur.

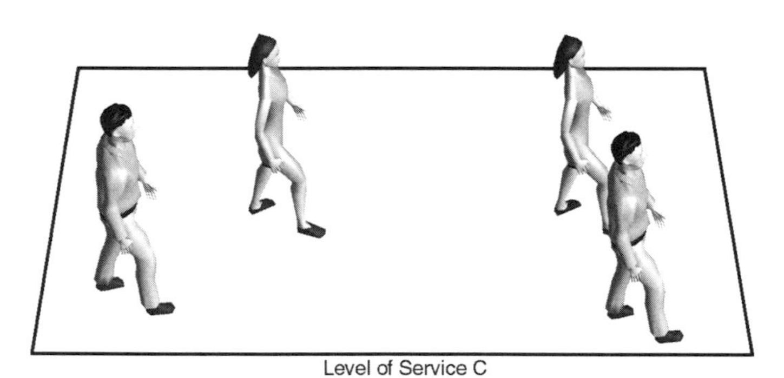

Walkways: Level of Service C

Average Pedestrian Area Occupancy: 15-25 sq. ft. (1.4-2.3 m sq.) per person.
Average Flow Volume: 10-15 PFM.

At Walkway Level of Service C, freedom to select individual walking speed and freely pass other pedestrians is restricted. Where pedestrians cross movements and reverse flows exist, there is a high probability of conflict requiring frequent adjustment of speed and direction to avoid contact. Designs consistent with this Level of Service would represent reasonably fluid flow. However, considerable friction and interaction between pedestrians is likely to occur, particularly in multidirectional flow situations. Examples of this type of design would be heavily used transportation terminals, public buildings, or open spaces where severe peaking, combined with space restrictions, limit design flexibility.

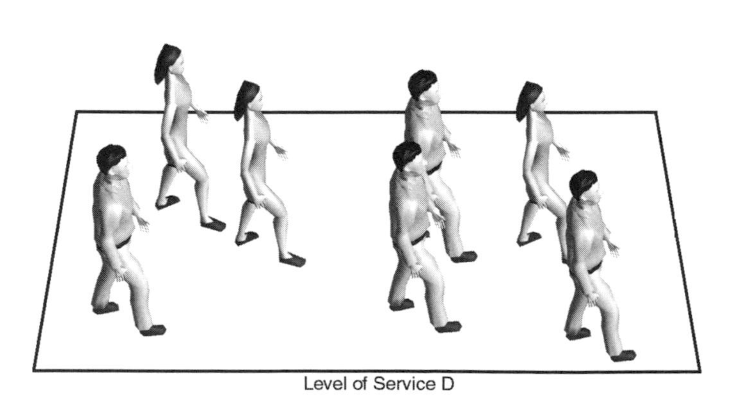

Fig. 15. Level of Service Standards of Walkways: Pedestrian flow volume and area relationships.

Walkways: Level of Service D

Average Pedestrian Area Occupancy: 10-15 sq. ft. (1 m-1.4 m sq.) per person.
Average Flow Volume: 15-20 PFM.

At Walkway Level of Service D, the majority of persons would have their normal walking speeds restricted and reduced, due to difficulties in bypassing slower moving pedestrians and avoiding conflicts. Pedestrians involved in reverse flow and crossing movements would be severally restricted, with the occurrence of multiple conflicts with others. Designs at this Level of Service would be representative of the most crowded public areas, where it is necessary to alter walking stride and direction continually to maintain reasonable forward progress. At this Level of Service there is some probability of intermittently reaching critical density, causing momentary stoppages of flow. Designs consistent with this Level of Service would represent only the most crowded public areas.

Walkways: Level of Service E

Average Pedestrian Area Occupancy: 5-10 sq. ft. (.46-1 m sq.) per person.

Average Flow Volume: 20-25 PFM.

At Walkway Level of Service E, virtually all pedestrians would have their normal walking speeds restricted, requiring frequent adjustments of gait. At the lower end of the range, forward progress would only be made by shuffling. Insufficient areas would be available to bypass slower moving pedestrians. Extreme difficulties would be experiences by pedestrians attempting reverse flow and cross flow movements. The design volume approaches the maximum attainable capacity of the walkway, with resulting frequent stoppages and interruptions of flow. This design range should only be acceptable for short peaks in the most crowded areas. This design level would occur naturally with a bulk arrival traffic patterns that immediately exceeds available capacity, and this is the only design situation for which it would be recommended. Examples would include sports stadium design or rail transit facilities where there may be a large but short term existing of passengers from a train. When this Level of Service is assumed for these design conditions, the limited adequacy of pedestrian holding areas at critical design sections, and all supplementary pedestrian facilities, must be carefully evaluated for amenity and safety.

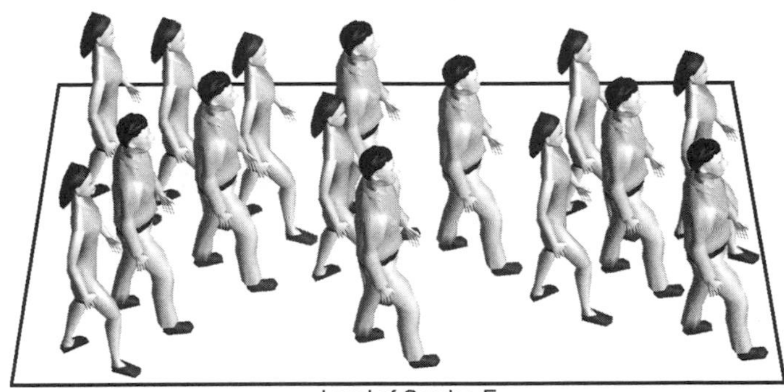

Walkways: Level of Service F

Average Pedestrian Area Occupancy: 5 sq. ft. (.46 m sq.) per person or less.

Average Flow Volume: Variable, up to 25 PFM.

At Walkway Level of Service F, all pedestrian walking speeds are extremely restricted. Forward progress can only be made by shuffling. There would be frequent, unavoidable contact with other pedestrians, and reverse or crossing movements would be virtually impossible. Traffic flow would be sporadic, with forward progress based on the movement of those in front. This Level of Service is representative of a loss of control and a complete breakdown in traffic flow. Pedestrian areas below 5 sq. ft. are more representative of queuing, rather than a traffic flow situation. This Level of Service is not recommended for walkway design.

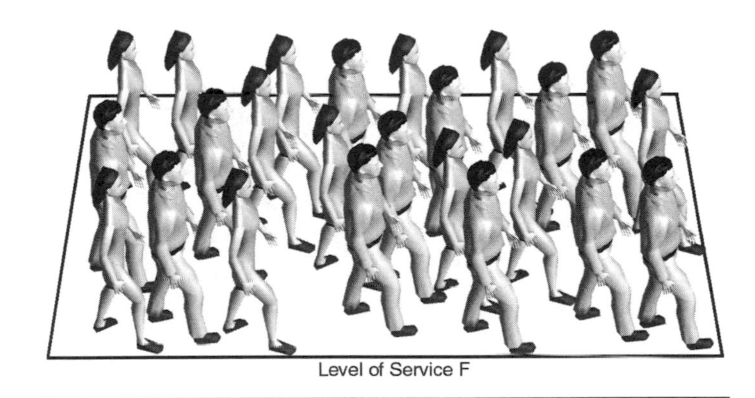

Fig. 15. cont. Level of Service Standards of Walkways: Pedestrian flow volume and area relationships.

Stairway Standards

When designing stairs, increased attention is required to avoid and eliminate safety hazards (Fig. 16). In addition to design judgment in evaluating the traffic patterns and peaking characteristics recommended for use of Walkway Standards, the following factors should be considered in stairway design:

- Stairs should be located so as to be readily visible and identifiable as a means of direct access to the levels they are designed to interconnect;
- Clear areas large enough to allow for queuing pedestrians should be provided at the approaches to all stairways;
- Stairs should be well lighted;
- Stair nosing, riser, tread, and railing configurations should be designed to assist human locomotion, particularly for the individuals with limited capacity;
- Riser heights should be kept below 7 in. (18 cm), to reduce energy expenditure and to increase traffic efficiency;

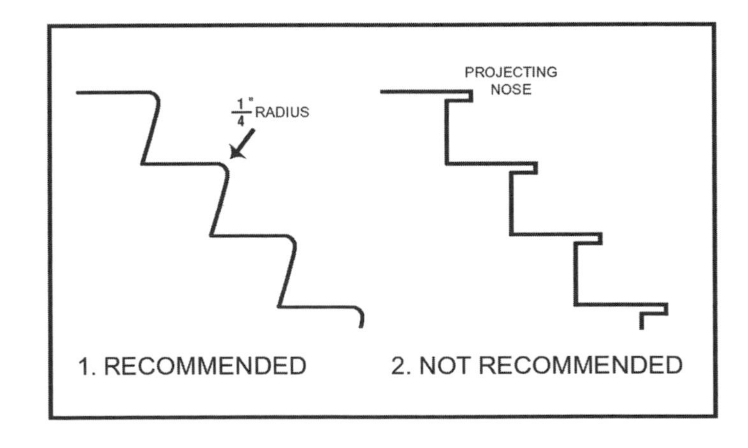

Fig. 16. Stair design. Rounded stair nosings are preferred. Stairs with projected noising cause locomotion difficulties.

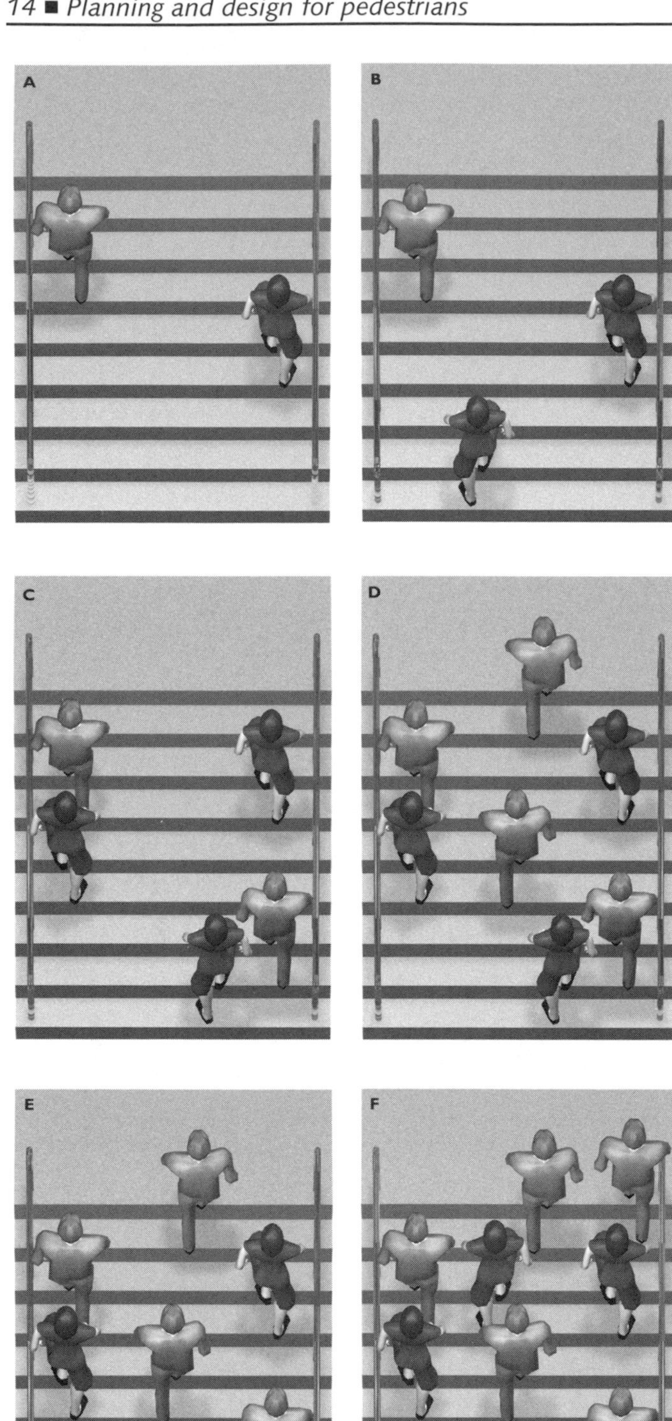

Fig. 17. Level of Service Standards of Stairways.

- When a stairway is placed directly within a corridor, the lower capacity of the stairway is the controlling factor in the design of the pedway section;
- Where minor, reverse flow traffic volumes frequently occur on a stair, the effective width of the stair for the major direction design flow should be reduced by a minimum of one traffic lane, or 30 in. (41 cm).

Stairway Levels of Service with pedestrian flow volume and area relationships are illustrated in Fig. 17 and Fig. 18.

Stairways: Level of Service A

Average Pedestrian Area Occupancy: 20 sq. ft. (1.9 m sq.) per person, or greater.
Average Flow Volume: 5 PFM, or less.

At Stairway Level of Service A, sufficient area is provided to freely select locomotion speed, and to bypass other slower moving pedestrians. No serious difficulties would be experienced with reverse traffic flows. Designs at this Level of Service would be consistent with public buildings or plazas that have no severe traffic peaks or space limitations.

Stairways: Level of Service B

Average Pedestrian Area Occupancy: 15-20 sq. ft. (1.4-1.9 m sq.) per person.
Average Flow Volume: 5-7 PFM.

At Stairway Level of Service B, representing a space approximately 5 treads long and 3 to 4 ft. (.9 to 1.2 m) wide, virtually all persons may freely select locomotion speeds. However, in the lower range of area occupancy, some difficulties would be experienced in passing slower moving pedestrians. Reverse flows would cause minor traffic conflicts. Designs at this Level of Service would be consistent with transportation terminals and public buildings that have recurrent peak demands and no serious space limitations.

Stairways: Level of Service C

Average Pedestrian Area Occupancy: 10-15 sq. ft. (.9-1.4 m sq.) per person.
Average Flow Volume: 7-10 PFM.

At Stairway Level of Service C, representing a space approximately 4 to 5 treads long and about 3 ft. (.9 m) wide, locomotion speeds would be restricted slightly, due to an inability to pass slower moving pedestrians. Minor reverse traffic flows would encounter some difficulties. Design at this Level of Service would be consistent with transportation terminals and public buildings with recurrent peak demands and some space restrictions.

Stairways: Level of Service D

Average Pedestrian Area Occupancy: 7-10 sq. ft. (.7-.9 m sq.) per person.
Average Flow Volume: 10-13 PFM.

At Stairway Level of Service D, representing a space approximately 3 to 4 treads long and 2 to 3 ft. (.6-.9 m) wide, locomotion speeds are restricted for the majority of persons, due to the limited open tread space and an inability to bypass slower moving pedestrians. Reverse flows would encounter significant difficulties and traffic conflicts. Designs at this Level of Service would be consistent with the more crowded public buildings and transportation terminals that are subjected to relatively severe peak demands.

Stairways: Level of Service E

Average Pedestrian Area Occupancy: 4-7 sq. ft. (.4-.7 m sq.) per person.
Average Flow Volume: 13-17 PFM.

At Stairway Level of Service E. representing a space approximately 2 to 4 tread lengths long and 2 ft. (61 cm) wide, the minimum possible area for locomotion on stairs, virtually all persons would have their normal locomotion speeds reduced, because of the minimum tread length space and inability to bypass others. Intermittent stoppages are likely to occur, as the critical pedestrian density is exceeded. Reverse traffic flows would experience serious conflicts. This Level of Service would only occur naturally with a bulk arrival traffic pattern that immediately exceeds available capacity. This is the only design situation for which it would be recommended. Examples would include sports stadiums or transit facilities where there is a large uncontrolled, short-term exodus of pedestrians.

Stairways: Level of Service F

Average Pedestrian Area Occupancy: 4 sq. ft. (.4 m sq.) per person, or less.
Average Flow Volume: Variation to 17 PFM.

At Stairway Level of Service F, representing a space approximately 1 to 2 tread lengths long and 2 ft. (61 cm) wide, there is a complete breakdown in traffic flow, with many stoppages. Forward progress would depend on movement of those in front. This Level of Service is not recommended for design.

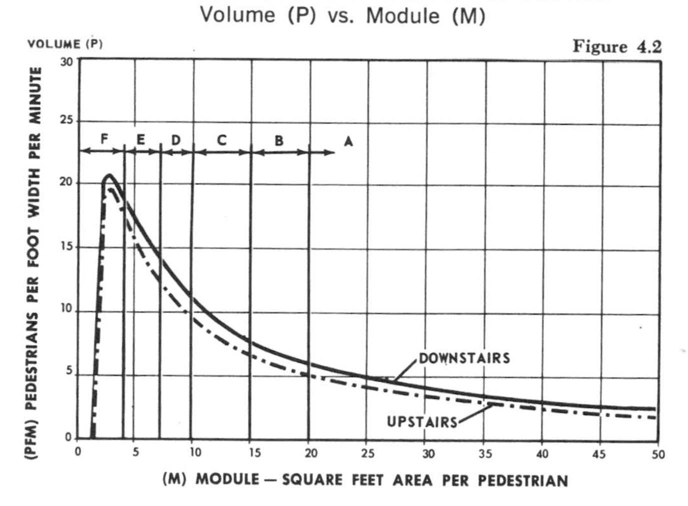

Fig. 18. Level of Service Standards of Stairways. Pedestrian flow volume and area relationships.

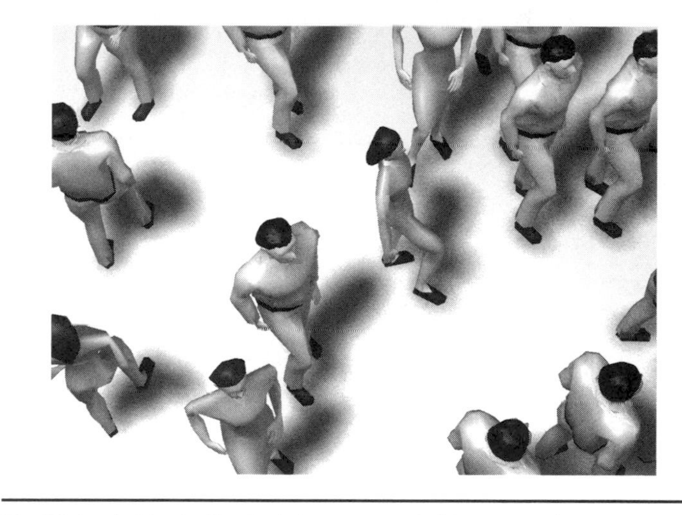

Fig. 19. Level of Service Standards (representative) of Queuing in the A to B range.

Queuing standards

Queuing Level of Service standards are based on human dimensions, personal space preferences and pedestrian mobility. The designer should not only apply queuing standards in areas designed primarily for pedestrian waiting, such as elevator and theater lobbies, but rather in other areas in which queuing is likely to result from service stoppages of inadequate capacity of pedestrian service facilities. Pedestrian holding areas on the approaches to stairs, or other critical areas, should also be designed to hold waiting pedestrians. Areas such as railway and bus platforms have critical pedestrian holding capacities, which, if exceeded, can cause persons to be injured by being pushed onto tracks or roadways. In addition to their ability to hold standees, queuing areas have different internal circulation requirements, based on the type of use. For examples, an airport baggage claim area must be capable of holding persons waiting for baggage as well as those moving out of the area with baggage. Queuing Levels of Service that are as dense as Level D are not recommended, except as a temporary exigency, such as traffic crosswalks. Levels E density is not acceptable except for elevators. Level F is unacceptable in any designed usage. Illustrations representative of Queuing Levels of Service from A to D are shown in Figs. 19 and 20.

Queuing: Level of Service A—Free Circulation Zone

Average Pedestrian Area Occupancy: 13 sq. ft. (1.2 m sq.) per person, or more.
Average Inter-person spacing: 4 ft. (1.2 m) or more.

At Queuing Level of Service A, space is provided for standing and free circulation through the queuing area without disturbing others. Examples would include better-designed passenger concourse areas and baggage claim areas.

Queuing: Level of Service B—Restricted Circulation Zone

Average Pedestrian Area Occupancy: 10-13 sq. ft. (.9-1.2 m sq.) per person.
Average Inter-person spacing: 3-1/2 to 4 ft. (1.1 to 1.2 m).

At Queuing Level of Service B, space is provided for standing and restricted circulation through the queue without disturbing others. Examples would include railroad platforms and passenger concourse areas.

Queuing: Level of Service C—Personal Comfort Zone

Average Pedestrian Area Occupancy: 7-10 sq. ft. (.6-.9 m sq.) per person.
Average Inter-person spacing: 3 to 3-1/2 ft. (.9 to 1.1 m).

At Queuing Level of Service C, space is provided for standing and restricted circulation through the queuing area by disturbing others. It is within the range of the personal comfort body buffer zone established by psychological experiments. Examples would include ordered queue ticket selling areas and elevator lobbies.

Box. A:
Principles of safety and security in public places

The following are based upon principles established for City of Nottingham, England (Source: Wekerle, Gerda and Carolyn Whitman, *Safe Cities: Guidelines for Planning, Design and Management*. New York: Van Nostrand Reinhold 1995)

1 Design for pedestrians to move about in well-lit, wide-circulation routes that reflect existing patterns of movement.

This principle asserts that well-used and vibrant streets are essential for safety. Measures that enhance street activity, frequency and intensity of use also encourage safety. Measures that detract from street life may increase risks, such as overhead walkways and underground malls.

2 Consider safety of people and property together rather than separately.

In the traditional Neighborhood Watch and Business Watch approaches, it is sometimes assumed that the concern is about robbery rather than personal safety. Street crime, assault and harassment—less visible than broken windows and stolen property—have higher economic and social costs.

3 Use opportunities for enhancing natural surveillance.

Responsible business owners, neighbors and strangers are the best defenses against crime. Urban plazas, parks, pathways and parking lots should be designed not only to allow people to see and be seen but also be provided with call boxes and graphics clearly indicating ways of helping yourself or getting help.

4 Provide good maintenance.

Good maintenance is crucial for lasting design improvements. Adding lighting does no good if not frequently cleaned or if burned out bulbs are not immediately replaced. A broken fence no longer fulfills its function. Critical maintenance items should be designed for easy access for cleaning, replacement and repair.

5 Make sure solutions to one problem don't cause another.

Safety is part of the integrated design objectives to improve urban quality and public amenity. Will the energy-saving benefit of reducing outdoor lighting impact the level of safety? Will fencing a playground create an entrapment zone? Active, carefully designed, well lit and well-maintained urban places help address all such concerns.

6 Involve neighborhoods in public safety planning.

Local business owners, residents, and community leaders will know local crime and nuisance patterns that pose safety risks and should be invited to serve a significant role in creating a comprehensive urban design plan for safety and security.

Fig. 20. Level of Service Standards (representative) of Queuing in the C to D range.

Queuing: Level of Service D—No Touch Zone

Average Pedestrian Area Occupancy: 3-7 sq. ft. (.28 to .6 m sq.) per person.
Average Inter-person spacing: 2 to 3 ft. (.6 to.9 m)

At Queuing Level of Service D, space is provided for standing without personal contact with others, but circulation through the queuing area is severally restricted, and forward movement is only possible as a group. Examples would include motorstair queuing areas, pedestrian safety islands, or holding areas at crosswalks. Based on psychological experiments, this level of area occupancy is not recommended for long-term waiting periods.

Queuing: Level of Service E—Touch Zone

Average Pedestrian Area Occupancy: 2-3 sq. ft. (.19 to .28 m sq.) per person.
Average Inter-person spacing: 2 ft. (.6 m) or less.

At Queuing Level of Service E, space is provided for standing but personal contact with others is unavoidable. Circulation within the queuing area is not possible. This level of area occupancy can only be sustained for short periods of time without physical and psychological discomfort. The only recommended application would be for elevator occupancy.

Queuing: Level of Service F—The Body Ellipse

Average Pedestrian Area Occupancy: 2 sq. ft. (.19 m sq.) per person, or less.
Average Inter-person spacing: Close contact with surrounding persons.

At Queuing Level of Service F, space is approximately equivalent to the area of the human body. Standing is possible, but close unavoidable contact with surrounding standees causes physical and psychological discomfort. No movement is possible, and in large crowds the potential for panic exists. ■

REFERENCES

Fruin, John J. *Pedestrian Planning and Design.* New York: Metropolitan Association of Urban Designers and Environmental Planners, Inc., 1971, reprinted in 1987. Mobile, AL. Elevator World Publications.

"Pedestrian Circulation" in Charles W. Harris and Nicholas T. Dines *Time-Saver Standards for Landscape Architecture*. New York: McGraw-Hill 1998.

Roberto Brambilla and Gianni Longo

Summary
This article offers guidelines and checklists for design of pedestrian streets, created as traffic-free zones. The term "traffic-free zone" defines areas where motor vehicle traffic is banned or curtailed and priority given to pedestrians, often served by public transportation. A pedestrian street can attract a diversity and range of activities, drawing all ages and social strata, tourists and residents into the public life of the city.

Key words
central business district, citizen participation, pedestrian planning, survey methods, traffic calming, traffic-free zone, way finding

PHOTO: Donald Watson

Pedestrian street in Boulder, Colorado

Pedestrian zones: a design guide

Beginning in the early 1960s, urban writers such as Lewis Mumford, Bernard Rudofsky, Jane Jacobs and William H. Whyte, among others, advocated for the potential of returning pedestrian street life to cities engulfed by traffic. Studies by sociologists Herbert Gans and Nathan Glazer, among others, pointed out the failure of early urban renewal attempts. European pedestrian experiments provided examples for a new approach to renewal, appropriate for preservation and revitalization of historic urban centers.

A pedestrian street often features amenities for relaxation, entertainment or broader set of activities. The space may be landscaped with trees, flowers and water. Fountains can be a focus of activity for children as well as adults. Street lighting is usually proportioned to pedestrian scale, with variations for dramatic effect and for safety. Amphitheater or speaker platforms are often included in a street's design to encourage use of the street as a public place. Benches and tables provide places for socializing. Design of pedestrian precincts described in this article include:

- **Pedestrian districts.** Characterized by eliminating vehicular traffic over a portion of a city and considered as a unit for architectural, historic or commercial reasons. Many European cities have adopted this type of traffic-free zoning because it suits the physical conditions of historic central areas.

- **Pedestrian streets.** Pedestrian streets are individual streets from which traffic has been eliminated. Emergency vehicles, however, have access, and service and delivery trucks are often allowed during restricted hours. The term "pedestrian street" is synonymous with "pedestrian mall," a term that was first used in North America to describe traffic-free zones, such as a central street.

Many pedestrian streets have been built as an omnibus solution to all of a city's problems, and in such cases they have usually failed. Unless a traffic-free zone is conceived in the context of an overall city effort to solve its problems, it cannot succeed for long. Furthermore, the project must be designed at a time and for a place where it can function as designed. Some pedestrian streets have simply been built too late, after stores and people have already deserted the downtown. In such cases, eliminating cars is unlikely to turn the tide.

1 ESTABLISHING DESIGN PERFORMANCE CRITERIA

Especially as one involves municipal, business and community groups, it is important to establish clear design objectives that are responsive to the requirements of the stakeholder community. Design objectives should be stated in terms of performance criteria to be used as a basis for evaluation of the design proposals.

In formulating design objectives within a community, the experience of another city, while familiar to the community, may not be applicable to every situation and to the proposed project context.

Credits: This article is based on "Handbook for Pedestrian Action," Columbia University Center for Advanced Research in Urban and Environmental Affairs sponsored by the United States Department of Housing and Urban Development, National Endowment for the Arts and the U.S. President's Council on Environmental Quality. 1977.

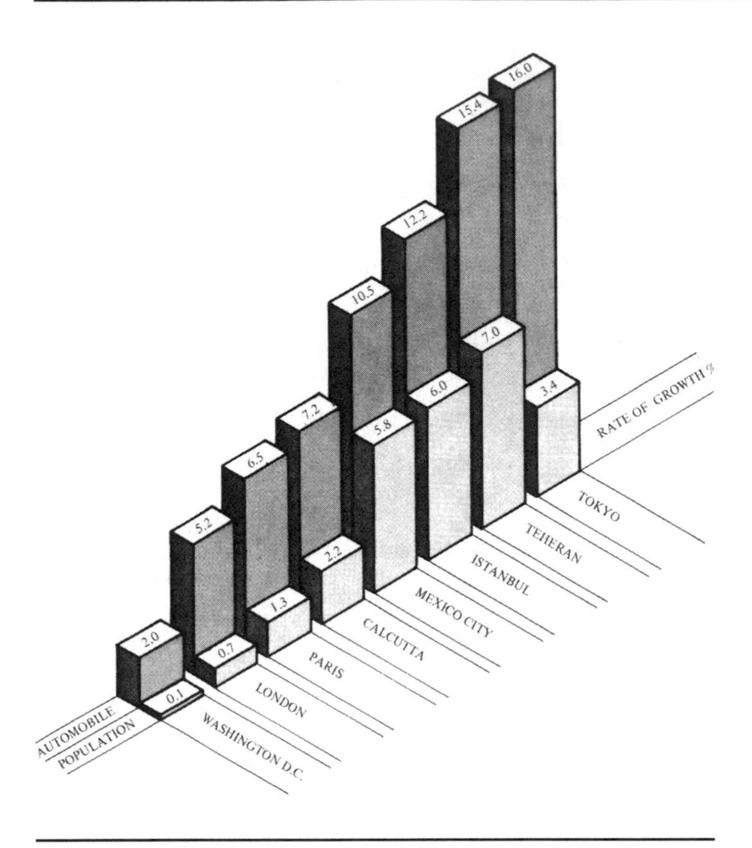

Annual growth rate of population and private automobiles in selected cities 1960–1970. Data obtained from: World Bank Sector Policy Paper, Urban Transport (Washington: World Bank, May, 1975).

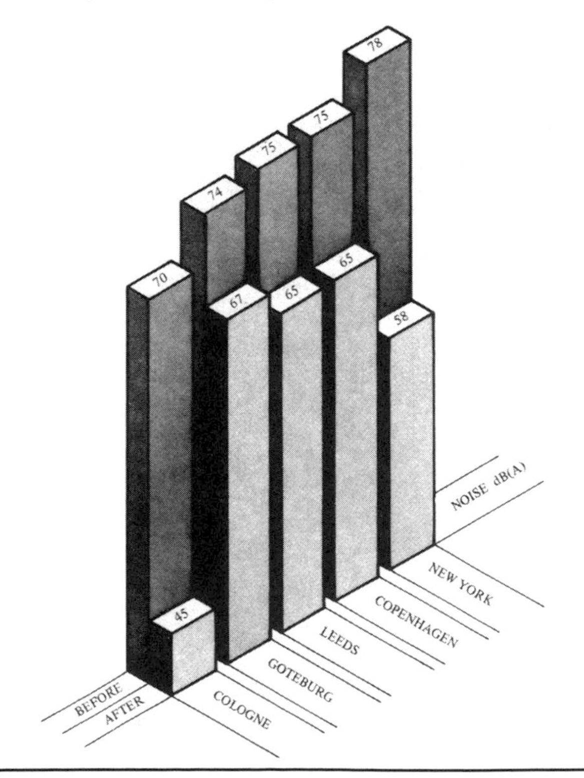

Noise pollution levels before and after the banning of vehicular traffic from city streets. Data obtained from: *Pedestrian Streets*, op. cit. *Streets for People*, op. cit.

The design team can help to determine which proposals are feasible under which circumstances and to map out alternative. The design must support the intended activities in the pedestrian street community, not the reverse, since the ultimate goal of design is to translate functional, economic, social and environmental objectives into concrete terms.

Grouping a list of daily activities into a reasonable number of categories can facilitate design proposals. The following divisions provide a workable model:

- **Functional objectives:** operation and use of the street, its accessibilty, mobility, maintenance and services.
- **Improved mobility:** for people, goods, service and emergency vehicles, and access for maintenance of the utility system.
- **Improved access by car/or public transit:** This may involve updating street or highway patterns, modifying existing bus and/or subway lines, or adding a few more facilities. Signs, lighting, and traffic information may need improvements. The project should also include adequate parking. Improvements in connections between parking, public transit and pedestrian areas can replace the loss of immediate parking spaces.
- **Social and community improvement:** Participation in, as well as the observation of, activities such as resting, talking, playing sports or games, celebrating festivities and events, monitoring children at play, babysitting, promoting educational and cultural performances.
- **Economic improvement:** meeting the needs and requirements of income producing functions related to office and commercial activities.
- **Environmental improvement:** improving the quality of the physical environment by preserving and restoring historic landmarks, providing green areas, flowers and trees, fountains and ponds and enhancing the visual aspect of the overall street scene.

These criteria are discussed in detail in the remainder of this article.

2 TRAFFIC MANAGEMENT

A pedestrian street discourages if not eliminates the use of private vehicles. A pedestrian street must be seen as part of a traffic system plan, that is, the success of the pedestrian zone depends on a larger, overall traffic plan that address the urban-wide transportation needs. The plan must address potential problems of increased traffic and parking on adjacent streets. In many cities, such as Portland, Oregon, the downtown pedestrian zones are supported by an affordable, efficient and attractive public transportation

It must provide for goods delivery and servicing. A common variation, as in Boulder, Colorado, is to maintain traffic on streets perpendicular to the pedestrian street, with limited and traffic-calmed crossings. A variety of strategies have been adopted to permit deliveries to stores, which face onto pedestrian areas. Rear servicing has been the most successful method. In cases where rear service has proved impractical, delivery access within a pedestrian area has been permitted on a restricted basis.

As an assist to pedestrian movement within a large pedestrian precinct, people movers have been employed frequently in the United States. Pomona, Fresno and Sacramento, in California, provide colorful, electrically powered vehicles for shoppers' convenience.

Improving public transportation is fundamental to the success of a pedestrian zone. Once traffic restrictions are initiated in an urban center, public transportation provides the easiest access for everyone to shopping and business areas. Transitways are a compromise between a completely traffic-free district and the need for a shoppers transportation system on the pedestrian street.

Another alternative of public transportation has been "para-transit" systems, such as car pools and subscription car and bus services. Para-transit options also include demand-response services, such as shared taxis and dial-a-ride service.

In a number of North American cities, pedestrian streets have been created to provide better accessibility to downtown stores and offices. Therefore, emphasis has been placed on reorganizing traffic patterns, rather than reducing the number of automobiles. The point of traffic-free zoning is not to eliminate the automobile, but to separate vehicular and pedestrian movement. In order to accomplish this, most cities have reduced, as much as possible, the number of streets crossing the pedestrian zone, and have rerouted cross-town traffic so that major streets do not cut through pedestrian areas.

In order to compete with the convenience of suburban shopping centers, cities must provide parking near to and accessible to and from the pedestrian precinct. Parking is usually a problem in downtown areas, and the creation of pedestrian zones adds to the problem by eliminating a number of spaces. The solution has been to build garages with direct entrance onto pedestrian areas. Part of the success of Nicollet Street in Minneapolis has been due to the creation of fringe parking areas that greatly improved downtown accessibility.

3 ECONOMIC REVITALIZATION

A pedestrian area has the potential for helping central business districts to remain viable commercially through a combination of improved retail trade, new investors and new development. To compete with regional shopping centers, downtown commercial districts must offer a wide range of shopping opportunities to customers and present a vital economic image.

There are times however when the best of planning, public interest and good will cannot sustain the economic gains of a traffic-free zone. The key to success seem to be timeliness. In some cases, the trend toward decay is irreversible. In such cases, even a well designed, easily accessible pedestrian district fails.

Though merchants are the first to benefit from a street's success, they do not always support traffic-free zones. They sometimes oppose car limitations, fearing a loss of customers, or else are concerned with vandalism and changes in the type of shoppers. One of the most common results of conversion to pedestrian streets is a change in the type and quality of shops. Luxury and specialty stores seem to stay and thrive, bringing regional and national attention to the central shopping

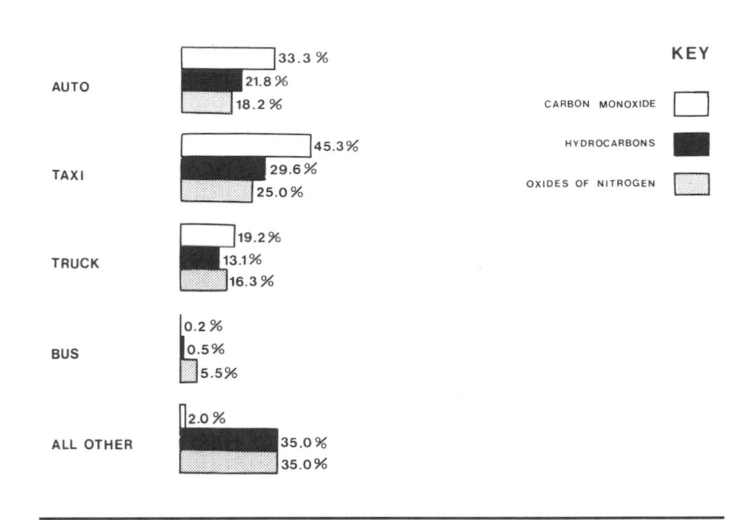

Sources of air pollution in midtown. The above data were prepared by New York City's Office of Midtown Planning and Development, during the battle for the Madison Avenue Mall. Source: Department of Air Resources.

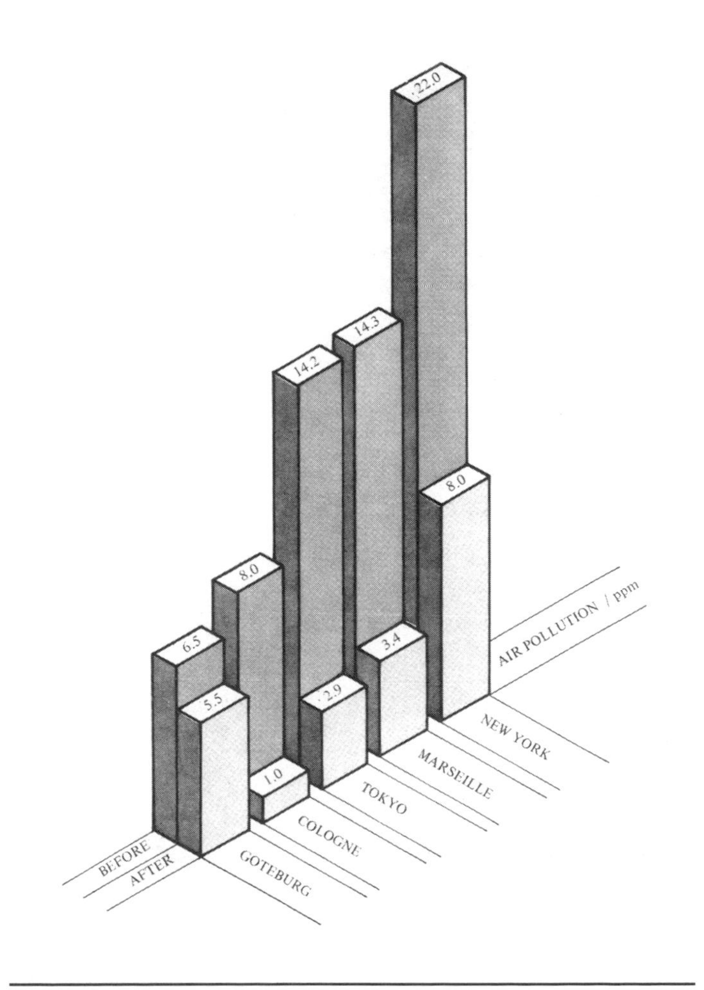

Air pollution levels before and after the banning of vehicular traffic from city streets.

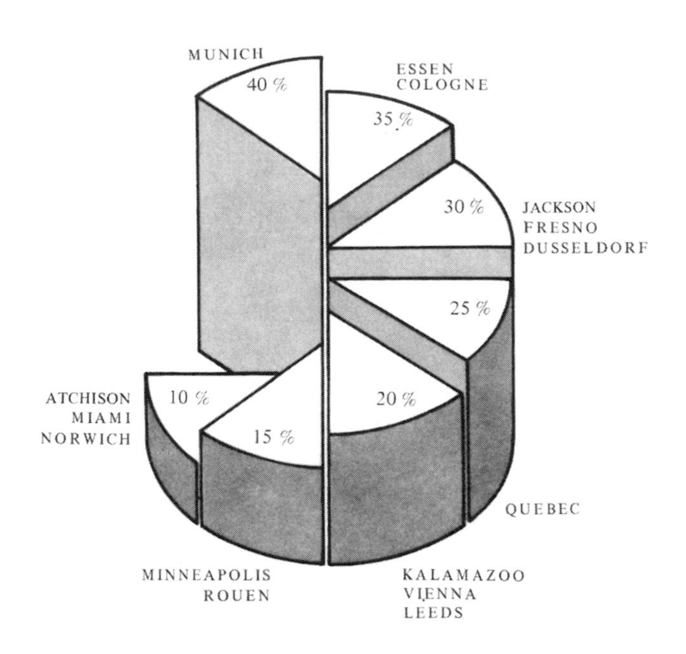

Retail sales increase in European and North American pedestrian districts. Data obtained from: Laurence Alexander, op.cit. *Streets for People* (Paris: OECD, 1974).

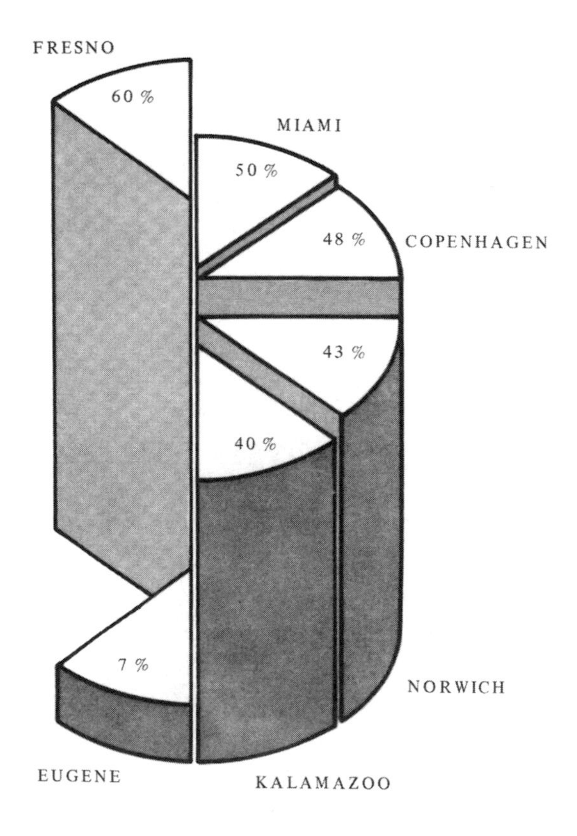

Increased pedestrian flow in traffic-free zones. Data obtained from: Laurence A. Alexander, *Downtown Malls,* an annual review, vol.1 (New York: Downtown Research and Development Center, 1975).

areas. When shopping and sightseeing are combined in a traffic-free area, tourism profits.

Ideally, a traffic-free zone should be used during the night as well as day if the maximum economic potential is to be realized. In order to stimulate round-the-clock activity, recreational facilities, including theaters, restaurants and movie houses are also typically included in the plans. The introduction or reintroduction of housing near traffic-free zones should be given priority.

4 ENVIRONMENTAL IMPROVEMENT

In the United States, concern about urban air pollution led to the passage of the Clean Air Act of 1970. Mobile air pollutant sources account for roughly 95 percent of carbon monoxide emissions, 65 percent of hydrocarbons, 40 percent of nitrogen oxides, 15 percent of suspended particulates, and 50 percent of photochemical smog formed by atmospheric reaction of hydrocarbons and nitrogen oxides.

Given the truck and car traffic's negative impact on the air-quality and noise level of the urban environment, traffic-free zones has been seen as a useful tool to reduce pollution, as well as to improve the appearance of downtown areas and the preservation and enhancement of historic districts.

Two factors influence urban air quality: street level pollution, and upper-level pollution. Upper-level pollution, less concentrated but distributed throughout the city, comes from vehicles as well as stationary sources and is rapidly dispelled in most cities. Street level pollution, on the other hand, is highly concentrated and local. It is not so easily dispelled, especially in enclosed urban spaces. Along with encouraging nonpolluting forms of transit, such as electric vehicles, the most effective way to clean the air is to limit automobile traffic.

The human experience and physical image of downtown is seriously affected by the presence of cars. Pedestrians don't need large conspicuous signs to draw their attention when the distraction of traffic isn't hindering their ability to perceive their environment. Coordinated and subdued graphics help to create visual and sensory coherence. Particularly in pedestrian zones that seek to encourage tourists and new visitors, there is a need and opportunity to utilize way-finding and interpretive graphics, just as one would for a venue that experience a large number of first-time visitors.

Creating a pedestrian-friendly streetscape is among the first and often least-cost options for creating the new appearance of a pedestrian zone. Repaving streets to sidewalk level has the effect of eliminating accessibility barriers established by a separation of road, curb and sidewalk, while adding a sense of formal unity to the entire street. Pedestrian-oriented lighting helps to restore a more intimate and natural scale to converted streets. Landscaped islands, sculpture and linear parks set into an existing right-of-way are elements in the design of traffic-free zones. They too provide a cost-effective means to enhance the appearance and functional benefits of plants and trees to urban areas.

In both Europe and North America, the desire to introduce human scale and amenities into urban centers has become a conscious aspect of the effort in shaping the design of traffic-free areas. A coherence of appropriate street furniture—including benches, waste and recycling collectors, fountains and transit stops—have been introduced in pedestrian precincts and streets, which cities can use to bring character to the downtown environment.

Pedestrian zoning is part of the restoration of historic districts. In numerous cities, exemplified by European examples, traffic-free zoning has been a conservation measure to restore the unity of the historic urban fabric. Elimination of cars is often required, in order to mitigate air-borne pollution and vibrations that destroy physical structures.

5 SOCIAL BENEFITS

Pedestrian zoning gives the greatest range to pedestrian activities; an environment free from the restraints imposed by traffic in order to offer unhampered movement, social amenities and comfort to walkers. As density increases in cities, allocation of open space decreases to the point where pedestrians are squeezed into the leftover spaces between traffic and buildings. There are many methods for locating usable space to create an environment for pedestrian activity, improve the social image of the city, protect the pedestrian and reduce vehicular accidents.

Providing socially responsive places in crowded downtown areas is an important service to an urban population. Visitors and residents alike are affected in their image of the city by the evidence and accessibility of pedestrian activities. Towns in which the streets are populated both day and night seem welcoming, especially for visitors whose perceptions of a city's character are determined by its street activities.

Safety is improved by pedestrian zoning. Traffic-free zones are often created for the purpose of reducing traffic accidents. Pedestrian deaths caused by automobiles represent 20 percent of all traffic fatalities in the United States and 30 percent in Europe. This makes safety in downtown areas an important issue. The increase of hours and intensity of use, along with improved lighting and accessibility help to create an atmosphere of security, which discourages crime.

6 COMMUNITY PARTICIPATION

Traffic-free zoning proposals have often generated controversy. Potential opponents may include those most directly affected by the project, such as merchants on the site who fear a loss of business if the project fails; taxpayers who object to public funds being spent on a pedestrian zone; taxi drivers, truckers and private bus companies which see traffic-free zoning as a threat to their interests; automotive firms which tend to object to any street being closed off. City traffic planning agencies often resist the idea of having the present street system disrupted. Financial interests, such as investors in suburban shopping malls, may represent competing proposals.

The success of a traffic-free project can depend upon a group's capacity to involve all stakeholders and interested individuals in the planning processes, sharing of actual (not token) decision making. Those affected by pedestrian projects should always have the opportunity to participate in the decisions that influence their various roles in the community.

Public liaison and promoting a traffic-free zone can be as important as the resulting improvements. Often, after years of decline and congestion

This type of questionnaire can help take the pulse of one's community.

Dear Neighbor:

Some of your neighbors on the block are interested in improving it by making it more attractive. Your answers to the following questions are to let us know if you are interested in the idea, as well as what you feel should be done and how much you'd be willing to do to improve the block. We'll get back to you with the results of the questionnaire shortly.

Thank you!

1. How long have you lived on the block?____
2. Do you like living on the block?____
3. Do you rent or own the place where you live?____
4. Do you work?____ If so, what do you do?____
5. Age____ Male or female____
6. How many people live with you?____
7. What are their ages and sexes? ____
8. Do you find the block attractive-looking?____ unattractive?____ neither?____
9. Do you ever use the street as a place to chat, sit in the sun, play games? Please describe briefly the ways in which you use your street.____
10. If additions were made on the street, which of the following would you like to see:
    a) additional lighting____
    b) banners____
    c) benches or seats____
    d) community bulletin board____
    e) more attractive paving____
    f) painted wall murals____
    g) fresh paint on fences, building fronts, etc.____
    h) trees and plants____
11. If a vacant lot on the block or nearby is available, would you like to see it cleaned up, decorated and turned into:
    a) a small park____
    b) a playground____
    c) a garden____
12. If street changes are made, should they be designed especially to serve any of these groups: children____ elderly people____ young adults____ adults____ teenagers____ everyone____
13. Where on the block would you like to see changes? At the ends____ (which one? or both?)____ in the middle____ elsewhere____ (please say where)____
14. Do you think the kinds of changes suggested in this questionnaire would be popular or unpopular on the block?____ Please give the reasons for your answer:____
15. Would you help to make the street more attractive by contributing time?____ money?____
16. Do you have a skill that would be useful in making changes on the block, such as carpentry, masonry, painting, sewing, designing, working with electricity?____
17. Would you be willing to volunteer this skill – using your own talents and perhaps teaching or supervising others – to help improve the block?____
18. Are you in a profession whose services might be of use to a block association (such as law, insurance, communication)?____
19. Would you be willing to contribute time and advice to help your block association in a professional capacity?____
20. Do you have any suggestions for improving the block?____

Source: Mary Grozier and Richard Roberts, *New York's City Streets.* Council on the Environment of New York City: New York

in a downtown area, people have forgotten its potential and are not aware that a pedestrian project could benefit them. Local merchants, fearing change, imagine a loss of retail trade. It is often necessary to stimulate their imaginations in a direction that has lain dormant for years, or has never existed at all.

Special events are among the best ways to promote the pedestrian zone. Through games, happenings, star attractions, contests and prizes, a traffic-free district can be publicized before and after completion. The events should be clearly identified with the site, either by name, location or sponsor, and preferably all three.

By way of summary, the following list of reasons to create traffic-free zones provides some a useful summary of the above points:

- **To attract people.** This creates more opportunities for shopping, socializing, business and fun, along with increased financial sources for both the citizens and municipality.

- **To provide a sense of place that strengthens community identity and pride.** This improves community relations and reduces feelings of alienation, while creating a place for all people to congregate.

- **To reduce noise and air pollution.**

- **To provide a safe and attractive environment.** A pedestrian zone is part of the community's public environment, accessible to people of all ages and abilities.

- **To improve the visual environment.** Signs, lights, spaces, colors and textures can be designed to relate to the person on foot, rather than to the person on wheels.

- **To promote urban conservation** through planting and landscaping and cultural conservation including preservation, building restoration and renewal.

- **To increase property values** and, consequently, the city's revenue from real estate taxes.

- **To invite walking and to permit special rights-of-way to be reserved for bicycles and public transportation vehicles.** This improves mobility through the city center and helps save energy.

- **To decrease the number of motor vehicle related accidents,** saving lives, police work and judicial time.

- **To promote citizen participation** in the planning and implementation of the pedestrian area, as an instrument for public education and engagement in urban life.

7 PLANNING SURVEYS

Planning a traffic-free zone is a complex process. Potential problems and possible solutions, as well as the results from extensive analysis of local physical, economic and social resources must be properly synthesized and evaluated. The survey should identify all of the social, economic and physical factors that might influence a pedestrian project: it should document relationships, as well as recorded statistics.

The planning survey is a guide to a street's feasibility. Many of the data can be determined easily and informally, at least for preliminary guidance. Even when results are incomplete or not conclusive, they chart a general course for further study and advocacy efforts. The survey has three objectives:

- It provides information about the city and the potential impact of a pedestrian project, so that sponsors and designers can base their decisions upon a concrete assessment of community realities.

- It provides tools for sponsors and designers so that they can undertake the project effectively and gain the financial and political support necessary for a full-scale feasibility study.

- It provides information, which may be useful in the full-scale professional feasibility study. This information, brought together at minimal cost, can also help to narrow and define professional participation in later stages of the project.

To aid the planning survey process, the following summary provides analysis checklists for the significant factors involved in design of a pedestrian area:

Natural environment

Although the residents of a city are aware of its general climate and outdoor comfort levels, objective data on this subject is useful to the urban advocate and designer. Climate data is available from local airports agricultural stations, as well as published sources.

- **Temperature and humidity:** profile of comfortable outdoor weather, range and variation of temperatures.

- **Precipitation:** amount and frequency of average rainfall and snow.

- **Wind:** force, direction, frequency and effect.

- **Solar access:** with reference to the prospective pedestrian street, the sun exposure (documented by orientation and shadow patterns on ground areas and facades.

- **Landscape:** documentation of successful native plant and tree varieties.

Built environment

Existing development patterns provide the starting point for design proposals. What has already been built along the street route establishes a setting for the project, its infrastructure and definition of its initial amenities or restrictions. This portion of the survey is an assemblage of basic data about the buildings and other characteristics of the street site. The information can be assembled from maps, photography, observation or local interviews, and city's planning records.

- **Buildings and open space:** approximate ratio and general patterns of use.

- **Type, scale, configuration, and materials:** architectural appearance of the area and its specific design properties.

- **Function:** approximate number and ratio of residential, office, commercial, institutional, recreational, governmental and retail space.

- **Ownership:** description of property ownership.

- **Outdoor services and amenities:** existing benches, telephone booths, kiosks, fountains, restrooms, planters and other such elements of the street site.

Circulation patterns

Basic traffic patterns, route capacities, congestion figures and overall concerns about mobility and accessibility within the proposed pedestrian area need to be quantified as part of the advocacy process, since the advocate is frequently called upon to define a project in these terms. Data is usually available from local transportation planning agencies, although additional statistical research in the field may be required.

- **Patterns:** peak and off-peak traffic flow statistics and density figures. Average speeds at varying times during the day and night for the full range of transportation modes.

- **Actors:** description of pedestrians, bicycles, cars, trucks, taxis, subways, trolleys, service vehicles and so forth.

- **Traffic accidents (documented from police reports):** Data if available helps to describe conflicts of pedestrians, bicyclists and vehicles.

- **Facilities:** inventory and capacity of the street system (roads, tracks, sidewalks, plazas, and parking spaces).

- **Emergency:** existing and potential road access on the pedestrian route for delivery and servicing purposes, including ambulance, fire, and police vehicles.

Utility network

Streets are often proposed for the oldest parts of the downtown areas, where utilities may be the most outdated and in need of repair. The proposed project often provides an incentive for improvements. Its design must allow access to utility networks for maintenance and operation, so it is important to know the location of utilities in the proposed pedestrian area and how they are reached, above or below ground level. Data may be obtained from the utility companies and the city agencies.

- **Gas, steam and fuel oil:** supply routes and equipment, access for maintenance and service, system condition.

- **Water, electricity, telephone, sewers:** maps of existing conditions and service requirements, along with planned improvements. Normally documented on city maps.

Physical conditions and trends

Since the pedestrian street tends to improve the physical conditions in a given urban area, the current conditions within a proposed pedestrian zone provide important arguments for the project. This information should be quantified and made available for future professional requirements but is also useful for the formulation of arguments in direct dealings with the community.

- **Pollution:** sources of air-, water-, noise pollution with notes if conditions improving or deteriorating.

- **Planned construction:** plans for new buildings or other development in the proposed pedestrian street site area.

- **Condition of existing structures:** assessment of condition of buildings, roads or transportation facilities, utilities. An exterior assessment often provides data sufficient for initial planning.

- **Streetscape:** general features and appearance of the existing street, its landscaping and furniture, as well as sidewalk curbs and street surface conditions.

- **Building appearance:** including those of special architectural or cultural interest. (Detailed assessment of structural conditions may be required subsequently.)

Demographics

It is important to identify the potential users of the proposed street site, since they will either be served or displaced by the street. While a new pedestrian zone should attract new elements to the area, it should also serve the legitimate needs of its current inhabitants, and should indicate the categories of people not already served by or drawn to the specific projects. A knowledge of their background and interests can establish a framework within which to provide additional facilities intended to draw them back into the city. This information can be obtained from direct observation and interviews. Social service agencies and other municipal offices can also provide data.

- **Age:** Children with parents, school children, teenagers, young, middle-aged or elderly adults frequent the street.

- **Income range of users:** present and potential (helpful for market projections.)

Community conditions and trends

Streets often gain their greatest support in areas where urban decay threatens the status quo. The survey should focus on the existing social conditions and trends that the street may be able to improve. If conditions and outlook are already bright, a proposal must be formulated to add substantially to the area's image or environment. Local observation and interviews are again the best survey approach. The capacity of local stakeholders to be involved significantly should also be assessed, since they have the largest commitment to the long-term improvement of their area. There may be wide-ranging opinion from support to opposition. No matter how far apart the different stakeholders may appear to be at first, some basis of mutual agreement can be sought, such as reduction of crime, improvement of community quality and improved investment opportunities.

- **Crime:** increasing or decreasing vandalism, violence, prostitution, robbery, and related incidents in the area.

- **Stakeholder capacity:** capacity for local leadership to represent and involve interested organizations, sponsors and participants, including businesses, banks, civic and faith-based organizations, neighborhood associations, and other advocates of community stakeholders.

Commercial and market conditions

Commercial operations in the pedestrian area may be the group most directly affected by the street's design, construction and financing. Their interests should be determined as early as possible in the advocacy process. Door to door checks, or talks with the local Chamber of Commerce will yield useful information.

- **Role of local commercial interests** in the central business district.
- **Regional market statistics for city and region.**
- **Local market analysis:** types of shops possible on designated pedestrian route, size, old or new, service or amenity oriented, owner- or tenant-operated. Quality of service and merchandise.
- **Ownership trends:** rate of turnover, buildings owner-operated, locally owned, or controlled by absentee landlords.
- **Commercial and residential activity:** present and projected, in areas of the proposed pedestrian project, including types and sizes of business and residences.
- **Industrial:** A great deal of industrial activity is unlikely to function on the street, but some light industry may survive there. Adjacent delivery and pick-up networks could make introduction of a pedestrian street difficult. Since manufacturing and other industrial processes rely on low rents and overhead to increase their profit margins, many may be threatened by the increased property values and tax rates a street could bring.
- **Business profile detail** (appropriate where commercial and market analysis is critical to success of the project): number, size and types of businesses, number of employees, of what skills, income and interests, hours of operation, service requirements and logistics that may conflict with or aid pedestrian activities.

Transportation

Available transportation facilities and use contribute to the economic health of an area and should be included in this survey. Government transportation agencies typically have data on existing or planned facilities and services. Deficiencies can be accounted for in the street proposal.

- Population served (determine "service area" by the radius of typical travel time and distance of users and visitors that the pedestrian street might attract.)
- Means of transport and routes, frequency and capacity, including bus, rail, minibus and related stops and connections.
- Bicycles, including connections with existing or prospective bikeways and/or greenways.
- Parking, present and projected.

Real estate conditions and trends

The feasibility of a pedestrian street project is most often based on the positive output of a cost-benefit analysis. This analysis largely depends on the evaluation of local economics. If the conditions are good, the street can maintain and accelerate the present trend. If they are decaying, the plan must propose to reverse and ameliorate the situation.

- Real estate trends
- Sales tax trends
- Vacancy rates in stores
- Housing occupancy and density
- Investment patterns

Administrative and legal resources

Preliminary analysis can serve as an introduction to laws, practices, policies and precedents that apply to pedestrian zoning. In many cases appropriate legal or administrative provisions for a street will not exist and must be improvised or created through legislative authority. A brief survey should reveal precedents and applicable regulations. Data can be found through municipality planning, transportation, or building agencies.

- Specific street legislation and regulations
- Zoning requirements and procedures
- Building codes and other applicable regulations
- Taxes and assessments: property taxes, frontage, street benefits to sales surcharge

In Box A and B on the pages that follow, checklists are provided for Central Business District assessment. These may provide a convenient supplement to the survey data items outlined above. ■

Box A: CBD PROBLEM CHECKLIST

1. GENERAL APPEARANCE
   .... a. Cluttered, unattractive entrances to CBD.
   .... b. Lack of landscape plantings and green spaces.
   .... c. Dirty, littered streets, sidewalks and alleys.
   .... d. Visual chaos of poles, signs and wires.
   .... e. Lack of design harmony among buildings.
   .... f. Lack of views, vistas, and visual focal points.
2. BUILDINGS
   .... a. Poorly maintained exterior appearance.
   .... b. Drab, uninteresting interiors.
   .... c. Functionally obsolete size and shape.
   .... d. Vacant upper stories.
   .... e. Dirty, cluttered rear entrances.
   .... f. Inharmonious remodeling.
   .... g. Absentee ownerships.
3. SIGNS
   .... a. Excessively large.
   .... b. Overhang public right-of-way.
   .... c. Poorly maintained.
   .... d. Gaudy, garish and ugly.
   .... e. Difficult to read.
   .... f. Poorly designed.
   .... g. Inharmonious with building architecture.
4. STREETS AND ALLEYS
   .... a. Too narrow for traffic and parking needs.
   .... b. Poor surface condition.
   .... c. Inadequate storm drainage.
   .... d. Lack proper markings and directional signs.
   .... e. Rough railroad crossings.
   .... f. Dirty and littered.
5. TRAFFIC
   .... a. Congested, slow moving.
   .... b. Inconvenient circulation pattern.
   .... c. Turning conflicts of intersections.
   .... d. Loading zone conflicts.
   .... e. Poor access routes to CBD.
   .... f. Through traffic conflicts.
   .... g. Excessive truck traffic
6. PARKING
   .... a. Insufficient number of spaces.
   .... b. On-street spaces conflict with traffic.
   .... c. Unattractive, poorly designed lots.
   .... d. Inconvenient location.
   .... e. Dirty, muddy or rough surface.
   .... f. Poorly lighted.
   .... g. Slow turn-over.
   .... h. Employees use prime customer spaces.
   .... i. Spaces too small, difficult to use.
   .... j. Obsolete fee structure
7. PEDESTRIAN FACILITIES
   .... a. Rough, broken sidewalks.
   .... b. High curbs.
   .... c. Pedestrian-automobile conflicts.
   .... d. Dark side streets.
   .... e. Unattractive routes between stores and parking areas.
   .... f. Lack of benches, fountains, rest rooms, phones, trash containers, and information centers.
   .... g. No protection from inclement weather.
   .... h. Narrow sidewalks.
   .... i. Excessive noise, dust, or objectionable odors.
8. LAND USE
   .... a. Lack of major shopping store.
   .... b. Non-commercial dead spots in shopping frontage.
   .... c. Excessive vacant buildings and land.
   .... d. Lack of room for expansion.
   .... e. Objectionable uses that produce noise, dust, odors, smoke, or traffic conflicts.
   .... f. Lack of compact, convenient retail core.
   .... g. Outlying retail uses compete with CBD.
9. UTILITIES
   .... a. Water system old and undersized.
   .... b. Inadequate fire demand storage, pressure, or hydrants.
   .... c. Sanitary sewer old and undersized.
   .... d. Poor storm drainage.
   .... e. Tangled mess of overhead wires.
   .... f. Inadequate, unattractive street lighting.
   .... g. Streets continually torn up for repairs.
10. MERCHANDISING AND CUSTOMER RELATIONS
    .... a. Limited selection and variety.
    .... b. Lack competitive pricing.
    .... c. Poor quality.
    .... d. Unattractive window and interior displays.
    .... e. Lack prompt and courteous attention to customers.
    .... f. Lack of product knowledge.
    .... g. Irresponsible service and maintenance practices.
11. COMMUNITY ATTITUDE
    .... a. No community concern or pride in CBD.
    .... b. Businessmen not interested in improving their stores.
    .... c. No public-private cooperation.
    .... d. It's too late to save the CBD.
    .... e. No imagination.
    .... f. Nobody wants to spend money.
    .... g. Don't need any help.
12. PLANNING ACTIVITY
    .... a. No active planning agency.
    .... b. No plans prepared for CBD improvement.
    .... c. Plan prepared, but gathering dust.
    .... d. Inadequate or no zoning ordinance.
    .... e. Inadequate or no construction codes.
    .... f. Inadequate or no sign regulations.
13. AREA TRENDS
    .... a. New regional shopping center within 60 miles.
    .... b. New highway connection to major cities.
    .... c. Declining rural population.
    .... d. Neighboring community has improved CBD.
    .... e. Unstable employment base.
    .... f. Workers commuting farther to larger cities.
14. OTHER
    .... a. ....
    .... b. ....
    .... c. ....
    .... d. ....

Source: Central Business District Improvement Manual for Iowa Communities (Des Moines, Iowa: Division of Municipal Affairs, Office for Planning and Programming, 1971)

Box B: A CHECKLIST OF POTENTIAL CBD ASSETS

1. GENERAL APPEARANCE
.... a. Interesting skyline if clutter eliminated.
.... b. Open spaces could be landscaped.
.... c. Vistas could be created to certain major buildings.
.... d. Unifying design concept possible among buildings.
.... e. Unnecessary poles, signs and wires could easily be removed.

2. BUILDINGS
.... a. Buildings with interesting architectural details.
.... b. Buildings of historical value.
.... c. Rear entrances could be improved.
.... d. Vacancies allow space for expansion.
.... e. Most buildings in sound structural condition.

3. SIGNS
.... a. Obsolete signs easily removed.
.... b. Interesting old signs to restore.
.... c. Sign panels harmonizing with building possible.
.... d. Flush mounted wall signs could be easily viewed.
.... e. Uniform "under-canopy" signs could be used.

4. STREETS AND ALLEYS
.... a. Wide main street right-of-way.
.... b. Pavement width could be reduced.
.... c. Traffic markings and signs easily replaced.
.... d. Regular street clean-up program could be initiated.
.... e. Street resufacing could be programmed.

5. TRAFFIC
.... a. Traffic could circulate around main shopping street.
.... b. One-way movements could be eliminated.
.... c. Some turning movements could be eliminated.
.... d. Through traffic could bypass CBD.
.... e. Better locations possible for loading zones or bus stops.

6. PARKING
.... a. Some on-street spaces could be eliminated.
.... b. Locations available for more off-street parking.
.... c. Existing parking lot design and layout could be improved.
.... d. Space available for landscaping and screening.
.... e. Employees could park in other locations than customer spaces.

7. PEDESTRIAN FACILITIES
.... a. Sidewalks could be widened.
.... b. Spaces for benches, fountains and restrooms could be created.
.... c. Landscaped arcade through mid-block to parking area possible.
.... d. Mall or semi-mall could be developed.
.... e. Canopy could be installed along entire street.

8. LAND USE
.... a. Reasonably compact shopping core.
.... b. Some non-CBD uses would be willing to relocate.
.... c. No major outlying commercial uses yet.
.... d. Good variety of retail uses.
.... e. Adjacent land available for expansion.

9. UTILITIES
.... a. Water and sewer systems feasible to improve.
.... b. Overhead wires could be relocated or placed underground.
.... c. Street lighting ready soon for replacement.
.... d. Storm drainage can be improved.
.... e. Utility companies cooperative and interested in community.

10. MERCHANDISING AND CUSTOMER RELATIONS
.... a. Good market available for quality goods and services.
.... b. Experienced and knowledgeable businessmen.
.... c. Active Chamber of Commerce.

11. COMMUNITY ATTITUDE
.... a. Strong community spirit.
.... b. Have worked for educational, recreational and medical service improvements in past.
..... c. Progressive local government.
.... d. Government and business will cooperate.
.... e. Take pride in high quality of local improvements.

12. PLANNING ACTIVITY
.... a. Active planning agency and planning program.
.... b. Good Zoning Ordinance.
.... c. CBD Plan has been discussed.
.... d. Public officials recognized value of professional assistance.
.... e. Funds available for continuing planning program.

13. AREA TRENDS
.... a. Trade area population has increased spendable income.
.... b. Larger farms require more goods and services.
.... c. Our community is center for governmental, educational or medical services.
.... d. Good recreational or tourism potential.
.... e. Metropolitan population decentralizing into our trade area.

14. OTHER
.... a. ....
.... b. ....
.... c. ....
.... d. ....

Source: Central Business District Improvement Manual for Iowa Communities (Des Moines, Iowa: Division of Municipal Affairs, Office for Planning and Programming, 1971)

Roberto Brambilla and Gianni Longo

Summary

European experience in design of pedestrian precincts is relevant to urban designers for many reasons: they illustrate a variety of goals and diversity of urban scale of a broad cross section of European experiments. Their design goals range from strictly functional ones dealing with traffic control strategies, to humanistic ones dealing with conversation of the urban fabric and improvement of residential conditions in central areas.

Key words

European cities, pedestrian zones, pedestrian planning strategies, traffic control

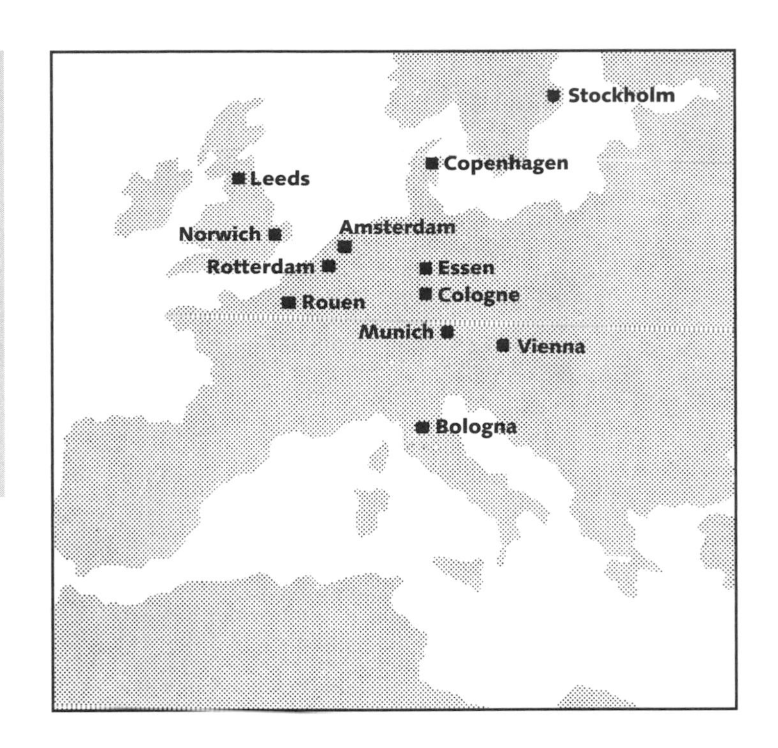

Pedestrian precincts: twelve European cities

In the past forty years, the cores of European cities have experienced an unprecedented revival, which is affecting not only their appearance, but also their functioning. Residential neighborhoods, historic districts and large groups of buildings have been restored. City streets have been freed from traffic and returned to pedestrians. Tourists and residents alike have made these new public spaces the focus of their activities.

As a conservation measure, traffic-free zoning has been introduced in numerous European cities to restore the unity of their historic urban fabric. Banning vehicular traffic has proved most effective in reducing noise and air pollution levels, and in reducing the damage caused to historic buildings by continuous vibrations and emissions from traffic. Economics are an important factor in pedestrian zoning. The restoration of housing stock, the reuse of abandoned buildings as community centers and the recycling of streets into pedestrian spaces have all been results of economic priorities. Tourism has profited enomously by combining shopping and sightseeing in traffic free areas.

STOCKHOLM
KEY PLAN – Stockholm's central business district:
1 The Torg
2 Drottninggatan
3 Biblioteksgatan
4 Bryggargatan

1 STOCKHOLM

While most pedestrian streets in Europe are glamorous but short, Stockholm's pedestrian system extends throughout the entire city.

Credits: This article is based on the study "Rediscovery of the Pedestrian: 12 European Cities" published by Columbia University Center for Advanced Research in Urban and Environmental Affairs 1976 and sponsored by the United States Department of Housing and Urban Development, National Endowment for the Arts and the U.S. President's Council on Environmental Quality.

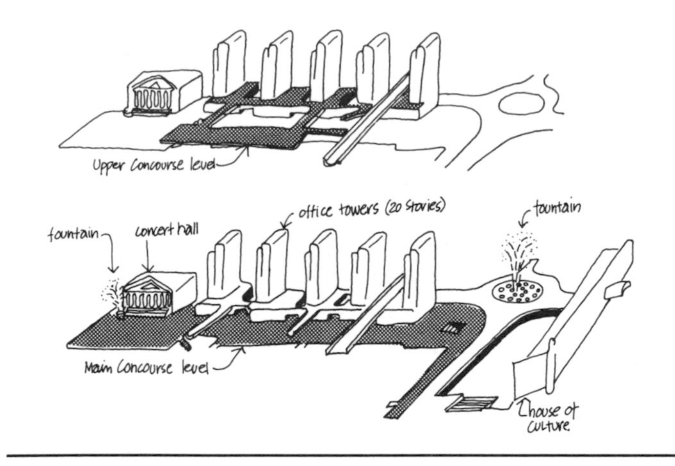

Vertical separation of pedestrian movement was included in the design of the Torg in Stockholm, Sweden.

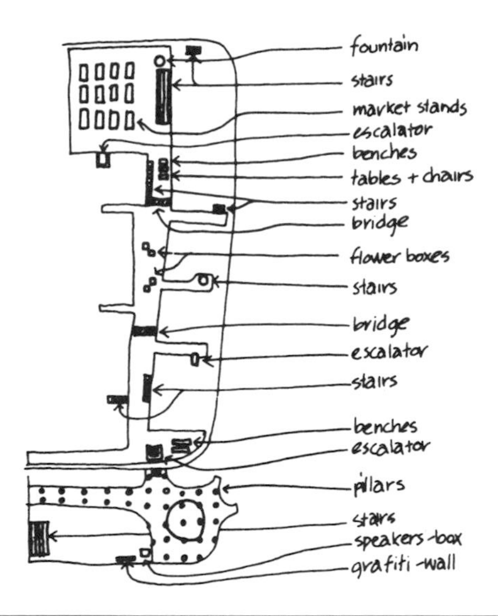

Facilitating equipment. Vertical separation of pedestrian movement was included in the design of the Torg in Stockholm, Sweden.

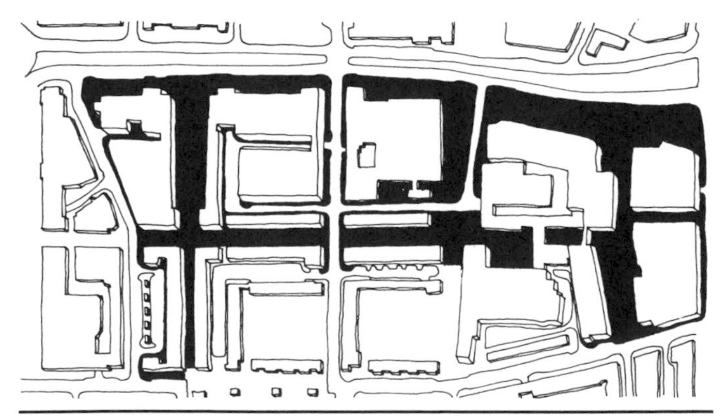

Profile of Functions in Lijnbaan

Shops	1,470,000 sq. feet
Offices	720,000 sq. feet
Professional Offices	210,000 sq. feet
Hotels, Restaurants and Cafes	140,000 sq. feet

In the mall there are almost 10,000 theater and movie seats, 1,000 hotel beds and 1,279 housing units. All of these contribute to a vigorous street life.

The major traffic free area in downtown Stockholm is the Torg, a multiblock mix of office towers, commercial and cultural buildings, planned in 1946 and largely completed by 1962. And in 1972 a major experiment was initiated in two residential areas—Ostermalm, close to the downtown, and Aspudden, a suburb built in 1910. These particular experiments were aimed at eliminating all through traffic from residential districts by segregating it to peripheral roads, thus transforming the interior streets into de facto walkways.

City profile

Stockholm, the capital and largest city in Sweden was founded in the middle of the 13th century and officially recognized as the capital in 1436. The medieval nucleus of the city remains largely intact. Today the street network and many of the buildings from that period still exist. The modern central business district developed within the so-called "stone town," a district which dates back to the 17th century.

Design features

Hard-edged in design and with little greenery, the Torg is nevertheless lively. There have been very few changes since it was conceived in the 1950s, and the hard, glossy materials reflect the aesthetic of that decade. A series of recurring elements conveys a sense of cohesion and modernity. Pedestrian bridges connect terraces and wide stairways connect the two levels. Uniformly spaced suspended light fixtures in the center of the pedestrian street give a sense of scale. In general, the dark paving of the pedestrian walkway and the building materials of the Torg have been used skillfully. Connected directly to the lobbies of the office towers, the second level of the Torg consists of roof gardens, which lead to more small commercial stores.

Traffic-free downtown

Stockholm's traffic-free experiments have largely been initiated by merchants. Their attitude toward pedestrian zoning, therefore, has usually been positive. The Chalmers Institute of Technology in Gotherberg carried out a survey of the country's pedestrian streets. Businessmen on twenty four of thirty streets surveyed were 75–100% in favor of traffic bans; on five streets, 50–75% of the merchants approved; and on one street only 25–50% were satisfied.

Because of the complicated interrelation of factors affecting sales volume, it is hard to determine exactly how influential pedestrian zoning has been. Numerous studies, however, do reflect a satisfactory level of sales on Sweden's pedestrian streets, which coincides with the attitude of shopkeepers. A survey made by the Swedish Retail Federation, which took into account 110 companies located on traffic free streets throughout the country, recorded either stable or improved conditions for 90% of the business establishments, with the remainder showing a decrease.

2 ROTTERDAM

Lijnbaan, in Rotterdam's central commercial district, was one of the first completely pedestrianized shopping areas in the world. Rotterdam's entire central district was destroyed by German air raids, and the establishment of the city's traffic-free area was part of overall postwar construction. Two-story shops line the linear street, and squares at both ends each contain large department stores. Lijnbaan has served as a model for numerous pedestrian malls around the world. Its design features have been adopted in many European pedestrian schemes.

City profile

Rotterdam is the second largest city in the Netherlands. The old definition of it as a "harbor with a city attached" no longer corresponds to the physiognomy of this modern industrial city. However, the port remains the city's primary economic strength, and its development is closely related to the city's growth. When the Ruhr region became a major industrial center during the first half of the 19th century, Rotterdam increased in importance, establishing itself as the principal port for the region. This had inevitable consequences for the entire city. In 1850, Rotterdam had a population of 100,000, By the year 1900 this figure had tripled. After the Second World War, the port of Rotterdam expanded rapidly to meet the increasing demands of Europe's growing economy. Today, it is among largest harbors in the world.

Lijnbaan

Lijnbaan was designed as a single unit and completed in 1955. The Lijnbaan concourse is surrounded by two-story buildings. These are architecturally unimpressive but create the proper pedestrian scale. Organization of different kinds of commercial establishments together in one place determined the distribution and functional arrangement of its buildings, which share a number of centralized services.

Department stores face the entrance to the large square at each end of the mall. Small retail stores, restaurants and open-air cafes line the pedestrian concourse, contributing significantly to its lively atmosphere. Groups of high-rise residential buildings located on each side of the street were designed to increase the relatively low real estate value of commercial parcels. They play a very important role in Lijnbaan, providing street activity twenty-four hours a day in this primarily commercial development. The mall is widely used by residents of the towers as an extension of their recreational space.

Design features

All aspects of Lijnbaan's design have been controlled to give a feeling of unity. The paving is uniform throughout the project, and bridges connecting the two sides of the mall are a welcome addition in rainy weather. Flower beds are interspersed with seating to lend color to the area. The large department stores in the squares at each end contrast with the uniform height of the buildings along Lijnbaan's mall, which derive some variety from balconies placed at different levels. These are used by musicians during the warm weather. Thus, Lijnbaan maintains the characteristics of a neighborhood in spite of being the center of a highly commercial district.

ROTTERDAM

The city center in 1933 (top). The city center, after the debris from bombings had been removed (middle). Partially damaged buildings were restored. The city center in 1970 (bottom). Visible in the lower right corner is the Lijnbaan shopping and housing complex.

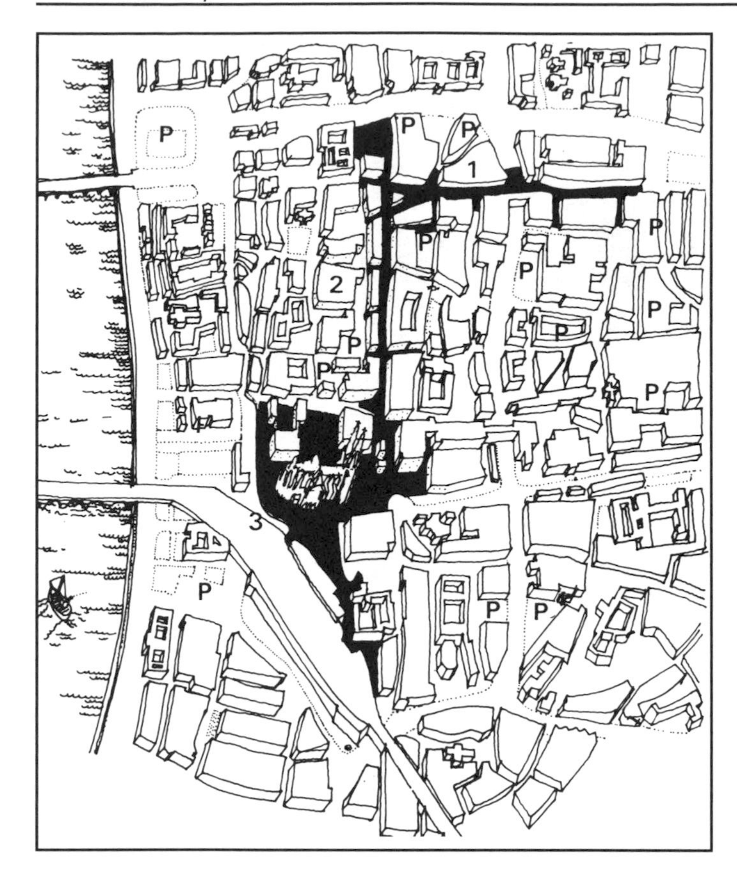

COLOGNE
KEY PLAN
1 Shildergasse
2 Hohe Strasse
3 The Domplatze

The plan also indicates the parking lots and garages constructed immediately after the war to absorb the predicted increase in automobile traffic.

3 COLOGNE

During the Second World War, Allied bombing destroyed ninety percent of Cologne. After the war, reconstruction efforts were implemented in two separate but inter-related phases. Redesign of the city's circulation network was planned to accommodate the expected increase in the number of private motor vehicles. The city's older central section was rebuilt with land use modifications that provided a framework for the introduction of a major pedestrian spine.

Cologne's pedestrian zone extends from the city's world famous cathedral to the Neumarket, the most popular meeting place in the city. The train station and the bus terminal constitute the other poles of the district, and people move between them at all times of day. The system is like a conveyor belt, funneling shoppers and tourists back and forth through the heart of the city; only around the Cathedral does the stream of people slow down. Neither Hohe Strasse nor Shildergasse provide many pedestrian amenities. Nevertheless, they are among the most successful commercial trafficfree streets in Europe.

City profile

Cologne was founded by the Romans in the 3rd century. It grew rapidly in the Middle Ages because of its strategic position where the Rhine flows into the North Sea. Goods from England, France and the Low Countries passed through the city on their way to Eastern Europe. Famous for the variety and beauty of its Romanesque churches and its Gothic cathedral, today Cologne is the fourth largest city in Germany. The city has regained its pre-war prominence as Germany's major Rhine port.

Hohe Strasse and Shildergasse—Design features

Hohe Strasse and Shildergasse are extremely different in their physical and commercial makeup. Hohe Strasse is narrow, reflecting its ancient Roman character. It is lined with medium- to high-priced boutiques and jewelers that give the street an air of elegance. Its small cafes and fast food restaurants cater to the several million tourists and visitors who come to Cologne each year. Shildergasse on the other hand is wide, with large department stores along most of its length.

The design of both streets is based on their commercial nature. As a consequence, pedestrian amenities such as places for people to relax are not provided. The linear configuration of the streets and their function as connectors of the city's major attractions contribute to the "conveyor belt" effect of this pedestrian system.

Environmental impact

The traffic bans on Hohe Strasse and Shildergasse caused a number of side streets to become de facto walkways, greatly improving the overall environmental conditions of the old city.

Cologne's planners were among the few that took before and after readings of noise and air pollution in the pedestrianized area. Air pollution levels dropped from 8 parts per million (ppm) of carbon monoxide to 1pm. Noise levels were reduced by 15 decibels, equivalent to cutting the original noise level in half. These results were much publicized at the time and have constituted a strong argument for introducing pedestrian zones in Germany and Europe.

The Domplatze

Cologne's pedestrian system has been enriched by the recent renovations to the square around the city's famous Cathedral. Directly connected to the Hohe Strasse, the square is also linked to the pedestrian concourse of the railroad station via bridges and escalators, and to the traffic-free promenades along the riverfronts by a pedestrian bridge over the Rhine. Seats, trees and open-air cafes have been installed, and the square provides a place to rest and view the celebrated Dome.

4 COPENHAGEN

By the early 1950s, it became evident to Copenhagen's planners that the increasing number of private automobiles could not be absorbed by the narrow streets in the city's historic core. In addition to causing unbearable congestion, motor vehicle traffic was causing environmental deterioration and affecting the social and economic potential of the downtown area. After experimenting with a number of traffic control strategies, city planners proposed pedestrian zoning as a solution.

In Danish, strøget means "to stroll," a favorite Danish custom. Strøget is also the name for Copenhagen's pedestrian zone, the first in Denmark. Strøget clearly identifies Copenhagen's downtown as a pleasant place to spend time. It attracts people from all over, as well as outside, the country. Strøget's success can be measured by the fact that Copenhagen's pedestrian system has continued to expand, with the goal of eliminating traffic from the entire central business district.

City profile

The population and scale of Copenhagen was limited by the boundaries of the 15th century walls until 1850, when the military cordon protecting the city opened up. The medieval street patterns remained, but fires had destroyed most of the old buildings and new ones had been built. By the end of the 19th century, the population had jumped to 451,488 persons, and the shape and growth of the city had changed significantly. Workers had moved into the city and the Industrial Revolution made its impact. Speculative housing projects were built along the city's perimeter with intense development occurring 1870 and 1890. It is possible to read the phases of this development in the three concentric rings of high density housing that start in the medieval center and move outward.

Strøget

The original Strøget consisted of three contiguous streets, running from the Town Hall Square to Kings New Square. An experimental traffic ban was initiated in 1962. In 1964 the area was declared a permanent pedestrian zone.

Design features

The atmosphere throughout the pedestrian system is pleasant, even though its design is quite simple. The paving is gray, with dark midstripes indicating service routes (open to delivery vehicles from 4:00 a.m. to 11:00 a.m.). Street signs are modest. It is left to the pedestrians to provide color and liveliness. Shops give careful attention to their window displays, which are designed to capture the interest of strollers. Unexpected alleyways and arcades provide variety in the scale of the area.

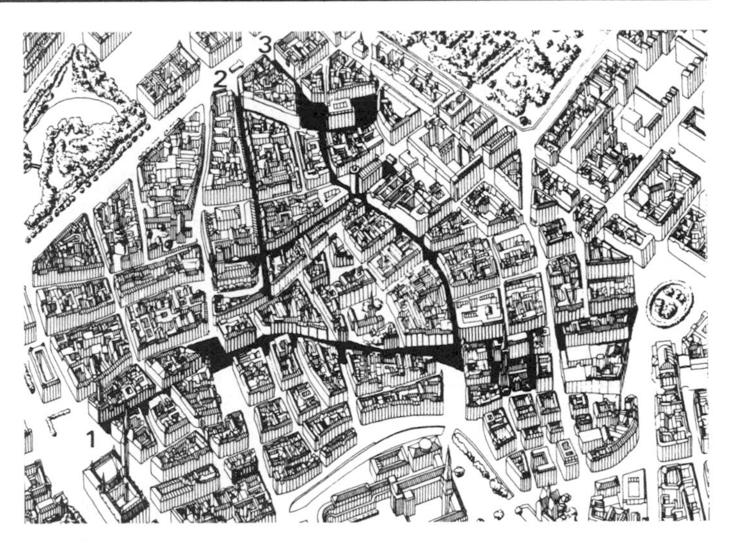

COPENHAGEN
KEY PLAN
1 Strøget
2 Fiolstraede
3 Kobmagergade

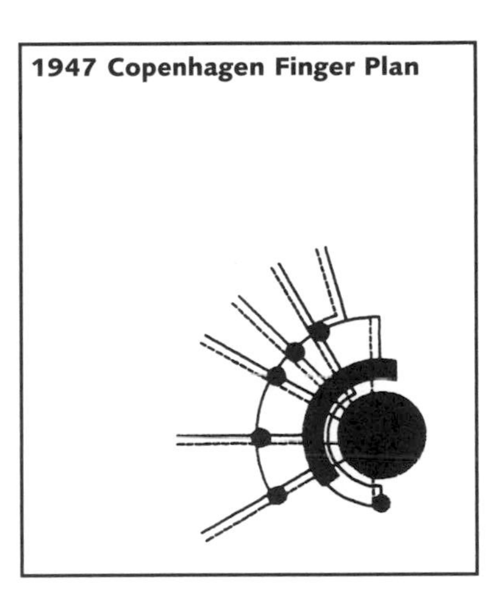

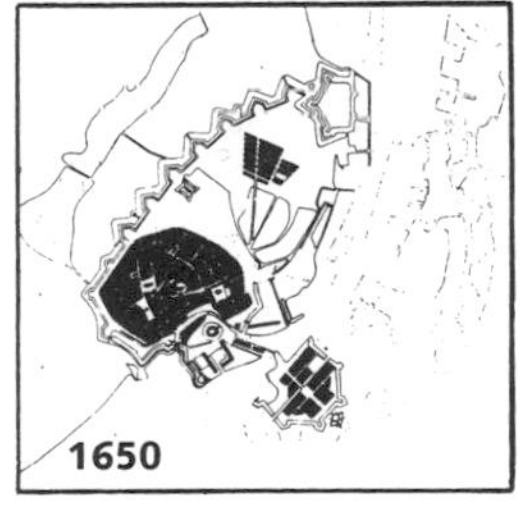

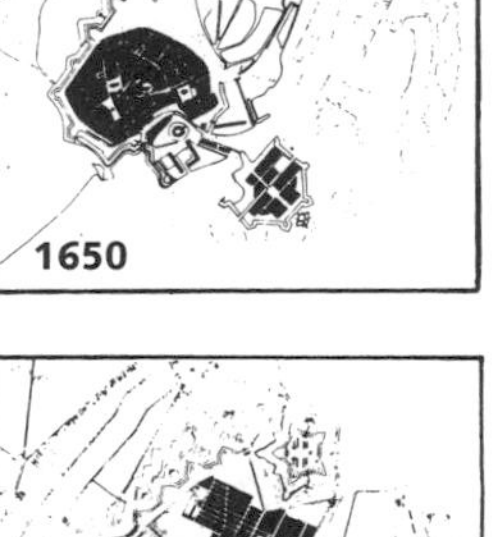

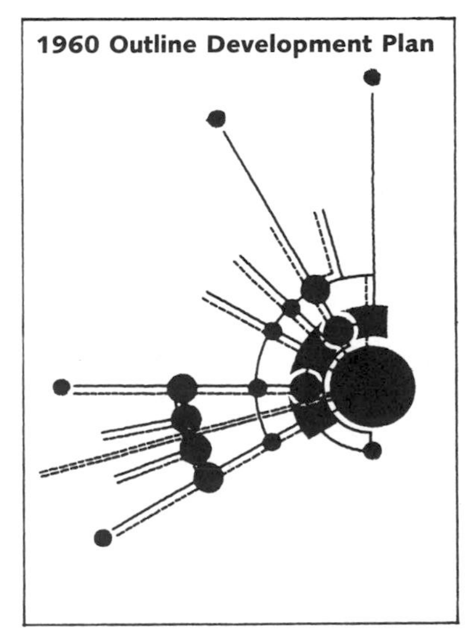

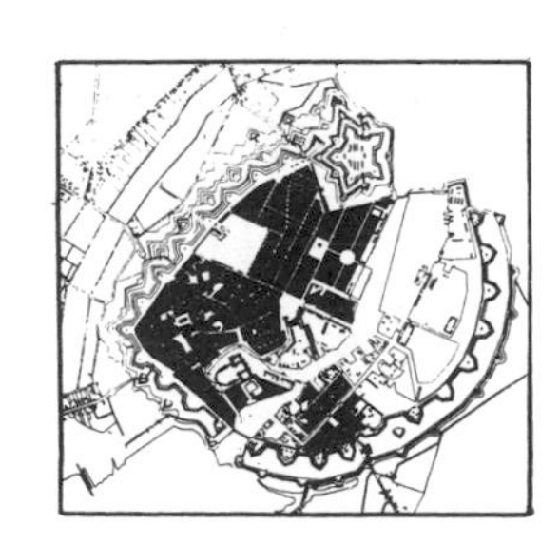

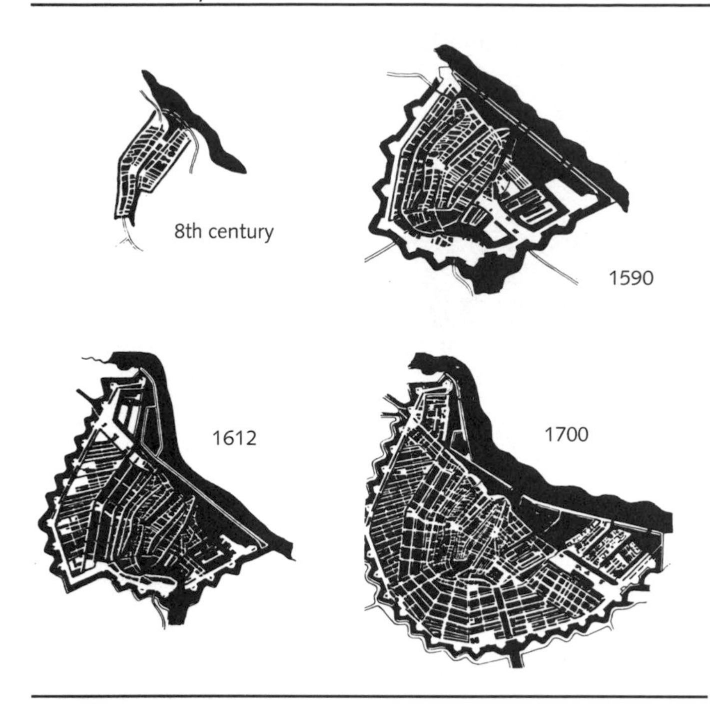

The expansion of the city core within the protective walls.

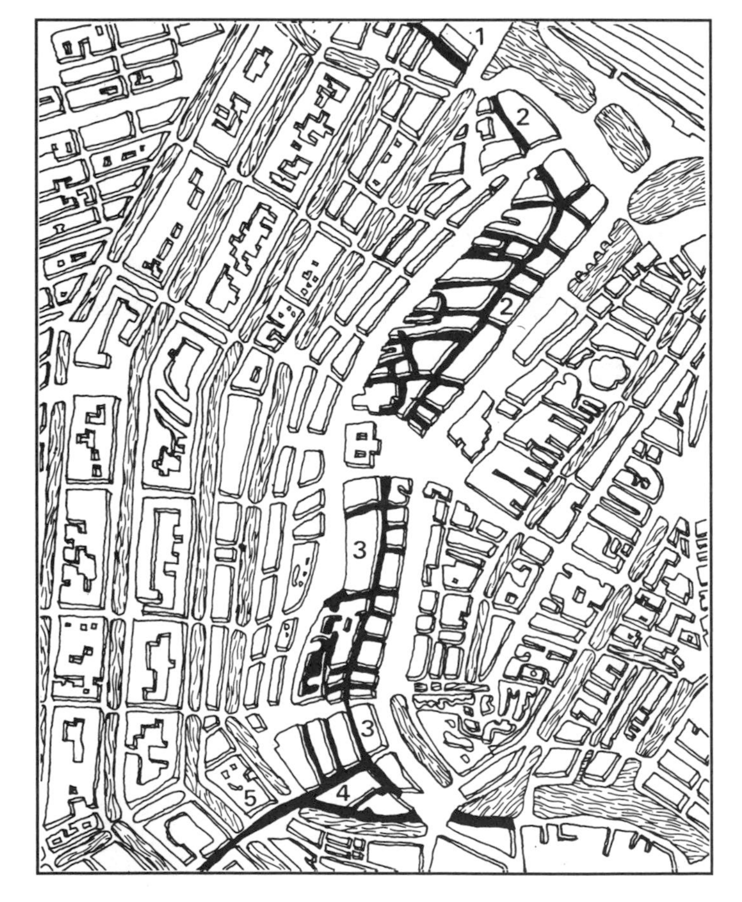

AMSTERDAM

KEY PLAN
1 Haarlemmerstraat
2 Nieuvendijk
3 Kalverstraat
4 Heiligeveg

The pedestrian routes themselves offer little chance to sit and relax. They are quite narrow and there are few benches or seating areas. But the singular thrust of the streets is relieved by open plaza areas, often highlighted by cultural and historic landmarks. Here, there are benches and opportunities to meet and talk or watch informal entertainment.

The whole system of pedestrian zones benefits from the urban fabric of the city, which retains a medieval character, although the buildings date mostly from the 18th and 19th centuries. Pedestrians throughout the city will frequently turn a corner and suddenly find fountains, historic buildings, or well delineated towers (Copenhagen is the City of Beautiful Towers). The narrow streets form a spider web across the city, with plazas and open spaces at many of the intersections.

Copenhagen planning strategies emphasize the administration and improvement of existing urban resources, rather than the capital creation of new ones, especially in conjunction when extending the pedestrian system.

5 AMSTERDAM

Amsterdam's pedestrian streets were introduced in an attempt to cure traffic problems, which developed in the city after the Second World War. Congestion was especially severe in the old part of the city, where vehicular streets were wedged between canals and bridges, and where the particular nature of the subsoil made construction of underpasses, parking garages and underground public transit systems impossible. Amsterdam is one of the best-preserved larger European cities, but is relatively young. Founded about 700 years ago, the city had 210,000 inhabitants by the 17th century, making it the fourth largest city in Europe at that time.

Amsterdam's organized growth began as early as 1612, when the first official plan for the city was adopted. Known as the Three Canal Plan, it quadrupled the area of the city and allowed for future population growth without need to rebuild fortifications, which had to be created on landfill in Holland. This plan provided sufficient land for expansion until 1874. During those three and a half centuries the city took on its characteristic appearance, which is visible today: a series of concentric canals, which have the triple function of defense, transportation and drainage. The city's policy of rationalized growth was carried over into the first half of this century beginning with the 1901 Housing Act. This was followed by a series of plans in 1906, 1907 and 1917.

Central pedestrian district

Traffic congestion in the central commercial streets seriously jeopardized the retail economy of the area. Pedestrian zoning was introduced on a major scale in an effort to reverse this trend. Early in 1960, Kalverstraat and Nieuwendijk were officially closed to traffic during shopping hours, but it was not until eleven years later that they assumed their present appearance. In 1971, they were repaved, and further restrictions were placed on the delivery of goods. The streets are the core of Amsterdam's pedestrian system.

Neither Kalverstraat nor Nieuwendijk has any street furniture, display cases or special lighting fixtures. However, this lack of amenities is not noticed in the crush of people, mostly tourists, on the streets during the day. After nightfall the pedestrian streets are not as busy, but they are far from deserted. Small shopkeepers leave their windows lighted until late at night, making the evening hours an ideal time for browsing.

In 1971, Leidsestraat was transformed into a pedestrian street, but its streetcar line was retained. This street, once one of the most elegant in central Amsterdam, had lost most of its glamour over the years. At the same time the square in front of the Municipal Theater was converted from a parking lot into a raised café terrace; and the pedestrian zone was extended along the southern side of the Singel, a street famous for its permanent flower market. In March of 1973, motor traffic was banned from Haarlemmerstraat, which completed a ribbon of pedestrian streets encircling the center of Amsterdam. As a consequence, all the streets which cross this ribbon became walkways.

6 BOLOGNA

Bologna's traffic-free zones were part of an overall effort to revitalize the historic core and control the city's territorial growth. However, the pedestrian district itself was implemented exclusively on the basis of its humanistic and cultural benefits. "Comprehensive conservation" of the physical and social fabric of central Bologna has been the guiding concept of the administration of this northern Italian city. Bologna's pedestrian streets were planned within this local context, creating a setting for community life in the city core. Contrary to other pedestrian efforts, Bologna's planners were not particularly concerned with improving retail economics. Nevertheless, shops witnessed increases in sales due to the traffic free streets.

City profile

There has been very little change in Bologna's layout since its foundations. The original Roman grid of 200 B.C. is still visible. One of the major provincial centers of the Empire, Bologna maintained its importance in the early Middle Ages when it became a trading center between the indigenous population and invading Visigoths, Huns and Longobards. The University of Bologna, which became one of the most famous centers of learning in Europe, was founded in the 11th century.

In the mid-1960s, people began leaving the city, attracted by new peripheral housing. Within the city, traffic congestion was becoming more dense, and environmental conditions were deteriorating. Bologna's social and cultural domination of the region was slipping away as a direct result of population loss. It was evident that remedial action was necessary.

In 1972, the Via Calderini and Via dell'Inferno were closed. Traffic on streets perpendicular to these thoroughfares was eliminated. After agreement had been reached with shopkeepers, the streets around one of Bologna's characteristic commercial blocks were permanently closed to traffic in 1973. Banning traffic from these areas benefited the entire city center, since the number of cars entering it was reduced greatly.

Creation of the pedestrian "island," as residents call it, was a fundamental step in Bologna's transformation. It is a basic element in the new way of life the city can offer its residents. The unexpectedly overwhelming success of the pedestrian zone has encouraged the city to continue clearing streets and plazas of both parked and moving vehicles, and limiting access to all but resident traffic in selected areas. As the organizations and citizen committees in various neighborhoods became more efficient, pedestrian streets begin to emerge in residential districts.

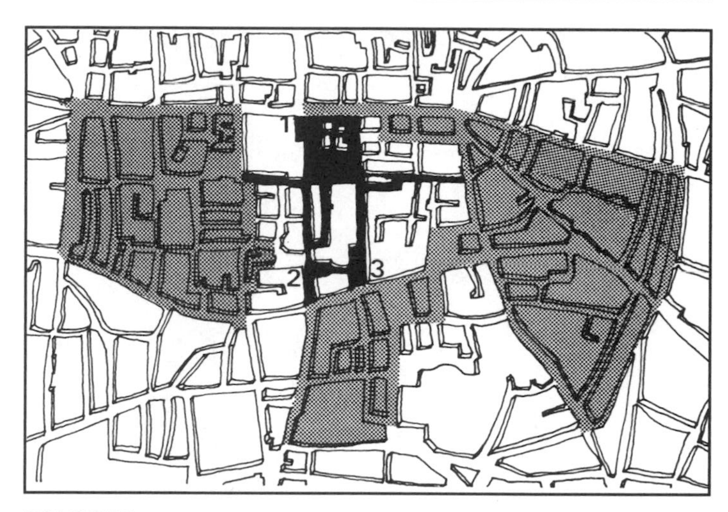

BOLOGNA
KEY PLAN
1 Piazza Maggiore
2 Via Massimo D'azeglio
3 Piazza del Nettuno

The shaded areas indicate the so-called Blue Zones, where parking is prohibited and access limited.

One year after the center of Bologna was converted into a pedestrian zone, a poll was taken among the city's residents. 43% of the people interviewed from all over the metropolitan area strongly identified the center of the city with the Piazza Maggiore. When asked what image they associated with the piazza, 28% identified it as a monumental space which is part of a unified urban complex. However, 52% identified it as an important place for social interaction. Some 15% of those interviewed saw it as a place for adult activities such as political demonstrations and festivals. 23% of the people viewed the Piazza Maggiore as being populated by pigeons for children to play with and people to feed. On the other hand, the Piazza Maggiore as a theater or setting for collective events was very real to the people interviewed.

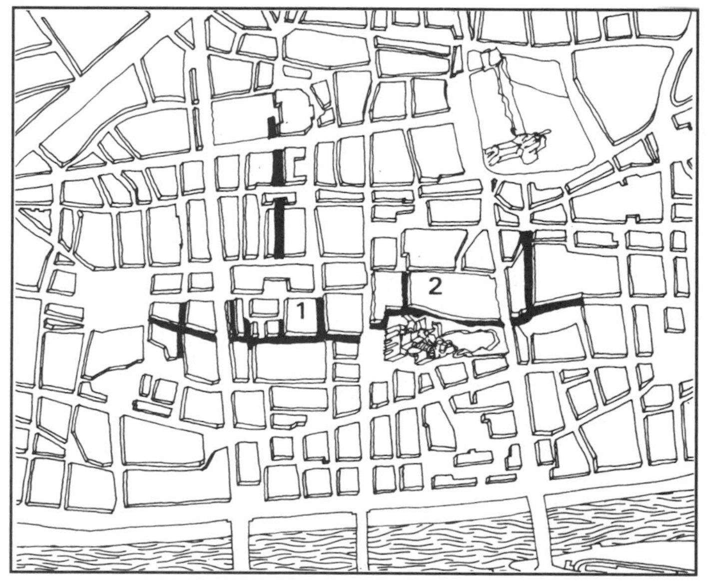

ROUEN
KEY PLAN
1 Rue du Gros Horloge
2 The cathedral district

The Gros Horloge gives the city's major commercial street its name.

The Malraux Law. Named after Andre Malraux, the Cultural Affairs Minister, the law was passed by the national government to protect not only isolated monuments but entire neighborhoods—secteurs sauvegardes or landmark districts. This legislation not only prevents demolition for slum clearance and urban renewal, but also calls for careful preservation and restoration of worthwhile elements on the site. Ugly annexes, remodeled facades, garish signs and storefronts are removed to reveal the original building beneath the 19th and 20th century excrescences.

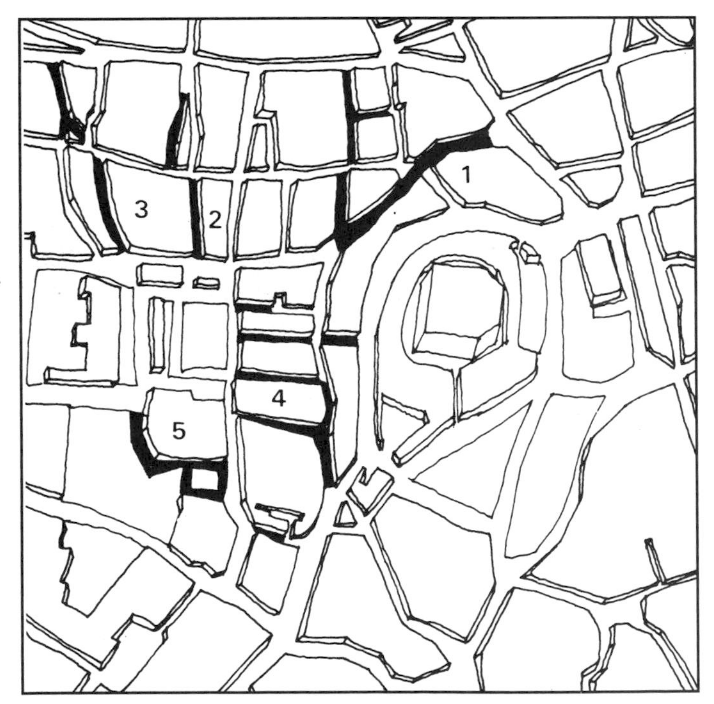

NORWICH
KEY PLAN
1 London Street
2 Dove Street
3 Lower Goat Lane
4 White Lion Street
5 Hay Hill

7 ROUEN

Rouen leads France in creation of pedestrian streets. It was the first French town to initiate a comprehensive pedestrian system as part of a plan aimed at reversing a trend which was making the city a mere economic suburb of Paris, some 135 km. (84 mi.) away. The city's initial traffic-free experiment was located along a major commercial street. The pedestrian system was later expanded to ease the flow of tourists around the Cathedral. Pedestrian zoning was implemented in both areas in order to revitalize the city's central area, improve the quality of street life and restore the prestigious historic buildings.

City profile

Founded during the Gallo Roman period, the city grew along two axes at right angles to one other that intersect near the site on which the Cathedral was built. As the capital of Normandy, Rouen has a rich history. Joan of Arc was burned at the stake in the Market Square. In the 19th century, Ruskin and Victor Hugo made it an important stop on the Romantic itinerary of medieval architecture in France.

Rouen sustained heavy bombing during the Second World War, and much of its core had to be rebuilt. As a result, this area now contains both narrow, medieval streets and wide modern thoroughfares In spite of the severe bomb damage in the war, Rouen still offered many architectural attractions. Tourism became an important part of its economy. Most of the historic and architectural landmarks are protected by the 1962 Malraux Law.

Rue Du Gros Horloge

In 1970, general traffic was banned on the Rue du Gros Horloge, between Rue Jeanne d'Arc and the Cathedral Square. This pedestrian zone—only 7.3 meters (24 ft.) wide and 266 meters (875 ft.) in length—is a showcase of medieval half-timbered shops and town-houses. The street forms a vista leading to a Renaissance gate with a clock, the Gros Horloge, set above its archway. The Rue de Gros Horloge was a pioneer effort to accustom the public to pedestrian zoning. Although the size of the first phase of the project seems too small as a model case, it had remarkable impact at both local and national levels in France.

As a low-cost strategy to develop and protect the center, the pedestrian experiment has been an immense success. No major new construction or urban renewal was required. The principal changes were readjusting flow on the four central bridges to equalize traffic capacity and creating surface parking for the city-owned land. The Rue du Gros Horloge itself was recycled. That is, most of the materials were salvaged from the street and used for paving, planters and benches.

8 NORWICH

Norwich was the first city in Great Britain to eliminate traffic permanently from a central commercial street. Cars were banned on London Street because they constituted a hazard to the crowds of shoppers that overflowed the narrow sidewalks. Initially, the ban was temporary, during which time numerous discussions were held with merchants along the street to assess the impact of the experiment. The permanent conversion of the street took place only after the traffic ban proved advantageous to citizens and businesses alike.

Norwich is an example of the possibility of unifying the entire commercial district with a network of pedestrian streets, in order to transform the core area into a totally car free environment. In conjunction with the pedestrian program, the city with help from private contributors launched a massive effort to restore and rehabilitate historic buildings in the core area.

City profile

Established nearly 1200 years ago, Norwich is the largest city in East Anglia, a flat agricultural region with one of the highest population growth rates in Great Britain. By the 11th century, it was one of the largest towns in England with an economy based on its markets. The city served as an important inland port for the textile industry. In addition, it was an important religious center.

Aspects of the district have changed dramatically since it has become a pedestrian area. Restaurant owners note that people request tables near street windows to watch the passing crowd, and in some instances tables have been put outside in the Continental fashion. Merchants have seen their sales increase from five to twenty percent, and window displays are no longer soiled by auto exhaust. Building owners on London Street have redecorated their stores according to a color and design scheme established by an advisory committee. Graphic controls unify all signs in the area. About one-third of the buildings are redecorated every year, giving the street a continuously fresh appearance.

Mini-buses still operate in the city's central commercial district, moving shoppers to and from the major transportation terminals.

The historic district

To protect the residential zone around Norwich's 12th century Cathedral, and the Cathedral itself, from structural damage and air pollution, a second pedestrian system was created in that quarter of town. As a complement to the large pedestrian zone and restoration program, Norwich has encouraged more housing in the center of the city in order to increase the kinds and amounts of activity in that area.

9 LEEDS

Leeds, one of England's major industrial cities, implemented the country's largest pedestrian zone in the short period of four years. The introduction of pedestrian streets was part of the city's vigorous efforts to insure the efficient functioning of Leeds' commercial district, which is the major retail center for over 3,000,000 people. Unlike most British cities, the central area of Leeds is not a complex interweaving of streets and buildings from various historic periods. The city is a product of the Industrial Revolution and was planned with precisely defined residential, commercial and industrial districts which do not overlap. Even though the commercial district was self-contained, traffic-free zoning required careful planning in order to draw the area's disparate elements.

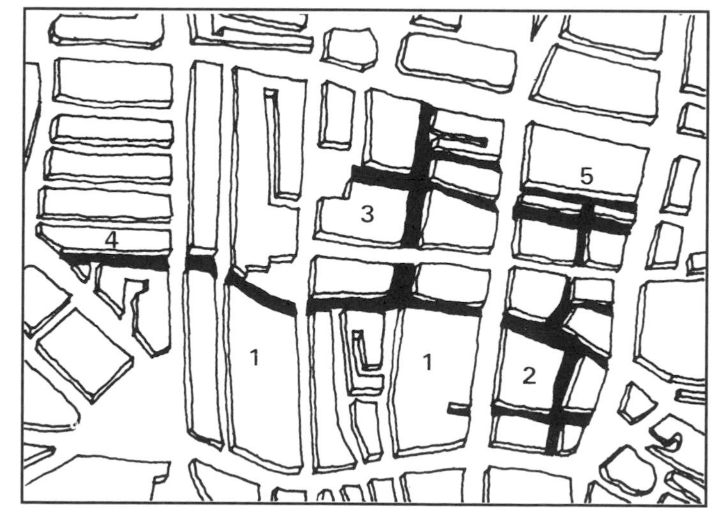

LEEDS
KEY PLAN
1 Commercial Street
2 Kirk-gate
3 Lands Lane
4 Bond Street
5 Queen Victoria Street

City profile

Leeds is a major clothing manufacture and wholesale center in Great Britain. It is located in central Yorkshire where woolen industries were introduced by emigrating Flemish weavers in the 14th century.

In the 19th century, the city became a major center of the woolen trade because of abundant waterpower derived from the River Aire. Clothing manufacture because the dominant industry.

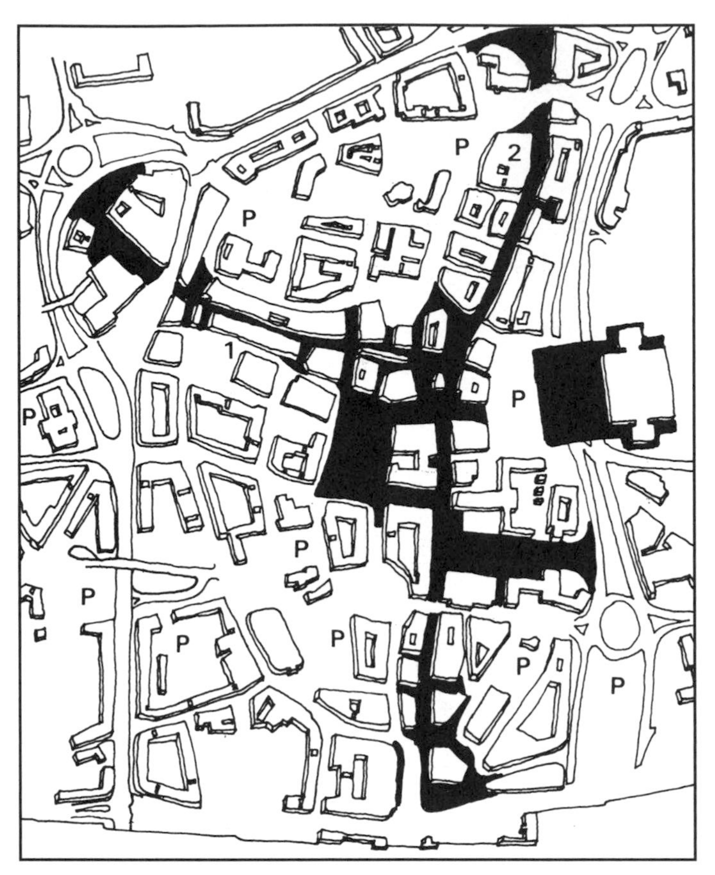

ESSEN
KEY PLAN
1 Limbecker Strasse
2 Kettwiger Strasse

The availability of parking spaces is also indicated.

As the woolen mills expanded in the second half of the nineteenth century, the city grew along a transportation system typical of that era. A radial road system was created from the center outward. The customary row housing of Victorian England grew up along the system. Freight moved mainly by water and rail lines. Today, the northern bank of the River Aire is primarily commercial with the market and shops to the east and offices and warehouses to the west. The south bank of the river is given over to industry, with the railway station and City Square at the intersection of these three zones.

The Leeds Approach

According to the Leeds Town and Country Planning Act, "Two-stage closure is possible: An experimental closure may be introduced for up to eighteen months where an authority is unsure of its success: a permanent closure may follow this or be done initially. Traders' motoring organizations and police are first consulted. This is followed by an order published in local papers. A period of twenty-one days is allowed for objections. These are considered by the Planning Committee, which may decide to revise the plan, hold an inquiry or proceed. Permissions to proceed is sought from the Secretary of State who is informed of the objections to the closure, and he has the power to make the order official. Only when permanent closure is obtained can permanent features such as trees be planted. Until then only temporary landscaping is allowed, as the space is still technically a road. Powers are available to control the times for delivery vans to enter the street."

The Pedestrian Network

Within the overall framework of traffic control and large-scale renewal, the city banned motor vehicles completely from several major shopping streets. The elimination of cars from the core resulted in a renewal of the area's charm, reduction of pollution levels, and a strengthening of Leeds' role as a regional shopping center. During the initial stages of its traffic-free experiment, the city provided a small bus that moved shoppers through the pedestrian zone. This service was discontinued due to objections from shoppers themselves. A mini-bus still operates around the pedestrian network, and connects it with the central rail and bus stations.

10 ESSEN

Essen, along with Leeds, England, has one of Europe's most extensive pedestrian systems. Both Essen and Leeds are centers of large industrial regions, and, in both cities, traffic-free zoning was introduced to rationalize the traffic conditions as well as to improve the retail trade in highly specialized commercial and business areas.

City profile

Though thoroughly modern, the city was founded in the 9th century. The cathedral dates from 852 AD when Bishop Altfrid of Hildesheim laid the foundations for it and the Convent of Asnidi. Some 1,000 years later, this is still the heart of the city. The medieval town grew into a kidney shaped configuration defined by a ring of fortified walls. Today, a road has replaced the walls, and the city has expanded far beyond its medieval quarter.

Pedestrian system

The pedestrian system of central Essen is based on two major streets that cross in the heart of the city, Limbecker Strasse and Kettwiger Strasse. They measure, respectively, 55 meters (1,800 ft.) and 914 meters (3,000 ft.) in length. Pedestrian walkways fan out in all directions from these two streets, so that every point within the central business district can be conveniently reached on foot. The two main streets differ sharply in appearance and character. Limbecker averages 8.2 meters (27 ft.) in width and is lined with retail shops. It has only one restaurant and no places of entertainment. Consequently, it is busy in the day, but practically dead at night. Kettwiger, on the other hand, is 18 meters (60 ft.) wide and lined with fashionable shops, restaurants and cinemas that attract patrons far into the evening hours.

11 MUNICH

Munich's pedestrian system symbolically and physically binds together the medieval heart of the city. The Frauenkirche, New City Hall, shops and stores along the Marien Platz, Kaufingerstrasse and Neuhauserstrasse are a single and unified pedestrian scheme, which succinctly expresses the governmental, religious and commercial functions of the city to its citizens and visitors.

City profile

Munich was founded in the early 12th century but was predated by a monastery from which is derives its name. The city is famous for its university and museum. It is also well known for its industrial output of optical and precision instruments, heavy machinery and beer. During the Second World war, Munich was severely damaged. The scars have remained visible, although medieval and Renaissance buildings still stand in the city's central core. Reconstruction began immediately after the war, and as Germany's economic recovery picked up, the population increased at the rate of 40,000 people per year. The inner city population, however, has decreased due partially to the lack of housing in the center of Munich.

Pedestrian scheme

Munich's first pedestrian street was created between two major medieval gates: the Karlstor on the west side and the reconstructed Rathaus gate on the east side. The city's entire traffic-free system consists of this central spine connecting the two gates, with a series of smaller branches extending to the side streets, which have been transformed into vehicle free dead ends. There are several plazas within the boundaries of the pedestrianized area. At only one point is the major pedestrian street intersected by automobile traffic.

Planners were concerned about the apparently great width of the central spine, which is at one point 18 meters (60 ft.) wide. This concern became purely academic once the street was open because of the sheer number of people who used it. Actually, the width of the main pedestrian thoroughfare is a pleasing contrast to the surrounding narrow side streets. To some degree, this openness is complemented by existing medieval porticos, joined to modern arcades to create a pleasant shopping ambiance on rainy days.

The historic buildings in the pedestrian area have been restored and cleaned. They are floodlit at night. The street has been further enhanced with seven fountains, some of which are new and large enough to

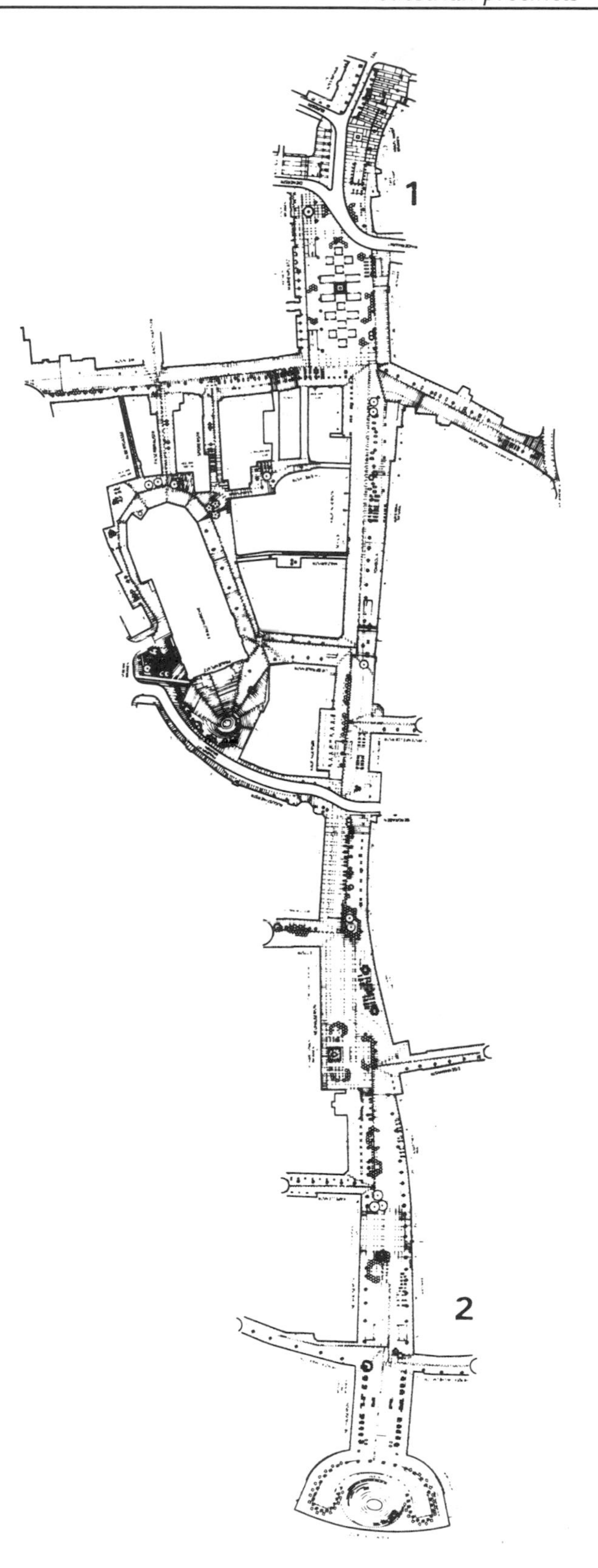

MUNICH
KEY PLAN
1 Kaufingerstrasse
2 Neuhauserstrasse

In Munich, the stackable chairs that people can arrange as they please are one of the major innovations in pedestrian design.

provide seating. Others, formerly in different parts of the city, were moved into the traffic-free district.

Design features

Stackable chairs are perhaps the single most innovative touch in this pedestrian scheme. People can move the chairs about as they please and group them for special events. Octagonal concrete flower planters are distributed along the street. People often sit on them though they are not comfortable. Planters are filled with seasonable flowering plants throughout the year. Trees have been planted in several places.

Relatively short streetlights were placed down the center of each pedestrian street to give a sense of scale. Freestanding glass display cases were also installed. However, merchants found that shoppers preferred looking into actual shop windows, as has been true in other pedestrian zones. The city has profited enormously from its pedestrian district. The old core contains not only civic monuments, but also, more importantly, stores, hotels, restaurants and cafes. The pedestrian zone, together with these attractions, generates a constant and varied street life, day and night.

12 VIENNA

Vienna presents a remarkable dilemma: it must either convert the whole inner city into a pedestrian zone or abandon the concept. This situation is due to two major factors. Residences, shops and offices are mixed together to the degree that it is impossible to define a clear cut commercial district. In addition, the core layout is such that fragmented pedestrian zoning would create an impossible traffic jam on surrounding streets.

As a compromise between totally closing streets in the inner city or leaving cars and pedestrians to fight it out, the city decided to experiment with a temporary closing of its most famous shopping street, Karntner Strasse, for the Christmas season of 1971. Using "Street Happenings," temporary furniture and other low-cost devises, the city actively involved tourists and residents in the pedestrian environment. Subsequently, this street was permanently repaved and furnished.

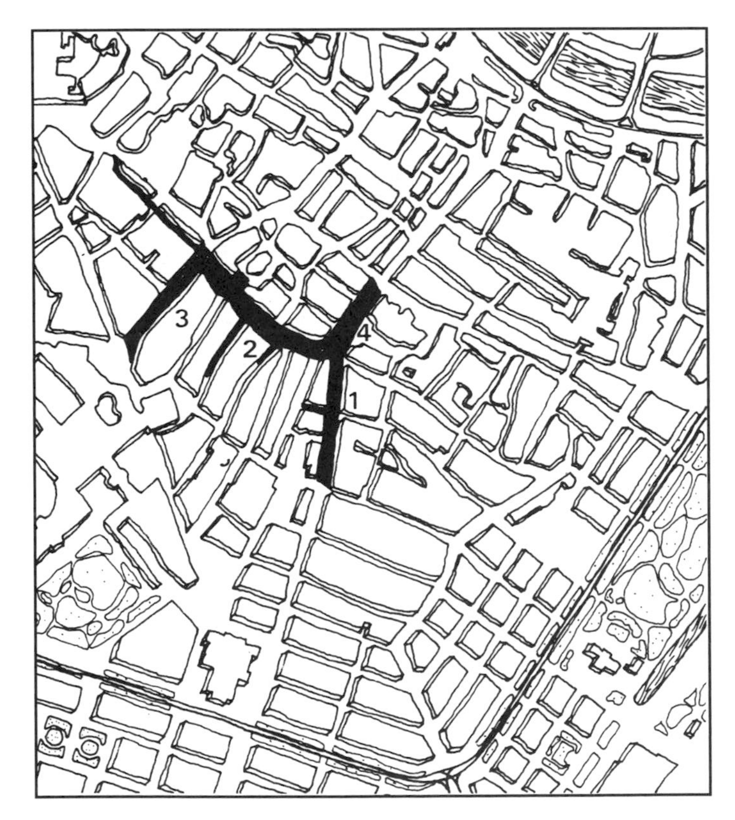

VIENNA
KEY PLAN
1 Karnterstrasse
2 Graben
3 Kohlmarkt
4 Stefanplatz

City profile

Today, Vienna is the capital of Austria, but for most of its history it was the capital of vast empires. The city dominated the trade and politics of central Europe through its position on the Danube. From 806 to 1558 AD, it was the capital of the Holy Roman Empire and capital of the Austro-Hungarian Empire until 1918. Then it became "an elephant in a backyard," as Winston Churchill put it.

In 1860 the medieval fortified walls surrounding Vienna were torn down and replaced by a boulevard, the famous Ring, which was landscaped with trees and surrounded by public parks and gardens. It became the most fashionable part of the city. About a mile beyond the Ringstrasse, a second, outer line of fortifications, the Linienwall, was torn down in 1890 and the Gurtel, German for "belt," became the second ring boulevard. As part of this belt, a rapid transit system was constructed which connects the railroad stations. The surrounding area was rapidly built up as a dense residential zone.

Planning strategies

In December of 1971, the firm of Victor Gruen International presented the city with recommendations for saving Vienna from the attack of automobile traffic. Gruen's report also strongly emphasized improving the ring road system created in the 19th century.

To implement a complete pedestrian zone in central Vienna, the city decided to build a new subway system rather than rely on buses for rapid transit.

The street happenings

When the idea of converting major streets in the inner city to walkways was first aired, merchants were divided in their opinions about it. In a poll taken two months before the Karntner Strasse experiment, only 52 percent of the merchants thought it was a good idea. That two percent edge decided the future of pedestrian streets in Vienna.

A series of "street happenings" for the Christmas Promenade, as the experiment was called, transformed the previously congested vehicular streets into exciting, if unusual, pedestrian walkways. The program, which was very inexpensive, included the following:

- Illuminating all the facades on the street with floodlights mounted on wooden towers.
- Installing loudspeakers for music, information and announcements of events. No advertising was allowed.
- Putting up a tent in the street for small dramatic and musical events.
- Placing colorful balls, twelve feet in diameter, for pedestrians to move around and play with.
- Setting up a "School for Walking" created to stimulate a series of unusual situations—undulating floors, soft surfaces, hanging plastic strips and rotating rollers—to which the pedestrians were to react.
- Special events were also scheduled in the pedestrian area and traditional Christmas decorations were used to create a festive atmosphere.

Merchants were far from pleased with these street happenings and during the course of the experiment a number of the attractions were removed. In particular, the large balls and the "School for Walking" generated the most disapproval. However, the program was successful in that it focused attention on the possibilities inherent in the pedestrian street, and only a month after the experiment began it was given an unlimited extension of time.

Because creation of pedestrian zones in Vienna has been an all-or- nothing proposition, the city has undertaken a far more ambitious program than most European cities. When the new subway system is completed and adequate parking is created on the Rings, the city's historic core will become a unified pedestrian zone. ■

The School for Walking

Graben

Clare Cooper Marcus with Carolyn Francis
and Rob Russell

Summary
To create an urban plaza, urban designers must know what kind to make, where to place it, how to design it to provide a humanizing contribution to urban life. Contemporary plaza types include the street plaza, corporate foyer, the urban oasis, the transit foyer, and the grand public place. Design recommendations on all these types are reviewed including location, size, visual complexity, activities, microclimate, boundaries, circulation, seating, planting, public art, paving, and related amenities.

Key words
environmental design research, microclimate, outdoor space, pedestrian, plaza, public art, seating, sidewalk, street, universal design

Town square, Malmö, Sweden

Urban plazas: design review checklist

1 TYPES OF DOWNTOWN PLAZAS

Downtown spaces can be categorized in many ways: by size, use, relationship to street, style, predominant function, archi tectural form, location, and so on. This article is concerned with the interplay of form and use, how the physical environment influences activities, socialization, or simply repose. The classification is based on a mix of form and use, moving from smallest to largest in size. The typology is not necessarily exhaustive; rather it is presented as a starting point for design thinking about downtown plaza areas.

The typology is an attempt to make some sense of the varied categories of downtown open space in cities. It can be applied to most cities as a basis for:

- understanding the variety of spaces described as *urban plazas*,
- categorizing plaza spaces in a specific city, and
- developing local guidelines for specific plaza types.

The street plaza

The street plaza is a small portion of public open space immediately adjacent to the sidewalk and closely connected to the street. It sometimes is a widening of the sidewalk itself or an extension of it under an arcade. Such spaces are generally used for brief periods of sitting, waiting, and watching. They tend to be used more by men than by women.

PHOTO: Robert Russell

Fig. 1. The seating edge. A seating wall facing Sansome Street, San Francisco, often used by construction workers and bicycle messengers.

- **The seating edge:** A seating height wall of stepped edge to a sidewalk (Fig. 1.)
- **The widened sidewalk:** A widened portion of the sidewalk that is furnished with seating blocks, steps, or bollards. Used primarily for viewing passersby (Fig. 2.)

Credits: This article is abstracted by Clare Cooper Marcus with Carolyn Francis and Rob Russell from *People Places: Design Guidelines for Urban Open Space*, Cooper Marcus and Francis, eds. 1998. (See References for complete citation.)

Fig. 2. The impressive forecourt. The Federal Office building, Seattle, has a plaza intended primarily as a walking route to an important building, with trees, sculpture, but few places to sit.

Fig. 3. The stage set. Security Pacific Plaza, Los Angeles, seems to be a stage set for a play that never happens.

PHOTO: Michael McKinley

Fig. 4. The widened sidewalk. Crocker Plaza, one of the most highly used plazas in San Francisco, is little more than sets of steps looking onto adjacent sidewalks.

- **The bus waiting place:** A portion of the sidewalk at a bus stop, sometimes furnished with a bench, shelter, kiosk, or litter container.

- **The pedestrian link:** An outdoor passage or alley that connects two blocks or, sometimes, two plazas.

- **The corner sun pocket:** A building footprint that is designed to open up a small plaza where two streets meet and where there is access to sun during the peak lunchtime period.

- **The arcade plaza:** A sidewalk that is widened by means of an extension under a building overhang.

The corporate foyer

The corporate foyer is part of a new, generally high-rise building complex. Its principle function is to provide an attractive, often elegant entry and image for its corporate sponsor. It is usually privately owned but accessible to the public. It is sometimes locked after business hours.

- **The decorative porch:** A small decorative entry, sometimes planted or supplied with seating or a water feature.

- **The impressive forecourt:** A larger entry plaza, often finished in expensive materials (marble, travertine) and sometimes designed to discourage any use by passing through (Fig. 3.)

- **The stage set:** A very large corporate plaza flanked by an impressive tall building that it helps frame it. The plaza is primarily a stage set with buildings as a backdrop (Fig. 4.)

The urban oasis

The urban oasis is a type of plaza that is more heavily planted, has a garden or park image, and is partially secluded from the street. Its location and design deliberately set this place apart from the noise and activity of the city. It is often popular for lunchtime eating, reading, socializing, and it is the one category that tends to attract more women than men, or at least equal proportions of each. The urban oasis has a quiet reflective quality.

- **The outdoor lunch plaza:** A plaza separated from the street by a level change or a pierced wall and furnished for comfortable lunchtime use.

- **The garden oasis:** A small plaza, often enclosed and secluded from the street, whose high density and variety of planting conveys a garden image (Fig. 5.)

- **The roof garden:** A rooftop area developed as a garden setting for sitting, walking, and viewing.

The transit foyer

The transit foyer is a type of plaza space created for easy access in and out of heavily used public transit terminals.

- **The subway entry place**: A place for passing through, waiting, meeting, and watching. It sometimes becomes a favorite hangout for a particular group (*e.g.* teens) who can reach this place by public transit.

• **The bus terminal:** Where many city bus lines converge and many commuters arrive and leave the city center each day. It is primarily a space to move through, but it sometimes attracts vendors of newspapers, flowers, light snacks, and the like.

The grand public place

The grand public space comes closes to our image of the old-world town square or piazza. When located near a diversity of land uses (office, retail, warehouse, transit) it tends to attract users from a greater distance and in greater variety than do other plazas. Such a plaza is often big and flexible enough to host brown-bag lunch crowds; outdoor cafes; passers through; and the occasional concerts, art shows, exhibits, and rallies. It is usually a public area owned and is often considered "the heart of the city."

• **The city plaza:** An area predominately hard surfaced, centrally located, and highly visible. It is often the setting for programmed events such as concerts, performances, and political rallies.

• **The city square:** A centrally located, often historic place where major thoroughfares intersect. Unlike many other kinds of plazas, it is not attached to a particular building; rather, it often encompasses one or more complete city blocks and is usually bounded by streets

PHOTO: Robert Russell

Fig. 5. The garden oasis. The TransAmerica Redwood Park, San Francisco, is a small through-block space with trees, lawn, fountain, and ample seating that is often used very much for lunch-hour relaxation.

PHOTO: Jennifer Webber

Fig. 6. The best-used plazas are often those that are easily accessible to a variety of people. Justin Herman Plaza in San Francisco is within easy walking distance of corporate office buildings, downtown stores and housing, and tourist and convention hotels.

PHOTO: Robert Russell

Fig. 7. Casual or "secondary" seating on steps, walls, mounds, and planters does not appear empty when people are not present.

2 DESIGN RECOMMENDATIONS AND CHECKLIST

Location

The best locations are those that attract a variety of users and both active and passive uses (Figs. 6 and 7.) A study of the effect of context on the use of five downtown Minneapolis plazas found that the most frequently used plaza was in the area of greatest land use diversity, where office and retail districts overlapped.

In determining whether or not a new plaza would be an asset in the proposed location, designer and client should ask the following questions:

1. Does the analysis of nearby public open space indicate that a proposed new space will be welcomed and used?

2. Have the client and designer determined for which functions the plaza should be designed? For example, as a visual setback for a building, transition zone, lunchtime relaxation, bus waiting, sidewalk cafés, displays or exhibits, performances, or mid-block pedestrian thoroughfare?

3. Have the correlations between block location and type of space been considered, either in choosing a location for the plaza within an entire planned development, or in determining how best to configure. and detail a particularly located plaza? For example, has the high-use potential of a corner location at grade been considered, or the oasis potential of a mid-block cul-de-sac location?

4. Assuming a catchment area of 900 ft. (274 m), will a currently unserved population be served by the proposed development?

5. Are there many workers in the catchment area, to ensure a lunchtime clientele?

Fig. 8. Walking and sitting are two activities accommodated in this plaza in the French Quarter of New Orleans.

PHOTO: Jennifer Webber

Fig. 9. Men tend to predominate in up-front, on-display locations in urban plazas.

6. Is the plaza located where a diversity of people can use it, for example, workers, tourists, and shoppers?

7. Does the location of the plaza tie into an existing or proposed pedestrian system for downtown?

8. Does the local climate warrant providing a plaza? If an outdoor space can be used for less than three months of the year, an additional public indoor space should be considered.

Size

It is difficult to make recommendations regarding size, as every location and context is different. Kevin Lynch suggested that dimensions of 40 ft. (12 m) appear intimate in scale; up to 80 ft. (24 m) is still a pleasant human scale; and that most of the successful enclosed squares of the past have not exceeded 450 ft. (137 m) in smaller dimension (Lynch 1971). Gehl proposed a maximum dimension of 230 to 330 ft. (70 to 100 m), as this is the maximum distance for being able to see events. This might be combined with the maximum distance for being able to see facial expressions, 65 to 80 ft. (20 to 25 m).

These dimensional guidelines suggest:

9. Given that every location and context is different, have the suggestions by Lynch and Gehl been considered in regard to limiting plaza dimensions?

Visual complexity

In a study of downtown plazas in Vancouver, Canada by Joardar and Neill (1978, p. 488), the authors observe that, "Both density and variety as opposed to sparseness and repetition appeared to be perceptually important," suggesting:

10. Does the design incorporate a wide variety of forms, colors, and textures—fountains, sculptures different places to sit, nooks and corners, plants and shrubs, changes in level?

11. If a complex view from the plaza is possible, has the design capitalized on it?

Uses and activities

A number of questions should be considered to ensure a successful program of uses and activities in each plaza. (Fig. 8).

12. Has the plaza been designed to accommodate either lingerers or passers-through? or if both functions are to be included, have distinct subareas of the plaza been provided to avoid conflict?

13. If people are encouraged to take shortcuts through the plaza, have barriers between the sidewalk and plaza been eliminated, including grade changes?

14. To encourage people to stop and linger in the plaza, have dense furnishings, attractive focal elements, and defined edges been used? If concerts, rallies, and so on are anticipated, have unimpeded open areas been provided?

15. Does the plaza design address the differences between men's predominant preference for a "front yard" experience—public, interactive—and many women's desire for a relaxed and secure "backyard" experience (Fig. 9)?

16. To minimize vandalism and the presence of "undesirables" (or to render them inconspicuous in the crowd), has the plaza been designed to encourage heavy use, rather than by "hardening" the design?

Microclimate

Climatic considerations involve thinking carefully about sunlight, temperature, glare, wind, precipitation, and overall comfort. Comfortable outdoor conditions can be modified by bioclimatic design, including shading, planting and evaporative cooling in summer and solar exposure and wind protection in winter (Fig. 10). Easy indoor/outdoor access to and from conditioned indoor public spaces should also be considered for greater variablity.

17. Is the plaza sited to receive maximum, year-round sunshine, providing sun exposed areas and shaded areas, especially where people may sit at different times?

18. Where the summers are very hot, is shade provided by means of vegetation, canopies, trellises, and so on?

19. As a city policy, is building height and mass controlled to preserve and enhance solar access, that is, sunlight reaching public open spaces?

20. Does the plaza site have mid-day (lunchtime) temperatures for at least three months of the year in the outdoor comfort range? (between 50°F and 80°F (10°C and 26.6°C) as a function of temperature, sun, wind, and clothing).

21. For those months when mid-day temperatures average are in the outdoor comfort range, have sun-shade patterns been calculated to predict where seating areas should be located?

22. Does glare from adjacent buildings create unpleasant visual and/or temperature conditions in the plaza?

23. Can reflected light off adjacent buildings be used to brighten the plaza's shadowed areas?

24. Have local wind patterns been evaluated for the plaza site? Will windiness lead to nonuse, particularly in cities with marginally hot summers?

Boundaries

A plaza should be perceived as a distinct place, and yet must be visible and functionally accessible to passersby. Exposure to adjacent sidewalks is essential; a successful plaza has one of preferably two sides exposed to public rights-of-way, the more likely that they are to feel invited into it; thus, an extension of plaza planting onto the sidewalk may imply to passersby that they are already in the plaza. Even a minor barrier or level change can considerably reduce the number of passersby who enter and use a plaza (Pushkarev and Zupan, 1975).

Fig. 10. The microclimate of an outdoor-indoor garden terrace adjacent to Tucson Airport is modified by shading, planting and evaporative cooling.

PHOTO: Michael McKinley

Fig. 11. The facade of the Citicorp Atrium is imposing and seemingly closed, but in fact, there are entries to a semioutdoor plaza between the columns.

Fig. 12. Different people need different forms of seating: a young office worker in Justin Herman Plaza, San Francisco, relaxing on a wave shaped concrete ledge.

Fig. 13. Some of the most popular public seating places are steps and walls alongside busy streets and sidewalks, depicted here in Cambridge, Massachusetts.

Fig. 14. Privacy while sitting: Flower-filled planters on a seating ziggurat at 101 California Street, San Francisco, permit singles or pairs of people to find semiprivate niches.

25. Do boundaries such as paving changes or planting define the plaza as a distinct space from the sidewalk, without rendering the plaza visually or functionally inaccessible to passersby?

26. Does the plaza have at least two sides exposed to public rights-of-way, unless it is intended to function as an oasis?

27. Have plaza design features such as plantings been extended into the public right-of-way, to draw attention to the plaza?

28. Can any needed grade change between the plaza and sidewalk be kept below three feet?

29. Have the visual and functional transitions between the plaza and adjacent buildings been considered? Has the personal space of either the plaza users or the building users been violated by placing seating, tables, or desks too close on either side of windows or doors?

30. Do ground-level building uses enliven the plaza, incorporating retail stores and cafés rather than offices or blank walls?

31. Is outdoor café seating available, in attractive colors to draw people in?

32. Have the plaza's edges been designed with many nooks and corners, to provide a variety of seating and viewing opportunities?

33. If it is a large plaza, has it been divided into subspaces to provide a variety of experiential settings for users?

34. Have such features as grade changes, planting diversity, and seating arrangement been used to create subareas?

Subspaces

With the exception of plazas specifically designed for large public gatherings, markets, or rallies, large plazas should be divided into subspaces, that is, smaller and recognizable units, to encourage use. Careful consideration should be given to the boundaries of subspaces and to their size (Fig. 11.)

35. Are subspaces separated from one another without creating a sense of isolation for users?

36. Are subspaces large enough so that users entering an area will not feel as though they are intruding if someone is already in that space?

37. Are subspaces scaled so that a person will not feel intimidated or alienated sitting there alone or with few others present?

Circulation

The principal use of many plazas is by pedestrians entering and leaving nearby buildings. Regardless of local weather, the aesthetics of the plaza, or anything else, people will take the shortest and straightest route between the sidewalk (bus stop, car drop-off, intersection) and the nearest building entry. A necessary analysis in plaza design is predicting the route by which people will flow in and out of a building, to ensure an unimpeded path for their movement.

38. Has the plaza been designed to mesh with, or enhance, existing downtown circulation patterns?

39. Are plazas linked by a system of safe pedestrian walkways, malls, street closures, and the like, to encourage walking?

40. Has thought been given to predicting the direct routes between sidewalk and building entries that people will take at rush hours?

41. Does the plaza layout also allow easy access to a café, bank, or retail establishment peripheral to the plaza; access to seating or viewing areas; and opportunities for shortcuts or pleasant walk-throughs?

42. If there is a need or desire to guide pedestrian flows, have physical barriers such as walls, planters, bollards, or distinct changes in level or texture been used to do so, rather than color or pattern changes in paving, which have been shown to be ineffective?

43. Does the plaza design allow for the tendency of pedestrians to walk in the center of spaces and sitters to gravitate to the edges of spaces?

44. Does the plaza accommodate the needs for the disabled, the elderly, parents with strollers, and vendors with carts? Do ramps parallel stairs whenever possible, or at least allow access to every level?

Seating

Even when zoning regulations encourage the provision of more downtown public open space; there is not necessarily a parallel increase in places to sit (Figs. 12 and 16). Perhaps the most detailed evaluation of outdoor seating behavior, William Whyte's study of Manhattan plazas, reported: "After three months of checking our various factors—such as sun angles, size of spaces, nearness to transit—we came to a spectacular conclusion: people sit most where there are places to sit. Other things matter too, food, fountains, tables, sunlight, shade, trees—but this simplest of amenities, a place to sit, is far and away the most important element in plaza use." (Whyte 1974, pp. 30). To design for the importance of seating:

45. Does the design recognize that seating is the most important element in encouraging plaza use?

46. Does the seating meet the needs of varying types of sitters commonly found in most plazas?

47. Has seating been placed in those locations that are sunny during lunch hours, or, in very hot locations, in the shade?

48. Does the plaza seating reflect the fact that sitters are drawn to locations where they can see other people passing by?

49. Has secondary seating (mounds of grass, steps with a view, seating walls, retaining walls that allow sitting) been incorporated into the plaza design, to increase overall seating capacity without creating a "sea of benches" that might intimidate potential users when sparsely populated?

50. Is there at least as much primary as secondary seating in the plaza?

Fig. 15. A kiosk next to ziggurat seating at 101 California Street, San Francisco expands further the range of seating and privacy options.

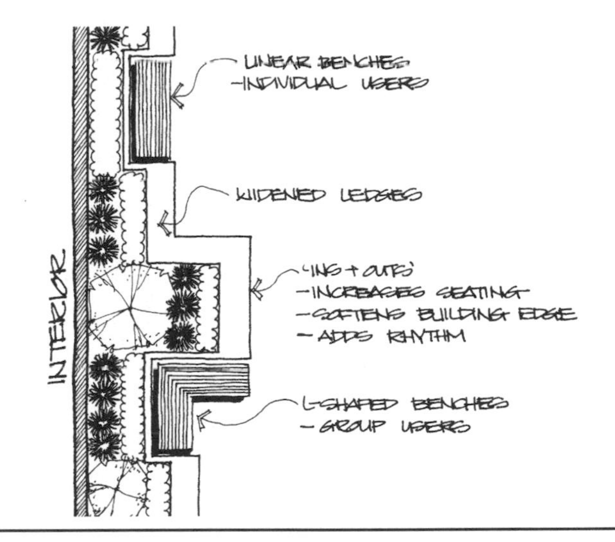

Fig. 16. The more articulated the edge of a plaza is, the more edge sitters can sit on it.

PHOTO: Robert Russell

Fig. 17. Redwood trees in this small San Francisco plaza park (Trans-America Redwood Park) surrounded by skyscrapers help screen out adjacent buildings and create an oasis-like setting.

Fig. 18. A café may draw people to a plaza above street level. This café at the First Interstate Bank Centers in Seattle offers access by stairs and by escalator.

Fig. 19. The contemporary street sculpture in Malmö, Sweden, brings art to a level where people can touch and enjoy it.

51. Are elements intended as secondary seating (with the exception of lawns) within the optimal sixteen-to-thirty-inch height range?

52. Have wooden benches been given high priority, and do they include those that are three by six feet and backless, for flexible use?

53. Is some seating linear (benches, steps or ledges) or circular and outward facing to allow people to sit close to strangers without the need for eye contact or interaction?

54. Are there wide, backless benches, right-angle arrangements, and movable chairs and tables to accommodate groups?

55. Has seating been located to allow a range of choices, from sunny to shady?

56. Has a sense of privacy been created for some of the seating, through the placement of planters or other designed elements?

57. Have a variety of seating orientations been included to allow water views, distant views, views of entertainers, foliage views, views of passersby?

58. Have seating materials been used that seem "warm" such as wood, and have those been avoided that seem "cold" (concrete, metal, stone) or that even look as though they might damage clothing if sat on?

59. Has the appropriate amount of seating been provided with reference to standard guidelines? *e.g.*, Project for Public Space recommends one linear foot of seating per 30 sq. ft. of plaza area (1 linear m of seating per 9 sq. m of plaza area). The San Francisco Downtown Plan guideline is a one-to-one ratio of one linear foot (or meter) of seating for each linear foot (or meter) of plaza perimeter.

Planting

The variety and quality of textural, color, massing, aural, and olfactory effects created by a careful planting plan can add immeasurable to the plaza's use (Fig. 17).

60. Has a variety of planting been used to heighten and enliven the users' perception of change in color, light, ground slope, smells, sounds, and textures?

61. Have feathery leaved, quasi-open trees been selected where a see-through effect to other subareas is desirable?

62. If the plaza must be sunken, have trees been planted that will soon grow above sidewalk level?

63. Have open canopy trees been selected for windy plazas, to reduce potential damage associated with dense foliage and high winds?

64. Have a variety of annuals, perennials, shrubs, and tress been selected for their color and fragrance?

65. Has the eventual height and mass of mature plants been considered, in regards to views, shade, and maintenance?

66. Have tree plantings been used that screen out adjacent building walls but, if necessary, still allow light to reach building windows?

67. Is there adequate seating so that people are not forced to sit in planted areas, thus damaging the vegetation? Are planter seat walls wide enough to prevent users from sitting in planted areas?

68. Do lawns vary the plaza's overall character and encourage picnicking, sleeping, reading, sunbathing, sprawling, and other casual activities?

69. Is the lawn area raised or sloped to improve seating and viewing opportunities, and has it avoided creating a vast "prairie" expanse in favor of smaller more intimate areas?

Level changes

Careful consideration of aesthetic and psychological effects, the perils of sunken plazas, the wise use of people attractors in sunken plazas, and the use of raised plazas are all important variables in the use of level changes in plazas (Fig. 18).

70. Have some modest but observable changes in level been included in the plaza design, to create smaller subareas?

71. Have level changes been considered as a means to separate seating areas and circulation?

72. If level changes are used, has a visual connection between levels been maintained?

73. Where level changes are incorporated, have ramps been provided to allow access for disabled people, those with baby strollers, and so forth?

74. Is there an elevated vantage point, with a wall or railing to lean on while watching people?

75. Have dramatic grade changes between plaza and sidewalk (either up or down) been avoided, as such plazas will be underused?

76. If a plaza must be sunken more than slightly, has some eye-catching feature been included to encourage people to enter?

77. If a plaza must be raised more than slightly, has planting been used to "announce" its presence and draw people upward?

Public art and sculpture

The authors of Livable Cities propose excellent criteria for evaluating art in public places that, "should make a positive contribution to the life of the city, and to the well-being of its inhabitants. . .[public art] should generously give the public some positive benefit—delight, amenity, fantasy, joy, sociability—in a word, a sense of well-being" (Crowhurst-Lennard and Lennard 1987, pp. 89, 90) (Figs. 19 and 21).

78. If public art has been included in the plaza design, will it be able to create a sense of joy and delight, stimulate play and creativity, and promote communication among viewers?

Fig. 20. Lovejoy Plaza fountain in a Portland, Oregon. Lawrence Halprin, Landscape Architect. Plaza encourages interaction with the water.

Fig. 21. Sculpture for different moods in Seattle: a tranquil composition in a bank plaza by Isamu Noguchi.

79. Can people interact with any planned public art—touch it, climb on it, move it, play in it?

80. Is the art likely to "speak" to a large proportion of the public, rather than an elite few?

81. Has a fountain or other water feature been included in the plaza design, for its visual and aural attraction?

82. Does the sound of the fountain screen out traffic noise?

83. Is the fountain in scale with plaza space?

84. Will wind cause water spray to blow, thereby rendering the sitting areas unusable? If so, is a gardener or plaza manager available to adjust the fountain?

85. Has a fountain been designed to be "hands on," so that plaza visitors can interact with it?

86. Have the costs of running the fountain been calculated to ensure that the fountain can be operated?

87. If sculptural elements are to be used in the plaza, will they be scaled to the plaza itself?

88. Is some of the sculpture experiential, that is, can people sit around it, climb on it, or alter its shape?

89. Is the sculpture located so as not to impede plaza circulation patterns and sight lines?

90. Has the sculpture been located off center, to avoid creating the impression that the plaza is merely a background for it?

Paving

To even a casual observer of people in public places, it becomes apparent that people seek to move from any two origins to destination points from A to B in as direct a line as possible—establishing so called "desire lines" and most apparent for example by pathways tracked in the snow. All major circulation routes must accommodate this principle, or people will take shortcuts across lawns or even planting to get to where they want to go, as directly as possible. Surfaces that most people avoid (and so can be used to channel movement) are large-sized gravel and cobbles. Women are more likely to avoid them than are men. A change in surface that is readily apparent to the feet and eyes, such as the transition from sidewalk paving to brick, can define a plaza as a separate place without discouraging entry.

91. Do major circulation routes follow the plaza users' principal "desire lines"?

92. If a design's intention is to channel pedestrian movement, have cobbles or large gravel been used where walking needs to be discouraged?

93. Has a change in paving been used to signify the transition from sidewalk to plaza, without discouraging entry?

Food

William Whyte observes that a Manhattan plaza with a food kiosk or outdoor restaurant is much more likely to attract users than is one without such features (Whyte 1980). This is more than a local custom (Fig. 22.)

94. Are any food services available in and next to the plaza, such as food vendors, a food kiosk, or an indoor-outdoor café?

95. Are there comfortable places to sit and eat either a bag lunch or food bought from a vendor?

96. Have drinking fountains, restrooms, and telephones been provided to augment the facilities for eating, as one would find in a restaurant?

97. Have enough trash/recycling containers been distributed around the plaza to prevent littering of food wrappers and containers?

Programs

Designers customarily "sign off" the design project when the construction is completed. But in the case of urban plazas, as in most other designed spaces, subsequent management of the plaza is crucial to its success or demise. The provision of food in a plaza is one crucial element in its success; another is the provision of programs that will attract a variety of activities and uses.

98. Do the plaza's management policies encourage special events, such as temporary exhibits, concerts, and theatrical events?

99. Does the plaza design include a functional stage area that can be used for sitting, eating lunch, and other activities during non performance periods?

100. Is the stage situated to avoid undue disruption to pedestrian circulation and to avoid making the audience face directly into the sun?

101. Will movable chairs be provided for the audience, and is there storage nearby for such chairs when not in use?

102. Are there places on the plaza to post event schedules and notices, so that they will be readily visible to plaza users?

103. Is there some method available to announce an event—decorations, banners?

104. Is there a place for temporary concessions to set up on event days?

Vendors

Small items have been sold from stalls, wagons, handcarts, and kiosks in city plazas since cities first evolved. But when retail districts developed and especially since department stores sprang up, vending began to be viewed as detrimental to commercial districts. Merchants saw it as unfair competition; city officials worried about health codes and congestion. Since the 1960s, at least in United States, there has been a remarkable reversal of this attitude in many downtown areas.

Merchants began to see that vending certain types of goods in particular locations can increase the popularity of retail areas, enliven the environment of a plaza or a sidewalk, and provide security.

105. Has the plaza been designed to accommodate vendors, whose presence will add to the vitality of the space, provide a measure of security, and often increase the popularity of surrounding retail outlets?

106. Have vendors been considered especially for plazas that are already popular for lunchtime use, poorly used and in need of something to draw users, and/or sidewalk or transit plazas with many pedestrians?

107. Does the plaza include an area that can be used for a farmers' market?

108. When providing for a market or vendors, could a colorful, fabric "roof" be provided for that area, to draw attention to the facility, provide shelter and shade, and contrast with the scale of downtown buildings?

109. Has the area for vendors or market been situated so as to be easily accessible and highly visible, yet not impede regular plaza circulation?

Information and signs

Although permanent employees in any building soon find their way about even if there are no directional signs, the occasional visitor or new employee may become disoriented without this information.

110. Is the name of the building clearly displayed and well lit after dark?

111. Is the main entrance to the building obvious and accessible?

112. After entering the building, is an information/reception desk immediately visible, or are there at least clear signs to one?

113. Are there signs directing visitors to elevators, restrooms, telephones, and cafeteria or coffee shop?

114. On leaving the building, are there clear signs indicating the way to transit stops, taxi ranks, and nearby streets?

115. Has provision of a simple, clear map of the neighborhood been considered?

Universal design

Universal design is an overriding design awareness and thus provides a helpful summary and overview of all design elements (Fig. 23). Universal design is the sensitivity and commitment to design for all people of all ages and of all abilities. Universal design responds to legally mandated requirements of accessibility, but goes well beyond minimum provisions by design details that do not marginalize in any way. The Center for Universal Design, North Carolina State University, suggests "Principles of Universal Design" that apply to design of urban plazas (Salamen and Ostroff 1997; also *cf.* article by Edward Steinfeld, this volume).

Fig. 22. Eating at outdoor cafes is universally popular in the summer months. In Vancouver, Canada, the glassed-in structure to the left offers a unique facility—public indoor seating with tables for brown baggers who want to eat inside.

PHOTO: John P. Salmen

Fig. 23. "Rest Seat" designed by Donnelly Design as prototype for public seating. Rest Seat is intended to make sitting and rising from a seated position easier and safer by providing extended armrests and a seating surface that is pitched slightly forward and higher than conventional seating.

116. Equitable use. The design is useful and accessible to people with diverse abilities.

117. Flexibility in use. The design accommodates a wide range of individual preferences and abilities.

118. Simple and intuitive use. Use of the design is easy to understand, regardless of the user's experience, knowledge, language skills or current concentration level.

119. Perceptible information. The design communicates necessary information effectively to the user, regardless of ambient condition or the user's sensory abilities.

120. Tolerance for error. The design minimizes hazards and the adverse consequences of accidental or unintended actions.

Fig. 24. Paley Park, a vest pocket park in New York City.

121. Low physical effort. The design can be used efficiently and comfortably and with a minimum of fatigue.

122. Size and space for approach and use. Appropriate size and space is provided for approach, reach, manipulation and use, regardless of user's body size, posture or mobility.

Maintenance and amenities

Finally, in any public space, people will care for an environment if they see that management cares (Fig. 24).

123. Will there be adequate staff to maintain plantings, so that lawns are green and trimmed, dead flowers are removed, and so forth? If there is some question about the availability of maintenance, an effort should be made to use attractive yet low-maintenance planting.

124. Are there enough litter containers and a collection schedule that will prevent their overflowing?

125. Will lawns, as well as shrubs and flowers in planters that double as seats, be watered on a schedule that will leave them dry and usable during lunchtime? ■

REFERENCES

Cooper Marcus, Clare and Carolyn Francis, eds. 1998. *People Places: Design Guidelines for Urban Open Space,* NY: Van Nostrand Reinhold. 2nd Edition, pp. 13–84.

Crowhurst-Lennard, Suzanne H., and Henry L. Lennard. 1987. *Livable Cities - People and Places: Socialized design principles for the future of the city.* Southhampton, NY: Gondolier Press.

Gehl, Jan. 1987. *Life between buildings: using public space.* NY: Van Nostrand Reinhold.

Joardan, S.P., and J.W. Neill. 1978. "The subtle differences in configuration of small public spaces." *Landscape Architecture* 68(11): 487–491.

Lynch, Kevin. 1976. *Site Planning* (second edition) Cambridge: MIT Press.

Projects for Public Spaces. 1984. *Managing downtown public spaces.* Chicago: Planners Press.

Pushkarev, Boris, and Jeffrey Zupan. 1975. *Urban space for pedestrians.* Cambridge: MIT Press.

Rudofsky, Bernard. 1969. *Streets for People: A primer for Americans.* Garden City, NY: Anchor Press/Doubleday.

Salmen, John P. S., and Elaine Ostroff. Universal Design and Accessible Design. 1997. *Time-Saver Standards for Architectural Design Data.* Donald Watson, ed. Seventh Edition. New York: McGraw-Hill.

Sitte, Camillo. 1945. *The art of building cities: City building according to artistic fundamentals.* [First published in German, 1889; trans. Charles Stewart]. New York: Reinhold.

Whyte, William H. 1980. *The Social Life of Small Urban Spaces.* Washington DC: Conservation Foundation.

Whyte, William H. 1974. "The best street life in the world." *New York Magazine*, July 15, pp. 26–33.

Clare Cooper Marcus

Summary

Shared outdoor space has many precedents: monastic cloistered gardens, collegial courts of Oxford, grassy courtyards of Hampstead Garden Suburb, common greens of Radburn and courtyard clusters of contemporary live-work lofts. Documented by postoccupancy evaluation, shared outdoor spaces are a vital component of the neighborhood landscape. They provide an important transition space between the privacy of the dwelling and neighborhood parks, streets, and town centers. Their role is undervalued in current urban design theory and practice and needs to be understood for their benefits as effective settings for local social life and the safe play.

Key words

children's environments, *cul-de-sac*, housing cooperatives, green landscape, new urbanism, postoccupancy evaluation, roadway, semiprivate space, shared outdoor space, *woonerf*

Oxford University College Quadrangle

Shared outdoor space

1 SHARED OUTDOOR SPACE: A FORGOTTEN URBAN TYPE

In contemporary Western cities, one can identify four broad categories of outdoor space. The first is private space, owned by an individual and accessible only to the owners, or to those they invite onto their land. In this category would be yards of private homes, private estates, and so on. The second is generally called public space—areas such as parks and streets that are publicly owned and accessible to all. A third category consists of spaces such as corporate plazas or a university campus, which are privately owned but accessible to the general public. The final category, needed a term of art proposed here as "shared outdoor space," is owned by a group and is usually accessible only to members of that group. Such spaces include community gardens, the common landscaped areas of condominium developments, clustered housing, assisted living facilities, and co-housing, as well as historic precedents such as Radburn, Baldwin Hills, Sunnyside Gardens and numerous others found in most long-established cities (Figs. 1 and 2.)

This last category might be further described as falling into two types: places where a member of the general public may freely wander through (although they may be responded to as an outsider) and strictly exclusive places such as private clubs, golf courses, and gated communities where it is not possible to gain access unless a resident or member. For purposes here, the term "shared outdoor space" will be applied to the first of the above two subcategories. This article presents an overview of the qualities of shared outdoor spaces, documented by observation and postoccupancy evaluation, that provide evidence of how they work successfully and why it is important to focus on this neglected element of the urban design vocabulary and the neighborhood landscape.

Fig. 1. Historic example of shared outdoor space: Bungalow Court, 1920, Berkeley, California

Many contemporary cities are expanding, becoming denser, and throughout the U.S. and perhaps abroad increasingly inhabited by mobile populations. Lengthy commutes, often by all adults in family households, leaves little time for fostering close ties with neighbors. Fear of crime (even if not justified by actual crime statistics) makes

Credits: Photos accompanying this article are by the author.

Fig. 2. Shared outdoor space that is neither private (belonging to an individual) nor public: Vancouver, Canada

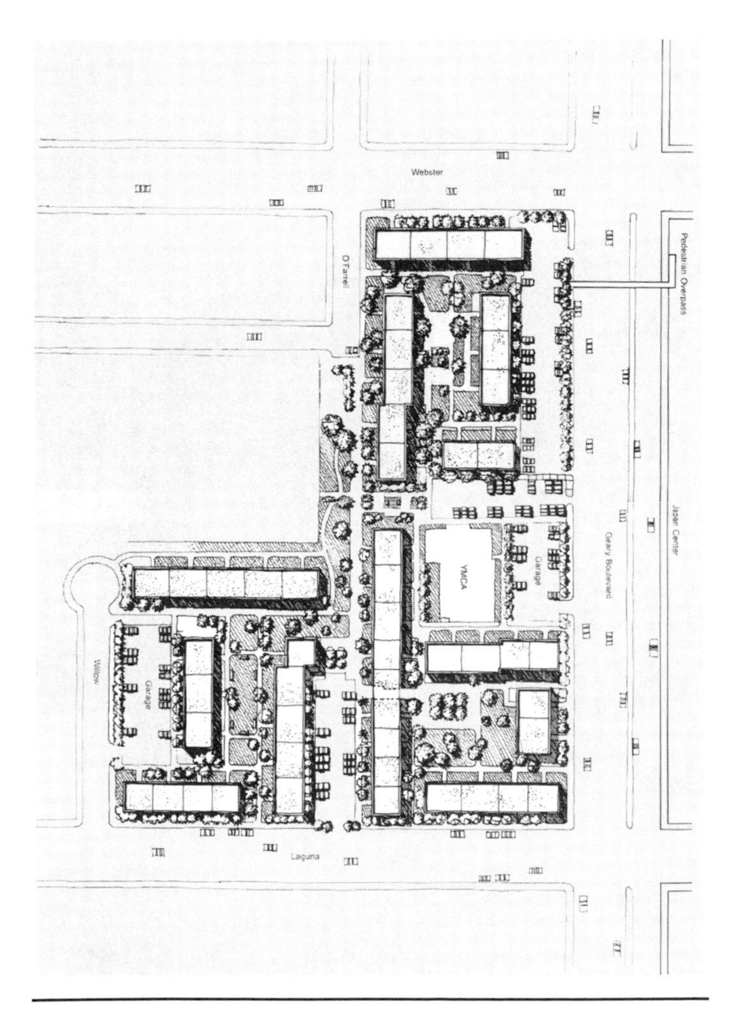

Fig. 3. St. Francis Square, San Francisco: A middle income-housing co-op with shared courtyards. Drawing courtesy of the Bruner Foundation.

parents leery of allowing their children to spend time alone on public sidewalks or to walk alone to neighborhood parks. As a result of an aging population, fewer people have the energy to maintain private yards; condominium and retirement communities seem attractive choices, both because of access to shared outdoor space (which residents don't have to maintain), and because of the potential for a greater sense of community. Parents raising young children, as well as some "empty nesters" and telecommuters, are attracted to innovative developments such as cohousing because of the potential of close neighborhood ties and safe pedestrian spaces for neighborly interaction and children's play. Because of these societal developments it seems appropriate at this time to look at shared outdoor space—what it has to offer, how it is designed, how it is used—and the pitfalls of confusing it with public outdoor space.

This article will examine four examples of shared outdoor space:

- Southside Park, Sacramento, Calif., a cohousing community established in 1993.
- Cherry Hill, an affordable housing facility in Petaluma, Calif., opened in 1992.
- Village Homes, an initially controversial subdivision in Davis, Calif., completed in1981.
- St. Francis Square, San Francisco, a mid-density housing co-op, dating from 1964.

In each case, areas of shared landscaped space are heavily used by residents (particularly children); residents make decisions about their use and upkeep; and their presence and social significance add greatly to a sense of place and identity shared by residents. Evidence for the use and significance of these neighborhood landscapes will be drawn from systematic post-occupancy evaluation (POE) studies in the case of Cherry Hill and St. Francis Square, and from more informal observations of Southside Park and Village Homes.

It is important that urban designers, landscape architects and architects recognize the distinctions between truly public spaces and shared outdoor landscapes, each with its own particular design attributes. It is especially important that this discussion occur when "traditional neighborhood" and New Urbanist (NU) planning principles are widely promoted, often without sufficient recognition of this valuable form of outdoor space. In many built examples of suburban developments, many espousing New Urbanism principles, emphasis is placed on large, public parks, while shared outdoor spaces (as defined in this paper) are not in evidence.

2 SHARED OUTDOOR SPACE IN FOUR COMMUNITIES

St. Francis Square

St. Francis Square, completed in 1964, is a 299-unit middle-income co-op in the Western Addition district of San Francisco. It was the first of many similar medium-density, garden apartment plans built during the urban renewal era of the 1960s and 1970s. The client of St. Francis Square (the Pension Fund of the ILWU) challenged the designers (Robert Marquis, Claude Stoller, and Lawrence Halprin) to create a safe, green, quiet community that would provide an option for middle-income families who wanted to raise their children in the city (Figs. 3–6.)

The designers created a pedestrian-oriented site plan with parking on the periphery of the three-block site and three-story apartment buildings facing inwards onto three thoughtfully landscaped courtyards. The three courtyards, each serving 100 households, became the shared outdoor space of the development and were critical to the strong sense of community that quickly developed at St. Francis Square and that has been characteristic of the development ever since.

This shared outdoor space, which is owned and maintained by the co-op, is critical to the community in a number of ways. It provides a green, quiet outlook with trees screening the view of nearby apartments, thereby reducing perceived density. It provides an attractive, safe landscape for children's play, with grassy slopes, pathways and play equipment—all within sight and calling distance of home. Sitting outside with a small child, walking home from one of the three shared laundries or from a parked vehicle, and many other daily activities are opportunities for adult residents to see each other and stop for a chat. The courtyards at St. Francis Square became, in effect, the family backyard writ large. If these spaces had been the equivalent of a public park, accessible to all, it is unlikely that they could have supported the strong sense of community that exists at St. Francis Square. Parents would not allow their children to play outside alone and residents would be less likely to help maintain the space, challenge a stranger, or come out to help a neighbor in need.

The author conducted a POE in 1969–70, and its findings were confirmed and expanded by one year's participant observation when living there with her family (1973–74). Subsequent site visits, plus conversations with current management, confirm the continued relevancy of basic findings of almost thirty years ago. The shared outdoor space at St. Francis Square is highly valued and well used by residents because:

- Narrow entries between buildings clearly mark the passage from the public space of street and sidewalk into the shared space of landscaped courtyards
- The courtyards are human-scaled (c. 150 x 150 ft. (46 x 46 m) with a height-to-width ratio circa1:6)
- The courtyards are bounded by the units they serve and almost all units have views into the outdoor space (facilitating child supervision)
- Considerable attention and budget was expended on the landscape elements of this shared space so that it is highly usable and attractive for adults and children, and has stood the test of time
- There are clear boundaries in the form of fenced patios, and/or "Keep off" landscaping between the private spaces of apartments, their outdoor space, and the shared space of the courtyards
- There is easy access from apartments and patios into the courtyards

The success of St. Francis Square has served as a model for many comparable programs in San Francisco and elsewhere. There is always a waiting list of people wanting to move in.

Fig. 4. Walk-up apartments look out onto shared landscaped space (St. Francis Square).

Fig. 5. Clear distinctions between private interior space, private patios, and shared outdoor space. Residents add to the public realm by planting outside their patio fences (St. Francis Square).

Fig. 6. Sensitively designed shared outdoor space ensures that children can play within sight and calling distance of home (St. Francis Square).

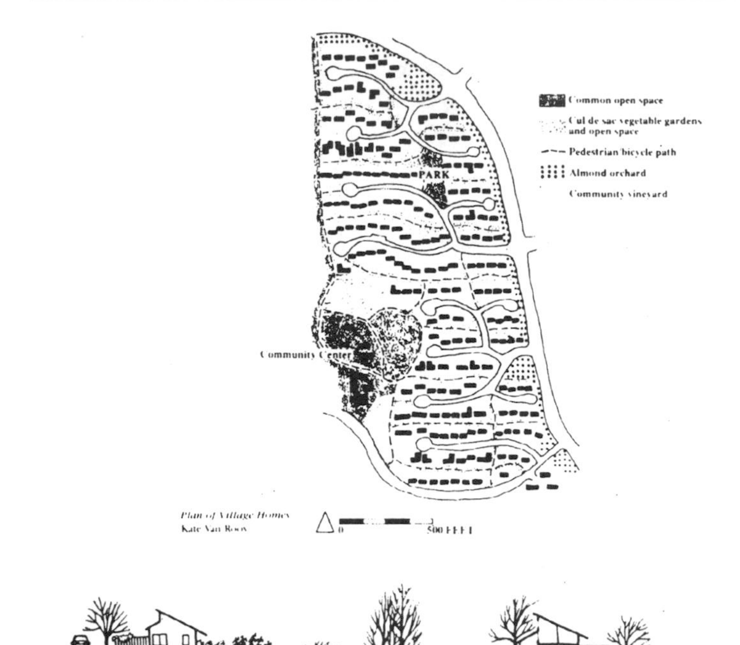

Fig. 7. Village Homes is a 240-unit visionary neighborhood with *cul-de-sac* and shared outdoor space in Davis, California.

Fig. 8. Narrow, tree-lined culs-de-sac limit traffic flow, reduce summer heat, and provide locales for neighborly encounters (Village Homes).

Fig. 9. Footpaths lines with fruit trees between the back patios of houses (Village Homes).

Village Homes

Village Homes is a 240-unit visionary neighborhood created and designed by Michael and Judy Corbett in 1981 on the outskirts of the university town of Davis, California. A recent book by the Corbetts (2000) documents the story of Village Homes from its beginnings as a "hippie subdivision," initially eschewed by banks and the local real estate industry, to its current status as "the most desirable neighborhood in Davis." At the heart of Village Homes' success, both aesthetically and socially, is the use of shared outdoor space as the core structure of the neighborhood's design. This space consists of *culs-de-sac* access and a central common green. The long, narrow, tree-shaded dead-end streets keep the neighborhood cooler in summer, save money on infrastructure, eliminate through traffic, and create quiet and safe spaces for children to play and neighbors to meet. An extensive pedestrian green at the heart of the neighborhood includes spaces for ball games and picnics, community-owned gardens, a vineyard and orchard, and drainage swales taking the place of storm sewers, reducing summer irrigation costs by one-third and providing environments for wildlife and children's exploratory play (Figs. 7–10.)

This attractive environment, though accessible to outsiders bicycling and walking, is definitely not a public park. Bounded by the inward-facing residences at Village Homes, it provides a green heart to the neighborhood, a safe and interesting area for children's play and adult exercise, and a strong sense of identity lacking in nearby grid-pattern subdivisions. A survey quoted in the Corbetts' book indicates that residents of Village Homes know, on average, 40 neighbors and have three or four close friends in the neighborhood. In comparison, a nearby standard subdivision, residents know an average of 17 neighbors and have one friend in the neighborhood.

Cherry Hill

Cherry Hill is a twenty-nine unit development of townhouses for low- and moderate-income families with children in Petaluma, a small town north of San Francisco in Sonoma County. The client of the project was Burbank Housing Development Corporation, a nonprofit developer. Residents first took occupancy in January 1992 (Figs. 11–14.)

A major goal of the site plan was to provide a safe environment for the many children expected to live there and to support a sense of

Fig. 10. A common green used for ball games, picnics, and neighborhood gatherings (Village Homes).

community among the residents. The Project Manager had read about woonerf, or residential precincts in Holland and other countries of Northern Europe commonly employed as a means of calming traffic (Cooper Marcus and Sarkissian 1986). When he asked where he could go to see one—"The council needs to be convinced"—and was told the nearest was in Holland, he took the risk of pressing the designers (woman-owned firm Morse and Cleaver) to go ahead with the *woonerf* idea. The resulting site plan consists of a narrow 22–ft. ((6.7 m) access road that creates a one-way loop around a central green. Off the loop road are four paved courtyards permitting cars to drive up to each house and creating safe hard-surface play areas. Similar to the European experience, cars and pedestrians coexist safely without sidewalks since cars drive very slowly as they enter the development, their speed regulated by the narrow roadway, speed bumps, and dead-ends of the loop-plus-courtyards. Unlike the standard grid, no one enters Cherry Hill except residents or their visitors.

A POE study conducted by architecture graduate students confirmed the success of these design decisions in April 1993 under the direction of this author. Methods consisted of interviews with 17 of the 29 households, and seven and a half hours of observation and activity mapping of the use of the shared outdoor spaces. 88% of those interviewed in the sample socialized with other families in their immediate courtyard. 64% socialized with families elsewhere in Cherry Hill. Asked where they were most likely to "bump into people they knew and stop for a chat," 94% cited their courtyard, 38% the central green, and 30% the street. 88% reported they would recognize a stranger walking in Cherry Hill and two-thirds were very satisfied with the site plan. Reasons cited for satisfaction were that it is safe for children: it is close, intimate, simple, homey, convenient, and encourages community. The 18% "not satisfied" thought that the courtyards created a "fish bowl" effect, that the whole site was too tight, and that the central green needed to be more visible. When asked how they would rate the sense of community at Cherry Hill, 71% rated it "strong" or "very strong."

On visiting Cherry Hill today and reviewing the aggregate activity maps of observations in 1993, it is clear that the dead-end *cul-de-sac* site plan, inspired by the *woonerf* model, has been highly successful in facilitating a strong sense of community and children's play. During daylight, nonschool hours, children are all over the neighborhood:

Fig. 12. Common green, Cherry Hill.

Fig. 13. A *cul-de-sac* type site plan ensures that traffic is limited and streets as well can be used for play (Cherry Hill).

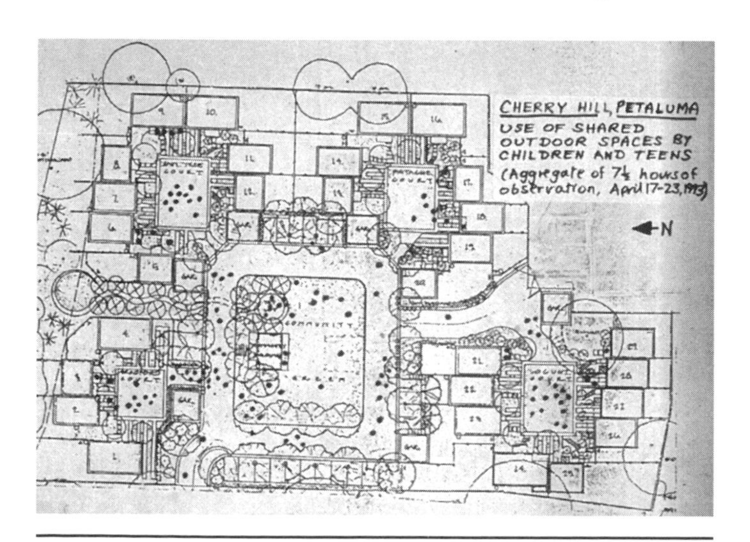

Fig. 11. Use of shared outdoor space by children and teens at an affordable housing development, Cherry Hill, Petaluma, California.

Fig. 14. Houses are clustered around small paved courts (on left, and distance), well used for play and neighborly meetings (Cherry Hill).

Legend

B	Bike storage
C	Compost
CH	Common house
K	Kid's play equipment
R	Recycling
V	Vegetable garden
W	Workshop

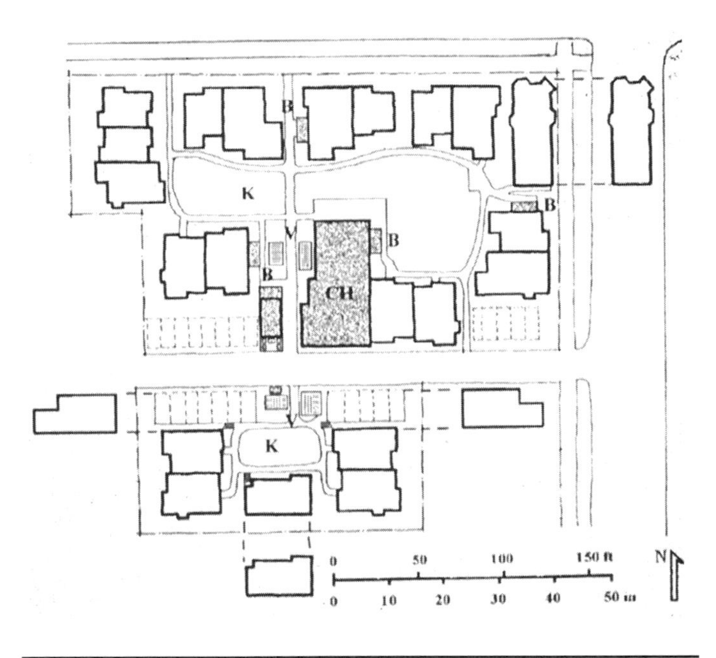

Fig. 15. Site plan of Southside Park cohousing community, Sacramento, California.

rollerblading, rolling on the grassy slope, going round the loop on scooters, watching adults working on their cars, clustered around an ice cream truck. That parents perceive the roadway as safe is confirmed by the fact that they have formally designated two sections of it for games—four square and basketball. Virtually the only cars entering the site are those of residents, and it is *their* children who are at play, so they are extra careful. It seems reasonable to assume that were this a standard grid pattern neighborhood with through traffic and no shared outdoor space, there would be less of a sense of community, and much less outdoor play. Significantly, when asked if their children watched less TV after moving to Cherry Hill, 50% said that they did, the choice to spend time outdoors and with nearby playmates being more attractive than the flickering screen. (The other 50% responded either that they had no TV, or that their children watched about as often as they had in their previous residence.)

Southside Park

Southside Park cohousing is a twenty-five unit urban "in-fill" development in inner city Sacramento, California. Mogavero Notestine and Associates designed it in consultation with the sixty-seven residents (forty adults and twenty-seven children). It was completed in 1993 with fourteen market-rate, six moderate-income, and five low-income condominiums. The site plan fits into the existing grid, with most of the houses clustered around a common green on the interior of the block, and the remainder (two renovated "Victorians" and several new units) in a smaller cluster across an alley (Figs. 15–18.)

Front porches mark the house entries from the street; back porches and patios look out onto the common green. Residents eat meals together several times a week in the 2,500 sq. ft. (71 sq. m) common house. While there has been no systematic POE of Southside Park, many casual visits by this author confirm what the residents and the designers hoped for: that its many children are attracted to play in the safe, enclosed common outdoor space comprising lawns, pathways and a play equipment area; and that adult residents frequently meet while outdoors with their children, using the common laundry, working in the raised garden beds, or walking back and forth to parking and the common house. As at St. Francis Square, Village Homes, and Cherry Hill, the sense of community and the range of children's

Fig. 16. Houses face onto neighborhood streets and back onto shared outdoor space (Southside Park).

Fig. 17. A play area in the shared outdoor space is a focus for children's activities (Southside Park).

outdoor play opportunities at Southside Park are due largely to a site plan that curbs traffic flow and creates a central pedestrian green space which is shared by residents and is not generally accessible to the public. Ironically, contrary to some design assumptions, the street-facing porches at Southside Park are used by residents when they are seeking privacy, since the shared outdoor space on the interior of the block is such a social space.

3 CHARACTERISTICS OF SUCCESSFUL SHARED OUTDOOR SPACE

Post-occupancy evaluation and related studies of these four sites (*cf.* Cooper Marcus and Sarkissian 1986) indicate how effective shared outdoor space can be as a significant component of the neighborhood landscape if the following criteria are met: (Figs. 19–21)

- The shared outdoor space is bounded by the dwellings it serves and is clearly not a public park
- Entry points into the space from a public street or sidewalk are designed so that it is clear that one is entering a setting which is not public space
- Its dimensions and the height-to-width ratio of buildings to outdoor space create a human-scaled setting
- There are clear boundaries and easy access between what is private (dwelling unit, patio, yard) and what is shared

As much care is focused on the layout, circulation patterns, planting plan, furnishings, and lighting of the shared outdoor space as is normally focused on the dwelling interiors. In particular, the design needs to focus on children (play equipment, paths for wheeled vehicles, and areas for exploratory play), as research shows that children will comprise more than 80% of users of such spaces if they are designed with the above criteria in mind.

These details are critical. It is the lack of many or all of these characteristics that accounts for the failings of the shared outdoor spaces

Fig. 19. Shared outdoor space is most successful when it is bounded by the dwellings it serves and is clearly not a public park. (Muir Commons co-housing, Davis, California.)

Fig. 20. Entry into shared outdoor space from a public street is via a narrow opening—as here on right. This helps to define the space so it is not misconstrued as a public park (Southside Park, Sacramento).

Fig. 18. Patios and decks look onto the shared outdoor space, facilitating parents to oversee children's activities and permitting casual meetings between neighbors.

Fig. 21. A human-scaled setting is created when the ratio of the height of the buildings to the width of the shared outdoor space is carefully considered (Housing Co-op, Vancouver, Canada).

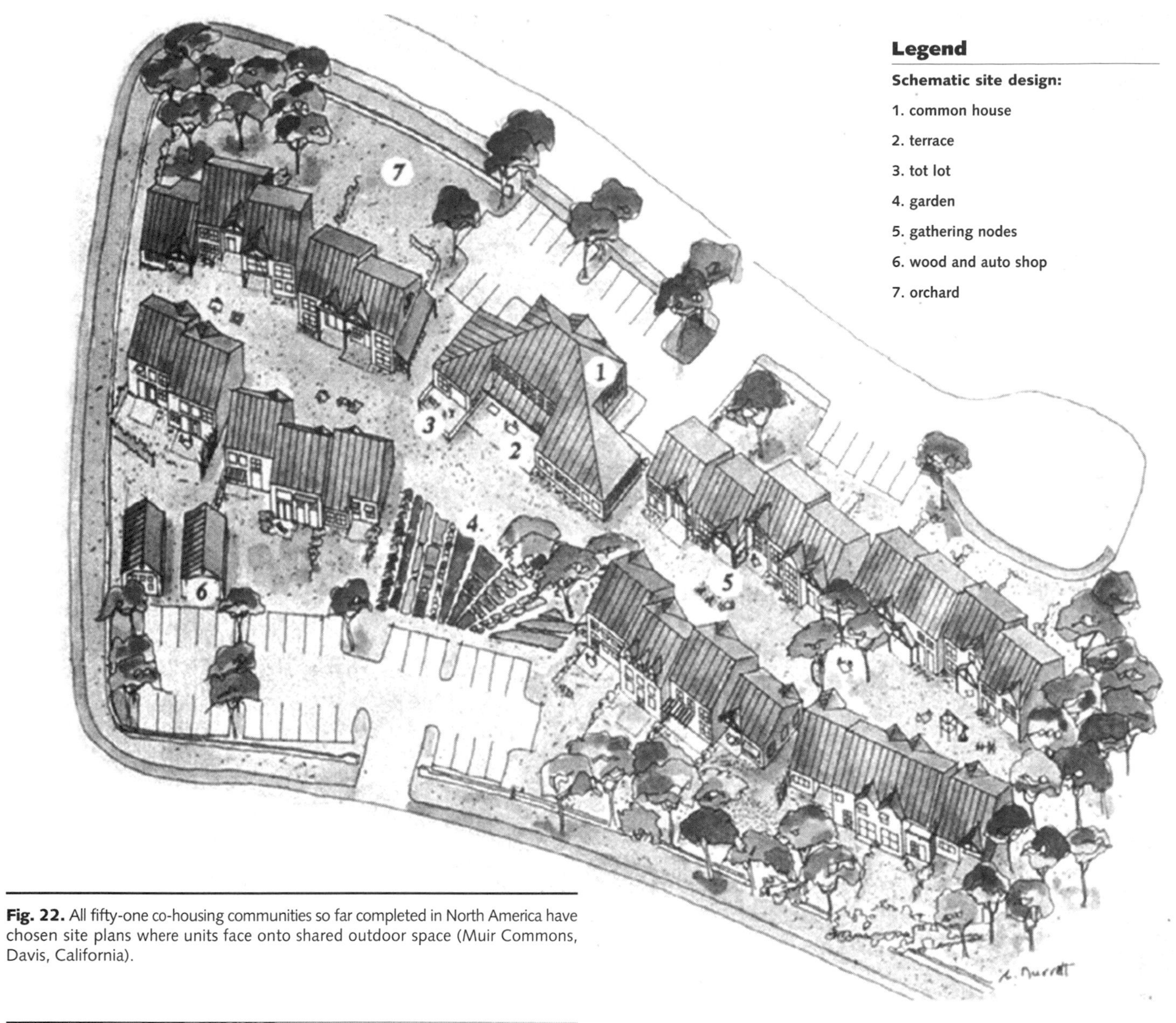

Fig. 22. All fifty-one co-housing communities so far completed in North America have chosen site plans where units face onto shared outdoor space (Muir Commons, Davis, California).

Fig. 23. It is essential that there are clear boundaries between what is private and what is shared (St. Francis Square).

Fig. 24. It is essential that shared outdoor space be designed for children who will comprise 80% or more of the users in family housing. A poor example—how not to do it (Affordable Housing, San Francisco).

of many postwar public housing projects, and many suburban Planned Unit Developments of the 1960s. Unfortunately, some design critics remarking that such spaces often became poorly maintained "no-man's lands," have concluded (incorrectly in this author's opinion) that they could never work (Coleman 1985, Duany *et al.* 2000). There is ample contrary evidence that, appropriately designed, not only do shared outdoor spaces "work," but also are actively desired by people who are able to exercise choice over how they live. Most if not all of the co-housing communities completed or under construction in North America (approximately 70 cases at the time of this writing) have chosen site plans where units face onto shared outdoor space (*Co-housing* 2000)(Figs. 22–24).

Additionally, many consumer-preference surveys indicate that suburban house buyers prefer houses that look onto *culs-de-sac*, rather than those that look onto through streets. In a national study conducted by American LIVES (1996), for example, 80% of a sample of prospective homebuyers preferred houses facing onto shared *culs-de-sac*, as opposed to those facing through streets, since they were safer for children and created a more neighborly context for adults. In market research focus groups for the planning of the NU community of Celebration, Florida, the overwhelming majority wanted *culs-de-sac*. Two national surveys published in Urban Land (February 1997) show a preference for *cul-de-sac* neighborhoods over "through street" layouts (Figs. 25–28).

The advantages of the *cul-de-sac* over the through street are that:

- They are quieter and safer for children;
- They provide potential for more neighborly interaction;
- There is a greater sense of privacy;
- Residents have a greater ability to distinguish neighbors from strangers; and,
- There are generally lower burglary and vandalism rates.

As in landscaped shared outdoor space, such settings are viewed by parents as safe for even small children's play. Neighborhood parks, while suitable for older children and teens, are rarely used by small children unless accompanied by a parent. As the development of "mixer courts" in the United Kingdom indicates, *culs-de-sac* can be attractively landscaped spaces and need not be the hot-asphalt "blobs" which characterize these settings in many American suburbs.

4 BENEFITS OF SHARED OUTDOOR SPACE

There is a further reason why designers should look seriously at communities consisting of clusters of individuals and families ranging from approximately 25 to 250 households. People are not being *forced* to live in such groupings; they are choosing to live there, and in the case of co-housing, working long and hard to bring them into being. What is happening, this author speculates, is a yearning for a community of neighbors whose size is such that one can recognize everyone; whose form is such that numerous casual encounters occur each day (the "compost" of community); and whose sense of ownership and control is such that subtle changes and modifications can be made to the shared environment as community needs change. People have a need to relate to a group larger than the family unit, but smaller than

Fig. 25. Many consumer preference surveys show that consumers prefer *culs-de-sac* over through streets—but they don't have to look like this.

Fig. 26. British variation on the *cul-de-sac*, termed a "mixer court"; asphalt replaced by attractive paved court which permits auto-access and meets fire-safety regulations (Newcastle, U.K.).

Fig. 27. The same mixer-court from inside, looking back at the entry.

Fig. 28. People need to relate to a group that is larger than the family, but smaller than a planner-designated neighborhood. Preparations for a party in shared outdoor space of a Canadian co-op (Vancouver, Canada).

Fig. 29. Residents sharing outdoor space immediately outside their homes are more likely to help make changes. Numerous modifications to the shared outdoor space at St. Francis Square over its 40-year history help sustain a strong sense of community.

a planner-designated neighborhood. It is interesting that there is no generally accepted name for this type and size of group (in other cultures, it might be a tribe, a clan, a large extended family). Nor do we have a term for the outdoor space it shares, which has been variously termed semipublic, semiprivate, communal, shared, or cluster. This lack of commonly agreed-upon definitions has perhaps made it more difficult to recognize this type of group, and to design space they may share.

Social scientists and commentators on contemporary U.S. society have noted an increasing withdrawal from public life, a rise in shyness, a fear of the "other" and the emergence of gated communities (Blakely and Snyder 1999). This suggests the need for a category of space and a form of grouping that provides a transition zone between complete privacy and complete publicness.

There are more demographic, economic, and psychological reasons why the creation of a community with shared outdoor space is particularly appropriate at this time. With increasing numbers of families where all adults are employed, the presence of a safe and interesting communal play space *right outside* the house is particularly attractive. This is true for single-parent families as well. The potential sociability of a traffic-free, green area at the heart of a community (the shared outdoor space) is appealing not only to those with children, but to the increasing number of single-person households (both young and elderly).

Finally, the presence of shared outdoor space whose management and maintenance may be shared by the group allows for people to have some control over their environment and to mold it to their needs as time goes by. Residents of St. Francis Square, for example, have made numerous changes to the shared outdoor areas of their community as needs have changed during the almost 40-year history of the co-op. Outside the private dwelling of its associated private outdoor space, there are relatively few opportunities for small groups to have some sense of accomplishment through hands-on manipulation of the local environment (Fig. 29). Evidence from interviews in communities with shared outdoor space indicates that this "working together" on the shared near-home environment provides a profound sense of shared responsibility and community (Cooper Marcus 1971, Cooper Marcus 1992).

5 THE SIGNIFICANCE OF A GREEN LANDSCAPE IN SHARED OUTDOOR SPACE

A series of studies conducted by researchers at the University of Illinois provide compelling evidence for the importance of shared outdoor spaces with grass and trees, versus comparable spaces that are hard-surfaced.

In a study of shared outdoor space and the effects of trees on preference ratings and sense of safety, a sample of residents in Robert Taylor Homes, a Chicago public housing project, were shown photo simulations of trees and grass added to a bleak courtyard adjacent to their apartments. "Residents' responses to the current condition of this outdoor space could hardly be worse" (Kuo *et al.* 1998, p. 42). Not only did they dislike its appearance, they also felt unsafe there. "Contrary to predictions made by law enforcement officials and some housing managers, residents' responses indicate that basic landscaping would be very welcome" (Kuo *et al.* 1998, p. 55). The more trees and grass depicted in the courtyard, the more residents liked it and the safer they said they would feel in it. The maximum density shown was twenty-two trees per acre, and it was made clear that the trees were high branching and that no shrubs would block views. Residents were so enthusiastic the planting trees in their courtyard, they were very willing to participate in the greening process itself. "A recurrent topic of concern among participants was how to ensure that any trees planted would survive; when asked how this could be done, the reply was straightforward: 'You've got to involve everybody' " (Kuo *et al.* 1998, p. 50).

A subsequent study in Ida B. Wells, a nearby low-rise housing project determined that residents involved in greening activities (planting flowers or caring for plants and trees in shared outdoor spaces) experienced stronger neighboring ties and a greater sense of community, felt a greater sense of ownership over these spaces, picked up the litter more often, and felt they had greater control over what happened in such spaces, than did residents not engaged in greening activities (Brunson, Kuo, and Sullivan 2002).

Earlier studies by these same researchers revealed other benefits of tree-filled courtyards with trees. The amount of time residents spent in equal-sized common spaces in two inner-city neighborhoods was strongly predicted by the presence, location, and number of trees (Coley *et al.* 1997). The closer trees were to residential buildings, the more people spent time outside near them. They reported:

> *"The historical structures of respected 'old heads'—adults in the neighborhood who watched out for, disciplined, and befriended children—are largely absent in today's ghettoes, while the prevalence of single-parent families and absent and unemployed male figures has contributed to lower levels of child supervision. . .The results of this study indicate that trees draw mixed groups of children and adults outdoors together. It is likely that the presence of adults both increases the children's supervision and also increases their opportunities to interact personally with adults in their neighborhood" (p. 490).*

It is important to assign adequate budgets to the planting and landscaping of such settings. A study comparing outdoor activities in low- and high-vegetation courtyards at the Ida B. Wells public housing project in Chicago found that in the relatively barren spaces, levels of play and access to adults were approximately half as much as those found in spaces with more grass and trees, and the incidence of creative play was significantly lower in the former. A study in Robert Taylor Homes found that "the more vegetation there was in a shared common space, the stronger the neighborhood ties near that space—compared to residents living adjacent to relatively barren spaces, individuals living adjacent to greener common spaces had more social activities and more visitors, knew more of their neighbors, reported their neighbors were more concerned with helping and supporting one another, and had stronger feelings of belonging (Kuo *et al.* 1998b, p. 843).

The authors concluded that shared green spaces may be especially important in providing settings for informal social contact in neighborhoods. This is particularly so in many public housing projects, where neighbors have a great deal in common (in terms of ethnicity, socioeconomic status and presence of children at home), and where many residents are severely limited in terms of mobility (lack of transport and low income) and are thus constrained in their access to both places and people outside the neighborhood (Kuo *et al.* 1998b, p. 845). Indeed, it seems likely that availability of informal social contacts in shared outdoor space may be critically important in low-income communities where residents have few options other than nearby neighbors for social support and sharing of resources.

Another intriguing conclusion of the above study is that, because greenery in common spaces leads to more use and stronger social networks, and because neighbors with strong social ties are more effective at exercising control over unwanted behaviors, "the greening of these spaces may yield surprising indirect benefits in the form of lower levels of crime" (Kuo *et al.* 1998b, p. 848).

6 DESIGN AND POLICY IMPLICATIONS

The design research literature on site planning (Gehl 1987; Cooper Marcus and Sarkissian1986) provides a methodology for objective assessment and rigorously established findings of how spaces work for people in their dynamic social and environmental design context. These and above-referenced examples offer ample evidence suggest that shared outdoor space provides a setting for casual social interaction; for strengthening social networks and a sense of community; for children's play; and for enhancing a sense of responsibility and safety in the neighborhood. These findings are particularly pertinent in lower-income settings where residents may not be able to sustain wider social networks or take their children to areas of public recreation.

Many public housing projects across the United States are undergoing modernization and rehabilitation. It is critical that these redesigns take into account the crucial importance of shared outdoor space. The provisions of defined private outdoor space and nearby public parks are important, but not sufficient.

Many planned unit developments, including those espousing tenets of New Urbanism (NU) seem to obviate the use of *culs-de-sac* and instead favor the grid and public parks bounded by through streets. While such parks are important in neighborhood design, the social significance of shared outdoor space that is not fully public is overlooked. In recent NU literature, the only reference to space shared by a group yet not fully public is the alley, a design device to ensure that curb cuts and garages do not mar the streetscape and that houses can be sited closer together (Fig. 30). While these goals are laudable, it seems unlikely that a sense of local identity can be facilitated as much by these utilitarian passageways as it is by the provision of common

Fig. 30. The alley: Shared outdoor space in a New Urbanist development, Santa Fe, New Mexico

greens bounded by the units they serve. NU-inspired building codes are increasingly being adopted. While many of their humanizing principles are positive, some types of site plans seem to be legislated out of existence, such as those at St. Francis Square, Village Homes, Cherry Hill, and Southside Park, apparently because they are based on the *cul-de-sac* model of shared outdoor space, the very reasons as noted that they succeed.

It is especially important that design guidelines of housing projects be examined in light of the above conclusions. Proposing that neighborhoods be designed or re-designed with outdoor space designated as either private (yards, patios) or public (streets, sidewalks, parks) will not provide the critically important setting of shared outdoor space for child and adult social and play activities close to home. Just because such settings in public housing and Planned Unit Developments were poorly understood, inadequately designed, and underused in the past does not justify throwing out this significant category of outdoor space in contemporary medium- and high-density housing. Designers should attend to what residents who use and value such space have to tell and should provide shared outdoor settings as green oases in newly-built and re-designed urban and suburban neighborhoods. ■

REFERENCES

American LIVES Inc. 1996. 1995 *New Urbanism Study*. San Francisco: American LIVES Inc.

Anonymous. 1993. *Post-occupancy Evaluation of Cherry Hill*, Petaluma, CA. Berkeley: unpublished.

Blakely, Edward, and Mary Gail Snyder. 1999. *Fortress America: Gated communities in the United States*. Washington, DC: Brookings Institution Press.

Brunson, Liesette, Frances E. Kuo, and William C. Sullivan. 2002. Sowing the seeds of community: Greening and gardening in inner-city neighborhoods. *American Journal of Community Psychology*.

Coley, Rebekah Levine, Frances E. Kuo, and William C. Sullivan. 1997. Where does community grow? The social context created by nature in urban public housing. *Environment and Behavior* 29(4): 468-494.

Coleman, Alice. 1985. *Utopia on trial: Vision and reality in planned housing*. London: H. Shipman.

Cooper Marcus, Clare. 1992. *Architecture and a Sense of Community: The case of co-housing*. Unpublished manuscript, Berkeley, Calif.

________. 1971. St. Francis Square: Attitudes of its residents. *AIA Journal*.

________. 1970. *Resident attitudes towards the environment at St. Francis Square, San Francisco: A summary of the initial findings*. Institute of Urban and Regional Development, University of California Working Paper No. 126.

Cooper Marcus, Clare, and Wendy Sarkissian. 1986. *Housing as if people matter: Site design guidelines for medium-density family housing*. Berkeley: University of California.

Corbett, Judy, and Michael Corbett. 2000. *Designing sustainable communities: Learning from Village Homes*. Washington, DC: Island Press.

Duany, Andres, Elizabeth Plater-Zyberk, and Jeff Speck. 2000. *Suburban nation: The rise of sprawl and the decline of the American Dream*. New York: North Point Press.

Frantz, D., and C. Collins. *Celebration, U.S.A*. 1999. New York: Henry Holt.

Fulton, William. 1996. *The new urbanism: Hope or hype for American communities?* Cambridge: Lincoln Institute of Land Policy.

Gehl, Jan. 1987. *Life between buildings: Using public space*. New York: Van Nostrand Reinhold.

Kuo, Frances E., Magdalena Bacaico, and William C. Sullivan. 1998. Transforming inner-city landscapes: trees, sense of safety, and preference. *Environment and Behavior* 30(1): 28-59.

Kuo, Frances E., William C. Sullivan, Rebekah Levine Coley, and Liesette Brunson. 1998b. Fertile ground for community: Inner-city neighborhood common spaces. *American Journal of Community Psychology* 26(6): 823-851.

Kweon, Byoung-suk, William C. Sullivan, and Angela R. Wiley. 1998. Green common spaces and the social integration of inner-city older adults. *Environment and Behavior* 30(6): 832-959.

Taylor, Andrea Faber, Angela Wiley, Frances E. Kuo, and William C. Sullivan. 1998. Growing up in the inner city: Green spaces as places to grow. *Environment and Behavior* 30(1): 3-27.

[Various]. 2000. *Co-housing: The Journal of the Co-housing Network* 13 (1).

Susan Goltsman

Summary

A quality play and learning environment is more than a collection of play equipment. The entire site with all its elements—from vegetation to storage—can be a play and learning resource for children with and without disabilities. This article discusses how to create a universally designed play area that integrates the needs and abilities of all children into the design. An integrated play area allows children with disabilities to participate in play experiences that other children take for granted. Included are general elements and guidelines as well as specific performance criteria for a variety of play settings.

Key words

accessibility, community participation, fences, gardens, gradients, outdoor play, play equipment, ramps, sand play, universal design, water play

Universal design in outdoor play areas

Play is more than just having fun. It is a process through which children develop their physical, mental, and social skills. It is value-laden and culturally based. In the past, most play experiences occurred in unstructured child-chosen places. Children with disabilities, depending on the type and severity of the disability and the attitudes of their parents, generally have less access to these free-range play settings found around the neighborhood. They also have very limited choices within most structured play settings. It is possible to create well-designed play areas that successfully integrate the needs of children with and without disabilities. The key is to provide diverse physical and social environments so that children with disabilities are a part of the overall play experience.

1 DESIGNING AN INTEGRATED PLAY AREA

A well-designed play area that integrates children with and without disabilities consists of a range of settings carefully layered onto a site. They contain one or more of the following elements: entrances, pathways, fences and enclosures, signage, play equipment, game areas, land forms and topography, trees and vegetation, gardens, animal habitats, water play, sand play, loose parts, gathering places, stage areas, storage, and ground covering and safety surfacing. In any play area design, each play setting varies in importance, depending on community values, site constraints, and location. The way these elements are used will also determine the degree of accessibility and integration possible in that environment. However, in designing a play space of any size, the full range of settings should be considered.

Having a diversity of play settings and opportunities within the settings is key to integration and access in a play area. Since play to be developmental must be a challenging, not every part of the environment should be physically accessible to every user. Therefore, a play area must support a range of challenges, both mentally and physically. Physical challenge within the play area must be part of a progression of challenges that promote an individual's skill.

On the other hand, the play area's social experience must be accessible to all. Unlike a physical challenge, which to be developmental must be "earned" through a child's efforts, the opportunity for a social interaction should be easy. Social integration is the basic reason why a play area must be accessible to children of all abilities. If the play area truly serves the range of children who use is, then it is considered universally designed.

Creating a universally designed play area requires integrating the needs and abilities of all children into the design of play areas. The diversity of both physical and social environments is the key to accommodating the variety of users in a play area. Obtaining physical diversity means placing a broad range of challenges within the play setting. Such an environment will allow more children to participate, make choices, take on challenges, develop skills, and, most importantly, play together.

Credits: This article is reprinted with permission from Preiser, Wolfgang F. E., and Elaine Ostroff, editors, (2001) *Universal Design Handbook*. New York: McGraw Hill. Photos are courtesy of Moore Iacofano Goltsman (MIG).

Social diversity is linked to physical diversity. Contact between children of different abilities will naturally increase in play areas that are open to a wider spectrum of users. This interaction is particularly critical for children with functional limitations, who are so often denied these social experiences.

Elements of an integrated play area

- **Consider the many ways in which children with disabilities can interact.** When arranging the play area, integrate accessible play equipment with the rest of the play setting. Placing less challenging activities directly next to those requiring greater physical ability will encourage interaction across all ability levels.

- **Provide an accessible route that connects every activity area and every accessible play component in the play setting.** (A play component is defined as an item that provides an opportunity for play; it can be a single piece of equipment or part of a larger composite structure.) Even though not every play component will be physically accessible to everyone, simply enabling all children to be near the action provides opportunities and choices, and promotes the possibility of communication with others. This is a major step towards integration.

- **Ensure that at least one of each kind of play component on the ground is accessible and usable by children with mobility impairments.** Likewise, at least half of the play components elevated above ground should be accessible. Access onto and off of equipment can be provided with ramps, transfer platforms (automated lifts for the wheel-chair bound), or other appropriate methods of access. Remember that ramps and transfer systems can also serve as physical challenges and should be designed so that they add to the diversity of the environment.

- **Make sand play accessible by providing raised sand areas or installing a transfer system into the sand.** Note that raised sand areas, which provide space underneath for wheelchairs, provide severely limited play experiences because the necessary clearance keeps the sand depth at a minimum. Providing a transfer system into the sand area will allow users to enjoy full-body sand play.

- **Make portions of gathering places accessible to promote social interaction.** These areas are important areas of interaction and allow groups of people to play, eat, watch, socialize, and congregate. Include accessible seating—such as benches without backrests and arm supports—so people of varying abilities can sit together.

- **Don't forget safety guidelines, which outline important parameters such as head entrapments, safety surfacing; use zones.** At times, however, provisions for safety and accessibility can conflict. For example, a raised sand shelf could be considered hazardous because the shelf is more than twenty inches off the ground. If you strictly followed the safety requirements, you would need to construct a non-climbable enclosure along the edge of the shelf, which would defeat the whole purpose of the design. In such cases, seek solutions that provide other means of access that mitigate the safety hazard, such as installing rubber safety surfacing on the ground below the shelf.

2 PERFORMANCE CRITERIA FOR INCLUSIVE PLAY SETTINGS

A universally designed play setting is not "high tech." It is "design tech." In order to accommodate the needs of children with varied abilities, the overall design or individual components need to be reconsidered so that there is an inclusive system. The focus should be based on good anthropometric data, user-based design guidelines and performance criteria. The accessible components that are created, such as transfer systems for manufactured equipment, are not stigmatizing by their appearance.

The options available for creating a universally designed play area are all based on low technology, innovative thinking, and problem solving. In some instances, certain design solutions, especially new ones, may require fabrication and manufacturing in a state-of-the-art factory. Ideas should be shared in the manufacture of play components. Users, designers, and manufacturers working together will advance this relatively new area of environmental design.

The following are performance criteria for creating accessibility and integration in play areas:

Accessible route of travel within a play setting

Because play is primarily a social experience, accessible routes through a play setting must connect all types of activities. A path that connects the accessible play elements within a play setting is essential. Without this connection, children with disabilities can too easily find themselves isolated from their friends without disabilities. The accessible route within a play setting avoids problems caused by circulation design flaws and promotes social interaction.

A good play setting has many routes through the space. A route itself may be the play experience. Pathways, for example, can be a play element in themselves, supporting wheeled toys, running games, and exploration. To be accessible, pathways require firm and stable surfaces and correct grades (one up to 20 percent) and cross slopes (2 percent or less). The quality of the pathway system sets the tone for the environment. Pathways can be wide with small branches, long and straight or circuitous and meandering. Each creates different play behaviors and experiences. To provide the range of challenges necessary for a variety of developmentally appropriate play experiences, minimum routes or auxiliary pathways through a play experience are exempt from the strict requirements of the primary accessible route of travel.

Criteria for accessible route design:

- An accessible route to and from the intended use of the different activities within the play area setting must be provided.

- The accessible route should be a minimum of sixty inches wide but can be adjusted down to thirty-six inches if it is in conjunction with a bench or play activity.

- The cross slope of the accessible route of travel shall not exceed 1:50.

- The slope of the accessible route of travel should not exceed 1:20.

- If a slope exceeds 1:20 it is a ramp. A ramp on the accessible route of travel on the ground plane should not exceed a slope of 1:16.

- If the accessible route of travel is adjacent to loose fill material or there is a drop-off, then the edge of the pathway should be treated to protect a wheelchair from falling off the route and into the loose fill material. This can be accomplished by beveling the edge with a slope that does not exceed thirty percent (a vertical raised edge would create a trip hazard for walking children). If this route is within the use zone of the play equipment, the path and the edge treatment should be made of safety surfacing.

- Changes in level along the path should not exceed 1/2 in. (1.27 cm).

- Where egress from an accessible play activity occurs in loose fill surface which is not firm, stable, and slip-resistant, a means to the point of access for that play activity shall be provided. The surfacing material should not splinter, scrape, puncture, or abrade the skin when being crawled upon.

Play equipment

Most equipment settings stimulate large muscle activity and kinesthetic experience, but they can also support nonphysical aspects of child development. Equipment can provide opportunities to experience height and can serve as landmarks to assist orientation and wayfinding. They may also become rendezvous spots, stimulate social interaction, and provide hideaways in hiding and chasing games. Small, semienclosed spaces support dramatic play, and seating, shelves, and tables encourage social play (Fig. 1).

Selection criteria for manufactured play equipment for a recreation area:

- Properly selected equipment can support the development of creativity and cooperation, especially structures that incorporate sand and water play. Play structures can be converted to other temporary uses like stage settings; loose parts such as backdrops or banners can be strung from and attached to the equipment for special events and dramatic play activities. Equipment settings must be designed as part of a comprehensive multipurpose play environment. Isolated pieces of equipment are ineffective.

- Equipment should be properly sited, selected, and installed over appropriate shock-absorbing surfaces. Procedures and standards for equipment purchase, installation, and maintenance must be developed. A diligent safety inspection program must be implemented.

- There are a number of well-documented safety issues related to manufactured play equipment: falls, entrapments, protrusions, collisions, and splinters. All of these issues must be addressed in the design, maintenance, and supervision of play equipment.

- Equipment should be accessible, but must be designed primarily for children, not wheelchairs. Transfer points should always be marked both visually and tactually. The most significant aspect of making a piece of equipment accessible is to understand that children with disabilities need many of the same challenges as children without disabilities. Using synthetic surfacing can provide access to, under, and through the equipment for children who use wheelchairs.

Fig. 1. A play village that provides access through and around makes the dramatic play social experience available to all.

RECENT U.S. LEGISLATION

Laws recently enacted in the United States to assure and promote the rights of persons with disabilities have now created market demand and social acceptance of an all-inclusive, universally designed play environment. In 1998, the Architectural and Transportation Barriers Compliance Board created a committee to amend the Americans with Disabilities Act Accessibility Guidelines (ADAAG) by adding a section for play areas. The guidelines were developed by a regulatory negotiation committee composed of representatives from a variety of interest groups. The final guidelines proposed for rule (ADAAG 36 CFR Part 1191) have gone through public review and await publication in the Federal Register. The requirements of this rule will become part of the enforceable ADA guidelines once adopted by the Department of Justice.

Fig. 2. Water sprays installed below a rubber surface provide water play for everyone.

Fig. 3. Giant nautilus sand play element.

Getting in the center of action may be as important as climbing to the highest point for some children.

- Play equipment provides opportunities for integration, especially when programmed with other activities. Play settings should be exciting and attractive for parents as well as children—adults accompany children to the park or playground more often today than in the past. It is equally important to design for parents using wheelchairs who accompany their able-bodied children.

Play equipment is designed to provide a physical challenge as well as social interaction. For equipment to be appropriate for different skill levels, it must be graduated in its challenge opportunities. Access up to, onto, through, and off equipment should also provide a variety of challenge levels appropriate to the age of the intended users.

Water play areas

Water in all its forms is a universal play material because it can be manipulated in so many ways. One can splash it, pour it, use it to float objects, and mix it with dirt to form "magic potions." Permanent or temporary, the multisensory quality of water makes a substantial contribution to child development. Water settings include: a hose in the sandpit, puddles, ponds, drinking fountains, bubblers, sprinklers, sprays, cascades, pools, and a dew-covered leaf (Figs. 2 and 3).

If water play is provided, a part of the play area must be wheelchair accessible. If the water source is manipulated by children, then it must be usable by all children. If loose parts such as buckets are provided and children have access to the equipment storage, then the storage must be usable by all children. When water is provided for play, the following dimensions apply:

- **Forward reach:** 36–20 in. (91–50 cm).
- **Side reach:** 36–20 in. (91–50 cm).
- **Clear space:** 36 x 55 in. (91 x 140 cm). The clear space should be located at the part of the water play area where the most water play will occur. If the water source is part of the active play area and children turn the water on and off, then it must be accessible. If the water source is part of a spray pool, the area under the spray should be accessible. Accessibility should involve the dimensions for both clear space and reach.
- **Clearance ranges:** top height to access water 30 in. (76 cm) maximum, under clearance 27 in. minimum (69 cm).

Sand play areas

Children will play in dirt wherever they find it. Using "props" such as a few twigs, a small plastic toy, or a few stones, children can create an imaginary world in the dirt, around the roots of a tree or in a raised planter. The sandbox is a refined and sanitized version of dirt play. It works best if it retains dirt's play qualities. Provide small, intimate group spaces, adequate play surfaces, and access to water and other small play props.

If a sand play area is provided, part of it must be accessible. Important elements are clear floor space, maneuvering room, reach and clearance ranges, and operating mechanisms for control of sand flow.

When products such as buckets and shovels will likely be used in the sand play area, storage places should be at accessible reach range.

Raised sand play is a very limiting play experience because of the way a raised area must be constructed. To provide a place for the wheelchair under the sand shelf, there is very little depth of sand available for play. Therefore a raised sand area by itself is not a substitution for full body sand play. All sand play opportunities on the site should be usable by all children to the greatest extent possible.

If the sand area is designed to allow children to play inside it, a place within the sand play area should be provided where a participant can rest or lean against a firm, stationary back support in close proximity to the main activity area. Back support can be provided by any vertical surface that is a minimum height of 1 in. (2.54 cm) and a minimum width of 6 in. (15 cm), depending on the size of the child. Back support can be a boulder, a log, or a post that is holding up a shade structure. A transfer system into a sand area may also be necessary if the area is large and contains a variety of sand activities. A transfer system would be appropriate if there are no areas of raised sand play in the primary activity area, or if the sand area is over 100 square feet (9.29 sq. meters) and the raised sand area would tend to isolate accessible sand play activities.

When raised sand is provided, the following clearance ranges apply:

- **Top height to sand:** 30–34 (76–86 cm) maximum
- **Under clearance:** 27 in. (69 cm) minimum
- **Side Reach:** 36–20 in. (91–50 cm).
- **Forward reach:** 36–20 in. (91–50 cm).
- **Clear space for wheelchair:** 36 x 55 in. (91 x 140 cm).

Depending on site conditions and the amount of sand play, shade may be required. If shade is required by site conditions, it may be provided through a variety of means, such as trees, tents, umbrellas, and structures. This advisory requirement for shade is based on the site context, program, and users. Some shade in or around sand is usually desirable.

Gathering places

To support social development and cooperation, children need comfortable gathering places. Parents and play leaders need comfortable places for sitting, socializing, and supervising.

If gathering places are provided, a portion of them should be accessible. A gathering place contains fixed elements to support play, eating, watching, talking, or assembling for a programmed activity. Gathering places should serve people of all ages (Fig. 4).

- **Seating:** A variety of seat choices should be made available. At least 50 percent of fixed benches should have no backs or arms.
- **Tables:** Where tables are provided, a variety of sizes and seating arrangements should be provided.

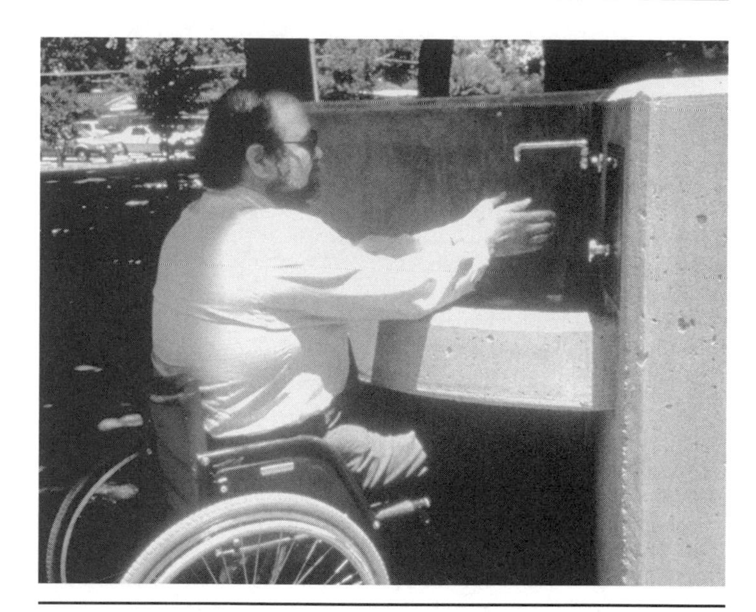

Fig. 4. Accessible wash-up sink serves people of all ages and abilities.

- **Game tables:** Game tables provide a place for two to four people to play board games. Where fewer than five game tables are provided, a minimum of one four-sided game table should include an accessible space on one side.
- **Storage:** If storage is supplied as part of a gathering area and is used by children, accessible shelves and hooks should be a maximum of thirty-six inches above the ground. The amount of storage is dependent upon program requirements.
- **Shade:** Shade may be desirable for gathering areas where people will be participating in activities over a long period of time. Shade can be provided through a variety of means such as trees, canopies, or trellises, depending on site context.

Garden Settings

An effective play-and-learn activity, gardens allow children to interact with nature and with each other. Garden beds and tools can be adapted for use by children with disabilities. Scent gardens are attractive to all children and are of special appeal to children with low vision.

Gardens in play areas are primarily used to enable planting, tending, studying, and harvesting vegetation. Depending on the type and height of plantings, planter boxes may require a raised area for access or a transfer point. A garden must provide a minimum of one accessible garden plot.

Raised Gardens

If a raised area is provided, then:

- The raised area should be located as part of the main garden area. The program determines the amount of raised area, but a minimum of 10% of the garden should be raised.

- The edge should be raised above the ground surface to a minimum of 20 in. (51 cm) and a maximum of 30 in. (76 cm).
- Garden growing area should require access either by side or by forward reach twelve to thirty-six inches above ground.
- **Transfer systems:** If children are required to sit in the dirt to garden, a transfer point should be provided which enables a participant to transfer into the garden.
- **Potting and maintenance Areas:** Potting and preparation areas should require access either by forward or side reach. The amount of area to be made accessible depends on the program. At least one workstation for potting should be made accessible.
- **Storage:** Storage areas for the garden should provide access for children who use wheelchairs. Hooks and shelves should be a maximum of 36 in. off the ground.
- **Circulation:** Aisles with widths 36–44 in. (91–112 cm) around the garden should be provided on a main aisle so a child using a wheelchair or walker can get to the garden. This larger aisle width of 48–60 in. (122–152 cm) should also provide access to the accessible gardening spaces.

Vegetation, trees, and landforms

Vegetation, trees, and topography are important features in a play setting. These features should be integrated into the flow of play activities and spaces, or they can be play features in themselves.

Landforms help children explore movement through space and provide for varied circulation. Topographic variety stimulates fantasy play, orientation skills, hide-and-seek games, viewing, rolling, climbing, sliding, and jumping. "Summit" points must accommodate wheelchairs and provide support for other disabling conditions.

Trees and vegetation comprise one of the most ignored topics in the design of play environments. They are two of the most important elements for integration, however, because everyone can enjoy and share them. Vegetation stimulates exploratory behavior, fantasy, and imagination. It is a major source of play props, including leaves, flowers, fruits, nuts, seeds, and sticks. It allows children to learn about the environment through direct experience.

Designers and program providers should integrate plants into play settings rather than creating separate "nature areas." For children with physical disabilities, the experience of being in trees can be replicated by providing trees that a wheelchair can roll into or under. You can create an accessible mini-forest by planting small trees or large branching bushes.

If vegetation, trees and/or landforms are used as a feature, a means should be provided for access up to and around the feature. Tree grates and other site furniture that support or protect the feature must also be selected so as not to entrap wheels, canes, crutch tips, etc.

Animal habitats

Contact with wildlife and domestic animals stimulates a caring and responsible attitude towards other living things, provides a therapeutic effect and offers many learning opportunities. Play areas can provide opportunities to care for or observe domestic animals. Existing or created habitats of butterflies and other insects, aquatic life, birds, and small animals should be protected. Planting appropriate vegetation will attract insects and birds.

Entrances and Signage

Entrances are transition zones that help orient, inform, and introduce users to the site. They are places for congregating and for displaying information. Not all play areas, though, have defined entrances. Sometimes entry to a play area can be provided from all directions.

Signs can be permanent or temporary, informative or playful. Expressive and informative displays use walls, floors, ground surfaces, structures, ceilings, sky wires, and roof lines on or near a play area to hang, suspend, and fly materials for art and education. Signage is a visual, tactile, or auditory means of conveying information and it must communicate a message of "ALL Users Welcome." You can use appropriate heights, depths, colors, pictures, and tactile qualities to make signs accessible. Signs should primarily communicate graphically. Talking signs are also effective.

Fences, enclosures, and barriers

An enclosure is a primary means of differentiating and defining the child's environment. For example, fences can double back on themselves to provide small social settings. Fences, enclosures, and barriers protect fragile environments, define pathways, enclose activity areas, and designate social settings. Low fences could be play elements and should be considered as such. The entrance to an enclosure should be clearly visible and wide enough for wheelchair passage with widths of 36 in. (91 cm) minimum, 48 in. (122 cm) preferable.

To fully support child development and integration of children with and without disabilities, a risk management program and professional play leadership should augment a well-designed play environment. Part of the play leader's job is to set up, manipulate, and modify the physical environment to facilitate creative activity. Since the environment can either support or hamper play, designers must learn how they can best empower children through design. Compensating for an inadequate environment is a drain on the leader's time and energy.

3 CASE STUDY: IBACH PARK

Ibach Park is a 19.8-acre (8-hectare) neighborhood park, owned and operated by the City of Tualatin Parks and Recreation Department and was designed by Moore Iacofano Goltsman (MIG). The design and development of the park reflects the benefits desired by community residents who were active participants in the design process. Park features include a 3/4 acre (.3035 hectare) play area that celebrates the rich heritage of Tualatin from prehistory to early settlement days: a preteen play area, a soccer field, a softball field, tennis courts, a basketball court, picnic areas, an open turf area, the Hedges Creek greenway trail development and restoration, interpretive signage, restrooms, and parking.

Design process

The design goals were:

- The design is developed with input from the local community.

- The play area integrates children of varying abilities.
- Create a play environment that encourages social and intellectual development.

To ensure that as many community members as possible would have an opportunity to contribute their ideas, visions, and goals to the site design, a series of five workshops were held. Participants included residents, members of the Tualatin Parks and Recreation Advisory Board, and members of the Tualatin Disabilities Advisory Board. As part of the public participation process, a design charrette was held to involve residents in identifying the park's future benefits. From these public meetings, a design concept emerged that was related to the site's archaeological context. This concept was then overlaid with the functional requirements for a neighborhood park, including active and passive recreational opportunities. Other overall goals for the park included safety, security, accessibility, environmental preservation and education, and ease of maintenance (Fig. 5).

Fig. 5. An interactive design process allows the community to help make design trade-offs.

The design team worked closely with maintenance staff during the design process to reduce operation costs. Some features designed to reduce costs include special court surfacing on tennis and basketball courts; timed release of water for water play features; the use of synthetic safety surfacing in the play area; plant selection; and the design of mow strips to facilitate equipment use. Restroom doors have timers that can be set automatically to lock and unlock, again reducing personnel costs. Sand-based softball and soccer fields are designed for year-round use with complete underground drainage and irrigation to increase playing time and user safety.

Elements of the design

Children's play area

Based on community members input and the results of the community design charrette, the design team developed an interpretive design concept for the Ibach Park play area. The accessible play area explores the rich heritage of Tualatin, from prehistory to early European settlement days. Children learn about the history of Tualatin through interactive play, reenacting historic events in play settings. The historic play area was designed to meet the developmental needs of preschool and school-age children, and to allow children with and without disabilities to have equal access to integrated play opportunities. The final design provides play area users with an interactive tour through the history of Tualatin, allowing people to experience a bit of the City's past.

The play area design includes three distinct areas reflecting significant historic periods: the Prehistory area, the Native American area, and the Early European Settler area. A water course of real and simulated water runs through the entire play area, symbolizing the historic impact the Tualatin River has had on the life of the city and its people. In areas where real water was not feasible, the river flow is continued, represented by blue synthetic safety surfacing. A segment of the river is a water play element that allows children to turn on the river flow by pressing against a bollard. The water flows into the riverbed, which is at the ground level on one side and at a raised height on the other side to allow water play while using a wheelchair or while standing (Fig. 6).

Fig. 6. Water play channel at Ibach Park provides wheelchair access on one side.

Fig. 7. Water element, mastodon ribcage and sand play at Ibach Park. The transfer platform (left) and missing mastodon rib (right) facilitate access (MIG).

Prehistory area

The Prehistory area begins at a high, rocky area, simulating a mountain (such as Mt. Hood, which can be viewed from this site). It contains a child-sized version of the meteor that landed on earth millions of years ago and was carried west to Tualatin by surging waters of the Bretz Floods over 15,000 years ago. In addition, an archaeological dig allows children to unearth fossil rocks and fern prints, and explore a giant mastodon ribcage, providing them with a wonderful sense of discovery. Children not only see what happens when a layer of earth is scraped away to discover fossils, but they can develop a better sense of the natural world and the history of life. The mastodon ribcage also provides an excellent climbing structure. One missing rib permits wheelchair access (Fig. 7).

Native American area

Replicas of Native American petroglyphs signal the change from prehistory to Tualatin's Native American history. The Native American area contains a circle of drums that allow children to create music. Each drum is at a different height so that children can play either standing or sitting down. A dugout canoe, cut out to accommodate wheelchairs, allows children to experience how the Native Americans might have traveled down the river to catch fish or look for a better hunting ground.

Early European settler area

This area contains provides an opportunity to learn what it was like for the early settlers of the region. Young children can play in log cabin-style play houses, prepare breakfast at the sand tables, harness a team of horses to pull a covered wagon, ride a ferry across the Tualatin River, or care for the family milk cow.

Children can view the play environment from the swing area, which is entirely covered with accessible safety surfacing and includes a bucket swing for young children or children with disabilities. Located at a distance from the play area for younger children is a separate area that allows both active play and "hanging out" space for youth ages 10 to 14. Both play areas meet state-of-art safety standards.

The final design for the Ibach Park play area symbolizes the collective memory of generations of Tualatin inhabitants. It translates local history into an interactive educational resource, providing play and learning opportunities through direct interaction with the natural and built environments. The result is a diversity of play opportunities and environments that allow children of all abilities to discover, learn, and have fun. ■

REFERENCES

American Society for Testing and Materials. 1995. *Standard Consumer Safety Performance Specification for Playground Equipment in Public Use.* West Conshohocken, PA: ASTM.

Center for Accessible Housing. 1992. *The Recommendations for Accessibility Standards for Children's Environments.* Raleigh, NC: North Carolina State University.

International Association for the Child's Right to Play: www. ncsu.edu/ipa

Moore, R.C. 1993. *Plants for Play: A Plant Selection Guide for Children's Outdoor Environments.* Berkeley, CA: MIG Communications.

Moore, R.C., D.S. Iacofano, and S. M. Goltsman. 1992. *Play For All Guidelines: Planning, Design, and Management of Outdoor Play Settings For All Children.* Berkeley, CA: MIG Communications.

Moore, R.C., and H.H. Wong. 1997. *Natural Learning: Creating Environments for Rediscovering Nature's Way of Teaching.* Berkeley, CA: MIG Communications.

PLAE, Inc., 1993. *Universal Access to Outdoor Recreation Areas: A Design Guide.* Berkeley, CA: MIG Communications.

Preiser, Wolfgang F. E., and Elaine Ostroff, editors. 2001. *Universal Design Handbook.* New York: McGraw-Hill.

U.S. Architectural and Transportation Barriers Compliance Board, 1991. *Americans with Disabilities Act Accessibility Guidelines (ADAAG): Accessibility Requirements for New Construction and Alternation of Buildings and Facilities Covered by the ADA.* Washington, DC: Access Board. (See also www.access-board.gov)

U.S. Architectural and Transportation Barriers Compliance Board, 1994. *Recommendations for Accessibility Guidelines: Recreational Facilities and Outdoor Developed Areas.* Washington, DC: Access Board.

7 • URBAN DESIGN DETAILS

Edward Steinfeld

Summary

This article reviews the basic principles of universal design for transport systems. It describes the opportunity for universal design, planning for social equity, and the necessity for seamless continuity of access for all, across time and in multiple locations. Strategies for way finding, negotiating level changes and long distances, vehicle loading, ticketing and security are also described as components of the universal design of transport systems.

Key words

accessibility, airports, signage, transit terminal, transport design, universal design, way finding

The bus transit terminals at Curitiba, Brazil.

Universal design for urban transportation

1 THE OPPORTUNITY FOR UNIVERSAL DESIGN

Universal design, also called "design for all," "transgenerational design" and "inclusive design," is a philosophy that has replaced the term "accessible design" as the paradigm of design with consideration for disability and aging (Box A). There is considerable difference between the two approaches. In accessible design, the goal is to provide access for people with disabilities, usually according to code required minimum rules. In universal design, the goal is to improve usability and accessibility for all people, including those with limitations of function. Universal design should improve convenience for everyone, not just people with disabilities and the elderly. Design of mass transport systems is one of the areas where the value of universal design is most obvious.

Advanced transportation systems have transformed our cities from a geography characterized by central cities with outlying suburbs and towns dependent upon them into the "deconcentrated" urban geography of the multicentered regional network (see, for example, Gottdiener, 1994). While the central city still has great economic and cultural significance, it now has competition from suburban development. In some parts of the world, the major cities cannot even be viewed as distinct entities anymore. High-speed transportation systems often make it possible for journeys between destinations in different cities hundreds of miles apart to be more convenient and faster than journeys within each city itself.

The rapid evolution of high-speed mass transportation systems—including airways, rail systems and ships—provides both challenges and opportunities for universal design. On one hand, older transportation systems can be very difficult and expensive to change. For example, subway systems like those in New York, London and Paris include deep tunnels primarily served by stairways and escalators. Surface rail systems, like Japan Railways, one of the most sophisticated in the world, have systemwide features like differences between train floor and loading platform heights that create major obstacles (Kanbayashi, 1999).

On the other hand, throughout the world, governments are recognizing the importance of good transportation. In many countries, the emphasis of urban transportation planning has shifted from highway infrastructure to mass transport systems. New mass transit projects are conceived regularly as part of showcase urban developments as in Barcelona prior to the Olympics. Cities such as Sao Paulo in Brazil are rapidly expanding their public transportation systems to overcome the massive congestion caused by unplanned growth. In the U.S., the prototypical automobile-oriented country, the environmental problems caused by overreliance on automobiles and fossil fuel consumption has led to an expansion of public mass transportation systems like those in Los Angeles, Washington DC, Atlanta and the Bay Area of Northern California. Worldwide, new airports and airport expansion projects are being built and often include connections to rail systems at "multimodal" transportation hubs.

These developments offer opportunities to make strategic improvements in the universal design and accessibility of infrastructure and insure that the entire population will benefit from more efficient and easier to use transportation. But, most importantly, it is critical that new systems do not discriminate against underrepresented groups such as people with disabilities. And, ways should be sought to

Credits: This article and the graphics are based in part on Steinfeld, E. (2001) "Universal design in mass transportation" in Preiser, W.F.E. and E. Ostroff, *Universal Design Handbook*, New York: McGraw Hill.

Principles of Universal Design

1 **Equitable use.** The design is useful and marketable to people with diverse abilities.

2 **Flexibility in use.** The design accommodates a wide range of individual preferences and abilities.

3 **Simple and intuitive use.** Use of the design is easy to understand, regardless of the user's experience, knowledge, language skills or current concentration level.

4 **Perceptible information.** The design communicates necessary information effectively to the user, regardless of ambient condition or the user's sensory abilities.

5 **Tolerance for error.** The design minimizes hazards and the adverse consquences of accidental or unintended actions.

6 **Low physical effort.** The design can be used efficiently and comfortably and with a minimum of fatigue.

7 **Size and space for approach and use.** Appropriate size and space is provided for approach, reach, manipulation and use, regardless of user's body size, posture or mobility.

Source: Center for Universal Design, North Carolina State University

upgrade the older infrastructure and vehicle stock to insure that the benefits of regional and global transportation are available to all.

Providing for social equity

Transportation planning is motivated by many goals, the most significant being economic and political. At the broadest level, transportation planning has implications for social justice and equity. The major challenge for universal design in transportation planning is at the level of policy—the commitment to serve all citizens with an efficient and affordable system that brings people where they want to go when they want to go there. This commitment has major economic and social benefits for a society over the long run.

In the United States, the financial investment in infrastructure that supports private transportation far surpasses the investment for public transportation infrastructure. Since so few localities have comprehensive mass transportation systems, efforts to increase funding for urban mass transportation do not obtain strong support in comparison to policies that support private automobile systems. Thus, inner city low-income residents are, in general, served by systems that are inconvenient, intermittent and often do not bring their passengers where new jobs are located. Similar problems are common in developing countries where the lack of affordable and convenient mass transportation can be a significant impediment to employment, particularly for women with small children. In the future, the aging of the population will be increasing concern for equity in urban planning and design. The older generation may be particularly disadvantaged by lack of affordable mass transport, especially in places like American suburbs where loss of ability to drive creates limiting conditions on mobility.

Providing for seamless continuity

Continuity of service in space and time is a key attribute of a universal transportation system. A system cannot really be universally accessible if one can only reach a limited number of destinations or if only one bus or station in twenty is usable. Continuity of accessible features should be provided through every link in the travel chain: the path from home to transit stop or station, access to stop or station, transfer points, the final stop or station and the path from there to the destination. Seamless continuity means that all these links in the chain are fully accessible and that no detours or extra links have to be added because of barriers that restrict access. The level of universal design available is clearly dependent on the weakest link in the system. For a wheelchair user, one curb at a crosswalk might be as significant as a huge stairway or lack of an elevator where there is a major level change.

Many cities with older rail transit systems have upgraded strategic stations or made new stations accessible without addressing the accessibility of existing stations. Cities with bus systems often attempt to solve the problem of inaccessibility by serving each route with a limited number of accessible buses, usually equipped with lifts. Neither strategy is successful because they cannot provide seamless continuity. Accessible buses are not available when needed or it is impossible to reach one's destination because the station is not accessible.

Universal design solutions develop seamless continuity over time by developing a strategic plan for upgrading existing rail stations as well as by providing full accessibility to new stations. Where it is impossible to upgrade existing stations to a fully accessible level, alternatives routes can be developed with an accessible bus system. Bus systems can be fully upgraded over time so that all buses are accessible. Beyond the vehicles and stations, the pedestrian paths to the stations and bus stops must also be fully accessible to insure continuity in the travel chain. In older historical city centers, this can be a difficult problem to solve due to the presence of very high curbs and cobblestone or other walking surfaces that are difficult to use. However, there are many examples of good solutions to these problems. For example, a strip of smooth paved concrete walkway can be provided down the center of a cobblestone or brick sidewalk. Curb ramps can be located parallel to the street where curbs are extremely high. Paving reconstruction programs can include raising the elevation of roadways at intersections to reduce the height of curbs at crossings.

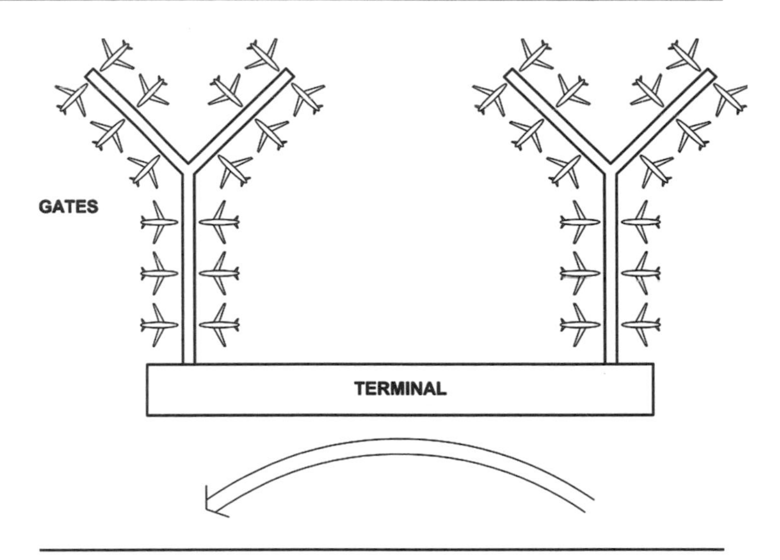

Fig. 1. Tree-like terminal plan.

2 STRATEGIES FOR THE UNIVERSAL DESIGN OF TRANSIT

Way finding assistance

In a complex transportation system, finding stations, finding ticketing areas and finding the vehicle one wants and the services one needs are perhaps the most important concerns of the traveler. Most travelers can understand the basic organization of terminals with their common spatial syntax of entry area, ticketing area, main circulation halls and gate areas. But, way finding becomes difficult in the more complex terminals, especially multimodal transportation centers.

A terminal that reduces the number of decision points for the traveler between the entry area and the gate areas reduces the potential for decision-making error. For example, consider an airport terminal with a tree-like structure as in the older parts of O'Hare International Airport in Chicago (Fig. 1). As travelers proceed deeper into the system they must make a series of decisions about which branch of each fork to select. Only the links encountered are visible to them. If a mistake is made, it becomes hard to orient oneself to the whole and recover from the wrong choice. Moreover, recovering from an error that leads one to the wrong branch can be very time-consuming and increase physical effort substantially because one has to retrace the path back down the trunk and then back out the new branch. Although the overall system is easy to comprehend, actual experience of the plan is not the same as viewing a diagram of the whole. The newer United Airlines terminal at Chicago O'Hare, on the other hand, divided up the gates into two separate structures (Fig. 2). The only choice is to decide from which of the two one's flight leaves. All the gates in each section are relatively close to one another and visible to the traveler from one strategic point. Going the wrong way is not as serious a mistake in this layout. A moving walkway connects the two terminals reducing the burden of travel from one to the other.

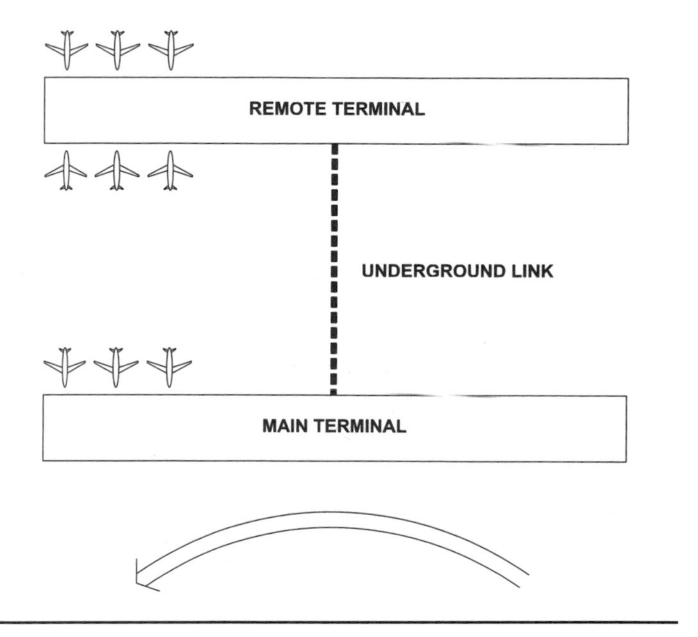

Fig. 2. Island terminal plan.

Making the system understandable is a key contribution that urban design can make to improve usability of transportation systems. Stations and stops should have a unique and recognizable appearance. There needs to be a strong graphic identity for the system and also a means to distinguish routes and different types of vehicles (*e.g.*, express and local).

The overall design of transportation centers and the relationship between the surrounding context and the terminal can make a significant difference for all travelers. Urban design issues are very important in exposing and highlighting the operation of the transportation system. The following urban design strategies can be effective:

Fig. 3. Talking sign system above the door (black square) provides visually impaired people with a verbal message in a form that includes directionality.

Fig. 4. Talking sign system on the loading platform.

- Build terminals above ground wherever possible.
- Provide views to the surrounding context for orientation.
- If terminals have to be below grade, develop strong landmarks within the terminal for orientation—light courts and atria are two good approaches.
- Insure that terminals have circulation that provides direct routes from the key origins and destinations.
- Connect older structures and new ones with open spaces that can become gathering places and that expose the new and old.

In the street and in complex transportation centers, a guidance system is needed to help people overcome the complexity of the environment. Signs and symbols should be legible and easily understood for the sighted traveler. To assist visually impaired people find their way around in the outdoor pedestrian environment, guide tracks can be introduced on walkways and used as a decorative device as well as a functional assist. Audible beacons can also be added to street crossings to assist the visually impaired traveler. These beacons can be annoying to people who work or live in the surrounding areas. This problem can be avoided if the signals are broadcast on a radio frequency; people who rely on them should be given free receivers. Pedestrian control signaling systems can improve safety for all people. They are especially useful in the few countries where vehicles drive on the left side and strangers can easily make the mistake of crossing when looking the wrong way. Transit stops can have unique markers on guide strips or special beacons.

Some cities are adopting nonvisual messaging technology that allows the visually impaired traveler to find destinations without the need for tactile and Braille signage (Figs. 3 and 4). The traveler who needs them can obtain a receiver that scans the environment for transmitters that continuously broadcast voice messages over an infrared band. These signs have directionality and serve as beacons. As wireless web and personal digital assistant technology develops, there will be other approaches that can be adopted to accomplish similar goals for a broader population. In urban design of transportation, it is important to consider adoption of information systems that can benefit not only people with visual impairments but also the general population. In the near future, a small palm sized computer could serve as a way finding guide through complex terminals and city districts. Not only might this work visually but also by using speech, providing options for the traveler's specific needs and preferences. These systems will not only provide way finding information but also information like scheduling, status of departures and arrivals, location of amenities and other useful information for the traveler. Moreover, they will allow people who speak different languages to access the same information as the native speakers.

Negotiating level changes in terminals

A key aspect of mobility in terminals is negotiating level changes. Level changes are often necessary to efficiently process arriving and departing passengers and their baggage in large terminals. Even in small buildings and individual transit stops, the boarding platforms are often above or below ground, out of the way of surface traffic. Rail transportation presents the most difficult problems with level changes. Terminals that serve intersecting rail lines must have more than one level in order to allow one line to pass over the other.

Level changes within transportation terminals are accommodated the same way they would be in any multilevel structure, with elevators, stairs and ramps. In terminals where access is restricted to some zones, however, for either fare control or security reasons, providing an accessible path of travel that does not require one to walk up or down stairs is complicated by the need to provide accessibility accommodations on either side of the security perimeter.

One of the most serious problems faced by people with mobility impairments is being forced to rely on assistance to negotiate changes in level or circumvent security perimeters to use accessible routes. Forced dependency results in inconvenience, embarrassment, anger and exposure to injury by poorly trained attendants (Kanbayachi, 1999). This is especially the case when people who use wheelchairs must use an entirely different circulation path than other travelers. Careful design can reduce the need for duplicate elevators or ramps, for example, a system that keeps all passengers at the same level once they are within the secure perimeter. It should be noted that the same strategies adopted to assist people with mobility impairments also are beneficial to many other travelers. They help everyone with luggage, people with infants in strollers, frail older people, small children and women in the late stages of pregnancy as well. From the perspective of management, universal design reduces the need for special training for all new staff, concerns about back injuries caused by lifting passengers and it also improves productivity because workers do not have to leave their posts as often to assist an individual traveler.

Fig. 5. Ferry terminal ramp system.

The geometry of rail terminals is governed by the tracks. Access to them requires underpasses or overpasses to reach any platforms not adjacent to the entry hallway. The level changes need to be at least one complete story. Since large terminals have many tracks, providing wheelchair access to platforms can be very involved. Each platform has to have an elevator or ramp access. To minimize elevators, a transverse link can be provided that is served by both an elevator and escalators or stairs. From this link, ramps can feed each platform for all travelers. But, such a system has to be planned into the design from the early stages. The two critical factors to be addressed in planning are the horizontal length required for the ramps and the difference in elevation between the transverse link and the loading platforms. Without considering these factors in the basic planning of the terminal, there may not be enough room available for ramps and an elevator for each platform will be required. This circulation problem is less severe if the rail terminal is at the end of the line or on a spur. In such a case, access to each platform can be provided from the end of the tracks at the same level and an overpass or underpass is not necessary. However, most railway terminals have to traverse the path of the train lines.

The following design strategies are useful for planning terminals to accommodate people who cannot use stairs easily:

- Plan terminals to reduce level changes to minimum.
- Provide accessibility for wheelchairs, carriages, bicycles and other personal wheeled mobility devices to all vehicle-loading zones.
- Maintain accessible circulation within and without the security perimeter.

Figs. 6 and 7. Raised platforms for light rail work with ramps to provide access.

Fig. 8. "Stairless" bus.

Negotiating long distances in terminals

Regardless of how well a terminal is planned, the sheer size of large terminals results in the need to traverse long distances. Some prototypical terminal designs are more compact than others; for example, compare a terminal that clusters gates together in a node to one that spreads the gates out in long fingers. In Reagan International Airport in Washington, DC, the plan has a long pedestrian hallway with the gates all distributed in fingers off the base; the passenger who has to rush from one gate to another in different fingers has a long route to travel in a short time. In large terminals like this, it is often impossible to avoid excessively long routes. Most new terminals provide assistance through mechanical means to reduce the burden of these distances and increase the speed of traffic flow through the terminal. At many airports, planning several smaller terminals linked by an automated rapid transit system has reduced the distance between gates. Each remote terminal is an island that can have gates on all sides. This reduces overall corridor length. At Newark International Airport and at Chicago O'Hare International, new transit links have been added to the original multiterminal complexes. A major construction project is underway to link terminals at Kennedy Airport to the metropolitan rapid transit system in the New York City area.

The following urban design strategies to reduce the effort required to move around in terminals:

- Plan the location and the overall area of terminals to minimize the distance of trips within them.
- Integrate transit systems into terminal design and locate transit stations as close as possible to the exit of terminals if not right inside the terminal.
- Provide automated walkways wherever there are long pedestrian paths in terminals and to transit stations, parking and rental cars.
- Plan terminals to separate courtesy carts from pedestrian traffic; in large airports include a transit link that connects all terminals and transit stations.

Vehicle loading

One of the most difficult issues in universal design of transportation systems is accommodating level changes to board and exit vehicles. This is an urban design issue because in planning transit systems and improvements, it is critical to avoid the mistakes of the past and insure compatibility between station design and vehicle design. Once in place, it is extremely difficult to overcome barriers caused by level changes. There are two basic accommodation strategies that can be used separately or in tandem. The first focuses on design of the vehicle and the second on design of the transit station or terminal. In addition to accommodating the level change, there is also a need to protect people from falls and other safety hazards in the loading area especially when there is a loading platform. A uniform design approach throughout a system eliminates surprises for the traveler, especially travelers who have visual impairments.

Raising the entire secure area of transportation terminals off the ground to the same level as vehicle floors is a good universal design strategy. It requires passengers to negotiate only two level changes at the end and the beginning of the trip. Moreover, this strategy benefits all riders by increasing safety and convenience and improving service response by reducing the time of loading and unloading a vehicle. Finally, it makes it easier to maintain security in the system since no one can get on vehicles unless they are on a platform. All entries to platforms can be controlled easily. Typically, platform loading is used only in rail systems but bus systems can also benefit from the use of this strategy. In Curitiba, Brazil, all express bus terminals and stops are raised off the ground and are accessible by ramp or lift as well as stairs. Once riders enter the secure area, they stay at the higher level and never have to use stairs or any other vertical circulation during their trip. All vehicles are loaded with great ease from the platform. In Buffalo, New York, a variation of this strategy was used for a light rail rapid transit system that has both underground and aboveground stations (Figs. 6 and 7). In the below ground stations, trains are loaded directly from a platform as in all subways. All the above ground stations have a small raised platform level with the vehicle floor and they all have ramps.

Consistency and continuity across the system and over time is important, so that standards have to be established for both rolling stock and terminal construction. In existing systems where there are inconsistencies between levels, vehicles will have to be fitted with lifts or small ramps. The latter is preferable because all passengers can use them. But, where there are great differences in levels, lifts may be needed because ramps would be either too long to fit on the platforms or too steep to negotiate. A small ramp (drawbridge) can be used to overcome small level differences between vehicle and platform.

One major obstacle to insuring continuity in the travel chain in rail transport is the need to coordinate different companies. In a continent like Europe where crossing international borders is common, the need to insure compatibility of systems across international borders requires a high level of coordination. But there are also intra-urban issues because there may be different companies serving the same stations, *e.g.*, commuter rail and rapid transit. And there may be different types of vehicles coming into the same stations or platforms (on different sides), *e.g.*, light rail and subway systems.

Many systems—especially transit and shuttle buses but also trolleys—must load and unload on street surfaces. There are several technologies available to reduce or compensate for the level change between vehicle and street/walkway surface. The best approaches reduce the difference in height to a minimum. A good example is the "low floor" bus that has a lower floor level than conventional buses and a "kneeling" hydraulic feature that brings the floor level within one step of the ground when activated. Such vehicles reduce the difficulty of entering and exiting significantly. This helps children, parents with children, older people and people with limitations of mobility. It also helps people with visual impairments because it reduces the complexity of entering and exiting a strange vehicle. A wide doorway and handholds add another level of convenience.

The low floor/kneeling bus does not eliminate the step up entirely so it is not sufficient to provide a truly accessible vehicle. The most common solution to this problem is the wheelchair lift. Several types of wheelchair lifts are available for transit buses. They include lifts at the front or rear doors, lifts that slide out from under the bus floor and lifts that are integrated into the stairs of a bus and fold out when the lift is needed. Another solution that works particularly well is the "stairless bus" (Fig. 8). This bus combines the low floor and kneeling features with a ramp that serves all passengers. Compared to the lifts, this solution is a far better example of universal design. Lifts only benefit wheelchair users. They are time-consuming to operate and expensive to purchase and maintain. The person using the lift becomes a spectacle. Some riders and drivers may even become annoyed with the extra time required to load or unload someone with the lift, reinforcing the stigma associated with disability. With the stairless bus, however, all passengers enter the same way when the ramp is used. No one becomes a spectacle and there is no delay caused by the need to accommodate a person with a disability. Where the ramp can be extended onto a curb, the incline is eliminated as well, adding great convenience for all passengers.

Avoiding falls off loading platforms is a major safety concern for individuals with visual impairments and children. There are several methods that can be used to protect the traveler from falling. One is the use of a gate and barrier system. This is by far the safest strategy. However, it constrains the location of where vehicles can stop to load and unload. In Curitiba, Brazil, the express bus stations are constructed of a prefabricated plastic tube system in a circular section that protects all passengers from falling off the platform. Bus trains have automated systems that open the doors on both buses and stations when sensors on the platform detect that the bus is in the proper position. At that time, travelers are free to load and unload.

The most secure system for protecting waiting passengers at the platform edge is a physical barrier along the entire platform. In Curitiba, the tube stations provide such a barrier. They have sliding door openings where the doors to the vehicle will be when the buses stop at the stations. In new subway stations guard rails and even complete glazed enclosures with automated sliding doors are being introduced. Not only do such barriers protect people with visual impairments but they also protect the general population from being pushed off the platform and they prevent suicides by jumping in front of trains. Current technology supports precise docking of vehicles that enables this approach to platform safety.

Ticketing and security

Urban designers can have an influence on the equipment used in transit systems. Such systems are often part of street furniture and integrated into kiosks and other structures. For many riders, disabled or not, the most annoying parts of a transit system to use are the ticketing and security areas. This is especially the case for first time users who do not understand how it works. The problems are compounded if the rider has a visual disability or does not read or speak the local language. In many cases, there are unusual rules that have to be followed with significant penalties for transgressors. In others, ticket prices vary based on the length of the trip and are difficult to determine without prior experience with the system. In most cases, the rider must purchase a ticket or token and then use it in a fare gate. This can often require the use of two different machines. Security systems at airports and other terminals require complex maneuvering under great social pressure, especially at rush hours. At the destination, users often find that the ticket they purchased is not sufficient to leave the system and need to add fare or upgrade tickets to exit. And, transfers to other lines and vehicle types often have to be obtained. Universal design can help to reduce the stress and increase convenience for all riders.

The operation of fare and ticketing machines is often incomprehensible, even if one reads the local language. But, through design that provides strong "affordances" and utilizes intuitive methods of operation (Norman, 1988), the steps for using these machines can be conveyed easily without complicated instructions. To be usable by people with visual impairments, fare and ticketing machines should be standardized in design and include tactile or audible cues to their operation.

The most direct way to increase usability of ticketing systems is to simplify the task of purchasing tickets. For example, many systems separate the change machine from the ticket purchase machine. While it may be useful to have separate change machines for convenience in purchasing food or beverages, there is no reason to separate those functions for purchasing tickets. Ticketing machines can also provide change. This eliminates a source of congestion and reduces the number of tasks necessary to use the system. A second strategy is to combine the ticket machine with the access gate. For example, money can be used to get access instead of a token or fare card. Fare cards could be issued as money is inserted in a combined ticketing/access gate machine.

Additional simplification can be achieved by the use of fare cards that one can purchase for varying denominations and that can be used throughout a system, even different modes of transport. Passengers only have to pay once to buy the card and can use it for many trips until it runs out. Day, week or month passes eliminate the need to obtain transfers so such options greatly facilitate the use of transit systems. Finally, the actual task of buying a fare card or token can be simplified through the use of swipe card systems for purchasing tickets with credit cards or sensor systems with prepaid or automated credit card billing like those now used on highways.

Ideally, aside from local buses or light rail vehicles that stop on the street, the only time one should have to pay is when entering the security perimeter of a transit system. Once in the system, there should be a continuity of path within the perimeter without having to pass through and exit and re-enter. This may not be possible, however, when changing modes, *e.g.*, from train to bus, or lines. In these cases, inter-modal passes or transfers should be available to avoid the need to pay again. Transfer dissemination should be obvious and evident to the user. In many cases, one has to ask the driver of a vehicle for a transfer. This can be impossible or difficult for people who have communication limitations.

Universal design strategies for ticketing systems include:

- Provide intuitive fail-safe ticket purchasing machines.
- Simplify the ticketing system by reducing steps and offering optional payment plans and methods.
- Reduce the number of payments to a minimum, preferably to one.
- Fare and ticket collection machines should be designed for use without having to read and without vision.

3 CONCLUSION

Universal design in transportation systems is in its infancy. While many of the lessons learned in the general design of accessible buildings are applicable to transportation terminals, stations and stops, there are many unique concerns that cannot be addressed by existing guidelines and standards. For example, current guidelines and standards do not provide any guidance on access to dynamic scheduling information. Furthermore, many of the universal design concerns are beyond issues of minimal accommodation for disability like inability to read in the local language. Moreover, high technology applications are emerging that could increase the options available for universal design.

As the demand for transportation at all scales continues increasing, the issues identified here are likely to become more and more important for urban designers. Research is needed both to identify the severity of problems and priorities of the traveling public and also to address the gap in information needed for design decisions. It is important that this research give attention to the differences and similarities between travelers with consideration to age, disability, citizens, and visitors. It is also important that groups like bicycle users, parents with small children and others who use wheeled mobility devices be included as part of the research population. The value of including universal design features in transportation systems will be increased if research can demonstrate their widespread value to all travelers. But, the specific needs of people with disabilities need to be identified and prioritized as well so that they do not become neglected in the pursuit of greater convenience and usability for the broader population. It is likely that those specific needs can be addressed in ways that will benefit everyone. Some examples include better information on scheduling changes in terminals and greater accessibility of vehicles.

The challenge of universal design in transportation is not only to eliminate discrimination in access but also to insure social integration in use. While the existing mass transport infrastructure is difficult to change, particularly old underground systems in deep tunnels and interurban rail systems, there are many opportunities for practicing universal design in developing countries and the rapidly growing suburbs of major cities.

Finally, it is critical to remember that reducing environmental pollution and depletion of dwindling fossil fuel resources are key goals of mass transport planning. Increasing convenience in using public transit for all can is not only a universal design strategy but also it is a more general planning strategy for attracting commuters away from automobiless. ■

REFERENCES

Gottdiener, M. 1985. *The Social Production of Urban Space.* Austin: University of Texas Press.

Kanbayashi, A. 1999. "Accessibility for the disabled." *Japan Railway and Transport Review,* No. 20.

Norman, D. 1988. *The Design of Everyday Things.* New York: Doubleday.

Salmen, John P. S., and Elaine Ostroff (1997). "Universal design and accessible design." Watson, Donald, ed., *Time-Saver Standards for Architectural Design Data.* New York: McGraw-Hill.

Steinfeld, E., 2001. "Universal design in mass transportation." Preiser, W.F.E., and E. Ostroff, eds., *Universal Design Handbook,* New York: McGraw Hill.

Walter M. Kulash and Ian M. Lockwood

Summary
Traffic calming, a rapidly growing traffic engineering practice, consists of physical measures that slow motorists to a "desired" speed. Traffic calming measures can be added to existing streets or incorporated into new or rebuilt streets. Traffic calming measures include changes in vertical surface, narrowings, and lateral shifts. Frequently, traffic-calming sites include combinations of the above, along with streetscaping.

This article lists the basic types of traffic calming measures, establishes their technical basis, and provides dimensions and advice for implementation.

Key words
automobile, bulbout, *chicanes*, livable communities, neighborhood, pedestrian, roundabouts, speed control, street, streetscape, traffic calming, traffic circle

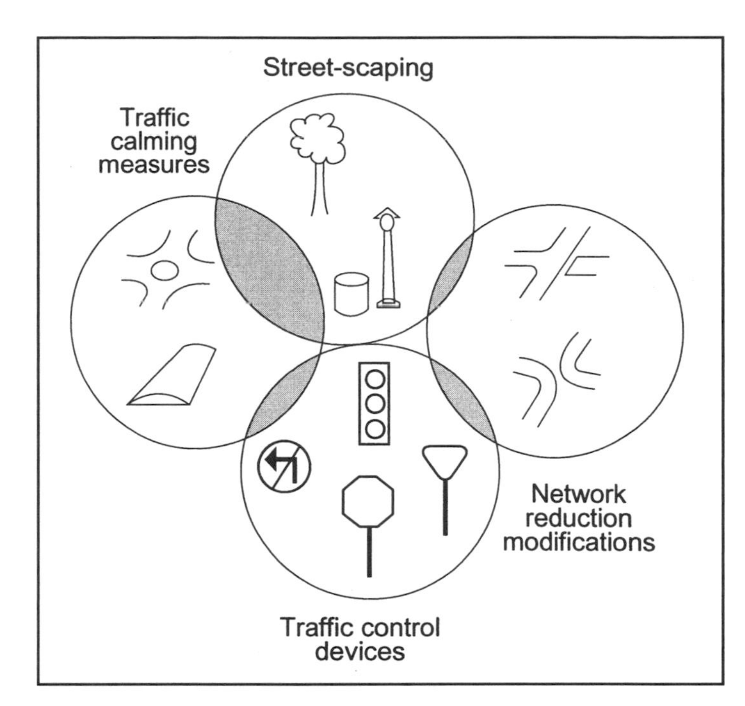

Fig. 1. Relationships of street elements.

Traffic calming

1 OVERVIEW

Traffic calming involves physical measures that:

- reduce the negative effects of motor vehicle use,
- alter driver behavior,
- improve the conditions for nonmotorized street users.

Typically, traffic calming slows motorists to a "desired" speed and develops the street(s) in a context-sensitive way to meet the goals and objectives of the community (*e.g.*, homeowners, business owners, etc.) Traffic calming can be accomplished by:

- Retrofitting the existing streets with regularly spaced measures, and/or,
- Rebuilding the streets to include new cross-sections.

Traffic calming is becoming an increasingly important part of the effort for cities, towns, and villages to become safer and increasingly livable, economically successful, and sustainable. Traffic calming is popular and accepted throughout the world and is now acceptable in North America at the local, county, state, and federal levels on all types of streets with pedestrian activity. Traffic calming has a myriad of applications in urban areas, but is also an option for rural towns and villages where the rural highway enters the town and becomes the main street.

Traffic control devices (*e.g.*, signs, signals, and pavement markings), often support, but do not in themselves, constitute traffic calming measures (Fig. 1). Similarly, with streetscaping elements. Route modification measures (*e.g.*, street closures, partial closures, one-way streets, and turn prohibitions) or network reducing measures should never be considered traffic-calming measures.

Desired speeds

Traffic calming measures reduce the motor vehicle speeds to "desired" speeds (*i.e.*, the speeds that the community wants); typically 20 mph (32 kph) or less for residential streets, and 25 to 30 mph (40 to 48 kph) on commercial streets, collector streets, and arterial streets. The desired speed and the design speed (*i.e.*, the speed used for engineering design purposes) are the same. The intent is that the design of the street self-enforces the desired speed without the need for constant police enforcement.

Categories of measures

Categories of traffic calming measures include vertical measures, narrowings; and lateral shifts and deflections. Most traffic calming sites incorporate combinations of the above and related streetscaping.

Collision severity and vehicle speed

The number of collisions on traffic-calmed streets is about 50% lower than on comparable conventionally designed streets, due to lower speeds, shorter stopping distances and wider driver fields of vision. Furthermore, the rate of injury accidents decreases exponentially as

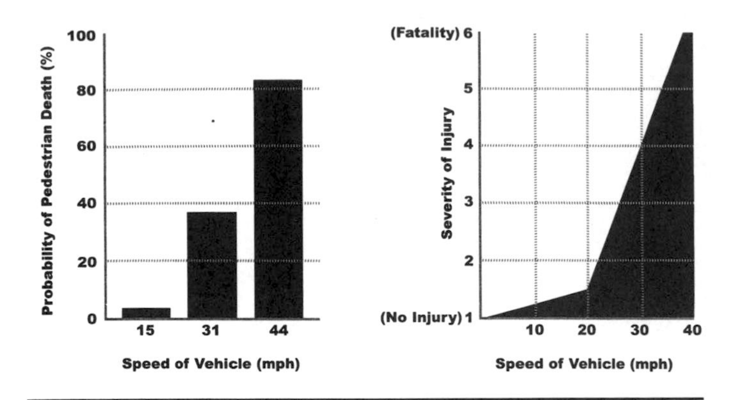

Fig. 2. Pedestrian safety. (Source: Limpert 1994)

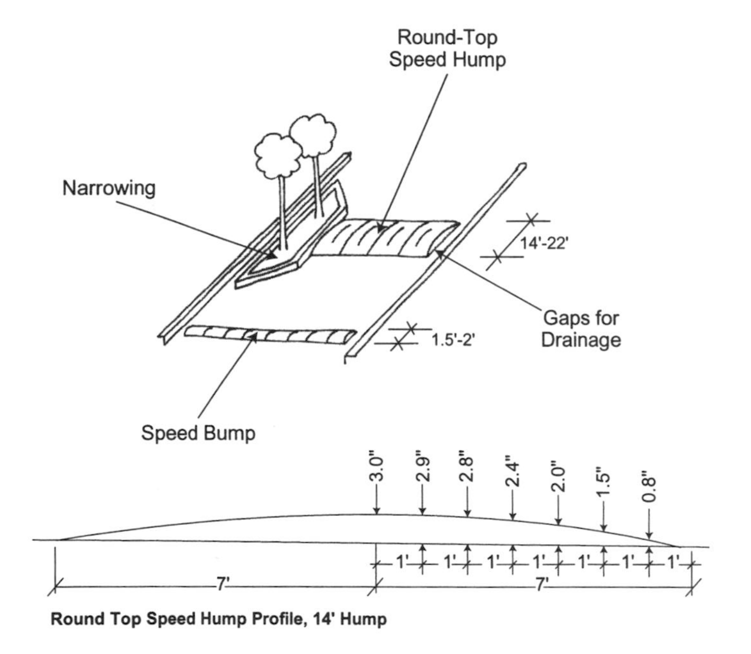

Fig. 3. Round-top speed hump and speed bump.

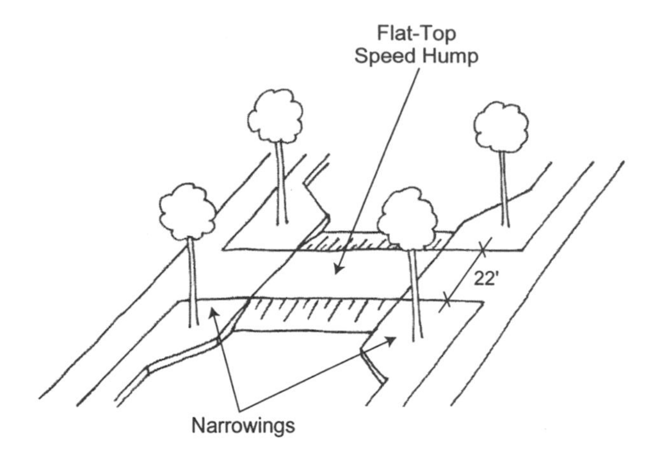

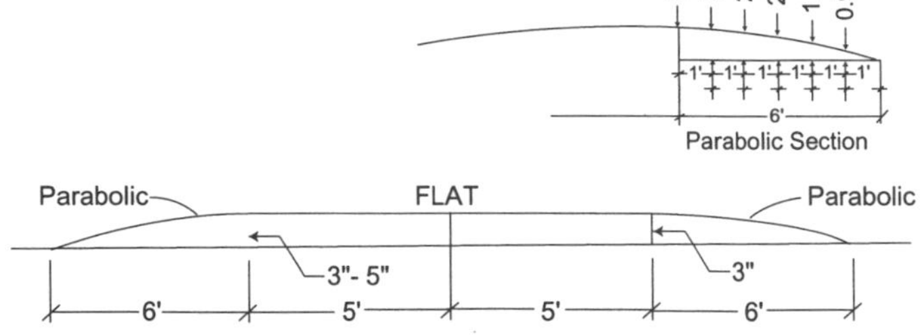

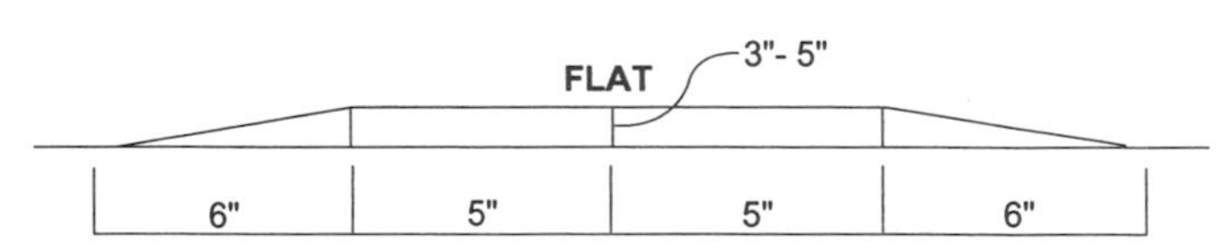

Fig. 4. Flat-top speed humps.

motor vehicle speeds decrease. This more-than-proportional decrease reflects the exponential decrease in kinetic energy of the motor vehicle as speeds decrease. (Fig. 2).

Pedestrian crossing and vehicle speed

The ability for pedestrians to safety cross the street increase disproportionally as vehicle speeds decrease. The space between vehicles (gap) considered minimum for safe crossing by pedestrians will decrease, due to their (correct) understanding of vehicle stopping distance and driver manners. At speeds of 25 mph (40 kph) or less, drivers are likely to allow pedestrians to cross, while over that they are not. The impact is therefore double-edged: better gaps for crossing, and, at the same time, better behaved and more courteous drivers. The result is a dramatic increase in the ability and comfort of crossing a traffic stream. This safety is further compounded by greatly reduced probability of injury in collisions occurring at lower speeds.

2 VERTICAL MEASURES

Drivers slow down for vertical measures due to the uncomfortable ride at higher speeds. Performance varies depending on the measures' slopes, heights, and profiles, as well as the spacing between of the measures along the street. The effect of vertical measures is stronger on trucks and buses, compared with automobiles.

Round-top speed humps

Unlike speed bumps, which are about 4 in. (10 cm) high and about 18 in. (46 cm) long, round top-speeds humps (Fig. 3) are 3–4 in. (7.6–10.2 cm) high and 12 to 14 ft. (3.66 to 4.27 m) long. In addition, the humps are much more effective on speeding and are legally acceptable on public streets. The round-top speed humps come in a variety of profiles and sizes, depending on what the designer wishes to achieve. A common profile is the parabolic crown (Fig. 3).

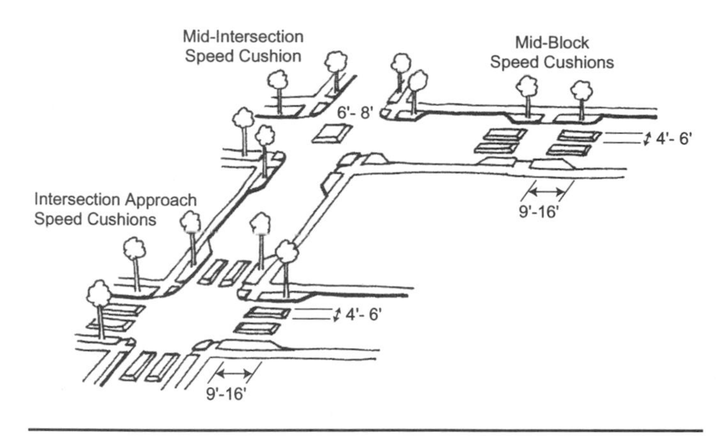

Fig. 5. Speed cushions.

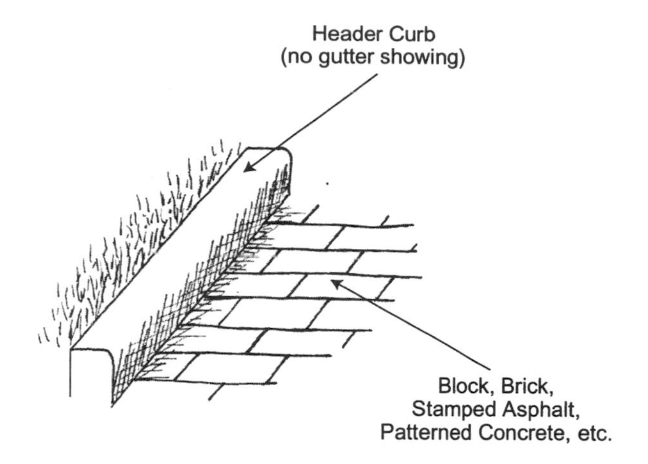

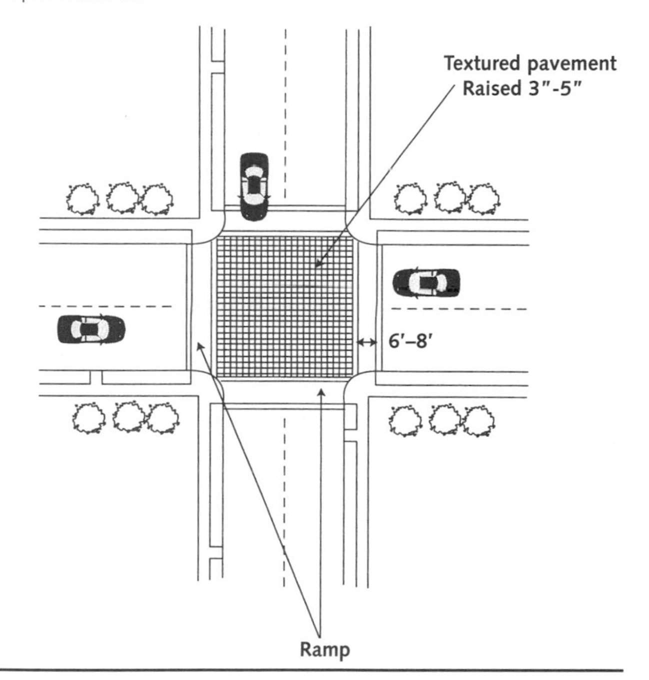

Fig. 6. Raised intersections.

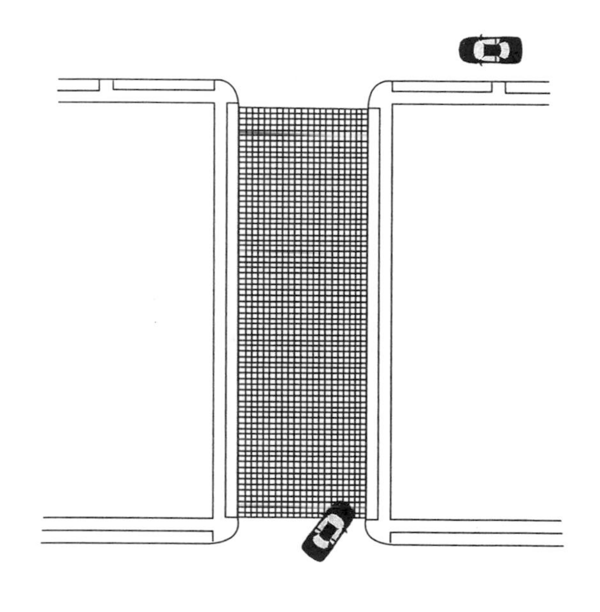

Fig. 7. Textured surface.

Construction materials vary from smooth asphalt, to printed and stained asphalt, to stamped concrete, to concrete paver stones. Gaps are normally left between the ends of the hump and the curbs to allow water to drain past the hump. Frequently, humps are used in combination with other traffic calming measures such as bulbouts, pinch points, or medians.

Flat-top speed humps

Flat-top speed humps (Fig. 4) are similar to round-top speed humps, but have a flat (not rounded) top. This flat top can include a pedestrian crossing. The ramps to/from the flat top vary in profile and length. Typical ramp lengths are six feet with typical flat tops of 10 ft. (3.01 m) for a total length of approximately 22 ft. (6.7 m). A simple straight ramp is now widely used. Flat top speed humps make safe pedestrian crossings because:

- Drivers need to slow down at the crossing,
- The measures are conspicuous so that drivers know to expect pedestrians,
- Pedestrians are effectively four to six inches taller than they would be otherwise so that they can see and be seen easier.

Frequently, flat-top speed humps are used in conjunction with bulbouts on both sides to make the crossing more conspicuous, provide a safe vantage point for pedestrians watching for cars, to prevent parking near or on the crossing, to shorten the crossing distance, and to further slow down drivers.

Speed cushions

Speed cushions are small, flat-top humps that do not extend fully across the street (Fig. 5). Typically, they affect only one pair (*i.e.*, right side or left side) of the vehicle wheels. Speed cushions can permit bike lanes and storm drainage to continue unhindered at the original street grade.

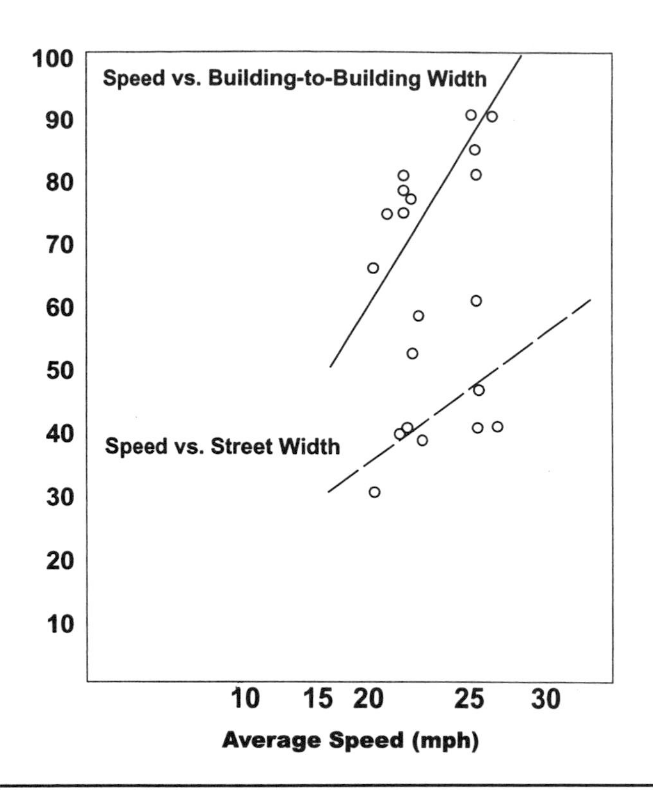

Fig. 8. Speed vs. paved width and building setbacks. (Source: Smith Appleyard 1981)

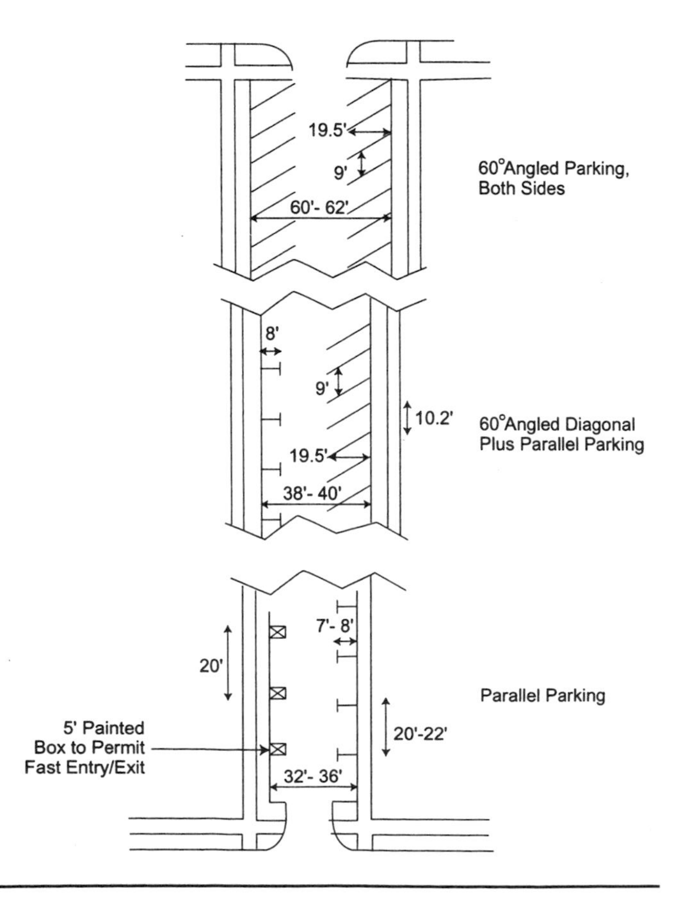

Fig. 9. On-street parking.

Raised intersections

Raised intersections (Fig. 6) involve raising the whole intersection surface to sidewalk height and providing motor vehicle ramps on the approaches. They enjoy the same benefits for pedestrian accommodation as the flat-top speed humps. They too can be used in conjunction with bulbouts.

Textured surfaces

Drivers tend to drive faster on smooth streets than on comparable, but rough streets. Changing the texture of the street (Fig. 7) using bricks, concrete pavers, cobblestones, or stamped pavement, will help drivers slow down. Texture alone will likely be insufficient and can result in increased tire noise. Therefore, it is more effective to combine texture with other traffic calming measures.

Rumble strips should not be used as a traffic calming measure, but rather in their customary role as a warning for something likely to be unexpected by the driver, such as a T-intersection or a rural stop sign. The noise that they cause is problematical in or near neighborhoods. Further, rumble strips lose their effectiveness when used as traffic calming measures, since they do not compel a reduced speed, and drivers learn to simply ignore them.

3 NARROWINGS

While there is no definitive published guideline relating street width and design speed, there is a widespread understanding that narrow street width reduces speed. The width between buildings has a similar effect (Fig. 8).

On-street parking

On-street parking (Fig. 9) slows motor vehicle speeds by narrowing the travel lanes. This narrowing is particularly effective because of the height of the parked cars and the articulation (irregular appearance) of the enclosure that the parked cars provide. Further, the occasional parking maneuvers of slowing or stopping cars are a frequent reminder, to motorists, of the other users of the street. Beyond its immediate traffic calming effect, on-street parking greatly improves the pedestrian qualities of the street, by putting a barrier of parked cars between the sidewalk and moving vehicles. Pedestrian benefits are increased through the use of bulbouts, which result in more sidewalk space and shorter crosswalks at intersections.

Typical types of on-street parking include parallel parking and diagonal parking (Fig. 9). These types may be combined as desired, with parallel and diagonal patterns on opposite sides of the street, or alternating on the same side of the street to create or accentuate lateral shifts.

A desirable complement to on-street parking is the intersection bulbout, which defines and shields the parking, as well as provides a better street corner for pedestrians. Mid-block bulbouts also define the parking areas, as well as providing pedestrian crosswalks, transit stops and places for trees. By regularly placing bulbouts, a continuous street tree appearance can be gained.

Parking along the medians on divided streets is an inexpensive and effective way to reallocate excess pavement width. If only one parking lane can be accommodated along the median, it can be alternated along either side of the median. Though only one row of trees can be accommodated in the median, the appearance of a double row of trees can then be created.

"Back-in/head-out" diagonal parking (Fig. 10) is superior to conventional "head-in/back-out" diagonal parking. Both types of diagonal parking have common dimensions, but the back-in/head-out is superior for safety reasons due to better visibility when leaving. This is particularly important on busy streets or where drivers find their views blocked by large vehicles, tinted windows, etc., in adjacent vehicles in the case of head-in/back-out angled parking. In other words, drivers do not back blindly into an active travel lane. The back-in maneuver is simpler than a parallel parking maneuver. Furthermore, with back-in/head-out parking, the open doors of the parked vehicle block pedestrian access to the travel lane and guide pedestrians to the sidewalk, which is a safety benefit, particularly for children. Further, back-in/head-out parking puts most cargo loading (into trunks, tailgates) on the curb, rather than in the street.

Bicycle lanes

Adding the on-street bicycle lane (Fig. 11) somewhat narrows the traveled way for motor vehicles, while providing for an otherwise-neglected mode of travel. Bicycle lane dimensions are now standardized throughout the U.S. in AASHTO Bicycle Facilities Guidelines.

Where parallel parking is present, the bicycle lane is striped between the parking lane and the motor vehicle lane. When a bicycle lane is adjacent to diagonal parking, it is preferable to use back-in/head-out angled parking for safety reasons. In this way, exiting motorists do not back blindly into the bicycle lane, but rather back up while entering the parking stall with far greater viability.

A much more sophisticated way of accommodating the bicycle lane is the protected bicycle lane with the bicycle lane adjacent to the sidewalk. Patterned after widespread European experience, this design locates a one-way bicycle lane (in the direction of motor vehicular traffic) adjacent to the sidewalk, at the same level as the sidewalk (see Klien and Lardner "Urban Bikeways," earlier in this volume). At intersection approaches and departures, where parking is prohibited anyway, the bicycle lane transitions horizontally to the side of the curb, and vertically down to street pavement level. Through the intersection, the bike lane is adjacent to the motor vehicular through lane as in a conventional in-street bicycle lane. On the far side of the intersection, the pattern is reversed, with the bicycle lane again transitioning horizontally to adjacent to the sidewalk, and vertically to the same grade as the sidewalk. The paving material for the bicycle lane is usually red in color or marked to distinguish it from the rest of the street.

Clearly, this arrangement narrows most of the street, while providing separate facilities for cyclists. Furthermore, the street is effectively wider at the intersections for the turning needs of larger vehicles. With the possible exception of expert riders, the needs of the majority of cyclists are helped by protected bicycle lanes when compared to the conventional, completely in-street bike lane. Vulnerable cyclists feel safer. There are far less, or no, car door-opening problems. Cyclists can access shops, bicycle parking, etc. without having to maneuver between parked cars. In-line skaters and other nonmotorized, wheeled street users, increasingly expected to operate as cyclists, are well served by the protected bicycle lane.

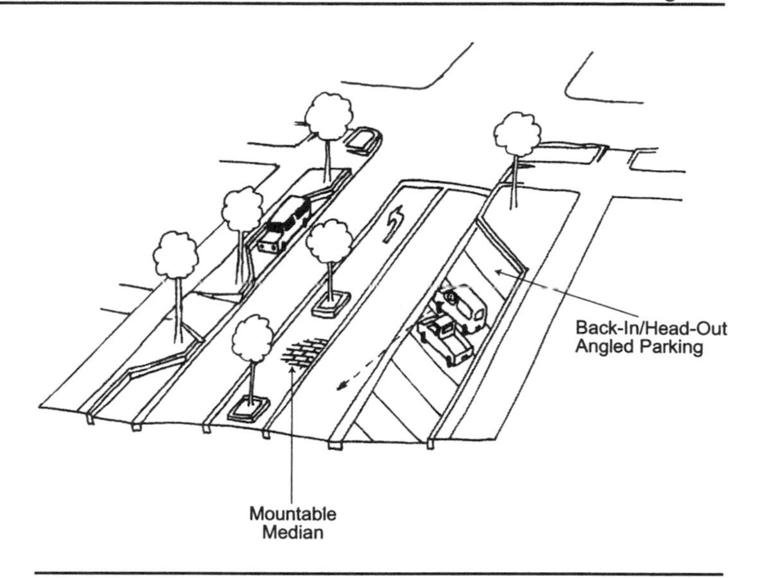

Fig. 10. Back-in / head-out angled parking (shown with median).

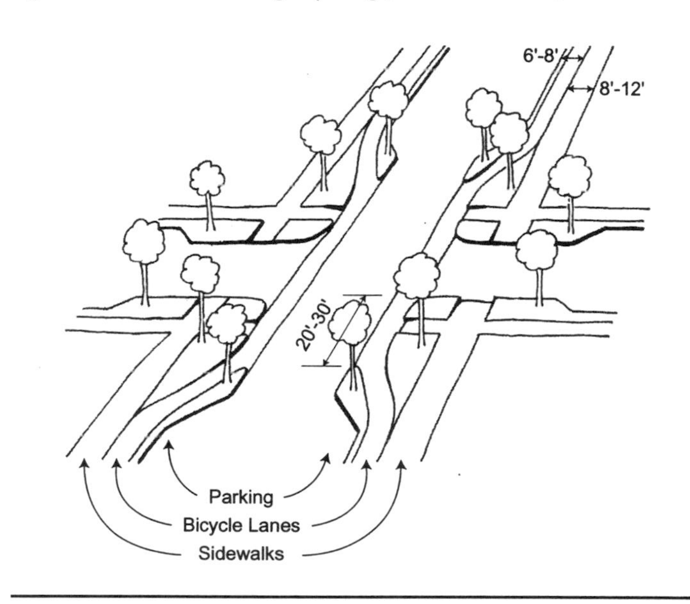

Fig. 11. On-street bicycle lane.

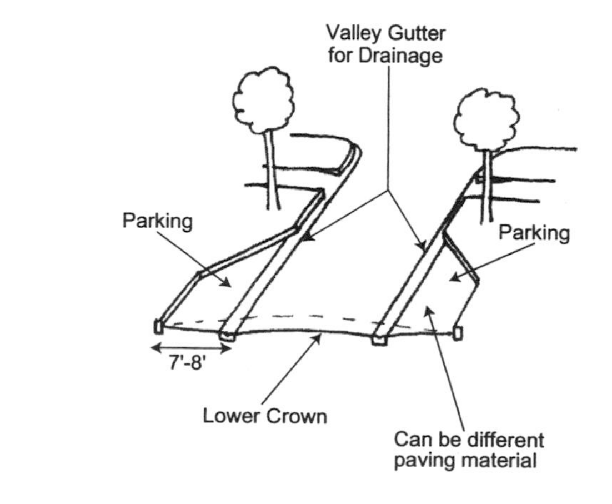

Fig. 12. Valley gutter.

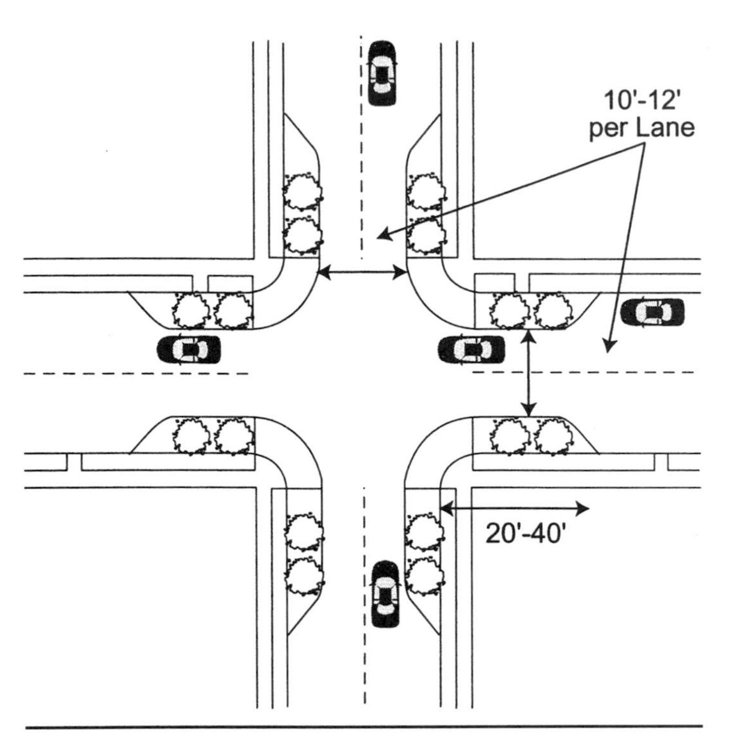

Fig. 13. Intersection bulbouts.

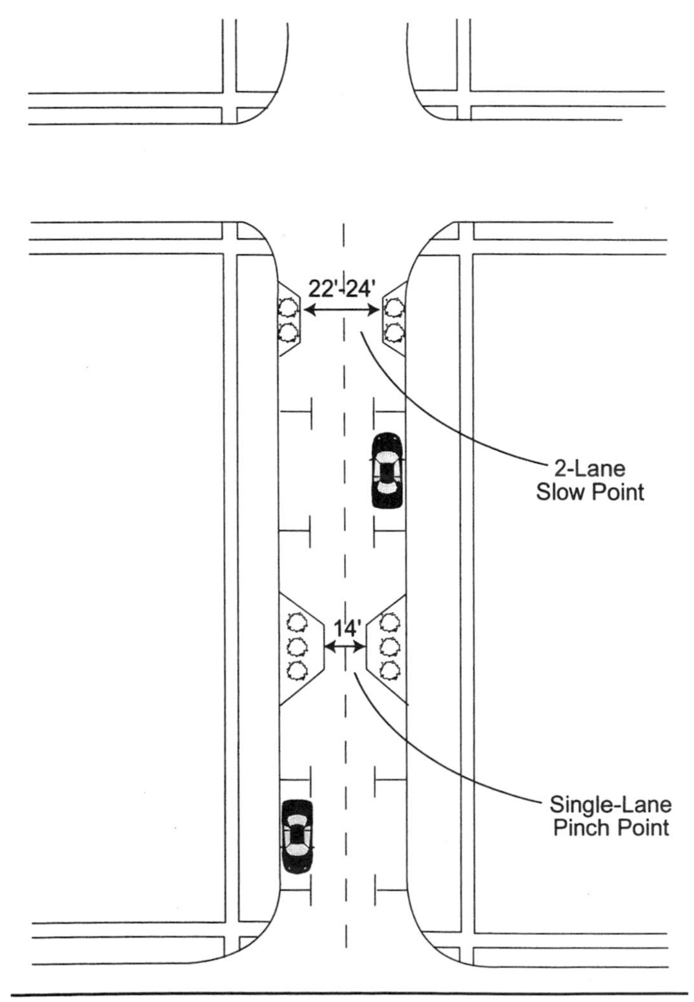

Fig. 14. Mid-block bulbouts.

Changing the street's cross-slope profile (valley gutters)

Rebuilding a street (or building a new street) with the drainage along a valley gutter between the parking and the traveled lanes (Fig. 12) serves a number of traffic calming functions. The valley gutter, a gently sloping V-shaped band, breaks the cross-section profile of the street, and consequently narrows the appearance of the street. This apparent narrowing is further emphasized when different paving materials or colors are used for the travel lanes (typically asphalt or red brick), the valley gutter (typically concrete) and the parking lanes (gray brick, textured pavement, *etc.*). The valley gutter design combines a number of subtle advantages. Because there are four cross slopes (rather than two), the centerline of the street is noticeably lower than on a conventionally crowned street. This reduces the elevation of the motor vehicles relative to the pedestrians on the sidewalk. Though seemingly a small and subtle change, it makes an important difference to the relationship between pedestrians and drivers. On a street with bulbouts (corner, mid-block or both), the valley gutter puts the drainage along the seam between edge of bulbout and edge of traveled lane, thereby circumventing the challenge of storm water being trapped at bulbouts on a conventionally crowned street. For similar reasons, the accumulation of debris along the curb is far less problematical than for a conventionally crown street with bulbouts. Yet another advantage of the valley gutter street is the possibility of paving the parking lane, or parts of it, in pervious materials that are advantageous to street trees (particularly in tight urban quarters) and that reduce storm runoff.

Narrowing at intersections (bulbouts)

Bulbouts (narrowing the street at an intersection) are versatile and are the most widely used traffic calming measure. One or both of the intersecting streets can be bulged out. The width of the roadway is typically no wider than necessary to accommodate the through lanes allocating 10-12 ft. (3.1-3.7 m) per lane. Where a left-turn lane is warranted, an additional 10-11 ft. (3.1-3.4 m) is added to the width of the opening. On low-volume streets, the opening through the bulbout can be reduced to a single lane of traffic with a 12-14 ft. (3.7-4.3 m) curb-to-curb face, thereby constituting an "intersection pinch-point." (Fig. 13)

Besides calming traffic, bulbouts define the on-street parking area, shield the ends of the parking zone from moving traffic, and discourage drivers from using the parking lane for overtaking other vehicles.

Bulbouts usually permit the stop bar on the approach lanes to be moved further into the intersection, which reduces the intersection size and signal clearance intervals, and sometimes solves problems with inadequate sight distance triangle from vehicles approaching from the side street. Bulbouts enlarge the corner radius, thereby frequently resolving issues of "wasted" space (unusable by either vehicles or pedestrians) near the corners.

Intersection bulbouts invite a wide variety of complementing street design measures, some of them further calming traffic. The shorter pedestrian crossing distances resulting from bulbout intersections invite the use of textured crosswalks, raised crosswalks and raised intersections. The bulbouts should be elongated to provide enough room for a vertical element such as a tree. Where a tree is impossible for sight triangle reasons on the side street approach, the increase in sidewalk area provides good locations for planters, information kiosks, benches and a wide variety of other street furniture. The additional sidewalk space is a useful mitigation measure where recent "build to" ordinances are requiring new buildings to locate close to the street, thereby putting a premium on sidewalk space.

Bulbouts can be an extraordinary efficient traffic calming measure. The construction at a single corner affects traffic in all directions on both intersecting streets, defines and shields parking, complements pedestrian measures, provides more sidewalk and can help augment an entrance measure for a district, neighborhood or corridor.

Mid-block bulbouts

Bulbouts at mid-block locations (*i.e.*, away from intersections) are a close counterpart of the intersection bulbout. They define parking, narrow the appearance of the pavement, prevent vehicle overtaking, and can be complemented by pedestrian crosswalks (Fig. 14). Typically, the width of the opening between mid-block bulbouts is keyed to that of intersection bulbouts. Typically, a single lane of traffic in each direction is permitted. For low volume streets, a single lane "pinch point" of 12-14 ft. (3.7-4.3 m) wide can be formed by mid-block bulbouts.

Because they are not part of the sidewalk (as are corner bulbouts), mid-block bulbouts that are not used as pedestrian crossings can leave the existing curb and gutter intact, so that no drainage changes need to be made. Further economy can be gained by bordering the mid-block bulbouts with simple extruded curb, placed on top of existing pavement (as in commercial parking lots). Because the bulbouts are in a parking, not a driving lane, the lighter duty construction of the extruded curb is often adequate.

A series of regularly spaced mid-block bulbouts containing street trees forms an attractive enclosure for the street, further reducing motor vehicle speeds. The additional space gained by a bulbout is frequently the difference between a proper location for a tree and an otherwise cramped location too close to the building fronts.

The same features that complement corner bulbouts also combine well with mid-block bulbouts. These include a valley gutter cross-section for the street, different paving material in the parking lane, pedestrian crosswalks, textured pedestrian crosswalks and raised pedestrian crosswalks. Further, bus stops and shelters can be placed on the mid-block bulbouts, so that bus drivers do not have to leave and reenter the travel lane, parking spaces are not consumed in taper lengths, and the sidewalks are not obstructed or cramped by bus shelters.

Medians

Short sections of median which do not block side street or property access and which reduce the street cross-section to a single lane in each direction (Fig. 15), are a highly effective traffic calming measure. They narrow the street and, when planted with trees, narrow the apparent width of the street even more. They help discourage overtaking on two-lane streets as well as on streets with center lanes. They also help define and shield left-turn lanes. They are also frequently used as pedestrian refuges.

On the other hand, longer sections of median, cutting off access to properties and local streets are not considered traffic calming, but as a means to deny access. Long medians that leave more than one travel lane in each direction (typically for emergency pull-over reasons) may even increase speeds.

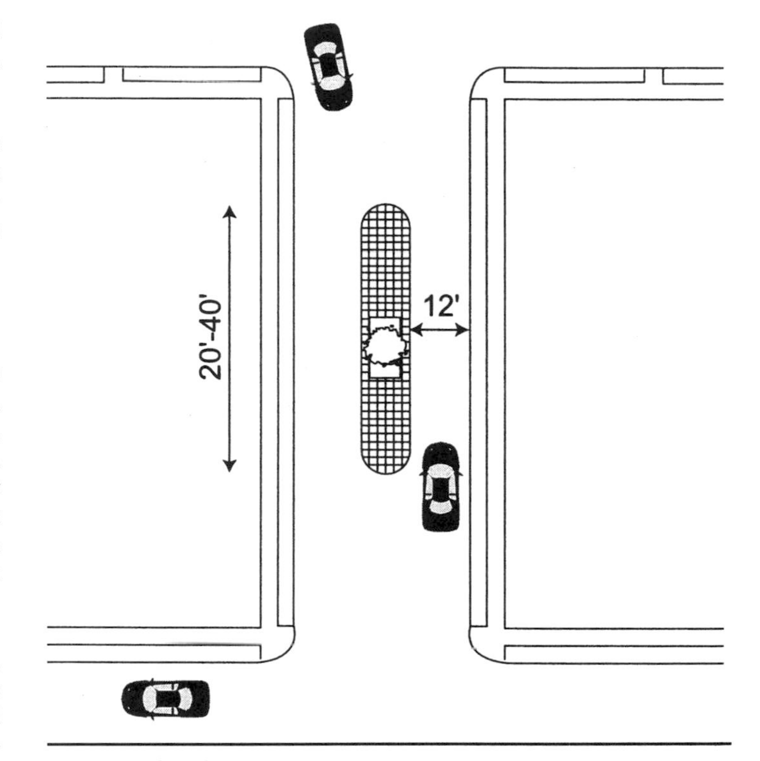

Fig. 15. Median islands.

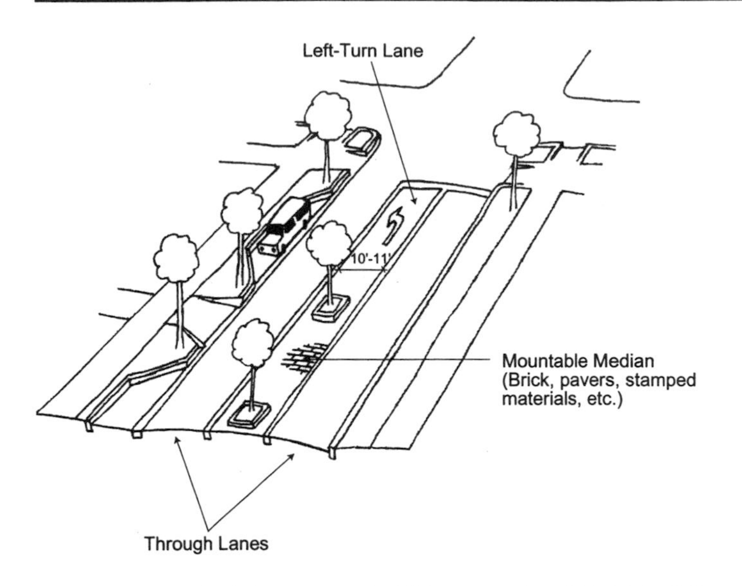

Fig. 16. Mountable medians.

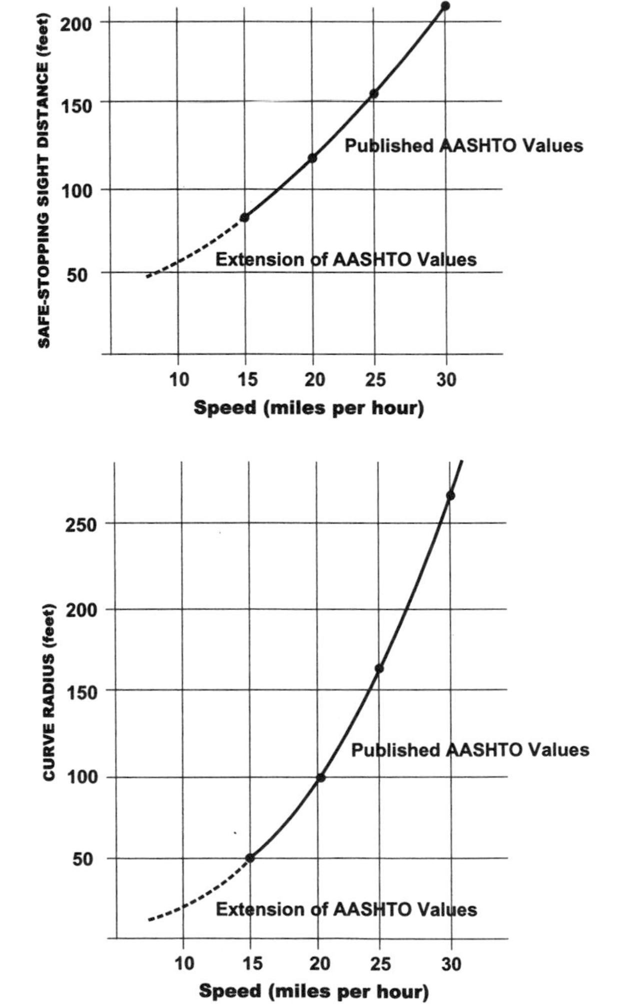

Fig. 17. Speed-related design elements (Source: AASHTO 1994).

Long medians that are traversable and mountable (Fig. 16) are an exception. The median is built out of a material that contrasts the travel lanes and is flush with the travel lanes to make it easily mountable. Mountable medians have fairly regular trees placed along them to prevent them from being used for a travel lane and to help provide an increased sense of narrowing. Care is needed to ensure that no driveways are blocked by the placement of a tree. The remaining lane in each direction can be one-lane wide because the mountable median can be used for emergency pullover purposes. At intersections, mountable medians can end at the intersection to provide a pedestrian refuge, or they can end away from the intersection to provide a left-turn lane.

4 LATERAL SHIFTS AND DEFLECTION

Deflecting the path of the vehicle reduces the design speed (and therefore calms traffic) in two ways: firstly, by introducing a curve of small radius into the driving path and secondly, by reducing the driver's sight distance of the road ahead. These two speed-related design elements are covered in AASHTO *A Policy on Geometric Design of Highways and Streets* (AASHTO *Green Book*) for speeds of 20 miles per hour (32 kph) and above (Fig. 17). For speeds, under 20 mph, extrapolate downward.

On new and rebuilt roads, deflection can be gained through curving of the centerline of the road. On existing roads, deflection is gained by varying the vehicle path within the available pavement or, more ambitiously, within the available right-of-way. While deflection at mid-block locations is a long-standing traffic calming measure, perhaps the most popular application of deflection is now at intersections, in the form of mini-traffic circles and roundabouts.

A further guideline for deflection measures is that a reasonable design vehicle—usually the "single unit" or Suburban Utility Vehicle (SUV), or a tractor-trailer with a wheelbase of 40 ft. (12.2 m) or WB40—be accommodated. The turning template for the SUV is given in Fig. 18.

Mid-block deflection

At mid-block locations, deflection can be gained by medians in the center of the street, by bulbouts from the side of the street, and combinations of both (Fig. 19). For example medians can be used in conjunction with bulbouts and on-street parking prior to and/or after the median to create lateral shifts in the street. Deflection can also be gained from arranging parking in an asymmetrical method, so that the traveled lane shifts from side to side as motorists proceed along it.

Mid-block deflectors, placed on minor streets near their intersections with major streets, perform two important functions:

- They help block the view down the minor street from the major, thus reducing its appeal from a cut-through perspective, and, conversely,
- They screen the view of the major street from the minor street, which is often appealing to those on the minor street depending on the aesthetic quality of the major streets.

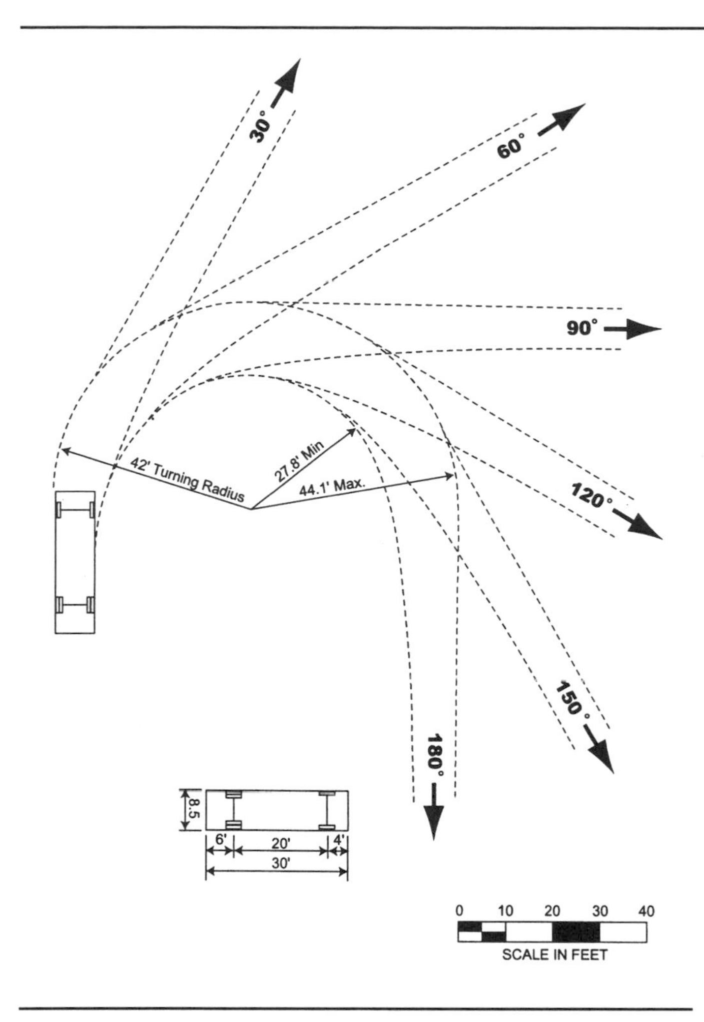

Fig. 18. Single-unit truck turning template. (Source: AASHTO 1994)

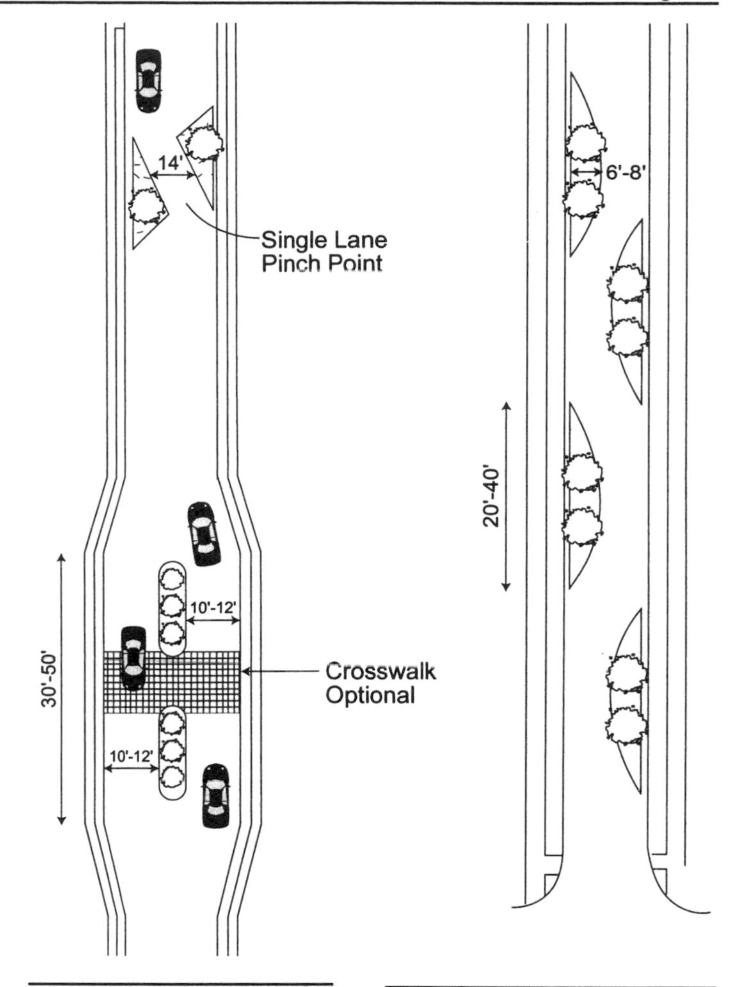

Fig. 19. Mid-block deflectors.

Fig. 20. *Chicanes.*

Chicanes

Chicanes resemble S-curves and require drivers to turn left-right-left or right-left-right along a street (Fig. 20). These can be achieved by employing two or three narrowings that alternate sides of the street or by rebuilding the section of street. Care must be taken to ensure that drivers have to deflect while going through the chicane. *Chicanes* result in the loss of on-street parking in the vicinity of the measure. Vertical landscaping on chicanes is important to make them conspicuous.

Intersection deflections

Intersection deflections can be achieved through intersections, by shifting the lanes and/or angling the lanes at the approaches (Fig. 21). Care needs to be taken so that drivers understand that a shift is taking place. On streets with parking on one side, intersection deflections can be used to switch parking form one side of the street to the other while minimizing any loss in the number of parking spaces due to transition areas.

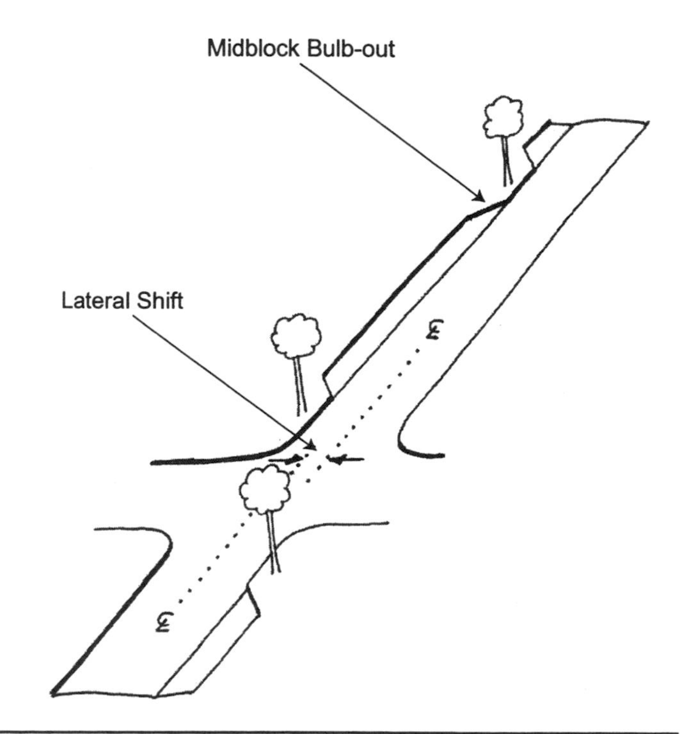

Fig. 21. Lateral shift through intersection.

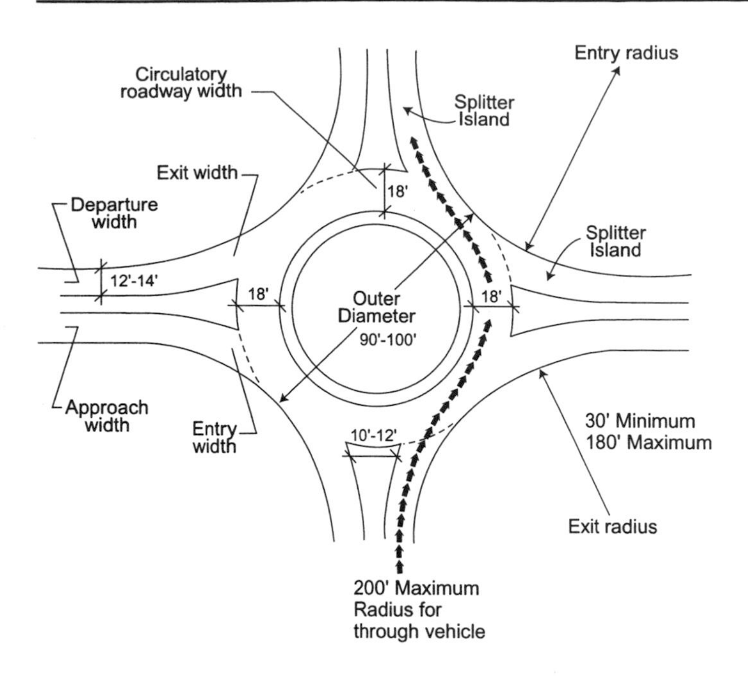

Fig. 22. Roundabout.

"SW" Street Width	"R" Curb Radius	"OS" Offset	"D" Circle Diameter	"OW" Opening Width
20	15'	5.5'	9'	16'
20	20'	4.5'	11'	18'
25	15'	5.0'	15'	17 '
25	20'	4.5'	16'	18'
30	15'	5.0'	20'	17'
30	20'	4.0'	22'	19'

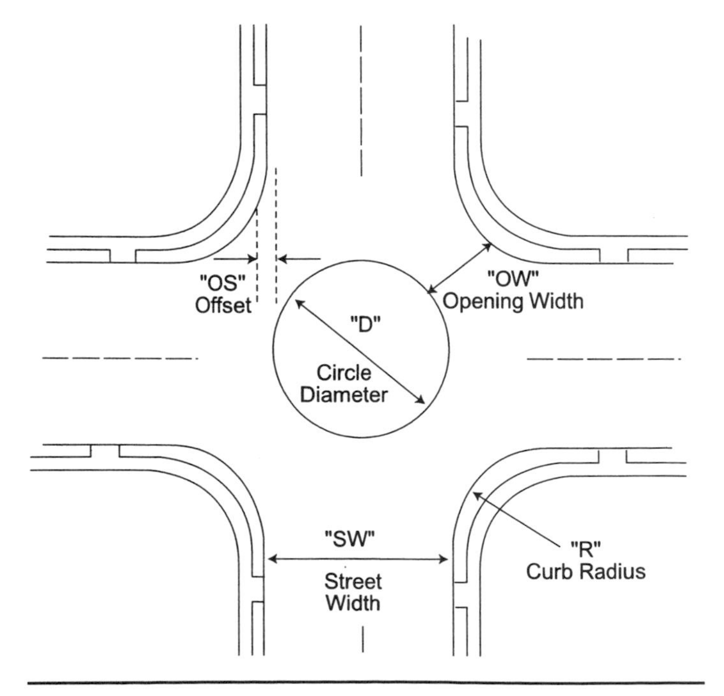

Fig. 23. Traffic mini-circle.

Roundabouts

Roundabouts (Fig. 22) are both a traffic calming measure and a highly efficient intersection design. They calm traffic by introducing three successive reverse curves of short radius: to the right to enter the circle, to the left to go around the circle, and to the right to exit the circle. Many roundabouts further calm traffic by reducing the sight distance for oncoming motorists, with trees or man-made features in the center of the roundabout.

The roundabout is a highly efficient intersection design because it instantly assigns, and constantly reassigns, the right-of-way to each newly arriving vehicle. This rapid assignment of right-of-way to the entering vehicle differs sharply from the signalized intersection, in which motorists must not only wait for their green phase, but also usually wait for a queue of stored vehicles to clear ahead of them.

The size of the central island largely determines the operating speed of vehicles. Small islands cause little deflection, and therefore provide little speed reduction. At the other extreme, overly large islands provide a large turning radius, and therefore little reduction in speed.

Splitter islands are typically used at roundabouts that would otherwise be signalized intersections, because:

- They reduce the disparity in speed between circulating and incoming vehicles,
- They eliminate the possibility of right-angles collisions,
- They increase the car-carrying capacity of the intersection while providing a refuge for pedestrians.

Mini-traffic circles

A mini-traffic circle (Fig. 23), a circular island in the center of an existing intersection, is a highly effective traffic calming measure, dealing with two streets at once. It is often confused with the roundabout, which it resembles. The mini-traffic circle and roundabout, however, are distinctly different measures. Mini-traffic circles can, and almost always are, considerably smaller, since they don't need to merge traffic in a circular fashion, but rather depend on ordinary traffic control devices (*e.g.*, stop and yield signs) to assign right-of-way. The smaller size of the mini-traffic circle often permits it to be placed within the existing curbs, minimizing the cost of the installation.

Mini-traffic circles are usually preferable to roundabouts on lower-volume streets. As volumes increase through the 2,000-5,000 ADT range, roundabouts may become preferable for their vehicular capacity. Above this range, it is highly likely that a roundabout is the preferred solution.

At a mini-traffic circle, automobiles proceed around in a counterclockwise fashion, in the same manner as at a roundabout. Trucks larger than pickup trucks—*i.e.*, the 30-ft. (9.14 m) single unit truck or larger—can make left turns in front of the mini-traffic circle, after yielding right-of-way.

There is no minimum size for mini-traffic circles. However, the amount of deflection diminishes as size decreases, as well as the ability to have plantings or built features that serve to reduce the sight distance.

Mini-traffic circles can be used at T-intersections as well (Fig. 24).

They are designed the same way as at a four-approach intersection except that there are only three approaches. These are relatively rare because of the effect on the top of the T in terms of right-of-way needs or the effect on landscaping or the sidewalk. The "impeller" offers a deflection option for T-intersections and is designed in a similar way as a mini-traffic circle, except that half of the circle is located to the top of the T and the other half stays in place, but with a break in it for left turns.

The mid-block mini- traffic circle, though also seldom used, is a completely appropriate traffic calming measure. Motorists must make three successive turns in opposite directions, a key factor in assuring a reduction in speed. It is designed the same way as at an intersection, except it effectively has only two approaches, one from each side.

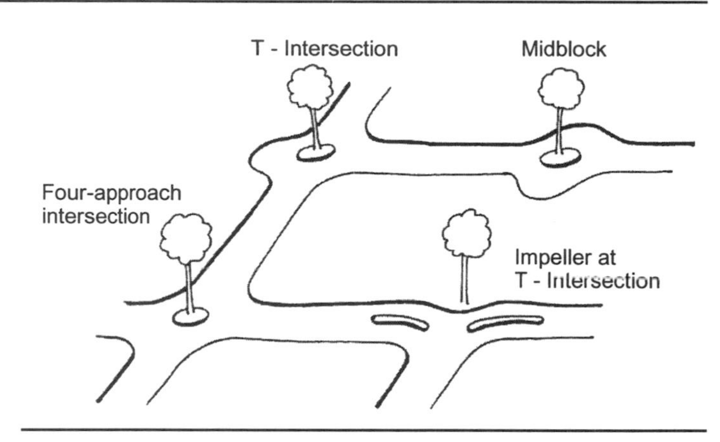

Fig. 24. Mini-traffic circle, "T" intersection and midblock.

Traffic circles

A traffic circle is a circular block of land surrounded by a one-way street flowing in a counterclockwise direction. The land area is usually large enough to accommodate a park or civic building. A traffic circle is not a traffic calming measure itself but is simply a different arrangement of the block structure and street network to create an interesting site. All the streets intersecting the traffic circle usually intersect at right angles forming T-intersections. Occasionally, four-approach intersections exist if a driveway is located inside the circle. The circle and its intersections can incorporate traffic calming measures in a similar fashion to regular orthogonal intersections and streets.

Entrance ways

Entranceways (Fig. 25) have several roles. First they announce to those who are arriving that they are entering a traffic-calmed area and that it is time to slow down and pay special driving attention. Well-designed entranceways also identify the area or district. Second, the height and bulk of the entranceway feature can reduce both the real and perceived width of the street, thereby lowering driving speeds. Other actions, most often-textured pavement, speed tables and single-lane "pinch points," are particularly complementary to an entranceway. The entranceway can also be combined with a transit shelter, weather shelter, information kiosk, and signposts.

Street closures and partial closures, including or not including gates, are not traffic calming measures and should never be installed.

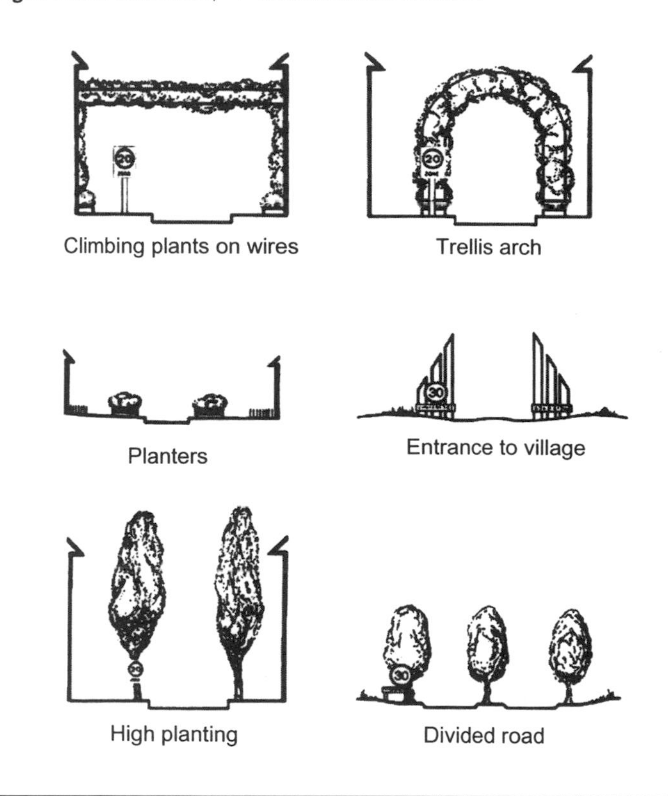

Fig. 25. Entrances and gateways.

6 PROGRAMMING OF TRAFFIC CALMING MEASURES

Number and spacing of traffic calming measures

Continuous cross-section changes (*e.g.*, on-street parking, bicycle lanes, continuous narrowing, street trees) can be used without limit despite the length of the street, subject only to the ability to fund them. However, intermittent traffic calming measures, requiring driver response (turning, slowing) at specific locations, should be limited so that drivers are subject to no more than 8–12 measures on any possible trip that they make within the area. A greater intensity of measures is overly annoying.

A well-designed area-wide traffic-calming plan (Fig. 26) will use continuous cross-section actions on "framework" streets (*i.e.*, longer streets that would require more that 8–12 measures in succession). The occasional measure that requires a further driver response is

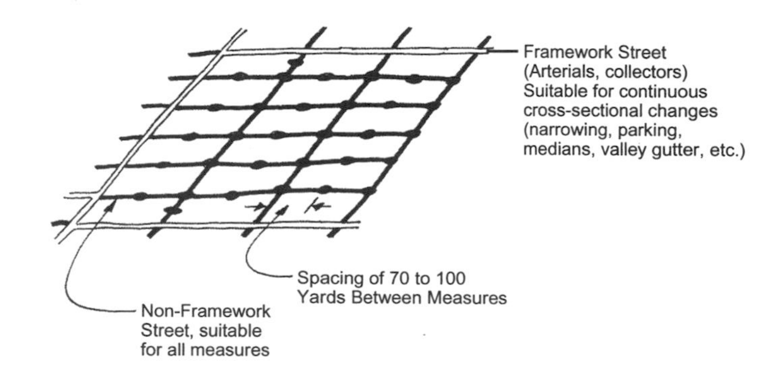

Fig. 26. Area-wide traffic calming.

appropriate even on framework streets when a major pedestrian generator is involved like a school or park. On the "non-framework" streets, locating measures at a spacing of 75–100 yards (21–28 m) per measure is typical to achieve desired speeds of 20 mph (32 kph). The best approach is to plan measures in the obvious places first (*e.g.,* at pedestrian generators, at pedestrian crossings, etc.) and then fill in to achieve the correct spacing.

Guidelines for allocating traffic calming measures

Allocating traffic calming measures on the basis of numerical scores ("warrants") is not advisable. Invariably, warrants focus on the few easily measured motor vehicle traffic characteristics (for example, speeds, volumes, and collisions) furnishing an incomplete and frequently misleading view of the need for traffic calming. The really important factors in neighborhood, district, or corridor design; such as its historical character, age makeup, types of businesses, overall vitality, construction type, proximity of buildings to the street, walkability, aesthetics, etc. is difficult to capture in a numerical warrant. Further, the funds and efforts spent in gathering the data to support warrants drains the traffic calming program of funds. The use of numerical scores or warrants based on motor vehicle speeds, volumes and collisions gives the allocation exercise a guise of objectivity due to its inherent, false assumption that all else is equal. The circumstances for each potential project are as dramatically diverse as the involved streets, land uses, and populations, as well as the community's goals, objectives, policies and priorities. The one-size-fits-all allocation is therefore arbitrary and fifers greatly in its applicability. Applying judgment is a better approach because then the aforementioned variety of factors can be considered for each situation.

Requiring petitions (from residents, property owners, businesses, etc.) as a prerequisite for traffic calming is highly inadvisable. Petitions voluntarily produced are useable input to any traffic-calming program, but requiring such petitions is highly inadvisable. Mandatory petitions are expensive and distracting, and costly to both funding and management attention of the traffic calming program. Worse, petitions are frequently a divisive, rather than cohesive factor in opinion making. The outcome can vary greatly, depending on the wording of the petition and on who is collecting signatures. Worst of all, mandatory petitions makes it more onerous for the professional transportation practitioner to design a context-sensitive, safe, traffic calmed street.

Rather than focusing on counterproductive measures such as warrants and petitions, the allocation of traffic calming effort should follow from street design deficiencies. Important deficiencies include excessively wide streets, absence of on-street parking, lack of crosswalks, lack of sidewalks, absence of street trees, highway-type (for example cobra-head) rather than pedestrian-scale street lighting, lengthy traffic signal cycles, absence of pedestrian crossing signals, poor aesthetics, poor retail activity, etc.

Defending against tort liability claims

Successful large tort liability claims (*i.e.,* jury awards) typically involve a driver "victimized" by a fault in the road design or operation, and suffering severe injury. Consequently, the vast majority of large-settlement tort claims hinge on factors that victimize drivers in high-speed collisions, such as malfunctioning traffic control devices (signals, stop signs, *etc.*) or misaligned pavement and pavement edges. The low-speed environment of traffic calming and superior safety record, on the other hand, removes both of these apparent requisites of major tort actions, specifically: a victim eliciting sympathy from a jury and severe bodily injury. Exposure to tort liability claims arising from traffic calming measures can be managed through simple planning and management actions:

- Have a written statement of rationale for traffic calming, adopted by local government (City Council, County Commission, and so forth). Compelling rationales are, for example, neighborhood preservation, Main Street business viability, school and campus safety, historical district design, and the preservation of the environment. A clear statement of a rationale defends against the traffic-calming program being attacked as "capricious."
- Do normal monitoring for the safety and maintenance. This monitoring need not be anything beyond the normal local collision reporting system, with some further effort made to clearly identify those locations that contain traffic calming measures.
- Address the known safety problems in a systematic manner, spending available funds in a cost-effective manner. This principle has long been a defense against highway tort claims of all types, since there is never enough funding to cure all known safety defects and there can therefore be no reasonable expectation of doing so. ■

REFERENCES

American Association of State Highway and Transportation Officials (AASHTO). nd. *Bicycle Facilities Guidelines.* Washington, DC: AASHTO.

American Association of State Highway and Transportation Officials (AASHTO). 1994. *A Policy on Geometric Design of Highways and Streets* (the AASHTO *Green Book*) Washington, DC: AASHTO.

Devon County Council. 1991, *Traffic Calming Guidelines.* Exeter, Great Britain: Smart Print.

Ewing, Reid. August 1999, *Traffic Calming: State of the Practice.* Washington, DC: Institute of Transportation Engineers, Publication IR-098.

Limpert, R. 1994. *Motor Vehicle Accident Reconstruction and Cause Analysis.* Charlottesville, VA: Michie Company.

Lockwood, Ian M. July 1997. *ITE Traffic Calming Definition.* Washington, DC: *ITE Journal.*

Smith, D. T., and D. Appleyard. 1981. *Improving the Residential Street Environment—Final Report.* Washington, DC: Federal Highway Administration.

Mark C. Childs

Summary
Places where people can park at one location stop, leave their cars and carry out multiple tasks on foot are identified as having both high social and economic value. The points of arrival and departure, the parking lots, parking garages, and street parking are critical design elements in promoting efficiency and amenity of urban life. The patterns presented in this article support the forming of such public space through the proper design of parking facilities.

Keywords
accessible parking, automobile, commercial vehicles, curb, dimensions, driveway, grades, parking, queuing, stall, traffic

PHOTO: Donald Watson

Santa Barbara, CA municipal parking structure subsumed within downtown mixed zone commercial/cultural district fabric.

Parking and circulation dimensions

The automobile plays a vital role in providing mobility and accessibility to a great variety of places. In the United States, cars are advertised and virtually identified with the idea of freedom itself. Yet we need to stop building unnecessary roads, to reduce the number of trips generated each day, and to redirect urban design efforts toward enhancing the quality of life at the beginning and at the end of every trip.

The origins and destinations of car trips must become places of serious consideration for designers. The following tables and diagrams offer minimum, maximum and ideal standards for various aspects of sparking lot design.

Comment on parking standards

Standards are not a substitute for thought. They must be used within the whole context of the design. A few of the most common misuses of standards follow:

The minimum as maximum. When a minimum dimension is given, it does not mean it is the best dimension, only that it is the least one can use in most cases without significant problems. Often a few more inches make the difference between OK and great.

Standards as design. The direct application of standards is not the end of design. The opportunities to make parking lots into great places will not arise from the simple application of standards. See Childs (1999) for ideas and design patterns for going beyond the standards.

Standards as the unmitigated truth. Standards are informed judgments meant to be generally applicable. If the particular application is significantly different, do some research to understand the basis of the standard and then you may modify the standard to better fit your situation.

It's not interesting unless I break the standards. Good standards represent experience about what makes something safe and gracious, and minimums often imply the dimension that just barely makes an action feasible. Simply "breaking the rules" could easily be making things worse.

My standard trumps your judgment. Pedestrians have lost space to auto traffic in many cases simply because the auto-traffic advocates have adopted standards, while pedestrian advocates used judgment tailored to the case.

Escalating standards. In 1929, recommended stall widths were 7 ft. (2.13 m) and under. Today a stall width less than this is unacceptable. Parking stall dimensions are based on vehicle sizes. As traffic lanes and parking spaces have been made larger to accommodate the largest vehicles, it has become easier to operate large vehicles. This encourages more people to buy larger vehicles (other things being equal) and, in

Credits: This article is adapted by the author from Chapter 13, *Parking Spaces: A Design, Implementation and Use Manual*, New York: McGraw-Hill 1999. Stefanos Polyzoides's preface is excerpted in introductory paragraphs of this article.

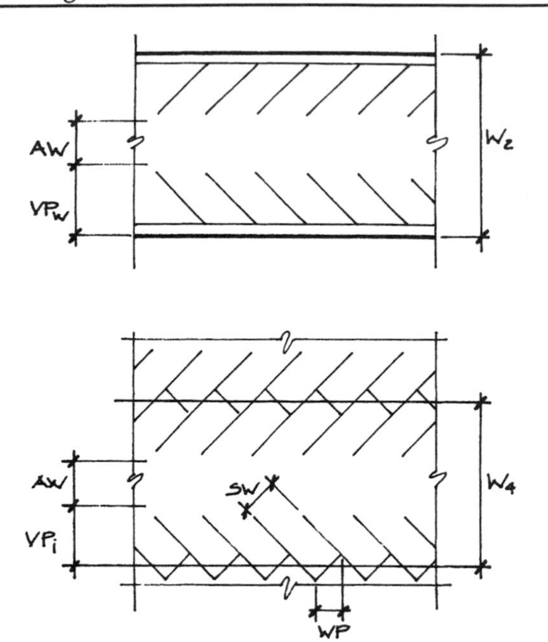

Fig. 1. Module variables.

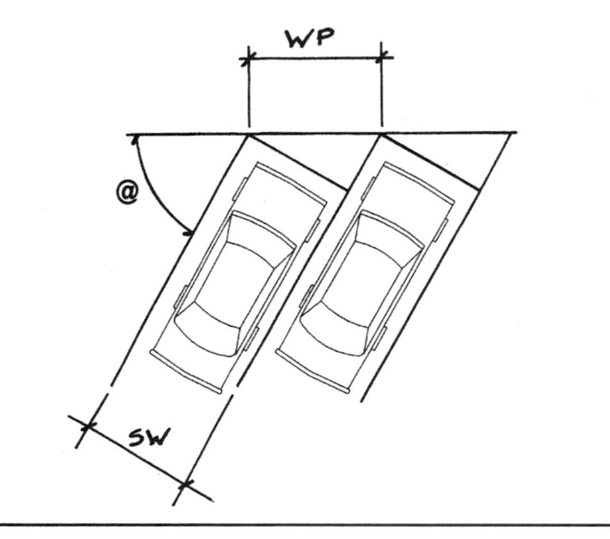

Fig. 2. Stall widths.

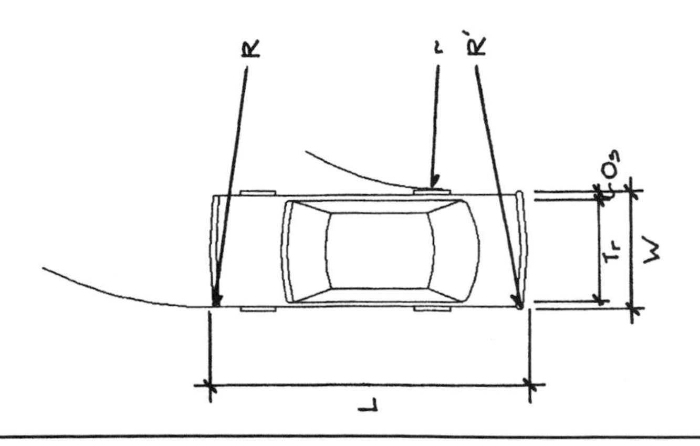

Fig. 3. Vehicle design variables.

response, street dimensions increase. Part of the reason European parking stalls are smaller than stalls in the United States is that respect for and investment in preautomobile cities has limited the growth in parking stall size and thus limited the growth in car size.

NOTE: Dimension tables are grouped at the end of the article.

1 PARKING LOTS

Bay & stall dimensions

There are a number of different approaches to the sizing and dimensioning of parking facilities. Table 1, Stall and Module Dimensions, incorporates a service approach that allows a designer to vary dimensions based on the expected use of the lot. The tabulation assumes a mix of small and large cars close to the 1995 national average. To use Table 1, three initial decisions must be made:

1. The appropriate stall width based on the expected turnover of the lot (Table 2),
2. If and how small car stalls will be provided (see below), and
3. The angle of parking. Slightly smaller dimensions are recommended in certain conditions in a (reduced) level-of-service approach (Chrest, Smith & Bhuyan 1996). (Figs. 1-3)

Table 5 lists some special conditions for vehicle stall dimensions. Parking for other vehicles should be based on the vehicle's length, width and turning radii (Tables 3 and 4).

Small car stalls

It is estimated that in 1995, half the cars on the road in the United States were small cars, 5'–9" wide by 14'–11" long (1.75 by 4.55 m) or smaller (Chrest, Smith & Bhuyan 1996, p. 30). Many local regulations allow a specified percentage of stalls to be designated for small or compact cars. However, the average difference in size between small and large cars has been decreasing so that it is often confusing for motorists to decide if they are driving a small or standard car. Small car stalls are ideally designated in three different conditions:

- In employee parking lots, stalls for compact cars may constitute up to 40% of the general stalls. This standard supplies compact stalls for 80% of the expected number of small cars based upon the 1995 U.S. average. When a survey has been conducted and if there is reason to believe that the number of small cars varies from the average, then the supply of compact stalls should be adjusted. In order to build smaller parking lots, businesses could offer rewards to employees who drive compacts. 100 small car stalls and aisles use about 5,300 sq. ft. (~493 sq. m) less land than would 100 standard stalls.
- At stadiums or other lots in which someone directs traffic to parking areas, cars may be sorted by size. Often stall markings are not used in these conditions to preserve flexibility. Designers should calculate the number of vehicles that may be parked under various expected conditions (e.g., compacts only, mixed, large only) and provide bays that have as little waste space under each condition as possible. A 68 ft. (20.7 m) long bay, for example, provides nine 7.5 ft. (2.3 m) wide stalls or eight 8.5 ft. (2.59 m) wide stalls.

- In general service lots, up to 40% of the stalls may be for compacts (adjusted for local conditions). Alternatively, very large cars could be given special stalls. By removing them from the mix of cars, the general stalls and aisles could be made smaller. For example, if 5 to 10% of stalls were designated for large cars, a stall width two sizes smaller could be used for the remainder of the lot, and aisles could be approximately 6 in. (~15 cm) smaller. The large size cars, classes 10 and 11, made up 7% of the market in 1993. (Chrest, Smith & Bhuyan 1996, p. 33). These large car stalls should be the least convenient so that smaller cars do not occupy them. Since there is limited experience with this approach, a traffic engineer should be consulted to tailor dimensions to expected local conditions.

Lanes

Table 6 shows widths for various traffic lanes. All of these dimensions are for public streets. Local regulations and standards often specify similar dimensions for parking lot lanes. The Uniform Fire Code requires that fire lanes are a minimum of 20 ft. (6.09 m) wide, have a height clearance of at least 13.5 ft. (4.11 m), have a 40 ft. (11.98 m) minimum radius on curves, and a maximum dead-end length of 150 ft. (46.3 m).

Driveway location

The existing street and pedestrian network must be evaluated before designing a parking lot to determine the best location for entrances and exits. Local codes normally limit the size, number and location of access points. Consideration in locating access points include:

- Avoid crossing busy pedestrian routes.
- Minimize the number and size of curb cuts to reduce conflicts with pedestrian and street traffic.
- Integrate driveways into the street system. Place entrances off of alleys when possible; otherwise integrate the driveways with intersections. If neither if these is feasible place driveways far away from intersections to avoid turning conflicts. Entrances must be "upstream" from exits.
- Avoid left turns across traffic if possible (provide turn lanes where volume is significant).
- Keep internal traffic flow simple.
- Avoid requiring cars to back up onto sidewalks in order to exit a parking stall.
- Entry control devices require at least a two-space off-street reservoir. (See below to calculate queuing for large lots).

Driveway width

Driveway width should be kept to a minimum to limit sidewalk curb-cut lengths and to minimize pavement. However, sufficient width must be given to accommodate traffic flow. See Table 7.

Number of driveway lanes

For large lots, the standard method for determining the number of lanes entering or exiting a lot is based on the highest demand for the lot during any 15 minute period. The method is given below, but some thought may be given to the use of this standard. For example, the following charts indicate that we should design stadium parking lots so that they can empty in 30 minutes. Perhaps adjacent merchants would prefer that people are induced to stay a bit longer. Adjacent roads could be a bit smaller and cheaper if the crowd was metered out more slowly, and perhaps more people would use transit and the lot could be smaller. Limited exit capacity may be an adjunct to increased parking fees for jurisdictions wishing to reduce auto travel in downtowns.

The rate at which cars may enter a lot is affected by the angle of approach to the lot (in order of ease: straight approach, left turn to entrance, right turn to entrance), and drivers' familiarity with the lot (*e.g.*, commuters are more familiar with a lot than are tourists). Rates up to 1000 cars per lane per hour are possible. Entrance fee and control booths reduce the entrance rate by 50 to 83%. Attendants can park 8 to 16 cars per hour per attendant.

The standard formula for the number of lanes is N =(SxR)/(PxU). N is number of lanes, S is number of stalls, R is the percent of lot capacity moving at peak hour (Table 8). P is the peak hour factor (Table 9), and U is the design capacity of the lane (Table 10).

Queuing

Entrance routes for lots at which people tend to arrive at the same time (*e.g.* employee parking, special events) should have off-street queuing space to avoid having cars block the sidewalk and street. Alternatively, work hours can be staggered and special events can have incentives for arriving early. The formula to estimate the number of cars queued is: L= i2/(1 – i) where L = number of vehicles waiting, i = peak hour volume/ maximum service rate (Tables 8 and 10) (Weant & Levinson 1990, 186–187.)

Grades

Table 11 shows maximum recommended grades within parking lots. Care should be taken to minimize the grades of routes used by the mobility impaired. Ramps may have a maximum slope of 1 in 12 over a maximum length of 30 ft. (9.14 m) between landings. (Fig. 4)

End islands

In lots with more than 200 cars or difficult conditions, islands should be created at the end of bays to assure sight distances, provide adequate turning radii, protect vehicles at the end of parking bays and limit parking encroachment into cross aisles. End islands also help delineate circulation and may be used to plant trees, pile snow or store grocery carts, and provide places to place light poles, fire hydrants, low signage and other equipment. Striped pavement in and of itself is not very effective in discouraging use of these areas for auto parking, and drivers turning into an aisle often cut the corner, putting themselves on the wrong side of the aisle. Curbs or bollards are often needed to establish the end island.

The character of these islands is a significant design feature. They line the cross street and act as gateways to each bay of parking. Fig. 5 shows typical end islands. The size and geometry of the islands should be adjusted for particular conditions. Design considerations include:

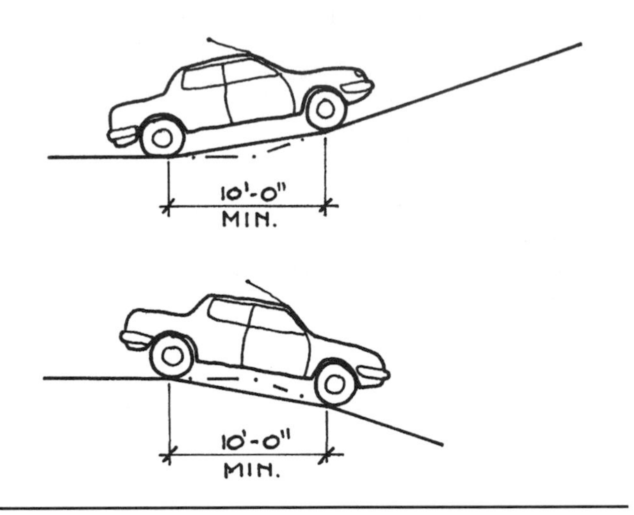

Fig. 4. Slopes at grade changes. Change grade by a maximum of 10° increments with 10 ft. (3.05 m) minimum between changes of grade.

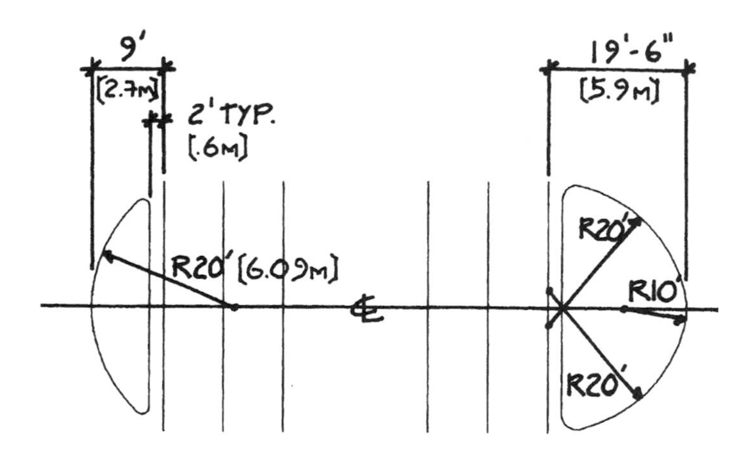

Fig. 5. Diagram of typical end islands.

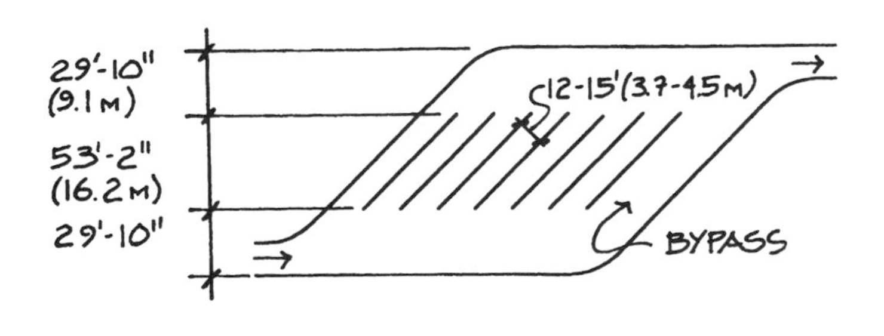

Fig. 6. Commercial truck parking layout.

- A typical car has an inside turning radius of about 15 ft. (4.57 m) and end islands with smaller radii may cause cars to "turn wide" into the opposing lane.
- Care should be taken with the placement of the radii to avoid exposing the end of an adjacent car and its disembarking passengers to "clipping" from turning cars.
- Cars parking next to the island need a walking aisle of at least 24 in. (61 cm) between them and any landscaping.

Adequate sight distance increases with the speed of cars and when the road is curved. For example, a car making a right turn onto a 20 mph (32 kph) aisle (typical small lot speeds) needs to be able to see about 250 ft. (~76 m) ahead; turning onto a 35 mph (56 kph) aisle, the driver needs to see about 550 ft. (168 m) ahead to judge the safety of making the turn (Stover & Koepke 1989, Table 1). Thus the depth of the island should increase next to higher speed or curved aisles/roads.

End islands may also be increased in size to provide space for trees or other amenities.

Stall markings

Marking stalls aids in the even distribution of vehicles within a lot and is necessary for stalls for the handicapped. Painted or applied stripes are typical. Street reflectors offer higher visibility, and can be felt if a tire crosses them. Changes in pavement color and material can also help designate stalls. Painted or applied stall stripes are usually 4 in. (~10 cm) wide and about 6 in. (~15 cm) shorter than the stall to encourage drivers to pull fully into the stall. Reflective and ridged stripes are available that provide increased visibility and a tactile presence to pedestrians and drivers.

There is disagreement about the effectiveness of double stripes to encourage centering of cars. A study by Paul Box (1994) of shopping center, motel and office building parking lots found no significant difference in parking efficiency between single and double stripes and thus recommends avoiding the expense of paired lines. However, for stalls for the mobility impaired, possibly grocery store lots, and in other conditions where the usable width of the between-doors aisle is large and critical for pedestrian access, the aisle should be marked with two stripes and crosshatching.

The design manuals of the 1950s suggest that stall marks be continued up walls to help drivers locate the stall. Careful placement of parking meters, trees, bollards and other vertical elements can provide the same function.

Overhangs, wheel stops and bumpers

Wheel stops, curbs, bumpers and bollards are used to protect structures, limit parking encroachment onto pedestrian pathways or planting areas, and signal to drivers when they have fully pulled into a stall. Sometimes they are used to prevent driving across rows of empty stalls because this cutting across is perceived as a hazardous activity. Backing up from parking stalls, however, generates more parking lot accidents (Box 1981), and thus cars should not be prevented from pulling through stalls when possible.

The Institute of Transportation Engineers discourages wheel stops in parking lots because they may present a tripping hazard, interfere with snow plowing, and trap trash (ITE 1990). Wheel stops should only be used to protect a wall, column or other item at a location where pedestrians are unlikely to walk. Where wheel stops are used, an overhang of about 2.5 ft. (76 cm) for 90 degree front-in parking and about 4.5 ft. (1.37 m) for 90-degree back-in should be provided beyond the wheel stop. Concrete and recycled plastic wheel stops come in a range of sizes. Typically a 6 ft. (~1.8 m) stop is used for each stall, and 6 in. (~15 cm) is the maximum height for a wheel stop. Wheel stops should not be shared between adjacent stalls because it creates confusion and blocks pedestrian circulation in the aisle.

Curbs are often used in lieu of wheel stops. Curbs present a smaller tripping hazard than wheel stops because the action of stepping up or down is easier than the action of stepping over and because they appear at an expected location. Curbs along walkways should be designed as steps with rounded, skid-resistant and contrasting-color nosings. Curbs require the same overhang setbacks as wheel stops.

Some vehicles overhang their wheels farther than the overhang setbacks recommended above and cars occasionally ride up or over curbs and wheel stops. Thus curbs and wheel stops should not be used when positive limitation is required. Bollards at least 36 in. (~ 90 cm) tall or bumpers can be used to protect buildings, trees, fire hydrants or other structures within harm's way.

2 BICYCLE PARKING

Some cities require that places to park and secure bicycles be provided whenever automobile parking is required. For example, Albuquerque requires that buildings provide a number of bicycle stalls equal to 10% of the number of required car stalls with a minimum of three bike stalls.

Bicycle stalls require either a lockable enclosure or secure stationary racks which support the frame and to which the frame and both wheels can be locked. A space with minimum dimensions of 6 ft. (1.83 m) long, 2.5 ft. (.76 m) wide and with a minimum overhead clearance of 7.5 feet (2.28 m) is necessary for each stall. An access aisle 5 ft. (1.52 m) wide minimum in needed beside or between each row of bike stalls.

Ideally, bike stalls should be covered to protect the bikes from rain and the heat of the sun. The parking area should be well lighted, close to the entrance of the building, e.g., not more than 50 ft. (15.2 m) and visible to passersby and/or building occupants. Bikes may also be parked in secured areas within a building.

The State of Oregon Department of Transportation developed a set of questions to help evaluate bike-parking racks (ODOT 1992) including:

- Can the rack support children and adults sitting, standing or jumping on it without bending? Is the rack safe for children to play near?
- Can rack capacity be expanded as needs increase?
- Is there adequate lighting?
- Does the rack allow a bike to be easily and securely locked to it without damage to the bike? Will the bike stand up even if bumped?

3 COMMERCIAL TRUCK PARKING

There are an estimated 185,000 commercial truck stalls near interstates at truck stops. On a typical night, 90% of these spaces are occupied, and a study by the Federal Highway Administration (1996) estimates that currently another 28,400 stalls are needed to make certain that exhausted truck drivers have a safe and secure place to pull off the highway.

This same report advises that public highway rest stops are not a direct substitute for private truck stops. The rest stops are used primarily during the day for short rests, while the truck stops provide longer nighttime parking. Significant factors in the preference for truck stops are the added security of the truck stop, and the services offered.

Fig. 6 shows the preferred layout of stalls for single trailer trucks. This pattern is up to 70% more land efficient than parallel parking (Federal Highway Administration 1996, p. xxiii). More than half of truck stop lots currently have no markings, providing the flexibility to accommodate double and triple trailer rigs but complicating parking maneuvers.

4 LOTS ACCESSIBLE TO THE MOBILITY IMPAIRED

The first case brought into Federal Court that resulted in a civil penalty under Title III of the Americans with Disabilities Act (ADA) was for failure to make parking accessible. Parking is a critical element of accessibility.

The ADA is a civil rights law. The Department of Justice has the charge of enforcing the law; people who believe they have been discriminated against may sue the property owner. The guidelines issued by the government are not a building code subject to state or local approval or variances.

There are current efforts to create regulations that if followed, will serve to show that the design meets the intent of the law. The information in this book was compiled from publications of the Architectural and Transportation Barriers Compliance Board (ATBCB) and other sources as noted. It should be viewed as the opinion of the author, informed by the material quoted. The law and best practices continue to evolve. The designer should review materials and conditions that apply directly to the project at hand. Lots owned by government agencies generally follow Title II rules that are usually more stringent than the Title III rules for privately owned lots discussed below.

Required number of accessible stalls

Whenever parking is supplied, no matter how the total amount of parking was determined, a portion of the stalls must be accessible to people with mobility impairment (hereafter called "accessible stalls"). Local codes may exceed the Federal requirement for required number of accessible stalls shown in Table 12. The more stringent rule governs. When a facility has more than one lot, the required number of stalls is determined lot by lot. In employee or contract lots, accessible stalls must be provided but "accessible spaces may be used by persons without disabilities when they are not needed by (persons) with disabilities" (ATBCB 1994). When the use of a facility, *e.g.* a senior center, indicates that more accessible stalls are needed than are required by Table 12 criteria, a study should be conducted to determine an adequate supply of accessible stalls.

Location of stalls

The location of accessible stalls must give mobility-impaired persons preferential treatment in terms of access and must not discriminate against them in terms of amenities (e.g. if the general stalls have hail protection canopies, the accessible stalls must also). The shorthand rule is that accessible stalls should be located with the shortest possible route to the entrance(s). Relevant U.S. regulations include:

> *Accessible parking spaces serving a particular building shall be located on the shortest route of travel from adjacent parking to an accessible entrance. In parking facilities that do not serve a particular building, accessible parking shall be located on the shortest accessible route of travel to an accessible pedestrian entrance of the parking facility. In buildings with multiple accessible entrances with adjacent parking, accessible parking shall be dispersed and located closest to the accessible entrances.* (ADAAG 4.6.2)

> *Accessible spaces can be provided in other lots or locations, or in the case of parking garages, on one level only when equal or greater access is provided in terms of proximity to an accessible entrance, cost and convenience. . . .The minimum number of spaces must still be determined separately for each lot.* (ATBCB 1994)

Van-accessible stalls

See Table 5 for required dimensions of van-accessible stalls and Table 12 for the required number of stalls. These stalls must be marked with "van-accessible," but this does not restrict the stall to use by vans (ATBCB 1994).

Notes on the layout of stalls

Two accessible stalls may share an access aisle. However, this should only be done when the stalls are at 90 degrees and allow both front-in and back-in parking. [Title II modification to ADAAG, 14.2.6(1)(b)]

Curb ramps or other obstructions may not be within the stall's access aisle, but may begin at the curb face when vehicles overhang a curb (Chrest, Smith & Bhuyan 1996, pp. 212.)

Car overhang may not obstruct the clear width of a sidewalk access route. Wheel stops and/or a reinforced signpost may help limit car overhang.

Signage

"Accessible parking spaces shall be designated as reserved by a sign showing the symbol of accessibility. . .(Van) spaces. . .shall have an additional sign "van-accessible" mounted below the symbol of accessibility." (ADAAG 4.6.4)

ADAAG requires that the sign not be obscured by a car or parked van. Centering the sign on the access aisle may help its visibility.

Equipment

Equipment such as parking meters, automated teller machines, pay stations, and ticket dispensers must have accessible controls. Most such equipment is now designed with operating mechanisms that are considered accessible, and the designer's major role is to place the controls at a proper level and to provide clear access to the controls.

For example, ADAAG 14.2(2) controls the design and placement of parking meters in public rights-of-way:

> *Parking meter controls shall be 42 in (1065mm) maximum above finished public sidewalk. . . and where parking meters serve accessible parking spaces, a stable, firm, and slip-resistant clear ground space a minimum of 30 in. by 48 in. (76 cm by 122 cm), shall be provided at the controls. . .parking meters shall be located at or near the head or foot of the parking space so as not to interfere with the operation of a side lift or a passenger side transfer.*

Existing lots

Bulletin #6 (AATBCB 1994) suggests that existing lots must be made accessible when it is possible to do so:

> *ADAAG established minimum requirements for new construction or alterations. However, existing facilities not being altered may be subject to requirements for access. Title III of the ADA, which covers the private sector, requires the removal of barriers in places of public accommodation where it is 'readily achievable' to do so. This requirement is addressed by regulations issued by the Department of Justice. Under these regulations, barrier removal must comply with ADAAG requirements to the extent that is readily achievable to do so. For example, when restriping a parking lot to provide accessible spaces, if it is not readily achievable to provide the full number of accessible spaces required by ADAAG, a lesser number may be provided. The requirement to remove barriers, however, remains a continuing obligation; what is not readily achievable at one point may become achievable in the future.*

When alternations are made (e.g., realigning striping or resurfacing, but not routine maintenance) whatever is altered must be made accessible unless technically infeasible (ADAAG 4.1.6 (1)) and improvements to the path of travel to the lot must be made, up to a cost equal to 20% of the project budget (Title II, Rule 36.403 (f)).

Passenger-loading zones

There must be at least one passenger-loading zone for the mobility impaired whenever designated loading zones are provided. There must be an access aisle at least 5 ft. wide by 20 ft. long (1.53 by 6.10 m) adjacent and parallel to the vehicle pull-up space. A clear height of 9.5 ft. (2.90 m) is required at the loading zone and along the vehicle route, to, from and within the zone. The vehicle space and the access aisle shall be level with surface slopes not exceeding 1:50 (2%) in all directions (ADAAG 4.6.6). Neither curb ramps nor street furniture may occupy the access aisle space (ATBCB 1994).

Curb parking and loading zones

Table 13 gives dimensions for accessible curb stalls, bus stalls, and stalls for other vehicles. Local regulations may require different dimensions.

6 CIRCULATIONS PATTERNS

The circulation though a parking lot is critical to its performance. Overall goals are to create a simple, legible route that allows drivers to easily circulate past all available stalls on their way in, and past as few stalls as possible on their way out. It is prudent to avoid dead-end aisles when possible, although even in high turnover lots, parking cul-de-sacs for 10 to 12 cars may function well. Circulation patterns should be reviewed to remove "choke" points—places that many cars must pass through. Additionally, stalls for the mobility impaired need to be placed as near to entrances as possible, fire lanes often are required by local codes, and conflicts between cars and between pedestrians should be minimized. Usually there are trade-offs between these goals. For the layout of vast lots, see the Urban Land Institute's Parking Requirements for Shopping Centers (ULI 1982). Following are some possible layouts for smaller lots. In general, any of these patterns is more efficient if the aisles run the long dimension of the lot.

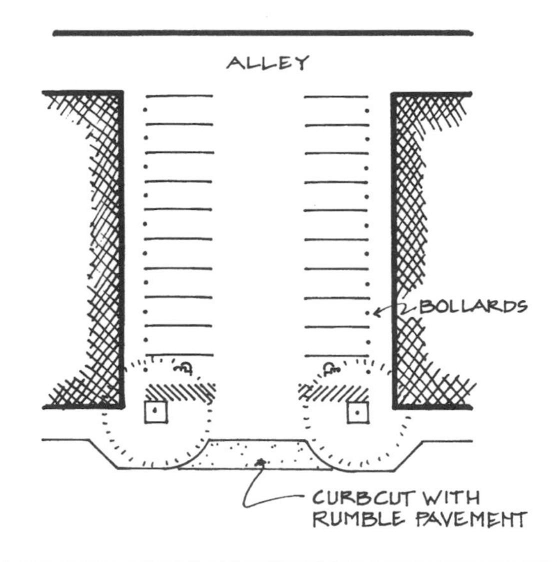

Fig. 7. Two-way bay.

The parking row

The most common unit of parking is the two-way aisle with 90-degree stalls and exits at both ends. (Fig. 7)

Advantages:

- Accessible stalls are easily incorporated into the design;
- The wide aisles increase separation between cars and pedestrians in the aisle;
- The two-way aisles allows drivers to exit efficiently;
- Does not require aisle directional signs and markings.

Disadvantages:

- Two-way traffic may increase the conflict between pedestrians and cars;
- This pattern cannot be fitted into all constrained sites.

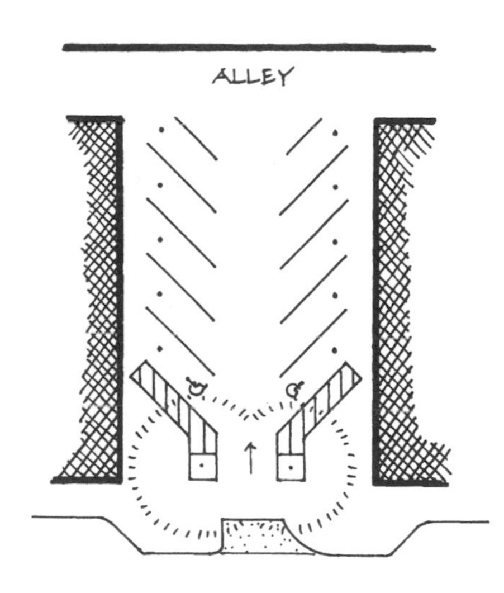

Fig. 8. One-way slot.

One-way slot (Fig. 8)

Advantages:

- This pattern can be fitted into narrow sites;
- One-way entrance and exit simplifies circulation and reduces required curb cut width;
- Pedestrian/car conflict reduced;
- Angle parking is perceived as the easiest in which to park.

Disadvantages:

- Drivers cannot recirculate within the lot;
- Signage of one-way entrance and exit required;
- Stalled or slow vehicle blocks entire system;
- Requires an alley or must run block face to block face.

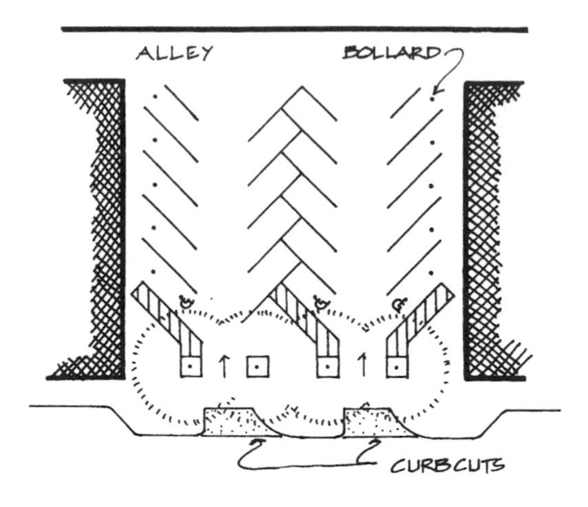

Fig. 9. Herringbone pattern.

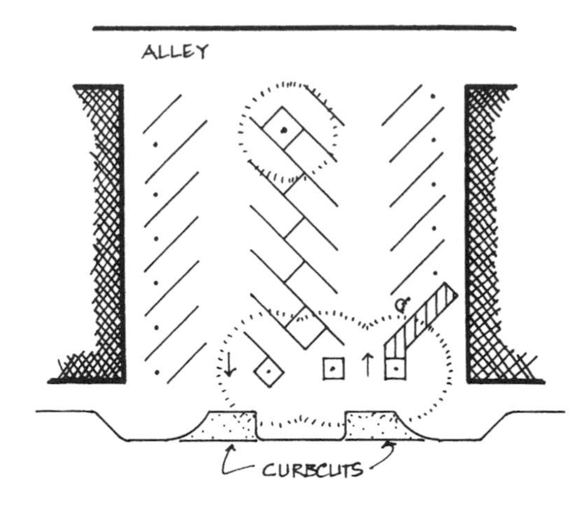

Fig. 10. One-way look.

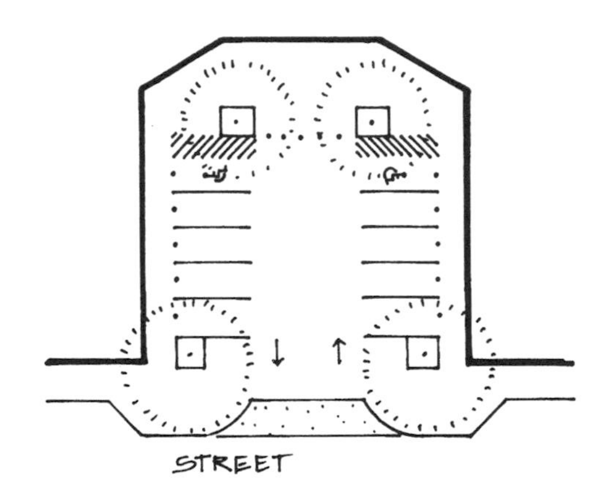

Fig. 11. Dead-end lots.

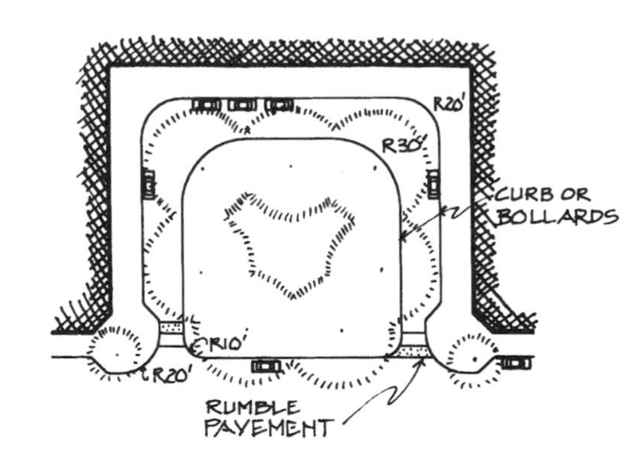

Fig. 12. Drop-off turnaround.

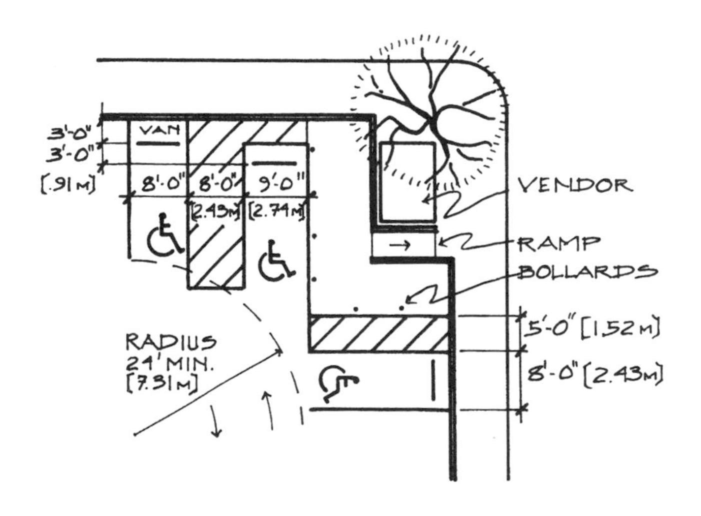

Fig. 13. Accessible corners (after Chrest, Smith and Bhuyan 1996).

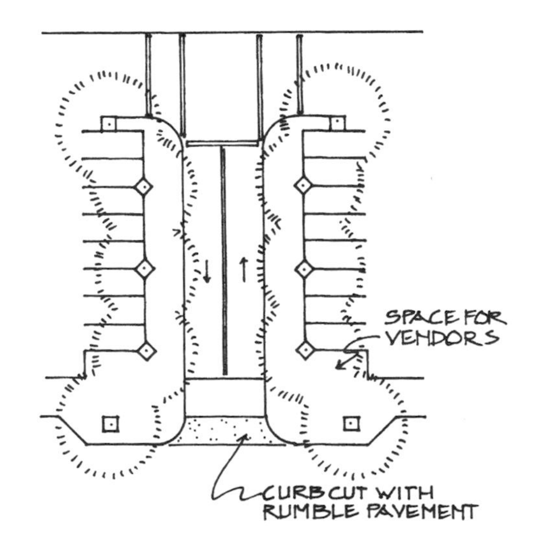

Fig. 14. Entrance aisle.

Herringbone (Fig. 9)

The one-way slot can be expanded into multiple one-way bays with the herringbone pattern. The advantages and disadvantages are similar to the one-way slot except that because cars cannot recirculate within the lot and cannot pass by all the stalls, there may be significant inefficiency in parking.

One-way loop (Fig. 10)

The advantages and disadvantages of the one-way loop are similar to the one-way slot. The exit should be "downstream" of the entrance. A cross aisle can be added to allow cars to recirculate within the lot.

Dead-end lots (Fig.11)

Dead-end lots should be limited to 10–12 cars for public parking and 40 cars for low-turnover employee or contract parking and should have back-out stubs at the dead end.

Advantages:

- All stalls are along edges allowing pedestrians to avoid crossing traffic;
- With a small lot the curbcut width can be minimal.

Disadvantages:

- The size must be limited to reduce conflicts due to excessive turnover.

Drop-off turnabouts (Fig. 12)

Advantages:

- Removes the activity of dropping-off and picking-up from the street;
- Allows architectural design of the place of arrival and departure.

Disadvantages:

- Can be less time- and land-efficient than on-street drop-offs.

Accessible corners (Fig. 13)

- The corners of parking lots can provide good access routes to accessible stalls.
- The advantages and disadvantages are dependent on the layout of the entire lot and must be evaluated on a case-by-case basis.

Entrance aisle (Fig. 14)

Lots that are likely to generate significant traffic, such as grocery store lots or other high-turnover, large lots, should have entrance aisles. These aisles provide a place to slow down cars as they enter the car commons and space for cars waiting to exit the lot. They can also provide direct pedestrian paths that do not cross parking areas. ■

REFERENCES

ATBCB (Architectural and Transportation Barriers Compliance Board) *Americans with Disabilities Act Accessibilities Guidelines for Buildings and Facilities; Final Guidelines.* (ADAAG) 36 CFR Part 1191. July 26, 1991.

ATBCB 1994. *Bulletin #6: Parking*

Box, Paul C. 1981. "Parking Lot Accident Characteristics." *ITE Journal*, December 1981.

Box, Paul C. 1992. "Parking and terminals." *Traffic Engineering Handbook*, fourth ed., Englewood Cliffs, NJ: Prentice-Hall.

Box, Paul C. 1994. "Effect of Single vs. Double Line Parking Stall." *ITE Journal*, May 1994.

Burrage, Robert H., and Edward G. Mogren. 1957. *Parking*. Saugatuck, CT: Eno Foundation for Highway Traffic Control.

Childs, Mark C. 1999. *Parking Spaces: A design, implementation and use manual.* New York: McGraw Hill.

Chrest, Anthony P., Mary S. Smith, and Sam Bhuyan, 1996. *Parking Structures: Planning, Design, Construction, Maintenance and Repair.* New York: Chapman & Hall.

Dare, Charles E. "Consideration of Special Purpose Vehicles in Parking Lot Design," *ITE Journal,* May 1985.

Federal Highway Administration. 1996. *Commercial Driver Rest and Parking Requirements.* Report no. FHWA-MC-96-0010.

ITE (Institute of Transportation Engineers) 1990. Committee 5D-8, *Guidelines for Parking Facility Location and Design.* Washington, DC.

ODOT (Oregon's Department of Transportation) 1992. *Bicycle Parking Facilities.*

Ricker, Edmund R. 1957. *Traffic Design of Parking Garages.* Westport, CT: Eno Foundation for Highway Traffic Control.

Stover, Vergil G., and, Frank J. Koepke. 1989. "End Islands as an Element of Site Design," *ITE Journal,* November 1989.

ULI (Urban Land Institute). 1999. *Parking Requirements for Shopping Centers.* 2nd Edition. Washington, DC: Urban Land Institute.

Untermann, Richard K. 1984. *Accommodating the Pedestrian: Adapting Towns and Neighborhoods for Walking and Bicycling.* New York: Van Nostrand Reinhold.

Weant, Robert A., and Herbert S. Levinson. 1990. *Parking.* Westport, CT: Eno Foundation for Highway Traffic Control.

Table 1. Stall and Module Dimensions

		Stall Width		Stall Depth Parallel to Aisle		Aisle Width *Min.* (AW) ft	Minimum Modules	
		Par. to Car (Sw) ft	Par. to Aisle (WP) ft	to Wall (VPw) ft	to Interlock (VPi) ft		Wall to Wall (W_2) ft	Interlock (W_4) ft
Two-Way Aisle								
	90°							
Mix	G	10.00	10.0	18.4	18.4	24.0	60.8	60.8
	A	9.00	9.0	18.4	18.4	24.0	60.8	60.8
	B	8.75	8.8	18.4	18.4	24.0	60.8	60.8
	C	8.50	8.5	18.4	18.4	24.0	60.8	60.8
	D	8.25	8.3	18.4	18.4	24.0	60.8	60.8
Small	A	8.00	8.0	15.1	15.1	22.3	52.4	52.4
	B	7.75	7.8	15.1	15.1	22.3	52.4	52.4
	C	7.50	7.5	15.1	15.1	22.3	52.4	52.4
	D	7.25	7.3	15.1	15.1	22.3	52.4	52.4
Large		9.00	9.0	18.7	18.7	24.0	61.3	61.3
One-Way Aisles								
	75°					(+2 ft min. for 2-way aisle—see note)		
Mix	G	10.00	10.4	19.4	18.6	20.8	59.6	58.0
	A	9.00	9.3	19.4	18.6	21.0	59.9	58.2
	B	8.75	9.1	19.4	18.6	21.1	60.0	58.3
	C	8.50	8.8	19.4	18.6	21.2	60.0	58.4
	D	8.25	8.5	19.4	18.6	21.2	60.1	58.4
Small	A	8.00	8.3	16.2	15.4	20.0	52.5	50.8
	B	7.75	8.0	16.2	15.4	20.1	52.5	50.9
	C	7.50	7.8	16.2	15.4	20.1	52.6	50.9
	D	7.25	7.5	16.2	15.4	20.2	52.7	51.0
Large		9.00	9.3	19.7	18.9	20.9	60.4	58.7
	70°							
Mix	G	10.00	10.6	19.5	18.4	18.2	57.2	55.0
	A	9.00	9.6	19.5	18.4	18.6	57.5	55.3
	B	8.75	9.3	19.5	18.4	18.7	57.6	55.4
	C	8.50	9.0	19.5	18.4	18.7	57.7	55.5
	D	8.25	8.8	19.5	18.4	18.8	57.8	55.6
Small	A	8.00	8.5	16.4	15.3	17.9	50.6	48.4
	B	7.75	8.2	16.4	15.3	18.0	50.7	48.5
	C	7.50	8.0	16.4	15.3	18.0	50.8	48.6
	D	7.25	7.7	16.4	15.3	18.1	50.9	48.7
Large		9.00	9.6	19.8	18.7	19.1	58.7	56.4

Table 1. Stall and Module Dimensions (cont.)

		Stall Width		Stall Depth Parallel to Aisle			Minimum Modules	
		Par. to Car (Sw) ft	Par. to Aisle (WP) ft	to Wall (VPw) ft	to Interlock (VPi) ft	Aisle Width *Min.* (AW) ft	Wall to Wall (W_2) ft	Interlock (W_4) ft
One-Way Aisles (*Cont.*)								
	65°							
Mix	G	10.00	11.0	19.4	18.0	15.7	54.4	51.7
	A	9.00	9.9	19.4	18.0	16.1	54.9	52.2
	B	8.75	9.7	19.4	18.0	16.2	55.0	52.3
	C	8.50	9.4	19.4	18.0	16.3	55.1	52.4
	D	8.25	9.1	19.4	18.0	16.4	55.2	52.5
Small	A	8.00	8.8	16.4	15.0	15.7	48.5	45.8
	B	7.75	8.6	16.4	15.0	15.8	48.6	45.9
	C	7.50	8.3	16.4	15.0	15.9	48.7	46.0
	D	7.25	8.0	16.4	15.0	16.0	48.8	46.1
Large		9.00	9.9	19.7	18.3	16.6	56.0	53.2
	60°							
Mix	G	10.00	11.5	19.1	17.5	13.2	51.4	48.2
	A	9.00	10.4	19.1	17.5	13.7	51.9	48.7
	B	8.75	10.1	19.1	17.5	13.8	52.0	48.8
	C	8.50	9.8	19.1	17.5	13.9	52.2	49.0
	D	8.25	9.5	19.1	17.5	14.0	52.3	49.1
Small	A	8.00	9.2	16.3	14.7	13.6	46.1	42.9
	B	7.75	8.9	16.3	14.7	13.7	46.2	43.0
	C	7.50	8.7	16.3	14.7	13.8	46.4	43.2
	D	7.25	8.4	16.3	14.7	13.9	46.5	43.3
Large		9.00	10.4	19.5	17.8	14.1	53.1	49.8
	55°							
Mix	G	10.00	12.2	18.7	16.9	11.7	49.2	45.6
	A	9.00	11.0	18.7	16.9	11.2	48.7	45.1
	B	8.75	10.7	18.7	16.9	11.4	48.9	45.2
	C	8.50	10.4	18.7	16.9	11.5	49.0	45.4
	D	8.25	10.1	18.7	16.9	11.7	49.2	45.5
Small	A	8.00	9.8	16.0	14.2	11.5	43.5	39.8
	B	7.75	9.5	16.0	14.2	11.6	43.7	40.0
	C	7.50	9.2	16.0	14.2	11.7	43.8	40.1
	D	7.25	8.9	16.0	14.2	11.9	44.0	40.3
Large		9.00	11.0	19.1	17.2	11.7	49.9	46.1

Table 1. Stall and Module Dimensions (cont.)

		Stall Width		Stall Depth Parallel to Aisle		Aisle	Minimum Modules	
		Par. to Car (Sw) ft	Par. to Aisle (WP) ft	to Wall (VPw) ft	to Interlock (VPi) ft	Width *Min.* (AW) ft	Wall to Wall (W_2) ft	Interlock (W_4) ft
One-Way Aisles (*Cont.*)								
	50°							
Mix	G	10.00	13.1	18.2	16.2	11.4	47.8	43.7
	A	9.00	11.7	18.2	16.2	11.0	47.4	43.3
	B	8.75	11.4	18.2	16.2	11.0	47.4	43.3
	C	8.50	11.1	18.2	16.2	11.0	47.4	43.3
	D	8.25	10.8	18.2	16.2	11.0	47.4	43.3
Small	A	8.00	10.4	15.7	13.6	11.0	42.4	38.2
	B	7.75	10.1	15.7	13.6	11.0	42.4	38.2
	C	7.50	9.8	15.7	13.6	11.0	42.4	38.2
	D	7.25	9.5	15.7	13.6	11.0	42.4	38.2
Large		9.00	11.7	18.6	16.4	11.3	48.4	44.1
	45°							
Mix	G	10.00	14.1	17.5	15.3	11.0	46.1	41.5
	A	9.00	12.7	17.5	15.3	11.0	46.1	41.5
	B	8.75	12.4	17.5	15.3	11.0	46.1	41.5
	C	8.50	12.0	17.5	15.3	11.0	46.1	41.5
	D	8.25	11.7	17.5	15.3	11.0	46.1	41.5
Small	A	8.00	11.3	15.2	12.9	11.0	41.4	36.9
	B	7.75	11.0	15.2	12.9	11.0	41.4	36.9
	C	7.50	10.6	15.2	12.9	11.0	41.4	36.9
	D	7.25	10.3	15.2	12.9	11.0	41.4	36.9
Large		9.00	12.7	17.9	15.6	11.0	46.8	42.1

1. Figures 2 and 3 define dimensions used in Table 1.
2. Stalls angled between 90 and 60° confuse whether the aisle is one-way or two-way. Do not use angles between 90 and 75°. Some designers advocate using 75° stalls with two-way aisles because the right-hand-side parking maneuver is easier into an angled stall. However, making a left-hand turn to park in a 75° stall is difficult. A minimum of 22 feet is necessary for two-way aisles, and 24 to 25 feet allows ample walking space and occasional left-hand parking. Stalls at angles between 45 and 0° (parallel parking) are not generally advisable because they are often space-inefficient and confusing. The formula used to calculate Table 1 is given in Table 3; if special circumstances warrant, stall dimensions or angles other than those given may be calculated.
3. In the interest of legibility, metric dimensions are not given. The dimension in meters may be derived by dividing feet by 3.28. However, it may be better to recalculate, using Table 3 and appropriate metric stall dimensions.
4. Stall stripes are often painted 6 to 10 inches shorter than the stall depth to encourage drivers to fully pull into the stall.
5. The chart uses a minimum aisle width of 11 feet. This dimension is minimally sufficient to allow passage of cars and pedestrians. In high-turnover or special situations such as lots primarily serving the elderly or children, a pedestrian walkway and/or a wider aisle should be provided.
6. Sources: Adapted and recalculated from Ricker 1957; Weant and Levinson 1990; Box 1992; and Chrest, Smith, and Bhuyan 1996.
7. See Table 4 and Figure 4 for the design vehicles and clearances used to calculate this table.

Table 2 Turnover Categories and Stall Widths

Name		Stall Width Parallel to Car	Application
Mix*	G	10.00 ft (3.05 m)	Grocery stores and others that use shopping carts†
	A	9.00 ft (2.74 m)	Very high turnover rates—post office, convenience stores Areas with a high percentage (>20%) of passenger trucks
	B	8.75 ft (2.67 m)	High turnover rates—general retail
	C	8.50 ft (2.59 m)	Medium turnover rates—airport, hospitals, residential
	D	8.25 ft (2.52 m)	Low turnover rates—employee parking
Small	A	8.00 ft (2.44 m)	Same as categories above
	B	7.75 ft (2.36 m)	
	C	7.50 ft (2.29 m)	
	D	7.25 ft (2.21 m)	

* The mix of cars includes about 40% small cars, based on 1995 U.S. averages.

† There are a variety of grocery cart models ranging in width from approximately 18 to 32 inches. I recommend "hairpin" double stripes at least 2 feet wide for lots that use grocery carts. This 2-foot width is included in the 10-foot centerline-to-centerline dimensions.

Table 3 Formulas for Parking Stall and Module Dimensions

	Formula
Sw =	Given
WP =	Sw/sin @
VPw =	W cos @ + (L + cb) sin @
VPi =	.5Wcos @ + (L + cb) sin @
AW =	For lots with @ < critical angle, $AW = R' + c - \sin @[(r - Os)^2 - (r - Os - i + c)^2]^{.5} - \cos @(r + tr + Os - Sw)$ For lots with @ > critical angle, $AW = R' + c + \sin @[(R^2 - (r + tr + Os + i - c)^2]^{.5} - \cos @(r + tr + Os + Sw)$
W_2 =	AW + 2VPw
W_4 =	AW + 2Vpi

Critical Angle

Mixed Stalls		**Small Car Stalls**		**Large Car Stalls**	
Stall Width	**Critical Angle**	**Stall Width**	**Critical Angle**	**Stall Width**	**Critical Angle**
10.00	57.6	8.00	51.6	9.0	55.0
9.00	54.8	7.75	50.7		
8.75	54.0	7.50	49.8		
8.50	53.2	7.25	48.8		
8.25	52.4				

Critical angle = $\text{arccot} \{[(R^2 - (r + tr + Os + i - c)^2)^{.5} + ((r - Os)^2 - (r - Os - i + c)^2)^{.5}]/2Sw\}$.

Sw = stall width parallel to car; WP = stall width parallel to aisle; VPw = stall depth from wall parallel to aisle; VPi = stall depth from interlock parallel to aisle; AW = aisle width min.; W2 = module wall to wall; W4 = module interlock to interlock; @ = stall angle; W = design vehicle width; L = design vehicle length; cb = curb clearance .5 ft; R ft = design vehicle wall-to-wall rear radius; c = car clearance; Os = design vehicle side overhang; r = design vehicle curb-to-curb rear radius; i = walkway between cars; tr = design vehicle rear width.

For 90° lots, use 85% of the AW value given by the aisle width formula (see Weant and Levinson 1990, 160).

Aisle width formula from Edmund R. Ricker, *Traffic Design of Parking Garages,* Eno Foundation, 1957.

Table 4 Design Vehicle Dimensions

Type	**Mix**	**Small Car**	**Large**	**Passenger truck**
Length (L) inches	215 (5.46 m)	175 (4.45 m)	218 (5.54 m)	212 (5.39 m)
Width (W) inches	77 (1.96 m)	66 (1.68 m)	80 (2.03 m)	80 (2.03 m)
Wall-to-wall front radius (R) feet	20.5 (6.25 m)	18 (5.49 m)	20.75 (6.33 m)	
Wall-to-wall rear radius (R′) feet	17.4 (5.31 m)	15 (4.57 m)	17.5 (5.34 m)	
Curb-to-curb rear radius (r) feet	12 (3.66 m)	9.6 (2.93 m)	12.25 (3.73 m)	
Rear width (tr) feet	5.1 (1.55 m)	4.6 (1.4 m)	5.08 (1.55 m)	
Side overhang (Os) feet	0.63 (.192 m)	0.46 (.14 m)	0.75 (.228 m)	0.66 (.201 m)
Walkway between cars (i) feet	2 (.61 m)	2 (.61 m)	2 (.61 m)	2.85 (.87 m)
Car clearance (c) feet	1.5 (.46 m)	1.5 (.46 m)	1.5 (.46 m)	

SOURCES: Weant and Levinson 1990, 157; passenger truck data from Dare 1985.

Table 5 Stall Dimensions for Special Conditions

	Width	Length	Clr. Height
Designated large	9 ft (2.74 m)	18.5–20 ft (5.79–6.1 m)	
Passenger truck[1]	9 ft (2.74 m)	18.5 ft (5.64 m)	
Handicap car[2]	8 ft (2.44 m) + 5 ft (1.52 m) aisle	17.5 ft (5.34 m)	
Handicap van[2]	8 ft (2.44 m) + 8 ft aisle	17.5 ft (5.34 m)	8.16 ft (2.49 m)
Universal[3] (handicap car or van)	11 ft (3.35 m) + 5 ft (1.52 m) aisle		8.16 ft (2.49 m)
Valet[4]	7.5 ft (2.29 m)	17 ft (5.18 m)	
Europe typical[5]	7.83–8.16 ft (2.39–2.49 m)	15.58–16.42 ft (4.75–5 m)	
Bicycle[6]	2.5 ft (.76 m)	6 ft (1.83 m)	
Motorcycle[6]	3.33 ft (1.02 m)	7 ft (2.13 m)	

SOURCES: [1] Dare 1985.
[2] ADAAG.
[3] Bulletin #6.
[4] Burrage 1957, 242.
[5] Hunnicutt 1982, 650.
[6] Weant and Levinson 1990, 167.

Table 6 Lane Widths

Type	Width	Notes
Fire lane	20 ft (6.1 m) min. typ.	
Curb parking lane	6–8 ft (2.03 m)	10 ft (3.05 m) when also a traffic lane
Parking + traffic lane	18 ft (5.49 m) min.	
No-parking one-way	10 ft (3.05 m) min.	
No-parking two-way	16 ft (4.88 m) min.	

Table 7 Driveway Widths

Type	Entry Lanes	Exit Lanes	Total Width	Corner Radii
Commercial[1]				
Typical	1	1	22–30 ft (6.71–9.15 m)	15 ft (4.57 m)
Large volume	1 @ 14–16 ft (4.29–4.88 m)	2 @ 10–11 ft (3.05–3.35 m)	34–36 ft (10.37–10.98 m)	15 ft (4.57 m)
Very high volume	2 @ 10–11 ft (3.05–3.35 m)	2 @ 10–11 ft (3.05–3.35 m)	40–44 ft (12.20–13.41 m)	20 ft (6.09 m)
Residential[2]				
Driveway for 1 car			8 ft (2.44 m) max.	
Driveway for 2 to 10 cars			12 ft (3.66 m) max.	

SOURCES: [1]Box 1992, 206.
[2]Untermann 1984, 202.

Table 8 Design Percent of Lot Capacity Moving at Peak Hour (R)

	Morning Peak		Afternoon Peak	
Land Use	**In**	**Out**	**In**	**Out**
Office	40–70%	5–15%	5–20%	40–70%
Medical office	40–60%	50–80%	60–80%	60–90%
General retail	20–50%	30–60%	30–60%	30–60%
Convenience retail	80–150%	80–150%	80–150%	80–150%
Hospital visitors	30–40%	40–50%	40–60%	50–75%
Hospital employees	60–75%	5–10%	10–15%	60–75%
Residential	5–10%	30–50%	30–50%	10–30%
Hotel/motel	30–50%	50–80%	30–60%	10–30%
	Before event		After event	
Special event	80–100%		85–200%	

SOURCE: Adapted from Chrest, Smith, and Bhuyan, 1996.

Table 9 Peak Hour Factors (PHF)

Condition	PHF
Special events	.5–.65
Single lane	.75 max.
Two lanes	.85 max.
Three lanes	.95 max.

SOURCE: Adapted from Weant and Levinson, 1990.

Table 10 Design Capacity for Lanes (*U*)

	Entrance Vehicles/Hour/Lane		Exit Vehicles/Hour/Lane	
Control device	**Design Rate**	**Max. Rate**	**Design Rate**	**Max. Rate**
No stop	800	1,050	375	475
Automatic ticket dispenser	525	650		
Push-button ticket dispenser	450	525		
Coin/token-operated gate	150	200	150	200
Fixed fee to cashier with gate	200	250	200	250
Fixed fee to cashier with no gate	250	350	250	350
Variable fee to cashier			150	200
Validated ticket to operator			300	350
Machine-read ticket			375	425
Coded-card reader	350	400	350	400
Proximity card reader	500	550	500	550

SOURCE: Adapted from Weant and Levinson, 1990, 186.
NOTE: This table assumes an easy or straight approach with no sharp turns.

Table 11 Grades in Parking Lots

Grade	Condition
6% max.	Continuous slope in parking lot
12% max., 30 ft long	Nonparking automobile ramps with pedestrians allowed
15% max.	Nonparking automobile ramps with signs banning pedestrians
>6% change	A vertical curve transition is required;
1% min. / 2% rec.	Slope to drain asphalt
.5% min. / 2% rec.	Slope to drain concrete
2% (1:50) max.	Slope within accessible stalls

SOURCE: Adapted from Chrest, Smith, and Bhuyan, 1996; ITE, 1982; and Untermann, 1984.

Table 12 Required Number of Accessible Stalls

General Case

Total in Parking Lot	Required Minimum Number of Accessible Spaces	Source
1–25	1	ADAAG 4.1.3 (5) (a)
26–50	2	
51–75	3	
76–100	4	
101–150	5	
151–200	6	
201–300	7	
301–400	8	
401–500	9	
501–1000	2% of total	
1001 and over	20 + 1 per 100 over 1000	

No. of Accessible Spaces	Required Minimum Number of Van-Accessible Spaces	Source
1–8	1	ADAAG 4.1.3 (5) (a)
9–16	2	
17–24	3	
25–32	4	
33 and over	1 additional van-accessible per 8 accessible spaces	

Special Cases

Place	Requirement	Source
Medical outpatient units	10% of total stalls in lots serving visitors and patients.	ADAAG 4.1.2 (5) (d) and Bulletin #6
Medical units that specialize in persons with mobility impairments	20% of total stalls in lots serving visitors and patients.	ADAAG 4.1.2 (5) (d) and Bulletin #6
Valet parking	No stalls required. However, an accessible loading zone is required, and it is strongly recommended that self-park stalls be provided.	Bulletin #6

Table 13 Curb Parking

	Stall Length	Other	Source
Accessible loading	22 ft min. (6.71 m)	Platform 5 ft (1.53 m) wide, 20 ft (6.10 m) long, 9.5 ft (2.90 m) clear height	ADAAG 4.6.5 and 4.6.6
Truck loading	30–60 ft (9.15–18.29 m)	Add truck length per additional truck	Weant and Levinson
Drop-offs/taxi	50 ft (15.24 m)	Add 25 ft (7.62 m) per additional vehicle	Weant and Levinson
Paired (length per pair)	44–50 ft (13.41–15.24 m)	20 ft (6.10 m) stalls	Hunnicutt, 666 Hunnicutt, 666
Compact	19 ft (5.79 m)		Hunnicutt, 666
End stall	20 ft (6.10 m)		Hunnicutt, 666
Interior stall	22 ft–24 ft (6.71–7.32 m)		Hunnicutt, 666

Length of Curbside Bus Loading Zones

	Wheel Position from Curb			
	6 Inches (~15 cm)	1 Foot (~30.5 cm)	One 40-ft Bus (~12.2 m)	Additional per Bus
Upstream of intersection	L + 85 ft +(~25.9 m)	L + 65 ft +(~19.8 m)	105–125 ft (~32–38 m)	L + 5 ft +(~1.5 m)
Downstream of intersection				
Street width >39 ft (~11.9 m)	L + 55 ft +(~16.8 m)	L + 40 ft +(~12.2 m)	80–95 ft (~24.4–29 m)	L
Street width 32–39 ft (~9.75–11.9 m)	L + 70 ft +(~21.3 m)	L + 55 ft +(~16.8 m)	95–110 ft (~29–33.5 m)	L
Midblock				
Street width >39 ft (~11.9 m)	L + 135 ft +(~41.2 m)	L + 100 ft +(~30.5 m)	140–175 ft (~42.7–53.4 m)	L
Street width 32–39 ft (~9.75–11.9 m)	L + 150 ft +(~45.7 m)	L + 115 ft +(~35 m)	155–190 ft (~47.25–58 m)	L

Source for bus loading: Adapted from Homburger and Quinby, 1982. "Urban Transit" in *Transportation and Traffic Engineering Handbook*, 2nd edition.
L = length of bus.

Lynda H. Schneekloth, Editor

Summary

The green infrastructure of a city consists of those parts that contribute to the natural processes of keeping the water and the air clean and recycling of waste. It includes the parks and wild lands, stream corridors, utility corridors and vacant regenerating sites. These fragments of city property, if considered as a single system similar to transportation or waste treatment, offer opportunities for keeping our cities clean and for providing recreational space.

Key words

arboriculture, bioswale, conservation corridor, ecoroof, greenwalls, heat island, infrastructure, parks, porous paving, retaining wall, vegetation

Urban green infrastructure

Cities are a complex interaction of the natural and built elements. In order to maintain some degree of balance in the natural systems, we have to introduce engineered systems to create and transport energy, to remove and process wastes, to control storm runoff, and so on. This article presents details that reflect a rethinking in our conventional engineering responses, seeking ways to work more closely with natural processes to resolve some of the deficiencies and excesses that come with urban living. This approach is described by the term "green infrastructure."

1 GREEN INFRASTRUCTURE

The "green infrastructure" of a city is comprised of natural and designed systems and elements of the city that function in ways analogous to natural processes in managing air, water, microclimatic and energy resources. The most obvious part of this infrastructure are trees, open spaces of vacant lots, lawns and parks, and stream corridors, that is, all places that have water-pervious surfaces and/or soil to support plant material. Because it imitates natural systems, "green structure" is holistic and includes waterways and microclimatic systems that vegetation, land and water bodies create—essentially those parts of the urban system that are ecologically based (Fig. 1).

The green infrastructure performs ecological, recreational and aesthetic functions in the city. It improves the quality of the urban environment, provides access to natural habitats, avoids damage to the built form, and, in general, keeps all of us healthy. Moreover, wise use and expansion of green structure is cost-efficient at both the individual home/business level and for the municipality.

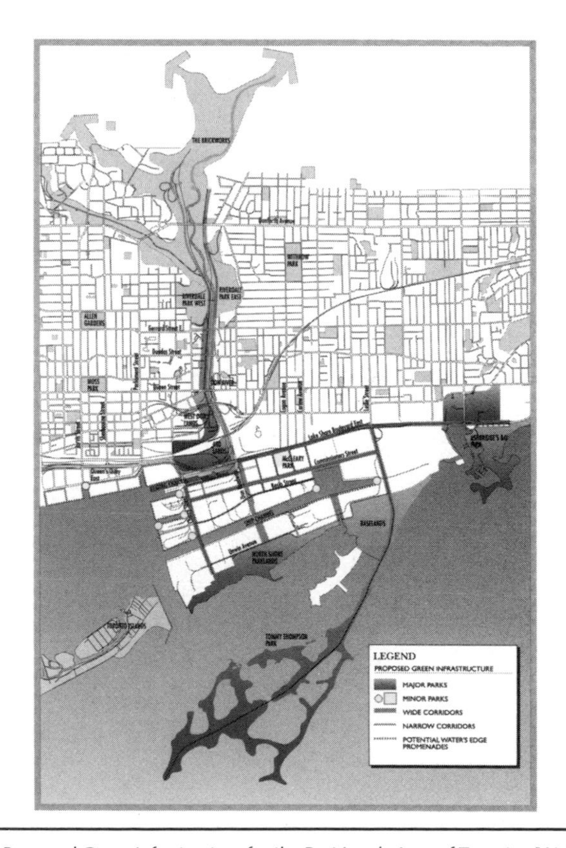

Fig. 1. Proposed Green Infrastructure for the Port Lands Area of Toronto. (Waterfront Regeneration Trust)

Credits: Material for this article is excerpted by permission from Hough, Benson and Evenson (1987) and Thompson and Sorvig (2000), cited in the References. Photos are by Lynda Schneekloth unless otherwise noted.

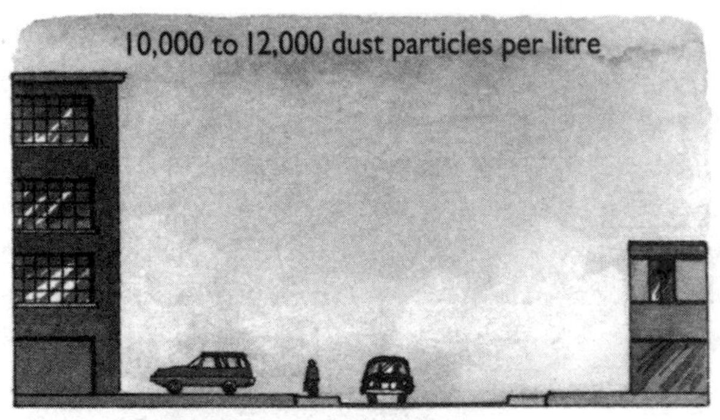

Fig. 2. Air quality benefits provided by trees. (Waterfront Regeneration Trust)

Air Quality Improvement

Vegetation reduces air pollution as it filters dust particles and pollutants attached to them (Fig. 2). For example, the 1993 Chicago Urban Forest Climate Project, a study of the cleansing function of Chicago's forest, found that the forest "removed an estimated 17 tons of carbon monoxide, 93 tons of sulfur dioxide, 98 tons of nitrogen dioxide, 210 tons of ozone, and 223 tons of fine particulate matter." (Hough, Benson and Evenson 1987). Trees also absorb carbon. In this study it was estimated that the 11% forest cover sequestered 155,000 tons of carbon each year.

Microclimate modification

Nonporous urban surfaces absorb and hold heat during warm weather, contributing to the "heat island effect," wherein temperatures can be between 8–10% hotter than the surrounding countryside. Relatively minor green projects can make a significant difference in both pollution control and heat reduction. An EPA Urban Heat Island Mitigation Initiative study of Los Angeles suggested that increasing green space by five percent in Los Angeles and replacing dark roofs and asphalt with lighter surfaces including green roofs, could lower overall temperatures by 4°F—resulting in significant energy savings and 10% less smog. In other words, the green infrastructure of a city is a natural air conditioner (*cf.* Lawrence Berkeley Laboratory 2002). The greater its coverage and canopy, the greater the benefits.

If strategically planted, trees serve as windbreaks, in part by lifting strong seasonal winds up and over the leeward structures and by breaking down strong wind patterns, important in winter cities where the chill factor greatly increases the sense of discomfort and the use of the outdoor space. Further, winter wind is responsible for drifting snow patterns. Vegetation can serve as shelterbelts if property designed, controlling where snow will accumulate.

Stormwater management

One of the most important benefits of the green infrastructure is in naturalizing the hydrological cycles in a city. The hard surfaces of the urban fabric increase the intensity of the run-off and the amount of pollutants in urban waters. Instead of water soaking into the ground, it travels quickly into storm drainage systems that flow into rivers and streams, causing increased flooding and erosion. The green fabric, on the other hands, absorbs the water at the source, recharging the groundwater, filtering pollutants, and slowing down the energy of water travel. This improves water quality and, as a nonstructural approach, is cost-effective.

Fig. 3. Tifft Nature Preserve is a constructed/preserved area immediately adjacent to downtown Buffalo. It is comprised of a wetland that dates back to the ice ages, reconstructed industrial slips, and a covered municipal landfill.

Biodiversity

The urban environment is home to more than human beings, and one might argue that the more we encourage wildlife in the city at appropriate places, the more varied and enriched will be the quality of daily life. A rich variety of birds and animals is an indicator of a healthy environment. Wildlife in the city moves through riparian corridors along rivers and streams and large parks that have areas of native vegetation. The health of these habitats, however, depends to a large degree on their size and connectivity—one of the reasons for doing a green inventory of a city to locate significant areas for wildlife habitat, sanctuaries and corridor links between natural conservation systems.

Recreational opportunities

One of the most visible and important functions of the green structure is for recreation, *e.g.*, in addition to using riparian corridors as flood and erosion control and habitat links, these are prime areas for bicycle trails and nature hikes. Major parks, with large and diverse ecological systems, provide parks for active recreation and sports fields, but also for passive recreation, bird watching and school field trips for science classes. Each part of the urban green fabric should be considered as a multi-use structure (Figs. 3–5).

Green infrastructure inventory

To take advantage of urban green infrastructure benefits, an inventory documents where the green spaces of the city exist, what ecological systems are there, and how they are connected. Land use categories assist in researching the green structure, but their definitions and ownership patterns are often inadequate to generate a picture of the green systems as an infrastructure.

Land use categories that contribute to the green infrastructure include parks, waterways, cemeteries, church and school open spaces, city farms and community gardens, utility corridors, rail lines, quarries, vacant lands and even brownfields, which need to be identified and hopefully remediated. Small city lots, street trees and backyards are also important. Street trees are an important part of the urban forest, but because of their limited space, do not provide all of the functions of a green space. Backyards, as well, are very important parts but because they are private and out of view are difficult to inventory.

An aerial survey provides a basis of an inventory and can be combined with windshield surveys and the generalized land use analysis. The rough identification of such spaces does not necessarily yield information on the types or quality of those environments, but gives a preliminary indication of green areas and watersheds that run through the urban fabric.

Regulatory provisions exist for the preservation of green space such as setbacks, easements, special review districts, public ownership, and so on. With an inventory, it is possible to change certain planning, design, and construction patterns to increase the green structure at the urban and site scale. One of the main opportunities is in the management of urban waters such as stormwater management and stream corridor restorations.

Two complementary urban design strategies—less paving and more vegetation—help increase the viability of the green infrastructure.

2 REDUCING PAVING

Hardscape paving has been implicated in a wide range of ecological problems. Most paving materials create surface stability by excluding water from the soil, and this impermeability causes a number of difficulties. Soil absorbs rainfall and nurtures flora, fauna, and humans, but impervious surfaces increase runoff, causing erosion and flooding, depleting soil water, and contributing to siltation and water pollution.

Modern construction has created such vast nonporous areas that many communities are being forced to limit the creation of new impervious surfaces. This hardening of the landscape results in a net

Fig. 4. Native Plants are found throughout the city along waterways and rail lines. To preserve these plants it is often necessary to provide maintenance initially to remove aggressive and invasive species.

Fig. 5. Aerial photograph of a section of the Buffalo River that runs through the center of Buffalo demonstrating the relationship between residential areas, railroad corridors, a wetland and riparian habitat surrounded by regenerating industrial brownfields. All of these are a part of the city's green structure.

Fig. 6. A gravel parking lot in an area without heavy traffic supplements a more impervious surface parking area.

Fig. 7. A dense urban parking area uses diverse paving types, some of which allow water to filter into the ground.

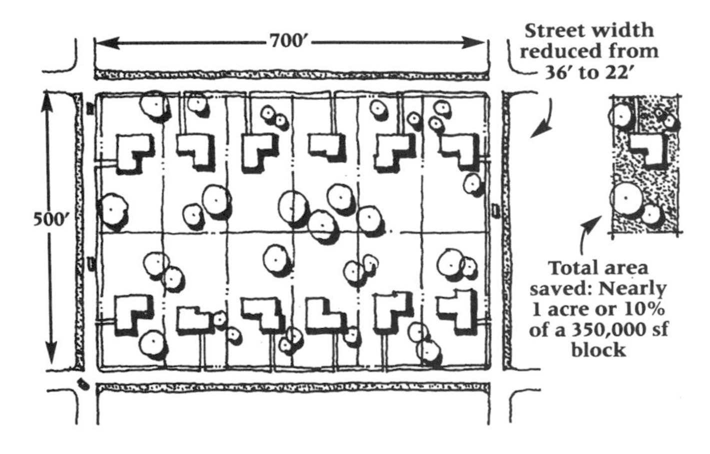

Fig. 8. One acre of developable or protected land is saved by reducing street width from thirty-six feet to twenty-two feet around an eight-acre city block. (Thompson and Sorvig)

reduction in the biologically productive surface of the earth as areas of pavement replace cornfields, meadows, forest, or desert. Moreover, paving consumes non-renewable resources both in building the lots and in the fuel required to truck the materials to the site. Asphalt, the material for most parking lots, is a complex mix of hydrocarbons, the mixing and application of which is an air-polluting act in itself.

Site planning policies can help to avoid unnecessary paving. Some opportunities include:

- **Density zoning.** Local policy that uses overall density (a number of units per acre, or a percentage of acreage devoted to structures) works better than minimum lot sizes, because it allows flexible adaptation to site topography.

- **Cluster development.** Placing several buildings together surrounded by open space, rather than each in the center of its separate lot, can greatly reduce infrastructure costs, including paving.

- **Combined land uses.** Zoning that allows residences and workplaces to coexist makes walking, biking, or public transit much easier for workers. This is often a matter of removing barriers to coexistence from existing zoning laws.

- **Impervious surface limits.** Set a maximum percentage of the site area that can be impervious. This must include both paved and roofed areas, existing and new. Where this level is set to 10% or lower, streams and other hydrological features of the area can be considered protected. Above 10%, impacts are serious enough to require mitigation; and where 30% of the area is impervious, degradation of the ecosystem is almost inevitable. In urban areas already far over this threshold, incentive programs for reducing impervious cover can be effective.

- **Street width limits.** Oversized roads also have negative effects on traffic safety and diminish the quality of life for communities through which they pass. Current research shows that the real cause of most accidents, and especially of serious injury accidents, is speed itself, and that wide, straight, flat roadways encourage drivers to speed.

- **Planted islands in turnarounds.** Paving the center of a turnaround is of no use to drivers and can be replaced by permeable, planted surfaces as a matter of policy.

- **Pollutant collector.** These include paving at gas stations, car washes, dumpster pads, and other point sources of concentrated pollutants. Isolating runoff from contaminated sources keeps the flow on ordinary streets much cleaner.

- **Storm drain inlet labeling.** Knowledge of where pavement runoff goes can decrease public dumping of pollutants onto pavement and into drains.

Parking areas are often designed to meet a single "peak day" (or even peak hour) projection. To change this standard practice will require code changes. Therefore, the ongoing goal of reducing the total amount of new parking is largely a task for local government.

One problem with conventional construction of parking lots is to require clearing the site of trees and leveling the landforms. A design approach that is appropriate for forested and other sensitive sites is to scatter the parking throughout the site. Not only does this approach require much less cutting of trees and disturbance of the forest floor, it allows stormwater from the slender roadway and scattered parking spaces to run directly onto the woodland floor (there are no curbs or gutters) and soak in.

Furthermore, not all parking areas need to be paved. In fact, many lightly used parking lots would be better (from a water-quality standpoint) if they were surfaced with a more permeable material like gravel. *No ground surface should be any more impervious than necessary.* Even if the paving itself remains hard, approaches like eliminating gutters and curbs, infiltrating stormwater in planted areas, or using porous pavement can legitimately be thought of as "softer" than conventional engineering (Figs. 6–10).

Bioswales

Beyond (or instead of) the curb, install grassed or vegetated areas called *bioswales*—linear, planted drainage channels (Fig. 10). A typical bioswale moves stormwater runoff as slowly as possible along a gentle incline, keeping the rain on the site as long as possible and allowing it to soak into the ground—contrary to conventional engineering practice. At the lowest point of the swale, there is usually a raised drain inlet that empties any overflow (during particularly heavy storms) into the nearest waterway. Along with the infiltrating function, bioswales cleanse runoff via their plants and soil microbes.

Bioswales function particularly well in parking lots, which generate runoff laden with pollutants that drip from cars and collect on the parking lot surface. In addition to their decontamination functions, these bioswales create a place for lush plantings amid the parked cars. Wetland plants—cattails, bulrushes, yellow iris, and others, mostly natives—slow the water while biologically breaking down pollutants. Contaminants that escape this gauntlet are captured in the topsoil, where soil microorganisms attack them. Thus filtered, the stormwater seeps through the subsoil into the underlying water table.

Porous paving materials

Another way to decrease the stormwater impacts of paving is to make the pavement more permeable so that infiltration occurs through the surface of the paving itself (Fig. 11).

• **Porous Asphalt and Porous Concrete.** These are materials similar to those that go by a variety of names: no-fines paving, pervious paving, permeable paving, and *percrete* (for "percolating concrete"). Stone aggregate is held together with either asphalt or Portland cement as a binder; some high-tech versions have used epoxies to bind the stone. Porous paving is strong enough for parking pedestrian use, and some road surfaces. In addition to its ecological advantages, porous paving can save construction, real estate, and maintenance costs, because it doubles as a stormwater system and eliminates storm drains. It may be twelve to thirty-eight percent cheaper overall.

A second, greater savings occurs where a porous paving reservoir system substitutes for open stormwater detention, retention, or infiltration basins. The land area otherwise required for the basins is freed for other uses. A third advantage is reduced maintenance costs, particularly where snow removal is significant. Snow that falls on porous paving tends to melt quickly and drain into the pores. The size and depth of the reservoir must be designed to fit site conditions: soil permeability, slope, and the local design storm.

Fig. 9. Scattered parking preserves trees creating more pleasant parking area and retaining the good services of the trees.

Fig. 10. A bioswale filters, slows, and infiltrates runoff from parking. The raised grate overflows to the storm sewer only in very heavy storms. (Thompson and Sorvig)

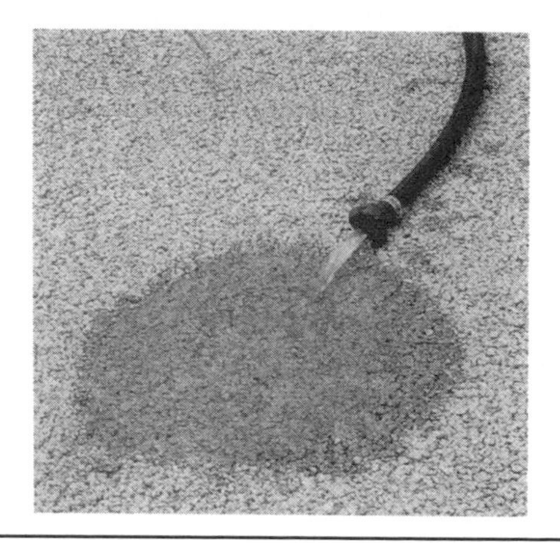

Fig. 11. Porous concrete, shown here, and porous asphalt support vehicles but permit water to infiltrate rapidly and snow to melt quickly. (Thompson and Sorvig)

Fig. 12. Grasscret™ pavers are used in residential parking in Aarhus, Denmark.

Fig. 13. Unit pavers at Cranbrook.

Fig. 14. In its undisturbed condition, the shape of a tree resembles a wine glass set on a plate. (Moll & Young)

• **Grass pavers.** Grassed paving systems allow turf grass to grow through an open cell of concrete or plastic that transfers the weight of vehicles to an underlying base course. According to one manufacturer's study, every 1,000 square feet of grass paving infiltrates nearly 7,000 gallons per 10 inches of rainfall, which would otherwise be runoff; converts enough carbon dioxide to oxygen to supply 22 adults for a year; provides cooling equivalent to 1.7 tons of air conditioning annually; and, in the case of one manufacturer, recycles more than 400 pounds of plastic in the product itself. Grassed paving is somewhat limited in its applications because grass will not survive constant daily traffic.

Three general types of grassed paving systems exist: poured-in-plastic systems such as Bormanite's Grasscret™ consist of steel reinforced concrete; precast concrete pavers that like poured-in-place concrete provide rigid structural support; and a large number of the available systems as recycled plastic pavers (Figs. 12 and 13).

• **Unit Pavers.** Another potentially permeable surface uses unit pavers (pavers set as individual pieces, rather than a continuous sheet like poured concrete). They must be laid on sand, crushed stone, stone screenings, or some other permeable material. Because the percolation actually takes place in the joints between the pavers, the width and the material of the joints become critical.

- Use wide joints.
- Use thicker pavers to compensate for loss of rigidity, if necessary.
- Use permeable joint-filler materials.
- After initial installation, settling of the paving occurs; brush in more coarse joint-filler materials, rather than allow finer debris to accumulate and block the pore space.
- Where possible, leave joints lower than the walking surface.
- Make the base course beneath the pavers as coarse as possible to prevent water being retained in the surface layer.
- Do not compact the base course excessively.

• **Cool asphalt with plants reflective surfaces.** Conventional parking lots, as noted earlier, are a major contributor to the heat islands in American cities. There are two ways of reducing the heat increases from paving. The first is to plant shade trees. The second approach is to increase the reflectivity—measured by the reflectivity index of albedo—of pavement and thus reduce its heat absorption capacity. Asphalt can be lightened in several ways. One is to specify that the mix include light colored stone, both aggregate and fines. Color coating asphalt developed in large part as a decorative system but has promising environmental possibilities. These can give asphalt a surface of almost any color, and when light colors are used, they will make paving less heat-absorptive.

3 URBAN VEGETATION

A corollary to the principle of minimizing urban hardscape is to replace impervious surfaces with more water absorbing materials. One of the most obvious is vegetation. This section will discuss some opportunities regarding vegetation.

Trees

Trees are the most obvious part of the urban landscape, both in their presence and in their absence (Figs. 14–16). There has been resurgence in tree planting in urban environments, many of which were devastated by the Dutch Elm disease. Moll and Young (1992) summarize the economic worth of a tree. "A single tree would provide this much dollar value benefit for one year: air conditioning, $73; controlling erosion and stormwater, $75; wildlife shelter, $75; and controlling air pollution, $50. The total is $273 a year. Compounding this amount for fifty years at 5%, the grand total is $57,151."

Moreover, urban trees are more valuable than their country cousins because they enhance real estate. In some cases, the value of a urban lot can increase by twenty percent if it has trees, and on average the increased value is between five and seven percent. Looked at nationally, this can add an extra $5000 per lot. In the same way that trees create a value for lots, they add value to neighborhoods in general—the American dream is to live on a tree-lined street.

Landscape plants represent a significant financial investment, whether purchased from a nursery, transplanted, or protected on-site. Healthy plants, and the construction that keeps them that way, are essential to the functional, ecological, and aesthetic success of a built landscape.

Tree planting standards

Inadequate planting structures, particularly ones with too little soil volume, are the leading cause of an epidemic of urban street tree deaths. The average life span of urban trees has been estimated to be as short as two years, and few experts give them longer than ten years. These are trees that could live fifty years or more in suburban settings or in the wild.

A widely accepted *minimum* is 300 cu. ft. (1.4 cu m), that is, a pit 10 ft. square x 3 ft. deep (approx. 3 m square x 1 m deep). This is much more than many street trees ever get, yet it is truly adequate only for trees whose mature trunk Diameter-at-Breast-Height (DBH) is less than 6 in. (15 cm). For a 24 in. (60 cm) DBH tree, about 1,500 cu. ft. (about 43 cu. m) of soil are recommended—a pit about 22 x 22 ft. by 3 ft. deep (about 6.7 m square x 1 m deep). (Increased *depth* is not of much value to most plants, since root growth stays mainly in the top foot of soil.) This relationship between the amount of tree canopy and the root-supporting soil volume is critical in determining long-term tree health. Available root volume may be even less than it appears at the surface. Utility lines frequently run through tree pits; steam lines are particularly lethal, but all utility lines steal space from already inadequate root volume.

New standards of arboriculture have introduced two methods of designing street-tree pits: the "continuous trench" structure and the "root path trench."

• **For the continuous trench structure**, soil under pavement is deliberately compacted for engineering support of the sidewalk or traffic lane. This subsoil creates a wall around the conventional pit, often as hard as concrete. The continuous trench stretches from tree to tree, under paving strengthened by reinforcement. This design greatly increases the soil volume available to each tree. It also requires slightly different details of sidewalk construction, which any experienced contractor can readily learn. Variations on the design are used for plazas, sidewalk plantings, and other urban situations.

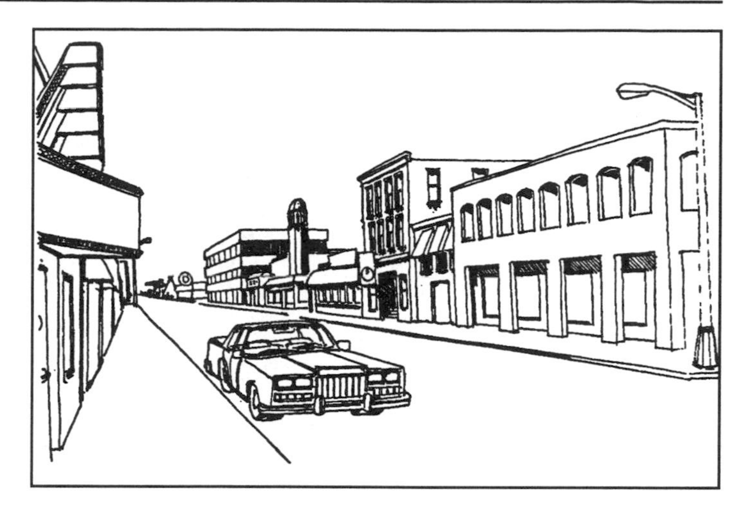

Fig. 15. The same urban street with and without trees. (Moll & Young)

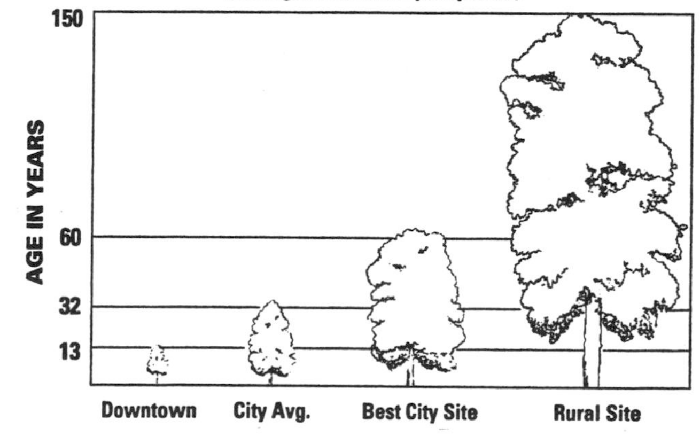

Fig. 16. Average life of street trees. (Moll & Young)

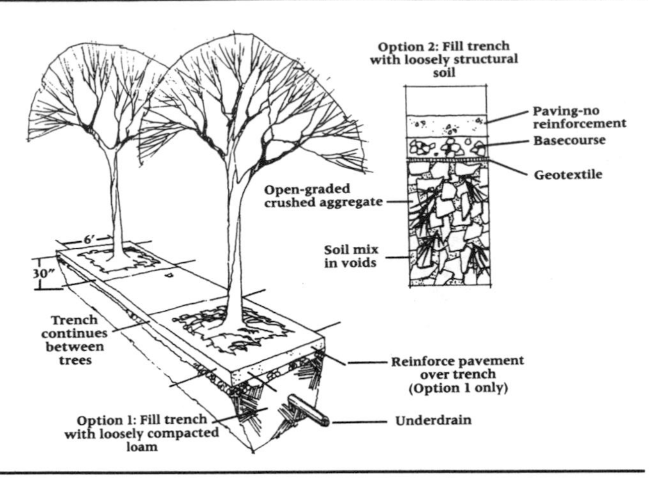

Fig. 17. A diagram of the continuous trench-planting pit that greatly facilitates the survival rate of trees because it permits sufficient root space. (Thompson and Sorvig)

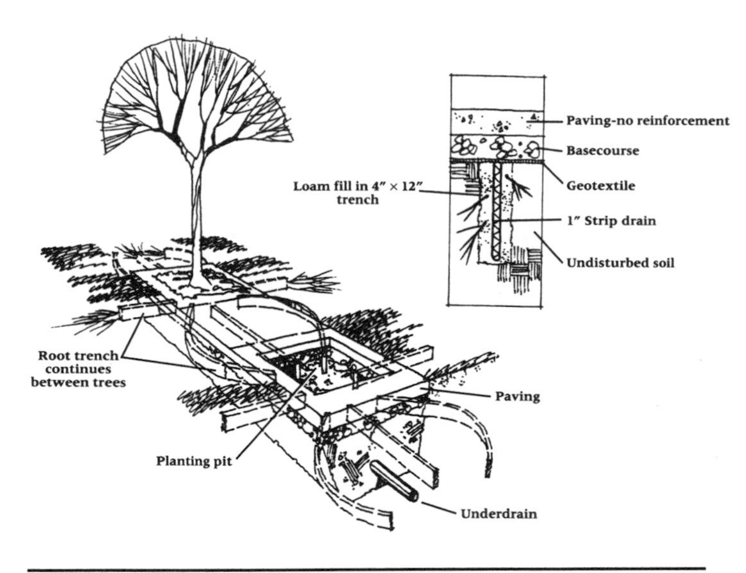

Fig. 18. A diagram of the root path trench pit that requires less excavation than continuous trenches, but still provides air and water "paths" that lead root growth. (Thompson and Sorvig)

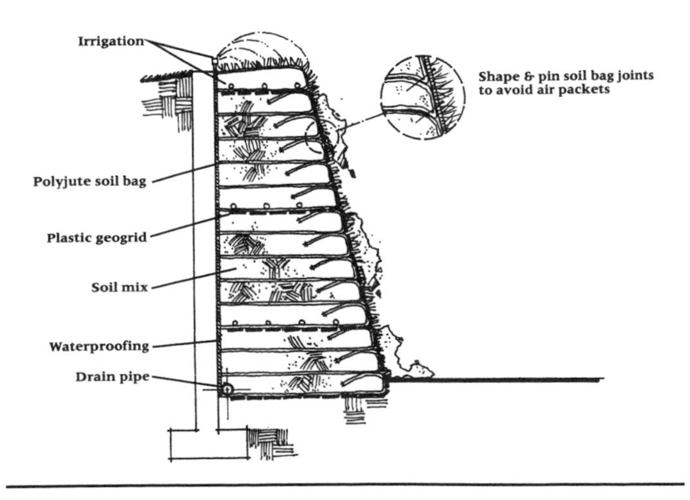

Fig. 19. A greenwall against a structure, designed for zoo use by CLR Design. (Thompson and Sorvig)

- **The root path trench system** leads the roots out of the pit in small radial trenches, about four inches wide by twelve inches deep. Each trench contains a drainlike product, a plastic "waffle" core wrapped in geotextile, which brings both water and air through the length of the trench. Surrounded by good planting soil, this air and water source provides the conditions roots need to grow, and thus the roots follow the trench. The surrounding soil must be good enough for them to spread eventually but does not need to be replaced wholesale with planting mix.

The standard details for planting trees and shrubs have changed. In particular, the recommended size and shape of planting holes have changed. An older standard of "twice the width of the root ball" is now considered a minimum. In good soils, a shallow pit just 6 in. (15 cm) deeper than the root ball, but at least 6 ft. (1.8 m) wide, is now preferred. In poor, clayey, or compacted soils, the minimum width of the pit increases dramatically, to 15 to 20 ft. (4.6 to 6 m) wide.

The bottom and sides of the pit must be roughened, so that existing and filled soil will bind together. Clay soils, in particular, will "glaze" when dug, making a surface that stops roots as effectively as a pot and creating "virtual container" conditions. Soil for filling the pit should be amended with compost or other organic matter up to about five percent by weight. Making the soil too rich can discourage the roots from leaving the pit. Many experts now believe that staking and wrapping the trunk of trees should be avoided. Ideally, the tree should be oriented in the same direction it was growing in the nursery.

Caution about planters, raised beds, and containers

Any plant grown in a container or planter is under more stress than is the same species planted in the ground. The limited soil volume in the container tends to dry out, heat up, or freeze quickly and can easily become waterlogged or deficient in nutrients. Containers are most often set on hard surfaces, which stress the plant with excess heat. These stresses make container plants particularly hard to sustain, whether in a plaza or a roof garden. Subjected to extremes of climate, plants that are *also* stressed by undersized planting structures seldom survive (Figs. 17 and 18).

Greenwalls

Greenwalls offer effective alternatives to conventional landscape retaining walls of cast-in-place concrete, metal, or wood (Fig. 19). A vegetated surface suits many functions and aesthetic preferences: it deadens and diffuses noise, makes graffiti impossible, cuts heat and glare, holds or slows rainwater, traps air pollutants, and processes carbon dioxide, while providing food and shelter for wildlife. Most greenwalls use small, light elements, installed without heavy equipment. Many require reduced materials, no formwork, and for some types no footings, saving money and resources. Most deal flexibly with unstable soils, settling, and deflection—even earthquakes. Careful attention to irrigation and microclimate is richly repaid.

The choices for greenwall structures include:

- **Block** – engineered with gaps where plants root through the wall.
- **Crib wall** – concrete or wood elements stacked log-cabin style. A related stackable unit looks like giant jacks from a child's game.

- **Frame** – interlocking circle-or diamond-shaped units stacked like masonry (mostly in Europe and Japan). Also used flat to "blanket" water channels. For parking, Grass-crete is a similar concept.
- **Trough** – stackable soil filled tubs (retaining or freestanding).
- **Gabion** –wire baskets filled with stones to provide a strong but permeable wall or dam.
- **Mesh** – like minigabions, holding a thin layer of soil to a surface.
- **Cell** – flexible, strong honeycombs filled with soil. Closely related are plastic turf support systems like Grasspave.
- **Sandbag** – geotextiles wrapped around soil, formally called "vegetated geogrid."

Bioengineering weaves together living woody plants for structural strength; inert materials are usually secondary. Greenwalls contrast with other closely related methods in one important way. They derive their strength *primarily* from the inert part of the system; planting protects the surface and adds some strength. Both have advantages and are often combined (Fig. 20).

Greenwalls are as effective as conventional structures for slope retention in almost all situations. Planted surfaces offer habitat, making greenwalls much more attractive for many sites over and above their ecological advantages as plant cover. As a general rule, any structure used to stabilize a slope should not be monolithic but rather an open system around and through which plants may root and establish themselves.

The simplest plantable retaining structure is a *dry stone wall*, usually lower than six ft. (1.8 m) in height, constructed against the toe of a slope in a single course one rock wide. The wall is built by stacking local stones on top of one another. For taller structures, *gabions* are an alternative to stone walls.

A somewhat more sophisticated retaining structure is a *crib wall*—an open-faced, interlocking structure of wood or concrete beams assembled log-cabin style and embedded in the slope. Normally, crib walls are battered to improve stability, although vertical crib walls may be specified with appropriate foundations. The beauty of the crib wall is that there are ample openings between the beams through which plants grow. Where walls must be extremely high or nearly vertical, concrete may be a better choice.

Any wall can be draped with trailing or climbing plants, especially if terraced. A true greenwall has plants growing *on its surface*, which requires soil spaces. There are two ways of achieving this: leaving out blocks in each course, or rounding the corners of each block.

Flexible soil support systems

An entirely different concept for greenwalls relies on flexible materials rather than masonry to make soil stand upright. Mesh, honeycomb, or fabric, these flexible materials are filled with soil. The weight of the soil prevents the support material from moving, and in turn, the support keeps the soil from slumping.

Naturalistic greenwalls suit zoos or historic themes. In other settings, greenwalls could be ornamental, patterned with colored sedums or

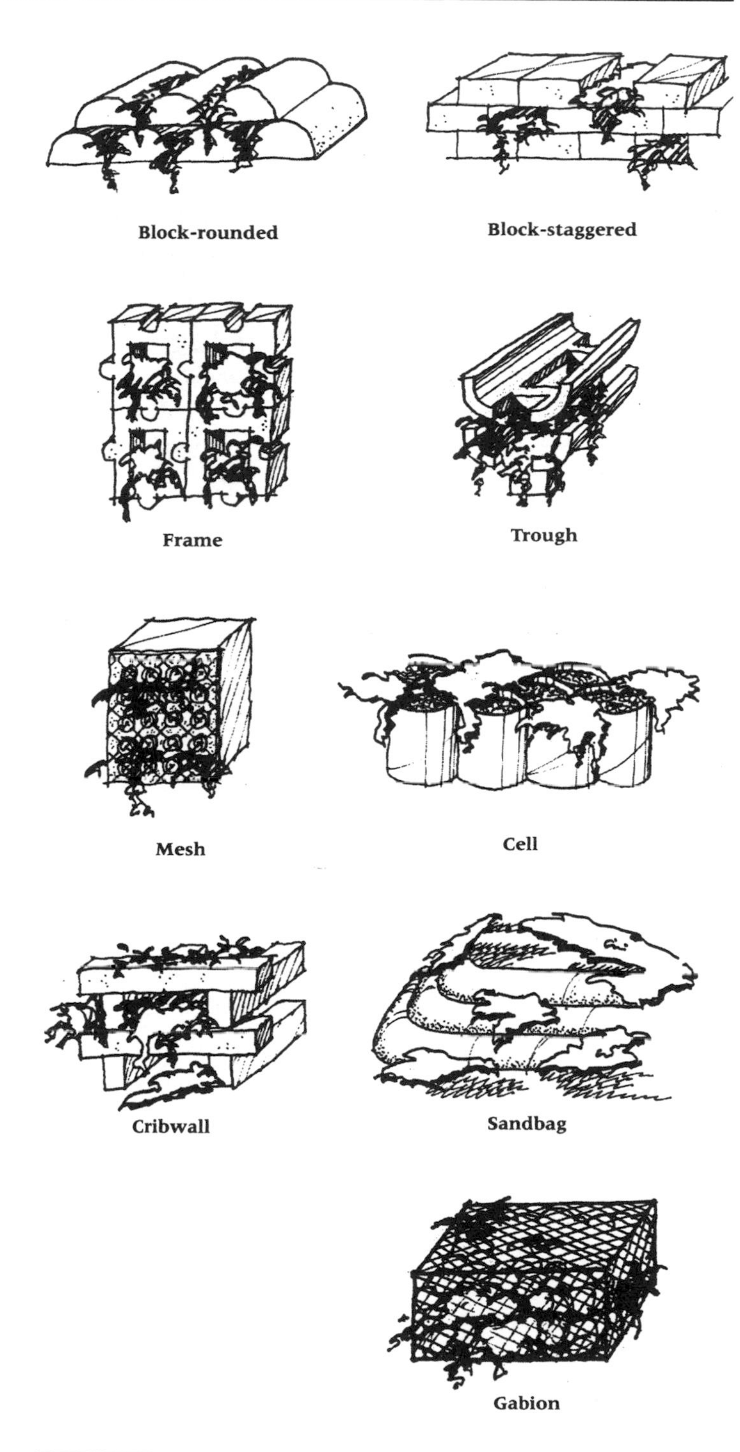

Fig. 20. Greenwalls combine bioengineering with a variety of hard structures: several basic concepts are diagrammed above. (Thompson and Sorvig)

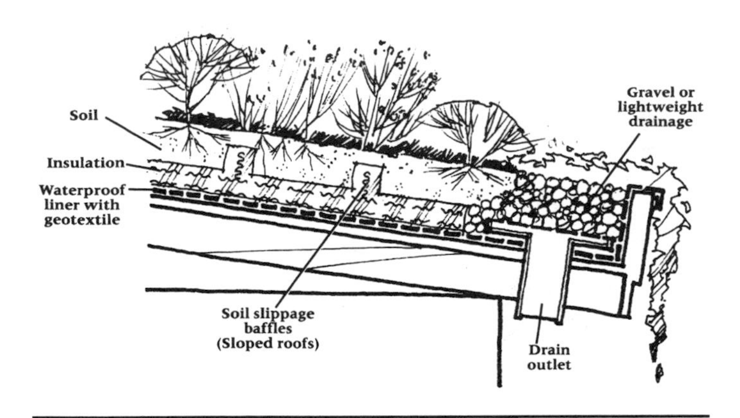

Fig. 21. Unlike conventional roof gardens, ecoroofs are light enough to retrofit on existing structures. (Thompson and Sorvig)

Table 1. Ecoroof Costs (Based on European experience)

Materials	$/sq ft
Fleece layers	$0.45
Root protection mats	$0.74
Waterproof seal	$0.74
Soil mixture	$0.60
Plants and seed	$0.30
Total	$2.83*
Additional Option	
Heat insulation	$1.49
Drainage layer	$0.52

*Plus incidentals: sealants, clamps connectors, edge lumber, etc. Contractor fees not included.

Source: Beckman *et al.* 1997

blooming displays. One limitation: any fabric-reinforced system relies on the weight of fill and needs a wide footprint. For this reason, they are best used where the slope is fill or to cover built walls.

Cellular containment

Honeycombs of heavy polyethylene sheet are shipped folded flat, expanding when pulled like crepe-paper holiday decorations. Once staked at the edges, the expanded cell sheet is strong enough to walk on as workers fill cells with soil. Each cell is about 8 in. (20 cm) square, available in 2- to 8-in. (5 to 20 cm) depths.

A single layer of cells can be laid over an existing slope for stabilization; filled with gravel, it substitutes for paving. Cells are available with perforated sidewalks used to stabilize stream crossings. To make a greenwall, cell sheets are laid horizontally on top of one another, stepping upward as steeply as a 1-to-4 horizontal-to-vertical slope. All but the edge cells are filled with gravel; edge cells, exposed by the stepping structure, are filled with planting soil. The polyethylene edge of each layer remains exposed but is quickly covered by plants.

Designing greenwalls

Design considerations for greenwalls include:

- Microclimate on any vertical surface depends on compass orientation and is usually severe – hot/sunny, cold/shady, or alternating daily.
- Irrigation can be sprayed onto the wall, channeled down from the top, or (using drippers) run on or behind the face.
- Soil mix and plant selection are critical.
- Especially if the greenwall covers a building, plan scrupulously for maintenance of the underlying structure.
- Costs are often twenty-five to fifty percent less than cast-in-place concrete, but they can only truly be compared design by design.
- Be sure to plan for maintenance during plant establishment.

Ecoroofs

Every contemporary city has, in the words of Toronto environmental designer and author Michael Hough, "hundreds of acres of rooftops that for the most part lie desolate and forgotten." Hough's description is true even of economically vibrant cities: at ground level, they are lively, but at roof level, lifeless. Conventional roofs are impervious to water and exposed to high winds; they cause severe microclimates by absorbing or reflecting heat.

Roof gardens—at least as conventionally seen as rare exceptions and thus of minimal effect in most urban areas—do not adequately address the problem of sterile roof expanses. Conventional roof gardens are typically prohibitive to place on existing roofs, due to the added weight of earth required.

There are alternatives to conventional roof gardens that address the issues of low maintenance, hardiness and light weight. Variously known as *ecoroofs*, *green roofs*, or *extensive* roof gardens (as opposed to conventional *intensive* roof gardens), this new model has emerged from the crowded cities of northern Europe. Ecoroofs typically cover the entire roof of a building with a continuous layer of growing medium, as thin as 50 millimeters (about 2 in.), that supports low-maintenance vegetation. In concept they are a lightweight modern version of the sod roof, a centuries-old tradition in Scandinavia. They are not intended to be walked upon and generally do not feature pedestrian access.

Ecoroofs require little additional load-bearing capacity and may be retrofitted to many existing buildings with no modification in their structural systems (Fig. 21). They do not require flat roofs as do conventional roof gardens but may be installed on roofs with slopes up to thirty degrees from the horizontal. In contrast to conventional roof gardens, whose gardenesque plantings require irrigation, fertilization, and intensive maintenance, ecoroofs typically require little or no irrigation (at least in temperate climates) and no fertilizer. This results from specification for each location of a very densely rooted layer of turf grasses and plant mix appropriate to the regional microclimate of the installation. The thick matting tends to all but exclude intrusion of windblown "weeds."

The requirements of an ecoroof are relatively modest, yet the environmental benefits are considerable:

- Improves the building's thermal insulation.
- Reduces the urban "heat island" effect, by absorbing less heat.
- Produces oxygen, absorbs carbon dioxide, and filters air pollution.
- Stores carbon.
- Provides wildlife habitat, especially for birds.
- Absorbs up to 75% of rain falling on it, thus slowing storm water runoff.
- Ecoroof materials and approaches.

Ecoroof underlayments are similar to conventional roof gardens. Like other roof gardens, ecoroofs feature the following layers: a waterproof membrane, a layer of insulation, a drainage layer, and the growing medium, sometimes referred to as the "substrate." In their soil conditions ecoroofs differ markedly from conventional roof gardens, which rely on significant depths of high-quality soil. Ecoroofs generally make do with poor and relatively thin growing medium, adequate for wildflowers. This microenvironment requires plants that do well in the thin, nutrient-poor substrate such as xerophytic plants. This reduces the need for irrigation; in fact, many authorities stress that irrigation is unnecessary for ecoroofs unless there is an extended dry spell. If irrigation is a must, consider use of greywater, treated effluent, or rainwater harvesting.

Native plants

There has been a resurgent interest in the use of "native" plant material in the United States for everything from street trees to parklands and as a replacement for the ubiquitous lawn. Part of this strategy is the preservation of existing, intact ecosystems found in the city, and the restoration of native plant communities wherever possible.

Although there is much controversy about an exact definition of *native plants*, the general idea is to use plants that grow sustainably in a region either because they have been there for a very long time, co-evolving with insects, soils, animals and other plants, or that have "naturalized," that is, have found a niche to occupy. Native plants are more sustainable because they are adapted to local conditions, and they attract a diversity of birds, butterflies and animals because they are a known food source.

Native plants are effective in formal plantings such as street trees and green walls, but their most appropriate uses are as communities of plants, ecosystems such as the xeric community recommended for the ecoroof, wetlands and wetland forests for stream bank stabilization, and appropriate restoration of diverse urban ecosystems.

One of the most interesting opportunities for the use of native plant communities is as a replacement for the lawn. There are 20,000,000 acres of lawns in residential settings in the U.S., each requiring high levels of maintenance, fertilizers, pesticides, water and fossil fuels. And besides, they are biological deserts. Bormann, Balmori and Geballe (1993) argue that lawns are a significant environmental hazard:

- A lawnmower pollutes as much in one hour as does driving an automobile for 350 miles.
- 60,000 to 70,000 severe accidents result from lawnmowers.
- 580,000,000 gallons of gasoline are used for lawnmowers.
- 30–60% of urban fresh water is used for watering lawns.
- $5,250,000,000 is spent on fossil fuel-derived fertilizers for U.S. lawns.
- 67,000,000 pounds of synthetic pesticides are used on U.S. lawns.
- $700,000,000 is spent on pesticides for U.S. lawns.

The EPA website on native plants reports on a study by Applied Ecological Services of larger properties. It estimates that over a 20-year period, the cumulative cost of maintaining a prairie or a wetland totals $3,000 per acre versus $20,000 per acre for nonnative turf grasses. As a general practice, there are many advantages and few disadvantages in using native plants. They may be, however, more difficult to specify because they are not readily available at standard nurseries (although the frequent specification of them usually increases their availability), and the restoration of native plant communities requires expertise not usually available as a standard landscape service. However, they are cost-effective and environmentally sustainable.

Beyond parks

The green structure of a city includes its municipal parks, but is much more. Parks require high levels of maintenance because of our perception of the "aesthetic" of parkland, and because of this simple maintenance of the existing parks often stress municipal budgets. But throughout every city are pockets and systems of green that are fortuitous and naturally evolving, areas that we perceive as "vacant lots" and wastelands (Fig. 22). All of these areas are part of the green structure of the city, forming an infrastructure that contributes to the environmental health of a region. Plants "can perform a number of functions better than any technological equivalent yet invented, notably in bioengineering and in cleaning up air, water, and soil pollutants. Because of their essential role in making life possible, as well as the social and financial costs of raising them, cultivated plants are too valuable to abuse" (Thompson and Sorvig 2000, p. 129). ■

Fig. 22. Even small urban lots can use native plants. An urban residential home in Buffalo has a restored WNY edge community in the front yard.

REFERENCES

Beckman, Stephanie, *et al.* 1997. *Greening our Cities: An Analysis of the Benefits and Barriers Associated with Green Roofs.* Portland: Oregon State University.

Bormann, F. Herbert, Diana Balmori, Gordon T. Geballe. 1993. *Redesigning the American Lawn.* New Haven: Yale University Press.

Hauer, R.J., R.W. Miller, and D.M. Ouimet. 1994. Street Tree Decline and Construction Damage. *Journal of Arboriculture 20* (2): 94–97.

Hough, Michael. 1984. *City Form and Natural Process.* New York: Van Nostrand Reinhold Co.

Hough, Michael, Beth Benson and Jeff Evenson. 1987. *Greening The Toronto Port Land.* Toronto: Waterfront Regeneration Trust.

Lawrence Berkeley Laboratory. 2002. Heat Island Mitigation Studies. Environmental Energy Technologies Division. Berkeley, CA: Lawrence Berkley National Laboratory. http://eetd.lbl.gov/Heatisland/CoolRoofs/

Moll, Gary, and Stanley Young. 1992. *Growing Greener Cities.* Los Angeles: Living Planet Press.

Spirn, Anne Whiston. 1984. *The Granite Garden: Urban Nature and Human Design.* New York: Basic Books.

Thompson, J. William and Kim Sorvig. 2000. *Sustainable Landscape Construction: A Guide to Green Building Outdoors.* Washington, DC: Island Press.

Table 2: Comparison of Costs and Values of Landscape Plantings (Source: Thompson and Sorvig).

Service, Value, or Cost	Amount	Notes
Purchase or replace nursery stock	\$25 to \$750	Varies regionally; based on informal survey of nurseries.
Cost to install and establish one tree	\$75 to \$3,000	Through second year; based on CTLA rule of thumb, 2-3 x initial cost of tree.
Annual maintenance investment, one tree	\$0 to \$75	Informal estimate of likely costs.
Oversize replacement (>6")	9"= \$955 to \$5,725	CTLA, \$15 to 90 per sq. inch of trunk cross-section area.
Oxygen production, one mature tree	\$32,000	Mich. Forestry
Air pollution control, one mature tree	\$62,000	Mich. Forestry
Water cycling and purification one mature tree	\$37,500	Mich. Forestry
Erosion control, one mature tree	\$32,000	Mich. Forestry
Energy saving (heating and cooling 40% adjacent structure), one mature tree	\$26,000	50 years x Annual.
Insurance limit for one tree under ordinary property-owner policy	\$500	Informal survey of several policies.
Litigation value of one tree	\$15,000	1981 Arlington, VA U.S. Tax destroyed court case on record.
Annual losses of trees caused by Construction in Milwaukee, WI	\$800,000	Hauer, Miller, and Ouimet 1994, p. 94–97.
Annual energy savings of entire	\$4,000,000,000	Rowan Rountree, USFS, urban forest cited at http://www.treelink.org

Lynda H. Schneekloth, editor

Summary
This article presents details of natural and built watersheds, stormwater management, and streamside corridor restoration as ways to improve water quality for aquifer recharge and fresh water sources for consumption, recreation and habitat rejuvenation. Urban watershed and systems require design and management to become beneficial resources for urban life and its natural systems. Rainwater discharge, rivers and streams systems are often poorly designed, channelized and misused as sewers and waste repositories. Environmentally sensitive urban watershed design can be a major resource for communities if managed from a watershed perspective.

Key words
aquifer, bioretention, ecological restoration, hydrology, pollution, stormwater, stream corridor, water quality, watershed

PHOTO: Lynda Schneekloth

Constructed wetland along the Buffalo River.

Urban waterways

All cities are located in specific watersheds and receive some amount of rainfall each year. Many cities are situated at the mouth of rivers or along rivers because of the early dependence upon water transportation and waterpower. All water that passes through a city must somehow be managed so that good quantities and quality of water are available for consumption and to avoid adverse effects such as flooding. Well-protected and managed waterways are an asset to a community and can bring recreational and ecological benefits to the citizens. All landforms, building and urban pavement, vegetation and surface soil geology serve as "sinks, "catchments" and "filters" of rain- and stormwater runoff. How this water flow is designed and managed has a direct bearing on the priceless resource of the subsurface aquifer, the vital water storage that determines the health and sustainability of any human settlement.

A watershed is an area of land that drains water into a common outlet or body of water. All buildings and urban structures have a "watershed address," that is, it shares and sometimes determines the flow of surface water within its regional geology. This can be very large scale, such as the Mississippi River watershed or Hudson River watershed (Fig. 1). Watershed can also be medium or small scale characterized only by small streams and vernal ponds. A watershed approach to managing urban waters attempts to protect waters, people, health and property through a holistic, dispersed, and diverse approach. This article reviews design criteria and details of a watershed-based approach to urban design, focusing upon details of stormwater management and stream corridor restoration.

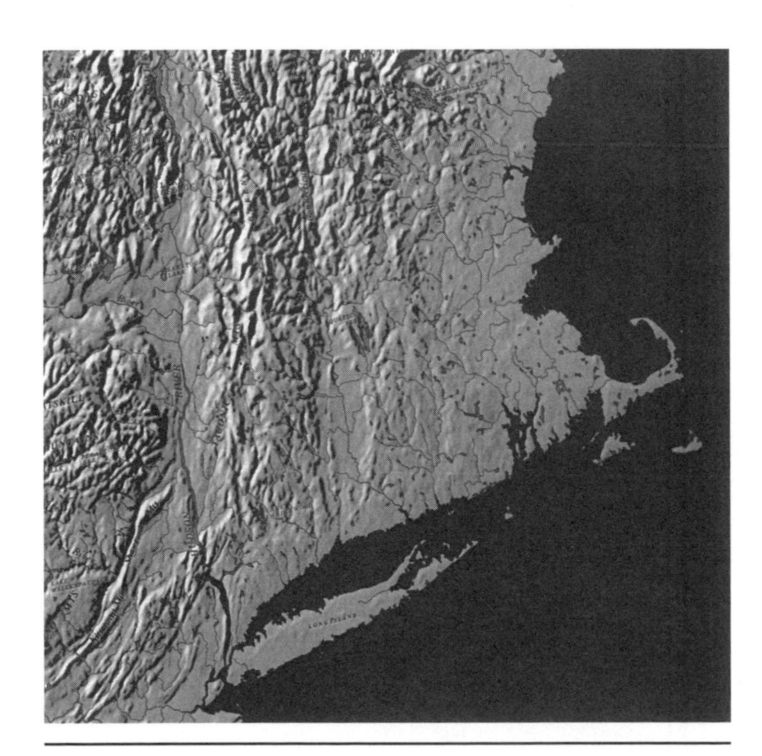

Fig. 1. Northeastern United States watersheds (excerpt from *Landforms and Drainage of the 48 States*, reproduced courtesy of Raven Maps and Images <Ravenmaps.com>

Credits: Material for this article is by permission from Clayto (1988), The Stormwater Center, and the Federal Interagency Stream Restoration Working Group (1998), cited in the References.

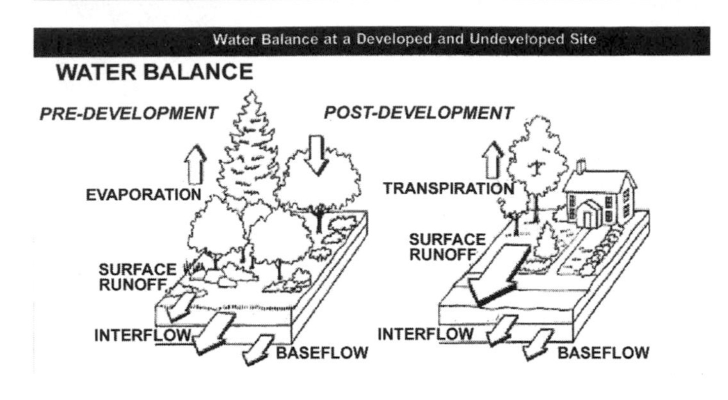

Fig. 2. Water balance at a developed and undeveloped site. (Clayto 1988)

1 STORMWATER MANAGEMENT

Urban development has a profound influence on the quality of local streams because it dramatically alters the local hydrologic cycle. Most often undertaken without regard to ecological requirement, trees that had intercepted rainfall are removed, and natural depressions that had temporarily ponded water are graded to a uniform slope. The spongy humus layer of the forest floor that had absorbed rainfall is scraped off, eroded or severely compacted. Having lost its natural storage capacity, a cleared and graded site can no longer prevent rainfall from being rapidly converted into stormwater runoff (Fig. 2).

The situation typically worsens after construction. Rooftops, roads, parking lots, driveways and other impervious surfaces no longer allow rainfall to soak into the ground. Consequently, most rainfall is directly converted into stormwater runoff, increasing the volume of stormwater runoff, and as a result, the erosion of soil that accompanies heavy runoff. For example, a parking lot can produce sixteen times more stormwater runoff each year than a meadow of identical area.

Without natural or designed catchments and porous surfaces, increases in storm water runoff are most often too much for the existing capacity of a drainage system to handle. As a result, the drainage system is often "improved" to rapidly collect runoff and quickly convey it away (using curb and gutter, enclosed storm sewers, and lined channels). The stormwater runoff is subsequently discharged to downstream waters, such as streams, reservoirs, lakes or estuaries. The damage caused by such poor stormwater management includes declining water quality, diminishing groundwater recharge, flooding, and erosion. In addition, when stormwaters carry the pollutants and chemicals of paved, manicured lawn or chemically managed agricultural fields, the stormwater deposits heavy doses of pollutants or nutrients beyond the natural capacity of healthy waterways.

Because planners, architects and engineers are responsible for design of systems by which urban waterways are handled, it is important to understand how hydrology, water quality and urban design are related. The consequences of stormwater mismanagement and design approaches for best management practices for treating stormwater are described below.

Declining water quality

Impervious surfaces accumulate pollutants deposited from the atmosphere, leaked from vehicles, or wind-blown in from adjacent areas. During storm events, these pollutants quickly wash off, and are rapidly delivered to downstream waters. Urban runoff has elevated concentrations of both phosphorus and nitrogen, which can enrich streams, lakes, reservoirs and estuaries (known as eutrophication). Excess nutrients promote algal growth that blocks sunlight from reaching underwater grasses and depletes oxygen in bottom waters. Urban runoff has been identified as a key and controllable source of nutrients.

Both suspended and deposited sediments can have adverse effects on aquatic life in streams, lakes and estuaries. Sediments also transport other attached pollutants. Sources of sediment include washoff of particles that are deposited on impervious surfaces and the erosion of streambanks and construction sites.

Organic matter, washed from impervious surfaces during storms, can present a problem in slower moving downstream waters. As organic matter decomposes, it can deplete dissolved oxygen in lakes and tidal waters. Low levels of oxygen in the water can have an adverse impact on aquatic life. Also, bacteria levels in stormwater runoff routinely exceed public health standards for water contact recreation. Stormwater runoff can also lead to the closure of adjacent shellfish beds and swimming beaches and may increase the cost of treating drinking water at water supply reservoirs.

Other pollutants include hydrocarbons, from vehicles that leak oil and grease; trace metals that can be toxic to aquatic life at certain concentrations; pesticide accumulations that approach or exceed toxicity thresholds for aquatic life; and chlorides/salts that are applied to roads and parking lots in winter at much higher concentrations than many freshwater organisms can tolerate. Impervious surfaces may increase temperature in receiving waters, adversely impacting aquatic life that requires cold and cool water conditions.

Finally, considerable quantities of trash and debris are washed through the storm drain networks. The trash and debris accumulate in streams and lakes and detract from their natural beauty.

Diminishing groundwater recharge and quality

Groundwater is a critical water resource. Not only do many people depend on groundwater for their drinking water, but also the health of many aquatic systems is dependent on its steady discharge (Fig. 3). The slow infiltration of rainfall through the soil layer is essential for replenishing groundwater. The amount of rainfall that recharges groundwater varies, depending on the slope, soil and vegetation.

Urban land uses and activities degrade groundwater quality if stormwater runoff is directed into the soil without adequate treatment. Certain land uses and activities are known to produce higher loads of metals and toxic chemicals, very typical of waterways that have been or are still used for industry. Stormwater runoff should never be infiltrated into the soil if a site has any history of soil contamination.

Degradation of stream channels

Stormwater runoff is a powerful force that influences the geometry of streams. After development, both the frequency and magnitude of storm flows increase dramatically (Fig. 4). Consequently, urban stream channels experience more bankfull and subbankfull flow events each year than they had prior to development.

As a result, the bed and bank of a stream are both exposed to highly erosive flows more frequently and for longer periods. Streams typically respond to this change by increasing their cross-sectional area to handle the more frequent and erosive flows either by channel widening or down cutting, or both. The stream enters a highly unstable phase, and experiences severe streambank erosion and habitat degradation. In this phase, the stream often experiences some of the following changes:

- Rapid stream widening
- Increased streambank quality (through sediment deposition and embedding of the substrate)
- Loss of pool/riffle structure in the stream channel
- Degradation of stream habitat structure
- Creation of fish barriers by culverts and other stream crossings.

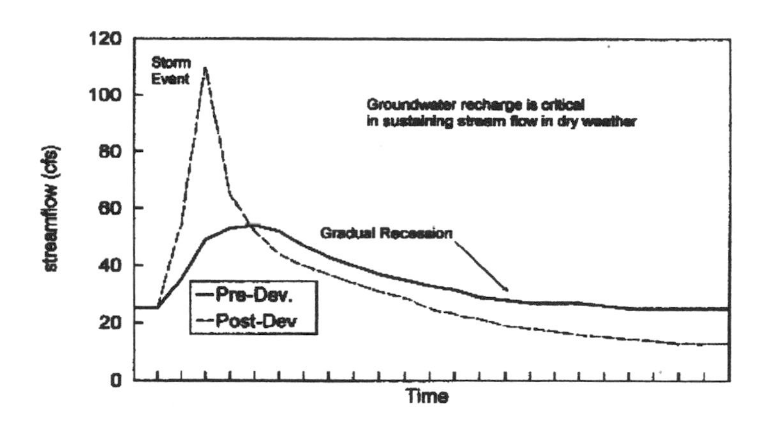

Fig. 3. Graph illustrating the decline in streamflow due to diminished groundwater recharge. (Clayto 1988)

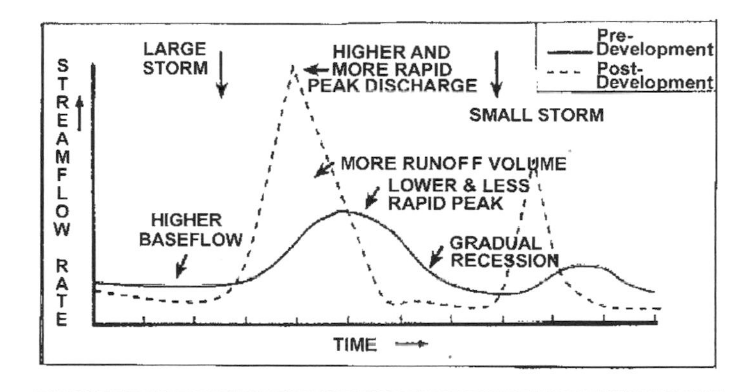

Fig. 4. Graph illustrating the change in streamflow following development. (Clayto 1988)

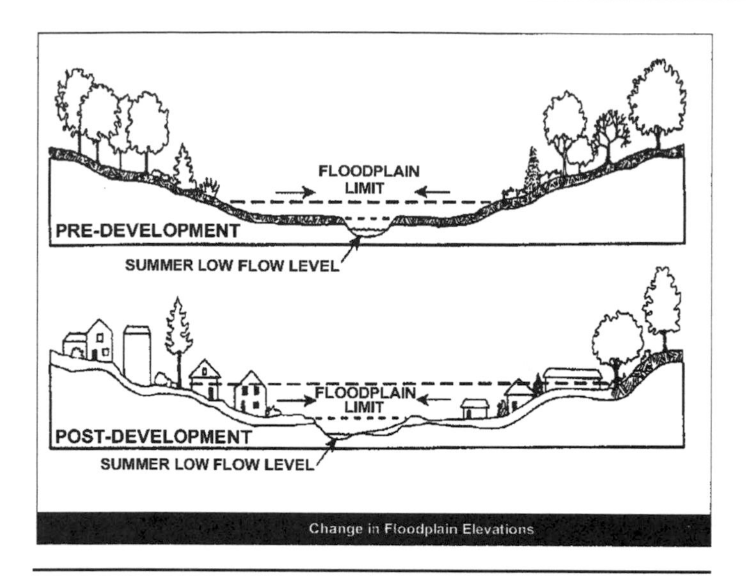

Fig. 5. Predevelopment and postdevelopment floodplain elevations. (Clayto 1988)

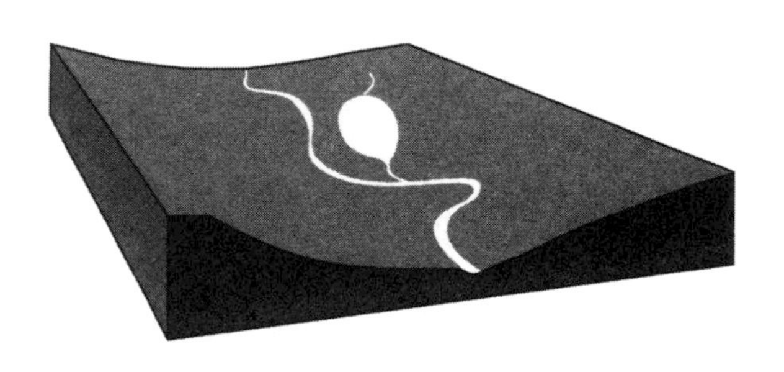

Fig. 6. Sediment Basins can help preserve water quality and provide erosion control downstream by trapping some larger particles. Barriers can be employed in conjunction with excavated pools, constructed across a drainage way or off-stream and connected to the stream by a flow diversion channel to trap and store waterborne sediment and debris. (FISR 1988)

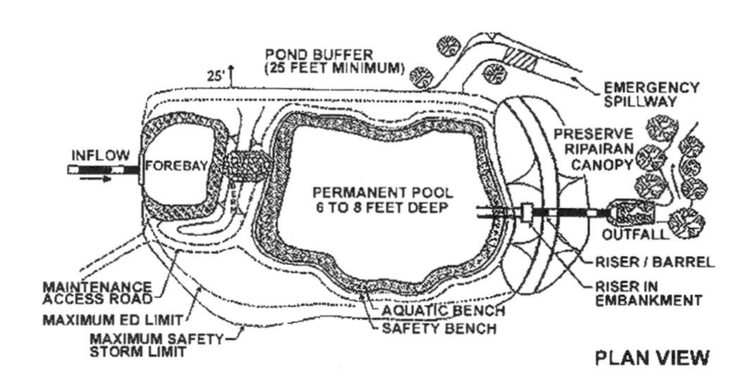

Fig. 7. Example of "Wet Extended Detention (ED) pond that provides water retention for flood control and water quality protection. (Stormwater Center)

The decline in the physical habitat of the stream, coupled with lower base flows and higher stormwater pollutant loads, has a severe impact on the aquatic community.

Flooding

Waterflow events that exceed the capacity of the stream channel spill out into the adjacent flood plain. These are termed "overbank" floods, and can damage property and downstream drainage structures. Overbank floods are ranked in terms of their statistical return frequency. For example, a flood that has a fifty percent chance of occurring in any given year is termed a "two year" flood. Similarly, a flood that has a ten percent chance of occurring in any given year is terms a "ten-year flood." Urban development increases the peak discharge rate associated with a given design storm, because impervious surfaces generate greater runoff volumes and drainage systems deliver it more rapidly to a stream (Fig. 5).

Floodplain expansion

The level areas bordering streams and rivers are known as floodplains. Operationally, the floodplain is usually defined as the land area within the limits of the 100-year storm flow water elevation. The 100-year storm has a 1% chance of occurring in any given year. These floods can be very destructive, and can pose a threat to property and human life.

Floodplains are natural flood storage areas and help to attenuate downstream flooding. Floodplains are very important habitat areas, encompassing riparian forests, wetlands, and wildlife corridors. Consequently, many jurisdictions restrict or even prohibit new development within the 100-year floodplain to prevent flood hazards and conserve habitats.

As with overbank floods, development sharply increases the peak discharge rate associated with the 100-year design storm. As a consequence, the elevation of a stream's 100-year floodplain becomes higher and the boundaries of its floodplain expand.

Best practices for managing stormwater

There are many ways of managing stormwater, including regional landuse planning and conservation of wetlands, aquatic buffers, stream corridor management (including erosion and sediment control), better site planning, and on-site storm water treatment practices that improve water quality (Fig. 6). Some of these are large-scale efforts, and others very specific on-site design and management ideas.

One innovative plan is the City of Vancouver's Downspout Disconnection Program. Vancouver is limiting the amount of water going into the stormwater system by collecting it at the site of individual homes and businesses through the reintroduction of rainbarrels or underground storage. The diversion of water from the storm drains reduces the overload on the wastewater treatment plant, particularly during storms, and helps limit the number of times raw sewage enters the rivers (an undesirable consequence of the combined sewer systems currently existing in many U.S. cities).

Stormwater treatment practices to improve water quality

There are many practices that help store or hold water and release it slowly into the storm drains and/or river and streams. Detention ponds on roofs and parking lots, for example, hold water and release it slowly. But these do not improve water quality. The following set of Stormwater Treatment Practices (STPs) meet water quality goals and have demonstrated the capability to remove 80% of the Total Suspended Solids (TSS). The examples shown are for purposes of illustration only and each application should be designed for a site's specific situation (*cf.* Storm Water Center website listed in References).

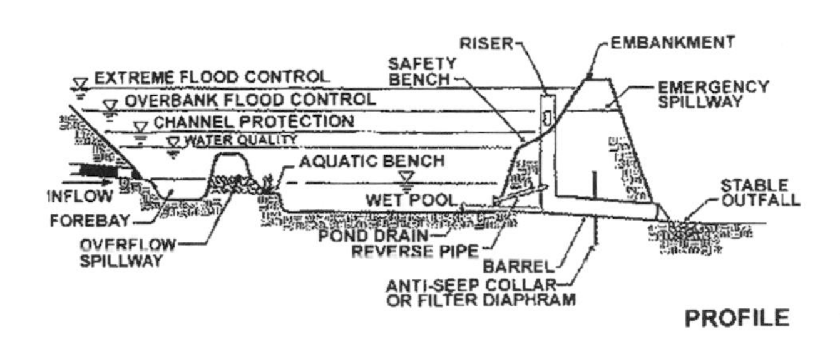

Fig. 8. Cross-section of ED pond. (Stormwater Center)

Stormwater ponds

Stormwater ponds are a combination of a permanent pool, extended detention or shallow marsh equivalent. Design variants include: Micropool Extended Detention Pond, Wet Pond, Wet Extended Detention Pond, Multiple Pond System, and "Pocket" Pond (Figs. 7 and 8).

Wetlands

Stormwater wetlands create shallow constructed marsh areas to treat urban stormwater and often incorporate small permanent pools and/or extended detention storage to achieve full water quality. Design variants include: Shallow Wetland, Extended Detention Shallow Wetland, Pond/Wetland System, and "Pocket" Wetland (Figs. 9 and 10).

- **Wetland Feasibility Criteria:** A water balance should be performed to demonstrate that a stormwater wetland could withstand a significant drought at summer evaporation rates without completely drawing down.
- **Wetland Conveyance Criteria:** Flowpaths from the inflow points to the outflow points of stormwater wetlands should be maximized. A minimum flowpath of 2:1 (length to relative width) should be provided across the stormwater wetland.
- **Wetland Pretreatment Criteria:** Sediment regulation is critical to sustain stormwater wetlands. Consequently, a forebay should be located at the inlet, and a micropool should be located at the outlet.
- **Wetland Treatment Criteria:** The surface area of the entire stormwater wetland should be at least one percent of the contributing drainage area (1.5% for shallow marsh design).
- **Wetland Landscaping Criteria:** A landscaping plan should be provided that indicates the methods used to establish and maintain wetland coverage. Minimum elements of a plan include: delineation of pondscaping zones, selection of corresponding plant species, planting plan, sequence for preparing wetland bed (including soil amendments, if needed) and sources of plant material.

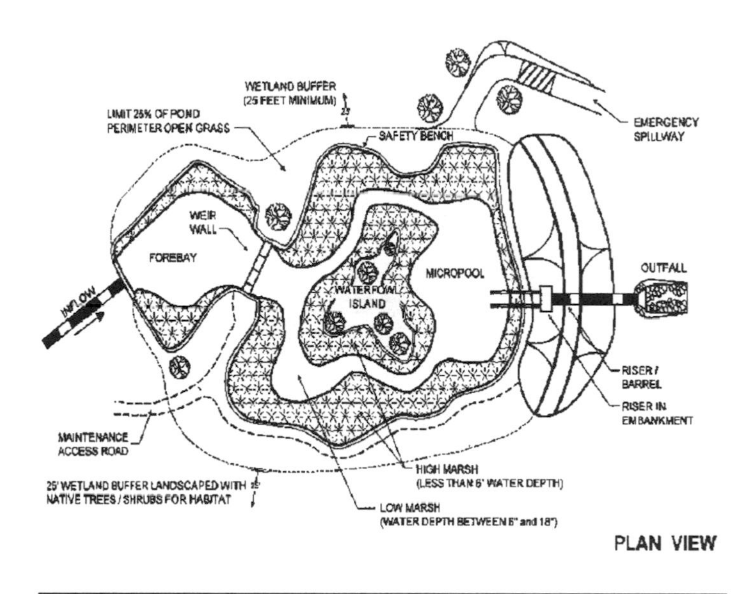

Fig. 9. Example of designed wetland that meets both water retention and water quality control. (Stormwater Center)

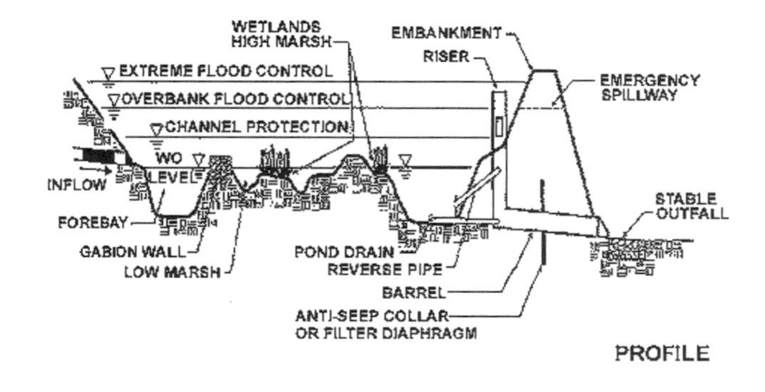

Fig. 10. Cross-section of designed wetland.

Infiltration systems

Stormwater infiltration practices capture and temporarily store the water before allowing it to infiltrate into the soil. Design variants include: Infiltration Trench and Infiltration Basin.

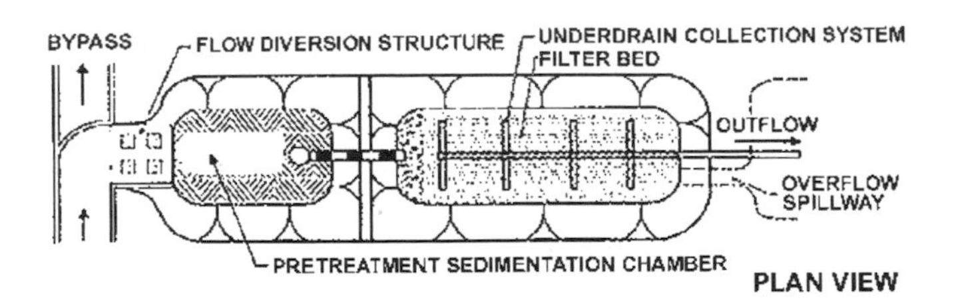

Fig. 11. Example of Sandfilter design that treats stormwater run-off. (Stormwater Center)

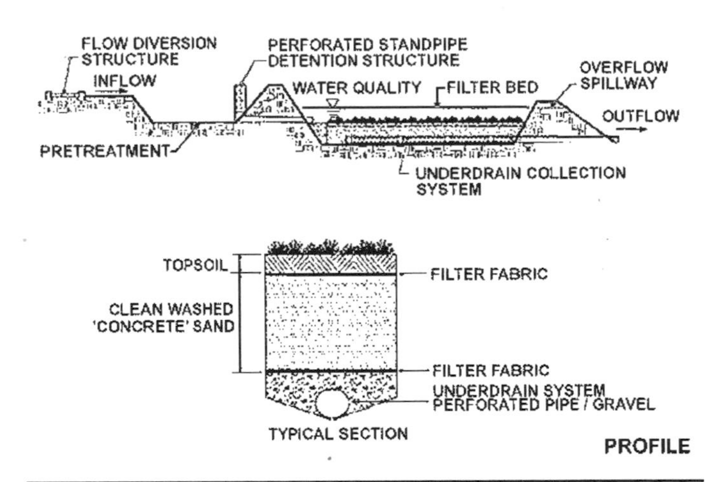

Fig. 12. Cross-section of sand filter design.

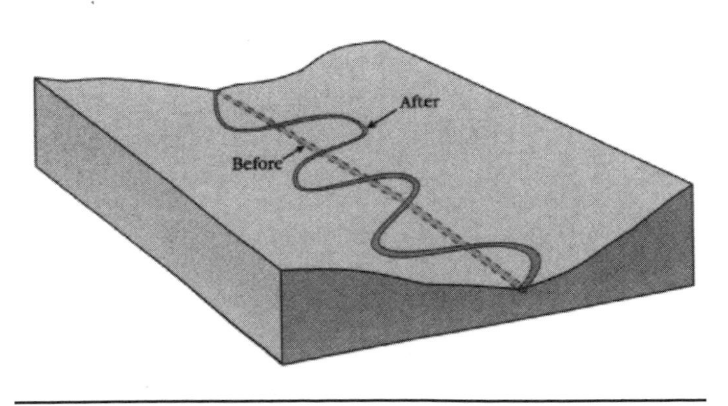

Fig. 13. Stream Meander Restoration reintroduces the natural dynamics to an engineered stream that will improve channel stability, habitat quality, aesthetics, and other stream corridor functions or values. (FISR 1988)

Filtering systems

Stormwater filtering systems capture and temporarily store the water, and pass it through a filter bed of sand, organic matter, soil or other media. Filtered runoff may be collected and returned to the conveyance system, or allowed to partially exfiltrate into the soil. Design variants include:

Surface Sand Filter, Underground Sand Filter, Perimeter Sand Filter, Organic Filter and Bioretention.

Figs. 11 and 12 illustrate the design of a sand filter to treat stormwater runoff. The filter provides recharge and treats the water quality volume.

- **Filtering Feasibility Criteria:** Most stormwater filters normally require two to six feet of head. The perimeter sand filter can be designed to function with as little as one foot of head. The maximum contributing area to an individual stormwater filtering system is usually less than 10 acres (4 ha). Sand and organic filtering systems are generally applied to land uses with a high percentage of impervious surfaces. Sites with imperviousness less than 75% will require full sedimentation pretreatment techniques.

- **Filtering Landscaping Criteria:** A dense and vigorous vegetative cover should be established over the contributing pervious drainage areas before runoff can be accepted into the facility.

Open channel systems

Open channel systems are vegetated open channels explicitly designed to capture and treat the water within dry or wet cells formed by checkdams or other means. Design variants include: Dry Swale, Wet Swale, and Grass Channels—often called bioswales when plants are added that have very good water absorptive properties.

2 STREAM CORRIDOR RESTORATION

The United States has more than 3.5 million miles (5.6 million km) of rivers and streams that, along with closely associated floodplain and upland areas, comprise corridors of great economic, social, cultural, and environmental value. These corridors are complex ecosystems that include the land, plants, animals, and network of streams within them. They perform ecological functions such as modulating streamflow, storing water, removing harmful materials from water, and providing habitat for aquatic and terrestrial plants and animals (Fig. 13).

Definitions: Restoration, rehabilitation and reclamation

- **Restoration** is reestablishment of the structure and function of ecosystems (National Research Council 1992). Ecological restoration is the process of returning an ecosystem as closely as possible to predisturbance conditions and functions. Implicit in this definition is that ecosystems are naturally dynamic. It is therefore not possible to recreate a system exactly. The restoration process reestablishes the general structure, function, and dynamic but self-sustaining behavior of the ecosystem.

- **Rehabilitation** makes the land useful again after a disturbance. It involves the recovery of ecosystem functions and processes in a degraded habitat (Dunster and Dunster 1996). Rehabilitation does not necessarily reestablish the predisturbance condition, but does involve establishing geological and hydrologically stable landscapes that support the natural ecosystem mosaic.

- **Reclamation** is a series of activities intended to change the biophysical capacity of an ecosystem. The resulting ecosystem is different from the one existing prior to recovery (Dunster and Dunster 1996). The term implies adapting wild or natural resources to serve a utilitarian human purpose such as the conversion of riparian or wetland ecosystems to agricultural, industrial, or urban uses.

- **Restoration** differs from rehabilitation and reclamation in that restoration is a holistic process not achieved through the isolated manipulation of individual elements. While restoration aims to return an ecosystem to a former natural condition, rehabilitation and reclamation imply putting a landscape to a new or altered use to serve a particular human.

Change in hydrology after urbanization

The hydrology of urban streams changes as sites are cleared and impervious cover such as rooftops, roadways, parking lots, sidewalks, and driveways replaces natural vegetation. One of the consequences is that more of a stream's annual flow is delivered as stormwater runoff rather than the baseflow. Depending on the proportion of watershed impervious cover, the annual volume of stormwater runoff can increase by up to sixteen times that for natural areas (Schueler 1995). In addition, since impervious cover prevents rainfall from infiltrating into the soil, less flow is available to recharge ground water. Therefore, during extended periods without rainfall, baseflow levels are often reduced in urban streams (Simmons and Reynolds 1982). (Fig. 14)

Storm runoff moves more rapidly over smooth, hard pavement than over natural vegetation. As a result, the rising limbs of storm hydrographs become steeper and higher in urbanizing areas. Recession limbs also decline more steeply in urban streams.

Designing urban stream buffers

The edges of waterways are among the most critical zones in the hydrological cycle. How water edges are designed and managed makes a great difference in water quality. Landscape practices that replace natural vegetation along stream and river edges and instead install manicured lawnscape do damage to waterway quality, by eliminating the natural filtration and holding capacity that more substantial vegetation can provide. Informed of the critical role of streamside vegetation, many municipalities are enacting guidelines and regulations to address waterway corridors, also referred to as "streamside buffer zones."

Streamside buffers, as might be provided by dense vegetation, wildflowers, bushes and woodland, create a rich variety of environmental roles:

- Vegetation filters and cleanses stormwater runoff,
- Ground vegetation slows the rate of stormwater flow, reducing erosion.
- Vegetation and marshy soils absorb rising waters, reducing flooding.
- Established root systems secure river embankments, protected against erosion.
- Overhanging branches shade water edges, providing cooler water for aquatic species (fish, amphibians).

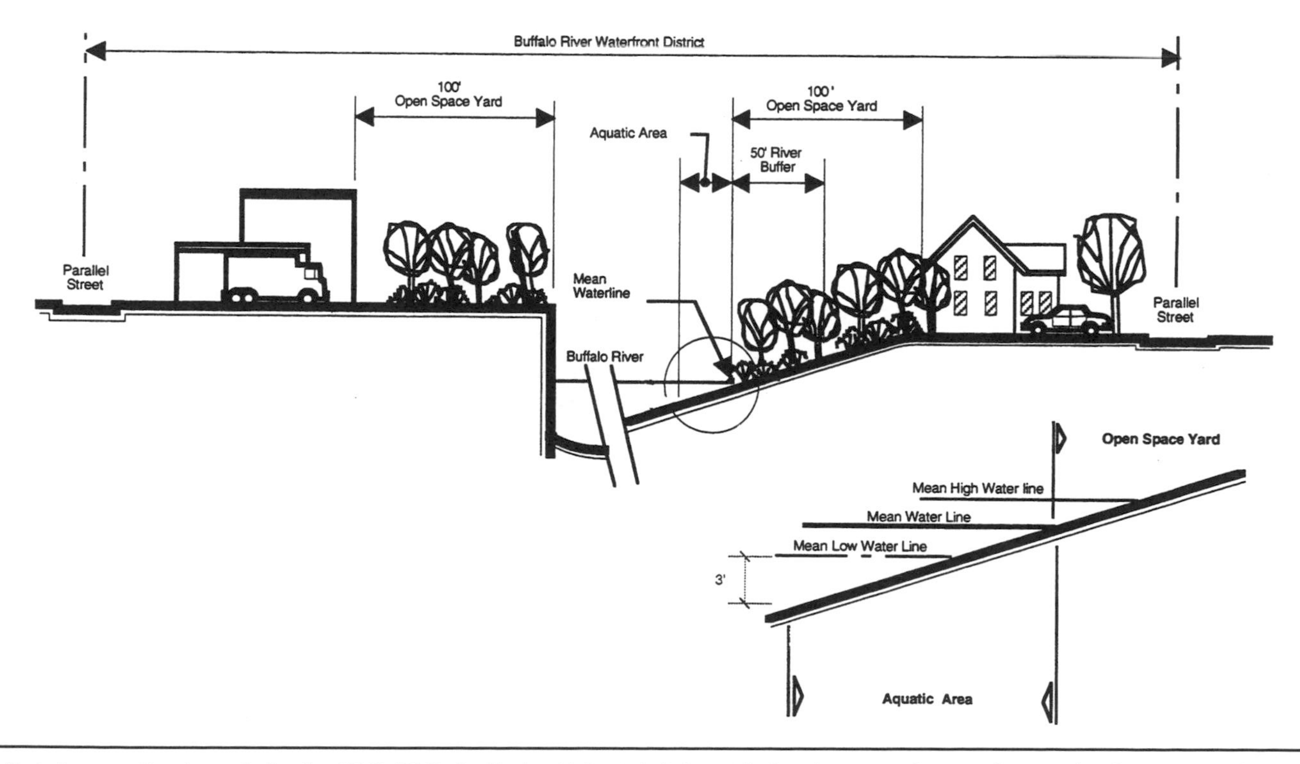

Fig. 14. Typical cross section demonstrating the 100 ft. (30.5 m) setback guideline to include a 50 ft. densely vegetated zone and a second 50 ft. (15.24 m) of open space to preserve the water quality along the urban river. (Friends of the Buffalo River 1996)

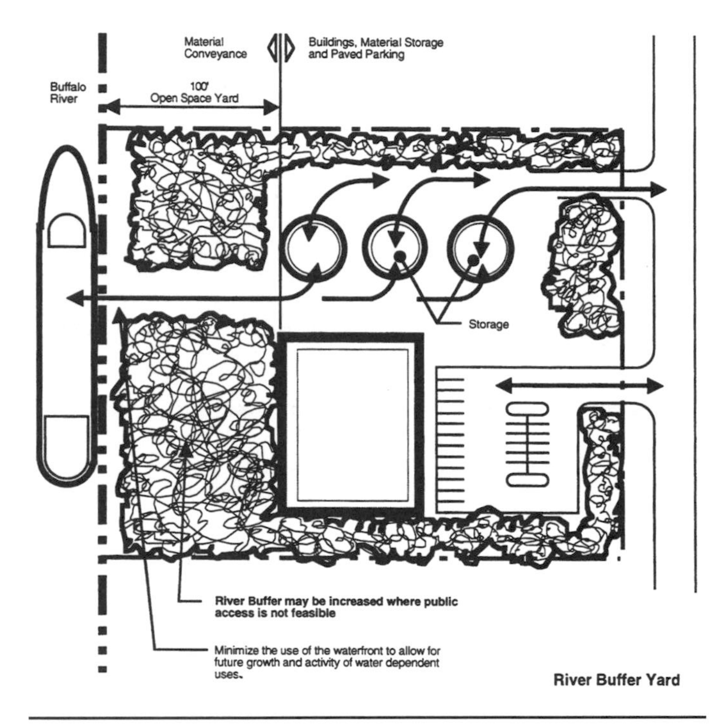

Fig. 15. Diagram of 100 ft. (30.5 m) open space yard as could be developed for industrial and/or commercial use along the Buffalo River. (Friends of the Buffalo River 1996)

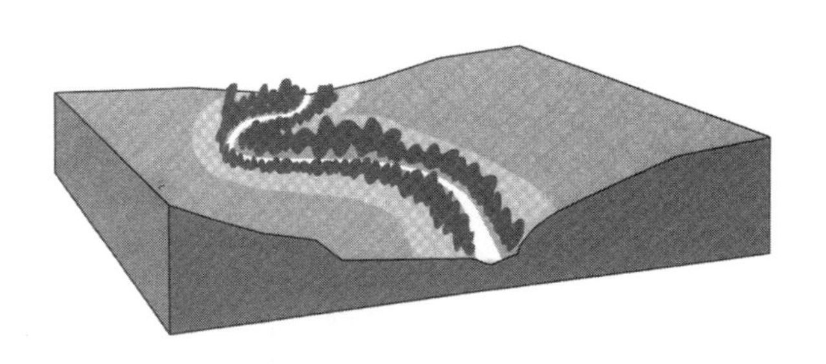

Fig. 16. Riparian Forest Buffers demonstrates the use of streamside vegetation to lower water temperatures, provide a source of detritus and large woody debris, improve habitat, and reduce sediment, organic materials, nutrients, pesticides and other pollutants migrating to the stream. (FISR 1988)

- Vegetation provides protective cover for water-edge insects, bird life and small animal habitat.
- Overhanging branches, reeds, grasses and root systems along shallow water edges provide cover for small fishes (spawning zones).

The ability of an urban stream buffer to realize these benefits depends on how stream buffers are planned, designed, and maintained. Subject to specific circumstances defined by the criteria below, vegetation should be maintained within the buffer zone that is appropriate to the local ecology. Manicured lawns and agriculture/animal husbandry that utilize should be strictly prohibited and/or their runoff subject to retention and cleansing.

The following design criteria govern how to design and manage a stream buffer.

Criterion 1: Minimum Total Buffer Width

Most local buffer criteria require that development be set back a fixed and uniform distance from the stream channel. Recommended urban stream buffers range from 20 to 200 ft. (6 to 60 m) in width from each side of the stream, with an average of 100 ft. (30 m) (Schueler 1995). A minimum width of at least 100 ft. is recommended to provide adequate stream protection.

Criterion 2: Three-Zone Buffer System

Effective urban stream buffers have three lateral zones—streamside, middle core, and outer zone. Each zone performs a different function, and has a different width, vegetative target and management scheme. The streamside zone protects the physical and ecological integrity of the stream ecosystem. The vegetative target is mature riparian forest that can provide shade, leaf litter, woody debris, and erosion protection to the stream. The middle zone extends from the outward boundary of the streamside zone, and varies in width, depending on stream order, the extent of the 100-year floodplain, adjacent steep slopes, and protected wetland areas. Its key functions are to provide further distance between upland development and the stream. The vegetative target for this zone is also mature forest, but some clearing may be allowed for stormwater management, access, and recreational uses (Fig. 15).

The outer zone is the "buffer's buffer," an additional 25 ft. (7.6 m) setback from the outward edge of the middle zone to the nearest permanent structure. In most instances, it is a residential backyard. The vegetative target for the outer zone is usually turf or lawn, although the property owner is encouraged to plant trees and shrubs, and thus increase the total width of the buffer. Very few uses are restricted in this zone. Indeed, gardening, compost piles, yard wastes, and other common residential activities often will occur in the outer zone.

Criterion 3: Predevelopment Vegetative Target

The ultimate vegetative target for urban stream buffers should be specified as the predevelopment riparian plant community—usually mature forest. Notable exceptions include prairie streams of the U.S. Midwest, or arroyos of the arid West, that may have a grass or shrub cover in the riparian zone. In general, the vegetative target should be based on the natural vegetative community present in the floodplain, as determined from reference riparian zones. Turfgrass is allowed for the outer zone of the buffer (Fig. 16).

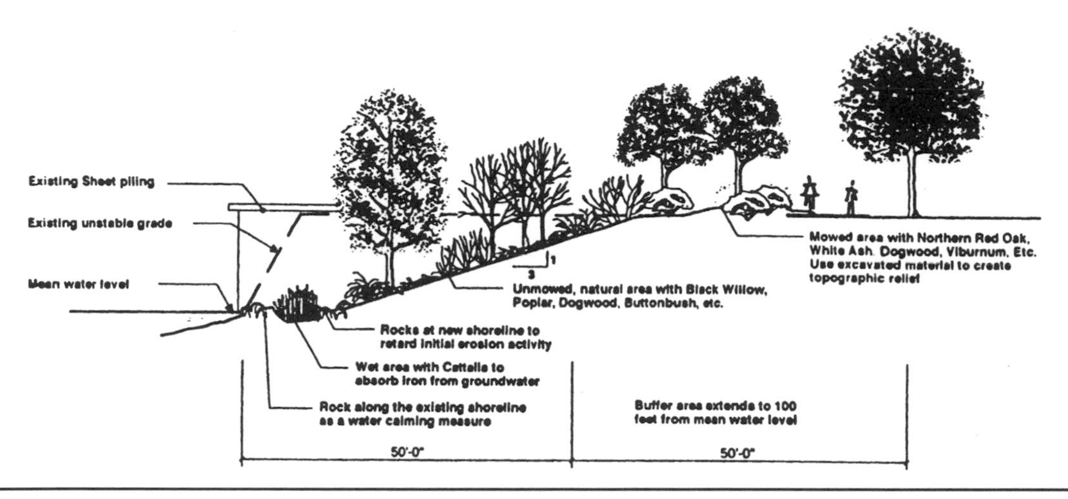

Fig. 17. Shoreline restoration that includes the removal of steep engineered slope to a stable, vegetated slope within the 100-ft. (30 m) open space yard. (Friends of the Buffalo River 1996)

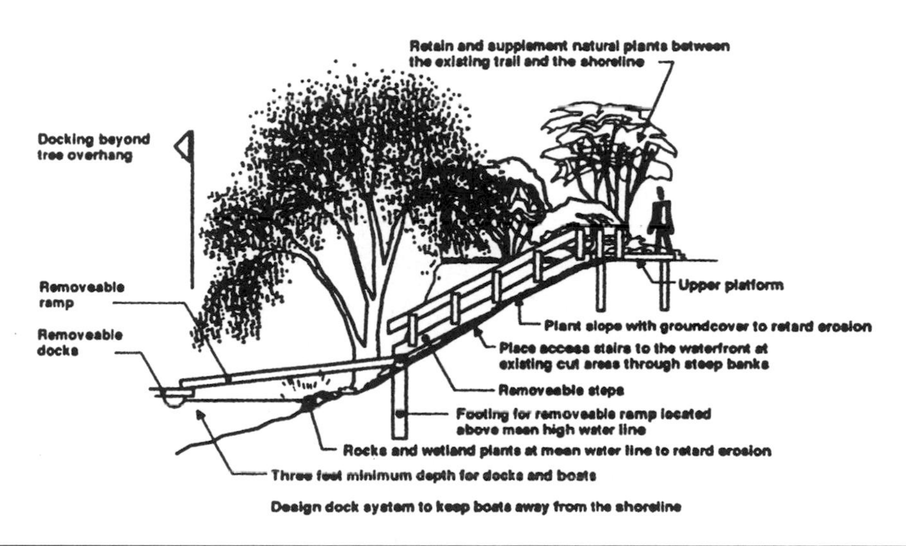

Fig. 18. Shoreline restoration where slopes are steep and in heavily used areas. (Friends of the Buffalo River 1996)

Criterion 4: Buffer Expansion and Contraction

Many communities require the minimum width of the buffer to be expanded under certain conditions. Specifically, the average width of the middle zone can be expanded to include:

- The full extent of the 100-year floodplain
- All undevelopable steep slopes (greater than twenty-five percent)
- Steep slopes (five to twenty-five percent slope, at four additional feet of slope per one percent increment of slope above five percent)
- Any adjacent delineated wetlands or critical habitats

Criterion 5: Buffer Delineation

When delineating the boundaries of a buffer, criteria should be developed to address the following (Fig. 17):

- At what mapping scale will streams be defined?
- Where does the stream begin and the buffer end?
- From what point should the inner edge of the buffer be measured?

Criterion 6: Buffer Crossings

Major objectives for stream buffers are to maintain an unbroken corridor of riparian forest and to allow for upstream and downstream fish passage in the stream network. It is not always possible to try to meet these goals. Provision must be made for linear forms of development that must cross the stream or the buffer, such as roads, bridges, fairways, underground utilities, enclosed storm drains or outfall channels.

Criterion 7: Stormwater Runoff

Buffers can be an important component of stormwater treatment at a development site. They cannot treat all the stormwater runoff generated within a watershed. Generally, a buffer system can only treat runoff from less than 10% of the contributing watershed to the stream. Therefore, some kind of structural Best Management Practice must be installed to treat the quantity and quality of storm water runoff from the remaining 90% of the watershed (Fig. 18).

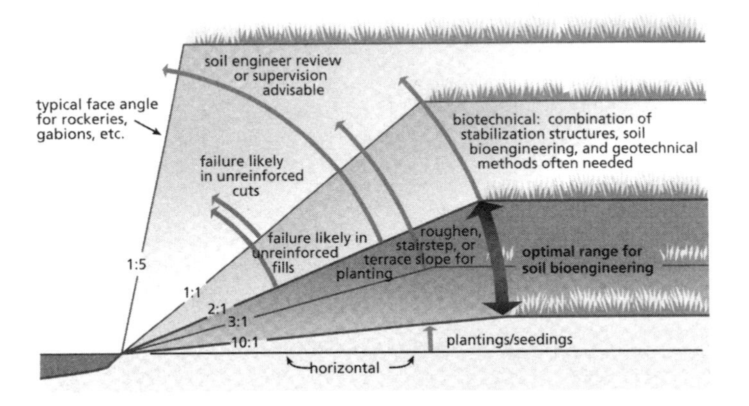

Fig. 19. Diagram of cuts and fills along shorelines and recommendations for type of treatment depending on the slope gradient. (FISR 1988)

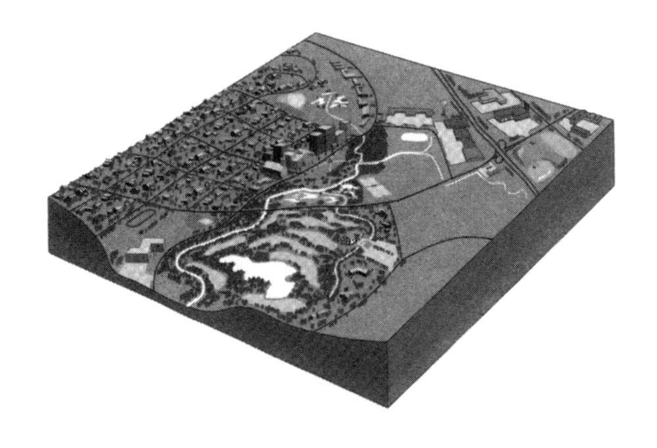

Fig. 20. Best Management Practices in Urban Environments includes individual and systematic approaches designed to offset, reduce, or protect against the impacts of urban development and urban activities on the stream corridor. (FISR 1988)

Criterion 8: Buffers During Plan Review and Construction

The limits and uses of the stream buffer systems should be well defined during each stage of the development process, from initial plan review through construction, with appropriate standards of detail (Fig.19).

Criterion 9: Buffer Education and Enforcement

The future integrity of a buffer system requires a strong education and enforcement program. Thus, it is important to make the buffer "visible" to the community, and to encourage greater buffer awareness and stewardship among adjacent residents. Several simple steps can be taken:

- Mark the buffer boundaries with permanent signs that describe allowable uses.
- Educate buffer owners about the benefits and uses of the buffer with pamphlets, stream walks, and meetings with homeowners associations.
- Ensure that new owners are fully informed about buffer limits/uses when property is sold or transferred.
- Engage residents in a buffer stewardship program that includes reforestation and backyard "streamside buffer" programs.
- Conduct annual buffer walks (ideally these are clean-ups) to raise community awareness and to check on encroachment.

Criterion 10: Buffer Flexibility

In most regions of the country, a 100-ft. (30 m) buffer will take about 5% of the total land area in any given watershed out of use or production. While this constitutes a relatively modest land reserve at the watershed scale, it can be a significant hardship for a landowner whose property is adjacent to a stream. Many communities are legitimately concerned that stream buffer requirements could represent an uncompensated "taking" of private property. These concerns can be eliminated if a community incorporates several simple measures to ensure fairness and flexibility when administering its buffer program. As a general rule, the intent of the buffer program is to modify the location of development in relation to the stream but not its overall intensity. Some flexible measures in the buffer ordinance include:

- Maintaining buffers in private ownership
- Buffer averaging
- Density compensation
- Variances
- Conservation easements

Best management practices in urban areas

Best Management Practices (BMPs) are individual or systematic approaches designed to offset, reduce, or protect against the impacts of urban development and urban activities on the stream corridor. The following are presented as examples of the many techniques that are being used in support of stream corridor restoration. The examples are conceptual and contain little design guidance. Therefore, all restoration techniques require further design before implementation (Fig. 20).

Applications and effectiveness

- Used to improve and/or restore ecological functions that have been impaired by urban activities.
- Needs to be integrated with BMPs on other lands in the landscape to assure that stream restoration is applied along the entire stream corridor to the extent possible.
- The use of individual urban BMPs should be coordinated with an overall plan for restoring the stream system.
- Urban sites are highly variable and have a high potential for disturbance.
- BMPs may include: extended detention dry basins, wet ponds, constructed wetlands, oil-water separators, vegetated swales, filter strips, infiltration basins and trenches, porous pavement, and urban forestry.

Bank shaping and planting

Bank shaping and planting involves regrading streambanks to a stable slope, placing topsoil and other materials needed for sustaining plant growth, and selecting, installing and establishing appropriate plant species (Fig. 21).

Applications and effectiveness

- Most successful on streambanks where moderate erosion and channel migration are anticipated.
- Reinforcement at the toe of the embankment is often needed.
- Enhances conditions for colonization of native species.
- Used in conjunction with other protective practices where flow velocities exceed the tolerance range for available plants, and where erosion occurs below base flows.
- Streambank soil materials, probable groundwater fluctuation, and bank loading conditions are factors for determining appropriate slope conditions.
- Slope stability analyses are recommended.

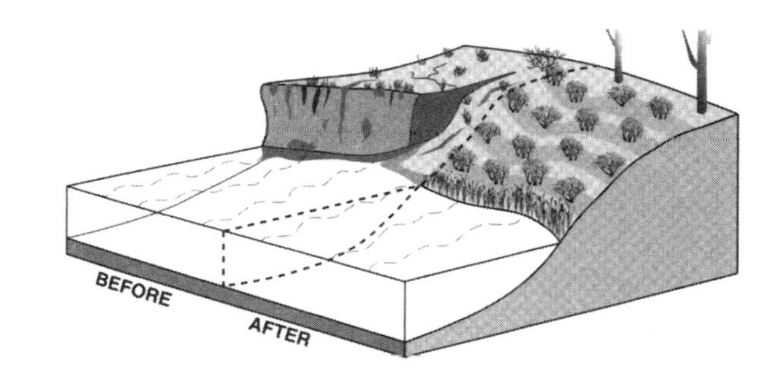

Fig. 21. Bank Shaping and Planting, through regrading of streambanks, creates a stable slope. Topsoil and other materials are needed to sustain appropriate plant growth. (FISR 1988)

Brush mattresses

Brush mattresses are a combination of live stakes, live facines, and branch cuttings installed to cover and physically protect streambanks, eventually to sprout and establish numerous individual plants (Fig. 22).

Applications and effectiveness

- Form an immediate protective cover over the streambank.
- Capture sediment during flood flows.
- Provide opportunities for rooting of the cuttings over the stream-bank.
- Rapidly restores riparian vegetation and streamside habitat.
- Enhance conditions for colonization of native vegetation.
- Limited to the slope above base flow levels.

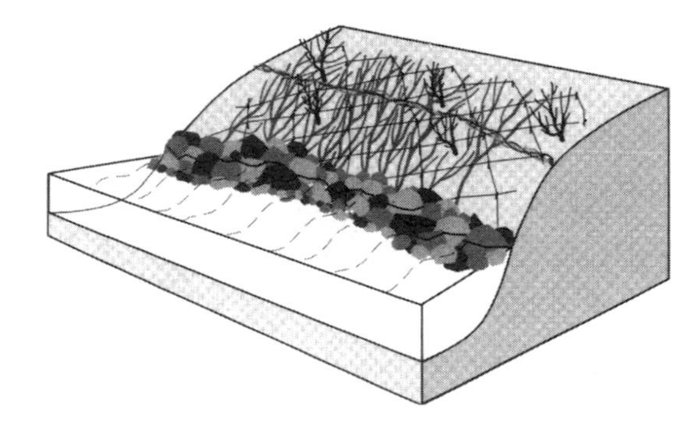

Fig. 22. Brush Mattresses use a combination of live stakes, live facines, and branch cuttings installed to cover and physically protect streambanks. Eventually these sprout and establish numerous individual plants that protect the streambed. (FISR 1988)

- Toe protection is required where toe scour is anticipated.
- Appropriate where exposed streambanks are threatened by high flows prior to vegetation establishment.
- Should not be used on slopes that are experiencing mass movement or other slope instability.

Coconut fiber roll

Coconut fiber rolls are cylindrical structures composed of coconut husk fibers bound together with twine woven from coconut material to protect slopes from erosion while trapping sediment, which encourages plant growth within the fiber roll (Fig. 23).

Applications and effectiveness

- Most commonly available in twelve-inch diameter by twenty-foot lengths.
- Typically staked near the toe of the streambank with dormant cuttings and rooted plants inserted into slits into the rolls.
- Appropriate where moderate toe stabilization is required in conjunction with restoration of the streambank and the sensitivity of the site allows for only minor disturbance.
- Provide an excellent medium for promoting plant growth at the water's edge.
- Not appropriate for sites with high velocity flows or large ice build-up.
- Flexibility for molding to the existing curvature of the streambank.
- Requires little site disturbance.
- The rolls are buoyant and require secure anchoring.
- Can be expensive.
- An effective life of six to ten years.
- Should, where appropriate, be used with soil bioengineering systems and vegetative plantings to stabilize the upper bank and ensure a regenerative source of streamside vegetation.
- Enhances conditions for colonization of native vegetation.

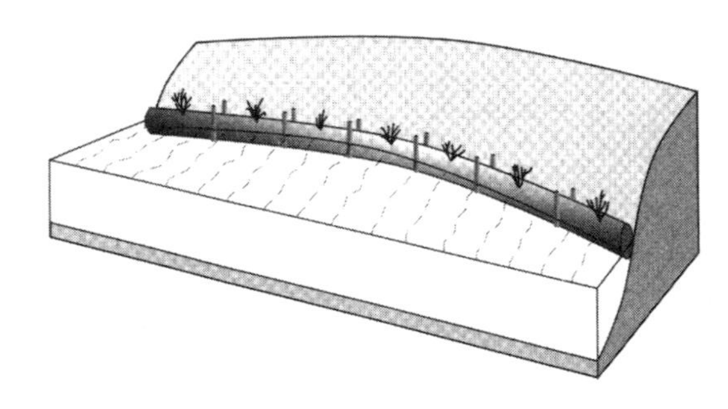

Fig. 23. Coconut Fiber Rolls are cylindrical structures composed of coconut husk fibers bound together with twine woven from coconut material. These protect slopes from erosion while trapping sediment and encourage plant growth within the fiber roll. (FISR 1988)

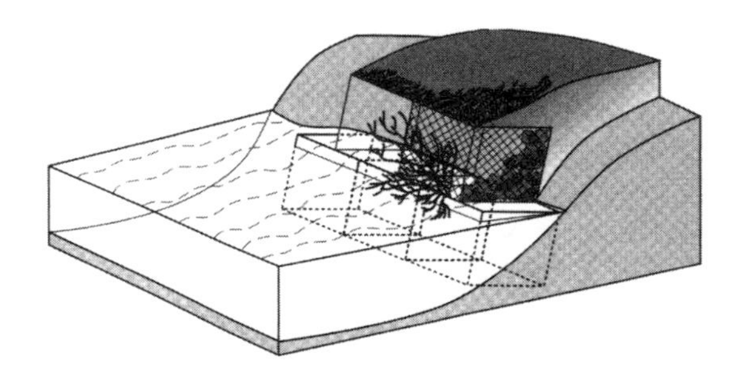

Fig. 24. Vegetated Gabions are wire-mesh; rectangular baskets filled with small to medium size rock and soil and laced together to form a structure toe or sidewall. Live branch cuttings are placed on each consecutive layer between the rock-filled baskets to take root, consolidate the structure and bind it to the slope. (FISR 1988)

Vegetated gabions

Vegetated gabions are wire-mesh; rectangular baskets filled with small to medium size rock and soil and laced together to form a structural toe or sidewall. Live branch cuttings are placed on each consecutive layer between the rock filled baskets to take root, consolidate the structure, and bind it to the slope.

Applications and effectiveness

- Useful for protecting steep slopes where scouring or undercutting is occurring or there are heavy loading conditions.

- Can be a cost-effective solution where some form of structural solution is needed and other materials are not readily available or must be brought in from distant sources.
- Useful when design requires rock size greater than what is locally available.
- Effective where bank slope is steep and requires moderate structural support.
- Appropriate at the base of a slope where a low toe wall is needed to stabilize the slope and reduce slope steepness.
- Will not resist large, lateral earth stresses.
- Should, where appropriate, be used with soil bioengineering systems and vegetative plantings to stabilize the upper bank and ensure a regenerative source of streambank vegetation.
- Require a stable foundation.
- Are expensive to install and replace.
- Appropriate where channel side slopes must be steeper than appropriate for riprap or other material, or where channel toe protection is needed, but rock riprap of the desired size is not readily available.
- Are available in vinyl coated wire as well as galvanized steel to improve durability.
- Not appropriate in heavy bedload streams or those with severe ice action because of serious abrasion damage potential.

Live stakes

Live stakes are live, woody cuttings which are tamped into the soil to root, grow and create a living root mat that stabilizes the soil by reinforcing and binding soil particles together, and by extracting excess soil moisture (Figs. 25, 26).

Applications and effectiveness

- Effective where site conditions are uncomplicated, construction time is limited, and an inexpensive method is needed.
- Appropriate for repair of small earth slips and slumps that are frequently wet.
- Can be used to stake down surface erosion control materials.
- Stabilize intervening areas between other soil bioengineering techniques.
- Rapidly restores riparian vegetation and streamside habitat.
- Should, where appropriate, be used with other soil bioengineering systems and vegetative plantings.
- Enhance conditions for colonization of vegetation from the surrounding plant community.
- Requires toe protection where toe scour is anticipated.

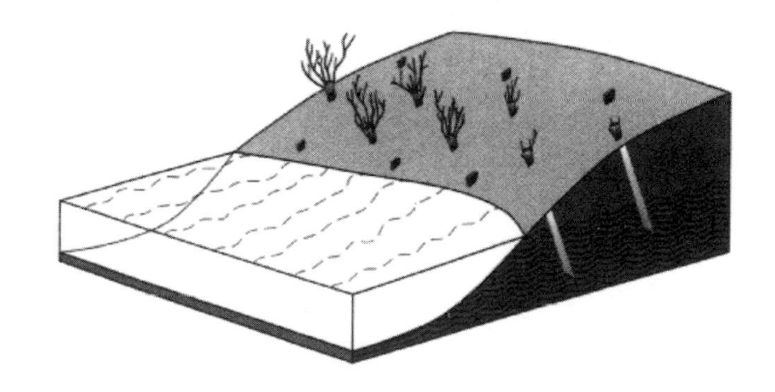

Fig. 25. Live Stakes are woody cuttings, which are tamped into the soil to root, grow and create a living root mat that stabilizes the soil. It reinforces and binds soil particles together and extracts excess soil moisture. (FISR 1988)

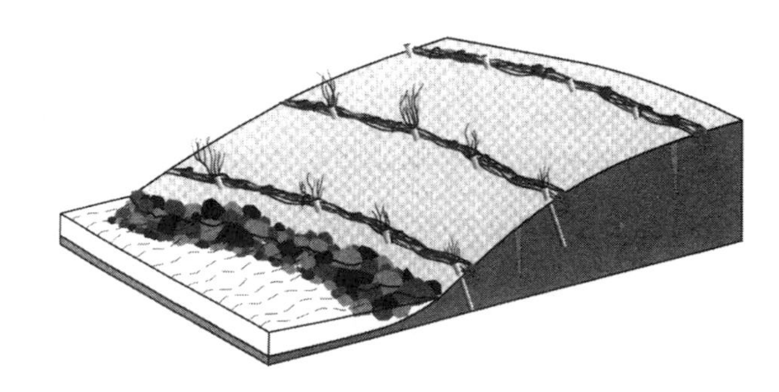

Fig. 26. Live Facines are dormant branch cuttings bound together into long sausage-like, cylindrical bundles and places in shallow trenches on slopes to reduce erosion and shallow sliding. (FISR 1988)

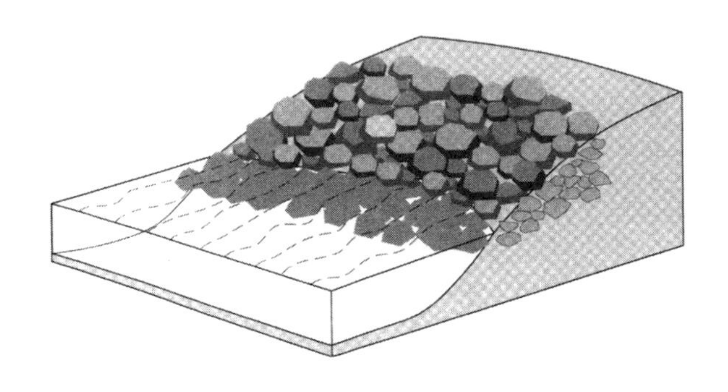

Fig. 27. Riprap is a blanket of appropriately sized stones extending from the toe of slope to a height needed for long term durability. (FISR 1988)

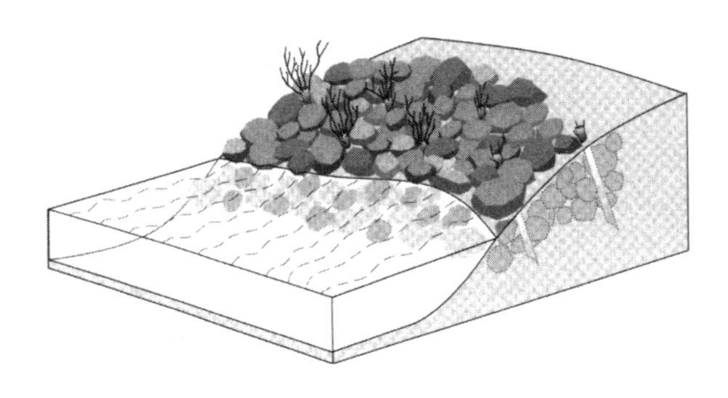

Fig. 28. Joint Plantings are live stakes tamped into joints or openings between rock which have been previously installed on a slope or while rock is being placed on the slope face. (FISR 1988)

Riprap

Riprap is a blanket of appropriately sized stones extending from the toe of slope to a height needed for long-term durability (Fig. 27).

Applications and effectiveness

- Can be vegetated—see joint plantings (Fig. 28).
- Appropriate where long-term durability is needed, design discharge are high, there is a significant threat to life or high value property, or there is no practical way to otherwise incorporate vegetation into the design.
- Should, where appropriate, be used with soil bioengineering systems and vegetative plantings to stabilize the upper bank and ensure a regenerative source of streambank vegetation.
- Flexible and not impaired by slight movement from settlement or other adjustments.
- Should not be placed to an elevation above which vegetative or soil bioengineering systems are an appropriate alternative.
- Commonly used form of bank protection.
- Can be expensive if materials are not locally available. ■

REFERENCES

Center for Watershed Protection, Silver Spring, MD. <center@cwp.org>

City of Vancouver Downspout Disconnection Program, <http://www.city farmer.org/downspout96.html>

Clayto, Richard.1988 (Entry #181). *Why Stormwater Matters.* Maryland Stormwater Design Manual. http://www.stormwatercenter.net.

Dunster, J., and Dunster K., 1996. *Dictionary of natural resource management.* Vancover: University of British Columbia.

FISR (Federal Interagency Stream Restoration) Working Group. 1998. *Stream Corridor Restoration: Principles, Processes and Practices Manual.* Washington, DC: National Engineering Handbook, Part 653. GPO No. 0120-A; SuDocs No. A 57.6/2:EN3PT.653. <http://www.usda.gov/stream_restoration>

Friends of the Buffalo River, 1996. *Buffalo River Greenway Plan and Design Guidelines.* Buffalo, NY: Friends of the Buffalo River.

National Research Council (NRC). 1992. *Restoration of aquatic ecosystems: science, technology, and public policy.* Washington, DC: National Academy Press.

Schueler, T. 1995. "The importance of imperviousness." *Watershed Protection Techniques (1)* 3: 100-111.

Simmons, D., and Reynolds, R. 1982. Effects of urbanization on baseflow of selected southshore streams, Long Island, NY. *Water Resources Bulletin* 181 (5): 797-805.

Stormwater Center Website <http:///www.stormwater center.net/ Manual Builder>

Donald Watson, editor

Summary
Ease of way finding is a requirement of any plan organization of the physical environment. The urban designer's designation of circulation systems must be coordinated with the design and placement of informational graphics. This article presents an overview of guidelines and standards for graphic communication in urban settings, including pedestrian way finding, street graphics and interpretive planning, by which visitor information and orientation graphics are coordinated with architecture and planning.

Key words
circulation, evaluation, graphics, pedestrian, interpretation, maps, motorist, orientation, planning, signage, way finding

PHOTO: Donald Watson

Graphic communication, way finding and interpretive planning

1 WAY-FINDING

Way finding can be described as spatial problem solving by a pedestrian traveler or motorist in determining one's destination, distance and path. Pedestrians in a hurry or otherwise focused upon a planned destination will follow the *perceived path of least effort* (even if it may be actually longer). Clear circulation and consistent and easily read signage should be easily seen and understood amidst conflicting or distracting sights or sounds. In all cases, clarity of direction from the viewer's changing perspective is the guiding criterion in designing circulation for ease of orientation and way finding.

Easily maneuvered and distinct circulation paths in public places should be provided—especially where there are different purposes, speeds and streams of pedestrian traffic— along with appropriate "step-asides" to step out of the rush of pedestrian traffic. This is especially helpful at points of transition, such as entryways, where *cul-de-sac* seating areas and/or orientation places are needed for both safety and convenience.

A clear circulation system that is free of conflicting traffic should be evident to all, especially to first time visitors. Graphic signs and symbols help to provide consistency, redundancy and reassurance to overlay and support way finding,

Kevin Lynch (1960) discusses the importance of way finding in urban contexts in *The Image of the City*. Studies reported by Rachel and Stephen Kaplan (1998) indicate the value of orientation and way finding, alongside the negative impacts of being lost or disoriented in a variety of circumstances. They describe how when people feel oriented and self-assured, they can find their way around. Their openness to explore an area is increased..."Making it easy for visitors to acquire knowledge of their situation will contribute to the quality of their experiences."

Design for way finding should provide:

- **Reassurance** about orientation and way finding communicated directly in the way that a facility or urban setting is designed. Circulation pathways and graphic communication should be clear, direct and obvious to a first time visitor.

- **Visitor's understanding** of the spatial organization of a setting, including locations of likely destinations and routes to them, should be provided by introductory orientation stations and maps located at significant decision points.

Passini (1984) presents a methodology for designing circulation and graphic systems for way finding in urban places (Box A). He defines *spatial orientation* as the ability to orient oneself in space, and

Credits: Selected illustrations, text and data cited in this article are from following key references: *With People in Mind* by Kaplan, Kaplan and Ryna (1998), *Wayfinding in Architecture* by Romedi Passini (1984) and *Street Graphics* by Ewald and Mandelker (1971). Additional citations are listed in the References.

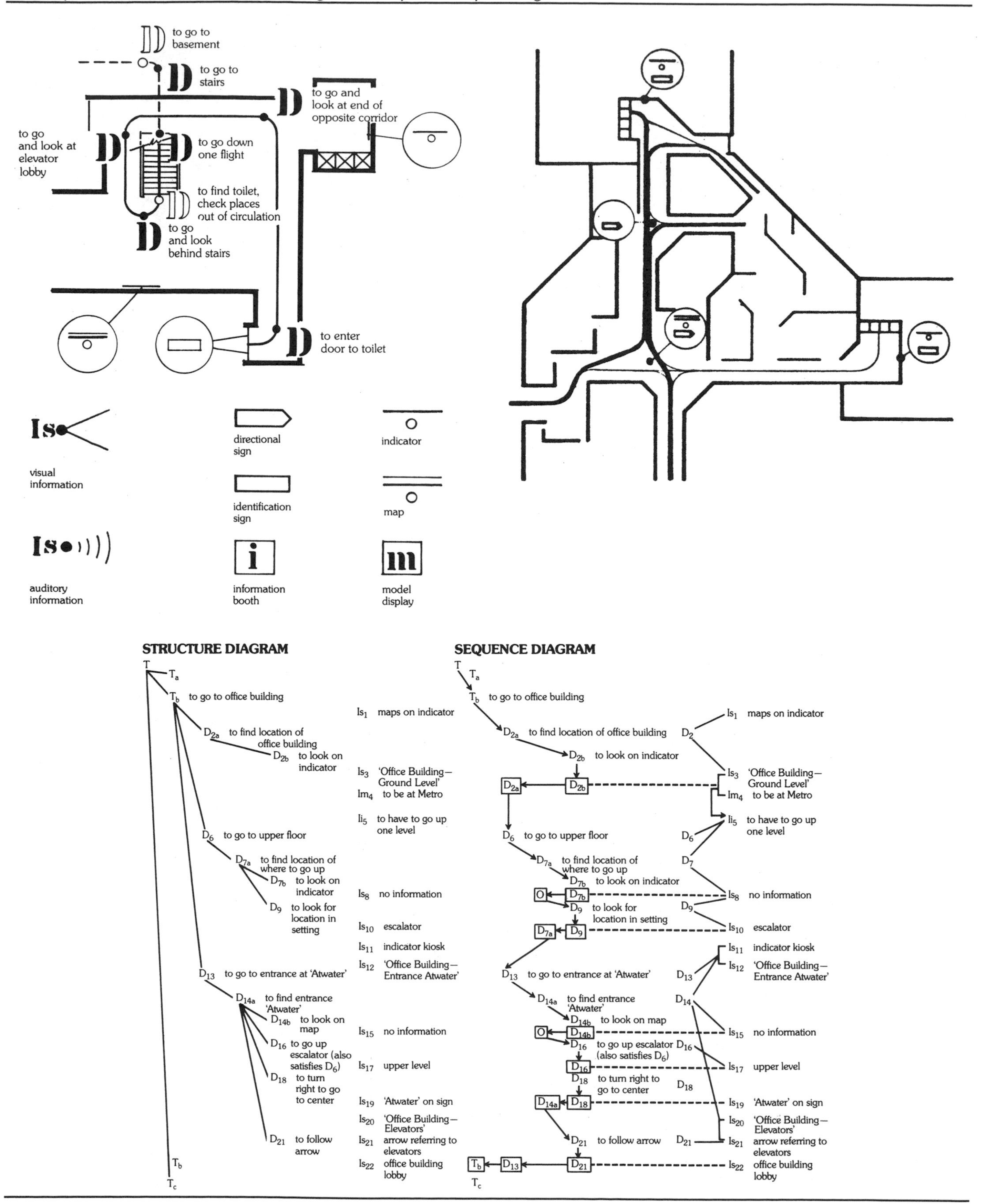

Box A: Plan and circulation path analysis establishes each decision point, for which ease of way finding should be mutually reinforced by design of the circulation and by directional and identification graphics. (Passini 1984)

way finding as the ability "to determine what to do in order to reach a destination." Disorientation results if one is lost, that is, without a "mental map" or a spatial representation of where one is *and* also without the ability to take action to determine one's location. He defines three purposes of way finding signs: direction (where to go), identification (identifying object, function or destination), and reassurance (repetitive signs that help reassure the pedestrian).

Passini recommends that way finding be evident in planning every public environment, mapped in spatial terms as a sequence of decision points in the circulation, with each major decision point provided with consistent graphic signage. A way finding design is thus planned and designed as a *circulation decision plan*, to which Passini gives the term *Ariadne's Thread*, after the Greek myth of Theseus's escape from the Minotaur, retracing the thread given by Ariadne.

Mapping

Way finding can be enhanced by posted maps, portable maps, and signs that are well designed. If not done well, maps can contribute to confusion, rather than clarity. (Figs. 1-6)

- Maps should identify key decision points in a visitor's circulation path (Fig. 1).
- Maps should be simplified and instantly understood "for the mind's eye" in graphic representation (Figs. 2-4).
- Maps should contain key label and symbol information "where one needs it," rather than with complex legends.
- Maps and orientation signs should display information from the viewer's perspective, utilizing north arrows if possible for consistent orientation (Fig. 5).
- Maps and orientation signage should be "prototyped" and tested as mock-ups before installation.

Fig. 1. Maps should emphasize main highlights and path structure.

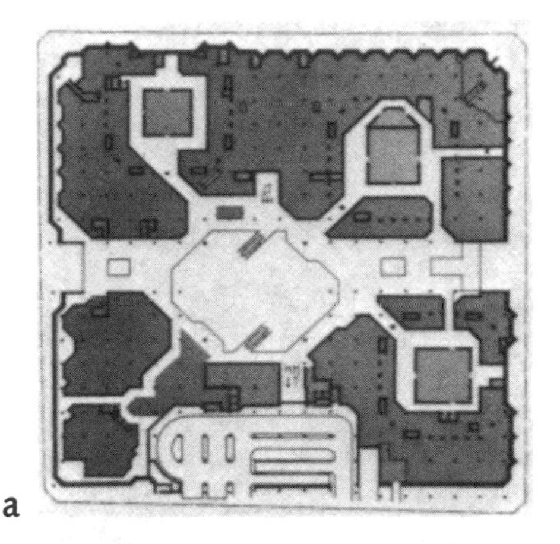

a

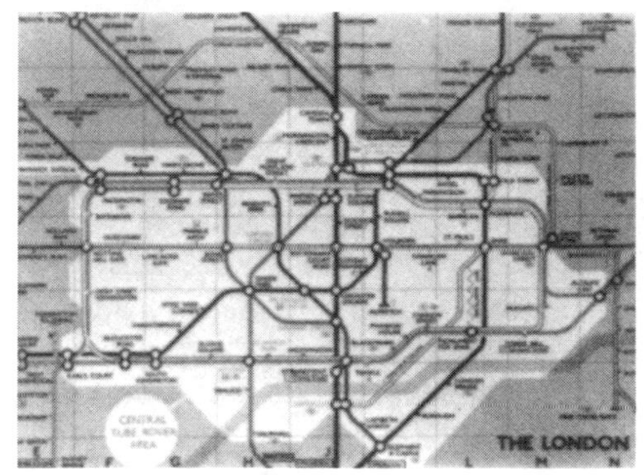

b

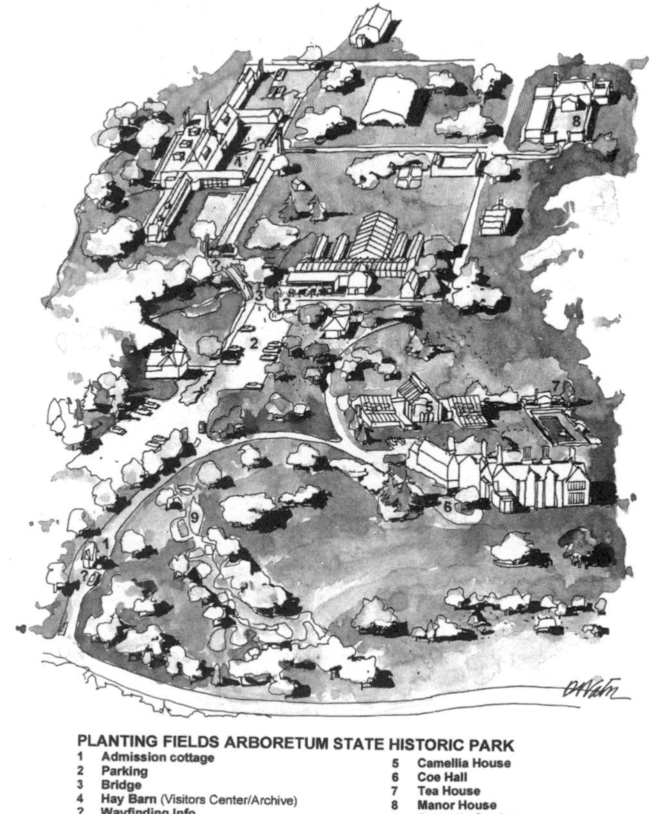

c

Fig. 2. Mapping choices (a) true-to-scale plan (b) diagrammatic, and (c) pictorial axonometric. (EarthRise Initiatives)

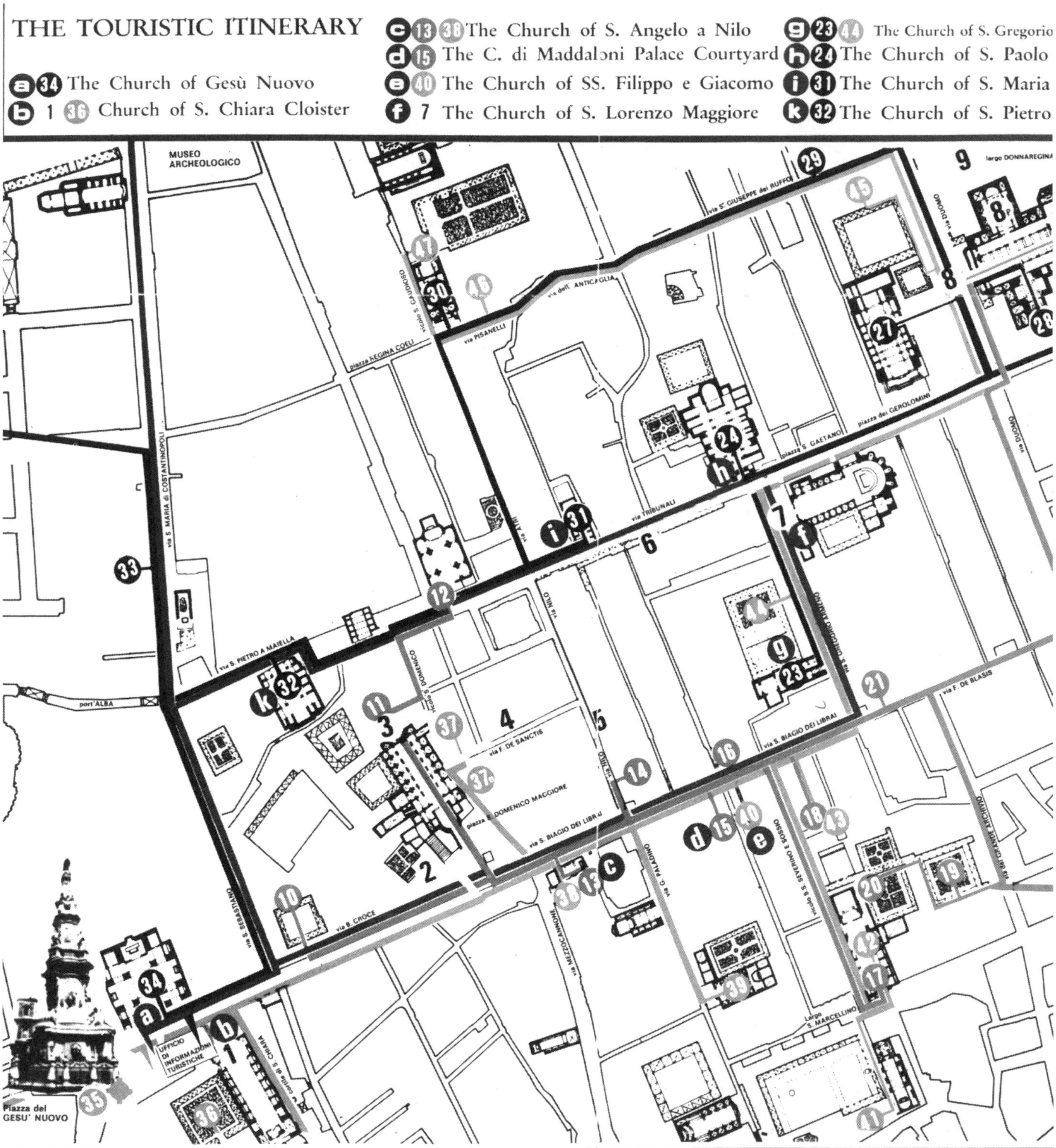

THE TOURISTIC ITINERARY

a 34 The Church of Gesù Nuovo
b 1 36 Church of S. Chiara Cloister
c 13 38 The Church of S. Angelo a Nilo
d 15 The C. di Maddaloni Palace Courtyard
e 40 The Church of SS. Filippo e Giacomo
f 7 The Church of S. Lorenzo Maggiore
g 23 44 The Church of S. Gregorio
h 24 The Church of S. Paolo
i 31 The Church of S. Maria
k 32 The Church of S. Pietro

Fig. 3. Extract of diagrammatic, true-to-scale plan of Naples, Italy Old City, a stratified multilayered map. Itineraries are distinguished by color and symbol for tours of Baroque, Rococo, Medieval, and Renaissance sites. (Courtesy: Azienda Autonoma di Soggiorno Cura e Turismo, Napoli)

Fig. 4. Precise and true-to-scale axonometric of Acropolis provides easily visualized orientation and way finding. Drawing by N. Gouvoussis reproduced by permission Gouvoussis Publishers.

Fig. 5. At Albany, New York Visitor's Center, way finding signage includes maps oriented to the perspective of the visitor.

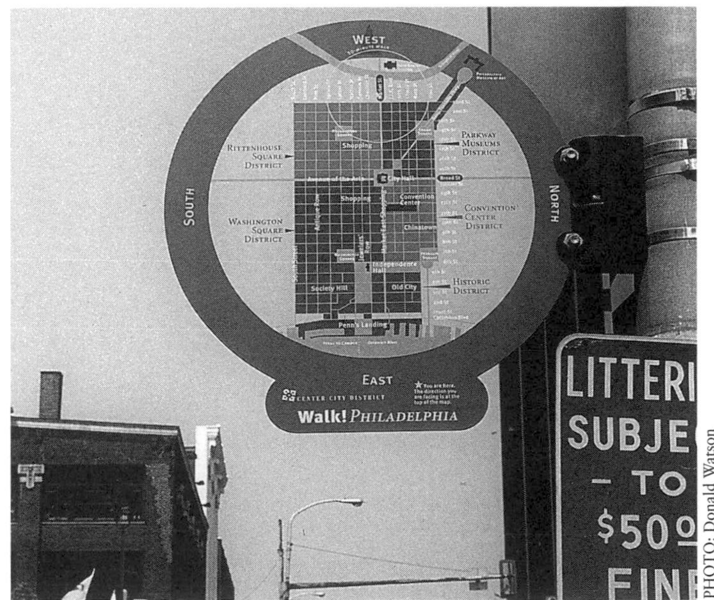

Fig. 6. Philadelphia street graphics utilize its memorable downtown plan for orientation and way finding.

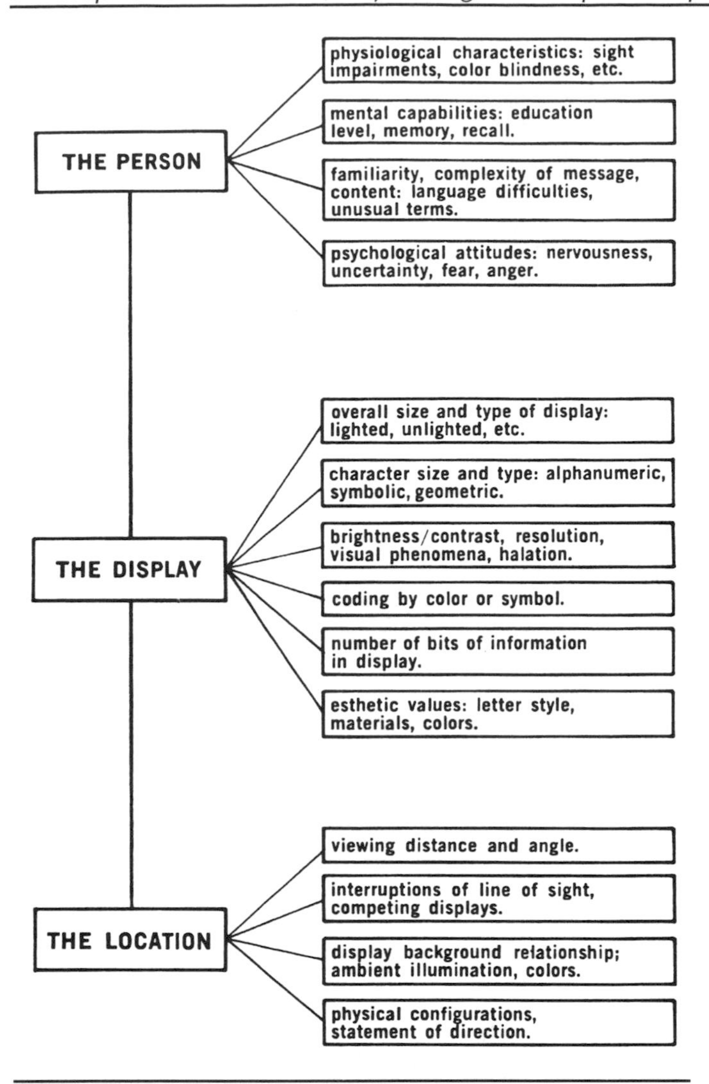
THE PERSON
physiological characteristics: sight impairments, color blindness, etc.
mental capabilities: education level, memory, recall.
familiarity, complexity of message, content: language difficulties, unusual terms.
psychological attitudes: nervousness, uncertainty, fear, anger.
THE DISPLAY
overall size and type of display: lighted, unlighted, etc.
character size and type: alphanumeric, symbolic, geometric.
brightness/contrast, resolution, visual phenomena, halation.
coding by color or symbol.
number of bits of information in display.
esthetic values: letter style, materials, colors.
THE LOCATION
viewing distance and angle.
interruptions of line of sight, competing displays.
display background relationship; ambient illumination, colors.
physical configurations, statement of direction.

Fig. 7. Elements of visual communication by signs. (Fruin 1971)

PHOTO: Donald Watson

Fig. 8. Way finding and information signage reinforced by lighting. Eero Saarinen's TWA Terminal, JFK Airport, New York.

Directional signs for pedestrians

John J. Fruin (1971) discusses directional signs as visual aids in way finding. (Also see his article on *Pedestrian Planning* earlier in this volume). Fruin describes how signage helps to "confirm the pedestrian's expectations of a space, communicating supplemental information and assistance." He details the three basic elements of signing as the person or viewer, the display or sign itself, and the location of the sign. (Fig. 7).

The attributes of good signing can be compared to factors that enhance human short-term memory, essentially: "to introduce, repeat, repeat, and summarize," ideally via some easily remembered graphic. Short-term memory is improved by rehearsal, or in the case of signage, by introductions to the terms and conventions of the signing system, including:

- use of short, familiar, and consistent terms,
- avoidance of the need to translate the sign message into other units, terms or meanings.
- use of repetition to improve continuity and consistency of sign format and means of presentation.

Different degrees of learning are required for understanding signs. Complicated, color-coded signing systems can confuse the pedestrian with insufficient time, or with visual impairment or learning disabilities. An estimate 8% of the adult population is subject to some extent of color blindness, so that color-coding alone should not be relied upon for key way finding graphics. Many others have functional illiteracy or the inability to understand complicated instructions. The display or sign itself should be easily recognized as a directional sign.

Sign location can be a significant factor in clarity of the communication. Way finding signage should have clear, unobstructed lines of sign, and good viewing angles. Competing visual displays, or a confused background, or reflective glare off the sign face all significantly reduce its value.

Lighting should reinforce the directional lighting, ideally be illuminating directional signage to be distinctive in the visual field, with backlit signs providing nearly ideal clarity and illumination of the message. In high circulation spaces, such as transit terminals, lighting should reinforce the direction of flow (Fig. 8).

Exit signs selected by the designer for emergency evacuation directions should be evaluated for legibility in smoke conditions. *The Lighting Handbook* (IESNA 1998) provides detailed standards and requirements of emergency exit signage applicable to enclosed or semi-enclosed structures and spaces.

Letter sizes recommended for pedestrian signs

For directional signs legible to the pedestrian, a rule-of-thumb for directional sign lettering height recommended by Fruin is a letter-size minimum of 2 in. (5.08 cm), plus an additional 1 in. (2.54 cm) of letter height for each 25 ft. (7.6 m) of viewing distance. Thus to be legible from 100 ft. (30.5 m), a pedestrian sign would be 4 in. (10 cm) high. More specific dimensional guidelines for signs that have interpretive content beyond orientation and way finding are indicated in Table 1. (Ham 1992, p. 266)

Table 1. Viewing distances and minimum heights of letters for informational and interpretive signs. Sizes given are minimum for each viewing distance. (Ham 1992)

Type of Text	1 to 4 feet	4 to 6 feet	30 feet	60 feet
Titles	2cm (3/4") >72 pt.	2.5cm (1") >96 pt.	10cm (4") 348 pt.	15cm (6") >567 pt.
Headings	1.3cm (1/2") >48 pt.	2cm (3/4") >72 pt.	8cm (3") 288 pt.	13cm (5") >480 pt.
Body text	0.6cm (1/4") >24 pt.	1.3cm (1/2") >48 pt.	6cm (2") 192 pt.	10cm (4") >384 pt.
Captions and specimen labels	0.5cm (3/16") >18 pt.	.6cm (1/4") > 24 pt.	N/A	N/A

2 STREET GRAPHICS

Street graphics as defined by Ewald and Mandelker (1971) is "graphic communication along streets and highways, including symbols and letters as they appear on signs, billboards, storefronts, marquees, canopies, and all other stationary visual media." They recommend and propose design standards for a comprehensive street graphics "system" to help order *all* means of visual communication seen from a public right-of-way. Such standards—difficult to implement due to the myriad of private and public property owners involved in exterior urban settings—are nevertheless necessary to achieve a coherent visual environment.

Comprehensive street graphics have been controlled in designated zones, such as pedestrian zones, historic districts, and commercial and industrial areas of cities. In the United States this control is most often accomplished through private agreements among key business interests rather than by public regulation, or in districts largely controlled by a single owner. In addition, a few municipalities—such as Carmel and Ojai, California, Aspen, Colorado, Colonial Williamsburg, and Georgetown, Washington, DC—have adopted public regulations to reinforce the desire to project a distinctive and coherent community image.

In the terms of Ewald and Mandelker, the primary function of street graphics is "to index the environment, that is, to tell people where they can find what." They argue that:

> *Street graphics affect public safety and can make a significant contribution to the life-style of a community. They are an important expression of a community's culture. They can announce, inform, delight, stimulate, or designate. Most of all street graphics should clearly index. Using the environment for selling may be considered as a legitimate but special privilege to be granted only by official action of the community at particular, specific locations.* [Ewald and Mandelker 1971, p. 4]

The discussion that follows, excerpted from this reference, describes the role of street graphics to create effective communication between people and the urban environment.

Street signs

Ewald and Mandelker recommend against the widespread use of white condensed capital letters often used on a green background for street name signs, "First of all, all capital condensed letters are difficult to read, and it seems axiomatic that street signs should be designed

a

b

Fig. 9. Applying street graphic system (a) before and (b) after. (Ewald and Mandelker 1971)

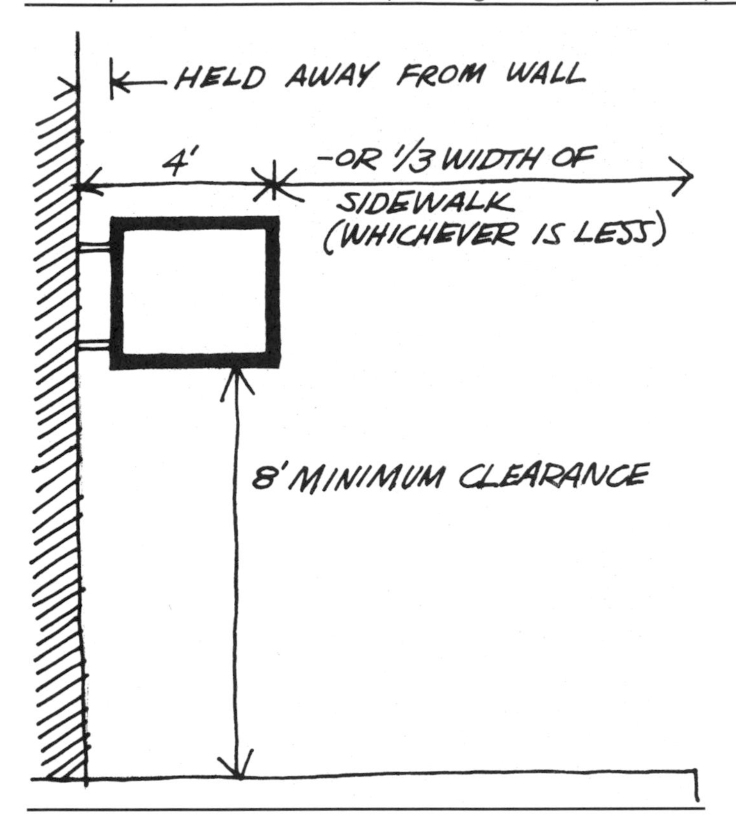

Fig. 10. Guidelines for graphics that project over sidewalks. (Ewald and Mandelker 1971)

for maximum legibility in the conditions under which they are most frequently seen—from a moving vehicle." [p. 6]. They recommend that the more legible combination of upper and lowercase letters with an "x" height (ratio of lowercase to capital letter height) of two-thirds be used for street signs, and that the size of letters and the total area of signs be determined on the basis of road widths and traffic speeds (Figs. 10–13).

A consistent approach to street numbers should be required as an element of street signage, at least 3 in. (7.6 cm) high. The size should of course be increased if necessary for legibility in the circumstances in which they are seen. In any streetscape, similar placement and location of address numbers adds to clarity and ease of identification.

Perception

In all but exclusively pedestrian zones, the prime audience for most street graphics is the motorist, whose orientation and control of speed and direction are crucial for their own as well as other's safety. For this reason alone, it makes sense to approach the problem of street graphic design from the perception point of view of the motorist in a car or other moving vehicle, in which case, the urban graphic designer should have foremost in mind that:

- Seeing takes time.
- There are limits to the detail that can be discerned (as well as the scope of the visual field that is observed).
- Contrast is of vital significance.

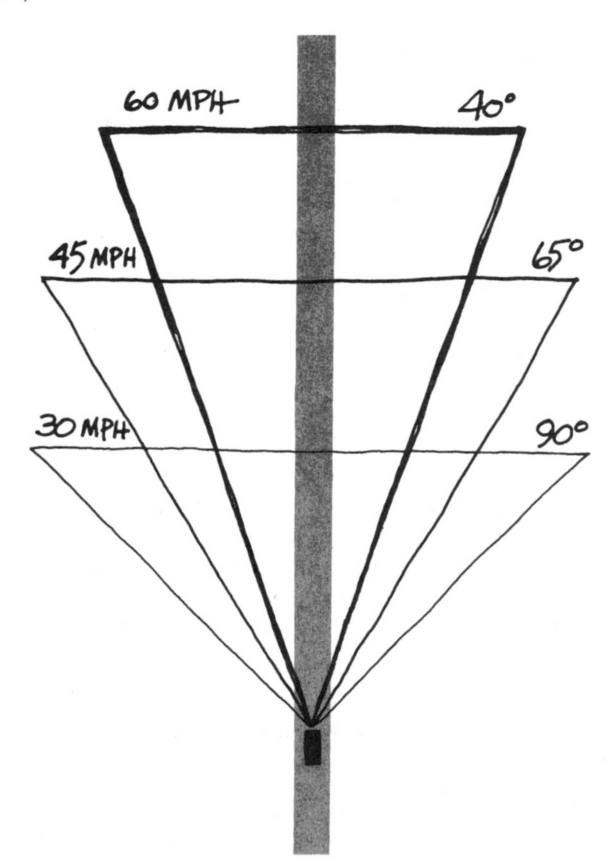

Fig. 11. Angle of Vision. (Ewald and Mandelker 1971)

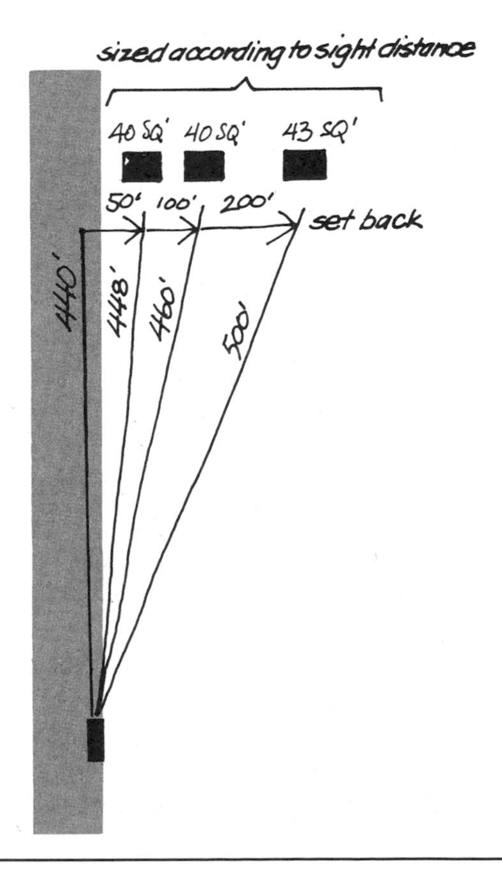

Fig. 12. Typical setback of graphic sign makes little difference in the size of graphic required. (Ewald and Mandelker 1971)

Analysis of how people see from a moving car is well established. The classic reference is *Human Limitations in Automobile Driving* by J. R. Hamilton and Louis L. Thurstone (1937). Tunnard and Pushkarev (1963), summarize their findings on perception from a moving automobile as follows:

As driving speed increases,

1. The driver's concentration increases.

2. The driver's point of concentration recedes. The driver's eyes feel their way ahead of the wheels.

- At 25 mi. per hr. (32 KpH), the natural focusing point lies approximately 600 ft. (180 m) ahead of the car;

- At 45 mi. per hr. (72 KpH) it lies some 1200 ft. (366 m) ahead of the car.

3. The driver's peripheral vision decreases.

4. Foreground details begin to fade.

- At 40 mi. per hr. (64 KpH), the closest point of clear vision lies about 80 ft. (15.24 m) ahead of the car.

- At 60 mi. per hr. (97 KpH), the driver can see clearly only that detail which lies within an area 110 to 1400 ft. (34 m) ahead of the car and within a forward viewing angle of 40°. Since at that speed, the distance between 110 to 1,400 ft. is covered in less than 15 seconds, it follows that elaborate detail in highway graphics is totally meaningless, if not distracting.

5. The driver's perception of space and speed deteriorates, and the driver's judgment becomes more dependent on visual clues picked up along the highway.

Visual responses are not instantaneous. It takes the human eye 0.1 to 0.3 second to fixate on an object, provided the eye and the object are in a relatively fixed position with respect to one another. Moreover, it takes the eye about one second to change focus from the dashboard to some detail on the road ahead.

One of the strongest visual stimuli is a flashing light. When seen out of the corner of the eye, a flashing light will cause the eyes to swing involuntarily to focus on it—a delightful sensation at the slow pace of Fremont Avenue in Las Vegas, or Piccadilly Circus in London, but dangerous under high-speed driving conditions. Similarly, when an object is difficult to identify, the eye fixates on it longer and jumps back to view it repeatedly.

Lettering sizes for signs for motorists

For directional signs legible to the motorist, Ewald and Mandelker recommend "1 in. of increased height for every 50 ft. of viewing distance" (2.54 cm for every 15 m), a rule-of-thumb stated differently than Fruin's recommendation for pedestrian signs, but with the same resulting sizes. Design guidelines for motorists' signage especially emphasizes the need for a readable typeface and good contrast between the letter and its background. Other important aspects of legibility include:

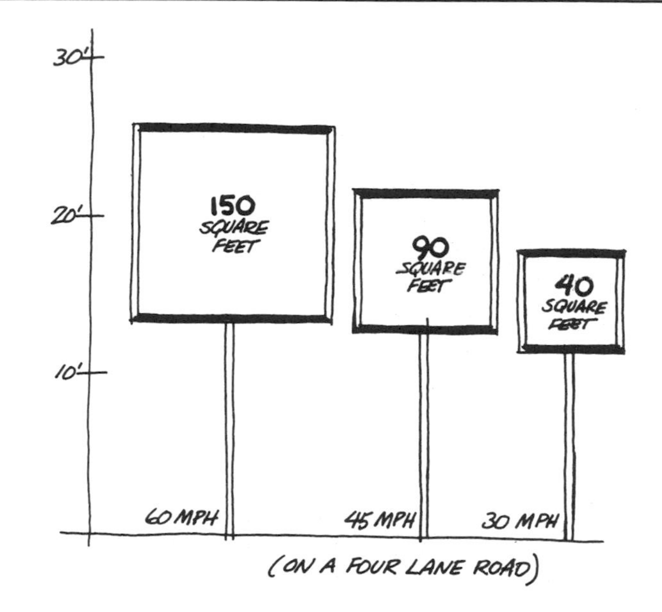

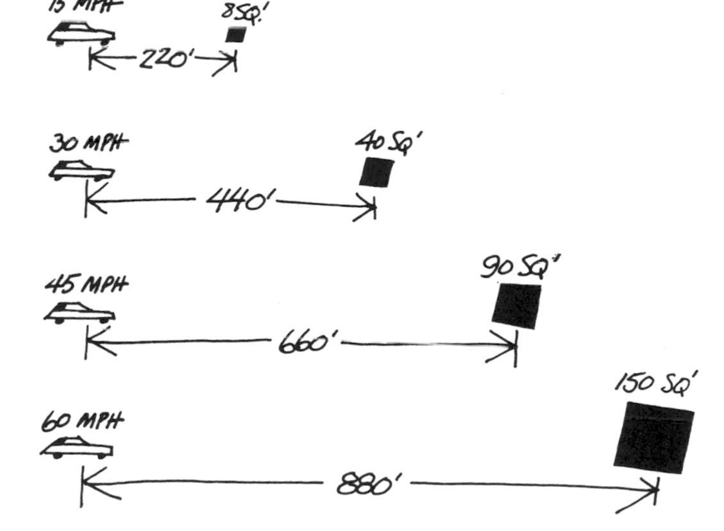

Fig. 13. Size of graphic legible to motorists depends upon roadway speed and number of lanes. (Ewald and Mandelker 1971)

Table 2. Operational summary of proposed system of street graphics (Ewald and Mandelker 1971)

Determinant	Designation	Basis of Control
Identity which street graphic is to communicate	Individual proprietor and/or community	Good design, within the framework of the other three determinants
Type of activity to which street graphic pertains	Commercial, Industrial, Institutional activities	Appropriateness
Character of surrounding area in which street graphic is displayed	Commercial/Industrial/ Institutional/Residential/ Rural Specially designated areas (historic, scenic, commercial plazas)	Compatibility
How the street graphic is seen	Pedestrian or motorist at various speeds; 2-4-6 lanes of traffic, expressways	Legibility

KEY MESSAGE	EDUCATIONAL CONTENT	EXAMPLES	OUTCOME EVALUATION	PROGRAM ELEMENTS
1 ***Albany Pine Bush Preserve*** ***A globally unique and endangered landscape...*** ***a place like no other on earth...***	• Ecology at the macro-level. • Once a >25,000, now >6,000 ac. • Glacial, geologic history, glacial lake, sand, wind dunes, barrens, fire & vernal ponds. • Natural landscape, threatened by inappropriate development and fire suppression. • Partners & volunteers now work to protect & manage the Pine Bush. *Emphasis is on "appreciation" and interactive engagement with the Pine Bush Preserve environment/landscape.*	• Views from the road and trails; special settings that inspire wonder and appreciation for the Pine Bush past, present and future. • Sun court: sundial, geologic sculpture, seasonal fountain (rainbow in summer, ice forms in winter) • Views of restored dunes, pitch pines, vernal ponds, & scrub oak barrens, native wildflowers & fire. • Role of fire and forest history (general macro-level).	• *% of general public that knows the Albany Pine Bush as globally unique and special.* • *number of visitors or "members" and % of visitors who can identify key Pine Bush images.* • *survey(s) of trail & facility users that demonstrate an incurred % of users, program participants and volunteers.* • *public recognition, e.g. news items and notices.*	**(A) Logo and brochures.** **(B) Trails/Interpretive settings.** **(C) Sun court.** **(D) Restored pine bush.** **(E) Orientation theater.**
2 ***Albany Pine Bush Preserve*** ***Discover the details...*** ***rare plants and animals in nature.***	• Ecology at the micro-level, using magnifying glass & microscope. • Discover the beauty & complexity of nature. • Trees, plants, insects, amphibians, birds, mammals, co-dependencies. • Indigenous and invasive species. *Emphasis is on details & stories, nature study, natural sciences and biology, developed and told by educational partners.*	• Trail stations & exhibits at the Center & trail-heads to introduce rare species and tell the story of codependence within the Preserve. • Settings on trails that inspire appreciation of interdependence of nature. • Introductory self-study and guided tours. • Interpretive programs with graphics to reinforce modules. • Special topics based on interests of staff, volunteers & partners.	• *prototyping* • *user evaluations of tours, programs, & visitors which demonstrate % that know APB is home to multiple rare & endangered species, including Karner Blue butterfly.* • *program attendance* • *educational performance measures* • *income from programs* • *gift shop sales*	**(F) Maps/Trail head signs.** **(B) Trailheads, trails/Interpretive settings.** **(G) Outdoor classroom.** **(H) Greenhouse & garden landscaping.** **(I) Biodiversity Lab.** **(J) Gift Shop as an interpretive station.** **(K) Program modules.**

billboards & advertising
mailer
maps
WEB page print out
directional signs (arriving and departing)
exit salutation
1 Entry sign & logo
legal residential building lot set aside
(future) manager's residence
buffer
overflow parking (placed within landscape & fields)
parking (dispersed)
buffer
Restored Pine Bush
(future) community recreation & education wing
Picnic & Outdoor Events restored landscape
Event Pavilion
canteen/ restrooms
arrival/ drop-off plaza
2 Gateway orientation
2a sun court "sun, sky and water"
3e Naturalist's outdoor classroom
greeting & exit comments
2b maps & models
5c gift shop
5a orient' theatre
5 APBPC Center
5b Public Forum
Educational resources
Research archive
staff offices
Oxbridge Room
3a green house
4c Field study burn area
4 Woodland trail 60 min. +
4b Vernal pool classroom
4a Trailhead
numbered interpretive trail
3d program theme
"we can't manage without you"
burn management
3c program theme
"right here in our own backyard"
living history
3 Discovery & Sensory Trail 20 min.
3b program theme
"it's all in the details"
ecological history & restoration
"no place like this on earth"
Restored Pine Bush butterfly gardens & winter bird feeders
(future) cafeteria terrace
service & recycling
EarthRise Initiatives Interpretive Planning
key
First Phase Construction
Interpretive sequence
Fund-raising ventures
PROGRAM ELEMENTS

Box B: (a) matrix of theme, and key messages matched (moving horizontally) to interpretive program elements. (b) Extract of storyboard that places the themes and message content in circulation diagram. From Interpretive Plan of Albany, New York , Pine Bush. Earthrise Initiatives.

- Spacing between the individual letters making up a word,
- Thickness of the letters,
- General layout,
- Contrast. It is obvious enough white letters on a black background are more legible than white letters on a pale grey background. Colors, too, contrast with one another in ways that affect the legibility of words.

The amount of information that people can receive, process, and remember is strictly limited: the upper limit may be considered to be "five plus or minus two," that is, not more than seven items, however such items are measured. For purposes of the street graphics approach discussed here, an item is one syllable, symbol, abbreviation, broken plane, or discontinuous odd shape.

A summary of the street graphic system proposed by Ewald and Mandelker is summarized in Table 2.

3 PLANNING FOR INTERPRETATION

For many locations, providing orientation for visitors through interpretive media is combined with way finding. The terms *interpretation* and *interpretive planning* describe a comprehensive approach to design of communication media and their setting to inform people about places.

Interpretive planning creates a framework for signs, exhibits and related learning opportunities presented as part of a visitor's experience of a place. The interpretive approach to public communication can be traced to the early 20th c. work of Enos Mills (1880–1922) —a naturalist and author who contributed to the National Park Service Act of 1916 and established Rocky Mountain National Park—and of Freeman Tilden (1883–1980), a nature writer and consultant to the U.S. Park Service. Tilden defined interpretation as. . . ."an educational activity, which aims to reveal meanings and relationships through the use of original objects, by firsthand experience, and by illustrative media, rather than simply to communicate factual information." (Tilden 1957)

Developed over the past thirty years, interpretive planning coordinates graphic design, circulation, orientation and way finding, as well as educational signs and exhibits based on consistent presentation of thematic qualities. The art of interpretation is evident in design of educational media in public parks, museums and other visitor venues, where the quality of visitor experience is a predominant criterion. The methods of interpretive planning are also applicable to orientation and way finding in urban contexts. For examples of early steps in the process of formulating the script of an interpretive plan, see Box B.

Beck and Cable (1998, p.10) offer a list of guiding principles for interpretation based on writings of Enos Mills and Freeman Tilden. Although intended for designers of interpretive programs for recreational parks and nature preserves, they are also useful as a checklist for orientation and interpretive media for urban settings.

1. Interpretation should relate the subject content to the lives of visitors.

2. Interpretation goes beyond providing factual information, should reveal deeper meaning and promote inquiry and discovery.

3. The interpretive presentation should be designed as a story that informs and enlightens.

4. The purpose of the interpretive story is to inspire and to provoke visitors to broaden their horizons.

5. Interpretation should present a complete theme or thesis and address the whole person.

6. Interpretation for children, teenagers, and seniors, when these comprise uniform groups, should follow fundamentally different approaches for various learning styles and abilities.

7. Every place has a history. Interpretation can bring the past alive.

8. Media technology can reveal the world in exciting new ways if done with foresight and care.

9. Interpretative planning should concern itself with the quantity and quality (selection and accuracy) of information presented. Avoid long text.

10. Interpretation should be based on familiar communication techniques, developing knowledge and experience in both interpreters and visitors.

11. Interpretive writing should address what visitors would like to know. Use visitor interviews and surveys to test interpretive media.

12. The interpretive program must be capable of attracting broad community, volunteer, and organizational support and be subject to change and improvement.

13. Interpretation should instill the ability to appreciate and respect and preserve the environment and culture it represents.

14. Interpretation can promote positive and engaging experience of place though coordinated program and facility design, utilizing visual sequence for an unfolding story.

15. The ability to convey passion and to instill curiosity and interest in the subject content is an essential ingredient for effective interpretation.

An interpretive planning framework is best defined by an overriding theme, comprised of a set of easily communicated messages and program elements that may convey various subthemes of the educational and recreational experience.

Visitor orientation and learning is essentially self-directed and self-motivated. Many if not most visits to public places for tourism are primarily for recreation, with education being secondary. Interpretive planning therefore utilizes the principles of discovery learning, recognizing the innate curiosity and also the great variety of learning styles of individuals of all ages, interests and abilities.

An interpretive plan is constantly evaluated by monitoring the visitor experience and requires careful testing and feedback, especially when

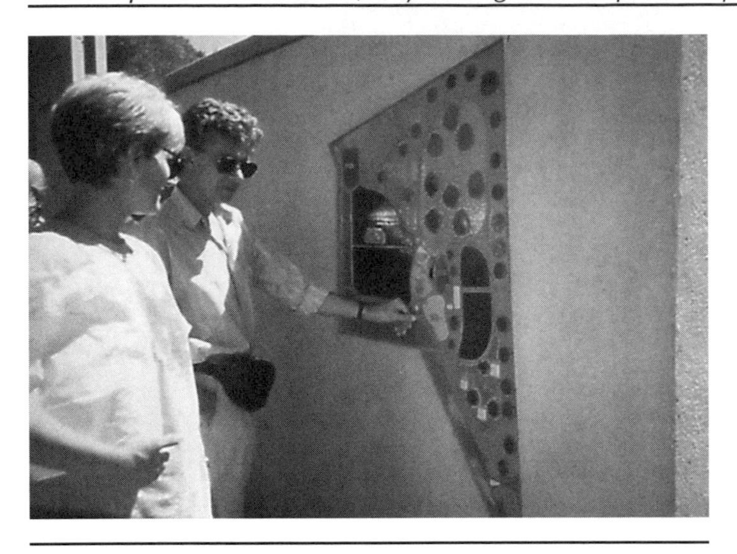

Fig. 14. Signage wall with tile raised tactile and visual map for visually impaired, combined with chimes at destination points. Flood Park, San Mateo County, California. Moore, Iacofano, Goltsman.

Fig. 15. Talking signs, San Francisco, CA. Individual identifying a bus shelter from a distance with range of 50 ft. (15 m) with infrared way finding receiver. Designer: Bill Crandall with Smith Kettlewell Rehabilitation.

the purpose of interpretive design is to achieve stated objectives based upon visitor expectations and capacities.

The importance of universal design. Universal design is an approach to designing interpretive programs and facilities to be fully accessible and enjoyable by people of all ages and abilities. Universal design seeks out opportunities: *e.g.*, indoor and outdoor gardens and botanical displays offer opportunities to create "sensory gardens" which can be enjoyed (and also interpreted) by the visually impaired through the design of fragrance and sounds (Figs. 14 and 15).

Terms frequently used in interpretive planning

An interpretive plan consists of:

- Overall theme based on the vision and goals of public outreach.
- Key messages, sub themes of the interpretive/educational curriculum.
- Elements or media for communicating program content (publications, signs, exhibits, landscape, buildings, trails)
- Outcomes to be achieved through the interpretive and educational program.

Overall theme. Usually stated in a succinct and memorable phrase, the theme encapsulates and communicates the vision of the interpretive program, derived from the organizational mission. The mission and the theme help to convey core values, goals and aspirations of the organization in a compelling and memorable way.

Subthemes and key messages. The key subthemes—often definable as messages and/or topic content of the interpretive (educational) programs—provide focus and coherence to all communications.

Interpretive elements are the physical objects of the media and materials by which the interpretation is conveyed. They typically may include a Web page about the site, brochures, signs, maps and walk books, exhibits and the landscape and architectural settings that reinforce the interpretive messages and values. These may be redundant and "multileveled," that is, presented in different formats to respond to the diversity of learning styles of people of all ages and abilities.

Outcomes define the intended or actual results in terms of the interpretive plan elements, obtained from an unbiased process of evaluation. These specific outcomes (such as the intention to convey a particular educational topic awareness) should build to a larger overall outcome.

Interpretive planning is used to orient visitors in public places designed for recreation and education, such as public parks, nature preserves, visitor centers, museums, and similar outdoor educational venues. (*Cf.* References, under National Association for Interpretation). Interpretive planning is based on the design of messages, as a story or thread of experience of a site.

Evaluation of outcomes

Interpretive planning is guided by evaluation, based on explicit objectives. Major graphic systems and exhibit should be tested as mock-ups and prototypes. The following is a protocol of evaluation commonly used in museum exhibits and equally applicable to outdoor and indoor graphics and interpretive settings.

Evaluation methods:

A. Front-end evaluation (predesign)

- Define the learning objectives of the exhibit
- Define the audience (age and/or learning group)
- Establish "predesign" guidelines by questionnaire, interview and "focus groups"
- Define the guidelines for preliminary exhibit design form and content, including "expected outcomes," and, to the extent possible, how these outcomes will be measured.

B. Formative evaluation (preinstallation)

- Prepare various mock-up designs and/or learning settings
- Pretest various mock-ups with the audience ("target" learning groups) including focus groups and interviews. (Evaluation is carried out by "third-party" evaluator (*e.g.*, different than the sponsor and/or designer).
- Document findings, including outcomes and measurements.
- Change the design as appropriate, including ways to continuously improve the exhibit/educational setting.

C. Summative evaluation(s) (postinstallation)

- Establish a means to continually evaluate and improve the effectiveness of the exhibit/learning setting. ■

Fig.16 Interpretive graphic sculptures. Bridge Square, Minneapolis, Minnesota Federal Reserve Bank plaza. The site's history is depicted in five bronze-cast medallions of three-dimensional models indicating civic improvements, starting with the Great Waterfall, the savannas and forests of the site prior to development through various phases of urban design and development from 1847 to the modern era. Graphic designer: Deborah Beckett, HOK Planning Group.

REFERENCES

Beck, Larry, and Ted Cable. 1998. *Interpretation for the 21st Century: Fifteen guiding principles for interpreting nature and culture.* Champaign, IL: Sagamore Publishing.

Ewald, William R., Jr. and Daniel R. Mandelker. 1971. *Street Graphics: a concept and a system.* Washington, DC: American Society of Landscape Architecture Foundation.

Fruin, John J. 1971, 1987. *Pedestrian planning and design.* Mobile, AL: Elevator Press.

Hamilton, J.R., and Louis L. Thurstone. 1937. *Safe driving: Human Limitations in Automobile Driving.* Toronto: Doubleday.

Ham, Sam H.1992. *Environmental Interpretation.* Golden, CO: North American Press.

Illuminating Engineering Society of North America (IESNA), 1998. *Lighting Handbook: Reference & Application.* Mark Rea, ed., New York: Illuminating Engineering Society of North America. <www.iesna.org>

Kaplan, Rachael, Stephen Kaplan and Robert Ryna (1998). *With People in Mind: design and management of everyday nature.* Washington, DC: Island Press.

Lynch, Kevin. 1960. *The Image of the City.* Cambridge, MA: MIT Press.

National Association for Interpretation. Fort Collins, CO. <www.interpnet.com>

Passini, Romedi. 1984. *Wayfinding in Architecture.* New York: Van Nostrand Reinhold.

Tilden, Freeman. 1957, 1977. *Interpreting Our Heritage. University of North Carolina.* Raleigh, North Carolina.

Tunnard, Christopher, and Boris Pushkarev. 1963. *Man-Made America: Chaos or Control: an inquiry into selected problems of design in the urbanized landscape.* New Haven: Yale University Press.

Anne Whiston Spirn

Summary
This article summarizes the factors that influence air quality at street level. It identifies potential problem areas and describes urban design strategies to reduce either the level of air pollutants or human exposure to them. A classification analysis allows urban designers to classify air quality design opportunities and constraints

Key words
air flow, dispersion models, emission levels, environmental quality, health, pollution, radiation inversions, sinks, wind shadow

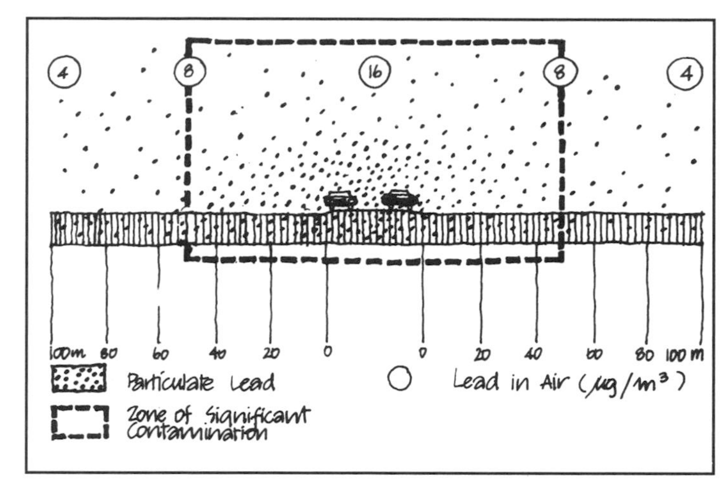

Fig. 1. In flat, open terrain under calm conditions, air pollution levels are highest adjacent to the road and decrease with distance from it. This diagram illustrates atmospheric levels of particulate lead generated by a traffic volume of 25,000 cars per day (adapted from Smith 1976).

Better air quality at street level: strategies for urban design

The street is the busiest outdoor open space in the city. People who live and work in urban areas spend much of their time driving, walking, sitting, and playing outdoors. For policemen, street vendors, and taxi drivers, the street itself is a workplace. Unfortunately, the street is also the source for much of the city's air pollution. In Boston, for example, transportation accounts for ninety-four percent of the carbon monoxide emitted within the city. In certain street-side locales, carbon monoxide concentration may exceed national safety standards for one-hour exposure, the length of a normal lunch break. Improving air quality at street level must be an objective of any comprehensive street improvement program.

The degree to which pollutants emitted at street level are concentrated or dispersed is greatly affected by the form of the city. Urban design, especially if integrated with other measures, could play an important role in dispersing street level air pollutants. At the very least, urban design can help in limiting human exposure in areas where pollutants are highly concentrated.

1 STREET-LEVEL AIR POLLUTION: DISTRIBUTION AND EXPOSURE

Transportation-related air pollutants and their spatial and temporal distribution differ greatly from spot to spot within a city. They are a function of commuting patterns, traffic volume and speed, meteorological conditions, the topography of urban form, and the materials of which the city is composed. Pollution levels may be as much as six to ten times higher in one place than another. The application of effective urban design measures to reduce exposure to street-level air pollution depends on an understanding of these factors, both individually and in combination.

Emission levels

The quality of pollutants emitted by motor vehicles is directly proportional to traffic volume and speed. Vehicles emit fewer pollutants at steady speeds and greater quantities in stop-and-go traffic or while idling at stoplights.

Proximity to the roadway

In flat, open terrain under calm conditions, air pollution concentrations tend to be highest adjacent to the road and to decrease with distance from it (Fig. 1).

Air circulation: winds and breezes

Wind patterns vary from city to city, influenced by regional climate and physiographic setting. The degree to which winds penetrate the city at street level in influenced by the orientation and continuity of open spaces, their dimension and shape, and the topography of

Credits: The work presented in this chapter was performed as part of research supported by the Boston Redevelopment Authority and the National Endowment for the Arts. An earlier version of this article appears in Moudon, 1987 (See References).

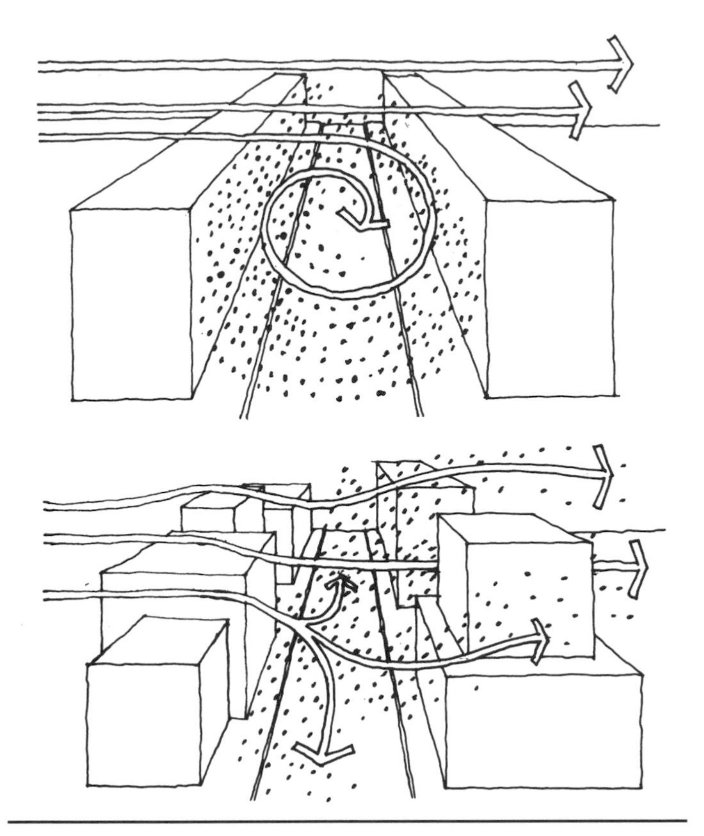

Fig. 2. Street canyons lined with buildings of similar height, oriented perpendicular to the wind direction (top), tend to have poorer air circulation than street canyons that are lined with buildings of different heights and interspersed with open areas (bottom).

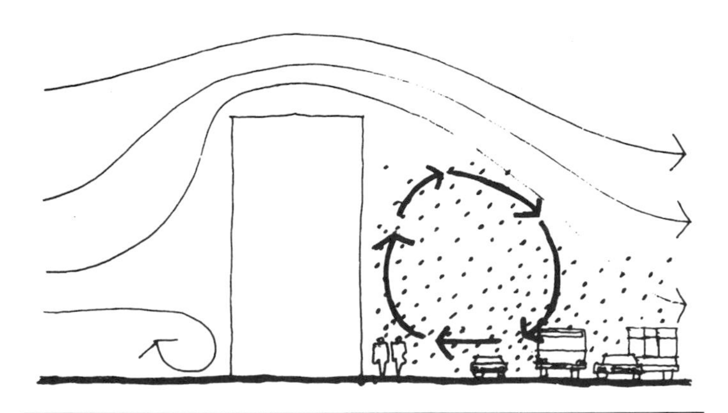

Fig. 3. Wind shadows have reduced air circulation and pollutants emitted with them may tend to build up.

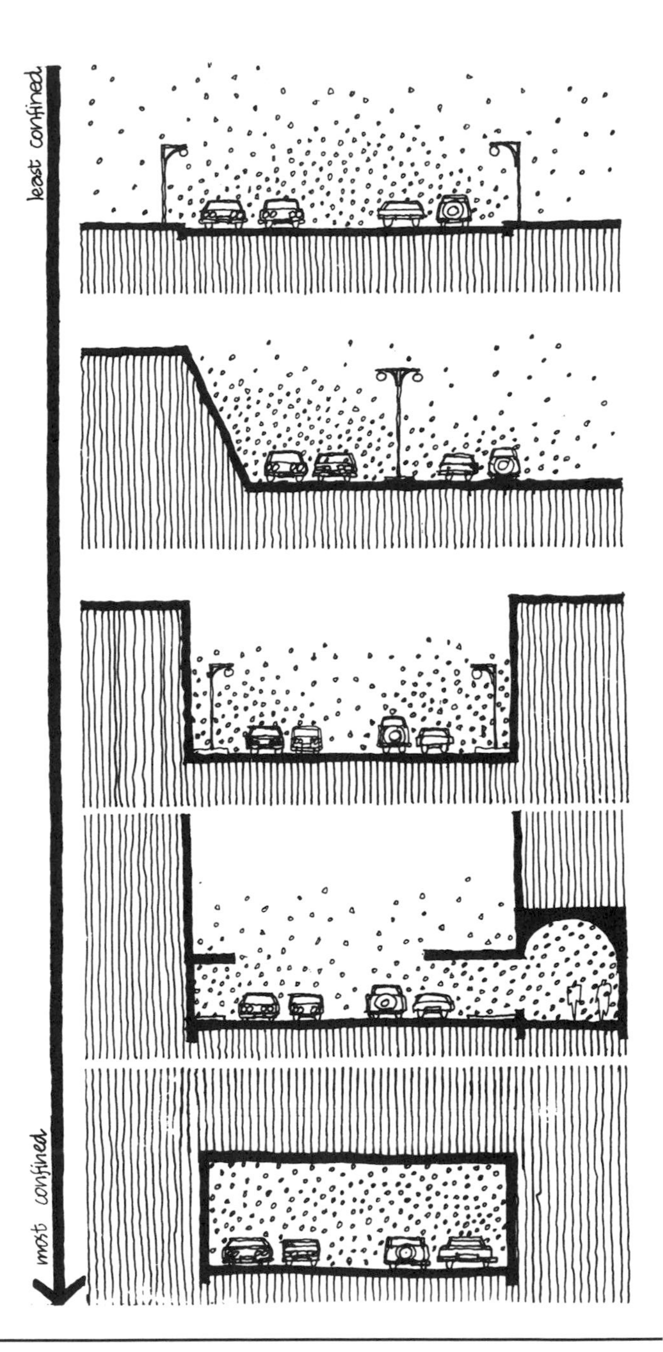

Fig. 4. The more enclosed a space, the more likely the accumulation of pollutants. (Adapted from Kurtzweg 1973.)

buildings. Street canyons, for example, may channel winds or prevent them from reaching street level, depending on their shape, dimension, and orientation (Fig. 2).

Within the city, wind speed and direction are altered as air flows around an obstacle (be it a building, a wall, or a grove of trees), creating zones of accelerated wind speeds and of reduced air circulation. Pollutants emitted within a well-ventilated situation may be quickly dispersed, whereas pollutants trapped with the wakes of buildings can become concentrated (Fig. 3). Patterns of air movement around an obstacle are influenced by its dimensions, orientation, shape, porosity, and surface roughness. The higher and wider the obstacle, for example, the larger the wind shadow it casts.

Air circulation: inversions

Inversions inhibit vertical mixing of air, thus slowing the dispersion of air pollutants. The frequency and duration of citywide inversions are influenced by regional weather patterns. However, local, micro-inversions can also occur as a function of physiography. On calm and clear nights, cool air will tend to flow into valley bottoms where it may trap pollutants until the midmorning sun warms the air near the surface.

Air circulation: spatial confinement

The more enclosed or confined a space is by a building, walls, embankments, or canopies adjacent to or over the roadway, the less opportunity pollutants within that space have to disperse (Fig. 4).

Sinks

Under some conditions, plants may act as "sinks" for certain pollutants. Large, densely planted groves of trees and shrubs, for example, may effectively filter particulates (Fig. 5).

Pollution-sensitive users and activities

Certain individuals, for example, young children and the elderly, are especially vulnerable to air pollution. The places they use are henceforth described as pollution-sensitive. Activities that entail frequent or prolonged use of a place are also described with this phrase.

Table 1 summarizes locations prone to air pollution as a function of the preceding factors. This checklist is both a descriptive and a prescriptive tool: it permits the analysis of a place in terms of factors that influence emissions, dispersion, and pollution-sensitivity. Table 1 enables the urban designer to predict where potential problems and opportunities are likely to result from certain combinations of factors that are likely to arise. By tabulating the overlap of conditions in a single location, for example, one can identify and then empirically evaluate, the potential "hot spots," that is, localized high pollution zones. The checklist can be employed to classify design situations. There are ranked from 1 to 6, whereby Class 1 indicates low or no emissions and with good air circulation and where Class 6 indicates high emissions and with poor air circulation. Each class entails a choice of strategies for reducing human exposure to street-level air pollution.

Table 1. Places prone or sensitive to street-level air pollution: a checklist

HIGH EMISSIONS

- Major street or highway (especially with parking lanes)
- Busy intersection
- Taxi stand
- Bus depot
- Parking garage entrance and exhaust vents
- Tunnel entrance and exhaust vents

PROXIMITY TO THE ROAD

- Median strip
- Traffic island
- Curbside
- Roadside strip

POOR AIR CIRCULATION

WINDS
- Wind shadows at leeward base of buildings
- Short street canyons blocked at both ends
- Long street canyons perpendicular to prevailing winds
- Urban districts with narrow, irregular street patterns

BREEZES
- Locations upwind of pollution source
- Windward base of obstruction to breeze

INVERSIONS
- Valley bottom
- Street canyon bottom

POOR AIR CIRCULATION

SPACIAL ENCLOSURE
- High, narrow street canyons
- Streetside arcade
- Bus shelter
- Interior atrium
- Tunnel
- Parking garage

POLLUTION-SENSITIVE USES

- Playgrounds and schoolyards
- Sports fields
- Sitting areas used either frequently or for a long period of time by same individuals
- Sitting areas used by elderly or convalescents
- Outdoor cafés
- Bus stops
- Building entrances, windows, and intake vents

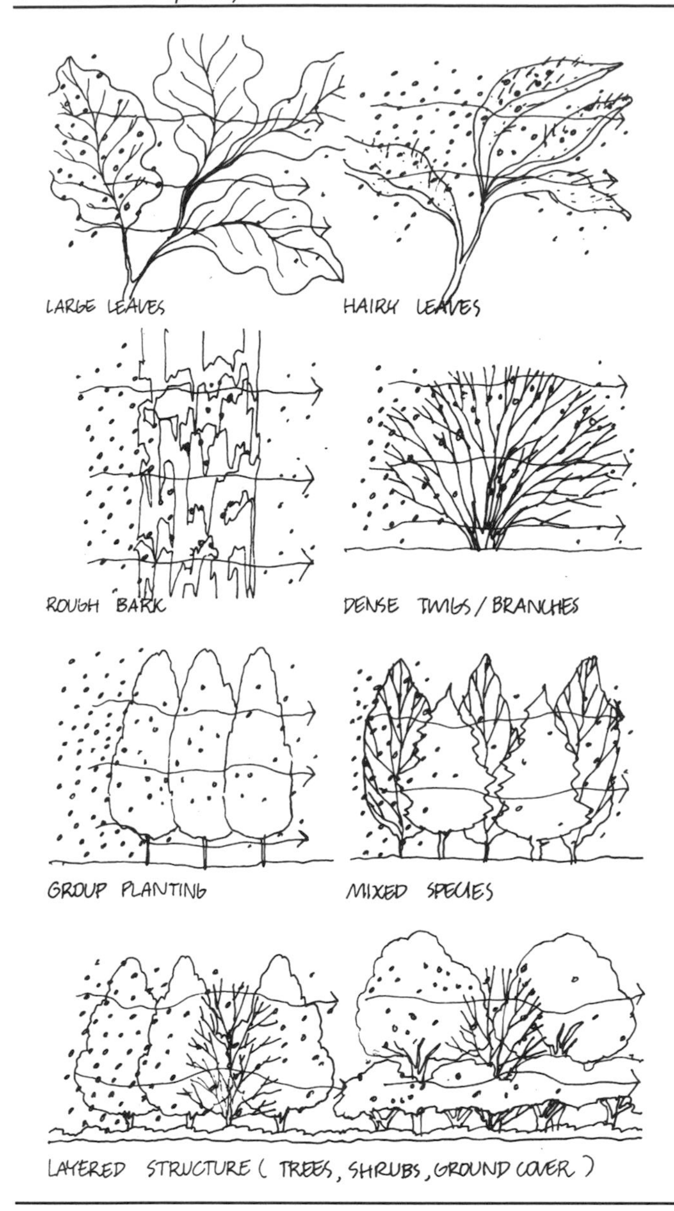

Fig. 5. Plants as sinks for particulates. (Adapted from Smith and Staskawicz 1977.)

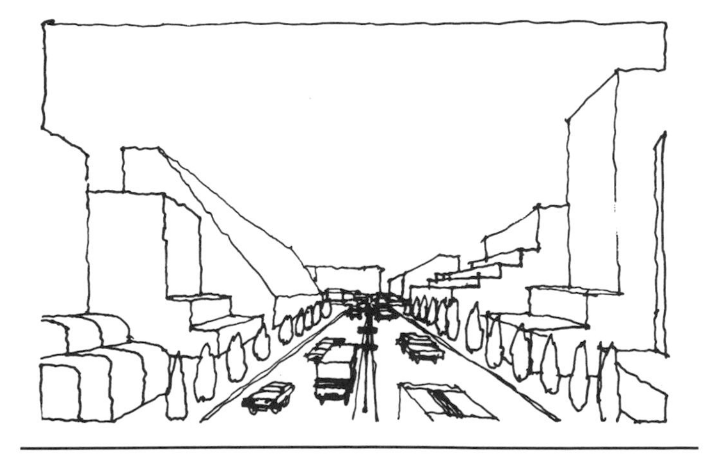

Fig. 6. To promote air circulation in street canyons, step buildings back from the street, increase openings and vary building heights.

2 URBAN DESIGN STRATEGIES

There are four basic strategies for addressing street level air pollution: prevention or reduction of emissions; enhancement of air circulation; removal of pollutants from the air, and protection of pollution-sensitive uses (see Table 3). Not all strategies are equally appropriate for every design situation. Table 2 identifies those applicable to each class. For Class 2, for example, the concern is poor air circulation, and the strategy should be to improve air circulation. For Class 3, the main concern is proximity to the road, and the design strategy should therefore be aimed at filtering pollutants and separating pollution-sensitive uses from the road, either through distance or a barrier. All four design strategies are appropriate for Class 6, which is characterized by a combination of high emissions, proximity to the road, and poor air circulation.

Strategy I: Preventing or reducing emissions

Various measures have been implemented to prevent or reduce emissions. Removing the source of emissions—whether it be the motor vehicle itself or a use like a taxi stand that may generate high emissions—is the most effective way to reduce or eliminate transportation-related air pollution in a particular location. Although elimination of the motor vehicle is rarely feasible, it may be the most attractive option under certain circumstances. Prime candidates are urban districts with narrow streets, relatively even building heights, and long blocks with few openings or with a highly irregular street pattern. These characteristics offer little opportunity for natural dispersion of street level air pollutants and therefore may be good candidates for conversion to pedestrian zones with restricted vehicular access. This was done in the center of Bologna, Italy, for example, where narrow streets lined by buildings with deep arcades trapped vehicle emissions. Bologna designated the core of this area as a pedestrian zone and a larger area as a zone with restricted vehicular access. (See article 6.7, "Pedestrian precincts: twelve European cities.")

A network of pedestrian streets provide another alternative. By linking a series of streets in dense urban districts, a convenient pedestrian route can be created through the city with minimal disruption to vehicular traffic. Even if the elimination of traffic from a district or street is not feasible, emissions at a particular spot may be substantially reduced by relocating high-emission uses, such as taxi stands, bus depots, loading zones, and parking garage entrances. Care should be taken to site such sources in well-ventilated locations, away from pollution-sensitive activities.

Strategy 2: Enhancing air circulation

Promoting the penetration of winds and breezes, inhibiting the formation and duration of local inversions, and avoiding spatial confinement can enhance air circulation.

It is difficult, even impractical, to create a design or plan that responds equally well to all wind directions. One must therefore determine the most important wind directions in a given city and take note of the city's seasonal variations (if any). Wind analysis requires understanding prevailing microclimatic wind data (available from airport and/or other city data) and local microclimatic effects determined by local building forms (documented by local observation and/or wind tunnel studies using models). A maximally effective design to enhance air circulation at any given spot will be framed within the larger context of street, district, and city. Failure to account for this larger context may nullify efforts at the local scale.

Major open spaces can funnel winds into the city. Highways and parkways, as well as stream valleys and linear parks, are potential ground-level wind channels. Continuous corridors, oriented to channel wind from desired directions into and through the city should be created. Large open areas may also bring winds to ground level in densely built districts.

Street canyons can be designed to promote ventilation through the use of the following strategies: widening the canyon; stepping buildings back from the street; increasing openings in the canyon; introducing uneven building heights or buildings of pyramidal or irregular shape; and designing intersections to deflect winds into the street canyon. Varying building setbacks to create a rough, irregular profile may have the additional benefit of reducing channelization of winds, a condition that might otherwise cause pedestrian discomfort (Fig. 6).

One can promote air circulation around the base of buildings and other obstacles by orienting them to pose minimum surface area to the wind. This can be accomplished through the use of openings or porous building material to permit airflow, while reducing wind speeds, and through the placement of another obstacle nearby to deflect and direct winds into areas where air might otherwise be stagnant (Fig. 7).

The duration of micro-inversions in enclosed low spots (called radiation inversions) can be shortened if early morning sun is permitted to penetrate at ground level and if the flow of breezes is maintained through the city. Street canyons, especially when located in a topographic low spot, should be designed to permit sun to reach the ground relatively early in the morning. In streets oriented north-south, the street canyon can be widened, building heights lowered, and buildings either set back or stepped back from the east side of the street.

Spatial confinement should be avoided in locations with relatively high air pollution emissions, such as heavily traveled streets. The degree of spatial confinement in street canyons can be reduced through widening the canyon, increasing openings in the street canyon, lowering building heights, and setting or stepping buildings back from the street. Street-side shelters should be designed with open sides or high canopies to reduce spatial confinement (Fig. 8).

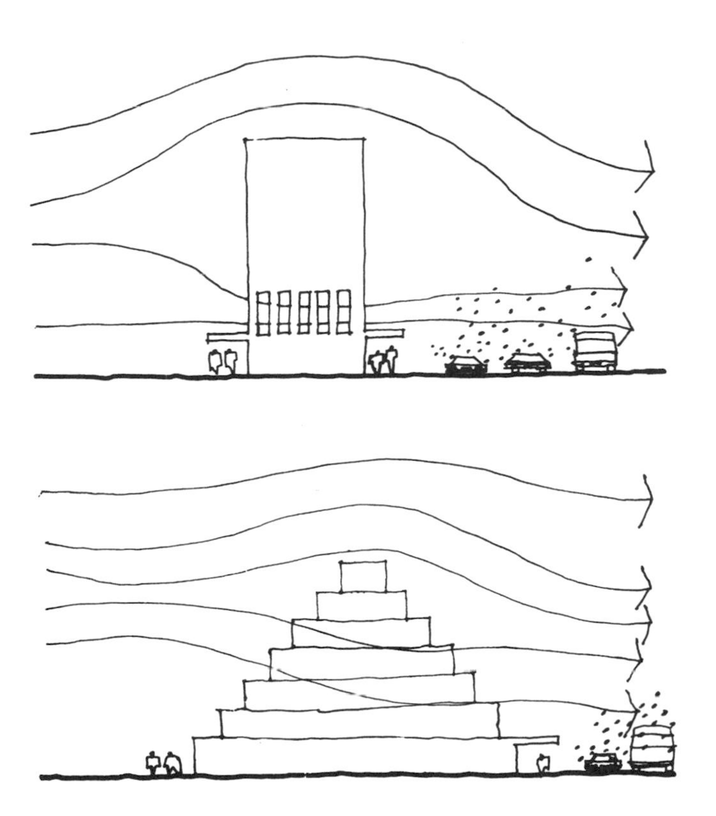

Fig. 7. To reduce wind shadow at the base of a building (top), design buildings with a pyramidal shape or with openings that permit air flow (bottom).

Fig. 8. To promote air circulation in street-side arcades design them with high canopies.

Strategy 3: Removing pollutants from the air

Whenever possible, the roadside zone should be landscaped to filter the air. The effectiveness of such "filters" can be enhanced greatly by design and accomplished with relatively low cost means. Major streets and highways should be accorded sufficient breadth to accommodate a dense, layered arrangement of plants (although street trees planted in a single row, or highway rights-of-way, that consist of grassy meadows with scattered trees, will have an insignificant impact as filters). Highway rights-of-way with a sloped embankment covered by woodland (leaving a mowed strip adjacent to the road for safety) will not only filter the air, but will cost less to maintain than grass (Fig. 9).

Strategy 4: Protecting pollution-sensitive uses

Pollution-sensitive uses and activity areas—such as playground and schoolyards, sitting areas, outdoor cafes, and building entrances and intake vents—should be set back from the road in areas of good air

Table 2. Street-level air pollution design solutions

	OPPORTUNITY FOR POLLUTION-SENSITIVE USES	CONSTRAINT FOR POLLUTION-SENSITIVE USES	APPROPRIATE DESIGN STRATEGIES (SEE TABLE 3)
CLASS 1 · Low or no emissions · Outside roadside zone · Good air circulation	■		
CLASS 2 · Low or no emissions · Outside roadside zone · Poor air circulation	■		2A–C 2D
CLASS 3 · Low-to-moderate emissions · Within roadside zone · Good air circulation		■	3A 4B–C
CLASS 4 · Low-to-moderate emissions · Within roadside zone · Poor air circulation		■	2A–D 3A 4A–D
CLASS 5 · High emissions · Within roadside zone · Good air circulation		■	1A 3A 4A–D
CLASS 6 · High emissions · Within roadside zone · Poor air circulation		■	1A 2A-D 3A 4A–D

Table 3. Strategies to reduce human exposure

STRATEGY 1
PREVENTING OR REDUCING EMISSIONS

A. Remove source of emissions
   1. Designate pedestrian districts
   2. Designate pedestrian streets
   3. Remove high-emission uses
B. Prevent release of pollutants
C. Improve engine efficiency
D. Improve traffic flow
E. Reduce number of vehicles
F. Reduce peak emissions

STRATEGY 2
ENHANCING AIR CIRCULATION

A. Promote penetration of winds and breezes
   1. Design major open spaces to enhance the penetration of winds into the city
   2. Design street canyons to promote the penetration of winds
   3. Promote air circulation around the base of buildings and other obstacles
B. Inhibit inversions
   1. Design street canyons to enhance solar access
C. Avoid spatial confinement
   1. Design street canyons to reduce spatial confinement
   2. Design street-side shelters to reduce spatial confinement

STRATEGY 3
REMOVING POLLUTANTS FROM THE AIR

A. Plant landscape filters
   1. Design the landscape of major streets to filter particulate pollutants
   2. Design the landscape within highway rights-of-way to filter particulate pollutants

STRATEGY 4
PROTECTING POLLUTION-SENSITIVE USES

A. Locate pollution-sensitive uses outside high-pollution zones
   1. Locate pollution-sensitive uses in areas of good air circulation, with low to moderate emissions, upwind of major pollution sources
   2. Maintain distance between pollution-sensitive uses and major streets and highways
   3. Discourage the use of high-pollution zones for pollution-sensitive uses
B. Locate high-emission uses judiciously
   1. Site high-emission uses away from existing pollution-sensitive uses
   2. Site high-emission uses in areas with good air circulation
   3. Site high-emission uses downwind of existing pollution-sensitive uses
C. Buffer pollution-sensitive uses from high-emission uses
   1. Use barriers to separate pollution-sensitive and pollution-generating uses
D. Regulate time of use to separate pollution-sensitive and pollution-generating uses
   1. Establish vehicle-free time zones during intensive pedestrian use
   2. Limit traffic during times of limited air circulation

circulation. New uses that may generate high emissions, such as parking garages and major streets and including building ventilation exhausts, should be located downwind and away from existing pollution-sensitive uses in such areas with good air circulation (Fig. 10).

Pollution sensitive uses must sometimes be located in zones of relatively high emissions; however, uses such as sitting areas and outdoor cafes can be protected through buffers, for example, walls, planters, and berms. Planted berms will not only provide a barrier, but can also be designed to filter particulate pollutants (Fig. 11).

Occupation of the same space by pollution-sensitive and pollution generating uses is sometimes unavoidable. In such cases, temporal separation may be the ideal way to reduce pedestrian exposure to air pollution.

3 COMPREHENSIVE APPROACH TO STREET-LEVEL AIR QUALITY

In the seventeenth century, the celebrated diarist John Evelyn proposed a comprehensive plan to alleviate air pollution in London. Evelyn's plan, employed the four basic urban design strategies described above. Evelyn recommended the prohibition of high-sulphur coal, the relocation of polluting land uses such as tanneries from central London to outlying areas downwind, and the planting of entire square blocks with trees and flowers to sweeten the air. Rarely since that time has there been so comprehensive a proposal to improve urban air quality. Many programs in the United States today focus on reducing emissions and ignore the potential of other strategies. If an air quality program is to be effective, it must integrate multiple strategies and coordinate efforts both at the citywide and street-corner scales (Fig. 12). The specifics of such a program will vary from city to city, depending on climatic setting, transportation patterns, and existing urban form. What works in one city may be inappropriate or impractical in another. Whatever its particularities, however, a plan to address air pollution should always acknowledge the issues of energy conservation and climatic comfort. Increased air pollution is often the by-product of profligate energy use. The promotion of climatic comfort in outdoor spaces may contribute to decreased air circulation and an accretion of air pollutants: as, for example, when sidewalks and building entrances are protected by deep arcades with vehicular access, or when winds that might ventilate a space are reflected from it.

NOTES

1. Dispersion models, both physical and mathematical, have been developed to predict where, when, and to what extent air pollution concentrations will occur. Many, but not all, of the variables described in this chapter are incorporated into most mathematical and physical models.

2. The most reliable way to predict airflow through the complex aerodynamic spaces and surfaces of a city is to test a scale model in a wind tunnel, manipulating the form of buildings and open areas to achieve adequate air circulation.

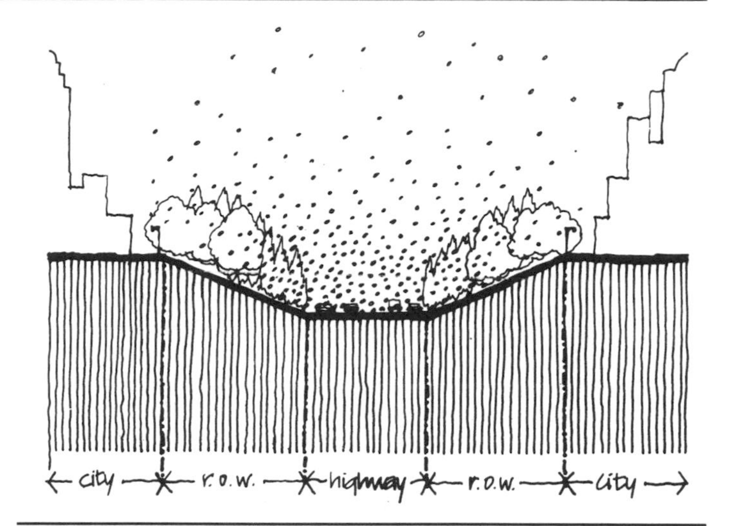

Fig. 9. Sloped highway embankments and woodland growth help filter pollutants from the air.

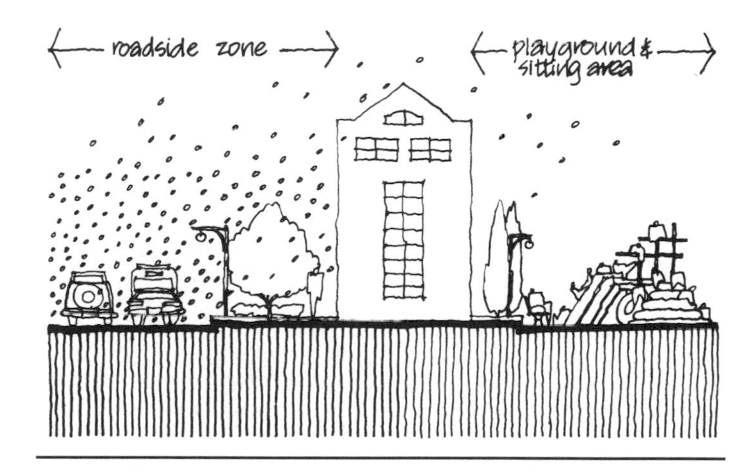

Fig. 10. Pollution-sensitive uses could be located away from high-emission zones: a concept showing sitting areas and playground set apart from major street and highway.

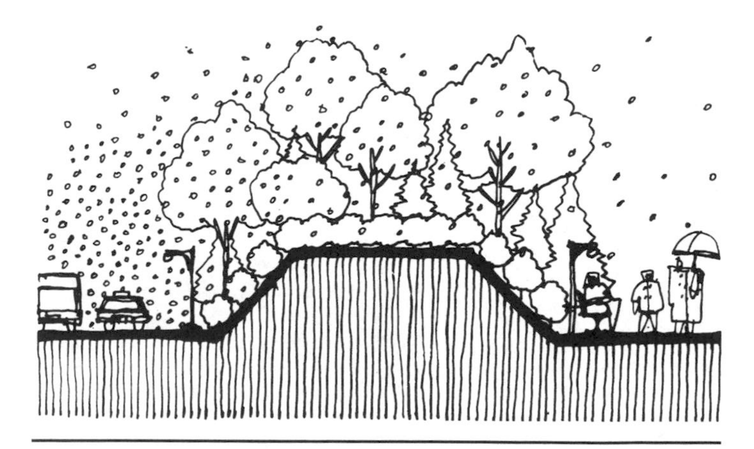

Fig. 11. Pollution-sensitive uses could be buffered from high-emission uses: a concept showing uses separated by planting berm as a vertical and horizontal barrier.

REFERENCES

Moudon, Anne Vernez. 1987. *Public Streets for Public Use.* New York: Van Nostrand Reinhold.

Smith, W. H. 1976. "Lead contamination of the Road's Ecosystem." *Journal of the Air Pollution Control Association* 26:753–66.

Smith, W. H., and B. J. Staskawicz. 1977. "Removal of Atmospheric Particles by Leaves and Twigs of Urban Trees. Some Preliminary Observations and Assessment of Research Needs." *Environmental Management* 1: 317–30.

Spirn, A. W. 1984. *The Granite Garden: Urban Nature Human Design.* New York: Basic Books.

———. 1986. *Air Quality at Street Level Strategies Urban Design.* Cambridge: Harvard Graduate School Design.

———, and W. G. Batchelor. 1985. "Street Level Pollution and Urban Form: A Review of Recent Literature." Cambridge: Harvard Graduate School of Design.

a.

c.

b.

d.

Fig. 12. A comprehensive approach to improving air quality through urban design in Stuttgart, Federal Republic of Germany: (a) aerial view of the city showing open space system design to funnel cool, clean air into and through the city (the strip of park in the center is the Schlossgarten); (b) hillside canyon with public staircase designed to funnel air off the hillsides into downtown; (c) Schlossgarten, the downtown park designed both for recreation and for improving the flow of air through the city; (d) major downtown highway with pedestrian areas buffered by dense landscaping of trees and shrubs. Photo: Jane S. Katz.

James P. Cowan

Summary

Noise negatively affects more people in urban areas than any other environmental stressor. The most common environmental noise sources to which people are exposed are highway and rail vehicles, aircraft, industrial plants, and ventilation equipment on buildings. These sources have been associated with reduced real estate value and degrading the quality of urban life worldwide. With proper planning of communities near highways, airports, rail lines, and industrial facilities, these effects can be significantly reduced. This article discusses the practical options to minimize noise impacts on urban communities, along with relevant regulation limits for reference.

Key words

acoustics, acoustical buffer, berming, decibels, noise control, noise regulation

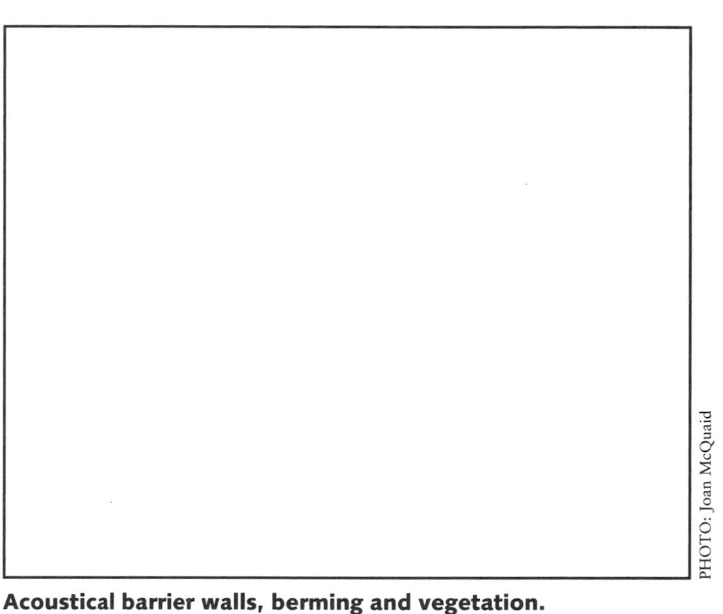

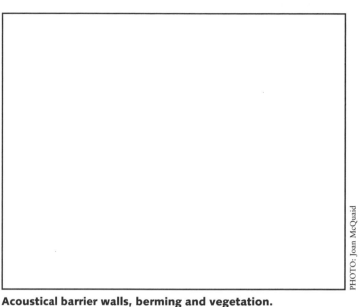

PHOTO: Joan McQuaid

Acoustical barrier walls, berming and vegetation.

Acoustic considerations for urban design

1 ENVIRONMENTAL NOISE FUNDAMENTALS

Terminology

The sound that we hear is typically described in terms of sound pressure levels. Sound pressure levels rate the pressure fluctuations that stimulate our sense of hearing. Because of the nature of the human hearing mechanism and its reaction to these pressure fluctuations, it is most practical to describe sound pressure fluctuations in terms of *decibels*. Decibels are based on a logarithmic scale, similar to that used to rate earthquakes (the Richter scale), and therefore do not follow the same mathematical rules as most of other rating systems. Our ears respond differently to changes in pitch or frequency. This variation in response has been programmed into a filtering system that is used in most sound measuring equipment. Decibels that have been filtered in this manner (to take account for human frequency sensitivity) are known as A-weighted decibels and are denoted as dBA or dB(A). Typical dBA levels that one would encounter range from 0 dBA at which sounds can just be detected by a person without hearing damage, to 120 dBA at which most people will feel pain from the sound exposure. Table 1 shows common sound sources with their approximate dBA sound levels for reference.

Most people can just notice a change when sound pressure levels rise or fall by a factor of 3 dBA. A dBA increase of 3 also corresponds to a doubling of sound power. Therefore, two sound sources, each generating a sound pressure level of 60 dBA at a certain location, would combine to generate a total sound pressure level of 63 dBA at the same location when both sources are operating. An addition of 10 dBA would sound as if a sound source doubled in loudness. For example, a 70 dBA sound source would sound twice as loud as a 60 dBA source. Alternately, a 60 dBA sound source would sound half as loud as a 70 dBA source. Because of the logarithmic nature of the decibel scale, it would take ten sources of equal intensity to generate a 10 dBA increase over the sound generated by one source.

Sound levels are usually rated not only by their volume but also by their duration. The most common ratings for sound levels over extended periods of time are Leq and Ldn. Leq is the equivalent sound level, which is an energy average of sound over a specified period of time. Since sound levels are continuously changing, the Leq provides a single number rating for the sound over specified durations. The Federal Highway Administration requires the use of a 1-hour Leq as the basis for state Departments of Transportation to evaluate the potential impact of a new or expanded highway. Other agencies use the Ldn as a yardstick for land use compatibility with sound sources. The Ldn is the day-night equivalent level, which is a 24-hour Leq value with the stipulation that all sound events occurring between 10:00 PM and 7:00 AM have 10 dBA added to them to compensate for the extra sensitivity of sounds occurring during normal sleeping hours. Many federal agencies (such as the Federal Aviation Administration, the Federal Transit Administration, and the Department of Housing and Urban Development) use Ldn in their restrictions. Many municipalities regulate noise in terms of instantaneous maximum levels that cannot be exceeded. For reference, the Occupational Safety and Health Administration (OSHA) sets limits to avoid hearing loss starting at an 8-hour time-weighted average of 90 dBA.

Table 1. Typical Sound Pressure Levels for Various Environments in dBA

dBA Level	Perception of Loudness	Sound Source
20	1/8 as loud as 50 dBA	Broadcast studio
30	1/4 as loud as 50 dBA	Rural area at night
40	1/2 as loud as 50 dBA	Suburban area at night
50	Reference level	Suburban area during the day
60	2 times louder than 50 dBA	Urban area during the day
70	4 times louder than 50 dBA	Passing cars at 25 ft. (7.62 m)
80	8 times louder than 50 dBA	Highway vehicles at 50 ft. (15.24 m)
90	16 times louder than 50 dBA	Passing truck or bus at 25 ft. (7.62 m)
100	32 times louder than 50 dBA	Passing subway train from platform

Table 2. Noise Reduction Options

Control at the Source	Control in the Path	Control at the Listener
Maintenance	Enclose source	Relocate listener
Avoid resonance	Install barrier	Enclose listener
Relocate source	Install proper muffler	Have listener use hearing protection
Remove unnecessary sources	Install absorptive treatment	Add masking sound at listener
Use quieter model	Isolate vibrations	Add masking sound at listener
Redesign source to be quieter	Active noise control	Add masking sound at listener

Outdoor sound travel

The urban designer deals primarily with the outdoor sound environment, except for occasionally ensuring that intrusive sounds do not penetrate building exteriors. Sound travel outdoors, especially over distances greater than 200 to 300 ft. (60 to 90 m) from a sound source, is highly dependent on weather conditions. The atmospheric conditions that affect sound travel most significantly are temperature variations, wind currents, and humidity. In terms of temperature variations, sound waves generally bend toward cooler temperatures. For example, with all other weather conditions remaining unchanged, on a typical summer's afternoon temperature decreases with increasing altitude. In this case, sound waves tend to bend upward and generate what is known as a shadow zone. If you are in a shadow zone, you may be able to see a sound source at a distance but not hear it since shadow zones can decrease sound levels by up to 20 dBA at distances greater than 500 ft. (150 m) from a sound source. Shadow zones can also be set up when sound is traveling against the wind.

These conditions are reversed when temperatures are cooler close to the ground than at higher elevations, as would be the case early in the morning or over a calm body of water. In these cases, sound waves tend to bend toward the ground, bounce off reflective ground surfaces, and travel farther than expected. This is why sound is said to "carry" well over water. Sound also tends to travel farther than expected when it is traveling with the wind. (Figs. 1 and 2)

An acoustic condition unique to downtown urban areas is the urban canyon effect. In this case, tall buildings having flat, acoustically reflective exterior surfaces are parallel to each other on either side of a street. Sound generated between the buildings reflects many times between the building exteriors and can, in many cases, travel along the lengths of entire streets with minimal reduction. This effect can be minimized by specifying exterior finishes for buildings on at least one side of the street to be acoustically absorptive.

2 ENVIRONMENTAL NOISE REDUCTION METHODS

Noise by definition is unwanted sound. It is derived from the word "nausea," emphasizing our unhealthy associations with noise. Most generally, noise can be controlled at its source, in the path between the source and the listener, or at the listener. Table 2 summarizes the general options available. If the noise can be controlled at its source, it is unnecessary to consider the path or listener locations. Likewise, if the noise can be controlled in the path between the source and listener, it is unnecessary to consider the listener's location for noise control measures.

The options for noise control at the source are generally self-explanatory. Although they are the preferred noise control options, they are often impractical logistically or economically. Most often, noise control options are limited to the path between the source and the listener and at the listener. Given many misconceptions about these options, it is useful to discuss some of them further.

Enclosures

Enclosures can be effective at reducing noise levels, as long as they are designed properly. Consider the following points when designing noise enclosures:

The enclosure must completely surround the noise source, having no air gaps. Air gaps can significantly compromise the noise reduction effectiveness of partitions. Think of waterproofing. If water can leak through a partition, so can noise. An enclosure with any side open is not an enclosure but a barrier, and the noise reduction effectiveness of barriers is limited to 15 dBA, independent of the barrier material. Enclosures, on the other hand, can provide up to 70 dBA of reduction if properly designed.

The enclosure must be isolated from the ground or any structural members of a building. An enclosure covering the sides and top of a noise source but having the bottom open (since the source is sitting on the ground) can compromise its effectiveness for several reasons. First, the chances of the sides of the enclosure perfectly sealing to the ground are slim, and therefore, air gaps would result. Second, vibrations will be carried along the ground since the source is in direct contact with it. The only way to reduce these vibrations is to vibrationally isolate the source from the ground using tuned springs (appropriate for the source), pads, or the bottom of a multilayered enclosure.

The enclosure should not be comprised of only sound absorptive material. Sound absorptive material can be effective in reducing noise when it is used as part of a multilayered enclosure (on the inside); however, absorptive material on its own is not effective in reducing noise. The main purpose of absorptive material is to control reflections within spaces, not to control sound transmission out of spaces.

The enclosure must consider that some sources require ventilation. This cannot translate to leaving a simple opening in the enclosure without severely compromising the noise control effectiveness of the enclosure. Ventilation systems must be developed that minimize noise transmission. Simple louvers do not provide adequate noise reduction.

The enclosure should be constructed using multilayered construction for maximum efficiency. Doubling the mass of an enclosure would add 6 dB to its noise reduction effectiveness. This can easily lead to excessive weight for an effective homogeneous enclosure. Multilayered enclosures can add more than 20 dB of effectiveness under similar space requirements to massive enclosures with a fraction of the weight.

Barriers

A barrier is contrasted from an enclosure by it being open to the air on at least one side. Because of diffraction, in which sound waves bend around barriers (see Fig. 3), noise barriers are limited to 15 dBA of noise reduction capability, independent of the material. It is also important to have no air spaces within or under the barriers, since this will compromise their already limited effectiveness. This is why what

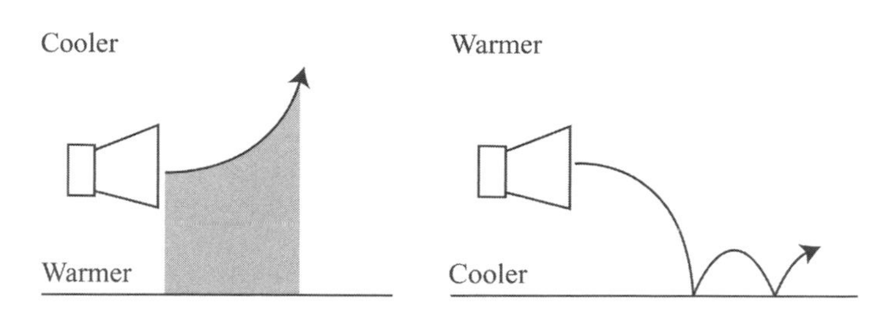

Fig. 1. Sound travel in temperature variations.

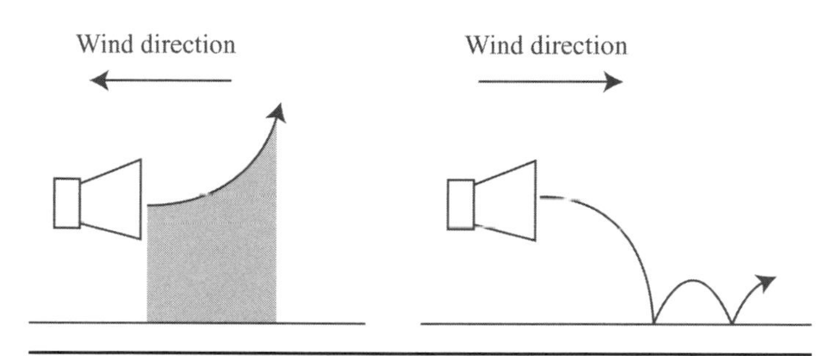

Fig. 2. Sound travel in wind currents.

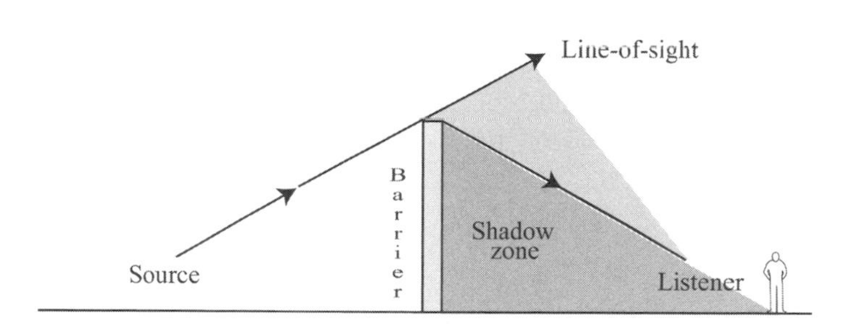

Fig. 3. The diffraction of sound over a barrier.

is typically called a "fence" does not function effectively as a noise barrier. A berm or hill will provide a barrier effect while preserving the natural landscape.

The noise reduction effectiveness of barriers is typically rated by the insertion loss (denoted IL). IL is the simple reduction in sound pressure level, at a specific location, with a barrier in place. In other words, IL is the difference between conditions with and without a barrier.

To provide any insertion loss, a barrier must break the line-of-sight between the sound source and listener. In other words, if you can see a sound source on the other side of a barrier, that barrier is providing no sound reduction (from that source) for you. Breaking this line-of-sight typically provides a minimum IL of 3 to 5 dBA, with IL increasing as one goes farther into the shadow zone of the barrier.

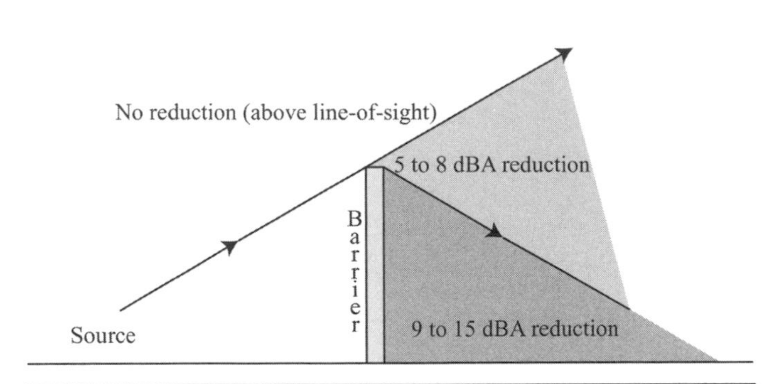

Fig. 4. The noise reduction effectiveness for sound traveling over a barrier (cross-sectional view).

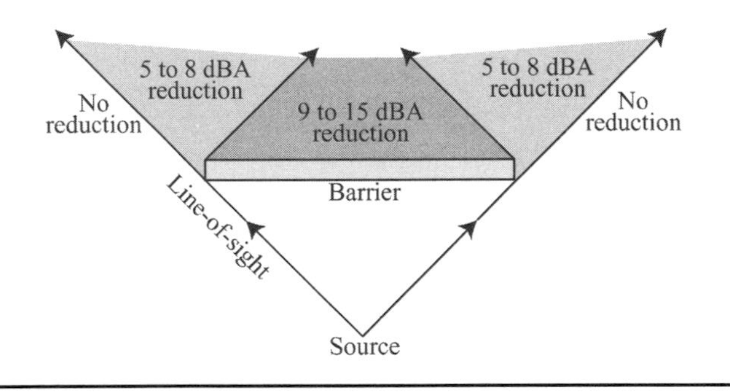

Fig. 5. The noise reduction effectiveness for sound traveling around a barrier (top view).

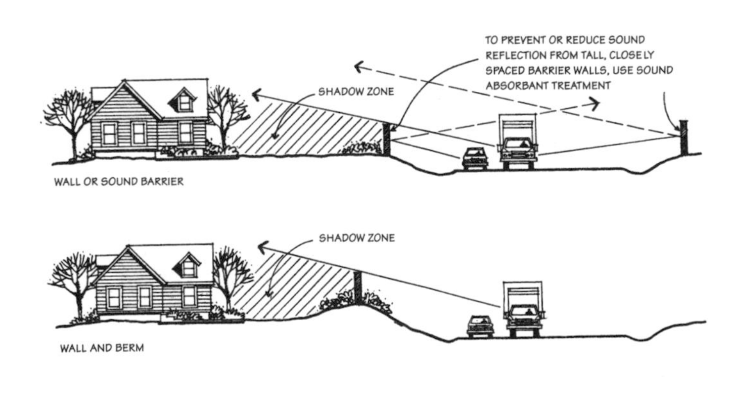

Fig. 6. Barrier wall or combination of wall and berm. (Source: Harris and Dines 1998, pp. 660–9.)

Noise barriers can be effective at reducing noise levels within 200 ft. (60 m) of a sound source but not beyond that distance. That effectiveness is limited, however, by the phenomenon of diffraction. Independent of the material, barriers provide a maximum of 15 dBA of noise reduction to a listener. The most important design aspects of the barrier are that it is solid, it can stand up to the elements, and it breaks the line-of-sight between the source and listener. Any air gaps will compromise a barrier's already limited effectiveness. Figs. 4–6 show the noise reduction effectiveness of outdoor barriers.

It is often thought that trees or other types of vegetation between a source and listener will provide a barrier effect. However, studies confirm that vegetation has minimal effect on reducing noise, unless it is in the form a dense forested area of evergreens more than 100 ft. (30 m) thick. The only natural design that will serve as an effective noise barrier is a berm or hill. A natural or man-made topographical depression that causes the line-of-sight to a sound source can be just as effective for noise control as a barrier.

Rows of buildings between a sound source and listener can provide as much noise reduction as a barrier as long as the buildings cover at least 70% of the street length. Less than a 50% coverage will provide little, if any, noise shielding.

When noise barriers are on both sides of a roadway and parallel to each other, they can create an environment similar to that of an urban canyon. This can cause up to a 3 dBA degradation in the effectiveness of the barriers. A simple solution to this would be to cover the parallel barrier faces with an acoustically absorptive material.

Mufflers

Mufflers are devices that are inserted in the path of ductwork or piping with the specific intention of reducing sound traveling through that conduit. They are often effective for controlling noise from stacks on rooftop ventilation equipment. The effectiveness of mufflers is typically rated using insertion loss. Mufflers must be designed for each purpose to preserve the required pressure characteristics. For that reason, each muffler is unique to its installation.

Vibration Isolation

Mechanical equipment can generate vibrations that can travel through a building's structural members to affect remote locations within a building. It is therefore prudent to isolate any heavy equipment from any structural members of buildings. This can be accomplished by mounting the equipment on springs, pads, or inertia blocks; however, a specialist trained in vibration analysis should perform the selection of specific isolating equipment. The main reason for this is that each vibration isolation device is tuned to a specific frequency range. If this is not matched properly with the treated equipment, the devices can amplify the vibrations and cause more of a problem than would have occurred without any treatment.

Active noise control

Passive noise control involves all of the noise control methods discussed so far in this article, in which the sound field is not directly altered. Active noise control involves electronically altering the character of the sound wave to reduce its level. In this case, a microphone measures the noise and a processor generates a mirror image of (180∞ out of phase from) that source. This mirror image is then reproduced by a loudspeaker in the path of the original sound. This new sound cancels enough of the original signal to reduce levels by up to 40 dB in the appropriate circumstances. Although this is a very powerful noise control tool, active noise control is only practical in local environments and for tonal frequencies below 500 Hz. Ventilation ducts and industrial plant stacks are ideal candidates for active noise control systems. This is because they are enclosed environments and because their dominant noise is often low frequency pure tones (associated with fan characteristics).

Buffer zones

Space is one aspect of the outdoor environment that is often available, although sometimes at a premium in urban areas. Sound generally dissipates at a rate of 3 to 6 dBA per doubling of distance from a source within 200 to 300 ft, (60 to 90 m) of that source. Its decay rate beyond that is highly variable depending on the atmospheric (mainly temperature variations, wind currents, and humidity) and terrain conditions between the source and listener. However, sound levels generally decrease with increasing distance from a source. Therefore, the greater distance that can be placed between an objectionable sound source and a listener, the better.

It is best to avoid placing an objectionable sound source near a still body of water that lies between the source and a listener because temperature effects will cause the sound to travel across the body of water with little reduction. Although wind currents are constantly changing, it is best to avoid locating a noise-sensitive building (such as a residence, house of worship, health care facility, or school) in the prevailing downwind direction of a noise source.

Masking

Masking systems are composed of natural or electronic components that add sound to an environment to cover or mask objectionable sounds. They work best when they blend with the environment to the point at which they go unnoticed. Acceptable sounds outdoors include more natural sounds, such as running water or rustling leaves. Outdoor fountains are not only effective in masking sound, but they can also add aesthetically to an area. Any other natural masking sounds would have to be added electronically using weather-proofed loudspeakers.

3 NOISE REGULATIONS

It is important for the urban designer to understand noise regulations to avoid misunderstandings about limitations and responsibilities. This discussion has been divided into the categories of federal, state, and local levels to establish the responsibilities of the different governmental levels.

Federal

The Federal Highway Administration (FHWA) has promulgated noise regulations, in Title 23 Code of Federal Regulations Part 772 (*Procedures for Abatement of Highway Traffic Noise and Construction Noise*), that establish limits for the consideration of noise mitigation for new or expanded highway projects. These limits, known as noise abatement criteria, are in terms of land use and 1-hour Leq values that cannot be approached or exceeded. The definition of "approach or exceed" is deferred to each State's Department of Transportation. The most common of these noise abatement criteria is 67 dBA, used for residential communities. In addition to the absolute limit of 67 dBA, the FHWA regulations also have a provision that noise levels generated by a new highway project do not "substantially increase" noise levels to a noise-sensitive area (such as a residential community). The definition of this term is also deferred to each State's Department of Transportation.

The FHWA offers literature to the planning community to assist them in dealing effectively with noise issues. *Guidelines for Considering Noise in Land Use Planning and Control* is an interagency report published in 1980 as a joint effort of the U.S. Department of Transportation, Department of Defense, Environmental Protection Agency, Veterans Administration, and Department of Housing and Urban Development. Another valuable document available through the FHWA is *The Audible Landscape: A Manual for Highway Noise and Land Use*, published by the FHWA in 1976. The former document deals with planning for all transportation noise sources while the latter deals with only highway noise. Although these documents have not been written recently, the principles and examples listed in the documents are as valuable today as they were when they were written.

The Federal Aviation Administration (FAA) regulates aircraft noise abatement for the aircraft themselves and for their noise exposures in communities. The Federal Aviation Regulation (FAR) Part 150, entitled *Airport Noise Compatibility Planning* (issued in 1989) establishes a standardized noise and land use compatibility program including plotting noise contour maps around airports. Noise abatement, usually in the form of upgrading windows, doors, and attic insulation, are considered for homes exposed to aircraft-generated Ldn levels of 65 dBA or more.

The U.S. Department of Housing and Urban Development (HUD) issued their noise regulations in 1979, which is published in the Federal Register as Title 24 Code of Federal Regulations Part 51B. This sets limits for acceptability of funding for HUD-assisted residential projects. These limits range from an unmitigated background sound level of 65 dBA Ldn to extra sound attenuation in buildings exposed to levels between 65 and 75 dBA Ldn. HUD will not fund projects in areas greater than 75 dBA Ldn.

The Federal Transit Administration (FTA) uses noise criteria listed in *Transit Noise and Vibration Impact Assessment* (U.S. DOT report number DOT-T-95-16 published in 1995). These criteria are based on the increase in sound levels caused by a transit (light rail or bus) project ranging from 0 to 10 dBA depending on the existing background sound level. For example, no increase is permitted if the background levels are 75 dBA while a 10 dBA increase is permitted if the background levels are 40 dBA. These values are in terms of either hourly Leq or Ldn, depending on whether or not there is sensitivity to nighttime sounds. Hourly Leq is used if there is no nighttime sensitivity and Ldn is used if there is. The Federal Railroad Administration has recently adopted noise limits for interstate freight rail similar to those imposed by the FTA.

State

As is mentioned above, each State's Department of Transportation is required to interpret the "approach or exceed" and "substantial increase" phrases in the FHWA regulations. Depending on the state, "approach or exceed" has been defined as from 1 to 3 dBA of the 67 dBA hourly Leq limit. Therefore, depending on the state, the limit for noise abatement consideration ranges from 64 to 66 dBA. A "substantial increase" in noise levels ranges from 6 to 15 dBA, depending on the state. Only the Departments of Transportation deal with highway-generated noise.

Some states regulate nontransportation noise sources with statutes that are similar to local noise ordinances.

Local

Local municipalities, through their zoning and administrative codes, can restrict the development of residential and other noise-sensitive land uses in the vicinity of highways, airports, rail lines, or industrial facilities.

4 PLANNING ISSUES

Noise control design options

As is mentioned above, noise control options for urban communities are generally limited to constructing noise barriers or berms and providing buffer zones between noise sources and residential communities to have as large a distance as possible between the sources and the affected communities (Fig. 6). If outdoor activities would not be at issue, sound control designs can be incorporated into the residential buildings facing the noise sources. These designs include installing windows and doors that would be more efficient in sound reduction than typical designs and designing building facades without windows or doors facing noise sources.

Responsibility of enforcement

Federal agencies assume no responsibility in enforcing noise issues. The FAA and FTA evaluate proposed new projects or expansions to existing facilities and consider noise abatement when their impact criteria are exceeded. State Departments of Transportation evaluate the potential noise impacts of new or expanded highways on noise-sensitive areas. If the FHWA criteria are predicted to be approached or exceeded, or if a substantial noise increase is predicted at an existing noise-sensitive location as a result of a proposed highway construction project, the State Department of Transportation has the responsibility to consider noise abatement (based on its established procedure for approving the construction of noise abatement measures). This only deals, however, with state and interstate highways. Local roads are the responsibility of each local municipality. Except for transportation sources, noise code enforcement is the responsibility of local municipalities.

Responsibility of funding

New state highway construction is funded by each State's Department of Transportation. The FAA funds residential sound insulation programs through the administration of each airport. Transit authorities typically fund any noise abatement for transit projects. New interstate highway construction is usually funded by a combination of federal and state money. It is this money that is also used for the construction of noise abatement measures, if approved by the agencies involved. Funding for local roads is the responsibility of each local municipality. Noise abatement designs can be funded by the municipality or the developers of communities, depending on the local administrative codes.

Public involvement

The public is the ultimate client of designers, developers, planners, and municipal officials. An educated public provides the greatest opportunity for cooperation and resolution to noise-related disputes in the most effective ways. Noise problems are also most effectively controlled when they are recognized and resolved in the design process of any community. It is the responsibility of the design community to be educated about noise issues as they relate to urban design. It is then the design community's responsibility to educate the public about these noise issues before problems arise. In that way, all parties will understand the issues at hand and will be able to make informed, effective decisions in the design process. ■

REFERENCES

Cowan, James P. 1994. *Handbook of Environmental Acoustics*. New York: John Wiley & Sons.

Harris, Charles W., and Nicholas T. Dines. 1998. *Time-Saver Standards for Landscape Architecture*. 2nd edition. New York: McGraw-Hill.

Harris, Cyril M. 1991. *Handbook of Acoustical Measurements and Noise Control*. New York: McGraw-Hill.

Walter A. Cooper

Summary

The vast majority of information produced in the world today exists only in electronic or multimedia form. Communities without adequate on-line access to this information will increasingly fall behind in commercial, educational and cultural development. Urban planners must provide for an appropriate telecommunications infrastructure to transport, store, process and disseminate electronic information.

Key words

antenna, central office, communications, local loop, microwave, network, optical fiber, regulations, satellite, site design, telecommunications, wireless

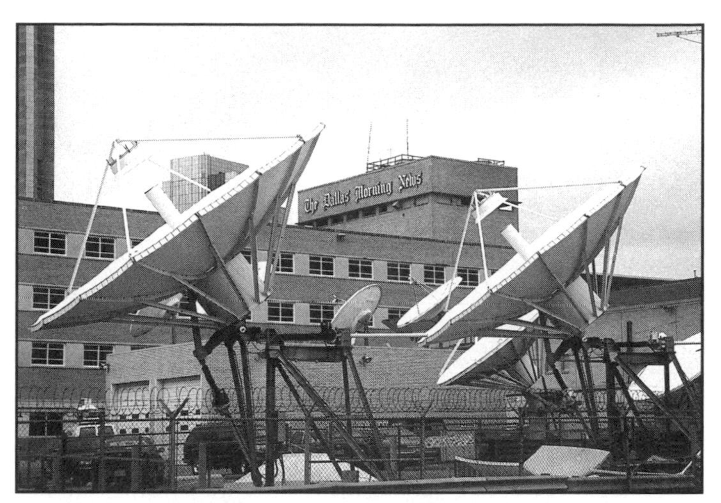

Satellites and microwave communications support news operations in a major metropolitan area.

Urban telecommunications infrastructure

Introduction

Urban planners and architects play an increasingly critical role in planning and design of telecommunications infrastructure. In the past, this responsibility may have been left mainly to the local telecommunications company. The continuing deployment and availability of an adequate infrastructure was more or less taken for granted by those outside the telecommunications industry. Any enhancements to the basic infrastructure were treated as amenities to attract "high-tech" development. Today, access to advanced telecommunications services can no longer be considered just an amenity. University of California researchers at Berkeley estimate that the world now produces 1 to 2 exabytes (billions of gigabytes, or approximately 8 to 16 x 10^{18} bits) of new information each year. Only .003 percent of this data is printed on paper. The rest exists only in electronic or multimedia form (Lyman and Varian 2000). The economic, social, cultural and political viability of the community depends on efficient and cost-effective, broadband access to this electronic data.

The infrastructure for producing, processing, transporting and storing all this information is a vast collection of electronic and computer networks connected by fiber optic strands, microwave links, metallic cables, switching hardware and millions of lines of software code that collectively operates as an enormous global network. Deregulation of the telecommunications industry has layered new dimensions of capability and complexity onto the traditional information infrastructure. The breakup of the Bell System in 1984, the commercialization of the Internet (the Worldwide Web) in the 1990s and the proliferation of wireless services are but a few key milestones in a sweeping process of de facto and regulatory change in the telecommunications landscape.

More recently, the Telecommunications Act of 1996 and its implementing regulations have accelerated the process and greatly changed the landscape of the telecommunications industry. Key objectives or the Telecommunications Act were to:

- Spur innovation and competition in the industry
- Remove regulatory barriers to entry into telecommunications markets
- Reduce the price of services
- Bring the benefits of advanced telecommunications services to all Americans

There are now over 2,000 competitors in a telecommunications market that was once dominated by AT&T and a handful of smaller regional telephone companies.

Now electronic information is produced or disseminated by Internet Service Providers (ISPs) and a myriad of corporations, educational institutions, entertainment companies, governments and other entities. Many businesses and institutions are removing applications such as

Credits: Photos are by the author unless otherwise noted.

accounting, payroll and human resources from their office file servers and turning them over to Applications Service Providers (ASPs) whose facilities may be hundreds or thousands of miles away. The information is transported and delivered to the user via public and private telecommunications networks.

Typically, these networks are operated by long haul and local telecommunications companies. The long-haul carriers are generally referred to as Interexchange Carriers (IXCs) and the local carriers are called Local Exchange Carriers (LECs). LECs are further divided into Incumbent LECs (or ILECs, the former regulated Bell System companies) and Competitive LECs (CLECs). Continuing deregulation of the industry is blurring the distinction between local and inter-city carriers. Many companies now provide both local and long haul facilities. The Internet, to cite one example, is one of the most visible and fastest growing, elements of this exploding global infrastructure. It is estimated that the number of host computers on the Internet grew from 376,000 in 1990 to over 122 million as of May 2001 (Telcordia 2001). Internet traffic doubles every 90 days.

While these changes have brought new services, lower prices and great technological innovation to the nation's communities, they have also complicated the process of urban planning and design. It is no longer sufficient to rely on "the Telephone Company" to plan, design and deploy the telecommunications infrastructure needed by today's and tomorrow's communities. There is no longer a single telecommunications entity with this responsibility. At the same time, the same competitive environment and market forces that produced new services and lower prices have also led to an uneven distribution of resources. Studies by the Federal Communications Commission have confirmed that many lower-tier urban areas, rural communities and inner city areas are being left behind in the deployment of advanced telecommunications services (FCC 2000.) Unless governments, planners and community groups proactively plan for the development and enhancement of the urban telecommunications infrastructure, this trend is likely to continue. Communities on the wrong side of the "digital divide" will fall further and further behind in an increasingly on-line society.

A good example of proactive planning is Berkshire Connect. Leaders in the Berkshire region of western Massachusetts saw the area as greatly disadvantaged in access to advanced telecommunications services at affordable prices. Planners and business leaders identified and aggregated demand and demonstrated the economic feasibility of providing advanced services in the region. As a result, the telecommunications industry made several million dollars of private investment in Berkshire County and the area now enjoys ready access to advanced services at prices comparable to those paid in the major metropolitan centers (FCC 2000.)

Rules, regulations and standards

The activities of the ISPs, ASPs, IXCs, ILECs, CLECs and other service providers are governed at the federal, state and local levels. Telecommunications companies, especially the ILECs, are still subject to considerable regulation, which is embodied mainly in Title 47 of the Code of Federal Regulations (the FCC Rules). States have similar regulations. Local authorities exercise oversight of telecommunications companies mainly through code enforcement, planning and zoning regulations and various franchise arrangements. With some notable exceptions, local authorities may exercise a great deal of control over the siting, construction and operation of telecommunications facilities.

Technical standards are established partly by government regulation and partly by national, international and industry standards bodies. The International Telecommunications Union (ITU) sets standards to ensure compatibility of national networks. The most prominent U.S. bodies are the:

- Federal Communications Commission (FCC)
- American National Standards Institute (ANSI)
- Institute of Electrical and Electronic Engineers (IEEE)
- Telecommunications Industry Association (TIA)
- Electronic Industry Alliance (EIA)

Another important source of technical standards is the facility design guidelines of Telcordia™ Technologies (formerly Bell Communications Research, Inc. – Bellcore), the research and development arm of the former Bell operating companies.

Wireless telecommunications facilities are subject to additional FCC and Federal Aviation Administration (FAA) regulations and technical standards and usually require an operating license from the FCC. Structural and wind loading standards for antenna structures and towers are subject to state and local building codes and ANSI/EIA/TIA-222-E.

Broadband services

Broadband communication service is essential for effective access to advanced electronic data and multimedia resources. The FCC definition of a broadband service is one capable of providing a data transfer rate of at least 200 kilobits per second (Kb/s) in both directions. Many businesses and commercial interests consider 512 Kb/s or even 1.5 megabits per second (Mb/s) to be the bare minimum standard for broadband communications. Much higher rates (millions to billions of bits per second) are common in many commercial, scientific and academic networks.

A typical modem connection on the Public Switched Telephone Network (PSTN) can support up to about 56 kilobits per second (Kb/s) and is thus not considered broadband by any definition. Even a 128 Kb/s Integrated Services Digital Network (ISDN) connection, once considered a very advanced digital service, does not qualify.

Digital Subscriber Line Service (DSL), originally intended as a vehicle for telephone companies to offer video services to residential customers, is now widely used to deliver broadband data service at various speeds up to about 6 Mb/s, but usually much less. DSL is offered to residential and small business customers by both ILECs and CLECs. DSL is delivered over the existing copper cable "local loop" that normally provides conventional phone service. Because of the technical limitations of the local loop, there is a trade-off between the data rate and the maximum distance the DSL subscriber can be from the service provider's central office. This can be a serious limitation, especially in less densely populated areas. Data speed decreases with distance out to a maximum of about 12,000 to 18,000 cable-feet. Beyond this distance, DSL cannot provide broadband service, even if the local loop is in perfect condition.

Cable modem service is similar to DSL in that it is aimed at residential and small business customers. The difference is that it takes advantage of the inherently broadband capabilities of the coaxial cable used to deliver cable TV (CATV). Unlike the PSTN local loop, CATV infrastructure is not ubiquitous. Many residences and most businesses are not wired for CATV service and thus, cannot subscribe to cable modem service. Another limitation of cable modem technology is that it shares bandwidth among all users on the same cable loop. Data speed, while often higher than that available with DSL, decreases as more customers use the network. Because of this, CATV providers will usually not guarantee a specific data rate.

As noted above, Basic Rate ISDN is not a broadband service. It is possible, however to combine several IDSN lines using an "inverse multiplexer" to create a broadband connection at multiples of 128 Kb/s. This is often used to support relatively inexpensive video teleconferencing. There is a broadband version of ISDN called Primary Rate ISDN which provides approximately 1.5 Mb/s service, although it is not widely deployed in the United States.

T-1 service is more commonly used than Primary Rate ISDN. It also provides 1.5 Mb/s and is often used by medium sized businesses for voice, data, video and/or Internet access. T-1 service is part of a hierarchy of digital services originally developed in the 1960s to connect telephone company switching centers (Interexchange Facilities). This hierarchy can provide data rates of approximately 1.5 Mb/s (T-1), 6.3 Mb/s (T-2) and 45 Mb/s (T-3). Collectively, these services are referred to as "T-carrier." T-1 and ISDN can be delivered over the copper local loop with distance limitations similar to those of DSL. For longer distances, T-carrier and ISDN are delivered over optical fiber, coaxial cable (decreasingly), microwave and satellite facilities. ISDN and T-carrier services provide a fixed amount of bandwidth between two points. They cannot expand or contract their capacity in accordance with communications traffic demand. If they are sized to handle peak loads, they may have idle capacity most of the time. The idle bandwidth is not available to other users of the network and the circuits cannot be "turned off" when not in use. This is very inefficient, especially for some types of data and video traffic.

A service called Synchronous Optical Network (SONET) has been developed to overcome some of the limitations of T-carrier. SONET provides more flexibility and much higher data rates than T-carrier. SONET bandwidth increments are referred to as "OC" (for optical carrier) and typically range from OC-1 (51.8 Mb/s) to OC-192 (9, 953 Mb/s) and beyond. SONET is usually delivered on optical fiber and can be accommodated on digital microwave radio. Table 1 compares T-carrier and SONET services.

Table 1. Typical T-1 and SONET Data Rates

T-Carrier	SONET	Data Rate (Megabits per second)	Equivalent Number of 64 Kb/s Channels
T-1	--	1.544	24
T-2	--	6.312	96
T-3	--	44.736	672
--	OC-1	51.84	672
--	OC-3	155.52	1,344
--	OC-12	622.08	5,376
--	OC-48	2,488	21,504
--	OC-192	9,953	86,016

Network switching

A factor limiting the efficiency of conventional networks is circuit switching, the traditional method for connecting telephone calls. Circuit switching provides a dedicated physical connection among two or more points on the network for the duration of the phone call or data exchange. The bandwidth is unavailable for any other use during that period, even though the circuit is idle during pauses in the conversation or data transfer. This idle time is typically more than 50 percent of the total connection time. Modem-based data communication compounds this problem. People surfing the Web, for example, may occupy a circuit path for hours at a time in a network that was designed for telephone calls lasting only a few minutes each.

Packet switching was developed to make networks more efficient. The original ARPANET (now the Internet) made use of a packet technology called Transmission Control Protocol/Internet Protocol (TCP/IP, or simply IP). Frame Relay is another early packet switching method. It was designed for the PSTN, but is declining in use. A more recent version of PSTN packet switching is Asynchronous Transfer Mode (ATM). Both IP and ATM greatly improve the efficiency of networks by breaking all voice, data or video traffic down into equal size "packets" of digitized data. The packets are sent over the network using all available bandwidth and multiple paths. Packets are prioritized so that those that can tolerate the least delay (say, a voice conversation) are handled before lower priority traffic (say, a batch data transfer). Full use is made of available network bandwidth by interleaving of multiple customers' traffic over the same facilities. Once the packets reach the destination, they are disassembled and the data contents are arranged in the correct order. The switching and packet assembly-disassembly process is so fast that it can appear to be instantaneous to the customer. Both IP and ATM can provide "virtual permanent connections" (VPN) to emulate the dedicated connections of T-carrier facilities without the associated bandwidth penalties.

IP was originally designed to handle only data. It does not provide the quality of service expected for voice communications or videoteleconferencing sessions. ATM was designed to handle voice, data and video, but at the price of a higher bandwidth "overhead." While both IP and ATM networks are widely deployed today, it is expected that the two technologies will evolve into a single standard similar to the latest IP model (IP Version 6). Multi-Protocol Label Switching (MPLS) is compatible with both IP and ATM and may be the vehicle for the eventual merger or intersnetworking of the two standards.

Collocation facilities

The trend away from circuit switching and dedicated T-carrier networks toward "connectionless" IP and ATM packet networks and the proliferation service providers have important implications for urban planners and designers. Whereas traditional networks are highly centralized and hierarchical, packet networks are more distributed and peer-oriented. Thus, the major resources and facilities of traditional networks tend to be concentrated in major commercial centers. Early Internet activity and other advanced information services tended to conform to the model of the traditional networks. This led to significant development, in the 1990s, of "telco hotels" and "collocation centers" in the Tier One cities. These collocation centers also provide a convenient and economical means for multiple service providers, ILECs and CLECs to interconnect their networks.

PHOTO: STCC

Fig. 1. Springfield Technology Park is a successful high technology redevelopment project.

Now, networks are becoming more distributed to take advantage of packet technology and to move information resources closer to the customer. This reduces "latency" in the network and reduces bandwidth demands on the long haul networks. It also means that the facilities (collocation centers) to house the network components, file servers, ISPs and ASPs must be accommodated in Second and Third Tier Metropolitan Statistical Areas (MSAs) and even rural areas. Increasingly, these collocation facilities provide a hub for urban economic development.

High-technology enterprises and start-up businesses often cluster around these facilities to take advantage of inexpensive access to broadband services. Thus, collocation centers are frequently seen as catalysts for sustainable development in urban enterprise zones. One example of the successful application of this concept is Springfield Technology Park in Springfield, MA. This 500,000 sq. ft. (46,450 sq. m) facility was established in November 1996 in buildings that once housed part of the historic Springfield Armory. The area had become economically depressed with the closing of the Armory and other light manufacturing businesses. The facility is affiliated with the adjacent Springfield Technical Community College (STCC) and now houses more than 10 telecommunications companies, related technology businesses, a small business incubator and a telecommunications development and training center. By all accounts, it and the STCC have led a booming economic revival in the area (STCC 2001) (Fig. 1).

Siting of collocation facilities presents new challenges for the planner. These are 24/7, mission critical facilities. They require abundant and reliable electrical power, fuel storage, security and proximity to major telecommunications rights-of-way (ROW). In the past, abandoned industrial facilities were often converted to this use. Their location in or near major cities and along rail and highway rights-of-way (often used for distribution of optical fiber networks) was favorable for this use. Disadvantages include potential cleanup costs, limited expandability and limited flexibility for tenants equipment layouts. Because of this, there is a trend toward purpose-built collocation facilities, especially in the lower-tier communities. In addition to the zoning, land use, environmental, logistics and other evaluation factors typically considered for site development, additional generic site and facility considerations can be classified into several categories. For the site in general these include:

- Risks from natural and man-made hazards
- Utility services adequate to sustain power-intensive, mission-critical operations on a 24/7 basis
- Access to telecommunications networks, facilities and ROW
- Ability to locate equipment in or near Incumbent Local Exchange Carrier central offices
- Appropriate workforce and customer base
- Convenient highway and airport access
- On-site or nearby food, entertainment, lodging and shopping
- Fuel and water storage for sustained operations during utility failures

Table 2. Typical Collocation Center Tenant Requirements

Type of Tenant	Likely Uses	Space Needs (1000 USF)	Security Needs	24/7 Operation	Connected Watts/USF	Supplemental Cooling / Back-up Power	Fire Protection	Population Density per 1000 USF
CLEC/IXC	Large switch	5-15	Very High	Yes	40-60	Yes/yes	Gas or Pre Action	1
CLEC/IXC	Network Equip.	1-5	Very High	Yes	40-60	Yes/yes	Gas or Pre Action	1
CLEC/IXC	Network Ops Center	1-2	Very High	Yes	10-15	Yes/yes	Pre Action	10
ASP	Server Farm	1-25	High	Yes	25-40	Yes/yes	Pre Action	2
Large ISP/ Web Host	Server Farm & Network Equip.	1-25	High	Yes	25-40	Yes/yes	Pre Action or Wet	2
Small ISP	Servers & Network Equip.	Cage or Cabinet	Medium	Yes	20-30	Yes/yes (shared)	Pre action or Wet	Very Low or None
Data Warehouse	Server Farm	1-25	Very High	Yes	25-40	Yes/yes	Gas or Pre Action	1
Software Development	Server Farm/Work Stations	1-40	Medium	No	8-12	Yes/yes (shared)	Pre action or Wet	10
Call Center	Switch/ Work Stations	1-40	Medium	2-3 Shifts	6-8	Yes/UPS	Wet	15
Multi Media/ Computer Animation	Server Farm/ Work Stations	1-40	Medium	1-2 Shifts	8-12	Yes/UPS	Pre Action or Wet	1
Incubator Company	Work Stations	< 1-2	Medium to Low	No	4-6	No/no	Wet	10
Legacy Business	Servers & Work Stations	1-40 +	Medium to Low	No	4-6	No/yes (shared)	Wet	10

It is important to note that collocation centers not only need access to the long haul networks that connect them to other cities and regions, they need to connect to their local customers as well. Service providers most often deliver advanced services to their customers over the ILECs' "local loop" infrastructure. The collocation center must be near an ILEC central office in order to connect to the local loop. This "last mile" of the network infrastructure is usually its greatest bottleneck. It was never designed to provide the capacity needed for the more advanced telecommunications services. Technologies like DSL can squeeze more bandwidth out of the local loop, but there are limitations. In many cases, the local loop must be upgraded with optical fiber or bypassed using wireless technology. ROW must be provided for upgraded local distribution.

Collocation centers are often thought of as "telco" or "switch" sites, yet they often house a variety of uses. These can include ASPs, data warehouses, call centers, ISP "server farms," and many similar uses. Table 2 summarizes typical facility planning considerations for a multi-use collocation facility.

Wired networks

Telecommunications networks are tied together with metallic or optical (glass) fiber cables. In the local loop, the most common medium is paired conductor copper cable. This medium is a legacy of the PSTN. Besides voice and modem-based data, copper can transport DSL or T-1 to provide broadband data service and even video service. There are serious limitations, however, as noted above. Coaxial cable is another metallic medium. It is deployed mainly for CATV networks and other video transmission. Its use is declining in favor of optical fiber.

PHOTO: Courtesy Carlon Telecom Systems.

Fig. 2. A cable plow places optical fiber duct in a single operation.

PHOTO: Courtesy Carlon Telecom Systems.

Fig. 3. Telecommunications duct being placed in a trench.

Since the late 1970s, optical fiber has become the medium of choice in long haul networks and is finding its way into the local loop, especially in more densely populated urban areas. Compared with copper cable, optical fiber provides enormous capacity. A single pair of hair-thin fiber strands can transport information over great distances at up to 20 billion bits per second (20 gigabits/s). New technologies like wavelength division multiplexing (WDM) can expand this capacity even further by sending different wavelengths ("colors") of laser light over the same strands. Fifty-five wavelengths can be multiplexed onto a single fiber giving it a capacity of 1.1 trillion bits per second. This is the equivalent of about 15 million voice channels or enough capacity to transmit the entire contents of the Encyclopedia Britannica 20,000 times in one second (Nellist and Gilbert 1999). Higher rates of multiplexing have been demonstrated recently.

Unlike copper cables, optical fibers and the pulses of light they transmit are impervious to electromagnetic interference. Thus, optical fiber can be installed on electric transmission towers or in electrical duct bank, the latter being a useful feature in congested urban streets. At least one manufacturer makes composite optical fiber/power cable for this purpose. Optical fiber can also be installed in telecommunications duct bank, directly buried or installed on utility poles like conventional metallic telecommunications cable. (Fig. 2)

If old copper cables can be removed from existing telecommunications ductbank, many times their capacity in optical fiber cable can be installed in the vacated space. A single conventional 4-inch telecommunications conduit can easily accommodate at over 2,000 strands of fiber. Conduits can be subducted for multiple carriers. PVC conduit works well for optical fiber cable because of its low pulling friction. Buried fiber routes need to be planned carefully, however, to avoid excessive bends between manholes or other pulling points. Bends should be sweeping bends with a minimum 40-ft. (12.2 m) radius, if possible. Bends of less than this radius must be manufactured bends only and will shorten distances between pulling points. Access points include manholes (sometimes called maintenance holes-abbreviated MH), hand holds, aboveground pedestals and controlled environment vaults (CEVs). CEVs are similar to large MHs and contain electronic network equipment, power and air conditioning.

MHs should be linear (Type "A") with all cable direction changes outside the MH. If this is not possible, type "L," "T" or "J" MHs may be used to accommodate cable bends. As with substandard conduit bends, this condition will shorten allowable pulling distances and make installation and future cable additions more difficult.

Railroad ROW and the shoulders of limited access highways make good pathways for buried optical fiber cable. Conduits can be placed in trenches or plowed in. The nature of these ROW eliminates many issues with grade and stream crossings. In urban streets, appropriate use of lateral boring and similar techniques will minimize the need for trenching. Overhead cable placement requires careful attention to span lengths, pole attachment and vertical separation of services. Underground cable placement may be more cost-effective in many cases. (Fig. 3)

An emerging technology for optical fiber infrastructure is Airblown Fiber (ABF). With ABF, empty plastic tubes are buried in the ground and the fiber strands are later blown into the tubes using a special machine. Aboveground pedestals are used instead of MH or hand holes. It is only necessary to access the infrastructure at the pedestal access points. If fiber strands need to be replaced for any reason, the old strands are blown out and new ones are blown in. The cost of materials for this technology is relatively more costly than for conventional optical fiber infrastructure, but installation can be faster, reducing labor costs. Once the ABF infrastructure is in place, future additions of changes of fiber are very convenient. ABF as also been used successfully in renovation of historic buildings. If ABF tubes are placed within the building infrastructure during renovation, it is possible to effect future optical fiber installations or modifications without disturbing the building fabric. (Fig. 4)

Optical fiber has not found widespread use in the local loop because of cost. The cable itself is not expensive, but the electronic and optical devices needed to activate ("light") the cable are costly. Economies of scale offset this cost in long haul, interexchange applications and even in high usage local loop situations (office parks, campuses and in urban centers with large office buildings). Even so, only about 5% of office buildings in the United States are currently connected by optical fiber.

New techniques like Passive Optical Network (PON) are gradually overcoming the cost issue and making optical fiber local loop feasible for less densely populated areas. PON, as the name implies, makes use of passive components in lieu of the more expensive electro-optical components of conventional optical fiber networks. This trade-off reduces the distance and bandwidth capability of the optical fiber but still provides much greater capability than the conventional copper local loop. PON can be implemented using conventional fiber construction or ABF. As individual customers and small businesses demand more advanced services, the return on investment for PON becomes more attractive.

PHOTO: Courtesy Sumitomo Electric Company

Fig. 4. An air blown fiber tube assembly.

Wireless networks

Beginning with the advent of cellular phone service in the 1980s there has been an increasing proliferation of wireless facilities and services. These include mobile services such as cellular (and its functional equivalents), "wireless e-mail," paging and growing number of fixed wireless services. The latter category includes point-to-multipoint networks like Local Multipoint Distribution Service (LMDS) Multichannel Multipoint Distribution Service (MMDS) and point-to-point services like microwave. Satellite services can be either point-to-point or point-to-multipoint and can cover a much larger geographic area than the other wireless services. Emerging digital broadcast standards have triggered the construction of numerous new television broadcast towers. Unlicensed microwave and infrared devices are also available, but potential interference problems with unlicensed technology and the short range of infrared limit their applicability to urban telecommunications needs. A brief description of the most common wireless networks follows. (Fig. 5)

Fig. 5. Broadcast and microwave towers dominate the skyline of a community.

In the U.S. marketplace, mobile services include conventional cellular service and three functionally equivalent services. These are Broadband Personal Communications Service (PCS), Narrowband PCS and Enhanced Specialized Mobile Radio Service (ESMR). Licensing policies permit up to six providers of these functionally equivalent services within a single market. Originally conceived as mobile or portable telephones, cellular and its equivalents are becoming mobile data services as well. They can connect to laptop computers to provide data connections. The more advanced generations of devices (so-called G2.5 or G3) units can send and receive electronic mail messages and can display data and graphics from the Internet and other on-line sources. There is an increasing convergence between

Fig. 6. A typical cellular tower.

Fig. 7. Roof-mounted microwave antennas provide point-to-point broadband communications in an office park.

portable phones and other hand-held personal digital assistant (PDA) devices. ESMR provides a two-way radio function in addition to phone and data services.

As the name implies, cellular (and equivalent) services work by breaking a geographic area down into a number of relative small, contiguous cells. Near the geographic center of each cell is a base station, usually mounted on a tower. This base station is connected to the wired telecommunications network and has the capability to connect a mobile station within its cell to the PSTN and to other networks. If the mobile unit passes out of the cell, the network can "handoff" the unit to the next cell without dropping the call. In practice, a typical cellular or PCS cell has a radius of roughly 2 to 10 miles, which can vary greatly depending on the antenna height, operating frequency, terrain, morphology, and subscriber density. High-density subscriber environments require smaller and more numerous cells because the call handling capacity of each is limited. As cells become saturated with calling volume, additional cells must be constructed. (Fig. 6)

The cellular model permits relatively reliable coverage because of the limited area served by the base station and, more important, it permits the re-use of the assigned radio frequencies in noncontiguous cells. This greatly increases the number of customers that can be served with the relatively limited radio frequency spectrum available for this purpose. The disadvantage of this arrangement is the number or base stations required within a geographic area. Whereas early cellular installations used relatively tall towers and large cells, the trend is toward smaller, more numerous cells and base station antennas mounted closer to the ground. In densely populated urban areas, cellular antennas are usually mounted on buildings and individual cells may cover only a city block or two. With minor differences, PCS and ESMR work in the same manner as cellular.

There are satellite-based mobile phone systems such as Iridium, but these do not appear likely to replace conventional ground-based cellular, PCS and ESMR systems. Satellite systems are expensive to use, require large, rather cumbersome handsets and do not work in areas that are "shadowed" from the satellites. To date, such systems have failed to capture a large market share and their long-term economic viability is yet to be demonstrated.

While cellular, PCS and ESMR are described as mobile services, there has been a growing use of "mobile" phones to replace wired phones, especially in residential areas. Aggressive pricing and promotion by service providers is accelerating this trend.

Fixed wireless services include microwave, satellite, LMDS and MMDS. Point-to-point Microwave has been in use for more than sixty years and was once the mainstay of the transcontinental telecommunications network. Optical fiber has largely replaced long haul microwave, although microwave is still extensively used in areas of rough terrain and in more sparsely populated regions. Modern digital microwave can carry SONET signals at data rates up to about 1,000 Mb/s over distances of up to 35 mi. (56.3 km) per "hop." Repeater stations can extend this distance to hundreds or thousands of miles. Microwave is also used to replace the local loop, providing high capacity connections to office parks, individual office buildings, campuses and other concentrations of network customers. (Fig. 7)

Fixed satellite services work on the same principal as microwave, but cover a much larger area because the "repeater station" is located in a satellite 22,000 mi. (35,400 km) above the earth. This allows the

Table 3. PCS System Buildout Requirements
(Source: FCC Rules: 47CFR24.5 October 1, 2000)

Type of Licensee	Operating Frequency Range (MHz)	Minimum coverage requirement within 5 years of license grant	Minimum coverage requirement within 10 years of license grant
Narrowband PCS (Nationwide Licensee)	901-941	An area of 750,000 square kilometers or 37.5% of the US population	A area of 1,500,000 square kilometers or 75% of the US population
Narrowband PCS (Regional Licensee)	901-941	An area of 150,000 square kilometers or 37.5% of the regional population	An area of 300,000 square kilometers or 75% of the regional population
Narrowband PCS (MTA Licensee)	901-941	An area of 75,000 square kilometers or 25% of the licensed area or 37.5% of the licensed area population	An area of 150,000 square kilometers or 50% of the licensed area or 75% of the licensed area population
Broadband PCS (30 MHz Licensee)	1,850-2,200	1/3 of the licensed area population	2/3 of the licensed area population
Broadband PCS (10 MHz Licensee)	1,850-2,200	1/4 of the licensed area population	No requirement

Note 1: As an alternative, Narrowband PCS licensees may satisfy buildout requirements by demonstrating that they provide substantial service to their licensed areas within ten years of license grant. Substantial service is defined as service that is sound, favorable, and substantially about a level of mediocre service that would barely warrant renewal.

Note 2: Broadband PCS licensees must serve the required portion of the population with a signal level sufficient to provide adequate service.

repeater to "see" and communicate with earth stations anywhere within an enormous geographic area. Limitations of satellite service are relatively high cost, limited capacity and signal delay. It takes about .25 seconds for the radio signal to travel to the satellite and back. This makes voice conversation difficult and causes problems with certain data transfer protocols. On the other hand, satellites are ideal for the distribution of one-way television programming and other data over a wide area.

Two-way satellite connections are available for Internet and other data access. The upstream data is at a slower speed than the downstream data. The dish antenna points the subscriber data at the appropriate satellite. Uplink speeds range from 50Kb/s to 150Kb/s. Downlink speeds range from about 150 Kbps to more than 1,200 Mb/s, depending on factors such as the capacity of the server, and the sizes of download files, and the amount of Internet traffic. While the uplink data speeds do not qualify as broadband, satellite systems offer an alternative in rural areas where Digital Subscriber Line (DSL) and cable modems are not available.

LMDS and MMDS are newer fixed wireless services capable of providing broadband connections over relatively short distances, *e.g.*, approximately 2 to 20 mi. (3.2 to 32.2 km). Like DSL, LMDS and MMDS were originally designed to distribute TV signals in competition with CATV operators ("wireless cable"). LMDS operates at a higher frequency than MMDS and, because the signal is easily reflected, may provide coverage even without direct line of sight. MMDS has greater range and can serve a larger area if line of sight is maintained. Like DSL, these technologies have been adapted for two-way local data transport. LMDS and MMDS are cellular in nature like cellular telephony, but are fixed, rather than mobile services. LMDS and MMDS require a base station near the center of the cell and the customers. LMDS and MMDS typically provide data transmission at rates of up to 30-50 Mb/s.

The siting of wireless towers and facilities, especially cellular and PCS facilities, has become a vexing issue for many local planning authorities. National policy strongly encourages the rapid deployment of wireless services, including multiple competitive, but functionally equivalent services, within designated markets. Licensees must serve a significant portion of their license areas within specified five- and ten-year "buildout" periods. The buildout rules are confusing to some planners because they are not consistent among the functionally equivalent services. Cellular service providers, for example must serve a specific geographic area or risk loss of the unserved portions to their competitors. An area is considered "served" if a certain signal level is maintained within the area, even though an unspecified number of coverage "holes" may exist within the coverage area. The rules are

Fig. 8. Cellular antennas can be mounted on power transmission towers.

Fig. 9. Cellular antenna tower disguised as a tree.

more complex for PCS service providers. They must serve a portion of a licensed area or a percentage of the population within that area or risk license forfeiture. As Table 3 shows, the rules vary for different types of PCS licensees. As complex and inconsistent as these rules may seem, it is important for the planner to recognize that no cellular or PCS licensee is required by regulations to provide universal or 100% coverage. There is considerable latitude in making antenna siting decisions.

In December 2000, over 104,000 cellular base stations were supplying service to more than 115,000,000 subscribers in the United States, a growth of 27.2% over the previous year (CTIA 2001). In addition, newer services like LMDS and MMDS are entering some markets for the first time and many broadcasters are constructng new transmission towers to accommodate the new digital TV standards. The speed and intensity of wireless facility deployment has taken some communities by surprise.

Besides the usual concerns related to the appropriate placement of any facility in the community, planners and local officials are faced with new concerns unique to wireless technology. Chief among these concerns are aesthetics and public health and safety. There has also been confusion and concern about how much local control over these facilities the Telecommunications Act permits. The Telecommunications Act speaks specifically about Personal Wireless Facilities (cellular and related networks). The Act specifically preserves the authority of local authorities to ". . . regulate the placement, construction and modification of Personal Wireless Facilities." (TCA 1996). Further, there is nothing in the Act to prevent local authorities from:

- Protecting navigable airspace from tall structures.
- Requiring several PWFs to co-locate or use existing structures, if feasible.
- Minimizing the size and community impact of PWFs consistent with need.
- Confirming PWF compliance with applicable standards and regulations.

In particular, local authorities have several tools at their disposal for ensuring that PWFs are constructed in a manner acceptable to the community. For example, Federal Aviation Regulations provide for determining whether a proposed tall structure would constitute an obstruction or hazard to air navigation. It is not well known, however, that the FAA has no power to prevent the construction of such structures. The FAA can only require obstruction painting or lighting, but cannot otherwise regulate the construction of such structures. The FAA relies entirely on local authorities to exercise this control through their zoning powers. Local authorities may limit the construction of structures that would have an adverse impact on air commerce. Structures less than 200 ft. (61 m) above ground level generally do not have an impact on air navigation unless they are located near an airport.

Local officials can also require that PWFs and other wireless facilities conform to the aesthetic requirements of the community. The visual impact of antennas can be minimized by mounting them on or within existing structures. Roof-mounted equipment can be set back to minimize visibility from street level. Wall-mounted antennas can be painted to match the surfaces they are mounted on. Antennas can be mounted behind radio transparent material specifically designed to conceal them without degrading their performance. These materials come in a variety of colors and textures to blend with existing buildings and other structures.

Antennas and antenna towers can also be disguised to look like other structures (silos, flag poles, clock towers or trees) or can be mounted on existing lamp standards, electric transmission towers or billboards. (Figs. 8 and 9)

Even if antennas are not concealed or camouflaged, their impact can be reduced by using natural materials (laminated wood monopole towers in lieu of steel), by using "cross-polarized' antennas (fewer antenna elements, mounted close to the structure) and by limiting tower height to the minimum necessary.

The Act does contain three limitations on communities' right to regulate PWFs however. Local authorities may not:

- Unreasonably discriminate against providers of functionally equivalent services.
- Prohibit or have the effect of prohibiting the provision of Personal Wireless Services.
- Regulate PWFs based on the environmental effects of radio-frequency emissions to the extent such facilities comply with the FCC regulations regarding radio frequency emissions.

Given the widespread perceptions about the possible health effects of radio frequency radiation from cell towers, other PWFs and broadcast facilities, it is prudent for planners and local authorities to satisfy themselves that existing or proposed wireless facilities do, in fact, comply with radio frequency FCC radio frequency power density standards. This is not only important for peace of mind within the community, but because in most cases, the FCC exempts wireless service providers from demonstrating compliance at the federal level. A very useful guide for making this determination is FCC OET Bulletin 65, which provides simplified formulas and charts for this purpose. Obviously, if an installation were found not to comply with the regulations, there would be ample grounds for local planning or zoning authorities to require modification or deny approval for construction. The following table summarizes the maximum allowable levels of radio frequency power density that a wireless facility may produce in locations accessible by the general public and by workers, respectively. The values given for power density are milliwatts per square centimeter (mW/cm2) and are frequency dependent. For example, at cellular and ESMR frequencies (around 800 MHz) the public exposure limit is about 0.53 mW/cm2. For PCS it is about 1.0 and for FM and TV broadcast towers it is 0.2. (Table 4)

Federal regulations do not require PWF licensees to provide perfect coverage of their assigned areas. Regulations recognize that 100% wireless coverage over a large geographic area would be cost-prohibitive if not technically impractical. For this reason, the regulations permit significant "coverage holes" within a licensee's area. The degree of coverage licensees must provide within specified buildout periods is summarized in Table 3 above. Planners should recognize this when making base station siting decisions.

Table 4. Limits for Maximum Permissible Exposure
(Source: FCC OET Bulletin No. 65, August 1997)

A — Limits for Occupational/Controlled Exposure

Frequency (f) Range (MHz)	Power Density (mW/cm^2)	Averaging Time (Minutes)
0.3-3.0	100	6
3.0-30	900/f	6
30-300	1.0	6
300-1,500	f/300	6
1,500-100,000	5.0	6

B — Limits for General Population/Uncontrolled Exposure

Frequency (f) Range (MHz)	Power Density (mW/cm^2)	Averaging Time (Minutes)
0.3-3.0	100	30
3.0-30	180/f	30
30-300	0.2	30
300-1,500	f/1500	30
1,500-100,000	1.0	30

Note 1: Occupational/controlled exposure limits apply in situations where persons are exposed as a consequence of their employment provided those persons are fully aware of the potential for exposure and can exercise control over their exposure.

Note 2: General Population/uncontrolled exposures apply in situations in which the general public may be exposed, or where persons that are exposed as a consequence of their employment may not be fully aware of the potential for exposure or can not exercise control over their exposure.

A very useful tool to assist planners in making wireless facility siting decisions is radio frequency propagation modeling software. Using one of several predictive propagation models and a digital terrain map, it is relatively simple to estimate the quality and extent of coverage that can be achieved from a given site for cellular, PCS, TV broadcast, or any other wireless service. Various sites can be tested and compared and alternatives such as antenna heights can be investigated. There are several propagation software packages. They vary in cost and sophistication, but all can produce reasonably accurate "coverage maps" for a proposed installation.

Conclusions

Information is the lifeblood of every community, business and institution. The way this information is produced, processed, delivered and stored has been changing dramatically in the last fifteen years and the pace of this change is likely to accelerate. Market forces and government regulations alone will not assure the appropriate distribution of information services to all communities. Urban planners and designers, governments and businesses must take a proactive role in planning, designing, developing and maintaining the appropriate infrastructure and facilities. ■

REFERENCES

ANSI/EIA/TIA-222-E. 1991. *Structural Standards for Steel Antenna Towers and Antenna Supporting Structures.* Washington: Electronic Industries Alliance.

ANSI/EIA/TIA-758. 1999. *Customer-Owned Outside Plant Standard.* Washington: Electronic Industries Alliance.

BISCI. 1998. *Telecommunications Distribution Methods Manual,* 8th Edition. Tampa, FL: Building Industry Consulting Service International.

BISCI. 2001. *Customer-Owned Outside Plant Design Manual,* 2nd Edition. Tampa, FL: Building Industry Consulting Service International.

CTIA. 2001. "Semi-Annual Wireless Industry Survey." Washington: Cellular Telecommunications & Internet Association. http://www.wow-com/industry/stats.

FCC. 2000. *Deployment of Advanced Telecommunications Capability: Second Report.* Washington: Federal Communications Commission.

Freeman, Roger L. 1985. *Reference Manual for Telecommunications Engineering.* New York: John Wiley & Sons.

Lyman, Peter, and Hal R. Varian. 2000. "How Much Information." School of Information Management and Systems, University of California at Berkeley. http://www.sims.berkeley.edu/how-much-info.

Nellist, John G., and Elliott M. Gilbert. 1999. *Understanding Modern Telecommunications.* Norwood, MA: Artech House, Inc.

OET Bulletin 65. 1997. *Evaluating Compliance with FCC Guidelines for Human Exposure to Radiofrequency Electromagnetic Fields.* Washington. Federal Communications Commission.

Rainer, George. 1990. *Understanding Infrastructure.* New York: John Wiley & Sons.

Rappaport, Theodore S. 1996. *Wireless Communications, Principles & Practice.* Upper Saddle River, NJ: Prentice Hall PTR.

STCC. 2001. "STCC Technology Park." Springfield: Springfield Technical Community College. http://www.stcc.mass.edu/techpark.

Telcordia. 2001. "Internet Sizer." Telcordia™ Technologies. Piscataway, NJ: http://www.netsizer.com

Title 14. Code of Federal Regulations. 2001. *Aeronautics and Space.* Washington: Federal Aviation Administration.

Title 47. Code of Federal Regulations. 2001. *Telecommunications.* Washington: Federal Communications Commission.

Nancy Clanton, editor

Summary
This article presents an overview of urban outdoor lighting design criteria. Included are general principles, characteristics of various lamps, and definitions of technical terms associated with lighting design. Recommendations emphasize appropriateness of lighting design to specific conditions of a site, and use of minimal appropriate lighting to reduce energy cost and to prevent light pollution and "light trespass" effects.

Key words
color, design criteria, glare, illumination levels, landscape lighting, light pollution, lighting controls, luminaire selection, pedestrian, photometric chart, safety

Nighttime aerial photo of Los Angeles, CA.

Urban outdoor lighting

The objectives of outdoor lighting design include:

- Improve the legibility of critical nodes, landmarks, and circulation and activity zones in the landscape;
- Facilitate the safe movement of pedestrians and vehicles, promoting a more secure environment, and minimizing the potential for personal harm and damage to property; and
- Help to reveal the salient features of a site by layering the light with soft ambient light and key accent lighting at a desired intensity in order to encourage nighttime use.
- Provide a environmentally responsible lighting that does not overlight or produce glare to improve light quality and to minimize light pollution and light trespass effect.

1 GENERAL DESIGN PRINCIPLES

Key points: General design principles

1. Subtle but recognizable distinctions can be made between major and minor roads, paths, and use areas by varying the distribution and brightness of the light and by varying the height, spacing and color lams.
2. Clear lighting patterns reinforce the direction of circulation, delineate intersections, and provide a visual cue to what conditions lie ahead.
3. Glare from exposed light sources is a major safety concern. Luminaire location and mounting height, fixture type, and lamp intensity must be carefully selected to optimize light distribution and minimize glare.
4. Security is not necessarily enhanced by increasing illuminance levels on the ground. Consider peripheral lighting, vertical illuminance levels and good color-rendering sources as well.
5. Color differentiation, unobtrusive illumination of background spaces, and bright illumination of objects of interest are common approaches for articulating landscape character.

Nighttime visibility and function (lighting quality not quantity)

Too often, *lighting quantity* or lighting levels are used for design instead of *lighting quality*. *Lighting quality* involves contrast, brightness adaptation, glare and light source color.

Increasing contrast will increase visibility. An example of poor contrast would be a person in dark clothing against a dark landscape or façade.

Credits: This article is an update of "Outdoor Lighting, in Charles W. Harris and Nicholas T. Dines, editors. 1998. *Time-Saver Standards for Landscape Architecture* and Illuminating Engineering Society of North America (IESNA) Outdoor Environmental Lighting Committee. See References for full citations.

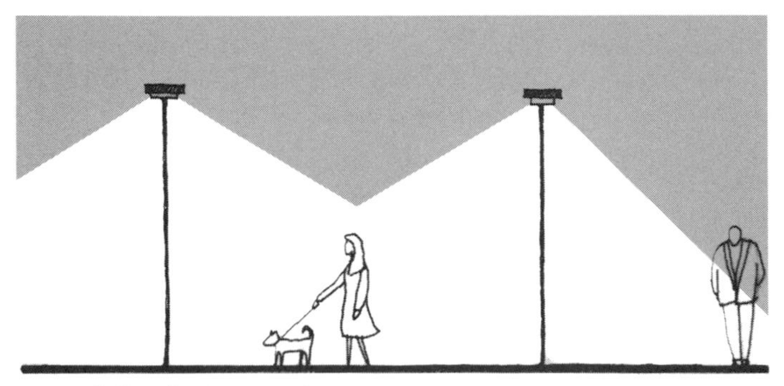

vertical distribution overlap

Fig. 1. Pathway and pedestrian walkway lighting.

If the façade is lighted, objects are easier to see. Our eyes adjust to the brightest object in our field of view. This adjustment of our eyes is referred to as brightness adaptation. If an object is very bright, like uncontrolled light from a floodlight, everything else in the immediate surrounding area appears relatively dark, making it harder to detect object details.

Glare is usually caused by uncontrolled light emitted from unshielded luminaires. An example of this is unshielded wall pack fixtures or floodlights located on a building façade. These situations can be easily avoided with proper equipment selection, location, aiming and shielding.

Light source color is another key to low light level visibility. Our night vision is very sensitive to short wavelength light (blue and green light), resulting in crisp and clear vision, especially in our peripheral vision. Reaction time and color recognition under low light levels is far superior with white light sources like metal halide, fluorescent, and inductive lamps.

Safety and security

Safety involves providing light on hazards so that they are detected with sufficient reaction time. Hazards may include pedestrian path and vehicle intersections, crosswalks, stairs and ramps. The lighting system, along with other site design elements, must provide visual information to assist users in avoiding such things as a collision or loss of bearings.

Security is often referred to as the "perception of safety." Providing for security involves lighting potentially hazardous locations and situations. For example, an increase in reaction time can give potential crime victims a better chance to change direction, find refuge, or call for help. Lighting can also act as a deterrent by increasing the visibility in an area of concern. Lighting is required in many secured areas to ensure that no encroachment goes unnoticed. However, it should be noted that an increase in the number of people in an area will be a more effective deterrent of crime than an increase in light level.

Darkness, together with unfamiliar surroundings, can incite strong feelings of insecurity. To provide a sense of security, possible hiding places and dense shadows should be minimized by the placement of appropriate light fixtures.

Pedestrian walkway lights: Walkway lights should have enough peripheral distribution to illuminate the immediate surroundings. Vertical light distribution over walkway areas should cover or overlap at a height of 2.13 m (7 ft.) so that visual recognition of other pedestrians is maintained. When the pedestrian's sense of security is a primary consideration, low mounting height with close spacing and a vertical illumination pattern may be the most effective approach. (Fig. 1)

Surveillance: For surveillance needs, lighting requirements should permit the detection of suspicious movement rather than provide for the recognition of definitive details. For the same expenditure of light energy, it is often more effective to light backgrounds, thereby generating silhouettes, than to light the foreground (*e.g.*, lighting the vertical face of a building instead of its horizontal foreground. It is also desirable to highlight entrances and to direct lighting away from points of surveillance. (Fig. 2)

peripheral lighting

undesirable

desirable

Fig. 2. Lighting for surveillance and security.

Vandalism: The best way to reduce the vandalism of light fixtures is to use fixtures that are durable enough to withstand abuse, or to place them out of reach. An alternative solution may be to use hardware that is less expensive to replace.

Atmosphere and character

A consistency of design expression can be achieved by identifying the common elements in a landscape that give it character, and then using similar approaches to their lighting. The clarity with which an object is perceived is influenced by its context.

Background: Exterior spaces should have a well-defined sense of background. Background spaces should be illuminated as unobtrusively as possible to meet the functional needs of safe circulation and protecting people and property. Whenever possible, these needs should be accommodated with peripheral lighting from the walkways, signage, entrances, and other elements relevant to the definition of the space.

Foreground: Foreground spaces or objects may be major elements and should be treated accordingly. Foreground spaces should utilize local lighting that produces maximum focus, minimum distractions, and no glare. Objects of interest and activities can be brightly illuminated while the background produces only minimal distraction.

Illumination of objects (shape accentuation): The direction of the light source is important for perception of three-dimensional objects. The ability to perceive volumetric form is influenced by the gradient of light and shadow falling on the object. Uniformly distributed, diffused light results in poorly rendered shadows; one must then rely upon outline and color in order to perceive the shape and form of the object. Conversely, a single point source will produce maximum shadows but may also minimize the perception of details. (Fig. 3)

Usually, the best way to illuminate standing objects is with a combination of both types of lighting. One source should accentuate shape and form by contrasting the surface with sharp shadows while the other source provides fill-lighting for details.

Color perception: Differences in lamplight color are often used with great effect in public lighting to color code roadways or to clearly delineate one area from another. As the general illumination level rises in a given situation, preference usually shifts away from a warm appearance toward the cool range. Accurate color rendition will aid recognition and improve the perception of outdoor environments. This is especially important at the pedestrian scale, where the color contrast of paving and landscape materials is often subtle. Fig. 4 illustrates the Color Rendering Index (CRI) defined below in Section 4, Terminology.

Environmentally responsible lighting

Environmentally responsible lighting includes minimizing light trespass and lighting pollution, and using minimal energy through lighting equipment selection and operation. The publication Lighting for Exterior Environments (IESNA 1999) provides lighting design criteria to limit light pollution and light trespass.

Light trespass is sometimes referred to as the "light shining in my window" syndrome. Usual culprits are unshielded floodlights, high wattage pedestrian lights, wall packs and other unshielded luminaires that are improperly located and poorly aimed. Given increasing public

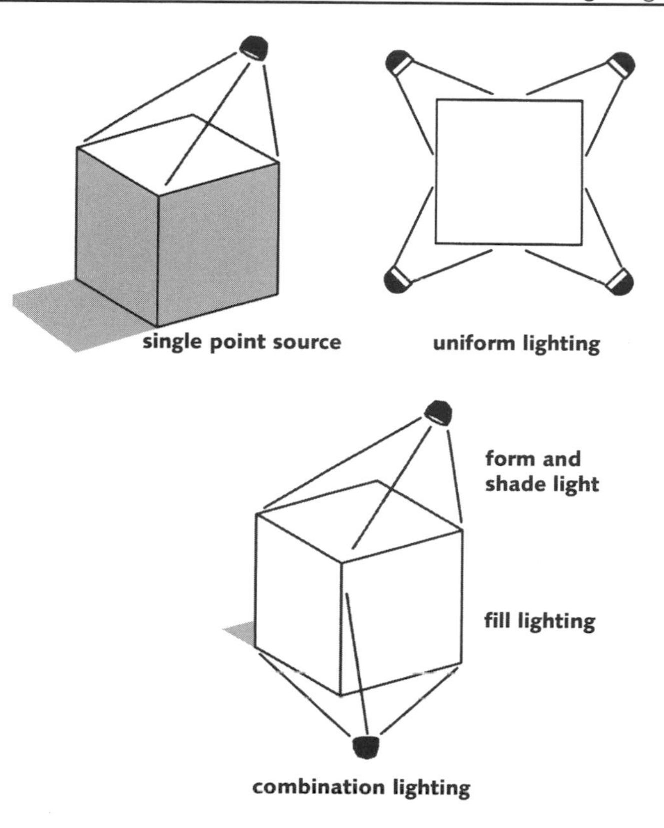

Fig. 3. Shape accentuation.

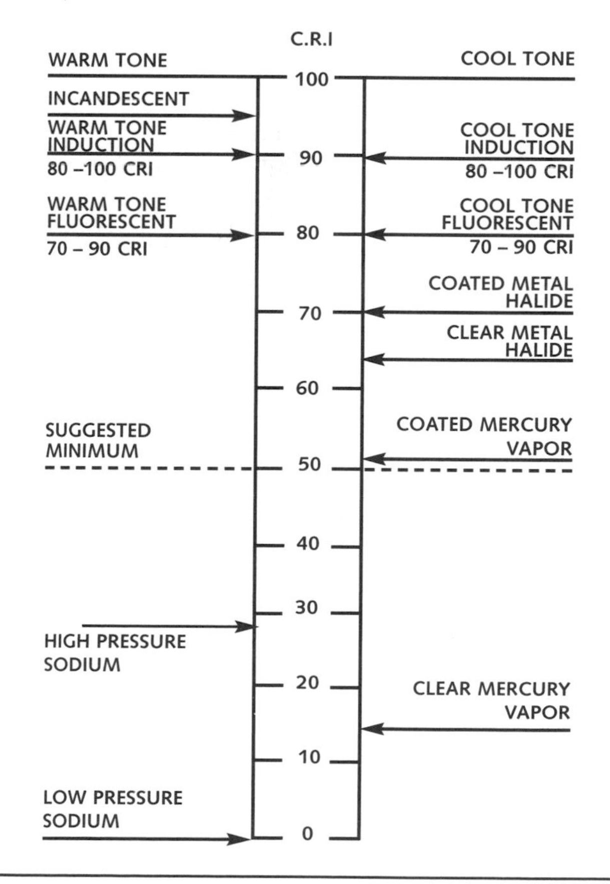

Fig. 4. Color Rendering Index (CRI).

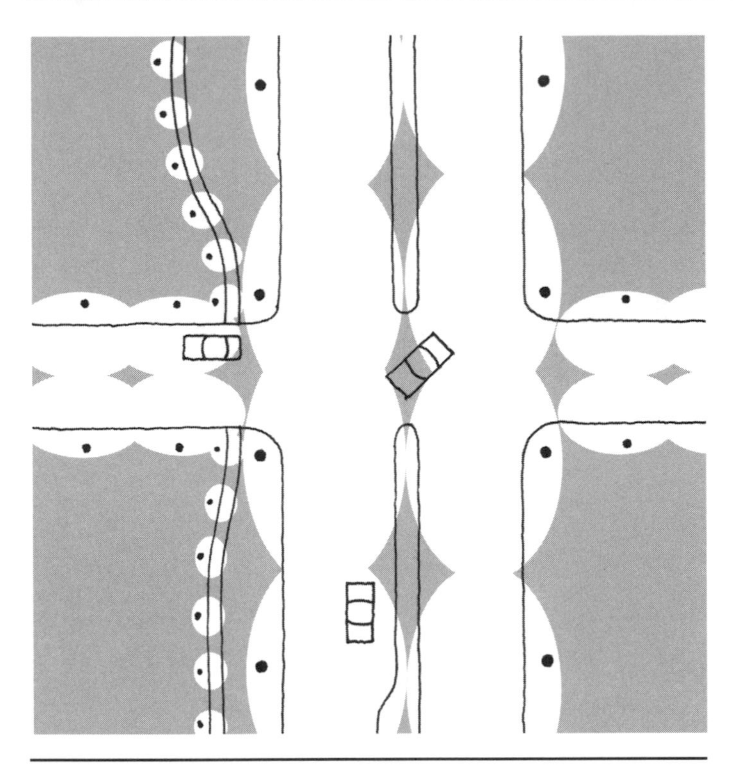

Fig. 5. Major and minor roads are distinct and pedestrian routes defined.

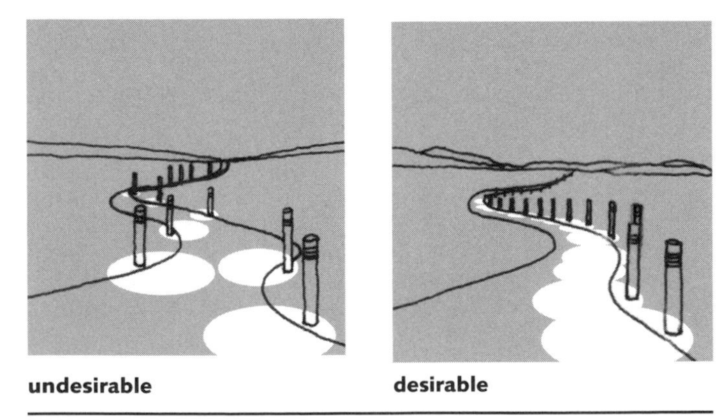

Fig. 6. Clear lighting patterns.

awareness and legal recourses, light trespass should be vigilantly avoided by careful lighting design and specification. Light trespass can be minimized with careful equipment selection, proper location, and proper aiming and shielding.

Light pollution is uncontrolled light that travels into the atmosphere. This light is wasted energy and creates a "sky glow." Unshielded luminaires and excessively high lighting levels cause light pollution. High wattage luminaires with poor visual shielding are increasingly not permitted. Excessive light levels with high amounts of reflected light that result from improper lighting design and installation are an increasing public concern. Use low wattage, shielded luminaires that are properly located and aimed.

Specific exterior light design criteria

The key to quality exterior lighting is to place light only where it is needed, without causing glare. By not wasting light, smaller lamp wattages can be utilized to achieve superior effects. The most important result is improved visibility. Another by-product is reduced energy usage and improved maintenance. Design criteria include basics such as lighting levels (illuminance), uniformity, and brightness balance (luminance), as well as recommendations for reducing glare, light trespass, and light pollution.

Deciding what to light

In some circumstances, it may be equally as important to determine what not to light as to determine what to light. Light can help guide people through a site or campus. Lighting certain commonly used nighttime paths will encourage safe movement from destination to destination. Other areas can be left dark to discourage use. It is important to provide a smooth transition between lighted areas and nonlighted areas.

Orientation

One purpose of outdoor lighting is way finding. As motorists and pedestrians weave their way throughout our nighttime built environment, lighting can help orient, guide and aid in visual tasks.

Layers of light

Outdoor environments that require lighting should be softly lighted to provide a pleasant ambient level. Overlighting should be avoided since it creates an imbalance between a site and adjacent streets and properties. Additional accent and task lighting can be added to guide people through an area and add visual interest.

Lighting hierarchy

Driver and pedestrian orientation can be aided by providing a hierarchy of lighting effects that correspond to the different zones and uses of a site. For instance, subtle but recognizable distinctions can be made between major and minor roads, paths, and use areas by varying the distribution and brightness of the light and by varying the height, spacing, and color of lamps. Attaining high levels of illumination along circulation routes does not have to be a prime consideration in outdoor lighting. If a clear and consistent system is provided, low levels may be adequate for safe circulation. (Fig. 5)

Clear lighting patterns

Clear optical guidance can be provided with the alignment of light fixtures positioned in consistent, recognizable, and unambiguous patterns. A staggered layout of road and pathway lights tends to obscure rather than reinforce the direction of circulation and the location of intersections (Fig. 6).

Conflict areas

Conflict areas such as intersections, walkway convergence, and pedestrian crossings should be highlighted in a manner such that pedestrians, bicyclists, skateboarders and other moving objects are easily seen and detected in plenty of time.

Changes in terrain

Stairs, ramps, bridges, change in walking surfaces and other "changes in terrain" need to be highlighted in such a manner that defines the shape and texture of potential tripping hazards. An example is to light stairs so that shadows clearly define the stairway without obscuring treads and risers.

Placement of luminaires

Spacing, height, and distribution of luminaires should avoid foliage shadows, provide uniformity, and vertical surface illumination. High mounting and wide spacing of fixtures may result in disruptions to the illumination pattern due to tree shadows (Fig. 7 top). Lower mounting heights and closer spacing between fixtures may create a more uniform distribution of light promoting the pedestrian's sense of security (Fig. 7 bottom).

Minimizing glare

Glare is major inhibitor of good visibility and can be produced by any scale of luminaire. Glare is produced by bright light sources in your field of view. This may include lamp, reflector or lens brightness. When luminaires are aimed towards you, the glare increases. When luminaires are aimed down, glare is greatly reduced. Luminaire apparent brightness as produced by a lens, can appear too bright if the light source is too powerful. Luminaires with full cut-off distributions aim the light in a downward manner. This distribution type is ideal for area and roadway lighting with higher lumen output. Attention to reducing and eliminating glare helps at the same time to minimizing and eliminating light trespass and night sky pollution (Fig. 8).

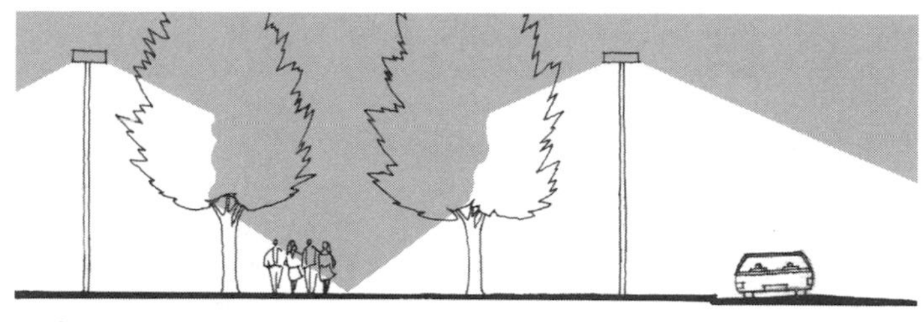

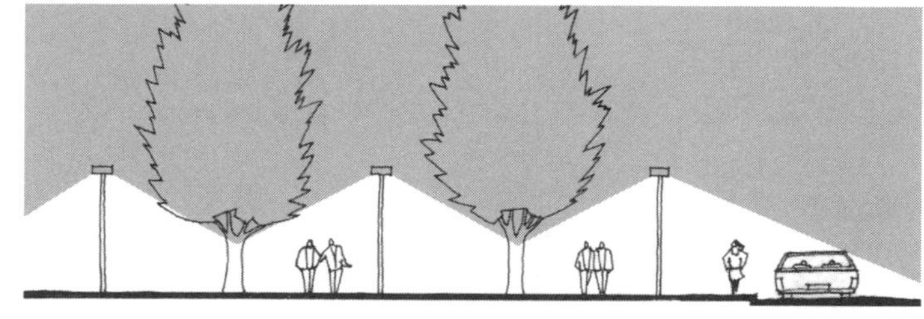

Fig. 7. Luminaire mounting heights. Low-mount fixtures provide better uniformity and illumination of vertical surfaces.

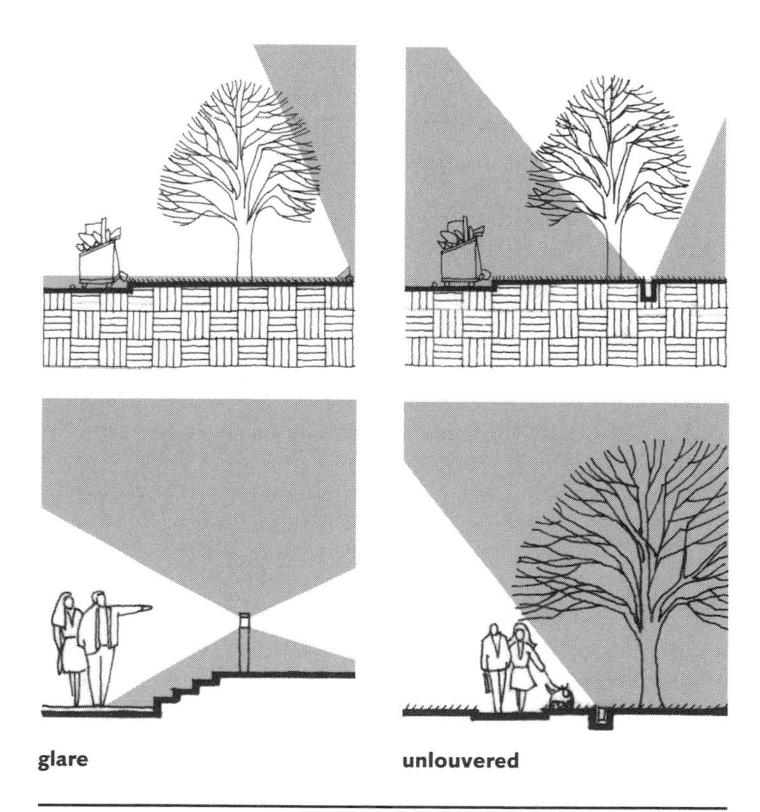

Fig. 8. Careless placement creates glare, light trespass and night sky pollution.

2 LAMP CHARACTERISTICS AND LIGHT DISTRIBUTION

Key points: Lamp characteristics

Selection of a lamp involves trade-offs between lamp size, optical control, efficacy, appearance, color temperature, color rendition, lamp life, costs, and maintenance (Table 1).

1. A variety of lamp types are commercially available. Selection of a lamp involves trade-offs between lamp size, optical control, efficacy, appearance, color temperature, color rendition, lamp life, costs and maintenance.

2. Illumination data for lighting fixtures are illustrated by photometric charts provided by the manufacturer. These charts illustrate light patterns on horizontal and vertical planes.

3. Uniformity of illumination is described by a ratio of light intensity values in lux (footcandles). A low ration appears more evenly lit and uniform, while a high ratio displays distinct and contrasting values.

4. Light fixtures can be broadly categorized into four main types based on size and design purpose: Low-level landscape lights, intermediate-height pedestrian lights, parking lot and roadway lights, and high-mast lights.

5. Levels of illumination are related to types of uses and other characteristics within use areas. Subject to new research findings and to consensus-based standard setting, standards are subject to change. Caution should be exercised when referencing previously published levels of illumination. See below, "Note regarding recommended levels of illumination."

Incandescent lamps: Incandescent lamps have superior color rendition and a warm white appearance. The disadvantage of a short lamp life can be overcome by the use of a rugged traffic signal lamp rated at 8000 hours nominal life or by undervoltaging the circuits to extend the life. Incandescent lamps have the lowest efficacy of all the lamps. However, they are inexpensive and the small filament permits good optical control. The most effective application for incandescent lamps is for residential accent lighting. Using a directional lamp such as an MR-16, can produce accent lighting on an object or landscape. This light source can also be easily dimmed. If used commercially, its short life and high energy cost may be difficult to justify.

Fluorescent lamps: Although compact shapes are becoming more widely used, most fluorescent lamps are long and linear, making optical control very difficult. They tend to produce glare unless they are well baffled. They have a good color rendition, whitish appearance, and superior life. Although they have good efficacy, their light output may be severely diminished by very cold weather. Fluorescent lamps, especially compact fluorescents, are excellent choice for decorative wall sconces and low-level pedestrian lighting. Their instant "on" capability, low wattage, low glare and long life make them an ideal lamping solution for many decorative luminaires.

Induction lamps: Induction lamps are relatively new, high-frequency sources that have no filaments or electrodes, but rather use the electromagnetic spectrum to directly energize a phosphor coating on the bulb. These lamps have a light quality similar to fluorescent in a 90-115 mm (3 1/2-4 1/2 in.) spherical lamp envelope, but with a significantly longer average rated life (up to 100,000 hours), and with almost no sensitivity to ambient temperature.

Mercury vapor lamps: Mercury vapor (MV) lamps have good efficacy, excellent life, but are the least energy efficient of HID lamps. Strong in the blue-green end of the color spectrum, the lamp is popular for foliage lighting but is otherwise not acceptable for color rendering properties. For these reasons, it is recommended that metal halide lamps be considered in lieu of mercury vapor lamps.

Metal halide lamps: Metal halide (MH) lamps offer superior optical control and color rendition, being an excellent white light source. Their efficacy is very high for outdoor lighting tasks, especially when peripheral detection is important. Since metal halide comes in a variety of color temperatures from 3000K (similar to an incandescent) to cool 5000K (similar to moon light), the installation can be designed for different effects. Currently, the metal halide lamp has a shorter life than other HID sources, yet the life and lumen maintenance are being improved constantly.

High-pressure sodium lamps: High-pressure sodium (HPS) lamps have good efficacy for on-axis tasks. Their efficacy drops when peripheral detection is important. There popularity in recent years accounts for their preserved high efficacy, and low maintenance. The orange-yellow appearance and a mediocre color rendition also work

Table 1. SUMMARY OF LIGHT CHARACTERISTICS

Lamp	Wattage range, M (ft)	Efficacy, lumen/watt*	Average life, hrs	Apparent color	Color rendering	Initial cost of equipment
Incandescent	3-300 (10-1000)	10-25	750-2000	Warm white	Best overall	Low
Fluorescent	4.5-64.5 (15-215)	40-80	7500-15,000	Warm to cool white	Very Good	Medium
Induction	16.5-25.5 (55-85)	63-70	100,000	White	Very Good	High
[1] Mercury vapor (deluxe white)	12-300 (40-1000)	25-60	24,000	Cool white	Good	Medium
Metal halide	52.5-450 (175-1500)	65-105	7500-20,000	Cool white	Good	High
High-pressure sodium (STP)	10.5-300 (35-1000)	60-120	—	Orange-yellow	Poor	High
[1] 'White' high-pressure sodium	45-75 (150-250)	75-80	—	Warm white	Very good	Very high
[2] Low-pressure sodium	5.4-54 (18-180)	70-150 -	—	Intense yellow	Very poor	High

*Includes ballast losses
Notes: [1] not recommended; [2] not recommended except for locations near observatories

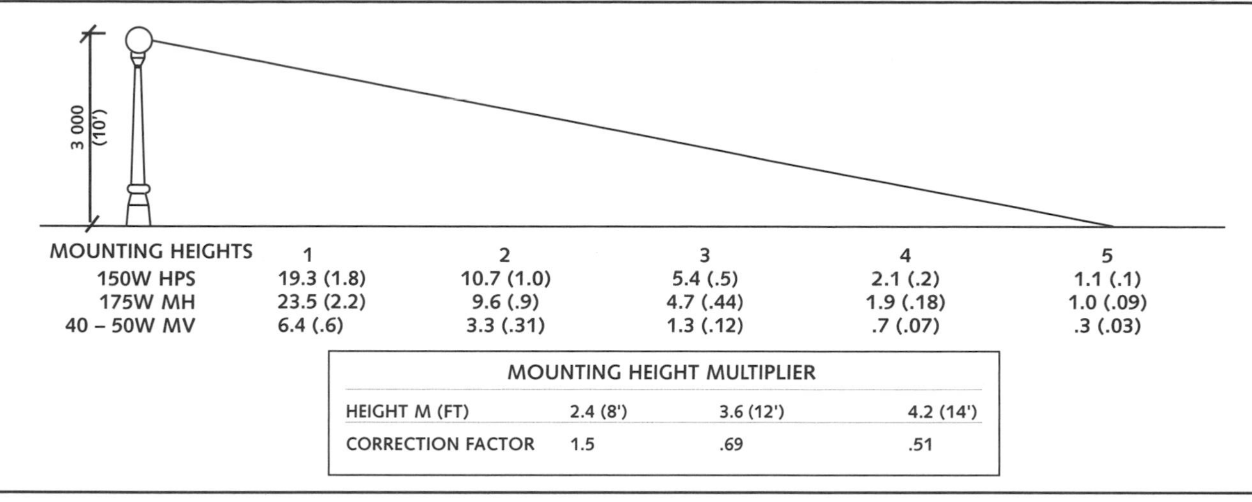

3 000
(10')
MOUNTING HEIGHTS
150W HPS
175W MH
40 – 50W MV
1
19.3 (1.8)
23.5 (2.2)
6.4 (.6)
2
10.7 (1.0)
9.6 (.9)
3.3 (.31)
3
5.4 (.5)
4.7 (.44)
1.3 (.12)
4
2.1 (.2)
1.9 (.18)
.7 (.07)
5
1.1 (.1)
1.0 (.09)
.3 (.03)
MOUNTING HEIGHT MULTIPLIER
HEIGHT M (FT) 2.4 (8') 3.6 (12') 4.2 (14')
CORRECTION FACTOR 1.5 .69 .51

Fig. 9. Typical photometric chart for roadway, walkway or area lighting. Lux (footcandle) levels displayed are for a mounting height of 3 m (~10 ft.). For other mounting heights, use the multiplier listed in the inset. Once minimum illumination levels are identified, fixture spacing is determined by multiplying the number of corresponding mounting heights by two.

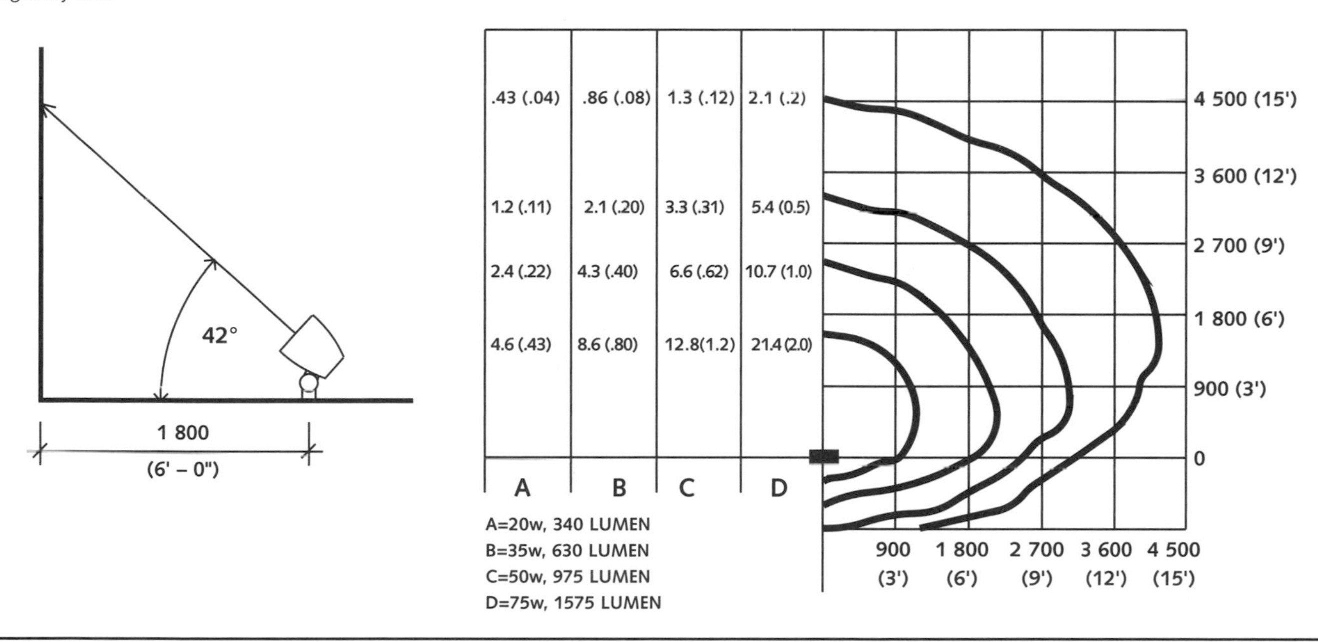

Fig. 10. Typical photometric chart for directional lighting, displayed in lux (footcandles). Illumination levels listed are based on a typical half 42° aiming angle.

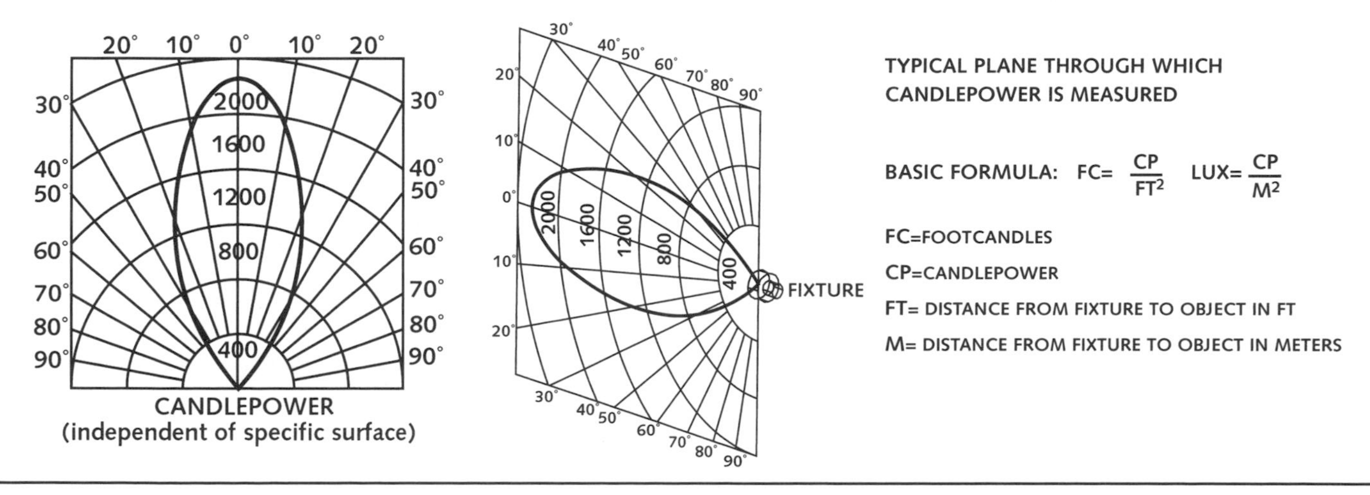

Fig. 11. Typical photometric chart for directional lighting, displayed in *candela*. Maximum *candela* in this example is at 0° (2,200 candlepowers). Conversion to lux can be calculated by the indicated formula.

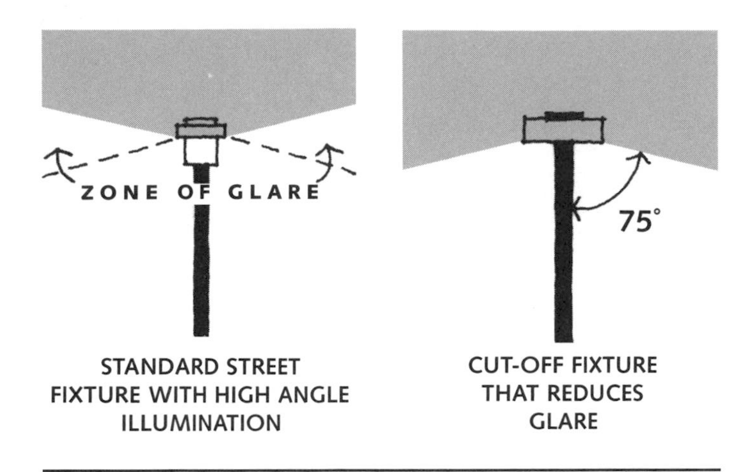

Fig. 12. Cutoff light distribution. Consideration of cutoff is critical in minimizing glare, light trespass and night sky light pollution.

Table 2. UNIFORMITY RATIOS

UNIFORMITY RATIO		
Average, lux (fc)	Minimum, lux (fc)	Visual description of illuminated field
21.4 (2)	10.7 (1)	Just a visible difference in light intensities
32.3 (3)*	10.7 (1)*	The high values of the field are twice as bright as the low values
43.0 (4)†	10.7 (1)†	—
107.6 (10)	10.7 (1)	Very distinct focal highlights; spotty

* Average and minimum uniformity ratios usually recommended for roads.

† Average and minimum uniformity ratios usually recommended for walkways.

against this lamp. It rarely enhances foliage colors because of deficiencies at the blue-green end of the color spectrum.

'White' high-pressure sodium lamps: These lamps provide excellent color rendition in a warm tone similar to incandescent. Efficacy is sacrificed to obtain improved color. The resulting characteristics are a cross between metal halide and incandescent lamps. These lamps are very expensive and operate similar to metal halide lamps.

Low-pressure sodium lamps: Low-pressure sodium (LPS) lamps have high lumen ratings, but very low lumen effectiveness multipliers. A large arc tube results in poor optical control, but the lamp has superior life. Since the lamp is monochromatic, its color rendering properties are very poor. But it is this monochromatic property that makes it an ideal lamp source for near observatories, since astronomers need only filter out one narrow bandwidth of light.

Light distribution

Horizontal and vertical distribution: Horizontal illumination is especially important along the ground plane where changes in grade occur. However, a considerable portion of the night environment is perceived through direct and silhouette lighting of vertical objects and surfaces. Both patterns should be carefully coordinated in developing a successful lighting scheme.

Illumination data for outdoor lighting fixtures are illustrated by the manufacturers' photometric charts. These charts illustrate the actual light patterns and intensity levels on horizontal and vertical planes. Fig. 9 illustrates basic photometric data for walkway, road or area lighting. Lux (footcandle) measurements are given for horizontal distances based on mounting height and type of fixture. General spacing is usually 4 times mounting height in parking lots and 5 to 6 times the mounting height for walkways. If continuous lighting along a walkway is not desired, locate the pedestrian lights at walkway intersections or other possible conflict areas.

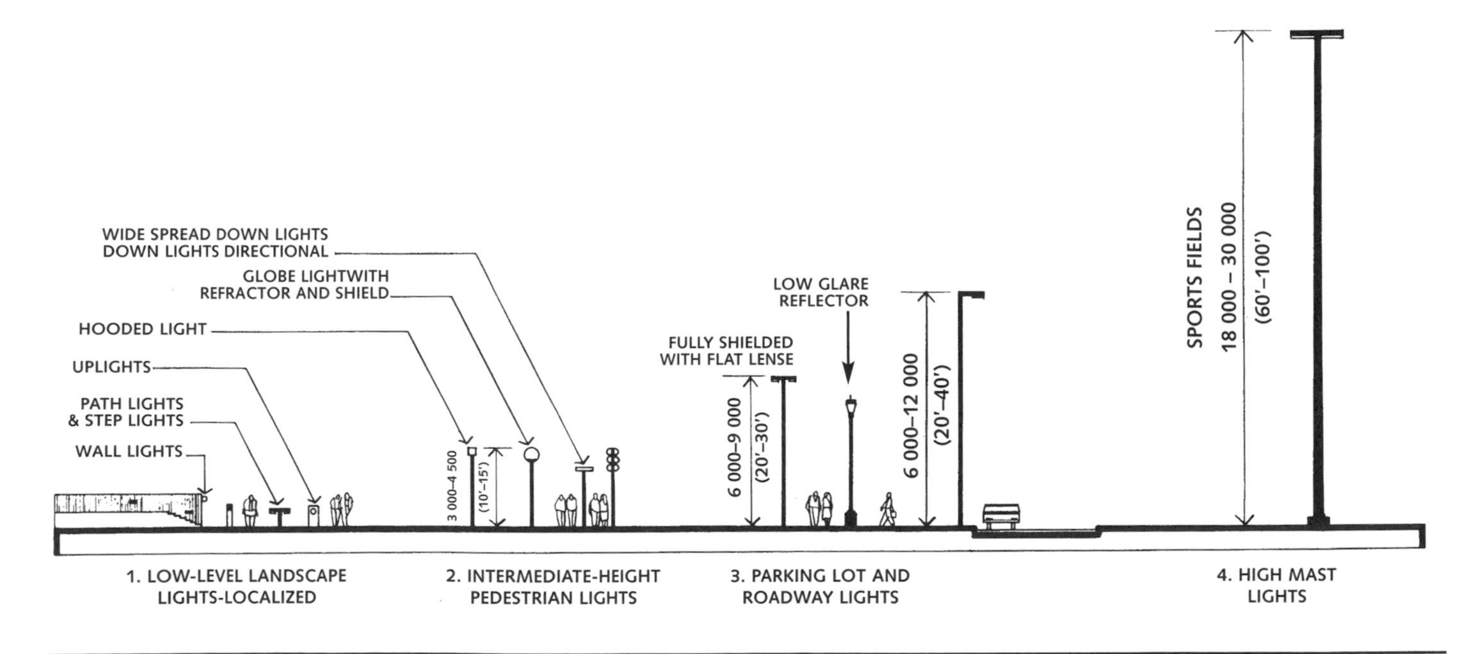

Fig. 13. Categories of light fixtures.

Fig. 10 illustrates photometric data for directional lighting, typically used for wall or signage applications. The aiming angle is commonly specified, and the fixture type is selected based on distance and illumination requirements.

A third type of photometric chart shows the distribution of candlepower, or intensity, in various directions, independent of any specific surface (Fig. 11). The maximum value within any given point on the distribution curve can be converted into lux (footcandles) with the formula shown within Fig. 11. This chart is occasionally used when determining light intensity and the angle of distribution for accent lighting (uplighting of trees, and floodlighting).

Basic light distribution patterns

Low-level path lights: These fixtures produce circular patterns of light that are symmetrical around the light center. Whenever fixtures produce a light pattern that is symmetrical, only one-half is shown, permitting maximum size and accuracy, as in Figs. 9.

Wall or sign lights: These fixtures are nearly always used to light vertical surfaces. Therefore, photometrics are presented on a vertical plane with the fixture set at an optimum distance (which varies per fixture) and backlit from the plane, as in Fig. 10. For a long wall or sign, the spacing from fixture to fixture can be determined by overlapping curves until the minimum acceptable light level between fixtures is established.

Accent lights: Photometric or adjustable accent lights, where the aiming angle and distance to lighted objects can vary, must be expressed in terms of the light output from the source rather than of the light falling upon the object. Candlepower is measured on a typical plane through the fixture and is charted in curve form, as in Fig. 11.

Uniformity

The uniformity of an illuminated field can be described with a ratio of light intensity values in lux (footcandles). The uniformity ratio typically compares the average illumination with the minimum footcandle value of a particular field. A low ratio appears more evenly lit and very uniform. The opposite is true for a high ratio where the two values are wide apart, resulting in a field that has distinct and contrasting values. Refer to Table 2 for a general description of different uniformity ratios.

Photometric distribution types

Many kinds of light fixtures greatly reduce glare by restricting high-angle light to not more than 75° above nadir, e.g., measured from the perpendicular (Fig. 12). Cut-off consideration is critical in minimizing glare, light trespass and light pollution. The following criteria are the general industry classifications of degrees of high-angle cutoff:

1. **Noncutoff:** unrestricted high-angle illumination.
2. **Semicutoff:** not more than 5% of peak intensity radiating above 90° and 20% of peak intensity above 80°.
3. **Cutoff:** not more than 2-1/2% of peak intensity radiating above 90° and 10% of peak intensity above 80°.
4. **Full cut-off:** Zero candela intensity occurs at an angle of 90° above nadir, and at all greater angles from nadir. Additionally, the candela per 1000 lamp lumens does not numerically exceed 100 (10%) at a vertical angle of 80° above nadir. This applies to all lateral angles around the luminaire.

Categories of light fixtures

Various categories of light fixtures commonly used in outdoor lighting situations are described below (Fig. 13).

Low-level landscape lights

Typical characteristics include:

1. Heights usually less than 1.83 m (6 ft.) but sometimes up to 3.05 m (10 ft.).
2. Lamps may be incandescent, compact fluorescent, induction, metal halide. Mercury vapor lamps or high-pressure sodium lamps are not recommended (see "Lamp characteristics" above).
3. Low-wattage capabilities, with limited intensities.
4. Substantial variety, with some sizes and shapes fitting within modules of finish materials (brick, etc.).
5. Finite light patterns, with directing capabilities.
6. Light sources are usually below eye level, so glare must be controlled.
7. Low maintenance requirements but high susceptibility to vandalism.
8. Use low wattage directional lamps for accent lighting with internal shielding louvers.

Intermediate-height pedestrian lights

Typical characteristics include:

1. Average heights of 3.05 to 3.66 m (10 to 12 ft.).
2. Lamps can be compact fluorescent, induction, or metal halide. Incandescent, mercury vapor, or high-pressure sodium not recommended.
3. Substantial variety of fixtures and respective lighting patterns.
4. Used in or around pedestrian pavements, and considered pedestrian in scale.
5. Lower fixture mounting heights are susceptible to vandalism and damage from lawn-mowers.

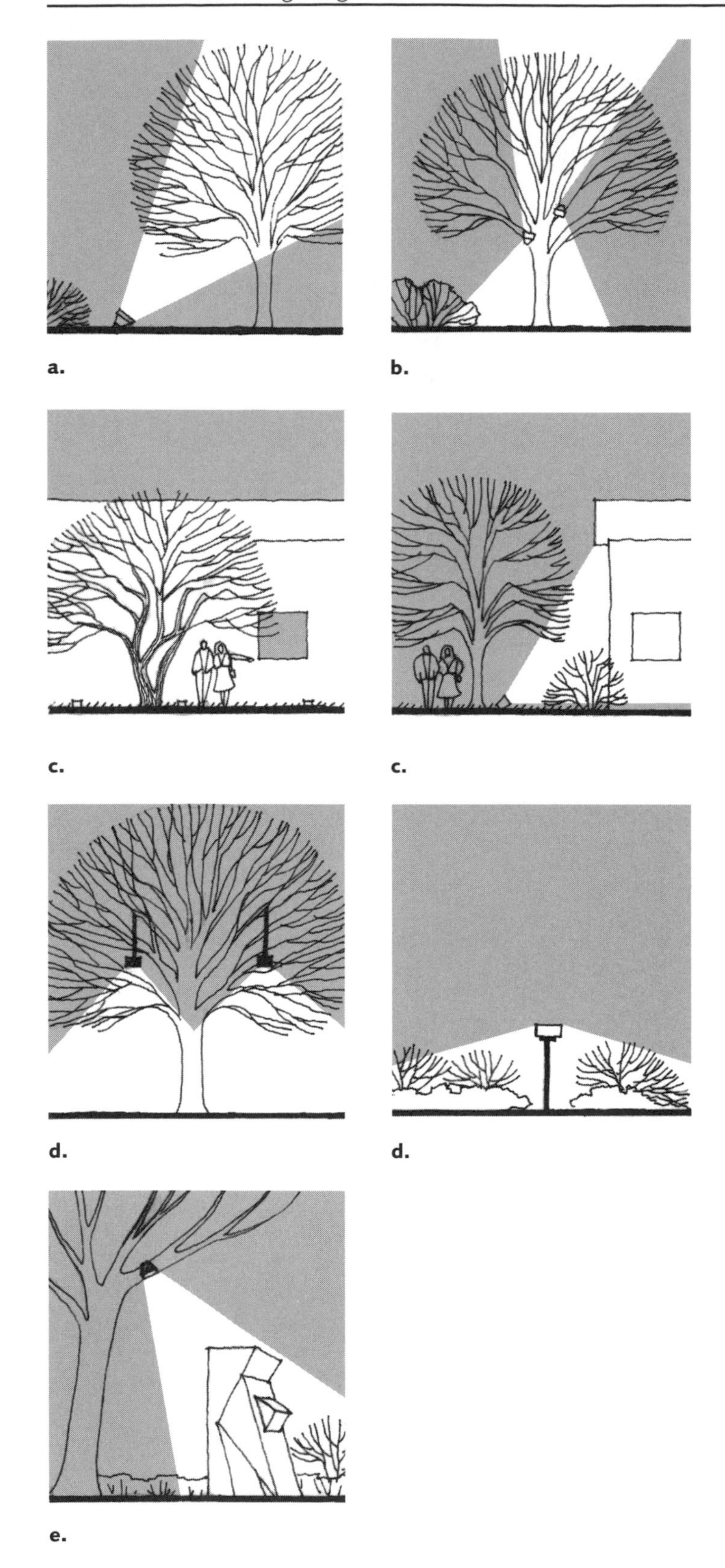

Fig. 14. Landscape lighting effects. (a) Uplighting (directional viewing). (b) Moonlighting. (c) Silhouette lighting. (d) Spread lighting. (e) Spotlighting.

Parking lot and roadway lights

Typical characteristics include:

1. Average heights of 6.1 to 12.2 m (20 to 40 ft.).
2. Lamps can be mercury vapor, metal halide induction, or high-pressure sodium.
3. Used to light streets, parking lots, and recreational, commercial, and industrial areas.

High-mast lights

Typical characteristics include:

1. Average heights of 18.3 to 30.5 m (60 to 100 ft.).
2. Lamps can be metal halide or high-pressure sodium.
3. Used for large parking lots, highway interchanges, and sports fields.
4. Fixtures must be lowered to be maintained.

Landscape lighting effects

Lighting effects most frequently used in outdoor lighting situations are described below (Fig. 14). Minimize the quantity of light. Use as low wattage fixtures as possible especially in metal halide. Light only those selected landscape and features to provide safety, way finding, and aesthetic highlighting.

Uplighting for directional viewing: When a lighted object can be seen from one direction only, above-grade accent lights can be used. To prevent glare, fixtures should be aimed away from observers and, if possible, concealed to keep the landscape uncluttered. (Fig. 14a)

Uplighting for all-around viewing: If the lighted object can be seen from any direction, then adjustable well lights with louvers should be considered. With newly planted trees, place uplights as close as possible to the outside of the root ball. Placing fixtures midway between trees is rarely satisfactory. The light typically misses the trunk and most foliage and is a common cause of light pollution. It is particularly unsuccessful if trees are deciduous, especially during the winter stage. Leaves, rain-splashed dirt, snow and ice limit the practicality of ground level uplights.

Moonlighting: The effect of moonlight filtering through the trees is another pleasing outdoor lighting technique. Up-and-down lighting is used to create this effect, which requires that fixtures be carefully placed in trees. Ground lighting is accented by shadows from leaves and branches. (Fig. 14b)

Silhouette lighting: Trees and shrubs with interesting branching structure can be dramatically expressed when silhouetted against a wall or building facade. Such lighting also provides additional security near the building. (Fig. 14c)

Spread lighting: Spread lights produce circular patterns of illumination for general area lighting. They are effective for groundcovers, low shrubs, walks, and steps. However, to take full advantage of the light throw, fixtures should be kept to open areas so that shrubbery does not restrict light distribution. The overhead spread light provides additional height and throw. When used in eating or recreational areas, several fixtures should be used to soften shadows while creating a uniform lighting effect. (Fig. 14d)

Spotlighting: Special objects such as statues, sculpture, or specimen shrubs can be lighted with well-shielded fixtures using spot lamps. By mounting adjustable overhead in trees or nearby structures, glare and fixture distraction can be eliminated. If ground-mounted fixtures are used, they should be concealed with shrubbery and louvered. (Fig. 14e)

Path lighting: Path lights are essentially spread lights at a lower height. In areas where other landscape lighting is used, a high degree of light shielding is necessary for path lights. This prevents the glare that inhibits a full view of the surrounding landscape. If no other outdoor lighting is used in the immediate area, less-shielded path lights may be acceptable. These fixtures illuminate the path and some of the surrounding landscape as well, but there remains the possibility that the glare will be disruptive.

Low voltage systems

Low voltage lighting systems offer an alternative to more energy-consumptive 120-volt systems. These systems work particularly well in informal, small-scale, residential settings. In addition to reduced energy usage, low voltage systems offer safe and easy installation, longer lamp life (up to twice as long as 120-volt lamps), small fixtures that can be hidden in the landscape, and very low light levels to achieve a variety of lighting effects. Low voltage systems are not typically appropriate for larger projects requiring significant levels of illumination, or public sites where durability is a concern. Low voltage systems include a transformer, cable, connectors, and the fixtures. (Fig. 15)

Transformers are required to convert standard 120-volt output to the proper operating voltage. They must provide adequate power to accommodate all fixtures proposed, including anticipated expansion. The transformer's wattage rating must be equal to or greater than the wattage of all fixtures combined. Large or diverse lighting schemes may require multiple transformers. Transformers may use manual, automatic timer, or photocell systems that detect light levels to turn power on and off for the system. Current photocell technology is suitable only for small-scale residential applications due to reliability concerns.

Systems typically use 12-gauge low voltage cable feeds with 18-gauge fixture wire, unless otherwise specified by the manufacturer. Plastic connectors are used to join each fixture to the cable feed. A wide variety of low voltage fixtures are available to achieve various lighting effects.

Controls

Exterior lighting controls are extremely important from saving energy to minimizing light trespass and light pollution. In addition, controls can create different scenes in the lighting design.

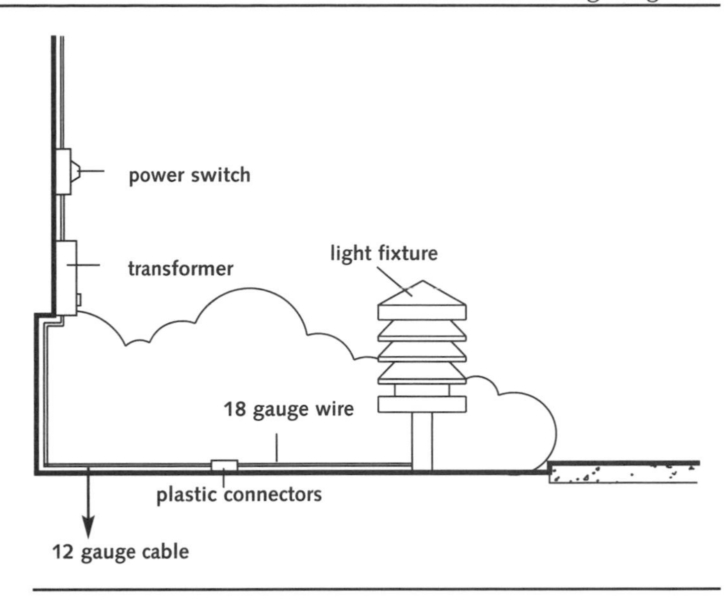

Fig. 15. Low voltage lighting system.

Dusk to Dawn: With a photocell, lights come on at dusk and off at dawn. Even though this type of control is the most common, it is discouraged because it uses the greatest amount of energy and has the greatest environmental impact.

Dusk to Off: A photocell/timeclock combination is a great alternative to the dusk to dawn scenario. Lights come on at dusk, but are able to turn off at a specified time in the evening.

Motion sensors: Motion sensors are an excellent alternative to standard lighting controls. The lighting only turns "on" when people are in the area. Motion sensors are best applied at entrances and walkway lighting that is used occasionally. They may also be used to alert security personnel that someone is in the area.

Satellite/FM Control: When a community or a campus wishes to control the lighting as a whole, satellite or FM controls work very well. With a centralized signal, different lighted areas can be turned on/off or dimmed depending on the need.

3 NOTE REGARDING RECOMMENDED LEVELS OF ILLUMINATION

The IESNA Lighting Handbook (IESNA 1998) discusses in depth that the level of illumination is only one of many criteria required to design outdoor lighting. Other criteria such as glare, shadows, and object identification are often more important than lighting levels. Too often, only the lighting level is used to design a project and could result in a substandard design. Additionally, recommended lighting levels are [as of this writing] subject to change, based upon new research on task performance. Designers are advised to refer only to current data. Rather than following quantitative minimum lighting levels, outdoor lighting design should be guided by the requirements to provide lighting that is attractive, energy-efficient and provides for safety and security, while minimizing or eliminating glare, light trespass and night-sky light pollution.

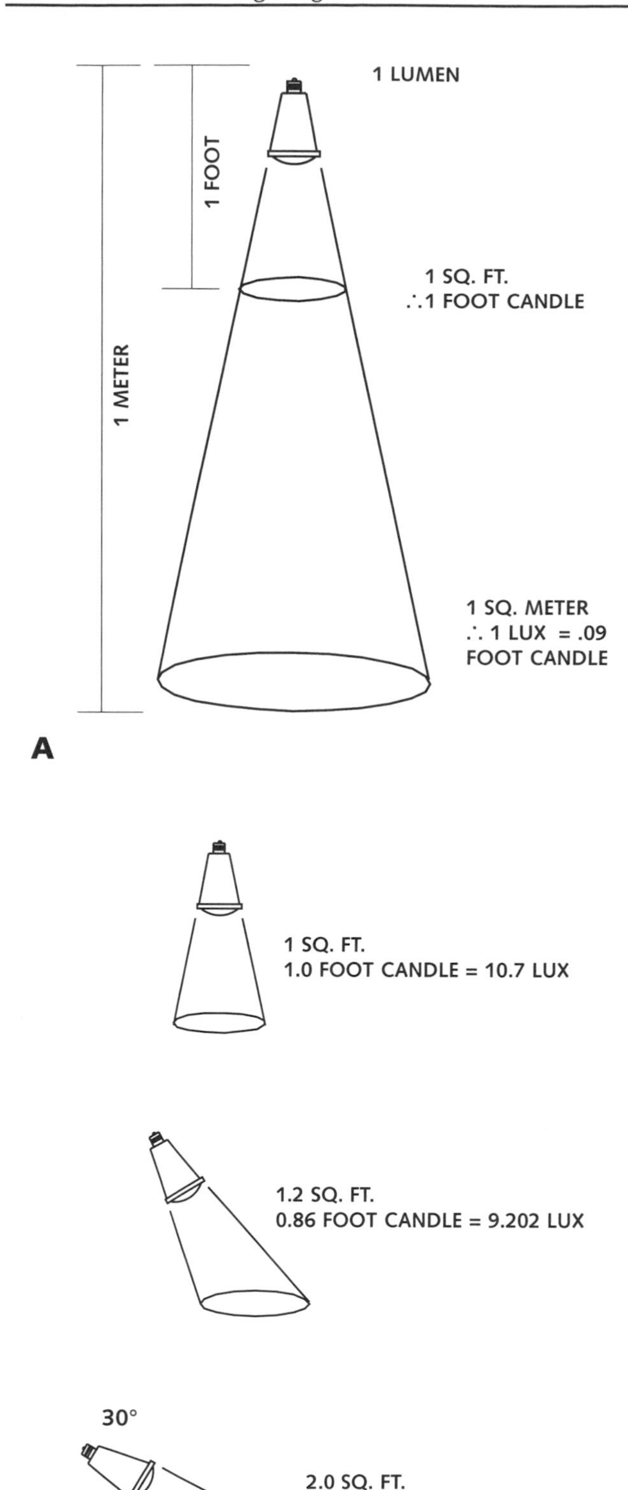

Fig. 16. (a) Lux and footcandle illustration. (b) Incident illumination.

4 TERMINOLOGY

Lumen: A quantitative unit of measurement referring to the total amount of light energy emitted by a light source, without regard to the direction of its distribution.

Footcandle (fc): A U.S. unit of measurement referring to incident light. Footcandles can be derived from lumens (1 fc = 1 lumen/sq. ft.) or candelas (fc = candelas/distance sq.).

Lux (lx): The International Standard (SI) measure of incident light. A value of one lux is equal to one lumen uniformly distributed over an area of one square meter (10.7 lx = 1 fc) (Fig. 16).

Candlepower: The unit of intensity of a light source in a specific direction often referred to as Candela. One candela directed perpendicularly to a surface one foot away generates one footcandle of light.

Illuminance: Incident light, or light striking a surface (Fig. 16). Angle of incidence or incident angle is measured from the plane of the surface that is illuminated.

Luminance: Light leaving a surface, whether due to the surface's reflectance, or because it is the surface of a light-emitting object (like a light bulb). Luminance is the measurable form of brightness, which is a subjective sensation.

Efficacy: A measure of how efficiently a lamp converts electric power (Watts) into light energy (lumens) without regard to the effectiveness of its illumination. It should not be assumed that a lamp that has high efficacy will give better illumination than a less efficient lamp (Fig. 17). Recent research has shown that lamp sources with a higher blue light content (similar to metal halide, fluorescent and induction lamps), are more effective at night, especially in detecting objects in our peripheral vision. As a result, lumen ratings of lamp sources with a white light distribution may have a higher lumen rating than other lamp sources (such as height pressure sodium or low pressure sodium) when used for exterior lighting. "Lumen Effectiveness Multipliers" may be applied to lamp ratings when calculating lamp efficacy.

Light depreciation: Lamp output (lumens) will depreciate over its effective life. Illumination will be reduced further due to an accumulation of dirt and grime on the lamp and fixture. Adjustments should be made to compensate for this depreciation when determining the average values of illumination maintained over time. A maintenance factor of 50 to 70 percent is common for outdoor applications. New installations are routinely designed to deliver 1–1/2 to 2 times as much illumination as needed, to sustain this maintained output over the anticipated life of the lamp.

Color: Two measures used to describe the color characteristics of lamps are (1) the apparent color and (2) the color-rendering index.

Apparent color of a light source is given by the color temperature. Various index numbers are used to rank sources on a scale that range from warm to cool in appearance. Preference for one or another is a matter of taste and usually varies with the context of the application and with the illumination level. Warm tones tend to be favored when illumination is low and cooler tones are preferred under high lighting levels.

Color rendering index (CRI) is a measurement of the degree to which object colors are faithfully rendered. This scale ranges from 0 to 100 and is a reasonable approximation of color rendering accuracy. CRI is completely independent of whether a light source casts the object in a warm or cool tone. The CRI graph shows the ranking of the major outdoor light sources (Fig. 4). As a general guideline, a minimum CRI of 50 is suggested to attain a reasonably faithful or natural color rendition. Lamps ranked significantly below this are judged to cause visible distortions to appearance.

Glare: A point or surface of luminance that is above one's current state of adaptation. The human visual system can comfortably see in light levels ranging from starlight to noonday sun, but cannot do so over this entire range at the same time. We adapt to one limited range or another, and perceive glare as any brightness above our current state of adaptation. Disability glare impairs visibility and is primarily a physiological phenomenon; *e.g.*, the nighttime glare from an oncoming vehicle's headlights can momentarily blind a driver's perception of the road ahead.

Discomfort glare does not impair visibility but is primarily a psychological phenomenon or an annoyance that may produce fatigue if it continues over an extended period of time.

Obnoxious glare: When glare results in complaints from adjacent property owners or from users, then the lighting installation may produce obnoxious glare. Care should be given to limit luminaire brightness and overlighting to prevent this occurrence.

Cutoff light distribution: A term used in reference to the optical design of some fixture types. Proper placement of lamps and the use of carefully aligned reflectors, can effectively eliminate intense high-angle light. Most cutoff designs severely restrict fixture intensities above 75° from nadir, that is, within 15° of horizontal. ■

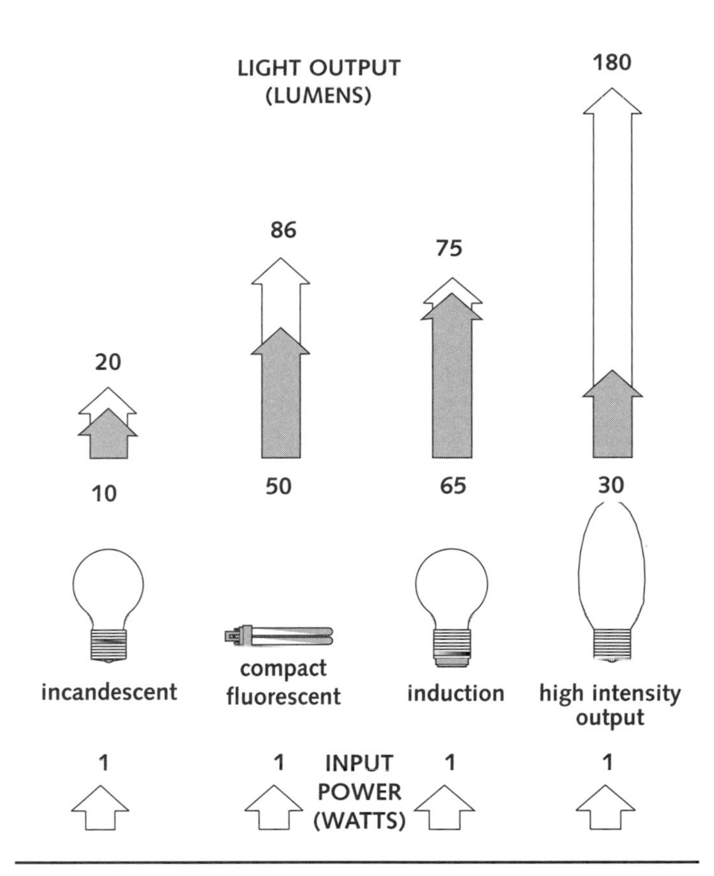

Fig. 17. Lamp efficacy.

REFERENCES

Harris, Charles W., and Nicholas T. Dines, editors. 1998. *Time-Saver Standards for Landscape Architecture.* 2nd ed. New York: McGraw-Hill.

Illuminating Engineering Society of North America (IESNA) 1998. *Lighting Handbook: Reference & Application.* Mark Rea, ed., New York: Illuminating Engineering Society of North America. www.iesna.org

Illuminating Engineering Society of North America (IESNA) 1999. *Lighting for Exterior Environments. RP-33.* New York: Illuminating Engineering Society of North America.

Illuminating Engineering Society of North America (IESNA) 2000. "Addressing Obtrusive Light (Urban Sky Glow and Light Trespass) in Conjunction with Roadway Lighting" Technical Memorandum TM-10-00. New York: Illuminating Engineering Society of North America.

8 • CASE STUDIES IN URBAN DESIGN

Robert G. Shibley

Summary

Since its inception in 1987, the Rudy Bruner Award for Urban Excellence (RBA) has been dedicated to discovering and celebrating places that are distinguished not only by their quality design, but by their social, economic, and contextual contributions to their urban environments. Looking beyond form, the Award asks how projects fit into and improve their urban context on many levels. While this does not negate the importance of good design, it does alter the boundaries of what good design means, placing new emphasis on the complex processes of collaboration required to create and sustain an excellent urban place.

Key words

case study, evaluation, placemaking, process

Pike Place Market in Seattle, Washington.

Case Studies in Urban Design

1 WHAT IS THE RUDY BRUNER AWARD FOR URBAN EXCELLENCE?

Beginning in1987 the Rudy Bruner Award has recognized forty-two winners from across the country that have exemplified urban excellence—places that arrived at good design through a thoughtful process powered by varied and often deeply felt values. The criteria for entering the competition are intentionally broad, and the winners have been a diverse group that has made very different contributions to America's cities. Many represent new models of urban placemaking guided by creative visions that have successfully challenged conventional wisdom. Most are products of hard-won collaborations between very different people with very different agendas. All have lent vitality to the neighborhoods in which they are located.

The Rudy Bruner Award highlights the complex process of urban placemaking, emphasizing the many elements beyond good design that produce urban excellence. Through studying the development and functioning of these projects, urban designers are able to discover creative ways to respond to some of our cities' most intractable problems.

The Bruner Award also serves as a national forum on the nature of urban excellence. Rather than approaching projects with a set of preconceived standards, the Award seeks to flesh out its ideals through winners' own demonstrations of excellence. In essence, the Award hopes to find and understand, rather than to dictate, urban excellence. To ensure a lively process not dominated by any one perspective, each Selection Committee brings together a distinguished panel representing several different kinds of urban expertise, including architects, landscape architects, planners, developers, community organizers, financiers, and the mayor of a major metropolitan area.

As the Selection Committee discusses the project applications, they consider a wide variety of questions: What kinds of places make neighborhoods and cities better places to live, work, and play? How did these places come into being? What visions powered their creation? How did these visions become a reality? What obstacles had to be overcome? What makes a place important in its urban context? In this way the Selection Committee explores the dynamic nature of urban excellence and contributes to a broader understanding of the critical urban issues of the day.

Credits: This article is based on introductions to the seven publications of the Bruner Foundation on Rudy Bruner Award Winners since 1987 all by Simeon Bruner as well as on Shibley, R. (1995) "Redefining Excellence in the Urban Environment: The Rudy Bruner Award Program" in Schneekloth, Lynda. and Robert Shibley, (1995) *Placemaking: The Art and Practice of Building Communities,* (New York: John Wiley and Sons), and on Shibley, R. (2000) "Learning about Urban Excellence" by the author as published in Shibley, R. *et. al.* (2000) in *Commitment to Place: Urban Excellence and Community*, (Boston: Bruner Foundation).

Because urban excellence is a lived experience, not an abstract quality defined at a conference table, Bruner Foundation staff visit each selected finalist for two or three days, exploring the project and pursuing questions raised by the Selection Committee. These site visit teams serve as the Committee's eyes and ears. They tour all parts of the project, interview between 15 and 25 key participants (including "unofficial" community participants and project opponents), take photographs, and observe patterns of use.

The project organizers arrange some of these activities, but the site visit teams also pursue their own agenda in order to investigate the Selection Committee's questions and concerns. Afterwards, the teams present their findings to the Selection Committee. With the team leadership on hand to answer additional questions, the Committee then debates the merits of each project to decide upon a winner. In doing so, they explore the issues facing urban areas and come to a deeper understanding of the kinds of processes and places that embody urban excellence.

The Seattle skyline and home for Pike Place Market.

1987 Gold Medal

Pike Market Place, Seattle, WA

A farmers market, with owner operated small businesses, low, middle & upper income residences, and a social service network.

Selection Committee:

Clare Cooper Marcus, Department of Landscape Architecture, University of California at Berkeley, CA
Vernon George, President, Hammer, Silver & George Assoc.,Washington D.C. Cresswoth Lander, Director, Department of Human and Community Development, Tucson, AZ
George Latimer, Mayor, St. Paul, MN
Theodore Liebman, Liebman Melting Partnership, New York, NY
William Whyte, American Conservation Association, New York, NY

2 REDEFINING EXCELLENCE IN THE URBAN ENVIRONMENT

Cities are more than a collection of buildings. They are home to human communities who live, work, and relate to each other and their urban settings in an endless variety of ways. Urban places therefore sit within many contexts, at many different levels of "urban fabric": they relate stylistically to their material surroundings; they contribute to patterns of economic activity; they are positioned within the flow of pedestrian and vehicular traffic; they occupy a niche within an urban ecosystem of water, energy, and waste production and disposal; and, perhaps most importantly, they play a role in forming and sustaining urban communities. They are, in other words, more than skillfully constructed relationships among construction materials, environmental features and spatial definitions. They are parts of the social and ecological landscape of a city as well.

For an urban intervention to be a successful "fit" and for it to do the work it is intended to do, therefore, it must be tailored to respond effectively to all levels of its urban landscape. Much of what urban designers do relates to this necessity. Formal design methodologies such as figure ground studies measure the fit, workability, and impact of a project within those elements of the urban fabric traditionally recognized as the domain of professional urban designers. But such tools cannot measure some of the most important social dynamics.

Process does matter. It frames the success or failure in achieving the cooperation needed to advance the project. It guards or fails to guard against doing damage to existing social and physical conditions even as it strives to advance such circumstances. The Bruner archive of success stories offers clear testimony that "knowing what to do doesn't get it done."

But, one might protest, urban designers are trained as experts in design and formal evaluation; isn't it someone else's job to navigate the social and political processes of getting places built? While this perspective may seem hard-headed (or, alternately, romantically artistic), in reality it means relinquishing planner and designer voices in the processes that will fundamentally shape the urban place they are trying to design. The process of creating places influences the human relationships they house, both between people and places and between people and communities in places. As a result, there are

The Tenant Interim Lease Program in New York. New York brought thousands of units of housing back to life under interim lease arrangements leading to cooperative ownership by low income residents.

New York City's Greenmarket (below) enlivens the street life of the city as it delivers farm produce to inner city neighborhoods and delights children.

Pioneer Square in Portland, Oregon.

1989 Gold Medals

Tenant Interim Lease Program, New York, NY

The creation of tenant self-managed cooperatives through an interim lease and renovation program.

The 1972 Portland Downtown Plan, Portland, OR

Office, retail, housing, open space, recreational/cultural, transportation results from a new downtown plan.

Selection Committee:

Mary Decker, First National Bank of Chicago, IL
George E. Hartman, Jr., Hartman-Cox Architects, Washington, D.C.
David Lawrence, Sr. Vice President, Gerald Hines Interests, New York, NY
Joseph Riley, Mayor of Charleston, Charleston, NC
Anna Whinston Spirn, Department of Landscape Architecture, University of Penn, Philadelphia, PA
Aaron Zaretsky, Grove Arcade Public Market Foundation, Asheville, NC

1991 Gold Medal

Greenmarket, New York, NY

Fresh farm products grown in New York delivered through street vending to inner-city neighborhoods.

Selection Committee:

Gwendolyn Clemons, Director of Planning & Development for Cook County IL, Chicago, IL
Lawrence Halprin, Lawrence Halprin Studio, San Francisco, CA
Tony Hiss, formerly with The New Yorker, New York, NY
Joseph McNeely, President, The Development Training Institute, Baltimore, MD
Adele Naude Santos, Adele Naude Santos Architects, Philadelphia, PA
Vincent Schoemehl, Mayor, St. Louis, MO

values embedded in places, and in the process of constructing places, that shape how a place will function as an element in the social landscape. And while some of the value of an urban place results from its formal architectural qualities, as a part of a living city its true contribution lies in the way it sustains the healthy functioning of urban communities. The most beautiful building in the world will be adjudged a failure if no one uses it, or worse, if it damages social networks and activities that previously functioned well.

To manage the creation of a successful urban place, then, requires attention to three interrelated aspects of design: place, process, and values. To speak of process is to acknowledge that urban places are lived in, over time, before, after, and even at the same time as they are constructed, reconstructed, or otherwise "planned." The design and construction of a specific project is merely one chapter in the long history and future of a place. But it is not an insignificant or timeless chapter: lives still move forward during the planning and implementation process. The relationships established during this time influence how people will feel about—and ultimately how they will use—the place being created. These relationships, in turn, depend upon the values that inform the process. Does the process make an effort to affirm and build upon existing urban fabric, or is it solely focused on future potential? Does it respond to needs as defined by the communities who use a place, or does it unfold according to a formal logic unrelated to the specific urban context? Clearly, decisions about such values will shape the kind of place ultimately created.

To illustrate the impossibility of separating place, process, and values in the practice of urban design, consider three of the Bruner Award's Gold Medal Winners: the 1972 Portland (Oregon) Downtown Plan, Pike Place Market in Seattle, and Yerba Buena Gardens in San Francisco. In each case, a city was faced with a deteriorating area within its center. The first efforts to address these "blighted" districts came in the 1950s and 1960s during the heyday of urban renewal. Confident that professional expertise could lead them to renewal, Portland, Seattle, and San Francisco brought in their own versions

The New Communities Corporation converted this nursing home into a 99 unit senior citizen apartment complex. In represents one of several interventions in the poorer sections of Newark.

Single room occupancy units on Times Square.

1993 Gold Medals

New Community Corporation, Newark, NJ

A housing, day care, social services, employment training, health care, and commercial revitalization program.

Harbor Point, Boston, MA

A renovated public housing project (Columbia Point) involving its conversion to a mixed income rental community.

Selection Committee:

Sara Bode, President, Honey Tree Learning Center, Chicago, former Mayor, Oak Park, IL
Denise Fairchild, Director, Local Initiative Support Corporation, Washington DC
Harvey Gantt, FAIA, Architect and Urban Designer,former Mayor, Charlotte, NC
Ed McNamara, Director, Neighborhood Partnership Fund, Portland, OR
Frank Sanchis, Vice President, National Trust for Historic Preservation, Washington D.C.
Robert Sommer, Ph.D., environmental psychologist, Prof. of psychology, Director, Center Consumer Research, University of California at Davis, CA

1997 Gold Medal

Times Square

The Times Square has meticulously restored historic 652 room hotel to provide permanent single room occupancy units at affordable rents.

Robert Curvin, Vice President for Communications, Ford Foundation, New York, NY
Roberta Feldman, Ph.D., Co-Director, City Design Center, University of Illinois, Chicago, IL
Susan Rice, former Sr. Vice President of Fleet Bank, New York, NY
Hon. Kurt L. Schmoke, Mayor of Baltimore, MD
Robert M. Weinberg, President, Market Place Development, Boston, MA

of the "best and the brightest" to fix their downtowns. Sure of their own wisdom, these urban designers—often in consultation with business leaders—assessed the needs of the downtowns and produced gripping and often grand new visions for what their future should look like. Often, they featured high-rise buildings designed to expand or stimulate financial districts, linked to parking garages or tourist attractions to draw in visitors from outside the city center.

It is important to recognize that these plans and urban design proposals worked according to their own logic. They were products of excellent professional technique applied to a set of problems and design guidelines, and, for the most part, they were brilliant solutions to the given set of conditions. This apparent excellence may help explain why their proponents were so flabbergasted when they all failed, utterly, ignominiously, and amidst harsh public rancor. Why this failure? A variety of reasons attended each project, but one common thread emerged: in each instance, city residents and communities were excluded from decision making about the places that they lived and worked. Alienated from the process of "fixing" the city, they interpreted the projects as hostile invasions into their home territories, making undesirable changes in the service of someone else's goals.

At the most basic level, then, the process of these three projects affected the eventual design, in that none of them were ever built. The processes proved alienating and ineffective because the values that informed them were exclusive ones: the people and communities that should naturally have served as their strongest base of support (after all, who wouldn't want their declining neighborhood revived?) had been systematically—and sometimes hostilely—prevented from participating. It was only after residents in the three cities organized themselves and forcibly inserted themselves into the design process, that these projects began to take on the excellence that characterizes them today.

In all three cases, inclusion of affected stakeholders was legally mandated by court rulings or political victories. The open, public

The Yerba Buena Gardens complex in San Francisco, California south of Market Street.

Courtyard, Philadelphia, PA. Village of the Arts complex.

1999 Gold Medal

Yerba Buena Gardens, San Francisco, CA

A complex of cultural, residential, open space, convention center, children oriented, entertainment, commercial and entertainment-retail developments.

Selection Committee:

Curits M. Davis, AIA, John Hancock Mutual Life Insurance Co., Boston, MA
Lawrence Goldman, President/CEO, New Jersey Performing Arts Center, Newark, NJ
Min Kantrowitz, AICP, March, President, Min Kantrowitz and Assoc., Albuquerque, NM
Rick Lowe, Founding Director, Project Row Houses, Houston, TX
Frieda Molina, MCP, Manpower Development Resources Corp., San Francisco, CA
Hon. Tom Murphy, Mayor, Pittsburgh, PA

2001 Gold Medal

Village of Arts and Humanities, Philadelphia, Pennsylvania

An arts-based community organization developing a vital and wholesome urban village.

Craig Barton, AIA, RBGC Associates, and University of Virginia, Charlottesville, VA
John Bok, Esq.,Foley, Hoag, and Eliot, LLP, Boston, MA
Rosanne Haggerty, Executive Director, Common Ground, HDFC Inc., New York, NY
Allan B. Jacobs, Professor, University of California at Berkeley, Berkeley, CA
Gail R. Shibley, Asst. Secretary for Public Affairs, US Department of Labor, Washington, DC
Hon. Wellington Webb, Mayor, Denver, CO

processes that then unfolded produced changes in priorities that resulted in radically changed designs: the fortress-like cultural enclave designed for Yerba Buena, for example, became a permeable set of open spaces inviting locals to mix with tourists in the new multi-use cultural district. Seattle decided that the best way to revive the Pike Place Market area was to save the market itself rather than bulldozing and replacing all but a little of it. And Portland developed a downtown plan that emphasized public transit and public open spaces over more parking and high rise development.

In each case, the values embedded in the process of creation led to very different places, both in terms of design and in terms of how they were ultimately used by city residents and tourists. The excellence recognized by the Rudy Bruner Award had much to do with the healthy communities these places have fostered, and the special relationships these communities have developed with their places. Residents of all three cities came to care about their places as they cared for them, and the results were easy to see. Not only had residents helped shape the initial design projects, but also they remained organized and aware stewards of their places. The creation of their excellent downtown areas had not, therefore, ended when the first projects were completed. New dangers to central city vitality were successfully fended off, and new opportunities eagerly seized. Without these organized community advocates, Yerba Buena, Pike Place, and downtown Portland would undeniably be different places today.

3 LEARNING ABOUT URBAN EXCELLENCE: THE CASE STUDY METHODOLOGY

If urban design is to include a broader effort to address the many levels upon which urban fabric operates, urban designers must be equipped with tools that allow them to consistently evaluate and work with the social as well as the material aspects of design. The Rudy Bruner Award's mission has, in part, been to identify and make available a variety of successful ideas and approaches.

This set of tools, taken together as a coherent philosophy of "placemaking," addresses not only the physical construction and maintenance of places, but also the quality of relationships between people and places as well as among the communities that make and use places. Its central premise is that making and caring for places—becoming committed to them—can be the basis for making and caring for communities. Excellence is achieved when good design comes out of a process that is sensitive to (and inclusive of) the human relationships in which it develops, and that, like a good guest, leaves those relationships better than it found them.

Because the social dynamics of each urban intervention is unique, placemaking cannot be "operationalized" as a set of specific design imperatives to be applied universally to every context. Rather, each instance must reflect the specific communities, places, needs, and opportunities attached to a particular urban location. How, then, can we teach about the tools and strategies experienced placemakers have developed to produce successful places? Traditional formal techniques of design evaluation are inadequate. Instead, we must turn to case studies: to stories. Stories acknowledge that social processes are in fact processes—they happen over time, and how they happen influences where they go. And since humans experience their own lives as interlocking sets of stories, narratives are the best way to characterize the complex social dynamics that attend an urban intervention.

For example, for the three Gold Medalists discussed above, public opposition to the original projects materialized because residents began to believe that city planners did not have their best interests at heart. Citizens constructed a narrative whereby developers and city hall were conspiring to wreck the places they lived and worked. And this narrative, appearing nowhere on the best figure ground study, had actual power in the design and construction of Seattle, Portland, and San Francisco. To understand the power of that narrative, and how it was transformed to one of collaboration, trust, pride, and progress, we must examine the project itself as a story.

Placemaking brings the stories—the essence of human experience—back into urban design. Urban planners and designers are responsible for the kinds of stories they participate in. Whether or not they choose to pay attention to the social dynamics of their efforts, those efforts will have a social impact. They will become parts of the ongoing narrative of urban life. And, as often as not, it is the realm of that social narrative that holds the success or failure of an urban project. Placemaking is thus more than a moral ideal; it is a practical recipe for success. In fleshing out what placemaking means in practice, the following case studies make this practicality abundantly clear. ■

REFERENCES

Farbstein, Jay, and Richard Wener, 1992. *1991 Rudy Bruner Award for Excellence in the Urban Environment: Connections: Creating Urban Excellence.* Boston: Bruner Foundation, Inc.

________. 1993. *1993 Rudy Bruner Award for Excellence in the Urban Environment: Rebuilding Communities: Recreating Urban Excellence.* Boston: Bruner Foundation, Inc.

________. 1996. *1995 Rudy Bruner Award for Excellence in the Urban Environment: Building Coalitions for Urban Excellence.* Boston: Bruner Foundation, Inc.

________. 1998. *1997 Rudy Bruner Award for Excellence in the Urban Environment:Visions of Urban Excellence.* Boston: Bruner Foundation, Inc.

Langdon, Philip, with Robert Shibley and Polly Welch. 1990. *Urban Excellence.* New York: Van Nostrand Reinhold.

Peirce, Neal, and Robert Guskind. 1993. *Breakthroughs: Recreating the American City.* New Brunswick, New Jersey: Center for Urban Policy Research, Rutgers, State University of New Jersey.

Schneekloth, L., and Robert Shibley 1995. *Placemaking: The Art and Practice of Building Communities.* NY: John Wiley and Sons.

Shibley, Robert, with Emily Axelrod, Jay Farbstein, and Richard Wener. 2000. *1999 Rudy Bruner Award for Excellence in the Urban Environment: Commitment to Place: Urban Excellence and Community.* Boston: Bruner Foundation, Inc.

For related web sites, see:

www.buffalo.edu/libraries/projects/bruner/

www.urbandesignproject.org

www.brunerfoundation.org

Robert Shibley and David Herzberg

Summary
Portland's 1972 Downtown Plan was recognized by the 1989 Rudy Bruner Award Selection Committee because of its mix of inclusive planning and design, its physical adaptability to new conditions over time, and the redefinition of transit, parking, waterfront, and public space issues into the problem of access. City residents' participation in the planning process helped them connect to the result of their efforts and shape the vision of the downtown. The foundations laid in the Downtown Plan were adaptable enough to nest well with the later Central City Plan. Finally, the clean air act violations in the City of the early seventies necessitated a reduction in automotive use leading planners to define what had been seen as parking problems as problems of access. This has led to an excellent transit system, more in downtown residential development, and a more bikeable city than the one prior to 1972.

Key words
consensus, democracy, downtown, placemaking, public access, transit, values

Fig. 1. Portland, Oregon with Mt. Hood in the background.

The 1972 Portland Downtown Plan

1 FROM PARKING GARAGE TO MASTER PLAN

By all rights, the centerpiece of downtown Portland ought to be a 12-story parking garage. The year was 1971, and the once-lovely central city had been declining for years. Neal Peirce and Robert Guskind's Bruner case book recounts a familiar story: suburban growth, the appearance of a major shopping mall outside the downtown, parking problems, a precipitous fall in the number of housing units, a steadily worsening self-image—all conspiring to draw people and resources out of the heart of the city.

A flurry of new development in the late 1960s produced several high-rise buildings that did little to reinvigorate the area. Just south of the downtown, the city-run Portland Development Corporation bulldozed a "blighted and economically isolated neighborhood" to make way for offices and business services (Peirce and Guskind, p. 72). Meanwhile, department store executives, anxious to jump-start sagging sales, embraced the suggestion of an out-of-town developer to replace a small two-story garage with a huge parking structure. "All retailers want parking within one block of the hosiery counter," an observer explained; "that's their conception of how far a woman will walk" (Peirce and Guskind, p. 56).

Of all the blows to their civic pride, it was the garage that pushed Portlanders over the top. The symbolism could not have been worse: the block selected for the new parking garage had formerly been home to the Portland Hotel, a glittering, million-dollar Victorian castle that had catered to the nation's most prominent cultural and political elites for more than a half century beginning in 1891. Few city residents forgot (or forgave) the decision in 1951 to raze this icon of Portland's national status and replace it with a two-story parking garage, a blue-and-gold affair operated by Union Oil and (report Peirce and Guskind) "ranked among the Northwest's ugliest urban structures" (p. 56). Thus, when developers raised the idea of putting an even bigger garage in, "citizen reaction was instant—and negative. The city's planning commission found citizens carrying anti-garage signs at its hearing. The city council decided the political price was too high, and the garage was a dead letter" (Peirce and Guskind, p. 56).

Credits: This chapter is based on Peirce, Neal and Robert Guskind, 1993 "Portland's 1972 Downtown Plan: Rebirth of the Public City," in *Breakthroughs: Re-Creating the American City.* New Brunswick, NJ: Center for Urban Policy Research, Rutgers, State University of New Jersey and Boston: the Bruner Foundation. It is also based on Shibley, Robert and Polly Welch's *"Portland's 1972 Downtown Plan in the 1989 Rudy Bruner Award for Excellence in the Urban Environment Site Visit Report,"* an unpublished technical report for the Bruner Foundation.

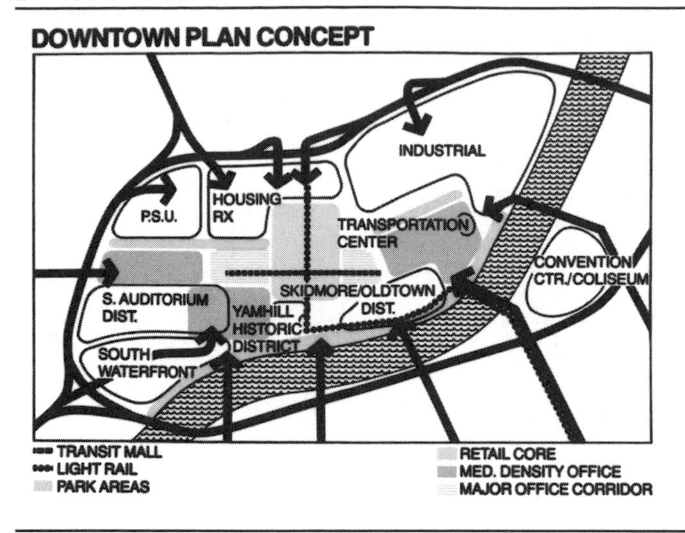

Fig. 2. The 1972 Portland Downtown Plan concept is represented with a simple diagram.

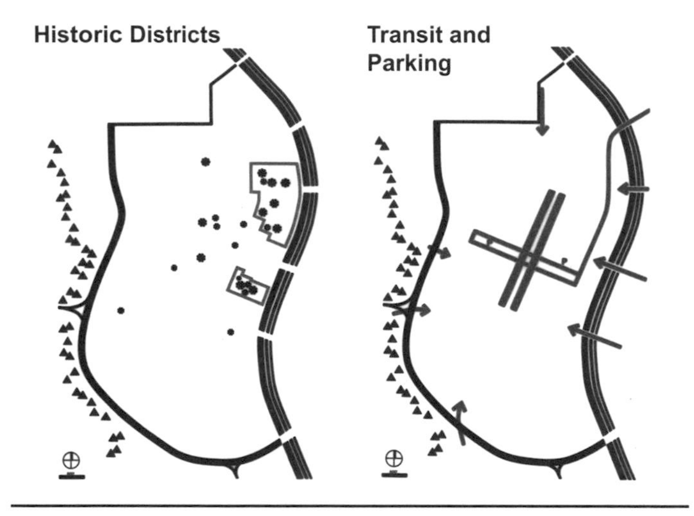

Fig. 3. The plan further illustrated specific land use ideas in diagrams like these on the location of the historic districts and on the approach to transit and parking.

Fig. 4. The Yam Hill Historic District is an example of the strong citizen support for the historic preservation.

Hoping to put the failed garage behind them, Portland business leaders opened a dialogue with Mayor Terry Shrunk in the hopes of addressing their parking needs with a more comprehensive "Downtown Plan." The dialogue quickly led to the creation of a semi-official "Portland Improvement Committee" composed of twelve top business leaders and advised by a variety of high-profile city and private planners. But the passions unleashed by the ill-fated garage were not to be calmed so easily.

As Peirce and Guskind tell it, "a number of vocal citizens, including civically concerned architects and neighborhood activists, expressed alarm about business interests' dominant role. In response to those concerns, Mayor Shrunk decided to appoint a broadly based citizens committee. This Citizens Advisory Committee (CAC) had seventeen members including representatives from the general public, from business, and from environmental and arts groups. It divided itself up into subcommittees on such topics as transportation and parking, housing, retailing, and the future of the city's waterfront—and then let any citizen of Portland join any one of those subcommittees" (p. 57).

"Over the course of fourteen months of intensive activity and public participation," Peirce and Guskind continue, "the CAC worked closely with the professional planning team to frame the city's 1972 Downtown Plan goals and guidelines. A project office was opened downtown, a kind of 'neutral turf' neither the territory of city hall nor the business interests. Throughout this highly collaborative process, with every camp from the major businesses to city hall to architects and neighborhood voices intimately involved, a far-ranging plan evolved. In contrast to traditional, highly prescriptive master plans, the initial document was basic and 'doable,' not highly detailed. It focused less on specific solutions than general guidelines for approaching each issue area" (p. 58).

Generally, the ruling guideline might best be described as planning for "public access." This meant access to the downtown through creatively expanded public transportation, with an 11-block transit mall linked to light rail; access to housing with a full range of new or preserved units; pedestrian access to the waterfront via a public park rather than a multilane roadway; public access to central open spaces in the form of the grand Pioneer Courthouse Square (on the proposed garage site); and so forth. Guided by this goal of designing for people, Portland systematically re-created the welcoming atmosphere it had perfected in the first half of the 20th century.

2 THE SYMBOLIC ECONOMY OF A PARKING GARAGE

It may sound foolish to give credit for Portland's rebirth to a failed parking garage. How could such a simple structure—and one that was never even built—play such a major role in a decades-long urban success story? The answer is simple: the parking garage was more than just a building. Like all other urban spaces, real or proposed, it reverberated on all levels of urban fabric. When they protested the new garage, Portlanders were not angered by the prospect of more parking spaces. They did not have a theoretical opposition to buildings that housed such parking spaces. Rather, they were incensed by the narrative of decline and exclusion embodied by that specific project, at that specific location. The social reality of their once open and accessible city had been slowly changing under an invasion of high rises, garages, and bulldozers. The latest decision by a remote private entity to again remake a place felt by many Portlanders to "belong" to the city as a whole—and to be representative of what the city should be like—finally pushed residents to demand a voice in the process of dealing with the downtown's decline.

Thus it was not the failure of the garage-as-building—the architect's garage—that initiated Portland's award-winning 1972 Downtown Plan. Rather, it was the parking garage as a social and cultural entity. It was the successful opposition to the idea of the garage, and the kind of city it embodied, that infused the design process with vitality and direction. The garage did not support the kind of urban life city residents wanted, and it symbolically trampled upon a precious civic memory. Unaware—or at least uncaring—of the complex of activities and meanings associated with the site, city planners had focused only on the building and its immediate economic impact. Only when the full spectrum of social as well as physical and economic realities came under consideration was success a real possibility. To the extent that "the parking garage that never was" helped make this obvious to the city, business leaders, and (most importantly) city residents, it is indeed correct to claim that this apparently humble example of modern architecture was responsible for the rebirth of a major American city.

3 PORTLAND'S 1972 DOWNTOWN PLAN TODAY

- **A planning process:** "The Bruner award," Peirce and Guskind note, "went officially for the process by which Portland gathered broad community consensus and set its vision for two decades of remarkable inner-city planning and progress" (p. 52).
- **A series of public improvements.**
- **Public transportation:** A handsome 11-block transit mall linked to light rail and punctuated by vintage trolleys.
- **Open public space:** The centrally located Pioneer Courthouse Square is a great brick plaza easily accessible from surrounding streets and buildings.
- **A harborside park:** The six-lane Harbor Drive insulating the Willamette River from the city was closed and replaced by Tom McCall Waterfront Park.
- **Public art:** A plethora of fountains, greenery, and mischievous public art designed to be enjoyed by ordinary residents as well as art critics.
- **Distinctive architecture:** A preservationist ethic has maintained some historic buildings, while some new buildings—*e.g.*, the Justice and Portland Buildings—have brought nationally notable architecture to the city.
- **Housing:** A full range of housing preserved or newly constructed.

4 LESSONS FROM PORTLAND'S 1972 DOWNTOWN PLAN

The city's new "friendlier" feel and streetscape have made it a popular place to be again: the downtown is lively at night and so popular that the new light rail system initially carried more passengers into the city Saturdays than on weekdays. Underlying this transformation,

Fig. 5. The modern bus station design works well with the historic Pioneer Courthouse.

Fig. 6. Portland continues to make very good use of its historic building stock.

Fig. 7. The approach to public transit included developing and coordinating light rail, vintage trolley, and an eleven-block transit mall in service of access downtown.

Fig. 8. The transit mall downtown Portland. The transit system provides riders with closed circuit television system providing riders up to the minute arrival and departure information in thirty-one stops.

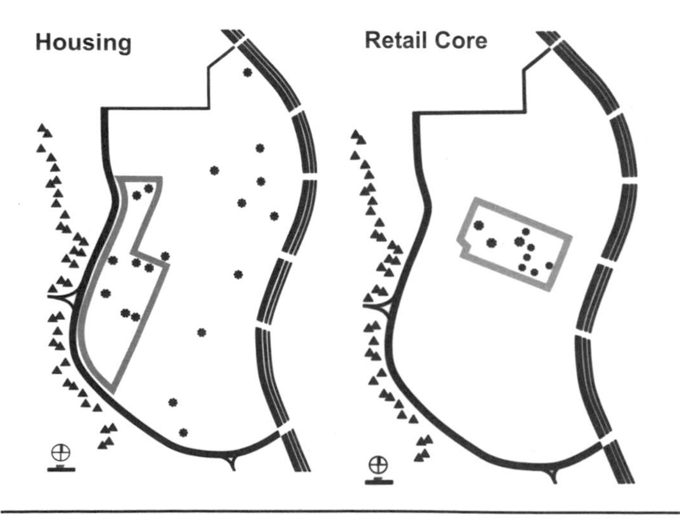

Fig. 9. The housing and retail core diagrams from Portland's downtown plan.

Peirce and Guskind explain, are a number of concrete achievements. The "public access" program's success can be seen in a massive increase in transit trips into the downtown from 79,000 daily in 1975 to 128,000 in 1985. Transit's share of commuter trips went up from 28 to 52 percent. Meanwhile, the downtown retail district increased its penetration of the regional retail market from 7 percent to more than 30 percent of dollar volume sales—the diametric opposite of the trends almost everywhere else in America.

Unsurprisingly, the expansion of business activity has been remarkable: total downtown retail space rose to 5 million square feet; department stores returned to the central city; and, perhaps most importantly, downtown employment grew by 30,000. Total assessed property value has increased 382 percent. Overall, the $500 million of public investment made between 1972 and 1986 leveraged $1.7 billion of private investment in the downtown. Guided by a publicly supported vision and enjoying considerable positive momentum, Portland has used its Downtown Plan to rebuild a lively, well-loved central city.

Portland's 1972 Downtown Plan forces us to broaden our understanding of excellent urban "places." How can a plan—and a plan with few specifics at that—be a place? Only if we acknowledge that urban places exist as social artifacts with histories and futures does it become possible to recognize the importance of the process of planning, construction, and maintenance in the makeup of an urban place. By awarding not the City of Portland itself but the design process that produced the successful environment, the Rudy Bruner Award sought to emphasize the fundamental interconnection between place, process, and values. Portland's downtown is indeed an excellent place; but what makes it more excellent than, say, a central city focused around a twelve-story parking garage is the social programming that produced it and, more importantly, uses it and makes it such a lively place. If city residents had not participated in the planning, they would not be as connected to the result as they obviously are.

Part of this is purely structural: the city would simply not have been designed to be so open and welcoming to its residents and suburban visitors. The key here is the vital link between benefits proposed at the drafting table, and real world uses. On the drafting table, a twelve-story garage offered certain attractive benefits; but because these benefits took for granted the social behavior of city residents in unrealistic ways, they never materialized. The same parking garage might have been a smashing success at a different location, where increased parking was truly the key issue for stakeholders. Portland reveals that, at least in some instances, the best way to find out what the key issues really are is also the most straightforward: you ask.

The Rudy Bruner Award Selection Committee identified three key interrelated elements of Portland's 1972 Downtown Plan that contributed to its success: its inclusive planning process; its adaptability; and its conception of "public access."

Inclusive planning

When business leaders or city hall tried to go it alone in responding to Portland's decline, they produced narrow and shortsighted solutions catering to only one element of the city's rich and complex urban ecology. A parking garage, for example, aimed to jump-start the economy by persuading more women to buy hosiery downtown. Ultimately, this kind of thinking angered and energized opposition in the very people for whom they were supposedly planning. The story of the garage warns us to interrogate our initial goals very carefully. If those goals include having city residents use a place in certain ways, it only makes sense to test the abstractions of the drafting table with real input from the people who will ultimately live in and use a place.

Everyone necessary for the success of a place should be included in the process of planning it

To renew its downtown, Portland needed more than enough parking spaces—it needed to repair the damaged relationship between its residents and its central city. Parking was a problem, to be sure, but a larger problem was the fading desire of city residents to be downtown. One element of fixing this problem was to construct new attractions downtown, such as the Pioneer Courthouse Square, the public artworks, and the waterside park.

But perhaps even more crucial was the forging of ties between citizens and city that resulted from residents' participation in the process of creating those attractions. At three crucial moments, grassroots organizations exercised a key influence on the plan's development: first, in opposing the garage; second, in rejecting a "Mount Hood Freeway" proposal in favor of a light rail system; and third, in successfully crusading for a waterside park to replace Harbor Drive rather than broadening it to ten lanes as proposed. Beyond making for a higher quality downtown, each of these instances increased Portlanders' emotional investment in the city and heightened their awareness of their own role in its success or failure. The process of building a "public city" actually helped create the "public" to use it.

Energetic public dissent is an opportunity to invite new collaborators

Portland's Downtown Plan succeeded in part because of the key decision to create the Citizens' Action Committee. Facing an organized and oppositional public, Terry Shrunk, the mayor responsible for some of the initial "bulldozer"-style renewal efforts, decided to cooperate rather than continue to fight. By inviting any Portland resident to join the CAC's subcommittees, he took advantage of the outpouring of passion and energy that, ironically, his own decisions had helped to unleash. Later, when the Pioneer Courthouse Square project appeared doomed, a citizens' fund-raising committee was able to marshal public support by selling personalized bricks at $15 a piece, raising $1.5 million through this and other similar efforts. The lesson? It is easy to get people to talk about grand urban projects, but far harder to find people willing to devote time and energy to doing something about it. Peoples' lives are full. Ignoring an opportunity to engage them in the massive challenge of urban planning would indeed be a waste.

Inclusiveness means vesting real authority and responsibility

Public participation in the Portland Downtown Plan was no token affair. At key moments in the plan's evolution—both as it was conceived and as it was implemented over the next two decades—city residents shaped fundamental elements of its vision and values. Sometimes this influence was nerve-wracking to business leaders or planning professionals, such as when Portlanders called for abolishing Harbor Drive, a major six-lane thoroughfare, or demanding that a public open space rather than a parking garage sit at the heart of the city. That these innovations succeeded is not proof that residents are the true repositories of urban planning wisdom, or that professionals and business leaders are obstacles to progress. Rather, Portland's

Fig. 10. Cornerstone Housing on the waterfront.

Fig. 11. A key to the revitalization of downtown Portland was an aggressive program of new and rehabilitated housing throughout the downtown.

Fig. 12. Portland enjoys several department stores in the downtown and has experienced substantial growth in its retail share of the regional market since implementing its 1972 downtown plan.

Fig. 13. Development downtown Portland attends to sustaining and creating new public space as a way of encouraging private development as well as making a livable city.

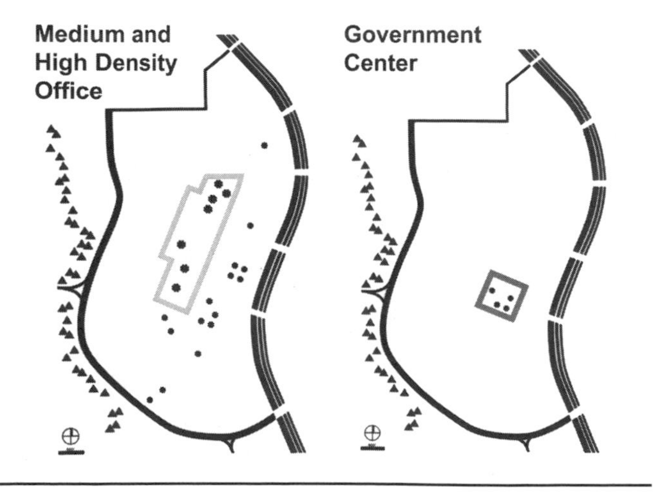

Fig. 14. Diagrams on medium and high-density office locations and the government center.

success reflects the ability of these groups to work together: residents supplied the vision—"public access," say, over more cars and parking—and then urban experts worked within that vision to come up with a successful plan. Neither one could have done it alone.

Adaptability

In recognizing Portland's 1972 Downtown Plan, the Rudy Bruner Award included as elements of the plan's excellence a number of developments not included or even foreseen by its authors: the light rail system, for example, and the Tom McCall Waterfront Park. Why should such unexpected additions reflect on the excellence of the original plan? Because the plan established design guidelines based on values and vision rather than specific details for projects. By defining what the downtown should be like, but not determining exactly how the city should attain that goal, the document built in the kind of adaptability necessary to survive the twists and turns of urban history. Relying on the continued participation and vigilance of residents, sharpened by the organizing power of the formal plan, Portland approached its renewal decades with a set of principles in mind and an open eye for opportunities to achieving them. It is a testament to the power of the original plan that new developments like the park harmonize so well with the spirit of the original document and with other elements already in place. By not closing itself off to new possibilities, the plan kept itself open to achieving excellence over the long term.

Vision, not design, should be the crux of a plan

The goal of most urban interventions is to achieve certain social, economic, political, and cultural effects. Generally, even the most cleverly and beautifully designed building will be considered a failure if the social programming for which it is intended never materializes. Portland's 1972 Downtown Plan acknowledged this ranking of priorities by establishing values like "public access" rather than specific plans as the heart of its vision. The goal was to facilitate travel in and around the downtown. Broadened in this way to include any solution, not just more parking, this goal allowed serious consideration of public transportation, which proved the most attractive option given the air pollution problems faced by Portland at that time.

The next step was to construct public transportation so as to encourage travel, again in keeping with the basic vision. One result was the handsome architecture and ubiquitous public art of the central transit mall. Another was "the world's first closed-circuit television system to provide riders at each of the thirty-one stops with around-the-clock arrival and departure information plus route information" (Peirce and Guskind, p. 60). Each of these developments was a creative response to the central imperative of the plan, which was to increase access.

Planning is not over once construction has begun

Clearly, Portland has benefited from the interpretation of the 1972 document as an ongoing set of guiding principles rather than a one-shot effort to improve the downtown. Seizing unexpected opportunities in the name of the original vision has resulted in some of the most successful elements of the plan, like the Waterfront Park, Pioneer Square, and the light rail system. In each case, downtown planning was assumed to be a constant affair, never completed and always following the values as established through stakeholder collaboration.

Maintain the collaborations that produced the original vision to oversee its implementation and maintenance

Notably, even after the initial success of public organizations in helping to create a more acceptable downtown plan in 1972, public outcry and opposition remained an important element of future developments. The Waterfront Park, for example, only came into being because of citizen pressure to remove Harbor Drive over the protests of planning professionals. Pioneer Courthouse Square required Portlanders' money as well as their active political support. These instances do not reflect a weakness in the 1972 Downtown Plan. Rather, they highlight the necessity for continued vigilance over the implementation of a project. True collaboration in urban planning rarely comes about because of basic generosity on the part of all stakeholders; usually, it is the happy result of an otherwise stalemated conflict. Only when stakeholders have forcibly carved a place for themselves at the negotiation table can they be sure to maintain that place and meaningfully exercise their voice in the planning process. And just as planning does not end with construction, neither does the need for this enforced collaboration.

Fig. 15. Portland City Hall helps to anchor the government district. Michael Graves, FAIA Architect.

Inclusive places

Not every individual urban place is intended to be open and public. Portland's experience, however, strongly argues in favor of the idea that cities themselves ought to be. Guided by the central principle of "public access," Portland's 1972 Downtown Plan led to the formation of a welcoming, friendly place that successfully attracted residents and visitors. Here the link between the social values and processes inherent in the plan and the ultimate physical design become evident. The inclusive and evolving planning process produced an inclusive place adaptable enough to respond to changes and new opportunities.

Fig. 16. The medium and high density office core was to step down to the water and to the rest of downtown.

Public spaces should be publicly accessible

The 1972 plan called for reviving the downtown not just architecturally and economically, but socially—the goal was to rebuild the downtown as the lively heart of a major city, a place people wanted to go. To accomplish this, all new development was designed to be literally or metaphorically welcoming. The public transit system allowed people to get to the city, while its attractive design carried the clear message that transit riders were valued. Pioneer Courthouse Square reversed the dynamic of many central cities in the 1980s with inviting open public space rather than, as Peirce and Guskind put it, "mammoth private structures walling off street life" (p. 65). Waterfront Park replaced a major obstacle between the public and the Willamette River, allowing greater access to one of the city's natural advantages. The public artworks were deliberately designed to be enjoyed by art critics and local children alike. And the efforts to preserve or construct housing were about access at its most basic sense: living space. Together, these elements created the possibility of an again-lively downtown, a possibility that retail establishments and other entrepreneurs eagerly took advantage of. To attract people, one must design for them.

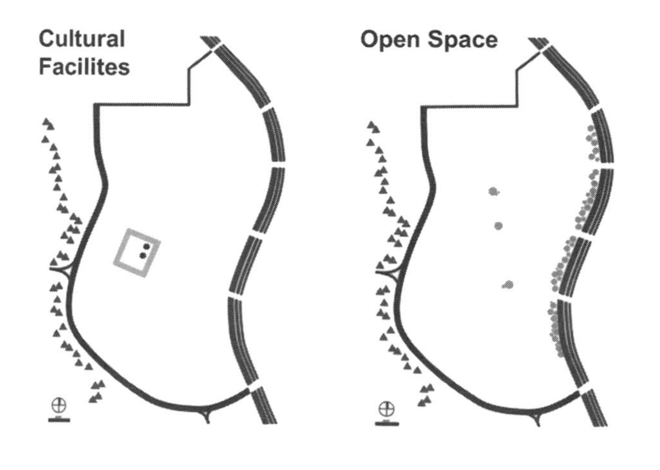

Fig. 17. The diagrams on cultural facilities and open space.

Process and values always influence design

It is no accident that an inclusive process ruled by the value of "access" created a downtown centered on open public spaces. Every planning initiative carries with it a set of values and a process for implementing those values. Consider the garage again: the process was decision making by a developer and retailers. The value was increasing retail activity by making automobile access to the downtown easier. The resulting place would have been a parking garage. The building reflected the

Fig. 18. A 1958 image of the problem of access across Harbor Drive on the Portland Waterfront. (Courtesy of the Oregon Historical Society, O.H. 57776 as published in *Breakthroughs: Re-creating The American City,* p.67)

Fig. 19. The transformation of Harbor Drive into Governor Tom McCall Waterfront Park never created the predicted traffic snarls.

Fig. 20. The transit mall is both an efficient element of the transportation system in Portland and a good public space.

Fig. 21. Downtown Portland has a tradition of public space including Lawrence Halprin's Lovejoy Fountain. Lawrence Halprin, Landscape Architect.

voices and values involved in the planning process. To see the 1972 Downtown Plan as "political" and the garage as a simple building misses the basic fact that every urban intervention reflects the agendas and values permitted to influence the planning process. Failing to engage the public in planning is not refusing to get involved in the social and political dynamics of a project; it is, in fact, making important social and political decisions about who will or will not be involved in shaping the final product. Whatever those decisions are, they should be made consciously, and with an awareness of how they can affect the design and, ultimately, the success of the project.

6 CONCLUSIONS

The story of Portland's 1972 Downtown Plan makes it clear that much of what takes place in urban planning happens in social, political, and cultural arenas, even though the ultimate product is a structure or structures in the physical world. Urban places can be elegant, beautiful, even works of genius in architecture and landscape, but they only achieve true excellence through their relationship with the people who use them, and the relationships between people that they foster.

Portland was able to reverse the decline of its central city because it added a key new player to the table of urban experts: the passionate and active citizens who were "experts" on what they wanted out of the city being designed for them. By incorporating them into the process, Portland strengthened their commitment to the city, which ended up being a key element in the downtown's rebirth. In return, Portland residents got the kind of city they wanted, one that welcomed them in their desire to participate in it. This invigorated relationship between people and their place also provided a continuing oversight that helped to avoid backtracking and sharpened the city's ability to take advantage of new opportunities as they arose.

Ultimately, Portland's 1972 Downtown Plan succeeded because it recognized that urban renewal is about more than rebuilding decaying building stock. It is about rebuilding the urban communities that are the lifeblood of a city. And, clearly, one important way to strengthen urban communities and confirm their commitment to their places is to involve them in the process of renewing the physical infrastructure of the city. This involvement itself helps enact a virtuous circle of self-reinforcing dynamics, exactly the opposite of the vicious circle that led from one high rise to another as residents fled the downtown. Acknowledging the importance of such social dynamics allows planners to employ their expertise in the most effective manner possible, rather than wasting it on projects that only tangentially meet (if they do not directly oppose) real social needs. ■

Fig. 22. Pioneer Courthouse Square, Portland Oregon; (top) In 1971, the site of the future Pioneer Court House Square was a two story parking garage with plans in place for a twelve-story parking garage. This garage plan was a catalyst for vocal citizens and others to approach downtown development planning in an entirely new way (bottom).

Fig. 23. The forecourt and bus stops in front of Pioneer Courthouse Square.

Fig. 24. 303 housing units at McCormick Pier.

REFERENCES

For further reading on the Portland planning experience see:

Abbott, Carl. 1983 *Portland: Planning, Politics and Growth in the Twentieth Century City.* Lincoln: U. of Nebraska Press.

_________ 1991 "Urban Design in Portland, Oregon as policy and process, 1960–1989," *Planning Perspectives* 6: 1-18.

_________ 1994 "The Oregon Planning Style," in Planning the Oregon Way: A Twenty Year Evaluation. Carl Abbott, Deborah Howe and Sy Adler, eds. Corvallis: Oregon State U. Press.

Clary, Bruce. 1986 "A Framework for Citizen Participation: Portland's Office of Neighborhood Associations." *Management Information Service Report.* Washington: International City Management Association, vol. 18, no. 9 pp. 1–13.

Hovey, Bradshaw. 1998 "Building the City, Structuring Change: Portland's Implicit Utopian Project," *Utopian Studies*, St. Louis, Mo.: Society for Utopian Studies.

Ehrenhalt, Alan. 1997 "The Great Wall of Portland," *Governing,* 10 (May) 8:20–24.

Oliver, Gordon 1994 "Portland Revs Up for Action," *Planning,* 60 Chicago: American Planning Association, 8:8–12.

Robert Shibley and David Herzberg

Summary

Yerba Buena's success as an urban intervention owes a great deal to the inclusive process that created it. Such inclusiveness was not the initial approach; it was forced upon San Francisco's Redevelopment Agency by citizen lawsuits. The lawsuits made progress more difficult and negotiated, but they also ensured that progress would be more responsive to the needs of the people and organizations that would live and work in the project area. For Yerba Buena, the inclusive approach transformed an apparent mess of social, legal, and economic dilemmas into a successful process that spoke to the needs of diverse constituencies who are now proud to call the project "home." In awarding Yerba Buena as its 1999 winner, the Rudy Bruner Award Selection Committee identified three characteristics: inclusive leadership, creative use of conflict, and commitment to place.

Key words

conflict, cultural district, democracy, inclusiveness, placemaking, process, public open space, values.

A representation of Yerba Buena Gardens showing the project in the context of the surrounding San Francisco skyline.

Yerba Buena Gardens, San Francisco, CA

1 A HISTORY OF THE PROJECT THAT WASN'T

Yerba Buena Center should have been a successful urban intervention. The plans, drafted with brilliance and precision in the 1960s by Kenzo Tange, Gerald M. McCue and Associates, Lawrence Halprin, and Mario Ciampi, were cutting-edge exemplars of the urban design wisdom of their day. The goal was to reverse the decay that postwar economic changes had visited upon the "South of Market" neighborhood adjacent to San Francisco's financial district. The plans called for extruding the financial district into several blocks across Market Street with fortress-like ring of office towers protectively encircling a central sports stadium and convention center. Backed by the might of one of the nation's first Redevelopment Agencies and the money of a top-flight developer, the project seemed destined for a glorious future.

Yerba Buena Center, however, did not succeed. It was nearly a complete failure. Indeed, within five years of its inception, the project had essentially been abandoned amidst an avalanche of citizen opposition and lawsuits. For when the Redevelopment Agency put forward its plans, it had ignored a crucial fact: several thousand people, nearly all of them single, elderly, male, and poor, still called the South of Market area home. Little concerned over such "bums," the Agency initiated mass dislocations while planning only a few hundred units of new housing to replace thousands of lost units. As demolitions continued, citizen opposition to the project grew increasingly intense and sophisticated. Resistance culminated in a 1970 lawsuit filed by Tenants and Owners in Opposition to Redevelopment (TOOR) that successfully forced the Redevelopment Agency to acknowledge its legal obligation to provide housing for displaced residents.

Citizen anger and distrust had reached such a fever pitch that it took years for the Redevelopment Agency to regain credibility. It was only through the intervention of popular San Francisco Mayor George Moscone that Yerba Buena was reborn. Moscone appointed a committee to study the area and produce a consensus design vision, explicitly encouraging citizen input through public hearings and discussions. This process affirmed some of the central components

Credits: This chapter is based on "Yerba Buena Gardens" in Shibley, Robert with Emily Axelrod, Jay Farbstein, and Richard Wener. 2000. *1999 Rudy Bruner Award for Urban Excellence: Commitment to Place: Urban Excellence and Community.* Boston: Bruner Foundation, Inc.

Fig. 1. A 1908 photo of the newly rebuilt Mission Street in the vicinity of Yerba Buena Gardens. (California Historical Society, San Francisco, in Bloomfield, p. 385)

Fig. 2. A 1961 photo of Third and Mission, looking south on Mission. (Photo Bill Nichols, courtesy Call-Bulletin Collection, San Francisco History Room, San Francisco Public Library in Bloomfield, p. 391.) The area surrounding Yerba Buena Gardens illustrates how very different the project site could have been without the urban renewal and displacement of thousands of residents. Some critics will never forgive the clearance of the old building fabric.

of the old vision—mixed commercial use, convention center, and public garden—but also added essential new elements, such as subsidized low-income housing, public and cultural as well as a retail center, and other amenities to be located atop an underground convention center expansion. The garden was no longer to be surrounded by office towers but rather open to neighborhood.

Given a second opportunity to revitalize the South of Market neighborhood, the Redevelopment Agency studiously avoided its earlier mistakes. For the most part, this meant paying careful attention to the social realities of the place it still hoped to see one day as Yerba Buena. An inclusive process designed to give a meaningful voice to all stakeholders was anchored by a series of creative partnerships that vested responsibility as well as authority in multiple locations.

This approach did more than lower the risk of lawsuits; it changed the nature of the project. Today's new-look "Yerba Buena Gardens" is a successful downtown district. It still meets the original design criteria, boasting world-class architecture, internationally recognized cultural institutions, and bustling commercial venues. But instead of being a "fortress" enclosure protected against the people who live south of Market Street, it is a thriving and diverse residential neighborhood that incorporates the energy and authenticity of both its new and long-term residents. This change in spatial form reflects more than an updated set of urban design guidelines—it is the physical embodiment of a different kind of process powered by a different set of values.

2 ANATOMY OF A FAILURE

The first Yerba Buena did not fail because of anything visible from the drafting board. Indeed, even today the original plans for Yerba Buena Center still hold a kind of grand artistic appeal, and the logic behind the original program and design still makes sense. The trouble is, what "makes sense" economically as well as on the drafting board does not always "make sense" in the real world of cities. Urban design is not just a matter of economic development, design and landscapes. It is a process that takes place in human communities who often care a great deal about where they live and work. Urban planners and designers must understand how their work fits into this social context just as they must understand the best ways to work with different spatial, typological, and material opportunities.

Unfortunately, human relationships are not as easy to manipulate as the physical media of the urban designer or even the economics of development. No figure-ground study can capture the intricate dynamics of a living city block or neighborhood. And no single strategy can be applied blindly to all contexts. Each urban intervention takes place within a unique context that demands unique solutions and practices. Case studies like this one of Yerba Buena can, however, reveal certain guiding principles that maximize a project's chances of success. In doing so, this case suggests, these principles also alter in subtle but fundamental ways what "success" in urban design means.

3 YERBA BUENA GARDENS TODAY

Public space

- A 5.5-acre central garden linked to an extensive open space network.

Commercial development

- **Convention Center:** an original "big box" building and a two-part underground expansion.
- **Hotels:** a Marriott, W Hotel, Four Seasons Hotel and Tower, and a Carpenter and Company project. These are mostly high-end facilities, some with condominiums and commercial venues at ground level.
- **Entertainment:** the centerpiece is Sony's "Metreon," its first large-scale urban entertainment center.

Museums and cultural facilities

- **Center for the Arts:** two buildings designed by James Polshek (the theater) and Fumihiko Maki (the galleries and forum).
- **San Francisco Museum of Modern Art**, designed by Mario Botta.
- **Children's Center for Technology and the Arts** ("Zeum"): a studio for technology and the arts targeting older youth.
- **Other youth facilities:** a historic carousel, bowling alley, ice skating rink, child care center, and children's garden.

Housing

- **TODCO homes:** four housing projects providing units for 1700 low-income, disabled, and elderly residents.
- **Other housing:** a mix ranging from SRO projects to 1500 market rate units (20% subsidized low-mod) to the high-end condos at the hotels.

4 LESSONS FROM YERBA BUENA GARDENS

Economic activity South of Market has increased astronomically and land values have continued to rise. The growth in hotel space, combined with the drawing power of the convention center, regional and international art venues, and the Metreon have given a major boost to the tourism industry. While there is clear evidence of gentrification, low-income housing providers have successfully put an end to mass displacement. The area has a diverse mix of rich and poor, and enjoys broad ethnic diversity. The children's facilities, open public space, cultural facilities, and other amenities are well used by locals as well as visitors. All these elements are underwritten and stabilized by an interlocking network of politically aware neighborhood organizations who care about Yerba Buena and are determined to work together to secure its best interests.

While Yerba Buena Gardens features the work of very well-known architects, the Bruner selection committee made it clear that their award was not, fundamentally, about that fact. Instead, the committee stressed that Yerba Buena's success as an urban intervention owes a great deal to the placemaking process that created it, ungainly and conflict-ridden as it was. Indeed, without such obstacles such as the TOOR lawsuit, key elements of the project would never have come about.

Yerba Buena's development process opened up to input from a diversity of constituencies with a stake in the South of Market area. Such

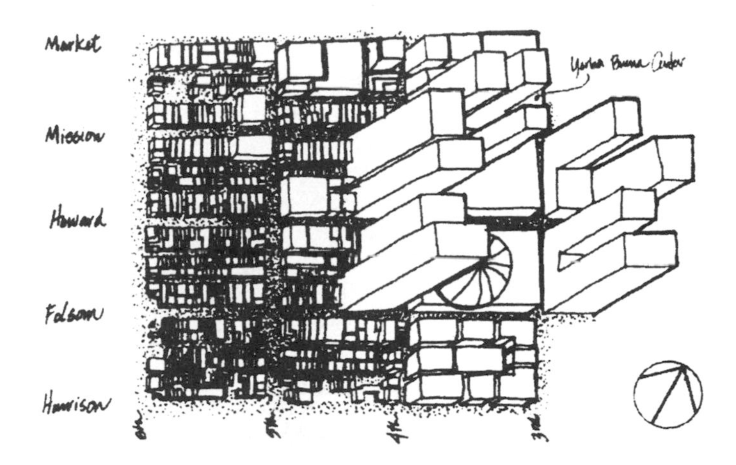

Fig. 3 Early conceptions of Yerba Buena Gardens envisioned a complex of high-rise buildings in almost a fortress fashion. (Drawing reprinted with permission, *Places* from Cheryl Parker's "Making a 21st Century Neighborhood," vol. 10 #1, 1995)

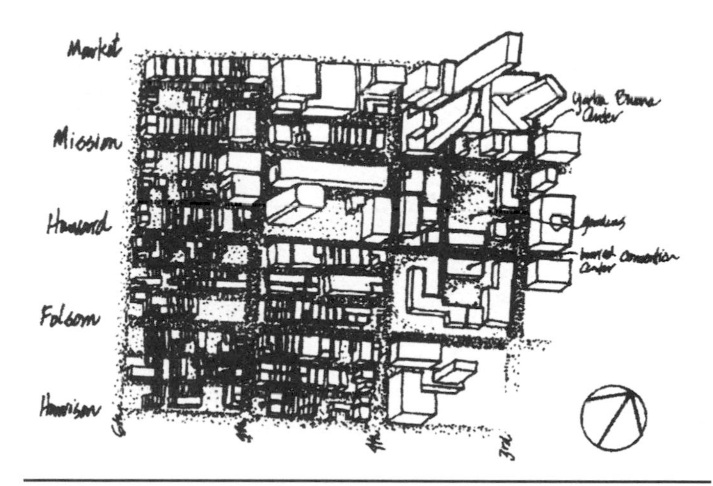

Fig. 4 Over the years the Yerba Buena Gardens master plan softened its edges, opening it to the surrounding neighborhood. (Drawing reprinted with permission, *Places* from Cheryl Parker's "Making a 21st Century Neighborhood," vol. 10 #1, 1995)

Fig. 5 A pylon at the entrance to the complex at Yerba Buena sets the ground rules for the use of the park consistent with all of the San Francisco Park Code.

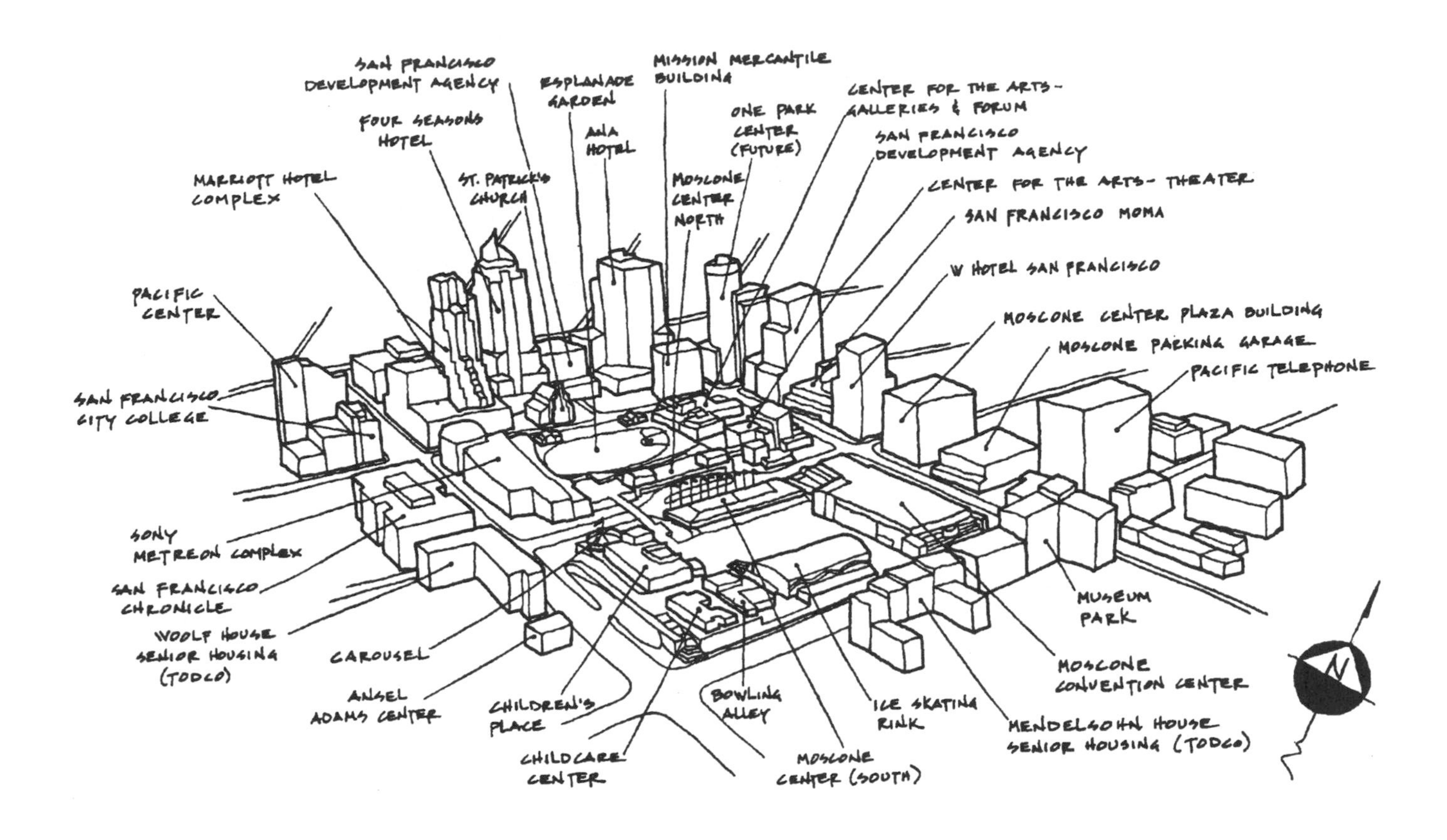

Fig. 7. A very rich and diverse mix of uses contributes to the financial, economic and social success of Yerba Buena Gardens. (Graphic from Shibley *et. al.*, pg. 8.)

inclusiveness was hardly a preferred strategy to begin with; it was forced upon the Redevelopment Agency by the lawsuits. If it made progress more difficult and negotiated, however, it also made progress more responsive to the needs of the people and organizations that would live and work in the project area. And since it is those people, ultimately, who make or break an urban place, their ownership in the process and final product mattered a great deal. For Yerba Buena, the inclusive approach transformed an apparent mess of social, legal, and economic dilemmas into a successful process that spoke to the needs of diverse constituencies who are now proud to call the project "home."

How did this project succeed in embracing the challenging and messy work of inclusive planning? In recognizing Yerba Buena as its 1999 winner, the Rudy Bruner Award Selection Committee identified three main characteristics: inclusive leadership, creative use of conflict, and commitment to place.

Inclusive leadership

As its history suggests, no single towering "urban pioneer" dominates the Yerba Buena story. Yet real leadership of the highest quality has been a crucial element of this project's success. In fact, it is precisely the difficult and contested conditions present in Yerba Buena that are most conducive to good leadership. For when no single player can dominate the process, no one has the power to insist upon a plan that ignores others' needs. Every player must willingly buy into the game. This is necessary, but it is also desirable—the goal, after all, is to create a place where every player will live and work together. Under these conditions, good leadership means escaping the zero-sum game by fostering in each stakeholder the faith that their individual agendas will be strengthened, not weakened, by incorporating others' agendas into a richer shared vision.

Avoid zero-sum games

Along every step of the path from the "urban fortress" of what was called Yerba Buena Center to the friendlier Yerba Buena Gardens, the Redevelopment Agency and other project leaders made deals that added to, rather than watered down, the central vision. For example, the Agency helped transform one-time foe TOOR into a nonprofit development corporation (TODCO) responsible for building and managing much of the court-mandated low-income housing. It recognized TOOR's strength and commitment to the project and put it to work. In another instance, the Agency bowed to the demands of local organizations for youth programming far beyond initial conceptions (a bowling alley and recreational ice rink hardly conform to conventionally defined "highest and best land-use" that was defined by early conceptions of Yerba Buena Center). Again, the deal making became a device for incorporating new energy and commitment into the broader vision. Good leadership meant recognizing that these

Fig. 8. The Mendohlson House senior housing operated by TODCOR.

Fig. 9. An example of low-income housing adjacent to Yerba Buena Gardens.

were not instances of choosing between two agendas, but were rather opportunities to enrich the overall project by incorporating both of them. Sally B. Woodbridge (1995) writes in *Progressive Architecture* that the area still looks cobbled together, in part because of the almost 30 years of false starts for the development. This is contrasted with Clifford Pearson's 1999 review for *Architectural Record* that focused on the Metreon describing it as a successful part of the way the very successful arts complex is "kept busy around the clock."

Lead with vision, not by fiat

Throughout the project, the Redevelopment Agency persuaded others to take significant risks in the service of a vision of what Yerba Buena could one day be. In a sense, it was the power of the vision as much as the Agency's political and economic might that did the persuading. One of the most important of the Agency's strategies in this regard involved financial and design "exactions" agreed to by potential developers to maintain the project's funding and design priorities. Many rents, for example, are tied to profits, and some rent payments go to support the central Gardens and the arts complex. Some buildings, like Sony's Metreon and the 3rd and Mission Street project, were redesigned at significant cost to ensure proper architectural "fit" (Sony also built a Tai Chi park at the behest of local seniors). All developers were required to turn to locals as a "first source" employee pool. Beyond the exactions, other examples include the luring of SFMoMA across Market Street and the placement of the convention center expansion underground. In each case, the Agency worked with other players to assure the common good was realized through the shared vision.

Share responsibility for the project among all stakeholders

As it negotiated its way through these compromises, the Redevelopment Agency crafted an array of unusual and creative public-private partnerships. The creation of TODCO was one. Another was what the Bruner Site Visit Team came to call "client construction." For several of the key facilities such as the Center for the Arts and Zeum, the Agency reserved space, earmarked operating funds, and actually began building long before there was a client to run the finished projects. Rather than selecting an existing organization to take over the facility, the Agency went to the community to develop its own institutions during construction, and then, after literally growing them to fit their tasks, encouraged them to form private nonprofit agencies. By sharing real power with groups such as TODCO and the arts institutions, the Agency also spread responsibility for the project among all stakeholders.

Be ready to be persuaded as well as to persuade

The Redevelopment Agency showed good leadership in its ability to "sell" its vision of Yerba Buena to other players in the private sector. Just as valuable, however, was the Agency's willingness to embrace this

Fig. 10. Gardeners work in plots provided by the San Francisco Redevelopment Authority as part of the low and moderate-income housing provided in the immediate area surrounding Yerba Buena Gardens.

Fig. 11. Even when fully programmed there are opportunities for passive recreation in the gardens.

Fig. 12. To the left is Fumihiko Maki's Center for the Arts Galleries and Forum and John Ploshek's performance facility is to the right. Both facilities create a framed view of Mario Botta's San Francisco Museum of Modern Art from the Garden.

process in reverse—to "buy" others' visions for Yerba Buena and thus enrich the whole project. From the public garden to the youth advocates' bowling alley, such additions have been crucial to the Agency's success.

The uses of conflict

The conflicts in Yerba Buena have ultimately served the purpose of communication. While battling over relocation and struggling over what to include in the project, the various players felt each other out. They became familiar with the aspirations and reasoning of the other stakeholders. And, finally, they reached a point where lines of communication had been opened and compromise had become possible—if everyone was willing to make the basic decision to abandon the fight and work together. It may be strange to call a lawsuit a means of communication, but given the outcome, that is exactly what it turned out to be.

Allow conflict to add value to the project

Yerba Buena's conflicts delayed and even halted development at times, but in return they significantly enriched the overall project. As a result of the TOOR lawsuit, for example, Yerba Buena is not the sterile financial district or precious cultural precinct it might have been. Instead, it is a richer and more authentic place whose own residential population frees it somewhat from a total dependence upon tourists and external investment. Because of strong community presence at the negotiation table, even the high-profile arts facilities are open to locals, creating opportunities for neighborhood artists while conferring vibrancy to the cultural institutions.

Another gift of conflict, the hard-won youth programming, has added to Yerba Buena's unique appeal in a nation full of urban projects where young people have been "designed out" as unprofitable security risks. And of course the central open park proved to be a key attraction flexible enough to draw the wide diversity of institutions that have now gathered around it. Plans, and planners, have their own inertia; oftentimes, conflict is an opportunity to test old assumptions and accept new ideas.

Conflict can strengthen the stakeholders who will help create and manage a place

Yerba Buena's conflict-ridden history has also produced less tangible benefits, the most important of which are the host of community organizations that have come to embody the neighborhood's new social and political self-awareness. With the central stroke of hiring its most implacable foe (TOOR) as housing coordinator, the Agency set a tone that has permeated the project ever since. Community organizations like TODCO and the South of Market Foundation are all part of a broader group, the Yerba Buena Alliance, which since 1991 has provided a space where all area stakeholders can come together to promote the neighborhood through media, community relations and community outreach. Since time and change never stops in a city, the continuing vigilance of these organizations is necessary to ensure Yerba Buena's continued success.

Commitment to place

What enabled the various players to get past their combativeness, to lay down their arms and move forward together? Conflict does not in every instance become communication. Some of Yerba Buena's good fortune may be due to the long-term stewardship of Helen Sause, whose personal trustworthiness became apparent over time. The most important factor in turning conflict into communication has been the third key element of Yerba Buena's success: an overarching commitment to place that kept the messy, inefficient, and often downright ugly process going for nearly five decades.

Commitment to a place means commitment to its history and for what it is as well as what it could be

The Redevelopment Agency's initial dream of an "urban fortress" was founded upon the idea of completely replacing the existing South of Market area and its residents. More than any other single factor, it was this utter disregard for the existing neighborhood that spelled doom for the first Yerba Buena. Only when the Agency had embraced its legal obligation to consider the current population could it work together with other stakeholders in an atmosphere of good faith. And only in such an atmosphere could the riskiest decisions be made. The central public gardens were designed to address the needs of current neighborhood residents, children from all over the bay area, and a vibrant arts community, as well as office workers over lunch. As such, they were designed to attract future investment in a manner more robust than the more modest mix of users originally proposed. Many elements of the youth programming, now a much-heralded success, originally served as examples of the project's commitment to the existing community and its needs. The creation of TODCO was a similar story: it placed the needs of the neighborhood over the integrity of the original plan. Yerba Buena's future potential, ultimately, only made sense—only became possible—when envisioned in the context of its history.

Commitment to a place means commitment to realizing its full potential, not simply preserving what exists at all costs

The economic and tourist potential of the South of Market neighborhood could no more be ignored than could the area's current residents. Both were facts, both were the realities of the place. With its eye on the future, the Agency risked the expense of hiring top-flight architects like James Stewart Polshek, and Fumihiko Maki. It lured the SFMoMA with architect Mario Botta to the "wrong" side of Market Street, once a formidable urban boundary line. And, creatively wielding the reality of a golden future, it leveraged exactions from private developers that continue to support the arts, garden, and youth programs. The rich mix of users and uses at Yerba Buena Gardens make good business sense.

The depth of shared commitment to this neighborhood is revealed, in part, by the willingness of the Redevelopment Agency's community partners to go along with this aspect of the plan. The private developers, for example, agreed to the exactions; SFMoMA agreed to be lured; TOOR chose to join up even though the project would radically alter the character of the neighborhood. Youth organizations demanded access, not an end to development. Without this shared commitment to achieve Yerba Buena's potential without abandoning its history, the project may never have happened—and would in any case be very different from the highly successful urban mix it is today.

6 ANATOMY OF A SUCCESS

Taken together, Yerba Buena's many excellent qualities can be understood as part of the unified notion of "placemaking."

Fig. 13. The historic carousel relocated from San Francisco's Playland-at-the-Beach is an important part of the intergenerational mix of activity at the Gardens.

Fig. 14. The SONY Metreon is a retail and entertainment center facing the Gardens. It was redesigned following a public design review to allow additional light into the gardens through additional roof glazing.

Fig 15. The Zeum is a children's arts facility where teens from all over San Francisco experiment with video and digital arts. Here is a class gathered in the courtyard adjacent to the Zeum entrance.

Fig.16. Dancing in front of the Martin Luther King Memorial facing the garden.

Fig. 17. The Church is one of the last remaining structures in the site area that survived the bulldozers of the 1960 urban renewal.

Placemaking refers not only to the physical construction and maintenance of places, but also to the quality of relationships between people and places as well as among the communities that make and use places. Its central premise is that making and caring for places—becoming committed to them—can be the basis for making and caring for communities. Excellence is achieved when good design comes out of a process that is sensitive to (and inclusive of) the human relationships in which it develops, and that, like a good guest, leaves those relationships better than it found them.

The complex evolution of Yerba Buena's nearly 50-year history reveals not only the desirability but the practicality of placemaking. After the failure of the initial top-down, grand urban renewal project, Yerba Buena only began to move forward after a more inclusive and democratic process was implemented. The resulting neighborhood holds true to the best of the original vision but is, by any standard, vastly superior to its initial formulation. The courage to make the difficult transition from one type of process to another stemmed from three important elements of the Yerba Buena project that any urban intervention would do well to emulate: inclusive leadership willing to share responsibility (and credit) as necessary; an ability to learn from and grow through conflict rather than just to survive it; and an overarching commitment to a place, for what it could be but also for its history and for what it is. ■

REFERENCES

Bloomfield, Anne B. 1996 "A History of the California Society's New Mission Street Neighborhood" in *California History*, Winter 1995/1996, Vol. LXXIV No. 4.

Hartman, Chester. 1974. *Yerba Buena: Land Grab and Community Resistance in San Francisco*. San Francisco: Glide Publications.

Parker, Cheryl. 1995. "Making a 21st century neighborhood" *Places 10* (1): p. 36–45.

Pearson, Clifford A. 1999 "Metreon, San Francisco, California: What's a shopping and entertainment center doing in an arts complex? Keeping the place busy around the clock," *Architectural Record* (vol. 10, pp. 154–160).

Woodbridge, Sally B. 1995 "When good urban places go awry" (Yerba Buena Gardens, San Francisco), *Progressive Architecture* (vol. 76, pp. 60–67).

Robert Shibley and David Herzberg

Summary

Historic preservation can be a two-edged sword: it profitably requires you to take stock of what already exists before proceeding to plan anew, but it can also hobble needed efforts to address current realities when they demand fundamental change. Seattle's Pike Place Market story represents a successful balancing of the past and future in a way that unleashed new creative energies. Pike Place also reinforces the necessity of viewing urban places on many levels at once: not just buildings, but an economy; not just an economy, but also a unique "social ecology" of farmers, shoppers, entrepreneurs, and residents. There are three elements to Pike Place's renewal process: harnessing the power of citizens' commitment to place, employing "indigenous" solutions suitable for unique local conditions, and preserving the complex "social ecology" responsible for the market's excellence in the first place.

Key words

context-specific design, democracy, historic preservation, placemaking, process, public markets, social ecology, social values

Pike Place Market looking east from Elliot Bay against the Seattle skyline.

Pike Place Market, Seattle, Washington

1 THE UN-MAKING AND MAKING OF AN EXCELLENT URBAN PLACE

For the first half of the 20th century, Seattle's Pike Place Market was by any standards an excellent urban place. As Langdon (1990) explains, the market's origins lay in the early 20th century, during the Progressive Era. Following the reformist ethic of the day, the City council designated a public space in the downtown area where farmers could sell produce directly to consumers with no middlemen siphoning money from the transaction.

From the first motley collection of fewer than a dozen farm wagons on August 17, 1907, the market grew to encompass hundreds of farmers and thousands of shoppers. Its evident popularity spurred demand for an ongoing series of physical improvements undertaken over several years, including roofed buildings and paved roadways. As it evolved, however, it remained true to its origins as a way for farmers and consumers to save money: whether financed by the City or by private developers, the design emphasized "openness, ease of circulation, simplicity, and economy, without expensive decoration that would repel cost-conscious customers" (Langdon, p. 22). No grand urban castle or dreamscape, the market nonetheless earned an important niche in the economic and social ecology of Seattle and its surrounding farmlands.

The urban landscape is never static, however, and in the space of a single decade the thriving market fell upon hard times. Perhaps the most significant event in this decade of ruin was the internment of Americans of Japanese descent during World War II. By the early 1940s, Langdon notes, "a large proportion of the market's farmers were Japanese, and their absence was sorely felt" (p. 23). Most were unable to recover their farms after the war. Meanwhile, even when peace returned, larger currents continued to eat away at the market's foundation: suburbs lured people away from the city, technological advances in refrigeration improved the produce in supermarkets, and new "agri-business" giants began selling directly to supermarket chains. In the decade between 1939 and 1949, the market virtually disappeared, as the number of licensed farmers shrank from 515 to 53.

"As Pike Place Public Market and its surroundings became shabbier," Langdon continues, "the city government began to consider taking a more active, dramatic role in shaping the market area. As early as 1950, there was a proposal by at least one city planning commissioner to demolish the market and replace it with a seven-story parking

Credit: This chapter is based on "Pike Place Public Market" in Langdon, Philip with Robert Shibley and Polly Welch. 1990. *Urban Excellence.* New York: Van Nostrand Reinhold.

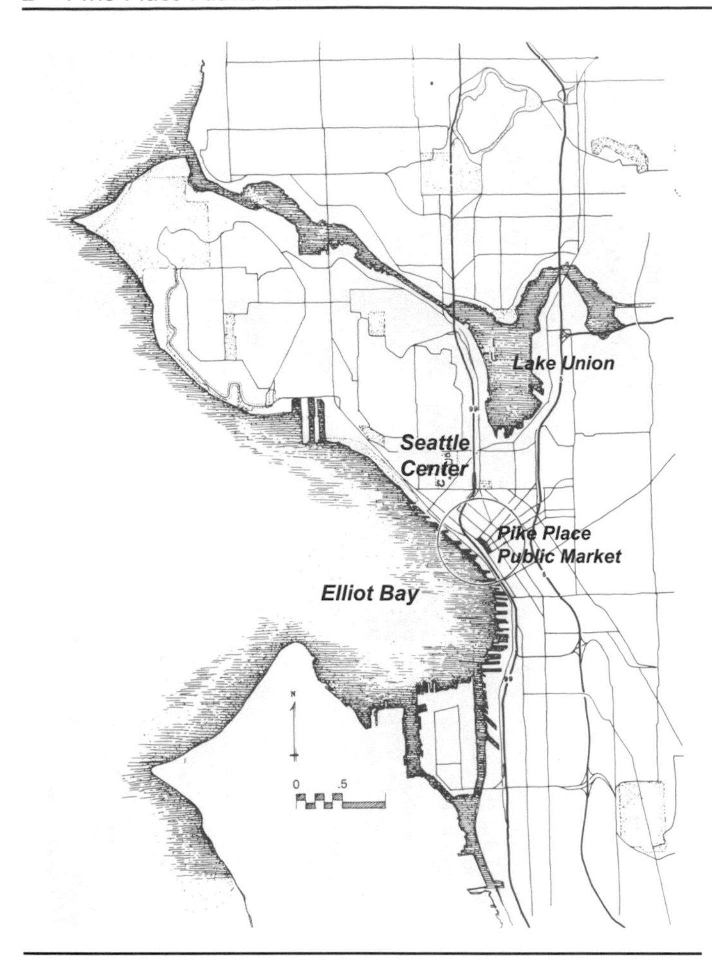

Pike Place on Elliot Bay south of Lake Union and the Seattle Center and just north of Pioneer Square and the King Dome.

The rummage shop is right next to the sliver shop in this sketch of the market by Victor Steinbrueck (from Steinbrueck,1978).

garage" (p. 23). While this particular plan was never adopted, the slash-and-burn "urban renewal" logic that it represented still held sway in most thinking about the market. The city ultimately decided to seek federal urban renewal funds to "give Pike Place a radically different character as part of its 1963 Downtown Plan," which, among other things, called for plunking a high-rise building in the center of the market (p. 23).

At this moment, it appeared that an unhappy confluence of socio-demographic shifts and urban design thinking had destroyed a once vital part of Seattle. But Pike Place still had one important asset that had not been drained away during its low years: a network of people who cared about its existence and who were committed to preserving and reviving it as a social and architectural institution. Victor Steinbrueck, a professor of architecture at the University of Washington and the market's patron saint, organized the first of many grass-roots volunteer organizations to "save" the market from the bulldozers—and from its own decline predating the urban renewal threat.

These citizens' groups focused their energies on reviving the complex "social ecology" supported by the market as well as preserving the organically grown beehive of buildings and sheds. In 1971 they successfully changed the terms of the debate with a referendum that established a Pike Place historic district to be overseen by a Market Historical Commission. The Commission had a mandate to preserve both buildings and uses "deemed to have architectural, cultural, economic, and historical value" (Langdon, p. 26).

Over the next decades, the market reversed its earlier decline, owing in no small part to the willingness of all players to get on board the project once it had been defined by the referendum. Structural and architectural investment have been accompanied by a rebirth in business and cultural activities. A "Preservation and Development Authority," appointed by the Mayor, manages a careful balance between preserving the market's authenticity and historic mission, and ensuring that the market remains a financial success.

One important consequence of Pike Place's commitment to maintaining its historic "social ecology" has been the evolution of the area into much more than simply a market. An array of social services and affordable housing, for example, complement regulations ensuring that goods sold still serve locals as well as visitors. By balancing these various concerns, Pike Place Public Market has become a place well-loved (and fiercely protected) by those who use it, whether they use it as a place to sell wares, a place to shop, a place to live, or simply as a familiar icon representing Seattle's unique urban identity.

To renew or rebuild?

If Steinbrueck had been a lone advocate, representative of nothing more compelling than his own nostalgia for days gone by, the Pike Place Public Market would now be a parking lot, or a complex of high rises, or some other very different kind of place—and deservedly so. The Pike Place story is about historic preservation, but not the museum kind. The market was not "preserved" as a monument to past greatness. Rather, Steinbrueck's vision succeeded because it connected with the present, living needs and future aspirations of real people and real opportunities in the existing Seattle economy. Preservation can be a two-edged sword: it profitably requires you to take stock of what already exists before proceeding to plan anew, but it can also hobble needed efforts to address current realities when they demand fundamental change. The Pike Place story represents a successful balancing of the past and future in a way that unleashed new creative energies.

Pike Place also reinforces the necessity of viewing urban places on many levels at once: not just buildings, but an economy; not just an economy, but also a "social ecology" of farmers, shoppers, entrepreneurs, and residents. And, significantly, not just any buildings, economy, and social ecology, but a place-specific unique environment so attuned to its location in Seattle that it serves as one of a handful of civic emblems.

Pike Place is east of Seattle's downtown in a 7.5 acre historic district.

Postcard of the "Sanitary Market" in the early 1920s (from the Pike Place Archives).

2 PIKE PLACE PUBLIC MARKET IN 1987

A public market: Pike Place "favors person-to-person sales of hard-to-find goods, including ethnic and seasonal products; of goods that involve light manufacturing processes that are interesting to watch; of those catering to the pedestrian or offered in a natural state, rather than prepackaged; and of those bringing together people of varied backgrounds" (Langdon, p. 48). No chains are permitted, and a "robin hood rent structure" encourages small enterprises. Farmers are subsidized with low rental rates and prime selling locations. Regulations also ensure a supply of reasonably priced food for lower-income city residents.

- **A historic district:** The "humble but not plain" (Langdon, p. 39) architecture of the market has been preserved to impart a sense of the city's history.
- **A social services provider:** Includes a senior center, child care center, community clinic, food bank, and affordable housing for low-income residents and seniors.
- **A tourist attraction:** In addition to everyday fare, the market offers high-end goods and "food theater" for the visitor. High-end housing units have also become a part of the mix.

3 LESSONS FROM PIKE PLACE PUBLIC MARKET

Employment increased 66% between 1973 and 1982, from 1,428 to 2,370. A long-term decline in number of low- and moderate-income units in the downtown area was stabilized in 1978. Meanwhile, higher-end residencies have begun to collect in a once-stigmatized neighborhood. In all, Pike Place is an economically successful and healthily diverse urban place, serving a variety of constituencies—poor and wealthy locals, tourists, gourmets, and farmers—and fostering pride and community self-awareness among them.

Unlike many projects submitted to the Rudy Bruner Award for consideration, Pike Place was not the result of an organized planning process. Nor did it primarily represent the efforts of one or more central designers. Rather, it grew up "organically," with parts added piecemeal as the need (or opportunity) for them surfaced. The original market's excellence can be attributed in large part to this "organic" evolution, with the whole neatly adapted to its social and economic uses because the one grew out of the other.

The excellence of the present-day market was different: it resulted from a conscious process that recognized the virtues of what had already been created and sought to revive them. No single architectural vision, no perfect design plan, could capture what had been excellent

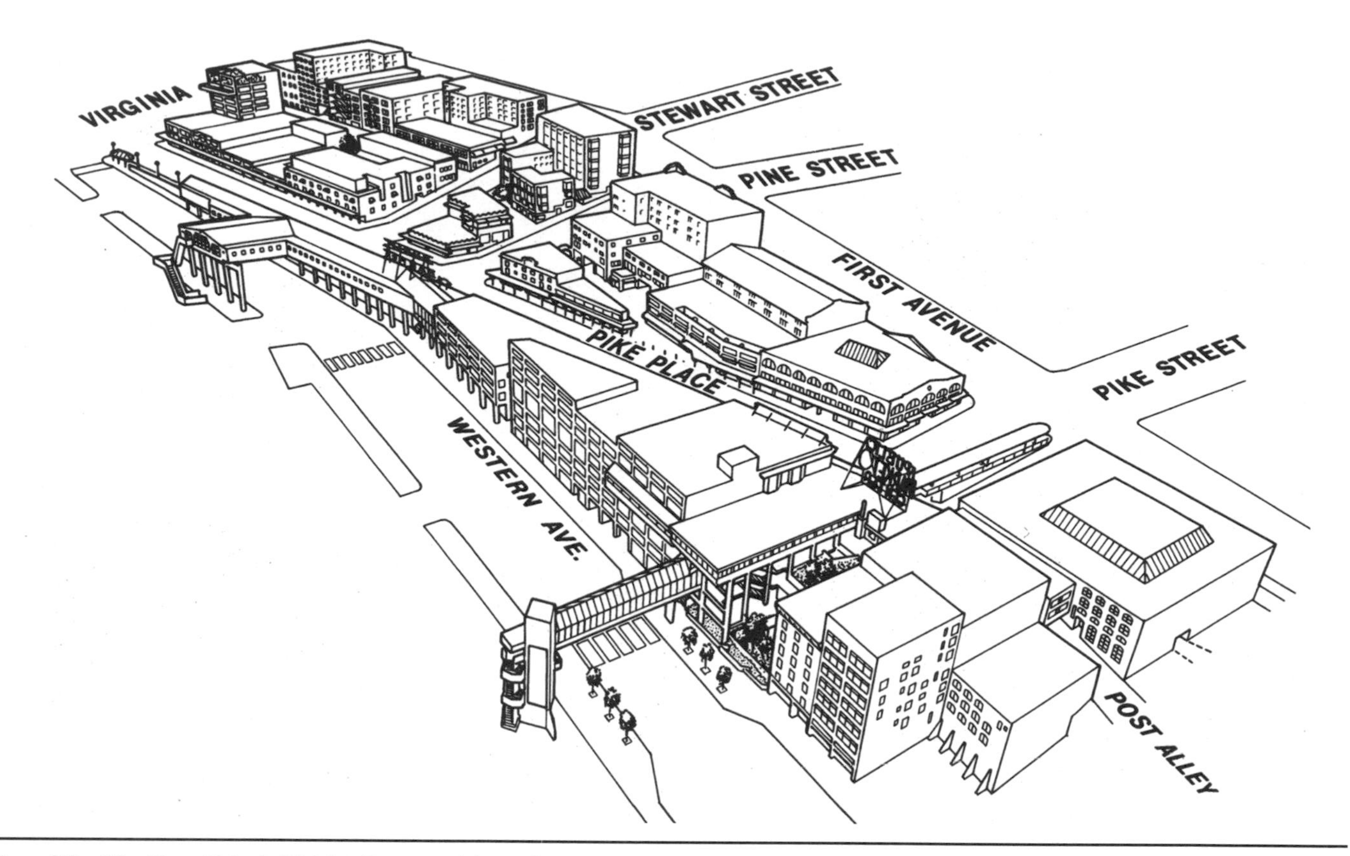

Drawing of the Pike Place Historic District (Courtesy of the Market Foundation).

Postcard of Pike Place from the 1920s (from the Pike Place Archives).

Sketch of the Corner Market (Courtesy of the Market Foundation).

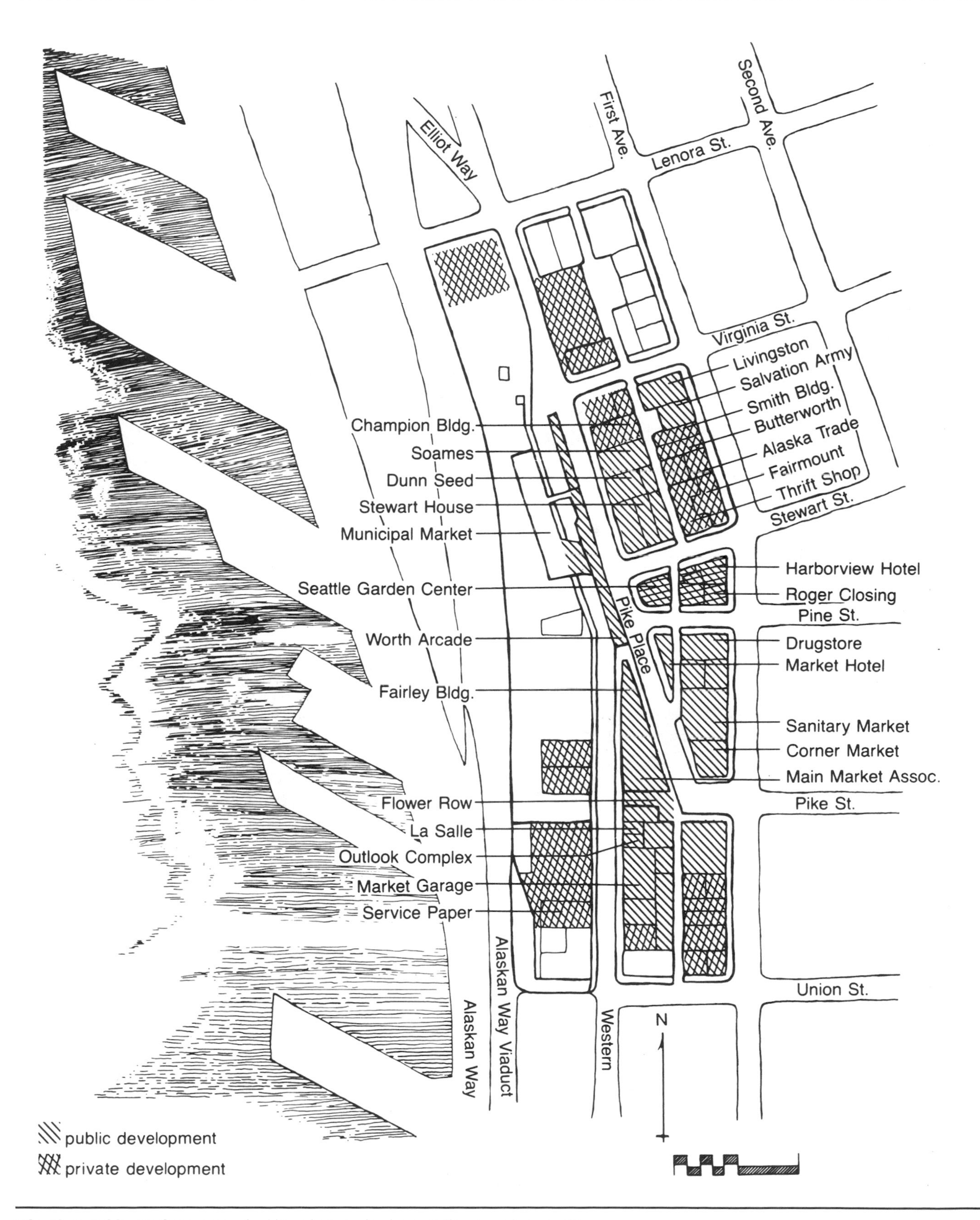

Pike Place Public Market is a mix of public and private development (from Langdon, pg. 52).

Table 1. Distribution of leasable market space in 1986.

250,000 sq. ft. Leasable Market Space	
Food	26%
Restaurant	17%
Used Goods	15%
Other	33%
Total	100%

Table 2. 800 housing units were built in the Market from 1975–1986.

800 Housing Units 1975 - 1986	
Single Room Occupancy	5%
Low Income	57%
Moderate income	11%
Condominium	27%
Total	100%

Table 3. Market Foundation programs at Pike Place Market, 1986.

Other Programs	
Senior Center	45,000 visits/year
Health Center	21,000 visits/ year
Food Bank	$1.6 million worth of food with $80,000/year
Child Care	100 families served annually

Table 4. Space changes at Pike Place Market between 1974–1983.

Selected Changes in Space Use 1974 - 1983		
Total sq. ft.	1,122,500	703,400
Wholesale	23%	3%
Retail	16%	56%
Office/service	15%	32%

about the market in its heyday; that excellence lay in the relationships that had developed between a sometimes ungainly place and the people who used and inhabited it. Only by acknowledging those relationships, and placing them at the center of the renewal process, was Seattle able to succeed in capitalizing on its market inheritance.

The Rudy Bruner Award Selection Committee identified three key interrelated elements to this renewal process: harnessing the power of citizens' commitment to place, employing "indigenous" solutions suitable for unique local conditions, and preserving the complex "social ecology" responsible for the market's excellence in the first place.

Citizens' commitment to place

Without public support for Steinbrueck's vision of a preserved and revived market, Pike Place would not exist today. The market's success has depended fundamentally on this public support, from the initial referendum that created it to the basic decision to shop there and keep the money flowing. As often happens, this public support has turned out to be a two-way street: not only has it engendered an excellent urban place, it has also fostered community self-awareness and organizations that serve locals in a variety of other ways as well.

Use public energy to get things done

The first thing public opposition achieved was time to think, as redevelopment plans were delayed. Market preservationists came to recognize the value of the "social ecology" as well as the buildings, while developers paused long enough to register economic changes that undercut some of their own design reasoning. But opposition itself was never the end goal of the market's energized constituencies, and public pressure soon led to the 1971 referendum, which lit a fire under a new kind of redevelopment spirit. Citizen opposition could have been considered a roadblock, an obstacle to development. But for Pike Place it was ultimately recognized as an engine for development.

One of the key moments in Pike Place's history was when opponents of preservation (including the Mayor) acknowledged their defeat in 1971 and harnessed public energy to power and oversee the restructuring of the market. The historical half of the Preservation and Development Committee, understaffed and underfunded, nonetheless attracted enough energetic and capable volunteers to provide a counterweight to purely economic decision making. Fundraising activities similarly traded upon the voluntary support of the public. Rather than wasting time, resources, and caring in an endless battle with preservationists, the Mayor and other pro-development forces acknowledged the depth of citizen feeling about the market and helped preservationists put it to work in constructive ways.

Use public energy in meaningful, not just symbolic, ways

Symbolism is important; no doubt "owning" a monogrammed ceramic floor tile in Pike Place is psychologically important in motivating people to donate money to the market. But Pike Place has put peoples' commitment to work in more meaningful ways as well. Committees depend upon volunteers to make crucial decisions about what can be sold or built in the market. The Market Foundation, a social services umbrella organization, helps raise funds and also ensures that the social fabric surrounding the market remains healthy and well serviced according to the needs of its populations. The Merchants Association, Public Development Authority, Historic District, City Hall, the Market

Foundation, and the Friends of the Market (Steinbrueck's advocacy organization) similarly keep power distributed among a wide number of constituents through both formal and informal systems of checks and balances built into the governance of the Market.

While such a diffuse power structure can be awkward and inefficient, it also prevents ill-considered planning mistakes; for example, when a parking garage and office building complex was proposed for the edge of the market, an organized public uproar forced a reconsideration. The garage was ultimately built, but without the office and retail; instead, the garage sits beneath three stories of congregate housing for low-income elderly people.

"Indigenous" solutions

One of the remarkable things about Pike Place is that the urban market revival was not a predictable urban solution, imported from other cities. It was an indigenous solution, developed by local people who looked at what they had and figured out how to use it to maximum advantage. The best solutions to urban issues, Langdon observes, "rarely are stock ideas grafted onto local terrain" (p. 29).

For Pike Place, and other urban projects, this means a twofold effort: affirming and building on what already exists and who already lives there; and striving towards a vision of the place and people's unique potential. Balancing these two efforts is not a zero sum game, with each concession to history necessitating a sacrifice of future potential. Rather, there is a large "sweet spot" in the middle where these two dynamics feed on each other, producing a vision of the future charged and energized by a recognition of what is already there to work with. One major advantage of this is that such a project usually garners far more support among existing residents and other constituencies, because it builds on what they already value.

A rich mix of retail occurs in the market from newsstands to salmon.

Use your own strengths

At the time that urban renewal mavens proposed bulldozing and rebuilding the area, Pike Place was home to a public market tradition stretching back more than half a century. The generic parking garage or high-rise building no doubt satisfied a purely formalistic evaluation of the needs of Seattle's downtown district, but it paid no heed to the social potential built into Pike Place over its decades as a public market. Unsurprisingly, the 1963 Downtown Plan found many opponents and relatively few friends. One cannot say for certain that it would have failed, but certainly a far safer bet was to acknowledge and build upon the existing infrastructure of the public market and its well-organized and committed supporters. By capitalizing on what it already had, both in buildings and in people, Seattle turned what might have been an eyesore and a public burden into a unique and famed city landmark.

Preserve what is already working

One of the most compelling of Pike Place's many stories is that of the architects and contractors who worked on the market's various decaying buildings. To address the biggest project, an L-shaped complex at the heart of the market, Langdon notes that "architects spent up to a full year walking through the market, learning how its multiple levels were actually built and creating drawings of the existing structure on which the renovations could be based" (p. 36). The strange, conglomerate architecture of the market was to be preserved, not simply for historic reasons, but because it worked, giving the mar-

The Public Market Center sign and clock are important symbols of Pike Place Market.

The 1963 downtown plan proposed a high-rise building over the street towering over the old sanitary market. The proposal was defeated in a public referendum (sketch from the Pike Place Archives).

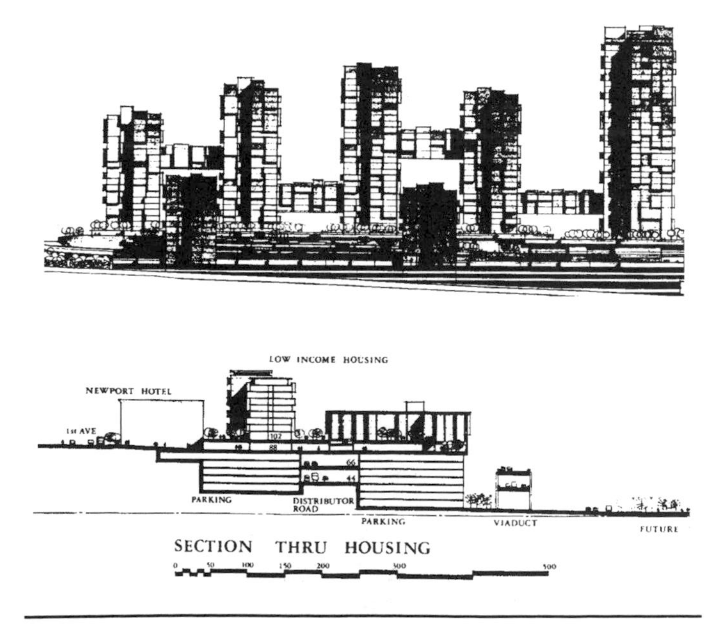

The 1963 downtown plan for the Market included several tall structures and a high concentration of low income housing.

ket its unique feel and inviting repeated exploratory visits over and above routine shopping.

Such an ethic also functioned in the decision to carefully manage the economic activity of the market. To preserve Pike Place's historic character—the character that engendered such devotion among its users and that was therefore to some degree responsible for its success —the management prohibits chain businesses, subsidizes farmers and small entrepreneurs with lower rents, and encourages "food theater" and other entertainment that dramatizes connections between consumers and producers.

Pursue future potential with no apologies

There is no entrance fee to the Pike Place Market; it is not a re-created village along the lines of Old Salem in Massachusetts. If it is a tourist and gourmet attraction, it is also a living, working market full of people making a living or purchasing needed items. Without this authenticity, preserving the market would make far less sense. The challenge for Pike Place was to remain true to its roots while capitalizing as much as possible on its potential for becoming a market suitable to Seattle's new postwar realities. Supermarket chains and suburbs were not going to disappear because the public wanted to preserve the historic market. But this seeming conflict was actually an opportunity: an opportunity to "sell" the very authenticity that people so valued in the market.

This occurred in a variety of ways, from the literal selling of ceramic floor tiles for fundraising; to the use of "Rachel," a life-sized brass piggy bank for donations to the market; and the ubiquitous market slogan "meet the producer." In perhaps the most important manifestation of this strategy, tourist-oriented and gourmet items were introduced into the mix of products for sale at the market. Significantly, this was done within the context of the market's larger aims: in a "robin hood" rent structure, high-end businesses subsidized farmers and smaller entrepreneurs, for example.

In each case, Pike Place's potential was envisioned as an outgrowth of the value that was already there. Thus even though gentrification has accompanied the market's rebirth, it has not been sudden and catastrophic, and there has been time to put in place institutions to prevent any radical upending of the area's diversity.

"Social ecologies"

Despite efforts to remain true to the market's historic architecture, the most significant thread connecting the new Pike Place to the original is the "social ecology" of interconnected and interdependent people linked together by the market. Pike Place reveals that a successful urban place is more than its buildings and its bottom line; it lies rather in the social relationships fostered among people and between people and a place. What, after all, would Pike Place be without the unique relationships it has fostered? A poor excuse for a supermarket. But instead, it is a vital, healthy, diverse place where all the most valued dynamics of urban life unfold.

"Social Ecology" is a variable, not a constant

If a designer preoccupies him- or herself solely with the buildings, assuming that the social ecology will simply take care of itself, a rude surprise might be the result. Nor can one rely entirely on universal design principles tailored to foster certain types of social activity.

No one, for example, would propose to build a new building with the meandering features of the Pike Place market. And yet, without that unique structure, the complex collection of social networks that have routed themselves through the market would likely fall apart. The design is part of the community narrative that helps sustain the healthy social ecology.

This is not to say that conventional design wisdom has no role to play; as the market was reworked to bring it up to code and to expand the available space, professional design principles were used at every juncture. They were just used within the context of design decisions made for the purpose of maintaining the social ecology. They were a means, not an end in themselves. Pike Place succeeds in large part because of its continued and vigilant preoccupation with developing and maintaining the social ecology that inhabits and enlivens it.

Planning a place can strengthen communities

If Pike Place benefited from public support, a variety of Seattle communities have similarly benefited from the market. Organizations like the Market Foundation help maintain the market, but also serve as institutionalized voices for communities of low income and elderly residents who formerly had little or no political muscle. In coming together to preserve their place or social program, they also re-created themselves as more tightly knit and self-aware communities, able to continue the process of holding on to what currently works while looking towards the future. In other words, this was a self-reinforcing cycle: in preserving the "social ecology" of the market, the market itself was renewed and revitalized, which in turn renewed and revitalized the social ecology.

Victor Steinbrueck's sketches of the market illustrate the character of the unique structures and passages (from Steinbrueck 1978).

4 IN A CITY, BUILDING IS ALWAYS REBUILDING

Pike Place Public Market is, in some ways, a unique story of re-creating a historic urban landmark. But every urban place presents similar dynamics, since every intervention in urban fabric takes place at a location where something already exists, whatever it may be. Every urban project, in some sense, is built on top of a historic location, and carries the possibility of taking advantage of what already exists. Obviously, not every project will do so as explicitly as Pike Place, but this story does reveal how a project can aim at the future even by such resolutely backward-looking strategies as historic preservation. The key, both for Pike Place and for other projects, was the understanding of preservation as a matter not just of buildings but of social ecology as well.

Ultimately, the Pike Place story argues for the importance of understanding urban design and planning as a social as well as an architectural activity. The logic of the project was at heart almost entirely social, from the valuing of the market in the first place, to the preservation of the unusual architecture, to the systematic efforts to maintain the "authenticity" of the market experience. Quality architecture had its place, certainly: reconstructing the original design of the market was no easy task. But without the social logic, this would have been an incomprehensible waste of time and effort. The "place" of Pike Place Public Market is only excellent because of the values that informed the process of restructuring it: harnessing the power of people's commitment to their places; designing "indigenous" solutions that value an existing place and its potential; and designing for the needs of the social ecology that inhabits and animates a place. ■

Rachel the pig is one of the many ways the Market Foundation raises money to support its housing, daycare, senior center, and food bank supports to the Market. (Langdon *et. al.* pg. 33)

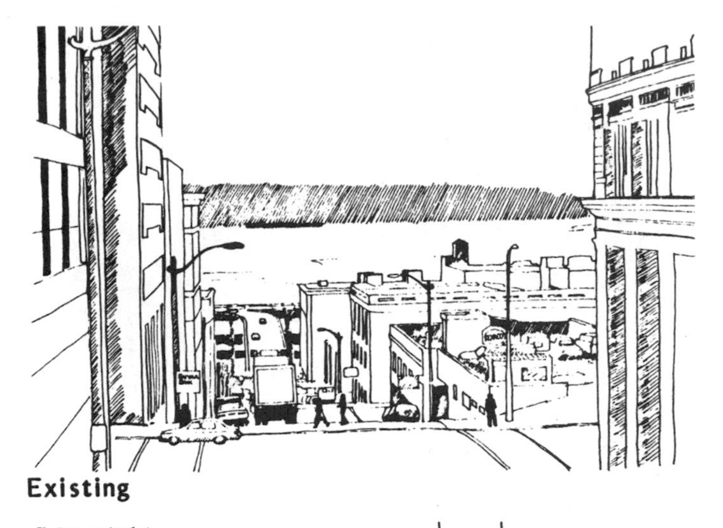

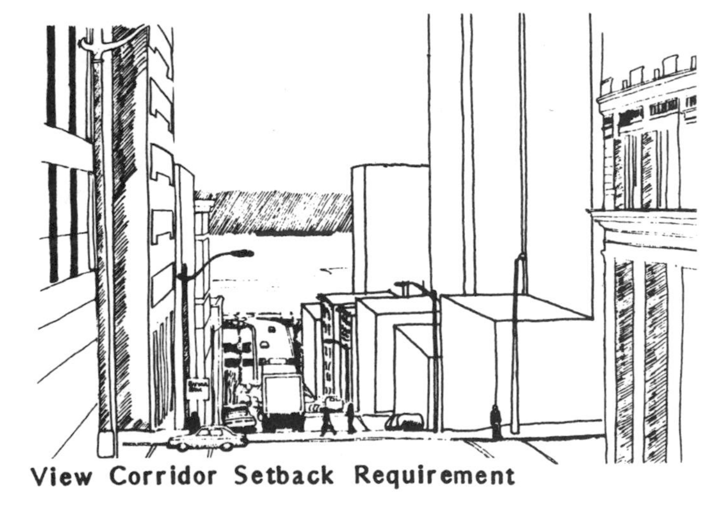

View corridor guidelines protect the views approaching Pike Place and the rest of the waterfront by relating height and setback elevations. (City of Seattle Land Use and Transportation Project. 1984)

REFERENCES

City of Seattle Land Use and Transportation Project. 1984. Mayor's Recommended Land Use & Transportation Plan for Downtown Seattle, Seattle, WA: City of Seattle.

Farbstein, Jay, Robert Shibley, Polly Welch, and Richard Wener with Emily H. Axelrod. 1998. *Sustaining Urban Excellence: Learning from the Rudy Bruner Award for Urban Excellence 1987-1993*, Boston, MA: Bruner Foundation, Inc.

Fox, Nancy. 1989. "Case Study: Pike Place Market, Seattle." In Bernard J. Frieden and Lynne B. Sagalyn (eds.). *Downtown, Inc.: How America Rebuilds Cities*. Cambridge, MA: MIT Press.

Spitzer, Theodore, and Hilary Baum. 1995. *Public Markets and Community Revitalization*. Urban Land Institute and Project for Public Spaces, Inc.

Steinbrueck, Victor. 1978. *Market Sketchbook*. Seattle and London: University of Washington Press.

Robert Shibley, Emily Axelrod, Jay Farbstein, Polly Welch, and Richard Wener

Summary

The three common threads of excellent urban projects are: first, successful placemakers have understood that making urban places is a long process extending both before and after the construction of a particular project. Second, inclusive processes that rely on democratic values tend to produce places more attuned to the full spectrum of needs and possibilities inherent in a place. Finally, while excellent places have tended to follow the general contours of the first two threads, they have been powerfully shaped by the unique circumstances and opportunities in which they unfolded.

Key words

community, democratic places, funding, maintenance, placemaking, process, transitions, values

The Southwest Corridor in Boston had its origins in community opposition to a highway proposal that would have further divided their neighborhoods. Neal Peirce and Robert Guskind (1993) refer to the Southwest Corridor as, ". . .landmark testimony to how "officialdom" and citizens, poor and rich, can work together. . ."

Making and sustaining excellent urban places

1 PLACE, PROCESS AND VALUES

As the preceding articles have made clear, the Rudy Bruner Award's trinity of "place, process, and values" is inseparable. Excellent design reflects the values physically embedded by the process of creation. Inclusive processes, powered by democratic values, produce certain types of places. These places may share some formal qualities, such as openness and permeability to the public, but then again, they may not: an inclusive process by definition opens itself up to the unique contextual dynamics of a particular physical and social environment, meaning that the end product will be uniquely tailored to "fit." In short, it would be artificial and misleading to discuss the design qualities of an excellent place without also discussing the process and values necessary to bring about such places. They stand in relation to each other, and lose their meaning when considered separately.

This inseparability does not mean that categorical or analytical thinking is impossible with regards to placemaking. Indeed, while the decisions to be made in a particular project are invariably difficult and hard fought, the practical terms of placemaking are relatively simple and easy to digest. If we cannot say with certainty that every excellent place should have this or that material or spatial characteristic, we can identify threads of place-process-value that have commonly occurred in the excellent places awarded by the Bruner Foundation. That is to say, certain types of processes, powered by certain types of values, tend to produce certain types of places—whether those be successful processes and values producing successful places, or inadequate processes informed by limiting values that result in failed or even never-built projects.

The three common threads of excellent urban projects are: first, successful placemakers understand that making urban places is a long process extending both before and after the construction of a particular project. A design should address a place's ongoing narrative, addressing existing problems or opportunities but also remaining flexible enough to meet the needs and opportunities of the future. Second, inclusive processes that rely on democratic values tend to produce places more attuned to the full spectrum of needs and possibilities inherent in a place. The active participation of a place's constituencies not only makes the full and vital use of a place more likely, but also fosters community coherence and self-awareness so that the place's natural caretakers become more motivated and competent to maintain it over time. Finally, excellent places are often powerfully shaped by the unique circumstances and opportunities in which they unfold. This has especially been true in two of the more traditional and fundamentally important aspects of professional urban design practice: funding and design. A variety of creative public and private funding strategies emerge from this attention to local context. Similarly, excellent designs typically present "indigenous" solutions carefully crafted to meet local conditions.

Credits: This essay is partially based on Farbstein, Jay, Robert Shibley, Polly Welch, and Richard Wener, with Emily Axelrod. 1998. *Sustaining Urban Excellence: Learning from the Rudy Bruner Award for Urban Excellence, 1987–1993*. Boston: Bruner Foundation, Inc. and Robert Shibley and Polly Welch. 1990. "Excellence in Urban Development: Building on Conflicts and Agreements," in *Urban Land, Sept., Vol. 49, No. 9.*

Fig. 1. Decking the Southwest Corridor created the conditions for new sites for economic development in abutting neighborhoods. It also knit the fabric of the neighborhoods back together with recreation and transit station developments.

Fig. 2. Transit stations became key nodes in Boston's Southwest Corridor. Above is the park entrance to the Back Bay station.

Fig. 3. Park gates are integrated with the pumping station ventilators on the Southwest Corridor.

2 TRANSITIONS: PLACES AS PROCESSES

Acknowledging that places evolve over time—that places are also processes—opens up a project to the participation of real and existing communities, strengthening those communities and tightening the "fit" between people and their places.

How do the people who care for good places know whether or not to change them? If changes are required, how do these same people know when and how to effect the changes? Once changes are made, how are the day-to-day activities in the places used to inform decisions about future changes, maintenance requirements, and the best uses of the place? The changes implied in these questions are easiest to understand when they involve physical adjustments to the place through new construction, renovation, or even simple repairs. A more common and less well-understood variety of change involves enabling subtle shifts in how constituents perceive the places, how the succession of leadership occurs, or how the rules of place modification and management evolve.

Knowing how to change is knowing how to mediate relationships between people and between people and place, ensuring constructive transitions in perceptions, rules constraining use patterns, design standards, and leadership. Just as an architect selects building materials based on knowledge of their durability and flexibility over time, an urban designer must establish design processes sturdy enough to meet new needs and opportunities as people live and work together in the "completed" place.

From self-conscious to un-self-conscious places

Urban interventions often occur in so-called "blighted" areas, where deteriorating building stock and declining economic activity are accompanied by a fading or already moribund community. Successful projects transform problems such as these into opportunities, allowing the crisis to define a discrete place with a clear vision responsive to recognizable needs. These "natural" boundary conditions specify what is within the project and what is excluded, and becomes a powerful narrative energizing and making possible project implementation. Indeed, defining such clear boundary conditions may well be the first step in nurturing the self-aware communities necessary to care for urban places.

Truly excellent places come about when such "self-conscious" projects are designed to be flexible enough to evolve over time into "un-self-conscious" places integrated with their surroundings and defined, not by the logic of the urban intervention, but by the natural (and fluid) boundaries of interlocking urban communities.

Such transitions are exemplified in the experiences of two projects with similarities, the Southwest Corridor in Boston and the Radial Re-use Plan in Lincoln, Nebraska. Both projects originated in public protests against planned but unbuilt highways whose attendant disinvestment had "blighted" swaths of urban fabric. These blighted areas, clearly defined and easily recognizable, served to focus the energies of local communities eager to rescue their ailing neighborhoods.

In Boston, the "people before highways" actions resulted in massive public investment in linear parks, a multimodal transportation system, and attendant infrastructure and community development in what became known as the Southwest Corridor. Once the revitalization project had succeeded, however, the notion of a clearly demarcated

"Corridor" had served its purpose. The area was ready to be re-integrated with adjacent neighborhoods, with new boundaries reflecting the social fabric of the region rather than the design parameters of the original project. The "old" corridor project became a part of the multimodal transit system serving the Boston region and, except for historical records of this development prominently displayed within its old boundary, the Southwest Corridor is frankly difficult to observe as one place or concept. Its prominence as a focus for organization continues to dim in favor of smaller acts of daily life and development within each community.

In Lincoln, local communities called for, and got, a new bike path and accompanying neighborhood and community development projects in the old highway right-of-way. Activists rallied their efforts around the very conditions that had allowed this part of the city to be ignored for two decades: a major university enjoying the benefits of "soft" land on its borders which allowed for inexpensive expansion, and a city council constructed entirely of at-large members which enabled this area of the city to be underrepresented in governance and infrastructure investment decisions.

Over time, however, the project changed from a clearly defined Radial Reuse Plan to a far less self-conscious component of a regional recreation path, and dozens of smaller scale neighborhood and community development projects. The abstraction of a monolithic plan with a fixed boundary has given way to discrete acts of neighborhood advocacy and development that no longer depend on the old project definition. In many respects the metaphor of the project might be that of reweaving torn fabric: before the reweaving, the fabric had a clearly defined hole. After the reweaving, one should no longer be able to identify the location of the damage.

Sometimes the transition from self-conscious to un-self-conscious is not about project boundaries, but is rather about project constituency and goal orientation. For example, an initial and often stated goal of the conversion of the Columbia Point housing project in Boston into the Harbor Point mixed income development was to create an environment where the former residents of Columbia Point would be fully integrated into the mainstream of Harbor Point life. As years passed, however, residents became bored with or even hostile to continuing efforts to measure "progress" on their integration. It has become self-defeating for them to be identified as targets for integration rather than as community members pursuing the daily tasks of living and relating to neighbors. The original project definition focuses attention on mixed income living, and that very attention now may actually get in the way of full realization of the goal. The complexity of constructing healthy relationships is not necessarily improved by continuously calling attention to differences in income level.

The origins of a project are necessarily self-conscious. The need to identify and define a problem concrete enough to grapple with, and to marshal enough support and energy to move forward, both require a relatively straightforward narrative with a clear urban "protagonist." But after project construction, a place may or may not be well served by sustaining the same level of self-consciousness. In order to effect a successful transition between project development and sustaining a project, think through these transitions clearly ahead of time and explore the long-term implications of defining project boundaries and goals for organizational support systems.

Fig. 4. The Southwest Corridor under construction literally illustrates the linking of abutting neighborhoods.

Fig. 5. The cover on the corridor creates opportunities for recreation trails, play areas, community gardening, transit stations, and easy access through the park to neighborhood facilities in abutting neighborhoods.

Fig. 6. Radial Re-use Plan in Lincoln, Nebraska, is a modest program compared to the Southwest Corridor, it was also a citizen confrontation of transportation planning that led to reconnecting neighborhoods through a linear park. Decades of disinvestment were reversed after plans for a highway north of O Street in Lincoln were finally laid to rest.

Fig. 7. The Brooklyn-Queens Greenway unifies the Olmsted and Moses park systems by providing links that make the full 40-mile system free of automobiles.

Fig. 8. The failed Columbia Point public housing project was converted to mixed income "luxury apartments" at Harbor Point.

Some projects, of course, need to sustain their distinct identity. Significant preservation efforts, for example, or projects devoted to specific ethnic identities often benefit from consistently asserting their differences. Those involved in the lives of projects must locate their efforts on a continuum between self-conscious and un-self-conscious. On occasion, they will need to embrace the tension between states, allowing projects to be one or the other as necessary.

The evolution of rules and goals

Managing places often involves adjusting intentions and operation to meet new demands, clientele, and surrounding circumstances. As the nature of the human and economic environment around a project evolves, the project itself needs to adapt so that it can continue to serve as an excellent home for human activity. Again it is the successful relationships among people and between people and their places that define excellence, and it is the job of those who manage a place to find a way to sustain these relationships over time.

The Ocean Drive renovations along a street of historic structures in Miami Beach have grown from an initial eighteen modest renovations to fifty-nine substantial ones. As the project grew, its nature changed: increasingly stringent historic preservation standards led to higher costs and more exclusive properties. To the surprise of all involved, the area witnessed a dramatic upsurge in economic value and activity. In response to these new and unforeseen circumstances, the original goal of including both middle and high-income residents evolved into an intentional drive for gentrification, resulting in significant increases in city tax revenue. Although seizing this opportunity meant sacrificing other goals, the painful decision reflected real new possibilities that had changed the project's basic narrative.

Rule changes as well as goal changes seem to be the norm in many of the projects recognized by the Rudy Bruner Award. The shifting financial circumstances of Pike Place Market, for example, have again opened up the question of balance between tourist-oriented craft sales and the farmers' market. Success is allowing another public market, Greenmarket in New York City, to press for higher percentages of local goods in their sales. The economic pressure on the Saint Francis Square cooperative housing project in San Francisco and its surrounding neighborhood may eventually shift the cooperative's goal from sustaining the complex as a cooperative lifestyle choice to selling it as an economic bonanza. By contrast, the Park at Post Office Square in Boston holds steadfast to its original goals and operating principles because "they are still the right ones." And even in Ocean Drive, the campaign to preserve historic property still remains. However, what was "preservation for preservation sake" has been transformed into "preservation as a means to economic ends."

Shifting project goals to meet new circumstances is a critical element of sustaining good places, so long as the changes are managed equitably. "Project success" is defined by all those affected by the project. It is political and contextual. The limits to tolerance need to be negotiated as circumstances change. In cases like Ocean Drive, stating the goal of mixed-income occupancy in absolute terms may well have reduced the level and quality of historic preservation activity and would certainly have reduced the tax revenues to the city. It is difficult but important to frame project goals in a manner that enable them to evolve based on new information. Sometimes this will mean redefining what constitutes error in the interests of service to a broader public good.

Leadership succession

Leadership is central in transitions such as moving from a self-conscious to un-self-conscious project, evolving from one goal to another, or shifting from one set of rules to another, and other changes that influence places. Sometimes the leadership in project origination is able to see and respond to the need for change as it occurs, but often such changes require a restructuring of leadership. These are often fragile transitions, which, if handled poorly, can destroy the spirit of a project or place. Much as at the beginning of projects, these crucial moments will produce changes that reflect the values and process with which they are implemented.

The abrupt departure of inspirational leader and founder Jose Galura of the Fairmount Health Center in North Philadelphia, for example, was concurrent with the exposure of serious financial difficulties from which the organization is only now beginning to recover. Other projects face similarly risky challenges. Will the New Community Corporation, so successful in revitalizing a central ward in Newark, lose its effectiveness or veer from its central values when Msgr. William Linder finally retires? Will the block club leaders in Cincinnati's historic Betts-Longworth neighborhood be able to share power with new tenants and continue the development of the area?

The Bruner archives offer some guidelines for the types of organizational structures that best accommodate these difficult transitions. Saint Francis Square and Pike Place Market have robust systems of democratic checks and balances that do not vest too much power in any one source. As a result they are both conservative and stable entities. Other organizations where power is centralized, like the Casa Rita women's homeless shelter in the Bronx, the New Community Corporation, Harbor Point, and even in Betts-Longworth at the community based organization level, are more volatile and face potentially difficult transitions in the future. New Community is clearly paying a lot of attention to this issue of transition and its board of directors has been heavily involved in transition planning. Harbor Point is addressing it through a stiff hierarchy that favors the economic implications of decisions. Often in the origins of a project a key person is central to a good start. Retaining the commitment to continuation, however, may require the leadership to be more diffused and shared in a manner that promotes commitment to continued maintenance.

Transitions in the lives of projects might be the neglected stepchild of policy action in government, in assessment criteria by award selection committees, and in the formation of project goals by private developers and public officials. We are often too quick to frame absolute criteria for success, using a "bottom line" mentality that does not allow for changing circumstances and an ongoing stance of critical investigation into what makes a place successful. The challenge is how to balance this relativistic perspective with the perceived need to control development and force it to meet predetermined ends. Acknowledging that there is a tension between these two approaches and embracing the tension as part of the management process may be the best approach to sustaining good places.

Fig. 9. Columbia Point prior to its becoming Harbor Point in Boston was a very self-conscious effort to integrate a range of incomes in the development and avoid any displacement.

Fig. 10. The adaptive re-use of public housing into a mixed race and income community at Harbor Point models a move from 20% occupancy in Columbia Point to 98% in the Harbor Point of 1996 and a shift from 100% low income in Columbia Point to 70% market rate and 30% subsidized low income units.

Fig. 11. Ocean Drive, in Miami Beach, Florida now boasts 59 substantial renovations of their art deco hotel structures breathing new life into the economy of the city.

Fig. 12. Part of the Ocean Drive project involved widening the streets even as automobile traffic increased. Above left is the 1990 restoration of the Waldorf Hotel and to the right is the Palace Grill, see as renovated in 1990.

Fig. 13. The transformation of north Philadelphia Neighborhoods like the one above was part of the goal of the Philadelphia Health Services organization as it extended its reach into the neighborhood with Fairmount Health Center.

Fig. 14. The site of the new Fairmount Health Center chosen by Jose Galura was an old auto dealership that was converted to a "re-manufacturing" warehouse situated on contested turf between Hispanic and African American communities.

3 PARTICIPATORY DEMOCRACY: BUILDING PLACES, BUILDING COMMUNITIES

Inclusive processes that incorporate the energy and capabilities of existing communities strengthen both places and communities.

Many stories about urban placemaking are potent examples of people engaging the essential drama of democracy—making decisions that reflect the will and best interests of a pluralist nation. Often these decisions have been about seizing opportunities to redress the social, political, and economic inequities of the urban renewal decades. The Bruner projects provide a wide range of opportunities to reflect upon the benefits and pitfalls of participatory democracy that offer alternatives to the planning practices that created urban renewal.

About a third of the Bruner finalists had a strong citizen constituency from the start, usually a small group of aggrieved citizens organizing themselves to voice opposition or to take public action on a problem. Some of these continue to be managed and operated by the original citizen organization. Other projects engaged one or more constituencies in a limited way to inform the process, or sought the advice of advocates for a particular constituency. A few projects were conceived, designed, and developed by single individuals. Although they are not without advisory boards and public review, decision-making power is clearly held by the leader. It is noteworthy that in the history of the award, there have been no finalists where citizen participation failed or the recommendations of a participatory process were ignored in placemaking. This might be attributable to the self-selection inherent in asking applicants to reflect on values and process as ingredients in their projects' success.

The role of organized communities in sustaining places

When a group of people perceive a shared problem and organize themselves to take action in relation to a situation or place, they are often recognized by public decision makers as a force to be reckoned with, giving credibility to their network and mission. When the group is included in the placemaking process the potential power of its constituents is strengthened by building their political knowledge and expertise. These empowered constituencies or individual members have become valuable civic resources in subsequent placemaking.

The regional constituencies for open space preservation who were included in the design of the Post Office Square in Boston, for example, gained credibility and attracted important allies from the phenomenal success of the park. That engagement, in turn, increased the value of their advocacy for additional open space within the city. Veterans of the Southwest Corridor have been faced with nearly endless opportunities to utilize the skills in community advocacy developed during Corridor design meetings. A number of empty development parcels and the promised relocation of a transit line still engage the stalwarts in the Roxbury community in Boston. Almost all of the public housing adjacent to the Corridor has undergone major redevelopment over the last decade, involving in planning and design decisions some of the same residents who had represented their community's interests in the Corridor process.

In Portland, the city is engaged in its third, even more ambitious, growth planning effort in two decades. The current process sustains the Portland Downtown Plan's tradition of inclusive values and democratic process by virtue of continued participation of citizens devoted to sustained involvement. The success of the first round has doubtlessly contributed to sustaining the interest and energy of Portland residents.

The role of real inclusiveness in building communities and community leaders

For many people, involvement in saving or making a special urban place is their first experience in having an impact on something seemingly immutable. Beyond the immediate feeling of empowerment, many participants thereby discover a role that brings new dimensions and skills to their lives. In these transformative experiences citizens find a voice, become knowledgeable and confident, and possibly even emerge as leaders. In a sense, the making of a place serves as a training ground for the communities and community leaders who will use it, providing opportunities to develop organizational skills and expertise that benefit both the project and the people.

The women on public assistance at Harbor Point in Boston, for example, perceived themselves as powerless because they were living in some of the most hopeless public housing in the country. When they organized to reject HUD's funding of a quick fix, they realized that they could positively affect change. They sought out a private development/management team who recognized that the opinions, energy, and tenacity of the residents would make a better project.

The Tenant Interim Lease Program in New York has promoted residents cooperatively buying the in rem apartment buildings in which they live. It has a constituency of residents who have voluntarily learned accounting, maintenance, and management skills to support their newly formed co-ops, and who have thereby greatly enhanced employment opportunities. The program is predicated on building the capacity of housing residents to own and manage their apartment buildings. In addition to its real estate rescues, it has encouraged people with little sense of power and self-worth to develop organizational leadership skills. Some of those residents have used this training to develop a new business or start other ventures beneficial to the neighborhood.

In the Radial Re-use Plan in Lincoln, NE, two young activists who helped organize their neighborhoods to challenge the city's plan for a highway were instrumental in getting the makeup of the city council revised from members-at-large to district representation. They themselves were elected to the Council during the decade that it took to get the highway right-of-way redesigned as a community park and bikeway. In Newark, Debra Channeyfield declared her candidacy for city council, urging New Communities Corporation to take more of a community organizing role and looking for their support. As daughter of one of NCC's founders and a product of its child care program, she had maintained an active role in the organization's operations.

Between demonstrating its benefits to Stowe and researching everything published on bikeways, Anne Lusk, who single-handedly promoted and developed support for the Stowe Recreation Path in Vermont, has become a nationally recognized expert, sitting on national advisory councils and helping other bike-path efforts get off the ground around the country. Rita Zimmer, who founded the Casa Rita transitional house for homeless women and children in the South Bronx, purposefully filled her eighteen-person board with individuals who could lend their professional skills to her mission and who would

Fig. 15. The new Fairmount Health Center was a 1989 RBA winner, cited for its contribution as a symbol of pride in the new community, a new start for very distressed neighborhoods and a place of quality health care in a high-need area.

Fig. 16. The interior of Fairmount Health Center is meant to be as open to the public as the street outside.

Fig. 17. Casa Rita, a shelter for homeless women and children in the South Bronx, New York was a 1987 award winner. Strong leadership from the board at Women in Need was essential to the start-up and to sustaining this award winner into a change of program involving the adoption of substance abusing women with children.

Fig. 18. The Park at Post Office Square in Boston, Massachusetts was a blighted aboveground parking garage before it became an urban park with space for fourteen hundred cars below it.

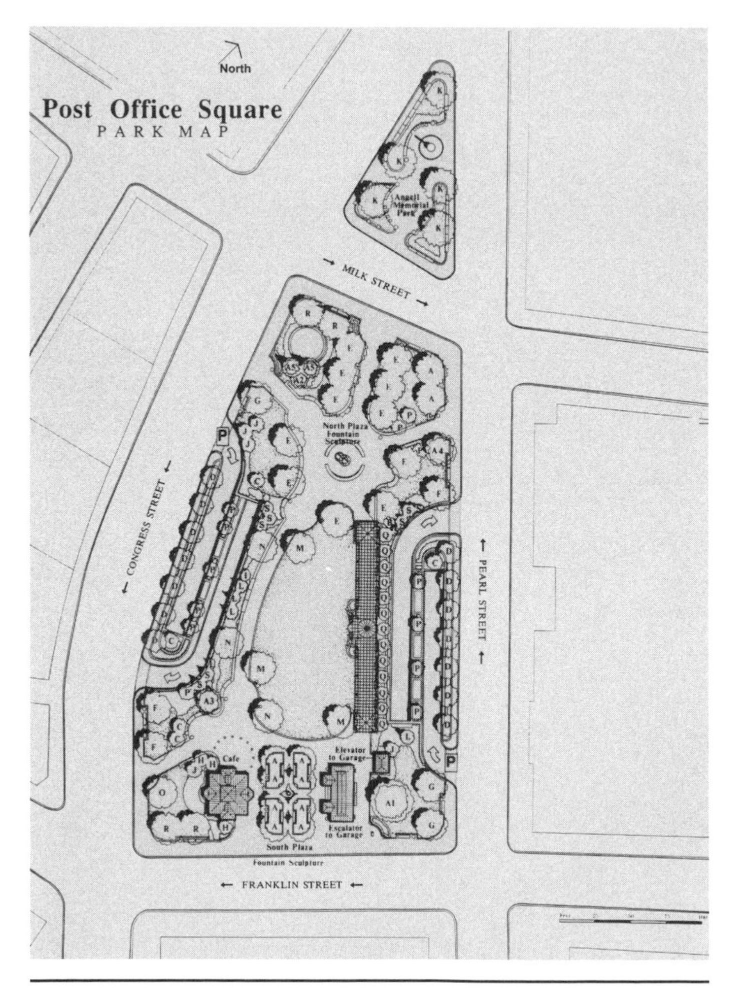

Fig. 19. The Park at Post Office Square enjoyed the support of a large regional constituency for open space as it stayed the course for development with a clear vision and a strong core leadership group.

welcome the opportunity to learn how to be a productive board member as well as become more familiar with the issues of homelessness. Once the building was occupied, she added women residents to the board as well.

The role of participation in sustaining places

Historically, in the United States, the planning and design of urban places has been the domain of politicians, commissions of people with vested interests, private benefactors, large institutions, and professionals working for public agencies. The interests and needs of urban communities and constituencies not represented in this traditional decision structure were not considered important enough to merit a seat at the table. Changes in social and political awareness have now carved a place for all interested parties. But having a seat at the table is not sufficient. It needs to be supported by a process that allows all aspects of democracy to be enacted, including the right to dissent; a process in which power is negotiated and shared, and the outcomes reflect, even if they do not fully satisfy, all interests. The greater the stake people have in the design process and product, the more likely it is that they will maintain the place over time, and the more competent they will be to apply their placemaking experience to other related problems that arise.

The participation process in the Southwest Corridor was so thorough, the impact on decision making so visible, and the outcomes so understandable that it convinced most people who came in touch with it that it was the best way to plan public places. Both the staff and the citizens took that belief and that expectation to many other public projects in Boston and elsewhere over the following years. The Corridor also provides evidence that citizens value the control and accountability that participation provides, and seek to extend it to managing and maintaining the places that have been created. Neighborhood groups along the corridor spent many months trying to persuade the Metropolitan District Commission (MDC) to allow them to maintain the parkland above the transit system. The current arrangement lets local residents participate in park maintenance using MDC lawn mowers.

Portland's Downtown Plan spawned a participatory process that has been brought into play for each of the subsequent planning efforts, the Central City Plan and the 2020 Regional Plan. Citizens have come to expect active involvement and responsive urban design decisions. This can be seen in another Portland Bruner winner, the West Clinton Action Plan. When REACH, the local community development corporation associated with the West Clinton neighborhood, initiated a bottom-up process for revitalizing the neighborhood, the city's community development department funded the plan and adopted the action planning model for use across the city.

Three cooperatively owned and managed housing developments—Cabrillo Village in Saticoy, California, St. Francis Square in San Francisco, and the Tenant Interim Lease Program in New York—suggest that shared ownership has many of the same benefits as participatory planning. In each case, the residents show a high degree of commitment to maintaining the development and have started to address more difficult problems like evicting dangerous residents.

Tailoring participation to suit project conditions

A formal citizen participation process is not always the most efficient way to accomplish placemaking. While law or ethics often requires it, private developers usually weigh its benefits against its cost in time and energy. Seeking input from constituencies at the inception of a project can minimize the possibility of contentious meetings and challenges later on, but not all projects seem to need representative public input. In some cases, touching base with selected constituencies is adequate. Making decisions about participation requires a sense of balance, between process and action on one hand, and between "local" and "expert" knowledge on the other.

Striking a balance between process and action requires different approaches for different projects. New Communities Corporation in Newark sometimes jumps into situations with only minimal planning because the opportunity could vanish with publicity and process. The Brooklyn-Queens Greenway Project in New York worries that a process that has not yet led to many visible outcomes will not be able to sustain the level of participation that it needs to ensure long-range success. At Harbor Point, some residents speak of the decision-making process as being "like watching sausage get made," while others embrace the process as the key to "empowerment and care giving."

The 1972 Portland Downtown Plan was so participation-oriented that some people felt "meeting-ed to death." Citizen participation has become something of a ritual in the city, however, and the process has changed from citizen members actually drafting and redrafting guidelines to staff doing the work from citizen input and review. The Southwest Corridor project addressed this issue of action by developing an intrinsic reward for reaching group consensus: when public meetings reached consensus the project director made decisions on the spot. When consensus did not emerge, the differing opinions were forwarded to the transit agency for their consideration.

Melding citizen and expert input can be difficult. Planning and design professionals may not concur with the opinions of the participants in the planning process. The Southwest Corridor is a case in point. As the project director pointed out, residents will not participate if they think people from the outside will come in and tell them what to do. The many consulting firms working on this gigantic project had to develop a new appreciation for process, since decision making was indeed shared with the citizenry. When the engineers were interviewed years later, some indicated that this experience had profoundly changed the way they have worked with clients on other jobs.

Some projects engaged in no formal participatory process, relying, in most cases, on their knowledge or the knowledge of a consultant to determine what would be appropriate for place users. Ocean Drive, Miami Beach, Florida; Park at Post Office Square, Boston; Quality Hill, Kansas, MO; and Roslindale Village Main Street, Boston, were all clearly driven by the goal of making a successful place that would also generate economic return. For the most part, market analysis replaced grassroots input. None of these places appear to be missing essential elements that might have emerged through citizen participation, nor have any of them faced angry challenges from disenfranchised residents. What they may not have captured, however, is a loyal constituency to sustain them in hard times. Time will tell.

Fig. 20. The new garage at Post Office Square handles more cars, with greater safety and efficiency than the aboveground facility it replaced, and it gives back quality open space.

Fig. 21. The fountain at Post Office Square is a popular lunch spot for nearby workers.

Fig. 22. The New York City Greenmarket supports 97% of all its costs from fees charged to farmer participants.

Democratic places

The Bruner finalists reflect democratic values not only in the process that led to their development but also in the designs of the places. The Park at Post Office Square and the Stowe Recreation Path, for example, are designed to invite the public onto or through private land. Post Office Square "feels democratic" to its users because secretaries and bank presidents share the space on an equal footing. In Stowe a similar classlessness exists when summer tourists share the bike path with local residents. Stowe property owners signed easements so that the path could cross their land. As Anne Lusk acquired each critical easement, observers wondered whether every one of the other landowners would agree. A single holdout could ruin the concept of the undertaking. This, too, is symbolic of what participatory democracy is about: the fragile interplay and essential connection between private interests and common good.

Some places act as a "commons"—everyone shares in the responsibility to use a place without diminishing its common benefit by overuse or abuse. St. Francis Square has shared courtyards and private yards. Residents saw the court as symbolic of their community, representing the cooperative governance in physical form, and also providing a place for residents to gather to play with the community children or chat with their neighbors. A housing administrator in San Francisco pointed out that most of the housing developments built more recently have no common space other than sidewalks and seem much less cohesive as communities.

In almost every one of the Bruner stories, people developed a stake in their places through some form of participation. Some chose not to follow their input through the democratic "sausage machine," while others made sure they could track the trade-offs and decisions at every juncture. It appears that people felt ongoing support for their place and its evolution either way.

Fig. 23. One aspect of the process of participation in Boston's Southwest Corridor project successfully involved abutting neighborhood groups in a review of approaches to construction of the tunnel. The consultation resulted in a decision to use a slurry wall system.

Many of the projects represent urban placemaking strategies at a time when citizen involvement was coming of age and city residents were eager to have a role in shaping their environment. Participation at the grass roots level seemed to promise a more inclusive process and has provided more responsive and equitable places for many urban dwellers. The Bruner finalists reflect many levels of involvement and a range of values related to citizen input. Virtually every project, however, has been and continues to be the beneficiary of constituency support. The places that were born from constituency efforts seem to place the greatest value on maintaining this broad connection to community. Those places are actively sustained, as well.

Citizens benefit in many ways from participating in making urban places, not the least of which is the sense of empowerment it brings. Membership in a constituency for an urban place also gives people access to community experiences and consideration of the common good beyond individual needs. It remains to be seen whether people will make time in their lives to continue the tradition of citizen participation in civic decisions.

4 PLANNING FOR REAL PLACES: DESIGN THAT FITS

Inclusive thinking about ongoing processes creates opportunities to employ innovative strategies and designs uniquely suitable for a particular urban location.

Funding that fits

Every project needs funding, and it is in facing this reality that planners' idealism can be compromised. But placemakers attuned to the unique challenges and opportunities presented by their projects often find creative ways to leverage funding out of their commitment to the values already inherent in a place and its communities. When the development process is an inclusive or democratic one, the energized constituents can provide more than volunteer labor or planning suggestions — they often prove willing to contribute financially as well. Alternately, when a project is clearly benefiting from the commitment and approval of its constituencies, it becomes much more attractive to potential sources of funding. Ever conservative, money prefers to hop on moving bandwagons if possible.

Some of the most successful of these funding efforts were successful because they were linked to the fundamental values and inclusive process of a place. Perhaps the best example of this is Yerba Buena Gardens, where the city's Redevelopment Authority lured private developers to the site by selling their commitment to transforming the project into a bustling public space. The evident support of the well-organized local residents who had forced this transformation, combined with the vision of a pedestrian oasis of cultural facilities and open parks in the heart of a major urban downtown, made Yerba Buena seem a sure bet to investors like Sony.

Other projects used similar context-specific tactics. The National AIDS Memorial Grove in San Francisco, for example, sold spaces for names to be carved in the park's hardscape elements, an appropriate gesture for a place intended to memorialize loved ones lost to the epidemic. ARTScorpsLA, in Los Angeles, enacted its principle of public land as "common turf" by simply seizing and making use of derelict—but privately owned—land in the barrio. Pike Place Market presented, among other things, a life-sized "piggy bank" to solicit private contributions from shoppers—a successful effort to play on the market's theme of farm products "fresh from the producer."

Not all fundraising can be so neatly incorporated into a project's identity. Many projects succeeded by carefully weaving public and private money and energies. In Casa Rita, for example, Rita Zimmer clearly provided the central impetus through Women in Need, but also relied heavily on government funding and private gifts. The New Community Corporation similarly grew out of a private nonprofit, organized by Monsignor William Linder and Mary Smith as a response to the tremendous physical and social damage that occurred during the 1967 Newark riots. It grew by filling niches left void by city government, supplying housing, health care, education, job training, and more, using government and foundation grants where possible, and creating its own income sources where not.

The Tenant Interim Lease Program is the product of New York City regulations and funding, even though it runs to a significant degree on the organizing efforts of the Urban Homestead Assistance Board and the labor and energy of the tenants, who work to make their formerly rental apartment buildings into cooperatively owned ones. The Betts-Longworth Historic District in Cincinnati was created and promoted by planners working for the city, a consortium of banks made loans, and not-for-profit and for-profit private developers built housing.

Where possible, successful projects have established income streams largely or totally from private sources that are sufficient to support

Fig. 24. Work crews are part of what makes the National Aids Memorial Garden in San Francisco work. The regular schedule of volunteer workdays gives participants a special relationship to the grove and among themselves as it furthers the goals of the memorial.

Fig. 25. Reconstructed migrant labor shacks at Cabrillo Village now serve as permanent homes.

Fig. 26. The Stowe Recreation Path creates a large public commons for the community through donated easements along a creek.

Fig. 27. The Stowe Recreation Path creates places of beauty for everyone to enjoy. Both the place and the process of creating it are dedicated to connecting private interests of abutting landholders to the public good.

the development and ongoing maintenance. Evidence suggests that a disproportionate reliance on government action and funding may be hazardous to the long-term sustainability of a project. Under these circumstances, a change in policy or in administration can place survival of a dependent effort in peril. Beyond Homelessness in San Francisco, for example, was a housing plan closely associated with the goals of one mayoral administration; the next incoming mayor felt no mandate to push for those particular solutions to homelessness.

To avoid such dependence, the Park at Post Office Square took advantage of the scarce availability of parking in Boston's financial district. The funds they received for the right to buy parking spaces, and loans based on parking revenues, supported creation of the large underground parking lot and the beautiful park on top. Now, after several years of operation, the parking operation turns a profit, which will eventually help support other (public) parks in the city. Greenmarket similarly traded on its ability to bring farm fresh produce to New York City locations, funding 97% of its operation out of fees charged to farmer-participants. Ocean Drive exploited its historic art deco buildings and prime beachfront acreage to direct private money towards the goal of preservation and development.

Most community development efforts, however, deal with situations in which such valuable and salable resources are not available. Often they are in the least marketable neighborhoods of a city. An innovative strategy that may have more general applicability is the idea of mixing profit and not-for-profit activities — creating funds that provide an internal source of subsidies for other inherently less profitable or even money-draining activities. New Community Corporation, for example, has actively sought to create profitable enterprises to support its many nonprofit operations. The most successful to date has been the Pathmark Shopping Center that it owns jointly with the Pathmark Corporation. The supermarket has become the most profitable of the chain's New Jersey stores, and its profits help fund a variety of other programs. New Community also runs fast food and upscale restaurants and has developed a factory for modular housing.

Public and private partnerships

Given the complexity of the social problems addressed by Bruner finalists and the overlapping concerns and jurisdictions involved, it is not surprising that few of the projects were conceived and created by one single organization or sector of society. The clearest and most enduring story of the Bruner Awards may indeed be the highly collaborative nature of successful urban projects, often involving people from many different kinds of organizations and backgrounds. In particular, a wide and complex range of partnerships between public and private actors has characterized most of the projects. Some of these have primarily related to funding, but there are other activities and kinds of support provided by the public sector that can also make or break community development efforts.

Often these are far from dramatic, and basically constitute the ongoing, "public housekeeping" functions of advice, regulation, and oversight. The Fairmount Health Center in Philadelphia, for example, forged a special relationship to career staffers when the U.S. Public Health Service (PHS) intervened in a crisis. Regulatory requirements of the PHS, and routine visits for oversight of those regulations, led to the discovery and eventual correction of severe management and fiscal problems. The PHS staffers played a clear and positive role in saving the Center from extinction. Since that time, the PHS personnel, who feel pride in the recovery, go out of their way to communicate both information and support for the agency's new leadership.

Government employees also routinely work with their for-profit and not-for-profit partners to set strategies and then help advocate for these policies with other government agencies and at higher governmental levels. The ongoing coalition orchestrating Ocean Drive is a good example: the business community and the preservationists have not co-opted the city's planning department, but instead share interests with it. City professionals routinely advocate within city government for responsiveness to the Drive's needs. The joint creation and honing of governmental regulations permits the developmental agenda to bloom. Another example is provided by the high regard state workers hold for Casa Rita's personalized approach to housing the homeless. Policy makers were influenced by Casa Rita's design in formulating their policy for "tier two" transitional housing programs, although they utilized it in a much diluted form.

Government agency staffers can also, by their public and private actions, help affirm and validate the values and ideological orientation of a project. Early on in the story of the Radial Re-use Project, city employees emerged as allies of the newly assertive neighborhood groups. Other illustrations of these shared values can be seen in the case of a San Francisco city department executive who has chosen to live at St. Francis Square, by the employees of the City of Portland who are visible as members of community development boards implementing Portland's Downtown Plan, by Cincinnati municipal workers who live in Betts-Longworth, and by New York City transportation department workers who regularly cycle along the Brooklyn-Queens Greenway to work.

In some cases, governmental bodies have even moved to formally share statutory decision making with a non-profit. The allocation of the very significant interstate transportation grants (ISTEA) is currently the domain of a fourteen-person committee in New York State. Ten of the members represent government entities, while the other four are from nonprofits, two of whom are Brooklyn-Queens Greenway activists. This is a tangible manifestation of the close alliance of local community groups and public servants. Greenway veterans speak with marked respect and gratitude for New York City government professionals who "knew everything" and shared their knowledge.

Democratic places

Places that win a Bruner Award for Urban Excellence are places in harmony and balance with the values, processes, and social outcomes supported by their environment. While the Bruner Award is not focused on urban design, urban excellence cannot avoid incorporating appropriate design. At the minimum, excellence requires that the physical environment not inhibit, restrain, detract, or (at worst) destroy the social environment. At best, excellent projects are wonderful places in every sense of the word, with spatial and physical qualities that express, symbolize, and support the social processes they contain.

This kind of design success in the urban environment can come out of a variety of processes. Capable or even noted architects, landscape architects, and planners designed some projects; others have benefited from the thoughtful, caring actions of citizens. Bruner Selection Committees have identified at least three elements of excellent design: the use and preservation of the value inhering in the existing physical

Fig. 28. The shared public space at St. Francis Square in San Francisco, California is symbolic of their community, represents the cooperative governance in physical form, and also provides a place for public gatherings and children play areas.

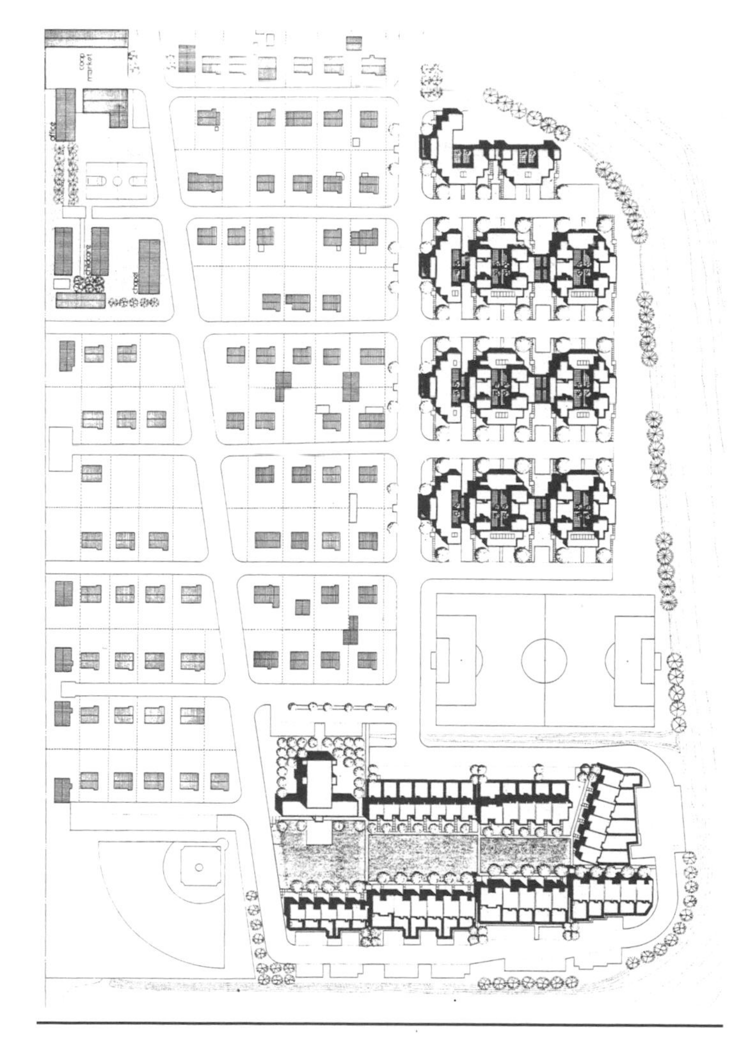

Fig. 29. The site plan for Cabrillo Village illustrates several "democratic spaces." The grid streets of the old camp and new four-plex units, the common green of the row house development, and the basketball court related to the community building which links all three residential areas.

environment of a place; the accessibility and openness of public spaces; and a commitment to the highest quality of design and construction.

Preservation contributes something vitally important to the urban fabric and to urban life. Preserving and adapting older buildings provides a sense of continuity of experience and memory, which helps us to understand and relate to the city. It sustains "social ecologies" and builds meaning to urban life. Parkside Preservation in Philadelphia, for example, resuscitated an almost impossibly moribund downtown neighborhood and community with its investment in historic preservation of the buildings along Parkside Avenue. Pike Place similarly revived a complex network of social activities by salvaging Seattle's historic farmers' market. Quality Hill in Kansas City, MO, refurbished a number of historic buildings, restoring housing and creating space for recreation and offices. Betts-Longworth reversed policies that had nearly eliminated a historic district near downtown Cincinnati. In each case, a project to revitalize ailing urban places drew energy from historic narratives associated with those places.

The accessibility of public spaces directly reflects the question of how values become encoded into places. When a process is opened up in meaningful ways to affected constituencies, the diversity of voices involved tends to push the design towards openness, so that the needs of many different communities can be addressed. The synergistic potential of such openness is clearest in projects that developed to revolve around a central open public space, like Yerba Buena Gardens, the Park at Post Office Square, the Pioneer Courthouse Square Project of the Portland Downtown Plan, the Stowe Recreation Path, the art-parks created by ARTScorpsLA, the Southwest Corridor, the Radial Re-use Project, Saint Francis Square, the Brooklyn-Queens Greenway, and others. But open public space is only one aspect of accessibility.

Other examples include the commitment to public transportation in the Portland Downtown Plan; the openness of the cultural institutions in Yerba Buena to locals, and the inclusion of children's facilities in the project; ARTScorpsLA's willingness to allow anyone, including gang members, to participate in the creation of its public artworks; and so forth. In these cases, and many others, inclusive processes installed long-standing avenues for public access to the physical and social environment of the place.

Excellent design is a fundamental principle of the Bruner Awards. Indeed, the valuing of a place and its communities embodied in the process of placemaking would be undermined if the final product were shabby or just minimally acceptable. Excellent design is the capstone of a process wherein communities demonstrate and act upon their caring for their places. On a more prosaic level, it is a crucial feature in a project's ability to leverage its process and values into financial and political support for maintaining the place over time.

Oftentimes, this excellence in design is fairly conventional, reflecting the activities of talented professionals. Yerba Buena, for example, became a showcase for skilled practitioners like Mario Botta, Fumihiko Maki, James Stewart Polshek, and I.M. Pei. The Portland Public Market in Maine was designed entirely by a professional with extensive experience in planning public markets. Historic preservation projects like Parkside Preservation, Ocean Drive, Betts, Longworth, Quality Hill, and others relied on the architectural feats of past professionals. In some instances, such as in ARTScorpsLA, design professionals were purposefully kept in an advisory position so that even the physical elements of the project would reflect the active participation of constituents. While such participation may be most appropriate for the small and relatively simple art parks produced by ARTScorpsLA, the notion of constituents influencing design outcome is a common one in Bruner winners: Portland's Downtown Plan provided for public art designed to be enjoyed by laypeople, as was Yerba Buena Gardens' memorial to Martin Luther King, Jr. In each case, the commitment to serving the real needs of urban communities fundamentally shaped project design. And in each case, those needs were best served by the creation of places better than strictly necessary.

In summary

The importance of local context for the design and development of a project is a key theme in Bruner winners, but this is not to suggest that a successful project can be a model for others in different locations. The public market template established at Pike Place, for example, has been used to good effect in New York City's Greenmarket and Portland, Maine's Portland Public Market. In each case, though, even where a programming or design concept was imported from elsewhere, it was closely and carefully refashioned to suit the needs—and the opportunities—of a local place. Portland's market, for example, was designed to make use of local building materials, and developed a "food theater" dynamic intended to dramatize the importance of Maine's ailing small farming heritage. The array of public open spaces lodged at the heart of so many projects is remarkable. Consider the grand squares like the Pioneer Courthouse in Portland and the Park at Post Office Square in Boston, the natural pathways and bike lanes in Stowe and New York City, and the courtyards of St. Francis Square as well as the gardens in the National Aids Memorial Grove in San Francisco.

Just as design has reflected the imperatives of local conditions, so have funding strategies and efforts to build partnerships between the public and private sectors. Many places have managed to leverage their profit potential into support for locally desirable goals, like the spectacular success of Yerba Buena Gardens in providing amenities for low-income seniors and community youth by attracting massive investment by international-grade developers. Other projects with more modest profit potential have creatively negotiated public funding, or persuaded an involved public to contribute their own money. And nearly all projects have pieced together collaborations with the government or private sector to ensure the best possible exploitation of opportunities.

The advantages extracted from difficult or even disastrous conditions; the creative financing plucked out of seemingly parched economic landscapes; the effective cooperation between stubborn activists, profit-driven private companies, and bureaucratic public agencies—all of these stem from the basic placemaking decision to know an urban place before remaking it, to learn the existing relationships between people and between people and their places, and to build upon existing value insofar as possible. The most important step a placemaker makes, in a certain sense, is the one not made: it is the conscious decision not to see an urban intervention as if it had no history or context. ■

Index

D

E

H

I

M

N

O

S

T